P9-EER-987

Extensive and Varied Exercise Sets	An abundant collection of exercises is included in an exercise set at the end of each section. Exercises are organized within categories. Your instructor will usually provide guidance on which exercises to work. The exercises in the first category, Practice Exercises, follow the same order as the section's worked examples.	The paral[illegible] Exercises [illegible] you refer to the worked examples and use them as models for solving these problems.
Practice Plus Problems	This category of exercises contains more challenging problems that often require you to combine several skills or concepts.	It is important to dig in and develop your problem-solving skills. Practice Plus Exercises provide you with ample opportunity to do so.

3 Review for Quizzes and Tests

Feature	Description	Benefit
Mid-Chapter Check Points	At approximately the midway point in the chapter, an integrated set of review exercises allows you to review the skills and concepts you learned separately over several sections.	By combining exercises from the first half of the chapter, the Mid-Chapter Check Points give a comprehensive review before you move on to the material in the remainder of the chapter.
Chapter Review Grids	Each chapter contains a review chart that summarizes the definitions and concepts in every section of the chapter. Examples that illustrate these key concepts are also included in the chart.	Review this chart and you'll know the most important material in the chapter!
Chapter Review Exercises	A comprehensive collection of review exercises for each of the chapter's sections follows the review grid.	Practice makes perfect. These exercises contain the most significant problems for each of the chapter's sections.
Chapter Tests	Each chapter contains a practice test with approximately 25 problems that cover the important concepts in the chapter. Take the practice test, check your answers, and then watch the Chapter Test Prep Video CD to see worked-out solutions for any exercises you miss.	You can use the chapter test to determine whether you have mastered the material covered in the chapter.
Chapter Test Prep Video CDs	These video CDs found at the back of your text contain worked-out solutions to every exercise in each chapter test.	The videos let you review any exercises you miss on the chapter test.
Cumulative Review Exercises	Beginning with Chapter 2, each chapter concludes with a comprehensive collection of mixed cumulative review exercises. These exercises combine problems from previous chapters and the present chapter, providing an ongoing cumulative review.	Ever forget what you've learned? These exercises ensure that you are not forgetting anything as you move forward.

More Tools for Your Mathematics Success

Student Study Pack

Get math help when YOU need it! The **Student Study Pack** provides you with the ultimate set of support resources to go along with your text. Packaged at no charge with a new textbook, the **Student Study Pack** contains these invaluable tools:

- **Student Solutions Manual**

  A printed manual containing full solutions to odd-numbered textbook exercises.

- **Prentice Hall Math Tutor Center**

  Tutors provide one-on-one tutoring for any problem with an answer at the back of the book. You can contact the Tutor Center via a toll-free phone number, fax, or email.

- **CD-ROM Lecture Series**

  A comprehensive set of textbook-specific CD-ROMs containing short video clips of the textbook objectives being reviewed and key examples being solved.

Tutorial and Homework Options

MYMATHLAB

MyMathLab, packaged at no charge with a new textbook, is a complete online multimedia resource to help you succeed in learning. **MyMathLab** features:

- The entire textbook online.
- Problem-solving video clips and practice exercises, correlated to the examples and exercises in the text.
- Online tutorial exercises with guided examples.
- Online homework and tests.
- Generates a personalized study plan based on your test results.
- Tracks all online homework, tests, and tutorial work you do in **MyMathLab** gradebook.

Introductory and Intermediate Algebra for College Students

Second Edition

Introductory and Intermediate Algebra for College Students

Robert Blitzer

Miami-Dade College

PEARSON
Prentice Hall

Upper Saddle River, New Jersey 07458

Library of Congress Cataloging-in-Publication Data

Blitzer, Robert.
Introductory and intermediate algebra for college students / Robert Blitzer.—2nd ed.
p. cm.
Includes indexes.
ISBN 0-13-149259-4
1. Algebra—Textbooks. I. Title.

QA154.3.B5845 2006
512.9—dc22 2004063813

Senior Acquisitions Editor: *Paul Murphy*
Editor in Chief: *Christine Hoag*
Project Manager: *Liz Covello*
Production Editor: *Prepare, Inc.*
Assistant Managing Editor: *Bayani Mendoza de Leon*
Senior Managing Editor: *Linda Mihatov Behrens*
Executive Managing Editor: *Kathleen Schiaparelli*
Media Project Manager: *Audra J. Walsh*
Media Production Editor: *Donna Crilly*
Managing Editor, Digital Supplements: *Nicole M. Jackson*
Assistant Manufacturing Manager/Buyer: *Michael Bell*
Executive Marketing Manager: *Eilish Collins Main*
Marketing Assistant: *Rebecca Alimena*
Director of Marketing: *Patrice Jones*
Development Editor: *Don Gecewicz*
Editor in Chief, Development: *Carol Trueheart*
Editorial Assistant: *Mary Burket*
Assistant Editor/Print Supplements Editor: *Christina Simoneau*
Art Director: *Maureen Eide*
Interior Design: *Running River Design*
Cover Art and Designer: *Koala Bear Design*
Art Editor: *Thomas Benfatti*
Director of Creative Services: *Paul Belfanti*
Director, Image Resource Center: *Melinda Reo*
Manager, Rights and Permissions: *Zina Arabia*
Manager, Visual Research: *Beth Brenzel*
Image Permission Coordinator: *Debbie Hewitson*
Photo Researcher: *Elaine Soares; Beaura K. Ringrose*
Art Studios: *Scientific Illustrators; Laserwords*
Compositor: *Prepare, Inc.*

PEARSON Prentice Hall

© 2006, 2002 Pearson Education, Inc.
Pearson Prentice Hall
Pearson Education, Inc.
Upper Saddle River, New Jersey 07458

All rights reserved. No part of this book may be reproduced, in any form or by any means, without written permission from the publisher.

Pearson Prentice Hall™ is a trademark of Pearson Education, Inc.

Printed in the United States of America
10 9 8 7 6 5 4 3 2 1

ISBN 0-13-149259-4

Pearson Education LTD., *London*
Pearson Education Australia PTY, Limited, *Sydney*
Pearson Education Singapore, Pte. Ltd
Pearson Education North Asia Ltd, *Hong Kong*
Pearson Education Canada, Ltd., *Toronto*
Pearson Educación de Mexico, S.A. de C.V.
Pearson Education—Japan, *Tokyo*
Pearson Education Malaysia, Pte. Ltd

CONTENTS

CHAPTER 3 LINEAR EQUATIONS IN TWO VARIABLES 184

CHAPTER 4 SYSTEMS OF LINEAR EQUATIONS 245

CHAPTER 5 EXPONENTS AND POLYNOMIALS 305

CHAPTER 6 FACTORING POLYNOMIALS 382

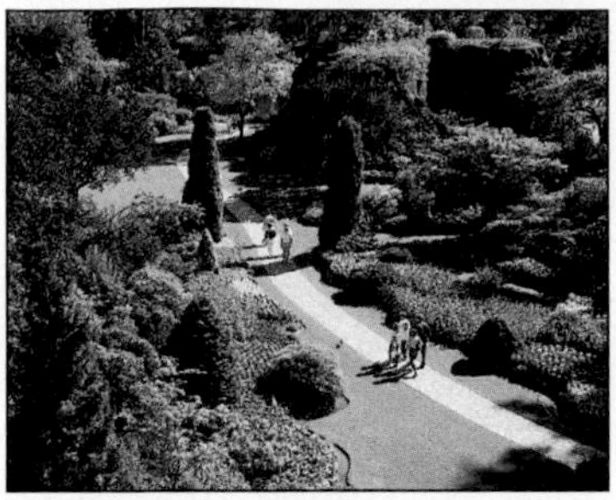

CHAPTER 7 RATIONAL EXPRESSIONS 441

CHAPTER 8 BASICS OF FUNCTIONS 530

CHAPTER 9 INEQUALITIES AND PROBLEM SOLVING 578

CHAPTER 10 RADICALS, RADICAL FUNCTIONS, AND RATIONAL EXPONENTS 637

CHAPTER 11 QUADRATIC EQUATIONS AND FUNCTIONS 717

CHAPTER 12 Exponential and Logarithmic Functions 798

CHAPTER 13 Conic Sections and Systems of Nonlinear Equations 872

CHAPTER 14 Sequences, Series, and Probability 933

PREFACE

Introductory and Intermediate Algebra for College Students, Second Edition, provides comprehensive, in-depth coverage of the topics required in a course combining the study of introductory and intermediate algebra. The book is written for college students who have no previous experience in algebra and for those who need a review of basic algebraic concepts before moving on to intermediate algebra. I wrote the book to help diverse students, with different backgrounds and career plans, to succeed in a combined introductory and intermediate algebra course. *Introductory and Intermediate Algebra for College Students*, Second Edition, has two primary goals:

1. To help students acquire a solid foundation in the skills and concepts of introductory and intermediate algebra, without the repetition of topics in two separate texts.

2. To show students how algebra can model and solve authentic real-world problems.

One major obstacle in the way of achieving these goals is the fact that very few students actually read their textbook. This has been a regular source of frustration for me and for my colleagues in the classroom. Anecdotal evidence gathered over years highlights two basic reasons why students do not take advantage of their textbook:

- "I'll never use this information."
- "I can't follow the explanations."

I've written every page of the Second Edition with the intent of eliminating these two objections. The ideas and tools I've used to do so are described in the features that follow. These features and their benefits are highlighted for the student in "A Brief Guide to Getting the Most from This Book" that appears inside the front cover.

What's New in the Second Edition?

I believe students and instructors will welcome the following new features:

- **Practice Plus Exercises.** More challenging practice exercises that often require students to combine several skills or concepts have been added to the exercise sets. The 750 Practice Plus Exercises in the Second Edition, averaging 10 of these exercises per exercise set, provide instructors with the option of creating assignments that take practice exercises to a more challenging level than in the previous edition.
- **Mid-Chapter Check Points.** At approximately the midway point in each chapter, an integrated set of review exercises allows students to review and assimilate the skills and concepts they learned separately over several sections. The 285 exercises that make up the Mid-Chapter Check Points, averaging 20 exercises per check point, are of a mixed nature, requiring students to discriminate which concepts or skills to apply. The Mid-Chapter Check Points should help students bring together the different objectives covered in the first half of the chapter before they move on to the material in the remainder of the chapter.

- **New Applications and Real-World Data.** I researched hundreds of books, magazines, almanacs, and online data sites to prepare the Second Edition. As a result, many new, innovative applications, supported by data that extend as far up to the present as possible, appear in 318 new application exercises and examples.

- **Over 2000 New Examples and Exercises.** In addition to the 750 Practice Plus Exercises, the 285 Mid-Chapter Check Points, and the 318 new application exercises, the Second Edition contains 736 new exercises that appear in the various categories of the exercise sets.

- **Increased Study Tip Boxes.** The book's Study Tip boxes offer suggestions for problem solving, point out common errors to avoid, and provide informal hints and suggestions. These invaluable hints appear in greater abundance in the Second Edition.

- **Increased Use of Function Notation in Equation Solving.** Because functions are the core of Chapters 8 through 14 (the intermediate algebra portion of the course), students are repeatedly shown how functions relate to equations and graphs. Many of the new practice exercises in the second half of the book use function notation in solving both equations and inequalities.

- **Graphical and Numerical Approaches to Problems.** Although the use of graphing utilities is optional, students can look at the new side-by-side features in Using Technology boxes and, with the assistance of the explanatory voice balloons, begin to understand both graphical and numerical approaches to problems even if they are not using a graphing utility in the course.

- **Chapter Test Prep Video CDs.** Packaged at the back of the text, these video CDs provide students with step-by-step solutions for each of the exercises in the book's chapter tests.

What Content and Organizational Changes Have Been Made to the Second Edition?

- Section 1.1 (Fractions) contains new discussions on mixed numbers and improper fractions, as well as the use of prime factorizations in reducing fractions and finding LCDs.

- Section 2.4 (Formulas and Percents) includes new discussions on percent increase, percent decrease, and ways percents are frequently used incorrectly.

- Section 5.6 (Long Division of Polynomials; Synthetic Division) includes long division problems with nonlinear divisors.

- Section 7.6 (Solving Rational Equations) contains a new discussion on solving formulas with rational expressions for a variable.

- The Mid-Textbook Check Point (Are You Prepared for Intermediate Algebra?) allows students to review introductory algebra topics before starting the intermediate algebra portion of the book. Presented in the format of the new Mid-Chapter Check Points and appearing immediately before Chapter 8 (rather than as an appendix in the back of the book, as in the previous edition), the Mid-Textbook Check Point gives students a fast way to review and practice prerequisite skills needed in intermediate algebra.

- Chapter 8 (Basics of Functions), the book's transitional chapter into intermediate algebra, is now devoted exclusively to functions. An integrated graphing functional approach is emphasized throughout the remainder of the book. Because both functions and systems of linear equations in three variables were included in this chapter in the previous edition, the following organizational changes were made:
  - Systems of linear equations in three variables are discussed in the final section of Chapter 4 (Systems of Linear Equations).
  - Matrix solutions to linear systems and Cramer's Rule, topics frequently covered in college algebra, are included in separate appendices.
  - Composite and inverse functions, appearing in the chapter on exponential and logarithmic functions in the previous edition, are discussed in the final section of Chapter 8. Instructors who wish to include these topics in basics of functions can now do so, even if exponential and logarithmic functions are not covered in the course.
- Section 8.1 (Introduction to Functions) provides a new introduction to relations and functions using *Celebrity Jeopardy*'s highest earners and their winnings. Also included is a new discussion of identifying the domain and the range of a function from its graph.
- Section 11.1 (The Square Root Property and Completing the Square; Distance and Midpoint Formulas) now includes a discussion of the distance between two points in rectangular coordinates, as well as the midpoint formula. Distance and midpoint formulas, appearing in the chapter on conic sections in the previous edition, follow naturally from the square root property and can now be included in the course even if conic sections are not covered.
- Section 11.5 (Polynomial and Rational Inequalities) is a reworked and expanded discussion of quadratic and rational inequalities from the previous edition. Solution procedures are now developed and organized around polynomial functions, rational functions, and their graphs.
- Section 12.5 (Exponential Growth and Decay; Modeling Data) contains new examples involving choosing models for data before technology is used to obtain these models.
- Section 14.4 (The Binomial Theorem) now gives the formula for $(r + 1)$st term, rather than the rth term, of the expansion of $(a + b)^n$. Many students find the formula for the $(r + 1)$st term easier to work with when finding a particular term in a binomial expansion.
- Section 14.6 (Probability) contains new probability examples using tables with real-world data.

What Familiar Features Have Been Retained in the Second Edition?

The features described below that helped make the First Edition so popular continue in the Second Edition.

- **Detailed Worked-Out Examples.** Each worked example is titled, making clear the purpose of the example. Examples are clearly written and provide students with detailed step-by-step solutions. No steps are omitted and each step is thoroughly explained to the right of the mathematics.
- **Check Point Examples.** Each example is followed by a similar matched problem, called a Check Point, offering students the opportunity to test their understanding of the example by working a similar exercise. The answers to the Check Points are provided in the answer section.

- **Explanatory Voice Balloons.** Voice balloons are used in a variety of ways to demystify mathematics. They translate algebraic ideas into everyday English, help clarify problem-solving procedures, present alternative ways of understanding concepts, and connect problem solving to concepts students have already learned.

- **Extensive and Varied Exercise Sets.** An abundant collection of exercises is included in an exercise set at the end of each section. Exercises are organized within seven category types: Practice Exercises, Practice Plus Exercises, Application Exercises, Writing in Mathematics, Critical Thinking Exercises, Technology Exercises, and Review Exercises. This format makes it easy to create well-rounded homework assignments. The order of the practice exercises is exactly the same as the order of the section's worked examples. This parallel order enables students to refer to the titled examples and their detailed explanations to achieve success working the practice exercises.

- **Chapter-Opening and Section-Opening Scenarios.** Every chapter and every section open with a scenario presenting a unique application of algebra in students' lives outside the classroom. These scenarios are revisited in the course of the chapter or section in an example, discussion, or exercise.

- **Section Objectives.** Learning objectives open every section. These objectives help students recognize and focus on the section's most important ideas. The objectives are stated in the margin at their point of use.

- **Chapter Review Grids.** Each chapter contains a review chart that summarizes the definitions and concepts in every section of the chapter. Examples that illustrate these key concepts are also included in the chart.

- **End-of-Chapter Materials.** A comprehensive collection of review exercises for each of the chapter's sections follows the review grid. This is followed by a chapter test that enables students to test their understanding of the material covered in the chapter. Beginning with Chapter 2, each chapter concludes with a comprehensive collection of mixed cumulative review exercises.

- **Graphing.** Chapter 1 contains an introduction to graphing, a topic that is integrated throughout the book. Line, bar, circle, and rectangular coordinate graphs that use real data appear in nearly every section and exercise set. Many examples and exercises use graphs to explore relationships between data and to provide ways of visualizing a problem's solution.

- **Geometric Problem Solving.** Chapter 2 contains a section on problem solving in geometry that teaches geometric concepts that are important to a student's understanding of algebra. There is a frequent emphasis on problem solving in geometric situations, as well as on geometric models that allow students to visualize algebraic formulas.

- **Thorough, Yet Optional Technology.** Although the use of graphing utilities is optional, they are utilized in Using Technology boxes to enable students to visualize and gain numerical insight into algebraic concepts. The use of graphing utilities is also reinforced in the technology exercises appearing in the exercise sets for those who want this option. With the book's early introduction to graphing, students can look at the calculator screens in the Using Technology boxes and gain an increased understanding of an example's solution even if they are not using a graphing utility in the course.

- **Study Tips.** Study Tip boxes appear throughout the book.

- **Enrichment Essays.** These discussions provide historical, interdisciplinary, and otherwise interesting connections to the algebra under study, showing students that math is an interesting and dynamic discipline.
- **Discovery.** Discover for Yourself boxes, found throughout the text, encourage students to further explore algebraic concepts. These explorations are optional and their omission does not interfere with the continuity of the topic under consideration.
- **Chapter Projects.** At the end of each chapter is a collaborative activity that gives students the opportunity to work cooperatively as they think and talk about mathematics. Additional group projects can be found in the *Instructor's Resource Manual*. Many of these exercises should result in interesting group discussions.

Resources for the Instructor

Print

Annotated Instructor's Edition (ISBN: 0-13-149260-8)

- Answers appear in place on the same text page as exercises or in the Graphing Answer Section
- Answers to all exercises in the exercise sets, Mid-Chapter Check Points, Chapter Reviews, Chapter Tests, and Cumulative Reviews

Instructor's Solutions Manual (ISBN: 0-13-192178-9)

- Detailed step-by-step solutions to the even-numbered section exercises
- Solutions to every exercise (odd and even) in the Mid-Chapter Check Points, Chapter Reviews, Chapter Tests, and Cumulative Reviews
- Solution methods reflect those emphasized in the text.

Instructor's Resource Manual with Tests (ISBN: 0-13-192177-0)

- Six test forms per chapter—3 free response, 3 multiple choice
- Two *Cumulative Tests* for all even-numbered chapters
- Two *Final Exams*
- Answers to all test items
- *Mini-Lectures* for each section with brief lectures including key learning objectives, classroom examples, and teaching notes
- Additional *Activities*, two per chapter, providing short group activities in a convenient ready-to-use handout format
- *Skill Builders* providing an enhanced worksheet for each text section, including concept rules, explained examples, and extra problems for students
- Twenty *Additional Exercises* per section for added test exercises or worksheets

Media

Lab Pack CD Lecture Series (ISBN: 0-13-192180-0)

- Organized by section, *Lab Pack CD Lecture Series* contains problem-solving techniques and examples from the textbook.

- Step-by-step solutions to selected exercises from each textbook section marked with a video icon

TestGen (ISBN: 0-13-192181-9)

- Windows and Macintosh compatible
- Algorithmically driven, text-specific testing program covering all objectives of the text
- Chapter Test file for each chapter provides algorithms specific to exercises in each *Chapter Test* from the text.
- Edit and add your own questions with the built-in question editor, which allows you to create graphs, import graphics, and insert math notation.
- Create a nearly unlimited number of tests and worksheets, as well as assorted reports and summaries.
- Networkable for administering tests and capturing grades online, or on a local area network

MyMathLab
www.mymathlab.com
An all-in-one, online tutorial, homework, assessment, and course management tool with the following features:

- Rich and flexible set of course materials, featuring free-response exercises algorithmically generated for unlimited practice and mastery
- Entire textbook online with links to multimedia resources including video clips, practice exercises, and animations that are correlated to the textbook examples and exercises
- Homework and test managers to select and assign online exercises correlated directly to the text
- A personalized Study Plan generated based on student test results. The Study Plan links directly to unlimited tutorial exercises for the areas students need to study and retest, so they can practice until they have mastered the skills and concepts.
- *MyMathLab Gradebook* allows you to track all of the online homework, tests, and tutorial work while providing grade control.
- Import *TestGen* tests

MathXL®
www.mathxl.com
A powerful online homework, tutorial, and assessment system that allows instructors to:

- Create, edit, and assign online homework and tests using algorithmically generated exercises correlated at the objective level to the textbook
- Track student work in *MathXL*'s online gradebook

Resources for the Student

Student Solutions Manual | CD/Video | PH Math/Tutor Center | MathXL Tutorials on CD | MathXL® | MyMathLab | Interactmath.com

Student Study Pack (ISBN: 0-13-154936-7)

Get math help when YOU need it! Available at no charge when packaged with a new text, *Student Study Pack* provides the ultimate set of support resources to go along with the text. The *Student Study Pack* includes the *Student Solutions Manual*, access to the *Prentice Hall Math Tutor Center*, and the *CD Lecture Videos*.

Print

Student Solutions Manual (ISBN: 0-13-192179-7)

- Solutions to all odd-numbered section exercises
- Solutions to every exercise (odd and even) in the Mid-Chapter Check Points, Chapter Reviews, Chapter Tests, and Cumulative Reviews
- Solution methods reflect those emphasized in the text.

Media

MyMathLab

An all-in-one, online tutorial, homework, assessment, course management tool with the following student features:

- The entire textbook online with links to multimedia resources including video clips, practice exercises, and animations correlated to the textbook examples and exercises
- Online tutorial, homework, and tests
- A personalized Study Plan based on student test results. The Study Plan links directly to unlimited tutorial exercises for the areas students need to study and retest, so they can practice until they have mastered the skills and concepts.

MathXL®

A powerful online homework, tutorial, and assessment system that allows students to:

- Take chapter tests and receive a personalized study plan based on their test results
- See diagnosed weaknesses and link directly to tutorial exercises for the objectives they need to study and retest
- Access supplemental animations and video clips directly from selected exercises

www.InterActMath.com

- Have the power of the *MathXL®* text-specific tutorial exercises available for unlimited practice online, without an access code, and without tracking capabilities

MathXL® Tutorials on CD (ISBN: 0-13-192176-2)

An interactive tutorial that provides:

- Algorithmically generated practice exercises correlated at the objective level
- Practice exercises accompanied by an example and a guided solution
- Tutorial video clips within the exercise to help students visualize concepts
- Easy-to-use tracking of student activity and scores and printed summaries of students' progress

Chapter Test Prep Video CDs (ISBN: 0-13-192174-6)

- Provide step-by-step video solutions to each exercise in each Chapter Test in the textbook
- Packaged at no charge with the text, inside the back cover

PH Tutor Center (ISBN: 0-13-064604-0)

- Tutorial support via phone, fax, or email bundled at no charge with a new text, or purchased separately with a used book
- Staffed by developmental math faculty
- Available 5 days a week, 7 hours a day

Acknowledgments

An enormous benefit of authoring a successful series is the broad-based feedback I receive from the students, dedicated users, and reviewers. Every change to this edition is the result of their thoughtful comments and suggestions. I would like to express my appreciation to all the reviewers, whose collective insights form the backbone of this revision. In particular, I would like to thank:

Howard Anderson	*Skagit Valley College*
John Anderson	*Illinois Valley Community College*
Michael H. Andreoli	*Miami Dade College–North Campus*
Hien Bui	*Hillsborough Community College*
Warren J. Burch	*Brevard Community College*
Alice Burstein	*Middlesex Community College*
Edie Carter	*Amarillo College*
Thomas B. Clark	*Trident Technical College*
Sandra Pryor Clarkson	*Hunter College*
Bettyann Daley	*University of Delaware*
Robert A. Davies	*Cuyahoga Community College*
Paige Davis	*Lurleen B. Wallace Community College*
Ben Divers, Jr.	*Ferrum College*
Irene Doo	*Austin Community College*
Charles C. Edgar	*Onondaga Community College*
Rhoderick Fleming	*Wake Technical Community College*
Susan Forman	*Bronx Community College*
Donna Gerken	*Miami-Dade College*
Marion K. Glasby	*Anne Arundel Community College*
Sue Glascoe	*Mesa Community College*
Jay Graening	*University of Arkansas*
Robert B. Hafer	*Brevard Community College*
Mary Lou Hammond	*Spokane Community College*
Donald Herrick	*Northern Illinois University*
Beth Hooper	*Golden West College*
Tracy Hoy	*College of Lake County*
Judy Kasabian	*Lansing Community College*
Gary Kersting	*North Central Michigan College*
Gary Knippenberg	*Lansing Community College*
Mary Kochler	*Cuyahoga Community College*
Kristi Laird	*Jackson Community College*
Jennifer Lempke	*North Central Michigan College*
Sandy Lofstock	*St. Petersburg College*
Hank Martel	*Broward Community College*
Diana Martelly	*Miami-Dade College*
John Robert Martin	*Tarrant County Junior College*
Mikal McDowell	*Cedar Valley College*
Irwin Metviner	*State University of New York at Old Westbury*

Terri Moser	*Austin Community College*
Robert Musselman	*California State University, Fresno*
Kamilia Nemri	*Spokane Community College*
Allen R. Newhart	*Parkersburg Community College*
Steve O'Donnell	*Rogue Community College*
Jeff Parent	*Oakland Community College*
Kate Rozsa	*Mesa Community College*
Scott W. Satake	*Eastern Washington University*
Mike Schramm	*Indian River Community College*
Kathy Shepard	*Monroe County Community College*
Gayle Smith	*Lane Community College*
Linda Smoke	*Central Michigan University*
Dick Spangler	*Tacoma Community College*
Janette Summers	*University of Arkansas*
Robert Thornton	*Loyola University*
Lucy C. Thrower	*Francis Marion College*
Andrew Walker	*North Seattle Community College*
Margaret Williamson	*Milwaukee Area Technical College*
Roberta Yellott	*McNeese State University*
Marilyn Zopp	*McHenry County College*

Special thanks to Professors Doreen Kelly and Kate Rozsa at Mesa Community College for your work on the custom edition that resulted in many of the practice plus exercises and many of the new exercises in the Second Edition. Thank you, Brad Davis, consummate mathematician and proofreader, for preparing the answer and graphing answer sections, as well as for your invaluable edits that kept me at my mathematical and syntactical best.

Additional acknowledgments are extended to Paul Murphy, senior acquisitions editor; Liz Covello, project manager; Barbara Mack, production editor; Linda Martino full service editor from Prepare; and Bayani Mendoza de Leon, assistant managing editor. Lastly, my thanks to the staff at Prentice Hall for their continuing support: Eilish Main, executive marketing manager; Patrice Jones, director of marketing; Chris Hoag, editor-in-chief; and to the entire Prentice Hall sales force for their confidence and enthusiasm about my books.

I hope that my love for learning, as well as my respect for the diversity of students I have taught and learned from over the years, is apparent throughout this new edition. By connecting algebra to the whole spectrum of learning, it is my intent to show students that their world is profoundly mathematical, and indeed, π is in the sky.

Robert Blitzer

TO THE STUDENT

I've written this book so that you can learn about the power of algebra and how it relates directly to your life outside the classroom. All concepts are carefully explained, important definitions and procedures are set off in boxes, and worked-out examples that present solutions in a step-by-step manner appear in every section. Each example is followed by a similar matched problem, called a Check Point, for you to try so that you can actively participate in the learning process as you read the book. (Answers to all Check Points appear in the back of the book.) Study Tips offer hints and suggestions and often point out common errors to avoid. A great deal of attention has been given to applying algebra to your life to make your learning experience both interesting and relevant.

As you begin your studies, I would like to offer some specific suggestions for using this book and for being successful in this course:

1. Read the book. Read each section with pen (or pencil) in hand. Move through the worked-out examples with great care. These examples provide a model for doing exercises in the exercise sets. As you proceed through the reading, do not give up if you do not understand every single word. Things will become clearer as you read on and see how various procedures are applied to specific worked-out examples.

2. Work problems every day and check your answers. The way to learn mathematics is by doing mathematics, which means working the Check Points and assigned exercises in the exercise sets. The more exercises you work, the better you will understand the material.

3. Review for quizzes and tests. After completing a chapter, study the chapter review chart, work the exercises in the Chapter Review, and work the exercises in the Chapter Test. Answers to all these exercises are given in the back of the book.

The methods that I've used to help you read the book, work the problems, and review for tests are described in "A Brief Guide to Getting the Most from This Book" that appears inside the front cover. Spend a few minutes reviewing the guide to familiarize yourself with the book's features and their benefits.

4. Use the resources available with this book. Additional resources to aid your study are described following the guide to getting the most from your book. These resources include a Solutions Manual, a Chapter Test Prep Video CD, MyMathLab, an online version of the book with links to multimedia resources, MathXL®, an online homework, tutorial, and assessment system of the text, and tutorial support at no charge at the PH Tutor Center.

5. Attend all lectures. No book is intended to be a substitute for valuable insights and interactions that occur in the classroom. In addition to arriving for lecture on time and being prepared, you will find it useful to read the section before it is covered in lecture. This will give you a clear idea of the new material that will be discussed.

I wrote this book in Point Reyes National Seashore, 40 miles north of San Francisco. The park consists of 75,000 acres with miles of pristine surf-washed beaches, forested ridges, and bays bordered by white cliffs. It was my hope to convey the beauty and excitement of mathematics using nature's unspoiled beauty as a source of inspiration and creativity. Enjoy the pages that follow as you empower yourself with the algebra needed to succeed in college, your career, and in your life.

Regards,
Bob
Robert Blitzer

ABOUT THE AUTHOR

Bob Blitzer is a native of Manhattan and received a Bachelor of Arts degree with dual majors in mathematics and psychology (minor: English literature) from the City College of New York. His unusual combination of academic interests led him toward a Master of Arts in mathematics from the University of Miami and a doctorate in behavioral sciences from Nova University. Bob is most energized by teaching mathematics and has taught a variety of mathematics courses at Miami-Dade College for nearly 30 years. He has received numerous teaching awards, including Innovator of the Year from the League for Innovations in the Community College, and was among the first group of recipients at Miami-Dade College for an endowed chair based on excellence in the classroom. In addition to *Introductory and Intermediate Algebra for College Students*, Bob has written *Introductory Algebra for College Students, Intermediate Algebra for College Students*, *Essentials of Intermediate Algebra for College Students*, *Essentials of Introductory and Intermediate Algebra for College Students, Algebra for College Students, Thinking Mathematically, College Algebra, Algebra and Trigonometry*, and *Precalculus*, all published by Prentice Hall.

Sitting in the biology department office, you overhear two of the professors discussing the possible adult heights of their respective children. Looking at the blackboard that they've been writing on, you see that there are formulas that can estimate the height a child will attain as an adult. If the child is x years old and h inches tall, that child's adult height, H, in inches, is approximated by one of the following formulas:

Girls:

$$H = \frac{h}{0.00028x^3 - 0.0071x^2 + 0.0926x + 0.3524}$$

Boys:

$$H = \frac{h}{0.00011x^3 - 0.0032x^2 + 0.0604x + 0.3796}.$$

You will use one of these formulas in Exercise 118 on page 91.

CHAPTER 1

The Real Number System

When you encounter formulas with symbols such as those that appear in the estimation of a child's adult height, don't panic! You are already familiar with many symbols—the smiley face, the peace symbol, the heart symbol, the dollar sign, and even symbols on your calculator or computer. In this chapter, you will become familiar with the special symbolic notation of algebra. You will see that the language of algebra describes your world and holds the power to solve many of its problems.

SECTION 1.1 FRACTIONS

Objectives

1. Convert between mixed numbers and improper fractions.
2. Write the prime factorization of a composite number.
3. Reduce or simplify fractions.
4. Multiply fractions.
5. Divide fractions.
6. Add and subtract fractions with identical denominators.
7. Add and subtract fractions with unlike denominators.
8. Solve problems involving fractions.

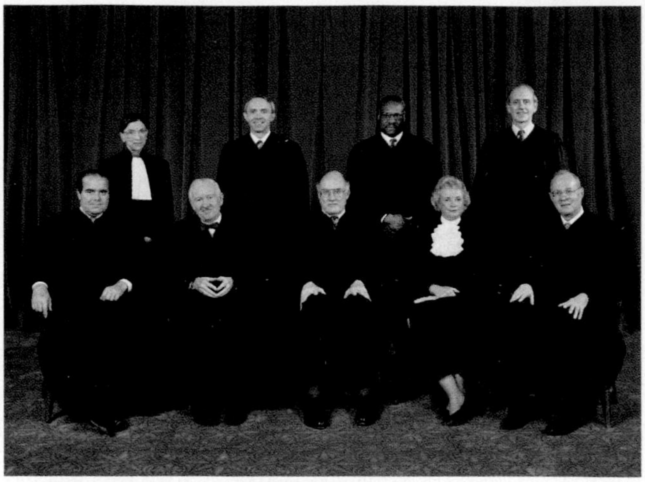

The Nine Justices of the U.S. Supreme Court

"If I were asked where I place the American aristocracy, I should reply without hesitation that it occupies the judicial bench and bar. Scarcely any political question arises in the United States that is not resolved, sooner or later, into a judicial question."

Alexis de Tocqueville, *Democracy in America*

President Jimmy Carter had no opportunity to make an appointment to the Supreme Court. However, he selected more African Americans, Hispanics, and women for the lower federal courts than all other prior presidents combined—40 women, 37 African Americans, and 16 Hispanics. The graph in Figure 1.1 shows the number of female and minority appointments to federal judgeships for Presidents Carter, Reagan, Bush Senior, and Clinton.

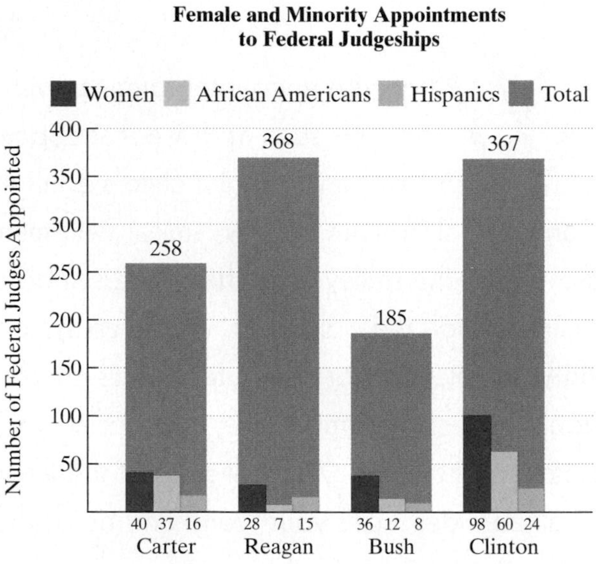

FIGURE 1.1

Source: James Burns et al., *Government by the People*, Prentice Hall, 2002

We can use the numbers shown in Figure 1.1 to form various fractions. For example, the graph indicates that President Carter appointed a total of 258 federal judges. Of the 258 judges, there were 40 women. We can say that the *fraction* of women appointed to federal judgeships under Carter was $\frac{40}{258}$. Can you see how this fraction is used to refer to part (40 women) of a whole (258 total appointments)?

In a fraction, the number that is written above the fraction bar is called the **numerator**. The number below the fraction bar is called the **denominator**.

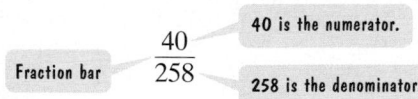

$$\frac{40}{258}$$

40 is the numerator.

Fraction bar

258 is the denominator.

The numbers 40 and 258 are examples of *natural numbers*. The **natural numbers** are the numbers that we use for counting:

$$1, 2, 3, 4, 5, \ldots.$$

The three dots after the 5 indicate that the list continues in the same manner without ending.

Fractions appear throughout algebra. We are frequently given numerical information that involves fractions. In this section, we present a brief review of operations with fractions that we will use in algebra.

1 Convert between mixed numbers and improper fractions.

Mixed Numbers and Improper Fractions A **mixed number** consists of the addition of a natural number and a fraction, expressed without the use of an addition sign. Here is an example of a mixed number:

$$3\frac{4}{5}.$$

The natural number is 3 and the fraction is $\frac{4}{5}$. $3\frac{4}{5}$ means $3 + \frac{4}{5}$.

An **improper fraction** is a fraction whose numerator is greater than its denominator. An example of an improper fraction is $\frac{19}{5}$.

The mixed number $3\frac{4}{5}$ can be converted to the improper fraction $\frac{19}{5}$ using the following procedure:

CONVERTING A MIXED NUMBER TO AN IMPROPER FRACTION

1. Multiply the denominator of the fraction by the natural number and add the numerator to this product.
2. Place the result from step 1 over the denominator in the mixed number.

EXAMPLE 1 Converting from Mixed Number to Improper Fraction

Convert $3\frac{4}{5}$ to an improper fraction.

SOLUTION

$$3\frac{4}{5} = \frac{5 \cdot 3 + 4}{5}$$

Multiply the denominator by the natural number and add the numerator.

$$= \frac{15 + 4}{5} = \frac{19}{5}$$

Place the result over the mixed number's denominator.

■

STUDY TIP

After working each Check Point, check your answer in the answer section before continuing your reading.

✔ **CHECK POINT 1** Convert $2\frac{5}{8}$ to an improper fraction.

An improper fraction can be converted to a mixed number using the following procedure:

CONVERTING AN IMPROPER FRACTION TO A MIXED NUMBER

1. Divide the denominator into the numerator. Record the quotient (the result of the division) and the remainder.
2. Write the mixed number using the following form:

$$\text{quotient}\,\frac{\text{remainder}}{\text{original denominator}}.$$

natural number part — fraction part

EXAMPLE 2 Converting from an Improper Fraction to a Mixed Number

Convert $\frac{42}{5}$ to a mixed number.

SOLUTION

Step 1. **Divide the denominator into the numerator.**

$$\begin{array}{r} 8 \\ 5\overline{)42} \\ \underline{40} \\ 2 \end{array}$$

quotient (8)

remainder (2)

Step 2. **Write the mixed number using quotient $\frac{\textbf{remainder}}{\textbf{original denominator}}$.** Thus,

$$\frac{42}{5} = 8\frac{2}{5}.$$

quotient (8); remainder (2); original denominator (5)

■

CHECK POINT 2 Convert $\frac{5}{3}$ to a mixed number.

STUDY TIP

In applied problems, answers are usually expressed as mixed numbers, which many people find more meaningful than improper fractions. Improper fractions are often easier to work with when performing operations with fractions.

Using Variables to Express Fractions Algebra uses letters, such as a and b, to represent numbers. Such letters are called **variables**. Throughout this section, we will use variables to express fractions. These fractions can be represented by the *variable expression*

$$\frac{a}{b}.$$

In this expression, the variable a represents any natural number and the variable b also represents any natural number. Can you see why $\frac{a}{b}$ is called a variable expression? Its value varies with the choice of a and the choice of b. Here are some examples:

Choose $a = 5$ and $b = 3$. $\quad \frac{a}{b} = \frac{5}{3}$ $\qquad$ Choose $a = 3$ and $b = 5$. $\quad \frac{a}{b} = \frac{3}{5}$ $\qquad$ Choose $a = 1$ and $b = 100$. $\quad \frac{a}{b} = \frac{1}{100}$.

2 Write the prime factorization of a composite number.

Factors and Prime Factorizations In the section opener, we saw that the fraction of women appointed to federal judgeships under President Carter was $\frac{40}{258}$. As you looked at this fraction, did you attempt to *simplify* it by *reducing it to its lowest terms*? Before we simplify fractions, let's first review how to factor the natural numbers that make up the numerator and the denominator.

To **factor** a natural number means to write it as a multiplication of two or more natural numbers. For example, 21 can be factored as $7 \cdot 3$. In the statement $7 \cdot 3 = 21$, 7 and 3 are called the **factors** and 21 is the **product**.

$$7 \cdot 3 = 21$$

7 is a factor of 21. 3 is a factor of 21. The product of 7 and 3 is 21.

STUDY TIP

Although we use a raised dot to indicate multiplication, as in $7 \cdot 3$, there are other ways to represent the product of 7 and 3:

7×3
$7(3)$
$(7)3$
$(7)(3)$.

Are 7 and 3 the only factors of 21? The answer is no because 21 can also be factored as $1 \cdot 21$. Thus, 1 and 21 are also factors of 21. The factors of 21 are 1, 3, 7, and 21.

Unlike the number 21, some natural numbers have only two factors: the number itself and 1. For example, the number 7 has only two factors: 7 (the number itself) and 1. The only way to factor 7 is $1 \cdot 7$ or, equivalently, $7 \cdot 1$. For this reason, 7 is called a *prime number*.

PRIME NUMBERS A **prime number** is a natural number greater than 1 that has only itself and 1 as factors.

The first ten prime numbers are

$$2, 3, 5, 7, 11, 13, 17, 19, 23, \text{ and } 29.$$

Can you see why the natural number 15 is not in this list? In addition to having 15 and 1 as factors ($15 = 1 \cdot 15$), it also has factors of 3 and 5 ($15 = 3 \cdot 5$). The number 15 is an example of a *composite number*.

STUDY TIP

The number 1 is the only natural number that is neither a prime number nor a composite number.

COMPOSITE NUMBERS A **composite number** is a natural number greater than 1 that is not a prime number.

Every composite number can be expressed as the product of prime numbers. For example, the composite number 45 can be expressed as

$$45 = 3 \cdot 3 \cdot 5.$$

This product contains only prime numbers: 3 and 5.

Expressing a composite number as the product of prime numbers is called the **prime factorization** of that composite number. The prime factorization of 45 is $3 \cdot 3 \cdot 5$. The order in which we write these factors does not matter. This means that

$$45 = 3 \cdot 3 \cdot 5 \quad \text{or} \quad 45 = 5 \cdot 3 \cdot 3 \quad \text{or} \quad 45 = 3 \cdot 5 \cdot 3.$$

To find the prime factorization of a composite number, begin by selecting any two numbers whose product is the number to be factored. If one or both of the factors are not prime numbers, continue to factor each composite number. Stop when all numbers are prime.

EXAMPLE 3 Prime Factorization of a Composite Number

Find the prime factorization of 100.

SOLUTION Begin by selecting any two numbers whose product is 100. Here is one possibility:

$$100 = 4 \cdot 25.$$

Because the factors 4 and 25 are not prime, we factor each of these composite numbers.

$$\begin{aligned} 100 &= 4 \cdot 25 && \text{This is our first factorization.} \\ &= 2 \cdot 2 \cdot 5 \cdot 5 && \text{Factor 4 and 25.} \end{aligned}$$

Notice that 2 and 5 are both prime. The prime factorization of 100 is $2 \cdot 2 \cdot 5 \cdot 5$. ■

✔ CHECK POINT 3 Find the prime factorization of 36.

3 Reduce or simplify fractions.

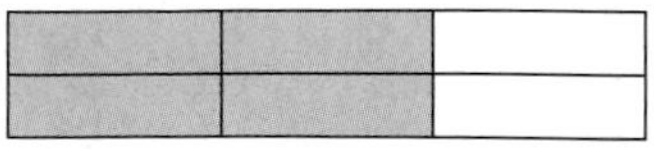

FIGURE 1.2

Reducing Fractions Two fractions are **equivalent** if they represent the same value. Writing a fraction as an equivalent fraction with a smaller denominator is called **reducing a fraction**. A fraction is **reduced to its lowest terms** when the numerator and denominator have no common factors other than 1.

Look at the rectangle in Figure 1.2. Can you see that it is divided into 6 equal parts? Of these 6 parts, 4 of the parts are red. Thus, $\frac{4}{6}$ of the rectangle is red.

The rectangle in Figure 1.2 is also divided into 3 equal stacks and 2 of the stacks are red. Thus, $\frac{2}{3}$ of the rectangle is red. Because both $\frac{4}{6}$ and $\frac{2}{3}$ of the rectangle are red, we can conclude that $\frac{4}{6}$ and $\frac{2}{3}$ are equivalent fractions.

How can we show that $\frac{4}{6} = \frac{2}{3}$ without using Figure 1.2? Prime factorizations of 4 and 6 play an important role in the process. So does the **Fundamental Principle of Fractions**.

FUNDAMENTAL PRINCIPLE OF FRACTIONS In words: The value of a fraction does not change if the numerator and denominator are divided (or multiplied) by the same nonzero number.

In algebraic language: If $\frac{a}{b}$ is a fraction and c is a nonzero number, then

$$\frac{a \cdot c}{b \cdot c} = \frac{a}{b}.$$

We use prime factorizations and the Fundamental Principle to reduce $\frac{4}{6}$ to its lowest terms as follows:

$$\frac{4}{6} = \frac{2 \cdot 2}{3 \cdot 2} = \frac{2}{3}$$

Write prime factorizations of 4 and 6.

Divide the numerator and the denominator by the common prime factor, 2.

Here is a procedure for writing a fraction in lowest terms:

REDUCING A FRACTION TO ITS LOWEST TERMS

1. Write the prime factorization of the numerator and the denominator.
2. Divide the numerator and the denominator by the greatest common factor, the product of all factors common to both.

Division lines can be used to show dividing out a fraction's numerator and denominator by common factors:

$$\frac{4}{6} = \frac{2 \cdot \cancel{2}}{3 \cdot \cancel{2}} = \frac{2}{3}.$$

STUDY TIP

When reducing a fraction to its lowest terms, only factors that are common to the numerator and the denominator can be divided out. **If you have not factored** and expressed the numerator and denominator in terms of multiplication, **do not divide out.**

Correct:

$$\frac{2 \cdot \cancel{2}}{3 \cdot \cancel{2}} = \frac{2}{3}$$

Incorrect:

$$\frac{2 + \cancel{2}}{3 + \cancel{2}} = \frac{2}{3}$$

Note that $\frac{2+2}{3+2} = \frac{4}{5}$, not $\frac{2}{3}$.

EXAMPLE 4 Reducing Fractions

Reduce each fraction to its lowest terms:

a. $\frac{6}{14}$ **b.** $\frac{15}{75}$ **c.** $\frac{25}{11}$ **d.** $\frac{11}{33}$.

SOLUTION For each fraction, begin with the prime factorization of the numerator and the denominator.

a. $\frac{6}{14} = \frac{3 \cdot \cancel{2}}{7 \cdot \cancel{2}} = \frac{3}{7}$ 2 is the greatest common factor of 6 and 14. Divide the numerator and the denominator by 2.

Including 1 as a factor is helpful when all other factors can be divided out.

b. $\frac{15}{75} = \frac{3 \cdot 5}{3 \cdot 25} = \frac{1 \cdot \cancel{3} \cdot \cancel{5}}{\cancel{3} \cdot \cancel{5} \cdot 5} = \frac{1}{5}$ $3 \cdot 5$, or 15, is the greatest common factor of 15 and 75. Divide the numerator and the denominator by $3 \cdot 5$.

c. $\frac{25}{11} = \frac{5 \cdot 5}{1 \cdot 11}$

Because 11 and 25 share no common factor (other than 1), $\frac{11}{25}$ is already reduced to its lowest terms.

d. $\frac{11}{33} = \frac{1 \cdot \cancel{11}}{3 \cdot \cancel{11}} = \frac{1}{3}$ 11 is the greatest common factor of 11 and 33. Divide the numerator and denominator by 11.

When reducing fractions, it may not always be necessary to write prime factorizations. In some cases, you can use inspection to find the greatest common factor of the numerator and the denominator. For example, when reducing $\frac{15}{75}$, you can use 15 rather than $3 \cdot 5$:

$$\frac{15}{75} = \frac{1 \cdot \cancel{15}}{5 \cdot \cancel{15}} = \frac{1}{5}.$$

CHECK POINT 4 Reduce each fraction to its lowest terms:

a. $\frac{10}{15}$ **b.** $\frac{42}{24}$ **c.** $\frac{13}{15}$ **d.** $\frac{9}{45}$.

4 Multiply fractions.

Multiplying Fractions The result of multiplying two fractions is called their **product**.

MULTIPLYING FRACTIONS In words: The product of two or more fractions is the product of their numerators divided by the product of their denominators.

In algebraic language: If $\frac{a}{b}$ and $\frac{c}{d}$ are fractions, then

$$\frac{a}{b} \cdot \frac{c}{d} = \frac{a \cdot c}{b \cdot d}.$$

Here is an example that illustrates the rule in the previous box:

$$\frac{3}{8}\cdot\frac{5}{11}=\frac{3\cdot 5}{8\cdot 11}=\frac{15}{88}.$$

The product of $\frac{3}{8}$ and $\frac{5}{11}$ is $\frac{15}{88}$.

Multiply numerators and multiply denominators.

EXAMPLE 5 Multiplying Fractions

Multiply. If possible, reduce the product to its lowest terms:

a. $\frac{3}{7}\cdot\frac{2}{5}$ **b.** $5\cdot\frac{7}{12}$ **c.** $\frac{2}{3}\cdot\frac{9}{4}$ **d.** $\left(3\frac{2}{3}\right)\left(1\frac{1}{4}\right)$.

SOLUTION

a. $\frac{3}{7}\cdot\frac{2}{5}=\frac{3\cdot 2}{7\cdot 5}=\frac{6}{35}$ Multiply numerators and multiply denominators.

b. $5\cdot\frac{7}{12}=\frac{5}{1}\cdot\frac{7}{12}=\frac{5\cdot 7}{1\cdot 12}=\frac{35}{12}$ or $2\frac{11}{12}$ Write 5 as $\frac{5}{1}$. Then multiply numerators and multiply denominators.

c. $\frac{2}{3}\cdot\frac{9}{4}=\frac{2\cdot 9}{3\cdot 4}=\frac{18}{12}=\frac{3\cdot \cancel{6}}{2\cdot \cancel{6}}=\frac{3}{2}$ or $1\frac{1}{2}$

Simplify $\frac{18}{12}$; 6 is the greatest common factor of 18 and 12.

d. $\left(3\frac{2}{3}\right)\left(1\frac{1}{4}\right)=\frac{11}{3}\cdot\frac{5}{4}=\frac{11\cdot 5}{3\cdot 4}=\frac{55}{12}$ or $4\frac{7}{12}$ ■

CHECK POINT 5 Multiply. If possible, reduce the product to its lowest terms:

a. $\frac{4}{11}\cdot\frac{2}{3}$ **b.** $6\cdot\frac{3}{5}$

c. $\frac{3}{7}\cdot\frac{2}{3}$ **d.** $\left(3\frac{2}{5}\right)\left(1\frac{1}{2}\right)$.

STUDY TIP

You can divide numerators and denominators by common factors before performing multiplication. Then multiply the remaining factors in the numerators and multiply the remaining factors in the denominators. For example,

$$\frac{7}{15}\cdot\frac{20}{21}=\frac{\cancel{7}\cdot 1}{\cancel{5}\cdot 3}\cdot\frac{\cancel{5}\cdot 4}{\cancel{7}\cdot 3}=\frac{1\cdot 4}{3\cdot 3}=\frac{4}{9}.$$

7 is the greatest common factor of 7 and 21.

5 is the greatest common factor of 15 and 20.

Study Tip (continued)

The divisions involving the common factors, 7 and 5, are often shown as follows:

$$\frac{7}{15}\cdot\frac{20}{21}=\frac{\overset{1}{\cancel{7}}}{\underset{3}{\cancel{15}}}\cdot\frac{\overset{4}{\cancel{20}}}{\underset{3}{\cancel{21}}}=\frac{1\cdot 4}{3\cdot 3}=\frac{4}{9}.$$

Divide by 7. Divide by 5.

---

5 Divide fractions.

Dividing Fractions The result of dividing two fractions is called their **quotient**. A geometric figure is useful for developing a process for determining the quotient of two fractions.

Consider the division

$$\frac{4}{5}\div\frac{1}{10}.$$

We want to know how many $\frac{1}{10}$'s are in $\frac{4}{5}$. We can use Figure 1.3 to find this quotient. The rectangle is divided into fifths. The dashed lines further divide the rectangle into tenths.

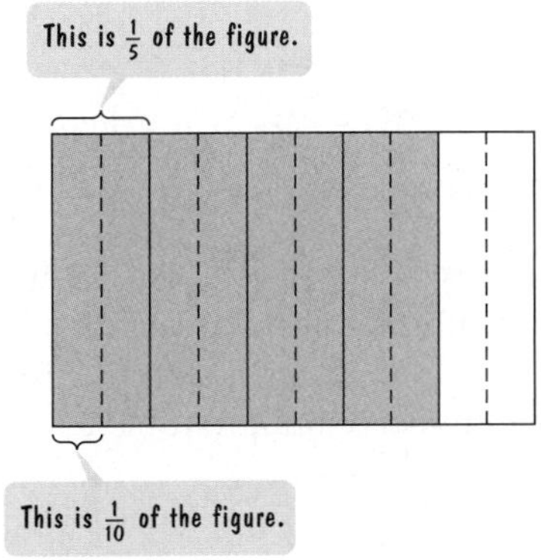
This is $\frac{1}{5}$ of the figure.
This is $\frac{1}{10}$ of the figure.

FIGURE 1.3

Figure 1.3 shows that $\frac{4}{5}$ of the rectangle is red. How many $\frac{1}{10}$'s of the rectangle does this include? Can you see that this includes eight of the $\frac{1}{10}$ pieces? Thus, there are eight $\frac{1}{10}$'s in $\frac{4}{5}$:

$$\frac{4}{5}\div\frac{1}{10}=8.$$

We can obtain the quotient 8 in the following way:

$$\frac{4}{5}\div\frac{1}{10}=\frac{4}{5}\cdot\frac{10}{1}=\frac{4\cdot 10}{5\cdot 1}=\frac{40}{5}=8.$$

Change the division to multiplication. Invert the divisor, $\frac{1}{10}$.

By inverting the divisor, $\frac{1}{10}$, and obtaining $\frac{10}{1}$, we are writing the divisor's *reciprocal*. Two fractions are **reciprocals** of each other if their product is 1. Thus, $\frac{1}{10}$ and $\frac{10}{1}$ are reciprocals because $\frac{1}{10}\cdot\frac{10}{1}=1$.

Generalizing from the result shown above and using the word *reciprocal*, we obtain the following rule for dividing fractions:

DIVIDING FRACTIONS In words: The quotient of two fractions is the first fraction multiplied by the reciprocal of the second fraction.

In algebraic language: If $\frac{a}{b}$ and $\frac{c}{d}$ are fractions and $\frac{c}{d}$ is not 0, then

$$\frac{a}{b}\div\frac{c}{d}=\frac{a}{b}\cdot\frac{d}{c}.$$

Change division to multiplication. Invert $\frac{c}{d}$ and write its reciprocal.

EXAMPLE 6 Dividing Fractions

Divide: **a.** $\frac{2}{3} \div \frac{7}{15}$ **b.** $\frac{3}{4} \div 5$ **c.** $4\frac{3}{4} \div 1\frac{1}{2}$.

SOLUTION

a. $\frac{2}{3} \div \frac{7}{15} = \frac{2}{3} \cdot \frac{15}{7} = \frac{2 \cdot 15}{3 \cdot 7} = \frac{30}{21} = \frac{10 \cdot \cancel{3}}{7 \cdot \cancel{3}} = \frac{10}{7} \text{ or } 1\frac{3}{7}$

Change division to multiplication. Invert $\frac{7}{15}$ and write its reciprocal. Simplify: 3 is the greatest common factor of 30 and 21.

b. $\frac{3}{4} \div 5 = \frac{3}{4} \div \frac{5}{1} = \frac{3}{4} \cdot \frac{1}{5} = \frac{3 \cdot 1}{4 \cdot 5} = \frac{3}{20}$

Change division to multiplication. Invert $\frac{5}{1}$ and write its reciprocal.

c. $4\frac{3}{4} \div 1\frac{1}{2} = \frac{19}{4} \div \frac{3}{2} = \frac{19}{4} \cdot \frac{2}{3} = \frac{19 \cdot 2}{4 \cdot 3} = \frac{38}{12} = \frac{19 \cdot \cancel{2}}{6 \cdot \cancel{2}} = \frac{19}{6} \text{ or } 3\frac{1}{6}$ ■

CHECK POINT 6 Divide:

a. $\frac{5}{4} \div \frac{3}{8}$ **b.** $\frac{2}{3} \div 3$ **c.** $3\frac{3}{8} \div 2\frac{1}{4}$

6 Add and subtract fractions with identical denominators.

Adding and Subtracting Fractions with Identical Denominators The result of adding two fractions is called their **sum**. The result of subtracting two fractions is called their **difference**. A geometric figure is useful for developing a process for determining the sum or difference of two fractions with identical denominators.

Consider the addition

$$\frac{3}{7} + \frac{2}{7}.$$

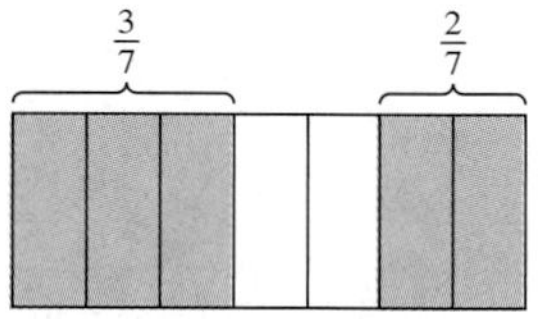

FIGURE 1.4

We can use Figure 1.4 to find this sum. The rectangle is divided into sevenths. On the left, $\frac{3}{7}$ of the rectangle is red. On the right, $\frac{2}{7}$ of the rectangle is red. Including both the left and the right, a total of $\frac{5}{7}$ of the rectangle is red. Thus,

$$\frac{3}{7} + \frac{2}{7} = \frac{5}{7}.$$

We can obtain the sum $\frac{5}{7}$ in the following way:

$$\frac{3}{7} + \frac{2}{7} = \frac{3 + 2}{7} = \frac{5}{7}.$$

Add numerators and put this result over the common denominator.

Generalizing from this result gives us the following rule:

ADDING AND SUBTRACTING FRACTIONS WITH IDENTICAL DENOMINATORS

In words: The sum or difference of two fractions with identical denominators is the sum or difference of their numerators over the common denominator.

In algebraic language: If $\frac{a}{b}$ and $\frac{c}{b}$ are fractions, then

$$\frac{a}{b} + \frac{c}{b} = \frac{a + c}{b} \quad \text{and} \quad \frac{a}{b} - \frac{c}{b} = \frac{a - c}{b}.$$

EXAMPLE 7 Adding and Subtracting Fractions with Identical Denominators

Perform the indicated operations:

a. $\frac{3}{11}+\frac{4}{11}$ **b.** $\frac{11}{12}-\frac{5}{12}$ **c.** $5\frac{1}{4}-2\frac{3}{4}$.

SOLUTION

a. $\frac{3}{11}+\frac{4}{11}=\frac{3+4}{11}=\frac{7}{11}$

b. $\frac{11}{12}-\frac{5}{12}=\frac{11-5}{12}=\frac{6}{12}=\frac{1\cdot \cancel{6}}{2\cdot \cancel{6}}=\frac{1}{2}$

c. $5\frac{1}{4}-2\frac{3}{4}=\frac{21}{4}-\frac{11}{4}=\frac{21-11}{4}=\frac{10}{4}=\frac{\cancel{2}\cdot 5}{\cancel{2}\cdot 2}=\frac{5}{2}$ or $2\frac{1}{2}$ ■

CHECK POINT 7 Perform the indicated operations:

a. $\frac{2}{11}+\frac{3}{11}$ **b.** $\frac{5}{6}-\frac{1}{6}$ **c.** $3\frac{3}{8}-1\frac{1}{8}$.

7 Add and subtract fractions with unlike denominators.

Adding and Subtracting Fractions with Unlike Denominators How do we add or subtract fractions with different denominators? We must first rewrite them as equivalent fractions with the same denominator. We do this by using the Fundamental Principle of Fractions: The value of a fraction does not change if the numerator and the denominator are multiplied by the same nonzero number. Thus, if $\frac{a}{b}$ is a fraction and c is a nonzero number, then

$$\frac{a}{b}=\frac{a\cdot c}{b\cdot c}.$$

EXAMPLE 8 Writing an Equivalent Fraction

Write $\frac{3}{4}$ as an equivalent fraction with a denominator of 16.

SOLUTION To obtain a denominator of 16, we must multiply the denominator of the given fraction, $\frac{3}{4}$, by 4. So that we do not change the value of the fraction, we also multiply the numerator by 4.

$$\frac{3}{4}=\frac{3\cdot 4}{4\cdot 4}=\frac{12}{16}$$ ■

CHECK POINT 8 Write $\frac{2}{3}$ as an equivalent fraction with a denominator of 21.

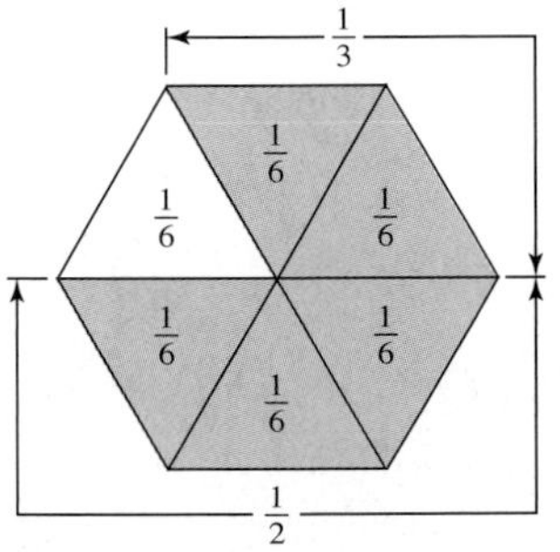

FIGURE 1.5 $\frac{1}{2}+\frac{1}{3}=\frac{5}{6}$

Equivalent fractions can be used to add fractions with different denominators, such as $\frac{1}{2}$ and $\frac{1}{3}$. Figure 1.5 indicates that the sum of half the whole figure and one-third of the whole figure results in 5 parts out of 6, or $\frac{5}{6}$, of the figure. Thus,

$$\frac{1}{2}+\frac{1}{3}=\frac{5}{6}.$$

We can obtain the sum $\frac{5}{6}$ if we rewrite each fraction as an equivalent fraction with a denominator of 6.

$$\frac{1}{2}+\frac{1}{3}=\frac{1\cdot 3}{2\cdot 3}+\frac{1\cdot 2}{3\cdot 2}$$ Rewrite each fraction as an equivalent fraction with a denominator of 6.

$$=\frac{3}{6}+\frac{2}{6}$$ Perform the multiplications. We now have a common denominator.

$$=\frac{3+2}{6}$$ Add the numerators and place this sum over the common denominator.

$$=\frac{5}{6}$$ Perform the addition.

When adding $\frac{1}{2}$ and $\frac{1}{3}$, there are many common denominators that we can use, such as 6, 12, 18, and so on. The given denominators, 2 and 3, divide into all of these numbers. However, the denominator 6 is the smallest number that 2 and 3 divide into. For this reason, 6 is called the *least common denominator*, abbreviated LCD.

ADDING AND SUBTRACTING FRACTIONS WITH UNLIKE DENOMINATORS

1. Rewrite the fractions as equivalent fractions with the least common denominator.
2. Add or subtract the numerators, putting this result over the common denominator.

EXAMPLE 9 Adding and Subtracting Fractions with Unlike Denominators

Perform the indicated operation: **a.** $\frac{1}{5}+\frac{3}{4}$ **b.** $\frac{3}{4}-\frac{1}{6}$ **c.** $2\frac{7}{15}-1\frac{4}{5}$.

SOLUTION

a. Just by looking, can you can tell that the smallest number divisible by both 5 and 4 is 20? Thus, the least common denominator for the denominators 5 and 4 is 20. We rewrite both fractions as equivalent fractions with the least common denominator, 20.

$$\frac{1}{5}+\frac{3}{4}=\frac{1\cdot 4}{5\cdot 4}+\frac{3\cdot 5}{4\cdot 5}$$ To obtain denominators of 20, multiply the numerator and denominator of the first fraction by 4 and the second fraction by 5.

$$=\frac{4}{20}+\frac{15}{20}$$ Perform the multiplications.

$$=\frac{4+15}{20}$$ Add the numerators and put this sum over the least common denominator.

$$=\frac{19}{20}$$ Perform the addition.

DISCOVER FOR YOURSELF

Try Example 9(a), $\frac{1}{5}+\frac{3}{4}$, using a common denominator of 40. Because both 5 and 4 divide into 40, 40 is a common denominator, although not the *least* common denominator. Describe what happens. What is the advantage of using the least common denominator?

b. By looking, can you tell that the smallest number divisible by both 4 and 6 is 12? Thus, the least common denominator for the denominators 4 and 6 is 12. We rewrite both fractions as equivalent fractions with the least common denominator, 12.

$$\frac{3}{4}-\frac{1}{6}=\frac{3\cdot 3}{4\cdot 3}-\frac{1\cdot 2}{6\cdot 2}$$ To obtain denominators of 12, multiply the numerator and denominator of the first fraction by 3 and the second fraction by 2.

$$=\frac{9}{12}-\frac{2}{12}$$ Perform the multiplications.

$$=\frac{9-2}{12}$$ Subtract the numerators and put this difference over the least common denominator.

$$=\frac{7}{12}$$ Perform the subtraction.

c. $2\frac{7}{15} - 1\frac{4}{5} = \frac{37}{15} - \frac{9}{5}$ Convert each mixed number to an improper fraction.

The smallest number divisible by both 15 and 5 is 15. Thus, the least common denominator for the denominators 15 and 5 is 15. Because the first fraction already has a denominator of 15, we only have to rewrite the second fraction.

$= \frac{37}{15} - \frac{9 \cdot 3}{5 \cdot 3}$ To obtain denominators of 15, multiply the numerator and denominator of the second fraction by 3.

$= \frac{37}{15} - \frac{27}{15}$ Perform the multiplications.

$= \frac{37 - 27}{15}$ Subtract the numerators and put this difference over the common denominator.

$= \frac{10}{15}$ Perform the subtraction.

$= \frac{2 \cdot \cancel{5}}{3 \cdot \cancel{5}} = \frac{2}{3}$ Reduce to lowest terms. ■

CHECK POINT 9 Perform the indicated operation:

a. $\frac{1}{2} + \frac{3}{5}$ **b.** $\frac{4}{3} - \frac{3}{4}$ **c.** $3\frac{1}{6} - 1\frac{11}{12}$.

EXAMPLE 10 Using Prime Factorizations to Find the LCD

Perform the indicated operation: $\frac{1}{15} + \frac{7}{24}$.

SOLUTION We need to first find the least common denominator. Using inspection, it is difficult to determine the smallest number divisible by both 15 and 24. We will use their prime factorizations to find the least common denominator:

$$15 = 5 \cdot 3 \quad \text{and} \quad 24 = 8 \cdot 3 = 2 \cdot 2 \cdot 2 \cdot 3.$$

The different prime factors are 5, 3, and 2. The least common denominator is obtained by using the greatest number of times each factor appears in any prime factorization. Because 5 and 3 appear as prime factors and 2 is a factor of 24 three times, the least common denominator is

$$5 \cdot 3 \cdot 2 \cdot 2 \cdot 2 = 5 \cdot 3 \cdot 8 = 120.$$

Now we can rewrite both fractions as equivalent fractions with the least common denominator, 120.

$\frac{1}{15} + \frac{7}{24} = \frac{1 \cdot 8}{15 \cdot 8} + \frac{7 \cdot 5}{24 \cdot 5}$ To obtain denominators of 120, multiply the numerator and denominator of the first fraction by 8 and the second fraction by 5.

$= \frac{8}{120} + \frac{35}{120}$ Perform the multiplications.

$= \frac{8 + 35}{120}$ Add the numerators and put this sum over the least common denominator.

$= \frac{43}{120}$ Perform the addition. ■

CHECK POINT 10 Perform the indicated operation: $\frac{3}{10} + \frac{7}{12}$.

8 Solve problems involving fractions.

Applications You can hardly pick up a newspaper without being bombarded with graphs that describe how we live, what we want, and even how we are likely to die. Understanding fractions and their operations is often helpful in interpreting the information given in these graphs.

Circle graphs, also called **pie charts**, show how a quantity is divided into parts. Circle graphs are divided into pieces, called **sectors**. The sizes of the sectors show the relative sizes of the categories. Figure 1.6 is an example of a typical circle graph. The graph shows the breakdown of the number of countries in the world that are free, partly free, or not free.

World's Countries by Status of Freedom

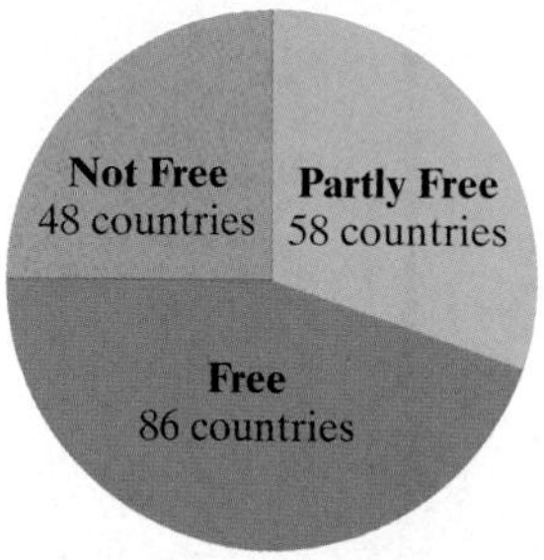

FIGURE 1.6

Source: Larry Berman and Bruce Murphy, *Approaching Democracy*, 4th edition, Prentice Hall, 2003

EXAMPLE 11 Using a Circle Graph to Interpret Information

Use Figure 1.6 to determine the fraction of the world's countries that are not free. Reduce the answer to its lowest terms.

SOLUTION The sector on the upper left shows that there are 48 countries that are not free. By adding the numbers in the three sectors, we see that there are a total of

$$48 + 58 + 86$$

or 192 countries.

Fraction of nonfree countries

$$= \frac{\text{number of countries that are not free}}{\text{total number of countries}}$$

$$= \frac{48}{192} = \frac{2 \cdot 24}{2 \cdot 96} = \frac{2 \cdot 2 \cdot 12}{2 \cdot 2 \cdot 48} = \frac{\cancel{2} \cdot \cancel{2} \cdot \cancel{12} \cdot 1}{\cancel{2} \cdot \cancel{2} \cdot \cancel{12} \cdot 4} = \frac{1}{4}$$

The fraction of the world's countries that are not free is $\frac{1}{4}$. ■

✔ **CHECK POINT 11** Use Figure 1.6 to determine the fraction of the world's countries that are partly free. Reduce the answer to its lowest terms.

The graph in Figure 1.6 can show the fraction of countries, rather than the actual number of countries, in each of the three sectors. If a circle graph shows a fraction within each of its sectors, the sum of the fractional parts must be 1, representing one whole circle.

EXAMPLE 12 Miss America Hair Color

The Miss America title has been awarded, with some breaks, since 1921. The graph in Figure 1.7 might disprove the often-quoted claim that "gentlemen prefer blondes." What fraction of Miss America titleholders have been blondes?

SOLUTION We begin by finding the fraction of Miss Americas who have not been blondes. Add the fractions of brunettes, $\frac{7}{10}$, and redheads, $\frac{3}{50}$.

Miss America Hair Color 1921–2003

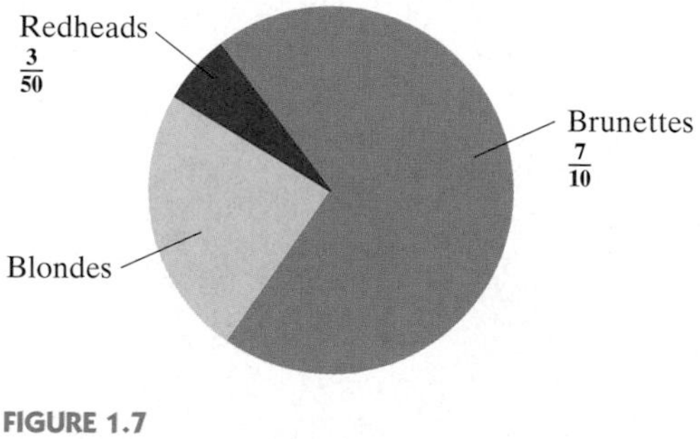

FIGURE 1.7

Source: Ben Schott, *Schott's Miscellany*, Bloomsbury, 2003

$$\frac{7}{10}+\frac{3}{50}$$ The least common denominator is 50.

$$=\frac{7\cdot 5}{10\cdot 5}+\frac{3}{50}$$ To obtain denominators of 50, multiply the numerator and denominator of the first fraction by 5.

$$=\frac{35}{50}+\frac{3}{50}$$ Perform the multiplications.

$$=\frac{35+3}{50}=\frac{38}{50}$$ Add the numerators and put this sum over the least common denominator.

$$=\frac{19\cdot \not{2}}{25\cdot \not{2}}=\frac{19}{25}$$ Reduce to lowest terms.

Thus, $\frac{19}{25}$ of the Miss Americas have not been blondes. The fractions representing all three hair colors in Figure 1.7 must add up to 1. Therefore, the fraction of Miss Americas who have been blondes can be found by subtracting $\frac{19}{25}$ from 1:

$$1-\frac{19}{25}=\frac{25}{25}-\frac{19}{25}=\frac{25-19}{25}=\frac{6}{25}.$$

We see that $\frac{6}{25}$ of Miss America titleholders have been blondes. ■

CHECK POINT 12 At a workshop on enhancing creativity, $\frac{1}{4}$ of the participants are musicians, $\frac{2}{5}$ are artists, $\frac{1}{10}$ are actors, and the remaining participants are writers. Find the fraction of people at the workshop who are writers.

1.1 EXERCISE SET

Student Solutions Manual | CD/Video | PH Math/Tutor Center | MathXL Tutorials on CD | MathXL® | MyMathLab | Interactmath.com

Practice Exercises

In Exercises 1–6, convert each mixed number to an improper fraction.

1. $2\frac{3}{8}$ **2.** $2\frac{7}{9}$ **3.** $7\frac{3}{5}$

4. $6\frac{2}{5}$ **5.** $8\frac{7}{16}$ **6.** $9\frac{5}{16}$

In Exercises 7–12, convert each improper fraction to a mixed number.

7. $\frac{23}{5}$ **8.** $\frac{47}{8}$ **9.** $\frac{76}{9}$

10. $\frac{59}{9}$ **11.** $\frac{711}{20}$ **12.** $\frac{788}{25}$

In Exercises 13–28, identify each natural number as prime or composite. If the number is composite, find its prime factorization.

13. 22 **14.** 15 **15.** 20

16. 75 **17.** 37 **18.** 23

19. 36 **20.** 100 **21.** 140

22. 110 **23.** 79 **24.** 83

25. 81 **26.** 64

27. 240 **28.** 360

In Exercises 29–40, simplify each fraction by reducing it to its lowest terms.

29. $\frac{10}{16}$ **30.** $\frac{8}{14}$ **31.** $\frac{15}{18}$ **32.** $\frac{18}{45}$

33. $\frac{35}{50}$ **34.** $\frac{45}{50}$ **35.** $\frac{32}{80}$ **36.** $\frac{75}{80}$

37. $\frac{44}{50}$ **38.** $\frac{38}{50}$ **39.** $\frac{120}{86}$ **40.** $\frac{116}{86}$

In Exercises 41–90, perform the indicated operation. Where possible, reduce the answer to its lowest terms.

41. $\frac{2}{5}\cdot\frac{1}{3}$ **42.** $\frac{3}{7}\cdot\frac{1}{4}$ **43.** $\frac{3}{8}\cdot\frac{7}{11}$

44. $\frac{5}{8}\cdot\frac{3}{11}$ **45.** $9\cdot\frac{4}{7}$ **46.** $8\cdot\frac{3}{7}$

47. $\frac{1}{10}\cdot\frac{5}{6}$ **48.** $\frac{1}{8}\cdot\frac{2}{3}$ **49.** $\frac{5}{4}\cdot\frac{6}{7}$

50. $\frac{7}{4}\cdot\frac{6}{11}$ **51.** $\left(3\frac{3}{4}\right)\left(1\frac{3}{5}\right)$

52. $\left(2\frac{4}{5}\right)\left(1\frac{1}{4}\right)$ **53.** $\frac{5}{4}\div\frac{4}{3}$

54. $\frac{7}{8}\div\frac{2}{3}$ **55.** $\frac{18}{5}\div 2$ **56.** $\frac{12}{7}\div 3$

57. $2\div\frac{18}{5}$ **58.** $3\div\frac{12}{7}$ **59.** $\frac{3}{4}\div\frac{1}{4}$

60. $\frac{3}{7}\div\frac{1}{7}$ **61.** $\frac{7}{6}\div\frac{5}{3}$ **62.** $\frac{7}{4}\div\frac{3}{8}$

63. $\frac{1}{14}\div\frac{1}{7}$ **64.** $\frac{1}{8}\div\frac{1}{4}$ **65.** $6\frac{3}{5}\div 1\frac{1}{10}$

66. $1\frac{3}{4}\div 2\frac{5}{8}$ **67.** $\frac{2}{11}+\frac{4}{11}$ **68.** $\frac{5}{13}+\frac{2}{13}$

69. $\frac{7}{12}+\frac{1}{12}$
70. $\frac{5}{16}+\frac{1}{16}$
71. $\frac{5}{8}+\frac{5}{8}$
72. $\frac{3}{8}+\frac{3}{8}$
73. $\frac{7}{12}-\frac{5}{12}$
74. $\frac{13}{18}-\frac{5}{18}$
75. $\frac{16}{7}-\frac{2}{7}$
76. $\frac{17}{5}-\frac{2}{5}$
77. $\frac{1}{2}+\frac{1}{5}$
78. $\frac{1}{3}+\frac{1}{5}$
79. $\frac{3}{4}+\frac{3}{20}$
80. $\frac{2}{5}+\frac{2}{15}$
81. $\frac{3}{8}+\frac{5}{12}$
82. $\frac{3}{10}+\frac{2}{15}$
83. $\frac{11}{18}-\frac{2}{9}$
84. $\frac{17}{18}-\frac{4}{9}$
85. $\frac{4}{3}-\frac{3}{4}$
86. $\frac{3}{2}-\frac{2}{3}$
87. $\frac{7}{10}-\frac{3}{16}$
88. $\frac{7}{30}-\frac{5}{24}$
89. $3\frac{3}{4}-2\frac{1}{3}$
90. $3\frac{2}{3}-2\frac{1}{2}$

Practice Plus

In Exercises 91–94, perform the indicated operation. Write the answer as a variable expression.

91. $\frac{3}{4}\cdot\frac{a}{5}$
92. $\frac{2}{3}\div\frac{a}{7}$
93. $\frac{11}{x}+\frac{9}{x}$
94. $\frac{10}{y}-\frac{6}{y}$

In Exercises 95–96, perform the indicated operations. Begin by performing operations in parentheses.

95. $\left(\frac{1}{2}-\frac{1}{3}\right)\div\frac{5}{8}$
96. $\left(\frac{1}{2}+\frac{1}{4}\right)\div\left(\frac{1}{2}+\frac{1}{3}\right)$

Application Exercises

97. The circle graph shows the results of a survey in which American adults were asked the following question:

If you got conflicting or different reports of the same news story from television, newspapers, radio, or magazines, which would you be most inclined to believe?

The circle graph shows the fraction of American adults finding each medium the most believable.

Most Believable News Source

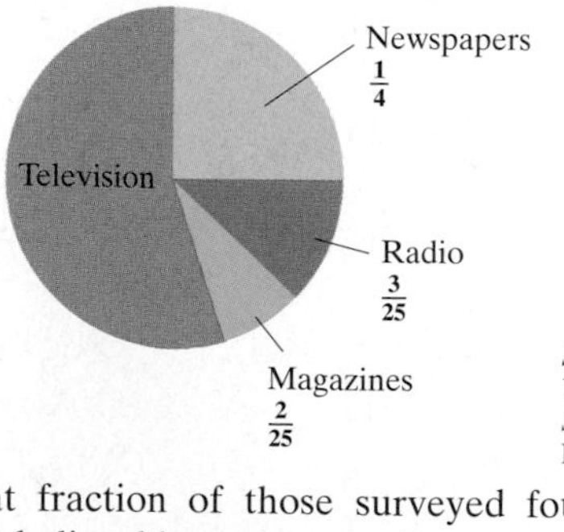

Source: Thomas R. Dye, *Who's Running America? Seventh Edition*, Prentice Hall, 2002

a. What fraction of those surveyed found television the most believable news source?

b. If 2000 people were surveyed, how many found radio the most believable news source?

98. The circle graph shows the results of one of the questions in a telephone poll taken for *Time*/CNN during 2002's summer of corporate scandals.

Which of the following best describes your view of the recent accounting scandals involving major American corporations?

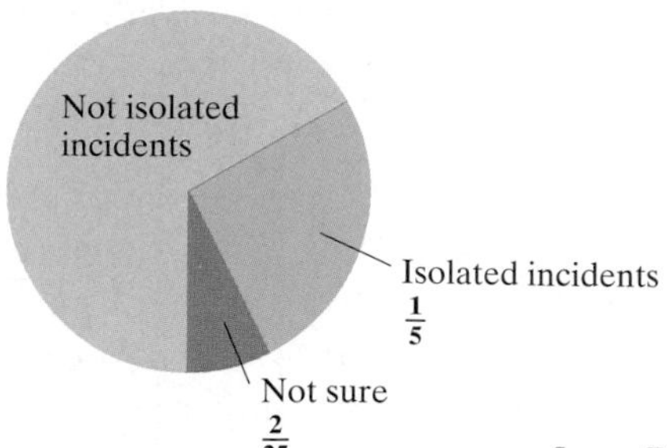

Source: *Time*/CNN

a. What fraction of those surveyed replied that the accounting scandals were not isolated incidents?

b. If 2000 people were surveyed, how many were not sure if the accounting scandals were isolated incidents?

The ratio of a to b can be expressed by the fraction $\frac{a}{b}$. In Exercises 99–100, use the natural numbers shown in Figure 1.1 on page 2 to find the indicated ratio. Reduce each fraction to its lowest terms.

99. the ratio of Hispanic judges to female judges appointed by Carter

100. the ratio of African-American judges to female judges appointed by Reagan

A common application of fractions involves preparing food for a different number of servings than what the recipe calls for. The amount of each ingredient can be found as follows:

Amount of ingredient needed

$$=\frac{\textit{desired serving size}}{\textit{recipe serving size}}\times\textit{ingredient amount in the recipe.}$$

Use this information to solve Exercises 101–102. Give answers as mixed numbers.

101. A chocolate-chip cookie recipe for five dozen cookies requires $\frac{3}{4}$ cup sugar. If you want to make eight dozen cookies, how much sugar is needed?

102. A mix for eight servings of instant potatoes requires $2\frac{2}{3}$ cups of water. If you want to make six servings, how much water is needed?

103. If you walk $\frac{3}{4}$ mile and then jog $\frac{2}{5}$ mile, what is the total distance covered? How much farther did you walk than jog?

104. Some companies pay people extra when they work more than a regular 40-hour work week. The overtime pay is often $1\frac{1}{2}$ times the regular hourly rate. This is called *time and a half.* A summer job for students pays \$12 per hour and offers time and a half for the hours worked over 40. If a student works 46 hours during one week, what is the student's total pay before taxes?

105. The legend of a map indicates that 1 inch = 16 miles. If the distance on the map between two cities is $2\frac{3}{8}$ inches, how far apart are the cities?

Writing in Mathematics

Writing about mathematics will help you to learn mathematics. For all writing exercises in this book, use complete sentences to respond to the questions. Some writing exercises can be answered in a sentence; others require a paragraph or two. You can decide how much you need to write as long as your writing clearly and directly answers the question in the exercise. Standard references such as a dictionary and a thesaurus should be helpful.

106. Explain how to convert a mixed number to an improper fraction and give an example.

107. Explain how to convert an improper fraction to a mixed number and give an example.

108. Describe the difference between a prime number and a composite number.

109. What is meant by the prime factorization of a composite number?

110. What is the Fundamental Principle of Fractions?

111. Explain how to reduce a fraction to its lowest terms. Give an example with your explanation.

112. Explain how to multiply fractions and give an example.

113. Explain how to divide fractions and give an example.

114. Describe how to add or subtract fractions with identical denominators. Provide an example with your description.

115. Explain how to add fractions with different denominators. Use $\frac{5}{6} + \frac{1}{2}$ as an example.

116. Explain what is wrong with this statement. "If you'd like to save some money, I'll be happy to sell you my computer system for only $\frac{3}{2}$ of the price I originally paid for it."

Critical Thinking Exercises

117. Which one of the following is true?

a. $\frac{1}{2} + \frac{1}{5} = \frac{2}{7}$ **b.** $\frac{2+6}{2} = \frac{\cancel{2}+6}{\cancel{2}} = 6$ **c.** $\frac{1}{2} \div 4 = 2$

d. Every fraction has infinitely many equivalent fractions.

118. Shown below is a short excerpt from "The Star-Spangled Banner." The time is $\frac{3}{4}$, which means that each measure must contain notes that add up to $\frac{3}{4}$. The values of the different notes tell musicians how long to hold each note.

𝅝 $= 1$ 𝅗𝅥 $= \frac{1}{2}$ ♩ $= \frac{1}{4}$ ♪ $= \frac{1}{8}$

Use vertical lines to divide this line of "The Star-Spangled Banner" into measures.

SECTION 1.2

Objectives

1. Define the sets that make up the real numbers.
2. Graph numbers on a number line.
3. Express rational numbers as decimals.
4. Classify numbers as belonging to one or more sets of the real numbers.
5. Understand and use inequality symbols.
6. Find the absolute value of a real number.

THE REAL NUMBERS

The U.N. building is designed with three golden rectangles.

The United Nations Building in New York was designed to represent its mission of promoting world harmony. Viewed from the front, the building looks like three rectangles stacked upon each other. In each rectangle, the ratio of the width to height is $\sqrt{5} + 1$ to 2, approximately 1.618 to 1. The ancient Greeks believed that such a rectangle, called a **golden rectangle**, was the most visually pleasing of all rectangles.

The ratio 1.618 to 1 is approximate because $\sqrt{5}$ is an irrational number, a special kind of real number. Irrational? Real? Let's make sense of all this by describing the kinds of numbers you will encounter in this course.

1 Define the sets that make up the real numbers.

Natural Numbers and Whole Numbers Before we describe the set of real numbers, let's be sure you are familiar with some basic ideas about sets. A **set** is a collection of objects whose contents can be clearly determined. The objects in a set are called the **elements** of the set. For example, the set of numbers used for counting can be represented by

$$\{1, 2, 3, 4, 5, \dots\}.$$

The braces, $\{\}$, indicate that we are representing a set. This form of representing a set uses commas to separate the elements of the set. Remember that the three dots after the 5 indicate that there is no final element and that the listing goes on forever.

We have seen that the set of numbers used for counting is called the set of **natural numbers**. When we combine the number 0 with the natural numbers, we obtain the set of **whole numbers**.

NATURAL NUMBERS AND WHOLE NUMBERS

The set of **natural numbers** is $\{1, 2, 3, 4, 5, \dots\}$.

The set of **whole numbers** is $\{0, 1, 2, 3, 4, 5, \dots\}$.

Integers and the Number Line The whole numbers do not allow us to describe certain everyday situations. For example, if the balance in your checking account is \$30 and you write a check for \$35, your checking account is overdrawn by \$5. We can write this as -5, read *negative* 5. The set consisting of the natural numbers, 0, and the negatives of the natural numbers is called the set of **integers**.

INTEGERS The set of **integers** is

$$\{\underbrace{\dots, -4, -3, -2, -1}_{\text{Negative integers}}, 0, \underbrace{1, 2, 3, 4, \dots}_{\text{Positive integers}}\}.$$

Notice that the term **positive integers** is another name for the natural numbers. The positive integers can be written in two ways:

1. Use a "+" sign. For example, +4 is "positive four."
2. Do not write any sign. For example, 4 is assumed to be "positive four."

EXAMPLE 1 Practical Examples of Negative Integers

Write a negative integer that describes each of the following situations:

a. A debt of \$10

b. The shore surrounding the Dead Sea is 1312 feet below sea level.

SOLUTION

a. A debt of \$10 can be expressed by the negative integer -10 (negative ten).

b. The shore surrounding the Dead Sea is 1312 feet below sea level, expressed as -1312. ■

CHECK POINT 1 Write a negative integer that describes each of the following situations:

a. A debt of \$500

b. Death Valley, the lowest point in North America, is 282 feet below sea level.

2 Graph numbers on a number line.

The **number line** is a graph we use to visualize the set of integers, as well as sets of other numbers. The number line is shown in Figure 1.8.

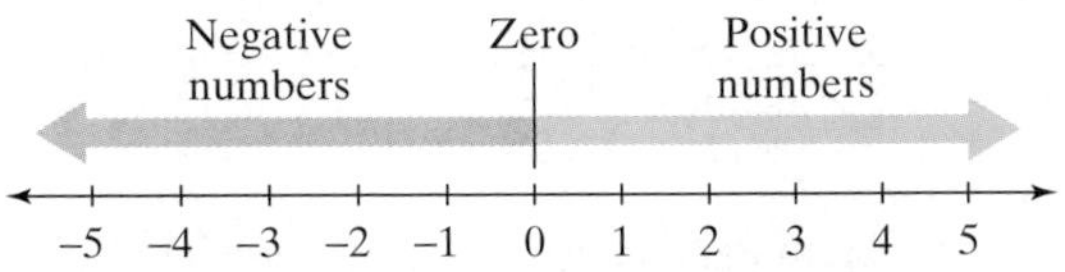

FIGURE 1.8 The number line

The number line extends indefinitely in both directions, shown by the arrows on the left and the right. Zero separates the positive numbers from the negative numbers on the number line. The positive integers are located to the right of 0, and the negative integers are located to the left of 0. Zero is neither positive nor negative. For every positive integer on a number line, there is a corresponding negative integer on the opposite side of 0.

Integers are graphed on a number line by placing a dot at the correct location for each number.

EXAMPLE 2 Graphing Integers on a Number Line

Graph: **a.** -3 **b.** 4 **c.** 0.

SOLUTION Place a dot at the correct location for each integer.

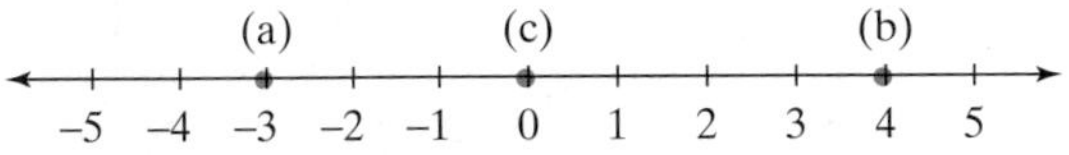

■

✔ **CHECK POINT 2** Graph: **a.** -4 **b.** 0 **c.** 3.

Rational Numbers If two integers are added, subtracted, or multiplied, the result is always another integer. This, however, is not always the case with division. For example, 10 divided by 5 is the integer 2. By contrast, 5 divided by 10 is $\frac{1}{2}$, and $\frac{1}{2}$ is not an integer. To permit divisions such as $\frac{5}{10}$, we enlarge the set of integers, calling the new collection the *rational numbers*. The set of **rational numbers** consists of all the numbers that can be expressed as a quotient of two integers, with the denominator not 0.

THE RATIONAL NUMBERS The set of **rational numbers** is the set of all numbers that can be expressed in the form $\frac{a}{b}$, where a and b are integers and b is not equal to 0, written $b \neq 0$. The integer a is called the **numerator** and the integer b is called the **denominator**.

Here are two examples of rational numbers:

STUDY TIP

In Section 1.7, you will learn that a negative number divided by a positive number gives a negative result. Thus, $\frac{-3}{4}$ can also be written as $-\frac{3}{4}$.

- $\frac{1}{2}$
  - a, the integer forming the numerator, is 1.
  - b, the integer forming the denominator, is 2.
- $\frac{-3}{4}$
  - a, the integer forming the numerator, is −3.
  - b, the integer forming the denominator, is 4.

Is the integer 5 another example of a rational number? Yes. The integer 5 can be written with a denominator of 1.

$$5 = \frac{5}{1}$$

- a, the integer forming the numerator, is 5.
- b, the integer forming the denominator, is 1.

All integers are also rational numbers because they can be written with a denominator of 1.

How can we express a negative mixed number, such as $-2\frac{3}{4}$, in the form $\frac{a}{b}$? Copy the negative sign and then follow the procedure discussed in the previous section:

$$-2\frac{3}{4} = -\frac{4\cdot 2+3}{4} = -\frac{8+3}{4} = -\frac{11}{4} = \frac{-11}{4}.$$

$a = -11$

$b = 4$

Copy the negative sign from step to step and convert $2\frac{3}{4}$ to an improper fraction.

Rational numbers are graphed on a number line by placing a dot at the correct location for each number.

EXAMPLE 3 Graphing Rational Numbers on a Number Line

Graph: **a.** $\frac{7}{2}$ **b.** -4.6.

SOLUTION Place a dot at the correct location for each rational number.

a. Because $\frac{7}{2} = 3\frac{1}{2}$, its graph is midway between 3 and 4.

b. Because $-4.6 = -4\frac{6}{10}$, its graph is $\frac{6}{10}$ of a unit to the left of -4.

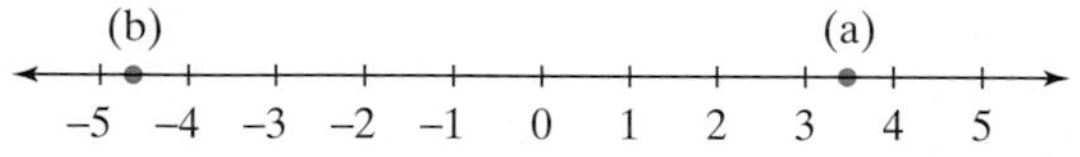

CHECK POINT 3 Graph: **a.** $\frac{9}{2}$ **b.** -1.2.

3 Express rational numbers as decimals.

Every rational number can be expressed as a fraction or a decimal. To express the fraction $\frac{a}{b}$ as a decimal, divide the denominator, b, into the numerator, a.

USING TECHNOLOGY

Calculators and Decimals

Given a rational number, $\frac{a}{b}$, you can express it as a decimal by entering

a [÷] b

into a calculator. In the case of a repeating decimal, the calculator rounds off in the final decimal place displayed.

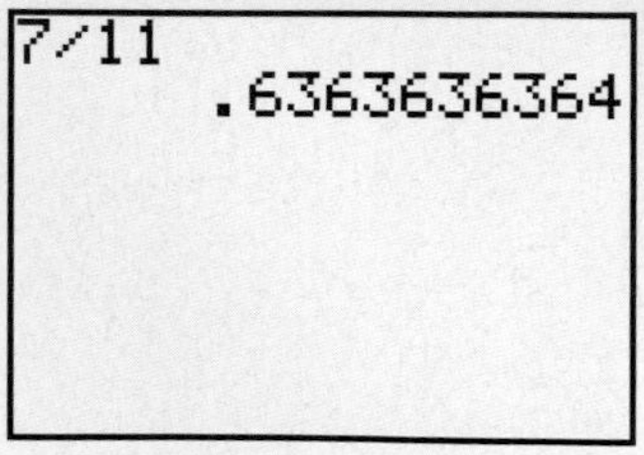

This graphing calculator screen for $\frac{7}{11}$ shows

.6363636364

rather than

.6363636363 ....

EXAMPLE 4 Expressing Rational Numbers as Decimals

Express each rational number as a decimal: **a.** $\frac{5}{8}$ **b.** $\frac{7}{11}$.

SOLUTION In each case, divide the denominator into the numerator.

a.
```
   0.625
8)5.000
  4 8
    20
    16
     40
     40
      0
```

b.
```
     0.6363...
11)7.0000...
   6 6
     40
     33
      70
      66
       40
       33
        70
        ⋮
```

In Example 4, the decimal for $\frac{5}{8}$, namely 0.625, stops and is called a **terminating decimal**. Other examples of terminating decimals are

$$\frac{1}{4} = 0.25, \quad \frac{2}{5} = 0.4, \quad \text{and} \quad \frac{7}{8} = 0.875.$$

By contrast, the division process for $\frac{7}{11}$ results in $0.6363\ldots$, with the digits 63 repeating over and over indefinitely. To indicate this, write a bar over the digits that repeat. Thus,

$$\frac{7}{11} = 0.\overline{63}.$$

The decimal for $\frac{7}{11}$, $0.\overline{63}$, is called a **repeating decimal**. Other examples of repeating decimals are

$$\frac{1}{3} = 0.333\ldots = 0.\overline{3} \quad \text{and} \quad \frac{2}{3} = 0.666\ldots = 0.\overline{6}.$$

RATIONAL NUMBERS AND DECIMALS Any rational number can be expressed as a decimal. The resulting decimal will either terminate (stop), or it will have a digit that repeats or a block of digits that repeat.

USING TECHNOLOGY

You can obtain decimal approximations for irrational numbers using a calculator. For example, to approximate $\sqrt{2}$, use the following keystrokes:

Many Scientific Calculators

2 [√] or 2 [2nd / INV] [x^2]

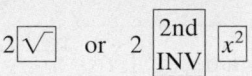

Many Graphing Calculators

[√] 2 [ENTER]

The display may read 1.41421356237, although your calculator may display more or fewer digits. Between which two integers would you graph $\sqrt{2}$ on a number line?

CHECK POINT 4 Express each rational number as a decimal:

a. $\frac{3}{8}$ **b.** $\frac{5}{11}$.

Irrational Numbers Can you think of a number that, when written in decimal form, neither terminates nor repeats? An example of such a number is $\sqrt{2}$ (read: "the square root of 2"). The number $\sqrt{2}$ is a number that can be multiplied by itself to obtain 2. No terminating or repeating decimal can be multiplied by itself to get 2. However, some approximations come close to 2.

- 1.4 is an approximation of $\sqrt{2}$:
$$1.4 \times 1.4 = 1.96.$$
- 1.41 is an approximation of $\sqrt{2}$:
$$1.41 \times 1.41 = 1.9881.$$
- 1.4142 is an approximation of $\sqrt{2}$:
$$1.4142 \times 1.4142 = 1.99996164.$$

Can you see how each approximation in the list is getting better? This is because the products are getting closer and closer to 2.

The number $\sqrt{2}$, whose decimal representation does not come to an end and does not have a block of repeating digits, is an example of an **irrational number**.

THE IRRATIONAL NUMBERS Any number that can be represented on the number line that is not a rational number is called an **irrational number**. Thus, the set of irrational numbers is the set of numbers whose decimal representations are neither terminating nor repeating.

ENRICHMENT ESSAY

The Best and Worst of π

In 1999, two Japanese mathematicians used two different computer programs to calculate π to over 206 billion digits. The calculations took the computer 43 hours!

The most inaccurate version of π came from the 1897 General Assembly of Indiana. Bill No. 246 stated that "π was by law 4."

Perhaps the best known of all the irrational numbers is π (pi). This irrational number represents the distance around a circle (its circumference) divided by the diameter of the circle. In the *Star Trek* episode "Wolf in the Fold," Spock foils an evil computer by telling it to "compute the last digit in the value of π." Because π is an irrational number, there is no last digit in its decimal representation:

$$\pi = 3.14159265358979323846264338327950\ldots.$$

Because irrational numbers cannot be represented by decimals that come to an end, mathematicians use symbols such as $\sqrt{2}$, $\sqrt{3}$, and π to represent these numbers. However, **not all square roots are irrational**. For example, $\sqrt{25} = 5$ because 5 multiplied by itself is 25. Thus, $\sqrt{25}$ is a natural number, a whole number, an integer, and a rational number $\left(\sqrt{25} = \frac{5}{1}\right)$.

4 Classify numbers as belonging to one or more sets of the real numbers.

The Set of Real Numbers All numbers that can be represented by points on the number line are called **real numbers**. Thus, the set of real numbers is formed by combining the rational numbers and the irrational numbers. Every real number is either rational or irrational.

The sets that make up the real numbers are summarized in Table 1.1. Notice the use of the symbol $\approx$ in the examples of irrational numbers. The symbol $\approx$ means "is approximately equal to."

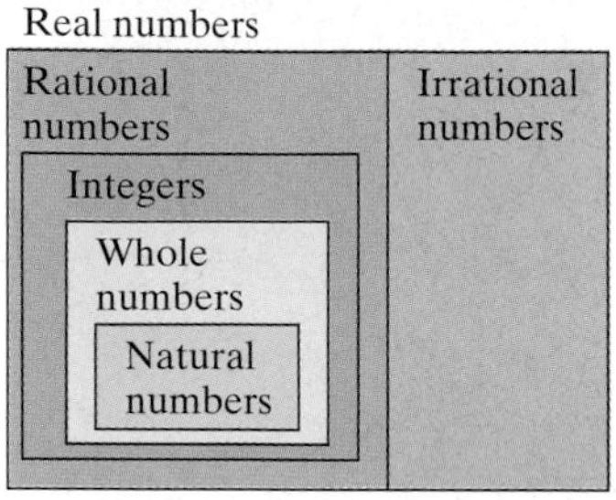

This diagram shows that every real number is rational or irrational.

Table 1.1 The Sets that Make Up the Real Numbers

Name	Description	Examples
Natural numbers	$\{1, 2, 3, 4, 5, \ldots\}$ These numbers are used for counting.	$2, 3, 5, 17$
Whole numbers	$\{0, 1, 2, 3, 4, 5, \ldots\}$ The set of whole numbers is formed by adding 0 to the set of natural numbers.	$0, 2, 3, 5, 17$
Integers	$\{\ldots, -5, -4, -3, -2, -1, 0, 1, 2, 3, 4, 5, \ldots\}$ The set of integers is formed by adding negatives of the natural numbers to the set of whole numbers.	$-17, -5, -3, -2, 0,$ $2, 3, 5, 17$
Rational numbers	The set of rational numbers is the set of all numbers that can be expressed in the form $\frac{a}{b}$, where a and b are integers and b is not equal to 0, written $b \neq 0$. Rational numbers can be expressed as terminating or repeating decimals.	$-17 = \frac{-17}{1}, -5 = \frac{-5}{1}, -3, -2,$ $0, 2, 3, 5, 17,$ $\frac{2}{5} = 0.4,$ $\frac{-2}{3} = -0.6666\cdots = -0.\overline{6}$
Irrational numbers	The set of irrational numbers is the set of all numbers whose decimal representations are neither terminating nor repeating. Irrational numbers cannot be expressed as a quotient of integers.	$\sqrt{2} \approx 1.414214$ $-\sqrt{3} \approx -1.73205$ $\pi \approx 3.142$ $-\frac{\pi}{2} \approx -1.571$

EXAMPLE 5 Classifying Real Numbers

Consider the following set of numbers:

$$\left\{-7, -\frac{3}{4}, 0, 0.\overline{6}, \sqrt{5}, \pi, 7.3, \sqrt{81}\right\}.$$

List the numbers in the set that are

a. natural numbers. **b.** whole numbers. **c.** integers.
d. rational numbers. **e.** irrational numbers. **f.** real numbers.

SOLUTION

a. Natural numbers: The natural numbers are the numbers used for counting. The only natural number in the set is $\sqrt{81}$ because $\sqrt{81} = 9$. (9 multiplied by itself is 81.)

b. Whole numbers: The whole numbers consist of the natural numbers and 0. The elements of the set that are whole numbers are 0 and $\sqrt{81}$.

c. Integers: The integers consist of the natural numbers, 0, and the negatives of the natural numbers. The elements of the set that are integers are $\sqrt{81}$, 0, and -7.

d. Rational numbers: All numbers in the set that can be expressed as the quotient of integers are rational numbers. These include $-7\left(-7 = \frac{-7}{1}\right)$, $-\frac{3}{4}$, $0\left(0 = \frac{0}{1}\right)$, and $\sqrt{81}\left(\sqrt{81} = \frac{9}{1}\right)$. Furthermore, all numbers in the set that are terminating or repeating decimals are also rational numbers. These include $0.\overline{6}$ and 7.3.

e. Irrational numbers: The irrational numbers in the set are $\sqrt{5}(\sqrt{5} \approx 2.236)$ and $\pi(\pi \approx 3.14)$. Both $\sqrt{5}$ and π are only approximately equal to 2.236 and 3.14, respectively. In decimal form, $\sqrt{5}$ and π neither terminate nor have blocks of repeating digits.

f. Real numbers: All the numbers in the given set are real numbers. ■

CHECK POINT 5 Consider the following set of numbers:

$$\left\{-9, -1.3, 0, 0.\overline{3}, \frac{\pi}{2}, \sqrt{9}, \sqrt{10}\right\}.$$

List the numbers in the set that are

a. natural numbers. **b.** whole numbers.

c. integers. **d.** rational numbers.

e. irrational numbers. **f.** real numbers.

5 Understand and use inequality symbols.

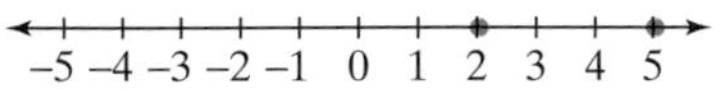

FIGURE 1.9

Ordering the Real Numbers On the real number line, the real numbers increase from left to right. The lesser of two real numbers is the one farther to the left on a number line. The greater of two real numbers is the one farther to the right on a number line.

Look at the number line in Figure 1.9. The integers 2 and 5 are graphed. Observe that 2 is to the left of 5 on the number line. This means that 2 is less than 5:

$2 < 5$: 2 is less than 5 because 2 is to the *left* of 5 on the number line.

In Figure 1.9, we can also observe that 5 is to the right of 2 on the number line. This means that 5 is greater than 2.

$5 > 2$: 5 is greater than 2 because 5 is to the *right* of 2 on the number line.

The symbols $<$ and $>$ are called **inequality symbols.** These symbols always point to the lesser of the two real numbers when the inequality is true.

2 is less than 5. — $2 < 5$ — The symbol points to 2, the lesser number.

5 is greater than 2. — $5 > 2$ — The symbol points to 2, the lesser number.

EXAMPLE 6 Using Inequality Symbols

Insert either $<$ or $>$ in the shaded area between each pair of numbers to make a true statement:

a. 3 ☐ 17 **b.** −4.5 ☐ 1.2 **c.** −5 ☐ −83 **d.** $\frac{4}{5}$ ☐ $\frac{2}{3}$.

SOLUTION In each case, mentally compare the graph of the first number to the graph of the second number. If the first number is to the left of the second number, insert the symbol $<$ for "is less than." If the first number is to the right of the second number, insert the symbol $>$ for "is greater than."

a. Compare the graphs of 3 and 17 on the number line. Because 3 is to the left of 17, this means that 3 is less than 17: $3 < 17$.

b. Compare the graphs of -4.5 and 1.2. Because -4.5 is to the left of 1.2, this means that -4.5 is less than 1.2: $-4.5 < 1.2$.

c. Compare the graphs of -5 and -83. Because -5 is to the right of -83, this means that -5 is greater than -83: $-5 > -83$.

d. Compare the graphs of $\frac{4}{5}$ and $\frac{2}{3}$. To do so, convert to decimal notation or use a common denominator. Using decimal notation, $\frac{4}{5} = 0.8$ and $\frac{2}{3} = 0.\overline{6}$. Because 0.8 is to the right of $0.\overline{6}$, this means that $\frac{4}{5}$ is greater than $\frac{2}{3}$: $\frac{4}{5} > \frac{2}{3}$. ■

✔ CHECK POINT 6 Insert either $<$ or $>$ in the shaded area between each pair of numbers to make a true statement:

a. $14 \quad 5$ **b.** $-5.4 \quad 2.3$ **c.** $-19 \quad -6$ **d.** $\frac{1}{4} \quad \frac{1}{2}$.

The symbols $<$ and $>$ may be combined with an equal sign, as shown in the table.

Symbols	Meaning	Example	Explanation
$a \le b$	a is less than or equal to b.	$3 \le 7$ $7 \le 7$	Because $3 < 7$ Because $7 = 7$
$b \ge a$	b is greater than or equal to a.	$7 \ge 3$ $-5 \ge -5$	Because $7 > 3$ Because $-5 = -5$

When using the symbol $\le$ (is less than or equal to), the inequality is a true statement if either the $<$ part or the $=$ part is true. When using the symbol $\ge$ (is greater than or equal to), the inequality is a true statement if either the $>$ part or the $=$ part is true.

EXAMPLE 7 Using Inequality Symbols

Determine whether each inequality is true or false:

a. $-7 \le 4$ **b.** $-7 \le -7$ **c.** $-9 \ge 6$.

SOLUTION

a. $-7 \le 4$ is true because $-7 < 4$ is true.

b. $-7 \le -7$ is true because $-7 = -7$ is true.

c. $-9 \ge 6$ is false because neither $-9 > 6$ nor $-9 = 6$ is true. ■

✔ CHECK POINT 7 Determine whether each inequality is true or false:

a. $-2 \le 3$ **b.** $-2 \ge -2$ **c.** $-4 \ge 1$.

6 Find the absolute value of a real number.

Absolute Value Suppose you are feeling a bit lazy. Instead of your usual jog, you decide to use your finger to stroll along the number line. You start at 0 and end the finger walk at -5. You have covered a distance of 5 units.

Absolute value describes distance from 0 on a number line. If a represents a real number, the symbol $|a|$ represents its absolute value, read "the absolute value of a." In terms of your walk, we can write

$$|-5| = 5.$$

The absolute value of -5 is 5 because -5 is 5 units from 0 on the number line.

ABSOLUTE VALUE The absolute value of a real number a, denoted by $|a|$, is the distance from 0 to a on the number line. Because absolute value describes a distance, it is never negative.

EXAMPLE 8 Finding Absolute Value

Find the absolute value: **a.** $|-3|$ **b.** $|5|$ **c.** $|0|$.

SOLUTION The solution is illustrated in Figure 1.10.

a. $|-3| = 3$ *The absolute value of -3 is 3 because -3 is 3 units from 0.*

b. $|5| = 5$ *5 is 5 units from 0.*

c. $|0| = 0$ *0 is 0 units from itself.*

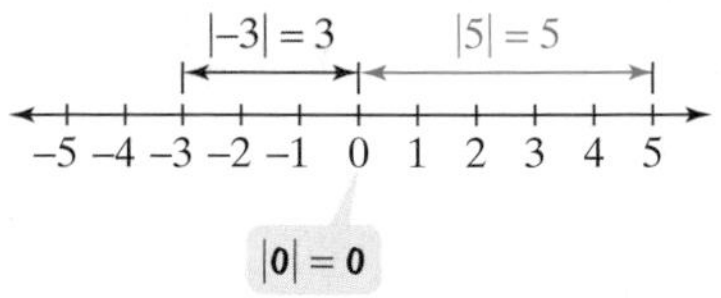

FIGURE 1.10 Absolute value describes distance from 0 on a number line.

Can you see that the absolute value of a real number is either positive or zero? Zero is the only real number whose absolute value is 0: $|0| = 0$. **The absolute value of any other real number is always positive.**

CHECK POINT 8 Find the absolute value:

a. $|-4|$ **b.** $|6|$ **c.** $|-\sqrt{2}|$.

1.2 EXERCISE SET

Student Solutions Manual CD/Video PH Math/Tutor Center MathXL Tutorials on CD MathXL® MyMathLab Interactmath.com

Practice Exercises

In Exercises 1–8, write a positive or negative integer that describes each situation.

1. Meteorology: 20° below zero
2. Navigation: 65 feet above sea level
3. Health: A gain of 8 pounds
4. Economics: A loss of $12,500.00
5. Banking: A withdrawal of $3000.00
6. Physics: An automobile slowing down at a rate of 3 meters per second each second.
7. Economics: A budget deficit of 4 billion dollars
8. Football: A 14-yard loss

In Exercises 9–20, start by drawing a number line that shows integers from -5 to 5. Then graph each real number on your number line.

9. 2 **10.** 5 **11.** -5 **12.** -2 **13.** $3\frac{1}{2}$ **14.** $2\frac{1}{4}$

15. $\frac{11}{3}$ **16.** $\frac{7}{3}$ **17.** -1.8 **18.** -3.4 **19.** $-\frac{16}{5}$ **20.** $-\frac{11}{5}$

In Exercises 21–32, express each rational number as a decimal.

21. $\frac{3}{4}$ **22.** $\frac{3}{5}$ **23.** $\frac{7}{20}$

24. $\frac{3}{20}$ **25.** $\frac{7}{8}$ **26.** $\frac{5}{16}$

27. $\frac{9}{11}$ **28.** $\frac{3}{11}$ **29.** $-\frac{1}{2}$

30. $-\frac{1}{4}$ **31.** $-\frac{5}{6}$ **32.** $-\frac{7}{6}$

In Exercises 33–36, list all numbers from the given set that are: **a.** *natural numbers,* **b.** *whole numbers,* **c.** *integers,* **d.** *rational numbers,* **e.** *irrational numbers,* **f.** *real numbers.*

33. $\{-9, -\frac{4}{5}, 0, 0.25, \sqrt{3}, 9.2, \sqrt{100}\}$

34. $\{-7, -0.\overline{6}, 0, \sqrt{49}, \sqrt{50}\}$

35. $\left\{-11, -\frac{5}{6}, 0, 0.75, \sqrt{5}, \pi, \sqrt{64}\right\}$

36. $\{-5, -0.\overline{3}, 0, \sqrt{2}, \sqrt{4}\}$

37. Give an example of a whole number that is not a natural number.

38. Give an example of an integer that is not a whole number.

39. Give an example of a rational number that is not an integer.

40. Give an example of a rational number that is not a natural number.

41. Give an example of a number that is an integer, a whole number, and a natural number.

42. Give an example of a number that is a rational number, an integer, and a real number.

43. Give an example of a number that is an irrational number and a real number.

44. Give an example of a number that is a real number, but not an irrational number.

In Exercises 45–62, insert either < or > in the shaded area between each pair of numbers to make a true statement.

45. $\frac{1}{2} \;\square\; 2$

46. $4 \;\square\; -3$

47. $3 \;\square\; -\frac{5}{2}$

48. $3 \;\square\; \frac{3}{2}$

49. $-4 \;\square\; -6$

50. $-\frac{5}{2} \;\square\; -\frac{5}{3}$

51. $-2.5 \;\square\; 1.5$

52. $-1.25 \;\square\; -0.5$

53. $-\frac{3}{4} \;\square\; -\frac{5}{4}$

54. $0 \;\square\; -\frac{1}{2}$

55. $-4.5 \;\square\; 3$

56. $-5.5 \;\square\; 2.5$

57. $\sqrt{2} \;\square\; 1.5$

58. $\sqrt{3} \;\square\; 2$

59. $0.\overline{3} \;\square\; 0.3$

60. $0.6 \;\square\; 0.\overline{6}$

61. $-\pi \;\square\; -3.5$

62. $-\frac{\pi}{2} \;\square\; -2.3$

In Exercises 63–70, determine whether each inequality is true or false.

63. $-5 \geq -13$

64. $-5 \leq -8$

65. $-9 \geq -9$

66. $-14 \leq -14$

67. $0 \geq -6$

68. $0 \geq -13$

69. $-17 \geq 6$

70. $-14 \geq 8$

In Exercises 71–78, find each absolute value.

71. $|6|$

72. $|3|$

73. $|-7|$

74. $|-9|$

75. $\left|\frac{5}{6}\right|$

76. $\left|\frac{4}{5}\right|$

77. $|-\sqrt{11}|$

78. $|-\sqrt{29}|$

Practice Plus

In Exercises 79–86, insert either <, >, or = in the shaded area to make a true statement.

79. $|-6| \;\square\; |-3|$

80. $|-20| \;\square\; |-50|$

81. $\left|\frac{3}{5}\right| \;\square\; |-0.6|$

82. $\left|\frac{5}{2}\right| \;\square\; |-2.5|$

83. $\frac{30}{40} - \frac{3}{4} \;\square\; \frac{14}{15} \cdot \frac{15}{14}$

84. $\frac{17}{18} \cdot \frac{18}{17} \;\square\; \frac{50}{60} - \frac{5}{6}$

85. $\frac{8}{13} \div \frac{8}{13} \;\square\; |-1|$

86. $|-2| \;\square\; \frac{4}{17} \div \frac{4}{17}$

Application Exercises

Temperatures sometimes fall below zero. A combination of low temperature and wind makes it feel colder than the actual temperature. The table shows how cold it feels when low temperatures are combined with different wind speeds.

Windchill

Wind	**Temperature (°F)**											
(mph)	**35**	**30**	**25**	**20**	**15**	**10**	**5**	**0**	**−5**	**−10**	**−15**	**−20**
5	31	25	19	13	7	−1	−5	−11	−16	−22	−28	−34
10	27	21	15	9	3	−4	−10	−16	−22	−28	−35	−41
15	25	19	13	6	0	−7	−13	−19	−26	−32	−39	−45
20	24	17	11	4	−2	−9	−15	−22	−29	−35	−42	−48
25	23	16	9	3	−4	−11	−17	−24	−31	−37	−44	−51

Source: National Weather Service

Use the information from the table to solve Exercises 87–88.

87. Write a negative integer that indicates how cold the temperature feels when the temperature is 15° Fahrenheit and the wind is blowing at 20 miles per hour.

88. Write a negative integer that indicates how cold the temperature feels when the temperature is 10° Fahrenheit and the wind is blowing at 15 miles per hour.

The following table shows the amount of money, in millions of dollars, collected and spent by the U.S. government from 1997 through 2002.

Year	Money Collected	Money Spent
1997	1,579,300	1,601,200
1998	1,721,800	1,652,600
1999	1,827,500	1,703,000
2000	2,025,200	1,788,800
2001	1,991,000	1,863,900
2002	1,946,100	2,052,300

Money is expressed in millions of dollars.
Source: Office of Management and Budget

Use the information from the table to solve Exercises 89–90.

89. List the years for which money collected $<$ money spent. Was there a budget surplus or deficit in these years?

90. List the years for which money collected $>$ money spent. Was there a budget surplus or deficit in these years?

Writing in Mathematics

Writing about mathematics will help you to learn mathematics. For all writing exercises in this book, use complete sentences to respond to the questions. Some writing exercises can be answered in a sentence; others require a paragraph or two. You can decide how much you need to write as long as your writing clearly and directly answers the question in the exercise. Standard references such as a dictionary and a thesaurus should be helpful.

91. What is a set?

92. What are the natural numbers?

93. What are the whole numbers?

94. What are the integers?

95. How does the set of integers differ from the set of whole numbers?

96. Describe how to graph a number on the number line.

97. What is a rational number?

98. Explain how to express $\frac{3}{8}$ as a decimal.

99. Describe the difference between a rational number and an irrational number.

100. If you are given two real numbers, explain how to determine which one is the lesser.

101. Describe what is meant by the absolute value of a number. Give an example with your explanation.

102. Give an example of an everyday situation that can be described using integers but not using whole numbers.

Critical Thinking Exercises

103. Which one of the following statements is true?

a. Every rational number is an integer.
b. Some whole numbers are not integers.
c. Some rational numbers are not positive.
d. Irrational numbers cannot be negative.

104. Which one of the following statements is true?

a. $\sqrt{36}$ is an irrational number.
b. Some real numbers are not rational numbers.
c. Some integers are not rational numbers.
d. All whole numbers are positive.

105. Answer this question without using a calculator. Between which two consecutive integers is $-\sqrt{47}$?

106. We have used a code to describe an activity that is important to your success in school and at work. Here is the coded message:

$(20, 18)$ $(-8, -12)$ $(0, 9)$ $(-14, -17)$
$(0, 11)$ $(-17, 9)$ $(-16, 14)$ $(-22, 7)$.

Read across the two rows. Each number pair represents one letter of the coded message. The first number pair represents the first letter, the second pair, the second letter, and so on. Use the following table to decode the message.

First Decoding	Second Decoding					
Use the greater of the two numbers in each pair. If the greater number is negative, then use its absolute value.	0 = blank space		1 = A	2 = B	3 = C	4 = D
	5 = E	6 = F	7 = G	8 = H	9 = I	10 = J
	11 = K	12 = L	13 = M	14 = N	15 = O	16 = P
	17 = Q	18 = R	19 = S	20 = T	21 = U	22 = V
	23 = W	24 = X	25 = Y	26 = Z		

Technology Exercises

In Exercises 107–110, use a calculator to find a decimal approximation for each irrational number, correct to three decimal places. Between which two integers should you graph each of these numbers on the number line?

107. $\sqrt{3}$

108. $-\sqrt{12}$

109. $1 - \sqrt{2}$

110. $2 - \sqrt{5}$

SECTION 1.3

Objectives

1. Plot ordered pairs in the rectangular coordinate system.
2. Find coordinates of points in the rectangular coordinate system.
3. Interpret information given by line graphs.
4. Interpret information given by bar graphs.

ORDERED PAIRS AND GRAPHS

S.S. Archives/Shooting Star International. © All rights reserved.

The beginning of the seventeenth century was a time of innovative ideas and enormous intellectual progress in Europe. English theatergoers enjoyed a succession of exciting new plays by Shakespeare. William Harvey proposed the radical notion that the heart was a pump for blood rather than the center of emotion. Galileo, with his new-fangled invention called the telescope, supported the theory of Polish astronomer Copernicus that the sun, not the Earth, was the center of the solar system. Monteverdi was writing the world's first grand operas. French mathematicians Pascal and Fermat invented a new field of mathematics called probability theory.

Into this arena of intellectual electricity stepped French aristocrat René Descartes (1596–1650). Descartes, propelled by the creativity surrounding him, developed a new branch of mathematics that brought together algebra and geometry in a unified way—a way that visualized numbers as points on a graph, equations as geometric figures, and geometric figures as equations. This new branch of mathematics, called *analytic geometry*, established Descartes as one of the founders of modern thought and among the most original mathematicians and philosophers of any age. We begin this section by looking at Descartes's deceptively simple idea, called the **rectangular coordinate system** or (in his honor) the **Cartesian coordinate system**.

1 Plot ordered pairs in the rectangular coordinate system.

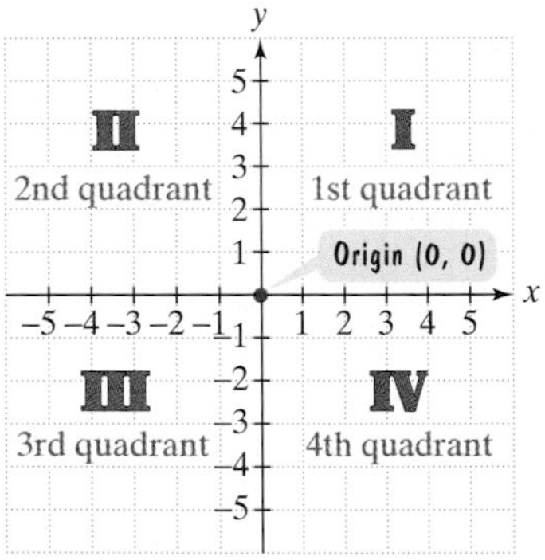

FIGURE 1.11 The rectangular coordinate system

Points and Ordered Pairs Descartes used two number lines that intersect at right angles at their zero points, as shown in Figure 1.11. The horizontal number line is the ***x*-axis**. The vertical number line is the ***y*-axis**. The point of intersection of these axes is their zero points, called the **origin**. Positive numbers are shown to the right and above the origin. Negative numbers are shown to the left and below the origin. The axes divide the plane into four quarters, called **quadrants**. The points located on the axes are not in any quadrant.

Each point in the rectangular coordinate system corresponds to an **ordered pair** of real numbers, (x, y). Examples of such pairs are $(4, 2)$ and $(-5, -3)$. The first number in each pair, called the ***x*-coordinate**, denotes the distance and direction from the origin along the x-axis. The second number, called the ***y*-coordinate**, denotes vertical distance and direction along a line parallel to the y-axis or along the y-axis itself.

Figure 1.12 shows how we **plot**, or locate, the points corresponding to the ordered pairs $(4, 2)$ and $(-5, -3)$. We plot $(4, 2)$ by going 4 units from 0 to the right along the x-axis. Then we go 2 units up parallel to the y-axis. We plot $(-5, -3)$ by going 5 units from 0 to the left along the x-axis and 3 units down parallel to the y-axis. The phrase "the point corresponding to the ordered pair $(-5, -3)$" is often abbreviated as "the point $(-5, -3)$."

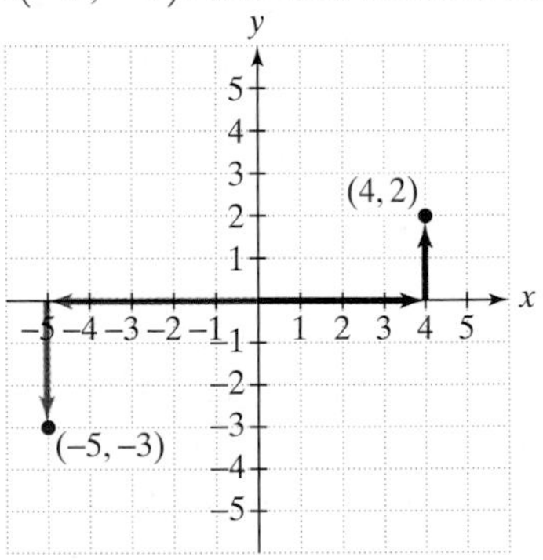

FIGURE 1.12 Plotting (4, 2) and (−5, −3)

EXAMPLE 1 Plotting Points in the Rectangular Coordinate System

Plot the points: $A(-3, 5)$, $B(2\ -4)$, $C(5, 0)$, $D(-5, -3)$, $E(0, 4)$ and $F(0, 0)$.

SOLUTION See Figure 1.13. We move from the origin and plot the points in the following way:

$A(-3, 5)$:	3 units left, 5 units up
$B(2, -4)$:	2 units right, 4 units down
$C(5, 0)$:	5 units right, 0 units up or down
$D(-5, -3)$:	5 units left, 3 units down
$E(0, 4)$:	0 units right or left, 4 units up
$F(0, 0)$:	0 units right or left, 0 units up or down

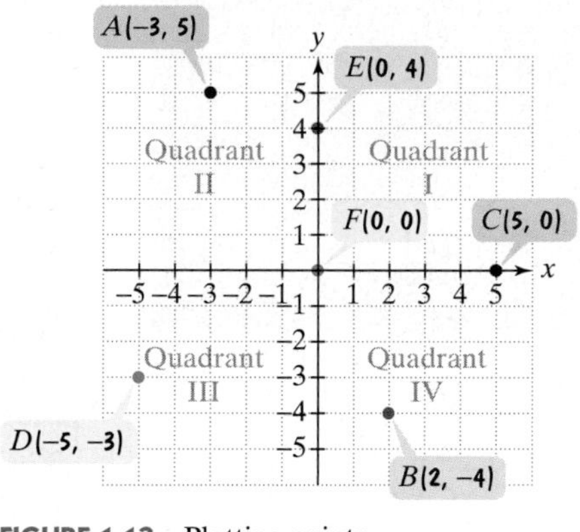

FIGURE 1.13 Plotting points

The phrase *ordered pair* is used because **order is important**. For example, the points (2, 5) and (5, 2) are not the same. To plot (2, 5), move 2 units right and 5 units up. To plot (5, 2), move 5 units right and 2 units up. The points (2, 5) and (5, 2) are in different locations. **The order in which coordinates appear makes a difference in a point's location.**

CHECK POINT 1 Plot the points: $A(-2, 4)$, $B(4, -2)$, $C(-3, 0)$, and $D(0, -3)$.

2 Find coordinates of points in the rectangular coordinate system.

In the rectangular coordinate system, each ordered pair corresponds to exactly one point. Example 2 illustrates that each point in the rectangular coordinate system corresponds to exactly one ordered pair.

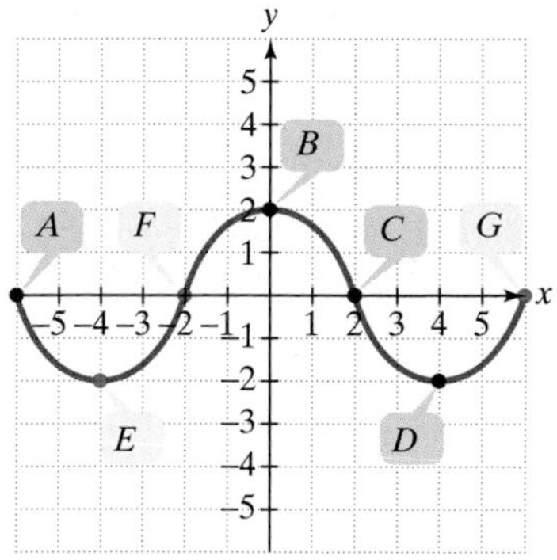

FIGURE 1.14 Finding coordinates of points

EXAMPLE 2 Finding Coordinates of Points

Determine the coordinates of points *A, B, C,* and *D* shown in Figure 1.14.

SOLUTION

Point	Position from the Origin	Coordinates
A	6 units left, 0 units up or down	$(-6, 0)$
B	0 units right or left, 2 units up	$(0, 2)$
C	2 units right, 0 units up or down	$(2, 0)$
D	4 units right, 2 units down	$(4, -2)$

CHECK POINT 2 Determine the coordinates of points *E, F,* and *G* shown in Figure 1.14.

The rectangular coordinate system lets us visualize relationships between two quantities, as shown in the next example.

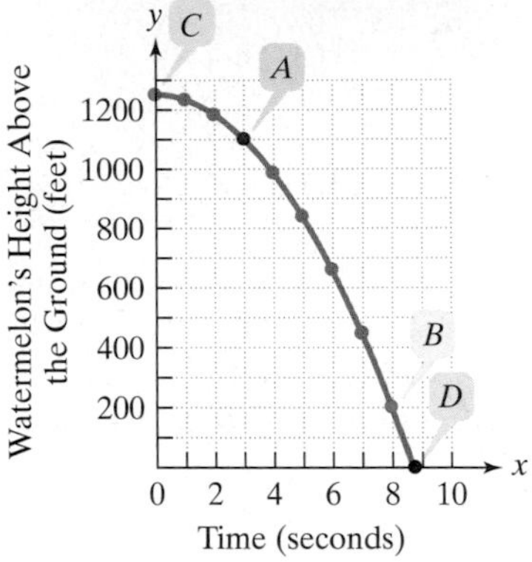

FIGURE 1.15

EXAMPLE 3 An Application of the Rectangular Coordinate System

Comedian David Letterman has been known to drop watermelons out of windows and watch them fall. Suppose that he drops a watermelon from the observation deck at the top of New York's Empire State Building. (Yes, the sidewalk on the ground below the building has been cleared!) Letterman drops the melon and watches it glide past the side of the building, smiling at the instant it splatters on the ground. The points in Figure 1.15 show the height of the melon above the ground at different times. Find the coordinates of point A. Then interpret the coordinates in terms of the information given.

SOLUTION Let's take a few minutes to look at the rectangular coordinate system. The x-axis represents the time, in seconds, that the watermelon is in the air. Each mark on the x-axis represents one second. The y-axis represents the melon's height, in feet, above the ground. Each mark on the y-axis represents 100 feet. Because time and height are not negative, Figure 1.15 shows only the first quadrant of the rectangular coordinate system and its boundary.

Now let us find the coordinates of point A. Point A is 3 units right and 1100 units up from the origin. Thus, the coordinates of point A are (3, 1100). This means that after 3 seconds, the watermelon is 1100 feet above the ground. ■

✔ **CHECK POINT 3** Use Figure 1.15 to find the coordinates of point B. Then interpret the coordinates in terms of the information given.

Take another look at Figure 1.15. How can we find the coordinates of points C and D? It appears that we must estimate one or more of these coordinates and arrive at reasonable approximations.

EXAMPLE 4 Estimating Coordinates of a Point

Use Figure 1.15 to estimate the coordinates of point C. Then interpret the coordinates in terms of the information given.

SOLUTION Point C is 0 units right from the origin. Thus, its x-coordinate is 0. How far up from the origin is point C? Its distance up is approximately midway between 1200 and 1300. Thus, its y-coordinate is approximately 1250. A reasonable estimate is that the coordinates of point C are (0, 1250). This means that after 0 seconds, or at the very instant Letterman dropped the watermelon, it was approximately 1250 feet above the ground. Equivalently, the melon was dropped from a height of approximately 1250 feet. ■

✔ **CHECK POINT 4** Use Figure 1.15 to estimate the coordinates of point D. Then interpret the coordinates in terms of the information given.

STUDY TIP

Any point on the x-axis has a y-coordinate of 0. Any point on the y-axis has an x-coordinate of 0.

Throughout your study of algebra, you will see how the rectangular coordinate system is used to create graphs that show relationships between quantities. Magazines and newspapers also use graphs to display information. In the remainder of this section, we will discuss how to interpret line and bar graphs.

3 Interpret information given by line graphs.

Line Graphs **Line graphs** are often used to illustrate trends over time. Some measure of time, such as months or years, frequently appears on the horizontal axis. Amounts are generally listed on the vertical axis. Points are drawn to represent the given information. The graph is formed by connecting the points with line segments.

Figure 1.16 is an example of a typical line graph. The graph shows the average age at which women in the United States married for the first time from 1890 through 2002. The years are listed on the horizontal axis and the ages are listed on the vertical axis. The symbol ⸗ on the vertical axis shows that there is a break in values between 0 and 20. Thus, the first tick mark on the vertical axis represents an average age of 20.

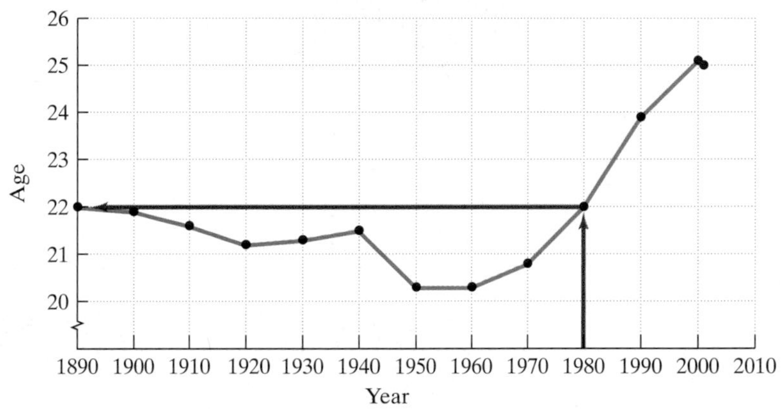

FIGURE 1.16

Source: U.S. Census Bureau

A line graph displays information in the first quadrant of a rectangular coordinate system. By identifying points on line graphs and their coordinates, you can interpret specific information given by the graph.

For example, the red lines in Figure 1.16 show how to find the average age at which women married for the first time in 1980.

STEP 1 Locate 1980 on the horizontal axis.

STEP 2 Locate the point above 1980.

STEP 3 Read across to the corresponding age on the vertical axis.

The age is 22. Thus, in 1980, women in the United States married for the first time at an average age of 22.

EXAMPLE 5 Using a Line Graph

The line graph in Figure 1.17 shows the fraction of federal prisoners in the United States sentenced for drug offenses from 1970 through 2001.

a. For the period shown, estimate the maximum fraction of federal prisoners sentenced for drug offenses. When did this occur?

b. Table 1.2 shows the number, in thousands, of federal prisoners in the United States for four selected years. Estimate the number of federal prisoners sentenced for drug offenses for the year in part (a).

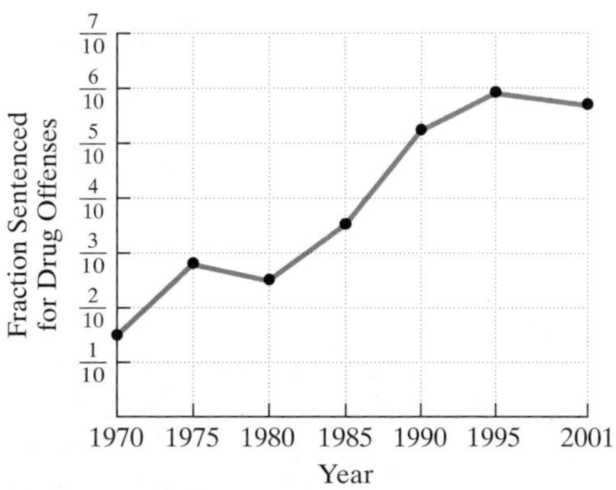

FIGURE 1.17

Source: Frank Schmalleger, *Criminal Justice Today*, 7th Edition, Prentice Hall, 2003

Table 1.2
Number, in Thousands, of U.S. Federal Prisoners

Year	Federal Prisoners
1990	60
1995	90
2000	130
2001	140

Source: Bureau of Justice Statistics

SOLUTION

a. The maximum fraction of federal prisoners sentenced for drug offenses can be found by locating the highest point on the graph. This point lies above 1995 on the horizontal axis. Read across to the corresponding fraction on the vertical axis. The number falls slightly below $\frac{6}{10}$. It appears that $\frac{6}{10}$, or $\frac{3}{5}$, is a reasonable estimate. Thus, the maximum fraction of federal prisoners sentenced for drug offenses is approximately $\frac{3}{5}$. This occurred in 1995.

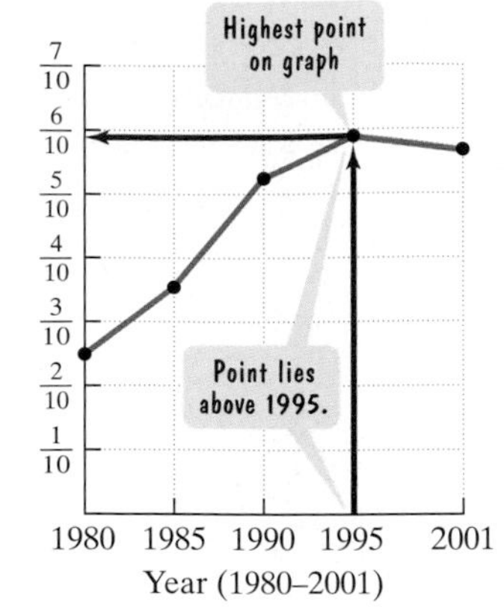

b. Table 1.2 shows that there were 90 thousand federal prisoners in 1995. The number sentenced for drug offenses can be determined as follows:

$\frac{3}{5}$ — of — the total prison population

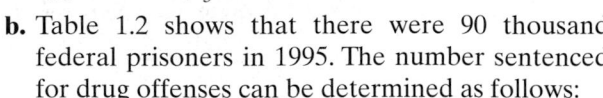

$$\text{Number sentenced for drug offenses} = \frac{3}{5} \cdot 90 = \frac{3 \cdot 90}{5} = \frac{270}{5} = 54.$$

In 1995, approximately 54 thousand federal prisoners were sentenced for drug offenses. ■

CHECK POINT 5 Use Figure 1.17 and Table 1.2 to estimate the number of federal prisoners sentenced for drug offenses in 1990.

4 Interpret information given by bar graphs.

Bar Graphs **Bar graphs** are convenient for showing comparisons among items. The bars may be vertical or horizontal, and their heights or lengths are used to show the amounts of different items. Figure 1.18 is an example of a typical bar graph. The graph shows the number of cars per 100 people in ten selected countries.

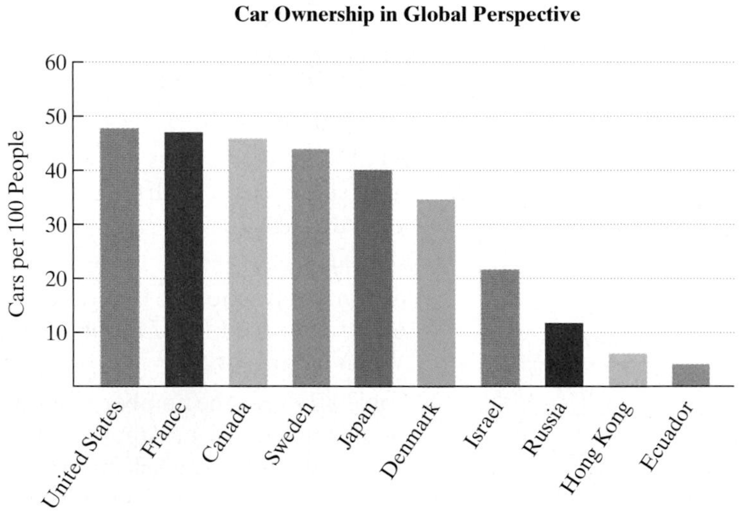

FIGURE 1.18

Source: The World Bank

EXAMPLE 6 Using a Bar Graph

Using Figure 1.18, answer the following:

a. Estimate the number of cars per 100 people in the United States.

b. Which countries have fewer than 20 cars per 100 people?

SOLUTION

a. To estimate the number of cars per 100 people in the United States, look at the top of the bar representing the United States and then read the cars-per-100-people scale on the vertical axis. The bar extends more than midway between 40 and 50, but is less than 50 by a few units. It appears that 48 is a reasonable estimate. Thus, there are approximately 48 cars per 100 people in the United States.

b. To find which countries have fewer than 20 cars per 100 people, we first locate the 20 mark on the vertical axis. Then we look for bars whose heights do not exceed 20. There are three such bars, namely, the three bars located in the right portion of the graph. The names of the countries below these bars show that Russia, Hong Kong, and Ecuador have fewer than 20 cars per 100 people. ■

CHECK POINT 6 Using Figure 1.18, answer the following:

a. Estimate the number of cars per 100 people in Israel.

b. Which countries have more than 40 cars per 100 people?

1.3 EXERCISE SET

Student Solutions Manual | CD/Video | PH Math/Tutor Center | MathXL Tutorials on CD | MathXL® | MyMathLab | Interactmath.com

Practice Exercises

In Exercises 1–8, plot the given point in a rectangular coordinate system. Indicate in which quadrant each point lies.

1. $(3, 5)$ **2.** $(5, 3)$ **3.** $(-5, 1)$

4. $(1, -5)$ **5.** $(-3, -1)$ **6.** $(-1, -3)$

7. $(6, -3.5)$ **8.** $(-3.5, 6)$

In Exercises 9–24, plot the given point in a rectangular coordinate system.

9. $(-3, -3)$ **10.** $(-5, -5)$ **11.** $(-2, 0)$

12. $(-5, 0)$ **13.** $(0, 2)$ **14.** $(0, 5)$

15. $(0, -3)$ **16.** $(0, -5)$ **17.** $\left(\frac{5}{2}, \frac{7}{2}\right)$

18. $\left(\frac{7}{2}, \frac{5}{2}\right)$ **19.** $\left(-5, \frac{3}{2}\right)$ **20.** $\left(-\frac{9}{2}, -4\right)$

21. $(0, 0)$ **22.** $\left(-\frac{5}{2}, 0\right)$ **23.** $\left(0, -\frac{5}{2}\right)$

24. $\left(0, \frac{7}{2}\right)$

In Exercises 25–32, give the ordered pairs that correspond to the points labeled in the figure.

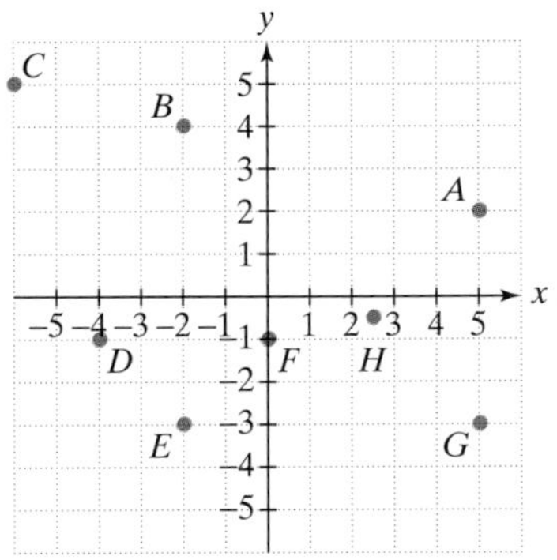

25. *A* **26.** *B* **27.** *C*

28. *D* **29.** *E* **30.** *F*

31. *G* **32.** *H*

33. In which quadrants are the *y*-coordinates positive?

34. In which quadrants are the *x*-coordinates negative?

35. In which quadrants do the *x*-coordinates and the *y*-coordinates have the same sign?

36. In which quadrants do the x-coordinates and the y-coordinates have opposite signs?

Practice Plus

37. Graph four points in at least two different quadrants such that the absolute value of the x-coordinate is the y-coordinate.

38. Graph four points in at least two different quadrants such that the absolute value of the y-coordinate is the x-coordinate.

39. The point $A(-2, y)$ lies on the line that connects $(-3, 5)$ and $(1, 3)$. Estimate the y-coordinate of point A.

40. The point $A(-1, y)$ lies on the line that connects $(-4, -1)$ and $(2, -4)$. Estimate the y-coordinate of point A.

Application Exercises

A football is thrown by a quarterback to a receiver. The points in the figure show the height of the football, in feet, above the ground in terms of its distance, in yards, from the quarterback. Use this information to solve Exercises 41–46.

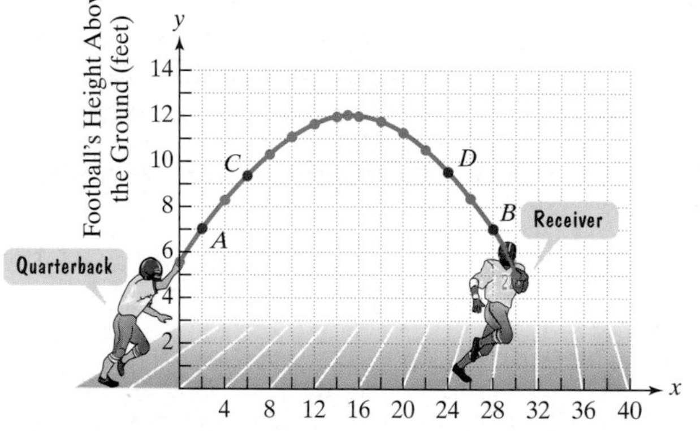

41. Find the coordinates of point A. Then interpret the coordinates in terms of the information given.

42. Find the coordinates of point B. Then interpret the coordinates in terms of the information given.

43. Estimate the coordinates of point C.

44. Estimate the coordinates of point D.

45. What is the football's maximum height? What is its distance from the quarterback when it reaches its maximum height?

46. What is the football's height when it is caught by the receiver? What is the receiver's distance from the quarterback when he catches the football?

Afghanistan accounts for 76% of the world's illegal opium production. Opium-poppy cultivation nets big money in a country where most people earn less than $1 per day. (Source: Newsweek) The line graph shows opium-poppy cultivation, in thousands of acres, in Afghanistan from 1990 through 2002. Use the graph to solve Exercises 47–52.

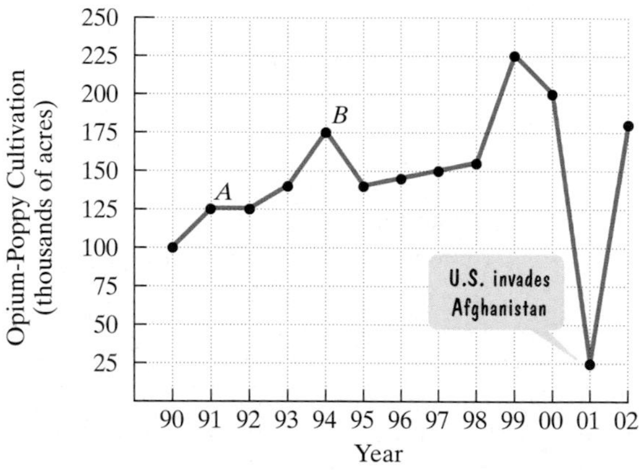

Source: U.N. Office on Drugs and Crime

47. What are the coordinates of point A? What does this mean in terms of the information given by the graph?

48. What are the coordinates of point B? What does this mean in terms of the information given by the graph?

49. For the period shown, when did opium cultivation reach a minimum? How many thousands of acres were used to cultivate the illegal crop?

50. For the period shown, when did opium cultivation reach a maximum? How many thousands of acres were used to cultivate the illegal crop?

51. Between which two years did opium cultivation not change?

52. Between which two years did opium cultivation increase at the greatest rate? What is a reasonable estimate of the increase, in thousands of acres, used to cultivate the illegal crop during this period?

The line graphs at the top of the next page show that oil consumption in the United States has continued to increase as domestic production of crude oil has been dropping. Use the information shown by the two line graphs to solve Exercises 53–60.

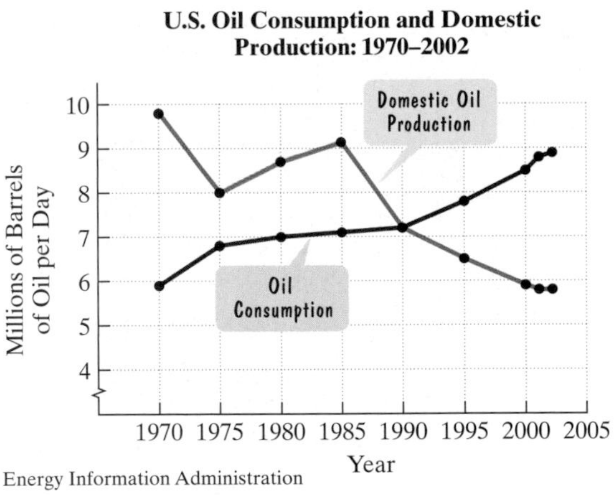

Source: Energy Information Administration

53. Find an estimate for oil consumption in 1985.

54. Find an estimate for domestic oil production in 1985.

55. In which year were oil consumption and domestic production the same? Estimate the millions of barrels of oil consumed and produced per day during that year.

56. For which years did oil consumption exceed domestic production?

57. For the period shown, when did domestic oil production reach a maximum? What is a reasonable estimate for oil production during that year?

58. For the period shown, when did oil consumption reach a maximum? What is a reasonable estimate for oil consumption during that year?

59. Find an estimate for the difference between production and consumption in 1980.

60. Find an estimate for the difference between consumption and production in 2000.

You can't judge the quality of a movie by its domestic box-office gross, but you can't keep making movies that don't make money. The bar graph shows that actor Kevin Costner needs a hit. Use the information in the graph to solve Exercises 61–64.

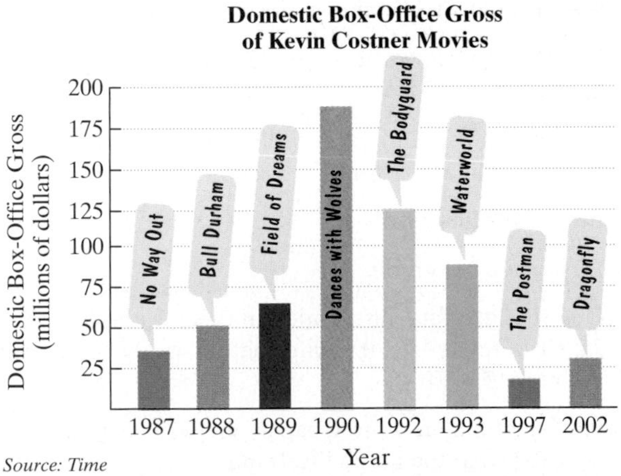

Source: Time

61. Estimate the box-office gross, in millions of dollars, of *The Bodyguard*.

62. Estimate the box-office gross, in millions of dollars, of *Waterworld*.

63. Which movies have a box-office gross of less than $50 million?

64. Which movies have a box-office gross of more than $100 million?

The bar graph shows life expectancy in the United States by year of birth. Use the graph to solve Exercises 65–68.

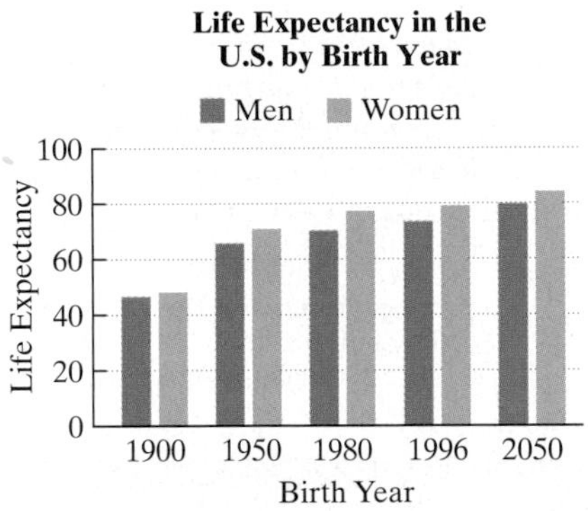

Source: U.S. Census Bureau

65. Estimate the life expectancy for men born in 1900.

66. Estimate the life expectancy for women born in 2050.

67. By approximately how many more years can women born in 1996 expect to live as compared to men born in 1950?

68. For which genders and for which birth years does life expectancy exceed 40 years but is at most 60 years?

The bar graph shows the annual median income (the income in the middle of the list when all of the groups' incomes are arranged in order) for six groups, organized by gender and race. Use the information in the graph to solve Exercises 69–72.

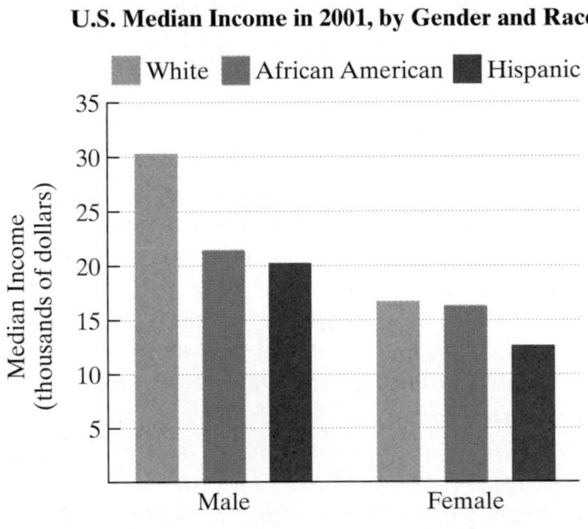

Source: U.S. Census Bureau

(Refer to the graph on the previous page)

69. Estimate the median income, in thousands of dollars, for African-American men.

70. Estimate the median income, in thousands of dollars, for white women.

71. Estimate the difference, in thousands of dollars, between median income of Hispanic men and Hispanic women.

72. Estimate the difference, in thousands of dollars, between the median income of African-American men and African-American women.

Candidates for president of the United States keep jumping into the race earlier and earlier. The bar graph shows the average number of days before the nominating convention that candidates declared their intention to run for the office. Use the information in the graph to solve Exercises 73–74.

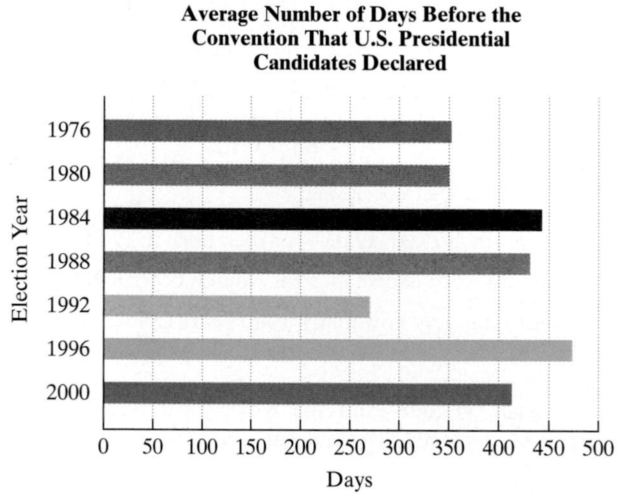

Source: *Newsweek*

73. In 1992, Clinton's 284 days, a short campaign, didn't cost him the win. Estimate the average number of days before the convention that candidates declared in 1992. By approximately how many days did Clinton's declaration exceed the 1992 average?

74. Typical of candidates today, in 2000, Bush declared 405 days prior to the convention. Estimate the average number of days before the convention that candidates declared in 2000. By approximately how many days was Bush's declaration less than the 2000 average?

Despite various programs over the past three decades, oil from other countries accounts for an ever-greater share of energy in the United States. The bar graph in the next column shows the fraction of oil used in the United States that was obtained from other countries for five selected years. In Exercises 75–78, use the information shown by the graph to write the year or years that fit the given description.

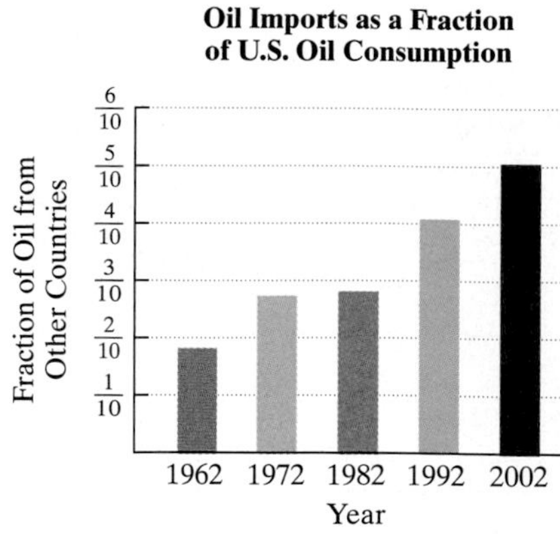

Source: Energy Information Administration

75. The United States obtained more than $\frac{1}{2}$ of its oil from other countries.

76. The United States obtained less than $\frac{1}{5}$ of its oil from other countries.

77. The fraction of imported oil exceeded $\frac{1}{5}$ but was at most $\frac{1}{2}$.

78. The fraction of imported oil exceeded $\frac{2}{5}$ but was at most $\frac{1}{2}$.

79. In 2002, $\frac{1}{10}$ of U.S. imported oil was from Saudi Arabia. Use the information shown in the graph to estimate the fraction of oil used in the United States in 2002 that was imported from Saudi Arabia.

80. In 2002, $\frac{2}{25}$ of U.S. imported oil was from Venezuela. Use the information shown in the graph to estimate the fraction of oil used in the United States in 2002 that was imported from Venezuela.

Writing in Mathematics

81. What is the rectangular coordinate system?

82. Explain how to plot a point in the rectangular coordinate system. Give an example with your explanation.

83. Explain why $(5, -2)$ and $(-2, 5)$ do not represent the same point.

84. Explain how to find the coordinates of a point in the rectangular coordinate system.

85. Describe a line graph.

86. Describe a bar graph.

87. Describe how the information in the bar graph for Exercises 75–80 is related to the information in the line graphs for Exercises 53–60.

88. Find a graph in a newspaper, magazine, or almanac and describe what the graph illustrates.

Critical Thinking Exercises

89. **a.** Graph each of the following sets of points in a separate rectangular coordinate system:

$$A = \{(-3, 9), (-2, 4), (-1, 1), (0, 0), (1, 1), (2, 4), (3, 9)\}$$
$$B = \{(9, -3), (4, -2), (1, -1), (0, 0), (1, 1), (4, 2), (9, 3)\}.$$

b. In which set (A or B) is each x-coordinate associated with exactly one y-coordinate? In which set are x-coordinates associated with more than one y-coordinate? Describe how these differences can be seen from your graphs in part (a).

90. **a.** Graph each of the following points:

$$\left(1, \frac{1}{2}\right), (2, 1), \left(3, \frac{3}{2}\right), (4, 2).$$

Parts (b)–(d) can be answered by changing the sign of one or both coordinates of the points in part (a).

b. What must be done to the coordinates so that the resulting graph is a mirror-image reflection about the y-axis of your graph in part (a)?

c. What must be done to the coordinates so that the resulting graph is a mirror-image reflection about the x-axis of your graph in part (a)?

d. What must be done to the coordinates so that the resulting graph is a straight-line extension of your graph in part (a)?

Review Exercises

From here on, each exercise set will contain three review exercises. It is essential to review previously covered topics to improve your understanding of the topics and to help maintain your mastery of the material. If you are not certain how to solve a review exercise, turn to the section and the illustrative example given in parentheses at the end of each exercise.

91. Add: $\frac{3}{4} + \frac{2}{5}$. (Section 1.1, Example 9)

92. Insert < or > in the shaded area to make a true statement: $-\frac{1}{4}$ ▢ 0. (Section 1.2, Example 6)

93. Find the absolute value: $|-5.83|$. (Section 1.2, Example 8)

SECTION 1.4

Objectives

1. Evaluate algebraic expressions.
2. Understand and use the vocabulary of algebraic expressions.
3. Use commutative properties.
4. Use associative properties.
5. Use distributive properties.
6. Combine like terms.
7. Simplify algebraic expressions.
8. Use algebraic expressions that model reality.

BASIC RULES OF ALGEBRA

Algebraic Expressions Feeling attractive with a suntan that gives you a "healthy glow"? Think again. Direct sunlight is known to promote skin cancer. Although sunscreens protect you from burning, dermatologists are concerned with the long-term damage that results from the sun even without sunburn.

We have seen that algebra uses letters, such as x and y, to represent numbers. Recall that such letters are called **variables**. For example, we can let x represent the number of minutes that a person can stay in the sun without burning with no sunscreen. With a number 6 sunscreen, exposure time without burning is six times as long, or 6 times x. This can be written $6 \cdot x$, but it is usually expressed as $6x$. Placing a number and a letter next to one another indicates multiplication.

Notice that $6x$ combines the number 6 and the variable x using the operation of multiplication. A combination of variables and numbers using the operations of addition, subtraction, multiplication, or division, as well as powers or roots, is called an **algebraic expression**. Here are some examples of algebraic expressions:

$$x + 6, \quad x - 6, \quad 6x, \quad \frac{x}{6}, \quad 3x + 5, \quad \sqrt{x} + 7.$$

1 Evaluate algebraic expressions.

Evaluating Algebraic Expressions We can replace a variable that appears in an algebraic expression by a number. We are **substituting** the number for the variable. The process is called **evaluating the expression**. For example, we can evaluate $6x$ (from the sunscreen example) for $x = 15$. We substitute 15 for x. We obtain $6 \cdot 15$, or 90. This means if you can stay in the sun for 15 minutes without burning when you don't put on any lotion, then with a number 6 lotion, you can "cook" for 90 minutes without burning.

Many algebraic expressions involve more than one operation. The order in which we add, subtract, multiply, and divide is important. In Section 1.8, we will discuss the rules for the order in which operations should be done. For now, follow this order:

1. Perform calculations within parentheses first.
2. Perform multiplication before addition.

EXAMPLE 1 Evaluating an Algebraic Expression

The algebraic expression $2.35x + 179.5$ describes the population of the United States, in millions, x years after 1960. Evaluate the expression for $x = 40$. Describe what the answer means in practical terms.

SOLUTION We begin by substituting 40 for x. Because $x = 40$, we will be finding the U.S. population 40 years after 1960, in the year 2000.

$2.35x + 179.5$

Replace x with 40.

$= 2.35(40) + 179.5$

$= 94 + 179.5$ — Perform the multiplication: $2.35(40) = 94$.

$= 273.5$ — Perform the addition.

According to the given algebraic expression, in 2000, the population of the United States was 273.5 million. ■

According to the U.S. Bureau of the Census, in 2000 the population of the United States was 281.4 million. Notice that the algebraic expression in Example 1 provides an approximate, but not an exact, description of the actual population. Many algebraic expressions approximately describe some aspect of reality, and we say that they *model* reality.

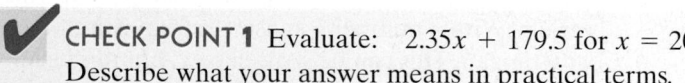

✔ **CHECK POINT 1** Evaluate: $2.35x + 179.5$ for $x = 20$.
Describe what your answer means in practical terms.

2 Understand and use the vocabulary of algebraic expressions.

The Vocabulary of Algebraic Expressions We have seen that an algebraic expression combines numbers and variables. Here is another example of an algebraic expression:

$$7x + 3.$$

The **terms** of an algebraic expression are those parts that are separated by addition. For example, the algebraic expression $7x + 3$ contains two terms, namely $7x$ and 3. Notice that a term is a number, a variable, or a number multiplied by one or more variables.

The numerical part of a term is called its **coefficient**. In the term $7x$, the 7 is the coefficient. If a term containing one or more variables is written without a coefficient, the coefficient is understood to be 1. Thus, x means $1x$ and ab means $1ab$.

A term that consists of just a number is called a **constant term**. The constant term of $7x + 3$ is 3.

The parts of each term that are multiplied are called the **factors of the term**. The factors of the term $7x$ are 7 and x.

Like terms are terms that have exactly the same variable factors. Here are two examples of like terms:

$7x$ and $3x$ — These terms have the same variable factor, x.

$4y$ and $9y$. — These terms have the same variable factor, y.

By contrast, here are some examples of terms that are not like terms. These terms do not have the same variable factor.

$7x$ and 3 — The variable factor of the first term is x. The second term has no variable factor.

$7x$ and $3y$ — The variable factor of the first term is x. The variable factor of the second term is y.

Constant terms are like terms. Thus, the constant terms 7 and -12 are like terms.

EXAMPLE 2 Using the Vocabulary of Algebraic Expressions

Use the algebraic expression

$$4x + 7 + 5x$$

to answer the following questions:

a. How many terms are there in the algebraic expression?
b. What is the coefficient of the first term?
c. What is the constant term?
d. What are the like terms in the algebraic expression?

SOLUTION

a. Because terms are separated by addition, the algebraic expression $4x + 7 + 5x$ contains three terms.

$$4x + 7 + 5x$$

First term — Second term — Third term

b. The coefficient of the first term, $4x$, is 4.
c. The constant term in $4x + 7 + 5x$ is 7.
d. The like terms in $4x + 7 + 5x$ are $4x$ and $5x$. These terms have the same variable factor, x. ■

CHECK POINT 2 Use the algebraic expression $6x + 2x + 11$ and answer each of the four questions in Example 2.

Equivalent Algebraic Expressions In Example 2, we considered the algebraic expression

$$4x + 7 + 5x.$$

Let's compare this expression with a second algebraic expression

$$9x + 7.$$

Evaluate each expression for some choice of x. We will select $x = 2$.

$4x + 7 + 5x$	$9x + 7$
Replace x with 2.	Replace x with 2.
$= 4 \cdot 2 + 7 + 5 \cdot 2$	$= 9 \cdot 2 + 7$
$= 8 + 7 + 10$	$= 18 + 7$
$= 25$	$= 25$

DISCOVER FOR YOURSELF

Show that $4x + 7 + 5x$ and $9x + 7$ have the same value when $x = 10$.

Both algebraic expressions have the same value when $x = 2$. Regardless of what number you select for x, the algebraic expressions $4x + 7 + 5x$ and $9x + 7$ will have the same value. These expressions are called *equivalent algebraic expressions*. Two algebraic expressions that have the same value for all replacements are called **equivalent algebraic expressions**. Because $4x + 7 + 5x$ and $9x + 7$ are equivalent algebraic expressions, we write

$$4x + 7 + 5x = 9x + 7.$$

Properties of Real Numbers and Algebraic Expressions We now turn to basic properties, or rules, that you know from past experiences in working with whole numbers and fractions. These properties will be extended to include all real numbers and algebraic expressions. We will give each property a name so that we can refer to it throughout the study of algebra.

3 Use commutative properties.

The Commutative Properties The addition or multiplication of two real numbers can be done in any order. For example, $3 + 5 = 5 + 3$ and $3 \cdot 5 = 5 \cdot 3$. Changing the order does not change the answer of a sum or a product. These facts are called **commutative properties**.

STUDY TIP

The commutative property does not hold for subtraction or division.

$$6 - 1 \neq 1 - 6$$
$$8 \div 4 \neq 4 \div 8$$

THE COMMUTATIVE PROPERTIES Let a, b, and c represent real numbers, variables, or algebraic expressions.

Commutative Property of Addition

$$a + b = b + a$$

Changing order when adding does not affect the sum.

Commutative Property of Multiplication

$$ab = ba$$

Changing order when multiplying does not affect the product.

EXAMPLE 3 Using the Commutative Properties

Use the commutative properties to write an algebraic expression equivalent to each of the following: **a.** $y + 6$ **b.** $5x$.

SOLUTION

a. By the commutative property of addition, an algebraic expression equivalent to $y + 6$ is $6 + y$. Thus,

$$y + 6 = 6 + y.$$

b. By the commutative property of multiplication, an algebraic expression equivalent to $5x$ is $x5$. Thus,

$$5x = x5.$$

■

CHECK POINT 3 Use the commutative properties to write an algebraic expression equivalent to each of the following: **a.** $x + 14$ **b.** $7y$.

EXAMPLE 4 Using the Commutative Properties

Write an algebraic expression equivalent to $13x + 8$ using

a. the commutative property of addition.
b. the commutative property of multiplication.

SOLUTION

a. By the commutative property of addition, we change the order of the terms being added. This means that an algebraic expression equivalent to $13x + 8$ is $8 + 13x$:

$$13x + 8 = 8 + 13x.$$

b. By the commutative property of multiplication, we change the order of the factors being multiplied. This means that an algebraic expression equivalent to $13x + 8$ is $x13 + 8$:

$$13x + 8 = x13 + 8.$$

■

CHECK POINT 4 Write an algebraic expression equivalent to $5x + 17$ using

a. the commutative property of addition.
b. the commutative property of multiplication.

ENRICHMENT ESSAY

Commutative Words and Sentences

The commutative property states that a change in order produces no change in the answer. The words and sentences listed here suggest a characteristic of the commutative property; they read the same from left to right and from right to left!

- dad
- repaper
- never odd or even
- Go deliver a dare, vile dog!
- May a moody baby doom a yam?
- Madam, in Eden I'm Adam.
- Ma is a nun, as I am.
- A man, a plan, a canal: Panama
- Are we not drawn onward, we few, drawn onward to new era?

4 Use associative properties.

The Associative Properties Parentheses indicate groupings. We perform operations within the parentheses first. For example,

$$(2 + 5) + 10 = 7 + 10 = 17$$

and

$$2 + (5 + 10) = 2 + 15 = 17.$$

In general, the way in which three numbers are grouped does not change their sum. It also does not change their product. These facts are called the **associative properties**.

STUDY TIP

The associative property does not hold for subtraction or division.

$$(6 - 3) - 1 \neq 6 - (3 - 1)$$
$$(8 \div 4) \div 2 \neq 8 \div (4 \div 2)$$

THE ASSOCIATIVE PROPERTIES Let a, b, and c represent real numbers, variables, or algebraic expressions.

Associative Property of Addition

$$(a + b) + c = a + (b + c)$$

Changing grouping when adding does not affect the sum.

Associative Property of Multiplication

$$(ab)c = a(bc)$$

Changing grouping when multiplying does not affect the product.

The associative properties can be used to simplify some algebraic expressions by removing the parentheses.

EXAMPLE 5 Simplifying Using the Associative Properties

Simplify: **a.** $3 + (8 + x)$ **b.** $8(4x)$.

SOLUTION

a. $3 + (8 + x)$ — This is the given algebraic expression.

$= (3 + 8) + x$ — Use the associative property of addition to group the first two numbers.

$= 11 + x$ — Add within parentheses.

Using the commutative property of addition, this simplified algebraic expression can also be written as $x + 11$.

b. $8(4x)$ — This is the given algebraic expression.

$= (8 \cdot 4)x$ — Use the associative property of multiplication to group the first two numbers.

$= 32x$ — Multiply within parentheses.

We can use the commutative property of multiplication and write this simplified algebraic expression as $x32$ or $x \cdot 32$. However, it is customary to express a term with its coefficient on the left. Thus, we use $32x$ as the simplified form of the algebraic expression. ■

CHECK POINT 5 Simplify:

a. $8 + (12 + x)$ **b.** $6(5x)$.

ENRICHMENT ESSAY

The Associative Property and the English Language

In the English language, sentences can take on different meanings depending on the way the words are associated with commas. Here are two examples.

- *Do not break your bread or roll in your soup.*
  Do not break your bread, or roll in your soup.
- *Woman, without her man, is nothing.*
  Woman, without her, man is nothing.

The next example involves the use of both basic properties to simplify an algebraic expression.

EXAMPLE 6 Using the Commutative and Associative Properties

Simplify: $7 + (x + 2)$.

SOLUTION

$7 + (x + 2)$ — This is the given algebraic expression.

$= 7 + (2 + x)$ — Use the commutative property to change the order of the addition.

$= (7 + 2) + x$ — Use the associative property to group the first two numbers.

$= 9 + x$ — Add within parentheses.

STUDY TIP

Commutative: changes *order*
Associative: changes *grouping*

Using the commutative property of addition, an equivalent algebraic expression is $x + 9$. ■

CHECK POINT 6 Simplify: $8 + (x + 4)$.

5 Use distributive properties.

The Distributive Properties The **distributive property** involves both multiplication and addition. The property shows how to multiply the sum of two numbers by a third number. Consider, for example, $4(7 + 3)$, which can be calculated in two ways. One way is to perform the addition within the grouping symbols and then multiply:

$$4(7 + 3) = 4(10) = 40.$$

The other way is to *distribute* the multiplication by 4 over the addition by first multiplying each number within the parentheses by 4 and then adding:

$$4(7 + 3) = 4 \cdot 7 + 4 \cdot 3 = 28 + 12 = 40.$$

The result in both cases is 40. Thus,

$$4(7 + 3) = 4 \cdot 7 + 4 \cdot 3.$$ Multiplication distributes over addition.

The distributive property allows us to rewrite the product of a number and a sum as the sum of two products.

THE DISTRIBUTIVE PROPERTY Let a, b, and c represent real numbers, variables, or algebraic expressions.

$$a(b + c) = ab + ac$$

Multiplication distributes over addition.

STUDY TIP

Do not confuse the distributive property with the associative property of multiplication.

Distributive:

$$4(5 + x) = 4 \cdot 5 + 4x = 20 + 4x$$

Associative:

$$4(5 \cdot x) = (4 \cdot 5)x = 20x$$

EXAMPLE 7 Using the Distributive Property

Multiply: $6(x + 4)$.

SOLUTION Multiply *each term* inside the parentheses, x and 4, by the multiplier outside, 6.

$6(x + 4) = 6x + 6 \cdot 4$ Use the distributive property to remove parentheses.

$= 6x + 24$ Multiply: $6 \cdot 4 = 24$. ■

CHECK POINT 7 Multiply: $5(x + 3)$.

EXAMPLE 8 Using the Distributive Property

Multiply: $5(3y + 7)$.

SOLUTION Multiply *each term* inside the parentheses, $3y$ and 7, by the multiplier outside, 5.

$5(3y + 7) = 5 \cdot 3y + 5 \cdot 7$ Use the distributive property to remove parentheses.

$= 15y + 35$ Multiply. Use the associative property of multiplication to find $5 \cdot 3y$: $5 \cdot 3y = (5 \cdot 3)y = 15y$. ■

CHECK POINT 8 Multiply: $6(4y + 7)$.

STUDY TIP

When using the distributive property to remove parentheses, be sure to multiply *each term* inside the parentheses by the multiplier outside.

Incorrect!

~~$5(3y + 7) = 5 \cdot 3y + 7$~~
~~$= 15y + 7$~~

7 must also be multiplied by 5.

Table 1.3 shows a number of other forms of the distributive property.

Table 1.3 Other Forms of the Distributive Property

Property	Meaning	Example
$a(b - c) = ab - ac$	Multiplication distributes over subtraction.	$5(4x - 3) = 5 \cdot 4x - 5 \cdot 3$ $= 20x - 15$
$a(b + c + d) = ab + ac + ad$	Multiplication distributes over three or more terms in parentheses.	$4(x + 10 + 3y)$ $= 4x + 4 \cdot 10 + 4 \cdot 3y$ $= 4x + 40 + 12y$
$(b + c)a = ba + ca$	Multiplication on the right distributes over addition (or subtraction).	$(x + 7)9 = x \cdot 9 + 7 \cdot 9$ $= 9x + 63$

6 Combine like terms.

Combining Like Terms The distributive property

$$a(b + c) = ab + ac$$

lets us add and subtract like terms. To do this, we will usually apply the property in the form

$$ax + bx = (a + b)x$$

and then combine a and b. For example,

$$3x + 7x = (3 + 7)x = 10x.$$

This process is called **combining like terms**.

EXAMPLE 9 Combining Like Terms

Combine like terms: **a.** $4x + 15x$ **b.** $7a - 2a$.

SOLUTION

a. $4x + 15x$ These are like terms because 4x and 15x have identical variable factors.
$= (4 + 15)x$ Apply the distributive property.
$= 19x$ Add within parentheses.

b. $7a - 2a$ These are like terms because 7a and 2a have identical variable factors.
$= (7 - 2)a$ Apply the distributive property.
$= 5a$ Subtract within parentheses. ■

DISCOVER FOR YOURSELF

Can you think of a fast method that will immediately give each result in Example 9? Describe the method.

✔ **CHECK POINT 9** Combine like terms: **a.** $7x + 3x$ **b.** $9a - 4a$.

When combining like terms, you may find yourself leaving out the details of the distributive property. For example, you may simply write

$$7x + 3x = 10x.$$

It might be useful to think along these lines: Seven things plus three of the (same) things give ten of those things. To add like terms, add the coefficients and copy the common variable.

COMBINING LIKE TERMS MENTALLY

1. Add or subtract the coefficients of the terms.
2. Use the result of step 1 as the coefficient of the term's variable factor.

When an expression contains three or more terms, use the commutative and associative properties to group like terms. Then combine the like terms.

EXAMPLE 10 Grouping and Combining Like Terms

Simplify: **a.** $7x + 5 + 3x + 8$ **b.** $4x + 7y + 2x + 3y$.

SOLUTION

a. $7x + 5 + 3x + 8$

$= (7x + 3x) + (5 + 8)$ Rearrange terms and group the like terms using the commutative and associative properties. This step is often done mentally.

$= 10x + 13$ Combine like terms: $7x + 3x = 10x$. Combine constant terms: $5 + 8 = 13$.

b. $4x + 7y + 2x + 3y$

$= (4x + 2x) + (7y + 3y)$ Group like terms.

$= 6x + 10y$ Combine like terms by adding coefficients and keeping the variable factor. ■

CHECK POINT 10 Simplify:

a. $8x + 7 + 10x + 3$ **b.** $9x + 6y + 5x + 2y$.

7 Simplify algebraic expressions.

Simplifying Algebraic Expressions An algebraic expression is **simplified** when parentheses have been removed and like terms have been combined.

SIMPLIFYING ALGEBRAIC EXPRESSIONS

1. Use the distributive property to remove parentheses.
2. Rearrange terms and group like terms using commutative and associative properties. This step may be done mentally.
3. Combine like terms by combining the coefficients of the terms and keeping the same variable factor.

EXAMPLE 11 Simplifying an Algebraic Expression

Simplify: $5(3x + 7) + 6x$.

SOLUTION

$$
\begin{aligned}
&5(3x + 7) + 6x \\
&= 5 \cdot 3x + 5 \cdot 7 + 6x && \text{Use the distributive property to remove the parentheses.} \\
&= 15x + 35 + 6x && \text{Multiply.} \\
&= (15x + 6x) + 35 && \text{Group like terms.} \\
&= 21x + 35 && \text{Combine like terms.}
\end{aligned}
$$

DISCOVER FOR YOURSELF

Substitute 10 for x in both $5(3x + 7) + 6x$ and $21x + 35$. Do you get the same answer in each case? Which form of the expression is easier to work with?

✔ **CHECK POINT 11** Simplify: $7(2x + 3) + 11x$.

EXAMPLE 12 Simplifying an Algebraic Expression

Simplify: $6(2x + 4y) + 10(4x + 3y)$.

SOLUTION

$$
\begin{aligned}
&6(2x + 4y) + 10(4x + 3y) \\
&= 6 \cdot 2x + 6 \cdot 4y + 10 \cdot 4x + 10 \cdot 3y && \text{Use the distributive property to remove the parentheses.} \\
&= 12x + 24y + 40x + 30y && \text{Multiply.} \\
&= (12x + 40x) + (24y + 30y) && \text{Group like terms.} \\
&= 52x + 54y && \text{Combine like terms.}
\end{aligned}
$$

✔ **CHECK POINT 12** Simplify: $7(4x + 3y) + 2(5x + y)$.

8 Use algebraic expressions that model reality.

Applications Have you had a good workout lately? The next example involves an algebraic expression that approximately describes, or models, your optimum heart rate during exercise.

EXAMPLE 13 Modeling Optimum Heart Rate

The optimum heart rate is the rate that a person should achieve during exercise for the exercise to be most beneficial. The algebraic expression

$$0.6(220 - a)$$

describes a person's optimum heart rate, in beats per minute, where a represents the age of the person in years.

a. Use the distributive property to rewrite the algebraic expression without parentheses.

b. Use each form of the algebraic expression to determine the optimum heart rate for a 20-year-old runner.

SOLUTION

a.
$$
\begin{aligned}
0.6(220 - a) &= 0.6(220) - 0.6a && \text{Use the distributive property to remove parentheses.} \\
&= 132 - 0.6a && \text{Multiply: } 0.6(220) = 132.
\end{aligned}
$$

b. To determine the optimum heart rate for a 20-year-old runner, substitute 20 for a in each form of the algebraic expression.

Using $0.6(220 - a)$:	**Using $132 - 0.6a$:**
$0.6(220 - 20)$	$132 - 0.6(20)$
$= 0.6(200)$	$= 132 - 12$
$= 120$	$= 120$

Both forms of the algebraic expression indicate that the optimum heart rate for a 20-year-old runner is 120 beats per minute.

CHECK POINT 13 Use each form of the algebraic expression in Example 13 to determine the optimum heart rate for a 40-year-old runner.

1.4 EXERCISE SET

Student Solutions Manual · CD/Video · PH Math/Tutor Center · MathXL Tutorials on CD · MathXL® · MyMathLab · Interactmath.com

Practice Exercises

In Exercises 1–10, evaluate each algebraic expression for the given value of the variable.

1. $x + 11; x = 4$
2. $x - 7; x = 11$
3. $8x; x = 10$
4. $12x; x = 5$
5. $5x + 6; x = 8$
6. $6x + 9; x = 4$
7. $7(x + 3); x = 2$
8. $8(x + 4); x = 3$
9. $\frac{5}{9}(F - 32); F = 77$
10. $\frac{5}{9}(F - 32); F = 50$

In Exercises 11–16, an algebraic expression is given. Use each expression to answer the following questions.

a. *How many terms are there in the algebraic expression?*
b. *What is the numerical coefficient of the first term?*
c. *What is the constant term?*
d. *Does the algebraic expression contain like terms? If so, what are the like terms?*

11. $3x + 5$
12. $9x + 4$
13. $x + 2 + 5x$
14. $x + 6 + 7x$
15. $4y + 1 + 3x$
16. $8y + 1 + 10x$

In Exercises 17–24, use the commutative property of addition to write an equivalent algebraic expression.

17. $y + 4$
18. $x + 7$
19. $5 + 3x$
20. $4 + 9x$
21. $4x + 5y$
22. $10x + 9y$
23. $5(x + 3)$
24. $6(x + 4)$

In Exercises 25–32, use the commutative property of multiplication to write an equivalent algebraic expression.

25. $9x$
26. $8x$
27. $x + y6$
28. $x + y7$
29. $7x + 23$
30. $13x + 11$
31. $5(x + 3)$
32. $6(x + 4)$

In Exercises 33–36, use an associative property to rewrite each algebraic expression. Once the grouping has been changed, simplify the resulting algebraic expression.

33. $7 + (5 + x)$
34. $9 + (3 + x)$
35. $7(4x)$
36. $8(5x)$

In Exercises 37–56, use a distributive property to rewrite each algebraic expression without parentheses.

37. $3(x + 5)$
38. $4(x + 6)$
39. $8(2x + 3)$
40. $9(2x + 5)$
41. $\frac{1}{3}(12 + 6r)$
42. $\frac{1}{4}(12 + 8r)$
43. $5(x + y)$
44. $7(x + y)$
45. $3(x - 2)$
46. $4(x - 5)$
47. $2(4x - 5)$
48. $6(3x - 2)$
49. $\frac{1}{2}(5x - 12)$
50. $\frac{1}{3}(7x - 21)$
51. $(2x + 7)4$
52. $(5x + 3)6$
53. $6(x + 3 + 2y)$
54. $7(2x + 4 + y)$
55. $5(3x - 2 + 4y)$
56. $4(5x - 3 + 7y)$

In Exercises 57–74, simplify each algebraic expression.

57. $7x + 10x$

58. $5x + 13x$

59. $11a - 3a$

60. $14b - 5b$

61. $3 + (x + 11)$

62. $7 + (x + 10)$

63. $5y + 3 + 6y$

64. $8y + 7 + 10y$

65. $2x + 5 + 7x - 4$

66. $7x + 8 + 2x - 3$

67. $11a + 12 + 3a + 2$

68. $13a + 15 + 2a + 11$

69. $5(3x + 2) - 4$

70. $2(5x + 4) - 3$

71. $12 + 5(3x - 2)$

72. $14 + 2(5x - 1)$

73. $7(3a + 2b) + 5(4a + 2b)$

74. $11(6a + 3b) + 4(12a + 5b)$

Practice Plus

In Exercises 75–78, name the property illustrated by each true statement.

75. $6x + (2y + 7y) = 6x + (7y + 2y)$

76. $6x + (2y + 7y) = (2y + 7y) + 6x$

77. $6x + (2y + 7y) = (6x + 2y) + 7y$

78. $(6x)4 = 4(6x)$

In Exercises 79–82, determine if each statement is true or false. Do not use a calculator.

79. $468(787 + 289) = 787 + 289(468)$

80. $468(787 + 289) = 787(468) + 289(468)$

81. $58 \cdot 9 + 32 \cdot 9 = (58 + 32) \cdot 9$

82. $58 \cdot 9 \cdot 32 \cdot 9 = (58 \cdot 32) \cdot 9$

Application Exercises

83. Suppose you can stay in the sun for x minutes without burning when you don't put on any lotion. The algebraic expression $15x$ describes how long you can tan without burning with a number 15 lotion. Evaluate the algebraic expression for $x = 20$. Describe what the answer means in practical terms.

84. Suppose that the cost of an item, excluding tax, is x dollars. The algebraic expression $0.06x$ describes the sales tax on that item. Evaluate the algebraic expression for $x = 400$. Describe what the answer means in practical terms.

The algebraic expression

$$405x + 5565$$

gives an approximate description of credit-card debt per U.S. household x years after 1994. The actual debt per household is shown in the bar graph at the top of the next column. Use this information to solve Exercises 85–86.

Credit-Card Debt per U.S. Household

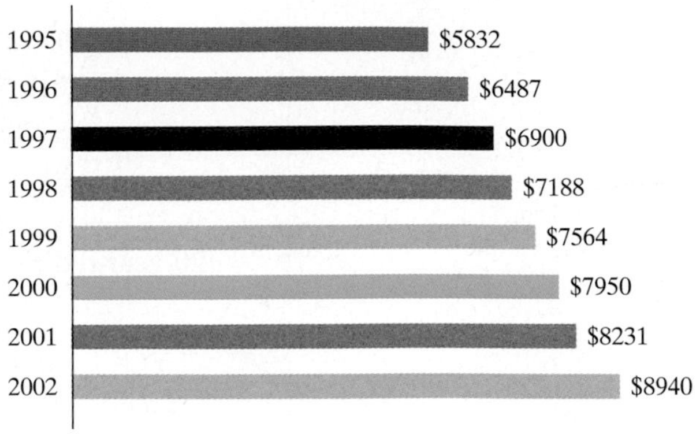

Source: Cardweb.com, Inc.

85. Evaluate the algebraic expression for $x = 8$. Describe what the answer means in practical terms. How well does the algebraic expression model the actual data for the appropriate year shown in the bar graph?

86. Evaluate the algebraic expression for $x = 7$. Describe what the answer means in practical terms. How well does the algebraic expression model the actual data for the appropriate year shown in the bar graph?

87. Testosterone is the hormone of choice for hundreds of thousands of middle-aged American men determined to stave off symptoms of andropause, the male version of menopause. The algebraic expression

$$2(0.18x + 0.01) + 0.02x$$

gives an approximate description of the millions of prescriptions for testosterone in the United States x years after 1997. The line graph indicates the actual number of testosterone prescriptions.

Testosterone Prescriptions in the U.S.

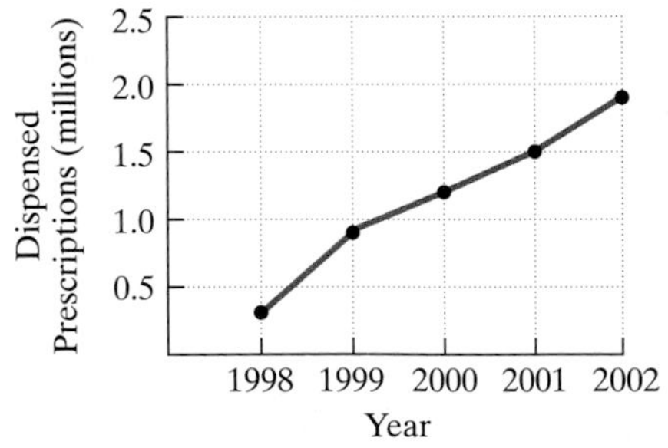

Source: NDC HEALTH

a. Simplify the algebraic expression.

b. Use the simplified algebraic expression to find the number of testosterone prescriptions in 2002. How well does the algebraic expression model the data shown in the line graph for 2002?

88. The equivalent algebraic expressions

$$\frac{DA + D}{24} \quad \text{and} \quad \frac{D(A + 1)}{24}$$

describe the drug dosage for children between the ages of 2 and 13. In each algebraic expression, D stands for an adult dose and A represents the child's age. If an adult dose of ibuprofen is 200 milligrams, what is the proper dose for a 12-year-old child? Use both forms of the algebraic expressions to answer the question. Which form is easier to use?

Writing in Mathematics

89. What is an algebraic expression? Provide an example with your description.

90. What does it mean to evaluate an algebraic expression? Provide an example with your description.

91. What is a term? Provide an example with your description.

92. What are like terms? Provide an example with your description.

93. What are equivalent algebraic expressions?

94. State a commutative property and give an example.

95. State an associative property and give an example.

96. State a distributive property and give an example.

97. Explain how to add like terms. Give an example.

98. What does it mean to simplify an algebraic expression?

99. An algebra student incorrectly used the distributive property and wrote $3(5x + 7) = 15x + 7$. If you were that student's teacher, what would you say to help the student avoid this kind of error?

100. You can transpose the letters in the word "conversation" to form the phrase "voices rant on." From "total abstainers" we can form "sit not at ale bars." What two algebraic properties do each of these transpositions (called anagrams) remind you of? Explain your answer.

Critical Thinking Exercises

101. Which one of the following statements is true?

a. Subtraction is a commutative operation.

b. $(24 \div 6) \div 2 = 24 \div (6 \div 2)$

c. $7y + 3y = (7 + 3)y$ for any value of y

d. $2x + 5 = 5x + 2$

102. Which one of the following statements is true?

a. $a + (bc) = (a + b)(a + c)$ In words, addition can be distributed over multiplication.

b. $4(x + 3) = 4x + 3$

c. Not every algebraic expression can be simplified.

d. Like terms contain the same coefficients.

103. A business that manufactures small alarm clocks has weekly fixed costs of \$5000. The average cost per clock for the business to manufacture x clocks is described by

$$\frac{0.5x + 5000}{x}.$$

a. Find the average cost when $x = 100$, 1000, and 10,000.

b. Like all other businesses, the alarm clock manufacturer must make a profit. To do this, each clock must be sold for at least 50¢ more than what it costs to manufacture. Due to competition from a larger company, the clocks can be sold for \$1.50 each and no more. Our small manufacturer can only produce 2000 clocks weekly. Does this business have much of a future? Explain.

Review Exercises

104. Express $\frac{4}{9}$ as a decimal. (Section 1.2, Example 4)

105. Plot $(-3, -1)$ in a rectangular coordinate system. (Section 1.3, Example 1)

106. Divide: $\frac{3}{7} \div \frac{15}{7}$. (Section 1.1, Example 6)

MID-CHAPTER CHECK POINT

CHAPTER 1

What You Know: We reviewed operations with fractions. We defined the real numbers and represented them as points on a number line. We used the rectangular coordinate system to represent ordered pairs of real numbers. We saw how information is displayed in line and bar graphs. Finally, we introduced some basic rules of algebra and used the commutative, associative, and distributive properties to simplify algebraic expressions.

In Exercises 1–9, perform the indicated operation or simplify the given expression. Where possible, reduce fractional answers to lowest terms.

1. $15a + 14 + 9a - 13$

2. $\frac{7}{10} - \frac{8}{15}$

3. $\frac{2}{3} \cdot \frac{3}{4}$

4. $7(9x + 3) + \frac{1}{3}(6x - 15)$

5. $\frac{5}{22} + \frac{5}{33}$

6. $\frac{3}{5} \div \frac{9}{10}$

7. $\frac{23}{105} - \frac{2}{105}$

8. $2\frac{7}{9} \div 3$

9. $5\frac{2}{9} - 3\frac{1}{6}$

10. Plot $\left(3, -5\frac{1}{2}\right)$ in a rectangular coordinate system. In which quadrant does the point lie?

In Exercises 11–13, rewrite $5(x + 3)$ as an equivalent expression using the given property.

11. the commutative property of multiplication

12. the commutative property of addition

13. the distributive property

Use the bar graph to solve Exercises 14–15.

Percentage of Americans Ages 12 and Older Listening to Internet Radio

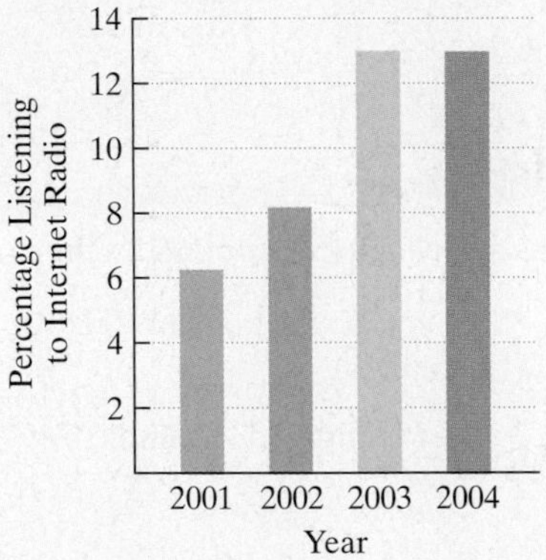

Source: Edison Media Research

14. Find an estimate for the difference between the percentage of Americans listening to Internet radio in 2004 and in 2001.

15. For which years shown did the percentage of Americans listening to Internet radio exceed 7%?

16. Insert either < or > in the shaded area to make a true statement:

$$-8000 \quad\square\quad -8\frac{1}{4}.$$

17. List all the rational numbers in this set:

$$\left\{-11, -\frac{3}{7}, 0, 0.45, \sqrt{23}, \sqrt{25}\right\}.$$

In Exercises 18–19, evaluate each algebraic expression for the given value of the variable.

18. $10x + 7; x = \frac{3}{5}$

19. $8(x - 2); x = \frac{5}{2}$

Use the graphs to solve Exercises 20–22.

Grades of U.S. Undergraduate College Students

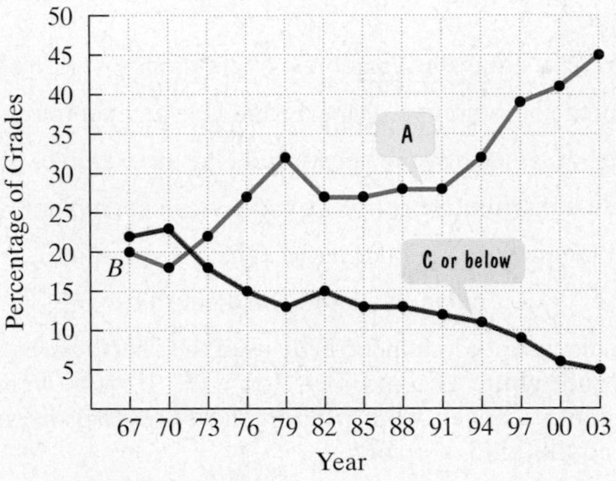

Source: UCLA Higher Education Research Institute

20. What are the coordinates of point B? What does this mean in terms of the information given by the graph?

21. For the period shown, when did the percentage of grades consisting of C or below reach a maximum? What is a reasonable estimate of the percentage of grades in this category for that year?

22. Find the difference between the percentage of A's in 2003 and the percentage of C's or below in 2003.

23. Find the absolute value: $|-19.3|$.

24. Express $\frac{1}{11}$ as a decimal.

25. Rewrite without parentheses:

$$8(7x - 10 + 3y).$$

SECTION 1.5

Objectives

1. Add numbers with a number line.
2. Find sums using identity and inverse properties.
3. Add numbers without a number line.
4. Use addition rules to simplify algebraic expressions.
5. Solve applied problems using a series of additions.

ADDITION OF REAL NUMBERS

It has not been a good day! First, you lost your wallet with \$30 in it. Then, to get through the day, you borrowed \$10, which you somehow misplaced. Your loss of \$30 followed by a loss of \$10 is an overall loss of \$40. This can be written

$$-30 + (-10) = -40.$$

The result of adding two or more numbers is called the **sum** of the numbers. The sum of -30 and -10 is -40. You can think of gains and losses of money to find sums. For example, to find $17 + (-13)$, think of a gain of \$17 followed by a loss of \$13. There is an overall gain of \$4. Thus, $17 + (-13) = 4$. In the same way, to find $-17 + 13$, think of a loss of \$17 followed by a gain of \$13. There is an overall loss of \$4, so $-17 + 13 = -4$.

1 Add numbers with a number line.

Adding with a Number Line We use the number line to help picture the addition of real numbers. Here is the procedure for finding $a + b$, the sum of a and b, using the number line:

USING THE NUMBER LINE TO FIND A SUM Let a and b represent real numbers. To find $a + b$ using a number line,

1. Start at a.
2. **a.** If b is **positive**, move b units to the **right**.
   b. If b is **negative**, move b units to the **left**.
   c. If b is **0**, **stay** at a.
3. The number where we finish on the number line represents the sum of a and b.

This procedure is illustrated in Examples 1 and 2. Think of moving to the right as a gain and moving to the left as a loss.

EXAMPLE 1 Adding Real Numbers Using a Number Line

Find the sum using a number line:

$$3 + (-5).$$

SOLUTION We find $3 + (-5)$ using the number line in Figure 1.19.

Step 1. We consider 3 to be the first number, represented by a in the preceding box. We start at a, or 3.

Step 2. We consider -5 to be the second number, represented by b. Because this number is negative, we move 5 units to the left.

Step 3. We finish at -2 on the number line. The number where we finish represents the sum of 3 and -5. Thus,

$$3 + (-5) = -2.$$

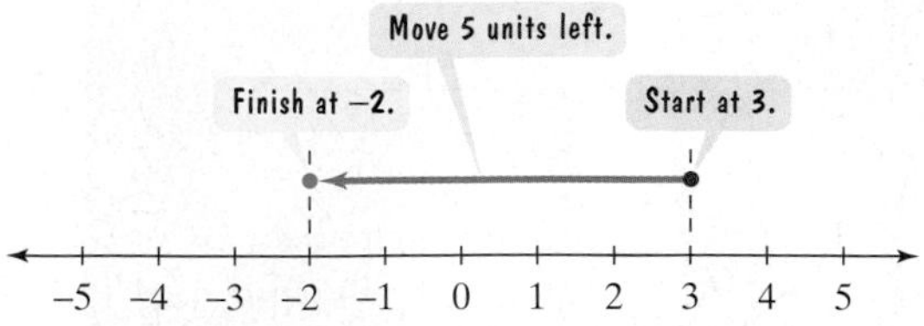

FIGURE 1.19 $3 + (-5) = -2$

Observe that if there is a gain of \$3 followed by a loss of \$5, there is an overall loss of \$2. ■

CHECK POINT 1 Find the sum using a number line:

$$4 + (-7).$$

EXAMPLE 2 Adding Real Numbers Using a Number Line

Find each sum using a number line: **a.** $-3 + (-4)$ **b.** $-6 + 2$.

SOLUTION

a. To find $-3 + (-4)$, start at -3. Move 4 units to the left. We finish at -7. Thus,

$$-3 + (-4) = -7.$$

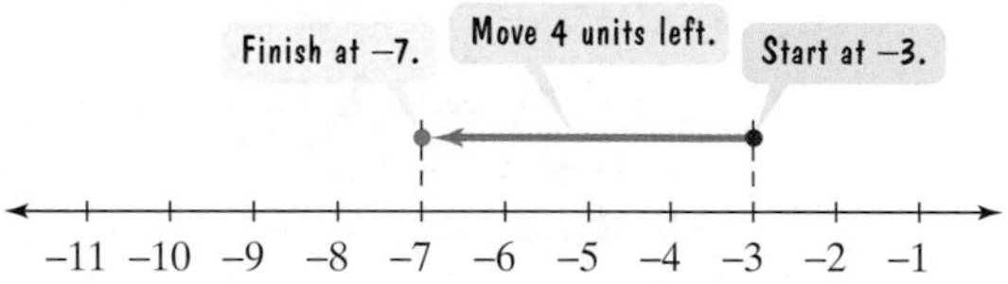

Observe that if there is a loss of \$3 followed by a loss of \$4, there is an overall loss of \$7.

b. To find $-6 + 2$, start at -6. Move 2 units to the right because 2 is positive. We finish at -4. Thus,

$$-6 + 2 = -4.$$

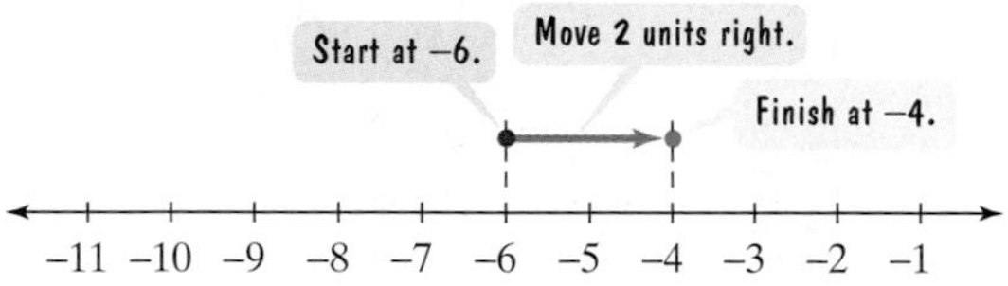

Observe that if there is a loss of \$6 followed by a gain of \$2, there is an overall loss of \$4. ■

CHECK POINT 2 Find each sum using a number line:

a. $-1 + (-3)$ **b.** $-5 + 3$.

2 Find sums using identity and inverse properties.

The Number Line and Properties of Addition The number line can be used to picture some useful properties of addition. For example, let's see what happens if we add two numbers with different signs but the same absolute value. Two such numbers are 3 and −3. To find $3 + (-3)$ on a number line, we start at 3 and move 3 units to the left. We finish at 0. Thus,

$$3 + (-3) = 0.$$

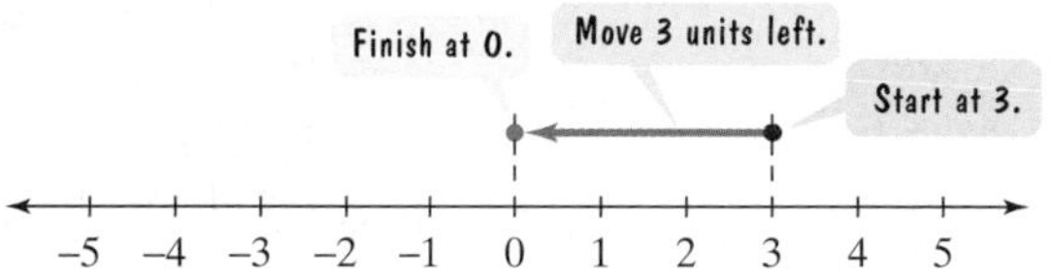

Numbers that are opposites, such as 3 and −3, are called *additive inverses*. **Additive inverses** are pairs of real numbers that are the same number of units from zero on the number line, but are on opposite sides of zero. Thus, −3 is the additive inverse of 3, and 5 is the additive inverse of −5. The additive inverse of 0 is 0. Other additive inverses come in pairs.

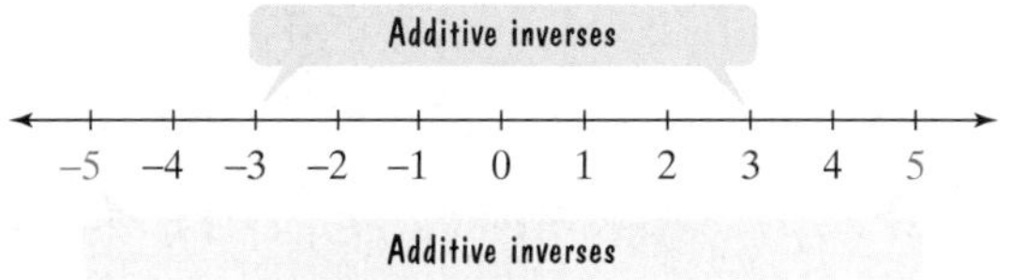

In general, the sum of any real number, denoted by a, and its additive inverse, denoted by $-a$, is zero:

$$a + (-a) = 0.$$

This property is called the **inverse property of addition**. In Section 1.4, we discussed the commutative and associative properties of addition. We now add two additional properties to our previous list, shown in Table 1.4.

Table 1.4 Identity and Inverse Properties of Addition

Let a be a real number, a variable, or an algebraic expression.

Property	Meaning	Examples
Identity Property of Addition	Zero can be deleted from a sum. $a + 0 = a$ $0 + a = a$	• $4 + 0 = 4$ • $-3x + 0 = -3x$ • $0 + (5a + b) = 5a + b$
Inverse Property of Addition	The sum of a real number and its additive inverse gives 0, the additive identity. $a + (-a) = 0$ $(-a) + a = 0$	• $6 + (-6) = 0$ • $3x + (-3x) = 0$ • $[-(2y + 1)] + (2y + 1) = 0$

3 Add numbers without a number line.

Adding without a Number Line Now that we can picture the addition of real numbers, we look at two rules for using absolute value to add signed numbers.

ADDING TWO NUMBERS WITH THE SAME SIGN

1. Add the absolute values.
2. Use the common sign as the sign of the sum.

STUDY TIP

The sum of two positive numbers is always positive.
The sum of two negative numbers is always negative.

EXAMPLE 3 Adding Real Numbers

Add without using a number line:

a. $-11 + (-15)$ **b.** $-0.2 + (-0.8)$ **c.** $-\frac{3}{4} + \left(-\frac{1}{2}\right)$.

SOLUTION In each part of this example, we are adding numbers with the same sign.

a. $-11 + (-15) = -26$

Add absolute values: $11 + 15 = 26$.

Use the common sign.

b. $-0.2 + (-0.8) = -1$

Add absolute values: $0.2 + 0.8 = 1.0$ or 1.

Use the common sign.

c. $-\frac{3}{4} + \left(-\frac{1}{2}\right) = -\frac{5}{4}$

Add absolute values: $\frac{3}{4} + \frac{1}{2} = \frac{3}{4} + \frac{2}{4} = \frac{5}{4}$.

Use the common sign.

■

CHECK POINT 3 Add without using a number line:

a. $-10 + (-25)$ **b.** $-0.3 + (-1.2)$ **c.** $-\frac{2}{3} + \left(-\frac{1}{6}\right)$.

We also use absolute value to add two real numbers with different signs.

USING TECHNOLOGY

You can use a calculator to add signed numbers. Here are the keystrokes for finding $-11 + (-15)$:

Scientific Calculator

11 [+/-] [+] 15 [+/-] [=]

Graphing Calculator

[(−)] 11 [+] [(−)] 15 [ENTER]

Here are the keystrokes for finding $-13 + 4$:

Scientific Calculator

13 [+/-] [+] 4 [=]

Graphing Calculator

[(−)] 13 [+] 4 [ENTER]

ADDING TWO NUMBERS WITH DIFFERENT SIGNS

1. Subtract the smaller absolute value from the greater absolute value.
2. Use the sign of the number with the greater absolute value as the sign of the sum.

EXAMPLE 4 Adding Real Numbers

Add without using a number line: **a.** $-13 + 4$ **b.** $-0.2 + 0.8$ **c.** $-\frac{3}{4} + \frac{1}{2}$.

SOLUTION In each part of this example, we are adding numbers with different signs.

a. $-13 + 4 = -9$

Subtract absolute values: $13 - 4 = 9$.

Use the sign of the number with the greater absolute value.

b. $-0.2 + 0.8 = 0.6$

Subtract absolute values: $0.8 - 0.2 = 0.6$.

Use the sign of the number with the greater absolute value. The sign is assumed to be positive.

c. $-\frac{3}{4} + \frac{1}{2} = -\frac{1}{4}$

Subtract absolute values: $\frac{3}{4} - \frac{1}{2} = \frac{3}{4} - \frac{2}{4} = \frac{1}{4}$.

Use the sign of the number with the greater absolute value.

■

STUDY TIP

The sum of two numbers with different signs may be positive or negative. Keep in mind that the sign of the sum is the sign of the number with the greater absolute value.

CHECK POINT 4 Add without using a number line:

a. $-15 + 2$ **b.** $-0.4 + 1.6$ **c.** $-\frac{2}{3} + \frac{1}{6}$.

4 Use addition rules to simplify algebraic expressions.

Algebraic Expressions The rules for adding real numbers can be used to simplify certain algebraic expressions.

EXAMPLE 5 Simplifying Algebraic Expressions

Simplify:

a. $-11x + 7x$

b. $7y + (-12z) + (-9y) + 15z$

c. $3(8 - 7x) + 5(2x - 4)$.

SOLUTION

a. $-11x + 7x$ The given algebraic expression has two like terms. $-11x$ and $7x$ have identical variable factors.

$= (-11 + 7)x$ Apply the distributive property.

$= -4x$ Add within parentheses: $-11 + 7 = -4$.

b. $7y + (-12z) + (-9y) + 15z$ The colors indicate that there are two pairs of like terms.

$= 7y + (-9y) + (-12z) + 15z$ Arrange like terms so that they are next to one another.

$= [7 + (-9)]y + [(-12) + 15]z$ Apply the distributive property.

$= -2y + 3z$ Add within the grouping symbols: $7 + (-9) = -2$ and $-12 + 15 = 3$.

c. $3(8 - 7x) + 5(2x - 4)$

$= 3 \cdot 8 - 3 \cdot 7x + 5 \cdot 2x - 5 \cdot 4$ Use the distributive property to remove the parentheses.

$= 24 - 21x + 10x - 20$ Multiply.

$= (-21x + 10x) + (24 - 20)$ Group like terms.

$= (-21 + 10)x + (24 - 20)$ Apply the distributive property.

$= -11x + 4$ Perform operations within parentheses: $-21 + 10 = -11$ and $24 - 20 = 4$. ■

CHECK POINT 5 Simplify:

a. $-20x + 3x$ **b.** $3y + (-10z) + (-10y) + 16z$

c. $4(10 - 8x) + 4(3x - 5)$.

5 Solve applied problems using a series of additions.

Applications Positive and negative numbers are used in everyday life to represent such things as gains and losses in the stock market, rising and falling temperatures, deposits and withdrawals on bank statements, and ascending and descending motion. Positive and negative numbers are used to solve applied problems involving a series of additions.

One way to add a series of positive and negative numbers is to use the commutative and associative properties.

- Add all the positive numbers.
- Add all the negative numbers.
- Add the sums obtained in the first two steps.

The next example illustrates this idea.

EXAMPLE 6 An Application of Adding Signed Numbers

A glider was towed 1000 meters into the air and then let go. It descended 70 meters into a thermal (rising bubble of warm air), which took it up 2100 meters. At this point it dropped 230 meters into a second thermal. Then it rose 1200 meters. What was its altitude at that point?

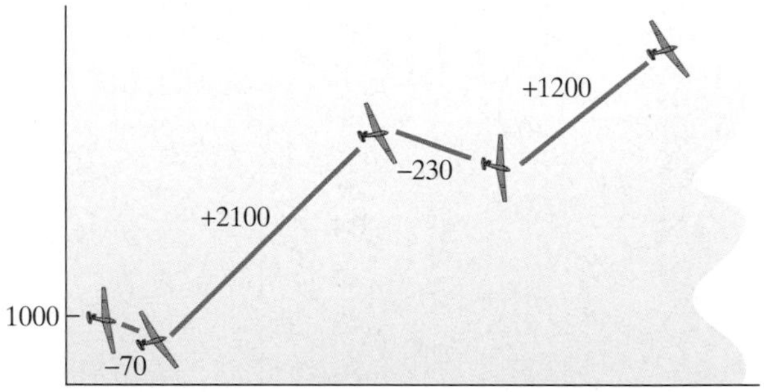

SOLUTION We use the problem's conditions to write a sum. The altitude of the glider is expressed by the following sum:

Towed to 1000 meters	then	Descended 70 meters	then	Taken up 2100 meters	then	Dropped 230 meters	then	Rose 1200 meters
1000	+	(−70)	+	2100	+	(−230)	+	1200.

$1000 + (-70) + 2100 + (-230) + 1200$	This is the sum arising from the problem's conditions.
$= (1000 + 2100 + 1200) + [(-70) + (-230)]$	Use the commutative and associative properties to group the positive and negative numbers.
$= 4300 + (-300)$	Add the positive numbers. Add the negative numbers.
$= 4000$	Add the results.

The altitude of the glider is 4000 meters. ■

DISCOVER FOR YOURSELF

Try working Example 6 by adding from left to right. You should still obtain 4000 for the sum. Which method do you find easier?

✔ **CHECK POINT 6** The water level of a reservoir is measured over a five-month period. During this time, the level rose 2 feet, then fell 4 feet, then rose 1 foot, then fell 5 feet, and then rose 3 feet. What was the change in the water level at the end of the five months?

1.5 EXERCISE SET

Student Solutions Manual | CD/Video | PH Math/Tutor Center | MathXL Tutorials on CD | MathXL® | MyMathLab | Interactmath.com

Practice Exercises

In Exercises 1–8, find each sum using a number line.

1. $7 + (-3)$ **2.** $7 + (-2)$ **3.** $-2 + (-5)$

4. $-1 + (-5)$ **5.** $-6 + 2$ **6.** $-8 + 3$

7. $3 + (-3)$ **8.** $5 + (-5)$

In Exercises 9–46, find each sum without the use of a number line.

9. $-7 + 0$ **10.** $-5 + 0$

11. $30 + (-30)$ **12.** $15 + (-15)$

13. $-30 + (-30)$ **14.** $-15 + (-15)$

15. $-8 + (-10)$ **16.** $-4 + (-6)$

17. $-0.4 + (-0.9)$ **18.** $-1.5 + (-5.3)$

19. $-\frac{7}{10} + \left(-\frac{3}{10}\right)$ **20.** $-\frac{7}{8} + \left(-\frac{1}{8}\right)$

21. $-9 + 4$ **22.** $-7 + 3$

23. $12 + (-8)$ **24.** $13 + (-5)$

25. $6 + (-9)$ **26.** $3 + (-11)$

27. $-3.6 + 2.1$ **28.** $-6.3 + 5.2$

29. $-3.6 + (-2.1)$ **30.** $-6.3 + (-5.2)$

31. $\frac{9}{10} + \left(-\frac{3}{5}\right)$ **32.** $\frac{7}{10} + \left(-\frac{2}{5}\right)$

33. $-\frac{5}{8} + \frac{3}{4}$ **34.** $-\frac{5}{6} + \frac{1}{3}$

35. $-\frac{3}{7} + \left(-\frac{4}{5}\right)$ **36.** $-\frac{3}{8} + \left(-\frac{2}{3}\right)$

37. $4 + (-7) + (-5)$ **38.** $10 + (-3) + (-8)$

39. $85 + (-15) + (-20) + 12$

40. $60 + (-50) + (-30) + 25$

41. $17 + (-4) + 2 + 3 + (-10)$

42. $19 + (-5) + 1 + 8 + (-13)$

43. $-45 + \left(-\frac{3}{7}\right) + 25 + \left(-\frac{4}{7}\right)$

44. $-50 + \left(-\frac{7}{9}\right) + 35 + \left(-\frac{11}{9}\right)$

45. $3.5 + (-45) + (-8.4) + 72$

46. $6.4 + (-35) + (-2.6) + 14$

In Exercises 47–64, simplify each algebraic expression.

47. $-10x + 2x$ **48.** $-19x + 10x$

49. $25y + (-12y)$ **50.** $26y + (-14y)$

51. $-8a + (-15a)$ **52.** $-9a + (-13a)$

53. $-4 + 7x + 5 + (-13x)$

54. $-5 + 8x + 3 + (-16x)$

55. $7b + 2 + (-b) + (-6)$

56. $10b + 7 + (-b) + (-15)$

57. $7x + (-5y) + (-9x) + 2y$

58. $13x + (-9y) + (-11x) + 3y$

59. $4(5x - 3) + 6$

60. $5(2x - 3) + 4$

61. $8(3 - 4y) + 35y$

62. $7(5 - 3y) + 25y$

63. $6(2 - 9a) + 7(3a + 5)$

64. $8(3 - 7a) + 4(2a + 3)$

Practice Plus

In Exercises 65–68, find each sum.

65. $|-3 + (-5)| + |2 + (-6)|$

66. $|4 + (-11)| + |-3 + (-4)|$

67. $-20 + [-|15 + (-25)|]$

68. $-25 + [-|18 + (-26)|]$

In Exercises 69–70, insert either $<$, $>$, or $=$ in the shaded area to make a true statement.

69. $6 + [2 + (-13)] \;\square\; -3 + [4 + (-8)]$

70. $[(-8) + (-6)] + 10 \;\square\; -8 + [9 + (-2)]$

Application Exercises

Solve Exercises 71–80 by writing a sum of signed numbers and adding.

71. The greatest temperature variation recorded in a day is 100 degrees in Browning, Montana, on January 23, 1916. The low temperature was −56°F. What was the high temperature?

72. In Spearfish, South Dakota, on January 22, 1943, the temperature rose 49 degrees in two minutes. If the initial temperature was −4°F, what was the high temperature?

73. The Dead Sea is the lowest elevation on earth, 1312 feet below sea level. What is the elevation of a person standing 712 feet above the Dead Sea?

74. Lake Assal in Africa is 512 feet below sea level. What is the elevation of a person standing 642 feet above Lake Assal?

75. The temperature at 8:00 A.M. was −7°F. By noon it had risen 15°F, but by 4:00 P.M. it had fallen 5°F. What was the temperature at 4:00 P.M.?

76. On three successive plays, a football team lost 15 yards, gained 13 yards, and then lost 4 yards. What was the team's total gain or loss for the three plays?

77. A football team started with the football at the 27-yard line, advancing toward the center of the field (the 50-yard line). Four successive plays resulted in a 4-yard gain, a 2-yard loss, an 8-yard gain, and a 12-yard loss. What was the location of the football at the end of the fourth play?

78. The water level of a reservoir is measured over a five-month period. At the beginning, the level is 20 feet. During this time, the level rose 3 feet, then fell 2 feet, then fell 1 foot, then fell 4 feet, and then rose 2 feet. What is the reservoir's water level at the end of the five months?

79. The bar graph shows the budget surplus or deficit for the United States government from 1999 through 2003.

U.S. Government Budget Surplus/Deficit

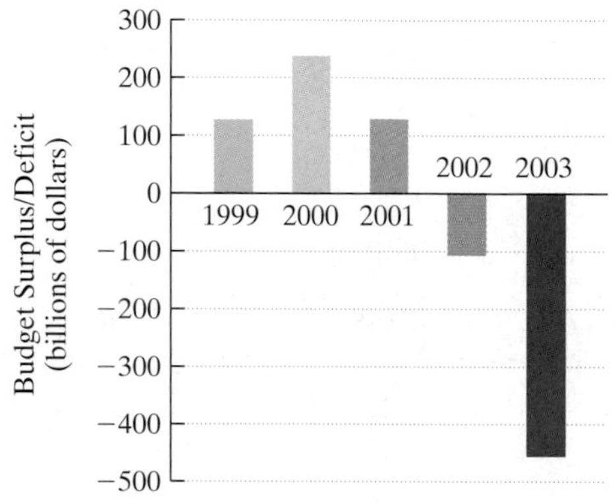

Source: Budget of the U.S. Government

In 1999, the U.S. government had a budget surplus of \$126 billion. This surplus then changed as follows:

2000: increased by \$110 billion

2001: decreased by \$109 billion

2002: decreased by \$233 billion

2003: decreased by \$349 billion.

What was the government's budget surplus or deficit in 2003?

80. According to *Newsweek*, the Ford Motor Company "swung from most profitable carmaker to biggest loser." The bar graph in the next column shows Ford's net profits and losses (the sum of income and expenses) from 1998 through 2002. In 1998, Ford's net profit was \$22 billion. This profit then changed as follows:

1999: decreased by \$16 billion

2000: decreased by \$3 billion

2001: decreased by \$8 billion

2002: increased by \$4 billion.

What was the company's net profit or loss in 2002?

Net Profits and Losses of the Ford Motor Company

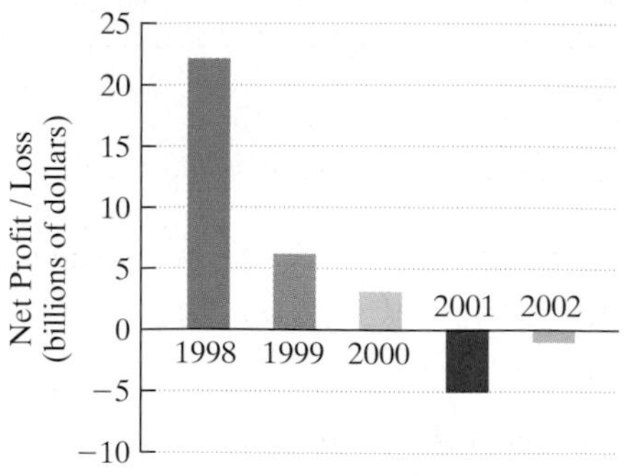

Source: Newsweek

Writing in Mathematics

81. Explain how to add two numbers with a number line. Provide an example with your explanation.

82. What are additive inverses?

83. Describe how the inverse property of addition

$$a + (-a) = 0$$

can be shown on a number line.

84. Without using a number line, describe how to add two numbers with the same sign. Give an example.

85. Without using a number line, describe how to add two numbers with different signs. Give an example.

86. Write a problem that can be solved by finding the sum of at least three numbers, some positive and some negative. Then explain how to solve the problem.

87. Without a calculator, you can add numbers using a number line, using absolute value, or using gains and losses. Which method do you find most helpful? Why is this so?

Critical Thinking Exercises

88. Which one of the following statements is true?

a. The sum of a positive number and a negative number is a negative number.

b. $|-9 + 2| = 9 + 2$

c. If two numbers are both positive or both negative, then the absolute value of their sum equals the sum of their absolute values.

d. $\frac{3}{4} + \left(-\frac{3}{5}\right) = -\frac{3}{20}$

89. Which one of the following statements is true?

a. The sum of a positive number and a negative number is a positive number.

b. If one number is positive and the other negative, then the absolute value of their sum equals the sum of their absolute values.

c. $\frac{3}{4} + \left(-\frac{2}{3}\right) = -\frac{1}{12}$

d. The sum of zero and a negative number is always a negative number.

In Exercises 90–91, find the missing term.

90. $5x + \_ + (-11x) + (-6y) = -6x + 2y$

91. $\_ + 11x + (-3y) + 3x = 7(2x - 3y)$

Technology Exercises

92. Use a calculator to verify any five of the sums that you found in Exercises 17–46.

93. Use a calculator to verify any three of the answers that you obtained in Application Exercises 71–80.

Review Exercises

94. Determine whether this inequality is true or false: $19 \geq -18$. (Section 1.2, Example 7)

95. Consider the set

$$\{-6, -\pi, 0, 0.\bar{7}, \sqrt{3}, \sqrt{4}\}.$$

List all numbers from the set that are **a.** natural numbers, **b.** whole numbers, **c.** integers, **d.** rational numbers, **e.** irrational numbers, **f.** real numbers. (Section 1.2, Example 5)

96. Plot $\left(\frac{7}{2}, -\frac{5}{2}\right)$ in a rectangular coordinate system. In which quadrant does the point lie? (Section 1.3, Example 1)

SECTION 1.6

Objectives

1. Subtract real numbers.
2. Simplify a series of additions and subtractions.
3. Use the definition of subtraction to identify terms.
4. Use the subtraction definition to simplify algebraic expressions.
5. Solve problems involving subtraction.

SUBTRACTION OF REAL NUMBERS

People are going to live longer in the 21st century. This will put added pressure on the Social Security and Medicare systems. How insecure is Social Security's future? In this section, we use subtraction of real numbers to numerically describe one aspect of the insecurity.

The Meaning of Subtraction Time for a new computer! Your favorite model, which normally sells for \$1500, has an incredible price reduction of \$600. The computer's reduced price, \$900, can be expressed in two ways:

$$1500 - 600 = 900 \quad \text{or} \quad 1500 + (-600) = 900.$$

This means that

$$1500 - 600 = 1500 + (-600).$$

To subtract 600 from 1500, we add 1500 and the additive inverse of 600. Generalizing from this situation, we define subtraction as follows:

DEFINITION OF SUBTRACTION For all real numbers a and b,

$$a - b = a + (-b).$$

In words: To subtract b from a, add the additive inverse of b to a. The result of subtraction is called the **difference**.

1 Subtract real numbers.

A Procedure for Subtracting Real Numbers The definition of subtraction gives us a procedure for subtracting real numbers.

SUBTRACTING REAL NUMBERS

1. Change the subtraction operation to addition.
2. Change the sign of the number being subtracted.
3. Add, using one of the rules for adding numbers with the same sign or different signs.

USING TECHNOLOGY

You can use a calculator to subtract signed numbers. Here are the keystrokes for finding $5 - (-6)$:

Scientific Calculator

5 [−] 6 [+/−] [=]

Graphing Calculator

5 [−] [(−)] 6 [ENTER]

Here are the keystrokes for finding $-9 - (-3)$:

Scientific Calculator

9 [+/−] [−] 3 [+/−] [=]

Graphing Calculator

[(−)] 9 [−] [(−)] 3 [ENTER]

Don't confuse the subtraction key on a graphing calculator, [−], with the sign change or additive inverse key, [(−)]. What happens if you do?

EXAMPLE 1 Using the Definition of Subtraction

Subtract: **a.** $7 - 10$ **b.** $5 - (-6)$ **c.** $-9 - (-3)$.

SOLUTION

a. $7 - 10 = 7 + (-10) = -3$

Change the subtraction to addition. Replace 10 with its additive inverse.

b. $5 - (-6) = 5 + 6 = 11$

Change the subtraction to addition. Replace −6 with its additive inverse.

c. $-9 - (-3) = -9 + 3 = -6$

Change the subtraction to addition. Replace −3 with its additive inverse. ■

CHECK POINT 1 Subtract:

a. $3 - 11$ **b.** $4 - (-5)$ **c.** $-7 - (-2)$.

The definition of subtraction can be applied to real numbers that are not integers.

EXAMPLE 2 Using the Definition of Subtraction

Subtract. **a.** $-5.2 - (-11.4)$ **b.** $-\frac{3}{4} - \frac{2}{3}$ **c.** $4\pi - (-9\pi)$.

SOLUTION

a. $-5.2 - (-11.4) = -5.2 + 11.4 = 6.2$

Change the subtraction to addition. Replace −11.4 with its additive inverse.

b. $-\frac{3}{4} - \frac{2}{3} = -\frac{3}{4} + \left(-\frac{2}{3}\right) = -\frac{9}{12} + \left(-\frac{8}{12}\right) = -\frac{17}{12}$

Change the subtraction to addition. Replace $\frac{2}{3}$ with its additive inverse.

c. $4\pi - (-9\pi) = 4\pi + 9\pi = (4 + 9)\pi = 13\pi$

Change the subtraction to addition. Replace -9π with its additive inverse. ■

Reading the symbol "−" can be a bit tricky. The way you read it depends on where it appears. For example,

$$-5.2 - (-11.4)$$

is read "negative five point two minus negative eleven point four." Read parts (b) and (c) of Example 2 aloud. When is "−" read "negative" and when is it read "minus"?

✔ **CHECK POINT 2** Subtract:

a. $-3.4 - (-12.6)$ **b.** $-\frac{3}{5} - \frac{1}{3}$ **c.** $5\pi - (-2\pi)$.

2 Simplify a series of additions and subtractions.

Problems Containing a Series of Additions and Subtractions In some problems, several additions and subtractions occur together. We begin by converting all subtractions to additions of additive inverses, or opposites.

SIMPLIFYING A SERIES OF ADDITIONS AND SUBTRACTIONS

1. Change all subtractions to additions of additive inverses.
2. Group and then add all the positive numbers.
3. Group and then add all the negative numbers.
4. Add the results of steps 2 and 3.

EXAMPLE 3 Simplifying a Series of Additions and Subtractions

Simplify: $7 - (-5) - 11 - (-6) - 19$.

SOLUTION

$7 - (-5) - 11 - (-6) - 19$

$= 7 + 5 + (-11) + 6 + (-19)$ Write subtractions as additions of additive inverses.

$= (7 + 5 + 6) + [(-11) + (-19)]$ Group the positive numbers. Group the negative numbers.

$= 18 + (-30)$ Add the positive numbers. Add the negative numbers.

$= -12$ Add the results. ■

✔ **CHECK POINT 3** Simplify: $10 - (-12) - 4 - (-3) - 6$.

3 Use the definition of subtraction to identify terms.

Subtraction and Algebraic Expressions We know that the terms of an algebraic expression are separated by addition signs. Let's use this idea to identify the terms of the following algebraic expression:

$$9x - 4y - 5.$$

Because terms are separated by addition, we rewrite the algebraic expression as additions of additive inverses, or opposites. Thus,

$$9x - 4y - 5 = 9x + (-4y) + (-5).$$

The three terms of the algebraic expression are $9x$, $-4y$, and -5.

EXAMPLE 4 **Using the Definition of Subtraction to Identify Terms**

Identify the terms of the algebraic expression:

$$2xy - 13y - 6.$$

SOLUTION Rewrite the algebraic expression as additions of additive inverses.

$$2xy - 13y - 6 = \underbrace{2xy}_{\text{First term}} + \underbrace{(-13y)}_{\text{Second term}} + \underbrace{(-6)}_{\text{Third term}}$$

Because terms are separated by addition, the terms are $2xy$, $-13y$, and -6. ■

CHECK POINT 4 Identify the terms of the algebraic expression:

$$-6 + 4a - 7ab.$$

4 Use the subtraction definition to simplify algebraic expressions.

The procedure for subtracting real numbers can be used to simplify certain algebraic expressions that involve subtraction.

EXAMPLE 5 **Simplifying Algebraic Expressions**

Simplify: **a.** $2 + 3x - 8x$ **b.** $-4x - 9y - 2x + 12y$.

SOLUTION

a. $2 + 3x - 8x$ This is the given algebraic expression.

$= 2 + 3x + (-8x)$ Write the subtraction as the addition of an additive inverse.

$= 2 + [3 + (-8)]x$ Apply the distributive property.

$= 2 + (-5x)$ Add within the grouping symbols

$= 2 - 5x$ Be concise and express as subtraction.

b. $-4x - 9y - 2x + 12y$ This is the given algebraic expression.

$= -4x + (-9y) + (-2x) + 12y$ Write the subtractions as the additions of additive inverses.

$= -4x + (-2x) + (-9y) + 12y$ Arrange like terms so that they are next to one another.

$= [-4 + (-2)]x + (-9 + 12)y$ Apply the distributive property.

$= -6x + 3y$ Add within the grouping symbols. ■

STUDY TIP

You can think of gains and losses of money to work the distributive property mentally:

- $3x - 8x = -5x$ A gain of 3 dollars followed by a loss of 8 of those dollars is a net loss of 5 of those dollars.
- $-9y + 12y = 3y$ A loss of 9 dollars followed by a gain of 12 of those dollars is a net gain of 3 of those dollars.

CHECK POINT 5 Simplify:

a. $4 + 2x - 9x$ **b.** $-3x - 10y - 6x + 14y$.

5 Solve problems involving subtraction.

Applications Subtraction is used to solve problems in which the word "difference" appears. The difference between real numbers a and b is expressed as $a - b$.

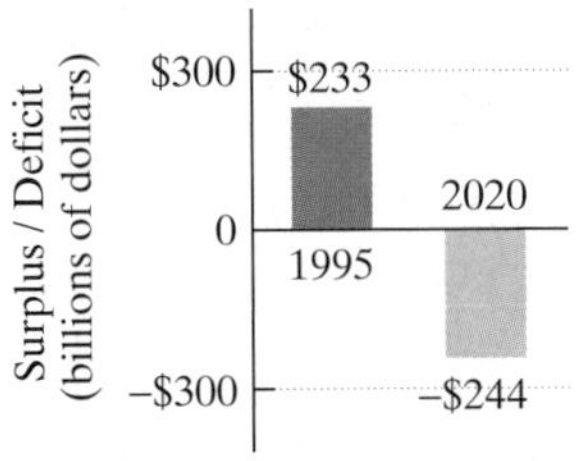

FIGURE 1.20 Social Security Annual Surplus/Deficit

Source: U.S. Office of Management and Budget

EXAMPLE 6 An Application of Subtraction Using the Word "Difference"

The bar graph in Figure 1.20 shows that in 1995, Social Security had an annual cash surplus of $233 billion. By 2020, this amount is expected to be a negative number—a deficit of $244 billion. What is the difference between the 1995 surplus and the projected 2020 deficit?

SOLUTION

The difference | is | the 1995 surplus | minus | the 2020 deficit.

$$= 233 - (-244)$$
$$= 233 + 244 = 477$$

The difference between the 1995 surplus and the projected 2020 deficit is $477 billion. ■

CHECK POINT 6 The peak of Mount Everest is 8848 meters above sea level. The Marianas Trench, on the floor of the Pacific Ocean, is 10,915 meters below sea level. What is the difference in elevation between the peak of Mount Everest and the Marianas Trench?

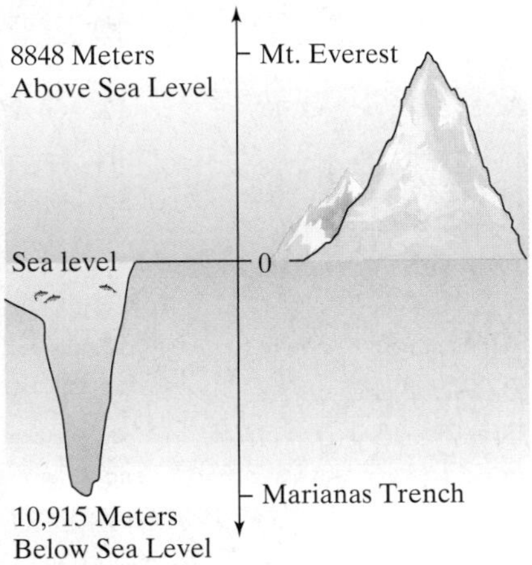

1.6 EXERCISE SET

Student Solutions Manual | CD/Video | PH Math/Tutor Center | MathXL Tutorials on CD | MathXL® | MyMathLab | Interactmath.com

Practice Exercises

1. Consider the subtraction $5 - 12$.
   a. Find the additive inverse, or opposite, of 12.
   b. Rewrite the subtraction as the addition of the additive inverse of 12.
2. Consider the subtraction $4 - 10$.
   a. Find the additive inverse, or opposite, of 10.
   b. Rewrite the subtraction as the addition of the additive inverse of 10.
3. Consider the subtraction $5 - (-7)$.
   a. Find the additive inverse, or opposite, of -7.
   b. Rewrite the subtraction as the addition of the additive inverse of -7.
4. Consider the subtraction $2 - (-8)$.
   a. Find the additive inverse, or opposite, of -8.
   b. Rewrite the subtraction as the addition of the additive inverse of -8.

In Exercises 5–50, perform the indicated subtraction.

5. $14 - 8$
6. $15 - 2$
7. $8 - 14$
8. $2 - 15$
9. $3 - (-20)$
10. $5 - (-17)$
11. $-7 - (-18)$
12. $-5 - (-19)$
13. $-13 - (-2)$
14. $-21 - (-3)$
15. $-21 - 17$
16. $-29 - 21$

17. $-45 - (-45)$
18. $-65 - (-65)$
19. $23 - 23$
20. $26 - 26$
21. $13 - (-13)$
22. $15 - (-15)$
23. $0 - 13$
24. $0 - 15$
25. $0 - (-13)$
26. $0 - (-15)$
27. $\frac{3}{7} - \frac{5}{7}$
28. $\frac{4}{9} - \frac{7}{9}$
29. $\frac{1}{5} - \left(-\frac{3}{5}\right)$
30. $\frac{1}{7} - \left(-\frac{3}{7}\right)$
31. $-\frac{4}{5} - \frac{1}{5}$
32. $-\frac{4}{9} - \frac{1}{9}$
33. $-\frac{4}{5} - \left(-\frac{1}{5}\right)$
34. $-\frac{4}{9} - \left(-\frac{1}{9}\right)$
35. $\frac{1}{2} - \left(-\frac{1}{4}\right)$
36. $\frac{2}{5} - \left(-\frac{1}{10}\right)$
37. $\frac{1}{2} - \frac{1}{4}$
38. $\frac{2}{5} - \frac{1}{10}$
39. $9.8 - 2.2$
40. $5.7 - 3.3$
41. $-3.1 - (-1.1)$
42. $-4.6 - (-1.1)$
43. $1.3 - (-1.3)$
44. $1.4 - (-1.4)$
45. $-2.06 - (-2.06)$
46. $-3.47 - (-3.47)$
47. $5\pi - 2\pi$
48. $9\pi - 7\pi$
49. $3\pi - (-10\pi)$
50. $4\pi - (-12\pi)$

In Exercises 51–68, simplify each series of additions and subtractions.

51. $13 - 2 - (-8)$
52. $14 - 3 - (-7)$
53. $9 - 8 + 3 - 7$
54. $8 - 2 + 5 - 13$
55. $-6 - 2 + 3 - 10$
56. $-9 - 5 + 4 - 17$
57. $-10 - (-5) + 7 - 2$
58. $-6 - (-3) + 8 - 11$
59. $-23 - 11 - (-7) + (-25)$
60. $-19 - 8 - (-6) + (-21)$
61. $-823 - 146 - 50 - (-832)$
62. $-726 - 422 - 921 - (-816)$
63. $1 - \frac{2}{3} - \left(-\frac{5}{6}\right)$
64. $2 - \frac{3}{4} - \left(-\frac{7}{8}\right)$
65. $-0.16 - 5.2 - (-0.87)$
66. $-1.9 - 3 - (-0.26)$
67. $-\frac{3}{4} - \frac{1}{4} - \left(-\frac{5}{8}\right)$
68. $-\frac{1}{2} - \frac{2}{3} - \left(-\frac{1}{3}\right)$

In Exercises 69–72, identify the terms in each algebraic expression.

69. $-3x - 8y$
70. $-9a - 4b$
71. $12x - 5xy - 4$
72. $8a - 7ab - 13$

In Exercises 73–84, simplify each algebraic expression.

73. $3x - 9x$
74. $2x - 10x$
75. $4 + 7y - 17y$
76. $5 + 9y - 29y$
77. $2a + 5 - 9a$
78. $3a + 7 - 11a$
79. $4 - 6b - 8 - 3b$
80. $5 - 7b - 13 - 4b$
81. $13 - (-7x) + 4x - (-11)$
82. $15 - (-3x) + 8x - (-10)$
83. $-5x - 10y - 3x + 13y$
84. $-6x - 9y - 4x + 15y$

Practice Plus

In Exercises 85–90, find the value of each expression.

85. $-|-9 - (-6)| - (-12)$
86. $-|-8 - (-2)| - (-6)$
87. $\frac{5}{8} - \left(\frac{1}{2} - \frac{3}{4}\right)$
88. $\frac{9}{10} - \left(\frac{1}{4} - \frac{7}{10}\right)$
89. $|-9 - (-3 + 7)| - |-17 - (-2)|$
90. $|24 - (-16)| - |-51 - (-31 + 2)|$

Application Exercises

91. The peak of Mount Kilimanjaro, the highest point in Africa, is 19,321 feet above sea level. Qattara Depression, Egypt, one of the lowest points in Africa, is 436 feet below sea level. What is the difference in elevation between the peak of Mount Kilimanjaro and the Qattara Depression?

92. The peak of Mount Whitney is 14,494 feet above sea level. Mount Whitney can be seen directly above Death Valley, which is 282 feet below sea level. What is the difference in elevation between these geographic locations?

The bar graph shows the occupations with the greatest projected growth and the greatest projected losses from 2000 through 2010. Use the graph to solve Exercises 93–96.

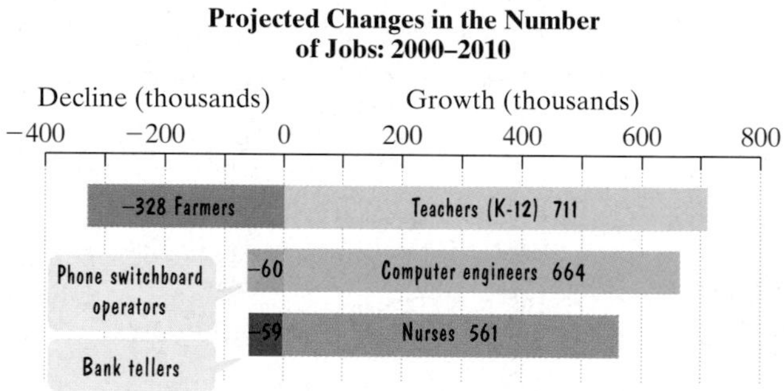

Source: U.S. Bureau of Labor Statistics

93. What is the difference, in thousands of jobs, between the projected increase in teachers' jobs and the projected decrease in farmers' jobs?

94. What is the difference, in thousands of jobs, between the projected increase in computer engineering jobs and the projected decrease in jobs for phone switchboard operators?

95. By how many thousands of jobs does the decline for phone switchboard operators exceed the decline for farmers?

96. By how many thousands of jobs does the decline for bank tellers exceed the decline for farmers?

Do you enjoy cold weather? If so, try Fairbanks, Alaska. The average daily low temperature for each month in Fairbanks is shown in the bar graph. Use the graph to solve Exercises 97–100.

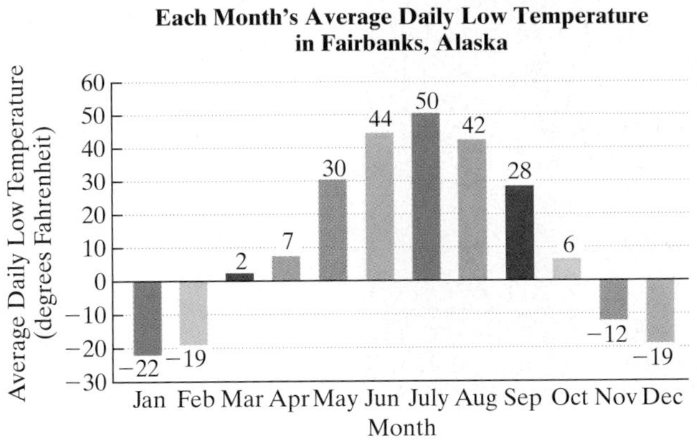

Source: The Weather Channel Enterprises, Inc.

97. What is the difference between the average daily low temperatures for March and February?

98. What is the difference between the average daily low temperatures for October and November?

99. How many degrees warmer is February's average low temperature than January's average low temperature?

100. How many degrees warmer is November's average low temperature than December's average low temperature?

When a person receives a drug injected into a muscle, the concentration of the drug in the body depends on the time elapsed since the injection. The points in the rectangular system show the concentration of the drug, measured in milligrams per 100 milliliters, from the time the drug is injected until 13 hours later. Use this information to solve Exercises 101–106.

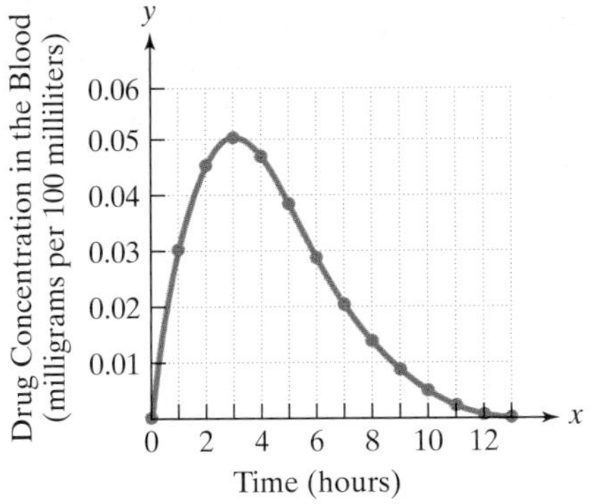

101. What is the drug's maximum concentration and when does this occur?

102. What happens by the end of 13 hours?

103. What is the approximate difference between the drug's concentration 4 hours after it was injected and 1 hour after it was injected?

104. What is the approximate difference between the drug's concentration 4 hours after it was injected and 7 hours after it was injected?

105. When is the drug's concentration increasing?

106. When is the drug's concentration decreasing?

Writing in Mathematics

107. Explain how to subtract real numbers.

108. How is $4 - (-2)$ read?

109. Explain how to simplify a series of additions and subtractions. Provide an example with your explanation.

110. Explain how to find the terms of the algebraic expression $5x - 2y - 7$.

111. Write a problem that can be solved by finding the difference between two numbers. At least one of the numbers should be negative. Then explain how to solve the problem.

Critical Thinking Exercises

112. Which one of the following statements is true?

a. If a and b are negative numbers, then $a - b$ is a negative number.

b. $7 - (-2) = 5$

c. The difference between 0 and a negative number is always a positive number.

d. None of the given statements is true.

113. The golden age of Athens culminated in 212 B.C. and the golden age of India culminated in A.D. 500. Determine the number of years that elapsed between these dates. (*Note:* When the calendar was reformed, the number 0 had not been invented. There was no year 0 and the year A.D. 1 followed the year 1 B.C. Calculate the difference between the years in the usual way and then use this added bit of information to modify your answer.)

114. Find the value:

$$-1 + 2 - 3 + 4 - 5 + 6 - \cdots - 99 + 100.$$

Technology Exercises

115. Use a calculator to verify any five of the differences that you found in Exercises 5–46.

116. Use a calculator to verify any three of the answers that you found in Exercises 51–68.

Review Exercises

117. Graph on a number line: -4.5. (Section 1.2; Example 3)

118. Use the commutative property of addition to write an equivalent algebraic expression: $10(a + 4)$. (Section 1.4; Example 4)

119. Give an example of an integer that is not a natural number (Section 1.2; Example 5)

SECTION 1.7

Objectives

1 Multiply real numbers.
2 Multiply more than two real numbers.
3 Find multiplicative inverses.
4 Use the definition of division.
5 Divide real numbers.
6 Simplify algebraic expressions involving multiplication.
7 Use algebraic expressions that model reality.

MULTIPLICATION AND DIVISION OF REAL NUMBERS

Technology is now promising to bring light, fast, and beautiful wheelchairs to millions of disabled people. The cost of manufacturing these radically different wheelchairs can be modeled by an algebraic expression containing division. In this section, we will see how this algebraic expression illustrates that low prices are possible with high production levels, urgently needed in this situation. There are more than half a billion people with disabilities in developing countries; an estimated 20 million need wheelchairs right now.

Multiplying Real Numbers Suppose that things go from bad to worse for Social Security, and the projected \$244 billion deficit in 2020 triples by the end of the twenty-first century. The new deficit is

$$3(-244) = (-244) + (-244) + (-244) = -732$$

or \$732 billion. Thus,

$$3(-244) = -732.$$

The result of the multiplication, −732, is called the **product** of 3 and −244. The numbers being multiplied, 3 and −244, are called the **factors** of the product.

Rules for multiplying real numbers are described in terms of absolute value. For example, $3(-244) = -732$ illustrates that the product of numbers with different signs is found by multiplying their absolute values. The product is negative.

$$3(-244) = -732$$

Multiply absolute values: $|3| \cdot |-244| = 3 \cdot 244 = 732$.

Factors have different signs and the product is negative.

1 Multiply real numbers.

The following rules are used to determine the sign of the product of two numbers:

THE PRODUCT OF TWO REAL NUMBERS

- The product of two real numbers with **different signs** is found by multiplying their absolute values. The product is **negative**.
- The product of two real numbers with the **same sign** is found by multiplying their absolute values. The product is **positive**.
- The product of 0 and any real number is 0. Thus, for any real number a,

$$a \cdot 0 = 0 \quad \text{and} \quad 0 \cdot a = 0.$$

EXAMPLE 1 Multiplying Real Numbers

Multiply:

a. $6(-3)$ **b.** $-\frac{1}{5}\cdot\frac{2}{3}$ **c.** $(-9)(-10)$ **d.** $(-1.4)(-2)$ **e.** $(-372)(0)$.

SOLUTION

a. $6(-3) = -18$

Multiply absolute values: $6 \cdot 3 = 18$.

Different signs: negative product

b. $-\frac{1}{5}\cdot\frac{2}{3} = -\frac{2}{15}$

Multiply absolute values: $\frac{1}{5}\cdot\frac{2}{3} = \frac{1\cdot 2}{5\cdot 3} = \frac{2}{15}$.

Different signs: negative product

c. $(-9)(-10) = 90$

Multiply absolute values: $9 \cdot 10 = 90$.

Same sign: positive product

d. $(-1.4)(-2) = 2.8$

Multiply absolute values: $(1.4)(2) = 2.8$.

Same sign: positive product

e. $(-372)(0) = 0$

The product of 0 and any real number is 0: $a \cdot 0 = 0$. ■

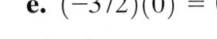

✔ CHECK POINT 1 Multiply: **a.** $8(-5)$ **b.** $-\frac{1}{3}\cdot\frac{4}{7}$ **c.** $(-12)(-3)$ **d.** $(-1.1)(-5)$ **e.** $(-543)(0)$.

2 Multiply more than two real numbers.

Multiplying More Than Two Numbers How do we perform more than one multiplication, such as

$$-4(-3)(-2)?$$

Because of the associative and commutative properties, we can order and group the numbers in any manner. Each pair of negative numbers will produce a positive product. Thus, the product of an even number of negative numbers is always positive. By contrast, the product of an odd number of negative numbers is always negative.

$$-4(-3)(-2) = -24$$

Multiply absolute values: $4 \cdot 3 \cdot 2 = 24$.

Odd number of negative numbers (three): negative product

MULTIPLYING MORE THAN TWO NUMBERS

1. Assuming that no factor is zero,
   - The product of an **even** number of **negative numbers** is **positive**.
   - The product of an **odd** number of **negative numbers** is **negative**.

   The multiplication is performed by multiplying the absolute values of the given numbers.
2. If any factor is 0, the product is 0.

EXAMPLE 2 Multiplying More Than Two Numbers

Multiply: **a.** $(-3)(-1)(2)(-2)$ **b.** $(-1)(-2)(-2)(3)(-4)$.

SOLUTION

a. $(-3)(-1)(2)(-2) = -12$

Multiply absolute values: $3 \cdot 1 \cdot 2 \cdot 2 = 12$.

Odd number of negative numbers (three): negative product

b. $(-1)(-2)(-2)(3)(-4) = 48$

Multiply absolute values: $1 \cdot 2 \cdot 2 \cdot 3 \cdot 4 = 48$.

Even number of negative numbers (four): positive product

■

CHECK POINT 2 Multiply:

a. $(-2)(3)(-1)(4)$ **b.** $(-1)(-3)(2)(-1)(5)$.

Is it always necessary to count the number of negative factors when multiplying more than two numbers? No. If any factor is 0, you can immediately write 0 for the product. For example,

$$(-37)(423)(0)(-55)(-3.7) = 0.$$

If any factor is 0, the product is 0.

3 Find multiplicative inverses.

The Meaning of Division The result of dividing the real number a by the nonzero real number b is called the **quotient** of a and b. We can write this quotient as $a \div b$ or $\frac{a}{b}$.

We know that subtraction is defined in terms of addition of an additive inverse, or opposite:

$$a - b = a + (-b).$$

In a similar way, we can define division in terms of multiplication. For example, the quotient of 8 and 2 can be written as multiplication:

$$8 \div 2 = 8 \cdot \frac{1}{2}.$$

We call $\frac{1}{2}$ the *multiplicative inverse*, or *reciprocal*, of 2. Two numbers whose product is 1 are called **multiplicative inverses** or **reciprocals** of each other. Thus, the multiplicative inverse of 2 is $\frac{1}{2}$ and the multiplicative inverse of $\frac{1}{2}$ is 2 because $2 \cdot \frac{1}{2} = 1$.

EXAMPLE 3 Finding Multiplicative Inverses

Find the multiplicative inverse of each number:

a. 5 **b.** $\frac{1}{3}$ **c.** -4 **d.** $-\frac{4}{5}$.

SOLUTION

a. The multiplicative inverse of 5 is $\frac{1}{5}$ because $5 \cdot \frac{1}{5} = 1$.

b. The multiplicative inverse of $\frac{1}{3}$ is 3 because $\frac{1}{3} \cdot 3 = 1$.

c. The multiplicative inverse of -4 is $-\frac{1}{4}$ because $(-4)\left(-\frac{1}{4}\right) = 1$.

d. The multiplicative inverse of $-\frac{4}{5}$ is $-\frac{5}{4}$ because $\left(-\frac{4}{5}\right)\left(-\frac{5}{4}\right) = 1$. ■

CHECK POINT 3 Find the multiplicative inverse of each number:

a. 7 **b.** $\frac{1}{8}$ **c.** -6 **d.** $-\frac{7}{13}$.

Can you think of a real number that has no multiplicative inverse? The number **0 has no multiplicative inverse** because 0 multiplied by any number is never 1, but always 0.

We now define division in terms of multiplication by a multiplicative inverse.

4 Use the definition of division.

DEFINITION OF DIVISION If a and b are real numbers and b is not 0, then the quotient of a and b is defined as

$$a \div b = a \cdot \frac{1}{b}.$$

In words: The quotient of two real numbers is the product of the first number and the multiplicative inverse of the second number.

USING TECHNOLOGY

You can use a calculator to multiply and divide signed numbers. Here are the keystrokes for finding

$$(-173)(-256):$$

Scientific Calculator

173 [+/−] [×] 256 [+/−] [=]

Graphing Calculator

[(−)] 173 [×] [(−)] 256 [ENTER]

The number 44288 should be displayed.

Division is performed in the same manner, using [÷] instead of [×]. What happens when you divide by 0? Try entering

8 [÷] 0

and pressing [=] or [ENTER].

EXAMPLE 4 Using the Definition of Division

Use the definition of division to find each quotient: **a.** $-15 \div 3$ **b.** $\frac{-20}{-4}$.

SOLUTION

a. $-15 \div 3 = -15 \cdot \frac{1}{3} = -5$

Change the division to multiplication. Replace 3 with its multiplicative inverse.

b. $\frac{-20}{-4} = -20 \cdot \left(-\frac{1}{4}\right) = 5$

Change the division to multiplication. Replace −4 with its multiplicative inverse. ■

CHECK POINT 4 Use the definition of division to find each quotient:

a. $-28 \div 7$ **b.** $\frac{-16}{-2}$.

5 Divide real numbers.

A Procedure for Dividing Real Numbers Because the quotient $a \div b$ is defined as the product $a \cdot \frac{1}{b}$, the sign rules for dividing numbers are the same as the sign rules for multiplying them.

THE QUOTIENT OF TWO REAL NUMBERS

- The quotient of two real numbers with **different signs** is found by dividing their absolute values. The quotient is **negative**.
- The quotient of two real numbers with the **same sign** is found by dividing their absolute values. The quotient is **positive**.
- Division of a nonzero number by zero is undefined.
- Any nonzero number divided into 0 is 0.

EXAMPLE 5 Dividing Real Numbers

Divide: **a.** $\frac{8}{-2}$ **b.** $-\frac{3}{4} \div \left(-\frac{5}{9}\right)$ **c.** $\frac{-20.8}{4}$ **d.** $\frac{0}{-7}$.

SOLUTION

a. $\frac{8}{-2} = -4$

Divide absolute values: $\frac{8}{2} = 4$.

Different signs: negative quotient

b. $-\frac{3}{4} \div \left(-\frac{5}{9}\right) = \frac{27}{20}$

Divide absolute values: $\frac{3}{4} \div \frac{5}{9} = \frac{3}{4} \cdot \frac{9}{5} = \frac{27}{20}$.

Same sign: positive quotient

c. $\frac{-20.8}{4} = -5.2$

Divide absolute values: $4\overline{)20.8}$ = 5.2.

Different signs: negative quotient

d. $\frac{0}{-7} = 0$

Any nonzero number divided into 0 is 0.

STUDY TIP

Here's another reason why division by zero is undefined. We know that

$$\frac{12}{4} = 3 \text{ because } 3 \cdot 4 = 12.$$

Now think about $\frac{-7}{0}$. If

$$\frac{-7}{0} = ? \text{ then } ? \cdot 0 = -7.$$

However, any real number multiplied by 0 is 0 and not −7. No matter what number we replace by ? in $? \cdot 0 = -7$, we can never obtain −7.

Can you see why $\frac{0}{-7}$ must be 0? The definition of division tells us that

$$\frac{0}{-7} = 0 \cdot \left(-\frac{1}{7}\right)$$

and the product of 0 and any real number is 0. By contrast, the definition of division does not allow for division by 0 because 0 does not have a multiplicative inverse. It is incorrect to write

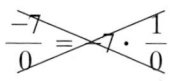

$$\frac{-7}{0} = -7 \cdot \frac{1}{0}.$$

(crossed out) 0 does not have a multiplicative inverse.

Division by zero is not allowed or not defined. Thus, $\frac{-7}{0}$ does not represent a real number. A real number can never have a denominator of 0.

CHECK POINT **5** Divide:

a. $\frac{-32}{-4}$ **b.** $-\frac{2}{3} \div \frac{5}{4}$ **c.** $\frac{21.9}{-3}$ **d.** $\frac{0}{-5}$.

6 Simplify algebraic expressions involving multiplication.

Multiplication and Algebraic Expressions In Section 1.4, we discussed the commutative and associative properties of multiplication. We also know that multiplication distributes over addition and subtraction. We now add some additional properties to our previous list (Table 1.5). These properties are frequently helpful in simplifying algebraic expressions.

Table 1.5 Additional Properties of Multiplication

Let *a* be a real number, a variable, or an algebraic expression.

Property	Meaning	Examples
Identity Property of Multiplication	1 can be deleted from a product. $a \cdot 1 = a$ $1 \cdot a = a$	• $\sqrt{3} \cdot 1 = \sqrt{3}$ • $1x = x$ • $1(2x + 3) = 2x + 3$
Inverse Property of Multiplication	If a is not 0: $a \cdot \frac{1}{a} = 1$ $\frac{1}{a} \cdot a = 1$ The product of a nonzero number and its multiplicative inverse, or reciprocal, gives 1, the multiplicative identity.	• $6 \cdot \frac{1}{6} = 1$ • $3x \cdot \frac{1}{3x} = 1$ (x is not 0.) • $\frac{1}{(y-2)} \cdot (y - 2) = 1$ (y is not 2.)
Multiplication Property of −1	Negative 1 times a is the additive inverse, or opposite, of a. $-1 \cdot a = -a$ $a(-1) = -a$	• $-1 \cdot \sqrt{3} = -\sqrt{3}$ • $-1\left(-\frac{3}{4}\right) = \frac{3}{4}$ • $-1x = -x$ • $-(x + 4) = -1(x + 4) = -x - 4$
Double Negative Property	The additive inverse of $-a$ is a. $-(-a) = a$	• $-(-4) = 4$ • $-(-6y) = 6y$

In the preceding table, we used two steps to remove the parentheses from $-(x + 4)$. First, we used the multiplication property of -1.

$$-(x + 4) = -1(x + 4)$$

Then we used the distributive property, distributing -1 to each term in parentheses.

$$-1(x + 4) = (-1)x + (-1)4 = -x + (-4) = -x - 4$$

There is a fast way to obtain $-(x + 4) = -x - 4$ in just one step.

NEGATIVE SIGNS AND PARENTHESES If a negative sign precedes parentheses, remove the parentheses and change the sign of every term within the parentheses.

Here are some examples that illustrate this method.

$$-(11x + 5) = -11x - 5$$
$$-(11x - 5) = -11x + 5$$
$$-(-11x + 5) = 11x - 5$$
$$-(-11x - 5) = 11x + 5$$

EXAMPLE 6 Simplifying Algebraic Expressions

Simplify: **a.** $-2(3x)$ **b.** $6x + x$ **c.** $8a - 9a$ **d.** $-3(2x - 5)$ **e.** $-(3y - 8)$.

SOLUTION We will show all steps in the solution process. However, you probably are working many of these steps mentally.

DISCOVER FOR YOURSELF

Verify each simplification in Example 6 by substituting 5 for the variable. The value of the given expression should be the same as the value of the simplified expression. Which expression is easier to evaluate?

a. $-2(3x)$ This is the given algebraic expression.
$= (-2 \cdot 3)x$ Use the associative property and group the first two numbers.
$= -6x$ Numbers with opposite signs have a negative product.

b. $6x + x$ This is the given algebraic expression.
$= 6x + 1x$ Use the multiplication property of 1.
$= (6 + 1)x$ Apply the distributive property.
$= 7x$ Add within parentheses.

c. $8a - 9a$ This is the given algebraic expression.
$= (8 - 9)a$ Apply the distributive property.
$= -1a$ Subtract within parentheses: $8 - 9 = 8 + (-9) = -1$.
$= -a$ Apply the multiplication property of -1.

d. $-3(2x - 5)$ This is the given algebraic expression.
$= -3(2x) - (-3) \cdot (5)$ Apply the distributive property.
$= -6x - (-15)$ Multiply.
$= -6x + 15$ Subtraction is the addition of an additive inverse.

e. $-(3y - 8)$ This is the given algebraic expression.
$= -3y + 8$ Remove parentheses by changing the sign of every term inside the parentheses. ■

CHECK POINT 6 Simplify: **a.** $-4(5x)$ **b.** $9x + x$ **c.** $13b - 14b$ **d.** $-7(3x - 4)$ **e.** $-(7y - 6)$.

Before turning to applications, let's try one additional example involving simplification.

EXAMPLE 7 Simplifying an Algebraic Expression

Simplify: $5(2y - 9) - (9y - 8)$.

SOLUTION

$5(2y - 9) - (9y - 8)$ This is the given algebraic expression.
$= 5 \cdot 2y - 5 \cdot 9 - (9y - 8)$ Apply the distributive property over the first parentheses.
$= 10y - 45 - (9y - 8)$ Multiply.
$= 10y - 45 - 9y + 8$ Remove the second parentheses by changing the sign of each term within parentheses.
$= (10y - 9y) + (-45 + 8)$ Group like terms.
$= 1y + (-37)$ Combine like terms. For the variable terms, $10y - 9y = 10y + (-9y) = [10 + (-9)]y = 1y$.
$= y + (-37)$ Use the multiplication property of 1: $1y = y$.
$= y - 37$ Express addition of an additive inverse as subtraction. ■

CHECK POINT 7 Simplify: $4(3y - 7) - (13y - 2)$

A Summary of Operations with Real Numbers Operations with real numbers are summarized in Table 1.6.

Table 1.6 Summary of Operations with Real Numbers

Signs of Numbers	Addition	Subtraction	Multiplication	Division
Both Numbers Are Positive	Sum Is Always Positive	Difference May Be Either Positive or Negative	Product Is Always Positive	Quotient Is Always Positive
Examples 8 and 2 2 and 8	$8 + 2 = 10$ $2 + 8 = 10$	$8 - 2 = 6$ $2 - 8 = -6$	$8 \cdot 2 = 16$ $2 \cdot 8 = 16$	$8 \div 2 = 4$ $2 \div 8 = \frac{1}{4}$
One Number Is Positive and the Other Number Is Negative	Sum May Be Either Positive or Negative	Difference May Be Either Positive or Negative	Product Is Always Negative	Quotient Is Always Negative
Examples 8 and −2 −8 and 2	$8 + (-2) = 6$ $-8 + 2 = -6$	$8 - (-2) = 10$ $-8 - 2 = -10$	$8(-2) = -16$ $-8(2) = -16$	$8 \div (-2) = -4$ $-8 \div 2 = -4$
Both Numbers Are Negative	Sum Is Always Negative	Difference May Be Either Positive or Negative	Product Is Always Positive	Quotient Is Always Positive
Examples −8 and −2 −2 and −8	$-8 + (-2) = -10$ $-2 + (-8) = -10$	$-8 - (-2) = -6$ $-2 - (-8) = 6$	$-8(-2) = 16$ $-2(-8) = 16$	$-8 \div (-2) = 4$ $-2 \div (-8) = \frac{1}{4}$

7 Use algebraic expressions that model reality.

Applications Algebraic expressions that model reality frequently contain division.

EXAMPLE 8 Average Cost of Producing a Wheelchair

A company that manufactures wheelchairs has monthly fixed costs of $500,000. The average cost per wheelchair for the company to manufacture x wheelchairs per month is modeled by the algebraic expression

$$\frac{400x + 500{,}000}{x}.$$

Find the average cost per wheelchair for the company to manufacture

a. 10,000 wheelchairs per month. **b.** 50,000 wheelchairs per month.

c. 100,000 wheelchairs per month.

What happens to the average cost per wheelchair as the production level increases?

SOLUTION

a. We are interested in the average cost per wheelchair for the company if 10,000 wheelchairs are manufactured per month. Because x represents the number of wheelchairs manufactured per month, we substitute 10,000 for x in the given algebraic expression.

$$\frac{400x + 500{,}000}{x} = \frac{400(10{,}000) + 500{,}000}{10{,}000} = \frac{4{,}000{,}000 + 500{,}000}{10{,}000}$$

$$= \frac{4{,}500{,}000}{10{,}000} = 450$$

The average cost per wheelchair of producing 10,000 wheelchairs per month is \$450.

b. Now, 50,000 wheelchairs are manufactured per month. We find the average cost per wheelchair by substituting 50,000 for x in the given algebraic expression.

$$\frac{400x + 500{,}000}{x} = \frac{400(50{,}000) + 500{,}000}{50{,}000} = \frac{20{,}000{,}000 + 500{,}000}{50{,}000}$$

$$= \frac{20{,}500{,}000}{50{,}000} = 410$$

The average cost per wheelchair of producing 50,000 wheelchairs per month is \$410.

c. Finally, the production level has increased to 100,000 wheelchairs per month. We find the average cost per wheelchair for the company by substituting 100,000 for x in the given algebraic expression.

$$\frac{400x + 500{,}000}{x} = \frac{400(100{,}000) + 500{,}000}{100{,}000} = \frac{40{,}000{,}000 + 500{,}000}{100{,}000}$$

$$= \frac{40{,}500{,}000}{100{,}000} = 405$$

The average cost per wheelchair of producing 100,000 wheelchairs per month is \$405.

As the production level increases, the average cost of producing each wheelchair decreases. This illustrates the difficulty with small businesses. It is nearly impossible to have competitively low prices when production levels are low. ■

The points in the rectangular coordinate system in Figure 1.21 show the relationship between production level and cost. The x-axis represents the number of wheelchairs produced per month. The y-axis represents the average cost per wheelchair for the company. The symbol ⌇ on the y-axis indicates that there is a break in the values of y between 0 and 400. Thus, the values of y begin at 400.

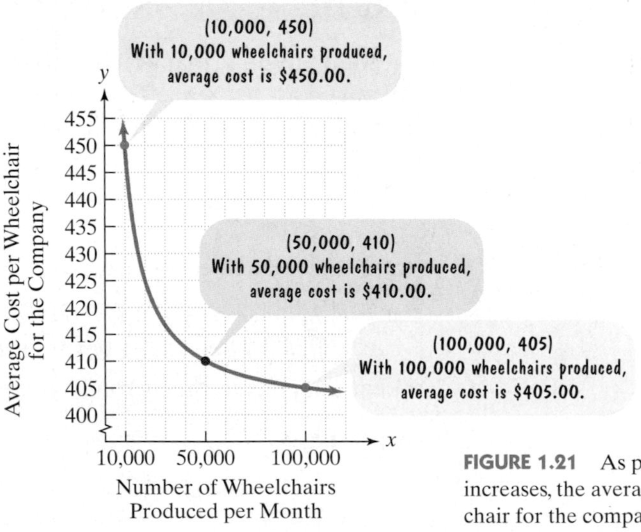

FIGURE 1.21 As production level increases, the average cost per wheelchair for the company decreases.

The three points with the voice balloons illustrate our computations in Example 8. The points in Figure 1.21 are falling from left to right. Can you see how this shows that the company's cost per wheelchair is decreasing as their production level increases?

CHECK POINT 8 A company that manufactures running shoes has weekly fixed costs of \$300,000. The average cost per pair of running shoes for the company to manufacture x pairs per week is modeled by the algebraic expression

$$\frac{30x + 300{,}000}{x}.$$

Find the average cost per pair of running shoes for the company to manufacture

a. 1000 pairs per week. **b.** 10,000 pairs per week.

c. 100,000 pairs per week.

1.7 EXERCISE SET

Student Solutions Manual | CD/Video | PH Math/Tutor Center | MathXL Tutorials on CD | MathXL® | MyMathLab | Interactmath.com

Practice Exercises

In Exercises 1–34, perform the indicated multiplication.

1. $5(-9)$
2. $10(-7)$
3. $(-8)(-3)$
4. $(-9)(-5)$
5. $(-3)(7)$
6. $(-4)(8)$
7. $(-19)(-1)$
8. $(-11)(-1)$
9. $0(-19)$
10. $0(-11)$
11. $\frac{1}{2}(-24)$
12. $\frac{1}{3}(-21)$
13. $\left(-\frac{3}{4}\right)(-12)$
14. $\left(-\frac{4}{5}\right)(-30)$
15. $-\frac{3}{5}\cdot\left(-\frac{4}{7}\right)$
16. $-\frac{5}{7}\cdot\left(-\frac{3}{8}\right)$
17. $-\frac{7}{9}\cdot\frac{2}{3}$
18. $-\frac{5}{11}\cdot\frac{2}{7}$
19. $3(-1.2)$
20. $4(-1.2)$
21. $-0.2(-0.6)$
22. $-0.3(-0.7)$
23. $(-5)(-2)(3)$
24. $(-6)(-3)(10)$
25. $(-4)(-3)(-1)(6)$
26. $(-2)(-7)(-1)(3)$
27. $-2(-3)(-4)(-1)$
28. $-3(-2)(-5)(-1)$
29. $(-3)(-3)(-3)$
30. $(-4)(-4)(-4)$
31. $5(-3)(-1)(2)(3)$
32. $2(-5)(-2)(3)(1)$
33. $(-8)(-4)(0)(-17)(-6)$
34. $(-9)(-12)(-18)(0)(-3)$

In Exercises 35–42, find the multiplicative inverse of each number.

35. 4
36. 3
37. $\frac{1}{5}$
38. $\frac{1}{7}$
39. -10
40. -12
41. $-\frac{2}{5}$
42. $-\frac{4}{9}$

In Exercises 43–46,

a. *Rewrite the division as multiplication involving a multiplicative inverse.*

b. *Use the multiplication from part (a) to find the given quotient.*

43. $-32 \div 4$
44. $-18 \div 6$
45. $\frac{-60}{-5}$
46. $\frac{-30}{-5}$

In Exercises 47–76, perform the indicated division or state that the expression is undefined.

47. $\frac{12}{-4}$
48. $\frac{40}{-5}$
49. $\frac{-21}{3}$
50. $\frac{-60}{6}$
51. $\frac{-90}{-3}$
52. $\frac{-66}{-6}$
53. $\frac{0}{-7}$
54. $\frac{0}{-8}$
55. $\frac{7}{0}$
56. $\frac{-8}{0}$
57. $-15 \div 3$
58. $-80 \div 8$
59. $120 \div (-10)$
60. $130 \div (-10)$
61. $(-180) \div (-30)$
62. $(-150) \div (-25)$
63. $0 \div (-4)$
64. $0 \div (-10)$
65. $-4 \div 0$
66. $-10 \div 0$
67. $\frac{-12.9}{3}$
68. $\frac{-21.6}{3}$

69. $-\frac{1}{2} \div \left(-\frac{3}{5}\right)$

70. $-\frac{1}{2} \div \left(-\frac{7}{9}\right)$

71. $-\frac{14}{9} \div \frac{7}{8}$

72. $-\frac{5}{16} \div \frac{25}{8}$

73. $\frac{1}{3} \div \left(-\frac{1}{3}\right)$

74. $\frac{1}{5} \div \left(-\frac{1}{5}\right)$

75. $6 \div \left(-\frac{2}{5}\right)$

76. $8 \div \left(-\frac{2}{9}\right)$

In Exercises 77–96, simplify each algebraic expression.

77. $-5(2x)$

78. $-9(3x)$

79. $-4\left(-\frac{3}{4}y\right)$

80. $-5\left(-\frac{3}{5}y\right)$

81. $8x + x$

82. $12x + x$

83. $-5x + x$

84. $-6x + x$

85. $6b - 7b$

86. $12b - 13b$

87. $-y + 4y$

88. $-y + 9y$

89. $-4(2x - 3)$

90. $-3(4x - 5)$

91. $-3(-2x + 4)$

92. $-4(-3x + 2)$

93. $-(2y - 5)$

94. $-(3y - 1)$

95. $4(2y - 3) - (7y + 2)$

96. $5(3y - 1) - (14y - 2)$

Practice Plus

In Exercises 97–104, write a numerical expression for each phrase. Then simplify the numerical expression by performing the given operations.

97. 8 added to the product of 4 and −10

98. 14 added to the product of 3 and −15

99. The product of −9 and −3, decreased by −2

100. The product of −6 and −4, decreased by −5

101. The quotient of −18 and the sum of −15 and 12

102. The quotient of −25 and the sum of −21 and 16

103. The difference between −6 and the quotient of 12 and −4

104. The difference between −11 and the quotient of 20 and −5

Application Exercises

The graph shows the millions of welfare recipients in the United States who received cash assistance from 1994 through 2001. The data can be modeled by the algebraic expression

$$-1.4t + 14.7,$$

which represents the number of welfare recipients, in millions, t years after 1994. Use this algebraic expression to solve Exercises 105–106.

Welfare Recipients in the U.S.

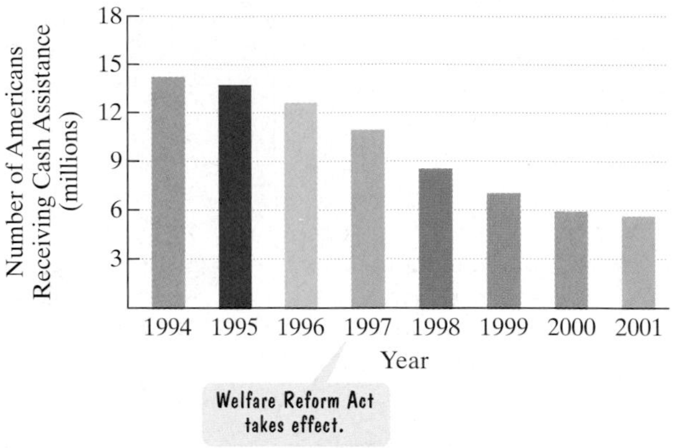

Source: Department of Health and Human Services

105. According to the model, how many welfare recipients were there in 2000? How well does the algebraic expression model the actual data for 2000 shown in the bar graph?

106. According to the model, how many welfare recipients were there in 1997? How well does the algebraic expression model the actual data for 1997 shown in the bar graph?

In an experiment on memory, students in a language class are asked to memorize 40 vocabulary words in Latin, a language with which the students are not familiar. After studying the words for one day, students are tested each day after to see how many words they remember. The class average is taken and the results are graphed as points in the rectangular coordinate system. Use the points shown to solve Exercises 107–108.

Average Number of Words Remembered over Time

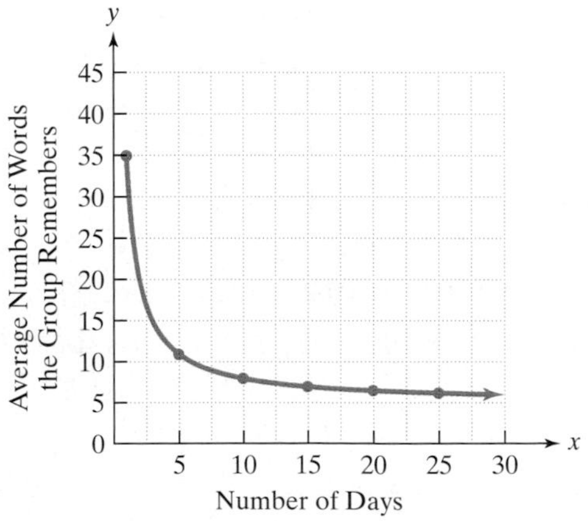

107. a. Find a reasonable estimate of the number of Latin words remembered after 5 days.

b. The algebraic expression

$$\frac{5x + 30}{x}$$

models the number of Latin words remembered by the students after x days. Use this expression to find the number of Latin words remembered after 5 days. How does this compare with your estimate from part (a)?

108. a. Find a reasonable estimate of the number of Latin words remembered after 15 days.

b. The algebraic expression

$$\frac{5x + 30}{x}$$

models the number of Latin words remembered by the students after x days. Use this expression to find the number of Latin words remembered after 15 days. How does this compare with your estimate from part (a)?

In Palo Alto, California, a government agency ordered computer-related companies to contribute to a pool of money to clean up underground water supplies. (The companies had stored toxic chemicals in leaking underground containers.) The algebraic expression

$$\frac{200x}{100 - x}$$

models the cost, in tens of thousands of dollars, for removing x percent of the contaminants. Use this algebraic expression to solve Exercises 109–110.

109. a. Substitute 50 for x and find the cost, in tens of thousands of dollars, for removing 50% of the contaminants.

b. Find the cost, in tens of thousands of dollars, for removing 80% of the contaminants.

c. Describe what is happening to the cost of the cleanup as the percentage of contaminant removed increases.

110. a. Substitute 60 for x and find the cost, in tens of thousands of dollars, for removing 60% of the contaminants.

b. Find the cost, in tens of thousands of dollars, for removing 90% of the contaminants.

c. Describe what is happening to the cost of the cleanup as the percentage of contaminants removed increases.

Writing in Mathematics

111. Explain how to multiply two real numbers. Provide examples with your explanation.

112. Explain how to determine the sign of a product that involves more than two numbers.

113. Explain how to find the multiplicative inverse of a number.

114. Why is it that 0 has no multiplicative inverse?

115. Explain how to divide real numbers.

116. Why is division by zero undefined?

117. Explain how to simplify an algebraic expression in which a negative sign precedes parentheses.

118. A politician promises to do "whatever it takes" to seize "one hundred percent" of all illegal drugs that enter the country. Suppose that the cost, in millions of dollars, for seizing x percent of the illegal drugs entering the country is modeled by the algebraic expression

$$\frac{130x}{100 - x}.$$

According to this algebraic expression, can the politician keep his or her promise? Explain your answer.

Critical Thinking Exercises

119. Which one of the following statements is true?

a. Multiplying a negative number by a nonnegative number will always give a negative number.

b. The product of two negative numbers is always a positive number.

c. The product of -3 and 4 is 12.

d. The product of real numbers a and b is not always equal to the product of real numbers b and a.

120. Which one of the following statements is true?

a. The product of two negative numbers is sometimes a negative number.

b. Both the addition and the multiplication of two negative numbers result in a positive number.

c. $\left(-\frac{1}{2}\right)\left(-\frac{1}{2}\right) = \frac{1}{4}$

d. Reversing the order of the two factors in a product results in a different answer.

In Exercises 121–124, write an algebraic expression for the given English phrase.

121. The value, in cents, of x nickels

122. The distance covered by a car traveling at 50 miles per hour for x hours

123. The monthly salary, in dollars, for a person earning x dollars per year

124. The fraction of people in a room who are women if there are 40 women and x men in the room

Technology Exercises

125. Use a calculator to verify any five of the products that you found in Exercises 1–34.

126. Use a calculator to verify any five of the quotients that you found in Exercises 47–76.

127. Simplify using a calculator:

$$0.3(4.7x - 5.9) - 0.07(3.8x - 61).$$

128. Use your calculator to attempt to find the quotient of -3 and 0. Describe what happens. Does the same thing occur when finding the quotient of 0 and -3? Explain the difference. Finally, what happens when you enter the quotient of 0 and itself?

Review Exercises

In Exercises 129–131, perform the indicated operation.

129. $-6 + (-3)$ (Section 1.5, Example 3)

130. $-6 - (-3)$ (Section 1.6, Example 1)

131. $-6 \div (-3)$ (Section 1.7, Example 4)

SECTION 1.8

Objectives

1 Evaluate exponential expressions.

2 Simplify algebraic expressions with exponents.

3 Use the order of operations agreement.

4 Evaluate formulas.

EXPONENTS, ORDER OF OPERATIONS, AND MATHEMATICAL MODELS

It's been another one of those days! Traffic is really backed up on the highway. Finally, you see the source of the traffic jam—a minor fender-bender. Still stuck in traffic, you notice that the driver appears to be quite young. This might seem like a strange observation. After all, what does a driver's age have to do with his or her chance of getting into an accident? In this section, we see how algebra describes your world, including a relationship between age and numbers of car accidents.

1 Evaluate exponential expressions.

Natural Number Exponents Although people do a great deal of talking, the total output since the beginning of gabble to the present day, including all baby talk, love songs, and congressional debates, only amounts to about 10 million billion words. This can be expressed as 16 factors of 10, or 10^{16} words.

Exponents such as 2, 3, 4, and so on are used to indicate repeated multiplication. For example,

$$10^2 = 10 \cdot 10 = 100,$$

$$10^3 = 10 \cdot 10 \cdot 10 = 1000, \quad 10^4 = 10 \cdot 10 \cdot 10 \cdot 10 = 10{,}000.$$

The 10 that is repeated when multiplying is called the **base**. The small numbers above and to the right of the base are called **exponents** or **powers**. The exponent tells the number of times the base is to be used when multiplying. In 10^3, the base is 10 and the exponent is 3.

Any number with an exponent of 1 is the number itself. Thus, $10^1 = 10$.

USING TECHNOLOGY

You can use a calculator to evaluate exponential expressions. For example, to evaluate 5^3, press the following keys:

Many Scientific Calculators

5 [y^x] 3 [=]

Many Graphing Calculators

5 [∧] 3 [ENTER]

Although calculators have special keys to evaluate powers of ten and squaring bases, you can always use one of the sequences shown here.

Multiplications that are expressed in exponential notation are read as follows:

10^1: "ten to the first power"
10^2: "ten to the second power" or "ten squared"
10^3: "ten to the third power" or "ten cubed"
10^4: "ten to the fourth power"
10^5: "ten to the fifth power"
etc.

Any real number can be used as the base. Thus,

$$7^2 = 7 \cdot 7 = 49 \text{ and } (-3)^4 = (-3)(-3)(-3)(-3) = 81.$$

The bases are 7 and -3, respectively. Do not confuse $(-3)^4$ and -3^4.

$$-3^4 = -(3 \cdot 3 \cdot 3 \cdot 3) = -81$$

The negative is not taken to the power because it is not inside parentheses.

An exponent applies only to a base. A negative sign is not part of a base unless it appears in parentheses.

EXAMPLE 1 Evaluating Exponential Expressions

Evaluate: **a.** 4^2 **b.** $(-5)^3$ **c.** $(-2)^4$ **d.** -2^4.

SOLUTION

Exponent is 2.

a. $4^2 = 4 \cdot 4 = 16$ — The exponent indicates that the base is used as a factor two times.

Base is 4.

We read $4^2 = 16$ as "4 to the second power is 16" or "4 squared is 16."

Exponent is 3.

b. $(-5)^3 = (-5)(-5)(-5)$ — The exponent indicates that the base is used as a factor three times.

Base is -5.

$= -125$ — An odd number of negative factors yields a negative product.

We read $(-5)^3 = -125$ as "the number negative 5 to the third power is negative 125" or "negative 5 cubed is negative 125."

Exponent is 4.

c. $(-2)^4 = (-2)(-2)(-2)(-2)$ — The exponent indicates the base is used as a factor four times.

Base is -2.

$= 16$ — An even number of negative factors yields a positive product.

We read $(-2)^4 = 16$ as "the number negative 2 to the fourth power is 16."

Exponent is 4.

d. $-2^4 = -(2 \cdot 2 \cdot 2 \cdot 2)$ The negative is not inside parentheses and is not taken to the fourth power.

Base is 2.

$= -16$ Multiply the twos and copy the negative.

We read $-2^4 = -16$ as "the negative of 2 raised to the fourth power is negative 16" or "the opposite, or additive inverse, of 2 raised to the fourth power is negative 16." ■

CHECK POINT 1 Evaluate:

a. 6^2 **b.** $(-4)^3$ **c.** $(-1)^4$ **d.** -1^4.

The formal algebraic definition of a natural number exponent summarizes our discussion:

DEFINITION OF A NATURAL NUMBER EXPONENT If b is a real number and n is a natural number,

$$b^n = \underbrace{b \cdot b \cdot b \cdot \ldots \cdot b}_{b \text{ appears as a factor } n \text{ times.}}$$

(Exponent: n; Base: b)

b^n is read "the nth power of b" or "b to the nth power." Thus, the nth power of b is defined as the product of n factors of b. The expression b^n is called an **exponential expression**.

Furthermore, $b^1 = b$.

2 Simplify algebraic expressions with exponents.

Exponents and Algebraic Expressions The distributive property can be used to simplify certain algebraic expressions that contain exponents. For example, we can use the distributive property to combine like terms in the algebraic expression $4x^2 + 6x^2$:

$$4x^2 + 6x^2 = (4 + 6)x^2 = 10x^2.$$

First term with variable factor x^2; Second term with variable factor x^2; The common variable factor is x^2.

EXAMPLE 2 Simplifying Algebraic Expressions

Simplify, if possible: **a.** $7x^3 + 2x^3$ **b.** $5x^2 + x^2$ **c.** $3x^2 + 4x^3$.

SOLUTION

a. $7x^3 + 2x^3$ There are two like terms with the same variable factor, namely x^3.

$= (7 + 2)x^3$ Apply the distributive property.

$= 9x^3$ Add within parentheses.

STUDY TIP

When adding algebraic expressions, if you have like terms you add only the numerical coefficients—not the exponents. **Exponents are never added when the operation is addition.** Avoid these common errors.

- ~~$7x^3 + 2x^3 = 9x^6$~~
- ~~$5x^2 + x^2 = 6x^4$~~
- ~~$3x^2 + 4x^3 = 7x^5$~~

INCORRECT

b. $5x^2 + x^2$ — There are two like terms with the same variable factor, namely x^2.

$= 5x^2 + 1x^2$ — Use the multiplication property of 1.

$= (5 + 1)x^2$ — Apply the distributive property.

$= 6x^2$ — Add within parentheses.

c. $3x^2 + 4x^3$ cannot be simplified. The terms $3x^2$ and $4x^3$ are not like terms because they have different variable factors, namely x^2 and x^3. ■

CHECK POINT 2 Simplify, if possible:

a. $16x^2 + 5x^2$ **b.** $7x^3 + x^3$ **c.** $10x^2 + 8x^3$.

3 Use the order of operations agreement.

Order of Operations Suppose that you want to find the value of $3 + 7 \cdot 5$. Which procedure shown is correct?

$$3 + 7 \cdot 5 = 3 + 35 = 38 \quad \text{or} \quad 3 + 7 \cdot 5 = 10 \cdot 5 = 50$$

If you know the answer, you probably know certain rules, called the **order of operations**, to make sure that there is only one correct answer. One of these rules states that if a problem contains no parentheses or other grouping symbols, perform multiplication before addition. Thus, the procedure on the left is correct because the multiplication of 7 and 5 is done first. Then the addition is performed. The correct answer is 38.

Some problems contain grouping symbols, such as parentheses, ( ); brackets, [ ]; braces, { }; absolute value symbols, | |; or fraction bars. These grouping symbols tell us what to do first. Here are two examples:

- $(3 + 7) \cdot 5 = 10 \cdot 5 = 50$

First, perform operations in grouping symbols.

- $8|6 - 16| = 8|-10| = 8 \cdot 10 = 80.$

Here are the rules for determining the order in which operations should be performed:

STUDY TIP

Here's a sentence to help remember the order of operations: "Please excuse my dear Aunt Sally."

Please	**P**arentheses
Excuse	**E**xponents
{**M**y	{**M**ultiplication
Dear}	**D**ivision}
{**A**unt	{**A**ddition
Sally}	**S**ubtraction}

ORDER OF OPERATIONS

1. Perform all operations within grouping symbols.
2. Evaluate all exponential expressions.
3. Do all multiplications and divisions in the order in which they occur, working from left to right.
4. Finally, do all additions and subtractions in the order in which they occur, working from left to right.

In the third step, be sure to do all multiplications and divisions *as they occur* from left to right. For example,

$$8 \div 4 \cdot 2 = 2 \cdot 2 = 4$$ Do the division first because it occurs first.

$$8 \cdot 4 \div 2 = 32 \div 2 = 16.$$ Do the multiplication first because it occurs first.

EXAMPLE 3 Using the Order of Operations

Simplify: $18 + 2 \cdot 3 - 10$.

SOLUTION There are no grouping symbols or exponential expressions. In cases like this, we multiply and divide before adding and subtracting.

$$18 + 2 \cdot 3 - 10 = 18 + 6 - 10$$ Multiply: $2 \cdot 3 = 6$.

$$= 24 - 10$$ Add and subtract from left to right: $18 + 6 = 24$.

$$= 14$$ Subtract: $24 - 10 = 14$. ■

✔ **CHECK POINT 3** Simplify: $20 + 4 \cdot 3 - 17$.

EXAMPLE 4 Using the Order of Operations

Simplify: $6^2 - 24 \div 2^2 \cdot 3 - 1$.

SOLUTION There are no grouping symbols. Thus, we begin by evaluating exponential expressions. Then we multiply or divide. Finally, we add or subtract.

$$6^2 - 24 \div 2^2 \cdot 3 - 1$$

$$= 36 - 24 \div 4 \cdot 3 - 1$$ Evaluate exponential expressions: $6^2 = 6 \cdot 6 = 36$ and $2^2 = 2 \cdot 2 = 4$.

$$= 36 - 6 \cdot 3 - 1$$ Perform the multiplications and divisions from left to right. Start with $24 \div 4 = 6$.

$$= 36 - 18 - 1$$ Now do the multiplication: $6 \cdot 3 = 18$.

$$= 18 - 1$$ Finally, perform the subtraction from left to right: $36 - 18 = 18$.

$$= 17$$ Complete the subtraction: $18 - 1 = 17$. ■

✔ **CHECK POINT 4** Simplify: $7^2 - 48 \div 4^2 \cdot 5 - 2$.

EXAMPLE 5 Using the Order of Operations

Simplify: **a.** $(2 \cdot 5)^2$ **b.** $2 \cdot 5^2$.

SOLUTION

a. Because $(2 \cdot 5)^2$ contains grouping symbols, namely parentheses, we perform the operation within parentheses first.

$$(2 \cdot 5)^2 = 10^2$$ Multiply within parentheses: $2 \cdot 5 = 10$.

$$= 100$$ Evaluate the exponential expression: $10^2 = 10 \cdot 10 = 100$.

b. Because $2 \cdot 5^2$ does not contain grouping symbols, we begin by evaluating the exponential expression.

$$2 \cdot 5^2 = 2 \cdot 25$$ Evaluate the exponential expression: $5^2 = 5 \cdot 5 = 25$.

$$= 50$$ Now do the multiplication: $2 \cdot 25 = 50$. ■

✔ **CHECK POINT 5** Simplify: **a.** $(3 \cdot 2)^2$ **b.** $3 \cdot 2^2$.

EXAMPLE 6 Using the Order of Operations

Simplify: $\left(\frac{1}{2}\right)^3 - \left(\frac{1}{2} - \frac{3}{4}\right)^2(-4)$.

SOLUTION Because grouping symbols appear, we perform the operation within parentheses first.

$$\left(\frac{1}{2}\right)^3 - \left(\frac{1}{2} - \frac{3}{4}\right)^2(-4)$$

$$= \left(\frac{1}{2}\right)^3 - \left(-\frac{1}{4}\right)^2(-4)$$ Work inside parentheses first: $\frac{1}{2} - \frac{3}{4} = \frac{2}{4} - \frac{3}{4} = \frac{2}{4} + \left(-\frac{3}{4}\right) = -\frac{1}{4}$.

$$= \frac{1}{8} - \frac{1}{16}(-4)$$ Evaluate exponential expressions: $\left(\frac{1}{2}\right)^3 = \frac{1}{2}\cdot\frac{1}{2}\cdot\frac{1}{2} = \frac{1}{8}$ and $\left(-\frac{1}{4}\right)^2 = \left(-\frac{1}{4}\right)\left(-\frac{1}{4}\right) = \frac{1}{16}$.

$$= \frac{1}{8} - \left(-\frac{1}{4}\right)$$ Multiply: $\frac{1}{16}\cdot\left(\frac{-4}{1}\right) = -\frac{4}{16} = -\frac{1}{4}$.

$$= \frac{3}{8}$$ Subtract: $\frac{1}{8} - \left(-\frac{1}{4}\right) = \frac{1}{8} + \frac{1}{4} = \frac{1}{8} + \frac{2}{8} = \frac{3}{8}$. ■

CHECK POINT 6 Simplify: $\left(-\frac{1}{2}\right)^2 - \left(\frac{7}{10} - \frac{8}{15}\right)^2(-18)$.

Some expressions contain many grouping symbols. An example of such an expression is $2[5(4 - 7) + 9]$. The grouping symbols are the parentheses and the brackets.

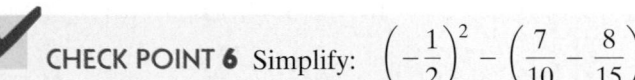

The parentheses, the innermost grouping symbols, group $4 - 7$.

$$2[5(4 - 7) + 9]$$

The brackets, the outermost grouping symbols, group $5(4 - 7) + 9$.

When combinations of grouping symbols appear, **perform operations within the innermost grouping symbols first**. Then work to the outside, performing operations within the outermost grouping symbols.

EXAMPLE 7 Using the Order of Operations

Simplify: $2[5(4 - 7) + 9]$.

SOLUTION

$$2[5(4 - 7) + 9]$$

$$= 2[5(-3) + 9]$$ Work inside parentheses first: $4 - 7 = 4 + (-7) = -3$.

$$= 2[-15 + 9]$$ Work inside brackets and multiply: $5(-3) = -15$.

$$= 2[-6]$$ Add inside brackets: $-15 + 9 = -6$. The resulting problem can also be expressed as $2(-6)$.

$$= -12$$ Multiply: $2[-6] = -12$. ■

Parentheses can be used for both innermost and outermost grouping symbols. For example, the expression $2[5(4-7)+9]$ can also be written $2(5(4-7)+9)$. However, too many parentheses can be confusing. The use of both parentheses and brackets makes it easier to identify inner and outer groupings.

CHECK POINT 7 Simplify: $4[3(6-11)+5]$.

EXAMPLE 8 Using the Order of Operations

Simplify: $18 \div 6 + 4[5 + 2(8-10)^3]$.

SOLUTION

$18 \div 6 + 4[5 + 2(8-10)^3]$

$= 18 \div 6 + 4[5 + 2(-2)^3]$ — Work inside parentheses first: $8 - 10 = 8 + (-10) = -2$.

$= 18 \div 6 + 4[5 + 2(-8)]$ — Work inside brackets and evaluate the exponential expression: $(-2)^3 = (-2)(-2)(-2) = -8$.

$= 18 \div 6 + 4[5 + (-16)]$ — Work inside brackets and multiply: $2(-8) = -16$.

$= 18 \div 6 + 4[-11]$ — Work inside brackets and add: $5 + (-16) = -11$.

$= 3 + 4[-11]$ — Perform the multiplications and divisions from left to right. Start with $18 \div 6 = 3$.

$= 3 + (-44)$ — Now do the multiplication: $4(-11) = -44$.

$= -41$ — Finally, perform the addition: $3 + (-44) = -41$. ■

CHECK POINT 8 Simplify: $25 \div 5 + 3[4 + 2(7-9)^3]$.

Fraction bars are grouping symbols that separate expressions into two parts, the numerator and the denominator. Consider, for example,

$$\frac{2(3-12) + 6 \cdot 4}{2^4 + 1}.$$

The fraction bar is the grouping symbol.

The numerator is one part of the expression.

The denominator is the other part of the expression.

We can use brackets instead of the fraction bar. An equivalent expression is

$$[2(3-12) + 6 \cdot 4] \div [2^4 + 1].$$

The grouping suggests a method for simplifying expressions with fraction bars as grouping symbols:

- Simplify the numerator.
- Simplify the denominator.
- If possible, simplify the fraction.

EXAMPLE 9 Using the Order of Operations

Simplify: $\dfrac{2(3-12)+6\cdot 4}{2^4+1}$.

SOLUTION

$$\frac{2(3-12)+6\cdot 4}{2^4+1}$$

$$=\frac{2(-9)+6\cdot 4}{16+1}$$ Work inside parentheses in the numerator: $3-12=3+(-12)=-9$. Evaluate the exponential expression in the denominator: $2^4=2\cdot 2\cdot 2\cdot 2=16$.

$$=\frac{-18+24}{16+1}$$ Multiply in the numerator: $2(-9)=-18$ and $6\cdot 4=24$.

$$=\frac{6}{17}$$ Perform the addition in the numerator and the denominator.

CHECK POINT 9 Simplify: $\dfrac{5(4-9)+10\cdot 3}{2^3-1}$.

EXAMPLE 10 Using the Order of Operations

Evaluate: $-x^2-7x$ for $x=-2$.

SOLUTION We begin by substituting -2 for each occurrence of x in the algebraic expression. Then we use the order of operations to evaluate the expression.

$$-x^2-7x$$

Replace x with -2.

$$=-(-2)^2-7(-2)$$

$$=-4-7(-2)$$ Evaluate the exponential expression: $(-2)^2=(-2)(-2)=4$.

$$=-4-(-14)$$ Multiply: $7(-2)=-14$.

$$=10$$ Subtract: $-4-(-14)=-4+14=10$.

CHECK POINT 10 Evaluate: $-x^2-4x$ for $x=-5$.

STUDY TIP

The evaluation in Example 10 might be easier by replacing each x with parentheses. Then place the value of x in each parentheses:

$$-x^2-7x$$
$$=-(\quad)^2-7(\quad)$$

Place the value of x here.

Some algebraic expressions contain two sets of grouping symbols. Using the order of operations, grouping symbols are removed from innermost (parentheses) to outermost (brackets).

EXAMPLE 11 Simplifying an Algebraic Expression

Simplify: $18x^2+4-[6(x^2-2)+5]$.

SOLUTION

$18x^2 + 4 - [6(x^2 - 2) + 5]$

$= 18x^2 + 4 - [6x^2 - 12 + 5]$ — Use the distributive property to remove parentheses: $6(x^2 - 2) = 6x^2 - 6 \cdot 2 = 6x^2 - 12$.

$= 18x^2 + 4 - [6x^2 - 7]$ — Add inside brackets: $-12 + 5 = -7$.

$= 18x^2 + 4 - 6x^2 + 7$ — Remove brackets by changing the sign of each term within brackets.

$= (18x^2 - 6x^2) + 4 + 7$ — Group like terms.

$= 12x^2 + 11$ — Combine like terms. ■

CHECK POINT 11 Simplify: $14x^2 + 5 - [7(x^2 - 2) + 4]$.

4 Evaluate formulas.

Applications: Formulas and Mathematical Models One aim of algebra is to provide a compact, symbolic description of the world. These descriptions involve the use of *formulas*. A **formula** is a statement of equality that uses letters to express a relationship between two or more variables. For example, one variety of crickets chirps faster as the temperature rises. You can calculate the temperature by counting the number of times a cricket chirps per minute and applying the following formula:

$$T = 0.3n + 40.$$

In the formula, T is the temperature, in degrees Fahrenheit, and n is the number of cricket chirps per minute. We can use this formula to determine the temperature if you are sitting on your porch and count 80 chirps per minute. Here is how to do so:

$T = 0.3n + 40$ — This is the given formula.

$T = 0.3(80) + 40$ — Substitute 80 for n.

$T = 24 + 40$ — Multiply: $0.3(80) = 24$.

$T = 64.$ — Add.

When there are 80 cricket chirps per minute, the temperature is 64 degrees.

The process of finding formulas to describe real-world phenomena is called **mathematical modeling**. Such formulas, together with the meaning assigned to the variables, are called **mathematical models**. We often say that these formulas model, or describe, the relationship among the variables.

In creating mathematical models, we strive for both accuracy and simplicity. For example, the formula $T = 0.3n + 40$ is relatively simple to use. However, you should not get upset if you count 80 cricket chirps and the actual temperature is 62 degrees, rather than 64 degrees, as predicted by the formula. Many mathematical models give an approximate, rather than an exact, description of the relationship between variables.

Sometimes a mathematical model gives an estimate that is not a good approximation or is extended to include values of the variable that do not make sense. In these cases, we say that **model breakdown** has occurred. Here is an example:

Use the mathematical model $T = 0.3n + 40$ with $n = 1200$ (1200 cricket chirps per minute).

$$T = 0.3(1200) + 40 = 360 + 40 = 400$$

At 400° F, forget about 1200 chirps per minute! At this temperature, the cricket would "cook" and, alas, all chirping would cease.

EXAMPLE 12 Car Accidents and Age

The mathematical model

$$N = 0.4x^2 - 36x + 1000$$

approximates the number of accidents, N, per 50 million miles driven, for drivers who are x years old. The formula applies to drivers ages 16 through 74, inclusive. How many accidents, per 50 million miles driven, are there for 20-year-old drivers?

SOLUTION In the mathematical model, x represents the age of the driver. We are interested in 20-year-old drivers. Thus, we substitute 20 for each occurrence of x. Then we use the order of operations to find N, the number of accidents per 50 million miles driven.

$N = 0.4x^2 - 36x + 1000$ — This is the given mathematical model.

$N = 0.4(20)^2 - 36(20) + 1000$ — Replace each occurrence of x with 20.

$N = 0.4(400) - 36(20) + 1000$ — Evaluate the exponential expression: $(20)^2 = (20)(20) = 400$.

$N = 160 - 720 + 1000$ — Multiply from left to right: $0.4(400) = 160$ and $36(20) = 720$.

$N = -560 + 1000$ — Perform additions and subtractions from left to right. Subtract: $160 - 720 = 160 + (-720) = -560$.

$N = 440$ — Add: $-560 + 1000 = 440$.

Thus, 20-year-old drivers have 440 accidents per 50 million miles driven. ■

STUDY TIP

To find

$$160 - 720 + 1000$$
$$= 160 + (-720) + 1000,$$

you can also first add the positive numbers

$$160 + 1000 = 1160$$

and then add the negative number to that result:

$$1160 + (-720) = 440.$$

How do the number of accidents for 20-year-old drivers compare to, say, the number for 45-year-old drivers? The answer is given by the points in the rectangular coordinate system in Figure 1.22. The x-coordinate of each point represents the driver's age. The y-coordinate represents the number of accidents per 50 million miles driven.

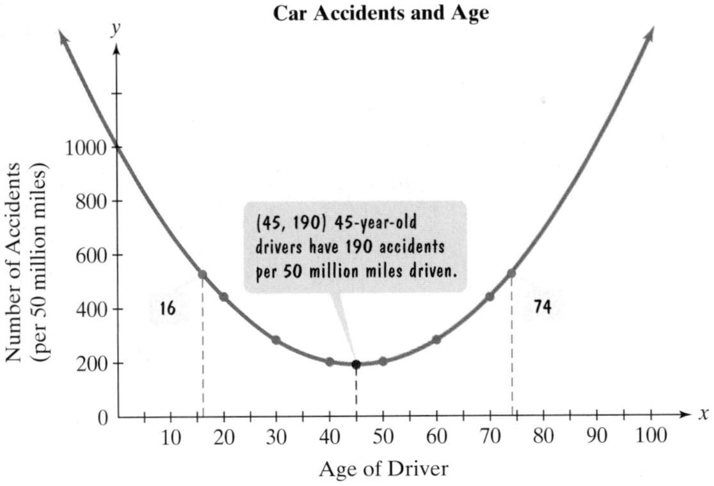

FIGURE 1.22

Can you see that the point (45, 190) is lower than any of the other points? In practical terms, this indicates that 45-year-olds have the least number of car accidents, 190 per 50 million miles driven. Drivers both younger and older than 45 have more accidents per 50 million miles driven.

CHECK POINT 12 Use the mathematical model described in Example 12, $N = 0.4x^2 - 36x + 1000$, to answer this question: How many accidents, per 50 million miles driven, are there for 40-year-old drivers?

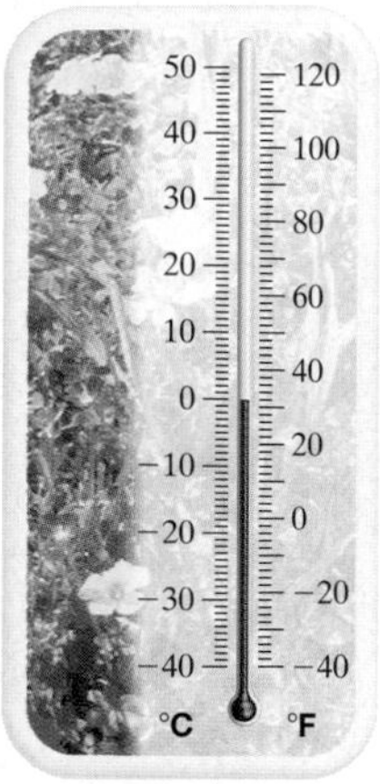

FIGURE 1.23 The Celsius scale is on the left and the Fahrenheit scale is on the right.

Some formulas give an exact, rather than an approximate, relationship between variables. For example, Figure 1.23 shows temperatures on the Celsius scale and on the Fahrenheit scale. The formula

$$C = \frac{5}{9}(F - 32)$$

expresses an exact relationship between Fahrenheit temperature, F, and Celsius temperature, C.

The Formula	What the Formula Tells Us
$C = \frac{5}{9}(F - 32)$	If 32 is subtracted from the Fahrenheit temperature, $F - 32$, and this difference is multiplied by $\frac{5}{9}$, the resulting product, $\frac{5}{9}(F - 32)$, gives the Celsius temperature.

EXAMPLE 13 Converting from Fahrenheit to Celsius

The temperature on a warm spring day is 77°F. Use the formula $C = \frac{5}{9}(F - 32)$ to find the equivalent temperature on the Celsius scale.

SOLUTION Because the temperature is 77°F, we substitute 77 for F in the given formula. Then we use the order of operations to find the value of C.

$C = \frac{5}{9}(F - 32)$ This is the given formula.

$C = \frac{5}{9}(77 - 32)$ Replace F with 77.

$C = \frac{5}{9}(45)$ Work inside parentheses first: $77 - 32 = 45$.

$C = 25$ Multiply: $\frac{5}{9}(45) = \frac{5}{\cancel{9}_1} \cdot \frac{\cancel{45}^5}{1} = \frac{25}{1} = 25$.

Thus, 77°F is equivalent to 25°C. ■

CHECK POINT 13 The temperature on a warm summer day is 86°F. Use the formula $C = \frac{5}{9}(F - 32)$ to find the equivalent temperature on the Celsius scale.

1.8 EXERCISE SET

Student Solutions Manual | CD/Video | PH Math/Tutor Center | MathXL Tutorials on CD | MathXL® | MyMathLab | Interactmath.com

Practice Exercises

In Exercises 1–14, evaluate each exponential expression.

1. 9^2 **2.** 3^2 **3.** 4^3

4. 6^3 **5.** $(-4)^2$ **6.** $(-10)^2$

7. $(-4)^3$ **8.** $(-10)^3$ **9.** $(-5)^4$

10. $(-1)^6$ **11.** -5^4 **12.** -1^6

13. -10^2 **14.** -8^2

In Exercises 15–28, simplify each algebraic expression, or explain why the expression cannot be simplified.

15. $7x^2 + 12x^2$ **16.** $6x^2 + 18x^2$

17. $10x^3 + 5x^3$ **18.** $14x^3 + 8x^3$

19. $8x^4 + x^4$ **20.** $14x^4 + x^4$

21. $26x^2 - 27x^2$

22. $29x^2 - 30x^2$

23. $27x^3 - 26x^3$

24. $30x^3 - 29x^3$

25. $5x^2 + 5x^3$

26. $8x^2 + 8x^3$

27. $16x^2 - 16x^2$

28. $34x^2 - x^2$

In Exercises 29–72, use the order of operations to simplify each expression.

29. $7 + 6 \cdot 3$

30. $3 + 4 \cdot 5$

31. $45 \div 5 \cdot 3$

32. $40 \div 4 \cdot 2$

33. $6 \cdot 8 \div 4$

34. $8 \cdot 6 \div 2$

35. $14 - 2 \cdot 6 + 3$

36. $36 - 12 \div 4 + 2$

37. $8^2 - 16 \div 2^2 \cdot 4 - 3$

38. $10^2 - 100 \div 5^2 \cdot 2 - 1$

39. $3(-2)^2 - 4(-3)^2$

40. $5(-3)^2 - 2(-4)^2$

41. $(4 \cdot 5)^2 - 4 \cdot 5^2$

42. $(3 \cdot 5)^2 - 3 \cdot 5^2$

43. $(2 - 6)^2 - (3 - 7)^2$

44. $(4 - 6)^2 - (5 - 9)^2$

45. $6(3 - 5)^3 - 2(1 - 3)^3$

46. $-3(-6 + 8)^3 - 5(-3 + 5)^3$

47. $[2(6 - 2)]^2$

48. $[3(4 - 6)]^3$

49. $2[5 + 2(9 - 4)]$

50. $3[4 + 3(10 - 8)]$

51. $[7 + 3(2^3 - 1)] \div 21$

52. $[11 - 4(2 - 3^3)] \div 37$

53. $\dfrac{10 + 8}{5^2 - 4^2}$

54. $\dfrac{6^2 - 4^2}{2 - (-8)}$

55. $\dfrac{37 + 15 \div (-3)}{2^4}$

56. $\dfrac{22 + 20 \div (-5)}{3^2}$

57. $\dfrac{(-11)(-4) + 2(-7)}{7 - (-3)}$

58. $\dfrac{-5(7 - 2) - 3(4 - 7)}{-13 - (-5)}$

59. $4|10 - (8 - 20)|$

60. $6|7 - 4 \cdot 3|$

61. $8(-10) + |4(-5)|$

62. $4(-15) + |3(-10)|$

63. $-2^2 + 4[16 \div (3 - 5)]$

64. $-3^2 + 2[20 \div (7 - 11)]$

65. $24 \div \dfrac{3^2}{8 - 5} - (-6)$

66. $30 \div \dfrac{5^2}{7 - 12} - (-9)$

67. $\dfrac{\frac{1}{4} - \frac{1}{2}}{\frac{1}{3}}$

68. $\dfrac{\frac{3}{5} - \frac{7}{10}}{\frac{1}{2}}$

69. $-\dfrac{9}{4}\left(\dfrac{1}{2}\right) + \dfrac{3}{4} \div \dfrac{5}{6}$

70. $\left[-\dfrac{4}{7} - \left(-\dfrac{2}{5}\right)\right]\left[-\dfrac{3}{8} + \left(-\dfrac{1}{9}\right)\right]$

71. $\dfrac{\frac{7}{9} - 3}{\frac{5}{6}} \div \dfrac{3}{2} + \dfrac{3}{4}$

72. $\dfrac{\frac{17}{25}}{\frac{3}{5} - 4} \div \dfrac{1}{5} + \dfrac{1}{2}$

In Exercises 73–80, evaluate each algebraic expression for the given value of the variable.

73. $x^2 + 5x; x = 3$

74. $x^2 - 2x; x = 6$

75. $3x^2 - 8x; x = -2$

76. $4x^2 - 2x; x = -3$

77. $-x^2 - 10x; x = -1$

78. $-x^2 - 14x; x = -1$

79. $\dfrac{6y - 4y^2}{y^2 - 15}; y = 5$

80. $\dfrac{3y - 2y^2}{y(y - 2)}; y = 5$

In Exercises 81–88, simplify each algebraic expression by removing parentheses and brackets.

81. $3[5(x - 2) + 1]$

82. $4[6(x - 3) + 1]$

83. $3[6 - (y + 1)]$

84. $5[2 - (y + 3)]$

85. $7 - 4[3 - (4y - 5)]$

86. $6 - 5[8 - (2y - 4)]$

87. $2(3x^2 - 5) - [4(2x^2 - 1) + 3]$

88. $4(6x^2 - 3) - [2(5x^2 - 1) + 1]$

Practice Plus

In Exercises 89–92, express each sentence as a single numerical expression. Then use the order of operations to simplify the expression.

89. Cube −2. Subtract this exponential expression from −10.

90. Cube −5. Subtract this exponential expression from −100.

91. Subtract 10 from 7. Multiply this difference by 2. Square this product.

92. Subtract 11 from 9. Multiply this difference by 2. Raise this product to the fourth power.

In Exercises 93–96, let x represent the number. Express each sentence as a single algebraic expression. Then simplify the expression.

93. Multiply a number by 5. Add 8 to this product. Subtract this sum from the number.

94. Multiply a number by 3. Add 9 to this product. Subtract this sum from the number.

95. Cube a number. Subtract 4 from this exponential expression. Multiply this difference by 5.

96. Cube a number. Subtract 6 from this exponential expression. Multiply this difference by 4.

Application Exercises

Medical researchers have found that the desirable heart rate, R, in beats per minute, for beneficial exercise is approximated by the mathematical models

$$R = 165 - 0.75A \qquad \text{for men}$$
$$R = 143 - 0.65A \qquad \text{for women}$$

where A is the person's age. Use these mathematical models to solve Exercises 97–98.

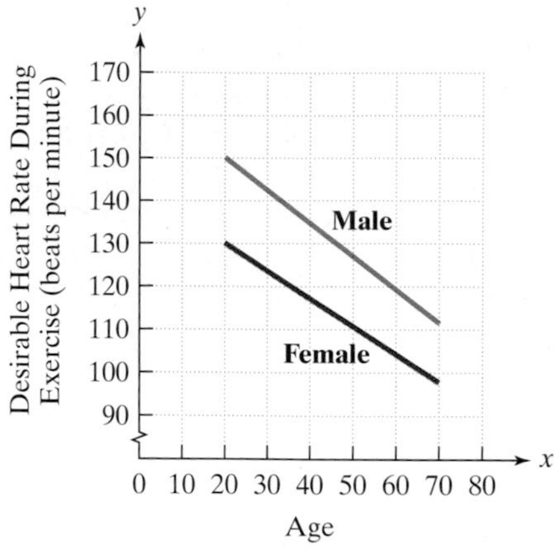

97. What is the desirable heart rate during exercise for a 40-year-old man? Identify your computation as an appropriate point in the rectangular coordinate system.

98. What is the desirable heart rate during exercise for a 40-year-old woman? Identify your computation as an appropriate point in the rectangular coordinate system.

The bar graph shows the cost of Medicare, in billions of dollars, through 2005. The data can be modeled by the formula

$$N = 1.2x^2 + 15.2x + 181.4,$$

where N represents Medicare spending, in billions of dollars, x years after 1995. Use this formula to solve Exercises 99–100.

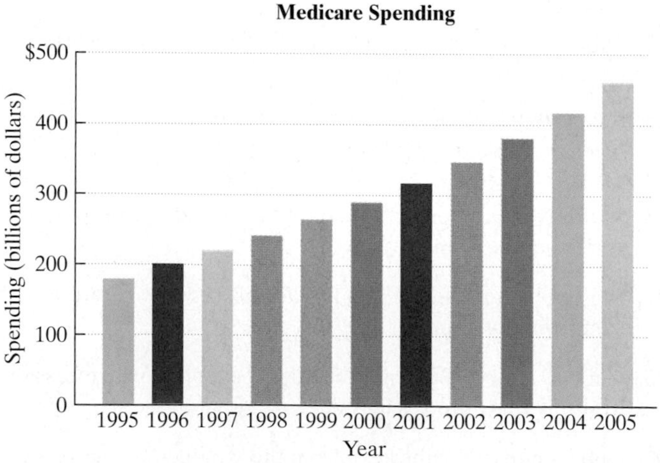

Source: Congressional Budget Office

99. According to the formula, what was the cost of Medicare, in billions of dollars, in 2000? How well does the formula describe the cost for that year shown by the bar graph?

100. According to the formula, what was the cost of Medicare, in billions of dollars, in 2005? How well does the formula describe the cost for that year shown by the bar graph?

Bariatrics is the field of medicine that deals with the overweight. Bariatric surgery closes off a large part of the stomach. As a result, patients eat less and have a diminished appetite. Celebrities like pop singer Carnie Wilson and the Today show's weatherman Al Roker have become no-longer-larger-than-life walking billboards for the operation. The bar graph shows the number of bariatric surgeries from 1992 through 2002.

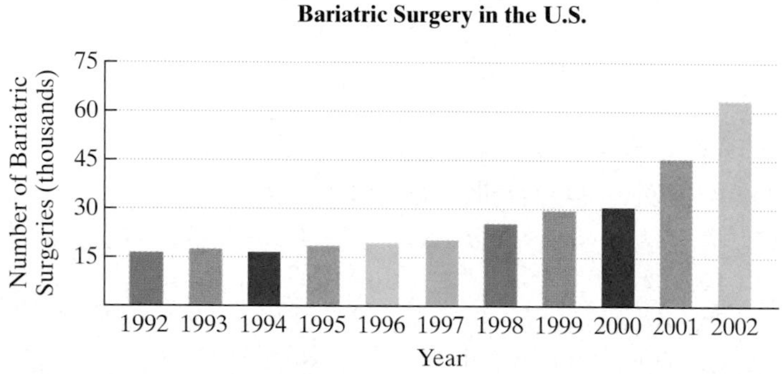

Source: American Society for Bariatric Surgery

The formula

$$N = 0.12x^3 - x^2 + 3x + 15$$

models the number of bariatric surgeries, N, in thousands, x years after 1992. Use this formula to solve Exercises 101–102.

101. According to the formula, how many bariatric surgeries were performed in 2002? How well does the formula model the data shown in the bar graph?

102. According to the formula, how many bariatric surgeries were performed in 1994? How well does the formula model the data shown in the bar graph?

The formula

$$C = \frac{5}{9}(F - 32)$$

expresses the relationship between Fahrenheit temperature, F, and Celsius temperature, C. In Exercises 103–106, use the formula to convert the given Fahrenheit temperature to its equivalent temperature on the Celsius scale.

103. 68°F

104. 41°F

105. −22°F

106. −31°F

Writing in Mathematics

107. Describe what it means to raise a number to a power. In your description, include a discussion of the difference between -5^2 and $(-5)^2$.

108. Explain how to simplify $4x^2 + 6x^2$. Why is the sum not equal to $10x^4$?

109. Why is the order of operations agreement needed?

110. What is a formula?

111. The formula $F = \frac{9}{5}C + 32$ expresses the relationship between Celsius temperature, C, and Fahrenheit temperature, F. You'll be leaving the cold of winter for a vacation to Hawaii. CNN International reports a temperature in Hawaii of 30°C. Should you pack a winter coat? Use the formula to explain your answer.

Critical Thinking Exercises

112. Which one of the following is true?

a. If x is -3, then the value of $-3x - 9$ is -18.

b. The algebraic expression $\frac{6x + 6}{x + 1}$ cannot have the same value when two different replacements are made for x such as $x = -3$ and $x = 2$.

c. A miniature version of a space shuttle is an example of a mathematical model.

d. The value of $\frac{|3 - 7| - 2^3}{(-2)(-3)}$ is the fraction that results when $\frac{1}{3}$ is subtracted from $-\frac{1}{3}$.

113. Simplify: $\frac{1}{4} - 6(2 + 8) \div \left(-\frac{1}{3}\right)\left(-\frac{1}{9}\right)$.

Grouping symbols can be inserted into $4 + 3 \cdot 7 - 4$ *so that the resulting value is 45. By placing parentheses around the addition we obtain*

$$(4 + 3) \cdot 7 - 4 = 7 \cdot 7 - 4 = 49 - 4 = 45.$$

In Exercises 114–115, insert parentheses in each expression so that the resulting value is 45.

114. $2 \cdot 3 + 3 \cdot 5$

115. $2 \cdot 5 - \frac{1}{2} \cdot 10 \cdot 9$

Technology Exercises

The United States has more people in prison, as well as more people in prison per capita, than any other western industrialized nation. The bar graph shows the number of inmates in U.S. state and federal prisons in nine selected years from 1985 through 2001.

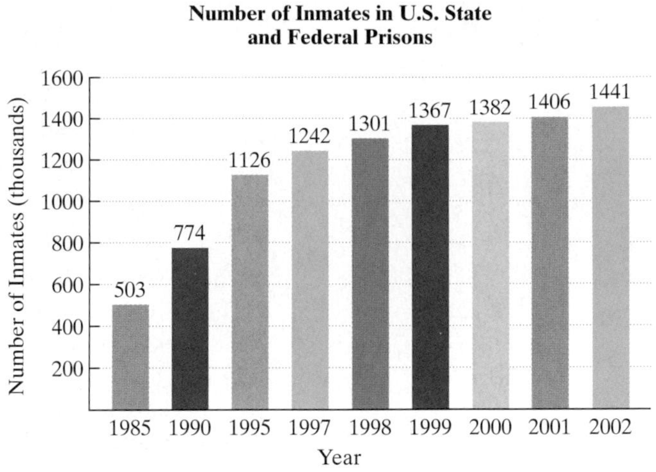

Source: U.S. Justice Department

The data in the graph can be modeled by each of the following formulas:

$$N = 59.5x + 505.12$$
$$N = -0.619x^2 + 69.52x + 485.28$$
$$N = -0.239x^3 + 5.288x^2 + 33.28x + 503.47.$$

In each formula, x represents the number of years after 1985 and N represents the number of inmates, in thousands. Use this information to solve Exercises 116–117.

116. Use a calculator to determine which of the three formulas serves as the best model for the inmate population in 2001.

117. Use a calculator to determine which, if any, of the three formulas serves as the best model for the inmate population for the period shown by the graph.

118. In this exercise, use a calculator and one of the formulas that estimates the adult height of a child given in the chapter introduction on page 1. Predict the adult height, H, in inches using the actual height of a child whose age, x, and height, h, in inches, you know.

Review Exercises

119. Simplify: $-8 - 2 - (-5) + 11$ (Section 1.6, Example 3)

120. Multiply: $-4(-1)(-3)(2)$. (Section 1.7, Example 2)

121. Give an example of a real number that is not an irrational number. (Section 1.2, Example 5).

GROUP PROJECT

CHAPTER 1

One measure of physical fitness is your *resting heart rate*. Generally speaking, the more fit you are, the lower your resting heart rate. The best time to take this measurement is when you first awaken in the morning, before you get out of bed. Lie on your back with no body parts crossed and take your pulse in your neck or wrist. Use your index and second fingers and count your pulse beat for one full minute to get your resting heart rate. A resting heart rate under 48 to 57 indicates high fitness, 58 to 62, above average fitness, 63 to 70, average fitness, 71 to 82, below average fitness, and 83 or more, low fitness.

Another measure of physical fitness is your percentage of body fat. You can estimate your body fat using the following formulas:

For men: Body fat $= -98.42 + 4.15w - 0.082b$

For women: Body fat $= -76.76 + 4.15w - 0.082b$

where $w =$ waist measurement, in inches, and $b =$ total body weight, in pounds. Then divide your body fat by your total weight to get your body fat percentage. For men, less than 15% is considered athletic, 25% about average. For women, less than 22% is considered athletic, 30% about average.

Each group member should bring his or her age, resting heart rate, and body fat percentage to the group. Using the data, the group should create three graphs.

a. Create a graph that shows age and resting heart rate for group members.

b. Create a graph that shows age and body fat percentage for group members.

c. Create a graph that shows resting heart rate and body fat percentage for group members.

For each graph, select the style (line or bar) that is most appropriate.

CHAPTER 1 SUMMARY

Definitions and Concepts	Examples
Section 1.1 Fractions	
Mixed Numbers and Improper Fractions A mixed number consists of the addition of a natural number (1, 2, 3, ...) and a fraction, expressed without the use of an addition sign. An improper fraction has a numerator that is greater than its denominator. To convert a mixed number to an improper fraction, multiply the denominator by the natural number and add the numerator. Then place this result over the original denominator.	Convert $5\frac{3}{7}$ to an improper fraction. $5\frac{3}{7} = \frac{7 \cdot 5 + 3}{7} = \frac{35 + 3}{7} = \frac{38}{7}$
To convert an improper fraction to a mixed number, divide the denominator into the numerator and write the mixed number using $\text{quotient}\frac{\text{remainder}}{\text{original denominator}}$.	Convert $\frac{14}{3}$ to a mixed number. $\frac{14}{3} = 4\frac{2}{3}$ $\quad 3\overline{)14}$ quotient 4, 12, remainder 2
A prime number is a natural number greater than 1 that has only itself and 1 as factors. A composite number is a natural number greater than 1 that is not a prime number. The prime factorization of a composite number means to express the composite number as the product of prime numbers.	Find the prime factorization: $60 = 6 \cdot 10$ $= 2 \cdot 3 \cdot 2 \cdot 5$

Definitions and Concepts	Examples
Section 1.1 Fractions (continued)	
A fraction is reduced to its lowest terms when the numerator and denominator have no common factors other than 1. To reduce a fraction to its lowest terms, divide both the numerator and the denominator by their greatest common factor. The greatest common factor can be found by inspection or prime factorizations of the numerator and the denominator.	Reduce to lowest terms: $\frac{8}{14} = \frac{\cancel{2} \cdot 4}{\cancel{2} \cdot 7} = \frac{4}{7}$
Multiplying Fractions The product of two or more fractions is the product of their numerators divided by the product of their denominators.	Multiply: $\frac{2}{7} \cdot \frac{5}{9} = \frac{2 \cdot 5}{7 \cdot 9} = \frac{10}{63}$
Dividing Fractions The quotient of two fractions is the first multiplied by the reciprocal (or multiplicative inverse) of the second.	Divide: $\frac{4}{9} \div \frac{3}{7} = \frac{4}{9} \cdot \frac{7}{3} = \frac{4 \cdot 7}{9 \cdot 3} = \frac{28}{27}$
Adding and Subtracting Fractions with Identical Denominators Add or subtract numerators. Put this result over the common denominator.	Subtract: $\frac{5}{8} - \frac{3}{8} = \frac{5-3}{8} = \frac{2}{8} = \frac{\cancel{2} \cdot 1}{\cancel{2} \cdot 4} = \frac{1}{4}$
Adding and Subtracting Fractions with Unlike Denominators Rewrite the fractions as equivalent fractions with the least common denominator. Then add or subtract numerators, putting this result over the common denominator.	Add: $\frac{3}{8} + \frac{5}{12} = \frac{3}{8} \cdot \frac{3}{3} + \frac{5}{12} \cdot \frac{2}{2}$ The LCD is 24. $= \frac{9}{24} + \frac{10}{24} = \frac{19}{24}$
Section 1.2 The Real Numbers	
A set is a collection of objects, called elements, whose contents can be clearly determined.	$\{a, b, c\}$
A line used to visualize numbers is called a number line.	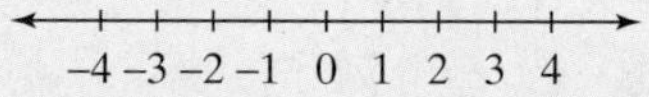
Real Numbers: the set of all numbers that can be represented by points on the number line **The Sets That Make Up the Real Numbers** • Natural Numbers: $\{1, 2, 3, 4, \ldots\}$ • Whole Numbers: $\{0, 1, 2, 3, 4, \ldots\}$ • Integers: $\{\ldots, -3, -2, -1, 0, 1, 2, 3, \ldots\}$ • Rational Numbers: the set of numbers that can be expressed as the quotient of an integer and a nonzero integer; can be expressed as terminating or repeating decimals • Irrational Numbers: the set of numbers that cannot be expressed as the quotient of integers; decimal representations neither terminate nor repeat.	Given the set $\left\{-1.4, 0, 0.\overline{7}, \frac{9}{10}, \sqrt{2}, \sqrt{4}\right\}$ list the • natural numbers: $\sqrt{4}$, or 2 • whole numbers: 0, $\sqrt{4}$ • rational numbers: $-1.4, 0, 0.\overline{7}, \frac{9}{10}, \sqrt{4}$ • irrational numbers: $\sqrt{2}$ • real numbers: $-1.4, 0, 0.\overline{7}, \frac{9}{10}, \sqrt{2}, \sqrt{4}$

Definitions and Concepts	Examples

Section 1.2 The Real Numbers (continued)

For any two real numbers, a and b, a is less than b if a is to the left of b on the number line.

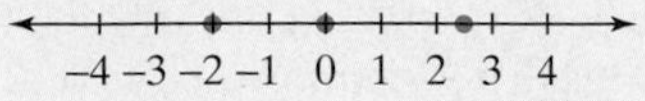

Inequality Symbols

$<$: is less than $>$: is greater than

$\leq$: is less than or equal to $\geq$: is greater than or equal to

$-2 < 0$ $0 > -2$

$0 < 2.5$ $2.5 > 0$

The absolute value of a, written $|a|$, is the distance from 0 to a on the number line.

$|4| = 4$ $|0| = 0$ $|-6| = 6$

Section 1.3 Ordered Pairs and Graphs

The rectangular coordinate system consists of a horizontal number line, the x-axis, and a vertical number line, the y-axis, intersecting at their zero points, the origin. Each point in the system corresponds to an ordered pair of real numbers (x, y). The first number in the pair is the x-coordinate; the second number is the y-coordinate.

Plot: $(2, 3), (-5, 4), (-4, -3), (5, -2)$.

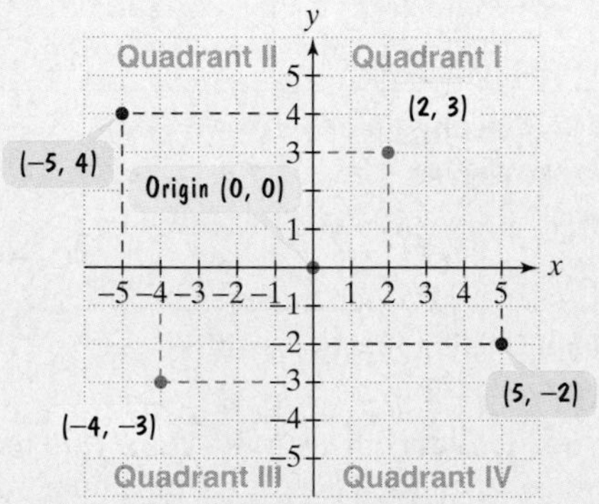

Information is often displayed using line graphs, bar graphs, and circle graphs. Line graphs are often used to illustrate trends over time. Bar graphs are convenient for showing comparisons among items.

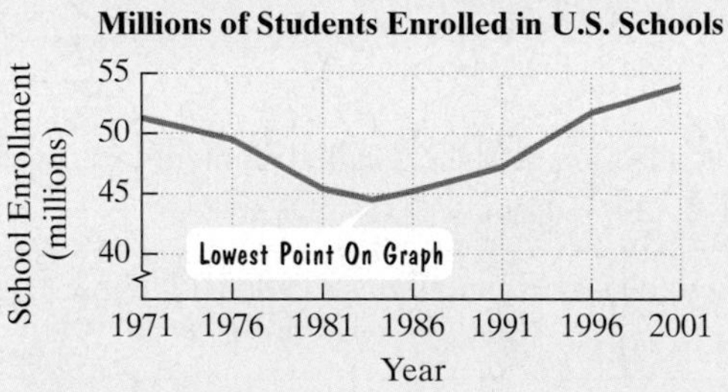

Source: National Education Association

In which year did enrollment reach a minimum? 1984
Estimate enrollment for that year: 44.5 million.

Section 1.4 Basic Rules of Algebra

A letter used to represent a number is called a variable. An algebraic expression is a combination of variables, numbers, and operation symbols. Terms are separated by addition. The parts of each term that are multiplied are its factors. Like terms have the same variable factors raised to the same powers. To evaluate an algebraic expression, substitute a given number for the variable and simplify.

Evaluate: $6(x + 3) + 4x$ when $x = 5$.

Replace x with 5.

$= 6(5 + 3) + 4 \cdot 5$

$= 6(8) + 4 \cdot 5$

$= 48 + 20$

$= 68$

Definitions and Concepts	Examples

Section 1.4 Basic Rules of Algebra (continued)

Properties of Real Numbers and Algebraic Expressions

- Commutative Properties:

$$a + b = b + a$$
$$ab = ba$$

- Associative Properties:

$$(a + b) + c = a + (b + c)$$
$$(ab)c = a(bc)$$

- Distributive Properties:

$$a(b + c) = ab + ac$$
$$(b + c)a = ba + ca$$
$$a(b - c) = ab - ac$$
$$(b - c)a = ba - ca$$
$$a(b + c + d) = ab + ac + ad$$

Commutative of Addition:

$$5x + 4 = 4 + 5x$$

Commutative of Multiplication:

$$5x + 4 = x5 + 4$$

Associative of Addition:

$$6 + (4 + x) = (6 + 4) + x = 10 + x$$

Associative of Multiplication:

$$7(10x) = (7 \cdot 10)x = 70x$$

Distributive:

$$8(x + 5 + 4y) = 8x + 40 + 32y$$

Distributive to Combine Like Terms:

$$8x + 12x = (8 + 12)x = 20x$$

Simplifying Algebraic Expressions

Use the distributive property to remove grouping symbols. Then combine like terms.

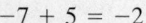

$$4(5x + 7) + 13x$$
$$= 20x + 28 + 13x$$
$$= (20x + 13x) + 28$$
$$= 33x + 28$$

Section 1.5 Addition of Real Numbers

Sums on a Number Line

To find $a + b$, the sum of a and b, on a number line, start at a. If b is positive, move b units to the right. If b is negative, move b units to the left. If b is 0, stay at a. The number where we finish on the number line represents $a + b$.

$$-7 + 5 = -2$$

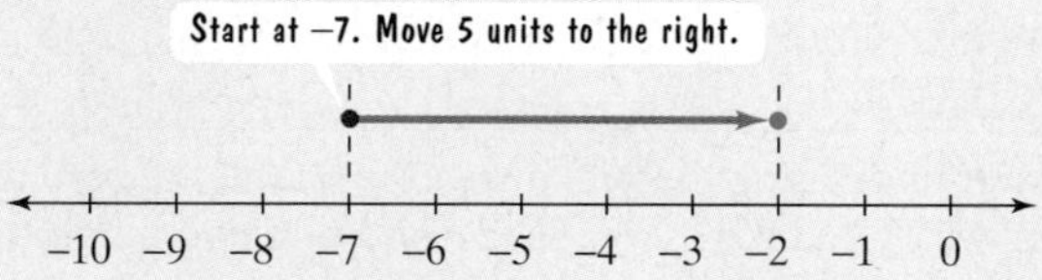

Additive inverses are pairs of real numbers that are the same number of units from zero on the number line, but on opposite sides of zero.

- Identity Property of Addition:

$$a + 0 = 0 \qquad 0 + a = a$$

- Inverse Property of Addition:

$$a + (-a) = 0 \qquad (-a) + a = 0$$

The additive inverse (or opposite) of 4 is −4.
The additive inverse of −1.7 is 1.7.

Identity Property of Addition:

$$4x + 0 = 4x$$

Inverse Property of Addition:

$$4x + (-4x) = 0$$

Addition without a Number Line

To add two numbers with the same sign, add their absolute values and use their common sign. To add two numbers with different signs, subtract the smaller absolute value from the greater absolute value and use the sign of the number with the greater absolute value.

Add:

$$10 + 4 = 14$$
$$-4 + (-6) = -10$$
$$-30 + 5 = -25$$
$$12 + (-8) = 4$$

To add a series of positive and negative numbers, add all the positive numbers and add all the negative numbers. Then add the resulting positive and negative sums.

$$5 + (-3) + (-7) + 2$$
$$= (5 + 2) + [(-3) + (-7)]$$
$$= 7 + (-10)$$
$$= -3$$

Definitions and Concepts	Examples
Section 1.6 Subtraction of Real Numbers	
To subtract b from a, add the additive inverse of b to a: $$a - b = a + (-b).$$ The result is called the difference between a and b.	Subtract: $-7 - (-5) = -7 + 5 = -2$ $-\frac{3}{4} - \frac{1}{2} = -\frac{3}{4} + \left(-\frac{1}{2}\right)$ $= -\frac{3}{4} + \left(-\frac{2}{4}\right) = -\frac{5}{4}$
To simplify a series of additions and subtractions, change all subtractions to additions of additive inverses. Then use the procedure for adding a series of positive and negative numbers.	Simplify: $-6 - 2 - (-3) + 10$ $= -6 + (-2) + 3 + 10$ $= -8 + 13$ $= 5$
Section 1.7 Multiplication and Division of Real Numbers	
The result of multiplying a and b, ab, is called the product of a and b. If the two numbers have different signs, the product is negative. If the two numbers have the same sign, the product is positive. If either number is 0, the product is 0.	Multiply: $-5(-10) = 50$ $\frac{3}{4}\left(-\frac{5}{7}\right) = -\frac{3}{4}\cdot\frac{5}{7} = -\frac{15}{28}$
Assuming that no number is 0, the product of an even number of negative numbers is positive. The product of an odd number of negative numbers is negative. If any number is 0, the product is 0.	Multiply: $(-3)(-2)(-1)(-4) = 24$ $(-3)(2)(-1)(-4) = -24$
The result of dividing the real number a by the nonzero real number b is called the quotient of a and b. If two numbers have different signs, their quotient is negative. If two numbers have the same sign, their quotient is positive. Division by zero is undefined.	Divide: $\frac{21}{-3} = -7$ $-\frac{1}{3} \div (-3) = \frac{1}{3}\cdot\frac{1}{3} = \frac{1}{9}$
Two numbers whose product is 1 are called multiplicative inverses or reciprocals of each other. The number 0 has no multiplicative inverse. • Identity Property of Multiplication $a \cdot 1 = a \qquad 1 \cdot a = a$ • Inverse Property of Multiplication If a is not 0: $a \cdot \frac{1}{a} = 1 \qquad \frac{1}{a} \cdot a = 1$ • Multiplication Property of -1 $-1a = -a \qquad a(-1) = -a$ • Double Negative Property $-(-a) = a$	The multiplicative inverse of 4 is $\frac{1}{4}$. The multiplicative inverse of $-\frac{1}{3}$ is -3. Simplify: $1x = x$ $7x \cdot \frac{1}{7x} = 1$ $4x - 5x = -1x = -x$ $-(-7y) = 7y$
If a negative sign precedes parentheses, remove parentheses and change the sign of every term within parentheses.	Simplify: $-(7x - 3y + 2) = -7x + 3y - 2$

Definitions and Concepts	Examples
Section 1.8 Exponents, Order of Operations, and Mathematical Models	
If b is a real number and n is a natural number, b^n, the nth power of b, is the product of n factors of b. Furthermore, $b^1 = b$.	Evaluate: $8^2 = 8 \cdot 8 = 64$ $(-5)^3 = (-5)(-5)(-5) = -125$
Order of Operations **1.** Perform operations within grouping symbols, starting with the innermost grouping symbols. **2.** Evaluate exponential expressions. **3.** Multiply and divide in order from left to right. **4.** Add and subtract in order from left to right.	Simplify: $5(4-6)^2 - 2(1-3)^3$ $= 5(-2)^2 - 2(-2)^3$ $= 5(4) - 2(-8)$ $= 20 - (-16)$ $= 20 + 16 = 36$
Some algebraic expressions contain two sets of grouping symbols: parentheses, the inner grouping symbols, and brackets, the outer grouping symbols. To simplify such expressions, use the order of operations and remove grouping symbols from innermost (parentheses) to outermost (brackets).	Simplify: $5 - 3[2(x+1) - 7]$ $= 5 - 3[2x + 2 - 7]$ $= 5 - 3[2x - 5]$ $= 5 - 6x + 15$ $= -6x + 20$
A formula is a statement of equality that uses letters to express a relationship between two or more variables. Formulas that describe real-world phenomena are called mathematical models. Many mathematical models give an approximate, rather than an exact, description of the relationship between variables.	The formula $N = 0.4x^2 + 0.5$ models the millions of people, N, in the United States using cable modems x years after 1996.

CHAPTER 1 REVIEW EXERCISES

1.1 *In Exercises 1–2, convert each mixed number to an improper fraction.*

1. $3\frac{2}{7}$ **2.** $5\frac{9}{11}$

In Exercises 3–4, convert each improper fraction to a mixed number.

3. $\frac{17}{9}$ **4.** $\frac{27}{5}$

In Exercises 5–7, identify each natural number as prime or composite. If the number is composite, find its prime factorization.

5. 60 **6.** 63 **7.** 67

In Exercises 8–9, simplify each fraction by reducing it to its lowest terms.

8. $\frac{15}{33}$ **9.** $\frac{40}{75}$

In Exercises 10–15, perform the indicated operation. Where possible, reduce the answer to its lowest terms.

10. $\frac{3}{5} \cdot \frac{7}{10}$ **11.** $\frac{4}{5} \div \frac{3}{10}$ **12.** $1\frac{2}{3} \div 6\frac{2}{3}$

13. $\frac{2}{9} + \frac{4}{9}$ **14.** $\frac{5}{6} + \frac{7}{9}$ **15.** $\frac{3}{4} - \frac{2}{15}$

16. The gas tank of a car is filled to its capacity. The first day, $\frac{1}{4}$ of the tank's gas is used for travel. The second day, $\frac{1}{3}$ of the tank's original amount of gas is used for travel. What fraction of the tank is filled with gas at the end of the second day?

1.2 *In Exercises 17–18, graph each real number on a number line.*

17. -2.5 **18.** $4\frac{3}{4}$

In Exercises 19–20, express each rational number as a decimal.

19. $\frac{5}{8}$ **20.** $\frac{3}{11}$

21. Consider the set

$$\left\{-17, -\frac{9}{13}, 0, 0.75, \sqrt{2}, \pi, \sqrt{81}\right\}.$$

List all numbers from the set that are: **a.** natural numbers, **b.** whole numbers, **c.** integers, **d.** rational numbers, **e.** irrational numbers, **f.** real numbers.

22. Give an example of an integer that is not a natural number.

23. Give an example of a rational number that is not an integer.

24. Give an example of a real number that is not a rational number.

In Exercises 25–28, insert either < or > in the shaded area between each pair of numbers to make a true statement.

25. $-93 \;\square\; 17$ **26.** $-2 \;\square\; -200$

27. $0 \;\square\; -\frac{1}{3}$ **28.** $-\frac{1}{4} \;\square\; -\frac{1}{5}$

In Exercises 29–30, determine whether each inequality is true or false.

29. $-13 \geq -11$ **30.** $-126 \leq -126$

In Exercises 31–32, find each absolute value.

31. $|-58|$ **32.** $|2.75|$

1.3 *In Exercises 33–36, plot the given point in a rectangular coordinate system. Indicate in which quadrant each point lies.*

33. $(1, -5)$ **34.** $(4, -3)$

35. $\left(\frac{7}{2}, \frac{3}{2}\right)$ **36.** $(-5, 2)$

37. Give the ordered pairs that correspond to the points labeled in the figure.

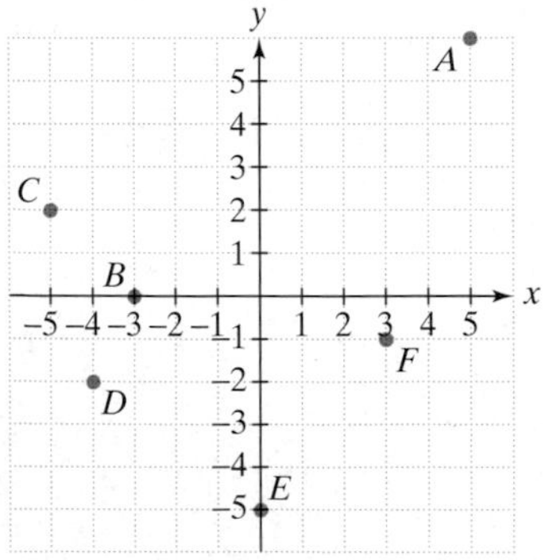

We live in an era of democratic aspiration. The number of democracies worldwide is on the rise. The line graph shows the number of democracies worldwide, in four-year periods, from 1973 through 2001, including 2002. Use the graph to solve Exercises 38–42.

Tracking Democracy: Number of Democracies Worldwide

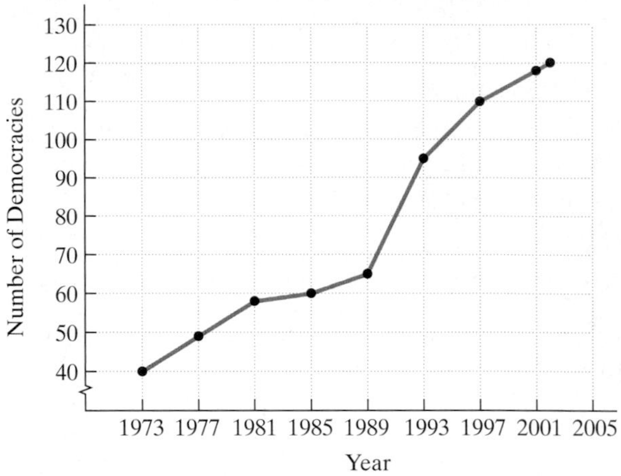

Source: The Freedom House

38. Find an estimate for the number of democracies in 1989.

39. How many more democracies were there in 2002 than in 1973?

40. In which four-year period did the number of democracies increase at the greatest rate?

41. In which four-year period did the number of democracies increase at the slowest rate?

42. In which year were there 49 democracies?

The United States has the highest rate of television ownership in the world. The bar graph shows the seven countries in the world with the greatest number of televisions per 100 people. Use the graph to solve Exercises 43–44.

Television Ownership in Global Perspective

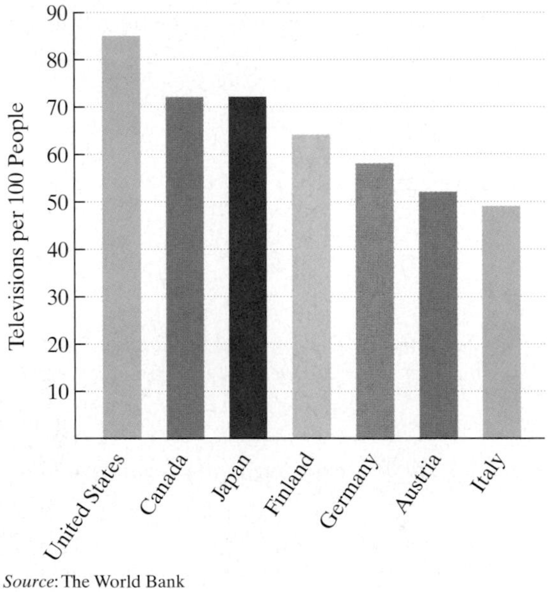

Source: The World Bank

43. Estimate the number of televisions per 100 people in the United States.

44. Which countries have more than 50 but fewer than 70 televisions per 100 people?

1.4 *In Exercises 45–46, evaluate each algebraic expression for the given value of the variable.*

45. $7x + 3;\ x = 10$

46. $5(x - 4);\ x = 12$

47. Use the commutative property of addition to write an equivalent algebraic expression: $7 + 13y$.

48. Use the commutative property of multiplication to write an equivalent algebraic expression: $9(x + 7)$.

In Exercises 49–50, use an associative property to rewrite each algebraic expression. Then simplify the resulting algebraic expression.

49. $6 + (4 + y)$

50. $7(10x)$

51. Use the distributive property to rewrite without parentheses: $6(4x - 2 + 5y)$.

In Exercises 52–53, simplify each algebraic expression.

52. $4a + 9 + 3a - 7$

53. $6(3x + 4) + 5(2x - 1)$

54. Suppose that a store is selling all computers at 25% off the regular price. If x is the regular price, the algebraic expression $x - 0.25x$ describes the sale price. Evaluate the expression when $x = 2400$. Describe what the answer means in practical terms.

1.5

55. Use a number line to find the sum: $-6 + 8$.

In Exercises 56–58, find each sum without the use of a number line.

56. $8 + (-11)$

57. $-\frac{3}{4} + \frac{1}{5}$

58. $7 + (-5) + (-13) + 4$

In Exercises 59–60, simplify each algebraic expression.

59. $8x + (-6y) + (-12x) + 11y$

60. $10(4 - 3y) + 28y$

61. The Dead Sea is the lowest elevation on Earth, 1312 feet below sea level. If a person is standing 512 feet above the Dead Sea, what is that person's elevation?

62. The water level of a reservoir is measured over a five-month period. At the beginning, the level is 25 feet. During this time, the level fell 3 feet, then rose 2 feet, then rose 1 foot, then fell 4 feet, and then rose 2 feet. What is the reservoir's water level at the end of the five months?

1.6

63. Rewrite $9 - 13$ as the addition of an additive inverse.

In Exercises 64–66, perform the indicated subtraction.

64. $-9 - (-13)$

65. $-\frac{7}{10} - \frac{1}{2}$

66. $-3.6 - (-2.1)$

In Exercises 67–68, simplify each series of additions and subtractions.

67. $-7 - (-5) + 11 - 16$

68. $-25 - 4 - (-10) + 16$

69. Simplify: $3 - 6a - 8 - 2a$.

70. What is the difference in elevation between a plane flying 26,500 feet above sea level and a submarine traveling 650 feet below sea level?

1.7 *In Exercises 71–73, perform the indicated multiplication.*

71. $-7(-12)$

72. $\frac{3}{5}\left(-\frac{5}{11}\right)$

73. $5(-3)(-2)(-4)$

In Exercises 74–76, perform the indicated division or state that the expression is undefined.

74. $\frac{45}{-5}$

75. $-17 \div 0$

76. $-\frac{4}{5} \div \left(-\frac{2}{5}\right)$

In Exercises 77–78, simplify each algebraic expression.

77. $-4\left(-\frac{3}{4}x\right)$

78. $-3(2x - 1) - (4 - 5x)$

1.8 *In Exercises 79–81, evaluate each exponential expression.*

79. $(-6)^2$

80. -6^2

81. $(-2)^5$

In Exercises 82–83, simplify each algebraic expression, or explain why the expression cannot be simplified.

82. $4x^3 + 2x^3$

83. $4x^3 + 4x^2$

In Exercises 84–92, use the order of operations to simplify each expression.

84. $-40 \div 5 \cdot 2$

85. $-6 + (-2) \cdot 5$

86. $6 - 4(-3 + 2)$

87. $28 \div (2 - 4^2)$

88. $36 - 24 \div 4 \cdot 3 - 1$

89. $-8[-4 - 5(-3)]$

90. $\dfrac{6(-10 + 3)}{2(-15) - 9(-3)}$

91. $\left(\frac{1}{2} + \frac{1}{3}\right) \div \left(\frac{1}{4} - \frac{3}{8}\right)$

92. $\frac{1}{2} - \frac{2}{3} \div \frac{5}{9} + \frac{3}{10}$

In Exercises 93–94, evaluate each algebraic expression for the given value of the variable.

93. $x^2 - 2x + 3; x = -1$

94. $-x^2 - 7x; x = -2$

In Exercises 95–96, simplify each algebraic expression.

95. $4[7(a - 1) + 2]$

96. $-6[4 - (y + 2)]$

On the average, infant girls weigh 7 pounds at birth and gain 1.5 pounds for each month for the first six months. The formula

$$W = 1.5x + 7$$

models a baby girl's weight, W, in pounds, after x months, where x is less than or equal to 6. Use the formula to solve Exercises 97–98.

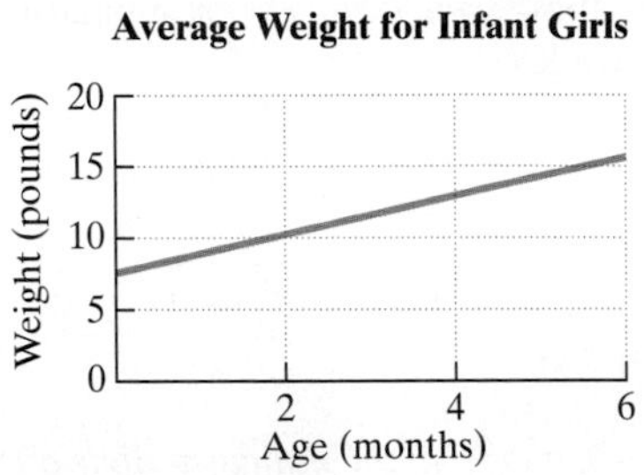

97. What does an infant girl weigh after four months? Identify your computation as an appropriate point on the line graph.

98. What does an infant girl weigh after six months? Identify your computation as an appropriate point on the line graph.

Among people under 40, opinions are split on downloading music: 45% think it's stealing, while 46% think it's not. (Source: Newsweek, September 22, 2003) Downloading and music CD prices that remained high through 2003 resulted in the decline of CD sales shown in the bar graph. The data can be modeled by the formula

$$N = -26x^2 + 143x + 740,$$

where N represents the number of music CD sales, in millions, x years after 1997. Use the formula to solve Exercises 99–100.

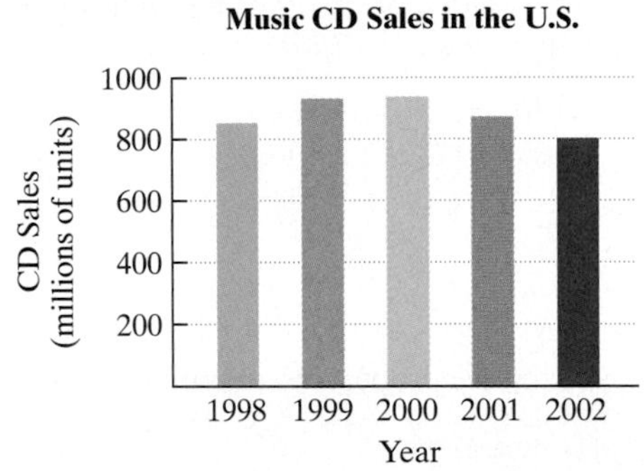

Source: RIAA

99. According to the formula, how many millions of CDs were sold in 2002? How well does the formula describe sales for that year shown by the bar graph?

100. What do you notice about CD sales from 1998 through 2002 from the bar graph that is not obvious by looking at the formula?

CHAPTER 1 TEST

Remember to use your Chapter Test Prep Video CD to see the worked-out solutions to the test questions you want to review.

In Exercises 1–10, perform the indicated operation or operations.

1. $1.4 - (-2.6)$

2. $-9 + 3 + (-11) + 6$

3. $3(-17)$

4. $\left(-\frac{3}{7}\right) \div \left(-\frac{15}{7}\right)$

5. $\left(3\frac{1}{3}\right)\left(-1\frac{3}{4}\right)$

6. $-50 \div 10$

7. $-6 - (5 - 12)$

8. $(-3)(-4) \div (7 - 10)$

9. $(6 - 8)^2(5 - 7)^3$

10. $\dfrac{3(-2) - 2(2)}{-2(8 - 3)}$

In Exercises 11–13, simplify each algebraic expression.

11. $11x - (7x - 4)$

12. $5(3x - 4y) - (2x - y)$

13. $6 - 2[3(x + 1) - 5]$

14. List all the rational numbers in this set.

$$\left\{-7, -\frac{4}{5}, 0, 0.25, \sqrt{3}, \sqrt{4}, \frac{22}{7}, \pi\right\}$$

15. Insert either < or > in the shaded area to make a true statement: $-1 \quad -100$.

16. Find the absolute value: $|-12.8|$.

17. Plot $(-4, 3)$ in a rectangular coordinate system. In which quadrant does the point lie?

18. Find the coordinates of point A in the figure.

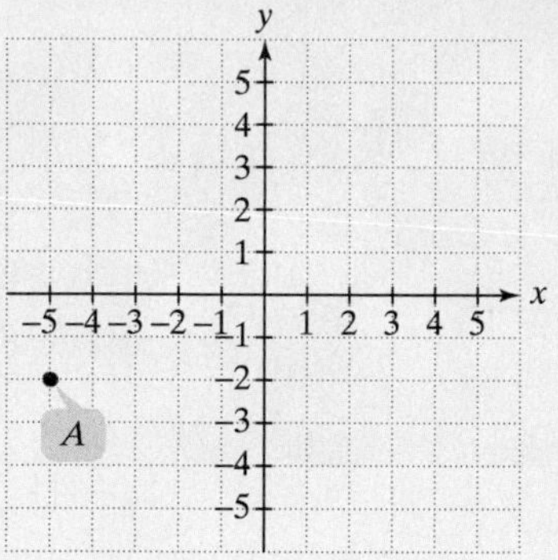

In Exercises 19–20, evaluate each algebraic expression for the given value of the variable.

19. $5(x - 7); x = 4$

20. $x^2 - 5x; x = -10$

21. Use the commutative property of addition to write an equivalent algebraic expression: $2(x + 3)$.

22. Use the associative property of multiplication to rewrite $-6(4x)$. Then simplify the expression.

23. Use the distributive property to rewrite without parentheses: $7(5x - 1 + 2y)$.

Tule elk are introduced into a newly acquired habitat. The points in the rectangular coordinate system show the elk population every ten years. Use this information to solve Exercises 24–25.

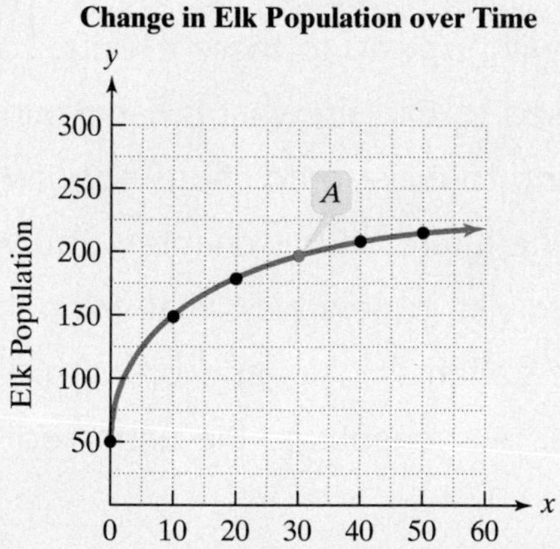

24. Find the coordinates of point A. Then interpret the coordinates in terms of the information given.

25. How many elk were introduced into the habitat?

26. Use the bar graph to find a reasonable estimate for the millions of DVDs sold in the United States in 2002.

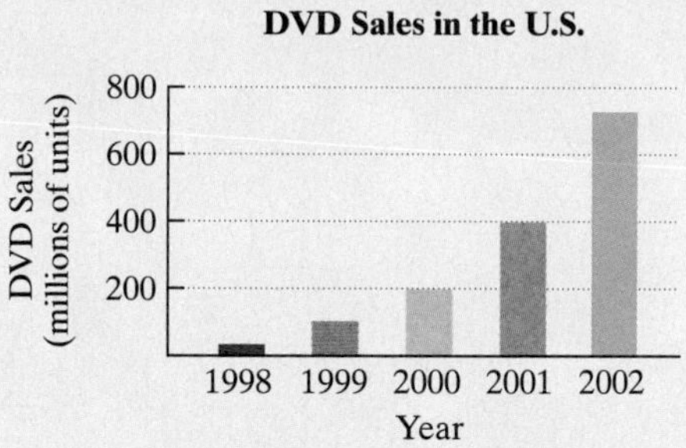

Source: International Recording Media Association

27. The formula

$$T = 3(A - 20)^2 \div 50 + 10$$

describes the average running time, T, in seconds, for a person who is A years old to run the 100-yard dash. How long does it take a 30-year-old runner to run the 100-yard dash?

28. What is the difference in elevation between a plane flying 16,200 feet above sea level and a submarine traveling 830 feet below sea level?

29. Use the graph to find a reasonable estimate, in thousands of dollars, for the average price of an existing single-family home in the United States in 1999.

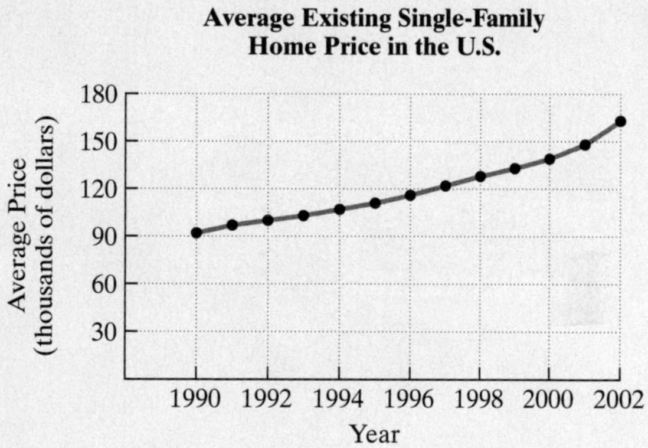

Source: National Association of Realtors

30. The data in Exercise 29 can be modeled by the formula

$$N = 0.3x^2 + 1.2x + 92.7,$$

where N represents the average price of an existing single-family home, in thousands of dollars, x years after 1989. Use the formula to find the average price of such a home in 1999.

CHAPTER 2

Reading your psychology textbook, you were intrigued by a study that found undergraduate students with an average or high sense of humor reported few increases in depression over time. By contrast, students with a low sense of humor were more likely to become depressed in response to negative life events. Is there a way to model these variables and use formulas to predict how we will respond to difficult life events?

This problem appears as Exercises 75–76 in Exercise Set 2.3.

Linear Equations and Inequalities in One Variable

The belief that humor and laughter can have positive benefits on our lives is not new. The Bible tells us, "A merry heart doeth good like a medicine, but a broken spirit drieth the bones" (Proverbs 17:22). Algebra can be used to model the influence that humor plays in our responses to negative life events. The resulting formulas predict how low- and high-humor college students will respond to these events. Formulas can be used to explain what is happening in the present and to make predictions about what might occur in the future. In this chapter, you will learn to use formulas in new ways that will help you to recognize patterns, logic, and order in a world that can appear chaotic to the untrained eye.

SECTION 2.1

THE ADDITION PROPERTY OF EQUALITY

Objectives

1. Check whether a number is a solution to an equation.
2. Use the addition property of equality to solve equations.
3. Solve applied problems using formulas.

Unfortunately, many of us have been fined for driving over the speed limit. The amount of the fine depends on how fast we are speeding. Suppose that a highway has a speed limit of 60 miles per hour. The amount that speeders are fined, F, is described by the formula

$$F = 10x - 600,$$

where x is the speed in miles per hour. A friend, whom we shall call Leadfoot, borrows your car and returns a few hours later with a $400 speeding fine. Leadfoot is furious, protesting that the car was barely driven over the speed limit. Should you believe Leadfoot?

To decide if Leadfoot is telling the truth, use the formula

$$F = 10x - 600.$$

Leadfoot was fined $400, so substitute 400 for F:

$$400 = 10x - 600.$$

Now you need to find the value for x. This variable represents Leadfoot's speed, which resulted in the $400 fine.

The use of algebra in a variety of everyday applications often leads to an *equation*. An **equation** is a statement that two algebraic expressions are equal. Thus, $400 = 10x - 600$ is an example of an equation. The equal sign divides the equation into two parts, the left side and the right side:

$$\underbrace{\boxed{400}}_{\text{Left side}} = \underbrace{\boxed{10x - 600}}_{\text{Right side}}$$

The two sides of an equation can be reversed, so we can express this equation as

$$10x - 600 = 400.$$

The form of this equation is $ax + b = c$, with $a = 10$, $b = -600$, and $c = 400$. Any equation in this form is called a **linear equation in one variable**. The exponent on the variable in such an equation is 1.

In the next three sections, we will study how to solve linear equations. **Solving an equation** is the process of finding the number (or numbers) that make the equation a true statement. These numbers are called the **solutions**, or **roots**, of the equation, and we say that they **satisfy** the equation.

1 Check whether a number is a solution to an equation.

Checking Whether a Number Is a Solution to an Equation A proposed solution to an equation can be checked by substituting that number for each occurrence of the variable in the equation. If the substitution results in a true statement, the number is a solution. If the substitution results in a false statement, the number is not a solution.

EXAMPLE 1 Checking Proposed Solutions (Is Leadfoot Telling the Truth?)

Consider the equation

$$10x - 600 = 400.$$

(Remember that x represents Leadfoot's speed that resulted in the \$400 fine.) Determine whether

a. 60 is a solution. **b.** 100 is a solution.

SOLUTION

a. To determine whether 60 is a solution to the equation, we substitute 60 for x.

$10x - 600 = 400$ This is the given equation.

$10(60) - 600 \stackrel{?}{=} 400$ Substitute 60 for x. The question mark over the equal sign indicates that we do not know yet if the statement is true.

$600 - 600 \stackrel{?}{=} 400$ Multiply: $10(60) = 600$.

$0 = 400$ Subtract: $600 - 600 = 0$.

This statement is false.

Because the check results in a false statement, we conclude that 60 is not a solution to the given equation. (Leadfoot was not doing 60 miles per hour in your borrowed car.)

b. To determine whether 100 is a solution to the equation, we substitute 100 for x.

$10x - 600 = 400$ This is the given equation.

$10(100) - 600 \stackrel{?}{=} 400$ Substitute 100 for x.

$1000 - 600 \stackrel{?}{=} 400$ Multiply: $10(100) = 1000$.

$400 = 400$ Subtract: $1000 - 600 = 400$.

This statement is true.

Because the check results in a true statement, we conclude that 100 is a solution to the given equation. Thus, 100 satisfies the equation. (Leadfoot was doing an outrageous 100 miles per hour, and lied with the claim that your car was barely driven over the speed limit.) ■

✔ **CHECK POINT 1** Consider the equation $5x - 3 = 17$. Determine whether

a. 3 is a solution. **b.** 4 is a solution.

2 Use the addition property of equality to solve equations.

Using the Addition Property of Equality to Solve Equations Consider the equation

$$x = 11.$$

By inspection, we can see that the solution to this equation is 11. If we substitute 11 for x, we obtain the true statement $11 = 11$.

Now consider the equation

$$x - 3 = 8.$$

If we substitute 11 for x, we obtain $11 - 3 \stackrel{?}{=} 8$. Subtracting on the left side, we get the true statement $8 = 8$.

STUDY TIP

An equation is like a balanced scale—balanced because its two sides are equal. To maintain this balance, whatever is done to one side must also be done to the other side.

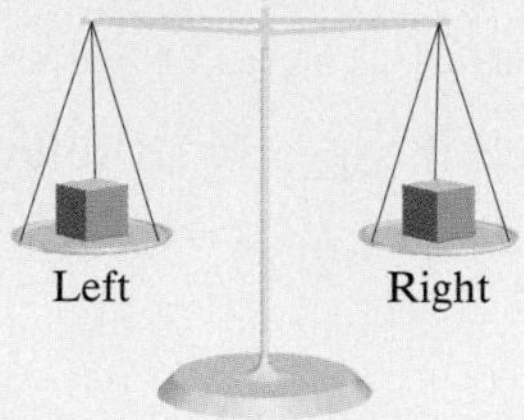

Consider $x - 3 = 8$.

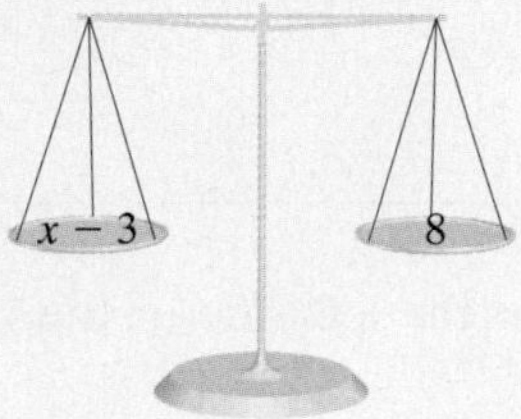

The scale is balanced if the left and right sides are equal.

Add 3 to the left side.

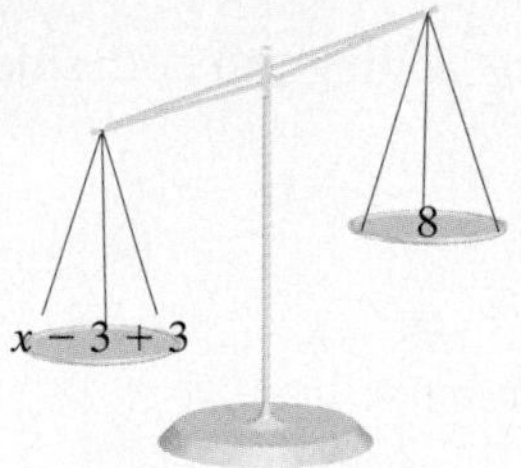

Keep the scale balanced by adding 3 to the right side.

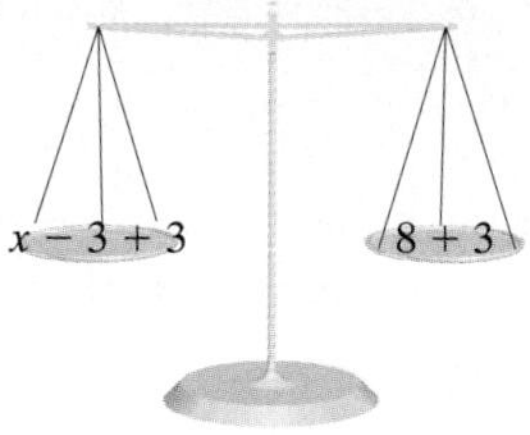

Thus, $x = 11$.

The equations $x - 3 = 8$ and $x = 11$ both have the same solution, namely 11, and are called *equivalent equations*. **Equivalent equations** are equations that have the same solution.

The idea in solving a linear equation is to get an equivalent equation with the variable (the letter) by itself on one side of the equal sign and a number by itself on the other side. For example, consider the equation $x - 3 = 8$. To get x by itself on the left side, add 3 to the left side, because $x - 3 + 3$ gives $x + 0$, or just x. You must then add 3 to the right side also. By doing this, we are using the **addition property of equality**.

THE ADDITION PROPERTY OF EQUALITY The same real number (or algebraic expression) may be added to both sides of an equation without changing the equation's solution. This can be expressed symbolically as follows:

$$\text{If } a = b, \text{ then } a + c = b + c.$$

EXAMPLE 2 Solving an Equation Using the Addition Property

Solve the equation: $x - 3 = 8$.

SOLUTION We can isolate the variable, x, by adding 3 to both sides of the equation.

$x - 3 = 8$	This is the given equation.
$x - 3 + 3 = 8 + 3$	Add 3 to both sides.
$x + 0 = 11$	This step is often done mentally and not listed.
$x = 11$	

By inspection, we can see that the solution to $x = 11$ is 11. To check this proposed solution, replace x with 11 in the original equation.

Check

$x - 3 = 8$	This is the original equation.
$11 - 3 \stackrel{?}{=} 8$	Substitute 11 for x.
$8 = 8$	Subtract: $11 - 3 = 8$.

This statement is true.

Because the check results in a true statement, we conclude that the solution to the given equation is 11. ■

The set of an equation's solutions is called its **solution set**. Thus, the solution set of the equation in Example 2 is $\{11\}$. The solution can be expressed as 11 or, using set notation, $\{11\}$. However, do not write the solution as $x = 11$. **The solution of an equation should not be given as an equivalent equation.**

CHECK POINT 2 Solve the equation and check your proposed solution:

$$x - 5 = 12.$$

When we use the addition property of equality, we add the same number to both sides of an equation. We know that subtraction is the addition of an additive inverse. Thus, the addition property also lets us subtract the same number from both sides of an equation without changing the equation's solution.

EXAMPLE 3 Subtracting the Same Number from Both Sides

Solve and check: $z + 1.4 = 2.06$.

SOLUTION

$z + 1.4 = 2.06$ — This is the given equation.

$z + 1.4 - 1.4 = 2.06 - 1.4$ — Subtract 1.4 from both sides. This is equivalent to adding -1.4 to both sides.

$z = 0.66$ — Subtracting 1.4 from both sides eliminates 1.4 on the left.

Can you see that the solution to $z = 0.66$ is 0.66? To check this proposed solution, replace z with 0.66 in the original equation.

Check

$z + 1.4 = 2.06$ — This is the original equation.

$0.66 + 1.4 \stackrel{?}{=} 2.06$ — Substitute 0.66 for z.

$2.06 = 2.06$ — This statement is true.

This true statement indicates that the solution is 0.66, or the solution set is $\{0.66\}$. ■

✔ CHECK POINT 3 Solve and check: $z + 2.8 = 5.09$.

When isolating the variable, we can isolate it on either the left side or the right side of an equation.

EXAMPLE 4 Isolating the Variable on the Right

Solve and check: $-\frac{1}{2} = x - \frac{2}{3}$.

SOLUTION We can isolate the variable, x, on the right side by adding $\frac{2}{3}$ to both sides of the equation.

$-\frac{1}{2} = x - \frac{2}{3}$ — This is the given equation.

$-\frac{1}{2} + \frac{2}{3} = x - \frac{2}{3} + \frac{2}{3}$ — Add $\frac{2}{3}$ to both sides, isolating x on the right.

$-\frac{3}{6} + \frac{4}{6} = x$ — Rewrite each fraction as an equivalent fraction with a denominator of 6: $-\frac{1}{2} + \frac{2}{3} = -\frac{1}{2}\cdot\frac{3}{3} + \frac{2}{3}\cdot\frac{2}{2} = -\frac{3}{6} + \frac{4}{6}$.

$\frac{1}{6} = x$ — Add on the left side: $-\frac{3}{6} + \frac{4}{6} = \frac{-3 + 4}{6} = \frac{1}{6}$.

Take a moment to check the proposed solution, $\frac{1}{6}$. Substitute $\frac{1}{6}$ for x in the original equation. You should obtain $-\frac{1}{2} = -\frac{1}{2}$. This true statement indicates that the solution is $\frac{1}{6}$, or the solution set is $\left\{\frac{1}{6}\right\}$. ■

STUDY TIP

The equations $a = b$ and $b = a$ have the same meaning. If you prefer, you can solve

$$-\frac{1}{2} = x - \frac{2}{3}$$

by reversing the two sides and solving

$$x - \frac{2}{3} = -\frac{1}{2}.$$

✔ CHECK POINT 4 Solve and check: $-\frac{1}{2} = x - \frac{3}{4}$.

In Example 5, we combine like terms before using the addition property.

EXAMPLE 5 Combining Like Terms before Using the Addition Property

Solve and check: $5y + 3 - 4y - 8 = 6 + 9$.

SOLUTION

$5y + 3 - 4y - 8 = 6 + 9$	This is the given equation.
$y - 5 = 15$	Combine like terms: $5y - 4y = y$, $3 - 8 = -5$, and $6 + 9 = 15$.
$y - 5 + 5 = 15 + 5$	Add 5 to both sides.
$y = 20$	

To check the proposed solution, 20, replace y with 20 in the original equation.

Check

$5y + 3 - 4y - 8 = 6 + 9$	Be sure to use the original equation and not the simplified form in the second step above. (Why?)
$5(20) + 3 - 4(20) - 8 \stackrel{?}{=} 6 + 9$	Substitute 20 for y.
$100 + 3 - 80 - 8 \stackrel{?}{=} 6 + 9$	Multiply on the left.
$103 - 88 \stackrel{?}{=} 6 + 9$	Combine positive numbers and combine negative numbers on the left.
$15 = 15$	This statment is true.

This true statement verifies that the solution is 20, or the solution set is $\{20\}$. ■

CHECK POINT 5 Solve and check: $8y + 7 - 7y - 10 = 6 + 4$.

Adding and Subtracting Variable Terms on Both Sides of an Equation In some equations, variable terms appear on both sides. Here is an example:

$$4x = 7 + 3x.$$

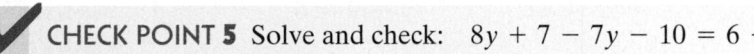

A variable term, $4x$, is on the left side. A variable term, $3x$, is on the right side.

Our goal is to isolate all the variable terms on one side of the equation. We can use the addition property of equality to do this. The property allows us to add or subtract the same variable term on both sides of an equation without changing the solution. Let's see how we can use this idea to solve $4x = 7 + 3x$.

EXAMPLE 6 Using the Addition Property to Isolate Variable Terms

Solve and check: $4x = 7 + 3x$.

SOLUTION In the given equation, variable terms appear on both sides. We can isolate them on one side by subtracting $3x$ from both sides of the equation.

$4x = 7 + 3x$	This is the given equation.
$4x - 3x = 7 + 3x - 3x$	Subtract $3x$ from both sides and isolate variable terms on the left.
$x = 7$	Subtracting $3x$ from both sides eliminates $3x$ on the right. On the left, $4x - 3x = 1x = x$.

To check the proposed solution, 7, replace x with 7 in the original equation.

Check

$4x = 7 + 3x$	Use the original equation.
$4(7) \stackrel{?}{=} 7 + 3(7)$	Substitute 7 for x.
$28 \stackrel{?}{=} 7 + 21$	Multiply: $4(7) = 28$ and $3(7) = 21$.
$28 = 28$	This statement is true.

This true statement verifies that the solution is 7, or the solution set is $\{7\}$. ■

✔ CHECK POINT 6 Solve and check: $7x = 12 + 6x$.

EXAMPLE 7 Solving an Equation by Isolating the Variable

Solve and check: $3y - 9 = 2y + 6$.

SOLUTION Our goal is to isolate variable terms on one side and constant terms on the other side. Let's begin by isolating the variable on the left.

$3y - 9 = 2y + 6$	This is the given equation.
$3y - 2y - 9 = 2y - 2y + 6$	Isolate the variable terms on the left by subtracting $2y$ from both sides.
$y - 9 = 6$	Subtracting $2y$ from both sides eliminates $2y$ on the right. On the left, $3y - 2y = 1y = y$.

Now we isolate the constant terms on the right by adding 9 to both sides.

$y - 9 + 9 = 6 + 9$	Add 9 to both sides.
$y = 15$	

Check

$3y - 9 = 2y + 6$	Use the original equation.
$3(15) - 9 \stackrel{?}{=} 2(15) + 6$	Substitute 15 for y.
$45 - 9 \stackrel{?}{=} 30 + 6$	Multiply: $3(15) = 45$ and $2(15) = 30$.
$36 = 36$	This statement is true.

The solution is 15, or the solution set is $\{15\}$. ■

✔ CHECK POINT 7 Solve and check: $3x - 6 = 2x + 5$.

3 Solve applied problems using formulas.

Applications Our next example shows how the addition property of equality can be used to find the value of a variable in a mathematical model.

EXAMPLE 8 An Application: Vocabulary and Age

There is a relationship between the number of words in a child's vocabulary, V, and the child's age, A, in months, for ages between 15 and 50 months, inclusive. This relationship can be modeled by the formula

$$V + 900 = 60A.$$

Use the formula to find the vocabulary of a child at the age of 30 months.

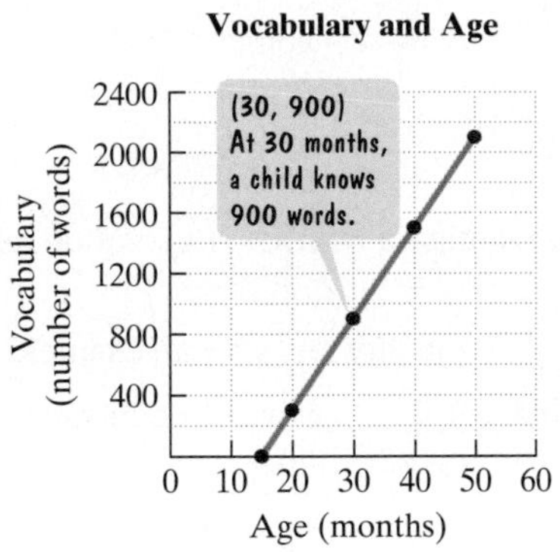

FIGURE 2.1

SOLUTION In the formula, A represents the child's age, in months. We are interested in a 30-month-old child. Thus, we substitute 30 for A. Then we use the addition property of equality to find V, the number of words in the child's vocabulary.

$$V + 900 = 60A \quad \text{This is the given formula.}$$
$$V + 900 = 60(30) \quad \text{Substitute 30 for } A.$$
$$V + 900 = 1800 \quad \text{Multiply: } 60(30) = 1800.$$
$$V + 900 - 900 = 1800 - 900 \quad \text{Subtract 900 from both sides and solve for } V.$$
$$V = 900$$

At the age of 30 months, a child has a vocabulary of 900 words. ■

The points in the rectangular coordinate system in Figure 2.1 allow us to "see" the formula $V + 900 = 60A$. The first coordinate of each point represents the child's age, in months. The second coordinate represents the child's vocabulary. The points are rising steadily from left to right. This shows that a typical child's vocabulary is steadily increasing with age.

CHECK POINT 8 Use the formula $V + 900 = 60A$ to find the vocabulary of a child at the age of 50 months.

2.1 EXERCISE SET

Student Solutions Manual | CD/Video | PH Math/Tutor Center | MathXL Tutorials on CD | MathXL® | MyMathLab | Interactmath.com

Practice Exercises

Solve each equation in Exercises 1–44 using the addition property of equality. Be sure to check your proposed solutions.

1. $x - 4 = 19$

2. $y - 5 = -18$

3. $z + 8 = -12$

4. $z + 13 = -15$

5. $-2 = x + 14$

6. $-13 = x + 11$

7. $-17 = y - 5$

8. $-21 = y - 4$

9. $7 + z = 11$

10. $18 + z = 14$

11. $-6 + y = -17$

12. $-8 + y = -29$

13. $x + \frac{1}{3} = \frac{7}{3}$

14. $x + \frac{7}{8} = \frac{9}{8}$

15. $t + \frac{5}{6} = -\frac{7}{12}$

16. $t + \frac{2}{3} = -\frac{7}{6}$

17. $x - \frac{3}{4} = \frac{9}{2}$

18. $x - \frac{3}{5} = \frac{7}{10}$

19. $-\frac{1}{5} + y = -\frac{3}{4}$

20. $-\frac{1}{8} + y = -\frac{1}{4}$

21. $3.2 + x = 7.5$

22. $-2.7 + w = -5.3$

23. $x + \frac{3}{4} = -\frac{9}{2}$

24. $r + \frac{3}{5} = -\frac{7}{10}$

25. $5 = -13 + y$

26. $-11 = 8 + x$

27. $-\frac{3}{5} = -\frac{3}{2} + s$

28. $\frac{7}{3} = -\frac{5}{2} + z$

29. $830 + y = 520$

30. $-90 + t = -35$

31. $r + 3.7 = 8$

32. $x + 10.6 = -9$

33. $-3.7 + m = -3.7$

34. $y + \frac{7}{11} = \frac{7}{11}$

35. $6y + 3 - 5y = 14$

36. $-3x - 5 + 4x = 9$

37. $7 - 5x + 8 + 2x + 4x - 3 = 2 + 3 \cdot 5$

38. $13 - 3r + 2 + 6r - 2r - 1 = 3 + 2 \cdot 9$

39. $7y + 4 = 6y - 9$

40. $4r - 3 = 5 + 3r$

41. $12 - 6x = 18 - 7x$

42. $20 - 7s = 26 - 8s$

43. $4x + 2 = 3(x - 6) + 8$

44. $7x + 3 = 6(x - 1) + 9$

Practice Plus

The equations in Exercises 45–48 contain small geometric figures that represent real numbers. Use the addition property of equality to isolate x on one side of the equation and the geometric figures on the other side.

45. $x - \square = \triangle$

46. $x + \square = \triangle$

47. $2x + \triangle = 3x + \square$

48. $6x - \triangle = 7x - \square$

In Exercises 49–52, use the given information to write an equation. Let x represent the number described in each exercise. Then solve the equation and find the number.

49. If 12 is subtracted from a number, the result is -2. Find the number.

50. If 23 is subtracted from a number, the result is -8. Find the number.

51. The difference between $\frac{2}{5}$ of a number and 8 is $\frac{7}{5}$ of that number. Find the number.

52. The difference between 3 and $\frac{2}{7}$ of a number is $\frac{5}{7}$ of that number. Find the number.

Application Exercises

Formulas frequently appear in the business world. For example, the cost, C, of an item (the price paid by a retailer) plus the markup, M, on that item (the retailer's profit) equals the selling price, S, of the item. The formula is

$$C + M = S.$$

Use the formula to solve Exercises 53–54.

53. The selling price of a computer is \$1850. If the markup on the computer is \$150, find the cost to the retailer for the computer.

54. The selling price of a television is \$650. If the cost to the retailer for the television is \$520, find the markup.

The formula

$$d + 525{,}000 = 5000c$$

models the relationship between the annual number of deaths in the United States from heart disease, d, and the average cholesterol level, c, of blood. (Cholesterol level, c, is expressed in milligrams per deciliter of blood.) Use this formula to solve Exercises 55–56.

55. The average cholesterol level for people in the United States is 210. According to the formula, how many deaths per year from heart disease can be expected with this cholesterol level?

56. Suppose that the average cholesterol level for people in the United States could be reduced to 180. Determine the number of annual deaths from heart disease that can be expected with this reduced cholesterol level.

Statins, used to control high cholesterol, are the top-selling class of prescription drugs in the United States. The bar graph shows statin sales, in billions of dollars, from 1998 through 2002. The data can be modeled by the formula

$$S - 1.6x = 5.8,$$

where S is statin sales, in billions of dollars, x years after 1998. Use this formula to solve Exercises 57–58.

Statin Sales in the U.S.

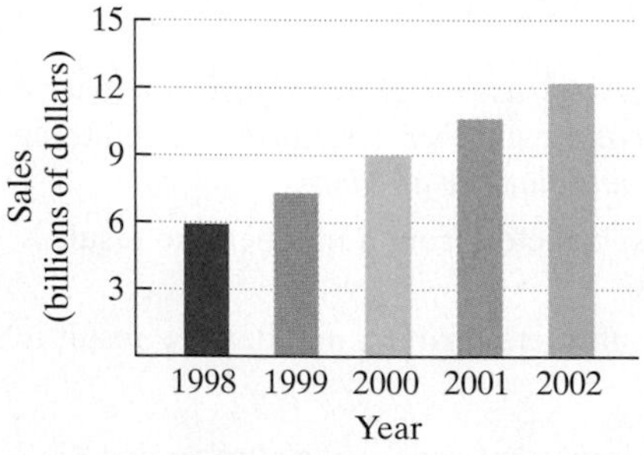

Source: Lawrence Berkeley National Laboratory

57. According to the formula, what will Americans spend on statins in 2005?

58. According to the formula, what will Americans spend on statins in 2006?

Writing in Mathematics

59. What is an equation?

60. Explain how to determine whether a number is a solution to an equation.

61. State the addition property of equality and give an example.

62. Explain why $x + 2 = 9$ and $x + 2 = -6$ are not equivalent equations.

63. What is the difference between solving an equation such as

$$5y + 3 - 4y - 8 = 6 + 9$$

and simplifying an algebraic expression such as

$$5y + 3 - 4y - 8?$$

If there is a difference, which topic should be taught first? Why?

Critical Thinking Exercises

64. Which one of the following statements is true?

a. If $y - a = -b$, then $y = a + b$.

b. If $y + 7 = 0$, then $y = 7$.

c. The solution to $4 - x = -3x$ is -2.

d. If 7 is added on one side of an equation, then it should be subtracted on the other side.

65. Write an equation with a negative solution that can be solved by adding 100 to both sides.

Technology Exercises

Use a calculator to solve each equation in Exercises 66–67.

66. $x - 7.0463 = -9.2714$

67. $6.9825 = 4.2296 + y$

Review Exercises

68. Plot $(-3, 1)$ in rectangular coordinates. In which quadrant does the point lie? (Section 1.3, Example 1)

69. Simplify: $-16 - 8 \div 4 \cdot (-2)$. (Section 1.8, Example 4)

70. Simplify: $3[7x - 2(5x - 1)]$. (Section 1.8, Example 11)

SECTION 2.2

Objectives

1 Use the multiplication property of equality to solve equations.

2 Solve equations in the form $-x = c$.

3 Use the addition and multiplication properties to solve equations.

4 Solve applied problems using formulas.

THE MULTIPLICATION PROPERTY OF EQUALITY

Could you live to be 125? The number of Americans ages 100 or older could approach 850,000 by 2050. Some scientists predict that by 2100, our descendants could live to be 200 years of age. In this section, we will see how a formula can be used to make these kinds of predictions as we turn to a new property for solving linear equations.

1 Use the multiplication property of equality to solve equations.

Using the Multiplication Property of Equality to Solve Equations Can the addition property of equality be used to solve every linear equation in one variable? No. For example, consider the equation

$$\frac{x}{5} = 9.$$

We cannot isolate the variable x by adding or subtracting 5 on both sides. To get x by itself on the left side, multiply the left side by 5:

$$5 \cdot \frac{x}{5} = \left(5 \cdot \frac{1}{5}\right)x = 1x = x.$$

5 is the multiplicative inverse of $\frac{1}{5}$.

You must then multiply the right side by 5 also. By doing this, we are using the **multiplication property of equality**.

THE MULTIPLICATION PROPERTY OF EQUALITY The same nonzero real number (or algebraic expression) may multiply both sides of an equation without changing the solution. This can be expressed symbolically as follows:

If $a = b$ and $c \neq 0$, then $ac = bc$.

EXAMPLE 1 Solving an Equation Using the Multiplication Property

Solve the equation: $\frac{x}{5} = 9$.

SOLUTION We can isolate the variable, x, by multiplying both sides of the equation by 5.

$$\frac{x}{5} = 9 \quad \text{This is the given equation.}$$

$$5 \cdot \frac{x}{5} = 5 \cdot 9 \quad \text{Multiply both sides by 5.}$$

$$1x = 45 \quad \text{Simplify.}$$

$$x = 45 \quad 1x = x$$

By substituting 45 for x in the original equation, we obtain the true statement $9 = 9$. This verifies that the solution is 45, or the solution set is $\{45\}$. ■

CHECK POINT 1 Solve the equation: $\frac{x}{3} = 12$.

When we use the multiplication property of equality, we multiply both sides of an equation by the same nonzero number. We know that division is multiplication by a multiplicative inverse. Thus, the multiplication property also lets us divide both sides of an equation by a nonzero number without changing the solution.

EXAMPLE 2 Dividing Both Sides by the Same Nonzero Number

Solve: **a.** $6x = 30$ **b.** $-7y = 56$ **c.** $-18.9 = 3z$.

SOLUTION In each equation, the variable is multiplied by a number. We can isolate the variable by dividing both sides of the equation by that number.

a. $6x = 30$ This is the given equation.

$\frac{6x}{6} = \frac{30}{6}$ Divide both sides by 6.

$1x = 5$ Simplify.

$x = 5$ $1x = x$

By substituting 5 for x in the original equation, we obtain the true statement $30 = 30$. The solution is 5, or the solution set is $\{5\}$.

b. $-7y = 56$ This is the given equation.

$\frac{-7y}{-7} = \frac{56}{-7}$ Divide both sides by -7.

$1y = -8$ Simplify.

$y = -8$ $1y = y$

By substituting -8 for y in the original equation, we obtain the true statement $56 = 56$. The solution is -8, or the solution set is $\{-8\}$.

c. $-18.9 = 3z$ This is the given equation.

$\frac{-18.9}{3} = \frac{3z}{3}$ Divide both sides by 3.

$-6.3 = 1z$ Simplify.

$-6.3 = z$ $1z = z$

By substituting -6.3 for z in the original equation, we obtain the true statement $-18.9 = -18.9$. The solution is -6.3, or the solution set is $\{-6.3\}$. ■

CHECK POINT 2 Solve:

a. $4x = 84$ **b.** $-11y = 44$ **c.** $-15.5 = 5z$.

Some equations have a variable term with a fractional coefficient. Here is an example:

$$\frac{3}{4}y = 12.$$

The coefficient of the term $\frac{3}{4}y$ is $\frac{3}{4}$.

To isolate the variable, multiply both sides of the equation by the multiplicative inverse of the fraction. For the equation $\frac{3}{4}y = 12$, the multiplicative inverse of $\frac{3}{4}$ is $\frac{4}{3}$. Thus, we solve $\frac{3}{4}y = 12$ by multiplying both sides by $\frac{4}{3}$.

EXAMPLE 3 Using the Multiplication Property to Eliminate a Fractional Coefficient

Solve: **a.** $\frac{3}{4}y = 12$ **b.** $9 = -\frac{3}{5}x$.

SOLUTION

a.

$$\frac{3}{4}y = 12$$ This is the given equation.

$$\frac{4}{3}\left(\frac{3}{4}y\right) = \frac{4}{3} \cdot 12$$ Multiply both sides by $\frac{4}{3}$, the multiplicative inverse of $\frac{3}{4}$.

$$1y = 16$$ On the left, $\frac{4}{3}\left(\frac{3}{4}y\right) = \left(\frac{4}{3} \cdot \frac{3}{4}\right)y = 1y$. On the right, $\frac{4}{3} \cdot \frac{12}{1} = \frac{48}{3} = 16$.

$$y = 16$$ $1y = y$

By substituting 16 for y in the original equation, we obtain the true statement $12 = 12$. The solution is 16, or the solution set is $\{16\}$.

b.

$$9 = -\frac{3}{5}x$$ This is the given equation.

$$-\frac{5}{3} \cdot 9 = -\frac{5}{3}\left(-\frac{3}{5}x\right)$$ Multiply both sides by $-\frac{5}{3}$, the multiplicative inverse of $-\frac{3}{5}$.

$$-15 = 1x$$ Simplify.

$$-15 = x$$ $1x = x$

By substituting -15 for x in the original equation, we obtain the true statement $9 = 9$. The solution is -15, or the solution set is $\{-15\}$. ■

STUDY TIP

The equation

$$9 = -\frac{3}{5}x$$

can be expressed as

$$9 = -\frac{3x}{5}$$

or

$$9 = \frac{-3x}{5}.$$

CHECK POINT 3 Solve: **a.** $\frac{2}{3}y = 16$ **b.** $28 = -\frac{7}{4}x$.

2 Solve equations in the form $-x = c$.

Equations and Coefficients of −1 How do we solve an equation in the form $-x = c$, such as $-x = 4$? Because the equation means $-1x = 4$, we have not yet obtained a solution. The solution of an equation is obtained from the form $x =$ some number. The equation $-x = 4$ is not yet in this form. We still need to isolate x. We can do this by multiplying or dividing both sides of the equation by -1. We will multiply by -1.

EXAMPLE 4 Solving Equations in the Form $-x = c$

Solve: **a.** $-x = 4$ **b.** $-x = -7$.

SOLUTION We multiply both sides of each equation by -1. This will isolate x on the left side.

a.

$$-x = 4$$ This is the given equation.

$$-1x = 4$$ Rewrite $-x$ as $-1x$.

$$(-1)(-1x) = (-1)(4)$$ Multiply both sides by -1.

$$1x = -4$$ On the left, $(-1)(-1) = 1$. On the right, $(-1)(4) = -4$.

$$x = -4$$ $1x = x$

Check

$$-x = 4 \quad \text{This is the original equation.}$$
$$-(-4) \stackrel{?}{=} 4 \quad \text{Substitute } -4 \text{ for } x.$$
$$4 = 4 \quad -(-a) = a, \text{ so } -(-4) = 4.$$

This true statement indicates that the solution is −4, or the solution set is $\{-4\}$.

b.

$$-x = -7 \quad \text{This is the given equation.}$$
$$-1x = -7 \quad \text{Rewrite } -x \text{ as } -1x.$$
$$(-1)(-1x) = (-1)(-7) \quad \text{Multiply both sides by } -1.$$
$$1x = 7 \quad (-1)(-1) = 1 \text{ and } (-1)(-7) = 7.$$
$$x = 7 \quad 1x = x$$

By substituting 7 for x in the original equation, we obtain the true statement $-7 = -7$. The solution is 7, or the solution set is $\{7\}$. ■

STUDY TIP

If $-x = c$, then the equation's solution is the additive inverse (the opposite) of c. For example, the solution of $-x = -7$ is the additive inverse of -7, which is 7.

✔ CHECK POINT 4 Solve: **a.** $-x = 5$ **b.** $-x = -3$.

3 Use the addition and multiplication properties to solve equations.

Equations Requiring Both the Addition and Multiplication Properties When an equation does not contain fractions, we will often use the addition property of equality before the multiplication property of equality. Our overall goal is to isolate the variable with a coefficient of 1 on either the left or right side of the equation.

Here is the procedure that we will be using to solve the equations in the next three examples:

- Use the addition property of equality to isolate the variable term.
- Use the multiplication property of equality to isolate the variable.

EXAMPLE 5 Using Both the Addition and Multiplication Properties

Solve: $3x + 1 = 7$.

SOLUTION We begin by isolating the variable term, $3x$, subtracting 1 from both sides. Then we isolate the variable, x, by dividing both sides by 3.

- **Use the addition property of equality to isolate the variable term.**

$$3x + 1 = 7 \quad \text{This is the given equation.}$$
$$3x + 1 - 1 = 7 - 1 \quad \text{Use the addition property, subtracting 1 from both sides.}$$
$$3x = 6 \quad \text{Simplify.}$$

- **Use the multiplication property of equality to isolate the variable.**

$$\frac{3x}{3} = \frac{6}{3} \quad \text{Divide both sides by 3.}$$
$$x = 2 \quad \text{Simplify.}$$

By substituting 2 for x in the original equation, we obtain the true statement $7 = 7$. The solution is 2, or the solution set is $\{2\}$. ■

✔ CHECK POINT 5 Solve: $4x + 3 = 27$.

EXAMPLE 6 Using Both the Addition and Multiplication Properties

Solve: $-2y - 28 = 4.$

SOLUTION We begin by isolating the variable term, $-2y$, adding 28 to both sides. Then we isolate the variable, y, by dividing both sides by -2.

- **Use the addition property of equality to isolate the variable term.**

$$-2y - 28 = 4 \quad \text{This is the given equation.}$$
$$-2y - 28 + 28 = 4 + 28 \quad \text{Use the addition property, adding 28 to both sides.}$$
$$-2y = 32 \quad \text{Simplify.}$$

- **Use the multiplication property of equality to isolate the variable.**

$$\frac{-2y}{-2} = \frac{32}{-2} \quad \text{Divide both sides by } -2.$$
$$y = -16 \quad \text{Simplify.}$$

Take a moment to substitute -16 for y in the given equation. Do you obtain the true statement $4 = 4$? The solution is -16, or the solution set is $\{-16\}$. ■

✔ CHECK POINT 6 Solve: $-4y - 15 = 25.$

EXAMPLE 7 Using Both the Addition and Multiplication Properties

Solve: $3x - 14 = -2x + 6.$

SOLUTION We will use the addition property to collect all terms involving x on the left and all numerical terms on the right. Then we will isolate the variable, x, by dividing both sides by its coefficient.

- **Use the addition property of equality to isolate the variable term.**

$$3x - 14 = -2x + 6 \quad \text{This is the given equation.}$$
$$3x + 2x - 14 = -2x + 2x + 6 \quad \text{Add } 2x \text{ to both sides.}$$
$$5x - 14 = 6 \quad \text{Simplify.}$$
$$5x - 14 + 14 = 6 + 14 \quad \text{Add 14 to both sides.}$$
$$5x = 20 \quad \text{Simplify. The variable term, } 5x\text{, is isolated on the left. The numerical term, 20, is isolated on the right.}$$

- **Use the multiplication property of equality to isolate the variable.**

$$\frac{5x}{5} = \frac{20}{5} \quad \text{Divide both sides by 5.}$$
$$x = 4 \quad \text{Simplify.}$$

Check

$$3x - 14 = -2x + 6 \quad \text{Use the original equation.}$$
$$3(4) - 14 \stackrel{?}{=} -2(4) + 6 \quad \text{Substitute the proposed solution, 4, for } x.$$
$$12 - 14 \stackrel{?}{=} -8 + 6 \quad \text{Multiply.}$$
$$-2 = -2 \quad \text{Simplify.}$$

The true statement $-2 = -2$ verifies that the solution is 4, or the solution set is $\{4\}$. ■

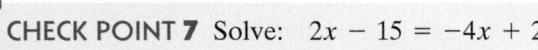

✔ CHECK POINT 7 Solve: $2x - 15 = -4x + 21.$

4 Solve applied problems using formulas.

Applications Your life expectancy is related to the year when you were born. The bar graph in Figure 2.2 shows life expectancy in the United States by year of birth. The data for U.S. women shown by the green bars can be modeled by the formula

$$E = 0.18t + 71,$$

where E is the life expectancy for women born t years after 1950.

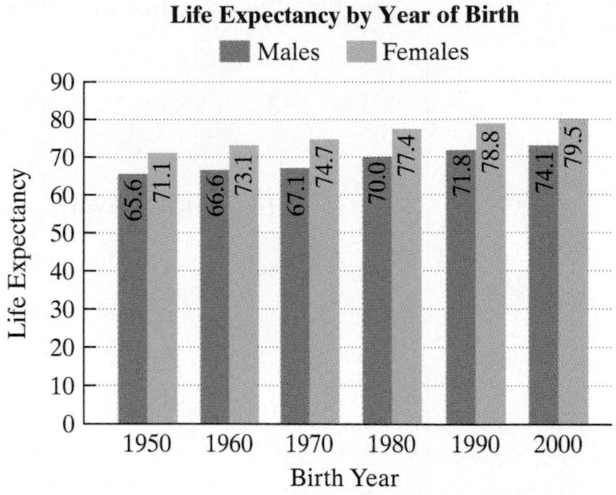

FIGURE 2.2
Source: U.S. Bureau of the Census

EXAMPLE 8 Using the Formula For Life Expectancy

Use the formula

$$E = 0.18t + 71$$

to determine the year of birth for which U.S. women can expect to live 87.2 years.

SOLUTION We are interested in a life expectancy of 87.2 years. We substitute 87.2 for E in the formula and solve for t. Keep in mind that t represents birth years *after* 1950.

$$E = 0.18t + 71$$ This is the given formula.

$$87.2 = 0.18t + 71$$ Replace E with 87.2.

Our goal is to isolate t.

$$87.2 - 71 = 0.18t + 71 - 71$$ Isolate the term containing t by subtracting 71 from both sides.

$$16.2 = 0.18t$$ Simplify.

$$\frac{16.2}{0.18} = \frac{0.18t}{0.18}$$ Divide both sides by 0.18.

$$90 = t$$ Simplify: $0.18\overline{)16.20}$ = 90.

The formula indicates that U.S. women born 90 years after 1950, or in 2040, can expect to live 87.2 years. ■

✔ **CHECK POINT 8** The data for U.S. men shown by the blue bars in Figure 2.2 can be modeled by the formula $E = 0.16t + 65$, where E represents life expectancy for men born t years after 1950. Determine the year of birth for which U.S. men can expect to live 76.2 years.

2.2 EXERCISE SET

Student Solutions Manual | CD/Video | PH Math/Tutor Center | MathXL Tutorials on CD | MathXL® | MyMathLab | Interactmath.com

Practice Exercises

Solve each equation in Exercises 1–28 using the multiplication property of equality. Be sure to check your proposed solutions.

1. $\frac{x}{6} = 5$ **2.** $\frac{x}{7} = 4$

3. $\frac{x}{-3} = 11$ **4.** $\frac{x}{-5} = 8$

5. $5y = 35$ **6.** $6y = 42$

7. $-7y = 63$ **8.** $-4y = 32$

9. $-28 = 8z$ **10.** $-36 = 8z$

11. $-18 = -3z$ **12.** $-54 = -9z$

13. $-8x = 6$ **14.** $-8x = 4$

15. $17y = 0$ **16.** $-16y = 0$

17. $\frac{2}{3}y = 12$ **18.** $\frac{3}{4}y = 15$

19. $28 = -\frac{7}{2}x$ **20.** $20 = -\frac{5}{8}x$

21. $-x = 17$ **22.** $-x = 23$

23. $-47 = -y$ **24.** $-51 = -y$

25. $-\frac{x}{5} = -9$ **26.** $-\frac{x}{5} = -1$

27. $2x - 12x = 50$ **28.** $8x - 3x = -45$

Solve each equation in Exercises 29–54 using both the addition and multiplication properties of equality. Check proposed solutions.

29. $2x + 1 = 11$ **30.** $2x + 5 = 13$

31. $2x - 3 = 9$ **32.** $3x - 2 = 9$

33. $-2y + 5 = 7$ **34.** $-3y + 4 = 13$

35. $-3y - 7 = -1$ **36.** $-2y - 5 = 7$

37. $12 = 4z + 3$ **38.** $14 = 5z - 21$

39. $-x - 3 = 3$ **40.** $-x - 5 = 5$

41. $6y = 2y - 12$ **42.** $8y = 3y - 10$

43. $3z = -2z - 15$ **44.** $2z = -4z + 18$

45. $-5x = -2x - 12$ **46.** $-7x = -3x - 8$

47. $8y + 4 = 2y - 5$ **48.** $5y + 6 = 3y - 6$

49. $6z - 5 = z + 5$ **50.** $6z - 3 = z + 2$

51. $6x + 14 = 2x - 2$ **52.** $9x + 2 = 6x - 4$

53. $-3y - 1 = 5 - 2y$ **54.** $-3y - 2 = -5 - 4y$

Practice Plus

The equations in Exercises 55–58 contain small geometric figures that represent real numbers. Use the multiplication property of equality to isolate x on one side of the equation and the geometric figures on the other side.

55. $\frac{x}{\square} = \triangle$ **56.** $\triangle = \square x$

57. $\triangle = -x$ **58.** $\frac{-x}{\square} = \triangle$

In Exercises 59–62, use the given information to write an equation. Let x represent the number described in each exercise. Then solve the equation and find the number.

59. If a number is multiplied by 6, the result is 10. Find the number.

60. If a number is multiplied by -6, the result is 20. Find the number.

61. If a number is divided by -9, the result is 5. Find the number.

62. If a number is divided by -7, the result is 8. Find the number.

Application Exercises

The formula

$$M = \frac{n}{5}$$

models your distance, M, in miles, from a lightning strike in a thunderstorm if it takes n seconds to hear thunder after seeing the lightning. Use this formula to solve Exercises 63–64.

63. If you are 2 miles away from the lightning flash, how long will it take the sound of thunder to reach you?

64. If you are 3 miles away from the lightning flash, how long will it take the sound of thunder to reach you?

The Mach number is a measurement of speed, named after the man who suggested it, Ernst Mach (1838–1916). The formula

$$M = \frac{A}{740}$$

indicates that the speed of an aircraft, A, in miles per hour, divided by the speed of sound, approximately 740 miles per hour, results in the Mach number, M. Use the formula to determine the speed, in miles per hour, of the aircrafts in Exercises 65–66. (Note: When an aircraft's speed increases beyond Mach 1, it is said to have broken the sound barrier.)

65.

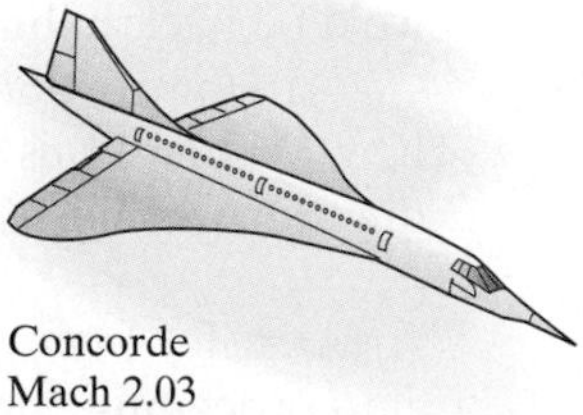

Concorde
Mach 2.03
Speed = ?

66.

SR-71 Blackbird
Mach 3.3
Speed = ?

67. The formula $P = -0.5d + 100$ models the percentage, P, of lost hikers found in search and rescue missions when members of the search team walk parallel to one another separated by a distance of d yards. If a search and rescue team finds 70% of lost hikers, substitute 70 for P in the formula and find the parallel distance of separation between members of the search party.

68. The formula $M = 420x + 720$ models the data for the amount of money lost to credit card fraud worldwide, M, in millions of dollars, x years after 1989. In which year did losses amount to 4080 million dollars?

Writing in Mathematics

69. State the multiplication property of equality and give an example.

70. Explain how to solve the equation $-x = -50$.

71. Explain how to solve the equation $2x + 8 = 5x - 3$.

72. What might occur in the future to cause life expectancy to exceed the predictions made by the models in Example 8 and Check Point 8 on page 116?

Critical Thinking Exercises

73. Which one of the following statements is true?

a. If $7x = 21$, then $x = 21 - 7$.

b. If $3x - 4 = 16$, then $3x = 12$.

c. If $3x + 7 = 0$, then $x = \frac{7}{3}$.

d. The solution to $6x = 0$ is not a natural number.

In Exercises 74–75, write an equation with the given characteristics.

74. The solution is a positive integer and the equation can be solved by dividing both sides by -60.

75. The solution is a negative integer and the equation can be solved by multiplying both sides by $\frac{4}{5}$.

Technology Exercises

Solve each equation in Exercises 76–77. Use a calculator to help with the arithmetic. Check your solution using the calculator.

76. $3.7x - 19.46 = -9.988$

77. $-72.8y - 14.6 = -455.43 - 4.98y$

Review Exercises

78. Evaluate: $(-10)^2$. (Section 1.8, Example 1)

79. Evaluate: -10^2. (Section 1.8, Example 1)

80. Evaluate $x^3 - 4x$ for $x = -1$. (Section 1.8, Example 10)

SECTION 2.3

SOLVING LINEAR EQUATIONS

Objectives

1. Solve linear equations.
2. Solve linear equations containing fractions.
3. Identify equations with no solution or infinitely many solutions.
4. Solve applied problems using formulas.

Yes, we overindulged, but it was delicious. Now if we could just grow a few inches taller, we'd be back in line with our recommended weights.

In this section, we will see how algebra models the relationship between weight and height. To use this mathematical model, it would be helpful to have a systematic procedure for solving linear equations. We open the section with such a procedure.

1 Solve linear equations.

A Step-By-Step Procedure for Solving Linear Equations Here is a step-by-step procedure for solving a linear equation in one variable. Not all of these steps are necessary to solve every equation.

SOLVING A LINEAR EQUATION

1. Simplify the algebraic expression on each side.
2. Collect all the variable terms on one side and all the constant terms on the other side.
3. Isolate the variable and solve.
4. Check the proposed solution in the original equation.

EXAMPLE 1 Solving a Linear Equation

Solve and check: $2x - 8x + 40 = 13 - 3x - 3$.

SOLUTION

Step 1. **Simplify the algebraic expression on each side.**

$2x - 8x + 40 = 13 - 3x - 3$	This is the given equation.
$-6x + 40 = 10 - 3x$	Combine like terms: $2x - 8x = -6x$ and $13 - 3 = 10$.

Step 2. **Collect variable terms on one side and constant terms on the other side.** We will collect variable terms on the left by adding $3x$ to both sides. We will collect the numbers on the right by subtracting 40 from both sides.

$-6x + 40 + 3x = 10 - 3x + 3x$	Add $3x$ to both sides.
$-3x + 40 = 10$	Simplify: $-6x + 3x = -3x$.
$-3x + 40 - 40 = 10 - 40$	Subtract 40 from both sides.
$-3x = -30$	Simplify.

Step 3. **Isolate the variable and solve.** We isolate the variable, x, by dividing both sides by -3.

$\frac{-3x}{-3} = \frac{-30}{-3}$	Divide both sides by -3.
$x = 10$	Simplify.

Step 4. **Check the proposed solution in the original equation.** Substitute 10 for x in the original equation.

$2x - 8x + 40 = 13 - 3x - 3$	This is the original equation.
$2 \cdot 10 - 8 \cdot 10 + 40 \stackrel{?}{=} 13 - 3 \cdot 10 - 3$	Substitute 10 for x.
$20 - 80 + 40 \stackrel{?}{=} 13 - 30 - 3$	Perform the indicated multiplications.
$-60 + 40 \stackrel{?}{=} -17 - 3$	Subtract: $20 - 80 = -60$ and $13 - 30 = -17$.
$-20 = -20$	Simplify.

By substituting 10 for x in the original equation, we obtain the true statement $-20 = -20$. This verifies that the solution is 10, or the solution set is $\{10\}$. ■

DISCOVER FOR YOURSELF

Solve the equation in Example 1 by collecting terms with the variable on the right and numbers on the left. What do you observe?

✔ **CHECK POINT 1** Solve and check: $-7x + 25 + 3x = 16 - 2x - 3$.

EXAMPLE 2 Solving a Linear Equation

Solve and check: $5x = 8(x + 3)$.

SOLUTION

Step 1. **Simplify the algebraic expression on each side.** Use the distributive property to remove parentheses on the right.

$5x = 8(x + 3)$	This is the given equation.
$5x = 8x + 24$	Use the distributive property.

Step 2. **Collect variable terms on one side and constant terms on the other side.** We will work with $5x = 8x + 24$ and collect variable terms on the left by subtracting $8x$ from both sides. The only constant term, 24, is already on the right.

$$5x - 8x = 8x + 24 - 8x \quad \text{Subtract } 8x \text{ from both sides.}$$
$$-3x = 24 \quad \text{Simplify: } 5x - 8x = -3x.$$

Step 3. **Isolate the variable and solve.** We isolate the variable, x, by dividing both sides by -3.

$$\frac{-3x}{-3} = \frac{24}{-3} \quad \text{Divide both sides by } -3.$$
$$x = -8 \quad \text{Simplify.}$$

Step 4. **Check the proposed solution in the original equation.** Substitute -8 for x in the original equation.

$$5x = 8(x + 3) \quad \text{This is the original equation.}$$
$$5(-8) \stackrel{?}{=} 8(-8 + 3) \quad \text{Substitute } -8 \text{ for } x.$$
$$5(-8) \stackrel{?}{=} 8(-5) \quad \text{Perform the addition in parentheses: } -8 + 3 = -5.$$
$$-40 = -40 \quad \text{Multiply.}$$

The true statement $-40 = -40$ verifies that -8 is the solution, or the solution set is $\{-8\}$. ■

CHECK POINT 2 Solve and check: $8x = 2(x + 6)$.

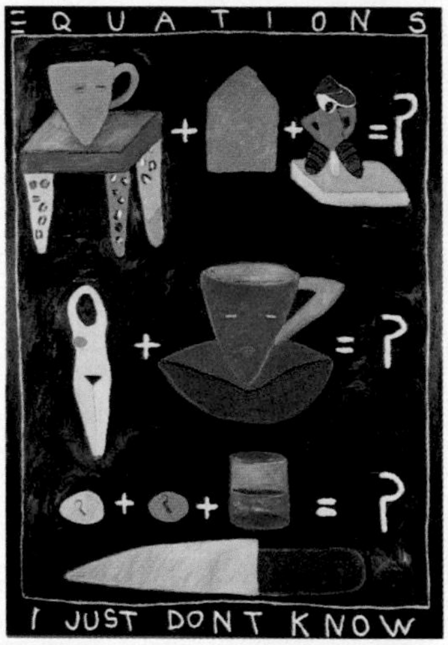

The compact, symbolic notation of algebra enables us to use a clear step-by-step method for solving equations, designed to avoid the confusion shown in the painting.

Squeak Carnwath *Equations* 1981, oil on cotton canvas 96 in. h × 72 in. w.

EXAMPLE 3 Solving a Linear Equation

Solve and check: $2(x - 3) - 17 = 13 - 3(x + 2)$.

SOLUTION

Step 1. **Simplify the algebraic expression on each side.**

Do not begin with 13 – 3. Multiplication (the distributive property) is applied before subtraction.

$$2(x - 3) - 17 = 13 - 3(x + 2) \quad \text{This is the given equation.}$$
$$2x - 6 - 17 = 13 - 3x - 6 \quad \text{Use the distributive property.}$$
$$2x - 23 = -3x + 7 \quad \text{Combine like terms.}$$

Step 2. **Collect variable terms on one side and constant terms on the other side.** We will collect variable terms on the left by adding $3x$ to both sides. We will collect the numbers on the right by adding 23 to both sides.

$$2x - 23 + 3x = -3x + 7 + 3x \quad \text{Add } 3x \text{ to both sides.}$$
$$5x - 23 = 7 \quad \text{Simplify: } 2x + 3x = 5x.$$
$$5x - 23 + 23 = 7 + 23 \quad \text{Add 23 to both sides.}$$
$$5x = 30 \quad \text{Simplify.}$$

Step 3. **Isolate the variable and solve.** We isolate the variable, x, by dividing both sides by 5.

$$\frac{5x}{5} = \frac{30}{5} \quad \text{Divide both sides by 5.}$$
$$x = 6 \quad \text{Simplify.}$$

Step 4. Check the proposed solution in the original equation. Substitute 6 for x in the original equation.

$$2(x-3)-17=13-3(x+2) \quad \text{This is the original equation.}$$
$$2(6-3)-17\stackrel{?}{=}13-3(6+2) \quad \text{Substitute 6 for x.}$$
$$2(3)-17\stackrel{?}{=}13-3(8) \quad \text{Simplify inside parentheses.}$$
$$6-17\stackrel{?}{=}13-24 \quad \text{Multiply.}$$
$$-11=-11 \quad \text{Subtract.}$$

The true statement $-11 = -11$ verifies that 6 is the solution, or the solution set is $\{6\}$. ■

CHECK POINT 3 Solve and check: $4(2x+1)-29=3(2x-5)$.

2 Solve linear equations containing fractions.

Linear Equations with Fractions Equations are easier to solve when they do not contain fractions. How do we remove fractions from an equation? We begin by multiplying both sides of the equation by the least common denominator of any fractions in the equation. The least common denominator is the smallest number that all denominators will divide into. Multiplying every term on both sides of the equation by the least common denominator will eliminate the fractions in the equation. Example 4 shows how we "clear an equation of fractions."

EXAMPLE 4 Solving a Linear Equation Involving Fractions

Solve and check: $\frac{3x}{2}=\frac{x}{5}-\frac{39}{5}$.

SOLUTION The denominators are 2, 5, and 5. The smallest number that is divisible by 2, 5, and 5 is 10. We begin by multiplying both sides of the equation by 10, the least common denominator.

$$\frac{3x}{2}=\frac{x}{5}-\frac{39}{5} \quad \text{This is the given equation.}$$
$$10\cdot\frac{3x}{2}=10\left(\frac{x}{5}-\frac{39}{5}\right) \quad \text{Multiply both sides by 10.}$$
$$10\cdot\frac{3x}{2}=10\cdot\frac{x}{5}-10\cdot\frac{39}{5} \quad \text{Use the distributive property. Be sure to multiply all terms by 10.}$$
$$\overset{5}{\cancel{10}}\cdot\frac{3x}{\underset{1}{\cancel{2}}}=\overset{2}{\cancel{10}}\cdot\frac{x}{\underset{1}{\cancel{5}}}-\overset{2}{\cancel{10}}\cdot\frac{39}{\underset{1}{\cancel{5}}} \quad \text{Divide out common factors in the multiplications.}$$
$$15x=2x-78 \quad \text{Complete the multiplications. The fractions are now cleared.}$$

At this point, we have an equation similar to those we previously have solved. Collect the variable terms on one side and the constant terms on the other side.

$$15x-2x=2x-2x-78 \quad \text{Subtract 2x to get the variable terms on the left.}$$
$$13x=-78 \quad \text{Simplify.}$$

Isolate x by dividing both sides by 13.

$$\frac{13x}{13}=\frac{-78}{13} \quad \text{Divide both sides by 13.}$$
$$x=-6 \quad \text{Simplify.}$$

Check the proposed solution. Substitute -6 for x in the original equation. You should obtain $-9 = -9$. This true statement verifies that the solution is -6, or the solution set is $\{-6\}$. ■

CHECK POINT 4 Solve and check: $\frac{x}{4} = \frac{2x}{3} + \frac{5}{6}$.

3 Identify equations with no solution or infinitely many solutions.

Linear Equations with No Solution or Infinitely Many Solutions Thus far, each equation that we have solved has had a single solution. However, some equations are not true for even one real number. Such an equation is called an **inconsistent equation**, or a **contradiction**. Here is an example of such an equation:

$$x = x + 4.$$

There is no number that is equal to itself plus 4. This equation has no solution. Its solution set is written either as

$$\{\ \} \quad \text{or} \quad \varnothing.$$

These symbols stand for the empty set, a set with no elements.

An equation that is true for all real numbers is called an **identity**. An example of an identity is

$$x + 3 = x + 2 + 1.$$

Every number plus 3 is equal to that number plus 2 plus 1. Every real number is a solution to this equation. Thus, the solution set to this equation is the set of all real numbers. This set is written as $\mathbb{R}$. In Section 2.7, you will learn a second notation for representing the set of all real numbers.

If you attempt to solve an equation with no solution, you will eliminate the variable and obtain a false statement, such as $2 = 5$. If you attempt to solve an equation that is true for every real number, you will eliminate the variable and obtain a true statement, such as $4 = 4$.

EXAMPLE 5 Solving a Linear Equation

Solve: $2x + 6 = 2(x + 4)$.

SOLUTION

$2x + 6 = 2(x + 4)$	This is the given equation.
$2x + 6 = 2x + 8$	Use the distributive property.
$2x + 6 - 2x = 2x + 8 - 2x$	Subtract 2x from both sides.
$6 = 8$	Simplify.

Keep reading. $6 = 8$ is not the solution.

The original equation is equivalent to the false statement $6 = 8$, which is false for every value of x. The equation is inconsistent and has no solution. You can express this by writing "no solution" or using one of the symbols for the empty set, { } or ∅. ■

CHECK POINT 5 Solve: $3x + 7 = 3(x + 1)$.

EXAMPLE 6 Solving a Linear Equation

Solve: $-3x + 5 + 5x = 4x - 2x + 5$.

SOLUTION

$-3x + 5 + 5x = 4x - 2x + 5$	This is the given equation.
$2x + 5 = 2x + 5$	Combine like terms: $-3x + 5x = 2x$ and $4x - 2x = 2x$.
$2x + 5 - 2x = 2x + 5 - 2x$	Subtract 2x from both sides.
$5 = 5$	Simplify.

Keep reading. $5 = 5$ is not the solution.

STUDY TIP

If you are concerned by the vocabulary of equation types, keep in mind that there are three possible situations. We can state these situations informally as follows:

1. x = a real number — conditional equation
2. x = all real numbers — identity
3. x = no real numbers. — inconsistent equation

The original equation is equivalent to the true statement $5 = 5$, which is true for every value of x. The equation is an identity and all real numbers are solutions. You can express this by writing "all real numbers" or using the notation $\mathbb{R}$ for the set of all real numbers. Try substituting any number of your choice for x in the original equation. You will obtain a true statement. ■

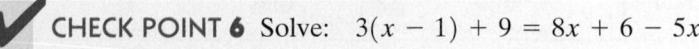

✔ **CHECK POINT 6** Solve: $3(x - 1) + 9 = 8x + 6 - 5x$.

An equation that is not an identity, but that is true for at least one real number, is called a **conditional equation**. The equation $2x + 3 = 17$ is an example of a conditional equation. The equation is not an identity and is true only if x is 7.

4 Solve applied problems using formulas.

Applications The next example shows how our procedure for solving equations with fractions can be used to find the value of a variable in a mathematical model.

EXAMPLE 7 Modeling Weight and Height

The formula

$$\frac{W}{2} - 3H = 53$$

models the recommended weight, W, in pounds, for a male, where H represents the man's height, in inches, over 5 feet. What is the recommended weight for a man who is 6 feet, 3 inches tall?

SOLUTION Keep in mind that H represents height in inches *above 5 feet.* A man who is 6 feet, 3 inches tall is 1 foot, 3 inches above 5 feet. Because 12 inches = 1 foot, he is $12 + 3$, or 15 inches, above 5 feet tall. To find his recommended weight, we substitute 15 for H in the formula and solve for W:

$$\frac{W}{2} - 3H = 53 \quad \text{This is the given formula.}$$

$$\frac{W}{2} - 3 \cdot 15 = 53 \quad \text{Substitute 15 for } H.$$

$$\frac{W}{2} - 45 = 53 \quad \text{Multiply: } 3 \cdot 15 = 45.$$

Multiply both sides of the equation by 2, the least common denominator:

$$2\left(\frac{W}{2} - 45\right) = 2 \cdot 53$$

$$2 \cdot \frac{W}{2} - 2 \cdot 45 = 2 \cdot 53 \quad \text{Use the distributive property.}$$

$$W - 90 = 106 \quad \text{Multiply: } 2 \cdot \frac{W}{2} = \frac{2}{1} \cdot \frac{W}{2} = \frac{2W}{2} = 1W = W.$$

$$W - 90 + 90 = 106 + 90 \quad \text{Add 90 to both sides.}$$

$$W = 196 \quad \text{Simplify.}$$

The recommended weight for a man whose height is 6 feet, 3 inches is 196 pounds. ■

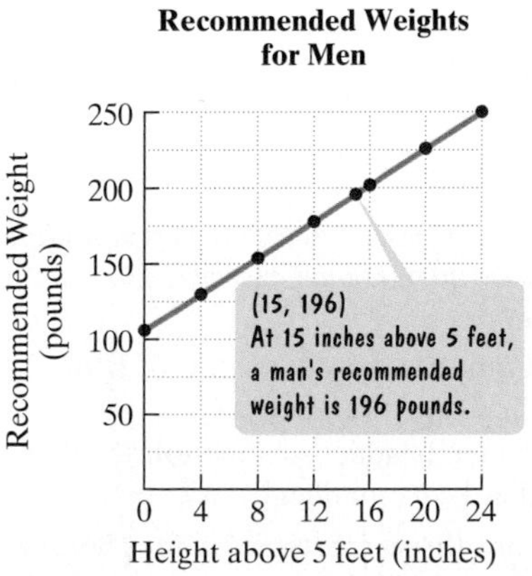

FIGURE 2.3

The points in the rectangular coordinate system in Figure 2.3 allow us to "see" the formula $\frac{W}{2} - 3H = 53$. The first coordinate of each point represents a man's height, in inches, above 5 feet. The second coordinate represents that man's recommended weight, in pounds. The points are rising steadily from left to right. This shows that the recommended weight increases as height increases.

CHECK POINT 7 Use the formula $\frac{W}{2} - 3H = 53$ to find the recommended weight for a man who is 5 feet, 3 inches tall.

2.3 EXERCISE SET

Student Solutions Manual | CD/Video | PH Math/Tutor Center | MathXL Tutorials on CD | MathXL® | MyMathLab | Interactmath.com

Practice Exercises

In Exercises 1–30, solve each equation. Be sure to check your proposed solution by substituting it for the variable in the original equation.

1. $5x + 3x - 4x = 10 + 2$
2. $4x + 8x - 2x = 20 - 15$
3. $4x - 9x + 22 = 3x + 30$
4. $3x + 2x + 64 = 40 - 7x$
5. $3x + 6 - x = 8 + 3x - 6$
6. $3x + 2 - x = 6 + 3x - 8$
7. $4(x + 1) = 20$
8. $3(x - 2) = -6$
9. $7(2x - 1) = 42$
10. $4(2x - 3) = 32$
11. $38 = 30 - 2(x - 1)$
12. $20 = 44 - 8(2 - x)$
13. $2(4z + 3) - 8 = 46$
14. $3(3z + 5) - 7 = 89$
15. $6x - (3x + 10) = 14$
16. $5x - (2x + 14) = 10$
17. $5(2x + 1) = 12x - 3$
18. $3(x + 2) = x + 30$
19. $3(5 - x) = 4(2x + 1)$
20. $3(3x - 1) = 4(3 + 3x)$
21. $8(y + 2) = 2(3y + 4)$
22. $8(y + 3) = 3(2y + 12)$
23. $3(x + 1) = 7(x - 2) - 3$
24. $5x - 4(x + 9) = 2x - 3$
25. $5(2x - 8) - 2 = 5(x - 3) + 3$
26. $7(3x - 2) + 5 = 6(2x - 1) + 24$
27. $6 = -4(1 - x) + 3(x + 1)$
28. $100 = -(x - 1) + 4(x - 6)$
29. $10(z + 4) - 4(z - 2) = 3(z - 1) + 2(z - 3)$
30. $-2(z - 4) - (3z - 2) = -2 - (6z - 2)$

Solve and check each equation in Exercises 31–46. Begin your work by rewriting each equation without fractions.

31. $\frac{x}{5} - 4 = -6$
32. $\frac{x}{2} + 13 = -22$
33. $\frac{2x}{3} - 5 = 7$
34. $\frac{3x}{4} - 9 = -6$
35. $\frac{2y}{3} - \frac{3}{4} = \frac{5}{12}$
36. $\frac{3y}{4} - \frac{2}{3} = \frac{7}{12}$
37. $\frac{x}{3} + \frac{x}{2} = \frac{5}{6}$
38. $\frac{x}{4} - \frac{x}{5} = 1$
39. $20 - \frac{z}{3} = \frac{z}{2}$
40. $\frac{z}{5} - \frac{1}{2} = \frac{z}{6}$
41. $\frac{y}{3} + \frac{2}{5} = \frac{y}{5} - \frac{2}{5}$
42. $\frac{y}{12} + \frac{1}{6} = \frac{y}{2} - \frac{1}{4}$
43. $\frac{3x}{4} - 3 = \frac{x}{2} + 2$
44. $\frac{3x}{5} - \frac{2}{5} = \frac{x}{3} + \frac{2}{5}$
45. $\frac{x-3}{5} - 1 = \frac{x-5}{4}$
46. $\frac{x-2}{3} - 4 = \frac{x+1}{4}$

In Exercises 47–64, solve each equation. Identify equations that have no solution, or equations that are true for all real numbers.

47. $3x - 7 = 3(x + 1)$
48. $2(x - 5) = 2x + 10$
49. $2(x + 4) = 4x + 5 - 2x + 3$
50. $3(x - 1) = 8x + 6 - 5x - 9$
51. $7 + 2(3x - 5) = 8 - 3(2x + 1)$
52. $2 + 3(2x - 7) = 9 - 4(3x + 1)$
53. $4x + 1 - 5x = 5 - (x + 4)$
54. $5x - 5 = 3x - 7 + 2(x + 1)$
55. $4(x + 2) + 1 = 7x - 3(x - 2)$
56. $5x - 3(x + 1) = 2(x + 3) - 5$
57. $3 - x = 2x + 3$
58. $5 - x = 4x + 5$
59. $\frac{x}{3} + 2 = \frac{x}{3}$
60. $\frac{x}{4} + 3 = \frac{x}{4}$
61. $\frac{x}{2} - \frac{x}{4} + 4 = x + 4$
62. $\frac{x}{2} + \frac{2x}{3} + 3 = x + 3$
63. $\frac{2}{3}x = 2 - \frac{5}{6}x$
64. $\frac{2}{3}x = \frac{1}{4}x - 8$

Practice Plus

The equations in Exercises 65–66 contain small figures (□, △, and \$) that represent real numbers. Use properties of equality to isolate x on one side of the equation and the small figures on the other side.

65. $\frac{x}{\square} + \triangle = \$$
66. $\frac{x}{\square} - \triangle = -\$$
67. If $\frac{x}{5} - 2 = \frac{x}{3}$, evaluate $x^2 - x$.
68. If $\frac{3x}{2} + \frac{3x}{4} = \frac{x}{4} - 4$, evaluate $x^2 - x$.

In Exercises 69–72, use the given information to write an equation. Let x represent the number described in each exercise. Then solve the equation and find the number.

69. When one-third of a number is added to one-fifth of the number, the sum is 16. What is the number?
70. When two-fifths of a number is added to one-fourth of the number, the sum is 13. What is the number?
71. When 3 is subtracted from three-fourths of a number, the result is equal to one-half of the number. What is the number?

72. When 30 is subtracted from seven-eighths of a number, the result is equal to one-half of the number. What is the number?

Application Exercises

In Massachusetts, speeding fines are determined by the formula

$$F = 10(x - 65) + 50,$$

where F is the cost, in dollars, of the fine if a person is caught driving x miles per hour. Use this formula to solve Exercises 73–74.

73. If a fine comes to \$250, how fast was that person speeding?

74. If a fine comes to \$400, how fast was that person speeding?

The graph indicates a relationship between sense of humor and depression. Persons with a low sense of humor have higher levels of depression in response to negative life events than those with a high sense of humor.

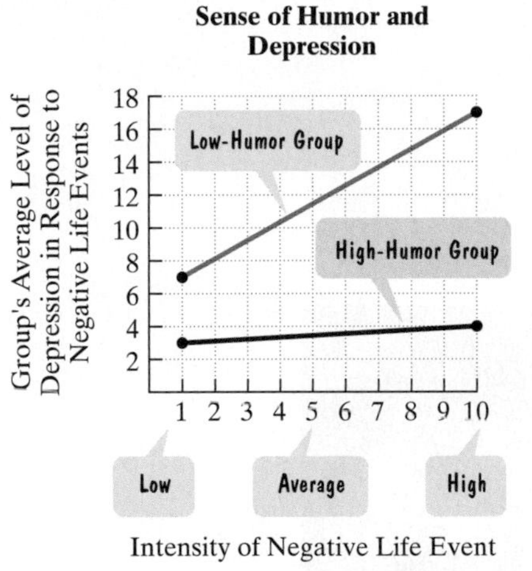

Source: Steven Davis and Joseph Palladino, *Psychology Third Edition*, Prentice Hall, 2003.

The data can be modeled by the formulas

$$D = \frac{10}{9}N + \frac{53}{9} \quad \text{low humor}$$

$$D = \frac{1}{9}N + \frac{26}{9}, \quad \text{high humor}$$

where N is the intensity of a negative life event (from 1, low, to 10, high) and D is the level of depression in response to that event. Use these formulas to solve Exercises 75–76.

75. If the low-humor group averages a level of depression of 10 in response to a negative life event, what is the intensity of that event? How is the solution shown on the line graph?

76. If the high-humor group averages a level of depression of 3.5, or $\frac{7}{2}$, in response to a negative life event, what is the intensity of that event? How is the solution shown on the line graph?

The formula

$$p = 15 + \frac{5d}{11}$$

describes the pressure of sea water, p, in pounds per square foot, at a depth of d feet below the surface. Use the formula to solve Exercises 77–78.

77. The record depth for breath-held diving, by Francisco Ferreras (Cuba) off Grand Bahama Island, on November 14, 1993, involved pressure of 201 pounds per square foot. To what depth did Ferreras descend on this ill-advised venture? (He was underwater for 2 minutes and 9 seconds!)

78. At what depth is the pressure 20 pounds per square foot?

Writing in Mathematics

79. In your own words, describe how to solve a linear equation.

80. Explain how to solve a linear equation containing fractions.

81. Suppose that you solve $\frac{x}{5} - \frac{x}{2} = 1$ by multiplying both sides by 20, rather than the least common denominator of 5 and 2 (namely, 10). Describe what happens. If you get the correct solution, why do you think we clear the equation of fractions by multiplying by the *least* common denominator?

82. Suppose you are an algebra teacher grading the following solution on an examination:

Solve: $-3(x - 6) = 2 - x$

Solution:
$$-3x - 18 = 2 - x$$
$$-2x - 18 = 2$$
$$-2x = -16$$
$$x = 8.$$

You should note that 8 checks, and the solution is 8. The student who worked the problem therefore wants full credit. Can you find any errors in the solution? If full credit is 10 points, how many points should you give the student? Justify your position.

Critical Thinking Exercises

83. Which one of the following statements is true?

a. The equation $3(x + 4) = 3(4 + x)$ has precisely one solution.

b. The equation $2y + 5 = 0$ is equivalent to $2y = 5$.

c. If $2 - 3y = 11$ and the solution to the equation is substituted into $y^2 + 2y - 3$, a number results that is neither positive nor negative.

d. The equation $x + \frac{1}{3} = \frac{1}{2}$ is equivalent to $x + 2 = 3$.

84. A woman's height, h, is related to the length of her femur, f (the bone from the knee to the hip socket), by the formula $f = 0.432h - 10.44$. Both h and f are measured in inches. A partial skeleton is found of a woman in which the femur is 16 inches long. Police find the skeleton in an area where a woman slightly over 5 feet tall has been missing for over a year. Can the partial skeleton be that of the missing woman? Explain.

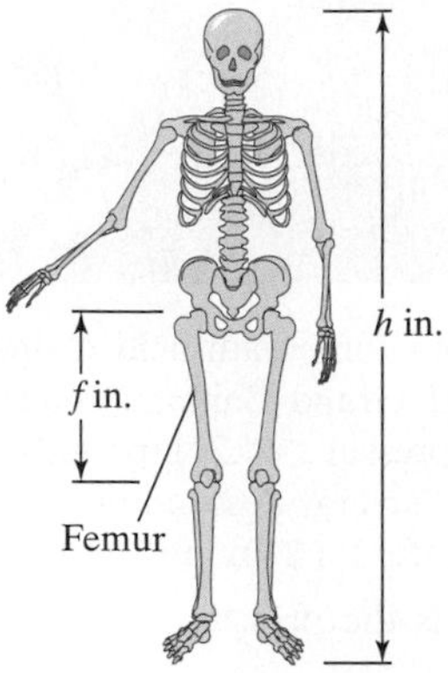

Solve each equation in Exercises 85–86.

85. $\frac{2x - 3}{9} + \frac{x - 3}{2} = \frac{x + 5}{6} - 1$

86. $2(3x + 4) = 3x + 2[3(x - 1) + 2]$

Technology Exercises

Solve each equation in Exercises 87–88. Use a calculator to help with the arithmetic. Check your solution using the calculator.

87. $2.24y - 9.28 = 5.74y + 5.42$

88. $4.8y + 32.5 = 124.8 - 9.4y$

Review Exercises

In Exercises 89–91, insert either < or > in the box between each pair of numbers to make a true statement.

89. $-24 \square -20$ (Section 1.2, Example 6)

90. $-\frac{1}{3} \square -\frac{1}{5}$ (Section 1.2, Example 6)

91. Simplify: $-9 - 11 + 7 - (-3)$. (Section 1.6, Example 3)

SECTION 2.4

FORMULAS AND PERCENTS

Objectives

1. Solve a formula for a variable.
2. Express a decimal as a percent.
3. Express a percent as a decimal.
4. Use the percent formula.
5. Solve applied problems involving percent change.

"And, if elected, it is my solemn pledge to cut your taxes by 10% for each of my first three years in office, for a total cut of 30%."

Did you know that one of the most common ways that you are given numerical information is with percents? In this section, you will learn to use a formula that will help you to understand percents, enabling you to make sense of the politician's promise.

1 Solve a formula for a variable.

Solving a Formula for One of Its Variables We know that solving an equation is the process of finding the number (or numbers) that make the equation a true statement. All of the equations we have solved contained only one letter, such as x or y.

By contrast, the formulas we have seen contain two or more letters, representing two or more variables. Here is an example:

$$C \quad + \quad M \quad = \quad S.$$

A retailer's cost of an item | plus | the retailer's markup (profit) | equals | the items's selling price.

We say that this formula is solved for the variable S because S is alone on one side of the equation and the other side does not contain an S.

Solving a formula for a variable means rewriting the formula so that the variable is isolated on one side of the equation. It does not mean obtaining a numerical value for that variable.

The addition and multiplication properties of equality are used to solve a formula for one of its variables. Consider the retailer's formula, $C + M = S$. How do we solve this formula for C? Use the addition property to isolate C by subtracting M from both sides:

We need to isolate C.

$$C + M = S \qquad \text{This is the given formula.}$$
$$C + M - M = S - M \qquad \text{Subtract } M \text{ from both sides.}$$
$$C = S - M. \qquad \text{Simplify.}$$

Solved for C, the formula $C = S - M$ tells us that the cost of an item for a retailer is the item's selling price minus its markup.

To solve a formula for one of its variables, treat that variable as if it were the only variable in the equation. Think of the other variables as if they were numbers. Use the addition property of equality to isolate all terms with the specified variable on one side of the equation and all terms without the specified variable on the other side. Then use the multiplication property of equality to get the specified variable alone.

Our first example involves the formula for the area of a rectangle. The **area of a two-dimensional figure** is the number of square units it takes to fill the interior of the figure. A **square unit** is a square, each of whose sides is one unit in length, as illustrated in Figure 2.4. The figure shows that there are 12 square units contained within the rectangle. The area of the rectangle is 12 square units. Notice that the area can be determined in the following manner:

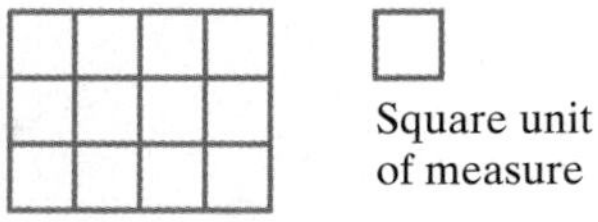

FIGURE 2.4

Across | Down

$$4 \text{ units} \cdot 3 \text{ units} = 4 \cdot 3 \text{ units} \cdot \text{units} = 12 \text{ square units}.$$

The area of a rectangle is the product of the distance across, its length, and the distance down, its width.

AREA OF A RECTANGLE The area, A, of a rectangle with length l and width w is given by the formula

$$A = lw.$$

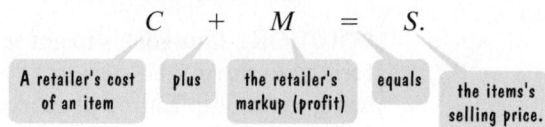

EXAMPLE 1 Solving a Formula for a Variable

Solve the formula $A = lw$ for w.

SOLUTION Our goal is to get w by itself on one side of the formula. There is only one term with w, lw, and it is already isolated on the right side. We isolate w on the right by using the multiplication property of equality and dividing both sides by l.

We need to isolate w.

$$A = lw \quad \text{This is the given formula.}$$

$$\frac{A}{l} = \frac{lw}{l} \quad \text{Isolate } w \text{ by dividing both sides by } l.$$

$$\frac{A}{l} = w \quad \text{Simplify: } \frac{lw}{l} = 1w = w.$$

The formula solved for w is $\frac{A}{l} = w$ or $w = \frac{A}{l}$. Thus, the area of a rectangle divided by its length is equal to its width. ■

CHECK POINT 1 Solve the formula $A = lw$ for l.

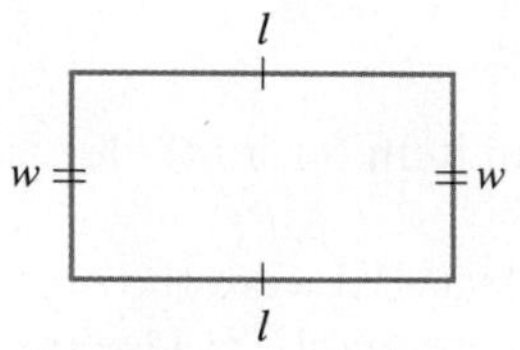

FIGURE 2.5 A rectangle with length l and width w

The perimeter, P, of a two-dimensional figure is the sum of the lengths of its sides. Perimeter is measured in linear units, such as inches, feet, yards, meters, or kilometers.

Example 2 involves the perimeter of a rectangle. Because perimeter is the sum of the lengths of the sides, the perimeter of the rectangle shown in Figure 2.5 is $l + w + l + w$. This can be expressed as

$$P = 2l + 2w.$$

PERIMETER OF A RECTANGLE The perimeter, P, of a rectangle with length l and width w is given by the formula

$$P = 2l + 2w.$$

The perimeter of a rectangle is the sum of twice the length and twice the width.

EXAMPLE 2 Solving a Formula for a Variable

Solve the formula $2l + 2w = P$ for w.

SOLUTION First, isolate $2w$ on the left by subtracting $2l$ from both sides. Then solve for w by dividing both sides by 2.

We need to isolate w.

$$2l + 2w = P \quad \text{This is the given formula.}$$

$$2l - 2l + 2w = P - 2l \quad \text{Isolate } 2w \text{ by subtracting } 2l \text{ from both sides.}$$

$$2w = P - 2l \quad \text{Simplify.}$$

$$\frac{2w}{2} = \frac{P - 2l}{2} \quad \text{Isolate } w \text{ by dividing both sides by 2.}$$

$$w = \frac{P - 2l}{2} \quad \text{Simplify.}$$

■

CHECK POINT 2 Solve the formula $2l + 2w = P$ for l.

EXAMPLE 3 Solving a Formula for a Variable

The total price of an article purchased on a monthly deferred payment plan is described by the following formula:

$$T = D + pm.$$

In this formula, T is the total price, D is the down payment, p is the amount of the monthly payment, and m is the number of payments. Solve the formula for p.

SOLUTION First, isolate pm on the right by subtracting D from both sides. Then isolate p from pm by dividing both sides of the formula by m.

We need to isolate p.

$T = D + pm$	This is the given formula. We want p alone.
$T - D = D - D + pm$	Isolate pm by subtracting D from both sides.
$T - D = pm$	Simplify.
$\frac{T - D}{m} = \frac{pm}{m}$	Now isolate p by dividing both sides by m.
$\frac{T - D}{m} = p$	Simplify: $\frac{pm}{m} = \frac{p\cancel{m}^{1}}{\cancel{m}_{1}} = p \cdot 1 = p.$

■

CHECK POINT 3 Solve the formula $T = D + pm$ for m.

The next example has a formula that contains a fraction. To solve for a variable in a formula involving fractions, we begin by multiplying both sides by the least common denominator of all fractions in the formula. This will eliminate the fractions. Then we solve for the specified variable.

EXAMPLE 4 Solving a Formula Containing a Fraction for a Variable

Solve the formula $\frac{W}{2} - 3H = 53$ for W.

SOLUTION Do you remember seeing this formula in the last section? It models the recommended weight, W, for a male, where H represents the man's height, in inches, over 5 feet. We begin by multiplying both sides of the formula by 2 to eliminate the fraction. Then we isolate the variable W.

$\frac{W}{2} - 3H = 53$	This is the given formula.
$2\left(\frac{W}{2} - 3H\right) = 2 \cdot 53$	Multiply both sides by 2.
$2 \cdot \frac{W}{2} - 2 \cdot 3H = 2 \cdot 53$	Use the distributive property.

We need to isolate W.

$$W - 6H = 106 \qquad \text{Simplify.}$$

$$W - 6H + 6H = 106 + 6H \qquad \text{Isolate } W \text{ by adding } 6H \text{ to both sides.}$$

$$W = 106 + 6H \qquad \text{Simplify.}$$

This form of the formula makes it easy to find a man's recommended weight, W, if we know his height, H, in inches, over 5 feet. ■

CHECK POINT 4 Solve for x: $\frac{x}{3} - 4y = 5$.

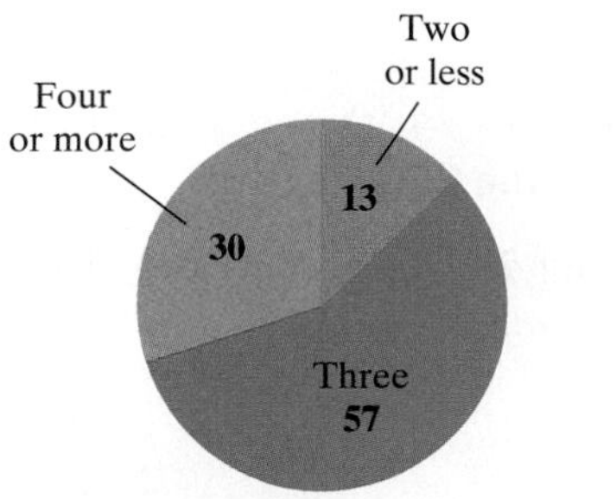

FIGURE 2.6 Number of bedrooms in privately owned single-family U.S. houses per 100 houses.

Source: U.S. Census Bureau and HUD

Before turning to a formula involving percent, let's review some of the basics of percent.

Basics of Percent **Percents** are the result of expressing numbers as a part of 100. The word *percent* means *per hundred*. For example, the circle graph in Figure 2.6 shows that 57 out of every 100 single-family homes have three bedrooms. Thus, $\frac{57}{100} = 57\%$, indicating that 57% of the houses have three bedrooms. The percent sign, %, is used to indicate the number of parts out of one hundred parts.

By definition, $57\% = \frac{57}{100}$. We can express the fraction $\frac{57}{100}$ in decimal notation as 0.57. Thus, $0.57 = 57\%$. Here is a general procedure for expressing a decimal number as a percent:

2 Express a decimal as a percent.

EXPRESSING A DECIMAL NUMBER AS A PERCENT

1. Move the decimal point two places to the right.
2. Attach a percent sign.

EXAMPLE 5 Expressing a Decimal as a Percent

Express 0.47 as a percent.

SOLUTION

Move decimal point two places right.

0.47 % ← Attach a percent sign.

Thus, $0.47 = 47\%$. ■

CHECK POINT 5 Express 0.023 as a percent.

We reverse the procedure of Example 5 to express a percent as a decimal number.

3 Express a percent as a decimal.

EXPRESSING A PERCENT AS A DECIMAL NUMBER

1. Move the decimal point two places to the left.
2. Remove the percent sign.

STUDY TIP

Dictionaries indicate that the word *percentage* has the same meaning as the word *percent*. Use the word that sounds best in the circumstance.

EXAMPLE 6 Expressing Percents as Decimals

Express each percent as a decimal: **a.** 19% **b.** 180%.

SOLUTION Use the two steps in the box.

a.

$$19\% = 19.\% = 0.19\%$$

The percent sign is removed.

The decimal point starts at the far right.

The decimal point is moved two places to the left.

Thus, 19% = 0.19.

b. 180% = 1.80% = 1.80 or 1.8. ■

CHECK POINT 6 Express each percent as a decimal:

a. 67% **b.** 250%.

4 Use the percent formula.

A Formula Involving Percent Percents are useful in comparing two numbers. To compare the number A to the number B using a percent P, the following formula is used:

A is P percent of B.

$$A = P \cdot B.$$

In the formula

$$A = PB$$

B = the base number, P = the percent (in decimal form), and A = the number compared to B.

There are three basic types of percent problems that can be solved using the percent formula

$$A = PB. \quad \text{A is P percent of B.}$$

Question	**Given**	**Percent Formula**
What is P percent of B?	P and B	Solve for A.
A is P percent of what?	A and P	Solve for B.
A is what percent of B?	A and B	Solve for P.

Let's look at an example of each type of problem.

EXAMPLE 7 Using the Percent Formula: What Is P Percent of B?

What is 8% of 20?

SOLUTION We use the formula $A = PB$: A is P percent of B. We are interested in finding the quantity A in this formula.

What is 8% of 20?

$$A = 0.08 \cdot 20 \quad \text{Express 8\% as 0.08.}$$

$$A = 1.6$$

Multiply: .08 × 20 = 1.60

Thus, 1.6 is 8% of 20. The answer is 1.6. ■

CHECK POINT 7 What is 9% of 50?

EXAMPLE 8 Using the Percent Formula: A Is P Percent of What?

4 is 25% of what?

SOLUTION We use the formula $A = PB$: A is P percent of B. We are interested in finding the quantity B in this formula.

4	is	25%	of	what?
4	=	0.25	·	B

$$4 = 0.25 \cdot B \quad \text{Express 25\% as 0.25.}$$

$$\frac{4}{0.25} = \frac{0.25B}{0.25} \quad \text{Divide both sides by 0.25.}$$

$$16 = B \quad \text{Simplify: } 0.25\overline{)4.00} = 16.$$

Thus, 4 is 25% of 16. The answer is 16. ■

CHECK POINT 8 9 is 60% of what?

EXAMPLE 9 Using the Percent Formula: A Is What Percent of B?

1.3 is what percent of 26?

SOLUTION We use the formula $A = PB$: A is P percent of B. We are interested in finding the quantity P in this formula.

1.3	is	what percent	of	26?
1.3	=	P	·	26

$$1.3 = P \cdot 26$$

$$\frac{1.3}{26} = \frac{P \cdot 26}{26} \quad \text{Divide both sides by 26.}$$

$$0.05 = P \quad \text{Simplify: } 26\overline{)1.30} = 0.05.$$

We change 0.05 to a percent by moving the decimal point two places to the right and adding a percent sign: $0.05 = 5\%$. Thus, 1.3 is 5% of 26. The answer is 5%. ■

CHECK POINT 9 18 is what percent of 50?

5 Solve applied problems involving percent change.

Applications Percents are used for comparing changes, such as increases or decreases in sales, population, prices, and production. If a quantity changes, its percent increase or percent decrease can be determined by asking the following question:

The change is what percent of the original amount?

The question is answered using the percent formula as follows:

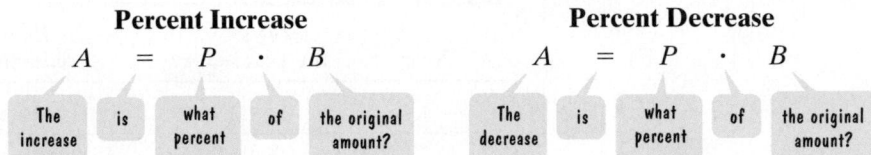

Percent Increase

$$A = P \cdot B$$

The increase | is | what percent | of | the original amount?

Percent Decrease

$$A = P \cdot B$$

The decrease | is | what percent | of | the original amount?

EXAMPLE 10 Finding Percent Decrease

A jacket regularly sells for \$135.00 The sale price is \$60.75. Find the percent decrease in the jacket's price.

SOLUTION The percent decrease in price can be determined by asking the following question:

The price decrease is what percent of the original price (\$135.00)?

The price decrease is the difference between the original price and the sale price (\$60.75):

$$\$135.00 - \$60.75 = \$74.25.$$

Now we use the percent formula to find the percent decrease.

The price decrease | is | what percent | of | the original price?

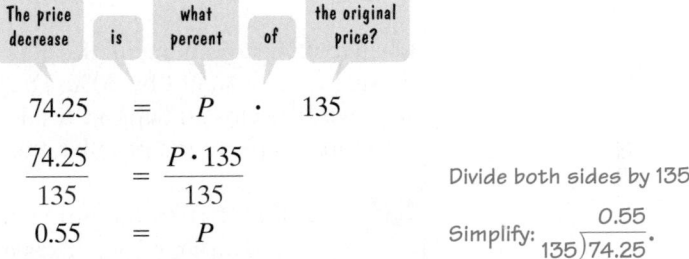

$$74.25 = P \cdot 135$$

$$\frac{74.25}{135} = \frac{P \cdot 135}{135}$$ Divide both sides by 135.

$$0.55 = P$$ Simplify: $135\overline{)74.25}$ = 0.55.

We change 0.55 to a percent by moving the decimal point two places to the right and adding a percent sign: 0.55 = 55%. Thus, the percent decrease in the jacket's price is 55%. ■

✔ **CHECK POINT 10** A television regularly sells for \$940. The sale price is \$611. Find the percent decrease in the television's price.

In our next example, we look at one of the many ways that percent can be used incorrectly.

EXAMPLE 11 Promises of a Politician

A politician states, "If you elect me to office, I promise to cut your taxes for each of my first three years in office by 10% each year, for a total reduction of 30%." Evaluate the accuracy of the politician's statement.

SOLUTION To make things simple, let's assume that a taxpayer paid \$100 in taxes in the year previous to the politician's election. A 10% reduction during year 1 is 10% of \$100.

$$10\% \text{ of previous year's tax} = 10\% \text{ of } \$100 = 0.10 \cdot \$100 = \$10$$

With a 10% reduction the first year, the taxpayer will pay only \$100 − \$10, or \$90, in taxes during the politician's first year in office.

The table on the next page shows how we calculate the new, reduced tax for each of the first three years in office.

Year	Tax paid the year before	10% reduction	Taxes paid this year
1	\$100	$0.10 \cdot \$100 = \$10$	$\$100 - \$10 = \$90$
2	\$ 90	$0.10 \cdot \$90 = \$9$	$\$90 - \$9 = \$81$
3	\$ 81	$0.10 \cdot \$81 = \$8.10$	$\$81 - \$8.10 = \$72.90$

The tax reduction is the amount originally paid, \$100.00, minus the amount paid during the politician's third year in office, \$72.90:

$$\$100.00 - \$72.90 = \$27.10.$$

Now we use the percent formula to determine the percent decrease in taxes over the three years.

The tax decrease | is | what percent | of | the original tax?

$$27.1 = P \cdot 100$$

$$\frac{27.1}{100} = \frac{P \cdot 100}{100} \qquad \text{Divide both sides by 100.}$$

$$0.271 = P \qquad \text{Simplify.}$$

Change 0.271 to a percent: $0.271 = 27.1\%$. The percent decrease is 27.1%. The taxes decline by 27.1%, not by 30%. The politician is ill-informed in saying that three consecutive 10% cuts add up to a total tax cut of 30%. In our calculation, which serves as a counterexample to the promise, the total tax cut is only 27.1%. ■

CHECK POINT 11 Suppose you paid \$1200 in taxes. During year 1, taxes decrease by 20%. During year 2, taxes increase by 20%.

a. What do you pay in taxes for year 2?

b. How do your taxes for year 2 compare with what you originally paid, namely \$1200? If the taxes are not the same, find the percent increase or decrease.

2.4 EXERCISE SET

Student Solutions Manual | CD/Video | PH Math/Tutor Center | MathXL Tutorials on CD | MathXL® | MyMathLab | Interactmath.com

Practice Exercises

In Exercises 1–26, solve each formula for the specified variable. Do you recognize the formula? If so, what does it describe?

1. $d = rt$ for r

2. $d = rt$ for t

3. $I = Prt$ for P

4. $I = Prt$ for r

5. $C = 2\pi r$ for r

6. $C = \pi d$ for d

7. $E = mc^2$ for m

8. $V = \pi r^2 h$ for h

9. $y = mx + b$ for m

10. $y = mx + b$ for x

11. $T = D + pm$ for p

12. $P = C + MC$ for M

13. $A = \frac{1}{2}bh$ for b

14. $A = \frac{1}{2}bh$ for h

15. $M = \frac{n}{5}$ for n

16. $M = \frac{A}{740}$ for A

17. $\frac{c}{2} + 80 = 2F$ for c

18. $p = 15 + \frac{5d}{11}$ for d

19. $A = \frac{1}{2}(a + b)$ for a

20. $A = \frac{1}{2}(a + b)$ for b

21. $S = P + Prt$ for r

22. $S = P + Prt$ for t

23. $A = \frac{1}{2}h(a + b)$ for b

24. $A = \frac{1}{2}h(a + b)$ for a

25. $Ax + By = C$ for x

26. $Ax + By = C$ for y

In Exercises 27–34, express each decimal as a percent.

27. 0.89
28. 0.16
29. 0.002
30. 0.008
31. 4.78
32. 5.38
33. 100
34. 85

In Exercises 35–44, express each percent as a decimal.

35. 27%
36. 83%
37. 63.4%
38. 2.15%
39. 170%
40. 360%
41. 3%
42. 8%
43. $\frac{1}{2}\%$
44. $\frac{1}{4}\%$

Use the percent formula, $A = PB$: A is P percent of B, to solve Exercises 45–56.

45. What is 3% of 200?
46. What is 8% of 300?
47. What is 18% of 40?
48. What is 16% of 90?
49. 3 is 60% of what?
50. 8 is 40% of what?
51. 24% of what number is 40.8?
52. 32% of what number is 51.2?
53. 3 is what percent of 15?
54. 18 is what percent of 90?
55. What percent of 2.5 is 0.3?
56. What percent of 7.5 is 0.6?
57. If 5 is increased to 8, the increase is what percent of the original number?
58. If 5 is increased to 9, the increase is what percent of the original number?
59. If 4 is decreased to 1, the decrease is what percent of the original number?
60. If 8 is decreased to 6, the decrease is what percent of the original number?

Practice Plus

In Exercises 61–68, solve each equation for x.

61. $y = (a + b)x$

62. $y = (a - b)x$

63. $y = (a - b)x + 5$

64. $y = (a + b)x - 8$

65. $y = cx + dx$

66. $y = cx - dx$

67. $y = Ax - Bx - C$

68. $y = Ax + Bx + C$

Application Exercises

69. The average, or mean, A, of three exam grades, x, y, and z, is given by the formula

$$A = \frac{x + y + z}{3}.$$

   a. Solve the formula for z.

   b. Use the formula in part (a) to solve this problem. On your first two exams, your grades are 86% and 88%: $x = 86$ and $y = 88$. What must you get on the third exam to have an average of 90%?

70. The average, or mean, A, of four exam grades, x, y, z, and w, is given by the formula

$$A = \frac{x + y + z + w}{4}.$$

   a. Solve the formula for w.

   b. Use the formula in part (a) to solve this problem. On your first three exams, your grades are 76%, 78%, and 79%: $x = 76$, $y = 78$, and $z = 79$. What must you get on the fourth exam to have an average of 80%?

71. If you are traveling in your car at an average rate of r miles per hour for t hours, then the distance, d, in miles, that you travel is described by the formula $d = rt$: distance equals rate times time.

   a. Solve the formula for t.

   b. Use the formula in part (a) to find the time that you travel if you cover a distance of 100 miles at an average rate of 40 miles per hour.

72. The formula $F = \frac{9}{5}C + 32$ expresses the relationship between Celsius temperature, C, and Fahrenheit temperature, F.

   a. Solve the formula for C.

   b. Use the formula from part (a) to find the equivalent Celsius temperature for a Fahrenheit temperature of 59°.

A recent Time/CNN telephone poll included never-married single women between the ages of 18 and 49 and never-married single men between the ages of 18 and 49. The circle graphs show the results for one of the questions in the poll. Use this information to solve Exercises 73–74.

If You Couldn't Find the Perfect Mate, Would You Marry Someone Else?

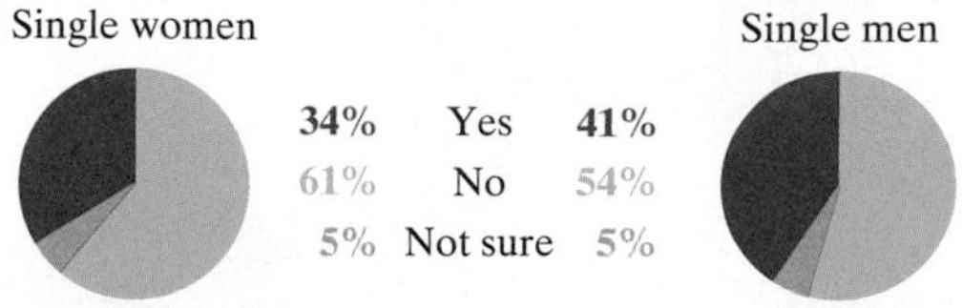

Source: Time, August 28, 2000

73. There were 1200 single women who participated in the poll. How many stated they would marry someone other than the perfect mate?

74. There were 1200 single men who participated in the poll. How many stated they would marry someone other than the perfect mate?

75. In 2002, the leading cause of death in the United States was heart disease, resulting in 710,760 deaths. If 30% of all deaths were caused by heart disease, find the total number of deaths in the United States in 2002.
(*Source*: Department of Health and Human Services)

76. In 2001, 4% of the Hispanic population in the United States was Cuban. If the Cuban-American population at that time was approximately 1.4 million, what was the approximate U.S. Hispanic population in 2001?

In 2002, there were approximately 12,000 hate crimes reported to the FBI in the United States. The circle graph shows the breakdown of this total number. Use this information to solve Exercises 77–78.

Motivation for 12,000 U.S. Hate-Crime Incidents

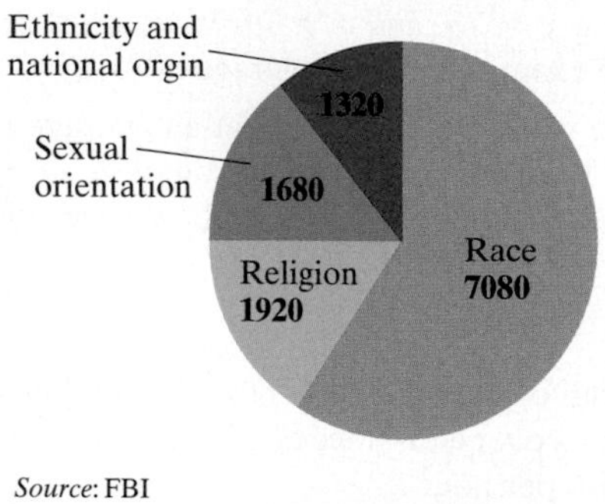

Source: FBI

77. What percent of hate-crime incidents were motivated by race?

78. What percent of hate-crime incidents were motivated by sexual orientation?

79. A charity has raised \$7500, with a goal of raising \$60,000. What percent of the goal has been raised?

80. A charity has raised \$225,000, with a goal of raising \$500,000. What percent of the goal has been raised?

81. A restaurant bill came to \$60. If 15% of this bill is left as a tip, how much was the tip?

82. If income tax is \$3502 plus 28% of taxable income over \$23,000, how much is the income tax on a taxable income of \$35,000?

83. Suppose that the local sales tax rate is 6% and you buy a car for \$16,800.

a. How much tax is due?

b. What is the car's total cost?

84. Suppose that the local sales tax rate is 7% and you buy a graphing calculator for \$96.

a. How much tax is due?

b. What is the calculator's total cost?

85. An exercise machine with an original price of \$860 is on sale at 12% off.

a. What is the discount amount?

b. What is the exercise machine's sale price?

86. A dictionary that normally sells for \$16.50 is on sale at 40% off.

a. What is the discount amount?

b. What is the dictionary's sale price?

87. A sofa regularly sells for \$840. The sale price is \$714. Find the percent decrease in the sofa's price.

88. A fax machine regularly sells for \$380. The sale price is \$266. Find the percent decrease in the machine's price.

89. Suppose that you put \$10,000 in a rather risky investment recommended by your financial advisor. During the first year, your investment decreases by 30% of its original value. During the second year, your investment increases by 40% of its first-year value. Your advisor tells you that there must have been a 10% overall increase of your original \$10,000 investment. Is your financial advisor using percentages properly? If not, what is your actual percent gain or loss of your original \$10,000 investment?

90. The price of a color printer is reduced by 30% of its original price. When it still does not sell, its price is reduced by 20% of the reduced price. The salesperson informs you that there has been a total reduction of 50%. Is the salesperson using percentages properly? If not, what is the actual percent reduction from the original price?

Writing in Mathematics

91. Explain what it means to solve a formula for a variable.

92. What is a percent?

93. Describe how to express a decimal number as a percent and give an example.

94. Describe how to express a percent as a decimal number and give an example.

95. What does the percent formula, $A = PB$, describe? Give an example of how the formula is used.

96. Describe one way in which you use percents in your daily life.

Critical Thinking Exercises

97. Which one of the following statements is true?

a. If $ax + b = 0$, then $x = \frac{b}{a}$.

b. If $A = lw$, then $w = \frac{l}{A}$.

c. If $A = \frac{1}{2}bh$, then $b = \frac{A}{2h}$.

d. Solving $x - y = -7$ for y gives $y = x + 7$.

98. In psychology, an intelligence quotient, Q, also called IQ, is measured by the formula

$$Q = \frac{100M}{C},$$

where M = mental age and C = chronological age. Solve the formula for C.

99. The height, h, in feet, of water in a fountain is described by the formula

$$h = -16t^2 + 64t$$

and the velocity, v, in feet per second, of water in the fountain is described by $v = -32t + 64$. Find the time when the water's velocity is 16 feet per second, and then find the water's height at that time.

Review Exercises

100. Solve and check: $5x + 20 = 8x - 16$. (Section 2.2, Example 7)

101. Solve and check: $5(2y - 3) - 1 = 4(6 + 2y)$. (Section 2.3, Example 3)

102. Simplify: $x - 0.3x$. (Section 1.4, Example 9)

MID-CHAPTER CHECK POINT

CHAPTER 2

What You Know: We learned a step-by-step procedure for solving linear equations, including equations with fractions. We saw that some linear equations have no solution, whereas others have all real numbers as solutions. We used the addition and multiplication properties of equality to solve formulas for variables. Finally, we worked with the percent formula $A = PB$: A is P percent of B.

1. Solve: $\frac{x}{2} = 12 - \frac{x}{4}$.

2. Solve: $5x - 42 = -57$.

3. Solve for C: $H = \frac{EC}{825}$

4. What is 6% of 140?

5. Solve: $\frac{-x}{10} = -3$.

6. Solve: $1 - 3(y - 5) = 4(2 - 3y)$.

7. Solve for r: $S = 2\pi rh$

8. 12 is 30% of what?

9. Solve: $\frac{3y}{5} + \frac{y}{2} = \frac{5y}{4} - 3$.

10. Solve: $5z + 7 = 6(z - 2) - 4(2z - 3)$

11. Solve for x: $Ax - By = C$.

12. Solve: $6y + 7 + 3y = 3(3y - 1)$

13. The formula $D = 0.12x + 5.44$ models the number of children in the United States with physical disabilities, D, in millions, x years after 2000. According to this model, in which year will there be 6.4 million children in the United States with physical disabilities?

14. Solve: $10\left(\frac{1}{2}x + 3\right) = 10\left(\frac{3}{5}x - 1\right)$

15. 50 is what percent of 400?

16. Solve: $\frac{3(m + 2)}{4} = 2m + 3$.

17. If 40 is increased to 50, the increase is what percent of the original number?

18. Solve: $12w - 4 + 8w - 4 = 4(5w - 2)$.

SECTION 2.5

Objectives

1 Translate English phrases into algebraic expressions.

2 Solve algebraic word problems using linear equations.

AN INTRODUCTION TO PROBLEM SOLVING

You started college with your best friend. This semester, you and your friend are taking two classes together. However, your friend often misses class and is not doing the necessary homework between classes to succeed. What can you say to your friend, who values your advice and who is in danger of flunking out of college if things continue on their present course?

Some problems have many plans for finding an answer. To solve your friend's problem, or any problem for that matter, we need to understand the problem fully, devise a plan for solving it, and then carry out the plan. However, problem solving in algebra is easier than solving the many problematic situations encountered in everyday life. Why? Algebra provides a step-by-step strategy for solving problems. As you become familiar with this strategy, you will learn to solve a wide variety of problems.

A Strategy for Solving Word Problems Using Equations Problem solving is an important part of algebra. The problems in this book are presented in English. We must translate from the ordinary language of English into the language of algebraic equations. To translate, however, we must understand the English prose and be familiar with the forms of algebraic language. Here are some general steps we will follow in solving word problems:

STRATEGY FOR SOLVING WORD PROBLEMS

Step 1 Read the problem carefully. Attempt to state the problem in your own words and state what the problem is looking for. Let x (or any variable) represent one of the unknown quantities in the problem.

Step 2 If necessary, write expressions for any other unknown quantities in the problem in terms of x.

Step 3 Write an equation in x that describes the conditions of the problem.

Step 4 Solve the equation and answer the problem's question.

Step 5 Check the solution *in the original wording* of the problem, not in the equation obtained from the words.

Take great care with step 1. Reading a word problem is not the same as reading a newspaper. Reading the problem involves slowly working your way through its parts, making notes on what is given, and perhaps rereading the problem a few times. Only at this point should you let x represent one of the quantities.

1 Translate English phrases into algebraic expressions.

The most difficult step in this process is step 3 because it involves translating verbal conditions into an algebraic equation. Translations of some commonly used English phrases are listed in Table 2.1. We choose to use x to represent the variable, but we can use any letter.

STUDY TIP

Cover the right column in Table 2.1 with a sheet of paper and attempt to formulate the algebraic expression in the column on your own. Then slide the paper down and check your answer. Work through the entire table in this manner.

Table 2.1 Algebraic Translations of English Phrases

English Phrase	Algebraic Expression
Addition	
The sum of a number and 7	$x + 7$
Five more than a number; a number plus 5	$x + 5$
A number increased by 6; 6 added to a number	$x + 6$
Subtraction	
A number minus 4	$x - 4$
A number decreased by 5	$x - 5$
A number subtracted from 8	$8 - x$
The difference between a number and 6	$x - 6$
The difference between 6 and a number	$6 - x$
Seven less than a number	$x - 7$
Seven minus a number	$7 - x$
Nine fewer than a number	$x - 9$
Multiplication	
Five times a number	$5x$
The product of 3 and a number	$3x$
Two-thirds of a number (used with fractions)	$\frac{2}{3}x$
Seventy-five percent of a number (used with decimals)	$0.75x$
Thirteen multiplied by a number	$13x$
A number multiplied by 13	$13x$
Twice a number	$2x$
Division	
A number divided by 3	$\frac{x}{3}$
The quotient of 7 and a number	$\frac{7}{x}$
The quotient of a number and 7	$\frac{x}{7}$
The reciprocal of a number	$\frac{1}{x}$
More than one operation	
The sum of twice a number and 7	$2x + 7$
Twice the sum of a number and 7	$2(x + 7)$
Three times the sum of 1 and twice a number	$3(1 + 2x)$
Nine subtracted from 8 times a number	$8x - 9$
Twenty-five percent of the sum of 3 times a number and 14	$0.25(3x + 14)$
Seven times a number, increased by 24	$7x + 24$
Seven times the sum of a number and 24	$7(x + 24)$

STUDY TIP

Here are three similar English phrases that have very different translations:

7 minus 10: $7 - 10$

7 less than 10: $10 - 7$

7 is less than 10: $7 < 10$.

Think carefully about what is expressed in English before you translate into the language of algebra.

EXAMPLE 1 Translating English Phrases into Algebraic Expressions

Write each English phrase as an algebraic expression. Let x represent the number.

a. Six subtracted from 5 times a number

b. The quotient of 9 and a number, decreased by 4 times the number

SOLUTION

a. Six subtracted from | 5 times a number

$5x - 6$

The algebraic expression for "six subtracted from 5 times a number" is $5x - 6$.

b.

The quotient of 9 and a number,	decreased by	4 times the number
$\frac{9}{x}$	$-$	$4x$

The algebraic expression for "the quotient of 9 and a number, decreased by 4 times the number" is $\frac{9}{x} - 4x$. ■

CHECK POINT 1 Write each English phrase as an algebraic expression. Let x represent the number.

a. Four times a number, increased by 6

b. The quotient of a number decreased by 4 and 9.

2 Solve algebraic word problems using linear equations.

Applying the Strategy for Solving Word Problems Now that we've practiced writing algebraic expressions for English phrases, let's apply our five-step strategy for solving word problems.

EXAMPLE 2 Solving a Word Problem

Nine subtracted from eight times a number is 39. Find the number.

SOLUTION

Step 1. **Let x represent one of the quantities.** Because we are asked to find a number, let

$$x = \text{the number.}$$

Step 2. **Represent other quantities in terms of x.** There are no other unknown quantities to find, so we can skip this step.

Step 3. **Write an equation in x that describes the conditions.**

Nine subtracted from	eight times a number	is	39.
-9	$8x$	$=$	39

$$8x \quad -9 \quad = \quad 39$$

Step 4. **Solve the equation and answer the question.**

$$8x - 9 = 39 \quad \text{This is the equation for the problem's conditions.}$$

$$8x - 9 + 9 = 39 + 9 \quad \text{Add 9 to both sides.}$$

$$8x = 48 \quad \text{Simplify.}$$

$$\frac{8x}{8} = \frac{48}{8} \quad \text{Divide both sides by 8.}$$

$$x = 6 \quad \text{Simplify.}$$

The number is 6.

Step 5. **Check the proposed solution in the original wording of the problem.** "Nine subtracted from eight times a number is 39." The proposed number is 6. Eight times 6 is $8 \cdot 6$, or 48. Nine subtracted from 48 is $48 - 9$, or 39. The proposed solution checks in the problem's wording, verifying that the number is 6. ■

CHECK POINT 2 Four subtracted from six times a number is 68. Find the number.

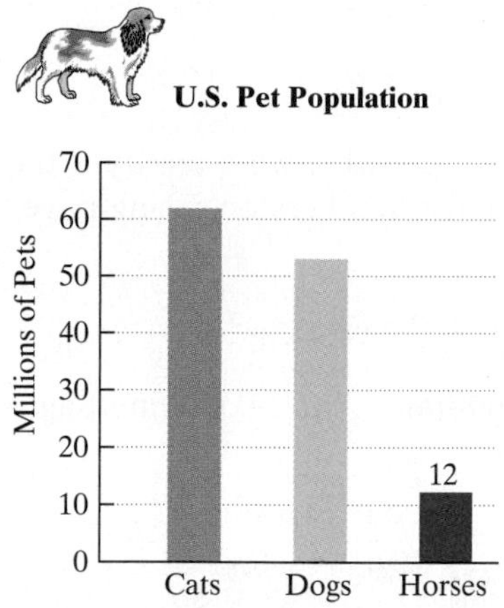

Americans spend more than $30 billion per year on their pets, from MRIs to spas to doggy diapers.

Source: American Veterinary Medical Association

EXAMPLE 3 Pet Population

Americans love their pets. The number of cats in the United States exceeds the number of dogs by 7.5 million. The number of cats and dogs combined is 114.7 million. Determine the number of dogs and cats in the United States.

SOLUTION

Step 1. **Let x represent one of the quantities.** We know something about the number of cats: the cat population exceeds the dog population by 7.5 million. This means that there are 7.5 million more cats than dogs. We will let

x = the number (in millions) of dogs in the United States.

Step 2. **Represent other quantities in terms of x.** The other unknown quantity is the number of cats. Because there are 7.5 million more cats than dogs, let

$x + 7.5$ = the number (in millions) of cats in the United States.

Step 3. **Write an equation in x that describes the conditions.** The number of cats and dogs combined is 114.7 million.

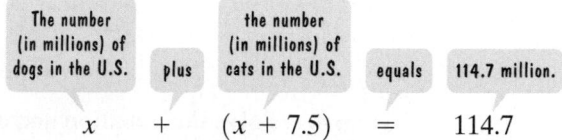

The number (in millions) of dogs in the U.S.	plus	the number (in millions) of cats in the U.S.	equals	114.7 million.
x	$+$	$(x + 7.5)$	$=$	114.7

Step 4. **Solve the equation and answer the question.**

$x + (x + 7.5) = 114.7$ — This is the equation specified by the conditions of the problem.

$2x + 7.5 = 114.7$ — Regroup and combine like terms on the left side.

$2x + 7.5 - 7.5 = 114.7 - 7.5$ — Subtract 7.5 from both sides.

$2x = 107.2$ — Simplify.

$\frac{2x}{2} = \frac{107.2}{2}$ — Divide both sides by 2.

$x = 53.6$ — Simplify.

Because x represents the number (in millions) of dogs, there are 53.6 million dogs in the United States. Because $x + 7.5$ represents the number (in millions) of cats, there are $53.6 + 7.5$, or 61.1 million cats in the United States.

Step 5. **Check the proposed solution in the original wording of the problem.** The problem states that the number of cats and dogs combined is 114.7 million. By adding 53.6 million, the dog population, and 61.1 million, the cat population, we do, indeed, obtain a sum of 114.7 million. ■

✔ **CHECK POINT 3** Americans go to great lengths to treat sick dogs. Specialized veterinary care can be costly. The average cost of a magnetic resonance imaging (MRI) scan exceeds the average cost of acupuncture by $463. Combined, the two procedures cost $687. Determine the average cost for each of these veterinary procedures.
(*Source*: American Veterinary Medical Association)

EXAMPLE 4 Consecutive Integers

Two pages that face each other in a book have 145 as the sum of their page numbers. What are the page numbers?

SOLUTION

Step 1. **Let x represent one of the quantities.** We will let

x = the page number of the page on the left.

Step 2. **Represent other quantities in terms of x.** The other unknown quantity is the page number of the facing page on the right. Page numbers on facing pages are consecutive integers. Thus,

$x + 1$ = the page number of the page on the right.

Step 3. **Write an equation in x that describes the conditions.** The two facing pages have 145 as the sum of their page numbers.

The page number on the left	plus	the page number on the right	equals	145.
x	$+$	$(x + 1)$	$=$	145

Step 4. **Solve the equation and answer the question.**

$x + (x + 1) = 145$	This is the equation for the problem's conditions.
$2x + 1 = 145$	Regroup and combine like terms.
$2x + 1 - 1 = 145 - 1$	Subtract 1 from both sides.
$2x = 144$	Simplify.
$\frac{2x}{2} = \frac{144}{2}$	Divide both sides by 2.
$x = 72$	Simplify.

Thus,

the page number on the left $= x = 72$

and

the page number on the right $= x + 1 = 72 + 1 = 73$.

The page numbers are 72 and 73.

Step 5. **Check the proposed solution in the original wording of the problem.** The problem states that the sum of the page numbers on the facing pages is 145. By adding 72, the page number on the left, and 73, the page number on the right, we do, indeed, obtain a sum of 145. ■

✔ **CHECK POINT 4** Two pages that face each other in a book have 193 as the sum of their page numbers. What are the page numbers?

Example 4 and Check Point 4 involved consecutive integers. By contrast, some word problems involve consecutive odd integers, such as 5, 7, and 9. Other word problems involve consecutive even integers, such as 6, 8, and 10. When working with consecutive even or consecutive odd integers, we must continually add 2 to move from one integer to the next successive integer in the list.

Table 2.2 should be helpful in solving consecutive integer problems.

Table 2.2 Consecutive Integers

English Phrase	Algebraic Expressions	Example
Two consecutive integers	$x, x + 1$	13, 14
Three consecutive integers	$x, x + 1, x + 2$	−8, −7, −6
Two consecutive even integers	$x, x + 2$	40, 42
Two consecutive odd integers	$x, x + 2$	−37, −35
Three consecutive even integers	$x, x + 2, x + 4$	30, 32, 34
Three consecutive odd integers	$x, x + 2, x + 4$	9, 11, 13

EXAMPLE 5 Renting a Car

Rent-a-Heap Agency charges $125 per week plus $0.20 per mile to rent a small car. How many miles can you travel for $335?

SOLUTION

Step 1. **Let x represent one of the quantities.** Because we are asked to find the number of miles we can travel for $335, let

$$x = \text{the number of miles.}$$

Step 2. **Represent other quantities in terms of x.** There are no other unknown quantities to find, so we can skip this step.

Step 3. **Write an equation in x that describes the conditions.** Before writing the equation, let us consider a few specific values for the number of miles traveled. The rental charge is $125 plus $0.20 for each mile.

3 miles: The rental charge is \$125 + \$0.20(3).
30 miles: The rental charge is \$125 + \$0.20(30).
100 miles: The rental charge is \$125 + \$0.20(100).
x miles: The rental charge is \$125 + \$0.20x.

The weekly charge of $125	plus	the charge of $0.20 per mile for x miles	equals	the total $335 rental charge.
125	+	$0.20x$	=	335

Step 4. **Solve the equation and answer the question.**

$125 + 0.20x = 335$ — This is the equation specified by the conditions of the problem.

$125 + 0.20x - 125 = 335 - 125$ — Subtract 125 from both sides.

$0.20x = 210$ — Simplify.

$\dfrac{0.20x}{0.20} = \dfrac{210}{0.20}$ — Divide both sides by 0.20.

$x = 1050$ — Simplify.

You can travel 1050 miles for $335.

Step 5. **Check the proposed solution in the original wording of the problem.** Traveling 1050 miles should result in a total rental charge of $335. The mileage charge of $0.20 per mile is

$$\$0.20(1050) = \$210.$$

Adding this to the $125 weekly charge gives a total rental charge of

$$\$125 + \$210 = \$335.$$

Because this results in the given rental charge of $335, this verifies that you can travel 1050 miles. ■

CHECK POINT 5 A taxi charges $2.00 to turn on the meter plus $0.25 for each eighth of a mile. If you have $10.00, how many eighths of a mile can you go? How many miles is that?

We will be using the formula for the perimeter of a rectangle, $P = 2l + 2w$, in our next example. Twice the rectangle's length plus twice the rectangle's width is its perimeter.

EXAMPLE 6 Finding the Dimensions of a Soccer Field

A rectangular soccer field is twice as long as it is wide. If the perimeter of a soccer field is 300 yards, what are the field's dimensions?

SOLUTION

Step 1. **Let x represent one of the quantities.** We know something about the length; the field is twice as long as it is wide. We will let

$$x = \text{the width.}$$

Step 2. **Represent other quantities in terms of x.** Because the field is twice as long as it is wide, let

$$2x = \text{the length.}$$

Figure 2.7 illustrates the soccer field and its dimensions.

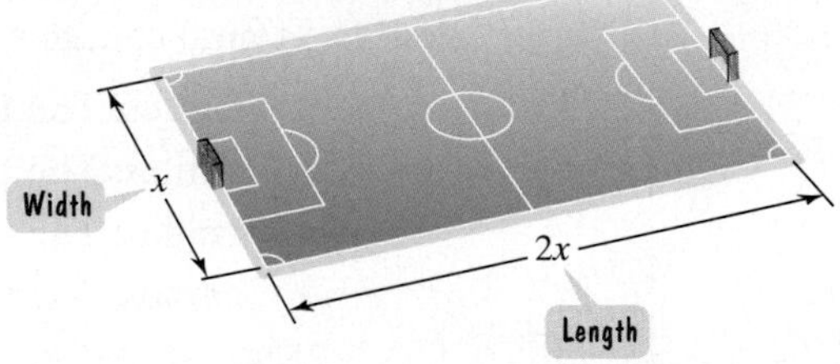

FIGURE 2.7

Step 3. **Write an equation in x that describes the conditions.** Because the perimeter of a soccer field is 300 yards,

Twice the length	plus	twice the width	is	the perimeter.
$2 \cdot 2x$	$+$	$2 \cdot x$	$=$	$300.$

Step 4. **Solve the equation and answer the question.**

$$2 \cdot 2x + 2 \cdot x = 300 \quad \text{This is the equation for the problem's conditions.}$$

$$4x + 2x = 300 \quad \text{Multiply.}$$

$$6x = 300 \quad \text{Combine like terms.}$$

$$\frac{6x}{6} = \frac{300}{6} \quad \text{Divide both sides by 6.}$$

$$x = 50 \quad \text{Simplify.}$$

Thus,

$$\text{Width} = x = 50$$

$$\text{Length} = 2x = 2(50) = 100.$$

The dimensions of a soccer field are 50 yards by 100 yards.

Step 5. **Check the proposed solution in the original wording of the problem.** The perimeter of the soccer field using the dimensions that we found is 2(50 yards) + 2(100 yards) = 100 yards + 200 yards, or 300 yards. Because the problem's wording tells us that the perimeter is 300 yards, our dimensions are correct. ■

✔ **CHECK POINT 6** A rectangular swimming pool is three times as long as it is wide. If the perimeter of the pool is 320 feet, what are the pool's dimensions?

EXAMPLE 7 A Price Reduction

Your local computer store is having a sale. After a 30% price reduction, you purchase a new computer for $980. What was the computer's price before the reduction?

SOLUTION

Step 1. **Let x represent one of the quantities.** We will let x = the original price of the computer before the reduction.

Step 2. **Represent other quantities in terms of x.** There are no other unknown quantities to find, so we can skip this step.

Step 3. **Write an equation in x that describes the conditions.** The computer's original price minus the 30% reduction is the reduced price, $980.

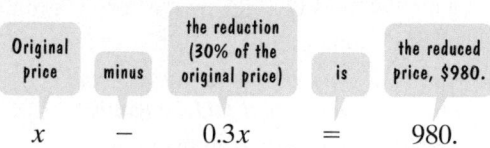

Original price	minus	the reduction (30% of the original price)	is	the reduced price, $980.
x	$-$	$0.3x$	$=$	$980.$

STUDY TIP

Observe that the original price, x, reduced by 30% is $x - 0.3x$ and *not* $x - 0.3$.

Step 4. **Solve the equation and answer the question.**

$$x - 0.3x = 980 \quad \text{This is the equation for the problem's conditions.}$$

$$0.7x = 980 \quad \text{Combine like terms: } x - 0.3x = 1x - 0.3x = 0.7x.$$

$$\frac{0.7x}{0.7} = \frac{980}{0.7} \quad \text{Divide both sides by 0.7.}$$

$$x = 1400 \quad \text{Simplify: } 0.7\overline{)980.0} = 1400.$$

The computer's price before the reduction was $1400.

Step 5. **Check the proposed solution in the original wording of the problem.** The price before the reduction, $1400, minus the reduction in price should equal the reduced price given in the original wording, $980. The reduction in price is equal to 30% of the price before the reduction, $1400. To find the reduction, we multiply the decimal equivalent of 30%, 0.30 or 0.3, by the original price, $1400:

$$30\% \text{ of } \$1400 = (0.3)(\$1400) = \$420.$$

Now we can determine whether the calculation for the price before the reduction, $1400, minus the reduction, $420, is equal to the reduced price given in the problem, $980. We subtract:

$$\$1400 - \$420 = \$980.$$

This verifies that the price of the computer before the reduction was $1400. ■

✔ **CHECK POINT 7** After a 40% price reduction, an exercise machine sold for $564. What was the exercise machine's price before this reduction?

2.5 EXERCISE SET

Student Solutions Manual | CD/Video | PH Math/Tutor Center | MathXL Tutorials on CD | MathXL® | MyMathLab | Interactmath.com

Practice Exercises

In Exercises 1–14, let x represent the number. Write each English phrase as an algebraic expression.

1. The sum of a number and 7
2. A number increased by 15
3. A number subtracted from 25
4. 46 less than a number
5. 9 decreased by 4 times a number
6. 15 less than the product of 8 and a number
7. The quotient of 83 and a number
8. The quotient of a number and 83
9. The sum of twice a number and 40
10. Twice the sum of a number and 40
11. 93 subtracted from 9 times a number
12. The quotient of 13 and a number, decreased by 7 times the number
13. Eight times the sum of a number and 14
14. Nine times the difference of a number and 5

In Exercises 15–34, let x represent the number. Use the given conditions to write an equation. Solve the equation and find the number.

15. A number increased by 60 is equal to 410. Find the number.
16. The sum of a number and 43 is 107. Find the number.
17. A number decreased by 23 is equal to 214. Find the number.
18. The difference between a number and 17 is 96. Find the number.
19. The product of 7 and a number is 126. Find the number.
20. The product of 8 and a number is 272. Find the number.
21. The quotient of a number and 19 is 5. Find the number.
22. The quotient of a number and 14 is 8. Find the number.
23. The sum of four and twice a number is 56. Find the number.
24. The sum of five and three times a number is 59. Find the number.
25. Seven subtracted from five times a number is 178. Find the number.
26. Eight subtracted from six times a number is 298. Find the number.
27. A number increased by 5 is two times the number. Find the number.
28. A number increased by 12 is four times the number. Find the number.
29. Twice the sum of four and a number is 36. Find the number.
30. Three times the sum of five and a number is 48. Find the number.
31. Nine times a number is 30 more than three times that number. Find the number.
32. Five more than four times a number is that number increased by 35. Find the number.
33. If the quotient of three times a number and five is increased by four, the result is 34. Find the number.
34. If the quotient of three times a number and four is decreased by three, the result is nine. Find the number.

Application Exercises

In Exercises 35–62, use the five-step strategy to solve each problem.

35. Two of the most expensive movies ever made were *Titanic* and *Waterworld*. The cost to make *Titanic* exceeded the cost to make *Waterworld* by $25 million. The combined cost to make the two movies was $375 million. Find the cost of making each of these movies.

Paramount Pictures Corporation, Inc.

36. Each day, the number of births in the world exceeds the number of deaths by 229 thousand. The combined number of births and deaths is 521 thousand. Determine the number of births and the number of deaths per day.

Daily Growth of World Population

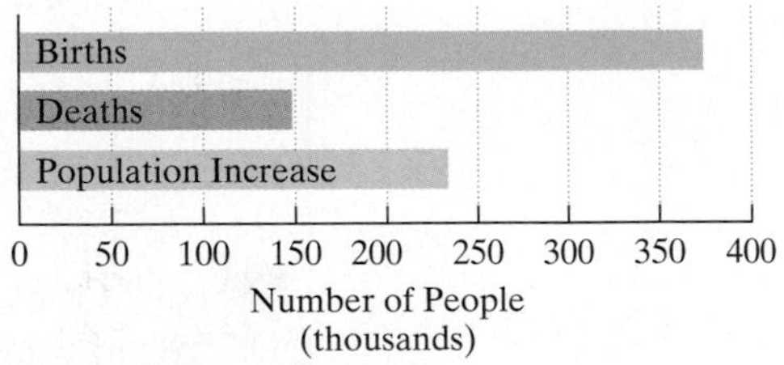

Source: "Population Update" 2000

37. The circle graph shows the political ideology of U.S. college freshmen. The percentage of liberals exceeds twice that of conservatives by 4.4%. Using the displayed percents, it can be shown that liberals and conservatives combined account for 57.2% of college freshmen. Find the percentage of liberals and the percentage of conservatives.

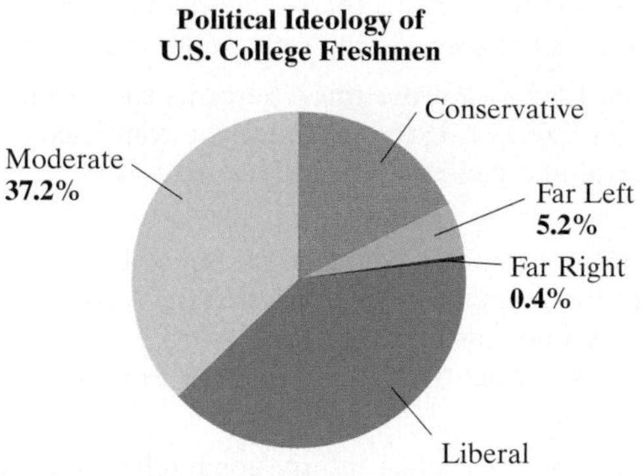

Source: The Chronicle of Higher Education

38. Commuters in one-third of the largest cities in the United States spend more than 40 hours per year, equivalent to one work week, sitting in traffic. The bar graph shows the number of hours in traffic per year for the average motorist in ten cities. The average motorist in Los Angeles spends 32 hours less than twice that of the average motorist in Miami stuck in traffic each year. In the two cities combined, 139 hours are spent by the average motorist per year in traffic. How many hours are wasted in traffic by the average motorist in Los Angeles and Miami?

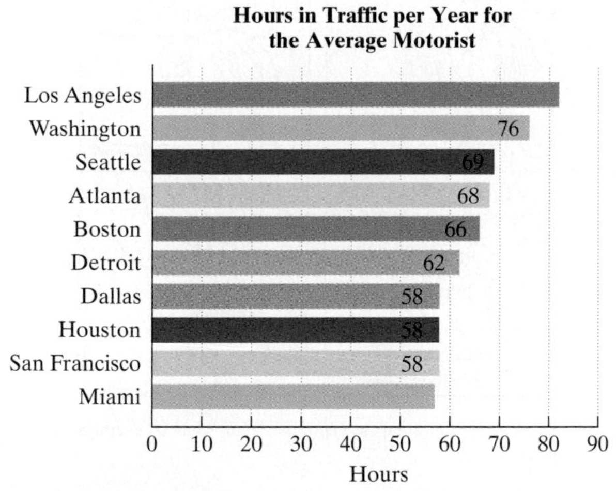

Source: Texas Transportation Institute

39. The sum of the page numbers on the facing pages of a book is 629. What are the page numbers?

40. The sum of the page numbers on the facing pages of a book is 525. What are the page numbers?

41. The highest-grossing North American concert tour was the Rolling Stones (1994), followed closely by Bruce Springsteen (2003). Combined, the two tours grossed $241 million. When expressed in millions, the earnings for the tours are consecutive integers. Find the gross, in millions, for the Rolling Stones tour and the Springsteen tour. (*Source*: *Rolling Stone*)

42. The first Super Bowl was played between the Green Bay Packers and the Kansas City Chiefs in 1967. Only once, in 1991, were the winning and losing scores in the Super Bowl consecutive integers. If the sum of the scores was 39, what were the scores?

43. Find two consecutive even integers whose sum is 66.

44. Find two consecutive odd integers whose sum is 72.

45. A car rental agency charges $200 per week plus $0.15 per mile to rent a car. How many miles can you travel in one week for $320?

46. A car rental agency charges $180 per week plus $0.25 per mile to rent a car. How many miles can you travel in one week for $395?

47. The average weight for female infants at birth is 7 pounds, with a monthly weight gain of 1.5 pounds. After how many months does a baby girl weigh 16 pounds?

48. The total revenue from Indian casinos in the United States has been increasing at approximately $2.2 billion per year. In 2002, Indian casinos reported a combined revenue of $12.7 billion. In which year will total revenue reach $28.1 billion? (*Source*: National Indian Gaming Commission)

49. A rectangular field is four times as long as it is wide. If the perimeter of the field is 500 yards, what are the field's dimensions?

50. A rectangular field is five times as long as it is wide. If the perimeter of the field is 288 yards, what are the field's dimensions?

51. An American football field is a rectangle with a perimeter of 1040 feet. The length is 200 feet more than the width. Find the width and length of the rectangular field.

52. A basketball court is a rectangle with a perimeter of 86 meters. The length is 13 meters more than the width. Find the width and length of the basketball court.

53. A bookcase is to be constructed as shown in the figure. The length is to be 3 times the height. If 60 feet of lumber is available for the entire unit, find the length and height of the bookcase.

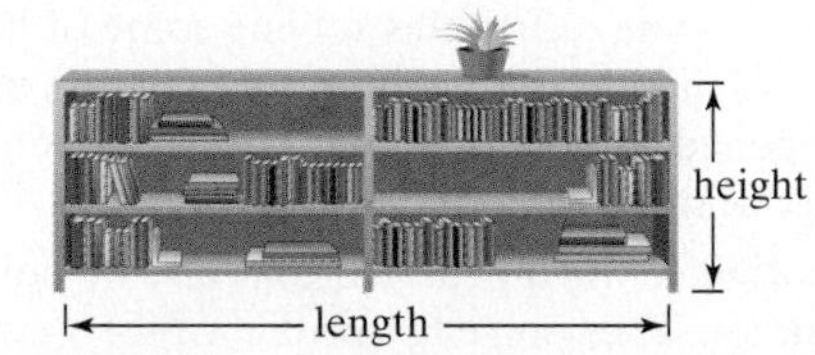

54. The height of the bookcase in the figure is 3 feet longer than the length of a shelf. If 18 feet of lumber is available for the entire unit, find the length and height of the unit.

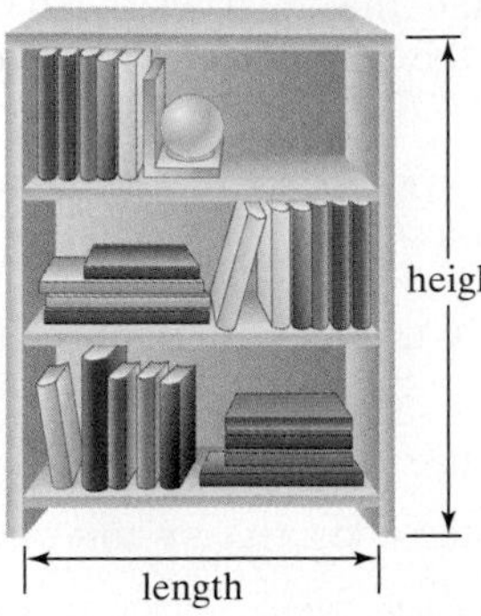

55. After a 20% reduction, you purchase a television for \$320. What was the television's price before the reduction?

56. After a 30% reduction, you purchase a VCR for \$98. What was the VCR's price before the reduction?

57. The average yearly earnings of pharmacists increased by 30% from 2001 to 2002. If salaries averaged \$87,100 in 2002, what was the average salary in 2001?
(*Source*: Bureau of Labor Statistics)

58. The average yearly earnings of physical education teachers increased by 40% from 2001 to 2002. If salaries averaged \$63,000 in 2002, what was the average salary in 2001?
(*Source*: Bureau of Labor Statistics)

59. Including 6% sales tax, a car sold for \$15,370. Find the price of the car before the tax was added.

60. Including 8% sales tax, a bed-and-breakfast inn charges \$172.80 per night. Find the inn's nightly cost before the tax is added.

61. An automobile repair shop charged a customer \$448, listing \$63 for parts and the remainder for labor. If the cost of labor is \$35 per hour, how many hours of labor did it take to repair the car?

62. A repair bill on a sailboat came to \$1603, including \$532 for parts and the remainder for labor. If the cost of labor is \$63 per hour, how many hours of labor did it take to repair the sailboat?

Writing in Mathematics

63. In your own words, describe a step-by-step approach for solving algebraic word problems.

64. Many students find solving linear equations much easier than solving algebraic word problems. Discuss some of the reasons why this is the case.

65. Did you have some difficulties solving some of the problems that were assigned in this exercise set? Discuss what you did if this happened to you. Did your course of action enhance your ability to solve algebraic word problems?

66. Write an original word problem that can be solved using a linear equation. Then solve the problem.

Critical Thinking Exercises

67. Which English statement given below is correctly translated into an algebraic equation?

a. Ten pounds less than Bill's weight (x) equals 160 pounds: $10 - x = 160$.

b. Four more than five times a number (x) is one less than six times that number: $5x + 4 = 1 - 6x$.

c. Seven is three more than some number (x): $7 + 3 = x$.

d. None of the above is correctly translated.

68. Explain how to use the three percents shown on the circle graph in Exercise 37 to determine the combined percentage of liberal and conservative college freshmen.

69. An HMO pamphlet contains the following recommended weight for women: "Give yourself 100 pounds for the first 5 feet plus 5 pounds for every inch over 5 feet tall." Using this description, which height corresponds to an ideal weight of 135 pounds?

70. The rate for a particular international telephone call is \$0.55 for the first minute and \$0.40 for each additional minute. Determine the length of a call that costs \$6.95.

71. In a film, the actor Charles Coburn plays an elderly "uncle" character criticized for marrying a woman when he is 3 times her age. He wittily replies, "Ah, but in 20 years time I shall only be twice her age." How old is the "uncle" and the woman?

72. Answer the question in the following *Peanuts* cartoon strip. (*Note*: You may not use the answer given in the cartoon!)

PEANUTS reprinted by permission of United Features Syndicate, Inc.

Review Exercises

73. Solve and check: $\frac{4}{5}x = -16$. (Section 2.2, Example 3)

74. Solve and check: $6(y - 1) + 7 = 9y - y + 1$. (Section 2.3, Example 3)

75. Solve for w: $V = \frac{1}{3}lwh$. (Section 2.4, Example 4)

SECTION 2.6

Objectives

1. Solve problems using formulas for perimeter and area.
2. Solve problems using formulas for a circle's area and circumference.
3. Solve problems using formulas for volume.
4. Solve problems involving the angles of a triangle.
5. Solve problems involving complementary and supplementary angles.

PROBLEM SOLVING IN GEOMETRY

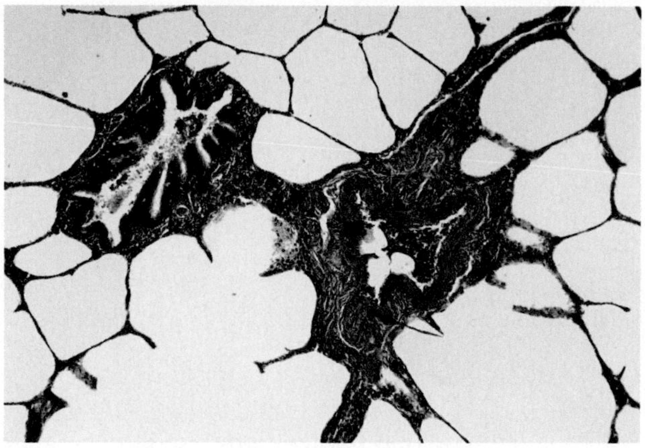

A portion of the human lung magnified 160 times

Geometry is about the space you live in and the shapes that surround you. You're even made of it. The human lung consists of nearly 300 spherical air sacs, geometrically designed to provide the greatest surface area within the limited volume of our bodies. Viewed in this way, geometry becomes an intimate experience.

For thousands of years, people have studied geometry in some form to obtain a better understanding of the world in which they live. A study of the shape of your world will provide you with many practical applications that will help to increase your problem-solving skills.

1 Solve problems using formulas for perimeter and area.

Geometric Formulas for Perimeter and Area Solving geometry problems often requires using basic geometric formulas. Formulas for perimeter and area are summarized in Table 2.3. Remember that perimeter is measured in linear units, such as feet or meters, and area is measured in square units, such as square feet, ft^2, or square meters, m^2.

Table 2.3 Common Formulas for Perimeter and Area

Square	Rectangle	Triangle	Trapezoid
$A = s^2$ $P = 4s$	$A = lw$ $P = 2l + 2w$	$A = \frac{1}{2}bh$	$A = \frac{1}{2}h(a + b)$

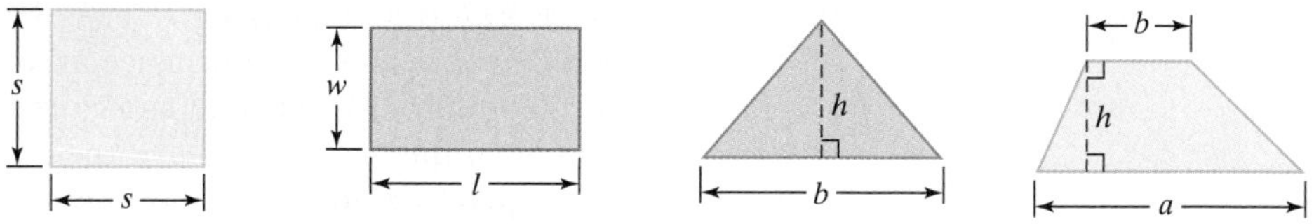

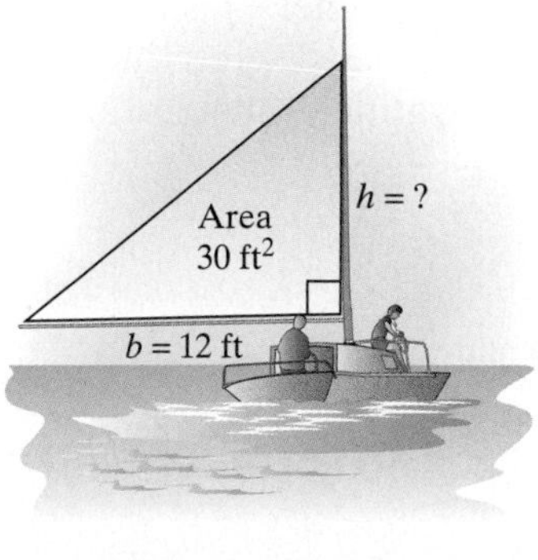

FIGURE 2.8 Finding the height of a triangular sail

EXAMPLE 1 Using the Formula for the Area of a Triangle

A sailboat has a triangular sail with an area of 30 square feet and a base that is 12 feet long. (See Figure 2.8.) Find the height of the sail.

SOLUTION We begin with the formula for the area of a triangle given in Table 2.3.

$A = \frac{1}{2}bh$	The area of a triangle is $\frac{1}{2}$ the product of its base and height.
$30 = \frac{1}{2}(12)h$	Substitute 30 for A and 12 for b.
$30 = 6h$	Simplify.
$\frac{30}{6} = \frac{6b}{6}$	Divide both sides by 6.
$5 = h$	Simplify.

The height of the sail is 5 feet.

Check

The area is $A = \frac{1}{2}bh = \frac{1}{2}(12 \text{ feet})(5 \text{ feet}) = 30$ square feet. ■

✔ CHECK POINT **1** A sailboat has a triangular sail with an area of 24 square feet and a base that is 4 feet long. Find the height of the sail.

2 Solve problems using formulas for a circle's area and circumference.

The point at which a pebble hits a flat surface of water becomes the center of a number of circular ripples.

Geometric Formulas for Circumference and Area of a Circle It's a good idea to know your way around a circle. Clocks, angles, maps, and compasses are based on circles. Circles occur everywhere in nature: in ripples on water, patterns on a butterfly's wings, and cross sections of trees. Some consider the circle to be the most pleasing of all shapes.

A **circle** is a set of points in the plane equally distant from a given point, its center. Figure 2.9 shows two circles. A **radius** (plural: radii), r, is a line segment from the center to any point on the circle. For a given circle, all radii have the same length. A **diameter**, d, is a line segment through the center whose endpoints both lie on the circle. For a given circle, all diameters have the same length. In any circle, **the length of a diameter is twice the length of a radius.**

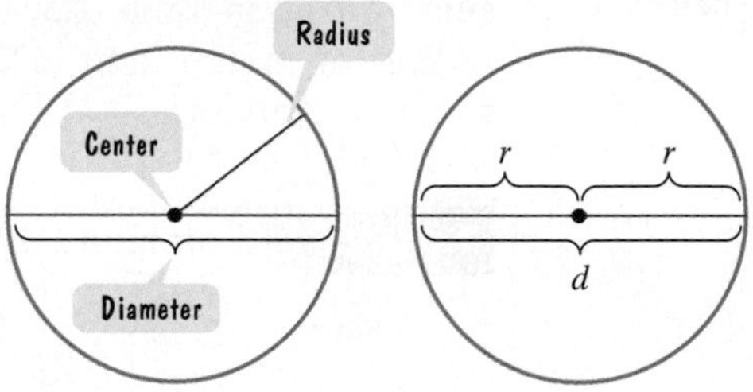

FIGURE 2.9

The words *radius* and *diameter* refer to both the line segments in Figure 2.9 as well as their linear measures. The distance around a circle (its perimeter) is called its **circumference.** Formulas for the area and circumference of a circle are given in terms of π and appear in Table 2.4. We have seen that π is an irrational number and is approximately equal to 3.14.

Table 2.4 Formulas for Circles

Circle	Area	Circumference
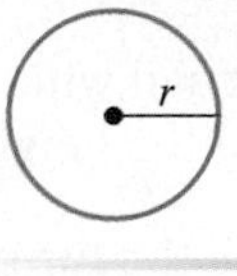	$A = \pi r^2$	$C = 2\pi r$

When computing a circle's area or circumference by hand, round π to 3.14. When using a calculator, use the $\boxed{\pi}$ key, which gives the value of π rounded to approximately 11 decimal places. In either case, calculations involving π give approximate answers. These answers can vary slightly depending on how π is rounded. The symbol $\approx$ (is approximately equal to) will be written in these calculations.

EXAMPLE 2 Finding the Area and Circumference of a Circle

Find the area and circumference of a circle with a diameter measuring 20 inches.

SOLUTION The radius is half the diameter, so $r = \frac{20}{2} = 10$ inches.

$A = \pi r^2$	$C = 2\pi r$	*Use the formulas for area and circumference of a circle.*
$A = \pi(10)^2$	$C = 2\pi(10)$	*Substitute 10 for r.*
$A = 100\pi$	$C = 20\pi$	

The area of the circle is 100π square inches and the circumference is 20π inches. Using the fact that $\pi \approx 3.14$, the area is approximately 100(3.14), or 314 square inches. The circumference is approximately 20(3.14), or 62.8 inches. ■

USING TECHNOLOGY

You can use your calculator to obtain the price per square inch for each pizza in Example 3. The price per square inch for the large pizza, $\frac{15}{64\pi}$, is approximated by one of the following keystrokes:

Many Scientific Calculators

15 $\boxed{\div}$ $\boxed{(}$ 64 $\boxed{\times}$ $\boxed{\pi}$ $\boxed{)}$ $\boxed{=}$

Many Graphing Calculators

15 $\boxed{\div}$ $\boxed{(}$ 64 $\boxed{\pi}$ $\boxed{)}$ $\boxed{\text{ENTER}}$

CHECK POINT 2 The diameter of a circular landing pad for helicopters is 40 feet. Find the area and circumference of the landing pad. Express answers in terms of π. Then round answers to the nearest square foot and foot, respectively.

EXAMPLE 3 Problem Solving Using the Formula for a Circle's Area

Which one of the following is the better buy: a large pizza with a 16-inch diameter for \$15.00 or a medium pizza with an 8-inch diameter for \$7.50?

SOLUTION The better buy is the pizza with the lower price per square inch. The radius of the large pizza is $\frac{1}{2} \cdot 16$ inches, or 8 inches, and the radius of the medium pizza is $\frac{1}{2} \cdot 8$ inches, or 4 inches. The area of the surface of each circular pizza is determined using the formula for the area of a circle.

$$\text{Large pizza: } A = \pi r^2 = \pi(8 \text{ in.})^2 = 64\pi \text{ in.}^2 \approx 201 \text{ in.}^2$$
$$\text{Medium pizza: } A = \pi r^2 = \pi(4 \text{ in.})^2 = 16\pi \text{ in.}^2 \approx 50 \text{ in.}^2$$

For each pizza, the price per square inch is found by dividing the price by the area:

$$\text{Price per square inch for large pizza} = \frac{\$15.00}{64\pi \text{ in.}^2} \approx \frac{\$15.00}{201 \text{ in.}^2} \approx \frac{\$0.07}{\text{in.}^2}$$

$$\text{Price per square inch for medium pizza} = \frac{\$7.50}{16\pi \text{ in.}^2} \approx \frac{\$7.50}{50 \text{ in.}^2} = \frac{\$0.15}{\text{in.}^2}.$$

The large pizza costs approximately \$0.07 per square inch and the medium pizza costs approximately \$0.15 per square inch. Thus, the large pizza is the better buy. ■

In Example 3, did you at first think that the price per square inch would be the same for the large and the medium pizzas? After all, the radius of the large pizza is twice that of the medium pizza, and the cost of the large is twice that of the medium. However, the large pizza's area, 64π square inches, is *four times the area* of the medium pizza, 16π square inches. Doubling the radius of a circle increases its area by four times the original amount.

> **CHECK POINT 3** Which one of the following is the better buy: a large pizza with an 18-inch diameter for \$20.00 or a medium pizza with a 14-inch diameter for \$14.00?

3 Solve problems using formulas for volume.

Geometric Formulas for Volume A shoe box and a basketball are examples of three-dimensional figures. **Volume** refers to the amount of space occupied by such a figure. To measure this space, we begin by selecting a cubic unit. One such cubic unit, 1 cubic centimeter (cm^3), is shown in Figure 2.10.

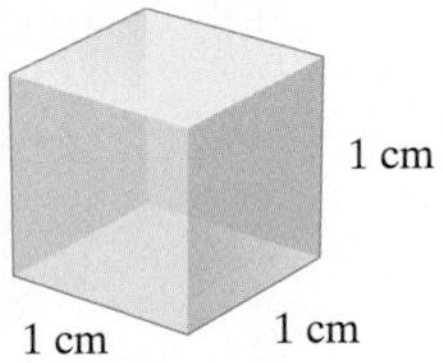

FIGURE 2.10

The edges of a cube all have the same length. Other cubic units used to measure volume include 1 cubic inch ($in.^3$) and 1 cubic foot (ft^3). The volume of a solid is the number of cubic units that can be contained in the solid.

Formulas for volumes of three-dimensional figures are given in Table 2.5.

Table 2.5 Common Formulas for Volume

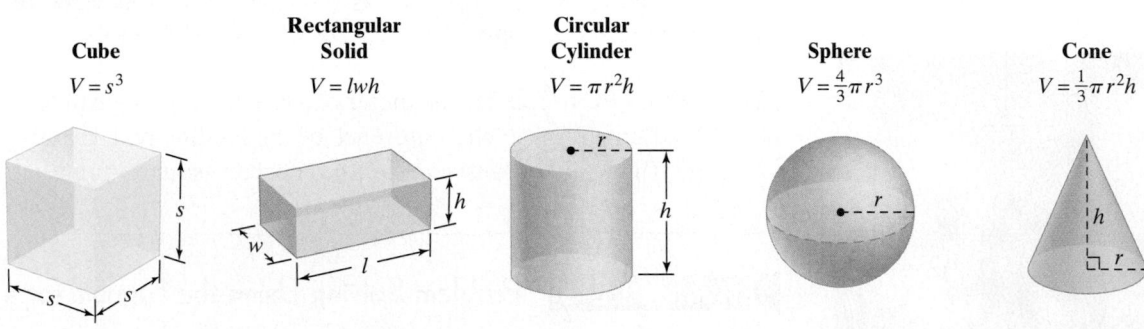

Cube	Rectangular Solid	Circular Cylinder	Sphere	Cone
$V = s^3$	$V = lwh$	$V = \pi r^2 h$	$V = \frac{4}{3}\pi r^3$	$V = \frac{1}{3}\pi r^2 h$

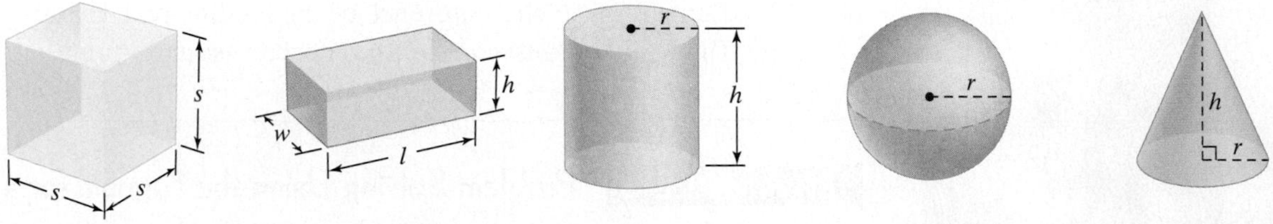

EXAMPLE 4 Using the Formula for the Volume of a Cylinder

A cylinder with a radius of 2 inches and a height of 6 inches has its radius doubled. (See Figure 2.11.) How many times greater is the volume of the larger cylinder than the volume of the smaller cylinder?

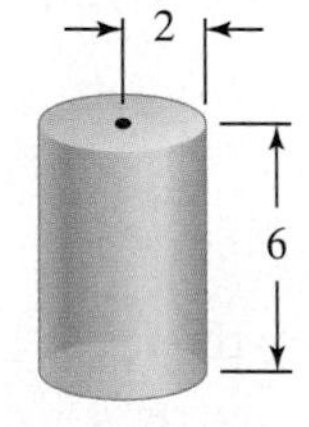

Radius: 2 inches
Height: 6 inches

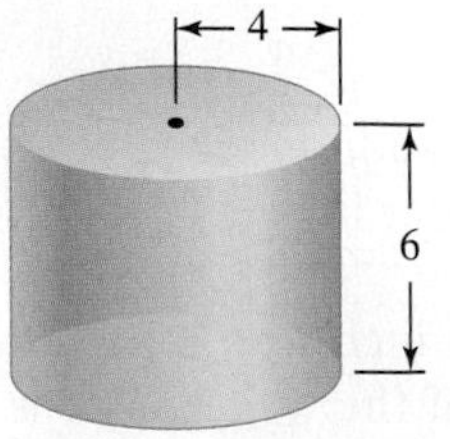

Radius: 4 inches
Height: 6 inches

FIGURE 2.11 Doubling a cylinder's radius

SOLUTION We begin with the formula for the volume of a cylinder given in Table 2.5. Find the volume of the smaller cylinder and the volume of the larger cylinder. To compare the volumes, divide the volume of the larger cylinder by the volume of the smaller cylinder.

$$V = \pi r^2 h$$ *Use the formula for the volume of a cylinder.*

Radius is doubled.

$$V_{\text{Smaller}} = \pi(2)^2(6) \quad V_{\text{Larger}} = \pi(4)^2(6)$$ *Substitute the given values.*

$$V_{\text{Smaller}} = \pi(4)(6) \quad V_{\text{Larger}} = \pi(16)(6)$$

$$V_{\text{Smaller}} = 24\pi \quad V_{\text{Larger}} = 96\pi$$

The volume of the smaller cylinder is 24π cubic inches. The volume of the larger cylinder is 96π cubic inches. We use division to compare the volumes:

$$\frac{V_{\text{Larger}}}{V_{\text{Smaller}}} = \frac{96\pi}{24\pi} = \frac{4}{1}.$$

Thus, the volume of the larger cylinder is 4 times the volume of the smaller cylinder. ■

CHECK POINT 4 A cylinder with a radius of 3 inches and a height of 5 inches has its height doubled. How many times greater is the volume of the larger cylinder than the volume of the smaller cylinder?

EXAMPLE 5 Applying Volume Formulas

An ice cream cone is 5 inches deep and has a radius of 1 inch. A spherical scoop of ice cream also has a radius of 1 inch. (See Figure 2.12.) If the ice cream melts into the cone, will it overflow?

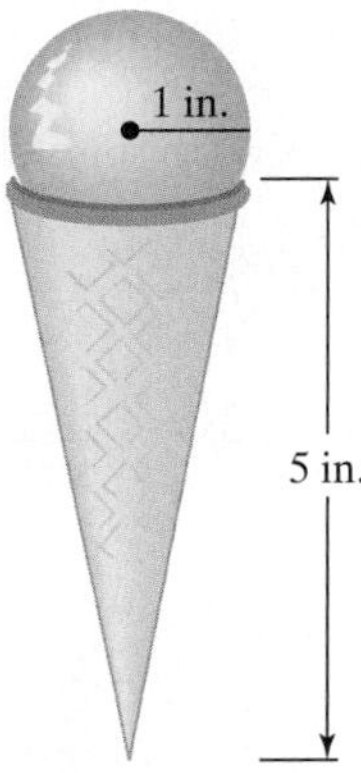

FIGURE 2.12

SOLUTION The ice cream will overflow if the volume of the ice cream, a sphere, is greater than the volume of the cone. Find the volume of each.

$$V_{\text{cone}} = \frac{1}{3}\pi r^2 h = \frac{1}{3}\pi(1\text{ in.})^2 \cdot 5\text{ in.} = \frac{5\pi}{3}\text{ in.}^3 \approx 5\text{ in.}^3$$

$$V_{\text{sphere}} = \frac{4}{3}\pi r^3 = \frac{4}{3}\pi(1\text{ in.})^3 = \frac{4\pi}{3}\text{ in.}^3 \approx 4\text{ in.}^3$$

The volume of the spherical scoop of ice cream is less than the volume of the cone, so there will be no overflow. ■

CHECK POINT 5 A basketball has a radius of 4.5 inches. If the ball is filled with 350 cubic inches of air, is this enough air to fill it completely?

4 Solve problems involving the angles of a triangle.

The Angles of a Triangle The hour hand of a clock moves from 12 to 2. The hour hand suggests a **ray**, a part of a line that has only one endpoint and extends forever in the opposite direction. An *angle* is formed as the ray in Figure 2.13 rotates from 12 to 2.

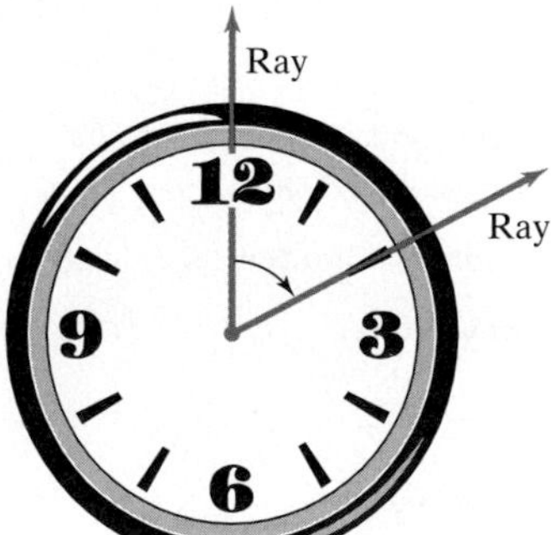

FIGURE 2.13 Clock with rays rotating to form an angle

An **angle**, symbolized ∡, is made up of two rays that have a common endpoint. Figure 2.14 shows an angle. The common endpoint, B in the figure, is called the **vertex** of the angle. The two rays that form the angle are called its **sides**. The four ways of naming the angle are shown to the right of Figure 2.14.

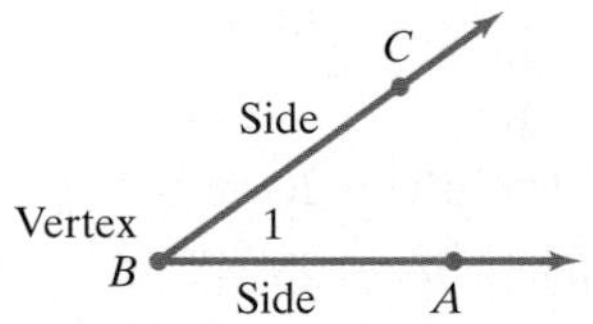

Naming the Angle

∡1 ∡B ∡ABC ∡CBA

Vertex alone — Vertex letter in the middle

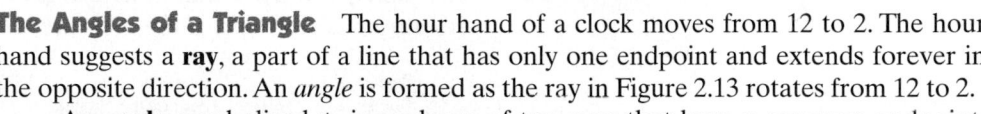

FIGURE 2.14 An angle: two rays with a common endpoint

One way to measure angles is in **degrees**, symbolized by a small, raised circle °. Think of the hour hand of a clock. From 12 noon to 12 midnight, the hour hand moves around in a complete circle. By definition, the ray has rotated through 360 degrees, or 360°. Using 360° as the amount of rotation of a ray back onto itself, a degree, 1°, is $\frac{1}{360}$ of a complete rotation.

Our next problem is based on the relationship among the three angles of any triangle.

THE ANGLES OF A TRIANGLE

The sum of the measures of the three angles of any triangle is 180°.

EXAMPLE 6 Angles of a Triangle

In a triangle, the measure of the first angle is twice the measure of the second angle. The measure of the third angle is 20° less than the second angle. What is the measure of each angle?

SOLUTION

Step 1. Let x represent one of the quantities. Let

$$x = \text{the measure of the second angle.}$$

Step 2. Represent other quantities in terms of x. The measure of the first angle is twice the measure of the second angle. Thus, let

$$2x = \text{the measure of the first angle.}$$

The measure of the third angle is 20° less than the second angle. Thus, let

$$x - 20 = \text{the measure of the third angle.}$$

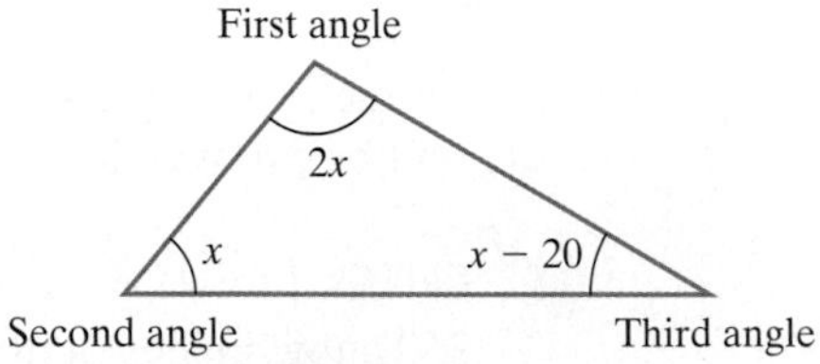

Step 3. Write an equation in x that describes the conditions. Because we are working with a triangle, the sum of the measures of its three angles is 180°.

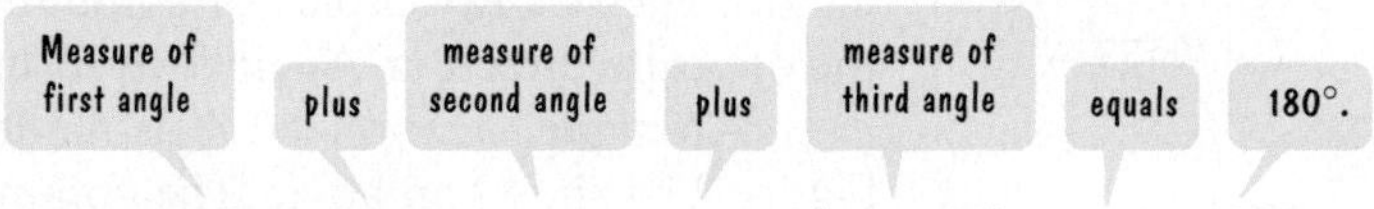

$$2x \quad + \quad x \quad + \quad (x - 20) \quad = \quad 180$$

Step 4. Solve the equation and answer the question.

$$2x + x + (x - 20) = 180$$ This is the equation that describes the sum of the measures of the angles.

$$4x - 20 = 180$$ Regroup and combine like terms.

$$4x - 20 + 20 = 180 + 20$$ Add 20 to both sides.

$$4x = 200$$ Simplify.

$$\frac{4x}{4} = \frac{200}{4}$$ Divide both sides by 4.

$$x = 50$$ Simplify.

Measure of first angle $= 2x = 2 \cdot 50 = 100$

Measure of second angle $= x = 50$

Measure of third angle $= x - 20 = 50 - 20 = 30$

The angles measure 100°, 50°, and 30°.

Step 5. Check the proposed solution in the original wording of the problem. The problem tells us that we are working with a triangle's angles. Thus, the sum of the measures should be 180°. Adding the three measures, we obtain $100° + 50° + 30°$, giving the required sum of 180°. ■

✔ **CHECK POINT 6** In a triangle, the measure of the first angle is three times the measure of the second angle. The measure of the third angle is 20° less than the second angle. What is the measure of each angle?

5 Solve problems involving complementary and supplementary angles.

Complementary and Supplementary Angles Two angles with measures having a sum of 90° are called **complementary angles**. For example, angles measuring 70° and 20° are complementary angles because $70° + 20° = 90°$. For angles such as those measuring 70° and 20°, each angle is a **complement** of the other: The 70° angle is the complement of the 20° angle and the 20° angle is the complement of the 70° angle. The measure of the complement can be found by subtracting the angle's measure from 90°. For example, we can find the complement of a 25° angle by subtracting 25° from 90°: $90° - 25° = 65°$. Thus, an angle measuring 65° is the complement of one measuring 25°.

Two angles with measures having a sum of 180° are called **supplementary angles**. For example, angles measuring 110° and 70° are supplementary angles because $110° + 70° = 180°$. For angles such as those measuring 110° and 70°, each angle is a **supplement** of the other: The 110° angle is the supplement of the 70° angle and the 70° angle is the supplement of the 110° angle. The measure of the supplement can be found by subtracting the angle's measure from 180°. For example, we can find the supplement of a 25° angle by subtracting 25° from 180°: $180° - 25° = 155°$. Thus, an angle measuring 155° is the supplement of one measuring 25°.

ALGEBRAIC EXPRESSIONS FOR COMPLEMENTS AND SUPPLEMENTS

Measure of an angle: x

Measure of the angle's complement: $90 - x$

Measure of the angle's supplement: $180 - x$

EXAMPLE 7 Angle Measures and Complements

The measure of an angle is 40° less than four times the measure of its complement. What is the angle's measure?

SOLUTION

Step 1. **Let x represent one of the quantities.** Let

$$x = \text{the measure of the angle.}$$

Step 2. **Represent other unknown quantities in terms of x.** Because this problem involves an angle and its complement, let

$$90 - x = \text{the measure of the complement.}$$

Step 3. **Write an equation in x that describes the conditions.**

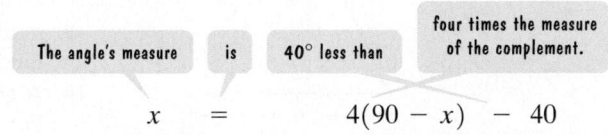

The angle's measure | is | 40° less than | four times the measure of the complement.

$$x = 4(90 - x) - 40$$

Step 4. Solve the equation and answer the question.

$$x = 4(90 - x) - 40 \quad \text{This is the equation that describes the problem's conditions.}$$
$$x = 360 - 4x - 40 \quad \text{Use the distributive property.}$$
$$x = 320 - 4x \quad \text{Simplify: } 360 - 40 = 320.$$
$$x + 4x = 320 - 4x + 4x \quad \text{Add 4x to both sides.}$$
$$5x = 320 \quad \text{Simplify.}$$
$$\frac{5x}{5} = \frac{320}{5} \quad \text{Divide both sides by 5.}$$
$$x = 64 \quad \text{Simplify.}$$

The angle measures 64°.

Step 5. Check the proposed solution in the original wording of the problem. The measure of the complement is $90° - 64° = 26°$. Four times the measure of the complement is $4 \cdot 26°$, or 104°. The angle's measure, 64°, is 40° less than 104°: $104° - 40° = 64°$. As specified by the problem's wording, the angle's measure is 40° less than four times the measure of its complement. ■

CHECK POINT 7 The measure of an angle is twice the measure of its complement. What is the angle's measure?

2.6 EXERCISE SET

Student Solutions Manual | CD/Video | PH Math/Tutor Center | MathXL Tutorials on CD | MathXL® | MyMathLab | Interactmath.com

Practice Exercises

Use the formulas for perimeter and area in Table 2.3 on page 149 to solve Exercises 1–12.

In Exercises 1–2, find the perimeter and area of each rectangle.

1.

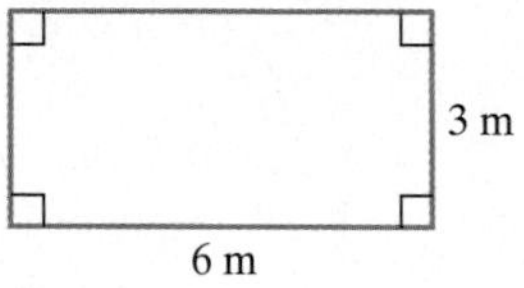

2.

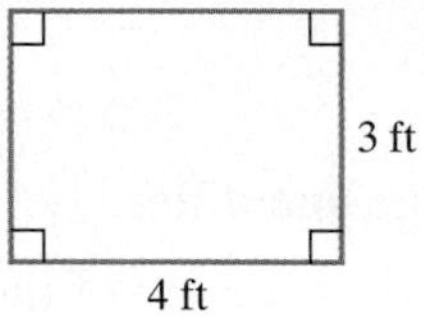

In Exercises 3–4, find the area of each triangle.

3.

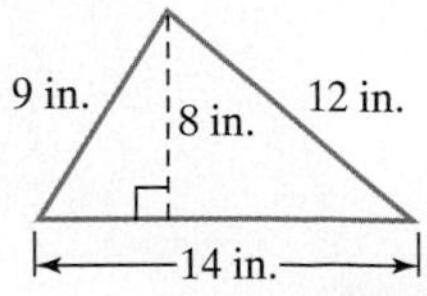

4.

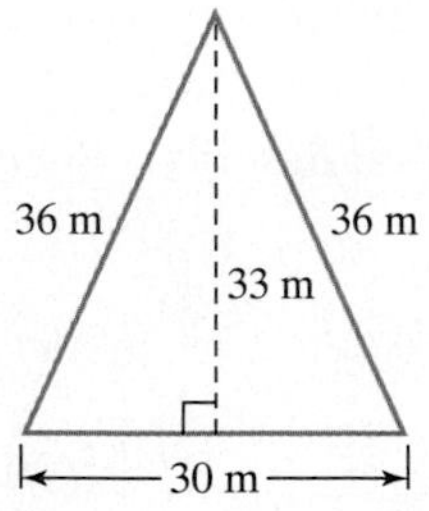

In Exercises 5–6, find the area of each trapezoid.

5.

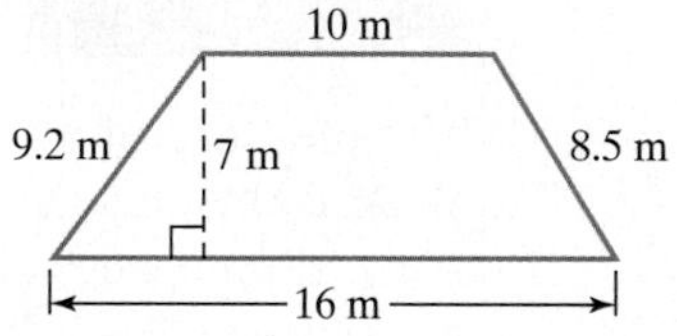

6.

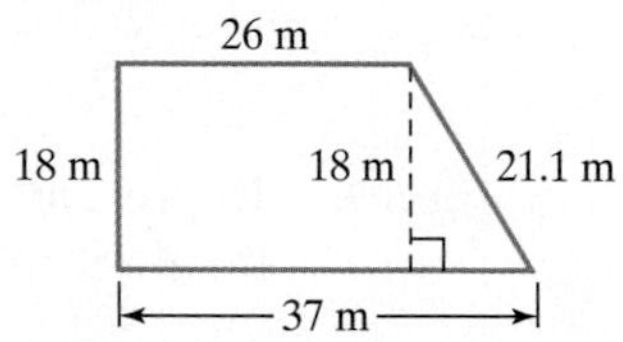

7. A rectangular swimming pool has a width of 25 feet and an area of 1250 square feet. What is the pool's length?

8. A rectangular swimming pool has a width of 35 feet and an area of 2450 square feet. What is the pool's length?

9. A triangle has a base of 5 feet and an area of 20 square feet. Find the triangle's height.

10. A triangle has a base of 6 feet and an area of 30 square feet. Find the triangle's height.

11. A rectangle has a width of 44 centimeters and a perimeter of 188 centimeters. What is the rectangle's length?

12. A rectangle has a width of 46 centimeters and a perimeter of 208 centimeters. What is the rectangle's length?

Use the formulas for the area and circumference of a circle in Table 2.4 on page 150 to solve Exercises 13–18.

In Exercises 13–16, find the area and circumference of each circle. Express answers in terms of π. Then round to the nearest whole number.

13.

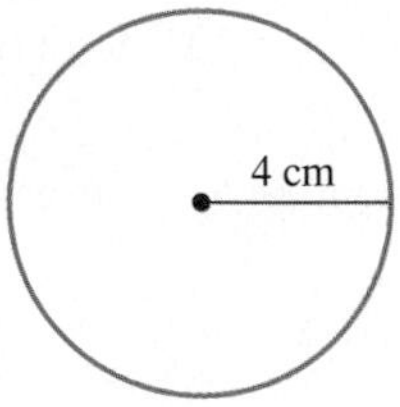

14.

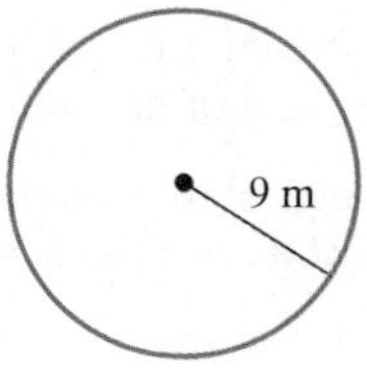

15.

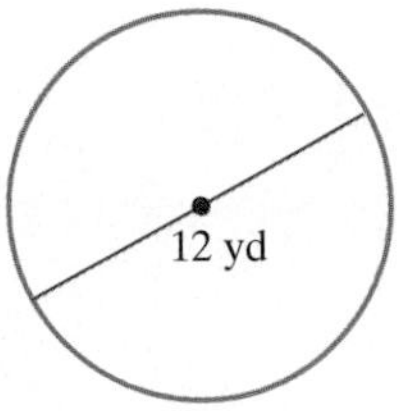

16.

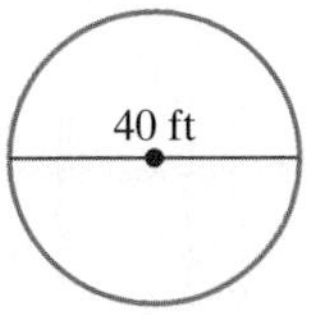

17. The circumference of a circle is 14π inches. Find the circle's radius and diameter.

18. The circumference of a circle is 16π inches. Find the circle's radius and diameter.

Use the formulas for volume in Table 2.5 on page 152 to solve Exercises 19–30.

In Exercises 19–26, find the volume of each figure. Where applicable, express answers in terms of π. Then round to the nearest whole number.

19.

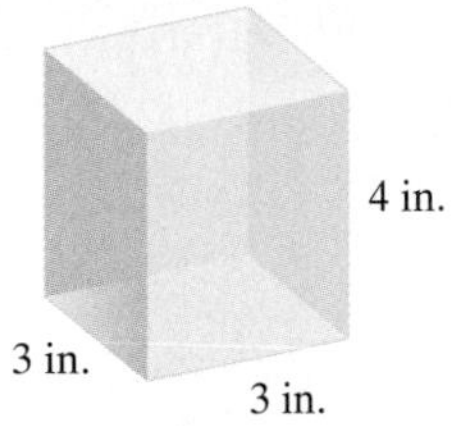

20.

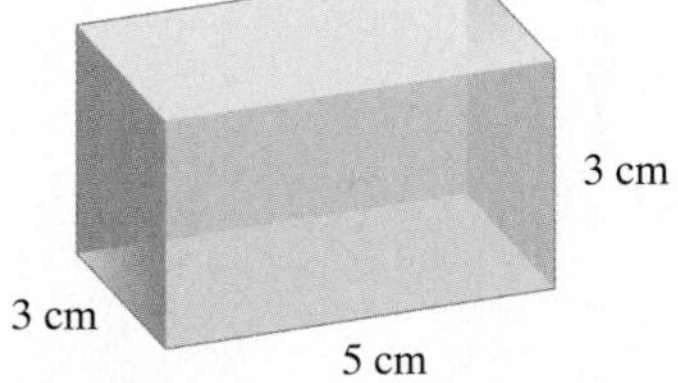

21.

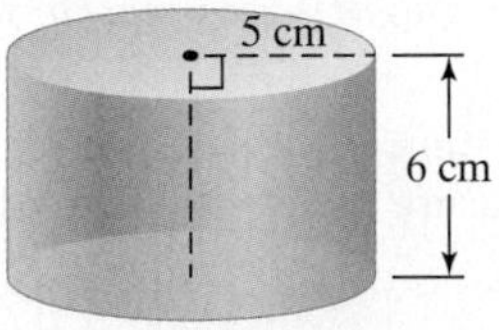

22.

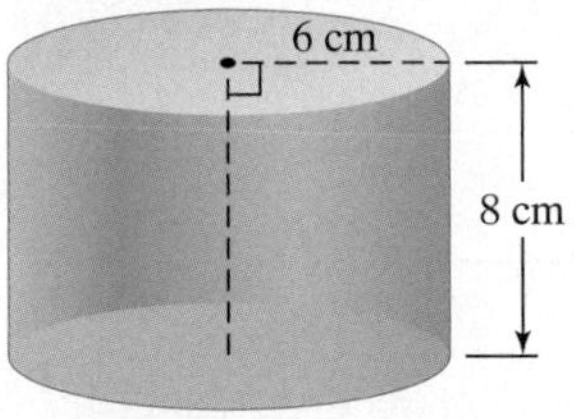

23.

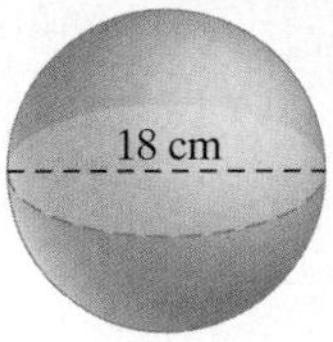

24.

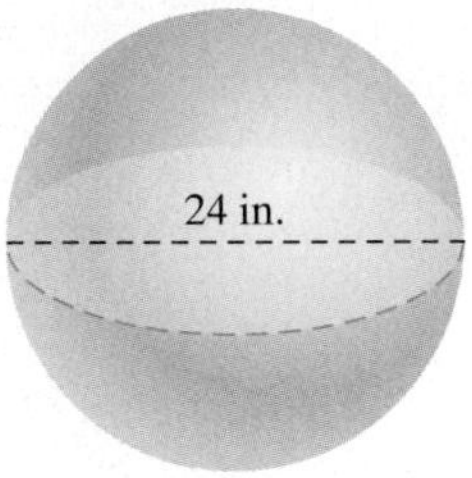

25.

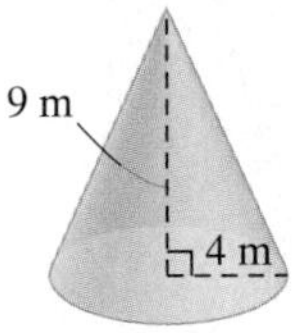

26.

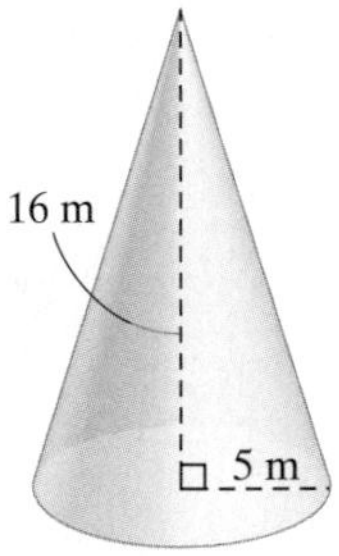

27. Solve the formula for the volume of a circular cylinder for h.

28. Solve the formula for the volume of a cone for h.

29. A cylinder with radius 3 inches and height 4 inches has its radius tripled. How many times greater is the volume of the larger cylinder than the smaller cylinder?

30. A cylinder with radius 2 inches and height 3 inches has its radius quadrupled. How many times greater is the volume of the larger cylinder than the smaller cylinder?

Use the relationship among the three angles of any triangle to solve Exercises 31–36.

31. Two angles of a triangle have the same measure and the third angle is 30° greater than the measure of the other two. Find the measure of each angle.

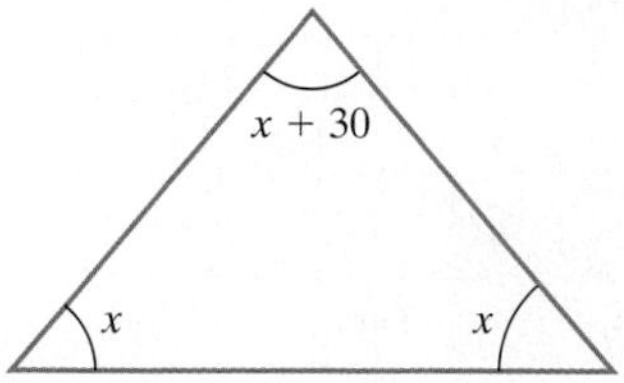

32. One angle of a triangle is three times as large as another. The measure of the third angle is 40° more than that of the smallest angle. Find the measure of each angle.

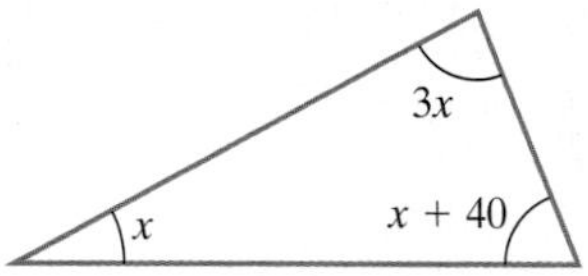

Find the measure of each angle whose degree measure is represented in terms of x in the triangles in Exercises 33–34.

33.

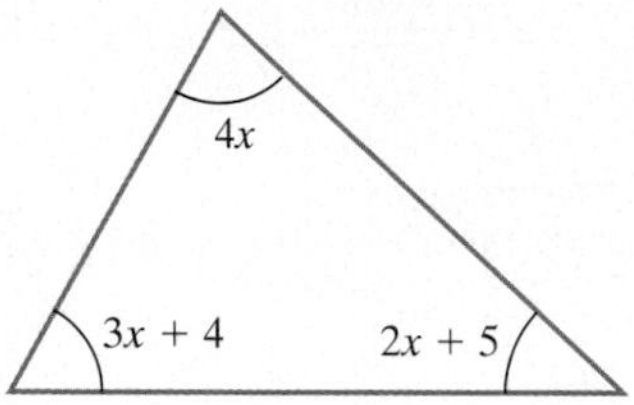

34.

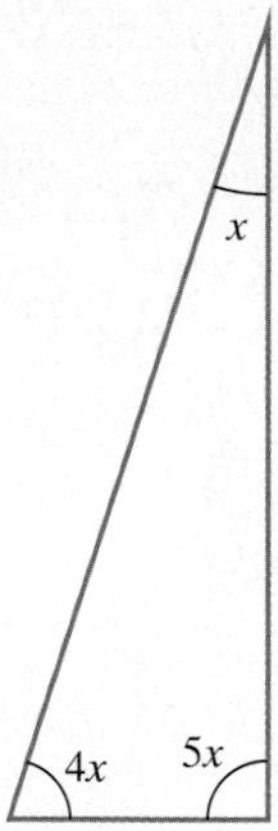

35. One angle of a triangle is twice as large as another. The measure of the third angle is 20° more than that of the smallest angle. Find the measure of each angle.

36. One angle of a triangle is three times as large as another. The measure of the third angle is 30° greater than that of the smallest angle. Find the measure of each angle.

In Exercises 37–40, find the measure of the complement of each angle.

37. 58° **38.** 41° **39.** 88° **40.** 2°

In Exercises 41–44, find the measure of the supplement of each angle.

41. 132° **42.** 93° **43.** 90° **44.** 179.5°

In Exercises 45–50, use the five-step problem-solving strategy to find the measure of the angle described.

45. The angle's measure is 60° more than that of its complement.

46. The angle's measure is 78° less than that of its complement.

47. The angle's measure is three times that of its supplement.

48. The angle's measure is 16° more than triple that of its supplement.

49. The measure of the angle's supplement is 10° more than three times that of its complement.

50. The measure of the angle's supplement is 52° more than twice that of its complement.

Practice Plus

In Exercises 51–53, find the area of each figure.

51.

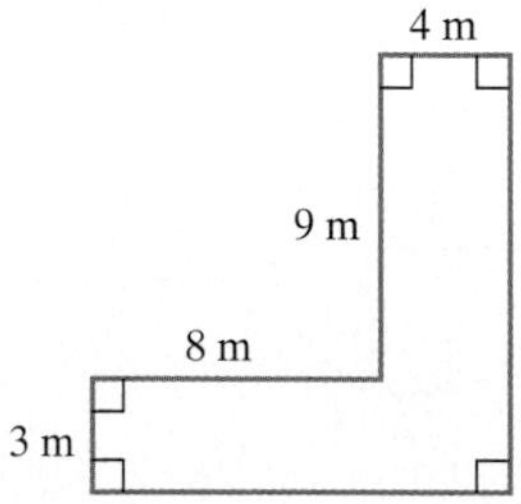

52.

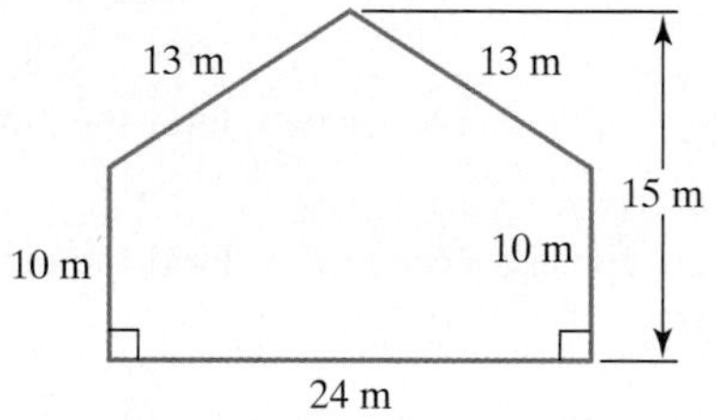

53.

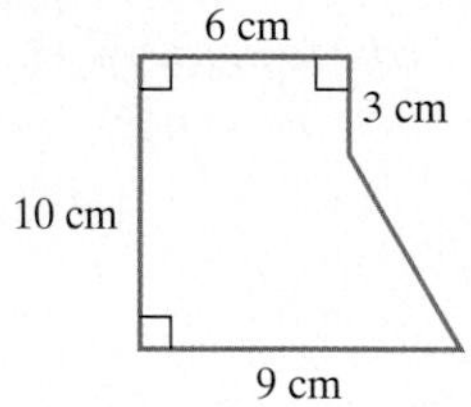

54. Find the area of the shaded region in terms of π.

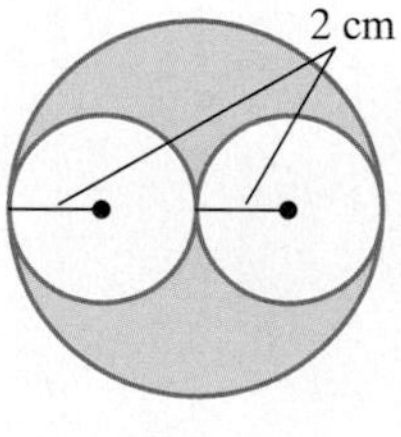

In Exercises 55–56, find the volume of the darkly shaded region. In Exercise 55, use the fact that the volume of a pyramid is $\frac{1}{3}$ the volume of a rectangular solid having the same base and the same height. In Exercise 56, express the answer in terms of π.

55.

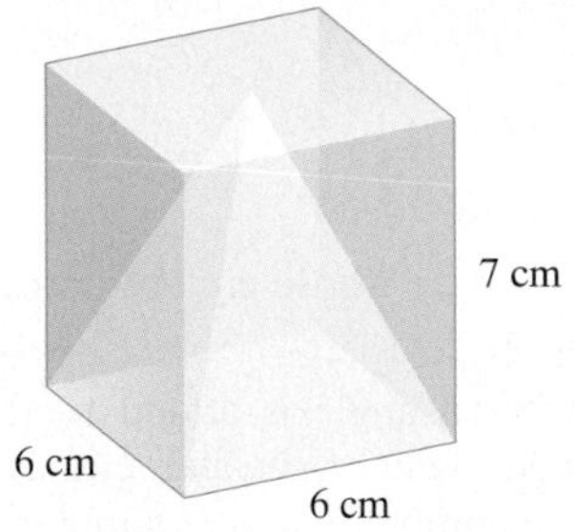

56.

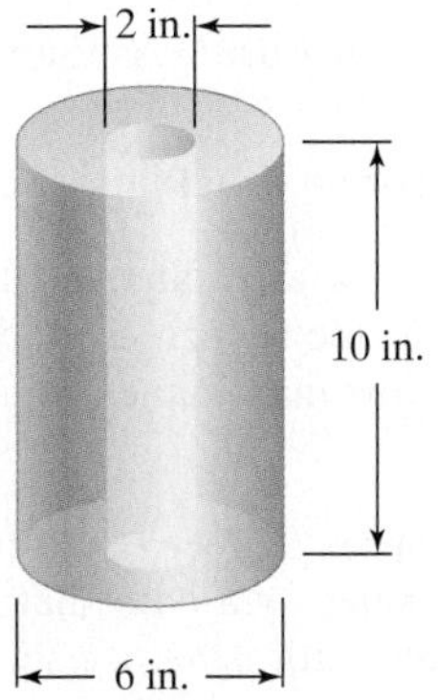

Application Exercises

Use the formulas for perimeter and area in Table 2.3 on page 149 to solve Exercises 57–58.

57. Taxpayers with an office in their home may deduct a percentage of their home-related expenses. This percentage is based on the ratio of the office's area to the area of the home. A taxpayer with an office in a 2200-square-foot home maintains a 20-foot by 16-foot office. If the yearly electricity bills for the home come to $4800, how much of this is deductible?

58. The lot in the figure shown, except for the house, shed, and driveway, is lawn. One bag of lawn fertilizer costs $25.00 and covers 4000 square feet.

a. Determine the minimum number of bags of fertilizer needed for the lawn.

b. Find the total cost of the fertilizer.

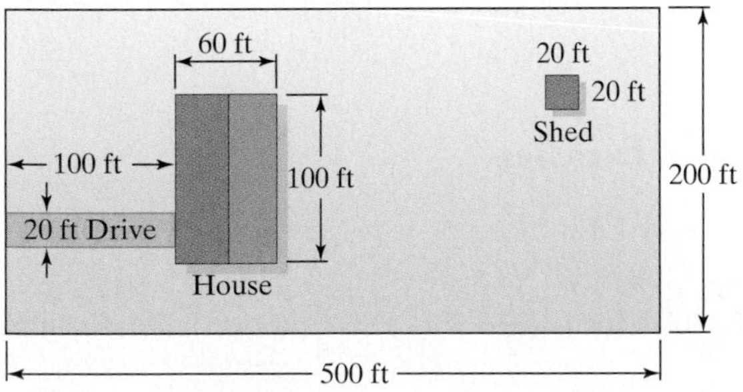

Use the formulas for the area and the circumference of a circle in Table 2.4 on page 150 to solve Exercises 59–64. Round all circumference and area calculations to the nearest whole number.

59. Which one of the following is a better buy: a large pizza with a 14-inch diameter for $12.00 or a medium pizza with a 7-inch diameter for $5.00?

60. Which one of the following is a better buy: a large pizza with a 16-inch diameter for $12.00 or two small pizzas, each with a 10-inch diameter, for $12.00?

61. If asphalt pavement costs $0.80 per square foot, find the cost to pave the circular road in the figure shown. Round to the nearest dollar.

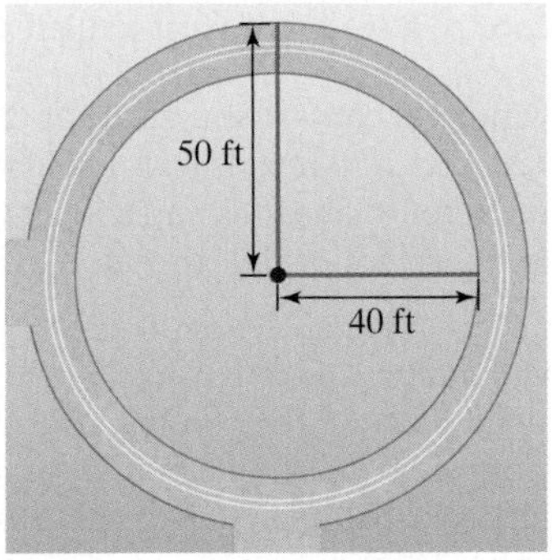

62. Hardwood flooring costs $10.00 per square foot. How much will it cost (to the nearest dollar) to cover the dance floor shown in the figure with hardwood flooring?

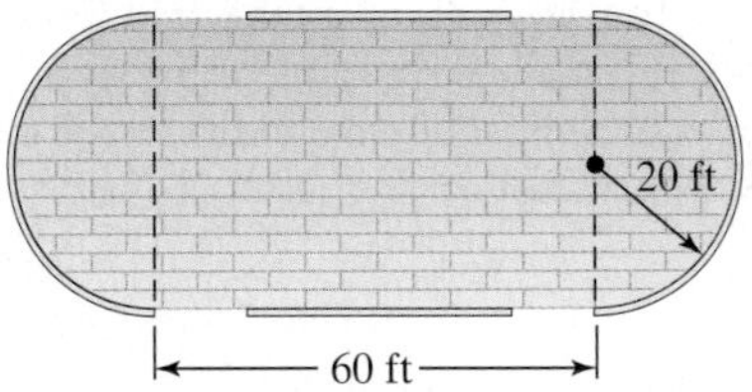

63. A glass window is to be placed in a house. The window consists of a rectangle, 6 feet high by 3 feet wide, with a semicircle at the top. Approximately how many feet of stripping, to the nearest tenth of a foot, will be needed to frame the window?

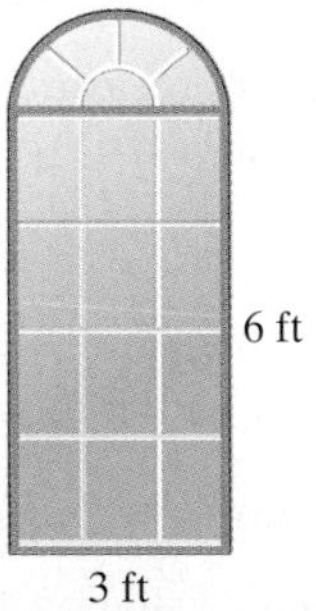

64. How many plants spaced every 6 inches are needed to surround a circular garden with a 30-foot radius?

Use the formulas for volume in Table 2.5 on page 152 to solve Exercises 65–69. When necessary, round all volume calculations to the nearest whole number.

65. A water reservoir is shaped like a rectangular solid with a base that is 50 yards by 30 yards, and a vertical height of 20 yards. At the start of a three-month period of no rain, the reservoir was completely full. At the end of this period, the height of the water was down to 6 yards. How much water was used in the three-month period?

66. A building contractor is to dig a foundation 4 yards long, 3 yards wide, and 2 yards deep for a toll booth's foundation. The contractor pays $10 per load for trucks to remove the dirt. Each truck holds 6 cubic yards. What is the cost to the contractor to have all the dirt hauled away?

67. Two cylindrical cans of soup sell for the same price. One can has a diameter of 6 inches and a height of 5 inches. The other has a diameter of 5 inches and a height of 6 inches. Which can contains more soup and, therefore, is the better buy?

68. The tunnel under the English Channel that connects England and France is one of the world's longest tunnels. The Chunnel, as it is known, consists of three separate tunnels built side by side. Each is a half-cylinder that is 50,000 meters long and 4 meters high. How many cubic meters of dirt had to be removed to build the Chunnel?

69. You are about to sue your contractor who promised to install a water tank that holds 500 gallons of water. You know that 500 gallons is the capacity of a tank that holds 67 cubic feet. The cylindrical tank has a radius of 3 feet and a height of 2 feet 4 inches. Does the evidence indicate you can win the case against the contractor if it goes to court?

Writing in Mathematics

70. Using words only, describe how to find the area of a triangle.

71. Describe the difference between the following problems: How much fencing is needed to enclose a garden? How much fertilizer is needed for the garden?

72. Describe how volume is measured. Explain why linear or square units cannot be used.

73. What is an angle?

74. If the measures of two angles of a triangle are known, explain how to find the measure of the third angle.

75. Can a triangle contain two 90° angles? Explain your answer.

76. What are complementary angles? Describe how to find the measure of an angle's complement.

77. What are supplementary angles? Describe how to find the measure of an angle's supplement?

78. Describe an application of a geometric formula involving area or volume.

79. Write and solve an original problem involving the measures of the three angles of a triangle.

Critical Thinking Exercises

80. Which one of the following is true?

a. It is not possible to have a circle whose circumference is numerically equal to its area.

b. When the measure of a given angle is added to three times the measure of its complement, the sum equals the sum of the measures of the complement and supplement of the angle.

c. The complement of an angle that measures less than 90° is an angle that measures more than 90°.

d. Two complementary angles cannot be equal in measure.

81. Suppose you know the cost for building a rectangular deck measuring 8 feet by 10 feet. If you decide to increase the dimensions to 12 feet by 15 feet, by how many times will the cost increase?

82. A rectangular swimming pool measures 14 feet by 30 feet. The pool is surrounded on all four sides by a path that is 3 feet wide. If the cost to resurface the path is $2 per square foot, what is the total cost of resurfacing the path?

83. What happens to the volume of a sphere if its radius is doubled?

84. A scale model of a car is constructed so that its length, width, and height are each $\frac{1}{10}$ the length, width, and height of the actual car. By how many times does the volume of the car exceed its scale model?

85. Find the measure of the angle of inclination, denoted by x in the figure, for the road leading to the bridge.

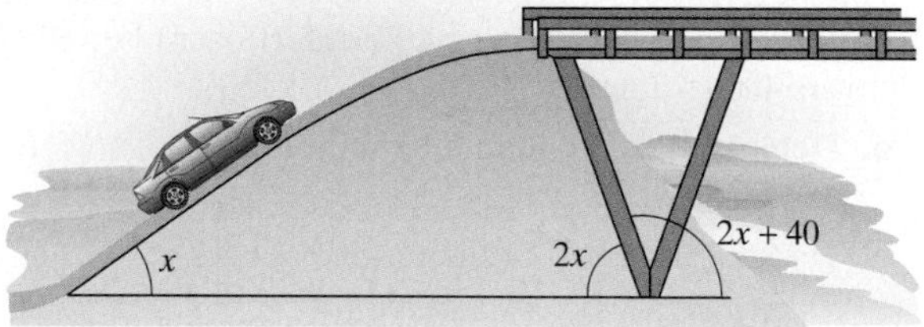

Review Exercises

86. Solve for s: $P = 2s + b$. (Section 2.4, Example 3)

87. Solve for x: $\frac{x}{2} + 7 = 13 - \frac{x}{4}$. (Section 2.3, Example 4)

88. Simplify: $[3(12 \div 2^2 - 3)^2]^2$. (Section 1.8, Example 8)

SECTION 2.7

Objectives

1. Graph the solutions of an inequality on a number line.
2. Use set-builder notation.
3. Understand properties used to solve linear inequalities.
4. Solve linear inequalities.
5. Identify inequalities with no solution or infinitely many solutions.
6. Solve problems using linear inequalities.

SOLVING LINEAR INEQUALITIES

Do you remember Rent-a-Heap, the car rental company that charged \$125 per week plus \$0.20 per mile to rent a small car? In Example 5 on page 143 we asked the question: How many miles can you travel for \$335? We let x represent the number of miles and set up a linear equation as follows:

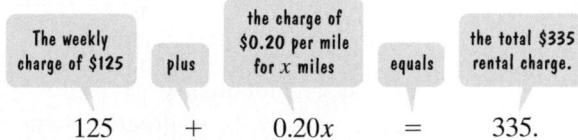

The weekly charge of \$125	plus	the charge of \$0.20 per mile for x miles	equals	the total \$335 rental charge.
125	$+$	$0.20x$	$=$	335.

Because we are limited by how much money we can spend on everything from buying clothing to renting a car, it is also possible to ask: How many miles can you travel if you can spend *at most* \$335? We again let x represent the number of miles. Spending *at most* \$335 means that the amount spent on the weekly rental must be *less than or equal to* \$335:

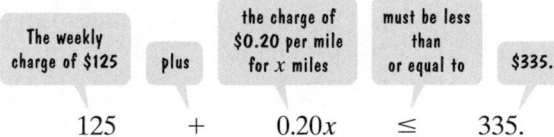

The weekly charge of \$125	plus	the charge of \$0.20 per mile for x miles	must be less than or equal to	\$335.
125	$+$	$0.20x$	$\le$	335.

Using the commutative property of addition, we can express this inequality as

$$0.20x + 125 \le 335.$$

The form of this inequality is $ax + b \le c$, with $a = 0.20$, $b = 125$, and $c = 335$. Any inequality in this form is called a **linear inequality in one variable**. The symbol between $ax + b$ and c can be $\le$ (is less than or equal to), $<$ (is less than), $\ge$ (is greater than or equal to), or $>$ (is greater than). The greatest exponent on the variable in such an inequality is 1.

In this section, we will study how to solve linear inequalities such as $0.20x + 125 \le 335$. **Solving an inequality** is the process of finding the set of numbers that will make the inequality a true statement. These numbers are called the **solutions** of the inequality, and we say that they **satisfy** the inequality. The set of all solutions is called the **solution set** of the inequality. We begin by discussing how to graph and how to represent these solution sets.

1 Graph the solutions of an inequality on a number line.

Graphs of Inequalities There are infinitely many solutions to the inequality $x < 3$, namely, all real numbers that are less than 3. Although we cannot list all the solutions, we can make a drawing on a number line that represents these solutions. Such a drawing is called the **graph of the inequality**.

Graphs of solutions to linear inequalities are shown on a number line by shading all points representing numbers that are solutions. *Open dots* (∘) indicate endpoints that are *not solutions* and *closed dots* (•) indicate endpoints that *are solutions.*

EXAMPLE 1 Graphing Inequalities

Graph the solutions of each inequality: **a.** $x < 3$ **b.** $x \geq -1$ **c.** $-1 < x \leq 3$.

SOLUTION

a. The solutions of $x < 3$ are all real numbers that are less than 3. They are graphed on a number line by shading all points to the left of 3. The open dot at 3 indicates that 3 is not a solution, but numbers such as 2.9999 and 2.6 are. The arrow shows that the graph extends indefinitely to the left.

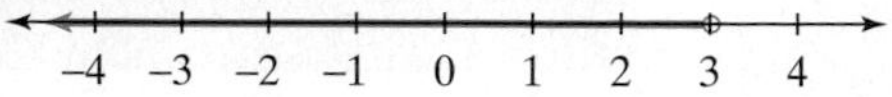

b. The solutions of $x \geq -1$ are all real numbers that are greater than or equal to -1. We shade all points to the right of -1 and the point for -1 itself. The closed dot at -1 shows that -1 is a solution of the given inequality. The arrow shows that the graph extends indefinitely to the right.

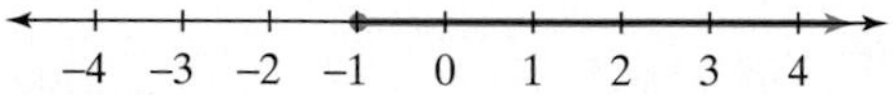

c. The inequality $-1 < x \leq 3$ is read "-1 is less than x *and* x is less than or equal to 3," or "x is greater than -1 *and* less than or equal to 3." The solutions of $-1 < x \leq 3$ are all real numbers between -1 and 3, not including -1 but including 3. The open dot at -1 indicates that -1 is not a solution. The closed dot at 3 shows that 3 is a solution. Shading indicates the other solutions.

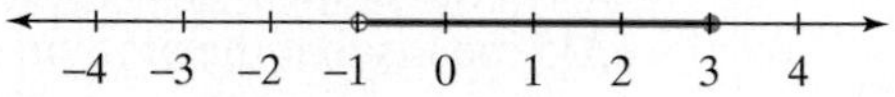

■

STUDY TIP

Because an inequality symbol points to the smaller number, $x < 3$ (x is less than 3) may be expressed as $3 > x$ (3 is greater than x).

✔ **CHECK POINT 1** Graph the solutions of each inequality:

a. $x < 4$ **b.** $x \geq -2$ **c.** $-4 \leq x < 1$.

2 Use set-builder notation.

Solution Sets The solutions of $x < 3$ are all real numbers that are less than 3. We can use the set concept introduced in Chapter 1 and state that the solution is the *set of all real numbers less than 3.* We use **set-builder notation** to write the solution set of $x < 3$ as

$$\{x|x < 3\}.$$

We read this as "the set of all x such that x is less than 3." Solutions of inequalities should be expressed in set-builder notation.

3 Understand properties used to solve linear inequalities.

Properties Used to Solve Linear Inequalities Back to our question: How many miles can you drive your Rent-a-Heap car if you can spend at most \$335 per week? We answer the question by solving

$$0.20x + 125 \leq 335$$

for x. The solution procedure is nearly identical to that for solving

$$0.20x + 125 = 335.$$

Our goal is to get x by itself on the left side. We do this by subtracting 125 from both sides to isolate $0.20x$:

$0.20x + 125 \leq 335$	This is the given inequality.
$0.20x + 125 - 125 \leq 335 - 125$	Subtract 125 from both sides.
$0.20x \leq 210.$	Simplify.

Finally, we isolate x from $0.20x$ by dividing both sides of the inequality by 0.20:

$$\frac{0.20x}{0.20} \leq \frac{210}{0.20} \quad \text{Divide both sides by 0.20.}$$

$$x \leq 1050. \quad \text{Simplify.}$$

With at most \$335 per week to spend, you can travel at most 1050 miles.

We started with the inequality $0.20x + 125 \leq 335$ and obtained the inequality $x \leq 1050$ in the final step. Both of these inequalities have the same solution set, namely $\{x | x \leq 1050\}$. Inequalities such as these, with the same solution set, are said to be **equivalent**.

We isolated x from $0.20x$ by dividing both sides of $0.20x \leq 210$ by 0.20, a positive number. Let's see what happens if we divide both sides of an inequality by a negative number. Consider the inequality $10 < 14$. Divide both 10 and 14 by -2:

$$\frac{10}{-2} = -5 \quad \text{and} \quad \frac{14}{-2} = -7.$$

Because -5 lies to the right of -7 on the number line, -5 is greater than -7:

$$-5 > -7.$$

Notice that the direction of the inequality symbol is reversed:

$$10 < 14$$
$$\updownarrow$$
$$-5 > -7$$

Dividing by -2 changes the direction of the inequality symbol.

In general, **when we multiply or divide both sides of an inequality by a negative number, the direction of the inequality symbol is reversed**. When we reverse the direction of the inequality symbol, we say that we change the *sense* of the inequality.

We can isolate a variable in a linear inequality the same way we can isolate a variable in a linear equation. The following properties are used to create equivalent inequalities:

STUDY TIP

English phrases such as "at least" and "at most" can be represented by inequalities.

English Sentence	Inequality
x is at least 5.	$x \geq 5$
x is at most 5.	$x \leq 5$
x is between 5 and 7.	$5 < x < 7$
x is no more than 5.	$x \leq 5$
x is no less than 5.	$x \geq 5$

Properties of Inequalities

Property	The Property in Words	Example
The Addition Property of Inequality If $a < b$, then $a + c < b + c$. If $a < b$, then $a - c < b - c$.	If the same quantity is added to or subtracted from both sides of an inequality, the resulting inequality is equivalent to the original one.	$2x + 3 < 7$ Subtract 3: $2x + 3 - 3 < 7 - 3$ Simplify: $2x < 4$
The Positive Multiplication Property of Inequality If $a < b$ and c is positive, then $ac < bc$. If $a < b$ and c is positive, then $\frac{a}{c} < \frac{b}{c}$.	If we multiply or divide both sides of an inequality by the same positive quantity, the resulting inequality is equivalent to the original one.	$2x < 4$ Divide by 2: $\frac{2x}{2} < \frac{4}{2}$ Simplify: $x < 2$
The Negative Multiplication Property of Inequality If $a < b$ and c is negative, then $ac > bc$. If $a < b$ and c is negative, then $\frac{a}{c} > \frac{b}{c}$.	If we multiply or divide both sides of an inequality by the same negative quantity and reverse the direction of the inequality symbol, the resulting inequality is equivalent to the original one.	$-4x < 20$ Divide by -4 and reverse the sense of the inequality: $\frac{-4x}{-4} > \frac{20}{-4}$ Simplify: $x > -5$

4 Solve linear inequalities.

Solving Linear Inequalities Involving Only One Property of Inequality If you can solve a linear equation, it is likely that you can solve a linear inequality. Why? The procedure for solving linear inequalities is nearly the same as the procedure for solving linear equations, with one important exception: **When multiplying or dividing by a negative number, reverse the direction of the inequality symbol, changing the sense of the inequality.**

DISCOVER FOR YOURSELF

Can you check all solutions to Example 2 in the given inequality? Is a partial check possible? Select a real number that is less than 5 and show that it satisfies $x + 3 < 8$.

EXAMPLE 2 Solving a Linear Inequality

Solve and graph the solution set on a number line:

$$x + 3 < 8.$$

SOLUTION Our goal is to isolate x. We can do this by using the addition property, subtracting 3 from both sides.

$x + 3 < 8$	This is the given inequality.
$x + 3 - 3 < 8 - 3$	Subtract 3 from both sides.
$x < 5$	Simplify.

The solution set consists of all real numbers that are less than 5. We express this in set-builder notation as

$\{x \mid x < 5\}$. This is read "the set of all x such that x is less than 5."

The graph of the solution set is shown as follows:

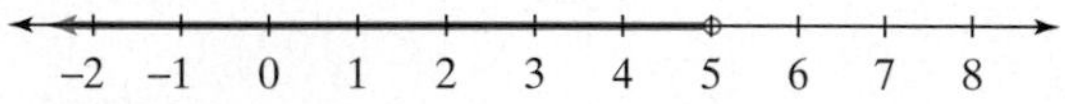

■

CHECK POINT 2 Solve and graph the solution set on a number line:

$$x + 6 < 9.$$

EXAMPLE 3 Solving a Linear Inequality

Solve and graph the solution set on a number line:

$$4x - 1 \geq 3x - 6.$$

SOLUTION Our goal is to isolate all terms involving x on one side and all numerical terms on the other side, exactly as we did when solving equations. Let's begin by using the addition property to isolate variable terms on the left.

$4x - 1 \geq 3x - 6$	This is the given inequality.
$4x - 3x - 1 \geq 3x - 3x - 6$	Subtract 3x from both sides.
$x - 1 \geq -6$	Simplify.

Now we isolate the numerical terms on the right. Use the addition property and add 1 to both sides.

$x - 1 + 1 \geq -6 + 1$	Add 1 to both sides.
$x \geq -5$	Simplify.

The solution set consists of all real numbers that are greater than or equal to −5. We express this in set-builder notation as

$\{x \mid x \geq -5\}$. This is read "the set of all x such that x is greater than or equal to −5."

The graph of the solution set is shown as follows:

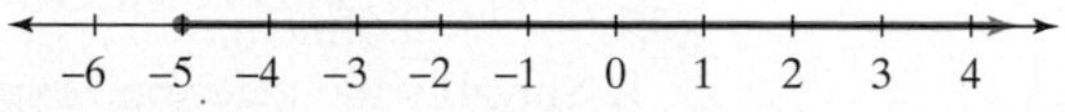

■

CHECK POINT 3 Solve and graph the solution set on a number line:

$$8x - 2 \geq 7x - 4.$$

We solved the inequalities in Examples 2 and 3 using the addition property of inequality. Now let's practice using the multiplication property of inequality. Do not forget to reverse the direction of the inequality symbol when multiplying or dividing both sides by a negative number.

EXAMPLE 4 Solving Linear Inequalities

Solve and graph the solution set on a number line: **a.** $\frac{1}{3}x < 5$ **b.** $-3x < 21$.

SOLUTION In each case, our goal is to isolate x. In the first inequality, this is accomplished by multiplying both sides by 3. In the second inequality, we can do this by dividing both sides by -3.

a. $\frac{1}{3}x < 5$ This is the given inequality.

$3 \cdot \frac{1}{3}x < 3 \cdot 5$ Isolate x by multiplying by 3 on both sides. The symbol $<$ stays the same because we are multiplying both sides by a positive number.

$x < 15$ Simplify.

The solution set is $\{x | x < 15\}$. The graph of the solution set is shown as follows:

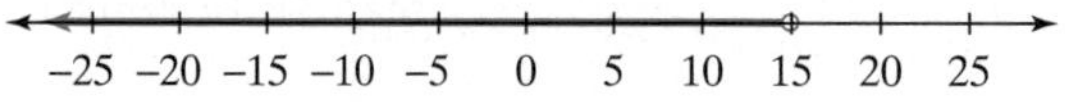

b. $-3x < 21$ This is the given inequality.

$\frac{-3x}{-3} > \frac{21}{-3}$ Isolate x by dividing by -3 on both sides. The symbol $<$ must be reversed because we are dividing both sides by a negative number.

$x > -7$ Simplify.

The solution set is $\{x | x > -7\}$. The graph of the solution set is shown as follows:

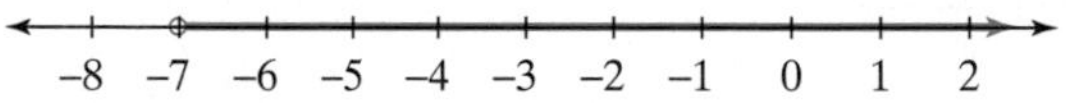

■

CHECK POINT 4 Solve and graph the solution set on a number line:

a. $\frac{1}{4}x < 2$ **b.** $-6x < 18$.

Inequalities Requiring Both the Addition and Multiplication Properties If an inequality does not contain fractions, it can be solved using the following procedure. Notice, again, how similar this procedure is to the procedure for solving an equation.

SOLVING A LINEAR INEQUALITY

1. Simplify the algebraic expression on each side.
2. Use the addition property of inequality to collect all the variable terms on one side and all the constant terms on the other side.
3. Use the multiplication property of inequality to isolate the variable and solve. Reverse the sense of the inequality when multiplying or dividing both sides by a negative number.
4. Express the solution set in set-builder notation and graph the solution set on a number line.

EXAMPLE 5 Solving a Linear Inequality

Solve and graph the solution set on a number line:

$$4y - 7 \geq 5.$$

SOLUTION

Step 1. **Simplify each side.** Because each side is already simplified, we can skip this step.

Step 2. **Collect variable terms on one side and constant terms on the other side.** The variable term, $4y$, is already on the left. We will collect constant terms on the right by adding 7 to both sides.

$$4y - 7 \geq 5 \quad \text{This is the given inequality.}$$

$$4y - 7 + 7 \geq 5 + 7 \quad \text{Add 7 to both sides.}$$

$$4y \geq 12 \quad \text{Simplify.}$$

Step 3. **Isolate the variable and solve.** We isolate the variable, y, by dividing both sides by 4. Because we are dividing by a positive number, we do not reverse the inequality symbol.

$$\frac{4y}{4} \geq \frac{12}{4} \quad \text{Divide both sides by 4.}$$

$$y \geq 3 \quad \text{Simplify.}$$

Step 4. **Express the solution set in set-builder notation and graph the set on a number line.** The solution set consists of all real numbers that are greater than or equal to 3, expressed in set-builder notation as $\{y|y \geq 3\}$. The graph of the solution set is shown as follows:

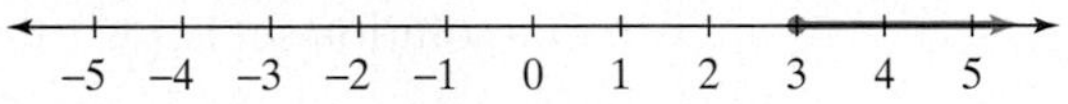

■

STUDY TIP

It is possible to perform a partial check for an inequality. Select one number from the solution set. Substitute that number into the original inequality and perform the resulting computations. You should obtain a true statement.

✔ **CHECK POINT 5** Solve and graph the solution set on a number line:

$$5y - 3 \geq 17.$$

EXAMPLE 6 Solving a Linear Inequality

Solve and graph the solution set on a number line:

$$7x + 15 \geq 13x + 51.$$

STUDY TIP

You can solve

$$7x + 15 \geq 13x + 51$$

by isolating x on the right side. Subtract $7x$ from both sides:

$$7x + 15 - 7x \geq 13x + 51 - 7x$$
$$15 \geq 6x + 51.$$

Now subtract 51 from both sides:

$$15 - 51 \geq 6x + 51 - 51$$
$$-36 \geq 6x.$$

Finally, divide both sides by 6:

$$\frac{-36}{6} \geq \frac{6x}{6}$$
$$-6 \geq x.$$

This last inequality means the same thing as

$$x \leq -6.$$

SOLUTION

Step 1. **Simplify each side.** Because each side is already simplified, we can skip this step.

Step 2. **Collect variable terms on one side and constant terms on the other side.** We will collect variable terms on the left and constant terms on the right.

$7x + 15 \geq 13x + 51$	This is the given inequality.
$7x + 15 - 13x \geq 13x + 51 - 13x$	Subtract 13x from both sides.
$-6x + 15 \geq 51$	Simplify.
$-6x + 15 - 15 \geq 51 - 15$	Subtract 15 from both sides.
$-6x \geq 36$	Simplify.

Step 3. **Isolate the variable and solve.** We isolate the variable, x, by dividing both sides by -6. Because we are dividing by a negative number, we must reverse the inequality symbol.

$\frac{-6x}{-6} \leq \frac{36}{-6}$	Divide both sides by −6 and reverse the sense of the inequality.
$x \leq -6$	Simplify.

Step 4. **Express the solution set in set-builder notation and graph the set on a number line.** The solution set consists of all real numbers that are less than or equal to -6, expressed in set-builder notation as $\{x \mid x \leq -6\}$. The graph of the solution set is shown as follows:

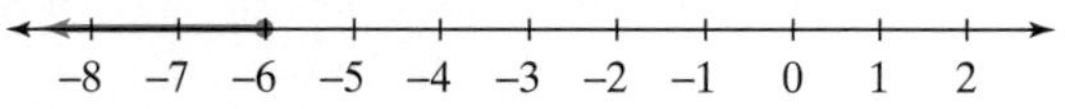

■

CHECK POINT 6 Solve and graph the solution set: $6 - 3x \leq 5x - 2$.

EXAMPLE 7 Solving a Linear Inequality

Solve and graph the solution set on a number line:

$$2(x - 3) + 5x \leq 8(x - 1).$$

SOLUTION

Step 1. **Simplify each side.** We use the distributive property to remove parentheses. Then we combine like terms.

$2(x - 3) + 5x \leq 8(x - 1)$	This is the given inequality.
$2x - 6 + 5x \leq 8x - 8$	Use the distributive property.
$7x - 6 \leq 8x - 8$	Add like terms on the left.

Step 2. **Collect variable terms on one side and constant terms on the other side.** We will collect variable terms on the left and constant terms on the right.

$7x - 8x - 6 \leq 8x - 8x - 8$	Subtract 8x from both sides.
$-x - 6 \leq -8$	Simplify.
$-x - 6 + 6 \leq -8 + 6$	Add 6 to both sides.
$-x \leq -2$	Simplify.

Step 3. **Isolate the variable and solve.** To isolate x in $-x \leq -2$, we must eliminate the negative sign in front of the x. Because $-x$ means $-1x$, we can do this by multiplying (or dividing) both sides of the inequality by -1. We are multiplying by a negative number. Thus, we must reverse the inequality symbol.

$$(-1)(-x) \geq (-1)(-2)$$ Multiply both sides of $-x \leq -2$ by -1 and reverse the sense of the inequality.

$$x \geq 2$$ Simplify.

Step 4. **Express the solution set in set-builder notation and graph the set on a number line.** The solution set consists of all real numbers that are greater than or equal to 2, expressed in set-builder notation as $\{x|x \geq 2\}$. The graph of the solution set is shown as follows:

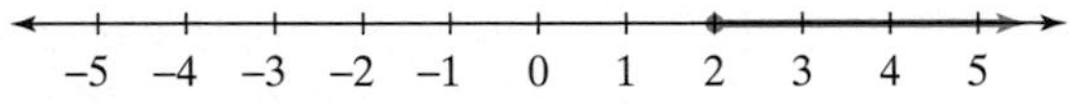

■

✔ CHECK POINT **7** Solve and graph the solution set:

$$2(x - 3) - 1 \leq 3(x + 2) - 14.$$

5 Identify inequalities with no solution or infinitely many solutions.

Inequalities with Unusual Solution Sets We have seen that some equations have no solution. This is also true for some inequalities. An example of such an inequality is

$$x > x + 1.$$

There is no number that is greater than itself plus 1. This inequality has no solution. Its solution set is $\emptyset$, the empty set.

By contrast, some inequalities are true for all real numbers. An example of such an inequality is

$$x < x + 1.$$

Every real number is less than itself plus 1. The solution set is $\{x|x \text{ is a real number}\}$ or $\mathbb{R}$.

If you attempt to solve an inequality that has no solution, you will eliminate the variable and obtain a false statement, such as $0 > 1$. If you attempt to solve an inequality that is true for all real numbers, you will eliminate the variable and obtain a true statement, such as $0 < 1$.

EXAMPLE 8 Solving a Linear Inequality

Solve: $3(x + 1) > 3x + 5$.

SOLUTION

$$3(x + 1) > 3x + 5$$ This is the given inequality.

$$3x + 3 > 3x + 5$$ Apply the distributive property.

$$3x + 3 - 3x > 3x + 5 - 3x$$ Subtract $3x$ from both sides.

Keep reading. $3 > 5$ is not the solution.

$$3 > 5$$ Simplify.

The original inequality is equivalent to the false statement $3 > 5$, which is false for every value of x. The inequality has no solution. The solution set is $\emptyset$, the empty set. ■

CHECK POINT 8 Solve: $4(x + 2) > 4x + 15$.

EXAMPLE 9 Solving a Linear Inequality

Solve: $2(x + 5) \le 5x - 3x + 14$.

SOLUTION

$2(x + 5) \le 5x - 3x + 14$	This is the given inequality.
$2x + 10 \le 5x - 3x + 14$	Apply the distributive property.
$2x + 10 \le 2x + 14$	Combine like terms.
$2x + 10 - 2x \le 2x + 14 - 2x$	Subtract 2x from both sides.
$10 \le 14$	Simplify.

Keep reading. $10 \le 14$ is not the solution.

The original inequality is equivalent to the true statement $10 \le 14$, which is true for every value of x. The solution is the set of all real numbers, written

$$\{x | x \text{ is a real number}\} \text{ or } \mathbb{R}.$$ ■

CHECK POINT 9 Solve: $3(x + 1) \ge 2x + 1 + x$.

6 Solve problems using linear inequalities.

Applications As you know, different professors may use different grading systems to determine your final course grade. Some professors require a final examination; others do not. In our next example, a final exam is required *and* it also counts as two grades.

EXAMPLE 10 An Application: Final Course Grade

To earn an A in a course, you must have a final average of at least 90%. On the first four examinations, you have grades of 86%, 88%, 92%, and 84%. If the final examination counts as two grades, what must you get on the final to earn an A in the course?

SOLUTION We will use our five-step strategy for solving algebraic word problems.

Steps 1 and 2. **Represent unknown quantities in terms of x.** Let x = your grade on the final examination.

Step 3. **Write an inequality in x that describes the conditions.** The average of the six grades is found by adding the grades and dividing the sum by 6.

$$\text{Average} = \frac{86 + 88 + 92 + 84 + x + x}{6}$$

Because the final counts as two grades, the x (your grade on the final examination) is added twice. This is also why the sum is divided by 6.

To get an A, your average must be at least 90. This means that your average must be greater than or equal to 90.

Your average | must be greater than or equal to | 90.

$$\frac{86 + 88 + 92 + 84 + x + x}{6} \ge 90$$

Step 4. Solve the inequality and answer the problem's question.

$$\frac{86 + 88 + 92 + 84 + x + x}{6} \geq 90$$
This is the inequality for the given conditions.

$$\frac{350 + 2x}{6} \geq 90$$
Combine like terms in the numerator.

$$6\left(\frac{350 + 2x}{6}\right) \geq 6(90)$$
Multiply both sides by 6, clearing the fraction.

$$350 + 2x \geq 540$$
Multiply.

$$350 + 2x - 350 \geq 540 - 350$$
Subtract 350 from both sides.

$$2x \geq 190$$
Simplify.

$$\frac{2x}{2} \geq \frac{190}{2}$$
Divide both sides by 2.

$$x \geq 95$$
Simplify.

You must get at least 95% on the final examination to earn an A in the course.

Step 5. Check. We can perform a partial check by computing the average with any grade that is at least 95. We will use 96. If you get 96% on the final examination, your average is

$$\frac{86 + 88 + 92 + 84 + 96 + 96}{6} = \frac{542}{6} = 90\frac{1}{3}.$$

Because $90\frac{1}{3} > 90$, you earn an A in the course. ■

CHECK POINT 10 To earn a B in a course, you must have a final average of at least 80%. On the first three examinations, you have grades of 82%, 74%, and 78%. If the final examination counts as two grades, what must you get on the final to earn a B in the course?

2.7 EXERCISE SET

Student Solutions Manual | CD/Video | PH Math/Tutor Center | MathXL Tutorials on CD | MathXL® | MyMathLab | Interactmath.com

Practice Exercises

In Exercises 1–12, graph the solutions of each inequality on a number line.

1. $x > 5$

2. $x > -3$

3. $x < -2$

4. $x < 0$

5. $x \geq -4$

6. $x \geq -6$

7. $x \leq 4.5$

8. $x \leq 7.5$

9. $-2 < x \leq 6$

10. $-3 \leq x < 6$

11. $-1 < x < 3$

12. $-2 \leq x \leq 0$

Describe each graph in Exercises 13–18 using set-builder notation.

13. 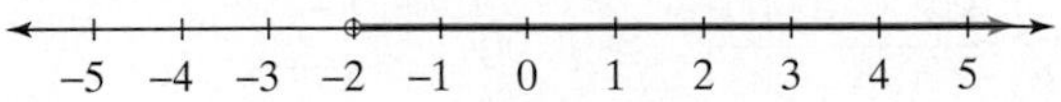

14. 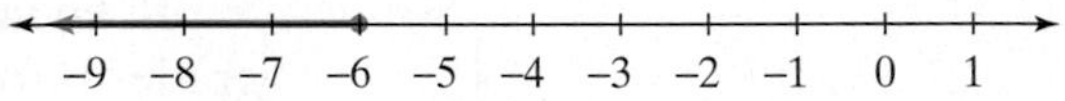

15. 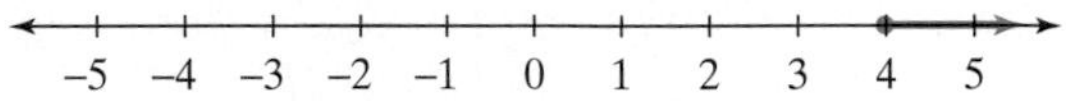

16. 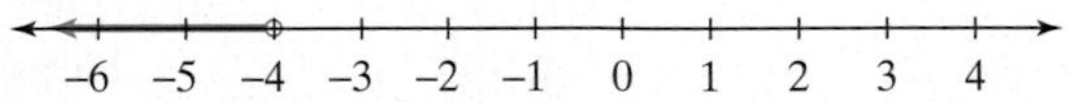

17. 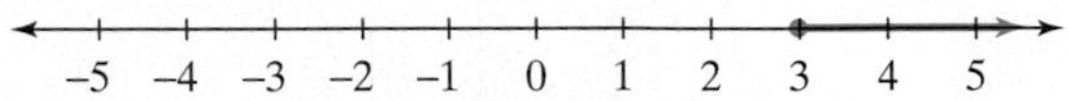

18. 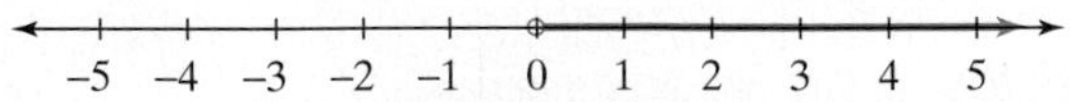

Use the addition property of inequality to solve each inequality in Exercises 19–36. Express the solution set in set-builder notation and graph the set on a number line.

19. $x - 3 > 4$

20. $x + 1 < 6$

21. $x + 4 \leq 10$

22. $x - 5 \geq 2$

23. $y - 2 < 0$

24. $y + 3 \geq 0$

25. $3x + 4 \leq 2x + 7$

26. $2x + 9 \leq x + 2$

27. $5x - 9 < 4x + 7$

28. $3x - 8 < 2x + 11$

29. $7x - 7 > 6x - 3$

30. $8x - 9 > 7x - 3$

31. $x - \frac{2}{3} > \frac{1}{2}$

32. $x - \frac{1}{3} \geq \frac{5}{6}$

33. $y + \frac{7}{8} \leq \frac{1}{2}$

34. $y + \frac{1}{3} \leq \frac{3}{4}$

35. $-15y + 13 > 13 - 16y$

36. $-12y + 17 > 20 - 13y$

Use the multiplication property of inequality to solve each inequality in Exercises 37–54. Express the solution set in set-builder notation and graph the set on a number line.

37. $\frac{1}{2}x < 4$

38. $\frac{1}{2}x > 3$

39. $\frac{x}{3} > -2$

40. $\frac{x}{4} < -1$

41. $4x < 20$

42. $6x < 18$

43. $3x \geq -21$

44. $7x \geq -56$

45. $-3x < 15$

46. $-7x > 21$

47. $-3x \geq 15$

48. $-7x \leq 21$

49. $-16x > -48$

50. $-20x > -140$

51. $-4y \leq \frac{1}{2}$

52. $-2y \leq \frac{1}{2}$

53. $-x < 4$

54. $-x > -3$

Use both the addition and multiplication properties of inequality to solve each inequality in Exercises 55–78. Express the solution set in set-builder notation and graph the set on a number line.

55. $2x - 3 > 7$

56. $3x + 2 \leq 14$

57. $3x + 3 < 18$

58. $8x - 4 > 12$

59. $3 - 7x \leq 17$

60. $5 - 3x \geq 20$

61. $-2x - 3 < 3$

62. $-3x + 14 < 5$

63. $5 - x \leq 1$

64. $3 - x \geq -3$

65. $2x - 5 > -x + 6$

66. $6x - 2 \geq 4x + 6$

67. $2y - 5 < 5y - 11$

68. $4y - 7 > 9y - 2$

69. $3(2y - 1) < 9$

70. $4(2y - 1) > 12$

71. $3(x + 1) - 5 < 2x + 1$

72. $4(x + 1) + 2 \geq 3x + 6$

73. $8x + 3 > 3(2x + 1) - x + 5$

74. $7 - 2(x - 4) < 5(1 - 2x)$

75. $\frac{x}{3} - 2 \geq 1$

76. $\frac{x}{4} - 3 \geq 1$

77. $1 - \frac{x}{2} > 4$

78. $1 - \frac{x}{2} < 5$

In Exercises 79–88, solve each inequality. Identify inequalities that have no solution, or inequalities that are true for all real numbers.

79. $4x - 4 < 4(x - 5)$

80. $3x - 5 < 3(x - 2)$

81. $x + 3 < x + 7$

82. $x + 4 < x + 10$

83. $7x \leq 7(x - 2)$

84. $3x + 1 \leq 3(x - 2)$

85. $2(x + 3) > 2x + 1$

86. $5(x + 4) > 5x + 10$

87. $5x - 4 \leq 4(x - 1)$

88. $6x - 3 \leq 3(x - 1)$

Practice Plus

In Exercises 89–92, use properties of inequality to rewrite each inequality so that x is isolated on one side.

89. $3x + a > b$

90. $-2x - a \leq b$

91. $y \leq mx + b$ and $m < 0$

92. $y > mx + b$ and $m > 0$

We know that $|x|$ represents the distance from 0 to x on a number line. In Exercises 93–96, use each sentence to describe all possible locations of x on a number line. Then rewrite the given sentence as an inequality involving $|x|$.

93. The distance from 0 to x on a number line is less than 2.

94. The distance from 0 to x on a number line is less than 3.

95. The distance from 0 to x on a number line is greater than 2.

96. The distance from 0 to x on a number line is greater than 3.

Application Exercises

The bar graph shows the percentage of wages full-time workers pay in income tax in eight selected countries. (The percents shown are averages for single earners without children.) Let x represent the percentage of wages workers pay in income tax. In Exercises 97–102, write the name of the country or countries described by the given inequality.

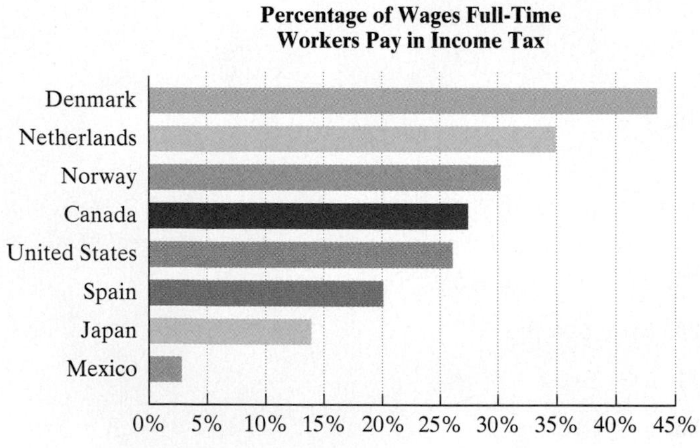

Source: *The Washington Post*

97. $x \geq 30\%$

98. $x > 30\%$

99. $x < 20\%$

100. $x \leq 20\%$

101. $25\% \leq x < 40\%$

102. $5\% < x \leq 25\%$

The line graph shows the declining consumption of cigarettes in the United States. The data shown by the graph can be modeled by

$$N = 550 - 9x,$$

where N is the number of cigarettes consumed, in billions, x years after 1988. Use this formula to solve Exercises 103–104.

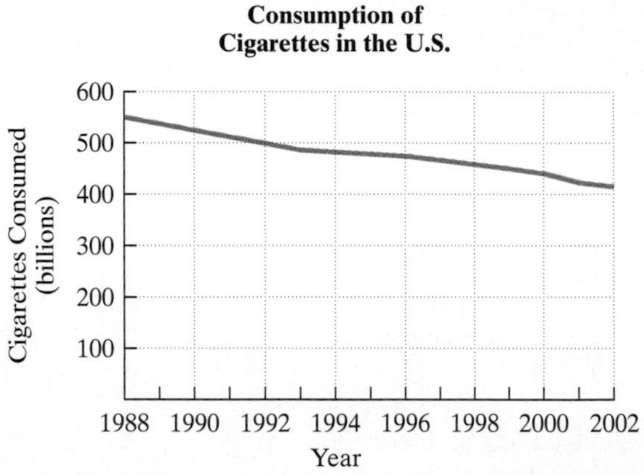

Source: Economic Research Service, USDA

103. Describe how many years after 1988 cigarette consumption will be less than 370 billion cigarettes each year. Which years are included in your description?

104. Describe how many years after 1988 cigarette consumption will be less than 325 billion cigarettes each year. Which years are included in your description?

105. On two examinations, you have grades of 86 and 88. There is an optional final examination, which counts as one grade. You decide to take the final in order to get a course grade of A, meaning a final average of at least 90.

a. What must you get on the final to earn an A in the course?

b. By taking the final, if you do poorly, you might risk the B that you have in the course based on the first two exam grades. If your final average is less than 80, you will lose your B in the course. Describe the grades on the final that will cause this to happen.

106. On three examinations, you have grades of 88, 78, and 86. There is still a final examination, which counts as one grade.

a. In order to get an A, your average must be at least 90. If you get 100 on the final, compute your average and determine if an A in the course is possible.

b. To earn a B in the course, you must have a final average of at least 80. What must you get on the final to earn a B in the course?

107. A car can be rented from Continental Rental for $80 per week plus 25 cents for each mile driven. How many miles can you travel if you can spend at most $400 for the week?

108. A car can be rented from Basic Rental for $60 per week plus 50 cents for each mile driven. How many miles can you travel if you can spend at most $600 for the week?

109. An elevator at a construction site has a maximum capacity of 3000 pounds. If the elevator operator weighs 245 pounds and each cement bag weighs 95 pounds, how many bags of cement can be safely lifted on the elevator in one trip?

110. An elevator at a construction site has a maximum capacity of 2800 pounds. If the elevator operator weighs 265 pounds and each cement bag weighs 65 pounds, how many bags of cement can be safely lifted on the elevator in one trip?

Writing in Mathematics

111. When graphing the solutions of an inequality, what is the difference between an open dot and a closed dot?

112. When solving an inequality, when is it necessary to change the direction of the inequality symbol? Give an example.

113. Describe ways in which solving a linear inequality is similar to solving a linear equation.

114. Describe ways in which solving a linear inequality is different from solving a linear equation.

115. Using current trends, future costs of Medicare can be modeled by $C = 18x + 250$, where x represents the number of years after 2000 and C represents the cost of Medicare, in billions of dollars. Use the formula to write a word problem that can be solved using a linear inequality. Then solve the problem.

Critical Thinking Exercises

116. Which one of the following statements is true?

a. The inequality $x - 3 > 0$ is equivalent to $x < 3$.

b. The statement "x is at most 5" is written $x < 5$.

c. The inequality $-4x < -20$ is equivalent to $x > -5$.

d. The statement "the sum of x and 6% of x is at least 80" is written $x + 0.06x \geq 80$.

117. A car can be rented from Basic Rental for $260 per week with no extra charge for mileage. Continental charges $80 per week plus 25 cents for each mile driven to rent the same car. How many miles should be driven in a week to make the rental cost for Basic Rental a better deal than Continental's?

118. Membership in a fitness club costs $500 yearly plus $1 per hour spent working out. A competing club charges $440 yearly plus $1.75 per hour for use of their equipment. How many hours must a person work out yearly to make membership in the first club cheaper than membership in the second club?

Technology Exercises

Solve each inequality in Exercises 119–120. Use a calculator to help with the arithmetic.

119. $1.45 - 7.23x > -1.442$

120. $126.8 - 9.4y \leq 4.8y + 34.5$

Review Exercises

121. 8 is 40% of what number? (Section 2.4, Example 8)

122. The length of a rectangle exceeds the width by 5 inches. The perimeter is 34 inches. What are the rectangle's dimensions? (Section 2.5, Example 6)

123. Solve and check: $5x + 16 = 3(x + 8)$. (Section 2.3, Example 2)

GROUP PROJECT

CHAPTER 2

One of the best ways to learn how to *solve* a word problem in algebra is to *design* word problems of your own. Creating a word problem makes you very aware of precisely how much information is needed to solve the problem. You must also focus on the best way to present information to a reader and on how much information to give. As you write your problem, you gain skills that will help you solve problems created by others.

The group should design five different word problems that can be solved using an algebraic equation. All of the problems should be on different topics. For example, the group should not have more than one problem on finding a number. The group should turn in both the problems and their algebraic solutions.

CHAPTER 2 SUMMARY

Definitions and Concepts	Examples
Section 2.1 The Addition Property of Equality	
A linear equation in one variable can be written in the form $ax + b = c$, where a is not zero.	$3x + 7 = 9$ is a linear equation.
Equivalent equations have the same solution.	$2x - 4 = 6$, $2x = 10$, and $x = 5$ are equivalent equations.
The Addition Property of Equality Adding the same number (or algebraic expression) to or subtracting the same number (or algebraic expression) from both sides of an equation does not change its solution.	• $x - 3 = 8$ $x - 3 + 3 = 8 + 3$ $x = 11$ • $x + 4 = 10$ $x + 4 - 4 = 10 - 4$ $x = 6$
Section 2.2 The Multiplication Property of Equality	
The Multiplication Property of Equality Multiplying both sides or dividing both sides of an equation by the same nonzero real number (or algebraic expression) does not change the solution.	• $\frac{x}{-5} = 6$ $-5\left(\frac{x}{-5}\right) = -5(6)$ $x = -30$ • $-50 = -5y$ $\frac{-50}{-5} = \frac{-5y}{-5}$ $10 = y$
Equations and Coefficients of −1 If $-x = c$, multiply both sides by -1 to solve for x. The solution is the additive inverse of c.	$-x = -12$ $(-1)(-x) = (-1)(-12)$ $x = 12$
Using the Addition and Multiplication Properties If an equation does not contain fractions, • Use the addition property to isolate the variable term. • Use the multiplication property to isolate the variable.	$-2x - 5 = 11$ $-2x - 5 + 5 = 11 + 5$ $-2x = 16$ $\frac{-2x}{-2} = \frac{16}{-2}$ $x = -8$
Section 2.3 Solving Linear Equations	
Solving a Linear Equation **1.** Simplify each side.	Solve: $7 - 4(x - 1) = x + 1.$ $7 - 4x + 4 = x + 1$ $-4x + 11 = x + 1$
2. Collect all the variable terms on one side and all the constant terms on the other side.	$-4x - x + 11 = x - x + 1$ $-5x + 11 = 1$ $-5x + 11 - 11 = 1 - 11$ $-5x = -10$

Definitions and Concepts	Examples
Section 2.3 Solving Linear Equations (continued)	
3. Isolate the variable and solve. (If the variable is eliminated and a false statement results, the inconsistent equation has no solution. If a true statement results, all real numbers are solutions of the identity.)	$\frac{-5x}{-5} = \frac{-10}{-5}$ $x = 2$
4. Check the proposed solution in the original equation.	$7 - 4(x - 1) = x + 1$ $7 - 4(2 - 1) \stackrel{?}{=} 2 + 1$ $7 - 4(1) \stackrel{?}{=} 2 + 1$ $7 - 4 \stackrel{?}{=} 2 + 1$ $3 = 3$, true The solution is 2, or the solution set is {2}.
Equations Containing Fractions Multiply both sides (all terms) by the least common denominator. This clears the equation of fractions.	$\frac{x}{5} + \frac{1}{2} = \frac{x}{2} - 1$ $10\left(\frac{x}{5} + \frac{1}{2}\right) = 10\left(\frac{x}{2} - 1\right)$ $10 \cdot \frac{x}{5} + 10 \cdot \frac{1}{2} = 10 \cdot \frac{x}{2} - 10 \cdot 1$ $2x + 5 = 5x - 10$ $-3x = -15$ $x = 5$ The solution is 5, or the solution set is {5}.
Types of Equations An equation that is true for all real numbers, ℝ, is called an identity. When solving an identity, the variable is eliminated and a true statement, such as $3 = 3$, results. An equation that is not true for even one real number is called an inconsistent equation. A false statement, such as $3 = 7$, results when solving such an equation, whose solution set is ∅, the empty set. A conditional equation is not an identity, but is true for at least one real number.	Solve: $4x + 5 = 4(x + 2)$ $4x + 5 = 4x + 8$ $5 = 8$, false The inconsistent equation has no solution: ∅. Solve: $5x - 4 = 5(x + 1) - 9$ $5x - 4 = 5x + 5 - 9$ $5x - 4 = 5x - 4$ $-4 = -4$, true All real number satisfy the identity: ℝ.
Section 2.4 Formulas and Percents	
To solve a formula for one of its variables, use the steps for solving a linear equation and isolate the specified variable on one side of the equation.	Solve for l: $w = \frac{P - 2l}{2}$ $2w = 2\left(\frac{P - 2l}{2}\right)$ $2w = P - 2l$ $2w - P = P - P - 2l$ $2w - P = -2l$ $\frac{2w - P}{-2} = \frac{-2l}{-2}$ $\frac{2w - P}{-2} = l$

Definitions and Concepts	Examples
Section 2.4 Formulas and Percents (continued)	
The word *percent* means *per hundred*. The symbol % denotes percent.	$47\% = \frac{47}{100}$ $3\% = \frac{3}{100}$
To express a decimal as a percent, move the decimal point two places to the right and add a percent sign.	$0.37 = 37\%$ $0.006 = 0.6\%$
To express a percent as a decimal, move the decimal point two places to the left and remove the percent sign.	$250\% = 250\% = 2.5$ $4\% = 04.\% = 0.04$
A Formula Involving Percent *A* is *P* percent of *B*. $A = P \cdot B$ In the formula $A = PB$, P is expressed as a decimal.	• What is 5% of 20? $A = 0.05 \cdot 20$ $A = 1$ Thus, 1 is 5% of 20. • 6 is 30% of what? $6 = 0.3 \cdot B$ $\frac{6}{0.3} = B$ $20 = B$ Thus, 6 is 30% of 20. • 33 is what percent of 75? $33 = P \cdot 75$ $\frac{33}{75} = P$ $P = 0.44 = 44\%$ Thus, 33 is 44% of 75.
Section 2.5 An Introduction to Problem Solving	
Strategy for Solving Word Problems Step 1 Let x represent one of the quantities.	The length of a rectangle exceeds the width by 3 inches. The perimeter is 26 inches. What are the rectangle's dimensions? Let x = the width.
Step 2 Represent other quantities in terms of x.	$x + 3$ = the length
Step 3 Write an equation that describes the conditions.	Twice length plus twice width is perimeter. $2(x + 3) + 2x = 26$

Definitions and Concepts	Examples
Section 2.5 An Introduction to Problem Solving (continued)	
Step 4 Solve the equation and answer the question.	$2(x + 3) + 2x = 26$ $2x + 6 + 2x = 26$ $4x + 6 = 26$ $4x = 20$ $x = 5$ The width (x) is 5 inches and the length $(x + 3)$ is $5 + 3$, or 8 inches.
Step 5 Check the proposed solution in the original wording of the problem.	Perimeter $= 2(5 \text{ in.}) + 2(8 \text{ in.})$ $= 10 \text{ in.} + 16 \text{ in.} = 26 \text{ in.}$ This checks with the given perimeter.
Section 2.6 Problem Solving in Geometry	
Solving geometry problems often requires using basic geometric formulas. Formulas for perimeter, area, circumference, and volume are given in Tables 2.3 (page 149), 2.4 (page 150), and 2.5 (page 152) in Section 2.6.	A sailboat's triangular sail has an area of 24 ft^2 and a base of 8 ft. Find its height. $A = \frac{1}{2}bh$ $24 = \frac{1}{2}(8)h$ $24 = 4h$ $6 = h$ The sail's height is 6 ft.
The sum of the measures of the three angles of any triangle is 180°.	In a triangle, the first angle measures 3 times the second and the third measures 40° less than the second. Find each angle's measure. Second angle $= x$ First angle $= 3x$ Third angle $= x - 40$ Sum of measures is 180°. $x + 3x + (x - 40) = 180$ $5x - 40 = 180$ $5x = 220$ $x = 44$ The angles measure $x = 44$, $3x = 3 \cdot 44 = 132$, and $x - 40 = 44 - 40 = 4$. The angles measure 44°, 132°, and 4°.
Two complementary angles have measures whose sum is 90°. Two supplementary angles have measures whose sum is 180°. If an angle measures x, its complement measures $90 - x$, and its supplement measures $180 - x$.	An angle measures five times its complement. Find the angle's measure. $x =$ angle's measure $90 - x =$ measure of complement $x = 5(90 - x)$ $x = 450 - 5x$ $6x = 450$ $x = 75$ The angle measures 75°.

Definitions and Concepts	Examples
Section 2.7 Solving Linear Inequalities	
A linear inequality in one variable can be written in one of these forms: $ax + b < c \quad ax + b \le c$ $ax + b > c \quad ax + b \ge c$ where a is not 0.	$3x + 6 > 12$ is a linear inequality.
Set-Builder Notation and Graphs $\{x \mid a < x < b\}$ $\{x \mid a \le x \le b\}$ $\{x \mid a \le x < b\}$ $\{x \mid x > b\}$ $\{x \mid x \le a\}$ 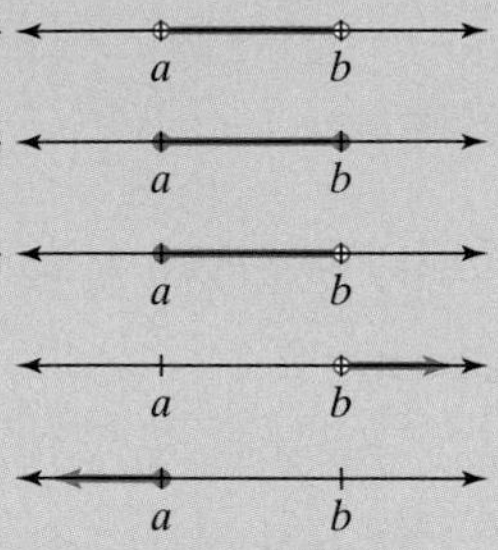 	• Graph the solutions of $x < 4$. 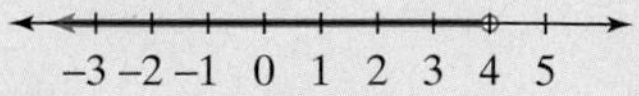 • Graph the solutions of $-2 < x \le 1$. 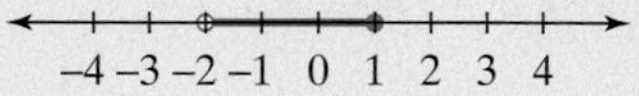
The Addition Property of Inequality Adding the same number to or subtracting the same number from both sides of an inequality does not change the solutions.	$x + 3 < 8$ $x + 3 - 3 < 8 - 3$ $x < 5$
The Positive Multiplication Property of Inequality Multiplying or dividing both sides of an inequality by the same positive number does not change the solutions.	$\frac{x}{6} \ge 5$ $6 \cdot \frac{x}{6} \ge 6 \cdot 5$ $x \ge 30$
The Negative Multiplication Property of Inequality Multiplying or dividing both sides of an inequality by the same negative number and reversing the direction of the inequality sign does not change the solutions.	$-3x \le 12$ $\frac{-3x}{-3} \ge \frac{12}{-3}$ $x \ge -4$
Solving Linear Inequalities Use the procedure for solving linear equations. When multiplying or dividing by a negative number, reverse the direction of the inequality symbol. Express the solution set in set-builder notation and graph the set on a number line. If the variable is eliminated and a false statement results, the inequality has no solution. The solution set is $\emptyset$, the empty set. If a true statement results, the solution is the set of all real numbers: $\{x \mid x \text{ is a real number}\}$ or $\mathbb{R}$.	Solve: $x + 4 \ge 6x - 16$ $x + 4 - 6x \ge 6x - 16 - 6x$ $-5x + 4 \ge -16$ $-5x + 4 - 4 \ge -16 - 4$ $-5x \ge -20$ $\frac{-5x}{-5} \le \frac{-20}{-5}$ $x \le 4$ $\{x \mid x \le 4\}$ 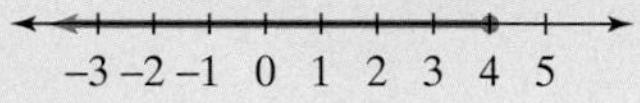

CHAPTER 2 REVIEW EXERCISES

2.1 *Solve each equation in Exercises 1–5 using the addition property of equality. Be sure to check proposed solutions.*

1. $x - 10 = 22$

2. $-14 = y + 8$

3. $7z - 3 = 6z + 9$

4. $4(x + 3) = 3x - 10$

5. $6x - 3x - 9 + 1 = -5x + 7x - 3$

2.2 *Solve each equation in Exercises 6–13 using the multiplication property of equality. Be sure to check proposed solutions.*

6. $\frac{x}{8} = 10$

7. $\frac{y}{-8} = 7$

8. $7z = 77$

9. $-36 = -9y$

10. $\frac{3}{5}x = -9$

11. $30 = -\frac{5}{2}y$

12. $-x = 25$

13. $\frac{-x}{10} = -1$

Solve each equation in Exercises 14–18 using both the addition and multiplication properties of equality. Check proposed solutions.

14. $4x + 9 = 33$

15. $-3y - 2 = 13$

16. $5z + 20 = 3z$

17. $5x - 3 = x + 5$

18. $3 - 2x = 9 - 8x$

19. The formula $F = 1.2x + 21.6$ models the average family income, F, in thousands of dollars, for Puerto Ricans x years after 1990. How many years after 1990 is the average family income expected to reach \$40.8 thousand? In which year is this expected to occur?

2.3 *Solve and check each equation in Exercises 20–28.*

20. $5x + 9 - 7x + 6 = x + 18$

21. $3(x + 4) = 5x - 12$

22. $1 - 2(6 - y) = 3y + 2$

23. $2(x - 4) + 3(x + 5) = 2x - 2$

24. $-2(y - 4) - (3y - 2) = -2 - (6y - 2)$

25. $\frac{2x}{3} = \frac{x}{6} + 1$

26. $\frac{x}{2} - \frac{1}{10} = \frac{x}{5} + \frac{1}{2}$

27. $3(8x - 1) = 6(5 + 4x)$

28. $4(2x - 3) + 4 = 8x - 8$

29. The optimum heart rate that a person should achieve during exercise for the exercise to be most beneficial is modeled by $r = 0.6(220 - a)$, where a represents a person's age and r represents that person's optimum heart rate, in beats per minute. If the optimum heart rate is 120 beats per minute, how old is that person?

2.4 *In Exercises 30–34, solve each formula for the specified variable.*

30. $I = Pr$ for r

31. $V = \frac{1}{3}Bh$ for h

32. $P = 2l + 2w$ for w

33. $A = \frac{B + C}{2}$ for B

34. $T = D + pm$ for m

In Exercises 35–36, express each decimal as a percent.

35. 0.72

36. 0.0035

In Exercises 37–39, express each percent as a decimal.

37. 65%

38. 150%

39. 3%

40. What is 8% of 120?

41. 90 is 45% of what?

42. 36 is what percent of 75?

43. If 6 is increased to 12, the increase is what percent of the original number?

44. If 5 is decreased to 3, the decrease is what percent of the original number?

45. A college that had 40 students for each lecture course increased the number to 45 students. What is the percent increase in the number of students in a lecture course?

46. Consider the following statement:

> My portfolio fell 10% last year, but then it rose 10% this year, so at least I recouped my losses.

Is this statement true? In particular, suppose you invested \$10,000 in the stock market last year. How much money would be left in your portfolio with a 10% fall and then a 10% rise? If there is a loss, what is the percent decrease, to the nearest tenth of a percent, in your portfolio?

47. The radius is one of two bones that connect the elbow and the wrist. The formula $r = \frac{h}{7}$ models the length of a woman's radius, r, in inches, and her height, h, in inches.

a. Solve the formula for h.

b. Use the formula in part (a) to find a woman's height if her radius is 9 inches long.

48. Every day, the average U.S. household uses 91 gallons of water flushing toilets. The circle graph on the next page shows that this represents 26% of the total number of gallons of water used per day. How many gallons of water does the average U.S. household use per day?

Where U.S. Households Use Water

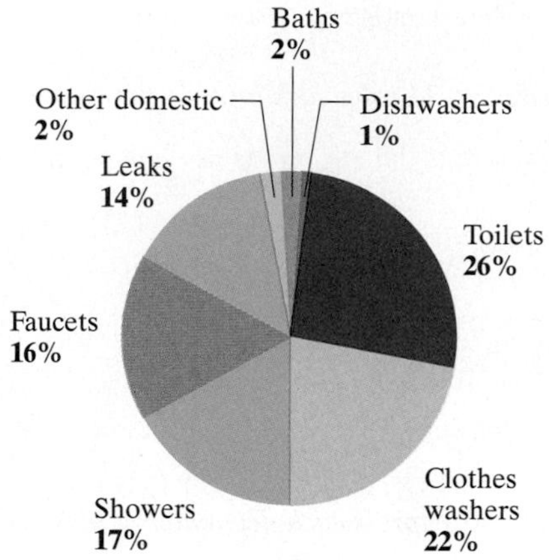

Source: American Water Works Association

2.5 *In Exercises 49–56, use the five-step strategy to solve each problem.*

49. Six times a number, decreased by 20, is four times the number. Find the number.

50. On average, the number of unhealthy air days per year in Los Angeles exceeds three times that of New York City by 48 days. If Los Angeles and New York combined have 268 unhealthy air days per year, determine the number of unhealthy days for the two cities.
(*Source*: Environmental Protection Agency)

51. Two pages that face each other in a book have 93 as the sum of their page numbers. What are the page numbers?

52. The two female artists in the United States with the most platinum albums are Barbra Streisand followed by Madonna. (A platinum album represents one album sold per 266 people.) The number of platinum albums by these two singers form consecutive odd integers. Combined, they have 96 platinum albums. Determine the number of platinum albums by Streisand and the number of platinum albums by Madonna.

53. In 2003, the average weekly salary for workers in the United States was \$612. If this amount is increasing by \$15 yearly, in how many years after 2003 will the average salary reach \$747? In which year will that be?

54. A bank's total monthly charge for a checking account is \$6 plus \$0.05 per check. If your total monthly charge is \$6.90, how many checks did you write during that month?

55. A rectangular field is three times as long as it is wide. If the perimeter of the field is 400 yards, what are the field's dimensions?

56. After a 25% reduction, you purchase a table for \$180. What was the table's price before the reduction?

2.6 *Use a formula for area to find the area of each figure in Exercises 57–59.*

57.

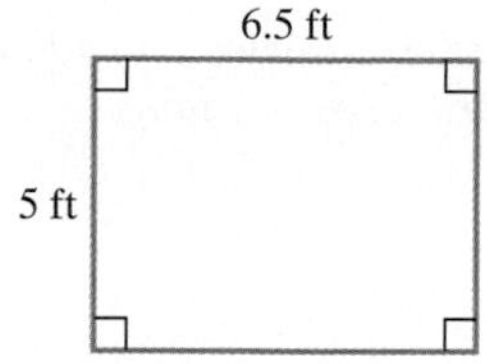

58.

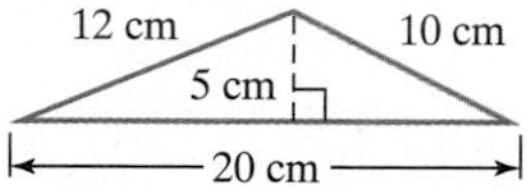

59.

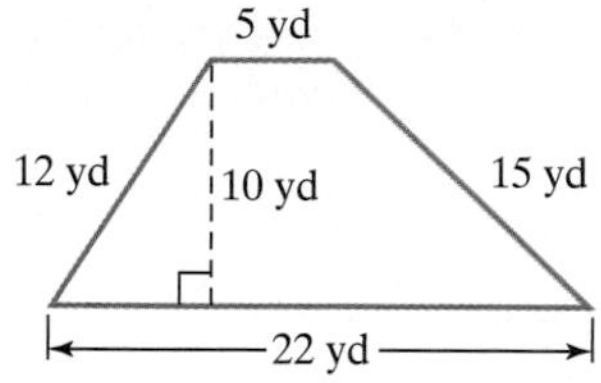

60. Find the circumference and the area of a circle with a diameter of 20 meters. Round answers to the nearest whole number.

61. A sailboat has a triangular sail with an area of 42 square feet and a base that measures 14 feet. Find the height of the sail.

62. A rectangular kitchen floor measures 12 feet by 15 feet. A stove on the floor has a rectangular base measuring 3 feet by 4 feet, and a refrigerator covers a rectangular area of the floor measuring 3 feet by 4 feet. How many square feet of tile will be needed to cover the kitchen floor not counting the area used by the stove and the refrigerator?

63. A yard that is to be covered with mats of grass is shaped like a trapezoid. The bases are 80 feet and 100 feet, and the height is 60 feet. What is the cost of putting the grass mats on the yard if the landscaper charges \$0.35 per square foot?

64. Which one of the following is a better buy: a medium pizza with a 14-inch diameter for \$6.00 or two small pizzas, each with an 8-inch diameter, for \$6.00?

Use a formula for volume to find the volume of each figure in Exercises 65–67. Where applicable, express answers in terms of π. Then round to the nearest whole number.

65.

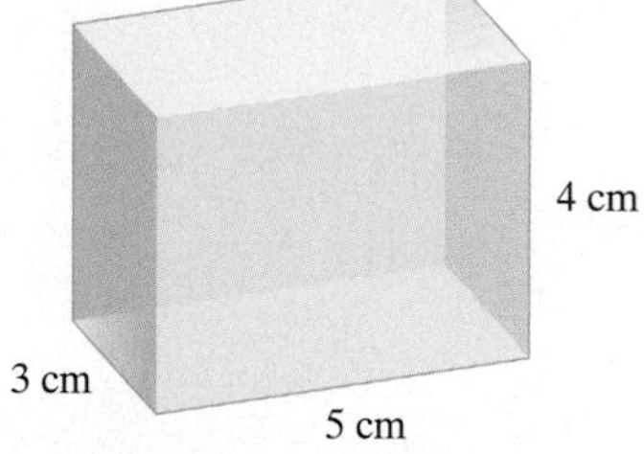

66.

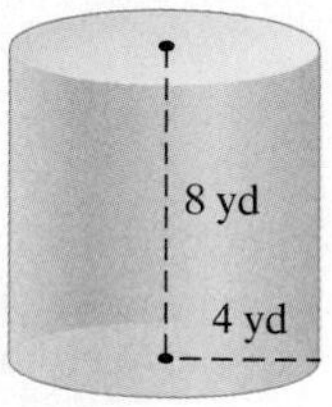

67.

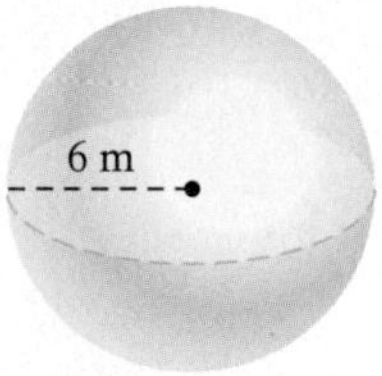

68. A train is being loaded with freight containers. Each box is 8 meters long, 4 meters wide, and 3 meters high. If there are 50 freight containers, how much space is needed?

69. A cylindrical fish tank has a diameter of 6 feet and a height of 3 feet. How many tropical fish can be put in the tank if each fish needs 5 cubic feet of water?

70. Find the measure of each angle of the triangle shown in the figure.

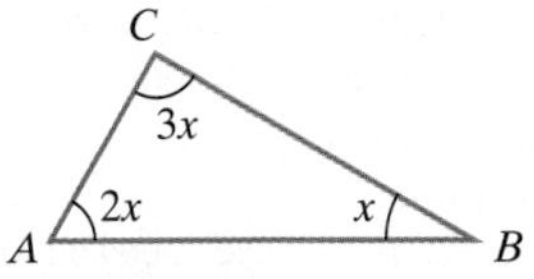

71. In a triangle, the measure of the first angle is 15° more than twice the measure of the second angle. The measure of the third angle exceeds that of the second angle by 25°. What is the measure of each angle?

72. Find the measure of the complement of a 57° angle.

73. Find the measure of the supplement of a 75° angle.

74. How many degrees are there in an angle that measures 25° more than the measure of its complement?

75. The measure of the supplement of an angle is 45° less than four times the measure of the angle. Find the measure of the angle and its supplement.

2.7 *In Exercises 76–77, graph the solutions of each inequality on a number line.*

76. $x < -1$

77. $-2 < x \le 4$

Describe each graph in Exercises 78–79 using set-builder notation.

78.

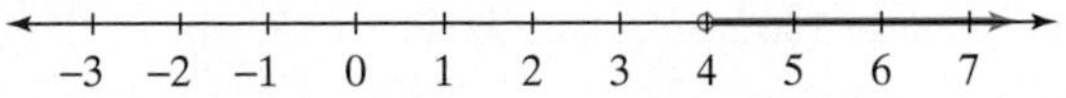

79.

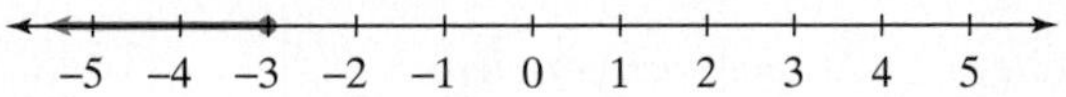

Solve each inequality in Exercises 80–87. Express the solution set in set-builder notation and graph the set on a number line. If the inequality has no solution or is true for all real numbers, so state. It is not necessary to graph solution sets for these inequalities.

80. $2x - 5 < 3$

81. $\frac{x}{2} > -4$

82. $3 - 5x \le 18$

83. $4x + 6 < 5x$

84. $6x - 10 \ge 2(x + 3)$

85. $4x + 3(2x - 7) \le x - 3$

86. $2(2x + 4) > 4(x + 2) - 6$

87. $-2(x - 4) \le 3x + 1 - 5x$

88. To pass a course, a student must have an average on three examinations of at least 60. If a student scores 42 and 74 on the first two tests, what must be earned on the third test to pass the course?

89. A long distance telephone service charges 10¢ for the first minute and 5¢ for each minute thereafter. The cost, C, in cents, for a call lasting x minutes is modeled by the formula

$$C = 10 + 5(x - 1).$$

How many minutes can you talk on the phone if you do not want the cost to exceed \$5, or 500¢?

CHAPTER 2 TEST

Remember to use your Chapter Test Prep Video CD to see the worked-out solutions to the test questions you want to review.

In Exercises 1–6, solve each equation.

1. $4x - 5 = 13$

2. $12x + 4 = 7x - 21$

3. $8 - 5(x - 2) = x + 26$

4. $3(2y - 4) = 9 - 3(y + 1)$

5. $\frac{3}{4}x = -15$

6. $\frac{x}{10} + \frac{1}{3} = \frac{x}{5} + \frac{1}{2}$

7. The formula $N = 2.4x + 180$ models U.S. population, N, in millions x years after 1960. How many years after 1960 is the U.S. population expected to reach 324 million? In which year is this expected to occur?

In Exercises 8–9, solve each formula for the specified variable.

8. $V = \pi r^2 h$ for h

9. $l = \frac{P - 2w}{2}$ for w

10. What is 6% of 140?

11. 120 is 80% of what?

12. 12 is what percent of 240?

In Exercises 13–17, solve each problem.

13. The product of 5 and a number, decreased by 9, is 306. What is the number?

14. In New York City, a fitness trainer earns \$22,870 more per year than a preschool teacher. The yearly average salaries for fitness trainers and preschool teachers combined are \$79,030. Determine the average yearly salary of a fitness trainer and a preschool teacher in New York City. (*Source*: *Time*, April 14, 2003)

15. A long-distance telephone plan has a monthly fee of \$15.00 and a rate of \$0.05 per minute. How many minutes can you chat long distance in a month for a total cost, including the \$15.00, of \$45.00?

16. A rectangular field is twice as long as it is wide. If the perimeter of the field is 450 yards, what are the field's dimensions?

17. After a 20% reduction, you purchase a new Stephen King novel for \$28. What was the book's price before the reduction?

In Exercises 18–19, find the area of each figure.

18.

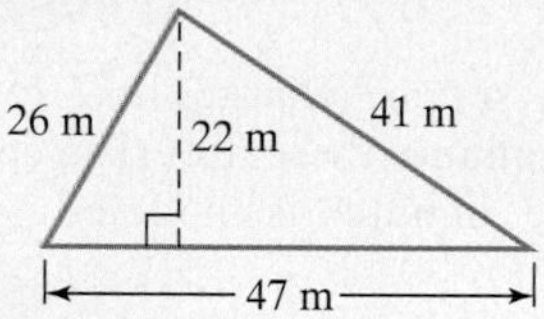

19.

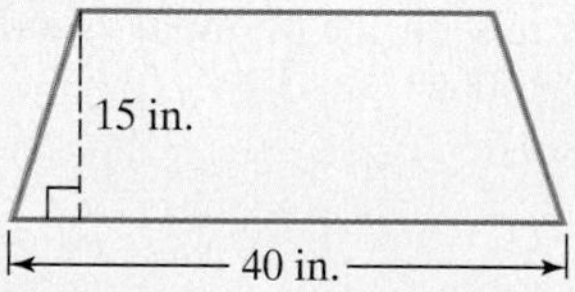

In Exercises 20–21, find the volume of each figure. Where applicable, express answers in terms of π. Then round to the nearest whole number.

20.

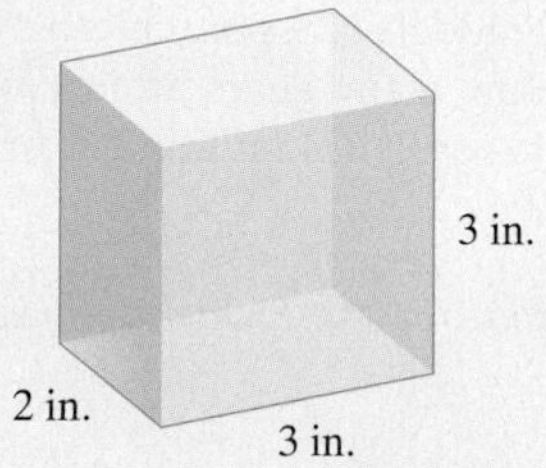

21.

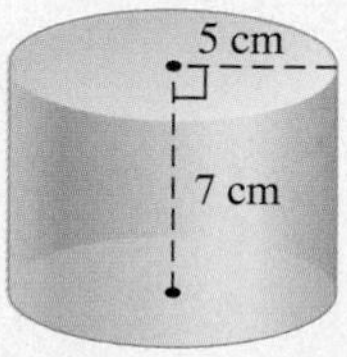

22. What will it cost to cover a rectangular floor measuring 40 feet by 50 feet with square tiles that measure 2 feet on each side if a package of 10 tiles costs \$13 per package?

23. A sailboat has a triangular sail with an area of 56 square feet and a base that measures 8 feet. Find the height of the sail.

24. In a triangle, the measure of the first angle is three times that of the second angle. The measure of the third angle is 30° less than the measure of the second angle. What is the measure of each angle?

25. How many degrees are there in an angle that measures 16° more than the measure of its complement?

In Exercises 26–27, graph the solutions of each inequality on a number line.

26. $x > -2$

27. $-4 \le x < 1$

28. Use set-builder notation to describe the following graph.

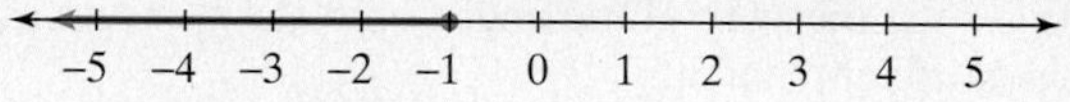

Solve each inequality in Exercises 29–31. Express the solution set in set-builder notation and graph the set on a number line.

29. $\frac{x}{2} < -3$

30. $6 - 9x \ge 33$

31. $4x - 2 > 2(x + 6)$

32. A student has grades on three examinations of 76, 80, and 72. What must the student earn on a fourth examination to have an average of at least 80?

33. The length of a rectangle is 20 inches. For what widths is the perimeter greater than 56 inches?

CUMULATIVE REVIEW EXERCISES (CHAPTERS 1–2)

In Exercises 1–3, perform the indicated operation or operations.

1. $-8 - (12 - 16)$ **2.** $(-3)(-2) + (-2)(4)$

3. $(8 - 10)^3(7 - 11)^2$

4. Simplify: $2 - 5[x + 3(x + 7)]$.

5. List all the rational numbers in this set:

$$\left\{-4, -\frac{1}{3}, 0, \sqrt{2}, \sqrt{4}, \frac{\pi}{2}, 1063\right\}.$$

6. Plot $(-2, -1)$ in a rectangular coordinate system. In which quadrant does the point lie?

7. Insert either $<$ or $>$ in the box to make a true statement:

$$-10{,}000 \square -2.$$

8. Use the distributive property to rewrite without parentheses:

$$6(4x - 1 - 5y).$$

The graph shows the unemployment rate in the United States from 1990 through mid-2003. Use the information in the graph to solve Exercises 9–10.

U.S. Unemployment Rate

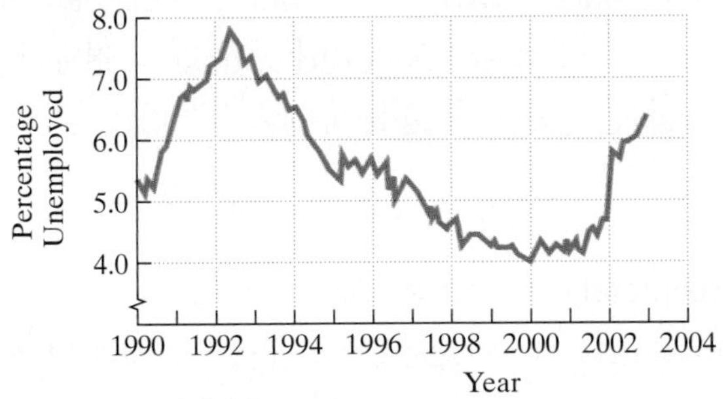

Source: Bureau of Labor Statistics

9. For the period shown, in which year was the unemployment rate at a minimum? What percentage of the work force was unemployed in that year?

10. For the period shown, during which year did the unemployment rate reach a maximum? Estimate the percentage of the work force unemployed, to the nearest tenth of a percent, at that time.

In Exercises 11–12, solve each equation.

11. $5 - 6(x + 2) = x - 14$

12. $\frac{x}{5} - 2 = \frac{x}{3}$

13. Solve for A: $V = \frac{1}{3}Ah$.

14. 48 is 30% of what?

15. The length of a rectangular parking lot is 10 yards less than twice its width. If the perimeter of the lot is 400 yards, what are its dimensions?

16. A gas station owner makes a profit of 40 cents per gallon of gasoline sold. How many gallons of gasoline must be sold in a year to make a profit of \$30,000 from gasoline sales?

17. Graph the solution set of $-2 < x \leq 3$ on a number line.

Solve each inequality in Exercises 18–19. Express the solution set in set-builder notation and graph the set on a number line.

18. $3 - 3x > 12$

19. $5 - 2(3 - x) \leq 2(2x + 5) + 1$

20. You take a summer job selling medical supplies. You are paid \$600 per month plus 4% of the sales price of all the supplies you sell. If you want to earn more than \$2500 per month, what value of medical supplies must you sell?

CHAPTER 3

Although the world's attention is often on large endangered mammals from Africa to the oceans, the United States is home to more than 395 endangered animals and close to 600 endangered plants.

This discussion is developed algebraically in Exercises 59 in Exercise Set 3.4.

Linear Equations in Two Variables

In the United States, the Department of the Interior classifies species as endangered or threatened. Endangered species are those in danger of extinction. Threatened species are those likely to become endangered species within the foreseeable future. The formula

$$E = 10x + 166$$

approximately describes the number of endangered animal species in the United States, represented by E, x years after 1980. This equation in two variables tells of ten new animal species per year classified by the federal government as endangered.

In this chapter, you will learn graphing and modeling methods for situations involving two variables. With these methods, you will be able to create formulas that model the data of your world.

SECTION 3.1

Objectives

1. Determine whether an ordered pair is a solution of an equation.
2. Find solutions of an equation in two variables.
3. Use point plotting to graph linear equations.
4. Use point plotting to graph other kinds of equations.
5. Use graphs of linear equations to solve problems.

GRAPHING EQUATIONS IN TWO VARIABLES

A picture, as they say, is worth a thousand words. Have you seen pictures of gas-guzzling cars from the 1950s, with their huge fins and overstated designs? The worst year for automobile fuel efficiency was 1958, when U.S. cars averaged a dismal 12.4 miles per gallon.

There is a formula that models fuel efficiency of U.S. cars over time. The formula is

$$y = 0.0075x^2 - 0.2672x + 14.8.$$

The variable x represents the number of years after 1940. The variable y represents the average number of miles per gallon for U.S. automobiles. Looking at the formula does not make it obvious that 1958, 18 years after 1940, was the worst year for fuel efficiency. However, if we could somehow make a picture of the formula, such as the one shown in Figure 3.1, the lowest point on the picture would reveal approximately 1958 as the year in which gas-guzzling cars averaged less than 13 miles per gallon. The shape of the graph also shows decreasing fuel efficiency from 1940 through 1958 and increasing fuel efficiency after 1958. In this chapter, we will be making pictures of equations. We can use these pictures to visualize the behavior of the variables in the equation.

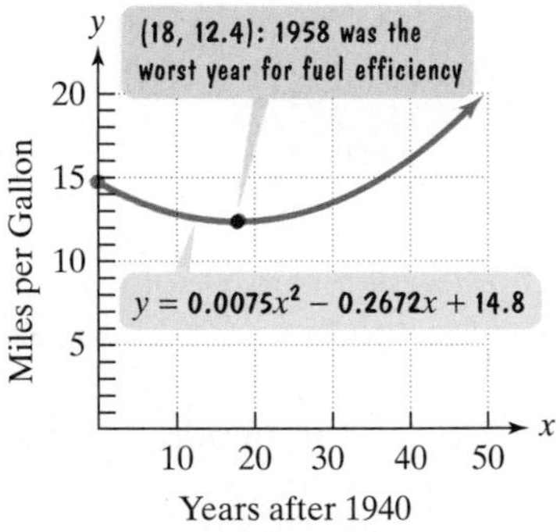

FIGURE 3.1 A picture of a formula

Solutions of Equations The rectangular coordinate system allows us to visualize relationships between two variables by connecting any equation in two variables with a geometric figure. Consider, for example, the following equation in two variables:

$$x + y = 10.$$

The sum of two numbers, x and y, is 10.

Many pairs of numbers fit the description in the voice balloon, such as $x = 1$ and $y = 9$, or $x = 3$ and $y = 7$. The phrase "$x = 1$ and $y = 9$" is abbreviated using the ordered pair $(1, 9)$. Similarly, the phrase "$x = 3$ and $y = 7$" is abbreviated using the ordered pair $(3, 7)$.

1 Determine whether an ordered pair is a solution of an equation.

A **solution of an equation in two variables**, x and y, is an ordered pair of real numbers with the following property: When the x-coordinate is substituted for x and the y-coordinate is substituted for y in the equation, we obtain a true statement. For example, $(1, 9)$ is a solution of the equation $x + y = 10$. When 1 is substituted for x and 9 is substituted for y, we obtain the true statement $1 + 9 = 10$, or $10 = 10$. Because there are infinitely many pairs of numbers that have a sum of 10, the equation $x + y = 10$ has infinitely many solutions. Each ordered-pair solution is said to **satisfy** the equation. Thus, $(1, 9)$ satisfies the equation $x + y = 10$.

EXAMPLE 1 Deciding Whether an Ordered Pair Satisfies an Equation

Determine whether each ordered pair is a solution of the equation

$$x - 4y = 14:$$

a. $(2, -3)$ **b.** $(12, 1)$.

SOLUTION

a. To determine whether $(2, -3)$ is a solution of the equation, we substitute 2 for x and -3 for y.

$x - 4y = 14$ — This is the given equation.

$2 - 4(-3) \stackrel{?}{=} 14$ — Substitute 2 for x and −3 for y.

$2 - (-12) \stackrel{?}{=} 14$ — Multiply: $4(-3) = -12$.

This statement is true. $14 = 14$ — Subtract: $2 - (-12) = 2 + 12 = 14$.

Because we obtain a true statement, we conclude that $(2, -3)$ is a solution of the equation $x - 4y = 14$. Thus, $(2, -3)$ satisfies the equation.

b. To determine whether $(12, 1)$ is a solution of the equation, we substitute 12 for x and 1 for y.

$x - 4y = 14$ — This is the given equation.

$12 - 4(1) \stackrel{?}{=} 14$ — Substitute 12 for x and 1 for y.

$12 - 4 \stackrel{?}{=} 14$ — Multiply: $4(1) = 4$.

This statement is false. $8 = 14$ — Subtract: $12 - 4 = 8$.

Because we obtain a false statement, we conclude that $(12, 1)$ is not a solution of $x - 4y = 14$. The ordered pair $(12, 1)$ does not satisfy the equation. ■

CHECK POINT 1 Determine whether each ordered pair is a solution of the equation $x - 3y = 9$:

a. $(3, -2)$ **b.** $(-2, 3)$.

In this chapter, we will use x and y to represent the variables of an equation in two variables. However, any two letters can be used. Solutions are still ordered pairs. The first number in an ordered pair usually replaces the variable that occurs first alphabetically. The second number in an ordered pair usually replaces the variable that occurs last alphabetically.

2 Find solutions of an equation in two variables.

How do we find ordered pairs that are solutions of an equation in two variables, x and y?

- Select a value for one of the variables.
- Substitute that value into the equation and find the corresponding value of the other variable.
- Use the values of the two variables to form an ordered pair (x, y). This pair is a solution of the equation.

EXAMPLE 2 Finding Solutions of an Equation

Find five solutions of

$$y = 2x - 1.$$

Select integers for x, starting with -2 and ending with 2.

SOLUTION We organize the process of finding solutions in the following table of values.

Start with these values of x.

Substitute x into $y = 2x - 1$ and compute y.

Use values for x and y to form an ordered-pair solution.

Any numbers can be selected for x. There is nothing special about integers from -2 to 2, inclusive. We chose these values to include two negative numbers, 0, and two positive numbers. We also wanted to keep the resulting computations for y relatively simple.

x	$y = 2x - 1$	(x, y)
-2	$y = 2(-2) - 1 = -4 - 1 = -5$	$(-2, -5)$
-1	$y = 2(-1) - 1 = -2 - 1 = -3$	$(-1, -3)$
0	$y = 2 \cdot 0 - 1 = 0 - 1 = -1$	$(0, -1)$
1	$y = 2 \cdot 1 - 1 = 2 - 1 = 1$	$(1, 1)$
2	$y = 2 \cdot 2 - 1 = 4 - 1 = 3$	$(2, 3)$

Look at the ordered pairs in the last column. Five solutions of $y = 2x - 1$ are $(-2, -5)$, $(-1, -3)$, $(0, -1)$, $(1, 1)$, and $(2, 3)$. ■

✔ **CHECK POINT 2** Find five solutions of $y = 3x + 2$. Select integers for x, starting with -2 and ending with 2.

3 Use point plotting to graph linear equations.

Graphing Linear Equations in the Form $y = mx + b$ In Example 2, we found five solutions of $y = 2x - 1$. We can generate as many ordered-pair solutions as desired to $y = 2x - 1$ by substituting numbers for x and then finding the corresponding values for y. The **graph of the equation** is the set of all points whose coordinates satisfy the equation.

One method for graphing an equation such as $y = 2x - 1$ is the **point-plotting method**.

THE POINT-PLOTTING METHOD FOR GRAPHING AN EQUATION IN TWO VARIABLES

1. Find several ordered pairs that are solutions of the equation.
2. Plot these ordered pairs as points in the rectangular coordinate system.
3. Connect the points with a smooth curve or line.

EXAMPLE 3 Graphing an Equation Using the Point-Plotting Method

Graph the equation: $y = 3x$.

SOLUTION

Step 1. **Find several ordered pairs that are solutions of the equation.** Because there are infinitely many solutions, we cannot list them all. To find some solutions of the equation, we select integers for x, starting with -2 and ending with 2.

Start with these values of x. | Substitute x into $y = 3x$ and compute y. | These are some solutions of $y = 3x$.

x	$y = 3x$	(x, y)
-2	$y = 3(-2) = -6$	$(-2, -6)$
-1	$y = 3(-1) = -3$	$(-1, -3)$
0	$y = 3 \cdot 0 = 0$	$(0, 0)$
1	$y = 3 \cdot 1 = 3$	$(1, 3)$
2	$y = 3 \cdot 2 = 6$	$(2, 6)$

Step 2. **Plot these ordered pairs as points in the rectangular coordinate system.** The five ordered pairs in the table of values are plotted in Figure 3.2(a).

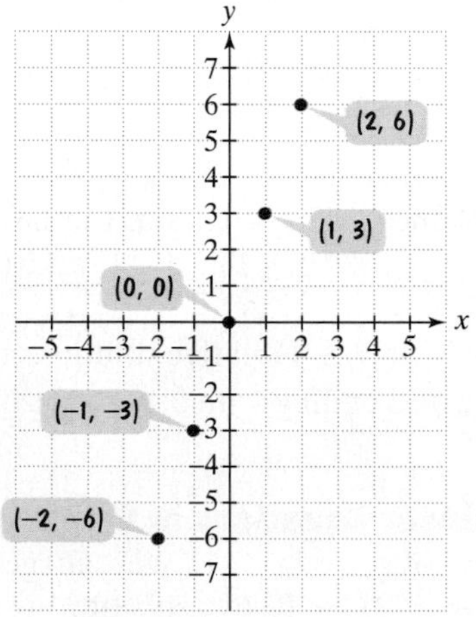

FIGURE 3.2 (a) Some solutions of $y = 3x$ plotted as points

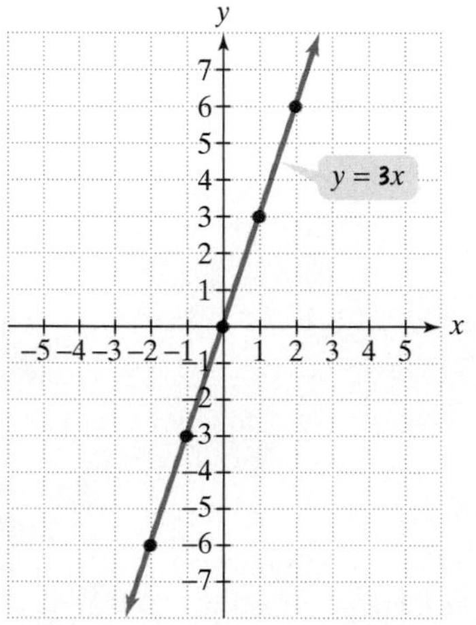

FIGURE 3.2 (b) The graph of $y = 3x$

Step 3. **Connect the points with a smooth curve or line.** The points lie along a straight line. The graph of $y = 3x$ is shown in Figure 3.2(b). The arrows on both ends of the line indicate that it extends indefinitely in both directions. ■

CHECK POINT 3 Graph the equation: $y = 2x$.

Equations like $y = 3x$ and $y = 2x$ are called **linear equations in two variables** because the graph of each equation is a line. Any equation that can be written in the form $y = mx + b$, where m and b are constants, is a linear equation in two variables. Here are examples of linear equations in two variables:

$$y = 3x \qquad\qquad y = 3x - 2$$

$$\text{or} \quad y = 3x + 0 \qquad\qquad \text{or} \quad y = 3x + (-2)$$

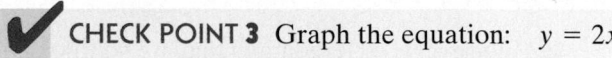

This is in the form of $y = mx + b$ with $m = 3$ and $b = 0$. | This is in the form of $y = mx + b$ with $m = 3$ and $b = -2$.

Can you guess how the graph of the linear equation $y = 3x - 2$ compares with the graph of $y = 3x$? In Example 3, we graphed $y = 3x$. Now, let's graph the equation $y = 3x - 2$.

EXAMPLE 4 Graphing a Linear Equation in Two Variables

Graph the equation: $y = 3x - 2$.

SOLUTION

Step 1. **Find several ordered pairs that are solutions of the equation.** To find some solutions, we select integers for x, starting with -2 and ending with 2.

Start with x. Compute y. Form the ordered pair (x, y).

x	$y = 3x - 2$	(x, y)
-2	$y = 3(-2) - 2 = -6 - 2 = -8$	$(-2, -8)$
-1	$y = 3(-1) - 2 = -3 - 2 = -5$	$(-1, -5)$
0	$y = 3 \cdot 0 - 2 = 0 - 2 = -2$	$(0, -2)$
1	$y = 3 \cdot 1 - 2 = 3 - 2 = 1$	$(1, 1)$
2	$y = 3 \cdot 2 - 2 = 6 - 2 = 4$	$(2, 4)$

Step 2. **Plot these ordered pairs as points in the rectangular coordinate system.** The five ordered pairs in the table of values are plotted in Figure 3.3(a).

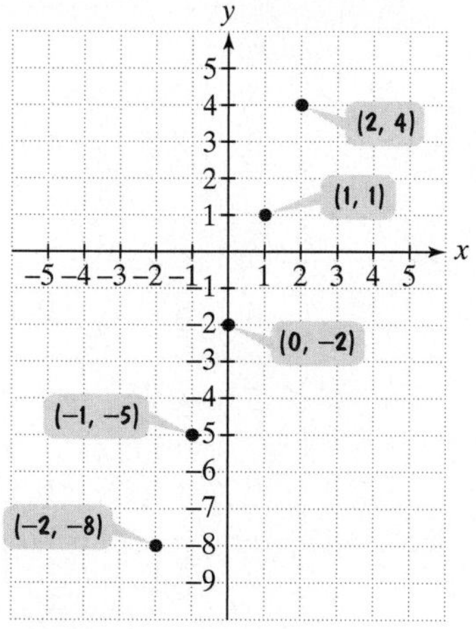

FIGURE 3.3 (a) Some solutions of $y = 3x - 2$ plotted as points

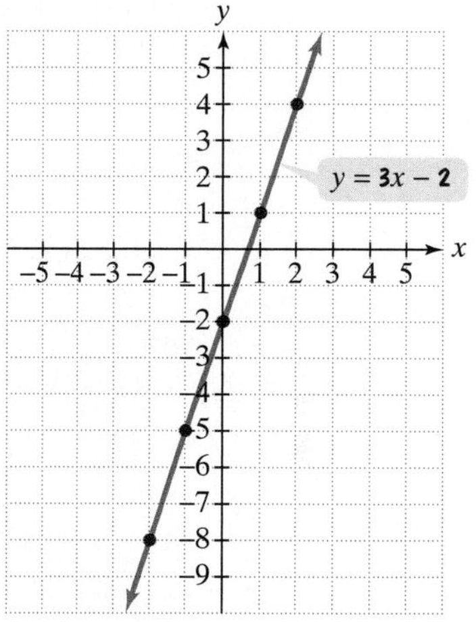

FIGURE 3.3 (b) The graph of $y = 3x - 2$

Step 3. **Connect the points with a smooth curve or line.** The points lie along a straight line. The graph of $y = 3x - 2$ is shown in Figure 3.3(b). ■

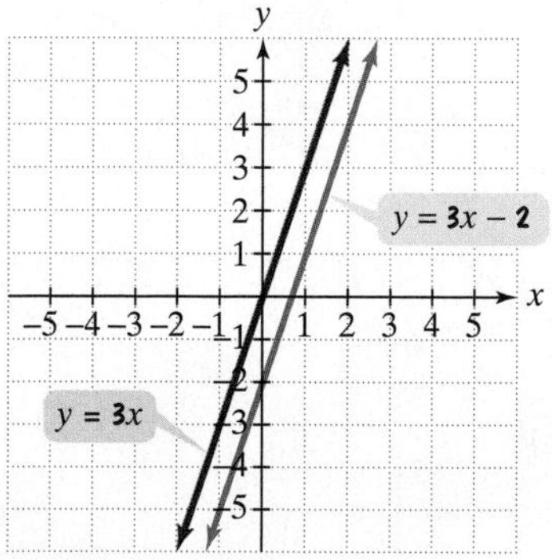

FIGURE 3.4

Now we are ready to compare the graphs of $y = 3x - 2$ and $y = 3x$. The graphs of both linear equations are shown in the same rectangular coordinate system in Figure 3.4. Can you see that the blue graph of $y = 3x - 2$ is parallel to the red graph of $y = 3x$ and shifted 2 units down? Instead of crossing the y-axis at $(0, 0)$, the graph now crosses the y-axis at $(0, -2)$.

COMPARING GRAPHS OF LINEAR EQUATIONS If the value of m does not change,

- The graph of $y = mx + b$ is the graph of $y = mx$ shifted b units up when b is a positive number.
- The graph of $y = mx + b$ is the graph of $y = mx$ shifted b units down when b is a negative number.

CHECK POINT 4 Graph the equation: $y = 2x - 2$.

EXAMPLE 5 Graphing a Linear Equation in Two Variables

Graph the equation: $y = \frac{2}{3}x + 1$.

SOLUTION

Step 1. **Find several ordered pairs that are solutions of the equation.** Notice that m, the coefficient of x, is $\frac{2}{3}$. When m is a fraction, we will select values of x that are multiples of the denominator. In this way, we can avoid values of y that are fractions. Because the denominator of $\frac{2}{3}$ is 3, we select multiples of 3 for x. Let's use −6, −3, 0, 3, and 6.

Start with multiples of 3 for x. | Compute y. | Form the ordered pair (x, y).

x	$y = \frac{2}{3}x + 1$	(x, y)
−6	$y = \frac{2}{3}(-6) + 1 = -4 + 1 = -3$	(−6, −3)
−3	$y = \frac{2}{3}(-3) + 1 = -2 + 1 = -1$	(−3, −1)
0	$y = \frac{2}{3} \cdot 0 + 1 = 0 + 1 = 1$	(0, 1)
3	$y = \frac{2}{3} \cdot 3 + 1 = 2 + 1 = 3$	(3, 3)
6	$y = \frac{2}{3} \cdot 6 + 1 = 4 + 1 = 5$	(6, 5)

Step 2. **Plot these ordered pairs as points in the rectangular coordinate system.** The five ordered pairs in the table of values are plotted in Figure 3.5.

Step 3. **Connect the points with a smooth curve or line.** The points lie along a straight line. The graph of $y = \frac{2}{3}x + 1$ is shown in Figure 3.5.

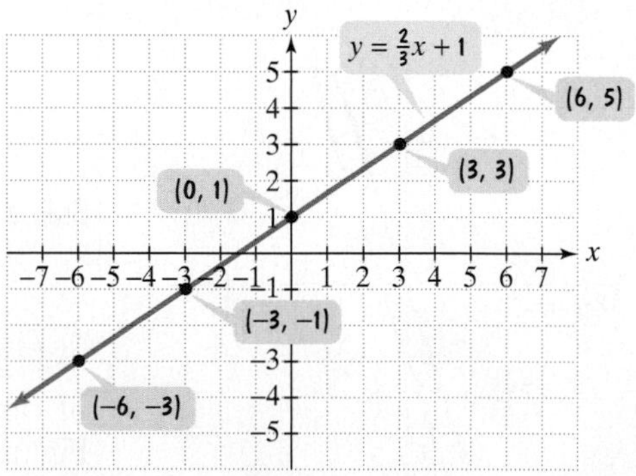

FIGURE 3.5 The graph of $y = \frac{2}{3}x + 1$

CHECK POINT 5 Graph the equation: $y = \frac{1}{2}x + 2$.

4 Use point plotting to graph other kinds of equations.

Graphing Nonlinear Equations in Two Variables

Look at the picture of this gymnast. He has created a perfect balance in which the two halves of his body are mirror images of each other. Is it possible for graphs to have mirrorlike qualities? Yes. Although our next graph is not a straight line, we can obtain its cuplike U-shape using the point-plotting method for graphing an equation in two variables.

EXAMPLE 6 Graphing a Nonlinear Equation in Two Variables

Graph the equation: $y = x^2 - 4$.

SOLUTION The given equation involves two variables, x and y. However, because the variable x is squared, it is not a linear equation in two variables.

$$y = x^2 - 4$$

This is not in the form $y = mx + b$ because x is squared.

Although the graph is not a line, it is still a picture of all the ordered-pair solutions of $y = x^2 - 4$. Thus, we can use the point-plotting method to obtain the graph.

Step 1. **Find several ordered pairs that are solutions of the equation.** To find some solutions, we select integers for x, starting with -3 and ending with 3.

Start with x. Compute y. Form the ordered pair (x, y).

x	$y = x^2 - 4$	(x,y)
-3	$y = (-3)^2 - 4 = 9 - 4 = 5$	$(-3, 5)$
-2	$y = (-2)^2 - 4 = 4 - 4 = 0$	$(-2, 0)$
-1	$y = (-1)^2 - 4 = 1 - 4 = -3$	$(-1, -3)$
0	$y = 0^2 - 4 = 0 - 4 = -4$	$(0, -4)$
1	$y = 1^2 - 4 = 1 - 4 = -3$	$(1, -3)$
2	$y = 2^2 - 4 = 4 - 4 = 0$	$(2, 0)$
3	$y = 3^2 - 4 = 9 - 4 = 5$	$(3, 5)$

Step 2. **Plot these ordered pairs as points in the rectangular coordinate system.** The seven ordered pairs in the table of values are plotted in Figure 3.6(a) on the next page.

STUDY TIP

If the graph of an equation is not a straight line, use more solutions than when graphing lines. These extra solutions are needed to get a better general idea of the graph's shape.

Step 3. **Connect the points with a smooth curve.** The seven points are joined with a smooth curve in Figure 3.6(b). The graph of $y = x^2 - 4$ is a curve where the part of the graph to the right of the y-axis is a reflection of the part to the left of it, and vice versa. The arrows on both ends of the curve indicate that it extends indefinitely in both directions.

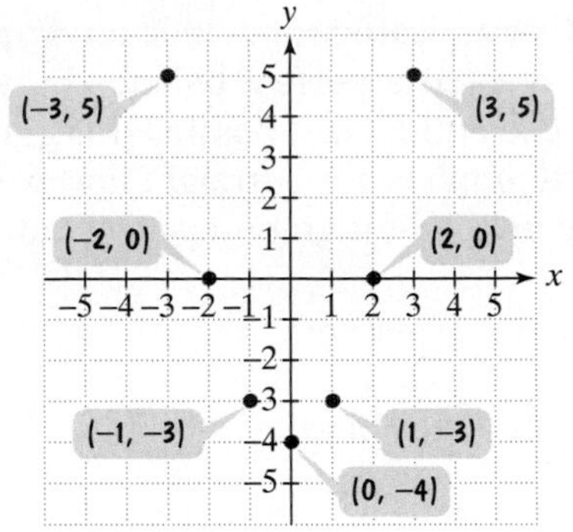

FIGURE 3.6 (a) Some solutions of $y = x^2 - 4$ plotted as points

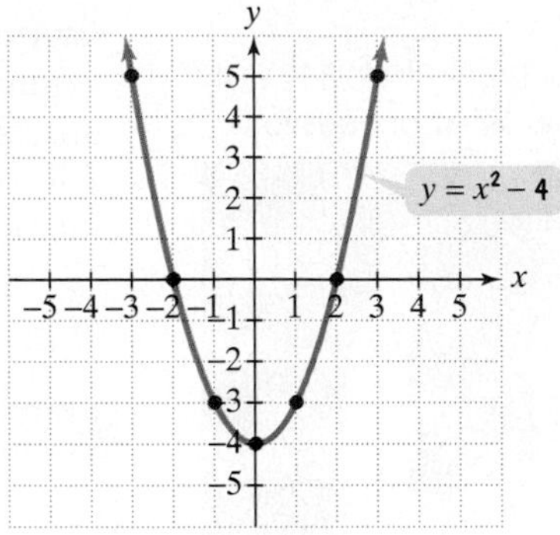

FIGURE 3.6 (b) The graph of $y = x^2 - 4$

CHECK POINT 6 Graph the equation: $y = x^2 - 1$. Select integers for x, starting with -3 and ending with 3.

5 Use graphs of linear equations to solve problems.

Applications Part of the beauty of the rectangular coordinate system is that it allows us to "see" mathematical formulas and visualize the solution to a problem. This idea is demonstrated in Example 7.

EXAMPLE 7 An Application Using Graphs of Linear Equations

The toll to a bridge costs \$2.50. Commuters who use the bridge frequently have the option of purchasing a monthly coupon book for \$21.00. With the coupon book, the toll is reduced to \$1.00. The monthly cost, y, of using the bridge x times can be described by the following formulas:

Without the coupon book:

$y = 2.50x$ — The monthly cost, y, is \$2.50 times the number of times, x, that the bridge is used.

With the coupon book:

$y = 21 + 1 \cdot x$ — The monthly cost, y, is \$21 for the book plus \$1 times the number of times, x, that the bridge is used.

$y = 21 + x$

a. Let $x = 0, 2, 4, 10, 12, 14$, and 16. Make a table of values for each linear equation showing seven solutions for the equation.

b. Graph the equations in the same rectangular coordinate system.

c. What are the coordinates of the intersection point for the two graphs? Interpret the coordinates in practical terms.

SOLUTION

a. Tables of values showing seven solutions for each equation follow.

Without the Coupon Book

x	$y = 2.5x$	(x, y)
0	$y = 2.5(0) = 0$	(0, 0)
2	$y = 2.5(2) = 5$	(2, 5)
4	$y = 2.5(4) = 10$	(4, 10)
10	$y = 2.5(10) = 25$	(10, 25)
12	$y = 2.5(12) = 30$	(12, 30)
14	$y = 2.5(14) = 35$	(14, 35)
16	$y = 2.5(16) = 40$	(16, 40)

With the Coupon Book

x	$y = 21 + x$	(x, y)
0	$y = 21 + 0 = 21$	(0, 21)
2	$y = 21 + 2 = 23$	(2, 23)
4	$y = 21 + 4 = 25$	(4, 25)
10	$y = 21 + 10 = 31$	(10, 31)
12	$y = 21 + 12 = 33$	(12, 33)
14	$y = 21 + 14 = 35$	(14, 35)
16	$y = 21 + 16 = 37$	(16, 37)

b. Now we are ready to graph the two equations. Because the x- and y-coordinates are nonnegative, it is only necessary to use the origin, the positive portions of the x- and y-axes, and the first quadrant of the rectangular coordinate system. The x-coordinates in our tables begin at 0 and end at 16. We will let each tick mark on the x-axis represent two units. However, the y-coordinates in our tables begin at 0 and get as large as 40 in the formula that describes the monthly cost without the coupon book. So that our y-axis does not get too long, we will let each tick mark on the y-axis represent five units. Using this setup and the two tables of values, we construct the graphs of $y = 2.5x$ and $y = 21 + x$, shown in Figure 3.7.

c. The graphs intersect at (14, 35). This means that if the bridge is used 14 times in a month, the total monthly cost without the coupon book is the same as the total monthly cost with the coupon book, namely \$35.

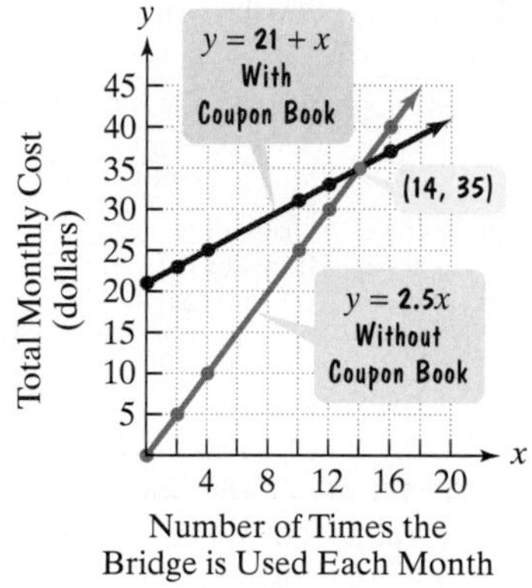

FIGURE 3.7 Options for a toll

In Figure 3.7, look at the two graphs to the right of the intersection point (14, 35). The red graph of $y = 21 + x$ lies below the blue graph of $y = 2.5x$. This means that if the bridge is used more than 14 times in a month ($x > 14$), the (red) monthly cost, y, with the coupon book is cheaper than the (blue) monthly cost, y, without the coupon book.

CHECK POINT 7 The toll to a bridge costs \$2.00. If you use the bridge x times in a month, the monthly cost, y, is $y = 2x$. With a \$10 coupon book, the toll is reduced to \$1.00. The monthly cost, y, of using the bridge x times in a month with the coupon book is $y = 10 + x$.

a. Let $x = 0, 2, 4, 6, 8, 10$, and 12. Make tables of values showing seven solutions of $y = 2x$ and seven solutions of $y = 10 + x$.

b. Graph the equations in the same rectangular coordinate system.

c. What are the coordinates of the intersection point for the two graphs? Interpret the coordinates in practical terms.

ENRICHMENT ESSAY

Mathematical Blossom

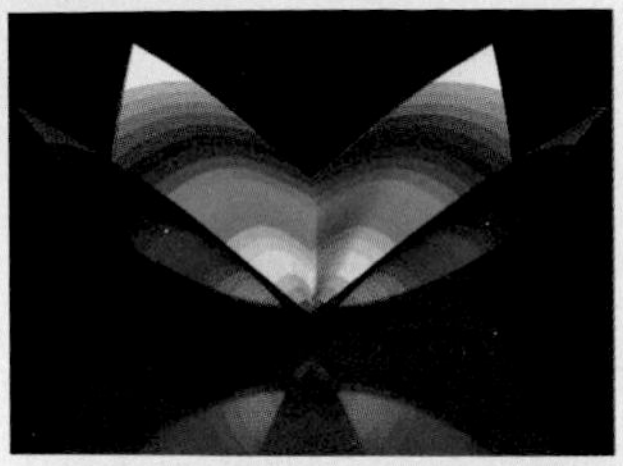

Graph of an equation in a three-dimensional Cartesian plane

This picture is a graph in a three-dimensional rectangular coordinate system. For many, the picture is more interesting than its equation:

$$z = (|x| - |y|)^2 + \frac{2|xy|}{\sqrt{x^2 + y^2}}.$$

Sometimes it's pleasant to simply get the "feel" of an equation by seeing its picture. Turning equations into visual images hints at the beauty that lies within mathematics.

USING TECHNOLOGY

Graphing calculators or graphing software packages for computers are referred to as **graphing utilities** or graphers. A graphing utility is a powerful tool that quickly generates the graph of an equation in two variables. Figure 3.8 shows two such graphs for the equations in Examples 3 and 4.

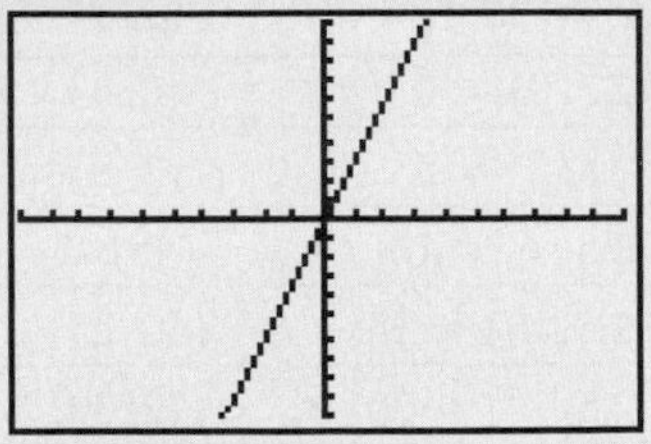

FIGURE 3.8(a) The graph of $y = 3x$

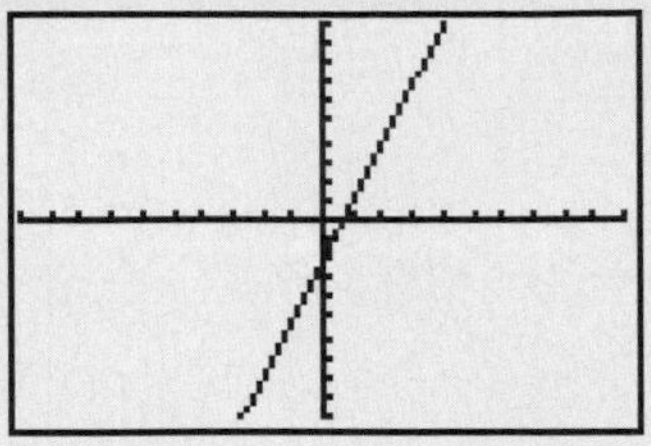

FIGURE 3.8(b) The graph of $y = 3x - 2$

What differences do you notice between these graphs and the graphs that we drew by hand? They do seem a bit "jittery." Arrows do not appear on both ends of the graphs. Furthermore, numbers are not given along the axes. For both graphs in Figure 3.8, the x-axis extends from -10 to 10 and the y-axis also extends from -10 to 10. The distance represented by each consecutive tick mark is one unit. We say that the **viewing rectangle**, or the **viewing window**, is $[-10, 10, 1]$ by $[-10, 10, 1]$.

$[-10,$ 10, 1] by $[-10,$ 10, 1].

- -10: The minimum x-value along the x-axis is -10.
- 10: The maximum x-value along the x-axis is 10.
- 1: The scale on the x-axis is 1 unit per tick mark.
- -10: The minimum y-value along the y-axis is -10.
- 10: The maximum y-value along the y-axis is 10.
- 1: The scale on the y-axis is 1 unit per tick mark.

To graph an equation in x and y using a graphing utility, enter the equation and specify the size of the viewing rectangle. The size of the viewing rectangle sets minimum and maximum values for both the x- and y-axes. Enter these values, as well as the values between consecutive tick marks, on the respective axes. The $[-10, 10, 1]$ by $[-10, 10, 1]$ viewing rectangle used in Figure 3.8 is called the **standard viewing rectangle**.

On most graphing utilities, the display screen is two-thirds as high as it is wide. By using a square setting, you can equally space the x and y tick marks. (This does not occur in the standard viewing rectangle.) Graphing utilities can also *zoom in* and *zoom out*. When you zoom in, you see a smaller portion of the graph, but you do so in greater detail. When you zoom out, you see a larger portion of the graph. Thus, zooming out may help you to develop a better understanding of the overall character of the graph. With practice, you will become more comfortable with graphing equations in two variables using your graphing utility. You will also develop a better sense of the size of the viewing rectangle that will reveal needed information about a particular graph.

3.1 EXERCISE SET

Student Solutions Manual | CD/Video | PH Math/Tutor Center | MathXL Tutorials on CD | MathXL® | MyMathLab | Interactmath.com

Practice Exercises

In Exercises 1–12, determine whether each ordered pair is a solution of the given equation.

1. $y = 3x \quad (2, 3), (3, 2), (-4, -12)$

2. $y = 4x \quad (3, 12), (12, 3), (-5, -20)$

3. $y = -4x \quad (-5, -20), (0, 0), (9, -36)$

4. $y = -3x \quad (-5, 15), (0, 0), (7, -21)$

5. $y = 2x + 6 \quad (0, 6), (-3, 0), (2, -2)$

6. $y = 8 - 4x \quad (8, 0), (16, -2), (3, -4)$

7. $3x + 5y = 15$ $(-5, 6), (0, 5), (10, -3)$

8. $2x - 5y = 0$ $(-2, 0), (-10, 6), (5, 0)$

9. $x + 3y = 0$ $(0, 0), \left(1, \frac{1}{3}\right), \left(2, -\frac{2}{3}\right)$

10. $x + 5y = 0$ $(0, 0), \left(1, \frac{1}{5}\right), \left(2, -\frac{2}{5}\right)$

11. $x - 4 = 0$ $(4, 7), (3, 4), (0, -4)$

12. $y + 2 = 0$ $(0, 2), (2, 0), (0, -2)$

In Exercises 13–20, find five solutions of each equation. Select integers for x, starting with −2 and ending with 2. Organize your work in a table of values.

13. $y = 12x$
14. $y = 14x$
15. $y = -10x$
16. $y = -20x$
17. $y = 8x - 5$
18. $y = 6x - 4$
19. $y = -3x + 7$
20. $y = -5x + 9$

In Exercises 21–44, graph each linear equation in two variables. Find at least five solutions in your table of values for each equation.

21. $y = x$
22. $y = x + 1$
23. $y = x - 1$
24. $y = x - 2$
25. $y = 2x + 1$
26. $y = 2x - 1$
27. $y = -x + 2$
28. $y = -x + 3$
29. $y = -3x - 1$
30. $y = -3x - 2$
31. $y = \frac{1}{2}x$
32. $y = -\frac{1}{2}x$
33. $y = -\frac{1}{4}x$
34. $y = \frac{1}{4}x$
35. $y = \frac{1}{3}x + 1$
36. $y = \frac{1}{3}x - 1$
37. $y = -\frac{3}{2}x + 1$
38. $y = -\frac{3}{2}x + 2$
39. $y = -\frac{5}{2}x - 1$
40. $y = -\frac{5}{2}x + 1$
41. $y = x + \frac{1}{2}$
42. $y = x - \frac{1}{2}$
43. $y = 4$, or $y = 0x + 4$
44. $y = 3$, or $y = 0x + 3$

Graph each equation in Exercises 45–50. Find seven solutions in your table of values for each equation by using integers for x, starting with −3 and ending with 3.

45. $y = x^2$
46. $y = x^2 - 2$
47. $y = x^2 + 1$
48. $y = x^2 + 2$
49. $y = 4 - x^2$
50. $y = 9 - x^2$

Practice Plus

In Exercises 51–54, write each sentence as a linear equation in two variables. Then graph the equation.

51. The y-variable is 3 more than the x-variable.
52. The y-variable exceeds the x-variable by 4.
53. The y-variable exceeds twice the x-variable by 5.
54. The y-variable is 2 less than 3 times the x-variable.
55. At the beginning of a semester, a student purchased eight pens and six pads for a total cost of \$14.50.
    a. If x represents the cost of one pen and y represents the cost of one pad, write an equation in two variables that reflects the given conditions.
    b. If pads cost \$0.75 each, find the cost of one pen.
56. A nursery offers a package of three small orange trees and four small grapefruit trees for \$22.
    a. If x represents the cost of one orange tree and y represents the cost of one grapefruit tree, write an equation in two variables that reflects the given conditions.
    b. If a grapefruit tree costs \$2.50, find the cost of an orange tree.

Application Exercises

The graph shows the percentage of divorced Americans, by race.

Divorced Americans, by Race

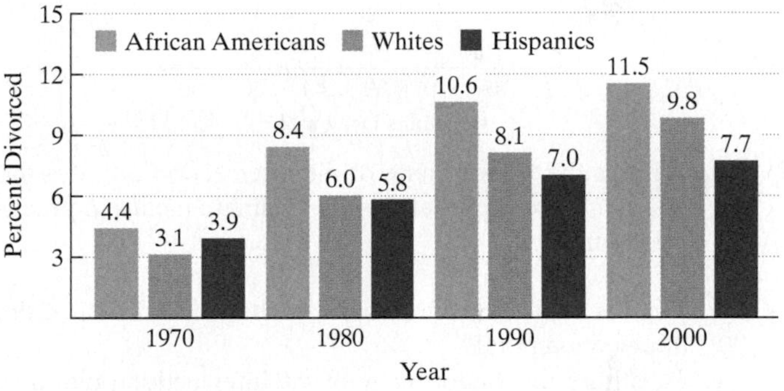

Source: U.S. Census Bureau

The data can be modeled by linear equations in two variables:

African Americans: $y = 0.24x + 5.2$

Whites: $y = 0.22x + 3.4$

Hispanics: $y = 0.13x + 4.2$

In each model, x is the number of years after 1970 and y is the percentage of divorced Americans in the racial group. Use this information to solve Exercises 57–58.

57. a. Find four solutions of the linear equation that models the data for African Americans. Use 0, 10, 20, and 30 for x. Organize your work in a table of values. How well does the linear equation model the data shown in the bar graph?

b. Repeat part (a) for the linear equation that models the data for whites.

c. Repeat part (a) for the linear equation that models the data for Hispanics.

58. **a.** Use the tables of values from Exercise 57 to graph any one of the three linear equations.

b. If you have not yet done so, extend your graph to include $x = 40$. Use the line to predict the percentage of divorced Americans, rounded to the nearest tenth of a percent, in this group in 2010.

59. A rental company charges \$40.00 a day plus \$0.35 per mile to rent a moving truck. The total cost, y, for a day's rental if x miles are driven is described by $y = 40 + 0.35x$. A second company charges \$36.00 a day plus \$0.45 per mile, so the daily cost, y, if x miles are driven is described by $y = 36 + 0.45x$. The graphs of the two equations are shown in the same rectangular coordinate system.

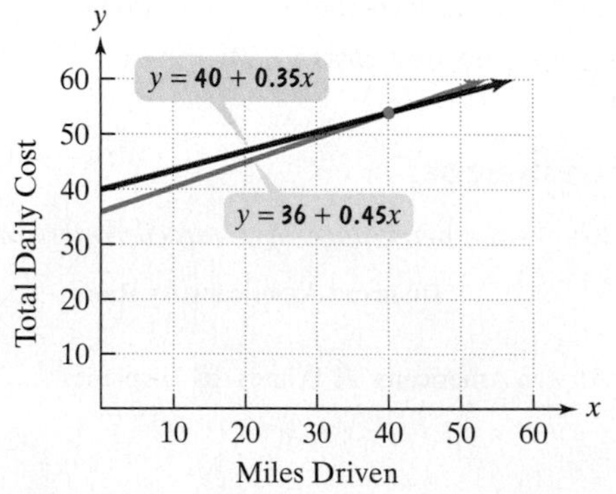

a. What is the x-coordinate of the intersection point of the graphs? Describe what this x-coordinate means in practical terms.

b. What is a reasonable estimate for the y-coordinate of the intersection point?

c. Substitute the x-coordinate of the intersection point into each of the equations and find the corresponding value for y. Describe what this value represents in practical terms. How close is this value to your estimate from part (b)?

60. The linear equation in two variables $y = 166x + 1781$ models the cost, y, in tuition and fees per year, of a four-year public college x years after 1990.

a. Find five solutions of $y = 166x + 1781$. Use 0, 5, 10, 15, and 20 for x. Organize your work in a table of values.

b. Use the solutions in part (a) to graph $y = 166x + 1781$. What does the shape of the graph indicate about the cost of a four-year public college?

61. The linear equation in two variables $y = 50x + 30{,}000$ models the total weekly cost, y, in dollars, for a business that manufactures x racing bicycles each week. The equation indicates that the business has weekly fixed costs of \$30,000 plus a cost of \$50 to manufacture each bicycle.

a. Find five solutions of $y = 50x + 30{,}000$. Use 0, 10, 20, 30, and 40 for x. Organize your work in a table of values.

b. Use the solutions in part (a) to graph $y = 50x + 30{,}000$.

Writing in Mathematics

62. How do you determine whether an ordered pair is a solution of an equation in two variables, x and y?

63. Explain how to find ordered pairs that are solutions of an equation in two variables, x and y.

64. What is the graph of an equation?

65. Explain how to graph an equation in two variables in the rectangular coordinate system.

Critical Thinking Exercises

66. Which one of the following is true?

a. The graph of $y = 3x + 1$ is parallel to the graph of $y = 2x$, but shifted up 1 unit.

b. The graph of any equation in the form $y = mx + b$ passes through the point $(0, b)$.

c. The ordered pair $(3, 4)$ satisfies the equation

$$2y - 3x = -6.$$

d. If $(2, 5)$ satisfies an equation, then $(5, 2)$ also satisfies the equation.

Graph each equation in Exercises 67–68. Find seven solutions in your table of values by using integers for x, starting with −3 and ending with 3.

67. $y = |x|$ **68.** $y = |x| + 1$

69. Although the level of air pollution varies from day to day and from hour to hour, during the summer the level of air pollution depends on the time of day. The equation in two variables

$$y = 0.1x^2 - 0.4x + 0.6$$

describes the level of air pollution, in parts per million (ppm), where x corresponds to the number of hours after 9 A.M.

a. Find six solutions of the equation. Select integers for x, starting with 0 and ending with 5.

b. Researchers have determined that a level of 0.3 ppm or more of pollutants in the air can be hazardous to your health. Based on the six solutions in part (a), at what times between 9 A.M. and 2 P.M. should runners exercise to avoid unsafe air?

Technology Exercises

Use a graphing utility to graph each equation in Exercises 70–73 in a standard viewing rectangle. Then use the TRACE *feature to trace along the line and find the coordinates of two points.*

70. $y = 2x - 1$ **71.** $y = -3x + 2$

72. $y = \frac{1}{2}x$ **73.** $y = \frac{3}{4}x - 2$

74. Use a graphing utility to verify any five of your hand-drawn graphs in Exercises 21–44. Use an appropriate viewing rectangle and the ZOOM SQUARE feature to make the graph look like the one you drew by hand.

75. The linear equation $y = 2.4x + 180$ models U.S. population, y, in millions, x years after 1960. Use a graphing utility to graph the equation in a $[0, 90, 10]$ by $[0, 500, 100]$ viewing rectangle. What does the shape of the graph indicate about changing U.S. population over time?

Review Exercises

76. Solve: $3x + 5 = 4(2x - 3) + 7$. (Section 2.3, Example 3)

77. Simplify: $3(1 - 2 \cdot 5) - (-28)$. (Section 1.8, Example 7)

78. Solve for h: $V = \frac{1}{3}Ah$. (Section 2.4, Example 4)

SECTION 3.2

Objectives

1 Use a graph to identify intercepts.

2 Graph a linear equation in two variables using intercepts.

3 Graph horizontal or vertical lines.

GRAPHING LINEAR EQUATIONS USING INTERCEPTS

Despite the nutritional disaster floating within the cans, the international sale of carbonated soft drinks continues to flourish. A can of Coca-Cola is sold every six seconds throughout the world.

The carbonated soft drinks market in the United States appears to be losing its fizz. As consumption of carbonated soft drinks declines, there is an emerging market for alternative beverages, including juices, teas, bottled waters, and sports drinks. The line graphs in Figure 3.9 show the projected U.S. beverage market share for carbonated and noncarbonated alternative beverages through 2010.

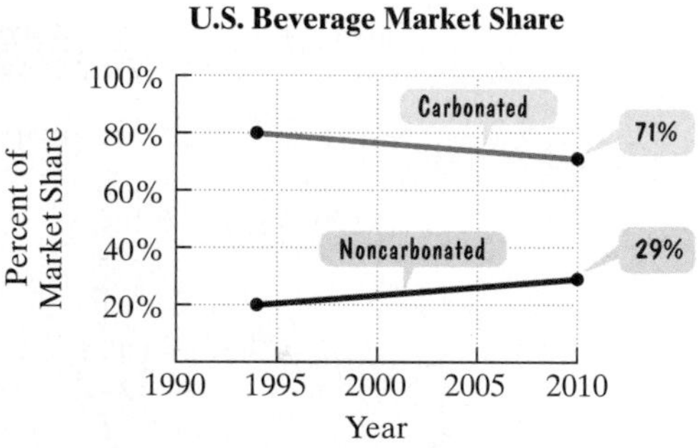

FIGURE 3.9

Source: Beverage Digest

The decreasing market share for carbonated beverages can be modeled by

$$y = -0.57x + 80.$$

The variable x represents the number of years after 1994. The variable y represents the market share of carbonated beverages, measured as a percentage of all sales.

There is another way that we can write the equation $y = -0.57x + 80$. We will collect the x- and y-terms on the left side. This is done by adding $0.57x$ to both sides:

$$0.57x + y = 80.$$

The form of this equation is $Ax + By = C$.

$$0.57x + y = 80$$

A, the coefficient of x, is 0.57. B, the coefficient of y, is 1. C, the constant on the right, is 80.

All equations of the form $Ax + By = C$ are straight lines when graphed as long as A and B are not both zero. To graph linear equations of this form, we will use two important points: the *intercepts*.

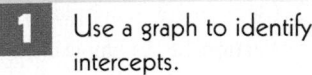

1 Use a graph to identify intercepts.

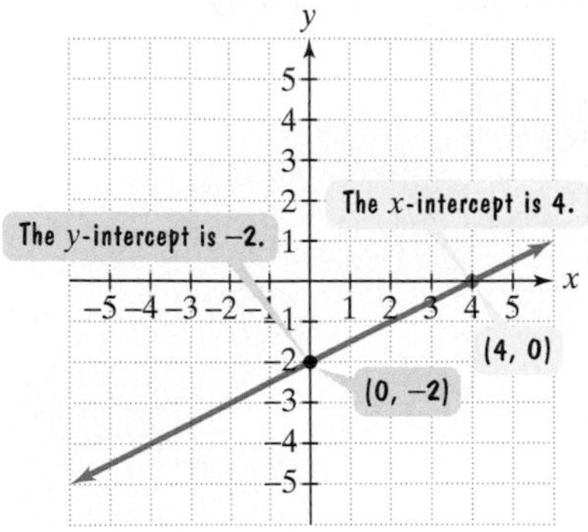

FIGURE 3.10 The graph of $2x - 4y = 8$

Intercepts An ***x*-intercept** of a graph is the x-coordinate of a point where the graph intersects the x-axis. For example, look at the graph of $2x - 4y = 8$ in Figure 3.10. The graph crosses the x-axis at $(4, 0)$. Thus, the x-intercept is 4. **The *y*-coordinate corresponding to an *x*-intercept is always zero.**

A ***y*-intercept** of a graph is the y-coordinate of a point where the graph intersects the y-axis. The graph of $2x - 4y = 8$ in Figure 3.10 shows that the graph crosses the y-axis at $(0, -2)$. Thus, the y-intercept is -2. **The *x*-coordinate corresponding to a *y*-intercept is always zero.**

STUDY TIP

Mathematicians tend to use two ways to describe intercepts. Did you notice that we are using single numbers? If a graph's x-intercept is a, it passes through the point $(a, 0)$. If a graph's y-intercept is b, it passes through the point $(0, b)$.

Some books state that the x-intercept is the *point* $(a, 0)$ and the x-intercept is *at a* on the x-axis. Similarly, the y-intercept is the *point* $(0, b)$ and the y-intercept is *at b* on the y-axis. In these descriptions, the intercepts are the actual points where a graph crosses the axes.

Although we'll describe intercepts as single numbers, we'll immediately state the point on the x- or y-axis that the graph passes through. Here's the important thing to keep in mind:

x-intercept: The corresponding y is 0.
y-intercept: The corresponding x is 0.

EXAMPLE 1 Identifying Intercepts

Identify the x- and y-intercepts.

a.

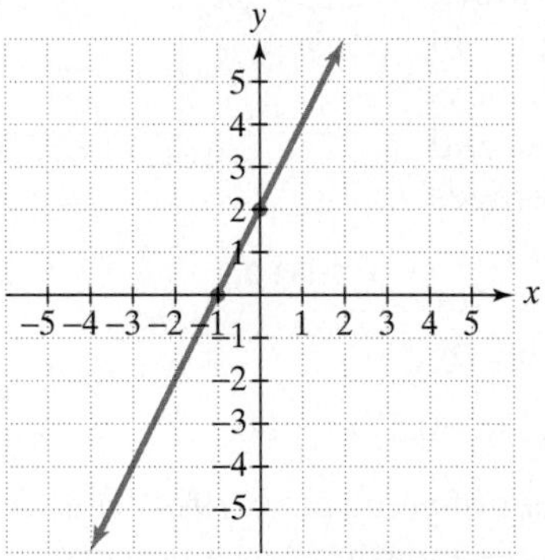

b.

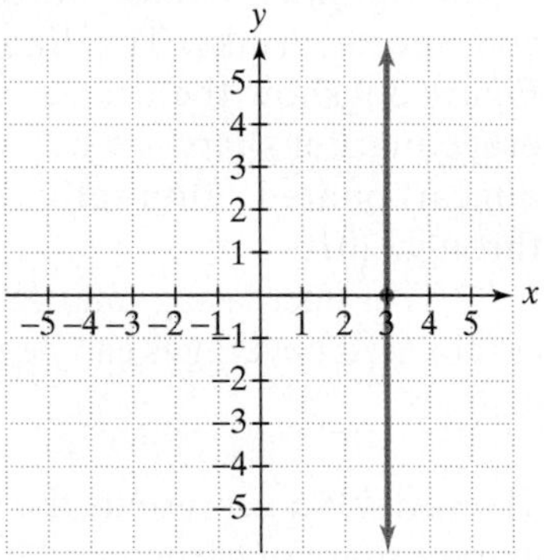

c.

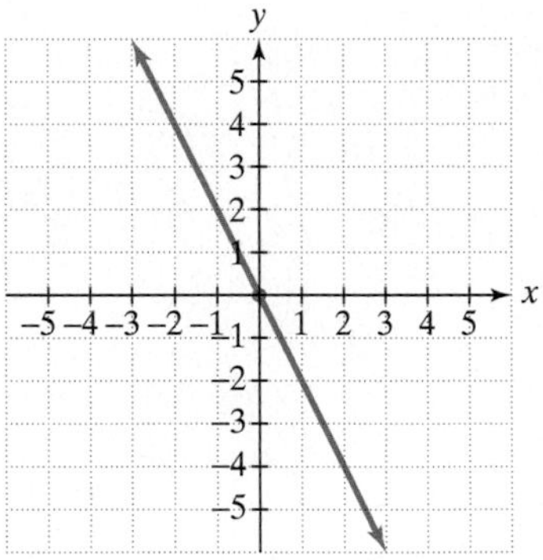

SOLUTION

a. The graph crosses the x-axis at $(-1, 0)$. Thus, the x-intercept is -1. The graph crosses the y-axis at $(0, 2)$. Thus, the y-intercept is 2.

b. The graph crosses the x-axis at $(3, 0)$, so the x-intercept is 3. This vertical line does not cross the y-axis. Thus, there is no y-intercept.

c. This graph crosses the x- and y-axes at the same point, the origin. Because the graph crosses both axes at $(0, 0)$, the x-intercept is 0 and the y-intercept is 0. ■

CHECK POINT 1 Identify the x- and y-intercepts.

a.

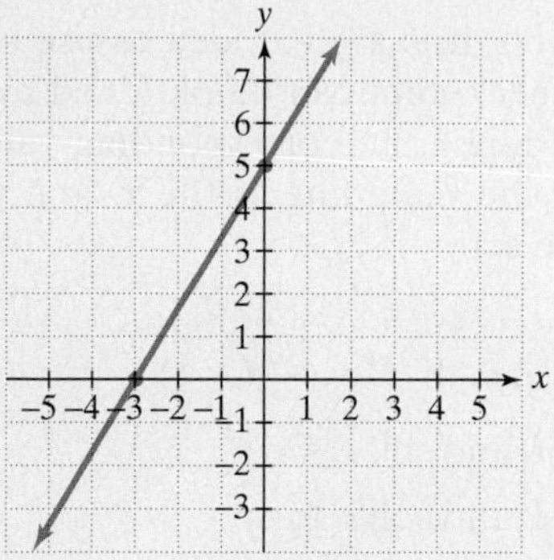

b.

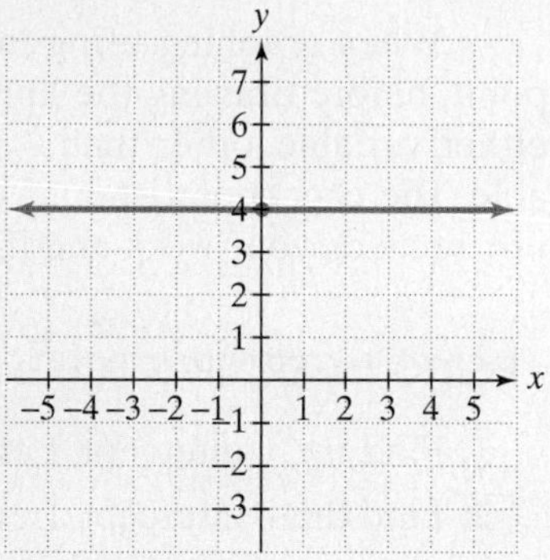

c.

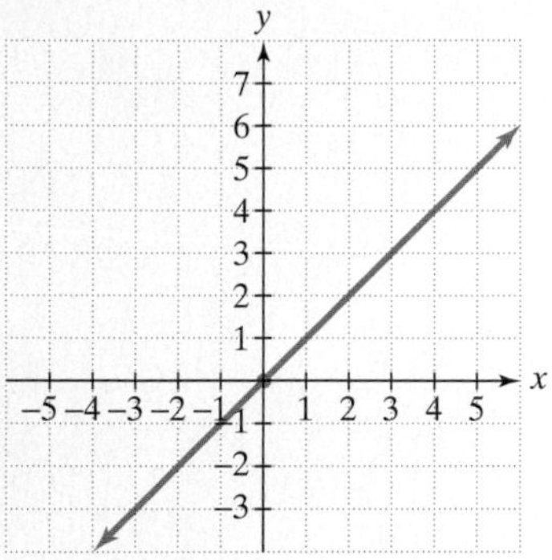

2 Graph a linear equation in two variables using intercepts.

Graphing Using Intercepts An equation of the form $Ax + By = C$, where A, B, and C are integers, is called the **standard form** of the equation of a line. The equation can be graphed by finding the x- and y-intercepts, plotting the intercepts, and drawing a straight line through these points. How do we find the intercepts of a line, given its equation? Because the y-coordinate of the x-intercept is 0, to find the x-intercept,

- Substitute 0 for y in the equation.
- Solve for x.

EXAMPLE 2 Finding the x-Intercept

Find the x-intercept of the graph of $3x - 4y = 24$.

SOLUTION To find the x-intercept, let $y = 0$ and solve for x.

$3x - 4y = 24$	This is the given equation.
$3x - 4 \cdot 0 = 24$	Let $y = 0$.
$3x = 24$	Simplify: $4 \cdot 0 = 0$ and $3x - 0 = 3x$.
$x = 8$	Divide both sides by 3.

The x-intercept is 8. The graph of $3x - 4y = 24$ passes through the point $(8, 0)$. ■

CHECK POINT 2 Find the x-intercept of the graph of $4x - 3y = 12$.

Because the x-coordinate of the y-intercept is 0, to find the y-intercept,

- Substitute 0 for x in the equation.
- Solve for y.

EXAMPLE 3 Finding the y-Intercept

Find the y-intercept of the graph of $3x - 4y = 24$.

SOLUTION To find the y-intercept, let $x = 0$ and solve for y.

$3x - 4y = 24$	This is the given equation.
$3 \cdot 0 - 4y = 24$	Let $x = 0$.
$-4y = 24$	Simplify: $3 \cdot 0 = 0$ and $0 - 4y = -4y$.
$y = -6$	Divide both sides by -4.

The y-intercept is -6. The graph of $3x - 4y = 24$ passes through the point $(0, -6)$. ■

CHECK POINT 3 Find the y-intercept of the graph of $4x - 3y = 12$.

When graphing using intercepts, it is a good idea to use a third point, a checkpoint, before drawing the line. A checkpoint can be obtained by selecting a value for either variable, other than 0, and finding the corresponding value for the other variable. The checkpoint should lie on the same line as the x- and y-intercepts. If it does not, recheck your work and find the error.

USING INTERCEPTS TO GRAPH $Ax + By = C$

1. Find the x-intercept. Let $y = 0$ and solve for x.
2. Find the y-intercept. Let $x = 0$ and solve for y.
3. Find a checkpoint, a third ordered-pair solution.
4. Graph the equation by drawing a line through the three points.

EXAMPLE 4 Using Intercepts to Graph a Linear Equation

Graph: $3x + 2y = 6$.

SOLUTION

Step 1. **Find the x-intercept. Let $y = 0$ and solve for x.**

$$3x + 2 \cdot 0 = 6$$
$$3x = 6$$
$$x = 2$$

The x-intercept is 2, so the line passes through $(2, 0)$.

Step 2. **Find the y-intercept. Let $x = 0$ and solve for y.**

$$3 \cdot 0 + 2y = 6$$
$$2y = 6$$
$$y = 3$$

The y-intercept is 3, so the line passes through $(0, 3)$.

Step 3. **Find a checkpoint, a third ordered-pair solution.** For our checkpoint, we will let $x = 1$ and find the corresponding value for y.

$$3x + 2y = 6 \quad \text{This is the given equation.}$$
$$3 \cdot 1 + 2y = 6 \quad \text{Substitute 1 for x.}$$
$$3 + 2y = 6 \quad \text{Simplify.}$$
$$2y = 3 \quad \text{Subtract 3 from both sides.}$$
$$y = \frac{3}{2} \quad \text{Divide both sides by 2.}$$

The checkpoint is the ordered pair $\left(1, \frac{3}{2}\right)$, or $(1, 1.5)$.

Step 4. **Graph the equation by drawing a line through the three points.** The three points in Figure 3.11 lie along the same line. Drawing a line through the three points results in the graph of $3x + 2y = 6$. ■

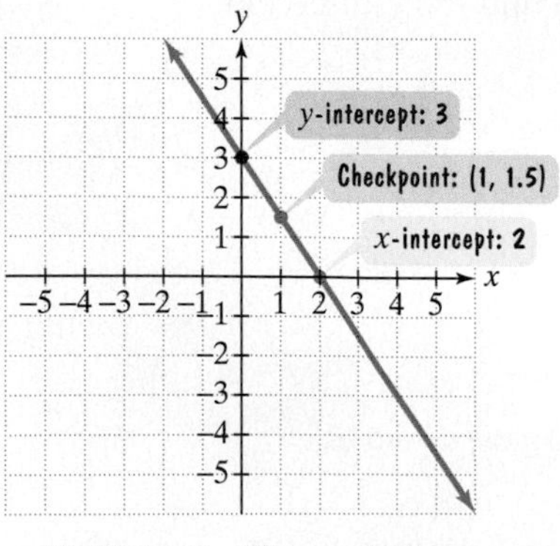

FIGURE 3.11 The graph of $3x + 2y = 6$

USING TECHNOLOGY

You can use a graphing utility to graph equations of the form $Ax + By = C$. Begin by solving the equation for y. For example, to graph $3x + 2y = 6$, solve the equation for y.

$$3x + 2y = 6 \quad \text{This is the equation to be graphed.}$$
$$3x - 3x + 2y = -3x + 6 \quad \text{Subtract } 3x \text{ from both sides.}$$
$$2y = -3x + 6 \quad \text{Simplify.}$$
$$\frac{2y}{2} = \frac{-3x + 6}{2} \quad \text{Divide both sides by 2.}$$
$$y = -\frac{3}{2}x + 3 \quad \text{Simplify.}$$

This is the equation to enter into your graphing utility. The graph of $y = -\frac{3}{2}x + 3$ or, equivalently, $3x + 2y = 6$, is shown below in a $[-6, 6, 1]$ by $[-6, 6, 1]$ viewing rectangle.

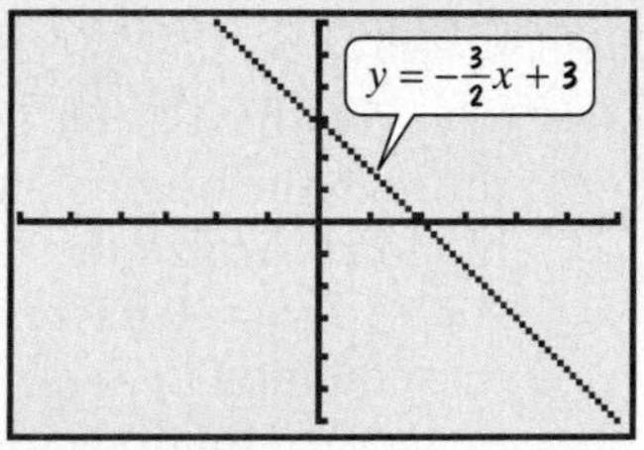

✔ CHECK POINT 4 Graph: $2x + 3y = 6$.

EXAMPLE 5 Using Intercepts to Graph a Linear Equation

Graph: $2x - y = 4$.

SOLUTION

Step 1. Find the x-intercept. Let $y = 0$ and solve for x.

$$2x - 0 = 4$$
$$2x = 4$$
$$x = 2$$

The x-intercept is 2, so the line passes through $(2, 0)$.

Step 2. Find the y-intercept. Let $x = 0$ and solve for y.

$$2 \cdot 0 - y = 4$$
$$-y = 4$$
$$y = -4$$

The y-intercept is -4, so the line passes through $(0, -4)$.

Step 3. Find a checkpoint, a third ordered-pair solution. For our checkpoint, we will let $x = 1$ and find the corresponding value for y.

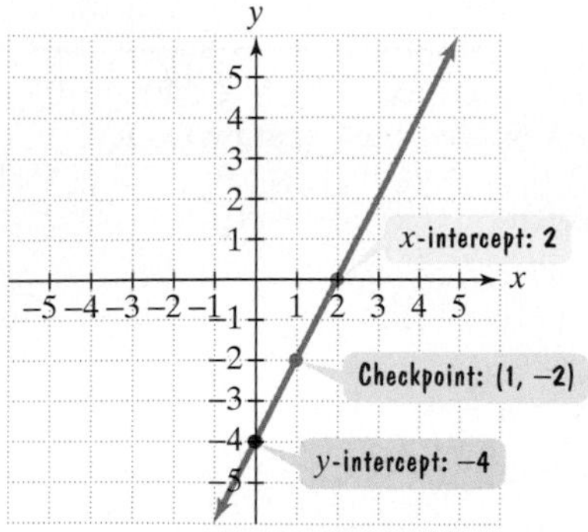

FIGURE 3.12 The graph of $2x - y = 4$

$2x - y = 4$	This is the given equation.
$2 \cdot 1 - y = 4$	Substitute 1 for x.
$2 - y = 4$	Simplify.
$-y = 2$	Subtract 2 from both sides.
$y = -2$	Multiply (or divide) both sides by −1.

The checkpoint is $(1, -2)$.

Step 4. **Graph the equation by drawing a line through the three points.** The three points in Figure 3.12 lie along the same line. Drawing a line through the three points results in the graph of $2x - y = 4$. ■

✔ CHECK POINT 5 Graph: $x - 2y = 4$.

We have seen that not all lines have two different intercepts. Some lines pass through the origin. Thus, they have an x-intercept of 0 and a y-intercept of 0. Is it possible to recognize these lines by their equations? Yes. **The graph of the linear equation $Ax + By = 0$ passes through the origin.** Notice that the constant on the right side of this equation is 0.

An equation of the form $Ax + By = 0$ can be graphed by using the origin as one point on the line. Find two other points by finding two other solutions of the equation. Select values for either variable, other than 0, and find the corresponding values for the other variable.

EXAMPLE 6 Graphing a Linear Equation of the Form $Ax + By = 0$

Graph: $x + 2y = 0$.

SOLUTION Because the constant on the right is 0, the graph passes through the origin. The x- and y-intercepts are both 0. Remember that we are using two points and a checkpoint to determine a line. Thus, we still want to find two other points. Let $y = -1$ to find a second ordered-pair solution. Let $y = 1$ to find a third ordered-pair (checkpoint) solution.

Let $y = -1$.	Let $y = 1$.
$x + 2y = 0$	$x + 2y = 0$
$x + 2(-1) = 0$	$x + 2 \cdot 1 = 0$
$x + (-2) = 0$	$x + 2 = 0$
$x = 2$	$x = -2$

The solutions are $(2, -1)$ and $(-2, 1)$. Plot these two points, as well as the origin—that is, $(0, 0)$. The three points in Figure 3.13 lie along the same line. Drawing a line through the three points results in the graph of $x + 2y = 0$. ■

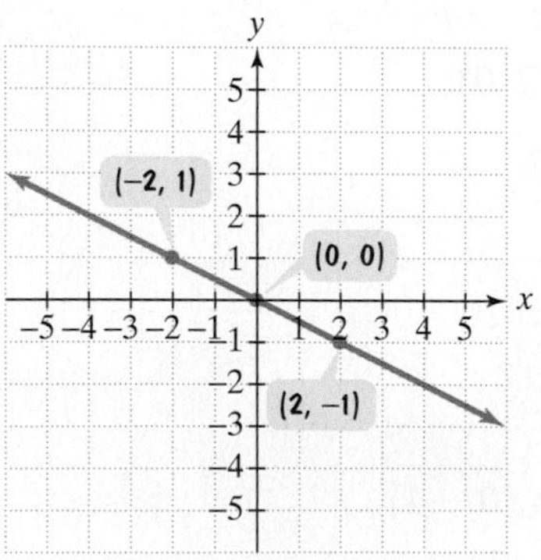

FIGURE 3.13 The graph of $x + 2y = 0$

✔ CHECK POINT 6 Graph: $x + 3y = 0$.

3 Graph horizontal or vertical lines.

Equations of Horizontal and Vertical Lines Some things change very little. For example, from 1985 to the present, the number of Americans participating in downhill skiing has remained relatively constant, indicated by the graph shown in Figure 3.14. Shown in the figure is a horizontal line that passes through or near most of the data points.

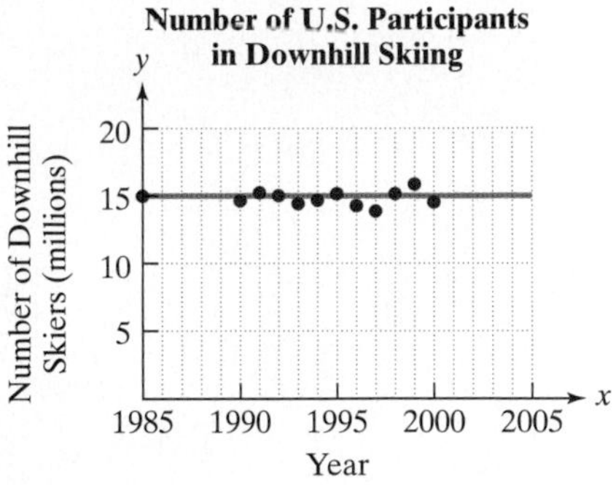

FIGURE 3.14

Source: National Ski Areas Association

We can use the horizontal line in Figure 3.14 to write an equation that reasonably models the data. The y-intercept of the line is 15, so the graph passes through $(0, 15)$. Furthermore, all points on the line have a value of y that is always 15. Thus, an equation that models the number of participants in downhill skiing for the period shown is

$$y = 15.$$

The popularity of downhill skiing has remained relatively constant in the United States at approximately 15 million participants each year.

The equation $y = 15$ can be expressed as $0x + 1y = 15$. We know that the graph of any equation of the form $Ax + By = C$ is a line as long as A and B are not both zero. The graph of $y = 15$ suggests that when A is zero, the line is horizontal.

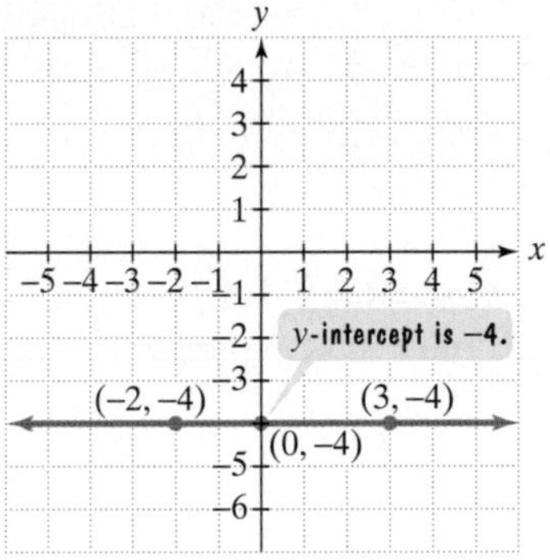

FIGURE 3.15 The graph of $y = -4$

EXAMPLE 7 Graphing a Horizontal Line

Graph the linear equation: $y = -4$.

SOLUTION All ordered pairs that are solutions of $y = -4$ have a value of y that is always -4. Any value can be used for x. Let's select three of the possible values for x: -2, 0, and 3. Using these values of x, three ordered pairs that are solutions of $y = -4$ are $(-2, -4)$, $(0, -4)$, and $(3, -4)$. Plot each of these points. Drawing a line that passes through the three points gives the horizontal line shown in Figure 3.15. ■

CHECK POINT 7 Graph the linear equation: $y = 3$.

Next, let's see what we can discover about the graph of an equation of the form $Ax + By = C$ when B is zero.

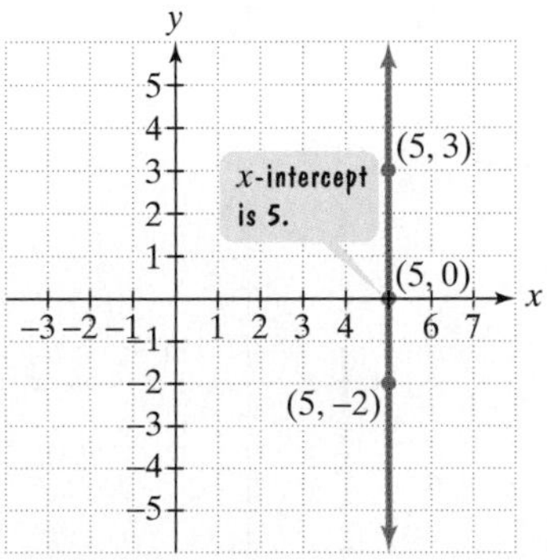

FIGURE 3.16 The graph $x = 5$

EXAMPLE 8 Graphing a Vertical Line

Graph the linear equation: $x = 5$.

SOLUTION All ordered pairs that are solutions of $x = 5$ have a value of x that is always 5. Any value can be used for y. Let's select three of the possible values for y: -2, 0, and 3. Using these values of y, three ordered pairs that are solutions of $x = 5$ are $(5, -2)$, $(5, 0)$, and $(5, 3)$. Drawing a line that passes through the three points gives the vertical line shown in Figure 3.16. ■

STUDY TIP

Do not confuse two-dimensional graphing and one-dimensional graphing of $x = 5$. The graph of $x = 5$ in a two-dimensional rectangular coordinate system is the vertical line in Figure 3.16 on page 203. By contrast, the graph of $x = 5$ on a one-dimensional number line representing values of x is a single point at 5:

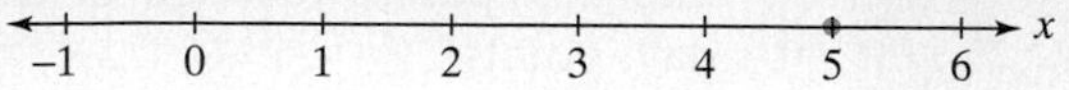

CHECK POINT 8 Graph the linear equation: $x = -2$.

HORIZONTAL AND VERTICAL LINES The graph of a linear equation in one variable is a horizontal or vertical line.

The graph of $y = b$ is a horizontal line. The y-intercept is b.

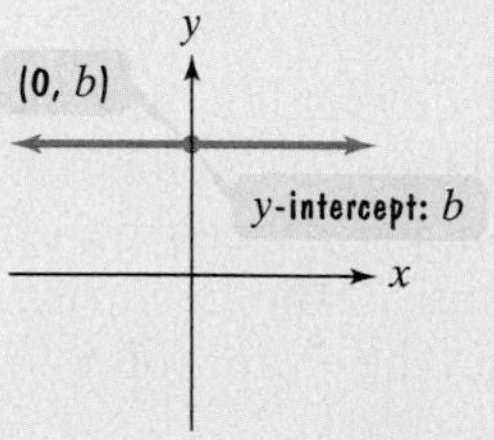

The graph of $x = a$ is a vertical line. The x-intercept is a.

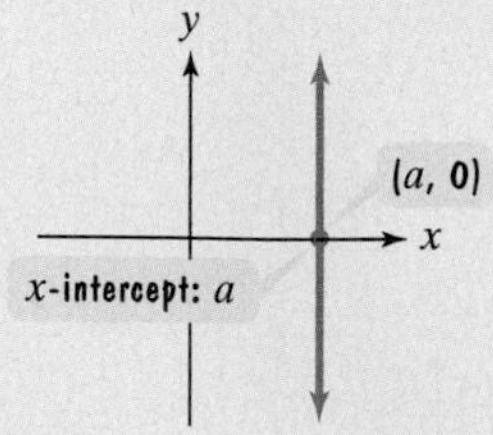

3.2 EXERCISE SET

Student Solutions Manual | CD/Video | PH Math/Tutor Center | MathXL Tutorials on CD | MathXL® | MyMathLab | Interactmath.com

Practice Exercises

In Exercises 1–8, use the graph to identify the

a. *x-intercept, or state that there is no x-intercept;* **b.** *y-intercept, or state that there is no y-intercept.*

1.

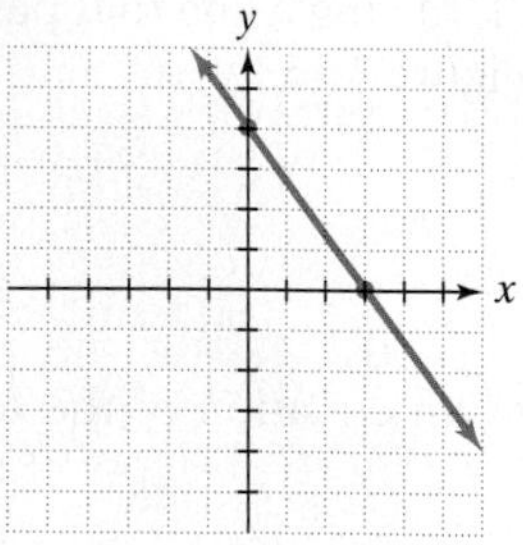

2.

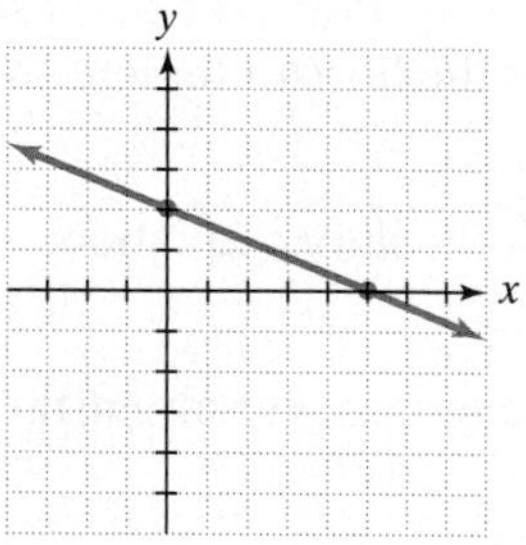

3.

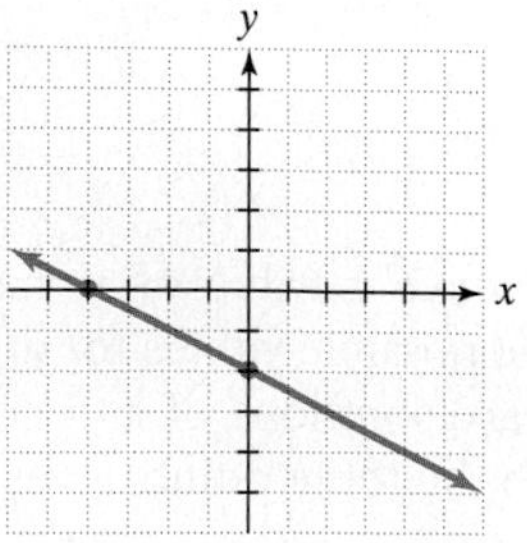

4.

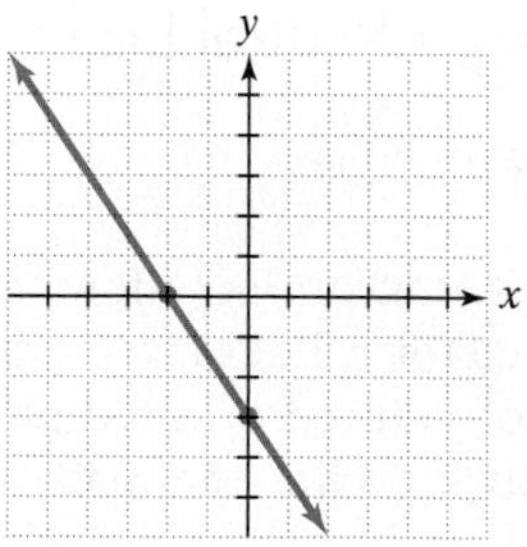

5.

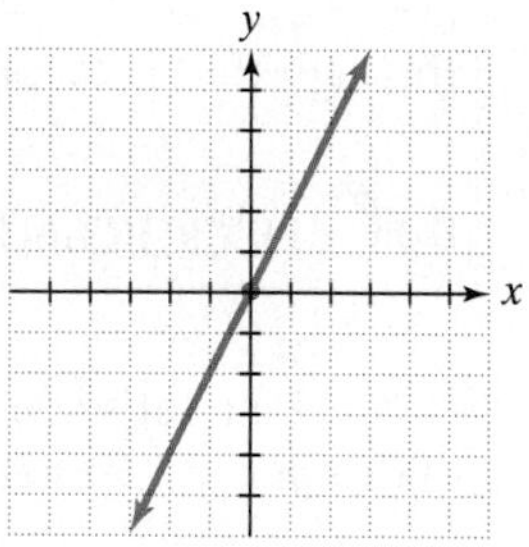

6.

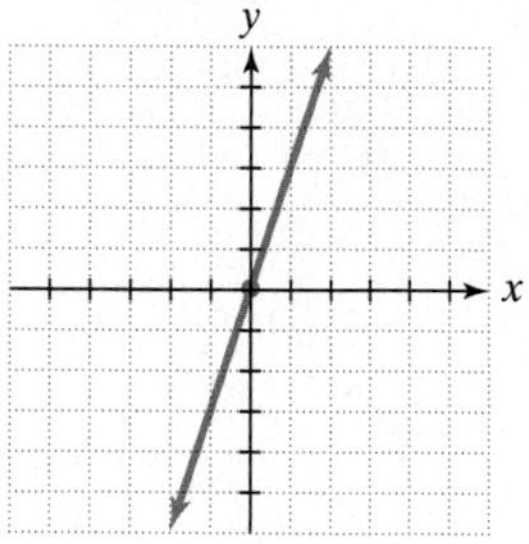

7.

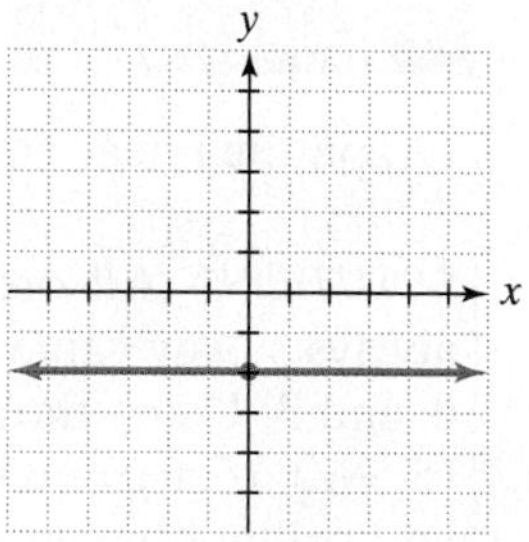

8.

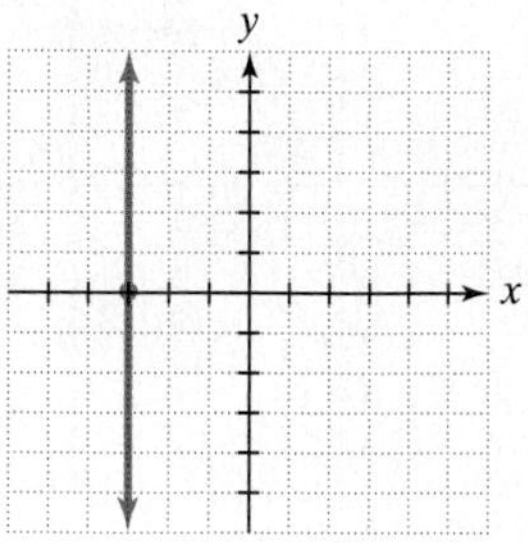

In Exercises 9–18, find the x-intercept and the y-intercept of the graph of each equation. Do not graph the equation.

9. $2x + 5y = 20$

10. $2x + 6y = 30$

11. $2x - 3y = 15$

12. $4x - 5y = 10$

13. $-x + 3y = -8$

14. $-x + 3y = -10$

15. $7x - 9y = 0$

16. $8x - 11y = 0$

17. $2x = 3y - 11$

18. $2x = 4y - 13$

In Exercises 19–40, use intercepts and a checkpoint to graph each equation.

19. $x + y = 5$

20. $x + y = 6$

21. $x + 3y = 6$

22. $2x + y = 4$

23. $6x - 9y = 18$

24. $6x - 2y = 12$

25. $-x + 4y = 6$

26. $-x + 3y = 10$

27. $2x - y = 7$

28. $2x - y = 5$

29. $3x = 5y - 15$

30. $2x = 3y + 6$

31. $25y = 100 - 50x$

32. $10y = 60 - 40x$

33. $2x - 8y = 12$

34. $3x - 6y = 15$

35. $x + 2y = 0$

36. $2x + y = 0$

37. $y - 3x = 0$

38. $y - 4x = 0$

39. $2x - 3y = -11$

40. $3x - 2y = -7$

In Exercises 41–46, write an equation for each graph.

41.

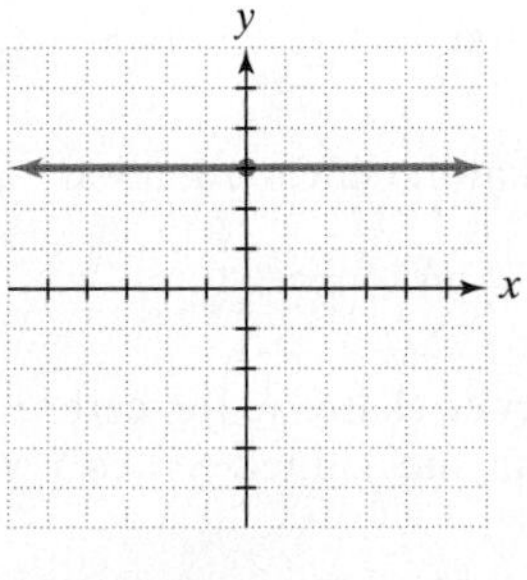

42.

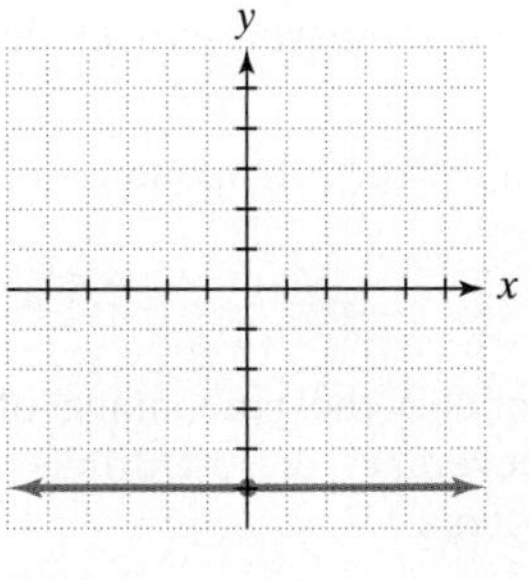

43.

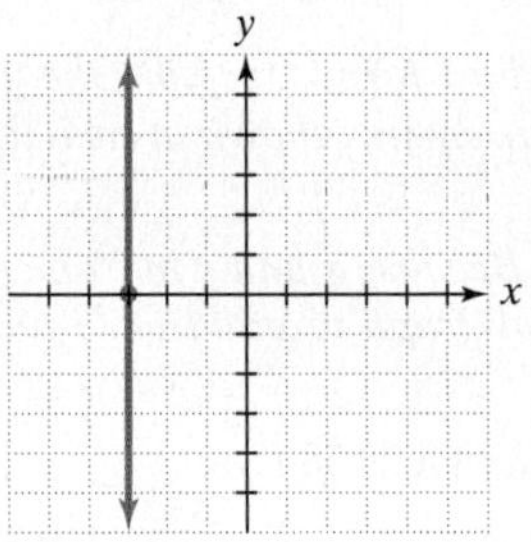

44.

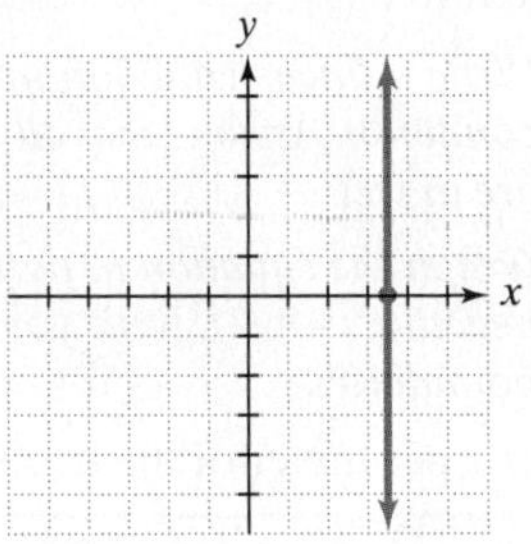

45.

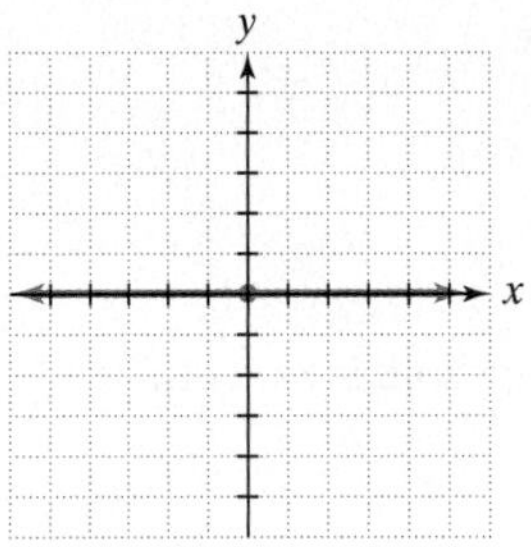

46.

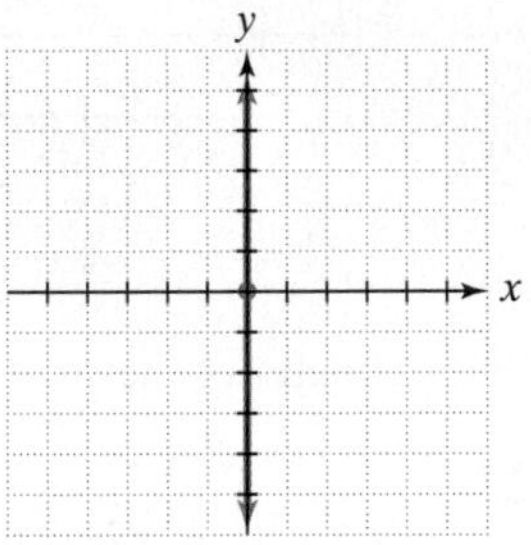

In Exercises 47–62, graph each equation.

47. $y = 4$

48. $y = 2$

49. $y = -2$

50. $y = -3$

51. $x = 2$

52. $x = 4$

53. $x + 1 = 0$

54. $x + 5 = 0$

55. $y - 3.5 = 0$

56. $y - 2.5 = 0$

57. $x = 0$

58. $y = 0$

59. $3y = 9$

60. $5y = 20$

61. $12 - 3x = 0$

62. $12 - 4x = 0$

Practice Plus

In Exercises 63–68, match each equation with one of the graphs shown in Exercises 1–8.

63. $3x + 2y = -6$

64. $x + 2y = -4$

65. $y = -2$

66. $x = -3$

67. $4x + 3y = 12$

68. $2x + 5y = 10$

In Exercises 69–70,

a. *Write a linear equation in standard form satisfying the given condition. Assume that all measurements shown in each figure are in feet.*

b. *Graph the equation in part (a). Because x and y must be nonnegative (why?), limit your final graph to quadrant I and its boundaries.*

69. The perimeter of the larger rectangle is 58 feet.

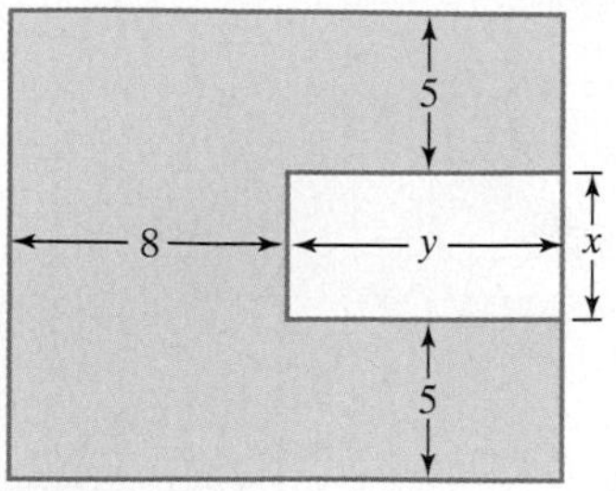

70. The perimeter of the shaded trapezoid is 84 feet.

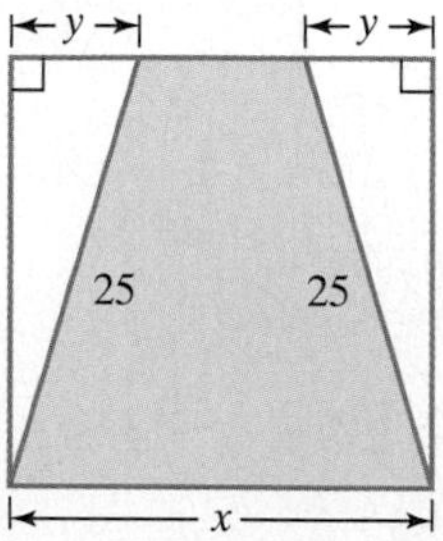

Application Exercises

The flight of a vulture is observed for 30 seconds. The graph shows the vulture's height, in meters, during this period of time. Use the graph to solve Exercises 71–75.

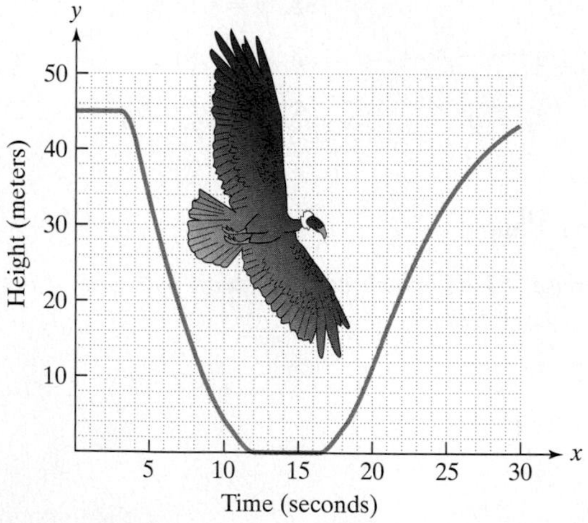

71. During which period of time is the vulture's height decreasing?

72. During which period of time is the vulture's height increasing?

73. What is the y-intercept? What does this mean about the vulture's height at the beginning of the observation?

74. During the first three seconds of observation, the vulture's flight is graphed as a horizontal line. Write the equation of the line. What does this mean about the vulture's flight pattern during this time?

75. Use integers to write five x-intercepts of the graph. What is the vulture doing during these times?

Too late for that flu shot now! It's only 8 A.M. and you're feeling lousy. Fascinated by the way that algebra models the world (your author is projecting a bit here), you decide to construct a graph showing your body temperature from 8 A.M. through 3 P.M. You decide to let x represent the number of hours after 8 A.M. and y represent your temperature at time x. The graph is shown. Use it to solve Exercises 76–80.

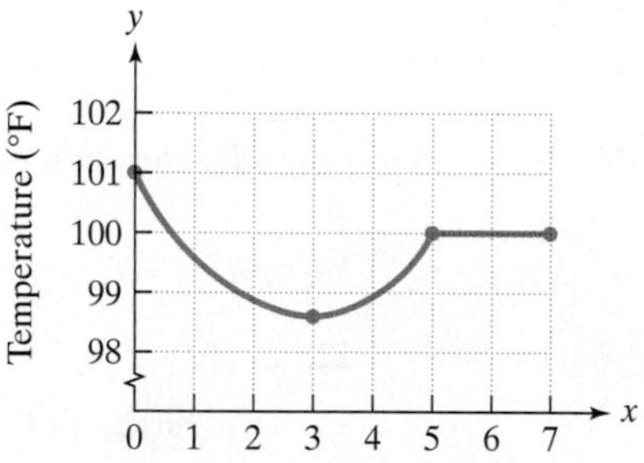

76. What is the y-intercept? What does this mean about your temperature at 8 A.M.?

77. During which period of time is your temperature decreasing?

78. Estimate your minimum temperature during the time period shown. How many hours after 8 A.M. does this occur? At what time does this occur?

79. During which period of time is your temperature increasing?

80. From five hours after 8 A.M. until seven hours after 8 A.M., your temperature is graphed as a portion of a horizontal line. Write the equation of the line. What does this mean about your temperature over this period of time?

81. In the section opener, we saw that the linear equation

$$0.57x + y = 80$$

models the percentage of market share, y, for carbonated beverages of soft drinks sold in the United States x years after 1994.

a. Find the y-intercept of the equation's graph. Describe what this means in terms of the variables in the equation.

b. How well does your description in part (a) model the appropriate portion of the data shown in Figure 3.9 on page 197?

c. Find the x-intercept of the equation's graph. Round to the nearest whole number. Describe what this means in terms of the variables in the equation. Does this seem reasonable or has model breakdown occurred?

82. As shown in the bar graph, the percentage of people in the United States satisfied with their lives remains relatively constant for all age groups. If x represents a person's age and y represents the percentage of people satisfied with their lives at that age, write an equation that reasonably models the data.

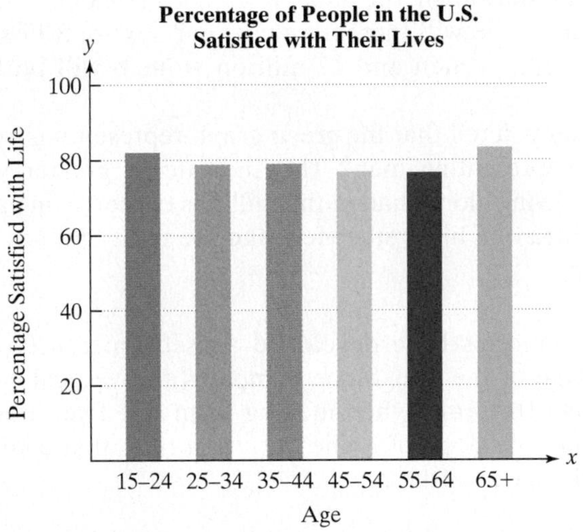

Source: *Culture Shift in Advanced Industrial Society*, Princeton University Press

Writing in Mathematics

83. What is an x-intercept of a graph?

84. What is a y-intercept of a graph?

85. If you are given an equation of the form $Ax + By = C$, explain how to find the x-intercept.

86. If you are given an equation of the form $Ax + By = C$, explain how to find the y-intercept.

87. Explain how to graph $Ax + By = C$ if C is not equal to zero.

88. Explain how to graph a linear equation of the form $Ax + By = 0$.

89. How many points are needed to graph a line? How many should actually be used? Explain.

90. Describe the graph of $y = 200$.

91. Describe the graph of $x = -100$.

92. We saw that the number of skiers in the United States has remained constant over time. Exercise 82 showed that the percentage of people satisfied with their lives remains constant for all age groups. Give another example of a real-world phenomenon that has remained relatively constant. Try writing an equation that models this phenomenon.

Critical Thinking Exercises

93. Write the equation of the line passing through the point $(5, 6)$ and parallel to the line whose equation is $y = -1$.

In Exercises 94–95, find the coefficients that must be placed in each shaded area so that the equation's graph will be a line with the specified intercepts.

94. $\square x + \square y = 10$; x-intercept $= 5$; y-intercept $= 2$

95. $\square x + \square y = 12$; x-intercept $= -2$; y-intercept $= 4$

Technology Exercises

96. Use a graphing utility to verify any five of your hand-drawn graphs in Exercises 19–40. Solve the equation for y before entering it.

In Exercises 97–100, use a graphing utility to graph each equation. You will need to solve the equation for y before entering it. Use the equation displayed on the screen to identify the x-intercept and the y-intercept.

97. $2x + y = 4$

98. $3x - y = 9$

99. $2x + 3y = 30$

100. $4x - 2y = -40$

Review Exercises

101. Find the absolute value: $|-13.4|$. (Section 1.2, Example 8)

102. Simplify: $7x - (3x - 5)$. (Section 1.7, Example 7)

103. Graph: $-2 \le x < 4$. (Section 2.7, Example 1)

SECTION 3.3

Objectives

1 Compute a line's slope.

2 Use slope to show that lines are parallel.

3 Calculate rate of change in applied situations.

SLOPE

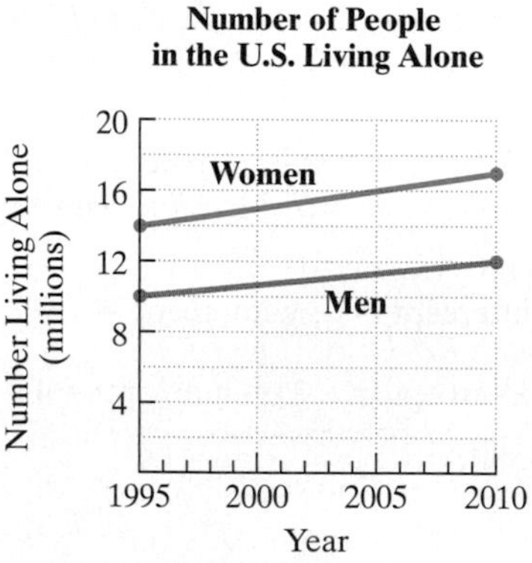

FIGURE 3.17

Source: Forrester Research

A best guess at the look of our nation in the next decades indicates that the number of men and women living alone will increase each year. Figure 3.17 shows that by 2010, approximately 12 million men and 17 million women will be living alone.

By looking at Figure 3.17, can you tell that the green graph representing women is steeper than the blue graph representing men? This indicates a greater yearly increase in the millions of women living alone than in the millions of men living alone. In this section, we will study the idea of a line's steepness and see what that has to do with how its variables are changing.

The Slope of a Line Mathematicians have developed a useful measure of the steepness of a line, called the **slope** of the line. Slope compares the vertical change (the **rise**) to the horizontal change (the **run**) when moving from one fixed point to another along the line. To calculate the slope of a line, we use a ratio that compares the change in y (the rise) to the change in x (the run).

DEFINITION OF SLOPE The **slope** of the line through the distinct points (x_1, y_1) and (x_2, y_2) is

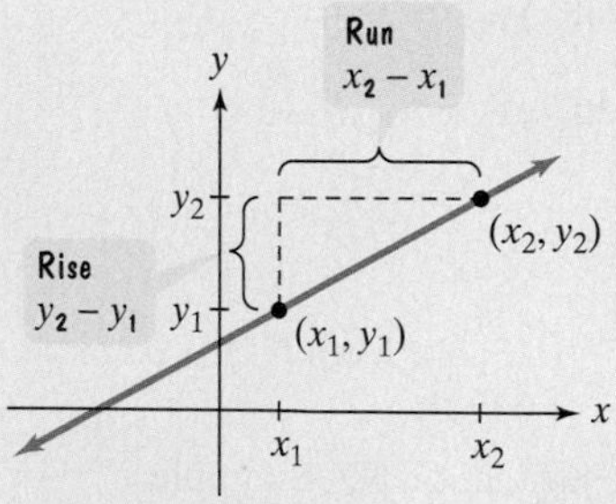

$$\frac{\text{Change in } y}{\text{Change in } x} = \frac{\text{Rise}}{\text{Run}} = \frac{y_2 - y_1}{x_2 - x_1},$$

where $x_2 - x_1 \neq 0$.

It is common notation to let the letter m represent the slope of a line. The letter m is used because it is the first letter of the French verb *monter*, meaning to rise, or to ascend.

1 Compute a line's slope.

EXAMPLE 1 Using the Definition of Slope

Find the slope of the line passing through each pair of points:

a. $(-3, -1)$ and $(-2, 4)$ **b.** $(-3, 4)$ and $(2, -2)$.

SOLUTION

a. Let $(x_1, y_1) = (-3, -1)$ and $(x_2, y_2) = (-2, 4)$. We obtain the slope as follows:

$$m = \frac{\text{Change in } y}{\text{Change in } x} = \frac{y_2 - y_1}{x_2 - x_1} = \frac{4 - (-1)}{-2 - (-3)} = \frac{5}{1} = 5.$$

The situation is illustrated in Figure 3.18(a). The slope of the line is 5, indicating that there is a vertical change, a rise, of 5 units for each horizontal change, a run, of 1 unit. The slope is positive and the line rises from left to right.

STUDY TIP

When computing slope, it makes no difference which point you call (x_1, y_1) and which point you call (x_2, y_2). If we let $(x_1, y_1) = (-2, 4)$ and $(x_2, y_2) = (-3, -1)$, the slope is still 5:

$$m = \frac{\text{Change in } y}{\text{Change in } x} = \frac{y_2 - y_1}{x_2 - x_1} = \frac{-1 - 4}{-3 - (-2)} = \frac{-5}{-1} = 5.$$

However, you should not subtract in one order in the numerator $(y_2 - y_1)$ and then in a different order in the denominator $(x_1 - x_2)$. The slope is *not* -5:

$$\frac{-1 - 4}{-2 - (-3)} = \frac{-5}{1} = -5. \quad \text{Incorrect}$$

b. We can let $(x_1, y_1) = (-3, 4)$ and $(x_2, y_2) = (2, -2)$. The slope of the line shown in Figure 3.18(b) is computed as follows:

$$m = \frac{\text{Change in } y}{\text{Change in } x} = \frac{y_2 - y_1}{x_2 - x_1} = \frac{-2 - 4}{2 - (-3)} = \frac{-6}{5} = -\frac{6}{5}.$$

The slope of the line is $-\frac{6}{5}$. For every vertical change of -6 units (6 units down), there is a corresponding horizontal change of 5 units. The slope is negative and the line falls from left to right.

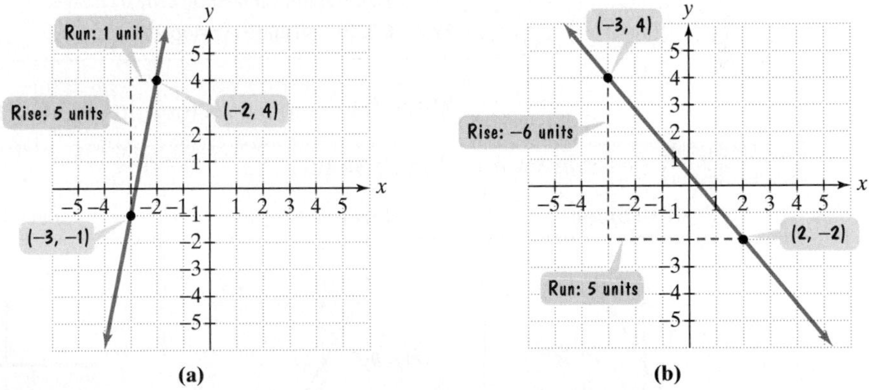

FIGURE 3.18 Visualizing slope

■

CHECK POINT 1 Find the slope of the line passing through each pair of points:

a. $(-3, 4)$ and $(-4, -2)$ **b.** $(4, -2)$ and $(-1, 5)$.

ENRICHMENT ESSAY

Slope and the Streets of San Francisco

San Francisco's Filbert Street has a slope of 0.613, meaning that for every horizontal distance of 100 feet, the street ascends 61.3 feet vertically. With its 31.5° angle of inclination, the street is too steep to pave and is only accessible by wooden stairs.

EXAMPLE 2 Using the Definition of Slope for Horizontal and Vertical Lines

Find the slope of the line passing through each pair of points:

a. (5, 4) and (3, 4) **b.** (2, 5) and (2, 1).

SOLUTION

a. Let $(x_1, y_1) = (5, 4)$ and $(x_2, y_2) = (3, 4)$. We obtain the slope as follows:

$$m = \frac{\text{Change in } y}{\text{Change in } x} = \frac{y_2 - y_1}{x_2 - x_1} = \frac{4 - 4}{3 - 5} = \frac{0}{-2} = 0.$$

The situation is illustrated in Figure 3.19(a). Can you see that the line is horizontal? Because any two points on a horizontal line have the same y-coordinate, these lines neither rise nor fall from left to right. The change in y, $y_2 - y_1$, is always zero. Thus, **the slope of any horizontal line is zero.**

b. We can let $(x_1, y_1) = (2, 5)$ and $(x_2, y_2) = (2, 1)$. Figure 3.19(b) shows that these points are on a vertical line. We attempt to compute the slope as follows:

$$m = \frac{\text{Change in } y}{\text{Change in } x} = \frac{1 - 5}{2 - 2} = \frac{-4}{0}$$

Division by zero is undefined.

Because division by zero is undefined, the slope of the vertical line in Figure 3.19(b) is undefined. In general, **the slope of any vertical line is undefined.**

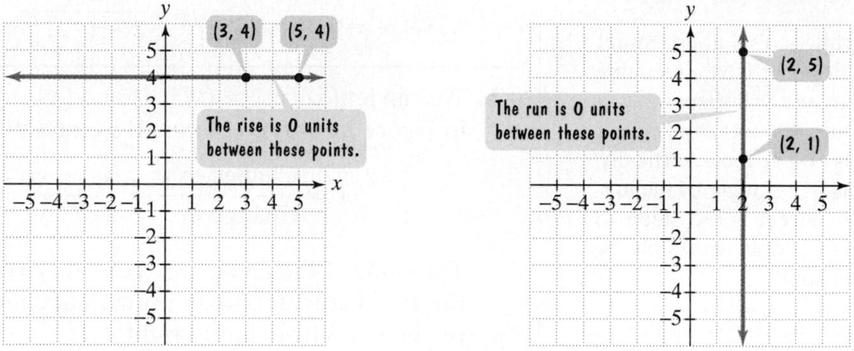

(a) Horizontal lines have no vertical change. (b) Vertical lines have no horizontal change.

FIGURE 3.19 Visualizing slope

Table 3.1 summarizes four possibilities for the slope of a line.

Table 3.1 Possibilities for a Line's Slope

Positive Slope	Negative Slope	Zero Slope	Undefined Slope
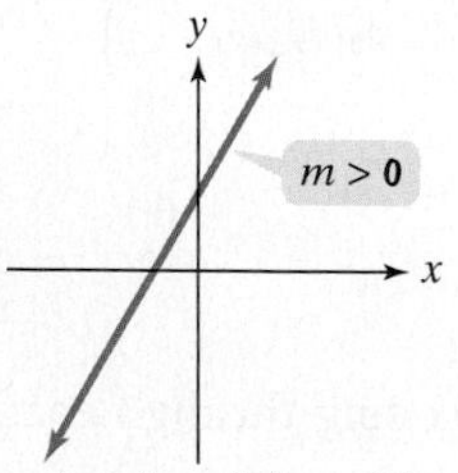	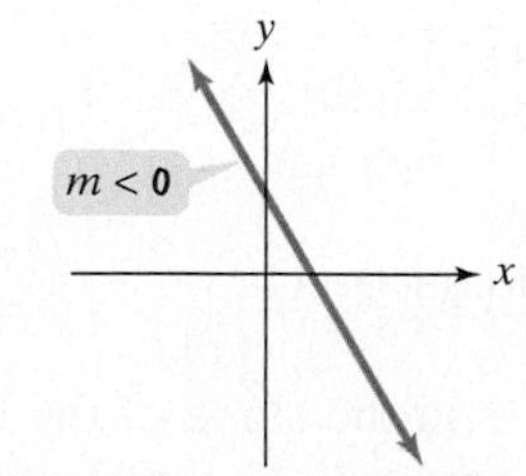	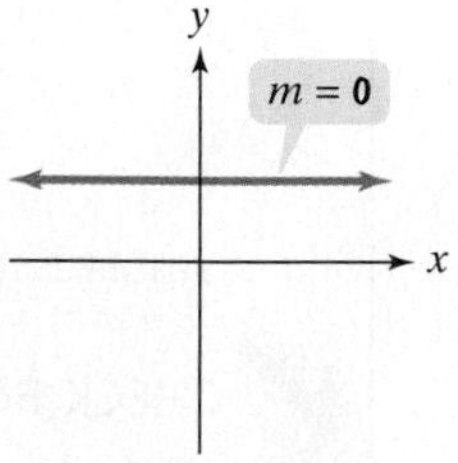	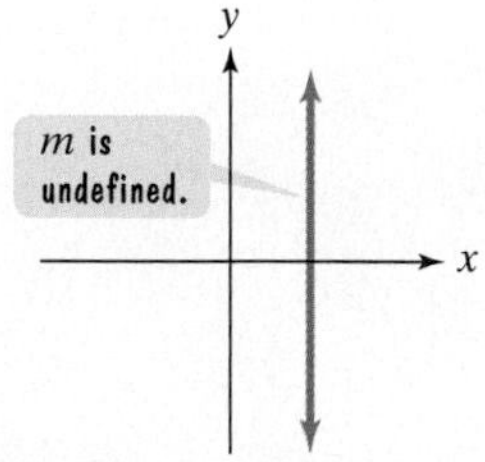
Line rises from left to right.	Line falls from left to right.	Line is horizontal.	Line is vertical.

CHECK POINT 2 Find the slope of the line passing through each pair of points or state that the slope is undefined:

a. (6, 5) and (2, 5) **b.** (1, 6) and (1, 4).

2 Use slope to show that lines are parallel.

Slope and Parallel Lines Two nonintersecting lines that lie in the same plane are **parallel**. If two lines do not intersect, the ratio of the vertical change to the horizontal change is the same for each line. Because two parallel lines have the same "steepness," they must have the same slope.

SLOPE AND PARALLEL LINES

1. If two nonvertical lines are parallel, then they have the same slope.
2. If two distinct nonvertical lines have the same slope, then they are parallel.
3. Two distinct vertical lines, each with undefined slope, are parallel.

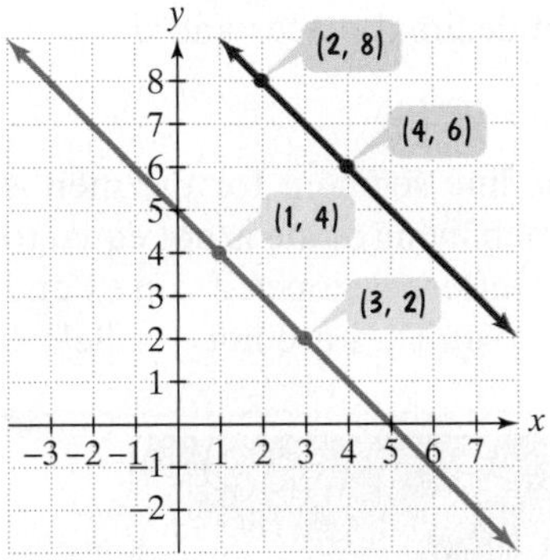

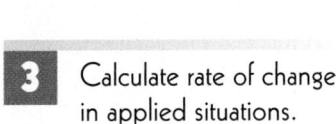

FIGURE 3.20 Using slope to show that lines are parallel

EXAMPLE 3 Using Slope to Show That Lines Are Parallel

Show that the line passing through (1, 4) and (3, 2) is parallel to the line passing through (2, 8) and (4, 6).

SOLUTION The situation is illustrated in Figure 3.20. The lines certainly look like they are parallel. Let's use equal slopes to confirm this fact. For each line, we compute the ratio of the difference in y-coordinates to the difference in x-coordinates. Be sure to subtract the coordinates in the same order.

Slope of the line through (1, 4) and (3, 2) is

$$\frac{4-2}{1-3} = \frac{2}{-2} = -1.$$

Slope of the line through (2, 8) and (4, 6) is

$$\frac{8-6}{2-4} = \frac{2}{-2} = -1.$$

Because the slopes are equal, the lines are parallel. ■

CHECK POINT 3 Show that the line passing through (4, 2) and (6, 6) is parallel to the line passing through (0, −2) and (1, 0).

3 Calculate rate of change in applied situations.

Slope as Rate of Change Slope is defined as the ratio of a change in y to a corresponding change in x. It tells how fast y is changing with respect to x. Thus, the slope of a line represents its rate of change.

Our next example shows how slope can be interpreted as a rate of change in an applied situation. When calculating slope in applied problems, keep track of the units in the numerator and the denominator.

EXAMPLE 4 Slope as a Rate of Change

The line graphs for the number of women and men living alone are shown again in Figure 3.21 at the top of the next page. Find the slope of the line segment for the women. Describe what this slope represents.

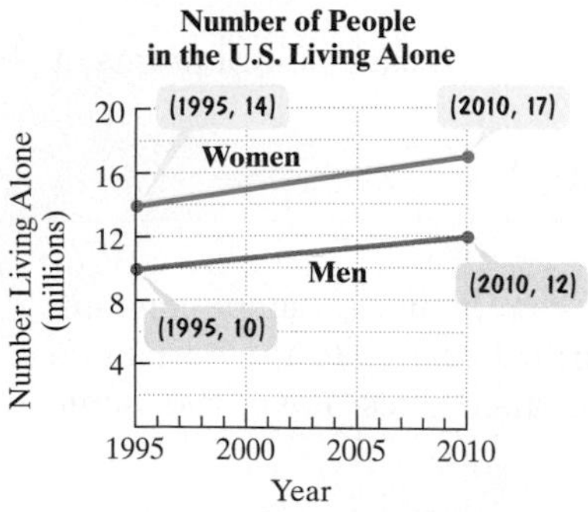

FIGURE 3.21
Source: Forrester Research

SOLUTION We let x represent a year and y the number of women living alone in that year. The two points shown on the line segment for women have the following coordinates:

$$(1995, 14) \quad \text{and} \quad (2010, 17).$$

In 1995, 14 million U.S. women lived alone.

In 2010, 17 million U.S. women are projected to live alone.

Now we compute the slope:

$$m = \frac{\text{Change in } y}{\text{Change in } x} = \frac{17 - 14}{2010 - 1995}$$

$$= \frac{3}{15} = \frac{1}{5} = \frac{0.2 \text{ million people}}{\text{year}}.$$

The unit in the numerator is *million women.*

The unit in the denominator is *year.*

The slope indicates that the number of U.S. women living alone is projected to increase by 0.2 million each year. The rate of change is 0.2 million women per year. ■

✔ **CHECK POINT 4** Use the graph in Example 4 to find the slope of the line segment for the men. Express the slope correct to two decimal places and describe what it represents.

In Check Point 4, did you find that the slope of the line segment for the men is different from that of the women? The rate of change for men living alone is not equal to the rate of change for women living alone. Because of these different slopes, if you extend the line segments in Figure 3.21, the resulting lines will intersect. They are not parallel.

ENRICHMENT ESSAY

Railroads and Highways

The steepest part of Mt. Washington Cog Railway in New Hampshire has a 37% grade. This is equivalent to a slope of $\frac{37}{100}$. For every horizontal change of 100 feet, the railroad ascends 37 feet vertically. Engineers denote slope by grade, expressing slope as a percentage.

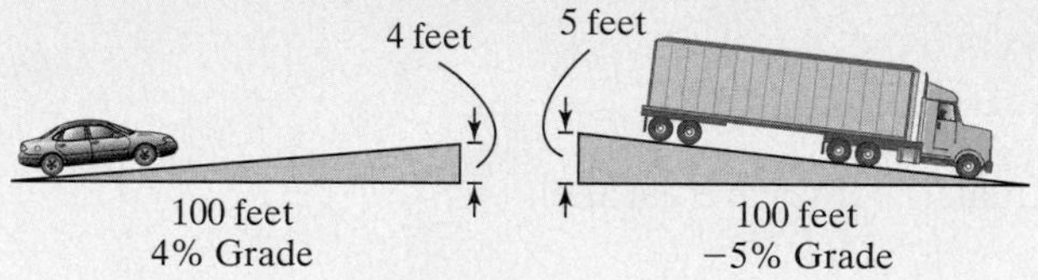

Railroad grades are usually less than 2%, although in the mountains they may go as high as 4%. The grade of the Mt. Washington Cog Railway is phenomenal, making it necessary for locomotives to *push* single cars up its steepest part.

A Mount Washington Cog Railway locomotive pushing a single car up the steepest part of the railroad. The locomotive is about 120 years old.

3.3 EXERCISE SET

Student Solutions Manual | CD/Video | PH Math/Tutor Center | MathXL Tutorials on CD | MathXL® | MyMathLab | Interactmath.com

Practice Exercises

In Exercises 1–10, find the slope of the line passing through each pair of points or state that the slope is undefined. Then indicate whether the line through the points rises, falls, is horizontal, or is vertical.

1. $(4, 7)$ and $(8, 10)$

2. $(2, 1)$ and $(3, 4)$

3. $(-2, 1)$ and $(2, 2)$

4. $(-1, 3)$ and $(2, 4)$

5. $(4, -2)$ and $(3, -2)$

6. $(4, -1)$ and $(3, -1)$

7. $(-2, 4)$ and $(-1, -1)$

8. $(6, -4)$ and $(4, -2)$

9. $(5, 3)$ and $(5, -2)$

10. $(3, -4)$ and $(3, 5)$

In Exercises 11–22, find the slope of each line, or state that the slope is undefined.

11.

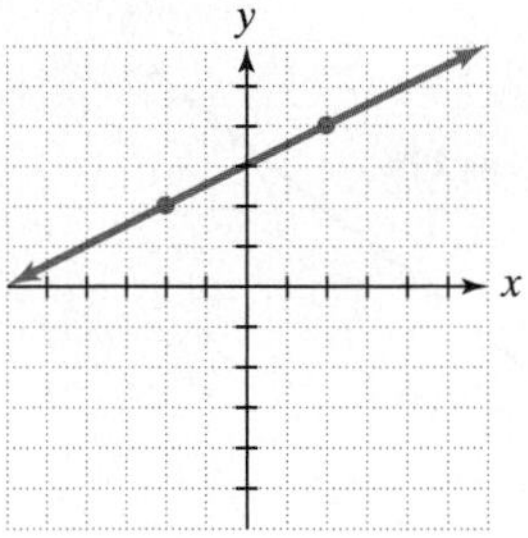

12.

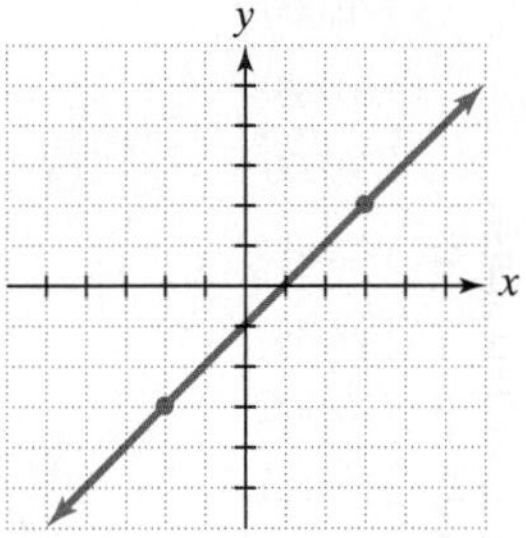

13.

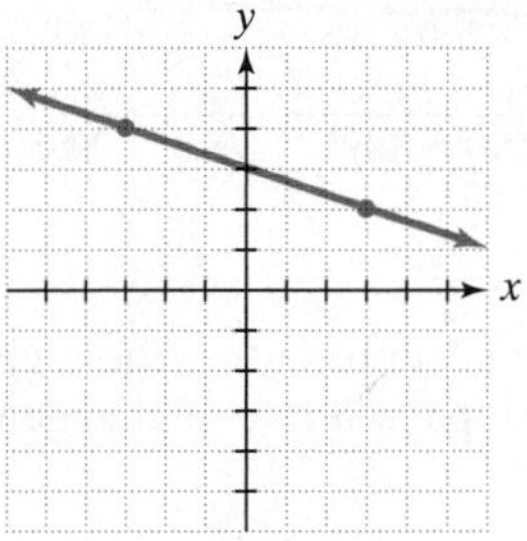

14.

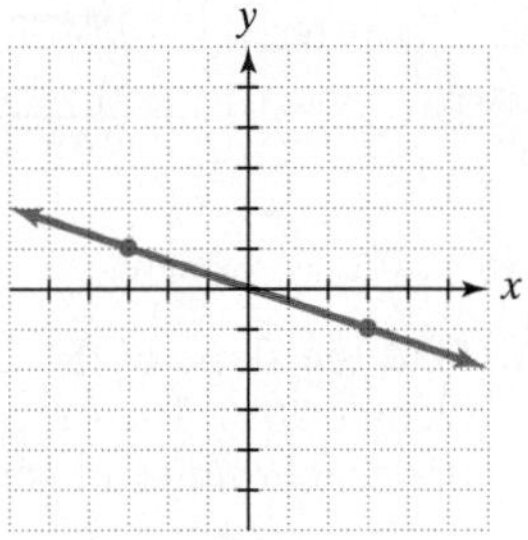

15.

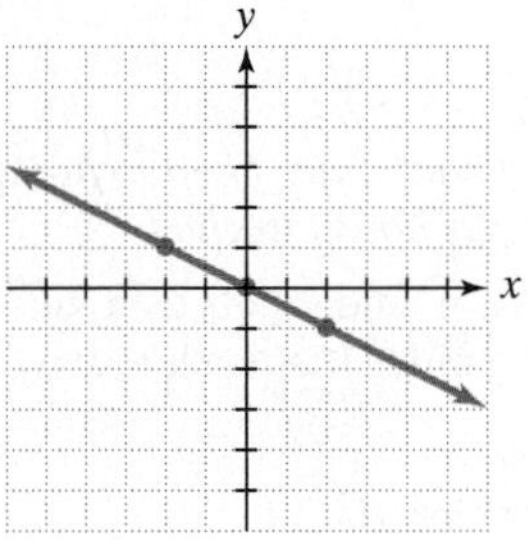

16.

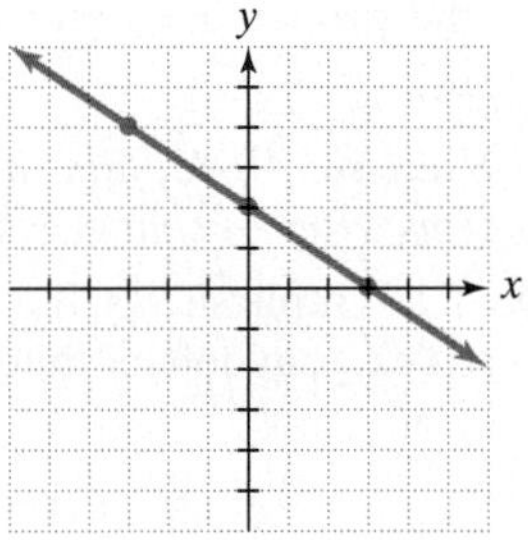

17.

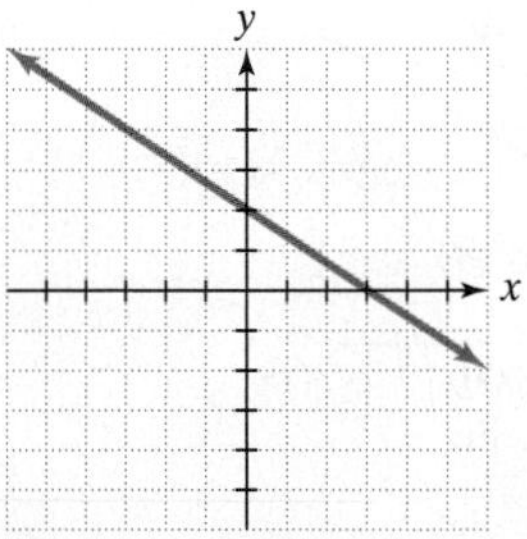

18.

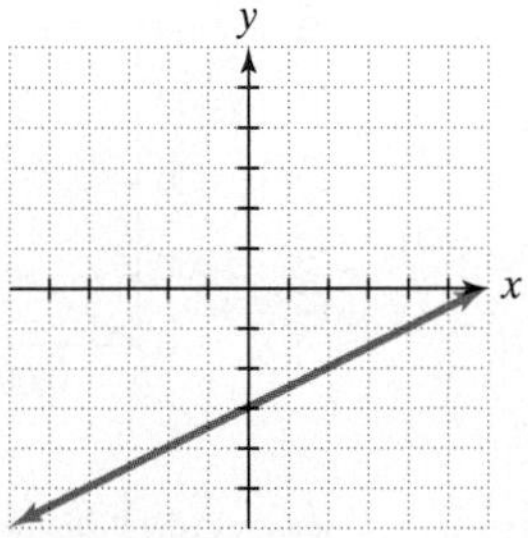

19.

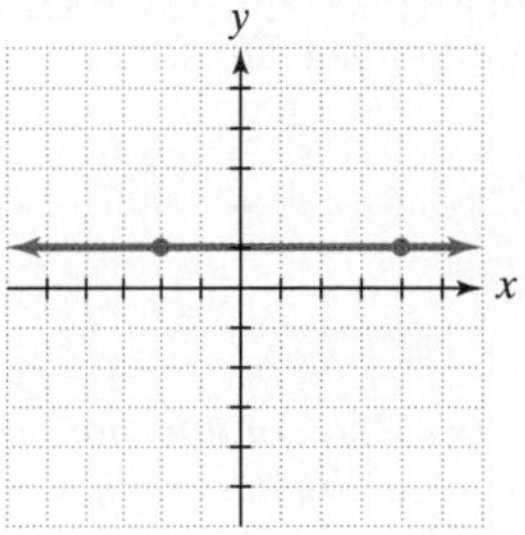

20.

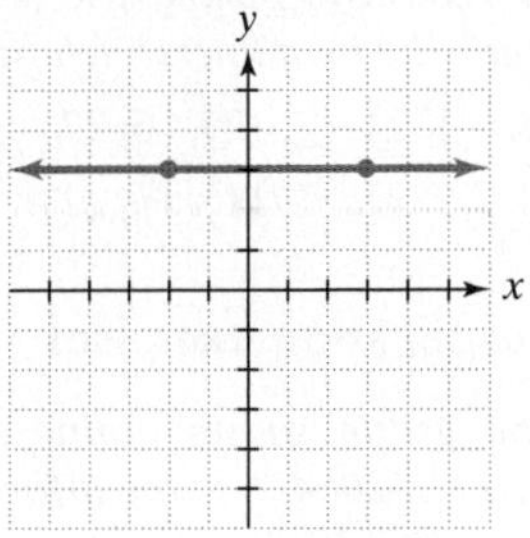

21.

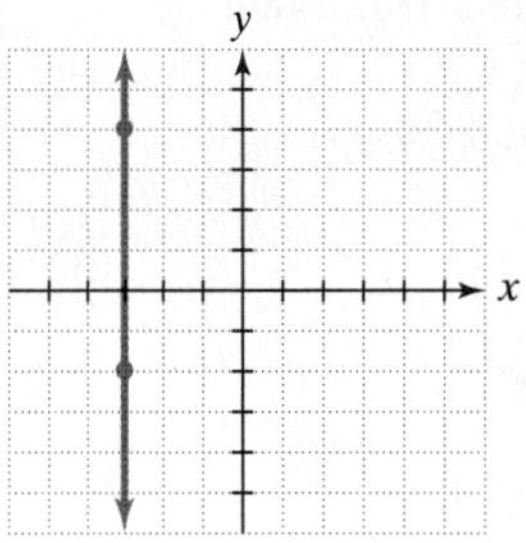

22.

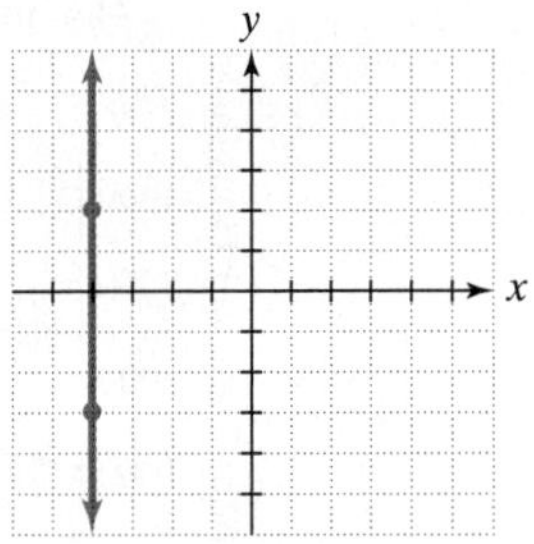

In Exercises 23–26, determine whether the distinct lines through each pair of points are parallel.

23. $(-2, 0)$ and $(0, 6)$; $(1, 8)$ and $(0, 5)$

24. $(2, 4)$ and $(6, 1)$; $(-3, 1)$ and $(1, -2)$

25. $(0, 3)$ and $(1, 5)$; $(-1, 7)$ and $(1, 10)$

26. $(-7, 6)$ and $(0, 4)$; $(-9, -3)$ and $(1, 5)$

Practice Plus

27. On the same set of axes, draw lines passing through the origin with slopes -1, $-\frac{1}{2}$, 0, $\frac{1}{3}$, and 2.

28. On the same set of axes, draw lines with y-intercept 4 and slopes -1, $-\frac{1}{2}$, 0, $\frac{1}{3}$, and 2.

Use slopes to solve Exercises 29–30.

29. Show that the points whose coordinates are $(-3, -3)$, $(2, -5)$, $(5, -1)$, and $(0, 1)$ are the vertices of a four-sided figure whose opposite sides are parallel. (Such a figure is called a *parallelogram*.)

30. Show that the points whose coordinates are $(-3, 6)$, $(2, -3)$, $(11, 2)$, and $(6, 11)$ are the vertices of a four-sided figure whose opposite sides are parallel.

31. The line passing through $(5, y)$ and $(1, 0)$ is parallel to the line joining $(2, 3)$ and $(-2, 1)$. Find y.

32. The line passing through $(1, y)$ and $(7, 12)$ is parallel to the line joining $(-3, 4)$ and $(-5, -2)$. Find y.

In Exercises 33–34, use slopes to determine whether the given points lie on a line. (Such points are said to be collinear.)

33. $(1, 2), (3, 12), (6, 27)$

34. $(1, 3), (3, 11), (6, 22)$

Application Exercises

The graph shows online shopping per U.S. online household through 2004. Use the information provided by the graph to solve Exercises 35–36.

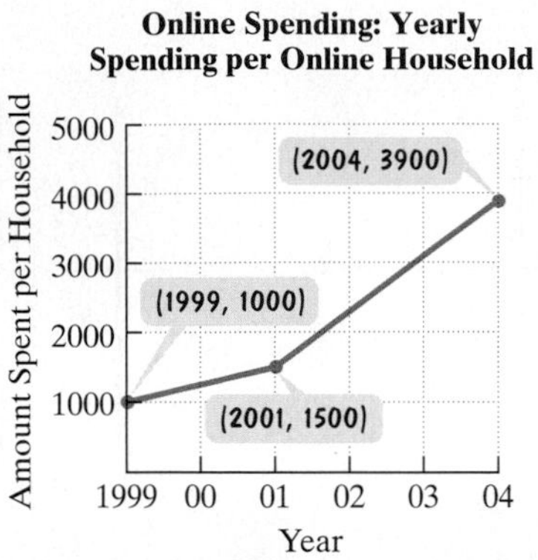

Source: Forrester Research

35. Find the slope of the line passing through (1999, 1000) and (2001, 1500). What does this represent in terms of the increase in online shopping per year?

36. Find the slope of the line passing through (2001, 1500) and (2004, 3900). What does this represent in terms of the increase in online shopping per year?

The graphs approximate the percentage of full-time police officers in the United States, by gender, y, for year x, where $1972 \le x \le 2002$. *Use this information to solve Exercises 37–38.*

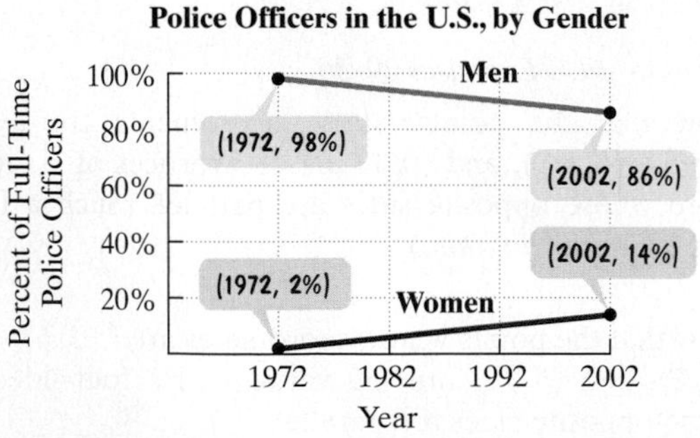

Source: National Center for Women and Policing

37. Find the slope of the line segment representing men. Then express the slope as a rate of change with the proper units attached.

38. Find the slope of the line segment representing women. Then express the slope as a rate of change with the proper units attached.

The graphs approximate the mean, or average, income for well-to-do U.S. families (the highest 20%) and the lowest-income U.S. families (the lowest 20%), y, for year x, where $1980 \le x \le 2002$. *Use this information to solve Exercises 39–40.*

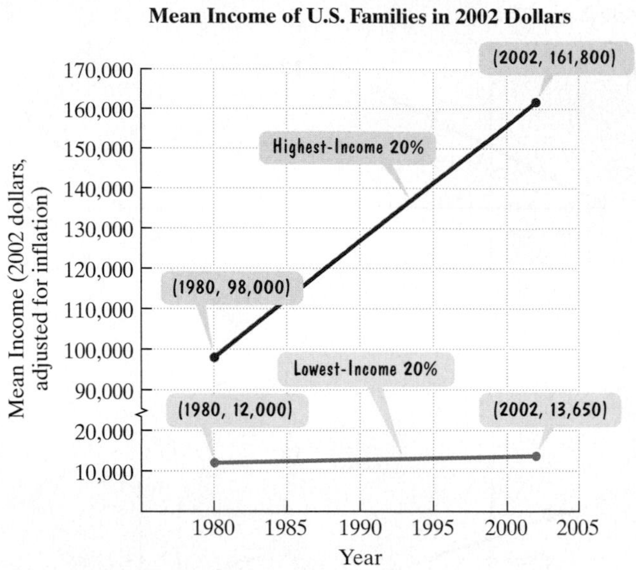

Source: U.S. Census Bureau

39. Find the slope of the line segment representing the highest-income 20%. Then express the slope as a rate of change with the proper units attached.

40. Find the slope of the line segment representing the lowest-income 20%. Then express the slope as a rate of change with the proper units attached.

In Exercises 41–42, find the slope of each line. Then express the slope as a rate of change with the proper units attached.

41. The graph shows the cost to own and operate a full-size truck, y, in dollars, in terms of the miles the truck is driven, x.

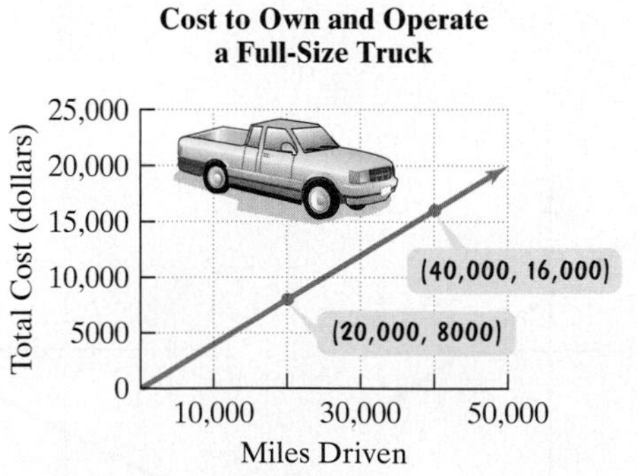

Source: Federal Highway Administration

42. The graph shows the cost to own and operate a compact car, y, in dollars, in terms of the miles the car is driven, x.

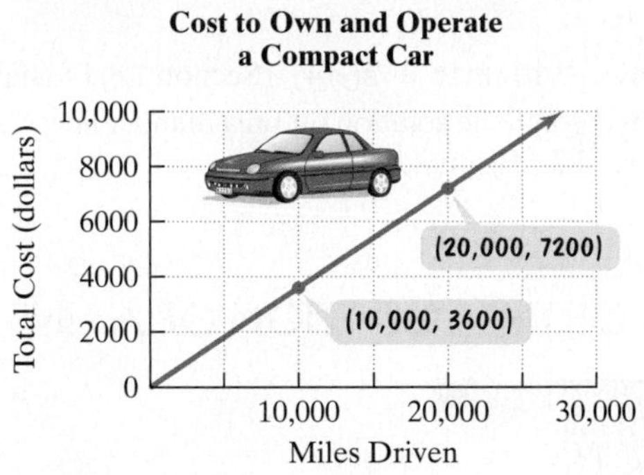

Source: Federal Highway Administration

The pitch of a roof refers to the absolute value of its slope. In Exercises 43–44, find the pitch of each roof shown.

43.

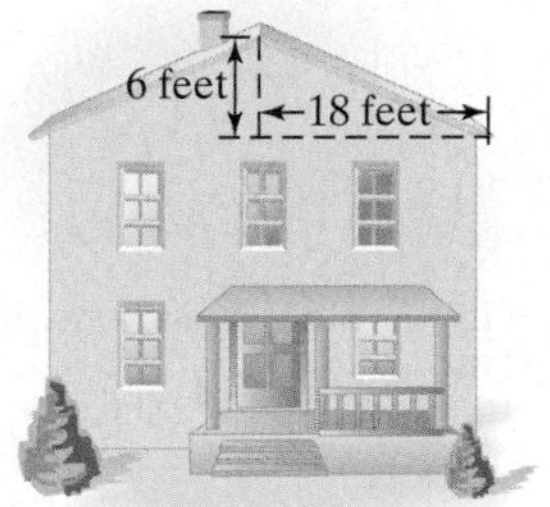

44.

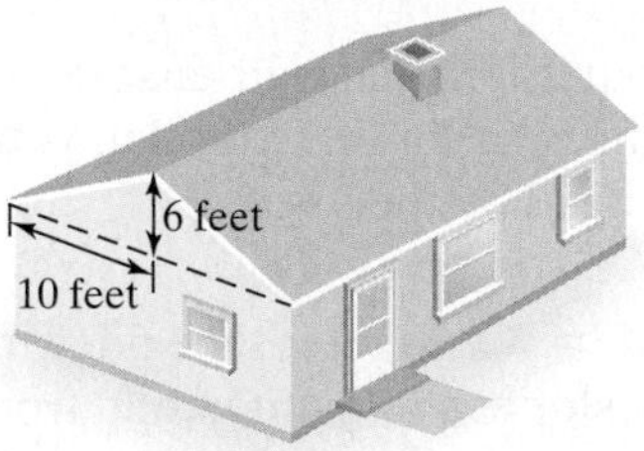

The grade of a road or ramp refers to its slope expressed as a percent. Use this information to solve Exercises 45–46.

45. Construction laws are very specific when it comes to access ramps for the disabled. Every vertical rise of 1 foot requires a horizontal run of 12 feet. What is the grade of such a ramp? Round to the nearest tenth of a percent.

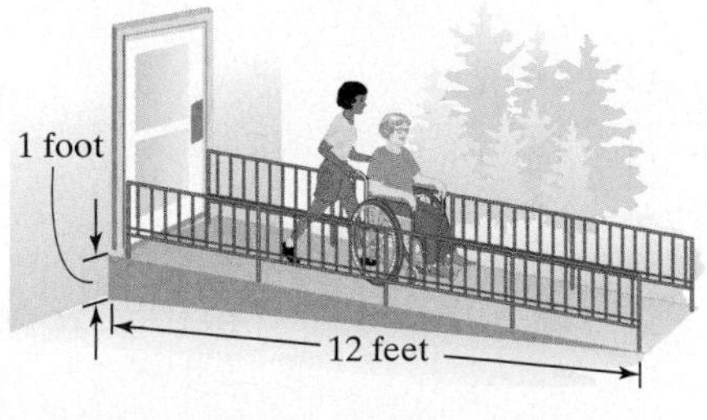

46. A college campus goes beyond the standards described in Exercise 45. All wheelchair ramps on campus are designed so that every vertical rise of 1 foot is accompanied by a horizontal run of 14 feet. What is the grade of such a ramp? Round to the nearest tenth of a percent.

Writing in Mathematics

47. What is the slope of a line?

48. Describe how to calculate the slope of a line passing through two points.

49. What does it mean if the slope of a line is zero?

50. What does it mean if the slope of a line is undefined?

51. If two lines are parallel, describe the relationship between their slopes.

52. Look back at the graph for Exercises 35–36. Do you think that the line through the points corresponding to the year 2001 and the year 2004 will model online spending per online household in the year 2040? Explain your answer.

Critical Thinking Exercises

53. Which one of the following is true?

a. Slope is run divided by rise.

b. The line through $(2, 2)$ and the origin has slope 1.

c. A line with slope 3 can be parallel to a line with slope -3.

d. The line through $(3, 1)$ and $(3, -5)$ has zero slope.

In Exercises 54–55, use the figure shown to make the indicated list.

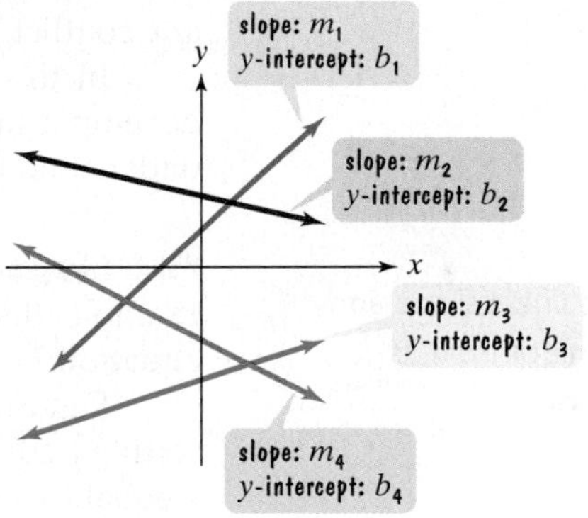

54. List the slopes m_1, m_2, m_3, and m_4 in order of decreasing size.

55. List the y-intercepts b_1, b_2, b_3, and b_4 in order of decreasing size.

Technology Exercises

Use a graphing utility to graph each equation in Exercises 56–59. Then use the TRACE *feature to trace along the line and find the coordinates of two points. Use these points to compute the line's slope.*

56. $y = 2x + 4$

57. $y = -3x + 6$

58. $y = -\frac{1}{2}x - 5$

59. $y = \frac{3}{4}x - 2$

60. In Exercises 56–59, compare the slope that you found with the line's equation. What relationship do you observe between the line's slope and one of the constants in the equation?

Review Exercises

61. A 36-inch board is cut into two pieces. One piece is twice as long as the other. How long are the pieces? (Section 2.5, Example 3)

62. Simplify: $-10 + 16 \div 2(-4)$. (Section 1.8, Example 4)

63. Solve and graph the solution set on a number line: $2x - 3 \leq 5$. (Section 2.7, Example 5)

SECTION 3.4

THE SLOPE-INTERCEPT FORM OF THE EQUATION OF A LINE

Objectives

1 Find a line's slope and y-intercept from its equation.

2 Graph lines in slope-intercept form.

3 Use slope and y-intercept to graph $Ax + By = C$.

4 Use slope and y-intercept to model data.

Children in Kampala, Uganda, discuss HIV/AIDS issues.

AIDS, famine, malaria, and civil wars have all taken their toll on Africa. In 2003. Secretary of State Colin Powell declared, "HIV is now more destructive than any army, any conflict, any weapon of mass destruction."

In this section, you will study the form of an equation that will enable you to develop a model for the growth of HIV/AIDS in Africa. Using this model, you will understand the need to fight AIDS on a continent ravaged by the virus.

1 Find a line's slope and y-intercept from its equation.

The Slope-Intercept Form of the Equation of a Line Let's begin with an example that shows how easy it is to find a line's slope and y-intercept from its equation.

Figure 3.22 shows the graph of $y = 2x + 4$. Verify that the x-intercept is -2 by setting y equal to 0 and solving for x. Similarly, verify that the y-intercept is 4 by setting x equal to 0 and solving for y.

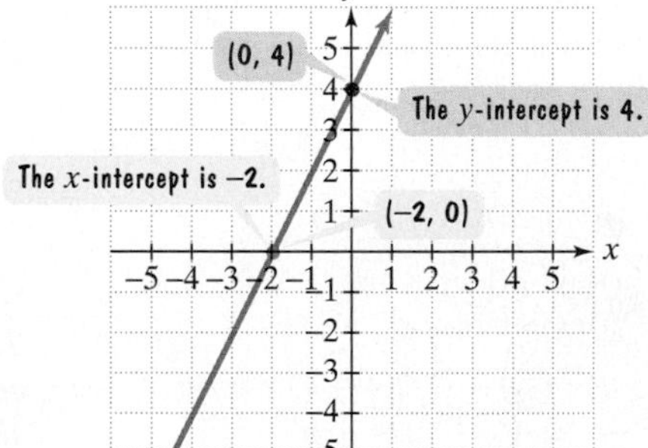

FIGURE 3.22 The graph of $y = 2x + 4$

Now that we have two points on the line, $(-2, 0)$ and $(0, 4)$, we can calculate the slope of the graph of $y = 2x + 4$.

$$\text{Slope} = \frac{\text{Change in } y}{\text{Change in } x}$$

$$= \frac{4 - 0}{0 - (-2)} = \frac{4}{2} = 2$$

We see that the slope of the line is 2, the same as the coefficient of x in the equation $y = 2x + 4$. The y-intercept is 4, the same as the constant in the equation $y = 2x + 4$.

$$y = 2x + 4$$

The slope is 2. The y-intercept is 4.

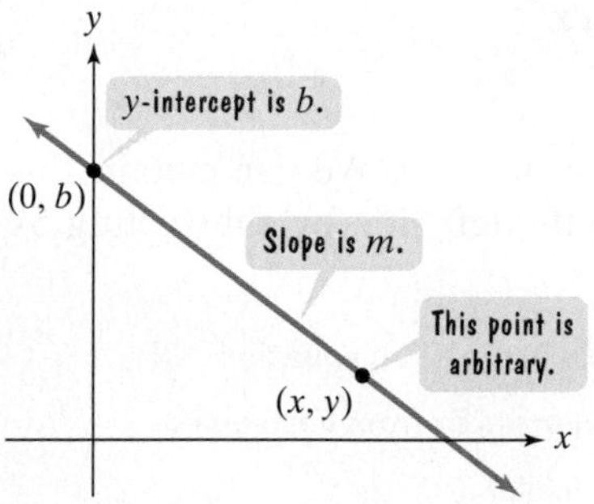

FIGURE 3.23 A line with slope m and y-intercept b

It is not merely a coincidence that the x-coefficient is the line's slope and the constant term is the y-intercept. Let's find the equation of any nonvertical line with slope m and y-intercept b. Because the y-intercept is b, the point $(0, b)$ lies on the line. Now, let (x, y) represent any other point on the line, shown in Figure 3.23. Keep in mind that the point (x, y) is arbitrary and is not in one fixed position. By contrast, the point $(0, b)$ is fixed.

Regardless of where the point (x, y) is located, the steepness of the line in Figure 3.23 remains the same. Thus, the ratio for slope stays a constant m. This means that for all points along the line

$$m = \frac{\text{Change in } y}{\text{Change in } x} = \frac{y - b}{x - 0} = \frac{y - b}{x}.$$

We can clear the fraction by multiplying both sides by x, the least common denominator.

$$m = \frac{y - b}{x}$$ This is the slope of the line in Figure 3.23.

$$mx = \frac{y - b}{x} \cdot x$$ Multiply both sides by x.

$$mx = y - b$$ Simplify: $\frac{y - b}{\cancel{x}} \cdot \cancel{x} = y - b$.

$$mx + b = y - b + b$$ Add b to both sides and solve for y.

$$mx + b = y$$ Simplify.

Now, if we reverse the two sides, we obtain the **slope-intercept form** of the equation of a line.

> SLOPE-INTERCEPT FORM OF THE EQUATION OF A LINE The **slope-intercept equation** of a nonvertical line with slope m and y-intercept b is
>
> $$y = mx + b.$$

Thus, if a line's equation is written with y isolated on one side, the x-coefficient is the line's slope and the constant term is the y-intercept.

EXAMPLE 1 Finding a Line's Slope and y-Intercept from Its Equation

Find the slope and the y-intercept of the line with the given equation:

a. $y = 2x - 4$ **b.** $y = \frac{1}{2}x + 2$ **c.** $5x + y = 4$.

STUDY TIP

The variables in $y = mx + b$ vary in different ways. The variables for slope, m, and y-intercept, b, vary from one line's equation to another. However, they remain constant in the equation of a single line. By contrast, the variables x and y represent the infinitely many points, (x, y), on a single line. Thus, these variables vary in both the equation of a single line, as well as from one equation to another.

SOLUTION

a. We write $y = 2x - 4$ as $y = 2x + (-4)$. The slope is the x-coefficient and the y-intercept is the constant term.

$$y = 2x + (-4)$$

The slope is 2. The y-intercept is -4.

b. The equation $y = \frac{1}{2}x + 2$ is in the form $y = mx + b$. We can find the slope, m, by identifying the coefficient of x. We can find the y-intercept, b, by identifying the constant term.

$$y = \frac{1}{2}x + 2$$

The slope is $\frac{1}{2}$. The y-intercept is 2.

c. The equation $5x + y = 4$ is not in the form $y = mx + b$. We can obtain this form by isolating y on one side. We isolate y on the left side by subtracting $5x$ from both sides.

$$5x + y = 4 \quad \text{This is the given equation.}$$
$$5x - 5x + y = -5x + 4 \quad \text{Subtract 5x from both sides.}$$
$$y = -5x + 4 \quad \text{Simplify.}$$

Now, the equation is in the form $y = mx + b$. The slope is the coefficient of x and the y-intercept is the constant term.

$$y = -5x + 4$$

The slope is -5. The y-intercept is 4.

■

CHECK POINT 1 Find the slope and the y-intercept of the line with the given equation:

a. $y = 5x - 3$ **b.** $y = \frac{2}{3}x + 4$

c. $7x + y = 6$.

2 Graph lines in slope-intercept form.

Graphing $y = mx + b$ by Using the Slope and y-Intercept If a line's equation is written with y isolated on one side, we can use the y-intercept and the slope to obtain its graph.

GRAPHING $y = mx + b$ BY USING THE SLOPE AND y-INTERCEPT

1. Plot the point containing the y-intercept on the y-axis. This is the point $(0, b)$.
2. Obtain a second point using the slope, m. Write m as a fraction, and use rise over run, starting at the point containing the y-intercept, to plot this point.
3. Use a straightedge to draw a line through the two points. Draw arrowheads at the ends of the line to show that the line continues indefinitely in both directions.

EXAMPLE 2 Graphing by Using the Slope and y-Intercept

Graph the line whose equation is $y = 4x - 3$.

SOLUTION We write $y = 4x - 3$ in the form $y = mx + b$.

$$y = 4x + (-3)$$

The slope is 4. The y-intercept is -3.

Now that we have identified the slope and the y-intercept, we use the three steps in the box to graph the equation.

Step 1. **Plot the point containing the y-intercept on the y-axis.** The y-intercept is -3. We plot the point $(0, -3)$, shown in Figure 3.24(a).

Step 2. **Obtain a second point using the slope, m. Write m as a fraction, and use rise over run, starting at the point containing the y-intercept, to plot this point.** The slope, 4, written as a fraction is $\frac{4}{1}$.

$$m = \frac{4}{1} = \frac{\text{Rise}}{\text{Run}}$$

We plot the second point on the line by starting at $(0, -3)$, the first point. Based on the slope, we move 4 units *up* (the rise) and 1 unit to the *right* (the run). This puts us at a second point on the line, $(1, 1)$, shown in Figure 3.24(b).

Step 3. **Use a straightedge to draw a line through the two points.** The graph of $y = 4x - 3$ is shown in Figure 3.24(c).

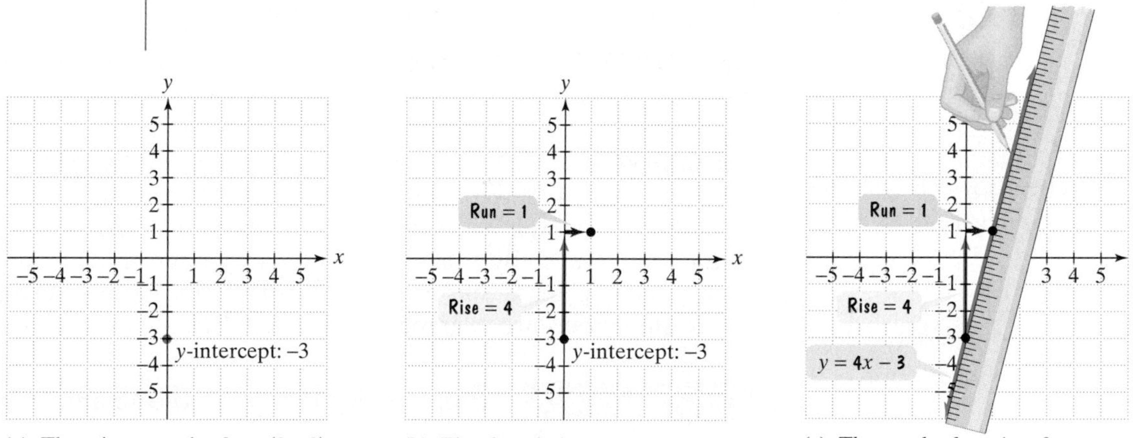

(a) The y-intercept is –3, so (0, –3) is a point on the line.

(b) The slope is 4.

(c) The graph of $y = 4x - 3$

FIGURE 3.24 Graphing $y = 4x - 3$ using the y-intercept and slope

■

✔ CHECK POINT 2 Graph the line whose equation is $y = 3x - 2$.

EXAMPLE 3 Graphing by Using the Slope and y-Intercept

Graph the line whose equation is $y = \frac{2}{3}x + 2$.

SOLUTION The equation of the line, $y = \frac{2}{3}x + 2$, is in the form $y = mx + b$. We can find the slope, m, by identifying the coefficient of x. We can find the y-intercept, b, by identifying the constant term.

$$y = \frac{2}{3}x + 2$$

The slope is $\frac{2}{3}$.

The y-intercept is 2.

Now that we have identified the slope and the y-intercept, we use the three-step procedure to graph the equation.

Step 1. **Plot the point containing the y-intercept on the y-axis.** The y-intercept is 2. We plot $(0, 2)$, shown in Figure 3.25.

Step 2. **Obtain a second point using the slope, m. Write m as a fraction, and use rise over run, starting at the point containing the y-intercept, to plot this point.** The slope, $\frac{2}{3}$, is already written as a fraction.

$$m = \frac{2}{3} = \frac{\text{Rise}}{\text{Run}}$$

We plot the second point on the line by starting at $(0, 2)$, the first point. Based on the slope, we move 2 units *up* (the rise) and 3 units to the *right* (the run). This puts us at a second point on the line, $(3, 4)$, shown in Figure 3.25.

Step 3. **Use a straightedge to draw a line through the two points.** The graph of $y = \frac{2}{3}x + 2$ is shown in Figure 3.25. ■

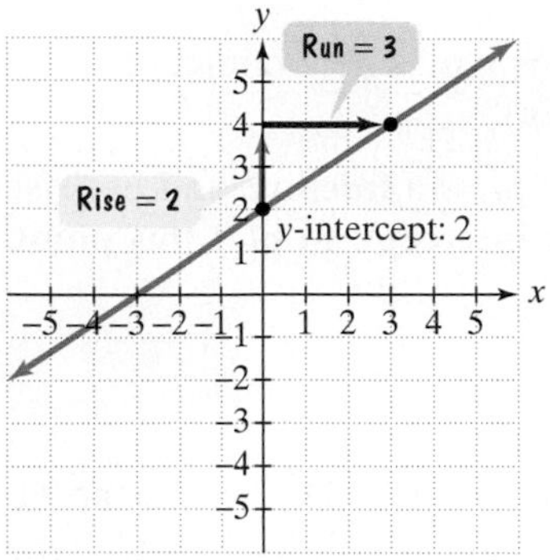

FIGURE 3.25 The graph of $y = \frac{2}{3}x + 2$

CHECK POINT 3 Graph the line whose equation is $y = \frac{3}{5}x + 1$.

3 Use slope and y-intercept to graph $Ax + By = C$.

Graphing $Ax + By = C$ by Using the Slope and y-Intercept Earlier in this chapter, we considered linear equations of the form $Ax + By = C$. We used x- and y-intercepts, as well as checkpoints, to graph these equations. It is also possible to obtain the graphs by using the slope and y-intercept. To do this, begin by solving $Ax + By = C$ for y. This will put the equation in slope-intercept form. Then use the three-step procedure to graph the equation. This is illustrated in Example 4.

EXAMPLE 4 Graphing by Using the Slope and y-Intercept

Graph the linear equation $2x + 5y = 0$ by using the slope and y-intercept.

SOLUTION We put the equation in slope-intercept form by solving for y.

$$2x + 5y = 0 \quad \text{This is the given equation.}$$

$$2x - 2x + 5y = -2x + 0 \quad \text{Subtract } 2x \text{ from both sides.}$$

$$5y = -2x + 0 \quad \text{Simplify.}$$

$$\frac{5y}{5} = \frac{-2x + 0}{5} \quad \text{Divide both sides by 5.}$$

$$y = \frac{-2x}{5} + \frac{0}{5} \quad \text{Divide each term in the numerator by 5.}$$

$$y = -\frac{2}{5}x + 0 \quad \text{Simplify.}$$

Now that the equation is in slope-intercept form, we can use the slope and y-intercept to obtain its graph. Examine the slope-intercept form:

$$y = -\frac{2}{5}x + 0.$$

slope: $-\frac{2}{5}$ — y-intercept: 0

Note that the slope is $-\frac{2}{5}$ and the y-intercept is 0. Use the y-intercept to plot $(0, 0)$ on the y-axis. Then locate a second point by using the slope.

$$m = -\frac{2}{5} = \frac{-2}{5} = \frac{\text{Rise}}{\text{Run}}$$

Because the rise is -2 and the run is 5, move *down* 2 units and to the *right* 5 units, starting at the point $(0, 0)$. This puts us at a second point on the line, $(5, -2)$. The graph of $2x + 5y = 0$ is the line drawn through these points, shown in Figure 3.26. ■

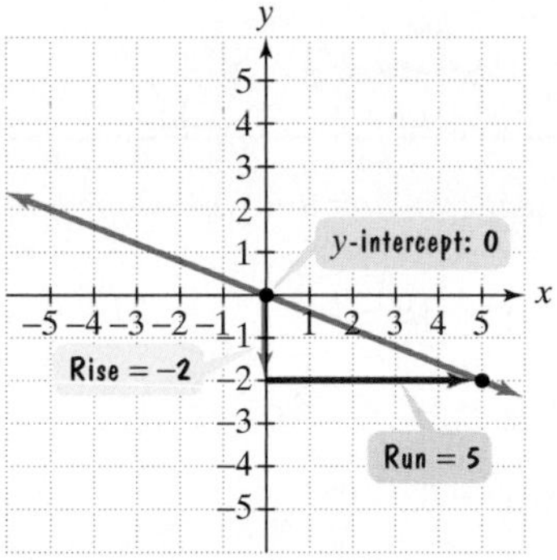

FIGURE 3.26 The graph of $2x + 5y = 0$, or $y = -\frac{2}{5}x + 0$

DISCOVER FOR YOURSELF

Obtain a second point in Example 4 by writing the slope as follows:

$$m = \frac{2}{-5} = \frac{\text{Rise}}{\text{Run}}$$

$-\frac{2}{5}$ can be expressed as $\frac{-2}{5}$ or $\frac{2}{-5}$.

Obtain a second point in Figure 3.26 by moving *up* 2 units and to the *left* 5 units, starting at $(0, 0)$. What do you observe once you graph the line?

CHECK POINT 4 Graph the linear equation $3x + 4y = 0$ by using the slope and y-intercept.

4 Use slope and y-intercept to model data.

Modeling with the Slope-Intercept Form of the Equation of a Line If an equation in slope-intercept form models some physical situation, then the slope and y-intercept have physical interpretations. For the equation $y = mx + b$, the y-intercept, b, tells us what is happening to y when x is 0. If x represents time, the y-intercept describes the value of y at the beginning, or when time equals 0. The slope represents the rate of change in y per unit change in x.

Let's see how we can use these ideas to develop a model for data. In the previous section, we looked at line graphs for the number of U.S. men and women living alone, repeated in Figure 3.27. Let's develop a model for the data for women living alone. We let x represent the number of years after 1995. At the beginning of our data, or 0 years after 1995, 14 million women lived alone. Thus, $b = 14$. In Example 4 in the previous section, we found that $m = 0.2$ (rate of change is 0.2 million women per year). An equation of the form $y = mx + b$ that models the data is

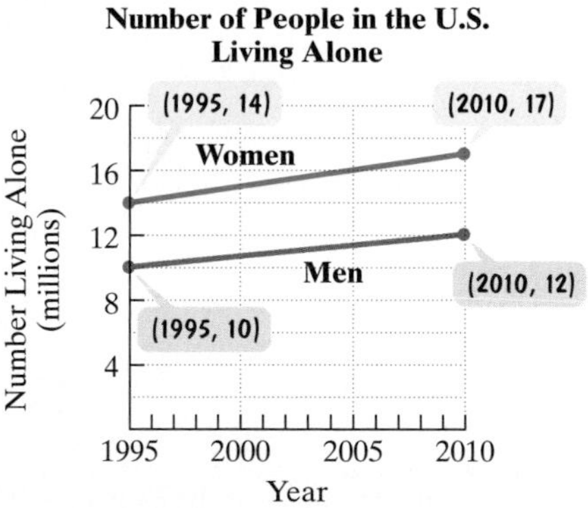

FIGURE 3.27

$$y = 0.2x + 14,$$

where y is the number, in millions, of U.S. women living alone x years after 1995.

3.4 EXERCISE SET

Student Solutions Manual | CD/Video | PH Math/Tutor Center | MathXL Tutorials on CD | MathXL® | MyMathLab | Interactmath.com

Practice Exercises

In Exercises 1–12, find the slope and the y-intercept of the line with the given equation.

1. $y = 3x + 2$

2. $y = 9x + 4$

3. $y = 3x - 5$

4. $y = 4x - 2$

5. $y = -\frac{1}{2}x + 5$

6. $y = -\frac{3}{4}x + 6$

7. $y = 7x$

8. $y = 10x$

9. $y = 10$

10. $y = 7$

11. $y = 4 - x$

12. $y = 5 - x$

In Exercises 13–26, begin by solving the linear equation for y. This will put the equation in slope-intercept form. Then find the slope and the y-intercept of the line with this equation.

13. $-5x + y = 7$

14. $-9x + y = 5$

15. $x + y = 6$

16. $x + y = 8$

17. $6x + y = 0$

18. $8x + y = 0$

19. $3y = 6x$

20. $3y = -9x$

21. $2x + 7y = 0$

22. $2x + 9y = 0$

23. $3x + 2y = 3$

24. $4x + 3y = 4$

25. $3x - 4y = 12$

26. $5x - 2y = 10$

In Exercises 27–38, graph each linear equation using the slope and y-intercept.

27. $y = 2x + 4$

28. $y = 3x + 1$

29. $y = -3x + 5$

30. $y = -2x + 4$

31. $y = \frac{1}{2}x + 1$

32. $y = \frac{1}{3}x + 2$

33. $y = \frac{2}{3}x - 5$

34. $y = \frac{3}{4}x - 4$

35. $y = -\frac{3}{4}x + 2$

36. $y = -\frac{2}{3}x + 4$

37. $y = -\frac{5}{3}x$

38. $y = -\frac{4}{3}x$

In Exercises 39–46,

a. *Put the equation in slope-intercept form by solving for y.*

b. *Identify the slope and the y-intercept.*

c. *Use the slope and y-intercept to graph the equation.*

39. $3x + y = 0$

40. $2x + y = 0$

41. $3y = 4x$

42. $4y = 5x$

43. $2x + y = 3$

44. $3x + y = 4$

45. $7x + 2y = 14$

46. $5x + 3y = 15$

In Exercises 47–52, graph both linear equations in the same rectangular coordinate system. If the lines are parallel, explain why.

47. $y = 3x + 1$
$y = 3x - 3$

48. $y = 2x + 4$
$y = 2x - 3$

49. $y = -3x + 2$
$y = 3x + 2$

50. $y = -2x + 1$
$y = 2x + 1$

51. $x - 2y = 2$
$2x - 4y = 3$

52. $x - 3y = 9$
$3x - 9y = 18$

Practice Plus

In Exercises 53–58, write an equation in the form $y = mx + b$ of the line that is described.

53. The y-intercept is 5 and the line is parallel to the line whose equation is $3x + y = 6$.

54. The y-intercept is -4 and the line is parallel to the line whose equation is $2x + y = 8$.

55. The line has the same y-intercept as the line whose equation is $16y = 8x + 32$ and is parallel to the line whose equation is $3x + 3y = 9$.

56. The line has the same y-intercept as the line whose equation is $2y = 6x + 8$ and is parallel to the line whose equation is $4x + 4y = 20$.

57. The line rises from left to right. It passes through the origin and a second point with equal x- and y-coordinates.

58. The line falls from left to right. It passes through the origin and a second point with opposite x- and y-coordinates.

Application Exercises

59. The formula $E = 10x + 166$ models the number of endangered animal species, E, in the United States x years after 1980.

a. What is the slope of this model? What does it represent in this situation?

b. What is the y-intercept of this model? What does it represent in this situation?

c. Use the model to find the number of endangered animal species in the United States in 1980, 1985, 1990, 1995, 2000, and 2003. How well does the model describe the actual data shown in the bar graph?

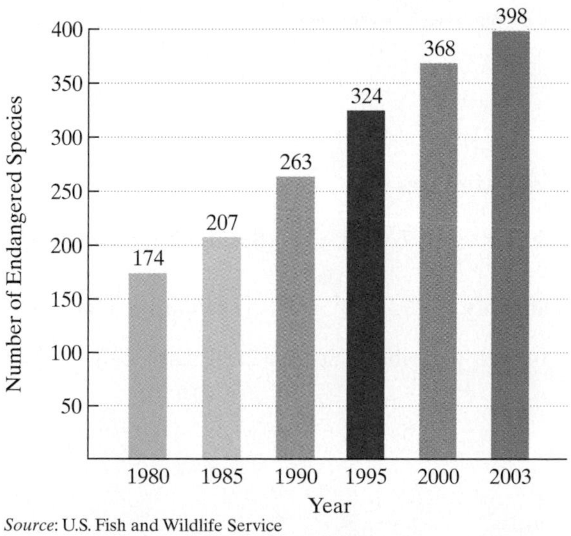

Source: U.S. Fish and Wildlife Service

60. A salesperson receives a fixed salary plus a percentage of all sales. The linear equation $y = 0.05x + 500$ describes the weekly salary, y, in dollars, in terms of weekly sales, x, also in dollars.

a. Use the formula to find the weekly salary for sales of \$0, \$1, \$2, \$3, \$4, \$5, \$100, and \$1000.

b. What is the slope of this equation? What does it represent in this situation?

c. What is the y-intercept of this equation? What does it represent in this situation?

61. In 2002, approximately 29 million people, including 3 million children, were living with HIV/AIDS in sub-Saharan Africa. The region accounted for 70% of the world's new HIV infections. The bar graph shows the number of people, in millions, living with the virus from 1997 through 2002. The graph in the rectangular coordinate system approximates the data.

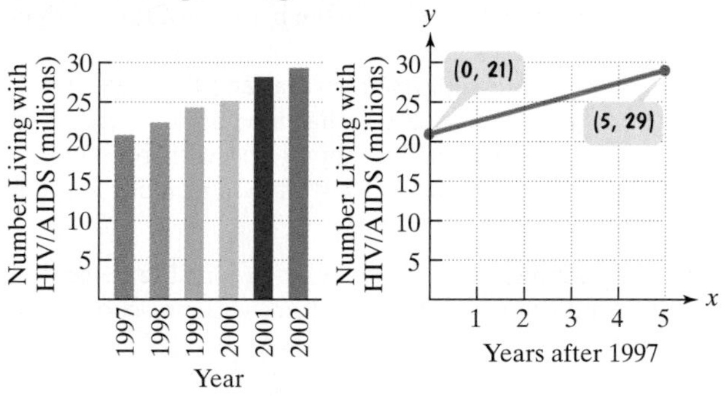

Source: U.N. AIDS

a. According to the rectangular coordinate graph, what is the y-intercept? Describe what this represents in this situation.

b. Use the coordinates of the two points shown to compute the slope. What does this mean in terms of rate of change?

c. Use the y-intercept from part (a) and the slope from part (b) to write an equation that models the number of people, in millions, living with HIV in sub-Saharan Africa, y, x years after 1997. Write your model in the form $y = mx + b$.

d. If worldwide efforts to fight AIDS in sub-Saharan Africa are not successful, use your model from part (c) to predict the number of people, in millions, who will be living with the virus in 2006.

62. The graph shows that the percentage of married women in the U.S. labor force who have children under 6 has been increasing steadily since 1960.

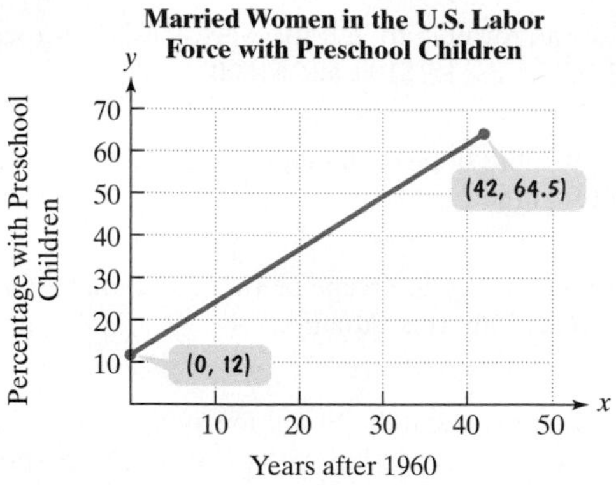

Source: James M. Henslin, *Essentials of Sociology*, Allyn and Bacon, 2002

a. According to the graph, what is the y-intercept? Describe what this represents in this situation.

b. Use the coordinates of the two points shown to compute the slope. What does this mean in terms of rate of change?

c. Use the y-intercept from part (a) and the slope from part (b) to write an equation that models the percentage of married women in the labor force with preschool children, y, x years after 1960. Write your model in the form $y = mx + b$.

d. Use your model from part (c) to predict the percentage of married women in the labor force with preschool children in 2006.

Writing in Mathematics

63. Describe how to find the slope and the y-intercept of a line whose equation is given.

64. Describe how to graph a line using the slope and y-intercept. Provide an original example with your description.

65. A formula in the form $y = mx + b$ models the cost, y, of a four-year college x years after 2003. Would you expect m to be positive, negative, or zero? Explain your answer.

Critical Thinking Exercises

66. Which one of the following is true?

a. The equation $y = mx + b$ shows that no line can have a y-intercept that is numerically equal to its slope.

b. Every line in the rectangular coordinate system has an equation that can be expressed in slope-intercept form.

c. The line $3x + 2y = 5$ has slope $-\frac{3}{2}$.

d. The line $2y = 3x + 7$ has a y-intercept of 7.

67. The relationship between Celsius temperature, C, and Fahrenheit temperature, F, can be described by a linear equation in the form $F = mC + b$. The graph of this equation contains the point $(0, 32)$: Water freezes at 0°C or at 32°F. The line also contains the point $(100, 212)$: Water boils at 100°C or at 212°F. Write the linear equation expressing Fahrenheit temperature in terms of Celsius temperature.

68. The graph below indicates that lower fertility rates (the number of births per woman) are related to the percentage of the population using contraceptives. A line that best fits the data is shown. Estimate the y-intercept and the slope of this line. Then write the line's slope-intercept equation. Use the equation to find the number of births per woman if 90% of married women of child-bearing age used contraceptives.

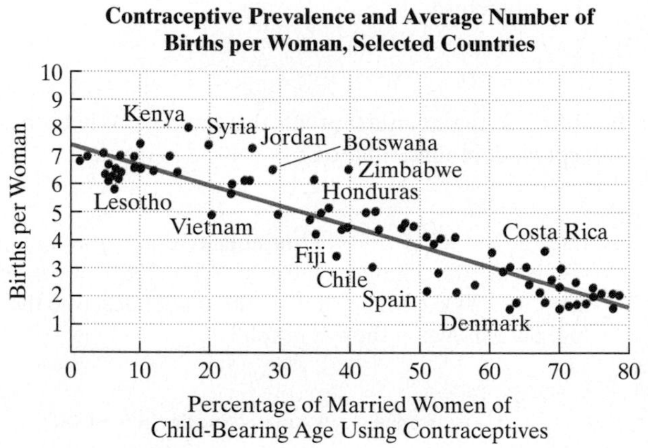

Source: Population Reference Bureau

Review Exercises

69. Solve: $\frac{x}{2} + 7 = 13 - \frac{x}{4}$. (Section 2.3, Example 4)

70. Simplify: $3(12 \div 2^2 - 3)^2$. (Section 1.8, Example 6)

71. 14 is 25% of what? (Section 2.4, Example 8)

MID-CHAPTER CHECK POINT

CHAPTER 3

What You Know: We learned to graph equations in two variables using point plotting, as well as a variety of other techniques. We used intercepts and a checkpoint to graph linear equations in the form $Ax + By = C$. We saw that the graph of a linear equation in one variable is a horizontal or a vertical line: $y = b$ graphs as a horizontal line and $x = a$ graphs as a vertical line. We determined a line's steepness, or rate of change, by computing its slope and we saw that lines with the same slope are parallel. Finally, we learned to graph linear equations in slope-intercept form, $y = mx + b$, using the slope, m, and the y-intercept, b.

In Exercises 1–3, use each graph to determine

a. *the x-intercept, or state that there is no x-intercept.*
b. *the y-intercept, or state that there is no y-intercept.*
c. *the line's slope, or state that the slope is undefined.*

1.

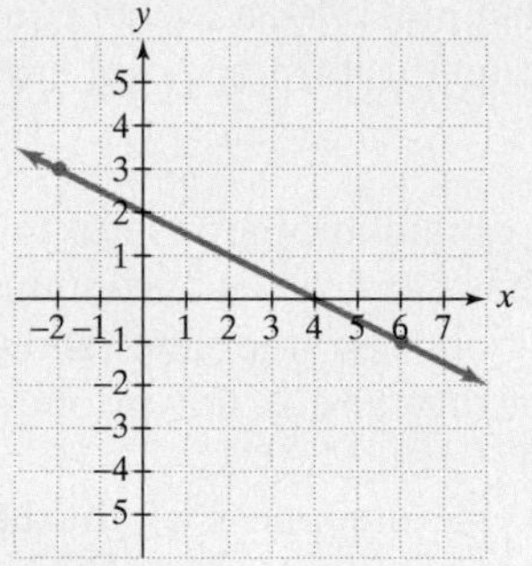

2.

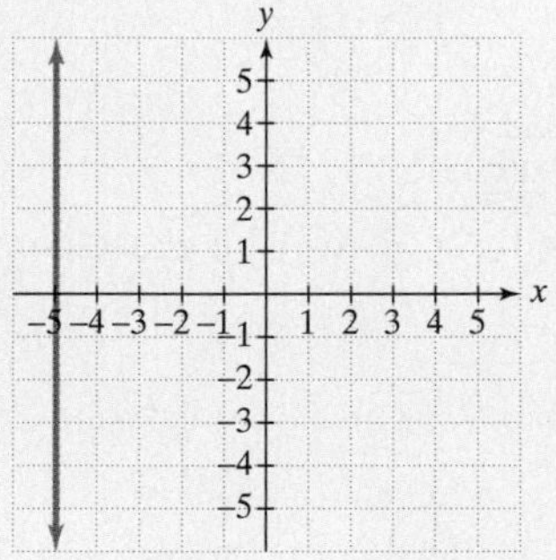

3.

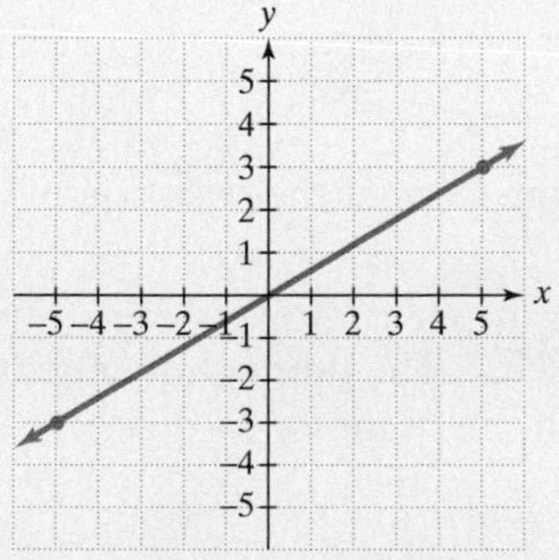

In Exercises 4–15, graph each equation in a rectangular coordinate system.

4. $y = -2x$	**5.** $y = -2$	**6.** $x + y = -2$
7. $y = \frac{1}{3}x - 2$	**8.** $x = 3.5$	**9.** $4x - 2y = 8$
10. $y = 3x + 2$	**11.** $3x + y = 0$	**12.** $y = x^2 - 4$
13. $y = x - 4$	**14.** $5y = -3x$	**15.** $5y = 20$

16. Find the slope and the y-intercept of the line whose equation is $5x - 2y = 10$.

17. Determine whether the line through $(2, -4)$ and $(7, 0)$ is parallel to a second line through $(-4, 2)$ and $(1, 6)$.

18. The graph shows the percentage of U.S. colleges that offered distance learning by computer for selected years from 1995 through 2002.

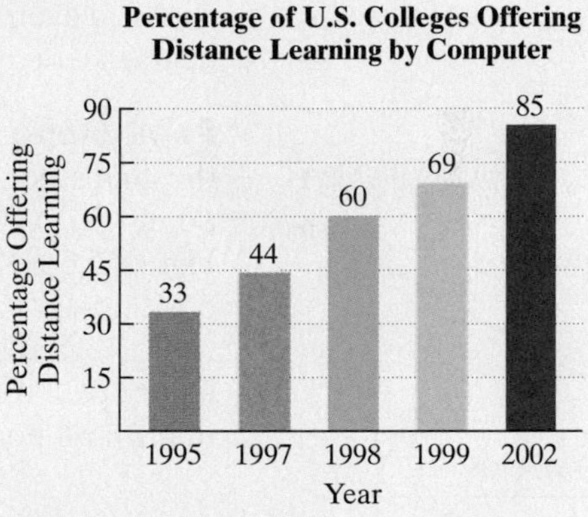

Source: International Data Corporation

The data for the years 1995 through 2002 can be modeled by the linear equation

$$y = 7.8x + 33$$

where x is the number of years after 1995 and y is the percentage of U.S. colleges offering distance learning.

a. Find the y-intercept of this model.
b. Describe what this y-intercept represents.
c. Find the slope of this model.
d. Describe the meaning of this slope as a rate of change.

SECTION 3.5

Objectives

1. Use the point-slope form to write equations of a line.
2. Find slopes and equations of parallel and perpendicular lines.
3. Model data with linear equations and make predictions.

THE POINT-SLOPE FORM OF THE EQUATION OF A LINE

Surprised by the number of people smoking cigarettes in movies and television shows made in the 1940s and 1950s? At that time, there was little awareness of the relationship between tobacco use and numerous diseases. Cigarette smoking was seen as a healthy way to relax and help digest a hearty meal. Then, in 1964, a linear equation changed everything. To understand the mathematics behind this turning point in public health, we explore another form of a line's equation.

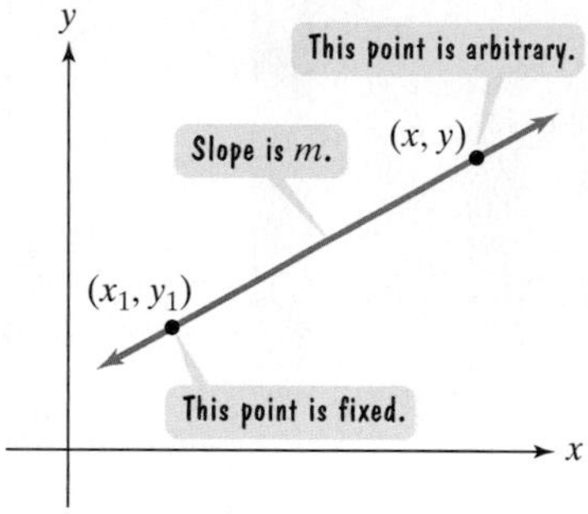

FIGURE 3.28 A line passing through (x_1, y_1) with slope m

Point-Slope Form We can use the slope of a line to obtain another useful form of the line's equation. Consider a nonvertical line that has slope m and contains the point (x_1, y_1). Now, let (x, y) represent any other point on the line, shown in Figure 3.28. Keep in mind that the point (x, y) is arbitrary and is not in one fixed position. By contrast, the point (x_1, y_1) is fixed.

Regardless of where the point (x, y) is located, the steepness of the line in Figure 3.28 remains the same. Thus, the ratio for slope stays a constant m. This means that for all points along the line,

$$m = \frac{\text{Change in } y}{\text{Change in } x} = \frac{y - y_1}{x - x_1}.$$

We can clear the fraction by multiplying both sides by $x - x_1$, the least common denominator.

$$m = \frac{y - y_1}{x - x_1}$$ This is the slope of the line in Figure 3.28.

$$m(x - x_1) = \frac{y - y_1}{x - x_1} \cdot (x - x_1)$$ Multiply both sides by $x - x_1$.

$$m(x - x_1) = y - y_1$$ Simplify: $\frac{y - y_1}{\cancel{x - x_1}} \cdot \cancel{x - x_1} = y - y_1$

Now, if we reverse the two sides, we obtain the **point-slope form** of the equation of a line.

POINT-SLOPE FORM OF THE EQUATION OF A LINE The **point-slope equation** of a nonvertical line with slope m that passes through the point (x_1, y_1) is

$$y - y_1 = m(x - x_1).$$

For example, the point-slope equation of the line passing through (1, 4) with slope 2 ($m = 2$) is

$$y - 4 = 2(x - 1).$$

1 Use the point-slope form to write equations of a line.

Using the Point-Slope Form to Write a Line's Equation If we know the slope of a line and a point not containing the y-intercept through which the line passes, the point-slope form is the equation that we should use. Once we have obtained this equation, it is customary to solve for y and write the equation in slope-intercept form. Examples 1 and 2 illustrate these ideas.

EXAMPLE 1 Writing the Point-Slope Form and the Slope-Intercept Form

Write the point-slope form and the slope-intercept form of the equation of the line with slope 4 that passes through the point $(-1, 3)$.

SOLUTION We begin with the point-slope equation of a line with $m = 4$, $x_1 = -1$, and $y_1 = 3$.

$$y - y_1 = m(x - x_1)$$ This is the point-slope form of the equation.

$$y - 3 = 4[x - (-1)]$$ Substitute the given values.

$$y - 3 = 4(x + 1)$$ We now have the point-slope form of the equation of the given line.

Now we solve this equation for y and write an equivalent equation in slope-intercept form ($y = mx + b$).

$$y - 3 = 4(x + 1)$$ This is the point-slope equation.

$$y - 3 = 4x + 4$$ Use the distributive property.

$$y = 4x + 7$$ Add 3 to both sides.

The slope-intercept form of the line's equation is $y = 4x + 7$. ■

CHECK POINT 1 Write the point-slope form and the slope-intercept form of the equation of the line with slope 6 that passes through the point $(2, -5)$.

EXAMPLE 2 Writing the Point-Slope Form and the Slope-Intercept Form

A line passes through the points $(4, -3)$ and $(-2, 6)$. (See Figure 3.29.) Find the equation of the line

a. in point-slope form. **b.** in slope-intercept form.

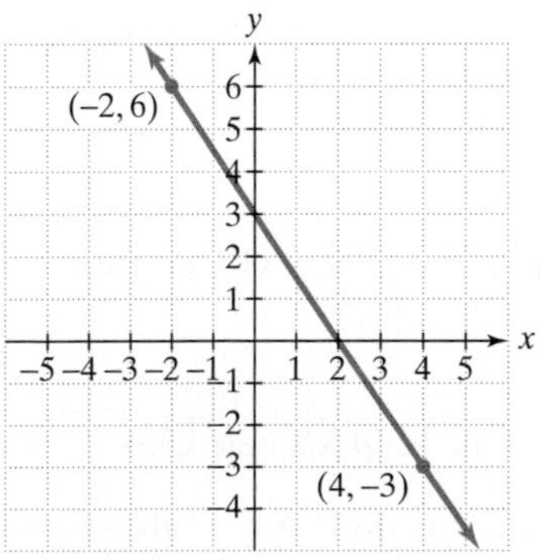

FIGURE 3.29

SOLUTION

a. To use the point-slope form, we need to find the slope. The slope is the change in the y-coordinates divided by the corresponding change in the x-coordinates.

$$m = \frac{6 - (-3)}{-2 - 4} = \frac{9}{-6} = -\frac{3}{2}$$ This is the definition of slope using $(4, -3)$ and $(-2, 6)$.

We can take either point on the line, $(-2, 6)$ or $(4, -3)$, to be (x_1, y_1). Let's use $(x_1, y_1) = (4, -3)$. Now, we are ready to write the point-slope equation.

$$y - y_1 = m(x - x_1) \qquad \text{This is the point-slope form of the equation.}$$

$$y - (-3) = -\frac{3}{2}(x - 4) \qquad \text{Substitute: } (x_1, y_1) = (4, -3) \text{ and } m = -\frac{3}{2}.$$

$$y + 3 = -\frac{3}{2}(x - 4) \qquad \text{Simplify.}$$

This equation is the point-slope form of the equation of the line shown in Figure 3.29 on page 227.

b. Now, we solve this equation for y and write an equivalent equation in slope-intercept form $(y = mx + b)$.

$$y + 3 = -\frac{3}{2}(x - 4) \qquad \text{This is the point-slope equation.}$$

$$y + 3 = -\frac{3}{2}x + 6 \qquad \text{Use the distributive property.}$$

$$y = -\frac{3}{2}x + 3 \qquad \text{Subtract 3 from both sides.}$$

This equation is the slope-intercept form of the equation of the line shown in Figure 3.29 on page 227. ■

DISCOVER FOR YOURSELF

If you are given two points on a line, you can use either point for (x_1, y_1) when you write its point-slope equation. Rework Example 2 using $(-2, 6)$ for (x_1, y_1). Once you solve for y, you should obtain the same slope-intercept equation as the one shown in the last line of the solution to Example 2.

✔ **CHECK POINT 2** A line passes through the points $(-2, -1)$ and $(-1, -6)$. Find the equation of the line

a. in point-slope form. **b.** in slope-intercept form.

In Examples 1 and 2, we eventually write a line's equation in slope-intercept form? But where do we start our work?

Starting with $y = mx + b$	**Starting with $y - y_1 = m(x - x_1)$**
Begin with the slope-intercept form if you know **1.** The slope of the line and the y-intercept.	Begin with the point-slope form if you know **1.** The slope of the line and a point on the line not containing the y-intercept or **2.** Two points on the line, neither of which contains the y-intercept.

2 Find slopes and equations of parallel and perpendicular lines.

Parallel and Perpendicular Lines The next example uses the fact that parallel lines have the same slope.

EXAMPLE 3 Writing Equations of a Line Parallel to a Given Line

Write an equation of the line passing through $(-3, 1)$ and parallel to the line whose equation is $y = 2x + 1$. Express the equation in point-slope form and slope-intercept form.

SOLUTION The situation is illustrated in Figure 3.30. We are looking for the equation of the red line shown on the left. How do we obtain this equation? Notice that the line passes through the point $(-3, 1)$. Using the point-slope form of the line's equation, we have $x_1 = -3$ and $y_1 = 1$.

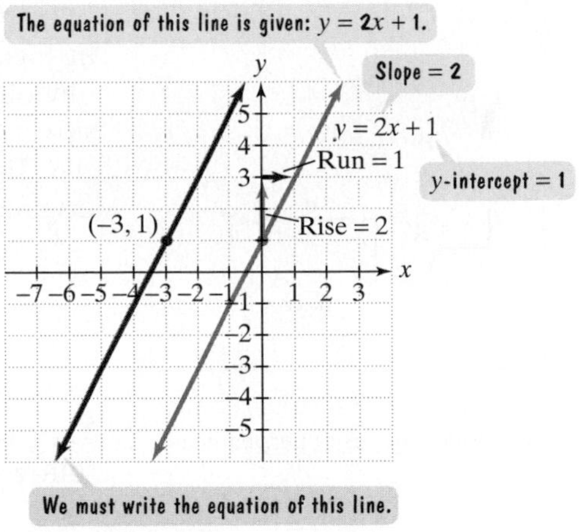
The equation of this line is given: y = 2x + 1.
Slope = 2
y = 2x + 1
Run = 1
y-intercept = 1
(−3, 1)
Rise = 2
y
x
We must write the equation of this line.
FIGURE 3.30

$$y - y_1 = m(x - x_1)$$

$y_1 = 1$ $x_1 = -3$

Now the only thing missing from the equation is m, the slope of the red line. Do we know anything about the slope of either line in Figure 3.30? The answer is yes; we know the slope of the dark blue line on the right, whose equation is given.

$$y = 2x + 1$$

The slope of the blue line on the right in Figure 3.30 is 2.

Parallel lines have the same slope. Because the slope of the blue line is 2, the slope of the red line, the line whose equation we must write, is also 2: $m = 2$. We now have values for x_1, y_1, and m for the red line.

$$y - y_1 = m(x - x_1)$$

$y_1 = 1$ $m = 2$ $x_1 = -3$

The point-slope form of the red line's equation is

$$y - 1 = 2[x - (-3)] \text{ or}$$

$$y - 1 = 2(x + 3).$$

Solving for y, we obtain the slope-intercept form of the equation.

$y - 1 = 2x + 6$ Apply the distributive property.

$y = 2x + 7$ Add 1 to both sides. This is the slope-intercept form, $y = mx + b$, of the equation. ■

CHECK POINT 3 Write an equation of the line passing through $(-2, 5)$ and parallel to the line whose equation is $y = 3x + 1$. Express the equation in point-slope form and slope-intercept form.

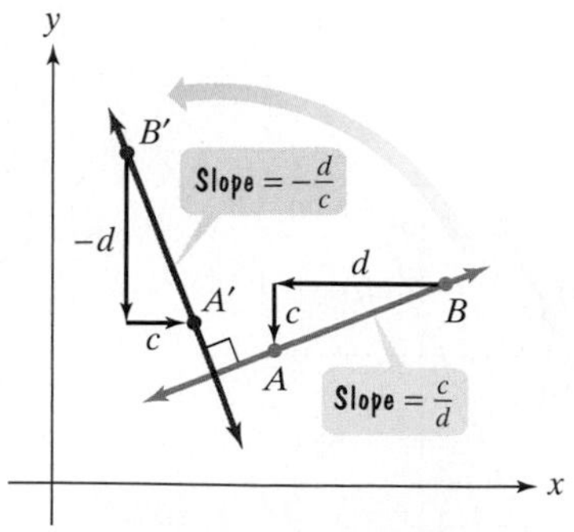

FIGURE 3.31 Slopes of perpendicular lines

Two lines that intersect at a right angle (90°) are said to be **perpendicular**, shown in Figure 3.31. The relationship between the slopes of perpendicular lines is not as obvious as the relationship between parallel lines. Figure 3.31 shows line AB, with slope $\frac{c}{d}$. Rotate line AB through 90° to the left to obtain line $A'B'$, perpendicular to line AB. The figure indicates that the rise and the run of the new line are reversed from the original line, but the rise is now negative. This means that the slope of the new line is $-\frac{d}{c}$. Notice that the product of the slopes of the two perpendicular lines is -1:

$$\left(\frac{c}{d}\right)\left(-\frac{d}{c}\right) = -1.$$

This relationship holds for all perpendicular lines and is summarized in the following box:

SLOPE AND PERPENDICULAR LINES

1. If two nonvertical lines are perpendicular, then the product of their slopes is -1.
2. If the product of the slopes of two lines is -1, then the lines are perpendicular.
3. A horizontal line having zero slope is perpendicular to a vertical line having undefined slope.

An equivalent way of stating this relationship is to say that one line is perpendicular to another line if its slope is the *negative reciprocal* of the slope of the other. For example, if a line has slope 5, any line having slope $-\frac{1}{5}$ is perpendicular to it. Similarly, if a line has slope $-\frac{3}{4}$, any line having slope $\frac{4}{3}$ is perpendicular to it.

EXAMPLE 4 Finding the Slope of a Line Perpendicular to a Given Line

Find the slope of any line that is perpendicular to the line whose equation is $x + 4y - 8 = 0$.

SOLUTION We begin by writing the equation of the given line in slope-intercept form. Solve for y.

$$x + 4y - 8 = 0$$ This is the given equation.

$$4y = -x + 8$$ To isolate the y-term, subtract x and add 8 on both sides.

$$y = -\frac{1}{4}x + 2$$ Divide both sides by 4.

Slope is $-\frac{1}{4}$.

The given line has slope $-\frac{1}{4}$. Any line perpendicular to this line has a slope that is the negative reciprocal of $-\frac{1}{4}$. Thus, the slope of any perpendicular line is 4. ■

✔ **CHECK POINT 4** Find the slope of any line that is perpendicular to the line whose equation is $x + 3y - 12 = 0$.

3 Model data with linear equations and make predictions.

USING TECHNOLOGY

You can use a graphing utility to obtain a model for a scatter plot in which the data points fall on or near a straight line. The line that best fits the data is called the **regression line**. After entering the data in Figure 3.32(b), a graphing utility displays a scatter plot of the data and the regression line.

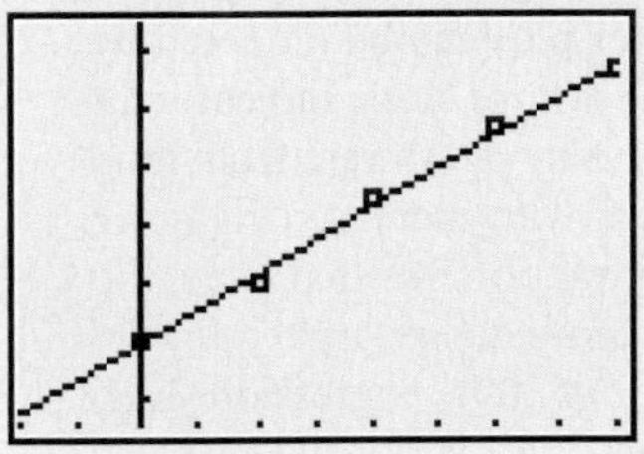

[−10, 40, 5] by [25, 39, 2]

Also displayed is the regression line's equation.

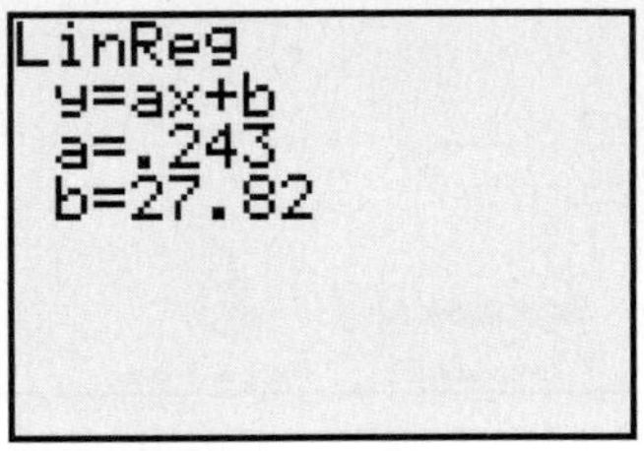

Applications Linear equations are useful for modeling data that fall on or near a line. For example, the bar graph in Figure 3.32(a) gives the median age of the U.S. population in the indicated year. (The median age is the age in the middle when all the ages of the U.S. population are arranged from youngest to oldest.) The data are displayed as a set of five points in a rectangular coordinate system in Figure 3.32(b).

The Graying of America: Median Age of the U.S. Population

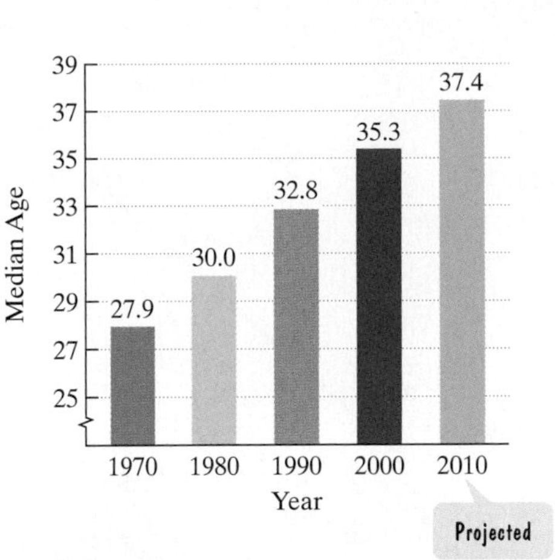

FIGURE 3.32(a)

Source: U.S. Census Bureau

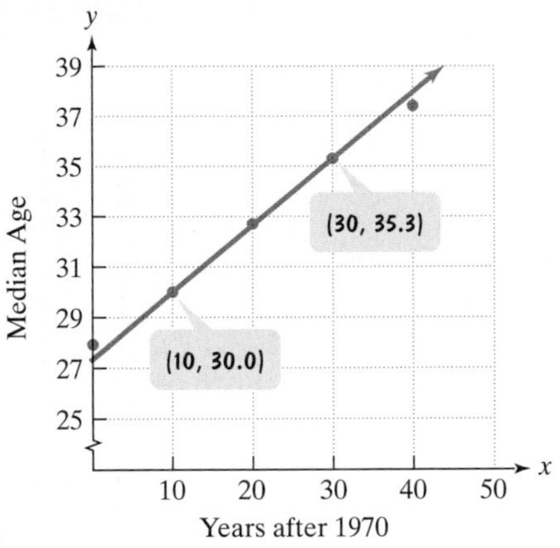

FIGURE 3.32(b)

A set of points representing data is called a **scatter plot**. Also shown on the scatter plot in Figure 3.32(b) is a line that passes through or near the five points. By writing the equation of this line, we can obtain a model of the data and make predictions about the median age of the U.S. population in the future.

EXAMPLE 5 Modeling the Graying of America

Write the slope-intercept equation of the line shown in Figure 3.32(b). Use the equation to predict the median age of the U.S. population in 2020.

SOLUTION The line in Figure 3.32(b) passes through (10, 30.0) and (30, 35.3). We start by finding its slope.

$$m = \frac{\text{Change in } y}{\text{Change in } x} = \frac{35.3 - 30.0}{30 - 10} = \frac{5.3}{20} = 0.265$$

The slope indicates that each year the median age of the U.S. population is increasing by 0.265 years.

Now, we write the line's slope-intercept equation.

$y - y_1 = m(x - x_1)$ — Begin with the point-slope form.

$y - 30.0 = 0.265(x - 10)$ — Either ordered pair can be (x_1, y_1). Let $(x_1, y_1) = (10, 30.0)$. From above, $m = 0.265$.

$y - 30.0 = 0.265x - 2.65$ — Apply the distributive property.

$y = 0.265x + 27.35$ — Add 30 to both sides and solve for y.

A linear equation that models the median age of the U.S. population, y, x years after 1970 is

$$y = 0.265x + 27.35.$$

Now, let's use this equation to predict the median age in 2020. Because 2020 is 50 years after 1970, substitute 50 for x and compute y.

$$y = 0.265(50) + 27.35 = 40.6$$

Our model predicts that the median age of the U.S. population in 2020 will be 40.6. ■

CHECK POINT 5 Use the data points (10, 30.0) and (20, 32.8) from Figure 3.32(b) on page 231 to write a slope-intercept equation that models the median age of the U.S. population x years after 1970. Use this model to predict the median age in 2020.

ENRICHMENT ESSAY

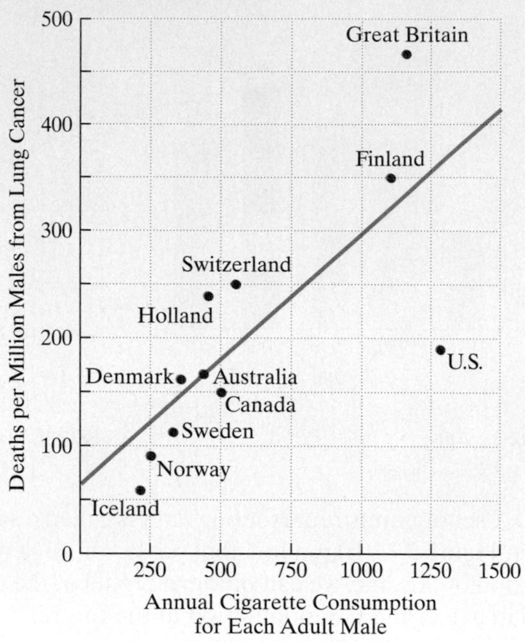

Source: *Smoking and Health*, Washington, D.C., 1964

Cigarettes and Lung Cancer

This scatter plot shows a relationship between cigarette consumption among males and deaths due to lung cancer per million males. The data are from 11 countries and date back to a 1964 report by the U.S. Surgeon General. The scatter plot can be modeled by a line whose slope indicates an increasing death rate from lung cancer with increased cigarette consumption. At that time, the tobacco industry argued that in spite of this regression line, tobacco use is not the cause of cancer. Recent data do, indeed, show a causal effect between tobacco use and numerous diseases.

3.5 EXERCISE SET

Student Solutions Manual | CD/Video | PH Math/Tutor Center | MathXL Tutorials on CD | MathXL® | MyMathLab | Interactmath.com

Practice Exercises

Write the point-slope form of the line satisfying each of the conditions in Exercises 1–28. Then use the point-slope form of the equation to write the slope-intercept form of the equation.

1. Slope $= 3$, passing through $(2, 5)$

2. Slope $= 6$, passing through $(3, 1)$

3. Slope $= 5$, passing through $(-2, 6)$

4. Slope $= 7$, passing through $(-4, 9)$

5. Slope $= -8$, passing through $(-3, -2)$

6. Slope $= -4$, passing through $(-5, -2)$

7. Slope $= -12$, passing through $(-8, 0)$

8. Slope $= -11$, passing through $(0, -3)$

9. Slope $= -1$, passing through $\left(-\frac{1}{2}, -2\right)$

10. Slope $= -1$, passing through $\left(-4, -\frac{1}{4}\right)$

11. Slope $= \frac{1}{2}$, passing through the origin

12. Slope $= \frac{1}{3}$, passing through the origin

13. Slope $= -\frac{2}{3}$, passing through $(6, -2)$

14. Slope $= -\frac{3}{5}$, passing through $(10, -4)$

15. Passing through $(1, 2)$ and $(5, 10)$

16. Passing through $(3, 5)$ and $(8, 15)$

17. Passing through $(-3, 0)$ and $(0, 3)$

18. Passing through $(-2, 0)$ and $(0, 2)$

19. Passing through $(-3, -1)$ and $(2, 4)$

20. Passing through $(-2, -4)$ and $(1, -1)$

21. Passing through $(-4, -1)$ and $(3, 4)$

22. Passing through $(-6, 1)$ and $(2, -5)$

23. Passing through $(-3, -1)$ and $(4, -1)$

24. Passing through $(-2, -5)$ and $(6, -5)$

25. Passing through $(2, 4)$ with x-intercept $= -2$

26. Passing through $(1, -3)$ with x-intercept $= -1$

27. x-intercept $= -\frac{1}{2}$ and y-intercept $= 4$

28. x-intercept $= 4$ and y-intercept $= -2$

In Exercises 29–44, the equation of a line is given. Find the slope of a line that is **a.** *parallel to the line with the given equation; and* **b.** *perpendicular to the line with the given equation.*

29. $y = 5x$

30. $y = 3x$

31. $y = -7x$

32. $y = -9x$

33. $y = \frac{1}{2}x + 3$

34. $y = \frac{1}{4}x - 5$

35. $y = -\frac{2}{5}x - 1$

36. $y = -\frac{3}{7}x - 2$

37. $4x + y = 7$

38. $8x + y = 11$

39. $2x + 4y - 8 = 0$

40. $3x + 2y - 6 = 0$

41. $2x - 3y - 5 = 0$

42. $3x - 4y + 7 = 0$

43. $x = 6$

44. $y = 9$

In Exercises 45–48, write an equation for line L in point-slope form and slope-intercept form.

45.

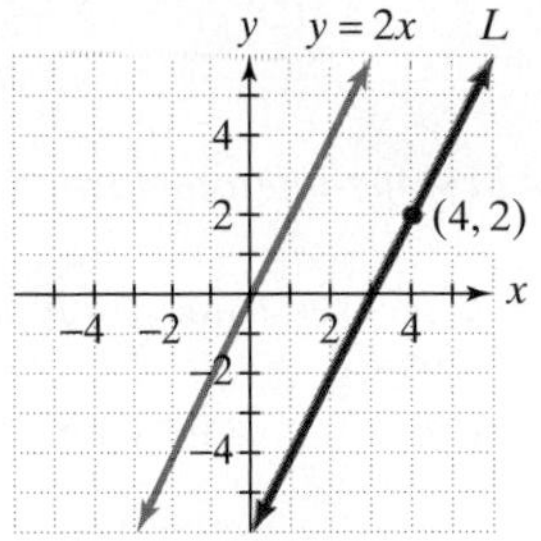

L is parallel to $y = 2x$.

46.

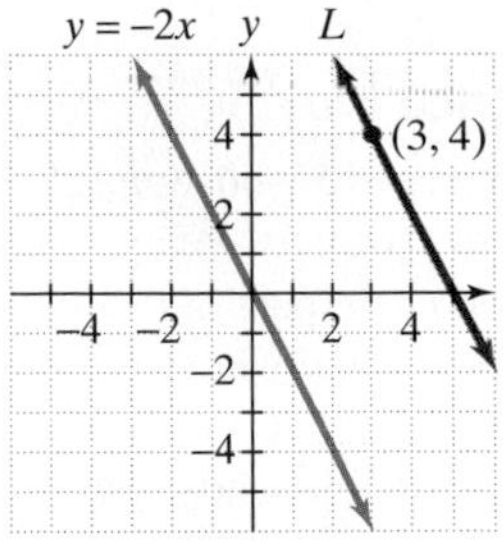

L is parallel to $y = -2x$.

47.

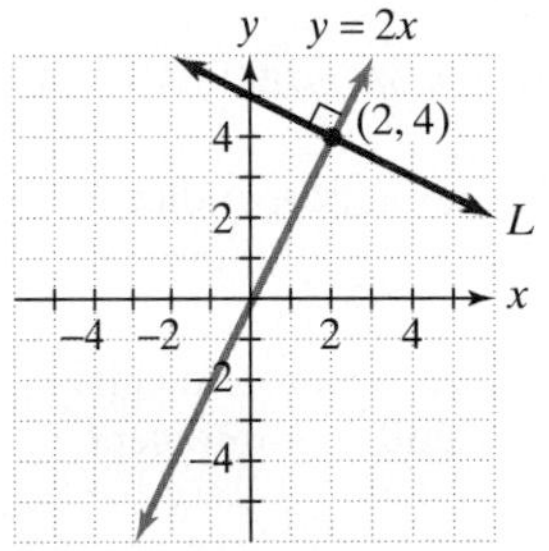

L is perpendicular to $y = 2x$.

48.

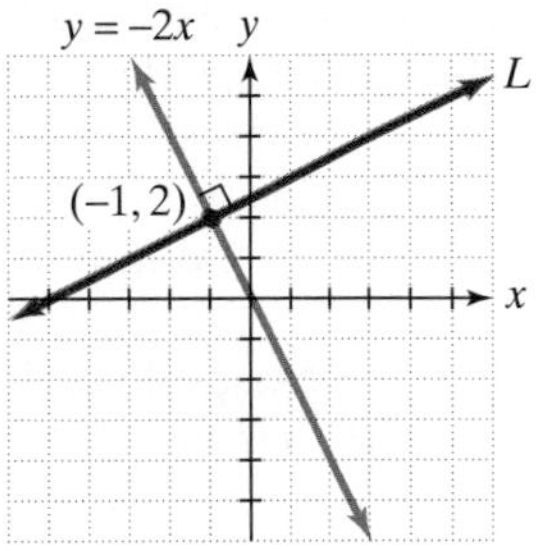

L is perpendicular to $y = -2x$.

In Exercises 49–56, use the given conditions to write an equation for each line in point-slope form and slope-intercept form.

49. Passing through $(-8, -10)$ and parallel to the line whose equation is $y = -4x + 3$

50. Passing through $(-2, -7)$ and parallel to the line whose equation is $y = -5x + 4$

51. Passing through $(2, -3)$ and perpendicular to the line whose equation is $y = \frac{1}{5}x + 6$

52. Passing through $(-4, 2)$ and perpendicular to the line whose equation is $y = \frac{1}{3}x + 7$

53. Passing through $(-2, 2)$ and parallel to the line whose equation is $2x - 3y - 7 = 0$

54. Passing through $(-1, 3)$ and parallel to the line whose equation is $3x - 2y - 5 = 0$

55. Passing through $(4, -7)$ and perpendicular to the line whose equation is $x - 2y - 3 = 0$

56. Passing through $(5, -9)$ and perpendicular to the line whose equation is $x + 7y - 12 = 0$

Practice Plus

In Exercises 57–66, write an equation in slope-intercept form of the line satisfying the given conditions.

57. The line passes through $(2, 4)$ and has the same y-intercept as the line whose equation is $x - 4y = 8$.

58. The line passes through $(2, 6)$ and has the same y-intercept as the line whose equation is $x - 3y = 18$.

59. The line has an x-intercept at -4 and is parallel to the line containing $(3, 1)$ and $(2, 6)$.

60. The line has an x-intercept at -6 and is parallel to the line containing $(4, -3)$ and $(2, 2)$.

61. The line passes through $(-1, 5)$ and is perpendicular to the line whose equation is $x = 6$.

62. The line passes through $(-2, 6)$ and is perpendicular to the line whose equation is $x = -4$.

63. The line passes through $(-6, 4)$ and is perpendicular to the line that has an x-intercept of 2 and a y-intercept of -4.

64. The line passes through $(-5, 6)$ and is perpendicular to the line that has an x-intercept of 3 and a y-intercept of -9.

65. The line is perpendicular to the line whose equation is $3x - 2y = 4$ and has the same y-intercept as this line.

66. The line is perpendicular to the line whose equation is $4x - y = 6$ and has the same y-intercept as this line.

67. What is the slope of a line that is parallel to the line whose equation is $Ax + By = C, B \neq 0$?

68. What is the slope of a line that is perpendicular to the line whose equation is $Ax + By = C, A \neq 0$ and $B \neq 0$?

Application Exercises

69. We seem to be fed up with being lectured at about our waistlines. The points in the graph show the average weight of American adults from 1990 through 2002. Also shown is a line that passes through or near the points.

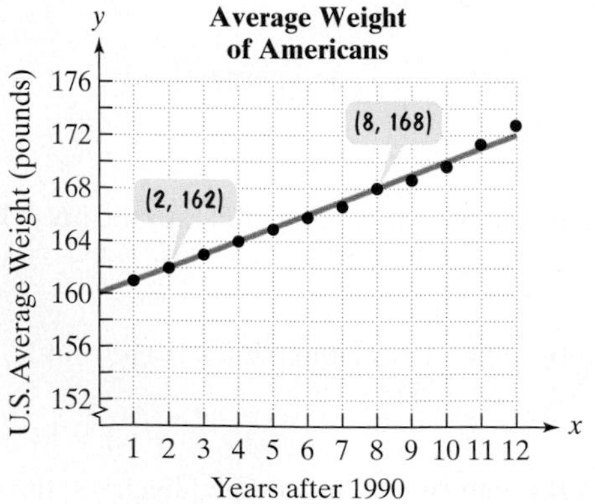

Source: Diabetes Care

a. Use the two points whose coordinates are shown by the voice balloons to find the point-slope equation of the line that models average weight of Americans, y, in pounds, x years after 1990.

b. Write the equation in part (a) in slope-intercept form.

c. Use the slope-intercept equation to predict the average weight of Americans in 2010.

70. Films may not be getting any better, but in this era of moviegoing, the number of screens available for new films and the classics has exploded. The points in the graph show the number of screens in the United States from 1995 through 2002. Also shown is a line that passes through or near the points.

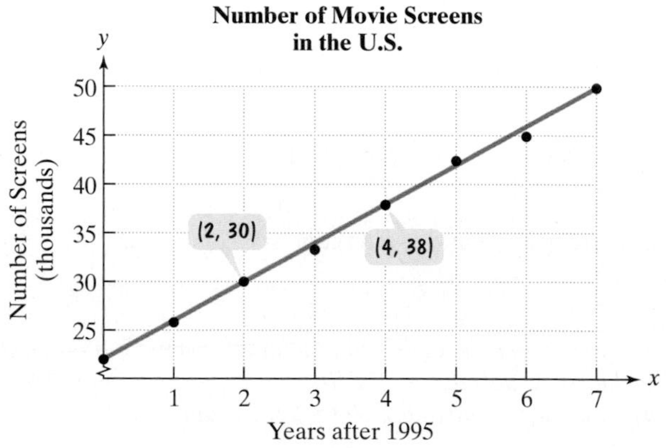

Source: Motion Picture Association of America

a. Use the two points whose coordinates are shown by the voice balloons to find the point-slope equation of the line that models the number of screens, y, in thousands, x years after 1995.

b. Write the equation in part (a) in slope-intercept form.

c. Use the slope-intercept equation to predict the number of screens, in thousands, in 2010.

71. Is there a relationship between education and prejudice? With increased education, does a person's level of prejudice tend to decrease? The scatter plot at the top of the next column shows ten data points, each representing the number of years of school completed and the score on a test measuring prejudice for each subject. Higher scores on this 1-to-10 test indicate greater prejudice. Also shown is the regression line, the line that best fits the data. Use two points on this line to write both its point-slope and slope-intercept equations. Then use the slope-intercept equation to predict the score on the prejudice test for a person with seven years of education.

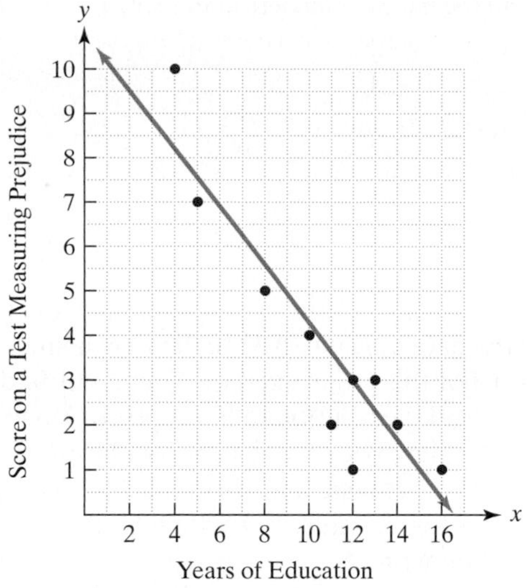

72. A business discovers a linear relationship between the number of shirts it can sell and the price per shirt. In particular, 20,000 shirts per week can be sold at \$19 each. Raising the price to \$55 causes the sales to fall to 2000 shirts per week. Write the point-slope and slope-intercept equations of the *demand line* through the ordered pairs

(20,000 shirts, \$19) and (2000 shirts, \$55).

Then determine the number of shirts that can be sold per week at \$50 each.

73. The scatter plot shows the average number of minutes each that 16 people exercise per week and the average number of headaches per month each person experiences.

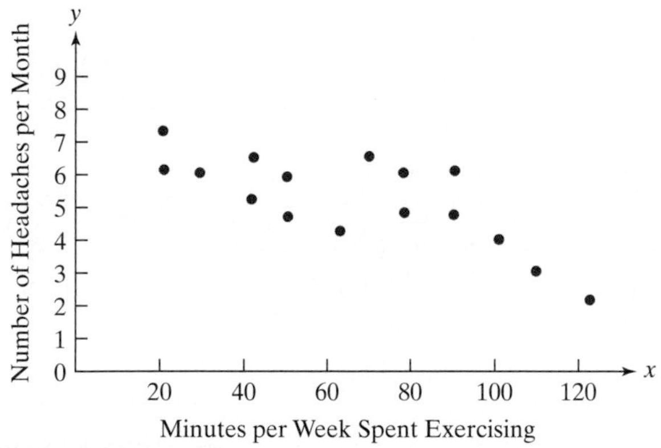

a. Draw a line that fits the data so that the spread of the data points around the line is as small as possible.

b. Use the coordinates of two points along your line to write its point-slope and slope-intercept equations.

c. Use the equation in part (b) to predict the number of headaches per month for a person exercising 130 minutes per week.

d. What general observation can you make about the number of headaches per month as the number of minutes per week spent exercising increases?

Writing in Mathematics

74. Describe how to write the equation of a line if its slope and a point along the line are known.

75. Describe how to write the equation of a line if two points along the line are known.

76. Take a second look at the scatter plot in Exercise 71, shown on the left. Although there is a relationship between education and prejudice, we cannot necessarily conclude that increased education causes a person's level of prejudice to decrease. Offer two or more possible explanations for the data in the scatter plot.

Critical Thinking Exercises

77. Which one of the following is true?

a. If a line has undefined slope, then it has no equation.

b. The line whose equation is $y - 3 = 7(x + 2)$ passes through $(-3, 2)$.

c. The point-slope form cannot be applied to the line through the points $(2, -5)$ and $(2, 6)$.

d. The slope of the line whose equation is $3x + y = 7$ is 3.

78. Excited about the success of celebrity stamps, post office officials were rumored to have put forth a plan to institute two new types of thermometers. On these new scales, °E represents degrees Elvis and °M represents degrees Madonna. If it is known that $40°E = 25°M$, $280°E = 125°M$, and degrees Elvis is linearly related to degrees Madonna, write an equation expressing E in terms of M.

Technology Exercises

79. Use a graphing utility to graph $y = 1.75x - 2$. Select the best viewing rectangle possible by experimenting with the range settings to show that the line's slope is $\frac{7}{4}$.

80. Use a graphing utility to graph the slope-intercept equation that you wrote in Exercise 72. Then select an appropriate range setting and use the TRACE feature to graphically show the number of shirts that can be sold at \$50 each.

81. **a.** Use the statistical menu of a graphing utility to enter the ten data points shown in the scatter plot in Exercise 71.

b. Use the DRAW menu and the scatter plot capability to draw a scatter plot of the data points like the one shown in Exercise 71.

c. Select the linear regression option. Use your utility to obtain values for a and b for the equation of the regression line, $y = ax + b$. You may also be given a *correlation coefficient*, r. Values of r close to 1 indicate that the points can be described by a linear relationship and the regression line has a positive slope. Values of r close to -1 indicate that the points can be described by a linear relationship and the regression line has a negative slope. Values of r close to 0 indicate no linear relationship between the variables.

d. Use the appropriate sequence (consult your manual) to graph the regression equation on top of the points in the scatter plot.

Review Exercises

82. How many sheets of paper, weighing 2 grams each, can be put in an envelope weighing 4 grams if the total weight must not exceed 29 grams? (Section 2.7, Example 8, and Section 2.5, Example 5)

83. List all the natural numbers in this set:

$$\left\{-2, 0, \frac{1}{2}, 1, \sqrt{3}, \sqrt{4}\right\}.$$

(Section 1.2, Example 5)

84. Use intercepts to graph $3x - 5y = 15$. (Section 3.2, Example 4)

GROUP PROJECT

CHAPTER 3

In Example 5 on page 231, we used the data in Figure 3.32 to develop a linear equation that modeled the graying of America. For this group exercise, you might find it helpful to pattern your work after Figure 3.32 and the solution to Example 5. Group members should begin by consulting an almanac, newspaper, magazine, or the Internet to find data that lie approximately on or near a straight line. Working by hand or using a graphing utility, construct a scatter plot for the data. If working by hand, draw a line that approximately fits the data and then write its equation. If using a graphing utility, obtain the equation of the regression line. Then use the equation of the line to make a prediction about what might happen in the future. Are there circumstances that might affect the accuracy of this prediction? List some of these circumstances.

CHAPTER 3 SUMMARY

Definitions and Concepts	Examples

Section 3.1 Graphing Equations in Two Variables

An ordered pair is a solution of an equation in two variables if replacing the variables by the coordinates of the ordered pair results in a true statement.

Is $(-1, 4)$ a solution of $2x + 5y = 18$?

$$2(-1) + 5 \cdot 4 \stackrel{?}{=} 18$$
$$-2 + 20 \stackrel{?}{=} 18$$
$$18 = 18, \text{ true}$$

Thus, $(-1, 4)$ is a solution.

One method for graphing an equation in two variables is point plotting. Find several ordered-pair solutions, plot them as points, and connect the points with a smooth curve or line.

Graph: $y = 2x + 1$.

x	$y = 2x + 1$	(x, y)
-2	$y = 2(-2) + 1 = -3$	$(-2, -3)$
-1	$y = 2(-1) + 1 = -1$	$(-1, -1)$
0	$y = 2 \cdot 0 + 1 = 1$	$(0, 1)$
1	$y = 2 \cdot 1 + 1 = 3$	$(1, 3)$
2	$y = 2 \cdot 2 + 1 = 5$	$(2, 5)$

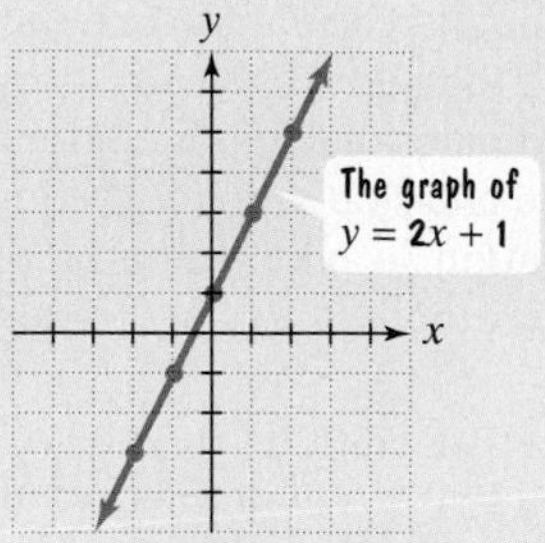

Definitions and Concepts	Examples

Section 3.2 Graphing Linear Equations Using Intercepts

If a graph intersects the x-axis at $(a, 0)$, then a is an x-intercept.
If a graph intersects the y-axis at $(0, b)$, then b is a y-intercept.

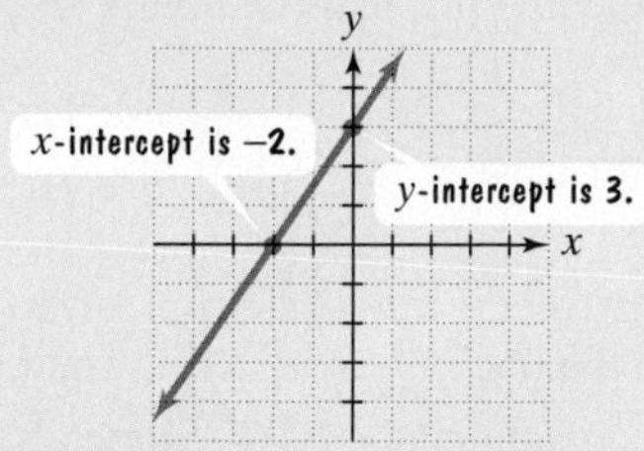

An equation of the form $Ax + By = C$, where A, B, and C are integers, is called the standard form of the equation of a line. The graph of $Ax + By = C$ is a line that can be obtained using intercepts. To find the x-intercept, let $y = 0$ and solve for x. To find the y-intercept, let $x = 0$ and solve for y. Find a checkpoint, a third ordered-pair solution. Graph the equation by drawing a line through the three points.

Graph using intercepts: $4x + 3y = 12$.

x-intercept: $4x = 12$
$x = 3$

y-intercept: $3y = 12$
$y = 4$

Checkpoint: Let $x = 2$.
$8 + 3y = 12$
$3y = 4$
$y = \frac{4}{3}$

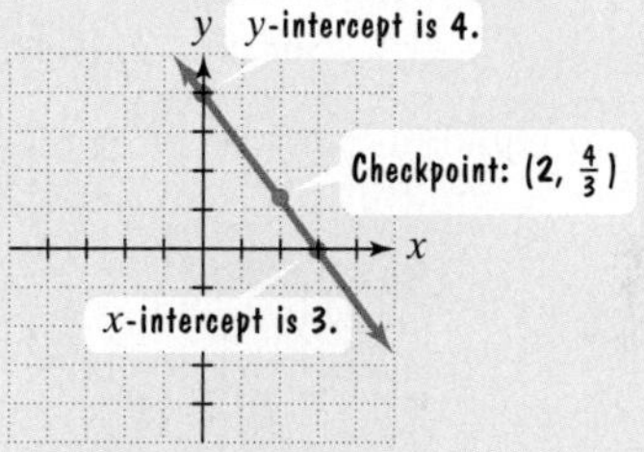

The graph of $Ax + By = 0$ is a line that passes through the origin. Find two other points by finding two other solutions of the equation. Graph the equation by drawing a line through the origin and these two points.

Graph: $x + 2y = 0$.

$x = 2$: $2 + 2y = 0$
$2y = -2$
$y = -1$

$y = 1$: $x + 2(1) = 0$
$x = -2$

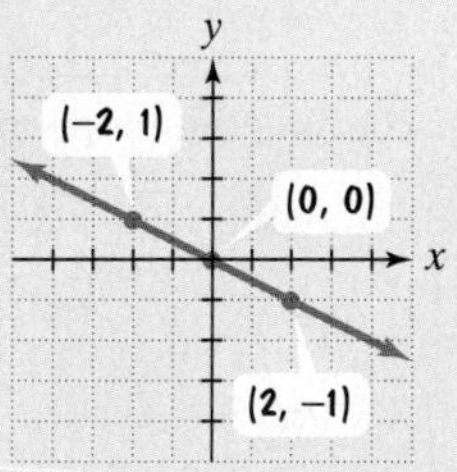

Horizontal and Vertical Lines
The graph of $y = b$ is a horizontal line. The y-intercept is b.
The graph of $x = a$ is a vertical line. The x-intercept is a.

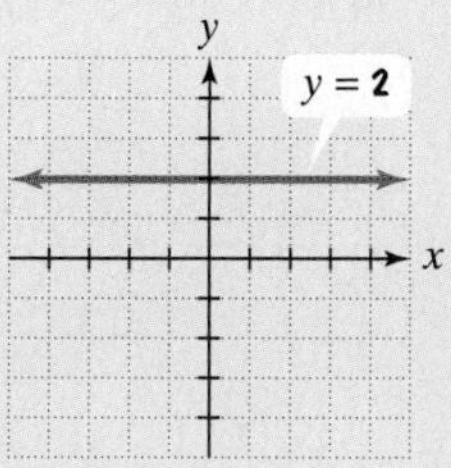

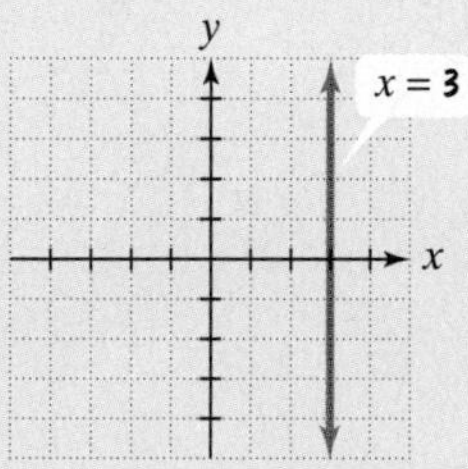

Definitions and Concepts	Examples

Section 3.3 Slope

The slope, m, of the line through the points (x_1, y_1) and (x_2, y_2) is

$$m = \frac{y_2 - y_1}{x_2 - x_1}, \quad x_2 - x_1 \neq 0.$$

If the slope is positive, the line rises from left to right. If the slope is negative, the line falls from left to right.

The slope of a horizontal line is 0. The slope of a vertical line is undefined.

If two distinct nonvertical lines have the same slope, then they are parallel.

Find the slope of the line passing through the points shown.

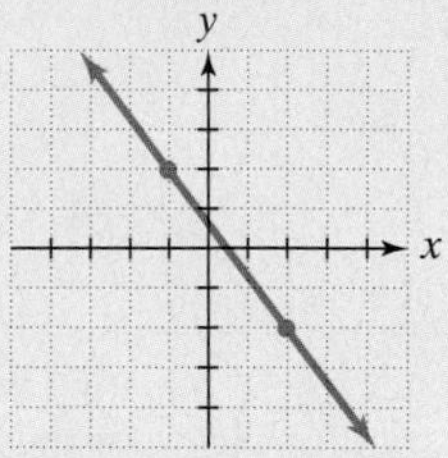

Let $(x_1, y_1) = (-1, 2)$ and $(x_2, y_2) = (2, -2)$.

$$m = \frac{y_2 - y_1}{x_2 - x_1} = \frac{-2 - 2}{2 - (-1)} = \frac{-4}{3} = -\frac{4}{3}$$

Section 3.4 The Slope-Intercept Form of the Equation of a Line

The slope-intercept equation of a nonvertical line with slope m and y-intercept b is

$$y = mx + b.$$

Find the slope and the y-intercept of the line with the given equation.

- $y = -2x + 5$

  Slope -2. y-intercept is 5.

- $2x + 3y = 9$ (Solve for y)

  $3y = -2x + 9$ Subtract $2x$.

  $y = -\frac{2}{3}x + 3$ Divide by 3.

  Slope is $-\frac{2}{3}$. y-intercept is 3.

Graphing $y = mx + b$ Using the Slope and y-Intercept

1. Plot the point containing the y-intercept on the y-axis. This is the point $(0, b)$.
2. Use the slope, m, to obtain a second point. Write m as a fraction, and use rise over run, starting at the point containing the y-intercept, to plot this point.
3. Graph the equation by drawing a line through the two points.

Graph: $y = -\frac{3}{4}x + 1$.

Slope is $-\frac{3}{4}$. y-intercept is 1.

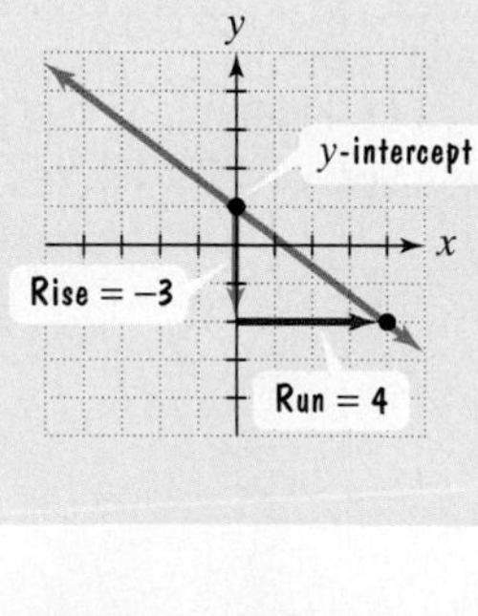

Definitions and Concepts	Examples
Section 3.5 The Point-Slope Form of the Equation of a Line	
The point-slope equation of a nonvertical line with slope m that passes through the point (x_1, y_1) is $$y - y_1 = m(x - x_1).$$	Slope $= -3$, passing through $(-1, 5)$ [$m = -3$; $x_1 = -1$; $y_1 = 5$] The line's point-slope equation is $$y - 5 = -3[x - (-1)].$$ Simplify: $$y - 5 = -3(x + 1).$$
To write the point-slope form of the line passing through two points, begin by using the points to compute the slope, m. Use either given point as (x_1, y_1) and write the point-slope equation: $$y - y_1 = m(x - x_1).$$ Solving this equation for y gives the slope-intercept form of the line's equation.	Write an equation in point-slope form and slope-intercept form of the line passing through $(-1, -3)$ and $(4, 2)$. $$m = \frac{2 - (-3)}{4 - (-1)} = \frac{2 + 3}{4 + 1} = \frac{5}{5} = 1$$ Using $(4, 2)$ as (x_1, y_1), the point-slope equation is $$y - 2 = 1(x - 4).$$ Solve for y to obtain the slope-intercept form. $$y = x - 2$$ Add 2 to both sides.
Nonvertical parallel lines have the same slope. If the product of the slopes of two lines is -1, then the lines are perpendicular. One line is perpendicular to another line if its slope is the negative reciprocal of the slope of the other.	Write point-slope and slope-intercept equations of the line passing through $(2, -1)$ [x_1; y_1] and perpendicular to $y = -\frac{1}{5}x + 6$. [slope] Slope, m, of perpendicular line is 5, the negative reciprocal of $-\frac{1}{5}$. $$y - (-1) = 5(x - 2)$$ Point-slope equation $$y + 1 = 5(x - 2)$$ $$y + 1 = 5x - 10$$ $$y = 5x - 11$$ Slope intercept equation

CHAPTER 3 REVIEW EXERCISES

3.1 *In Exercises 1–2, determine whether each ordered pair is a solution of the given equation.*

1. $y = 3x + 6 \quad (-3, 3), (0, 6), (1, 9)$

2. $3x - y = 12 \quad (0, 4), (4, 0), (-1, 15)$

In Exercises 3–4,

a. *Find five solutions of each equation. Organize your work in a table of values.*

b. *Use the five solutions in the table to graph each equation.*

3. $y = 2x - 3$
4. $y = \frac{1}{2}x + 1$
5. Graph the equation: $y = x^2 - 3$. Select integers for x, starting with -3 and ending with 3.
6. The linear equation in two variables $y = 5x - 41$ models the percentage of U.S. adults, y, with x years of education who are doing volunteer work.

a. Find four solutions of the equation. Use 10, 12, 14, and 16 for x. Organize your work in a table of values.

b. How well does the given equation model the data shown in the following table? Explain your answer.

Years of Education	10	12	14	16
Percentage Doing Volunteer Work	8.3%	18.8%	28.1%	38.4%

Source: U.S. Bureau of Labor

3.2 *In Exercises 7–9, use the graph to identify the*

a. *x-intercept, or state that there is no x-intercept.*

b. *y-intercept, or state that there is no y-intercept.*

7.

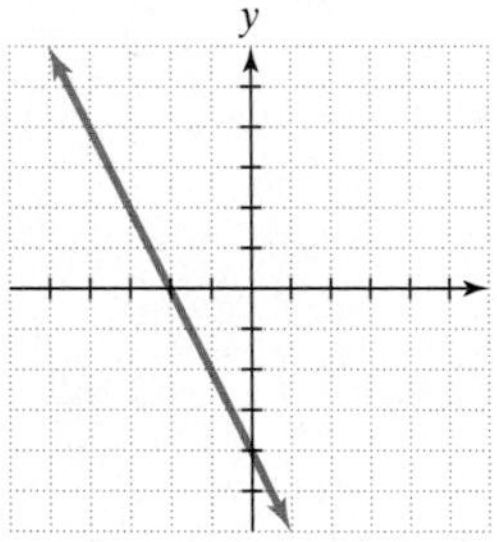

8.

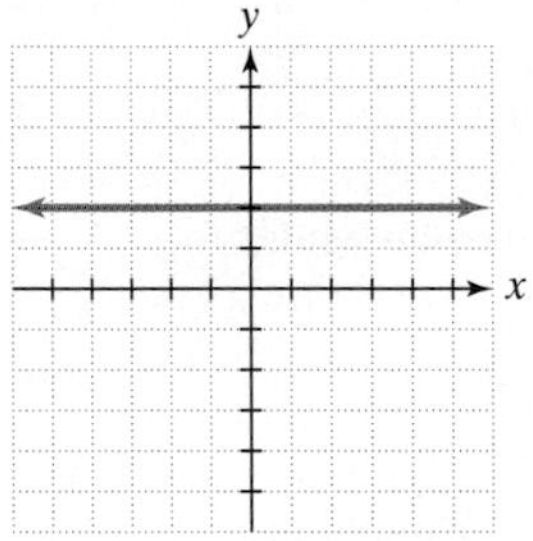

9.

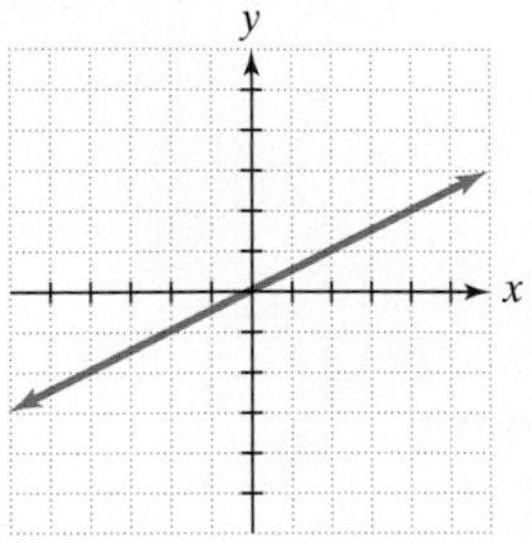

In Exercises 10–13, use intercepts to graph each equation.

10. $2x + y = 4$
11. $3x - 2y = 12$
12. $3x = 6 - 2y$
13. $3x - y = 0$

In Exercises 14–17, graph each equation.

14. $x = 3$
15. $y = -5$
16. $y + 3 = 5$
17. $2x = -8$
18. The graph shows the Fahrenheit temperature, y, x hours after noon.

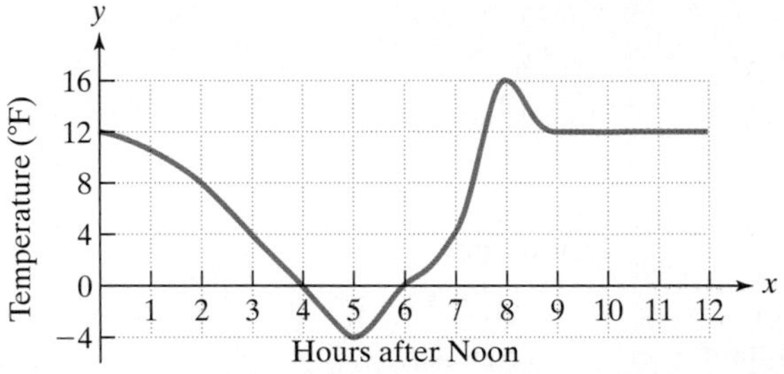

a. At what time did the minimum temperature occur? What is the minimum temperature?

b. At what time did the maximum temperature occur? What is the maximum temperature?

c. What are the x-intercepts? In terms of time and temperature, interpret the meaning of these intercepts.

d. What is the y-intercept? What does this mean in terms of time and temperature?

e. From 9 P.M. until midnight, the graph is shown as a horizontal line. What does this mean about the temperature over this period of time?

3.3 *In Exercises 19–22, calculate the slope of the line passing through the given points. If the slope is undefined, so state. Then indicate whether the line rises, falls, is horizontal, or is vertical.*

19. $(3, 2)$ and $(5, 1)$
20. $(-1, 2)$ and $(-3, -4)$
21. $(-3, 4)$ and $(6, 4)$
22. $(5, 3)$ and $(5, -3)$

In Exercises 23–26, find the slope of each line, or state that the slope is undefined.

23.

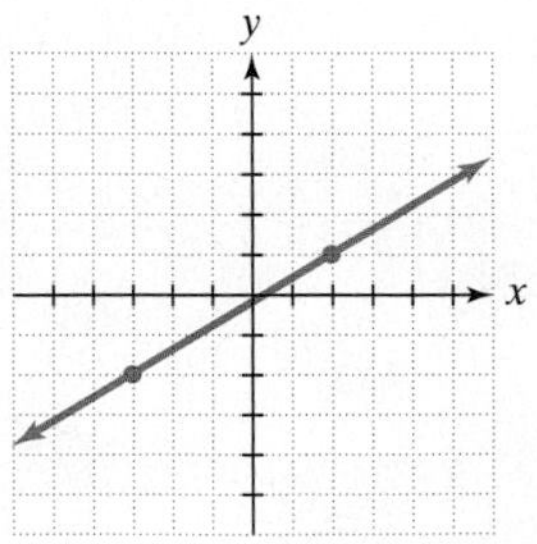

24.

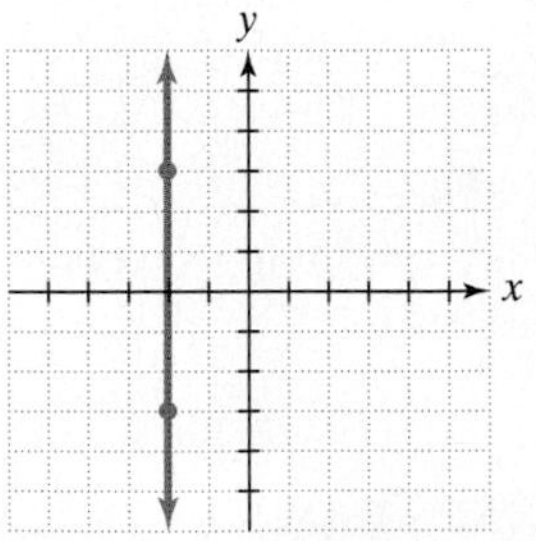

25.

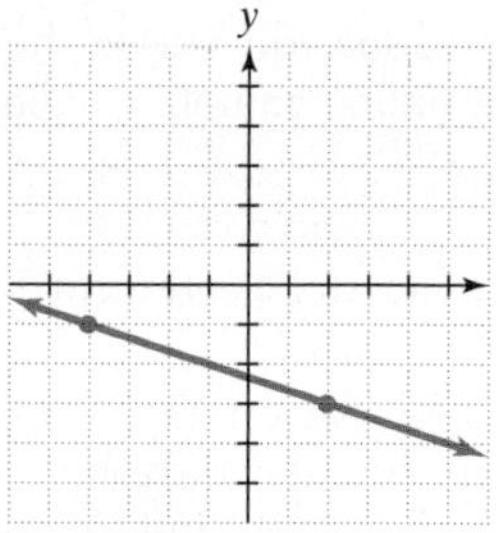

26.

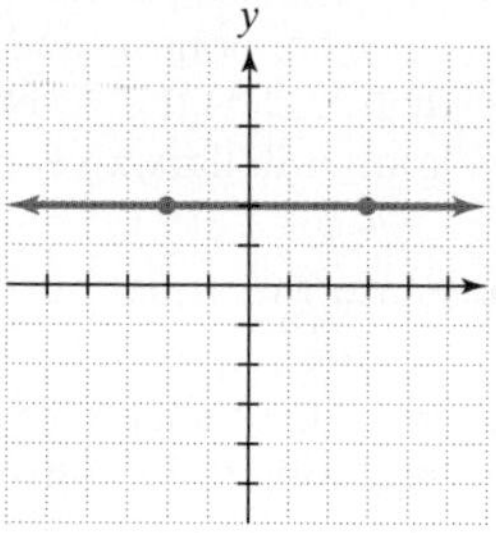

In Exercises 27–28, determine whether the distinct lines through each pair of points are parallel.

27. $(-1, -3)$ and $(2, -8)$
$(8, -7)$ and $(9, 10)$

28. $(5, 4)$ and $(9, 7)$
$(-6, 0)$ and $(-2, 3)$

29. The graph shows new AIDS diagnoses among the general U.S. population, y, for year x, where $1999 \leq x \leq 2003$.

New AIDS Diagnoses in the U.S.

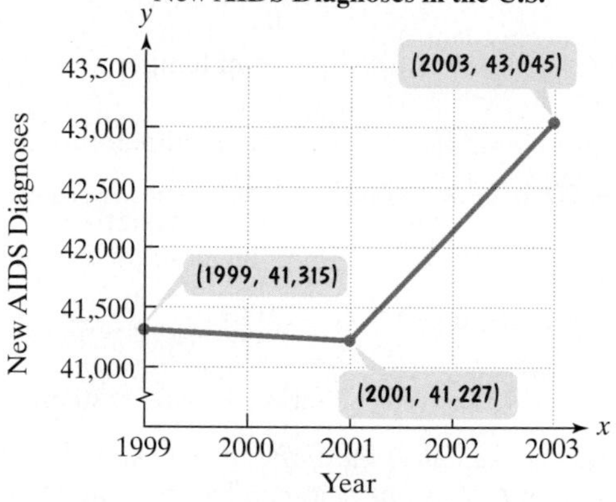

Source: Centers for Disease Control

a. Find the slope of the line passing through (1999, 41,315) and (2001, 41,227). Then express the slope as a rate of change with the proper units attached.

b. Find the slope of the line passing through (2001, 41,227) and (2003, 43,045). Then express the slope as a rate of change.

c. Draw a line passing through (1999, 41,315) and (2003, 43,045) and find its slope. Is the slope the average of the slopes of the lines that you found in parts (a) and (b)? Explain your answer.

3.4 *In Exercises 30–33, find the slope and the y-intercept of the line with the given equation.*

30. $y = 5x - 7$

31. $y = 6 - 4x$

32. $y = 3$

33. $2x + 3y = 6$

In Exercises 34–36, graph each linear equation using the slope and y-intercept.

34. $y = 2x - 4$

35. $y = \frac{1}{2}x - 1$

36. $y = -\frac{2}{3}x + 5$

In Exercises 37–38, write each equation in slope-intercept form. Then use the slope and y-intercept to graph the equation.

37. $y - 2x = 0$

38. $\frac{1}{3}x + y = 2$

39. Graph $y = -\frac{1}{2}x + 4$ and $y = -\frac{1}{2}x - 1$ in the same rectangular coordinate system. Are the lines parallel? If so, explain why.

40. The graph shows the average age of U.S. whites, African Americans, and Americans of Hispanic origin from 1990 through 2000.

Average Age of U.S. Whites, African Americans, and Hispanics

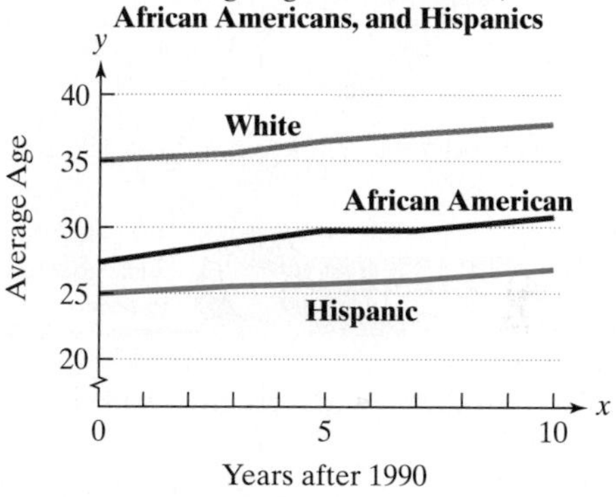

Source: U.S. Census Bureau

a. What is the smallest y-intercept? Describe what this represents in this situation.

b. The average age of the group with the greatest y-intercept was approximately 38 in 2000. Use the points $(0, 35)$ and $(10, 38)$ to compute the slope for this group. What does this mean about their average age for the period shown?

c. Use the slope from part (b) and the y-intercept for this group shown by the graph to write an equation that models the group's average age, y, x years after 1990. Write your model in $y = mx + b$ form.

d. Use your model from part (c) to predict the average age for the group in 2010.

3.5 *Write the point-slope form of the line satisfying the conditions in Exercises 41–44. Then use the point-slope form of the equation to write the slope-intercept form.*

41. Slope $= 6$, passing through $(-4, 7)$

42. Passing through $(3, 4)$ and $(2, 1)$

43. Passing through $(4, -7)$ and parallel to the line whose equation is $3x + y - 9 = 0$

44. Passing through $(-2, 6)$ and perpendicular to the line whose equation is $y = \frac{1}{3}x + 4$

45. You can click a mouse and bet the house. The points in the graph show the dizzying growth of online gambling. With more than 1800 sites, the industry has become the Web's biggest moneymaker.

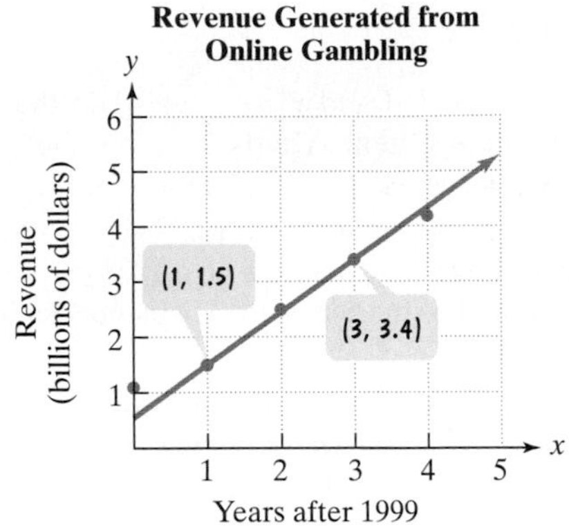

Source: Newsweek

a. Use the two points whose coordinates are shown by the voice balloons to find the point-slope equation of the line that models the revenue from online gambling, y, in billions of dollars, x years after 1999.

b. Write the equation in part (a) in slope-intercept form.

c. In 2003, nearly \$3.5 billion was lost on Internet bets, triggering a sharp backlash that threatened to shut down Internet wagering. If this crackdown on the industry is not successful, use your slope-intercept model to predict the billions of dollars in revenue from online gambling in 2009.

CHAPTER 3 TEST

Remember to use your Chapter Test Prep Video CD to see the worked-out solutions to the test questions you want to review.

1. Determine whether each ordered pair is a solution of $4x - 2y = 10$:

$$(0, -5), \quad (-2, 1), \quad (4, 3).$$

2. Find five solutions of $y = 3x + 1$. Organize your work in a table of values. Then use the five solutions in the table to graph the equation.

3. Graph: $y = x^2 - 1$. Select integers for x, starting with -3 and ending with 3.

4. Use the graph to identify the

a. x-intercept, or state that there is no x-intercept.

b. y-intercept, or state that there is no y-intercept.

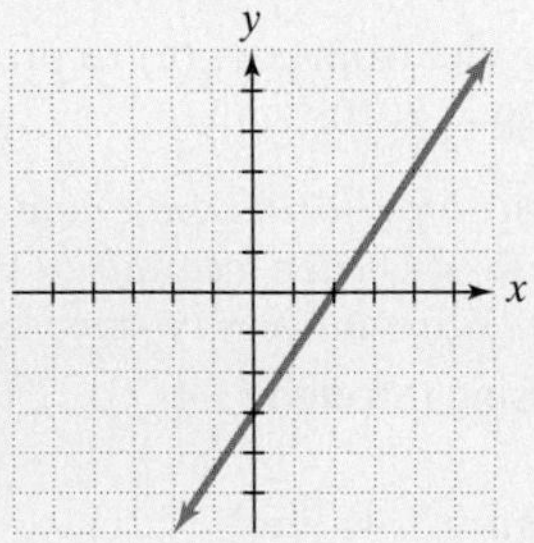

5. Use intercepts to graph $4x - 2y = -8$.

6. Graph $y = 4$ in a rectangular coordinate system.

In Exercises 7–8, calculate the slope of the line passing through the given points. If the slope is undefined, so state. Then indicate whether the line rises, falls, is horizontal, or is vertical.

7. $(-3, 4)$ and $(-5, -2)$

8. $(6, -1)$ and $(6, 3)$

9. Find the slope of the line in the figure shown or state that the slope is undefined.

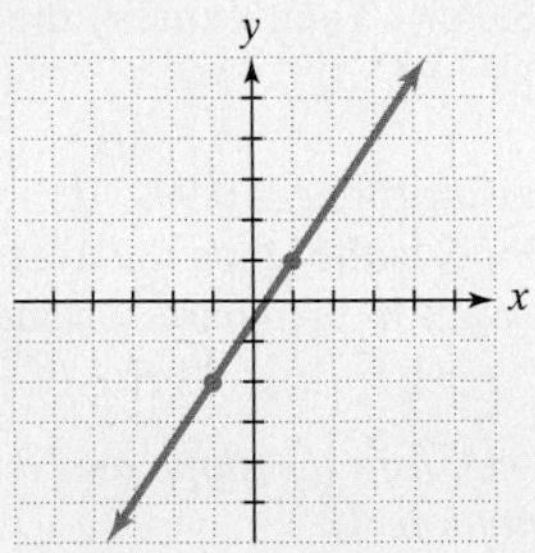

10. Determine whether the line through $(2, 4)$ and $(6, 1)$ is parallel to a second line through $(-3, 1)$ and $(1, -2)$.

In Exercises 11–12, find the slope and the y-intercept of the line with the given equation.

11. $y = -x + 10$

12. $2x + y = 6$

In Exercises 13–14, graph each linear equation using the slope and y-intercept.

13. $y = \frac{2}{3}x - 1$ **14.** $y = -2x + 3$

In Exercises 15–16, use the given conditions to write an equation for each line in point-slope form and slope-intercept form.

15. Slope $= -2$, passing through $(-1, 4)$

16. Passing through $(2, 1)$ and $(-1, -8)$

17. Passing through $(-2, 3)$ and perpendicular to the line whose equation is $y = -\frac{1}{2}x - 4$

18. Passing through $(6, -4)$ and parallel to the line whose equation is $x + 2y = 5$

19. The graph shows spending per pupil in public schools. Find the slope of the line passing through $(1970, 2100)$ and $(2000, 5280)$. Describe what the slope represents in this situation.

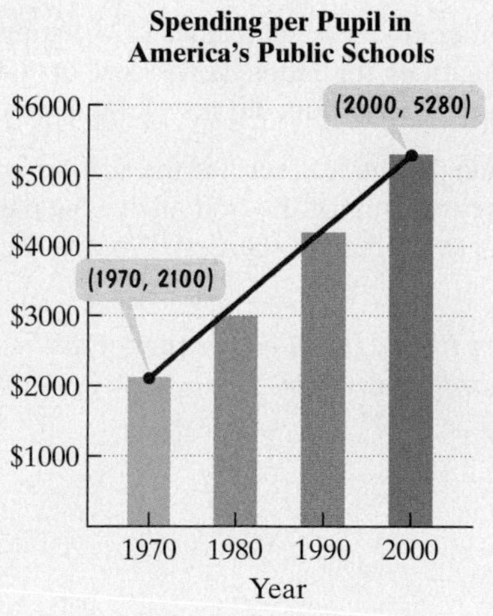

Source: National Education Association

20. The scatter plot shows the annual number of adult and child deaths worldwide due to AIDS from 1999 through 2003. Also shown is a line that models the data.

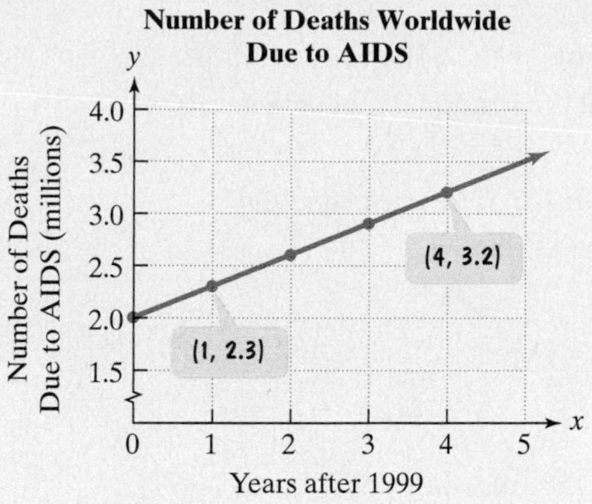

Source: World Health Organization

a. Use the two points whose coordinates are shown by the voice balloons to find the point-slope equation of the line that models the number of worldwide deaths due to AIDS, y, in millions, x years after 1999.

b. Write the equation in part (a) in slope-intercept form.

c. Use the linear equation to predict the number of worldwide deaths due to AIDS, in millions, in 2007.

CUMULATIVE REVIEW EXERCISES (CHAPTERS 1–3)

1. Perform the indicated operations:

$$\frac{10 - (-6)}{3^2 - (4 - 3)}.$$

2. Simplify: $6 - 2[3(x - 1) + 4]$.

3. List all the irrational numbers in this set: $\{-3, 0, 1, \sqrt{4}, \sqrt{5}, \frac{11}{2}\}$.

In Exercises 4–5, solve each equation.

4. $6(2x - 1) - 6 = 11x + 7$

5. $x - \frac{3}{4} = \frac{1}{2}$

6. Solve for x: $y = mx + b$.

7. 120 is 15% of what?

8. The formula $y = 4.5x - 46.7$ models the stopping distance, y, in feet, for a car traveling x miles per hour. If the stopping distance is 133.3 feet, how fast was the car traveling?

In Exercises 9–10, solve each inequality. Express the solution set in set-builder notation and graph the set on a number line.

9. $2 - 6x \geq 2(5 - x)$

10. $6(2 - x) > 12$

11. Evaluate $x^2 - 10x$ for $x = -3$.

12. Insert either $<$ or $>$ in the shaded area to make a true statement:

$$-2000 \quad \square \quad -3.$$

13. On February 8, the temperature in Manhattan at 10 P.M. was −4°F. By 3 A.M. the next day, the temperature had fallen 11°, but by noon the temperature increased by 21°. What was the temperature at noon?

14. The amount of money owed by doctors graduating from medical school is on the rise. The formula

$$D = 4x + 30$$

models the average debt, D, in thousands of dollars, of indebted medical-school graduates x years after 1985. Use the formula to determine in which year this debt will reach $150 thousand.

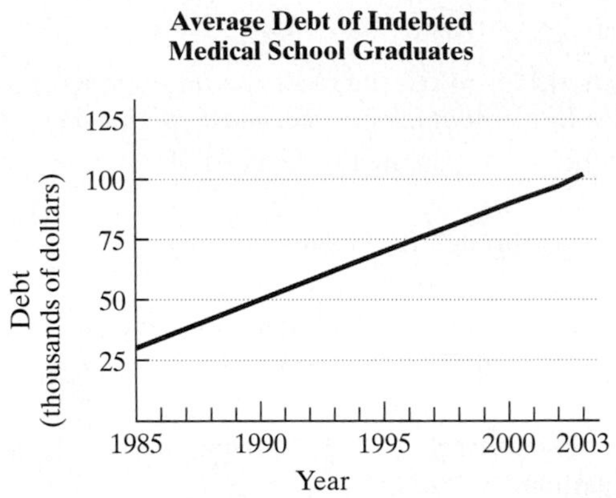

Source: American Medical Association

15. The length of a rectangular football field is 14 meters more than twice the width. If the perimeter is 346 meters, find the field's dimensions.

16. After a 10% weight loss, a person weighed 180 pounds. What was the weight before the loss?

17. A plumber charged a customer $228, listing $18 for parts and the remainder for labor. If the cost of the labor is $35 per hour, how many hours did the plumber work?

18. In a triangle, the measure of the second angle is 20° greater than the measure of the first angle. The measure of the third angle is twice that of the first. What is the measure of each angle?

In Exercises 19–20, graph each equation or inequality in the rectangular coordinate system.

19. $2x - y = 4$ **20.** $y = -4x + 3$

Mary Katherine Campbell,
Miss America 1922

Ericka Dunlap,
Miss America 2004

You are not a great fan of beauty pageants. However, as you were channel surfing, you tuned into the Miss America festivities. Is it your imagination, or does the icon of American beauty have that lean and hungry look?

This question is addressed in the Enrichment Essay "Missing America" in Section 4.1.

CHAPTER 4

Systems of Linear Equations

Television, movies, and magazines place great emphasis on physical beauty. Our culture emphasizes physical appearance to such an extent that it is a central factor in the perception and judgment of others. The modern emphasis on thinness as the ideal body shape has been suggested as a major cause of eating disorders among adolescent women. In this chapter, you will learn how systems of linear equations in two variables reveal the hidden patterns of your world, including a relationship between our changing cultural values of physical attractiveness and undernutrition.

SECTION 4.1

SOLVING SYSTEMS OF LINEAR EQUATIONS BY GRAPHING

Objectives

1. Decide whether an ordered pair is a solution of a linear system.
2. Solve systems of linear equations by graphing.
3. Use graphing to identify systems with no solution or infinitely many solutions.

Key West residents Brian Goss (left), George Wallace, and Michael Mooney (right) hold on to each other as they battle 90 mph winds along Houseboat Row in Key West, FL, on Friday, Sept. 25, 1998. The three had sought shelter behind a Key West hotel as Hurricane Georges descended on the Florida Keys, but were forced to seek other shelter when the storm conditions became too rough. Hundreds of people were killed by the storm when it swept through the Caribbean.

Problems ranging from scheduling airline flights to controlling traffic flow to routing phone calls over the nation's communication network often require solutions in a matter of moments. The solution to these real-world problems can involve solving thousands of equations having thousands of variables. AT&T's domestic long-distance network involves 800,000 variables! Meteorologists describing atmospheric conditions surrounding a hurricane must solve problems involving thousands of equations rapidly and efficiently. The difference between a two-hour warning and a two-day warning is a life-and-death issue for people in the path of one of nature's most destructive forces.

Although we will not be solving 800,000 equations with 800,000 variables, we will turn our attention to two equations with two variables, such as

$$\begin{aligned} 2x - 3y &= -4 \\ 2x + y &= 4. \end{aligned}$$

The methods that we consider for solving such problems provide the foundation for solving far more complex systems with many variables.

1 Decide whether an ordered pair is a solution of a linear system.

Systems of Linear Equations and Their Solutions We have seen that all equations in the form $Ax + By = C$ are straight lines when graphed. Two such equations, such as those listed above, are called a **system of linear equations** or a **linear system**. A **solution to a system of two linear equations in two variables** is an ordered pair that satisfies both equations in the system. For example, $(3, 4)$ satisfies the system

$$\begin{aligned} x + y &= 7 \qquad \text{(3 + 4 is, indeed, 7.)} \\ x - y &= -1. \qquad \text{(3 − 4 is, indeed, −1.)} \end{aligned}$$

Thus, $(3, 4)$ satisfies both equations and is a solution of the system. The solution can be described by saying that $x = 3$ and $y = 4$. The solution can also be described using set notation. The solution set of the system is $\{(3, 4)\}$—that is, the set consisting of the ordered pair $(3, 4)$.

A system of linear equations can have exactly one solution, no solution, or infinitely many solutions. We begin with systems having exactly one solution.

EXAMPLE 1 Determining Whether Ordered Pairs Are Solutions of a System

Consider the system:

$$x + 2y = 2$$
$$x - 2y = 6.$$

Determine if each ordered pair is a solution of the system:

a. $(4, -1)$ **b.** $(-4, 3)$.

SOLUTION

a. We begin by determining whether $(4, -1)$ is a solution. Because 4 is the x-coordinate and -1 is the y-coordinate of $(4, -1)$, we replace x with 4 and y with -1.

$$\begin{aligned} x + 2y &= 2 \\ 4 + 2(-1) &\stackrel{?}{=} 2 \\ 4 + (-2) &\stackrel{?}{=} 2 \\ 2 &= 2, \quad \text{true} \end{aligned} \qquad \begin{aligned} x - 2y &= 6 \\ 4 - 2(-1) &\stackrel{?}{=} 6 \\ 4 - (-2) &\stackrel{?}{=} 6 \\ 4 + 2 &\stackrel{?}{=} 6 \\ 6 &= 6, \quad \text{true} \end{aligned}$$

The pair $(4, -1)$ satisfies both equations: It makes each equation true. Thus, the ordered pair is a solution of the system.

b. To determine whether $(-4, 3)$ is a solution, we replace x with -4 and y with 3.

$$\begin{aligned} x + 2y &= 2 \\ -4 + 2 \cdot 3 &\stackrel{?}{=} 2 \\ -4 + 6 &\stackrel{?}{=} 2 \\ 2 &= 2, \quad \text{true} \end{aligned} \qquad \begin{aligned} x - 2y &= 6 \\ -4 - 2 \cdot 3 &\stackrel{?}{=} 6 \\ -4 - 6 &\stackrel{?}{=} 6 \\ -10 &= 6, \quad \text{false} \end{aligned}$$

The pair $(-4, 3)$ fails to satisfy *both* equations: It does not make both equations true. Thus, the ordered pair is not a solution of the system. ■

✔ **CHECK POINT 1** Consider the system:

$$2x - 3y = -4$$
$$2x + y = 4.$$

Determine if each ordered pair is a solution of the system:

a. $(1, 2)$ **b.** $(7, 6)$.

2 Solve systems of linear equations by graphing.

Solving Linear Systems by Graphing The solution of a system of linear equations can be found by graphing both of the equations in the same rectangular coordinate system. For a system with one solution, **the coordinates of the point of intersection give the system's solution**. For example, the system in Example 1,

$$x + 2y = 2$$
$$x - 2y = 6$$

is graphed in Figure 4.1. The solution of the system, $(4, -1)$, corresponds to the point of intersection of the lines.

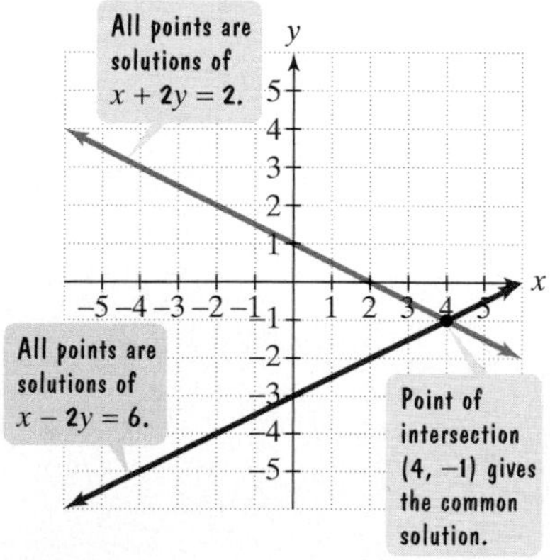

FIGURE 4.1 Visualizing a system's solution

SOLVING SYSTEMS OF TWO LINEAR EQUATIONS IN TWO VARIABLES, x AND y, BY GRAPHING

1. Graph the first equation.
2. Graph the second equation on the same axes.
3. If the lines representing the two graphs intersect at a point, determine the coordinates of this point of intersection. The ordered pair is the solution of the system.
4. Check the solution in both equations.

EXAMPLE 2 Solving a Linear System by Graphing

Solve by graphing:

$$\begin{aligned} 2x + 3y &= 6 \\ 2x + y &= -2. \end{aligned}$$

SOLUTION

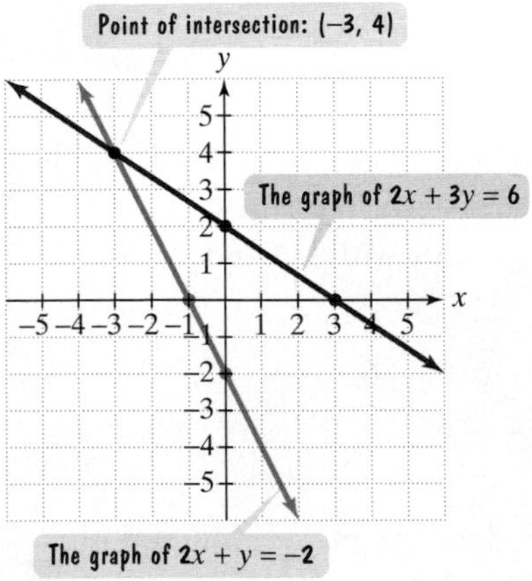

FIGURE 4.2

Step 1. **Graph the first equation.** We use intercepts to graph $2x + 3y = 6$.

x-intercept (Set $y = 0$.)	**y-intercept** (Set $x = 0$.)
$2x + 3 \cdot 0 = 6$	$2 \cdot 0 + 3y = 6$
$2x = 6$	$3y = 6$
$x = 3$	$y = 2$

The x-intercept is 3, so the line passes through $(3, 0)$. The y-intercept is 2, so the line passes through $(0, 2)$. The graph of $2x + 3y = 6$ is shown as the red line in Figure 4.2.

Step 2. **Graph the second equation on the same axes.** We use intercepts to graph $2x + y = -2$.

x-intercept (Set $y = 0$.)	**y-intercept** (Set $x = 0$.)
$2x + 0 = -2$	$2 \cdot 0 + y = -2$
$2x = -2$	$y = -2$
$x = -1$	

The x-intercept is -1, so the line passes through $(-1, 0)$. The y-intercept is -2, so the line passes through $(0, -2)$. The graph of $2x + y = -2$ is shown as the blue line in Figure 4.2.

Step 3. **Determine the coordinates of the intersection point. This ordered pair is the system's solution.** Using Figure 4.2, it appears that the lines intersect at $(-3, 4)$. The "apparent" solution of the system is $(-3, 4)$.

Step 4. **Check the solution in both equations.**

Check $(-3, 4)$ in $2x + 3y = 6$:	**Check $(-3, 4)$ in $2x + y = -2$:**
$2(-3) + 3 \cdot 4 \stackrel{?}{=} 6$	$2(-3) + 4 \stackrel{?}{=} -2$
$-6 + 12 \stackrel{?}{=} 6$	$-6 + 4 \stackrel{?}{=} -2$
$6 = 6$, true	$-2 = -2$, true

Because both equations are satisfied, $(-3, 4)$ is the solution of the system and $\{(-3, 4)\}$ is the solution set. ■

CHECK POINT 2 Solve by graphing:

$$\begin{aligned} 2x + y &= 6 \\ 2x - y &= -2. \end{aligned}$$

DISCOVER FOR YOURSELF

Must two lines intersect at exactly one point? Sketch two lines that have less than one intersection point. Now sketch two lines that have more than one intersection point. What does this say about each of these systems?

STUDY TIP

When solving linear systems by graphing, neatly drawn graphs are essential for determining points of intersection.

- Use rectangular coordinate graph paper.
- Use a ruler or straightedge.
- Use a pencil with a sharp point.

EXAMPLE 3 Solving a Linear System by Graphing

Solve by graphing:

$$y = -3x + 2$$
$$y = 5x - 6.$$

SOLUTION Each equation is in the form $y = mx + b$. Thus, we use the y-intercept, b, and the slope, m, to graph each line.

Step 1. Graph the first equation.

$$y = -3x + 2$$

The slope is −3. The y-intercept is 2.

The y-intercept is 2, so the line passes through $(0, 2)$. The slope is $-\frac{3}{1}$. Start at the y-intercept and move 3 units down (the rise) and 1 unit to the right (the run). The graph of $y = -3x + 2$ is shown as the red line in Figure 4.3.

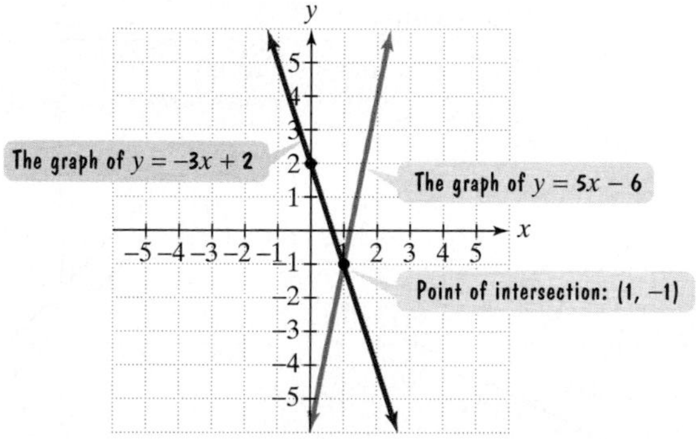

FIGURE 4.3

Step 2. Graph the second equation on the same axes.

$$y = 5x - 6$$

The slope is 5. The y-intercept is −6.

The y-intercept is −6, so the line passes through $(0, -6)$. The slope is $\frac{5}{1}$. Start at the y-intercept and move 5 units up (the rise) and 1 unit to the right (the run). The graph of $y = 5x - 6$ is shown as the blue line in Figure 4.3.

Step 3. Determine the coordinates of the intersection point. This ordered pair is the system's solution. Using Figure 4.3, it appears that the lines intersect at $(1, -1)$. The "apparent" solution of the system is $(1, -1)$.

Step 4. **Check the solution in both equations.**

Check (1, −1) in $y = -3x + 2$:	**Check (1, −1) in $y = 5x - 6$:**
$-1 \stackrel{?}{=} -3 \cdot 1 + 2$	$-1 \stackrel{?}{=} 5 \cdot 1 - 6$
$-1 \stackrel{?}{=} -3 + 2$	$-1 \stackrel{?}{=} 5 - 6$
$-1 = -1$, true	$-1 = -1$, true

Because both equations are satisfied, $(1, -1)$ is the solution and the system's solution set is $\{(1, -1)\}$. ■

CHECK POINT 3 Solve by graphing:

$$y = -x + 6$$
$$y = 3x - 6.$$

3 Use graphing to identify systems with no solution or infinitely many solutions.

Linear Systems Having No Solution or Infinitely Many Solutions We have seen that a system of linear equations in two variables represents a pair of lines. The lines either intersect, are parallel, or are identical. Thus, there are three possibilities for the number of solutions to a system of two linear equations.

THE NUMBER OF SOLUTIONS TO A SYSTEM OF TWO LINEAR EQUATIONS

The number of solutions to a system of two linear equations in two variables is given by one of the following. (See Figure 4.4.)

Number of Solutions	**What This Means Graphically**
Exactly one ordered-pair solution	The two lines intersect at one point.
No solution	The two lines are parallel.
Infinitely many solutions	The two lines are identical.

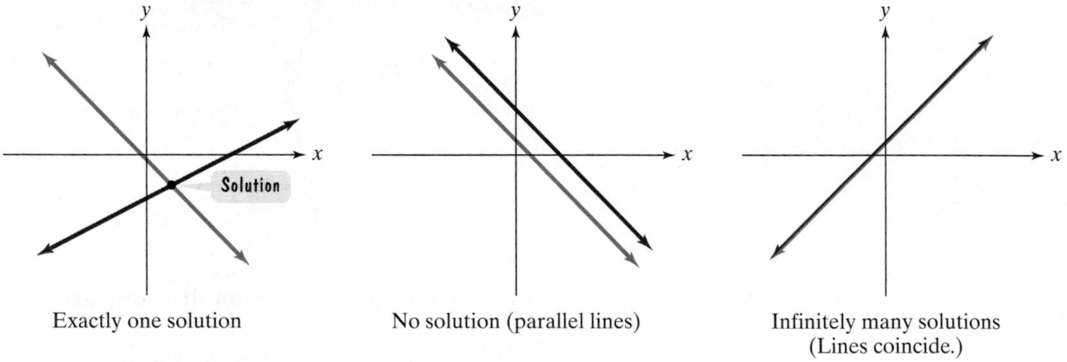

FIGURE 4.4 Possible graphs for a system of two linear equations in two variables

A linear system with no solution is called an **inconsistent system**. If you attempt to solve such a system by graphing, you will obtain two parallel lines. The solution set is the empty set, Ø.

EXAMPLE 4 A System with No Solution

Solve by graphing:

$$y = 2x - 1$$
$$y = 2x + 3.$$

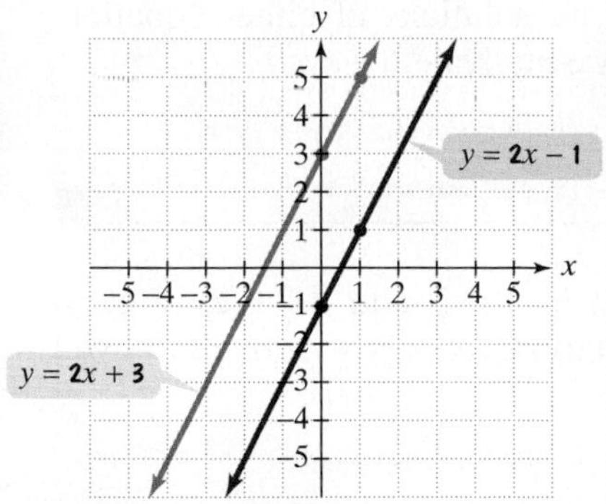

FIGURE 4.5 The graph of an inconsistent system

SOLUTION Compare the slopes and y-intercepts in the two equations.

The lines have the same slope, 2.

$$y = 2x - 1 \qquad y = 2x + 3$$

The lines have different y-intercepts, −1 and 3.

Figure 4.5 shows the graphs of the two equations. Because both equations have the same slope, 2, but different y-intercepts, the lines are parallel. The system is inconsistent and has no solution. The solution set is the empty set, ∅. ■

CHECK POINT 4 Solve by graphing:

$$y = 3x - 2$$
$$y = 3x + 1.$$

EXAMPLE 5 A System with Infinitely Many Solutions

Solve by graphing:

$$2x + y = 3$$
$$4x + 2y = 6.$$

SOLUTION We use intercepts to graph each equation.

• $2x + y = 3$

***x*-intercept**	***y*-intercept**
$2x + 0 = 3$	$2 \cdot 0 + y = 3$
$2x = 3$	$y = 3$
$x = \frac{3}{2}$	

Graph $(\frac{3}{2}, 0)$ and $(0, 3)$.

• $4x + 2y = 6$

***x*-intercept**	***y*-intercept**
$4x + 2 \cdot 0 = 6$	$4 \cdot 0 + 2y = 6$
$4x = 6$	$2y = 6$
$x = \frac{6}{4} = \frac{3}{2}$	$y = 3$

Graph $(\frac{3}{2}, 0)$ and $(0, 3)$.

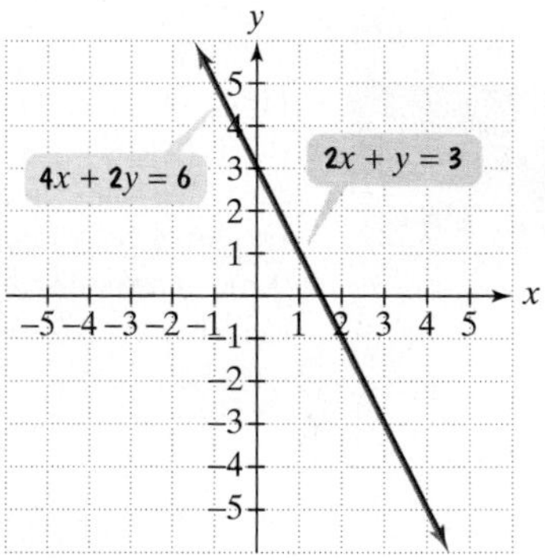

FIGURE 4.6 The graph of a system with infinitely many solutions

Both lines have the same x-intercept, $\frac{3}{2}$, or 1.5, and the same y-intercept, 3. Thus, the graphs of the two equations in the system are the same line, shown in Figure 4.6. The two equations have the same solutions. Any ordered pair that is a solution of one equation is a solution of the other, and, consequently, a solution of the system. The system has an infinite number of solutions, namely all points that are solutions of either equation. We express the solution set for the system in one of two equivalent ways:

$\{(x, y)|2x + y = 3\}$ The set of all ordered pairs (x, y) such that $2x + y = 3$

$\{(x, y)|4x + 2y = 6\}.$ The set of all ordered pairs (x, y) such that $4x + 2y = 6$

Take a second look at the two equations, $2x + y = 3$ and $4x + 2y = 6$, in Example 5. If you multiply both sides of the first equation, $2x + y = 3$, by 2, you will obtain the second equation, $4x + 2y = 6$.

$2x + y = 3$ This is the first equation in the system.

$2(2x + y) = 2 \cdot 3$ Multiply both sides by 2.

$2 \cdot 2x + 2y = 2 \cdot 3$ Use the distributive property.

This is the second equation in the system. $4x + 2y = 6$ Simplify.

Because $2x + y = 3$ and $4x + 2y = 6$ are different forms of the same equation, these equations are called *dependent equations*. In general, the equations in a linear system with infinitely many solutions are called **dependent equations**.

CHECK POINT **5** Solve by graphing:

$$x + y = 3$$
$$2x + 2y = 6.$$

ENRICHMENT ESSAY

Missing America

Here she is, Miss America, the icon of American beauty. Always thin, she is becoming more so. The scatter plot in the figure shows Miss America's body-mass index, a ratio comparing weight divided by the square of height. Two lines are also shown: a line that passes near the data points and a horizontal line representing the World Health Organization's cutoff point for undernutrition. The intersection point indicates that in approximately 1978, Miss America reached this cutoff. There she goes: If the trend continues, Miss America's body-mass index could reach zero in about 320 years.

Body-Mass Index of Miss America

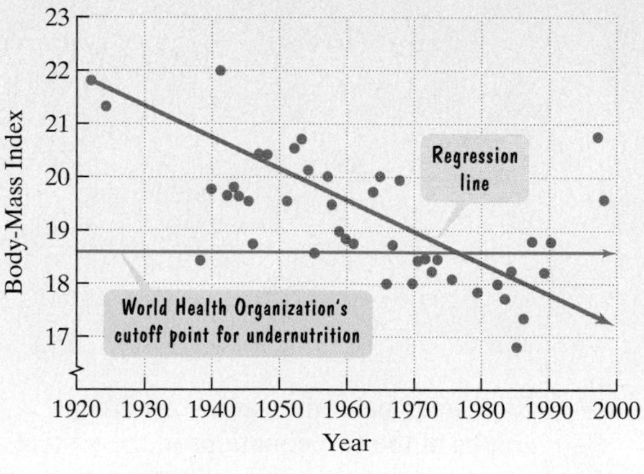

Source: Johns Hopkins School of Public Health

4.1 EXERCISE SET

Student Solutions Manual | CD/Video | PH Math/Tutor Center | MathXL Tutorials on CD | MathXL® | MyMathLab | Interactmath.com

Practice Exercises

In Exercises 1–10, determine whether the given ordered pair is a solution of the system.

1. $(2, -3)$
$2x + 3y = -5$
$7x - 3y = 23$

2. $(-2, -5)$
$6x - 2y = -2$
$3x + y = -11$

3. $\left(\frac{2}{3}, \frac{1}{9}\right)$
$x + 3y = 1$
$4x + 3y = 3$

4. $\left(\frac{7}{25}, -\frac{1}{25}\right)$
$4x + 3y = 1$
$3x - 4y = 1$

5. $(-5, 9)$
$5x + 3y = 2$
$x + 4y = 14$

6. $(10, 7)$
$6x - 5y = 25$
$4x + 15y = 13$

7. $(1400, 450)$
$x - 2y = 500$
$0.03x + 0.02y = 51$

8. $(200, 700)$
$-4x + y = -100$
$0.05x - 0.06y = -32$

9. $(8, 5)$
$5x - 4y = 20$
$3y = 2x + 1$

10. $(5, -2)$
$4x - 3y = 26$
$x = 15 - 5y$

In Exercises 11–42, solve each system by graphing. If there is no solution or an infinite number of solutions, so state, and use set notation to express solution sets.

11. $x + y = 6$
$x - y = 2$

12. $x + y = 2$
$x - y = 4$

13. $x + y = 1$
$y - x = 3$

14. $x + y = 4$
$y - x = 4$

15. $2x - 3y = 6$
$4x + 3y = 12$

16. $x + 2y = 2$
$x - y = 2$

17. $4x + y = 4$
$3x - y = 3$

18. $5x - y = 10$
$2x + y = 4$

19. $y = x + 5$
$y = -x + 3$

20. $y = x + 1$
$y = 3x - 1$

21. $y = 2x$
$y = -x + 6$

22. $y = 2x + 1$
$y = -2x - 3$

23. $y = -2x + 3$
$y = -x + 1$

24. $y = 3x - 4$
$y = -2x + 1$

25. $y = 2x - 1$
$y = 2x + 1$

26. $y = 3x - 1$
$y = 3x + 2$

27. $x + y = 4$
$x = -2$

28. $x + y = 6$
$y = -3$

29. $x - 2y = 4$
$2x - 4y = 8$

30. $2x + 3y = 6$
$4x + 6y = 12$

31. $y = 2x - 1$
$x - 2y = -4$

32. $y = -2x - 4$
$4x - 2y = 8$

33. $x + y = 5$
$2x + 2y = 12$

34. $x - y = 2$
$3x - 3y = -6$

35. $x - y = 0$
$y = x$

36. $2x - y = 0$
$y = 2x$

37. $x = 2$
$y = 4$

38. $x = 3$
$y = 5$

39. $x = 2$
$x = -1$

40. $x = 3$
$x = -2$

41. $y = 0$
$y = 4$

42. $y = 0$
$y = 5$

Practice Plus

Solve Exercises 43–50 without graphing. Find the slope and the y-intercept for the graph of each equation in the given system. Use this information (and not the equations' graphs) to determine if the system has no solution, one solution, or an infinite number of solutions.

43. $y = \frac{1}{2}x - 3$
$y = \frac{1}{2}x - 5$

44. $y = \frac{3}{4}x - 2$
$y = \frac{3}{4}x + 1$

45. $y = -\frac{1}{2}x + 4$
$3x - y = -4$

46. $y = -\frac{1}{4}x + 3$
$4x - y = -3$

47. $3x - y = 6$
$x = \frac{y}{3} + 2$

48. $2x - y = 4$
$x = \frac{y}{2} + 2$

49. $3x + y = 0$
$y = -3x + 1$

50. $2x + y = 0$
$y = -2x + 1$

Application Exercises

51. The graph shows the number of births in Massachusetts for women under 30 years old and women 30 years or older.

Number of Births in Massachusetts

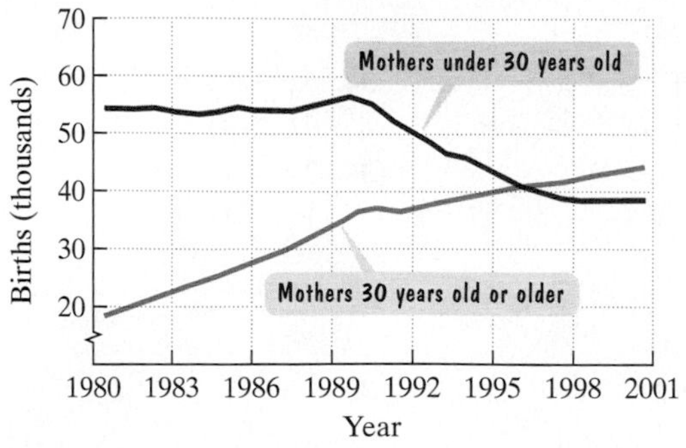

Source: Massachusetts Department of Public Health

a. Estimate the coordinates of the point of intersection. What does this mean in terms of the number of babies born to older mothers?

b. Describe what is happening to the number of births in Massachusetts to the right of the intersection point.

52. The figure shows scatter plots for the men's and women's winning times, in seconds, in the Olympic 100-meter freestyle swimming event. Also shown are lines that best fit the data, one for the men and one for the women.

Winning Times in the Olympic 100-Meter Freestyle Swimming Event

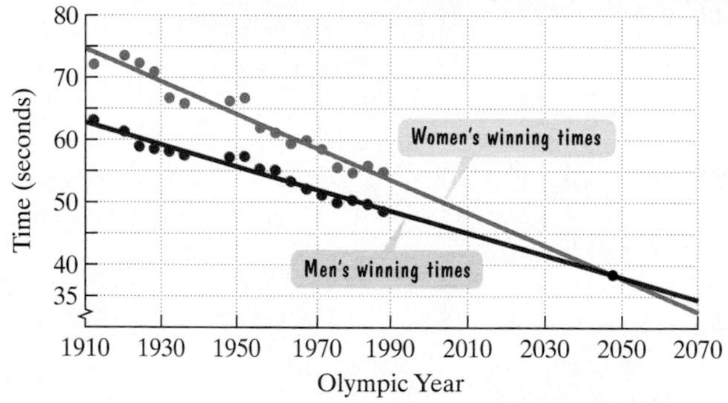

a. Estimate the coordinates of the point of intersection. What does this mean in terms of the women's time and the men's time?

b. Make a prediction about the swimming event in the Olympic years to the right of the intersection point.

Writing in Mathematics

53. What is a system of linear equations? Provide an example with your description.

54. What is a solution of a system of linear equations?

55. Explain how to determine if an ordered pair is a solution of a system of linear equations.

56. Explain how to solve a system of linear equations by graphing.

57. What is an inconsistent system? What happens if you attempt to solve such a system by graphing?

58. Explain how a linear system can have infinitely many solutions.

59. What are dependent equations? Provide an example with your description.

60. The following system models the winning times, y, in seconds, in the Olympic 500-meter speed skating event x years after 1970:

$$y = -0.19x + 43.7 \quad \text{Women}$$
$$y = -0.16x + 39.9. \quad \text{Men}$$

Use the slope of each model to explain why the system has a solution. What does this solution represent?

Critical Thinking Exercises

61. Which one of the following statements is true?

a. If a linear system has graphs with equal slopes, the system must be inconsistent.

b. If a linear system has graphs with equal y-intercepts, the system must have infinitely many solutions.

c. If a linear system has two distinct points that are solutions, then the graphs of the system's equations have equal slopes and equal y-intercepts.

d. It is possible for a linear system with one solution to have graphs with equal slopes.

62. Write a system of linear equations whose solution is $(5, 1)$. How many different systems are possible? Explain.

63. Write a system of equations with one solution, a system of equations with no solution, and a system of equations with infinitely many solutions. Explain how you were able to think of these systems.

64. Graph $y = x^2$ and $y = x + 2$ on the same axes. Find two ordered pairs that satisfy the system. Check that your answers satisfy both equations in the system.

Technology Exercises

65. Verify your solutions to any five exercises from Exercises 11 through 36 by using a graphing utility to graph the two equations in the system in the same viewing rectangle. After entering the two equations, one as y_1 and the other as y_2, and graphing them, use the TRACE and ZOOM features to find the coordinates of the intersection point. (It may first be necessary to solve the equation for y before entering it.) Many graphing utilities have a special INTERSECTION feature that displays the coordinates of the intersection point once the equations are graphed. Consult your manual.

Read Exercise 65. Then use a graphing utility to solve the systems in Exercises 66–73.

66. $y = 2x + 2$
$y = -2x + 6$

67. $y = -x + 5$
$y = x - 7$

68. $x + 2y = 2$
$x - y = 2$

69. $2x - 3y = 6$
$4x + 3y = 12$

70. $3x - y = 5$
$-5x + 2y = -10$

71. $2x - 3y = 7$
$3x + 5y = 1$

72. $y = \frac{1}{3}x + \frac{2}{3}$
$y = \frac{5}{7}x - 2$

73. $y = -\frac{1}{2}x + 2$
$y = \frac{3}{4}x + 7$

Review Exercises

In Exercises 74–76, perform the indicated operation.

74. $-3 + (-9)$ (Section 1.7, Table 1.6)

75. $-3 - (-9)$ (Section 1.7, Table 1.6)

76. $-3(-9)$ (Section 1.7, Table 1.6)

SECTION 4.2

Objectives

1 Solve linear systems by the substitution method.

2 Use the substitution method to identify systems with no solution or infinitely many solutions.

3 Solve problems using the substitution method.

SOLVING SYSTEMS OF LINEAR EQUATIONS BY THE SUBSTITUTION METHOD

Other than outrage, what is going on at the gas pumps? Is surging demand creating the increasing oil prices? Like all things in a free market economy, the price of a commodity is based on supply and demand. In this section, we use a second method for solving linear systems, the *substitution method*, to understand this economic phenomenon.

1 Solve linear systems by the substitution method.

Eliminating a Variable Using the Substitution Method Finding the solution of a linear system by graphing equations may not be easy to do. For example, a solution of $\left(-\frac{2}{3}, \frac{157}{29}\right)$ would be difficult to "see" as an intersection point on a graph.

Let's consider a method that does not depend on finding a system's solution visually: the substitution method. This method involves converting the system to one equation in one variable by an appropriate substitution.

EXAMPLE 1 Solving a System by Substitution

Solve by the substitution method:

$$y = -x - 1$$
$$4x - 3y = 24.$$

SOLUTION

Step 1. **Solve either of the equations for one variable in terms of the other.** This step has already been done for us. The first equation, $y = -x - 1$, is solved for y in terms of x.

Step 2. **Substitute the expression from step 1 into the other equation.** We substitute the expression $-x - 1$ for y into the other equation:

$$y = \boxed{-x - 1} \qquad 4x - 3\boxed{y} = 24. \qquad \text{Substitute } -x - 1 \text{ for } y.$$

This gives us an equation in one variable, namely

$$4x - 3(-x - 1) = 24.$$

The variable y has been eliminated.

Step 3. **Solve the resulting equation containing one variable.**

$4x - 3(-x - 1) = 24$	This is the equation containing one variable.
$4x + 3x + 3 = 24$	Apply the distributive property.
$7x + 3 = 24$	Combine like terms.
$7x = 21$	Subtract 3 from both sides.
$x = 3$	Divide both sides by 7.

Step 4. **Back-substitute the obtained value into the equation from step 1.** We now know that the x-coordinate of the solution is 3. To find the y-coordinate, we back-substitute the x-value into the equation from step 1,

$$y = -x - 1. \qquad \text{This is the equation from step 1.}$$

Substitute 3 for x.

$$y = -3 - 1$$

$$y = -4 \qquad \text{Simplify.}$$

With $x = 3$ and $y = -4$, the proposed solution is $(3, -4)$.

Step 5. **Check the proposed solution in both of the system's given equations.** Replace x with 3 and y with -4.

$$y = -x - 1 \qquad\qquad 4x - 3y = 24$$

$$-4 \stackrel{?}{=} -3 - 1 \qquad\qquad 4(3) - 3(-4) \stackrel{?}{=} 24$$

$$-4 = -4, \text{ true} \qquad\qquad 12 + 12 \stackrel{?}{=} 24$$

$$24 = 24, \text{ true}$$

The pair $(3, -4)$ satisfies both equations. The system's solution is $(3, -4)$ and the solution set is $\{(3, -4)\}$. ■

USING TECHNOLOGY

A graphing utility can be used to solve the system in Example 1. Graph each equation and use the intersection feature. The utility displays the solution $(3, -4)$ as $x = 3$, $y = -4$.

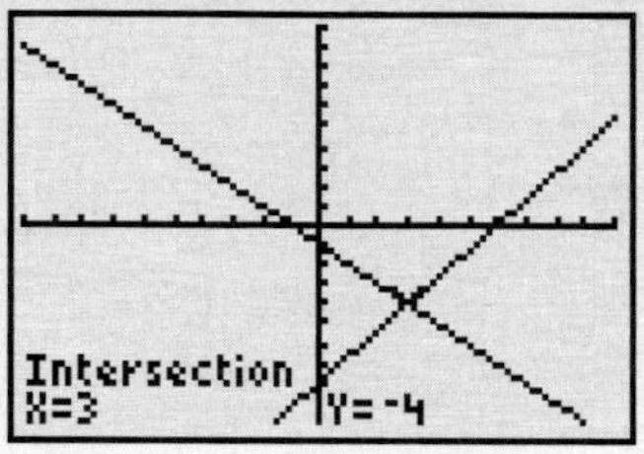

[−10, 10, 1] by [−10, 10, 1]

CHECK POINT 1 Solve by the substitution method:

$$y = 5x - 13$$

$$2x + 3y = 12.$$

Before considering additional examples, let's summarize the steps used in the substitution method.

STUDY TIP

In step 1, you can choose which variable to isolate in which equation. If possible, solve for a variable whose coefficient is 1 or -1 to avoid working with fractions.

SOLVING LINEAR SYSTEMS BY SUBSTITUTION

1. Solve either of the equations for one variable in terms of the other. (If one of the equations is already in this form, you can skip this step.)
2. Substitute the expression found in step 1 into the *other* equation. This will result in an equation in one variable.
3. Solve the equation containing one variable.
4. Back-substitute the value found in step 3 into the equation from step 1. Simplify and find the value of the remaining variable.
5. Check the proposed solution in both of the system's given equations.

EXAMPLE 2 Solving a System by Substitution

Solve by the substitution method:

$$\begin{aligned} 5x - 4y &= 9 \\ x - 2y &= -3. \end{aligned}$$

SOLUTION

Step 1. **Solve either of the equations for one variable in terms of the other.** We begin by isolating one of the variables in either of the equations. By solving for x in the second equation, which has a coefficient of 1, we can avoid fractions.

$x - 2y = -3$ This is the second equation in the given system.

$x = 2y - 3$ Solve for x by adding 2y to both sides.

Step 2. **Substitute the expression from step 1 into the other equation.** We substitute $2y - 3$ for x in the first equation.

$$x = \boxed{2y - 3} \qquad 5\boxed{x} - 4y = 9$$

This gives us an equation in one variable, namely

$$5(2y - 3) - 4y = 9.$$

The variable x has been eliminated.

Step 3. **Solve the resulting equation containing one variable.**

$5(2y - 3) - 4y = 9$ This is the equation containing one variable.

$10y - 15 - 4y = 9$ Apply the distributive property.

$6y - 15 = 9$ Combine like terms.

$6y = 24$ Add 15 to both sides.

$y = 4$ Divide both sides by 6.

Step 4. **Back-substitute the obtained value into the equation from step 1.** Now that we have the y-coordinate of the solution, we back-substitute 4 for y in the equation $x = 2y - 3$.

$x = 2y - 3$ Use the equation obtained in step 1.

$x = 2(4) - 3$ Substitute 4 for y.

$x = 8 - 3$ Multiply.

$x = 5$ Subtract.

With $x = 5$ and $y = 4$, the proposed solution is $(5, 4)$.

Step 5. **Check.** Take a moment to show that $(5, 4)$ satisfies both given equations. The solution is $(5, 4)$ and the solution set is $\{(5, 4)\}$. ■

STUDY TIP

Get into the habit of checking ordered-pair solutions in *both* equations of the system.

CHECK POINT 2 Solve by the substitution method:

$$\begin{aligned} 3x + 2y &= -1 \\ x - y &= 3. \end{aligned}$$

2 Use the substitution method to identify systems with no solution or infinitely many solutions.

USING TECHNOLOGY

A graphing utility was used to graph the equations in Example 3. The lines are parallel and have no point of intersection. This verifies that the system is inconsistent.

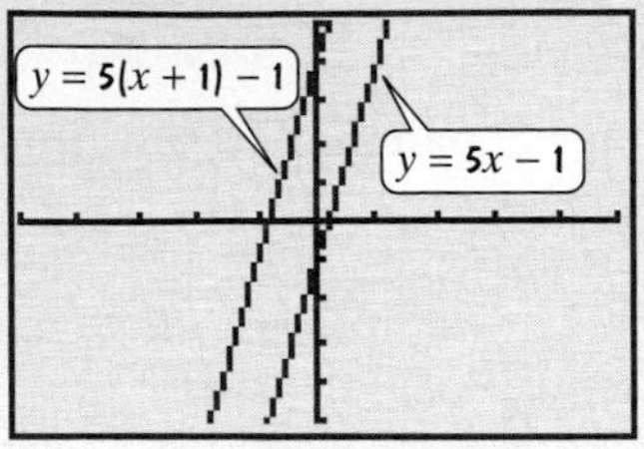

$[-5, 5, 1]$ by $[-5, 5, 1]$

The Substitution Method with Linear Systems Having No Solution or Infinitely Many Solutions Recall that a linear system with no solution is called an **inconsistent system**. If you attempt to solve such a system by substitution, you will eliminate both variables. A false statement such as $0 = 17$ will be the result.

EXAMPLE 3 Using the Substitution Method on an Inconsistent System

Solve the system:

$$y + 1 = 5(x + 1)$$
$$y = 5x - 1.$$

SOLUTION The variable y is isolated in the second equation. We use the substitution method and substitute the expression for y in the first equation.

$\boxed{y} + 1 = 5(x + 1) \quad y = \boxed{5x - 1}$ — Substitute $5x - 1$ for y.

$(5x - 1) + 1 = 5(x + 1)$ — This substitution gives an equation in one variable.

$5x = 5x + 5$ — Simplify on the left side. Use the distributive property on the right side.

$0 = 5$, false — Subtract $5x$ from both sides.

There are no values of x and y for which $0 = 5$.

The false statement $0 = 5$ indicates that the system is inconsistent and has no solution. The solution set is the empty set, $\varnothing$. ■

CHECK POINT 3 Solve the system:

$$3x + y = -5$$
$$y = -3x + 3.$$

Do you remember that the equations in a linear system with infinitely many solutions are called **dependent**? If you attempt to solve such a system by substitution, you will eliminate both variables. However, a true statement such as $5 = 5$ will be the result.

EXAMPLE 4 Using the Substitution Method on a System with Infinitely Many Solutions

Solve the system:

$$y = 3 - 2x$$
$$4x + 2y = 6.$$

SOLUTION The variable y is isolated in the first equation. We use the substitution method and substitute the expression for y in the second equation.

$y = \boxed{3 - 2x} \quad 4x + 2\boxed{y} = 6$ — Substitute $3 - 2x$ for y.

$4x + 2(3 - 2x) = 6$ — This substitution gives an equation in one variable.

$4x + 6 - 4x = 6$ — Apply the distributive property: $2(3 - 2x) = 2 \cdot 3 - 2 \cdot 2x = 6 - 4x$.

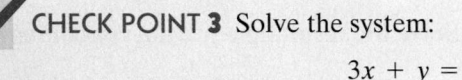

$6 = 6$, true — Simplify: $4x - 4x = 0$.

This true statement indicates that the system contains dependent equations and has infinitely many solutions. We express the solution set for the system in one of two equivalent ways:

$\{(x, y) | y = 3 - 2x\}$ The set of all ordered pairs (x, y) such that $y = 3 - 2x$

or $\{(x, y) | 4x + 2y = 6\}$. The set of all ordered pairs (x, y) such that $4x + 2y = 6$ ■

CHECK POINT 4 Solve the system:

$$y = 3x - 4$$
$$9x - 3y = 12.$$

3 Solve problems using the substitution method.

Applications An important application of systems of equations arises in connection with supply and demand. As the price of a product increases, the demand for that product decreases. However, at higher prices suppliers are willing to produce greater quantities of the product.

EXAMPLE 5 Supply and Demand Models

A chain of video stores specializes in cult films. The weekly demand and supply models for *The Rocky Horror Picture Show* are given by

$N = -13p + 760$ Demand model

$N = 2p + 430$ Supply model

in which p is the price of the video and N is the number of copies of the video sold or supplied each week to the chain of stores.

a. How many copies of the video can be sold and supplied at \$18 per copy?

b. Find the price at which supply and demand are equal. At this price, how many copies of *Rocky Horror* can be supplied and sold each week?

SOLUTION

a. To find how many copies of the video can be sold and supplied at \$18 per copy, we substitute 18 for p in the demand and supply models.

Demand Model	**Supply Model**
$N = -13p + 760$	$N = 2p + 430$
Substitute 18 for p.	Substitute 18 for p.
$N = -13 \cdot 18 + 760 = 526$	$N = 2 \cdot 18 + 430 = 466$

At \$18 per video, the chain can sell 526 copies of *Rocky Horror* in a week. The manufacturer is willing to supply 466 copies per week. This will result in a shortage of copies of the video. Under these conditions, the retail chain is likely to raise the price of the video.

b. We can find the price at which supply and demand are equal by solving the demand-supply linear system. We will use substitution, substituting $-13p + 760$ for N in the second equation.

$$N = \boxed{-13p + 760} \quad \boxed{N} = 2p + 430 \qquad \text{Substitute } -13p + 760 \text{ for } N.$$

$$-13p + 760 = 2p + 430 \qquad \text{The resulting equation contains only one variable.}$$

$$-15p + 760 = 430 \qquad \text{Subtract } 2p \text{ from both sides.}$$

$$-15p = -330 \qquad \text{Subtract 760 from both sides.}$$

$$p = 22 \qquad \text{Divide both sides by } -15.$$

The price at which supply and demand are equal is \$22 per video. To find the value of N, the number of videos supplied and sold weekly at this price, we back-substitute 22 for p into either the demand or the supply model. We'll use both models to make sure we get the same number in each case.

Demand Model	**Supply Model**
$N = -13p + 760$	$N = 2p + 430$
Substitute 22 for p.	Substitute 22 for p.
$N = -13 \cdot 22 + 760 = 474$	$N = 2 \cdot 22 + 430 = 474$

At a price of \$22 per video, 474 units of the video can be supplied and sold weekly. The intersection point, (22, 474), is shown in Figure 4.7.

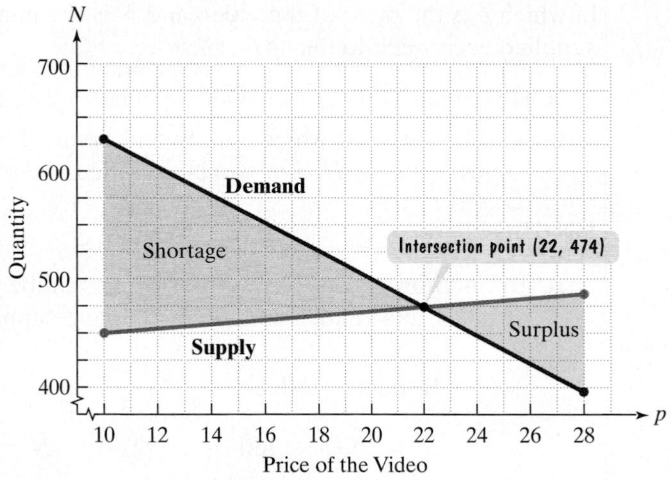

FIGURE 4.7 Priced at \$22 per video, 474 copies of the video can be supplied and sold weekly.

CHECK POINT 5 The demand for a product is modeled by $N = -20p + 1000$ and the supply for the product by $N = 5p + 250$. In these models, p is the price of the product and N is the number of units of the product supplied or sold weekly. At what price will supply equal demand? At that price, how many units of the product will be supplied and sold each week?

4.2 EXERCISE SET

Student Solutions Manual | CD/Video | PH Math/Tutor Center | MathXL Tutorials on CD | MathXL® | MyMathLab | Interactmath.com

Practice Exercises

In Exercises 1–32, solve each system by the substitution method. If there is no solution or an infinite number of solutions, so state, and use set notation to express solution sets.

1. $x + y = 4$
$y = 3x$

2. $x + y = 6$
$y = 2x$

3. $x + 3y = 8$
$y = 2x - 9$

4. $2x - 3y = -13$
$y = 2x + 7$

5. $x + 3y = 5$
$4x + 5y = 13$

6. $x + 2y = 5$
$2x - y = -15$

7. $2x - y = -5$
$x + 5y = 14$

8. $2x + 3y = 11$
$x - 4y = 0$

9. $2x - y = 3$
$5x - 2y = 10$

10. $-x + 3y = 10$
$2x + 8y = -6$

11. $-3x + y = -1$
$x - 2y = 4$

12. $-4x + y = -11$
$2x - 3y = 5$

13. $x = 9 - 2y$
$x + 2y = 13$

14. $6x + 2y = 7$
$y = 2 - 3x$

15. $y = 3x - 5$
$21x - 35 = 7y$

16. $9x - 3y = 12$
$y = 3x - 4$

17. $5x + 2y = 0$
$x - 3y = 0$

18. $4x + 3y = 0$
$2x - y = 0$

19. $2x - y = 6$
$3x + 2y = 5$

20. $2x - y = 4$
$3x - 5y = 2$

21. $2(x - 1) - y = -3$
$y = 2x + 3$

22. $x + y - 1 = 2(y - x)$
$y = 3x - 1$

23. $x = 2y + 9$
$x = 7y + 10$

24. $x = 5y - 3$
$x = 8y + 4$

25. $4x - y = 100$
$0.05x - 0.06y = -32$

26. $x + 6y = 8000$
$0.3x - 0.6y = 0$

27. $y = \frac{1}{3}x + \frac{2}{3}$
$y = \frac{5}{7}x - 2$

28. $y = -\frac{1}{2}x + 2$
$y = \frac{3}{4}x + 7$

29. $\frac{x}{6} - \frac{y}{2} = \frac{1}{3}$
$x + 2y = -3$

30. $\frac{x}{4} - \frac{y}{4} = -1$
$x + 4y = -9$

31. $2x - 3y = 8 - 2x$
$3x + 4y = x + 3y + 14$

32. $3x - 4y = x - y + 4$
$2x + 6y = 5y - 4$

Practice Plus

In Exercises 33–38, write a system of equations describing the given conditions. Then solve the system by the substitution method and find the two numbers.

33. The sum of two numbers is 81. One number is 41 more than the other. Find the numbers.

34. The sum of two numbers is 62. One number is 12 more than the other. Find the numbers.

35. The difference between two numbers is 5. Four times the larger number is 6 times the smaller number. Find the numbers.

36. The difference between two numbers is 25. Two times the larger number is 12 times the smaller number. Find the numbers.

37. The difference between two numbers is 1. The sum of the larger number and twice the smaller number is 7. Find the numbers.

38. The difference between two numbers is 5. The sum of the larger number and twice the smaller number is 14. Find the numbers.

In Exercises 39–40, multiply each equation in the system by an appropriate number so that the coefficients are integers. Then solve the system by the substitution method.

39. $0.7x - 0.1y = 0.6$
$0.8x - 0.3y = -0.8$

40. $1.25x - 0.01y = 4.5$
$0.5x - 0.02y = 1$

Application Exercises

41. At a price of p dollars per ticket, the number of tickets to a rock concert that can be sold is given by the demand model $N = -25p + 7500$. At a price of p dollars per ticket, the number of tickets that the concert's promoters are willing to make available is given by the supply model $N = 5p + 6000$.

a. How many tickets can be sold and supplied for $40 per ticket?

b. Find the ticket price at which supply and demand are equal. At this price, how many tickets will be supplied and sold?

42. The weekly demand and supply models for a particular brand of scientific calculator for a chain of stores are given by the demand model $N = -53p + 1600$ and the supply model $N = 75p + 320$. In these models, p is the price of the calculator and N is the number of calculators sold or supplied each week to the stores.

a. How many calculators can be sold and supplied at $12 per calculator?

b. Find the price at which supply and demand are equal. At this price, how many calculators of this type can be supplied and sold each week?

A business breaks even when the cost for running the business is equal to the money taken in by the business. In Exercises 43–44, determine how many units must be sold so that the business breaks even, experiencing neither loss nor profit.

43. A gasoline station has weekly costs and revenue (the money taken in by the station) that depend on the number of gallons of gasoline purchased and sold. If x gallons are purchased and sold, weekly costs are given by $y = 1.2x + 1080$ and weekly revenue by $y = 1.6x$. How many gallons of gasoline must be sold weekly for the station to break even?

44. An artist has monthly costs and revenue (the money taken in by the artist) that depend on the number of ceramic pieces produced and sold. If x ceramic pieces are produced and sold, monthly costs are given by $y = 4x + 2000$ and monthly revenue by $y = 9x$. How many ceramic pieces must be sold monthly for the artist to break even?

45. Infant mortality for African Americans is decreasing at a faster rate than it is for whites, shown by the graphs below. Infant mortality for African Americans can be modeled by $M = -0.41x + 22$ and for whites by $M = -0.18x + 10$. In both models, x is the number of years after 1980 and M is infant mortality, measured in deaths per 1000 live births. Use these models to project when infant mortality for African Americans and whites will be the same. What will be the infant mortality for both groups at that time?

Infant Mortality

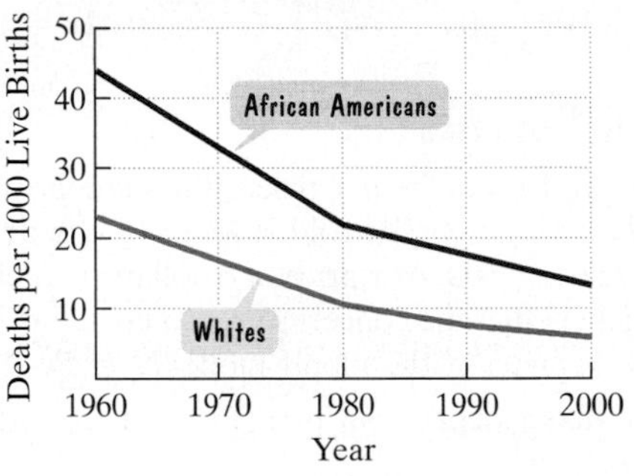

Source: National Center for Health Statistics

46. The equation $x + 10y = 2120$ models deaths from gunfire in the United States, y, in deaths per hundred thousand Americans, in year x. The equation $7x + 8y = 14{,}065$ models deaths from car accidents in the United States, y, in deaths per hundred thousand Americans, in year x. Solve the linear system formed by the two models. Then describe what the solution means in terms of the variables in the given models.

Writing in Mathematics

47. Describe a problem that might arise when solving a system of equations using graphing. Assume that both equations in the system have been graphed correctly and the system has exactly one solution.

48. Explain how to solve a system of equations using the substitution method. Use $y = 3 - 3x$ and $3x + 4y = 6$ to illustrate your explanation.

49. When using the substitution method, how can you tell if a system of linear equations has no solution?

50. When using the substitution method, how can you tell if a system of linear equations has infinitely many solutions?

51. The law of supply and demand states that, in a free market economy, a commodity tends to be sold at its equilibrium price. At this price, the amount that the seller will supply is the same amount that the consumer will buy. Explain how systems of equations can be used to determine the equilibrium price.

Critical Thinking Exercises

52. Which one of the following is true?

a. Solving an inconsistent system by substitution results in a true statement.

b. The line passing through the intersection of the graphs of $x + y = 4$ and $x - y = 0$ with slope $= 3$ has an equation given by $y - 2 = 3(x - 2)$.

c. Unlike the graphing method, where solutions cannot be seen, the substitution method provides a way to visualize solutions as intersection points.

d. To solve the system

$$\begin{aligned} 2x - y &= 5 \\ 3x + 4y &= 7 \end{aligned}$$

by substitution, replace y in the second equation with $5 - 2x$.

53. If $x = 3 - y - z$, $2x + y - z = -6$, and $3x - y + z = 11$, find the values for x, y, and z.

54. Find the value of m that makes

$$\begin{aligned} y &= mx + 3 \\ 5x - 2y &= 7 \end{aligned}$$

an inconsistent system.

Review Exercises

55. Graph: $4x + 6y = 12$. (Section 3.2, Example 4)

56. Solve: $4(x + 1) = 25 + 3(x - 3)$. (Section 2.3, Example 3)

57. List all the integers in this set: $\left\{-73, -\frac{2}{3}, 0, \frac{3}{1}, \frac{3}{2}, \frac{\pi}{1}\right\}$. (Section 1.2, Example 5)

SECTION 4.3

Objectives

1 Solve linear systems by the addition method.

2 Use the addition method to identify systems with no solution or infinitely many solutions.

3 Determine the most efficient method for solving a linear system.

SOLVING SYSTEMS OF LINEAR EQUATIONS BY THE ADDITION METHOD

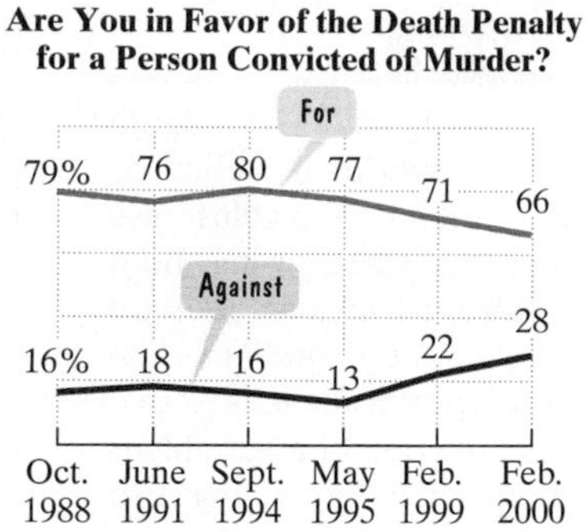

The graphs shown above are based on 543 adults polled nationally by *Newsweek*. If these trends continue, when will the percentage of Americans in favor of the death penalty be the same as the percentage of those who oppose it? The question can be answered by modeling the data with a system of linear equations and solving the system. However, the substitution method is not always the easiest way to solve linear systems. In this section we consider a third method for solving these systems.

1 Solve linear systems by the addition method.

Eliminating a Variable Using the Addition Method The substitution method is most useful if one of the given equations has an isolated variable. A third, and frequently the easiest, method for solving a linear system is the addition method. Like the substitution method, the addition method involves eliminating a variable and ultimately solving an equation containing only one variable. However, this time we eliminate a variable by adding the equations.

For example, consider the following system of equations:

$$3x - 4y = 11$$
$$-3x + 2y = -7.$$

When we add these two equations, the x-terms are eliminated. This occurs because the coefficients of the x-terms, 3 and -3, are opposites (additive inverses) of each other.

$$\begin{aligned} 3x - 4y &= 11 \\ -3x + 2y &= -7 \\ \hline \text{Add:} \quad -2y &= 4 \end{aligned}$$

The sum is an equation in one variable.

$$y = -2$$ Solve for y by dividing both sides by -2.

Now we can back-substitute -2 for y into one of the original equations to find x. It does not matter which equation you use; you will obtain the same value for x in either case. If we use either equation, we can show that $x = 1$ and the solution $(1, -2)$ satisfies both equations in the system.

When we use the addition method, we want to obtain two equations whose sum is an equation containing only one variable. The key step is to **obtain, for one of the variables, coefficients that differ only in sign**. To do this, we may need to multiply one or both equations by some nonzero number so that the coefficients of one of the variables, x or y, become opposites. Then when the two equations are added, this variable is eliminated.

EXAMPLE 1 Solving a System by the Addition Method

Solve by the addition method:

$$x + y = 4$$
$$x - y = 6.$$

SOLUTION The coefficients of y in the two equations, 1 and -1, differ only in sign. Therefore, by adding the two left sides and the two right sides, we can eliminate the y-terms.

$$\begin{aligned} x + y &= 4 \\ \underline{x - y} &= \underline{6} \\ \text{Add: } 2x + 0y &= 10 \\ 2x &= 10 \quad \text{Simplify.} \end{aligned}$$

Now y is eliminated and we can solve $2x = 10$ for x.

$$\begin{aligned} 2x &= 10 \\ x &= 5 \quad \text{Divide both sides by 2 and solve for } x. \end{aligned}$$

We back-substitute 5 for x into one of the original equations to find y. We will use both equations to show that we obtain the same value for y in either case.

Use the first equation:	**Use the second equation:**	
$x + y = 4$	$x - y = 6$	
$5 + y = 4$	$5 - y = 6$	Replace x with 5.
$y = -1.$	$-y = 1$	Solve for y.
	$y = -1.$	

Thus, $x = 5$ and $y = -1$. The proposed solution, $(5, -1)$, can be shown to satisfy both equations in the system. Consequently, the solution is $(5, -1)$ and the solution set is $\{(5, -1)\}$. ■

STUDY TIP

Although the addition method is also known as the elimination method, variables are eliminated when using both the substitution and addition methods. The name *addition method* specifically tells us that the elimination of a variable is accomplished by adding two equations.

✔ **CHECK POINT 1** Solve by the addition method:

$$x + y = 5$$
$$x - y = 9.$$

EXAMPLE 2 Solving a System by the Addition Method

Solve by the addition method:

$$3x - y = 11$$
$$2x + 5y = 13.$$

SOLUTION We must rewrite one or both equations in equivalent forms so that the coefficients of the same variable (either x or y) differ only in sign. Consider the terms in y in each equation, that is, $-1y$ and $5y$. To eliminate y, we can multiply each term of the first equation by 5 and then add the equations.

$$\begin{aligned} 3x - y = 11 &\xrightarrow{\text{Multiply by 5.}} & 15x - 5y &= 55 \\ 2x + 5y = 13 &\xrightarrow{\text{No change}} & \underline{2x + 5y} &= \underline{13} \\ & & \text{Add: } 17x + 0y &= 68 \quad \text{Simplify.} \\ & & 17x &= 68 \quad \text{Divide both sides by} \\ & & x &= 4 \quad \text{17 and solve for } x. \end{aligned}$$

Thus, $x = 4$. To find y, we back-substitute 4 for x into either one of the given equations. We'll use the second equation.

$2x + 5y = 13$	This is the second equation in the given system.
$2 \cdot 4 + 5y = 13$	Substitute 4 for x.
$8 + 5y = 13$	Multiply: $2 \cdot 4 = 8$.
$5y = 5$	Subtract 8 from both sides.
$y = 1$	Divide both sides by 5.

The solution is $(4, 1)$. Check to see that it satisfies both of the original equations in the system. The solution set is $\{(4, 1)\}$. ■

CHECK POINT 2 Solve by the addition method:

$$\begin{aligned} 4x - y &= 22 \\ 3x + 4y &= 26. \end{aligned}$$

Before considering additional examples, let's summarize the steps for solving linear systems by the addition method.

SOLVING LINEAR SYSTEMS BY ADDITION

1. If necessary, rewrite both equations in the form $Ax + By = C$.
2. If necessary, multiply either equation or both equations by appropriate nonzero numbers so that the sum of the x-coefficients or the sum of the y-coefficients is 0.
3. Add the equations in step 2. The sum is an equation in one variable.
4. Solve the equation in one variable.
5. Back-substitute the value obtained in step 4 into either of the given equations and solve for the other variable.
6. Check the solution in both of the original equations.

EXAMPLE 3 Solving a System by the Addition Method

Solve by the addition method:

$$\begin{aligned} 3x + 2y &= 48 \\ 9x - 8y &= -24. \end{aligned}$$

SOLUTION

Step 1. **Rewrite both equations in the form $Ax + By = C$.** Both equations are already in this form. Variable terms appear on the left and constants appear on the right.

Step 2. **If necessary, multiply either equation or both equations by appropriate numbers so that the sum of the x-coefficients or the sum of the y-coefficients is 0.** We can eliminate x or y. Let's eliminate x. Consider the terms in x in each equation, that is, $3x$ and $9x$. To eliminate x, we can multiply each term of the first equation by -3 and then add the equations.

$$\begin{aligned} 3x + 2y &= 48 \xrightarrow{\text{Multiply by } -3.} & -9x - 6y &= -144 \\ 9x - 8y &= -24 \xrightarrow{\text{No change}} & 9x - 8y &= -24 \end{aligned}$$

Step 3. **Add the equations.** Add: $\quad -14y = -168$

Step 4. **Solve the equation in one variable.** We solve $-14y = -168$ by dividing both sides by -14.

$$\frac{-14y}{-14} = \frac{-168}{-14} \quad \text{Divide both sides by } -14.$$

$$y = 12 \quad \text{Simplify.}$$

Step 5. **Back-substitute and find the value for the other variable.** We can back-substitute 12 for y into either one of the given equations. We'll use the first one.

$$3x + 2y = 48 \quad \text{This is the first equation in the given system.}$$

$$3x + 2(12) = 48 \quad \text{Substitute 12 for } y.$$

$$3x + 24 = 48 \quad \text{Multiply.}$$

$$3x = 24 \quad \text{Subtract 24 from both sides.}$$

$$x = 8 \quad \text{Divide both sides by 3.}$$

The solution is $(8, 12)$.

Step 6. **Check.** Take a few minutes to show that $(8, 12)$ satisfies both of the original equations in the system. The solution set is $\{(8, 12)\}$. ■

CHECK POINT 3 Solve by the addition method:

$$4x + 5y = 3$$
$$2x - 3y = 7.$$

Some linear systems have solutions that are not integers. If the value of one variable turns out to be a "messy" fraction, back-substitution might lead to cumbersome arithmetic. If this happens, you can return to the original system and use addition to find the value of the other variable.

EXAMPLE 4 Solving a System by the Addition Method

Solve by the addition method:

$$2x = 7y - 17$$
$$5y = 17 - 3x.$$

SOLUTION

Step 1. **Rewrite both equations in the form $Ax + By = C$.** We first arrange the system so that variable terms appear on the left and constants appear on the right. We obtain

$$2x - 7y = -17 \quad \text{Subtract 7y from both sides of the first equation.}$$

$$3x + 5y = 17. \quad \text{Add 3x to both sides of the second equation.}$$

Step 2. **If necessary, multiply either equation or both equations by appropriate numbers so that the sum of the x-coefficients or the sum of the y-coefficients is 0.** We can eliminate x or y. Let's eliminate x by multiplying the first equation by 3 and the second equation by -2.

$$2x - 7y = -17 \xrightarrow{\text{Multiply by 3.}} 6x - 21y = -51$$

$$3x + 5y = 17 \xrightarrow{\text{Multiply by } -2} \underline{-6x - 10y = -34}$$

Step 3. **Add the equations.** Add: $-31y = -85$

Step 4. **Solve the equation in one variable.** We solve $-31y = -85$ by dividing both sides by -31.

$$\frac{-31y}{-31} = \frac{-85}{-31} \quad \text{Divide both sides by } -31.$$

$$y = \frac{85}{31} \quad \text{Simplify.}$$

Step 5. **Back-substitute and find the value for the other variable.** Back-substitution of $\frac{85}{31}$ for y into either of the given equations results in cumbersome arithmetic. Instead, let's use the addition method on the given system in the form $Ax + By = C$ to find the value for x. Thus, we eliminate y by multiplying the first equation by 5 and the second equation by 7.

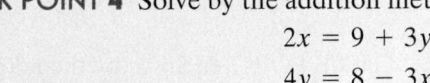

$$\begin{aligned} 2x - 7y &= -17 \xrightarrow{\text{Multiply by 5.}} & 10x - 35y &= -85 \\ 3x + 5y &= 17 \xrightarrow{\text{Multiply by 7.}} & \underline{21x + 35y} &= \underline{119} \\ & \text{Add:} & 31x &= 34 \\ & & x &= \frac{34}{31} \quad \text{Divide both sides by 31.} \end{aligned}$$

The solution is $\left(\frac{34}{31}, \frac{85}{31}\right)$.

Step 6. **Check.** For this system, a calculator is helpful in showing that $\left(\frac{34}{31}, \frac{85}{31}\right)$ satisfies both of the original equations in the system. The solution set is $\left\{\left(\frac{34}{31}, \frac{85}{31}\right)\right\}$. ■

CHECK POINT 4 Solve by the addition method:

$$2x = 9 + 3y$$
$$4y = 8 - 3x.$$

2 Use the addition method to identify systems with no solution or infinitely many solutions.

The Addition Method with Linear Systems Having No Solution or Infinitely Many Solutions As with the substitution method, if the addition method results in a false statement, the linear system is inconsistent and has no solution.

EXAMPLE 5 Using the Addition Method on an Inconsistent System

Solve the system:

$$4x + 6y = 12$$
$$6x + 9y = 12.$$

SOLUTION We can eliminate x or y. Let's eliminate x by multiplying the first equation by 3 and the second equation by -2.

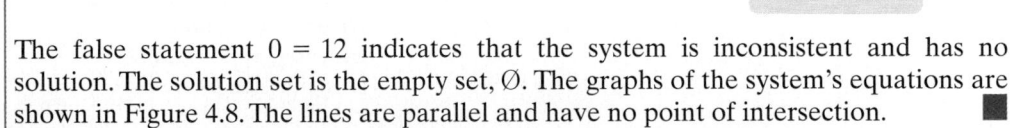

$$\begin{aligned} 4x + 6y &= 12 \xrightarrow{\text{Multiply by 3.}} & 12x + 18y &= 36 \\ 6x + 9y &= 12 \xrightarrow{\text{Multiply by } -2.} & \underline{-12x - 18y} &= \underline{-24} \\ & \text{Add:} & 0 &= 12 \end{aligned}$$

There are no values of x and y for which $0 = 12$.

The false statement $0 = 12$ indicates that the system is inconsistent and has no solution. The solution set is the empty set, $\varnothing$. The graphs of the system's equations are shown in Figure 4.8. The lines are parallel and have no point of intersection. ■

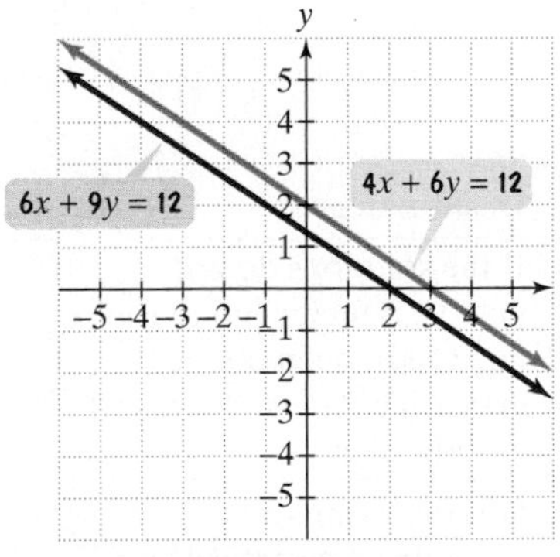

FIGURE 4.8 Visualizing the inconsistent system in Example 5

CHECK POINT 5 Solve the system:

$$x + 2y = 4$$
$$3x + 6y = 13.$$

If you use the addition method, how can you tell if a system has infinitely many solutions? As with the substitution method, you will eliminate both variables and obtain a true statement.

EXAMPLE 6 **Using the Addition Method on a System with Infinitely Many Solutions**

Solve by the addition method:

$$\begin{aligned} 2x - y &= 3 \\ -4x + 2y &= -6. \end{aligned}$$

SOLUTION We can eliminate y by multiplying the first equation by 2.

$$\begin{array}{rcr} 2x - y = 3 & \xrightarrow{\text{Multiply by 2.}} & 4x - 2y = 6 \\ -4x + 2y = -6 & \xrightarrow{\text{No change}} & \underline{-4x + 2y = -6} \\ & \text{Add:} & 0 = 0 \end{array}$$

The true statement $0 = 0$ indicates that the system contains dependent equations and has infinitely many solutions. Any ordered pair that satisfies the first equation also satisfies the second equation. We express the solution set for the system in one of two equivalent ways:

$$\{(x, y) | 2x - y = 3\} \text{ or } \{(x, y) | -4x + 2y = -6\}.$$

■

✔ CHECK POINT 6 Solve by the addition method:

$$\begin{aligned} x - 5y &= 7 \\ 3x - 15y &= 21. \end{aligned}$$

3 Determine the most efficient method for solving a linear system.

Comparing the Three Solution Methods The chart that follows compares the graphing, substitution, and addition methods for solving systems of linear equations. With increased practice, it becomes easier for you to select the best method for solving a particular linear system.

Comparing Solution Methods

Method	Advantages	Disadvantages
Graphing	You can see the solutions.	If the solutions do not involve integers or are too large to be seen on the graph, it's impossible to tell exactly what the solutions are.
Substitution	Gives exact solutions. Easy to use if a variable is on one side by itself.	Solutions cannot be seen. Can introduce extensive work with fractions when no variable has a coefficient of 1 or −1.
Addition	Gives exact solutions. Easy to use even if no variable has a coefficient of 1 or −1.	Solutions cannot be seen.

4.3 EXERCISE SET

Student Solutions Manual | CD/Video | PH Math/Tutor Center | MathXL Tutorials on CD | MathXL® | MyMathLab | Interactmath.com

Practice Exercises

In Exercises 1–44, solve each system by the addition method. If there is no solution or an infinite number of solutions, so state, and use set notation to express solution sets.

1. $x + y = -3$
$x - y = 11$

2. $x + y = 6$
$x - y = -2$

3. $2x + 3y = 6$
$2x - 3y = 6$

4. $3x + 2y = 14$
$3x - 2y = 10$

5. $x + 2y = 7$
$-x + 3y = 18$

6. $2x + y = -2$
$-2x - 3y = -6$

7. $5x - y = 14$
$-5x + 2y = -13$

8. $7x - 4y = 13$
$-7x + 6y = -11$

9. $3x + y = 7$
$2x - 5y = -1$

10. $3x - y = 11$
$2x + 5y = 13$

11. $x + 3y = 4$
$4x + 5y = 2$

12. $x + 2y = -1$
$4x - 5y = 22$

13. $-3x + 7y = 14$
$2x - y = -13$

14. $2x - 5y = -1$
$3x + y = 7$

15. $3x - 14y = 6$
$5x + 7y = 10$

16. $5x - 4y = 19$
$3x + 2y = 7$

17. $3x - 4y = 11$
$2x + 3y = -4$

18. $2x + 3y = -16$
$5x - 10y = 30$

19. $3x + 2y = -1$
$-2x + 7y = 9$

20. $5x + 3y = 27$
$7x - 2y = 13$

21. $3x = 2y + 7$
$5x = 2y + 13$

22. $9x = 25 + y$
$2y = 4 - 9x$

23. $2x = 3y - 4$
$-6x + 12y = 6$

24. $5x = 4y - 8$
$3x + 7y = 14$

25. $2x - y = 3$
$4x + 4y = -1$

26. $3x - y = 22$
$4x + 5y = -21$

27. $4x = 5 + 2y$
$2x + 3y = 4$

28. $3x = 4y + 1$
$4x + 3y = 1$

29. $3x - y = 1$
$3x - y = 2$

30. $4x - 9y = -2$
$-4x + 9y = -2$

31. $x + 3y = 2$
$3x + 9y = 6$

32. $4x - 2y = 2$
$2x - y = 1$

33. $7x - 3y = 4$
$-14x + 6y = -7$

34. $2x + 4y = 5$
$3x + 6y = 6$

35. $5x + y = 2$
$3x + y = 1$

36. $2x - 5y = -1$
$2x - y = 1$

37. $x = 5 - 3y$
$2x + 6y = 10$

38. $4x = 36 + 8y$
$3x - 6y = 27$

39. $4(3x - y) = 0$
$3(x + 3) = 10y$

40. $2(2x + 3y) = 0$
$7x = 3(2y + 3) + 2$

41. $x + y = 11$
$\frac{x}{5} + \frac{y}{7} = 1$

42. $x - y = -3$
$\frac{x}{9} - \frac{y}{7} = -1$

43. $\frac{4}{5}x - y = -1$
$\frac{2}{5}x + y = 1$

44. $\frac{x}{3} + y = 3$
$\frac{x}{2} - \frac{y}{4} = 1$

In Exercises 45–56, solve each system by the method of your choice. If there is no solution or an infinite number of solutions, so state, and use set notation to express solution sets. Explain why you selected one method over the other two.

45. $3x - 2y = 8$
$x = -2y$

46. $2x - y = 10$
$y = 3x$

47. $3x + 2y = -3$
$2x - 5y = 17$

48. $2x - 7y = 17$
$4x - 5y = 25$

49. $3x - 2y = 6$
$y = 3$

50. $2x + 3y = 7$
$x = 2$

51. $y = 2x + 1$
$y = 2x - 3$

52. $y = 2x + 4$
$y = 2x - 1$

53. $2(x + 2y) = 6$
$3(x + 2y - 3) = 0$

54. $2(x + y) = 4x + 1$
$3(x - y) = x + y - 3$

55. $3y = 2x$
$2x + 9y = 24$

56. $4y = -5x$
$5x + 8y = 20$

Practice Plus

In Exercises 57–64, solve each system or state that the system is inconsistent or dependent.

57. $\frac{3x}{5} + \frac{4y}{5} = 1$
$\frac{x}{4} - \frac{3y}{8} = -1$

58. $\frac{x}{3} - \frac{y}{2} = \frac{2}{3}$
$\frac{2x}{3} + y = \frac{4}{3}$

59. $5(x + 1) = 7(y + 1) - 7$
$6(x + 1) + 5 = 5(y + 1)$

60. $6x = 5(x + y + 3) - x$
$3(x - y) + 4y = 5(y + 1)$

61. $0.4x + y = 2.2$
$0.5x - 1.2y = 0.3$

62. $1.25x - 1.5y = 2$
$3.5x - 1.75y = 10.5$

63. $\frac{x}{2} = \frac{y+8}{3}$
$\frac{x+2}{2} = \frac{y+11}{3}$

64. $\frac{x}{2} = \frac{y+8}{4}$
$\frac{x+3}{2} = \frac{y+5}{4}$

Application Exercises

We opened this section with data from 1988 through 2000 showing the percentage of Americans for and against the death penalty for a person convicted of murder. The data can be modeled by the following system of equations:

$$13x + 12y = 992$$
The percent, y, in favor of the death penalty x years after 1988

$$-x + y = 16.$$
The percent, y, against the death penalty x years after 1988

Use this system to solve Exercises 65–66.

65. Use the addition method to determine in which year the percentage of Americans in favor of the death penalty will be the same as the percentage of Americans who oppose it. For that year, what percent will be for the death penalty and what percent will be against it?

66. Use the substitution method to solve Exercise 65.

Writing in Mathematics

67. Explain how to solve a system of equations using the addition method. Use $3x + 5y = -2$ and $2x + 3y = 0$ to illustrate your explanation.

68. When using the addition method, how can you tell if a system of linear equations has no solution?

69. When using the addition method, how can you tell if a system of linear equations has infinitely many solutions?

70. Take a second look at the data about the death penalty shown on page 263. Do you think that these trends will continue? Explain your answer.

71. The formula $3239x + 96y = 134{,}014$ models the number of daily evening newspapers, y, x years after 1980. The formula $-665x + 36y = 13{,}800$ models the number of daily morning newspapers, y, x years after 1980. What is the most efficient method for solving this system? Explain why. What does the solution mean in terms of the variables in the formulas? (It is not necessary to actually solve the system.)

Critical Thinking Exercises

72. Which one of the following statements is true?

a. If x can be eliminated by the addition method, y cannot be eliminated by using the original equations of the system.

b. If $Ax + 2y = 2$ and $2x + By = 10$ have graphs that intersect at $(2, -2)$, then $A = -3$ and $B = 3$.

c. The equations $y = x - 1$ and $x = y + 1$ are dependent.

d. If the two equations in a linear system are $5x - 3y = 7$ and $4x + 9y = 11$, multiplying the first equation by 4, the second by 5, and then adding equations will eliminate x.

73. Solve by expressing x and y in terms of a and b:

$$x - y = a$$
$$y = 2x + b.$$

74. The point of intersection of the graphs of the equations $Ax - 3y = 16$ and $3x + By = 7$ is $(5, -2)$. Find A and B.

Technology Exercises

75. Some graphing utilities can give the solution to a system of linear equations. (Consult your manual for details.) This capability is usually accessed with the SIMULT (simultaneous equations) feature. First, you must enter 2, for two equations in two variables. With each equation in $Ax + By = C$ form, you must then enter the coefficients for x and y and the constant term, one equation at a time. After entering all six numbers, press SOLVE. The solution will be displayed on the screen. (The x-value may be displayed as $x_1 =$ and the y-value as $x_2 =$.) Use this capability to verify the solution to any five of the exercises you solved in the practice exercises of this exercise set. Describe what happens when you use your graphing utility on a system with no solution or infinitely many solutions.

If your graphing utility has the feature described in Exercise 75, use it to solve each system in Exercises 76–78.

76. $\frac{1}{4}x - \frac{1}{4}y = -1$
$-3x + 7y = 8$

77. $x = 5y$
$2x - 3y = 7$

78. $0.6x + 0.08y = 4$
$3x + 2y = 4$

Review Exercises

79. For which number is 5 times the number equal to the number increased by 40? (Section 2.5, Example 2)

80. In which quadrant is $\left(-\frac{3}{2}, 15\right)$ located? (Section 1.3, Example 1)

81. Solve: $29{,}700 + 150x = 5000 + 1100x$. (Section 2.2, Example 7)

MID-CHAPTER CHECK POINT

CHAPTER 4

What You Know: We learned how to solve systems of linear equations by graphing, by the substitution method, and by the addition method. We saw that some systems, called inconsistent systems, have no solution, whereas other systems, called dependent systems, have infinitely many solutions.

In Exercises 1–3, solve each system by graphing.

1. $3x + 2y = 6$
$2x - y = 4$

2. $y = 2x - 1$
$y = 3x - 2$

3. $y = 2x - 1$
$6x - 3y = 12$

In Exercises 4–15, solve each system by the method of your choice.

4. $5x - 3y = 1$
$y = 3x - 7$

5. $6x + 5y = 7$
$3x - 7y = 13$

6. $x = \frac{y}{3} - 1$
$6x + y = 21$

7. $3x - 4y = 6$
$5x - 6y = 8$

8. $3x - 2y = 32$
$\frac{x}{5} + 3y = -1$

9. $x - y = 3$
$2x = 4 + 2y$

10. $x = 2(y - 5)$
$4x + 40 = y - 7$

11. $y = 3x - 2$
$y = 2x - 9$

12. $2x - 3y = 4$
$3x + 4y = 0$

13. $y - 2x = 7$
$4x = 2y - 14$

14. $4(x + 3) = 3y + 7$
$2(y - 5) = x + 5$

15. $\frac{x}{2} - \frac{y}{5} = 1$
$y - \frac{x}{3} = 8$

SECTION 4.4

Objectives

1. Solve problems using linear systems.
2. Solve simple interest problems.
3. Solve mixture problems.
4. Solve motion problems.

PROBLEM SOLVING USING SYSTEMS OF EQUATIONS

Enjoy chatting long distance on the phone? Telecommunication companies want your business. You can go online and get a list of their plans. Is there a monthly fee? What is the rate per minute? Does the plan involve a monthly minimum? In this section, you will learn to use systems of equations to select a plan that will save you the most money.

1 Solve problems using linear systems.

A Strategy for Solving Word Problems Using Systems of Equations When we solved problems in Chapters 2 and 3, we let x represent a quantity that was unknown. Problems in this section involve two unknown quantities. We will let x and

y represent these quantities. We then translate from the verbal conditions of the problem to a *system* of linear equations.

EXAMPLE 1 The World's Longest Snakes

The royal python and the anaconda are the world's longest snakes. The maximum length for each of these snakes is implied by the following description:

> Three royal pythons and two anacondas measure 161 feet. The royal python's length increased by triple the anaconda's length is 119 feet. Find the maximum length for each of these snakes.

SOLUTION

Step 1. Use variables to represent unknown quantities. Let x represent the royal python's length, in feet. Let y represent the anaconda's length, in feet.

Step 2. Write a system of equations describing the problem's conditions.

Three royal pythons	plus	two anacondas	measure	161 feet.
$3x$	$+$	$2y$	$=$	161

The royal python's length	increased by	triple the anaconda's length	is	119 feet.
x	$+$	$3y$	$=$	119

Step 3. Solve the system and answer the problem's question. The system

$$3x + 2y = 161$$
$$x + 3y = 119$$

can be solved by substitution or addition. Substitution works well because x in the second equation has a coefficient of 1. We can solve for x by subtracting $3y$ from both sides, thereby avoiding fractions. Addition also works well; if we multiply the second equation by -3, adding equations will eliminate x. We will use addition.

$$\begin{array}{rcl} 3x + 2y = 161 & \xrightarrow{\text{No change}} & 3x + 2y = 161 \\ x + 3y = 119 & \xrightarrow{\text{Multiply by } -3.} & \underline{-3x - 9y = -357} \\ & \text{Add:} & -7y = -196 \\ & & y = \dfrac{-196}{-7} = 28 \end{array}$$

Because y represents the anaconda's length, we see that the anaconda is 28 feet long. Now we can find x, the royal python's length. We do so by back-substituting 28 for y in either of the system's equations.

$x + 3y = 119$ We'll use the second equation.

$x + 3 \cdot 28 = 119$ Back-substitute 28 for y.

$x + 84 = 119$ Multiply: $3 \cdot 28 = 84$.

$x = 35$ Subtract 84 from both sides.

Because $x = 35$ and $y = 28$, the royal python is 35 feet long and the anaconda is 28 feet long.

Step 4. **Check the proposed answers in the original wording of the problem.** Three royal pythons and two anacondas should measure 161 feet:

$$3(35 \text{ ft}) + 2(28 \text{ ft}) = 105 \text{ ft} + 56 \text{ ft} = 161 \text{ ft}.$$

The royal python's length increased by triple the anaconda's length should be 119 feet:

$$35 \text{ ft} + 3(28 \text{ ft}) = 35 \text{ ft} + 84 \text{ ft} = 119 \text{ ft}.$$

This verifies that the royal python's and the anaconda's length are 35 feet and 28 feet, respectively. ■

CHECK POINT 1 The bustard and the condor are the world's heaviest flying bird and the world's heaviest bird of prey, respectively. The maximum weight for each of these birds is implied by the following description:

Two bustards and three condors weigh 173 pounds. The bustard's weight increased by double the condor's weight is 100 pounds. What is the maximum weight for each of these birds?

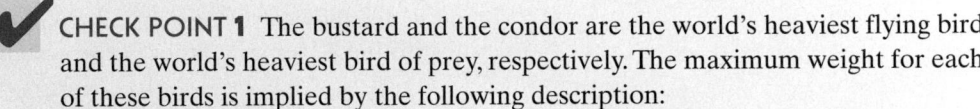

EXAMPLE 2 Cholesterol and Heart Disease

About 15 million hamburgers are eaten every day in the United States.

The verdict is in. After years of research, the nation's health experts agree that high cholesterol in the blood is a major contributor to heart disease. Thus, cholesterol intake should be limited to 300 milligrams or less each day. Fast foods provide a cholesterol carnival. All together, two McDonald's Quarter Pounders and three Burger King Whoppers with cheese contain 520 milligrams of cholesterol. Three Quarter Pounders and one Whopper with cheese exceed the suggested daily cholesterol intake by 53 milligrams. Determine the cholesterol content in each item.

SOLUTION

Step 1. **Use variables to represent the unknown quantities.** Let x represent the cholesterol content, in milligrams, of a Quarter Pounder. Let y represent the cholesterol content, in milligrams, of a Whopper with cheese.

Step 2. **Write a system of equations describing the problem's conditions.**

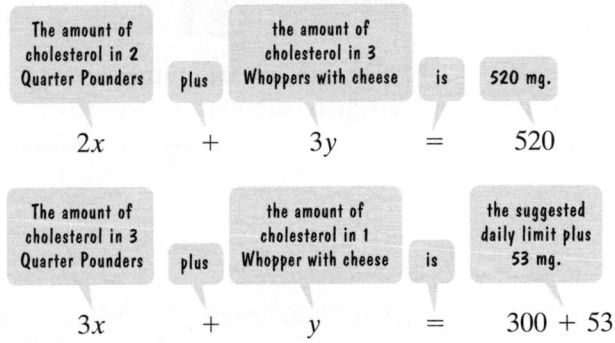

The amount of cholesterol in 2 Quarter Pounders	plus	the amount of cholesterol in 3 Whoppers with cheese	is	520 mg.
$2x$	$+$	$3y$	$=$	520

The amount of cholesterol in 3 Quarter Pounders	plus	the amount of cholesterol in 1 Whopper with cheese	is	the suggested daily limit plus 53 mg.
$3x$	$+$	y	$=$	$300 + 53$

Step 3. **Solve the system and answer the problem's question.** The system

$$2x + 3y = 520$$
$$3x + y = 353$$

can be solved by substitution or addition. We will use addition; if we multiply the second equation by -3, adding equations will eliminate y.

$$\begin{array}{rcl} 2x + 3y = 520 & \xrightarrow{\text{No change}} & 2x + 3y = 520 \\ 3x + y = 353 & \xrightarrow{\text{Multiply by } -3.} & \underline{-9x - 3y = -1059} \\ & \text{Add:} & -7x = -539 \\ & & x = \dfrac{-539}{-7} = 77 \end{array}$$

Because x represents the cholesterol content of a Quarter Pounder, we see that a Quarter Pounder contains 77 milligrams of cholesterol. Now we can find y, the cholesterol content of a Whopper with cheese. We do so by back-substituting 77 for x in either of the system's equations.

$3x + y = 353$ We'll use the second equation.

$3(77) + y = 353$ Back-substitute 77 for x.

$231 + y = 353$ Multiply.

$y = 122$ Subtract 231 from both sides.

Because $x = 77$ and $y = 122$, a Quarter Pounder contains 77 milligrams of cholesterol and a Whopper with cheese contains 122 milligrams of cholesterol.

Step 4. **Check the proposed answers in the original wording of the problem.** Two Quarter Pounders and three Whoppers with cheese contain

$$2(77 \text{ mg}) + 3(122 \text{ mg}) = 520 \text{ mg},$$

which checks with the given conditions. Furthermore, three Quarter Pounders and one Whopper with cheese contain

$$3(77 \text{ mg}) + 1(122 \text{ mg}) = 353 \text{ mg},$$

which does exceed the daily limit of 300 milligrams by 53 milligrams. ■

✔ CHECK POINT 2 How do the Quarter Pounder and Whopper with cheese measure up in the calorie department? Actually, not too well. Two Quarter Pounders and three Whoppers with cheese provide 2607 calories. Even combining one of each provides enough calories to bring tears to Jenny Craig's eyes—9 calories in excess of what is allowed on a 1000 calorie-a-day diet. Find the caloric content of each item.

EXAMPLE 3 Solar and Electric Heating Systems

The costs for two different kinds of heating systems for a three-bedroom home are given in the following table.

System	Cost to Install	Operating Cost/Year
Solar	\$29,700	\$150
Electric	\$5000	\$1100

After how many years will total costs for solar heating and electric heating be the same? What will be the cost at that time?

SOLUTION

Step 1. **Use variables to represent unknown quantities.** Let x represent the number of years the heating system is used. Let y represent the total cost for the heating system.

System	Cost to Install	Operating Cost/Year
Solar	$29,700	$150
Electric	$5000	$1100

Costs for heating systems, repeated

Step 2. Write a system of equations describing the problem's conditions.

Total cost for the solar system	equals	installation cost	plus	yearly operating cost	times	the number of years the system is used.
y	$=$	29,700	$+$	150	$\cdot$	x

Total cost for the electric system	equals	installation cost	plus	yearly operating cost	times	the number of years the system is used.
y	$=$	5000	$+$	1100	$\cdot$	x

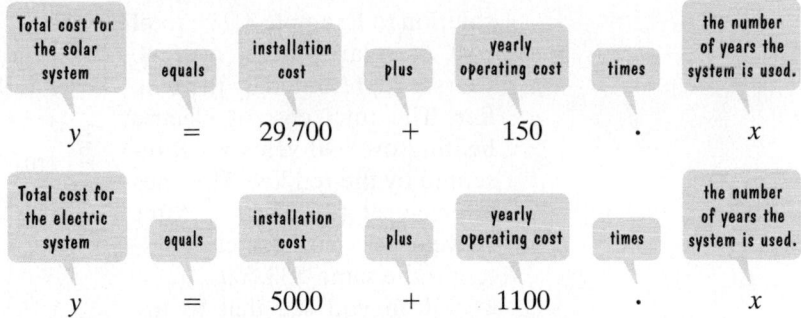

Step 3. Solve the system and answer the problem's question. We want to know after how many years the total costs for the two systems will be the same. We must solve the system

$$y = 29{,}700 + 150x$$
$$y = 5000 + 1100x.$$

Substitution works well because y is isolated in each equation.

$$y = \boxed{29{,}700 + 150x} \quad \boxed{y} = 5000 + 1100x$$ Substitute $29{,}700 + 150x$ for y.

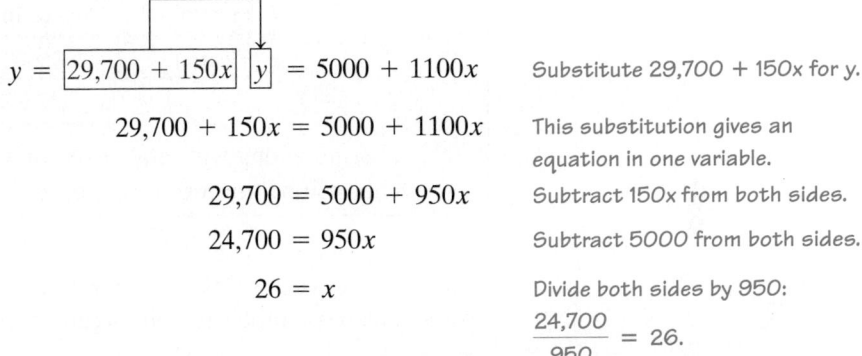

$$29{,}700 + 150x = 5000 + 1100x$$ This substitution gives an equation in one variable.

$$29{,}700 = 5000 + 950x$$ Subtract $150x$ from both sides.

$$24{,}700 = 950x$$ Subtract 5000 from both sides.

$$26 = x$$ Divide both sides by 950: $\frac{24{,}700}{950} = 26$.

Because x represents the number of years the heating system is used, we see that after 26 years, the total costs for the two systems will be the same. Now we can find y, the total cost. Back-substitute 26 for x in either of the system's equations. We will use the second equation, $y = 5000 + 1100x$.

$$y = 5000 + 1100 \cdot 26 = 5000 + 28{,}600 = 33{,}600$$

Because $x = 26$ and $y = 33{,}600$, after 26 years, the total costs for the two systems will be the same. The cost for each system at that time will be $33,600.

Step 4. Check the proposed answers in the original wording of the problem. Let's verify that after 26 years, the two systems will cost the same amount. The installation cost for the solar system is $29,700 and the yearly operating cost is $150. Thus, the total cost after 26 years is

$$\$29{,}700 + \$150(26) = \$29{,}700 + \$3900 = \$33{,}600.$$

The installation cost for the electric system is $5000 and the yearly operating cost is $1100. Thus, the total cost after 26 years is

$$\$5000 + \$1100(26) = \$5000 + \$28{,}600 = \$33{,}600.$$

This verifies that after 26 years the two systems will cost the same amount, $33,600. ■

The graphs in Figure 4.9 give us a way of visualizing the solution to Example 3. The total cost of solar heating over 40 years is represented by the blue line. The total cost of electric heating over 40 years is represented by the red line. The lines intersect at (26, 33,600): After 26 years, the cost for each system is the same, \$33,600.

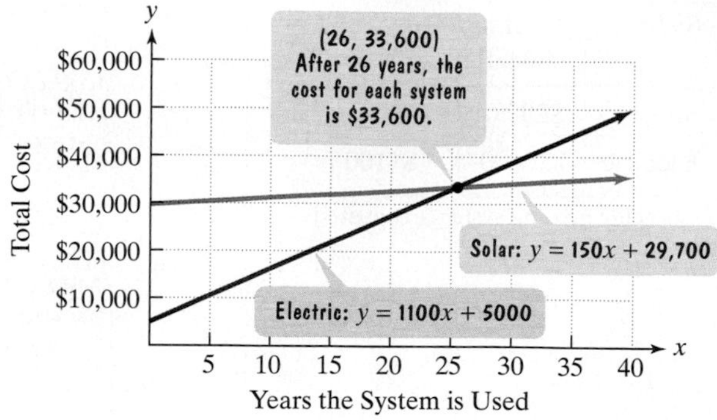

FIGURE 4.9 Visualizing total costs of solar and electric heating

Can you see that to the right of the intersection point, (26, 33,600), the blue graph representing solar costs lies below the red graph representing electric costs? Thus, after 26 years, or when $x > 26$, the cost for solar heating is less than the cost for electric heating.

CHECK POINT 3 Costs for two different kinds of heating systems for a three-bedroom house are given in the following table.

System	Cost to Install	Operating Cost/Year
Electric	\$5000	\$1100
Gas	\$12,000	\$700

After how long will total costs for electric heating and gas heating be the same? What will be the cost at that time?

Next, we will solve problems involving investments, mixtures, and motion with systems of equations. We will continue using our four-step problem-solving strategy. We will also use tables to help organize the information in the problems.

2 Solve simple interest problems.

Dual Investments with Simple Interest Simple interest involves interest calculated only on the amount of money that we invest, called the **principal**. The formula $I = Pr$ is used to find the simple interest, I, earned for one year when the principal, P, is invested at an annual interest rate, r. Dual investment problems involve different amounts of money in two or more investments, each paying a different rate.

EXAMPLE 4 Solving a Dual Investment Problem

Your grandmother needs your help. She has \$50,000 to invest. Part of this money is to be invested in noninsured bonds paying 15% annual interest. The rest of this money is to be invested in a government-insured certificate of deposit paying 7% annual interest. She told you that she requires \$6000 per year in extra income from both of these investments. How much money should be placed in each investment?

SOLUTION

Step 1. **Use variables to represent unknown quantities.**
Let x = the amount invested in the 15% noninsured bonds.
Let y = the amount invested in the 7% certificate of deposit.

Step 2. **Write a system of equations describing the problem's conditions.** Because Grandma has \$50,000 to invest,

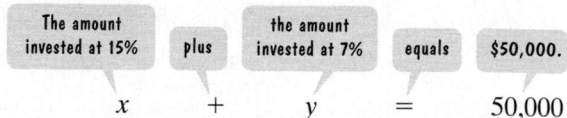

The amount invested at 15%	plus	the amount invested at 7%	equals	\$50,000.
x	$+$	y	$=$	50,000

Furthermore, Grandma requires \$6000 in total interest. We can use a table to organize the information in the problem and obtain a second equation.

	Principal (amount invested)	×	Interest rate	=	Interest earned
15% Investment	x		0.15		$0.15x$
7% Investment	y		0.07		$0.07y$

The interest for the two investments combined must be \$6000.

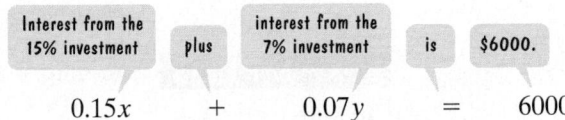

Interest from the 15% investment	plus	interest from the 7% investment	is	\$6000.
$0.15x$	$+$	$0.07y$	$=$	6000

Step 3. **Solve the system and answer the problem's question.** The system

$$x + y = 50{,}000$$
$$0.15x + 0.07y = 6000$$

can be solved by substitution or addition. Substitution works well because both variables in the first equation have coefficients of 1. Addition also works well; if we multiply the first equation by -0.15 or -0.07, adding equations will eliminate a variable. We will use addition.

$$\begin{array}{rcl} x + y = 50{,}000 & \xrightarrow{\text{Multiply by } -0.07.} & -0.07x - 0.07y = -3500 \\ 0.15x + 0.07y = 6000 & \xrightarrow{\text{No change}} & \underline{0.15x + 0.07y = 6000} \\ & \text{Add:} & 0.08x = 2500 \\ & & x = \dfrac{2500}{0.08} \\ & & x = 31{,}250 \end{array}$$

Because x represents the amount that should be invested at 15%, Grandma should place \$31,250 in 15% noninsured bonds. Now we can find y, the amount that she should place in the 7% certificate of deposit. We do so by back-substituting 31,250 for x in either of the system's equations.

$x + y = 50{,}000$ We'll use the first equation.

$31{,}250 + y = 50{,}000$ Back-substitute 31,250 for x.

$y = 18{,}750$ Subtract 31,250 from both sides.

Because $x = 31{,}250$ and $y = 18{,}750$, Grandma should invest \$31,250 at 15% and \$18,750 at 7%.

Step 4. **Check the proposed answers in the original wording of the problem.** Has Grandma invested \$50,000?

$$\$31{,}250 + \$18{,}750 = \$50{,}000$$

Yes, all her money was placed in the dual investments. Can she count on \$6000 interest? The interest earned on \$31,250 at 15% is (\$31,250)(0.15), or \$4687.50. The interest earned on \$18,750 at 7% is (\$18,750)(0.07), or \$1312.50. The total interest is \$4687.50 + \$1312.50, or \$6000, exactly as it should be. You've made your grandmother happy. (Now if you would just visit her more often ...) ■

CHECK POINT 4 You inherited \$5000 with the stipulation that for the first year the money had to be invested in two funds paying 9% and 11% annual interest. How much did you invest at each rate if the total interest earned for the year was \$487?

3 Solve mixture problems.

Problems Involving Mixtures Chemists and pharmacists often have to change the concentration of solutions and other mixtures. In these situations, the amount of a particular ingredient in the solution or mixture is expressed as a percentage of the total.

EXAMPLE 5 Solving a Mixture Problem

A chemist working on a flu vaccine needs to mix a 10% sodium-iodine solution with a 60% sodium-iodine solution to obtain 50 milliliters of a 30% sodium-iodine solution. How many milliliters of the 10% solution and of the 60% solution should be mixed?

SOLUTION

Step 1. **Use variables to represent unknown quantities.**
Let x = the number of milliliters of the 10% solution to be used in the mixture.
Let y = the number of milliliters of the 60% solution to be used in the mixture.

Step 2. **Write a system of equations describing the problem's conditions.** The situation is illustrated in Figure 4.10.

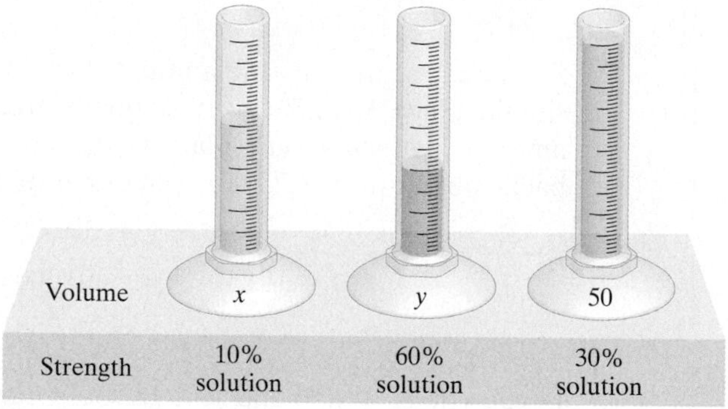

FIGURE 4.10

The chemist needs 50 milliliters of a 30% sodium-iodine solution. We form a table that shows the amount of sodium-iodine in each of the three solutions.

Solution	Number of milliliters	×	Percent of Sodium-Iodine	=	Amount of Sodium-Iodine
10% Solution	x		10% = 0.1		$0.1x$
60% Solution	y		60% = 0.6		$0.6y$
30% Mixture	50		30% = 0.3		$0.3(50) = 15$

The chemist needs to obtain a 50-milliliter mixture.

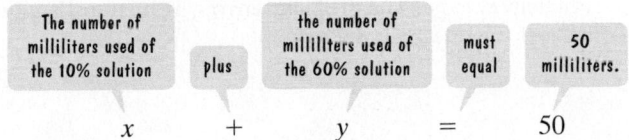

The number of milliliters used of the 10% solution	plus	the number of milliliters used of the 60% solution	must equal	50 milliliters.
x	$+$	y	$=$	50

The 50-milliliter mixture must be 30% sodium-iodine. The amount of sodium-iodine must be 30% of 50, or $(0.3)(50) = 15$ milliliters.

Amount of sodium-iodine in the 10% solution	plus	amount of sodium-iodine in the 60% solution	equals	amount of sodium-iodine in the mixture.
$0.1x$	$+$	$0.6y$	$=$	15

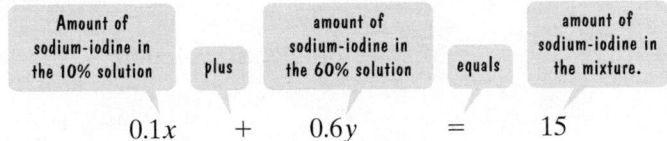

Step 3. **Solve the system and answer the problem's question.** The system

$$x + y = 50$$

$$0.1x + 0.6y = 15$$

can be solved by substitution or addition. Let's use substitution. The first equation can easily be solved for x or y. Solving for y, we obtain $y = 50 - x$.

$$y = \boxed{50 - x} \qquad 0.1x + 0.6\boxed{y} = 15$$

We substitute $50 - x$ for y in the second equation. This gives us an equation in one variable.

$$0.1x + 0.6(50 - x) = 15$$ This equation contains one variable, x.

$$0.1x + 30 - 0.6x = 15$$ Apply the distributive property.

$$-0.5x + 30 = 15$$ Combine like terms.

$$-0.5x = -15$$ Subtract 30 from both sides.

$$x = \frac{-15}{-0.5} = 30$$ Divide both sides by -0.5.

Back-substituting 30 for x in either of the system's equations ($x + y = 50$ is easier to use) gives $y = 20$. Because x represents the number of milliliters of the 10% solution and y the number of milliliters of the 60% solution, the chemist should mix 30 milliliters of the 10% solution with 20 milliliters of the 60% solution.

Step 4. **Check the proposed solution in the original wording of the problem.** The problem states that the chemist needs 50 milliliters of a 30% sodium-iodine solution. The amount of sodium-iodine in this mixture is 0.3(50), or 15 milliliters. The amount of sodium-iodine in 30 milliliters of the 10% solution is 0.1(30), or 3 milliliters. The amount of sodium-iodine in 20 milliliters of the 60% solution is $0.6(20) = 12$ milliliters. The amount of sodium-iodine in the two solutions used in the mixture is 3 milliliters + 12 milliliters, or 15 milliliters, exactly as it should be. ■

STUDY TIP

Problems involving dual investments and problems involving mixtures are both based on the same idea:
The total amount times the rate gives the amount.

Dual Investment Problems: principal · rate = interest

(principal: total money amount; rate: rate of interest; interest: amount of interest)

Percents are expressed as decimals in these equations.

Mixture Problems: solution · concentration = ingredient

(solution: total solution amount; concentration: percent of ingredient, or rate of concentration; ingredient: amount of ingredient)

Our dual investment problem involved mixing two investments. Our mixture problem involved mixing two liquids. The equations in these problems are obtained from similar conditions:

Dual Investment Problems

Interest from investment 1 + interest from investment 2 = amount of interest from mixed investments.

Mixture Problems

Ingredient amount in solution 1 + ingredient amount in solution 2 = amount of ingredient in mixture.

Being aware of the similarities between dual investment and mixture problems should make you a better problem solver in a variety of situations that involve mixtures.

✔ CHECK POINT **5** A chemist needs to mix a 12% acid solution with a 20% acid solution to obtain 160 ounces of a 15% acid solution. How many ounces of each of the acid solutions must be used?

4 Solve motion problems.

Problems Involving Motion Suppose that you ride your bike at an average speed of 12 miles per hour. What distance do you cover in 2 hours? Your distance is the product of your speed and the time that you travel:

$$\frac{12 \text{ miles}}{\cancel{\text{hour}}} \times 2 \cancel{\text{ hours}} = 24 \text{ miles}.$$

Your distance is 24 miles. Notice how the hour units cancel. The distance is expressed in miles.

In general, the distance covered by any moving body is the product of its average speed, or rate, and its time in motion:

A FORMULA FOR MOTION

$$d = rt$$

Distance equals rate times time.

Wind and water current have the effect of increasing or decreasing a traveler's rate.

STUDY TIP

It is not always necessary to use x and y to represent a problem's variables. Select letters that help you remember what the variables represent. For example, in Example 6, you may prefer using p and w rather than x and y:

p = plane's average rate in still air

w = wind's average rate.

EXAMPLE 6 Solving a Motion Problem

When a small airplane flies with the wind, it can travel 450 miles in 3 hours. When the same airplane flies in the opposite direction against the wind, it takes 5 hours to fly the same distance. Find the average rate of the plane in still air and the average rate of the wind.

SOLUTION

Step 1. **Use variables to represent unknown quantities.**
Let x = the average rate of the plane in still air.
Let y = the average rate of the wind.

Step 2. **Write a system of equations describing the problem's conditions.** As it travels with the wind, the plane's rate is increased. The net rate is its rate in still air, x, plus the rate of the wind, y, given by the expression $x + y$. As it travels against the wind, the plane's rate is decreased. The net rate is its rate in still air, x, minus the rate of the wind, y, given by the expression $x - y$. Here is a chart that summarizes the problem's information and includes the increased and decreased rates.

	Rate	×	Time	=	Distance
Trip with the Wind	$x + y$		3		$3(x + y)$
Trip against the Wind	$x - y$		5		$5(x - y)$

The problem states that the distance in each direction is 450 miles. We use this information to write our system of equations.

The distance of the trip with the wind | is | 450 miles.

$$3(x + y) = 450$$

The distance of the trip against the wind | is | 450 miles.

$$5(x - y) = 450$$

Step 3. **Solve the system and answer the problem's question.** We can simplify the system by dividing both sides of the equations by 3 and 5, respectively.

$$\begin{aligned} 3(x + y) &= 450 \xrightarrow{\text{Divide by 3.}} x + y = 150 \\ 5(x - y) &= 450 \xrightarrow{\text{Divide by 5.}} x - y = 90 \end{aligned}$$

Solve the system on the right by the addition method.

$$\begin{aligned} x + y &= 150 \\ x - y &= 90 \\ \text{Add: } 2x &= 240 \\ x &= 120 \quad \text{Divide both sides by 2.} \end{aligned}$$

Back-substituting 120 for x in either of the system's equations gives $y = 30$. Because $x = 120$ and $y = 30$, the average rate of the plane in still air is 120 miles per hour and the average rate of the wind is 30 miles per hour.

Step 4. **Check the proposed solution in the original wording of the problem.** Is the average rate of the plane in still air 120 miles per hour and the average rate of the wind 30 miles per hour? The problem states that the distance in each direction is 450 miles. The average rate of the plane with the wind is $120 + 30 = 150$ miles per hour. In 3 hours, it travels $150 \cdot 3$, or 450 miles, which checks with the stated condition. Furthermore, the average rate of the plane against the wind is $120 - 30 = 90$ miles per hour. In 5 hours, it travels $90 \cdot 5 = 450$ miles, which is the stated distance. ■

CHECK POINT 6 With the current, a motorboat can travel 84 miles in 2 hours. Against the current, the same trip takes 3 hours. Find the average rate of the boat in still water and the average rate of the current.

4.4 EXERCISE SET

Student Solutions Manual | CD/Video | PH Math/Tutor Center | MathXL Tutorials on CD | MathXL® | MyMathLab | Interactmath.com

Practice Exercises

In Exercises 1–4, let x represent one number and let y represent the other number. Use the given conditions to write a system of equations. Solve the system and find the numbers.

1. The sum of two numbers is 17. If one number is subtracted from the other, their difference is -3. Find the numbers.

2. The sum of two numbers is 5. If one number is subtracted from the other, their difference is 13. Find the numbers.

3. Three times a first number decreased by a second number is -1. The first number increased by twice the second number is 23. Find the numbers.

4. The sum of three times a first number and twice a second number is 43. If the second number is subtracted from twice the first number, the result is -4. Find the numbers.

Application Exercises

5. The graph in the next column makes Super Bowl Sunday look like a day of snack food bingeing in the United States. Combined, we wolf down 10.4 million pounds of potato chips and tortilla chips. The difference between consumption of potato chips and tortilla chips is 1.2 million pounds. How many millions of pounds of potato chips and tortilla chips are consumed on Super Bowl Sunday?

Millions of Pounds of Snack Food Consumed on Super Bowl Sunday

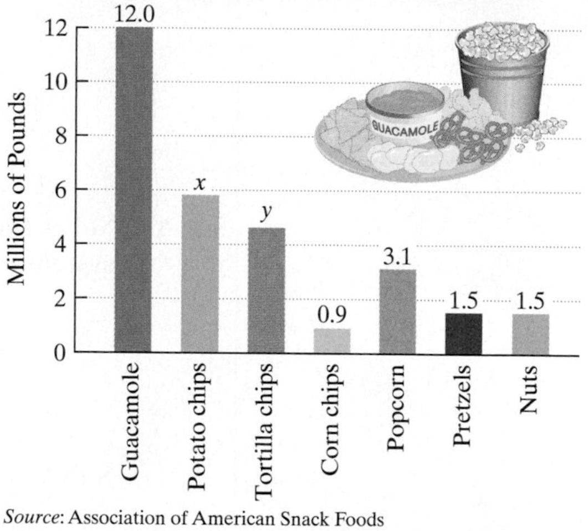

Source: Association of American Snack Foods

6. The graph at the top of the next page shows the places with the greatest number of documented fatal shark attacks from 1580 through 2002. Combined, 112 fatal attacks were recorded in the continental United States and South Africa. The difference between the number of attacks in the two countries was 26. How many documented fatal shark attacks were there in the continental United States and in South Africa from 1580 through 2002?

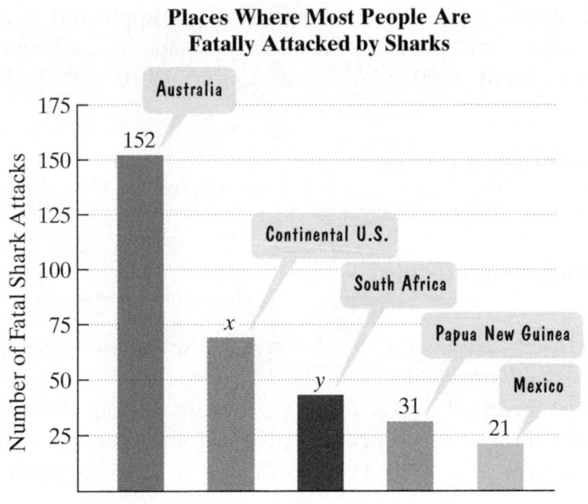

Source: International Shark Attack File

The graph shows the calories in some favorite fast foods. Use the information in Exercises 7–8 to find the exact caloric content of the specified foods.

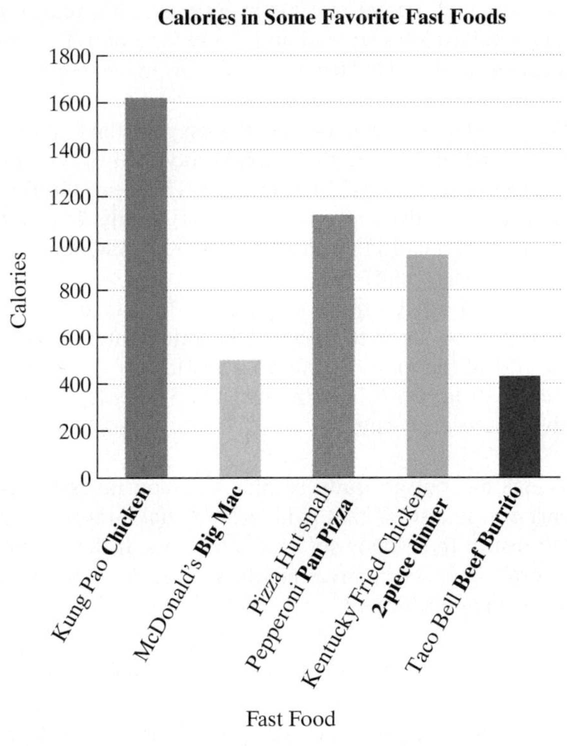

Source: Center for Science in the Public Interest

7. One pan pizza and two beef burritos provide 1980 calories. Two pan pizzas and one beef burrito provide 2670 calories. Find the caloric content of each item.

8. One Kung Pao chicken and two Big Macs provide 2620 calories. Two Kung Pao chickens and one Big Mac provide 3740 calories. Find the caloric content of each item.

9. Cholesterol intake should be limited to 300 mg or less each day. One serving of scrambled eggs from McDonalds and one Double Beef Whopper from Burger King exceed this intake by 241 mg. Two servings of scrambled eggs and three Double Beef Whoppers provide 1257 mg of cholesterol. Determine the cholesterol content in each item.

10. Two medium eggs and three cups of ice cream contain 701 milligrams of cholesterol. One medium egg and one cup of ice cream exceed the suggested daily cholesterol intake of 300 milligrams by 25 milligrams. Determine the cholesterol content in each item.

11. In a discount clothing store, all sweaters are sold at one fixed price and all shirts are sold at another fixed price. If one sweater and three shirts cost $42, while three sweaters and two shirts cost $56, find the price of one sweater and the price of one shirt.

12. A restaurant purchased eight tablecloths and five napkins for $106. A week later, a tablecloth and six napkins were bought for $24. Find the cost of one tablecloth and the cost of one napkin, assuming the same prices for both purchases.

13. You are choosing between two long-distance telephone plans. Plan A has a monthly fee of $20 with a charge of $0.05 per minute for all long-distance calls. Plan B has a monthly fee of $5 with a charge of $0.10 per minute for all long-distance calls.
   a. For how many minutes of long-distance calls will the costs for the two plans be the same? What will be the cost for each plan?
   b. If you make approximately 10 long-distance calls per month, each averaging 20 minutes, which plan should you select? Explain your answer.

14. You are choosing between two long-distance telephone plans. Plan A has a monthly fee of $15 with a charge of $0.08 per minute for all long-distance calls. Plan B has a monthly fee of $3 with a charge of $0.12 per minute for all long-distance calls.
   a. For how many minutes of long-distance calls will the costs for the two plans be the same? What will be the cost for each plan?
   b. If you make approximately 15 long-distance calls per month, each averaging 30 minutes, which plan should you select? Explain your answer.

15. You are choosing between two plans at a discount warehouse. Plan A offers an annual membership fee of $100 and you pay 80% of the manufacturer's recommended list price. Plan B offers an annual membership fee of $40 and you pay 90% of the manufacturer's recommended list price. How many dollars of merchandise would you have to purchase in a year to pay the same amount under both plans? What will be the cost for each plan?

16. You are choosing between two plans at a discount warehouse. Plan A offers an annual membership fee of $300 and you pay 70% of the manufacturer's recommended list price. Plan B offers an annual membership fee of $40 and you pay 90% of the manufacturer's recommended list price. How many dollars of merchandise would you have to purchase in a year to pay the same amount under both plans? What will be the cost for each plan?

The graphs show average weekly earnings of full-time wage and salary workers 25 and older, by educational attainment. Exercises 17–18 involve the information in these graphs.

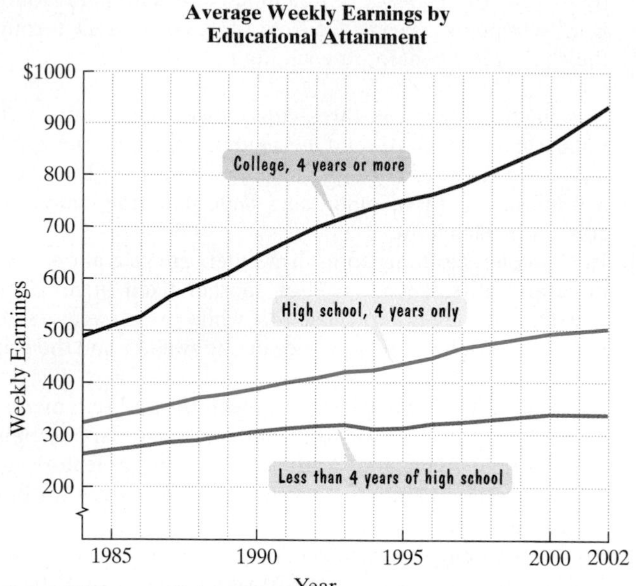

Source: U.S. Bureau of Labor Statistics

17. In 1985, college graduates averaged \$508 in weekly earnings. This amount has increased by approximately \$25 in weekly earnings per year. By contrast, in 1985, high school graduates averaged \$345 in weekly earnings. This amount has only increased by approximately \$9 in weekly earnings per year. How many years after 1985 will college graduates be earning twice the amount per week that high school graduates earn? In which year will this occur? What will be the weekly earnings for each group at that time?

18. In 1985, college graduates averaged \$508 in weekly earnings. This amount has increased by approximately \$25 in weekly earnings per year. By contrast, in 1985, people with less than four years of high school averaged \$270 in weekly earnings. This amount has only increased by approximately \$4 in weekly earnings per year. How many years after 1985 will college graduates be earning three times the amount per week that people with less than four years of high school earn? (Round to the nearest whole number.) In which year will this occur? What will be the weekly earnings for each group at that time?

19. Nutritional information for macaroni and broccoli is given in the table. How many servings of each would it take to get exactly 14 grams of protein and 48 grams of carbohydrates?

	Macaroni	Broccoli
Protein (grams/serving)	3	2
Carbohydrates (grams/serving)	16	4

20. The calorie-nutrient information for an apple and an avocado is given in the table. How many of each should be eaten to get exactly 1000 calories and 100 grams of carbohydrates?

	One Apple	One Avocado
Calories	100	350
Carbohydrates (grams)	24	14

Exercises 21–26 involve simple interest. Use the four-step strategy to solve each problem.

21. You invested \$7000 in two accounts paying 6% and 8% annual interest, respectively. If the total interest earned for the year was \$520, how much was invested at each rate?

22. You invested \$11,000 in stocks and bonds, paying 5% and 8% annual interest, respectively. If the total interest earned for the year was \$730, how much was invested in stocks and how much was invested in bonds?

23. You invested money in two funds. Last year, the first fund paid a dividend of 9% and the second a dividend of 3%, and you received a total of \$900 in interest. This year, the first fund paid a 10% dividend and the second only 1%, and you received a total of \$860 in interest. How much money did you invest in each fund?

24. You invested money in two funds. Last year, the first fund paid a dividend of 8% and the second a divident of 5%, and you received a total of \$1330 in interest. This year, the first fund paid a 12% dividend and the second only 2%, and you received a total of \$1500 in interest. How much money did you invest in each fund?

25. Things did not go quite as planned. You invested \$20,000, part of it in a stock that paid 12% annual interest. However, the rest of the money suffered a 5% loss. If the total annual income from both investments was \$1890, how much was invested at each rate?

26. Things did not go quite as planned. You invested \$30,000, part of it in a stock that paid 14% annual interest. However, the rest of the money suffered a 6% loss. If the total annual income from both investments was \$200, how much was invested at each rate?

Exercises 27–34 involve mixtures. Use the four-step strategy to solve each problem.

27. A wine company needs to blend a California wine with a 5% alcohol content and a French wine with a 9% alcohol content to obtain 200 gallons of wine with a 7% alcohol content. How many gallons of each kind of wine must be used?

28. A jeweler needs to mix an alloy with a 16% gold content and an alloy with a 28% gold content to obtain 32 ounces of a new alloy with a 25% gold content. How many ounces of each of the original alloys must be used?

29. For thousands of years, gold has been considered one of Earth's most precious metals. One hundred percent pure gold is 24-karat gold, which is too soft to be made into jewelry. In the United States, most gold jewelry is 14-karat gold, approximately 58% gold. If 18-karat gold is 75% gold and 12-karat gold is 50% gold, how much of each should be used to make a 14-karat gold bracelet weighing 300 grams?

30. In the "Peanuts" cartoon shown, solve the problem that is sending Peppermint Patty into an agitated state. How much cream and how much milk, to the nearest thousandth of a gallon, must be mixed together to obtain 50 gallons of cream that contains 12.5% butterfat?

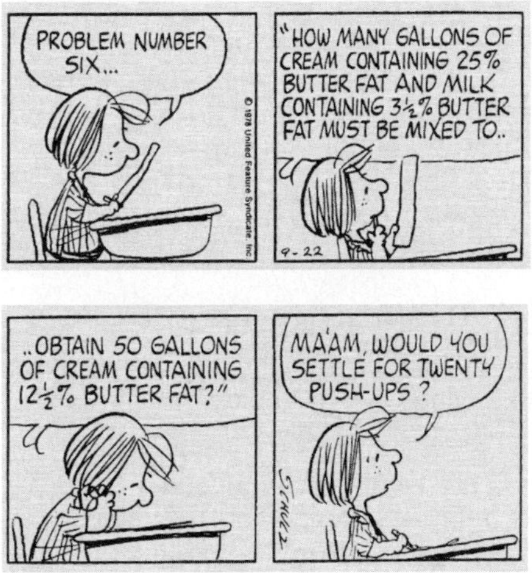

© 1978 United Media/United Feature Syndicate, Inc.

31. The manager of a candystand at a large multiplex cinema has a popular candy that sells for \$1.60 per pound. The manager notices a different candy worth \$2.10 per pound that is not selling well. The manager decides to form a mixture of both types of candy to help clear the inventory of the more expensive type. How many pounds of each kind of candy should be used to create a 75-pound mixture selling for \$1.90 per pound?

32. A grocer needs to mix raisins at \$2.00 per pound with granola at \$3.25 per pound to obtain 10 pounds of a mixture that costs \$2.50 per pound. How many pounds of raisins and how many pound of granola must be used?

33. A coin purse contains a mixture of 15 coins in nickels and dimes. The coins have a total value of \$1.10. Determine the number of nickels and the number of dimes in the purse.

34. A coin purse contains a mixture of 15 coins in dimes and quarters. The coins have a total value of \$3.30. Determine the number of dimes and the number of quarters in the purse.

Exercises 35–40 involve motion. Use the four-step strategy to solve each problem.

35. When a small plane flies with the wind, it can travel 800 miles in 5 hours. When the plane flies in the opposite direction, against the wind, it takes 8 hours to fly the same distance. Find the rate of the plane in still air and the rate of the wind.

36. When a plane flies with the wind, it can travel 4200 miles in 6 hours. When the plane flies in the opposite direction, against the wind, it takes 7 hours to fly the same distance. Find the rate of the plane in still air and the rate of the wind.

37. A boat's crew rowed 16 kilometers downstream, with the current, in 2 hours. The return trip upstream, against the current, covered the same distance, but took 4 hours. Find the crew's rowing rate in still water and the rate of the current.

38. A motorboat traveled 36 miles downstream, with the current, in 1.5 hours. The return trip upstream, against the current, covered the same distance, but took 2 hours. Find the boat's rate in still water and the rate of the current.

39. With the current, you can canoe 24 miles in 4 hours. Against the same current, you can canoe only $\frac{3}{4}$ of this distance in 6 hours. Find your rate in still water and the rate of the current.

40. With the current, you can row 24 miles in 3 hours. Against the same current, you can row only $\frac{2}{3}$ of this distance in 4 hours. Find your rowing rate in still water and the rate of the current.

Writing in Mathematics

41. Describe the conditions in a problem that enable it to be solved using a system of linear equations.

42. Write a word problem that can be solved by translating to a system of linear equations. Then solve the problem.

43. Exercises 13–16 involve using systems of linear equations to compare costs of long-distance telephone plans and plans at a discount warehouse. Describe another situation that involves choosing between two options that can be modeled and solved with a linear system.

Critical Thinking Exercises

44. A set of identical twins can only be recognized by the characteristic that one always tells the truth and the other always lies. One twin tells you of a lucky number pair: "When I multiply my first lucky number by 3 and my second lucky number by 6, the addition of the resulting numbers produces a sum of 12. When I add my first lucky number and twice my second lucky number, the sum is 5." Which twin is talking?

45. Tourist: "How many birds and lions do you have in your zoo?" Zookeeper: "There are 30 heads and 100 feet." Tourist: "I can't tell from that." Zookeeper: "Oh, yes, you can!" Can you? Find the number of each.

46. Find the measure of each angle whose degree measure is represented with a variable.

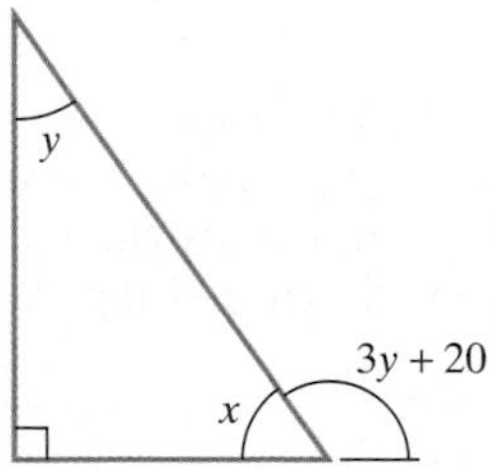

47. One apartment is directly above a second apartment. The resident living downstairs calls his neighbor living above him and states, "If one of you is willing to come downstairs, we'll have the same number of people in both apartments." The upstairs resident responds, "We're all too tired to move. Why don't one of you come up here? Then we'll have twice as many people up here as you've got down there." How many people are in each apartment?

48. In Lewis Carroll's *Through the Looking Glass*, the following dialogue takes place:

Tweedledum (to Tweedledee): The sum of your weight and twice mine is 361 pounds.

Tweedledee (to Tweedledum): Contrawise, the sum of your weight and twice mine is 362 pounds.

Find the weight of each of the two characters.

Technology Exercise

49. Select any two problems that you solved from Exercises 5–20. Use a graphing utility to graph the system of equations that you wrote for that problem. Then use the TRACE or INTERSECTION feature to show the point on the graphs that corresponds to the problem's solution.

Review Exercises

50. Find the slope of the line containing the points $(-6, 1)$ and $(2, -1)$ (Section 3.3, Example 1)

51. Add: $\frac{1}{5} + \left(-\frac{3}{4}\right)$. (Section 1.5, Example 4)

52. Graph: $y = x^2$. (Section 3.1, Example 6)

SECTION 4.5

SYSTEMS OF LINEAR EQUATIONS IN THREE VARIABLES

Objectives

1. Verify the solution of a system of linear equations in three variables.
2. Solve systems of linear equations in three variables.
3. Identify inconsistent and dependent systems.
4. Solve problems using systems in three variables.

All animals sleep, but the length of time they sleep varies widely: Cattle sleep for only a few minutes at a time. We humans seem to need more sleep than other animals, up to eight hours a day. Without enough sleep, we have difficulty concentrating, make mistakes in routine tasks, lose energy, and feel bad-tempered. There is a relationship between hours of sleep and death rate per year per 100,000 people. How many hours of sleep will put you in the group with the minimum death rate? In this section, you will learn how to solve systems of linear equations with more than two variables to answer this question.

1 Verify the solution of a system of linear equations in three variables.

Systems of Linear Equations in Three Variables and Their Solutions An equation such as $x + 2y - 3z = 9$ is called a *linear equation in three variables*. In general, any equation of the form

$$Ax + By + Cz = D,$$

where A, B, C, and D are real numbers such that A, B, and C are not all 0, is a **linear equation in three variables: x, y, and z.** The graph of this linear equation in three variables is a plane in three-dimensional space.

The process of solving a system of three linear equations in three variables is geometrically equivalent to finding the point of intersection (assuming that there is one) of three planes in space. (See Figure 4.11). A **solution** of a system of linear equations in three variables is an ordered triple of real numbers that satisfies all equations in the system. The **solution set** of the system is the set of all its solutions.

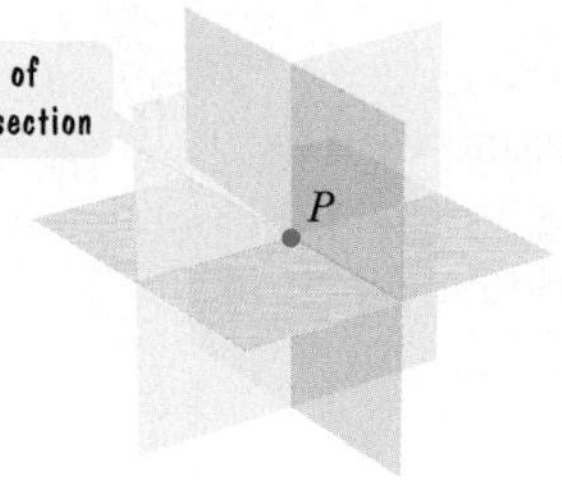

FIGURE 4.11

EXAMPLE 1 Determining Whether an Ordered Triple Satisfies a System

Show that the ordered triple $(-1, 2, -2)$ is a solution of the system:

$$\begin{aligned} x + 2y - 3z &= 9 \\ 2x - y + 2z &= -8 \\ -x + 3y - 4z &= 15. \end{aligned}$$

SOLUTION Because -1 is the x-coordinate, 2 is the y-coordinate, and -2 is the z-coordinate of $(-1, 2, -2)$, we replace x with -1, y with 2, and z with -2 in each of the three equations.

$$\begin{aligned} x + 2y - 3z &= 9 \\ -1 + 2(2) - 3(-2) &\stackrel{?}{=} 9 \\ -1 + 4 + 6 &\stackrel{?}{=} 9 \\ 9 &= 9, \quad \text{true} \end{aligned}$$

$$\begin{aligned} 2x - y + 2z &= -8 \\ 2(-1) - 2 + 2(-2) &\stackrel{?}{=} -8 \\ -2 - 2 - 4 &\stackrel{?}{=} -8 \\ -8 &= -8, \quad \text{true} \end{aligned}$$

$$\begin{aligned} -x + 3y - 4z &= 15 \\ -(-1) + 3(2) - 4(-2) &\stackrel{?}{=} 15 \\ 1 + 6 + 8 &\stackrel{?}{=} 15 \\ 15 &= 15, \quad \text{true} \end{aligned}$$

The ordered triple $(-1, 2, -2)$ satisfies the three equations: It makes each equation true. Thus, the ordered triple is a solution of the system. ■

CHECK POINT 1 Show that the ordered triple $(-1, -4, 5)$ is a solution of the system:

$$\begin{aligned} x - 2y + 3z &= 22 \\ 2x - 3y - z &= 5 \\ 3x + y - 5z &= -32. \end{aligned}$$

2 Solve systems of linear equations in three variables.

Solving Systems of Linear Equations in Three Variables by Eliminating Variables The method for solving a system of linear equations in three variables is similar to that used on systems of linear equations in two variables. We use addition to eliminate any variable, reducing the system to two equations in two variables. Once we obtain a system of two equations in two variables, we use addition or substitution to eliminate another variable. The result is a single equation in one variable. We solve this equation to get the value of the remaining variable. Other variable values are found by back-substitution.

STUDY TIP

It does not matter which variable you eliminate, as long as you do it in two different pairs of equations.

SOLVING LINEAR SYSTEMS IN THREE VARIABLES BY ELIMINATING VARIABLES

1. Reduce the system to two equations in two variables. This is usually accomplished by taking two different pairs of equations and using the addition method to eliminate the same variable from both pairs.
2. Solve the resulting system of two equations in two variables using addition or substitution. The result is an equation in one variable that gives the value of that variable.
3. Back-substitute the value of the variable found in step 2 into either of the equations in two variables to find the value of the second variable.
4. Use the values of the two variables from steps 2 and 3 to find the value of the third variable by back-substituting into one of the original equations.
5. Check the proposed solution in each of the original equations.

EXAMPLE 2 Solving a System in Three Variables

Solve the system:

$$\begin{aligned} 5x - 2y - 4z &= 3 && \text{Equation 1}\\ 3x + 3y + 2z &= -3 && \text{Equation 2}\\ -2x + 5y + 3z &= 3. && \text{Equation 3}\end{aligned}$$

SOLUTION There are many ways to proceed. Because our initial goal is to reduce the system to two equations in two variables, **the central idea is to take two different pairs of equations and eliminate the same variable from both pairs**.

Step 1. **Reduce the system to two equations in two variables.** We choose any two equations and use the addition method to eliminate a variable. Let's eliminate z from Equations 1 and 2. We do so by multiplying Equation 2 by 2. Then we add equations.

$$\begin{aligned} &(\text{Equation 1})\quad 5x - 2y - 4z = 3 \xrightarrow{\text{No change}} && 5x - 2y - 4z = 3\\ &(\text{Equation 2})\quad 3x + 3y + 2z = -3 \xrightarrow{\text{Multiply by 2.}} && \underline{6x + 6y + 4z = -6}\\ & && \text{Add: } 11x + 4y \quad = -3 \quad (\text{Equation 4})\end{aligned}$$

Now we must eliminate the *same* variable from another pair of equations. We can eliminate z from Equations 2 and 3. First, we multiply Equation 2 by -3. Next, we multiply Equation 3 by 2. Finally, we add equations.

$$\begin{aligned} &(\text{Equation 2})\quad 3x + 3y + 2z = -3 \xrightarrow{\text{Multiply by } -3.} && -9x - 9y - 6z = 9\\ &(\text{Equation 3})\quad -2x + 5y + 3z = 3 \xrightarrow{\text{Multiply by 2.}} && \underline{-4x + 10y + 6z = 6}\\ & && \text{Add: } -13x + y \quad = 15 \quad (\text{Equation 5})\end{aligned}$$

Equations 4 and 5 give us a system of two equations in two variables.

Step 2. **Solve the resulting system of two equations in two variables.** We will use the addition method to solve Equations 4 and 5 for x and y. To do so, we multiply Equation 5 by -4 and add this to Equation 4.

$$\begin{aligned} &(\text{Equation 4})\quad 11x + 4y = -3 \xrightarrow{\text{No change}} && 11x + 4y = -3\\ &(\text{Equation 5})\quad -13x + y = 15 \xrightarrow{\text{Multiply by } -4.} && \underline{52x - 4y = -60}\\ & && \text{Add: } 63x \quad = -63\\ & && x \quad = -1 \quad \text{Divide both sides by 63.}\end{aligned}$$

Step 3. **Use back-substitution in one of the equations in two variables to find the value of the second variable.** We back-substitute -1 for x in either Equation 4 or 5 to find the value of y. We will use Equation 5.

$$-13x + y = 15 \quad \text{Equation 5}$$
$$-13(-1) + y = 15 \quad \text{Substitute } -1 \text{ for } x.$$
$$13 + y = 15 \quad \text{Multiply.}$$
$$y = 2 \quad \text{Subtract 13 from both sides.}$$

Step 4. **Back-substitute the values found for two variables into one of the original equations to find the value of the third variable.** We can now use any one of the original equations and back-substitute the values of x and y to find the value for z. We will use Equation 2.

$$3x + 3y + 2z = -3 \quad \text{Equation 2}$$
$$3(-1) + 3(2) + 2z = -3 \quad \text{Substitute } -1 \text{ for } x \text{ and 2 for } y.$$
$$3 + 2z = -3 \quad \text{Multiply and then add: } 3(-1) + 3(2) = -3 + 6 = 3.$$
$$2z = -6 \quad \text{Subtract 3 from both sides.}$$
$$z = -3 \quad \text{Divide both sides by 2.}$$

With $x = -1$, $y = 2$, and $z = -3$, the proposed solution is the ordered triple $(-1, 2, -3)$.

Step 5. **Check.** Check the proposed solution, $(-1, 2, -3)$, by substituting the values for x, y, and z into each of the three original equations. These substitutions yield three true statements. Thus, the solution is $(-1, 2, -3)$ and the solution set is $\{(-1, 2, -3)\}$. ■

✔ CHECK POINT 2 Solve the system:

$$x + 4y - z = 20$$
$$3x + 2y + z = 8$$
$$2x - 3y + 2z = -16.$$

In some examples, one of the variables is already eliminated from a given equation. In this case, the same variable should be eliminated from the other two equations, thereby making it possible to omit one of the elimination steps. We illustrate this idea in Example 3.

EXAMPLE 3 Solving a System of Equations with a Missing Term

Solve the system:

$$x + z = 8 \quad \text{Equation 1}$$
$$x + y + 2z = 17 \quad \text{Equation 2}$$
$$x + 2y + z = 16. \quad \text{Equation 3}$$

$$\begin{aligned} x + \quad\; z &= 8 && \text{Equation 1} \\ x + y + 2z &= 17 && \text{Equation 2} \\ x + 2y + z &= 16 && \text{Equation 3} \end{aligned}$$

The given system (repeated)

SOLUTION

Step 1. **Reduce the system to two equations in two variables.** Because Equation 1 contains only x and z, we could eliminate y from Equations 2 and 3. This will give us two equations in x and z. To eliminate y from Equations 2 and 3, we multiply Equation 2 by -2 and add Equation 3.

$$\begin{aligned} &(\text{Equation 2}) \quad x + y + 2z = 17 \xrightarrow{\text{Multiply by } -2.} & -2x - 2y - 4z &= -34 \\ &(\text{Equation 3}) \quad x + 2y + z = 16 \xrightarrow{\text{No change}} & x + 2y + z &= 16 \\ &\text{Add:} & -x \qquad - 3z &= -18 \quad (\text{Equation 4}) \end{aligned}$$

Equation 4 and the given Equation 1 provide us with a system of two equations in two variables.

Step 2. **Solve the resulting system of two equations in two variables.** We will solve Equations 1 and 4 for x and z.

$$\begin{aligned} x + z &= 8 && \text{Equation 1} \\ -x - 3z &= -18 && \text{Equation 4} \\ \text{Add:} \quad -2z &= -10 \\ z &= 5 && \text{Divide both sides by } -2. \end{aligned}$$

Step 3. **Use back-substitution in one of the equations in two variables to find the value of the second variable.** To find x, we back-substitute 5 for z in either Equation 1 or 4. We will use Equation 1.

$$\begin{aligned} x + z &= 8 && \text{Equation 1} \\ x + 5 &= 8 && \text{Substitute 5 for } z. \\ x &= 3 && \text{Subtract 5 from both sides.} \end{aligned}$$

Step 4. **Back-substitute the values found for two variables into one of the original equations to find the value of the third variable.** To find y, we back-substitute 3 for x and 5 for z into Equation 2 or 3. We can't use Equation 1 because y is missing in this equation. We will use Equation 2.

$$\begin{aligned} x + y + 2z &= 17 && \text{Equation 2} \\ 3 + y + 2(5) &= 17 && \text{Substitute 3 for } x \text{ and 5 for } z. \\ y + 13 &= 17 && \text{Multiply and add.} \\ y &= 4 && \text{Subtract 13 from both sides.} \end{aligned}$$

We found that $z = 5$, $x = 3$, and $y = 4$. Thus, the proposed solution is the ordered triple $(3, 4, 5)$.

Step 5. Check. Substituting 3 for x, 4 for y, and 5 for z into each of the three original equations yields three true statements. Consequently, the solution is $(3, 4, 5)$ and the solution set is $\{(3, 4, 5)\}$. ■

CHECK POINT 3 Solve the system:

$$\begin{aligned} 2y - z &= 7 \\ x + 2y + z &= 17 \\ 2x - 3y + 2z &= -1. \end{aligned}$$

3 Identify inconsistent and dependent systems.

Inconsistent and Dependent Systems A system of linear equations in three variables represents three planes. The three planes need not intersect at one point. The planes may have no common point of intersection and represent an **inconsistent system** with no solution. Figure 4.12 illustrates some of the geometric possibilities for inconsistent systems.

Three planes are parallel with no common intersection point.

Two planes are parallel with no common intersection point.

Planes intersect two at a time. There is no intersection point common to all three planes.

FIGURE 4.12 Three planes may have no common point of intersection.

If you attempt to solve an inconsistent system algebraically, at some point in the solution process you will eliminate all three variables. A false statement, such as $0 = 17$, will be the result. For example, consider the system

$$\begin{aligned} 2x + 5y + z &= 12 &&\text{Equation 1}\\ x - 2y + 4z &= -10 &&\text{Equation 2}\\ -3x + 6y - 12z &= 20. &&\text{Equation 3} \end{aligned}$$

Suppose we reduce the system to two equations in two variables by eliminating x. To eliminate x in Equations 2 and 3, we multiply Equation 2 by 3 and add Equation 3:

$$\begin{array}{rcl} x - 2y + 4z = -10 & \xrightarrow{\text{Multiply by 3.}} & 3x - 6y + 12z = -30 \\ -3x + 6y - 12z = 20 & \xrightarrow{\text{No change}} & \underline{-3x + 6y - 12z = 20} \\ & \text{Add:} & 0 = -10 \end{array}$$

There are no values of x, y, and z for which $0 = -10$. The false statement $0 = -10$ indicates that the system is inconsistent and has no solution. The solution set is the empty set, $\varnothing$.

We have seen that a linear system that has at least one solution is called a **consistent system**. Planes that intersect at one point and planes that intersect at infinitely many points both represent consistent systems. Figure 4.13 illustrates planes that intersect at infinitely many points. The equations in these linear systems with infinitely many solutions are called **dependent**.

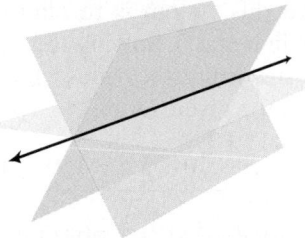

FIGURE 4.13 Three planes may intersect at infinitely many points.

The planes intersect along a common line.

The planes coincide.

If you attempt to solve a system with dependent equations algebraically, at some point in the solution process you will eliminate all three variables. A true statement, such as $0 = 0$, will be the result. If this occurs as you are solving a linear system, simply state that the equations are dependent.

4 Solve problems using systems in three variables.

Applications Systems of equations may allow us to find models for data without using a graphing utility. Three data points that do not lie on or near a line determine the graph of an equation of the form

$$y = ax^2 + bx + c, a \neq 0.$$

Such an equation has a cuplike U-shaped graph, making it ideal for modeling situations in which values of y are decreasing and then increasing.

The process of determining an equation whose graph contains given points is called **curve fitting**. In our next example, we fit the curve whose equation is $y = ax^2 + bx + c$ to three data points. Using a system of equations, we find values for a, b, and c.

EXAMPLE 4 Modeling Data Relating Sleep and Death Rate

In a study relating sleep and death rate, the following data were obtained. Use the equation $y = ax^2 + bx + c$ to model the data.

x (Average Number of Hours of Sleep)	y (Death Rate per Year per 100,000 Males)
4	1682
7	626
9	967

SOLUTION We need to find values for a, b, and c in $y = ax^2 + bx + c$. We can do so by solving a system of three linear equations in a, b, and c. We obtain the three equations by using the values of x and y from the data as follows:

$\boldsymbol{y = ax^2 + bx + c}$ *Use the quadratic function to model the data.*

When $x = 4$, $y = 1682$: $1682 = a \cdot 4^2 + b \cdot 4 + c$ or $16a + 4b + c = 1682$

When $x = 7$, $y = 626$: $626 = a \cdot 7^2 + b \cdot 7 + c$ or $49a + 7b + c = 626$

When $x = 9$, $y = 967$: $967 = a \cdot 9^2 + b \cdot 9 + c$ or $81a + 9b + c = 967.$

The easiest way to solve this system is to eliminate c from two pairs of equations, obtaining two equations in a and b. Solving this system gives $a = 104.5$, $b = -1501.5$, and $c = 6016$. We now substitute the values for a, b, and c into $y = ax^2 + bx + c$. The equation that models the given data is

$$y = 104.5x^2 - 1501.5x + 6016.$$

USING TECHNOLOGY

The graph of

$y = 104.5x^2 - 1501.5x + 6016$

is displayed in a $[3, 12, 1]$ by $[500, 2000, 100]$ viewing rectangle. The utility indicates that the lowest point on the graph is approximately $(7.2, 622.5)$. Men who average 7.2 hours of sleep are in the group with the lowest death rate, approximately 622.5 deaths per 100,000 males.

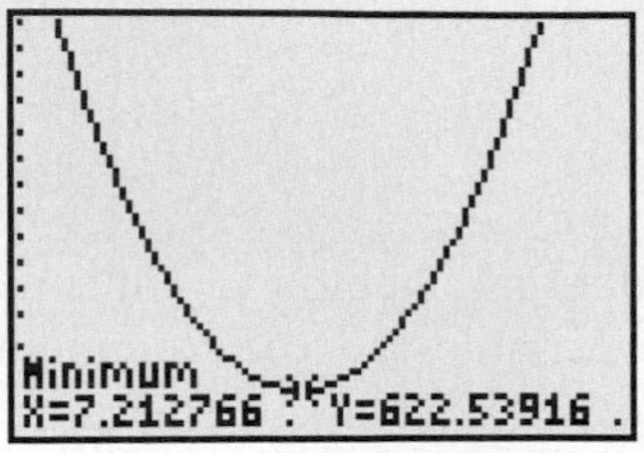

We can use the model that we obtained in Example 4 to find the death rate of males who average, say, 6 hours of sleep. Substitute 6 for x in $y = 104.5x^2 - 1501.5x + 6016$:

$$y = 104.5(6)^2 - 1501.5(6) + 6016 = 769.$$

According to the model, the death rate for males who average 6 hours of sleep is 769 deaths per 100,000 males.

CHECK POINT 4 Find the equation $y = ax^2 + bx + c$ whose graph passes through the points $(1, 4)$, $(2, 1)$, and $(3, 4)$.

Problems involving three unknowns can be solved using the same strategy for solving problems with two unknown quantities. You can let x, y, and z represent the unknown quantities. We then translate from the verbal conditions of the problem to a system of three equations in three variables. Problems of this type are included in the exercise set that follows.

4.5 EXERCISE SET

Student Solutions Manual | CD/Video | PH Math/Tutor Center | MathXL Tutorials on CD | MathXL® | MyMathLab | Interactmath.com

Practice Exercises

In Exercises 1–4, determine if the given ordered triple is a solution of the system.

1. $(2, -1, 3)$
$$\begin{aligned} x + y + z &= 4 \\ x - 2y - z &= 1 \\ 2x - y - z &= -1 \end{aligned}$$

2. $(5, -3, -2)$
$$\begin{aligned} x + y + z &= 0 \\ x + 2y - 3z &= 5 \\ 3x + 4y + 2z &= -1 \end{aligned}$$

3. $(4, 1, 2)$
$$\begin{aligned} x - 2y &= 2 \\ 2x + 3y &= 11 \\ y - 4z &= -7 \end{aligned}$$

4. $(-1, 3, 2)$
$$\begin{aligned} x - 2z &= -5 \\ y - 3z &= -3 \\ 2x - z &= -4 \end{aligned}$$

Solve each system in Exercises 5–22. If there is no solution or if there are infinitely many solutions and a system's equations are dependent, so state.

5.
$$\begin{aligned} x + y + 2z &= 11 \\ x + y + 3z &= 14 \\ x + 2y - z &= 5 \end{aligned}$$

6.
$$\begin{aligned} 2x + y - 2z &= -1 \\ 3x - 3y - z &= 5 \\ x - 2y + 3z &= 6 \end{aligned}$$

7.
$$\begin{aligned} 4x - y + 2z &= 11 \\ x + 2y - z &= -1 \\ 2x + 2y - 3z &= -1 \end{aligned}$$

8.
$$\begin{aligned} x - y + 3z &= 8 \\ 3x + y - 2z &= -2 \\ 2x + 4y + z &= 0 \end{aligned}$$

9.
$$\begin{aligned} 3x + 2y - 3z &= -2 \\ 2x - 5y + 2z &= -2 \\ 4x - 3y + 4z &= 10 \end{aligned}$$

10.
$$\begin{aligned} 2x + 3y + 7z &= 13 \\ 3x + 2y - 5z &= -22 \\ 5x + 7y - 3z &= -28 \end{aligned}$$

11.
$$\begin{aligned} 2x - 4y + 3z &= 17 \\ x + 2y - z &= 0 \\ 4x - y - z &= 6 \end{aligned}$$

12.
$$\begin{aligned} x + z &= 3 \\ x + 2y - z &= 1 \\ 2x - y + z &= 3 \end{aligned}$$

13.
$$\begin{aligned} 2x + y &= 2 \\ x + y - z &= 4 \\ 3x + 2y + z &= 0 \end{aligned}$$

14.
$$\begin{aligned} x + 3y + 5z &= 20 \\ y - 4z &= -16 \\ 3x - 2y + 9z &= 36 \end{aligned}$$

15.
$$\begin{aligned} x + y &= -4 \\ y - z &= 1 \\ 2x + y + 3z &= -21 \end{aligned}$$

16.
$$\begin{aligned} x + y &= 4 \\ x + z &= 4 \\ y + z &= 4 \end{aligned}$$

17.
$$\begin{aligned} 2x + y + 2z &= 1 \\ 3x - y + z &= 2 \\ x - 2y - z &= 0 \end{aligned}$$

18.
$$\begin{aligned} 3x + 4y + 5z &= 8 \\ x - 2y + 3z &= -6 \\ 2x - 4y + 6z &= 8 \end{aligned}$$

19. $5x - 2y - 5z = 1$
$10x - 4y - 10z = 2$
$15x - 6y - 15z = 3$

20. $x + 2y + z = 4$
$3x - 4y + z = 4$
$6x - 8y + 2z = 8$

21. $3(2x + y) + 5z = -1$
$2(x - 3y + 4z) = -9$
$4(1 + x) = -3(z - 3y)$

22. $7z - 3 = 2(x - 3y)$
$5y + 3z - 7 = 4x$
$4 + 5z = 3(2x - y)$

In Exercises 23–26, find the equation $y = ax^2 + bx + c$ whose graph passes through the given points.

23. $(-1, 6), (1, 4) (2, 9)$

24. $(-2, 7), (1, -2), (2, 3)$

25. $(-1, -4), (1, -2), (2, 5)$

26. $(1, 3), (3, -1), (4, 0)$

In Exercises 27–28, let x represent the first number, y the second number, and z the third number. Use the given conditions to write a system of equations. Solve the system and find the numbers.

27. The sum of three numbers is 16. The sum of twice the first number, 3 times the second number, and 4 times the third number is 46. The difference between 5 times the first number and the second number is 31. Find the three numbers.

28. The following is known about three numbers: Three times the first number plus the second number plus twice the third number is 5. If 3 times the second number is subtracted from the sum of the first number and 3 times the third number, the result is 2. If the third number is subtracted from the sum of 2 times the first number and 3 times the second number, the result is 1. Find the numbers.

Practice Plus

Solve each system in Exercises 29–30.

29. $\frac{x+2}{6} - \frac{y+4}{3} + \frac{z}{2} = 0$

$\frac{x+1}{2} + \frac{y-1}{2} - \frac{z}{4} = \frac{9}{2}$

$\frac{x-5}{4} + \frac{y+1}{3} + \frac{z-2}{2} = \frac{19}{4}$

30. $\frac{x+3}{2} - \frac{y-1}{2} + \frac{z+2}{4} = \frac{3}{2}$

$\frac{x-5}{2} + \frac{y+1}{3} - \frac{z}{4} = -\frac{25}{6}$

$\frac{x-3}{4} - \frac{y+1}{2} + \frac{z-3}{2} = -\frac{5}{2}$

In Exercises 31–32, find the equation $y = ax^2 + bx + c$ whose graph is shown. Select three points whose coordinates appear to be integers.

31.

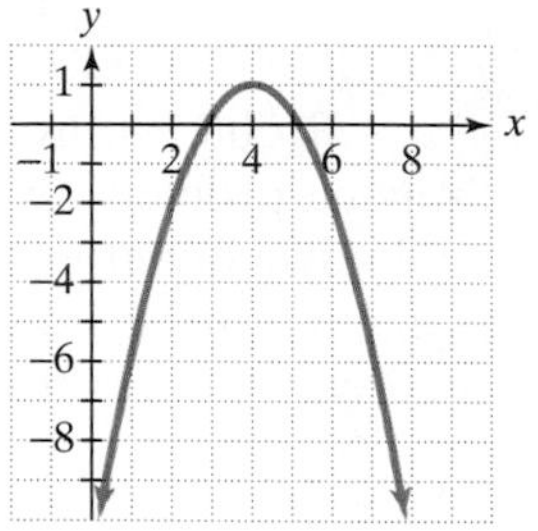

32.

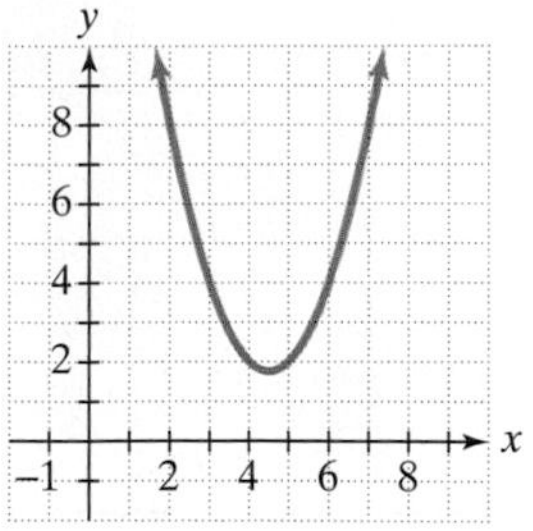

In Exercises 33–34, solve each system for (x, y, z) in terms of the nonzero constants a, b, and c.

33. $ax - by - 2cz = 21$
$ax + by + cz = 0$
$2ax - by + cz = 14$

34. $ax - by + 2cz = -4$
$ax + 3by - cz = 1$
$2ax + by + 3cz = 2$

Application Exercises

35. The bar graph shows the percentage of the U.S. population that was foreign-born from 1900 through 2002.

Percentage of U.S. Population That was Foreign-Born, 1900-2002

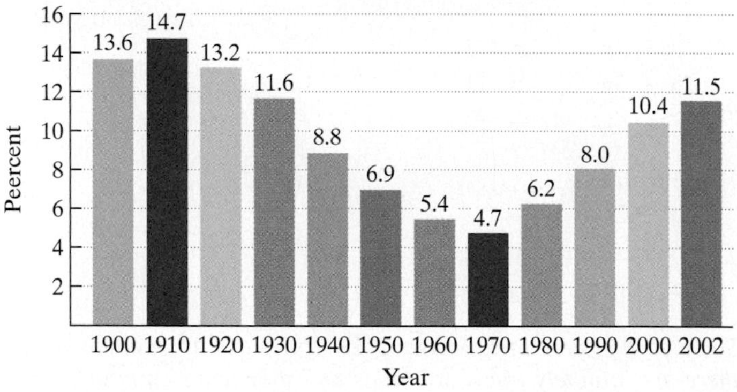

Source: U.S. Census Bureau

a. Write the data for 1900, 1970, and 2002 as ordered pairs (x, y), where x is the number of years after 1900 and y is the percentage of the U.S. population that was foreign-born in that year.

b. The three data points in part (a) can be modeled by the equation $y = ax^2 + bx + c$. Substitute each ordered pair into this equation, one ordered pair at a time, and write a system of linear equations in three variables that can be used to find values for a, b, and c. It is not necessary to solve the system.

36. The bar graph shows that the U.S. divorce rate increased from 1970 to 1985 and then decreased from 1985 to 2001.

U.S. Divorce Rates: Number of Divorces per 1000 People

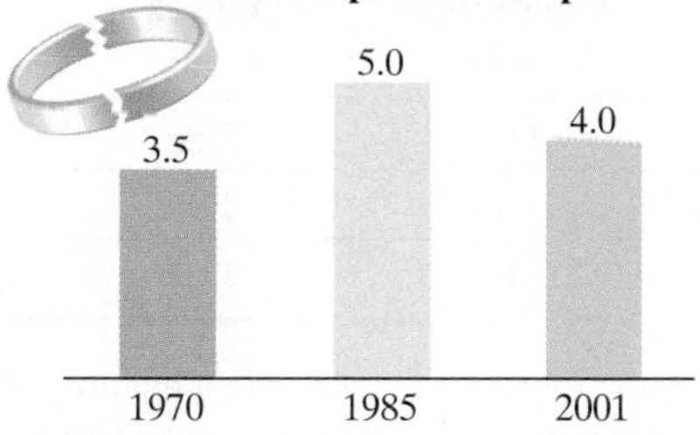

Source: U.S. Census Bureau

a. Write the data for 1970, 1985, and 2001 as ordered pairs (x, y), where x is the number of years after 1970 and y is that year's divorce rate.

b. The three data points in part (a) can be modeled by the equation $y = ax^2 + bx + c$. Substitute each ordered pair into this equation, one ordered pair at a time, and write a system of linear equations in three variables that can be used to find values for a, b, and c. It is not necessary to solve the system.

37. You throw a ball straight up from a rooftop. The ball misses the rooftop on its way down and eventually strikes the ground. A mathematical model can be used to describe the relationship for the ball's height above the ground, y, after x seconds. Consider the following data.

x, seconds after the ball is thrown	**y, ball's height, in feet, above the ground**
1	224
3	176
4	104

a. Find the equation $y = ax^2 + bx + c$ whose graph passes through the given points.

b. Use the equation in part (a) to find the value for y when $x = 5$. Describe what this means.

38. A mathematical model can be used to describe the relationship between the number of feet a car travels once the brakes are applied, y, and the number of seconds the car is in motion after the brakes are applied, x. A research firm collects the data shown at the top of the next column.

x, seconds in motion after brakes are applied	**y, feet car travels once the brakes are applied**
1	46
2	84
3	114

a. Find the equation $y = ax^2 + bx + c$ whose graph passes through the given points.

b. Use the equation in part (a) to find the value for y when $x = 6$. Describe what this means.

In Exercises 39–46, use the four-step strategy to solve each problem. Use x, y, and z to represent unknown quantities. Then translate from the verbal conditions of the problem to a system of three equations in three variables.

39. In current U.S. dollars, John D. Rockefeller's 1913 fortune of \$900 million would be worth about \$189 billion. The bar graph shows that Rockefeller is the wealthiest among the world's five richest people of all time. The combined estimated wealth, in current billions of U.S. dollars, of Andrew Carnegie, Cornelius Vanderbilt, and Bill Gates is \$256 billion. The difference between Carnegie's estimated wealth and Vanderbilt's is \$4 billion. The difference between Vanderbilt's estimated wealth and Gates's is \$36 billion. Find the estimated wealth, in current billions of U.S. dollars, of Carnegie, Vanderbilt, and Gates.

The Richest People of All Time Estimated Wealth, in Current Billions of U.S. Dollars

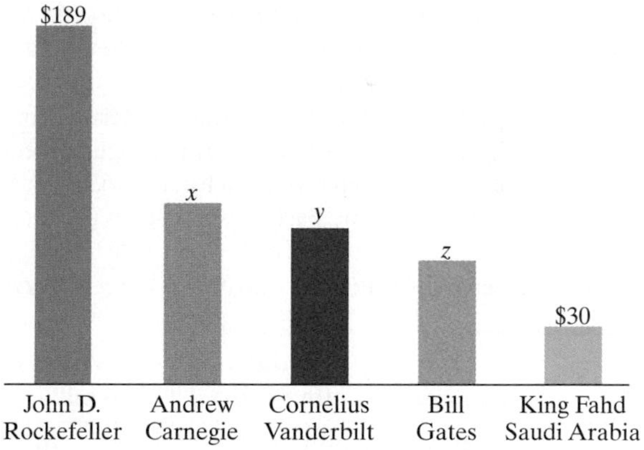

Source: Scholastic Book of World Records

40. The circle graph at the top of the next page indicates computers in use for the United States and the rest of the world. The percentage of the world's computers in Europe and Japan combined is 13% less than the percentage of the world's computers in the United States. If the percentage of the world's computers in Europe is doubled, it is only 3% more than the percentage of the world's computers in the United States. Find the percentage of the world's computers in the United States, Europe, and Japan.

Percentage of the World's Computers: U.S. and the World

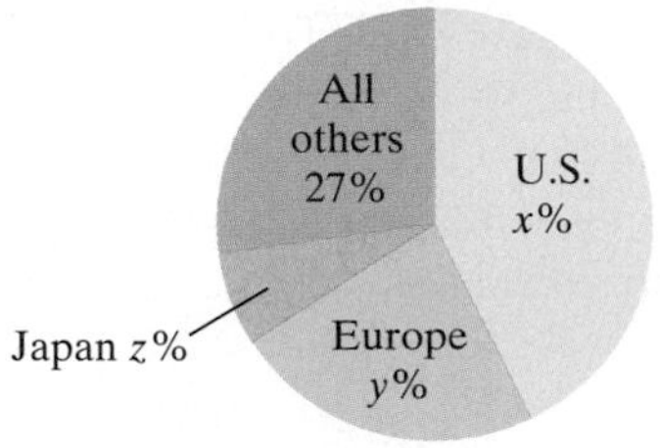

Source: Jupiter Communications

41. A person invested \$6700 for one year, part at 8%, part at 10%, and the remainder at 12%. The total annual income from these investments was \$716. The amount of money invested at 12% was \$300 more than the amount invested at 8% and 10% combined. Find the amount invested at each rate.

42. A person invested \$17,000 for one year, part at 10%, part at 12%, and the remainder at 15%. The total annual income from these investments was \$2110. The amount of money invested at 12% was \$1000 less than the amount invested at 10% and 15% combined. Find the amount invested at each rate.

43. At a college production of *Streetcar Named Desire*, 400 tickets were sold. The ticket prices were \$8, \$10, and \$12, and the total income from ticket sales was \$3700. How many tickets of each type were sold if the combined number of \$8 and \$10 tickets sold was 7 times the number of \$12 tickets sold?

44. A certain brand of razor blades comes in packages of 6, 12, and 24 blades, costing \$2, \$3, and \$4 per package, respectively. A store sold 12 packages containing a total of 162 razor blades and took in \$35. How many packages of each type were sold?

45. Three foods have the following nutritional content per ounce.

	Calories	Protein (in grams)	Vitamin C (in milligrams)
Food *A*	40	5	30
Food *B*	200	2	10
Food *C*	400	4	300

If a meal consisting of the three foods allows exactly 660 calories, 25 grams of protein, and 425 milligrams of vitamin C, how many ounces of each kind of food should be used?

46. A furniture company produces three types of desks: a children's model, an office model, and a deluxe model. Each desk is manufactured in three stages: cutting, construction, and finishing. The time requirements for each model and manufacturing stage are given in the following table.

	Children's model	Office model	Deluxe model
Cutting	2 hr	3 hr	2 hr
Construction	2 hr	1 hr	3 hr
Finishing	1 hr	1 hr	2 hr

Each week the company has available a maximum of 100 hours for cutting, 100 hours for construction, and 65 hours for finishing. If all available time must be used, how many of each type of desk should be produced each week?

Writing in Mathematics

47. What is a system of linear equations in three variables?

48. How do you determine whether a given ordered triple is a solution of a system in three variables?

49. Describe in general terms how to solve a system in three variables.

50. Describe what happens when using algebraic methods to solve an inconsistent system.

51. Describe what happens when using algebraic methods to solve a system with dependent equations.

52. AIDS is taking a deadly toll on southern Africa. Describe how to use the techniques that you learned in this section to obtain a model for African life span using projections with AIDS. Let x represent the number of years after 1985 and let y represent African life span in that year.

African Life Span

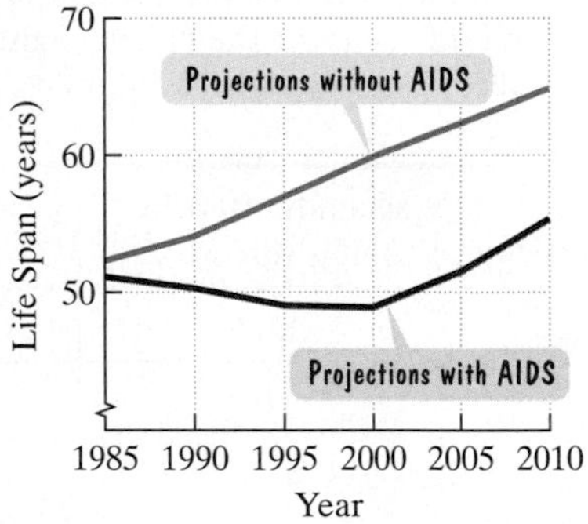

Source: United Nations

Technology Exercises

53. Does your graphing utility have a feature that allows you to solve linear systems by entering coefficients and constant terms? If so, use this feature to verify the solutions to any five exercises that you worked by hand from Exercises 5–16.

54. Verify your results in Exercises 23–26 by using a graphing utility to graph each equation that you obtained. Trace along the curve and convince yourself that the three points given in the exercise lie on the equation's graph.

Critical Thinking Exercises

55. Which one of the following is true?

a. The ordered triple (2, 15, 14) is the only solution of the equation $x + y - z = 3$.

b. The equation $x - y - z = -6$ is satisfied by $(2, -3, 5)$.

c. If two equations in a system are $x + y - z = 5$ and $2x + 2y - 2z = 7$, then the system must be inconsistent.

d. An equation with four variables, such as $x + 2y - 3z + 5w = 2$, cannot be satisfied by real numbers.

56. A modernistic painting consists of triangles, rectangles, and pentagons, all drawn so as to not overlap or share sides. Within each rectangle are drawn 2 red roses, and each pentagon contains 5 carnations. How many triangles, rectangles, and pentagons appear in the painting if the painting contains a total of 40 geometric figures, 153 sides of geometric figures, and 72 flowers?

57. In the following triangle, the degree measures of the three interior angles and two of the exterior angles are represented with variables. Find the measure of each interior angle.

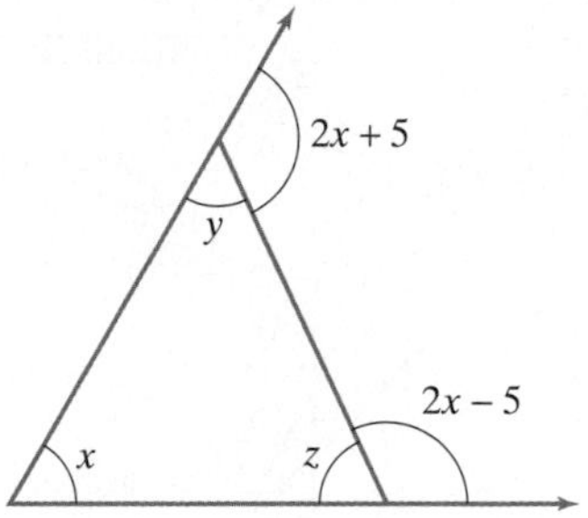

58. Two blocks of wood having the same length and width are placed on the top and bottom of a table, as shown in (a). Length A measures 32 centimeters. The blocks are rearranged as shown in (b). Length B measures 28 centimeters. Determine the height of the table.

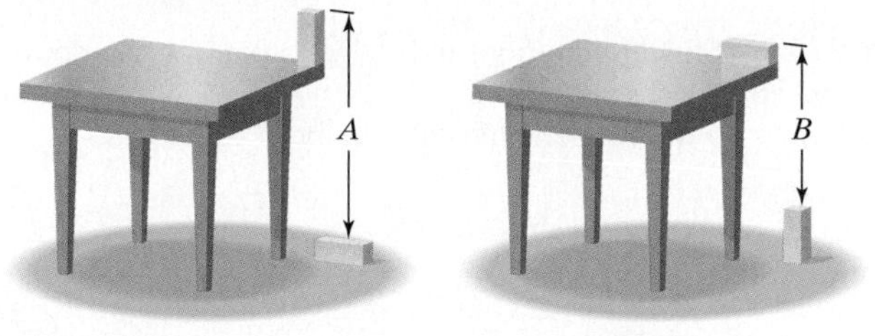

(a) **(b)**

Review Exercises

In Exercises 59–61, graph each equation.

59. $y = -\frac{3}{4}x + 3$ (Section 3.4, Example 3)

60. $-2x + y = 6$ (Section 3.2, Example 4)

61. $y = -5$ (Section 3.2, Example 7)

GROUP PROJECT

CHAPTER 4

Group members should go online and obtain a list of telecommunication companies that provide residential long-distance service. The list should contain the monthly fee, the monthly minimum, and the rate per minute for each service provider.

a. For each provider in the list, write an equation that describes the total monthly cost, y, of x minutes of long-distance phone calls.

b. Compare two of the plans. After how many minutes of long-distance calls will the costs for the two plans be the same? Solve a linear system to obtain your answer. What will be the cost for each plan?

c. Repeat part (b) for another two of the plans.

d. Each person should estimate the number of minutes he or she spends talking long distance each month. Group members should assist that person in selecting the plan that will save the most amount of money. Be sure to factor in the monthly minimum, if any, when choosing a plan. Whenever possible, use the equations that you wrote in part (a).

Caution: We've left something out! Your comparisons do not take into account the in-state rates of the plan. Furthermore, if you make international calls, international rates should also be one of your criteria in choosing a plan. Problem solving in real life can get fairly complicated.

CHAPTER 4 SUMMARY

Definitions and Concepts	Examples

Section 4.1 Solving Systems of Linear Equations by Graphing

A system of linear equations in two variables, x and y, consists of two equations of the form $Ax + By = C$. A solution is an ordered pair of numbers that satisfies both equations.

Determine whether $(3, -1)$ is a solution of

$$2x + 5y = 1$$
$$4x + y = 11.$$

Replace x with 3 and y with -1 in both equations.

$$2x + 5y = 1 \qquad 4x + y = 11$$
$$2 \cdot 3 + 5(-1) \stackrel{?}{=} 1 \qquad 4 \cdot 3 + (-1) \stackrel{?}{=} 11$$
$$6 + (-5) \stackrel{?}{=} 1 \qquad 12 + (-1) \stackrel{?}{=} 11$$
$$1 = 1, \text{ true} \qquad 11 = 11, \text{ true}$$

Thus, $(3, -1)$ is a solution of the system.

Using the graphing method, a solution of a linear system is a point common to the graphs of both equations in the system. If the graphs are parallel lines, the system has no solution and is called inconsistent. If the graphs are the same line, the system has infinitely many solutions. The equations are called dependent.

Solve by graphing: $2x + y = 4$

$$x + y = 2.$$

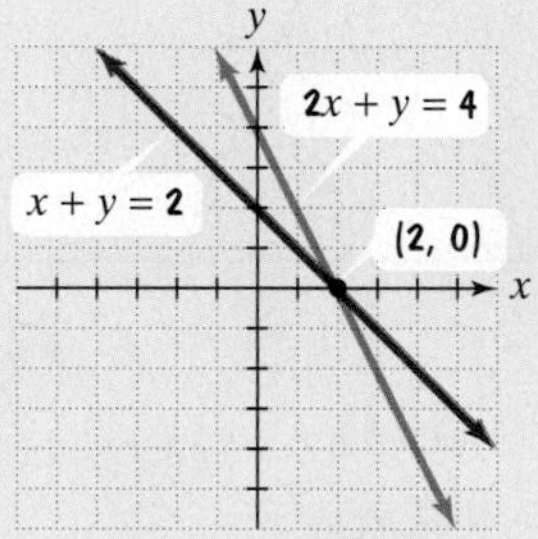

The solution is $(2, 0)$ and the solution set is $\{(2, 0)\}$.

Section 4.2 Solving Systems of Linear Equations by the Substitution Method

To solve a linear system by the substitution method,

1. Solve one equation for one variable in terms of the other.
2. Substitute the expression for that variable into the other equation. This will result in an equation in one variable.
3. Solve the equation in one variable.
4. Back-substitute the value of the variable found in step 3 in the equation from step 1. Simply and find the value of the remaining variable.
5. Check the proposed solution in both of the system's given equations.

If both variables are eliminated and a false statement results, the system has no solution. If both variables are eliminated and a true statement results, the system has infinitely many solutions.

Solve by the substitution method:

$$y = 2x + 3$$
$$7x - 5y = -18.$$

Substitute $2x + 3$ for y in the second equation.

$$7x - 5(2x + 3) = -18$$
$$7x - 10x - 15 = -18$$
$$-3x - 15 = -18$$
$$-3x = -3$$
$$x = 1$$

Find y. Substitute 1 for x in $y = 2x + 3$.

$$y = 2 \cdot 1 + 3 = 2 + 3 = 5$$

The solution, $(1, 5)$, checks. The solution set is $\{(1, 5)\}$.

Definitions and Concepts | Examples

Section 4.3 Solving Systems of Linear Equations by the Addition Method

To solve a linear system by the addition method,

1. Write equations in $Ax + By = C$ form.
2. Multiply one or both equations by nonzero numbers so that coefficients of a variable are opposites.
3. Add equations.
4. Solve the resulting equation for a variable.
5. Back-substitute the value of the variable into either original equation and find the value of the remaining variable.
6. Check the proposed solution in both the original equations.

If both variables are eliminated and a false statement results, the system has no solution. If both variables are eliminated and a true statement results, the system has infinitely many solutions.

Solve by the addition method:

$$\begin{aligned} 3x + y &= -11 \\ 6x - 2y &= -2. \end{aligned}$$

Eliminate y. Multiply both sides of the first equation by 2.

$$\begin{aligned} 6x + 2y &= -22 \\ \underline{6x - 2y} &= \underline{-2} \\ \text{Add: } 12x \quad &= -24 \\ x &= -2 \end{aligned}$$

Find y. Back-substitute -2 for x. We'll use the first equation.

$$\begin{aligned} 3(-2) + y &= -11 \\ -6 + y &= -11 \\ y &= -5 \end{aligned}$$

The solution, $(-2, -5)$, checks. The solution set is $\{(-2, -5)\}$.

Section 4.4 Problem Solving Using Systems of Equations

A Problem-Solving Strategy

1. Use variables, usually x and y, to represent unknown quantities.
2. Write a system of equations describing the problem's conditions.
3. Solve the system and answer the problem's question.
4. Check proposed answers in the problem's wording.

You invested \$14,000 in two stocks paying 7% and 9% interest. Total year-end interest was \$1180. How much was invested at each rate?

Let x = amount invested at 7% and
y = amount invested at 9%.

amount invested at 7% | amount invested at 9%

$$x + y = 14{,}000$$

interest from 7% investment | interest from 9% investment

$$0.07x + 0.09y = 1180$$

Solving by substitution or addition, $x = 4000$ and $y = 10{,}000$. Thus, \$4000 was invested at 7% and \$10,000 at 9%.

Section 4.5 Systems of Linear Equations in Three Variables

A system of linear equations in three variables, x, y, and z, consists of three equations of the form $Ax + By + Cz = D$. The solution set is the set consisting of the ordered triple that satisfies all three equations. The solution represents the point of intersection of three planes in space.

Is $(2, -1, 3)$ a solution of

$$\begin{aligned} 3x + 5y - 2z &= -5 \\ 2x + 3y - z &= -2 \\ 2x + 4y + 6z &= 18? \end{aligned}$$

Replace x with 2, y with -1, and z with 3. Using the first equation, we obtain:

$$\begin{aligned} 3 \cdot 2 + 5(-1) - 2(3) &\stackrel{?}{=} -5 \\ 6 - 5 - 6 &\stackrel{?}{=} -5 \\ -5 &= -5, \quad \text{true} \end{aligned}$$

The ordered triple $(2, -1, 3)$ satisfies the first equation. In a similar manner, it satisfies the other two equations and is a solution.

Definitions and Concepts	Examples

Section 4.5 Systems of Linear Equations in Three Variables (continued)

To solve a linear system in three variables by eliminating variables,

1. Reduce the system to two equations in two variables.
2. Solve the resulting system of two equations in two variables.
3. Use back-substitution in one of the equations in two variables to find the value of the second variable.
4. Back-substitute the values for two variables into one of the original equations to find the value of the third variable.
5. Check.

If all variables are eliminated and a false statement results, the system is inconsistent and has no solution. If a true statement results, the system contains dependent equations and has infinitely many solutions.

Solve:

$$2x + 3y - 2z = 0 \quad \boxed{1}$$
$$x + 2y - z = 1 \quad \boxed{2}$$
$$3x - y + z = -15. \quad \boxed{3}$$

Add equations $\boxed{2}$ and $\boxed{3}$ to eliminate z.

$$4x + y = -14 \quad \boxed{4}$$

Eliminate z again. Multiply equation $\boxed{3}$ by 2 and add to equation $\boxed{1}$.

$$8x + y = -30 \quad \boxed{5}$$

Multiply equation $\boxed{4}$ by -1 and add to equation $\boxed{5}$.

$$-4x - y = 14$$
$$\underline{8x + y = -30}$$
$$\text{Add: } 4x = -16$$
$$x = -4$$

Substitute -4 for x in $\boxed{4}$.

$$4(-4) + y = -14$$
$$y = 2$$

Substitute -4 for x and 2 for y in $\boxed{3}$.

$$3(-4) - 2 + z = -15$$
$$-14 + z = -15$$
$$z = -1$$

Checking verifies that $(-4, 2, -1)$ is the solution and $\{(-4, 2, -1)\}$ is the solution set.

Curve Fitting

Curve fitting is determining an equation whose graph contains given points. Three points that do not lie on a line determine the graph of an equation in the form

$$y = ax^2 + bx + c.$$

Use the three given points to create a system of three equations. Solve the system to find a, b, and c.

Find the equation $y = ax^2 + bx + c$ whose graph passes through the points $(-1, 2)$, $(1, 8)$, and $(2, 14)$.

Use $y = ax^2 + bx + c$.

When $x = -1, y = 2$: $\quad 2 = a(-1)^2 + b(-1) + c$

When $x = 1, y = 8$: $\quad 8 = a \cdot 1^2 + b \cdot 1 + c$

When $x = 2, y = 14$: $\quad 14 = a \cdot 2^2 + b \cdot 2 + c$

Solving,

$$a - b + c = 2$$
$$a + b + c = 8$$
$$4a + 2b + c = 14,$$

$a = 1, b = 3$, and $c = 4$. The equation, $y = ax^2 + bx + c$, is $y = x^2 + 3x + 4$.

CHAPTER 4 REVIEW EXERCISES

4.1 *In Exercises 1–2, determine whether the given ordered pair is a solution of the system.*

1. $(1, -5)$
$4x - y = 9$
$2x + 3y = -13$

2. $(-5, 2)$
$2x + 3y = -4$
$x - 4y = -10$

3. Does the graphing-utility screen show the solution for the following system? Explain.

$$x + y = 2$$
$$2x + y = -5$$

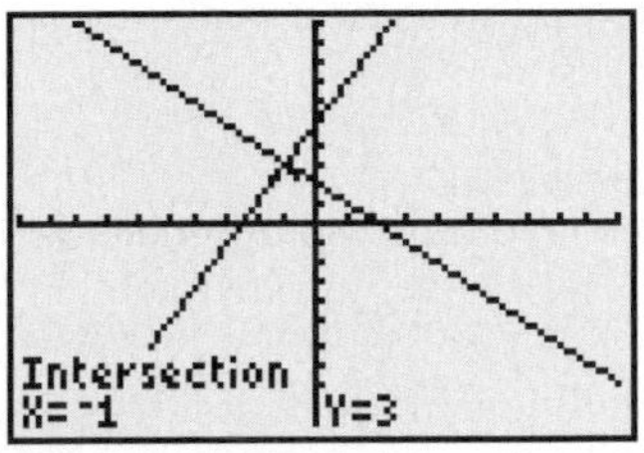

In Exercises 4–14, solve each system by graphing. If there is no solution or an infinite number of solutions, so state, and use set notation to express solution sets.

4. $x + y = 2$
$x - y = 6$

5. $2x - 3y = 12$
$-2x + y = -8$

6. $3x + 2y = 6$
$3x - 2y = 6$

7. $y = \frac{1}{2}x$
$y = 2x - 3$

8. $x + 2y = 2$
$y = x - 5$

9. $x + 2y = 8$
$3x + 6y = 12$

10. $2x - 4y = 8$
$x - 2y = 4$

11. $y = 3x - 1$
$y = 3x + 2$

12. $x - y = 4$
$x = -2$

13. $x = 2$
$y = 5$

14. $x = 2$
$x = 5$

4.2 *In Exercises 15–23, solve each system by the substitution method. If there is no solution or an infinite number of solutions, so state, and use set notation to express solution sets.*

15. $2x - 3y = 7$
$y = 3x - 7$

16. $2x - y = 6$
$x = 13 - 2y$

17. $2x - 5y = 1$
$3x + y = -7$

18. $3x + 4y = -13$
$5y - x = -21$

19. $y = 39 - 3x$
$y = 2x - 61$

20. $4x + y = 5$
$12x + 3y = 15$

21. $4x - 2y = 10$
$y = 2x + 3$

22. $x - 4 = 0$
$9x - 2y = 0$

23. $8y = 4x$
$7x + 2y = -8$

24. The weekly demand and supply models for the video *Titanic* at a chain of stores that sells videos are given by the demand model $N = -60p + 1000$ and the supply model $N = 4p + 200$, in which p is the price of the video and N is the number of videos sold or supplied each week to the chain of stores. Find the price at which supply and demand are equal. At this price, how many copies of *Titanic* can be supplied and sold each week?

4.3 *In Exercises 25–35, solve each system by the addition method. If there is no solution or an infinite number of solutions, so state, and use set notation to express solution sets.*

25. $x + y = 6$
$2x + y = 8$

26. $3x - 4y = 1$
$12x - y = -11$

27. $3x - 7y = 13$
$6x + 5y = 7$

28. $8x - 4y = 16$
$4x + 5y = 22$

29. $5x - 2y = 8$
$3x - 5y = 1$

30. $2x + 7y = 0$
$7x + 2y = 0$

31. $x + 3y = -4$
$3x + 2y = 3$

32. $2x + y = 5$
$2x + y = 7$

33. $3x - 4y = -1$
$-6x + 8y = 2$

34. $2x = 8y + 24$
$3x + 5y = 2$

35. $5x - 7y = 2$
$3x = 4y$

In Exercises 36–41, solve each system by the method of your choice. If there is no solution or an infinite number of solutions, so state, and use set notation to express solution sets.

36. $3x + 4y = -8$
$2x + 3y = -5$

37. $6x + 8y = 39$
$y = 2x - 2$

38. $x + 2y = 7$
$2x + y = 8$

39. $y = 2x - 3$
$y = -2x - 1$

40. $3x - 6y = 7$
$3x = 6y$

41. $y - 7 = 0$
$7x - 3y = 0$

4.4

42. The bar graph shows the five countries with the longest healthy life expectancy at birth. Combined, people in Japan and Switzerland can expect to spend 146.4 years in good health. The difference between healthy life expectancy between these two countries is 0.8 years. Find the healthy life expectancy at birth in Japan and Switzerland.

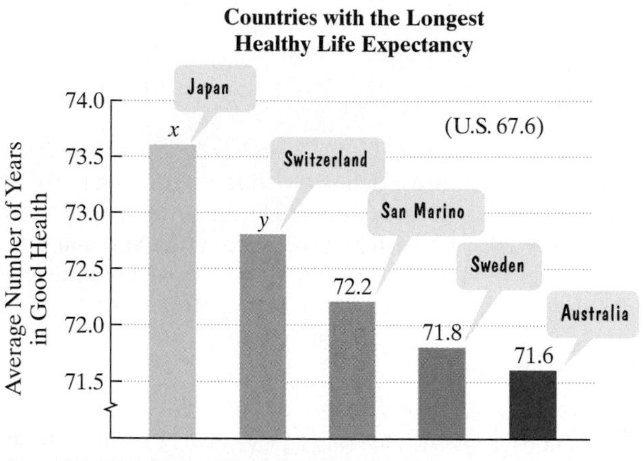

Source: World Health Organization

43. The gorilla and orangutan are the heaviest of the world's apes. Two gorillas and three orangutans weigh 1465 pounds. A gorilla's weight increased by twice an orangutan's weight is 815 pounds. Find the weight for each of these primates.

44. Health experts agree that cholesterol intake should be limited to 300 milligrams or less each day. Three ounces of shrimp and 2 ounces of scallops contain 156 milligrams of cholesterol. Five ounces of shrimp and 3 ounces of scallops contain 45 milligrams of cholesterol less than the suggested maximum daily intake. Determine the cholesterol content in an ounce of each item.

45. The perimeter of a table tennis top is 28 feet. The difference between 4 times the length and 3 times the width is 21 feet. Find the dimensions.

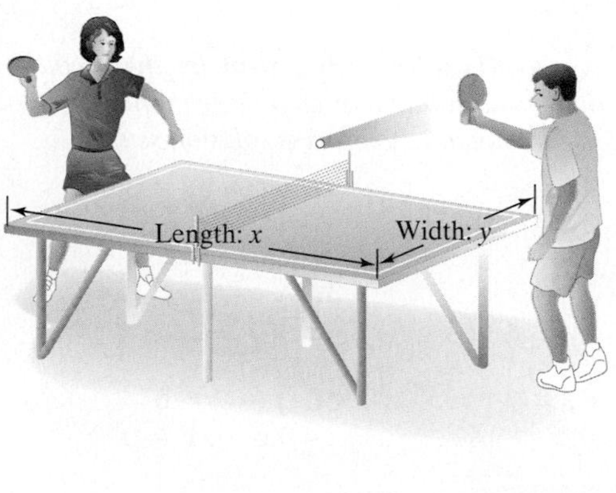

46. A travel agent offers two package vacation plans. The first plan costs \$360 and includes 3 days at a hotel and a rental car for 2 days. The second plan costs \$500 and includes 4 days at a hotel and a rental car for 3 days. The daily charge for the room is the same under each plan, as is the daily charge for the car. Find the cost per day for the room and for the car.

47. You are choosing between two long-distance telephone plans. One plan has a monthly fee of \$15 with a charge of \$0.05 per minute for all long-distance calls. The other plan has a monthly fee of \$10 with a charge of \$0.075 per minute for all long-distance calls. For how many minutes of long-distance calls will the costs for the two plans be the same? What will be the cost for each plan?

48. You invested \$9000 in two funds paying 4% and 7% annual interest, respectively. At the end of the year, the total interest from these investments was \$555. How much was invested at each rate?

49. A chemist needs to mix a solution that is 34% silver nitrate with one that is 4% silver nitrate to obtain 100 milliliters of a mixture that is 7% silver nitrate. How many milliliters of each of the solutions must be used?

50. When a plane flies with the wind, it can travel 2160 miles in 3 hours. When the plane flies in the opposite direction, against the wind, it takes 4 hours to fly the same distance. Find the rate of the plane in still air and the rate of the wind.

4.5

51. Is $(-3, -2, 5)$ a solution of the system

$$\begin{aligned} x + y + z &= 0 \\ 2x - 3y + z &= 5 \\ 4x + 2y + 4z &= 3? \end{aligned}$$

Solve each system in Exercises 52–54 by eliminating variables using the addition method. If there is no solution or if there are infinitely many solutions and a system's equations are dependent, so state.

52. $\begin{aligned} 2x - y + z &= 1 \\ 3x - 3y + 4z &= 5 \\ 4x - 2y + 3z &= 4 \end{aligned}$

53. $\begin{aligned} x + 2y - z &= 5 \\ 2x - y + 3z &= 0 \\ 2y + z &= 1 \end{aligned}$

54. $\begin{aligned} 3x - 4y + 4z &= 7 \\ x - y - 2z &= 2 \\ 2x - 3y + 6z &= 5 \end{aligned}$

55. Find the equation $y = ax^2 + bx + c$ whose graph passes through the points $(1, 4)$, $(3, 20)$, and $(-2, 25)$.

56. The graph shows the average yearly cost for veterinary care, per animal, in the United States. The combined yearly cost for a dog, a horse, and a cat is \$307. The difference between the cost for a dog and a horse is \$32, and the difference between the cost for a horse and a cat is \$16. Find the average yearly cost for veterinary care per dog, per horse, and per cat.

Average Yearly Cost for Veterinary Care, per Animal, in the U.S.

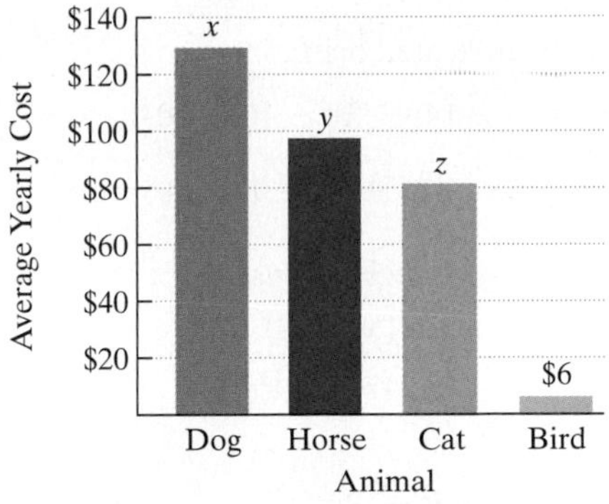

Source: American Veterinary Medical Association

CHAPTER 4 TEST

Remember to use your Chapter Test Prep Video CD to see the worked-out solutions to the test questions you want to review.

In Exercises 1–2, determine whether the given ordered pair is a solution of the system.

1. $(5, -5)$
$2x + y = 5$
$x + 3y = -10$

2. $(-3, 2)$
$x + 5y = 7$
$3x - 4y = 1$

In Exercises 3–4, solve each system by graphing. If there is no solution or an infinite number of solutions, so state, and use set notation to express solution sets.

3. $x + y = 6$
$4x - y = 4$

4. $2x + y = 8$
$y = 3x - 2$

In Exercises 5–7, solve each system by the substitution method. If there is no solution or an infinite number of solutions, so state, and use set notation to express solution sets.

5. $x = y + 4$
$3x + 7y = -18$

6. $2x - y = 7$
$3x + 2y = 0$

7. $2x - 4y = 3$
$x = 2y + 4$

In Exercises 8–10, solve each system by the addition method. If there is no solution or an infinite number of solutions, so state, and use set notation to express solution sets.

8. $2x + y = 2$
$4x - y = -8$

9. $2x + 3y = 1$
$3x + 2y = -6$

10. $3x - 2y = 2$
$-9x + 6y = -6$

11. According to the U.S. Census Bureau, the two most popular female first names in the United States are Mary and Patricia. Combined, these two names account for 3.7% of all female first names. The difference between the percentage of women named Mary and the percentage of women named Patricia is 1.5%. What percentage of all first names in the United States are Mary and what percentage are Patricia?

12. You are choosing between two long-distance telephone plans. One plan has a monthly fee of \$15 with a charge of \$0.05 per minute. The other plan has a monthly fee of \$5 with a charge of \$0.07 per minute. For how many minutes of long-distance calls will the costs for the two plans be the same? What will be the cost for each plan?

13. You invested \$9000 in two funds paying 6% and 7% annual interest, respectively. At the end of the year, the total interest from these investments was \$610. How much was invested at each rate?

14. You need to mix a 6% peroxide solution with a 9% peroxide solution to obtain 36 ounces of an 8% peroxide solution. How many ounces of each of the solutions must be used?

15. A paddleboat on the Mississippi River travels 48 miles downstream, with the current, in 3 hours. The return trip, against the current, takes the paddleboat 4 hours. Find the boat's rate in still water and the rate of the current.

16. Solve by eliminating variables using the addition method:

$$x + y + z = 6$$
$$3x + 4y - 7z = 1$$
$$2x - y + 3z = 5.$$

CUMULATIVE REVIEW EXERCISES (CHAPTERS 1–4)

1. Perform the indicated operations:

$$-14 - [18 - (6 - 10)].$$

2. Simplify: $6(3x - 2) - (x - 1)$.

In Exercises 3–4, solve each equation.

3. $17(x + 3) = 13 + 4(x - 10)$

4. $\frac{x}{4} - 1 = \frac{x}{5}$

5. Solve for t: $A = P + Prt$.

6. Solve and graph the solution set on a number line: $2x - 5 < 5x - 11$.

In Exercises 7–9, graph each equation in the rectangular coordinate system.

7. $x - 3y = 6$

8. $y = 4 - x^2$

9. $y = -\frac{3}{5}x + 2$

In Exercises 10–11, solve each linear system.

10. $3x - 4y = 8$
$4x + 5y = -10$

11. $2x - 3y = 9$
$y = 4x - 8$

12. Find the slope of the line passing through $(5, -6)$ and $(6, -5)$.

13. Write the point-slope form and the slope-intercept form of the equation of the line passing through $(-1, 6)$ with slope $= -4$.

14. The area of a triangle is 80 square feet. Find the height if the base is 16 feet.

15. If 10 pens and 15 pads cost \$26, and 5 of the same pens and 10 of the same pads cost \$16, find the cost of a pen and a pad.

16. List all the integers in this set:

$$\left\{-93, -\frac{7}{3}, 0, \sqrt{3}, \frac{7}{1}, \sqrt{100}\right\}.$$

The graphs show the percentage of U.S. households with one computer and multiple computers. Use the information provided by the graphs to solve Exercises 17–20.

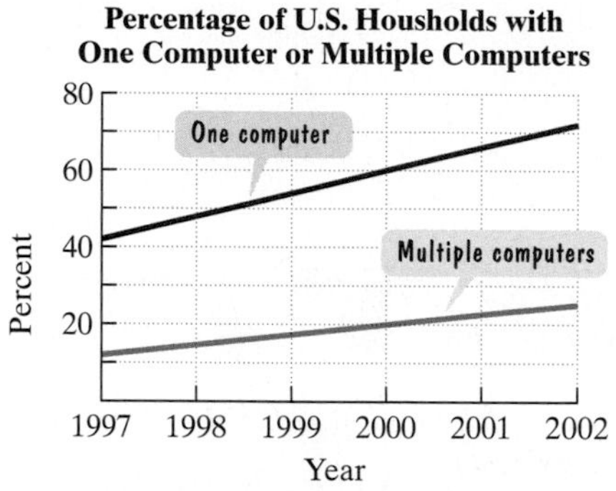

Source: Forrester Research. Inc.

17. What percentage of U.S. households had multiple computers in 2000?

18. Which graph has the greater slope? What does this mean in terms of the variables in this situation?

19. In 1997, 42% of U.S. households had one computer. This is increasing by approximately 6% per year. If this trend continues, in how many years after 1997 will 90% of U.S. households have one computer? In which year will that be?

20. The formula $y = \frac{8}{3}x + 12$ models the percentage of U.S. households, y, with multiple computers x years after 1997. Use the formula to find in which year 52% of U.S. households will have multiple computers.

One of the joys of your life is your dog, your very special buddy. Lately, however, you've noticed that your companion is slowing down a bit. He is now 8 years old and you wonder how this translates into human years. You remember something about every year of a dog's life being equal to seven years for a human. Is there a more accurate description?

This question is addressed in Exercises 103–106 in Exercise Set 5.1.

CHAPTER 5

Exponents and Polynomials

There is a formula that models the age in human years, y, of a dog that is x years old:

$$y = -0.001618x^4 + 0.077326x^3 - 1.2367x^2 + 11.460x + 2.914.$$

The algebraic expression on the right side of the formula contains variables to powers that are whole numbers and is an example of a polynomial. Much of what we do in algebra involves operations with polynomials. In this chapter, we study these operations, as well as the many applications of polynomials.

SECTION 5.1

Objectives

1 Understand the vocabulary used to describe polynomials.

2 Add polynomials.

3 Subtract polynomials.

4 Use mathematical models that contain polynomials.

ADDING AND SUBTRACTING POLYNOMIALS

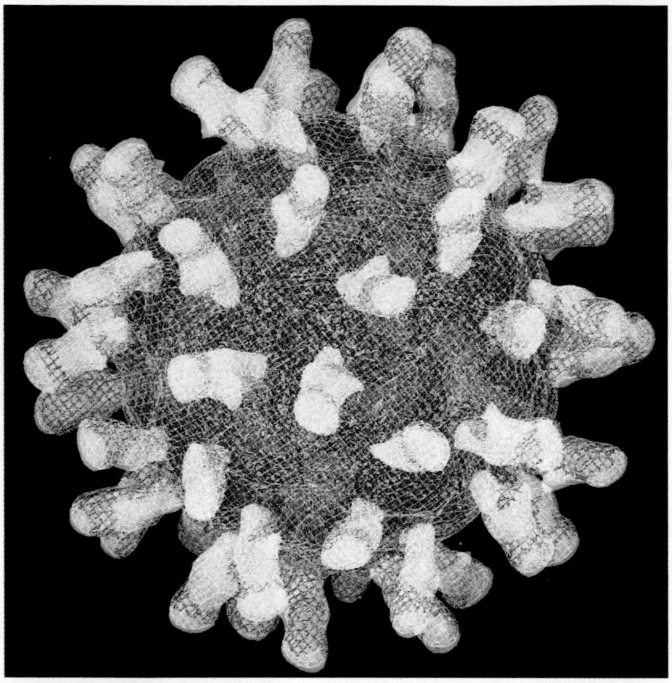

This computer-simulated model of the common cold virus was developed by researchers at Purdue University. Their discovery of how the virus infects human cells could lead to more effective treatment for the illness.

Runny nose? Sneezing? You are probably familiar with the unpleasant onset of a cold. We "catch cold" when the cold virus enters our bodies, where it multiplies. Fortunately, at a certain point the virus begins to die. The algebraic expression $-0.75x^4 + 3x^3 + 5$ describes the billions of viral particles in our bodies after x days of invasion. The expression enables mathematicians to determine the day on which there is a maximum number of viral particles and, consequently, the day we feel sickest.

The algebraic expression $-0.75x^4 + 3x^3 + 5$ is an example of a *polynomial*. A **polynomial** is a single term or the sum of two or more terms containing variables in the numerator with whole-number exponents. This particular polynomial contains three terms. Equations containing polynomials are used in such diverse areas as science, business, medicine, psychology, and sociology. In this section, we present basic ideas about polynomials. We then use our knowledge of combining like terms to find sums and differences of polynomials.

1 Understand the vocabulary used to describe polynomials.

Describing Polynomials Consider the polynomial

$$7x^3 - 9x^2 + 13x - 6.$$

We can express this polynomial as

$$7x^3 + (-9x^2) + 13x + (-6).$$

The polynomial contains four terms. It is customary to write the terms in the order of descending powers of the variables. This is the **standard form** of a polynomial.

We begin this chapter by limiting our discussion to polynomials containing only one variable. Each term of such a polynomial in x is of the form ax^n. The **degree** of ax^n is n. For example, the degree of the term $7x^3$ is 3.

STUDY TIP

We can express 0 in many ways, including $0x$, $0x^2$, and $0x^3$. It is impossible to assign a unique exponent to the variable. This is why 0 has no defined degree.

THE DEGREE OF ax^n If $a \neq 0$, the degree of ax^n is n. The degree of a nonzero constant term is 0. The constant 0 has no defined degree.

Here is an example of a polynomial and the degree of each of its four terms:

$$6x^4 - 3x^3 - 2x - 5.$$

degree 4 | degree 3 | degree 1 | degree of nonzero constant: 0

Notice that the exponent on x for the term $2x$, meaning $2x^1$, is understood to be 1. For this reason, the degree of $2x$ is 1.

A polynomial which when simplified has exactly one term is called a **monomial**. A **binomial** is a polynomial that has two terms, each with a different exponent. A **trinomial** is a polynomial with three terms, each with a different exponent. Polynomials with four or more terms have no special names.

The degree of a polynomial is the highest degree of all the terms of the polynomial. For example, $4x^2 + 3x$ is a binomial of degree 2 because the degree of the first term is 2, and the degree of the other term is less than 2. Also, $7x^5 - 2x^2 + 4$ is a trinomial of degree 5 because the degree of the first term is 5, and the degrees of the other terms are less than 5.

Up to now, we have used x to represent the variable in a polynomial. However, any letter can be used. For example,

- $7x^5 - 3x^3 + 8$ is a polynomial (in x) of degree 5. Because there are three terms, the polynomial is a trinomial.
- $6y^3 + 4y^2 - y + 3$ is a polynomial (in y) of degree 3. Because there are four terms, the polynomial has no special name.
- $z^7 + \sqrt{2}$ is a polynomial (in z) of degree 7. Because there are two terms, the polynomial is a binomial.

2 Add polynomials.

Adding Polynomials Recall that *like terms* are terms containing exactly the same variables to the same powers. Polynomials are added by combining like terms. For example, we can add the monomials $-9x^3$ and $13x^3$ as follows:

$$-9x^3 + 13x^3 = (-9 + 13)x^3 = 4x^3.$$

These like terms both contain x to the third power. | Add coefficients and keep the same variable factor, x^3.

EXAMPLE 1 Adding Polynomials

Add: $(-9x^3 + 7x^2 - 5x + 3) + (13x^3 + 2x^2 - 8x - 6)$.

SOLUTION The like terms are $-9x^3$ and $13x^3$, containing the same variable to the same power (x^3), as well as $7x^2$ and $2x^2$ (both contain x^2), $-5x$ and $-8x$ (both contain x), and the constant terms 3 and -6. We begin by grouping these pairs of like terms.

USING TECHNOLOGY

We can use a graphing utility to check the results of polynomial addition and subtraction. For example, try graphing

$$y_1 = (-9x^3 + 7x^2 - 5x + 3) + (13x^3 + 2x^2 - 8x - 6)$$

and

$$y_2 = 4x^3 + 9x^2 - 13x - 3$$

on the same screen, as shown below. (You may need to experiment with the viewing rectangle so that your graph does not appear to be cut off.) Because both graphs are identical, this shows that

$$y_1 = y_2$$

or that

$$(-9x^3 + 7x^2 - 5x + 3) + (13x^3 + 2x^2 - 8x - 6) = 4x^3 + 9x^2 - 13x - 3.$$

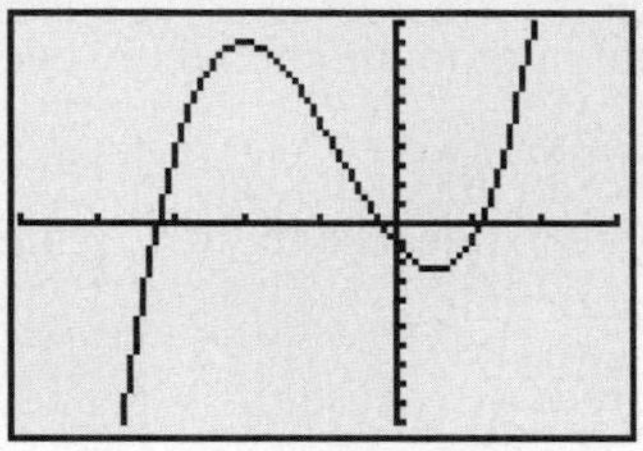

$[-5, 3, 1]$ by $[-30, 30, 3]$

$$(-9x^3 + 7x^2 - 5x + 3) + (13x^3 + 2x^2 - 8x - 6)$$

$$= (-9x^3 + 13x^3) + (7x^2 + 2x^2) + (-5x - 8x) + (3 - 6) \quad \text{Group like terms.}$$

$$= 4x^3 + 9x^2 + (-13x) + (-3) \quad \text{Combine like terms.}$$

$$= 4x^3 + 9x^2 - 13x - 3$$ ■

CHECK POINT 1 Add: $(-11x^3 + 7x^2 - 11x - 5) + (16x^3 - 3x^2 + 3x - 15)$.

Polynomials can be added by arranging like terms in columns. Then combine like terms, column by column.

EXAMPLE 2 Adding Polynomials Vertically

Add: $(-9x^3 + 7x^2 - 5x + 3) + (13x^3 + 2x^2 - 8x - 6)$.

SOLUTION

$$\begin{array}{rrrr} -9x^3 & 7x^2 & -5x & 3 \\ \underline{13x^3} & \underline{2x^2} & \underline{-8x} & \underline{-6} \\ 4x^3 & 9x^2 & -13x & -3 \end{array}$$

We consider each term separately and write like terms in columns.

Add, column by column.

Now add the four sums together:

$$4x^3 + 9x^2 + (-13x) + (-3) = 4x^3 + 9x^2 - 13x - 3.$$

This is the same answer that we found in Example 1. ■

CHECK POINT 2 Add the polynomials in Check Point 1 using a vertical format. Begin by arranging like terms in columns.

3 Subtract polynomials.

Subtracting Polynomials We subtract real numbers by adding the opposite, or additive inverse, of the number being subtracted. For example,

$$8 - 3 = 8 + (-3) = 5.$$

Subtraction of polynomials also involves opposites. If the sum of two polynomials is 0, the polynomials are **opposites**, or **additive inverses**, of each other. Here is an example:

$$(4x^2 - 6x - 7) + (-4x^2 + 6x + 7) = 0.$$

The opposite of $4x^2 - 6x - 7$ is $-4x^2 + 6x + 7$, and vice-versa.

Observe that the opposite of $4x^2 - 6x - 7$ can be obtained by changing the sign of each of its coefficients:

Polynomial		**Opposite**
$4x^2 - 6x - 7$	Change 4 to −4, change −6 to 6, and change −7 to 7.	$-4x^2 + 6x + 7$

In general, **the opposite of a polynomial is that polynomial with the sign of every coefficient changed**. Just as we did with real numbers, we subtract one polynomial from another by adding the opposite of the polynomial being subtracted.

SUBTRACTING POLYNOMIALS To subtract two polynomials, add the first polynomial and the opposite of the polynomial being subtracted.

EXAMPLE 3 Subtracting Polynomials

Subtract: $(7x^2 + 3x - 4) - (4x^2 - 6x - 7)$.

SOLUTION

$$(7x^2 + 3x - 4) - (4x^2 - 6x - 7)$$

Change the sign of each coefficient.

$= (7x^2 + 3x - 4) + (-4x^2 + 6x + 7)$ Add the opposite of the polynomial being subtracted.

$= (7x^2 - 4x^2) + (3x + 6x) + (-4 + 7)$ Group like terms.

$= 3x^2 + 9x + 3$ Combine like terms. ■

CHECK POINT 3 Subtract: $(9x^2 + 7x - 2) - (2x^2 - 4x - 6)$.

STUDY TIP

Be careful of the order in Example 4. For example, subtracting 2 from 5 means $5 - 2$. In general, subtracting B from A means $A - B$. The order of the resulting algebraic expression is not the same as the order in English.

EXAMPLE 4 Subtracting Polynomials

Subtract $2x^3 - 6x^2 - 3x + 9$ from $7x^3 - 8x^2 + 9x - 6$.

SOLUTION

$$(7x^3 - 8x^2 + 9x - 6) - (2x^3 - 6x^2 - 3x + 9)$$

Change the sign of each coefficient.

$= (7x^3 - 8x^2 + 9x - 6) + (-2x^3 + 6x^2 + 3x - 9)$ Add the opposite of the polynomial being subtracted.

$= (7x^3 - 2x^3) + (-8x^2 + 6x^2)$

$+ (9x + 3x) + (-6 - 9)$ Group like terms.

$= 5x^3 + (-2x^2) + 12x + (-15)$ Combine like terms.

$= 5x^3 - 2x^2 + 12x - 15$ ■

CHECK POINT 4 Subtract $3x^3 - 8x^2 - 5x + 6$ from $10x^3 - 5x^2 + 7x - 2$.

Subtraction can also be performed in columns.

EXAMPLE 5 Subtracting Polynomials Vertically

Use the method of subtracting by columns to find

$$(12y^3 - 9y^2 - 11y - 3) - (4y^3 - 5y + 8).$$

SOLUTION Arrange like terms in columns.

$$\begin{array}{rrrr} 12y^3 & -\ 9y^2 & -\ 11y & -\ 3 \\ -(4y^3 & & -\ 5y & +\ 8) \\ \hline \end{array}$$

Leave space for the missing term.

Add the opposite of the polynomial being subtracted.

$$\begin{array}{rrrr} 12y^3 & -\ 9y^2 & -\ 11y & -\ 3 \\ +\ -4y^3 & & +\ 5y & -\ 8 \\ \hline 8y^3 & -\ 9y^2 & -\ 6y & -\ 11 \end{array}$$

Change the sign of each coefficient of $4y^3 - 5y + 8$.

Combine like terms.

CHECK POINT 5 Use the method of subtracting by columns to find

$$(8y^3 - 10y^2 - 14y - 2) - (5y^3 - 3y + 6).$$

4 Use mathematical models that contain polynomials.

Applications Polynomials often appear in formulas that model real-world situations.

EXAMPLE 6 An Application: Death Rate

The formula

$$y = 0.036x^2 - 2.8x + 58.14$$

models the number of deaths per year per thousand people, y, for people who are x years old, $40 \le x \le 60$. Approximately how many people per thousand who are 50 years old die each year?

SOLUTION Because we are interested in people who are 50 years old, substitute 50 for x in the formula's polynomial.

$y = 0.036x^2 - 2.8x + 58.14$ This is the given formula.

$y = 0.036(50)^2 - 2.8(50) + 58.14$ Substitute 50 for x.

$y = 0.036(2500) - 2.8(50) + 58.14$ Evaluate the exponential expression: $50^2 = 50 \cdot 50 = 2500$.

$y = 90 - 140 + 58.14$ Perform the multiplications.

$y = 8.14 \approx 8$ Simplify.

Approximately 8 people per thousand who are 50 years old die each year.

We can use point plotting or a graphing utility to graph formulas that contain polynomials. These graphs contain only rounded curves with no sharp corners. For example, the graph of $y = 0.036x^2 - 2.8x + 58.14$, the formula from Example 6 that models the number of deaths per thousand, is shown in Figure 5.1. Our work in Example 6 can be visualized as a point on the curve.

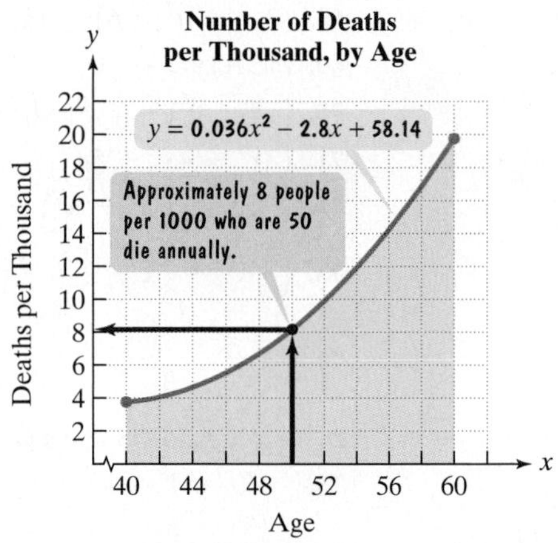

FIGURE 5.1 The graph of a formula containing a polynomial

> **CHECK POINT 6** Use the formula $y = 0.036x^2 - 2.8x + 58.14$ to answer this question: Approximately how many people per thousand who are 40 years old die annually? Identify your solution as a point on the curve in Figure 5.1.

5.1 EXERCISE SET

Student Solutions Manual | CD/Video | PH Math/Tutor Center | MathXL Tutorials on CD | MathXL® | MyMathLab | Interactmath.com

Practice Exercises

In Exercises 1–16, identify each polynomial as a monomial, a binomial, or a trinomial. Give the degree of the polynomial.

1. $3x + 7$

2. $5x - 2$

3. $x^3 - 2x$

4. $x^5 - 7x$

5. $8x^2$

6. $10x^2$

7. 5

8. 9

9. $x^2 - 3x + 4$

10. $x^2 - 9x + 2$

11. $7y^2 - 9y^4 + 5$

12. $3y^2 - 14y^5 + 6$

13. $15x - 7x^3$

14. $9x - 5x^3$

15. $-9y^{23}$

16. $-11y^{26}$

In Exercises 17–38, add the polynomials.

17. $(9x + 8) + (-17x + 5)$

18. $(8x - 5) + (-13x + 9)$

19. $(4x^2 + 6x - 7) + (8x^2 + 9x - 2)$

20. $(11x^2 + 7x - 4) + (27x^2 + 10x - 20)$

21. $(7x^2 - 11x) + (3x^2 - x)$

22. $(-3x^2 + x) + (4x^2 + 8x)$

23. $(4x^2 - 6x + 12) + (x^2 + 3x + 1)$

24. $(-7x^2 + 8x + 3) + (2x^2 + x + 8)$

25. $(4y^3 + 7y - 5) + (10y^2 - 6y + 3)$

26. $(2y^3 + 3y + 10) + (3y^2 + 5y - 22)$

27. $(2x^2 - 6x + 7) + (3x^3 - 3x)$

28. $(4x^3 + 5x + 13) + (-4x^2 + 22)$

29. $(4y^2 + 8y + 11) + (-2y^3 + 5y + 2)$

30. $(7y^3 + 5y - 1) + (2y^2 - 6y + 3)$

31. $(-2y^6 + 3y^4 - y^2) + (-y^6 + 5y^4 + 2y^2)$

32. $(7r^4 + 5r^2 + 2r) + (-18r^4 - 5r^2 - r)$

33. $\left(9x^3 - x^2 - x - \frac{1}{3}\right) + \left(x^3 + x^2 + x + \frac{4}{3}\right)$

34. $\left(12x^3 - x^2 - x + \frac{4}{3}\right) + \left(x^3 + x^2 + x - \frac{1}{3}\right)$

35. $\left(\frac{1}{5}x^4 + \frac{1}{3}x^3 + \frac{3}{8}x^2 + 6\right) + \left(-\frac{3}{5}x^4 + \frac{2}{3}x^3 - \frac{1}{2}x^2 - 6\right)$

36. $\left(\frac{2}{5}x^4 + \frac{2}{3}x^3 + \frac{5}{8}x^2 + 7\right) + \left(-\frac{4}{5}x^4 + \frac{1}{3}x^3 - \frac{1}{4}x^2 - 7\right)$

37. $(0.03x^5 - 0.1x^3 + x + 0.03) + (-0.02x^5 + x^4 - 0.7x + 0.3)$

38. $(0.06x^5 - 0.2x^3 + x + 0.05) + (-0.04x^5 + 2x^4 - 0.8x + 0.5)$

In Exercises 39–54, use a vertical format to add the polynomials.

39.
$$\begin{array}{r} 5y^3 - 7y^2 \\ \underline{6y^3 + 4y^2} \end{array}$$

40.
$$\begin{array}{r} 13x^4 - x^2 \\ \underline{7x^4 + 2x^2} \end{array}$$

41. $\begin{array}{r} 3x^2 - 7x + 4 \\ \underline{-5x^2 + 6x - 3} \end{array}$

42. $\begin{array}{r} 7x^2 - 5x - 6 \\ \underline{-9x^2 + 4x + 6} \end{array}$

43. $\begin{array}{r} \frac{1}{4}x^4 - \frac{2}{3}x^3 - 5 \\ \underline{-\frac{1}{2}x^4 + \frac{1}{5}x^3 + 4.7} \end{array}$

44. $\begin{array}{r} \frac{1}{3}x^9 - \frac{1}{5}x^5 - 2.7 \\ \underline{-\frac{3}{4}x^9 + \frac{2}{3}x^5 + 1} \end{array}$

45. $\begin{array}{r} y^3 + 5y^2 - 7y - 3 \\ \underline{-2y^3 + 3y^2 + 4y - 11} \end{array}$

46. $\begin{array}{r} y^3 + y^2 - 7y + 9 \\ \underline{-y^3 - 6y^2 - 8y + 11} \end{array}$

47. $\begin{array}{r} 4x^3 - 6x^2 + 5x - 7 \\ \underline{-9x^3 \qquad\quad - 4x + 3} \end{array}$

48. $\begin{array}{r} -4y^3 + 6y^2 - 8y + 11 \\ \underline{2y^3 \qquad\quad + 9y - 3} \end{array}$

49. $\begin{array}{l} 7x^4 - 3x^3 + x^2 \\ \underline{\qquad\qquad x^3 - x^2 + 4x - 2} \end{array}$

50. $\begin{array}{l} 7y^5 - 3y^3 + y^2 \\ \underline{\qquad\qquad 2y^3 - y^2 - 4y - 3} \end{array}$

51. $\begin{array}{r} 7x^2 - 9x + 3 \\ 4x^2 + 11x - 2 \\ \underline{-3x^2 + 5x - 6} \end{array}$

52. $\begin{array}{r} 7y^2 - 11y - 6 \\ 8y^2 + 3y + 4 \\ \underline{-9y^2 - 5y + 2} \end{array}$

53. $\begin{array}{r} 1.2x^3 - 3x^2 + 9.1 \\ 7.8x^3 - 3.1x^2 + 8 \\ \underline{1.2x^2 - 6} \end{array}$

54. $\begin{array}{r} 7.9x^3 - 6.8x^2 + 3.3 \\ 6.1x^3 - 2.2x^2 + 7 \\ \underline{4.3x^2 - 5} \end{array}$

In Exercises 55–74, subtract the polynomials.

55. $(x - 8) - (3x + 2)$
56. $(x - 2) - (7x + 9)$
57. $(x^2 - 5x - 3) - (6x^2 + 4x + 9)$
58. $(3x^2 - 8x - 2) - (11x^2 + 5x + 4)$
59. $(x^2 - 5x) - (6x^2 - 4x)$
60. $(3x^2 - 2x) - (5x^2 - 6x)$
61. $(x^2 - 8x - 9) - (5x^2 - 4x - 3)$
62. $(x^2 - 5x + 3) - (x^2 - 6x - 8)$
63. $(y - 8) - (3y - 2)$
64. $(y - 2) - (7y - 9)$
65. $(6y^3 + 2y^2 - y - 11) - (y^2 - 8y + 9)$
66. $(5y^3 + y^2 - 3y - 8) - (y^2 - 8y + 11)$
67. $(7n^3 - n^7 - 8) - (6n^3 - n^7 - 10)$
68. $(2n^2 - n^7 - 6) - (2n^3 - n^7 - 8)$
69. $(y^6 - y^3) - (y^2 - y)$
70. $(y^5 - y^3) - (y^4 - y^2)$
71. $(7x^4 + 4x^2 + 5x) - (-19x^4 - 5x^2 - x)$
72. $(-3x^6 + 3x^4 - x^2) - (-x^6 + 2x^4 + 2x^2)$
73. $\left(\frac{3}{7}x^3 - \frac{1}{5}x - \frac{1}{3}\right) - \left(-\frac{2}{7}x^3 + \frac{1}{4}x - \frac{1}{3}\right)$
74. $\left(\frac{3}{8}x^2 - \frac{1}{3}x - \frac{1}{4}\right) - \left(-\frac{1}{8}x^2 + \frac{1}{2}x - \frac{1}{4}\right)$

In Exercises 75–88, use a vertical format to subtract the polynomials.

75. $\begin{array}{r} 7x + 1 \\ \underline{-(3x - 5)} \end{array}$

76. $\begin{array}{r} 4x + 2 \\ \underline{-(3x - 5)} \end{array}$

77. $\begin{array}{r} 7x^2 - 3 \\ \underline{-(-3x^2 + 4)} \end{array}$

78. $\begin{array}{r} 9y^2 - 6 \\ \underline{-(-5y^2 + 2)} \end{array}$

79. $\begin{array}{r} 7y^2 - 5y + 2 \\ \underline{-(11y^2 + 2y - 3)} \end{array}$

80. $\begin{array}{r} 3x^5 - 5x^3 + 6 \\ \underline{-(7x^5 + 4x^3 - 2)} \end{array}$

81. $\begin{array}{r} 7x^3 + 5x^2 - 3 \\ \underline{-(-2x^3 - 6x^2 + 5)} \end{array}$

82. $\begin{array}{r} 3y^4 - 4y^2 + 7 \\ \underline{-(-5y^4 - 6y^2 - 13)} \end{array}$

83. $\begin{array}{r} 5y^3 + 6y^2 - 3y + 10 \\ \underline{-(6y^3 - 2y^2 - 4y - 4)} \end{array}$

84. $\begin{array}{r} 4y^3 + 5y^2 + 7y + 11 \\ \underline{-(-5y^3 + 6y^2 - 9y - 3)} \end{array}$

85. $\begin{array}{l} 7x^4 - 3x^3 + 2x^2 \\ \underline{-(\qquad - x^3 - x^2 + x - 2)} \end{array}$

86. $\begin{array}{l} 5y^6 - 3y^3 + 2y^2 \\ \underline{-(\qquad - y^3 - y^2 - y - 1)} \end{array}$

87. $\begin{array}{l} 0.07x^3 - 0.01x^2 + 0.02x \\ \underline{-(0.02x^3 - 0.03x^2 - \quad x)} \end{array}$

88. $\begin{array}{l} 0.04x^3 - 0.03x^2 + 0.05x \\ \underline{-(0.02x^3 - 0.06x^2 - \quad x)} \end{array}$

Practice Plus

In Exercises 89–92, perform the indicated operations.

89. $[(4x^2 + 7x - 5) - (2x^2 - 10x + 3)] - (x^2 + 5x - 8)$
90. $[(10x^3 - 5x^2 + 4x + 3) - (-3x^3 - 4x^2 + x)] - (7x^3 - 5x + 4)$
91. $[(4y^2 - 3y + 8) - (5y^2 + 7y - 4)] - [(8y^2 + 5y - 7) + (-10y^2 + 4y + 3)]$
92. $[(7y^2 - 4y + 2) - (12y^2 + 3y - 5)] - [(5y^2 - 2y - 8) + (-7y^2 + 10y - 13)]$
93. Subtract $x^3 - 2x^2 + 2$ from the sum of $4x^3 + x^2$ and $-x^3 + 7x - 3$.

94. Subtract $-3x^3 - 7x + 5$ from the sum of $2x^2 + 4x - 7$ and $-5x^3 - 2x - 3$.

95. Subtract $-y^2 + 7y^3$ from the difference between $-5 + y^2 + 4y^3$ and $-8 - y + 7y^3$. Express the answer in standard form.

96. Subtract $-2y^2 + 8y^3$ from the difference between $-6 + y^2 + 5y^3$ and $-12 - y + 13y^3$. Express the answer in standard form.

Application Exercises

97. The common cold is caused by a rhinovirus. The polynomial

$$-0.75x^4 + 3x^3 + 5$$

describes the billions of viral particles in our bodies after x days of invasion. Find the number of viral particles, in billions, after 0 days (the time of the cold's onset when we are still feeling well), 1 day, 2 days, 3 days, and 4 days. After how many days is the number of viral particles at a maximum and, consequently, the day we feel the sickest? By when should we feel completely better?

98. The polynomial $-0.02A^2 + 2A + 22$ is used by coaches to get athletes fired up so that they can perform well. The polynomial represents the performance level related to various levels of enthusiasm, from $A = 1$ (almost no enthusiasm) to $A = 100$ (maximum level of enthusiasm). Evaluate the polynomial when $A = 20$, $A = 50$, and $A = 80$. Describe what happens to performance as we get more and more fired up.

The graph shows cigarette consumption per U.S. adult from 1900 through 2001. The data from 1940 through 2001 can be modeled by the formula

$$y = -2.3x^2 + 135.3x + 2191,$$

where x represents years after 1940 and y represents cigarette consumption per U.S. adult. Use the formula to solve Exercises 99–100.

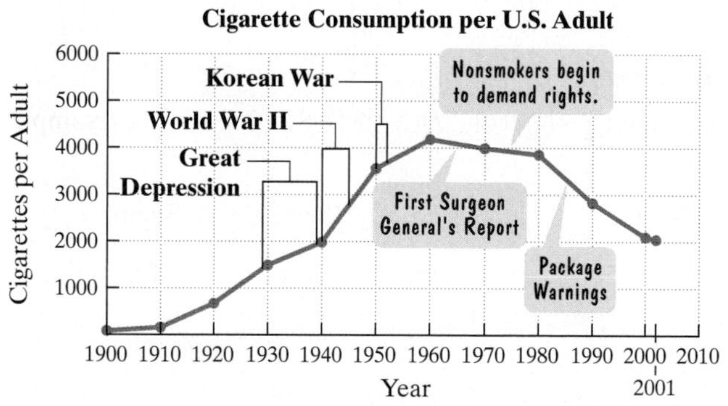

Source: U.S. Department of Health and Human Services

99. What was cigarette consumption per adult in 2000? How well does the formula model the actual data shown by the graph?

100. What was cigarette consumption per adult in 1980? How well does the formula model the actual data shown by the graph?

The wage gap is used to compare the status of women's earnings relative to men's. The wage gap is expressed as a percent and is calculated by dividing the median, or middlemost, annual earnings for women by the median annual earnings for men. The line graph shows the wage gap from 1960 through 2002. The data can be modeled by the formula

$$y = 0.012x^2 - 0.16x + 60,$$

where x represents years after 1960 and y represents median women's earnings as a percentage of median men's earnings. Use this information to solve Exercises 101–102.

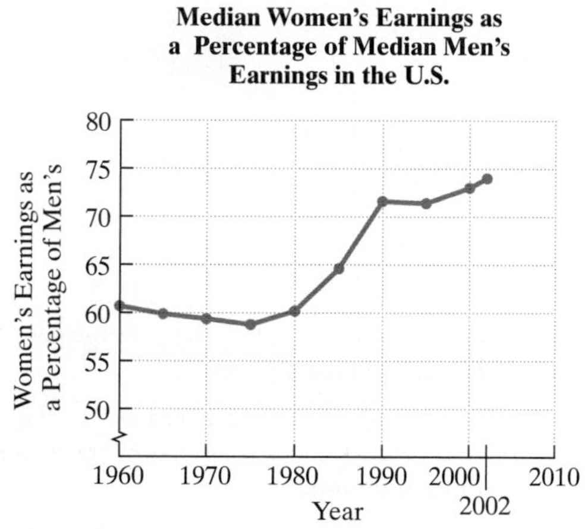

Source: U.S. Women's Bureau

101. a. Use the graph to estimate, to the nearest percent, women's earnings as a percentage of men's in 2000.

b. Use the mathematical model to find women's earnings as a percentage of men's in 2000.

c. In 2000, median annual earnings for U.S. women and men were \$27,355 and \$37,339, respectively. What were women's earnings as a percentage of men's? Use a calculator and round to the nearest tenth of a percent. How well do your answers in parts (a) and (b) model the actual data?

102. a. Use the graph to estimate, to the nearest percent, women's earnings as a percentage of men's in 1970.

b. Use the mathematical model to find women's earnings as a percentage of men's in 1970.

c. In 1970, median annual earnings for U.S. women and men were \$5440 and \$9184, respectively. What were women's earnings as a percentage of men's? Use a calculator and round to the nearest tenth of a percent. How well do your answers in parts (a) and (b) model the actual data?

The formula

$$y = -0.001618x^4 + 0.077326x^3 - 1.2367x^2 + 11.460x + 2.914$$

models the age in human years, y, of a dog that is x years old, where $x > 1$. The coefficients make it difficult to use this formula when computing by hand. However, a graph of the formula makes approximations possible. Use this information to solve Exercises 103–106.

Dog's Age in Human Years

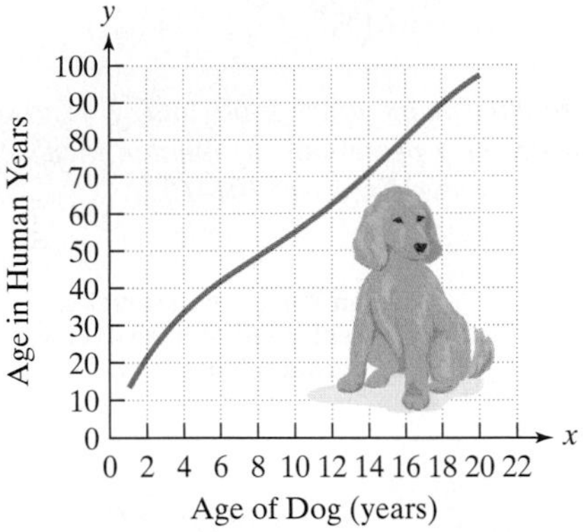

Source: U.C. Davis

103. a. If your dog is 6 years old, use the graph to estimate the equivalent age in human years.

b. Use the given mathematical model to verify your estimate in part (a). Use a calculator and round to the nearest tenth of a human year.

104. a. If your dog is 16 years old, use the graph to estimate the equivalent age in human years.

b. Use the given mathematical model to verify your estimate in part (a). Use a calculator and round to the nearest tenth of a human year.

105. If you are 25, use the graph to find the equivalent age for dogs.

106. If you are 45, use the graph to find the equivalent age for dogs.

Writing in Mathematics

107. What is a polynomial?

108. What is a monomial? Give an example with your explanation.

109. What is a binomial? Give an example with your explanation.

110. What is a trinomial? Give an example with your explanation.

111. What is the degree of a polynomial? Provide an example with your explanation.

112. Explain how to add polynomials.

113. Explain how to subtract polynomials.

114. A friend who is blind is having difficulty visualizing the relationship between age and deaths per thousand. Describe this relationship for your friend as age increases from 40 through 60. Use Figure 5.1 on page 311.

115. For Exercise 98, explain why performance levels do what they do as we get more and more fired up. If possible, describe an example of a time when you were too enthused and thus did poorly at something when you were hoping to do well.

Critical Thinking Exercises

116. Which one of the following is true?

a. In the polynomial $3x^2 - 5x + 13$, the coefficient of x is 5.

b. The degree of $3x^2 - 7x + 9x^3 + 5$ is 2.

c. $\frac{1}{5x^2} + \frac{1}{3x}$ is a binomial.

d. $(2x^2 - 8x + 6) - (x^2 - 3x + 5) = x^2 - 5x + 1$ for any value of x.

117. What polynomial must be subtracted from $5x^2 - 2x + 1$ so that the difference is $8x^2 - x + 3$?

118. The number of people who catch a cold t weeks after January 1 is $5t - 3t^2 + t^3$. The number of people who recover t weeks after January 1 is $t - t^2 + \frac{1}{3}t^3$. Write a polynomial in standard form for the number of people who are still ill with a cold t weeks after January 1.

119. Explain why it is not possible to add two polynomials of degree 3 and get a polynomial of degree 4.

Review Exercises

120. Simplify: $(-10)(-7) \div (1 - 8)$. (Section 1.8, Example 8)

121. Subtract: $-4.6 - (-10.2)$. (Section 1.6, Example 2)

122. Solve: $3(x - 2) = 9(x + 2)$. (Section 2.3, Example 3)

SECTION 5.2

Objectives

1. Use the product rule for exponents.
2. Use the power rule for exponents.
3. Use the products-to-powers rule.
4. Multiply monomials.
5. Multiply a monomial and a polynomial.
6. Multiply polynomials when neither is a monomial.

MULTIPLYING POLYNOMIALS

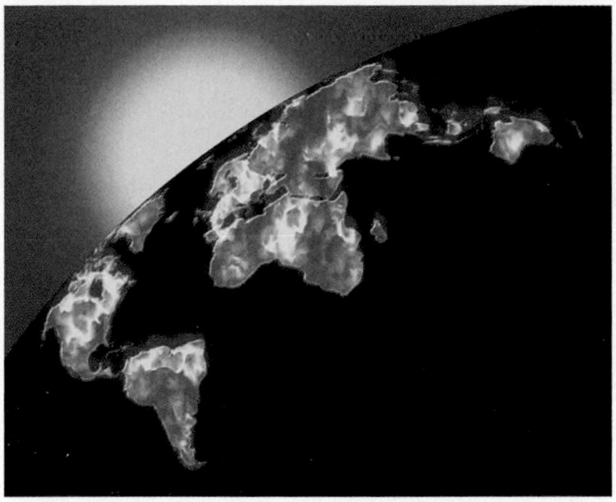

Recent advances in our understanding of climate have changed global warming from a subject for a disaster movie (the Statue of Liberty up to its chin in water) to a serious scientific and policy issue. Global warming appears to be related to the burning of fossil fuels, which adds carbon dioxide to the atmosphere. In the next few decades, we will see whether our use of fossil fuels will add enough carbon dioxide to the atmosphere to change it (and our climate) in significant ways. In this section's essay, you will see how a polynomial models trends in global warming through 2040. In the section itself, you will learn to multiply these algebraic expressions that play a significant role in modeling your world.

Before studying how polynomials are multiplied, we must develop some rules for working with exponents.

1 Use the product rule for exponents.

The Product Rule for Exponents We have seen that exponents are used to indicate repeated multiplication. For example, 2^4, where 2 is the base and 4 is the exponent, indicates that 2 occurs as a factor four times:

$$2^4 = 2 \cdot 2 \cdot 2 \cdot 2.$$

Now consider the multiplication of two exponential expressions, such as $2^4 \cdot 2^3$. We are multiplying 4 factors of 2 and 3 factors of 2. We have a total of 7 factors of 2:

4 factors of 2 | 3 factors of 2

$$2^4 \cdot 2^3 = (2 \cdot 2 \cdot 2 \cdot 2) \cdot (2 \cdot 2 \cdot 2)$$

Total: 7 factors of 2

Thus,

$$2^4 \cdot 2^3 = 2^7.$$

Caution: $2^4 \cdot 2^3$ is not equal to $2^{4 \cdot 3}$, or 2^{12}, as might be expected.

We can quickly find the exponent, 7, of the product by adding 4 and 3, the original exponents:

$$2^4 \cdot 2^3 = 2^{4+3} = 2^7.$$

This suggests the following rule:

THE PRODUCT RULE

$$b^m \cdot b^n = b^{m+n}$$

When multiplying exponential expressions with the same base, add the exponents. Use this sum as the exponent of the common base.

EXAMPLE 1 Using the Product Rule

Multiply each expression using the product rule:

a. $2^2 \cdot 2^3$ **b.** $x^7 \cdot x^9$ **c.** $y \cdot y^5$ **d.** $y^3 \cdot y^2 \cdot y^5$.

STUDY TIP

The product rule does not apply to exponential expressions with different bases:

- $x^7 \cdot y^9$, or x^7y^9, cannot be simplified.

SOLUTION

a. $2^2 \cdot 2^3 = 2^{2+3} = 2^5$ or 32

b. $x^7 \cdot x^9 = x^{7+9} = x^{16}$

c. $y \cdot y^5 = y^1 \cdot y^5 = y^{1+5} = y^6$

d. $y^3 \cdot y^2 \cdot y^5 = y^{3+2+5} = y^{10}$ ■

✔ **CHECK POINT 1** Multiply each expression using the product rule:

a. $2^2 \cdot 2^4$ **b.** $x^6 \cdot x^4$ **c.** $y \cdot y^7$ **d.** $y^4 \cdot y^3 \cdot y^2$.

2 Use the power rule for exponents.

The Power Rule for Exponents The next property of exponents applies when an exponential expression is raised to a power. Here is an example:

$$(3^2)^4.$$

The exponential expression 3^2 is raised to the fourth power.

There are 4 factors 3^2. Thus,

$$(3^2)^4 = 3^2 \cdot 3^2 \cdot 3^2 \cdot 3^2 = 3^{2+2+2+2} = 3^8.$$

Add exponents when multiplying with the same base.

We can obtain the answer, 3^8, by multiplying the exponents:

$$(3^2)^4 = 3^{2\cdot 4} = 3^8.$$

This suggests the following rule:

THE POWER RULE (POWERS TO POWERS)

$$(b^m)^n = b^{mn}$$

When an exponential expression is raised to a power, multiply the exponents. Place the product of the exponents on the base and remove the parentheses.

EXAMPLE 2 Using the Power Rule

Simplify each expression using the power rule:

a. $(2^3)^5$ **b.** $(x^6)^4$ **c.** $[(-3)^7]^5$.

STUDY TIP

Do not confuse the product and power rules. Note the following differences:

- $x^4 \cdot x^7 = x^{4+7} = x^{11}$
- $(x^4)^7 = x^{4 \cdot 7} = x^{28}$.

SOLUTION

a. $(2^3)^5 = 2^{3 \cdot 5} = 2^{15}$

b. $(x^6)^4 = x^{6 \cdot 4} = x^{24}$

c. $[(-3)^7]^5 = (-3)^{7 \cdot 5} = (-3)^{35}$ ■

CHECK POINT 2 Simplify each expression using the power rule:

a. $(3^4)^5$ **b.** $(x^9)^{10}$ **c.** $[(-5)^7]^3$.

3 Use the products-to-powers rule.

The Products-to-Powers Rule for Exponents The next property of exponents applies when we are raising a product to a power. Here is an example:

$$(2x)^4.$$

The product $2x$ is raised to the fourth power.

There are four factors of $2x$. Thus,

$$(2x)^4 = 2x \cdot 2x \cdot 2x \cdot 2x = 2 \cdot 2 \cdot 2 \cdot 2 \cdot x \cdot x \cdot x \cdot x = 2^4x^4.$$

We can obtain the answer, 2^4x^4, by raising each factor within the parentheses to the fourth power:

$$(2x)^4 = 2^4x^4.$$

This suggests the following rule:

PRODUCTS TO POWERS

$$(ab)^n = a^nb^n$$

When a product is raised to a power, raise each factor to the power.

EXAMPLE 3 Using the Products-to-Powers Rule

Simplify each expression using the products-to-powers rule:

a. $(5x)^3$ **b.** $(-2y^4)^5$.

SOLUTION

a. $(5x)^3 = 5^3x^3$ — Raise each factor to the third power.

$= 125x^3$ — $5^3 = 5 \cdot 5 \cdot 5 = 125$

b. $(-2y^4)^5 = (-2)^5(y^4)^5$ — Raise each factor to the fifth power.

$= (-2)^5y^{4 \cdot 5}$ — To raise an exponential expression to a power, multiply exponents: $(b^m)^n = b^{mn}$.

$= -32y^{20}$ — $(-2)^5 = (-2)(-2)(-2)(-2)(-2) = -32$ ■

CHECK POINT 3 Simplify each expression using the products-to-powers rule:

a. $(2x)^4$ **b.** $(-4y^2)^3$.

STUDY TIP

Try to avoid the following common errors that can occur when simplifying exponential expressions.

Correct	Incorrect	Description of Error
$b^3 \cdot b^4 = b^{3+4} = b^7$	$b^3 \cdot b^4 = b^{12}$	Exponents should be added, not multiplied.
$3^2 \cdot 3^4 = 3^{2+4} = 3^6$	$3^2 \cdot 3^4 = 9^{2+4} = 9^6$	The common base should be retained, not multiplied.
$(x^5)^3 = x^{5 \cdot 3} = x^{15}$	$(x^5)^3 = x^{5+3} = x^8$	Exponents should be multiplied, not added, when raising a power to a power.
$(4x)^3 = 4^3x^3 = 64x^3$	$(4x)^3 = 4x^3$	Both factors should be cubed.

4 Multiply monomials.

Multiplying Monomials Now that we have developed three properties of exponents, we are ready to turn to polynomial multiplication. We begin with the product of two monomials, such as $-8x^6$ and $5x^3$. This product is obtained by multiplying the coefficients, -8 and 5, and then multiplying the variables by using the product rule for exponents.

$$(-8x^6)(5x^3) = -8 \cdot 5 \cdot x^6 \cdot x^3 = -8 \cdot 5x^{6+3} = -40x^9$$

Multiply coefficients and add exponents.

STUDY TIP

Don't confuse adding and multiplying monomials.

Addition:

$$5x^4 + 6x^4 = 11x^4$$

Multiplication:

$$(5x^4)(6x^4) = (5 \cdot 6)(x^4 \cdot x^4) = 30x^{4+4} = 30x^8$$

Only like terms can be added or subtracted, but unlike terms may be multiplied.

Addition:

$5x^4 + 3x^2$ cannot be simplified.

Multiplication:

$$(5x^4)(3x^2) = (5 \cdot 3)(x^4 \cdot x^2) = 15x^{4+2} = 15x^6$$

MULTIPLYING MONOMIALS To multiply monomials with the same variable base, multiply the coefficients and then multiply the variables. Use the product rule for exponents to multiply the variables: Keep the variable and add the exponents.

EXAMPLE 4 Multiplying Monomials

Multiply: **a.** $(2x)(4x^2)$ **b.** $(-10x^6)(6x^{10})$.

SOLUTION

a. $(2x)(4x^2) = (2 \cdot 4)(x \cdot x^2)$ Multiply the coefficients and multiply the variables.

$= 8x^{1+2}$ Add exponents: $b^m \cdot b^n = b^{m+n}$.

$= 8x^3$ Simplify.

b. $(-10x^6)(6x^{10}) = (-10 \cdot 6)(x^6 \cdot x^{10})$ Multiply the coefficients and multiply the variables.

$= -60x^{6+10}$ Add exponents: $b^m \cdot b^n = b^{m+n}$.

$= -60x^{16}$ Simplify. ■

✔ CHECK POINT 4 Multiply: **a.** $(7x^2)(10x)$ **b.** $(-5x^4)(4x^5)$.

5 Multiply a monomial and a polynomial.

Multiplying a Monomial and a Polynomial That Is Not a Monomial We use the distributive property to multiply a monomial and a polynomial that is not a monomial. For example,

$$3x^2(2x^3 + 5x) = 3x^2 \cdot 2x^3 + 3x^2 \cdot 5x = 3 \cdot 2x^{2+3} + 3 \cdot 5x^{2+1} = 6x^5 + 15x^3.$$

Monomial Binomial Multiply coefficients and add exponents.

MULTIPLYING A MONOMIAL AND A POLYNOMIAL THAT IS NOT A MONOMIAL To multiply a monomial and a polynomial, use the distributive property to multiply each term of the polynomial by the monomial.

EXAMPLE 5 Multiplying a Monomial and a Polynomial

Multiply: **a.** $2x(x + 4)$ **b.** $3x^2(4x^3 - 5x + 2)$.

SOLUTION

a. $2x(x + 4) = 2x \cdot x + 2x \cdot 4$ Use the distributive property.

$= 2 \cdot 1x^{1+1} + 2 \cdot 4x$ To multiply the monomials, multiply coefficients and add exponents.

$= 2x^2 + 8x$ Simplify.

b. $3x^2(4x^3 - 5x + 2)$

$= 3x^2 \cdot 4x^3 - 3x^2 \cdot 5x + 3x^2 \cdot 2$ Use the distributive property.

$= 3 \cdot 4x^{2+3} - 3 \cdot 5x^{2+1} + 3 \cdot 2x^2$ To multiply the monomials, multiply coefficients and add exponents.

$= 12x^5 - 15x^3 + 6x^2$ Simplify.

Rectangles often make it possible to visualize polynomial multiplication. For example, Figure 5.2 shows a rectangle with length $2x$ and width $x + 4$. The area of the large rectangle is

$$2x(x + 4).$$

The sum of the areas of the two smaller rectangles is

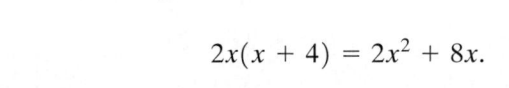

$$2x^2 + 8x.$$

Conclusion:

$$2x(x + 4) = 2x^2 + 8x.$$

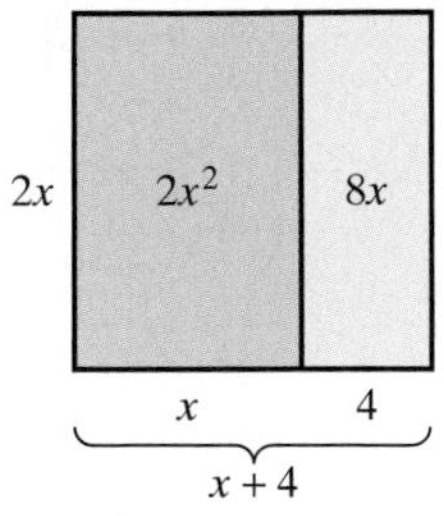

FIGURE 5.2

CHECK POINT 5 Multiply:

a. $3x(x + 5)$ **b.** $6x^2(5x^3 - 2x + 3)$.

6 Multiply polynomials when neither is a monomial.

Multiplying Polynomials When Neither Is a Monomial How do we multiply two polynomials if neither is a monomial? For example, consider

$$(2x + 3)(x^2 + 4x + 5).$$

Binomial — Trinomial

One way to perform this multiplication is to distribute $2x$ throughout the trinomial

$$2x(x^2 + 4x + 5)$$

and 3 throughout the trinomial

$$3(x^2 + 4x + 5).$$

Then combine the like terms that result. In general, the product of two polynomials is the polynomial obtained by multiplying each term of one polynomial by each term of the other polynomial and then combining like terms.

MULTIPLYING POLYNOMIALS WHEN NEITHER IS A MONOMIAL Multiply each term of one polynomial by each term of the other polynomial. Then combine like terms.

EXAMPLE 6 Multiplying Binomials

Multiply: **a.** $(x + 3)(x + 2)$ **b.** $(3x + 7)(2x - 4)$.

SOLUTION We begin by multiplying each term of the second binomial by each term of the first binomial.

a. $(x + 3)(x + 2)$

$= x(x + 2) + 3(x + 2)$ — Multiply the second binomial by each term of the first binomial.

$= x \cdot x + x \cdot 2 + 3 \cdot x + 3 \cdot 2$ — Use the distributive property.

$= x^2 + 2x + 3x + 6$ — Multiply. Note that $x \cdot x = x^1 \cdot x^1 = x^{1+1} = x^2$.

$= x^2 + 5x + 6$ — Combine like terms.

b. $(3x + 7)(2x - 4)$

$= 3x(2x - 4) + 7(2x - 4)$ — Multiply the second binomial by each term of the first binomial.

$= 3x \cdot 2x - 3x \cdot 4 + 7 \cdot 2x - 7 \cdot 4$ — Use the distributive property.

$= 6x^2 - 12x + 14x - 28$ — Multiply.

$= 6x^2 + 2x - 28$ — Combine like terms. ■

CHECK POINT 6 Multiply:

a. $(x + 4)(x + 5)$ **b.** $(5x + 3)(2x - 7)$.

STUDY TIP

You can visualize the polynomial multiplication in Example 6(a), using the rectangle with dimensions $x + 3$ and $x + 2$.

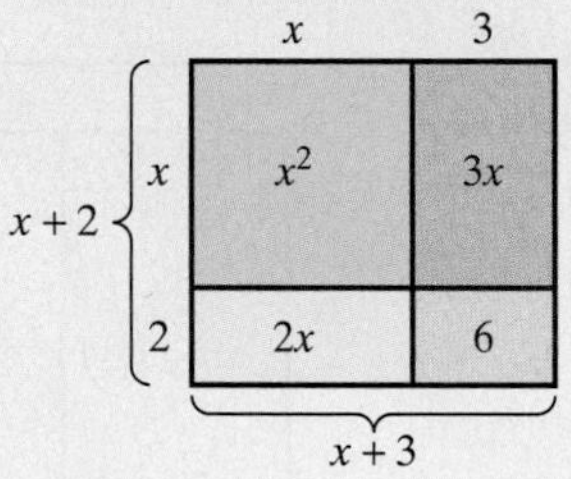

Area of large rectangle

$= (x + 3)(x + 2)$

Sum of areas of smaller rectangles

$= x^2 + 3x + 2x + 6$

$= x^2 + 5x + 6$

Conclusion:

$(x + 3)(x + 2) = x^2 + 5x + 6.$

USING TECHNOLOGY

A graphing utility can be used to see if a polynomial operation has been performed correctly. For example, to check

$$(x + 3)(x + 2) = x^2 + 5x + 6,$$

graph the left and right sides on the same screen, using

$$y_1 = (x + 3)(x + 2)$$

and

$$y_2 = x^2 + 5x + 6.$$

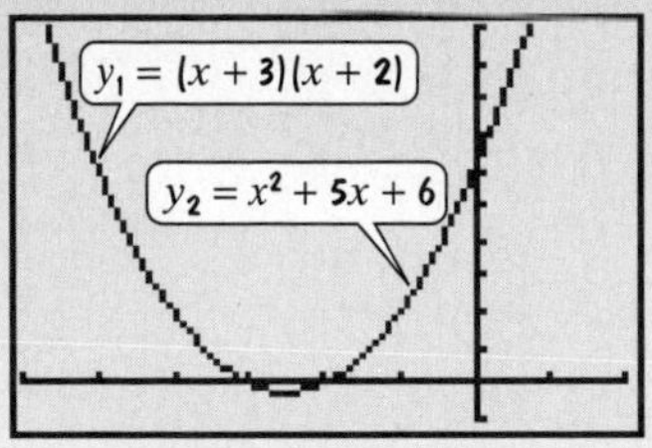

$[-6, 2, 1]$ by $[-1, 10, 1]$

As shown in the figure, both graphs are the same, verifying that the binomial multiplication was performed correctly.

EXAMPLE 7 Multiplying a Binomial and a Trinomial

Multiply: $(2x + 3)(x^2 + 4x + 5)$.

SOLUTION

$(2x + 3)(x^2 + 4x + 5)$

$= 2x(x^2 + 4x + 5) + 3(x^2 + 4x + 5)$ — Multiply the trinomial by each term of the binomial.

$= 2x \cdot x^2 + 2x \cdot 4x + 2x \cdot 5 + 3x^2 + 3 \cdot 4x + 3 \cdot 5$ — Use the distributive property.

$= 2x^3 + 8x^2 + 10x + 3x^2 + 12x + 15$ — Multiply monomials: Multiply coefficients and add exponents.

$= 2x^3 + 11x^2 + 22x + 15$ — Combine like terms: $8x^2 + 3x^2 = 11x^2$ and $10x + 12x = 22x$. ■

CHECK POINT 7 Multiply: $(5x + 2)(x^2 - 4x + 3)$.

Another method for solving Example 7 is to use a vertical format similar to that used for multiplying whole numbers.

$$\begin{array}{rrrr} & x^2 & + 4x & + 5 \\ & & 2x & + 3 \\ \hline & 3x^2 & + 12x & + 15 \\ 2x^3 & + 8x^2 & + 10x & \\ \hline 2x^3 & + 11x^2 & + 22x & + 15 \end{array}$$

Write like terms in the same column.

$3(x^2 + 4x + 5)$

$2x(x^2 + 4x + 5)$

Combine like terms.

EXAMPLE 8 Multiplying Polynomials Using a Vertical Format

Multiply: $(2x^2 - 3x)(5x^3 - 4x^2 + 7x)$.

SOLUTION To use the vertical format, it is most convenient to write the polynomial with the greatest number of terms in the top row.

$$\begin{array}{r} 5x^3 - 4x^2 + 7x \\ 2x^2 - 3x \\ \hline \end{array}$$

We now multiply each term in the top polynomial by the last term in the bottom polynomial.

$$\begin{array}{rrrr} & 5x^3 - & 4x^2 + & 7x \\ & & 2x^2 - & 3x \\ \hline & -15x^4 + & 12x^3 - & 21x^2 \end{array}$$

$-3x(5x^3 - 4x^2 + 7x)$

Then we multiply each term in the top polynomial by $2x^2$, the first term in the bottom polynomial. Like terms are placed in columns because the final step involves combining them.

$$\begin{array}{rrrrr} & & 5x^3 - & 4x^2 + & 7x \\ & & & 2x^2 - & 3x \\ \hline & -15x^4 + & 12x^3 & - 21x^2 & \\ 10x^5 - & 8x^4 + & 14x^3 & & \\ \hline 10x^5 - & 23x^4 + & 26x^3 & - 21x^2 & \end{array}$$

Write like terms in the same column.

$-3x(5x^3 - 4x^2 + 7x)$

$2x^2(5x^3 - 4x^2 + 7x)$

Combine like terms, which are lined up in columns. ■

CHECK POINT 8 Multiply using a vertical format: $(3x^2 - 2x)(2x^3 - 5x^2 + 4x)$.

ENRICHMENT ESSAY

Is It Hot in Here Or Is It Just Me?

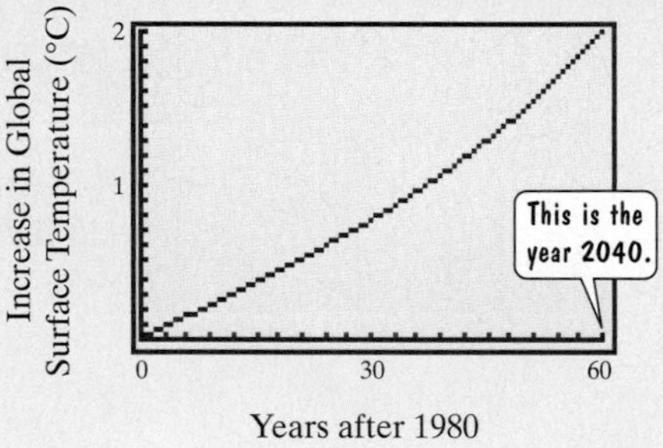

In the 1980s, a rising trend in global surface temperature was observed and the term "global warming" was coined. Scientists are more convinced than ever that burning coal, oil, and gas results in a buildup of gases and particles that trap heat and raise the planet's temperature. The average increase in global surface temperature, y, in degrees Celsius, x years after 1980 can be modeled by the polynomial formula

$$y = \frac{21}{5{,}000{,}000}x^3 - \frac{127}{1{,}000{,}000}x^2 + \frac{1293}{50{,}000}x.$$

The graph of this formula is shown above in a $[0, 60, 3]$ by $[0, 2, 0.1]$ viewing rectangle. The graph illustrates that the model predicts global warming will increase through the year 2040. Furthermore, the increasing steepness of the curve shows that global warming will increase at greater rates near the middle of the twenty-first century.

5.2 EXERCISE SET

Student Solutions Manual | CD/Video | PH Math/Tutor Center | MathXL Tutorials on CD | MathXL® | MyMathLab | Interactmath.com

Practice Exercises

In Exercises 1–8, multiply each expression using the product rule.

1. $x^{15} \cdot x^3$ **2.** $x^{12} \cdot x^4$

3. $y \cdot y^{11}$ **4.** $y \cdot y^{19}$

5. $x^2 \cdot x^6 \cdot x^3$ **6.** $x^4 \cdot x^3 \cdot x^5$

7. $7^9 \cdot 7^{10}$ **8.** $8^7 \cdot 8^{10}$

In Exercises 9–14, simplify each expression using the power rule.

9. $(6^9)^{10}$ **10.** $(6^7)^{10}$

11. $(x^{15})^3$ **12.** $(x^{12})^4$

13. $[(-20)^3]^3$ **14.** $[(-50)^4]^4$

In Exercises 15–24, simplify each expression using the products-to-powers rule.

15. $(2x)^3$ **16.** $(4x)^3$

17. $(-5x)^2$ **18.** $(-6x)^2$

19. $(4x^3)^2$ **20.** $(6x^3)^2$

21. $(-2y^6)^4$ **22.** $(-2y^5)^4$

23. $(-2x^7)^5$ **24.** $(-2x^{11})^5$

In Exercises 25–34, multiply the monomials.

25. $(7x)(2x)$ **26.** $(8x)(3x)$

27. $(6x)(4x^2)$ **28.** $(10x)(3x^2)$

29. $(-5y^4)(3y^3)$ **30.** $(-6y^4)(2y^3)$

31. $\left(-\frac{1}{2}a^3\right)\left(-\frac{1}{4}a^2\right)$ **32.** $\left(-\frac{1}{3}a^4\right)\left(-\frac{1}{2}a^2\right)$

33. $(2x^2)(-3x)(8x^4)$ **34.** $(3x^3)(-2x)(5x^6)$

In Exercises 35–54, find each product of the monomial and the polynomial.

35. $4x(x + 3)$ **36.** $6x(x + 5)$

37. $x(x - 3)$ **38.** $x(x - 7)$

39. $2x(x - 6)$ **40.** $3x(x - 5)$

41. $-4y(3y + 5)$

42. $-5y(6y + 7)$

43. $4x^2(x + 2)$ **44.** $5x^2(x + 6)$

45. $2y^2(y^2 + 3y)$ **46.** $4y^2(y^2 + 2y)$

47. $2y^2(3y^2 - 4y + 7)$

48. $4y^2(5y^2 - 6y + 3)$

49. $(3x^3 + 4x^2)(2x)$

50. $(4x^3 + 5x^2)(2x)$

51. $(x^2 + 5x - 3)(-2x)$

52. $(x^3 - 2x + 2)(-4x)$

53. $-3x^2(-4x^2 + x - 5)$

54. $-6x^2(3x^2 - 2x - 7)$

In Exercises 55–78, find each product. In each case, neither factor is a monomial.

55. $(x + 3)(x + 5)$

56. $(x + 4)(x + 6)$

57. $(2x + 1)(x + 4)$

58. $(2x + 5)(x + 3)$

59. $(x + 3)(x - 5)$

60. $(x + 4)(x - 6)$

61. $(x - 11)(x + 9)$

62. $(x - 12)(x + 8)$

63. $(2x - 5)(x + 4)$

64. $(3x - 4)(x + 5)$

65. $\left(\frac{1}{4}x + 4\right)\left(\frac{3}{4}x - 1\right)$

66. $\left(\frac{1}{5}x + 5\right)\left(\frac{3}{5}x - 1\right)$

67. $(x + 1)(x^2 + 2x + 3)$

68. $(x + 2)(x^2 + x + 5)$

69. $(y - 3)(y^2 - 3y + 4)$

70. $(y - 2)(y^2 - 4y + 3)$

71. $(2a - 3)(a^2 - 3a + 5)$

72. $(2a - 1)(a^2 - 4a + 3)$

73. $(x + 1)(x^3 + 2x^2 + 3x + 4)$

74. $(x + 1)(x^3 + 4x^2 + 7x + 3)$

75. $\left(x - \frac{1}{2}\right)(4x^3 - 2x^2 + 5x - 6)$

76. $\left(x - \frac{1}{3}\right)(3x^3 - 6x^2 + 5x - 9)$

77. $(x^2 + 2x + 1)(x^2 - x + 2)$

78. $(x^2 + 3x + 1)(x^2 - 2x - 1)$

In Exercises 79–92, use a vertical format to find each product.

79.
$$\begin{array}{r} x^2 - 5x + 3 \\ x + 8 \\ \hline \end{array}$$

80.
$$\begin{array}{r} x^2 - 7x + 9 \\ x + 4 \\ \hline \end{array}$$

81.
$$\begin{array}{r} x^2 - 3x + 9 \\ 2x - 3 \\ \hline \end{array}$$

82. $\begin{array}{r} y^2 - 5y + 3 \\ \underline{4y - 5} \end{array}$

83. $\begin{array}{r} 2x^3 + x^2 + 2x + 3 \\ \underline{x + 4} \end{array}$

84. $\begin{array}{r} 3y^3 + 2y^2 + y + 4 \\ \underline{y + 3} \end{array}$

85. $\begin{array}{r} 4z^3 - 2z^2 + 5z - 4 \\ \underline{3z - 2} \end{array}$

86. $\begin{array}{r} 5z^3 - 3z^2 + 4z - 3 \\ \underline{2z - 4} \end{array}$

87. $\begin{array}{r} 7x^3 - 5x^2 + 6x \\ \underline{3x^2 - 4x} \end{array}$

88. $\begin{array}{r} 9y^3 - 7y^2 + 5y \\ \underline{3y^2 + 5y} \end{array}$

89. $\begin{array}{r} 2y^5 - 3y^3 + y^2 - 2y + 3 \\ \underline{2y - 1} \end{array}$

90. $\begin{array}{r} n^4 - n^3 + n^2 - n + 1 \\ \underline{2n + 3} \end{array}$

91. $\begin{array}{r} x^2 + 7x - 3 \\ \underline{x^2 - x - 1} \end{array}$

92. $\begin{array}{r} x^2 + 6x - 4 \\ \underline{x^2 - x - 2} \end{array}$

Practice Plus

In Exercises 93–100, perform the indicated operations.

93. $(x + 4)(x - 5) - (x + 3)(x - 6)$

94. $(x + 5)(x - 6) - (x + 2)(x - 9)$

95. $4x^2(5x^3 + 3x - 2) - 5x^3(x^2 - 6)$

96. $3x^2(6x^3 + 2x - 3) - 4x^3(x^2 - 5)$

97. $(y + 1)(y^2 - y + 1) + (y - 1)(y^2 + y + 1)$

98. $(y + 1)(y^2 - y + 1) - (y - 1)(y^2 + y + 1)$

99. $(y + 6)^2 - (y - 2)^2$

100. $(y + 5)^2 - (y - 4)^2$

Application Exercises

101. Find a trinomial for the area of the rectangular rug shown here whose sides are $x + 5$ feet and $2x - 3$ feet:

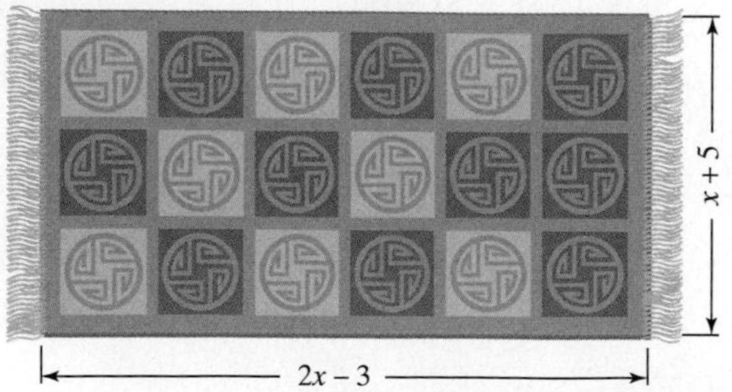

102. The base of a triangular sail is $4x$ feet and its height is $3x + 10$ feet. Write a binomial in terms of x for the area of the sail.

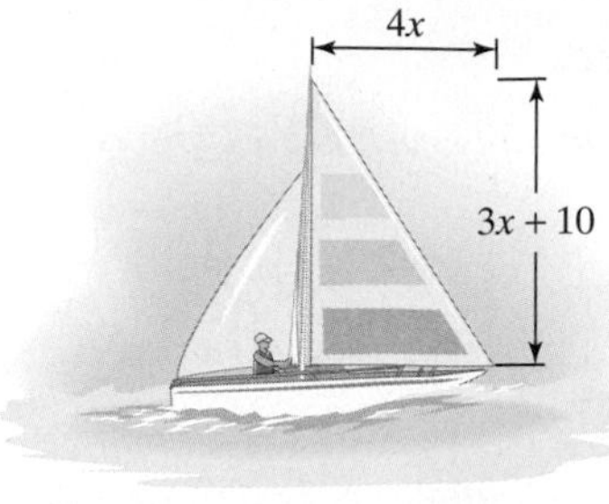

In Exercises 103–104,

a. *Express the area of the large rectangle as the product of two binomials.*

b. *Find the sum of the areas of the four smaller rectangles.*

c. *Use polynomial multiplication to show that your expressions for area in parts (a) and (b) are equal.*

103.

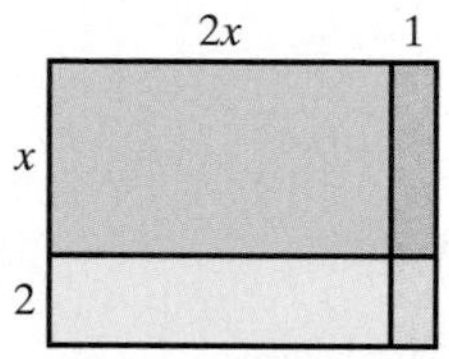

104.

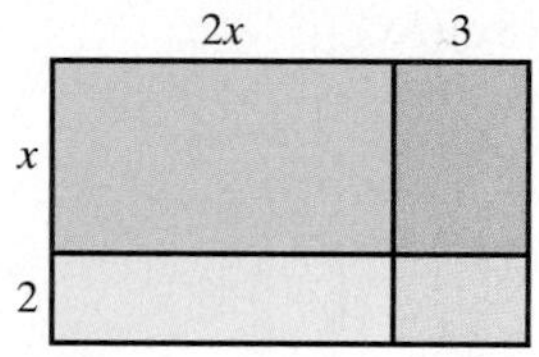

Writing in Mathematics

105. Explain the product rule for exponents. Use $2^3 \cdot 2^5$ in your explanation.

106. Explain the power rule for exponents. Use $(3^2)^4$ in your explanation.

107. Explain how to simplify an expression that involves a product raised to a power. Provide an example with your explanation.

108. Explain how to multiply monomials. Give an example.

109. Explain how to multiply a monomial and a polynomial that is not a monomial. Give an example.

110. Explain how to multiply polynomials when neither is a monomial. Give an example.

111. Explain the difference between performing these two operations:

$$2x^2 + 3x^2 \quad \text{and} \quad (2x^2)(3x^2).$$

112. Discuss situations in which a vertical format, rather than a horizontal format, is useful for multiplying polynomials.

113. Describe one change that might alter the prediction about global warming given by the model and graph on page 322.

Critical Thinking Exercises

114. Which one of the following is true?

a. $4x^3 \cdot 3x^4 = 12x^{12}$

b. $5x^2 \cdot 4x^6 = 9x^8$

c. $(y - 1)(y^2 + y + 1) = y^3 - 1$

d. Some polynomial multiplications can only be performed by using a vertical format.

115. Find a polynomial in descending powers of x representing the area of the shaded region.

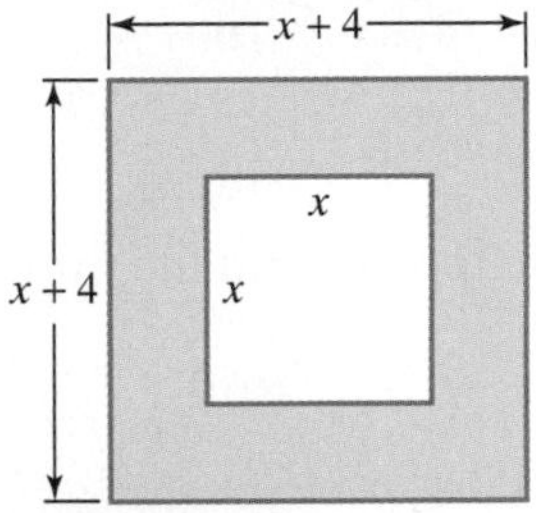

116. Find each of the products in parts (a)–(c).

a. $(x - 1)(x + 1)$

b. $(x - 1)(x^2 + x + 1)$

c. $(x - 1)(x^3 + x^2 + x + 1)$

d. Using the pattern found in parts (a)–(c), find $(x - 1)(x^4 + x^3 + x^2 + x + 1)$ without actually multiplying.

117. Find the missing factor.

$$(\_\_\_\_)\left(-\frac{1}{4}xy^3\right) = 2x^5y^3$$

Review Exercises

118. Solve: $4x - 7 > 9x - 2$. (Section 2.7, Example 6)

119. Graph $3x - 2y = 6$ using intercepts. (Section 3.2, Example 4)

120. Find the slope of the line passing through the points $(-2, 8)$ and $(1, 6)$. (Section 3.3, Example 1)

SECTION 5.3

Objectives

1 Use FOIL in polynomial multiplication.

2 Multiply the sum and difference of two terms.

3 Find the square of a binomial sum.

4 Find the square of a binomial difference.

SPECIAL PRODUCTS

Let's cut to the chase. Are there fast methods for finding products of polynomials? Yes. In this section, we use the distributive property to develop patterns that will let you multiply certain binomials quite rapidly.

1 Use FOIL in polynomial multiplication.

The Product of Two Binomials: FOIL Frequently, we need to find the product of two binomials. One way to perform this multiplication is to distribute each term in the first binomial through the second binomial. For example, we can find the product of the binomials $3x + 2$ and $4x + 5$ as follows:

$$(3x + 2)(4x + 5) = 3x(4x + 5) + 2(4x + 5)$$

Distribute $3x$ over $4x + 5$. Distribute 2 over $4x + 5$.

$$= 3x(4x) + 3x(5) + 2(4x) + 2(5)$$

$$= 12x^2 + 15x + 8x + 10.$$

We can also find the product of $3x + 2$ and $4x + 5$ using a method called FOIL, which is based on our work shown above. Any two binomials can be quickly multiplied by using the FOIL method, in which **F** represents the product of the **first** terms in each binomial, **O** represents the product of the **outside** terms, **I** represents the product of the **inside** terms, and **L** represents the product of the **last**, or second, terms in each binomial. For example, we can use the FOIL method to find the product of the binomials $3x + 2$ and $4x + 5$ as follows:

$$(3x + 2)(4x + 5) = 12x^2 + 15x + 8x + 10$$

(first, outside, inside, last; F: Product of First terms, O: Product of Outside terms, I: Product of Inside terms, L: Product of Last terms)

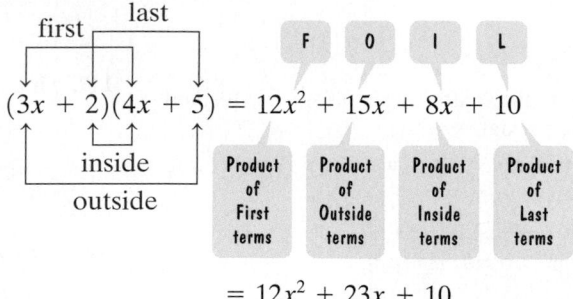

$$= 12x^2 + 23x + 10 \qquad \text{Combine like terms.}$$

In general, here's how to use the FOIL method to find the product of $ax + b$ and $cx + d$:

USING THE FOIL METHOD TO MULTIPLY BINOMIALS

$$(ax + b)(cx + d) = ax \cdot cx + ax \cdot d + b \cdot cx + b \cdot d$$

(first, outside, inside, last; F: Product of First terms, O: Product of Outside terms, I: Product of Inside terms, L: Product of Last terms)

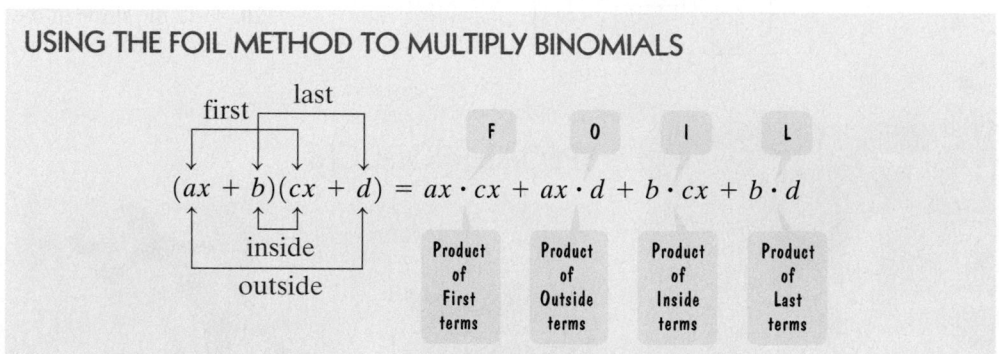

EXAMPLE 1 Using the FOIL Method

Multiply: $(x + 3)(x + 4)$.

SOLUTION

F: First terms $= x \cdot x = x^2 \quad (x + 3)(x + 4)$

O: Outside terms $= x \cdot 4 = 4x \quad (x + 3)(x + 4)$

I: Inside terms $= 3 \cdot x = 3x \quad (x + 3)(x + 4)$

L: Last terms $= 3 \cdot 4 = 12 \quad (x + 3)(x + 4)$

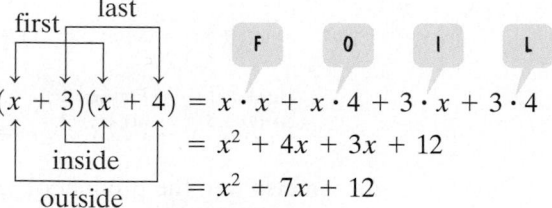

$$(x + 3)(x + 4) = x \cdot x + x \cdot 4 + 3 \cdot x + 3 \cdot 4$$
$$= x^2 + 4x + 3x + 12$$
$$= x^2 + 7x + 12 \qquad \text{Combine like terms.}$$

(first, outside, inside, last; F, O, I, L)

CHECK POINT 1 Multiply: $(x + 5)(x + 6)$.

EXAMPLE 2 Using the FOIL Method

Multiply: $(3x + 4)(5x - 3)$.

SOLUTION

first, last, inside, outside — F O I L

$$(3x + 4)(5x - 3) = 3x \cdot 5x + 3x(-3) + 4 \cdot 5x + 4(-3)$$
$$= 15x^2 - 9x + 20x - 12$$
$$= 15x^2 + 11x - 12 \quad \text{Combine like terms.}$$

■

CHECK POINT 2 Multiply: $(7x + 5)(4x - 3)$.

EXAMPLE 3 Using the FOIL Method

Multiply: $(2 - 5x)(3 - 4x)$.

SOLUTION

first, last, inside, outside — F O I L

$$(2 - 5x)(3 - 4x) = 2 \cdot 3 + 2(-4x) + (-5x)(3) + (-5x)(-4x)$$
$$= 6 - 8x - 15x + 20x^2$$
$$= 6 - 23x + 20x^2 \quad \text{Combine like terms.}$$

The product can also be expressed in standard form as $20x^2 - 23x + 6$. ■

CHECK POINT 3 Multiply: $(4 - 2x)(5 - 3x)$.

2 Multiply the sum and difference of two terms.

Multiplying the Sum and Difference of Two Terms We can use the FOIL method to multiply $A + B$ and $A - B$ as follows:

F O I L

$$(A + B)(A - B) = A^2 - AB + AB - B^2 = A^2 - B^2.$$

Notice that the outside and inside products have a sum of 0 and the terms cancel. The FOIL multiplication provides us with a quick rule for multiplying the sum and difference of two terms, referred to as a special-product formula.

THE PRODUCT OF THE SUM AND DIFFERENCE OF TWO TERMS

$$(A + B)(A - B) = A^2 - B^2$$

The product of the sum and the difference of the same two terms is the square of the first term minus the square of the second term.

EXAMPLE 4 Finding the Product of the Sum and Difference of Two Terms

Find each product by using the preceding rule:

a. $(4y + 3)(4y - 3)$ **b.** $(3x - 7)(3x + 7)$ **c.** $(5a^4 + 6)(5a^4 - 6)$.

SOLUTION Use the special-product formula shown.

$$(A + B)(A - B) = A^2 - B^2$$

First term squared − Second term squared = Product

a. $(4y + 3)(4y - 3) = (4y)^2 - 3^2 = 16y^2 - 9$

b. $(3x - 7)(3x + 7) = (3x)^2 - 7^2 = 9x^2 - 49$

c. $(5a^4 + 6)(5a^4 - 6) = (5a^4)^2 - 6^2 = 25a^8 - 36$ ■

CHECK POINT 4 Find each product:

a. $(7y + 8)(7y - 8)$ **b.** $(4x - 5)(4x + 5)$

c. $(2a^3 + 3)(2a^3 - 3)$.

3 Find the square of a binomial sum.

The Square of a Binomial Let's now find $(A + B)^2$, the square of a binomial sum. To do so, we begin with the FOIL method and look for a general rule.

$$(A + B)^2 = (A + B)(A + B) = A \cdot A + A \cdot B + A \cdot B + B \cdot B$$
$$= A^2 + 2AB + B^2$$

(F, O, I, L label the four products.)

This result implies the following rule, which is another example of a special-product formula:

THE SQUARE OF A BINOMIAL SUM

$$(A + B)^2 = A^2 + 2AB + B^2$$

The square of a binomial sum is first term squared plus 2 times the product of the terms plus last term squared.

STUDY TIP

Caution! The square of a sum is *not* the sum of the squares.

$(A + B)^2 \neq A^2 + B^2$ — The middle term $2AB$ is missing.

$(x + 3)^2 \neq x^2 + 9$ — Incorrect!

Show that $(x + 3)^2$ and $x^2 + 9$ are not equal by substituting 5 for x in each expression and simplifying.

EXAMPLE 5 Finding the Square of a Binomial Sum

Square each binomial using the preceding rule:

a. $(x + 3)^2$ **b.** $(3x + 7)^2$.

SOLUTION Use the special-product formula shown.

$$(A + B)^2 = A^2 + 2AB + B^2$$

	(First Term)^2	+	2 · Product of the Terms	+	(Last Term)^2	= Product
a. $(x + 3)^2 =$	x^2	+	$2 \cdot x \cdot 3$	+	3^2	$= x^2 + 6x + 9$
b. $(3x + 7)^2 =$	$(3x)^2$	+	$2(3x)(7)$	+	7^2	$= 9x^2 + 42x + 49$

■

✔ CHECK POINT 5 Square each binomial:

a. $(x + 10)^2$ **b.** $(5x + 4)^2$.

4 Find the square of a binomial difference.

Using the FOIL method on $(A - B)^2$, the square of a binomial difference, we obtain the following rule:

THE SQUARE OF A BINOMIAL DIFFERENCE

$$(A - B)^2 = A^2 - 2AB + B^2$$

The square of a binomial difference | is | first term squared | minus | 2 times the product of the terms | plus | last term squared.

EXAMPLE 6 Finding the Square of a Binomial Difference

Square each binomial using the preceding rule:

a. $(x - 4)^2$ **b.** $(5y - 6)^2$.

SOLUTION Use the special-product formula shown.

$$(A - B)^2 = A^2 - 2AB + B^2$$

	(First Term)^2	−	2 · Product of the Terms	+	(Last Term)^2	= Product
a. $(x - 4)^2 =$	x^2	−	$2 \cdot x \cdot 4$	+	4^2	$= x^2 - 8x + 16$
b. $(5y - 6)^2 =$	$(5y)^2$	−	$2(5y)(6)$	+	6^2	$= 25y^2 - 60y + 36$

■

✔ CHECK POINT 6 Square each binomial:

a. $(x - 9)^2$ **b.** $(7x - 3)^2$.

Figure 5.3 makes it possible to visualize the square of a binomial sum. The area of the large square is

$$(A + B)(A + B) \quad \text{or} \quad (A + B)^2.$$

The sum of the areas of the four smaller rectangles that make up the large square is

$$A^2 + AB + AB + B^2$$

or

$$A^2 + 2AB + B^2.$$

Conclusion:

$$(A + B)^2 = A^2 + 2AB + B^2.$$

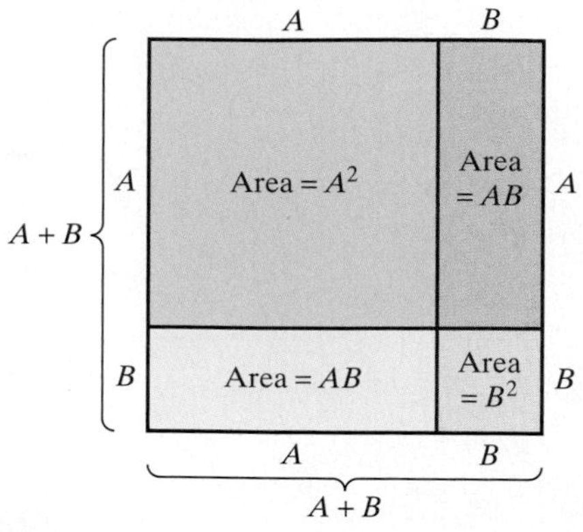

FIGURE 5.3

The following box summarizes the FOIL method and the three special products. The special products occur so frequently in algebra that it is convenient to memorize the form or pattern of these formulas.

FOIL and Special Products
Let *A*, *B*, *C*, and *D* be real numbers, variables, or algebraic expressions.

FOIL	***Example***
F O I L $(A + B)(C + D) = AC + AD + BC + BD$	F O I L $(2x + 3)(4x + 5) = (2x)(4x) + (2x)(5) + (3)(4x) + (3)(5)$ $= 8x^2 + 10x + 12x + 15$ $= 8x^2 + 22x + 15$
Sum and Difference of Two Terms	***Example***
$(A + B)(A - B) = A^2 - B^2$	$(2x + 3)(2x - 3) = (2x)^2 - 3^2$ $= 4x^2 - 9$
Square of a Binomial	***Example***
$(A + B)^2 = A^2 + 2AB + B^2$ $(A - B)^2 = A^2 - 2AB + B^2$	$(2x + 3)^2 = (2x)^2 + 2(2x)(3) + 3^2$ $= 4x^2 + 12x + 9$ $(2x - 3)^2 = (2x)^2 - 2(2x)(3) + 3^2$ $= 4x^2 - 12x + 9$

5.3 EXERCISE SET

Student Solutions Manual | CD/Video | PH Math/Tutor Center | MathXL Tutorials on CD | MathXL® | MyMathLab | Interactmath.com

Practice Exercises

In Exercises 1–24, use the FOIL method to find each product. Express the product in descending powers of the variable.

1. $(x + 4)(x + 6)$

2. $(x + 8)(x + 2)$

3. $(y - 7)(y + 3)$

4. $(y - 3)(y + 4)$

5. $(2x - 3)(x + 5)$

6. $(3x - 5)(x + 7)$
7. $(4y + 3)(y - 1)$
8. $(5y + 4)(y - 2)$
9. $(2x - 3)(5x + 3)$
10. $(2x - 5)(7x + 2)$
11. $(3y - 7)(4y - 5)$
12. $(4y - 5)(7y - 4)$
13. $(7 + 3x)(1 - 5x)$
14. $(2 + 5x)(1 - 4x)$
15. $(5 - 3y)(6 - 2y)$
16. $(7 - 2y)(10 - 3y)$
17. $(5x^2 - 4)(3x^2 - 7)$
18. $(7x^2 - 2)(3x^2 - 5)$
19. $(6x - 5)(2 - x)$
20. $(4x - 3)(2 - x)$
21. $(x + 5)(x^2 + 3)$
22. $(x + 4)(x^2 + 5)$
23. $(8x^3 + 3)(x^2 + 5)$
24. $(7x^3 + 5)(x^2 + 2)$

In Exercises 25–44, multiply using the rule for finding the product of the sum and difference of two terms.

25. $(x + 3)(x - 3)$
26. $(y + 5)(y - 5)$
27. $(3x + 2)(3x - 2)$
28. $(2x + 5)(2x - 5)$
29. $(3r - 4)(3r + 4)$
30. $(5z - 2)(5z + 2)$
31. $(3 + r)(3 - r)$
32. $(4 + s)(4 - s)$
33. $(5 - 7x)(5 + 7x)$
34. $(4 - 3y)(4 + 3y)$
35. $\left(2x + \frac{1}{2}\right)\left(2x - \frac{1}{2}\right)$
36. $\left(3y + \frac{1}{3}\right)\left(3y - \frac{1}{3}\right)$
37. $(y^2 + 1)(y^2 - 1)$
38. $(y^2 + 2)(y^2 - 2)$
39. $(r^3 + 2)(r^3 - 2)$
40. $(m^3 + 4)(m^3 - 4)$
41. $(1 - y^4)(1 + y^4)$
42. $(2 - s^5)(2 + s^5)$
43. $(x^{10} + 5)(x^{10} - 5)$
44. $(x^{12} + 3)(x^{12} - 3)$

In Exercises 45–62, multiply using the rule for the square of a binomial.

45. $(x + 2)^2$
46. $(x + 5)^2$
47. $(2x + 5)^2$
48. $(5x + 2)^2$
49. $(x - 3)^2$
50. $(x - 6)^2$
51. $(3y - 4)^2$
52. $(4y - 3)^2$
53. $(4x^2 - 1)^2$
54. $(5x^2 - 3)^2$
55. $(7 - 2x)^2$
56. $(9 - 5x)^2$
57. $\left(2x + \frac{1}{2}\right)^2$
58. $\left(3x + \frac{1}{3}\right)^2$
59. $\left(4y - \frac{1}{4}\right)^2$
60. $\left(2y - \frac{1}{2}\right)^2$
61. $(x^8 + 3)^2$
62. $(x^8 + 5)^2$

In Exercises 63–82, multiply using the method of your choice.

63. $(x - 1)(x^2 + x + 1)$
64. $(x + 1)(x^2 - x + 1)$
65. $(x - 1)^2$
66. $(x + 1)^2$
67. $(3y + 7)(3y - 7)$
68. $(4y + 9)(4y - 9)$
69. $3x^2(4x^2 + x + 9)$
70. $5x^2(7x^2 + x + 6)$
71. $(7y + 3)(10y - 4)$
72. $(8y + 3)(10y - 5)$
73. $(x^2 + 1)^2$
74. $(x^2 + 2)^2$
75. $(x^2 + 1)(x^2 + 2)$
76. $(x^2 + 2)(x^2 + 3)$

77. $(x^2 + 4)(x^2 - 4)$

78. $(x^2 + 5)(x^2 - 5)$

79. $(2 - 3x^5)^2$

80. $(2 - 3x^6)^2$

81. $\left(\frac{1}{4}x^2 + 12\right)\left(\frac{3}{4}x^2 - 8\right)$

82. $\left(\frac{1}{4}x^2 + 16\right)\left(\frac{3}{4}x^2 - 4\right)$

In Exercises 83–88, find the area of each shaded region. Write the answer as a polynomial in descending powers of x.

83.

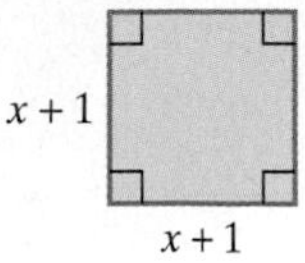

84.

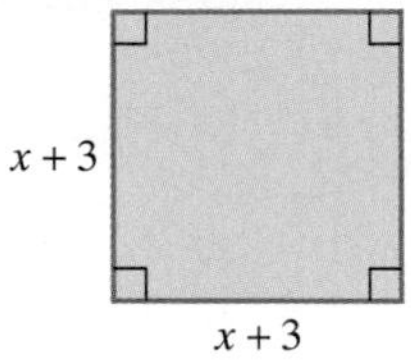

85.

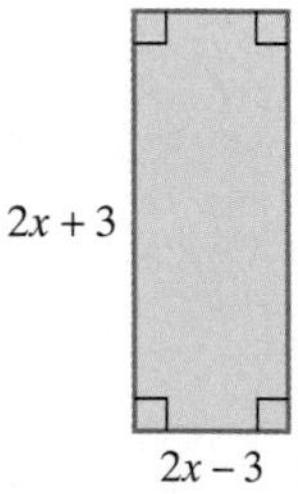

86.

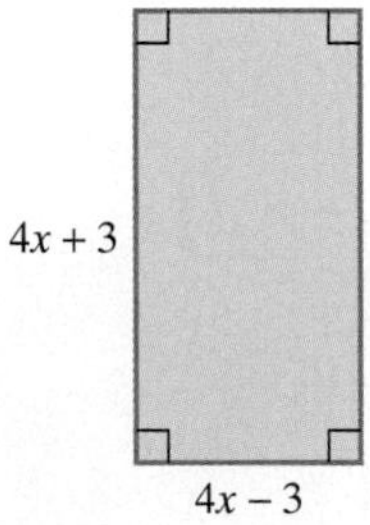

87.

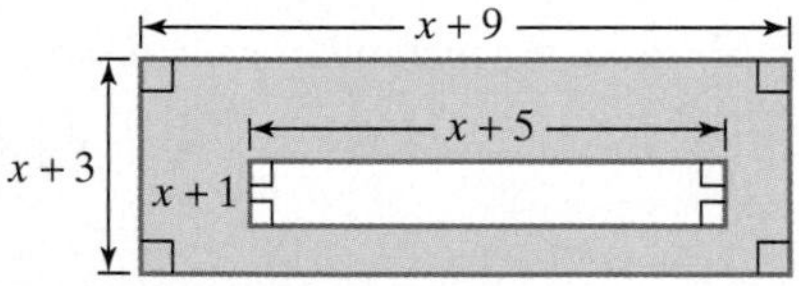

88.

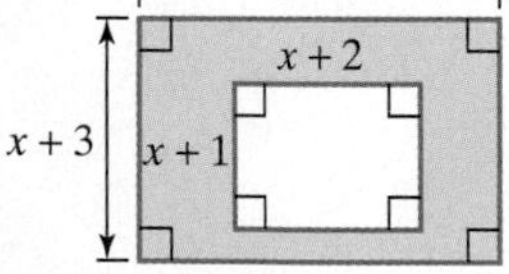

Practice Plus

In Exercises 89–96, multiply by the method of your choice.

89. $[(2x + 3)(2x - 3)]^2$

90. $[(3x + 2)(3x - 2)]^2$

91. $(4x^2 + 1)[(2x + 1)(2x - 1)]$

92. $(9x^2 + 1)[(3x + 1)(3x - 1)]$

93. $(x + 2)^3$

94. $(x + 4)^3$

95. $[(x + 3) - y][(x + 3) + y]$

96. $[(x + 5) - y][(x + 5) + y]$

Application Exercises

The square garden shown in the figure measures x yards on each side. The garden is to be expanded so that one side is increased by 2 yards and an adjacent side is increased by 1 yard. The graph shows the area of the expanded garden, y, in terms of the length of one of its original square sides, x. Use this information to solve Exercises 97–100.

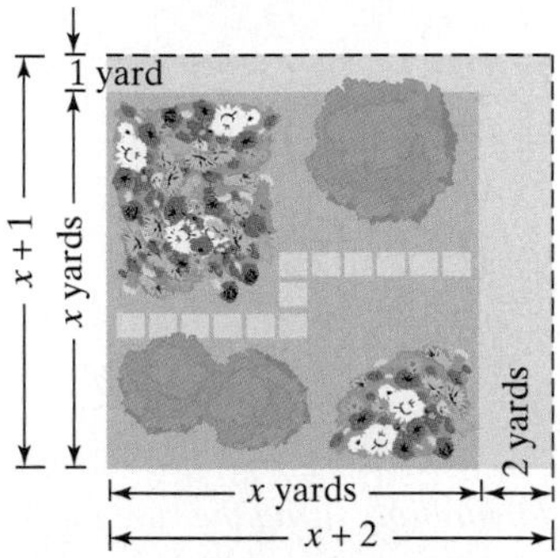

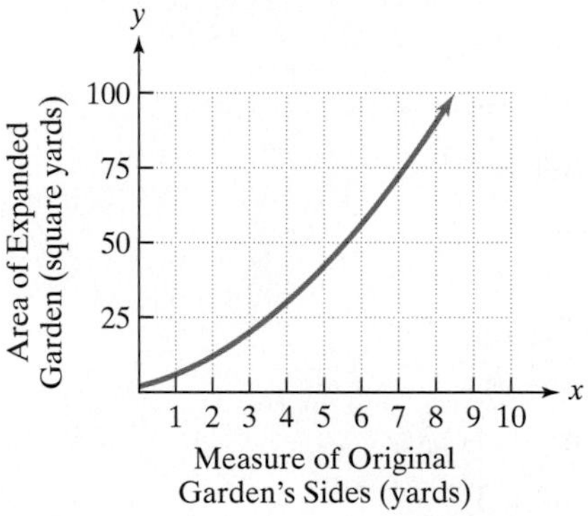

97. Write a product of two binomials that expresses the area of the larger garden.

98. Write a polynomial in descending powers of x that expresses the area of the larger garden.

99. If the original garden measures 6 yards on a side, use your expression from Exercise 97 to find the area of the larger garden. Then identify your solution as a point on the graph shown.

100. If the original garden measures 8 yards on a side, use your polynomial from Exercise 98 to find the area of the larger garden. Then identify your solution as a point on the graph shown.

The square painting in the figure measures x inches on each side. The painting is uniformly surrounded by a frame that measures 1 inch wide. Use this information to solve Exercises 101–102.

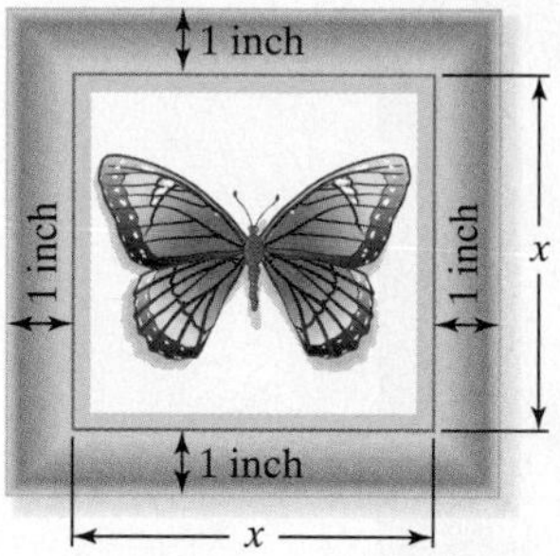

101. Write a polynomial in descending powers of x that expresses the area of the square that includes the painting and the frame.

102. Write an algebraic expression that describes the area of the frame. (*Hint:* The area of the frame is the area of the square that includes the painting and the frame minus the area of the painting.)

Writing in Mathematics

103. Explain how to multiply two binomials using the FOIL method. Give an example with your explanation.

104. Explain how to find the product of the sum and difference of two terms. Give an example with your explanation.

105. Explain how to square a binomial sum. Give an example with your explanation.

106. Explain how to square a binomial difference. Give an example with your explanation.

107. Explain why the graph for Exercises 97–100 is shown only in quadrant I.

Critical Thinking Exercises

108. Which one of the following is true?

a. $(3 + 4)^2 = 3^2 + 4^2$

b. $(2y + 7)^2 = 4y^2 + 28y + 49$

c. $(3x^2 + 2)(3x^2 - 2) = 9x^2 - 4$

d. $(x - 5)^2 = x^2 - 5x + 25$

109. What two binomials must be multiplied using the FOIL method to give a product of $x^2 - 8x - 20$?

110. Express the volume of the box as a polynomial in standard form.

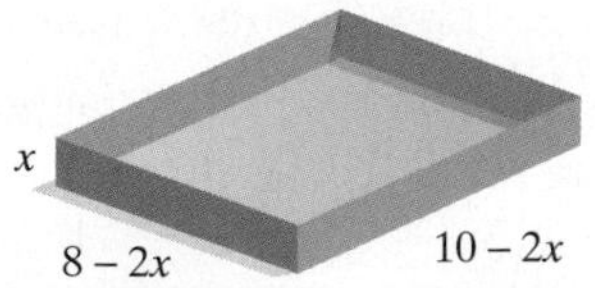

111. Express the area of the plane figure shown as a polynomial in standard form.

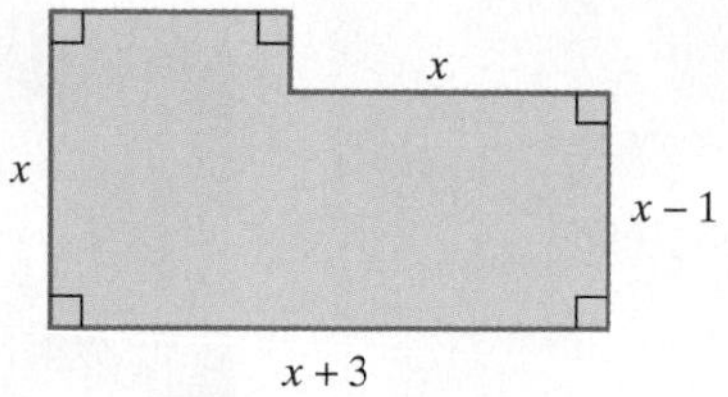

Technology Exercises

In Exercises 112–115, use a graphing utility to graph each side of the equation in the same viewing rectangle. (Call the left side y_1 and the right side y_2.) If the graphs coincide, verify that the multiplication has been performed correctly. If the graphs do not appear to coincide, this indicates that the multiplication is incorrect. In these exercises, correct the right side of the equation. Then graph the left side and the corrected right side to verify that the graphs coincide.

112. $(x + 1)^2 = x^2 + 1$; Use a $[-5, 5, 1]$ by $[0, 20, 1]$ viewing rectangle.

113. $(x + 2)^2 = x^2 + 2x + 4$; Use a $[-6, 5, 1]$ by $[0, 20, 1]$ viewing rectangle.

114. $(x + 1)(x - 1) = x^2 - 1$; Use a $[-6, 5, 1]$ by $[-2, 18, 1]$ viewing rectangle.

115. $(x - 2)(x + 2) + 4 = x^2$; Use a $[-6, 5, 1]$ by $[-2, 18, 1]$ viewing rectangle.

Review Exercises

In Exercises 116–118, solve each system by the method of your choice.

116. $2x + 3y = 1$

$y = 3x - 7$

(Section 4.2, Example 1)

117. $3x + 4y = 7$

$2x + 7y = 9$

(Section 4.3, Example 3)

118. $4x + y + 2z = 6$

$x + y + z = 1$

$x - 3y + z = 5$

(Section 4.5, Example 2)

SECTION 5.4

Objectives

1. Evaluate polynomials in several variables.
2. Understand the vocabulary of polynomials in two variables.
3. Add and subtract polynomials in several variables.
4. Multiply polynomials in several variables.

POLYNOMIALS IN SEVERAL VARIABLES

The next time you visit a lumberyard and go rummaging through piles of wood, think *polynomials*, although polynomials a bit different from those we have encountered so far. The construction industry uses a polynomial in two variables to determine the number of board feet that can be manufactured from a tree with a diameter of x inches and a length of y feet. This polynomial is

$$\frac{1}{4}x^2y - 2xy + 4y.$$

We call a polynomial containing two or more variables a **polynomial in several variables**. These polynomials can be evaluated, added, subtracted, and multiplied just like polynomials that contain only one variable.

1 Evaluate polynomials in several variables.

Evaluating a Polynomial in Several Variables Two steps can be used to evaluate a polynomial in several variables.

EVALUATING A POLYNOMIAL IN SEVERAL VARIABLES

1. Substitute the given value for each variable.
2. Perform the resulting computation using the order of operations.

EXAMPLE 1 Evaluating a Polynomial in Two Variables

Evaluate $2x^3y + xy^2 + 7x - 3$ for $x = -2$ and $y = 3$.

SOLUTION We begin by substituting -2 for x and 3 for y in the polynomial.

$2x^3y + xy^2 + 7x - 3$ — This is the given polynomial.

$= 2(-2)^3 \cdot 3 + (-2) \cdot 3^2 + 7(-2) - 3$ — Replace x with -2 and y with 3.

$= 2(-8) \cdot 3 + (-2) \cdot 9 + 7(-2) - 3$ — Evaluate exponential expressions: $(-2)^3 = (-2)(-2)(-2) = -8$ and $3^2 = 3 \cdot 3 = 9$.

$= -48 + (-18) + (-14) - 3$ — Perform the indicated multiplications.

$= -83$ — Add from left to right. ■

CHECK POINT 1 Evaluate $3x^3y + xy^2 + 5y + 6$ for $x = -1$ and $y = 5$.

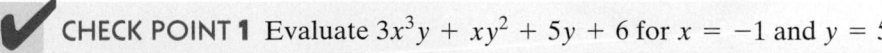

2 Understand the vocabulary of polynomials in two variables.

Describing Polynomials in Two Variables In this section, we will limit our discussion of polynomials in several variables to two variables.

In general, a **polynomial in two variables**, x and y, contains the sum of one or more monomials in the form ax^ny^m. The constant, a, is the **coefficient**. The exponents, n and m, represent whole numbers. The **degree** of the monomial ax^ny^m is $n + m$. We'll use the polynomial from the construction industry to illustrate these ideas.

The coefficients are $\frac{1}{4}$, -2, and 4.

$$\frac{1}{4}x^2y \quad - 2xy \quad + 4y$$

Degree of monomial: $2 + 1 = 3$ | Degree of monomial: $1 + 1 = 2$ | Degree of monomial: 1

The **degree of a polynomial in two variables** is the highest degree of all its terms. For the preceding polynomial, the degree is 3.

EXAMPLE 2 Using the Vocabulary of Polynomials

Determine the coefficient of each term, the degree of each term, and the degree of the polynomial:

$$7x^2y^3 - 17x^4y^2 + xy - 6y^2 + 9.$$

SOLUTION

Term	Coefficient	Degree (Sum of Exponents on the Variables)
$7x^2y^3$	7	$2 + 3 = 5$
$-17x^4y^2$	-17	$4 + 2 = 6$
xy	1	$1 + 1 = 2$
$-6y^2$	-6	2
9	9	0

Think of xy as $1x^1y^1$.

The degree of the polynomial is the highest degree of all its terms, which is 6. ■

✔ **CHECK POINT 2** Determine the coefficient of each term, the degree of each term, and the degree of the polynomial:

$$8x^4y^5 - 7x^3y^2 - x^2y - 5x + 11.$$

3 Add and subtract polynomials in several variables.

Adding and Subtracting Polynomials in Several Variables Polynomials in several variables are added by combining like terms. For example, we can add the monomials $-7xy^2$ and $13xy^2$ as follows:

$$-7xy^2 + 13xy^2 = (-7 + 13)xy^2 = 6xy^2.$$

These like terms both contain the variable factors x and y^2.

Add coefficients and keep the same variable factors, xy^2.

EXAMPLE 3 Adding Polynomials in Two Variables

Add: $(6xy^2 - 5xy + 7) + (9xy^2 + 2xy - 6)$.

SOLUTION

$$(6xy^2 - 5xy + 7) + (9xy^2 + 2xy - 6)$$
$$= (6xy^2 + 9xy^2) + (-5xy + 2xy) + (7 - 6) \quad \text{Group like terms.}$$
$$= 15xy^2 - 3xy + 1 \quad \text{Combine like terms by adding coefficients and keeping the same variable factors.}$$

CHECK POINT 3 Add: $(-8x^2y - 3xy + 6) + (10x^2y + 5xy - 10)$.

We subtract polynomials in two variables just as we did when subtracting polynomials in one variable. Add the first polynomial and the opposite of the polynomial being subtracted.

EXAMPLE 4 Subtracting Polynomials in Two Variables

Subtract:

$$(5x^3 - 9x^2y + 3xy^2 - 4) - (3x^3 - 6x^2y - 2xy^2 + 3).$$

SOLUTION

$$(5x^3 - 9x^2y + 3xy^2 - 4) - (3x^3 - 6x^2y - 2xy^2 + 3)$$

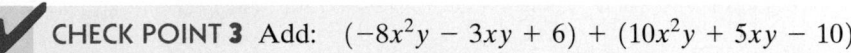

Change the sign of each coefficient.

$$= (5x^3 - 9x^2y + 3xy^2 - 4) + (-3x^3 + 6x^2y + 2xy^2 - 3) \quad \text{Add the opposite of the polynomial being subtracted.}$$
$$= (5x^3 - 3x^3) + (-9x^2y + 6x^2y) + (3xy^2 + 2xy^2) + (-4 - 3) \quad \text{Group like terms.}$$
$$= 2x^3 - 3x^2y + 5xy^2 - 7 \quad \text{Combine like terms by adding coefficients and keeping the same variable factors.}$$

CHECK POINT 4 Subtract: $(7x^3 - 10x^2y + 2xy^2 - 5) - (4x^3 - 12x^2y - 3xy^2 + 5)$.

4 Multiply polynomials in several variables.

Multiplying Polynomials in Several Variables The product of monomials forms the basis of polynomial multiplication. As with monomials in one variable, multiplication can be done mentally by multiplying coefficients and adding exponents on variables with the same base.

EXAMPLE 5 Multiplying Monomials

Multiply: $(7x^2y)(5x^3y^2)$.

SOLUTION

$$(7x^2y)(5x^3y^2)$$
$$= (7 \cdot 5)(x^2 \cdot x^3)(y \cdot y^2)$$ This regrouping can be worked mentally.
$$= 35x^{2+3}y^{1+2}$$ Multiply coefficients and add exponents on variables with same base.
$$= 35x^5y^3$$ Simplify. ■

CHECK POINT 5 Multiply: $(6xy^3)(10x^4y^2)$.

How do we multiply a monomial and a polynomial that is not a monomial? As we did with polynomials in one variable, multiply each term of the polynomial by the monomial.

EXAMPLE 6 Multiplying a Monomial and a Polynomial

Multiply: $3x^2y(4x^3y^2 - 6x^2y + 2)$.

SOLUTION

$$3x^2y(4x^3y^2 - 6x^2y + 2)$$
$$= 3x^2y \cdot 4x^3y^2 - 3x^2y \cdot 6x^2y + 3x^2y \cdot 2$$ Use the distributive property.
$$= 12x^{2+3}y^{1+2} - 18x^{2+2}y^{1+1} + 6x^2y$$ Multiply coefficients and add exponents on variables with the same base.
$$= 12x^5y^3 - 18x^4y^2 + 6x^2y$$ Simplify. ■

CHECK POINT 6 Multiply: $6xy^2(10x^4y^5 - 2x^2y + 3)$.

FOIL and the special-products formulas can be used to multiply polynomials in several variables.

EXAMPLE 7 Multiplying Polynomials in Two Variables

Multiply: **a.** $(x + 4y)(3x - 5y)$ **b.** $(5x + 3y)^2$.

SOLUTION We will perform the multiplication in part (a) using the FOIL method. We will multiply in part (b) using the formula for the square of a binomial, $(A + B)^2$.

a. $(x + 4y)(3x - 5y)$ Multiply these binomials using the FOIL method.

F O I L

$$= (x)(3x) + (x)(-5y) + (4y)(3x) + (4y)(-5y)$$
$$= 3x^2 - 5xy + 12xy - 20y^2$$
$$= 3x^2 + 7xy - 20y^2$$ Combine like terms.

$$(A + B)^2 = A^2 + 2 \cdot A \cdot B + B^2$$

b. $(5x + 3y)^2 = (5x)^2 + 2(5x)(3y) + (3y)^2$
$$= 25x^2 + 30xy + 9y^2$$ ■

CHECK POINT 7 Multiply:

a. $(7x - 6y)(3x - y)$ **b.** $(2x + 4y)^2$.

EXAMPLE 8 Multiplying Polynomials in Two Variables

Multiply: **a.** $(4x^2y + 3y)(4x^2y - 3y)$ **b.** $(x + y)(x^2 - xy + y^2)$.

SOLUTION We perform the multiplication in part (a) using the formula for the product of the sum and difference of two terms. We perform the multiplication in part (b) by multiplying each term of the trinomial, $x^2 - xy + y^2$, by x and y, respectively, and then adding like terms.

$$(A + B) \cdot (A - B) = A^2 - B^2$$

a. $(4x^2y + 3y)(4x^2y - 3y) = (4x^2y)^2 - (3y)^2$

$= 16x^4y^2 - 9y^2$

b. $(x + y)(x^2 - xy + y^2)$

$= x(x^2 - xy + y^2) + y(x^2 - xy + y^2)$ Multiply the trinomial by each term of the binomial.

$= x \cdot x^2 - x \cdot xy + x \cdot y^2 + y \cdot x^2 - y \cdot xy + y \cdot y^2$ Use the distributive property.

$= x^3 - x^2y + xy^2 + x^2y - xy^2 + y^3$ Add exponents on variables with the same base.

$= x^3 + y^3$ Combine like terms: $-x^2y + x^2y = 0$ and $xy^2 - xy^2 = 0$. ■

CHECK POINT 8 Multiply:

a. $(6xy^2 + 5x)(6xy^2 - 5x)$

b. $(x - y)(x^2 + xy + y^2)$.

5.4 EXERCISE SET

Student Solutions Manual | CD/Video | PH Math/Tutor Center | MathXL Tutorials on CD | MathXL® | MyMathLab | Interactmath.com

Practice Exercises

In Exercises 1–6, evaluate each polynomial for $x = 2$ and $y = -3$.

1. $x^2 + 2xy + y^2$

2. $x^2 + 3xy + y^2$

3. $xy^3 - xy + 1$

4. $x^3y - xy + 2$

5. $2x^2y - 5y + 3$

6. $3x^2y - 4y + 5$

In Exercises 7–8, determine the coefficient of each term, the degree of each term, and the degree of the polynomial.

7. $x^3y^2 - 5x^2y^7 + 6y^2 - 3$

8. $12x^4y - 5x^3y^7 - x^2 + 4$

In Exercises 9–20, add or subtract as indicated.

9. $(5x^2y - 3xy) + (2x^2y - xy)$

10. $(-2x^2y + xy) + (4x^2y + 7xy)$

11. $(4x^2y + 8xy + 11) + (-2x^2y + 5xy + 2)$

12. $(7x^2y + 5xy + 13) + (-3x^2y + 6xy + 4)$

13. $(7x^4y^2 - 5x^2y^2 + 3xy) + (-18x^4y^2 - 6x^2y^2 - xy)$

14. $(6x^4y^2 - 10x^2y^2 + 7xy) + (-12x^4y^2 - 3x^2y^2 - xy)$

15. $(x^3 + 7xy - 5y^2) - (6x^3 - xy + 4y^2)$

16. $(x^4 - 7xy - 5y^3) - (6x^4 - 3xy + 4y^3)$

17. $(3x^4y^2 + 5x^3y - 3y) - (2x^4y^2 - 3x^3y - 4y + 6x)$

18. $(5x^4y^2 + 6x^3y - 7y) - (3x^4y^2 - 5x^3y - 6y + 8x)$

19. $(x^3 - y^3) - (-4x^3 - x^2y + xy^2 + 3y^3)$

20. $(x^3 - y^3) - (-6x^3 + x^2y - xy^2 + 2y^3)$

21. Add: $\begin{array}{r} 5x^2y^2 - 4xy^2 + 6y^2 \\ \underline{-8x^2y^2 + 5xy^2 - y^2} \end{array}$

22. Add: $\begin{array}{r} 7a^2b^2 - 5ab^2 + 6b^2 \\ \underline{-10a^2b^2 + 6ab^2 + 6b^2} \end{array}$

23. Subtract: $\begin{array}{r} 3a^2b^4 - 5ab^2 + 7ab \\ \underline{-(-5a^2b^4 - 8ab^2 - ab)} \end{array}$

24. Subtract: $\begin{array}{r} 13x^2y^4 - 17xy^2 + xy \\ \underline{-(-7x^2y^4 - 8xy^2 - xy)} \end{array}$

25. Subtract $11x - 5y$ from the sum of $7x + 13y$ and $-26x + 19y$.

26. Subtract $23x - 5y$ from the sum of $6x + 15y$ and $x - 19y$.

In Exercises 27–76, find each product.

27. $(5x^2y)(8xy)$

28. $(10x^2y)(5xy)$

29. $(-8x^3y^4)(3x^2y^5)$

30. $(7x^4y^5)(-10x^7y^{11})$

31. $9xy(5x + 2y)$

32. $7xy(8x + 3y)$

33. $5xy^2(10x^2 - 3y)$

34. $6x^2y(5x^2 - 9y)$

35. $4ab^2(7a^2b^3 + 2ab)$

36. $2ab^2(20a^2b^3 + 11ab)$

37. $-b(a^2 - ab + b^2)$

38. $-b(a^3 - ab + b^3)$

39. $(x + 5y)(7x + 3y)$

40. $(x + 9y)(6x + 7y)$

41. $(x - 3y)(2x + 7y)$

42. $(3x - y)(2x + 5y)$

43. $(3xy - 1)(5xy + 2)$

44. $(7xy + 1)(2xy - 3)$

45. $(2x + 3y)^2$

46. $(2x + 5y)^2$

47. $(xy - 3)^2$

48. $(xy - 5)^2$

49. $(x^2 + y^2)^2$

50. $(2x^2 + y^2)^2$

51. $(x^2 - 2y^2)^2$

52. $(x^2 - y^2)^2$

53. $(3x + y)(3x - y)$

54. $(x + 5y)(x - 5y)$

55. $(ab + 1)(ab - 1)$

56. $(ab + 2)(ab - 2)$

57. $(x + y^2)(x - y^2)$

58. $(x^2 + y)(x^2 - y)$

59. $(3a^2b + a)(3a^2b - a)$

60. $(5a^2b + a)(5a^2b - a)$

61. $(3xy^2 - 4y)(3xy^2 + 4y)$

62. $(7xy^2 - 10y)(7xy^2 + 10y)$

63. $(a + b)(a^2 - b^2)$

64. $(a - b)(a^2 + b^2)$

65. $(x + y)(x^2 + 3xy + y^2)$

66. $(x + y)(x^2 + 5xy + y^2)$

67. $(x - y)(x^2 - 3xy + y^2)$

68. $(x - y)(x^2 - 4xy + y^2)$

69. $(xy + ab)(xy - ab)$

70. $(xy + ab^2)(xy - ab^2)$

71. $(x^2 + 1)(x^4y + x^2 + 1)$

72. $(x^2 + 1)(xy^4 + y^2 + 1)$

73. $(x^2y^2 - 3)^2$

74. $(x^2y^2 - 5)^2$

75. $(x + y + 1)(x + y - 1)$

76. $(x + y + 1)(x - y + 1)$

In Exercises 77–80 write a polynomial in two variables that describes the total area of each shaded region. Express each polynomial as the sum or difference of terms.

77.

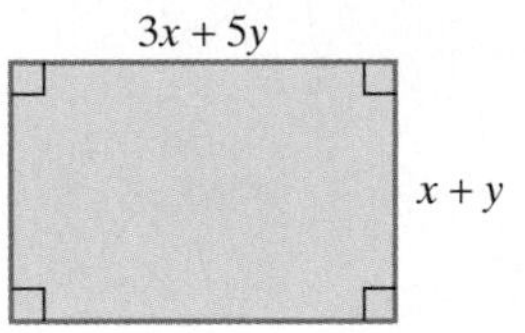

78.

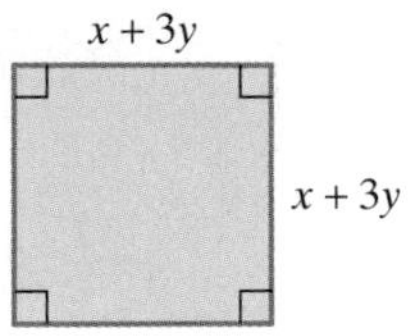

79.

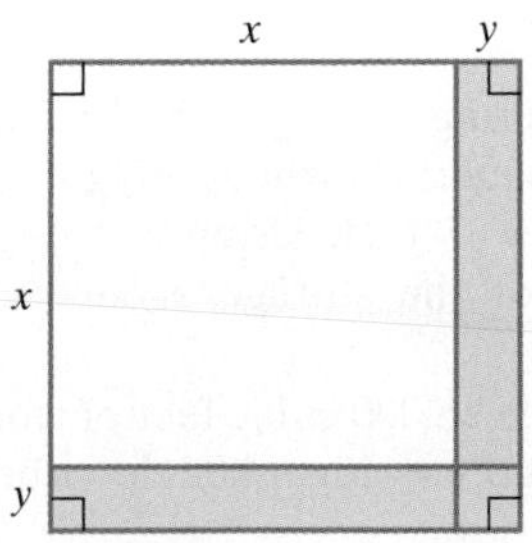

80.

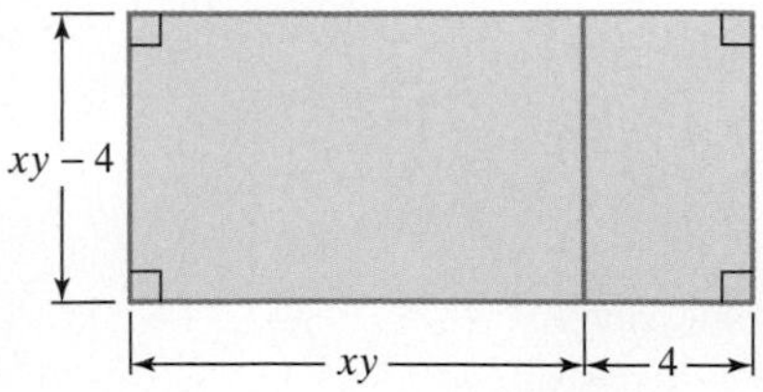

Practice Plus

In Exercises 81–86, find each product. As we said in the Section 5.3 opener, cut to the chase in each part of the polynomial multiplication: Use only the special-product formula for the sum and difference of two terms or the formulas for the square of a binomial.

81. $[(x^3y^3 + 1)(x^3y^3 - 1)]^2$

82. $[(1 - a^3b^3)(1 + a^3b^3)]^2$

83. $(xy - 3)^2(xy + 3)^2$ (Do not begin by squaring a binomial.)

84. $(ab - 4)^2(ab + 4)^2$ (Do not begin by squaring a binomial.)

85. $[x + y + z][x - (y + z)]$

86. $(a - b - c)(a + b + c)$

Application Exercises

87. The number of board feet, N, that can be manufactured from a tree with a diameter of x inches and a length of y feet is modeled by the formula

$$N = \frac{1}{4}x^2y - 2xy + 4y.$$

A building contractor estimates that 3000 board feet of lumber is needed for a job. The lumber company has just milled a fresh load of timber from 20 trees that averaged 10 inches in diameter and 16 feet in length. Is this enough to complete the job? If not, how many additional board feet of lumber is needed?

88. The storage shed shown in the figure has a volume given by the polynomial

$$2x^2y + \frac{1}{2}\pi x^2y.$$

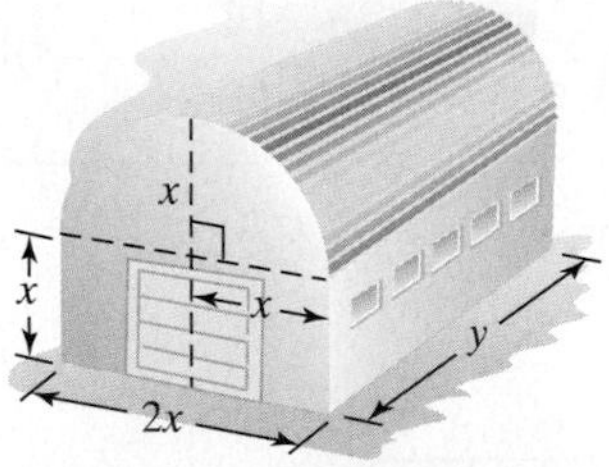

a. A small business is considering having a shed installed like the one shown in the figure. The shed's height, $2x$, is 26 feet and its length, y, is 27 feet. Using $x = 13$ and $y = 27$, find the volume of the storage shed.

b. The business requires at least 18,000 cubic feet of storage space. Should they construct the storage shed described in part (a)?

An object that is falling or vertically projected into the air has its height, in feet, above the ground given by

$$s = -16t^2 + v_0t + s_0,$$

where s is the height, in feet, v_0 is the original velocity of the object, in feet per second, t is the time the object is in motion, in seconds, and s_0 is the height, in feet, from which the object is dropped or projected. The figure shows that a ball is thrown straight up from a rooftop at an original velocity of 80 feet per second from a height of 96 feet. The ball misses the rooftop on its way down and eventually strikes the ground. Use the formula and this information to solve Exercises 89–91.

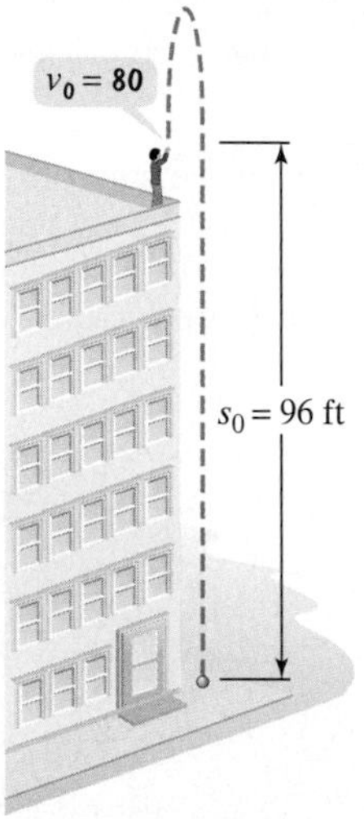

89. How high above the ground will the ball be 2 seconds after being thrown?

90. How high above the ground will the ball be 4 seconds after being thrown?

91. How high above the ground will the ball be 6 seconds after being thrown? Describe what this means in practical terms.

The graph visually displays the information about the thrown ball described in Exercises 89–91. The horizontal axis represents the ball's time in motion, in seconds. The vertical axis represents the ball's height above the ground, in feet. Use the graph to solve Exercises 92–97.

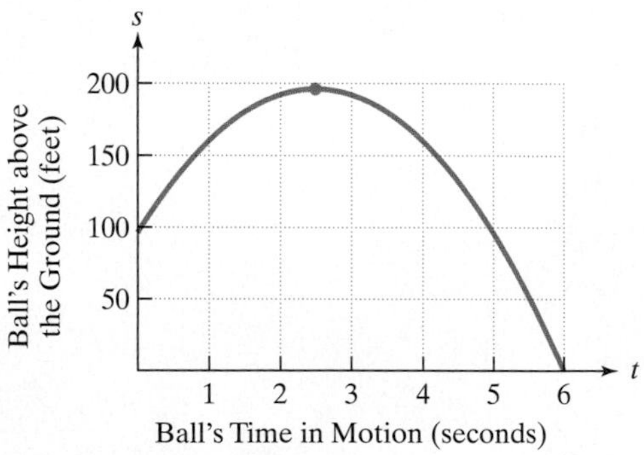

92. During which time period is the ball rising?

93. During which time period is the ball falling?

94. Identify your answer from Exercise 90 as a point on the graph.

95. Identify your answer from Exercise 89 as a point on the graph.

96. After how many seconds does the ball strike the ground?

97. After how many seconds does the ball reach its maximum height above the ground? What is a reasonable estimate of this maximum height?

Writing in Mathematics

98. What is a polynomial in two variables? Provide an example with your description.

99. Explain how to find the degree of a polynomial in two variables.

100. Suppose that you take up sky diving. Explain how to use the formula for Exercises 89–91 to determine your height above the ground at every instant of your fall.

Critical Thinking Exercises

101. Which one of the following is true?

a. The degree of $5x^{24} - 3x^{16}y^9 - 7xy^2 + 6$ is 24.

b. In the polynomial $4x^2y + x^3y^2 + 3x^2y^3 + 7y$, the term x^3y^2 has degree 5 and no numerical coefficient.

c. $(2x + 3 - 5y)(2x + 3 + 5y) = 4x^2 + 12x + 9 - 25y^2$

d. $(6x^2y - 7xy - 4) - (6x^2y + 7xy - 4) = 0$

In Exercises 102–103, find a polynomial in two variables that describes the area of the shaded region of each figure. Write the polynomial as the sum or difference of terms.

102.

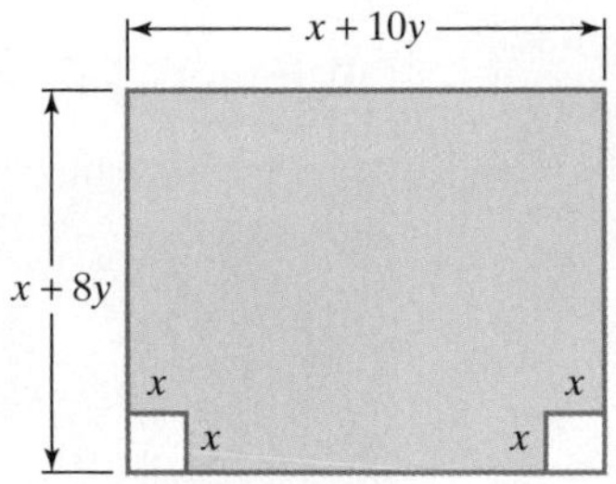

103.

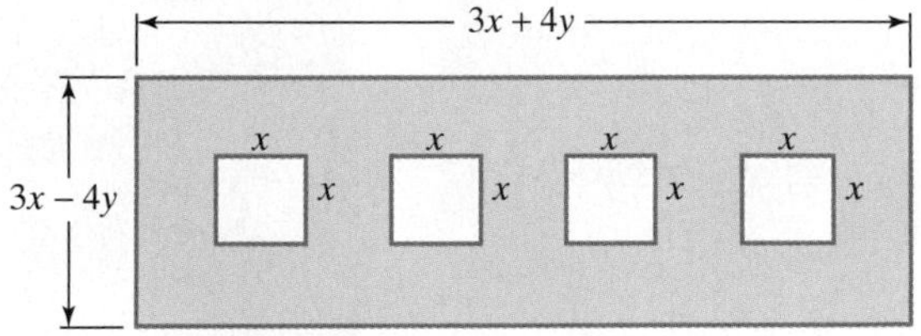

104. Use the formulas for the volume of a rectangular solid and a cylinder to derive the polynomial in Exercise 88 that describes the volume of the storage building.

Review Exercises

105. Solve for W: $R = \frac{L + 3W}{2}$. (Section 2.4, Example 4)

106. Subtract: $-6.4 - (-10.2)$. (Section 1.6, Example 2)

107. Write the point-slope and slope-intercept equations of a line passing through the point $(-2, 5)$ and parallel to the line whose equation is $3x - y = 9$. (Section 3.5, Example 3)

MID-CHAPTER CHECK POINT

CHAPTER 5

What You Know: We learned to add, subtract, and multiply polynomials. We used a number of fast methods for finding products of polynomials, including the FOIL method for multiplying binomials, a special-product formula for the product of the sum and difference of two terms $[(A+B)(A-B) = A^2 - B^2]$, and special-product formulas for squaring binomials $[(A+B)^2 = A^2 + 2AB + B^2;$ $(A - B)^2 = A^2 - 2AB + B^2]$. Finally, we applied all of these operations to polynomials in several variables.

In Exercises 1–21, perform the indicated operations.

1. $(11x^2y^3)(-5x^2y^3)$

2. $11x^2y^3 - 5x^2y^3$

3. $(3x + 5)(4x - 7)$

4. $(3x + 5) - (4x - 7)$

5. $(2x - 5)(x^2 - 3x + 1)$

6. $(2x - 5) + (x^2 - 3x + 1)$

7. $(8x - 3)^2$

8. $(-10x^4)(-7x^5)$

9. $(x^2 + 2)(x^2 - 2)$

10. $(x^2 + 2)^2$

11. $(9a - 10b)(2a + b)$

12. $7x^2(10x^3 - 2x + 3)$

13. $(3a^2b^3 - ab + 4b^2) - (-2a^2b^3 - 3ab + 5b^2)$

14. $2(3y - 5)(3y + 5)$

15. $(-9x^3 + 5x^2 - 2x + 7) + (11x^3 - 6x^2 + 3x - 7)$

16. $10x^2 - 8xy - 3(y^2 - xy)$

17. $(-2x^5 + x^4 - 3x + 10) - (2x^5 - 6x^4 + 7x - 13)$

18. $(x + 3y)(x^2 - 3xy + 9y^2)$

19. $(5x^4 + 4)(2x^3 - 1)$

20. $(y - 6z)^2$

21. $(2x + 3)(2x - 3) - (5x + 4)(5x - 4)$

SECTION 5.5

Objectives

1. Use the quotient rule for exponents.
2. Use the zero-exponent rule.
3. Use the quotients-to-powers rule.
4. Divide monomials.
5. Check polynomial division.
6. Divide a polynomial by a monomial.

DIVIDING POLYNOMIALS

To play the part of Charlie Chaplin, actor Robert Downey Jr. (1965–) learned to pantomime, speak two British dialects, and play left-handed tennis. His problems with substance abuse have fueled the debate over whether the illness should be handled by our criminal justice system or by health care professionals.

As you learn more mathematics, you will discover new ways to describe your world. Almost anything that you can think of involving variables can be modeled by a formula. For example, a polynomial models the annual number of drug convictions in the United States and another polynomial models drug arrests. By dividing the respective polynomials, we obtain an algebraic expression that describes the conviction rate for drug arrests.

In the next two sections, you will learn how to divide polynomials. Before turning to polynomial division, we must develop some additional rules for working with exponents.

1 Use the quotient rule for exponents.

The Quotient Rule for Exponents Consider the quotient of two exponential expressions, such as the quotient of 2^7 and 2^3. We are dividing 7 factors of 2 by 3 factors of 2. We are left with 4 factors of 2:

7 factors of 2

$$\frac{2^7}{2^3} = \frac{2\cdot2\cdot2\cdot2\cdot2\cdot2\cdot2}{2\cdot2\cdot2} = \frac{\cancel{2}\cdot\cancel{2}\cdot\cancel{2}\cdot2\cdot2\cdot2\cdot2}{\cancel{2}\cdot\cancel{2}\cdot\cancel{2}} = 2\cdot2\cdot2\cdot2$$

3 factors of 2 | Divide out pairs of factors: $\frac{2}{2} = 1$. | 4 factors of 2

Thus,

$$\frac{2^7}{2^3} = 2^4.$$

We can quickly find the exponent, 4, on the quotient by subtracting the original exponents:

$$\frac{2^7}{2^3} = 2^{7-3}.$$

This suggests the following rule:

THE QUOTIENT RULE

$$\frac{b^m}{b^n} = b^{m-n}, \quad b \neq 0$$

When dividing exponential expressions with the same nonzero base, subtract the exponent in the denominator from the exponent in the numerator. Use this difference as the exponent of the common base.

EXAMPLE 1 Using the Quotient Rule

Divide each expression using the quotient rule:

a. $\frac{2^8}{2^4}$ **b.** $\frac{x^{13}}{x^3}$ **c.** $\frac{y^{15}}{y}$.

SOLUTION

a. $\frac{2^8}{2^4} = 2^{8-4} = 2^4 \text{ or } 16$

b. $\frac{x^{13}}{x^3} = x^{13-3} = x^{10}$

c. $\frac{y^{15}}{y} = \frac{y^{15}}{y^1} = y^{15-1} = y^{14}$ ■

CHECK POINT 1 Divide each expression using the quotient rule:

a. $\frac{5^{12}}{5^4}$ **b.** $\frac{x^9}{x^2}$ **c.** $\frac{y^{20}}{y}$.

2 Use the zero-exponent rule.

Zero as an Exponent A nonzero base can be raised to the 0 power. The quotient rule can be used to help determine what zero as an exponent should mean. Consider the quotient of b^4 and b^4, where b is not zero. We can determine this quotient in two ways.

$$\frac{b^4}{b^4} = 1 \qquad \frac{b^4}{b^4} = b^{4-4} = b^0$$

Any nonzero expression divided by itself is 1.

Use the quotient rule and subtract exponents.

This means that b^0 must equal 1.

THE ZERO-EXPONENT RULE If b is any real number other than 0,

$$b^0 = 1.$$

EXAMPLE 2 Using the Zero-Exponent Rule

Use the zero-exponent rule to simplify each expression:

a. 7^0 **b.** $(-5)^0$ **c.** -5^0 **d.** $10x^0$ **e.** $(10x)^0$.

SOLUTION

a. $7^0 = 1$ Any nonzero number raised to the 0 power is 1.

b. $(-5)^0 = 1$ Any nonzero number raised to the 0 power is 1.

c. $-5^0 = -1$ $\qquad -5^0 = -(5^0) = -1$

Only 5 is raised to the 0 power.

d. $10x^0 = 10 \cdot 1 = 10$

Only x is raised to the 0 power.

e. $(10x)^0 = 1$

The entire expression, $10x$, is raised to the 0 power.

CHECK POINT 2 Use the zero-exponent rule to simplify each expression:

a. 14^0 **b.** $(-10)^0$ **c.** -10^0 **d.** $20x^0$ **e.** $(20x)^0$.

3 Use the quotients-to-powers rule.

The Quotients-to-Powers Rule for Exponents We have seen that when a product is raised to a power, we raise every factor in the product to the power:

$$(ab)^n = a^n b^n.$$

There is a similar property for raising a quotient to a power.

QUOTIENTS TO POWERS If a and b are real numbers and b is nonzero, then

$$\left(\frac{a}{b}\right)^n = \frac{a^n}{b^n}.$$

When a quotient is raised to a power, raise the numerator to the power and divide by the denominator raised to the power.

EXAMPLE 3 Using the Quotients-to-Powers Rule

Simplify each expression using the quotients-to-powers rule:

a. $\left(\frac{x}{4}\right)^2$ **b.** $\left(\frac{x^2}{5}\right)^3$ **c.** $\left(\frac{2a^3}{b^4}\right)^5$.

SOLUTION

a. $\left(\frac{x}{4}\right)^2 = \frac{x^2}{4^2} = \frac{x^2}{16}$ Square the numerator and the denominator.

b. $\left(\frac{x^2}{5}\right)^3 = \frac{(x^2)^3}{5^3} = \frac{x^{2\cdot 3}}{5\cdot 5\cdot 5} = \frac{x^6}{125}$ Cube the numerator and the denominator.

c. $\left(\frac{2a^3}{b^4}\right)^5 = \frac{(2a^3)^5}{(b^4)^5}$ Raise the numerator and the denominator to the fifth power.

$= \frac{2^5(a^3)^5}{(b^4)^5}$ Raise each factor in the numerator to the fifth power.

$= \frac{2^5 a^{3\cdot 5}}{b^{4\cdot 5}}$ To raise exponential expressions to powers, multiply exponents: $(b^m)^n = b^{mn}$.

$= \frac{32a^{15}}{b^{20}}$ Simplify.

CHECK POINT 3 Simplify each expression using the quotients-to-powers rule:

a. $\left(\frac{x}{5}\right)^2$ **b.** $\left(\frac{x^4}{2}\right)^3$ **c.** $\left(\frac{2a^{10}}{b^3}\right)^4$.

STUDY TIP

Try to avoid the following common errors that can occur when simplifying exponential expressions.

Correct	Incorrect	Description of Error
$\frac{2^{20}}{2^4} = 2^{20-4} = 2^{16}$	~~$\frac{2^{20}}{2^4} = 2^5$~~	Exponents should be subtracted, not divided.
$-8^0 = -1$	~~$-8^0 = 1$~~	Only 8 is raised to the 0 power.
$\left(\frac{x}{5}\right)^2 = \frac{x^2}{5^2} = \frac{x^2}{25}$	~~$\left(\frac{x}{5}\right)^2 = \frac{x^2}{5}$~~	The numerator and denominator must both be squared.

4 Divide monomials.

Dividing Monomials Now that we have developed three additional properties of exponents, we are ready to turn to polynomial division. We begin with the quotient of two monomials, such as $16x^{14}$ and $8x^2$. This quotient is obtained by dividing the coefficients, 16 and 8, and then dividing the variables using the quotient rule for exponents.

$$\frac{16x^{14}}{8x^2} = \frac{16}{8}x^{14-2} = 2x^{12}$$

Divide coefficients and subtract exponents.

DIVIDING MONOMIALS To divide monomials, divide the coefficients and then divide the variables. Use the quotient rule for exponents to divide the variables: Keep the variable and subtract the exponents.

EXAMPLE 4 Dividing Monomials

Divide: **a.** $\frac{-12x^8}{4x^2}$ **b.** $\frac{2x^3}{8x^3}$ **c.** $\frac{15x^5y^4}{3x^2y}$.

SOLUTION

a. $\frac{-12x^8}{4x^2} = \frac{-12}{4}x^{8-2} = -3x^6$

b. $\frac{2x^3}{8x^3} = \frac{2}{8}x^{3-3} = \frac{1}{4}x^0 = \frac{1}{4}\cdot 1 = \frac{1}{4}$

c. $\frac{15x^5y^4}{3x^2y} = \frac{15}{3}x^{5-2}y^{4-1} = 5x^3y^3$ ■

STUDY TIP

Look at the solution to Example 4(b). Rather than subtracting exponents for division that results in a 0 exponent, you might prefer to divide out x^3.

$$\frac{2\cancel{x^3}}{8\cancel{x^3}} = \frac{2}{8} = \frac{1}{4}$$

CHECK POINT 4 Divide:

a. $\frac{-20x^{12}}{10x^4}$ **b.** $\frac{3x^4}{15x^4}$ **c.** $\frac{9x^6y^5}{3xy^2}$.

5 Check polynomial division.

Checking Division of Polynomial Problems The answer to a division problem can be checked. For example, consider the following problem:

$$\frac{15x^5y^4}{3x^2y} = 5x^3y^3.$$

Dividend: the polynomial you are dividing into

Divisor: the polynomial you are dividing by

Quotient: the answer to your division problem

The quotient is correct if the product of the divisor and the quotient is the dividend. Is the quotient shown in the preceding equation correct?

$$(3x^2y)(5x^3y^3) = 3 \cdot 5x^{2+3}y^{1+3} = 15x^5y^4.$$

Divisor — Quotient — This is the dividend.

Because the product of the divisor and the quotient is the dividend, the answer to the division problem is correct.

> **CHECKING DIVISION OF POLYNOMIALS** To check a quotient in a division problem, multiply the divisor and the quotient. If this product is the dividend, the quotient is correct.

6 Divide a polynomial by a monomial.

Dividing a Polynomial That Is Not a Monomial by a Monomial To divide a polynomial by a monomial, we divide each term of the polynomial by the monomial. For example,

$$\frac{10x^8 + 15x^6}{5x^3} = \frac{10x^8}{5x^3} + \frac{15x^6}{5x^3} = \frac{10}{5}x^{8-3} + \frac{15}{5}x^{6-3} = 2x^5 + 3x^3.$$

polynomial dividend — monomial divisor — Divide the first term by $5x^3$. — Divide the second term by $5x^3$.

STUDY TIP

Try to avoid this common error:

Incorrect:

$$\frac{x^4 - \overset{1}{\cancel{x}}}{\underset{1}{\cancel{x}}} = \frac{x^4 - 1}{1} = x^4 - 1$$

Correct:

$$\frac{x^4 - x}{x} = \frac{x^4}{x} - \frac{x}{x}$$
$$= x^{4-1} - x^{1-1}$$
$$= x^3 - x^0$$
$$= x^3 - 1$$

Don't leave out the 1.

Is the quotient correct? Multiply the divisor and the quotient.

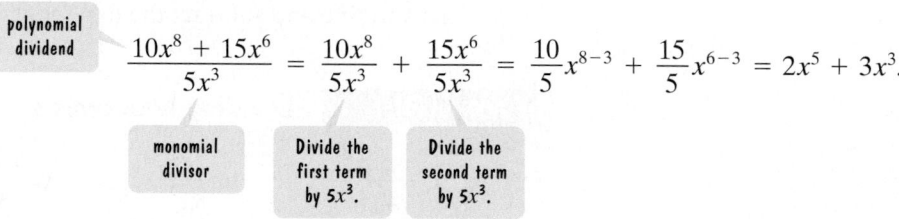

$$5x^3(2x^5 + 3x^3) = 5x^3 \cdot 2x^5 + 5x^3 \cdot 3x^3$$
$$= 5 \cdot 2x^{3+5} + 5 \cdot 3x^{3+3} = 10x^8 + 15x^6$$

Because this product gives the dividend, the quotient is correct.

> **DIVIDING A POLYNOMIAL THAT IS NOT A MONOMIAL BY A MONOMIAL** To divide a polynomial by a monomial, divide each term of the polynomial by the monomial.

EXAMPLE 5 Dividing a Polynomial by a Monomial

Find the quotient: $(-12x^8 + 4x^6 - 8x^3) \div 4x^2$.

SOLUTION

$$\frac{-12x^8 + 4x^6 - 8x^3}{4x^2}$$ Rewrite the division in a vertical format.

$$= \frac{-12x^8}{4x^2} + \frac{4x^6}{4x^2} - \frac{8x^3}{4x^2}$$ Divide each term of the polynomial by the monomial.

$$= \frac{-12}{4}x^{8-2} + \frac{4}{4}x^{6-2} - \frac{8}{4}x^{3-2}$$ Divide coefficients and subtract exponents.

$$= -3x^6 + x^4 - 2x$$ Simplify.

To check the answer, multiply the divisor and the quotient.

$$4x^2(-3x^6 + x^4 - 2x) = 4x^2(-3x^6) + 4x^2 \cdot x^4 - 4x^2(2x)$$
$$= 4(-3)x^{2+6} + 4x^{2+4} - 4 \cdot 2x^{2+1}$$
$$= -12x^8 + 4x^6 - 8x^3$$

Divisor: $4x^2$. Quotient: $-3x^6 + x^4 - 2x$. This is the dividend.

Because the product of the divisor and the quotient is the dividend, the answer—that is, the quotient—is correct. ■

CHECK POINT 5 Find the quotient: $(-15x^9 + 6x^5 - 9x^3) \div 3x^2$.

EXAMPLE 6 Dividing a Polynomial by a Monomial

Divide: $\dfrac{16x^5 - 9x^4 + 8x^3}{2x^3}$.

SOLUTION

$$\frac{16x^5 - 9x^4 + 8x^3}{2x^3}$$ This is the given polynomial division.

$$= \frac{16x^5}{2x^3} - \frac{9x^4}{2x^3} + \frac{8x^3}{2x^3}$$ Divide each term by $2x^3$.

$$= \frac{16}{2}x^{5-3} - \frac{9}{2}x^{4-3} + \frac{8}{2}x^{3-3}$$ Divide coefficients and subtract exponents. Did you immediately write the last term as 4?

$$= 8x^2 - \frac{9}{2}x + 4x^0$$ Simplify.

$$= 8x^2 - \frac{9}{2}x + 4$$ $x^0 = 1$, so $4x^0 = 4 \cdot 1 = 4$.

Check the answer by showing that the product of the divisor and the quotient is the dividend. ■

CHECK POINT 6 Divide: $\dfrac{25x^9 - 7x^4 + 10x^3}{5x^3}$.

EXAMPLE 7 **Dividing Polynomials in Two Variables**

Divide: $(15x^5y^4 - 3x^3y^2 + 9x^2y) \div 3x^2y$.

SOLUTION

$$\frac{15x^5y^4 - 3x^3y^2 + 9x^2y}{3x^2y}$$ Rewrite the division in a vertical format.

$$= \frac{15x^5y^4}{3x^2y} - \frac{3x^3y^2}{3x^2y} + \frac{9x^2y}{3x^2y}$$ Divide each term of the polynomial by the monomial.

$$= \frac{15}{3}x^{5-2}y^{4-1} - \frac{3}{3}x^{3-2}y^{2-1} + \frac{9}{3}x^{2-2}y^{1-1}$$ Divide coefficients and subtract exponents.

$$= 5x^3y^3 - xy + 3$$ Simplify.

Check the answer by showing that the product of the divisor and the quotient is the dividend. ■

CHECK POINT 7 Divide: $(18x^7y^6 - 6x^2y^3 + 60xy^2) \div 6xy^2$.

5.5 EXERCISE SET

Student Solutions Manual | CD/Video | PH Math/Tutor Center | MathXL Tutorials on CD | MathXL® | MyMathLab | Interactmath.com

Practice Exercises

In Exercises 1–10, divide each expression using the quotient rule. Express any numerical answers in exponential form.

1. $\frac{3^{20}}{3^5}$ **2.** $\frac{3^{30}}{3^{10}}$ **3.** $\frac{x^6}{x^2}$

4. $\frac{x^8}{x^4}$ **5.** $\frac{y^{13}}{y^5}$ **6.** $\frac{y^{19}}{y^6}$

7. $\frac{5^6 \cdot 2^8}{5^3 \cdot 2^4}$ **8.** $\frac{3^6 \cdot 2^8}{3^3 \cdot 2^4}$ **9.** $\frac{x^{100}y^{50}}{x^{25}y^{10}}$

10. $\frac{x^{200}y^{40}}{x^{25}y^{10}}$

In Exercises 11–24, use the zero-exponent rule to simplify each expression.

11. 2^0 **12.** 4^0 **13.** $(-2)^0$

14. $(-4)^0$ **15.** -2^0 **16.** -4^0

17. $100y^0$ **18.** $200y^0$ **19.** $(100y)^0$

20. $(200y)^0$ **21.** $-5^0 + (-5)^0$ **22.** $-6^0 + (-6)^0$

23. $-\pi^0 - (-\pi)^0$ **24.** $-\sqrt{3}^0 - (-\sqrt{3})^0$

In Exercises 25–36, simplify each expression using the quotients-to-powers rule. If possible, evaluate exponential expressions.

25. $\left(\frac{x}{3}\right)^2$ **26.** $\left(\frac{x}{5}\right)^2$

27. $\left(\frac{x^2}{4}\right)^3$ **28.** $\left(\frac{x^2}{3}\right)^3$

29. $\left(\frac{2x^3}{5}\right)^2$ **30.** $\left(\frac{3x^4}{7}\right)^2$

31. $\left(\frac{-4}{3a^3}\right)^3$ **32.** $\left(\frac{-5}{2a^3}\right)^3$

33. $\left(\frac{-2a^7}{b^4}\right)^5$ **34.** $\left(\frac{-2a^8}{b^3}\right)^5$

35. $\left(\frac{x^2y^3}{2z}\right)^4$ **36.** $\left(\frac{x^3y^2}{2z}\right)^4$

In Exercises 37–52, divide the monomials. Check each answer by showing that the product of the divisor and the quotient is the dividend.

37. $\frac{30x^{10}}{10x^5}$ **38.** $\frac{45x^{12}}{15x^4}$

39. $\frac{-8x^{22}}{4x^2}$ **40.** $\frac{-15x^{40}}{3x^4}$

41. $\frac{-9y^8}{18y^5}$ **42.** $\frac{-15y^{13}}{45y^9}$

43. $\frac{7y^{17}}{5y^5}$ **44.** $\frac{9y^{19}}{7y^{11}}$

45. $\frac{30x^7y^5}{5x^2y}$ **46.** $\frac{40x^9y^5}{2x^2y}$

47. $\frac{-18x^{14}y^2}{36x^2y^2}$ **48.** $\frac{-15x^{16}y^2}{45x^2y^2}$

49. $\frac{9x^{20}y^{20}}{7x^{20}y^{20}}$ **50.** $\frac{7x^{30}y^{30}}{15x^{30}y^{30}}$

51. $\dfrac{-5x^{10}y^{12}z^6}{50x^7y^3z^2}$ **52.** $\dfrac{-8x^{12}y^{10}z^4}{40x^2y^3z^2}$

In Exercises 53–78, divide the polynomial by the monomial. Check each answer by showing that the product of the divisor and the quotient is the dividend.

53. $\dfrac{10x^4 + 2x^3}{2}$ **54.** $\dfrac{20x^4 + 5x^3}{5}$

55. $\dfrac{14x^4 - 7x^3}{7x}$ **56.** $\dfrac{24x^4 - 8x^3}{8x}$

57. $\dfrac{y^7 - 9y^2 + y}{y}$ **58.** $\dfrac{y^8 - 11y^3 + y}{y}$

59. $\dfrac{24x^3 - 15x^2}{-3x}$ **60.** $\dfrac{10x^3 - 20x^2}{-5x}$

61. $\dfrac{18x^5 + 6x^4 + 9x^3}{3x^2}$

62. $\dfrac{18x^5 + 24x^4 + 12x^3}{6x^2}$

63. $\dfrac{12x^4 - 8x^3 + 40x^2}{4x}$

64. $\dfrac{49x^4 - 14x^3 + 70x^2}{-7x}$

65. $(4x^2 - 6x) \div x$

66. $(16y^2 - 8y) \div y$

67. $\dfrac{30z^3 + 10z^2}{-5z}$

68. $\dfrac{12y^4 - 42y^2}{-4y}$

69. $\dfrac{8x^3 + 6x^2 - 2x}{2x}$

70. $\dfrac{9x^3 + 12x^2 - 3x}{3x}$

71. $\dfrac{25x^7 - 15x^5 - 5x^4}{5x^3}$

72. $\dfrac{49x^7 - 28x^5 - 7x^4}{7x^3}$

73. $\dfrac{18x^7 - 9x^6 + 20x^5 - 10x^4}{-2x^4}$

74. $\dfrac{25x^8 - 50x^7 + 3x^6 - 40x^5}{-5x^5}$

75. $\dfrac{12x^2y^2 + 6x^2y - 15xy^2}{3xy}$

76. $\dfrac{18a^3b^2 - 9a^2b - 27ab^2}{9ab}$

77. $\dfrac{20x^7y^4 - 15x^3y^2 - 10x^2y}{-5x^2y}$

78. $\dfrac{8x^6y^3 - 12x^8y^2 - 4x^{14}y^6}{-4x^6y^2}$

Practice Plus

In Exercises 79–82, simplify each expression.

79. $\dfrac{2x^3(4x + 2) - 3x^2(2x - 4)}{2x^2}$

80. $\dfrac{6x^3(3x - 1) + 5x^2(6x - 3)}{3x^2}$

81. $\left(\dfrac{18x^2y^4}{9xy^2}\right) - \left(\dfrac{15x^5y^6}{5x^4y^4}\right)$

82. $\left(\dfrac{9x^3 + 6x^2}{3x}\right) - \left(\dfrac{12x^2y^2 - 4xy^2}{2xy^2}\right)$

83. Divide the sum of $(y + 5)^2$ and $(y + 5)(y - 5)$ by $2y$.

84. Divide the sum of $(y + 4)^2$ and $(y + 4)(y - 4)$ by $2y$.

In Exercises 85–86, the variable n in each exponent represents a natural number. Divide the polynomial by the monomial. Then use polynomial multiplication to check the quotient.

85. $\dfrac{12x^{15n} - 24x^{12n} + 8x^{3n}}{4x^{3n}}$

86. $\dfrac{35x^{10n} - 15x^{8n} + 25x^{2n}}{5x^{2n}}$

Application Exercises

87. The polynomial

$$28t^4 - 711t^3 + 5963t^2 - 1695t + 27{,}424$$

models the annual number of drug arrests in the United States t years after 1984. The polynomial

$$6t^4 - 207t^3 + 2128t^2 - 6622t + 15{,}220$$

models the annual number of drug convictions in the United States t years after 1984.

a. Write an algebraic expression that describes the conviction rate for drug arrests in the United States t years after 1984.

b. Can the polynomial division for the model in part (a) be performed using the methods that you learned in this section? Explain your answer.

88. The polynomial

$$0.0067t^2 + 2.56t + 250.39$$

models the U.S. population, in millions, t years after 1990. The polynomial

$$-35t^2 + 1160t + 10{,}890$$

models revenue, in millions of dollars, from video sales and rentals in the United States t years after 1990.

a. Write an algebraic expression that describes the average amount, in dollars, that each person in the United States spent on video sales and rentals t years after 1990.

b. Can the polynomial division for the model in part (a) be performed using the methods that you learned in this section? Explain your answer.

Writing in Mathematics

89. Explain the quotient rule for exponents. Use $\frac{3^6}{3^2}$ in your explanation.

90. Explain how to find any nonzero number to the 0 power.

91. Explain the difference between $(-7)^0$ and -7^0.

92. Explain how to simplify an expression that involves a quotient raised to a power. Provide an example with your explanation.

93. Explain how to divide monomials. Give an example.

94. Explain how to divide a polynomial that is not a monomial by a monomial. Give an example.

95. Are the expressions

$$\frac{12x^2 + 6x}{3x} \quad \text{and} \quad 4x + 2$$

equal for every value of x? Explain.

Critical Thinking Exercises

96. Which one of the following is true?

a. $x^{10} \div x^2 = x^5$ for all nonzero real numbers x.

b. $\frac{12x^3 - 6x}{2x} = 6x^2 - 6x$

c. $\frac{x^2 + x}{x} = x$

d. If a polynomial in x of degree 6 is divided by a monomial in x of degree 2, the degree of the quotient is 4.

97. What polynomial, when divided by $3x^2$, yields the trinomial $6x^6 - 9x^4 + 12x^2$ as a quotient?

In Exercises 98–99, find the missing coefficients and exponents designated by question marks.

98. $\frac{?x^8 - ?x^6}{3x^?} = 3x^5 - 4x^3$

99. $\frac{3x^{14} - 6x^{12} - ?x^7}{?x^?} = -x^7 + 2x^5 + 3$

Review Exercises

100. Find the absolute value: $|-20.3|$. (Section 1.2, Example 8)

101. Express $\frac{7}{8}$ as a decimal. (Section 1.2, Example 4)

102. Graph: $y = \frac{1}{3}x + 2$. (Section 3.4, Example 3)

SECTION 5.6

Objectives

1. Use long division to divide by a polynomial containing more than one term.
2. Divide polynomials using synthetic division.

LONG DIVISION OF POLYNOMIALS; SYNTHETIC DIVISION

For those of you who are dog lovers, you might still be thinking of the polynomial formula that models the age in human years, y, of a dog that is x years old, namely

$$y = -0.001618x^4 + 0.077326x^3 - 1.2367x^2 + 11.460x + 2.914.$$

Suppose that you are in your twenties, say 25. What is Fido's equivalent age? To answer this question, we must substitute 25 for y and solve the resulting polynomial equation for x:

$$25 = -0.001618x^4 + 0.077326x^3 - 1.2367x^2 + 11.460x + 2.914.$$

Don't panic! We won't be solving an equation as complicated as this—yet. You will learn to solve polynomial equations in which the highest power of the variable is 4 in more advanced algebra courses. Part of the method for solving such an equation involves the division of a polynomial by a binomial, such as

$$x + 3\overline{)x^2 + 10x + 21}.$$

Divisor has two terms and is a binomial.

The polynomial dividend has three terms and is a trinomial.

In this section, you will learn how to divide by a polynomial containing more than one term.

1 Use long division to divide by a polynomial containing more than one term.

The Steps in Dividing a Polynomial by a Binomial Dividing a polynomial by a binomial may remind you of long division.

DISCOVER FOR YOURSELF

Divide 3983 by 26 without the use of a calculator. Describe the process of the division using the four steps—*divide*, *multiply*, *subtract*, and *bring down*. What do you observe about this process? When does it come to an end?

When a divisor has more than one term, the four steps used to divide whole numbers—**divide, multiply, subtract, bring down the next term**—form the repetitive procedure for polynomial long division.

EXAMPLE 1 Dividing a Polynomial by a Binomial

Divide $x^2 + 10x + 21$ by $x + 3$.

SOLUTION The following steps illustrate how polynomial division is very similar to numerical division.

$$x + 3\overline{)x^2 + 10x + 21}$$

Arrange the terms of the dividend $(x^2 + 10x + 21)$ and the divisor $(x + 3)$ in descending powers of x.

$$\begin{array}{r} x \\ x + 3\overline{)x^2 + 10x + 21} \end{array}$$

Divide x^2 (the first term in the dividend) by x (the first term in the divisor): $\frac{x^2}{x} = x$. Align like terms.

$x(x + 3) = x^2 + 3x.$

$$\begin{array}{r} x \\ x + 3\overline{)x^2 + 10x + 21} \\ \underline{x^2 + 3x} \end{array}$$

Multiply each term in the divisor $(x + 3)$ by x, aligning terms of the product under like terms in the dividend.

$$\begin{array}{r} x \\ x + 3\overline{)x^2 + 10x + 21} \\ \underline{\ominus x^2 \ominus 3x} \\ 7x \end{array}$$

Change signs of the polynomial being subtracted.

Subtract $x^2 + 3x$ from $x^2 + 10x$ by changing the sign of each term in the lower expression and adding.

$$\begin{array}{r} x \\ x + 3\overline{)x^2 + 10x + 21} \\ \underline{x^2 + 3x} \\ 7x + 21 \end{array}$$

Bring down 21 from the original dividend and add algebraically to form a new dividend.

$$\begin{array}{r} x + 7 \\ x + 3\overline{)x^2 + 10x + 21} \\ \underline{x^2 + 3x} \\ 7x + 21 \end{array}$$

Find the second term of the quotient. Divide the first term of $7x + 21$ by x, the first term of the divisor: $\frac{7x}{x} = 7$.

$$\begin{array}{r}
x + 7 \\
x + 3\overline{)x^2 + 10x + 21} \\
\underline{x^2 + 3x} \\
7x + 21 \\
\underline{\ominus 7x \ominus 21} \\
0
\end{array}$$

$7(x + 3) = 7x + 21$

Remainder

Multiply the divisor $(x + 3)$ by 7, aligning under like terms in the new dividend. Then subtract to obtain the remainder of 0.

The quotient is $x + 7$ and the remainder is 0. We will not list a remainder of 0 in the answer. Thus,

$$\frac{x^2 + 10x + 21}{x + 3} = x + 7.$$

After performing polynomial long division, the answer can be checked. Find the product of the divisor and the quotient, and add the remainder. If the result is the dividend, the answer to the division problem is correct. For example, let's check our work in Example 1.

$$(x^2 + 10x + 21) \div (x + 3) = x + 7$$

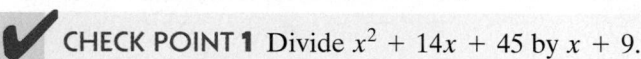

Dividend | Divisor | Quotient we wish to check

Multiply the divisor and the quotient and add the remainder, 0:

$$(x + 3)(x + 7) + 0 = x^2 + 7x + 3x + 21 + 0 = x^2 + 10x + 21$$

Divisor | Quotient | Remainder | This is the dividend.

Because we obtained the dividend, the quotient is correct.

CHECK POINT 1 Divide $x^2 + 14x + 45$ by $x + 9$.

Before considering additional examples, let's summarize the general procedure for dividing by a polynomial that contains more than one term.

LONG DIVISION OF POLYNOMIALS

1. **Arrange the terms** of both the dividend and the divisor in descending powers of any variable.
2. **Divide** the first term in the dividend by the first term in the divisor. The result is the first term of the quotient.
3. **Multiply** every term in the divisor by the first term in the quotient. Write the resulting product beneath the dividend with like terms lined up.
4. **Subtract** the product from the dividend.
5. **Bring down** the next term in the original dividend and write it next to the remainder to form a new dividend.
6. Use this new expression as the dividend and repeat this process until the remainder can no longer be divided. This will occur when the degree of the remainder (the highest exponent on a variable in the remainder) is less than the degree of the divisor.

In our next division, we will obtain a nonzero remainder.

EXAMPLE 2 Long Division of Polynomials

Divide: $\frac{7x - 9 - 4x^2 + 4x^3}{2x - 1}$.

SOLUTION We begin by writing the dividend in descending powers of x.

$$7x - 9 - 4x^2 + 4x^3 = 4x^3 - 4x^2 + 7x - 9$$

Think of 9 as $9x^0$. The powers descend from 3 to 0.

$$2x - 1\overline{)4x^3 - 4x^2 + 7x - 9}$$

This is the problem with the dividend in descending powers of x.

$$\begin{array}{r} 2x^2 \\ 2x - 1\overline{)4x^3 - 4x^2 + 7x - 9} \end{array}$$

Divide: $\frac{4x^3}{2x} = 2x^2$.

$$\begin{array}{r} 2x^2 \\ 2x - 1\overline{)4x^3 - 4x^2 + 7x - 9} \\ 4x^3 - 2x^2 \end{array}$$

$2x^2(2x - 1) = 4x^3 - 2x^2$

Multiply: $2x^2(2x - 1) = 4x^3 - 2x^2$.

$$\begin{array}{r} 2x^2 \\ 2x - 1\overline{)4x^3 - 4x^2 + 7x - 9} \\ \ominus 4x^3 \overset{\oplus}{-} 2x^2 \\ \hline - 2x^2 \end{array}$$

Change signs of the polynomial being subtracted.

Subtract: $4x^3 - 4x^2 - (4x^3 - 2x^2)$
$= 4x^3 - 4x^2 - 4x^3 + 2x^2$
$= -2x^2$.

$$\begin{array}{r} 2x^2 \\ 2x - 1\overline{)4x^3 - 4x^2 + 7x - 9} \\ 4x^3 - 2x^2 \downarrow \\ \hline - 2x^2 + 7x \end{array}$$

Bring down 7x. The new dividend is $-2x^2 + 7x$.

$$\begin{array}{r} 2x^2 - x \\ 2x - 1\overline{)4x^3 - 4x^2 + 7x - 9} \\ 4x^3 - 2x^2 \\ \hline - 2x^2 + 7x \end{array}$$

Divide: $\frac{-2x^2}{2x} = -x$.

$$\begin{array}{r} 2x^2 - x \\ 2x - 1\overline{)4x^3 - 4x^2 + 7x - 9} \\ 4x^3 - 2x^2 \\ \hline - 2x^2 + 7x \\ - 2x^2 + x \end{array}$$

$-x(2x - 1) = -2x^2 + x$

Multiply: $-x(2x - 1) = -2x^2 + x$.

$$\begin{array}{r} 2x^2 - x \\ 2x - 1\overline{)4x^3 - 4x^2 + 7x - 9} \\ 4x^3 - 2x^2 \\ \hline - 2x^2 + 7x \\ \overset{\oplus}{-} 2x^2 \overset{\ominus}{+} x \\ \hline 6x \end{array}$$

Subtract:
$-2x^2 + 7x - (-2x^2 + x)$
$= -2x^2 + 7x + 2x^2 - x$
$= 6x$.

$$
\begin{array}{r}
2x^2 - x \phantom{{}+ 3} \\
2x - 1 \overline{)\, 4x^3 - 4x^2 + 7x - 9} \\
\underline{4x^3 - 2x^2} \phantom{{}+ 7x - 9} \\
-2x^2 + 7x \phantom{{}- 9} \\
\underline{-2x^2 + x} \phantom{{}- 9} \\
6x - 9
\end{array}
$$

Bring down −9. The new dividend is 6x − 9.

$$
\begin{array}{r}
2x^2 - x + 3 \\
2x - 1 \overline{)\, 4x^3 - 4x^2 + 7x - 9} \\
\underline{4x^3 - 2x^2} \phantom{{}+ 7x - 9} \\
-2x^2 + 7x \phantom{{}- 9} \\
\underline{-2x^2 + x} \phantom{{}- 9} \\
6x - 9
\end{array}
$$

Divide: $\frac{6x}{2x} = 3.$

3(2x − 1) = 6x − 3

$$
\begin{array}{r}
2x^2 - x + 3 \\
2x - 1 \overline{)\, 4x^3 - 4x^2 + 7x - 9} \\
\underline{4x^3 - 2x^2} \phantom{{}+ 7x - 9} \\
-2x^2 + 7x \phantom{{}- 9} \\
\underline{-2x^2 + x} \phantom{{}- 9} \\
6x - 9 \\
\underline{6x - 3}
\end{array}
$$

Multiply: $3(2x - 1) = 6x - 3.$

$$
\begin{array}{r}
2x^2 - x + 3 \\
2x - 1 \overline{)\, 4x^3 - 4x^2 + 7x - 9} \\
\underline{4x^3 - 2x^2} \phantom{{}+ 7x - 9} \\
-2x^2 + 7x \phantom{{}- 9} \\
\underline{-2x^2 + x} \phantom{{}- 9} \\
6x - 9 \\
\underline{\overset{\ominus}{}6x \overset{\oplus}{-} 3} \\
-6
\end{array}
$$

Remainder

Subtract:
$6x - 9 - (6x - 3)$
$= 6x - 9 - 6x + 3$
$= -6$

The quotient is $2x^2 - x + 3$ and the remainder is -6. When there is a nonzero remainder, as in this example, list the quotient, plus the remainder above the divisor. Thus,

$$\frac{7x - 9 - 4x^2 + 4x^3}{2x - 1} = 2x^2 - x + 3 + \frac{-6}{2x - 1}$$

Remainder above divisor

Quotient

or

$$\frac{7x - 9 - 4x^2 + 4x^3}{2x - 1} = 2x^2 - x + 3 - \frac{6}{2x - 1}.$$

Check this result by showing that the product of the divisor and the quotient,

$$(2x - 1)(2x^2 - x + 3),$$

plus the remainder, -6, is the dividend, $7x - 9 - 4x^2 + 4x^3$. ■

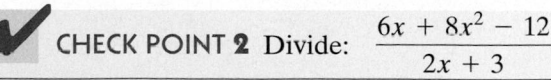

CHECK POINT 2 Divide: $\frac{6x + 8x^2 - 12}{2x + 3}$.

If a power of a variable is missing in a dividend, add that power of the variable with a coefficient of 0 and then divide. In this way, like terms will be aligned as you carry out the division.

EXAMPLE 3 Dividing a Polynomial with Missing Terms

Divide: $\dfrac{8x^3 - 1}{2x - 1}$.

SOLUTION We write the dividend, $8x^3 - 1$, as

$$8x^3 + 0x^2 + 0x - 1.$$

Use a coefficient of 0 with missing terms.

By doing this, we will keep all like terms aligned.

$4x^2(2x - 1) = 8x^3 - 4x^2$

$$\begin{array}{r} 4x^2 \\ 2x - 1 \overline{)\, 8x^3 + 0x^2 + 0x - 1} \\ \underline{\ominus 8x^3 \oplus 4x^2} \\ 4x^2 + 0x \end{array}$$

Divide $\left(\dfrac{8x^3}{2x} = 4x^2\right)$, multiply, subtract, and bring down the next term. The new dividend is $4x^2 + 0x$.

$2x(2x - 1) = 4x^2 - 2x$

$$\begin{array}{r} 4x^2 + 2x \\ 2x - 1 \overline{)\, 8x^3 + 0x^2 + 0x - 1} \\ \underline{8x^3 - 4x^2} \\ 4x^2 + 0x \\ \underline{\ominus 4x^2 \oplus 2x} \\ 2x - 1 \end{array}$$

Divide $\left(\dfrac{4x^2}{2x} = 2x\right)$, multiply $[2x(2x - 1) = 4x^2 - 2x]$, subtract, and bring down the next term. The new dividend is $2x - 1$.

$1(2x - 1) = 2x - 1$

$$\begin{array}{r} 4x^2 + 2x + 1 \\ 2x - 1 \overline{)\, 8x^3 + 0x^2 + 0x - 1} \\ \underline{8x^3 - 4x^2} \\ 4x^2 + 0x \\ \underline{4x^2 - 2x} \\ 2x - 1 \\ \underline{\ominus 2x \oplus 1} \\ 0 \end{array}$$

Divide $\left(\dfrac{2x}{2x} = 1\right)$, multiply $[1(2x - 1) = 2x - 1]$, and subtract. The remainder is 0.

Thus,

$$\frac{8x^3 - 1}{2x - 1} = 4x^2 + 2x + 1.$$

Check this result by showing that the product of the divisor and the quotient,

$$(2x - 1)(4x^2 + 2x + 1)$$

plus the remainder, 0, is the dividend, $8x^3 - 1$. ■

CHECK POINT 3 Divide: $\dfrac{x^3 - 1}{x - 1}$.

USING TECHNOLOGY

The graphs of $y_1 = \dfrac{8x^3 - 1}{2x - 1}$ and $y_2 = 4x^2 + 2x + 1$ are shown below.

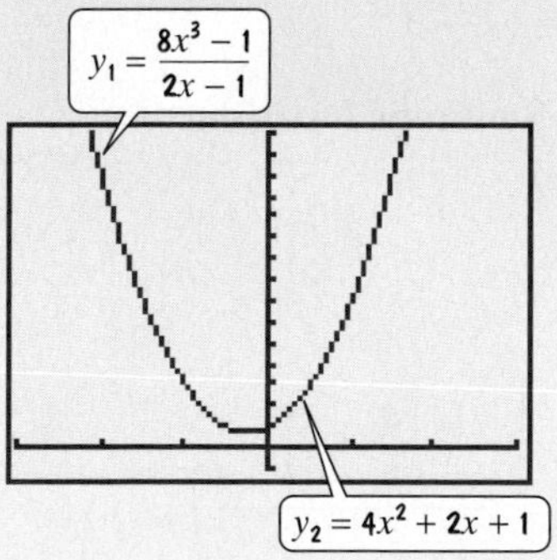

$[-3, 3, 1]$ by $[-1, 15, 1]$

The graphs coincide. Thus,

$$\frac{8x^3 - 1}{2x - 1} = 4x^2 + 2x + 1.$$

EXAMPLE 4 Long Division of Polynomials

Divide: $6x^4 + 5x^3 + 3x - 5$ by $3x^2 - 2x$.

SOLUTION We write the dividend, $6x^4 + 5x^3 + 3x - 5$, as $6x^4 + 5x^3 + 0x^2 + 3x - 5$ to keep all like terms aligned.

Multiply.

$$
\begin{array}{r|l}
 & 2x^2 + 3x + 2 \\
3x^2 - 2x & 6x^4 + 5x^3 + 0x^2 + 3x - 5 \\
 & \ominus 6x^4 \oplus\!- 4x^3 \\
 & \qquad 9x^3 + 0x^2 \\
 & \qquad \ominus 9x^3 \oplus\!- 6x^2 \\
 & \qquad\qquad 6x^2 + 3x \\
 & \qquad\qquad \ominus 6x^2 \oplus\!- 4x \\
 & \qquad\qquad\qquad 7x - 5
\end{array}
$$

$2x^2(3x^2 - 2x) = 6x^4 - 4x^3$

$3x(3x^2 - 2x) = 9x^3 - 6x^2$

$2(3x^2 - 2x) = 6x^2 - 4x$

Remainder

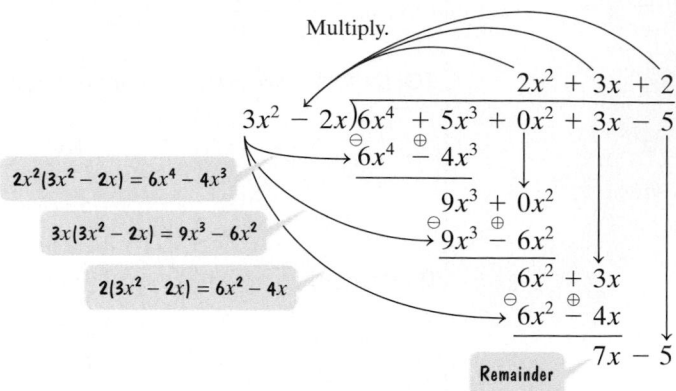

The division process is finished because the degree of $7x - 5$, which is 1, is less than the degree of the divisor $3x^2 - 2x$, which is 2. The answer is

$$\frac{6x^4 + 5x^3 + 3x - 5}{3x^2 - 2x} = 2x^2 + 3x + 2 + \frac{7x - 5}{3x^2 - 2x}.$$

■

CHECK POINT 4 Divide: $2x^4 + 3x^3 - 7x - 10$ by $x^2 - 2x$.

2 Divide polynomials using synthetic division.

Dividing Polynomials Using Synthetic Division We can use **synthetic division** to divide polynomials if the divisor is of the form $x - c$. This method provides a quotient more quickly than long division. Let's compare the two methods showing $x^3 + 4x^2 - 5x + 5$ divided by $x - 3$.

Long Division

Quotient

$$
\begin{array}{r|l}
 & x^2 + 7x + 16 \\
x - 3 & x^3 + 4x^2 - 5x + 5 \\
 & \ominus x^3 \oplus\!- 3x^2 \\
 & \qquad 7x^2 - 5x \\
 & \qquad \ominus 7x^2 \oplus\!- 21x \\
 & \qquad\qquad 16x + 5 \\
 & \qquad\qquad \ominus 16x \oplus\!- 48 \\
 & \qquad\qquad\qquad 53
\end{array}
$$

Dividend

Divisor $x - c$; $c = 3$

Remainder

Synthetic Division

$$
\begin{array}{r|rrrr}
3 & 1 & 4 & -5 & 5 \\
 & & 3 & 21 & 48 \\
\hline
 & 1 & 7 & 16 & 53
\end{array}
$$

Notice the relationship between the polynomials in the long division process and the numbers that appear in synthetic division.

These are the coefficients of the dividend $x^3 + 4x^2 - 5x + 5$.

The divisor is $x - 3$. This is 3, or c, in $x - c$.

$$\begin{array}{r|rrrr} 3 & 1 & 4 & -5 & 5 \\ & & 3 & 21 & 48 \\ \hline & 1 & 7 & 16 & 53 \end{array}$$

These are the coefficients of the quotient $x^2 + 7x + 16$.

This is the remainder.

Now let's look at the steps involved in synthetic division.

SYNTHETIC DIVISION To divide a polynomial by $x - c$:

Example

1. Arrange polynomials in descending powers, with a 0 coefficient for any missing term.

$$x - 3\overline{)x^3 + 4x^2 - 5x + 5}$$

2. Write c for the divisor, $x - c$. To the right, write the coefficients of the dividend.

$$\begin{array}{r|rrrr} 3 & 1 & 4 & -5 & 5 \end{array}$$

3. Write the leading coefficient of the dividend on the bottom row.

$$\begin{array}{r|rrrr} 3 & 1 & 4 & -5 & 5 \\ & \downarrow & \text{Bring down 1.} & & \\ \hline & 1 & & & \end{array}$$

4. Multiply c (in this case, 3) times the value just written on the bottom row. Write the product in the next column in the second row.

$$\begin{array}{r|rrrr} 3 & 1 & 4 & -5 & 5 \\ & & 3 & & \\ \hline & 1 & & & \end{array}$$

Multiply by 3: $3 \cdot 1 = 3$.

5. Add the values in this new column, writing the sum in the bottom row.

$$\begin{array}{r|rrrr} 3 & 1 & 4 & -5 & 5 \\ & & 3 & & \\ \hline & 1 & 7 & & \end{array}$$

Add.

6. Repeat this series of multiplications and additions until all columns are filled in.

$$\begin{array}{r|rrrr} 3 & 1 & 4 & -5 & 5 \\ & & 3 & 21 & \\ \hline & 1 & 7 & 16 & \end{array}$$

Add.

Multiply by 3: $3 \cdot 7 = 21$.

$$\begin{array}{r|rrrr} 3 & 1 & 4 & -5 & 5 \\ & & 3 & 21 & 48 \\ \hline & 1 & 7 & 16 & 53 \end{array}$$

Add.

Multiply by 3: $3 \cdot 16 = 48$.

7. Use the numbers in the last row to write the quotient, plus the remainder above the divisor. **The degree of the first term of the quotient is one less than the degree of the first term of the dividend.** The final value in this row is the remainder.

Written from 1 7 16 53 the last row of the synthetic division

$$\begin{array}{r} 1x^2 + 7x + 16 + \dfrac{53}{x - 3} \\ x - 3\overline{)x^3 + 4x^2 - 5x + 5} \end{array}$$

EXAMPLE 5 Using Synthetic Division

Use synthetic division to divide $5x^3 + 6x + 8$ by $x + 2$.

SOLUTION The divisor must be in the form $x - c$. Thus, we write $x + 2$ as $x - (-2)$. This means that $c = -2$. Writing a 0 coefficient for the missing x^2-term in the dividend, we can express the division as follows:

$$x - (-2)\overline{)5x^3 + 0x^2 + 6x + 8}.$$

Now we are ready to set up the problem so that we can use synthetic division.

Use the coefficients of the dividend $5x^3 + 0x^2 + 6x + 8$ in descending powers of x.

This is c in $x - (-2)$.

$$\underline{-2|}\ \ 5 \quad 0 \quad 6 \quad 8$$

We begin the synthetic division process by bringing down 5. This is followed by a series of multiplications and additions.

1. Bring down 5.

$$\begin{array}{r|rrrr} -2 & 5 & 0 & 6 & 8 \\ & & & & \\ \hline & 5 & & & \end{array}$$

2. Multiply: $-2(5) = -10$.

$$\begin{array}{r|rrrr} -2 & 5 & 0 & 6 & 8 \\ & & -10 & & \\ \hline & 5 & & & \end{array}$$

Multiply by −2.

3. Add: $0 + (-10) = -10$.

$$\begin{array}{r|rrrr} -2 & 5 & 0 & 6 & 8 \\ & & -10 & & \\ \hline & 5 & -10 & & \end{array}$$

Add.

4. Multiply: $-2(-10) = 20$.

$$\begin{array}{r|rrrr} -2 & 5 & 0 & 6 & 8 \\ & & -10 & 20 & \\ \hline & 5 & -10 & & \end{array}$$

Multiply by −2.

5. Add: $6 + 20 = 26$.

$$\begin{array}{r|rrrr} -2 & 5 & 0 & 6 & 8 \\ & & -10 & 20 & \\ \hline & 5 & -10 & 26 & \end{array}$$

Add.

6. Multiply: $-2(26) = -52$.

$$\begin{array}{r|rrrr} -2 & 5 & 0 & 6 & 8 \\ & & -10 & 20 & -52 \\ \hline & 5 & -10 & 26 & \end{array}$$

Multiply by −2.

7. Add: $8 + (-52) = -44$.

$$\begin{array}{r|rrrr} -2 & 5 & 0 & 6 & 8 \\ & & -10 & 20 & -52 \\ \hline & 5 & -10 & 26 & -44 \end{array}$$

Add.

The numbers in the last row represent the coefficients of the quotient and the remainder. The degree of the first term of the quotient is one less than that of the dividend. Because the degree of the dividend, $5x^3 + 6x + 8$, is 3, the degree of the quotient is 2. This means that the 5 in the last row represents $5x^2$.

$$\begin{array}{r|rrrr} -2 & 5 & 0 & 6 & 8 \\ & & -10 & 20 & -52 \\ \hline & 5 & -10 & 26 & -44 \end{array}$$

The quotient is $5x^2 - 10x + 26$.

The remainder is −44.

Thus,

$$\begin{array}{r} 5x^2 - 10x + 26 - \dfrac{44}{x+2} \\ x + 2\overline{)5x^3 + 6x + 8} \end{array}$$

■

CHECK POINT 5 Use synthetic division to divide

$$x^3 - 7x - 6 \text{ by } x + 2.$$

5.6 EXERCISE SET

Student Solutions Manual | CD/Video | PH Math/Tutor Center | MathXL Tutorials on CD | MathXL® | MyMathLab | Interactmath.com

Practice Exercises

In Exercises 1–40, divide as indicated. Check each answer by showing that the product of the divisor and the quotient, plus the remainder, is the dividend.

1. $\dfrac{x^2 + 6x + 8}{x + 2}$

2. $\dfrac{x^2 + 7x + 10}{x + 5}$

3. $\dfrac{2x^2 + x - 10}{x - 2}$

4. $\dfrac{2x^2 + 13x + 15}{x + 5}$

5. $\dfrac{x^2 - 5x + 6}{x - 3}$

6. $\dfrac{x^2 - 2x - 24}{x + 4}$

7. $\dfrac{2y^2 + 5y + 2}{y + 2}$

8. $\dfrac{2y^2 - 13y + 21}{y - 3}$

9. $\dfrac{x^2 - 3x + 4}{x + 2}$

10. $\dfrac{x^2 - 7x + 5}{x + 3}$

11. $\dfrac{5y + 10 + y^2}{y + 2}$

12. $\dfrac{-8y + y^2 - 9}{y - 3}$

13. $\dfrac{x^3 - 6x^2 + 7x - 2}{x - 1}$

14. $\dfrac{x^3 + 3x^2 + 5x + 3}{x + 1}$

15. $\dfrac{12y^2 - 20y + 3}{2y - 3}$

16. $\dfrac{4y^2 - 8y - 5}{2y + 1}$

17. $\dfrac{4a^2 + 4a - 3}{2a - 1}$

18. $\dfrac{2b^2 - 9b - 5}{2b + 1}$

19. $\dfrac{3y - y^2 + 2y^3 + 2}{2y + 1}$

20. $\dfrac{9y + 18 - 11y^2 + 12y^3}{4y + 3}$

21. $\dfrac{6x^2 - 5x - 30}{2x - 5}$

22. $\dfrac{4y^2 + 8y + 3}{2y - 1}$

23. $\dfrac{x^3 + 4x - 3}{x - 2}$

24. $\dfrac{x^3 + 2x^2 - 3}{x - 2}$

25. $\dfrac{4y^3 + 8y^2 + 5y + 9}{2y + 3}$

26. $\dfrac{2y^3 - y^2 + 3y + 2}{2y + 1}$

27. $\dfrac{6y^3 - 5y^2 + 5}{3y + 2}$

28. $\dfrac{4y^3 + 3y + 5}{2y - 3}$

29. $\dfrac{27x^3 - 1}{3x - 1}$

30. $\dfrac{8x^3 + 27}{2x + 3}$

31. $\dfrac{81 - 12y^3 + 54y^2 + y^4 - 108y}{y - 3}$

32. $\dfrac{8y^3 + y^4 + 16 + 32y + 24y^2}{y + 2}$

33. $\dfrac{4y^2 + 6y}{2y - 1}$

34. $\dfrac{10x^2 - 3x}{x + 3}$

35. $\dfrac{y^4 - 2y^2 + 5}{y - 1}$

36. $\dfrac{y^4 - 6y^2 + 3}{y - 1}$

37. $(4x^4 + 3x^3 + 4x^2 + 9x - 6) \div (x^2 + 3)$

38. $(3x^5 - x^3 + 4x^2 - 12x - 8) \div (x^2 - 2)$

39. $(15x^4 + 3x^3 + 4x^2 + 4) \div (3x^2 - 1)$

40. $(18x^4 + 9x^3 + 3x^2) \div (3x^2 + 1)$

In Exercises 41–58, divide using synthetic division. In the first two exercises, begin the process as shown.

41. $(2x^2 + x - 10) \div (x - 2)$ $\quad 2\rfloor \quad 2 \quad 1 \quad -10$

42. $(x^2 + x - 2) \div (x - 1)$ $\quad 1\rfloor \quad 1 \quad 1 \quad -2$

43. $(3x^2 + 7x - 20) \div (x + 5)$

44. $(5x^2 - 12x - 8) \div (x + 3)$

45. $(4x^3 - 3x^2 + 3x - 1) \div (x - 1)$

46. $(5x^3 - 6x^2 + 3x + 11) \div (x - 2)$

47. $(6x^5 - 2x^3 + 4x^2 - 3x + 1) \div (x - 2)$

48. $(x^5 + 4x^4 - 3x^2 + 2x + 3) \div (x - 3)$

49. $(x^2 - 5x - 5x^3 + x^4) \div (5 + x)$

50. $(x^2 - 6x - 6x^3 + x^4) \div (6 + x)$

51. $(3x^3 + 2x^2 - 4x + 1) \div \left(x - \frac{1}{3}\right)$

52. $(2x^4 - x^3 + 2x^2 - 3x + 1) \div \left(x - \frac{1}{2}\right)$

53. $\dfrac{x^5 + x^3 - 2}{x - 1}$

54. $\dfrac{x^7 + x^5 - 10x^3 + 12}{x + 2}$

55. $\dfrac{x^4 - 256}{x - 4}$

56. $\dfrac{x^7 - 128}{x - 2}$

57. $\dfrac{2x^5 - 3x^4 + x^3 - x^2 + 2x - 1}{x + 2}$

58. $\dfrac{x^5 - 2x^4 - x^3 + 3x^2 - x + 1}{x - 2}$

Practice Plus

In Exercises 59–68, divide as indicated.

59. $\dfrac{x^4 + y^4}{x + y}$

60. $\dfrac{x^5 + y^5}{x + y}$

61. $\dfrac{3x^4 + 5x^3 + 7x^2 + 3x - 2}{x^2 + x + 2}$

62. $\dfrac{x^4 - x^3 - 7x^2 - 7x - 2}{x^2 - 3x - 2}$

63. $\dfrac{4x^3 - 3x^2 + x + 1}{x^2 + x + 1}$

64. $\dfrac{x^4 - x^2 + 1}{x^2 + x + 1}$

65. $\dfrac{x^5 - 1}{x^2 - x + 2}$

66. $\dfrac{5x^5 - 7x^4 + 3x^3 - 20x^2 + 28x - 12}{x^3 - 4}$

67. $\dfrac{4x^3 - 7x^2y - 16xy^2 + 3y^3}{x - 3y}$

68. $\dfrac{12x^3 - 19x^2y + 13xy^2 - 10y^3}{4x - 5y}$

69. Divide the difference between $4x^3 + x^2 - 2x + 7$ and $3x^3 - 2x^2 - 7x + 4$ by $x + 1$.

70. Divide the difference between $4x^3 + 2x^2 - x - 1$ and $2x^3 - x^2 + 2x - 5$ by $x + 2$.

Application Exercises

71. Write a simplified polynomial that represents the length of the rectangle.

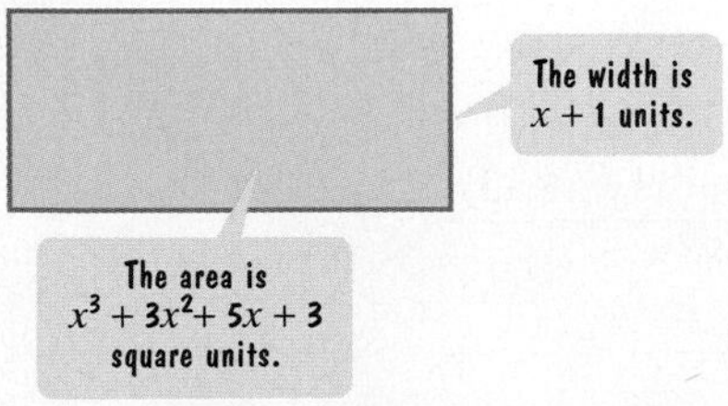

The width is $x + 1$ units.
The area is $x^3 + 3x^2 + 5x + 3$ square units.

72. Write a simplified polynomial that represents the measure of the base of the parallelogram.

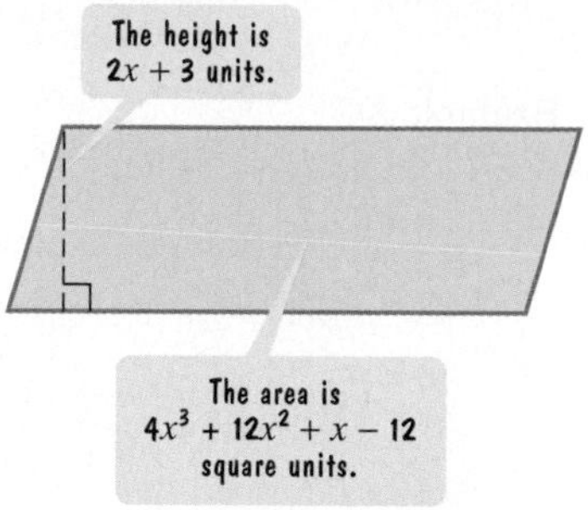
The height is
$2x + 3$ units.
The area is
$4x^3 + 12x^2 + x - 12$
square units.

You just signed a contract for a new job. The salary for the first year is \$30,000 and there is to be a percent increase in your salary each year. The algebraic expression

$$\frac{30{,}000x^n - 30{,}000}{x - 1}$$

describes your total salary over n years, where x is the sum of 1 and the yearly percent increase, expressed as a decimal. Use this information to solve Exercises 73–74.

73. a. Use the given expression and write a quotient of polynomials that describes your total salary over three years.

b. Simplify the expression in part (a) by performing the division.

c. Suppose you are to receive an increase of 5% per year. Thus, x is the sum of 1 and 0.05, or 1.05. Substitute 1.05 for x in the expression in part (a) as well as in the simplified form of the expression in part (b). Evaluate each expression. What is your total salary over the three-year period?

74. a. Use the given expression and write a quotient of polynomials that describes your total salary over four years.

b. Simplify the expression in part (a) by performing the division.

c. Suppose you are to receive an increase of 8% per year. Thus, x is the sum of 1 and 0.08, or 1.08. Substitute 1.08 for x in the expression in part (a) as well as in the simplified form of the expression in part (b). Evaluate each expression. What is your total salary over the four-year period?

Writing in Mathematics

75. In your own words, explain how to divide a polynomial by a binomial. Use $\frac{x^2 + 4}{x + 2}$ in your explanation.

76. When dividing a polynomial by a binomial, explain when to stop dividing.

77. After dividing a polynomial by a binomial, explain how to check the answer.

78. When dividing a binomial into a polynomial with missing terms, explain the advantage of writing the missing terms with zero coefficients.

Critical Thinking Exercises

79. Which one of the following is true?

a. If $4x^2 + 25x - 3$ is divided by $4x + 1$, the remainder is 9.

b. If polynomial division results in a remainder of zero, then the product of the divisor and the quotient is the dividend.

c. The degree of a polynomial is the power of the term that appears in the first position.

d. When a polynomial is divided by a binomial, the division process stops when the last term of the dividend is brought down.

80. When a certain polynomial is divided by $2x + 4$, the quotient is

$$x - 3 + \frac{17}{2x + 4}.$$

What is the polynomial?

81. Find the number k such that when $16x^2 - 2x + k$ is divided by $2x - 1$, the remainder is 0.

82. Describe the pattern that you observe in the following quotients and remainders.

$$\frac{x^3 - 1}{x + 1} = x^2 - x + 1 - \frac{2}{x + 1}$$

$$\frac{x^5 - 1}{x + 1} = x^4 - x^3 + x^2 - x + 1 - \frac{2}{x + 1}$$

Use this pattern to find $\frac{x^7 - 1}{x + 1}$. Verify your result by dividing.

Technology Exercises

In Exercises 83–87, use a graphing utility to determine whether the divisions have been performed correctly. Graph each side of the given equation in the same viewing rectangle. The graphs should coincide. If they do not, correct the expression on the right side by using polynomial division. Then use your graphing utility to show that the division has been performed correctly.

83. $\frac{x^2 - 4}{x - 2} = x + 2$

84. $\frac{x^2 - 25}{x - 5} = x - 5$

85. $\frac{2x^2 + 13x + 15}{x - 5} = 2x + 3$

86. $\frac{6x^2 + 16x + 8}{3x + 2} = 2x - 4$

87. $\frac{x^3 + 3x^2 + 5x + 3}{x + 1} = x^2 - 2x + 3$

Review Exercises

88. Solve the system:

$$7x - 6y = 17$$
$$3x + y = 18$$

(Section 4.3, Example 2)

89. What is 6% of 20? (Section 2.4, Example 7)

90. Solve: $\frac{x}{3} + \frac{2}{5} = \frac{x}{5} - \frac{2}{5}$ (Section 2.3, Example 4)

SECTION 5.7

NEGATIVE EXPONENTS AND SCIENTIFIC NOTATION

Objectives

1. Use the negative exponent rule.
2. Simplify exponential expressions.
3. Convert from scientific notation to decimal notation.
4. Convert from decimal notation to scientific notation.
5. Compute with scientific notation.
6. Solve applied problems with scientific notation.

You are listening to a discussion of the country's 6.8 trillion dollar deficit. It seems that this is a real problem, but then you realize that you don't really know what this number means. How can you look at this deficit in the proper perspective? If the national debt were evenly divided among all citizens of the country, how much would each citizen have to pay?

In the new millennium, literacy with numbers, called **numeracy**, will be a necessary skill for functioning in a meaningful way personally, professionally, and as a citizen. In this section, you will learn to use exponents to provide a way of putting large and small numbers into perspective.

1 Use the negative exponent rule.

Negative Integers as Exponents A nonzero base can be raised to a negative power. The quotient rule can be used to help determine what a negative integer as an exponent should mean. Consider the quotient of b^3 and b^5, where b is not zero. We can determine this quotient in two ways.

$$\frac{b^3}{b^5} = \frac{1 \cdot \cancel{b} \cdot \cancel{b} \cdot \cancel{b}}{\cancel{b} \cdot \cancel{b} \cdot \cancel{b} \cdot b \cdot b} = \frac{1}{b^2}$$

After dividing out pairs of factors, we have two factors of b in the denominator.

$$\frac{b^3}{b^5} = b^{3-5} = b^{-2}$$

Use the quotient rule and subtract exponents.

Notice that $\frac{b^3}{b^5}$ equals both b^{-2} and $\frac{1}{b^2}$. This means that b^{-2} must equal $\frac{1}{b^2}$. This example is a special case of the **negative exponent rule**.

THE NEGATIVE EXPONENT RULE If b is any real number other than 0 and n is a natural number, then

$$b^{-n} = \frac{1}{b^n}.$$

STUDY TIP

A negative exponent does not make the value of an expression negative. For example,

$$7^{-2} = \frac{1}{7^2} = \frac{1}{49}$$

is positive. Avoid these common errors:

Incorrect!

~~$7^{-2} = -7^2$~~

~~$7^{-2} = -\frac{1}{7^2}$~~

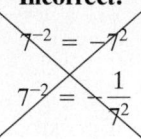

EXAMPLE 1 Using the Negative Exponent Rule

Use the negative exponent rule to write each expression with a positive exponent. Then simplify the expression.

a. 7^{-2} **b.** 4^{-3} **c.** $(-2)^{-4}$ **d.** -2^{-4} **e.** 5^{-1}

SOLUTION

a. $7^{-2} = \frac{1}{7^2} = \frac{1}{7 \cdot 7} = \frac{1}{49}$

b. $4^{-3} = \frac{1}{4^3} = \frac{1}{4 \cdot 4 \cdot 4} = \frac{1}{64}$

c. $(-2)^{-4} = \frac{1}{(-2)^4} = \frac{1}{(-2)(-2)(-2)(-2)} = \frac{1}{16}$

d. $-2^{-4} = -\frac{1}{2^4} = -\frac{1}{2 \cdot 2 \cdot 2 \cdot 2} = -\frac{1}{16}$

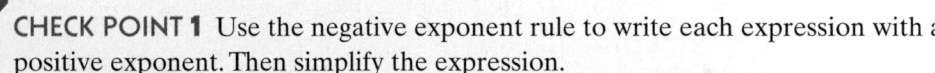

The negative is not inside parentheses and is not taken to the −4 power.

e. $5^{-1} = \frac{1}{5^1} = \frac{1}{5}$ ■

CHECK POINT 1 Use the negative exponent rule to write each expression with a positive exponent. Then simplify the expression.

a. 6^{-2} **b.** 5^{-3}

c. $(-3)^{-4}$ **d.** -3^{-4} **e.** 8^{-1}

Negative exponents can also appear in denominators. For example,

$$\frac{1}{2^{-10}} = \frac{1}{\frac{1}{2^{10}}} = 1 \div \frac{1}{2^{10}} = 1 \cdot \frac{2^{10}}{1} = 2^{10}.$$

In general, if a negative exponent appears in a denominator, an expression can be written with a positive exponent using

$$\frac{1}{b^{-n}} = b^n.$$

For example,

$$\frac{1}{2^{-3}} = 2^3 = 8 \quad \text{and} \quad \frac{1}{(-6)^{-2}} = (-6)^2 = 36.$$

Change only the sign of the exponent and not the sign of the base, −6.

NEGATIVE EXPONENTS IN NUMERATORS AND DENOMINATORS If b is any real number other than 0 and n is a natural number, then

$$b^{-n} = \frac{1}{b^n} \quad \text{and} \quad \frac{1}{b^{-n}} = b^n.$$

When a negative number appears as an exponent, switch the position of the base (from numerator to denominator or from denominator to numerator) and make the exponent positive. The sign of the base does not change.

EXAMPLE 2 Using Negative Exponents

Write each expression with positive exponents only. Then simplify, if possible.

a. $\frac{4^{-3}}{5^{-2}}$ **b.** $\left(\frac{3}{4}\right)^{-2}$ **c.** $\frac{1}{4x^{-3}}$ **d.** $\frac{x^{-5}}{y^{-1}}$

SOLUTION

a. $\frac{4^{-3}}{5^{-2}} = \frac{5^2}{4^3} = \frac{5 \cdot 5}{4 \cdot 4 \cdot 4} = \frac{25}{64}$

Switch the position of the bases and make the exponents positive.

b. $\left(\frac{3}{4}\right)^{-2} = \frac{3^{-2}}{4^{-2}} = \frac{4^2}{3^2} = \frac{4 \cdot 4}{3 \cdot 3} = \frac{16}{9}$

Switch the position of the bases and make the exponents positive.

c. $\frac{1}{4x^{-3}} = \frac{x^3}{4}$

Switch the position of the base and make the exponent positive. Note that only x is raised to the −3 power.

d. $\frac{x^{-5}}{y^{-1}} = \frac{y^1}{x^5} = \frac{y}{x^5}$

■

CHECK POINT 2 Write each expression with positive exponents only. Then simplify, if possible.

a. $\frac{2^{-3}}{7^{-2}}$ **b.** $\left(\frac{4}{5}\right)^{-2}$ **c.** $\frac{1}{7y^{-2}}$ **d.** $\frac{x^{-1}}{y^{-8}}$

2 Simplify exponential expressions.

Simplifying Exponential Expressions Properties of exponents are used to simplify exponential expressions. An exponential expression is **simplified** when

- No parentheses appear.
- No powers are raised to powers.
- Each base occurs only once.
- No negative or zero exponents appear.

SIMPLIFYING EXPONENTIAL EXPRESSIONS

	Example
1. If necessary, remove parentheses using $(ab)^n = a^n b^n$ or $\left(\frac{a}{b}\right)^n = \frac{a^n}{b^n}$.	$(xy)^3 = x^3y^3$
2. If necessary, simplify powers to powers using $(b^m)^n = b^{mn}$.	$(x^4)^3 = x^{4\cdot 3} = x^{12}$
3. If necessary, be sure that each base appears only once, using $b^m \cdot b^n = b^{m+n}$ or $\frac{b^m}{b^n} = b^{m-n}$.	$x^4 \cdot x^3 = x^{4+3} = x^7$
4. If necessary, rewrite exponential expressions with zero powers as 1 $(b^0 = 1)$. Furthermore, write the answer with positive exponents using $b^{-n} = \frac{1}{b^n}$ or $\frac{1}{b^{-n}} = b^n$.	$\frac{x^5}{x^8} = x^{5-8} = x^{-3} = \frac{1}{x^3}$

The following examples show how to simplify exponential expressions. In each example, assume that any variable in a denominator is not equal to zero.

STUDY TIP

There is often more than one way to simplify an exponential expression. For example, you may prefer to simplify Example 3 as follows:

$$x^{-9} \cdot x^4 = \frac{x^4}{x^9} = x^{4-9} = x^{-5} = \frac{1}{x^5}.$$

EXAMPLE 3 Simplifying an Exponential Expression

Simplify: $x^{-9} \cdot x^4$.

SOLUTION

$$x^{-9} \cdot x^4 = x^{-9+4} \qquad b^m \cdot b^n = b^{m+n}$$

$$= x^{-5} \qquad \text{The base, } x\text{, now appears only once.}$$

$$= \frac{1}{x^5} \qquad b^{-n} = \frac{1}{b^n}$$

■

CHECK POINT 3 Simplify: $x^{-12} \cdot x^2$.

EXAMPLE 4 Simplifying Exponential Expressions

Simplify:

a. $\frac{x^4}{x^9}$ **b.** $\frac{25x^6}{5x^8}$ **c.** $\frac{10y^7}{-2y^{10}}$.

SOLUTION

a. $\dfrac{x^4}{x^9} = x^{4-9} = x^{-5} = \dfrac{1}{x^5}$

b. $\dfrac{25x^6}{5x^8} = \dfrac{25}{5} \cdot \dfrac{x^6}{x^8} = 5x^{6-8} = 5x^{-2} = \dfrac{5}{x^2}$

c. $\dfrac{10y^7}{-2y^{10}} = \dfrac{10}{-2} \cdot \dfrac{y^7}{y^{10}} = -5y^{7-10} = -5y^{-3} = -\dfrac{5}{y^3}$ ■

CHECK POINT 4 Simplify:

a. $\dfrac{x^2}{x^{10}}$ **b.** $\dfrac{75x^3}{5x^9}$ **c.** $\dfrac{50y^8}{-25y^{14}}$.

EXAMPLE 5 Simplifying an Exponential Expression

Simplify: $\dfrac{(5x^3)^2}{x^{10}}$.

SOLUTION

$\dfrac{(5x^3)^2}{x^{10}} = \dfrac{5^2(x^3)^2}{x^{10}}$ Raise each factor in the product to the second power. Parentheses are removed using $(ab)^n = a^n b^n$.

$= \dfrac{5^2 x^{3 \cdot 2}}{x^{10}}$ Multiply powers to powers using $(b^m)^n = b^{mn}$.

$= \dfrac{25x^6}{x^{10}}$ Simplify.

$= 25x^{6-10}$ When dividing with the same base, subtract exponents: $\dfrac{b^m}{b^n} = b^{m-n}$.

$= 25x^{-4}$ Simplify. The base, x, now appears only once.

$= \dfrac{25}{x^4}$ Rewrite with a positive exponent using $b^{-n} = \dfrac{1}{b^n}$. ■

CHECK POINT 5 Simplify: $\dfrac{(6x^4)^2}{x^{11}}$.

EXAMPLE 6 Simplifying an Exponential Expression

Simplify: $\left(\dfrac{x^5}{x^2}\right)^{-3}$.

SOLUTION

Method 1. Remove parentheses first by raising the numerator and denominator to the -3 power.

$\left(\frac{x^5}{x^2}\right)^{-3} = \frac{(x^5)^{-3}}{(x^2)^{-3}}$ Use $\left(\frac{a}{b}\right)^n = \frac{a^n}{b^n}$ and raise the numerator and denominator to the -3 power.

$= \frac{x^{5(-3)}}{x^{2(-3)}}$ Multiply powers to powers using $(b^m)^n = b^{mn}$.

$= \frac{x^{-15}}{x^{-6}}$ Simplify.

$= x^{-15-(-6)}$ When dividing with the same base, subtract the exponent in the denominator from the exponent in the numerator: $\frac{b^m}{b^n} = b^{m-n}$.

$= x^{-9}$ Subtract: $-15 - (-6) = -15 + 6 = -9$. The base, x, now appears only once.

$= \frac{1}{x^9}$ Rewrite with a positive exponent using $b^{-n} = \frac{1}{b^n}$.

Method 2. First perform the division within the parentheses.

$\left(\frac{x^5}{x^2}\right)^{-3} = (x^{5-2})^{-3}$ Within parentheses, divide by subtracting exponents: $\frac{b^m}{b^n} = b^{m-n}$.

$= (x^3)^{-3}$ Simplify. The base, x, now appears only once.

$= x^{3(-3)}$ Multiply powers to powers: $(b^m)^n = b^{mn}$.

$= x^{-9}$ Simplify.

$= \frac{1}{x^9}$ Rewrite with a positive exponent using $b^{-n} = \frac{1}{b^n}$.

Which method do you prefer? ■

CHECK POINT 6 Simplify: $\left(\frac{x^8}{x^4}\right)^{-5}$.

Scientific Notation By the end of 2003, the national debt of the United States was about \$6.8 trillion. This is the amount of money the government has had to borrow over the years, mostly by selling bonds, because it has spent more than it has collected in taxes. A stack of \$1 bills equaling the national debt would rise to twice the distance from Earth to the moon. Because a trillion is 10^{12}, the national debt can be expressed as

$$6.8 \times 10^{12}.$$

The number 6.8×10^{12} is written in a form called *scientific notation.*

SCIENTIFIC NOTATION A positive number is written in **scientific notation** when it is expressed in the form

$$a \times 10^n,$$

where a is a number greater than or equal to 1 and less than 10 ($1 \le a < 10$) and n is an integer.

It is customary to use the multiplication symbol, $\times$, rather than a dot, when writing a number in scientific notation.

Here are two examples of numbers in scientific notation:

- Each day, 2.6×10^7 pounds of dust from the atmosphere settle on Earth.
- The length of the AIDS virus is 1.1×10^{-4} millimeter.

3 Convert from scientific notation to decimal notation.

We can use n, the exponent on the 10 in $a \times 10^n$, to change a number in scientific notation to decimal notation. If n is **positive**, move the decimal point in a to the **right** n places. If n is **negative**, move the decimal point in a to the **left** $|n|$ places.

EXAMPLE 7 Converting from Scientific to Decimal Notation

Write each number in decimal notation:

a. 2.6×10^7 **b.** 1.1×10^{-4}

SOLUTION In each case, we use the exponent on the 10 to move the decimal point. In part (a), the exponent is positive, so we move the decimal point to the right. In part (b), the exponent is negative, so we move the decimal point to the left.

a. $2.6 \times 10^7 = 26{,}000{,}000$

$n = 7$

Move the decimal point 7 places to the right.

b. $1.1 \times 10^{-4} = 0.00011$

$n = -4$

Move the decimal point $|-4|$ places, or 4 places, to the left.

■

CHECK POINT 7 Write each number in decimal notation:

a. 7.4×10^9 **b.** 3.017×10^{-6}.

4 Convert from decimal notation to scientific notation.

To convert a positive number from decimal notation to scientific notation, we reverse the procedure of Example 7.

CONVERTING FROM DECIMAL TO SCIENTIFIC NOTATION Write the number in the form $a \times 10^n$.

- Determine a, the numerical factor. Move the decimal point in the given number to obtain a number greater than or equal to 1 and less than 10.
- Determine n, the exponent on 10^n. The absolute value of n is the number of places the decimal point was moved. The exponent n is positive if the given number is greater than 10 and negative if the given number is between 0 and 1.

USING TECHNOLOGY

You can use your calculator's [EE] (enter exponent) or [EXP] key to convert from decimal to scientific notation. Here is how it's done for 0.000023:

Many Scientific Calculators

Keystrokes	Display
.000023 [EE] [=]	2.3 − 05

Many Graphing Calculators

Use the mode setting for scientific notation.

Keystrokes	Display
.000023 [ENTER]	2.3E−5

EXAMPLE 8 Converting from Decimal Notation to Scientific Notation

Write each number in scientific notation:

a. 4,600,000 **b.** 0.000023.

SOLUTION

a. $4{,}600{,}000 = 4.6 \times 10^6$

This number is greater than 10, so n is positive in $a \times 10^n$.

Move the decimal point in 4,600,000 to get $1 \le a < 10$.

The decimal point moved 6 places from 4,600,000 to 4.6.

b. $0.000023 = 2.3 \times 10^{-5}$

This number is less than 1, so n is negative in $a \times 10^n$.

Move the decimal point in 0.000023 to get $1 \le a < 10$.

The decimal point moved 5 places from 0.000023 to 2.3.

■

CHECK POINT 8 Write each number in scientific notation:

a. 7,410,000,000 **b.** 0.000000092.

5 Compute with scientific notation.

Computations with Scientific Notation Properties of exponents are used to perform computations with numbers that are expressed in scientific notation.

COMPUTATIONS WITH NUMBERS IN SCIENTIFIC NOTATION

Multiplication

$$(a \times 10^n) \times (b \times 10^m) = (a \times b) \times 10^{n+m}$$

Add the exponents on 10 and multiply the other parts of the numbers separately.

Division

$$\frac{a \times 10^n}{b \times 10^m} = \left(\frac{a}{b}\right) \times 10^{n-m}$$

Subtract the exponents on 10 and divide the other parts of the numbers separately.

Exponentiation

$$(a \times 10^n)^m = a^n \times 10^{nm}$$

Multiply exponents on 10 and raise the other part of the number to the power.

After the computation is completed, the answer may require an adjustment before it is back in scientific notation.

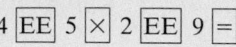

USING TECHNOLOGY

$(4 \times 10^5)(2 \times 10^9)$ On a Calculator:

Many Scientific Calculators

4 [EE] 5 [×] 2 [EE] 9 [=]

Display: 8. 14

Many Graphing Calculators

4 [EE] 5 [×] 2 [EE] 9 [ENTER]

Display: 8E14

EXAMPLE 9 Computations with Scientific Notation

Perform the indicated computations, writing the answers in scientific notation:

a. $(4 \times 10^5)(2 \times 10^9)$ **b.** $\dfrac{1.2 \times 10^6}{4.8 \times 10^{-3}}$ **c.** $(5 \times 10^{-4})^3$.

SOLUTION

a. $(4 \times 10^5)(2 \times 10^9) = (4 \times 2) \times (10^5 \times 10^9)$ Regroup.

$= 8 \times 10^{5+9}$ Add the exponents on 10 and multiply the other parts.

$= 8 \times 10^{14}$ Simplify.

b. $\dfrac{1.2 \times 10^6}{4.8 \times 10^{-3}} = \left(\dfrac{1.2}{4.8}\right) \times \left(\dfrac{10^6}{10^{-3}}\right)$ Regroup.

$= 0.25 \times 10^{6-(-3)}$ Subtract the exponents on 10 and divide the other parts.

$= 0.25 \times 10^9$ Simplify. Because 0.25 is not between 1 and 10, it must be written in scientific notation.

$= 2.5 \times 10^{-1} \times 10^9$ $0.25 = 2.5 \times 10^{-1}$

$= 2.5 \times 10^{-1+9}$ Add the exponents on 10.

$= 2.5 \times 10^8$ Simplify.

c. $(5 \times 10^{-4})^3 = 5^3 \times (10^{-4})^3$ $(ab)^n = a^n b^n$. Cube each factor in parentheses.

$= 125 \times 10^{-4 \cdot 3}$ Multiply the exponents and cube the other part of the number.

$= 125 \times 10^{-12}$ Simplify. 125 must be written in scientific notation.

$= 1.25 \times 10^2 \times 10^{-12}$ $125 = 1.25 \times 10^2$

$= 1.25 \times 10^{2+(-12)}$ Add the exponents on 10.

$= 1.25 \times 10^{-10}$ Simplify. ■

CHECK POINT 9 Perform the indicated computation, writing the answers in scientific notation:

a. $(3 \times 10^8)(2 \times 10^2)$ **b.** $\dfrac{8.4 \times 10^7}{4 \times 10^{-4}}$

c. $(4 \times 10^{-2})^3$.

6 Solve applied problems with scientific notation.

Applications: Putting Numbers in Perspective Due to tax cuts and spending increases, the United States began accumulating large deficits in the 1980s. To finance the deficit, the government had borrowed \$6.8 trillion as of the end of 2003. The graph in Figure 5.4 shows the national debt increasing over time.

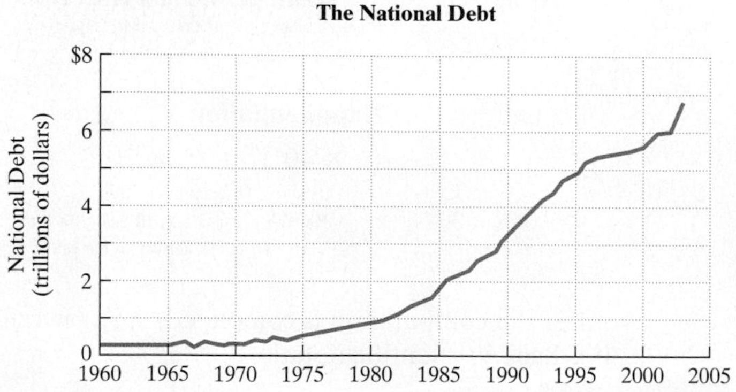

FIGURE 5.4

Source: Office of Management and Budget

Example 10 shows how we can use scientific notation to comprehend the meaning of a number such as 6.8 trillion.

EXAMPLE 10 The National Debt

As of November 2003, the national debt was \$6.8 trillion, or 6.8×10^{12} dollars. At that time, the U.S. population was approximately 290,000,000 (290 million), or 2.9×10^8. If the national debt was evenly divided among every individual in the United States, how much would each citizen have to pay?

SOLUTION The amount each citizen must pay is the total debt, 6.8×10^{12} dollars, divided by the number of citizens, 2.9×10^8.

$$\frac{6.8 \times 10^{12}}{2.9 \times 10^8} = \left(\frac{6.8}{2.9}\right) \times \left(\frac{10^{12}}{10^8}\right)$$
$$\approx 2.3 \times 10^{12-8}$$
$$= 2.3 \times 10^4$$
$$= 23{,}000$$

Every U.S. citizen would have to pay about \$23,000 to the federal government to pay off the national debt. ■

USING TECHNOLOGY

Here is the keystroke sequence for solving Example 10 using a calculator:

6.8 [EE] 12 [÷] 2.9 [EE] 8.

The quotient is displayed by pressing [=] on a scientific calculator and [ENTER] on a graphing calculator. The answer can be displayed in scientific or decimal notation. Consult your manual.

✔ **CHECK POINT 10** Approximately 2×10^4 people run in the New York City Marathon each year. Each runner runs a distance of 26 miles. Write the total distance covered by all the runners (assuming that each person completes the marathon) in scientific notation.

ENRICHMENT ESSAY

Earthquakes and Exponents

The earthquake that ripped through northern California on October 17, 1989, measured 7.1 on the Richter scale, killed more than 60 people, and injured more than 2400. Shown here is San Francisco's Marina district, where shock waves tossed houses off their foundations and into the street.

The Richter scale is misleading because it is not actually a 1 to 8, but rather a 1 to 10 million scale. Each level indicates a tenfold increase in magnitude from the previous level, making a 7.0 earthquake a million times greater than a 1.0 quake.

The following is a translation of the Richter scale:

Richter number (R)	**Magnitude** (10^{R-1})
1	$10^{1-1} = 10^0 = 1$
2	$10^{2-1} = 10^1 = 10$
3	$10^{3-1} = 10^2 = 100$
4	$10^{4-1} = 10^3 = 1000$
5	$10^{5-1} = 10^4 = 10{,}000$
6	$10^{6-1} = 10^5 = 100{,}000$
7	$10^{7-1} = 10^6 = 1{,}000{,}000$
8	$10^{8-1} = 10^7 = 10{,}000{,}000$

5.7 EXERCISE SET

Student Solutions Manual | CD/Video | PH Math/Tutor Center | MathXL Tutorials on CD | MathXL® | MyMathLab | Interactmath.com

Practice Exercises

In Exercises 1–28, write each expression with positive exponents only. Then simplify, if possible.

1. 8^{-2}
2. 9^{-2}
3. 5^{-3}
4. 4^{-3}
5. $(-6)^{-2}$
6. $(-7)^{-2}$
7. -6^{-2}
8. -7^{-2}
9. 4^{-1}
10. 6^{-1}
11. $2^{-1} + 3^{-1}$
12. $3^{-1} - 6^{-1}$
13. $\frac{1}{3^{-2}}$
14. $\frac{1}{4^{-3}}$
15. $\frac{1}{(-3)^{-2}}$
16. $\frac{1}{(-2)^{-2}}$
17. $\frac{2^{-3}}{8^{-2}}$
18. $\frac{4^{-3}}{2^{-2}}$
19. $\left(\frac{1}{4}\right)^{-2}$
20. $\left(\frac{1}{5}\right)^{-2}$
21. $\left(\frac{3}{5}\right)^{-3}$
22. $\left(\frac{3}{4}\right)^{-3}$
23. $\frac{1}{6x^{-5}}$
24. $\frac{1}{8x^{-6}}$
25. $\frac{x^{-8}}{y^{-1}}$
26. $\frac{x^{-12}}{y^{-1}}$
27. $\frac{3}{(-5)^{-3}}$
28. $\frac{4}{(-3)^{-3}}$

In Exercises 29–78, simplify each exponential expression. Assume that variables represent nonzero real numbers.

29. $x^{-8} \cdot x^3$
30. $x^{-11} \cdot x^5$
31. $(4x^{-5})(2x^2)$
32. $(5x^{-7})(3x^3)$
33. $\frac{x^3}{x^9}$
34. $\frac{x^5}{x^{12}}$
35. $\frac{y}{y^{100}}$
36. $\frac{y}{y^{50}}$
37. $\frac{30z^5}{10z^{10}}$
38. $\frac{45z^4}{15z^{12}}$
39. $\frac{-8x^3}{2x^7}$
40. $\frac{-15x^4}{3x^9}$
41. $\frac{-9a^5}{27a^8}$
42. $\frac{-15a^8}{45a^{13}}$
43. $\frac{7w^5}{5w^{13}}$
44. $\frac{7w^8}{9w^{14}}$
45. $\frac{x^3}{(x^4)^2}$
46. $\frac{x^5}{(x^3)^2}$
47. $\frac{y^{-3}}{(y^4)^2}$
48. $\frac{y^{-5}}{(y^3)^2}$
49. $\frac{(4x^3)^2}{x^8}$
50. $\frac{(5x^3)^2}{x^7}$
51. $\frac{(6y^4)^3}{y^{-5}}$
52. $\frac{(4y^5)^3}{y^{-4}}$
53. $\left(\frac{x^4}{x^2}\right)^{-3}$
54. $\left(\frac{x^6}{x^2}\right)^{-3}$
55. $\left(\frac{4x^5}{2x^2}\right)^{-4}$
56. $\left(\frac{6x^7}{2x^2}\right)^{-4}$
57. $(3x^{-1})^{-2}$
58. $(4x^{-1})^{-2}$
59. $(-2y^{-1})^{-3}$
60. $(-3y^{-1})^{-3}$
61. $\frac{2x^5 \cdot 3x^7}{15x^6}$
62. $\frac{3x^3 \cdot 5x^{14}}{20x^{14}}$
63. $(x^3)^5 \cdot x^{-7}$
64. $(x^4)^3 \cdot x^{-5}$
65. $(2y^3)^4 y^{-6}$
66. $(3y^4)^3 y^{-7}$
67. $\frac{(y^3)^4}{(y^2)^7}$
68. $\frac{(y^2)^5}{(y^3)^4}$
69. $(y^{10})^{-5}$
70. $(y^{20})^{-5}$
71. $(a^4b^5)^{-3}$
72. $(a^5b^3)^{-4}$
73. $(a^{-2}b^6)^{-4}$
74. $(a^{-7}b^2)^{-5}$
75. $\left(\frac{x^2}{2}\right)^{-2}$
76. $\left(\frac{x^2}{2}\right)^{-3}$
77. $\left(\frac{x^2}{y^3}\right)^{-3}$
78. $\left(\frac{x^3}{y^2}\right)^{-4}$

In Exercises 79–90, write each number in decimal notation without the use of exponents.

79. 8.7×10^2 **80.** 2.75×10^3

81. 9.23×10^5 **82.** 7.24×10^4

83. 3.4×10^0 **84.** 9.115×10^0

85. 7.9×10^{-1} **86.** 8.6×10^{-1}

87. 2.15×10^{-2} **88.** 3.14×10^{-2}

89. 7.86×10^{-4} **90.** 4.63×10^{-5}

In Exercises 91–106, write each number in scientific notation.

91. 32,400 **92.** 327,000

93. 220,000,000 **94.** 370,000,000,000

95. 713 **96.** 623

97. 6751 **98.** 9832

99. 0.0027 **100.** 0.00083

101. 0.0000202 **102.** 0.00000103

103. 0.005 **104.** 0.006

105. 3.14159 **106.** 2.71828

In Exercises 107–126, perform the indicated computations. Write the answers in scientific notation.

107. $(2 \times 10^3)(3 \times 10^2)$

108. $(3 \times 10^4)(3 \times 10^2)$

109. $(2 \times 10^5)(8 \times 10^3)$

110. $(4 \times 10^3)(5 \times 10^4)$

111. $\dfrac{12 \times 10^6}{4 \times 10^2}$ **112.** $\dfrac{20 \times 10^{20}}{10 \times 10^{10}}$

113. $\dfrac{15 \times 10^4}{5 \times 10^{-2}}$ **114.** $\dfrac{18 \times 10^2}{9 \times 10^{-3}}$

115. $\dfrac{15 \times 10^{-4}}{5 \times 10^2}$ **116.** $\dfrac{18 \times 10^{-2}}{9 \times 10^3}$

117. $\dfrac{180 \times 10^6}{2 \times 10^3}$ **118.** $\dfrac{180 \times 10^8}{2 \times 10^4}$

119. $\dfrac{3 \times 10^4}{12 \times 10^{-3}}$ **120.** $\dfrac{5 \times 10^2}{20 \times 10^{-3}}$

121. $(5 \times 10^2)^3$ **122.** $(4 \times 10^3)^2$

123. $(3 \times 10^{-2})^4$ **124.** $(2 \times 10^{-3})^5$

125. $(4 \times 10^6)^{-1}$ **126.** $(5 \times 10^4)^{-1}$

Practice Plus

In Exercises 127–134, simplify each exponential expression. Assume that variables represent nonzero real numbers.

127. $\dfrac{(x^{-2}y)^{-3}}{(x^2y^{-1})^3}$ **128.** $\dfrac{(xy^{-2})^{-2}}{(x^{-2}y)^{-3}}$

129. $(2x^{-3}yz^{-6})(2x)^{-5}$

130. $(3x^{-4}yz^{-7})(3x)^{-3}$

131. $\left(\dfrac{x^3y^4z^5}{x^{-3}y^{-4}z^{-5}}\right)^{-2}$

132. $\left(\dfrac{x^4y^5z^6}{x^{-4}y^{-5}z^{-6}}\right)^{-4}$

133. $\dfrac{(2^{-1}x^{-2}y^{-1})^{-2}(2x^{-4}y^3)^{-2}(16x^{-3}y^3)^0}{(2x^{-3}y^{-5})^2}$

134. $\dfrac{(2^{-1}x^{-3}y^{-1})^{-2}(2x^{-6}y^4)^{-2}(9x^3y^{-3})^0}{(2x^{-4}y^{-6})^2}$

In Exercises 135–138, perform the indicated computations. Express answers in scientific notation.

135. $(5 \times 10^3)(1.2 \times 10^{-4}) \div (2.4 \times 10^2)$

136. $(2 \times 10^2)(2.6 \times 10^{-3}) \div (4 \times 10^3)$

137. $\dfrac{(1.6 \times 10^4)(7.2 \times 10^{-3})}{(3.6 \times 10^8)(4 \times 10^{-3})}$

138. $\dfrac{(1.2 \times 10^6)(8.7 \times 10^{-2})}{(2.9 \times 10^6)(3 \times 10^{-3})}$

Application Exercises

In Exercises 139–142, rewrite the number in each statement in scientific notation.

139. King Mongkut of Siam (the king in the musical *The King and I*) had 9200 wives.

140. The top-selling music album of all time is "Their Greatest Hits" by the Eagles, selling 28,000,000 copies.

141. The volume of a bacterium is 0.0000000000000025 cubic meter.

142. Home computers can perform a multiplication in 0.00000000036 second.

In Exercises 143–146, use 10^6 for one million and 10^9 for one billion to rewrite the number in each statement in scientific notation.

143. In 2003, Americans consumed 600 million Big Macs at McDonald's.

144. In 2002, the United States imported 550 million barrels of crude oil from Saudi Arabia.

In Exercises 145–146, which appear on the next page, use the numbers displayed in the following graph.

Defense Spending: U.S. and 20 Next Top-Spending Nations

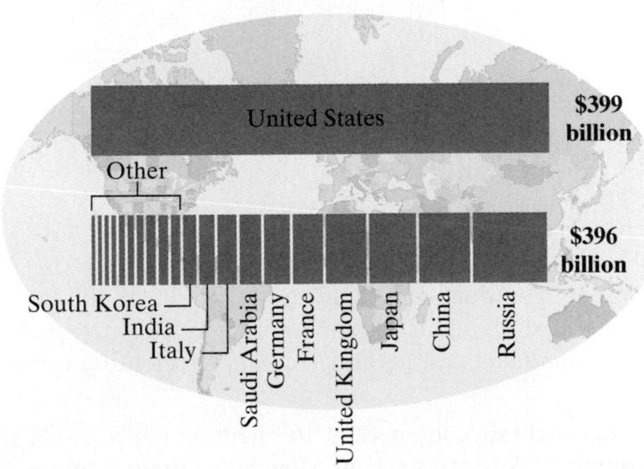

Source: U.S. Defense Department; Center for Defense Information

145. In 2003, the United States spent \$399 billion on defense.

146. In 2003, the United States spent more on defense than the 20 next top-spending nations combined. These nations spent \$396 billion on defense.

147. In 2002, the United States government spent approximately \$2 trillion, or 2×10^{12} dollars. Use the graph to write the amount that it spent on Social Security in scientific notation.

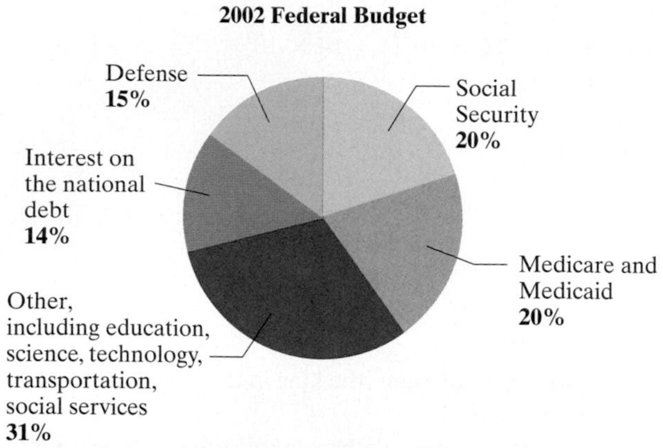

Source: U.S. Office of Management and Budget

148. Americans say they lead active lives, but for the 205 million of us ages 18 and older, walking is often as strenuous as it gets. Use the graph to write the number of Americans whose lifestyle is not very active. Express the answer in scientific notation.

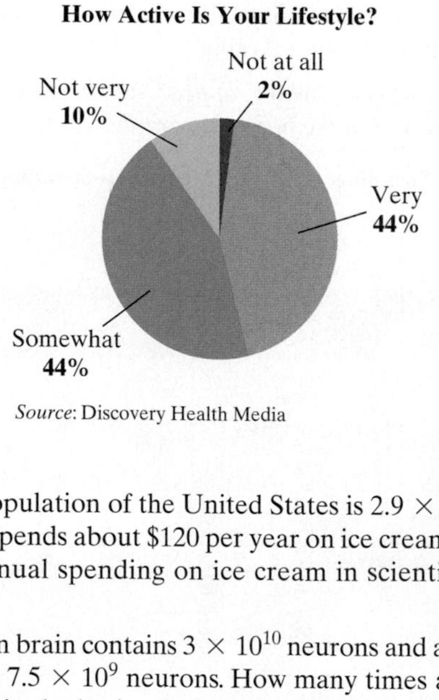

Source: Discovery Health Media

149. If the population of the United States is 2.9×10^8 and each person spends about \$120 per year on ice cream, express the total annual spending on ice cream in scientific notation.

150. A human brain contains 3×10^{10} neurons and a gorilla brain contains 7.5×10^9 neurons. How many times as many neurons are in the brain of a human as in the brain of a gorilla?

Use the motion formula $d = rt$, distance equals rate times time, and the fact that light travels at the rate of 1.86×10^5 *miles per second, to solve Exercises 151–152.*

151. If the moon is approximately 2.325×10^5 miles from Earth, how many seconds does it take moonlight to reach Earth?

152. If the sun is approximately 9.14×10^7 miles from Earth, how many seconds, to the nearest tenth of a second, does it take sunlight to reach Earth?

Writing in Mathematics

153. Explain the negative exponent rule and give an example.

154. How do you know if an exponential expression is simplified?

155. How do you know if a number is written in scientific notation?

156. Explain how to convert from scientific to decimal notation and give an example.

157. Explain how to convert from decimal to scientific notation and give an example.

158. Describe one advantage of expressing a number in scientific notation over decimal notation.

Critical Thinking Exercises

159. Which one of the following is true?

a. $4^{-2} < 4^{-3}$

b. $5^{-2} > 2^{-5}$

c. $(-2)^4 = 2^{-4}$

d. $5^2 \cdot 5^{-2} > 2^5 \cdot 2^{-5}$

160. Which one of the following is true?

a. $534.7 = 5.347 \times 10^3$

b. $\dfrac{8 \times 10^{30}}{4 \times 10^{-5}} = 2 \times 10^{25}$

c. $(7 \times 10^5) + (2 \times 10^{-3}) = 9 \times 10^2$

d. $(4 \times 10^3) + (3 \times 10^2) = 4.3 \times 10^3$

161. Give an example of a number where there is no advantage to using scientific notation instead of decimal notation. Explain why this is the case.

162. The mad Dr. Frankenstein has gathered enough bits and pieces (so to speak) for $2^{-1} + 2^{-2}$ of his creature-to-be. Write a fraction that represents the amount of his creature that must still be obtained.

Technology Exercises

163. Use a calculator in a fraction mode to check any five of your answers in Exercises 1–22.

164. Use a calculator to check any three of your answers in Exercises 79–90.

165. Use a calculator to check any three of your answers in Exercises 91–106.

166. Use a calculator with an [EE] or [EXP] key to check any four of your computations in Exercises 107–126. Display the result of the computation in scientific notation.

Review Exercises

167. Solve: $8 - 6x > 4x - 12$. (Section 2.7, Example 6)

168. Simplify: $24 \div 8 \cdot 3 + 28 \div (-7)$. (Section 1.8, Example 8)

169. List the whole numbers in this set:

$$\left\{-4, -\frac{1}{5}, 0, \pi, \sqrt{16}, \sqrt{17}\right\}.$$

(Section 1.2, Example 5)

GROUP PROJECT

CHAPTER 5

Putting Numbers into Perspective. A large number can be put into perspective by comparing it with another number. For example, we put the \$6.8 trillion national debt into perspective by comparing it to the number of U.S. citizens. The total distance covered by all the runners in the New York City Marathon (Check Point 10 on page 371) can be put into perspective by comparing this distance with, say, the distance from New York to San Francisco.

For this project, each group member should consult an almanac, a newspaper, or the World Wide Web to find a number greater than one million. Explain to other members of the group the context in which the large number is used. Express the number in scientific notation. Then put the number into perspective by comparing it with another number.

CHAPTER 5 SUMMARY

Section 5.1 Adding and Subtracting Polynomials

Definitions and Concepts	Examples
A polynomial is a single term or the sum of two or more terms containing variables in the numerator with whole number exponents. A monomial is a simplified polynomial with exactly one term; a binomial has exactly two terms; a trinomial has exactly three terms. The degree of a polynomial is the highest power of all the terms. The standard form of a polynomial is written in descending powers of the variable.	Polynomials Monomial: $2x^5$ Degree is 5. Binomial: $6x^3 + 5x$ Degree is 3. Trinomial: $7x + 4x^2 - 5$ Degree is 2.
To add polynomials, add like terms.	$(6x^3 + 5x^2 - 7x) + (-9x^3 + x^2 + 6x)$ $= (6x^3 - 9x^3) + (5x^2 + x^2) + (-7x + 6x)$ $= -3x^3 + 6x^2 - x$
The opposite, or additive inverse, of a polynomial is that polynomial with the sign of every coefficient changed. To subtract two polynomials, add the first polynomial and the opposite of the polynomial being subtracted.	$(5y^3 - 9y^2 - 4) - (3y^3 - 12y^2 - 5)$ $= (5y^3 - 9y^2 - 4) + (-3y^3 + 12y^2 + 5)$ $= (5y^3 - 3y^3) + (-9y^2 + 12y^2) + (-4 + 5)$ $= 2y^3 + 3y^2 + 1$

Definitions and Concepts	Examples
Section 5.2 Multiplying Polynomials	
Properties of Exponents Product Rule: $b^m \cdot b^n = b^{m+n}$ Power Rule: $(b^m)^n = b^{mn}$ Products to Powers: $(ab)^n = a^n b^n$	$x^3 \cdot x^8 = x^{3+8} = x^{11}$ $(x^3)^8 = x^{3 \cdot 8} = x^{24}$ $(-5x^2)^3 = (-5)^3(x^2)^3 = -125x^6$
To multiply monomials, multiply coefficients and add exponents.	$(-6x^4)(3x^{10}) = -6 \cdot 3x^{4+10} = -18x^{14}$
To multiply a monomial and a polynomial, multiply each term of the polynomial by the monomial.	$2x^4(3x^2 - 6x + 5)$ $= 2x^4 \cdot 3x^2 - 2x^4 \cdot 6x + 2x^4 \cdot 5$ $= 6x^6 - 12x^5 + 10x^4$
To multiply polynomials when neither is a monomial, multiply each term of one polynomial by each term of the other polynomial. Then combine like terms.	$(2x + 3)(5x^2 - 4x + 2)$ $= 2x(5x^2 - 4x + 2) + 3(5x^2 - 4x + 2)$ $= 10x^3 - 8x^2 + 4x + 15x^2 - 12x + 6$ $= 10x^3 + 7x^2 - 8x + 6$
Section 5.3 Special Products	
The FOIL method may be used when multiplying two binomials: First terms multiplied. Outside terms multiplied. Inside terms multiplied. Last terms multiplied.	F O I L $(3x + 7)(2x - 5) = 3x \cdot 2x + 3x(-5) + 7 \cdot 2x + 7(-5)$ $= 6x^2 - 15x + 14x - 35$ $= 6x^2 - x - 35$
The Product of the Sum and Difference of Two Terms $(A + B)(A - B) = A^2 - B^2$	$(4x + 7)(4x - 7) = (4x)^2 - 7^2$ $= 16x^2 - 49$
The Square of a Binomial Sum $(A + B)^2 = A^2 + 2AB + B^2$	$(x^2 + 6)^2 = (x^2)^2 + 2 \cdot x^2 \cdot 6 + 6^2$ $= x^4 + 12x^2 + 36$
The Square of a Binomial Difference $(A - B)^2 = A^2 - 2AB + B^2$	$(9x - 3)^2 = (9x)^2 - 2 \cdot 9x \cdot 3 + 3^2$ $= 81x^2 - 54x + 9$
Section 5.4 Polynomials in Several Variables	
To evaluate a polynomial in several variables, substitute the given value for each variable and perform the resulting computation.	Evaluate $4x^2y + 3xy - 2x$ for $x = -1$ and $y = -3$. $4x^2y + 3xy - 2x$ $= 4(-1)^2(-3) + 3(-1)(-3) - 2(-1)$ $= 4(1)(-3) + 3(-1)(-3) - 2(-1)$ $= -12 + 9 + 2 = -1$

Definitions and Concepts	Examples

Section 5.4 Polynomials in Several Variables (continued)

For a polynomial in two variables, the degree of a term is the sum of the exponents on its variables. The degree of the polynomial is the highest degree of all its terms.

$$7x^2y + 12x^4y^3 - 17x^5 + 6$$

degree: $2 + 1 = 3$ | degree: $4 + 3 = 7$ | degree: 5 | degree: 0

Degree of polynomial $= 7$

Polynomials in several variables are added, subtracted, and multiplied using the same rules for polynomials in one variable.

$$\begin{aligned}&(5x^2y^3 - xy + 4y^2) - (8x^2y^3 - 6xy - 2y^2)\\ &= (5x^2y^3 - xy + 4y^2) + (-8x^2y^3 + 6xy + 2y^2)\\ &= (5x^2y^3 - 8x^2y^3) + (-xy + 6xy) + (4y^2 + 2y^2)\\ &= -3x^2y^3 + 5xy + 6y^2\end{aligned}$$

F O I L

$$\begin{aligned}(3x - 2y)(x - y) &= 3x \cdot x + 3x(-y) + (-2y)x + (-2y)(-y)\\ &= 3x^2 - 3xy - 2xy + 2y^2\\ &= 3x^2 - 5xy + 2y^2\end{aligned}$$

Section 5.5 Dividing Polynomials

Additional Properties of Exponents

Quotient Rule: $\frac{b^m}{b^n} = b^{m-n}, \quad b \neq 0$

Zero-Exponent Rule: $b^0 = 1, \quad b \neq 0$

Quotients to Powers: $\left(\frac{a}{b}\right)^n = \frac{a^n}{b^n}, \quad b \neq 0$

$$\frac{x^{12}}{x^4} = x^{12-4} = x^8$$

$$(-3)^0 = 1 \qquad -3^0 = -(3^0) = -1$$

$$\left(\frac{y^2}{4}\right)^3 = \frac{(y^2)^3}{4^3} = \frac{y^{2\cdot3}}{4\cdot4\cdot4} = \frac{y^6}{64}$$

To divide monomials, divide coefficients and subtract exponents.

$$\frac{-40x^{40}}{20x^{20}} = \frac{-40}{20}x^{40-20} = -2x^{20}$$

To divide a polynomial by a monomial, divide each term of the polynomial by the monomial.

$$\begin{aligned}&\frac{8x^6 - 4x^3 + 10x}{2x}\\ &= \frac{8x^6}{2x} - \frac{4x^3}{2x} + \frac{10x}{2x}\\ &= 4x^{6-1} - 2x^{3-1} + 5x^{1-1} = 4x^5 - 2x^2 + 5\end{aligned}$$

Section 5.6 Long Division of Polynomials; Synthetic Division

To divide by a polynomial containing more than one term, begin by arranging all polynomials in descending powers of the variable. If a power of a variable is missing, add that power with a coefficient of 0. Repeat the four steps—divide, multiply, subtract, bring down the next term—until the degree of the remainder is less than the degree of the divisor.

Divide: $(2x^3 - x^2 - 7) \div (x - 2)$.

$$\begin{array}{r} 2x^2 + 3x + \ \ 6 \\ x - 2\overline{)2x^3 - \ \ x^2 + 0x - \ \ 7} \\ \underline{2x^3 - 4x^2} \qquad\qquad \\ 3x^2 + 0x \qquad \\ \underline{3x^2 - 6x} \qquad \\ 6x - \ \ 7 \\ \underline{6x - 12} \\ 5 \end{array}$$

The answer is $2x^2 + 3x + 6 + \frac{5}{x - 2}$.

Definitions and Concepts	Examples

Section 5.6 Long Division of Polynomials; Synthetic Division (continued)

Definitions and Concepts	Examples
A shortcut to long division, called synthetic division, can be used to divide a polynomial by a binomial of the form $x - c$.	Here is the division problem shown at the bottom of page 377 using synthetic division. $(2x^3 - x^2 - 7) \div (x - 2)$ Coefficients of the dividend, $2x^3 - x^2 + 0x - 7$ This is c in $x - c$. For $x - 2$, c is 2. $\underline{2}\rvert\ 2\ \ -1\ \ 0\ \ -7$ $\downarrow\ \ 4\ \ 6\ \ 12$ $2\ \ 3\ \ 6\ \ 5$ Coefficients of quotient; Remainder The answer is $2x^2 + 3x + 6 + \frac{5}{x-2}$.

Section 5.7 Negative Exponents and Scientific Notation

Definitions and Concepts	Examples
Negative Exponents in Numerators and Denominators If $b \neq 0$, $b^{-n} = \frac{1}{b^n}$ and $\frac{1}{b^{-n}} = b^n$.	$6^{-2} = \frac{1}{6^2} = \frac{1}{36}$ $\frac{1}{(-2)^{-4}} = (-2)^4 = 16$ $\left(\frac{2}{3}\right)^{-3} = \frac{2^{-3}}{3^{-3}} = \frac{3^3}{2^3} = \frac{27}{8}$
An exponential expression is simplified when • No parentheses appear. • No powers are raised to powers. • Each base occurs only once. • No negative or zero exponents appear.	Simplify: $\frac{(2x^4)^3}{x^{18}}$. $\frac{(2x^4)^3}{x^{18}} = \frac{2^3(x^4)^3}{x^{18}} = \frac{8x^{4\cdot 3}}{x^{18}} = \frac{8x^{12}}{x^{18}} = 8x^{12-18} = 8x^{-6} = \frac{8}{x^6}$
A positive number in scientific notation is expressed as $a \times 10^n$, where $1 \leq a < 10$ and n is an integer.	Write 2.9×10^{-3} in decimal notation. $2.9 \times 10^{-3} = .0029 = 0.0029$ Write 16,000 in scientific notation. $16{,}000 = 1.6 \times 10^4$
Use properties of exponents with base 10 $10^m \cdot 10^n = 10^{m+n}$, $\frac{10^m}{10^n} = 10^{m-n}$, and $(10^m)^n = 10^{mn}$ to perform computations with scientific notation.	$(5 \times 10^3)(4 \times 10^{-8})$ $= 5 \cdot 4 \times 10^{3-8}$ $= 20 \times 10^{-5}$ $= 2 \times 10^1 \times 10^{-5} = 2 \times 10^{-4}$

CHAPTER 5 REVIEW EXERCISES

5.1 *In Exercises 1–3, identify each polynomial as a monomial, binomial, or trinomial. Give the degree of the polynomial.*

1. $7x^4 + 9x$

2. $3x + 5x^2 - 2$

3. $16x$

In Exercises 4–8, add or subtract as indicated.

4. $(-6x^3 + 7x^2 - 9x + 3) + (14x^3 + 3x^2 - 11x - 7)$

5. $(9y^3 - 7y^2 + 5) + (4y^3 - y^2 + 7y - 10)$

6. $(5y^2 - y - 8) - (-6y^2 + 3y - 4)$
7. $(13x^4 - 8x^3 + 2x^2) - (5x^4 - 3x^3 + 2x^2 - 6)$

8. Subtract $x^4 + 7x^2 - 11x$ from $-13x^4 - 6x^2 + 5x$.

In Exercises 9–11, add or subtract as indicated.

9. Add. $\begin{array}{r} 7y^4 - 6y^3 + 4y^2 - 4y \\ \underline{\phantom{7y^4 - {}} y^3 - y^2 + 3y - 4} \end{array}$

10. Subtract. $\begin{array}{r} 7x^2 - 9x + 2 \\ \underline{-(4x^2 - 2x - 7)} \end{array}$

11. Subtract. $\begin{array}{r} 5x^3 - 6x^2 - 9x + 14 \\ \underline{-(-5x^3 + 3x^2 - 11x + 3)} \end{array}$

12. The polynomial $104.5x^2 - 1501.5x + 6016$ models the death rate per year per 100,000 men for men averaging x hours of sleep each night. Evaluate the polynomial when $x = 10$. Describe what the answer means in practical terms.

5.2 *In Exercises 13–17, simplify each expression.*

13. $x^{20} \cdot x^3$
14. $y \cdot y^5 \cdot y^8$
15. $(x^{20})^5$
16. $(10y)^2$
17. $(-4x^{10})^3$

In Exercises 18–26, find each product.

18. $(5x)(10x^3)$
19. $(-12y^7)(3y^4)$
20. $(-2x^5)(-3x^4)(5x^3)$
21. $7x(3x^2 + 9)$
22. $5x^3(4x^2 - 11x)$
23. $3y^2(-7y^2 + 3y - 6)$
24. $2y^5(8y^3 - 10y^2 + 1)$
25. $(x + 3)(x^2 - 5x + 2)$
26. $(3y - 2)(4y^2 + 3y - 5)$

In Exercises 27–28, use a vertical format to find each product.

27. $\begin{array}{r} y^2 - 4y + 7 \\ \underline{3y - 5} \end{array}$

28. $\begin{array}{r} 4x^3 - 2x^2 - 6x - 1 \\ \underline{2x + 3} \end{array}$

5.3 *In Exercises 29–41, find each product.*

29. $(x + 6)(x + 2)$
30. $(3y - 5)(2y + 1)$
31. $(4x^2 - 2)(x^2 - 3)$
32. $(5x + 4)(5x - 4)$
33. $(7 - 2y)(7 + 2y)$
34. $(y^2 + 1)(y^2 - 1)$
35. $(x + 3)^2$
36. $(3y + 4)^2$
37. $(y - 1)^2$
38. $(5y - 2)^2$
39. $(x^2 + 4)^2$
40. $(x^2 + 4)(x^2 - 4)$
41. $(x^2 + 4)(x^2 - 5)$
42. Write a polynomial in descending powers of x that represents the area of the shaded region.

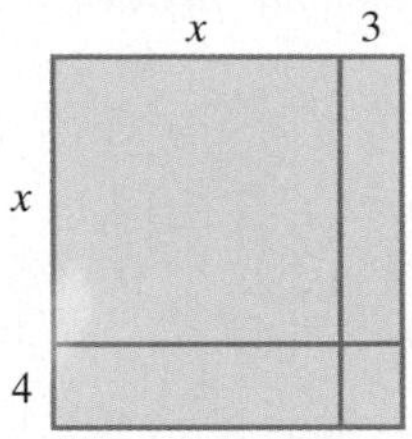

43. The parking garage shown in the figure below measures 30 yards by 20 yards. The length and the width are each increased by a fixed amount, x yards. Write a trinomial that describes the area of the expanded garage.

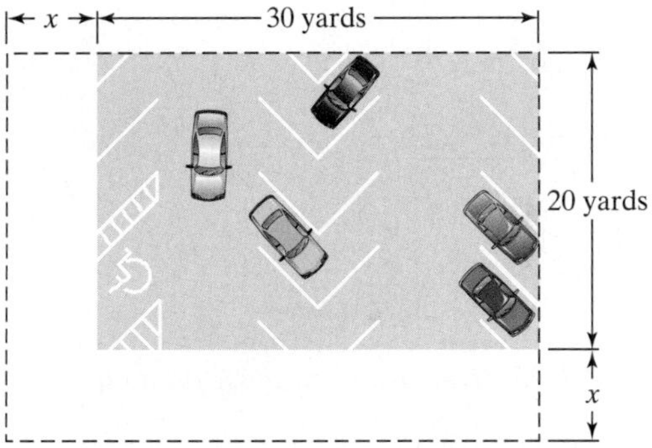

5.4

44. Evaluate $2x^3y - 4xy^2 + 5y + 6$ for $x = -1$ and $y = 2$.
45. Determine the coefficient of each term, the degree of each term, and the degree of the polynomial:

$4x^2y + 9x^3y^2 - 17x^4 - 12.$

In Exercises 46–55, perform the indicated operations.

46. $(7x^2 - 8xy + y^2) + (-8x^2 - 9xy + 4y^2)$

47. $(13x^3y^2 - 5x^2y - 9x^2) - (11x^3y^2 - 6x^2y - 3x^2 + 4)$

48. $(-7x^2y^3)(5x^4y^6)$
49. $5ab^2(3a^2b^3 - 4ab)$
50. $(x + 7y)(3x - 5y)$
51. $(4xy - 3)(9xy - 1)$
52. $(3x + 5y)^2$
53. $(xy - 7)^2$
54. $(7x + 4y)(7x - 4y)$
55. $(a - b)(a^2 + ab + b^2)$

5.5 *In Exercises 56–62, simplify each expression.*

56. $\frac{6^{40}}{6^{10}}$

57. $\frac{x^{18}}{x^3}$

58. $(-10)^0$

59. -10^0

60. $400x^0$

61. $\left(\frac{x^4}{2}\right)^3$

62. $\left(\frac{-3}{2y^6}\right)^4$

In Exercises 63–67, divide and check each answer.

63. $\frac{-15y^8}{3y^2}$

64. $\frac{40x^8y^6}{5xy^3}$

65. $\frac{18x^4 - 12x^2 + 36x}{6x}$

66. $\frac{30x^8 - 25x^7 - 40x^5}{-5x^3}$

67. $\frac{27x^3y^2 - 9x^2y - 18xy^2}{3xy}$

5.6 *In Exercises 68–71, divide and check each answer.*

68. $\frac{2x^2 + 3x - 14}{x - 2}$

69. $\frac{2x^3 - 5x^2 + 7x + 5}{2x + 1}$

70. $\frac{x^3 - 2x^2 - 33x - 7}{x - 7}$

71. $\frac{y^3 - 27}{y - 3}$

72. $(4x^4 + 6x^3 + 3x - 1) \div (2x^2 + 1)$

In Exercises 73–75, divide using synthetic division.

73. $(4x^3 - 3x^2 - 2x + 1) \div (x + 1)$

74. $(3x^4 - 2x^2 - 10x - 20) \div (x - 2)$

75. $(x^4 + 16) \div (x + 4)$

5.7 *In Exercises 76–80, write each expression with positive exponents only and then simplify.*

76. 7^{-2}

77. $(-4)^{-3}$

78. $2^{-1} + 4^{-1}$

79. $\frac{1}{5^{-2}}$

80. $\left(\frac{2}{5}\right)^{-3}$

In Exercises 81–89, simplify each exponential expression. Assume that variables in denominators do not equal zero.

81. $\frac{x^3}{x^9}$

82. $\frac{30y^6}{5y^8}$

83. $(5x^{-7})(6x^2)$

84. $\frac{x^4 \cdot x^{-2}}{x^{-6}}$

85. $\frac{(3y^3)^4}{y^{10}}$

86. $\frac{y^{-7}}{(y^4)^3}$

87. $(2x^{-1})^{-3}$

88. $\left(\frac{x^7}{x^4}\right)^{-2}$

89. $\frac{(y^3)^4}{(y^{-2})^4}$

In Exercises 90–92, write each number in decimal notation without the use of exponents.

90. 2.3×10^4

91. 1.76×10^{-3}

92. 9×10^{-1}

In Exercises 93–96, write each number in scientific notation.

93. 73,900,000

94. 0.00062

95. 0.38

96. 3.8

In Exercises 97–99, perform the indicated computation. Write the answers in scientific notation.

97. $(6 \times 10^{-3})(1.5 \times 10^6)$

98. $\frac{2 \times 10^2}{4 \times 10^{-3}}$

99. $(4 \times 10^{-2})^2$

100. A microsecond is 10^{-6} second and a nanosecond is 10^{-9} second. How many nanoseconds make a microsecond?

101. The world's population is approximately 6.3×10^9 people. Current projections double this population in 40 years. Write the population 40 years from now in scientific notation.

CHAPTER 5 TEST

Remember to use your Chapter Test Prep Video CD to see the worked-out solutions to the test questions you want to review.

1. Identify $9x + 6x^2 - 4$ as a monomial, binomial, or trinomial. Give the degree of the polynomial.

In Exercises 2–3, add or subtract as indicated.

2. $(7x^3 + 3x^2 - 5x - 11) + (6x^3 - 2x^2 + 4x - 13)$

3. $(9x^3 - 6x^2 - 11x - 4) - (4x^3 - 8x^2 - 13x + 5)$

In Exercises 4–10, find each product.

4. $(-7x^3)(5x^8)$

5. $6x^2(8x^3 - 5x - 2)$

6. $(3x + 2)(x^2 - 4x - 3)$

7. $(3y + 7)(2y - 9)$

8. $(7x + 5)(7x - 5)$

9. $(x^2 + 3)^2$

10. $(5x - 3)^2$

11. Evaluate $4x^2y + 5xy - 6x$ for $x = -2$ and $y = 3$.

In Exercises 12–14, perform the indicated operations.

12. $(8x^2y^3 - xy + 2y^2) - (6x^2y^3 - 4xy - 10y^2)$

13. $(3a - 7b)(4a + 5b)$

14. $(2x + 3y)^2$

In Exercises 15–17, divide and check each answer.

15. $\dfrac{-25x^{16}}{5x^4}$

16. $\dfrac{15x^4 - 10x^3 + 25x^2}{5x}$

17. $\dfrac{2x^3 - 3x^2 + 4x + 4}{2x + 1}$

18. Divide using synthetic division:

$$(3x^4 + 11x^3 - 20x^2 + 7x + 35) \div (x + 5).$$

In Exercises 19–20, write each expression with positive exponents only and then simplify.

19. 10^{-2} **20.** $\dfrac{1}{4^{-3}}$

In Exercises 21–26, simplify each expression.

21. $(-3x^2)^3$ **22.** $\dfrac{20x^3}{5x^8}$

23. $(-7x^{-8})(3x^2)$ **24.** $\dfrac{(2y^3)^4}{y^8}$

25. $(5x^{-4})^{-2}$ **26.** $\left(\dfrac{x^{10}}{x^5}\right)^{-3}$

27. Write 3.7×10^{-4} in decimal notation.

28. Write 7,600,000 in scientific notation.

In Exercises 29–30, perform the indicated computation. Write the answers in scientific notation.

29. $(4.1 \times 10^2)(3 \times 10^{-5})$

30. $\dfrac{8.4 \times 10^6}{4 \times 10^{-2}}$

31. Write a polynomial in descending powers of x that represents the area of the figure.

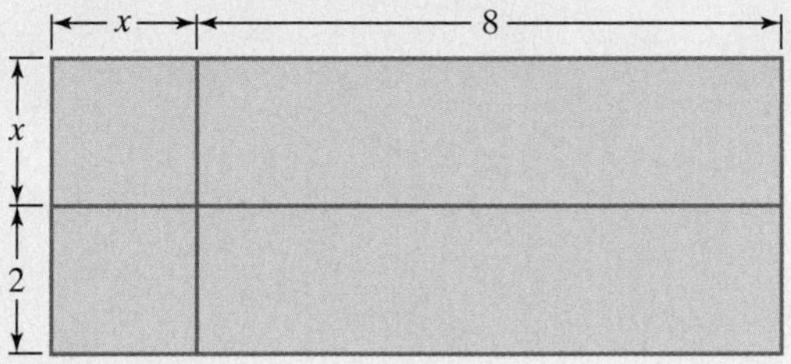

CUMULATIVE REVIEW EXERCISES (CHAPTERS 1–5)

In Exercises 1–2, perform the indicated operation or operations.

1. $(-7)(-5) \div (12 - 3)$ **2.** $(3 - 7)^2(9 - 11)^3$

3. What is the difference in elevation between a plane flying 14,300 feet above sea level and a submarine traveling 750 feet below sea level?

In Exercises 4–5, solve each equation.

4. $2(x + 3) + 2x = x + 4$

5. $\dfrac{x}{5} - \dfrac{1}{3} = \dfrac{x}{10} - \dfrac{1}{2}$

6. The length of a rectangular sign is 2 feet less than three times its width. If the perimeter of the sign is 28 feet, what are its dimensions?

7. Solve: $7 - 8x \le -6x - 5$. Express the solution set in set-builder notation and graph the solution set on a number line.

8. You invested \$6000 in two stocks paying 12% and 14% annual interest, respectively. At the end of the year, the total interest from these investments was \$772. How much was invested at each rate?

9. You need to mix a solution that is 70% antifreeze with one that is 30% antifreeze to obtain 20 liters of a mixture that is 60% antifreeze. How many liters of each of the solutions must be used?

10. Graph $y = -\frac{2}{5}x + 2$ using the slope and y-intercept.

11. Graph $x - 2y = 4$ using intercepts.

12. Find the slope of the line passing through the points $(-3, 2)$ and $(2, -4)$. Is the line rising, falling, horizontal, or vertical?

13. The slope of a line is -2 and the line passes through the point $(3, -1)$. Write the line's equation in point-slope form and slope-intercept form.

In Exercises 14–15, solve each system by the method of your choice.

14. $3x + 2y = 10$
$4x - 3y = -15$

15. $2x + 3y = -6$
$y = 3x - 13$

16. You are choosing between two long-distance telephone plans. One has a monthly fee of \$15 with a charge of \$0.05 per minute for all long-distance calls. The other plan has a monthly fee of \$5 with a charge of \$0.07 per minute for all long-distance calls. For how many minutes of long-distance calls will the costs for the two plans be the same? What will be the cost for each plan?

17. Write in scientific notation: 0.0024.

18. Subtract: $(9x^5 - 3x^3 + 2x - 7) - (6x^5 + 3x^3 - 7x - 9)$.

19. Divide: $\dfrac{x^3 + 3x^2 + 5x + 3}{x + 1}$.

20. Simplify: $\dfrac{(3x^2)^4}{x^{10}}$.

CHAPTER 6

Landscaped parks often form an oasis of greenery and calm in which to escape from the bustle of city life. Flowers, trees, ponds, and fountains provide a natural setting that mirrors the interest that many city dwellers have in their environment.

The role of polynomials in landscape design is explored in Exercises 84–85 in Exercise Set 6.6.

Factoring Polynomials

Have you ever thought about creating attractive and inviting home landscaping? Algebra and geometry play an important role in landscape design. In this chapter, you will see how rewriting a polynomial sum or difference in terms of multiplication can be used in the creation of horticultural masterpieces.

SECTION 6.1

Objectives

1. Factor monomials.
2. Find the greatest common factor.
3. Factor out the greatest common factor of a polynomial.
4. Factor by grouping.

THE GREATEST COMMON FACTOR AND FACTORING BY GROUPING

A two-year-old boy is asked, "Do you have a brother?" He answers, "Yes." "What is your brother's name?" "Tom." Asked if Tom has a brother, the two-year-old replies, "No." The child can go in the direction from self to brother, but he cannot reverse this direction and move from brother back to self.

As our intellects develop, we learn to reverse the direction of our thinking. Reversibility of thought is found throughout algebra. For example, we can multiply polynomials and show that

$$5x(2x + 3) = 10x^2 + 15x.$$

We can also reverse this process and express the resulting polynomial as

$$10x^2 + 15x = 5x(2x + 3).$$

Factoring a polynomial containing the sum of monomials means finding an equivalent expression that is a product.

Factoring $10x^2 + 15x$

Sum of monomials — Equivalent expression that is a product

$$10x^2 + 15x = 5x(2x + 3)$$

The factors of $10x^2 + 15x$ are $5x$ and $2x + 3$.

In this chapter, we will be factoring over the set of integers, meaning that the coefficients in the factors are integers. Polynomials that cannot be factored using integer coefficients are called **prime polynomials** over the set of integers.

1 Factor monomials.

Factoring Monomials Factoring a monomial means finding two monomials whose product gives the original monomial. For example, $30x^2$ can be factored in a number of different ways, such as

$30x^2 = (5x)(6x)$ The factors are $5x$ and $6x$.

$30x^2 = (15x)(2x)$ The factors are $15x$ and $2x$.

$30x^2 = (10x^2)(3)$ The factors are $10x^2$ and 3.

$30x^2 = (-6x)(-5x).$ The factors are $-6x$ and $-5x$.

Observe that each part of the factorization is called a *factor* of the given monomial.

DISCOVER FOR YOURSELF

Write three more ways of factoring the monomial $30x^2$.

2 Find the greatest common factor.

Factoring Out the Greatest Common Factor We use the distributive property to multiply a monomial and a polynomial of two or more terms. When we factor, we reverse this process, expressing the polynomial as a product.

Multiplication	**Factoring**
$a(b + c) = ab + ac$	$ab + ac = a(b + c)$

Here is a specific example:

Multiplication	**Factoring**
$5x(2x + 3)$	$10x^2 + 15x$
$= 5x \cdot 2x + 5x \cdot 3$	$= 5x \cdot 2x + 5x \cdot 3$
$= 10x^2 + 15x$	$= 5x(2x + 3)$.

In the process of finding an equivalent expression for $10x^2 + 15x$ that is a product, we used the fact that $5x$ is a factor of both $10x^2$ and $15x$. The factoring on the right shows that $5x$ is a *common factor* for all the terms of the binomial $10x^2 + 15x$.

In any factoring problem, the first step is to look for the *greatest common factor*. The **greatest common factor**, abbreviated GCF, is an expression of the highest degree that divides each term of the polynomial. Can you see that $5x$ is the greatest common factor of $10x^2 + 15x$? 5 is the greatest integer that divides 10 and 15. Furthermore, x is the greatest expression that divides x^2 and x.

The variable part of the greatest common factor always contains the smallest power of a variable that appears in all terms of the polynomial. For example, consider the polynomial

$$10x^2 + 15x.$$

x^1, or x, is the variable raised to the smallest exponent.

We see that x is the variable part of the greatest common factor, $5x$.

EXAMPLE 1 Finding the Greatest Common Factor

Find the greatest common factor of each list of terms:

a. $6x^3$ and $10x^2$ **b.** $15y^5$, $-9y^4$, and $27y^3$ **c.** x^5y^3, x^4y^4, and x^3y^2.

SOLUTION Use numerical coefficients to determine the coefficient of the GCF. Use variable factors to determine the variable factor of the GCF.

2 is the greatest integer that divides 6 and 10.

a. $6x^3$ and $10x^2$

x^2 is the variable raised to the smallest exponent.

We see that 2 is the coefficient of the GCF and x^2 is the variable factor of the GCF. Thus, the GCF of $6x^3$ and $10x^2$ is $2x^2$.

3 is the greatest integer that divides 15, −9, and 27.

b. $15y^5$, $-9y^4$, and $27y^3$

y^3 is the variable raised to the smallest exponent.

We see that 3 is the coefficient of the GCF and y^3 is the variable factor of the GCF. Thus, the GCF of $15y^5$, $-9y^4$, and $27y^3$ is $3y^3$.

x^3 is the variable, x, raised to the smallest exponent.

c. x^5y^3, x^4y^4, and x^3y^2

y^2 is the variable, y, raised to the smallest exponent.

Because all terms have coefficients of 1, 1 is the greatest integer that divides these coefficients. Thus, 1 is the coefficient of the GCF. The voice balloons show that x^3 and y^2 are the variable factors of the GCF. Thus, the GCF of x^5y^3, x^4y^4, and x^3y^2 is x^3y^2. ■

CHECK POINT 1 Find the greatest common factor of each list of terms:

a. $18x^3$ and $15x^2$ **b.** $-20x^2$, $12x^4$, and $40x^3$

c. x^4y, x^3y^2, and x^2y.

3 Factor out the greatest common factor of a polynomial.

When we factor a monomial from a polynomial, we determine the greatest common factor of all terms in the polynomial. Sometimes there may not be a GCF other than 1. When a GCF other than 1 exists, we use the following procedure:

FACTORING A MONOMIAL FROM A POLYNOMIAL

1. Determine the greatest common factor of all terms in the polynomial.
2. Express each term as the product of the GCF and its other factor.
3. Use the distributive property to factor out the GCF.

EXAMPLE 2 Factoring Out the Greatest Common Factor

Factor: $5x^2 + 30$.

SOLUTION The GCF of $5x^2$ and 30 is 5.

$$5x^2 + 30$$

$$= 5 \cdot x^2 + 5 \cdot 6$$ Express each term as the product of the GCF and its other factor.

$$= 5(x^2 + 6)$$ Factor out the GCF.

Because factoring reverses the process of multiplication, all factoring results can be checked by multiplying.

$$5(x^2 + 6) = 5 \cdot x^2 + 5 \cdot 6 = 5x^2 + 30$$

The factoring is correct because multiplication gives us the original polynomial. ■

CHECK POINT 2 Factor: $6x^2 + 18$.

STUDY TIP

When we express $5x^2 + 30$ as $5 \cdot x^2 + 5 \cdot 6$, we have factored the *terms* of the binomial, but not the binomial itself. The factorization of the binomial is not complete until we write

$$5(x^2 + 6).$$

Now we have expressed the binomial *as a product*.

EXAMPLE 3 Factoring Out the Greatest Common Factor

Factor: $18x^3 + 27x^2$.

SOLUTION We begin by determining the greatest common factor.

9 is the greatest integer that divides 18 and 27.

$18x^3$ and $27x^2$

x^2 is the variable raised to the smallest exponent.

The GCF of the two terms in the polynomial is $9x^2$.

$$18x^3 + 27x^2$$

$$= 9x^2(2x) + 9x^2(3)$$ Express each term as the product of the GCF and its other factor.

$$= 9x^2(2x + 3)$$ Factor out the GCF.

We can check this factorization by multiplying $9x^2$ and $2x + 3$, obtaining the original polynomial as the answer. ■

CHECK POINT 3 Factor: $25x^2 + 35x^3$.

DISCOVER FOR YOURSELF

What happens if you factor out $3x^2$ rather than $9x^2$ from $18x^3 + 27x^2$? Although $3x^2$ is a common factor of the two terms, it is not the *greatest* common factor. Remove $3x^2$ from $18x^3 + 27x^2$ and describe what happens with the second factor. Now factor again. Make the final result look like the factorization in Example 3. What is the advantage of factoring out the greatest common factor rather than just a common factor?

EXAMPLE 4 Factoring Out the Greatest Common Factor

Factor: $16x^5 - 12x^4 + 4x^3$.

SOLUTION First, determine the greatest common factor.

4 is the greatest integer that divides 16, −12, and 4.

$16x^5$, $-12x^4$, and $4x^3$

x^3 is the variable raised to the smallest exponent.

The GCF of the three terms of the polynomial is $4x^3$.

$$16x^5 - 12x^4 + 4x^3$$

$$= 4x^3 \cdot 4x^2 - 4x^3 \cdot 3x + 4x^3 \cdot 1$$ Express each term as the product of the GCF and its other factor.

$$= 4x^3(4x^2 - 3x + 1)$$ Factor out the GCF.

Don't leave out the 1. ■

CHECK POINT 4 Factor: $15x^5 + 12x^4 - 27x^3$.

EXAMPLE 5 Factoring Out the Greatest Common Factor

Factor: $27x^2y^3 - 9xy^2 + 81xy$.

SOLUTION First, determine the greatest common factor.

9 is the greatest integer that divides 27, −9, and 81.

$27x^2y^3$, $-9xy^2$, and $81xy$

The variables raised to the smallest exponents are x and y.

The GCF of the three terms of the polynomial is $9xy$.

$27x^2y^3 - 9xy^2 + 81xy$

$= 9xy \cdot 3xy^2 - 9xy \cdot y + 9xy \cdot 9$ — Express each term as the product of the GCF and its other factor.

$= 9xy(3xy^2 - y + 9)$ — Factor out the GCF. ■

CHECK POINT 5 Factor: $8x^3y^2 - 14x^2y + 2xy$.

4 Factor by grouping.

Factoring by Grouping Up to now, we have factored a monomial from a polynomial. By contrast, in our next example, the greatest common factor of the polynomial is a binomial.

EXAMPLE 6 Factoring Out the Greatest Common Binomial Factor

Factor:

a. $x^2(x + 3) + 5(x + 3)$ **b.** $x(y + 1) - 2(y + 1)$.

SOLUTION Let's identify the common binomial factor in each part of the problem.

$x^2(x + 3)$ and $5(x + 3)$ — The GCF, a binomial, is $x + 3$.

$x(y + 1)$ and $-2(y + 1)$ — The GCF, a binomial, is $y + 1$.

We factor out these common binomial factors as follows.

a. $x^2(x + 3) + 5(x + 3)$

$= (x + 3)x^2 + (x + 3)5$ — Express each term as the product of the GCF and its other factor, in that order. Hereafter, we omit this step.

$= (x + 3)(x^2 + 5)$ — Factor out the GCF, $x + 3$.

b. $x(y + 1) - 2(y + 1)$ — The GCF is $y + 1$.

$= (y + 1)(x - 2)$ — Factor out the GCF. ■

CHECK POINT 6 Factor:

a. $x^2(x + 1) + 7(x + 1)$ **b.** $x(y + 4) - 7(y + 4)$.

Some polynomials have only a greatest common factor of 1. However, by a suitable grouping of the terms, it still may be possible to factor. This process, called **factoring by grouping**, is illustrated in Example 7.

EXAMPLE 7 Factoring by Grouping

Factor: $x^3 + 4x^2 + 3x + 12$.

SOLUTION There is no factor other than 1 common to all terms. However, we can group terms that have a common factor:

$$\boxed{x^3 + 4x^2} + \boxed{3x + 12}.$$

Common factor is x^2.

Common factor is 3.

We now factor the given polynomial as follows:

$x^3 + 4x^2 + 3x + 12$

$= (x^3 + 4x^2) + (3x + 12)$ — *Group terms with common factors.*

$= x^2(x + 4) + 3(x + 4)$ — *Factor out the greatest common factor from the grouped terms. The remaining two terms have $x + 4$ as a common binomial factor.*

$= (x + 4)(x^2 + 3).$ — *Factor out the GCF, $x + 4$.*

Thus, $x^3 + 4x^2 + 3x + 12 = (x + 4)(x^2 + 3)$. Check the factorization by multiplying the right side of the equation using the FOIL method. Because the factorization is correct, you should obtain the original polynomial. ■

DISCOVER FOR YOURSELF

In Example 7, group the terms as follows:

$$(x^3 + 3x) + (4x^2 + 12).$$

Factor out the greatest common factor from each group and complete the factoring process. Describe what happens. What can you conclude?

✔ **CHECK POINT 7** Factor: $x^3 + 5x^2 + 2x + 10$.

FACTORING BY GROUPING

1. Group terms that have a common monomial factor. There will usually be two groups. Sometimes the terms must be rearranged.
2. Factor out the common monomial factor from each group.
3. Factor out the remaining common binomial factor (if one exists).

EXAMPLE 8 Factoring by Grouping

Factor: $xy + 5x - 4y - 20$.

SOLUTION There is no factor other than 1 common to all terms. However, we can group terms that have a common factor:

$$\boxed{xy + 5x} + \boxed{-4y - 20}.$$

Common factor is x: $xy + 5x = x(y + 5)$.

Use -4, rather than 4, as the common factor: $-4y - 20 = -4(y + 5)$. In this way, the common binomial factor, $y + 5$, appears.

The voice balloons illustrate that it is sometimes necessary to factor out a negative number from a grouping to obtain a common binomial factor for the two groupings. We now factor the given polynomial as follows:

$$xy + 5x - 4y - 20$$

$$= x(y + 5) - 4(y + 5) \quad \text{Factor } x \text{ and } -4\text{, respectively, from each grouping.}$$

$$= (y + 5)(x - 4). \quad \text{Factor out the GCF, } y + 5.$$

Thus, $xy + 5x - 4y - 20 = (y + 5)(x - 4)$. Using the commutative property of multiplication, the factorization can also be expressed as $(x - 4)(y + 5)$. Multiply these factors using the FOIL method to verify that, regardless of the order, these are the correct factors. ■

✔ **CHECK POINT 8** Factor: $xy + 3x - 5y - 15$.

6.1 EXERCISE SET

Student Solutions Manual | CD/Video | PH Math/Tutor Center | MathXL Tutorials on CD | MathXL® | MyMathLab | Interactmath.com

Practice Exercises

In Exercises 1–6, find three factorizations for each monomial.

1. $8x^3$

2. $20x^4$

3. $-12x^5$

4. $-15x^6$

5. $36x^4$

6. $27x^5$

In Exercises 7–18, find the greatest common factor of each list of terms.

7. 4 and $8x$

8. 5 and $15x$

9. $12x^2$ and $8x$

10. $20x^2$ and $15x$

11. $-2x^4$ and $6x^3$

12. $-3x^4$ and $6x^3$

13. $9y^5$, $18y^2$, and $-3y$

14. $10y^5$, $20y^2$, and $-5y$

15. xy, xy^2, and xy^3

16. x^2y, $3x^3y$, and $6x^2$

17. $16x^5y^4$, $8x^6y^3$, and $20x^4y^5$

18. $18x^5y^4$, $6x^6y^3$, and $12x^4y^5$

In Exercises 19–54, factor each polynomial using the greatest common factor. If there is no common factor other than 1 and the polynomial cannot be factored, so state.

19. $8x + 8$

20. $9x + 9$

21. $4y - 4$

22. $5y - 5$

23. $5x + 30$

24. $10x + 30$

25. $30x - 12$

26. $32x - 24$

27. $x^2 + 5x$

28. $x^2 + 6x$

29. $18y^2 + 12$

30. $20y^2 + 15$

31. $14x^3 + 21x^2$

32. $6x^3 + 15x^2$

33. $13y^2 - 25y$

34. $11y^2 - 30y$

35. $9y^4 + 27y^6$

36. $10y^4 + 15y^6$

37. $8x^2 - 4x^4$

38. $12x^2 - 4x^4$

39. $12y^2 + 16y - 8$

40. $15y^2 - 3y + 9$

41. $9x^4 + 18x^3 + 6x^2$

42. $32x^4 + 2x^3 + 8x^2$

43. $100y^5 - 50y^3 + 100y^2$

44. $26y^5 - 13y^3 + 39y^2$

45. $10x - 20x^2 + 5x^3$

46. $6x - 4x^2 + 2x^3$

47. $11x^2 - 23$

48. $12x^2 - 25$

49. $6x^3y^2 + 9xy$

50. $4x^2y^3 + 6xy$

51. $30x^2y^3 - 10xy^2 + 20xy$

52. $27x^2y^3 - 18xy^2 + 45x^2y$

53. $32x^3y^2 - 24x^3y - 16x^2y$

54. $18x^3y^2 - 12x^3y - 24x^2y$

In Exercises 55–66, factor each polynomial using the greatest common binomial factor.

55. $x(x + 5) + 3(x + 5)$

56. $x(x + 7) + 10(x + 7)$

57. $x(x + 2) - 4(x + 2)$

58. $x(x + 3) - 8(x + 3)$

59. $x(y + 6) - 7(y + 6)$

60. $x(y + 9) - 11(y + 9)$

61. $3x(x + y) - (x + y)$

62. $7x(x + y) - (x + y)$

63. $4x(3x + 1) + 3x + 1$

64. $5x(2x + 1) + 2x + 1$

65. $7x^2(5x + 4) + 5x + 4$

66. $9x^2(7x + 2) + 7x + 2$

In Exercises 67–84, factor by grouping.

67. $x^2 + 2x + 4x + 8$

68. $x^2 + 3x + 5x + 15$

69. $x^2 + 3x - 5x - 15$

70. $x^2 + 7x - 4x - 28$

71. $x^3 - 2x^2 + 5x - 10$

72. $x^3 - 3x^2 + 4x - 12$

73. $x^3 - x^2 + 2x - 2$

74. $x^3 + 6x^2 - 2x - 12$

75. $xy + 5x + 9y + 45$

76. $xy + 6x + 2y + 12$

77. $xy - x + 5y - 5$

78. $xy - x + 7y - 7$

79. $3x^2 - 6xy + 5xy - 10y^2$

80. $10x^2 - 12xy + 35xy - 42y^2$

81. $3x^3 - 2x^2 - 6x + 4$

82. $4x^3 - x^2 - 12x + 3$

83. $x^2 - ax - bx + ab$

84. $x^2 + ax + bx + ab$

Practice Plus

In Exercises 85–92, factor each polynomial.

85. $24x^3y^3z^3 + 30x^2y^2z + 18x^2yz^2$

86. $16x^2y^2z^2 + 32x^2yz^2 + 24x^2yz$

87. $x^3 - 4 + 3x^3y - 12y$

88. $x^3 - 5 + 2x^3y - 10y$

89. $4x^5(x + 1) - 6x^3(x + 1) - 8x^2(x + 1)$

90. $8x^5(x + 2) - 10x^3(x + 2) - 2x^2(x + 2)$

91. $3x^5 - 3x^4 + x^3 - x^2 + 5x - 5$

92. $7x^5 - 7x^4 + x^3 - x^2 + 3x - 3$

The figures for Exercises 93–94 show one or more circles drawn inside a square. Write a polynomial that represents the shaded area in each figure. Then factor the polynomial.

93.

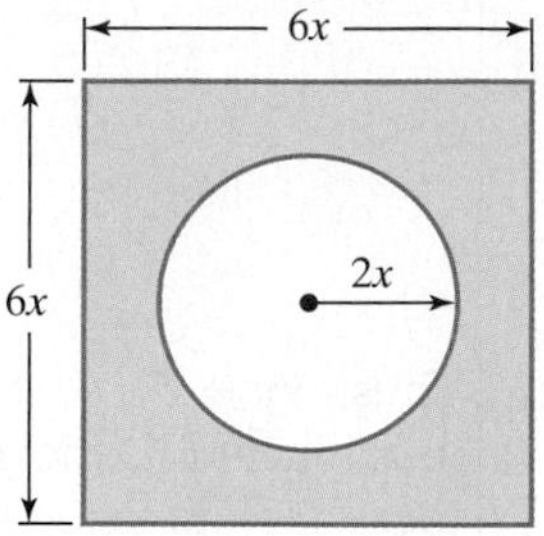

94.

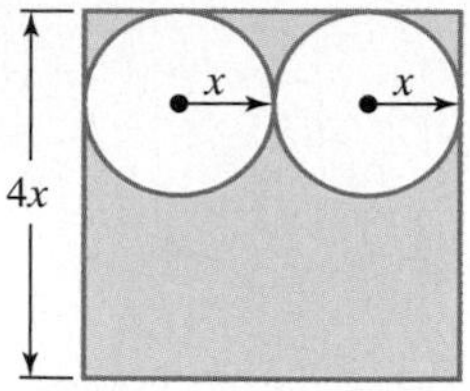

Application Exercises

95. An explosion causes debris to rise vertically with an initial velocity of 64 feet per second. The polynomial $64x - 16x^2$ describes the height of the debris above the ground, in feet, after x seconds.

a. Find the height of the debris after 3 seconds.

b. Factor the polynomial.

c. Use the factored form of the polynomial in part (b) to find the height of the debris after 3 seconds. Do you get the same answer as you did in part (a)? If so, does this prove that your factorization is correct? Explain.

96. An explosion causes debris to rise vertically with an initial velocity of 72 feet per second. The polynomial $72x - 16x^2$ describes the height of the debris above the ground, in feet, after x seconds.

a. Find the height of the debris after 4 seconds.

b. Factor the polynomial.

c. Use the factored form of the polynomial in part (b) to find the height of the debris after 4 seconds. Do you get the same answer as you did in part (a)? If so, does this prove that your factorization is correct? Explain.

In Exercises 97–98, write a polynomial for the length of each rectangle.

97.

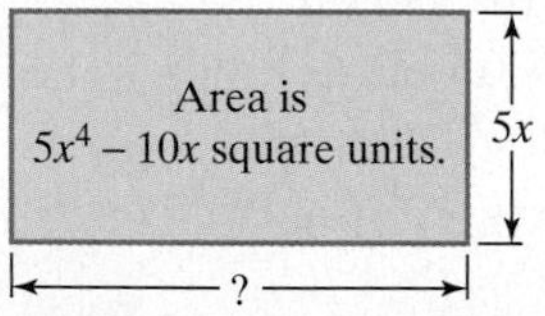

98.

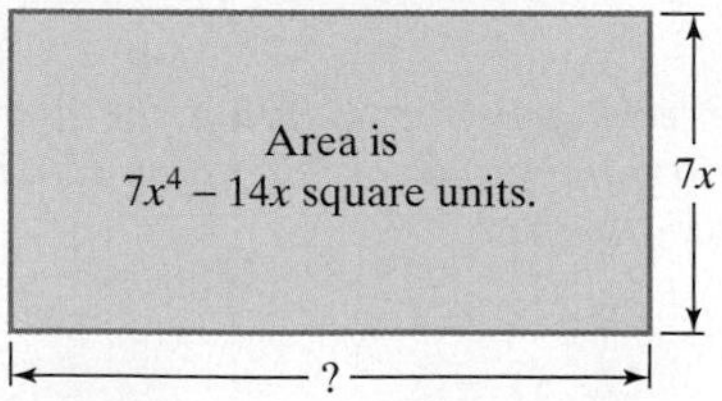

Writing in Mathematics

99. What is factoring?

100. What is a prime polynomial?

101. Explain how to find the greatest common factor of a list of terms. Give an example with your explanation.

102. Use an example and explain how to factor out the greatest common factor of a polynomial.

103. Suppose that a polynomial contains four terms and can be factored by grouping. Explain how to obtain the factorization.

104. Write a sentence that uses the word "factor" as a noun. Then write a sentence that uses the word "factor" as a verb.

Critical Thinking Exercises

105. Which one of the following is true?

a. Because a monomial contains one term, it follows that a monomial can be factored in precisely one way.

b. The GCF for $8x^3 - 16x^2$ is $8x$.

c. The integers 10 and 31 have no GCF.

d. $-4x^2 + 12x$ can be factored as $-4x(x - 3)$ or $4x(-x + 3)$.

106. Suppose you receive x dollars in January. Each month thereafter, you receive $100 more than you received the month before. Write a factored polynomial that describes the total dollar amount you receive from January through April.

In Exercises 107–108, write a polynomial that fits the given description. Do not use a polynomial that appears in this section or in the exercise set.

107. The polynomial has four terms and can be factored using a greatest common factor that has both a coefficient and a variable.

108. The polynomial has four terms and can be factored by grouping.

Technology Exercises

In Exercises 109–111, use a graphing utility to graph each side of the equation in the same viewing rectangle. Do the graphs coincide? If so, this means that the polynomial on the left side has been factored correctly. If not, factor the polynomial correctly and then use your graphing utility to verify the factorization.

109. $-3x - 6 = -3(x - 2)$

110. $x^2 - 2x + 5x - 10 = (x - 2)(x - 5)$

111. $x^2 + 2x + x + 2 = x(x + 2) + 1$

Review Exercises

112. Multiply: $(x + 7)(x + 10)$. (Section 5.3, Example 1)

113. Solve the system by graphing:

$$\begin{aligned} 2x - y &= -4 \\ x - 3y &= 3. \end{aligned}$$

(Section 4.1, Example 2)

114. Write the point-slope form of a line passing through $(-7, 2)$ and $(-4, 5)$. Then use the point-slope equation to write the slope-intercept equation. (Section 3.5, Example 2)

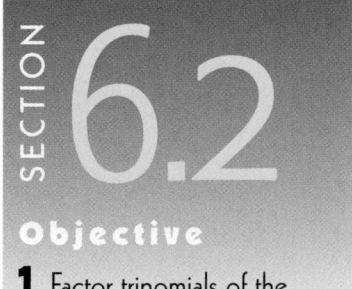

SECTION 6.2

Objective

1 Factor trinomials of the form $x^2 + bx + c$.

FACTORING TRINOMIALS WHOSE LEADING COEFFICIENT IS ONE

Not afraid of heights and cutting-edge excitement? How about skydiving? Behind your exhilarating experience is the world of algebra. After you jump from the airplane, your height above the ground at every instant of your fall can be described by a formula involving a variable that is squared. At a height of approximately 2000 feet, you'll need to open your parachute. How can you determine when you must do so?

The answer to this critical question involves using the factoring technique presented in this section. In Section 6.6, in which applications are discussed, this technique is applied to models involving the height of any free-falling object—in this case, you.

1 Factor trinomials of the form $x^2 + bx + c$.

A Strategy for Factoring $x^2 + bx + c$ In Section 5.3, we used the FOIL method to multiply two binomials. The product was often a trinomial. The following are some examples:

Factored Form		F O I L		Trinomial Form
$(x + 3)(x + 4)$	$=$	$x^2 + 4x + 3x + 12$	$=$	$x^2 + 7x + 12$
$(x - 3)(x - 4)$	$=$	$x^2 - 4x - 3x + 12$	$=$	$x^2 - 7x + 12$
$(x + 3)(x - 5)$	$=$	$x^2 - 5x + 3x - 15$	$=$	$x^2 - 2x - 15.$

Observe that each trinomial is of the form $x^2 + bx + c$, where the coefficient of the squared term is 1. Our goal in this section is to start with the trinomial form and, assuming that it is factorable, return to the factored form.

The first FOIL multiplication shown above indicates that $(x + 3)(x + 4) = x^2 + 7x + 12$. Let's reverse the sides of this equation:

$$x^2 + 7x + 12 = (x + 3)(x + 4).$$

We can make several important observations about the factors on the right side.

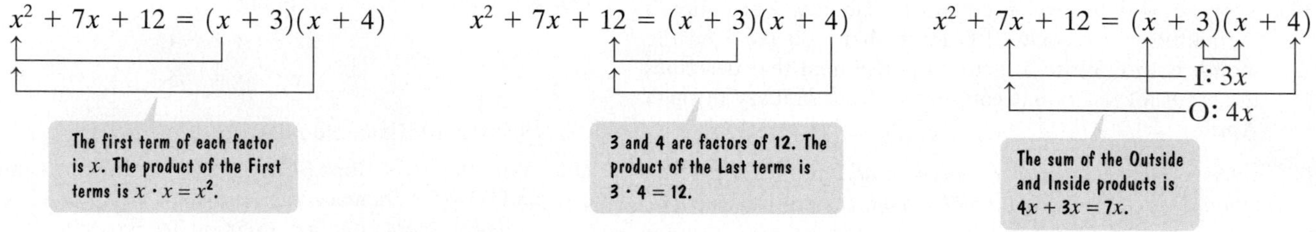
$x^2 + 7x + 12 = (x + 3)(x + 4)$
The first term of each factor is x. The product of the First terms is $x \cdot x = x^2$.
$x^2 + 7x + 12 = (x + 3)(x + 4)$
3 and 4 are factors of 12. The product of the Last terms is $3 \cdot 4 = 12$.
$x^2 + 7x + 12 = (x + 3)(x + 4)$
I: $3x$
O: $4x$
The sum of the Outside and Inside products is $4x + 3x = 7x$.

These observations provide us with a procedure for factoring $x^2 + bx + c$.

A STRATEGY FOR FACTORING $x^2 + bx + c$

1. Enter x as the first term of each factor.

$$(x \quad)(x \quad) = x^2 + bx + c$$

2. List pairs of factors of the constant c.
3. Try various combinations of these factors as the second term in each set of parentheses. Select the combination in which the sum of the Outside and Inside products is equal to bx.

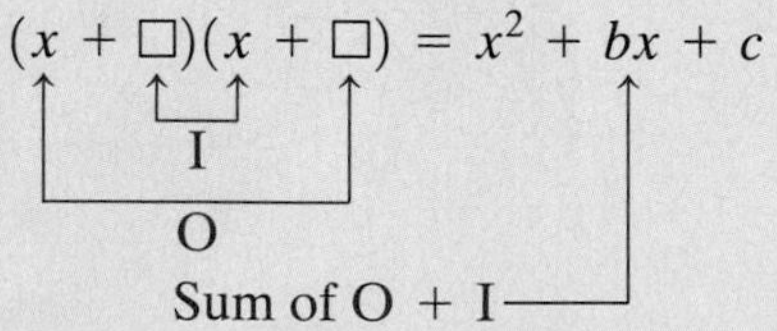
$(x + \square)(x + \square) = x^2 + bx + c$
I
O
Sum of O + I

4. Check your work by multiplying the factors using the FOIL method. You should obtain the original trinomial.

If none of the possible combinations yield an Outside product and an Inside product whose sum is equal to bx, the trinomial cannot be factored using integers and is called **prime** over the set of integers.

EXAMPLE 1 Factoring a Trinomial in $x^2 + bx + c$ Form

Factor: $x^2 + 6x + 8$.

SOLUTION

Step 1. **Enter x as the first term of each factor.**

$$x^2 + 6x + 8 = (x \quad)(x \quad)$$

To find the second term of each factor, we must find two integers whose product is 8 and whose sum is 6.

Step 2. **List pairs of factors of the constant, 8.**

Factors of 8	8, 1	4, 2	−8, −1	−4, −2

Step 3. **Try various combinations of these factors.** The correct factorization of $x^2 + 6x + 8$ is the one in which the sum of the Outside and Inside products is equal to $6x$. Here is a list of the possible factorizations:

Possible Factorizations of $x^2 + 6x + 8$	Sum of Outside and Inside Products (Should Equal $6x$)
$(x + 8)(x + 1)$	$x + 8x = 9x$
$(x + 4)(x + 2)$	$2x + 4x = 6x$
$(x - 8)(x - 1)$	$-x - 8x = -9x$
$(x - 4)(x - 2)$	$-2x - 4x = -6x$

This is the required middle term.

Thus, $x^2 + 6x + 8 = (x + 4)(x + 2)$.

Step 4. **Check this result by multiplying the right side using the FOIL method.** You should obtain the original trinomial. Because of the commutative property, we can also say that

$$x^2 + 6x + 8 = (x + 2)(x + 4).$$

USING TECHNOLOGY

If a polynomial contains one variable, a graphing utility can be used to check its factorization. For example, the factorization in Example 1

$$x^2 + 6x + 8 = (x + 4)(x + 2)$$

can be checked graphically or numerically.

Graphic Check

Use the GRAPH feature. Graph $y_1 = x^2 + 6x + 8$ and $y_2 = (x + 4)(x + 2)$ on the same screen. Because the graphs are identical, the factorization appears to be correct.

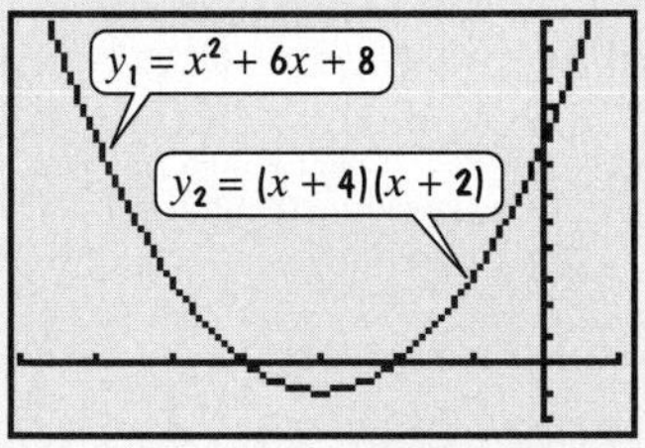

[−7, 1, 1] by [−2, 12, 1]

Numeric Check

Use the TABLE feature. Enter $y_1 = x^2 + 6x + 8$ and $y_2 = (x + 4)(x + 2)$ and press TABLE. Two columns of values are shown, one for y_1 and one for y_2. Because the corresponding values are equal regardless of how far up or down we scroll, the factorization is correct.

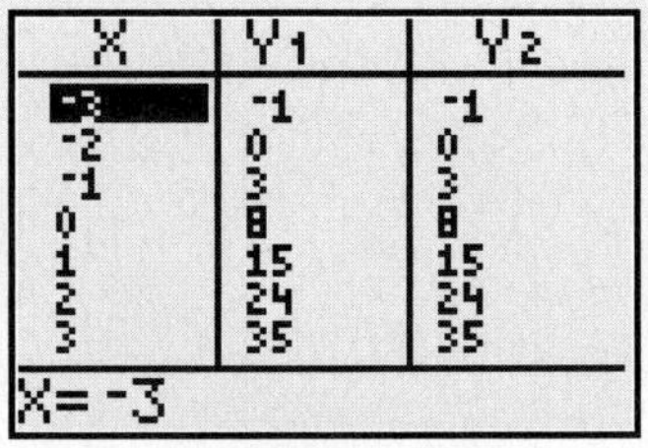

X	Y1	Y2
-3	-1	-1
-2	0	0
-1	3	3
0	8	8
1	15	15
2	24	24
3	35	35

X= -3

CHECK POINT 1 Factor: $x^2 + 5x + 6$.

EXAMPLE 2 Factoring a Trinomial in $x^2 + bx + c$ Form

Factor: $x^2 - 5x + 6$.

SOLUTION

Step 1. Enter x as the first term of each factor.

$$x^2 - 5x + 6 = (x \quad)(x \quad)$$

To find the second term of each factor, we must find two integers whose product is 6 and whose sum is -5.

Step 2. List pairs of factors of the constant, 6.

Factors of 6	6, 1	3, 2	−6, −1	−3, −2

Step 3. Try various combinations of these factors. The correct factorization of $x^2 - 5x + 6$ is the one in which the sum of the Outside and Inside products is equal to $-5x$. Here is a list of the possible factorizations:

Possible Factorizations of $x^2 - 5x + 6$	Sum of Outside and Inside Products (Should Equal $-5x$)
$(x + 6)(x + 1)$	$x + 6x = 7x$
$(x + 3)(x + 2)$	$2x + 3x = 5x$
$(x - 6)(x - 1)$	$-x - 6x = -7x$
$(x - 3)(x - 2)$	$-2x - 3x = -5x$

This is the required middle term.

Thus, $x^2 - 5x + 6 = (x - 3)(x - 2)$. Verify this result using the FOIL method. ■

STUDY TIP

To factor $x^2 + bx + c$ when c is positive, find two numbers with the same sign as the middle term.

$x^2 + 6x + 8 = (x + 2)(x + 4)$

Same signs

$x^2 - 5x + 6 = (x - 3)(x - 2)$

Same signs

In factoring a trinomial of the form $x^2 + bx + c$, you can speed things up by listing the factors of c and then finding their sums. We are interested in a sum of b. For example, in factoring $x^2 - 5x + 6$, we are interested in the factors of 6 whose sum is -5.

Factors of 6	6, 1	3, 2	−6, −1	−3, −2
Sum of Factors	7	5	−7	−5

This is the desired sum.

Thus, $x^2 - 5x + 6 = (x - 3)(x - 2)$.

CHECK POINT 2 Factor: $x^2 - 6x + 8$.

EXAMPLE 3 Factoring a Trinomial in $x^2 + bx + c$ Form

Factor: $x^2 + 2x - 35$.

SOLUTION

Step 1. Enter x as the first term of each factor.

$$x^2 + 2x - 35 = (x \quad)(x \quad)$$

To find the second term of each factor, we must find two integers whose product is -35 and whose sum is 2.

Step 2. List pairs of factors of the constant, −35.

Factors of −35	−35, 1	−7, 5	35, −1	7, −5

Step 3. Try various combinations of these factors. We are looking for the factors whose sum is 2.

Factors of −35	−35, 1	−7, 5	35, −1	7, −5
Sum of Factors	−34	−2	34	2

This is the desired sum.

Thus, $x^2 + 2x - 35 = (x + 7)(x - 5)$.

Step 4. Verify the factorization using the FOIL method.

$$(x + 7)(x - 5) = x^2 - 5x + 7x - 35 = x^2 + 2x - 35$$

Because the product of the factors is the original polynomial, the factorization is correct. ■

CHECK POINT 3 Factor: $x^2 + 3x - 10$.

EXAMPLE 4 Factoring a Trinomial Whose Leading Coefficient Is One

Factor: $y^2 - 2y - 99$.

SOLUTION

Step 1. Enter y as the first term of each factor.

$$y^2 - 2y - 99 = (y \quad)(y \quad)$$

To find the second term of each factor, we must find two integers whose product is −99 and whose sum is −2.

Step 2. List pairs of factors of the constant, −99.

Factors of −99	−99, 1	−11, 9	−33, 3	99, −1	11, −9	33, −3

Step 3. Try various combinations of these factors. We are interested in factors whose sum is −2.

Factors of −99	−99, 1	−11, 9	−33, 3	99, −1	11, −9	33, −3
Sum of Factors	−98	−2	−30	98	2	30

This is the desired sum.

Thus, $y^2 - 2y - 99 = (y - 11)(y + 9)$. Verify this result using the FOIL method. ■

CHECK POINT 4 Factor: $y^2 - 6y - 27$.

STUDY TIP

To factor $x^2 + bx + c$ when c is negative, find two numbers with opposite signs whose sum is the coefficient of the middle term.

$x^2 + 2x - 35 = (x + 7)(x - 5)$

Negative — Opposite signs

$y^2 - 2y - 99 = (y - 11)(y + 9)$

Negative — Opposite signs

EXAMPLE 5 Trying to Factor a Trinomial in $x^2 + bx + c$ Form

Factor: $x^2 + x - 5$.

SOLUTION

Step 1. **Enter x as the first term of each factor.**

$$x^2 + x - 5 = (x \quad\ )(x \quad\ )$$

To find the second term of each factor, we must find two integers whose product is -5 and whose sum is 1.

Steps 2 and 3. **List pairs of factors of the constant, -5, and try various combinations of these factors.** We are interested in factors whose sum is 1.

Factors of -5	$-5, 1$	$5, -1$
Sum of Factors	-4	4

No pair gives the desired sum, 1.

Because neither pair has a sum of 1, $x^2 + x - 5$ cannot be factored using integers. This trinomial is prime. ■

CHECK POINT 5 Factor: $x^2 + x - 7$.

EXAMPLE 6 Factoring a Trinomial in Two Variables

Factor: $x^2 - 5xy + 6y^2$.

SOLUTION

Step 1. **Enter x as the first term of each factor.** Because the last term of the trinomial contains y^2, the second term of each factor must contain y.

$$x^2 - 5xy + 6y^2 = (x \quad ?y)(x \quad ?y)$$

The question marks indicate that we are looking for the coefficients of y in each factor. To find these coefficients, we must find two integers whose product is 6 and whose sum is -5.

Steps 2 and 3. **List pairs of factors of the coefficient of the last term, 6, and try various combinations of these factors.** We are interested in factors whose sum is -5.

Factors of 6	$6, 1$	$3, 2$	$-6, -1$	$-3, -2$
Sum of Factors	7	5	-7	-5

This is the desired sum.

Thus, $x^2 - 5xy + 6y^2 = (x - 3y)(x - 2y)$.

Step 4. **Verify the factorization using the FOIL method.**

$$(x - 3y)(x - 2y) = x^2 - 2xy - 3xy + 6y^2 = x^2 - 5xy + 6y^2$$

Because the product of the factors is the original polynomial, the factorization is correct. ■

CHECK POINT 6 Factor: $x^2 - 4xy + 3y^2$.

Some polynomials can be factored using more than one technique. **Always begin by looking for a greatest common factor** and, if there is one, factor it out. A polynomial is **factored completely** when it is written as the product of prime polynomials.

EXAMPLE 7 Factoring Completely

Factor: $3x^3 - 15x^2 - 42x$.

SOLUTION The GCF of the three terms of the polynomial is $3x$. We begin by factoring out $3x$. Then we factor the remaining trinomial by the methods of this section.

$$3x^3 - 15x^2 - 42x$$

$$= 3x(x^2 - 5x - 14) \quad \text{Factor out the GCF.}$$

$$= 3x(x \quad)(x \quad) \quad \text{Begin factoring } x^2 - 5x - 14\text{. Find two integers whose product is } -14 \text{ and whose sum is } -5.$$

$$= 3x(x - 7)(x + 2) \quad \text{The integers are } -7 \text{ and } 2.$$

Thus,

$$3x^3 - 15x^2 - 42x = 3x(x - 7)(x + 2).$$

Be sure to include the GCF in the factorization.

How can we check this factorization? We will multiply the binomials using the FOIL method. Then use the distributive property and multiply each term of this product by $3x$. If the factorization is correct, we should obtain the original polynomial.

$$3x(x - 7)(x + 2) = 3x(x^2 + 2x - 7x - 14) = 3x(x^2 - 5x - 14) = 3x^3 - 15x^2 - 42x$$

Use the FOIL method on $(x - 7)(x + 2)$.

This is the original polynomial.

The factorization is correct. ■

CHECK POINT 7 Factor: $2x^3 + 6x^2 - 56x$.

6.2 EXERCISE SET

Student Solutions Manual | CD/Video | PH Math/Tutor Center | MathXL Tutorials on CD | MathXL® | MyMathLab | Interactmath.com

Practice Exercises

In Exercises 1–42, factor each trinomial, or state that the trinomial is prime. Check each factorization using FOIL multiplication.

1. $x^2 + 7x + 6$

2. $x^2 + 9x + 1$

3. $x^2 + 7x + 10$

4. $x^2 + 9x + 14$

5. $x^2 + 11x + 10$

6. $x^2 + 13x + 12$

7. $x^2 - 7x + 12$

8. $x^2 - 13x + 40$

9. $x^2 - 12x + 36$

10. $x^2 - 8x + 16$

11. $y^2 - 8y + 15$

12. $y^2 - 8y + 7$

13. $x^2 + 3x - 10$

14. $x^2 + 3x - 28$

15. $y^2 + 10y - 39$

16. $y^2 + 5y - 24$

17. $x^2 - 2x - 15$

18. $x^2 - 4x - 5$

19. $x^2 - 2x - 8$

20. $x^2 - 5x - 6$

21. $x^2 + 4x + 12$

22. $x^2 + 4x + 5$

23. $y^2 - 16y + 48$

24. $y^2 - 10y + 21$

25. $x^2 - 3x + 6$

26. $x^2 + 4x - 10$

27. $w^2 - 30w - 64$

28. $w^2 + 12w - 64$

29. $y^2 - 18y + 65$

30. $y^2 - 22y + 72$

31. $r^2 + 12r + 27$

32. $r^2 - 15r - 16$

33. $y^2 - 7y + 5$

34. $y^2 - 15y + 5$

35. $x^2 + 7xy + 6y^2$

36. $x^2 + 6xy + 8y^2$

37. $x^2 - 8xy + 15y^2$

38. $x^2 - 9xy + 14y^2$

39. $x^2 - 3xy - 18y^2$

40. $x^2 - xy - 30y^2$

41. $a^2 - 18ab + 45b^2$

42. $a^2 - 18ab + 80b^2$

In Exercises 43–66, factor completely.

43. $3x^2 + 15x + 18$

44. $3x^2 + 21x + 36$

45. $4y^2 - 4y - 8$

46. $3y^2 + 3y - 18$

47. $10x^2 - 40x - 600$

48. $2x^2 + 10x - 48$

49. $3x^2 - 33x + 54$

50. $2x^2 - 14x + 24$

51. $2r^3 + 6r^2 + 4r$

52. $2r^3 + 8r^2 + 6r$

53. $4x^3 + 12x^2 - 72x$

54. $3x^3 - 15x^2 + 18x$

55. $2r^3 + 8r^2 - 64r$

56. $3r^3 - 9r^2 - 54r$

57. $y^4 + 2y^3 - 80y^2$

58. $y^4 - 12y^3 + 35y^2$

59. $x^4 - 3x^3 - 10x^2$

60. $x^4 - 22x^3 + 120x^2$

61. $2w^4 - 26w^3 - 96w^2$

62. $3w^4 + 54w^3 + 135w^2$

63. $15xy^2 + 45xy - 60x$

64. $20x^2y - 100xy + 120y$

65. $x^5 + 3x^4y - 4x^3y^2$

66. $x^3y - 2x^2y^2 - 3xy^3$

Practice Plus

In Exercises 67–76, factor completely.

67. $2x^2y^2 - 32x^2yz + 30x^2z^2$

68. $2x^2y^2 - 30x^2yz + 28x^2z^2$

69. $(a + b)x^2 + (a + b)x - 20(a + b)$

70. $(a + b)x^2 - 13(a + b)x + 36(a + b)$

71. $x^2 + 0.5x + 0.06$

72. $x^2 - 0.5x - 0.06$

73. $x^2 - \frac{2}{5}x + \frac{1}{25}$

74. $x^2 + \frac{2}{3}x + \frac{1}{9}$

75. $-x^2 - 3x + 40$

76. $-x^2 - 4x + 45$

Application Exercises

77. You dive directly upward from a board that is 32 feet high. After t seconds, your height above the water is described by the polynomial

$$-16t^2 + 16t + 32.$$

a. Factor the polynomial completely. Begin by factoring -16 from each term.

b. Evaluate both the original polynomial and its factored form for $t = 2$. Do you get the same answer for each evaluation? Describe what this answer means.

78. You dive directly upward from a board that is 48 feet high. After t seconds, your height above the water is described by the polynomial

$$-16t^2 + 32t + 48.$$

a. Factor the polynomial completely. Begin by factoring -16 from each term.

b. Evaluate both the original polynomial and its factored form for $t = 3$. Do you get the same answer for each evaluation? Describe what this answer means.

Writing in Mathematics

79. Explain how to factor $x^2 + 8x + 15$.

80. Give two helpful suggestions for factoring $x^2 - 5x + 6$.

81. In factoring $x^2 + bx + c$, describe how the last terms in each factor are related to b and c.

82. Without actually factoring and without multiplying the given factors, explain why the following factorization is not correct:

$$x^2 + 46x + 513 = (x - 27)(x - 19).$$

Critical Thinking Exercises

83. Which one of the following is true?

a. A factor of $x^2 + x + 20$ is $x + 5$.

b. A trinomial can never have two identical factors.

c. A factor of $y^2 + 5y - 24$ is $y - 3$.

d. $x^2 + 4 = (x + 2)(x + 2)$

In Exercises 84–85, find all positive integers b so that the trinomial can be factored.

84. $x^2 + bx + 15$

85. $x^2 + 4x + b$

86. Factor: $x^{2n} + 20x^n + 99$.

87. Factor $x^3 + 3x^2 + 2x$. If x represents an integer, use the factorization to describe what the trinomial represents.

88. A box with no top is to be made from an 8-inch by 6-inch piece of metal by cutting identical squares from each corner and turning up the sides. (See the figure). The volume of the box is modeled by the polynomial $4x^3 - 28x^2 + 48x$. Factor the polynomial completely. Then use the dimensions given on the box and show that its volume is equivalent to the factorization that you obtain.

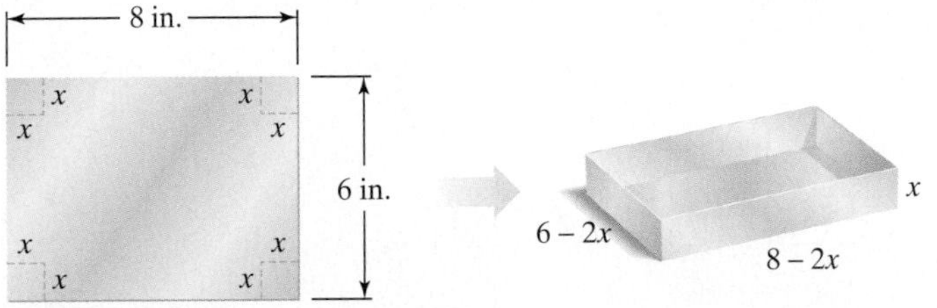

Technology Exercises

In Exercises 89–92, use the GRAPH *or* TABLE *feature of a graphing utility to determine if the polynomial on the left side of each equation has been correctly factored. If the graphs of y_1 and y_2 coincide, or if their corresponding table values are equal, this means that the polynomial on the left side has been correctly factored. If not, factor the trinomial correctly and then use your graphing utility to verify the factorization.*

89. $x^2 - 5x + 6 = (x - 2)(x - 3)$

90. $2x^2 + 2x - 12 = 2(x - 3)(x + 2)$

91. $x^2 - 2x + 1 = (x + 1)(x - 1)$

92. $2x^2 + 8x + 6 = (x + 3)(x + 1)$

Review Exercises

93. Multiply: $(2x + 3)(x - 2)$. (Section 5.3, Example 2)

94. Multiply: $(3x + 4)(3x + 1)$. (Section 5.3, Example 2)

95. Solve: $4(x - 2) = 3x + 5$. (Section 2.3, Example 2)

SECTION 6.3

FACTORING TRINOMIALS WHOSE LEADING COEFFICIENT IS NOT ONE

Objectives

1 Factor trinomials by trial and error.

2 Factor trinomials by grouping.

George Tooker, American, born 1920. "Farewell" 1966, egg tempera on gessoed masonite, 61×60.1 cm. (24.034×23.679 in.) P.967.76. Hood Museum of Art. Dartmouth College, Hanover, New Hampshire; gift of Pennington Haile, Class of 1924. ©George Tooker/DC Moore Gallery.

The special significance of the number 1 is reflected in our language. "One," "an," and "a" mean the same thing. The words "unit," "unity," "union," "unique," and "universal" are derived from the Latin word for "one." For the ancient Greeks, 1 was the indivisible unit from which all other numbers arose.

The Greeks' philosophy of 1 applies to our work in this section. Factoring trinomials whose leading coefficient is 1 is the basic technique from which other methods of factoring $ax^2 + bx + c$, where a is not equal to 1, follow.

1 Factor trinomials by trial and error.

Factoring by the Trial-and-Error Method How do we factor a trinomial such as $3x^2 - 20x + 28$? Notice that the leading coefficient is 3. We must find two binomials whose product is $3x^2 - 20x + 28$. The product of the First terms must be $3x^2$:

$$(3x \quad\;)(x \quad\;).$$

From this point on, the factoring strategy is exactly the same as the one we use to factor trinomials whose leading coefficient is 1.

A STRATEGY FOR FACTORING $ax^2 + bx + c$ Assume, for the moment, that there is no greatest common factor.

1. Find two First terms whose product is ax^2:

$$(\Box x + \quad\;)(\Box x + \quad\;) = ax^2 + bx + c.$$

2. Find two Last terms whose product is c:

$$(\Box x + \Box)(\Box x + \Box) = ax^2 + bx + c.$$

3. By trial and error, perform steps 1 and 2 until the sum of the Outside product and the Inside product is bx:

$$(\Box x + \Box)(\Box x + \Box) = ax^2 + bx + c.$$

I

O

Sum of O + I

If no such combinations exist, the polynomial is prime.

STUDY TIP

The *error* part of the factoring strategy plays an important role in the process. If you do not get the correct factorization the first time, this is not a bad thing. This error is often helpful in leading you to the correct factorization.

EXAMPLE 1 Factoring a Trinomial Whose Leading Coefficient Is Not One

Factor: $3x^2 - 20x + 28$.

SOLUTION

Step 1. Find two First terms whose product is $3x^2$.

$$3x^2 - 20x + 28 = (3x \quad\;)(x \quad\;)$$

Step 2. Find two Last terms whose product is 28. The number 28 has pairs of factors that are either both positive or both negative. Because the middle term, $-20x$, is negative, both factors must be negative. The negative factorizations of 28 are $-1(-28)$, $-2(-14)$, and $-4(-7)$.

Step 3. Try various combinations of these factors. The correct factorization of $3x^2 - 20x + 28$ is the one in which the sum of the Outside and Inside products is equal to $-20x$. Here is a list of the possible factorizations:

STUDY TIP

With practice, you will find that it is not necessary to list every possible factorization of the trinomial. As you practice factoring, you will be able to narrow down the list of possible factors to just a few. When it comes to factoring, practice makes perfect. (Sorry about the cliché.)

Possible Factorizations of $3x^2 - 20x + 28$	Sum of Outside and Inside Products (Should Equal $-20x$)
$(3x - 1)(x - 28)$	$-84x - x = -85x$
$(3x - 28)(x - 1)$	$-3x - 28x = -31x$
$(3x - 2)(x - 14)$	$-42x - 2x = -44x$
$(3x - 14)(x - 2)$	$-6x - 14x = -20x$
$(3x - 4)(x - 7)$	$-21x - 4x = -25x$
$(3x - 7)(x - 4)$	$-12x - 7x = -19x$

This is the required middle term.

Thus,

$$3x^2 - 20x + 28 = (3x - 14)(x - 2) \quad \text{or} \quad (x - 2)(3x - 14).$$

Show that this factorization is correct by multiplying the factors with the FOIL method. You should obtain the original trinomial. ■

CHECK POINT 1 Factor: $5x^2 - 14x + 8$.

EXAMPLE 2 Factoring a Trinomial Whose Leading Coefficient Is Not One

Factor: $8x^2 - 10x - 3$.

SOLUTION

Step 1. **Find two First terms whose product is $8x^2$.**

$$8x^2 - 10x - 3 \stackrel{?}{=} (8x \quad)(x \quad)$$
$$8x^2 - 10x - 3 \stackrel{?}{=} (4x \quad)(2x \quad)$$

Step 2. **Find two Last terms whose product is -3.** The possible factorizations are $1(-3)$ and $-1(3)$.

Step 3. **Try various combinations of these factors.** The correct factorization of $8x^2 - 10x - 3$ is the one in which the sum of the Outside and Inside products is equal to $-10x$. Here is a list of the possible factorizations:

Possible Factorizations of $8x^2 - 10x - 3$	Sum of Outside and Inside Products (Should Equal $-10x$)
$(8x + 1)(x - 3)$	$-24x + x = -23x$
$(8x - 3)(x + 1)$	$8x - 3x = 5x$
$(8x - 1)(x + 3)$	$24x - x = 23x$
$(8x + 3)(x - 1)$	$-8x + 3x = -5x$
$(4x + 1)(2x - 3)$	$-12x + 2x = -10x$
$(4x - 3)(2x + 1)$	$4x - 6x = -2x$
$(4x - 1)(2x + 3)$	$12x - 2x = 10x$
$(4x + 3)(2x - 1)$	$-4x + 6x = 2x$

This is the required middle term.

Thus,

$$8x^2 - 10x - 3 = (4x + 1)(2x - 3) \quad \text{or} \quad (2x - 3)(4x + 1).$$

Use FOIL multiplication to check either of these factorizations. ■

CHECK POINT 2 Factor: $6x^2 + 19x - 7$.

STUDY TIP

Here are some suggestions for reducing the list of possible factorizations for $ax^2 + bx + c$:

1. If b is relatively small, avoid the larger factors of a.
2. If c is positive, the signs in both binomial factors must match the sign of b.
3. If the trinomial has no common factor, no binomial factor can have a common factor.
4. Reversing the signs in the binomial factors reverses the sign of bx, the middle term.

EXAMPLE 3 Factoring a Trinomial in Two Variables

Factor: $2x^2 - 7xy + 3y^2$.

SOLUTION

Step 1. **Find two First terms whose product is $2x^2$.**

$$2x^2 - 7xy + 3y^2 = (2x \quad\quad)(x \quad\quad)$$

Step 2. **Find two Last terms whose product is $3y^2$.** The possible factorizations are $(y)(3y)$ and $(-y)(-3y)$.

Step 3. **Try various combinations of these factors.** The correct factorization of $2x^2 - 7xy + 3y^2$ is the one in which the sum of the Outside and Inside products is equal to $-7xy$. Here is a list of possible factorizations:

Possible Factorizations of $2x^2 - 7xy + 3y^2$	Sum of Outside and Inside Products (Should Equal $-7xy$)
$(2x + 3y)(x + y)$	$2xy + 3xy = 5xy$
$(2x + y)(x + 3y)$	$6xy + xy = 7xy$
$(2x - 3y)(x - y)$	$-2xy - 3xy = -5xy$
$(2x - y)(x - 3y)$	$-6xy - xy = -7xy$

This is the required middle term.

Thus,

$$2x^2 - 7xy + 3y^2 = (2x - y)(x - 3y) \quad \text{or} \quad (x - 3y)(2x - y).$$

Use FOIL multiplication to check either of these factorizations. ■

✔ **CHECK POINT 3** Factor: $3x^2 - 13xy + 4y^2$.

2 Factor trinomials by grouping.

Factoring by the Grouping Method A second method for factoring $ax^2 + bx + c$, $a \neq 0$, is called the **grouping method**. The method involves both trial and error, as well as grouping. The trial and error in factoring $ax^2 + bx + c$ depends on finding two numbers, p and q, for which $p + q = b$. Then we factor $ax^2 + px + qx + c$ by using grouping.

Let's see how this works by looking at our factorization in Example 2:

$$8x^2 - 10x - 3 = (2x - 3)(4x + 1).$$

If we multiply by using FOIL on the right, we obtain:

$$(2x - 3)(4x + 1) = 8x^2 + 2x - 12x - 3.$$

In this case, the desired numbers, p and q, are $p = 2$ and $q = -12$. Compare these numbers with ac and b in the given polynomial:

$$8x^2 - 10x - 3.$$

$ac = 8(-3) = -24$

$a = 8$ $\quad b = -10$ $\quad c = -3$

Can you see that p and q, 2 and -12, are factors of ac, or -24? Furthermore, p and q have a sum of b, namely -10. By expressing the middle term, $-10x$, in terms of p and q, we can factor by grouping as follows:

$$\begin{aligned} & 8x^2 - 10x - 3 \\ &= 8x^2 + (2x - 12x) - 3 && \text{Rewrite } -10x \text{ as } 2x - 12x. \\ &= (8x^2 + 2x) + (-12x - 3) && \text{Group terms.} \\ &= 2x(4x + 1) - 3(4x + 1) && \text{Factor from each group.} \\ &= (4x + 1)(2x - 3) && \text{Factor out the common binomial factor.} \end{aligned}$$

As we obtained in Example 2,

$$8x^2 - 10x - 3 = (4x + 1)(2x - 3).$$

Generalizing from this example, here's how to factor a trinomial by grouping:

FACTORING $ax^2 + bx + c$ USING GROUPING ($a \neq 1$)

1. Multiply the leading coefficient, a, and the constant, c.
2. Find the factors of ac whose sum is b.
3. Rewrite the middle term, bx, as a sum or difference using the factors from step 2.
4. Factor by grouping.

EXAMPLE 4 Factoring by Grouping

Factor by grouping: $2x^2 - x - 6$.

SOLUTION The trinomial is of the form $ax^2 + bx + c$.

$$2x^2 - x - 6$$

$a = 2$ $b = -1$ $c = -6$

Step 1. **Multiply the leading coefficient, a, and the constant, c.** Using $a = 2$ and $c = -6$,

$$ac = 2(-6) = -12.$$

Step 2. **Find the factors of ac whose sum is b.** We want the factors of -12 whose sum is b, or -1. The factors of -12 whose sum is -1 are -4 and 3.

Step 3. **Rewrite the middle term, $-x$, as a sum or difference using the factors from step 2, -4 and 3.**

$$2x^2 - x - 6 = 2x^2 - 4x + 3x - 6$$

Step 4. **Factor by grouping.**

$$\begin{aligned} &= (2x^2 - 4x) + (3x - 6) && \text{Group terms.} \\ &= 2x(x - 2) + 3(x - 2) && \text{Factor from each group.} \\ &= (x - 2)(2x + 3) && \text{Factor out the common binomial factor.} \end{aligned}$$

Thus,

$$2x^2 - x - 6 = (x - 2)(2x + 3) \quad \text{or} \quad (2x + 3)(x - 2).$$ ■

DISCOVER FOR YOURSELF

In step 2, we discovered that the desired numbers were -4 and 3, and we wrote $-x$ as $-4x + 3x$. What happens if we write $-x$ as $3x - 4x$? Use factoring by grouping on

$$\begin{aligned} & 2x^2 - x - 6 \\ &= 2x^2 + 3x - 4x - 6. \end{aligned}$$

Describe what happens.

✔ **CHECK POINT 4** Factor by grouping: $3x^2 - x - 10$.

EXAMPLE 5 Factoring by Grouping

Factor by grouping: $8x^2 - 22x + 5$.

SOLUTION The trinomial is of the form $ax^2 + bx + c$.

$$8x^2 - 22x + 5$$

$a = 8$ $b = -22$ $c = 5$

Step 1. **Multiply the leading coefficient, *a*, and the constant, *c*.** Using $a = 8$ and $c = 5$, $ac = 8 \cdot 5 = 40$.

Step 2. **Find the factors of *ac* whose sum is *b*.** We want the factors of 40 whose sum is b, or -22. The factors of 40 whose sum is -22 are -2 and -20.

Step 3. **Rewrite the middle term, $-22x$, as a sum or difference using the factors from step 2, -2 and -20.**

$$8x^2 - 22x + 5 = 8x^2 - 2x - 20x + 5$$

Step 4. **Factor by grouping.**

$= (8x^2 - 2x) + (-20x + 5)$ *Group terms.*

$= 2x(4x - 1) - 5(4x - 1)$ *Factor from each group.*

$= (4x - 1)(2x - 5)$ *Factor out the common binomial factor.*

Thus,

$$8x^2 - 22x + 5 = (4x - 1)(2x - 5) \quad \text{or} \quad (2x - 5)(4x - 1).$$

■

CHECK POINT 5 Factor by grouping: $8x^2 - 10x + 3$.

Factoring Completely Always begin the process of factoring a polynomial by looking for a greatest common factor. If there is one, **factor out the GCF first**. After doing this, you should attempt to factor the remaining trinomial by one of the methods presented in this section.

EXAMPLE 6 Factoring Completely

Factor completely: $15y^4 + 26y^3 + 7y^2$.

SOLUTION We will first factor out a common monomial factor from the polynomial and then factor the resulting trinomial by the methods of this section. The GCF of the three terms is y^2.

$15y^4 + 26y^3 + 7y^2 = y^2(15y^2 + 26y + 7)$ *Factor out the GCF.*

$= y^2(5y + 7)(3y + 1)$ *Factor $15y^2 + 26y + 7$ using trial and error or grouping.*

Thus,

$$15y^4 + 26y^3 + 7y^2 = y^2(5y + 7)(3y + 1) \quad \text{or} \quad y^2(3y + 1)(5y + 7).$$

Be sure to include the GCF, y^2, in the factorization.

■

CHECK POINT 6 Factor completely: $5y^4 + 13y^3 + 6y^2$.

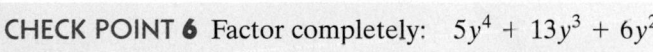

6.3 EXERCISE SET

Student Solutions Manual | CD/Video | PH Math/Tutor Center | MathXL Tutorials on CD | MathXL® | MyMathLab | Interactmath.com

Practice Exercises

In Exercises 1–58, use the method of your choice to factor each trinomial, or state that the trinomial is prime. Check each factorization using FOIL multiplication.

1. $2x^2 + 5x + 3$
2. $3x^2 + 5x + 2$
3. $3x^2 + 13x + 4$
4. $2x^2 + 7x + 3$
5. $2x^2 + 11x + 12$
6. $2x^2 + 19x + 35$
7. $5y^2 - 16y + 3$
8. $5y^2 - 17y + 6$
9. $3y^2 + y - 4$
10. $3y^2 - y - 4$
11. $3x^2 + 13x - 10$
12. $3x^2 + 14x - 5$
13. $3x^2 - 22x + 7$
14. $3x^2 - 10x + 7$
15. $5y^2 - 16y + 3$
16. $5y^2 - 8y + 3$
17. $3x^2 - 17x + 10$
18. $3x^2 - 25x - 28$
19. $6w^2 - 11w + 4$
20. $6w^2 - 17w + 12$
21. $8x^2 + 33x + 4$
22. $7x^2 + 43x + 6$
23. $5x^2 + 33x - 14$
24. $3x^2 + 22x - 16$
25. $14y^2 + 15y - 9$
26. $6y^2 + 7y - 24$
27. $6x^2 - 7x + 3$
28. $9x^2 + 3x + 2$
29. $25z^2 - 30z + 9$
30. $9z^2 + 12z + 4$
31. $15y^2 - y - 2$
32. $15y^2 + 13y - 2$
33. $5x^2 + 2x + 9$
34. $3x^2 - 5x + 1$
35. $10y^2 + 43y - 9$
36. $16y^2 - 46y + 15$
37. $8x^2 - 2x - 1$
38. $8x^2 - 22x + 5$
39. $9y^2 - 9y + 2$
40. $9y^2 + 5y - 4$
41. $20x^2 + 27x - 8$
42. $15x^2 - 19x + 6$
43. $2x^2 + 3xy + y^2$
44. $3x^2 + 4xy + y^2$
45. $3x^2 + 5xy + 2y^2$
46. $3x^2 + 11xy + 6y^2$
47. $2x^2 - 9xy + 9y^2$
48. $3x^2 + 5xy - 2y^2$
49. $6x^2 - 5xy - 6y^2$
50. $6x^2 - 7xy - 5y^2$
51. $15x^2 + 11xy - 14y^2$
52. $15x^2 - 31xy + 10y^2$
53. $2a^2 + 7ab + 5b^2$
54. $2a^2 + 5ab + 2b^2$
55. $15a^2 - ab - 6b^2$
56. $3a^2 - ab - 14b^2$
57. $12x^2 - 25xy + 12y^2$
58. $12x^2 + 7xy - 12y^2$

In Exercises 59–86, factor completely.

59. $4x^2 + 26x + 30$
60. $4x^2 - 18x - 10$
61. $9x^2 - 6x - 24$
62. $12x^2 - 33x + 21$
63. $4y^2 + 2y - 30$
64. $36y^2 + 6y - 12$
65. $9y^2 + 33y - 60$
66. $16y^2 - 16y - 12$
67. $3x^3 + 4x^2 + x$
68. $3x^3 + 14x^2 + 8x$
69. $2x^3 - 3x^2 - 5x$
70. $6x^3 + 4x^2 - 10x$
71. $9y^3 - 39y^2 + 12y$
72. $10y^3 + 12y^2 + 2y$
73. $60z^3 + 40z^2 + 5z$
74. $80z^3 + 80z^2 - 60z$
75. $15x^4 - 39x^3 + 18x^2$
76. $24x^4 + 10x^3 - 4x^2$
77. $10x^5 - 17x^4 + 3x^3$
78. $15x^5 - 2x^4 - x^3$
79. $6x^2 - 3xy - 18y^2$
80. $4x^2 + 14xy + 10y^2$
81. $12x^2 + 10xy - 8y^2$
82. $24x^2 + 3xy - 27y^2$
83. $8x^2y + 34xy - 84y$
84. $6x^2y - 2xy - 60y$
85. $12a^2b - 46ab^2 + 14b^3$
86. $12a^2b - 34ab^2 + 14b^3$

Practice Plus

In Exercises 87–90, factor completely.

87. $30(y+1)x^2 + 10(y+1)x - 20(y+1)$

88. $6(y+1)x^2 + 33(y+1)x + 15(y+1)$

89. $-32x^2y^4 + 20xy^4 + 12y^4$

90. $-10x^2y^4 + 14xy^4 + 12y^4$

91. **a.** Factor $2x^2 - 5x - 3$.

b. Use the factorization in part (a) to factor

$$2(y+1)^2 - 5(y+1) - 3.$$

Then simplify each factor.

92. **a.** Factor $3x^2 + 5x - 2$.

b. Use the factorization in part (a) to factor

$$3(y+1)^2 + 5(y+1) - 2.$$

Then simplify each factor.

93. Divide $3x^3 - 11x^2 + 12x - 4$ by $x - 2$. Use the quotient to factor $3x^3 - 11x^2 + 12x - 4$ completely.

94. Divide $2x^3 + x^2 - 13x + 6$ by $x - 2$. Use the quotient to factor $2x^3 + x^2 - 13x + 6$ completely.

Application Exercises

It is possible to construct geometric models for factorizations so that you can see the factoring. This idea is developed in Exercises 95–96.

95. Consider the following figure.

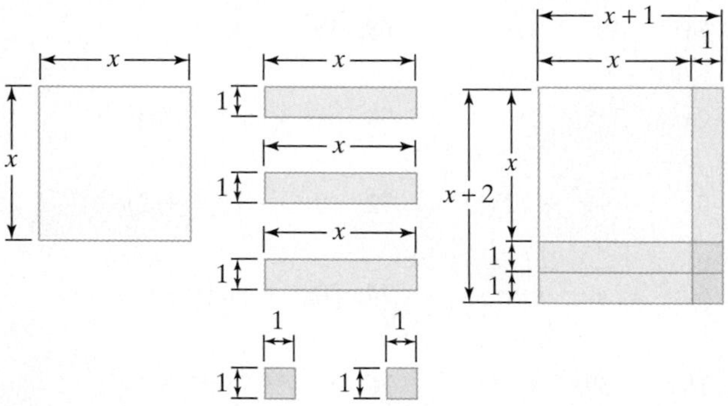

(a) (b)

a. Write a trinomial that expresses the sum of the areas of the six rectangular pieces shown in figure (a).

b. Express the area of the large rectangle in figure (b) as the product of two binomials.

c. Are the pieces in figures (a) and (b) the same? Set the expressions that you wrote in parts (a) and (b) equal to each other. What factorization is illustrated?

96. Copy the figure and cut out the six pieces. Use the pieces to create a geometric model for the factorization

$$2x^2 + 3x + 1 = (2x+1)(x+1)$$

by forming a large rectangle using all the pieces.

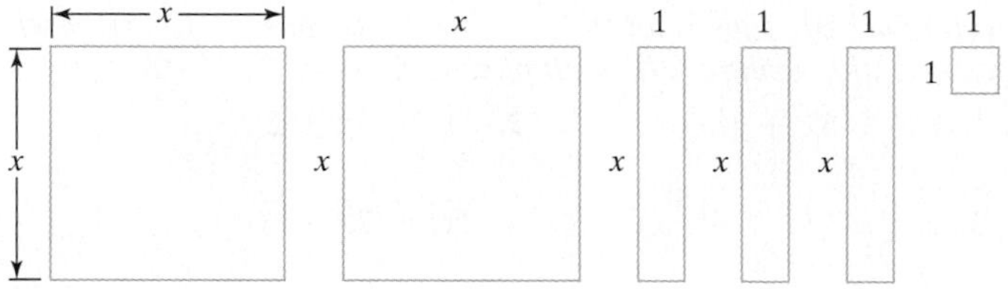

Writing in Mathematics

97. Explain how to factor $2x^2 - x - 1$.

98. Why is it a good idea to factor out the GCF first and then use other methods of factoring? Use $3x^2 - 18x + 15$ as an example. Discuss what happens if one first uses trial and error to factor as two binomials rather than first factoring out the GCF.

99. In factoring $3x^2 - 10x - 8$, a student lists $(3x-2)(x+4)$ as a possible factorization. Use FOIL multiplication to determine if this factorization is correct. If it is not correct, describe how the correct factorization can quickly be obtained using these factors.

100. Explain why $2x - 10$ cannot be one of the factors in the correct factorization of $6x^2 - 19x + 10$.

Critical Thinking Exercises

101. Which one of the following is true?

a. Once a GCF is factored from $18y^2 - 6y + 6$, the remaining trinomial factor is prime.

b. One factor of $12x^2 - 13x + 3$ is $4x + 3$.

c. One factor of $4y^2 - 11y - 3$ is $y + 3$.

d. The trinomial $3x^2 + 2x + 1$ has relatively small coefficients and therefore can be factored.

In Exercises 102–103, find all integers b so that the trinomial can be factored.

102. $3x^2 + bx + 2$

103. $2x^2 + bx + 3$

104. Factor: $3x^{10} - 4x^5 - 15$.

105. Factor: $2x^{2n} - 7x^n - 4$.

Review Exercises

In Exercises 106–108, perform the indicated operations.

106. $(9x+10)(9x-10)$ (Section 5.3, Example 4)

107. $(4x+5y)^2$ (Section 5.3, Example 5)

108. $(x+2)(x^2-2x+4)$ (Section 5.2, Example 7)

MID-CHAPTER CHECK POINT

CHAPTER 6

What You Know: We learned to factor out a polynomial's greatest common factor and to use grouping to factor polynomials with four terms. We factored polynomials with three terms, beginning with trinomials with leading coefficient 1 and moving on to $ax^2 + bx + c$, with $a \neq 1$. We saw that the factoring process should begin by looking for a GCF and, if there is one, factoring it out first.

In Exercises 1–12, factor completely, or state that the polynomial is prime.

1. $x^5 + x^4$
2. $x^2 + 7x - 18$
3. $x^2y^3 - x^2y^2 + x^2y$
4. $x^2 - 2x + 4$
5. $7x^2 - 22x + 3$
6. $x^3 + 5x^2 + 3x + 15$
7. $2x^3 - 11x^2 + 5x$
8. $xy - 7x - 4y + 28$
9. $x^2 - 17xy + 30y^2$
10. $25x^2 - 25x - 14$
11. $16x^2 - 70x + 24$
12. $3x^2 + 10xy + 7y^2$

SECTION 6.4

FACTORING SPECIAL FORMS

Objectives

1. Factor the difference of two squares.
2. Factor perfect square trinomials.
3. Factor the sum and difference of two cubes.

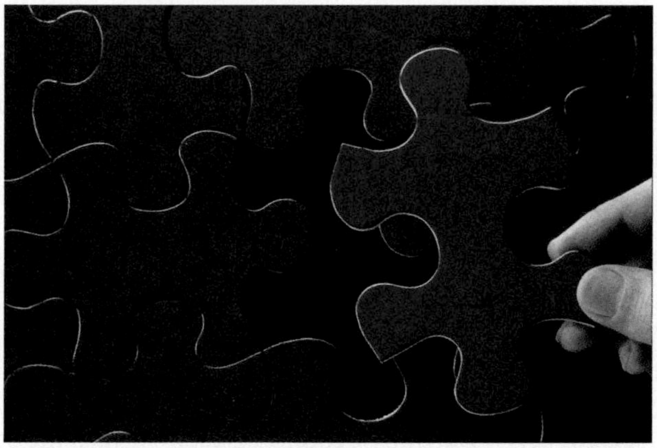

Do you enjoy solving puzzles? The process is a natural way to develop problem-solving skills that are important to every area of our lives. Engaging in problem solving for sheer pleasure releases chemicals in the brain that enhance our feeling of well-being. Perhaps this is why puzzles date back 12,000 years.

In this section, we develop factoring techniques by reversing the formulas for special products discussed in Chapter 5. These factorizations can be visualized by fitting pieces of a puzzle together to form rectangles.

1 Factor the difference of two squares.

Factoring the Difference of Two Squares A method for factoring the difference of two squares is obtained by reversing the special product for the sum and difference of two terms.

ENRICHMENT ESSAY

Visualizing the Factoring for the Difference of Two Squares

Yellow Shaded Area:

$A^2 - B^2$

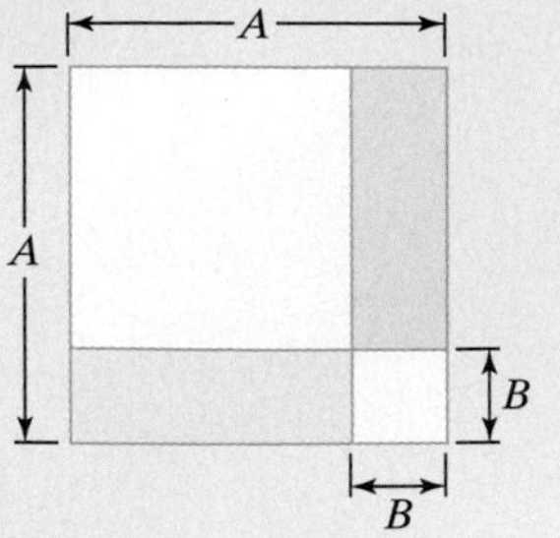

Yellow Shaded Area:

$(A + B)(A - B)$

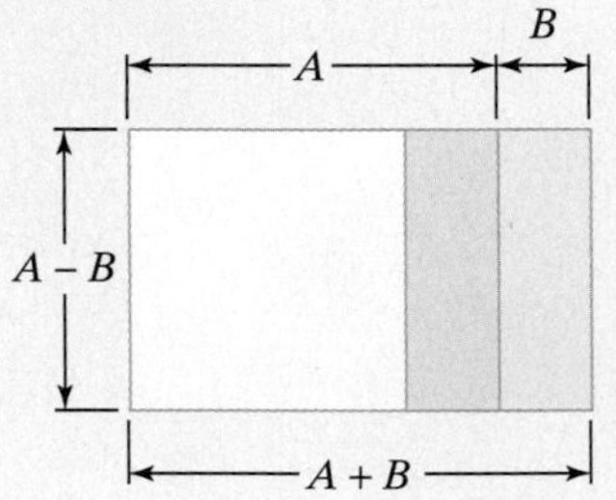

The yellow shaded areas are the same. Conclusion:

$A^2 - B^2 = (A + B)(A - B)$.

THE DIFFERENCE OF TWO SQUARES If A and B are real numbers, variables, or algebraic expressions, then

$$A^2 - B^2 = (A + B)(A - B).$$

In words: The difference of the squares of two terms factors as the product of a sum and a difference of those terms.

EXAMPLE 1 Factoring the Difference of Two Squares

Factor:

a. $x^2 - 4$ **b.** $81x^2 - 49$.

SOLUTION We must express each term as the square of some monomial. Then we use the formula for factoring $A^2 - B^2$.

a. $x^2 - 4 = x^2 - 2^2 = (x + 2)(x - 2)$

$A^2 - B^2 = (A + B)(A - B)$

b. $81x^2 - 49 = (9x)^2 - 7^2 = (9x + 7)(9x - 7)$ ■

CHECK POINT 1 Factor:

a. $x^2 - 81$ **b.** $36x^2 - 25$.

Can $x^2 - 5$ be factored using integers and the formula for factoring $A^2 - B^2$? No. The number 5 is not the square of an integer. Thus, $x^2 - 5$ is prime over the set of integers.

EXAMPLE 2 Factoring the Difference of Two Squares

Factor:

a. $9 - 16x^{10}$ **b.** $25x^2 - 4y^2$.

SOLUTION Begin by expressing each term as the square of some monomial. Then use the formula for factoring $A^2 - B^2$.

a. $9 - 16x^{10} = 3^2 - (4x^5)^2 = (3 + 4x^5)(3 - 4x^5)$

$A^2 - B^2 = (A + B)(A - B)$

b. $25x^2 - 4y^2 = (5x)^2 - (2y)^2 = (5x + 2y)(5x - 2y)$ ■

CHECK POINT 2 Factor:

a. $25 - 4x^{10}$ **b.** $100x^2 - 9y^2$.

When factoring, always check first for common factors. If there are common factors, factor out the GCF and then factor the resulting polynomial.

EXAMPLE 3 Factoring Out the GCF and Then Factoring the Difference of Two Squares

Factor:

a. $12x^3 - 3x$ **b.** $80 - 125x^2$.

SOLUTION

a. $12x^3 - 3x = 3x(4x^2 - 1) = 3x[(2x)^2 - 1^2] = 3x(2x + 1)(2x - 1)$

Factor out the GCF. $A^2 - B^2 = (A + B)(A - B)$

b. $80 - 125x^2 = 5(16 - 25x^2) = 5[4^2 - (5x)^2] = 5(4 + 5x)(4 - 5x)$

CHECK POINT 3 Factor:

a. $18x^3 - 2x$ **b.** $72 - 18x^2$.

We have seen that a polynomial is factored completely when it is written as the product of prime polynomials. To be sure that you have factored completely, check to see whether any of the factors can be factored.

EXAMPLE 4 A Repeated Factorization

Factor completely: $x^4 - 81$.

SOLUTION

$x^4 - 81 = (x^2)^2 - 9^2$ Express as the difference of two squares.

$= (x^2 + 9)(x^2 - 9)$ The factors are the sum and the difference of the expressions being squared.

$= (x^2 + 9)(x^2 - 3^2)$ The factor $x^2 - 9$ is the difference of two squares and can be factored.

$= (x^2 + 9)(x + 3)(x - 3)$ The factors of $x^2 - 9$ are the sum and the difference of the expressions being squared. ■

STUDY TIP

Factoring $x^4 - 81$ as

$$(x^2 + 9)(x^2 - 9)$$

is not a complete factorization. The second factor, $x^2 - 9$, is itself a difference of two squares and can be factored.

Are you tempted to further factor $x^2 + 9$, the sum of two squares, in Example 4? Resist the temptation! **The sum of two squares, $A^2 + B^2$, with no common factor other than 1 is a prime polynomial over the integers**.

CHECK POINT 4 Factor completely: $81x^4 - 16$.

2 Factor perfect square trinomials.

Factoring Perfect Square Trinomials Our next factoring technique is obtained by reversing the special products for squaring binomials. The trinomials that are factored using this technique are called **perfect square trinomials**.

ENRICHMENT ESSAY

Visualizing the Factoring for a Perfect Square Trinomial

Area:

$(A + B)^2$

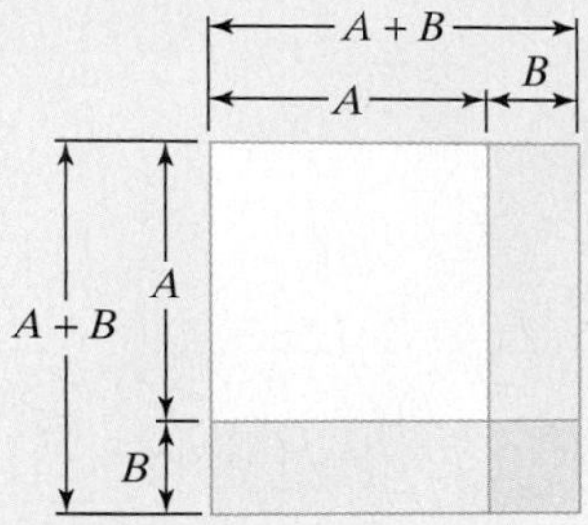

Sum of Areas:

$A^2 + 2AB + B^2$

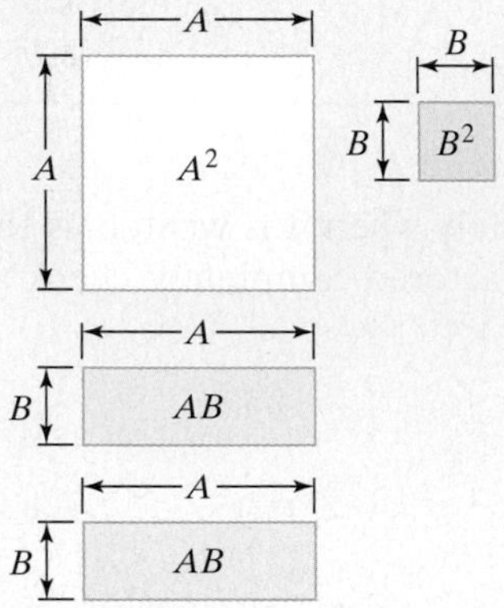

Conclusion:

$A^2 + 2AB + B^2 = (A + B)^2$

FACTORING PERFECT SQUARE TRINOMIALS Let A and B be real numbers, variables, or algebraic expressions.

1. $A^2 + 2AB + B^2 = (A + B)^2$
   Same sign
2. $A^2 - 2AB + B^2 = (A - B)^2$
   Same sign

The two items in the box show that perfect square trinomials come in two forms: one in which the middle term is positive and one in which the middle term is negative. Here's how to recognize a perfect square trinomial:

1. The first and last terms are squares of monomials or integers.
2. The middle term is twice the product of the expressions being squared in the first and last terms.

EXAMPLE 5 Factoring Perfect Square Trinomials

Factor:

a. $x^2 + 6x + 9$ **b.** $x^2 - 16x + 64$ **c.** $25x^2 - 60x + 36$.

SOLUTION

a. $x^2 + 6x + 9 = x^2 + 2 \cdot x \cdot 3 + 3^2 = (x + 3)^2$

$A^2 + 2AB + B^2 = (A + B)^2$

The middle term has a positive sign.

b. $x^2 - 16x + 64 = x^2 - 2 \cdot x \cdot 8 + 8^2 = (x - 8)^2$

$A^2 - 2AB + B^2 = (A - B)^2$

The middle term has a negative sign.

c. We suspect that $25x^2 - 60x + 36$ is a perfect square trinomial because $25x^2 = (5x)^2$ and $36 = 6^2$. The middle term can be expressed as twice the product of $5x$ and 6.

$$25x^2 - 60x + 36 = (5x)^2 - 2 \cdot 5x \cdot 6 + 6^2 = (5x - 6)^2$$

$A^2 - 2AB + B^2 = (A - B)^2$ ■

✔ CHECK POINT 5 Factor:

a. $x^2 + 14x + 49$ **b.** $x^2 - 6x + 9$

c. $16x^2 - 56x + 49$.

EXAMPLE 6 Factoring a Perfect Square Trinomial in Two Variables

Factor: $16x^2 + 40xy + 25y^2$.

SOLUTION Observe that $16x^2 = (4x)^2$, $25y^2 = (5y)^2$, and $40xy$ is twice the product of $4x$ and $5y$. Thus, we have a perfect square trinomial.

$$16x^2 + 40xy + 25y^2 = (4x)^2 + 2 \cdot 4x \cdot 5y + (5y)^2 = (4x + 5y)^2$$

$$A^2 \quad + \quad 2AB \quad + \quad B^2 \quad = \quad (A + B)^2$$

■

CHECK POINT 6 Factor: $4x^2 + 12xy + 9y^2$.

3 Factor the sum and difference of two cubes.

Factoring the Sum and Difference of Two Cubes We can use the following formulas to factor the sum or the difference of two cubes:

FACTORING THE SUM AND DIFFERENCE OF TWO CUBES

1. Factoring the Sum of Two Cubes

$$A^3 + B^3 = (A + B)(A^2 - AB + B^2)$$

Same sign Opposite signs

2. Factoring the Difference of Two Cubes

$$A^3 - B^3 = (A - B)(A^2 + AB + B^2)$$

Same sign Opposite signs

USING TECHNOLOGY

You can use a graphing utility to verify the factorization in Example 7. Enter

$$y_1 = x^3 + 8$$

and

$$y_2 = (x + 2)(x^2 - 2x + 4).$$

Graphic Check

The graphs are identical.

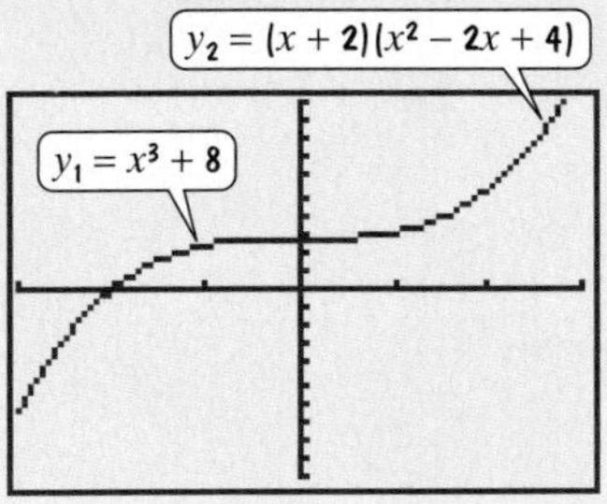

$[-3, 3, 1]$ by $[-30, 30, 3]$

Numeric Check

No matter how far up or down we scroll, $y_1 = y_2$.

X	Y1	Y2
-3	-19	-19
-2	0	0
-1	7	7
0	8	8
1	9	9
2	16	16
3	35	35

X= -3

EXAMPLE 7 Factoring the Sum of Two Cubes

Factor: $x^3 + 8$.

SOLUTION We must express each term as the cube of some monomial. Then we use the formula for factoring $A^3 + B^3$.

$$x^3 + 8 = x^3 + 2^3 = (x + 2)(x^2 - x \cdot 2 + 2^2) = (x + 2)(x^2 - 2x + 4)$$

$$A^3 \quad + \quad B^3 \quad = \quad (A + B) \quad (A^2 \quad - \quad AB \quad + \quad B^2)$$

■

CHECK POINT 7 Factor: $x^3 + 27$.

EXAMPLE 8 Factoring the Difference of Two Cubes

Factor: $27 - y^3$.

SOLUTION Express each term as the cube of some monomial. Then use the formula for factoring $A^3 - B^3$.

$$27 - y^3 = 3^3 - y^3 = (3 - y)(3^2 + 3y + y^2) = (3 - y)(9 + 3y + y^2)$$

$$A^3 - B^3 \quad = \quad (A \quad - \quad B) \quad (A^2 \quad + \quad AB \quad + \quad B^2)$$

■

CHECK POINT 8 Factor: $1 - y^3$.

EXAMPLE 9 Factoring the Sum of Two Cubes

Factor: $64x^3 + 125$.

SOLUTION Express each term as the cube of some monomial. Then use the formula for factoring $A^3 + B^3$.

$$64x^3 + 125 = (4x)^3 + 5^3 = (4x + 5)[(4x)^2 - (4x)(5) + 5^2]$$

$$A^3 + B^3 = (A + B)(A^2 - AB + B^2)$$

$$= (4x + 5)(16x^2 - 20x + 25)$$

■

CHECK POINT 9 Factor: $125x^3 + 8$.

6.4 EXERCISE SET

Student Solutions Manual | CD/Video | PH Math/Tutor Center | MathXL Tutorials on CD | MathXL® | MyMathLab | Interactmath.com

Practice Exercises

In Exercises 1–26, factor each difference of two squares.

1. $x^2 - 25$
2. $x^2 - 16$
3. $y^2 - 1$
4. $y^2 - 9$
5. $4x^2 - 9$
6. $9x^2 - 25$
7. $25 - x^2$
8. $16 - x^2$
9. $1 - 49x^2$
10. $1 - 64x^2$
11. $9 - 25y^2$
12. $16 - 49y^2$
13. $x^4 - 9$
14. $x^4 - 25$
15. $49y^4 - 16$
16. $49y^4 - 25$
17. $x^{10} - 9$
18. $x^{10} - 1$
19. $25x^2 - 16y^2$
20. $9x^2 - 25y^2$
21. $x^4 - y^{10}$
22. $x^{14} - y^4$
23. $x^4 - 16$
24. $x^4 - 1$
25. $16x^4 - 81$
26. $81x^4 - 1$

In Exercises 27–40, factor completely, or state that the polynomial is prime.

27. $2x^2 - 18$
28. $5x^2 - 45$
29. $2x^3 - 72x$
30. $2x^3 - 8x$
31. $x^2 + 36$
32. $x^2 + 4$
33. $3x^3 + 27x$
34. $3x^3 + 15x$
35. $18 - 2y^2$
36. $32 - 2y^2$
37. $3y^3 - 48y$
38. $3y^3 - 75y$
39. $18x^3 - 2x$
40. $20x^3 - 5x$

In Exercises 41–62, factor any perfect square trinomials, or state that the polynomial is prime.

41. $x^2 + 2x + 1$
42. $x^2 + 4x + 4$
43. $x^2 - 14x + 49$
44. $x^2 - 10x + 25$
45. $x^2 - 2x + 1$
46. $x^2 - 4x + 4$
47. $x^2 + 22x + 121$
48. $x^2 + 24x + 144$
49. $4x^2 + 4x + 1$
50. $9x^2 + 6x + 1$
51. $25y^2 - 10y + 1$
52. $64y^2 - 16y + 1$

53. $x^2 - 10x + 100$

54. $x^2 - 7x + 49$

55. $x^2 + 14xy + 49y^2$

56. $x^2 + 16xy + 64y^2$

57. $x^2 - 12xy + 36y^2$

58. $x^2 - 18xy + 81y^2$

59. $x^2 - 8xy + 64y^2$

60. $x^2 + 9xy + 16y^2$

61. $16x^2 - 40xy + 25y^2$

62. $9x^2 + 48xy + 64y^2$

In Exercises 63–70, factor completely.

63. $12x^2 - 12x + 3$

64. $18x^2 + 24x + 8$

65. $9x^3 + 6x^2 + x$

66. $25x^3 - 10x^2 + x$

67. $2y^2 - 4y + 2$

68. $2y^2 - 40y + 200$

69. $2y^3 + 28y^2 + 98y$

70. $50y^3 + 20y^2 + 2y$

In Exercises 71–88, factor using the formula for the sum or difference of two cubes.

71. $x^3 + 1$

72. $x^3 + 64$

73. $x^3 - 27$

74. $x^3 - 64$

75. $8y^3 - 1$

76. $27y^3 - 1$

77. $27x^3 + 8$

78. $125x^3 + 8$

79. $x^3y^3 - 64$

80. $x^3y^3 - 27$

81. $27y^4 + 8y$

82. $64y - y^4$

83. $54 - 16y^3$

84. $128 - 250y^3$

85. $64x^3 + 27y^3$

86. $8x^3 + 27y^3$

87. $125x^3 - 64y^3$

88. $125x^3 - y^3$

Practice Plus

In Exercises 89–96, factor completely.

89. $25x^2 - \frac{4}{49}$

90. $16x^2 - \frac{9}{25}$

91. $y^4 - \frac{y}{1000}$

92. $y^4 - \frac{y}{8}$

93. $0.25x - x^3$

94. $0.64x - x^3$

95. $(x + 1)^2 - 25$

96. $(x + 2)^2 - 49$

97. Divide $x^3 - x^2 - 5x - 3$ by $x - 3$. Use the quotient to factor $x^3 - x^2 - 5x - 3$ completely.

98. Divide $x^3 + 4x^2 - 3x - 18$ by $x - 2$. Use the quotient to factor $x^3 + 4x^2 - 3x - 18$ completely.

Application Exercises

In Exercises 99–102, find the formula for the area of the shaded region and express it in factored form.

99.

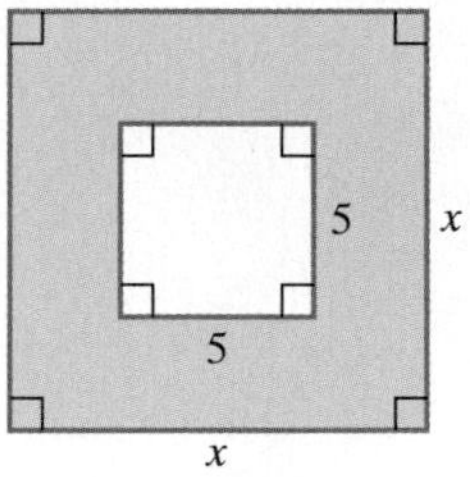

100.

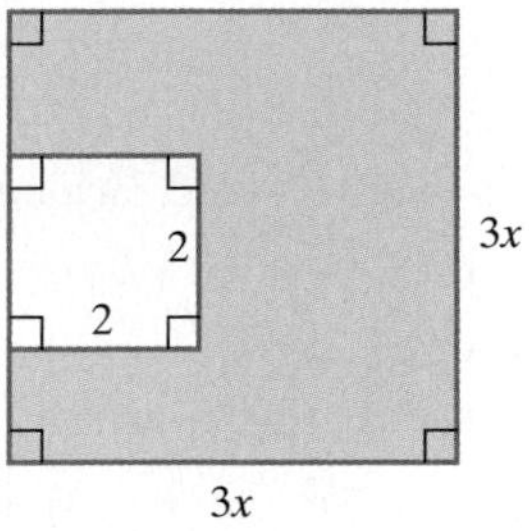

101.

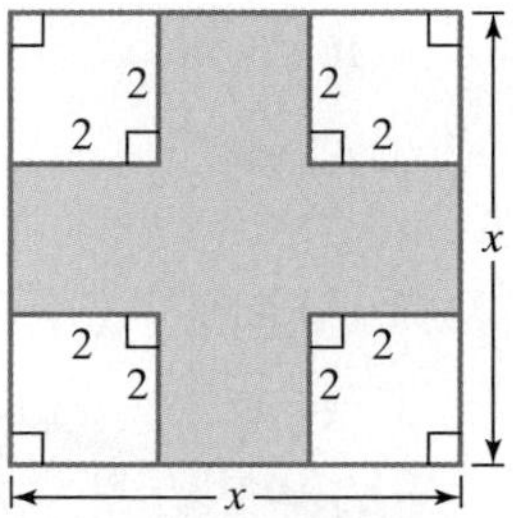

102.

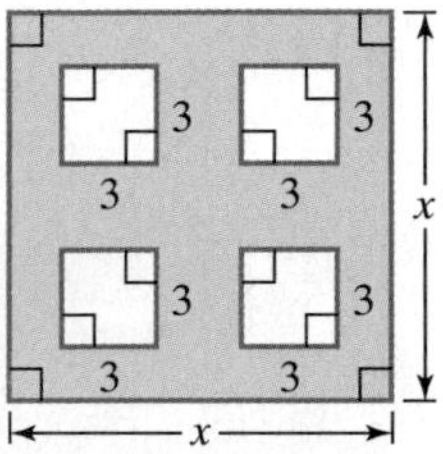

Writing in Mathematics

103. Explain how to factor the difference of two squares. Provide an example with your explanation.

104. What is a perfect square trinomial and how is it factored?

105. Explain why $x^2 - 1$ is factorable, but $x^2 + 1$ is not.

106. Explain how to factor $x^3 + 1$.

Critical Thinking Exercises

107. Which one of the following is true?

a. Because $x^2 - 25 = (x + 5)(x - 5)$, then $x^2 + 25 = (x - 5)(x + 5)$.

b. All perfect square trinomials are squares of binomials.

c. Any polynomial that is the sum of two squares is prime.

d. The polynomial $16x^2 + 20x + 25$ is a perfect square trinomial.

108. Where is the error in this "proof" that 2 = 0?

$a = b$	Suppose that a and b are any equal real numbers.
$a^2 = b^2$	Square both sides of the equation.
$a^2 - b^2 = 0$	Subtract b^2 from both sides.
$2(a^2 - b^2) = 2 \cdot 0$	Multiply both sides by 2.
$2(a^2 - b^2) = 0$	On the right side, $2 \cdot 0 = 0$.
$2(a + b)(a - b) = 0$	Factor $a^2 - b^2$.
$2(a + b) = 0$	Divide both sides by $a - b$.
$2 = 0$	Divide both sides by $a + b$.

In Exercises 109–112, factor each polynomial.

109. $x^2 - y^2 + 3x + 3y$

110. $x^{2n} - 25y^{2n}$

111. $4x^{2n} + 12x^n + 9$

112. $(x + 3)^2 - 2(x + 3) + 1$

In Exercises 113–114, find all integers k so that the trinomial is a perfect square trinomial.

113. $9x^2 + kx + 1$

114. $64x^2 - 16x + k$

Technology Exercises

In Exercises 115–118, use the GRAPH *or* TABLE *feature of a graphing utility to determine if the polynomial on the left side of each equation has been correctly factored. If the graphs of y_1 and y_2 coincide, or if their corresponding table values are equal, this means that the polynomial on the left side has been correctly factored. If not, factor the polynomial correctly and then use your graphing utility to verify the factorization.*

115. $4x^2 - 9 = (4x + 3)(4x - 3)$

116. $x^2 - 6x + 9 = (x - 3)^2$

117. $4x^2 - 4x + 1 = (4x - 1)^2$

118. $x^3 - 1 = (x - 1)(x^2 - x + 1)$

Review Exercises

119. Simplify: $(2x^2y^3)^4(5xy^2)$. (Section 5.7, Example 5)

120. Subtract: $(10x^2 - 5x + 2) - (14x^2 - 5x - 1)$. (Section 5.1, Example 3)

121. Divide: $\dfrac{6x^2 + 11x - 10}{3x - 2}$. (Section 5.6, Example 1)

SECTION 6.5

Objectives

1. Recognize the appropriate method for factoring a polynomial.
2. Use a general strategy for factoring polynomials.

A GENERAL FACTORING STRATEGY

Yogi Berra, catcher and renowned hitter for the New York Yankees (1946–1963), said it best: "If you don't know where you're going, you'll probably end up someplace else." When it comes to factoring, it's easy to know where you're going. Why? In this section, you will learn a step-by-step strategy that provides a plan and direction for solving factoring problems.

A Strategy for Factoring Polynomials It is important to practice factoring a wide variety of polynomials so that you can quickly select the appropriate technique. The polynomial is factored completely when all its polynomial factors, except possibly the monomial factors, are prime. Because of the commutative property, the order of the factors does not matter.

Here is a general strategy for factoring polynomials:

1 Recognize the appropriate method for factoring a polynomial.

A STRATEGY FOR FACTORING A POLYNOMIAL

1. If there is a common factor other than 1, factor out the GCF.
2. Determine the number of terms in the polynomial and try factoring as follows:
   a. If there are two terms, can the binomial be factored by one of the following special forms?

      Difference of two squares: $A^2 - B^2 = (A + B)(A - B)$

      Sum of two cubes: $A^3 + B^3 = (A + B)(A^2 - AB + B^2)$

      Difference of two cubes: $A^3 - B^3 = (A - B)(A^2 + AB + B^2)$

   b. If there are three terms, is the trinomial a perfect square trinomial? If so, factor by one of the following special forms:

      $$A^2 + 2AB + B^2 = (A + B)^2$$
      $$A^2 - 2AB + B^2 = (A - B)^2.$$

      If the trinomial is not a perfect square trinomial, try factoring by trial and error or grouping.

   c. If there are four or more terms, try factoring by grouping.
3. Check to see if any factors with more than one term in the factored polynomial can be factored further. If so, factor completely.
4. Check by multiplying.

2 Use a general strategy for factoring polynomials.

The following examples and those in the exercise set are similar to the previous factoring problems. One difference is that although these polynomials may be factored using the techniques we have studied in this chapter, each must be factored using at least two techniques. Also different is that these factorizations are not all of the same type. They are intentionally mixed to promote the development of a general factoring strategy.

EXAMPLE 1 Factoring a Polynomial

Factor: $4x^4 - 16x^2$.

SOLUTION

Step 1. **If there is a common factor, factor out the GCF.** Because $4x^2$ is common to both terms, we factor it out.

$$4x^4 - 16x^2 = 4x^2(x^2 - 4)$$ *Factor out the GCF.*

Step 2. **Determine the number of terms and factor accordingly.** The factor $x^2 - 4$ has two terms. It is the difference of two squares: $x^2 - 2^2$. We factor using the special form for the difference of two squares and rewrite the GCF.

$$4x^4 - 16x^2 = 4x^2(x + 2)(x - 2)$$ *Use $A^2 - B^2 = (A + B)(A - B)$ on $x^2 - 4$: $A = x$ and $B = 2$.*

Step 3. **Check to see if any factors with more than one term can be factored further.** No factor with more than one term can be factored further, so we have factored completely.

ENRICHMENT ESSAY

Weird Numbers

Mathematicians use the label **weird** to describe a number if

1. The sum of its factors, excluding the number itself, is greater than the number.
2. No partial collection of the factors adds up to the number.

The number 70 is weird. Its factors are 1, 2, 5, 7, 10, 14, and 35. The sum of these factors is 74, which is greater than 70. Two or more numbers in the list of factors cannot be added to obtain 70.

Weird numbers are rare. Below 10,000, the weird numbers are 70, 836, 4030, 5830, 7192, 7912, and 9272. It is not known whether an odd weird number exists.

Step 4. Check by multiplying.

$$4x^2(x+2)(x-2) = 4x^2(x^2-4) = 4x^4 - 16x^2$$

This is the original polynomial, so the factorization is correct. ■

CHECK POINT 1 Factor: $5x^4 - 45x^2$.

EXAMPLE 2 Factoring a Polynomial

Factor: $3x^2 - 6x - 45$.

SOLUTION

Step 1. If there is a common factor, factor out the GCF. Because 3 is common to all terms, we factor it out.

$$3x^2 - 6x - 45 = 3(x^2 - 2x - 15) \quad \text{Factor out the GCF.}$$

Step 2. Determine the number of terms and factor accordingly. The factor $x^2 - 2x - 15$ has three terms, but it is not a perfect square trinomial. We factor it using trial and error.

$$3x^2 - 6x - 45 = 3(x^2 - 2x - 15) = 3(x-5)(x+3)$$

Step 3. Check to see if factors can be factored further. In this case, they cannot, so we have factored completely.

Step 4. Check by multiplying.

$$3(x-5)(x+3) = 3(x^2 - 2x - 15) = 3x^2 - 6x - 45$$

FOIL

This is the original polynomial, so the factorization is correct. ■

CHECK POINT 2 Factor: $4x^2 - 16x - 48$.

EXAMPLE 3 Factoring a Polynomial

Factor: $7x^5 - 7x$.

SOLUTION

Step 1. If there is a common factor, factor out the GCF. Because $7x$ is common to both terms, we factor it out.

$$7x^5 - 7x = 7x(x^4 - 1) \quad \text{Factor out the GCF.}$$

Step 2. Determine the number of terms and factor accordingly. The factor $x^4 - 1$ has two terms. This binomial can be expressed as $(x^2)^2 - 1^2$, so it can be factored as the difference of two squares.

$$7x^5 - 7x = 7x(x^4 - 1) = 7x(x^2 + 1)(x^2 - 1)$$

Use $A^2 - B^2 = (A + B)(A - B)$ on $x^4 - 1$: $A = x^2$ and $B = 1$.

Step 3. Check to see if factors can be factored further. We note that $(x^2 - 1)$ is also the difference of two squares, $x^2 - 1^2$, so we continue factoring.

$$7x^5 - 7x = 7x(x^2 + 1)(x + 1)(x - 1)$$ *Factor $x^2 - 1$ as the difference of two squares.*

Step 4. Check by multiplying.

$$7x(x^2 + 1)(x + 1)(x - 1) = 7x(x^2 + 1)(x^2 - 1) = 7x(x^4 - 1) = 7x^5 - 7x$$

We obtain the original polynomial, so the factorization is correct. ■

CHECK POINT 3 Factor: $4x^5 - 64x$.

EXAMPLE 4 Factoring a Polynomial

Factor: $x^3 - 5x^2 - 4x + 20$.

SOLUTION

Step 1. If there is a common factor, factor out the GCF. Other than 1, there is no common factor.

Step 2. Determine the number of terms and factor accordingly. There are four terms. We try factoring by grouping.

$$x^3 - 5x^2 - 4x + 20$$

$$= (x^3 - 5x^2) + (-4x + 20)$$ *Group terms with common factors.*

$$= x^2(x - 5) - 4(x - 5)$$ *Factor from each group.*

$$= (x - 5)(x^2 - 4)$$ *Factor out the common binomial factor, $x - 5$.*

Step 3. Check to see if factors can be factored further. We note that $(x^2 - 4)$ is the difference of two squares, $x^2 - 2^2$, so we continue factoring.

$$x^3 - 5x^2 - 4x + 20 = (x - 5)(x + 2)(x - 2)$$ *Factor $x^2 - 4$ as the difference of two squares.*

We have factored completely because no factor with more than one term can be factored further.

Step 4. Check by multiplying.

F O I L

$$(x - 5)(x + 2)(x - 2) = (x - 5)(x^2 - 4) = x^3 - 4x - 5x^2 + 20$$

$$= x^3 - 5x^2 - 4x + 20$$

We obtain the original polynomial, so the factorization is correct. ■

USING TECHNOLOGY

You can use a graphing utility to check the factorization in Example 4. Enter the given polynomial and its complete factorization:

$$y_1 = x^3 - 5x^2 - 4x + 20 \quad \text{and} \quad y_2 = (x - 5)(x + 2)(x - 2).$$

Graphic Check

The graphs are identical.

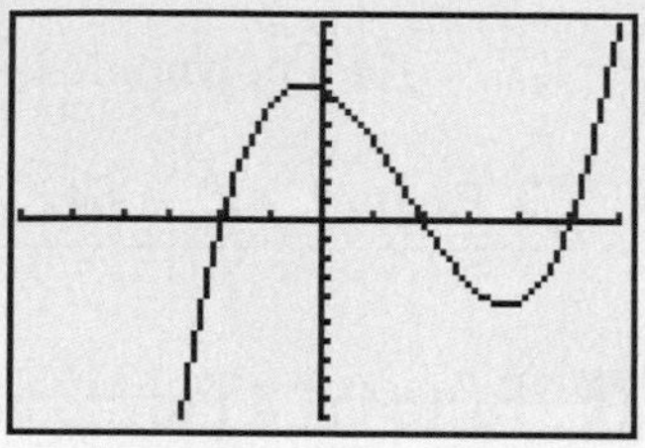

$[-6, 6, 1]$ by $[-30, 30, 3]$

Numeric Check

No matter how far up or down we scroll, $y_1 = y_2$.

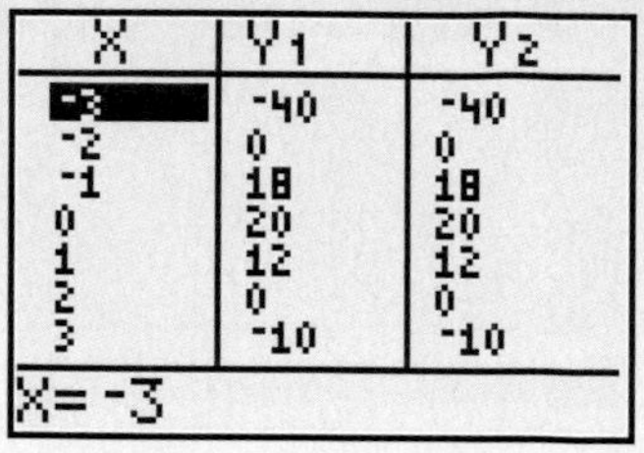

Caution: Keep in mind that a graphing utility cannot verify that a polynomial is factored completely. If you enter $(x - 5)(x^2 - 4)$ for y_2, which is not a complete factorization, you will still obtain identical graphs or tables with equal corresponding values.

✔ **CHECK POINT 4** Factor: $x^3 - 4x^2 - 9x + 36$.

EXAMPLE 5 Factoring a Polynomial

Factor: $2x^3 - 24x^2 + 72x$.

SOLUTION

Step 1. **If there is a common factor, factor out the GCF.** Because $2x$ is common to all terms, we factor it out.

$$2x^3 - 24x^2 + 72x = 2x(x^2 - 12x + 36) \qquad \text{Factor out the GCF.}$$

Step 2. **Determine the number of terms and factor accordingly.** The factor $x^2 - 12x + 36$ has three terms. Is it a perfect square trinomial? Yes. The first term, x^2, is the square of a monomial. The last term, 36, or 6^2, is the square of an integer. The middle term involves twice the product of x and 6. We factor using $A^2 - 2AB + B^2 = (A - B)^2$.

$$2x^3 - 24x^2 + 72x = 2x(x^2 - 12x + 36)$$

$$= 2x(x^2 - 2 \cdot x \cdot 6 + 6^2) \qquad \text{The second factor is a perfect square trinomial.}$$

$$A^2 - 2\,A\,B + B^2$$

$$= 2x(x - 6)^2 \qquad A^2 - 2AB + B^2 = (A - B)^2$$

Step 3. **Check to see if factors can be factored further.** In this problem, they cannot, so we have factored completely.

Step 4. Check by multiplying.

$$2x(x-6)^2 = 2x(x^2 - 12x + 36) = 2x^3 - 24x^2 + 72x$$

We obtain the original polynomial, so the factorization is correct. ■

CHECK POINT 5 Factor: $3x^3 - 30x^2 + 75x$.

EXAMPLE 6 Factoring a Polynomial

Factor: $3x^5 + 24x^2$.

SOLUTION

Step 1. If there is a common factor, factor out the GCF. Because $3x^2$ is common to both terms, we factor it out.

$$3x^5 + 24x^2 = 3x^2(x^3 + 8) \quad \text{Factor out the GCF.}$$

Step 2. Determine the number of terms and factor accordingly. The factor $x^3 + 8$ has two terms. This binomial can be expressed as $x^3 + 2^3$, so it can be factored as the sum of two cubes.

$$3x^5 + 24x^2 = 3x^2(\underbrace{x^3 + 2^3}_{A^3 + B^3}) \quad \text{Express } x^3 + 8 \text{ as the sum of two cubes.}$$

$$= 3x^2\underbrace{(x+2)(x^2 - 2x + 4)}_{(A+B)(A^2 - AB + B^2)} \quad \text{Factor the sum of two cubes.}$$

Step 3. Check to see if factors can be factored further. In this problem, they cannot, so we have factored completely.

Step 4. Check by multiplying.

$$3x^2(x+2)(x^2 - 2x + 4) = 3x^2[x(x^2 - 2x + 4) + 2(x^2 - 2x + 4)]$$

$$= 3x^2(x^3 - 2x^2 + 4x + 2x^2 - 4x + 8)$$

$$= 3x^2(x^3 + 8) = 3x^5 + 24x^2$$

We obtain the original polynomial, so the factorization is correct. ■

DISCOVER FOR YOURSELF

In Examples 1–6, substitute 1 for the variable in both the given polynomial and in its factored form. Evaluate each expression. What do you observe? Do this for a second value of the variable. Is this a complete check or only a partial check of the factorization? Explain.

CHECK POINT 6 Factor: $2x^5 + 54x^2$.

EXAMPLE 7 Factoring a Polynomial in Two Variables

Factor: $32x^4y - 2y^5$.

SOLUTION

Step 1. If there is a common factor, factor out the GCF. Because $2y$ is common to both terms, we factor it out.

$$32x^4y - 2y^5 = 2y(16x^4 - y^4) \quad \text{Factor out the GCF.}$$

Step 2. **Determine the number of terms and factor accordingly.** The factor $16x^4 - y^4$ has two terms. It is the difference of two squares: $(4x^2)^2 - (y^2)^2$. We factor using the special form for the difference of two squares.

$$32x^4y - 2y^5 = 2y[(4x^2)^2 - (y^2)^2]$$

$A^2 - B^2$ — Express $16x^4 - y^4$ as the difference of two squares.

$$= 2y(4x^2 + y^2)(4x^2 - y^2)$$

$(A + B)$ $(A - B)$ — $A^2 - B^2 = (A + B)(A - B)$

Step 3. **Check to see if factors can be factored further.** We note that the last factor, $4x^2 - y^2$, is also the difference of two squares, $(2x)^2 - y^2$, so we continue factoring.

$$32x^4y - 2y^5 = 2y(4x^2 + y^2)(2x + y)(2x - y)$$

Step 4. **Check by multiplying.** Multiply the factors in the factorization and verify that you obtain the original polynomial. ■

✔ CHECK POINT **7** Factor: $3x^4y - 48y^5$.

EXAMPLE 8 Factoring a Polynomial in Two Variables

Factor: $18x^3 + 48x^2y + 32xy^2$.

SOLUTION

Step 1. **If there is a common factor, factor out the GCF.** Because $2x$ is common to all terms, we factor it out.

$$18x^3 + 48x^2y + 32xy^2 = 2x(9x^2 + 24xy + 16y^2)$$

Step 2. **Determine the number of terms and factor accordingly.** The factor $9x^2 + 24xy + 16y^2$ has three terms. Is it a perfect square trinomial? Yes. The first term, $9x^2$, or $(3x)^2$, and the last term, $16y^2$, or $(4y)^2$, are squares of monomials. The middle term, $24xy$, is twice the product of $3x$ and $4y$. We factor using $A^2 + 2AB + B^2 = (A + B)^2$.

$$18x^3 + 48x^2y + 32xy^2 = 2x(9x^2 + 24xy + 16y^2)$$

$$= 2x[(3x)^2 + 2 \cdot 3x \cdot 4y + (4y)^2]$$

$A^2 + 2 \cdot A \cdot B + B^2$ — The second factor is a perfect square trinomial.

$$= 2x(3x + 4y)^2$$

$A^2 + 2AB + B^2 = (A + B)^2$

Step 3. **Check to see if factors can be factored further.** In this problem, they cannot, so we have factored completely.

Step 4. **Check by multiplication.** Multiply the factors in the factorization and verify that you obtain the original polynomial.

✔ CHECK POINT **8** Factor: $12x^3 + 36x^2y + 27xy^2$.

6.5 EXERCISE SET

Student Solutions Manual | CD/Video | PH Math/Tutor Center | MathXL Tutorials on CD | MathXL® | MyMathLab | Interactmath.com

Practice Exercises

In Exercises 1–62, factor completely, or state that the polynomial is prime. Check factorizations using multiplication or a graphing utility.

1. $5x^3 - 20x$ **2.** $4x^3 - 100x$

3. $7x^3 + 7x$ **4.** $6x^3 + 24x$

5. $5x^2 - 5x - 30$ **6.** $5x^2 - 15x - 50$

7. $2x^4 - 162$ **8.** $7x^4 - 7$

9. $x^3 + 2x^2 - 9x - 18$ **10.** $x^3 + 3x^2 - 25x - 75$

11. $3x^3 - 24x^2 + 48x$ **12.** $5x^3 - 20x^2 + 20x$

13. $2x^5 + 2x^2$ **14.** $2x^5 + 128x^2$

15. $6x^2 + 8x$ **16.** $21x^2 - 35x$

17. $2y^2 - 2y - 112$ **18.** $6x^2 - 6x - 12$

19. $7y^4 + 14y^3 + 7y^2$ **20.** $2y^4 + 28y^3 + 98y^2$

21. $y^2 + 8y - 16$ **22.** $y^2 - 18y - 81$

23. $16y^2 - 4y - 2$ **24.** $32y^2 + 4y - 6$

25. $r^2 - 25r$ **26.** $3r^2 - 27r$

27. $4w^2 + 8w - 5$ **28.** $35w^2 - 2w - 1$

29. $x^3 - 4x$ **30.** $9x^3 - 9x$

31. $x^2 + 64$ **32.** $y^2 + 36$

33. $9y^2 + 13y + 4$ **34.** $20y^2 + 12y + 1$

35. $y^3 + 2y^2 - 4y - 8$ **36.** $y^3 + 2y^2 - y - 2$

37. $16y^2 + 24y + 9$ **38.** $25y^2 + 20y + 4$

39. $4y^3 - 28y^2 + 40y$ **40.** $7y^3 - 21y^2 + 14y$

41. $y^5 - 81y$ **42.** $y^5 - 16y$

43. $20a^4 - 45a^2$ **44.** $48a^4 - 3a^2$

45. $12y^2 - 11y + 2$ **46.** $21x^2 - 25x - 4$

47. $9y^2 - 64$ **48.** $100y^2 - 49$

49. $9y^2 + 64$ **50.** $100y^2 + 49$

51. $2y^3 + 3y^2 - 50y - 75$ **52.** $12y^3 + 16y^2 - 3y - 4$

53. $2r^3 + 30r^2 - 68r$ **54.** $3r^3 - 27r^2 - 210r$

55. $8x^5 - 2x^3$ **56.** $y^9 - y^5$

57. $3x^2 + 243$ **58.** $27x^2 + 75$

59. $x^4 + 8x$ **60.** $x^4 + 27x$

61. $2y^5 - 2y^2$ **62.** $2y^5 - 128y^2$

Exercises 63–92 contain polynomials in several variables. Factor each polynomial completely and check using multiplication.

63. $6x^2 + 8xy$ **64.** $21x^2 - 35xy$

65. $xy - 7x + 3y - 21$ **66.** $xy - 5x + 2y - 10$

67. $x^2 - 3xy - 4y^2$ **68.** $x^2 - 4xy - 12y^2$

69. $72a^3b^2 + 12a^2 - 24a^4b^2$ **70.** $24a^4b + 60a^3b^2 + 150a^2b^3$

71. $3a^2 + 27ab + 54b^2$ **72.** $3a^2 + 15ab + 18b^2$

73. $48x^4y - 3x^2y$ **74.** $16a^3b^2 - 4ab^2$

75. $6a^2b + ab - 2b$ **76.** $16a^2 - 32ab + 12b^2$

77. $7x^5y - 7xy^5$ **78.** $3x^4y^2 - 3x^2y^2$

79. $10x^3y - 14x^2y^2 + 4xy^3$ **80.** $18x^3y + 57x^2y^2 + 30xy^3$

81. $2bx^2 + 44bx + 242b$ **82.** $3xz^2 - 72xz + 432x$

83. $15a^2 + 11ab - 14b^2$ **84.** $25a^2 + 25ab + 6b^2$

85. $36x^3y - 62x^2y^2 + 12xy^3$ **86.** $10a^4b^2 - 15a^3b^3 - 25a^2b^4$

87. $a^2y - b^2y - a^2x + b^2x$ **88.** $bx^2 - 4b + ax^2 - 4a$

89. $9ax^3 + 15ax^2 - 14ax$ **90.** $4ay^3 - 12ay^2 + 9ay$

91. $81x^4y - y^5$ **92.** $16x^4y - y^5$

Practice Plus

In Exercises 93–102, factor completely.

93. $10x^2(x + 1) - 7x(x + 1) - 6(x + 1)$

94. $12x^2(x - 1) - 4x(x - 1) - 5(x - 1)$

95. $6x^4 + 35x^2 - 6$ **96.** $7x^4 + 34x^2 - 5$

97. $(x - 7)^2 - 4a^2$ **98.** $(x - 6)^2 - 9a^2$

99. $x^2 + 8x + 16 - 25a^2$ **100.** $x^2 + 14x + 49 - 16a^2$

101. $y^7 + y$ **102.** $(y + 1)^3 + 1$

Application Exercises

103. A rock is dropped from the top of a 256-foot cliff. The height, in feet, of the rock above the water after t seconds is modeled by the polynomial $256 - 16t^2$. Factor this expression completely.

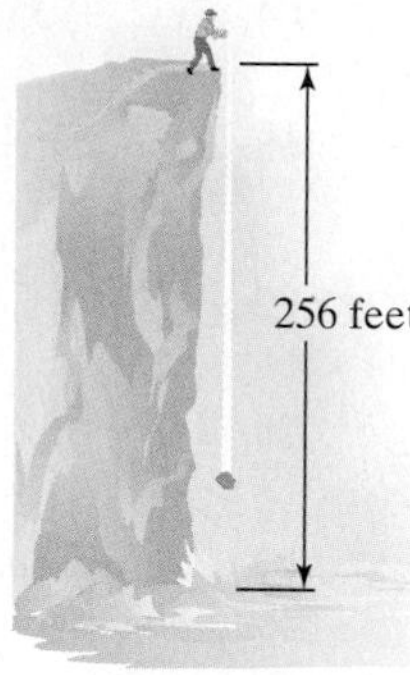

104. The building shown in the figure has a height represented by x feet. The building's base is a square and its volume is $x^3 - 60x^2 + 900x$ cubic feet. Express the building's dimensions in terms of x.

105. Express the area of the shaded ring shown in the figure in terms of π. Then factor this expression completely.

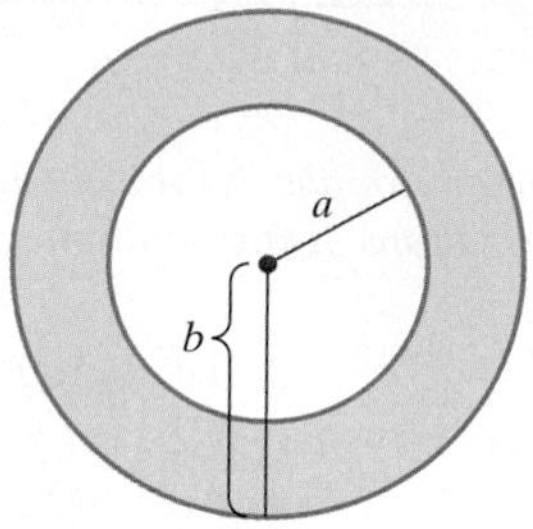

Writing in Mathematics

106. Describe a strategy that can be used to factor polynomials.

107. Describe some of the difficulties in factoring polynomials. What suggestions can you offer to overcome these difficulties?

108. You are about to take a great picture of fog rolling into San Francisco from the middle of the Golden Gate Bridge, 400 feet above the water. Whoops! You accidently lean too far over the safety rail and drop your camera. The height, in feet, of the camera after t seconds is modeled by the polynomial $400 - 16t^2$. The factored form of the polynomial is $16(5 + t)(5 - t)$. Describe something about your falling camera that is easier to see from the factored form, $16(5 + t)(5 - t)$, than from the form $400 - 16t^2$.

Critical Thinking Exercises

109. Which one of the following is true?

a. $x^2 - 9 = (x - 3)^2$ for any real number x.

b. The polynomial $4x^2 + 100$ is the sum of two squares and therefore cannot be factored.

c. If the general factoring strategy is used to factor a polynomial, at least two factorizations are necessary before the given polynomial is factored completely.

d. Once a common monomial factor is removed from $3xy^3 + 9xy^2 + 21xy$, the remaining trinomial factor cannot be factored further.

In Exercises 110–114, factor completely.

110. $3x^5 - 21x^3 - 54x$

111. $5y^5 - 5y^4 - 20y^3 + 20y^2$

112. $4x^4 - 9x^2 + 5$

113. $(x + 5)^2 - 20(x + 5) + 100$

114. $3x^{2n} - 27y^{2n}$

Technology Exercises

In Exercises 115–119, use the GRAPH *or* TABLE *feature of a graphing utility to determine if the polynomial on the left side of each equation has been correctly factored. If not, factor the polynomial correctly and then use your graphing utility to verify the factorization.*

115. $4x^2 - 12x + 9 = (4x - 3)^2$; $[-5, 5, 1]$ by $[0, 20, 1]$

116. $3x^3 - 12x^2 - 15x = 3x(x + 5)(x - 1)$; $[-5, 7, 1]$ by $[-80, 80, 10]$

117. $6x^2 + 10x - 4 = 2(3x - 1)(x + 2)$; $[-5, 5, 1]$ by $[-20, 20, 2]$

118. $x^4 - 16 = (x^2 + 4)(x + 2)(x - 2)$; $[-5, 5, 1]$ by $[-20, 20, 2]$

119. $2x^3 + 10x^2 - 2x - 10 = 2(x + 5)(x^2 + 1)$; $[-8, 4, 1]$ by $[-100, 100, 10]$

Review Exercises

120. Factor: $9x^2 - 16$. (Section 6.4, Example 1)

121. Graph using intercepts: $5x - 2y = 10$. (Section 3.2, Example 4)

122. The second angle of a triangle measures three times that of the first angle's measure. The third angle measures 80° more than the first. Find the measure of each angle. (Section 2.6, Example 6)

SECTION 6.6

Objectives

1 Use the zero-product principle.

2 Solve quadratic equations by factoring.

3 Solve problems with quadratic equations.

SOLVING QUADRATIC EQUATIONS BY FACTORING

The alligator, an endangered species, was the subject of a protection program at Florida's Everglades National Park. Park rangers used the formula

$$P = -10x^2 + 475x + 3500$$

to estimate the alligator population, P, after x years of the protection program. Their goal was to bring the population up to 7250. To find out how long the program had to be continued for this to happen, we substitute 7250 for P in the formula and solve for x:

$$7250 = -10x^2 + 475x + 3500.$$

Do you see how this equation differs from a linear equation? The highest exponent on x is 2. Solving such an equation involves finding the numbers that will make the equation a true statement. In this section, we use factoring to solve equations in the form $ax^2 + bx + c = 0$. We also look at applications of these equations.

The Standard Form of a Quadratic Equation We begin by defining a quadratic equation.

DEFINITION OF A QUADRATIC EQUATION A **quadratic equation** in x is an equation that can be written in the **standard form**

$$ax^2 + bx + c = 0,$$

where a, b, and c are real numbers, with $a \neq 0$. A quadratic equation in x is also called a **second-degree polynomial equation** in x.

Here is an example of a quadratic equation in standard form:

$$x^2 - 7x + 10 = 0.$$

$a = 1$ $b = -7$ $c = 10$

Use the zero-product principle.

Solving Quadratic Equations by Factoring We can factor the left side of the quadratic equation $x^2 - 7x + 10 = 0$. We obtain $(x - 5)(x - 2) = 0$. If a quadratic equation has zero on one side and a factored expression on the other side, it can be solved using the **zero-product principle**.

THE ZERO-PRODUCT PRINCIPLE If the product of two algebraic expressions is zero, then at least one of the factors is equal to zero.

If $AB = 0$, then $A = 0$ or $B = 0$.

For example, consider the equation $(x - 5)(x - 2) = 0$. According to the zero-product principle, this product can be zero only if at least one of the factors is zero. We set each individual factor equal to zero and solve each resulting equation for x.

$$(x - 5)(x - 2) = 0$$
$$x - 5 = 0 \quad \text{or} \quad x - 2 = 0$$
$$x = 5 \qquad\qquad x = 2$$

We can check each of the proposed solutions, 5 and 2, in the original quadratic equation, $x^2 - 7x + 10 = 0$. Substitute each one separately for x in the equation.

Check 5:

$$x^2 - 7x + 10 = 0$$
$$5^2 - 7 \cdot 5 + 10 \stackrel{?}{=} 0$$
$$25 - 35 + 10 \stackrel{?}{=} 0$$
$$0 = 0, \quad \text{true}$$

Check 2:

$$x^2 - 7x + 10 = 0$$
$$2^2 - 7 \cdot 2 + 10 \stackrel{?}{=} 0$$
$$4 - 14 + 10 \stackrel{?}{=} 0$$
$$0 = 0, \quad \text{true}$$

The resulting true statements indicate that the solutions are 5 and 2. Note that with a quadratic equation, we can have two solutions, compared to the linear equation that usually had one.

EXAMPLE 1 Using the Zero-Product Principle

Solve the equation: $(3x - 1)(x + 2) = 0$.

SOLUTION The product $(3x - 1)(x + 2)$ is equal to zero. By the zero-product principle, the only way that this product can be zero is if at least one of the factors is zero. Thus,

$$3x - 1 = 0 \quad \text{or} \quad x + 2 = 0.$$
$$3x = 1 \qquad\qquad x = -2 \qquad \text{Solve each equation for } x.$$
$$x = \frac{1}{3}$$

Because each linear equation has a solution, the original equation, $(3x - 1)(x + 2) = 0$, has two solutions, $\frac{1}{3}$ and -2. Check these solutions by substituting each one separately into the given equation. The equation's solution set is $\left\{-2, \frac{1}{3}\right\}$. ■

CHECK POINT 1 Solve the equation: $(2x + 1)(x - 4) = 0$.

2 Solve quadratic equations by factoring.

In Example 1 and Check Point 1, the given equations were in factored form. Here is a procedure for solving a quadratic equation when we must first do the factoring.

SOLVING A QUADRATIC EQUATION BY FACTORING

1. If necessary, rewrite the equation in the standard form $ax^2 + bx + c = 0$, moving all terms to one side, thereby obtaining zero on the other side.
2. Factor.
3. Apply the zero-product principle, setting each factor equal to zero.
4. Solve the equations in step 3.
5. Check the solutions in the original equation.

EXAMPLE 2 Solving a Quadratic Equation by Factoring

Solve: $2x^2 + 7x - 4 = 0$.

SOLUTION

Step 1. Move all terms to one side and obtain zero on the other side. All terms are already on the left and zero is on the other side, so we can skip this step.

Step 2. Factor.

$$2x^2 + 7x - 4 = 0$$
$$(2x - 1)(x + 4) = 0$$

Steps 3 and 4. Set each factor equal to zero and solve each resulting equation.

$$2x - 1 = 0 \quad \text{or} \quad x + 4 = 0$$
$$2x = 1 \qquad\qquad x = -4$$
$$x = \frac{1}{2}$$

Step 5. Check the solutions in the original equation.

Check $\frac{1}{2}$:

$$2x^2 + 7x - 4 = 0$$
$$2\left(\frac{1}{2}\right)^2 + 7\left(\frac{1}{2}\right) - 4 \stackrel{?}{=} 0$$
$$2\left(\frac{1}{4}\right) + 7\left(\frac{1}{2}\right) - 4 \stackrel{?}{=} 0$$
$$\frac{1}{2} + \frac{7}{2} - 4 \stackrel{?}{=} 0$$
$$4 - 4 \stackrel{?}{=} 0$$
$$0 = 0, \quad \text{true}$$

Check -4:

$$2x^2 + 7x - 4 = 0$$
$$2(-4)^2 + 7(-4) - 4 \stackrel{?}{=} 0$$
$$2(16) + 7(-4) - 4 \stackrel{?}{=} 0$$
$$32 + (-28) - 4 \stackrel{?}{=} 0$$
$$4 - 4 \stackrel{?}{=} 0$$
$$0 = 0, \quad \text{true}$$

The solutions are -4 and $\frac{1}{2}$, and the solution set is $\left\{-4, \frac{1}{2}\right\}$. ■

STUDY TIP

Do not confuse factoring a polynomial with solving a quadratic equation by factoring.

INCORRECT!

Factor: $2x^2 + 7x - 4$.

$(2x - 1)(x + 4)$

$2x - 1 = 0$ or $x + 4 = 0$

$x = \frac{1}{2}$ $\qquad x = -4$

✔ **CHECK POINT 2** Solve: $x^2 - 6x + 5 = 0$.

USING TECHNOLOGY

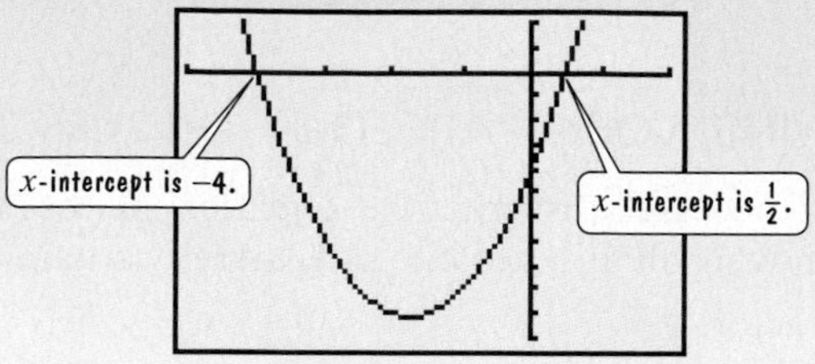

[−5, 2, 1] by [−11, 2, 1]

You can use a graphing utility to check the real number solutions of a quadratic equation. **The solutions of $ax^2 + bx + c = 0$ correspond to the x-intercepts for the graph of $y = ax^2 + bx + c$.** For example, to check the solutions of $2x^2 + 7x - 4 = 0$, graph $y = 2x^2 + 7x - 4$. The U-shaped, cuplike, graph is shown on the left. The x-intercepts are -4 and $\frac{1}{2}$, verifying -4 and $\frac{1}{2}$ as the solutions.

EXAMPLE 3 Solving a Quadratic Equation by Factoring

Solve: $3x^2 = 2x$.

SOLUTION

Step 1. **Move all terms to one side and obtain zero on the other side.** Subtract $2x$ from both sides and write the equation in standard form.

$$3x^2 - 2x = 2x - 2x$$
$$3x^2 - 2x = 0$$

Step 2. **Factor.** We factor out x from the two terms on the left side.

$$3x^2 - 2x = 0$$
$$x(3x - 2) = 0$$

Steps 3 and 4. **Set each factor equal to zero and solve the resulting equations.**

$$x = 0 \quad \text{or} \quad 3x - 2 = 0$$
$$3x = 2$$
$$x = \frac{2}{3}$$

Step 5. **Check the solutions in the original equation.**

Check 0:

$$3x^2 = 2x$$
$$3 \cdot 0^2 \stackrel{?}{=} 2 \cdot 0$$
$$0 = 0, \quad \text{true}$$

Check $\frac{2}{3}$:

$$3x^2 = 2x$$
$$3\left(\frac{2}{3}\right)^2 \stackrel{?}{=} 2\left(\frac{2}{3}\right)$$
$$3\left(\frac{4}{9}\right) \stackrel{?}{=} 2\left(\frac{2}{3}\right)$$
$$\frac{4}{3} = \frac{4}{3}, \quad \text{true}$$

The solutions are 0 and $\frac{2}{3}$, and the solution set is $\left\{0, \frac{2}{3}\right\}$. ■

STUDY TIP

Avoid dividing both sides of $3x^2 = 2x$ by x. You will obtain $3x = 2$ and, consequently, $x = \frac{2}{3}$. The other solution, 0, is lost. We can divide both sides of an equation by any *nonzero* real number. If x is zero, we lose the second solution.

CHECK POINT 3 Solve: $4x^2 = 2x$.

EXAMPLE 4 Solving a Quadratic Equation by Factoring

Solve: $x^2 = 6x - 9$.

SOLUTION

Step 1. **Move all terms to one side and obtain zero on the other side.** To obtain zero on the right, we subtract $6x$ and add 9 on both sides.

$$x^2 - 6x + 9 = 6x - 6x - 9 + 9$$
$$x^2 - 6x + 9 = 0$$

Step 2. **Factor.** The trinomial on the left side is a perfect square trinomial: $x^2 - 6x + 9 = x^2 - 2 \cdot x \cdot 3 + 3^2$. We factor using $A^2 - 2AB + B^2 = (A - B)^2$: $A = x$ and $B = 3$.

$$x^2 - 6x + 9 = 0$$
$$(x - 3)^2 = 0$$

Steps 3 and 4. **Set each factor equal to zero and solve the resulting equations.** Because both factors are the same, it is only necessary to set one of them equal to zero.

$$x - 3 = 0$$
$$x = 3$$

Step 5. **Check the solution in the original equation.**

Check 3:

$$x^2 = 6x - 9$$
$$3^2 \stackrel{?}{=} 6 \cdot 3 - 9$$
$$9 \stackrel{?}{=} 18 - 9$$
$$9 = 9, \quad \text{true}$$

The solution is 3 and the solution set is $\{3\}$. ■

USING TECHNOLOGY

The graph of $y = x^2 - 6x + 9$ is shown below. Notice that there is only one x-intercept, namely 3, verifying that the solution of

$$x^2 - 6x + 9 = 0$$

is 3.

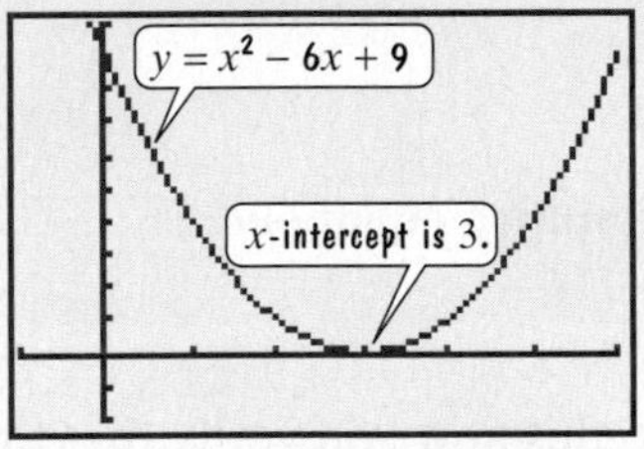

[−1, 6, 1] by [−2, 10, 1]

CHECK POINT 4 Solve: $x^2 = 10x - 25$.

EXAMPLE 5 Solving a Quadratic Equation by Factoring

Solve: $9x^2 = 16$.

SOLUTION

Step 1. **Move all terms to one side and obtain zero on the other side.** Subtract 16 from both sides and write the equation in standard form.

$$9x^2 - 16 = 16 - 16$$
$$9x^2 - 16 = 0$$

Step 2. **Factor.** The binomial on the left side is the difference of two squares: $9x^2 - 16 = (3x)^2 - 4^2$. We factor using $A^2 - B^2 = (A + B)(A - B)$: $A = 3x$ and $B = 4$.

$$9x^2 - 16 = 0$$
$$(3x + 4)(3x - 4) = 0$$

Steps 3 and 4. Set each factor equal to zero and solve the resulting equations. We use the zero-product principle to solve $(3x + 4)(3x - 4) = 0$.

$$3x + 4 = 0 \quad \text{or} \quad 3x - 4 = 0$$
$$3x = -4 \qquad\qquad 3x = 4$$
$$x = -\frac{4}{3} \qquad\qquad x = \frac{4}{3}$$

Step 5. Check the solutions in the original equation. Do this now and verify that the solutions of $9x^2 = 16$ are $-\frac{4}{3}$ and $\frac{4}{3}$. The equation's solution set is $\left\{-\frac{4}{3}, \frac{4}{3}\right\}$. ■

✔ CHECK POINT 5 Solve: $16x^2 = 25$.

EXAMPLE 6 Solving a Quadratic Equation by Factoring

Solve: $(x - 2)(x + 3) = 6$.

SOLUTION

Step 1. Move all terms to one side and obtain zero on the other side. We write the equation in standard form by multiplying out the product on the left side and then subtracting 6 from both sides.

$$(x - 2)(x + 3) = 6 \qquad \text{This is the given equation.}$$
$$x^2 + 3x - 2x - 6 = 6 \qquad \text{Use the FOIL method.}$$
$$x^2 + x - 6 = 6 \qquad \text{Simplify.}$$
$$x^2 + x - 6 - 6 = 6 - 6 \qquad \text{Subtract 6 from both sides.}$$
$$x^2 + x - 12 = 0 \qquad \text{Simplify.}$$

Step 2. Factor.

$$x^2 + x - 12 = 0$$
$$(x + 4)(x - 3) = 0$$

Steps 3 and 4. Set each factor equal to zero and solve the resulting equations.

$$x + 4 = 0 \quad \text{or} \quad x - 3 = 0$$
$$x = -4 \qquad\qquad x = 3$$

Step 5. Check the solutions in the original equation. Do this now and verify that the solutions are -4 and 3. The equation's solution set is $\{-4, 3\}$. ■

✔ CHECK POINT 6 Solve: $(x - 5)(x - 2) = 28$.

3 Solve problems with quadratic equations.

Applications of Quadratic Equations Solving quadratic equations by factoring can be used to answer questions about variables contained in mathematical models.

EXAMPLE 7 Modeling Motion

You throw a ball straight up from a rooftop 160 feet high with an initial speed of 48 feet per second. The formula

$$h = -16t^2 + 48t + 160$$

describes the ball's height above the ground, h, in feet, t seconds after you throw it. The ball misses the rooftop on its way down and eventually strikes the ground. The situation is illustrated in Figure 6.1. How long will it take for the ball to hit the ground?

FIGURE 6.1

SOLUTION The ball hits the ground when h, its height above the ground, is 0 feet. Thus, we substitute 0 for h in the given formula and solve for t.

$$h = -16t^2 + 48t + 160 \qquad \text{This is the formula that models the ball's height.}$$
$$0 = -16t^2 + 48t + 160 \qquad \text{Substitute 0 for } h.$$

It is easier to factor a trinomial with a positive leading coefficient. Thus, if a negative squared term appears in a quadratic equation, we make it positive by multiplying both sides of the equation by -1.

$$-1 \cdot 0 = -1(-16t^2 + 48t + 160)$$
$$0 = 16t^2 - 48t - 160$$

Do you see that each term on the right side of the equation changed sign? The left side of the equation remained zero. Now we continue to solve the equation.

$$16t^2 - 48t - 160 = 0$$ Reverse the two sides of the equation. This step is optional.

$$16(t^2 - 3t - 10) = 0$$ Factor out the GCF, 16.

Do not set the constant, 16, equal to zero: $16 \neq 0$.

$$16(t - 5)(t + 2) = 0$$ Factor the trinomial.

$$t - 5 = 0 \quad \text{or} \quad t + 2 = 0$$ Set each variable factor equal to 0.

$$t = 5 \qquad\qquad t = -2$$ Solve for t.

Because we begin describing the ball's height at $t = 0$, we discard the solution $t = -2$. The ball hits the ground after 5 seconds. ■

Figure 6.2 shows the graph of the formula $h = -16t^2 + 48t + 160$. The horizontal axis is labeled t, for the ball's time in motion. The vertical axis is labeled h, for the ball's height above the ground. Because time and height are both positive, the model is graphed in quadrant I only.

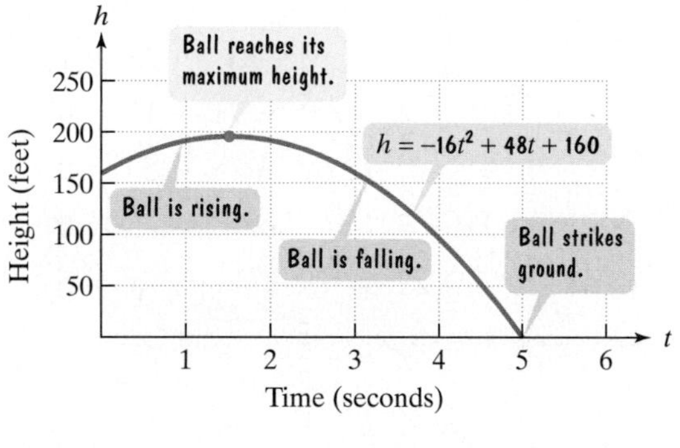

FIGURE 6.2

The graph visually shows what we discovered algebraically: The ball hits the ground after 5 seconds. The graph also reveals that the ball reaches its maximum height, nearly 200 feet, after 1.5 seconds. Then the ball begins to fall.

CHECK POINT 7 Use the formula $h = -16t^2 + 48t + 160$ to determine when the ball's height is 192 feet. Identify your solutions as points on the graph in Figure 6.2.

In our next example, we use our five-step strategy for solving word problems.

EXAMPLE 8 Solving a Problem About a Rectangle's Area

An architect is allowed no more than 15 square meters to add a small bedroom to a house. Because of the room's design in relationship to the existing structure, the width of its rectangular floor must be 7 meters less than two times the length. Find the precise length and width of the rectangular floor of maximum area that the architect is permitted.

SOLUTION

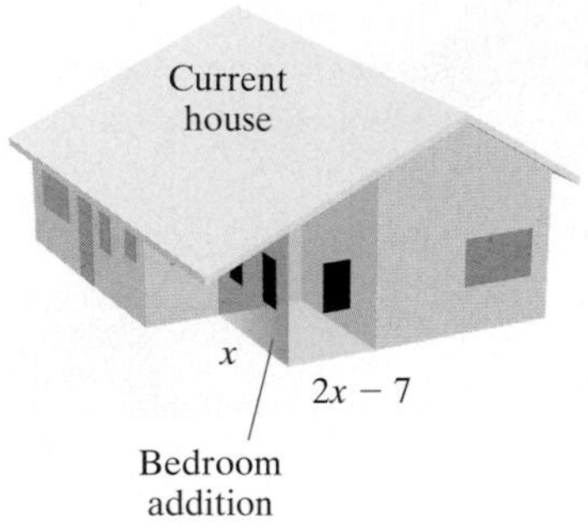

FIGURE 6.3

Step 1. **Let x represent one of the quantities.** We know something about the width: It must be 7 meters less than two times the length. We will let

$$x = \text{the length of the floor.}$$

Step 2. **Represent other quantities in terms of x.** Because the width must be 7 meters less than two times the length, let

$$2x - 7 = \text{the width of the floor.}$$

The problem is illustrated in Figure 6.3.

Step 3. **Write an equation that describes the conditions.** Because the architect is allowed no more than 15 square meters, an area of 15 square meters is the maximum area permitted. The area of a rectangle is the product of its length and its width.

Length of the floor | times | Width of the floor | is | the area.

$$x \cdot (2x - 7) = 15$$

Step 4. **Solve the equation and answer the question.**

$$x(2x - 7) = 15 \quad \text{This is the equation for the problem's conditions.}$$

$$2x^2 - 7x = 15 \quad \text{Use the distributive property.}$$

$$2x^2 - 7x - 15 = 0 \quad \text{Subtract 15 from both sides.}$$

$$(2x + 3)(x - 5) = 0 \quad \text{Factor.}$$

$$2x + 3 = 0 \quad \text{or} \quad x - 5 = 0 \quad \text{Set each factor equal to zero.}$$

$$2x = -3 \qquad x = 5 \quad \text{Solve the resulting equations.}$$

$$x = -\frac{3}{2}$$

A rectangle cannot have a negative length. Thus,

$$\text{Length} = x = 5$$
$$\text{Width} = 2x - 7 = 2 \cdot 5 - 7 = 10 - 7 = 3.$$

The architect is permitted a room of maximum area whose length is 5 meters and whose width is 3 meters.

Step 5. **Check the proposed solution in the original wording of the problem.** The area of the floor using the dimensions that we found is

$$A = lw = (5 \text{ meters})(3 \text{ meters}) = 15 \text{ square meters.}$$

Because the problem's wording tells us that the maximum area permitted is 15 square meters, our dimensions are correct. ■

CHECK POINT 8 The length of a rectangular sign is 3 feet longer than the width. If the sign's area is 54 square feet, find its length and width.

6.6 EXERCISE SET

Student Solutions Manual | CD/Video | PH Math/Tutor Center | MathXL Tutorials on CD | MathXL® | MyMathLab | Interactmath.com

Practice Exercises

In Exercises 1–8, solve each equation using the zero-product principle.

1. $x(x + 7) = 0$ **2.** $x(x - 3) = 0$

3. $(x - 6)(x + 4) = 0$ **4.** $(x - 3)(x + 8) = 0$

5. $(x - 9)(5x + 4) = 0$ **6.** $(x + 7)(3x - 2) = 0$

7. $10(x - 4)(2x + 9) = 0$ **8.** $8(x - 5)(3x + 11) = 0$

In Exercises 9–56, use factoring to solve each quadratic equation. Check by substitution or by using a graphing utility and identifying x-intercepts.

9. $x^2 + 8x + 15 = 0$ **10.** $x^2 + 5x + 6 = 0$

11. $x^2 - 2x - 15 = 0$ **12.** $x^2 + x - 42 = 0$

13. $x^2 - 4x = 21$ **14.** $x^2 + 7x = 18$

15. $x^2 + 9x = -8$ **16.** $x^2 - 11x = -10$

17. $x^2 + 4x = 0$ **18.** $x^2 - 6x = 0$

19. $x^2 - 5x = 0$ **20.** $x^2 + 3x = 0$

21. $x^2 = 4x$ **22.** $x^2 = 8x$

23. $2x^2 = 5x$ **24.** $3x^2 = 5x$

25. $3x^2 = -5x$ **26.** $2x^2 = -3x$

27. $x^2 + 4x + 4 = 0$ **28.** $x^2 + 6x + 9 = 0$

29. $x^2 = 12x - 36$ **30.** $x^2 = 14x - 49$

31. $4x^2 = 12x - 9$ **32.** $9x^2 = 30x - 25$

33. $2x^2 = 7x + 4$ **34.** $3x^2 = x + 4$

35. $5x^2 = 18 - x$ **36.** $3x^2 = 15 + 4x$

37. $x^2 - 49 = 0$ **38.** $x^2 - 25 = 0$

39. $4x^2 - 25 = 0$ **40.** $9x^2 - 100 = 0$

41. $81x^2 = 25$ **42.** $25x^2 = 49$

43. $x(x - 4) = 21$ **44.** $x(x - 3) = 18$

45. $4x(x + 1) = 15$

46. $x(3x + 8) = -5$

47. $(x - 1)(x + 4) = 14$

48. $(x - 3)(x + 8) = -30$

49. $(x + 1)(2x + 5) = -1$

50. $(x + 3)(3x + 5) = 7$

51. $y(y + 8) = 16(y - 1)$

52. $y(y + 9) = 4(2y + 5)$

53. $4y^2 + 20y + 25 = 0$

54. $4y^2 + 44y + 121 = 0$

55. $64w^2 = 48w - 9$ **56.** $25w^2 = 80w - 64$

Practice Plus

In Exercises 57–66, solve each equation and check your solutions.

57. $(x - 4)(x^2 + 5x + 6) = 0$

58. $(x - 5)(x^2 - 3x + 2) = 0$

59. $x^3 - 36x = 0$

60. $x^3 - 4x = 0$

61. $y^3 + 3y^2 + 2y = 0$

62. $y^3 + 2y^2 - 3y = 0$

63. $2(x - 4)^2 + x^2 = x(x + 50) - 46x$

64. $(x - 4)(x - 5) + (2x + 3)(x - 1) = x(2x - 25) - 13$

65. $(x - 2)^2 - 5(x - 2) + 6 = 0$

66. $(x - 3)^2 + 2(x - 3) - 8 = 0$

Application Exercises

A ball is thrown straight up from a rooftop 300 feet high. The formula

$$h = -16t^2 + 20t + 300$$

describes the ball's height above the ground, h, in feet, t seconds after it was thrown. The ball misses the rooftop on its way down and eventually strikes the ground. The graph of the formula is shown, with tick marks omitted along the horizontal axis. Use the formula to solve Exercises 67–69.

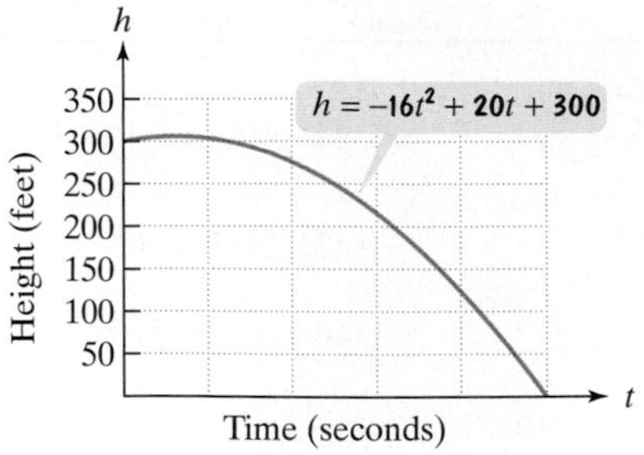

67. How long will it take for the ball to hit the ground? Use this information to provide tick marks with appropriate numbers along the horizontal axis in the figure shown.

68. When will the ball's height be 304 feet? Identify the solution as a point on the graph.

69. When will the ball's height be 276 feet? Identify the solution as a point on the graph.

An explosion causes debris to rise vertically with an initial speed of 72 feet per second. The formula

$$h = -16t^2 + 72t$$

describes the height of the debris above the ground, h, in feet, t seconds after the explosion. Use this information to solve Exercises 70–71.

70. How long will it take for the debris to hit the ground?

71. When will the debris be 32 feet above the ground?

The formula

$$N = 2x^2 + 22x + 320$$

models the number of inmates, N, in thousands, in U.S. state and federal prisons x years after 1980. The graph of the formula is shown in a [0, 20, 1] *by* [0, 1600, 100] *viewing rectangle. Use the formula to solve Exercises 72–73.*

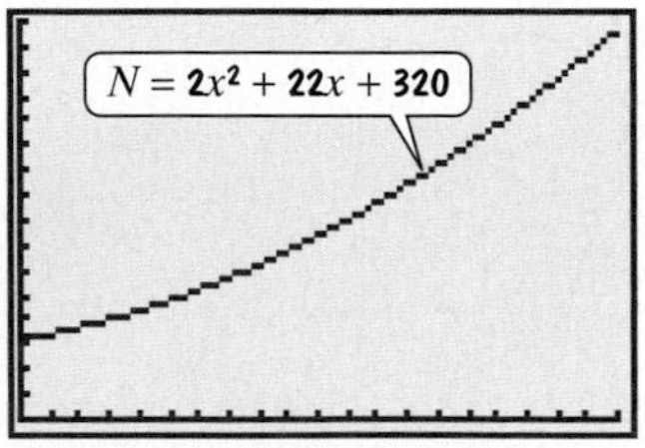

[0, 20, 1] by [0, 1600, 100]

72. In which year were there 740 thousand inmates in U.S. state and federal prisons? Identify the solution as a point on the graph shown.

73. In which year were there 1100 thousand inmates in U.S. state and federal prisons? Identify the solution as a point on the graph shown.

The alligator, an endangered species, is the subject of a protection program. The formula

$$P = -10x^2 + 475x + 3500$$

models the alligator population, P, after x years of the protection program, where $0 \le x \le 12$. *Use the formula to solve Exercises 74–75.*

74. After how long is the population up to 5990?

75. After how long is the population up to 7250?

The graph of the alligator population is shown over time. Use the graph to solve Exercises 76–77.

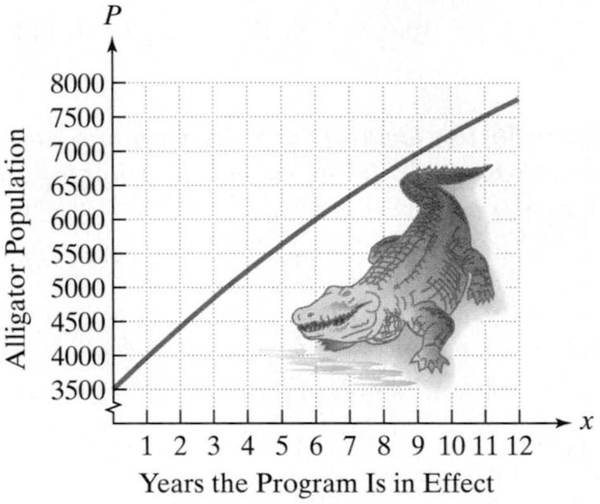

76. Identify your solution in Exercise 74 as a point on the graph.

77. Identify your solution in Exercise 75 as a point on the graph.

The formula

$$N = \frac{t^2 - t}{2}$$

describes the number of football games, N, that must be played in a league with t teams if each team is to play every other team once. Use this information to solve Exercises 78–79.

78. If a league has 36 games scheduled, how many teams belong to the league, assuming that each team plays every other team once?

79. If a league has 45 games scheduled, how many teams belong to the league, assuming that each team plays every other team once?

80. The length of a rectangular garden is 5 feet greater than the width. The area of the rectangle is 300 square feet. Find the length and the width.

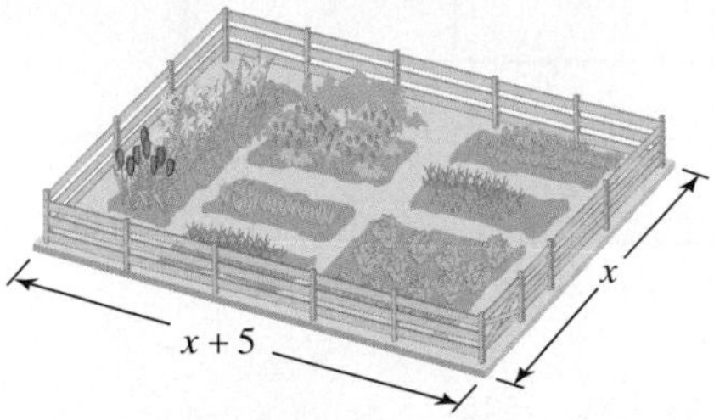

81. A rectangular parking lot has a length that is 3 yards greater than the width. The area of the parking lot is 180 square yards. Find the length and the width.

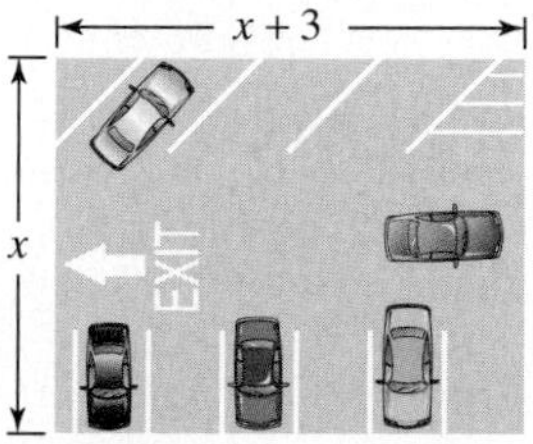

82. Each end of a glass prism is a triangle with a height that is 1 inch shorter than twice the base. If the area of the triangle is 60 square inches, how long are the base and height?

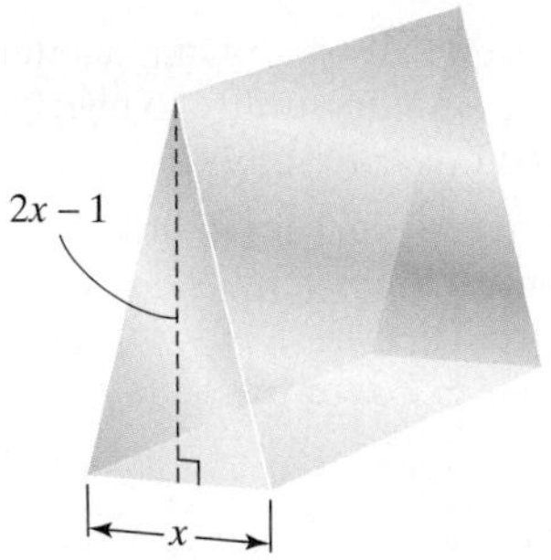

83. Great white sharks have triangular teeth with a height that is 1 centimeter longer than the base. If the area of one tooth is 15 square centimeters, find its base and height.

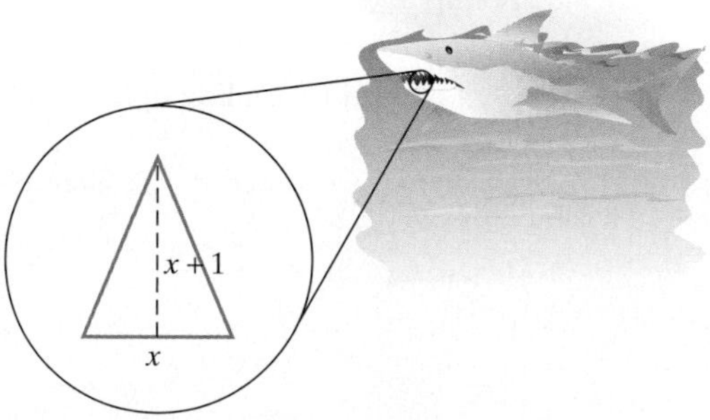

84. A vacant rectangular lot is being turned into a community vegetable garden measuring 15 meters by 12 meters. A path of uniform width is to surround the garden. If the area of the lot is 378 square meters, find the width of the path surrounding the garden.

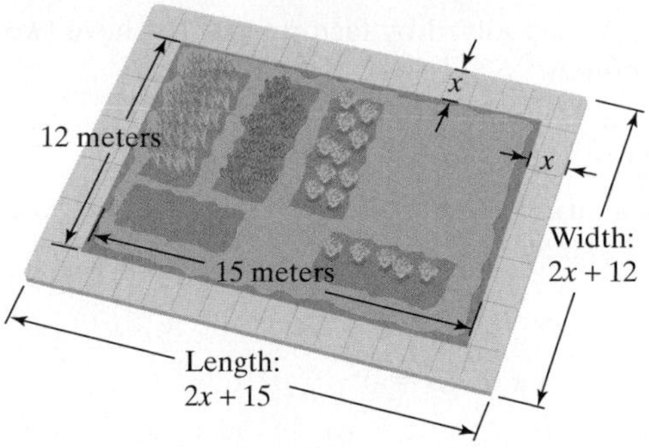

85. As part of a landscaping project, you put in a flower bed measuring 10 feet by 12 feet. You plan to surround the bed with a uniform border of low-growing plants.

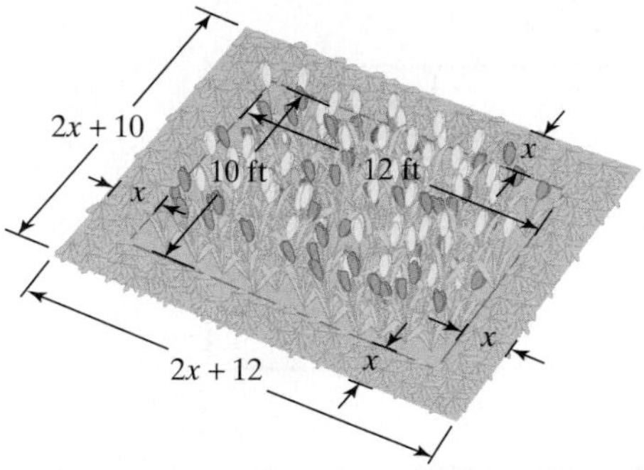

a. Write a polynomial that describes the area of the uniform border that surrounds your flower bed. (*Hint*: The area of the border is the area of the large rectangle shown in the figure minus the area of the flower bed.)

b. The low-growing plants surrounding the flower bed require 1 square foot each when mature. If you have 168 of these plants, how wide a strip around the flower bed should you prepare for the border?

Writing in Mathematics

86. What is a quadratic equation?

87. Explain how to solve $x^2 + 6x + 8 = 0$ using factoring and the zero-product principle.

88. If $(x + 2)(x - 4) = 0$ indicates that $x + 2 = 0$ or $x - 4 = 0$, explain why $(x + 2)(x - 4) = 6$ does not mean $x + 2 = 6$ or $x - 4 = 6$. Could we solve the equation using $x + 2 = 3$ and $x - 4 = 2$ because $3 \cdot 2 = 6$?

Critical Thinking Exercises

89. Which one of the following is true?

a. If $(x + 3)(x - 4) = 2$, then $x + 3 = 0$ or $x - 4 = 0$.

b. The solutions of the equation $4(x - 5)(x + 3) = 0$ are 4, 5, and -3.

c. Equations solved by factoring always have two different solutions.

d. Both 0 and $-\pi$ are solutions of the equation $x(x + \pi) = 0$.

90. Write a quadratic equation in standard form whose solutions are -3 and 5.

In Exercises 91–93, solve each equation.

91. $x^3 - x^2 - 16x + 16 = 0$

92. $3^{x^2-9x+20} = 1$

93. $(x^2 - 5x + 5)^3 = 1$

In Exercises 94–97, match each equation with its graph. The graphs are labeled (a) through (d).

94. $y = x^2 - x - 2$

95. $y = x^2 + x - 2$

96. $y = x^2 - 4$

97. $y = x^2 - 4x$

a.

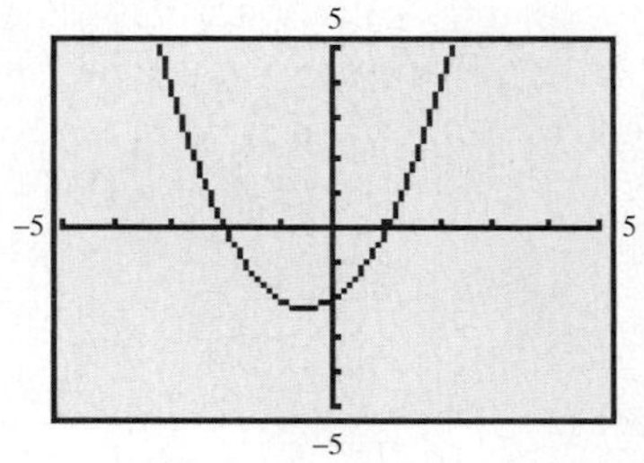

b.

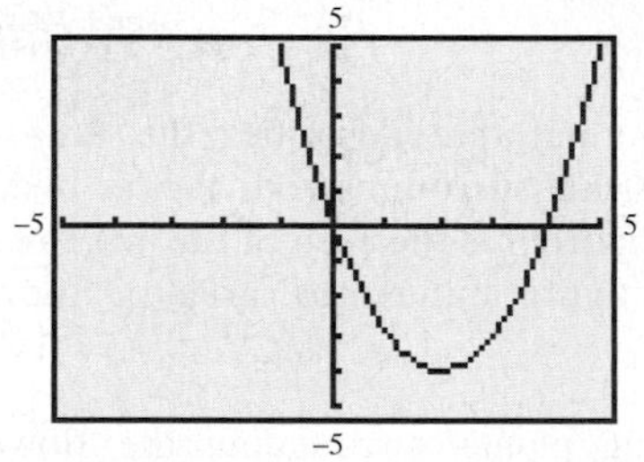

c.

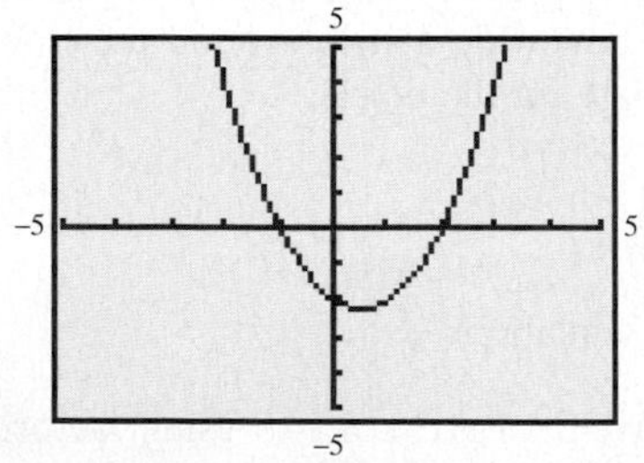

d.

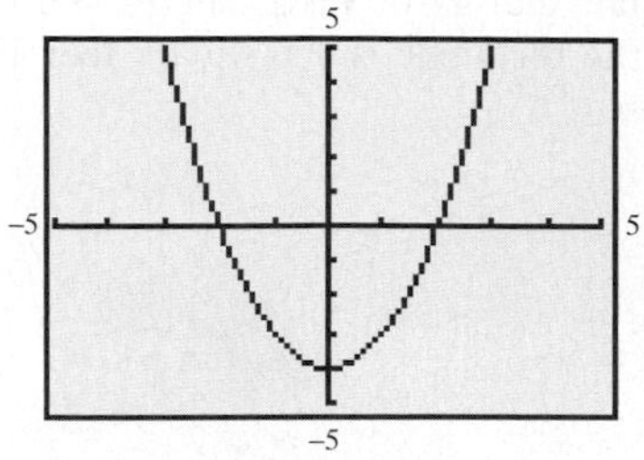

Technology Exercises

In Exercises 98–102, use the x-intercepts for the graph in a $[-10, 10, 1]$ by $[-13, 10, 1]$ viewing rectangle to solve the quadratic equation. Check by substitution.

98. Use the graph of $y = x^2 + 3x - 4$ to solve

$$x^2 + 3x - 4 = 0.$$

99. Use the graph of $y = x^2 + x - 6$ to solve

$$x^2 + x - 6 = 0.$$

100. Use the graph of $y = (x - 2)(x + 3) - 6$ to solve

$$(x - 2)(x + 3) - 6 = 0.$$

101. Use the graph of $y = x^2 - 2x + 1$ to solve

$$x^2 - 2x + 1 = 0.$$

102. Use the technique of identifying x-intercepts on a graph generated by a graphing utility to check any five equations that you solved in Exercises 9–56.

103. If you have access to a calculator that solves quadratic equations, consult the owner's manual to determine how to use this feature. Then use your calculator to solve any five of the equations in Exercises 9–56.

Review Exercises

104. Solve:

$$\begin{aligned} 4x + 2y - z &= 12 \\ 3x - y + z &= 6 \\ x + 3z &= 14 \end{aligned}$$

(Section 4.5, Example 3)

105. Simplify: $\left(\dfrac{8x^4}{4x^7}\right)^2$. (Section 5.7, Example 6)

106. Solve: $5x + 28 = 6 - 6x$. (Section 2.2, Example 7)

GROUP PROJECT

CHAPTER 6

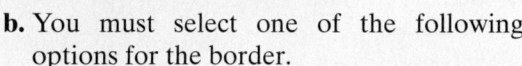

Group members are on the board of a condominium association. The condominium has just installed a small 35-foot-by-30-foot pool. Your job is to choose a material to surround the pool to create a border of uniform width.

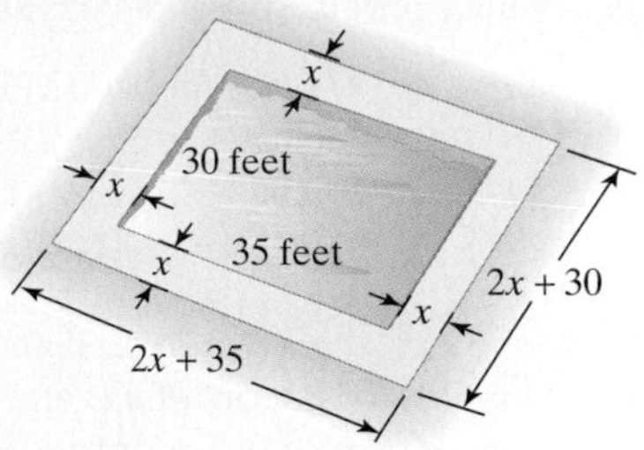

a. Begin by writing an algebraic expression for the area, in square feet, of the border around the pool. (*Hint*: The border's area is the combined area of the pool and border minus the area of the pool.)

b. You must select one of the following options for the border.

Options for the Border	Price
Cement	\$6 per square foot
Outdoor carpeting	\$5 per square foot plus \$10 per foot to install edging around the rectangular border
Brick	\$8 per square foot plus a \$60 charge for delivering the bricks

Write an algebraic expression for the cost of installing the border for each of these options.

c. You would like the border to be 5 feet wide. Use the algebraic expressions in part (b) to find the cost of the border for each of the three options.

d. You would prefer not to use cement. However, the condominium association is limited by a \$5000 budget. Given this limitation, approximately how wide can the border be using outdoor carpeting or brick? Which option should you select and why?

CHAPTER 6 SUMMARY

Definitions and Concepts	Examples
Section 6.1 The Greatest Common Factor and Factoring by Grouping	
Factoring a polynomial containing the sum of monomials means finding an equivalent expression that is a product. The greatest common factor, GCF, is an expression that divides every term of the polynomial. The variable part of the GCF contains the smallest power of a variable that appears in all terms of the polynomial.	Find the GCF of $16x^2y$, $20x^3y^2$, and $8x^2y^3$. The GCF of 16, 20, and 8 is 4. The GCF of x^2, x^3, and x^2 is x^2. The GCF of y, y^2, and y^3 is y. $\text{GCF} = 4 \cdot x^2 \cdot y = 4x^2y$
To factor a monomial from a polynomial, express each term as the product of the GCF and its other factor. Then use the distributive property to factor out the GCF.	$16x^2y + 20x^3y^2 + 8x^2y^3$ $= 4x^2y \cdot 4 + 4x^2y \cdot 5xy + 4x^2y \cdot 2y^2$ $= 4x^2y(4 + 5xy + 2y^2)$
To factor by grouping, factor out the GCF from each group. Then factor out the remaining common factor.	$xy + 5x - 3y - 15$ $= x(y + 5) - 3(y + 5)$ $= (y + 5)(x - 3)$

Definitions and Concepts	Examples
Section 6.2 Factoring Trinomials Whose Leading Coefficient Is One	
To factor a trinomial of the form $x^2 + bx + c$, find two numbers whose product is c and whose sum is b. The factorization is $(x + \text{one number})(x + \text{other number})$.	Factor: $x^2 + 9x + 20$. Find two numbers whose product is 20 and whose sum is 9. The numbers are 4 and 5. $x^2 + 9x + 20 = (x + 4)(x + 5)$
Section 6.3 Factoring Trinomials Whose Leading Coefficient Is Not One	
To factor $ax^2 + bx + c$ by trial and error, try various combinations of factors of ax^2 and c until a middle term of bx is obtained for the sum of outside and inside products.	Factor: $3x^2 + 7x - 6$. Factors of $3x^2$: $3x, x$ Factors of -6: 1 and -6, -1 and 6, 2 and -3, -2 and 3. A possible combination of these factors is $(3x - 2)(x + 3)$. Sum of outside and inside products should equal $7x$. $9x - 2x = 7x$ Thus, $3x^2 + 7x - 6 = (3x - 2)(x + 3)$.
To factor $ax^2 + bx + c$ by grouping, find the factors of ac whose sum is b. Write bx using these factors. Then factor by grouping.	Factor: $3x^2 + 7x - 6$. Find the factors of $3(-6)$, or -18, whose sum is 7. They are 9 and -2. $3x^2 + 7x - 6$ $= 3x^2 + 9x - 2x - 6$ $= 3x(x + 3) - 2(x + 3) = (x + 3)(3x - 2)$
Section 6.4 Factoring Special Forms	
The Difference of Two Squares $A^2 - B^2 = (A + B)(A - B)$	$9x^2 - 25y^2$ $= (3x)^2 - (5y)^2 = (3x + 5y)(3x - 5y)$
Perfect Square Trinomials $A^2 + 2AB + B^2 = (A + B)^2$ $A^2 - 2AB + B^2 = (A - B)^2$	$x^2 + 16x + 64 = x^2 + 2 \cdot x \cdot 8 + 8^2 = (x + 8)^2$ $25x^2 - 30x + 9 = (5x)^2 - 2 \cdot 5x \cdot 3 + 3^2 = (5x - 3)^2$
Sum and Difference of Cubes $A^3 + B^3 = (A + B)(A^2 - AB + B^2)$ $A^3 - B^3 = (A - B)(A^2 + AB + B^2)$	$8x^3 - 125 = (2x)^3 - 5^3$ $= (2x - 5)[(2x)^2 + 2x \cdot 5 + 5^2]$ $= (2x - 5)(4x^2 + 10x + 25)$

Definitions and Concepts	Examples
Section 6.5 A General Factoring Strategy	
A Factoring Strategy **1.** Factor out the GCF. **2. a.** If two terms, try $A^2 - B^2 = (A + B)(A - B)$ $A^3 + B^3 = (A + B)(A^2 - AB + B^2)$ $A^3 - B^3 = (A - B)(A^2 + AB + B^2)$. **b.** If three terms, try $A^2 + 2AB + B^2 = (A + B)^2$ $A^2 - 2AB + B^2 = (A - B)^2$. If not a perfect square trinomial, try trial and error or grouping. **c.** If four terms, try factoring by grouping. **3.** See if any factors can be factored further. **4.** Check by multiplying.	Factor: $2x^4 + 10x^3 - 8x^2 - 40x$. The GCF is $2x$. $2x^4 + 10x^3 - 8x^2 - 40x$ $= 2x(x^3 + 5x^2 - 4x - 20)$ Four terms: Try grouping. $= 2x[x^2(x + 5) - 4(x + 5)]$ $= 2x(x + 5)(x^2 - 4)$ This can be factored further. $= 2x(x + 5)(x + 2)(x - 2)$
Section 6.6 Solving Quadratic Equations by Factoring	
The Zero-Product Principle If $AB = 0$, then $A = 0$ or $B = 0$.	Solve: $(x - 6)(x + 10) = 0$ $x - 6 = 0$ or $x + 10 = 0$ $x = 6$ $\quad x = -10$ The solutions are -10 and 6, and the solution set is $\{-10, 6\}$.
A quadratic equation in x is an equation that can be written in the standard form $ax^2 + bx + c = 0, \quad a \neq 0.$ To solve by factoring, write the equation in standard form, factor, set each factor equal to zero, and solve each resulting equation. Check proposed solutions in the original equation.	Solve: $4x^2 + 9x = 9$. $4x^2 + 9x - 9 = 0$ $(4x - 3)(x + 3) = 0$ $4x - 3 = 0$ or $x + 3 = 0$ $x = \frac{3}{4}$ $\quad x = -3$ The solutions are -3 and $\frac{3}{4}$, and the solution set is $\left\{-3, \frac{3}{4}\right\}$.

CHAPTER 6 REVIEW EXERCISES

6.1 *In Exercises 1–5, factor each polynomial by using the greatest common factor. If there is no common factor other than 1 and the polynomial cannot be factored, so state.*

1. $30x - 45$

2. $12x^3 + 16x^2 - 400x$

3. $30x^4y + 15x^3y + 5x^2y$

4. $7(x + 3) - 2(x + 3)$

5. $7x^2(x + y) - (x + y)$

In Exercises 6–9, factor by grouping.

6. $x^3 + 3x^2 + 2x + 6$

7. $xy + y + 4x + 4$

8. $x^3 + 5x + x^2 + 5$

9. $xy + 4x - 2y - 8$

6.2 *In Exercises 10–17, factor completely, or state that the trinomial is prime.*

10. $x^2 - 3x + 2$

11. $x^2 - x - 20$

12. $x^2 + 19x + 48$

13. $x^2 - 6xy + 8y^2$

14. $x^2 + 5x - 9$

15. $x^2 + 16xy - 17y^2$

16. $3x^2 + 6x - 24$

17. $3x^3 - 36x^2 + 33x$

6.3 *In Exercises 18–26, factor completely, or state that the trinomial is prime.*

18. $3x^2 + 17x + 10$

19. $5y^2 - 17y + 6$

20. $4x^2 + 4x - 15$

21. $5y^2 + 11y + 4$

22. $8x^2 + 8x - 6$

23. $2x^3 + 7x^2 - 72x$

24. $12y^3 + 28y^2 + 8y$

25. $2x^2 - 7xy + 3y^2$

26. $5x^2 - 6xy - 8y^2$

6.4 *In Exercises 27–30, factor each difference of two squares completely.*

27. $4x^2 - 1$

28. $81 - 100y^2$

29. $25a^2 - 49b^2$

30. $z^4 - 16$

In Exercises 31–34, factor completely, or state that the polynomial is prime.

31. $2x^2 - 18$

32. $x^2 + 1$

33. $9x^3 - x$

34. $18xy^2 - 8x$

In Exercises 35–41, factor any perfect square trinomials, or state that the polynomial is prime.

35. $x^2 + 22x + 121$

36. $x^2 - 16x + 64$

37. $9y^2 + 48y + 64$

38. $16x^2 - 40x + 25$

39. $25x^2 + 15x + 9$

40. $36x^2 + 60xy + 25y^2$

41. $25x^2 - 40xy + 16y^2$

In Exercises 42–45, factor using the formula for the sum or difference of two cubes.

42. $x^3 - 27$

43. $64x^3 + 1$

44. $54x^3 - 16y^3$

45. $27x^3y + 8y$

In Exercises 46–47, find the formula for the area of the shaded region and express it in factored form.

46.

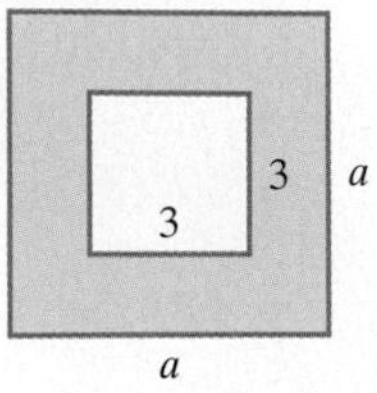

47.

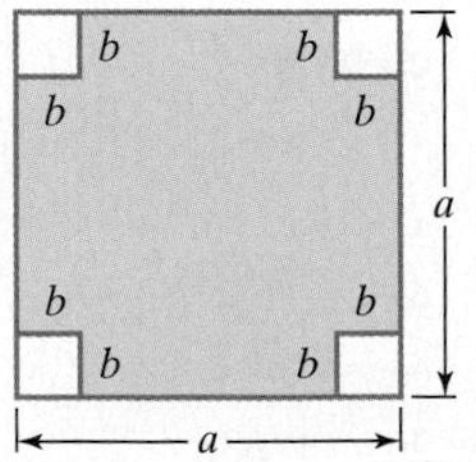

48. The figure shows a geometric interpretation of a factorization. Use the sum of the areas of the four pieces on the left and the area of the square on the right to write the factorization that is illustrated.

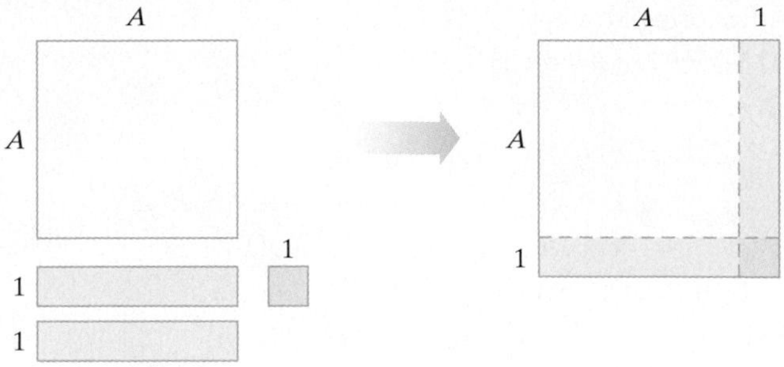

6.5 *In Exercises 49–81, factor completely, or state that the polynomial is prime.*

49. $x^3 - 8x^2 + 7x$

50. $10y^2 + 9y + 2$

51. $128 - 2y^2$

52. $9x^2 + 6x + 1$

53. $20x^7 - 36x^3$

54. $x^3 - 3x^2 - 9x + 27$

55. $y^2 + 16$

56. $2x^3 + 19x^2 + 35x$

57. $3x^3 - 30x^2 + 75x$

58. $3x^5 - 24x^2$

59. $4y^4 - 36y^2$

60. $5x^2 + 20x - 105$

61. $9x^2 + 8x - 3$

62. $10x^5 - 44x^4 + 16x^3$

63. $100y^2 - 49$

64. $9x^5 - 18x^4$

65. $x^4 - 1$

66. $2y^3 - 16$

67. $x^3 + 64$

68. $6x^2 + 11x - 10$

69. $3x^4 - 12x^2$

70. $x^2 - x - 90$

71. $25x^2 + 25xy + 6y^2$

72. $x^4 + 125x$

73. $32y^3 + 32y^2 + 6y$

74. $2y^2 - 16y + 32$

75. $x^2 - 2xy - 35y^2$

76. $x^2 + 7x + xy + 7y$

77. $9x^2 + 24xy + 16y^2$

78. $2x^4y - 2x^2y$

79. $100y^2 - 49z^2$

80. $x^2 + xy + y^2$

81. $3x^4y^2 - 12x^2y^4$

6.6 *In Exercises 82–83, solve each equation using the zero-product principle.*

82. $x(x - 12) = 0$

83. $3(x - 7)(4x + 9) = 0$

In Exercises 84–92, use factoring to solve each quadratic equation.

84. $x^2 + 5x - 14 = 0$

85. $5x^2 + 20x = 0$

86. $2x^2 + 15x = 8$

87. $x(x - 4) = 32$

88. $(x + 3)(x - 2) = 50$

89. $x^2 = 14x - 49$

90. $9x^2 = 100$

91. $3x^2 + 21x + 30 = 0$

92. $3x^2 = 22x - 7$

93. You dive from a board that is 32 feet above the water. The formula

$$h = -16t^2 + 16t + 32$$

describes your height above the water, h, in feet, t seconds after you dive. How long will it take you to hit the water?

94. The length of a rectangular sign is 3 feet longer than the width. If the sign has space for 40 square feet of advertising, find its length and its width.

95. The square lot shown here is being turned into a garden with a 3-meter path at one end. If the area of the garden is 88 square meters, find the dimensions of the square lot.

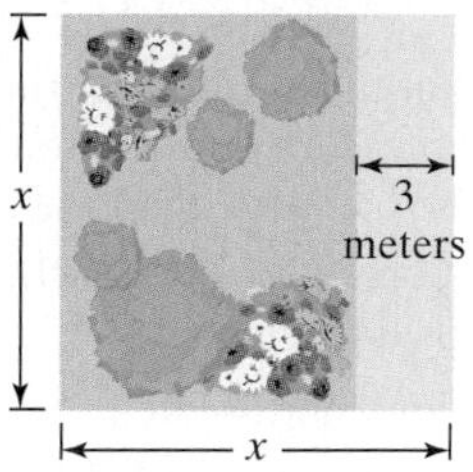

CHAPTER 6 TEST

Remember to use your Chapter Test Prep Video CD to see the worked-out solutions to the test questions you want to review.

In Exercises 1–21, factor completely, or state that the polynomial is prime.

1. $x^2 - 9x + 18$

2. $x^2 - 14x + 49$

3. $15y^4 - 35y^3 + 10y^2$

4. $x^3 + 2x^2 + 3x + 6$

5. $x^2 - 9x$

6. $x^3 + 6x^2 - 7x$

7. $14x^2 + 64x - 30$

8. $25x^2 - 9$

9. $x^3 + 8$

10. $x^2 - 4x - 21$

11. $x^2 + 4$

12. $6y^3 + 9y^2 + 3y$

13. $4y^2 - 36$

14. $16x^2 + 48x + 36$

15. $2x^4 - 32$

16. $36x^2 - 84x + 49$

17. $7x^2 - 50x + 7$

18. $x^3 + 2x^2 - 5x - 10$

19. $12y^3 - 12y^2 - 45y$

20. $y^3 - 125$

21. $5x^2 - 5xy - 30y^2$

In Exercises 22–27, solve each quadratic equation.

22. $x^2 + 2x - 24 = 0$

23. $3x^2 - 5x = 2$

24. $x(x - 6) = 16$

25. $6x^2 = 21x$

26. $16x^2 = 81$

27. $(5x + 4)(x - 1) = 2$

28. Find a formula for the area of the shaded region and express it in factored form.

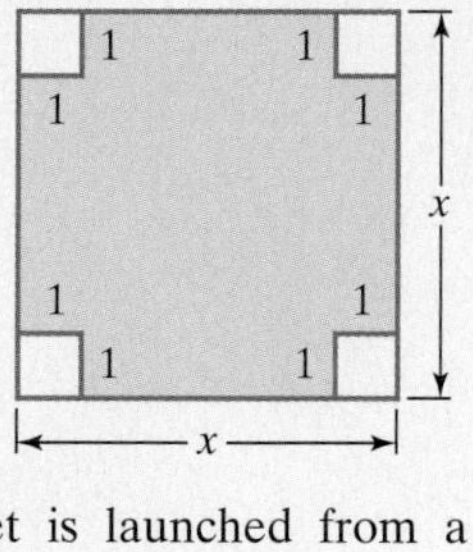

29. A model rocket is launched from a height of 96 feet. The formula

$$h = -16t^2 + 80t + 96$$

describes the rocket's height, h, in feet, t seconds after it was launched. How long will it take the rocket to reach the ground?

30. The length of a rectangular garden is 6 feet longer than its width. If the area of the garden is 55 square feet, find its length and its width.

CUMULATIVE REVIEW EXERCISES (CHAPTERS 1–6)

1. Simplify: $6[5 + 2(3 - 8) - 3]$.

2. Solve: $4(x - 2) = 2(x - 4) + 3x$.

3. Solve: $\frac{x}{2} - 1 = \frac{x}{3} + 1$.

4. Solve and express the solution set in set-builder notation. Graph the solution set on a number line.

$$5 - 5x > 2(5 - x) + 1$$

5. Find the measures of the angles of a triangle whose two base angles have equal measure and whose third angle is 10° less than three times the measure of a base angle.

6. A dinner for six people cost \$159, including a 6% tax. What was the dinner's cost before tax?

7. Graph using the slope and y-intercept: $y = -\frac{3}{5}x + 3$.

8. Write the point-slope form of the line passing through $(2, -4)$ and $(3, 1)$. Then use the point-slope form of the equation to write the slope-intercept equation.

9. Solve the system:

$$\begin{aligned} x - 2y + 2z &= 4 \\ 3x - y + 4z &= 4 \\ 2x + y - 3z &= 5. \end{aligned}$$

10. Solve the system:

$$\begin{aligned} 5x + 2y &= 13 \\ y &= 2x - 7. \end{aligned}$$

11. Solve the system:

$$\begin{aligned} 2x + 3y &= 5 \\ 3x - 2y &= -4. \end{aligned}$$

12. Subtract: $\frac{4}{5} - \frac{9}{8}$.

In Exercises 13–15, perform the indicated operations.

13. $\dfrac{6x^5 - 3x^4 + 9x^2 + 27x}{3x}$

14. $(3x - 5y)(2x + 9y)$

15. $\dfrac{6x^3 + 5x^2 - 34x + 13}{3x - 5}$

16. Write 0.0071 in scientific notation.

In Exercises 17–19, factor completely.

17. $3x^2 + 11x + 6$

18. $y^5 - 16y$

19. $4x^2 + 12x + 9$

20. The length of a rectangle is 2 feet greater than its width. If the rectangle's area is 24 square feet, find its dimensions.

At the start of the twenty-first century, we are plagued by questions about the environment. Will we run out of gas? How hot will it get? Will there be neighborhoods where the air is pristine? Can we make garbage disappear? Will there be any wilderness left? Which wild animals will become extinct? How much will it cost to clean up toxic wastes from our rivers so that they can safely provide food, recreation, and enjoyment of wildlife for the millions who live along and visit their shores?

The role of algebraic fractions in modeling environmental issues is introduced in Section 7.1 and developed in Example 6 in Section 7.6.

CHAPTER 7

Rational Expressions

When making decisions on public policies dealing with the environment, two important questions are

- What are the costs?
- What are the benefits?

Algebraic fractions play an important role in modeling the costs. By learning to work with these fractional expressions, you will gain new insights into phenomena as diverse as the dosage of drugs prescribed for children, inventory costs for a business, the cost of environmental cleanup, and even the shape of our heads.

SECTION 7.1

Objectives

1. Find numbers for which a rational expression is undefined.
2. Simplify rational expressions.
3. Solve applied problems involving rational expressions.

RATIONAL EXPRESSIONS AND THEIR SIMPLIFICATION

How do we describe the costs of reducing environmental pollution? We often use algebraic expressions involving quotients of polynomials. For example, the algebraic expression

$$\frac{250x}{100 - x}$$

describes the cost, in millions of dollars, to remove x percent of the pollutants that are discharged into a river. Removing a modest percentage of pollutants, say 40%, is far less costly than removing a substantially greater percentage, such as 95%. We see this by evaluating the algebraic expression for $x = 40$ and $x = 95$.

Evaluating $\frac{250x}{100 - x}$ for

$x = 40$: Cost is $\frac{250(40)}{100 - 40} \approx 167.$

$x = 95$: Cost is $\frac{250(95)}{100 - 95} = 4750.$

The cost increases from approximately \$167 million to a possibly prohibitive \$4750 million, or \$4.75 billion. Costs spiral upward as the percentage of removed pollutants increases.

Many algebraic expressions that describe costs of environmental projects are examples of *rational expressions*. In this section, we introduce rational expressions and their simplification.

DISCOVER FOR YOURSELF

What happens if you try substituting 100 for x in

$$\frac{250x}{100 - x}?$$

What does this tell you about the cost of cleaning up all of the river's pollutants?

1 Find numbers for which a rational expression is undefined.

Excluding Numbers from Rational Expressions A **rational expression** is the quotient of two polynomials. Some examples are

$$\frac{x-2}{4}, \quad \frac{4}{x-2}, \quad \frac{x}{x^2-1}, \quad \text{and} \quad \frac{x^2+1}{x^2+2x-3}.$$

Rational expressions indicate division, and division by zero is undefined. This means that **we must exclude any value or values of the variable that make a denominator zero**. For example, consider the rational expression

$$\frac{4}{x-2}.$$

When x is replaced with 2, the denominator is 0 and the expression is undefined.

$$\text{If } x = 2: \quad \frac{4}{x-2} = \frac{4}{2-2} = \frac{4}{0}$$

Division by zero is undefined.

Notice that if x is replaced by a number other than 2, such as 1, the expression is defined because the denominator is nonzero.

$$\text{If } x = 1: \quad \frac{4}{x-2} = \frac{4}{1-2} = \frac{4}{-1} = -4.$$

Thus, only 2 must be excluded as a replacement for x in the rational expression $\frac{4}{x-2}$.

USING TECHNOLOGY

We can use the TABLE feature of a graphing utility to verify our work with $\frac{4}{x-2}$. Enter

$$y_1 = 4 \; \boxed{\div} \; \boxed{(} \; \boxed{x} \; \boxed{-} \; 2 \; \boxed{)}$$

and press TABLE.

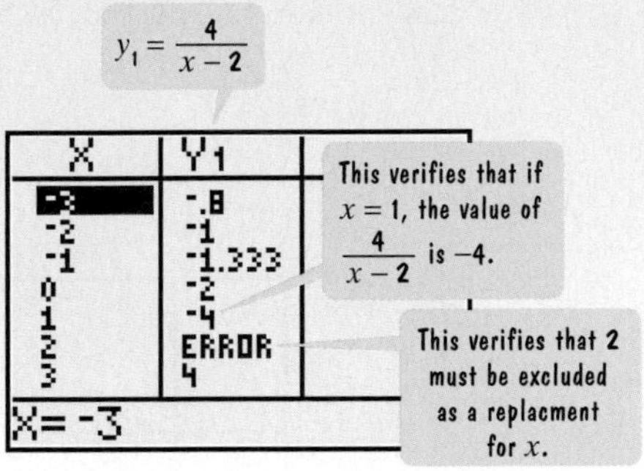
$y_1 = \frac{4}{x-2}$
X | Y1
-3 | -.8
-2 | -1
-1 | -1.333
0 | -2
1 | -4
2 | ERROR
3 | 4
X=-3
This verifies that if $x = 1$, the value of $\frac{4}{x-2}$ is -4.
This verifies that 2 must be excluded as a replacement for x.

EXCLUDING VALUES FROM RATIONAL EXPRESSIONS If a variable in a rational expression is replaced by a number that causes the denominator to be 0, that number must be excluded as a replacement for the variable. The rational expression is undefined at any value that produces a denominator of 0.

How do we determine the value or values of the variable for which a rational expression is undefined? Set the denominator equal to 0 and then solve the resulting equation for the variable.

EXAMPLE 1 Determining Numbers for Which Rational Expressions Are Undefined

Find all the numbers for which the rational expression is undefined:

a. $\frac{6x+12}{7x-28}$ **b.** $\frac{2x+6}{x^2+3x-10}$.

SOLUTION In each case, we set the denominator equal to 0 and solve.

$$\frac{6x+12}{7x-28} \qquad \frac{2x+6}{x^2+3x-10}$$

Exclude values of x that make these denominators 0.

a. $7x - 28 = 0$ — Set the denominator of $\frac{6x+12}{7x-28}$ equal to 0.

$7x = 28$ — Add 28 to both sides.

$x = 4$ — Divide both sides by 7.

Thus, $\frac{6x+12}{7x-28}$ is undefined for $x = 4$.

b. $x^2 + 3x - 10 = 0$ — Set the denominator of $\frac{2x+6}{x^2+3x-10}$ equal to 0.

$(x+5)(x-2) = 0$ — Factor.

$x + 5 = 0$ or $x - 2 = 0$ — Set each factor equal to 0.

$x = -5$ $\quad x = 2$ — Solve the resulting equations.

Thus, $\frac{2x+6}{x^2+3x-10}$ is undefined for $x = -5$ and $x = 2$.

USING TECHNOLOGY

When using a graphing utility to graph an equation containing a rational expression, you might not be pleased with the quality of the display. Compare these two graphs of $y = \frac{6x+12}{7x-28}$.

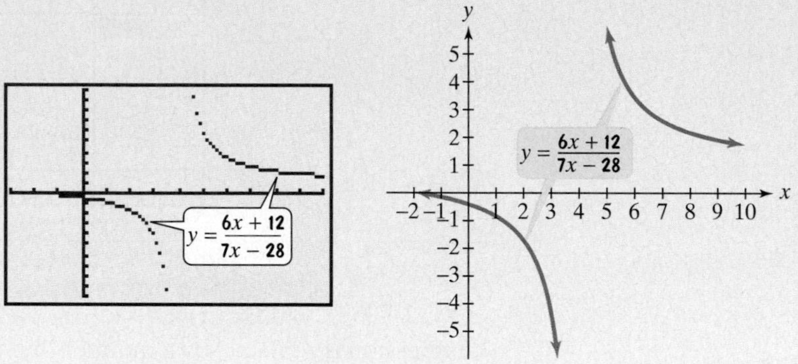

The graph on the left was obtained using the DOT mode in a $[-3, 10, 1]$ by $[-10, 10, 1]$ viewing rectangle. Examine the behavior of the graph near $x = 4$, the number for which the rational expression is undefined. The values of the rational expression are decreasing as the values of x get closer to 4 on the left and increasing as the values of x get closer to 4 on the right. However, there is no point on the graph corresponding to $x = 4$. Would you agree that this behavior is better illustrated in the hand-drawn graph on the right?

CHECK POINT 1 Find all the numbers for which the rational expression is undefined:

a. $\frac{7x-28}{8x-40}$ **b.** $\frac{8x-40}{x^2+3x-28}$.

Is every rational expression undefined for at least one number? No. Consider

$$\frac{x-2}{4}.$$

Because the denominator is not zero for any value of x, the rational expression is defined for all real numbers. Thus, it is not necessary to exclude any values for x.

2 Simplify rational expressions.

Simplifying Rational Expressions A rational expression is **simplified** if its numerator and denominator have no common factors other than 1 or -1. The following principle is used to simplify a rational expression:

FUNDAMENTAL PRINCIPLE OF RATIONAL EXPRESSIONS If P, Q, and R are polynomials, and Q and R are not 0,

$$\frac{PR}{QR} = \frac{P}{Q}.$$

As you read the Fundamental Principle, can you see why $\frac{PR}{QR}$ is not simplified? The numerator and denominator have a common factor, the polynomial R. By dividing the numerator and the denominator by the common factor, R, we obtain the simplified form $\frac{P}{Q}$. This is often shown as follows:

$$\frac{P\overset{1}{\cancel{R}}}{Q\underset{1}{\cancel{R}}} = \frac{P}{Q}.$$

Observe that $\frac{PR}{QR} = \frac{P}{Q} \cdot \frac{R}{R} = \frac{P}{Q} \cdot 1 = \frac{P}{Q}$.

The following procedure can be used to simplify rational expressions:

SIMPLIFYING RATIONAL EXPRESSIONS

1. Factor the numerator and the denominator completely.
2. Divide both the numerator and the denominator by any common factors.

EXAMPLE 2 Simplifying a Rational Expression

Simplify: $\frac{5x + 35}{20x}$.

SOLUTION

$$\frac{5x + 35}{20x} = \frac{5(x + 7)}{5 \cdot 4x}$$ Factor the numerator and denominator. Because the denominator is $20x$, $x \neq 0$.

$$= \frac{\overset{1}{\cancel{5}}(x + 7)}{\underset{1}{\cancel{5}} \cdot 4x}$$ Divide out the common factor of 5.

$$= \frac{x + 7}{4x}$$ ■

CHECK POINT 2 Simplify: $\frac{7x + 28}{21x}$.

EXAMPLE 3 Simplifying a Rational Expression

Simplify: $\dfrac{x^3 + x^2}{x + 1}$.

SOLUTION

$$\frac{x^3 + x^2}{x + 1} = \frac{x^2(x + 1)}{x + 1}$$

Factor the numerator. Because the denominator is $x + 1$, $x \neq -1$.

$$= \frac{x^2\overset{1}{\cancel{(x + 1)}}}{\underset{1}{\cancel{x + 1}}}$$

Divide out the common factor of $x + 1$.

$$= x^2$$

Simplifying a rational expression can change the numbers that make it undefined. For example, we just showed that

$$\frac{x^3 + x^2}{x + 1} = x^2.$$

This is undefined for $x = -1$.

This simplified form is defined for all real numbers.

Thus, to equate the two expressions, we must restrict the values for x in the simplified expression to exclude -1. We can write

$$\frac{x^3 + x^2}{x + 1} = x^2, \quad x \neq -1.$$

Hereafter, we will assume that the simplified rational expression is equal to the original rational expression for all real numbers except those for which either denominator is 0.

USING TECHNOLOGY

A graphing utility can be used to verify that

$$\frac{x^3 + x^2}{x + 1} = x^2, \quad x \neq -1.$$

Enter $y_1 = \dfrac{x^3 + x^2}{x + 1}$ and $y_2 = x^2$.

Graphic Check

The graphs of y_1 and y_2 appear to be identical. You can use the TRACE feature to trace y_1 and show that it is undefined for $x = -1$.

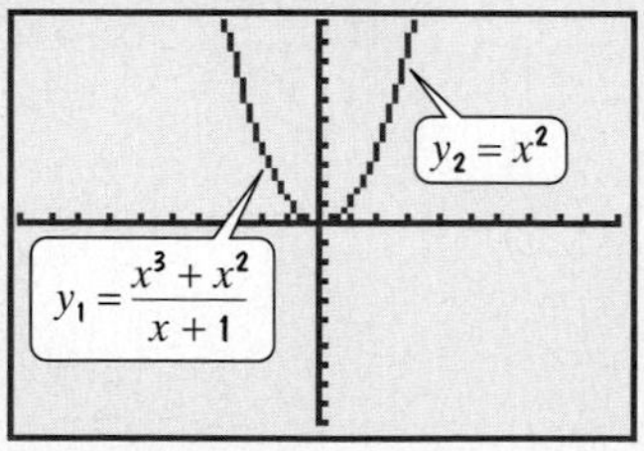

$[-10, 10, 1]$ by $[-10, 10, 1]$

Numeric Check

No matter how far up or down we scroll, if $x \neq -1$, $y_1 = y_2$. If $x = -1$, y_1 is undefined, although the value of y_2 is 1.

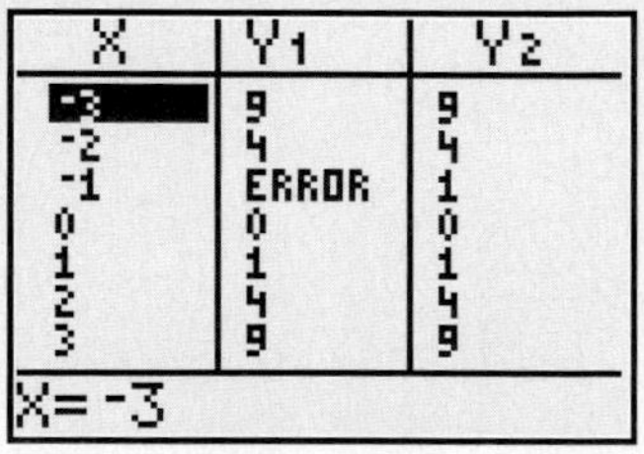

CHECK POINT 3 Simplify: $\dfrac{x^3 - x^2}{7x - 7}$.

EXAMPLE 4 Simplifying a Rational Expression

Simplify: $\frac{x^2 + 6x + 5}{x^2 - 25}$.

SOLUTION

$$\frac{x^2 + 6x + 5}{x^2 - 25} = \frac{(x+5)(x+1)}{(x+5)(x-5)}$$

Factor the numerator and denominator. Because the denominator is $(x + 5)(x - 5)$, $x \neq -5$ and $x \neq 5$.

$$= \frac{\overset{1}{\cancel{(x+5)}}(x+1)}{\underset{1}{\cancel{(x+5)}}(x-5)}$$

Divide out the common factor of $x + 5$.

$$= \frac{x+1}{x-5}$$

■

CHECK POINT 4 Simplify: $\frac{x^2 - 1}{x^2 + 2x + 1}$.

STUDY TIP

When simplifying rational expressions, only *factors* that are common to the *entire numerator* and the *entire denominator* can be divided out. **It is incorrect to divide out common terms from the numerator and denominator.**

Incorrect!

$$\frac{\overset{1}{\cancel{x}} + 5}{\underset{1}{\cancel{x}} + 7} = \frac{5}{7} \qquad \frac{x^2 - \overset{1}{\cancel{4}}}{\underset{1}{\cancel{4}}} = x^2 - 1 \qquad \frac{\overset{x}{\cancel{x^2}} - \overset{3}{\cancel{9}}}{\underset{1}{\cancel{x}} - \underset{1}{\cancel{3}}} = x - 3$$

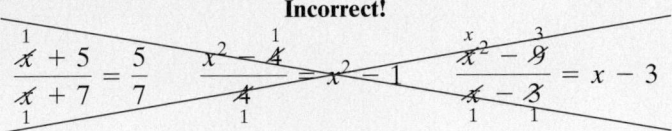

The first two expressions, $\frac{x+5}{x+7}$ and $\frac{x^2 - 4}{4}$, have no common factors in their numerators and denominators. Thus, these rational expressions are in simplified form. The rational expression $\frac{x^2 - 9}{x - 3}$ can be simplified as follows:

Correct

$$\frac{x^2 - 9}{x - 3} = \frac{(x+3)\overset{1}{\cancel{(x-3)}}}{\underset{1}{\cancel{x-3}}} = x + 3.$$

Divide out the common factor, $x - 3$.

Factors That Are Opposites How do we simplify rational expressions that contain factors in the numerator and denominator that are opposites, or additive inverses? Here is an example of such an expression:

$$\frac{x - 3}{3 - x}.$$

The numerator and denominator are opposites. They differ only in their signs.

Factor out -1 from either the numerator or the denominator. Then divide out the common factor.

$$\frac{x-3}{3-x} = \frac{-1(-x+3)}{3-x}$$

Factor -1 from the numerator. Notice how the sign of each term in the polynomial $x - 3$ changes.

$$= \frac{-1(3-x)}{3-x}$$

In the numerator, use the commutative property to rewrite $-x + 3$ as $3 - x$.

$$= \frac{-1\overset{1}{\cancel{(3-x)}}}{\underset{1}{\cancel{3-x}}}$$

Divide out the common factor of $3 - x$.

$$= -1$$

Our result, -1, suggests a useful property that is stated at the top of the next page.

SIMPLIFYING RATIONAL EXPRESSIONS WITH OPPOSITE FACTORS IN THE NUMERATOR AND DENOMINATOR The quotient of two polynomials that have opposite signs and are additive inverses is -1.

EXAMPLE 5 Simplifying a Rational Expression

Simplify: $\dfrac{4x^2 - 25}{15 - 6x}$.

SOLUTION

$$\frac{4x^2 - 25}{15 - 6x} = \frac{(2x + 5)(2x - 5)}{3(5 - 2x)}$$ Factor the numerator and denominator.

$$= \frac{(2x + 5)\overset{-1}{\cancel{(2x - 5)}}}{3\cancel{(5 - 2x)}}$$ The quotient of polynomials with opposite signs is -1.

$$= \frac{-(2x + 5)}{3} \text{ or } -\frac{2x + 5}{3} \text{ or } \frac{-2x - 5}{3}$$

Each of these forms is an acceptable answer. ■

3 Solve applied problems involving rational expressions.

✔ **CHECK POINT 5** Simplify: $\dfrac{9x^2 - 49}{28 - 12x}$.

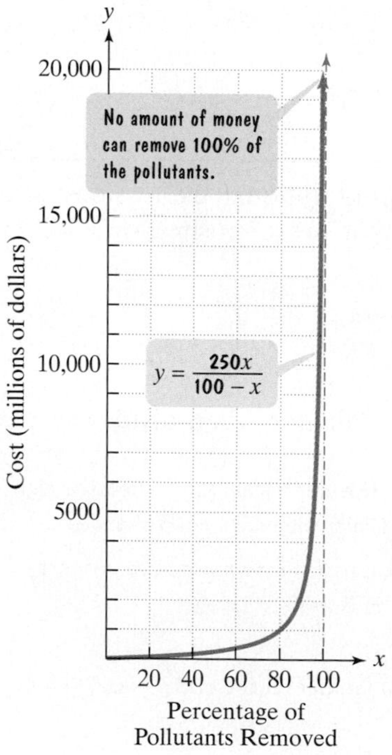

FIGURE 7.1

Applications The equation

$$y = \frac{250x}{100 - x}$$

models the cost, in millions of dollars, to remove x percent of the pollutants that are discharged into a river. This equation contains the rational expression that we looked at in the opening to this section. Do you remember how costs were spiraling upward as the percentage of removed pollutants increased?

Is it possible to clean up the river completely? To do this, we must remove 100% of the pollutants. The problem is that the rational expression is undefined for $x = 100$.

$$y = \frac{250x}{100 - x}$$ If $x = 100$, the value of the denominator is 0.

Notice how the graph of $y = \dfrac{250x}{100 - x}$, shown in Figure 7.1, approaches but never touches the dashed vertical line $x = 100$, our undefined value. The graph continues to rise more and more steeply, visually showing the escalating costs. By never touching the dashed vertical line, the graph illustrates that no amount of money will be enough to remove all pollutants from the river.

7.1 EXERCISE SET

Student Solutions Manual | CD/Video | PH Math/Tutor Center | MathXL Tutorials on CD | MathXL® | MyMathLab | Interactmath.com

Practice Exercises

In Exercises 1–20, find all numbers for which each rational expression is undefined. If the rational expression is defined for all real numbers, so state.

1. $\frac{5}{2x}$ **2.** $\frac{11}{3x}$

3. $\frac{x}{x-8}$ **4.** $\frac{x}{x-6}$

5. $\frac{13}{5x-20}$ **6.** $\frac{17}{6x-30}$

7. $\frac{x+3}{(x+9)(x-2)}$ **8.** $\frac{x+5}{(x+7)(x-9)}$

9. $\frac{4x}{(3x-17)(x+3)}$ **10.** $\frac{8x}{(4x-19)(x+2)}$

11. $\frac{x+5}{x^2+x-12}$ **12.** $\frac{7x-14}{x^2-9x+20}$

13. $\frac{x+5}{5}$ **14.** $\frac{x+7}{7}$

15. $\frac{y+3}{4y^2+y-3}$ **16.** $\frac{y+8}{6y^2-y-2}$

17. $\frac{y+5}{y^2-25}$ **18.** $\frac{y+7}{y^2-49}$

19. $\frac{5}{x^2+1}$ **20.** $\frac{8}{x^2+4}$

In Exercises 21–76, simplify each rational expression. If the rational expression cannot be simplified, so state.

21. $\frac{14x^2}{7x}$ **22.** $\frac{9x^2}{6x}$

23. $\frac{5x-15}{25}$ **24.** $\frac{7x+21}{49}$

25. $\frac{2x-8}{4x}$ **26.** $\frac{3x-9}{6x}$

27. $\frac{3}{3x-9}$ **28.** $\frac{12}{6x-18}$

29. $\frac{-15}{3x-9}$ **30.** $\frac{-21}{7x-14}$

31. $\frac{3x+9}{x+3}$ **32.** $\frac{5x-10}{x-2}$

33. $\frac{x+5}{x^2-25}$ **34.** $\frac{x+4}{x^2-16}$

35. $\frac{2y-10}{3y-15}$ **36.** $\frac{6y+18}{11y+33}$

37. $\frac{x+1}{x^2-2x-3}$ **38.** $\frac{x+2}{x^2-x-6}$

39. $\frac{4x-8}{x^2-4x+4}$ **40.** $\frac{x^2-12x+36}{4x-24}$

41. $\frac{y^2-3y+2}{y^2+7y-18}$ **42.** $\frac{y^2+5y+4}{y^2-4y-5}$

43. $\frac{2y^2-7y+3}{2y^2-5y+2}$ **44.** $\frac{3y^2+4y-4}{6y^2-y-2}$

45. $\frac{2x+3}{2x+5}$ **46.** $\frac{3x+7}{3x+10}$

47. $\frac{x^2+12x+36}{x^2-36}$ **48.** $\frac{x^2-14x+49}{x^2-49}$

49. $\frac{x^3-2x^2+x-2}{x-2}$ **50.** $\frac{x^3+4x^2-3x-12}{x+4}$

51. $\frac{x^3-8}{x-2}$ **52.** $\frac{x^3-125}{x^2-25}$

53. $\frac{(x-4)^2}{x^2-16}$ **54.** $\frac{(x+5)^2}{x^2-25}$

55. $\frac{x}{x+1}$ **56.** $\frac{x}{x+7}$

57. $\frac{x+4}{x^2+16}$ **58.** $\frac{x+5}{x^2+25}$

59. $\frac{x-5}{5-x}$ **60.** $\frac{x-7}{7-x}$

61. $\frac{2x-3}{3-2x}$ **62.** $\frac{5x-4}{4-5x}$

63. $\frac{x-5}{x+5}$ **64.** $\frac{x-7}{x+7}$

65. $\frac{4x-6}{3-2x}$ **66.** $\frac{9x-15}{5-3x}$

67. $\frac{4-6x}{3x^2-2x}$ **68.** $\frac{9-15x}{5x^2-3x}$

69. $\frac{x^2-1}{1-x}$ **70.** $\frac{x^2-4}{2-x}$

71. $\frac{y^2 - y - 12}{4 - y}$

72. $\frac{y^2 - 7y + 12}{3 - y}$

73. $\frac{x^2y - x^2}{x^3 - x^3y}$

74. $\frac{xy - 2x}{3y - 6}$

75. $\frac{x^2 + 2xy - 3y^2}{2x^2 + 5xy - 3y^2}$

76. $\frac{x^2 + 3xy - 10y^2}{3x^2 - 7xy + 2y^2}$

Practice Plus

In Exercises 77–84, simplify each rational expression.

77. $\frac{x^2 - 9x + 18}{x^3 - 27}$

78. $\frac{x^3 - 8}{x^2 + 2x - 8}$

79. $\frac{9 - y^2}{y^2 - 3(2y - 3)}$

80. $\frac{16 - y^2}{y(y - 8) + 16}$

81. $\frac{xy + 2y + 3x + 6}{x^2 + 5x + 6}$

82. $\frac{xy + 4y - 7x - 28}{x^2 + 11x + 28}$

83. $\frac{8x^2 + 4x + 2}{1 - 8x^3}$

84. $\frac{x^3 - 3x^2 + 9x}{x^3 + 27}$

Application Exercises

85. The rational expression

$$\frac{130x}{100 - x}$$

describes the cost, in millions of dollars, to inoculate x percent of the population against a particular strain of flu.

a. Evaluate the expression for $x = 40$, $x = 80$, and $x = 90$. Describe the meaning of each evaluation in terms of percentage inoculated and cost.

b. For what value of x is the expression undefined?

c. What happens to the cost as x approaches 100%? How can you interpret this observation?

86. The rational expression

$$\frac{60{,}000x}{100 - x}$$

describes the cost, in dollars, to remove x percent of the air pollutants in the smokestack emission of a utility company that burns coal to generate electricity.

a. Evaluate the expression for $x = 20$, $x = 50$, and $x = 80$. Describe the meaning of each evaluation in terms of percentage of pollutants removed and cost.

b. For what value of x is the expression undefined?

c. What happens to the cost as x approaches 100%? How can you interpret this observation?

Doctors use the rational expression

$$\frac{DA}{A + 12}$$

to determine the dosage of a drug prescribed for children. In this expression, A = the child's age and D = the adult dosage. Use the expression to solve Exercises 87–88.

87. If the normal adult dosage of medication is 1000 milligrams, what dosage should an 8-year-old child receive?

88. If the normal adult dosage of medication is 1000 milligrams, what dosage should a 4-year-old child receive?

89. A company that manufactures bicycles has costs given by the equation

$$C = \frac{100x + 100{,}000}{x}$$

in which x is the number of bicycles manufactured and C is the cost to manufacture each bicycle.

a. Find the cost per bicycle when manufacturing 500 bicycles.

b. Find the cost per bicycle when manufacturing 4000 bicycles.

c. Does the cost per bicycle increase or decrease as more bicycles are manufactured? Explain why this happens.

90. A company that manufactures small canoes has costs given by the equation

$$C = \frac{20x + 20{,}000}{x}$$

in which x is the number of canoes manufactured and C is the cost to manufacture each canoe.

a. Find the cost per canoe when manufacturing 100 canoes.

b. Find the cost per canoe when manufacturing 10,000 canoes.

c. Does the cost per canoe increase or decrease as more canoes are manufactured? Explain why this happens.

A drug is injected into a patient and the concentration of the drug in the bloodstream is monitored. The drug's concentration, y, in milligrams per liter, after x hours is modeled by

$$y = \frac{5x}{x^2 + 1}.$$

The graph of this equation, obtained with a graphing utility, is shown in the figure in a [0, 10, 1] *by* [0, 3, 1] *viewing rectangle. Use this information to solve Exercises 91–92.*

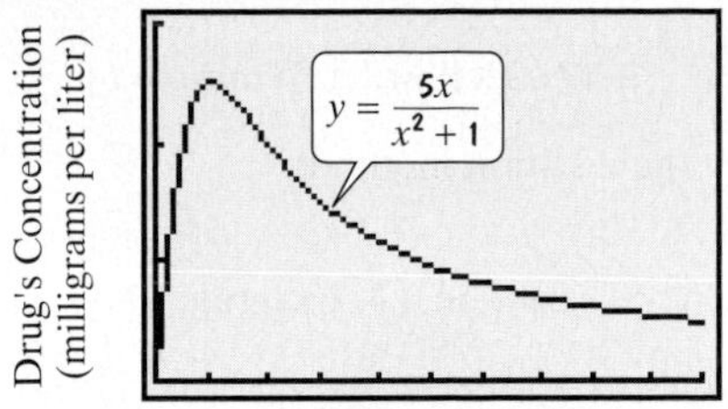

[0, 10, 1] by [0, 3, 1]

91. Use the equation to find the drug's concentration after 3 hours. Then identify the point on the equation's graph that conveys this information.

92. Use the graph of the equation to find after how many hours the drug reaches its maximum concentration. Then use the equation to find the drug's concentration at this time.

Body-mass index takes both weight and height into account when assessing whether an individual is underweight or overweight. The formula for body-mass index, BMI, *is*

$$\text{BMI} = \frac{703w}{h^2},$$

where w is weight, in pounds, and h is height, in inches. In adults, normal values for the BMI *are between 20 and 25, inclusive. Values below 20 indicate that an individual is underweight and values above 30 indicate that an individual is obese. Use this information to solve Exercises 93–94.*

93. Calculate the BMI, to the nearest tenth, for a 145-pound person who is 5 feet 10 inches tall. Is this person underweight?

94. Calculate the BMI, to the nearest tenth, for a 150-pound person who is 5 feet 6 inches tall. Is this person overweight?

95. The bar graph shows the total number of crimes in the United States, in millions, from 1995 through 2001.

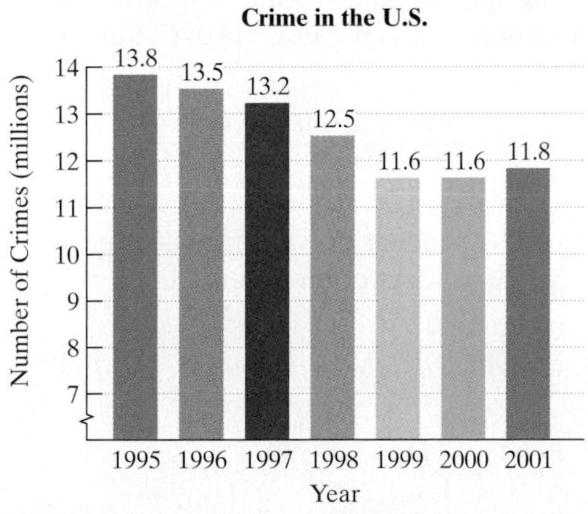

Source: FBI

The polynomial $3.7t + 257.4$ describes the U.S. population, in millions, t years after 1994. The polynomial $-0.4t + 14.2$ describes the number of crimes in the United States, in millions, t years after 1994.

a. Write a rational expression that describes the crime rate in the United States t years after 1994.

b. According to the rational expression in part (a), what was the crime rate in 2001? Round to two decimal places. How many crimes does this indicate per 100,000 inhabitants?

c. According to the FBI, there were 4161 crimes per 100,000 U.S. inhabitants in 2001. How well does the rational expression that you evaluated in part (b) model this number?

Writing in Mathematics

96. What is a rational expression? Give an example with your explanation.

97. Explain how to find the number or numbers, if any, for which a rational expression is undefined.

98. Explain how to simplify a rational expression.

99. Explain how to simplify a rational expression with opposite factors in the numerator and denominator.

100. A politician claims that each year the crime rate in the United States is decreasing. Explain how to use the polynomials in Exercise 95 to verify this claim.

101. Use the graph shown for Exercises 91–92 to write a description of the drug's concentration over time. In your description, try to convey as much information as possible that is displayed visually by the graph.

Critical Thinking Exercises

102. Which one of the following is true?

a. $\frac{x + 5}{x} = 5$ **b.** $\frac{x^2 + 3}{3} = x^2 + 1$

c. $\frac{3x + 9}{3x + 13} = \frac{9}{13}$

d. The expression $\frac{-3y - 6}{y + 2}$ reduces to the consecutive integer that follows -4.

103. Write a rational expression that cannot be simplified.

104. Write a rational expression that is undefined for $x = -4$.

105. Write a rational expression with $x^2 - x - 6$ in the numerator that can be simplified to $x - 3$.

Technology Exercises

In Exercises 106–109, use the GRAPH *or* TABLE *feature of a graphing utility to determine if the rational expression has been correctly simplified. If the simplification is wrong, correct it and then verify your answer using the graphing utility.*

106. $\frac{3x + 15}{x + 5} = 3, \quad x \neq -5$

107. $\dfrac{2x^2 - x - 1}{x - 1} = 2x^2 - 1, \ x \neq 1$

108. $\dfrac{x^2 - x}{x} = x^2 - 1, \quad x \neq 0$

109. Use a graphing utility to verify the graph in Figure 7.1 on page 448. TRACE along the graph as x approaches 100. What do you observe?

Review Exercises

110. Multiply: $\dfrac{5}{6} \cdot \dfrac{9}{25}$. (Section 1.1, Example 5)

111. Divide: $\dfrac{2}{3} \div 4$. (Section 1.1, Example 6)

112. Solve by the addition method:

$2x - 5y = -2$

$3x + 4y = 20$. (Section 4.3, Example 3)

SECTION 7.2

MULTIPLYING AND DIVIDING RATIONAL EXPRESSIONS

Objectives

1 Multiply rational expressions.

2 Divide rational expressions.

Highbrow wit in conjunction with lowbrow comedy characterize Stephen Sondheim's *A Funny Thing Happened on the Way to the Forum*. The musical is based on the plays of Plautus, comic dramatist of ancient Rome.

Your psychology class is learning various techniques to double what we remember over time. At the beginning of the course, students memorize 40 words in Latin, a language with which they are not familiar. The rational expression

$$\frac{5t + 30}{t}$$

models the class average for the number of words remembered after t days, where $t \geq 1$. If the techniques are successful, what will be the new memory model?

The new model can be found by multiplying the given rational expression by 2. In this section, you will see that we multiply rational expressions in the same way that we multiply rational numbers. Thus, we multiply numerators and multiply denominators. The rational expression for doubling what the class remembers over time is

$$\frac{2}{1} \cdot \frac{5t + 30}{t} = \frac{2(5t + 30)}{1 \cdot t} = \frac{2 \cdot 5t + 2 \cdot 30}{t} = \frac{10t + 60}{t}.$$

1 Multiply rational expressions.

Multiplying Rational Expressions The product of two rational expressions is the product of their numerators divided by the product of their denominators.

MULTIPLYING RATIONAL EXPRESSIONS If P, Q, R, and S are polynomials, where $Q \neq 0$ and $S \neq 0$, then

$$\frac{P}{Q} \cdot \frac{R}{S} = \frac{PR}{QS}.$$

EXAMPLE 1 Multiplying Rational Expressions

Multiply: $\frac{7}{x+3} \cdot \frac{x-2}{5}$.

SOLUTION

$$\frac{7}{x+3} \cdot \frac{x-2}{5} = \frac{7(x-2)}{(x+3)5}$$ Multiply numerators. Multiply denominators. ($x \neq -3$)

$$= \frac{7x-14}{5x+15}$$ ■

✔ **CHECK POINT 1** Multiply: $\frac{9}{x+4} \cdot \frac{x-5}{2}$.

Here is a step-by-step procedure for multiplying rational expressions. Before multiplying, divide out any factors common to both a numerator and a denominator.

MULTIPLYING RATIONAL EXPRESSIONS

1. Factor all numerators and denominators completely.
2. Divide numerators and denominators by common factors.
3. Multiply the remaining factors in the numerators and multiply the remaining factors in the denominators.

EXAMPLE 2 Multiplying Rational Expressions

Multiply: $\frac{x-3}{x+5} \cdot \frac{10x+50}{7x-21}$.

SOLUTION

$$\frac{x-3}{x+5} \cdot \frac{10x+50}{7x-21}$$

$$= \frac{x-3}{x+5} \cdot \frac{10(x+5)}{7(x-3)}$$ Factor as many numerators and denominators as possible.

$$= \frac{\overset{1}{\cancel{x-3}}}{\underset{1}{\cancel{x+5}}} \cdot \frac{10\overset{1}{\cancel{(x+5)}}}{7\underset{1}{\cancel{(x-3)}}}$$ Divide numerators and denominators by common factors.

$$= \frac{10}{7}$$ Multiply the remaining factors in the numerators and denominators. ■

✔ **CHECK POINT 2** Multiply: $\frac{x+4}{x-7} \cdot \frac{3x-21}{8x+32}$.

EXAMPLE 3 Multiplying Rational Expressions

Multiply: $\frac{x-7}{x-1} \cdot \frac{x^2-1}{3x-21}$.

SOLUTION

$$\frac{x-7}{x-1} \cdot \frac{x^2-1}{3x-21}$$

$$= \frac{x-7}{x-1} \cdot \frac{(x+1)(x-1)}{3(x-7)}$$ Factor as many numerators and denominators as possible.

$$= \frac{\overset{1}{\cancel{x-7}}}{\underset{1}{\cancel{x-1}}} \cdot \frac{(x+1)\overset{1}{\cancel{(x-1)}}}{3\underset{1}{\cancel{(x-7)}}}$$ Divide numerators and denominators by common factors.

$$= \frac{x+1}{3}$$ Multiply the remaining factors in the numerators and denominators. ■

CHECK POINT 3 Multiply: $\frac{x-5}{x-2} \cdot \frac{x^2-4}{9x-45}$.

EXAMPLE 4 Multiplying Rational Expressions

Multiply: $\frac{4x+8}{6x-3x^2} \cdot \frac{3x^2-4x-4}{9x^2-4}$.

SOLUTION

$$\frac{4x+8}{6x-3x^2} \cdot \frac{3x^2-4x-4}{9x^2-4}$$

$$= \frac{4(x+2)}{3x(2-x)} \cdot \frac{(3x+2)(x-2)}{(3x+2)(3x-2)}$$ Factor as many numerators and denominators as possible.

$$= \frac{4(x+2)}{3x\underset{1}{\cancel{(2-x)}}} \cdot \frac{\overset{1}{\cancel{(3x+2)}}\overset{-1}{\cancel{(x-2)}}}{\underset{1}{\cancel{(3x+2)}}(3x-2)}$$ Divide numerators and denominators by common factors. Because $2-x$ and $x-2$ have opposite signs, their quotient is -1.

$$= \frac{-4(x+2)}{3x(3x-2)} \text{ or } -\frac{4(x+2)}{3x(3x-2)}$$ Multiply the remaining factors in the numerators and denominators.

It is not necessary to carry out these multiplications.

■

CHECK POINT 4 Multiply: $\frac{5x+5}{7x-7x^2} \cdot \frac{2x^2+x-3}{4x^2-9}$.

2 Divide rational expressions.

Dividing Rational Expressions The quotient of two rational expressions is the product of the first expression and the multiplicative inverse, or reciprocal, of the second. The reciprocal is found by interchanging the numerator and the denominator.

DIVIDING RATIONAL EXPRESSIONS If P, Q, R, and S are polynomials, where $Q \neq 0$, $R \neq 0$, and $S \neq 0$, then

$$\frac{P}{Q} \div \frac{R}{S} = \frac{P}{Q} \cdot \frac{S}{R} = \frac{PS}{QR}.$$

Change division to multiplication.

Replace $\frac{R}{S}$ with its reciprocal by interchanging numerator and denominator.

Thus, **we find the quotient of two rational expressions by inverting the divisor and multiplying**. For example,

$$\frac{x}{7} \div \frac{6}{y} = \frac{x}{7} \cdot \frac{y}{6} = \frac{xy}{42}.$$

Change the division to multiplication.

Replace $\frac{6}{y}$ with its reciprocal by interchanging numerator and denominator.

STUDY TIP

When performing operations with rational expressions, if a rational expression is written without a denominator, it is helpful to write the expression with a denominator of 1. In Example 5, we wrote $x + 5$ as

$$\frac{x+5}{1}.$$

EXAMPLE 5 Dividing Rational Expressions

Divide: $(x + 5) \div \frac{x-2}{x+9}$.

SOLUTION

$$(x + 5) \div \frac{x-2}{x+9} = \frac{x+5}{1} \cdot \frac{x+9}{x-2}$$ Invert the divisor and multiply.

$$= \frac{(x+5)(x+9)}{x-2}$$ Multiply the factors in the numerators and denominators. We need not carry out the multiplication in the numerator. ■

✔ **CHECK POINT 5** Divide: $(x + 3) \div \frac{x-4}{x+7}$.

EXAMPLE 6 Dividing Rational Expressions

Divide: $\frac{x^2-2x-8}{x^2-9} \div \frac{x-4}{x+3}$.

SOLUTION

$$\frac{x^2-2x-8}{x^2-9} \div \frac{x-4}{x+3}$$

$$= \frac{x^2-2x-8}{x^2-9} \cdot \frac{x+3}{x-4}$$ Invert the divisor and multiply.

$$= \frac{(x-4)(x+2)}{(x+3)(x-3)} \cdot \frac{x+3}{x-4}$$ Factor as many numerators and denominators as possible.

$$= \frac{\overset{1}{\cancel{(x-4)}}(x+2)}{\underset{1}{\cancel{(x+3)}}(x-3)} \cdot \frac{\overset{1}{\cancel{(x+3)}}}{\underset{1}{\cancel{(x-4)}}}$$ Divide numerators and denominators by common factors.

$$= \frac{x+2}{x-3}$$ Multiply the remaining factors in the numerators and the denominators. ■

CHECK POINT 6 Divide: $\dfrac{x^2 + 5x + 6}{x^2 - 25} \div \dfrac{x + 2}{x + 5}$.

EXAMPLE 7 **Dividing Rational Expressions**

Divide: $\dfrac{y^2 + 7y + 12}{y^2 + 9} \div (7y^2 + 21y)$.

SOLUTION

$$\frac{y^2 + 7y + 12}{y^2 + 9} \div \frac{7y^2 + 21y}{1}$$ It is helpful to write the divisor with a denominator of 1.

$$= \frac{y^2 + 7y + 12}{y^2 + 9} \cdot \frac{1}{7y^2 + 21y}$$ Invert the divisor and multiply.

$$= \frac{(y + 4)(y + 3)}{y^2 + 9} \cdot \frac{1}{7y(y + 3)}$$ Factor as many numerators and denominators as possible.

$$= \frac{(y + 4)\overset{1}{\cancel{(y + 3)}}}{y^2 + 9} \cdot \frac{1}{7y\underset{1}{\cancel{(y + 3)}}}$$ Divide numerators and denominators by common factors.

$$= \frac{y + 4}{7y(y^2 + 9)}$$ Multiply the remaining factors in the numerators and the denominators. ■

CHECK POINT 7 Divide: $\dfrac{y^2 + 3y + 2}{y^2 + 1} \div (5y^2 + 10y)$.

7.2 EXERCISE SET

Student Solutions Manual | CD/Video | PH Math/Tutor Center | MathXL Tutorials on CD | MathXL® | MyMathLab | Interactmath.com

Practice Exercises

In Exercises 1–32, multiply as indicated.

1. $\dfrac{4}{x + 3} \cdot \dfrac{x - 5}{9}$

2. $\dfrac{8}{x - 2} \cdot \dfrac{x + 5}{3}$

3. $\dfrac{x}{3} \cdot \dfrac{12}{x + 5}$

4. $\dfrac{x}{5} \cdot \dfrac{30}{x - 4}$

5. $\dfrac{3}{x} \cdot \dfrac{4x}{15}$

6. $\dfrac{7}{x} \cdot \dfrac{5x}{35}$

7. $\dfrac{x - 3}{x + 5} \cdot \dfrac{4x + 20}{9x - 27}$

8. $\dfrac{x - 2}{x + 9} \cdot \dfrac{5x + 45}{2x - 4}$

9. $\dfrac{x^2 + 9x + 14}{x + 7} \cdot \dfrac{1}{x + 2}$

10. $\dfrac{x^2 + 9x + 18}{x + 6} \cdot \dfrac{1}{x + 3}$

11. $\dfrac{x^2 - 25}{x^2 - 3x - 10} \cdot \dfrac{x + 2}{x}$

12. $\dfrac{x^2 - 49}{x^2 - 4x - 21} \cdot \dfrac{x + 3}{x}$

13. $\dfrac{4y + 30}{y^2 - 3y} \cdot \dfrac{y - 3}{2y + 15}$

14. $\dfrac{9y + 21}{y^2 - 2y} \cdot \dfrac{y - 2}{3y + 7}$

15. $\dfrac{y^2 - 7y - 30}{y^2 - 6y - 40} \cdot \dfrac{2y^2 + 5y + 2}{2y^2 + 7y + 3}$

16. $\dfrac{3y^2 + 17y + 10}{3y^2 - 22y - 16} \cdot \dfrac{y^2 - 4y - 32}{y^2 - 8y - 48}$

17. $(y^2 - 9) \cdot \dfrac{4}{y - 3}$

18. $(y^2 - 16) \cdot \dfrac{3}{y - 4}$

19. $\dfrac{x^2 - 5x + 6}{x^2 - 2x - 3} \cdot \dfrac{x^2 - 1}{x^2 - 4}$

20. $\dfrac{x^2 + 5x + 6}{x^2 + x - 6} \cdot \dfrac{x^2 - 9}{x^2 - x - 6}$

21. $\dfrac{x^3 - 8}{x^2 - 4} \cdot \dfrac{x + 2}{3x}$

22. $\dfrac{x^2 + 6x + 9}{x^3 + 27} \cdot \dfrac{1}{x + 3}$

23. $\dfrac{(x - 2)^3}{(x - 1)^3} \cdot \dfrac{x^2 - 2x + 1}{x^2 - 4x + 4}$

24. $\dfrac{(x + 4)^3}{(x + 2)^3} \cdot \dfrac{x^2 + 4x + 4}{x^2 + 8x + 16}$

25. $\frac{6x+2}{x^2-1} \cdot \frac{1-x}{3x^2+x}$

26. $\frac{8x+2}{x^2-9} \cdot \frac{3-x}{4x^2+x}$

27. $\frac{25-y^2}{y^2-2y-35} \cdot \frac{y^2-8y-20}{y^2-3y-10}$

28. $\frac{2y}{3y-y^2} \cdot \frac{2y^2-9y+9}{8y-12}$

29. $\frac{x^2-y^2}{x} \cdot \frac{x^2+xy}{x+y}$

30. $\frac{4x-4y}{x} \cdot \frac{x^2+xy}{x^2-y^2}$

31. $\frac{x^2+2xy+y^2}{x^2-2xy+y^2} \cdot \frac{4x-4y}{3x+3y}$

32. $\frac{x^2-y^2}{x+y} \cdot \frac{x+2y}{2x^2-xy-y^2}$

In Exercises 33–64, divide as indicated.

33. $\frac{x}{7} \div \frac{5}{3}$

34. $\frac{x}{3} \div \frac{3}{8}$

35. $\frac{3}{x} \div \frac{12}{x}$

36. $\frac{x}{5} \div \frac{20}{x}$

37. $\frac{15}{x} \div \frac{3}{2x}$

38. $\frac{9}{x} \div \frac{3}{4x}$

39. $\frac{x+1}{3} \div \frac{3x+3}{7}$

40. $\frac{x+5}{7} \div \frac{4x+20}{9}$

41. $\frac{7}{x-5} \div \frac{28}{3x-15}$

42. $\frac{4}{x-6} \div \frac{40}{7x-42}$

43. $\frac{x^2-4}{x} \div \frac{x+2}{x-2}$

44. $\frac{x^2-4}{x-2} \div \frac{x+2}{4x-8}$

45. $(y^2-16) \div \frac{y^2+3y-4}{y^2+4}$

46. $(y^2+4y-5) \div \frac{y^2-25}{y+7}$

47. $\frac{y^2-y}{15} \div \frac{y-1}{5}$

48. $\frac{y^2-2y}{15} \div \frac{y-2}{5}$

49. $\frac{4x^2+10}{x-3} \div \frac{6x^2+15}{x^2-9}$

50. $\frac{x^2+x}{x^2-4} \div \frac{x^2-1}{x^2+5x+6}$

51. $\frac{x^2-25}{2x-2} \div \frac{x^2+10x+25}{x^2+4x-5}$

52. $\frac{x^2-4}{x^2+3x-10} \div \frac{x^2+5x+6}{x^2+8x+15}$

53. $\frac{y^3+y}{y^2-y} \div \frac{y^3-y^2}{y^2-2y+1}$

54. $\frac{3y^2-12}{y^2+4y+4} \div \frac{y^3-2y^2}{y^2+2y}$

55. $\frac{y^2+5y+4}{y^2+12y+32} \div \frac{y^2-12y+35}{y^2+3y-40}$

56. $\frac{y^2+4y-21}{y^2+3y-28} \div \frac{y^2+14y+48}{y^2+4y-32}$

57. $\frac{2y^2-128}{y^2+16y+64} \div \frac{y^2-6y-16}{3y^2+30y+48}$

58. $\frac{3y+12}{y^2+3y} \div \frac{y^2+y-12}{9y-y^3}$

59. $\frac{2x+2y}{3} \div \frac{x^2-y^2}{x-y}$

60. $\frac{5x+5y}{7} \div \frac{x^2-y^2}{x-y}$

61. $\frac{x^2-y^2}{8x^2-16xy+8y^2} \div \frac{4x-4y}{x+y}$

62. $\frac{4x^2-y^2}{x^2+4xy+4y^2} \div \frac{4x-2y}{3x+6y}$

63. $\frac{xy-y^2}{x^2+2x+1} \div \frac{2x^2+xy-3y^2}{2x^2+5xy+3y^2}$

64. $\frac{x^2-4y^2}{x^2+3xy+2y^2} \div \frac{x^2-4xy+4y^2}{x+y}$

Practice Plus

In Exercises 65–72, perform the indicated operation or operations.

65. $\left(\frac{y-2}{y^2-9y+18} \cdot \frac{y^2-4y-12}{y+2}\right) \div \frac{y^2-4}{y^2+5y+6}$

66. $\left(\frac{6y^2+31y+18}{3y^2-20y+12} \cdot \frac{2y^2-15y+18}{6y^2+35y+36}\right) \div \frac{2y^2-13y+15}{9y^2+15y+4}$

67. $\frac{3x^2+3x-60}{2x-8} \div \left(\frac{30x^2}{x^2-7x+10} \cdot \frac{x^3+3x^2-10x}{25x^3}\right)$

68. $\frac{5x^2-x}{3x+2} \div \left(\frac{6x^2+x-2}{10x^2+3x-1} \cdot \frac{2x^2-x-1}{2x^2-x}\right)$

69. $\frac{x^2+xz+xy+yz}{x-y} \div \frac{x+z}{x+y}$

70. $\frac{x^2-xz+xy-yz}{x-y} \div \frac{x-z}{y-x}$

71. $\frac{3xy+ay+3xb+ab}{9x^2-a^2} \div \frac{y^3+b^3}{6x-2a}$

72. $\frac{5xy-ay-5xb+ab}{25x^2-a^2} \div \frac{y^3-b^3}{15x+3a}$

Application Exercises

73. In the Section 7.1 opener, we used

$$\frac{250x}{100-x}$$

to describe the cost, in millions of dollars, to remove x percent of the pollutants that are discharged into the river. We were wrong. The cost will be half of what we originally anticipated. Write a rational expression that represents the reduced cost.

74. We originally thought that the cost, in dollars, to manufacture each of x bicycles was

$$\frac{100x + 100{,}000}{x}.$$

We were wrong. We can manufacture each bicycle at half of what we originally anticipated. Write a rational expression that represents the reduced cost.

Writing in Mathematics

75. Explain how to multiply rational expressions.

76. Explain how to divide rational expressions.

77. In dividing polynomials

$$\frac{P}{Q} \div \frac{R}{S},$$

why is it necessary to state that polynomial R is not equal to 0?

Critical Thinking Exercises

78. Which one of the following is true?

a. $5 \div x = \frac{1}{5} \cdot x$ for any nonzero number x.

b. $\frac{4}{x} \div \frac{x-2}{x} = \frac{4}{x-2}$ if $x \neq 0$ and $x \neq 2$.

c. $\frac{x-5}{6} \cdot \frac{3}{5-x} = \frac{1}{2}$ for any value of x except 5.

d. The quotient of two rational expressions can be found by dividing their numerators and dividing their denominators.

79. Find the missing polynomials: $\frac{\square}{\square} \cdot \frac{3x - 12}{2x} = \frac{3}{2}.$

80. Find the missing polynomials: $-\frac{1}{2x - 3} \div \frac{\square}{\square} = \frac{1}{3}.$

81. Divide:

$$\frac{9x^2 - y^2 + 15x - 5y}{3x^2 + xy + 5x} \div \frac{3x + y}{9x^3 + 6x^2y + xy^2}.$$

Technology Exercises

In Exercises 82–85, use the GRAPH *or* TABLE *feature of a graphing utility to determine if the multiplication or division has been performed correctly. If the answer is wrong, correct it and then verify your correction using the graphing utility.*

82. $\frac{x^2 + x}{3x} \cdot \frac{6x}{x + 1} = 2x$

83. $\frac{x^3 - 25x}{x^2 - 3x - 10} \cdot \frac{x + 2}{x} = x + 5$

84. $\frac{x^2 - 9}{x + 4} \div \frac{x - 3}{x + 4} = x - 3$

85. $(x - 5) \div \frac{2x^2 - 11x + 5}{4x^2 - 1} = 2x - 1$

Review Exercises

86. Solve: $2x + 3 < 3(x - 5)$. (Section 2.7, Example 6)

87. Factor completely: $3x^2 - 15x - 42$. (Section 6.5, Example 2)

88. Solve: $x(2x + 9) = 5$. (Section 6.6, Example 6)

SECTION 7.3

ADDING AND SUBTRACTING RATIONAL EXPRESSIONS WITH THE SAME DENOMINATOR

Objectives

1 Add rational expressions with the same denominator.

2 Subtract rational expressions with the same denominator.

3 Add and subtract rational expressions with opposite denominators.

Are you long, medium, or round? Your skull, that is. The varying shapes of the human skull create glorious diversity in the human species. By learning to add and subtract rational expressions with the same denominator, you will obtain an expression that models this diversity.

1 Add rational expressions with the same denominator.

Addition when Denominators Are the Same To add rational numbers having the same denominators, such as $\frac{2}{9}$ and $\frac{5}{9}$, we add the numerators and place the sum over the common denominator:

$$\frac{2}{9}+\frac{5}{9}=\frac{2+5}{9}=\frac{7}{9}.$$

We add rational expressions with the same denominator in an identical manner.

ADDING RATIONAL EXPRESSIONS WITH COMMON DENOMINATORS

If $\frac{P}{R}$ and $\frac{Q}{R}$ are rational expressions, then

$$\frac{P}{R}+\frac{Q}{R}=\frac{P+Q}{R}.$$

To add rational expressions with the same denominator, add numerators and place the sum over the common denominator. If possible, simplify the result.

USING TECHNOLOGY

The graphs of

$$y_1=\frac{2x-1}{3}+\frac{x+4}{3}$$

and

$$y_2=x+1$$

are the same line. Thus,

$$\frac{2x-1}{3}+\frac{x+4}{3}=x+1.$$

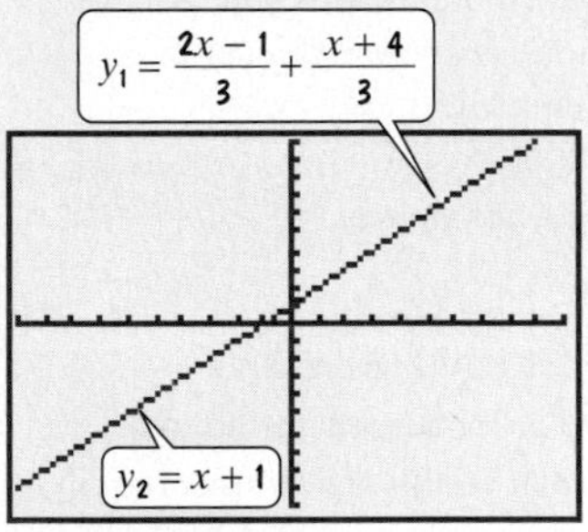

[−10, 10, 1] by [−10, 10, 1]

EXAMPLE 1 Adding Rational Expressions when Denominators Are the Same

Add: $\frac{2x-1}{3}+\frac{x+4}{3}$.

SOLUTION

$$\frac{2x-1}{3}+\frac{x+4}{3}=\frac{2x-1+x+4}{3}$$ Add numerators. Place this sum over the common denominator.

$$=\frac{3x+3}{3}$$ Combine like terms.

$$=\frac{\overset{1}{\cancel{3}}(x+1)}{\underset{1}{\cancel{3}}}$$ Factor and simplify.

$$=x+1$$ ■

✔ **CHECK POINT 1** Add: $\frac{3x-2}{5}+\frac{2x+12}{5}$.

EXAMPLE 2 Adding Rational Expressions when Denominators Are the Same

Add: $\frac{x^2}{x^2-9}+\frac{9-6x}{x^2-9}$.

SOLUTION

$$\frac{x^2}{x^2-9}+\frac{9-6x}{x^2-9}=\frac{x^2+9-6x}{x^2-9}$$ Add numerators. Place this sum over the common denominator.

$$=\frac{x^2-6x+9}{x^2-9}$$ Write the numerator in descending powers of x.

$$=\frac{(x-3)\overset{1}{\cancel{(x-3)}}}{(x+3)\underset{1}{\cancel{(x-3)}}}$$ Factor and simplify. What values of x are not permitted?

$$=\frac{x-3}{x+3}$$ ■

✔ **CHECK POINT 2** Add: $\dfrac{x^2}{x^2 - 25} + \dfrac{25 - 10x}{x^2 - 25}$.

2 Subtract rational expressions with the same denominator.

Subtraction when Denominators Are the Same The following box shows how to subtract rational expressions with the same denominator:

SUBTRACTING RATIONAL EXPRESSIONS WITH COMMON DENOMINATORS

If $\dfrac{P}{R}$ and $\dfrac{Q}{R}$ are rational expressions, then

$$\frac{P}{R} - \frac{Q}{R} = \frac{P - Q}{R}.$$

To subtract rational expressions with the same denominator, subtract numerators and place the difference over the common denominator. If possible, simplify the result.

EXAMPLE 3 Subtracting Rational Expressions when Denominators Are the Same

Subtract:

a. $\dfrac{2x + 3}{x + 1} - \dfrac{x}{x + 1}$ **b.** $\dfrac{5x + 1}{x^2 - 9} - \dfrac{4x - 2}{x^2 - 9}$.

SOLUTION

a. $\dfrac{2x + 3}{x + 1} - \dfrac{x}{x + 1} = \dfrac{2x + 3 - x}{x + 1}$ Subtract numerators. Place this difference over the common denominator.

$= \dfrac{x + 3}{x + 1}$ Combine like terms.

b. $\dfrac{5x + 1}{x^2 - 9} - \dfrac{4x - 2}{x^2 - 9} = \dfrac{5x + 1 - (4x - 2)}{x^2 - 9}$ Subtract numerators and include parentheses to indicate that both terms are subtracted. Place this difference over the common denominator.

$= \dfrac{5x + 1 - 4x + 2}{x^2 - 9}$ Remove parentheses and then change the sign of each term.

$= \dfrac{x + 3}{x^2 - 9}$ Combine like terms.

$= \dfrac{\overset{1}{\cancel{x + 3}}}{\underset{1}{\cancel{(x + 3)}}(x - 3)}$ Factor and simplify ($x \neq -3$ and $x \neq 3$).

$= \dfrac{1}{x - 3}$ ■

USING TECHNOLOGY

To check Example 3(b) numerically, enter

$$y_1 = \frac{5x + 1}{x^2 - 9} - \frac{4x - 2}{x^2 - 9}$$

$$y_2 = \frac{1}{x - 3}$$

and use the TABLE feature. If $x \neq -3$ and $x \neq 3$, no matter how far up or down we scroll, $y_1 = y_2$.

X	Y1	Y2
-3	ERROR	-.1667
-2	-.2	-.2
-1	-.25	-.25
0	-.3333	-.3333
1	-.5	-.5
2	-1	-1
3	ERROR	ERROR

X= -3

✔ **CHECK POINT 3** Subtract:

a. $\dfrac{4x + 5}{x + 7} - \dfrac{x}{x + 7}$ **b.** $\dfrac{3x^2 + 4x}{x - 1} - \dfrac{11x - 4}{x - 1}$.

STUDY TIP

When a numerator is being subtracted, be sure to **subtract every term in that expression.**

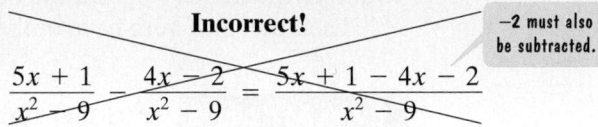

The − sign applies to the entire numerator, $4x - 2$.

Insert parentheses to indicate this.

The sign of every term of $4x - 2$ changes.

$$\frac{5x+1}{x^2-9} - \frac{4x-2}{x^2-9} = \frac{5x+1-(4x-2)}{x^2-9} = \frac{5x+1-4x+2}{x^2-9}$$

The entire numerator of the second rational expression must be subtracted. Avoid the common error of subtracting only the first term.

Incorrect!

-2 must also be subtracted.

$$\frac{5x+1}{x^2-9} - \frac{4x-2}{x^2-9} = \frac{5x+1-4x-2}{x^2-9}$$

EXAMPLE 4 Subtracting Rational Expressions when Denominators Are the Same

Subtract: $\dfrac{20y^2+5y+1}{6y^2+y-2} - \dfrac{8y^2-12y-5}{6y^2+y-2}$.

SOLUTION

$$\frac{20y^2+5y+1}{6y^2+y-2} - \frac{8y^2-12y-5}{6y^2+y-2}$$

Don't forget the parentheses.

$$= \frac{20y^2+5y+1-(8y^2-12y-5)}{6y^2+y-2}$$ Subtract numerators. Place this difference over the common denominator.

$$= \frac{20y^2+5y+1-8y^2+12y+5}{6y^2+y-2}$$ Remove parentheses and then change the sign of each term.

$$= \frac{(20y^2-8y^2)+(5y+12y)+(1+5)}{6y^2+y-2}$$ Group like terms. This step is usually performed mentally.

$$= \frac{12y^2+17y+6}{6y^2+y-2}$$ Combine like terms.

$$= \frac{\overset{1}{\cancel{(3y+2)}}(4y+3)}{\underset{1}{\cancel{(3y+2)}}(2y-1)}$$ Factor and simplify.

$$= \frac{4y+3}{2y-1}$$ ■

CHECK POINT 4 Subtract: $\dfrac{y^2+3y-6}{y^2-5y+4} - \dfrac{4y-4-2y^2}{y^2-5y+4}$.

3 Add and subtract rational expressions with opposite denominators.

Addition and Subtraction when Denominators Are Opposites How do we add or subtract rational expressions when denominators are opposites, or additive inverses? Here is an example of this type of addition problem:

$$\frac{x^2}{x-5} + \frac{4x+5}{5-x}.$$

These denominators are opposites. The differ only in their signs.

Multiply the numerator and the denominator of either of the rational expressions by -1. Then they will both have the same denominator.

EXAMPLE 5 Adding Rational Expressions when Denominators Are Opposites

Add: $\frac{x^2}{x-5} + \frac{4x+5}{5-x}$.

SOLUTION

$$\frac{x^2}{x-5} + \frac{4x+5}{5-x}$$

$$= \frac{x^2}{x-5} + \frac{(-1)}{(-1)} \cdot \frac{4x+5}{5-x}$$ Multiply the numerator and denominator of the second rational expression by -1.

$$= \frac{x^2}{x-5} + \frac{-4x-5}{-5+x}$$ Perform the multiplications by -1 by changing every term's sign.

$$= \frac{x^2}{x-5} + \frac{-4x-5}{x-5}$$ Rewrite $-5 + x$ as $x - 5$. Both rational expressions have the same denominator.

$$= \frac{x^2 + (-4x-5)}{x-5}$$ Add numerators. Place this sum over the common denominator.

$$= \frac{x^2 - 4x - 5}{x-5}$$ Remove parentheses.

$$= \frac{\overset{1}{\cancel{(x-5)}}(x+1)}{\underset{1}{\cancel{x-5}}}$$ Factor and simplify.

$$= x + 1$$

■

✔ **CHECK POINT 5** Add: $\frac{x^2}{x-7} + \frac{4x+21}{7-x}$.

ADDING AND SUBTRACTING RATIONAL EXPRESSIONS WITH OPPOSITE DENOMINATORS When one denominator is the additive inverse of the other, first multiply either rational expression by $\frac{-1}{-1}$ to obtain a common denominator.

EXAMPLE 6 Subtracting Rational Expressions when Denominators Are Opposites

Subtract: $\dfrac{5x - x^2}{x^2 - 4x - 3} - \dfrac{3x - x^2}{3 + 4x - x^2}$.

SOLUTION We note that $x^2 - 4x - 3$ and $3 + 4x - x^2$ are opposites. We multiply the second rational expression by $\frac{-1}{-1}$.

$$\frac{(-1)}{(-1)} \cdot \frac{3x - x^2}{3 + 4x - x^2} = \frac{-3x + x^2}{-3 - 4x + x^2}$$

Multiply the numerator and denominator by -1 by changing every term's sign.

$$= \frac{x^2 - 3x}{x^2 - 4x - 3}$$

Write the numerator and the denominator in descending powers of x.

We now return to the original subtraction problem.

$$\frac{5x - x^2}{x^2 - 4x - 3} - \frac{3x - x^2}{3 + 4x - x^2}$$

This is the given problem.

$$= \frac{5x - x^2}{x^2 - 4x - 3} - \frac{x^2 - 3x}{x^2 - 4x - 3}$$

Replace the second rational expression by the form obtained through multiplication by $\frac{-1}{-1}$.

$$= \frac{5x - x^2 - (x^2 - 3x)}{x^2 - 4x - 3}$$

Subtract numerators. Place this difference over the common denominator. Don't forget parentheses!

$$= \frac{5x - x^2 - x^2 + 3x}{x^2 - 4x - 3}$$

Remove parentheses and then change the sign of each term.

$$= \frac{-2x^2 + 8x}{x^2 - 4x - 3}$$

Combine like terms in the numerator. Although the numerator can be factored, further simplification is not possible. ■

CHECK POINT 6 Subtract: $\dfrac{7x - x^2}{x^2 - 2x - 9} - \dfrac{5x - 3x^2}{9 + 2x - x^2}$.

7.3 EXERCISE SET

Student Solutions Manual | CD/Video | PH Math/Tutor Center | MathXL Tutorials on CD | MathXL® | MyMathLab | Interactmath.com

Practice Exercises

In Exercises 1–38, add or subtract as indicated. Simplify the result, if possible.

1. $\dfrac{7x}{13} + \dfrac{2x}{13}$

2. $\dfrac{3x}{17} + \dfrac{8x}{17}$

3. $\dfrac{8x}{15} + \dfrac{x}{15}$

4. $\dfrac{9x}{24} + \dfrac{x}{24}$

5. $\dfrac{x - 3}{12} + \dfrac{5x + 21}{12}$

6. $\dfrac{x + 4}{9} + \dfrac{2x - 25}{9}$

7. $\dfrac{4}{x} + \dfrac{2}{x}$

8. $\dfrac{5}{x} + \dfrac{13}{x}$

9. $\dfrac{8}{9x} + \dfrac{13}{9x}$

10. $\dfrac{4}{9x} + \dfrac{11}{9x}$

11. $\dfrac{5}{x + 3} + \dfrac{4}{x + 3}$

12. $\dfrac{8}{x + 6} + \dfrac{10}{x + 6}$

13. $\dfrac{x}{x - 3} + \dfrac{4x + 5}{x - 3}$

14. $\dfrac{x}{x - 4} + \dfrac{9x + 7}{x - 4}$

15. $\dfrac{4x + 1}{6x + 5} + \dfrac{8x + 9}{6x + 5}$

16. $\dfrac{3x + 2}{3x + 4} + \dfrac{3x + 6}{3x + 4}$

17. $\frac{y^2+7y}{y^2-5y}+\frac{y^2-4y}{y^2-5y}$

18. $\frac{y^2-2y}{y^2+3y}+\frac{y^2+y}{y^2+3y}$

19. $\frac{4y-1}{5y^2}+\frac{3y+1}{5y^2}$

20. $\frac{y+2}{6y^3}+\frac{3y-2}{6y^3}$

21. $\frac{x^2-2}{x^2+x-2}+\frac{2x-x^2}{x^2+x-2}$

22. $\frac{x^2+9x}{4x^2-11x-3}+\frac{3x-5x^2}{4x^2-11x-3}$

23. $\frac{x^2-4x}{x^2-x-6}+\frac{4x-4}{x^2-x-6}$

24. $\frac{x}{2x+7}-\frac{2}{2x+7}$

25. $\frac{3x}{5x-4}-\frac{4}{5x-4}$

26. $\frac{x}{x-1}-\frac{1}{x-1}$

27. $\frac{4x}{4x-3}-\frac{3}{4x-3}$

28. $\frac{2y+1}{3y-7}-\frac{y+8}{3y-7}$

29. $\frac{14y}{7y+2}-\frac{7y-2}{7y+2}$

30. $\frac{2x+3}{3x-6}-\frac{3-x}{3x-6}$

31. $\frac{3x+1}{4x-2}-\frac{x+1}{4x-2}$

32. $\frac{x^3-3}{2x^4}-\frac{7x^3-3}{2x^4}$

33. $\frac{3y^2-1}{3y^3}-\frac{6y^2-1}{3y^3}$

34. $\frac{y^2+3y}{y^2+y-12}-\frac{y^2-12}{y^2+y-12}$

35. $\frac{4y^2+5}{9y^2-64}-\frac{y^2-y+29}{9y^2-64}$

36. $\frac{2y^2+6y+8}{y^2-16}-\frac{y^2-3y-12}{y^2-16}$

37. $\frac{6y^2+y}{2y^2-9y+9}-\frac{2y+9}{2y^2-9y+9}-\frac{4y-3}{2y^2-9y+9}$

38. $\frac{3y^2-2}{3y^2+10y-8}-\frac{y+10}{3y^2+10y-8}-\frac{y^2-6y}{3y^2+10y-8}$

In Exercises 39–64, denominators are additive inverses. Add or subtract as indicated. Simplify the result, if possible.

39. $\frac{4}{x-3}+\frac{2}{3-x}$

40. $\frac{6}{x-5}+\frac{2}{5-x}$

41. $\frac{6x+7}{x-6}+\frac{3x}{6-x}$

42. $\frac{6x+5}{x-2}+\frac{4x}{2-x}$

43. $\frac{5x-2}{3x-4}+\frac{2x-3}{4-3x}$

44. $\frac{9x-1}{7x-3}+\frac{6x-2}{3-7x}$

45. $\frac{x^2}{x-2}+\frac{4}{2-x}$

46. $\frac{x^2}{x-3}+\frac{9}{3-x}$

47. $\frac{y-3}{y^2-25}+\frac{y-3}{25-y^2}$

48. $\frac{y-7}{y^2-16}+\frac{7-y}{16-y^2}$

49. $\frac{6}{x-1}-\frac{5}{1-x}$

50. $\frac{10}{x-2}-\frac{6}{2-x}$

51. $\frac{10}{x+3}-\frac{2}{-x-3}$

52. $\frac{11}{x+7}-\frac{5}{-x-7}$

53. $\frac{y}{y-1}-\frac{1}{1-y}$

54. $\frac{y}{y-4}-\frac{4}{4-y}$

55. $\frac{3-x}{x-7}-\frac{2x-5}{7-x}$

56. $\frac{4-x}{x-9}-\frac{3x-8}{9-x}$

57. $\frac{x-2}{x^2-25}-\frac{x-2}{25-x^2}$

58. $\frac{x-8}{x^2-16}-\frac{x-8}{16-x^2}$

59. $\frac{x}{x-y}+\frac{y}{y-x}$

60. $\frac{2x-y}{x-y}+\frac{x-2y}{y-x}$

61. $\frac{2x}{x^2-y^2}+\frac{2y}{y^2-x^2}$

62. $\frac{2y}{x^2-y^2}+\frac{2x}{y^2-x^2}$

63. $\frac{x^2-2}{x^2+6x-7}+\frac{19-4x}{7-6x-x^2}$

64. $\frac{2x+3}{x^2-x-30}+\frac{x-2}{30+x-x^2}$

Practice Plus

In Exercises 65–72, perform the indicated operation or operations. Simplify the result, if possible.

65. $\frac{6b^2-10b}{16b^2-48b+27}+\frac{7b^2-20b}{16b^2-48b+27}-\frac{6b-3b^2}{16b^2-48b+27}$

66. $\frac{22b+15}{12b^2+52b-9}+\frac{30b-20}{12b^2+52b-9}-\frac{4-2b}{12b^2+52b-9}$

67. $\frac{2y}{y-5}-\left(\frac{2}{y-5}+\frac{y-2}{y-5}\right)$

68. $\frac{3x}{(x+1)^2}-\left[\frac{5x+1}{(x+1)^2}-\frac{3x+2}{(x+1)^2}\right]$

69. $\frac{b}{ac+ad-bc-bd}-\frac{a}{ac+ad-bc-bd}$

70. $\frac{y}{ax+bx-ay-by}-\frac{x}{ax+bx-ay-by}$

71. $\dfrac{(y-3)(y+2)}{(y+1)(y-4)} - \dfrac{(y+2)(y+3)}{(y+1)(4-y)} - \dfrac{(y+5)(y-1)}{(y+1)(4-y)}$

72. $\dfrac{(y+1)(2y-1)}{(y-2)(y-3)} + \dfrac{(y+2)(y-1)}{(y-2)(y-3)} - \dfrac{(y+5)(2y+1)}{(3-y)(2-y)}$

Application Exercises

73. Anthropologists and forensic scientists classify skulls using

$$\frac{L+60W}{L} - \frac{L-40W}{L},$$

where L is the skull's length and W is its width.

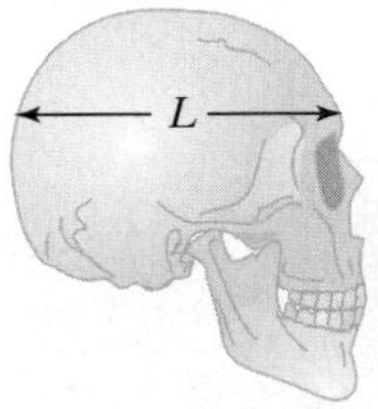

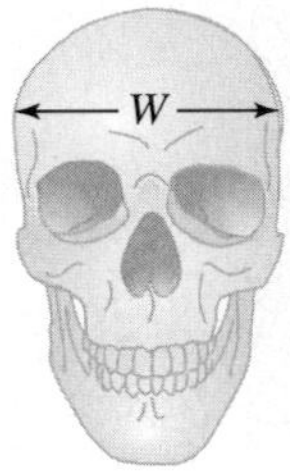

a. Express the classification as a single rational expression.

b. If the value of the rational expression in part (a) is less than 75, a skull is classified as long. A medium skull has a value between 75 and 80, and a round skull has a value over 80. Use your rational expression from part (a) to classify a skull that is 5 inches wide and 6 inches long.

74. The temperature, in degrees Fahrenheit, of a dessert placed in a freezer for t hours is modeled by

$$\frac{t+30}{t^2+4t+1} - \frac{t-50}{t^2+4t+1}.$$

a. Express the temperature as a single rational expression.

b. Use your rational expression from part (a) to find the temperature of the dessert, to the nearest hundredth of a degree, after 1 hour and after 2 hours.

In Exercises 75–76, find the perimeter of each rectangle.

75.

$\frac{5}{x+3}$ meters

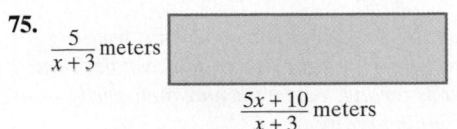

$\frac{5x+10}{x+3}$ meters

76.

$\frac{7}{x+4}$ inches

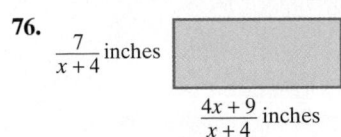

$\frac{4x+9}{x+4}$ inches

Writing in Mathematics

77. Explain how to add rational expressions when denominators are the same. Give an example with your explanation.

78. Explain how to subtract rational expressions when denominators are the same. Give an example with your explanation.

79. Describe two similarities between the following problems:

$$\frac{3}{8} + \frac{1}{8} \quad \text{and} \quad \frac{x}{x^2-1} + \frac{1}{x^2-1}.$$

80. Explain how to add rational expressions when denominators are opposites. Use an example to support your explanation.

Critical Thinking Exercises

81. Which one of the following is true?

a. The sum of two rational expressions with the same denominator can be found by adding numerators, adding denominators, and then simplifying.

b. $\dfrac{4}{b} - \dfrac{2}{-b} = -\dfrac{2}{b}$

c. The difference between two rational expressions with the same denominator can always be simplified.

d. $\dfrac{2x+1}{x-7} + \dfrac{3x+1}{x-7} - \dfrac{5x+2}{x-7} = 0$

In Exercises 82–83, perform the indicated operations. Simplify the result if possible.

82. $\left(\dfrac{3x-1}{x^2+5x-6} - \dfrac{2x-7}{x^2+5x-6}\right) \div \dfrac{x+2}{x^2-1}$

83. $\left(\dfrac{3x^2-4x+4}{3x^2+7x+2} - \dfrac{10x+9}{3x^2+7x+2}\right) \div \dfrac{x-5}{x^2-4}$

In Exercises 84–88, find the missing expression.

84. $\dfrac{2x}{x+3} + \dfrac{\square}{x+3} = \dfrac{4x+1}{x+3}$

85. $\dfrac{3x}{x+2} - \dfrac{\square}{x+2} = \dfrac{6-17x}{x+2}$

86. $\dfrac{6}{x-2} + \dfrac{\square}{2-x} = \dfrac{13}{x-2}$

87. $\dfrac{a^2}{a-4} - \dfrac{\square}{a-4} = a+3$

88. $\dfrac{3x}{x-5} + \dfrac{\square}{5-x} = \dfrac{7x+1}{x-5}$

Technology Exercises

In Exercises 89–91, use the GRAPH *or* TABLE *feature of a graphing utility to determine if the subtraction has been performed correctly. If the answer is wrong, correct it and then verify your correction using the graphing utility.*

89. $\frac{3x+6}{2} - \frac{x}{2} = x + 3$

90. $\frac{x^2+4x+3}{x+2} - \frac{5x+9}{x+2} = x - 2, x \neq -2$

91. $\frac{x^2-13}{x+4} - \frac{3}{x+4} = x + 4, x \neq -4$

Review Exercises

92. Subtract: $\frac{13}{15} - \frac{8}{45}$. (Section 1.1, Example 9)

93. Factor completely: $81x^4 - 1$. (Section 6.4, Example 4)

94. Divide: $\frac{3x^3 + 2x^2 - 26x - 15}{x+3}$. (Section 5.6, Example 2)

SECTION 7.4

Objectives

1 Find the least common denominator.

2 Add and subtract rational expressions with different denominators.

ADDING AND SUBTRACTING RATIONAL EXPRESSIONS WITH DIFFERENT DENOMINATORS

When my aunt asked how I liked my five-year-old nephew, I replied "medium rare." Unfortunately, my little joke did not get me out of baby sitting for the Dennis the Menace of our family. Now the little squirt doesn't want to go to bed because his head hurts. Does my aunt have any aspirin? What is the proper dosage for a child his age?

In this section's exercise set, you will use two formulas that model drug dosage for children. Before working with these models, we continue drawing on your experience from arithmetic to add and subtract rational expressions that have different denominators.

1 Find the least common denominator.

Finding the Least Common Denominator We can gain insight into adding rational expressions with different denominators by looking closely at what we do when adding fractions with different denominators. For example, suppose that we want to add $\frac{1}{2}$ and $\frac{2}{3}$. We must first write the fractions with the same denominator. We look for the smallest number that contains both 2 and 3 as factors. This number, 6, is then used as the *least common denominator*, or LCD.

The **least common denominator** of several rational expressions is a polynomial consisting of the product of all prime factors in the denominators, with each factor raised to the greatest power of its occurrence in any denominator.

FINDING THE LEAST COMMON DENOMINATOR

1. Factor each denominator completely.
2. List the factors of the first denominator.
3. Add to the list in step 2 any factors of the second denominator that do not appear in the list.
4. Form the product of each different factor from the list in step 3. This product is the least common denominator.

EXAMPLE 1 Finding the Least Common Denominator

Find the LCD of $\frac{7}{6x^2}$ and $\frac{2}{9x}$.

SOLUTION

Step 1. **Factor each denominator completely.**

$$6x^2 = 3 \cdot 2x^2 \quad (\text{or } 3 \cdot 2 \cdot x \cdot x)$$
$$9x = 3 \cdot 3x$$

Step 2. **List the factors of the first denominator.**

$$3, 2, x^2 \quad (\text{or } 3, 2, x, x)$$

Step 3. **Add any unlisted factors from the second denominator.** Two factors from $3 \cdot 3x$ are already in our list. These factors include x and one factor of 3. We add the other factor of 3 to our list. We have

$$3, 3, 2, x^2.$$

Step 4. **The least common denominator is the product of all factors in the final list.** Thus,

$$3 \cdot 3 \cdot 2x^2$$

or $18x^2$ is the least common denominator. ■

CHECK POINT 1 Find the LCD of $\frac{3}{10x^2}$ and $\frac{7}{15x}$.

EXAMPLE 2 Finding the Least Common Denominator

Find the LCD of $\frac{3}{x+1}$ and $\frac{5}{x-1}$.

SOLUTION

Step 1. **Factor each denominator completely.**

$$x + 1 = 1(x + 1)$$
$$x - 1 = 1(x - 1)$$

Step 2. **List the factors of the first denominator.**

$$1, x + 1$$

Step 3. **Add any unlisted factors from the second denominator.** We listed 1 and $x + 1$ as factors of the first denominator, $1(x + 1)$. The factors of the second denominator, $1(x - 1)$, include 1 and $x - 1$. One factor, 1, is already in our list, but the other factor, $x - 1$, is not. We add $x - 1$ to the list. We have

$$1, x + 1, x - 1.$$

Step 4. **The least common denominator is the product of all factors in the final list.** Thus,

$$1(x + 1)(x - 1)$$

or $(x + 1)(x - 1)$ is the least common denominator of $\frac{3}{x+1}$ and $\frac{5}{x-1}$. ■

✔ CHECK POINT 2 Find the LCD of $\frac{2}{x + 3}$ and $\frac{4}{x - 3}$.

EXAMPLE 3 Finding the Least Common Denominator

Find the LCD of

$$\frac{7}{5x^2 + 15x} \text{ and } \frac{9}{x^2 + 6x + 9}.$$

SOLUTION

Step 1. **Factor each denominator completely.**

$$5x^2 + 15x = 5x(x + 3)$$

$$x^2 + 6x + 9 = (x + 3)^2$$

Step 2. **List the factors of the first denominator.**

$$5, x, (x + 3)$$

Step 3. **Add any unlisted factors from the second denominator.** The second denominator is $(x + 3)^2$ or $(x + 3)(x + 3)$. One factor of $x + 3$ is already in our list, but the other factor is not. We add $x + 3$ to the list. We have

$$5, x, (x + 3), (x + 3).$$

Step 4. **The least common denominator is the product of all factors in the final list.** Thus,

$$5x(x + 3)(x + 3) \quad \text{or} \quad 5x(x + 3)^2$$

is the least common denominator. ■

✔ CHECK POINT 3 Find the LCD of $\frac{9}{7x^2 + 28x}$ and $\frac{11}{x^2 + 8x + 16}$.

2 Add and subtract rational expressions with different denominators.

Adding and Subtracting Rational Expressions with Different Denominators

Finding the least common denominator for two (or more) rational expressions is the first step needed to add or subtract the expressions. For example, to add $\frac{1}{2}$ and $\frac{2}{3}$, we first determine that the LCD is 6. Then we write each fraction in terms of the LCD.

$$\frac{1}{2} + \frac{2}{3} = \frac{1}{2} \cdot \frac{3}{3} + \frac{2}{3} \cdot \frac{2}{2}$$

Multiply the numerator and denominator of each fraction by whatever extra factors are required to form 6, the LCD.

$\frac{3}{3} = 1$ and $\frac{2}{2} = 1$. Multiplying by 1 does not change a fraction's value.

$$= \frac{3}{6} + \frac{4}{6}$$

$$= \frac{3+4}{6}$$

Add numerators. Place this sum over the LCD.

$$= \frac{7}{6}$$

We follow the same steps in adding or subtracting rational expressions with different denominators.

ADDING AND SUBTRACTING RATIONAL EXPRESSIONS THAT HAVE DIFFERENT DENOMINATORS

1. Find the LCD of the rational expressions.
2. Rewrite each rational expression as an equivalent expression whose denominator is the LCD. To do so, multiply the numerator and the denominator of each rational expression by any factor(s) needed to convert the denominator into the LCD.
3. Add or subtract numerators, placing the resulting expression over the LCD.
4. If possible, simplify the resulting rational expression.

EXAMPLE 4 Adding Rational Expressions with Different Denominators

Add: $\frac{7}{6x^2} + \frac{2}{9x}$.

SOLUTION

Step 1. **Find the least common denominator.** In Example 1, we found that the LCD for these rational expressions is $18x^2$.

Step 2. **Write equivalent expressions with the LCD as denominators.** We must rewrite each rational expression with a denominator of $18x^2$.

$$\frac{7}{6x^2} \cdot \frac{3}{3} = \frac{21}{18x^2} \qquad \frac{2}{9x} \cdot \frac{2x}{2x} = \frac{4x}{18x^2}$$

Multiply the numerator and denominator by 3 to get $18x^2$, the LCD.

Multiply the numerator and denominator by $2x$ to get $18x^2$, the LCD.

Because $\frac{3}{3} = 1$ and $\frac{2x}{2x} = 1$, we are not changing the value of either rational expression, only its appearance.

STUDY TIP

It is incorrect to add rational expressions by adding numerators and adding denominators. Avoid this common error.

Incorrect!

$$\frac{7}{6x^2} + \frac{2}{9x} = \frac{7+2}{6x^2+9x} = \frac{9}{6x^2+9x}$$

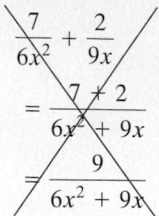

Now we are ready to perform the indicated addition.

$\frac{7}{6x^2} + \frac{2}{9x}$ This is the given problem. The LCD is $18x^2$.

$= \frac{7}{6x^2} \cdot \frac{3}{3} + \frac{2}{9x} \cdot \frac{2x}{2x}$ Write equivalent expressions with the LCD.

$= \frac{21}{18x^2} + \frac{4x}{18x^2}$

Steps 3 and 4. Add numerators, putting this sum over the LCD. Simplify if possible.

$$= \frac{21 + 4x}{18x^2} \quad \text{or} \quad \frac{4x + 21}{18x^2}$$

The numerator is prime and further simplification is not possible. ■

CHECK POINT 4 Add: $\frac{3}{10x^2} + \frac{7}{15x}$.

EXAMPLE 5 Adding Rational Expressions with Different Denominators

Add: $\frac{3}{x+1} + \frac{5}{x-1}$.

SOLUTION

Step 1. Find the least common denominator. The factors of the denominators are $x + 1$ and $x - 1$. In Example 2, we found that the LCD is $(x + 1)(x - 1)$.

Step 2. Write equivalent expressions with the LCD as denominators.

$\frac{3}{x+1} + \frac{5}{x-1}$

$= \frac{3(x-1)}{(x+1)(x-1)} + \frac{5(x+1)}{(x+1)(x-1)}$ Multiply each numerator and denominator by the extra factor required to form $(x + 1)(x - 1)$, the LCD.

Steps 3 and 4. Add numerators, putting this sum over the LCD. Simplify if possible.

$= \frac{3(x-1) + 5(x+1)}{(x+1)(x-1)}$

$= \frac{3x - 3 + 5x + 5}{(x+1)(x-1)}$ Use the distributive property to multiply and remove grouping symbols.

$= \frac{8x + 2}{(x+1)(x-1)}$ Combine like terms: $3x + 5x = 8x$ and $-3 + 5 = 2$. ■

We can factor 2 from the numerator of the answer in Example 5 to obtain

$$\frac{2(4x+1)}{(x+1)(x-1)}.$$

Because the numerator and denominator do not have any common factors, further simplification is not possible. In this section, unless there is a common factor in the numerator and denominator, we will leave an answer's numerator in unfactored form and the denominator in factored form.

CHECK POINT 5 Add: $\dfrac{2}{x+3} + \dfrac{4}{x-3}$.

EXAMPLE 6 Subtracting Rational Expressions with Different Denominators

Subtract: $\dfrac{x}{x+3} - 1$.

SOLUTION

Step 1. Find the least common denominator. We know that 1 means $\frac{1}{1}$. The factor of the first denominator is $x + 3$. Adding the factor of the second denominator, 1, the LCD is $1(x + 3)$ or $x + 3$.

Step 2. Write equivalent expressions with the LCD as denominators.

$$\frac{x}{x+3} - 1$$

$$= \frac{x}{x+3} - \frac{1}{1} \qquad \text{Write 1 as } \tfrac{1}{1}.$$

$$= \frac{x}{x+3} - \frac{1(x+3)}{1(x+3)}$$

Multiply the numerator and denominator of $\frac{1}{1}$ by the extra factor required to form $x + 3$, the LCD.

Steps 3 and 4. Subtract numerators, putting this difference over the LCD. Simplify if possible.

$$= \frac{x-(x+3)}{x+3}$$

$$= \frac{x-x-3}{x+3}$$

Remove parentheses and then change the sign of each term.

$$= \frac{-3}{x+3} \text{ or } -\frac{3}{x+3}$$

Simplify. ■

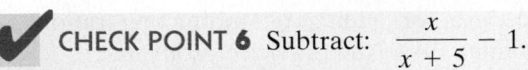

CHECK POINT 6 Subtract: $\dfrac{x}{x+5} - 1$.

EXAMPLE 7 Subtracting Rational Expressions with Different Denominators

Subtract: $\dfrac{y+2}{4y+16} - \dfrac{2}{y^2+4y}$.

SOLUTION

Step 1. **Find the least common denominator.** Start by factoring the denominators.

$$4y + 16 = 4(y + 4)$$
$$y^2 + 4y = y(y + 4)$$

The factors of the first denominator are 4 and $y + 4$. The only factor from the second denominator that is unlisted is y. Thus, the least common denominator is $4y(y + 4)$.

Step 2. **Write equivalent expressions with the LCD as denominators.**

$$\frac{y+2}{4y+16} - \frac{2}{y^2+4y}$$

$$= \frac{y+2}{4(y+4)} - \frac{2}{y(y+4)}$$ Factor denominators. The LCD is $4y(y + 4)$.

$$= \frac{(y+2)y}{4y(y+4)} - \frac{2 \cdot 4}{4y(y+4)}$$ Multiply each numerator and denominator by the extra factor required to form $4y(y + 4)$, the LCD.

Steps 3 and 4. **Subtract numerators, putting this difference over the LCD. Simplify if possible.**

$$= \frac{(y+2)y - 2 \cdot 4}{4y(y+4)}$$

$$= \frac{y^2 + 2y - 8}{4y(y+4)}$$ Use the distributive property: $(y + 2)y = y^2 + 2y$. Multiply: $2 \cdot 4 = 8$.

$$= \frac{\overset{1}{\cancel{(y+4)}}(y-2)}{4y\underset{1}{\cancel{(y+4)}}}$$ Factor and simplify.

$$= \frac{y-2}{4y}$$ ■

✔ **CHECK POINT 7** Subtract: $\dfrac{5}{y^2 - 5y} - \dfrac{y}{5y - 25}$.

In some situations, after factoring denominators, a factor in one denominator is the opposite of a factor in the other denominator. When this happens, we can use the following procedure:

ADDING AND SUBTRACTING RATIONAL EXPRESSIONS WHEN DENOMINATORS CONTAIN OPPOSITE FACTORS When one denominator contains the opposite factor of the other, first multiply either rational expression by $\frac{-1}{-1}$. Then apply the procedure for adding or subtracting rational expressions that have different denominators to the rewritten problem.

EXAMPLE 8 Adding Rational Expressions with Opposite Factors in the Denominators

Add: $\dfrac{x^2 - 2}{2x^2 - x - 3} + \dfrac{x - 2}{3 - 2x}$.

SOLUTION

Step 1. **Find the least common denominator.** Start by factoring the denominators.

$$2x^2 - x - 3 = (2x - 3)(x + 1)$$
$$3 - 2x = 1(3 - 2x)$$

Do you see that $2x - 3$ and $3 - 2x$ are opposite factors? Thus, we multiply either rational expression by $\frac{-1}{-1}$. We will use the second rational expression, resulting in $2x - 3$ in the denominator.

$$\frac{x^2 - 2}{2x^2 - x - 3} + \frac{x - 2}{3 - 2x}$$

$$= \frac{x^2 - 2}{(2x - 3)(x + 1)} + \frac{(-1)}{(-1)} \cdot \frac{x - 2}{3 - 2x}$$

Factor the first denominator. Multiply the second rational expression by $\frac{-1}{-1}$.

$$= \frac{x^2 - 2}{(2x - 3)(x + 1)} + \frac{-x + 2}{-3 + 2x}$$

Perform the multiplications by -1 by changing every term's sign.

$$= \frac{x^2 - 2}{(2x - 3)(x + 1)} + \frac{2 - x}{2x - 3}$$

The LCD of our rewritten addition problem is $(2x - 3)(x + 1)$.

Step 2. **Write equivalent expressions with the LCD as denominators.**

$$= \frac{x^2 - 2}{(2x - 3)(x + 1)} + \frac{(2 - x)(x + 1)}{(2x - 3)(x + 1)}$$

Multiply the numerator and denominator of the second rational expression by the extra factor required to form $(2x - 3)(x + 1)$, the LCD.

Steps 3 and 4. **Add numerators, putting this sum over the LCD. Simplify if possible.**

$$= \frac{x^2 - 2 + (2 - x)(x + 1)}{(2x - 3)(x + 1)}$$

$$= \frac{x^2 - 2 + 2x + 2 - x^2 - x}{(2x - 3)(x + 1)}$$

Use the FOIL method to multiply $(2 - x)(x + 1)$.

$$= \frac{(x^2 - x^2) + (2x - x) + (-2 + 2)}{(2x - 3)(x + 1)}$$

Group like terms.

$$= \frac{x}{(2x - 3)(x + 1)}$$

Combine like terms. ■

CHECK POINT 8 Add: $\dfrac{4x}{x^2 - 25} + \dfrac{3}{5 - x}$.

DISCOVER FOR YOURSELF

In Example 8, the denominators can be factored as follows:

$$2x^2 - x - 3 = (2x - 3)(x + 1)$$
$$3 - 2x = -1(2x - 3).$$

Using these factorizations, what is the LCD? Solve Example 8 by obtaining this LCD in each rational expression. Then combine the expressions. How does your solution compare with the one shown on the right?

7.4 EXERCISE SET

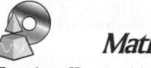

Student Solutions Manual | CD/Video | PH Math/Tutor Center | MathXL Tutorials on CD | MathXL® | MyMathLab | Interactmath.com

Practice Exercises

In Exercises 1–16, find the least common denominator of the rational expressions.

1. $\frac{7}{15x^2}$ and $\frac{13}{24x}$

2. $\frac{11}{25x^2}$ and $\frac{17}{35x}$

3. $\frac{8}{15x^2}$ and $\frac{5}{6x^5}$

4. $\frac{7}{15x^2}$ and $\frac{11}{24x^5}$

5. $\frac{4}{x-3}$ and $\frac{7}{x+1}$

6. $\frac{2}{x-5}$ and $\frac{3}{x+7}$

7. $\frac{5}{7(y+2)}$ and $\frac{10}{y}$

8. $\frac{8}{11(y+5)}$ and $\frac{12}{y}$

9. $\frac{17}{x+4}$ and $\frac{18}{x^2-16}$

10. $\frac{3}{x-6}$ and $\frac{4}{x^2-36}$

11. $\frac{8}{y^2-9}$ and $\frac{14}{y(y+3)}$

12. $\frac{14}{y^2-49}$ and $\frac{12}{y(y-7)}$

13. $\frac{7}{y^2-1}$ and $\frac{y}{y^2-2y+1}$

14. $\frac{9}{y^2-25}$ and $\frac{y}{y^2-10y+25}$

15. $\frac{3}{x^2-x-20}$ and $\frac{x}{2x^2+7x-4}$

16. $\frac{7}{x^2-5x-6}$ and $\frac{x}{x^2-4x-5}$

In Exercises 17–82, add or subtract as indicated. Simplify the result, if possible.

17. $\frac{3}{x}+\frac{5}{x^2}$

18. $\frac{4}{x}+\frac{8}{x^2}$

19. $\frac{2}{9x}+\frac{11}{6x}$

20. $\frac{5}{6x}+\frac{7}{8x}$

21. $\frac{4}{x}+\frac{7}{2x^2}$

22. $\frac{10}{x}+\frac{3}{5x^2}$

23. $6+\frac{1}{x}$

24. $3+\frac{1}{x}$

25. $\frac{2}{x}+9$

26. $\frac{7}{x}+4$

27. $\frac{x-1}{6}+\frac{x+2}{3}$

28. $\frac{x+3}{2}+\frac{x+5}{4}$

29. $\frac{4}{x}+\frac{3}{x-5}$

30. $\frac{3}{x}+\frac{4}{x-6}$

31. $\frac{2}{x-1}+\frac{3}{x+2}$

32. $\frac{3}{x-2}+\frac{4}{x+3}$

33. $\frac{2}{y+5}+\frac{3}{4y}$

34. $\frac{3}{y+1}+\frac{2}{3y}$

35. $\frac{x}{x+7}-1$

36. $\frac{x}{x+6}-1$

37. $\frac{7}{x+5}-\frac{4}{x-5}$

38. $\frac{8}{x+6}-\frac{2}{x-6}$

39. $\frac{2x}{x^2-16}+\frac{x}{x-4}$

40. $\frac{4x}{x^2-25}+\frac{x}{x+5}$

41. $\frac{5y}{y^2-9}-\frac{4}{y+3}$

42. $\frac{8y}{y^2-16}-\frac{5}{y+4}$

43. $\frac{7}{x-1}-\frac{3}{(x-1)^2}$

44. $\frac{5}{x+3}-\frac{2}{(x+3)^2}$

45. $\frac{3y}{4y-20}+\frac{9y}{6y-30}$

46. $\frac{4y}{5y-10}+\frac{3y}{10y-20}$

47. $\frac{y+4}{y}-\frac{y}{y+4}$

48. $\frac{y}{y-5}-\frac{y-5}{y}$

49. $\frac{2x+9}{x^2-7x+12}-\frac{2}{x-3}$

50. $\frac{3x+7}{x^2-5x+6}-\frac{3}{x-3}$

51. $\frac{3}{x^2-1}+\frac{4}{(x+1)^2}$

52. $\frac{6}{x^2 - 4} + \frac{2}{(x + 2)^2}$

53. $\frac{3x}{x^2 + 3x - 10} - \frac{2x}{x^2 + x - 6}$

54. $\frac{x}{x^2 - 2x - 24} - \frac{x}{x^2 - 7x + 6}$

55. $\frac{y}{y^2 + 2y + 1} + \frac{4}{y^2 + 5y + 4}$

56. $\frac{y}{y^2 + 5y + 6} + \frac{4}{y^2 - y - 6}$

57. $\frac{x - 5}{x + 3} + \frac{x + 3}{x - 5}$

58. $\frac{x - 7}{x + 4} + \frac{x + 4}{x - 7}$

59. $\frac{5}{2y^2 - 2y} - \frac{3}{2y - 2}$

60. $\frac{7}{5y^2 - 5y} - \frac{2}{5y - 5}$

61. $\frac{4x + 3}{x^2 - 9} - \frac{x + 1}{x - 3}$

62. $\frac{2x - 1}{x + 6} - \frac{6 - 5x}{x^2 - 36}$

63. $\frac{y^2 - 39}{y^2 + 3y - 10} - \frac{y - 7}{y - 2}$

64. $\frac{y^2 - 6}{y^2 + 9y + 18} - \frac{y - 4}{y + 6}$

65. $4 + \frac{1}{x - 3}$

66. $7 + \frac{1}{x - 5}$

67. $3 - \frac{3y}{y + 1}$

68. $7 - \frac{4y}{y + 5}$

69. $\frac{9x + 3}{x^2 - x - 6} + \frac{x}{3 - x}$

70. $\frac{x^2 + 9x}{x^2 - 2x - 3} + \frac{5}{3 - x}$

71. $\frac{x + 3}{x^2 + x - 2} - \frac{2}{x^2 - 1}$

72. $\frac{x}{x^2 - 10x + 25} - \frac{x - 4}{2x - 10}$

73. $\frac{y + 3}{5y^2} - \frac{y - 5}{15y}$

74. $\frac{y - 7}{3y^2} - \frac{y - 2}{12y}$

75. $\frac{x + 3}{3x + 6} + \frac{x}{4 - x^2}$

76. $\frac{x + 7}{4x + 12} + \frac{x}{9 - x^2}$

77. $\frac{y}{y^2 - 1} + \frac{2y}{y - y^2}$

78. $\frac{y}{y^2 - 1} + \frac{5y}{y - y^2}$

79. $\frac{x - 1}{x} + \frac{y + 1}{y}$

80. $\frac{x + 2}{y} + \frac{y - 2}{x}$

81. $\frac{3x}{x^2 - y^2} - \frac{2}{y - x}$

82. $\frac{7x}{x^2 - y^2} - \frac{3}{y - x}$

Practice Plus

In Exercises 83–92, perform the indicated operation or operations. Simplify the result, if possible.

83. $\frac{x + 6}{x^2 - 4} - \frac{x + 3}{x + 2} + \frac{x - 3}{x - 2}$

84. $\frac{x + 8}{x^2 - 9} - \frac{x + 2}{x + 3} + \frac{x - 2}{x - 3}$

85. $\frac{5}{x^2 - 25} + \frac{4}{x^2 - 11x + 30} - \frac{3}{x^2 - x - 30}$

86. $\frac{3}{x^2 - 49} + \frac{2}{x^2 - 15x + 56} - \frac{5}{x^2 - x - 56}$

87. $\frac{x + 6}{x^3 - 27} - \frac{x}{x^3 + 3x^2 + 9x}$

88. $\frac{x + 8}{x^3 - 8} - \frac{x}{x^3 + 2x^2 + 4x}$

89. $\frac{9y + 3}{y^2 - y - 6} + \frac{y}{3 - y} + \frac{y - 1}{y + 2}$

90. $\frac{7y - 2}{y^2 - y - 12} + \frac{2y}{4 - y} + \frac{y + 1}{y + 3}$

91. $\frac{3}{x^2 + 4xy + 3y^2} - \frac{5}{x^2 - 2xy - 3y^2} + \frac{2}{x^2 - 9y^2}$

92. $\frac{5}{x^2 + 3xy + 2y^2} - \frac{7}{x^2 - xy - 2y^2} + \frac{4}{x^2 - 4y^2}$

Application Exercises

Two formulas that approximate the dosage of a drug prescribed for children are

$$\text{Young's Rule: } C = \frac{DA}{A + 12}$$

$$\text{and Cowling's Rule: } C = \frac{D(A + 1)}{24}.$$

In each formula, A = the child's age, in years, D = an adult dosage, and C = the proper child's dosage. The formulas apply for ages 2 through 13, inclusive. Use the formulas to solve Exercises 93–96.

93. Use Young's rule to find the difference in a child's dosage for an 8-year-old child and a 3-year-old child. Express the answer as a single rational expression in terms of D. Then describe what your answer means in terms of the variables in the model.

94. Use Young's rule to find the difference in a child's dosage for a 10-year-old child and a 3-year-old child. Express the answer as a single rational expression in terms of D. Then describe what your answer means in terms of the variables in the model.

95. For a 12-year-old child, what is the difference in the dosage given by Cowling's rule and Young's rule? Express the answer as a single rational expression in terms of D. Then describe what your answer means in terms of the variables in the models.

96. Use Cowling's rule to find the difference in a child's dosage for a 12-year-old child and a 10-year-old child. Express the answer as a single rational expression in terms of D. Then describe what your answer means in terms of the variables in the model.

The graphs illustrate Young's rule and Cowling's rule when the dosage of a drug prescribed for an adult is 1000 milligrams. Use the graphs to solve Exercises 97–100.

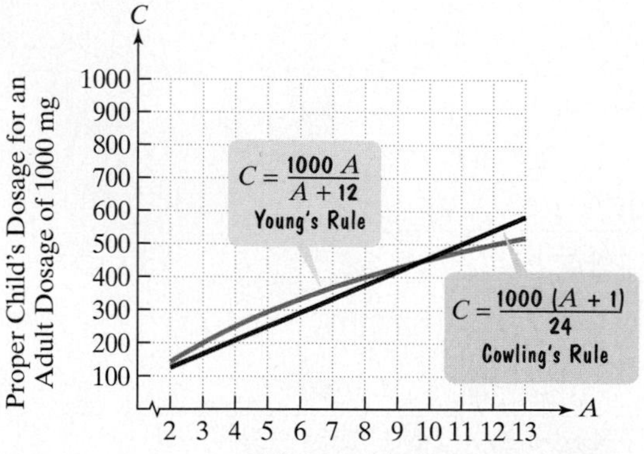

97. Does either formula consistently give a smaller dosage than the other? If so, which one?

98. Is there an age at which the dosage given by one formula becomes greater than the dosage given by the other? If so, what is a reasonable estimate of that age?

99. For what age under 11 is the difference in dosage given by the two formulas the greatest?

100. For what age over 11 is the difference in dosage given by the two formulas the greatest?

In Exercises 101–102, express the perimeter of each rectangle as a single rational expression.

101.

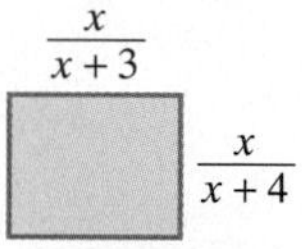

102.

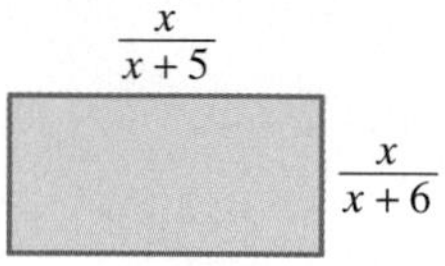

Writing in Mathematics

103. Explain how to find the least common denominator for denominators of $x^2 - 100$ and $x^2 - 20x + 100$.

104. Explain how to add rational expressions that have different denominators. Use $\frac{3}{x+5} + \frac{7}{x+2}$ in your explanation.

Explain the error in Exercises 105–106. Then rewrite the right side of the equation to correct the error that now exists.

105. $\frac{1}{x} + \frac{2}{5} = \frac{3}{x+5}$

106. $\frac{1}{x} + 7 = \frac{1}{x+7}$

107. The formulas in Exercises 93–96 relate the dosage of a drug prescribed for children to the child's age. Describe another factor that might be used when determining a child's dosage. Is this factor more or less important than age? Explain why.

Critical Thinking Exercises

108. Which one of the following is true?

a. $x - \frac{1}{5} = \frac{4}{5}x$

b. The LCD of $\frac{1}{x}$ and $\frac{2x}{x-1}$ is $x^2 - 1$.

c. $\frac{1}{x} + \frac{x}{1} = \frac{1}{\cancel{x}_1} + \frac{\cancel{x}^1}{1} = 1 + 1 = 2$

d. $\frac{2}{x} + 1 = \frac{2+x}{x}, x \neq 0$

In Exercises 109–110, perform the indicated operations. Simplify the result, if possible.

109. $\dfrac{y^2 + 5y + 4}{y^2 + 2y - 3} \cdot \dfrac{y^2 + y - 6}{y^2 + 2y - 3} - \dfrac{2}{y - 1}$

110. $\left(\dfrac{1}{x + h} - \dfrac{1}{x}\right) \div h$

In Exercises 111–112, find the missing rational expression.

111. $\dfrac{2}{x - 1} + \square = \dfrac{2x^2 + 3x - 1}{x^2(x - 1)}$

112. $\dfrac{4}{x - 2} - \square = \dfrac{2x + 8}{(x - 2)(x + 1)}$

Review Exercises

113. Multiply: $(3x + 5)(2x - 7)$. (Section 5.3, Example 2)

114. Graph: $3x - y = 3$. (Section 3.2, Example 5)

115. Write the slope-intercept form of the equation of the line passing through $(-3, -4)$ and $(1, 0)$. (Section 3.5, Example 2)

MID-CHAPTER CHECK POINT

CHAPTER 7

What You Know: We learned that it is necessary to exclude any value or values of a variable that make the denominator of a rational expression zero. We learned to simplify rational expressions by dividing the numerator and the denominator by common factors. We performed a variety of operations with rational expressions, including multiplication, division, addition, and subtraction.

1. Find all numbers for which $\dfrac{x^2 - 4}{x^2 - 2x - 8}$ is undefined.

In Exercises 2–4, simplify each rational expression.

2. $\dfrac{3x^2 - 7x + 2}{6x^2 + x - 1}$

3. $\dfrac{9 - 3y}{y^2 - 5y + 6}$

4. $\dfrac{16w^3 - 24w^2}{8w^4 - 12w^3}$

In Exercises 5–20, perform the indicated operations. Simplify the result, if possible.

5. $\dfrac{7x - 3}{x^2 + 3x - 4} - \dfrac{3x + 1}{x^2 + 3x - 4}$

6. $\dfrac{x + 2}{2x - 4} \cdot \dfrac{8}{x^2 - 4}$

7. $1 + \dfrac{7}{x - 2}$

8. $\dfrac{2x^2 + x - 1}{2x^2 - 7x + 3} \div \dfrac{x^2 - 3x - 4}{x^2 - x - 6}$

9. $\dfrac{1}{x^2 + 2x - 3} + \dfrac{1}{x^2 + 5x + 6}$

10. $\dfrac{17}{x - 5} + \dfrac{x + 8}{5 - x}$

11. $\dfrac{4y^2 - 1}{9y - 3y^2} \cdot \dfrac{y^2 - 7y + 12}{2y^2 - 7y - 4}$

12. $\dfrac{y}{y + 1} - \dfrac{2y}{y + 2}$

13. $\dfrac{w^2 + 6w + 5}{7w^2 - 63} \div \dfrac{w^2 + 10w + 25}{7w + 21}$

14. $\dfrac{2z}{z^2 - 9} - \dfrac{5}{z^2 + 4z + 3}$

15. $\dfrac{z + 2}{3z - 1} + \dfrac{5}{(3z - 1)^2}$

16. $\dfrac{8}{x^2 + 4x - 21} + \dfrac{3}{x + 7}$

17. $\dfrac{x^4 - 27x}{x^2 - 9} \cdot \dfrac{x + 3}{x^2 + 3x + 9}$

18. $\dfrac{x - 1}{x^2 - x - 2} - \dfrac{x + 2}{x^2 + 4x + 3}$

19. $\dfrac{x^2 - 2xy + y^2}{x + y} \div \dfrac{x^2 - xy}{5x + 5y}$

20. $\dfrac{5}{x + 5} + \dfrac{x}{x - 4} - \dfrac{11x - 8}{x^2 + x - 20}$

SECTION 7.5

Objectives

1 Simplify complex rational expressions by dividing.

2 Simplify complex rational expressions by multiplying by the LCD.

COMPLEX RATIONAL EXPRESSIONS

Do you drive to and from campus each day? If the one-way distance of your round-trip commute is d, then your average rate, or speed, is given by the expression

$$\frac{2d}{\frac{d}{r_1} + \frac{d}{r_2}}$$

in which r_1 and r_2 are your average rates on the outgoing and return trips, respectively. Do you notice anything unusual about this expression? It has two separate rational expressions in its denominator.

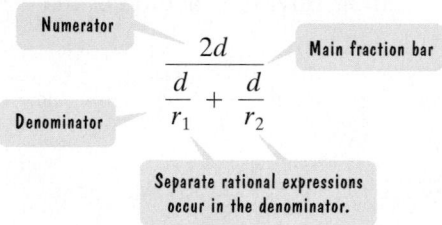

$$\frac{2d}{\frac{d}{r_1} + \frac{d}{r_2}}$$

Numerator

Main fraction bar

Denominator

Separate rational expressions occur in the denominator.

Complex rational expressions, also called **complex fractions**, have numerators or denominators containing one or more rational expressions. Here is another example of such an expression:

$$\frac{1 + \frac{1}{x}}{1 - \frac{1}{x}}.$$

Numerator

Main fraction bar

Denominator

Separate rational expressions occur in the numerator and denominator.

In this section, we study two methods for simplifying complex rational expressions.

1 Simplify complex rational expressions by dividing.

Simplifying by Rewriting Complex Rational Expressions as a Quotient of Two Rational Expressions One method for simplifying a complex rational expression is to combine its numerator into a single expression and combine its denominator into a single expression. Then perform the division by inverting the denominator and multiplying.

SIMPLIFYING A COMPLEX RATIONAL EXPRESSION BY DIVIDING

1. If necessary, add or subtract to get a single rational expression in the numerator.
2. If necessary, add or subtract to get a single rational expression in the denominator.
3. Perform the division indicated by the main fraction bar: Invert the denominator of the complex rational expression and multiply.
4. If possible, simplify.

The following examples illustrate the use of this first method:

EXAMPLE 1 Simplifying a Complex Rational Expression

Simplify:

$$\frac{\frac{1}{3}+\frac{2}{5}}{\frac{2}{5}-\frac{1}{3}}.$$

SOLUTION Let's first identify the parts of this complex rational expression.

Numerator

$$\frac{\frac{1}{3}+\frac{2}{5}}{\frac{2}{5}-\frac{1}{3}}$$

Main fraction bar

Denominator

Step 1. Add to get a single rational expression in the numerator.

$$\frac{1}{3}+\frac{2}{5}=\frac{1\cdot 5}{3\cdot 5}+\frac{2\cdot 3}{5\cdot 3}=\frac{5}{15}+\frac{6}{15}=\frac{11}{15}$$

The LCD is 3 · 5, or 15.

Step 2. Subtract to get a single rational expression in the denominator.

$$\frac{2}{5}-\frac{1}{3}=\frac{2\cdot 3}{5\cdot 3}-\frac{1\cdot 5}{3\cdot 5}=\frac{6}{15}-\frac{5}{15}=\frac{1}{15}$$

The LCD is 15.

Steps 3 and 4. Perform the division indicated by the main fraction bar: Invert and multiply. If possible, simplify.

$$\frac{\frac{1}{3}+\frac{2}{5}}{\frac{2}{5}-\frac{1}{3}}=\frac{\frac{11}{15}}{\frac{1}{15}}=\frac{11}{15}\cdot\frac{15}{1}=\frac{11}{\cancel{15}_{1}}\cdot\frac{\cancel{15}^{1}}{1}=11$$

Invert and multiply.

■

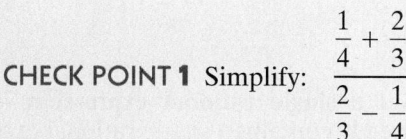

✔ **CHECK POINT 1** Simplify: $\frac{\frac{1}{4}+\frac{2}{3}}{\frac{2}{3}-\frac{1}{4}}$.

EXAMPLE 2 Simplifying a Complex Rational Expression

Simplify:

$$\frac{1+\frac{1}{x}}{1-\frac{1}{x}}.$$

SOLUTION

Step 1. Add to get a single rational expression in the numerator.

$$1+\frac{1}{x}=\frac{1}{1}+\frac{1}{x}=\frac{1\cdot x}{1\cdot x}+\frac{1}{x}=\frac{x}{x}+\frac{1}{x}=\frac{x+1}{x}$$

The LCD is $1 \cdot x$, or x.

Step 2. Subtract to get a single rational expression in the denominator.

$$1-\frac{1}{x}=\frac{1}{1}-\frac{1}{x}=\frac{1\cdot x}{1\cdot x}-\frac{1}{x}=\frac{x}{x}-\frac{1}{x}=\frac{x-1}{x}$$

The LCD is $1 \cdot x$, or x.

Steps 3 and 4. Perform the division indicated by the main fraction bar: Invert and multiply. If possible, simplify.

$$\frac{1+\frac{1}{x}}{1-\frac{1}{x}}=\frac{\frac{x+1}{x}}{\frac{x-1}{x}}=\frac{x+1}{x}\cdot\frac{x}{x-1}=\frac{x+1}{\cancel{x}_1}\cdot\frac{\cancel{x}^1}{x-1}=\frac{x+1}{x-1}$$

Invert and multiply.

■

CHECK POINT 2 Simplify: $\dfrac{2-\frac{1}{x}}{2+\frac{1}{x}}$.

EXAMPLE 3 Simplifying a Complex Rational Expression

Simplify:

$$\frac{\frac{1}{xy}}{\frac{1}{x}+\frac{1}{y}}.$$

SOLUTION

Step 1. Get a single rational expression in the numerator. The numerator, $\frac{1}{xy}$, already contains a single rational expression, so we can skip this step.

Step 2. Add to get a single rational expression in the denominator.

$$\frac{1}{x}+\frac{1}{y}=\frac{1\cdot y}{x\cdot y}+\frac{1\cdot x}{y\cdot x}=\frac{y}{xy}+\frac{x}{xy}=\frac{y+x}{xy}$$

The LCD is xy.

Steps 3 and 4. Perform the division indicated by the main fraction bar: Invert and multiply. If possible, simplify.

$$\frac{\frac{1}{xy}}{\frac{1}{x}+\frac{1}{y}}=\frac{\frac{1}{xy}}{\frac{y+x}{xy}}=\frac{1}{xy}\cdot\frac{xy}{y+x}=\frac{1}{\cancel{xy}_{1}}\cdot\frac{\cancel{xy}^{1}}{y+x}=\frac{1}{y+x}$$

Invert and multiply.

■

✔ CHECK POINT 3 Simplify: $\dfrac{\frac{1}{x}-\frac{1}{y}}{\frac{1}{xy}}$.

2 Simplify complex rational expressions by multiplying by the LCD.

Simplifying Complex Rational Expressions by Multiplying by the LCD A second method for simplifying a complex rational expression is to find the least common denominator of all the rational expressions in its numerator and denominator. Then multiply each term in its numerator and denominator by this least common denominator. Because we are multiplying by a form of 1, we will obtain an equivalent expression that does not contain fractions in its numerator or denominator.

SIMPLIFYING A COMPLEX RATIONAL EXPRESSION BY MULTIPLYING BY THE LCD

1. Find the LCD of all rational expressions within the complex rational expression.
2. Multiply both the numerator and the denominator of the complex rational expression by this LCD.
3. Use the distributive property and multiply each term in the numerator and denominator by this LCD. Simplify. No fractional expressions should remain in the numerator and denominator.
4. If possible, factor and simplify.

We now rework Examples 1, 2, and 3 using the method of multiplying by the LCD. Compare the two simplification methods to see if there is one method that you prefer.

EXAMPLE 4 Simplifying a Complex Rational Expression by the LCD Method

Simplify:

$$\frac{\frac{1}{3}+\frac{2}{5}}{\frac{2}{5}-\frac{1}{3}}.$$

SOLUTION The denominators in the complex rational expression are 3, 5, 5, and 3. The LCD is $3 \cdot 5$, or 15. Multiply both the numerator and denominator of the complex rational expression by 15.

$$\frac{\frac{1}{3}+\frac{2}{5}}{\frac{2}{5}-\frac{1}{3}} = \frac{15}{15} \cdot \frac{\left(\frac{1}{3}+\frac{2}{5}\right)}{\left(\frac{2}{5}-\frac{1}{3}\right)} = \frac{15 \cdot \frac{1}{3} + 15 \cdot \frac{2}{5}}{15 \cdot \frac{2}{5} - 15 \cdot \frac{1}{3}} = \frac{5+6}{6-5} = \frac{11}{1} = 11$$

$\frac{15}{15} = 1$, so we are not changing the complex fraction's value.

✔ **CHECK POINT 4** Simplify by the LCD method: $\dfrac{\frac{1}{4}+\frac{2}{3}}{\frac{2}{3}-\frac{1}{4}}$.

EXAMPLE 5 Simplifying a Complex Rational Expression by the LCD Method

Simplify:

$$\frac{1+\frac{1}{x}}{1-\frac{1}{x}}.$$

SOLUTION The denominators in the complex rational expression are $1, x, 1$, and x.

$$\frac{1+\frac{1}{x}}{1-\frac{1}{x}} = \frac{\frac{1}{1}+\frac{1}{x}}{\frac{1}{1}-\frac{1}{x}}$$

Denominators

Denominators

The LCD is $1 \cdot x$, or x. Multiply both the numerator and denominator of the complex rational expression by x.

$$\frac{1+\frac{1}{x}}{1-\frac{1}{x}} = \frac{x}{x} \cdot \frac{\left(1+\frac{1}{x}\right)}{\left(1-\frac{1}{x}\right)} = \frac{x \cdot 1 + x \cdot \frac{1}{x}}{x \cdot 1 - x \cdot \frac{1}{x}} = \frac{x+1}{x-1}$$

✔ **CHECK POINT 5** Simplify by the LCD method: $\dfrac{2-\frac{1}{x}}{2+\frac{1}{x}}$.

EXAMPLE 6 Simplifying a Complex Rational Expression by the LCD Method

Simplify:

$$\frac{\frac{1}{xy}}{\frac{1}{x}+\frac{1}{y}}.$$

SOLUTION The denominators in the complex rational expression are xy, x, and y. The LCD is xy. Multiply both the numerator and denominator of the complex rational expression by xy.

$$\frac{\frac{1}{xy}}{\frac{1}{x}+\frac{1}{y}} = \frac{xy}{xy}\cdot\frac{\left(\frac{1}{xy}\right)}{\left(\frac{1}{x}+\frac{1}{y}\right)} = \frac{xy\cdot\frac{1}{xy}}{xy\cdot\frac{1}{x}+xy\cdot\frac{1}{y}} = \frac{1}{y+x}.$$

■

✔ **CHECK POINT 6** Simplify by the LCD method: $\dfrac{\frac{1}{x}-\frac{1}{y}}{\frac{1}{xy}}$.

7.5 EXERCISE SET

Student Solutions Manual | CD/Video | PH Math/Tutor Center | MathXL Tutorials on CD | MathXL® | MyMathLab | Interactmath.com

Practice Exercises

In Exercises 1–40, simplify each complex rational expression by the method of your choice.

1. $\dfrac{\frac{1}{2}+\frac{1}{4}}{\frac{1}{2}+\frac{1}{3}}$

2. $\dfrac{\frac{1}{3}+\frac{1}{4}}{\frac{1}{3}+\frac{1}{6}}$

3. $\dfrac{5+\frac{2}{5}}{7-\frac{1}{10}}$

4. $\dfrac{1+\frac{3}{5}}{2-\frac{1}{4}}$

5. $\dfrac{\frac{2}{5}-\frac{1}{3}}{\frac{2}{3}-\frac{3}{4}}$

6. $\dfrac{\frac{1}{2}-\frac{1}{4}}{\frac{3}{8}+\frac{1}{16}}$

7. $\dfrac{\frac{3}{4}-x}{\frac{3}{4}+x}$

8. $\dfrac{\frac{2}{3}-x}{\frac{2}{3}+x}$

9. $\dfrac{7-\frac{2}{x}}{5+\frac{1}{x}}$

10. $\dfrac{8+\frac{3}{x}}{1-\frac{7}{x}}$

11. $\dfrac{2+\frac{3}{y}}{1-\frac{7}{y}}$

12. $\dfrac{4-\frac{7}{y}}{3-\frac{2}{y}}$

13. $\dfrac{\frac{1}{y}-\frac{3}{2}}{\frac{1}{y}+\frac{3}{4}}$

14. $\dfrac{\frac{1}{y}-\frac{3}{4}}{\frac{1}{y}+\frac{2}{3}}$

15. $\dfrac{\frac{x}{5}-\frac{5}{x}}{\frac{1}{5}+\frac{1}{x}}$

16. $\dfrac{\frac{3}{x}+\frac{x}{3}}{\frac{x}{3}-\frac{3}{x}}$

17. $\dfrac{1+\frac{1}{x}}{1-\frac{1}{x^2}}$

18. $\dfrac{1+\frac{2}{x}}{1-\frac{4}{x^2}}$

19. $\dfrac{\frac{1}{7}-\frac{1}{y}}{\frac{7-y}{7}}$

20. $\dfrac{\frac{1}{9}-\frac{1}{y}}{\frac{9-y}{9}}$

21. $\dfrac{x+\frac{2}{y}}{\frac{x}{y}}$

22. $\dfrac{x-\frac{2}{y}}{\frac{x}{y}}$

23. $\dfrac{\frac{1}{x}+\frac{1}{y}}{xy}$

24. $\dfrac{\frac{1}{x}+\frac{1}{y}}{x+y}$

25. $\dfrac{\frac{x}{y}+\frac{1}{x}}{\frac{y}{x}+\frac{1}{x}}$

26. $\dfrac{\frac{1}{x}+\frac{1}{y}}{\frac{1}{x}-\frac{1}{y}}$

27. $\dfrac{\frac{1}{y}+\frac{2}{y^2}}{\frac{2}{y}+1}$

28. $\dfrac{\frac{1}{y}+\frac{3}{y^2}}{\frac{3}{y}+1}$

29. $\dfrac{\frac{12}{x^2}-\frac{3}{x}}{\frac{15}{x}-\frac{9}{x^2}}$

30. $\dfrac{\frac{8}{x^2}-\frac{2}{x}}{\frac{10}{x}-\frac{6}{x^2}}$

31. $\dfrac{2+\frac{6}{y}}{1-\frac{9}{y^2}}$

32. $\dfrac{3+\frac{12}{y}}{1-\frac{16}{y^2}}$

33. $\dfrac{\frac{1}{x+2}}{1+\frac{1}{x+2}}$

34. $\dfrac{\frac{1}{x-2}}{1-\frac{1}{x-2}}$

35. $\dfrac{x-5+\frac{3}{x}}{x-7+\frac{2}{x}}$

36. $\dfrac{x+9-\frac{7}{x}}{x-6+\frac{4}{x}}$

37. $\dfrac{\frac{3}{xy^2}+\frac{2}{x^2y}}{\frac{1}{x^2y}+\frac{2}{xy^3}}$

38. $\dfrac{\frac{2}{x^3y}+\frac{5}{xy^4}}{\frac{5}{x^3y}-\frac{3}{xy}}$

39. $\dfrac{\frac{3}{x+1}-\frac{3}{x-1}}{\frac{5}{x^2-1}}$

40. $\dfrac{\frac{3}{x+2}-\frac{3}{x-2}}{\frac{5}{x^2-4}}$

Practice Plus

In Exercises 41–48, simplify each complex rational expression.

41. $\dfrac{\frac{6}{x^2+2x-15}-\frac{1}{x-3}}{\frac{1}{x+5}+1}$

42. $\dfrac{\frac{1}{x-2}-\frac{6}{x^2+3x-10}}{1+\frac{1}{x-2}}$

43. $\dfrac{y^{-1}-(y+5)^{-1}}{5}$

44. $\dfrac{y^{-1}-(y+2)^{-1}}{2}$

45. $\dfrac{1}{1-\frac{1}{x}}-1$

46. $\dfrac{1}{1-\frac{1}{x+1}}-1$

47. $\dfrac{1}{1+\frac{1}{1+\frac{1}{x}}}$

48. $\dfrac{1}{1+\frac{1}{1+\frac{1}{2}}}$

Application Exercises

49. The average rate on a round-trip commute having a one-way distance d is given by the complex rational expression

$$\frac{2d}{\frac{d}{r_1}+\frac{d}{r_2}}$$

in which r_1 and r_2 are the average rates on the outgoing and return trips, respectively. Simplify the expression. Then find your average rate if you drive to campus averaging 40 miles per hour and return home on the same route averaging 30 miles per hour.

50. If two electrical resistors with resistances R_1 and R_2 are connected in parallel (see the figure), then the total resistance in the circuit is given by the complex rational expression

$$\frac{1}{\frac{1}{R_1}+\frac{1}{R_2}}.$$

Simplify the expression. Then find the total resistance if $R_1 = 10$ ohms and $R_2 = 20$ ohms.

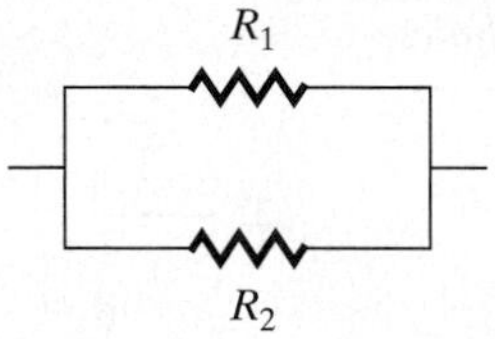

Writing in Mathematics

51. What is a complex rational expression? Give an example with your explanation.

52. Describe two ways to simplify $\dfrac{\frac{3}{x}+\frac{2}{x^2}}{\frac{1}{x^2}+\frac{2}{x}}$.

53. Which method do you prefer for simplifying complex rational expressions? Why?

Critical Thinking Exercises

54. Which one of the following is true?

a. The fraction $\dfrac{31{,}729{,}546}{72{,}578{,}112}$ is a complex rational expression.

b. $\dfrac{y-\frac{1}{2}}{y+\frac{3}{4}} = \dfrac{4y-2}{4y+3}$ for any value of y except $-\frac{3}{4}$.

c. $\dfrac{\frac{1}{4}-\frac{1}{3}}{\frac{1}{3}+\frac{1}{6}} = \frac{1}{12} \div \frac{3}{6} = \frac{1}{6}$

d. Some complex rational expressions cannot be simplified by both methods discussed in this section.

55. In one short sentence, five words or less, explain what

$$\frac{\frac{1}{x}+\frac{1}{x^2}+\frac{1}{x^3}}{\frac{1}{x^4}+\frac{1}{x^5}+\frac{1}{x^6}}$$

does to each number x.

In Exercises 56–57, simplify completely.

56. $\dfrac{2y}{2+\frac{2}{y}} + \dfrac{y}{1+\frac{1}{y}}$

57. $\dfrac{1+\frac{1}{y}-\frac{6}{y^2}}{1-\frac{5}{y}+\frac{6}{y^2}} - \dfrac{1-\frac{1}{y}}{1-\frac{2}{y}-\frac{3}{y^2}}$

Technology Exercises

In Exercises 58–60, use the GRAPH *or* TABLE *feature of a graphing utility to determine if the simplification is correct. If the answer is wrong, correct it and then verify your corrected simplification using the graphing utility.*

58. $\dfrac{x-\frac{1}{2x+1}}{1-\frac{x}{2x+1}} = 2x-1$

59. $\dfrac{\frac{1}{x}+1}{\frac{1}{x}} = 2$

60. $\dfrac{\frac{1}{x}+\frac{1}{3}}{\frac{1}{3x}} = x+\frac{1}{3}$

Review Exercises

61. Factor completely: $2x^3 - 20x^2 + 50x$. (Section 6.5, Example 2)

62. Solve: $2 - 3(x-2) = 5(x+5) - 1$. (Section 2.3, Example 3)

63. Multiply: $(x+y)(x^2 - xy + y^2)$. (Section 5.2, Example 7)

SECTION 7.6

SOLVING RATIONAL EQUATIONS

Objectives

1. Solve rational equations.
2. Solve problems involving formulas with rational expressions.
3. Solve a formula with a rational expression for a variable.

The time has come to clean up the river. Suppose that the government has committed \$375 million for this project. We know that

$$y = \frac{250x}{100-x}$$

models the cost, in millions of dollars, to remove x percent of the river's pollutants. What percentage of pollutants can be removed for \$375 million?

In order to determine the percentage, we use the given model. The government has committed \$375 million, so substitute 375 for y:

$$375 = \frac{250x}{100 - x} \quad \text{or} \quad \frac{250x}{100 - x} = 375.$$

The equation contains a rational expression.

Now we need to solve the equation and find the value for x. This variable represents the percentage of pollutants that can be removed for \$375 million.

A **rational**, or **fractional**, **equation** is an equation containing one or more rational expressions. The preceding equation is an example of a rational equation. Do you see that there is a variable in a denominator? This is a characteristic of many rational equations. In this section, you will learn a procedure for solving such equations.

1 Solve rational equations.

Solving Rational Equations We have seen that the LCD is used to add and subtract rational expressions. By contrast, when solving rational equations, **the LCD is used as a multiplier that clears an equation of fractions**.

USING TECHNOLOGY

We can use a graphing utility to verify the solution to Example 1. Graph each side of the equation:

$$y_1 = \frac{x}{4}$$

$$y_2 = \frac{1}{4} + \frac{x}{6}.$$

Trace along the lines or use the utility's intersection feature. The solution, as shown below, is the first coordinate of the point of intersection. Thus, the solution is 3.

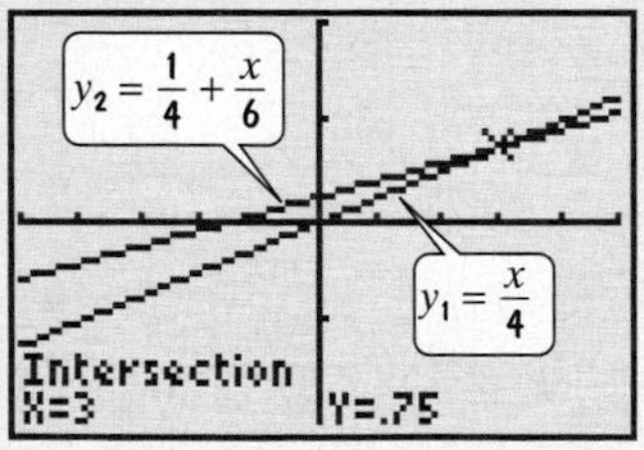

$[-5, 5, 1]$ by $[-2, 2, 1]$

EXAMPLE 1 Solving a Rational Equation

Solve: $\frac{x}{4} = \frac{1}{4} + \frac{x}{6}$.

SOLUTION The LCD of 4, 4, and 6 is 12. To clear the equation of fractions, we multiply both sides by 12.

$$\frac{x}{4} = \frac{1}{4} + \frac{x}{6} \qquad \text{This is the given equation.}$$

$$12\left(\frac{x}{4}\right) = 12\left(\frac{1}{4} + \frac{x}{6}\right) \qquad \text{Multiply both sides by 12, the LCD of all the fractions in the equation.}$$

$$12 \cdot \frac{x}{4} = 12 \cdot \frac{1}{4} + 12 \cdot \frac{x}{6} \qquad \text{Use the distributive property on the right side.}$$

$$3x = 3 + 2x \qquad \text{Simplify: } \frac{\overset{3}{\cancel{12}}}{1} \cdot \frac{x}{\cancel{4}} = 3x;\ \frac{\overset{3}{\cancel{12}}}{1} \cdot \frac{1}{\cancel{4}} = 3;\ \frac{\overset{2}{\cancel{12}}}{1} \cdot \frac{x}{\cancel{6}} = 2x.$$

$$x = 3 \qquad \text{Subtract } 2x \text{ from both sides.}$$

Substitute 3 for x in the original equation. You should obtain the true statement $\frac{3}{4} = \frac{3}{4}$. This verifies that the solution is 3 and the solution set is $\{3\}$. ■

CHECK POINT 1 Solve: $\frac{x}{6} = \frac{1}{6} + \frac{x}{8}$.

In Example 1, we solved a rational equation with constants in the denominators. Now, let's consider an equation such as

$$\frac{1}{x} = \frac{1}{5} + \frac{3}{2x}.$$

Can you see how this equation differs from the rational equation that we solved earlier? The variable, x, appears in two of the denominators. The procedure for solving this equation still involves multiplying each side by the least common denominator. However, we must avoid any values of the variable that make a denominator zero. For example, examine the denominators in the equation:

$$\frac{1}{x} = \frac{1}{5} + \frac{3}{2x}.$$

This denominator would equal zero if $x = 0$.

This denominator would equal zero if $x = 0$.

We see that x cannot equal zero. With this in mind, let's solve the equation.

EXAMPLE 2 Solving a Rational Equation

Solve: $\frac{1}{x} = \frac{1}{5} + \frac{3}{2x}$.

SOLUTION The denominators are x, 5, and $2x$. The least common denominator is $10x$. We begin by multiplying both sides of the equation by $10x$. We will also write the restriction that x cannot equal zero to the right of the equation.

$$\frac{1}{x} = \frac{1}{5} + \frac{3}{2x}, \quad x \neq 0$$ This is the given equation.

$$10x \cdot \frac{1}{x} = 10x\left(\frac{1}{5} + \frac{3}{2x}\right)$$ Multiply both sides by 10x.

$$10x \cdot \frac{1}{x} = 10x \cdot \frac{1}{5} + 10x \cdot \frac{3}{2x}$$ Use the distributive property. Be sure to multiply all terms by 10x.

$$\cancel{10x} \cdot \frac{1}{\cancel{x}} = \overset{2}{\cancel{10}}x \cdot \frac{1}{\underset{1}{\cancel{5}}} + \overset{5}{\cancel{10}}\cancel{x} \cdot \frac{3}{\underset{1}{\cancel{2x}}}$$ Divide out common factors in the multiplications.

$$10 = 2x + 15$$ Simplify.

Observe that the resulting equation,

$$10 = 2x + 15$$

is now cleared of fractions. With the variable term, $2x$, already on the right, we will collect constant terms on the left by subtracting 15 from both sides.

$$-5 = 2x$$ Subtract 15 from both sides.

$$-\frac{5}{2} = x$$ Divide both sides by 2.

We check our solution by substituting $-\frac{5}{2}$ into the original equation or by using a calculator. With a calculator, evaluate each side of the equation for $x = -\frac{5}{2}$, or for $x = -2.5$. Note that the original restriction that $x \neq 0$ is met. The solution is $-\frac{5}{2}$ and the solution set is $\left\{-\frac{5}{2}\right\}$. ■

✔ **CHECK POINT 2** Solve: $\frac{5}{2x} = \frac{17}{18} - \frac{1}{3x}$.

The following steps may be used to solve a rational equation:

SOLVING RATIONAL EQUATIONS

1. List restrictions on the variable. Avoid any values of the variable that make a denominator zero.
2. Clear the equation of fractions by multiplying both sides by the LCD of all rational expressions in the equation.
3. Solve the resulting equation.
4. Reject any proposed solution that is in the list of restrictions on the variable. Check other proposed solutions in the original equation.

EXAMPLE 3 Solving a Rational Equation

Solve: $x + \frac{1}{x} = \frac{5}{2}$.

SOLUTION

Step 1. **List restrictions on the variable.**

$$x + \frac{1}{x} = \frac{5}{2}$$

This denominator would equal 0 if $x = 0$.

The restriction is $x \neq 0$.

Step 2. **Multiply both sides by the LCD.** The denominators are x and 2. Thus, the LCD is $2x$. We multiply both sides by $2x$.

$$x + \frac{1}{x} = \frac{5}{2},\quad x \neq 0 \qquad \text{This is the given equation.}$$

$$2x\left(x + \frac{1}{x}\right) = 2x\left(\frac{5}{2}\right) \qquad \text{Multiply both sides by the LCD.}$$

$$2x \cdot x + 2x \cdot \frac{1}{x} = 2x \cdot \frac{5}{2} \qquad \text{Use the distributive property on the left side.}$$

$$2x^2 + 2 = 5x \qquad \text{Simplify.}$$

Step 3. **Solve the resulting equation.** Can you see that we have a quadratic equation? Write the equation in standard form and solve for x.

$$2x^2 - 5x + 2 = 0 \qquad \text{Subtract 5x from both sides.}$$

$$(2x - 1)(x - 2) = 0 \qquad \text{Factor.}$$

$$2x - 1 = 0 \quad \text{or} \quad x - 2 = 0 \qquad \text{Set each factor equal to 0.}$$

$$2x = 1 \qquad\qquad x = 2 \qquad \text{Solve the resulting equations.}$$

$$x = \frac{1}{2}$$

Step 4. **Check proposed solutions in the original equation.** The proposed solutions, $\frac{1}{2}$ and 2, are not part of the restriction that $x \neq 0$. Neither makes a denominator in the original equation equal to zero.

Check $\frac{1}{2}$:

$$x + \frac{1}{x} = \frac{5}{2}$$
$$\frac{1}{2} + \frac{1}{\frac{1}{2}} \stackrel{?}{=} \frac{5}{2}$$
$$\frac{1}{2} + 2 \stackrel{?}{=} \frac{5}{2}$$
$$\frac{1}{2} + \frac{4}{2} \stackrel{?}{=} \frac{5}{2}$$
$$\frac{5}{2} = \frac{5}{2}, \text{ true}$$

Check 2:

$$x + \frac{1}{x} = \frac{5}{2}$$
$$2 + \frac{1}{2} \stackrel{?}{=} \frac{5}{2}$$
$$\frac{4}{2} + \frac{1}{2} \stackrel{?}{=} \frac{5}{2}$$
$$\frac{5}{2} = \frac{5}{2}, \text{ true}$$

The solutions are $\frac{1}{2}$ and 2, and the solution set is $\left\{\frac{1}{2}, 2\right\}$. ■

✔ CHECK POINT 3 Solve: $x + \frac{6}{x} = -5$.

EXAMPLE 4 Solving a Rational Equation

Solve: $\frac{3x}{x^2 - 9} + \frac{1}{x - 3} = \frac{3}{x + 3}$.

SOLUTION

Step 1. List restrictions on the variable. By factoring denominators, it makes it easier to see values that make denominators zero.

$$\frac{3x}{(x + 3)(x - 3)} + \frac{1}{x - 3} = \frac{3}{x + 3}$$

This denominator is zero if $x = -3$ or $x = 3$.

This denominator is zero if $x = 3$.

This denominator is zero if $x = -3$.

The restrictions are $x \neq -3$ and $x \neq 3$.

Step 2. Multiply both sides by the LCD. The LCD is $(x + 3)(x - 3)$.

$$\frac{3x}{(x + 3)(x - 3)} + \frac{1}{x - 3} = \frac{3}{x + 3}, \quad x \neq -3, x \neq 3$$

This is the given equation with a denominator factored.

$$(x + 3)(x - 3)\left[\frac{3x}{(x + 3)(x - 3)} + \frac{1}{x - 3}\right] = (x + 3)(x - 3) \cdot \frac{3}{x + 3}$$

Multiply both sides by the LCD.

$$\cancel{(x + 3)}\,\cancel{(x - 3)} \cdot \frac{3x}{\cancel{(x + 3)}\,\cancel{(x - 3)}} + (x + 3)\cancel{(x - 3)} \cdot \frac{1}{\cancel{x - 3}}$$
$$= \cancel{(x + 3)}(x - 3) \cdot \frac{3}{\cancel{x + 3}}$$

Use the distributive property on the left side.

$$3x + (x + 3) = 3(x - 3)$$

Simplify.

Step 3. Solve the resulting equation.

$$3x + (x + 3) = 3(x - 3)$$ This is the equation cleared of fractions. Combine like terms on the left side.

$$4x + 3 = 3x - 9$$ Use the distributive property on the right side.

$$x + 3 = -9$$ Subtract 3x from both sides.

$$x = -12$$ Subtract 3 from both sides.

Step 4. Check proposed solutions in the original equation. The proposed solution, -12, is not part of the restriction that $x \neq -3$ and $x \neq 3$. Substitute -12 for x in the given equation and show that -12 is the solution. The equation's solution set is $\{-12\}$. ■

CHECK POINT 4 Solve: $\dfrac{11}{x^2 - 25} + \dfrac{4}{x + 5} = \dfrac{3}{x - 5}$.

EXAMPLE 5 Solving a Rational Equation

Solve: $\dfrac{8x}{x + 1} = 4 - \dfrac{8}{x + 1}$.

SOLUTION

Step 1. List restrictions on the variable.

$$\frac{8x}{x + 1} = 4 - \frac{8}{x + 1}$$

These denominators are zero if $x = -1$.

The restriction is $x \neq -1$.

Step 2. Multiply both sides by the LCD. The LCD is $x + 1$.

$$\frac{8x}{x + 1} = 4 - \frac{8}{x + 1}, \quad x \neq -1$$ This is the given equation.

$$(x + 1) \cdot \frac{8x}{x + 1} = (x + 1)\left[4 - \frac{8}{x + 1}\right]$$ Multiply both sides by the LCD.

$$\cancel{(x + 1)} \cdot \frac{8x}{\cancel{x + 1}} = (x + 1) \cdot 4 - \cancel{(x + 1)} \cdot \frac{8}{\cancel{x + 1}}$$ Use the distributive property on the right side.

$$8x = 4(x + 1) - 8$$ Simplify.

Step 3. Solve the resulting equation.

$$8x = 4(x + 1) - 8$$ This is the equation cleared of fractions.

$$8x = 4x + 4 - 8$$ Use the distributive property on the right side.

$$8x = 4x - 4$$ Simplify.

$$4x = -4$$ Subtract 4x from both sides.

$$x = -1$$ Divide both sides by 4.

Step 4. Check proposed solutions. The proposed solution, -1, is *not* a solution because of the restriction that $x \neq -1$. Notice that -1 makes both of the denominators zero in the original equation. There is *no solution to this equation.* The solution set is $\varnothing$, the empty set. ■

STUDY TIP

Reject any proposed solution that causes any denominator in a rational equation to equal 0.

CHECK POINT 5 Solve: $\dfrac{x}{x-3} = \dfrac{3}{x-3} + 9.$

STUDY TIP

It is important to distinguish between adding and subtracting rational expressions and solving rational equations. We *simplify* sums and differences of terms. On the other hand, we *solve* equations. This is shown in the following two problems, both with an LCD of $3x$.

Adding Rational Expressions	Solving Rational Equations
Simplify:	Solve:
$\dfrac{5}{3x} + \dfrac{3}{x}.$	$\dfrac{5}{3x} + \dfrac{3}{x} = 1.$
$= \dfrac{5}{3x} + \dfrac{3}{x} \cdot \dfrac{3}{3}$	$3x\left(\dfrac{5}{3x} + \dfrac{3}{x}\right) = 3x \cdot 1$
$= \dfrac{5}{3x} + \dfrac{9}{3x}$	$3x \cdot \dfrac{5}{3x} + 3x \cdot \dfrac{3}{x} = 3x$
$= \dfrac{5+9}{3x}$	$5 + 9 = 3x$
$= \dfrac{14}{3x}$	$14 = 3x$
	$\dfrac{14}{3} = x$

2 Solve problems involving formulas with rational expressions.

Applications of Rational Equations Rational equations can be solved to answer questions about variables contained in mathematical models.

EXAMPLE 6 A Government-Funded Cleanup

The formula

$$y = \frac{250x}{100 - x}$$

models the cost, y, in millions of dollars, to remove x percent of a river's pollutants. If the government commits \$375 million for this project, what percentage of pollutants can be removed?

SOLUTION Substitute 375 for y and solve the resulting rational equation for x.

$$375 = \frac{250x}{100 - x} \qquad \text{The LCD is } 100 - x.$$

$$(100 - x)375 = \cancel{(100 - x)} \cdot \frac{250x}{\cancel{100 - x}} \qquad \text{Multiply both sides by the LCD.}$$

$$375(100 - x) = 250x \qquad \text{Simplify.}$$

$$37{,}500 - 375x = 250x \qquad \text{Use the distributive property on the left side.}$$

$$37{,}500 = 625x \qquad \text{Add } 375x \text{ to both sides.}$$

$$\frac{37{,}500}{625} = \frac{625x}{625} \qquad \text{Divide both sides by 625.}$$

$$60 = x \qquad \text{Simplify.}$$

If the government spends \$375 million, 60% of the river's pollutants can be removed. ■

CHECK POINT 6 Use the model in Example 6 to answer this question: If government funding is increased to $750 million, what percentage of pollutants can be removed?

3 Solve a formula with a rational expression for a variable.

Solving a Formula for a Variable Formulas and mathematical models frequently contain rational expressions. We solve for a specified variable using the procedure for solving rational equations. The goal is to get the specified variable alone on one side of the equation. To do so, collect all terms with this variable on one side and all other terms on the other side. It is sometimes necessary to factor out the variable you are solving for.

EXAMPLE 7 Solving for a Variable in a Formula

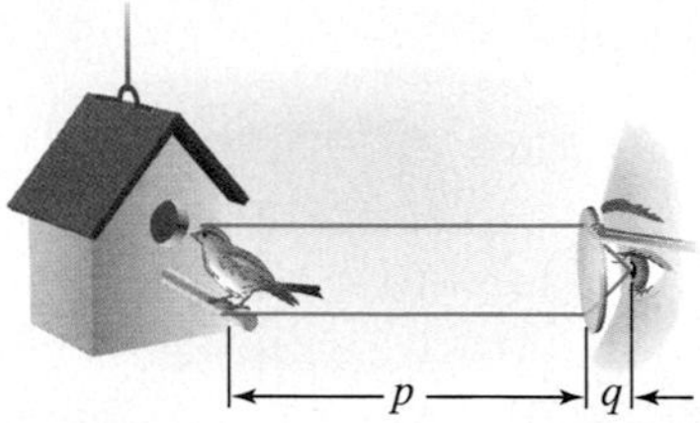

FIGURE 7.2

If you wear glasses, did you know that each lens has a measurement called its focal length, f? When an object is in focus, its distance from the lens, p, and the distance from the lens to your retina, q, satisfy the formula

$$\frac{1}{p} + \frac{1}{q} = \frac{1}{f}.$$

(See Figure 7.2.) Solve this formula for p.

SOLUTION Our goal is to isolate the variable p. We begin by multiplying both sides by the least common denominator, pqf, to clear the equation of fractions.

$$\frac{1}{p} + \frac{1}{q} = \frac{1}{f}$$ This is the given formula.

$$pqf\left(\frac{1}{p} + \frac{1}{q}\right) = pqf\left(\frac{1}{f}\right)$$ Multiply both sides by pqf, the LCD.

$$\cancel{p}qf\left(\frac{1}{\cancel{p}}\right) + p\cancel{q}f\left(\frac{1}{\cancel{q}}\right) = pq\cancel{f}\left(\frac{1}{\cancel{f}}\right)$$ Use the distributive property on the left side.

$$qf + pf = pq$$ Simplify.

Observe that the formula is now cleared of fractions. Collect terms with p, the specified variable, on one side of the equation. To do so, subtract pf from both sides.

$$qf + pf = pq$$ This is the equation cleared of fractions.

$$qf = pq - pf$$ Subtract pf from both sides.

$$qf = p(q - f)$$ Factor out p, the specified variable.

$$\frac{qf}{q - f} = \frac{p(q - f)}{q - f}$$ Divide both sides by $q - f$ and solve for p.

$$\frac{qf}{q - f} = p$$ Simplify. ■

CHECK POINT 7 Solve $\frac{1}{x} + \frac{1}{y} = \frac{1}{z}$ for x.

7.6 EXERCISE SET

Student Solutions Manual | CD/Video | PH Math/Tutor Center | MathXL Tutorials on CD | MathXL® | MyMathLab | Interactmath.com

Practice Exercises

In Exercises 1–44, solve each rational equation. If an equation has no solution, so state.

1. $\frac{x}{3} = \frac{x}{2} - 2$
2. $\frac{x}{5} = \frac{x}{6} + 1$
3. $\frac{4x}{3} = \frac{x}{18} - \frac{x}{6}$
4. $\frac{5x}{4} = \frac{x}{12} - \frac{x}{2}$
5. $2 - \frac{8}{x} = 6$
6. $1 - \frac{9}{x} = 4$
7. $\frac{2}{x} + \frac{1}{3} = \frac{4}{x}$
8. $\frac{5}{x} + \frac{1}{3} = \frac{6}{x}$
9. $\frac{2}{x} + 3 = \frac{5}{2x} + \frac{13}{4}$
10. $\frac{7}{2x} = \frac{5}{3x} + \frac{22}{3}$
11. $\frac{2}{3x} + \frac{1}{4} = \frac{11}{6x} - \frac{1}{3}$
12. $\frac{5}{2x} - \frac{8}{9} = \frac{1}{18} - \frac{1}{3x}$
13. $\frac{6}{x+3} = \frac{4}{x-3}$
14. $\frac{7}{x+1} = \frac{5}{x-3}$
15. $\frac{x-2}{2x} + 1 = \frac{x+1}{x}$
16. $\frac{7x-4}{5x} = \frac{9}{5} - \frac{4}{x}$
17. $x + \frac{6}{x} = -7$
18. $x + \frac{7}{x} = -8$
19. $\frac{x}{5} - \frac{5}{x} = 0$
20. $\frac{x}{4} - \frac{4}{x} = 0$
21. $x + \frac{3}{x} = \frac{12}{x}$
22. $x + \frac{3}{x} = \frac{19}{x}$
23. $\frac{4}{y} - \frac{y}{2} = \frac{7}{2}$
24. $\frac{4}{3y} - \frac{1}{3} = y$
25. $\frac{x-4}{x} = \frac{15}{x+4}$
26. $\frac{x-1}{2x+3} = \frac{6}{x-2}$
27. $\frac{1}{x-1} + 5 = \frac{11}{x-1}$
28. $\frac{3}{x+4} - 7 = \frac{-4}{x+4}$
29. $\frac{8y}{y+1} = 4 - \frac{8}{y+1}$
30. $\frac{2}{y-2} = \frac{y}{y-2} - 2$
31. $\frac{3}{x-1} + \frac{8}{x} = 3$
32. $\frac{2}{x-2} + \frac{4}{x} = 2$
33. $\frac{3y}{y-4} - 5 = \frac{12}{y-4}$
34. $\frac{10}{y+2} = 3 - \frac{5y}{y+2}$
35. $\frac{1}{x} + \frac{1}{x-3} = \frac{x-2}{x-3}$
36. $\frac{1}{x-1} + \frac{2}{x} = \frac{x}{x-1}$
37. $\frac{x+1}{3x+9} + \frac{x}{2x+6} = \frac{2}{4x+12}$
38. $\frac{3}{2y-2} + \frac{1}{2} = \frac{2}{y-1}$
39. $\frac{4y}{y^2-25} + \frac{2}{y-5} = \frac{1}{y+5}$
40. $\frac{1}{x+4} + \frac{1}{x-4} = \frac{22}{x^2-16}$
41. $\frac{1}{x-4} - \frac{5}{x+2} = \frac{6}{x^2-2x-8}$
42. $\frac{6}{x+3} - \frac{5}{x-2} = \frac{-20}{x^2+x-6}$
43. $\frac{2}{x+3} - \frac{2x+3}{x-1} = \frac{6x-5}{x^2+2x-3}$
44. $\frac{x-3}{x-2} + \frac{x+1}{x+3} = \frac{2x^2-15}{x^2+x-6}$

In Exercises 45–58, solve each formula for the specified variable.

45. $\frac{V_1}{V_2} = \frac{P_2}{P_1}$ for P_1 (chemistry)
46. $\frac{V_1}{V_2} = \frac{P_2}{P_1}$ for V_2 (chemistry)
47. $\frac{1}{p} + \frac{1}{q} = \frac{1}{f}$ for f (optics)
48. $\frac{1}{p} + \frac{1}{q} = \frac{1}{f}$ for q (optics)
49. $P = \frac{A}{1+r}$ for r (investment)
50. $S = \frac{a}{1-r}$ for r (mathematics)
51. $F = \frac{Gm_1m_2}{d^2}$ for m_1 (physics)
52. $F = \frac{Gm_1m_2}{d^2}$ for m_2 (physics)
53. $z = \frac{x - \bar{x}}{s}$ for x (statistics)
54. $z = \frac{x - \bar{x}}{s}$ for s (statistics)
55. $I = \frac{E}{R+r}$ for R (electronics)
56. $I = \frac{E}{R+r}$ for r (electronics)
57. $f = \frac{f_1f_2}{f_1+f_2}$ for f_1 (optics)
58. $f = \frac{f_1f_2}{f_1+f_2}$ for f_2 (optics)

Practice Plus

In Exercises 59–66, solve or simplify, whichever is appropriate.

59. $\dfrac{x^2 - 10}{x^2 - x - 20} = 1 + \dfrac{7}{x - 5}$

60. $\dfrac{x^2 + 4x - 2}{x^2 - 2x - 8} = 1 + \dfrac{4}{x - 4}$

61. $\dfrac{x^2 - 10}{x^2 - x - 20} - 1 - \dfrac{7}{x - 5}$

62. $\dfrac{x^2 + 4x - 2}{x^2 - 2x - 8} - 1 - \dfrac{4}{x - 4}$

63. $5y^{-2} + 1 = 6y^{-1}$

64. $3y^{-2} + 1 = 4y^{-1}$

65. $\dfrac{3}{y + 1} - \dfrac{1}{1 - y} = \dfrac{10}{y^2 - 1}$

66. $\dfrac{4}{y - 2} - \dfrac{1}{2 - y} = \dfrac{25}{y + 6}$

Application Exercises

A company that manufactures wheelchairs has fixed costs of \$500,000. The average cost per wheelchair, C, for the company to manufacture x wheelchairs per month is modeled by the formula

$$C = \frac{400x + 500{,}000}{x}.$$

Use this mathematical model to solve Exercises 67–68.

67. How many wheelchairs per month can be produced at an average cost of \$450 per wheelchair?

68. How many wheelchairs per month can be produced at an average cost of \$405 per wheelchair?

In Palo Alto, California, a government agency ordered computer-related companies to contribute to a pool of money to clean up underground water supplies. (The companies had stored toxic chemicals in leaking underground containers.) The formula

$$C = \frac{2x}{100 - x}$$

models the cost, C, in millions of dollars, for removing x percent of the contaminants. Use this mathematical model to solve Exercises 69–70.

69. What percentage of the contaminants can be removed for \$2 million?

70. What percentage of the contaminants can be removed for \$8 million?

We have seen that Young's rule

$$C = \frac{DA}{A + 12}$$

can be used to approximate the dosage of a drug prescribed for children. In this formula, A = the child's age, in years, D = an adult dosage, and C = the proper child's dosage. Use this formula to solve Exercises 71–72.

71. When the adult dosage is 1000 milligrams, a child is given 300 milligrams. What is that child's age? Round to the nearest year.

72. When the adult dosage is 1000 milligrams, a child is given 500 milligrams. What is that child's age?

A grocery store sells 4000 cases of canned soup per year. By averaging costs to purchase soup and to pay storage costs, the owner has determined that if x cases are ordered at a time, the yearly inventory cost, C, can be modeled by

$$C = \frac{10{,}000}{x} + 3x.$$

The graph of this model is shown below. Use this information to solve Exercises 73–74.

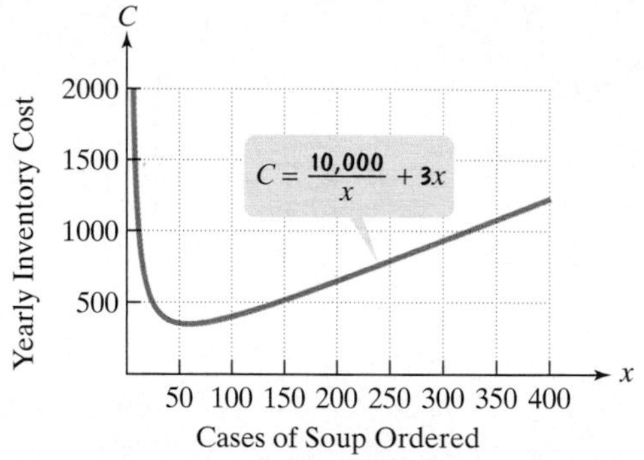

73. How many cases should be ordered at a time for yearly inventory costs to be \$350? Identify your solutions as points on the graph.

74. How many cases should be ordered at a time for yearly inventory costs to be \$790? Identify your solutions as points on the graph.

In baseball, a player's batting average is the total number of hits divided by the total number of times at bat. Use this information to solve Exercises 75–76.

75. A player has 12 hits after 40 times at bat. How many additional consecutive times must the player hit the ball to achieve a batting average of 0.440?

76. A player has eight hits after 50 times at bat. How many additional consecutive times must the player hit the ball to achieve a batting average of 0.250?

Writing in Mathematics

77. What is a rational equation?

78. Explain how to solve a rational equation.

79. Explain how to find restrictions on the variable in a rational equation.

80. Why should restrictions on the variable in a rational equation be listed before you begin solving the equation?

81. Describe similarities and differences between the procedures needed to solve the following problems:

Add: $\dfrac{2}{x} + \dfrac{3}{4}$ Solve: $\dfrac{2}{x} + \dfrac{3}{4} = 1.$

82. The equation

$$P = \frac{72{,}900}{100x^2 + 729}$$

models the percentage of people in the United States, P, who have x years of education and are unemployed. Use this model to write a problem that can be solved with a rational equation. It is not necessary to solve the problem.

Critical Thinking Exercises

83. Which one of the following is true?

a. $\frac{1}{x} + \frac{1}{6} = 6x\left(\frac{1}{x} + \frac{1}{6}\right) = 6 + x$

b. If a is any real number, the equation $\frac{a}{x} + 1 = \frac{a}{x}$ has no solution.

c. All real numbers satisfy the equation $\frac{3}{x} - \frac{1}{x} = \frac{2}{x}$.

d. To solve $\frac{5}{3x} + \frac{3}{x} = 1$, we must first add the rational expressions on the left side.

In Exercises 84–85, solve each rational equation.

84. $\frac{x + 1}{2x^2 - 11x + 5} = \frac{x - 7}{2x^2 + 9x - 5} - \frac{2x - 6}{x^2 - 25}$

85. $\left(\frac{x + 1}{x + 7}\right)^2 \div \left(\frac{x + 1}{x + 7}\right)^4 = 0.$

86. Find b so that the solution of

$$\frac{7x + 4}{b} + 13 = x$$

is -6.

Technology Exercises

In Exercises 87–89, use a graphing utility to solve each rational equation. Graph each side of the equation in the given viewing rectangle. The solution is the first coordinate of the point(s) of intersection. Check by direct substitution.

87. $\frac{x}{2} + \frac{x}{4} = 6$

$[-5, 10, 1]$ by $[-5, 10, 1]$

88. $\frac{50}{x} = 2x$

$[-10, 10, 1]$ by $[-20, 20, 2]$

89. $x + \frac{6}{x} = -5$

$[-10, 10, 1]$ by $[-10, 10, 1]$

Review Exercises

90. Factor completely: $x^4 + 2x^3 - 3x - 6$. (Section 6.1, Example 7)

91. Simplify: $(3x^2)(-4x^{-10})$. (Section 5.7, Example 3)

92. Simplify: $-5[4(x - 2) - 3]$. (Section 1.8, Example 11)

SECTION 7.7 APPLICATIONS USING RATIONAL EQUATIONS AND PROPORTIONS

Objectives

1. Solve problems involving motion.
2. Solve problems involving work.
3. Solve problems involving proportions.
4. Solve problems involving similar triangles.

The possibility of seeing a blue whale, the largest mammal ever to grace the earth, increases the excitement of gazing out over the ocean's swell of waves. Blue whales were hunted to near extinction in the last half of the nineteenth and the first half of the twentieth centuries. Using a method for estimating wildlife populations that we discuss in this section, by the mid-1960s it was determined that the world population of blue whales was less than 1000. This led the International Whaling Commission to prevent their extinction. A dramatic increase in blue whale sightings indicates an ongoing increase in their population and the success of the killing ban.

1 Solve problems involving motion.

Problems Involving Motion We have seen that the distance, d, covered by any moving body is the product of its average rate, r, and its time in motion, t: $d = rt$. Rational expressions appear in motion problems when the conditions of the problem involve the time traveled. We can obtain an expression for t, the time traveled, by dividing both sides of $d = rt$ by r.

$$d = rt \quad \text{Distance equals rate times time.}$$

$$\frac{d}{r} = \frac{rt}{r} \quad \text{Divide both sides by } r.$$

$$\frac{d}{r} = t \quad \text{Simplify.}$$

TIME IN MOTION

$$t = \frac{d}{r}$$

$$\text{Time traveled} = \frac{\text{Distance traveled}}{\text{Rate of travel}}$$

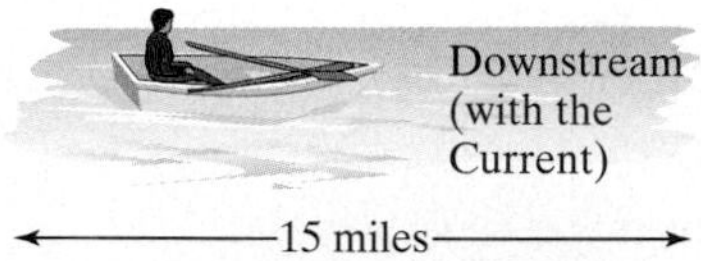

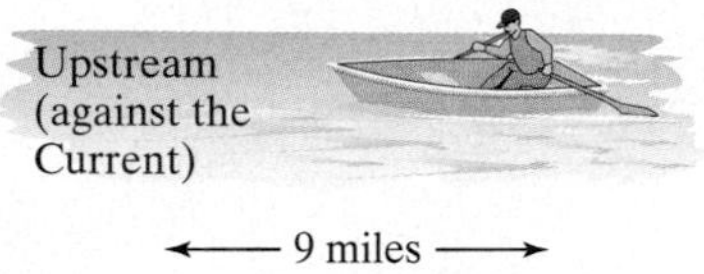

EXAMPLE 1 A Motion Problem Involving Time

In still water, your small boat averages 8 miles per hour. It takes you the same amount of time to travel 15 miles downstream, with the current, as 9 miles upstream, against the current. What is the rate of the water's current?

SOLUTION

Step 1. **Let x represent one of the quantities.** Let

$$x = \text{the rate of the current.}$$

Step 2. **Represent other quantities in terms of x.** We still need expressions for the rate of your boat with the current and the rate against the current. Traveling with the current, the boat's rate in still water, 8 miles per hour, is increased by the current's rate, x miles per hour. Thus,

$$8 + x = \text{the boat's rate with the current.}$$

Traveling against the current, the boat's rate in still water, 8 miles per hour, is decreased by the current's rate, x miles per hour. Thus,

$$8 - x = \text{the boat's rate against the current.}$$

Step 3. **Write an equation that describes the conditions.** By reading the problem again, we discover that the crucial idea is that the time spent going 15 miles with the current equals the time spent going 9 miles against the current. This information is summarized in the following table.

	Distance	Rate	Time = $\frac{\text{Distance}}{\text{Rate}}$
With the current	15	$8 + x$	$\frac{15}{8+x}$
Against the current	9	$8 - x$	$\frac{9}{8-x}$

These two times are equal.

We are now ready to write an equation that describes the problem's conditions.

The time spent going 15 miles with the current equals the time spent going 9 miles against the current.

$$\frac{15}{8+x} = \frac{9}{8-x}$$

Step 4. Solve the equation and answer the question.

$$\frac{15}{8+x} = \frac{9}{8-x}$$ This is the equation for the problem's conditions.

$$(8+x)(8-x)\cdot\frac{15}{8+x} = (8+x)(8-x)\cdot\frac{9}{8-x}$$ Multiply both sides by the LCD, $(8+x)(8-x)$.

$$15(8-x) = 9(8+x)$$ Simplify.

$$120 - 15x = 72 + 9x$$ Use the distributive property.

$$120 = 72 + 24x$$ Add 15x to both sides.

$$48 = 24x$$ Subtract 72 from both sides.

$$2 = x$$ Divide both sides by 24.

The rate of the water's current is 2 miles per hour.

Step 5. Check the proposed solution in the original wording of the problem. Does it take you the same amount of time to travel 15 miles downstream as 9 miles upstream if the current is 2 miles per hour? Keep in mind that your rate in still water is 8 miles per hour.

$$\text{Time required to travel 15 miles with the current} = \frac{\text{Distance}}{\text{Rate}} = \frac{15}{8+2} = \frac{15}{10} = 1\frac{1}{2}\text{ hours}$$

$$\text{Time required to travel 9 miles against the current} = \frac{\text{Distance}}{\text{Rate}} = \frac{9}{8-2} = \frac{9}{6} = 1\frac{1}{2}\text{ hours}$$

These times are the same, which checks with the original conditions of the problem. ■

✔ CHECK POINT **1** Forget the small boat! This time we have you canoeing on the Colorado River. In still water, your average canoeing rate is 3 miles per hour. It takes you the same amount of time to travel 10 miles downstream, with the current, as 2 miles upstream, against the current. What is the rate of the water's current?

2 Solve problems involving work.

Problems Involving Work You are thinking of designing your own Web site. You estimate that it will take 30 hours to do the job. In 1 hour, $\frac{1}{30}$ of the job is completed. In 2 hours, $\frac{2}{30}$, or $\frac{1}{15}$, of the job is completed. In 3 hours, the fractional part of the job done is $\frac{3}{30}$, or $\frac{1}{10}$. In x hours, the fractional part of the job that you can complete is $\frac{x}{30}$.

Your friend, who has experience developing Web sites, took 20 hours working on his own to design an impressive site. You wonder about the possibility of working together. How long would it take both of you to design your Web site?

Problems involving work usually have two people working together to complete a job. The amount of time it takes each person to do the job working alone is frequently known. The question deals with how long it will take both people working together to complete the job.

In work problems, **the number 1 represents one whole job completed**. For example, the completion of your Web site is represented by 1. Equations in work problems are based on the following condition:

Fractional part of the job done by the first person + fractional part of the job done by the second person = 1 (one whole job completed).

EXAMPLE 2 Solving a Problem Involving Work

You can design a Web site in 30 hours. Your friend can design the same site in 20 hours. How long will it take to design the Web site if you both work together?

SOLUTION

Step 1. **Let x represent one of the quantities.** Let x = the time, in hours, for you and your friend to design the Web site together.

Step 2. **Represent other quantities in terms of x.** Because there are no other unknown quantities, we can skip this step.

Step 3. **Write an equation that describes the conditions.** We construct a table to help find the fractional part of the task completed by you and your friend in x hours.

	Fractional part of job completed in 1 hour	Time working together	Fractional part of job completed in x hours
You	$\frac{1}{30}$	x	$\frac{x}{30}$
Your friend	$\frac{1}{20}$	x	$\frac{x}{20}$

You can design the site in 30 hours.

Your friend can design the site in 20 hours.

Fractional part of the job done by you + fractional part of the job done by your friend = one whole job.

$$\frac{x}{30} + \frac{x}{20} = 1$$

Step 4. **Solve the equation and answer the question.**

$$\frac{x}{30} + \frac{x}{20} = 1 \quad \text{This is the equation for the problem's conditions.}$$

$$60\left(\frac{x}{30} + \frac{x}{20}\right) = 60 \cdot 1 \quad \text{Multiply both sides by 60, the LCD.}$$

$$60 \cdot \frac{x}{30} + 60 \cdot \frac{x}{20} = 60$$ Use the distributive property on the left side.

$$2x + 3x = 60$$ Simplify: $\frac{\overset{2}{\cancel{60}}}{1} \cdot \frac{x}{\underset{1}{\cancel{30}}} = 2x$ and $\frac{\overset{3}{\cancel{60}}}{1} \cdot \frac{x}{\underset{1}{\cancel{20}}} = 3x$.

$$5x = 60$$ Combine like terms.

$$x = 12$$ Divide both sides by 5.

If you both work together, you can design your Web site in 12 hours.

Step 5. **Check the proposed solution in the original wording of the problem.** Will you both complete the job in 12 hours? Because you can design the site in 30 hours, in 12 hours, you can complete $\frac{12}{30}$, or $\frac{2}{5}$, of the job. Because your friend can design the site in 20 hours, in 12 hours, he can complete $\frac{12}{20}$, or $\frac{3}{5}$, of the job. Notice that $\frac{2}{5} + \frac{3}{5} = 1$, which represents the completion of the entire job, or one whole job. ■

STUDY TIP

Let

a = the time it takes person A to do a job working alone

b = the time it takes person B to do the same job working alone.

If x represents the time it takes for A and B to complete the entire job working together, then the situation can be modeled by the rational equation

$$\frac{x}{a} + \frac{x}{b} = 1.$$

✔ **CHECK POINT 2** One person can paint the outside of a house in 8 hours. A second person can do it in 4 hours. How long will it take them to do the job if they work together?

3 Solve problems involving proportions.

Problems Involving Proportions A **ratio** compares quantities by division. For example, this year's entering class at a medical school contains 60 women and 30 men. The ratio of women to men is $\frac{60}{30}$. We can express this ratio as a fraction reduced to lowest terms:

$$\frac{60}{30} = \frac{\cancel{30} \cdot 2}{\cancel{30} \cdot 1} = \frac{2}{1}.$$

This ratio can be expressed as 2:1, or 2 to 1.

A **proportion** is a statement that says that two ratios are equal. If the ratios are $\frac{a}{b}$ and $\frac{c}{d}$, then the proportion is

$$\frac{a}{b} = \frac{c}{d}.$$

We can clear this rational equation of fractions by multiplying both sides by bd, the least common denominator:

$$\frac{a}{b} = \frac{c}{d}$$ This is the given proportion.

$$bd \cdot \frac{a}{b} = bd \cdot \frac{c}{d}$$ Multiply both sides by bd ($b \neq 0$ and $d \neq 0$). Then simplify. On the left, $\frac{\cancel{b}d}{1} \cdot \frac{a}{\cancel{b}} = da = ad$. On the right, $\frac{b\cancel{d}}{1} \cdot \frac{c}{\cancel{d}} = bc$.

$$ad = bc$$

We see that the following principle is true for any proportion:

THE CROSS-PRODUCTS PRINCIPLE FOR PROPORTIONS

If $\frac{a}{b} = \frac{c}{d}$, then $ad = bc$. ($b \neq 0$ and $d \neq 0$)

The cross products ad and bc are equal.

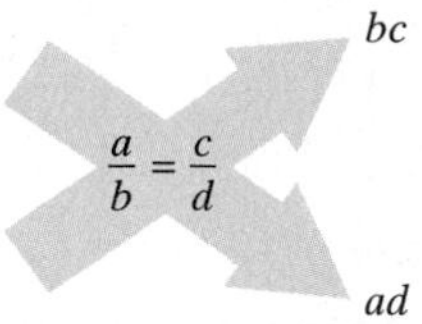

The cross-products principle: $ad = bc$

For example, if $\frac{2}{3} = \frac{6}{9}$, we see that $2 \cdot 9 = 3 \cdot 6$, or $18 = 18$.

Here is a procedure for solving problems involving proportions:

SOLVING APPLIED PROBLEMS USING PROPORTIONS

1. Read the problem and represent the unknown quantity by x (or any letter).
2. Set up a proportion by listing the given ratio on one side and the ratio with the unknown quantity on the other side. Each respective quantity should occupy the same corresponding position on each side of the proportion.
3. Drop units and apply the cross-products principle.
4. Solve for x and answer the question.

EXAMPLE 3 Applying Proportions: Calculating Taxes

The property tax on a house with an assessed value of \$65,000 is \$825. Determine the property tax on a house with an assessed value of \$180,000, assuming the same tax rate.

SOLUTION

Step 1. **Represent the unknown by x.** Let x = the tax on the \$180,000 house.

Step 2. **Set up a proportion.** We will set up a proportion comparing taxes to assessed value.

Tax on \$65,000 house / Assessed value (\$65,000) — equals — Tax on \$180,000 house / Assessed value (\$180,000)

Given ratio $\left\{ \frac{\$825}{\$65{,}000} \right. = \frac{\$x}{\$180{,}000}$ ← Unknown (numerator), ← Given quantity (denominator)

Step 3. **Drop the units and apply the cross-products principle.** We drop the dollar signs and begin to solve for x.

$$\frac{825}{65{,}000} = \frac{x}{180{,}000} \quad \text{This is the proportion for the problem's conditions.}$$

$$65{,}000x = (825)(180{,}000) \quad \text{Apply the cross-products principle.}$$

$$65{,}000x = 148{,}500{,}000 \quad \text{Multiply.}$$

Step 4. **Solve for x and answer the question.**

$$\frac{65{,}000x}{65{,}000} = \frac{148{,}500{,}000}{65{,}000} \quad \text{Divide both sides by 65,000.}$$

$$x \approx 2284.62 \quad \text{Round the value of } x \text{ to the nearest cent.}$$

The property tax on the \$180,000 house is approximately \$2284.62. ■

STUDY TIP

Here are three other correct proportions you can use in step 2:

- $\frac{\$65{,}000 \text{ value}}{\$825 \text{ tax}} = \frac{\$180{,}000 \text{ value}}{\$x \text{ tax}}$
- $\frac{\$65{,}000 \text{ value}}{\$180{,}000 \text{ value}} = \frac{\$825 \text{ tax}}{\$x \text{ tax}}$
- $\frac{\$180{,}000 \text{ value}}{\$65{,}000 \text{ value}} = \frac{\$x \text{ tax}}{\$825 \text{ tax}}$.

Each proportion gives the same cross product obtained in step 3.

CHECK POINT 3 The property tax on a house with an assessed value of \$45,000 is \$600. Determine the property tax on a house with an assessed value of \$112,500, assuming the same tax rate.

Sampling in Nature The method that was used to estimate the blue whale population described in the section opener is called the **capture-recapture method**. Because it is impossible to count each individual animal within a population, wildlife biologists randomly catch and tag a given number of animals. Sometime later they recapture a second sample of animals and count the number of recaptured tagged animals. The total size of the wildlife population is then estimated using the following proportion:

$$\frac{\text{Original number of tagged animals}}{\text{Total number of animals in the population}} = \frac{\text{Number of recaptured tagged animals}}{\text{Number of animals in second sample}}.$$

Initially unknown (x) ⟶ (Total number of animals in the population)

Known ratio (right side)

Although this is called the capture-recapture method, it is not necessary to recapture animals to observe whether or not they are tagged. This could be done from a distance, with binoculars for instance.

EXAMPLE 4 Applying Proportions: Estimating Wildlife Population

Wildlife biologists catch, tag, and then release 135 deer back into a wildlife refuge. Two weeks later they observe a sample of 140 deer, 30 of which are tagged. Assuming the ratio of tagged deer in the sample holds for all deer in the refuge, approximately how many deer are in the refuge?

SOLUTION

Step 1. **Represent the unknown by x.** Let x = the total number of deer in the refuge.

Step 2. **Set up a proportion.**

Original number of tagged deer / Total number of deer (Unknown ⟶ Total number of deer) equals Number of tagged deer in the observed sample / Total number of deer in the observed sample (Known ratio)

$$\frac{135}{x} = \frac{30}{140}$$

Steps 3 and 4. **Apply the cross-products principle, solve, and answer the question.**

$$\frac{135}{x} = \frac{30}{140}$$ This is the proportion for the problem's conditions.

$$(135)(140) = 30x$$ Apply the cross-products principle.

$$18{,}900 = 30x$$ Multiply.

$$\frac{18{,}900}{30} = \frac{30x}{30}$$ Divide both sides by 30.

$$630 = x$$ Simplify.

There are approximately 630 deer in the refuge. ■

CHECK POINT 4 Wildlife biologists catch, tag, and then release 120 deer back into a wildlife refuge. Two weeks later they observe a sample of 150 deer, 25 of which are tagged. Assuming the ratio of tagged deer in the sample holds for all deer in the refuge, approximately how many deer are in the refuge?

4 Solve problems involving similar triangles.

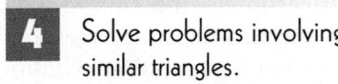

Pedestrian Crossing

Similar Triangles Shown in the margin is an international road sign. This sign is shaped just like the actual sign, although its size is smaller. Figures that have the same shape, but not the same size, are used in **scale drawings**. A scale drawing always pictures the exact shape of the object that the drawing represents. Architects, engineers, landscape gardeners, and interior decorators use scale drawings in planning their work.

Figures that have the same shape, but not necessarily the same size, are called **similar figures**. In Figure 7.3, triangles ABC and DEF are similar. Angles A and D measure the same number of degrees and are called **corresponding angles**. Angles C and F are corresponding angles, as are angles B and E. Angles with the same number of tick marks in Figure 7.3 are the corresponding angles.

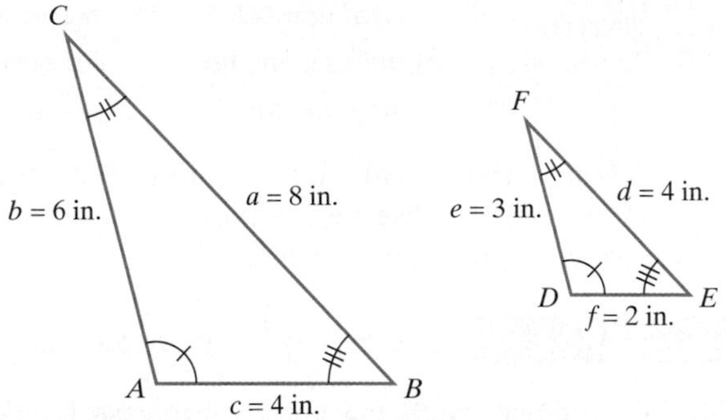

FIGURE 7.3

The sides opposite the corresponding angles are called **corresponding sides**. Although the measures of corresponding angles are equal, corresponding sides may or may not be the same length. For the triangles in Figure 7.3, each side in the smaller triangle is half the length of the corresponding side in the larger triangle.

The triangles in Figure 7.3 illustrate what it means to be **similar triangles**. **Corresponding angles have the same measure and the ratios of the lengths of the corresponding sides are equal**. For the triangles in Figure 7.3, each of these ratios is equal to 2:

$$\frac{a}{d} = \frac{8}{4} = 2 \qquad \frac{b}{e} = \frac{6}{3} = 2 \qquad \frac{c}{f} = \frac{4}{2} = 2.$$

In similar triangles, the lengths of the corresponding sides are proportional. Thus,

$$\frac{a}{d} = \frac{b}{e} = \frac{c}{f}.$$

If we know that two triangles are similar, we can set up a proportion to solve for the length of an unknown side.

EXAMPLE 5 Using Similar Triangles

The triangles in Figure 7.4 are similar. Find the missing length, x.

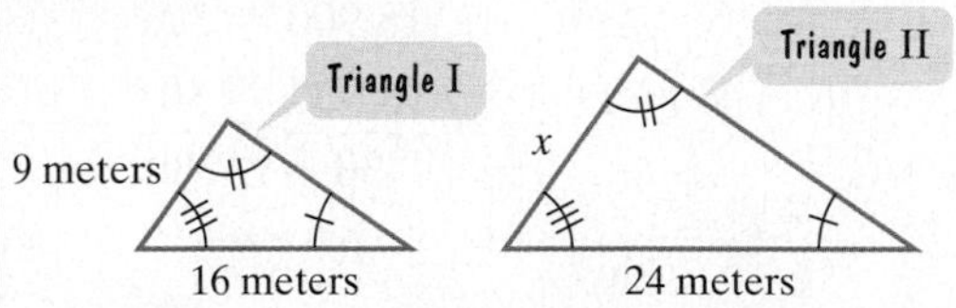

FIGURE 7.4

SOLUTION Because the triangles are similar, their corresponding sides are proportional.

Left side of Δ I. / Corresponding side on left of Δ II.

$$\frac{9}{x} = \frac{16}{24}$$

Bottom side of Δ I. / Corresponding side on bottom of Δ II.

We solve this rational equation by multiplying both sides by the LCD, $24x$. (You can also apply the cross-products principle for solving proportions.)

$$24x \cdot \frac{9}{x} = 24x \cdot \frac{16}{24} \qquad \text{Multiply both sides by the LCD, } 24x.$$

$$24 \cdot 9 = 16x \qquad \text{Simplify.}$$

$$216 = 16x \qquad \text{Multiply: } 24 \cdot 9 = 216.$$

$$13.5 = x \qquad \text{Divide both sides by 16.}$$

The missing length, x, is 13.5 meters. ■

CHECK POINT 5 The similar triangles in the figure are positioned so that they have the same orientation. Find the missing length, x.

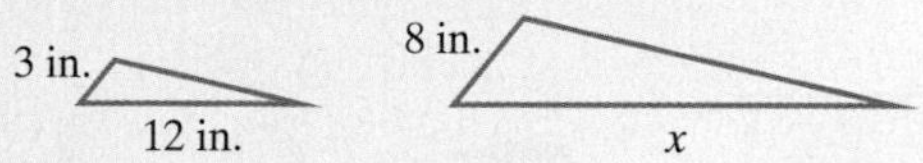

How can we quickly determine if two triangles are similar? **If the measures of two angles of one triangle are equal to those of two angles of a second triangle, then the two triangles are similar.** If the triangles are similar, then their corresponding sides are proportional.

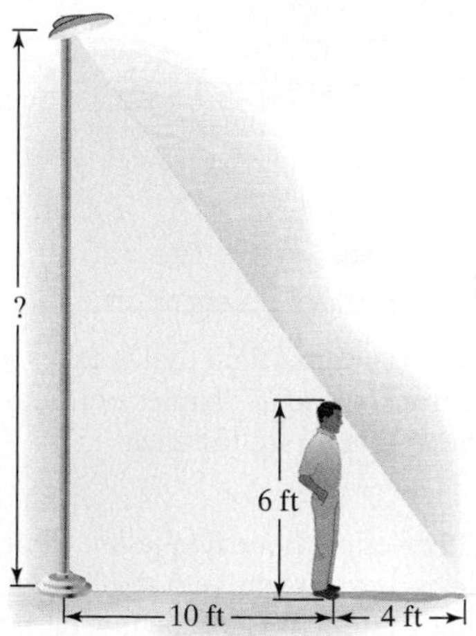

FIGURE 7.5

EXAMPLE 6 Problem Solving Using Similar Triangles

A man who is 6 feet tall is standing 10 feet from the base of a lamppost. (See Figure 7.5.) The man's shadow has a length of 4 feet. How tall is the lamppost?

SOLUTION The drawing in Figure 7.6 makes the similarity of the triangles easier to see. The large triangle with the lamppost on the left and the small triangle with the man on the left both contain 90° angles. They also share an angle. Thus, two angles of the large triangle are equal in measure to two angles of the small triangle. This means that the triangles are similar and their corresponding sides are proportional. We begin by letting x represent the height of the lamppost, in feet. Because corresponding sides of similar triangles are proportional,

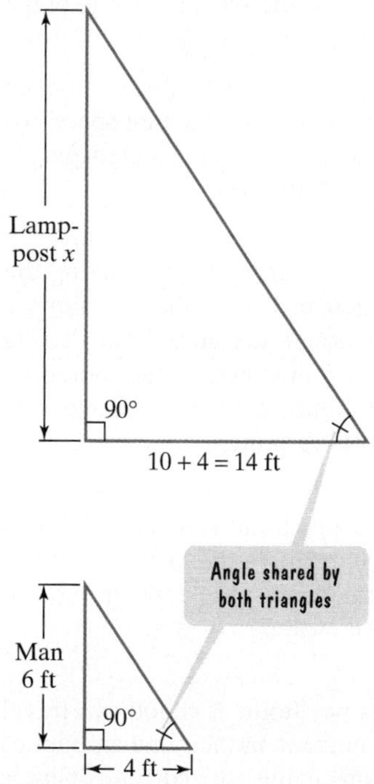

FIGURE 7.6

$$\frac{x}{6} = \frac{14}{4}.$$

Left side of big Δ. (x)
Corresponding side on left of small Δ. (6)
Bottom side of big Δ. (14)
Corresponding side on bottom of small Δ. (4)

We solve for x by multiplying both sides by the LCD, 12.

$$12 \cdot \frac{x}{6} = 12 \cdot \frac{14}{4} \qquad \text{Multiply both sides by the LCD, 12.}$$

$$2x = 42 \qquad \text{Simplify: } \frac{\overset{2}{\cancel{12}}}{1} \cdot \frac{x}{\underset{1}{\cancel{6}}} = 2x \text{ and } \frac{\overset{3}{\cancel{12}}}{1} \cdot \frac{14}{\underset{1}{\cancel{4}}} = 42.$$

$$x = 21 \qquad \text{Divide both sides by 2.}$$

The lamppost is 21 feet tall. ■

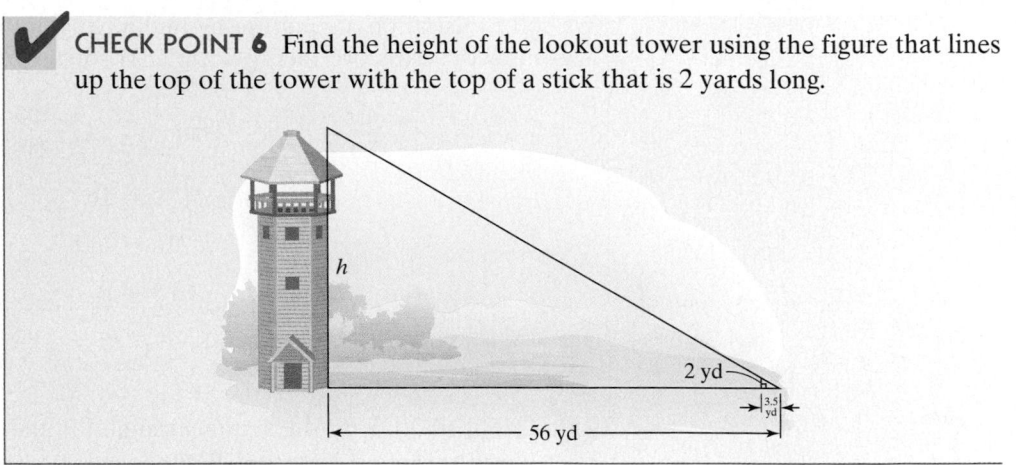

CHECK POINT 6 Find the height of the lookout tower using the figure that lines up the top of the tower with the top of a stick that is 2 yards long.

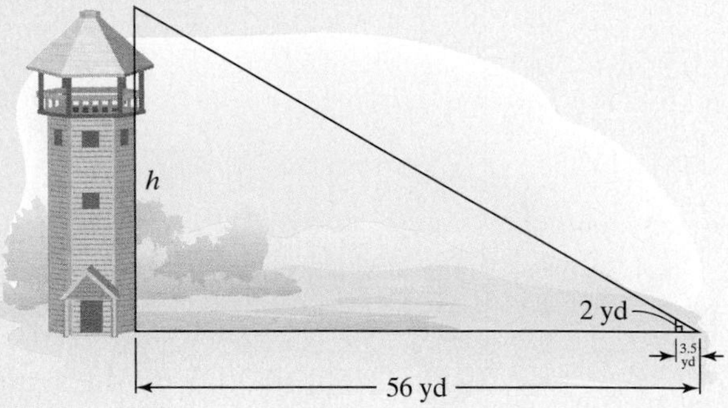

7.7 EXERCISE SET

Student Solutions Manual | CD/Video | PH Math/Tutor Center | MathXL Tutorials on CD | MathXL® | MyMathLab | Interactmath.com

Practice and Application Exercises

Use rational equations to solve Exercises 1–10. Each exercise is a problem involving motion.

1. How bad is the heavy traffic? You can walk 10 miles in the same time that it takes to travel 15 miles by car. If the car's rate is 3 miles per hour faster than your walking rate, find the average rate of each.

	Distance	Rate	Time $= \frac{\text{Distance}}{\text{Rate}}$
Walking	10	x	$\frac{10}{x}$
Car in Heavy Traffic	15	$x + 3$	$\frac{15}{x + 3}$

2. You can travel 40 miles on motorcycle in the same time that it takes to travel 15 miles on bicycle. If your motorcycle's rate is 20 miles per hour faster than your bicycle's, find the average rate for each.

	Distance	Rate	Time $= \frac{\text{Distance}}{\text{Rate}}$
Motorcycle	40	$x + 20$	$\frac{40}{x + 20}$
Bicycle	15	x	$\frac{15}{x}$

3. A jogger runs 4 miles per hour faster downhill than uphill. If the jogger can run 5 miles downhill in the same time that it takes to run 3 miles uphill, find the jogging rate in each direction.

4. A truck can travel 120 miles in the same time that it takes a car to travel 180 miles. If the truck's rate is 20 miles per hour slower than the car's, find the average rate for each.

5. In still water, a boat averages 15 miles per hour. It takes the same amount of time to travel 20 miles downstream, with the current, as 10 miles upstream, against the current. What is the rate of the water's current?

6. In still water, a boat averages 18 miles per hour. It takes the same amount of time to travel 33 miles downstream, with the current, as 21 miles upstream, against the current. What is the rate of the water's current?

7. As part of an exercise regimen, you walk 2 miles on an indoor track. Then you jog at twice your walking speed for another 2 miles. If the total time spent walking and jogging is 1 hour, find the walking and jogging rates.

8. The joys of the Pacific Coast! You drive 90 miles along the Pacific Coast Highway and then take a 5-mile run along a hiking trail in Point Reyes National Seashore. Your driving rate is nine times that of your running rate. If the total time for driving and running is 3 hours, find the average rate driving and the average rate running.

9. The water's current is 2 miles per hour. A boat can travel 6 miles downstream, with the current, in the same amount of time it travels 4 miles upstream, against the current. What is the boat's average rate in still water?

10. The water's current is 2 miles per hour. A canoe can travel 6 miles downstream, with the current, in the same amount of time it travels 2 miles upstream, against the current. What is the canoe's average rate in still water?

Use a rational equation to solve Exercises 11–16. Each exercise is a problem involving work.

11. You must leave for campus in 10 minutes or you will be late for class. Unfortunately, you are snowed in. You can shovel the driveway in 20 minutes and your brother claims he can do it in 15 minutes. If you shovel together, how long will it take to clear the driveway? Will this give you enough time before you have to leave?

12. You promised your parents that you would wash the family car. You have not started the job and they are due home in 16 minutes. You can wash the car in 40 minutes and your sister claims she can do it in 30 minutes. If you work together, how long will it take to do the job? Will this give you enough time before your parents return?

13. The MTV crew will arrive in one week and begin filming the city for *The Real World Kalamazoo*. The mayor is desperate to clean the city streets before filming begins. Two teams are available, one that requires 400 hours and one that requires 300 hours. If the teams work together, how long will it take to clean all of Kalamazoo's streets? Is this enough time before the cameras begin rolling?

14. A hurricane strikes and a rural area is without food or water. Three crews arrive. One can dispense needed supplies in 10 hours, a second in 15 hours, and a third in 20 hours. How long will it take all three crews working together to dispense food and water?

15. A pool can be filled by one pipe in 4 hours and by a second pipe in 6 hours. How long will it take using both pipes to fill the pool?

16. A pool can be filled by one pipe in 3 hours and by a second pipe in 6 hours. How long will it take using both pipes to fill the pool?

Use a proportion to solve each problem in Exercises 17–24.

17. The tax on a property with an assessed value of \$65,000 is \$720. Find the tax on a property with an assessed value of \$162,500.

18. The maintenance bill for a shopping center containing 180,000 square feet is \$45,000. What is the bill for a store in the center that is 4800 square feet?

19. St. Paul Island in Alaska has 12 fur seal rookeries (breeding places). In 1961, to estimate the fur seal pup population in the Gorbath rookery, 4963 fur seal pups were tagged in early August. In late August, a sample of 900 pups was observed and 218 of these were found to have been previously tagged. Estimate the total number of fur seal pups in this rookery.

20. To estimate the number of bass in a lake, wildlife biologists tagged 50 bass and released them in the lake. Later they netted 108 bass and found that 27 of them were tagged. Approximately how many bass are in the lake?

21. The ratio of monthly child support to a father's yearly income is 1:40. How much should a father earning \$38,000 annually pay in monthly child support?

22. The amount of garbage is proportional to the population. Dallas, Texas, has a population of 1.2 million and creates 38.4 million pounds of garbage each week. Find the amount of weekly garbage produced by New York City with a population of 8 million.

23. Height is proportional to foot length. A person whose foot length is 10 inches is 67 inches tall. In 1951, photos of large footprints were published. Some believed that these footprints were made by the "Abominable Snowman." Each footprint was 23 inches long. If indeed they belonged to the Abominable Snowman, how tall is the critter?

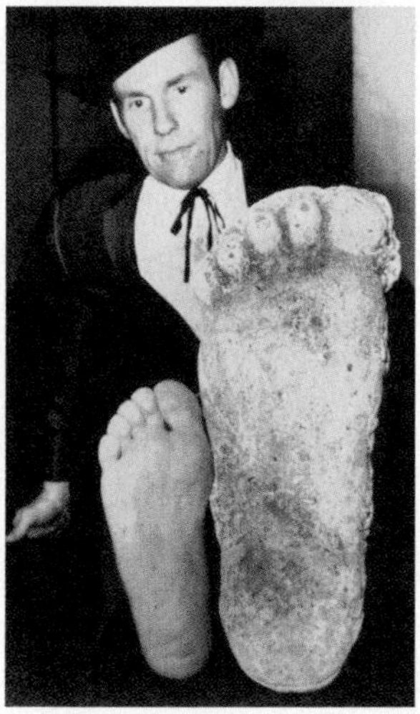

Roger Patterson comparing his foot with a plaster cast of a footprint of the purported "Bigfoot" that Mr. Patterson said he sighted in a California forest in 1967.

24. A person's hair length is proportional to the number of years it has been growing. After 2 years, a person's hair grows 8 inches. The longest moustache on record was grown by Kalyan Sain of India. Sain grew his moustache for 17 years. How long was it?

In Exercises 25–30, use similar triangles and the fact that corresponding sides are proportional to find the length of the side marked with an x.

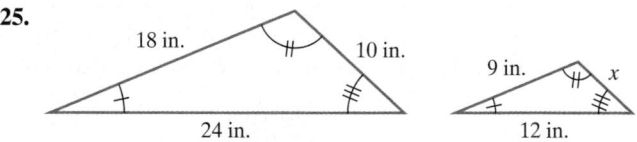

25.

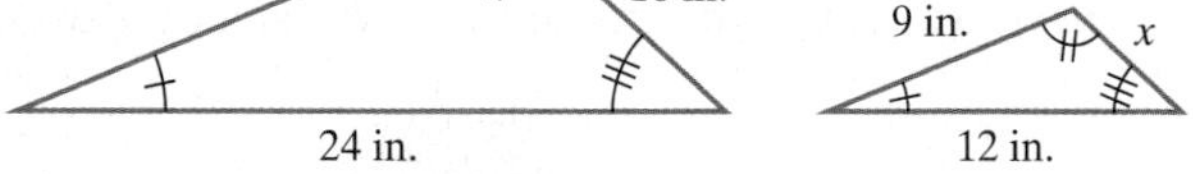

26.

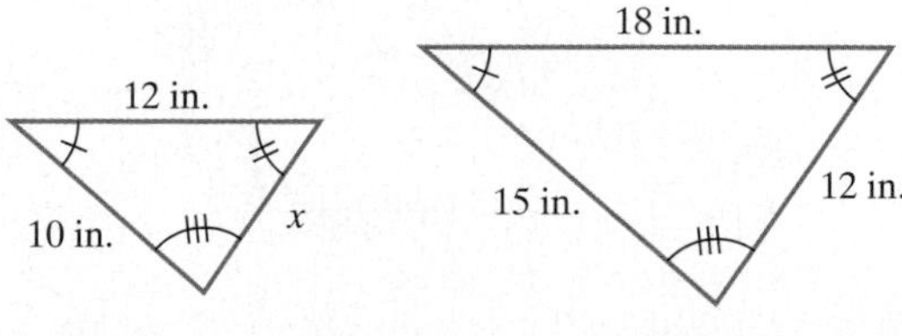

27.

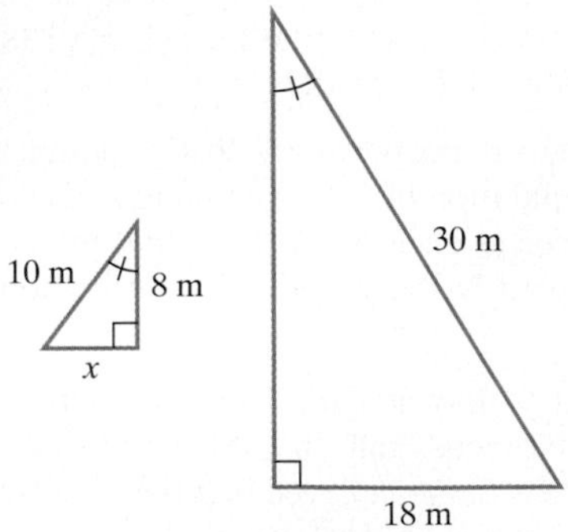

28.

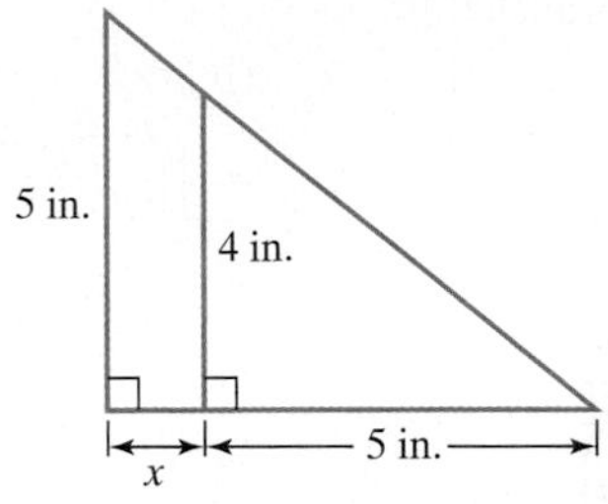

29.

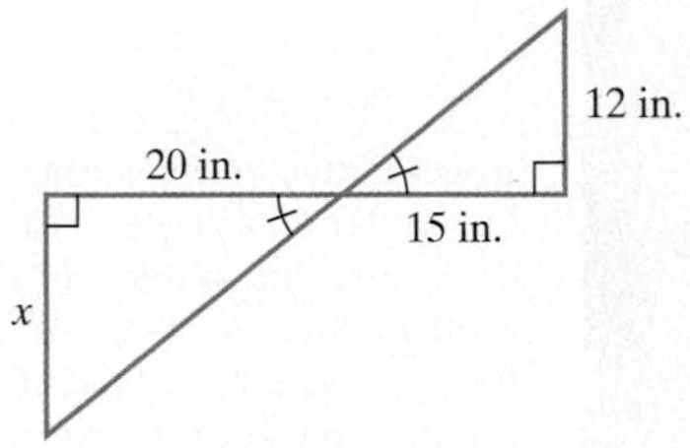

30.

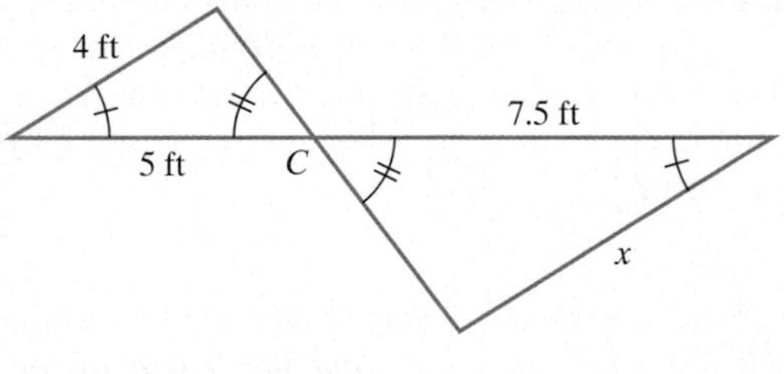

Use similar triangles to solve Exercises 31–32.

31. A tree casts a shadow 12 feet long. At the same time, a vertical rod 8 feet high casts a shadow 6 feet long. How tall is the tree?

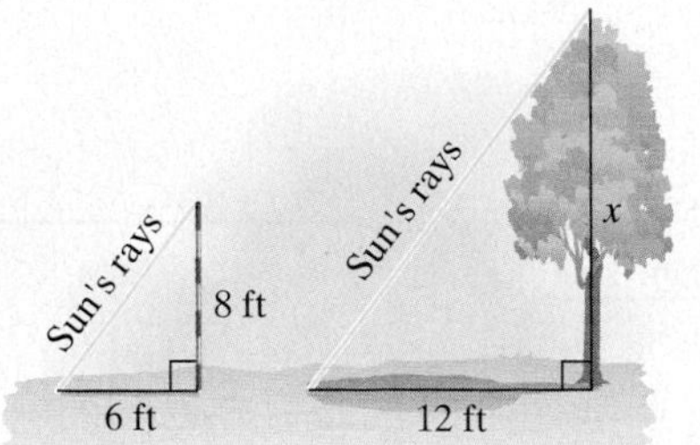

32. A person who is 5 feet tall is standing 80 feet from the base of a tree. The tree casts an 86-foot shadow. The person's shadow is 6 feet in length. What is the tree's height?

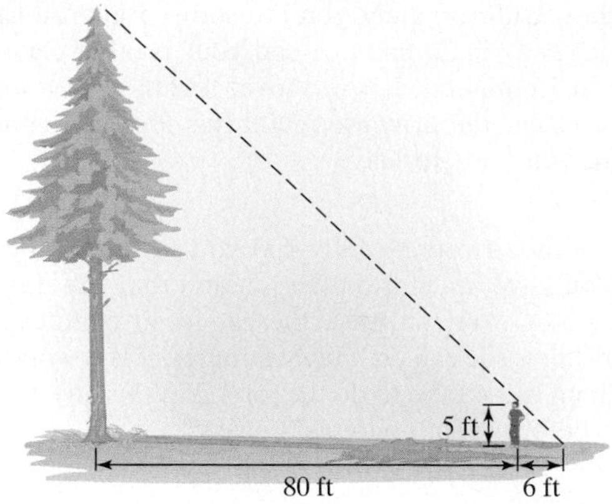

Writing in Mathematics

33. What is the relationship among time traveled, distance traveled, and rate of travel?

34. If you know how many hours it takes for you to do a job, explain how to find the fractional part of the job you can complete in x hours.

35. If you can do a job in 6 hours and your friend can do the same job in 3 hours, explain how to find how long it takes to complete the job working together. It is not necessary to solve the problem.

36. When two people work together to complete a job, describe one factor that can result in more or less time than the time given by the rational equations we have been using.

37. What is a proportion? Give an example with your description.

38. What are similar triangles?

39. If the ratio of the corresponding sides of two similar triangles is 1 to 1 $\left(\frac{1}{1}\right)$, what must be true about the triangles?

40. What are corresponding angles in similar triangles?

41. Describe how to identify the corresponding sides in similar triangles.

Critical Thinking Exercises

42. Two skiers begin skiing along a trail at the same time. The faster skier averages 9 miles per hour and the slower skier averages 6 miles per hour. The faster skier completes the trail $\frac{1}{4}$ hour before the slower skier. How long is the trail?

43. A snowstorm causes a bus driver to decrease the usual average rate along a 60-mile route by 15 miles per hour. As a result, the bus takes two hours longer than usual to complete the route. At what average rate does the bus usually cover the 60-mile route?

44. One pipe can fill a swimming pool in 2 hours, a second can fill the pool in 3 hours, and a third pipe can fill the pool in 4 hours. How many minutes, to the nearest minute, would it take to fill the pool with all three pipes operating?

45. Ben can prepare a company report in 3 hours. Shane can prepare the report in 4.2 hours. How long will it take them, working together, to prepare *four* company reports?

46. An experienced carpenter can panel a room 3 times faster than an apprentice can. Working together, they can panel the room in 6 hours. How long would it take each person working alone to do the job?

47. It normally takes 2 hours to fill a swimming pool. The pool has developed a slow leak. If the pool were full, it would take 10 hours for all the water to leak out. If the pool is empty, how long will it take to fill it?

48. Two investments have interest rates that differ by 1%. An investment for 1 year at the lower rate earns $175. The same principal invested for a year at the higher rate earns $200. What are the two interest rates?

Review Exercises

49. Factor: $25x^2 - 81$. (Section 6.4, Example 1)

50. Solve: $x^2 - 12x + 36 = 0$. (Section 6.6, Example 4)

51. Graph: $y = -\frac{2}{3}x + 4$. (Section 3.4, Example 3)

SECTION 7.8

Objectives

1 Solve direct variation problems.

2 Solve inverse variation problems.

3 Solve combined variation problems.

4 Solve problems involving joint variation.

MODELING USING VARIATION

Have you ever wondered how telecommunication companies estimate the number of phone calls expected per day between two cities? The formula

$$C = \frac{0.02P_1P_2}{d^2}$$

shows that the daily number of phone calls, C, increases as the populations of the cities, P_1 and P_2, in thousands, increase and decreases as the distance, d, between the cities increases.

Certain formulas occur so frequently in applied situations that they are given special names. Variation formulas show how one quantity changes in relation to other quantities. Quantities can vary *directly*, *inversely*, or *jointly*. In this section, we look at situations that can be modeled by each of these kinds of variation. And think of this. The next time you get one of those "all-circuits-are-busy" messages, you will be able to use a variation formula to estimate how many other callers you're competing with for those precious 5-cent minutes.

1 Solve direct variation problems.

Direct Variation When you swim underwater, the pressure in your ears depends on the depth at which you are swimming. The formula

$$p = 0.43d$$

describes the water pressure, p, in pounds per square inch, at a depth of d feet. We can use this linear function to determine the pressure in your ears at various depths:

If $d = 20$, $p = 0.43(20) = 8.6$. At a depth of 20 feet, water pressure is 8.6 pounds per square inch.

Doubling the depth doubles the pressure.

If $d = 40$, $p = 0.43(40) = 17.2$. At a depth of 40 feet, water pressure is 17.2 pounds per square inch.

Doubling the depth doubles the pressure.

If $d = 80$, $p = 0.43(80) = 34.4$. At a depth of 80 feet, water pressure is 34.4 pounds per square inch.

The formula $p = 0.43d$ illustrates that water pressure is a constant multiple of your underwater depth. If your depth is doubled, the pressure is doubled; if your depth is tripled, the pressure is triped; and so on. Because of this, the pressure in your ears is said to **vary directly** as your underwater depth. The **equation of variation** is

$$p = 0.43d.$$

Generalizing, we obtain the following statement:

> DIRECT VARIATION If a situation is described by an equation in the form
>
> $$y = kx$$
>
> where k is a constant, we say that ***y* varies directly as *x***. The number k is called the **constant of variation**.

Can you see that the direct variation equation, $y = kx$, is a special case of the linear equation $y = mx + b$? When $m = k$ and $b = 0$, $y = mx + b$ becomes $y = kx$. Thus, the slope of a direct variation equation is k, the constant of variation. Because b, the y-intercept, is 0, the graph of a direct variation equation is a line through the origin. This is illustrated in Figure 7.7, which shows the graph of $p = 0.43d$: Water pressure varies directly as depth.

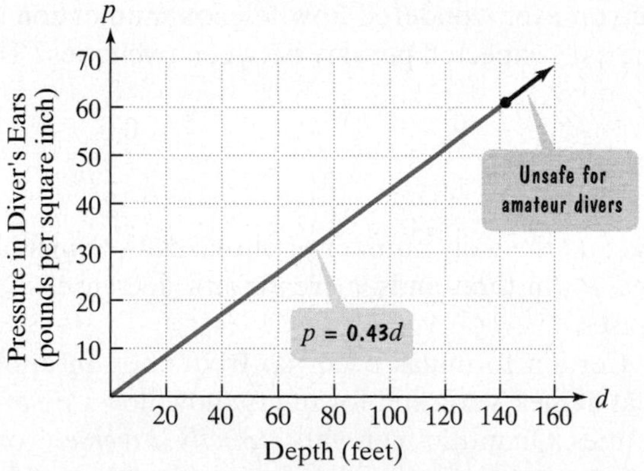

FIGURE 7.7 Water pressure at various depths

Problems involving direct variation can be solved using the following procedure. This procedure applies to direct variation problems, as well as to the other kinds of variation problems that we will discuss.

SOLVING VARIATION PROBLEMS

1. Write an equation that describes the given English statement.
2. Substitute the given pair of values into the equation in step 1 and find the value of k.
3. Substitute the value of k into the equation in step 1.
4. Use the equation from step 3 to answer the problem's question.

EXAMPLE 1 Solving a Direct Variation Problem

Many areas of Northern California depend on the snowpack of the Sierra Nevada mountain range for their water supply. The volume of water produced from melting snow varies directly as the volume of snow. Meteorologists have determined that 250 cubic centimeters of snow will melt to 28 cubic centimeters of water. How much water does 1200 cubic centimeters of melting snow produce?

SOLUTION

Step 1. **Write an equation.** We know that y varies directly as x is expressed as

$$y = kx.$$

By changing letters, we can write an equation that describes the following English statement: Volume of water, W, varies directly as volume of snow, S.

$$W = kS$$

Step 2. **Use the given values to find k.** We are told that 250 cubic centimeters of snow will melt to 28 cubic centimeters of water. Substitute 28 for W and 250 for S in the direct variation equation. Then solve for k.

$$W = kS$$ Volume of water varies directly as volume of melting snow.

$$28 = k(250)$$ 250 cubic centimeters of snow melt to 28 cubic centimeters of water.

$$\frac{28}{250} = \frac{k(250)}{250}$$ Divide both sides by 250.

$$0.112 = k$$ Simplify.

Step 3. **Substitute the value of k into the equation.**

$$W = kS$$ This is the equation from step 1.

$$W = 0.112S$$ Replace k, the constant of variation, with 0.112.

Step 4. **Answer the problem's question.** How much water does 1200 cubic centimeters of melting snow produce? Substitute 1200 for S in $W = 0.112S$ and solve for W.

$$W = 0.112S$$ Use the equation from step 3.

$$W = 0.112(1200)$$ Substitute 1200 for S.

$$W = 134.4$$ Multiply.

A snowpack measuring 1200 cubic centimeters will produce 134.4 cubic centimeters of water. ■

CHECK POINT 1 The number of gallons of water, W, used when taking a shower varies directly as the time, t, in minutes, in the shower. A shower lasting 5 minutes uses 30 gallons of water. How much water is used in a shower lasting 11 minutes?

The direct variation equation $y = kx$ is a linear equation. If $k > 0$, then the slope of the line is positive. Consequently, as x increases, y also increases.

2 Solve inverse variation problems.

Inverse Variation The distance from San Francisco to Los Angeles is 420 miles. The time that it takes to drive from San Francisco to Los Angeles depends on the rate at which one drives and is given by

$$\text{Time} = \frac{420}{\text{Rate}}.$$

For example, if you average 30 miles per hour, the time for the drive is

$$\text{Time} = \frac{420}{30} = 14,$$

or 14 hours. If you average 50 miles per hour, the time for the drive is

$$\text{Time} = \frac{420}{50} = 8.4,$$

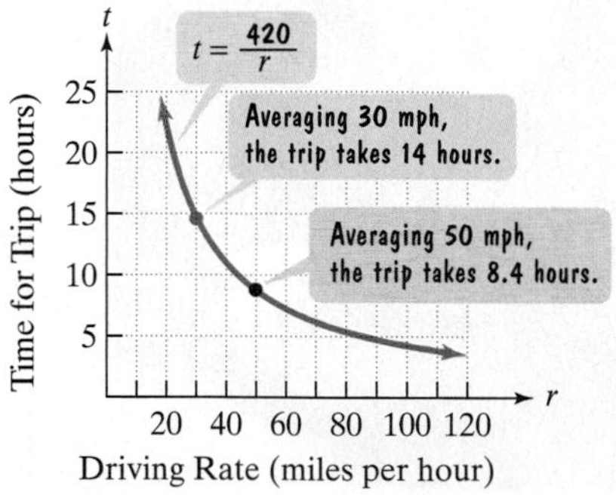

FIGURE 7.8

or 8.4 hours. As your rate (or speed) increases, the time for the trip decreases and vice versa. This is illustrated by the graph in Figure 7.8.

We can express the time for the San Francisco–Los Angeles trip using t for time and r for rate:

$$t = \frac{420}{r}.$$

This equation is an example of an **inverse variation** equation. Time, t, **varies inversely** as rate, r. When two quantities vary inversely, one quantity increases as the other decreases, and vice versa.

Generalizing, we obtain the following statement:

INVERSE VARIATION If a situation is described by an equation in the form

$$y = \frac{k}{x}$$

where k is a constant, we say that **y varies inversely as x.** The number k is called the **constant of variation.**

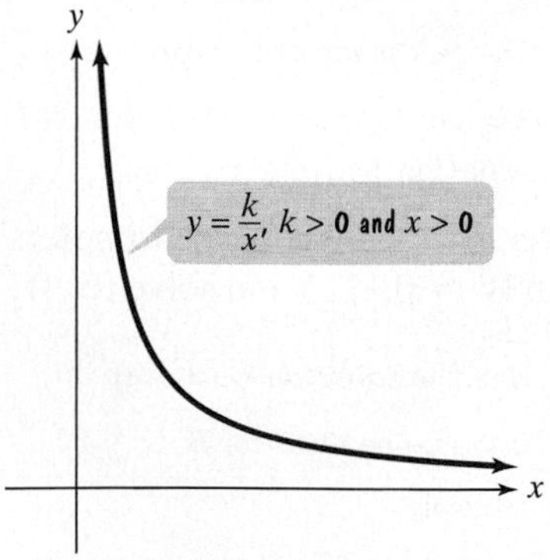

FIGURE 7.9 The graph of the inverse variation equation

Notice that the inverse variation equation

$$y = \frac{k}{x}$$

involves a rational expression, $\frac{k}{x}$. For $k > 0$ and $x > 0$, the graph of the equation takes on the shape shown in Figure 7.9. Under these conditions, as x increases, y decreases.

We use the same procedure to solve inverse variation problems as we did to solve direct variation problems. Example 2 illustrates this procedure.

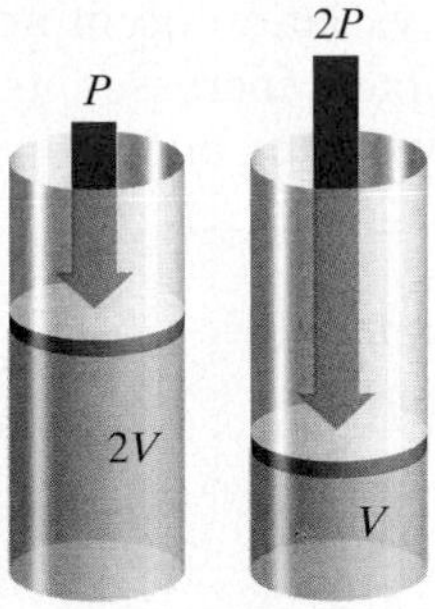

Doubling the pressure halves the volume.

EXAMPLE 2 Solving an Inverse Variation Problem

When you use a spray can and press the valve at the top, you decrease the pressure of the gas in the can. This decrease of pressure causes the volume of the gas in the can to increase. Because the gas needs more room than is provided in the can, it expands in spray form through the small hole near the valve. In general, if the temperature is constant, the pressure, P, of a gas in a container varies inversely as the volume, V, of the container. The pressure of a gas sample in a container whose volume is 8 cubic inches is 12 pounds per square inch. If the sample expands to a volume of 22 cubic inches, what is the new pressure of the gas?

SOLUTION

Step 1. **Write an equation.** We know that y varies inversely as x is expressed as

$$y = \frac{k}{x}.$$

By changing letters, we can write an equation that describes the following English statement: The pressure, P, of a gas in a container varies inversely as the volume, V.

$$P = \frac{k}{V}.$$

Step 2. **Use the given values to find k.** The pressure of a gas sample in a container whose volume is 8 cubic inches is 12 pounds per square inch. Substitute 12 for P and 8 for V in the inverse variation equation. Then solve for k.

$$P = \frac{k}{V}$$ Pressure varies inversely as volume.

$$12 = \frac{k}{8}$$ The pressure in an 8-cubic-inch container is 12 pounds per square inch.

$$12 \cdot 8 = \frac{k}{8} \cdot 8$$ Multiply both sides by 8.

$$96 = k$$ Simplify.

Step 3. **Substitute the value of k into the equation.**

$$P = \frac{k}{V}$$ Use the equation from step 1.

$$P = \frac{96}{V}$$ Replace k, the constant of variation, with 96.

Step 4. **Answer the problem's question.** We need to find the pressure when the volume expands to 22 cubic inches. Substitute 22 for V and solve for P.

$$P = \frac{96}{V} = \frac{96}{22} = 4\frac{4}{11}$$

When the volume is 22 cubic inches, the pressure of the gas is $4\frac{4}{11}$ pounds per square inch. ■

CHECK POINT 2 The length of a violin string varies inversely as the frequency of its vibrations. A violin string 8 inches long vibrates at a frequency of 640 cycles per second. What is the frequency of a 10-inch string?

3 Solve combined variation problems.

Combined Variation In **combined variation**, direct and inverse variation occur at the same time. For example, as the advertising budget, A, of a company increases, its monthly sales, S, also increase. Monthly sales vary directly as the advertising budget:

$$S = kA.$$

By contrast, as the price of the company's product, P, increases, its monthly sales, S, decrease. Monthly sales vary inversely as the price of the product:

$$S = \frac{k}{P}.$$

We can combine these two variation equations into one combined equation:

$$S = \frac{kA}{P}.$$

Monthly sales, S, vary directly as the advertising budget, A, and inversely as the price of the product, P.

The following example illustrates an application of combined variation:

EXAMPLE 3 Solving a Combined Variation Problem

The owners of Rollerblades Now determine that the monthly sales, S, of its skates vary directly as its advertising budget, A, and inversely as the price of the skates, P. When \$60,000 is spent on advertising and the price of the skates is \$40, the monthly sales are 12,000 pairs of rollerblades.

a. Write an equation of variation that describes this situation.

b. Determine monthly sales if the amount of the advertising budget is increased to \$70,000.

SOLUTION

a. Write an equation.

$$S = \frac{kA}{P}.$$

Translate "sales vary directly as the advertising budget and inversely as the skates' price."

Use the given values to find k.

$$12{,}000 = \frac{k(60{,}000)}{40}$$

When \$60,000 is spent on advertising ($A = 60{,}000$) and the price is \$40 ($P = 40$), monthly sales are 12,000 units ($S = 12{,}000$).

$$12{,}000 = k \cdot 1500$$

Divide 60,000 by 40.

$$\frac{12{,}000}{1500} = \frac{k \cdot 1500}{1500}$$

Divide both sides of the equation by 1500.

$$8 = k$$

Simplify.

Therefore, the equation of variation that describes monthly sales is

$$S = \frac{8A}{P}. \qquad \text{Substitute 8 for } k \text{ in } S = \frac{kA}{P}.$$

b. The advertising budget is increased to \$70,000, so $A = 70{,}000$. The skates' price is still \$40, so $P = 40$.

$$S = \frac{8A}{P} \qquad \text{This is the equation from part (a).}$$

$$S = \frac{8(70{,}000)}{40} \qquad \text{Substitute 70,000 for } A \text{ and 40 for } P.$$

$$S = 14{,}000 \qquad \text{Simplify.}$$

With a \$70,000 advertising budget and \$40 price, the company can expect to sell 14,000 pairs of rollerblades in a month (up from 12,000). ■

CHECK POINT 3 The number of minutes needed to solve an exercise set of variation problems varies directly as the number of problems and inversely as the number of people working to solve the problems. It takes 4 people 32 minutes to solve 16 problems. How many minutes will it take 8 people to solve 24 problems?

4 Solve problems involving joint variation.

Joint Variation Joint variation is a variation in which a variable varies directly as the product of two or more other variables. Thus, the equation $y = kxz$ is read "y varies jointly as x and z."

Joint variation plays a critical role in Isaac Newton's formula for gravitation:

$$F = G\frac{m_1 m_2}{d^2}.$$

The formula states that the force of gravitation, F, between two bodies varies jointly as the product of their masses, m_1 and m_2, and inversely as the square of the distance between them, d. (G is the gravitational constant.) The formula indicates that gravitational force exists between any two objects in the universe, increasing as the distance between the bodies decreases. One practical result is that the pull of the moon on the oceans is greater on the side of Earth closer to the moon. This gravitational imbalance is what produces tides.

EXAMPLE 4 Modeling Centrifugal Force

The centrifugal force, C, of a body moving in a circle varies jointly with the radius of the circular path, r, and the body's mass, m, and inversely with the square of the time, t, it takes to move about one full circle. A 6-gram body moving in a circle with radius 100 centimeters at a rate of 1 revolution in 2 seconds has a centrifugal force of 6000 dynes. Find the centrifugal force of an 18-gram body moving in a circle with radius 100 centimeters at a rate of 1 revolution in 3 seconds.

SOLUTION

$C = \dfrac{krm}{t^2}$	Translate "Centrifugal force, C, varies jointly with radius, r, and mass, m, and inversely with the square of time, t."
$6000 = \dfrac{k(100)(6)}{2^2}$	If $r = 100$, $m = 6$, and $t = 2$, then $C = 6000$.
$6000 = 150k$	Simplify.
$40 = k$	Divide both sides by 150 and solve for k.
$C = \dfrac{40rm}{t^2}$	Substitute 40 for k in the model for centrifugal force.
$C = \dfrac{40(100)(18)}{3^2}$	Find C when $r = 100$, $m = 18$, and $t = 3$.
$= 8000$	Simplify.

The centrifugal force is 8000 dynes. ■

CHECK POINT 4 The volume of a cone, V, varies jointly as its height, h, and the square of its radius r. A cone with a radius measuring 6 feet and a height measuring 10 feet has a volume of 120π cubic feet. Find the volume of a cone having a radius of 12 feet and a height of 2 feet.

7.8 EXERCISE SET

Student Solutions Manual | CD/Video | PH Math/Tutor Center | MathXL Tutorials on CD | MathXL® | MyMathLab | Interactmath.com

Practice Exercises

Use the four-step procedure for solving variation problems given on page 509 to solve Exercises 1–10.

1. y varies directly as x. $y = 65$ when $x = 5$. Find y when $x = 12$.
2. y varies directly as x. $y = 45$ when $x = 5$. Find y when $x = 13$.
3. y varies inversely as x. $y = 12$ when $x = 5$. Find y when $x = 2$.
4. y varies inversely as x. $y = 6$ when $x = 3$. Find y when $x = 9$.
5. y varies directly as x and inversely as the square of z. $y = 20$ when $x = 50$ and $z = 5$. Find y when $x = 3$ and $z = 6$.
6. a varies directly as b and inversely as the square of c. $a = 7$ when $b = 9$ and $c = 6$. Find a when $b = 4$ and $c = 8$.
7. y varies jointly as x and z. $y = 25$ when $x = 2$ and $z = 5$. Find y when $x = 8$ and $z = 12$.
8. C varies jointly as A and T. $C = 175$ when $A = 2100$ and $T = 4$. Find C when $A = 2400$ and $T = 6$.
9. y varies jointly as a and b and inversely as the square root of c. $y = 12$ when $a = 3, b = 2$, and $c = 25$. Find y when $a = 5, b = 3$, and $c = 9$.
10. y varies jointly as m and the square of n and inversely as p. $y = 15$ when $m = 2, n = 1$, and $p = 6$. Find y when $m = 3, n = 4$, and $p = 10$.

Practice Plus

In Exercises 11–20, write an equation that expresses each relationship. Then solve the equation for y.

11. x varies jointly as y and z.
12. x varies jointly as y and the square of z.
13. x varies directly as the cube of z and inversely as y.
14. x varies directly as the cube root of z and inversely as y.

15. x varies jointly as y and z and inversely as the square root of w.

16. x varies jointly as y and z and inversely as the square of w.

17. x varies jointly as z and the sum of y and w.

18. x varies jointly as z and the difference between y and w.

19. x varies directly as z and inversely as the difference between y and w.

20. x varies directly as z and inversely as the sum of y and w.

Application Exercises

Use the four-step procedure for solving variation problems given on page 509 to solve Exercises 21–36.

21. An alligator's tail length, T, varies directly as its body length, B. An alligator with a body length of 4 feet has a tail length of 3.6 feet. What is the tail length of an alligator whose body length is 6 feet?

22. An object's weight on the moon, M, varies directly as its weight on Earth, E. Neil Armstrong, the first person to step on the moon on July 20, 1969, weighed 360 pounds on Earth (with all of his equipment on) and 60 pounds on the moon. What is the moon weight of a person who weighs 186 pounds on Earth?

23. The height that a ball bounces varies directly as the height from which it was dropped. A tennis ball dropped from 12 inches bounces 8.4 inches. From what height was the tennis ball dropped if it bounces 56 inches?

24. The distance that a spring will stretch varies directly as the force applied to the spring. A force of 12 pounds is needed to stretch a spring 9 inches. What force is required to stretch the spring 15 inches?

25. If all men had identical body types, their weight would vary directly as the cube of their height. Shown at the top of the next column is Robert Wadlow, who reached a record height of 8 feet 11 inches (107 inches) before his death at age 22. If a man who is 5 feet 10 inches tall (70 inches) with the same body type as Mr. Wadlow weighs 170 pounds, what was Robert Wadlow's weight shortly before his death?

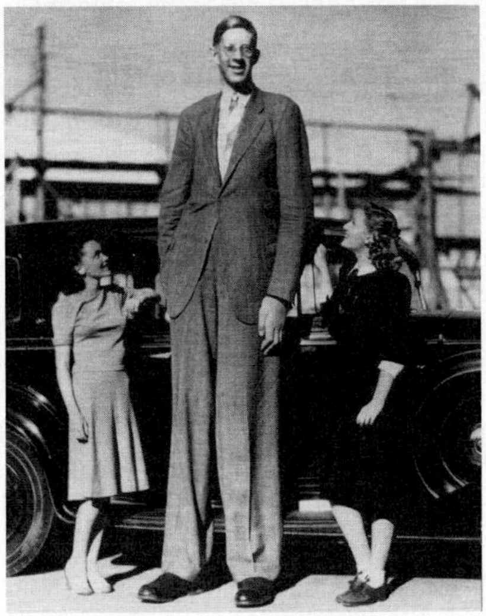

26. On a dry asphalt road, a car's stopping distance varies directly as the square of its speed. A car traveling at 45 miles per hour can stop in 67.5 feet. What is the stopping distance for a car traveling at 60 miles per hour?

27. The figure shows that a bicyclist tips the cycle when making a turn. The angle B, formed by the vertical direction and the bicycle, is called the banking angle. The banking angle varies inversely as the cycle's turning radius. When the turning radius is 4 feet, the banking angle is 28°. What is the banking angle when the turning radius is 3.5 feet?

28. The water temperature of the Pacific Ocean varies inversely as the water's depth. At a depth of 1000 meters, the water temperature is 4.4° Celsius. What is the water temperature at a depth of 5000 meters?

29. Radiation machines, used to treat tumors, produce an intensity of radiation that varies inversely as the square of the distance from the machine. At 3 meters, the radiation intensity is 62.5 milliroentgens per hour. What is the intensity at a distance of 2.5 meters?

30. The illumination provided by a car's headlight varies inversely as the square of the distance from the headlight. A car's headlight produces an illumination of 3.75 footcandles at a distance of 40 feet. What is the illumination when the distance is 50 feet?

31. Body-mass index, or BMI, takes both weight and height into account when assessing whether an individual is underweight or overweight. BMI varies directly as one's weight, in pounds, and inversely as the square of one's height, in inches. In adults, normal values for the BMI are between 20 and 25, inclusive. Values below 20 indicate that an individual is underweight and values above 30 indicate that an individual is obese. A person who weighs 180 pounds and is 5 feet, or 60 inches, tall has a BMI of 35.15. What is the BMI, to the nearest tenth, for a 170 pound person who is 5 feet 10 inches tall. Is this person overweight?

32. One's intelligence quotient, or IQ, varies directly as a person's mental age and inversely as that person's chronological age. A person with a mental age of 25 and a chronological age of 20 has an IQ of 125. What is the chronological age of a person with a mental age of 40 and an IQ of 80?

33. The heat loss of a glass window varies jointly as the window's area and the difference between the outside and inside temperatures. A window 3 feet wide by 6 feet long loses 1200 Btu per hour when the temperature outside is 20° colder than the temperature inside. Find the heat loss through a glass window that is 6 feet wide by 9 feet long when the temperature outside is 10° colder than the temperature inside.

34. Kinetic energy varies jointly as the mass and the square of the velocity. A mass of 8 grams and velocity of 3 centimeters per second has a kinetic energy of 36 ergs. Find the kinetic energy for a mass of 4 grams and velocity of 6 centimeters per second.

35. Sound intensity varies inversely as the square of the distance from the sound source. If you are in a movie theater and you change your seat to one that is twice as far from the speakers, how does the new sound intensity compare to that of your original seat?

36. Many people claim that as they get older, time seems to pass more quickly. Suppose that the perceived length of a period of time is inversely proportional to your age. How long will a year seem to be when you are three times as old as you are now?

37. The average number of daily phone calls, C, between two cities varies jointly as the product of their populations, P_1 and P_2, and inversely as the square of the distance, d, between them.

a. Write an equation that expresses this relationship.

b. The distance between San Francisco (population: 777,000) and Los Angeles (population: 3,695,000) is 420 miles. If the average number of daily phone calls between the cities is 326,000, find the value of k to two decimal places and write the equation of variation.

c. Memphis (population: 650,000) is 400 miles from New Orleans (population: 490,000). Find the average number of daily phone calls, to the nearest whole number, between these cities.

38. The force of wind blowing on a window positioned at a right angle to the direction of the wind varies jointly as the area of the window and the square of the wind's speed. It is known that a wind of 30 miles per hour blowing on a window measuring 4 feet by 5 feet exerts a force of 150 pounds. During a storm with winds of 60 miles per hour, should hurricane shutters be placed on a window that measures 3 feet by 4 feet and is capable of withstanding 300 pounds of force?

39. The table shows the values for the current, I, in an electric circuit and the resistance, R, of the circuit.

I (amperes)	0.5	1.0	1.5	2.0	2.5	3.0	4.0	5.0
R (ohms)	12	6.0	4.0	3.0	2.4	2.0	1.5	1.2

a. Graph the ordered pairs in the table of values, with values of I along the x-axis and values of R along the y-axis. Connect the eight points with a smooth curve.

b. Does current vary directly or inversely as resistance? Use your graph and explain how you arrived at your answer.

c. Write an equation of variation for I and R, using one of the ordered pairs in the table to find the constant of variation. Then use your variation equation to verify the other seven ordered pairs in the table.

Writing in Mathematics

40. What does it mean if two quantities vary directly?

41. In your own words, explain how to solve a variation problem.

42. What does it mean if two quantities vary inversely?

43. Explain what is meant by combined variation. Give an example with your explanation.

44. Explain what is meant by joint variation. Give an example with your explanation.

In Exercises 45–46, describe in words the variation shown by the given equation.

45. $z = \dfrac{k\sqrt{x}}{y^2}$

46. $z = kx^2\sqrt{y}$

47. We have seen that the daily number of phone calls between two cities varies jointly as their populations and inversely as the square of the distance between them. This model, used by telecommunication companies to estimate the line capacities needed among various cities, is called the *gravity model.* Compare the model to Newton's formula for gravitation on page 513 and describe why the name *gravity model* is appropriate.

Technology Exercise

48. Use a graphing utility to graph any three of the variation equations in Exercises 21–30. Then TRACE along each curve and identify the point that corresponds to the problem's solution.

Critical Thinking Exercises

49. In a hurricane, the wind pressure varies directly as the square of the wind velocity. If wind pressure is a measure of a hurricane's destructive capacity, what happens to this destructive power when the wind speed doubles?

50. The heat generated by a stove element varies directly as the square of the voltage and inversely as the resistance. If the voltage remains constant, what needs to be done to triple the amount of heat generated?

51. Galileo's telescope brought about revolutionary changes in astronomy. A comparable leap in our ability to observe the universe took place as a result of the Hubble Space Telescope. The space telescope can see stars and galaxies whose brightness is $\frac{1}{50}$ of the faintest objects now observable using ground-based telescopes. Use the fact that the brightness of a point source, such as a star, varies inversely as the square of its distance from an observer to show that the space telescope can see about seven times farther than a ground-based telescope.

Review Exercises

52. Solve:

$$8(2 - x) = -5x.$$

(Section 2.3, Example 2)

53. Divide:

$$\frac{27x^3 - 8}{3x + 2}.$$

(Section 5.6, Example 3)

54. Factor:

$$6x^3 - 6x^2 - 120x.$$

(Section 6.5, Example 2)

GROUP PROJECT

CHAPTER 7

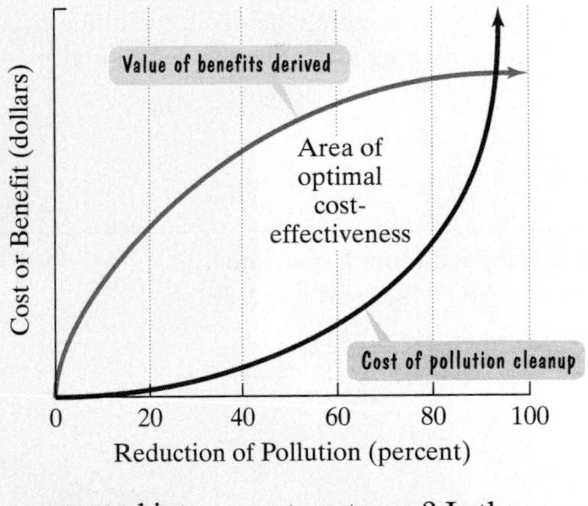

A cost-benefit analysis compares the estimated costs of a project with the benefits that will be achieved. Costs and benefits are given monetary values and compared using a benefit-cost ratio. As shown in the figure, a favorable ratio for a project means that the benefits outweigh the costs and the project is cost-effective. As a group, select an environmental project that was implemented in your area of the country. Research the cost and benefit graphs that resulted in the implementation of this project. How were the benefits converted into monetary terms? Is there an equation for either the cost model or the benefit model? Group members may need to interview members of environmental groups and businesses that were part of this project. You may wish to consult an environmental science textbook to find out more about cost-benefit analyses. After doing your research, the group should write or present a report explaining why the cost-benefit analysis resulted in the project's implementation.

CHAPTER 7 SUMMARY

Definitions and Concepts	Examples

Section 7.1 Rational Expressions and their Simplification

A rational expression is the quotient of two polynomials. To find values for which a rational expression is undefined, set the denominator equal to 0 and solve.

Find all numbers for which

$$\frac{7x}{x^2 - 3x - 4}$$

is undefined.

$$x^2 - 3x - 4 = 0$$
$$(x - 4)(x + 1) = 0$$
$$x - 4 = 0 \quad \text{or} \quad x + 1 = 0$$
$$x = 4 \qquad\qquad x = -1$$

Undefined at 4 and −1

To simplify a rational expression:

1. Factor the numerator and the denominator completely.
2. Divide the numerator and the denominator by any common factors.

If factors in the numerator and denominator are opposites, their quotient is −1.

Simplify: $\frac{3x + 18}{x^2 - 36}$.

$$\frac{3x + 18}{x^2 - 36} = \frac{3\overset{1}{\cancel{(x + 6)}}}{\underset{1}{\cancel{(x + 6)}}(x - 6)} = \frac{3}{x - 6}$$

Section 7.2 Multiplying and Dividing Rational Expressions

Multiplying Rational Expressions

1. Factor completely.
2. Divide numerators and denominators by common factors.
3. Multiply remaining factors in the numerators and multiply the remaining factors in the denominators.

$$\frac{x^2 + 3x - 10}{x^2 - 2x} \cdot \frac{x^2}{x^2 - 25}$$

$$= \frac{\overset{1}{\cancel{(x + 5)}}\overset{1}{\cancel{(x - 2)}}}{\underset{1}{\cancel{x}}\underset{1}{\cancel{(x - 2)}}} \cdot \frac{\overset{1}{\cancel{x}} \cdot x}{\underset{1}{\cancel{(x + 5)}}(x - 5)}$$

$$= \frac{x}{x - 5}$$

Dividing Rational Expressions

Invert the divisor and multiply.

$$\frac{3y + 3}{(y + 2)^2} \div \frac{y^2 - 1}{y + 2}$$

$$= \frac{3y + 3}{(y + 2)^2} \cdot \frac{y + 2}{y^2 - 1}$$

$$= \frac{3\overset{1}{\cancel{(y + 1)}}}{(y + 2)\underset{1}{\cancel{(y + 2)}}} \cdot \frac{\overset{1}{\cancel{(y + 2)}}}{\underset{1}{\cancel{(y + 1)}}(y - 1)}$$

$$= \frac{3}{(y + 2)(y - 1)}$$

Definitions and Concepts	Examples

Section 7.3 Adding and Subtracting Rational Expressions with the Same Denominator

To add or subtract rational expressions with the same denominator, add or subtract the numerators and place the result over the common denominator. If possible, simplify the resulting expression.

$$\frac{y^2 - 3y + 4}{y^2 + 8y + 15} - \frac{y^2 - 5y - 2}{y^2 + 8y + 15}$$
$$= \frac{y^2 - 3y + 4 - (y^2 - 5y - 2)}{y^2 + 8y + 15}$$
$$= \frac{y^2 - 3y + 4 - y^2 + 5y + 2}{y^2 + 8y + 15}$$
$$= \frac{2y + 6}{(y + 5)(y + 3)}$$
$$= \frac{2\overset{1}{\cancel{(y + 3)}}}{(y + 5)\underset{1}{\cancel{(y + 3)}}} = \frac{2}{y + 5}$$

To add or subtract rational expressions with opposite denominators, multiply either rational expression by $\frac{-1}{-1}$ to obtain a common denominator.

$$\frac{7}{x - 6} + \frac{x + 4}{6 - x}$$
$$= \frac{7}{x - 6} + \frac{(-1)}{(-1)} \cdot \frac{x + 4}{6 - x}$$
$$= \frac{7}{x + 6} + \frac{-x - 4}{x - 6}$$
$$= \frac{7 - x - 4}{x - 6} = \frac{3 - x}{x - 6}$$

Section 7.4 Adding and Subtracting Rational Expressions with Different Denominators

Finding the Least Common Denominator (LCD)

1. Factor denominators completely.
2. List factors of the first denominator.
3. Add to the list factors of the second denominator that are not already in the list.
4. The LCD is the product of factors in step 3.

Find the LCD of

$$\frac{x + 1}{2x - 2} \text{ and } \frac{2x}{x^2 + 2x - 3}.$$
$$2x - 2 = 2(x - 1)$$
$$x^2 + 2x - 3 = (x - 1)(x + 3)$$

Factors of first denominator: $2, x - 1$
Factors of second denominator not in the list: $x + 3$
LCD: $2(x - 1)(x + 3)$

Adding and Subtracting Rational Expressions with Different Denominators

1. Find the LCD.
2. Rewrite each rational expression as an equivalent expression with the LCD.
3. Add or subtract numerators, placing the resulting expression over the LCD.
4. If possible, simplify.

$$\frac{x + 1}{2x - 2} - \frac{2x}{x^2 + 2x - 3}$$
$$= \frac{x + 1}{2(x - 1)} - \frac{2x}{(x - 1)(x + 3)}$$

LCD is $2(x - 1)(x + 3)$.

$$= \frac{(x + 1)(x + 3)}{2(x - 1)(x + 3)} - \frac{2x \cdot 2}{2(x - 1)(x + 3)}$$
$$= \frac{x^2 + 4x + 3 - 4x}{2(x - 1)(x + 3)}$$
$$= \frac{x^2 + 3}{2(x - 1)(x + 3)}$$

Definitions and Concepts	Examples

Section 7.5 Complex Rational Expressions

Complex rational expressions have numerators or denominators containing one or more rational expressions. Complex rational expressions can be simplified by obtaining single expressions in the numerator and denominator and then dividing. They can also be simplified by multiplying the numerator and denominator by the LCD of all rational expressions within the complex rational expression.

Simplify by dividing: $\dfrac{\frac{1}{x}+5}{\frac{1}{x}-\frac{1}{3}}$.

$$=\frac{\frac{1}{x}+\frac{5x}{x}}{\frac{3}{3x}-\frac{x}{3x}}=\frac{\frac{1+5x}{x}}{\frac{3-x}{3x}}=\frac{1+5x}{\cancel{x}_1}\cdot\frac{3\cancel{x}^1}{3-x}$$

$$=\frac{3(1+5x)}{3-x}\quad\text{or}\quad\frac{3+15x}{3-x}$$

Simplify by the LCD method: $\dfrac{\frac{1}{x}+5}{\frac{1}{x}-\frac{1}{3}}$.

LCD is $3x$.

$$\frac{3x}{3x}\cdot\frac{\left(\frac{1}{x}+5\right)}{\left(\frac{1}{x}-\frac{1}{3}\right)}=\frac{3x\cdot\frac{1}{x}+3x\cdot 5}{3x\cdot\frac{1}{x}-3x\cdot\frac{1}{3}}$$

$$=\frac{3+15x}{3-x}$$

Section 7.6 Solving Rational Equations

A rational equation is an equation containing one or more rational expressions.

Solving Rational Equations

1. List restrictions on the variable.
2. Clear fractions by multiplying both sides by the LCD.
3. Solve the resulting equation.
4. Reject any proposed solution in the list of restrictions. Check other proposed solutions in the original equation.

Solve: $\dfrac{7x}{x^2-4}+\dfrac{5}{x-2}=\dfrac{2x}{x^2-4}$

$$\frac{7x}{(x+2)(x-2)}+\frac{5}{x-2}=\frac{2x}{(x+2)(x-2)}$$

Denominators would equal 0 if $x=-2$ or $x=2$. Restrictions: $x\neq-2$ and $x\neq 2$.

LCD is $(x+2)(x-2)$.

$$(x+2)(x-2)\left[\frac{7x}{(x+2)(x-2)}+\frac{5}{x-2}\right]=(x+2)(x-2)\cdot\frac{2x}{(x+2)(x-2)}$$

$$7x+5(x+2)=2x$$
$$7x+5x+10=2x$$
$$12x+10=2x$$
$$10=-10x$$
$$-1=x$$

The proposed solution, -1, is not part of the restriction $x\neq-2$ and $x\neq 2$. It checks. The solution is -1 and the solution set is $\{-1\}$.

Definitions and Concepts | Examples

Section 7.7 Applications Using Rational Equations and Proportions

Motion problems involving time are solved using

$$t = \frac{d}{r}.$$

$$\text{Time traveled} = \frac{\text{Distance traveled}}{\text{Rate of travel}}$$

It takes a cyclist who averages 16 miles per hour in still air the same time to travel 48 miles with the wind as 16 miles against the wind. What is the wind's rate?

x = wind's rate

$16 + x$ = cyclist's rate with wind

$16 - x$ = cyclist's rate against wind

	Distance	Rate	Time = $\frac{\text{Distance}}{\text{Rate}}$
With wind	48	$16 + x$	$\frac{48}{16 + x}$
Against wind	16	$16 - x$	$\frac{9}{16 - x}$

Two times are equal

$$\frac{48}{16 + x} = \frac{16}{16 - x}$$

$$(16 + x)(16 - x) \cdot \frac{48}{16 + x} = \frac{16}{16 - x} \cdot (16 + x)(16 - x)$$

$$48(16 - x) = 16(16 + x)$$

Solving this equation, $x = 8$.
The wind's rate is 8 miles per hour.

Work problems are solved using the following condition:

Fraction of job done by the first + fraction of job done by the second = 1.

One pipe fills a pool in 20 hours and a second pipe in 15 hours. How long will it take to fill the pool using both pipes?

x = time using both pipes

Fraction of pool filled by pipe 1 in x hours + fraction of pool filled by pipe 2 in x hours = 1.

$$\frac{x}{20} + \frac{x}{15} = 1$$

$$60\left(\frac{x}{20} + \frac{x}{15}\right) = 60 \cdot 1$$

$$3x + 4x = 60$$

$$7x = 60$$

$$x = \frac{60}{7} = 8\frac{4}{7} \text{ hours}$$

It will take $8\frac{4}{7}$ hours for both pipes to fill the pool.

Definitions and Concepts	Examples

Section 7.7 Applications Using Rational Equations and Proportions (continued)

A proportion is a statement in the form $\frac{a}{b} = \frac{c}{d}$. The cross-products principle states that if $\frac{a}{b} = \frac{c}{d}$, then $ad = bc$ ($b \neq 0$ and $d \neq 0$).

Solving Applied Problems Using Proportions

1. Read the problem and represent the unknown quantity by x (or any letter).
2. Set up a proportion by listing the given ratio on one side and the ratio with the unknown quantity on the other side.
3. Drop units and apply the cross-products principle.
4. Solve for x and answer the question.

30 elk are tagged and released. Sometime later, a sample of 80 elk are observed and 10 are tagged. How many elk are there?

x = number of elk

Tagged / Total

$$\frac{30}{x} = \frac{10}{80}$$

$$10x = 30 \cdot 80$$

$$10x = 2400$$

$$x = 240$$

There are 240 elk.

Similar triangles have the same shape, but not necessarily the same size. Corresponding angles have the same measure, and corresponding sides are proportional. If the measures of two angles of one triangle are equal to those of two angles of a second triangle, then the two triangles are similar.

Find x for these similar triangles.

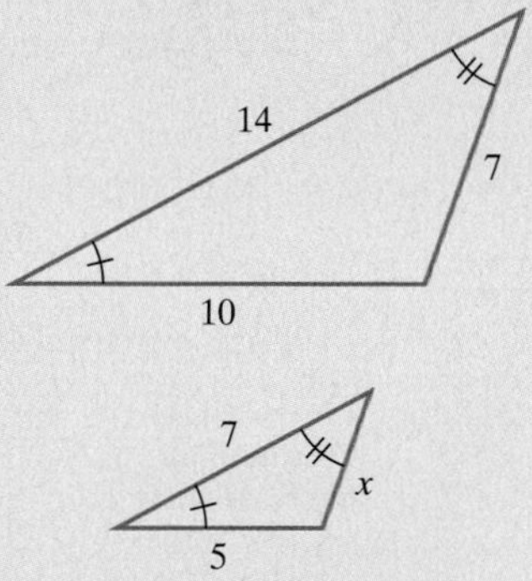

Corresponding sides are proportional:

$$\frac{7}{x} = \frac{10}{5}. \quad \left(\text{or } \frac{7}{x} = \frac{14}{7}\right)$$

$$5x \cdot \frac{7}{x} = \frac{10}{5} \cdot 5x$$

$$35 = 10x$$

$$x = \frac{35}{10} = 3.5$$

Definitions and Concepts	Examples

Section 7.8 Modeling Using Variation

English Statement	Equation
y varies directly as x.	$y = kx$
y varies directly as x^n.	$y = kx^n$
y varies inversely as x.	$y = \frac{k}{x}$
y varies inversely as x^n.	$y = \frac{k}{x^n}$
y varies directly as x and inversely as z.	$y = \frac{kx}{z}$
y varies jointly as x and z.	$y = kxz$

The time that it takes you to drive a certain distance varies inversely as your driving rate. Averaging 40 miles per hour, it takes you 10 hours to drive the distance. How long would the trip take averaging 50 miles per hour?

1. $t = \frac{k}{r}$ Time, t, varies inversely as rate, r.

2. It takes 10 hours at 40 miles per hour.

$$10 = \frac{k}{40}$$

$$k = 10(40) = 400$$

3. $t = \frac{400}{r}$

4. How long at 50 miles per hour? Substitute 50 for r.

$$t = \frac{400}{50} = 8$$

It takes 8 hours at 50 miles per hour.

CHAPTER 7 REVIEW EXERCISES

7.1 *In Exercises 1–4, find all numbers for which each rational expression is undefined. If the rational expression is defined for all real numbers, so state.*

1. $\frac{5x}{6x - 24}$

2. $\frac{x + 3}{(x - 2)(x + 5)}$

3. $\frac{x^2 + 3}{x^2 - 3x + 2}$

4. $\frac{7}{x^2 + 81}$

In Exercises 5–12, simplify each rational expression. If the rational expression cannot be simplified, so state.

5. $\frac{16x^2}{12x}$

6. $\frac{x^2 - 4}{x - 2}$

7. $\frac{x^3 + 2x^2}{x + 2}$

8. $\frac{x^2 + 3x - 18}{x^2 - 36}$

9. $\frac{x^2 - 4x - 5}{x^2 + 8x + 7}$

10. $\frac{y^2 + 2y}{y^2 + 4y + 4}$

11. $\frac{x^2}{x^2+4}$

12. $\frac{2x^2 - 18y^2}{3y - x}$

7.2 *In Exercises 13–17, multiply as indicated.*

13. $\frac{x^2 - 4}{12x} \cdot \frac{3x}{x + 2}$

14. $\frac{5x + 5}{6} \cdot \frac{3x}{x^2 + x}$

15. $\frac{x^2 + 6x + 9}{x^2 - 4} \cdot \frac{x - 2}{x + 3}$

16. $\frac{y^2 - 2y + 1}{y^2 - 1} \cdot \frac{2y^2 + y - 1}{5y - 5}$

17. $\frac{2y^2 + y - 3}{4y^2 - 9} \cdot \frac{3y + 3}{5y - 5y^2}$

In Exercises 18–22, divide as indicated.

18. $\frac{x^2 + x - 2}{10} \div \frac{2x + 4}{5}$

19. $\frac{6x + 2}{x^2 - 1} \div \frac{3x^2 + x}{x - 1}$

20. $\frac{1}{y^2 + 8y + 15} \div \frac{7}{y + 5}$

21. $\frac{y^2 + y - 42}{y - 3} \div \frac{y + 7}{(y - 3)^2}$

22. $\frac{8x + 8y}{x^2} \div \frac{x^2 - y^2}{x^2}$

7.3 *In Exercises 23–28, add or subtract as indicated. Simplify the result, if possible.*

23. $\frac{4x}{x+5} + \frac{20}{x+5}$

24. $\frac{8x-5}{3x-1} + \frac{4x+1}{3x-1}$

25. $\frac{3x^2+2x}{x-1} - \frac{10x-5}{x-1}$

26. $\frac{6y^2-4y}{2y-3} - \frac{12-3y}{2y-3}$

27. $\frac{x}{x-2} + \frac{x-4}{2-x}$

28. $\frac{x+5}{x-3} - \frac{x}{3-x}$

7.4 *In Exercises 29–31, find the least common denominator of the rational expressions.*

29. $\frac{7}{9x^3}$ and $\frac{5}{12x}$

30. $\frac{3}{x^2(x-1)}$ and $\frac{11}{x(x-1)^2}$

31. $\frac{x}{x^2+4x+3}$ and $\frac{17}{x^2+10x+21}$

In Exercises 32–42, add or subtract as indicated. Simplify the result, if possible.

32. $\frac{7}{3x} + \frac{5}{2x^2}$

33. $\frac{5}{x+1} + \frac{2}{x}$

34. $\frac{7}{x+3} + \frac{4}{(x+3)^2}$

35. $\frac{6y}{y^2-4} - \frac{3}{y+2}$

36. $\frac{y-1}{y^2-2y+1} - \frac{y+1}{y-1}$

37. $\frac{x+y}{y} - \frac{x-y}{x}$

38. $\frac{2x}{x^2+2x+1} + \frac{x}{x^2-1}$

39. $\frac{5x}{x+1} - \frac{2x}{1-x^2}$

40. $\frac{4}{x^2-x-6} - \frac{4}{x^2-4}$

41. $\frac{7}{x+3} + 2$

42. $\frac{2y-5}{6y+9} - \frac{4}{2y^2+3y}$

7.5 *In Exercises 43–47, simplify each complex rational expression.*

43. $\frac{\frac{1}{2}+\frac{3}{8}}{\frac{3}{4}-\frac{1}{2}}$

44. $\frac{\frac{1}{x}}{1-\frac{1}{x}}$

45. $\frac{\frac{1}{x}+\frac{1}{y}}{\frac{1}{xy}}$

46. $\frac{\frac{1}{x}-\frac{1}{2}}{\frac{1}{3}-\frac{x}{6}}$

47. $\frac{3+\frac{12}{x}}{1-\frac{16}{x^2}}$

7.6 *In Exercises 48–55, solve each rational equation. If an equation has no solution, so state.*

48. $\frac{3}{x} - \frac{1}{6} = \frac{1}{x}$

49. $\frac{3}{4x} = \frac{1}{x} + \frac{1}{4}$

50. $x + 5 = \frac{6}{x}$

51. $4 - \frac{x}{x+5} = \frac{5}{x+5}$

52. $\frac{2}{x-3} = \frac{4}{x+3} + \frac{8}{x^2-9}$

53. $\frac{2}{x} = \frac{2}{3} + \frac{x}{6}$

54. $\frac{13}{y-1} - 3 = \frac{1}{y-1}$

55. $\frac{1}{x+3} - \frac{1}{x-1} = \frac{x+1}{x^2+2x-3}$

56. Park rangers introduce 50 elk into a wildlife preserve. The formula

$$P = \frac{250(3t+5)}{t+25}$$

models the elk population, P, after t years. How many years will it take for the population to increase to 125 elk?

57. The formula

$$S = \frac{C}{1-r}$$

describes the selling price, S, of a product in terms of its cost to the retailer, C, and its markup, r, usually expressed as a percent. A small television cost a retailer \$140 and was sold for \$200. Find the markup. Express the answer as a percent.

In Exercises 58–62, solve each formula for the specified variable.

58. $P = \dfrac{R - C}{n}$ for C

59. $\dfrac{P_1V_1}{T_1} = \dfrac{P_2V_2}{T_2}$ for T_1

60. $T = \dfrac{A - P}{Pr}$ for P

61. $\dfrac{1}{R} = \dfrac{1}{R_1} + \dfrac{1}{R_2}$ for R

62. $I = \dfrac{nE}{R + nr}$ for n

7.7

63. In still water, a paddle boat averages 20 miles per hour. It takes the boat the same amount of time to travel 72 miles downstream, with the current, as 48 miles upstream, against the current. What is the rate of the water's current?

64. A car travels 60 miles in the same time that a car traveling 10 miles per hour faster travels 90 miles. What is the rate of each car?

65. A painter can paint a fence around a house in 6 hours. Working alone, the painter's apprentice can paint the same fence in 12 hours. How many hours would it take them to do the job if they worked together?

66. If a school board determines that there should be 3 teachers for every 50 students, how many teachers are needed for an enrollment of 5400 students?

67. To determine the number of trout in a lake, a conservationist catches 112 trout, tags them, and returns them to the lake. Later, 82 trout are caught, and 32 of them are found to be tagged. How many trout are in the lake?

68. The triangles shown in the figure are similar. Find the length of the side marked with an x.

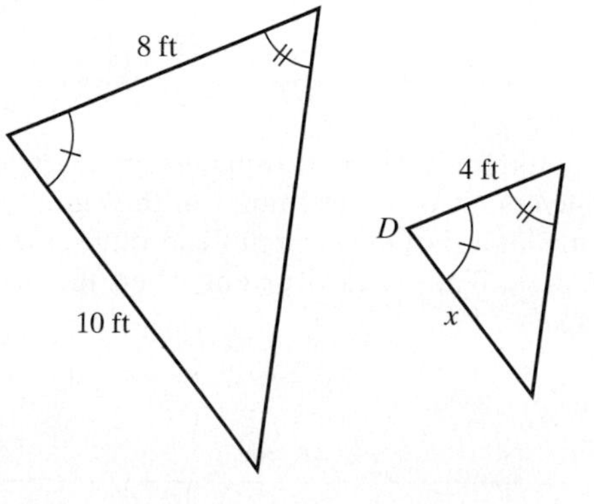

69. Find the height of the lamppost in the figure.

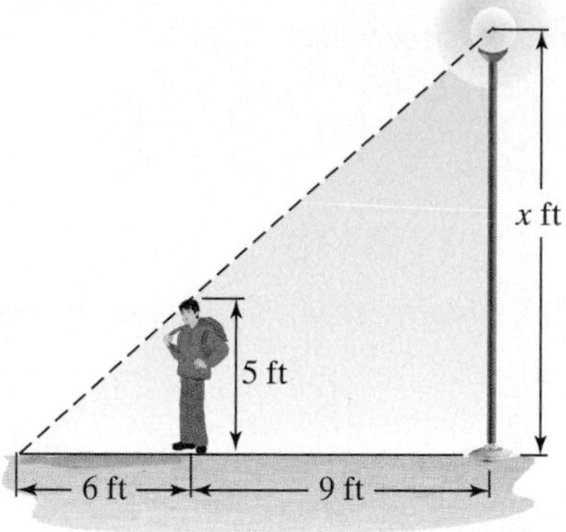

7.8 *Solve the variation problems in Exercises 70–75.*

70. A company's profit varies directly as the number of products it sells. The company makes a profit of $1175 on the sale of 25 products. What is the company's profit when it sells 105 products?

71. The distance that a body falls from rest varies directly as the square of the time of the fall. If skydivers fall 144 feet in 3 seconds, how far will they fall in 10 seconds?

72. The pitch of a musical tone varies inversely as its wavelength. A tone has a pitch of 660 vibrations per second and a wavelength of 1.6 feet. What is the pitch of a tone that has a wavelength of 2.4 feet?

73. The loudness of a stereo speaker, measured in decibels, varies inversely as the square of your distance from the speaker. When you are 8 feet from the speaker, the loudness is 28 decibels. What is the loudness when you are 4 feet from the speaker?

74. The time required to assemble computers varies directly as the number of computers assembled and inversely as the number of workers. If 30 computers can be assembled by 6 workers in 10 hours, how long would it take 5 workers to assemble 40 computers?

75. The volume of a pyramid varies jointly as its height and the area of its base. A pyramid with a height of 15 feet and a base with an area of 35 square feet has a volume of 175 cubic feet. Find the volume of a pyramid with a height of 20 feet and a base with an area of 120 square feet.

CHAPTER 7 TEST

Remember to use your Chapter Test Prep Video CD to see the worked-out solutions to the test questions you want to review.

1. Find all numbers for which

$$\frac{x+7}{x^2+5x-36}$$

is undefined.

In Exercises 2–3, simplify each rational expression.

2. $\dfrac{x^2+2x-3}{x^2-3x+2}$

3. $\dfrac{4x^2-20x}{x^2-4x-5}$

In Exercises 4–16, perform the indicated operations. Simplify the result, if possible.

4. $\dfrac{x^2-16}{10} \cdot \dfrac{5}{x+4}$

5. $\dfrac{x^2-7x+12}{x^2-4x} \cdot \dfrac{x^2}{x^2-9}$

6. $\dfrac{2x+8}{x-3} \div \dfrac{x^2+5x+4}{x^2-9}$

7. $\dfrac{5y+5}{(y-3)^2} \div \dfrac{y^2-1}{y-3}$

8. $\dfrac{2y^2+5}{y+3} + \dfrac{6y-5}{y+3}$

9. $\dfrac{y^2-2y+3}{y^2+7y+12} - \dfrac{y^2-4y-5}{y^2+7y+12}$

10. $\dfrac{x}{x+3} + \dfrac{5}{x-3}$

11. $\dfrac{2}{x^2-4x+3} + \dfrac{6}{x^2+x-2}$

12. $\dfrac{4}{x-3} + \dfrac{x+5}{3-x}$

13. $1 + \dfrac{3}{x-1}$

14. $\dfrac{2x+3}{x^2-7x+12} - \dfrac{2}{x-3}$

15. $\dfrac{8y}{y^2-16} - \dfrac{4}{y-4}$

16. $\dfrac{(x-y)^2}{x+y} \div \dfrac{x^2-xy}{3x+3y}$

In Exercises 17–18, simplify each complex rational expression.

17. $\dfrac{5+\dfrac{5}{x}}{2+\dfrac{1}{x}}$

18. $\dfrac{\dfrac{1}{x}-\dfrac{1}{y}}{\dfrac{1}{x}}$

In Exercises 19–21, solve each rational equation.

19. $\dfrac{5}{x} + \dfrac{2}{3} = 2 - \dfrac{2}{x} - \dfrac{1}{6}$

20. $\dfrac{3}{y+5} - 1 = \dfrac{4-y}{2y+10}$

21. $\dfrac{2}{x-1} = \dfrac{3}{x^2-1} + 1$

22. Solve: $R = \dfrac{as}{a+s}$ for a.

23. In still water, a boat averages 30 miles per hour. It takes the boat the same amount of time to travel 16 miles downstream, with the current, as 14 miles upstream, against the current. What is the rate of the water's current?

24. One pipe can fill a hot tub in 20 minutes and a second pipe can fill it in 30 minutes. If the hot tub is empty, how long will it take both pipes to fill it?

25. Park rangers catch, tag, and release 200 tule elk back into a wildlife refuge. Two weeks later they observe a sample of 150 elk, of which 5 are tagged. Assuming that the ratio of tagged elk in the sample holds for all elk in the refuge, how many elk are there in the park?

26. The triangles in the figure are similar. Find the length of the side marked with an x.

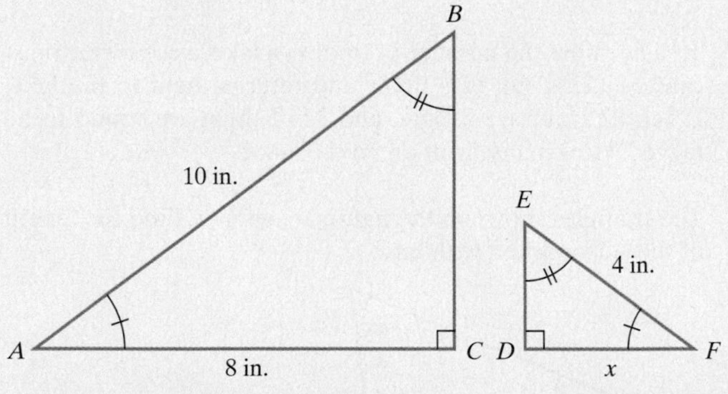

27. The amount of current flowing in an electrical circuit varies inversely as the resistance in the circuit. When the resistance in a particular circuit is 5 ohms, the current is 42 amperes. What is the current when the resistance is 4 ohms?

CUMULATIVE REVIEW EXERCISES (CHAPTERS 1–7)

In Exercises 1–6, solve each equation, inequality, or system of equations.

1. $2(x - 3) + 5x = 8(x - 1)$

2. $-3(2x - 4) > 2(6x - 12)$

3. $x^2 + 3x = 18$

4. $\dfrac{2x}{x^2 - 4} + \dfrac{1}{x - 2} = \dfrac{2}{x + 2}$

5. $y = 2x - 3$
$x + 2y = 9$

6. $3x + 2y = -2$
$-4x + 5y = 18$

In Exercises 7–9, graph each equation in a rectangular coordinate system.

7. $3x - 2y = 6$ **8.** $y = -2x + 3$ **9.** $y = -3$

In Exercises 10–12, simplify each expression.

10. $-21 - 16 - 3(2 - 8)$

11. $\left(\dfrac{4x^5}{2x^2}\right)^3$

12. $\dfrac{\frac{1}{x} - 2}{4 - \frac{1}{x}}$

In Exercises 13–15, factor completely.

13. $4x^2 - 13x + 3$

14. $4x^2 - 20x + 25$

15. $3x^2 - 75$

In Exercises 16–18, perform the indicated operations.

16. $(4x^2 - 3x + 2) - (5x^2 - 7x - 6)$

17. $\dfrac{-8x^6 + 12x^4 - 4x^2}{4x^2}$

18. $\dfrac{x + 6}{x - 2} + \dfrac{2x + 1}{x + 3}$

19. You invested \$4000, part at 5% and the remainder at 9% annual interest. At the end of the year, the total interest from these investments was \$311. How much was invested at each rate?

20. A 68-inch board is to be cut into two pieces. If one piece must be three times as long as the other, find the length of each piece.

MID-TEXTBOOK CHECK POINT Are You Prepared for Intermediate Algebra?

Algebra is cumulative. This means that **your performance in intermediate algebra depends heavily on the skills you acquired in introductory algebra**. Do you need a quick review of introductory algebra topics before starting the intermediate algebra portion of this book? This mid-textbook Check Point provides a fast way to review and practice the prerequisite skills needed in intermediate algebra.

STUDY TIP

You can quickly review the major topics of introductory algebra by studying the review grids for each of the first seven chapters in this book. Each chart summarizes the definitions and concepts in every section of the chapter. Examples that illustrate the key concepts are also included in the chart. The review charts, with worked-out examples, for each of the book's first seven chapters begin on pages 92, 174, 236, 298, 375, 435, and 518.

A Diagnostic Test for Your Introductory Algebra Skills. You can use the 36 exercises in this mid-textbook Check Point to test your understanding of introductory algebra topics. These exercises cover the fundamental algebra skills upon which the intermediate algebra portion of this book is based. Here are some suggestions for using these exercises as a diagnostic test:

1. Work through all 36 items at your own pace.
2. Use the answer section in the back of the book to check your work.
3. If your answer differs from that in the answer section or if you are not certain how to proceed with a particular item, turn to the section and the worked-out example given in parentheses at the end of each exercise. Study the step-by-step solution of the example that parallels the exercise and then try working the exercise again. If you feel that you need more assistance, study the entire section in which the example appears and work on a selected group of exercises in the exercise set for that section.

In Exercises 1–7, solve each equation, inequality, or system of equations.

1. $2 - 4(x + 2) = 5 - 3(2x + 1)$ (Section 2.3, Example 3)

2. $\frac{x}{2} - 3 = \frac{x}{5}$ (Section 2.3, Example 4)

3. $3x + 9 \geq 5(x - 1)$ (Section 2.7, Example 7)

4. $2x + 3y = 6$
$x + 2y = 5$ (Section 4.3, Example 2)

5. $3x - 2y = 1$
$y = 10 - 2x$ (Section 4.2, Example 1)

6. $\frac{3}{x + 5} - 1 = \frac{4 - x}{2x + 10}$ (Section 7.6, Example 4)

7. $x + \frac{6}{x} = -5$ (Section 7.6, Example 3)

In Exercises 8–16, perform the indicated operations. If possible, simplify the answer.

8. $\frac{12x^3}{3x^{12}}$ (Section 5.7, Example 4)

9. $4 \cdot 6 \div 2 \cdot 3 + (-5)$ (Section 1.8, Example 4)

10. $(6x^2 - 8x + 3) - (-4x^2 + x - 1)$ (Section 5.1, Example 3)

11. $(7x + 4)(3x - 5)$ (Section 5.3, Example 2)

12. $(5x - 2)^2$ (Section 5.3, Example 6)

13. $(x + y)(x^2 - xy + y^2)$ (Section 5.4, Example 8)

14. $\frac{x^2 + 6x + 8}{x^2} \div (3x^2 + 6x)$ (Section 7.2, Example 7)

15. $\frac{x}{x^2 + 2x - 3} - \frac{x}{x^2 - 5x + 4}$ (Section 7.4, Example 7)

16. $\frac{x - \frac{1}{5}}{5 - \frac{1}{x}}$ (Section 7.5, Examples 2 and 5)

In Exercises 17–22, factor completely.

17. $4x^2 - 49$ (Section 6.4, Example 1)

18. $x^3 + 3x^2 - x - 3$ (Section 6.5, Example 4)

19. $2x^2 + 8x - 42$ (Section 6.5, Example 2)

20. $x^5 - 16x$ (Section 6.4, Example 4)

21. $x^3 - 10x^2 + 25x$ (Section 6.4, Example 5)

22. $x^3 - 8$ (Section 6.4, Example 8)

In Exercises 23–25, graph each equation in a rectangular coordinate system.

23. $y = \frac{1}{3}x - 1$ (Section 3.4, Example 3)

24. $3x + 2y = -6$ (Section 3.2, Example 4)

25. $y = -2$ (Section 3.2, Example 7)

26. Find the slope of the line passing through the points $(-1, 3)$ and $(2, -3)$. (Section 3.3, Example 1)

27. Write the point-slope form of the equation of the line passing through the points $(1, 2)$ and $(3, 6)$. Then use the point-slope equation to write the slope-intercept form of the line's equation. (Section 3.5, Example 2)

In Exercises 28–36, use an equation or a system of equations to solve each problem.

28. Seven subtracted from five times a number is 208. Find the number. (Section 2.5, Example 2)

29. After a 20% reduction, a digital camera sold for \$256. What was the price before the reduction? (Section 2.5, Example 7)

30. A rectangular field is three times as long as it is wide. If the perimeter of the field is 400 yards, what are the field's dimensions? (Section 2.5, Example 6)

31. You invested \$20,000 in two accounts paying 7% and 9% annual interest, respectively. If the total interest earned for the year was \$1550, how much was invested at each rate? (Section 4.4, Example 4)

32. A chemist needs to mix a 40% acid solution with a 70% acid solution to obtain 12 liters of a 50% acid solution. How many liters of each solution should be used? (Section 4.4, Example 5)

33. A sailboat has a triangular sail with an area of 120 square feet and a base that is 15 feet long. Find the height of the sail. (Section 2.6, Example 1)

34. In a triangle, the measure of the first angle is 10° more than the measure of the second angle. The measure of the third angle is 20° more than four times that of the second angle. What is the measure of each angle? (Section 2.6, Example 6)

35. A salesperson works in the TV and stereo department of an electronics store. One day she sold 3 TVs and 4 stereos for \$2530. The next day, she sold 4 of the same TVs and 3 of the same stereos for \$2510. Find the price of a TV and a stereo. (Section 4.4, Example 1)

36. The length of a rectangle is 6 meters more than the width. The area is 55 square meters. Find the rectangle's dimensions. (Section 6.6, Example 8)

CHAPTER 8

'Tis the season and you've waited until the last minute to mail your holiday gifts. Your only option is overnight express mail. You realize that the cost of mailing a gift depends on its weight, but the mailing costs seem somewhat odd. Your packages that weigh 1.1 pounds, 1.5 pounds, and 2 pounds cost $15.75 each to send overnight. Packages that weigh 2.01 pounds and 3 pounds cost you $18.50 each. Finally, your heaviest gift is barely over 3 pounds and its mailing cost is $21.25. What sort of system is this in which costs increase by $2.75, stepping from $15.75 to $18.50 and from $18.50 to $21.25?

Graphs that ascend in steps are explored in Exercises 77–80 in Exercise Set 8.1.

Basics of Functions

The cost of mailing a package depends on its weight. The probability that you and another person in a room share the same birthday depends on the number of people in the room. In both of these situations, the relationship between variables can be illustrated with the notion of a *function*. Understanding this concept will give you a new perspective on many ordinary situations. Much of our work in the intermediate algebra portion of this book will be devoted to the important topic of functions and how they model your world.

SECTION 8.1

Objectives

1 Find the domain and range of a relation.
2 Determine whether a relation is a function.
3 Evaluate a function.
4 Use the vertical line test to identify functions.
5 Obtain information about a function from its graph.
6 Identify the domain and range of a function from its graph.

INTRODUCTION TO FUNCTIONS

Jerry Orbach	\$34,000
Charles Shaughnessy	\$31,800
Andy Richter	\$29,400
Norman Schwarzkopf	\$28,000
Jon Stewart	\$28,000

The answer: See the above list. The question: Who are *Celebrity Jeopardy's* five all-time highest earners? The list indicates a correspondence between the five all-time highest earners and their winnings. We can write this correspondence using a set of ordered pairs:

{(Orbach, \$34,000), (Shaughnessy, \$31,800), (Richter, \$29,400), (Schwarzkopf, \$28,000), (Stewart, \$28,000)}.

The mathematical term for a set of ordered pairs is a *relation*.

DEFINITION OF A RELATION A **relation** is any set of ordered pairs. The set of all first components of the ordered pairs is called the **domain** of the relation and the set of all second components is called the **range** of the relation.

1 Find the domain and range of a relation.

EXAMPLE 1 Finding the Domain and Range of a Relation

Find the domain and range of the relation:

{(Orbach, \$34,000), (Shaughnessy, \$31,800), (Richter, \$29,400), (Schwarzkopf, \$28,000), (Stewart, \$28,000)}.

SOLUTION The domain is the set of all first components. Thus, the domain is

{Orbach, Shaughnessy, Richter, Schwarzkopf, Stewart}.

The range is the set of all second components. Thus, the range is

{\$34,000, \$31,800, \$29,400, \$28,000}.

Although Schwarzkopf and Stewart both won \$28,000, it is not necessary to list \$28,000 twice.

■

CHECK POINT 1 Find the domain and the range of the relation:

$\{(5, 12.8), (10, 16.2), (15, 18.9), (20, 20.7), (25, 21.8)\}$.

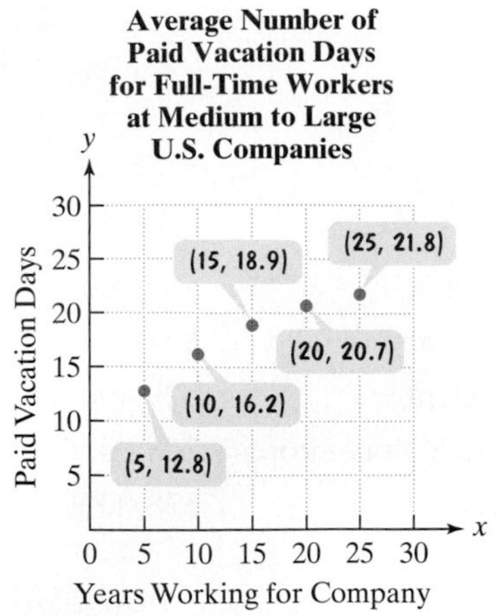

FIGURE 8.1 The graph of a relation showing a correspondence between years with a company and paid vacation days

Source: Bureau of Labor Statistics

As you worked Check Point 1, did you wonder if there was a rule that assigned the "inputs" in the domain to the "outputs" in the range? For example, for the ordered pair (15, 18.9), how does the output 18.9 depend on the input 15? Think paid vacation days! The first number in each ordered pair is the number of years that a full-time employee has been employed by a medium to large U.S. company. The second number is the average number of paid vacation days each year. Consider, for example, the ordered pair (15, 18.9).

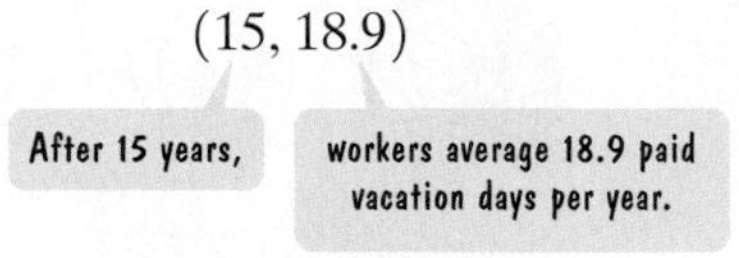

The relation in the vacation-days example can be pictured as follows:

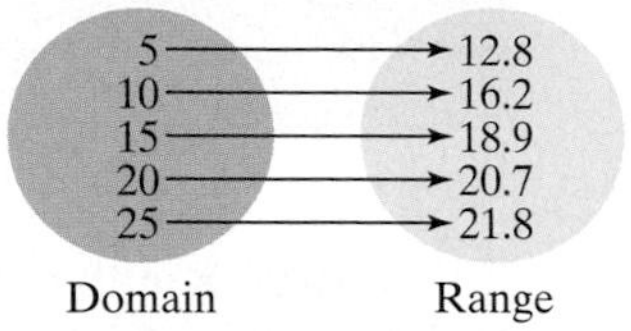

The five points in Figure 8.1 are another way to visually represent the relation.

2 Determine whether a relation is a function.

Jerry Orbach	$34,000
Charles Shaughnessy	$31,800
Andy Richter	$29,400
Norman Schwarzkopf	$28,000
Jon Stewart	$28,000

Functions Shown, again, in the margin are *Celebrity Jeopardy's* five all-time highest winners and their winnings. We've used this information to define two relations. Figure 8.2(a) shows a correspondence between winners and their winnings. Figure 8.2(b) shows a correspondence between winnings and winners.

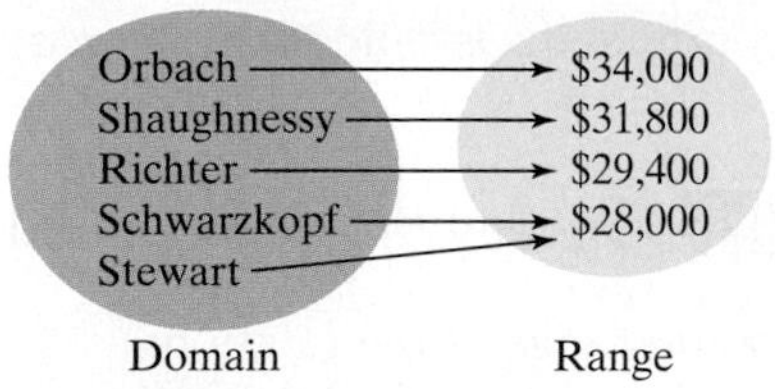

FIGURE 8.2 (a) Winners correspond to winnings.

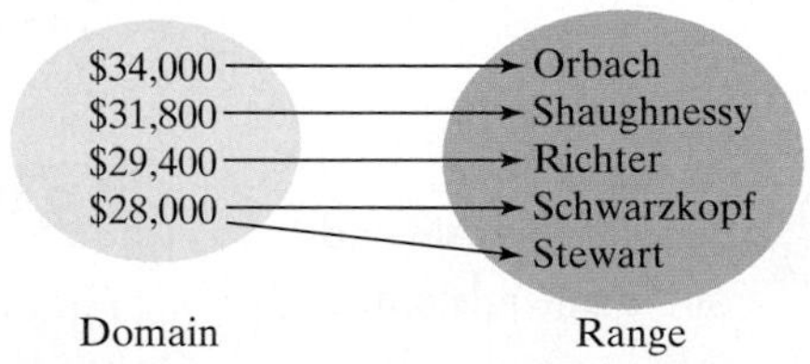

FIGURE 8.2 (b) Winnings correspond to winners.

A relation in which each member of the domain corresponds to exactly one member of the range is a **function**. Can you see that the relation in Figure 8.2(a) is a function? Each winner in the domain corresponds to exactly one winning amount in the range. If we know the winner, we can be sure of the amount won. Notice that more than one element in the domain can correspond to the same element in the range. (Schwarzkopf and Stewart both won $28,000.)

Is the relation in Figure 8.2(b) a function? Does each member of the domain correspond to precisely one member of the range? This relation is not a function because there is a member of the domain that corresponds to two different members of the range:

($28,000, Schwarzkopf) ($28,000, Stewart).

The member of the domain, $28,000, corresponds to both Schwarzkopf and Stewart in the range. If we know the amount won, $28,000, we cannot be sure of the winner. Because **a function is a relation in which no two ordered pairs have the same first**

component and different second components, the ordered pairs($28,000, Schwarzkopf) and ($28,000, Stewart) are not ordered pairs of a function.

Same first component

($28,000, Schwarzkopf) ($28,000, Stewart)

Different second components

> DEFINITION OF A FUNCTION A **function** is a correspondence from a first set, called the **domain**, to a second set, called the **range**, such that each element in the domain corresponds to *exactly one* element in the range.

Example 2 illustrates that not every correspondence between sets is a function.

EXAMPLE 2 Determining Whether a Relation Is a Function

Determine whether each relation is a function:

a. $\{(1, 5), (2, 5), (3, 7), (4, 8)\}$ **b.** $\{(5, 1), (5, 2), (7, 3), (8, 4)\}$.

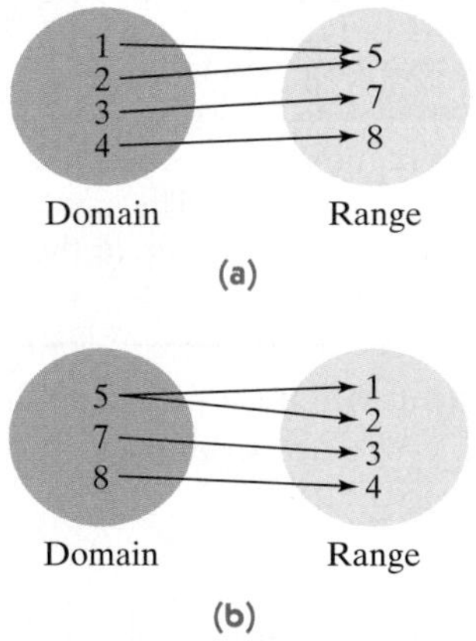

FIGURE 8.3

SOLUTION We begin by making a figure for each relation that shows the domain and the range (Figure 8.3).

a. Figure 8.3(a) shows that every element in the domain corresponds to exactly one element in the range. The element 1 in the domain corresponds to the element 5 in the range. Furthermore, 2 corresponds to 5, 3 corresponds to 7, and 4 corresponds to 8. No two ordered pairs in the given relation have the same first component and different second components. Thus, the relation is a function.

b. Figure 8.3(b) shows that 5 corresponds to both 1 and 2. If any element in the domain corresponds to more than one element in the range, the relation is not a function. This relation is not a function because two ordered pairs have the same first component and different second components.

Same first component

$(5, 1)$ $(5, 2)$

Different second components

STUDY TIP

If a relation is a function, reversing the components in each of its ordered pairs may result in a relation that is not a function.

Look at Figure 8.3(a) again. The fact that 1 and 2 in the domain correspond to the same number, 5, in the range does not violate the definition of a function. **A function can have two different first components with the same second component**. By contrast, a relation is not a function when two different ordered pairs have the same first component and different second components. Thus, the relation in Example 2(b) is not a function.

CHECK POINT 2 Determine whether each relation is a function:

a. $\{(1, 2), (3, 4), (5, 6), (5, 7)\}$

b. $\{(1, 2), (3, 4), (6, 5), (7, 5)\}$.

3 Evaluate a function.

Functions as Equations and Function Notation Functions are usually given in terms of equations rather than as sets of ordered pairs. For example, here is an equation that models paid vacation days each year as a function of years working for a company:

$$y = -0.016x^2 + 0.93x + 8.5.$$

The variable x represents years working for a company. The variable y represents the average number of vacation days each year. The variable y is a function of the variable x. For each value of x, there is one and only one value of y. The variable x is called the **independent variable** because it can be assigned any value from the domain. Thus, x can be assigned any positive integer representing years working for a company. The variable y is called the **dependent variable** because its value depends on x. Paid vacation days depend on years working for a company. The value of the dependent variable, y, is calculated after selecting a value for the independent variable, x.

If an equation in x and y gives one and only one value of y for each value of x, then the variable y is a function of the variable x. When an equation represents a function, the function is often named by a letter such as f, g, h, F, G, or H. Any letter can be used to name a function. Suppose that f names a function. Think of the domain as the set of the function's inputs and the range as the set of the function's outputs. As shown in Figure 8.4, the input is represented by x and the output by $f(x)$. The special notation $f(x)$, read "f of x" or "f at x," represents the **value of the function at the number x.**

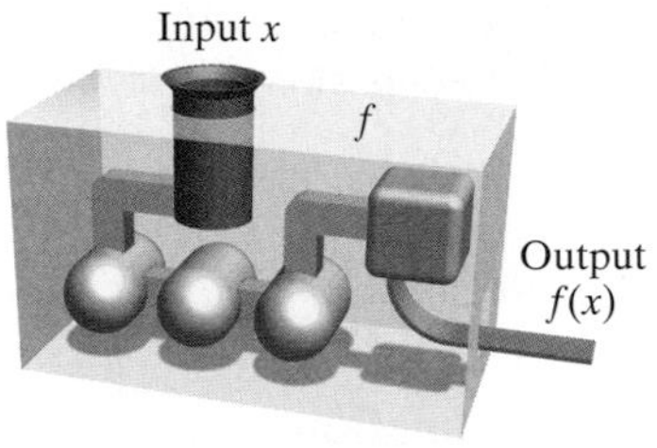

FIGURE 8.4 A "function machine" with inputs and outputs

Let's make this clearer by considering a specific example. We know that the equation

$$y = -0.016x^2 + 0.93x + 8.5$$

defines y as a function of x. We'll name the function f. Now, we can apply our new function notation.

Input: x Output: $f(x)$ Equation: $f(x) = -0.016x^2 + 0.93x + 8.5$

We read this equation as "f of x equals $-0.016x^2 + 0.93x + 8.5$."

$$x \quad f(x) \quad f(x) = -0.016x^2 + 0.93x + 8.5$$

STUDY TIP

The notation $f(x)$ does *not* mean "f times x." The notation describes the value of the function at x.

Suppose we are interested in finding $f(10)$, the function's output when the input is 10. To find the value of the function at 10, we substitute 10 for x. We are **evaluating the function** at 10.

$f(x) = -0.016x^2 + 0.93x + 8.5$	This is the given function.
$f(10) = -0.016(10)^2 + 0.93(10) + 8.5$	Replace each occurrence of x with 10.
$= -0.016(100) + 0.93(10) + 8.5$	Evaluate the exponential expression: $10^2 = 100$.
$= -1.6 + 9.3 + 8.5$	Perform the multiplications.
$= 16.2$	Add from left to right.

The statement $f(10) = 16.2$, read "f of 10 equals 16.2," tells us that the value of the function at 10 is 16.2. When the function's input is 10, its output is 16.2. (After 10 years, workers average 16.2 vacation days each year.) To find other function values, such as $f(15)$, $f(20)$, or $f(23)$, substitute the specified input values for x into the function's equation.

If a function is named f and x represents the independent variable, the notation $f(x)$ corresponds to the y-value for a given x. Thus,

$$f(x) = -0.016x^2 + 0.93x + 8.5 \quad \text{and} \quad y = -0.016x^2 + 0.93x + 8.5$$

define the same function. This function may be written as

$$y = f(x) = -0.016x^2 + 0.93x + 8.5.$$

USING TECHNOLOGY

Graphing utilities can be used to evaluate functions. The screens below show the evaluation of

$f(x) = -0.016x^2 + 0.93x + 8.5$

at 10 on a T1-83 Plus graphing calculator. The function f is named Y_1.

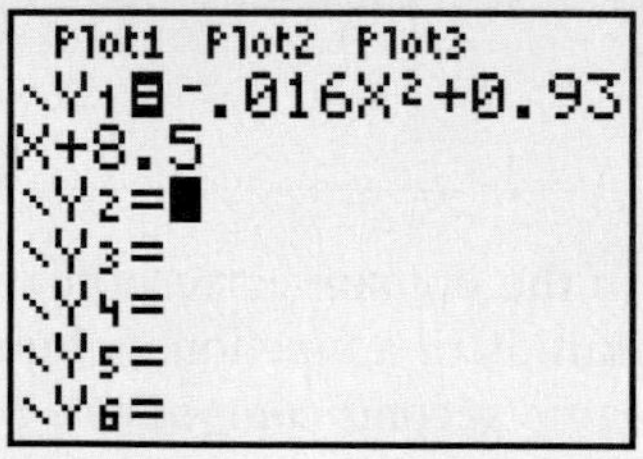

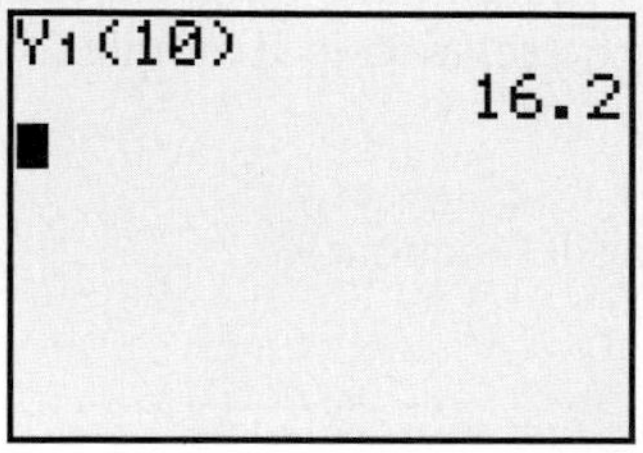

EXAMPLE 3 Using Function Notation

Find the indicated function value:

a. $f(4)$ for $f(x) = 2x + 3$

b. $g(-2)$ for $g(x) = 2x^2 - 1$

c. $h(-5)$ for $h(r) = r^3 - 2r^2 + 5$

d. $F(a + h)$ for $F(x) = 5x + 7$.

SOLUTION

a.

$$f(x) = 2x + 3$$ This is the given function.

$$f(4) = 2 \cdot 4 + 3$$ To find f of 4, replace x with 4.

$$= 8 + 3$$ Multiply: $2 \cdot 4 = 8$.

$$= 11$$ Add.

f of 4 is 11.

b.

$$g(x) = 2x^2 - 1$$ This is the given function.

$$g(-2) = 2(-2)^2 - 1$$ To find g of -2, replace x with -2.

$$= 2(4) - 1$$ Evaluate the exponential expression: $(-2)^2 = 4$.

$$= 8 - 1$$ Multiply: $2(4) = 8$.

$$= 7$$ Subtract.

g of -2 is 7.

c.

$$h(r) = r^3 - 2r^2 + 5$$ The function's name is h and r represents the independent variable.

$$h(-5) = (-5)^3 - 2(-5)^2 + 5$$ To find h of -5, replace each occurrence of r with -5.

$$= -125 - 2(25) + 5$$ Evaluate exponential expressions.

$$= -125 - 50 + 5$$ Multiply.

$$= -170$$ $-125 - 50 = -175$ and $-175 + 5 = -170$.

h of -5 is -170.

d.

$$F(x) = 5x + 7$$ This is the given function.

$$F(a + h) = 5(a + h) + 7$$ Replace x with $a + h$.

$$= 5a + 5h + 7$$ Apply the distributive property. ■

F of $a + h$ is $5a + 5h + 7$.

CHECK POINT 3 Find the indicated function value:

a. $f(6)$ for $f(x) = 4x + 5$

b. $g(-5)$ for $g(x) = 3x^2 - 10$

c. $h(-4)$ for $h(r) = r^2 - 7r + 2$

d. $F(a + h)$ for $F(x) = 6x + 9$.

STUDY TIP

Equations of the form $y = mx + b$, now expressed as $f(x) = mx + b$, were presented in Section 3.4. You can find a brief review of the slope-intercept form of the equation of a line in the Section 3.4 summary on page 238.

The functions in Check Point 3(a) and 3(d), $f(x) = 4x + 5$ and $F(x) = 6x + 9$, are examples of *linear functions*. **Linear functions** have equations of the form $f(x) = mx + b$. By contrast, the functions in Check Point 3(b) and 3(c), $g(x) = 3x^2 - 10$ and $h(r) = r^2 - 7r + 2$, are examples of *quadratic functions*. **Quadratic functions** have equations of the form $f(x) = ax^2 + bx + c$, $a \neq 0$.

4 Use the vertical line test to identify functions.

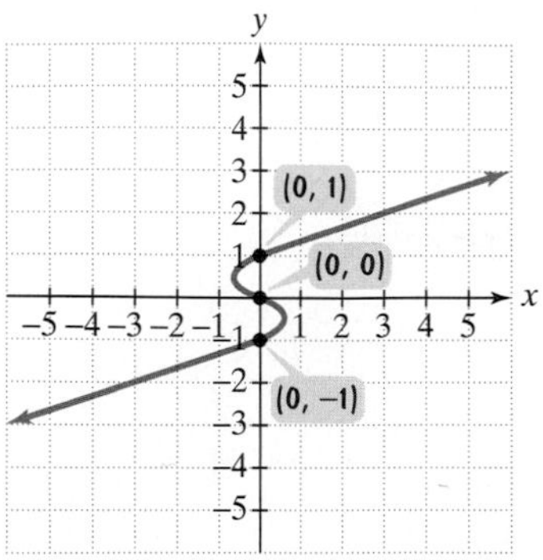

FIGURE 8.5 y is not a function of x because 0 is paired with three values of y, namely, 1, 0, and -1.

Graphs of Functions and the Vertical Line Test The **graph of a function** is the graph of its ordered pairs. For example, the graph of $f(x) = 3x + 1$ is the set of points (x, y) in the rectangular coordinate system satisfying the equation $y = 3x + 1$. Thus, the graph of f is a line with slope 3 and y-intercept 1. Similarly, the graph of $f(x) = x^2 + 3x + 5$ is the set of points (x, y) in the rectangular coordinate system satisfying the equation $y = x^2 + 3x + 5$.

Not every graph in the rectangular coordinate system is the graph of a function. The definition of a function specifies that no value of x can be paired with two or more different values of y. Consequently, if a graph contains two or more different points with the same first coordinate, the graph cannot represent a function. This is illustrated in Figure 8.5. Observe that points sharing a common first coordinate are vertically above or below each other.

This observation is the basis of a useful test for determining whether a graph defines y as a function of x. The test is called the **vertical line test**.

THE VERTICAL LINE TEST FOR FUNCTIONS If any vertical line intersects a graph in more than one point, the graph does not define y as a function of x.

EXAMPLE 4 Using the Vertical Line Test

Use the vertical line test to identify graphs in which y is a function of x.

a.

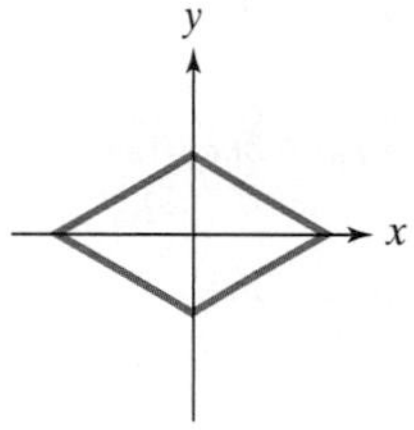

b.

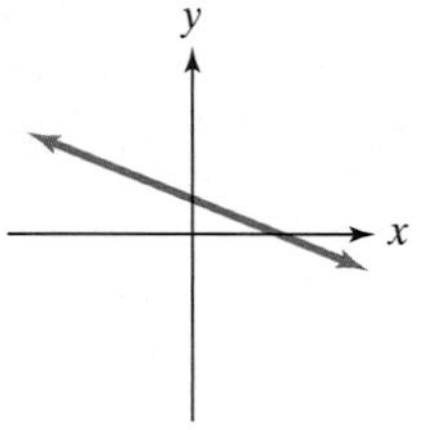

c.

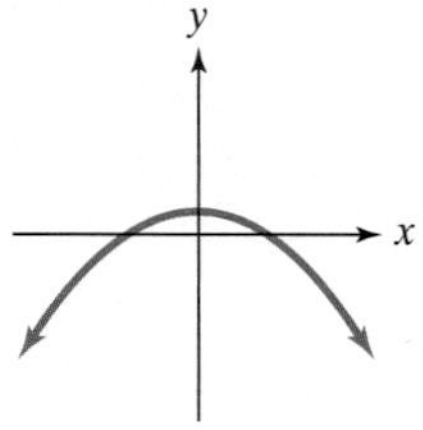

d.

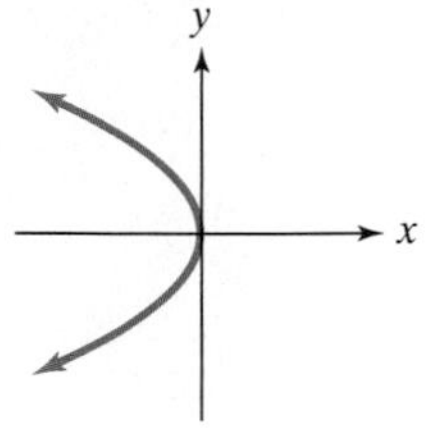

SOLUTION y is a function of x for the graphs in (b) and (c).

a.

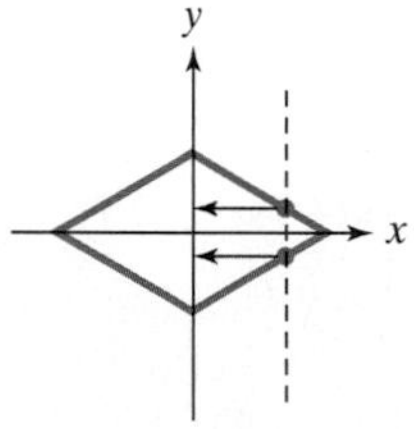

y is not a function of x. Two values of y correspond to one x-value.

b.

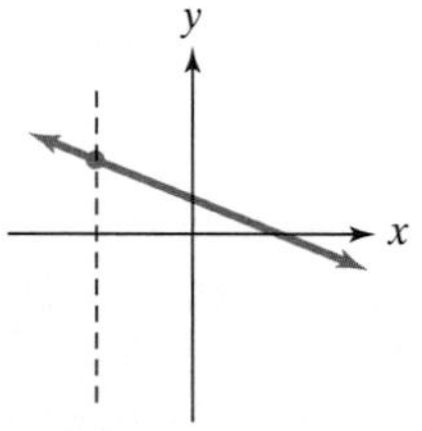

y is a function of x.

c.

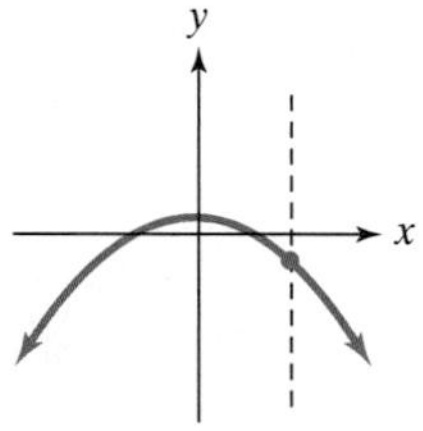

y is a function of x.

d.

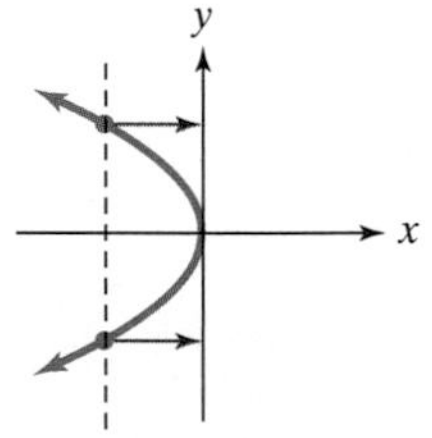

y is not a function of x. Two values of y correspond to one x-value. ■

CHECK POINT 4 Use the vertical line test to identify graphs in which y is a function of x.

a.

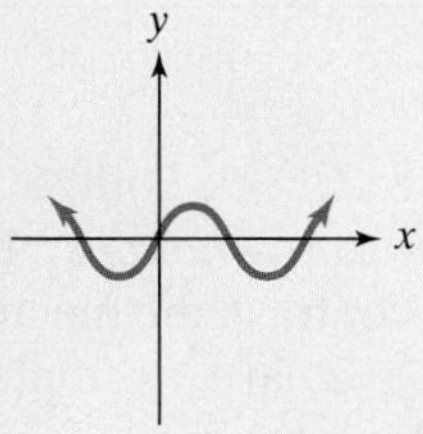

b.

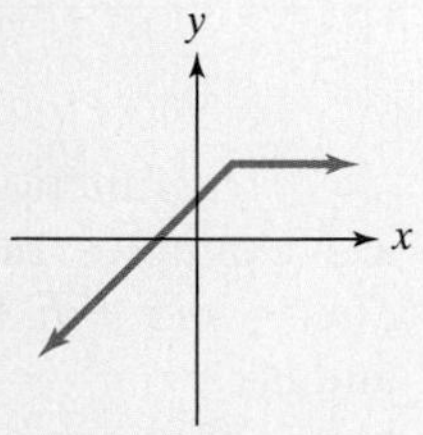

c.

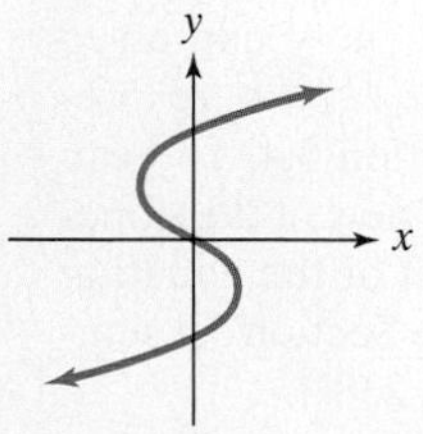

5 Obtain information about a function from its graph.

Obtaining Information from Graphs You can obtain information about a function from its graph. At the right or left of a graph, you will often find closed dots, open dots, or arrows.

- A closed dot indicates that the graph does not extend beyond this point and the point belongs to the graph.
- An open dot indicates that the graph does not extend beyond this point and the point does not belong to the graph.
- An arrow indicates that the graph extends indefinitely in the direction in which the arrow points.

EXAMPLE 5 Analyzing the Graph of a Function

The function

$$f(x) = -0.016x^2 + 0.93x + 8.5$$

models the average number of paid vacation days each year, $f(x)$, for full-time workers at medium to large U.S. companies after x years. The graph of f is shown in Figure 8.6.

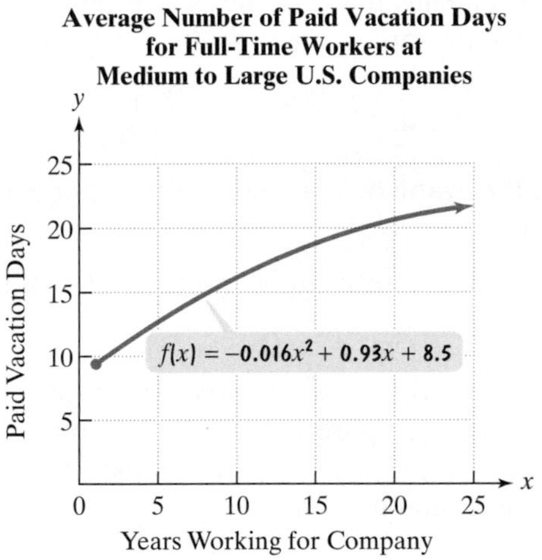

FIGURE 8.6

Source: Bureau of Labor Statistics.

a. Explain why f represents the graph of a function.

b. Use the graph to find a reasonable estimate of $f(5)$.

c. For what value of x is $f(x) = 20$?

d. Describe the general trend shown by the graph.

SOLUTION

a. No vertical line intersects the graph of f more than once. By the vertical line test, f represents the graph of a function.

b. To find $f(5)$, or f of 5, we locate 5 on the x-axis. The figure shows the point on the graph of f for which 5 is the first coordinate. From this point, we look to the y-axis to find the corresponding y-coordinate. A reasonable estimate of the y-coordinate is 13. Thus, $f(5) \approx 13$. After 5 years, a worker can expect approximately 13 paid vacation days.

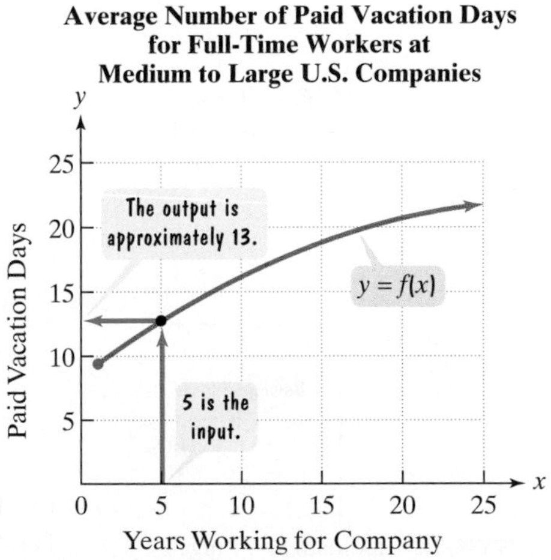

c. To find the value of x for which $f(x) = 20$, we locate 20 on the y-axis. The figure shows that there is one point on the graph of f for which 20 is the second coordinate. From this point, we look to the x-axis to find the corresponding x-coordinate. A reasonable estimate of the x-coordinate is 18. Thus, $f(x) = 20$ for $x \approx 18$. A worker with 20 paid vacation days has been with the company approximately 18 years.

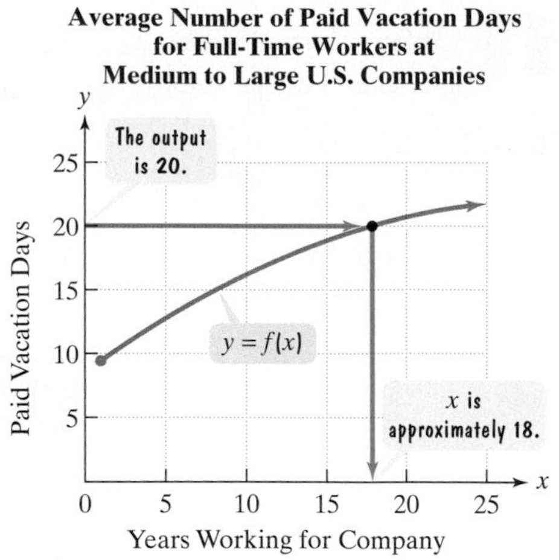

d. The graph of f is rising from left to right. This shows that paid vacation days increase as time with the company increases. However, the rate of increase is slowing down as the graph moves to the right. This means that the increase in paid vacation days takes place more slowly the longer an employee is with the company. ■

CHECK POINT 5

a. Use the graph of f in Figure 8.6 on page 537 to find a reasonable estimate of $f(10)$.

b. For what value of x is $f(x) = 15$? Round to the nearest whole number.

6 Identify the domain and range of a function from its graph.

Identifying Domain and Range from a Function's Graph Figure 8.7 illustrates how the graph of a function is used to determine the function's domain and its range.

Domain: set of inputs

Found on the x-axis

Range: set of outputs

Found on the y-axis

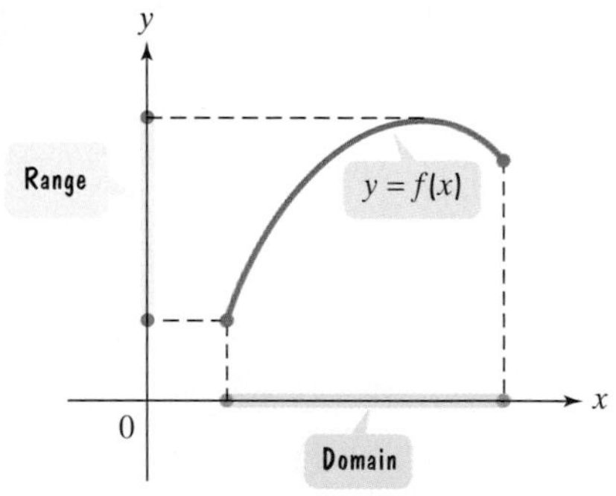

FIGURE 8.7 Domain and range of f

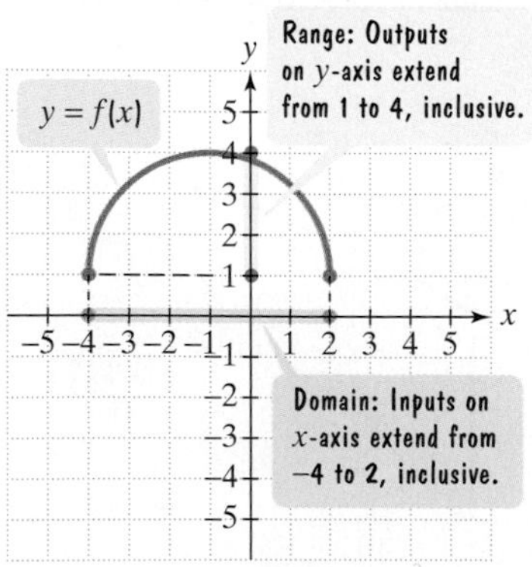

FIGURE 8.8 Domain and range of f

Let's apply these ideas to the graph of the function shown in Figure 8.8. To find the domain, look for all the inputs on the x-axis that correspond to points on the graph. Can you see that they extend from -4 to 2, inclusive? Using set-builder notation, the function's domain can be represented as follows:

$$\{x \mid -4 \leq x \leq 2\}.$$

The set of all x | such that | x is greater than or equal to -4 and less than or equal to 2.

To find the range, look for all the outputs on the y-axis that correspond to points on the graph. They extend from 1 to 4, inclusive. Using set-builder notation, the function's range can be represented as follows:

$$\{y \mid 1 \leq y \leq 4\}$$

The set of all y | such that | y is greater than or equal to 1 and less than or equal to 4.

EXAMPLE 6 Identifying the Domain and Range of a Function from Its Graph

Use the graph of each function to identify its domain and its range.

a.

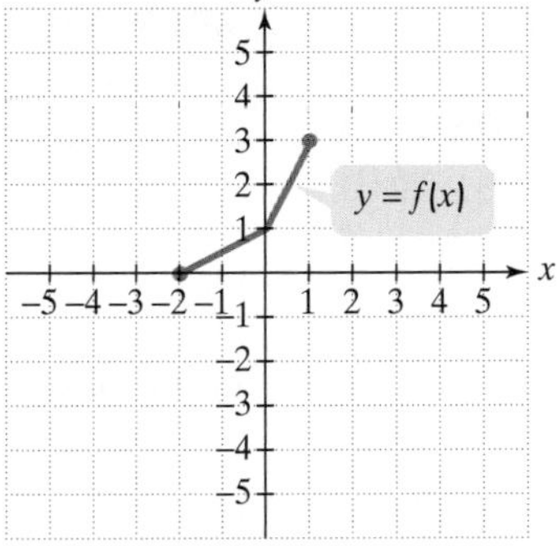
y = f(x)

b.

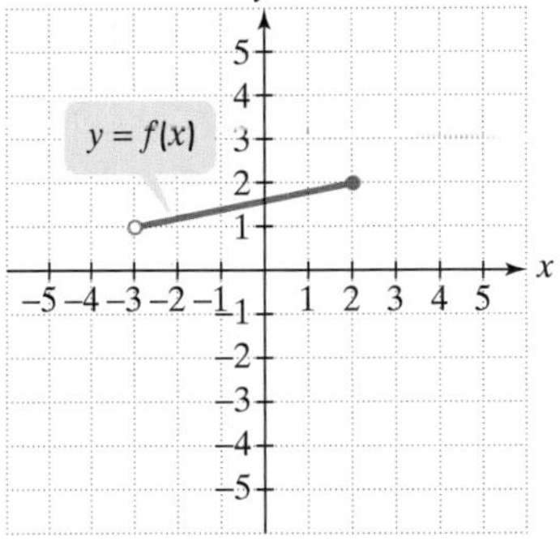
y = f(x)

c.

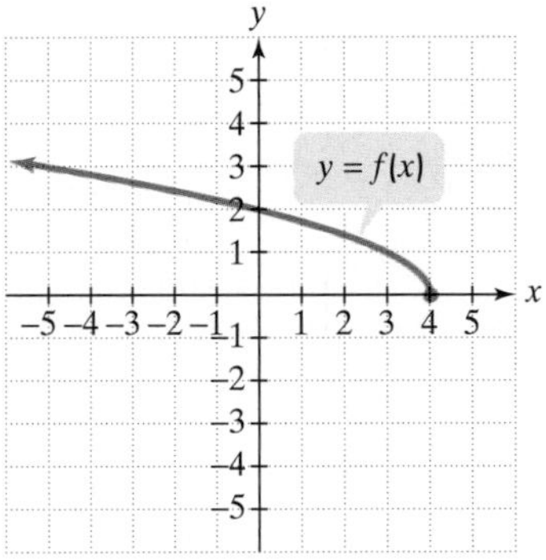
y = f(x)

d.

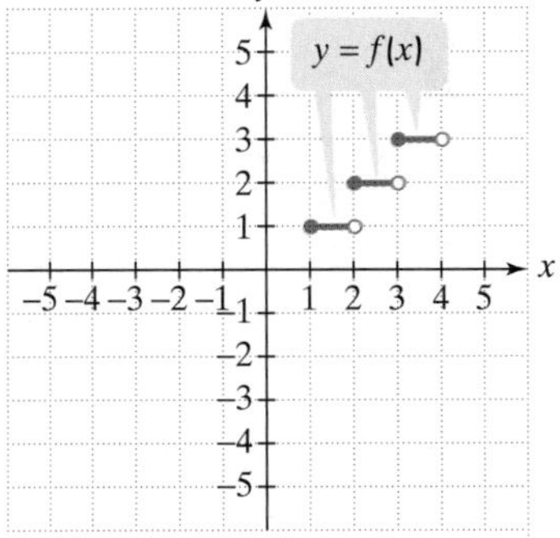
y = f(x)

SOLUTION For the graph of each function, the domain is highlighted in blue on the x-axis and the range is highlighted in green on the y-axis.

a.

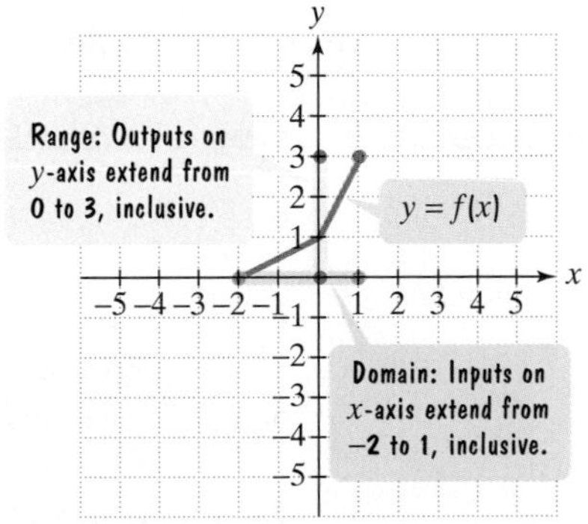
Range: Outputs on y-axis extend from 0 to 3, inclusive.
y = f(x)
Domain: Inputs on x-axis extend from −2 to 1, inclusive.

Domain $= \{x \mid -2 \leq x \leq 1\}$
Range $= \{y \mid 0 \leq y \leq 3\}$

b.

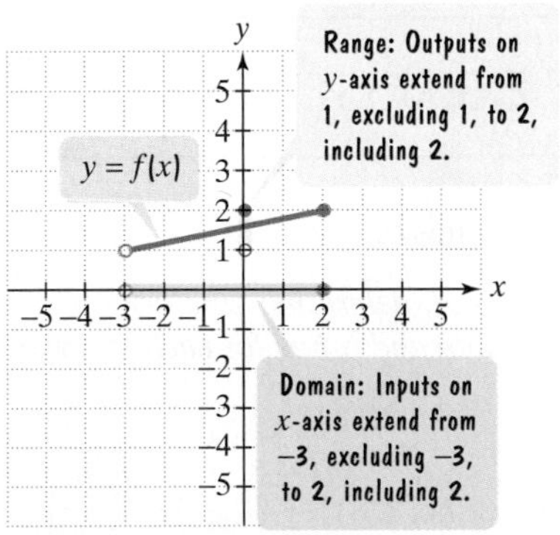
Range: Outputs on y-axis extend from 1, excluding 1, to 2, including 2.
y = f(x)
Domain: Inputs on x-axis extend from −3, excluding −3, to 2, including 2.

Domain $= \{x \mid -3 < x \leq 2\}$
Range $= \{y \mid 1 < y \leq 2\}$

c.

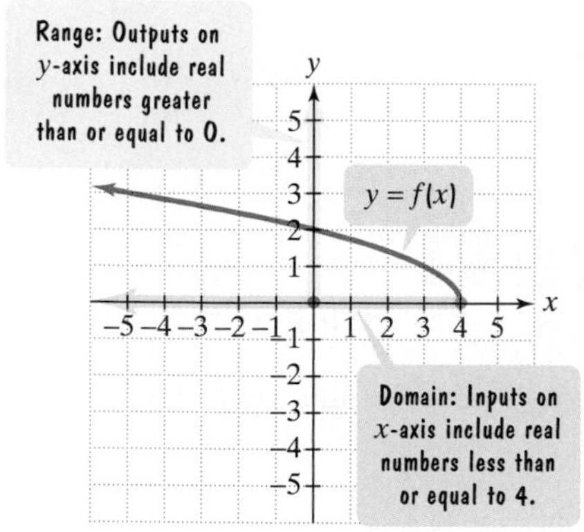
Range: Outputs on y-axis include real numbers greater than or equal to 0.
y = f(x)
Domain: Inputs on x-axis include real numbers less than or equal to 4.

Domain $= \{x \mid x \leq 4\}$
Range $= \{y \mid y \geq 0\}$

d.

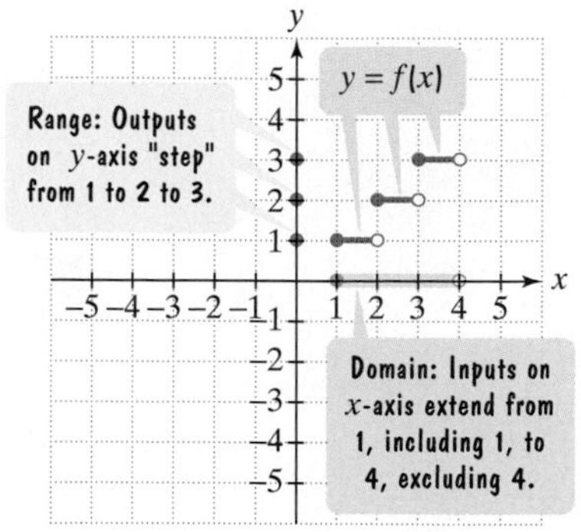
y = f(x)
Range: Outputs on y-axis "step" from 1 to 2 to 3.
Domain: Inputs on x-axis extend from 1, including 1, to 4, excluding 4.

Domain $= \{x \mid 1 \leq x < 4\}$
Range $= \{y \mid y = 1, 2, 3\}$ ■

CHECK POINT 6 Use the graph of each function to identify its domain and its range.

a.

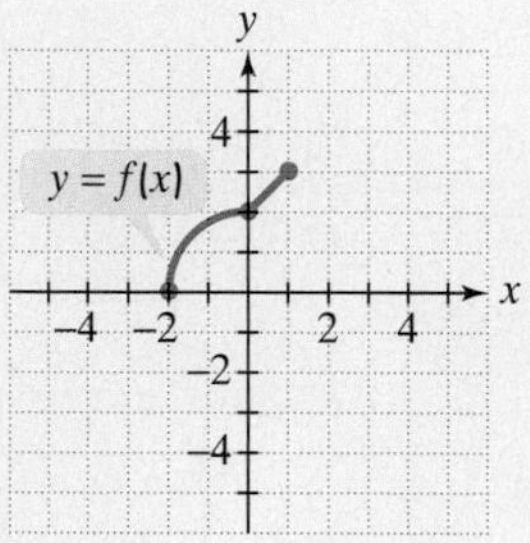
y = f(x)

b.

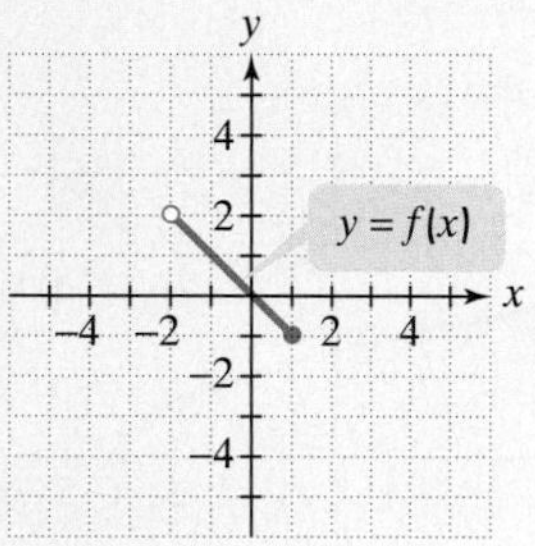
y = f(x)

c.

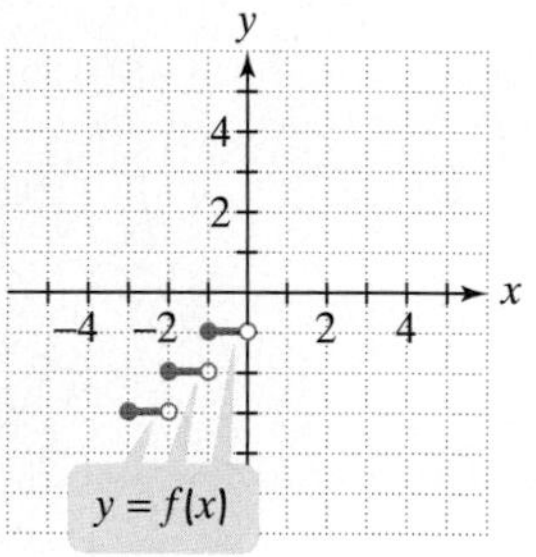
y = f(x)

8.1 EXERCISE SET

Student Solutions Manual | CD/Video | PH Math/Tutor Center | MathXL Tutorials on CD | MathXL® | MyMathLab | Interactmath.com

Practice Exercises

In Exercises 1–8, determine whether each relation is a function. Give the domain and range for each relation.

1. $\{(1, 2), (3, 4), (5, 5)\}$
2. $\{(4, 5), (6, 7), (8, 8)\}$
3. $\{(3, 4), (3, 5), (4, 4), (4, 5)\}$
4. $\{(5, 6), (5, 7), (6, 6), (6, 7)\}$
5. $\{(-3, -3), (-2, -2), (-1, -1), (0, 0)\}$
6. $\{(-7, -7), (-5, -5), (-3, -3), (0, 0)\}$
7. $\{(1, 4), (1, 5), (1, 6)\}$
8. $\{(4, 1), (5, 1), (6, 1)\}$

In Exercises 9–18, find the indicated function values.

9. $f(x) = x + 1$

a. $f(0)$ **b.** $f(5)$ **c.** $f(-8)$ **d.** $f(2a)$ **e.** $f(a + 2)$

10. $f(x) = x + 3$

a. $f(0)$ **b.** $f(5)$ **c.** $f(-8)$ **d.** $f(2a)$ **e.** $f(a + 2)$

11. $g(x) = 3x - 2$

a. $g(0)$ **b.** $g(-5)$ **c.** $g\left(\frac{2}{3}\right)$ **d.** $g(4b)$ **e.** $g(b + 4)$

12. $g(x) = 4x - 3$

a. $g(0)$ **b.** $g(-5)$ **c.** $g\left(\frac{3}{4}\right)$ **d.** $g(5b)$ **e.** $g(b + 5)$

13. $h(x) = 3x^2 + 5$

a. $h(0)$ **b.** $h(-1)$ **c.** $h(4)$ **d.** $h(-3)$ **e.** $h(4b)$

14. $h(x) = 2x^2 - 4$

a. $h(0)$ **b.** $h(-1)$ **c.** $h(5)$ **d.** $h(-3)$ **e.** $h(5b)$

15. $f(x) = 2x^2 + 3x - 1$

a. $f(0)$ **b.** $f(3)$ **c.** $f(-4)$ **d.** $f(b)$ **e.** $f(5a)$

16. $f(x) = 3x^2 + 4x - 2$

a. $f(0)$ **b.** $f(3)$ **c.** $f(-5)$ **d.** $f(b)$ **e.** $f(5a)$

17. $f(x) = \dfrac{2x - 3}{x - 4}$

a. $f(0)$ **b.** $f(3)$ **c.** $f(-4)$ **d.** $f(-5)$ **e.** $f(a + h)$

f. Why must 4 be excluded from the domain of f?

18. $f(x) = \dfrac{3x - 1}{x - 5}$

a. $f(0)$ **b.** $f(3)$ **c.** $f(-3)$ **d.** $f(10)$ **e.** $f(a + h)$

f. Why must 5 be excluded from the domain of f?

In Exercises 19–26, use the vertical line test to identify graphs in which y is a function of x.

19.

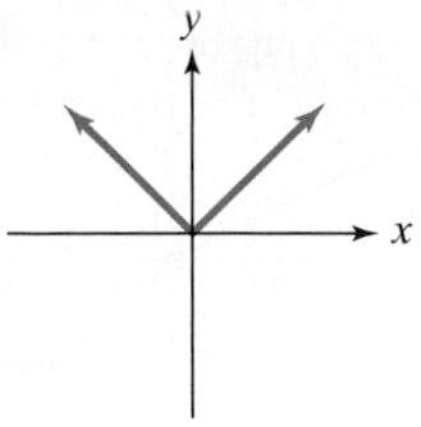

20.

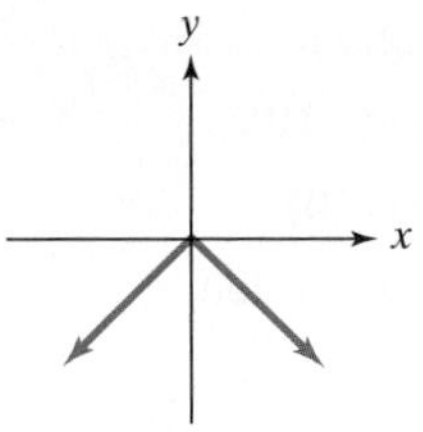

21.

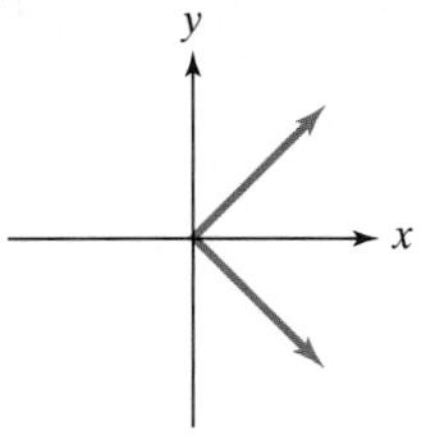

22.

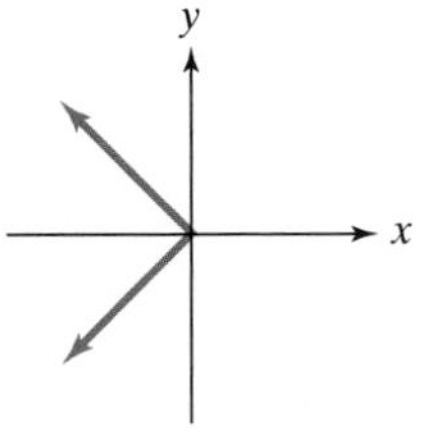

23.

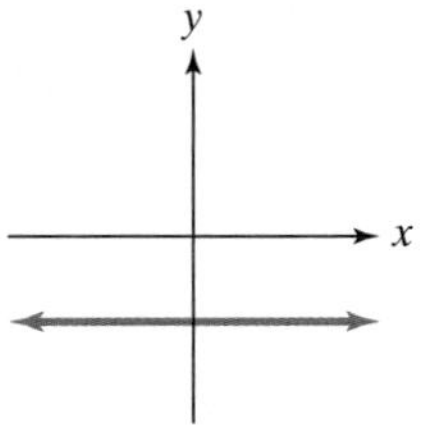

24.

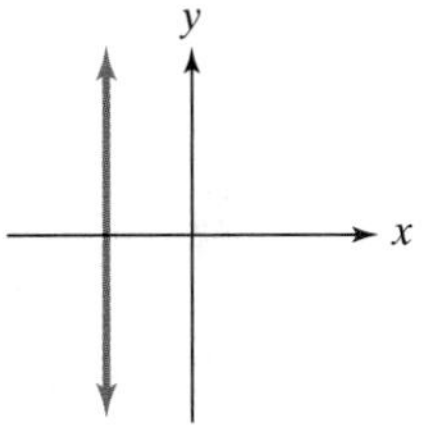

25.

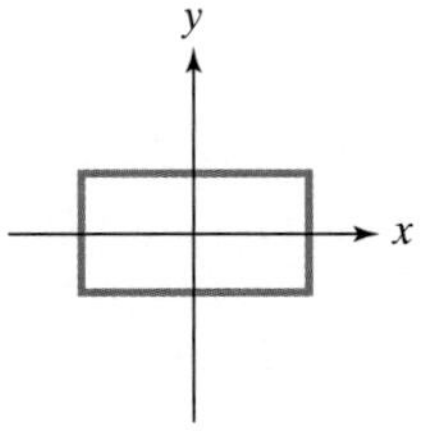

26.

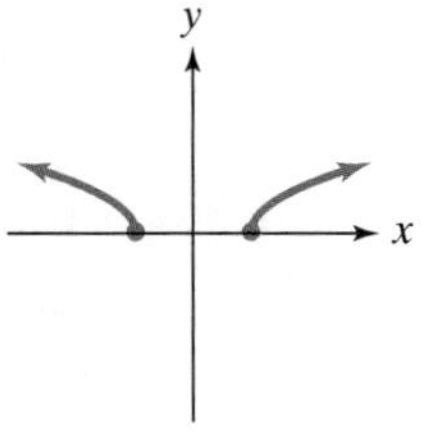

In Exercises 27–32, use the graph of f to find each indicated function value.

27. $f(-2)$

28. $f(2)$

29. $f(4)$

30. $f(-4)$

31. $f(-3)$

32. $f(-1)$

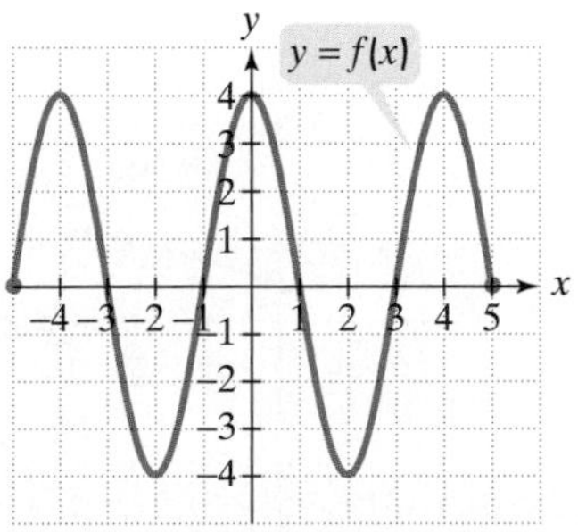

Use the graph of g to solve Exercises 33–38.

33. Find $g(-4)$.

34. Find $g(2)$.

35. Find $g(-10)$.

36. Find $g(10)$.

37. For what value of x is $g(x) = 1$?

38. For what value of x is $g(x) = -1$?

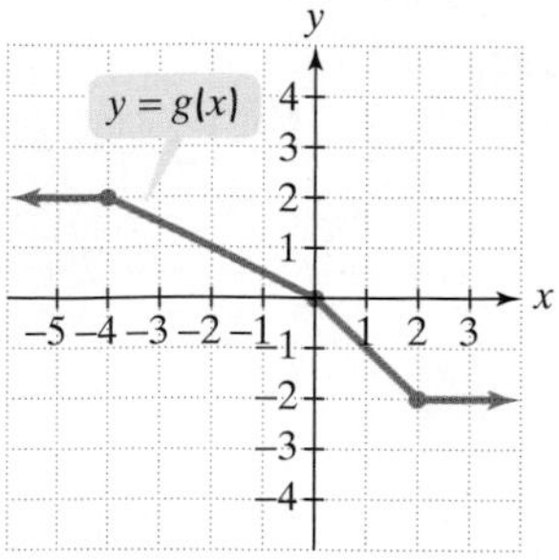

In Exercises 39–48, use the graph of each function to identify its domain and its range.

39.

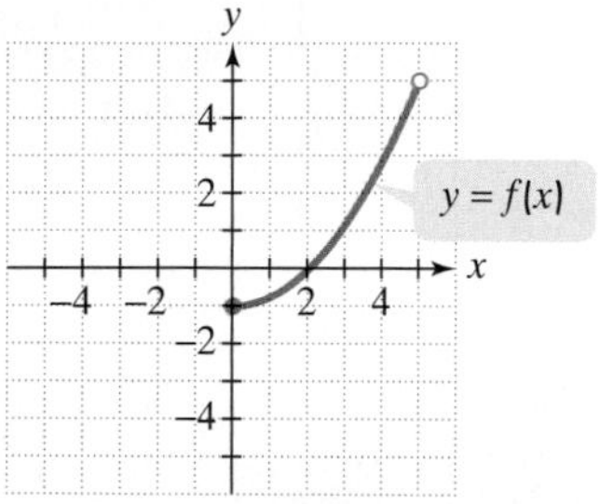

40.

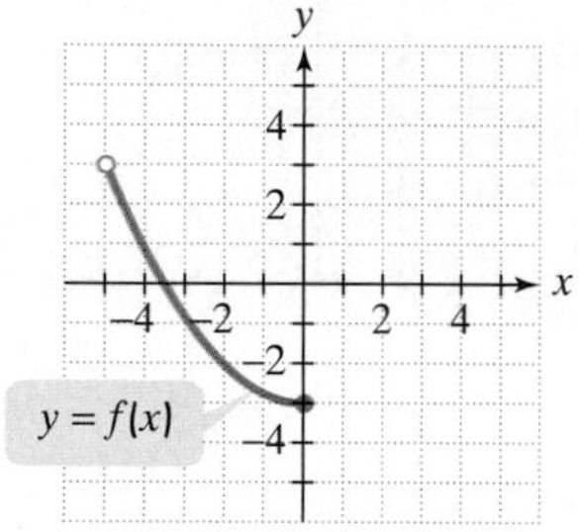

41.

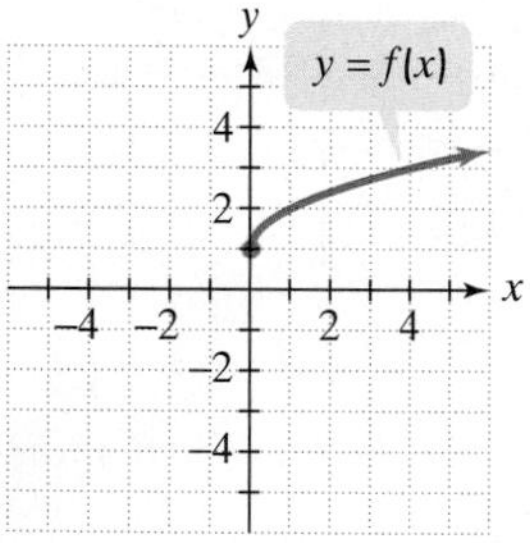

42.

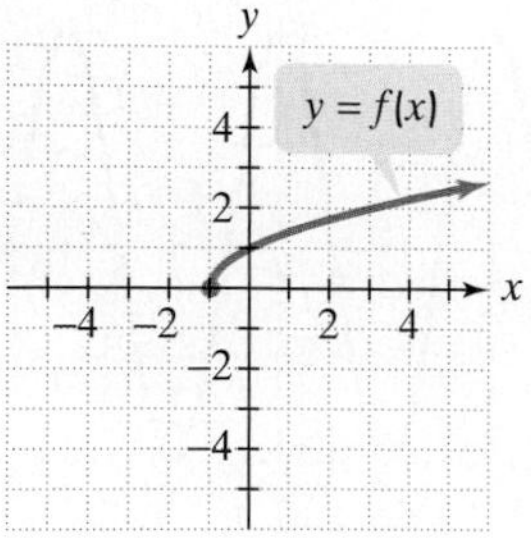

43.

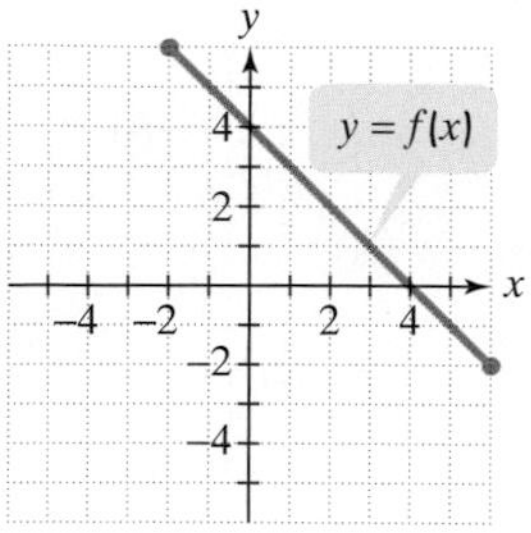

44.

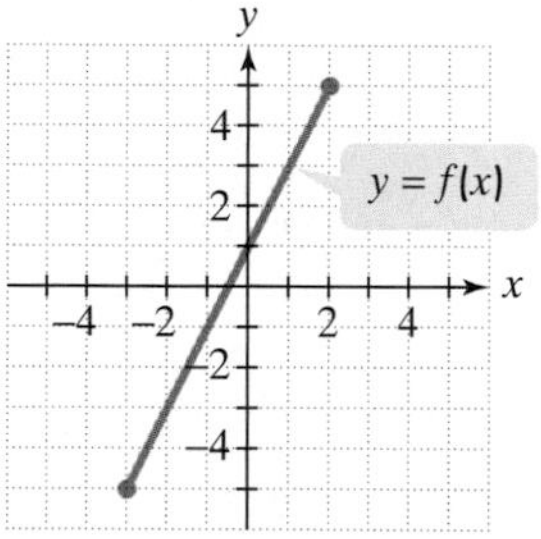

45.

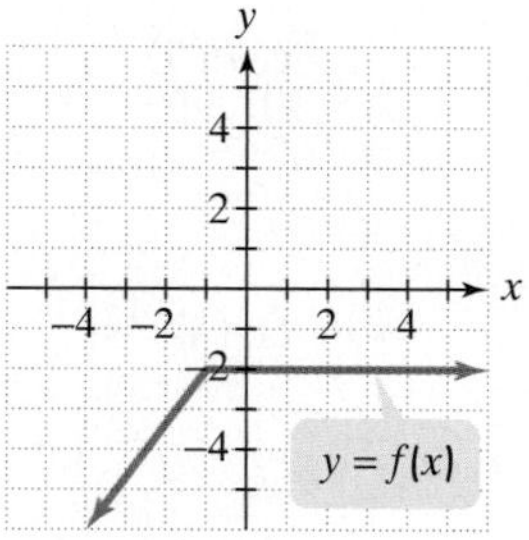

46.

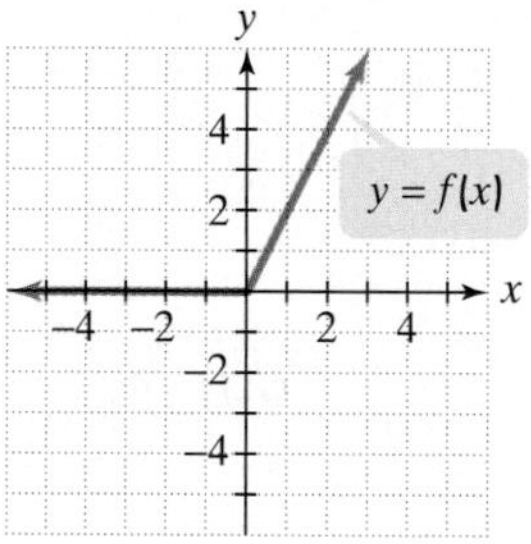

47.

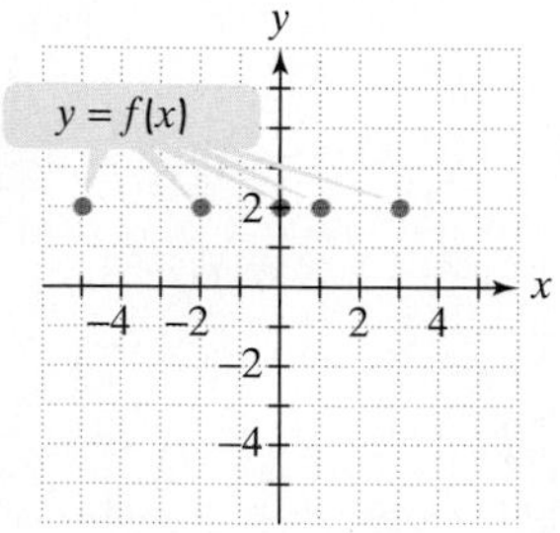

48.

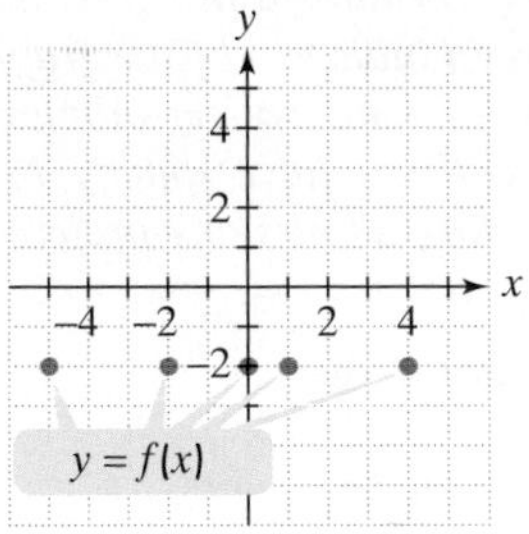

Practice Plus

In Exercises 49–50, let $f(x) = x^2 - x + 4$ *and* $g(x) = 3x - 5$.

49. Find $g(1)$ and $f(g(1))$.

50. Find $g(-1)$ and $f(g(-1))$.

In Exercises 51–52, let f and g be defined by the following table:

x	$f(x)$	$g(x)$
−2	6	0
−1	3	4
0	−1	1
1	−4	−3
2	0	−6

51. Find $\sqrt{f(-1) - f(0)} - [g(2)]^2 + f(-2) \div g(2) \cdot g(-1)$

52. Find $|f(1) - f(0)| - [g(1)]^2 + g(1) \div f(-1) \cdot g(2)$

In Exercises 53–54, find $f(-x) - f(x)$ *for the given function f. Then simplify the expression.*

53. $f(x) = x^3 + x - 5$

54. $f(x) = x^2 - 3x + 7$

In Exercises 55–56, each function is defined by two equations. The equation in the first row gives the output for negative numbers in the domain. The equation in the second row gives the output for nonnegative numbers in the domain. Find the indicated function values.

55. $f(x) = \begin{cases} 3x + 5 & \text{if } x < 0 \\ 4x + 7 & \text{if } x \geq 0 \end{cases}$

a. $f(-2)$ **b.** $f(0)$

c. $f(3)$ **d.** $f(-100) + f(100)$

56. $f(x) = \begin{cases} 6x - 1 & \text{if } x < 0 \\ 7x + 3 & \text{if } x \geq 0 \end{cases}$

a. $f(-3)$ **b.** $f(0)$

c. $f(4)$ **d.** $f(-100) + f(100)$

In Exercises 57–58, use the graph of each function to identify its domain and its range.

57.

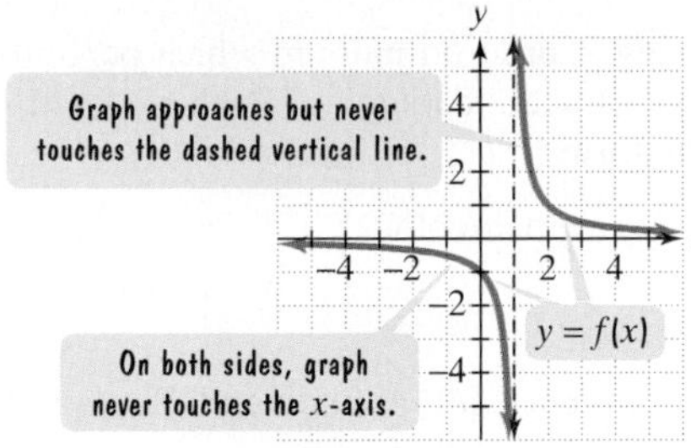

58.

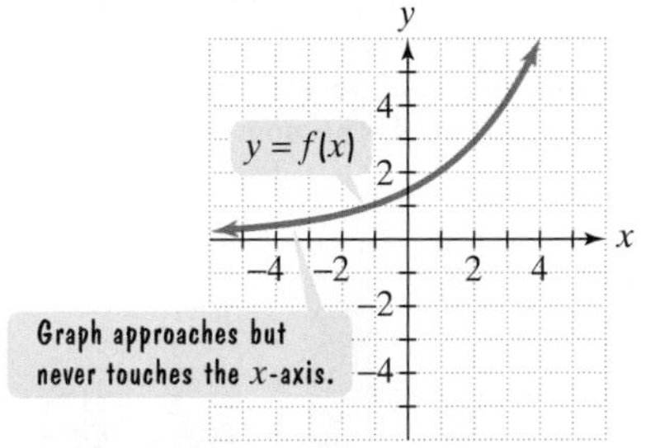

Application Exercises

59. The bar graph shows the breakdown of political ideologies in the United States.

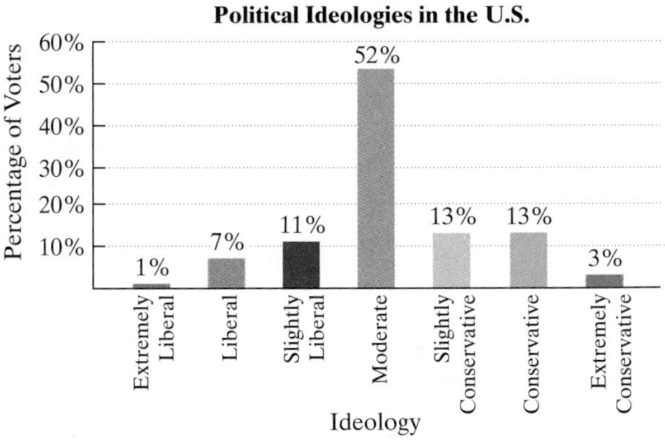

Source: Center for Political Studies, University of Michigan

a. Write a set of seven ordered pairs in which political ideologies correspond to percentages. Each ordered pair should be in the form

(ideology, percent).

Use EL, L, SL, M, SC, C, and EC to represent the respective ideologies from left to right.

b. Is the relation in part (a) a function? Explain your answer.

c. Write a set of seven ordered pairs in which percentages correspond to political ideologies. Each ordered pair should be in the form

(percent, ideology).

d. Is the relation in part (c) a function? Explain your answer.

60. The bar graph shows the percentage of teens in the United States who say they really need various technologies.

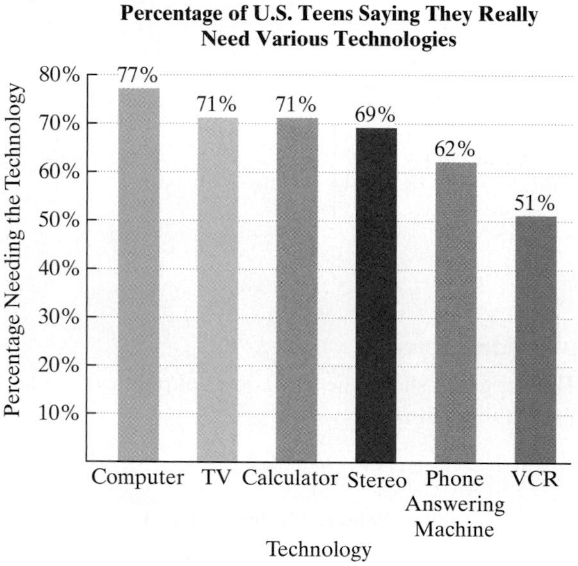

Source: Gallup Organization

a. Write a set of six ordered pairs in which technologies correspond to percentages. Each ordered pair should be in the form

(technology, percent).

b. Is the relation in part (a) a function? Explain your answer.

c. Write a set of six ordered pairs in which percentages correspond to technologies. Each ordered pair should be in the form

(percent, technology).

d. Is the relation in part (c) a function? Explain your answer.

The male minority? The graphs show enrollment in U.S. colleges, with projections through 2009. The trend indicated by the graphs is among the hottest topics of debate among college-admissions officers. Some private liberal arts colleges have quietly begun special efforts to recruit men—including admissions preferences for them.

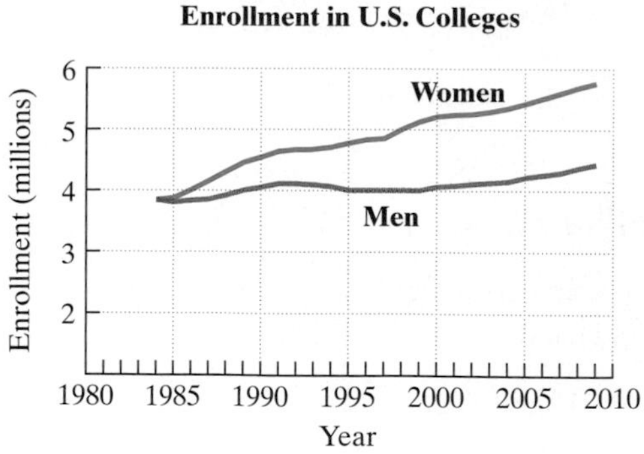

Source: Department of Education

The function

$$W(x) = 0.07x + 4.1$$

models the number of women, $W(x)$, in millions, enrolled in U.S. colleges x years after 1984. The function

$$M(x) = 0.01x + 3.9$$

models the number of men, $M(x)$, in millions, enrolled in U.S. colleges x years after 1984. Use these functions to solve Exercises 61–64.

61. Find and interpret $W(16)$. Identify this information as a point on the graph for women.

62. Find and interpret $M(16)$. Identify this information as a point on the graph for men.

63. Find and interpret $W(20) - M(20)$.

64. Find and interpret $W(25) - M(25)$.

The function

$$f(x) = 0.4x^2 - 36x + 1000$$

models the number of accidents, $f(x)$, per 50 million miles driven as a function of a driver's age, x, in years, where x includes drivers from ages 16 through 74, inclusive. The graph of f is shown. Use the equation for f to solve Exercises 65–68.

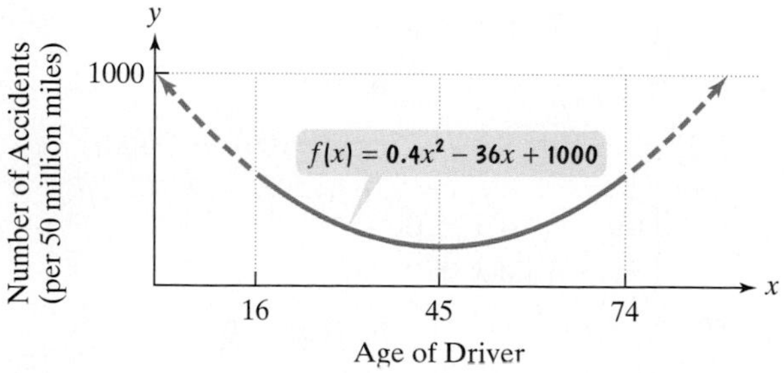

65. Find and interpret $f(20)$. Identify this information as a point on the graph of f.

66. Find and interpret $f(50)$. Identify this information as a point on the graph of f.

67. For what value of x does the graph reach its lowest point? Use the equation for f to find the minimum value of y. Describe the practical significance of this minimum value.

68. Use the graph to identify two different ages for which drivers have the same number of accidents. Use the equation for f to find the number of accidents for drivers at each of these ages.

The figure shows the percentage of Jewish Americans in the U.S. population, $f(x)$, x years after 1900. Use the graph to solve Exercises 69–76.

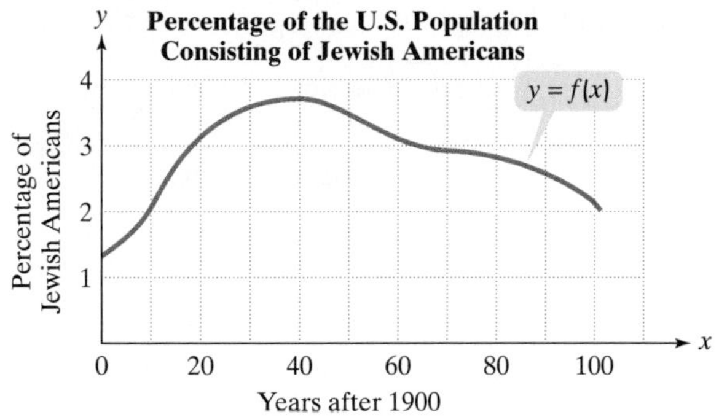

Source: American Jewish Yearbook

69. Use the graph to find a reasonable estimate of $f(60)$. What does this mean in terms of the variables in this situation?

70. Use the graph to find a reasonable estimate of $f(100)$. What does this mean in terms of the variables in this situation?

71. For what value or values of x is $f(x) = 3$? Round to the nearest year. What does this mean in terms of the variables in this situation?

72. For what value or values of x is $f(x) = 2.5$? Round to the nearest year. What does this mean in terms of the variables in this situation?

73. In which year did the percentage of Jewish Americans in the U.S. population reach a maximum? What is a reasonable estimate of the percentage for that year?

74. In which year was the percentage of Jewish Americans in the U.S. population at a minimum? What is a reasonable estimate of the percentage for that year?

75. Explain why f represents the graph of a function.

76. Describe the general trend shown by the graph.

The figure shows the cost of mailing a first-class letter, $f(x)$, as a function of its weight, x, in ounces. Use the graph to solve Exercises 77–80.

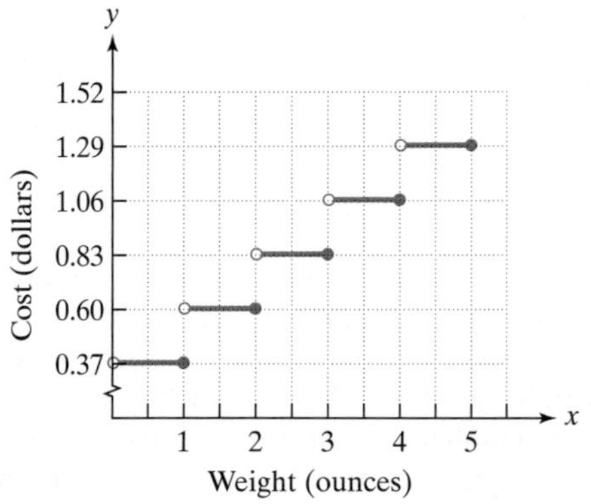

77. Find $f(3)$. What does this mean in terms of the variables in this situation?

78. Find $f(4)$. What does this mean in terms of the variables in this situation?

79. What is the cost of mailing a letter that weighs 1.5 ounces?

80. What is the cost of mailing a letter that weighs 1.8 ounces?

Writing in Mathematics

81. What is a relation? Describe what is meant by its domain and its range.

82. Explain how to determine whether a relation is a function. What is a function?

83. Does $f(x)$ mean f times x when referring to function f? If not, what does $f(x)$ mean? Provide an example with your explanation.

84. What is the graph of a function?

85. Explain how the vertical line test is used to determine whether a graph represents a function.

86. Explain how to identify the domain and range of a function from its graph.

87. For people filing a single return, federal income tax is a function of adjusted gross income because for each value of adjusted gross income there is a specific tax to be paid. By contrast, the price of a house is not a function of the lot size on which the house sits because houses on same-sized lots can sell for many different prices.

a. Describe an everyday situation between variables that is a function.

b. Describe an everyday situation between variables that is not a function.

88. Do you believe that the trend shown by the graphs for Exercises 61–64 should be reversed by providing admissions preferences for men? Explain your position on this issue.

Technology Exercise

89. The function

$$f(x) = -0.00002x^3 + 0.008x^2 - 0.3x + 6.95$$

models the number of annual physician visits, $f(x)$, by a person of age x. Graph the function in a $[0, 100, 5]$ by $[0, 40, 2]$ viewing rectangle. What does the shape of the graph indicate about the relationship between one's age and the number of annual physician visits? Use the TRACE or minimum function capability to find the coordinates of the minimum point on the graph of the function. What does this mean?

Critical Thinking Exercises

90. Which one of the following is true?

a. All relations are functions.

b. No two ordered pairs of a function can have the same second component and different first components.

c. The graph of every line is a function.

d. A horizontal line can intersect the graph of a function in more than one point.

91. If $f(x) = 3x + 7$, find $\dfrac{f(a+h) - f(a)}{h}$.

92. Give an example of a relation with the following characteristics: The relation is a function containing two ordered pairs. Reversing the components in each ordered pair results in a relation that is not a function.

93. If $f(x + y) = f(x) + f(y)$ and $f(1) = 3$, find $f(2)$, $f(3)$, and $f(4)$. Is $f(x + y) = f(x) + f(y)$ for all functions?

Review Exercises

94. Simplify: $24 \div 4[2 - (5 - 2)]^2 - 6$. (Section 1.8, Example 8)

95. Simplify: $\left(\dfrac{3x^2y^{-2}}{y^3}\right)^{-2}$. (Section 5.7, Example 6)

96. Solve: $\dfrac{x}{3} = \dfrac{3x}{5} + 4$. (Section 2.3, Example 4)

SECTION 8.2

Objectives

1 Find the domain of a function.

2 Find the sum of two functions.

3 Use the algebra of functions to combine functions and determine domains.

THE ALGEBRA OF FUNCTIONS

2006 Pontiac Solstice, the first two-seater from Pontiac in more than 20 years. Starting price: $30,000

America's big three automakers, GM, Ford, and Chrysler, know that consumers have always been willing to pay more for a cool design and a hot car. In 2004, Detroit gave consumers that opportunity, unleashing 40 new cars or updated models. The Big Three's discovery of the automobile in 2004 might seem odd, but from 1990 through 2003, Detroit focused much of its energy and money on SUVs and trucks, which commanded high prices and high profits, and saw less competition from imports. The line graphs in Figure 8.9 show the number, in millions, of cars and SUVs sold by the Big Three from 1990 through 2003. In this section, we will look at these data from the perspective of functions. By considering total sales of cars and SUVs, you will see that functions can be combined using procedures that will remind you of combining algebraic expressions.

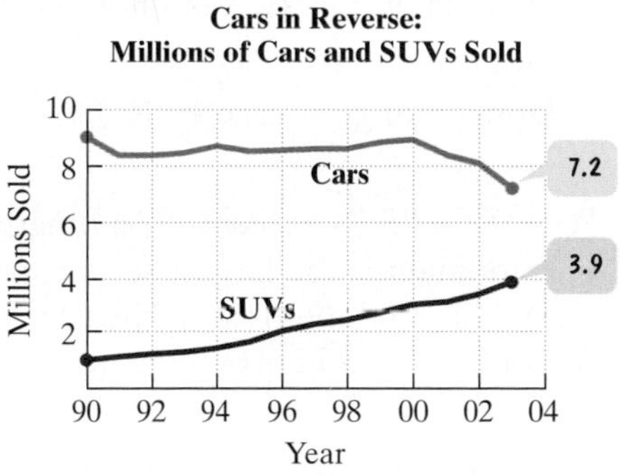

FIGURE 8.9 Sales by America's Big Three automakers

1 Find the domain of a function.

The Domain of a Function We begin with two functions that model the data in Figure 8.9.

$$C(x) = -0.14x + 9 \qquad S(x) = 0.22x + 1$$

Car sales, $C(x)$, in millions, x years after 1990

SUV sales, $S(x)$, in millions, x years after 1990

How far beyond 1990 should we extend these models? The trend in Figure 8.9 shows decreasing car sales and increasing SUV sales. Because the three automakers focused on SUVs from 1990 through 2003 and launched a fleet of new cars in 2004, the trend shown by the data changed in 2004. Thus, we should not extend the models beyond 2003. Because x represents the number of years after 1990,

$$\text{Domain of } C = \{x \mid x = 0, 1, 2, 3, \ldots, 13\}$$

and

$$\text{Domain of } S = \{x \mid x = 0, 1, 2, 3, \ldots, 13\}.$$

Functions that model data often have their domains explicitly given with the function's equation. However, for most functions, only an equation is given, and the domain is not specified. In cases like this, the domain of a function f is the largest set of real numbers for which the value of $f(x)$ is a real number. For example, consider the function

$$f(x) = \frac{1}{x - 3}.$$

Because division by 0 is undefined (and not a real number), the denominator, $x - 3$, cannot be 0. Thus, x cannot equal 3. The domain of the function consists of all real numbers other than 3, represented by

$$\text{Domain of } f = \{x \mid x \text{ is a real number and } x \neq 3\}.$$

USING TECHNOLOGY

You can graph a function and visually determine its domain. Look for inputs on the x-axis that correspond to points on the graph. For example, consider

$$g(x) = \sqrt{x}, \text{ or } y = \sqrt{x}.$$

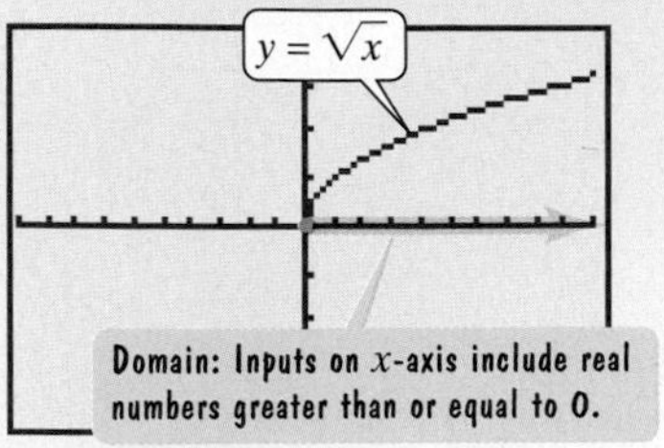

$y = \sqrt{x}$
Domain: Inputs on x-axis include real numbers greater than or equal to 0.

[−10, 10, 1] by [−4, 4, 1]

This verifies that $\{x|x \geq 0\}$ is the domain.

In Chapter 10, we will be studying square root functions such as

$$g(x) = \sqrt{x}.$$

The equation tells us to take the square root of x. Because only nonnegative numbers have square roots that are real numbers, the expression under the square root sign, x, must be nonnegative. Thus,

$$\text{Domain of } g = \{x|x \text{ is a nonnegative real number}\}.$$

Equivalently,

$$\text{Domain of } g = \{x|x \geq 0\}.$$

FINDING A FUNCTION'S DOMAIN If a function f does not model data or verbal conditions, its domain is the largest set of real numbers for which the value of $f(x)$ is a real number. Exclude from a function's domain real numbers that cause division by zero and real numbers that result in a square root of a negative number.

EXAMPLE 1 Finding the Domain of a Function

Find the domain of each function:

a. $f(x) = 3x + 2$ **b.** $g(x) = \dfrac{3x + 2}{x + 1}$.

SOLUTION

a. The function $f(x) = 3x + 2$ contains neither division nor a square root. For every real number, x, the algebraic expression $3x + 2$ is a real number. Thus, the domain of f is the set of all real numbers.

$$\text{Domain of } f = \{x|x \text{ is a real number}\}$$

b. The function $g(x) = \dfrac{3x + 2}{x + 1}$ contains division. Because division by 0 is undefined, we must exclude from the domain the value of x that causes $x + 1$ to be 0. Thus, x cannot equal -1.

$$\text{Domain of } g = \{x|x \text{ is a real number and } x \neq -1\}$$ ■

CHECK POINT 1 Find the domain of each function:

a. $f(x) = \dfrac{1}{2}x + 3$

b. $g(x) = \dfrac{7x + 4}{x + 5}$.

The Algebra of Functions We return to the functions that model millions of car and SUV sales from 1990 through 2003:

$$C(x) = -0.14x + 9 \qquad S(x) = 0.22x + 1.$$

Car sales, $C(x)$, in millions, x years after 1990

SUV sales, $S(x)$, in millions, x years after 1990

How can we use these functions to find total sales of cars and SUVs in 2003? Because 2003 is 13 years after 1990 and 13 is in the domain of each function, we need to find the sum of two function values:

$$C(13) + S(13).$$

Here is how it's done:

$$C(13) = -0.14(13) + 9 = 7.18 \qquad S(13) = 0.22(13) + 1 = 3.86$$

Substitute 13 for x in $C(x) = -0.14x + 9$. 7.18 million cars were sold in 2003. Substitute 13 for x in $S(x) = 0.22x + 1$. 3.86 million SUVs were sold in 2003.

$$C(13) + S(13) = 7.18 + 3.86 = 11.04.$$

Thus, a total of 11.04 million cars and SUVs were sold in 2003.

2 Find the sum of two functions.

There is a second way we can obtain this number. We can first add the functions C and S to obtain a new function, $C + S$. To do so, we add the terms to the right of the equal sign for $C(x)$ to the terms to the right of the equal sign for $S(x)$:

$$\begin{aligned}(C + S)(x) &= C(x) + S(x) \\ &= (-0.14x + 9) + (0.22x + 1) && \text{Add terms for } C(x) \text{ and } S(x). \\ &= 0.08x + 10. && \text{Combine like terms.}\end{aligned}$$

Thus,

$$(C + S)(x) = 0.08x + 10$$

Total car and SUV sales, in millions, x years after 1990

Do you see how we can use this new function to find total car and SUV sales in 2003? Substitute 13 for x in the equation for $C + S$:

$$(C + S)(13) = 0.08(13) + 10 = 11.04$$

Substitute 13 for x in $(C + S)(x) = 0.08x + 10$. As we found above, a total of 11.04 million cars and SUVs were sold in 2003.

The domain of the new function, $C + S$, consists of the numbers x that are in the domain of C and in the domain of S. Because both functions model data from 1990 through 2003,

$$\text{Domain of } C + S = \{x \mid x = 0, 1, 2, 3, \ldots, 13\}.$$

Here is a general definition for function addition:

THE SUM OF FUNCTIONS Let f and g be two functions. The **sum $f + g$** is the function defined by

$$(f + g)(x) = f(x) + g(x).$$

The domain of $f + g$ is the set of all real numbers that are common to the domain of f and the domain of g.

EXAMPLE 2 Finding the Sum of Two Functions

Let $f(x) = x^2 - 3$ and $g(x) = 4x + 5$. Find each at the following:

a. $(f + g)(x)$ **b.** $(f + g)(3)$.

SOLUTION

a. $(f + g)(x) = f(x) + g(x) = (x^2 - 3) + (4x + 5) = x^2 + 4x + 2$

Thus,

$$(f + g)(x) = x^2 + 4x + 2.$$

b. We find $(f + g)(3)$ by substituting 3 for x in the equation for $f + g$.

$$(f + g)(x) = x^2 + 4x + 2$$ This is the equation for $f + g$.

Substitute 3 for x.

$$(f + g)(3) = 3^2 + 4 \cdot 3 + 2 = 9 + 12 + 2 = 23$$ ■

✔ **CHECK POINT 2** Let $f(x) = 3x^2 + 4x - 1$ and $g(x) = 2x + 7$. Find each of the following:

a. $(f + g)(x)$ **b.** $(f + g)(4)$.

EXAMPLE 3 Adding Functions and Determining the Domain

Let $f(x) = \frac{4}{x}$ and $g(x) = \frac{3}{x + 2}$. Find each of the following:

a. $(f + g)(x)$ **b.** the domain of $f + g$.

SOLUTION

a. $(f + g)(x) = f(x) + g(x) = \frac{4}{x} + \frac{3}{x + 2}$

(In Chapter 7, we discussed how to add these rational expressions. In our study of the algebra of functions, we will leave these fractions in the form shown.)

b. The domain of $f + g$ is the set of all real numbers that are common to the domain of f and the domain of g. Thus, we must find the domains of f and g. We will do so for f first.

Note that $f(x) = \frac{4}{x}$ is a function involving division. Because division by 0 is undefined, x cannot equal 0.

$$\text{Domain of } f = \{x \mid x \text{ is a real number and } x \neq 0\}$$

The function $g(x) = \frac{3}{x + 2}$ is also a function involving division. Because division by 0 is undefined, x cannot equal -2.

$$\text{Domain of } g = \{x \mid x \text{ is a real number and } x \neq -2\}$$

To find $f(x) + g(x)$, x must be in both domains listed. Thus,

$$\text{Domain of } f + g = \{x \mid x \text{ is a real number and } x \neq 0 \text{ and } x \neq -2\}.$$ ■

✔ **CHECK POINT 3** Let $f(x) = \frac{5}{x}$ and $g(x) = \frac{7}{x - 8}$. Find each of the following:

a. $(f + g)(x)$ **b.** the domain of $f + g$.

3 Use the algebra of functions to combine functions and determine domains.

We can also combine functions using subtraction, multiplication, and division by performing operations with the algebraic expressions that appear on the right side of the equations. For example, the functions $f(x) = 2x$ and $g(x) = x - 1$ can be combined to form the difference, product, and quotient of f and g. Here's how it's done:

For each function, $f(x) = 2x$ and $g(x) = x - 1$.

Difference: $f - g$

$$(f - g)(x) = f(x) - g(x)$$
$$= 2x - (x - 1) = 2x - x + 1 = x + 1$$

Product: fg

$$(fg)(x) = f(x) \cdot g(x)$$
$$= 2x(x - 1) = 2x^2 - 2x$$

Quotient: $\frac{f}{g}$

$$\left(\frac{f}{g}\right)(x) = \frac{f(x)}{g(x)} = \frac{2x}{x - 1}, x \neq 1.$$

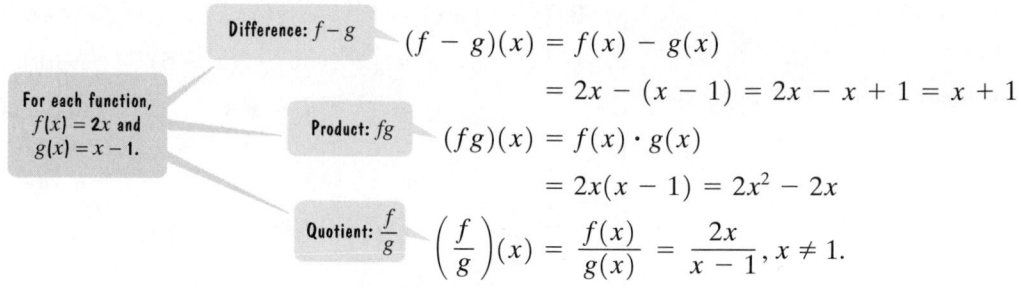

Just like the domain for $f + g$, the domain for each of these functions consists of all real numbers that are common to the domains of f and g. In the case of the quotient function $\frac{f(x)}{g(x)}$, we must remember not to divide by 0, so we add the further restriction that $g(x) \neq 0$.

The following definitions summarize our discussion:

THE ALGEBRA OF FUNCTIONS: SUM, DIFFERENCE, PRODUCT, AND QUOTIENT OF FUNCTIONS Let f and g be two functions. The **sum** $f + g$, the **difference** $f - g$, the **product** fg, and the **quotient** $\frac{f}{g}$ are functions whose domains are the set of all real numbers common to the domains of f and g, defined as follows:

1. Sum: $(f + g)(x) = f(x) + g(x)$
2. Difference: $(f - g)(x) = f(x) - g(x)$
3. Product: $(fg)(x) = f(x) \cdot g(x)$
4. Quotient: $\left(\frac{f}{g}\right)(x) = \frac{f(x)}{g(x)}$, provided $g(x) \neq 0$.

EXAMPLE 4 Using the Algebra of Functions

Let $f(x) = x^2 + x$ and $g(x) = x - 5$. Find each of the following:

a. $(f + g)(4)$

b. $(f - g)(x)$ and $(f - g)(-3)$

c. $(fg)(x)$ and $(fg)(-2)$

d. $\left(\frac{f}{g}\right)(x)$ and $\left(\frac{f}{g}\right)(7)$

SOLUTION

a. We can find $(f + g)(4)$ using $f(4)$ and $g(4)$.

$$f(x) = x^2 + x \qquad g(x) = x - 5$$
$$f(4) = 4^2 + 4 = 20 \qquad g(4) = 4 - 5 = -1$$

Thus,

$$(f + g)(4) = f(4) + g(4) = 20 + (-1) = 19.$$

We can also find $(f + g)(4)$ by first finding $(f + g)(x)$ and then substituting 4 for x:

$$(f + g)(x) = f(x) + g(x) \quad \text{This is the definition of the sum } f + g.$$
$$= (x^2 + x) + (x - 5) \quad \text{Substitute the given functions.}$$
$$= x^2 + 2x - 5. \quad \text{Simplify.}$$

Using $(f + g)(x) = x^2 + 2x - 5$. we have

$$(f + g)(4) = 4^2 + 2 \cdot 4 - 5 = 16 + 8 - 5 = 19.$$

$f(x) = x^2 + x$
$g(x) = x - 5$

The given functions, f and g (repeated)

STUDY TIP

Here are the details of the FOIL method we used to multiply the binomials: $(x^2 + x)(x - 5)$

$$\begin{aligned} &\quad\;\; \text{F} \qquad\quad \text{O} \qquad\quad \text{I} \qquad \text{L} \\ &= x^2 \cdot x + x^2(-5) + x \cdot x + x(-5) \\ &= x^3 - 5x^2 + x^2 - 5x \end{aligned}$$

Special product of polynomials are reviewed in the Section 5.3 summary on page 376.

b. $(f - g)(x) = f(x) - g(x)$ — This is the definition of the difference $f - g$.

$= (x^2 + x) - (x - 5)$ — Substitute the given functions.

$= x^2 + x - x + 5$ — Remove parentheses and change the sign of each term in the second set of parentheses.

$= x^2 + 5$ — Simplify.

Using $(f - g)(x) = x^2 + 5$, we have

$$(f - g)(-3) = (-3)^2 + 5 = 9 + 5 = 14.$$

c. $(fg)(x) = f(x) \cdot g(x)$ — This is the definition of the product fg.

$= (x^2 + x)(x - 5)$ — Substitute the given functions.

$= x^3 - 5x^2 + x^2 - 5x$ — Multiply using the FOIL method.

$= x^3 - 4x^2 - 5x$ — Combine like terms: $-5x^2 + x^2 = -4x^2$.

Using $(fg)(x) = x^3 - 4x^2 - 5x$, we have

$$\begin{aligned} (fg)(-2) &= (-2)^3 - 4(-2)^2 - 5(-2) \\ &= -8 - 4(4) - 5(-2) \quad \text{Evaluate exponential expressions.} \\ &= -8 - 16 + 10 = -14. \end{aligned}$$

We can also find $(fg)(-2)$ using the fact that

$$(fg)(-2) = f(-2) \cdot g(-2).$$

$f(x) = x^2 + x$ $\qquad$ $g(x) = x - 5$

$f(-2) = (-2)^2 + (-2) = 4 - 2 = 2$ $\qquad$ $g(-2) = -2 - 5 = -7$

Thus,

$$(fg)(-2) = f(-2) \cdot g(-2) = 2(-7) = -14.$$

d. $\left(\frac{f}{g}\right)(x) = \frac{f(x)}{g(x)}$ — This is the definition of the quotient $\frac{f}{g}$.

$= \frac{x^2 + x}{x - 5}$ — Substitute the given functions.

Using $\left(\frac{f}{g}\right)(x) = \frac{x^2 + x}{x - 5}$, we have

$$\left(\frac{f}{g}\right)(7) = \frac{7^2 + 7}{7 - 5} = \frac{56}{2} = 28.$$

■

CHECK POINT 4 Let $f(x) = x^2 - 2x$ and $g(x) = x + 3$. Find each of the following:

a. $(f + g)(5)$

b. $(f - g)(x)$ and $(f - g)(-1)$

c. $(fg)(x)$ and $(fg)(-4)$

d. $\left(\frac{f}{g}\right)(x)$ and $\left(\frac{f}{g}\right)(7)$

8.2 EXERCISE SET

Student Solutions Manual | CD/Video | PH Math/Tutor Center | MathXL Tutorials on CD | MathXL® | MyMathLab | Interactmath.com

Practice Exercises

In Exercises 1–10, find the domain of each function.

1. $f(x) = 3x + 5$

2. $f(x) = 4x + 7$

3. $g(x) = \frac{1}{x + 4}$

4. $g(x) = \frac{1}{x + 5}$

5. $f(x) = \frac{2x}{x - 3}$

6. $f(x) = \frac{4x}{x - 2}$

7. $g(x) = x + \frac{3}{5 - x}$

8. $g(x) = x + \frac{7}{6 - x}$

9. $f(x) = \frac{1}{x + 7} + \frac{3}{x - 9}$

10. $f(x) = \frac{1}{x + 8} + \frac{3}{x - 10}$

In Exercises 11–16, find **a.** $(f + g)(x)$ **b.** $(f + g)(5)$.

11. $f(x) = 3x + 1, g(x) = 2x - 6$

12. $f(x) = 4x + 2, g(x) = 2x - 9$

13. $f(x) = x - 5, g(x) = 3x^2$

14. $f(x) = x - 6, g(x) = 2x^2$

15. $f(x) = 2x^2 - x - 3, g(x) = x + 1$

16. $f(x) = 4x^2 - x - 3, g(x) = x + 1$

In Exercises 17–28, for each pair of functions, f and g, determine the domain of f + g.

17. $f(x) = 3x + 7, g(x) = 9x + 10$

18. $f(x) = 7x + 4, g(x) = 5x - 2$

19. $f(x) = 3x + 7, g(x) = \frac{2}{x - 5}$

20. $f(x) = 7x + 4, g(x) = \frac{2}{x - 6}$

21. $f(x) = \frac{1}{x}, g(x) = \frac{2}{x - 5}$

22. $f(x) = \frac{1}{x}, g(x) = \frac{2}{x - 6}$

23. $f(x) = \frac{8x}{x - 2}, g(x) = \frac{6}{x + 3}$

24. $f(x) = \frac{9x}{x - 4}, g(x) = \frac{7}{x + 8}$

25. $f(x) = \frac{8x}{x - 2}, g(x) = \frac{6}{2 - x}$

26. $f(x) = \frac{9x}{x - 4}, g(x) = \frac{7}{4 - x}$

27. $f(x) = x^2, g(x) = x^3$

28. $f(x) = x^2 + 1, g(x) = x^3 - 1$

In Exercises 29–48, let

$$f(x) = x^2 + 4x \quad \text{and} \quad g(x) = 2 - x.$$

Find each of the following.

29. $(f + g)(x)$ and $(f + g)(3)$

30. $(f + g)(x)$ and $(f + g)(4)$

31. $f(-2) + g(-2)$

32. $f(-3) + g(-3)$

33. $(f - g)(x)$ and $(f - g)(5)$

34. $(f - g)(x)$ and $(f - g)(6)$

35. $f(-2) - g(-2)$

36. $f(-3) - g(-3)$

37. $(fg)(x)$ and $(fg)(2)$

38. $(fg)(x)$ and $(fg)(3)$

39. $(fg)(5)$

40. $(fg)(6)$

41. $\left(\frac{f}{g}\right)(x)$ and $\left(\frac{f}{g}\right)(1)$

42. $\left(\frac{f}{g}\right)(x)$ and $\left(\frac{f}{g}\right)(3)$

43. $\left(\frac{f}{g}\right)(-1)$

44. $\left(\frac{f}{g}\right)(0)$

45. The domain of $f + g$

46. The domain of $f - g$

47. The domain of $\frac{f}{g}$

48. The domain of fg

Practice Plus

Use the graphs of f and g to solve Exercises 49–56.

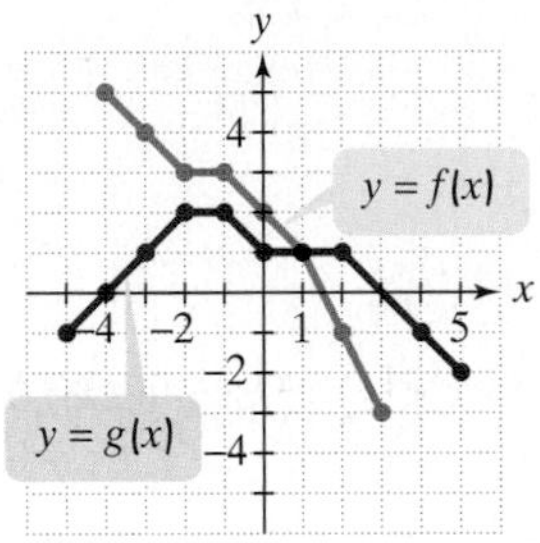

49. Find $(f + g)(-3)$.

50. Find $(g - f)(-2)$.

51. Find $(fg)(2)$.

52. Find $\left(\frac{g}{f}\right)(3)$.

53. Find the domain of $f + g$.

54. Find the domain of $\frac{f}{g}$.

55. Graph $f + g$.

56. Graph $f - g$.

Use the table defining f and g to solve Exercises 57–60.

x	$f(x)$	$g(x)$
−2	5	0
−1	3	−2
0	−2	4
1	−6	−3
2	0	1

57. Find $(f + g)(1) - (g - f)(-1)$.

58. Find $(f + g)(-1) - (g - f)(0)$.

59. Find $(fg)(-2) - \left[\left(\frac{f}{g}\right)(1)\right]^2$.

60. Find $(fg)(2) - \left[\left(\frac{g}{f}\right)(0)\right]^2$.

Application Exercises

The table shows the total number of births and the total number of deaths in the United States from 1995 through 2002.

Births and Deaths in the U.S.

Year	Births	Deaths
1995	3,899,589	2,312,132
1996	3,891,494	2,314,690
1997	3,880,894	2,314,245
1998	3,941,553	2,337,256
1999	3,959,417	2,391,399
2000	4,058,814	2,403,351
2001	4,025,933	2,416,425
2002	4,022,000	2,436,000

Source: Department of Health and Human Services

The data can be modeled by the following functions:

Number of births: $B(x) = 24{,}770x + 3{,}873{,}266$

Number of deaths: $D(x) = 20{,}205x + 2{,}294{,}970.$

In each function, x represents the number of years after 1995. Assume that the functions apply only to the years shown in the table. Use these functions to solve Exercises 61–64.

61. Find the domain of B.

62. Find the domain of D.

63. **a.** Find $(B - D)(x)$. What does this function represent?

b. Use the function in part (a) to find $(B - D)(6)$. What does this mean in terms of the U.S. population and to which year does this apply?

c. Use the data shown in the table to find $(B - D)(6)$. How well does the difference of functions used in part (b) model this number?

64. **a.** Find $(B - D)(x)$. What does this function represent?

b. Use the function in part (a) to find $(B - D)(2)$. What does this mean in terms of the U.S. population and to which year does this apply?

c. Use the data shown in the table to find $(B - D)(2)$. How well does the difference of functions used in part (b) model this number?

Consider the following functions:

$f(x)$ = population of the world's more-developed regions in year x

$g(x)$ = the population of the world's less-developed regions in year x

$h(x)$ = total world population in year x.

Use these functions and the graphs shown to answer Exercises 65–68.

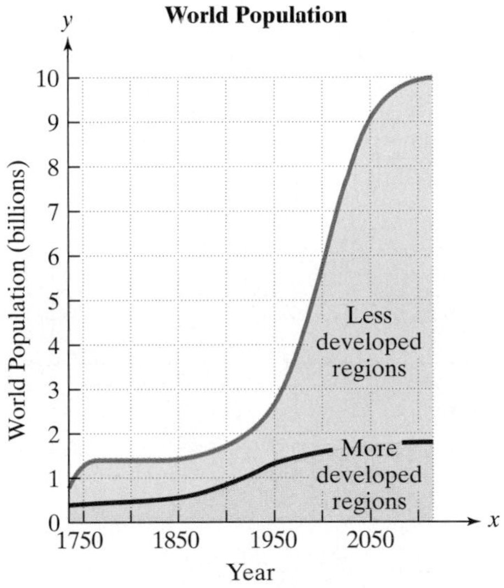

Source: Population Reference Bureau

65. What does the function $f + g$ represent?

66. What does the function $h - g$ represent?

67. Use the graph to estimate $(f + g)(2000)$.

68. Use the graph to estimate $(h - g)(2000)$.

69. A company that sells radios has yearly fixed costs of \$600,000. It costs the company \$45 to produce each radio. Each radio will sell for \$65. The company's costs and revenue are modeled by the following functions:

$C(x) = 600{,}000 + 45x$ This function models the company's costs.

$R(x) = 65x$ This function models the company's revenue.

Find and interpret $(R - C)(20{,}000)$, $(R - C)(30{,}000)$, and $(R - C)(40{,}000)$.

70. The function $f(t) = -0.14t^2 + 0.51t + 31.6$ models the U.S. population, $f(t)$, in millions, ages 65 and older, t years after 1990. The function $g(t) = 0.54t^2 + 12.64t + 107.1$ models the total yearly cost of Medicare, $g(t)$, in billions of dollars, t years after 1990.

a. What does the function $\frac{g}{f}$ represent?

b. Find and interpret $\frac{g}{f}(10)$.

Writing in Mathematics

71. If a function is defined by an equation, explain how to find its domain.

72. If equations for functions f and g are given, explain how to find $f + g$.

73. If the equations of two functions are given, explain how to obtain the quotient function and its domain.

74. If equations for functions f and g are given, describe two ways to find $(f - g)(3)$.

Technology Exercises

75. Graph the function $\frac{g}{f}$ from Exercise 70 in a $[0, 15, 1]$ by $[0, 60, 1]$ viewing rectangle. What does the shape of the graph indicate about the per capita costs of Medicare for the U.S. population ages 65 and over with increasing time?

In Exercises 76–79, graph each of the three functions in the same $[-10, 10, 1]$ by $[-10, 10, 1]$ viewing rectangle.

76. $y_1 = 2x + 3$
$y_2 = 2 - 2x$
$y_3 = y_1 + y_2$

77. $y_1 = x - 4$
$y_2 = 2x$
$y_3 = y_1 - y_2$

78. $y_1 = x$
$y_2 = x - 4$
$y_3 = y_1 \cdot y_2$

79. $y_1 = x^2 - 2x$
$y_2 = x$
$y_3 = \frac{y_1}{y_2}$

80. In Exercise 79, use the TRACE feature to trace along y_3. What happens at $x = 0$? Explain why this occurs.

Critical Thinking Exercises

81. Which one of the following is false?

a. If $(f + g)(a) = 0$, then $f(a)$ and $g(a)$ must be opposites, or additive inverses.

b. If $(f - g)(a) = 0$, then $f(a)$ and $g(a)$ must be equal.

c. If $\left(\frac{f}{g}\right)(a) = 0$, then $f(a)$ must be 0.

d. If $(fg)(a) = 0$, then $f(a)$ must be 0.

82. Use the graphs given in Exercises 65–68 to create a graph that shows the population, in billions, of less-developed regions from 1950 through 2050.

Review Exercises

83. Solve the system:

$$\begin{aligned} 11x + 4y &= -3 \\ -13x + y &= 15. \end{aligned}$$

(Section 4.3, Example 2)

84. Solve: $3(6 - x) = 3 - 2(x - 4)$. (Section 2.3, Example 3)

85. If $f(x) = 6x - 4$, find $f(b + 2)$. (Section 8.1, Example 3)

MID-CHAPTER CHECK POINT

CHAPTER 8

What You Know: We learned that a function is a relation in which no two ordered pairs have the same first component and different second components. We represented functions as equations and used function notation. We applied the vertical line test to identify graphs of functions. We determined the domain and range of a function from its graph, using inputs on the x-axis for the domain and outputs on the y-axis for the range. Finally, we developed an algebra of functions to combine functions and determine their domains.

In Exercises 1–6, determine whether each relation is a function. Give the domain and range for each relation.

1. $\{(2, 6), (1, 4), (2, -6)\}$

2. $\{(0, 1), (2, 1), (3, 4)\}$

3.

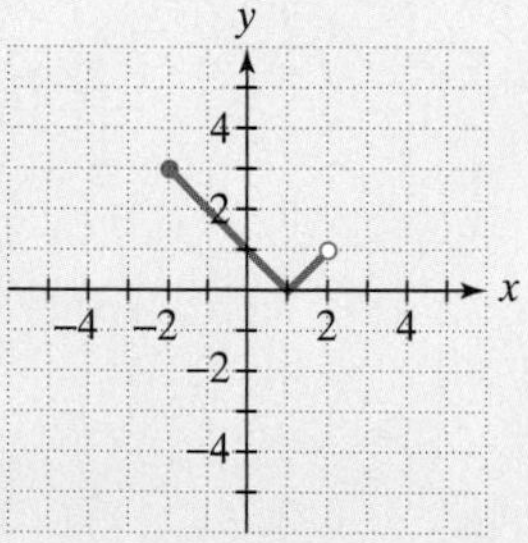

4.

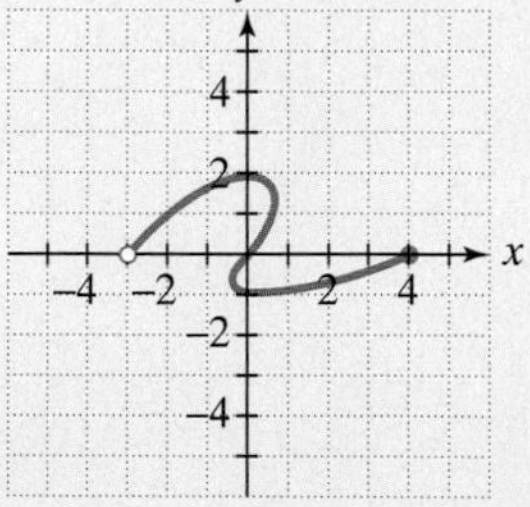

5.

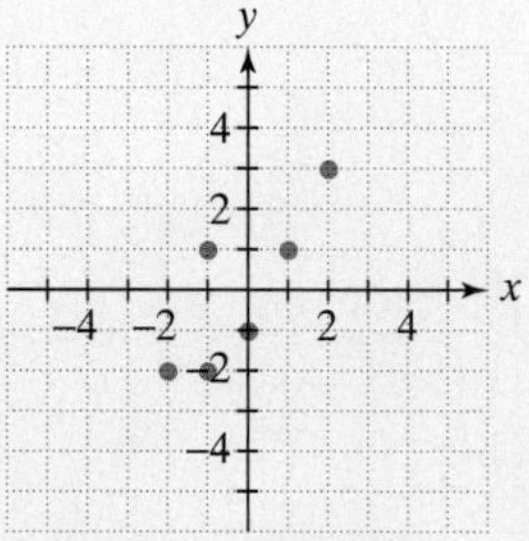

6.

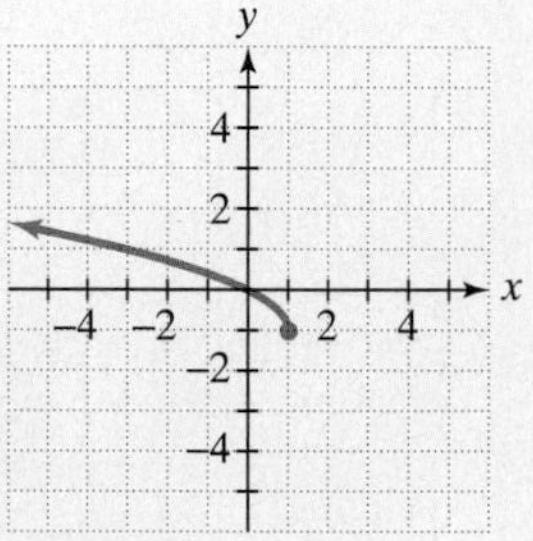

Use the graph of f to solve Exercises 7–12.

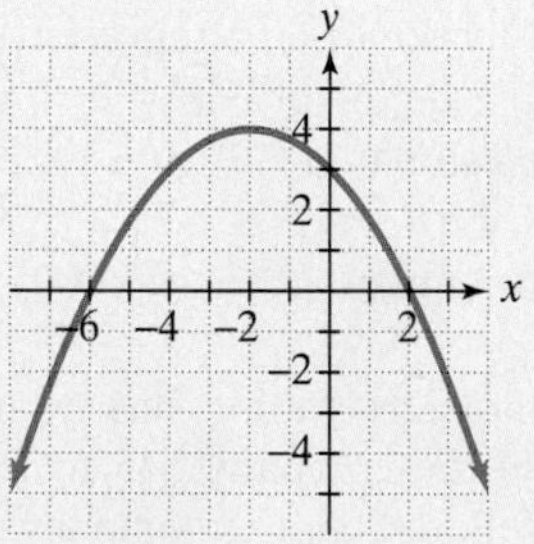

7. Explain why f represents the graph of a function.

8. Use the graph to find $f(-4)$.

9. For what value or values of x is $f(x) = 4$?

10. For what value or values of x is $f(x) = 0$?

11. Find the domain of f.

12. Find the range of f.

In Exercises 13–14, find the domain of each function.

13. $f(x) = (x + 2)(x - 2)$

14. $g(x) = \dfrac{1}{(x + 2)(x - 2)}$

In Exercises 15–22, let

$$f(x) = x^2 - 3x + 8 \text{ and } g(x) = -2x - 5.$$

Find each of the following:

15. $f(0) + g(-10)$

16. $f(-1) - g(3)$

17. $f(a) + g(a + 3)$

18. $(f + g)(x)$ and $(f + g)(-2)$

19. $(f - g)(x)$ and $(f - g)(5)$

20. $(fg)(x)$ and $(fg)(-1)$

21. $\left(\dfrac{f}{g}\right)(x)$ and $\left(\dfrac{f}{g}\right)(-4)$

22. The domain of $\dfrac{f}{g}$

SECTION 8.3

Objectives

1. Form composite functions.
2. Verify inverse functions.
3. Find the inverse of a function.
4. Use the horizontal line test to determine if a function has an inverse function.
5. Use the graph of a one-to-one function to graph its inverse function.

COMPOSITE AND INVERSE FUNCTIONS

You are an over-40 student returning to college. Through consistent hard work, you've earned a near-perfect grade point average, feeling more empowered by knowledge with each academic success. Unfortunately, when it comes to technology, you are at a disadvantage compared to younger students. There were no computers in the classroom during your high-school years, and you glaze over hearing students chatting about browsers, gigabytes, RAM, and DVDs. It's time to turn things around and become computer literate.

1 Form composite functions.

Composite Functions Luckily, your local computer store is having a sale. The models that are on sale cost either \$300 less than the regular price or 85% of the regular price. If x represents the computer's regular price, both discounts can be described with the following functions:

$$f(x) = x - 300 \qquad g(x) = 0.85x.$$

The computer is on sale for \$300 less than it's regular price.

The computer is on sale for 85% of its regular price.

At the store, you bargain with the salesperson. Eventually, she makes an offer you can't refuse. The sale price will be 85% of the regular price followed by a $300 reduction:

$$0.85x - 300.$$

85% of the regular price — followed by a $300 reduction

In terms of functions f and g, this offer can be obtained by taking the output of $g(x) = 0.85x$, namely $0.85x$, and using it as the input of f:

$$f(x) = x - 300$$

Replace x with $0.85x$, the output of $g(x) = 0.85x$.

$$f(0.85x) = 0.85x - 300.$$

Because $0.85x$ is $g(x)$, we can write this last equation as

$$f(g(x)) = 0.85x - 300.$$

We read this equation as "f of g of x is equal to $0.85x - 300$." We call $f(g(x))$ the **composition of the function f with g**, or a **composite function**. This composite function is written $f \circ g$. Thus,

$$(f \circ g)(x) = f(g(x)) = 0.85x - 300.$$

Like all functions, we can evaluate $f \circ g$ for a specified value of x in the function's domain. For example, here's how to find the value of this function at 1400:

$$(f \circ g)(x) = 0.85x - 300$$

This composite function describes the offer you cannot refuse.

Replace x with 1400.

$$(f \circ g)(1400) = 0.85(1400) - 300 = 1190 - 300 = 890.$$

This means that a computer that regularly sells for $1400 is on sale for $890 subject to both discounts. We can use a partial table of coordinates for each of the discount functions, g and f, to numerically verify this result.

Computer's regular price — 85% of the regular price

x	$g(x) = 0.85x$
1200	1020
1300	1105
1400	1190

85% of the regular price — $300 reduction

x	$f(x) = x - 300$
1020	720
1105	805
1190	890

Using these tables, we can find $(f \circ g)(1400)$:

$$(f \circ g)(1400) = f(g(1400)) = f(1190) = 890.$$

The table for g shows that $g(1400) = 1190$. — The table for f shows that $f(1190) = 890$.

This verifies that a computer that regularly sells for $1400 is on sale for $890 subject to both discounts.

Before you run out to buy a computer, let's generalize our discussion of the computer's double discount and define the composition of any two functions.

THE COMPOSITION OF FUNCTIONS The **composition of the function f with g** is denoted by $f \circ g$ and is defined by the equation

$$(f \circ g)(x) = f(g(x)).$$

The domain of the **composite function $f \circ g$** is the set of all x such that

1. x is in the domain of g and
2. $g(x)$ is in the domain of f.

The composition of f with g, $f \circ g$, is pictured as a machine with inputs and outputs in Figure 8.10. The diagram indicates that the output of g, or $g(x)$, becomes the input for "machine" f. If $g(x)$ is not in the domain of f, it cannot be input into machine f, and so $g(x)$ must be discarded.

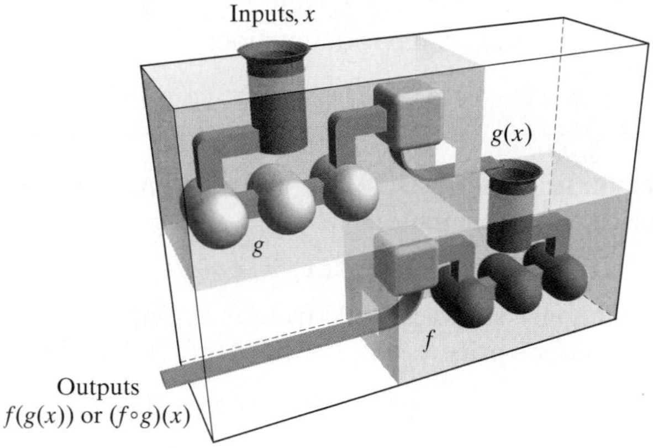

FIGURE 8.10 Inputting one function into a second function

EXAMPLE 1 Forming Composite Functions

Given $f(x) = 3x - 4$ and $g(x) = x^2 + 6$, find each of the following composite functions:

a. $(f \circ g)(x)$ **b.** $(g \circ f)(x)$.

SOLUTION

a. We begin with $(f \circ g)(x)$, the composition of f with g. Because $(f \circ g)(x)$ means $f(g(x))$, we must replace each occurrence of x in the equation for f with $g(x)$.

$$f(x) = 3x - 4$$ This is the given equation for f.

Replace x with $g(x)$.

$$(f \circ g)(x) = f(g(x)) = 3g(x) - 4$$

$$= 3(x^2 + 6) - 4$$ Because $g(x) = x^2 + 6$, replace $g(x)$ with $x^2 + 6$.

$$= 3x^2 + 18 - 4$$ Use the distributive property.

$$= 3x^2 + 14$$ Simplify.

Thus, $(f \circ g)(x) = 3x^2 + 14$.

b. Next, we find $(g \circ f)(x)$, the composition of g with f. Because $(g \circ f)(x)$ means $g(f(x))$, we must replace each occurrence of x in the equation for g with $f(x)$.

$$g(x) = x^2 + 6$$ This is the given equation for g.

Replace x with $f(x)$.

$$(g \circ f)(x) = g(f(x)) = (f(x))^2 + 6$$

$$= (3x - 4)^2 + 6$$ Because $f(x) = 3x - 4$, replace $f(x)$ with $3x - 4$.

$$= 9x^2 - 24x + 16 + 6$$ Use $(A - B)^2 = A^2 - 2AB + B^2$ to square $3x - 4$.

$$= 9x^2 - 24x + 22$$ Simplify.

Thus, $(g \circ f)(x) = 9x^2 - 24x + 22$. **Notice that $f \circ g$ is not the same function as $g \circ f$.** ■

✔ CHECK POINT **1** Given $f(x) = 5x + 6$ and $g(x) = x^2 - 1$, find each of the following composite functions:

a. $(f \circ g)(x)$ **b.** $(g \circ f)(x)$.

Inverse Functions Here are two functions that describe situations related to the price, x, of a computer:

$$f(x) = x - 300 \qquad g(x) = x + 300.$$

Function f subtracts \$300 from the computer's price and function g adds \$300 to the computer's price. Let's see what $f(g(x))$ does. Put $g(x)$ into f:

$$f(x) = x - 300$$ This is the given equation for f.

Replace x with $g(x)$.

$$f(g(x)) = g(x) - 300$$

$$= x + 300 - 300$$ Because $g(x) = x + 300$, replace $g(x)$ with $x + 300$.

$$= x.$$ This is the computer's original price.

By putting $g(x)$ into f and finding $f(g(x))$, we see that the computer's price, x, went through two changes: the first, an increase; the second, a decrease:

$$x + 300 - 300.$$

The final price of the computer, x, is identical to its starting price, x.

In general, if the changes made to x by function g are undone by the changes made by function f, then

$$f(g(x)) = x.$$

Assume, also, that this "undoing" takes place in the other direction:

$$g(f(x)) = x.$$

Under these conditions, we say that each function is the **inverse function** of the other. The fact that g is the inverse of f is expressed by renaming g as f^{-1}, read "f-inverse." For example, the inverse functions

$$f(x) = x - 300 \qquad g(x) = x + 300$$

are usually named as follows:

$$f(x) = x - 300 \qquad f^{-1}(x) = x + 300.$$

We can use partial tables of coordinates for f and f^{-1} to gain numerical insight into the relationship between a function and its inverse function.

Computer's regular price — $300 reduction

x	$f(x) = x - 300$
1200	900
1300	1000
1400	1100

Ordered pairs for f: (1200, 900), (1300, 1000), (1400, 1100)

Price with $300 reduction — $300 price increase

x	$f^{-1}(x) = x + 300$
900	1200
1000	1300
1100	1400

Ordered pairs for f^{-1}: (900, 1200), (1000, 1300), (1100, 1400)

The tables illustrate that if a function f is the set of ordered pairs (x, y), then the inverse, f^{-1}, is the set of ordered pairs (y, x). Using these tables, we can see how one function's changes to x are undone by the other function:

$$(f^{-1} \circ f)(1300) = f^{-1}(f(1300)) = f^{-1}(1000) = 1300.$$

The table for f shows that $f(1300) = 1000$.

The table for f^{-1} shows that $f^{-1}(1000) = 1300$.

The final price of the computer, $1300, is identical to its starting price, $1300.

With these ideas in mind, we present the formal definition of the inverse of a function:

STUDY TIP

The notation f^{-1} represents the inverse function of f. The -1 is *not* an exponent. The notation f^{-1} does *not* mean $\frac{1}{f}$:

$$f^{-1} \neq \frac{1}{f}.$$

DEFINITION OF THE INVERSE OF A FUNCTION Let f and g be two functions such that

$$f(g(x)) = x \quad \text{for every } x \text{ in the domain of } g$$

and

$$g(f(x)) = x \quad \text{for every } x \text{ in the domain of } f.$$

The function g is the **inverse of the function** f, and is denoted by f^{-1} (read "f-inverse"). Thus, $f(f^{-1}(x)) = x$ and $f^{-1}(f(x)) = x$. The domain of f is equal to the range of f^{-1}, and vice versa.

2 Verify inverse functions.

EXAMPLE 2 Verifying Inverse Functions

Show that each function is the inverse of the other:

$$f(x) = 5x \quad \text{and} \quad g(x) = \frac{x}{5}.$$

SOLUTION To show that f and g are inverses of each other, we must show that $f(g(x)) = x$ and $g(f(x)) = x$. We begin with $f(g(x))$.

$$f(x) = 5x$$ This is the given equation for f.

Replace x with $g(x)$.

$$f(g(x)) = 5g(x) = 5\left(\frac{x}{5}\right) = x$$ Because $g(x) = \frac{x}{5}$, replace $g(x)$ with $\frac{x}{5}$. Then simplify.

Next, we find $g(f(x))$.

$$g(x) = \frac{x}{5}$$ This is the given equation for g.

Replace x with $f(x)$.

$$g(f(x)) = \frac{f(x)}{5} = \frac{5x}{5} = x$$ Because $f(x) = 5x$, replace $f(x)$ with $5x$. Then simplify.

Because g is the inverse of f (and vice versa), we can use inverse notation and write

$$f(x) = 5x \quad \text{and} \quad f^{-1}(x) = \frac{x}{5}.$$

Notice how f^{-1} undoes the change produced by f: f changes x by multiplying by 5 and f^{-1} undoes this change by dividing by 5.

CHECK POINT 2 Show that each function is the inverse of the other:

$$f(x) = 7x \quad \text{and} \quad g(x) = \frac{x}{7}.$$

EXAMPLE 3 Verifying Inverse Functions

Show that each function is the inverse of the other:

$$f(x) = 3x + 2 \quad \text{and} \quad g(x) = \frac{x - 2}{3}.$$

SOLUTION To show that f and g are inverses of each other, we must show that $f(g(x)) = x$ and $g(f(x)) = x$. We begin with $f(g(x))$.

$$f(x) = 3x + 2$$ This is the equation for f.

Replace x with $g(x)$.

$$f(g(x)) = 3g(x) + 2 = 3\left(\frac{x - 2}{3}\right) + 2 = x - 2 + 2 = x$$

$g(x) = \frac{x - 2}{3}$

Next, we find $g(f(x))$.

$$g(x) = \frac{x - 2}{3}$$ This is the equation for g.

Replace x with $f(x)$.

$$g(f(x)) = \frac{f(x) - 2}{3} = \frac{(3x + 2) - 2}{3} = \frac{3x}{3} = x$$

$f(x) = 3x + 2$

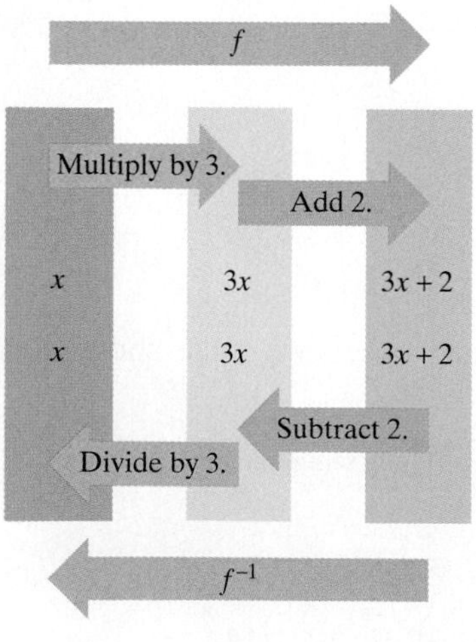

FIGURE 8.11 f^{-1} undoes the changes produced by f.

Because g is the inverse of f (and vice versa), we can use inverse notation and write

$$f(x) = 3x + 2 \quad \text{and} \quad f^{-1}(x) = \frac{x - 2}{3}.$$

Notice how f^{-1} undoes the changes produced by f: f changes x by *multiplying* by 3 and *adding* 2, and f^{-1} undoes this by *subtracting* 2 and *dividing* by 3. This "undoing" process is illustrated in Figure 8.11.

CHECK POINT 3 Show that each function is the inverse of the other:

$$f(x) = 4x - 7 \quad \text{and} \quad g(x) = \frac{x + 7}{4}.$$

3 Find the inverse of a function.

Finding the Inverse of a Function The definition of the inverse of a function tells us that the domain of f is equal to the range of f^{-1}, and vice versa. This means that if the function f is the set of ordered pairs (x, y), then the inverse of f is the set of ordered pairs (y, x). If a function is defined by an equation, we can obtain the equation for f^{-1}, the inverse of f, by interchanging the roles of x and y in the equation for the function f.

FINDING THE INVERSE OF A FUNCTION The equation for the inverse of a function f can be found as follows:

1. Replace $f(x)$ with y in the equation for $f(x)$.
2. Interchange x and y.
3. Solve for y. If this equation does not define y as a function of x, the function f does not have an inverse function and the procedure ends. If this equation does define y as a function of x, the function f has an inverse function.
4. If f has an inverse function, replace y in step 3 with $f^{-1}(x)$. We can verify our result by showing that $f(f^{-1}(x)) = x$ and $f^{-1}(f(x)) = x$.

STUDY TIP

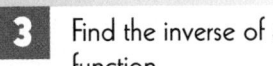

The procedure for finding a function's inverse uses a *switch-and-solve* strategy. Switch x and y, then solve for y.

EXAMPLE 4 Finding the Inverse of a Function

Find the inverse of $f(x) = 7x - 5$.

SOLUTION

Step 1. Replace $f(x)$ with y:

$$y = 7x - 5.$$

Step 2. Interchange x and y:

$$x = 7y - 5. \quad \text{This is the inverse function.}$$

Step 3. Solve for y:

$$x + 5 = 7y \quad \text{Add 5 to both sides.}$$

$$\frac{x + 5}{7} = y \quad \text{Divide both sides by 7.}$$

Step 4. Replace y with $f^{-1}(x)$:

$$f^{-1}(x) = \frac{x + 5}{7}. \quad \text{The equation is written with } f^{-1} \text{ on the left.}$$

Thus, the inverse of $f(x) = 7x - 5$ is $f^{-1}(x) = \frac{x + 5}{7}$.

The inverse function, f^{-1}, undoes the changes produced by f. f changes x by multiplying by 7 and subtracting 5. f^{-1} undoes this by adding 5 and dividing by 7. ■

DISCOVER FOR YOURSELF

In Example 4, we found that if $f(x) = 7x - 5$, then

$$f^{-1}(x) = \frac{x + 5}{7}.$$

Verify this result by showing that

$$f(f^{-1}(x)) = x$$

and

$$f^{-1}(f(x)) = x.$$

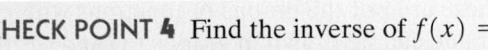

CHECK POINT 4 Find the inverse of $f(x) = 2x + 7$.

4 Use the horizontal line test to determine if a function has an inverse function.

The Horizontal Line Test and One-To-One Functions Some functions do not have inverses that are functions. For example, consider the function $f(x) = x^2$, or $y = x^2$. We can use a few of the solutions of $y = x^2$ to illustrate numerically that this function does not have an inverse:

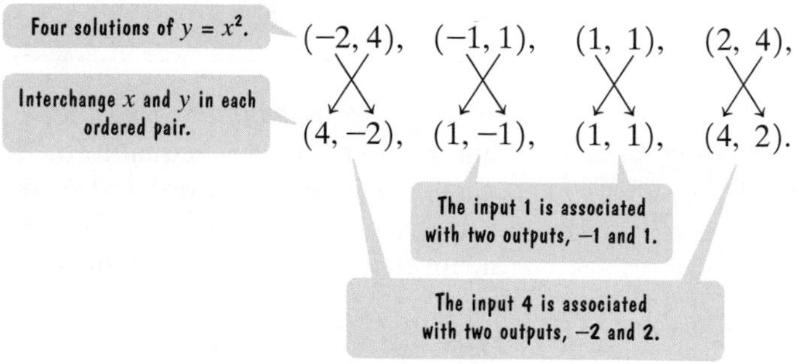

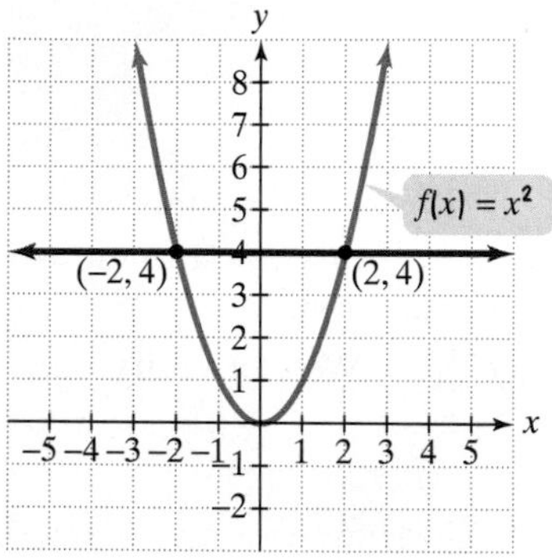

FIGURE 8.12 The horizontal line intersects the graph twice.

A function provides exactly one output for each input. Thus, the ordered pairs in the bottom row do not define a function.

Can we look at the graph of a function and tell if it represents a function with an inverse? Yes. The graph of the function $f(x) = x^2$ is shown in Figure 8.12. Four units above the x-axis, a horizontal line is drawn. This line intersects the graph at two of its points, $(-2, 4)$ and $(2, 4)$. Inverse functions have ordered pairs with the coordinates reversed. We just saw what happened when we interchanged x and y. We obtained $(4, -2)$ and $(4, 2)$, and these ordered pairs do not define a function.

If any horizontal line, such as the one in Figure 8.12, intersects a graph at two or more points, the set of these points will not define a function when their coordinates are reversed. This suggests the **horizontal line test** for inverse functions.

DISCOVER FOR YOURSELF

How might you restrict the domain of $f(x) = x^2$, graphed in Figure 8.12, so that the remaining portion of the graph passes the horizontal line test?

THE HORIZONTAL LINE TEST FOR INVERSE FUNCTIONS A function f has an inverse that is a function, f^{-1}, if there is no horizontal line that intersects the graph of the function f at more than one point.

EXAMPLE 5 Applying the Horizontal Line Test

Which of the following graphs represent functions that have inverse functions?

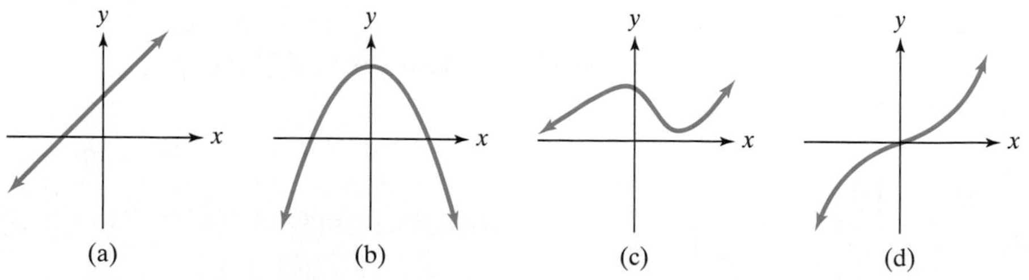

SOLUTION Notice that horizontal lines can be drawn in graphs (b) and (c) that intersect the graphs more than once. These graphs do not pass the horizontal line test. These are not the graphs of functions with inverse functions. By contrast, no horizontal line can be drawn in graphs (a) and (d) that intersects the graphs more than once.

These graphs pass the horizontal line test. Thus, the graphs in parts (a) and (d) represent functions that have inverse functions.

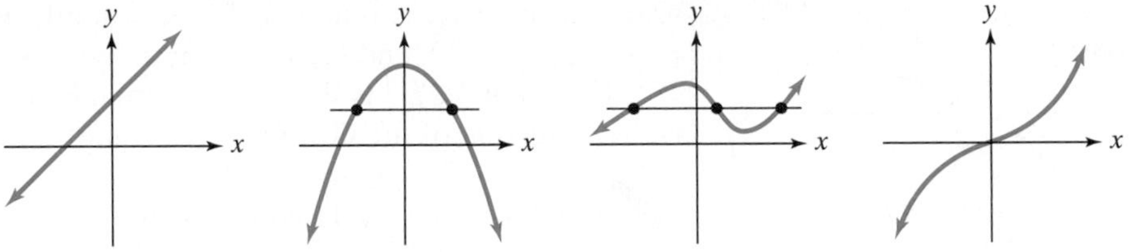

(a) Has an inverse function

(b) No inverse function

(c) No inverse function

(d) Has an inverse function

CHECK POINT 5 Which of the following graphs represent functions that have inverse functions?

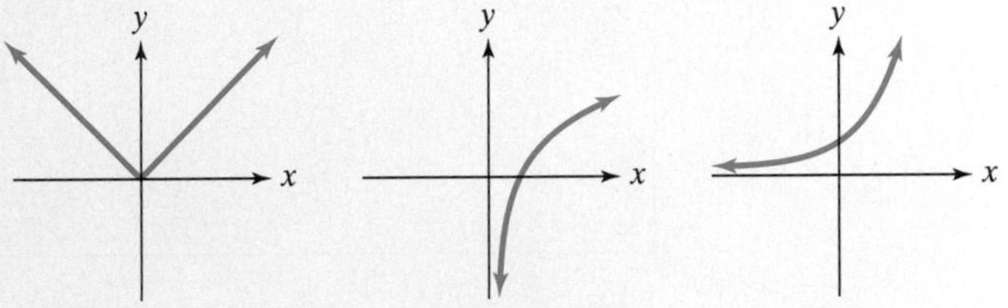

(a) (b) (c)

A function passes the horizontal line test when no two different ordered pairs have the same second component. This means that if $x_1 \neq x_2$, then $f(x_1) \neq f(x_2)$. Such a function is called a **one-to-one function**. Thus, a one-to-one function is a function in which no two different ordered pairs have the same second component. Only one-to-one functions have inverse functions. Any function that passes the horizontal line test is a one-to-one function. Any one-to-one function has a graph that passes the horizontal line test.

5 Use the graph of a one-to-one function to graph its inverse function.

Graphs of f and f^{-1} There is a relationship between the graph of a one-to-one function, f, and its inverse, f^{-1}. Because inverse functions have ordered pairs with the coordinates reversed, if the point (a, b) is on the graph of f, then the point (b, a) is on the graph of f^{-1}. The points (a, b) and (b, a) are symmetric with respect to the line $y = x$. Thus, **the graph of f^{-1} is a reflection of the graph of f about the line $y = x$.** This is illustrated in Figure 8.13.

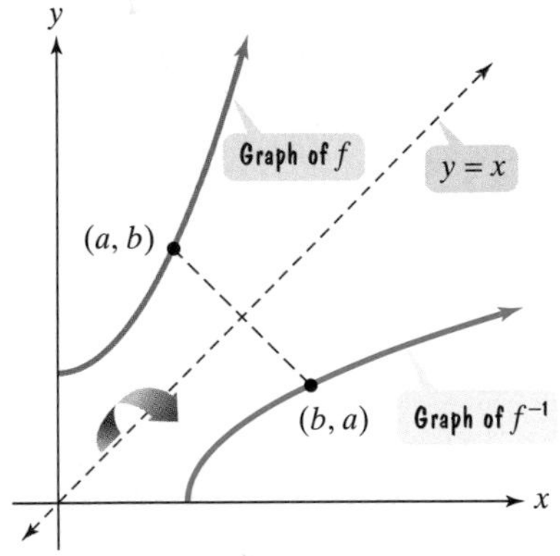

FIGURE 8.13 The graph of f^{-1} is a reflection of the graph of f about $y = x$.

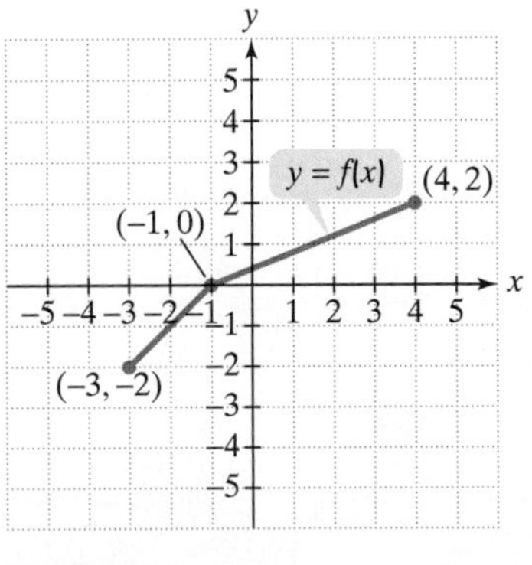

FIGURE 8.14

EXAMPLE 6 Graphing the Inverse Function

Use the graph of f in Figure 8.14 to draw the graph of its inverse function.

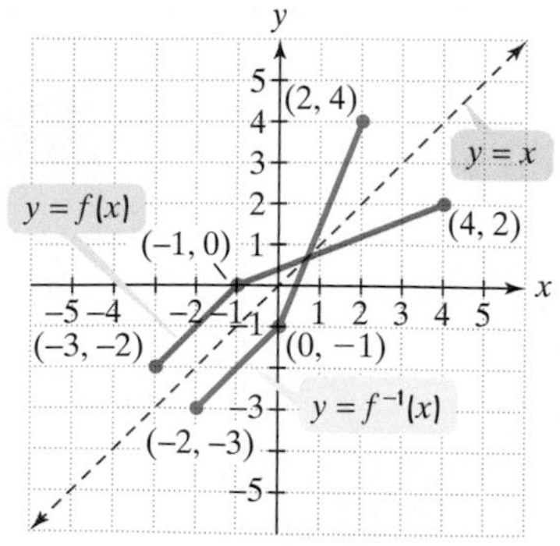

FIGURE 8.15 The graphs of f and f^{-1}

SOLUTION We begin by noting that no horizontal line intersects the graph of f, shown again in blue in Figure 8.15, at more than one point, so f does have an inverse function. Because the points $(-3, -2)$, $(-1, 0)$, and $(4, 2)$ are on the graph of f, the graph of the inverse function, f^{-1}, has points with these ordered pairs reversed. Thus, $(-2, -3)$, $(0, -1)$, and $(2, 4)$ are on the graph of f^{-1}. We can use these points to graph f^{-1}. The graph of f^{-1} is shown in green in Figure 8.15. Note that the green graph of f^{-1} is the reflection of the blue graph of f about the line $y = x$. ■

CHECK POINT 6 Use the graph of f in the figure below to draw the graph of its inverse function.

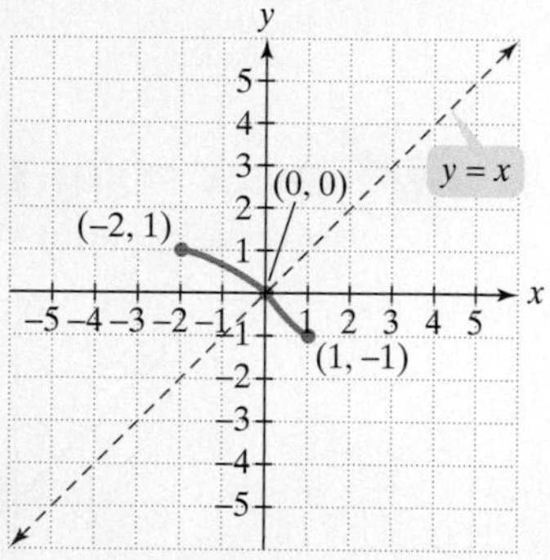

8.3 EXERCISE SET

Student Solutions Manual | CD/Video | PH Math/Tutor Center | MathXL Tutorials on CD | MathXL® | MyMathLab | Interactmath.com

Practice Exercises

In Exercises 1–14, find

a. $(f \circ g)(x)$;
b. $(g \circ f)(x)$;
c. $(f \circ g)(2)$.

1. $f(x) = 2x, \quad g(x) = x + 7$

2. $f(x) = 3x, \quad g(x) = x - 5$

3. $f(x) = x + 4, \quad g(x) = 2x + 1$

4. $f(x) = 5x + 2, \quad g(x) = 3x - 4$

5. $f(x) = 4x - 3, \quad g(x) = 5x^2 - 2$

6. $f(x) = 7x + 1, \quad g(x) = 2x^2 - 9$

7. $f(x) = x^2 + 2, \quad g(x) = x^2 - 2$

8. $f(x) = x^2 + 1, \quad g(x) = x^2 - 3$

9. $f(x) = \sqrt{x}, \quad g(x) = x - 1$

10. $f(x) = \sqrt{x}, \quad g(x) = x + 2$

11. $f(x) = 2x - 3, \quad g(x) = \dfrac{x + 3}{2}$

12. $f(x) = 6x - 3, \quad g(x) = \dfrac{x + 3}{6}$

13. $f(x) = \dfrac{1}{x}, \quad g(x) = \dfrac{1}{x}$

14. $f(x) = \dfrac{1}{x}, \quad g(x) = \dfrac{2}{x}$

In Exercises 15–24, find $f(g(x))$ and $g(f(x))$ and determine whether each pair of functions f and g are inverses of each other.

15. $f(x) = 4x$ and $g(x) = \dfrac{x}{4}$

16. $f(x) = 6x$ and $g(x) = \dfrac{x}{6}$

17. $f(x) = 3x + 8$ and $g(x) = \dfrac{x - 8}{3}$

18. $f(x) = 4x + 9$ and $g(x) = \dfrac{x - 9}{4}$

19. $f(x) = 5x - 9$ and $g(x) = \dfrac{x + 5}{9}$

20. $f(x) = 3x - 7$ and $g(x) = \dfrac{x + 3}{7}$

21. $f(x) = \dfrac{3}{x - 4}$ and $g(x) = \dfrac{3}{x} + 4$

22. $f(x) = \dfrac{2}{x - 5}$ and $g(x) = \dfrac{2}{x} + 5$

23. $f(x) = -x$ and $g(x) = -x$

24. $f(x) = -x$ and $g(x) = x$

The functions in Exercises 25–34 are all one-to-one. For each function,

a. *Find an equation for $f^{-1}(x)$, the inverse function.*

b. *Verify that your equation is correct by showing that $f(f^{-1}(x)) = x$ and $f^{-1}(f(x)) = x$.*

25. $f(x) = x + 3$

26. $f(x) = x + 5$

27. $f(x) = 2x$

28. $f(x) = 4x$

29. $f(x) = 2x + 3$

30. $f(x) = 3x - 1$

31. $f(x) = \dfrac{1}{x}$

32. $f(x) = \dfrac{2}{x}$

33. $f(x) = \dfrac{2x + 1}{x - 3}$

34. $f(x) = \dfrac{2x - 3}{x + 1}$

Which graphs in Exercises 35–40 represent functions that have inverse functions?

35.

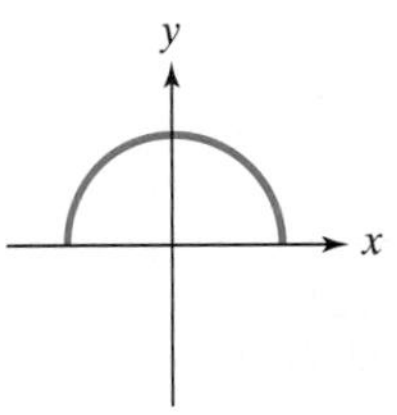

36.

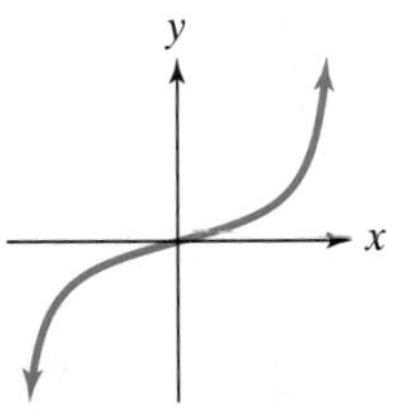

37.

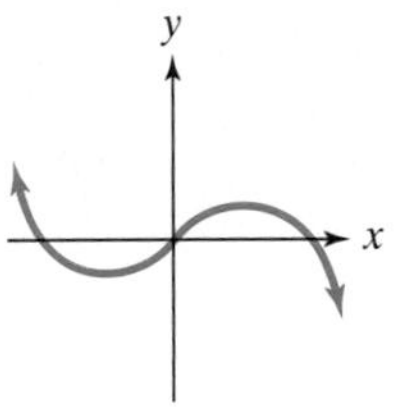

38.

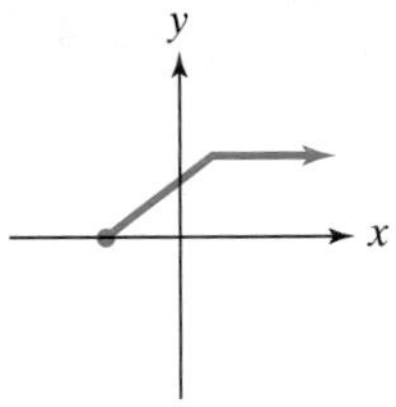

39.

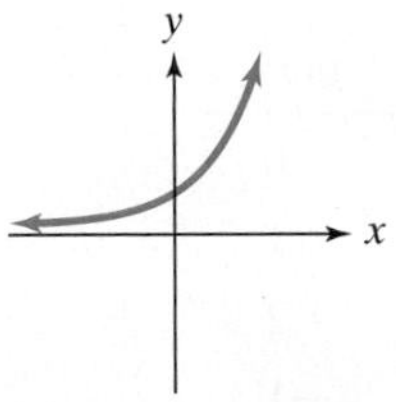

40.

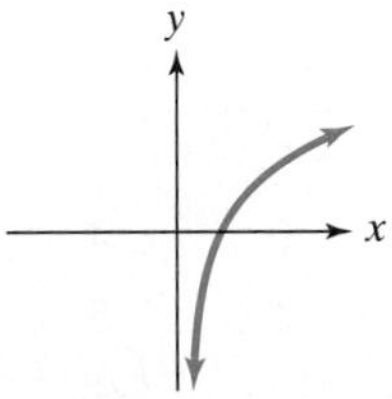

In Exercises 41–44, use the graph of f to draw the graph of its inverse function.

41.

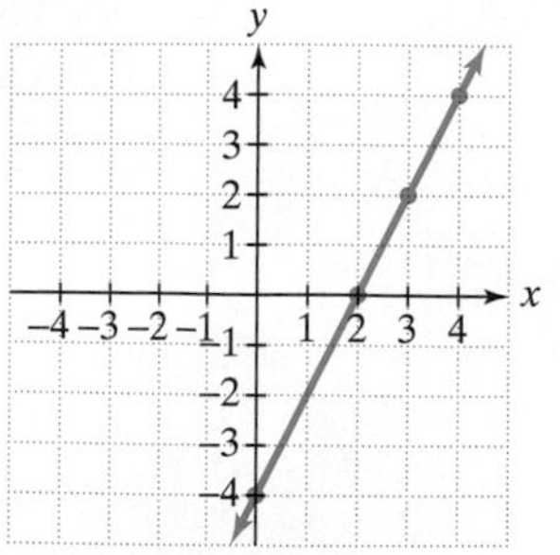

42.

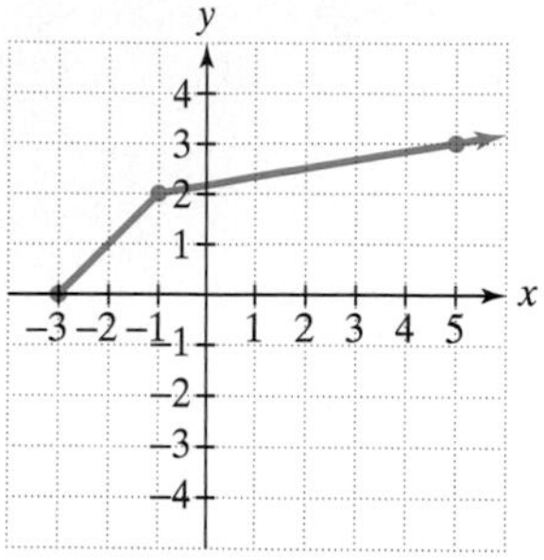

43.

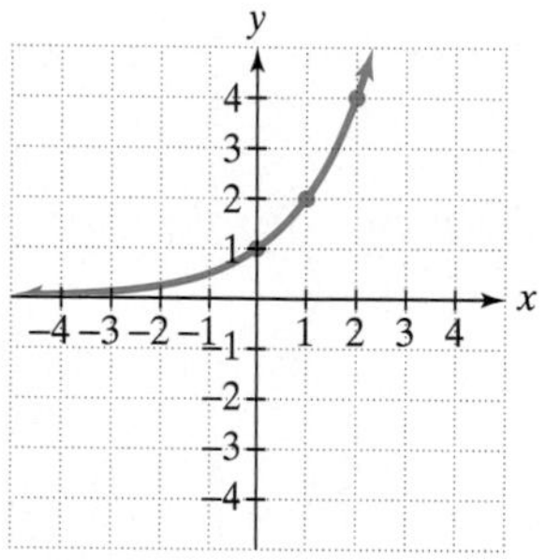

The graph approaches, but never touches the negative portion of the x-axis.

44.

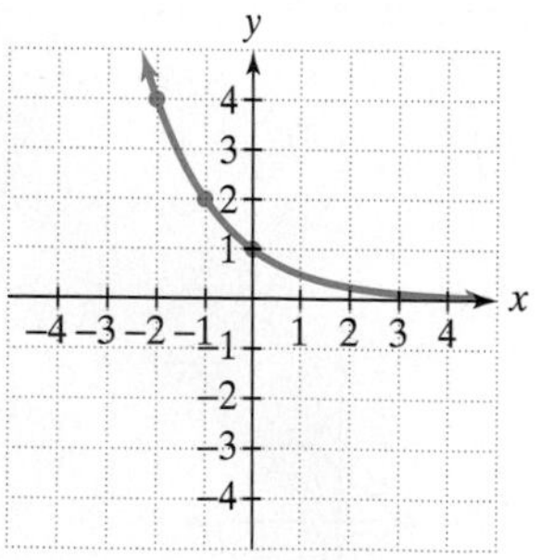

The graph approaches, but never touches the positive portion of the x-axis.

Practice Plus

In Exercises 45–50, f and g are defined by the following tables. Use the tables to evaluate each composite function.

x	$f(x)$
−1	1
0	4
1	5
2	−1

x	$g(x)$
−1	0
1	1
4	2
10	−1

45. $f(g(1))$

46. $f(g(4))$

47. $(g \circ f)(-1)$

48. $(g \circ f)(0)$

49. $f^{-1}(g(10))$

50. $f^{-1}(g(1))$

In Exercises 51–54, use the graphs of f and g to evaluate each composite function.

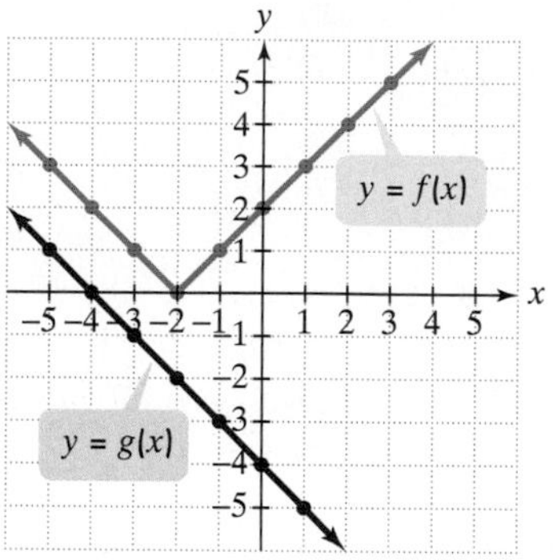

51. $(f \circ g)(-1)$

52. $(f \circ g)(1)$

53. $(g \circ f)(0)$

54. $(g \circ f)(-1)$

In Exercises 55–60, let

$$f(x) = 2x - 5$$
$$g(x) = 4x - 1$$
$$h(x) = x^2 + x + 2.$$

Evaluate the indicated function without finding an equation for the function.

55. $(f \circ g)(0)$

56. $(g \circ f)(0)$

57. $f^{-1}(1)$

58. $g^{-1}(7)$

59. $g(f[h(1)])$

60. $f(g[h(1)])$

Application Exercises

61. The regular price of a computer is x dollars. Let $f(x) = x - 400$ and $g(x) = 0.75x$.

a. Describe what the functions f and g model in terms of the price of the computer.

b. Find $(f \circ g)(x)$ and describe what this models in terms of the price of the computer.

c. Repeat part (b) for $(g \circ f)(x)$.

d. Which composite function models the greater discount on the computer, $f \circ g$ or $g \circ f$? Explain.

e. Find f^{-1} and describe what this models in terms of the price of the computer.

62. The regular price of a pair of jeans is x dollars. Let $f(x) = x - 5$ and $g(x) = 0.6x$.

a. Describe what functions f and g model in terms of the price of the jeans.

b. Find $(f \circ g)(x)$ and describe what this models in terms of the price of the jeans.

c. Repeat part (b) for $(g \circ f)(x)$.

d. Which composite function models the greater discount on the jeans, $f \circ g$ or $g \circ f$? Explain.

e. Find f^{-1} and describe what this models in terms of the price of the jeans.

In most societies, women say they prefer to marry men who are older than themselves, whereas men say they prefer women who are younger. Evolutionary psychologists attribute these preferences to female concern with a partner's material resources and male concern with a partner's fertility (Source: *Davie M. Buss, Psychological Inquiry, 6, 1–30). The graph shows the preferred age in a mate in five selected countries. Use the information in the graph to solve Exercises 63–64.*

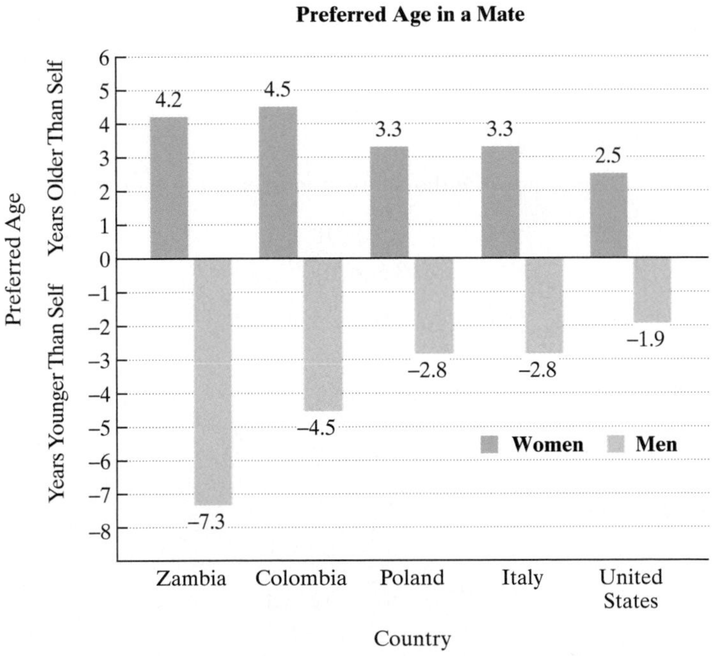

Source: Carole Wade and Carol Tavris, *Psychology*, 6th Edition, Prentice Hall, 2000

63. a. Consider a function, f, whose domain is the set of the five countries shown in the graph. Let the range be the set of the average number of years women in each of the respective countries prefer men who are older than themselves. Write function f as a set of ordered pairs.

b. Write the relation that is the inverse of f as a set of ordered pairs. Is this relation a function? Explain your answer.

64. a. Consider a function, f, whose domain is the set of the five countries shown in the graph. Let the range be the set of the average number of years men in each of the respective countries prefer women who are younger than themselves. Write f as a set of ordered pairs.

b. Write the relation that is the inverse of f as a set of ordered pairs. Is this relation a function? Explain your answer.

65. The graph represents the probability that two people in the same room share a birthday as a function of the number of people in the room. Call the function f.

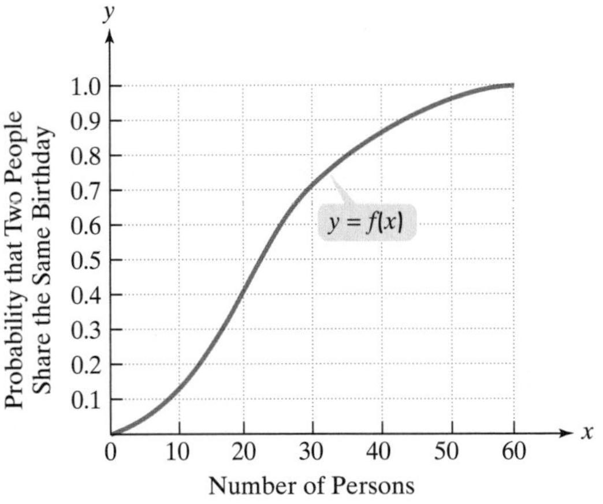

a. Explain why f has an inverse that is a function.

b. Describe in practical terms the meanings of $f^{-1}(0.25)$, $f^{-1}(0.5)$, and $f^{-1}(0.7)$.

66. The formula

$$y = f(x) = \frac{9}{5}x + 32$$

is used to convert from x degrees Celsius to y degrees Fahrenheit. The formula

$$y = g(x) = \frac{5}{9}(x - 32)$$

is used to convert from x degrees Fahrenheit to y degrees Celsius. Show that f and g are inverse functions.

Writing in Mathematics

67. Describe a procedure for finding $(f \circ g)(x)$.

68. Explain how to determine if two functions are inverses of each other.

69. Describe how to find the inverse of a one-to-one function.

70. What is the horizontal line test and what does it indicate?

71. Describe how to use the graph of a one-to-one function to draw the graph of its inverse function.

72. How can a graphing utility be used to visually determine if two functions are inverses of each other?

Technology Exercises

In Exercises 73–80, use a graphing utility to graph each function. Use the graph to determine whether the function has an inverse that is a function (that is, whether the function is one-to-one).

73. $f(x) = x^2 - 1$

74. $f(x) = \sqrt[3]{2 - x}$

75. $f(x) = \frac{x^3}{2}$

76. $f(x) = \frac{x^4}{4}$

77. $f(x) = |x - 2|$

78. $f(x) = (x - 1)^3$

79. $f(x) = -\sqrt{16 - x^2}$

80. $f(x) = x^3 + x + 1$

In Exercises 81–83, use a graphing utility to graph f and g in the same viewing rectangle. In addition, graph the line $y = x$ and visually determine if f and g are inverses.

81. $f(x) = 4x + 4, \quad g(x) = 0.25x - 1$

82. $f(x) = \frac{1}{x} + 2, \quad g(x) = \frac{1}{x - 2}$

83. $f(x) = \sqrt[3]{x} - 2, \quad g(x) = (x + 2)^3$

Critical Thinking Exercises

84. Which one of the following is true?

a. The inverse of $\{(1, 4), (2, 7)\}$ is $\{(2, 7), (1, 4)\}$.

b. The function $f(x) = 5$ is one-to-one.

c. If $f(x) = 3x$, then $f^{-1}(x) = \frac{1}{3x}$.

d. The domain of f is the same as the range of f^{-1}.

85. If $h(x) = \sqrt{3x^2 + 5}$, find functions f and g so that $h(x) = (f \circ g)(x)$.

86. If $f(x) = 3x$ and $g(x) = x + 5$, find $(f \circ g)^{-1}(x)$ and $(g^{-1} \circ f^{-1})(x)$.

87. Show that

$$f(x) = \frac{3x - 2}{5x - 3}$$

is its own inverse.

88. Consider the two functions defined by $f(x) = m_1x + b_1$ and $g(x) = m_2x + b_2$. Prove that the slope of the composite function of f with g is equal to the product of the slopes of the two functions.

Review Exercises

89. Divide and write the quotient in scientific notation:

$$\frac{4.3 \times 10^5}{8.6 \times 10^{-4}}.$$

(Section 5.7, Example 9)

90. Divide: $\frac{x^3 + 7x^2 - 2x + 3}{x - 2}$. (Section 5.6, Example 2)

91. Solve:

$$3x + 2y = 6$$
$$8x - 3y = 1.$$

(Section 4.3, Example 4)

GROUP PROJECT

CHAPTER 8

The bar graphs in Exercises 63 and 64 in Exercise Set 8.3 (see page 569) illustrate that if a relation is a function, reversing the components in each of its ordered pairs may result in a relation that is no longer a function. Group members should find examples of bar graphs, like the ones in Exercises 63 and 64, that illustrate this idea. Consult almanacs, newspapers, magazines, or the Internet. The group should select the graph with the most intriguing data. For the graph selected, write and solve a problem with four parts similar to Exercise 63 or 64.

CHAPTER 8 SUMMARY

Definitions and Concepts	Examples

Section 8.1 Introduction to Functions

Definitions and Concepts	Examples
A relation is any set of ordered pairs. The set of first components of the ordered pairs is the domain and the set of second components is the range. A function is a relation in which each member of the domain corresponds to exactly one member of the range. No two ordered pairs of a function can have the same first component and different second components.	The domain of the relation $\{(1, 2), (3, 4), (3, 7)\}$ is $\{1, 3\}$. The range is $\{2, 4, 7\}$. The relation is not a function: 3, in the domain, corresponds to both 4 and 7 in the range.
If a function is defined by an equation, the notation $f(x)$, read "f of x" or "f at x," describes the value of the function at the number, or input, x.	If $f(x) = 7x - 5$, then $f(a + 2) = 7(a + 2) - 5 = 7a + 14 - 5 = 7a + 9.$
The graph of a function is the graph of its ordered pairs. **The Vertical Line Test for Functions** If any vertical line intersects a graph in more than one point, the graph does not define y as a function of x. At the left or right of a function's graph, you will often find closed dots, open dots, or arrows. A closed dot shows that the graph ends and the point belongs to the graph. An open dot shows that the graph ends and the point does not belong to the graph. An arrow indicates that the graph extends indefinitely.	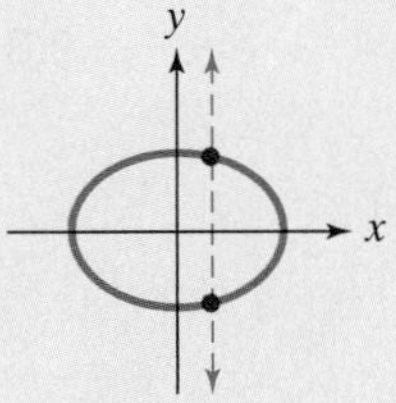 Not the graph of a function 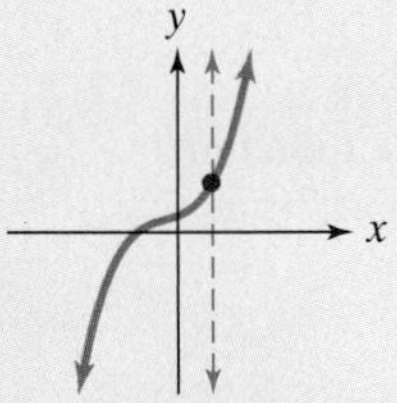 The graph of a function
The graph of a function can be used to determine the function's domain and its range. To find the domain, look for all the inputs on the x-axis that correspond to points on the graph. To find the range, look for all the outputs on the y-axis that correspond to points on the graph.	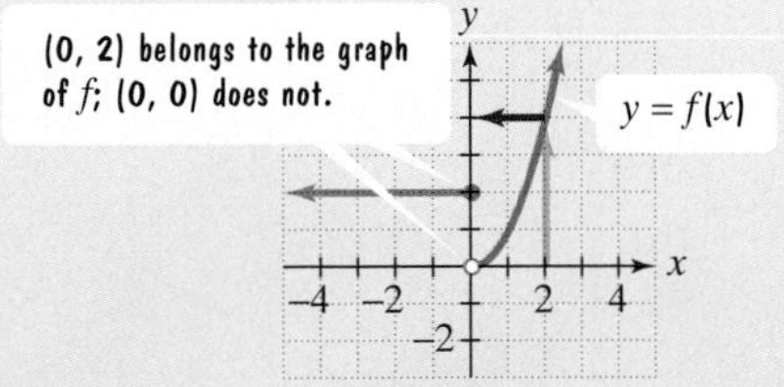 To find $f(2)$, locate 2 on the x-axis. The graph shows $f(2) = 4$. Domain of $f = \{x \mid x \text{ is a real number}\}$ Range of $f = \{y \mid y > 0\}$

Definitions and Concepts	Examples

Section 8.2 The Algebra of Functions

A Function's Domain
If a function f does not model data or verbal conditions, its domain is the largest set of real numbers for which the value of $f(x)$ is a real number. Exclude from a function's domain real numbers that cause division by zero and real numbers that result in a square root of a negative number.

$$f(x) = 7x + 13$$

Domain of $f = \{x|x \text{ is a real number}\}$

$$g(x) = \frac{7x}{12 - x}$$

Domain of $g = \{x|x \text{ is a real number and } x \neq 12\}$

The Algebra of Functions

Let f and g be two functions. The sum $f + g$, the difference $f - g$, the product fg, and the quotient $\frac{f}{g}$ are functions whose domains are the set of all real numbers common to the domains of f and g, defined as follows:

1. Sum: $(f + g)(x) = f(x) + g(x)$
2. Difference: $(f - g)(x) = f(x) - g(x)$
3. Product: $(fg)(x) = f(x) \cdot g(x)$
4. Quotient: $\left(\frac{f}{g}\right)(x) = \frac{f(x)}{g(x)}, g(x) \neq 0.$

Let $f(x) = x^2 + 2x$ and $g(x) = 4 - x$.

- $(f + g)(x) = (x^2 + 2x) + (4 - x) = x^2 + x + 4$

  $(f + g)(-2) = (-2)^2 + (-2) + 4 = 4 - 2 + 4 = 6$

- $(f - g)(x) = (x^2 + 2x) - (4 - x) = x^2 + 2x - 4 + x$
  $= x^2 + 3x - 4$

  $(f - g)(5) = 5^2 + 3 \cdot 5 - 4 = 25 + 15 - 4 = 36$

- $(fg)(x) = (x^2 + 2x)(4 - x) = 4x^2 - x^3 + 8x - 2x^2$
  $= -x^3 + 2x^2 + 8x$

  $(fg)(1) = -1^3 + 2 \cdot 1^2 + 8 \cdot 1 = -1 + 2 + 8 = 9$

- $\left(\frac{f}{g}\right)(x) = \frac{x^2 + 2x}{4 - x}, x \neq 4$

  $\left(\frac{f}{g}\right)(3) = \frac{3^2 + 2 \cdot 3}{4 - 3} = \frac{9 + 6}{1} = 15$

Section 8.3 Composite and Inverse Functions

Composite Functions
The composite function $f \circ g$ is defined by

$$(f \circ g)(x) = f(g(x)).$$

The composite function $g \circ f$ is defined by

$$(g \circ f)(x) = g(f(x)).$$

Let $f(x) = x^2 + x$ and $g(x) = 2x + 1$.

- $(f \circ g)(x) = f(g(x)) = (g(x))^2 + g(x)$

  Replace x with $g(x)$.

  $= (2x + 1)^2 + (2x + 1) = 4x^2 + 4x + 1 + 2x + 1$

  $= 4x^2 + 6x + 2$

- $(g \circ f)(x) = g(f(x)) = 2f(x) + 1$

  Replace x with $f(x)$.

  $= 2(x^2 + x) + 1 = 2x^2 + 2x + 1$

Definitions and Concepts	Examples

Section 8.3 Composite and Inverse Functions (continued)

Inverse Functions

If $f(g(x)) = x$ and $g(f(x)) = x$, function g is the inverse of function f, denoted f^{-1} and read "f inverse." The procedure for finding a function's inverse uses a switch-and-solve strategy. Switch x and y, then solve for y.

If $f(x) = 2x - 5$, find $f^{-1}(x)$.

$y = 2x - 5$ Replace $f(x)$ with y.

$x = 2y - 5$ Exchange x and y.

$x + 5 = 2y$ Solve for y.

$\frac{x+5}{2} = y$

$f^{-1}(x) = \frac{x+5}{2}$ Replace y with $f^{-1}(x)$.

The Horizontal Line Test for Inverse Functions

A function, f, has an inverse that is a function, f^{-1}, if there is no horizontal line that intersects the graph of f at more than one point. A one-to-one function is one in which no two different ordered pairs have the same second component. Only one-to-one functions have inverse functions. If the point (a, b) is on the graph of f, then the point (b, a) is on the graph of f^{-1}. The graph of f^{-1} is a reflection of the graph of f about the line $y = x$.

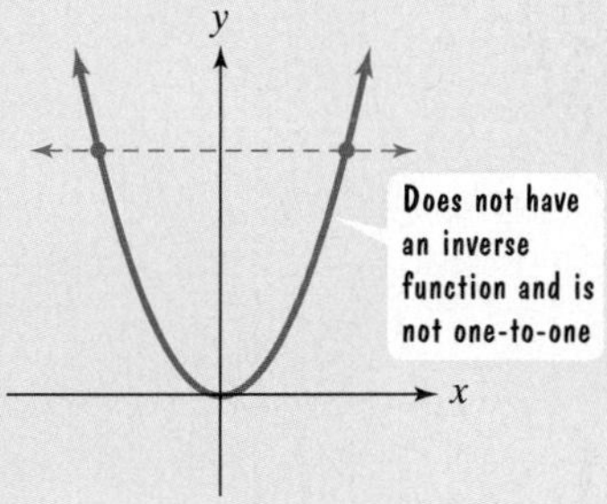

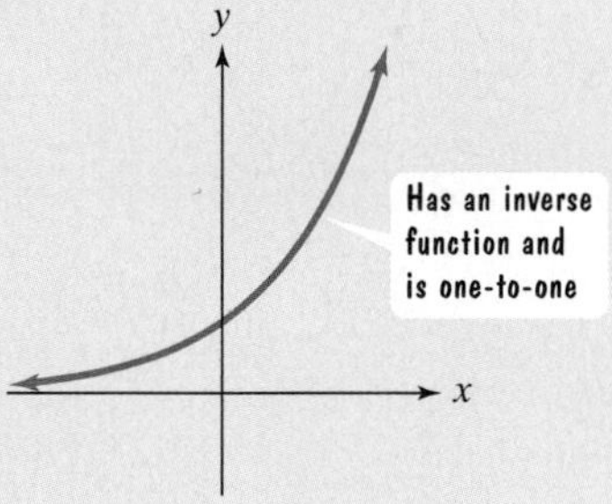

CHAPTER 8 REVIEW EXERCISES

8.1 *In Exercises 1–3, determine whether each relation is a function. Give the domain and range for each relation.*

1. $\{(3, 10), (4, 10), (5, 10)\}$

2. $\{(1, 12), (2, 100), (3, \pi), (4, -6)\}$

3. $\{(13, 14), (15, 16), (13, 17)\}$

In Exercises 4–5, find the indicated function values.

4. $f(x) = 7x - 5$

a. $f(0)$ **b.** $f(3)$ **c.** $f(-10)$

d. $f(2a)$ **e.** $f(a + 2)$

5. $g(x) = 3x^2 - 5x + 2$

a. $g(0)$ **b.** $g(5)$ **c.** $g(-4)$

d. $g(b)$ **e.** $g(4a)$

In Exercises 6–11, use the vertical line test to identify graphs in which y is a function of x.

6.

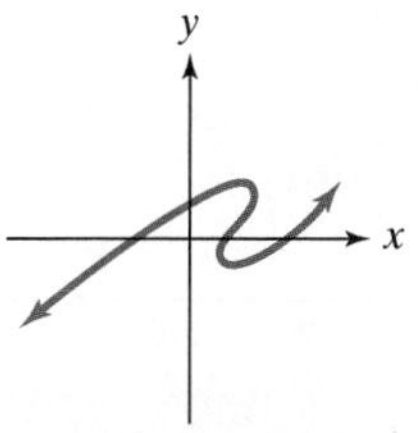

7.

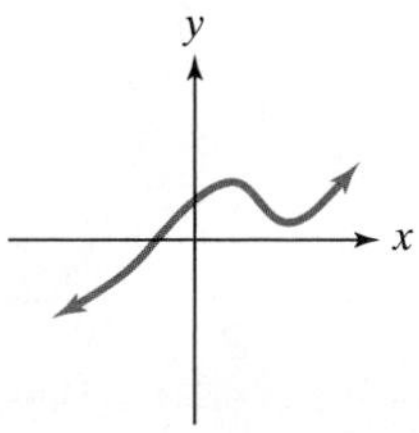

8.

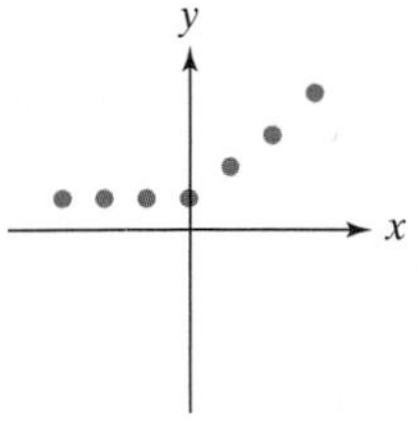

9.

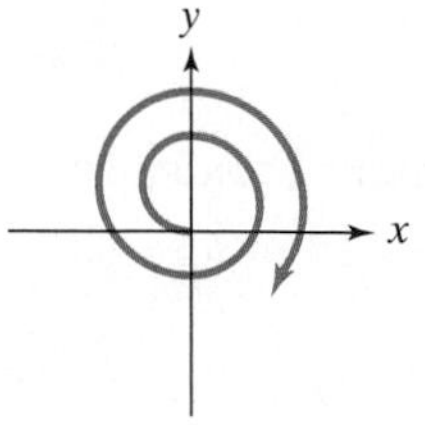

10.

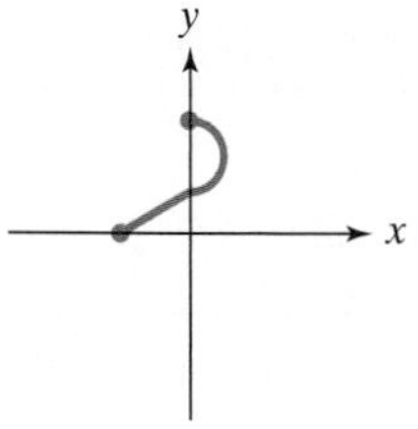

11.

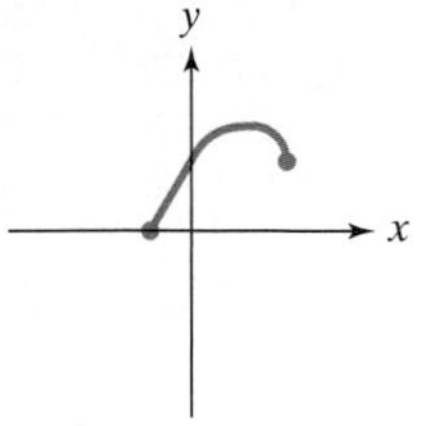

Use the graph of f to solve Exercises 12–16.

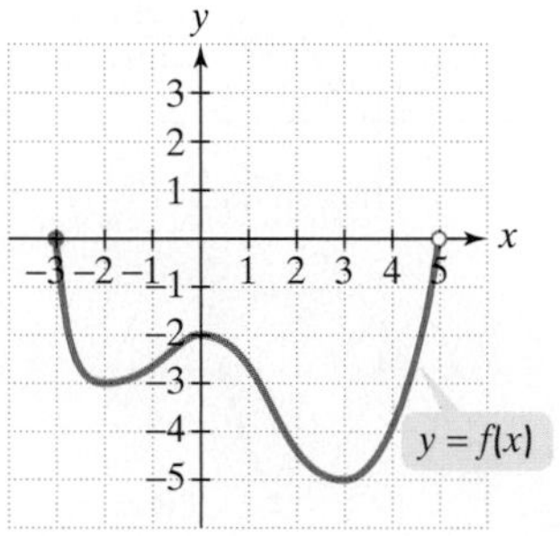

12. Find $f(-2)$.

13. Find $f(0)$.

14. For what value of x is $f(x) = -5$?

15. Find the domain of f.

16. Find the range of f.

17. The graph shows the height, in meters, of a vulture in terms of its time, in seconds, in flight.

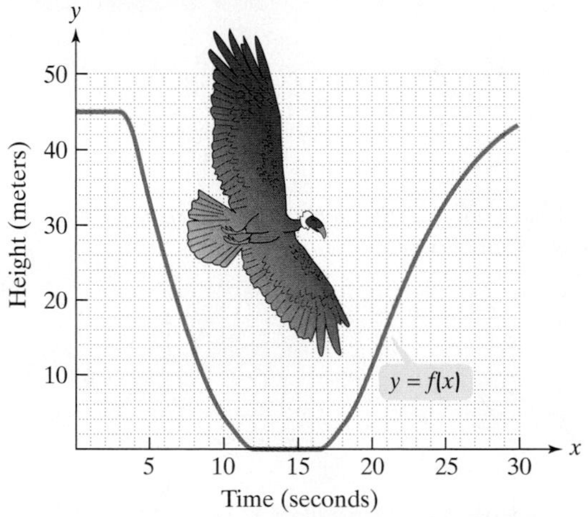

a. Use the graph to explain why the vulture's height is a function of its time in flight.

b. Find $f(15)$. Describe what this means in practical terms.

c. What is a reasonable estimate of the vulture's maximum height?

d. For what values of x is $f(x) = 20$? Describe what this means in practical terms.

e. Use the graph of the function to write a description of the vulture's flight.

8.2 *In Exercises 18–20, find the domain of each function.*

18. $f(x) = 7x - 3$

19. $g(x) = \dfrac{1}{x + 8}$

20. $f(x) = x + \dfrac{3x}{x - 5}$

In Exercises 21–22, find **a.** $(f + g)(x)$ *and* **b.** $(f + g)(3)$.

21. $f(x) = 4x - 5, \quad g(x) = 2x + 1$

22. $f(x) = 5x^2 - x + 4, \quad g(x) = x - 3$

In Exercises 23–24, for each pair of functions, f and g, determine the domain of $f + g$.

23. $f(x) = 3x + 4, \quad g(x) = \dfrac{5}{4 - x}$

24. $f(x) = \dfrac{7x}{x + 6}, \quad g(x) = \dfrac{4}{x + 1}$

In Exercises 25–32, let

$$f(x) = x^2 - 2x \quad \text{and} \quad g(x) = x - 5.$$

Find each of the following.

25. $(f + g)(x)$ and $(f + g)(-2)$

26. $f(3) + g(3)$

27. $(f - g)(x)$ and $(f - g)(1)$

28. $f(4) - g(4)$

29. $(fg)(x)$ and $(fg)(-3)$

30. $\left(\dfrac{f}{g}\right)(x)$ and $\left(\dfrac{f}{g}\right)(4)$

31. The domain of $f - g$

32. The domain of $\dfrac{f}{g}$

8.3 *In Exercises 33–34, find* **a.** $(f \circ g)(x)$; **b.** $(g \circ f)(x)$; **c.** $(f \circ g)(3)$.

33. $f(x) = x^2 + 3, \quad g(x) = 4x - 1$

34. $f(x) = \sqrt{x}, \quad g(x) = x + 1$

In Exercises 35–36, find $f(g(x))$ and $g(f(x))$ and determine whether each pair of functions f and g are inverses of each other.

35. $f(x) = \frac{3}{5}x + \frac{1}{2}$ and $g(x) = \frac{5}{3}x - 2$

36. $f(x) = 2 - 5x$ and $g(x) = \frac{2 - x}{5}$

The functions in Exercises 37–38 are all one-to-one. For each function,

a. Find an equation of $f^{-1}(x)$, the inverse function.

b. Verify that your equation is correct by showing that $f(f^{-1}(x)) = x$ and $f^{-1}(f(x)) = x$.

37. $f(x) = 4x - 3$

38. $f(x) = -\frac{1}{x}$

Which graphs in Exercises 39–42 represent functions that have inverse functions?

39.

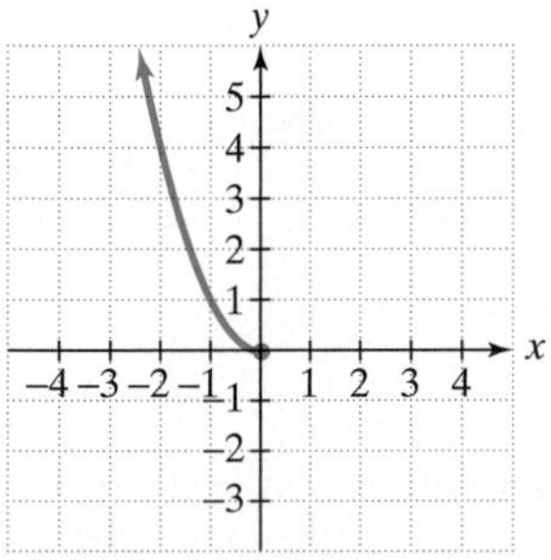

40.

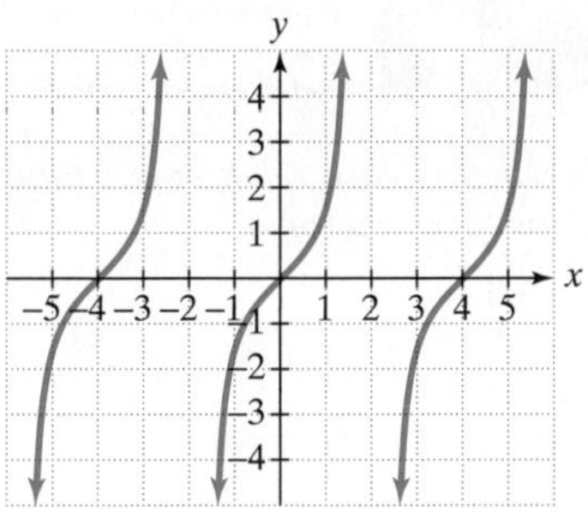

41.

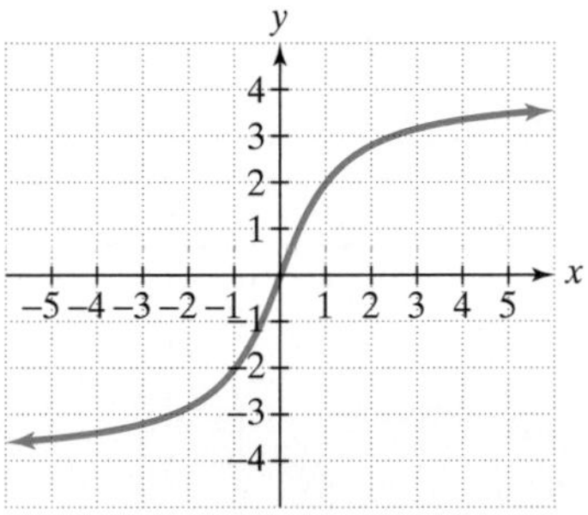

42.

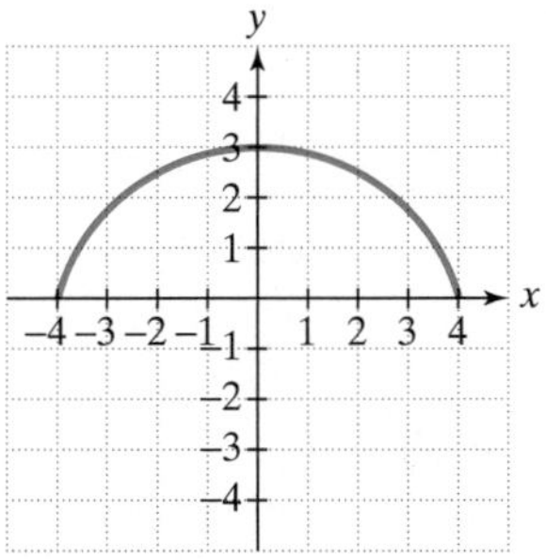

43. Use the graph of f in the figure shown to draw the graph of its inverse function.

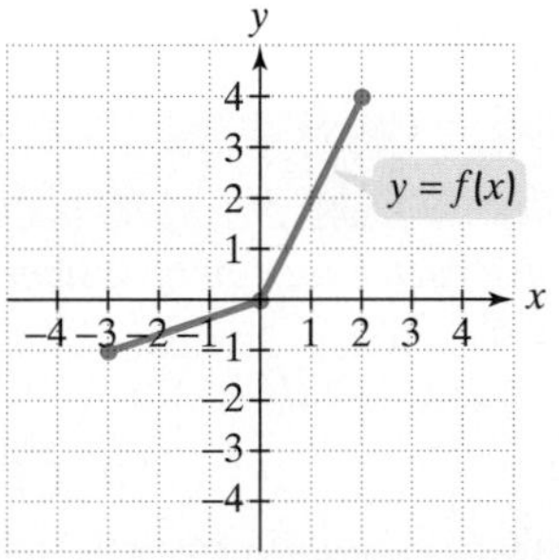

CHAPTER 8 TEST

Remember to use your Chapter Test Prep Video CD to see the worked-out solutions to the test questions you want to review.

In Exercises 1–2, determine whether each relation is a function. Give the domain and range for each relation.

1. $\{(1, 2), (3, 4), (5, 6), (6, 6)\}$

2. $\{(2, 1), (4, 3), (6, 5), (6, 6)\}$

3. If $f(x) = 3x - 2$, find $f(a + 4)$.

4. If $f(x) = 4x^2 - 3x + 6$, find $f(-2)$.

In Exercises 5–6, identify the graph or graphs in which y is a function of x.

5.

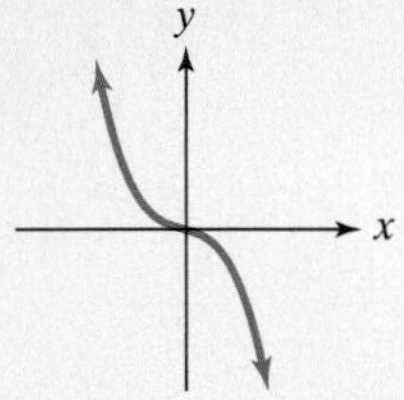

6.

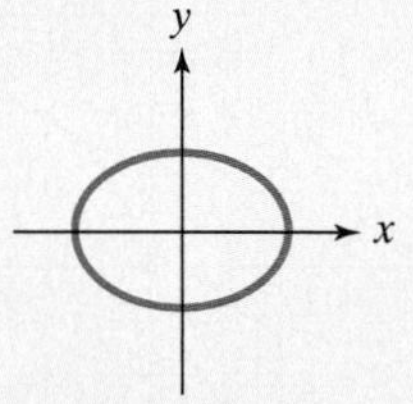

Use the graph of f to solve Exercises 7–10.

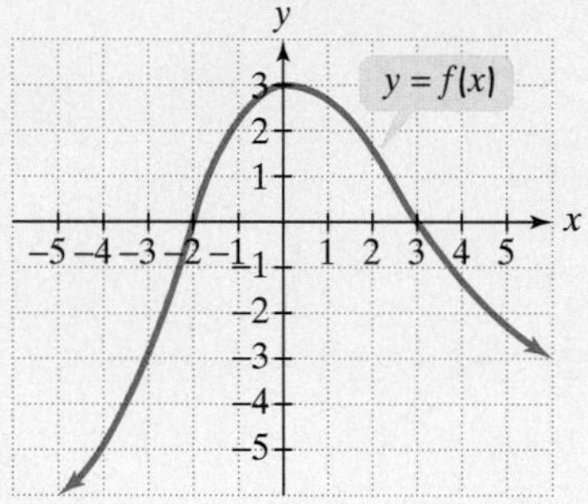

7. Find $f(6)$.

8. List two values of x for which $f(x) = 0$.

9. Find the domain of f.

10. Find the range of f.

11. Find the domain of $f(x) = \dfrac{6}{10 - x}$.

In Exercises 12–16, let

$$f(x) = x^2 + 4x \quad \textit{and} \quad g(x) = x + 2.$$

Find each of the following.

12. $(f + g)(x)$ and $(f + g)(3)$

13. $(f - g)(x)$ and $(f - g)(-1)$

14. $(fg)(x)$ and $(fg)(-5)$

15. $\left(\dfrac{f}{g}\right)(x)$ and $\left(\dfrac{f}{g}\right)(2)$

16. The domain of $\dfrac{f}{g}$

17. If $f(x) = x^2 + x$ and $g(x) = 3x - 1$, find $(f \circ g)(x)$ and $(g \circ f)(x)$.

18. If $f(x) = 5x - 7$, find $f^{-1}(x)$.

19. A function f models the amount given to charity as a function of income. The graph of f is shown in the figure.

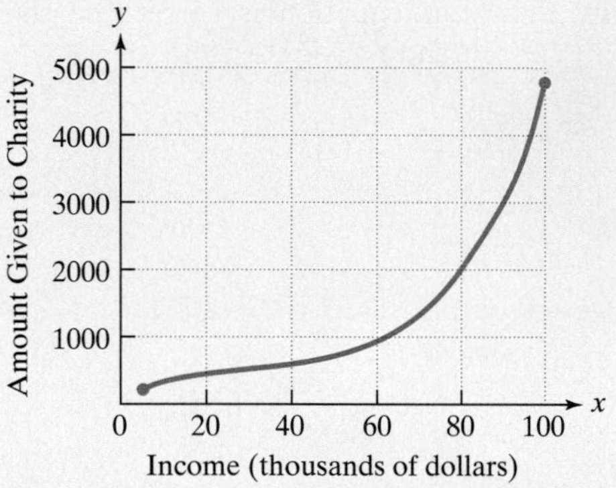

a. Explain why f has an inverse that is a function.

b. Find $f(80)$.

c. Describe in practical terms the meaning of $f^{-1}(2000)$.

CUMULATIVE REVIEW EXERCISES (CHAPTERS 1–8)

In Exercises 1–6, solve each equation or system of equations.

1. $2x + 3x - 5 + 7 = 10x + 3 - 6x - 4$

2. $2x^2 + 5x = 12$

3. $8x - 5y = -4$
$2x + 15y = -66$

4. $\dfrac{15}{x} - 4 = \dfrac{6}{x} + 3$

5. $-3x - 7 = 8$

6. If $f(x) = 2x^2 - 5x + 2$ and $g(x) = x^2 - 2x + 3$, find $(f - g)(x)$ and $(f - g)(3)$.

In Exercises 7–11, simplify each expression.

7. $\dfrac{8x^3}{-4x^7}$

8. $-8 - (-3) \cdot 4$

9. $\dfrac{\frac{1}{x} - \frac{1}{2}}{\frac{1}{3} - \frac{x}{6}}$

10. $\dfrac{4 - x^2}{3x^2 - 5x - 2}$

11. $-5 - (-8) - (4 - 6)$

In Exercises 12–13, factor completely.

12. $x^2 - 18x + 77$

13. $x^3 - 25x$

In Exercises 14–17, perform the indicated operations. If possible, simplify the answer.

14. $\dfrac{6x^3 - 19x^2 + 16x - 4}{x - 2}$

15. $(2x - 3)(4x^2 + 6x + 9)$

16. $\dfrac{3x}{x^2 + x - 2} - \dfrac{2}{x + 2}$

17. $\dfrac{5x^2 - 6x + 1}{x^2 - 1} \div \dfrac{16x^2 - 9}{4x^2 + 7x + 3}$

18. Solve the system:

$$\begin{aligned} x + 3y - z &= 5 \\ -x + 2y + 3z &= 13 \\ 2x - 5y - z &= -8. \end{aligned}$$

In Exercises 19–20, graph each equation in a rectangular coordinate system.

19. $2x - y = 4$

20. $y = -\dfrac{2}{3}x$

21. Is $\{(1, 5), (2, 5), (3, 5), (4, 5), (6, 5)\}$ a function? Give the relation's domain and range.

22. Find the slope of the line through $(-1, 5)$ and $(2, -3)$

23. Write the point-slope form of the equation of the line with slope 5, passing through $(-2, -3)$. Then use the point-slope equation to write the slope-intercept form of the line's equation.

24. Multiply and write the answer in scientific notation:

$$(7 \times 10^{-8})(3 \times 10^2).$$

25. Find the domain of $f(x) = \dfrac{1}{15 - x}$.

CHAPTER 9

Had a good workout lately? If so, could you tell if you were overdoing it or not pushing yourself hard enough?

Problems involving target heart rate ranges for various exercise goals appear as Exercises 59–62 in Exercise Set 9.4.

Inequalities and Problem Solving

Experts suggest a method for determining the heart rate you should maintain during aerobic activities, such as running, biking, or swimming. Subtract your age from 220; find 70% of that figure for the low end of the range and 80% for the high end. Your heart rate, in beats per minute, should stay between these two figures if you want to get the most cardiovascular benefit from your training.

The phrase "should stay between" indicates an algebraic inequality, a symbolic statement containing an inequality symbol such as $<$ (is less than). In this chapter, you will learn methods for modeling your world with inequalities. You will even see how inequalities are used to describe some of the most magnificent places in our nation's landscape.

SECTION 9.1

Objectives

1. Use interval notation.
2. Review how to solve linear inequalities.
3. Use linear inequalities to solve problems involving revenue, cost, and profit.

INTERVAL NOTATION AND BUSINESS APPLICATIONS USING LINEAR INEQUALITIES

Driving through your neighborhood, you see kids selling lemonade. Would it surprise you to know that this activity can be analyzed using linear inequalities? By doing so, you will view profit and loss in the business world in a new way. In this section, we use linear inequalities to solve problems and model business ventures.

1 Use interval notation.

Interval Notation Recall from Chapter 2 that any inequality in the form $ax + b \le c$ is called a **linear inequality in one variable**. The symbol between $ax + b$ and c can be $\le$ (is less than or equal to), $<$ (is less than), $\ge$ (is greater than or equal to), or $>$ (is greater than). The greatest exponent on the variable in such an inequality is 1.

Solving an inequality is the process of finding the set of numbers that make the inequality a true statement. These numbers are called the **solutions** of the inequality and we say that they **satisfy** the inequality. The set of all solutions is called the **solution set** of the inequality.

Graphs of solutions to linear inequalities are shown on a number line by shading all points representing numbers that are solutions. For example, the solutions of $x > 4$ are all real numbers that are greater than 4. They are graphed on a number line by shading all points to the right of 4. The open dot at 4 indicates that 4 is not a solution:

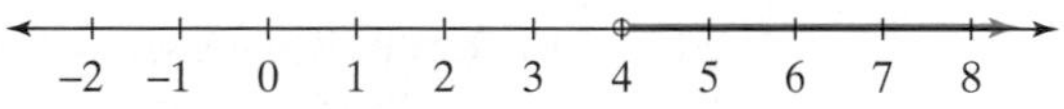

Throughout the remainder of this book, a convenient notation, called *interval notation*, will also be used to represent solution sets of inequalities. To help understand this notation, a different graphing notation will be used. Parentheses, rather than open dots, indicate endpoints that are not solutions. Thus, we graph the solutions of $x > 4$ as

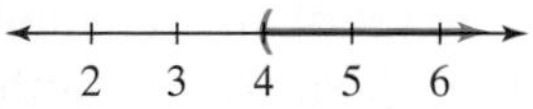

rather than as

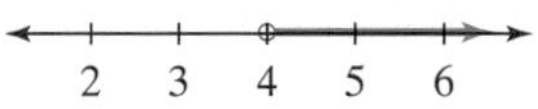

Square brackets, rather than closed dots, indicate endpoints that are solutions. Thus, we graph the solutions of $x \ge 4$ as

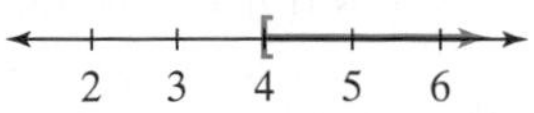

rather than as

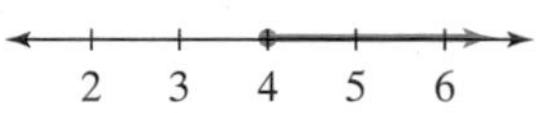

EXAMPLE 1 Graphing Inequalities

Graph the solutions of each inequality:

a. $x < 2$ **b.** $x \geq -3$ **c.** $-2 < x \leq 4$.

SOLUTION

a. The solutions of $x < 2$ are all real numbers that are less than 2. They are graphed on a number line by shading all points to the left of 2. The parenthesis at 2 indicates that 2 is not a solution, but numbers such as 1.9999 and 1.6 are. The arrow shows that the graph extends indefinitely to the left.

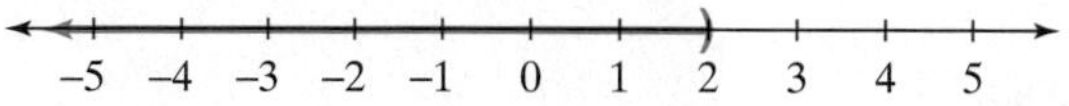

b. The solutions of $x \geq -3$ are all real numbers that are greater than or equal to -3. We shade all points to the right of -3 and the point for -3 itself. The bracket at -3 shows that -3 is a solution of the given inequality. The arrow shows that the graph extends indefinitely to the right.

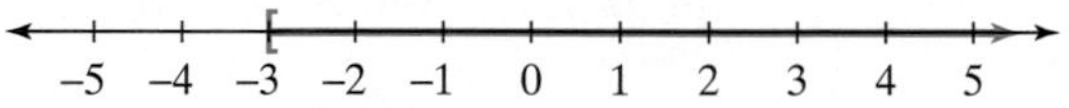

c. The inequality $-2 < x \leq 4$ is read "-2 is less than x *and* x is less than or equal to 4," or "x is greater than -2 *and* less than or equal to 4." The solutions of $-2 < x \leq 4$ are all real numbers between -2 and 4, not including -2 but including 4. The parenthesis at -2 indicates that -2 is not a solution. By contrast, the bracket at 4 shows that 4 is a solution. Shading indicates the other solutions.

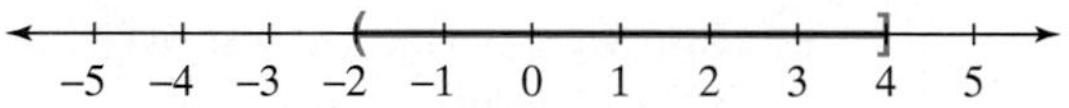

■

CHECK POINT 1 Graph the solutions of each inequality:

a. $x \leq 3$ **b.** $x > -5$ **c.** $-4 \leq x < 1$.

In Chapter 2, we used set-builder notation to represent solution sets of inequalities. Using this method, the solution set of $x > -4$ can be expressed as

$$\{x \mid x > -4\}.$$

The set of all x | such that | x is greater than -4.

We read this as "the set of all real numbers x such that x is greater than -4."

As previously mentioned, another method used to represent solution sets of inequalities is **interval notation**. Using this notation, the solution set of $x > -4$ is expressed as $(-4, \infty)$. The parenthesis at -4 indicates that -4 is not included in the interval. The infinity symbol, ∞, does not represent a real number. It indicates that the interval extends indefinitely to the right.

Table 9.1 lists nine possible types of intervals used to describe sets of real numbers.

Table 9.1 Intervals on the Real Number Line

Let a and b be real numbers such that $a < b$.

Interval Notation	Set-Builder Notation	Graph
(a, b)	$\{x \mid a < x < b\}$	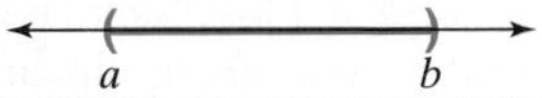
$[a, b]$	$\{x \mid a \le x \le b\}$	
$[a, b)$	$\{x \mid a \le x < b\}$	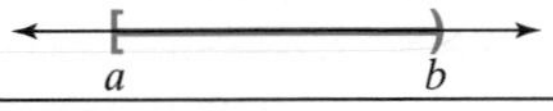
$(a, b]$	$\{x \mid a < x \le b\}$	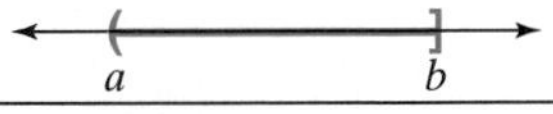
(a, ∞)	$\{x \mid x > a\}$	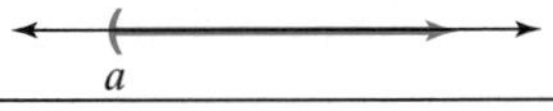
$[a, \infty)$	$\{x \mid x \ge a\}$	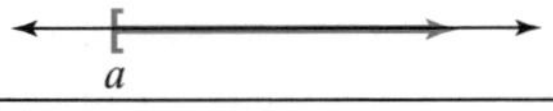
$(-\infty, b)$	$\{x \mid x < b\}$	
$(-\infty, b]$	$\{x \mid x \le b\}$	
$(-\infty, \infty)$	$\{x \mid x$ is a real number$\}$ or $\mathbb{R}$ (set of all real numbers)	

EXAMPLE 2 Intervals and Inequalities

Express each interval in set-builder notation and graph:

a. $(-1, 4]$ **b.** $[2.5, 4]$ **c.** $(-3, \infty)$.

SOLUTION

a. $(-1, 4] = \{x \mid -1 < x \le 4\}$

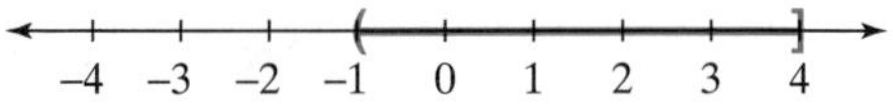

b. $[2.5, 4] = \{x \mid 2.5 \le x \le 4\}$

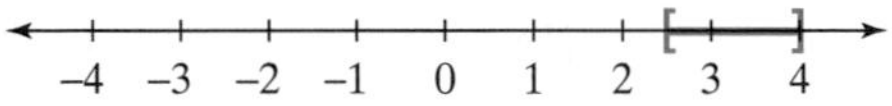

c. $(-3, \infty) = \{x \mid x > -3\}$

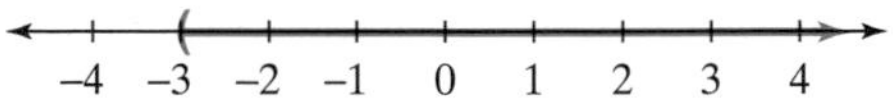

■

CHECK POINT 2 Express each interval in set-builder notation and graph:

a. $[-2, 5)$ **b.** $[1, 3.5]$ **c.** $(-\infty, -1)$.

2 Review how to solve linear inequalities.

STUDY TIP

If you need a more detailed presentation on solving linear inequalities, see Section 2.7, Examples 2 through 7, on pages 164 through 168.

Solving Linear Inequalities Recall from Chapter 2 the following procedure for solving a linear inequality. The only difference is that we will now use two notations, set-builder notation and interval notation, to express solution sets.

SOLVING A LINEAR INEQUALITY

1. Simplify the algebraic expression on each side.
2. Use the addition property of inequality to collect all the variable terms on one side and all the constant terms on the other side.
3. Use the multiplication property of inequality to isolate the variable and solve. Reverse the sense of the inequality when multiplying or dividing both sides by a negative number.
4. Express the solution set in set-builder or interval notation and graph the solution set on a number line.

EXAMPLE 3 Solving a Linear Inequality

Solve and graph the solution set on a number line:

$$2 - 12x \leq 7(1 - x).$$

SOLUTION

Step 1. **Simplify each side.** We use the distributive property to remove parentheses on the right side.

$$2 - 12x \leq 7(1 - x) \quad \text{This is the given inequality.}$$

$$2 - 12x \leq 7 - 7x \quad \text{Use the distributive property.}$$

Step 2. **Collect variable terms on one side and constant terms on the other side.** We will collect variable terms on the left and constant terms on the right.

$$2 - 12x + 7x \leq 7 - 7x + 7x \quad \text{Add 7x to both sides.}$$

$$2 - 5x \leq 7 \quad \text{Simplify.}$$

$$2 - 5x - 2 \leq 7 - 2 \quad \text{Subtract 2 from both sides.}$$

$$-5x \leq 5 \quad \text{Simplify.}$$

STUDY TIP

You can solve

$$2 - 12x \leq 7 - 7x$$

by isolating x on the right side. Add $12x$ to both sides:

$$2 - 12x + 12x \leq 7 - 7x + 12x$$

$$2 \leq 7 + 5x.$$

Now subtract 7 from both sides:

$$2 - 7 \leq 7 + 5x - 7$$

$$-5 \leq 5x.$$

Finally, divide both sides by 5:

$$\frac{-5}{5} \leq \frac{5x}{5}$$

$$-1 \leq x.$$

This last inequality means the same thing as

$$x \geq -1.$$

Step 3. **Isolate the variable and solve.** We isolate the variable, x, by dividing both sides by 5. Because we are dividing by a negative number, we must reverse the inequality symbol.

$$\frac{5x}{-5} \geq \frac{5}{-5} \quad \text{Divide both sides by } -5\text{. and reverse the sense of the inequality.}$$

$$x \geq -1 \quad \text{Simplify.}$$

Step 4. **Express the solution set in set-builder or interval notation and graph the set on a number line.** The solution set consists of all real numbers that are greater than or equal to -1, expressed in set-builder notation as $\{x | x \geq -1\}$. The interval notation for this solution set is $[-1, \infty)$. The graph of the solution set is shown as follows:

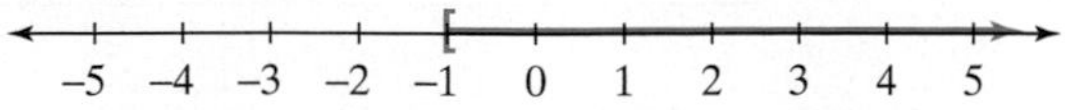

■

CHECK POINT 3 Solve and graph the solution set on a number line:

$$-12 - 8x \le 4(6 - x).$$

If an inequality contains fractions, begin by multiplying both sides by the least common denominator. This will clear the inequality of fractions.

EXAMPLE 4 Solving a Linear Inequality

Solve and graph the solution set on a number line:

$$\frac{x+3}{4} > \frac{x-2}{3} + \frac{1}{4}.$$

SOLUTION The denominators are 4, 3, and 4. The least common denominator is 12. We begin by multiplying both sides of the inequality by 12.

$$\frac{x+3}{4} > \frac{x-2}{3} + \frac{1}{4}$$ This is the given inequality.

$$12\left(\frac{x+3}{4}\right) > 12\left(\frac{x-2}{3} + \frac{1}{4}\right)$$ Multiply both sides by 12. Multiplying by a positive number preserves the sense of the inequality.

$$\frac{12}{1}\cdot\frac{x+3}{4} > \frac{12}{1}\cdot\frac{x-2}{3} + \frac{12}{1}\cdot\frac{1}{4}$$ Use the distributive property and multiply each term by 12.

$$\frac{\overset{3}{\cancel{12}}}{1}\cdot\frac{x+3}{\underset{1}{\cancel{4}}} > \frac{\overset{4}{\cancel{12}}}{1}\cdot\frac{x-2}{\underset{1}{\cancel{3}}} + \frac{\overset{3}{\cancel{12}}}{1}\cdot\frac{1}{\underset{1}{\cancel{4}}}$$ Divide out common factors in each multiplication.

$$3(x+3) > 4(x-2) + 3$$ The fractions are now cleared.

Now that fractions are cleared, we follow the four steps that we used in the previous example.

Step 1. **Simplify each side.**

$$3(x+3) > 4(x-2) + 3$$ This is the inequality with the fractions cleared.

$$3x + 9 > 4x - 8 + 3$$ Use the distributive property.

$$3x + 9 > 4x - 5$$ Simplify.

Step 2. **Collect variable terms on one side and constant terms on the other side.** We will collect variable terms on the left and constant terms on the right.

$$3x + 9 - 4x > 4x - 5 - 4x$$ Subtract 4x from both sides.

$$-x + 9 > -5$$ Simplify.

$$-x + 9 - 9 > -5 - 9$$ Subtract 9 from both sides.

$$-x > -14$$ Simplify.

Step 3. **Isolate the variable and solve.** To isolate x in $-x > -14$ we must eliminate the negative sign in front of the x. Because $-x$ means $-1x$, we can do this by multiplying (or dividing) both sides of the inequality by -1. We are multiplying by a negative number. Thus, we must reverse the inequality symbol.

$$(-1)(-x) > (-1)(-14)$$ Multiply both sides by -1 and reverse the sense of the inequality.

$$x > 14$$ Simplify.

Step 4. **Express the solution set in set-builder or interval notation and graph the set on a number line.** The solution set consists of all real numbers that are less than 14, expressed in set-builder notation as $\{x|x < 14\}$. The interval notation for this solution set is $(-\infty, 14)$. The graph of the solution set is shown as follows:

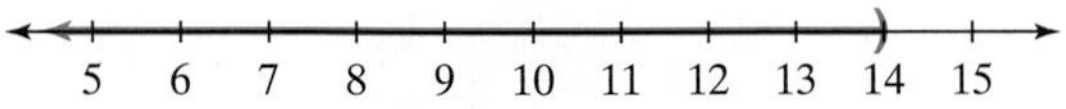

CHECK POINT 4 Solve and graph the solution set on a number line:

$$\frac{x-4}{2} > \frac{x-2}{3} + \frac{5}{6}.$$

3 Use linear inequalities to solve problems involving revenue, cost, and profit.

Functions of Business and Linear Inequalities As a young entrepreneur, did you ever try selling lemonade in your front yard? Suppose that you charged 55 cents for each cup and you sold 45 cups. Your **revenue** is your income from selling these 45 units, or $\$0.55(45) = \$24.75$. Your *revenue function* from selling x cups is

$$R(x) = 0.55x.$$

This is the unit price: 55¢ for each cup. This is the number of units sold.

For any business, the **revenue function**, $R(x)$, is the money generated by selling x units of the product:

$$R(x) = px.$$

Price per unit x units sold

Back to selling lemonade and energizing the neighborhood with white sugar: Is your revenue for the afternoon also your profit? No. We need to consider the cost of the business. You estimate that the lemons, white sugar, and bottled water cost 5 cents per cup. Furthermore, mommy dearest is charging you a \$10 rental fee for use of your (her?) front yard. Your *cost function* for selling x cups of lemonade is

$$C(x) = 10 + 0.05x.$$

This is your \$10 fixed cost. This is your variable cost: 5¢ for each cup produced.

For any business, the **cost function**, $C(x)$, is the cost of producing x units of the product:

$$C(x) = (\text{fixed cost}) + cx.$$

Cost per unit x units produced

The term on the right, cx, represents **variable cost**, because it varies based on the number of units produced. Thus, the cost function is the sum of the fixed cost and the variable cost.

REVENUE AND COST FUNCTIONS A company produces and sells x units of a product.

Revenue Function

$$R(x) = (\text{price per unit sold})x$$

Cost Function

$$C(x) = \text{fixed cost} + (\text{cost per unit produced})x$$

What does every entrepreneur, from a kid selling lemonade to Donald Trump, want to do? Generate profit, of course. The *profit* made is the money taken in, or the revenue, minus the money spent, or the cost. This relationship between revenue and cost allows us to define the *profit function*, $P(x)$.

THE PROFIT FUNCTION The profit, $P(x)$, generated after producing and selling x units of a product is given by the **profit function**

$$P(x) = R(x) - C(x)$$

where R and C are the revenue and cost functions, respectively.

Figure 9.1 shows the graphs of the revenue and cost functions for the lemonade business. Similar graphs and models apply no matter how small or large a business venture may be.

$$R(x) = 0.55x \qquad C(x) = 10 + 0.05x$$

Revenue is 55¢ times the number of cups sold.

Cost is \$10 plus 5¢ times the number of cups produced.

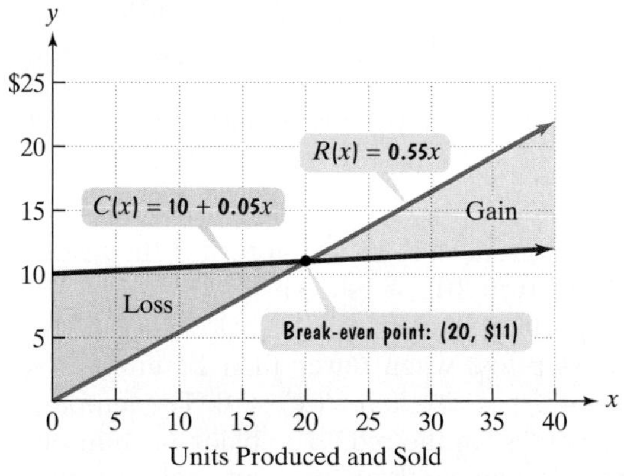

FIGURE 9.1

The lines intersect at the point $(20, 11)$. This means that when 20 cups are produced and sold, both cost and revenue are \$11. In business, this point of intersection is called the **break-even point**. At the break-even point, the money coming in is equal to

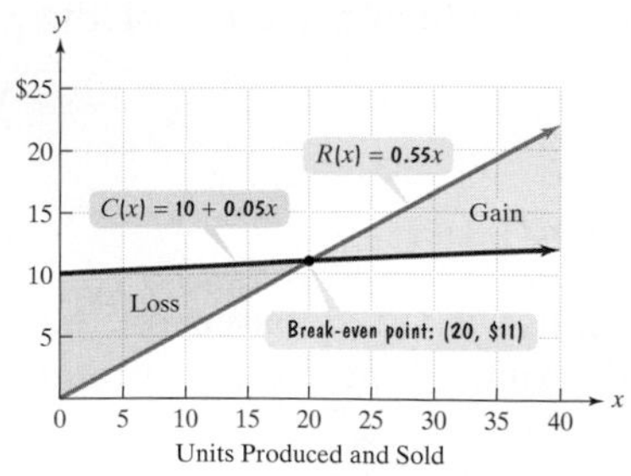

FIGURE 9.1 (repeated)

the money going out. Can you see what happens for x-values less than 20? The red cost graph is above the blue revenue graph: $C(x) > R(x)$. The cost is greater than the revenue and the business is losing money. Thus, if you sell fewer than 20 cups of lemonade, the result is a *loss*. By contrast, look at what happens for x-values greater than 20. The blue revenue graph is above the red cost graph: $R(x) > C(x)$. The revenue is greater than the cost and the business is making money. Thus, if you sell more than 20 cups of lemonade, the result is a *profit*.

EXAMPLE 5 Writing a Profit Function to Determine a Profit

a. Use the revenue and cost functions shown in Figure 9.1,

$$R(x) = 0.55x \quad \text{and} \quad C(x) = 10 + 0.05x,$$

to write the profit function for producing and selling x cups of lemonade.

b. Use and solve a linear inequality to answer this question: More than how many cups of lemonade must be sold to have a profit?

SOLUTION

a. The profit function is the difference between the revenue function and the cost function.

$$P(x) = R(x) - C(x) \qquad \text{This is the definition of the profit function.}$$
$$= 0.55x - (10 + 0.05x) \qquad \text{Substitute the given functions.}$$
$$= 0.55x - 10 - 0.05x \qquad \text{Distribute } -1 \text{ to each term in parentheses.}$$
$$= 0.50x - 10 \qquad \text{Simplify.}$$

The profit function is $P(x) = 0.50x - 10$.

b. A business makes money, or experiences a profit, when $P(x) > 0$. A business loses money, or experiences a loss, when $P(x) < 0$. Because we are interested in a profit, this occurs when $P(x) > 0$.

$$0.50x - 10 > 0 \qquad \text{This inequality models a profit.}$$
$$0.50x > 10 \qquad \text{Add 10 to both sides.}$$
$$\frac{0.50x}{0.50} > \frac{10}{0.50} \qquad \text{Divide both sides by 0.50. Division by a positive number does not reverse the sense of the inequality.}$$
$$x > 20 \qquad \text{Simplify.}$$

More than 20 cups of lemonade must be sold to have a profit.

DISCOVER FOR YOURSELF

A profit occurs when revenue exceeds cost:

$$R(x) > C(x).$$

Use $R(x) = 0.55x$ and $C(x) = 10 + 0.05x$ to verify that more than 20 cups of lemonade must be sold to have a profit.

The graph of the profit function, $P(x) = 0.50x - 10$, is shown in Figure 9.2. The red portion of the graph lies below the x-axis and shows a loss when fewer than 20 units are sold. Thus, if $x < 20$ then $P(x) < 0$. The lemonade business is "in the red." The black portion of the graph lies above the x-axis and shows a gain when more than 20 units are sold. Thus, if $x > 20$ then $P(x) > 0$. The lemonade business is "in the black."

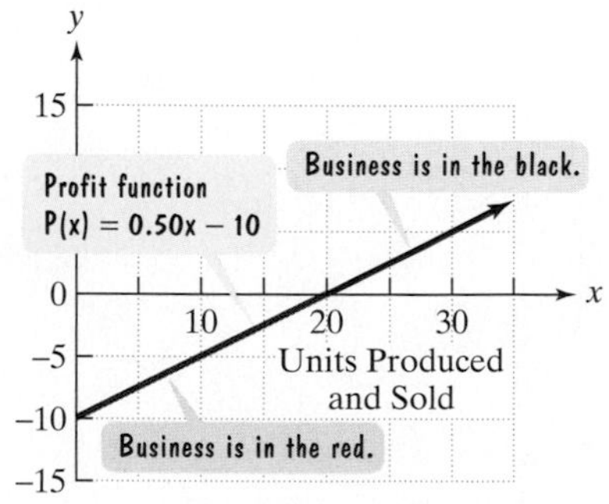

FIGURE 9.2 In the red and in the black

CHECK POINT 5

a. Use the revenue and cost functions

$$R(x) = 200x$$
$$C(x) = 160{,}000 + 75x$$

to write the profit function for producing and selling x units.

b. More than how many units must be produced and sold to have a profit?

EXAMPLE 6 Writing a Profit Function to Determine a Profit

Technology is now promising to bring light, fast, and beautiful wheelchairs to millions of disabled people. A company is planning to manufacture these radically different wheelchairs. Fixed cost will be $500,000 and it will cost $400 to produce each wheelchair. Each wheelchair will be sold for $600.

a. Write the cost function, C, of producing x wheelchairs.

b. Write the revenue function, R, from the sale of x wheelchairs.

c. Write the profit function, P, from producing and selling x wheelchairs.

d. More than how many wheelchairs must be produced and sold to have a profit?

Summary of Given Information
Fixed cost: $500,000
Cost to produce each wheelchair: $400
Selling price per wheelchair: $600

SOLUTION

a. The cost function is the sum of the fixed cost and variable cost.

Fixed cost of $500,000 | plus | Variable cost: $400 for each chair produced

$$C(x) = 500{,}000 + 400x$$

b. The revenue function is the money generated from the sale of x wheelchairs.

Revenue per chair, $600, times | the number of chairs sold

$$R(x) = 600x$$

c. The profit function is the difference between the revenue function and the cost function.

$P(x) = R(x) - C(x)$ — This is the definition of the profit function.

$= 600x - (500{,}000 + 400x)$ — Substitute the given functions.

$= 600x - 500{,}000 - 400x$ — Distribute -1 to each term in parentheses.

$= 200x - 500{,}000$ — Simplify.

The profit function is $P(x) = 200x - 500{,}000$.

USING TECHNOLOGY

The graphs of the wheelchair company's cost and revenue functions

$$C(x) = 500{,}000 + 400x$$

and $R(x) = 600x$

are shown in a

$$[0, 5000, 1000] \text{ by } [0, 3{,}000{,}000, 1{,}000{,}000]$$

viewing rectangle. To the right of the intersection point, the graph of the revenue function lies above that of the cost function. This confirms that producing and selling more than 2500 wheelchairs results in a profit.

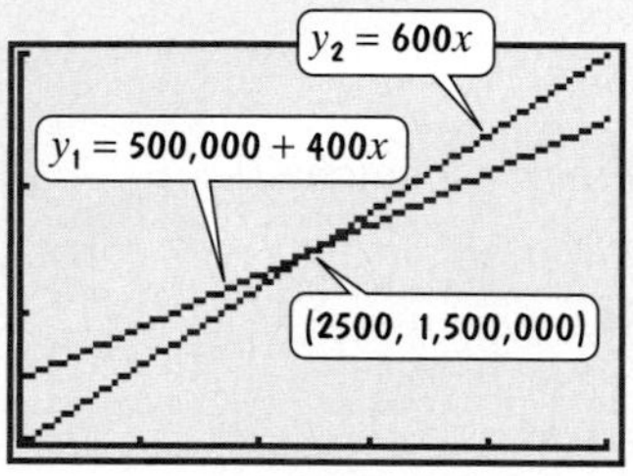

d. A profit occurs when $P(x) > 0$. We use the profit function $P(x) = 200x - 500{,}000$ from part (c).

$$200x - 500{,}000 > 0 \quad \text{Set the profit function greater than 0.}$$

$$200x > 500{,}000 \quad \text{Add 500,000 to both sides.}$$

$$\frac{200x}{200} > \frac{500{,}000}{200} \quad \text{Divide both sides by 200.}$$

$$x > 2500 \quad \text{Simplify.}$$

More than 2500 wheelchairs must be produced and sold to have a profit. ■

✔ **CHECK POINT 6** A company that manufactures running shoes has a fixed cost of \$300,000. Additionally, it costs \$30 to produce each pair of shoes. They are sold at \$80 per pair.

a. Write the cost function, C, of producing x pairs of running shoes.

b. Write the revenue function, R, from the sale of x pairs of running shoes.

c. Write the profit function, P, from producing and selling x pairs of running shoes.

d. More than how many pairs of running shoes must be produced and sold to have a profit?

9.1 EXERCISE SET

Student Solutions Manual | CD/Video | PH Math/Tutor Center | MathXL Tutorials on CD | MathXL® | MyMathLab | Interactmath.com

Practice Exercises

In Exercises 1–14, express each interval in set-builder notation and graph the interval on a number line.

1. $(1, 6]$ **2.** $(-2, 4]$
3. $[-5, 2)$ **4.** $[-4, 3)$
5. $[-3, 1]$ **6.** $[-2, 5]$
7. $(2, \infty)$ **8.** $(3, \infty)$
9. $[-3, \infty)$ **10.** $[-5, \infty)$
11. $(-\infty, 3)$ **12.** $(-\infty, 2)$
13. $(-\infty, 5.5)$ **14.** $(-\infty, 3.5]$

In Exercises 15–40, solve each linear inequality and graph the solution set on a number line.

15. $5x + 11 < 26$ **16.** $2x + 5 < 17$
17. $3x - 8 \ge 13$ **18.** $8x - 2 \ge 14$
19. $-9x \ge 36$ **20.** $-5x \le 30$
21. $8x - 11 \le 3x - 13$ **22.** $18x + 45 \le 12x - 8$
23. $4(x + 1) + 2 \ge 3x + 6$
24. $8x + 3 > 3(2x + 1) + x + 5$
25. $2x - 11 < -3(x + 2)$
26. $-4(x + 2) > 3x + 20$
27. $1 - (x + 3) \ge 4 - 2x$
28. $5(3 - x) \le 3x - 1$
29. $\frac{x}{4} - \frac{1}{2} \le \frac{x}{2} + 1$
30. $\frac{3x}{10} + 1 \ge \frac{1}{5} - \frac{x}{10}$
31. $1 - \frac{x}{2} > 4$

32. $7 - \frac{4}{5}x < \frac{3}{5}$

33. $\frac{x - 4}{6} \geq \frac{x - 2}{9} + \frac{5}{18}$

34. $\frac{4x - 3}{6} + 2 \geq \frac{2x - 1}{12}$

35. $7(y + 4) - 13 < 12 + 13(3 + y)$

36. $-3[7y - (2y - 3)] > -2(y + 1)$

37. $6 - \frac{2}{3}(3x - 12) \leq \frac{2}{5}(10x + 50)$

38. $\frac{2}{7}(7 - 21x) - 4 > 10 - \frac{3}{11}(11x - 11)$

39. $3[3(y + 5) + 8y + 7] + 5[3(y - 6) - 2(3y - 5)] < 2(4y + 3)$

40. $5[3(2 - 3y) - 2(5 - y)] - 6[5(y - 2) - 2(4y - 3)] < 3y + 19$

41. Let $f(x) = 3x + 2$ and $g(x) = 5x - 8$. Find all values of x for which $f(x) > g(x)$.

42. Let $f(x) = 2x - 9$ and $g(x) = 5x + 4$. Find all values of x for which $f(x) > g(x)$.

43. Let $f(x) = \frac{2}{5}(10x + 15)$ and $g(x) = \frac{1}{4}(8 - 12x)$. Find all values of x for which $g(x) \leq f(x)$.

44. Let $f(x) = \frac{3}{5}(10x - 15) + 9$ and $g(x) = \frac{3}{8}(16 - 8x) - 7$. Find all values of x for which $g(x) \leq f(x)$.

In Exercises 45–48, cost and revenue functions for producing and selling x units of a product are given. Cost and revenue are expressed in dollars.

a. *Write the profit function from producing and selling x units of the product.*

b. *More than how many units must be produced and sold to have a profit?*

45. $C(x) = 25{,}500 + 15x$
$R(x) = 32x$

46. $C(x) = 15{,}000 + 12x$
$R(x) = 32x$

47. $C(x) = 105x + 70{,}000$
$R(x) = 245x$

48. $C(x) = 1.2x + 1500$
$R(x) = 1.7x$

Practice Plus

In Exercises 49–50, solve each linear inequality and graph the solution set on a number line.

49. $2(x + 3) > 6 - \{4[x - (3x - 4) - x] + 4\}$

50. $3(4x - 6) < 4 - \{5x - [6x - (4x - (3x + 2))]\}$

In Exercises 51–52, write an inequality with x isolated on the left side that is equivalent to the given inequality.

51. $ax + b > c$; Assume $a < 0$.

52. $\frac{ax + b}{c} > b$; Assume $a > 0$ and $c < 0$.

In Exercises 53–54, use the graphs of y_1 and y_2 to solve each inequality.

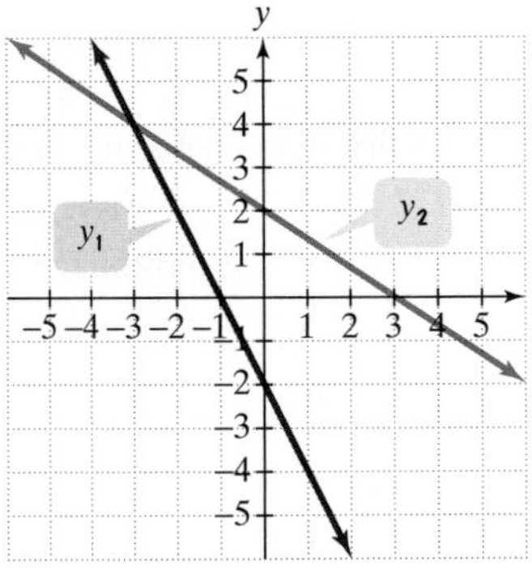

53. $y_1 \geq y_2$

54. $y_1 \leq y_2$

In Exercises 55–56, use the table of values for the linear functions y_1 and y_2 to solve each inequality.

X	Y1	Y2
-1.8	9.4	7.4
-1.7	9.1	7.6
-1.6	8.8	7.8
-1.5	8.5	8
-1.4	8.2	8.2
-1.3	7.9	8.4
-1.2	7.6	8.6

X=-1.8

55. $y_1 < y_2$

56. $y_1 > y_2$

Application Exercises

The graphs show that the three components of love, namely passion, intimacy, and commitment, progress differently over time. Passion peaks early in a relationship and then declines. By contrast, intimacy and commitment build gradually. Use the graphs to solve Exercises 57–64.

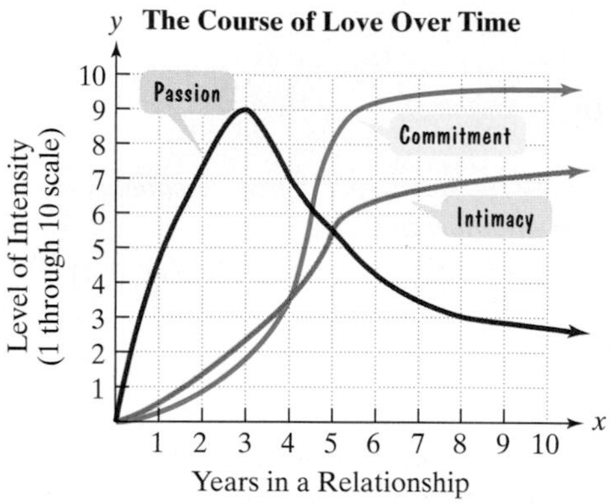

Source: R. J. Sternberg, A Triangular Theory of Love, *Psychological Review*, 93, 119–135.

57. Use interval notation to write an inequality that expresses for which years in a relationship intimacy is greater that commitment.

58. Use interval notation to write an inequality that expresses for which years in a relationship passion is greater than or equal to intimacy.

59. What is the relationship between passion and intimacy on the interval $[5, 7)$?

60. What is the relationship between intimacy and commitment on the interval $[4, 7)$?

61. What is the relationship between passion and commitment for $\{x|6 < x < 8\}$?

62. What is the relationship between passion and commitment for $\{x|7 < x < 9\}$?

63. What is the maximum level of intensity for passion? After how many years in a relationship does this occur?

64. After approximately how many years do levels of intensity for commitment exceed the maximum level of intensity for passion?

65. The percentage, P, of U.S. voters who used electronic voting systems, such as optical scans, in national elections can be modeled by the formula

$$P = 3.1x + 25.8,$$

where x is the number of years after 1994. In which years will more than 63% of U.S. voters use electronic systems?

66. The percentage, P, of U.S. voters who used punch cards or lever machines in national elections can be modeled by the formula

$$P = -2.5x + 63.1,$$

where x is the number of years after 1994. In which years will fewer than 38.1% of U.S. voters use punch cards or lever machines?

The Olympic 500-meter speed skating times have generally been decreasing over time. The formulas

$$W = -0.19t + 57 \quad \text{and} \quad M = -0.15t + 50$$

model the winning times, in seconds, for women, W, and men, M, t years after 1900. Use these models to solve Exercises 67–68.

67. Find values of t such that $W < M$. Describe what this means in terms of winning times.

68. Find values of t such that $W > M$. Describe what this means in terms of winning times.

Exercises 69–72 describe a number of business ventures. For each exercise,

a. *Write the cost function, C.*

b. *Write the revenue function, R.*

c. *Write the profit function, P.*

d. *More than how many units must be produced and sold to have a profit?*

69. A company that manufactures small canoes has a fixed cost of \$18,000. It costs \$20 to produce each canoe. The selling price is \$80 per canoe. (In solving this exercise, let x represent the number of canoes produced and sold.)

70. A company that manufactures bicycles has a fixed cost of \$100,000. It costs \$100 to produce each bicycle. The selling price is \$300 per bike. (In solving this exercise, let x represent the number of bicycles produced and sold.)

71. You invest in a new play. The cost includes an overhead of \$30,000, plus production costs of \$2500 per performance. A sold-out performance brings in \$3125. (In solving this exercise, let x represent the number of sold-out performances.)

72. You invested \$30,000 and started a business writing greeting cards. Supplies cost 2¢ per card and you are selling each card for 50¢. (In solving this exercise, let x represent the number of cards produced and sold.)

73. A company manufactures and sells blank audiocassette tapes. The weekly fixed cost is \$10,000 and it costs \$0.40 to produce each tape. The selling price is \$2.00 per tape. How many tapes must be produced and sold each week for the company to have a profit?

74. A company manufactures and sells personalized stationery. The weekly fixed cost is \$3000 and it costs \$3.00 to produce each package of stationery. The selling price is \$5.50 per package. How many packages of stationery must be produced and sold each week for the company to have a profit?

75. You are choosing between two long-distance telephone plans. Plan A has a monthly fee of \$15 with a charge of \$0.08 per minute for all long-distance calls. Plan B has a monthly fee of \$3 with a charge of \$0.12 per minute for all long-distance calls. How many minutes of long-distance calls in a month make plan A the better deal?

76. A city commission has proposed two tax bills. The first bill requires that a homeowner pay \$1800 plus 3% of the assessed home value in taxes. The second bill requires taxes of \$200 plus 8% of the assessed home value. What price range of home assessment would make the first bill a better deal?

Writing in Mathematics

77. When graphing the solutions of an inequality, what does a parenthesis signify? What does a bracket signify?

78. Describe a revenue function for a business venture.

79. Describe a cost function for a business venture. What are the two kinds of costs that are modeled by this function?

80. What is the profit function for a business venture and how is it determined?

81. If the profit function for a business venture is known, explain how to find how many units must be produced and sold to have a profit.

Technology Exercises

In Exercises 82–83, solve each inequality using a graphing utility. Graph each side separately. Then determine the values of x for which the graph on the left side lies above the graph on the right side.

82. $-3(x - 6) > 2x - 2$

83. $-2(x + 4) > 6x + 16$

84. Use a graphing utility's TABLE feature to verify your work in Exercises 82–83.

Use the same technique employed in Exercises 82–83 to solve each inequality in Exercises 85–86. In each case, what conclusion can you draw? What happens if you try solving the inequalities algebraically?

85. $12x - 10 > 2(x - 4) + 10x$

86. $2x + 3 > 3(2x - 4) - 4x$

87. A bank offers two checking account plans. Plan A has a base service charge of \$4.00 per month plus 10¢ per check. Plan B charges a base service charge of \$2.00 per month plus 15¢ per check.

a. Write models for the total monthly costs for each plan if x checks are written.

b. Use a graphing utility to graph the models in the same $[0, 50, 1]$ by $[0, 10, 1]$ viewing rectangle.

c. Use the graphs (and the intersection feature) to determine for what number of checks per month plan A will be better than plan B.

d. Verify the result of part (c) algebraically by solving an inequality.

Critical Thinking Exercises

88. Which one of the following is true?

a. The inequality $3x > 6$ is equivalent to $2 > x$.

b. The smallest real number in the solution set of $2x > 6$ is 4.

c. If x is at least 7, then $x > 7$.

d. The inequality $-3x > 6$ is equivalent to $-2 > x$.

89. What's wrong with this argument? Suppose x and y represent two real numbers, where $x > y$.

$2 > 1$	This is a true statement.
$2(y - x) > 1(y - x)$	Multiply both sides by $y - x$.
$2y - 2x > y - x$	Use the distributive property.
$y - 2x > -x$	Subtract y from both sides.
$y > x$	Add $2x$ to both sides.

The final inequality, $y > x$, is impossible because we were initially given $x > y$.

90. The figure shows a square with four identical equilateral triangles attached. If the perimeter of the figure is at least 24 centimeters, what are the possible values for k?

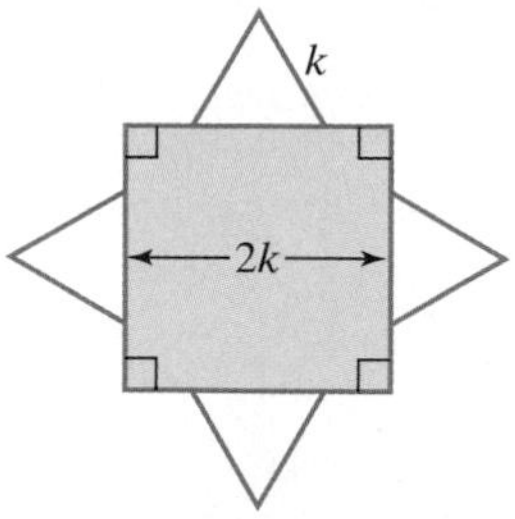

Review Exercises

91. If $f(x) = x^2 - 2x + 5$, find $f(-4)$. (Section 8.1, Example 3)

92. Solve the system:

$$\begin{aligned} 2x - y - z &= -3 \\ 3x - 2y - 2z &= -5 \\ -x + y + 2z &= 4. \end{aligned}$$

(Section 4.5, Example 2)

93. Factor: $25x^2 - 81$. (Section 6.4, Example 1)

SECTION 9.2

COMPOUND INEQUALITIES

Objectives

1. Find the intersection of two sets.
2. Solve compound inequalities involving *and*.
3. Find the union of two sets.
4. Solve compound inequalities involving *or*.

Boys and girls differ in their toy preferences. These differences are functions of our gender stereotypes—that is, our widely shared beliefs about males' and females' abilities, personality traits, and social behavior. Generally, boys have less leeway to play with "feminine" toys than girls do with "masculine" toys.

Which toys are requested by more than 40% of boys *and* more than 10% of girls? Which toys are requested by more than 40% of boys *or* more than 10% of girls? These questions are not the same. One involves inequalities joined by the word "and"; the other involves inequalities joined by the word "or." A **compound inequality** is formed by joining two inequalities with the word *and* or the word *or*.

Examples of Compound Inequalities

- $x - 3 < 5$ and $2x + 4 < 14$
- $3x - 5 \le 13$ or $5x + 2 > -3$

In this section, you will learn to solve compound inequalities. With this skill, you will be able to analyze childrens' toy preferences in the exercise set.

1 Find the intersection of two sets.

Compound Inequalities Involving *And* You need to determine whether there is sufficient support on campus to have a blood drive. You take a survey to obtain information, asking students two questions:

Would you be willing to donate blood?

Would you be willing to help serve a free breakfast to blood donors?

Part of your report involves the number of students willing to both donate blood *and* serve breakfast. You must focus on the set containing all students who are common to both the set of donors and the set of breakfast servers. This set is called the *intersection* of the two sets.

DEFINITION OF THE INTERSECTION OF SETS The **intersection** of sets A and B, written $A \cap B$, is the set of elements common to both set A **and** set B. This definition can be expressed in set-builder notation as follows:

$$A \cap B = \{x | x \in A \text{ AND } x \in B\}.$$

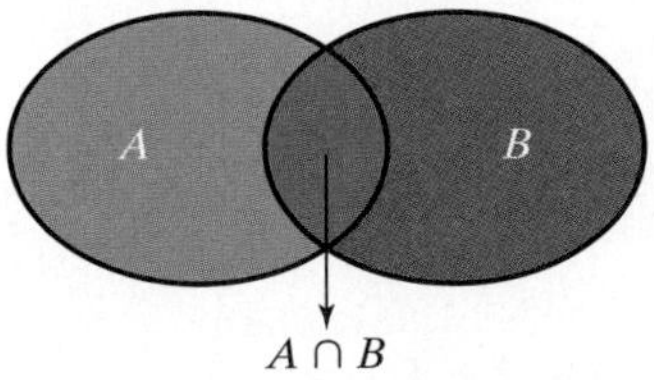

FIGURE 9.3 Picturing the intersection of two sets

Figure 9.3 shows a useful way of picturing the intersection of sets A and B. The figure indicates that $A \cap B$ contains those elements that belong to both A and B at the same time.

EXAMPLE 1 Finding the Intersection of Two Sets

Find the intersection: $\{7, 8, 9, 10, 11\} \cap \{6, 8, 10, 12\}$.

SOLUTION The elements common to $\{7, 8, 9, 10, 11\}$ and $\{6, 8, 10, 12\}$ are 8 and 10. Thus,

$$\{7, 8, 9, 10, 11\} \cap \{6, 8, 10, 12\} = \{8, 10\}.$$

CHECK POINT 1 Find the intersection: $\{3, 4, 5, 6, 7\} \cap \{3, 7, 8, 9\}$.

2 Solve compound inequalities involving *and*.

A number is a **solution of a compound inequality formed by the word *and*** if it is a solution of both inequalities. For example, the solution set of the compound inequality

$$x \leq 6 \quad \text{and} \quad x \geq 2$$

is the set of values of x that satisfy $x \leq 6$ and $x \geq 2$. Thus, the solution set is the intersection of the solution sets of the two inequalities.

What are the numbers that satisfy both $x \leq 6$ and $x \geq 2$? These numbers are easier to see if we graph the solution set to each inequality on a number line. These graphs are shown in Figure 9.4. The intersection is shown in the third graph.

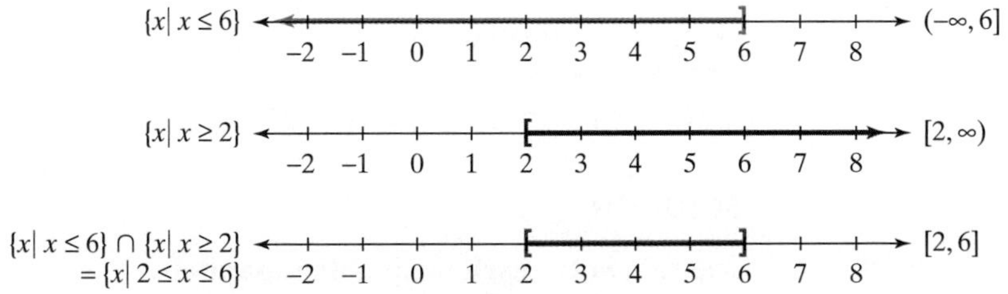

FIGURE 9.4 Numbers satisfying both $x \leq 6$ and $x \geq 2$

The numbers common to both sets are those that are less than or equal to 6 and greater than or equal to 2. This set is $\{x | 2 \leq x \leq 6\}$, or, in interval notation, $[2, 6]$.

Here is a procedure for finding the solution set of a compound inequality containing the word *and*.

SOLVING COMPOUND INEQUALITIES INVOLVING *AND*

1. Solve each inequality separately.
2. Graph the solution set to each inequality on a number line and take the intersection of these solution sets.

Keep in mind that this intersection appears as the portion of the number line that the two graphs have in common.

EXAMPLE 2 Solving a Compound Inequality with *And*

Solve: $x - 3 < 5$ and $2x + 4 < 14$.

SOLUTION

Step 1. Solve each inequality separately.

$$\begin{aligned} x - 3 &< 5 \quad \text{and} \quad & 2x + 4 &< 14 \\ x &< 8 & 2x &< 10 \\ & & x &< 5 \end{aligned}$$

Step 2. Take the intersection of the solution sets of the two inequalities. We graph the solution sets of $x < 8$ and $x < 5$. The intersection is shown in the third graph.

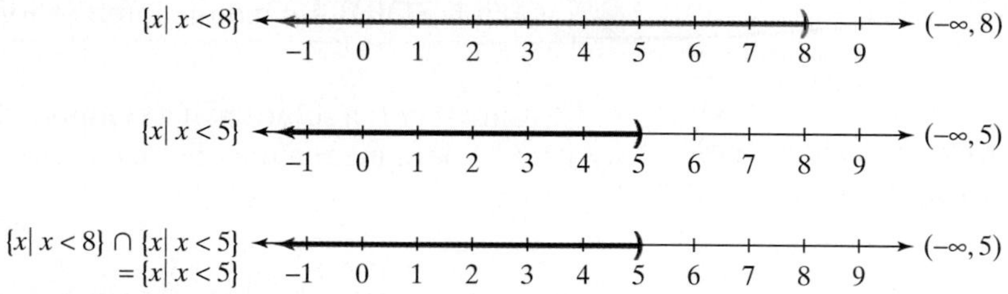

The numbers common to both sets are those that are less than 5. The solution set is $\{x|x < 5\}$, or, in interval notation, $(-\infty, 5)$. Take a moment to check that any number in $(-\infty, 5)$ satisfies both of the original inequalities. ■

CHECK POINT 2 Solve: $x + 2 < 5$ and $2x - 4 < -2$.

EXAMPLE 3 Solving a Compound Inequality with *And*

Solve: $2x - 7 > 3$ and $5x - 4 < 6$.

SOLUTION

Step 1. Solve each inequality separately.

$$\begin{aligned} 2x - 7 &> 3 \quad \text{and} \quad & 5x - 4 &< 6 \\ 2x &> 10 & 5x &< 10 \\ x &> 5 & x &< 2 \end{aligned}$$

Step 2. **Take the intersection of the solution sets of the two inequalities.** We graph the solution sets of $x > 5$ and $x < 2$. We use these graphs to find their intersection.

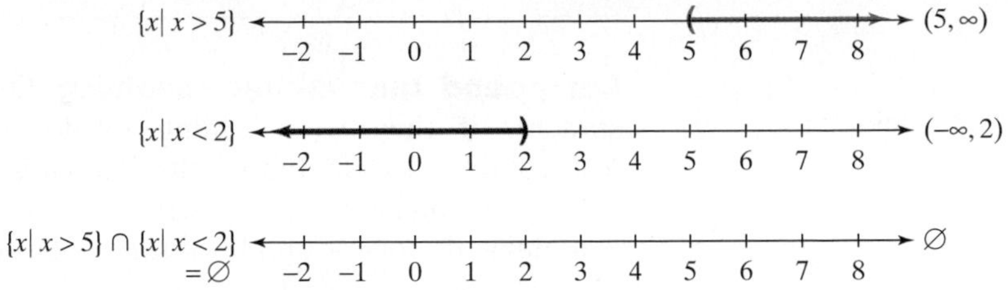

There is no number that is both greater than 5 and at the same time less than 2. Thus, the solution set is the empty set, ∅. ■

CHECK POINT 3 Solve: $4x - 5 > 7$ and $5x - 2 < 3$.

If $a < b$, the compound inequality

$$a < x \text{ and } x < b$$

can be written in the shorter form

$$a < x < b.$$

For example, the compound inequality

$$-3 < 2x + 1 \text{ and } 2x + 1 < 3$$

can be abbreviated

$$-3 < 2x + 1 < 3.$$

The word "and" does not appear when the inequality is written in the shorter form, although it is implied. The shorter form enables us to solve both inequalities at once. By performing the same operations on all three parts of the inequality, our goal is to **isolate *x* in the middle**.

EXAMPLE 4 Solving a Compound Inequality

Solve and graph the solution set:

$$-3 < 2x + 1 \le 3.$$

SOLUTION We would like to isolate x in the middle. We can do this by first subtracting 1 from all three parts of the compound inequality. Then we isolate x from $2x$ by dividing all three parts of the inequality by 2.

$-3 < 2x + 1 \le 3$	This is the given inequality.
$-3 - 1 < 2x + 1 - 1 \le 3 - 1$	Subtract 1 from all three parts.
$-4 < 2x \le 2$	Simplify.
$\frac{-4}{2} < \frac{2x}{2} \le \frac{2}{2}$	Divide each part by 2.
$-2 < x \le 1$	Simplify.

The solution set consists of all real numbers greater than -2 and less than or equal to 1, represented by $\{x \mid -2 < x \le 1\}$ in set-builder notation and $(-2, 1]$ in interval notation. The graph is shown as follows:

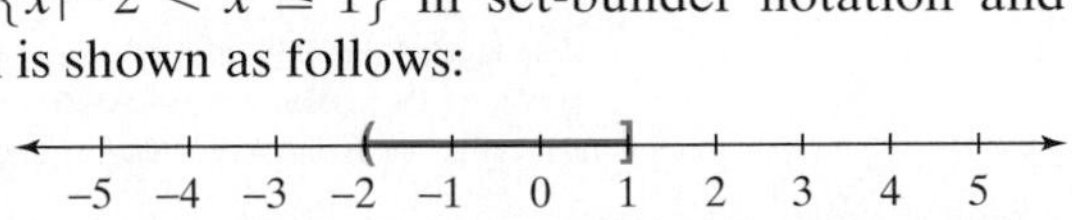

■

USING TECHNOLOGY

To check Example 4, graph each part of

$$-3 < 2x + 1 \le 3.$$

The figure shows that the graph of $y_2 = 2x + 1$ lies above the graph of $y_1 = -3$ and on or below the graph of $y_3 = 3$ when x is in the interval $(-2, 1]$.

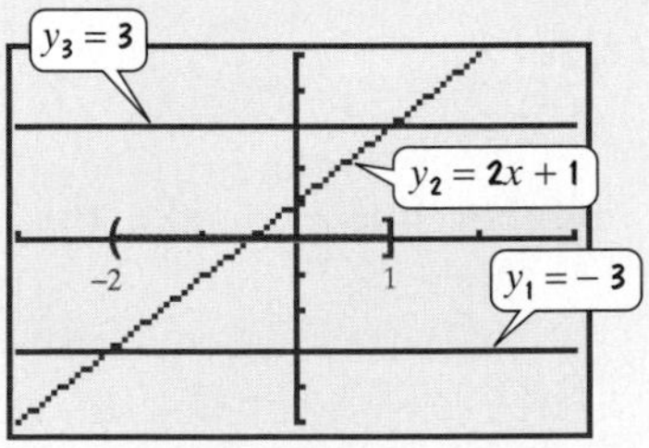

$[-3, 3, 1]$ by $[-5, 5, 1]$

CHECK POINT 4 Solve and graph the solution set: $1 \leq 2x + 3 < 11$.

3 Find the union of two sets.

Compound Inequalities Involving *Or* Earlier, we considered a survey asking students if they were willing to donate blood or serve breakfast to donors. You continue to sort the results. Part of your report involves the number of students willing to donate blood *or* serve breakfast *or* do both. You must focus on the set containing donors or breakfast servers or both. This set is called the *union* of two sets.

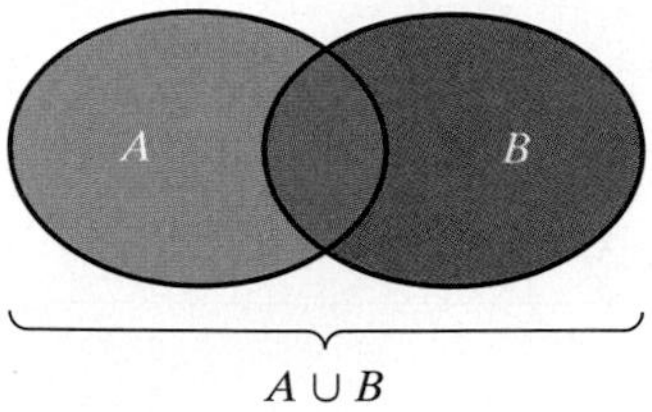

FIGURE 9.5 Picturing the union of two sets

DEFINITION OF THE UNION OF SETS The **union** of sets A and B, written $A \cup B$, is the set of elements that are members of set A **or** of set B or of both sets. This definition can be expressed in set-builder notation as follows:

$$A \cup B = \{x | x \in A \text{ OR } x \in B\}.$$

Figure 9.5 shows a useful way of picturing the union of sets A and B. The figure indicates that $A \cup B$ is formed by joining the sets together.

We can find the union of set A and set B by listing the elements of set A. Then, we include any elements of set B that have not already been listed. Enclose all elements that are listed with braces. This shows that the union of two sets is also a set.

STUDY TIP

When finding the union of two sets, do not list twice any elements that appear in both sets.

EXAMPLE 5 Finding the Union of Two Sets

Find the union: $\{7, 8, 9, 10, 11\} \cup \{6, 8, 10, 12\}$.

SOLUTION To find $\{7, 8, 9, 10, 11\} \cup \{6, 8, 10, 12\}$, start by listing all the elements from the first set, namely 7, 8, 9, 10, and 11. Now list all the elements from the second set that are not in the first set, namely 6 and 12. The union is the set consisting of all these elements. Thus,

$$\{7, 8, 9, 10, 11\} \cup \{6, 8, 10, 12\} = \{6, 7, 8, 9, 10, 11, 12\}.$$ ■

CHECK POINT 5 Find the union: $\{3, 4, 5, 6, 7\} \cup \{3, 7, 8, 9\}$.

4 Solve compound inequalities involving *or*.

A number is a **solution of a compound inequality formed by the word *or*** if it is a solution of either inequality. Thus, the solution set of a compound inequality formed by the word *or* is the union of the solution sets of the two inequalities.

SOLVING COMPOUND INEQUALITIES INVOLVING ***OR***

1. Solve each inequality separately.
2. Graph the solution set to each inequality on a number line and take the union of these solution sets.

Keep in mind that this union appears as the portion of the number line representing the total collection of numbers in the two graphs.

EXAMPLE 6 Solving a Compound Inequality with *Or*

Solve: $2x - 3 < 7$ or $35 - 4x \leq 3$.

SOLUTION

Step 1. Solve each inequality separately.

$$2x - 3 < 7 \quad \text{or} \quad 35 - 4x \leq 3$$
$$2x < 10 \qquad\qquad -4x \leq -32$$
$$x < 5 \qquad\qquad x \geq 8$$

Step 2. Take the union of the solution sets of the two inequalities. We graph the solution sets of $x < 5$ and $x \geq 8$. We use these graphs to find their union.

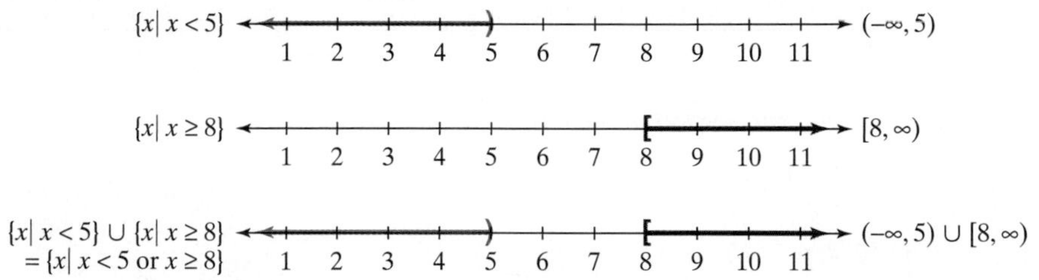

The solution set consists of all numbers that are less than 5 or greater than or equal to 8. The solution set is $\{x|x < 5 \text{ or } x \geq 8\}$, or, in interval notation, $(-\infty, 5) \cup [8, \infty)$. There is no shortcut way to express this union when interval notation is used. ■

CHECK POINT 6 Solve: $3x - 5 \leq -2$ or $10 - 2x < 4$.

EXAMPLE 7 Solving a Compound Inequality with *Or*

Solve: $3x - 5 \leq 13$ or $5x + 2 > -3$.

SOLUTION

Step 1. Solve each inequality separately.

$$3x - 5 \leq 13 \quad \text{or} \quad 5x + 2 > -3$$
$$3x \leq 18 \qquad\qquad 5x > -5$$
$$x \leq 6 \qquad\qquad x > -1$$

Step 2. Take the union of the solution sets of the two inequalities. We graph the solution sets of $x \leq 6$ and $x > -1$. We use these graphs to find their union.

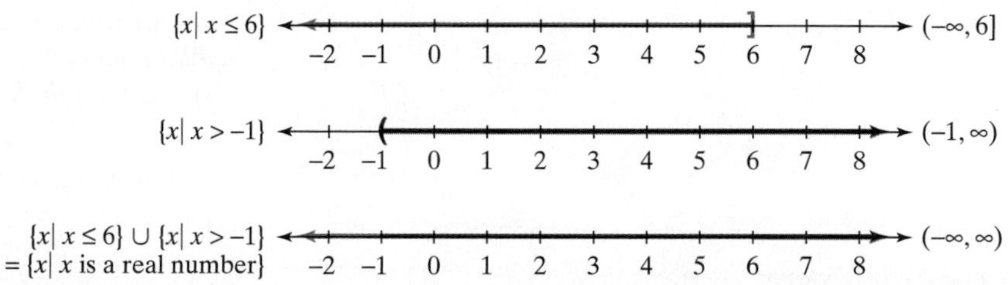

Because all real numbers are less than or equal to 6 or greater than -1, the union of the two sets fills the entire number line. Thus, the solution set is $\{x | x \text{ is a real number}\}$, or $\mathbb{R}$. The solution set in interval notation is $(-\infty, \infty)$. Any real number that you select will satisfy either or both of the original inequalities. ■

CHECK POINT 7 Solve: $2x + 5 \geq 3$ or $2x + 3 < 3$.

9.2 EXERCISE SET

Student Solutions Manual | CD/Video | PH Math/Tutor Center | MathXL Tutorials on CD | MathXL® | MyMathLab | Interactmath.com

Practice Exercises

In Exercises 1–6, find the intersection of the sets.

1. $\{1, 2, 3, 4\} \cap \{2, 4, 5\}$

2. $\{1, 3, 7\} \cap \{2, 3, 8\}$

3. $\{1, 3, 5, 7\} \cap \{2, 4, 6, 8, 10\}$

4. $\{0, 1, 3, 5\} \cap \{-5, -3, -1\}$

5. $\{a, b, c, d\} \cap \varnothing$

6. $\{w, y, z\} \cap \varnothing$

In Exercises 7–24, solve each compound inequality. Use graphs to show the solution set to each of the two given inequalities, as well as a third graph that shows the solution set of the compound inequality. Except for the empty set, express the solution set in both set-builder and interval notations.

7. $x > 3$ and $x > 6$

8. $x > 2$ and $x > 4$

9. $x \leq 5$ and $x \leq 1$

10. $x \leq 6$ and $x \leq 2$

11. $x < 2$ and $x \geq -1$

12. $x < 3$ and $x \geq -1$

13. $x > 2$ and $x < -1$

14. $x > 3$ and $x < -1$

15. $5x < -20$ and $3x > -18$

16. $3x \leq 15$ and $2x > -6$

17. $x - 4 \leq 2$ and $3x + 1 > -8$

18. $3x + 2 > -4$ and $2x - 1 < 5$

19. $2x > 5x - 15$ and $7x > 2x + 10$

20. $6 - 5x > 1 - 3x$ and $4x - 3 > x - 9$

21. $4(1 - x) < -6$ and $\frac{x - 7}{5} \leq -2$

22. $5(x - 2) > 15$ and $\frac{x - 6}{4} \leq -2$

23. $x - 1 \leq 7x - 1$ and $4x - 7 < 3 - x$

24. $2x + 1 > 4x - 3$ and $x - 1 \geq 3x + 5$

In Exercises 25–32, solve each inequality and graph the solution set on a number line. Express the solution set in both set-builder and interval notations.

25. $6 < x + 3 < 8$

26. $7 < x + 5 < 11$

27. $-3 \leq x - 2 < 1$

28. $-6 < x - 4 \leq 1$

29. $-11 < 2x - 1 \leq -5$

30. $3 \leq 4x - 3 < 19$

31. $-3 \leq \frac{2x}{3} - 5 < -1$

32. $-6 \leq \frac{x}{2} - 4 < -3$

In Exercises 33–38, find the union of the sets.

33. $\{1, 2, 3, 4\} \cup \{2, 4, 5\}$

34. $\{1, 3, 7, 8\} \cup \{2, 3, 8\}$

35. $\{1, 3, 5, 7\} \cup \{2, 4, 6, 8, 10\}$

36. $\{0, 1, 3, 5\} \cup \{2, 4, 6\}$

37. $\{a, e, i, o, u\} \cup \varnothing$

38. $\{e, m, p, t, y\} \cup \varnothing$

In Exercises 39–54, solve each compound inequality. Use graphs to show the solution set to each of the two given inequalities, as well as a third graph that shows the solution set of the compound inequality. Express the solution set in both set-builder and interval notations.

39. $x > 3$ or $x > 6$

40. $x > 2$ or $x > 4$

41. $x \leq 5$ or $x \leq 1$

42. $x \leq 6$ or $x \leq 2$
43. $x < 2$ or $x \geq -1$
44. $x < 3$ or $x \geq -1$
45. $x \geq 2$ or $x < -1$
46. $x \geq 3$ or $x < -1$
47. $3x > 12$ or $2x < -6$
48. $3x < 3$ or $2x > 10$
49. $3x + 2 \leq 5$ or $5x - 7 \geq 8$
50. $2x - 5 \leq -11$ or $5x + 1 \geq 6$
51. $4x + 3 < -1$ or $2x - 3 \geq -11$
52. $2x + 1 < 15$ or $3x - 4 \geq -1$
53. $-2x + 5 > 7$ or $-3x + 10 > 2x$
54. $16 - 3x \geq -8$ or $13 - x > 4x + 3$
55. Let $f(x) = 2x + 3$ and $g(x) = 3x - 1$. Find all values of x for which $f(x) \geq 5$ and $g(x) > 11$.
56. Let $f(x) = 4x + 5$ and $g(x) = 3x - 4$. Find all values of x for which $f(x) \geq 5$ and $g(x) \leq 2$.
57. Let $f(x) = 3x - 1$ and $g(x) = 4 - x$. Find all values of x for which $f(x) < -1$ or $g(x) < -2$.
58. Let $f(x) = 2x - 5$ and $g(x) = 3 - x$. Find all values of x for which $f(x) \geq 3$ or $g(x) < 0$.

Practice Plus

In Exercises 59–60, write an inequality with x isolated in the middle that is equivalent to the given inequality. Assume $a > 0, b > 0,$ and $c > 0$.

59. $-c < ax - b < c$
60. $-2 < \frac{ax - b}{c} < 2$

In Exercises 61–62, use the graphs of y_1, y_2, and y_3, to solve each compound inequality.

61. $-3 \leq 2x - 1 \leq 5$

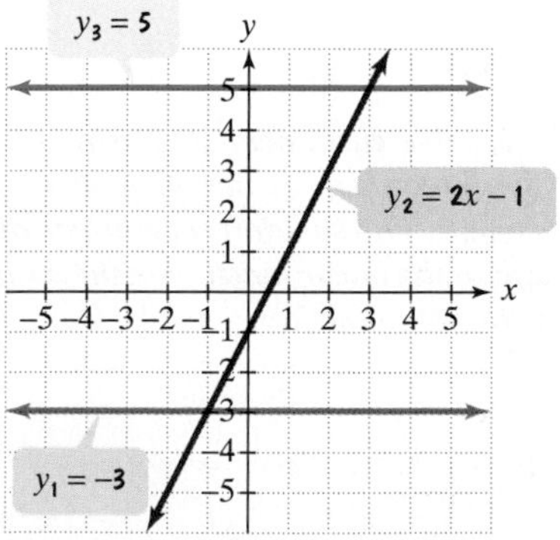

62. $x - 2 < 2x - 1 < x + 2$

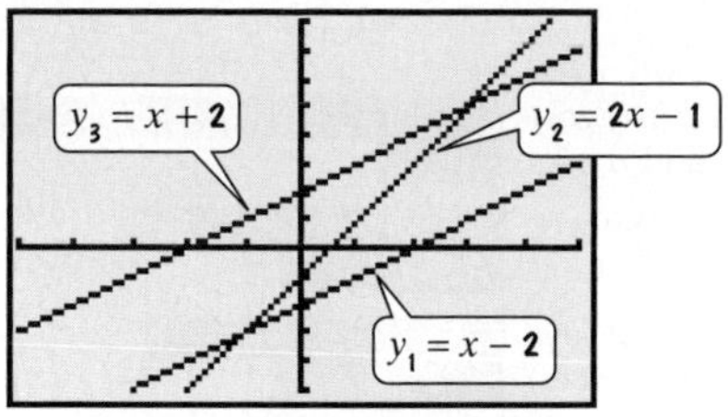

[−5, 5, 1] by [−5, 8, 1]

63. Solve $x - 2 < 2x - 1 < x + 2$, the inequality in Exercise 62, using algebraic methods. (*Hint*: Rewrite the inequality as $2x - 1 > x - 2$ and $2x - 1 < x + 2$.)
64. Use the hint given in Exercise 63 to solve $x \leq 3x - 10 \leq 2x$.

In Exercises 65–66, use the table to solve each inequality.

65. $-2 \leq 5x + 3 < 13$

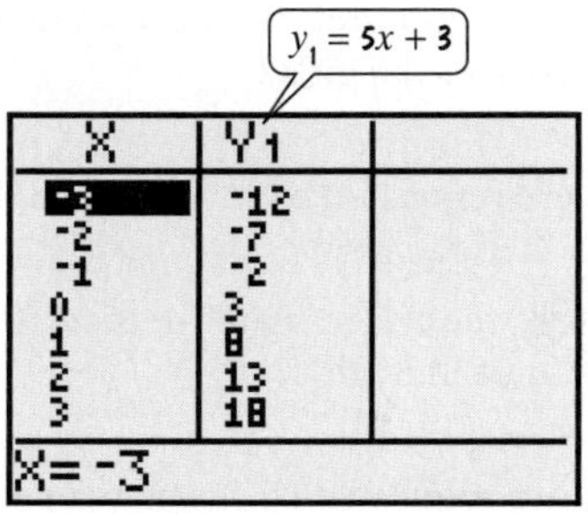

66. $-3 < 2x - 5 \leq 3$

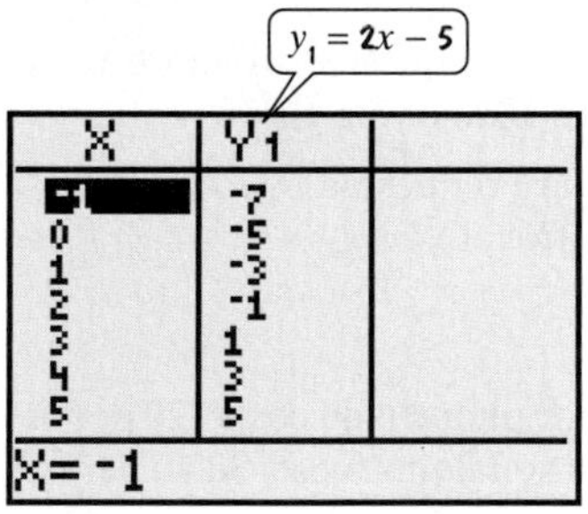

In Exercises 67–68, use the roster method to find the set of negative integers that are solutions of each inequality.

67. $5 - 4x \geq 1$ and $3 - 7x < 31$
68. $-5 < 3x + 4 \leq 16$

Application Exercises

As a result of cultural expectations about what is appropriate behavior for each gender, boys and girls differ substantially in their toy preferences. The graph on the next page shows the percentage of boys and girls asking for various types of toys in letters to Santa Claus. Use the information in the graph to solve Exercises 69–74.

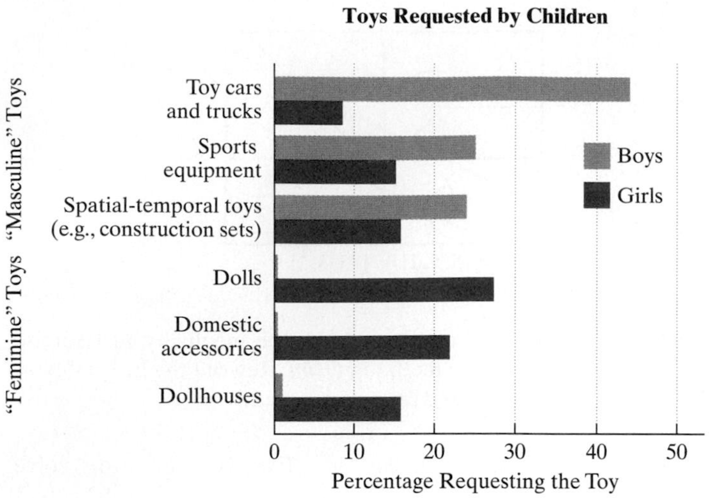

Source: Richardson, J. G., & Simpson, C. H. (1982). Children, gender and social structure: An analysis of the contents of letters to Santa Claus. *Child Development, 53*, 429–436.

69. Which toys were requested by more than 10% of the boys *and* less than 20% of the girls?

70. Which toys were requested by fewer than 5% of the boys *and* fewer than 20% of the girls?

71. Which toys were requested by more than 10% of the boys *or* less than 20% of the girls?

72. Which toys were requested by fewer than 5% of the boys *or* fewer than 20% of the girls?

73. Which toys were requested by more than 40% of the boys *and* more than 10% of the girls?

74. Which toys were requested by more than 40% of the boys *or* more than 10% of the girls?

75. A basic cellular phone plan costs \$20 per month for 60 calling minutes. Additional time costs \$0.40 per minute. The formula

$$C = 20 + 0.40(x - 60)$$

gives the monthly cost for this plan, C, for x calling minutes, where $x > 60$. How many calling minutes are possible for a monthly cost of at least \$28 and at most \$40?

76. The formula for converting Fahrenheit temperature, F, to Celsius temperature, C, is

$$C = \frac{5}{9}(F - 32).$$

If Celsius temperature ranges from 15° to 35°, inclusive, what is the range for the Fahrenheit temperature? Use interval notation to express this range.

77. On the first of four exams, your grades are 70, 75, 87, and 92. There is still one more exam, and you are hoping to earn a B in the course. This will occur if the average of your five exam grades is greater than or equal to 80 and less than 90. What range of grades on the fifth exam will result in earning a B? Use interval notation to express this range.

78. On the first of four exams, your grades are 82, 75, 80, and 90. There is still a final exam, and it counts as two grades. You are hoping to earn a B in the course. This will occur if the average of your six exam grades is greater than or equal to 80 and less than 90. What range of grades on the final exam will result in earning a B? Use interval notation to express this range.

79. The toll to a bridge is \$3.00. A three-month pass costs \$7.50 and reduces the toll to \$0.50. A six-month pass costs \$30 and permits crossing the bridge for no additional fee. How many crossings per three-month period does it take for the three-month pass to be the best deal?

80. Parts for an automobile repair cost \$175. The mechanic charges \$34 per hour. If you receive an estimate for at least \$226 and at most \$294 for fixing the car, what is the time interval that the mechanic will be working on the job?

Writing in Mathematics

81. Describe what is meant by the intersection of two sets. Give an example.

82. Explain how to solve a compound inequality involving "and."

83. Why is $1 < 2x + 3 < 9$ a compound inequality? Where are the two inequalities and what is the word that joins them?

84. Explain how to solve $1 < 2x + 3 < 9$.

85. Describe what is meant by the union of two sets. Give an example.

86. Explain how to solve a compound inequality involving "or."

87. How many Christmas trees are sold each holiday season? The function $T(x) = 0.9x + 32$ models the number of trees sold, $T(x)$, in millions, x years after 1990. Use the function to write a problem that can be solved using a compound inequality. Then solve the problem.

Technology Exercises

In Exercises 88–91, solve each inequality using a graphing utility. Graph each of the three parts of the inequality separately in the same viewing rectangle. The solution set consists of all values of x for which the graph of the linear function in the middle lies between the graphs of the constant functions on the left and the right.

88. $1 < x + 3 < 9$

89. $-1 < \frac{x + 4}{2} < 3$

90. $1 \le 4x - 7 \le 3$

91. $2 \le 4 - x \le 7$

Critical Thinking Exercises

92. Which one of the following is true?

a. $(-\infty, -1] \cap [-4, \infty) = [-4, -1]$

b. $(-\infty, 3) \cup (-\infty, -2) = (-\infty, -2)$

c. The union of two sets can never give the same result as the intersection of those sets.

d. The solution set of $x < a$ and $x > a$ is the set of all real numbers excluding a.

93. Solve and express the solution set in interval notation:

$$-7 \le 8 - 3x \le 20 \text{ and } -7 < 6x - 1 < 41.$$

The graphs of $f(x) = \sqrt{4 - x}$ and $g(x) = \sqrt{x + 1}$ are shown in a $[-3, 10, 1]$ by $[-2, 5, 1]$ viewing rectangle.

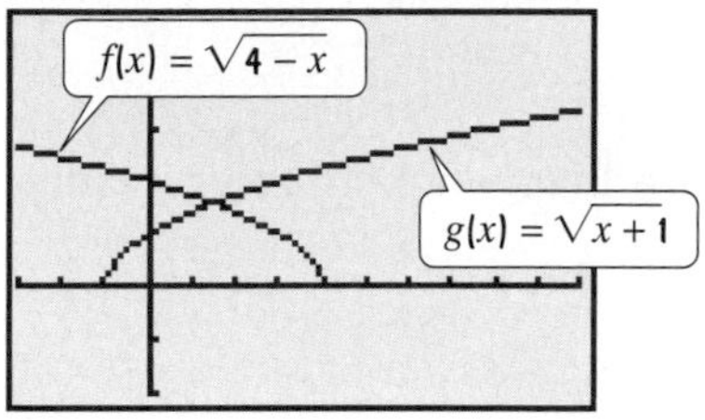

In Exercises 94–97, use the graphs and interval notation to express the domain of the given function.

94. The domain of f

95. The domain of g

96. The domain of $f + g$

97. The domain of $\frac{f}{g}$

98. At the end of the day, the change machine at a laundrette contained at least \$3.20 and at most \$5.45 in nickels, dimes, and quarters. There were 3 fewer dimes than twice the number of nickels and 2 more quarters than twice the number of nickels. What was the least possible number and the greatest possible number of nickels?

Review Exercises

99. If $f(x) = x^2 - 3x + 4$ and $g(x) = 2x - 5$, find $(g - f)(x)$ and $(g - f)(-1)$. (Section 8.2, Example 4)

100. Use function notation to write the equation of the line passing through (4, 2) and perpendicular to the line whose equation is $4x - 2y = 8$. (Section 3.5, Examples 3 and 4)

101. Simplify: $4 - [2(x - 4) - 5]$. (Section 1.8, Example 11)

SECTION 9.3

EQUATIONS AND INEQUALITIES INVOLVING ABSOLUTE VALUE

Objectives

1. Solve equations involving absolute value.
2. Solve inequalities involving absolute value.
3. Recognize absolute value inequalities with no solution or all real numbers as solutions.
4. Solve problems using absolute value inequalities.

*M*A*S*H* was set in the early 1950s during the Korean War. By the final episode, the show had lasted four times as long as the Korean War.

At the end of the twentieth century, there were 94 million households in the United States with television sets. The television program viewed by the greatest percentage of such households in that century was the final episode of *M*A*S*H*. Over 50 million American households watched this program.

Numerical information, such as the number of households watching a television program, is often given with a margin of error. Inequalities involving absolute value are used to describe errors in polling, as well as errors of measurement in manufacturing, engineering, science, and other fields. In this section, you will learn to solve equations and inequalities containing absolute value. With these skills, you will be able to analyze the percentage of households that watched the final episode of $M{*}A{*}S{*}H$.

1 Solve equations involving absolute value.

Equations Involving Absolute Value We have seen that the absolute value of a, denoted $|a|$, is the distance from 0 to a on a number line. Now consider **absolute value equations**, such as

$$|x| = 2.$$

This means that we must determine real numbers whose distance from the origin on a number line is 2. Figure 9.6 shows that there are two numbers such that $|x| = 2$, namely, 2 and -2. We write $x = 2$ or $x = -2$. This observation can be generalized as follows:

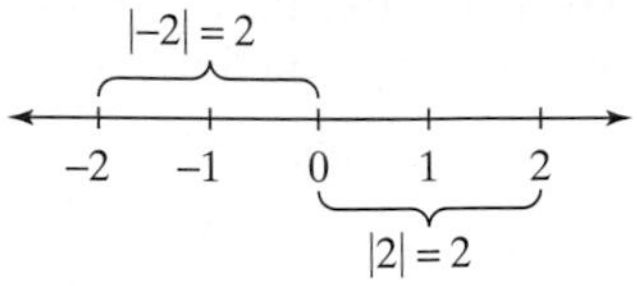

FIGURE 9.6 If $|x| = 2$, then $x = 2$ or $x = -2$.

REWRITING AN ABSOLUTE VALUE EQUATION WITHOUT ABSOLUTE VALUE BARS If c is a positive real number and X represents any algebraic expression, then $|X| = c$ is equivalent to $X = c$ or $X = -c$.

EXAMPLE 1 Solving an Equation Involving Absolute Value

Solve: $|2x - 3| = 11$.

SOLUTION

$|2x - 3| = 11$		This is the given equation.
$2x - 3 = 11$ or	$2x - 3 = -11$	Rewrite the equation without absolute value bars.
$2x = 14$	$2x = -8$	Add 3 to both sides of each equation.
$x = 7$	$x = -4$	Divide both sides of each equation by 2.

Check 7:	**Check −4:**	
$|2x - 3| = 11$	$|2x - 3| = 11$	This is the original equation.
$|2(7) - 3| \stackrel{?}{=} 11$	$|2(-4) - 3| \stackrel{?}{=} 11$	Substitute the proposed solutions.
$|14 - 3| \stackrel{?}{=} 11$	$|-8 - 3| \stackrel{?}{=} 11$	Perform operations inside the absolute value bars.
$|11| \stackrel{?}{=} 11$	$|-11| \stackrel{?}{=} 11$	
$11 = 11$, true	$11 = 11$, true	These true statements indicate that 7 and −4 are solutions.

The solutions are -4 and 7. We can also say that the solution set is $\{-4, 7\}$. ■

CHECK POINT 1 Solve: $|2x - 1| = 5$.

USING TECHNOLOGY

You can use a graphing utility to verify the solution set of an absolute value equation. Consider, for example,

$$|2x - 3| = 11.$$

Graph $y_1 = |2x - 3|$ and $y_2 = 11$. The graphs are shown in a $[-10, 10, 1]$ by $[-1, 15, 1]$ viewing rectangle. The x-coordinates of the intersection points are -4 and 7, verifying that $\{-4, 7\}$ is the solution set.

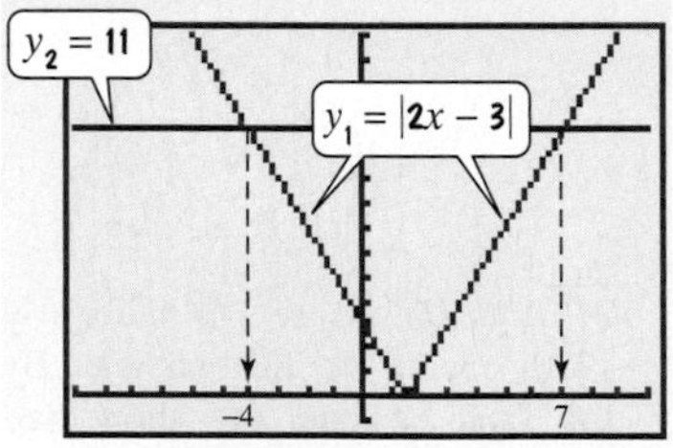

$[-10, 10, 1]$ by $[-1, 15, 1]$

The absolute value of a number is never negative. Thus, if X is an algebraic expression and c is a negative number, then $|X| = c$ has no solution. For example, the equation $|3x - 6| = -2$ has no solution because $|3x - 6|$ cannot be negative. The solution set is $\varnothing$, the empty set.

The absolute value of 0 is 0. Thus, if X is an algebraic expression and $|X| = 0$, the solution is found by solving $X = 0$. For example, the solution of $|x - 2| = 0$ is obtained by solving $x - 2 = 0$. The solution is 2 and the solution set is $\{2\}$.

To solve some absolute value equations, it is necessary to first isolate the expression containing the absolute value symbols. For example, consider the equation

$$3|2x - 3| - 8 = 25.$$

We need to isolate $|2x - 3|$.

How can we isolate $|2x - 3|$? Add 8 to both sides of the equation and then divide both sides by 3.

$$3|2x - 3| - 8 = 25 \quad \text{This is the given equation.}$$
$$3|2x - 3| = 33 \quad \text{Add 8 to both sides.}$$
$$|2x - 3| = 11 \quad \text{Divide both sides by 3.}$$

This results in the equation we solved in Example 1.

Some equations have two absolute value expressions, such as

$$|3x - 1| = |x + 5|.$$

These absolute value expressions are equal when the expressions inside the absolute value bars are equal to or opposites of each other.

REWRITING AN ABSOLUTE VALUE EQUATION WITH TWO ABSOLUTE VALUES WITHOUT ABSOLUTE VALUE BARS If $|X_1| = |X_2|$, then $X_1 = X_2$ or $X_1 = -X_2$.

EXAMPLE 2 Solving an Absolute Value Equation with Two Absolute Values

Solve: $|3x - 1| = |x + 5|$.

SOLUTION We rewrite the equation without absolute value bars.

$|X_1| = |X_2|$ means $X_1 = X_2$ or $X_1 = -X_2$

$$|3x - 1| = |x + 5| \quad \text{means} \quad 3x - 1 = x + 5 \quad \text{or} \quad 3x - 1 = -(x + 5).$$

We now solve the two equations that do not contain absolute value bars.

$$3x - 1 = x + 5 \quad \text{or} \quad 3x - 1 = -(x + 5)$$
$$2x - 1 = 5 \qquad 3x - 1 = -x - 5$$
$$2x = 6 \qquad 4x - 1 = -5$$
$$x = 3 \qquad 4x = -4$$
$$x = -1$$

Take a moment to complete the solution process by checking the two proposed solutions in the original equation. The solutions are -1 and 3, and the solution set is $\{-1, 3\}$. ■

CHECK POINT 2 Solve: $|2x - 7| = |x + 3|$.

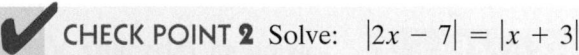

2 Solve inequalities involving absolute value.

Inequalities Involving Absolute Value Absolute value can also arise in inequalities. Consider, for example,

$$|x| < 2.$$

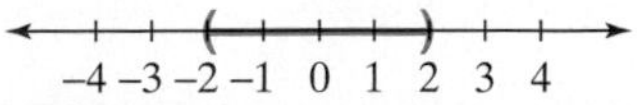

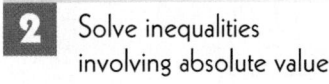

FIGURE 9.7 $|x| < 2$, so $-2 < x < 2$.

This means that the distance of x from 0 is *less than* 2, as shown in Figure 9.7. The interval shows values of x that lie less than 2 units from 0. Thus, x can lie between -2 and 2. That is, x is greater than -2 and less than 2: $-2 < x < 2$.

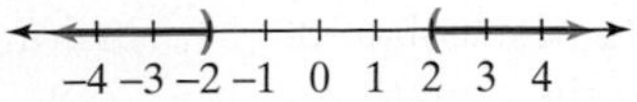

FIGURE 9.8 $|x| > 2$, so $x < -2$ or $x > 2$.

Some absolute value inequalities use the "greater than" symbol. For example, $|x| > 2$ means that the distance of x from 0 is *greater than* 2, as shown in Figure 9.8. Thus, x can be less than -2 *or* greater than 2: $x < -2$ or $x > 2$.

These observations suggest the following principles for solving inequalities with absolute value.

SOLVING AN ABSOLUTE VALUE INEQUALITY If X is an algebraic expression and c is a positive number,

1. The solutions of $|X| < c$ are the numbers that satisfy $-c < X < c$.
2. The solutions of $|X| > c$ are the numbers that satisfy $X < -c$ or $X > c$.

These rules are valid if $<$ is replaced by $\le$ and $>$ is replaced by $\ge$.

EXAMPLE 3 Solving an Absolute Value Inequality with <

Solve and graph the solution set on a number line: $|x - 4| < 3$.

SOLUTION We rewrite the inequality without absolute value bars.

$|X| < c$ **means** $-c < X < c$.

$$|x - 4| < 3 \quad \text{means} \quad -3 < x - 4 < 3.$$

We solve the compound inequality by adding 4 to all three parts.

$$-3 < x - 4 < 3$$
$$-3 + 4 < x - 4 + 4 < 3 + 4$$
$$1 < x < 7$$

The solution set is all real numbers greater than 1 and less than 7, denoted by $\{x \mid 1 < x < 7\}$ or $(1, 7)$. The graph of the solution set is shown as follows:

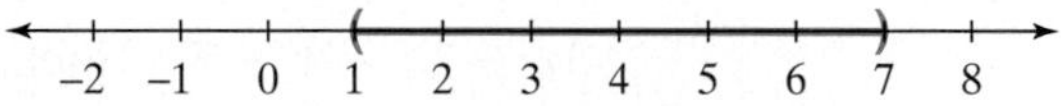

We can use the rectangular coordinate system to visualize the solution set of

$$|x - 4| < 3.$$

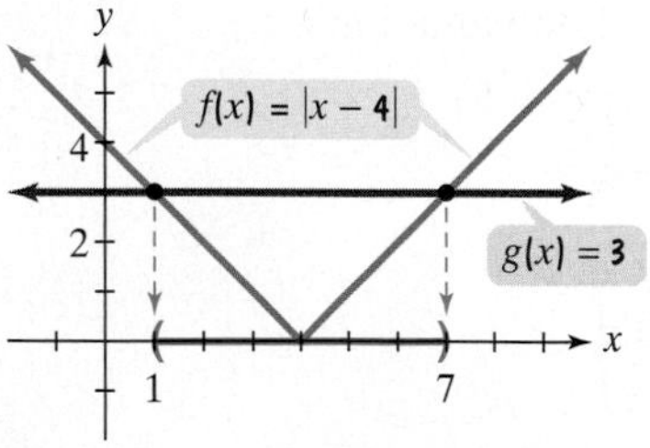

FIGURE 9.9 The solution set of $|x - 4| < 3$ is $(1, 7)$.

Figure 9.9 shows the graphs of $f(x) = |x - 4|$ and $g(x) = 3$. The solution set of $|x - 4| < 3$ consists of all values of x for which the blue graph of f lies below the red graph of g. These x-values make up the interval $(1, 7)$, which is the solution set.

CHECK POINT 3 Solve and graph the solution set on a number line: $|x - 2| < 5$.

EXAMPLE 4 Solving an Absolute Value Inequality with ≥

Solve and graph the solution set on a number line: $|2x + 3| \ge 5$.

SOLUTION We rewrite the inequality without absolute value bars.

$|X| \ge c$ **means** $X \le -c$ **or** $X \ge c$.

$$|2x + 3| \ge 5 \quad \text{means} \quad 2x + 3 \le -5 \quad \text{or} \quad 2x + 3 \ge 5.$$

We solve this compound inequality by solving each of these inequalities separately. Then we take the union of their solution sets.

$$2x + 3 \le -5 \quad \text{or} \quad 2x + 3 \ge 5$$ These are the inequalities without absolute value bars.

$$2x \le -8 \qquad\qquad 2x \ge 2$$ Subtract 3 from both sides.

$$x \le -4 \qquad\qquad x \ge 1$$ Divide both sides by 2.

The solution set consists of all numbers that are less than or equal to -4 or greater than or equal to 1. The solution set is $\{x | x \le -4 \text{ or } x \ge 1\}$, or, in interval notation, $(-\infty, -4] \cup [1, \infty)$. The graph of the solution set is shown as follows:

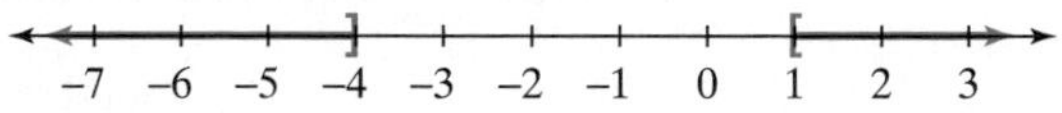
–7 –6 –5 –4 –3 –2 –1 0 1 2 3

✔ CHECK POINT 4 Solve and graph the solution set on a number line: $|2x - 5| \ge 3$.

3 Recognize absolute value inequalities with no solution or all real numbers as solutions.

Absolute Value Inequalities with Unusual Solution Sets If c is a positive number, the solutions of $|X| < c$ comprise a single interval. By contrast, the solutions of $|X| > c$ make up two intervals.

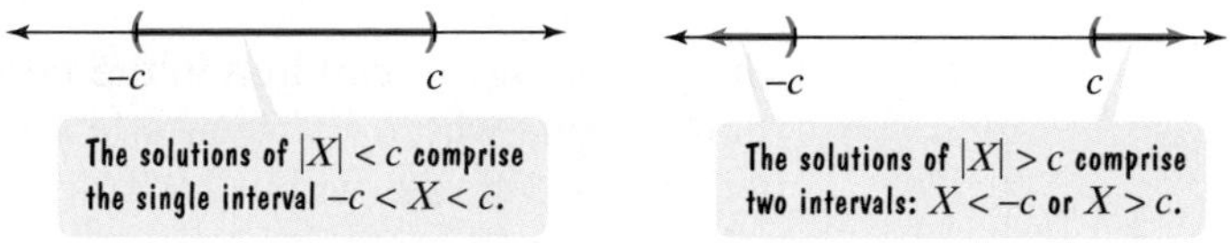
–c
c
The solutions of |X| < c comprise the single interval –c < X < c.
–c
c
The solutions of |X| > c comprise two intervals: X < –c or X > c.

Now let's see what happens to these inequalities if c is a negative number. Consider, for example, $|x| < -2$. Because $|x|$ always has a value that is greater than or equal to 0, there is no number whose absolute value is less than -2. The inequality $|x| < -2$ has no solution. The solution set is $\emptyset$.

Now consider the inequality $|x| > -2$. Because $|x|$ is never negative, all numbers have an absolute value that is greater than -2. All real numbers satisfy the inequality $|x| > -2$. The solution set is $(-\infty, \infty)$.

ABSOLUTE VALUE INEQUALITIES WITH UNUSUAL SOLUTION SETS If X is an algebraic expression and c is a negative number,

1. the inequality $|X| < c$ has no solution.
2. the inequality $|X| > c$ is true for all real numbers for which X is defined.

Table 9.2

What Is Bad about Being a Kid?

Kids Say	
Getting bossed around	17%
School, homework	15%
Can't do everything I want	11%
Chores	9%
Being grounded	9%

Source: Penn, Schoen, and Berland using 1172 interviews with children ages 6 to 14 from May 14 to June 1, 1999, Margin of error: ±2.9%

Note the margin of error.

Applications When you were between the ages of 6 and 14, how would you have responded to this question:

What is bad about being a kid?

In a random sample of 1172 children ages 6 to 14, 17% of the children responded, "Getting bossed around." The problem is that this is a single random sample. Do 17% of kids in the entire population of children ages 6 to 14 think that getting bossed around is a bad thing?

If you look at the results of a poll like the one in Table 9.2, you will observe that a **margin of error** is reported. The margin of error is ±2.9%. This means that the actual percentage of children who feel getting bossed around is a bad thing is at most 2.9% greater than or less than 17%. If x represents the percentage of children in the population who think that getting bossed around is a bad thing, then the poll's margin of error can be expressed as an absolute value inequality:

$$|x - 17| \le 2.9.$$

4 Solve problems using absolute value inequalities.

EXAMPLE 5 Analyzing a Poll's Margin of Error

The inequality

$$|x - 9| \le 2.9$$

describes the percentage of children in the population who think that being grounded is a bad thing about being a kid. (See Table 9.2.) Solve the inequality and interpret the solution.

SOLUTION We rewrite the inequality without absolute value bars.

$$|X| \le c \quad \text{means} \quad -c \le X \le c.$$

$$|x - 9| \le 2.9 \quad \text{means} \quad -2.9 \le x - 9 \le 2.9.$$

We solve the compound inequality by adding 9 to all three parts.

$$-2.9 \le x - 9 \le 2.9$$
$$-2.9 + 9 \le x - 9 + 9 \le 2.9 + 9$$
$$6.1 \le x \le 11.9$$

The percentage of children in the population who think that being grounded is a bad thing is somewhere between a low of 6.1% and a high of 11.9%. Notice that these percents are always within 2.9% above and below the given 9%, and that 2.9% is the poll's margin of error. ■

CHECK POINT 5 Solve the inequality:

$$|x - 11| \le 2.9.$$

Interpret the solution in terms of the information in Table 9.2 on page 605.

9.3 EXERCISE SET

Student Solutions Manual | CD/Video | PH Math/Tutor Center | MathXL Tutorials on CD | MathXL® | MyMathLab | Interactmath.com

Practice Exercises

In Exercises 1–38, find the solution set for each equation.

1. $|x| = 8$
2. $|x| = 6$
3. $|x - 2| = 7$
4. $|x + 1| = 5$
5. $|2x - 1| = 7$
6. $|2x - 3| = 11$
7. $\left|\frac{4x - 2}{3}\right| = 2$
8. $\left|\frac{3x - 1}{5}\right| = 1$
9. $|x| = -8$
10. $|x| = -6$
11. $|x + 3| = 0$
12. $|x + 2| = 0$
13. $2|y + 6| = 10$
14. $3|y + 5| = 12$
15. $3|2x - 1| = 21$
16. $2|3x - 2| = 14$
17. $|6y - 2| + 4 = 32$
18. $|3y - 1| + 10 = 25$
19. $7|5x| + 2 = 16$
20. $7|3x| + 2 = 16$
21. $|x + 1| + 5 = 3$
22. $|x + 1| + 6 = 2$
23. $|4y + 1| + 10 = 4$
24. $|3y - 2| + 8 = 1$
25. $|2x - 1| + 3 = 3$
26. $|3x - 2| + 4 = 4$
27. $|5x - 8| = |3x + 2|$
28. $|4x - 9| = |2x + 1|$
29. $|2x - 4| = |x - 1|$
30. $|6x| = |3x - 9|$

31. $|2x - 5| = |2x + 5|$

32. $|3x - 5| = |3x + 5|$

33. $|x - 3| = |5 - x|$

34. $|x - 3| = |6 - x|$

35. $|2y - 6| = |10 - 2y|$

36. $|4y + 3| = |4y + 5|$

37. $\left|\frac{2x}{3} - 2\right| = \left|\frac{x}{3} + 3\right|$

38. $\left|\frac{x}{2} - 2\right| = \left|x - \frac{1}{2}\right|$

In Exercises 39–70, solve and graph the solution set on a number line.

39. $|x| < 3$

40. $|x| < 5$

41. $|x - 2| < 1$

42. $|x - 1| < 5$

43. $|x + 2| \le 1$

44. $|x + 1| \le 5$

45. $|2x - 6| < 8$

46. $|3x + 5| < 17$

47. $|x| > 3$

48. $|x| > 5$

49. $|x + 3| > 1$

50. $|x - 2| > 5$

51. $|x - 4| \ge 2$

52. $|x - 3| \ge 4$

53. $|3x - 8| > 7$

54. $|5x - 2| > 13$

55. $|2(x - 1) + 4| \le 8$

56. $|3(x - 1) + 2| \le 20$

57. $\left|\frac{2x + 6}{3}\right| < 2$

58. $\left|\frac{3x - 3}{4}\right| < 6$

59. $\left|\frac{2x + 2}{4}\right| \ge 2$

60. $\left|\frac{3x - 3}{9}\right| \ge 1$

61. $\left|3 - \frac{2x}{3}\right| > 5$

62. $\left|3 - \frac{3x}{4}\right| > 9$

63. $|x - 2| < -1$

64. $|x - 3| < -2$

65. $|x + 6| > -10$

66. $|x + 4| > -12$

67. $|x + 2| + 9 \le 16$

68. $|x - 2| + 4 \le 5$

69. $2|2x - 3| + 10 > 12$

70. $3|2x - 1| + 2 > 8$

71. Let $f(x) = |5 - 4x|$. Find all values of x for which $f(x) = 11$.

72. Let $f(x) = |2 - 3x|$. Find all values of x for which $f(x) = 13$.

73. Let $f(x) = |3 - x|$ and $g(x) = |3x + 11|$. Find all values of x for which $f(x) = g(x)$.

74. Let $f(x) = |3x + 1|$ and $g(x) = |6x - 2|$. Find all values of x for which $f(x) = g(x)$.

75. Let $g(x) = |-1 + 3(x + 1)|$. Find all values of x for which $g(x) \le 5$.

76. Let $g(x) = |-3 + 4(x + 1)|$. Find all values of x for which $g(x) \le 3$.

77. Let $h(x) = |2x - 3| + 1$. Find all values of x for which $h(x) > 6$.

78. Let $h(x) = |2x - 4| - 6$. Find all values of x for which $h(x) > 18$.

Practice Plus

79. When 3 times a number is subtracted from 4, the absolute value of the difference is at least 5. Use interval notation to express the set of all numbers that satisfy this condition.

80. When 4 times a number is subtracted from 5, the absolute value of the difference is at most 13. Use interval notation to express the set of all numbers that satisfy this condition.

In Exercises 81–82, solve each inequality. Assume that $a > 0$ and $c > 0$. Use set-builder notation to express each solution set.

81. $|ax + b| < c$

82. $|ax + b| \ge c$

In Exercises 83–84, use the graph of $f(x) = |4 - x|$ to solve each equation or inequality.

83. $|4 - x| = 1$

84. $|4 - x| < 5$

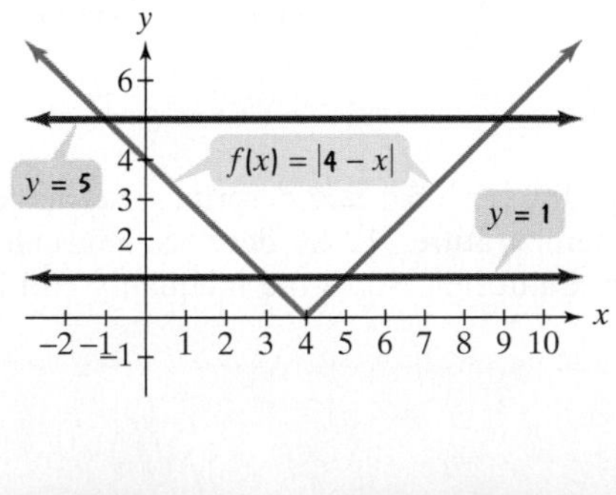

In Exercises 85–86, use the table to solve each inequality.

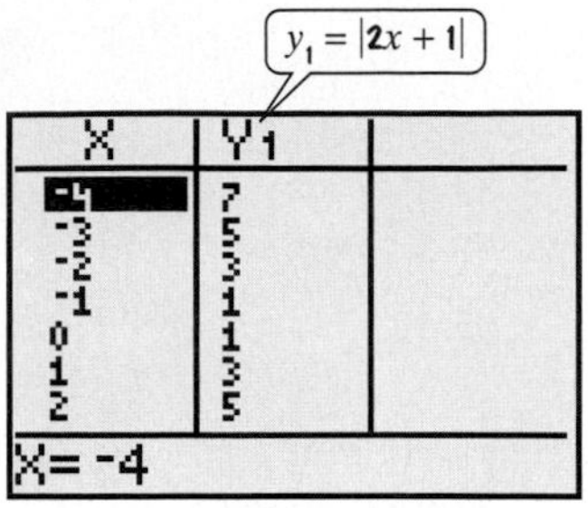

85. $|2x + 1| \leq 3$

86. $|2x + 1| \geq 3$

Application Exercises

The three television programs viewed by the greatest percentage of U.S. households in the twentieth century are shown in the table. The data are from a random survey of 4000 TV households by Nielsen Media Research. In Exercises 87–88, let x represent the actual viewing percentage in the U.S. population.

TV Programs with the Greatest U.S. Audience Viewing Percentage of the Twentieth Century

Program	Viewing Percentage in Survey
1. "M*A*S*H" Feb. 28, 1983	60.2%
2. "Dallas" Nov. 21, 1980	53.3%
3. "Roots" Part 8 Jan. 30, 1977	51.1%

Source: Nielsen Media Research

87. Solve the inequality: $|x - 60.2| \leq 1.6$. Interpret the solution in terms of the information in the table. What is the margin of error?

88. Solve the inequality: $|x - 51.1| \leq 1.6$. Interpret the solution in terms of the information in the table. What is the margin of error?

89. The inequality $|T - 57| \leq 7$ describes the range of monthly average temperature, T, in degrees Fahrenheit, for San Francisco, California. Solve the inequality and interpret the solution.

90. The inequality $|T - 50| \leq 22$ describes the range of monthly average temperature, T, in degrees Fahrenheit, for Albany, New York. Solve the inequality and interpret the solution.

The specifications for machine parts are given with tolerance limits that describe a range of measurements for which the part is acceptable. In Exercises 91–92, x represents the length of a machine part, in centimeters. The tolerance limit is 0.01 centimeter.

91. Solve: $|x - 8.6| \leq 0.01$. If the length of the machine part is supposed to be 8.6 centimeters, interpret the solution.

92. Solve: $|x - 9.4| \leq 0.01$. If the length of the machine part is supposed to be 9.4 centimeters, interpret the solution.

93. If a coin is tossed 100 times, we would expect approximately 50 of the outcomes to be heads. It can be demonstrated that a coin is unfair if h, the number of outcomes that result in heads, satisfies $\left|\frac{h - 50}{5}\right| \geq 1.645$. Describe the number of outcomes that determine an unfair coin that is tossed 100 times.

Writing in Mathematics

94. Explain how to solve an equation containing one absolute value expression.

95. Explain why the procedure that you described in Exercise 94 does not apply to the equation $|x - 5| = -3$. What is the solution set of this equation?

96. Describe how to solve an absolute value equation with two absolute values.

97. Describe how to solve an absolute value inequality involving the symbol $<$. Give an example.

98. Explain why the procedure that you described in Exercise 97 does not apply to the inequality $|x - 5| < -3$. What is the solution set of this inequality?

99. Describe how to solve an absolute value inequality involving the symbol $>$. Give an example.

100. Explain why the procedure that you described in Exercise 99 does not apply to the inequality $|x - 5| > -3$. What is the solution set of this inequality?

101. The final episode of *M*A*S*H* was viewed by more than 58% of U.S. television households. Is it likely that a popular television series in the twenty-first century will achieve a 58% market share? Explain your answer.

Technology Exercises

In Exercises 102–104, solve each equation using a graphing utility. Graph each side separately in the same viewing rectangle. The solutions are the x-coordinates of the intersection points.

102. $|x + 1| = 5$

103. $|3(x + 4)| = 12$

104. $|2x - 3| = |9 - 4x|$

In Exercises 105–107, solve each inequality using a graphing utility. Graph each side separately in the same viewing rectangle. The solution set consists of all values of x for which the the graph of the left side lies below the graph of the right side.

105. $|2x + 3| < 5$

106. $\left|\dfrac{2x - 1}{3}\right| < \dfrac{5}{3}$

107. $|x + 4| < -1$

In Exercises 108–110, solve each inequality using a graphing utility. Graph each side separately in the same viewing rectangle. The solution set consists of all values of x for which the graph of the left side lies above the graph of the right side.

108. $|2x - 1| > 7$

109. $|0.1x - 0.4| + 0.4 > 0.6$

110. $|x + 4| > -1$

111. Use a graphing utility to verify the solution sets for any five equations or inequalities that you solved by hand in Exercises 1–70.

Critical Thinking Exercises

112. Which one of the following is true?

a. All absolute value equations have two solutions.

b. The equation $|x| = -6$ is equivalent to $x = 6$ or $x = -6$.

c. We can rewrite the inequality $x > 5$ or $x < -5$ more compactly as $-5 < x < 5$.

d. Absolute value inequalities in the form $|ax + b| < c$ translate into *and* compound statements, which may be written as three-part inequalities.

113. Write an absolute value inequality for which the interval shown is the solution.

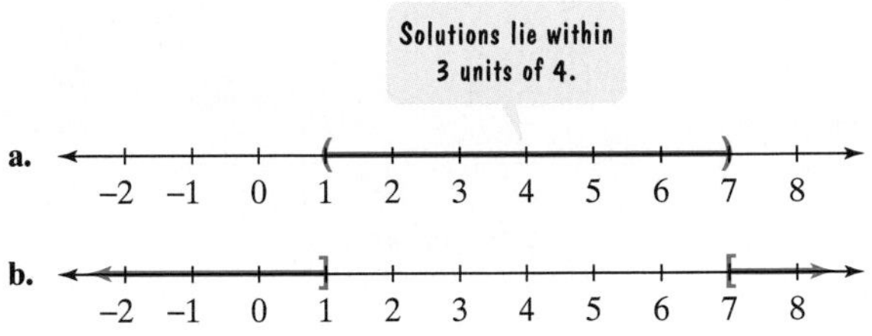

114. The percentage, p, of defective products manufactured by a company is given as $|p - 0.3\%| \le 0.2\%$. If 100,000 products are manufactured and the company offers a \$5 refund for each defective product, describe the company's cost for refunds.

115. Solve: $|2x + 5| = 3x + 4$.

Review Exercises

In Exercises 116–118, graph each linear function.

116. $3x - 5y = 15$ (Section 3.2, Example 5)

117. $f(x) = -\frac{2}{3}x$ or $y = -\frac{2}{3}x$ (Section 3.4, Example 4)

118. $f(x) = -2$ or $y = -2$ (Section 3.2, Example 7)

MID-CHAPTER CHECK POINT

CHAPTER 9

What You Know: We reviewed how to solve linear inequalities, expressing solution sets in set-builder and interval notations. We know that it is necessary to reverse the sense of an inequality when multiplying or dividing both sides by a negative number. We solved compound inequalities with *and* by finding the intersection of solution sets and with *or* by finding the union of solution sets. Finally, we solved equations and inequalities involving absolute value by carefully rewriting the given equation or inequality without absolute value bars. For positive values of c, we wrote $|X| = c$ as $X = c$ or $X = -c$, we wrote $|X| < c$ as $-c < X < c$, and we wrote $|X| > c$ as $X < -c$ or $X > c$.

In Exercises 1–17, solve each inequality or equation.

1. $4 - 3x \ge 12 - x$

2. $5 \le 2x - 1 < 9$

3. $|4x - 7| = 5$

4. $-10 - 3(2x + 1) > 8x + 1$

5. $2x + 7 < -11$ or $-3x - 2 < 13$

6. $|3x - 2| \le 4$

7. $|x + 5| = |5x - 8|$

8. $5 - 2x \ge 9$ and $5x + 3 > -17$

9. $3x - 2 > -8$ or $2x + 1 < 9$

10. $\frac{x}{2} + 3 \le \frac{x}{3} + \frac{5}{2}$

11. $\frac{2}{3}(6x - 9) + 4 > 5x + 1$

12. $|5x + 3| > 2$

13. $7 - \left|\frac{x}{2} + 2\right| \le 4$

14. $\frac{x + 3}{4} < \frac{1}{3}$

15. $5x + 1 \ge 4x - 2$ and $2x - 3 > 5$

16. $3 - |2x - 5| = -6$

17. $3 + |2x - 5| = -6$

18. A company that manufactures compact discs has fixed monthly overhead costs of \$60,000. Each disc costs \$0.18 to produce and sells for \$0.30.

a. Write the cost function, C, of producing x discs per month.

b. Write the revenue function, R, from the sale of x discs per month.

c. Write the profit function, P, from producing and selling x discs per month.

d. How many discs should be produced and sold each month for the company to have a profit of at least \$30,000?

19. A car rental agency rents a certain car for \$40 per day with unlimited mileage or \$24 per day plus \$0.20 per mile. How far can a customer drive this car per day for the \$24 option to cost no more than the unlimited mileage option?

20. To receive a B in a course, you must have an average of at least 80% but less than 90% on five exams. Your grades on the first four exams were 95%, 79%, 91%, and 86%. What range of grades on the fifth exam will result in a B for the course?

21. A retiree requires an annual income of at least \$9000 from an investment paying 7.5% annual interest. How much should the retiree invest to achieve the desired return?

SECTION 9.4

LINEAR INEQUALITIES IN TWO VARIABLES

Objectives

1 Graph a linear inequality in two variables.

2 Graph a system of linear inequalities.

3 Solve applied problems involving systems of inequalities.

This book was written in Point Reyes National Seashore, 40 miles north of San Francisco. The park consists of 75,000 acres with miles of pristine surf-washed beaches, forested ridges, and bays bordered by white cliffs.

Like your author, many people are kept inspired and energized surrounded by nature's unspoiled beauty. In this section, you will see how systems of inequalities model whether a region's natural beauty manifests itself in forests, grasslands, or deserts.

Linear Inequalities in Two Variables and Their Solutions We have seen that the graphs of equations in the form $Ax + By = C$ are straight lines. If we change the $=$ sign to $>$, $<$, $\geq$, or $\leq$, we obtain a **linear inequality in two variables.** Some examples of linear inequalities in two variables are $x + y > 2$, $3x - 5y \leq 15$, and $2x - y < 4$.

A **solution of an inequality in two variables,** x and y, is an ordered pair of real numbers with the following property: When the x-coordinate is substituted for x and the y-coordinate is substituted for y in the inequality, we obtain a true statement. For example, (3, 2) is a solution of the inequality $x + y > 1$. When 3 is substituted for x and 2 is substituted for y, we obtain the true statement $3 + 2 > 1$, or $5 > 1$. Because there are infinitely many pairs of numbers that have a sum greater than 1, the inequality $x + y > 1$ has infinitely many solutions. Each ordered-pair solution is said to **satisfy** the inequality. Thus, (3, 2) satisfies the inequality $x + y > 1$.

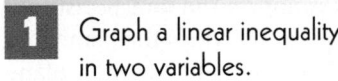

1 Graph a linear inequality in two variables.

The Graph of a Linear Inequality in Two Variables We know that the graph of an equation in two variables is the set of all points whose coordinates satisfy the equation. Similarly, the **graph of an inequality in two variables** is the set of all points whose coordinates satisfy the inequality.

Let's use Figure 9.10 to get an idea of what the graph of a linear inequality in two variables looks like. Part of the figure shows the graph of the linear equation $x + y = 2$. The line divides the points in the rectangular coordinate system into three sets. First, there is the set of points along the line, satisfying $x + y = 2$. Next, there is the set of points in the green region above the line. Points in the green region satisfy the linear inequality $x + y > 2$. Finally, there is the set of points in the purple region below the line. Points in the purple region satisfy the linear inequality $x + y < 2$.

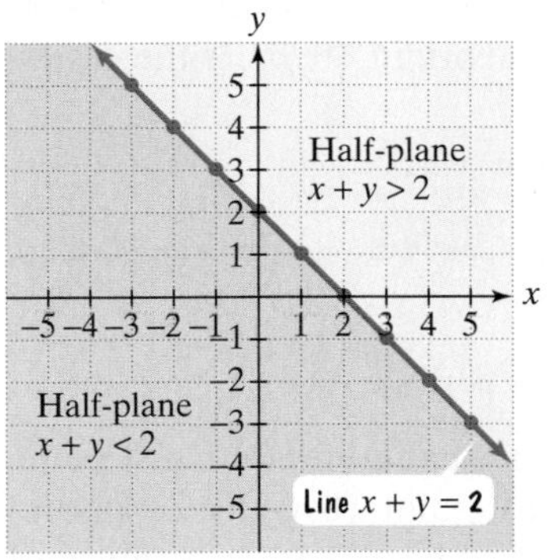

FIGURE 9.10

A **half-plane** is the set of all the points on one side of a line. In Figure 9.10, the green region is a half-plane. The purple region is also a half-plane. A half-plane is the graph of a linear inequality that involves $>$ or $<$. The graph of an inequality that involves $\geq$ or $\leq$ is a half-plane and a line. A solid line is used to show that a line is part of a graph. A dashed line is used to show that a line is not part of a graph.

GRAPHING A LINEAR INEQUALITY IN TWO VARIABLES

1. Replace the inequality symbol with an equal sign and graph the corresponding linear equation. Draw a solid line if the original inequality contains a $\leq$ or $\geq$ symbol. Draw a dashed line if the original inequality contains a $<$ or $>$ symbol.
2. Choose a test point in one of the half-planes that is not on the line. Substitute the coordinates of the test point into the inequality.
3. If a true statement results, shade the half-plane containing this test point. If a false statement results, shade the half-plane not containing this test point.

EXAMPLE 1 Graphing a Linear Inequality in Two Variables

Graph: $2x - 3y \geq 6$.

SOLUTION

Step 1. **Replace the inequality symbol by = and graph the linear equation.** We need to graph $2x - 3y = 6$. We can use intercepts to graph this line.

We set $y = 0$ to find the x-intercept:	**We set $x = 0$ to find the y-intercept:**
$2x - 3y = 6$	$2x - 3y = 6$
$2x - 3\cdot 0 = 6$	$2\cdot 0 - 3y = 6$
$2x = 6$	$-3y = 6$
$x = 3.$	$y = -2.$

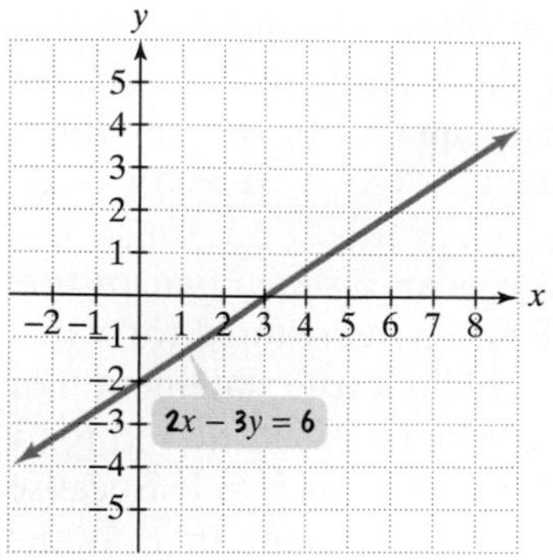

FIGURE 9.11 Preparing to graph $2x - 3y \geq 6$

The x-intercept is 3, so the line passes through $(3, 0)$. The y-intercept is -2, so the line passes through $(0, -2)$. Using the intercepts, the line is shown in Figure 9.11 as a solid line. This is because the inequality $2x - 3y \geq 6$ contains a $\geq$ symbol, in which equality is included.

Step 2. **Choose a test point in one of the half-planes that is not on the line. Substitute its coordinates into the inequality.** The line $2x - 3y = 6$ divides the plane into three parts—the line itself and two half-planes. The points in one half-plane satisfy $2x - 3y > 6$. The points in the other half-plane satisfy $2x - 3y < 6$. We need to find which half-plane belongs to the solution of $2x - 3y \geq 6$. To do so, we test a point from either half-plane. The origin, $(0, 0)$, is the easiest point to test.

$2x - 3y \geq 6$ — This is the given inequality.

$2\cdot 0 - 3\cdot 0 \overset{?}{\geq} 6$ — Test (0, 0) by substituting 0 for x and 0 for y.

$0 - 0 \overset{?}{\geq} 6$ — Multiply.

$0 \geq 6$ — This statement is false.

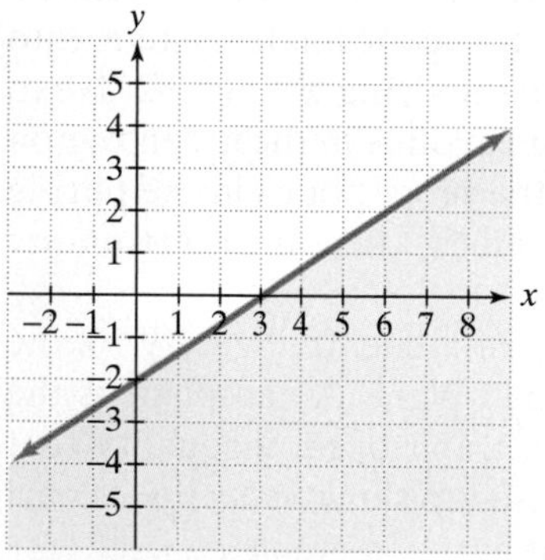

FIGURE 9.12 The graph of $2x - 3y \geq 6$

Step 3. **If a false statement results, shade the half-plane not containing the test point.** Because 0 is not greater than or equal to 6, the test point, $(0, 0)$, is not part of the solution set. Thus, the half-plane below the solid line $2x - 3y = 6$ is part of the solution set. The solution set is the line and the half-plane that does not contain the point $(0, 0)$, indicated by shading this half-plane. The graph is shown using green shading and a blue line in Figure 9.12. ■

✔ **CHECK POINT 1** Graph: $4x - 2y \geq 8$.

When graphing a linear inequality, test a point that lies in one of the half-planes and *not on the line dividing the half-planes*. The test point $(0, 0)$ is convenient because it is easy to calculate when 0 is substituted for each variable. However, if $(0, 0)$ lies on the dividing line and not in a half-plane, a different test point must be selected.

EXAMPLE 2 Graphing a Linear Inequality in Two Variables

Graph: $y > -\frac{2}{3}x$.

SOLUTION

Step 1. **Replace the inequality symbol by = and graph the linear equation.** Because we are interested in graphing $y > -\frac{2}{3}x$, we begin by graphing $y = -\frac{2}{3}x$. We can use the slope and the y-intercept to graph this linear function.

$$y = -\frac{2}{3}x + 0$$

Slope $= \frac{-2}{3} = \frac{\text{rise}}{\text{run}}$ $\qquad$ y-intercept $= 0$

The y-intercept is 0, so the line passes through $(0, 0)$. Using the y-intercept and the slope, the line is shown in Figure 9.13 as a dashed line. This is because the inequality $y > -\frac{2}{3}x$ contains a $>$ symbol, in which equality is not included.

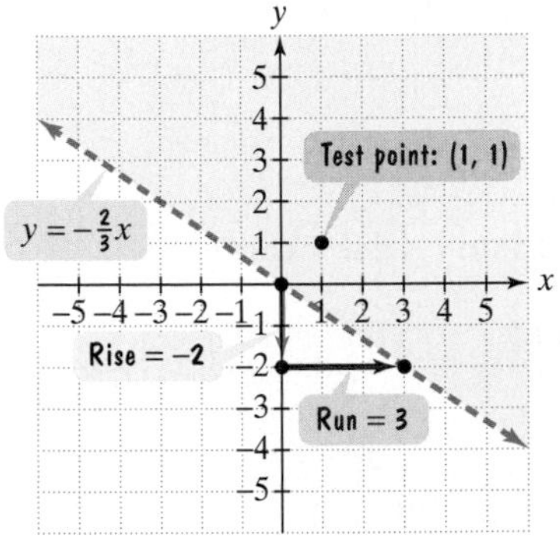

FIGURE 9.13 The graph of $y > -\frac{2}{3}x$

Step 2. **Choose a test point in one of the half-planes that is not on the line. Substitute its coordinates into the inequality.** We cannot use $(0, 0)$ as a test point because it lies on the line and not in a half-plane. Let's use $(1, 1)$, which lies in the half-plane above the line.

$$y > -\frac{2}{3}x \qquad \text{This is the given inequality.}$$

$$1 \overset{?}{>} -\frac{2}{3} \cdot 1 \qquad \text{Test } (1, 1) \text{ by substituting 1 for } x \text{ and 1 for } y.$$

$$1 > -\frac{2}{3} \qquad \text{This statement is true.}$$

Step 3. **If a true statement results, shade the half-plane containing the test point.** Because 1 is greater than $-\frac{2}{3}$, the test point $(1, 1)$ is part of the solution set. All the points on the same side of the line $y = -\frac{2}{3}x$ as the point $(1, 1)$ are members of the solution set. The solution set is the half-plane that contains the point $(1, 1)$, indicated by shading this half-plane. The graph is shown using green shading and a dashed blue line in Figure 9.13. ■

USING TECHNOLOGY

Most graphing utilities can graph inequalities in two variables with the SHADE feature. The procedure varies by model, so consult your manual. For most graphing utilities, you must first solve for y if it is not already isolated. The figure shows the graph of $y > -\frac{2}{3}x$. Most displays do not distinguish between dashed and solid boundary lines.

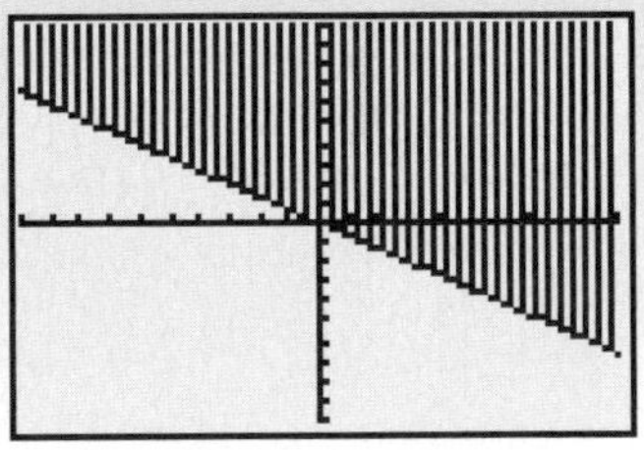

CHECK POINT 2 Graph: $y > -\frac{3}{4}x$.

Graphing Linear Inequalities without Using Test Points You can graph inequalities in the form $y > mx + b$ or $y < mx + b$ without using test points. The inequality symbol indicates which half-plane to shade.

- If $y > mx + b$, shade the half-plane above the line $y = mx + b$.
- If $y < mx + b$, shade the half-plane below the line $y = mx + b$.

> **STUDY TIP**
>
> Continue using test points to graph inequalities in the form $Ax + By > C$ or $Ax + By < C$. The graph of $Ax + By > C$ can lie above or below the line of $Ax + By = C$, depending on the value of B. The same comment applies to the graph of $Ax + By < C$.

It is also not necessary to use test points when graphing inequalities involving half-planes on one side of a vertical or a horizontal line.

For the Vertical Line $x = a$:

- If $x > a$, shade the half-plane to the right of $x = a$.
- If $x < a$, shade the half-plane to the left of $x = a$.

For the Horizontal Line $y = b$:

- If $y > b$, shade the half-plane above $y = b$.
- If $y < b$, shade the half-plane below $y = b$.

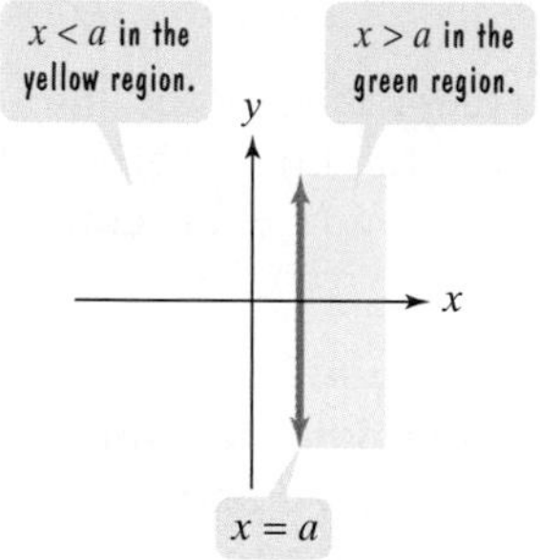
x < a in the yellow region.
x > a in the green region.
y
x
x = a

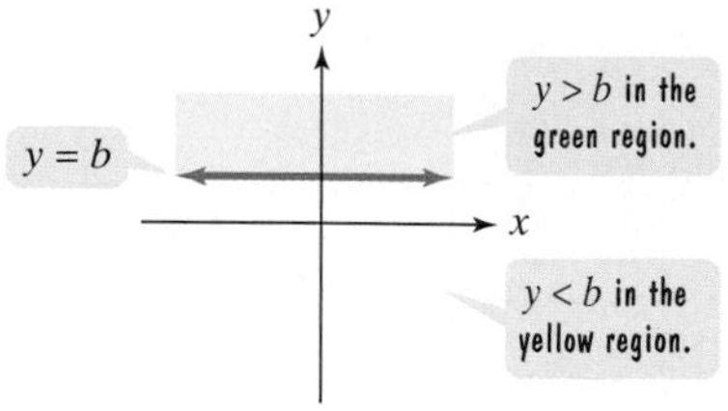
y
y > b in the green region.
y = b
x
y < b in the yellow region.

EXAMPLE 3 Graphing Inequalities without Using Test Points

Graph each inequality in a rectangular coordinate system:

a. $y \leq -3$ **b.** $x > 2$.

SOLUTION

a. $y \leq -3$

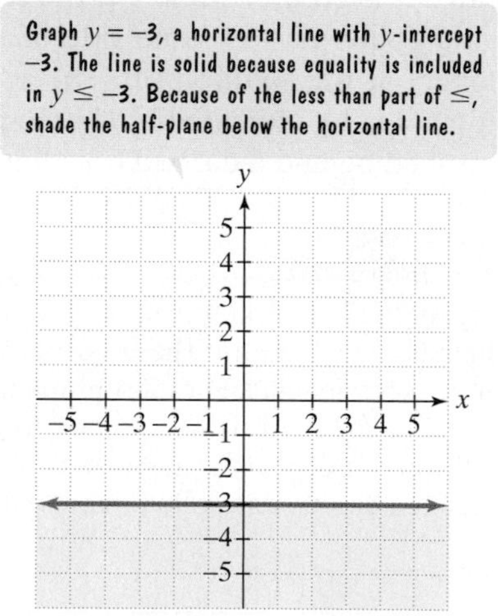
Graph y = −3, a horizontal line with y-intercept −3. The line is solid because equality is included in y ≤ −3. Because of the less than part of ≤, shade the half-plane below the horizontal line.
y
x
−5 −4 −3 −2 −1 1 2 3 4 5
5 4 3 2 1 −1 −2 −3 −4 −5

b. $x > 2$

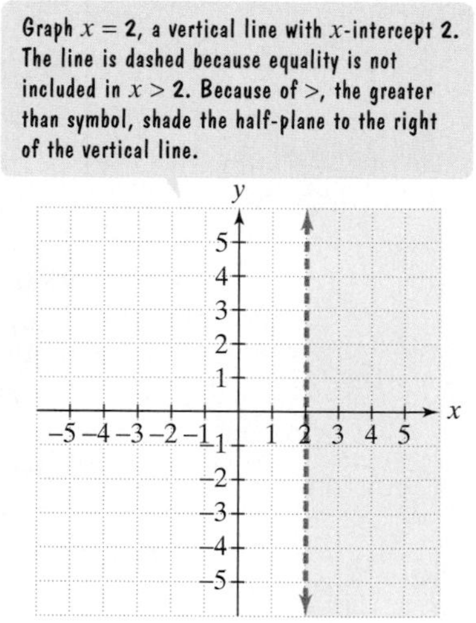
Graph x = 2, a vertical line with x-intercept 2. The line is dashed because equality is not included in x > 2. Because of >, the greater than symbol, shade the half-plane to the right of the vertical line.
y
x
−5 −4 −3 −2 −1 1 2 3 4 5
5 4 3 2 1 −1 −2 −3 −4 −5

■

✔ **CHECK POINT 3** Graph each inequality in a rectangular coordinate system:

a. $y > 1$ **b.** $x \leq -2$.

2 Graph a system of linear inequalities.

Systems of Linear Inequalities Just as two linear equations make up a system of linear equations, two or more linear inequalities make up a **system of linear inequalities**. Here is an example of a system of linear inequalities:

$$x - y < 1$$
$$2x + 3y \geq 12.$$

A **solution of a system of linear inequalities** in two variables is an ordered pair that satisfies each inequality in the system. The set of all such ordered pairs is the **solution set** of the system. Thus, to graph a system of inequalities in two variables, begin by graphing each individual inequality in the same rectangular coordinate system. Then find the region, if there is one, that is common to every graph in the system. This region of intersection gives a picture of the system's solution set.

EXAMPLE 4 Graphing a System of Linear Inequalities

Graph the solution set of the system:

$$x - y < 1$$
$$2x + 3y \geq 12.$$

SOLUTION Replacing each inequality symbol with an equal sign indicates that we need to graph $x - y = 1$ and $2x + 3y = 12$. We can use intercepts to graph these lines.

$x - y = 1$		$2x + 3y = 12$
x-intercept: $x - 0 = 1$	Set $y = 0$ in each equation.	x-intercept: $2x + 3 \cdot 0 = 12$
$x = 1$		$2x = 12$
The line passes through $(1, 0)$.		$x = 6$
		The line passes through $(6, 0)$.
y-intercept: $0 - y = 1$	Set $x = 0$ in each equation.	y-intercept: $2 \cdot 0 + 3y = 12$
$-y = 1$		$3y = 12$
$y = -1$		$y = 4$
The line passes through $(0, -1)$		The line passes through $(0, 4)$.

Now we are ready to graph the solution set of the system of linear inequalities.

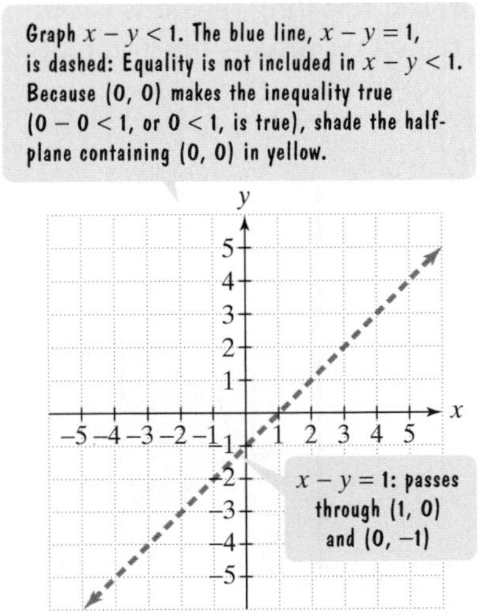

The graph of $x - y < 1$

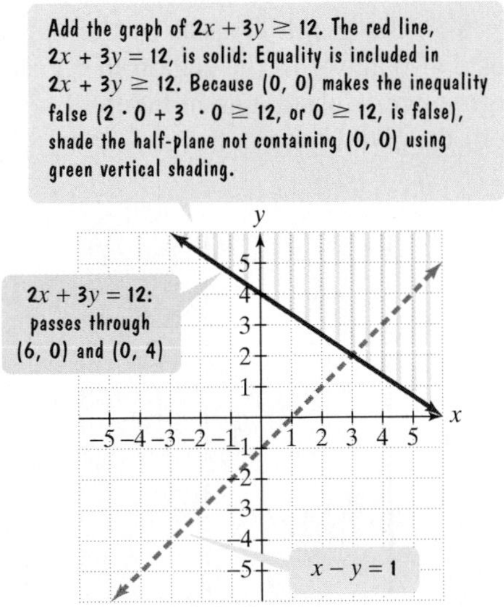

Adding the graph of $2x + 3y \geq 12$

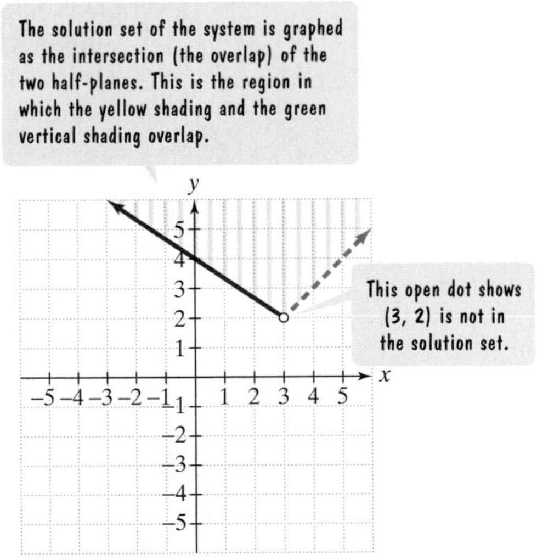
The solution set of the system is graphed as the intersection (the overlap) of the two half-planes. This is the region in which the yellow shading and the green vertical shading overlap.
y
x
This open dot shows (3, 2) is not in the solution set.

The graph of $x - y < 1$ and $2x + 3y \geq 12$

CHECK POINT 4 Graph the solution set of the system:

$$x - 3y < 6$$
$$2x + 3y \geq -6.$$

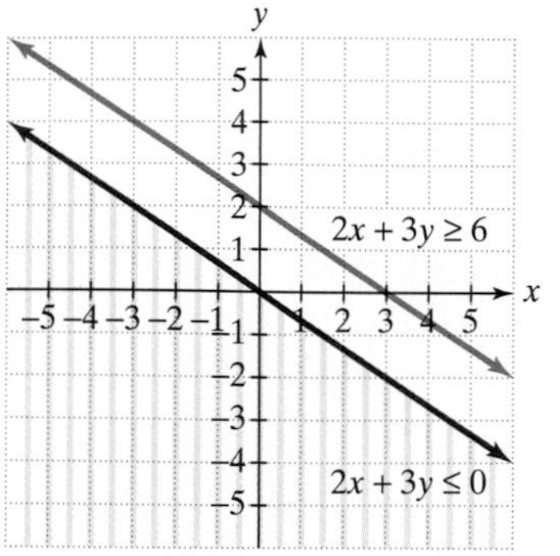

FIGURE 9.14 A system of inequalities with no solution

A system of inequalities has no solution if there are no points in the rectangular coordinate system that simultaneously satisfy each inequality in the system. For example, the system

$$2x + 3y \geq 6$$
$$2x + 3y \leq 0$$

whose separate graphs are shown in Figure 9.14 has no overlapping region. Thus, the system has no solution. The solution set is $\varnothing$, the empty set.

EXAMPLE 5 Graphing a System of Inequalities

Graph the solution set:

$$x - y < 2$$
$$-2 \leq x < 4$$
$$y < 3.$$

SOLUTION We begin by graphing $x - y < 2$, the first given inequality. The line $x - y = 2$ has an x-intercept of 2 and a y-intercept of -2. The test point $(0, 0)$ makes the inequality $x - y < 2$ true, and its graph is shown in Figure 9.15.

Now, let's consider the second given inequality, $-2 \leq x < 4$. Replacing the inequality symbols by $=$, we obtain $x = -2$ and $x = 4$, graphed as red vertical lines in Figure 9.16. The line of $x = 4$ is not included. Using $(0, 0)$ as a test point and substituting the x-coordinate, 0, into $-2 \leq x < 4$, we obtain the true statement $-2 \leq 0 < 4$. We therefore shade the region between the vertical lines. We must intersect this region with the yellow region in Figure 9.15. The resulting region is shown in yellow and green vertical shading in Figure 9.16.

Finally, let's consider the third given inequality, $y < 3$. Replacing the inequality symbol by $=$, we obtain $y = 3$, which graphs as a horizontal line. Because of the less than symbol in $y < 3$, the graph consists of the half-plane below the line $y = 3$. We must intersect this half-plane with the region in Figure 9.16. The resulting region is shown in yellow and green vertical shading in Figure 9.17. This region represents the graph of the solution set of the given system.

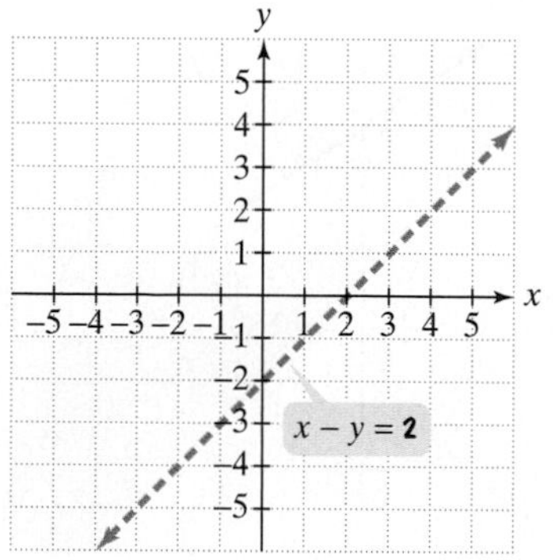

FIGURE 9.15 The graph of $x - y < 2$

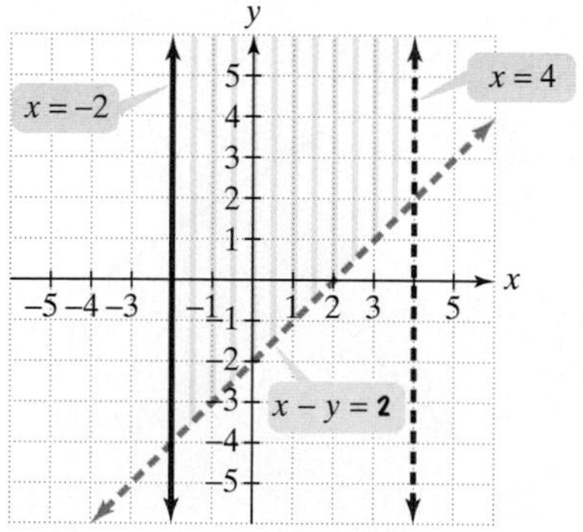

FIGURE 9.16 The graph of $x - y < 2$ and $-2 \leq x < 4$

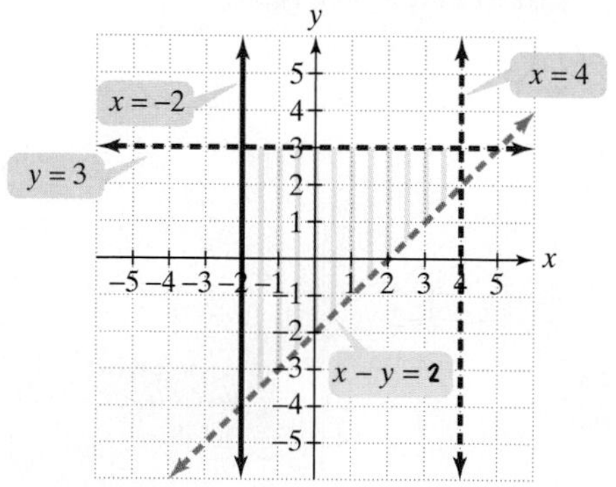

FIGURE 9.17 The graph of $x - y < 2$ and $-2 \leq x < 4$ and $y < 3$

CHECK POINT 5 Graph the solution set:

$$x + y < 2$$
$$-2 \le x < 1$$
$$y > -3.$$

3 Solve applied problems involving systems of inequalities.

Applications Temperature and precipitation affect whether or not trees and forests can grow. At certain levels of precipitation and temperature, only grasslands and deserts will exist. Figure 9.18 shows three kinds of regions—deserts, grasslands, and forests—that result from various ranges of temperature, T, and precipitation, P.

Systems of inequalities can be used to describe where forests, grasslands, and deserts occur. Because these regions occur when the average annual temperature, T, is 35°F or greater, each system contains the inequality $T \ge 35$.

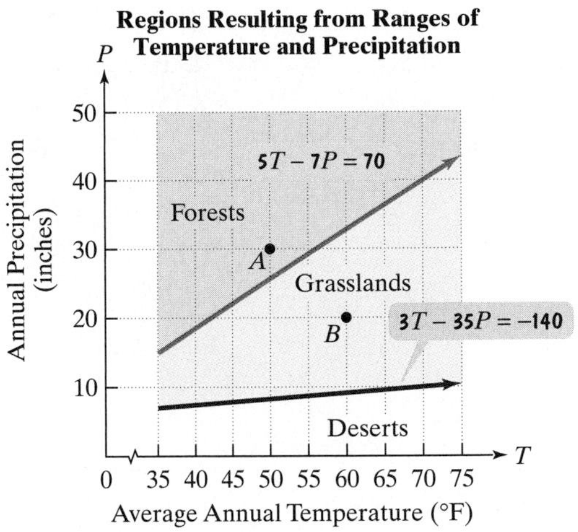

FIGURE 9.18

Source: A Miller and J. Thompson, *Elements of Meteorology*

Forests occur if	Grasslands occur if	Deserts occur if
$T \ge 35$	$T \ge 35$	$T \ge 35$
$5T - 7P < 70.$	$5T - 7P \ge 70$	$3T - 35P > -140.$
	$3T - 35P \le -140.$	

EXAMPLE 6 Forests and Systems of Inequalities

Show that point A in Figure 9.18 is a solution of the system of inequalities that describes where forests occur.

SOLUTION Point A has coordinates (50, 30). This means that if a region has an average annual temperature of 50°F and an average annual precipitation of 30 inches, a forest occurs. We can show that (50, 30) satisfies the system of inequalities for forests by substituting 50 for T and 30 for P in each inequality in the system.

$$T \ge 35 \qquad 5T - 7P < 70$$
$$50 \ge 35, \text{ true} \qquad 5 \cdot 50 - 7 \cdot 30 \stackrel{?}{<} 70$$
$$250 - 210 \stackrel{?}{<} 70$$
$$40 < 70, \text{ true}$$

The coordinates (50, 30) make each inequality true. Thus, (50, 30) satisfies the system for forests. ■

CHECK POINT 6 Show that point B in Figure 9.18 is a solution of the system of inequalities that describes where grasslands occur.

9.4 EXERCISE SET

Student Solutions Manual | CD/Video | PH Math/Tutor Center | MathXL Tutorials on CD | MathXL® | MyMathLab | Interactmath.com

Practice Exercises

In Exercises 1–22, graph each inequality.

1. $x + y \geq 3$

2. $x + y \geq 2$

3. $x - y < 5$

4. $x - y < 6$

5. $x + 2y > 4$

6. $2x + y > 6$

7. $3x - y \leq 6$

8. $x - 3y \leq 6$

9. $\frac{x}{2} + \frac{y}{3} < 1$

10. $\frac{x}{4} + \frac{y}{2} < 1$

11. $y > \frac{1}{3}x$

12. $y > \frac{1}{4}x$

13. $y \leq 3x + 2$

14. $y \leq 2x - 1$

15. $y < -\frac{1}{4}x$

16. $y < -\frac{1}{3}x$

17. $x \leq 2$

18. $x \leq -4$

19. $y > -4$

20. $y > -2$

21. $y \geq 0$

22. $x \leq 0$

In Exercises 23–46 graph the solution set of each system of inequalities or indicate that the system has no solution.

23. $\begin{aligned} 3x + 6y &\leq 6 \\ 2x + y &\leq 8 \end{aligned}$

24. $\begin{aligned} x - y &\geq 4 \\ x + y &\leq 6 \end{aligned}$

25. $\begin{aligned} 2x - 5y &\leq 10 \\ 3x - 2y &> 6 \end{aligned}$

26. $\begin{aligned} 2x - y &\leq 4 \\ 3x + 2y &> -6 \end{aligned}$

27. $\begin{aligned} y &> 2x - 3 \\ y &< -x + 6 \end{aligned}$

28. $\begin{aligned} y &< -2x + 4 \\ y &< x - 4 \end{aligned}$

29. $\begin{aligned} x + 2y &\leq 4 \\ y &\geq x - 3 \end{aligned}$

30. $\begin{aligned} x + y &\leq 4 \\ y &\geq 2x - 4 \end{aligned}$

31. $\begin{aligned} x &\leq 2 \\ y &\geq -1 \end{aligned}$

32. $\begin{aligned} x &\leq 3 \\ y &\leq -1 \end{aligned}$

33. $-2 \leq x < 5$

34. $-2 < y \leq 5$

35. $\begin{aligned} x - y &\leq 1 \\ x &\geq 2 \end{aligned}$

36. $\begin{aligned} 4x - 5y &\geq -20 \\ x &\geq -3 \end{aligned}$

37. $\begin{aligned} x + y &> 4 \\ x + y &< -1 \end{aligned}$

38. $\begin{aligned} x + y &> 3 \\ x + y &< -2 \end{aligned}$

39. $\begin{aligned} x + y &> 4 \\ x + y &> -1 \end{aligned}$

40. $\begin{aligned} x + y &> 3 \\ x + y &> -2 \end{aligned}$

41. $\begin{aligned} x - y &\leq 2 \\ x &\geq -2 \\ y &\leq 3 \end{aligned}$

42. $\begin{aligned} 3x + y &\leq 6 \\ x &\geq -2 \\ y &\leq 4 \end{aligned}$

43. $\begin{aligned} x &\geq 0 \\ y &\geq 0 \\ 2x + 5y &\leq 10 \\ 3x + 4y &\leq 12 \end{aligned}$

44. $\begin{aligned} x &\geq 0 \\ y &\geq 0 \\ 2x + y &\leq 4 \\ 2x - 3y &\leq 6 \end{aligned}$

45. $\begin{aligned} 3x + y &\leq 6 \\ 2x - y &\leq -1 \\ x &\geq -2 \\ y &\leq 4 \end{aligned}$

46. $\begin{aligned} 2x + y &\leq 6 \\ x + y &\geq 2 \\ 1 \leq x &\leq 2 \\ y &\leq 3 \end{aligned}$

Practice Plus

In Exercises 47–48, write each sentence as a linear inequality in two variables. Then graph the inequality.

47. The y-variable is at least 4 more than the product of -2 and the x-variable.

48. The y-variable is at least 2 more than the product of -3 and the x-variable.

In Exercises 49–50, write the given sentences as a system of linear inequalities in two variables. Then graph the system.

49. The sum of the x-variable and the y-variable is at most 4. The y-variable added to the product of 3 and the x-variable does not exceed 6.

50. The sum of the x-variable and the y-variable is at most 3. The y-variable added to the product of 4 and the x-variable does not exceed 6.

In Exercises 51–52, rewrite each inequality in the system without absolute value bars. Then graph the rewritten system in rectangular coordinates.

51. $\begin{aligned} |x| &\leq 2 \\ |y| &\leq 3 \end{aligned}$

52. $\begin{aligned} |x| &\leq 1 \\ |y| &\leq 2 \end{aligned}$

The graphs of solution sets of systems of inequalities involve finding the intersection of the solution sets of two or more inequalities. By contrast, in Exercises 53–54 you will be graphing the ***union*** *of the solution sets of two inequalities.*

53. Graph the union of $y > \frac{3}{2}x - 2$ and $y < 4$.

54. Graph the union of $x - y \geq -1$ and $5x - 2y \leq 10$.

Without graphing, in Exercises 55–58, determine if each system has no solution or infinitely many solutions.

55. $\begin{aligned} 3x + y &< 9 \\ 3x + y &> 9 \end{aligned}$

56. $\begin{aligned} 6x - y &\leq 24 \\ 6x - y &> 24 \end{aligned}$

57. $\begin{aligned} 3x + y &\leq 9 \\ 3x + y &\geq 9 \end{aligned}$

58. $\begin{aligned} 6x - y &\leq 24 \\ 6x - y &\geq 24 \end{aligned}$

Application Exercises

Maximum heart rate, H, in beats per minute is a function of age, a, modeled by the formula

$$H = 220 - a,$$

where $10 \le a \le 70$. The bar graph shows the target heart rate ranges for four types of exercise goals.

Target Heart Rate Ranges for Exercise Goals

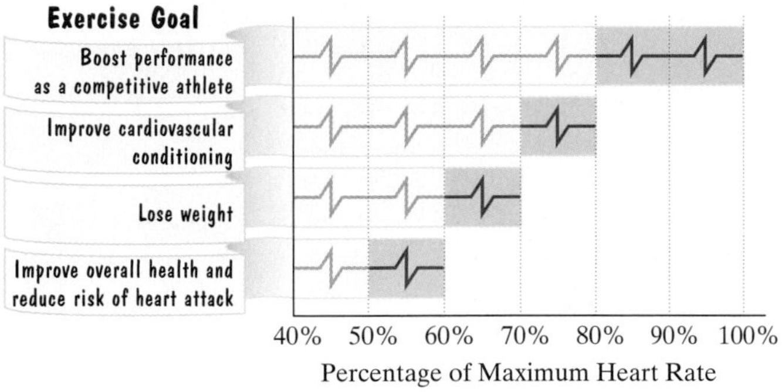

Source: Vitality

In Exercises 59–62, systems of inequalities will be used to model three of the target heart rate ranges shown in the bar graph. We begin with the target heart rate range for cardiovascular conditioning, modeled by the following system of inequalities:

$10 \le a \le 70$ — Heart rate ranges apply to ages 10 through 70, inclusive.

$H \ge 0.7(220 - a)$ — Target heart rate range is greater than or equal to 70% of maximum heart rate

$H \le 0.8(220 - a)$ — and less than or equal to 80% of maximum heart rate.

The graph of this system is shown in the figure. Use the graph to solve Exercises 59–60.

Target Heart Rate Range for Cardiovascular Conditioning

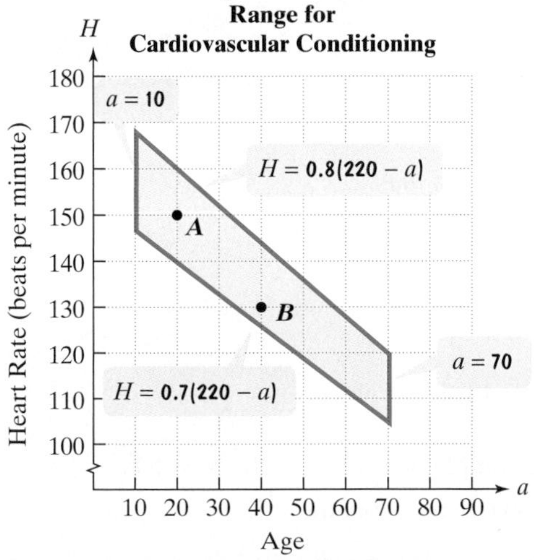

59. **a.** What are the coordinates of point A and what does this mean in terms of age and heart rate?

b. Show that point A is a solution of the system of inequalities.

60. **a.** What are the coordinates of point B and what does this mean in terms of age and heart rate?

b. Show that point B is a solution of the system of inequalities.

61. Write a system of inequalities that models the target heart rate range for the goal of losing weight.

62. Write a system of inequalities that models the target heart rate range for improving overall health.

63. On your next vacation, you will divide lodging between large resorts and small inns. Let x represent the number of nights spent in large resorts. Let y represent the number of nights spent in small inns.

a. Write a system of inequalities that models the following conditions:

> You want to stay at least 5 nights. At least one night should be spent at a large resort. Large resorts average \$200 per night and small inns average \$100 per night. Your budget permits no more than \$700 for lodging.

b. Graph the solution set of the system of inequalities in part (a).

c. Based on your graph in part (b), how many nights could you spend at a large resort and still stay within your budget?

64. **a.** An elevator can hold no more than 2000 pounds. If children average 80 pounds and adults average 160 pounds, write a system of inequalities that models when the elevator holding x children and y adults is overloaded.

b. Graph the solution set of the system of inequalities in part (a).

Writing in Mathematics

65. What is a linear inequality in two variables? Provide an example with your description.

66. How do you determine if an ordered pair is a solution of an inequality in two variables, x and y?

67. What is a half-plane?

68. What does a solid line mean in the graph of an inequality?

69. What does a dashed line mean in the graph of an inequality?

70. Explain how to graph $x - 2y < 4$.

71. What is a system of linear inequalities?

72. What is a solution of a system of linear inequalities?

73. Explain how to graph the solution set of a system of inequalities.

74. What does it mean if a system of linear inequalities has no solution?

Technology Exercises

Graphing utilities can be used to shade regions in the rectangular coordinate system, thereby graphing an inequality in two variables. Read the section of the user's manual for your graphing utility that describes how to shade a region. Then use your graphing utility to graph the inequalities in Exercises 75–78.

75. $y \leq 4x + 4$

76. $y \geq \frac{2}{3}x - 2$

77. $2x + y \leq 6$

78. $3x - 2y \geq 6$

79. Does your graphing utility have any limitations in terms of graphing inequalities? If so, what are they?

80. Use a graphing utility with a SHADE feature to verify any five of the graphs that you drew by hand in Exercises 1–22.

81. Use a graphing utility with a SHADE feature to verify any five of the graphs that you drew by hand for the systems in Exercises 23–46.

Critical Thinking Exercises

82. Write a linear inequality in two variables whose graph is shown.

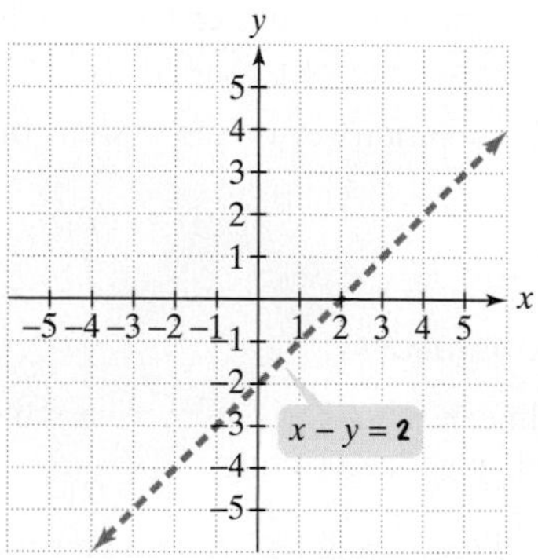

In Exercises 83–84, write a system of inequalities for each graph.

83.

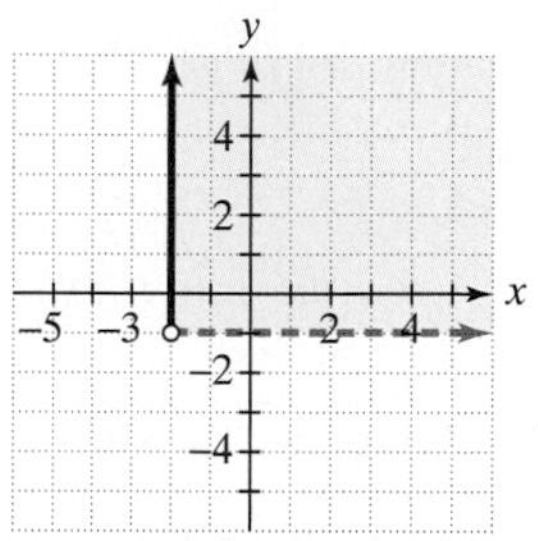

84.

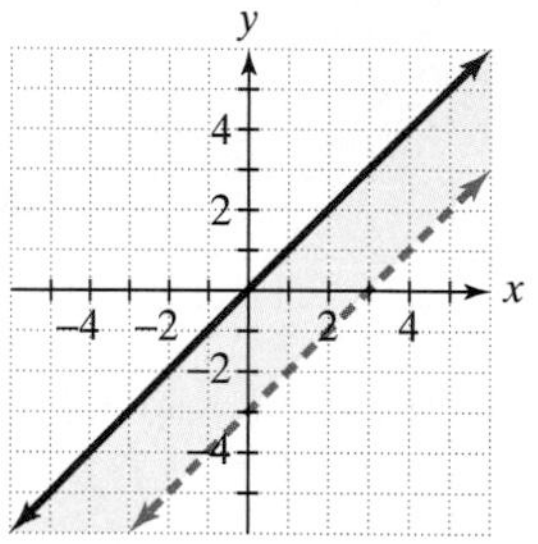

85. Write a system of inequalities whose solution set includes every point in the rectangular coordinate system.

86. Sketch the graph of the solution set for the following system of inequalities:

$$y \geq nx + b \ (n < 0, b > 0)$$

$$y \leq mx + b \ (m > 0, b > 0).$$

Review Exercises

87. Solve the system:

$$3x - y = 8$$

$$x - 5y = -2.$$

(Section 4.3, Example 2)

88. Solve by graphing:

$$y = 3x - 2$$

$$y = -2x + 8.$$

(Section 4.1, Example 3)

89. Factor completely: $2x^6 + 20x^5y + 50x^4y^2$.
(Section 6.5, Example 8)

SECTION 9.5

Objectives

1 Write an objective function describing a quantity that must be maximized or minimized.

2 Use inequalities to describe limitations in a situation.

3 Use linear programming to solve problems.

LINEAR PROGRAMMING

West Berlin children at Tempelhof Airport watch fleets of U.S. airplanes bringing in supplies to circumvent the Soviet blockade. The airlift began June 28, 1948 and continued for 15 months.

The Berlin Airlift (1948–1949) was an operation by the United States and Great Britain in response to military action by the former Soviet Union: Soviet troops closed all roads and rail lines between West Germany and Berlin, cutting off supply routes to the city. The Allies used a mathematical technique developed during World War II to maximize the amount of supplies transported. During the 15-month airlift, 278,228 flights provided basic necessities to blockaded Berlin, saving one of the world's great cities.

In this section, we will look at an important application of systems of linear inequalities. Such systems arise in **linear programming**, a method for solving problems in which a particular quantity that must be maximized or minimized is limited by other factors. Linear programming is one of the most widely used tools in management science. It helps businesses allocate resources to manufacture products in a way that will maximize profit. Linear programming accounts for more than 50% and perhaps as much as 90% of all computing time used for management decisions in business. The Allies used linear programming to save Berlin.

1 Write an objective function describing a quantity that must be maximized or minimized.

Objective Functions in Linear Programming Many problems involve quantities that must be maximized or minimized. Businesses are interested in maximizing profit. An operation in which bottled water and medical kits are shipped to earthquake victims needs to maximize the number of victims helped by this shipment. An **objective function** is an algebraic expression in two or more variables describing a quantity that must be maximized or minimized.

EXAMPLE 1 Writing an Objective Function

Bottled water and medical supplies are to be shipped to victims of an earthquake by plane. Each container of bottled water will serve 10 people and each medical kit will aid 6 people. If x represents the number of bottles of water to be shipped and y represents the number of medical kits, write the objective function that describes the number of people that can be helped.

SOLUTION Because each bottle of water serves 10 people and each medical kit aids 6 people, we have

The number of people helped	is	10 times the number of bottles of water	plus	6 times the number of medical kits.
	$=$	$10x$	$+$	$6y.$

Using z to represent the number of people helped, the objective function is

$$z = 10x + 6y.$$

Unlike the functions that we have seen so far, the objective function is an equation in three variables. For a value of x and a value of y, there is one and only one value of z. Thus, z is a function of x and y. ■

CHECK POINT 1 A company manufactures bookshelves and desks for computers. Let x represent the number of bookshelves manufactured daily and y the number of desks manufactured daily. The company's profits are \$25 per bookshelf and \$55 per desk. Write the objective function that describes the company's total daily profit, z, from x bookshelves and y desks. (Check Points 2 through 4 are related to this situation, so keep track of your answers.)

2 Use inequalities to describe limitations in a situation.

Constraints in Linear Programming Ideally, the number of earthquake victims helped in Example 1 should increase without restriction so that every victim receives water and medical kits. However, the planes that ship these supplies are subject to weight and volume restrictions. In linear programming problems, such restrictions are called **constraints**. Each constraint is expressed as a linear inequality. The list of constraints forms a system of linear inequalities.

EXAMPLE 2 Writing a Constraint

Each plane can carry no more than 80,000 pounds. The bottled water weighs 20 pounds per container and each medical kit weighs 10 pounds. Let x represent the number of bottles of water to be shipped and y the number of medical kits. Write an inequality that describes this constraint.

SOLUTION Because each plane can carry no more than 80,000 pounds, we have

The total weight of the water bottles	plus	the total weight of the medical kits	must be less than or equal to	80,000 pounds.
$20x$	$+$	$10y$	$\leq$	80,000.
Each bottle weighs 20 pounds.		Each kit weighs 10 pounds.		

The plane's weight constraint is described by the inequality

$$20x + 10y \leq 80{,}000.$$ ■

CHECK POINT 2 To maintain high quality, the company in Check Point 1 should not manufacture more than a total of 80 bookshelves and desks per day. Write an inequality that describes this constraint.

In addition to a weight constraint on its cargo, each plane has a limited amount of space in which to carry supplies. Example 3 demonstrates how to express this constraint.

EXAMPLE 3 Writing a Constraint

Planes can carry a total volume for supplies that does not exceed 6000 cubic feet. Each water bottle is 1 cubic foot and each medical kit also has a volume of 1 cubic foot. With x still representing the number of water bottles and y the number of medical kits, write an inequality that describes this second constraint.

SOLUTION Because each plane can carry a volume of supplies that does not exceed 6000 cubic feet, we have

The total volume of the water	plus	the total volume of the medical kits	must be less than or equal to	6000 cubic feet.
$1x$	$+$	$1y$	$\leq$	$6000.$
Each bottle is 1 cubic foot.		Each kit is 1 cubic foot.		

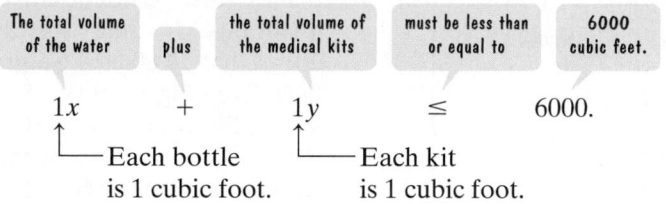

The plane's volume constraint is described by the inequality $x + y \leq 6000$. ■

In summary, here's what we have described so far in this aid-to-earthquake-victims situation:

$z = 10x + 6y$ — *This is the objective function describing the number of people helped with x bottles of water and y medical kits.*

$$20x + 10y \leq 80{,}000$$
$$x + y \leq 6000.$$

These are the constraints based on each plane's weight and volume limitations.

✔ **CHECK POINT 3** To meet customer demand, the company in Check Point 1 must manufacture between 30 and 80 bookshelves per day, inclusive. Furthermore, the company must manufacture at least 10 and no more than 30 desks per day. Write an inequality that describes each of these sentences. Then summarize what you have described about this company by writing the objective function for its profits, and the three constraints.

3 Use linear programming to solve problems.

Solving Problems with Linear Programming The problem in the earthquake situation described previously is to maximize the number of victims who can be helped, subject to the planes' weight and volume constraints. The process of solving this problem is called *linear programming*, based on a theorem that was proven during World War II.

SOLVING A LINEAR PROGRAMMING PROBLEM Let $z = ax + by$ be an objective function that depends on x and y. Furthermore, z is subject to a number of constraints on x and y. If a maximum or minimum value of z exists, it can be determined as follows:

1. Graph the system of inequalities representing the constraints.
2. Find the value of the objective function at each corner, or **vertex**, of the graphed region. The maximum and minimum of the objective function occur at one or more of the corner points.

EXAMPLE 4 Solving a Linear Programming Problem

Determine how many bottles of water and how many medical kits should be sent on each plane to maximize the number of earthquake victims who can be helped.

SOLUTION We must maximize $z = 10x + 6y$ subject to the following constraints:

$$20x + 10y \leq 80{,}000$$
$$x + y \leq 6000.$$

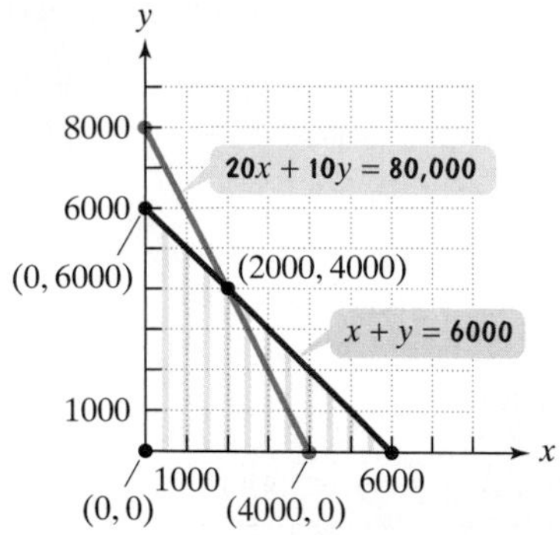

FIGURE 9.19 The region in quadrant I representing the constraints $20x + 10y \leq 80{,}000$ and $x + y \leq 6000$

Step 1. **Graph the system of inequalities representing the constraints.** Because x (the number of bottles of water per plane) and y (the number of medical kits per plane) must be nonnegative, we need to graph the system of inequalities in quadrant I and its boundary only. To graph the inequality $20x + 10y \leq 80{,}000$, we graph the equation $20x + 10y = 80{,}000$ as a solid blue line (Figure 9.19). Setting $y = 0$, the x-intercept is 4000 and setting $x = 0$, the y-intercept is 8000. Using $(0, 0)$ as a test point, the inequality is satisfied, so we shade below the blue line, as shown in yellow in Figure 9.19. Now we graph $x + y \leq 6000$ by first graphing $x + y = 6000$ as a solid red line. Setting $y = 0$, the x-intercept is 6000. Setting $x = 0$, the y-intercept is 6000. Using $(0, 0)$ as a test point, the inequality is satisfied, so we shade below the red line, as shown using green vertical shading in Figure 9.19.

We use the addition method to find where the lines $20x + 10y = 80{,}000$ and $x + y = 6000$ intersect.

$$\begin{array}{rcl} 20x + 10y = 80{,}000 & \xrightarrow{\text{No change}} & 20x + 10y = 80{,}000 \\ x + y = 6000 & \xrightarrow{\text{Multiply by } -10.} & \underline{-10x - 10y = -60{,}000} \\ & \text{Add:} & 10x = 20{,}000 \\ & & x = 2000 \end{array}$$

Back-substituting 2000 for x in $x + y = 6000$, we find $y = 4000$, so the intersection point is $(2000, 4000)$.

The system of inequalities representing the constraints is shown by the region in which the yellow shading and the green vertical shading overlap in Figure 9.19. The graph of the system of inequalities is shown again in Figure 9.20. The red and blue line segments are included in the graph.

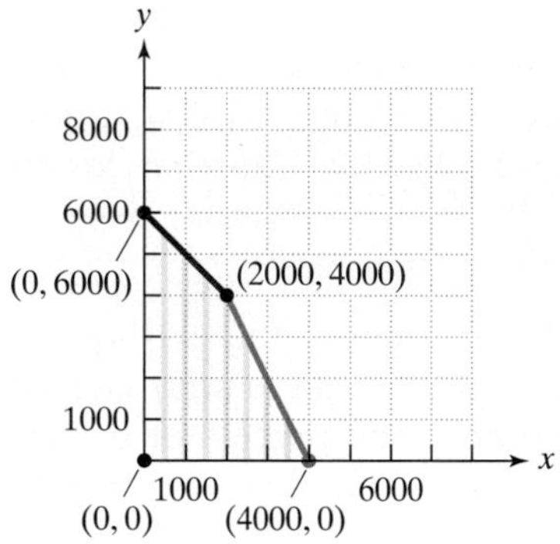

FIGURE 9.20

Step 2. **Find the value of the objective function at each corner of the graphed region. The maximum and minimum of the objective function occur at one or more of the corner points.** We must evaluate the objective function, $z = 10x + 6y$, at the four corners, or vertices, of the region in Figure 9.20.

Corner (x, y)	**Objective Function** $z = 10x + 6y$
$(0, 0)$	$z = 10(0) + 6(0) = 0$
$(4000, 0)$	$z = 10(4000) + 6(0) = 40{,}000$
$(2000, 4000)$	$z = 10(2000) + 6(4000) = 44{,}000$ ← maximum
$(0, 6000)$	$z = 10(0) + 6(6000) = 36{,}000$

Thus, the maximum value of z is 44,000 and this occurs when $x = 2000$ and $y = 4000$. In practical terms, this means that the maximum number of earthquake victims who can be helped with each plane shipment is 44,000. This can be accomplished by sending 2000 water bottles and 4000 medical kits per plane. ■

CHECK POINT 4 For the company in Check Points 1–3, how many bookshelves and how many desks should be manufactured per day to obtain maximum profit? What is the maximum daily profit?

EXAMPLE 5 Solving a Linear Programming Problem

Find the maximum value of the objective function

$$z = 2x + y$$

subject to the following constraints:

$$x \geq 0, y \geq 0$$
$$x + 2y \leq 5$$
$$x - y \leq 2.$$

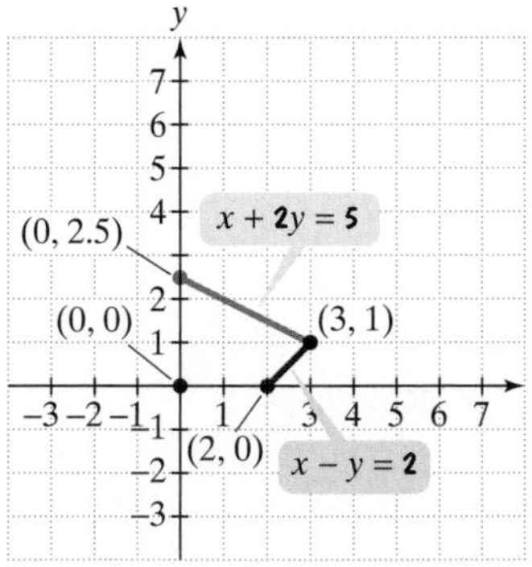

FIGURE 9.21 The graph of $x + 2y \leq 5$ and $x - y \leq 2$ in quadrant I

SOLUTION We begin by graphing the region in quadrant I ($x \geq 0$, $y \geq 0$) formed by the constraints. The graph is shown in Figure 9.21.

Now we evaluate the objective function at the four vertices of this region.

Objective function: $z = 2x + y$

At (0, 0): $z = 2 \cdot 0 + 0 = 0$

At (2, 0): $z = 2 \cdot 2 + 0 = 4$

At (3, 1): $z = 2 \cdot 3 + 1 = 7$ — Maximum value of z

At (0, 2.5): $z = 2 \cdot 0 + 2.5 = 2.5$

Thus, the maximum value of z is 7, and this occurs when $x = 3$ and $y = 1$. ■

We can see why the objective function in Example 5 has a maximum value that occurs at a vertex by solving the equation for y.

$$z = 2x + y$$ This is the objective function of Example 5.

$$y = -2x + z$$ Solve for y. Recall that the slope-intercept form of a line is $y = mx + b$.

Slope = −2 ; y-intercept = z

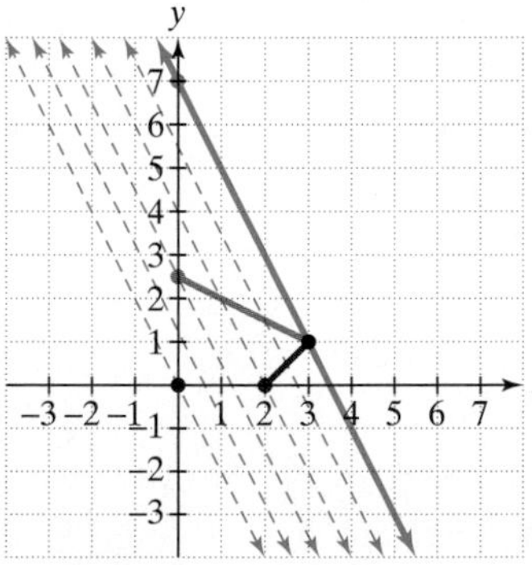

FIGURE 9.22 The line with slope −2 with the greatest y-intercept that intersects the shaded region passes through one of its vertices.

In this form, z represents the y-intercept of the objective function. The equation describes infinitely many parallel lines, each with slope −2. The process in linear programming involves finding the maximum z-value for all lines that intersect the region determined by the constraints. Of all the lines whose slope is −2, we're looking for the one with the greatest y-intercept that intersects the given region. As we see in Figure 9.22, such a line will pass through one (or possibly more) of the vertices of the region.

CHECK POINT 5 Find the maximum value of the objective function $z = 3x + 5y$ subject to the constraints $x \geq 0$, $y \geq 0$, $x + y \geq 1$, $x + y \leq 6$.

9.5 EXERCISE SET

Student Solutions Manual | CD/Video | PH Math/Tutor Center | MathXL Tutorials on CD | MathXL® | MyMathLab | Interactmath.com

Practice Exercises

In Exercises 1–4, find the value of the objective function at each corner of the graphed region. What is the maximum value of the objective function? What is the minimum value of the objective function?

1. Objective Function
$z = 5x + 6y$

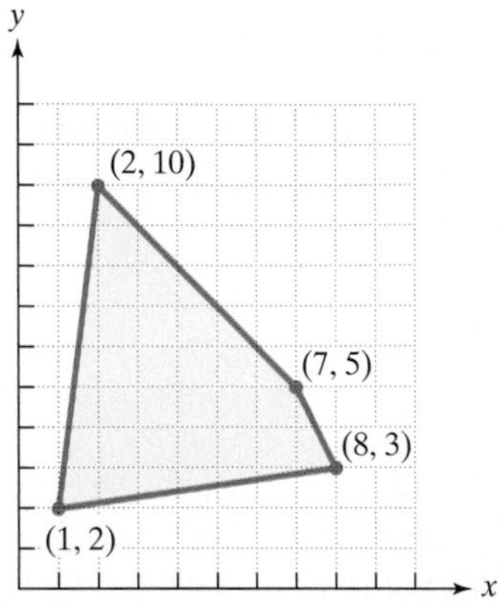

2. Objective Function
$z = 3x + 2y$

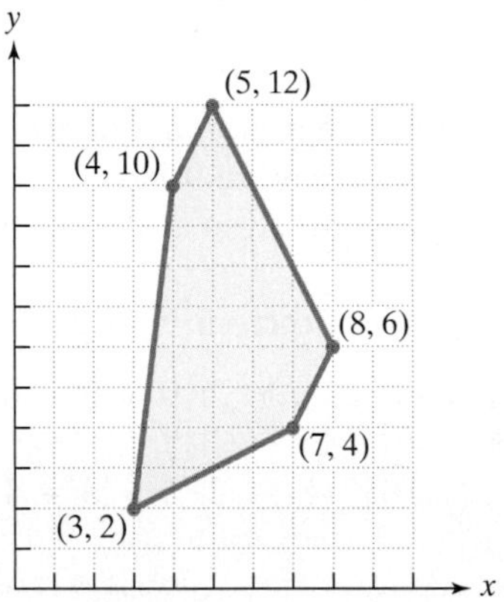

3. Objective Function
$z = 40x + 50y$

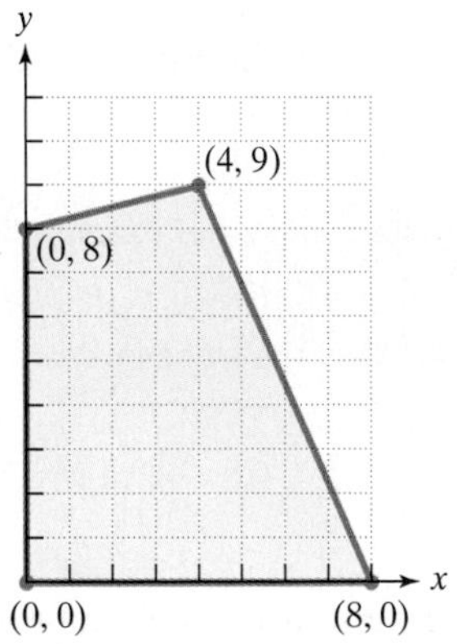

4. Objective Function
$z = 30x + 45y$

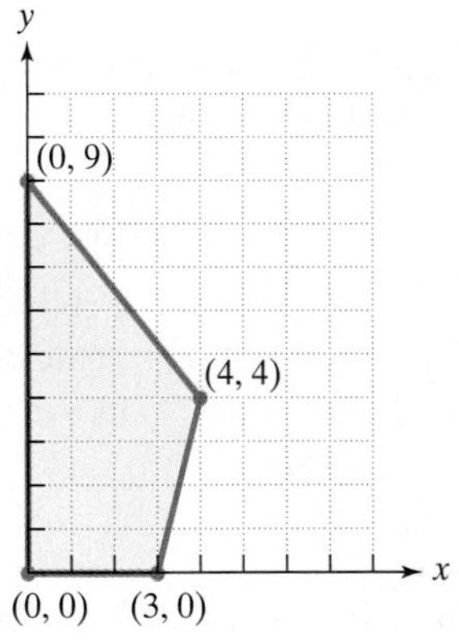

In Exercises 5–14, an objective function and a system of linear inequalities representing constraints are given.

a. *Graph the system of inequalities representing the constraints.*

b. *Find the value of the objective function at each corner of the graphed region.*

c. *Use the values in part (b) to determine the maximum value of the objective function and the values of x and y for which the maximum occurs.*

5. Objective Function $z = 3x + 2y$
Constraints
$x \geq 0, y \geq 0$
$2x + y \leq 8$
$x + y \geq 4$

6. Objective Function $z = 2x + 3y$
Constraints
$x \geq 0, y \geq 0$
$2x + y \leq 8$
$2x + 3y \leq 12$

7. Objective Function $z = 4x + y$
Constraints
$x \geq 0, y \geq 0$
$2x + 3y \leq 12$
$x + y \geq 3$

8. Objective Function $z = x + 6y$
Constraints
$x \geq 0, y \geq 0$
$2x + y \leq 10$
$x - 2y \geq -10$

9. Objective Function $z = 3x - 2y$
Constraints
$1 \leq x \leq 5$
$y \geq 2$
$x - y \geq -3$

10. Objective Function $z = 5x - 2y$
Constraints
$0 \leq x \leq 5$
$0 \leq y \leq 3$
$x + y \geq 2$

11. Objective Function $z = 4x + 2y$
Constraints
$x \geq 0, y \geq 0$
$2x + 3y \leq 12$
$3x + 2y \leq 12$
$x + y \geq 2$

12. Objective Function $z = 2x + 4y$
Constraints
$x \geq 0, y \geq 0$
$x + 3y \geq 6$
$x + y \geq 3$
$x + y \leq 9$

13. Objective Function $z = 10x + 12y$

Constraints
$$\begin{aligned} x \geq 0,\ y &\geq 0 \\ x + y &\leq 7 \\ 2x + y &\leq 10 \\ 2x + 3y &\leq 18 \end{aligned}$$

14. Objective Function $z = 5x + 6y$

Constraints
$$\begin{aligned} x \geq 0,\ y &\geq 0 \\ 2x + y &\geq 10 \\ x + 2y &\geq 10 \\ x + y &\leq 10 \end{aligned}$$

Application Exercises

15. A television manufacturer makes console and wide-screen televisions. The profit per unit is \$125 for the console televisions and \$200 for the wide-screen televisions.

a. Let $x =$ the number of consoles manufactured in a month and $y =$ the number of wide-screens manufactured in a month. Write the objective function that describes the total monthly profit.

b. The manufacturer is bound by the following constraints:

- Equipment in the factory allows for making at most 450 console televisions in one month.
- Equipment in the factory allows for making at most 200 wide-screen televisions in one month.
- The cost to the manufacturer per unit is \$600 for the console televisions and \$900 for the wide-screen televisions. Total monthly costs cannot exceed \$360,000.

Write a system of three inequalities that describes these constraints.

c. Graph the system of inequalities in part (b). Use only the first quadrant and its boundary, because x and y must both be nonnegative.

d. Evaluate the objective function for total monthly profit at each of the five vertices of the graphed region. [The vertices should occur at $(0, 0)$, $(0, 200)$, $(300, 200)$, $(450, 100)$, and $(450, 0)$.]

e. Complete the missing portions of this statement: The television manufacturer will make the greatest profit by manufacturing ___ console televisions each month and ___ wide-screen televisions each month. The maximum monthly profit is \$_____.

16. a. A student earns \$10 per hour for tutoring and \$7 per hour as a teacher's aid. Let $x =$ the number of hours each week spent tutoring and $y =$ the number of hours each week spent as a teacher's aid. Write the objective function that describes total weekly earnings.

b. The student is bound by the following constraints:

- To have enough time for studies, the student can work no more than 20 hours a week.
- The tutoring center requires that each tutor spend at least three hours a week tutoring.
- The tutoring center requires that each tutor spend no more than eight hours a week tutoring.

Write a system of three inequalities that describes these constraints.

c. Graph the system of inequalities in part (b). Use only the first quadrant and its boundary, because x and y are nonnegative.

d. Evaluate the objective function for total weekly earnings at each of the four vertices of the graphed region. [The vertices should occur at $(3, 0)$, $(8, 0)$, $(3, 17)$, and $(8, 12)$.]

e. Complete the missing portions of this statement: The student can earn the maximum amount per week by tutoring for _ hours per week and working as a teacher's aid for __ hours per week. The maximum amount that the student can earn each week is \$___.

Use the two steps for solving a linear programming problem, given in the box on page 623, to solve the problems in Exercises 17–23.

17. A manufacturer produces two models of mountain bicycles. The times (in hours) required for assembling and painting each model are given in the following table:

	Model A	Model B
Assembling	5	4
Painting	2	3

The maximum total weekly hours available in the assembly department and the paint department are 200 hours and 108 hours, respectively. The profits per unit are \$25 for model A and \$15 for model B. How many of each type should be produced to maximize profit?

18. A large institution is preparing lunch menus containing foods A and B. The specifications for the two foods are given in the following table:

Food	Units of Fat per Ounce	Units of Carbohydrates per Ounce	Units of Protein per Ounce
A	1	2	1
B	1	1	1

Each lunch must provide at least 6 units of fat per serving, no more than 7 units of protein, and at least 10 units of carbohydrates. The institution can purchase food A for \$0.12 per ounce and food B for \$0.08 per ounce. How many ounces of each food should a serving contain to meet the dietary requirements at the least cost?

19. Food and clothing are shipped to victims of a hurricane. Each carton of food will feed 12 people, while each carton of clothing will help 5 people. Each 20-cubic-foot box of food weighs 50 pounds and each 10-cubic-foot box of clothing weighs 20 pounds. The commercial carriers transporting food and clothing are bound by the following constraints:

- The total weight per carrier cannot exceed 19,000 pounds.
- The total volume must be less than 8000 cubic feet.

How many cartons of food and how many cartons of clothing should be sent with each plane shipment to maximize the number of people who can be helped?

20. On June 24, 1948, the former Soviet Union blocked all land and water routes through East Germany to Berlin. A gigantic airlift was organized using American and British planes to bring food, clothing, and other supplies to the more than 2 million people in West Berlin. The cargo capacity was 30,000 cubic feet for an American plane and 20,000 cubic feet for a British plane. To break the Soviet blockade, the Western Allies had to maximize cargo capacity, but were subject to the following restrictions:

- No more than 44 planes could be used.
- The larger American planes required 16 personnel per flight, double that of the requirement for the British planes. The total number of personnel available could not exceed 512.
- The cost of an American flight was \$9000 and the cost of a British flight was \$5000. Total weekly costs could not exceed \$300,000.

Find the number of American planes and the number of British planes that were used to maximize cargo capacity.

21. A theater is presenting a program on drinking and driving for students and their parents. The proceeds will be donated to a local alcohol information center. Admission is \$2.00 for parents and \$1.00 for students. However, the situation has two constraints: The theater can hold no more than 150 people and every two parents must bring at least one student. How many parents and students should attend to raise the maximum amount of money?

22. You are about to take a test that contains computation problems worth 6 points each and word problems worth 10 points each. You can do a computation problem in 2 minutes and a word problem in 4 minutes. You have 40 minutes to take the test and may answer no more than 12 problems. Assuming you answer all the problems attempted correctly, how many of each type of problem must you do to maximize your score? What is the maximum score?

23. In 1978, a ruling by the Civil Aeronautics Board allowed Federal Express to purchase larger aircraft. Federal Express's options included 20 Boeing 727s that United Airlines was retiring and/or the French-built Dassault Fanjet Falcon 20. To aid in their decision, executives at Federal Express analyzed the data at the top of the next column.

	Boeing 727	Falcon 20
Direct Operating Cost	\$1400 per hour	\$500 per hour
Payload	42,000 pounds	6000 pounds

Federal Express was faced with the following constraints:

- Hourly operating cost was limited to \$35,000.
- Total payload had to be at least 672,000 pounds.
- Only twenty 727s were available.

Given the constraints, how many of each kind of aircraft should Federal Express have purchased to maximize the number of aircraft?

Writing in Mathematics

24. What kinds of problems are solved using the linear programming method?

25. What is an objective function in a linear programming problem?

26. What is a constraint in a linear programming problem? How is a constraint represented?

27. In your own words, describe how to solve a linear programming problem.

28. Describe a situation in your life in which you would like to maximize something, but are limited by at least two constraints. Can linear programming be used in this situation? Explain your answer.

Critical Thinking Exercises

29. Suppose that you inherit \$10,000. The will states how you must invest the money. Some (or all) of the money must be invested in stocks and bonds. The requirements are that at least \$3000 be invested in bonds, with expected returns of \$0.08 per dollar, and at least \$2000 be invested in stocks, with expected returns of \$0.12 per dollar. Because the stocks are medium risk, the final stipulation requires that the investment in bonds should never be less than the investment in stocks. How should the money be invested so as to maximize your expected returns?

30. Consider the objective function $z = Ax + By$ ($A > 0$ and $B > 0$) subject to the following constraints: $2x + 3y \le 9$, $x - y \le 2$, $x \ge 0$, and $y \ge 0$. Prove that the objective function will have the same maximum value at the vertices $(3, 1)$ and $(0, 3)$ if $A = \frac{2}{3}B$.

Review Exercises

31. Solve: $x^2 - 12x + 36 = 0$
(Section 6.6, Example 4)

32. Divide: $\dfrac{1}{x^2 - 17x + 30} \div \dfrac{1}{x^2 + 7x - 18}$.
(Section 7.2, Example 6)

33. If $f(x) = x^3 + 2x^2 - 5x + 4$, find $f(-1)$.
(Section 8.1, Example 3)

GROUP PROJECT

CHAPTER 9

Each group member should research one situation that provides two different pricing options. These can involve areas such as public transportation options (with or without coupon books) or long-distance telephone plans or anything of interest. Be sure to bring in all the details for each option. At the group meeting, select the two pricing situations that are most interesting and relevant. Using each situation, write a word problem about selecting the better of the two options. The word problem should be one that can be solved using a linear inequality. The group should turn in the two problems and their solutions.

CHAPTER 9 SUMMARY

Definitions and Concepts	Examples

Section 9.1 Interval Notation and Business Applications Using Linear Inequalities

Definitions and Concepts:

A linear inequality in one variable can be written in the form $ax + b < c, ax + b \le c, ax + b > c$, or $ax + b \ge c$. The set of all numbers that make the inequality a true statement is its solution set. Graphs of solution sets are shown on a number line by shading all points representing numbers that are solutions. Parentheses indicate endpoints that are not solutions. Square brackets indicate endpoints that are solutions.

Examples:

- $(-2, 1] = \{x | -2 < x \le 1\}$

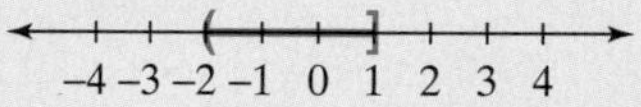

- $[-2, \infty) = \{x | x \ge -2\}$

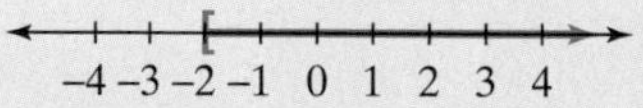

Definitions and Concepts:

Solving a Linear Inequality

1. Simplify each side.
2. Collect variable terms on one side and constant terms on the other side.
3. Isolate the variable and solve.

If an inequality is multiplied or divided by a negative number, the inequality symbol must be reversed.

Examples:

Solve: $2(x + 3) - 5x \le 15.$

$$2x + 6 - 5x \le 15$$

$$-3x + 6 \le 15$$

$$-3x \le 9$$

$$\frac{-3x}{-3} \ge \frac{9}{-3}$$

$$x \ge -3$$

Solution set: $\{x | x \ge -3\}$ or $[-3\ \infty)$

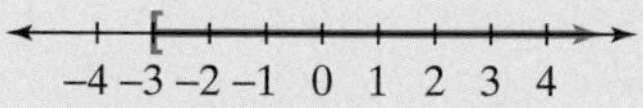

Definitions and Concepts	Examples

Section 9.1 Interval Notation and Business Applications Using Linear Inequalities (continued)

Functions of Business
A company produces and sells x units of a product.

Revenue Function

$$R(x) = (\text{price per unit sold})x$$

Cost Function

$$C(x) = \text{fixed cost} + (\text{cost per unit produced})x$$

Profit Function

$$P(x) = R(x) - C(x)$$

A business makes money, or has a profit, when $P(x) > 0$.
A business loses money, or has a loss, when $P(x) < 0$.

A company that manufactures lamps has a fixed cost of $80,000 and it costs $20 to produce each lamp. Lamps are sold for $70.

a. Write the cost function.

$$C(x) = 80{,}000 + 20x$$

Fixed cost

Variable cost: $20 per lamp

b. Write the revenue function.

$$R(x) = 70x$$

Revenue per lamp, $70, times number of lamps sold

c. Write the profit function.

$$P(x) = R(x) - C(x)$$
$$= 70x - (80{,}000 + 20x)$$
$$= 50x - 80{,}000$$

d. More than how many lamps must be produced and sold to have a profit?

Solve $P(x) > 0$.

$$50x - 80{,}000 > 0$$
$$50x > 80{,}000$$
$$x > 1600$$

More than 1600 lamps must be produced and sold to have a profit.

Section 9.2 Compound Inequalities

Intersection ($\cap$) and Union ($\cup$)
$A \cap B$ is the set of elements common to both set A and set B.
$A \cup B$ is the set of elements that are members of set A or set B or of both sets.

$\{1, 3, 5, 7\} \cap \{5, 7, 9, 11\} = \{5, 7\}$

$\{1, 3, 5, 7\} \cup \{5, 7, 9, 11\} = \{1, 3, 5, 7, 9, 11\}$

A compound inequality is formed by joining two inequalities with the word *and* or *or*. When the connecting word is *and*, graph each inequality separately and take the intersection of their solution sets.

Solve: $x + 1 > 3$ and $x + 4 \le 8$.

$x > 2$ and $x \le 4$

Solution set: $\{x|2 < x \le 4\}$ or $(2, 4]$

Definitions and Concepts	Examples

Section 9.2 Compound Inequalities (continued)

The compound inequality $a < x < b$ means $a < x$ and $x < b$. Solve by isolating the variable in the middle.

Solve: $-1 < \frac{2x+1}{3} \leq 2.$

$$-3 < 2x + 1 \leq 6 \quad \text{Multiply by 3.}$$

$$-4 < 2x \leq 5 \quad \text{Subtract 1.}$$

$$-2 < x \leq \frac{5}{2} \quad \text{Divide by 2.}$$

Solution set: $\left\{x \middle| -2 < x \leq \frac{5}{2}\right\}$ or $\left(-2, \frac{5}{2}\right]$

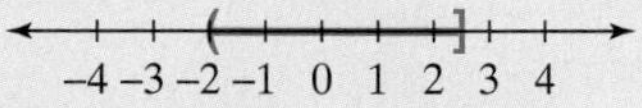

When the connecting word in a compound inequality is *or*, graph each inequality separately and take the union of their solution sets.

Solve: $x - 2 > -3$ or $2x \leq -6.$

$x > -1$ or $x \leq -3$

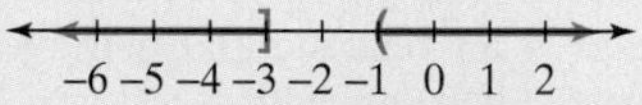

Solution set: $\{x | x \leq -3 \text{ or } x > -1\}$ or $(\infty, -3] \cup (-1, \infty)$

Section 9.3 Equations and Inequalities Involving Absolute Value

Absolute Value Equations

1. If $c > 0$, then $|X| = c$ means $X = c$ or $X = -c$.
2. If $c < 0$, then $|X| = c$ has no solution.
3. If $c = 0$, then $|X| = 0$ means $X = 0$.

Solve: $|2x - 7| = 3.$

$2x - 7 = 3$ or $2x - 7 = -3$

$2x = 10$ $\qquad$ $2x = 4$

$x = 5$ $\qquad$ $x = 2$

The solution set is $\{2, 5\}$.

Absolute Value Equations with Two Absolute Value Bars
If $|X_1| = |X_2|$, then $X_1 = X_2$ or $X_1 = -X_2$.

Solve: $|x - 6| = |2x + 1|.$

$x - 6 = 2x + 1$ or $x - 6 = -(2x + 1)$

$-x - 6 = 1$ $\qquad$ $x - 6 = -2x - 1$

$-x = 7$ $\qquad$ $3x - 6 = -1$

$x = -7$ $\qquad$ $3x = 5$

$$x = \frac{5}{3}$$

The solutions are -7 and $\frac{5}{3}$, and the solution set is $\left\{-7, \frac{5}{3}\right\}$.

If c is a positive number, then to solve $|X| < c$, solve the compound inequality $-c < X < c$. If c is negative, $|X| < c$ has no solution.

Solve: $|x - 4| < 3.$

$$-3 < x - 4 < 3$$

$$1 < x < 7$$

The solution set is $\{x | 1 < x < 7\}$ or $(1, 7)$.

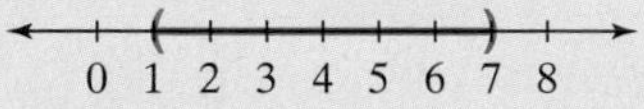

Definitions and Concepts	Examples

Section 9.3 Equations and Inequalities Involving Absolute Value (continued)

If c is a positive number, then to solve $|X| > c$, solve the compound inequality $X < -c$ *or* $X > c$. If c is negative, $|X| > c$ is true for all real numbers for which X is defined.

Solve: $\left|\frac{x}{3} - 1\right| \geq 2.$

$$\frac{x}{3} - 1 \leq -2 \quad \text{or} \quad \frac{x}{3} - 1 \geq 2.$$

$$x - 3 \leq -6 \quad \text{or} \quad x - 3 \geq 6 \qquad \text{Multiply by 3.}$$

$$x \leq -3 \quad \text{or} \quad x \geq 9 \qquad \text{Add 3.}$$

The solution set is $\{x|x \leq -3 \text{ or } x \geq 9\}$ or $(-\infty, -3] \cup [9, \infty)$.

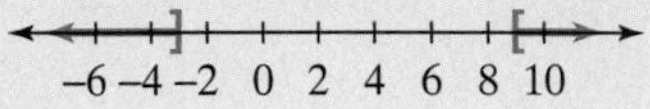

Section 9.4 Linear Inequalities in Two Variables

If the equal sign in $Ax + By = C$ is replaced with an inequality symbol, the result is a linear inequality in two variables. Its graph is the set of all points whose coordinates satisfy the inequality. To obtain the graph,

1. Replace the inequality symbol with an equal sign and graph the boundary line. Use a solid line for $\leq$ or $\geq$ and a dashed line for $<$ or $>$.
2. Choose a test point not on the line and substitute its coordinates into the inequality.
3. If a true statement results, shade the half-plane containing the test point. If a false statement results, shade the half-plane not containing the test point.

Graph: $x - 2y \leq 4$.

1. Graph $x - 2y = 4$. Use a solid line because the inequality symbol is $\leq$.
2. Test $(0, 0)$.

$$x - 2y \leq 4$$

$$0 - 2 \cdot 0 \overset{?}{\leq} 4$$

$$0 \leq 4, \qquad \text{true}$$

3. The inequality is true. Shade the half-plane containing $(0, 0)$.

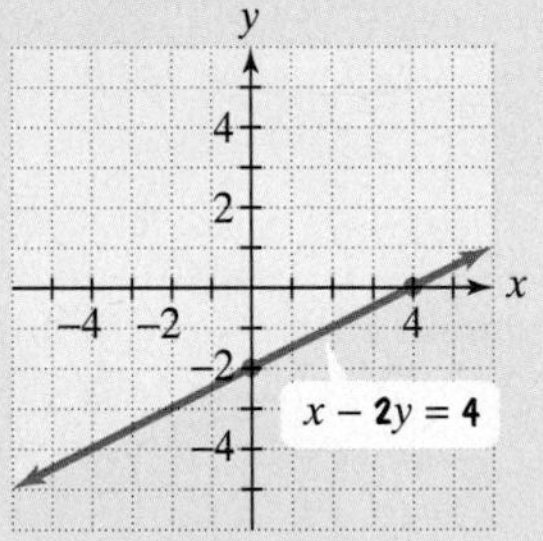

Two or more linear inequalities make up a system of linear inequalities. A solution is an ordered pair satisfying all inequalities in the system. To graph a system of inequalities, graph each inequality in the system. The overlapping region represents the solutions of the system.

Graph the solutions of the system:

$$y \leq -2x$$

$$x - y \geq 3.$$

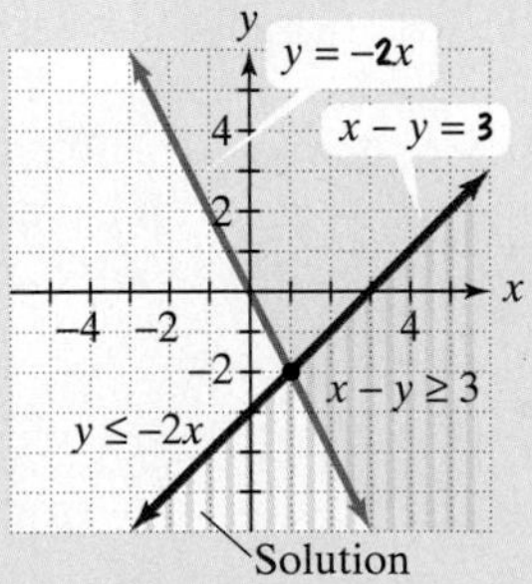

Definitions and Concepts	Examples

Section 9.5 Linear Programming

Linear programming is a method for solving problems in which a particular quantity that must be maximized or minimized is limited. An objective function is an algebraic expression in three variables describing a quantity that must be maximized or minimized. Constraints are restrictions, expressed as linear inequalities.

Solving a Linear Programming Problem

1. Graph the system of inequalities representing the constraints.
2. Find the value of the objective function at each corner, or vertex, of the graphed region. The maximum and minimum of the objective function occur at one or more vertices.

Find the maximum value of the objective function $z = 3x + 2y$ subject to the following constraints: $x \geq 0, y \geq 0, 2x + 3y \leq 18, 2x + y \leq 10$.

1. Graph the system of inequalities representing the constraints.

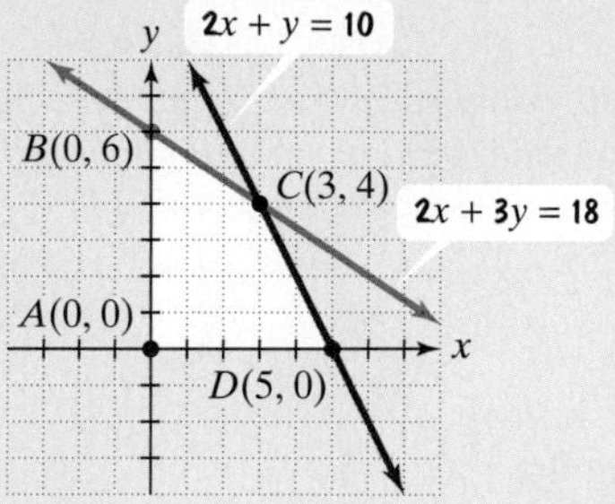

2. Evaluate the objective function at each vertex.

Vertex	$z = 3x + 2y$
$A(0, 0)$	$z = 3(0) + 2(0) = 0$
$B(0, 6)$	$z = 3(0) + 2(6) = 12$
$C(3, 4)$	$z = 3(3) + 2(4) = 17$
$D(5, 0)$	$z = 3(5) + 2(0) = 15$

The maximum value of the objective function is 17.

CHAPTER 9 REVIEW EXERCISES

9.1

In Exercises 1–3, express each interval in set-builder notation and graph the interval on a number line.

1. $(-2, 3]$

2. $[-1.5, 2]$

3. $[-1, \infty)$

In Exercises 4–8, solve each linear inequality and graph the solution set on a number line. Express the solution set in both set-builder and interval notations.

4. $-6x + 3 \leq 15$

5. $6x - 9 \geq -4x - 3$

6. $\frac{x}{3} - \frac{3}{4} - 1 > \frac{x}{2}$

7. $6x + 5 > -2(x - 3) - 25$

8. $3(2x - 1) - 2(x - 4) \geq 7 + 2(3 + 4x)$

9. The cost and revenue functions for producing and selling x units of a toaster oven are

$$C(x) = 40x + 357{,}000 \quad \text{and} \quad R(x) = 125x.$$

a. Write the profit function, P, from producing and selling x toaster ovens.

b. More than how many toaster ovens must be produced and sold to have profit?

Use this information to solve Exercises 10–13: A company is planning to produce and sell a new line of computers. The fixed cost will be \$360,000 and it will cost \$850 to produce each computer. Each computer will be sold for \$1150.

10. Write the cost function, C, of producing x computers.

11. Write the revenue function, R, from the sale of x computers.

12. Write the profit function, P, from producing and selling x computers.

13. More than how many computers must be produced and sold to have a profit?

14. A person can choose between two charges on a checking account. The first method involves a fixed cost of \$11 per month plus 6¢ for each check written. The second method involves a fixed cost of \$4 per month plus 20¢ for each check written. How many checks should be written to make the first method a better deal?

15. A salesperson earns \$500 per month plus a commission of 20% of sales. Describe the sales needed to receive a total income that exceeds \$3200 per month.

9.2 *In Exercises 16–19, let* $A = \{a, b, c\}$, $B = \{a, c, d, e\}$ *and* $C = \{a, d, f, g\}$. *Find the indicated set.*

16. $A \cap B$

17. $A \cap C$

18. $A \cup B$

19. $A \cup C$

In Exercises 20–30, solve each compound inequality. Except for the empty set, express the solution set in both set-builder and interval notations. Graph the solution set on a number line.

20. $x \le 3$ and $x < 6$

21. $x \le 3$ or $x < 6$

22. $-2x < -12$ and $x - 3 < 5$

23. $5x + 3 \le 18$ and $2x - 7 \le -5$

24. $2x - 5 > -1$ and $3x < 3$

25. $2x - 5 > -1$ or $3x < 3$

26. $x + 1 \le -3$ or $-4x + 3 < -5$

27. $5x - 2 \le -22$ or $-3x - 2 > 4$

28. $5x + 4 \ge -11$ or $1 - 4x \ge 9$

29. $-3 < x + 2 \le 4$

30. $-1 \le 4x + 2 \le 6$

31. On the first of four exams, your grades are 72, 73, 94, and 80. There is still one more exam, and you are hoping to earn a B in the course. This will occur if the average of your five exam grades is greater than or equal to 80 and less than 90. What range of grades on the fifth exam will result in receiving a B? Use interval notation to express this range.

9.3 *In Exercises 32–35, find the solution set for each equation.*

32. $|2x + 1| = 7$

33. $|3x + 2| = -5$

34. $2|x - 3| - 7 = 10$

35. $|4x - 3| = |7x + 9|$

In Exercises 36–39, solve and graph the solution set on a number line. Except for the empty set, express the solution set in both set-builder and interval notations.

36. $|2x + 3| \le 15$

37. $\left|\frac{2x + 6}{3}\right| > 2$

38. $|2x + 5| - 7 < -6$

39. $|2x - 3| + 4 \le -10$

40. Approximately 90% of the population sleeps h hours daily, where h is modeled by the inequality $|h - 6.5| \le 1$. Write a sentence describing the range for the number of hours that most people sleep. Do *not* use the phrase "absolute value" in your description.

9.4 *In Exercises 41–46, graph each inequality in a rectangular coordinate system.*

41. $3x - 4y > 12$

42. $x - 3y \le 6$

43. $y \le -\frac{1}{2}x + 2$

44. $y > \frac{3}{5}x$

45. $x \le 2$

46. $y > -3$

In Exercises 47–55, graph the solution set of each system of inequalities or indicate that the system has no solution.

47. $2x - y \le 4$
$x + y \ge 5$

48. $y < -x + 4$
$y > x - 4$

49. $-3 \le x < 5$

50. $-2 < y \le 6$

51. $x \ge 3$
$y \le 0$

52. $2x - y > -4$
$x \ge 0$

53. $x + y \le 6$
$y \ge 2x - 3$

54. $3x + 2y \ge 4$
$x - y \le 3$
$x \ge 0, y \ge 0$

55. $2x - y > 2$
$2x - y < -2$

9.5

56. Find the value of the objective function $z = 2x + 3y$ at each corner of the graphed region shown. What is the maximum value of the objective function? What is the minimum value of the objective function?

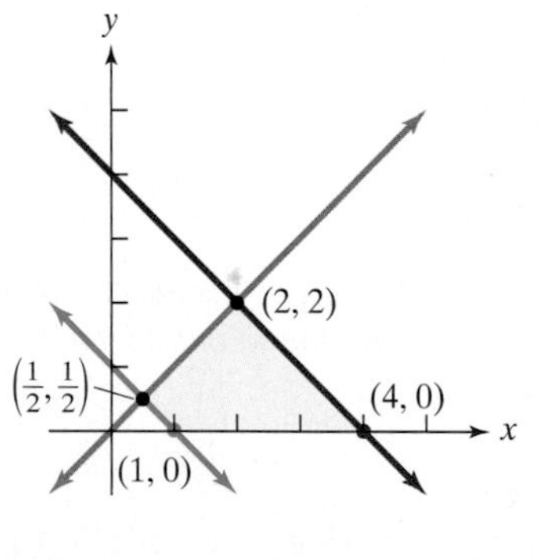

In Exercises 57–59, graph the region determined by the constraints. Then find the maximum value of the given objective function, subject to the constraints.

57. Objective Function $z = 2x + 3y$
Constraints
$x \ge 0, y \ge 0$
$x + y \le 8$
$3x + 2y \ge 6$

58. Objective Function $z = x + 4y$
Constraints $0 \le x \le 5, 0 \le y \le 7$
$x + y \ge 3$

59. Objective Function $z = 5x + 6y$
Constraints $x \ge 0, y \ge 0$
$y \le x$
$2x + y \le 12$
$2x + 3y \ge 6$

60. A paper manufacturing company converts wood pulp to writing paper and newsprint. The profit on a unit of writing paper is \$500 and the profit on a unit of newsprint is \$350.

a. Let x represent the number of units of writing paper produced daily. Let y represent the number of units of newsprint produced daily. Write the objective function that models total daily profit.

b. The manufacturer is bound by the following constraints:
- Equipment in the factory allows for making at most 200 units of paper (writing paper and newsprint) in a day.
- Regular customers require at least 10 units of writing paper and at least 80 units of newsprint daily.

Write a system of inequalities that models these constraints.

c. Graph the inequalities in part (b). Use only the first quadrant, because x and y must both be positive. (*Suggestion*: Let each unit along the x- and y-axes represent 20.)

d. Evaluate the objective profit function at each of the three vertices of the graphed region.

e. Complete the missing portions of this statement: The company will make the greatest profit by producing ___ units of writing paper and __ units of newsprint each day. The maximum daily profit is \$_____.

61. A manufacturer of lightweight tents makes two models whose specifications are given in the following table.

	Cutting Time per Tent	Assembly Time per Tent
Model A	0.9 hour	0.8 hour
Model B	1.8 hours	1.2 hours

Each month, the manufacturer has no more than 864 hours of labor available in the cutting department and at most 672 hours in the assembly division. The profits come to \$25 per tent for model A and \$40 per tent for model B. How many of each should be manufactured monthly to maximize the profit?

CHAPTER 9 TEST

Remember to use your Chapter Test Prep Video CD to see the worked-out solutions to the test questions you want to review.

In Exercises 1–2, express each interval in set-builder notation and graph the interval on a number line.

1. $[-3, 2)$ **2.** $(-\infty, -1]$

In Exercises 3–4, solve and graph the solution set on a number line. Express the solution set in both set-builder and interval notations.

3. $3(x + 4) \ge 5x - 12$

4. $\frac{x}{6} + \frac{1}{8} \le \frac{x}{2} - \frac{3}{4}$

5. A company is planning to manufacture computer desks. The fixed cost will be \$60,000 and it will cost \$200 to produce each desk. Each desk will be sold for \$450.

a. Write the cost function, C, of producing x desks.

b. Write the revenue function, R, from the sale of x desks.

c. Write the profit function, P, from producing and selling x desks.

d. More than how many desks must be produced and sold to have a profit?

6. Find the intersection: $\{2, 4, 6, 8, 10\} \cap \{4, 6, 12, 14\}$.

7. Find the union: $\{2, 4, 6, 8, 10\} \cup \{4, 6, 12, 14\}$.

In Exercises 8–12, solve each compound inequality. Except for the empty set, express the solution set in both set-builder and interval notations. Graph the solution set on a number line.

8. $2x + 4 < 2$ and $x - 3 > -5$

9. $x + 6 \ge 4$ and $2x + 3 \ge -2$

10. $2x - 3 < 5$ or $3x - 6 \le 4$

11. $x + 3 \le -1$ or $-4x + 3 < -5$

12. $-3 \le \frac{2x + 5}{3} < 6$

In Exercises 13–14, find the solution set for each equation.

13. $|5x + 3| = 7$

14. $|6x + 1| = |4x + 15|$

In Exercises 15–16, solve and graph the solution set on a number line. Express the solution set in both set-builder and interval notations.

15. $|2x - 1| < 7$

16. $|2x - 3| \ge 5$

17. The inequality $|b - 98.6| > 8$ describes a person's body temperature, b, in degrees Fahrenheit, when hyperthermia (extremely high body temperature) or hypothermia (extremely low body temperature) occurs. Solve the inequality and interpret the solution.

In Exercises 18–20, graph each inequality in a rectangular coordinate system.

18. $3x - 2y < 6$ **19.** $y \geq \frac{1}{2}x - 1$ **20.** $y \leq -1$

In Exercises 21–23, graph the solution set of each system of inequalities.

21. $x + y \geq 2$
$x - y \geq 4$

22. $3x + y \leq 9$
$2x + 3y \geq 6$
$x \geq 0, y \geq 0$

23. $-2 < x \leq 4$

24. Find the maximum value of the objective function $z = 3x + 5y$ subject to the following constraints: $x \geq 0, y \geq 0, x + y \leq 6, x \geq 2$.

25. A manufacturer makes two types of jet skis, regular and deluxe. The profit on a regular jet ski is \$200 and the profit on the deluxe model is \$250. To meet customer demand, the company must manufacture at least 50 regular jet skis per week and at least 75 deluxe models. To maintain high quality, the total number of both models of jet skis manufactured by the company should not exceed 150 per week. How many jet skis of each type should be manufactured per week to obtain maximum profit? What is the maximum weekly profit?

CUMULATIVE REVIEW EXERCISES (CHAPTERS 1–9)

In Exercises 1–2, solve each equation.

1. $5(x + 1) + 2 = x - 3(2x + 1)$

2. $\frac{2(x + 6)}{3} = 1 + \frac{4x - 7}{3}$

3. Simplify: $\frac{-10x^2y^4}{15x^7y^{-3}}$.

4. If $f(x) = x^2 - 3x + 4$, find $f(-3)$ and $f(2a)$.

5. If $f(x) = 3x^2 - 4x + 1$ and $g(x) = x^2 - 5x - 1$, find $(f - g)(x)$ and $(f - g)(2)$.

6. Use function notation to write the equation of the line passing through $(2, 3)$ and perpendicular to the line whose equation is $y = 2x - 3$.

In Exercises 7–10, graph each equation or inequality in a rectangular coordinate system.

7. $f(x) = 2x + 1$ **8.** $y > 2x$

9. $2x - y \geq 6$ **10.** $f(x) = -1$

11. Solve the system:

$$\begin{aligned} 3x - y + z &= -15 \\ x + 2y - z &= 1. \\ 2x + 3y - 2z &= 0 \end{aligned}$$

12. If $f(x) = \frac{x}{3} - 4$, find $f^{-1}(x)$.

13. If $f(x) = 3x^2 - 1$ and $g(x) = x + 2$, find $f(g(x))$ and $g(f(x))$.

14. A motel with 60 rooms charges \$90 per night for rooms with kitchen facilities and \$80 per night for rooms without kitchen facilities. When all rooms are occupied, the nightly revenue is \$5260. How many rooms of each kind are there?

15. Which of the following are functions?

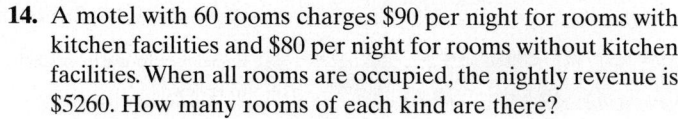

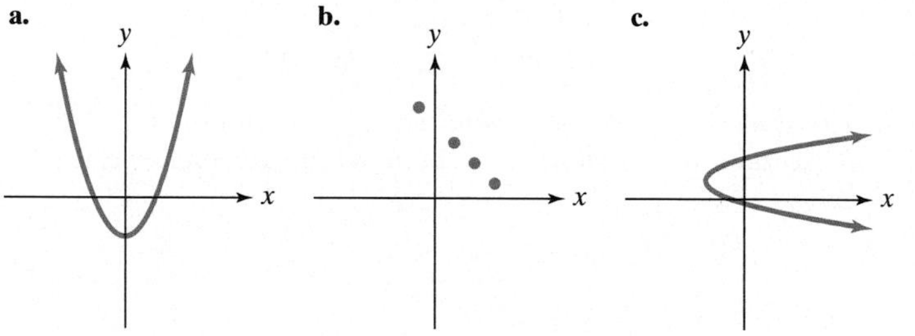

In Exercises 16–20, solve and graph the solution set on a number line. Express the solution set in both set-builder and interval notations.

16. $\frac{x}{4} - \frac{3}{4} - 1 \leq \frac{x}{2}$

17. $2x + 5 \leq 11$ and $-3x > 18$

18. $x - 4 \geq 1$ or $-3x + 1 \geq -5 - x$

19. $|2x + 3| \leq 17$

20. $|3x - 8| > 7$

You enjoy the single life. You have a terrific career, a support group of close friends, and your own home. Still, you feel the pressure to marry. Your parents belong to a generation that believes that you are not complete until you have found your "other half." They see your staying unattached as a "failure to marry." Perhaps you will someday enter into a permanent relationship, but at the moment you are satisfied with your life and career, and plan to stay single into the future.

This discussion is developed algebraically in the Section 10.6 opener and in Example 6 in Section 10.6.

CHAPTER 10

Radicals, Radical Functions, and Rational Exponents

An increase in the median age at which Americans first marry and the increased rate of divorce have contributed to the growth of our single population. Data indicate that the number of Americans living alone increased quite rapidly toward the end of the twentieth century. However, now this growth rate is beginning to slow down. In this chapter, you will see why radical functions with square roots are used to describe phenomena that are continuing to grow, but whose growth is leveling off. By learning about radicals and radical functions, you will have new algebraic tools for describing your world.

SECTION 10.1

Objectives

1. Evaluate square roots.
2. Evaluate square root functions.
3. Find the domain of square root functions.
4. Use models that are square root functions.
5. Simplify expressions of the form $\sqrt{a^2}$.
6. Evaluate cube root functions.
7. Simplify expressions of the form $\sqrt[3]{a^3}$.
8. Find even and odd roots.
9. Simplify expressions of the form $\sqrt[n]{a^n}$.

RADICAL EXPRESSIONS AND FUNCTIONS

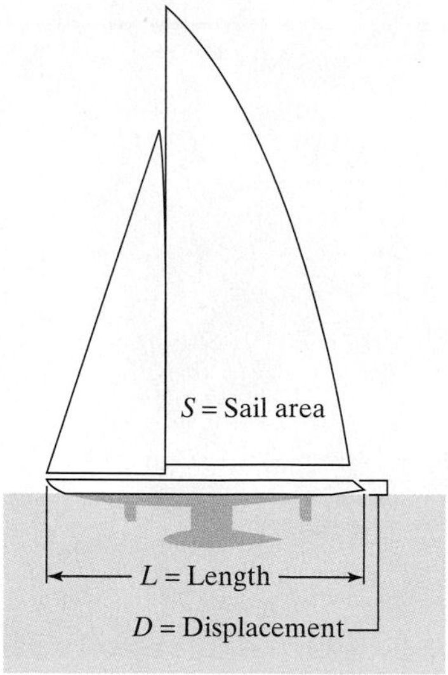

The America's Cup is the supreme event in ocean sailing. Competition is fierce and the costs are huge. Competitors look to mathematics to provide the critical innovation that can make the difference between winning and losing. The basic dimensions of competitors' yachts must satisfy an inequality containing square roots and cube roots:

$$L + 1.25\sqrt{S} - 9.8\sqrt[3]{D} \leq 16.296.$$

In the inequality, L is the yacht's length, in meters, S is its sail area, in square meters, and D is its displacement, in cubic meters.

In this section, we introduce a new category of expressions and functions that contain roots. You will see why square root functions are used to describe phenomena that are continuing to grow but whose growth is leveling off.

1 Evaluate square roots.

Square Roots From our earlier work with exponents, we are aware that the square of both 5 and -5 is 25:

$$5^2 = 25 \quad \text{and} \quad (-5)^2 = 25.$$

The reverse operation of squaring a number is finding the *square root* of the number. For example,

- One square root of 25 is 5 because $5^2 = 25$.
- Another square root of 25 is -5 because $(-5)^2 = 25$.

In general, **if $b^2 = a$, then b is a square root of a.**

The symbol $\sqrt{}$ is used to denote the *positive* or *principal square root* of a number. For example,

- $\sqrt{25} = 5$ because $5^2 = 25$ and 5 is positive.
- $\sqrt{100} = 10$ because $10^2 = 100$ and 10 is positive.

The symbol $\sqrt{}$ that we use to denote the principal square root is called a **radical sign**. The number under the radical sign is called the **radicand**. Together we refer to the radical sign and its radicand as a **radical expression**.

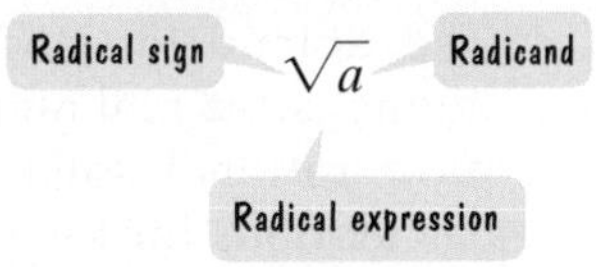

DEFINITION OF THE PRINCIPAL SQUARE ROOT If a is a nonnegative real number, the nonnegative number b such that $b^2 = a$, denoted by $b = \sqrt{a}$, is the **principal square root** of a.

The symbol $-\sqrt{}$ is used to denote the negative square root of a number. For example,

- $-\sqrt{25} = -5$ because $(-5)^2 = 25$ and -5 is negative.
- $-\sqrt{100} = -10$ because $(-10)^2 = 100$ and -10 is negative.

EXAMPLE 1 Evaluating Square Roots

Evaluate:

a. $\sqrt{81}$ **b.** $-\sqrt{9}$ **c.** $\sqrt{\frac{4}{49}}$

d. $\sqrt{0.0064}$ **e.** $\sqrt{36 + 64}$ **f.** $\sqrt{36} + \sqrt{64}$.

SOLUTION

a. $\sqrt{81} = 9$ — The principal square root of 81 is 9 because $9^2 = 81$.

b. $\sqrt{9} = -3$ — The negative square root of 9 is -3 because $(-3)^2 = 9$.

c. $\sqrt{\frac{4}{49}} = \frac{2}{7}$ — The principal square root of $\frac{4}{49}$ is $\frac{2}{7}$ because $\left(\frac{2}{7}\right)^2 = \frac{4}{49}$.

d. $\sqrt{0.0064} = 0.08$ — The principal square root of 0.0064 is 0.08 because $(0.08)^2 = (0.08)(0.08) = 0.0064$.

e. $\sqrt{36 + 64} = \sqrt{100}$ — Simplify the radicand.
$= 10$ — Take the principal square root of 100, which is 10.

f. $\sqrt{36} + \sqrt{64} = 6 + 8$ — $\sqrt{36} = 6$ because $6^2 = 36$. $\sqrt{64} = 8$
$= 14$ — because $8^2 = 64$. ■

STUDY TIP

In Example 1, parts (e) and (f), observe that $\sqrt{36 + 64}$ is not equal to $\sqrt{36} + \sqrt{64}$. In general,

$$\sqrt{a + b} \neq \sqrt{a} + \sqrt{b}$$

and

$$\sqrt{a - b} \neq \sqrt{a} - \sqrt{b}.$$

✔ CHECK POINT 1 Evaluate:

a. $\sqrt{64}$ **b.** $-\sqrt{49}$ **c.** $\sqrt{\frac{16}{25}}$

c. $\sqrt{0.0081}$ **e.** $\sqrt{9 + 16}$ **f.** $\sqrt{9} + \sqrt{16}$.

Let's see what happens to the radical expression $\sqrt{x}$ if x is a negative number. Is the square root of a negative number a real number? For example, consider $\sqrt{-25}$. Is there a real number whose square is -25? No. Thus, $\sqrt{-25}$ is not a real number. In general, **a square root of a negative number is not a real number**.

2 Evaluate square root functions.

Square Root Functions Because each nonnegative real number, x, has precisely one principal square root, $\sqrt{x}$, there is a **square root function** defined by

$$f(x) = \sqrt{x}.$$

The domain of this function is $[0, \infty)$. We can graph $f(x) = \sqrt{x}$ by selecting nonnegative real numbers for x. It is easiest to choose perfect squares, numbers that have rational square roots. Table 10.1 shows five such choices for x and the calculations for the corresponding outputs. We plot these ordered pairs as points in the rectangular coordinate system and connect the points with a smooth curve. The graph of $f(x) = \sqrt{x}$ is shown in Figure 10.1.

Table 10.1

x	$f(x) = \sqrt{x}$	(x, y) or $(x, f(x))$
0	$f(0) = \sqrt{0} = 0$	(0, 0)
1	$f(1) = \sqrt{1} = 1$	(1, 1)
4	$f(4) = \sqrt{4} = 2$	(4, 2)
9	$f(9) = \sqrt{9} = 3$	(9, 3)
16	$f(16) = \sqrt{16} = 4$	(16, 4)

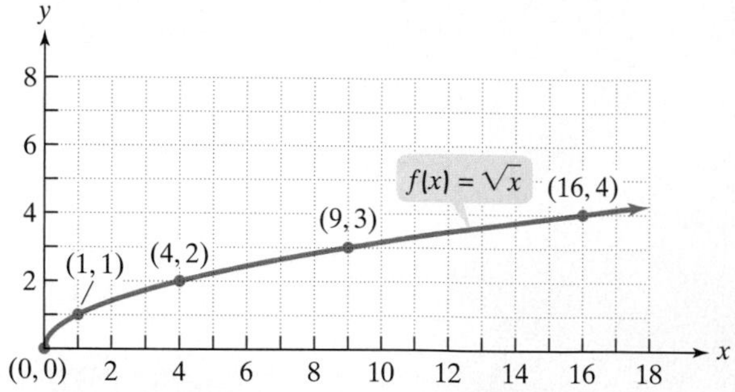

FIGURE 10.1 The graph of the square root function $f(x) = \sqrt{x}$

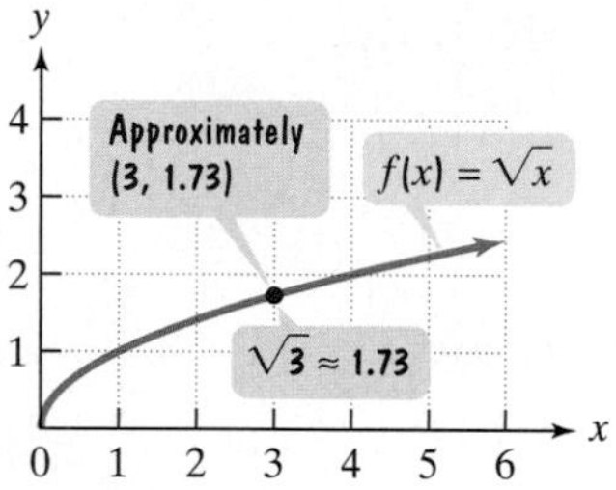

FIGURE 10.2 Visualizing $\sqrt{3}$ as a point on the graph of $f(x) = \sqrt{x}$

Is it possible to choose values of x for Table 10.1 that are not squares of integers, or perfect squares? Yes. For example, we can let $x = 3$. Thus, $f(3) = \sqrt{3}$. Because 3 is not a perfect square, $\sqrt{3}$ is an irrational number, one that cannot be expressed as a quotient of integers. We can use a calculator to find a decimal approximation of $\sqrt{3}$.

Many Scientific Calculators	**Many Graphing Calculators**
3 [√]	[√] 3 [ENTER]

Rounding the displayed number to two decimal places, $\sqrt{3} \approx 1.73$. This information is shown visually as a point, approximately (3, 1.73), on the graph of $f(x) = \sqrt{x}$ in Figure 10.2.

To evaluate a square root function, we use substitution, just as we did to evaluate other functions.

EXAMPLE 2 Evaluating Square Root Functions

For each function, find the indicated function value:

a. $f(x) = \sqrt{5x - 6}$; $f(2)$ **b.** $g(x) = -\sqrt{64 - 8x}$; $g(-3)$.

SOLUTION

a. $f(2) = \sqrt{5 \cdot 2 - 6}$ Substitute 2 for x in $f(x) = \sqrt{5x - 6}$.

$= \sqrt{4} = 2$ Simplify the radicand and take the square root.

b. $g(-3) = -\sqrt{64 - 8(-3)}$ Substitute -3 for x in $g(x) = -\sqrt{64 - 8x}$.

$= -\sqrt{88} \approx -9.38$ Simplify the radicand: $64 - 8(-3) = 64 - (-24) = 64 + 24 = 88$. Then use a calculator to approximate $\sqrt{88}$.

CHECK POINT 2 For each function, find the indicated function value:

a. $f(x) = \sqrt{12x - 20}$; $f(3)$

b. $g(x) = -\sqrt{9 - 3x}$; $g(-5)$.

3 Find the domain of square root functions.

We have seen that the domain of a function f is the largest set of real numbers for which the value of $f(x)$ is a real number. Because only nonnegative numbers have real square roots, the domain of a square root function is the set of real numbers for which the radicand is nonnegative.

EXAMPLE 3 Finding the Domain of a Square Root Function

Find the domain of

$$f(x) = \sqrt{3x + 12}.$$

SOLUTION The domain is the set of real numbers, x, for which the radicand, $3x + 12$, is nonnegative. We set the radicand greater than or equal to 0 and solve the resulting inequality.

$$3x + 12 \geq 0$$
$$3x \geq -12$$
$$x \geq -4$$

The domain of f is $\{x | x \geq -4\}$ or $[-4, \infty)$.

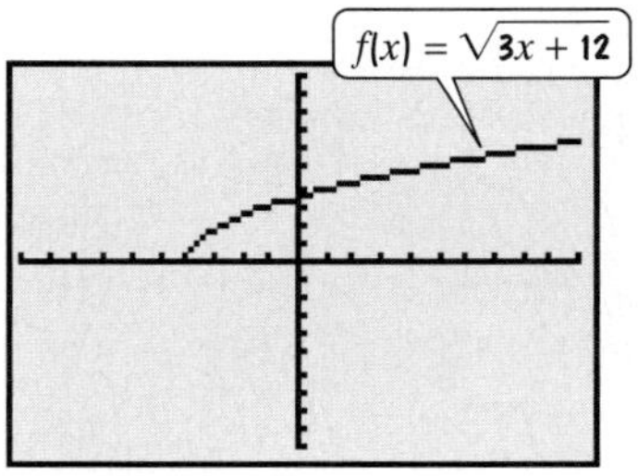

FIGURE 10.3

Figure 10.3 shows the graph of $f(x) = \sqrt{3x + 12}$ in a $[-10, 10, 1]$ by $[-10, 10, 1]$ viewing rectangle. The graph appears only for $x \geq -4$, verifying $[-4, \infty)$ as the domain. Can you see how the graph also illustrates this square root function's range? The graph only appears for nonnegative values of y. Thus, the range is $\{y | y \geq 0\}$ or $[0, \infty)$.

CHECK POINT 3 Find the domain of

$$f(x) = \sqrt{9x - 27}.$$

4 Use models that are square root functions.

The graph of the square root function $f(x) = \sqrt{x}$ is increasing from left to right. However, the rate of increase is slowing down as the graph moves to the right. This is why square root functions are often used to model growing phenomena with growth that is leveling off.

EXAMPLE 4 Modeling with a Square Root Function

The graph in Figure 10.4 shows the percentage of the U.S. population who used the Internet from 2000 through 2003. Although this percent grew from year to year, the growth was leveling off. The data can be modeled by the function

$$P(x) = 6.7\sqrt{x} + 44.4,$$

where $P(x)$ is the percentage of the U.S. population who used the Internet x years after 2000. According to the model, what percentage used the Internet in 2003? Round to the nearest tenth of a percent. How well does the model describe the actual data?

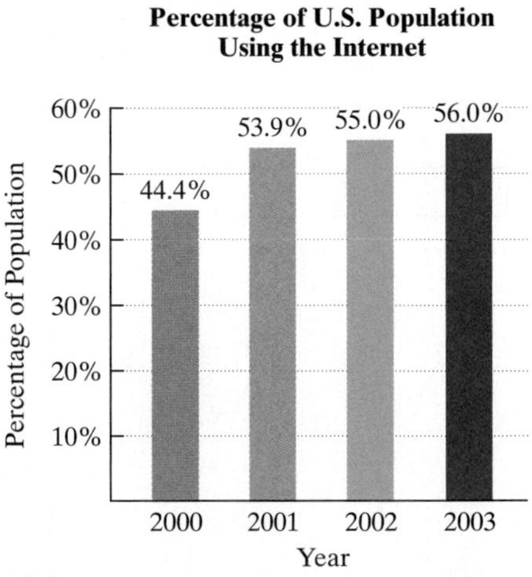

FIGURE 10.4

Source: U.S. Department of Commerce

SOLUTION Because 2003 is 3 years after 2000, we substitute 3 for x and evaluate the function at 3.

$$P(x) = 6.7\sqrt{x} + 44.4 \quad \text{Use the given function.}$$
$$P(3) = 6.7\sqrt{3} + 44.4 \quad \text{Substitute 3 for x.}$$
$$\approx 56.0 \quad \text{Use a calculator.}$$

The model indicates that approximately 56% of the U.S. population used the Internet in 2003. Figure 10.4 on the previous page shows 56%, so our model provides an excellent description of the actual data value for 2003. ■

CHECK POINT 4 Use the square root function in Example 4 to find the percentage of Internet users in 2002. Round to the nearest tenth of a percent. How well does the model describe the actual data?

5 Simplify expressions of the form $\sqrt{a^2}$.

Simplifying Expressions of the Form $\sqrt{a^2}$ You may think that $\sqrt{a^2} = a$. However, this is not necessarily true. Consider the following examples:

$$\sqrt{4^2} = \sqrt{16} = 4$$
$$\sqrt{(-4)^2} = \sqrt{16} = 4.$$

The result is not −4, but rather the absolute value of −4, or 4.

USING TECHNOLOGY

The graphs of

$f(x) = \sqrt{x^2}$ and $g(x) = |x|$

are shown in a $[-10, 10, 1]$ by $[-2, 10, 1]$ viewing rectangle. The graphs are the same. Thus,

$$\sqrt{x^2} = |x|.$$

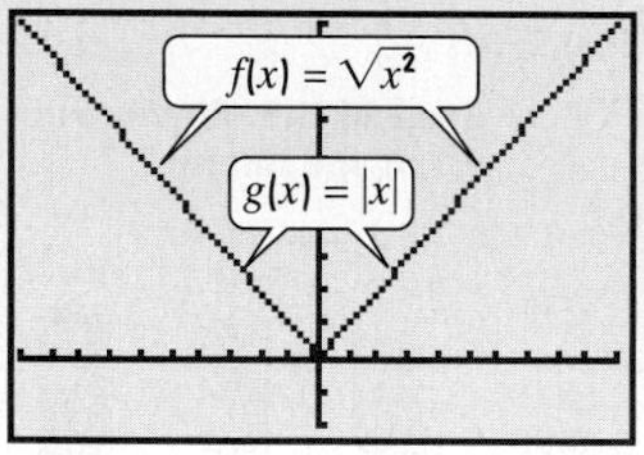

Here is a rule for simplifying expressions of the form $\sqrt{a^2}$:

SIMPLIFYING $\sqrt{a^2}$ For any real number a,

$$\sqrt{a^2} = |a|.$$

In words, the principal square root of a^2 is the absolute value of a.

EXAMPLE 5 Simplifying Radical Expressions

Simplify each expression:

a. $\sqrt{(-6)^2}$ **b.** $\sqrt{(x+5)^2}$ **c.** $\sqrt{25x^6}$ **d.** $\sqrt{x^2 - 4x + 4}$.

SOLUTION The principal square root of an expression squared is the absolute value of that expression. In parts (a) and (b), we are given squared radicands. In parts (c) and (d), it will first be necessary to express the radicand as an expression that is squared.

a. $\sqrt{(-6)^2} = |-6| = 6$

b. $\sqrt{(x+5)^2} = |x+5|$

c. To simplify $\sqrt{25x^6}$, first write $25x^6$ as an expression that is squared: $25x^6 = (5x^3)^2$. Then simplify.

$$\sqrt{25x^6} = \sqrt{(5x^3)^2} = |5x^3| \text{ or } 5|x^3|$$

d. To simplify $\sqrt{x^2 - 4x + 4}$, first write $x^2 - 4x + 4$ as an expression that is squared by factoring the perfect square trinomial: $x^2 - 4x + 4 = (x - 2)^2$. Then simplify.

$$\sqrt{x^2 - 4x + 4} = \sqrt{(x - 2)^2} = |x - 2|$$ ■

CHECK POINT 5 Simplify each expression:

a. $\sqrt{(-7)^2}$

b. $\sqrt{(x + 8)^2}$

c. $\sqrt{49x^{10}}$

d. $\sqrt{x^2 - 6x + 9}$.

In some situations, we are told that no radicands involve negative quantities raised to even powers. When the expression being squared is nonnegative, it is not necessary to use absolute value when simplifying $\sqrt{a^2}$. For example, assuming that no radicands contain negative quantities that are squared,

$$\sqrt{x^8} = \sqrt{(x^4)^2} = x^4$$

$$\sqrt{25x^2 + 10x + 1} = \sqrt{(5x + 1)^2} = 5x + 1.$$

6 Evaluate cube root functions.

Cube Roots and Cube Root Functions Finding the square root of a number reverses the process of squaring a number. Similarly, finding the cube root of a number reverses the process of cubing a number. For example, $2^3 = 8$, and so the cube root of 8 is 2. The notation that we use is $\sqrt[3]{8} = 2$.

DEFINITION OF THE CUBE ROOT OF A NUMBER The **cube root** of a real number a is written $\sqrt[3]{a}$.

$$\sqrt[3]{a} = b \quad \text{means that} \quad b^3 = a.$$

For example,

$$\sqrt[3]{64} = 4 \quad \text{because} \quad 4^3 = 64.$$

$$\sqrt[3]{-27} = -3 \quad \text{because} \quad (-3)^3 = -27.$$

In contrast to square roots, the cube root of a negative number is a real number. All real numbers have cube roots. The cube root of a positive number is positive. The cube root of a negative number is negative.

Because every real number, x, has precisely one cube root, $\sqrt[3]{x}$, there is a **cube root function** defined by

$$f(x) = \sqrt[3]{x}.$$

The domain of this function is the set of all real numbers. We can graph $f(x) = \sqrt[3]{x}$ by selecting perfect cubes, numbers that have rational cube roots, for x.

STUDY TIP

Some cube roots occur so frequently that you might want to memorize them.

$$\sqrt[3]{1} = 1$$
$$\sqrt[3]{8} = 2$$
$$\sqrt[3]{27} = 3$$
$$\sqrt[3]{64} = 4$$
$$\sqrt[3]{125} = 5$$
$$\sqrt[3]{216} = 6$$
$$\sqrt[3]{1000} = 10$$

Table 10.2 shows five such choices for x and the calculations for the corresponding outputs. We plot these ordered pairs as points in the rectangular coordinate system and connect the points with a smooth curve. The graph of $f(x) = \sqrt[3]{x}$ is shown in Figure 10.5.

Table 10.2

x	$f(x) = \sqrt[3]{x}$	(x, y) or $(x, f(x))$
-8	$f(-8) = \sqrt[3]{-8} = -2$	$(-8, -2)$
-1	$f(-1) = \sqrt[3]{-1} = -1$	$(-1, -1)$
0	$f(0) = \sqrt[3]{0} = 0$	$(0, 0)$
1	$f(1) = \sqrt[3]{1} = 1$	$(1, 1)$
8	$f(8) = \sqrt[3]{8} = 2$	$(8, 2)$

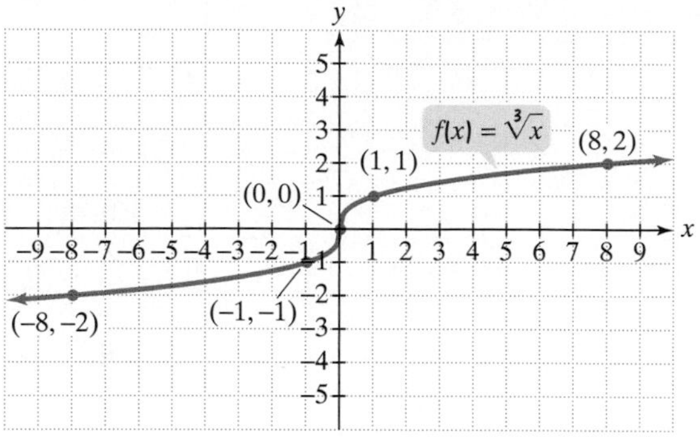

FIGURE 10.5 The graph of the cube root function $f(x) = \sqrt[3]{x}$

Notice that both the domain and the range of $f(x) = \sqrt[3]{x}$ are the set of all real numbers.

EXAMPLE 6 Evaluating Cube Root Functions

For each function, find the indicated function value:

a. $f(x) = \sqrt[3]{x - 2}$; $f(127)$ **b.** $g(x) = \sqrt[3]{8x - 8}$; $g(-7)$.

SOLUTION

a. $f(127) = \sqrt[3]{127 - 2}$ Substitute 127 for x in $f(x) = \sqrt[3]{x - 2}$.

$= \sqrt[3]{125}$ Simplify the radicand.

$= 5$ $\sqrt[3]{125} = 5$ because $5^3 = 125$.

b. $g(-7) = \sqrt[3]{8(-7) - 8}$ Substitute -7 for x in $g(x) = \sqrt[3]{8x - 8}$.

$= \sqrt[3]{-64}$ Simplify the radicand: $8(-7) - 8 = -56 - 8 = -64$.

$= -4$ $\sqrt[3]{-64} = -4$ because $(-4)^3 = -64$. ■

CHECK POINT 6 For each function, find the indicated function value:

a. $f(x) = \sqrt[3]{x - 6}$; $f(33)$

b. $g(x) = \sqrt[3]{2x + 2}$; $g(-5)$.

7 Simplify expressions of the form $\sqrt[3]{a^3}$.

Because the cube root of a positive number is positive and the cube root of a negative number is negative, absolute value is not needed to simplify expressions of the form $\sqrt[3]{a^3}$.

SIMPLIFYING $\sqrt[3]{a^3}$ For any real number a,

$$\sqrt[3]{a^3} = a.$$

In words, the cube root of any expression cubed is that expression.

EXAMPLE 7 Simplifying a Cube Root

Simplify: $\sqrt[3]{-64x^3}$.

SOLUTION Begin by expressing the radicand as an expression that is cubed: $-64x^3 = (-4x)^3$. Then simplify.

$$\sqrt[3]{-64x^3} = \sqrt[3]{(-4x)^3} = -4x$$

We can check our answer by cubing $-4x$:

$$(-4x)^3 = (-4)^3x^3 = -64x^3.$$

By obtaining the original radicand, we know that our simplification is correct. ■

CHECK POINT 7 Simplify: $\sqrt[3]{-27x^3}$.

8 Find even and odd roots.

Even and Odd *n*th Roots Up to this point, we have focused on square roots and cube roots. Other radical expressions have different roots. For example, the fifth root of a, written $\sqrt[5]{a}$, is the number b for which $b^5 = a$. Thus,

$$\sqrt[5]{32} = 2 \quad \text{because} \quad 2^5 = 2 \cdot 2 \cdot 2 \cdot 2 \cdot 2 = 32.$$

The radical expression $\sqrt[n]{a}$ represents the ***n*th root** of a. The number n is called the **index**. An index of 2 represents a square root and is not written. An index of 3 represents a cube root.

If the index n in $\sqrt[n]{a}$ is an odd number, a root is said to be an **odd root**. A cube root is an odd root. Other odd roots have the same characteristics as cube roots. Every real number has exactly one real root when n is odd. The (odd) nth root of a, $\sqrt[n]{a}$, is the number b for which $b^n = a$. An odd root of a positive number is positive and an odd root of a negative number is negative. For example,

$$\sqrt[5]{243} = 3 \quad \text{becausc} \quad 3^5 = 3 \cdot 3 \cdot 3 \cdot 3 \cdot 3 = 243$$

and $\sqrt[5]{-243} = -3$ because $(-3)^5 = (-3)(-3)(-3)(-3)(-3) = -243.$

STUDY TIP

Some higher even and odd roots occur so frequently that you might want to memorize them.

Fourth Roots	Fifth Roots
$\sqrt[4]{1} = 1$	$\sqrt[5]{1} = 1$
$\sqrt[4]{16} = 2$	$\sqrt[5]{32} = 2$
$\sqrt[4]{81} = 3$	$\sqrt[5]{243} = 3$
$\sqrt[4]{256} = 4$	
$\sqrt[4]{625} = 5$	

If the index n in $\sqrt[n]{a}$ is an even number, a root is said to be an **even root**. A square root is an even root. Other even roots have the same characteristics as square roots. Every positive real number has two real roots when n is even. One root is positive and one is negative. The positive root, called the **principal *n*th root** and represented by $\sqrt[n]{a}$, is the nonnegative number b for which $b^n = a$. The symbol $-\sqrt[n]{a}$ is used to denote the negative nth root. **An even root of a negative number is not a real number.**

EXAMPLE 8 Finding Even and Odd Roots

Find the indicated root, or state that the expression is not a real number:

a. $\sqrt[4]{81}$ **b.** $-\sqrt[4]{81}$ **c.** $\sqrt[4]{-81}$ **d.** $\sqrt[5]{-32}$.

SOLUTION

a. $\sqrt[4]{81} = 3$ The principal fourth root of 81 is 3 because $3^4 = 3 \cdot 3 \cdot 3 \cdot 3 = 81$.

b. $-\sqrt[4]{81} = -3$ The negative fourth root of 81 is -3 because $(-3)^4 = (-3)(-3)(-3)(-3) = 81$.

c. $\sqrt[4]{-81}$ is not a real number because the index, 4, is even and the radicand, -81, is negative. No real number can be raised to the fourth power to give a negative result such as -81. Real numbers to even powers can only result in nonnegative numbers.

d. $\sqrt[5]{-32} = -2$ because $(-2)^5 = (-2)(-2)(-2)(-2)(-2) = -32$. An odd root of a negative real number is always negative. ■

✔ **CHECK POINT 8** Find the indicated root, or state that the expression is not a real number:

a. $\sqrt[4]{16}$ **b.** $-\sqrt[4]{16}$

c. $\sqrt[4]{-16}$ **d.** $\sqrt[5]{-1}$.

9 Simplify expressions of the form $\sqrt[n]{a^n}$.

Simplifying Expressions of the Form $\sqrt[n]{a^n}$

We have seen that

$$\sqrt{a^2} = |a| \quad \text{and} \quad \sqrt[3]{a^3} = a.$$

Expressions of the form $\sqrt[n]{a^n}$ can be simplified in the same manner. Unless a is known to be nonnegative, absolute value notation is needed when n is even. When the index is odd, absolute value bars are not necessary.

SIMPLIFYING $\sqrt[n]{a^n}$ For any real number a,

1. If n is even, $\sqrt[n]{a^n} = |a|$.
2. If n is odd, $\sqrt[n]{a^n} = a$.

EXAMPLE 9 Simplifying Radical Expressions

Simplify:

a. $\sqrt[4]{(x-3)^4}$ **b.** $\sqrt[5]{(2x+7)^5}$ **c.** $\sqrt[6]{(-5)^6}$.

SOLUTION Each expression involves the nth root of a radicand raised to the nth power. Thus, each radical expression can be simplified. Absolute value bars are necessary in parts (a) and (c) because the index, n, is even.

a. $\sqrt[4]{(x-3)^4} = |x-3|$ $\qquad \sqrt[n]{a^n} = |a|$ if n is even.

b. $\sqrt[5]{(2x+7)^5} = 2x+7$ $\qquad \sqrt[n]{a^n} = a$ if n is odd.

c. $\sqrt[6]{(-5)^6} = |-5| = 5$ $\qquad \sqrt[n]{a^n} = |a|$ if n is even. ■

✔ **CHECK POINT 9** Simplify:

a. $\sqrt[4]{(x+6)^4}$

b. $\sqrt[5]{(3x-2)^5}$

c. $\sqrt[6]{(-8)^6}$.

10.1 EXERCISE SET

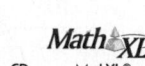

Student Solutions Manual | CD/Video | PH Math/Tutor Center | MathXL Tutorials on CD | MathXL® | MyMathLab | Interactmath.com

Practice Exercises

In Exercises 1–20, evaluate each expression, or state that the expression is not a real number.

1. $\sqrt{36}$
2. $\sqrt{16}$
3. $-\sqrt{36}$
4. $-\sqrt{16}$
5. $\sqrt{-36}$
6. $\sqrt{-16}$
7. $\sqrt{\frac{1}{25}}$
8. $\sqrt{\frac{1}{49}}$
9. $-\sqrt{\frac{9}{16}}$
10. $-\sqrt{\frac{4}{25}}$
11. $\sqrt{0.81}$
12. $\sqrt{0.49}$
13. $-\sqrt{0.04}$
14. $-\sqrt{0.64}$
15. $\sqrt{25 - 16}$
16. $\sqrt{144 + 25}$
17. $\sqrt{25} - \sqrt{16}$
18. $\sqrt{144} + \sqrt{25}$
19. $\sqrt{16 - 25}$
20. $\sqrt{25 - 144}$

In Exercises 21–26, find the indicated function values for each function. If necessary, round to two decimal places. If the function value is not a real number and does not exist, so state.

21. $f(x) = \sqrt{x - 2};\quad f(18), f(3), f(2), f(-2)$
22. $f(x) = \sqrt{x - 3};\quad f(28), f(4), f(3), f(-1)$
23. $g(x) = -\sqrt{2x + 3};\quad g(11), g(1), g(-1), g(-2)$
24. $g(x) = -\sqrt{2x + 1};\quad g(4), g(1), g\left(-\frac{1}{2}\right), g(-1)$
25. $h(x) = \sqrt{(x - 1)^2};\quad h(5), h(3), h(0), h(-5)$
26. $h(x) = \sqrt{(x - 2)^2};\quad h(5), h(3), h(0), h(-5)$

In Exercises 27–32, find the domain of each square root function. Then use the domain to match the radical function with its graph. [The graphs are labeled (a) through (f) and are shown in [−10, 10, 1] by [−10, 10, 1] viewing rectangles.]

27. $f(x) = \sqrt{x - 3}$
28. $f(x) = \sqrt{x + 2}$
29. $f(x) = \sqrt{3x + 15}$
30. $f(x) = \sqrt{3x - 15}$
31. $f(x) = \sqrt{6 - 2x}$
32. $f(x) = \sqrt{8 - 2x}$

a.

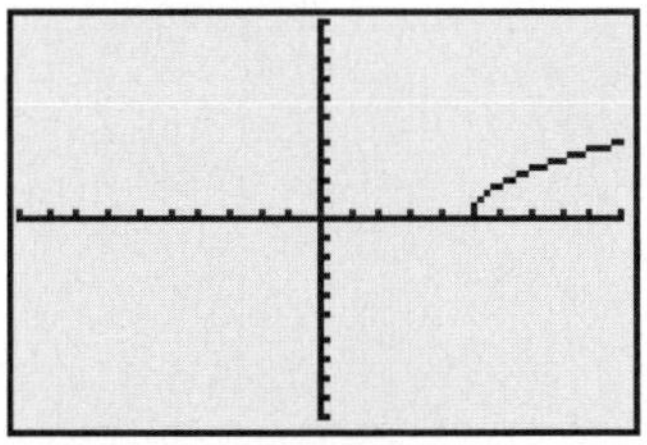

b.

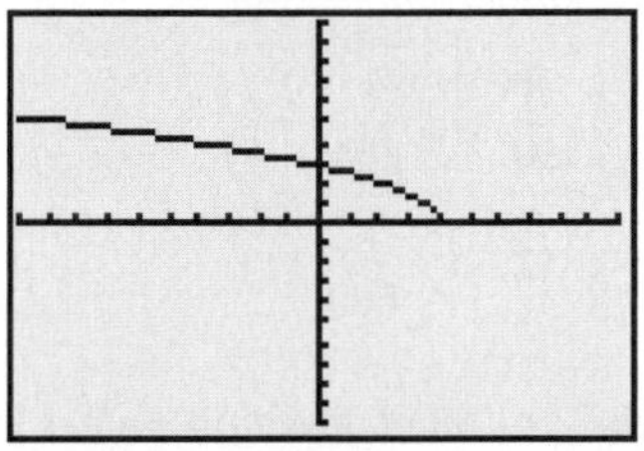

c.

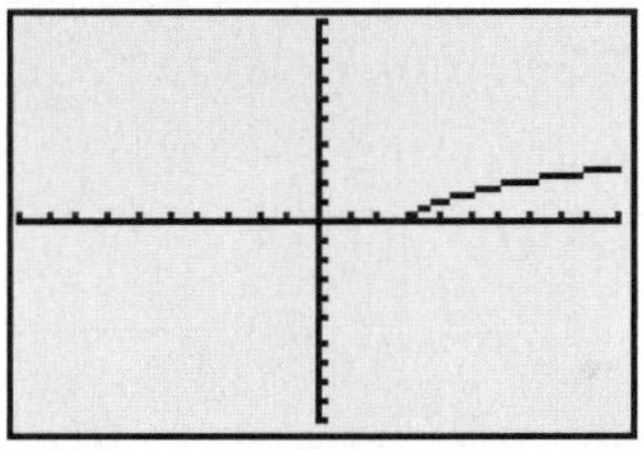

d.

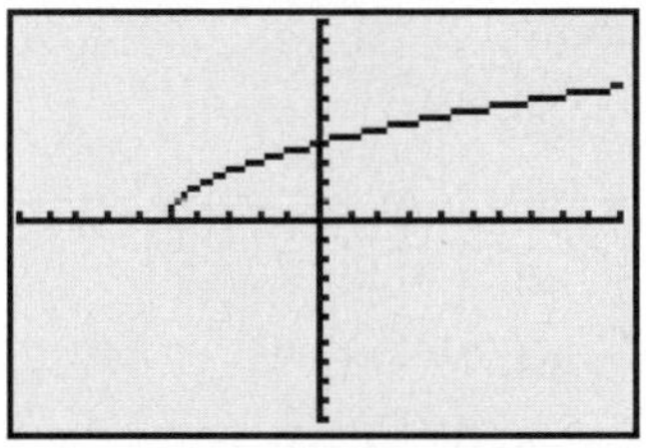

e.

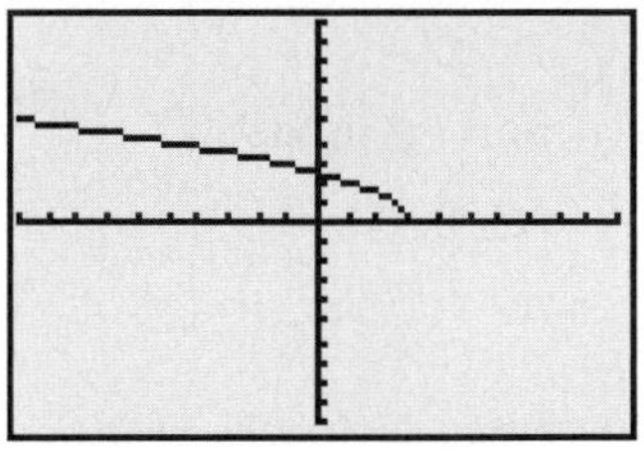

f.

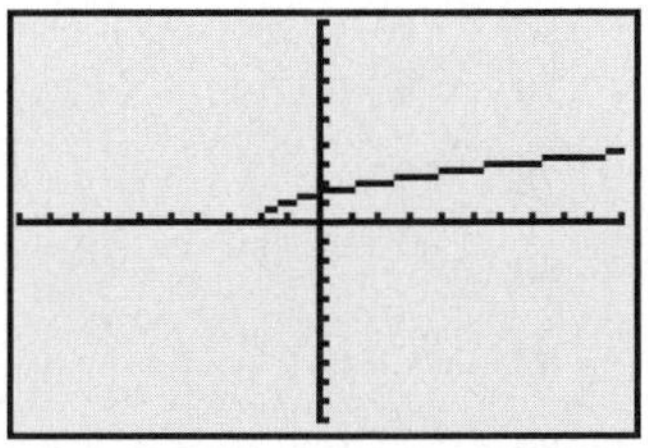

In Exercises 33–46, simplify each expression.

33. $\sqrt{5^2}$
34. $\sqrt{7^2}$

35. $\sqrt{(-4)^2}$ **36.** $\sqrt{(-10)^2}$

37. $\sqrt{(x-1)^2}$ **38.** $\sqrt{(x-2)^2}$

39. $\sqrt{36x^4}$ **40.** $\sqrt{81x^4}$

41. $-\sqrt{100x^6}$

42. $-\sqrt{49x^6}$

43. $\sqrt{x^2+12x+36}$

44. $\sqrt{x^2+14x+49}$

45. $-\sqrt{x^2-8x+16}$

46. $-\sqrt{x^2-10x+25}$

In Exercises 47–52, find each cube root.

47. $\sqrt[3]{27}$ **48.** $\sqrt[3]{64}$

49. $\sqrt[3]{-27}$ **50.** $\sqrt[3]{-64}$

51. $\sqrt[3]{\frac{1}{125}}$ **52.** $\sqrt[3]{\frac{1}{1000}}$

In Exercises 53–56, find the indicated function values for each function.

53. $f(x) = \sqrt[3]{x-1}$; $f(28), f(9), f(0), f(-63)$

54. $f(x) = \sqrt[3]{x-3}$; $f(30), f(11), f(2), f(-122)$

55. $g(x) = -\sqrt[3]{8x-8}$; $g(2), g(1), g(0)$

56. $g(x) = -\sqrt[3]{2x+1}$; $g(13), g(0), g(-63)$

In Exercises 57–74, find the indicated root, or state that the expression is not a real number.

57. $\sqrt[4]{1}$ **58.** $\sqrt[5]{1}$

59. $\sqrt[4]{16}$ **60.** $\sqrt[4]{81}$

61. $-\sqrt[4]{16}$ **62.** $-\sqrt[4]{81}$

63. $\sqrt[4]{-16}$ **64.** $\sqrt[4]{-81}$

65. $\sqrt[5]{-1}$ **66.** $\sqrt[7]{-1}$

67. $\sqrt[6]{-1}$ **68.** $\sqrt[8]{-1}$

69. $-\sqrt[4]{256}$ **70.** $-\sqrt[4]{10{,}000}$

71. $\sqrt[6]{64}$ **72.** $\sqrt[5]{32}$

73. $-\sqrt[5]{32}$ **74.** $-\sqrt[6]{64}$

In Exercises 75–88, simplify each expression. Include absolute value bars where necessary.

75. $\sqrt[3]{x^3}$ **76.** $\sqrt[5]{x^5}$

77. $\sqrt[4]{y^4}$ **78.** $\sqrt[6]{y^6}$

79. $\sqrt[3]{-8x^3}$ **80.** $\sqrt[3]{-125x^3}$

81. $\sqrt[3]{(-5)^3}$ **82.** $\sqrt[3]{(-6)^3}$

83. $\sqrt[4]{(-5)^4}$ **84.** $\sqrt[6]{(-6)^6}$

85. $\sqrt[4]{(x+3)^4}$ **86.** $\sqrt[4]{(x+5)^4}$

87. $\sqrt[5]{-32(x-1)^5}$

88. $\sqrt[5]{-32(x-2)^5}$

Practice Plus

In Exercises 89–92, complete each table and graph the given function. Identify the function's domain and range.

89. $f(x) = \sqrt{x} + 3$

x	$f(x) = \sqrt{x} + 3$
0	
1	
4	
9	

90. $f(x) = \sqrt{x} - 2$

x	$f(x) = \sqrt{x} - 2$
0	
1	
4	
9	

91. $f(x) = \sqrt{x-3}$

x	$f(x) = \sqrt{x-3}$
3	
4	
7	
12	

92. $f(x) = \sqrt{4-x}$

x	$f(x) = \sqrt{4-x}$
−5	
0	
3	
4	

In Exercises 93–96, find the domain of each function.

93. $f(x) = \dfrac{\sqrt[3]{x}}{\sqrt{30-2x}}$

94. $f(x) = \dfrac{\sqrt[3]{x}}{\sqrt{80-5x}}$

95. $f(x) = \dfrac{\sqrt{x-1}}{\sqrt{3-x}}$

96. $f(x) = \dfrac{\sqrt{x-2}}{\sqrt{7-x}}$

In Exercises 97–98, evaluate each expression.

97. $\sqrt[3]{\sqrt[4]{16} + \sqrt{625}}$

98. $\sqrt[3]{\sqrt{\sqrt{169} + \sqrt{9}} + \sqrt{\sqrt[3]{1000} + \sqrt[3]{216}}}$

Application Exercises

The table shows the median, or average, heights for boys of various ages in the United States, from birth through 60 months, or five years. The data can be modeled by the radical function

$$f(x) = 2.9\sqrt{x} + 20.1,$$

where $f(x)$ is the median height, in inches, of boys who are x months of age. Use the function to solve Exercises 99–100.

Boys' Median Heights

Age (months)	Height (inches)
0	20.5
6	27.0
12	30.8
18	32.9
24	35.0
36	37.5
48	40.8
60	43.4

Source: *The Portable Pediatrician for Parents*, by Laura Walther Nathanson, M.D., FAAP

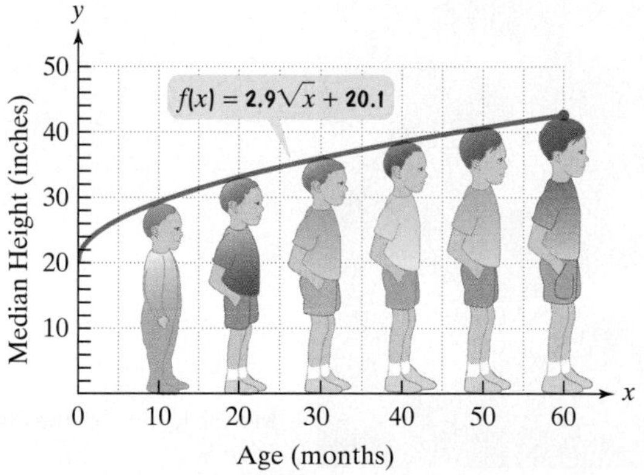

99. According to the model, what is the median height of boys who are 48 months, or four years, old? Use a calculator and round to the nearest tenth of an inch. How well does the model describe the actual median height shown in the table?

100. According to the model, what is the median height of boys who are 60 months, or five years, old? Use a calculator and round to the nearest tenth of an inch. How well does the model describe the actual median height shown in the table?

Police use the function $f(x) = \sqrt{20x}$ to estimate the speed of a car, $f(x)$, in miles per hour, based on the length, x, in feet, of its skid marks upon sudden braking on a dry asphalt road. Use the function to solve Exercises 101–102.

101. A motorist is involved in an accident. A police officer measures the car's skid marks to be 245 feet long. Estimate the speed at which the motorist was traveling before braking. If the posted speed limit is 50 miles per hour and the motorist tells the officer he was not speeding, should the officer believe him? Explain.

102. A motorist is involved in an accident. A police officer measures the car's skid marks to be 45 feet long. Estimate the speed at which the motorist was traveling before braking. If the posted speed limit is 35 miles per hour and the motorist tells the officer she was not speeding, should the officer believe her? Explain.

Writing in Mathematics

103. What are the square roots of 36? Explain why each of these numbers is a square root.

104. What does the symbol $\sqrt{}$ denote? Which of your answers in Exercise 103 is given by this symbol? Write the symbol needed to obtain the other answer.

105. Explain why $\sqrt{-1}$ is not a real number.

106. Explain how to find the domain of a square root function.

107. Explain how to simplify $\sqrt{a^2}$. Give an example with your explanation.

108. Explain why $\sqrt[3]{8}$ is 2. Then describe what is meant by the cube root of a real number.

109. Describe two differences between odd and even roots.

110. Explain how to simplify $\sqrt[n]{a^n}$ if n is even and if n is odd. Give examples with your explanations.

111. Explain the meaning of the words "radical," "radicand," and "index." Give an example with your explanation.

112. Describe the trend in a boy's growth from birth through five years, shown in the table for Exercises 99–100. Why is a square root function a useful model for the data?

113. The function $f(x) = 6.75\sqrt{x} + 12$ models the amount of new student loans, $f(x)$, in billions of dollars, x years after 1993. Write a word problem using the function. Then solve the problem.

Technology Exercises

114. Use a graphing utility to graph $y_1 = \sqrt{x}$, $y_2 = \sqrt{x+4}$, and $y_3 = \sqrt{x-3}$ in the same $[-5, 10, 1]$ by $[0, 6, 1]$ viewing rectangle. Describe one similarity and one difference that you observe among the graphs. Use the word "shift" in your response.

115. Use a graphing utility to graph $y = \sqrt{x}$, $y = \sqrt{x} + 4$, and $y = \sqrt{x} - 3$ in the same $[-1, 10, 1]$ by $[-10, 10, 1]$ viewing rectangle. Describe one similarity and one difference that you observe among the graphs.

116. Use a graphing utility to graph $f(x) = \sqrt{x}$, $g(x) = -\sqrt{x}$, $h(x) = \sqrt{-x}$, and $k(x) = -\sqrt{-x}$ in the same $[-10, 10, 1]$ by $[-4, 4, 1]$ viewing rectangle. Use the graphs to describe the domains and the ranges of functions *f, g, h*, and *k*.

117. Use a graphing utility to graph $y_1 = \sqrt{x^2}$ and $y_2 = -x$ in the same viewing rectangle.

a. For what values of x is $\sqrt{x^2} = -x$?

b. For what values of x is $\sqrt{x^2} \neq -x$?

Critical Thinking Exercises

118. Which one of the following is true?

a. The domain of $f(x) = \sqrt[3]{x-4}$ is $[4, \infty)$.

b. If n is odd and b is negative, then $\sqrt[n]{b}$ is not a real number.

c. The expression $\sqrt[n]{4}$ represents increasingly larger numbers for $n = 2, 3, 4, 5, 6$, and so on.

d. None of the above is true.

119. Write a function whose domain is $(-\infty, 5]$.

120. Let $f(x) = \sqrt{x-3}$ and $g(x) = \sqrt{x+1}$. Find the domain of $f + g$ and $\frac{f}{g}$.

121. Simplify: $\sqrt{(2x+3)^{10}}$.

In Exercises 122–123, graph each function by hand. Then describe the relationship between the function that you graphed and the graph of $f(x) = \sqrt{x}$.

122. $g(x) = \sqrt{x} + 2$

123. $h(x) = \sqrt{x+3}$

Review Exercises

124. Simplify: $3x - 2[x - 3(x + 5)]$. (Section 1.8, Example 11)

125. Simplify: $(-3x^{-4}y^3)^{-2}$. (Section 5.7, Example 6)

126. Solve: $|3x - 4| > 11$. (Section 9.3, Example 4)

SECTION 10.2

Objectives

1 Use the definition of $a^{\frac{1}{n}}$.

2 Use the definition of $a^{\frac{m}{n}}$.

3 Use the definition of $a^{-\frac{m}{n}}$.

4 Use models that contain rational exponents.

5 Simplify expressions with rational exponents.

6 Simplify radical expressions using rational exponents.

RATIONAL EXPONENTS

Marine iguanas of the Galápagos Islands

The Galápagos Islands are a chain of volcanic islands lying 600 miles west of Ecuador. They are famed for over 5000 species of plants and animals, including a rare flightless cormorant, marine iguanas, and giant tortoises weighing more than 600 pounds. Early in 2001, the plants and wildlife that live in the Galápagos were at risk from a massive oil spill that flooded 150,000 gallons of toxic fuel into one of the world's most fragile ecosystems. The long-term danger of the accident is that fuel sinking to the ocean floor will destroy algae that is vital to the food chain. Any imbalance in the food chain could threaten the rare Galápagos plant and animal species that have evolved for thousands of years in isolation with little human intervention.

At risk on these ecologically vulnerable islands are unique flora and fauna that helped to inspire Charles Darwin's theory of evolution. Darwin made an enormous collection of the islands' plant species. The function

$$f(x) = 29x^{\frac{1}{3}}$$

models the number of plant species, $f(x)$, on the various islands of the Galápagos in terms of the area, x, in square miles, of a particular island. But x to the *what* power? How can we interpret the information given by this function? In this section, we turn our attention to rational exponents such as $\frac{1}{3}$ and their relationship to roots of real numbers.

Defining Rational Exponents We define rational exponents so that their properties are the same as the properties for integer exponents. For example, suppose that $x = 7^{\frac{1}{3}}$. We know that exponents are multiplied when an exponential expression is raised to a power. For this to be true,

$$x^3 = \left(7^{\frac{1}{3}}\right)^3 = 7^{\frac{1}{3}\cdot 3} = 7^1 = 7.$$

We see that $x^3 = 7$. This means that x is the number whose cube is 7. Thus, $x = \sqrt[3]{7}$. Remember that we began with $x = 7^{\frac{1}{3}}$. This means that

$$7^{\frac{1}{3}} = \sqrt[3]{7}.$$

We can generalize this idea with the following definition:

1 Use the definition of $a^{\frac{1}{n}}$.

THE DEFINITION OF $a^{\frac{1}{n}}$ If $\sqrt[n]{a}$ represents a real number and $n \geq 2$ is an integer, then

$$a^{\frac{1}{n}} = \sqrt[n]{a}.$$

The denominator of the rational exponent is the radical's index.

If a is negative, n must be odd. If a is nonnegative, n can be any index.

EXAMPLE 1 Using the Definition of $a^{\frac{1}{n}}$

Use radical notation to rewrite each expression. Simplify, if possible:

a. $64^{\frac{1}{2}}$ **b.** $(-125)^{\frac{1}{3}}$ **c.** $(6x^2y)^{\frac{1}{5}}$.

SOLUTION

a. $64^{\frac{1}{2}} = \sqrt{64} = 8$

The denominator is the index.

b. $(-125)^{\frac{1}{3}} = \sqrt[3]{-125} = -5$

c. $(6x^2y)^{\frac{1}{5}} = \sqrt[5]{6x^2y}$

CHECK POINT 1 Use radical notation to rewrite each expression. Simplify, if possible:

a. $25^{\frac{1}{2}}$ **b.** $(-8)^{\frac{1}{3}}$ **c.** $(5xy^2)^{\frac{1}{4}}$.

In our next example, we begin with radical notation and rewrite the expression with rational exponents.

The radical's index becomes the exponent's denominator.

$$\sqrt[n]{a} = a^{\frac{1}{n}}$$

The radicand becomes the base.

EXAMPLE 2 Using the Definition of $a^{\frac{1}{n}}$

Rewrite with rational exponents:

a. $\sqrt[5]{13ab}$ **b.** $\sqrt[7]{\frac{xy^2}{17}}$.

SOLUTION Parentheses are needed to show that the entire radicand becomes the base.

a. $\sqrt[5]{13ab} = (13ab)^{\frac{1}{5}}$

The index is the exponent's denominator.

b. $\sqrt[7]{\frac{xy^2}{17}} = \left(\frac{xy^2}{17}\right)^{\frac{1}{7}}$ ■

CHECK POINT 2 Rewrite with rational exponents:

a. $\sqrt[4]{5xy}$ **b.** $\sqrt[5]{\frac{a^3b}{2}}$.

2 Use the definition of $a^{\frac{m}{n}}$.

Can rational exponents have numerators other than 1? The answer is yes. If the numerator is some other integer, we still want to multiply exponents when raising a power to a power. For this reason,

$$a^{\frac{2}{3}} = (a^{\frac{1}{3}})^2 \text{ and } a^{\frac{2}{3}} = (a^2)^{\frac{1}{3}}.$$

This means $(\sqrt[3]{a})^2$. This means $\sqrt[3]{a^2}$.

Thus,

$$a^{\frac{2}{3}} = (\sqrt[3]{a})^2 = \sqrt[3]{a^2}.$$

Do you see that the denominator, 3, of the rational exponent is the same as the index of the radical? The numerator, 2, of the rational exponent serves as an exponent in each of the two radical forms. We generalize these ideas with the following definition:

THE DEFINITION OF $a^{\frac{m}{n}}$ If $\sqrt[n]{a}$ represents a real number, $\frac{m}{n}$ is a positive rational number reduced to lowest terms, and $n \geq 2$ is an integer, then

$$a^{\frac{m}{n}} = (\sqrt[n]{a})^m$$

First take the nth root of a.

and

$$a^{\frac{m}{n}} = \sqrt[n]{a^m}.$$

First raise a to the m power.

The first form of the definition shown in the box involves taking the root first. This form is often preferable because smaller numbers are involved. Notice that the rational exponent consists of two parts, indicated by the following voice balloons:

The numerator is the exponent.

$$a^{\frac{m}{n}} = (\sqrt[n]{a})^m.$$

The denominator is the radical's index.

EXAMPLE 3 Using the Definition of $a^{\frac{m}{n}}$

Use radical notation to rewrite each expression and simplify:

a. $1000^{\frac{2}{3}}$ **b.** $16^{\frac{3}{2}}$ **c.** $-32^{\frac{3}{5}}$.

SOLUTION

a. $(1000)^{\frac{2}{3}} = (\sqrt[3]{1000})^2 = 10^2 = 100$

The denominator of $\frac{2}{3}$ is the root and the numerator is the exponent.

b. $16^{\frac{3}{2}} = (\sqrt{16})^3 = 4^3 = 64$

c. $-32^{\frac{3}{5}} = -(\sqrt[5]{32})^3 = -2^3 = -8$

The base is 32 and the negative sign is not affected by the exponent.

■

TECHNOLOGY

Here are the calculator keystroke sequences for $1000^{\frac{2}{3}}$:

Many Scientific Calculators

1000 [y^x] [(] 2 [÷] 3 [)] [=]

Many Graphing Calculators

1000 [^] [(] 2 [÷] 3 [)] [ENTER]

CHECK POINT 3 Use radical notation to rewrite each expression and simplify:

a. $8^{\frac{4}{3}}$ **b.** $25^{\frac{3}{2}}$ **c.** $-81^{\frac{3}{4}}$.

In our next example, we begin with radical notation and rewrite the expression with rational exponents. When changing from radical form to exponential form, the index becomes the denominator of the rational exponent.

EXAMPLE 4 Using the Definition of $a^{\frac{m}{n}}$

Rewrite with rational exponents:

a. $\sqrt[3]{7^5}$ **b.** $(\sqrt[4]{13xy})^9$.

SOLUTION

a. $\sqrt[3]{7^5} = 7^{\frac{5}{3}}$

The index is the exponent's denominator.

b. $(\sqrt[4]{13xy})^9 = (13xy)^{\frac{9}{4}}$ ■

CHECK POINT 4 Rewrite with rational exponents:

a. $\sqrt[3]{6^4}$ **b.** $(\sqrt[5]{2xy})^7$.

3 Use the definition of $a^{-\frac{m}{n}}$.

Can a rational exponent be negative? Yes. The way that negative rational exponents are defined is similar to the way that negative integer exponents are defined.

THE DEFINITION OF $a^{-\frac{m}{n}}$ If $a^{\frac{m}{n}}$ is a nonzero real number, then

$$a^{-\frac{m}{n}} = \frac{1}{a^{\frac{m}{n}}}.$$

EXAMPLE 5 Using the Definition of $a^{-\frac{m}{n}}$

Rewrite each expression with a positive exponent. Simplify, if possible:

a. $36^{-\frac{1}{2}}$ **b.** $125^{-\frac{1}{3}}$ **c.** $16^{-\frac{3}{4}}$ **d.** $(7xy)^{-\frac{4}{7}}$.

SOLUTION

a. $36^{-\frac{1}{2}} = \frac{1}{36^{\frac{1}{2}}} = \frac{1}{\sqrt{36}} = \frac{1}{6}$

b. $125^{-\frac{1}{3}} = \frac{1}{125^{\frac{1}{3}}} = \frac{1}{\sqrt[3]{125}} = \frac{1}{5}$

c. $16^{-\frac{3}{4}} = \frac{1}{16^{\frac{3}{4}}} = \frac{1}{(\sqrt[4]{16})^3} = \frac{1}{2^3} = \frac{1}{8}$

d. $(7xy)^{-\frac{4}{7}} = \frac{1}{(7xy)^{\frac{4}{7}}}$ ■

USING TECHNOLOGY

Here are the calculator keystroke sequences for $16^{-\frac{3}{4}}$:

Many Scientific Calculators

16 [y^x] [(] 3 [+/−] [÷] 4 [)] [=]

Many Graphing Calculators

16 [∧] [(] [(−)] 3 [÷] 4 [)] [ENTER]

CHECK POINT 5 Rewrite each expression with a positive exponent. Simplify, if possible:

a. $100^{-\frac{1}{2}}$ **b.** $8^{-\frac{1}{3}}$

c. $32^{-\frac{3}{5}}$ **d.** $(3xy)^{-\frac{5}{9}}$.

4 Use models that contain rational exponents.

Applications Now that you know the meaning of rational exponents, you can work with mathematical models that contain these exponents.

EXAMPLE 6 Calculating Windchill

The way that we perceive the temperature on a cold day depends on both air temperature and wind speed. The windchill is what the air temperature would have to be with no wind to achieve the same chilling effect on the skin. In 2002, the National Weather Service issued new windchill temperatures, shown in Table 10.3. (One reason for this new windchill index is that the wind speed is now calculated at 5 feet, the average height of the human body's face, rather than 33 feet, the height of the standard anemometer, an instrument that calculates wind speed.)

Table 10.3 New Windchill Temperature Index

Wind Speed (miles per hour)	Air Temperature (°F) 30	25	20	15	10	5	0	−5	−10	−15	−20	−25
5	25	19	13	7	1	−5	−11	−16	−22	−28	−34	−40
10	21	15	9	3	−4	−10	−16	−22	−28	−35	−41	−47
15	19	13	6	0	−7	−13	−19	−26	−32	−39	−45	−51
20	17	11	4	−2	−9	−15	−22	−29	−35	−42	−48	−55
25	16	9	3	−4	−11	−17	−24	−31	−37	−44	−51	−58
30	15	8	1	−5	−12	−19	−26	−33	−39	−46	−53	−60
35	14	7	0	−7	−14	−21	−27	−34	−41	−48	−55	−62
40	13	6	−1	−8	−15	−22	−29	−36	−43	−50	−57	−64
45	12	5	−2	−9	−16	−23	−30	−37	−44	−51	−58	−65
50	12	4	−3	−10	−17	−24	−31	−38	−45	−52	−60	−67
55	11	4	−3	−11	−18	−25	−32	−39	−46	−54	−61	−68
60	10	3	−4	−11	−19	−26	−33	−40	−48	−55	−62	−69

(Shaded) Frostbite occurs in 15 minutes or less.

Source: National Weather Service

The windchill temperatures shown in Table 10.3 can be calculated using

$$C = 35.74 + 0.6215t - 35.74v^{\frac{4}{25}} + 0.4275tv^{\frac{4}{25}}$$

in which C is the windchill, in degrees Fahrenheit, t is the air temperature, in degrees Fahrenheit, and v is the wind speed, in miles per hour. Use the formula to find the windchill temperature, to the nearest degree, when the air temperature is 25°F and the wind speed is 30 miles per hour.

SOLUTION Because we are interested in the windchill when the air temperature is 25°F and the wind speed is 30 miles per hour, substitute 25 for t and 30 for v. Then calculate C.

$$C = 35.74 + 0.6215t - 35.74v^{\frac{4}{25}} + 0.4275tv^{\frac{4}{25}}$$ This is the given formula.

$$C = 35.74 + 0.6215(25) - 35.74(30)^{\frac{4}{25}} + 0.4275(25)(30)^{\frac{4}{25}}$$ Substitute 25 for t and 30 for v.

$$\approx 8$$ Use a calculator.

When the air temperature is 25°F and the wind speed is 30 miles per hour, the windchill temperature is 8°. In other words, a 30-mile-per-hour wind makes 25° feel like 8°. ■

CHECK POINT 6 Use the windchill formula in Example 6 to find the windchill temperature, to the nearest degree, when the air temperature is 35°F and the wind speed is 15 miles per hour.

5 Simplify expressions with rational exponents.

Properties of Rational Exponents The same properties apply to rational exponents as to integer exponents. The following is a summary of these properties:

PROPERTIES OF RATIONAL EXPONENTS If m and n are rational exponents, and a and b are real numbers for which the following expressions are defined, then

1. $b^m \cdot b^n = b^{m+n}$ When multiplying exponential expressions with the same base, add the exponents. Use this sum as the exponent of the common base.
2. $\dfrac{b^m}{b^n} = b^{m-n}$ When dividing exponential expressions with the same base, subtract the exponents. Use this difference as the exponent of the common base.
3. $(b^m)^n = b^{mn}$ When an exponential expression is raised to a power, multiply the exponents. Place the product of the exponents on the base and remove the parentheses.
4. $(ab)^n = a^n b^n$ When a product is raised to a power, raise each factor to that power and multiply.
5. $\left(\dfrac{a}{b}\right)^n = \dfrac{a^n}{b^n}$ When a quotient is raised to a power, raise the numerator to that power and divide by the denominator to that power.

We can use these properties to simplify exponential expressions with rational exponents. As with integer exponents, an expression with rational exponents is **simplified** when:

- No parentheses appear.
- No powers are raised to powers.
- Each base occurs only once.
- No negative or zero exponents appear.

EXAMPLE 7 Simplifying Expressions with Rational Exponents

Simplify:

a. $6^{\frac{1}{7}} \cdot 6^{\frac{4}{7}}$ **b.** $\dfrac{32x^{\frac{1}{2}}}{16x^{\frac{3}{4}}}$ **c.** $\left(8.3^{\frac{3}{4}}\right)^{\frac{2}{3}}$ **d.** $\left(x^{-\frac{2}{5}}y^{\frac{1}{3}}\right)^{\frac{1}{2}}$.

SOLUTION

a. $6^{\frac{1}{7}} \cdot 6^{\frac{4}{7}} = 6^{\frac{1}{7}+\frac{4}{7}}$ To multiply with the same base, add exponents.

$= 6^{\frac{5}{7}}$ Simplify: $\frac{1}{7} + \frac{4}{7} = \frac{5}{7}$.

b. $\dfrac{32x^{\frac{1}{2}}}{16x^{\frac{3}{4}}} = \dfrac{32}{16}x^{\frac{1}{2}-\frac{3}{4}}$ Divide coefficients. To divide with the same base, subtract exponents.

$= 2x^{\frac{2}{4}-\frac{3}{4}}$ Write exponents in terms of the LCD, 4.

$= 2x^{-\frac{1}{4}}$ Subtract: $\frac{2}{4} - \frac{3}{4} = -\frac{1}{4}$.

$= \dfrac{2}{x^{\frac{1}{4}}}$ Rewrite with a positive exponent: $a^{-\frac{m}{n}} = \dfrac{1}{a^{\frac{m}{n}}}$.

c. $\left(8.3^{\frac{3}{4}}\right)^{\frac{2}{3}} = 8.3^{\left(\frac{3}{4}\right)\left(\frac{2}{3}\right)}$ — To raise a power to a power, multiply exponents.

$= 8.3^{\frac{1}{2}}$ — Multiply: $\frac{3}{4} \cdot \frac{2}{3} = \frac{6}{12} = \frac{1}{2}$.

d. $\left(x^{-\frac{2}{5}}y^{\frac{1}{3}}\right)^{\frac{1}{2}} = \left(x^{-\frac{2}{5}}\right)^{\frac{1}{2}}\left(y^{\frac{1}{3}}\right)^{\frac{1}{2}}$ — To raise a product to a power, raise each factor to the power.

$= x^{-\frac{1}{5}}y^{\frac{1}{6}}$ — Multiply: $-\frac{2}{5} \cdot \frac{1}{2} = -\frac{1}{5}$ and $\frac{1}{3} \cdot \frac{1}{2} = \frac{1}{6}$.

$= \dfrac{y^{\frac{1}{6}}}{x^{\frac{1}{5}}}$ — Rewrite with positive exponents. ■

CHECK POINT 7 Simplify:

a. $7^{\frac{1}{2}} \cdot 7^{\frac{1}{3}}$ **b.** $\dfrac{50x^{\frac{1}{3}}}{10x^{\frac{4}{3}}}$ **c.** $\left(9.1^{\frac{2}{5}}\right)^{\frac{3}{4}}$ **d.** $\left(x^{-\frac{3}{5}}y^{\frac{1}{4}}\right)^{\frac{1}{3}}$.

6 Simplify radical expressions using rational exponents.

Using Rational Exponents to Simplify Radical Expressions Some radical expressions can be simplified using rational exponents. We will use the following procedure:

SIMPLIFYING RADICAL EXPRESSIONS USING RATIONAL EXPONENTS

1. Rewrite each radical expression as an exponential expression with a rational exponent.
2. Simplify using properties of rational exponents.
3. Rewrite in radical notation if rational exponents still appear.

EXAMPLE 8 Simplifying Radical Expressions Using Rational Exponents

Use rational exponents to simplify:

a. $\sqrt[10]{x^5}$ **b.** $\sqrt[3]{27a^{15}}$ **c.** $\sqrt[4]{x^6y^2}$ **d.** $\sqrt{x} \cdot \sqrt[3]{x}$ **e.** $\sqrt[3]{\sqrt{x}}$.

SOLUTION

a. $\sqrt[10]{x^5} = x^{\frac{5}{10}}$ — Rewrite as an exponential expression.

$= x^{\frac{1}{2}}$ — Simplify the exponent.

$= \sqrt{x}$ — Rewrite in radical notation.

b. $\sqrt[3]{27a^{15}} = (27a^{15})^{\frac{1}{3}}$ — Rewrite as an exponential expression.

$= 27^{\frac{1}{3}}(a^{15})^{\frac{1}{3}}$ — Raise each factor in parentheses to the $\frac{1}{3}$ power.

$= \sqrt[3]{27} \cdot a^{15\left(\frac{1}{3}\right)}$ — To raise a power to a power, multiply exponents.

$= 3a^5$ — $\sqrt[3]{27} = 3$. Multiply exponents: $15 \cdot \frac{1}{3} = 5$.

c. $\sqrt[4]{x^6y^2} = (x^6y^2)^{\frac{1}{4}}$ Rewrite as an exponential expression.

$= (x^6)^{\frac{1}{4}}(y^2)^{\frac{1}{4}}$ Raise each factor in parentheses to the $\frac{1}{4}$ power.

$= x^{\frac{6}{4}}y^{\frac{2}{4}}$ To raise powers to powers, multiply.

$= x^{\frac{3}{2}}y^{\frac{1}{2}}$ Simplify.

$= (x^3y)^{\frac{1}{2}}$ $a^nb^n = (ab)^n$

$= \sqrt{x^3y}$ Rewrite in radical notation.

d. $\sqrt{x} \cdot \sqrt[3]{x} = x^{\frac{1}{2}} \cdot x^{\frac{1}{3}}$ Rewrite as exponential expressions.

$= x^{\frac{1}{2}+\frac{1}{3}}$ To multiply with the same base, add exponents.

$= x^{\frac{3}{6}+\frac{2}{6}}$ Write exponents in terms of the LCD, 6.

$= x^{\frac{5}{6}}$ Add: $\frac{3}{6} + \frac{2}{6} = \frac{5}{6}$.

$= \sqrt[6]{x^5}$ Rewrite in radical notation.

e. $\sqrt[3]{\sqrt{x}} = \sqrt[3]{x^{\frac{1}{2}}}$ Write the radicand as an exponential expression.

$= \left(x^{\frac{1}{2}}\right)^{\frac{1}{3}}$ Write the entire expression in exponential form.

$= x^{\frac{1}{6}}$ To raise powers to powers, multiply: $\frac{1}{2} \cdot \frac{1}{3} = \frac{1}{6}$.

$= \sqrt[6]{x}$ Rewrite in radical notation. ■

CHECK POINT 8 Use rational exponents to simplify:

a. $\sqrt[6]{x^3}$ **b.** $\sqrt[3]{8a^{12}}$ **c.** $\sqrt[8]{x^4y^2}$ **d.** $\dfrac{\sqrt{x}}{\sqrt[3]{x}}$ **e.** $\sqrt{\sqrt[3]{x}}$.

10.2 EXERCISE SET

Student Solutions Manual | CD/Video | PH Math/Tutor Center | MathXL Tutorials on CD | MathXL® | MyMathLab | Interactmath.com

Practice Exercises

In Exercises 1–20, use radical notation to rewrite each expression. Simplify, if possible.

1. $49^{\frac{1}{2}}$ **2.** $100^{\frac{1}{2}}$

3. $(-27)^{\frac{1}{3}}$ **4.** $(-64)^{\frac{1}{3}}$

5. $-16^{\frac{1}{4}}$ **6.** $-81^{\frac{1}{4}}$

7. $(xy)^{\frac{1}{3}}$ **8.** $(xy)^{\frac{1}{4}}$

9. $(2xy^3)^{\frac{1}{5}}$ **10.** $(3xy^4)^{\frac{1}{5}}$

11. $81^{\frac{3}{2}}$ **12.** $25^{\frac{3}{2}}$

13. $125^{\frac{2}{3}}$ **14.** $1000^{\frac{2}{3}}$

15. $(-32)^{\frac{3}{5}}$ **16.** $(-27)^{\frac{2}{3}}$

17. $27^{\frac{2}{3}} + 16^{\frac{3}{4}}$ **18.** $4^{\frac{5}{2}} - 8^{\frac{2}{3}}$

19. $(xy)^{\frac{4}{7}}$ **20.** $(xy)^{\frac{4}{9}}$

In Exercises 21–38, rewrite each expression with rational exponents.

21. $\sqrt{7}$ **22.** $\sqrt{13}$

23. $\sqrt[3]{5}$ **24.** $\sqrt[3]{6}$

25. $\sqrt[5]{11x}$ **26.** $\sqrt[5]{13x}$

27. $\sqrt{x^3}$ **28.** $\sqrt{x^5}$

29. $\sqrt[5]{x^3}$ **30.** $\sqrt[7]{x^4}$

31. $\sqrt[5]{x^2y}$ **32.** $\sqrt[7]{xy^3}$

33. $(\sqrt{19xy})^3$ **34.** $(\sqrt{11xy})^3$

35. $(\sqrt[6]{7xy^2})^5$ **36.** $(\sqrt[6]{9x^2y})^5$

37. $2x\sqrt[3]{y^2}$ **38.** $4x\sqrt[5]{y^2}$

In Exercises 39–54, rewrite each expression with a positive rational exponent. Simplify, if possible.

39. $49^{-\frac{1}{2}}$ **40.** $9^{-\frac{1}{2}}$

41. $27^{-\frac{1}{3}}$ **42.** $125^{-\frac{1}{3}}$

43. $16^{-\frac{3}{4}}$ **44.** $81^{-\frac{5}{4}}$

45. $8^{-\frac{2}{3}}$ **46.** $32^{-\frac{4}{5}}$

47. $\left(\frac{8}{27}\right)^{-\frac{1}{3}}$ **48.** $\left(\frac{8}{125}\right)^{-\frac{1}{3}}$

49. $(-64)^{-\frac{2}{3}}$ **50.** $(-8)^{-\frac{2}{3}}$

51. $(2xy)^{-\frac{7}{10}}$ **52.** $(4xy)^{-\frac{4}{7}}$

53. $5xz^{-\frac{1}{3}}$ **54.** $7xz^{-\frac{1}{4}}$

In Exercises 55–78, use properties of rational exponents to simplify each expression. Assume that all variables represent positive numbers.

55. $3^{\frac{3}{4}} \cdot 3^{\frac{1}{4}}$ **56.** $5^{\frac{2}{3}} \cdot 5^{\frac{1}{3}}$

57. $\dfrac{16^{\frac{3}{4}}}{16^{\frac{1}{4}}}$ **58.** $\dfrac{100^{\frac{3}{4}}}{100^{\frac{1}{4}}}$

59. $x^{\frac{1}{2}} \cdot x^{\frac{1}{3}}$ **60.** $x^{\frac{1}{2}} \cdot x^{\frac{2}{3}}$

61. $\dfrac{x^{\frac{4}{5}}}{x^{\frac{1}{5}}}$ **62.** $\dfrac{x^{\frac{3}{7}}}{x^{\frac{1}{7}}}$

63. $\dfrac{x^{\frac{1}{3}}}{x^{\frac{3}{4}}}$ **64.** $\dfrac{x^{\frac{1}{4}}}{x^{\frac{3}{5}}}$

65. $\left(5^{\frac{2}{3}}\right)^3$ **66.** $\left(3^{\frac{4}{5}}\right)^5$

67. $\left(y^{-\frac{2}{3}}\right)^{\frac{1}{4}}$ **68.** $\left(y^{-\frac{3}{4}}\right)^{\frac{1}{6}}$

69. $\left(2x^{\frac{1}{5}}\right)^5$ **70.** $\left(2x^{\frac{1}{4}}\right)^4$

71. $(25x^4y^6)^{\frac{1}{2}}$ **72.** $(125x^9y^6)^{\frac{1}{3}}$

73. $\left(x^{\frac{1}{2}}y^{-\frac{3}{5}}\right)^{\frac{1}{2}}$ **74.** $\left(x^{\frac{1}{4}}y^{-\frac{2}{5}}\right)^{\frac{1}{3}}$

75. $\dfrac{3^{\frac{1}{2}} \cdot 3^{\frac{3}{4}}}{3^{\frac{1}{4}}}$ **76.** $\dfrac{5^{\frac{3}{4}} \cdot 5^{\frac{1}{2}}}{5^{\frac{1}{4}}}$

77. $\dfrac{\left(3y^{\frac{1}{4}}\right)^3}{y^{\frac{1}{12}}}$ **78.** $\dfrac{\left(2y^{\frac{1}{5}}\right)^4}{y^{\frac{3}{10}}}$

In Exercises 79–112, use rational exponents to simplify each expression. If rational exponents appear after simplifying, write the answer in radical notation. Assume that all variables represent positive numbers.

79. $\sqrt[8]{x^2}$ **80.** $\sqrt[10]{x^2}$

81. $\sqrt[3]{8a^6}$ **82.** $\sqrt[3]{27a^{12}}$

83. $\sqrt[5]{x^{10}y^{15}}$ **84.** $\sqrt[5]{x^{15}y^{20}}$

85. $(\sqrt[3]{xy})^{18}$ **86.** $(\sqrt[3]{xy})^{21}$

87. $\sqrt[10]{(3y)^2}$ **88.** $\sqrt[12]{(3y)^2}$

89. $(\sqrt[6]{2a})^4$ **90.** $(\sqrt[8]{2a})^6$

91. $\sqrt[9]{x^6y^3}$ **92.** $\sqrt[4]{x^2y^6}$

93. $\sqrt{2} \cdot \sqrt[3]{2}$ **94.** $\sqrt{3} \cdot \sqrt[3]{3}$

95. $\sqrt[5]{x^2} \cdot \sqrt{x}$ **96.** $\sqrt[7]{x^2} \cdot \sqrt{x}$

97. $\sqrt[4]{a^2b} \cdot \sqrt[3]{ab}$ **98.** $\sqrt[6]{ab^2} \cdot \sqrt[3]{a^2b}$

99. $\dfrac{\sqrt[4]{x}}{\sqrt[5]{x}}$ **100.** $\dfrac{\sqrt[3]{x}}{\sqrt[4]{x}}$

101. $\dfrac{\sqrt[3]{y^2}}{\sqrt[6]{y}}$ **102.** $\dfrac{\sqrt[5]{y^2}}{\sqrt[10]{y^3}}$

103. $\sqrt[4]{\sqrt{x}}$ **104.** $\sqrt[5]{\sqrt{x}}$

105. $\sqrt{\sqrt{x^2y}}$ **106.** $\sqrt{\sqrt{xy^2}}$

107. $\sqrt[4]{\sqrt[3]{2x}}$ **108.** $\sqrt[5]{\sqrt[3]{2x}}$

109. $(\sqrt[4]{x^3y^5})^{12}$ **110.** $(\sqrt[5]{x^4y^2})^{20}$

111. $\dfrac{\sqrt[4]{a^5b^5}}{\sqrt{ab}}$ **112.** $\dfrac{\sqrt[4]{a^3b^3}}{\sqrt{ab}}$

Practice Plus

In Exercises 113–116, use the distributive property or the FOIL method to perform each multiplication.

113. $x^{\frac{1}{3}}\left(x^{\frac{1}{3}} - x^{\frac{2}{3}}\right)$

114. $x^{-\frac{1}{4}}\left(x^{\frac{9}{4}} - x^{\frac{1}{4}}\right)$

115. $\left(x^{\frac{1}{2}} - 3\right)\left(x^{\frac{1}{2}} + 5\right)$

116. $\left(x^{\frac{1}{3}} - 2\right)\left(x^{\frac{1}{3}} + 6\right)$

In Exercises 117–120, factor out the greatest common factor from each expression.

117. $6x^{\frac{1}{2}} + 2x^{\frac{3}{2}}$

118. $8x^{\frac{1}{4}} + 4x^{\frac{5}{4}}$

119. $15x^{\frac{1}{3}} - 60x$

120. $7x^{\frac{1}{3}} - 70x$

In Exercises 121–124, simplify each expression. Assume that all variables represent positive numbers.

121. $(49x^{-2}y^{4})^{-\frac{1}{2}}\left(xy^{\frac{1}{2}}\right)$

122. $(8x^{-6}y^{3})^{\frac{1}{3}}\left(x^{\frac{5}{6}}y^{-\frac{1}{3}}\right)^{6}$

123. $\left(\dfrac{x^{-\frac{5}{4}}y^{\frac{1}{3}}}{x^{-\frac{3}{4}}}\right)^{-6}$

124. $\left(\dfrac{x^{\frac{1}{2}}y^{-\frac{7}{4}}}{y^{-\frac{5}{4}}}\right)^{-4}$

Application Exercises

The Galápagos Islands, lying 600 miles west of Ecuador, are famed for their extraordinary wildlife. The function

$$f(x) = 29x^{\frac{1}{3}}$$

models the number of plant species, $f(x)$, on the various islands of the Galápagos chain in terms of the area, x, in square miles, of a particular island. Use the function to solve Exercises 125–126.

125. How many species of plants are on a Galápagos island that has an area of 8 square miles?

126. How many species of plants are on a Galápagos island that has an area of 27 square miles?

The function

$$f(x) = 70x^{\frac{3}{4}}$$

models the number of calories per day, $f(x)$, a person needs to maintain life in terms of that person's weight, x, in kilograms. (1 kilogram is approximately 2.2 pounds.) Use this model and a calculator to solve Exercises 127–128. Round answers to the nearest calorie.

127. How many calories per day does a person who weighs 80 kilograms (approximately 176 pounds) need to maintain life?

128. How many calories per day does a person who weighs 70 kilograms (approximately 154 pounds) need to maintain life?

We have seen that when the air temperature is t degrees Fahrenheit and the wind speed is v miles per hour, the windchill temperature, C, in degrees Fahrenheit, is given by the formula

$$C = 35.74 + 0.6215t - 35.74v^{\frac{4}{25}} + 0.4275tv^{\frac{4}{25}}.$$

Use this formula to solve Exercises 129–130.

129. **a.** Substitute 0 for t and write a function $C(v)$ that gives the windchill temperature as a function of wind speed for an air temperature of 0 °F.

b. Find and interpret $C(25)$. Use a calculator and round to the nearest degree.

c. Identify your solution to part (b) on the graph shown.

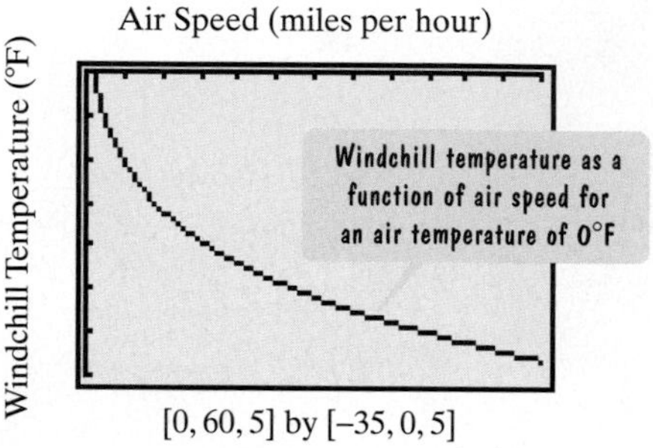

[0, 60, 5] by [−35, 0, 5]

130. **a.** Substitute 30 for t and write a function $C(v)$ that gives the windchill temperature as a function of wind speed for an air temperature of 30°F. Simplify the function's formula so that it contains exactly two terms.

b. Find and interpret $C(40)$. Use a calculator and round to the nearest degree.

c. Identify your solution to part (b) on the graph shown.

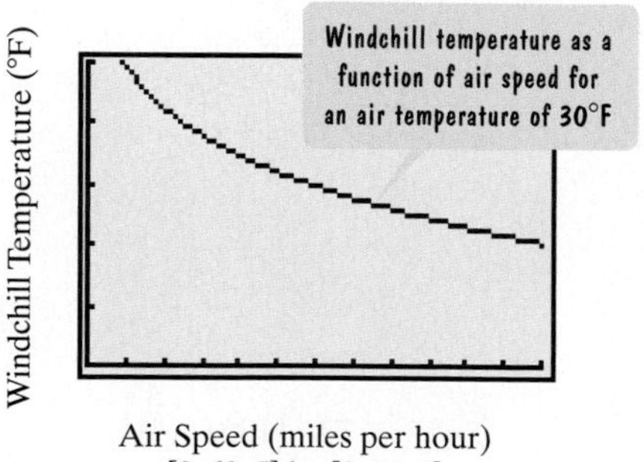

Air Speed (miles per hour)
[0, 60, 5] by [0, 25, 5]

Your job is to determine whether or not yachts are eligible for the America's Cup, the supreme event in ocean sailing. The basic dimensions of competitors' yachts must satisfy

$$L + 1.25\sqrt{S} - 9.8\sqrt[3]{D} \le 16.296,$$

where L is the yacht's length, in meters, S is its sail area, in square meters, and D is its displacement, in cubic meters. Use this information to solve Exercises 131–132.

131. **a.** Rewrite the inequality using rational exponents.

b. Use your calculator to determine if a yacht with length 20.85 meters, sail area 276.4 square meters, and displacement 18.55 cubic meters is eligible for the America's Cup.

132. **a.** Rewrite the inequality using rational exponents.

b. Use your calculator to determine if a yacht with length 22.85 meters, sail area 312.5 square meters, and displacement 22.34 cubic meters is eligible for the America's Cup.

Writing in Mathematics

133. What is the meaning of $a^{\frac{1}{n}}$? Give an example to support your explanation.

134. What is the meaning of $a^{\frac{m}{n}}$? Give an example.

135. What is the meaning of $a^{-\frac{m}{n}}$? Give an example.

136. Explain why $a^{\frac{1}{n}}$ is negative when n is odd and a is negative. What happens if n is even and a is negative? Why?

137. In simplifying $36^{\frac{3}{2}}$, is it better to use $a^{\frac{m}{n}} = \sqrt[n]{a^m}$ or $a^{\frac{m}{n}} = (\sqrt[n]{a})^m$? Explain.

138. How can you tell if an expression with rational exponents is simplified?

139. Explain how to simplify $\sqrt[3]{x} \cdot \sqrt{x}$.

140. Explain how to simplify $\sqrt[3]{\sqrt{x}}$.

Technology Exercises

141. Use a scientific or graphing calculator to verify your results in Exercises 15–18.

142. Use a scientific or graphing calculator to verify your results in Exercises 45–50.

Exercises 143–145 show a number of simplifications, not all of which are correct. Enter the left side of each equation as y_1 and the right side as y_2. Then use your graphing utility's TABLE *feature to determine if the simplification is correct. If it is not, correct the right side and use the* TABLE *feature to verify your simplification.*

143. $x^{\frac{3}{5}} \cdot x^{-\frac{1}{10}} = x^{\frac{1}{2}}$

144. $\left(x^{-\frac{1}{2}} \cdot x^{\frac{3}{4}}\right)^{-2} = x^{\frac{1}{2}}$

145. $\dfrac{x^{\frac{1}{4}}}{x^{\frac{1}{2}} \cdot x^{-\frac{3}{4}}} = \dfrac{1}{x^{\frac{1}{2}}}$

Critical Thinking Exercises

146. Which one of the following is true?

a. If n is odd, then $(-b)^{\frac{1}{n}} = -b^{\frac{1}{n}}$.

b. $(a + b)^{\frac{1}{n}} = a^{\frac{1}{n}} + b^{\frac{1}{n}}$

c. $8^{-\frac{2}{3}} = -4$

d. None of the above is true.

147. A mathematics professor recently purchased a birthday cake for her son with the inscription

$$\text{Happy } \left(2^{\frac{5}{2}} \cdot 2^{\frac{3}{4}} \div 2^{\frac{1}{4}}\right)\text{th Birthday.}$$

How old is the son?

148. The birthday boy in Exercise 147, excited by the inscription on the cake, tried to wolf down the whole thing. Professor Mom, concerned about the possible metamorphosis of her son into a blimp, exclaimed, "Hold on! It is your birthday, so why not take $\dfrac{8^{-\frac{4}{3}} + 2^{-2}}{16^{-\frac{3}{4}} + 2^{-1}}$ of the cake? I'll eat half of what's left over." How much of the cake did the professor eat?

149. Simplify: $\left[3 + \left(27^{\frac{2}{3}} + 32^{\frac{2}{5}}\right)\right]^{\frac{3}{2}} - 9^{\frac{1}{2}}$.

150. Find the domain of $f(x) = (x - 3)^{\frac{1}{2}}(x + 4)^{-\frac{1}{2}}$.

Review Exercises

151. Write the equation of the linear function whose graph passes through $(5, 1)$ and $(4, 3)$. (Section 3.5, Example 2)

152. Graph $y \le -\frac{3}{2}x + 3$. (Section 9.4, Example 2)

153. If $f(x) = 3x^2 - 5x + 4$, find $f(a + h)$. (Section 8.1, Example 3)

SECTION 10.3

Objectives

1. Use the product rule to multiply radicals.
2. Use factoring and the product rule to simplify radicals.
3. Multiply radicals and then simplify.

MULTIPLYING AND SIMPLIFYING RADICAL EXPRESSIONS

George Tooker (b. 1920) "Mirror II" 1963, egg tempera on gesso panel, 1968.4.

Gift of R. H. Donnelley Erdman (PA 1956). Addison Gallery of American Art, Phillips Academy, Andover, Massachusetts. All Rights Reserved.

A difficult challenge of middle adulthood is confronting the aging process. Middle-aged adults are forced to acknowledge their mortality as they witness the deaths of parents, colleagues, and friends. In addition, there are a number of physical transformations, including changes in vision that require glasses for reading, the onset of wrinkles and sagging skin, and a decrease in heart response. A change in heart response occurs fairly early; after 20, our hearts become less adept at accelerating in response to exercise. In this section, you will see how a radical function models changes in heart function throughout the aging process, as we turn to multiplying and simplifying radical expressions.

1 Use the product rule to multiply radicals.

The Product Rule for Radicals A rule for multiplying radicals can be generalized by comparing $\sqrt{25} \cdot \sqrt{4}$ and $\sqrt{25 \cdot 4}$. Notice that

$$\sqrt{25} \cdot \sqrt{4} = 5 \cdot 2 = 10 \quad \text{and} \quad \sqrt{25 \cdot 4} = \sqrt{100} = 10.$$

Because we obtain 10 in both situations, the original radical expressions must be equal. That is,

$$\sqrt{25} \cdot \sqrt{4} = \sqrt{25 \cdot 4}.$$

This result is a special case of the **product rule for radicals** that can be generalized as follows:

THE PRODUCT RULE FOR RADICALS If $\sqrt[n]{a}$ and $\sqrt[n]{b}$ are real numbers, then

$$\sqrt[n]{a} \cdot \sqrt[n]{b} = \sqrt[n]{ab}.$$

The product of two *n*th roots is the *n*th root of the product.

STUDY TIP

The product rule can be used only when the radicals have the same index. If indices differ, rational exponents can be used, as in $\sqrt{x}\cdot\sqrt[3]{x}$, which was Example 8(d) in the previous section.

EXAMPLE 1 Using the Product Rule for Radicals

Multiply:

a. $\sqrt{3}\cdot\sqrt{7}$ **b.** $\sqrt{x+7}\cdot\sqrt{x-7}$ **c.** $\sqrt[3]{7}\cdot\sqrt[3]{9}$ **d.** $\sqrt[8]{10x}\cdot\sqrt[8]{8x^4}$.

SOLUTION In each problem, the indices are the same. Thus, we multiply by multiplying the radicands.

a. $\sqrt{3}\cdot\sqrt{7} = \sqrt{3\cdot 7} = \sqrt{21}$

b. $\sqrt{x+7}\cdot\sqrt{x-7} = \sqrt{(x+7)(x-7)} = \sqrt{x^2-49}$

This is not equal to $\sqrt{x^2} - \sqrt{49}$.

c. $\sqrt[3]{7}\cdot\sqrt[3]{9} = \sqrt[3]{7\cdot 9} = \sqrt[3]{63}$

d. $\sqrt[8]{10x}\cdot\sqrt[8]{8x^4} = \sqrt[8]{10x\cdot 8x^4} = \sqrt[8]{80x^5}$ ■

CHECK POINT 1 Multiply:

a. $\sqrt{5}\cdot\sqrt{11}$ **b.** $\sqrt{x+4}\cdot\sqrt{x-4}$

c. $\sqrt[3]{6}\cdot\sqrt[3]{10}$ **d.** $\sqrt[7]{2x}\cdot\sqrt[7]{6x^3}$.

2 Use factoring and the product rule to simplify radicals.

Using Factoring and the Product Rule to Simplify Radicals A number that is the square of an integer is a **perfect square**. For example, 100 is a perfect square because $100 = 10^2$. A number is a **perfect cube** if it is the cube of an integer. Thus, 125 is a perfect cube because $125 = 5^3$. In general, a number is a **perfect *n*th power** if it is the nth power of an integer. Thus, p is a perfect nth power if there is an integer q such that $p = q^n$.

A radical of index n is **simplified** when its radicand has no factors other than 1 that are perfect nth powers. For example, $\sqrt{300}$ is not simplified because it can be expressed as $\sqrt{100\cdot 3}$ and 100 is a perfect square. We can use the product rule in the form

$$\sqrt[n]{ab} = \sqrt[n]{a}\cdot\sqrt[n]{b}$$

to simplify $\sqrt[n]{ab}$ when $\sqrt[n]{a}$ or $\sqrt[n]{b}$ is a perfect nth power. Consider $\sqrt{300}$. To simplify, we factor 300 so that one of its factors is the greatest perfect square possible.

$\sqrt{300} = \sqrt{100\cdot 3}$ — Factor 300. 100 is the greatest perfect square factor.

$= \sqrt{100}\cdot\sqrt{3}$ — Use the product rule: $\sqrt[n]{ab} = \sqrt[n]{a}\cdot\sqrt[n]{b}$.

$= 10\sqrt{3}$ — Write $\sqrt{100}$ as 10. We read $10\sqrt{3}$ as "ten times the square root of three."

USING TECHNOLOGY

You can use a calculator to provide numerical support that $\sqrt{300} = 10\sqrt{3}$. First find an approximation for $\sqrt{300}$:

300 [√] ≈ 17.32

or

[√] 300 [ENTER] ≈ 17.32.

Now find an approximation for $10\sqrt{3}$:

10 [×] 3 [√] ≈ 17.32

or

10 [√] 3 [ENTER] ≈ 17.32.

Correct to two decimal places,

$\sqrt{300} \approx 17.32$ and $10\sqrt{3} \approx 17.32$.

This verifies that

$$\sqrt{300} = 10\sqrt{3}.$$

Use this technique to support the numerical results for the answers in this section.

Caution: A simplified radical does not mean a decimal approximation.

SIMPLIFYING RADICAL EXPRESSIONS BY FACTORING A radical expression whose index is n is **simplified** when its radicand has no factors that are perfect nth powers. To simplify, use the following procedure:

1. Write the radicand as the product of two factors, one of which is the greatest perfect nth power.
2. Use the product rule to take the nth root of each factor.
3. Find the nth root of the perfect nth power.

EXAMPLE 2 Simplifying Radicals by Factoring

Simplify by factoring:

a. $\sqrt{75}$ **b.** $\sqrt[3]{54}$ **c.** $\sqrt[5]{64}$ **d.** $\sqrt{500xy^2}$.

SOLUTION

a. $\sqrt{75} = \sqrt{25 \cdot 3}$ — 25 is the greatest perfect square that is a factor of 75.

$= \sqrt{25} \cdot \sqrt{3}$ — Take the square root of each factor: $\sqrt[n]{ab} = \sqrt[n]{a} \cdot \sqrt[n]{b}$.

$= 5\sqrt{3}$ — Write $\sqrt{25}$ as 5.

b. $\sqrt[3]{54} = \sqrt[3]{27 \cdot 2}$ — 27 is the greatest perfect cube that is a factor of 54: $27 = 3^3$.

$= \sqrt[3]{27} \cdot \sqrt[3]{2}$ — Take the cube root of each factor: $\sqrt[n]{ab} = \sqrt[n]{a} \cdot \sqrt[n]{b}$.

$= 3\sqrt[3]{2}$ — Write $\sqrt[3]{27}$ as 3.

c. $\sqrt[5]{64} = \sqrt[5]{32 \cdot 2}$ — 32 is the greatest perfect fifth power that is a factor of 64: $32 = 2^5$.

$= \sqrt[5]{32} \cdot \sqrt[5]{2}$ — Take the fifth root of each factor: $\sqrt[n]{ab} = \sqrt[n]{a} \cdot \sqrt[n]{b}$.

$= 2\sqrt[5]{2}$ — Write $\sqrt[5]{32}$ as 2.

d. $\sqrt{500xy^2} = \sqrt{100y^2 \cdot 5x}$ — $100y^2$ is the greatest perfect square that is a factor of $500xy^2$: $100y^2 = (10y)^2$.

$= \sqrt{100y^2} \cdot \sqrt{5x}$ — Factor into two radicals.

$= 10|y|\sqrt{5x}$ — Take the square root of $100y^2$. ■

CHECK POINT 2 Simplify by factoring:

a. $\sqrt{80}$ **b.** $\sqrt[3]{40}$

c. $\sqrt[4]{32}$ **d.** $\sqrt{200x^2y}$.

EXAMPLE 3 Simplifying a Radical Function

If

$$f(x) = \sqrt{2x^2 + 4x + 2},$$

express the function, f, in simplified form.

SOLUTION Begin by factoring the radicand. The GCF is 2. Simplification is possible if we obtain a factor that is a perfect square.

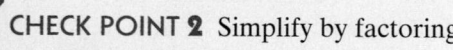

$f(x) = \sqrt{2x^2 + 4x + 2}$ — This is the given function.

$= \sqrt{2(x^2 + 2x + 1)}$ — Factor out the GCF.

$= \sqrt{2(x+1)^2}$ — Factor the perfect square trinomial: $A^2 + 2AB + B^2 = (A + B)^2$.

$= \sqrt{2} \cdot \sqrt{(x+1)^2}$ — Take the square root of each factor. The factor $(x + 1)^2$ is a perfect square.

$= \sqrt{2}\,|x + 1|$ — Take the square root of $(x + 1)^2$.

In simplified form,

$$f(x) = \sqrt{2}\,|x + 1|.$$

■

USING TECHNOLOGY

The graphs of

$$f(x) = \sqrt{2x^2 + 4x + 2}, \quad g(x) = \sqrt{2}|x + 1|, \quad \text{and} \quad h(x) = \sqrt{2}(x + 1)$$

are shown in three separate $[-5, 5, 1]$ by $[-5, 5, 1]$ viewing rectangles. The graphs in parts (a) and (b) are identical. This verifies that our simplification in Example 3 is correct: $\sqrt{2x^2 + 4x + 2} = \sqrt{2}|x + 1|$. Now compare the graphs in parts (a) and (c). Can you see that they are not the same? This illustrates the importance of not leaving out absolute value bars: $\sqrt{2x^2 + 4x + 2} \neq \sqrt{2}(x + 1)$.

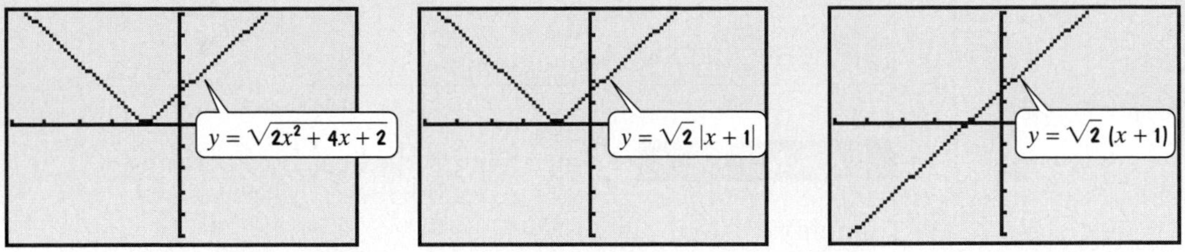

(a) (b) (c)

CHECK POINT 3 If $f(x) = \sqrt{3x^2 - 12x + 12}$, express the function, f, in simplified form.

For the remainder of this chapter, in situations that do not involve functions, we will **assume that no radicands involve negative quantities raised to even powers. Based upon this assumption, absolute value bars are not necessary when taking even roots.**

SIMPLIFYING WHEN VARIABLES TO EVEN POWERS IN A RADICAND ARE NONNEGATIVE QUANTITIES For any nonnegative real number a,

$$\sqrt[n]{a^n} = a.$$

In simplifying an nth root, how do we find variable factors in the radicand that are perfect nth powers? The **perfect nth powers have exponents that are divisible by n.** Simplification is possible by observation or by using rational exponents. Here are some examples:

- $\sqrt{x^6} = \sqrt{(x^3)^2} = x^3$ or $\sqrt{x^6} = (x^6)^{\frac{1}{2}} = x^3$

  6 is divisible by the index, 2. Thus, x^6 is a perfect square.

- $\sqrt[3]{y^{21}} = \sqrt[3]{(y^7)^3} = y^7$ or $\sqrt[3]{y^{21}} = (y^{21})^{\frac{1}{3}} = y^7$

  21 is divisible by the index, 3. Thus, y^{21} is a perfect cube.

- $\sqrt[6]{z^{24}} = \sqrt[6]{(z^4)^6} = z^4$ or $\sqrt[6]{z^{24}} = (z^{24})^{\frac{1}{6}} = z^4$.

  24 is divisible by the index, 6. Thus, z^{24} is a perfect 6th power.

EXAMPLE 4 Simplifying a Radical by Factoring

Simplify: $\sqrt{x^5y^{13}z^7}$.

SOLUTION We write the radicand as the product of the greatest perfect square factor and another factor. Because the index is 2, variables that have exponents that are

DISCOVER FOR YOURSELF

Square the answer in Example 4 and show that it is correct. If it is a square root, you should obtain the given radicand, $x^5y^{13}z^7$.

divisible by 2 are part of the perfect square factor. We use the greatest exponents that are divisible by 2.

$$\sqrt{x^5y^{13}z^7} = \sqrt{x^4 \cdot x \cdot y^{12} \cdot y \cdot z^6 \cdot z} \quad \text{Use the greatest even power of each variable.}$$

$$= \sqrt{(x^4y^{12}z^6)(xyz)} \quad \text{Group the perfect square factors.}$$

$$= \sqrt{x^4y^{12}z^6} \cdot \sqrt{xyz} \quad \text{Factor into two radicals.}$$

$$= x^2y^6z^3\sqrt{xyz} \quad \sqrt{x^4y^{12}z^6} = (x^4y^{12}z^6)^{\frac{1}{2}} = x^2y^6z^3$$

■

CHECK POINT 4 Simplify: $\sqrt{x^9y^{11}z^3}$.

EXAMPLE 5 Simplifying a Radical by Factoring

Simplify: $\sqrt[3]{32x^8y^{16}}$.

SOLUTION We write the radicand as the product of the greatest perfect cube factor and another factor. Because the index is 3, variables that have exponents that are divisible by 3 are part of the perfect cube factor. We use the greatest exponents that are divisible by 3.

$$\sqrt[3]{32x^8y^{16}} = \sqrt[3]{8 \cdot 4 \cdot x^6 \cdot x^2 \cdot y^{15} \cdot y} \quad \text{Identify perfect cube factors.}$$

$$= \sqrt[3]{(8x^6y^{15})(4x^2y)} \quad \text{Group the perfect cube factors.}$$

$$= \sqrt[3]{8x^6y^{15}} \cdot \sqrt[3]{4x^2y} \quad \text{Factor into two radicals.}$$

$$= 2x^2y^5\sqrt[3]{4x^2y} \quad \sqrt[3]{8} = 2 \text{ and } \sqrt[3]{x^6y^{15}} = (x^6y^{15})^{\frac{1}{3}} = x^2y^5.$$

■

CHECK POINT 5 Simplify: $\sqrt[3]{40x^{10}y^{14}}$.

EXAMPLE 6 Simplifying a Radical by Factoring

Simplify: $\sqrt[5]{64x^3y^7z^{29}}$.

SOLUTION We write the radicand as the product of the greatest perfect 5th power and another factor. Because the index is 5, variables that have exponents that are divisible by 5 are part of the perfect fifth factor. We use the greatest exponents that are divisible by 5.

$$\sqrt[5]{64x^3y^7z^{29}} = \sqrt[5]{32 \cdot 2 \cdot x^3 \cdot y^5 \cdot y^2 \cdot z^{25} \cdot z^4} \quad \text{Identify perfect fifth factors.}$$

$$= \sqrt[5]{(32y^5z^{25})(2x^3y^2z^4)} \quad \text{Group the perfect fifth factors.}$$

$$= \sqrt[5]{32y^5z^{25}} \cdot \sqrt[5]{2x^3y^2z^4} \quad \text{Factor into two radicals.}$$

$$= 2yz^5\sqrt[5]{2x^3y^2z^4} \quad \sqrt[5]{32} = 2 \text{ and } \sqrt[5]{y^5z^{25}} = (y^5z^{25})^{\frac{1}{5}} = yz^5.$$

■

CHECK POINT 6 Simplify: $\sqrt[5]{32x^{12}y^2z^8}$.

3 Multiply radicals and then simplify.

Multiplying and Simplifying Radicals We have seen how to use the product rule when multiplying radicals with the same index. Sometimes after multiplying, we can simplify the resulting radical.

EXAMPLE 7 Multiplying Radicals and Then Simplifying

Multiply and simplify:

a. $\sqrt{15}\cdot\sqrt{3}$ **b.** $7\sqrt[3]{4}\cdot 5\sqrt[3]{6}$ **c.** $\sqrt[4]{8x^3y^2}\cdot\sqrt[4]{8x^5y^3}$.

SOLUTION

a.
$$\sqrt{15}\cdot\sqrt{3}=\sqrt{15\cdot 3}$$ Use the product rule.
$$=\sqrt{45}=\sqrt{9\cdot 5}$$ 9 is the greatest perfect square factor of 45.
$$=\sqrt{9}\cdot\sqrt{5}=3\sqrt{5}$$

b.
$$7\sqrt[3]{4}\cdot 5\sqrt[3]{6}=35\sqrt[3]{4\cdot 6}$$ Use the product rule.
$$=35\sqrt[3]{24}=35\sqrt[3]{8\cdot 3}$$ 8 is the greatest perfect cube factor of 24.
$$=35\sqrt[3]{8}\cdot\sqrt[3]{3}=35\cdot 2\cdot\sqrt[3]{3}$$
$$=70\sqrt[3]{3}$$

c.
$$\sqrt[4]{8x^3y^2}\cdot\sqrt[4]{8x^5y^3}=\sqrt[4]{8x^3y^2\cdot 8x^5y^3}$$ Use the product rule.
$$=\sqrt[4]{64x^8y^5}$$ Multiply.
$$=\sqrt[4]{16\cdot 4\cdot x^8\cdot y^4\cdot y}$$ Identify perfect fourth factors.
$$=\sqrt[4]{(16x^8y^4)(4y)}$$ Group the perfect fourth factors.
$$=\sqrt[4]{16x^8y^4}\cdot\sqrt[4]{4y}$$ Factor into two radicals.
$$=2x^2y\sqrt[4]{4y}$$ $\sqrt[4]{16}=2$ and $\sqrt[4]{x^8y^4}=(x^8y^4)^{\frac{1}{4}}=x^2y$. ■

CHECK POINT 7 Multiply and simplify:

a. $\sqrt{6}\cdot\sqrt{2}$

b. $10\sqrt[3]{16}\cdot 5\sqrt[3]{2}$

c. $\sqrt[4]{4x^2y}\cdot\sqrt[4]{8x^6y^3}$.

10.3 EXERCISE SET

Student Solutions Manual | CD/Video | PH Math/Tutor Center | MathXL Tutorials on CD | MathXL® | MyMathLab | Interactmath.com

Practice Exercises

In Exercises 1–20, use the product rule to multiply.

1. $\sqrt{3}\cdot\sqrt{5}$

2. $\sqrt{7}\cdot\sqrt{5}$

3. $\sqrt[3]{2}\cdot\sqrt[3]{9}$

4. $\sqrt[3]{5}\cdot\sqrt[3]{4}$

5. $\sqrt[4]{11}\cdot\sqrt[4]{3}$

6. $\sqrt[5]{9}\cdot\sqrt[5]{3}$

7. $\sqrt{3x}\cdot\sqrt{11y}$

8. $\sqrt{5x}\cdot\sqrt{11y}$

9. $\sqrt[5]{6x^3}\cdot\sqrt[5]{4x}$

10. $\sqrt[4]{6x^2}\cdot\sqrt[4]{3x}$

11. $\sqrt{x+3}\cdot\sqrt{x-3}$

12. $\sqrt{x+6}\cdot\sqrt{x-6}$

13. $\sqrt[6]{x-4}\cdot\sqrt[6]{(x-4)^4}$

14. $\sqrt[6]{x-5}\cdot\sqrt[6]{(x-5)^4}$

15. $\sqrt{\frac{2x}{3}}\cdot\sqrt{\frac{3}{2}}$

16. $\sqrt{\frac{2x}{5}}\cdot\sqrt{\frac{5}{2}}$

17. $\sqrt[4]{\frac{x}{7}}\cdot\sqrt[4]{\frac{3}{y}}$

18. $\sqrt[4]{\frac{x}{3}}\cdot\sqrt[4]{\frac{7}{y}}$

19. $\sqrt[7]{7x^2y}\cdot\sqrt[7]{11x^3y^2}$

20. $\sqrt[9]{12x^2y^3}\cdot\sqrt[9]{3x^3y^4}$

In Exercises 21–32, simplify by factoring.

21. $\sqrt{50}$ **22.** $\sqrt{27}$

23. $\sqrt{45}$ **24.** $\sqrt{28}$

25. $\sqrt{75x}$ **26.** $\sqrt{40x}$

27. $\sqrt[3]{16}$ **28.** $\sqrt[3]{54}$

29. $\sqrt[3]{27x^3}$ **30.** $\sqrt[3]{250x^3}$

31. $\sqrt[3]{-16x^2y^3}$ **32.** $\sqrt[3]{-32x^2y^3}$

In Exercises 33–38, express the function, f, in simplified form. Assume that x can be any real number.

33. $f(x) = \sqrt{36(x+2)^2}$

34. $f(x) = \sqrt{81(x-2)^2}$

35. $f(x) = \sqrt[3]{32(x+2)^3}$

36. $f(x) = \sqrt[3]{48(x-2)^3}$

37. $f(x) = \sqrt{3x^2 - 6x + 3}$

38. $f(x) = \sqrt{5x^2 - 10x + 5}$

In Exercises 39–60, simplify by factoring. Assume that all variables in a radicand represent positive real numbers and no radicands involve negative quantities raised to even powers.

39. $\sqrt{x^7}$ **40.** $\sqrt{x^5}$

41. $\sqrt{x^8y^9}$ **42.** $\sqrt{x^6y^7}$

43. $\sqrt{48x^3}$ **44.** $\sqrt{40x^3}$

45. $\sqrt[3]{y^8}$ **46.** $\sqrt[3]{y^{11}}$

47. $\sqrt[3]{x^{14}y^3z}$ **48.** $\sqrt[3]{x^3y^{17}z^2}$

49. $\sqrt[3]{81x^8y^6}$ **50.** $\sqrt[3]{32x^9y^{17}}$

51. $\sqrt[3]{(x+y)^5}$

52. $\sqrt[3]{(x+y)^4}$

53. $\sqrt[5]{y^{17}}$ **54.** $\sqrt[5]{y^{18}}$

55. $\sqrt[5]{64x^6y^{17}}$ **56.** $\sqrt[5]{64x^7y^{16}}$

57. $\sqrt[4]{80x^{10}}$ **58.** $\sqrt[4]{96x^{11}}$

59. $\sqrt[4]{(x-3)^{10}}$

60. $\sqrt[4]{(x-2)^{14}}$

In Exercises 61–82, multiply and simplify. Assume that all variables in a radicand represent positive real numbers and no radicands involve negative quantities raised to even powers.

61. $\sqrt{12}\cdot\sqrt{2}$ **62.** $\sqrt{3}\cdot\sqrt{6}$

63. $\sqrt{5x}\cdot\sqrt{10y}$ **64.** $\sqrt{8x}\cdot\sqrt{10y}$

65. $\sqrt{12x}\cdot\sqrt{3x}$ **66.** $\sqrt{20x}\cdot\sqrt{5x}$

67. $\sqrt{50xy}\cdot\sqrt{4xy^2}$

68. $\sqrt{5xy}\cdot\sqrt{10xy^2}$

69. $2\sqrt{5}\cdot 3\sqrt{40}$ **70.** $3\sqrt{15}\cdot 5\sqrt{6}$

71. $\sqrt[3]{12}\cdot\sqrt[3]{4}$ **72.** $\sqrt[4]{4}\cdot\sqrt[4]{8}$

73. $\sqrt{5x^3}\cdot\sqrt{8x^2}$

74. $\sqrt{2x^7}\cdot\sqrt{12x^4}$

75. $\sqrt[3]{25x^4y^2}\cdot\sqrt[3]{5xy^{12}}$

76. $\sqrt[3]{6x^7y}\cdot\sqrt[3]{9x^4y^{12}}$

77. $\sqrt[4]{8x^2y^3z^6}\cdot\sqrt[4]{2x^4yz}$

78. $\sqrt[4]{4x^2y^3z^3}\cdot\sqrt[4]{8x^4yz^6}$

79. $\sqrt[5]{8x^4y^6z^2}\cdot\sqrt[5]{8xy^7z^4}$

80. $\sqrt[5]{8x^4y^3z^3}\cdot\sqrt[5]{8xy^9z^8}$

81. $\sqrt[3]{x-y}\cdot\sqrt[3]{(x-y)^7}$

82. $\sqrt[3]{x-6}\cdot\sqrt[3]{(x-6)^7}$

Practice Plus

In Exercises 83–92, simplify each expression. Assume that all variables in a radicand represent positive real numbers and no radicands involve negative quantities raised to even powers.

83. $-2x^2y\left(\sqrt[3]{54x^3y^7z^2}\right)$

84. $\dfrac{-x^2y^7}{2}\left(\sqrt[3]{-32x^4y^9z^7}\right)$

85. $-3y\left(\sqrt[5]{64x^3y^6}\right)$

86. $-4x^2y^7\left(\sqrt[5]{-32x^{11}y^{17}}\right)$

87. $\left(-2xy^2\sqrt{3x}\right)\left(xy\sqrt{6x}\right)$

88. $\left(-5x^2y^3z\sqrt{2xyz}\right)\left(-x^4z\sqrt{10xz}\right)$

89. $\left(2x^2y\sqrt[4]{8xy}\right)\left(-3xy^2\sqrt[4]{2x^2y^3}\right)$

90. $\left(5a^2b\sqrt[4]{8a^2b}\right)\left(4ab\sqrt[4]{4a^3b^2}\right)$

91. $\sqrt[5]{8x^4y^6}\left(\sqrt[5]{2xy^7} + \sqrt[5]{4x^6y^9}\right)$

92. $\sqrt[4]{2x^3y^2}\left(\sqrt[4]{6x^5y^6} + \sqrt[4]{8x^5y^7}\right)$

Application Exercises

The function

$$d(x) = \sqrt{\frac{3x}{2}}$$

models the distance, $d(x)$, in miles, that a person h feet high can see to the horizon. Use this function to solve Exercises 93–94.

93. The pool deck on a cruise ship is 72 feet above the water. How far can passengers on the pool deck see? Write the answer in simplified radical form. Then use the simplified radical form and a calculator to express the answer to the nearest tenth of a mile.

94. The captain of a cruise ship is on the star deck, which is 120 feet above the water. How far can the captain see? Write the answer in simplified radical form. Then use the simplified radical form and a calculator to express the answer to the nearest tenth of a mile.

Paleontologists use the function

$$W(x) = 4\sqrt{2x}$$

to estimate the walking speed of a dinosaur, $W(x)$, in feet per second, where x is the length, in feet, of the dinosaur's leg. The graph of W is shown in the figure. Use this information to solve Exercises 95–96.

Dinosaur Walking Speeds

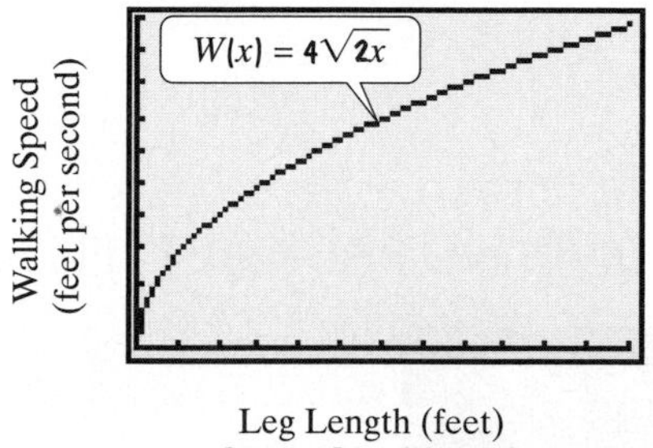

[0, 12, 1] by [0, 20, 2]

95. What is the walking speed of a dinosaur whose leg length is 6 feet? Use the function's equation and express the answer in simplified radical form. Then use the function's graph to estimate the answer to the nearest foot per second.

96. What is the walking speed of a dinosaur whose leg length is 10 feet? Use the function's equation and express the answer in simplified radical form. Then use the function's graph to estimate the answer to the nearest foot per second.

Your **cardiac index** *is your heart's output, in liters of blood per minute, divided by your body's surface area, in square meters. The cardiac index, $C(x)$, can be modeled by*

$$C(x) = \frac{7.644}{\sqrt[4]{x}}, \qquad 10 \le x \le 80,$$

where x is an individual's age, in years. The graph of the function is shown. Use the function to solve Exercises 97–98.

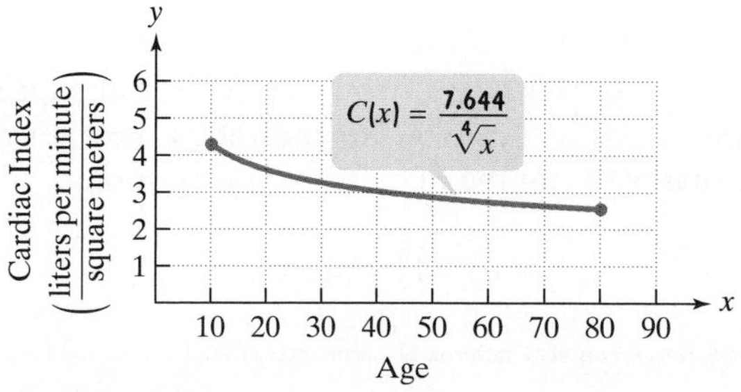

97. a. Find the cardiac index of a 32-year-old. Express the denominator in simplified radical form and reduce the fraction.

b. Use the form of the answer in part (a) and a calculator to express the cardiac index to the nearest hundredth. Identify your solution as a point on the graph.

98. a. Find the cardiac index of an 80-year-old. Express the denominator in simplified radical form and reduce the fraction.

b. Use the form of the answer in part (a) and a calculator to express the cardiac index to the nearest hundredth. Identify your solution as a point on the graph.

Writing in Mathematics

99. What is the product rule for radicals? Give an example to show how it is used.

100. Explain why $\sqrt{50}$ is not simplified. What do we mean when we say a radical expression is simplified?

101. In simplifying an nth root, explain how to find variable factors in the radicand that are perfect nth powers.

102. Without showing all the details, explain how to simplify $\sqrt[3]{16x^{14}}$.

103. As you get older, what would you expect to happen to your heart's output? Explain how this is shown in the graph for Exercises 97–98. Is this trend taking place progressively more rapidly or more slowly over the entire interval? What does this mean about this aspect of aging?

Technology Exercises

104. Use a calculator to provide numerical support for your simplifications in Exercises 21–24 and 27–28. In each case, find a decimal approximation for the given expression. Then find a decimal approximation for your simplified expression. The approximations should be the same.

In Exercises 105–108, determine if each simplification is correct by graphing the function on each side of the equation with your graphing utility. Use the given viewing rectangle. The graphs should be the same. If they are not, correct the right side of the equation and then use your graphing utility to verify the simplification.

105. $\sqrt{x^4} = x^2$
[0, 5, 1] by [0, 20, 1]

106. $\sqrt{8x^2} = 4x\sqrt{2}$
[−5, 5, 1] by [−5, 20, 1]

107. $\sqrt{3x^2 - 6x + 3} = (x-1)\sqrt{3}$
[−5, 5, 1] by [−5, 5, 1]

108. $\sqrt[3]{2x} \cdot \sqrt[3]{4x^2} = 4x$
[−10, 10, 1] by [−10, 10, 1]

Critical Thinking Exercises

109. Which one of the following is true?

a. $2\sqrt{5} \cdot 6\sqrt{5} = 12\sqrt{5}$

b. $\sqrt[3]{4} \cdot \sqrt[3]{4} = 4$

c. $\sqrt{12} = 2\sqrt{6}$

d. $\sqrt[3]{3^{15}} = 243$

110. If a number is tripled, what happens to its square root?

111. What must be done to a number so that its cube root is tripled?

112. If $f(x) = \sqrt[3]{2x}$ and $(fg)(x) = 2x$, find $g(x)$.

113. Graph $f(x) = \sqrt{(x-1)^2}$ by hand.

Review Exercises

114. Solve: $2x - 1 \le 21$ and $2x + 2 \ge 12$. (Section 9.2, Example 2)

115. Solve:

$$\begin{aligned} 5x + 2y &= 2 \\ 4x + 3y &= -4. \end{aligned}$$

(Section 4.3, Example 4)

116. Factor: $64x^3 - 27$. (Section 6.4, Example 8)

SECTION 10.4

ADDING, SUBTRACTING, AND DIVIDING RADICAL EXPRESSIONS

Objectives

1 Add and subtract radical expressions.

2 Use the quotient rule to simplify radical expressions.

3 Use the quotient rule to divide radical expressions.

The future is now: You have the opportunity to explore the cosmos in a starship traveling near the speed of light. The experience will enable you to understand the mysteries of the universe first hand, transporting you to unimagined levels of knowing and being. The down side: According to Einstein's theory of relativity, close to the speed of light, your aging rate relative to friends on Earth is nearly zero. You will return from your two-year journey to a futuristic world in which friends and loved ones are long dead. Do you explore space or stay here on Earth? In this section, in addition to learning to perform various operations with radicals, you will see how they model your return to a world where the people you knew have long since departed.

1 Add and subtract radical expressions

Adding and Subtracting Radical Expressions Two or more radical expressions that have the same indices and radicands are called **like radicals**. Like radicals are combined in exactly the same way that we combine like terms. For example,

$$7\sqrt{11} + 6\sqrt{11} = (7+6)\sqrt{11} = 13\sqrt{11}.$$

7 square roots of 11 plus 6 square roots of 11 result in 13 square roots of 11.

EXAMPLE 1 Adding and Subtracting Like Radicals

Simplify (add or subtract) by combining like radical terms:

a. $7\sqrt{2} + 8\sqrt{2}$ **b.** $\sqrt[3]{5} - 4x\sqrt[3]{5} + 8\sqrt[3]{5}$ **c.** $8\sqrt[6]{5x} - 5\sqrt[6]{5x} + 4\sqrt[3]{5x}$.

SOLUTION

a. $7\sqrt{2} + 8\sqrt{2}$

$= (7 + 8)\sqrt{2}$ Apply the distributive property.

$= 15\sqrt{2}$ Simplify.

b. $\sqrt[3]{5} - 4x\sqrt[3]{5} + 8\sqrt[3]{5}$

$= (1 - 4x + 8)\sqrt[3]{5}$ Apply the distributive property.

$= (9 - 4x)\sqrt[3]{5}$ Simplify.

c. $8\sqrt[6]{5x} - 5\sqrt[6]{5x} + 4\sqrt[3]{5x}$

$= (8 - 5)\sqrt[6]{5x} + 4\sqrt[3]{5x}$ Apply the distributive property to the two terms with like radicals.

$= 3\sqrt[6]{5x} + 4\sqrt[3]{5x}$ The indices, 6 and 3, differ. These are not like radicals and cannot be combined. ■

CHECK POINT 1 Simplify by combining like radical terms:

a. $8\sqrt{13} + 2\sqrt{13}$

b. $9\sqrt[3]{7} - 6x\sqrt[3]{7} + 12\sqrt[3]{7}$

c. $7\sqrt[4]{3x} - 2\sqrt[4]{3x} + 2\sqrt[3]{3x}$.

In some cases, radical expressions can be combined once they have been simplified. For example, to add $\sqrt{2}$ and $\sqrt{8}$, we can write $\sqrt{8}$ as $\sqrt{4 \cdot 2}$ because 4 is a perfect square factor of 8.

$$\sqrt{2} + \sqrt{8} = \sqrt{2} + \sqrt{4 \cdot 2} = 1\sqrt{2} + 2\sqrt{2} = (1 + 2)\sqrt{2} = 3\sqrt{2}$$

EXAMPLE 2 Combining Radicals That First Require Simplification

Simplify by combining like radical terms, if possible:

a. $7\sqrt{18} + 5\sqrt{8}$ **b.** $4\sqrt{27x} - 8\sqrt{12x}$ **c.** $7\sqrt{3} - 2\sqrt{5}$.

SOLUTION

a. $7\sqrt{18} + 5\sqrt{8}$

$= 7\sqrt{9 \cdot 2} + 5\sqrt{4 \cdot 2}$ Factor the radicands using the greatest perfect square factors.

$= 7\sqrt{9} \cdot \sqrt{2} + 5\sqrt{4} \cdot \sqrt{2}$ Take the square root of each factor.

$= 7 \cdot 3 \cdot \sqrt{2} + 5 \cdot 2 \cdot \sqrt{2}$ $\sqrt{9} = 3$ and $\sqrt{4} = 2$.

$= 21\sqrt{2} + 10\sqrt{2}$ Multiply.

$= (21 + 10)\sqrt{2}$ Apply the distributive property.

$= 31\sqrt{2}$ Simplify.

b. $4\sqrt{27x} - 8\sqrt{12x}$

$= 4\sqrt{9 \cdot 3x} - 8\sqrt{4 \cdot 3x}$ — Factor the radicands using the greatest perfect square factors.

$= 4\sqrt{9} \cdot \sqrt{3x} - 8\sqrt{4} \cdot \sqrt{3x}$ — Take the square root of each factor.

$= 4 \cdot 3 \cdot \sqrt{3x} - 8 \cdot 2 \cdot \sqrt{3x}$ — $\sqrt{9} = 3$ and $\sqrt{4} = 2$.

$= 12\sqrt{3x} - 16\sqrt{3x}$ — Multiply.

$= (12 - 16)\sqrt{3x}$ — Apply the distributive property.

$= -4\sqrt{3x}$ — Simplify.

c. $7\sqrt{3} - 2\sqrt{5}$ cannot be simplified. The radical expressions have different radicands, namely 3 and 5, and are not like terms. ■

CHECK POINT 2 Simplify by combining like radical terms, if possible:

a. $3\sqrt{20} + 5\sqrt{45}$

b. $3\sqrt{12x} - 6\sqrt{27x}$

c. $8\sqrt{5} - 6\sqrt{2}$.

EXAMPLE 3 Adding and Subtracting with Higher Indices

Simplify by combining like radical terms, if possible:

a. $2\sqrt[3]{16} - 4\sqrt[3]{54}$ **b.** $5\sqrt[3]{xy^2} + \sqrt[3]{8x^4y^5}$.

SOLUTION

a. $2\sqrt[3]{16} - 4\sqrt[3]{54}$

$= 2\sqrt[3]{8 \cdot 2} - 4\sqrt[3]{27 \cdot 2}$ — Factor the radicands using the greatest perfect cube factors.

$= 2\sqrt[3]{8} \cdot \sqrt[3]{2} - 4\sqrt[3]{27} \cdot \sqrt[3]{2}$ — Take the cube root of each factor.

$= 2 \cdot 2 \cdot \sqrt[3]{2} - 4 \cdot 3 \cdot \sqrt[3]{2}$ — $\sqrt[3]{8} = 2$ and $\sqrt[3]{27} = 3$.

$= 4\sqrt[3]{2} - 12\sqrt[3]{2}$ — Multiply.

$= (4 - 12)\sqrt[3]{2}$ — Apply the distributive property.

$= -8\sqrt[3]{2}$ — Simplify.

b. $5\sqrt[3]{xy^2} + \sqrt[3]{8x^4y^5}$

$= 5\sqrt[3]{xy^2} + \sqrt[3]{(8x^3y^3)xy^2}$ — Factor the second radicand using the greatest perfect cube factor.

$= 5\sqrt[3]{xy^2} + \sqrt[3]{8x^3y^3} \cdot \sqrt[3]{xy^2}$ — Take the cube root of each factor.

$= 5\sqrt[3]{xy^2} + 2xy\sqrt[3]{xy^2}$ — $\sqrt[3]{8} = 2$ and $\sqrt[3]{x^3y^3} = (x^3y^3)^{\frac{1}{3}} = xy$.

$= (5 + 2xy)\sqrt[3]{xy^2}$ — Apply the distributive property. ■

CHECK POINT 3 Simplify by combining like radical terms, if possible.

a. $3\sqrt[3]{24} - 5\sqrt[3]{81}$

b. $5\sqrt[3]{x^2y} + \sqrt[3]{27x^5y^4}$.

Dividing Radical Expressions We have seen that the root of a product is the product of the roots. The root of a quotient can also be expressed as the quotient of roots. Here is an example:

$$\sqrt{\frac{64}{4}} = \sqrt{16} = 4 \quad \text{and} \quad \frac{\sqrt{64}}{\sqrt{4}} = \frac{8}{2} = 4.$$

This expression is the square root of a quotient.

This expression is the quotient of two square roots.

The two procedures produce the same result, 4. This is a special case of the **quotient rule for radicals**.

2 Use the quotient rule to simplify radical expressions.

THE QUOTIENT RULE FOR RADICALS If $\sqrt[n]{a}$ and $\sqrt[n]{b}$ are real numbers and $b \neq 0$, then

$$\sqrt[n]{\frac{a}{b}} = \frac{\sqrt[n]{a}}{\sqrt[n]{b}}.$$

The nth root of a quotient is the quotient of the nth roots.

We know that a radical is simplified when its radicand has no factors other than 1 that are perfect nth powers. The quotient rule can be used to simplify some radicals. Keep in mind that all variables in radicands represent positive real numbers.

EXAMPLE 4 Using the Quotient Rule to Simplify Radicals

Simplify using the quotient rule:

a. $\sqrt[3]{\frac{16}{27}}$ **b.** $\sqrt{\frac{x^2}{25y^6}}$ **c.** $\sqrt[4]{\frac{7y^5}{x^{12}}}$.

SOLUTION We simplify each expression by taking the roots of the numerator and the denominator. Then we use factoring to simplify the resulting radicals, if possible.

a. $\sqrt[3]{\frac{16}{27}} = \frac{\sqrt[3]{16}}{\sqrt[3]{27}} = \frac{\sqrt[3]{8 \cdot 2}}{3} = \frac{\sqrt[3]{8} \cdot \sqrt[3]{2}}{3} = \frac{2\sqrt[3]{2}}{3}$

b. $\sqrt{\frac{x^2}{25y^6}} = \frac{\sqrt{x^2}}{\sqrt{25y^6}} = \frac{x}{5(y^6)^{\frac{1}{2}}} = \frac{x}{5y^3}$

Try to do this step mentally.

c. $\sqrt[4]{\frac{7y^5}{x^{12}}} = \frac{\sqrt[4]{7y^5}}{\sqrt[4]{x^{12}}} = \frac{\sqrt[4]{y^4 \cdot 7y}}{\sqrt[4]{x^{12}}} = \frac{y\sqrt[4]{7y}}{x^3}$ ■

CHECK POINT 4 Simplify using the quotient rule:

a. $\sqrt[3]{\frac{24}{125}}$ **b.** $\sqrt{\frac{9x^3}{y^{10}}}$ **c.** $\sqrt[3]{\frac{8y^7}{x^{12}}}$.

By reversing the two sides of the quotient rule, we obtain a procedure for dividing radical expressions.

3 Use the quotient rule to divide radical expressions.

DIVIDING RADICAL EXPRESSIONS If $\sqrt[n]{a}$ and $\sqrt[n]{b}$ are real numbers and $b \neq 0$, then

$$\frac{\sqrt[n]{a}}{\sqrt[n]{b}} = \sqrt[n]{\frac{a}{b}}.$$

To divide two radical expressions with the same index, divide the radicands and retain the common index.

EXAMPLE 5 Dividing Radical Expressions

Divide and, if possible, simplify:

a. $\frac{\sqrt{48x^3}}{\sqrt{6x}}$ **b.** $\frac{\sqrt{45xy}}{2\sqrt{5}}$ **c.** $\frac{\sqrt[3]{16x^5y^2}}{\sqrt[3]{2xy^{-1}}}$.

SOLUTION In each part of this problem, the indices in the numerator and the denominator are the same. Perform each division by dividing the radicands and retaining the common index.

a. $\frac{\sqrt{48x^3}}{\sqrt{6x}} = \sqrt{\frac{48x^3}{6x}} = \sqrt{8x^2} = \sqrt{4x^2 \cdot 2} = \sqrt{4x^2} \cdot \sqrt{2} = 2x\sqrt{2}$

b. $\frac{\sqrt{45xy}}{2\sqrt{5}} = \frac{1}{2} \cdot \sqrt{\frac{45xy}{5}} = \frac{1}{2} \cdot \sqrt{9xy} = \frac{1}{2} \cdot 3\sqrt{xy}$ or $\frac{3\sqrt{xy}}{2}$

c. $\frac{\sqrt[3]{16x^5y^2}}{\sqrt[3]{2xy^{-1}}} = \sqrt[3]{\frac{16x^5y^2}{2xy^{-1}}}$ Divide the radicands and retain the common index.

$= \sqrt[3]{8x^{5-1}y^{2-(-1)}}$ Divide factors in the radicand. Subtract exponents on common bases.

$= \sqrt[3]{8x^4y^3}$ Simplify.

$= \sqrt[3]{(8x^3y^3)x}$ Factor using the greatest perfect cube factor.

$= \sqrt[3]{8x^3y^3} \cdot \sqrt[3]{x}$ Factor into two radicals.

$= 2xy\sqrt[3]{x}$ Simplify. ■

CHECK POINT 5 Divide and, if possible, simplify:

a. $\frac{\sqrt{40x^5}}{\sqrt{2x}}$ **b.** $\frac{\sqrt{50xy}}{2\sqrt{2}}$

c. $\frac{\sqrt[3]{48x^7y}}{\sqrt[3]{6xy^{-2}}}$.

10.4 EXERCISE SET

Student Solutions Manual | CD/Video | PH Math/Tutor Center | MathXL Tutorials on CD | MathXL® | MyMathLab | Interactmath.com

Practice Exercises

In this exercise set, assume that all variables represent positive real numbers.

In Exercises 1–10, add or subtract as indicated.

1. $8\sqrt{5} + 3\sqrt{5}$
2. $7\sqrt{3} + 2\sqrt{3}$
3. $9\sqrt[3]{6} - 2\sqrt[3]{6}$
4. $9\sqrt[3]{7} - 4\sqrt[3]{7}$
5. $4\sqrt[5]{2} + 3\sqrt[5]{2} - 5\sqrt[5]{2}$
6. $6\sqrt[5]{3} + 2\sqrt[5]{3} - 3\sqrt[5]{3}$
7. $3\sqrt{13} - 2\sqrt{5} - 2\sqrt{13} + 4\sqrt{5}$
8. $8\sqrt{17} - 5\sqrt{19} - 6\sqrt{17} + 4\sqrt{19}$
9. $3\sqrt{5} - \sqrt[3]{x} + 4\sqrt{5} + 3\sqrt[3]{x}$
10. $6\sqrt{7} - \sqrt[3]{x} + 2\sqrt{7} + 5\sqrt[3]{x}$

In Exercises 11–28, add or subtract as indicated. You will need to simplify terms to identify the like radicals.

11. $\sqrt{3} + \sqrt{27}$
12. $\sqrt{5} + \sqrt{20}$
13. $7\sqrt{12} + \sqrt{75}$
14. $5\sqrt{12} + \sqrt{75}$
15. $3\sqrt{32x} - 2\sqrt{18x}$
16. $5\sqrt{45x} - 2\sqrt{20x}$
17. $5\sqrt[3]{16} + \sqrt[3]{54}$
18. $3\sqrt[3]{24} + \sqrt[3]{81}$
19. $3\sqrt{45x^3} + \sqrt{5x}$
20. $8\sqrt{45x^3} + \sqrt{5x}$
21. $\sqrt[3]{54xy^3} + y\sqrt[3]{128x}$
22. $\sqrt[3]{24xy^3} + y\sqrt[3]{81x}$
23. $\sqrt[3]{54x^4} - \sqrt[3]{16x}$
24. $\sqrt[3]{81x^4} - \sqrt[3]{24x}$
25. $\sqrt{9x - 18} + \sqrt{x - 2}$
26. $\sqrt{4x - 12} + \sqrt{x - 3}$
27. $2\sqrt[3]{x^4y^2} + 3x\sqrt[3]{xy^2}$
28. $4\sqrt[3]{x^4y^2} + 5x\sqrt[3]{xy^2}$

In Exercises 29–44, simplify using the quotient rule.

29. $\sqrt{\dfrac{11}{4}}$
30. $\sqrt{\dfrac{19}{25}}$
31. $\sqrt[3]{\dfrac{19}{27}}$
32. $\sqrt[3]{\dfrac{11}{64}}$
33. $\sqrt{\dfrac{x^2}{36y^8}}$
34. $\sqrt{\dfrac{x^2}{144y^{12}}}$
35. $\sqrt{\dfrac{8x^3}{25y^6}}$
36. $\sqrt{\dfrac{50x^3}{81y^8}}$
37. $\sqrt[3]{\dfrac{x^4}{8y^3}}$
38. $\sqrt[3]{\dfrac{x^5}{125y^3}}$
39. $\sqrt[3]{\dfrac{50x^8}{27y^{12}}}$
40. $\sqrt[3]{\dfrac{81x^8}{8y^{15}}}$
41. $\sqrt[4]{\dfrac{9y^6}{x^8}}$
42. $\sqrt[4]{\dfrac{13y^7}{x^{12}}}$
43. $\sqrt[5]{\dfrac{64x^{13}}{y^{20}}}$
44. $\sqrt[5]{\dfrac{64x^{14}}{y^{15}}}$

In Exercises 45–66, divide and, if possible, simplify.

45. $\dfrac{\sqrt{40}}{\sqrt{5}}$
46. $\dfrac{\sqrt{200}}{\sqrt{10}}$
47. $\dfrac{\sqrt[3]{48}}{\sqrt[3]{6}}$
48. $\dfrac{\sqrt[3]{54}}{\sqrt[3]{2}}$
49. $\dfrac{\sqrt{54x^3}}{\sqrt{6x}}$
50. $\dfrac{\sqrt{72x^3}}{\sqrt{2x}}$
51. $\dfrac{\sqrt{x^5y^3}}{\sqrt{xy}}$
52. $\dfrac{\sqrt{x^7y^6}}{\sqrt{x^3y^2}}$
53. $\dfrac{\sqrt{200x^3}}{\sqrt{10x^{-1}}}$
54. $\dfrac{\sqrt{500x^3}}{\sqrt{10x^{-1}}}$
55. $\dfrac{\sqrt{48a^8b^7}}{\sqrt{3a^{-2}b^{-3}}}$
56. $\dfrac{\sqrt{54a^7b^{11}}}{\sqrt{3a^{-4}b^{-2}}}$
57. $\dfrac{\sqrt{72xy}}{2\sqrt{2}}$
58. $\dfrac{\sqrt{50xy}}{2\sqrt{2}}$
59. $\dfrac{\sqrt[3]{24x^3y^5}}{\sqrt[3]{3y^2}}$
60. $\dfrac{\sqrt[3]{250x^5y^3}}{\sqrt[3]{2x^3}}$
61. $\dfrac{\sqrt[4]{32x^{10}y^8}}{\sqrt[4]{2x^2y^{-2}}}$
62. $\dfrac{\sqrt[5]{96x^{12}y^{11}}}{\sqrt[5]{3x^2y^{-2}}}$
63. $\dfrac{\sqrt[3]{x^2 + 5x + 6}}{\sqrt[3]{x + 2}}$
64. $\dfrac{\sqrt[3]{x^2 + 7x + 12}}{\sqrt[3]{x + 3}}$
65. $\dfrac{\sqrt[3]{a^3 + b^3}}{\sqrt[3]{a + b}}$
66. $\dfrac{\sqrt[3]{a^3 - b^3}}{\sqrt[3]{a - b}}$

Practice Plus

In Exercises 67–76, perform the indicated operations.

67. $\dfrac{\sqrt{32}}{5} + \dfrac{\sqrt{18}}{7}$
68. $\dfrac{\sqrt{27}}{2} + \dfrac{\sqrt{75}}{7}$
69. $3x\sqrt{8xy^2} - 5y\sqrt{32x^3} + \sqrt{18x^3y^2}$
70. $6x\sqrt{3xy^2} - 4x^2\sqrt{27xy} - 5\sqrt{75x^5y}$
71. $5\sqrt{2x^3} + \dfrac{30x^3\sqrt{24x^2}}{3x^2\sqrt{3x}}$
72. $7\sqrt{2x^3} + \dfrac{40x^3\sqrt{150x^2}}{5x^2\sqrt{3x}}$

73. $2x\sqrt{75xy} - \dfrac{\sqrt{81xy^2}}{\sqrt{3x^{-2}y}}$

74. $5\sqrt{8x^2y^3} - \dfrac{9x^2\sqrt{64y}}{3x\sqrt{2y^{-2}}}$

75. $\dfrac{15x^4\sqrt[3]{80x^3y^2}}{5x^3\sqrt[3]{2x^2y}} - \dfrac{75\sqrt[3]{5x^3y}}{25\sqrt[3]{x^{-1}}}$

76. $\dfrac{16x^4\sqrt[3]{48x^3y^2}}{8x^3\sqrt[3]{3x^2y}} - \dfrac{20\sqrt[3]{2x^3y}}{4\sqrt[3]{x^{-1}}}$

In Exercises 77–80, find $\left(\frac{f}{g}\right)(x)$ and the domain of $\left(\frac{f}{g}\right)$. Express each quotient function in simplified form.

77. $f(x) = \sqrt{48x^5}, \quad g(x) = \sqrt{3x^2}$

78. $f(x) = \sqrt{x^2 - 25}, \quad g(x) = \sqrt{x + 5}$

79. $f(x) = \sqrt[3]{32x^6}, \quad g(x) = \sqrt[3]{2x^2}$

80. $f(x) = \sqrt[3]{2x^6}, \quad g(x) = \sqrt[3]{16x}$

Application Exercises

What does travel in space have to do with radicals? Imagine that in the future we will be able to travel in starships at velocities approaching the speed of light (approximately 186,000 miles per second). According to Einstein's theory of relativity, time would pass more quickly on Earth than it would in the moving starship. The radical expression

$$R_f\frac{\sqrt{c^2 - v^2}}{\sqrt{c^2}}$$

gives the aging rate of an astronaut relative to the aging rate of a friend, R_f, on Earth. In the expression, v is the astronaut's velocity and c is the speed of light. Use the expression to solve Exercises 81–82. Imagine that you are the astronaut on the starship.

81. a. Use the quotient rule and simplify the expression that shows your aging rate relative to a friend on Earth. Working in a step-by-step manner, express your aging rate as

$$R_f\sqrt{1 - \left(\frac{v}{c}\right)^2}.$$

b. You are moving at velocities approaching the speed of light. Substitute c, the speed of light, for v in the simplified expression from part (a). Simplify completely. Close to the speed of light, what is your aging rate relative to a friend on Earth? What does this mean?

82. a. Use the quotient rule and simplify the expression that shows your aging rate relative to a friend on Earth. Working step-by-step, express your aging rate as

$$R_f\sqrt{1 - \left(\frac{v}{c}\right)^2}.$$

b. You are moving at 90% of the speed of light. Substitute $0.9c$ for v, your velocity, in the simplified expression from part (a). What is your aging rate, correct to two decimal places, relative to a friend on Earth? If you are gone for 44 weeks, approximately how many weeks have passed for your friend?

In Exercises 83–84, find the perimeter and area of each rectangle. Express answers in simplified radical form.

83.

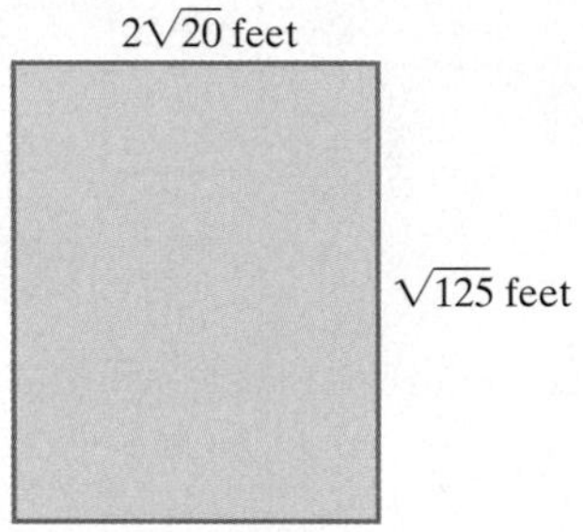

84.

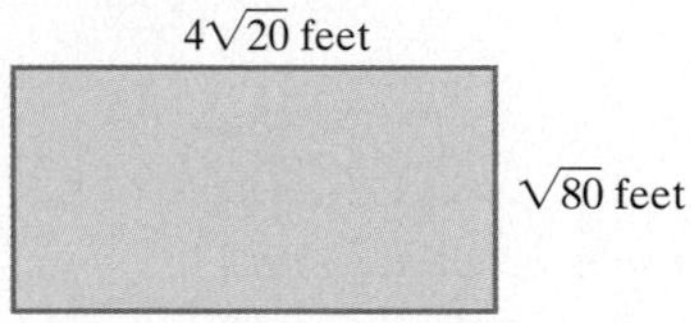

Writing in Mathematics

85. What are like radicals? Give an example with your explanation.

86. Explain how to add like radicals. Give an example with your explanation.

87. If only like radicals can be combined, why is it possible to add $\sqrt{2}$ and $\sqrt{8}$?

88. Explain how to simplify a radical expression using the quotient rule. Provide an example.

89. Explain how to divide radical expressions with the same index.

90. Answer the question posed in the section opener on page 670. What will you do: explore space or stay here on Earth? What are the reasons for your choice?

Technology Exercises

91. Use a calculator to provide numerical support to any four exercises that you worked from Exercises 1–66 that do not contain variables. Begin by finding a decimal approximation for the given expression. Then find a decimal approximation for your answer. The two decimal approximations should be the same.

In Exercises 92–94, determine if each operation is performed correctly by graphing the function on each side of the equation with your graphing utility. Use the given viewing rectangle. The graphs should be the same. If they are not, correct the right side of the equation and then use your graphing utility to verify the correction.

92. $\sqrt{4x} + \sqrt{9x} = 5\sqrt{x}$
$[0, 5, 1]$ by $[0, 10, 1]$

93. $\sqrt{16x} - \sqrt{9x} = \sqrt{7x}$
$[0, 5, 1]$ by $[0, 5, 1]$

94. $x\sqrt{8} + x\sqrt{2} = x\sqrt{10}$
$[-5, 5, 1]$ by $[-15, 15, 1]$

Critical Thinking Exercises

95. Which one of the following is true?

a. $\sqrt{5} + \sqrt{5} = \sqrt{10}$

b. $4\sqrt{3} + 5\sqrt{3} = 9\sqrt{6}$

c. If any two radical expressions are completely simplified, they can then be combined through addition or subtraction.

d. None of the above is true.

96. If an irrational number is decreased by $2\sqrt{18} - \sqrt{50}$, the result is $\sqrt{2}$. What is the number?

97. Simplify: $\dfrac{\sqrt{20}}{3} + \dfrac{\sqrt{45}}{4} - \sqrt{80}$.

98. Simplify: $\dfrac{6\sqrt{49xy}\,\sqrt{ab^2}}{7\sqrt{36x^{-3}y^{-5}}\,\sqrt{a^{-9}b^{-1}}}$.

Review Exercises

99. Solve: $2(3x - 1) - 4 = 2x - (6 - x)$.
(Section 2.3, Example 3)

100. Factor: $x^2 - 8xy + 12y^2$.
(Section 6.2, Example 6)

101. Add: $\dfrac{2}{x^2 + 5x + 6} + \dfrac{3x}{x^2 + 6x + 9}$.
(Section 7.4, Example 7)

MID-CHAPTER CHECK POINT

CHAPTER 10

What You Know: We learned to find roots of numbers. We saw that the domain of a square root function is the set of real numbers for which the radicand is nonnegative. We learned to simplify radical expressions, using $\sqrt[n]{a^n} = |a|$ if n is even and $\sqrt[n]{a^n} = a$ if n is odd. The definition $a^{\frac{m}{n}} = (\sqrt[n]{a})^m = \sqrt[n]{a^m}$ connected rational exponents and radicals. Finally, we performed various operations with radicals, including multiplication, addition, subtraction, and division.

In Exercises 1–23, simplify the given expression or perform the indicated operation(s) and, if possible, simplify. Assume that all variables represent positive real numbers.

1. $\sqrt{100} - \sqrt[3]{-27}$

2. $\sqrt{8x^5y^7}$

3. $3\sqrt[3]{4x^2} + 2\sqrt[3]{4x^2}$

4. $(3\sqrt[3]{4x^2})(2\sqrt[3]{4x^2})$

5. $27^{\frac{2}{3}} + (-32)^{\frac{3}{5}}$

6. $\left(64x^3y^{\frac{1}{4}}\right)^{\frac{1}{3}}$

7. $5\sqrt{27} - 4\sqrt{48}$

8. $\sqrt{\dfrac{500x^3}{4y^4}}$

9. $\dfrac{x}{\sqrt[4]{x}}$

10. $\sqrt[3]{54x^5}$

11. $\dfrac{\sqrt[3]{160}}{\sqrt[3]{2}}$

12. $\sqrt[5]{\dfrac{x^{10}}{y^{20}}}$

13. $\dfrac{\left(x^{\frac{2}{3}}\right)^2}{\left(x^{\frac{1}{4}}\right)^3}$

14. $\sqrt[6]{x^6y^4}$

15. $\sqrt[7]{(x - 2)^3} \cdot \sqrt[7]{(x - 2)^6}$

16. $\sqrt[4]{32x^{11}y^{17}}$

17. $4\sqrt[3]{16} + 2\sqrt[3]{54}$

18. $\dfrac{\sqrt[7]{x^4y^9}}{\sqrt[7]{x^{-5}y^7}}$

19. $(-125)^{-\frac{2}{3}}$

20. $\sqrt{2} \cdot \sqrt[3]{2}$

21. $\sqrt[3]{\dfrac{32x}{y^4}} \cdot \sqrt[3]{\dfrac{2x^2}{y^2}}$

22. $\sqrt{32xy^2} \cdot \sqrt{2x^3y^5}$

23. $4x\sqrt{6x^4y^3} - 7y\sqrt{24x^6y}$

In Exercises 24–25, find the domain of each function.

24. $f(x) = \sqrt{30 - 5x}$

25. $g(x) = \sqrt[3]{3x - 15}$

SECTION 10.5

MULTIPLYING WITH MORE THAN ONE TERM AND RATIONALIZING DENOMINATORS

Objectives

1. Multiply radicals with more than one term.
2. Use polynomial special products to multiply radicals.
3. Rationalize denominators containing one term.
4. Rationalize denominators containing two terms.
5. Rationalize numerators.

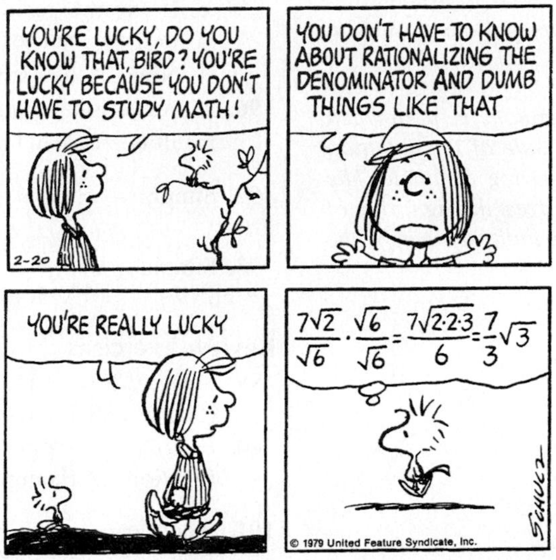

PEANUTS reprinted by permission of United Feature Syndicate, Inc.

The late Charles Schulz, creator of the "Peanuts" comic strip, transfixed 350 million readers worldwide with the joys and angst of his hapless Charlie Brown and Snoopy, a romantic self-deluded beagle. In 18,250 comic strips that spanned nearly 50 years, mathematics was often featured. Is the discussion of radicals shown above the real thing, or is it just an algebraic scam? By the time you complete this section on multiplying and dividing radicals, you will be able to decide.

1 Multiply radicals with more than one term.

Multiplying Radicals with More Than One Term Radical expressions with more than one term are multiplied in much the same way that polynomials with more than one term are multiplied. Example 1 uses the distributive property and the FOIL method to perform multiplications.

EXAMPLE 1 Multiplying Radicals

Multiply:

a. $\sqrt{7}(x+\sqrt{2})$ **b.** $\sqrt[3]{x}(\sqrt[3]{6}-\sqrt[3]{x^2})$ **c.** $(5\sqrt{2}+2\sqrt{3})(4\sqrt{2}-3\sqrt{3})$.

SOLUTION

a. $\sqrt{7}(x+\sqrt{2})$

$=\sqrt{7}\cdot x+\sqrt{7}\cdot\sqrt{2}$ Use the distributive property.

$=x\sqrt{7}+\sqrt{14}$ Multiply the radicals.

b. $\sqrt[3]{x}(\sqrt[3]{6}-\sqrt[3]{x^2})$

$=\sqrt[3]{x}\cdot\sqrt[3]{6}-\sqrt[3]{x}\cdot\sqrt[3]{x^2}$ Use the distributive property.

$=\sqrt[3]{6x}-\sqrt[3]{x^3}$ Multiply the radicals: $\sqrt[n]{a}\cdot\sqrt[n]{b}=\sqrt[n]{ab}$.

$=\sqrt[3]{6x}-x$ Simplify: $\sqrt[3]{x^3}=x$.

first, last, inside, outside

c. $(5\sqrt{2} + 2\sqrt{3})(4\sqrt{2} - 3\sqrt{3})$ Use the FOIL method.

F O I L

$= (5\sqrt{2})(4\sqrt{2}) + (5\sqrt{2})(-3\sqrt{3}) + (2\sqrt{3})(4\sqrt{2}) + (2\sqrt{3})(-3\sqrt{3})$

$= 20 \cdot 2 - 15\sqrt{6} + 8\sqrt{6} - 6 \cdot 3$ Multiply. Note that $\sqrt{2} \cdot \sqrt{2} = \sqrt{4} = 2$ and $\sqrt{3} \cdot \sqrt{3} = \sqrt{9} = 3$.

$= 40 - 15\sqrt{6} + 8\sqrt{6} - 18$ Complete the multiplications.

$= (40 - 18) + (-15\sqrt{6} + 8\sqrt{6})$ Group like terms. Try to do this step mentally.

$= 22 - 7\sqrt{6}$ Combine numerical terms and like radicals. ■

CHECK POINT 1 Multiply:

a. $\sqrt{6}(x + \sqrt{10})$

b. $\sqrt[3]{y}(\sqrt[3]{y^2} - \sqrt[3]{7})$

c. $(6\sqrt{5} + 3\sqrt{2})(2\sqrt{5} - 4\sqrt{2})$.

2 Use polynomial special products to multiply radicals.

Some radicals can be multiplied using the special products for multiplying polynomials.

STUDY TIP

Just to recap, special products for multiplying polynomials are summarized as follows:

Square of a Binomial:

$(A + B)^2 = A^2 + 2AB + B^2$

$(A - B)^2 = A^2 - 2AB + B^2$

Product of the Sum and Difference of Two Terms:

$(A + B)(A - B) = A^2 - B^2$.

EXAMPLE 2 Using Special Products to Multiply Radicals

Multiply:

a. $(\sqrt{3} + \sqrt{7})^2$ **b.** $(\sqrt{7} + \sqrt{3})(\sqrt{7} - \sqrt{3})$ **c.** $(\sqrt{a} - \sqrt{b})(\sqrt{a} + \sqrt{b})$.

SOLUTION Use the special-product formulas shown.

$(A + B)^2 = A^2 + 2 \cdot A \cdot B + B^2$

a. $(\sqrt{3} + \sqrt{7})^2 = (\sqrt{3})^2 + 2 \cdot \sqrt{3} \cdot \sqrt{7} + (\sqrt{7})^2$ Use the special product for $(A + B)^2$.

$= 3 + 2\sqrt{21} + 7$ Multiply the radicals.

$= 10 + 2\sqrt{21}$ Simplify.

$(A + B) \cdot (A - B) = A^2 - B^2$

b. $(\sqrt{7} + \sqrt{3})(\sqrt{7} - \sqrt{3}) = (\sqrt{7})^2 - (\sqrt{3})^2$ Use the special product for $(A + B)(A - B)$.

$= 7 - 3$ Simplify: $(\sqrt{a})^2 = a$.

$= 4$

$(A - B) \cdot (A + B) = A^2 - B^2$

c. $(\sqrt{a} - \sqrt{b})(\sqrt{a} + \sqrt{b}) = (\sqrt{a})^2 - (\sqrt{b})^2 = a - b$ ■

Radical expressions that involve the sum and difference of the same two terms are called **conjugates**. For example,

$$\sqrt{7}+\sqrt{3} \quad \text{and} \quad \sqrt{7}-\sqrt{3}$$

are conjugates of each other. Parts (b) and (c) of Example 2 illustrate that the product of two radical expressions need not be a radical expression:

$$(\sqrt{7}+\sqrt{3})(\sqrt{7}-\sqrt{3}) = 4$$
$$(\sqrt{a}-\sqrt{b})(\sqrt{a}+\sqrt{b}) = a-b.$$

The product of conjugates does not contain a radical.

Later in this section, we will use conjugates to simplify quotients.

CHECK POINT 2 Multiply:

a. $(\sqrt{5}+\sqrt{6})^2$

b. $(\sqrt{6}+\sqrt{5})(\sqrt{6}-\sqrt{5})$

c. $(\sqrt{a}-\sqrt{7})(\sqrt{a}+\sqrt{7})$.

3 Rationalize denominators containing one term.

Rationalizing Denominators Containing One Term You can use a calculator to compare the approximate values for $\frac{1}{\sqrt{3}}$ and $\frac{\sqrt{3}}{3}$. The two approximations are the same. This is not a coincidence:

$$\frac{1}{\sqrt{3}} = \frac{1}{\sqrt{3}}\cdot\boxed{\frac{\sqrt{3}}{\sqrt{3}}} = \frac{\sqrt{3}}{\sqrt{9}} = \frac{\sqrt{3}}{3}.$$

Any number divided by itself is 1. Multiplication by 1 does not change the value of $\frac{1}{\sqrt{3}}$.

This process involves rewriting a radical expression as an equivalent expression in which the denominator no longer contains any radicals. The process is called **rationalizing the denominator**. When the denominator contains a single radical with an nth root, **multiply the numerator and the denominator by a radical of index n that produces a perfect nth power in the denominator's radicand.**

EXAMPLE 3 Rationalizing Denominators

Rationalize each denominator:

a. $\frac{\sqrt{5}}{\sqrt{6}}$

b. $\sqrt[3]{\frac{7}{16}}$.

SOLUTION

a. If we multiply the numerator and the denominator of $\frac{\sqrt{5}}{\sqrt{6}}$ by $\sqrt{6}$, the denominator becomes $\sqrt{6}\cdot\sqrt{6} = \sqrt{36} = 6$. The denominator's radicand, 36, is a perfect square.

The denominator no longer contains a radical. Therefore, we multiply by 1, choosing $\frac{\sqrt{6}}{\sqrt{6}}$ for 1.

$$\frac{\sqrt{5}}{\sqrt{6}} = \frac{\sqrt{5}}{\sqrt{6}} \cdot \frac{\sqrt{6}}{\sqrt{6}}$$ Multiply the numerator and denominator by $\sqrt{6}$ to remove the radical in the denominator.

$$= \frac{\sqrt{30}}{\sqrt{36}}$$ Multiply numerators and multiply denominators. The denominator's radicand, 36, is a perfect square.

$$= \frac{\sqrt{30}}{6}$$ Simplify: $\sqrt{36} = 6$.

b. Using the quotient rule, we can express $\sqrt[3]{\frac{7}{16}}$ as $\frac{\sqrt[3]{7}}{\sqrt[3]{16}}$. We have cube roots, so we want the denominator's radicand to be a perfect cube. Right now, the denominator's radicand is 16 or 4^2. We know that $\sqrt[3]{4^3} = 4$. If we multiply the numerator and the denominator of $\frac{\sqrt[3]{7}}{\sqrt[3]{16}}$ by $\sqrt[3]{4}$, the denominator becomes

$$\sqrt[3]{16} \cdot \sqrt[3]{4} = \sqrt[3]{4^2} \cdot \sqrt[3]{4} = \sqrt[3]{4^3} = 4.$$

The denominator's radicand, 4^3, is a perfect cube. The denominator no longer contains a radical. Therefore, we multiply by 1, choosing $\frac{\sqrt[3]{4}}{\sqrt[3]{4}}$ for 1.

$$\sqrt[3]{\frac{7}{16}} = \frac{\sqrt[3]{7}}{\sqrt[3]{16}}$$ Use the quotient rule and rewrite as the quotient of radicals.

$$= \frac{\sqrt[3]{7}}{\sqrt[3]{4^2}}$$ Write the denominator's radicand as an exponential expression.

$$= \frac{\sqrt[3]{7}}{\sqrt[3]{4^2}} \cdot \frac{\sqrt[3]{4}}{\sqrt[3]{4}}$$ Multiply the numerator and denominator by $\sqrt[3]{4}$ to remove the radical in the denominator.

$$= \frac{\sqrt[3]{28}}{\sqrt[3]{4^3}}$$ Multiply numerators and denominators. The denominator's radicand, 4^3, is a perfect cube.

$$= \frac{\sqrt[3]{28}}{4}$$ Simplify: $\sqrt[3]{4^3} = 4$. ■

CHECK POINT 3 Rationalize each denominator:

a. $\frac{\sqrt{3}}{\sqrt{7}}$ **b.** $\sqrt[3]{\frac{2}{9}}$.

Example 3 showed that it is helpful to express the denominator's radicand using exponents. In this way, we can find the extra factor or factors needed to produce a perfect nth power. For example, suppose that $\sqrt[5]{8}$ appears in the denominator. We want a perfect fifth power. By expressing $\sqrt[5]{8}$ as $\sqrt[5]{2^3}$, we would multiply the numerator and denominator by $\sqrt[5]{2^2}$ because

$$\sqrt[5]{2^3} \cdot \sqrt[5]{2^2} = \sqrt[5]{2^5} = 2.$$

EXAMPLE 4 Rationalizing Denominators

Rationalize each denominator:

a. $\sqrt{\frac{3x}{5y}}$ **b.** $\frac{\sqrt[3]{x}}{\sqrt[3]{36y}}$ **c.** $\frac{10y}{\sqrt[5]{4x^3y}}$.

SOLUTION By examining each denominator, you can determine how to multiply by 1. Let's first look at the denominators. For the square root, we must produce exponents of 2 in the radicand. For the cube root, we need exponents of 3, and for the fifth root, we want exponents of 5.

- $\sqrt{5y}$
- $\sqrt[3]{36y}$ or $\sqrt[3]{6^2y}$
- $\sqrt[5]{4x^3y}$ or $\sqrt[5]{2^2x^3y}$

Multiply by $\sqrt{5y}$: $\sqrt{5y} \cdot \sqrt{5y} = \sqrt{25y^2} = 5y$.

Multiply by $\sqrt[3]{6y^2}$: $\sqrt[3]{6^2y} \cdot \sqrt[3]{6y^2} = \sqrt[3]{6^3y^3} = 6y$.

Multiply by $\sqrt[5]{2^3x^2y^4}$: $\sqrt[5]{2^2x^3y} \cdot \sqrt[5]{2^3x^2y^4} = \sqrt[5]{2^5x^5y^5} = 2xy$.

a. $$\sqrt{\frac{3x}{5y}} = \frac{\sqrt{3x}}{\sqrt{5y}} = \frac{\sqrt{3x}}{\sqrt{5y}} \cdot \frac{\sqrt{5y}}{\sqrt{5y}} = \frac{\sqrt{15xy}}{\sqrt{25y^2}} = \frac{\sqrt{15xy}}{5y}$$

Multiply by 1. $25y^2$ is a perfect square.

b. $$\frac{\sqrt[3]{x}}{\sqrt[3]{36y}} = \frac{\sqrt[3]{x}}{\sqrt[3]{6^2y}} = \frac{\sqrt[3]{x}}{\sqrt[3]{6^2y}} \cdot \frac{\sqrt[3]{6y^2}}{\sqrt[3]{6y^2}} = \frac{\sqrt[3]{6xy^2}}{\sqrt[3]{6^3y^3}} = \frac{\sqrt[3]{6xy^2}}{6y}$$

Multiply by 1. 6^3y^3 is a perfect cube.

c. $$\frac{10y}{\sqrt[5]{4x^3y}} = \frac{10y}{\sqrt[5]{2^2x^3y}} = \frac{10y}{\sqrt[5]{2^2x^3y}} \cdot \frac{\sqrt[5]{2^3x^2y^4}}{\sqrt[5]{2^3x^2y^4}}$$

Multiply by 1.

$$= \frac{10y\sqrt[5]{2^3x^2y^4}}{\sqrt[5]{2^5x^5y^5}} = \frac{10y\sqrt[5]{8x^2y^4}}{2xy} = \frac{5\sqrt[5]{8x^2y^4}}{x}$$

$2^5x^5y^5$ is a perfect 5th power. Simplify: Divide numerator and denominator by $2y$.

■

CHECK POINT 4 Rationalize each denominator:

a. $\sqrt{\frac{2x}{7y}}$

b. $\frac{\sqrt[3]{x}}{\sqrt[3]{9y}}$

c. $\frac{6x}{\sqrt[5]{8x^2y^4}}$.

4 Rationalize denominators containing two terms.

ENRICHMENT ESSAY

Golden Rectangles

The early Greeks believed that the most pleasing of all rectangles were **golden rectangles**, whose ratio of width to height is

$$\frac{w}{h} = \frac{2}{\sqrt{5} - 1}.$$

The Parthenon at Athens fits into a golden rectangle once the triangular pediment is reconstructed. By working Exercise 128 in the exercise set, you will rationalize the denominator of this rational expression and obtain a better understanding about the ratio of adjacent sides in all golden rectangles.

Rationalizing Denominators Containing Two Terms How can we rationalize a denominator if the denominator contains two terms with one or more square roots? **Multiply the numerator and the denominator by the conjugate of the denominator.** Here are three examples of such expressions:

- $\frac{8}{3\sqrt{2} + 4}$ — The conjugate of the denominator is $3\sqrt{2} - 4$.
- $\frac{2 + \sqrt{5}}{\sqrt{6} - \sqrt{3}}$ — The conjugate of the denominator is $\sqrt{6} + \sqrt{3}$.
- $\frac{h}{\sqrt{x + h} - \sqrt{x}}$ — The conjugate of the denominator is $\sqrt{x + h} + \sqrt{x}$.

The product of the denominator and its conjugate is found using the formula

$$(A + B)(A - B) = A^2 - B^2.$$

The simplified product will not contain a radical.

EXAMPLE 5 Rationalizing a Denominator Containing Two Terms

Rationalize the denominator: $\frac{8}{3\sqrt{2} + 4}$.

SOLUTION The conjugate of the denominator is $3\sqrt{2} - 4$. If we multiply the numerator and the denominator by $3\sqrt{2} - 4$, the simplified denominator will not contain a radical. Therefore, we multiply by 1, choosing $\frac{3\sqrt{2} - 4}{3\sqrt{2} - 4}$ for 1.

$$\frac{8}{3\sqrt{2} + 4} = \frac{8}{3\sqrt{2} + 4} \cdot \frac{3\sqrt{2} - 4}{3\sqrt{2} - 4}$$ Multiply by 1.

$$= \frac{8(3\sqrt{2} - 4)}{(3\sqrt{2})^2 - 4^2}$$ $(A + B)(A - B) = A^2 - B^2$

Leave the numerator in factored form. This helps simplify, if possible.

$$= \frac{8(3\sqrt{2} - 4)}{18 - 16}$$ $(3\sqrt{2})^2 = 9 \cdot 2 = 18$

$$= \frac{8(3\sqrt{2} - 4)}{2}$$ This expression can still be simplified.

$$= \frac{\overset{4}{\cancel{8}}(3\sqrt{2} - 4)}{\underset{1}{\cancel{2}}}$$ Divide the numerator and denominator by 2.

$$= 4(3\sqrt{2} - 4) \quad \text{or} \quad 12\sqrt{2} - 16$$ ■

CHECK POINT 5 Rationalize the denominator: $\frac{18}{2\sqrt{3} + 3}$.

EXAMPLE 6 Rationalizing a Denominator Containing Two Terms

Rationalize the denominator: $\dfrac{2+\sqrt{5}}{\sqrt{6}-\sqrt{3}}$.

SOLUTION Multiplication of both the numerator and denominator by $\sqrt{6}+\sqrt{3}$ will rationalize the denominator.

$$\frac{2+\sqrt{5}}{\sqrt{6}-\sqrt{3}} = \frac{2+\sqrt{5}}{\sqrt{6}-\sqrt{3}}\cdot\frac{\sqrt{6}+\sqrt{3}}{\sqrt{6}+\sqrt{3}}$$

Multiply by 1.

F O I L

$$= \frac{2\sqrt{6}+2\sqrt{3}+\sqrt{5}\cdot\sqrt{6}+\sqrt{5}\cdot\sqrt{3}}{(\sqrt{6})^2-(\sqrt{3})^2}$$

Use FOIL in the numerator and $(A-B)(A+B)=A^2-B^2$ in the denominator.

$$= \frac{2\sqrt{6}+2\sqrt{3}+\sqrt{30}+\sqrt{15}}{6-3}$$

$$= \frac{2\sqrt{6}+2\sqrt{3}+\sqrt{30}+\sqrt{15}}{3}$$

Further simplification is not possible. ■

CHECK POINT 6 Rationalize the denominator: $\dfrac{3+\sqrt{7}}{\sqrt{5}-\sqrt{2}}$.

5 Rationalize numerators.

Rationalizing Numerators We have seen that square root functions are often used to model growing phenomena with growth that is leveling off. Figure 10.6 shows a male's height as a function of his age. The pattern of his growth suggests modeling with a square root function.

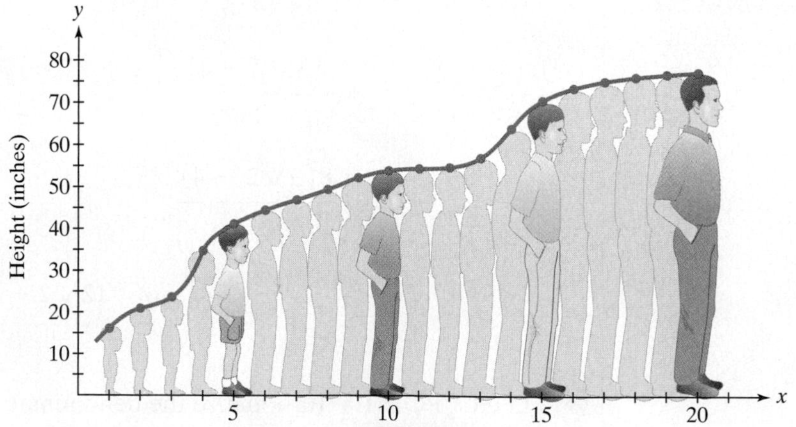

FIGURE 10.6

If we use $f(x) = \sqrt{x}$ to model height, $f(x)$, at age x, the expression

$$\frac{f(a+h)-f(a)}{h} = \frac{\sqrt{a+h}-\sqrt{a}}{h}$$

describes the man's average growth rate from age a to age $a + h$. Can you see that this expression is not defined if $h = 0$? However, to explore the man's average growth rates for successively shorter periods of time, we need to know what happens to the expression as h takes on values that get closer and closer to 0.

What happens to growth near the instant in time that the man is age a? The question is answered by first **rationalizing the numerator**. The procedure is similar to rationalizing the denominator. **To rationalize a numerator, multiply by 1 to eliminate the radical in the *numerator*.**

EXAMPLE 7 Rationalizing a Numerator

Rationalize the numerator:

$$\frac{\sqrt{a+h}-\sqrt{a}}{h}.$$

SOLUTION The conjugate of the numerator is $\sqrt{a+h}+\sqrt{a}$. If we multiply the numerator and the denominator by $\sqrt{a+h}+\sqrt{a}$, the simplified numerator will not contain radicals. Therefore, we multiply by 1, choosing $\frac{\sqrt{a+h}+\sqrt{a}}{\sqrt{a+h}+\sqrt{a}}$ for 1.

$$\frac{\sqrt{a+h}-\sqrt{a}}{h} = \frac{\sqrt{a+h}-\sqrt{a}}{h}\cdot\frac{\sqrt{a+h}+\sqrt{a}}{\sqrt{a+h}+\sqrt{a}}$$ Multiply by 1.

$$= \frac{\left(\sqrt{a+h}\right)^2-(\sqrt{a})^2}{h\left(\sqrt{a+h}+\sqrt{a}\right)}$$ $(A-B)(A+B) = A^2 - B^2$ Leave the denominator in factored form.

$$= \frac{a+h-a}{h\left(\sqrt{a+h}+\sqrt{a}\right)}$$ $\left(\sqrt{a+h}\right)^2 = a+h$ and $(\sqrt{a})^2 = a$.

$$= \frac{h}{h\left(\sqrt{a+h}+\sqrt{a}\right)}$$ Simplify the numerator.

$$= \frac{1}{\sqrt{a+h}+\sqrt{a}}$$ Simplify by dividing the numerator and the denominator by h.

ENRICHMENT ESSAY

Roller Coasters and Instantaneous Change

Roller coaster rides give you the opportunity to spend a few hair-raising minutes plunging hundreds of feet, accelerating to 80 miles per hour in seven seconds, and enduring vertical loops that turn you upside-down. By finding a function that models your distance above the ground, techniques such as rationalizing numerators help determine the instant when your velocity is the greatest. As you experience the glorious agony of the roller coaster, this is your moment of peak terror.

What happens to the rationalized expression in Example 7 as h is close to 0? The expression is close to

$$\frac{1}{\sqrt{a+0}+\sqrt{a}} = \frac{1}{\sqrt{a}+\sqrt{a}} = \frac{1}{2\sqrt{a}}.$$

The man's growth rate at one instant in time, age a, can be determined with this expression. This expression reveals how the square root function $f(x) = \sqrt{x}$ is changing, instant by instant.

CHECK POINT 7 Rationalize the numerator: $\frac{\sqrt{x+3}-\sqrt{x}}{3}$.

10.5 EXERCISE SET

Student Solutions Manual · CD/Video · PH Math/Tutor Center · MathXL Tutorials on CD · MathXL® · MyMathLab · Interactmath.com

Practice Exercises

In this exercise set, assume that all variables represent positive real numbers.

In Exercises 1–38, multiply as indicated. If possible, simplify any radical expressions that appear in the product.

1. $\sqrt{2}(x+\sqrt{7})$
2. $\sqrt{5}(x+\sqrt{3})$
3. $\sqrt{6}(7-\sqrt{6})$
4. $\sqrt{3}(5-\sqrt{3})$
5. $\sqrt{3}(4\sqrt{6}-2\sqrt{3})$
6. $\sqrt{6}(4\sqrt{6}-3\sqrt{2})$
7. $\sqrt[3]{2}(\sqrt[3]{6}+4\sqrt[3]{5})$
8. $\sqrt[3]{3}(\sqrt[3]{6}+7\sqrt[3]{4})$
9. $\sqrt[3]{x}(\sqrt[3]{16x^2}-\sqrt[3]{x})$
10. $\sqrt[3]{x}(\sqrt[3]{24x^2}-\sqrt[3]{x})$
11. $(5+\sqrt{2})(6+\sqrt{2})$
12. $(7+\sqrt{2})(8+\sqrt{2})$
13. $(6+\sqrt{5})(9-4\sqrt{5})$
14. $(4+\sqrt{5})(10-3\sqrt{5})$
15. $(6-3\sqrt{7})(2-5\sqrt{7})$
16. $(7-2\sqrt{7})(5-3\sqrt{7})$
17. $(\sqrt{2}+\sqrt{7})(\sqrt{3}+\sqrt{5})$
18. $(\sqrt{3}+\sqrt{2})(\sqrt{10}+\sqrt{11})$
19. $(\sqrt{2}-\sqrt{7})(\sqrt{3}-\sqrt{5})$
20. $(\sqrt{3}-\sqrt{2})(\sqrt{10}-\sqrt{11})$
21. $(3\sqrt{2}-4\sqrt{3})(2\sqrt{2}+5\sqrt{3})$
22. $(3\sqrt{5}-2\sqrt{3})(4\sqrt{5}+5\sqrt{3})$
23. $(\sqrt{3}+\sqrt{5})^2$
24. $(\sqrt{2}+\sqrt{7})^2$
25. $(\sqrt{3x}-\sqrt{y})^2$
26. $(\sqrt{2x}-\sqrt{y})^2$
27. $(\sqrt{5}+7)(\sqrt{5}-7)$
28. $(\sqrt{6}+2)(\sqrt{6}-2)$
29. $(2-5\sqrt{3})(2+5\sqrt{3})$
30. $(3-5\sqrt{2})(3+5\sqrt{2})$
31. $(3\sqrt{2}+2\sqrt{3})(3\sqrt{2}-2\sqrt{3})$
32. $(4\sqrt{3}+3\sqrt{2})(4\sqrt{3}-3\sqrt{2})$
33. $(3-\sqrt{x})(2-\sqrt{x})$
34. $(4-\sqrt{x})(3-\sqrt{x})$
35. $(\sqrt[3]{x}-4)(\sqrt[3]{x}+5)$
36. $(\sqrt[3]{x}-3)(\sqrt[3]{x}+7)$
37. $(x+\sqrt[3]{y^2})(2x-\sqrt[3]{y^2})$
38. $(x-\sqrt[5]{y^3})(2x+\sqrt[5]{y^3})$

In Exercises 39–64, rationalize each denominator.

39. $\dfrac{\sqrt{2}}{\sqrt{5}}$
40. $\dfrac{\sqrt{7}}{\sqrt{3}}$
41. $\sqrt{\dfrac{11}{x}}$
42. $\sqrt{\dfrac{6}{x}}$
43. $\dfrac{9}{\sqrt{3y}}$
44. $\dfrac{12}{\sqrt{3y}}$
45. $\dfrac{1}{\sqrt[3]{2}}$
46. $\dfrac{1}{\sqrt[3]{3}}$
47. $\dfrac{6}{\sqrt[3]{4}}$
48. $\dfrac{10}{\sqrt[3]{5}}$
49. $\sqrt[3]{\dfrac{2}{3}}$
50. $\sqrt[3]{\dfrac{3}{4}}$
51. $\dfrac{4}{\sqrt[3]{x}}$
52. $\dfrac{7}{\sqrt[3]{x}}$
53. $\sqrt[3]{\dfrac{2}{y^2}}$
54. $\sqrt[3]{\dfrac{5}{y^2}}$
55. $\dfrac{7}{\sqrt[3]{2x^2}}$
56. $\dfrac{10}{\sqrt[3]{4x^2}}$
57. $\sqrt[3]{\dfrac{2}{xy^2}}$
58. $\sqrt[3]{\dfrac{3}{xy^2}}$
59. $\dfrac{3}{\sqrt[4]{x}}$
60. $\dfrac{5}{\sqrt[4]{x}}$
61. $\dfrac{6}{\sqrt[5]{8x^3}}$
62. $\dfrac{10}{\sqrt[5]{16x^2}}$

63. $\frac{2x^2y}{\sqrt[5]{4x^2y^4}}$

64. $\frac{3xy^2}{\sqrt[5]{8xy^3}}$

In Exercises 65–74, simplify each radical expression and then rationalize the denominator.

65. $\frac{9}{\sqrt{3x^2y}}$

66. $\frac{25}{\sqrt{5x^2y}}$

67. $-\sqrt{\frac{75a^5}{b^3}}$

68. $-\sqrt{\frac{150a^3}{b^5}}$

69. $\sqrt{\frac{7m^2n^3}{14m^3n^2}}$

70. $\sqrt{\frac{5m^4n^6}{15m^3n^4}}$

71. $\frac{3}{\sqrt[4]{x^5y^3}}$

72. $\frac{5}{\sqrt[4]{x^2y^7}}$

73. $\frac{12}{\sqrt[3]{-8x^5y^8}}$

74. $\frac{15}{\sqrt[3]{-27x^4y^{11}}}$

In Exercises 75–92, rationalize each denominator. Simplify, if possible.

75. $\frac{8}{\sqrt{5}+2}$

76. $\frac{15}{\sqrt{6}+1}$

77. $\frac{13}{\sqrt{11}-3}$

78. $\frac{17}{\sqrt{10}-2}$

79. $\frac{6}{\sqrt{5}+\sqrt{3}}$

80. $\frac{12}{\sqrt{7}+\sqrt{3}}$

81. $\frac{\sqrt{a}}{\sqrt{a}-\sqrt{b}}$

82. $\frac{\sqrt{b}}{\sqrt{a}-\sqrt{b}}$

83. $\frac{25}{5\sqrt{2}-3\sqrt{5}}$

84. $\frac{35}{5\sqrt{2}-3\sqrt{5}}$

85. $\frac{\sqrt{5}+\sqrt{3}}{\sqrt{5}-\sqrt{3}}$

86. $\frac{\sqrt{11}-\sqrt{5}}{\sqrt{11}+\sqrt{5}}$

87. $\frac{\sqrt{x}+1}{\sqrt{x}+3}$

88. $\frac{\sqrt{x}-2}{\sqrt{x}-5}$

89. $\frac{5\sqrt{3}-3\sqrt{2}}{3\sqrt{2}-2\sqrt{3}}$

90. $\frac{2\sqrt{6}+\sqrt{5}}{3\sqrt{6}-\sqrt{5}}$

91. $\frac{2\sqrt{x}+\sqrt{y}}{\sqrt{y}-2\sqrt{x}}$

92. $\frac{3\sqrt{x}+\sqrt{y}}{\sqrt{y}-3\sqrt{x}}$

In Exercises 93–104, rationalize each numerator. Simplify, if possible.

93. $\sqrt{\frac{3}{2}}$

94. $\sqrt{\frac{5}{3}}$

95. $\frac{\sqrt[3]{4x}}{\sqrt[3]{y}}$

96. $\frac{\sqrt[3]{2x}}{\sqrt[3]{y}}$

97. $\frac{\sqrt{x}+3}{\sqrt{x}}$

98. $\frac{\sqrt{x}+4}{\sqrt{x}}$

99. $\frac{\sqrt{a}+\sqrt{b}}{\sqrt{a}-\sqrt{b}}$

100. $\frac{\sqrt{a}-\sqrt{b}}{\sqrt{a}+\sqrt{b}}$

101. $\frac{\sqrt{x+5}-\sqrt{x}}{5}$

102. $\frac{\sqrt{x+7}-\sqrt{x}}{7}$

103. $\frac{\sqrt{x}+\sqrt{y}}{x^2-y^2}$

104. $\frac{\sqrt{x}-\sqrt{y}}{x^2-y^2}$

Practice Plus

In Exercises 105–112, add or subtract as indicated. Begin by rationalizing denominators for all terms in which denominators contain radicals.

105. $\sqrt{2}+\frac{1}{\sqrt{2}}$

106. $\sqrt{5}+\frac{1}{\sqrt{5}}$

107. $\sqrt[3]{25}-\frac{15}{\sqrt[3]{5}}$

108. $\sqrt[4]{8}-\frac{20}{\sqrt[4]{2}}$

109. $\sqrt{6}-\sqrt{\frac{1}{6}}+\sqrt{\frac{2}{3}}$

110. $\sqrt{15}-\sqrt{\frac{5}{3}}+\sqrt{\frac{3}{5}}$

111. $\frac{2}{\sqrt{2}+\sqrt{3}}+\sqrt{75}-\sqrt{50}$

112. $\frac{5}{\sqrt{2}+\sqrt{7}}-2\sqrt{32}+\sqrt{28}$

113. Let $f(x) = x^2 - 6x - 4$. Find $f(3-\sqrt{13})$.

114. Let $f(x) = x^2 + 4x - 2$. Find $f(-2+\sqrt{6})$.

115. Let $f(x) = \sqrt{9+x}$. Find $f(3\sqrt{5}) \cdot f(-3\sqrt{5})$.

116. Let $f(x) = x^2$. Find $f(\sqrt{a+1}-\sqrt{a-1})$.

Application Exercises

The bar graph shows the percentage of U.S. households online from 1997 through 2003. The function

$$P(t) = 15.92\sqrt{t} + 19$$

models the percentage, $P(t)$, of U.S. households online t years after 1997. Use the function to solve Exercises 117–118.

Percentage of U.S. Households Online

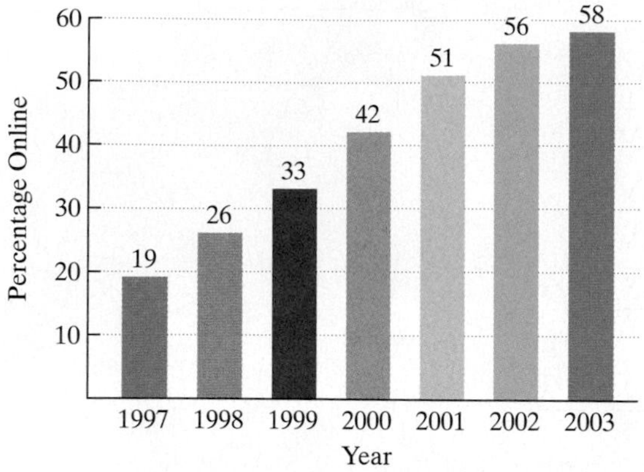

Source: U.S. Department of Commerce

117. Find $P(4)$ and describe what this means. Round to the nearest percent. How well does your answer model the actual data shown in the graph?

118. Find $P(6)$ and describe what this means. Use a calculator and round to the nearest percent. How well does your answer model the actual data shown in the graph?

In Exercises 119–120, use the data shown in the bar graph for the percentage of U.S. households online.

119. The average rate of change in the percentage of households online from 1997 through 2003 is

Change in percent

$$\frac{\text{percentage online in 2003} - \text{percentage online in 1997}}{2003 - 1997}$$

Change in time

Find the average yearly increase in the percentage of online households during this period.

120. The average rate of change in the percentage of households online from 2001 through 2003 is

Change in percent

$$\frac{\text{percentage online in 2003} - \text{percentage online in 2001}}{2003 - 2001}$$

Change in time

Find the average yearly increase in the percentage of online households during this period.

The algebraic expression

$$15.92\left(\frac{\sqrt{t+h} - \sqrt{t}}{h}\right)$$

models the average rate of change in the percentage of households online from t years after 1997 to $t + h$ years after 1997. Use the expression to solve Exercises 121–124.

121. Let $t = 0$ and $h = 6$. Evaluate the expression to find the average rate of change in the percentage of households online from 1997 through 2003—that is, between 0 years and 6 years after 1997. How well does your answer model the actual yearly increase in the percentage that you found in Exercise 119?

122. Let $t = 4$ and $h = 2$. Evaluate the expression to find the average rate of change in the percentage of households online from 2001 through 2003—that is, between 4 years and $4 + 2$, or 6, years after 1997. How well does your answer model the actual yearly increase in the percentage that you found in Exercise 120?

123. **a.** Rewrite the average rate of change as an equivalent expression by rationalizing the numerator of the rational expression in parentheses.

b. Substitute 0 for h in your expression from part (a) and simplify. The resulting expression gives the rate of change in the percentage of households online precisely t years after 1997.

c. Use the expression from part (b) to find the rate of change in the percentage of households online in 2003. Round to the nearest tenth of a percent.

124. Solve parts (a) and (b) of Exercise 123. Then find the rate of change in the percentage of households online in 2001. Round to the nearest tenth of a percent.

In Exercises 125–126, write expressions for the perimeter and area of each figure. Then simplify these expressions. Assume that all measures are given in inches.

125.

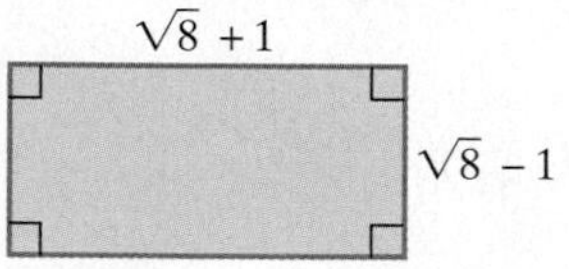

126.

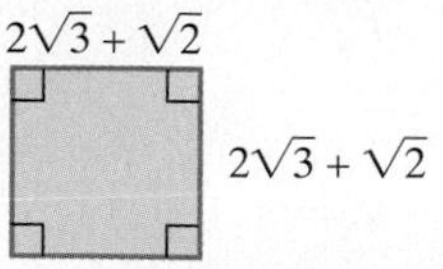

127. In the "Peanuts" cartoon shown in the section opener on page 678, Woodstock appears to be working steps mentally. Fill in the missing steps that show how to go from $\frac{7\sqrt{2\cdot 2\cdot 3}}{6}$ to $\frac{7}{3}\sqrt{3}$.

128. Rationalize the denominator of the golden ratio

$$\frac{2}{\sqrt{5}-1},$$

discussed in the enrichment essay on page 683. Then use a calculator and find the ratio of width to height, correct to the nearest hundredth, in golden rectangles.

Writing in Mathematics

129. Explain how to perform this multiplication: $\sqrt{2}(\sqrt{7}+\sqrt{10})$.

130. Explain how to perform this multiplication: $(2+\sqrt{3})(4+\sqrt{3})$.

131. Explain how to perform this multiplication: $(2+\sqrt{3})^2$.

132. What are conjugates? Give an example with your explanation.

133. Describe how to multiply conjugates.

134. Describe what it means to rationalize a denominator. Use both $\frac{1}{\sqrt{5}}$ and $\frac{1}{5+\sqrt{5}}$ in your explanation.

135. When a radical expression has its denominator rationalized, we change the denominator so that it no longer contains any radicals. Doesn't this change the value of the radical expression? Explain.

136. Square the real number $\frac{2}{\sqrt{3}}$. Observe that the radical is eliminated from the denominator. Explain whether this process is equivalent to rationalizing the denominator.

137. Describe the trend shown in the bar graph in Exercises 117–118 for the percentage of U.S. households online from 1997 through 2003. What explanation can you offer for this trend?

Technology Exercises

In Exercises 138–141, determine if each operation is performed correctly by graphing the function on each side of the equation with your graphing utility. Use the given viewing rectangle. The graphs should be the same. If they are not, correct the right side of the equation and then use your graphing utility to verify the correction.

138. $(\sqrt{x}-1)(\sqrt{x}-1)=x+1$

$[0, 5, 1]$ by $[-1, 2, 1]$

139. $(\sqrt{x}+2)(\sqrt{x}-2)=x^2-4$ for $x\geq 0$

$[0, 10, 1]$ by $[-10, 10, 1]$

140. $(\sqrt{x}+1)^2=x+1$

$[0, 8, 1]$ by $[0, 15, 1]$

141. $\frac{3}{\sqrt{x+3}-\sqrt{x}}=\sqrt{x+3}+\sqrt{x}$

$[0, 8, 1]$ by $[0, 6, 1]$

Critical Thinking Exercises

142. Which one of the following is true?

a. $\frac{\sqrt{3}+7}{\sqrt{3}-2}=-\frac{7}{2}$

b. $\frac{4}{\sqrt{x+y}}=\frac{4\sqrt{x-y}}{x-y}$

c. $\frac{4\sqrt{x}}{\sqrt{x}-y}=\frac{4x+4y\sqrt{x}}{x-y^2}$

d. $(\sqrt{x}-7)^2=x-49$

143. Solve:

$$7[(2x-5)-(x+1)]=\left(\sqrt{7}+2\right)\left(\sqrt{7}-2\right).$$

144. Simplify: $\left(\sqrt{2+\sqrt{3}}+\sqrt{2-\sqrt{3}}\right)^2$.

145. Rationalize the denominator: $\frac{1}{\sqrt{2}+\sqrt{3}+\sqrt{4}}$.

Review Exercises

146. Add: $\frac{2}{x-2}+\frac{3}{x^2-4}$.

(Section 7.4, Example 5)

147. Solve: $3x-4\leq 2$ and $4x+5\geq 5$ (Section 9.2, Example 2)

148. Is $\{(-1, 1), (1, 1), (-2, 4), (2, 4)\}$ a function? Is the inverse of this relation a function? (Section 8.1, Example 2, and Section 8.3, first paragraph on page 564)

SECTION 10.6

Objectives

1 Solve radical equations.

2 Use models that are radical functions to solve problems.

RADICAL EQUATIONS

A best guess at the look of our nation in the next decades indicates that the number of men and women living alone will increase each year. By 2010, approximately 28 million men and women are projected to be living alone. The function

$$f(x) = 2.6\sqrt{x} + 11$$

models the number of Americans living alone, $f(x)$, in millions, x years after 1970. How can we predict the year when, say, 29 million of us will live alone? Substitute 29 for $f(x)$ and solve for x:

$$29 = 2.6\sqrt{x} + 11.$$

The resulting equation contains a variable in the radicand and is called a *radical equation*. A **radical equation** is an equation in which the variable occurs in a square root, cube root, or any higher root. Some examples of radical equations are

$$\sqrt{2x+3} = 5, \quad \sqrt{3x+1} - \sqrt{x+4} = 1, \quad \text{and} \quad \sqrt[3]{3x-1} + 4 = 0.$$

Variables occur in radicands.

In this section, you will learn how to solve radical equations. Solving such equations will enable you to solve new kinds of problems using radical functions.

1 Solve radical equations.

Solving Radical Equations Consider the following radical equation:

$$\sqrt{x} = 9.$$

We solve the equation by squaring both sides:

Squaring both sides eliminates the square root.

$$(\sqrt{x})^2 = 9^2$$

$$x = 81.$$

The proposed solution, 81, can be checked in the original equation, $\sqrt{x} = 9$. Because $\sqrt{81} = 9$, the solution is 81 and the solution set is $\{81\}$.

In general, we solve radical equations with square roots by squaring both sides of the equation. We solve radical equations with nth roots by raising both sides of the equation to the nth power. Unfortunately, if n is even, all the solutions of the equation raised to the even power may not be solutions of the original equation. Consider, for example, the equation

$$x = 4.$$

If we square both sides, we obtain

$$x^2 = 16.$$

This new equation has two solutions, −4 and 4. By contrast, only 4 is a solution of the original equation, $x = 4$. For this reason, **when raising both sides of an equation to an even power, always check proposed solutions in the original equation**.

Here is a general method for solving radical equations with nth roots:

SOLVING RADICAL EQUATIONS CONTAINING nTH ROOTS

1. If necessary, arrange terms so that one radical is isolated on one side of the equation.
2. Raise both sides of the equation to the nth power to eliminate the nth root.
3. Solve the resulting equation. If this equation still contains radicals, repeat steps 1 and 2.
4. Check all proposed solutions in the original equation.

EXAMPLE 1 Solving a Radical Equation

Solve: $\sqrt{2x + 3} = 5$.

SOLUTION

Step 1. **Isolate a radical on one side.** The radical, $\sqrt{2x + 3}$, is already isolated on the left side of the equation, so we can skip this step.

Step 2. **Raise both sides to the *n*th power.** Because n, the index, is 2, we square both sides.

$$\sqrt{2x + 3} = 5 \quad \text{This is the given equation.}$$

$$\left(\sqrt{2x + 3}\right)^2 = 5^2 \quad \text{Square both sides to eliminate the radical.}$$

$$2x + 3 = 25 \quad \text{Simplify.}$$

Step 3. **Solve the resulting equation.**

$$2x + 3 = 25 \quad \text{The resulting equation is a linear equation.}$$

$$2x = 22 \quad \text{Subtract 3 from both sides.}$$

$$x = 11 \quad \text{Divide both sides by 2.}$$

Step 4. **Check the proposed, solution in the original equation.** Because both sides were raised to an even power, this check is essential.

Check 11:

$$\sqrt{2x + 3} = 5$$

$$\sqrt{2 \cdot 11 + 3} \stackrel{?}{=} 5$$

$$\sqrt{25} \stackrel{?}{=} 5$$

$$5 = 5, \quad \text{true}$$

The solution is 11 and the solution set is {11}. ■

CHECK POINT 1 Solve: $\sqrt{3x + 4} = 8$.

EXAMPLE 2 Solving a Radical Equation

Solve: $\sqrt{x-3}+6=5$.

SOLUTION

Step 1. **Isolate a radical on one side.** The radical, $\sqrt{x-3}$, can be isolated by subtracting 6 from both sides. We obtain

$$\sqrt{x-3}=-1.$$

A principal square root cannot be negative. This equation has no solution. Let's continue the solution procedure to see what happens.

Step 2. **Raise both sides to the *n*th power.** Because n, the index, is 2, we square both sides.

$$(\sqrt{x-3})^2=(-1)^2$$

$$x-3=1 \quad \text{Simplify.}$$

Step 3. **Solve the resulting equation.**

$$x-3=1 \quad \text{The resulting equation is a linear equation.}$$

$$x=4 \quad \text{Add 3 to both sides.}$$

Step 4. **Check the proposed solution in the original equation.**

Check 4:

$$\sqrt{x-3}+6=5$$

$$\sqrt{4-3}+6\stackrel{?}{=}5$$

$$\sqrt{1}+6\stackrel{?}{=}5$$

$$1+6\stackrel{?}{=}5$$

$$7=5, \quad \text{false}$$

This false statement indicates that 4 is not a solution. Thus, the equation has no solution. The solution set is Ø, the empty set.

Example 2 illustrates that extra solutions may be introduced when you raise both sides of a radical equation to an even power. Such solutions, which are not solutions of the given equation, are called **extraneous solutions**. Thus, 4 is an extraneous solution of $\sqrt{x-3}+6=5$.

CHECK POINT 2 Solve: $\sqrt{x-1}+7=2$.

EXAMPLE 3 Solving a Radical Equation

Solve: $x+\sqrt{26-11x}=4$.

SOLUTION

Step 1. **Isolate a radical on one side.** We isolate the radical, $\sqrt{26-11x}$, by subtracting x from both sides.

$$x+\sqrt{26-11x}=4 \quad \text{This is the given equation.}$$

$$\sqrt{26-11x}=4-x. \quad \text{Subtract } x \text{ from both sides.}$$

STUDY TIP

Be sure to square *both sides* of an equation. Do *not* square each term.

Correct:

$$\left(\sqrt{26-11}\right)^2 = (4-x)^2$$

Incorrect!

~~$\left(\sqrt{26-11}\right)^2 = 4^2 - x^2$~~

Step 2. Square both sides.

$$\left(\sqrt{26-11x}\right)^2 = (4-x)^2$$

$$26 - 11x = 16 - 8x + x^2$$ Simplify. Use the special-product formula $(A-B)^2 = A^2 - 2AB + B^2$ to square $4 - x$.

Step 3. Solve the resulting equation. Because of the x^2-term, the resulting equation is a quadratic equation. We need to write this quadratic equation in standard form. We can obtain zero on the left side by subtracting 26 and adding $11x$ on both sides.

$$26 - 26 - 11x + 11x = 16 - 26 - 8x + 11x + x^2$$

$$0 = x^2 + 3x - 10$$ Simplify.

$$0 = (x+5)(x-2)$$ Factor.

$x + 5 = 0$ or $x - 2 = 0$ Set each factor equal to zero.

$x = -5$ $\qquad$ $x = 2$ Solve for x.

Step 4. Check the proposed solutions in the original equation.

Check −5:

$$x + \sqrt{26-11x} = 4$$
$$-5 + \sqrt{26-11(-5)} \stackrel{?}{=} 4$$
$$-5 + \sqrt{81} \stackrel{?}{=} 4$$
$$-5 + 9 \stackrel{?}{=} 4$$
$$4 = 4,\ \text{true}$$

Check 2:

$$x + \sqrt{26-11x} = 4$$
$$2 + \sqrt{26-11\cdot 2} \stackrel{?}{=} 4$$
$$2 + \sqrt{4} \stackrel{?}{=} 4$$
$$2 + 2 \stackrel{?}{=} 4$$
$$4 = 4,\ \text{true}$$

The solutions are −5 and 2, and the solution set is $\{-5, 2\}$. ■

USING TECHNOLOGY

You can use a graphing utility to provide a graphic check that $\{-5, 2\}$ is the solution set of $x + \sqrt{26-11x} = 4$.

Use the given equation

$$x + \sqrt{26-11x} = 4.$$

Enter $y_1 = x + \sqrt{26-11x}$ in the $\boxed{y=}$ screen.

Enter $y_2 = 4$ in the $\boxed{y=}$ screen.

Display graphs for y_1 and y_2. The solutions are the x-coordinates of the intersection points. These x-coordinates are −5 and 2. This verifies $\{-5, 2\}$ as the solution set of $x + \sqrt{26-11x} = 4$.

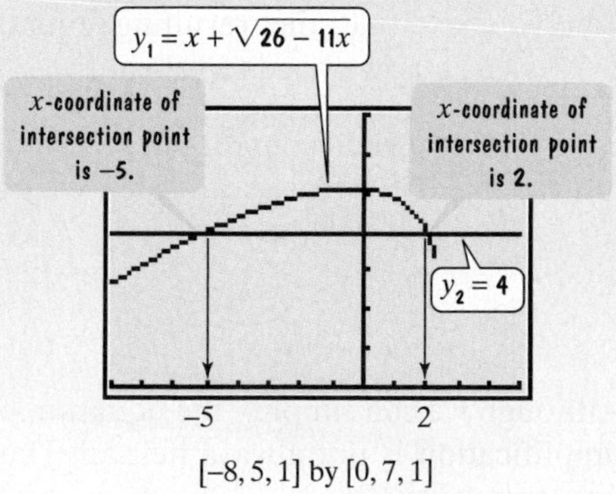

[−8, 5, 1] by [0, 7, 1]

Use the equivalent equation

$$x + \sqrt{26-11x} - 4 = 0.$$

Enter $y_1 = x + \sqrt{26-11x} - 4$ in the $\boxed{y=}$ screen.

Display the graph for y_1. The solutions are the x-intercepts. The x-intercepts are −5 and 2. This verifies $\{-5, 2\}$ as the solution set of $x + \sqrt{26-11x} - 4 = 0$.

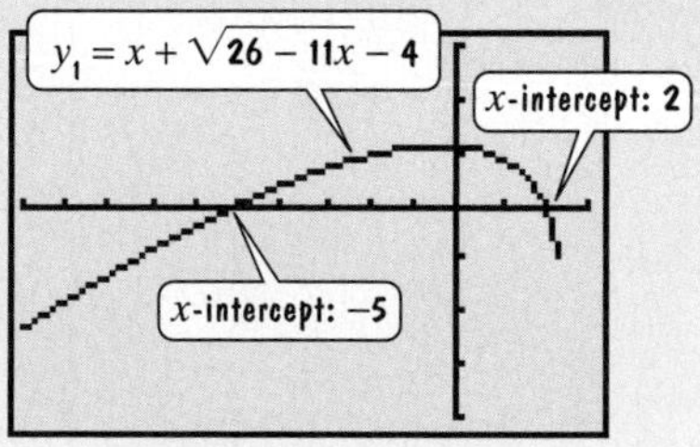

[−10, 3, 1] by [−4, 3, 1]

CHECK POINT 3 Solve: $\sqrt{6x+7} - x = 2$.

The solution of radical equations with two or more square root expressions involves isolating a radical, squaring both sides, and then repeating this process. Let's consider an equation containing two square root expressions.

EXAMPLE 4 Solving an Equation That Has Two Radicals

Solve: $\sqrt{3x+1} - \sqrt{x+4} = 1$.

SOLUTION

Step 1. Isolate a radical on one side. We can isolate the radical $\sqrt{3x+1}$ by adding $\sqrt{x+4}$ to both sides. We obtain

$$\sqrt{3x+1} = \sqrt{x+4} + 1.$$

Step 2. Square both sides.

$$\left(\sqrt{3x+1}\right)^2 = \left(\sqrt{x+4}+1\right)^2$$

Squaring the expression on the right side of the equation can be a bit tricky. We have to use the formula

$$(A+B)^2 = A^2 + 2AB + B^2.$$

Focusing on just the right side, here is how the squaring is done:

$$\underbrace{(A+B)^2}_{} = A^2 + 2 \cdot A \cdot B + B^2$$

$$\left(\sqrt{x+4}+1\right)^2 = \left(\sqrt{x+4}\right)^2 + 2 \cdot \sqrt{x+4} \cdot 1 + 1^2 = x + 4 + 2\sqrt{x+4} + 1.$$

Now let's return to squaring both sides.

$$\left(\sqrt{3x+1}\right)^2 = \left(\sqrt{x+4}+1\right)^2$$ Square both sides of the equation with an isolated radical.

$$3x + 1 = x + 4 + 2\sqrt{x+4} + 1$$ $\left(\sqrt{3x+1}\right)^2 = 3x + 1$; square the right side using the formula for $(A+B)^2$.

$$3x + 1 = x + 5 + 2\sqrt{x+4}$$ Combine numerical terms on the right side: $4 + 1 = 5$.

Can you see that the resulting equation still contains a radical, namely $\sqrt{x+4}$? Thus, we need to repeat the first two steps.

Repeat Step 1. Isolate a radical on one side. We isolate $2\sqrt{x+4}$, the radical term, by subtracting $x + 5$ from both sides. We obtain

$$3x + 1 = x + 5 + 2\sqrt{x+4}$$ This is the equation from our last step.

$$2x - 4 = 2\sqrt{x+4}.$$ Subtract x and subtract 5 from both sides.

Although we can simplify the equation by dividing both sides by 2, this sort of simplification is not always helpful. Thus, we will work with the equation in this form.

Repeat Step 2. Square both sides.

Be careful in squaring both sides. Use $(A - B)^2 = A^2 - 2AB + B^2$ to square the left side. Use $(AB)^2 = A^2B^2$ to square the right side.

$$(2x - 4)^2 = \left(2\sqrt{x + 4}\right)^2 \quad \text{Square both sides.}$$

$$4x^2 - 16x + 16 = 4(x + 4) \quad \text{Square both 2 and } \sqrt{x+4} \text{ on the right side.}$$

Step 3. Solve the resulting equation. We solve this quadratic equation by writing it in standard form.

$$4x^2 - 16x + 16 = 4x + 16 \quad \text{Use the distributive property.}$$
$$4x^2 - 20x = 0 \quad \text{Subtract } 4x + 16 \text{ from both sides.}$$
$$4x(x - 5) = 0 \quad \text{Factor.}$$
$$4x = 0 \quad \text{or} \quad x - 5 = 0 \quad \text{Set each factor equal to zero.}$$
$$x = 0 \qquad\qquad x = 5 \quad \text{Solve for } x.$$

Step 4. Check the proposed solutions in the original equation.

Check 0:	**Check 5:**
$\sqrt{3x+1} - \sqrt{x+4} = 1$	$\sqrt{3x+1} - \sqrt{x+4} = 1$
$\sqrt{3\cdot 0+1} - \sqrt{0+4} \stackrel{?}{=} 1$	$\sqrt{3\cdot 5+1} - \sqrt{5+4} \stackrel{?}{=} 1$
$\sqrt{1} - \sqrt{4} \stackrel{?}{=} 1$	$\sqrt{16} - \sqrt{9} \stackrel{?}{=} 1$
$1 - 2 \stackrel{?}{=} 1$	$4 - 3 \stackrel{?}{=} 1$
$-1 = 1$, false	$1 = 1$, true

The check indicates that 0 is not a solution. It is an extraneous solution brought about by squaring each side of the equation. The only solution is 5, and the solution set is $\{5\}$. ■

CHECK POINT 4 Solve: $\sqrt{x + 5} - \sqrt{x - 3} = 2$.

EXAMPLE 5 Solving a Radical Equation

Solve: $(3x - 1)^{\frac{1}{3}} + 4 = 0$.

SOLUTION Although we can rewrite the equation in radical form

$$\sqrt[3]{3x - 1} + 4 = 0,$$

it is not necessary to do so. Because the equation involves a cube root, we isolate the radical term—that is, the term with the rational exponent—and cube both sides.

$$(3x - 1)^{\frac{1}{3}} + 4 = 0 \quad \text{This is the given equation.}$$
$$(3x - 1)^{\frac{1}{3}} = -4 \quad \text{Subtract 4 from both sides and isolate the term with the rational exponent.}$$
$$\left[(3x - 1)^{\frac{1}{3}}\right]^3 = (-4)^3 \quad \text{Cube both sides.}$$
$$3x - 1 = -64 \quad \text{Multiply exponents on the left side and simplify.}$$
$$3x = -63 \quad \text{Add 1 to both sides.}$$
$$x = -21 \quad \text{Divide both sides by 3.}$$

Because both sides were raised to an odd power, it is not essential to check the proposed solution, -21. However, checking is always a good idea. Do so now and verify that -21 is the solution and the solution set is $\{-21\}$. ■

Example 5 illustrates that a radical equation with rational exponents can be solved by

1. isolating the expression with the rational exponent, and
2. raising both sides of the equation to a power that is the reciprocal of the rational exponent.

Keep in mind that it is essential to check when both sides have been raised to even powers. Thus, equations with rational exponents such as $\frac{1}{2}$ and $\frac{1}{4}$ must be checked.

CHECK POINT 5 Solve: $(2x-3)^{\frac{1}{3}}+3=0$.

2 Use models that are radical functions to solve problems.

Applications of Radical Equations Radical equations can be solved to answer questions about variables contained in radical functions.

EXAMPLE 6 Living Alone

The bar graph in Figure 10.7 shows the number of people in the United States living alone for selected years from 1970 through 2000. The function

$$f(x)=2.6\sqrt{x}+11$$

models the number of Americans living alone, $f(x)$, in millions, x years after 1970. According to the model, when will 29 million Americans live alone?

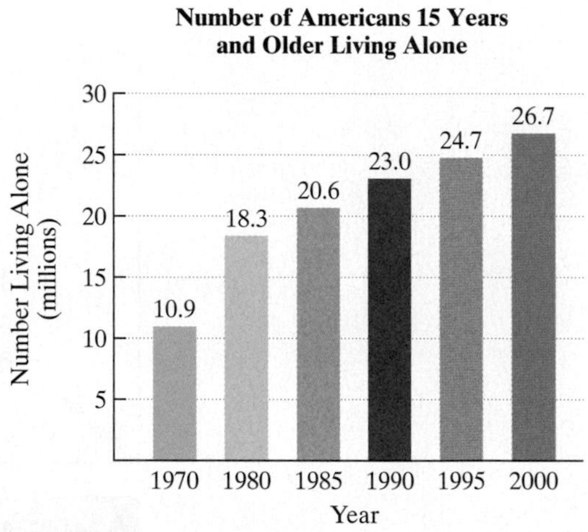

FIGURE 10.7
Source: U.S. Census Bureau

SOLUTION To find when 29 million Americans will live alone, substitute 29 for $f(x)$ in the given function. Then solve for x, the number of years after 1970.

$$f(x)=2.6\sqrt{x}+11 \quad \text{This is the given function.}$$

$$29=2.6\sqrt{x}+11 \quad \text{Substitute 29 for } f(x).$$

$$18=2.6\sqrt{x} \quad \text{Subtract 11 from both sides.}$$

$$\frac{18}{2.6}=\sqrt{x} \quad \text{Divide both sides by 2.6.}$$

$$\left(\frac{18}{2.6}\right)^2=(\sqrt{x})^2 \quad \text{Square both sides.}$$

$$48\approx x \quad \text{Use a calculator.}$$

The model indicates that 29 million Americans will live alone approximately 48 years after 1970. Because $1970+48=2018$, this is predicted to occur in 2018. ■

CHECK POINT 6 The function $f(x) = 2.9\sqrt{x} + 11$ serves as another model for the number of Americans living alone, $f(x)$, in millions, x years after 1970. According to this model, when will 29 million Americans live alone? Use a calculator and round to the nearest year. Does this model predict faster or slower growth in the number of people living alone than the function used in Example 6?

10.6 EXERCISE SET

Student Solutions Manual | CD/Video | PH Math/Tutor Center | MathXL Tutorials on CD | MathXL® | MyMathLab | Interactmath.com

Practice Exercises

In Exercises 1–38, solve each radical equation.

1. $\sqrt{3x - 2} = 4$

2. $\sqrt{5x - 1} = 8$

3. $\sqrt{5x - 4} - 9 = 0$

4. $\sqrt{3x - 2} - 5 = 0$

5. $\sqrt{3x + 7} + 10 = 4$

6. $\sqrt{2x + 5} + 11 = 6$

7. $x = \sqrt{7x + 8}$

8. $x = \sqrt{6x + 7}$

9. $\sqrt{5x + 1} = x + 1$

10. $\sqrt{2x + 1} = x - 7$

11. $x = \sqrt{2x - 2} + 1$

12. $x = \sqrt{3x + 7} - 3$

13. $x - 2\sqrt{x - 3} = 3$

14. $3x - \sqrt{3x + 7} = -5$

15. $\sqrt{2x - 5} = \sqrt{x + 4}$

16. $\sqrt{6x + 2} = \sqrt{5x + 3}$

17. $\sqrt[3]{2x + 11} = 3$

18. $\sqrt[3]{6x - 3} = 3$

19. $\sqrt[3]{2x - 6} - 4 = 0$

20. $\sqrt[3]{4x - 3} - 5 = 0$

21. $\sqrt{x - 7} = 7 - \sqrt{x}$

22. $\sqrt{x - 8} = \sqrt{x} - 2$

23. $\sqrt{x + 2} + \sqrt{x - 1} = 3$

24. $\sqrt{x - 4} + \sqrt{x + 4} = 4$

25. $2\sqrt{4x + 1} - 9 = x - 5$

26. $2\sqrt{x - 3} + 4 = x + 1$

27. $(2x + 3)^{\frac{1}{3}} + 4 = 6$

28. $(3x - 6)^{\frac{1}{3}} + 5 = 8$

29. $(3x + 1)^{\frac{1}{4}} + 7 = 9$

30. $(2x + 3)^{\frac{1}{4}} + 7 = 10$

31. $(x + 2)^{\frac{1}{2}} + 8 = 4$

32. $(x - 3)^{\frac{1}{2}} + 8 = 6$

33. $\sqrt{2x - 3} - \sqrt{x - 2} = 1$

34. $\sqrt{x + 2} + \sqrt{3x + 7} = 1$

35. $3x^{\frac{1}{3}} = (x^2 + 17x)^{\frac{1}{3}}$

36. $2(x - 1)^{\frac{1}{3}} = (x^2 + 2x)^{\frac{1}{3}}$

37. $(x + 8)^{\frac{1}{4}} = (2x)^{\frac{1}{4}}$

38. $(x - 2)^{\frac{1}{4}} = (3x - 8)^{\frac{1}{4}}$

Practice Plus

39. If $f(x) = x + \sqrt{x + 5}$, find all values of x for which $f(x) = 7$.

40. If $f(x) = x - \sqrt{x - 2}$, find all values of x for which $f(x) = 4$.

41. If $f(x) = (5x + 16)^{\frac{1}{3}}$ and $g(x) = (x - 12)^{\frac{1}{3}}$, find all values of x for which $f(x) = g(x)$.

42. If $f(x) = (9x + 2)^{\frac{1}{4}}$ and $g(x) = (5x + 18)^{\frac{1}{4}}$, find all values of x for which $f(x) = g(x)$.

In Exercises 43–46, solve each formula for the specified variable.

43. $r = \sqrt{\dfrac{3V}{\pi h}}$ for V

44. $r = \sqrt{\dfrac{A}{4\pi}}$ for A

45. $t = 2\pi\sqrt{\dfrac{l}{32}}$ for l

46. $v = \sqrt{\frac{FR}{m}}$ for m

47. If 5 times a number is decreased by 4, the principal square root of this difference is 2 less than the number. Find the number(s).

48. If a number is decreased by 3, the principal square root of this difference is 5 less than the number. Find the number(s).

In Exercises 49–50, find the x-intercept(s) of the graph of each function without graphing the function.

49. $f(x) = \sqrt{x + 16} - \sqrt{x} - 2$

50. $f(x) = \sqrt{2x - 3} - \sqrt{2x} + 1$

Application Exercises

In 2002, the average surface temperature on Earth was 57.9°F, approximately 1.4° higher than it was one hundred years ago. Worldwide temperatures have risen only 9°F since the end of the last ice age 12,000 years ago. Most climatologists are convinced that over the next one hundred years, global temperatures will continue to increase, possibly setting off a chain of devastating events beginning with a rise in sea levels worldwide and ending with the destruction of water supplies, forests, and agriculture in many parts of the world. The graph shows global annual average temperatures from 1880 through 2002, with projections from 2002 through 2100.

Global Annual Average Temperatures and Projections through 2100

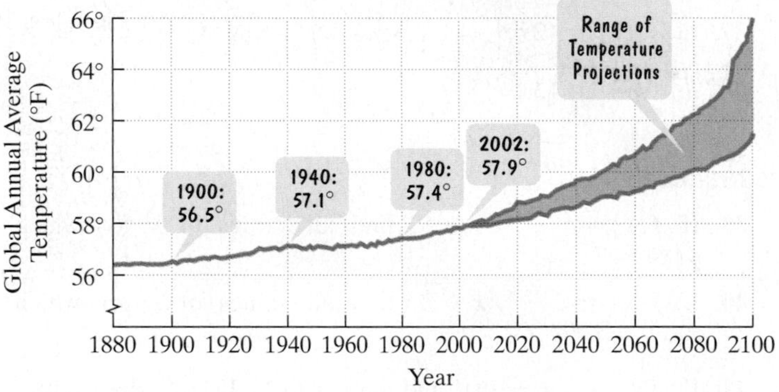

Source: National Oceanic and Atmospheric Administration

The temperature projections shown in the graph can be modeled by two functions:

$f(x) = 0.083x + 57.9$ — Models temperatures at the high end of the range

$g(x) = 0.36\sqrt{x} + 57.9.$ — Models temperatures at the low end of the range

In these functions, $f(x)$ and $g(x)$ describe projected global annual average temperatures, in degrees Fahrenheit, x years after 2002, where $0 \le x \le 98$. Use the functions to solve Exercises 51–54.

51. Use f and g to determine the temperatures at the high and low end of the range of projected global average temperatures for 2100. Round to the nearest tenth of a degree.

52. Use f and g to determine the temperatures at the high and low end of the range of projected global average temperatures for 2080. Round to the nearest tenth of a degree.

53. Use f and g to determine in which years the projected global average temperature will exceed the 2002 average of 57.9° by one degree. Round to the nearest year.

54. Use f and g to determine in which years the projected global average temperature will exceed the 2002 average of 57.9° by two degrees.

Out of a group of 50,000 births, the number of people, $f(x)$, surviving to age x is modeled by the function

$$f(x) = 5000\sqrt{100 - x}.$$

The graph of the function is shown. Use the function to solve Exercises 55–56.

Number of Survivors, by Age, from a Group of 50,000 Births

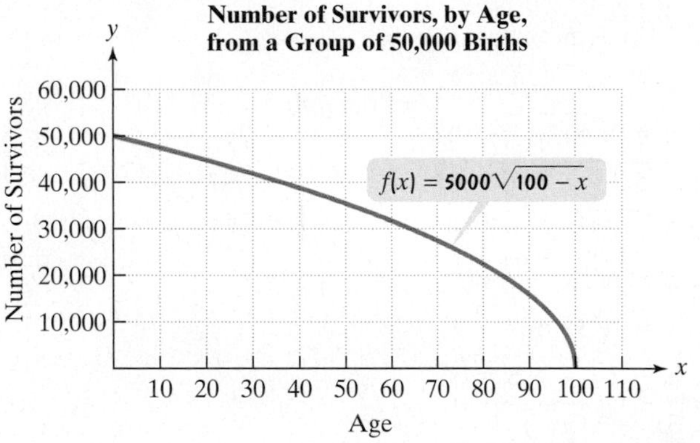

55. To what age will 40,000 people in the group survive? Identify the solution as a point on the graph of the function.

56. To what age will 35,000 people in the group survive? Identify the solution as a point on the graph of the function.

The function

$$f(x) = 29x^{\frac{1}{3}}$$

models the number of plant species, $f(x)$, on the islands of the Galápagos in terms of the area, x, in square miles, of a particular island. Use the function to solve Exercises 57–58.

57. What is the area of a Galápagos island that has 87 species of plants?

58. What is the area of a Galápagos island that has 58 species of plants?

For each planet in our solar system, its year is the time it takes the planet to revolve once around the sun. The function

$$f(x) = 0.2x^{\frac{3}{2}}$$

models the number of Earth days in a planet's year, $f(x)$, where x is the average distance of the planet from the sun, in millions of kilometers. Use the function to solve Exercises 59–60.

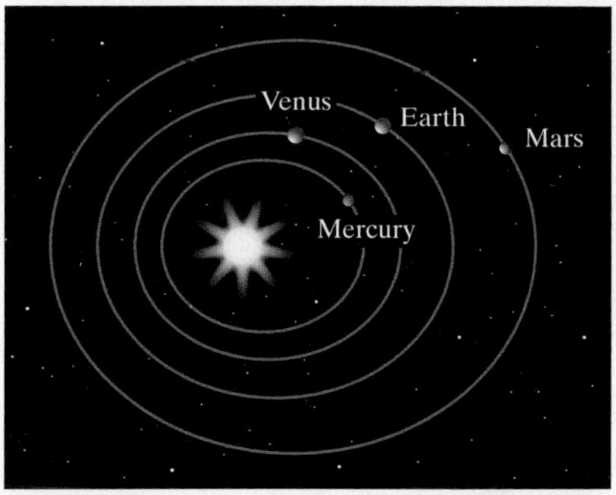

59. We, of course, have 365 Earth days in our year. What is the average distance of Earth from the sun? Use a calculator and round to the nearest million kilometers.

60. There are approximately 88 Earth days in the year of the planet Mercury. What is the average distance of Mercury from the sun? Use a calculator and round to the nearest million kilometers.

Writing in Mathematics

61. What is a radical equation?

62. In solving $\sqrt{2x-1} + 2 = x$, why is it a good idea to isolate the radical term? What if we don't do this and simply square each side? Describe what happens.

63. What is an extraneous solution to a radical equation?

64. Explain why $\sqrt{x} = -1$ has no solution.

65. Explain how to solve a radical equation with rational exponents.

66. The radical function in Example 6 on page 696 shows a gradual year-by-year increase in the number of Americans living alone. What explanations can you offer for this trend? Describe an event that might occur in the future that could cause this trend to change.

67. Describe the trend shown by the graph of f in Exercises 55–56. When is the rate of decrease most rapid? What does this mean about survival rate by age?

Technology Exercises

In Exercises 68–72, use a graphing utility to solve each radical equation. Graph each side of the equation in the given viewing rectangle. The equation's solution is given by the x-coordinate of the point(s) of intersection. Check by substitution.

68. $\sqrt{2x+2} = \sqrt{3x-5}$
$[-1, 10, 1]$ by $[-1, 5, 1]$

69. $\sqrt{x} + 3 = 5$
$[-1, 6, 1]$ by $[-1, 6, 1]$

70. $\sqrt{x^2+3} = x + 1$
$[-1, 6, 1]$ by $[-1, 6, 1]$

71. $4\sqrt{x} = x + 3$
$[-1, 10, 1]$ by $[-1, 14, 1]$

72. $\sqrt{x} + 4 = 2$
$[-2, 18, 1]$ by $[0, 10, 1]$

Critical Thinking Exercises

73. Which one of the following is true?

a. The first step in solving $\sqrt{x+6} = x + 2$ is to square both sides, obtaining $x + 6 = x^2 + 4$.

b. The equations $\sqrt{x+4} = -5$ and $x + 4 = 25$ have the same solution set.

c. If $T = 2\pi\sqrt{\frac{L}{32}}$, then $L = \frac{8T^2}{\pi^2}$.

d. The equation $\sqrt{x^2+9x+3} = -x$ has no solution because a principal square root is always nonnegative.

74. Find the length of the three sides of the right triangle shown in the figure.

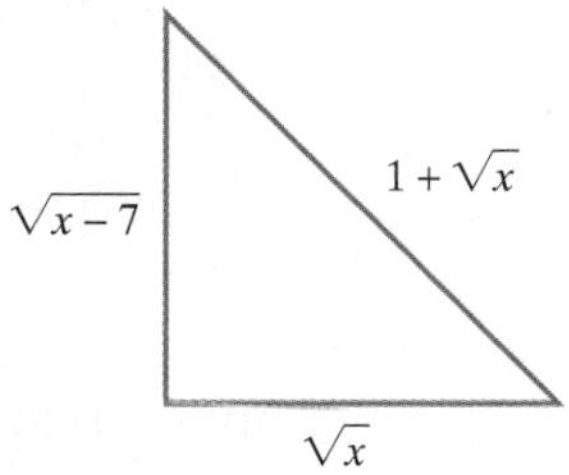

In Exercises 75–77, solve each equation.

75. $\sqrt[3]{x\sqrt{x}} = 9$

76. $\sqrt{\sqrt{x} + \sqrt{x+9}} = 3$

77. $(x-4)^{\frac{2}{3}} = 25$

Review Exercises

78. Divide using synthetic division:
$(4x^4 - 3x^3 + 2x^2 - x - 1) \div (x + 3)$.

(Section 5.6, Example 5)

79. Divide:

$$\frac{3x^2 - 12}{x^2 + 2x - 8} \div \frac{6x + 18}{x + 4}.$$

(Section 7.2, Example 6)

80. Factor: $y^2 - 6y + 9 - 25x^2$.

(Section 6.5, Example 3)

SECTION 10.7

Objectives

1. Express square roots of negative numbers in terms of i.
2. Add and subtract complex numbers.
3. Multiply complex numbers.
4. Divide complex numbers.
5. Simplify powers of i.

COMPLEX NUMBERS

© 2003 Roz Chast from Cartoonbank.com. All rights reserved.

Who is this kid warning us about our eyeballs turning black if we attempt to find the square root of -9? Don't believe what you hear on the street. Although square roots of negative numbers are not real numbers, they do play a significant role in algebra. In this section, we move beyond the real numbers and discuss square roots with negative radicands.

1 Express square roots of negative numbers in terms of i.

The Imaginary Unit i In Chapter 11 we will study equations whose solutions involve the square roots of negative numbers. Because the square of a real number is never negative, there is no real number x such that $x^2 = -1$. To provide a setting in which such equations have solutions, mathematicians invented an expanded system of numbers, the complex numbers. The *imaginary number i*, defined to be a solution of the equation $x^2 = -1$, is the basis of this new set.

THE IMAGINARY UNIT i The **imaginary unit i** is defined as

$$i = \sqrt{-1}, \quad \text{where} \quad i^2 = -1.$$

Using the imaginary unit i, we can express the square root of any negative number as a real multiple of i. For example,

$$\sqrt{-25} = \sqrt{25(-1)} = \sqrt{25}\sqrt{-1} = 5i.$$

We can check that $\sqrt{-25} = 5i$ by squaring $5i$ and obtaining -25.

$$(5i)^2 = 5^2i^2 = 25(-1) = -25$$

THE SQUARE ROOT OF A NEGATIVE NUMBER If b is a positive real number, then

$$\sqrt{-b} = \sqrt{b(-1)} = \sqrt{b}\sqrt{-1} = \sqrt{b}i \quad \text{or} \quad i\sqrt{b}.$$

EXAMPLE 1 Expressing Square Roots of Negative Numbers as Multiples of i

Write as a multiple of i:

a. $\sqrt{-9}$ **b.** $\sqrt{-3}$ **c.** $\sqrt{-80}$.

SOLUTION

a. $\sqrt{-9} = \sqrt{9(-1)} = \sqrt{9}\sqrt{-1} = 3i$

b. $\sqrt{-3} = \sqrt{3(-1)} = \sqrt{3}\sqrt{-1} = \sqrt{3}\,i$

c. $\sqrt{-80} = \sqrt{80(-1)} = \sqrt{80}\sqrt{-1} = \sqrt{16 \cdot 5}\sqrt{-1} = 4\sqrt{5}\,i$ ■

Be sure not to write i under the radical.

CHECK POINT 1 Write as a multiple of i:

a. $\sqrt{-64}$ **b.** $\sqrt{-11}$ **c.** $\sqrt{-48}$.

A new system of numbers, called *complex numbers*, is based on adding multiples of i, such as $5i$, to the real numbers.

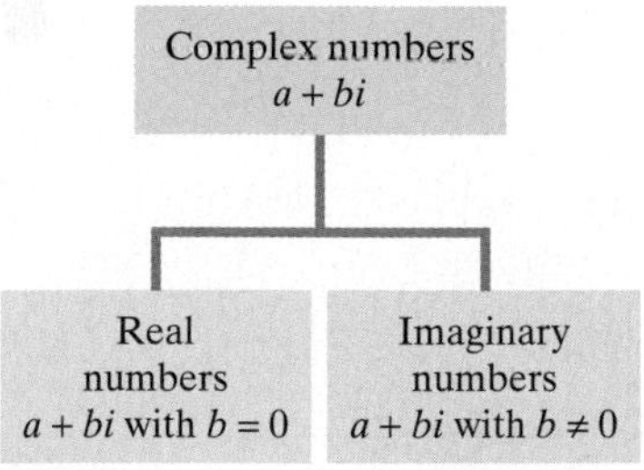

FIGURE 10.8 The complex number system

COMPLEX NUMBERS AND IMAGINARY NUMBERS The set of all numbers in the form

$$a + bi$$

with real numbers a and b, and i, the imaginary unit, is called the set of **complex numbers**. The real number a is called the **real part**, and the real number b is called the **imaginary part** of the complex number $a + bi$. If $b \neq 0$, then the complex number is called an **imaginary number** (Figure 10.8).

Here are some examples of complex numbers. Each number can be written in the form $a + bi$.

$-4 + 6i$ — a, the real part, is -4. b, the imaginary part, is 6.

$2i = 0 + 2i$ — a, the real part, is 0. b, the imaginary part, is 2.

$3 = 3 + 0i$ — a, the real part, is 3. b, the imaginary part, is 0.

Can you see that b, the imaginary part, is not zero in the first two complex numbers? Because $b \neq 0$, these complex numbers are imaginary numbers. By contrast, the imaginary part of the complex number on the right is zero. This complex number is not an imaginary number. The number 3, or $3 + 0i$, is a real number.

2 Add and subtract complex numbers.

Adding and Subtracting Complex Numbers The form of a complex number $a + bi$ is like the binomial $a + bx$. Consequently, we can add, subtract, and multiply complex numbers using the same methods we used for binomials, remembering that $i^2 = -1$.

ADDING AND SUBTRACTING COMPLEX NUMBERS

1. $(a + bi) + (c + di) = (a + c) + (b + d)i$
   In words, this says that you add complex numbers by adding their real parts, adding their imaginary parts, and expressing the sum as a complex number.
2. $(a + bi) - (c + di) = (a - c) + (b - d)i$
   In words, this says that you subtract complex numbers by subtracting their real parts, subtracting their imaginary parts, and expressing the difference as a complex number.

EXAMPLE 2 Adding and Subtracting Complex Numbers

Perform the indicated operations, writing the result in the form $a + bi$:

a. $(5 - 11i) + (7 + 4i)$ **b.** $(-5 + 7i) - (-11 - 6i)$.

SOLUTION

a. $(5 - 11i) + (7 + 4i)$

$= 5 - 11i + 7 + 4i$ Remove the parentheses.

$= 5 + 7 - 11i + 4i$ Group real and imaginary terms.

$= (5 + 7) + (-11 + 4)i$ Add real parts and add imaginary parts.

$= 12 - 7i$ Simplify.

b. $(-5 + 7i) - (-11 - 6i)$

$= -5 + 7i + 11 + 6i$ Remove the parentheses. Change signs of real and imaginary parts in the complex number being subtracted.

$= -5 + 11 + 7i + 6i$ Group real and imaginary terms.

$= (-5 + 11) + (7 + 6)i$ Add real parts and add imaginary parts.

$= 6 + 13i$ Simplify. ■

STUDY TIP

The following examples, using the same integers as in Example 2, show how operations with complex numbers are just like operations with polynomials.

a. $(5 - 11x) + (7 + 4x)$
$= 12 - 7x$

b. $(-5 + 7x) - (-11 - 6x)$
$= -5 + 7x + 11 + 6x$
$= 6 + 13x$

CHECK POINT 2 Add or subtract as indicated:

a. $(5 - 2i) + (3 + 3i)$

b. $(2 + 6i) - (12 - 4i)$.

3 Multiply complex numbers.

Multiplying Complex Numbers Multiplication of complex numbers is performed the same way as multiplication of polynomials, using the distributive property and the FOIL method. After completing the multiplication, we replace i^2 with -1. This idea is illustrated in the next example.

EXAMPLE 3 Multiplying Complex Numbers

Find the products:

a. $4i(3 - 5i)$ **b.** $(7 - 3i)(-2 - 5i)$.

SOLUTION

a. $4i(3 - 5i)$

$= 4i \cdot 3 - 4i \cdot 5i$ Distribute $4i$ throughout the parentheses.

$= 12i - 20i^2$ Multiply.

$= 12i - 20(-1)$ Replace i^2 with -1.

$= 20 + 12i$ Simplify to $12i + 20$ and write in $a + bi$ form.

b. $(7 - 3i)(-2 - 5i)$

F O I L

$$= -14 - 35i + 6i + 15i^2. \qquad \text{Use the FOIL method.}$$
$$= -14 - 35i + 6i + 15(-1) \qquad i^2 = -1$$
$$= -14 - 15 - 35i + 6i \qquad \text{Group real and imaginary terms.}$$
$$= -29 - 29i \qquad \text{Combine real and imaginary terms.}$$

■

CHECK POINT 3 Find the products:

a. $7i(2 - 9i)$ **b.** $(5 + 4i)(6 - 7i)$.

Consider the multiplication problem

$$5i \cdot 2i = 10i^2 = 10(-1) = -10.$$

This problem can also be given in terms of square roots of negative numbers:

$$\sqrt{-25} \cdot \sqrt{-4}.$$

Because the product rule for radicals only applies to real numbers, multiplying radicands is incorrect. **When performing operations with square roots of negative numbers, begin by expressing all square roots in terms of *i*.** Then perform the indicated operation.

CORRECT:

$$\sqrt{-25} \cdot \sqrt{-4} = \sqrt{25}\sqrt{-1} \cdot \sqrt{4}\sqrt{-1}$$
$$= 5i \cdot 2i$$
$$= 10i^2 = 10(-1) = -10$$

INCORRECT!

$$\sqrt{-25} \cdot \sqrt{-4} = \sqrt{(-25)(-4)}$$
$$= \sqrt{100}$$
$$= 10$$

EXAMPLE 4 Multiplying Square Roots of Negative Numbers

Multiply: $\sqrt{-3} \cdot \sqrt{-5}$.

SOLUTION

$$\sqrt{-3} \cdot \sqrt{-5} = \sqrt{3}\sqrt{-1} \cdot \sqrt{5}\sqrt{-1}$$
$$= \sqrt{3}i \cdot \sqrt{5}i \qquad \text{Express square roots in terms of } i.$$
$$= \sqrt{15}i^2 \qquad \sqrt{3} \cdot \sqrt{5} = \sqrt{15} \text{ and } i \cdot i = i^2.$$
$$= \sqrt{15}(-1) \qquad i^2 = -1$$
$$= -\sqrt{15}$$

■

CHECK POINT 4 Multiply: $\sqrt{-5} \cdot \sqrt{-7}$.

4 Divide complex numbers.

Conjugates and Division It is possible to multiply imaginary numbers and obtain a real number. Here is an example:

F O I L

$$(4 + 7i)(4 - 7i) = 16 - 28i + 28i - 49i^2$$
$$= 16 - 49i^2 = 16 - 49(-1) = 65.$$

Replace i^2 with -1.

ENRICHMENT ESSAY

Complex Numbers on a Postage Stamp

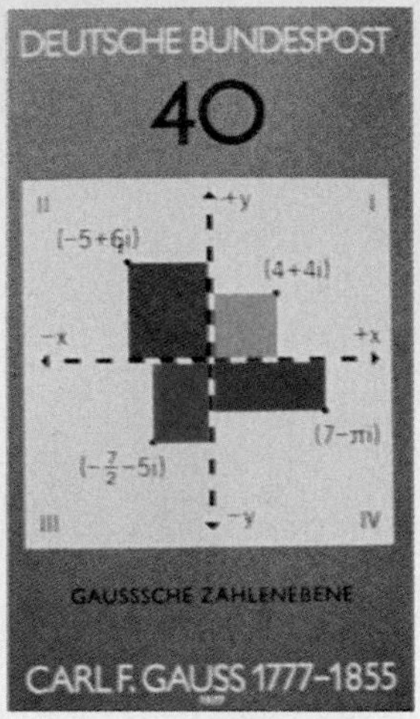

This stamp honors the work done by the German mathematician Carl Friedrich Gauss (1777–1855) with complex numbers. Gauss represented complex numbers as points in the plane.

You can also perform $(4 + 7i)(4 - 7i)$ using the formula

$$(A + B)(A - B) = A^2 - B^2.$$

A real number is obtained even faster:

$$(4 + 7i)(4 - 7i) = 4^2 - (7i)^2 = 16 - 49i^2 = 16 - 49(-1) = 65.$$

The **conjugate** of the complex number $a + bi$ is $a - bi$. The **conjugate** of the complex number $a - bi$ is $a + bi$. The multiplication problem that we just performed involved conjugates. The multiplication of conjugates always results in a real number:

$$(a + bi)(a - bi) = a^2 - (bi)^2 = a^2 - b^2i^2 = a^2 - b^2(-1) = a^2 + b^2.$$

The product eliminates i.

Conjugates are used to divide complex numbers. By multiplying the numerator and the denominator of the division by the conjugate of the denominator, you will obtain a real number in the denominator. Here are two examples of such divisions:

- $\dfrac{7 + 4i}{2 - 5i}$ — The conjugate of the denominator is $2 + 5i$.
- $\dfrac{5i - 4}{3i}$ or $\dfrac{5i - 4}{0 + 3i}$. — The conjugate of the denominator is $0 - 3i$, or $-3i$.

The procedure for dividing complex numbers, illustrated in Examples 5 and 6, should remind you of rationalizing denominators.

EXAMPLE 5 Using Conjugates to Divide Complex Numbers

Divide and simplify to the form $a + bi$:

$$\frac{7 + 4i}{2 - 5i}.$$

SOLUTION The conjugate of the denominator is $2 + 5i$. Multiplication of both the numerator and the denominator by $2 + 5i$ will eliminate i from the denominator.

$$\frac{7 + 4i}{2 - 5i} = \frac{7 + 4i}{2 - 5i} \cdot \frac{2 + 5i}{2 + 5i}$$ Multiply by 1.

F O I L

$$= \frac{14 + 35i + 8i + 20i^2}{2^2 - (5i)^2}$$ Use FOIL in the numerator and $(A - B)(A + B) = A^2 - B^2$ in the denominator.

$$= \frac{14 + 43i + 20i^2}{4 - 25i^2}$$ Simplify.

$$= \frac{14 + 43i + 20(-1)}{4 - 25(-1)}$$ $i^2 = -1$

$$= \frac{14 + 43i - 20}{4 + 25}$$ Perform the multiplications involving -1.

$$= \frac{-6 + 43i}{29}$$ Combine real terms in the numerator and denominator.

$$= -\frac{6}{29} + \frac{43}{29}i$$ Express the answer in the form $a + bi$.

■

CHECK POINT 5 Divide and simplify to the form $a + bi$:

$$\frac{6+2i}{4-3i}.$$

EXAMPLE 6 Using Conjugates to Divide Complex Numbers

Divide and simplify to the form $a + bi$:

$$\frac{5i-4}{3i}.$$

SOLUTION The conjugate of the denominator, $0 + 3i$, is $0 - 3i$. Multiplication of both the numerator and the denominator by $-3i$ will eliminate i from the denominator.

$$\frac{5i-4}{3i} = \frac{5i-4}{3i}\cdot\frac{-3i}{-3i}$$ Multiply by 1.

$$= \frac{-15i^2+12i}{-9i^2}$$ Multiply. Use the distributive property in the numerator.

$$= \frac{-15(-1)+12i}{-9(-1)}$$ $i^2 = -1$

$$= \frac{15+12i}{9}$$ Perform the multiplications involving -1.

$$= \frac{15}{9} + \frac{12}{9}i$$ Express the division in the form $a + bi$.

$$= \frac{5}{3} + \frac{4}{3}i$$ Simplify real and imaginary parts. ■

CHECK POINT 6 Divide and simplify to the form $a + bi$:

$$\frac{3-2i}{4i}.$$

5 Simplify powers of i.

Powers of i Cycles govern many aspects of life—heartbeats, sleep patterns, seasons, and tides all follow regular, predictable cycles. Surprisingly, so do powers of i. To see how this occurs, use the fact that $i^2 = -1$ and express each power of i in terms of i^2:

$$i$$
$$i^2 = -1$$
$$i^3 = i^2 \cdot i = (-1)i = -i$$
$$i^4 = (i^2)^2 = (-1)^2 = 1$$
$$i^5 = i^4 \cdot i = (i^2)^2 \cdot i = (-1)^2 \cdot i = i$$
$$i^6 = (i^2)^3 = (-1)^3 = -1.$$

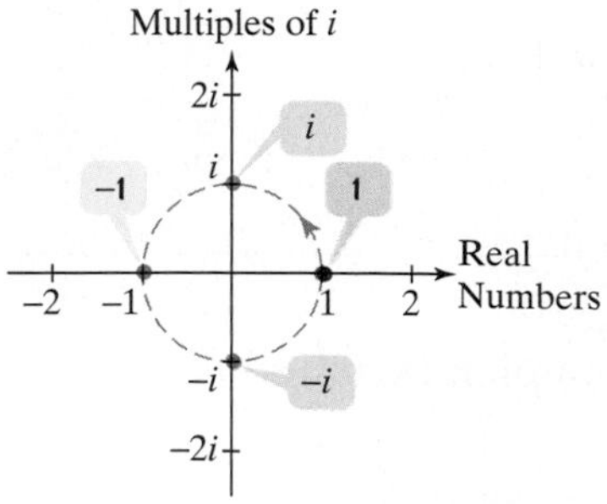

FIGURE 10.9

Can you see that the powers of i are cycling through the values i, -1, $-i$, and 1? The cycle is illustrated in Figure 10.9. In the figure, tick marks on the horizontal axis represent real numbers and tick marks along the vertical axis represent multiples of i. Using this representation, powers of i are equally spaced at 90° intervals on a circle having radius 1.

Here is a procedure for simplifying powers of i:

SIMPLIFYING POWERS OF i

1. Express the given power of i in terms of i^2.
2. Replace i^2 with -1 and simplify. Use the fact that -1 to an even power is 1 and -1 to an odd power is -1.

EXAMPLE 7 Simplifying Powers of i

Simplify:

a. i^{12} **b.** i^{39} **c.** i^{50}.

SOLUTION

a. $i^{12} = (i^2)^6 = (-1)^6 = 1$

b. $i^{39} = i^{38}i = (i^2)^{19}i = (-1)^{19}i = (-1)i = -i$

c. $i^{50} = (i^2)^{25} = (-1)^{25} = -1$ ■

CHECK POINT 7 Simplify:

a. i^{16} **b.** i^{25} **c.** i^{35}.

ENRICHMENT ESSAY

The Patterns of Chaos

R. F. Voss "29-Fold M-set Seahorse" computer-generated image

One of the new frontiers of mathematics suggests that there is an underlying order in things that appear to be random, such as the hiss and crackle of background noises as you tune a radio. Irregularities in the heartbeat, some of them severe enough to cause a heart attack, or irregularities in our sleeping patterns, such as insomnia, are examples of chaotic behavior. Chaos in the mathematical sense does not mean a complete lack of form or arrangement. In mathematics, chaos is used to describe something that appears to be random but is not actually random. The patterns of chaos appear in images like the one on the left, called the Mandelbrot set. Magnified portions of this image yield repetitions of the original structure, as well as new and unexpected patterns. The Mandelbrot set transforms the hidden structure of chaotic events into a source of wonder and inspiration.

The Mandelbrot set is made possible by opening up graphing to include complex numbers in the form $a + bi$. Although the details are beyond the scope of this text, the coordinate system that is used is shown in Figure 10.9. Plot certain complex numbers in this system, add color to the magnified boundary of the graph, and the patterns of chaos begin to appear.

10.7 EXERCISE SET

Student Solutions Manual · CD/Video · PH Math/Tutor Center · MathXL Tutorials on CD · MathXL® · MyMathLab · Interactmath.com

Practice Exercises

In Exercises 1–16, express each number in terms of i and simplify, if possible.

1. $\sqrt{-100}$
2. $\sqrt{-49}$
3. $\sqrt{-23}$
4. $\sqrt{-21}$
5. $\sqrt{-18}$
6. $\sqrt{-125}$
7. $\sqrt{-63}$
8. $\sqrt{-28}$
9. $-\sqrt{-108}$
10. $-\sqrt{-300}$
11. $5 + \sqrt{-36}$
12. $7 + \sqrt{-4}$
13. $15 + \sqrt{-3}$
14. $20 + \sqrt{-5}$
15. $-2 - \sqrt{-18}$
16. $-3 - \sqrt{-27}$

In Exercises 17–32, add or subtract as indicated. Write the result in the form a + bi.

17. $(3 + 2i) + (5 + i)$
18. $(6 + 5i) + (4 + 3i)$
19. $(7 + 2i) + (1 - 4i)$
20. $(-2 + 6i) + (4 - i)$
21. $(10 + 7i) - (5 + 4i)$
22. $(11 + 8i) - (2 + 5i)$
23. $(9 - 4i) - (10 + 3i)$
24. $(8 - 5i) - (6 + 2i)$
25. $(3 + 2i) - (5 - 7i)$
26. $(-7 + 5i) - (9 - 11i)$
27. $(-5 + 4i) - (-13 - 11i)$
28. $(-9 + 2i) - (-17 - 6i)$
29. $8i - (14 - 9i)$
30. $15i - (12 - 11i)$
31. $\left(2 + \sqrt{3}i\right) + \left(7 + 4\sqrt{3}i\right)$
32. $\left(4 + \sqrt{5}i\right) + \left(8 + 6\sqrt{5}i\right)$

In Exercises 33–62, find each product. Write imaginary results in the form a + bi.

33. $2i(5 + 3i)$
34. $5i(4 + 7i)$
35. $3i(7i - 5)$
36. $8i(4i - 3)$
37. $-7i(2 - 5i)$
38. $-6i(3 - 5i)$
39. $(3 + i)(4 + 5i)$
40. $(4 + i)(5 + 6i)$
41. $(7 - 5i)(2 - 3i)$
42. $(8 - 4i)(3 - 2i)$
43. $(6 - 3i)(-2 + 5i)$
44. $(7 - 2i)(-3 + 6i)$
45. $(3 + 5i)(3 - 5i)$
46. $(2 + 7i)(2 - 7i)$
47. $(-5 + 3i)(-5 - 3i)$
48. $(-4 + 2i)(-4 - 2i)$
49. $\left(3 - \sqrt{2}i\right)\left(3 + \sqrt{2}i\right)$
50. $\left(5 - \sqrt{3}i\right)\left(5 + \sqrt{3}i\right)$
51. $(2 + 3i)^2$
52. $(3 + 2i)^2$
53. $(5 - 2i)^2$
54. $(5 - 3i)^2$
55. $\sqrt{-7} \cdot \sqrt{-2}$
56. $\sqrt{-7} \cdot \sqrt{-3}$
57. $\sqrt{-9} \cdot \sqrt{-4}$
58. $\sqrt{-16} \cdot \sqrt{-4}$
59. $\sqrt{-7} \cdot \sqrt{-25}$
60. $\sqrt{-3} \cdot \sqrt{-36}$
61. $\sqrt{-8} \cdot \sqrt{-3}$
62. $\sqrt{-9} \cdot \sqrt{-5}$

In Exercises 63–84, divide and simplify to the form a + bi.

63. $\dfrac{2}{3 + i}$
64. $\dfrac{3}{4 + i}$
65. $\dfrac{2i}{1 + i}$
66. $\dfrac{5i}{2 + i}$
67. $\dfrac{7}{4 - 3i}$
68. $\dfrac{9}{1 - 2i}$
69. $\dfrac{6i}{3 - 2i}$
70. $\dfrac{5i}{2 - 3i}$
71. $\dfrac{1 + i}{1 - i}$
72. $\dfrac{1 - i}{1 + i}$
73. $\dfrac{2 - 3i}{3 + i}$
74. $\dfrac{2 + 3i}{3 - i}$
75. $\dfrac{5 - 2i}{3 + 2i}$
76. $\dfrac{6 - 3i}{4 + 2i}$
77. $\dfrac{4 + 5i}{3 - 7i}$
78. $\dfrac{5 - i}{3 - 2i}$
79. $\dfrac{7}{3i}$
80. $\dfrac{5}{7i}$
81. $\dfrac{8 - 5i}{2i}$
82. $\dfrac{3 + 4i}{5i}$
83. $\dfrac{4 + 7i}{-3i}$
84. $\dfrac{5 + i}{-4i}$

In Exercises 85–100, simplify each expression.

85. i^{10}
86. i^{14}
87. i^{11}
88. i^{15}
89. i^{22}
90. i^{46}
91. i^{200}
92. i^{400}
93. i^{17}
94. i^{21}
95. $(-i)^4$
96. $(-i)^6$
97. $(-i)^9$
98. $(-i)^{13}$
99. $i^{24} + i^2$
100. $i^{28} + i^{30}$

Practice Plus

In Exercises 101–108, perform the indicated operation(s) and write the result in the form a + bi.

101. $(2 - 3i)(1 - i) - (3 - i)(3 + i)$
102. $(8 + 9i)(2 - i) - (1 - i)(1 + i)$
103. $(2 + i)^2 - (3 - i)^2$
104. $(4 - i)^2 - (1 + 2i)^2$
105. $5\sqrt{-16} + 3\sqrt{-81}$
106. $5\sqrt{-8} + 3\sqrt{-18}$
107. $\dfrac{i^4 + i^{12}}{i^8 - i^7}$
108. $\dfrac{i^8 + i^{40}}{i^4 + i^3}$
109. Let $f(x) = x^2 - 2x + 2$. Find $f(1 + i)$.
110. Let $f(x) = x^2 - 2x + 5$. Find $f(1 - 2i)$.

In Exercises 111–114, simplify each evaluation to the form $a + bi$.

111. Let $f(x) = x - 3i$ and $g(x) = 4x + 2i$. Find $(fg)(-1)$.

112. Let $f(x) = 12x - i$ and $g(x) = 6x + 3i$. Find $(fg)\left(-\frac{1}{3}\right)$.

113. Let $f(x) = \dfrac{x^2 + 19}{2 - x}$. Find $f(3i)$.

114. Let $f(x) = \dfrac{x^2 + 11}{3 - x}$. Find $f(4i)$.

Application Exercises

Complex numbers are used in electronics to describe the current in an electric circuit. Ohm's law relates the current in a circuit, I, in amperes, the voltage of the circuit, E, in volts, and the resistance of the circuit, R, in ohms, by the formula $E = IR$. Use this formula to solve Exercises 115–116.

115. Find E, the voltage of a circuit, if $I = (4 - 5i)$ amperes and $R = (3 + 7i)$ ohms.

116. Find E, the voltage of a circuit, if $I = (2 - 3i)$ amperes and $R = (3 + 5i)$ ohms.

117. The mathematician Girolamo Cardano is credited with the first use (in 1545) of negative square roots in solving the now-famous problem, "Find two numbers whose sum is 10 and whose product is 40." Show that the complex numbers $5 + \sqrt{15}i$ and $5 - \sqrt{15}i$ satisfy the conditions of the problem. (Cardano did not use the symbolism $\sqrt{15}i$ or even $\sqrt{-15}$. He wrote R.m 15 for $\sqrt{-15}$, meaning "radix minus 15." He regarded the numbers 5 + R.m 15 and 5 – R.m 15 as "fictitious" or "ghost numbers," and considered the problem "manifestly impossible." But in a mathematically adventurous spirit, he exclaimed, "Nevertheless, we will operate.")

Writing in Mathematics

118. What is the imaginary unit i?

119. Explain how to write $\sqrt{-64}$ as a multiple of i.

120. What is a complex number? Explain when a complex number is a real number and when it is an imaginary number. Provide examples with your explanation.

121. Explain how to add complex numbers. Give an example.

122. Explain how to subtract complex numbers. Give an example.

123. Explain how to find the product of $2i$ and $5 + 3i$.

124. Explain how to find the product of $2i + 3$ and $5 + 3i$.

125. Explain how to find the product of $2i + 3$ and $2i - 3$.

126. Explain how to find the product of $\sqrt{-1}$ and $\sqrt{-4}$. Describe a common error in the multiplication that needs to be avoided.

127. What is the conjugate of $2 + 3i$? What happens when you multiply this complex number by its conjugate?

128. Explain how to divide complex numbers. Provide an example with your explanation.

129. Explain each of the three jokes in the cartoon on page 700.

130. A stand-up comedian uses algebra in some jokes, including one about a telephone recording that announces "You have just reached an imaginary number. Please multiply by i and dial again." Explain the joke.

Explain the errors in Exercises 131–132.

131. $\sqrt{-9} + \sqrt{-16} = \sqrt{-25} = \sqrt{25}i = 5i$

132. $\left(\sqrt{-9}\right)^2 = \sqrt{-9} \cdot \sqrt{-9} = \sqrt{81} = 9$

Critical Thinking Exercises

133. Which one of the following is true?

a. Some irrational numbers are not complex numbers.

b. $(3 + 7i)(3 - 7i)$ is an imaginary number.

c. $\dfrac{7 + 3i}{5 + 3i} = \dfrac{7}{5}$

d. In the complex number system, $x^2 + y^2$ (the sum of two squares) can be factored as $(x + yi)(x - yi)$.

In Exercises 134–136, perform the indicated operations and write the result in the form $a + bi$.

134. $\dfrac{4}{(2 + i)(3 - i)}$

135. $\dfrac{1 + i}{1 + 2i} + \dfrac{1 - i}{1 - 2i}$

136. $\dfrac{8}{1 + \frac{2}{i}}$

Review Exercises

137. Simplify:

$$\frac{\frac{x}{y^2} + \frac{1}{y}}{\frac{y}{x^2} + \frac{1}{x}}.$$

(Section 7.5, Example 3 or Example 6)

138. Solve for x: $\dfrac{1}{x} + \dfrac{1}{y} = \dfrac{1}{z}$.

(Section 7.6, Example 7)

139. If $f(x) = 2x^2 - x$ and $g(x) = x - 6$, find $(g - f)(3)$.

(Section 8.2, Example 4)

GROUP PROJECT

CHAPTER 10

Group members should consult an almanac, newspaper, magazine, or the Internet and return to the group with as much data as possible that show phenomena that are continuing to grow over time, but whose growth is leveling off. Select the five data sets that you find most intriguing. Let x represent the number of years after the first year in each data set. Model the data by hand using

$$f(x) = a\sqrt{x} + b.$$

Use the first and last data points to find values for a and b. The first data point corresponds to $x = 0$. Its second coordinate gives the value of b. To find a, substitute the second data point into $f(x) = a\sqrt{x} + b$, with the value that you obtained for b. Now solve the equation and find a. Substitute a and b into $f(x) = a\sqrt{x} + b$ to obtain a square root function that models each data set. Then use the function to make predictions about what might occur in the future. Are there circumstances that might affect the accuracy of the predictions? List some of these circumstances.

CHAPTER 10 SUMMARY

Section 10.1 Radical Expressions and Functions

Definitions and Concepts	Examples
If $b^2 = a$, then b is a square root of a. The principal square root of a, designated $\sqrt{a}$, is the nonnegative number satisfying $b^2 = a$. The negative square root of a is written $-\sqrt{a}$. A square root of a negative number is not a real number. A radical function in x is a function defined by an expression containing a root of x. The domain of a square root function is the set of real numbers for which the radicand is nonnegative.	Let $f(x) = \sqrt{6 - 2x}$. $f(-15) = \sqrt{6 - 2(-15)} = \sqrt{6 + 30} = \sqrt{36} = 6$ $f(5) = \sqrt{6 - 2 \cdot 5} = \sqrt{6 - 10} = \sqrt{-4}$, not a real number Domain of f: Set the radicand greater than or equal to zero. $6 - 2x \geq 0$ $-2x \geq -6$ $x \leq 3$ Domain of $f = \{x \mid x \leq 3\}$ or $(-\infty, 3]$.
The cube root of a real number a is written $\sqrt[3]{a}$. $\sqrt[3]{a} = b$ means that $b^3 = a$. The nth root of a real number a is written $\sqrt[n]{a}$. The number n is the index. Every real number has one root when n is odd. The odd nth root of a, $\sqrt[n]{a}$, is the number b for which $b^n = a$. Every positive real number has two real roots when n is even. An even root of a negative number is not a real number. If n is even, then $\sqrt[n]{a^n} = \|a\|$. If n is odd, then $\sqrt[n]{a^n} = a$.	• $\sqrt[3]{-8} = -2$ because $(-2)^3 = -8$. • $\sqrt[4]{-16}$ is not a real number. • $\sqrt{x^2 - 14x + 49} = \sqrt{(x - 7)^2} = \|x - 7\|$ • $\sqrt[3]{125(x + 6)^3} = 5(x + 6)$

Definitions and Concepts	Examples

Section 10.2 Rational Exponents

Definitions and Concepts	Examples
• $a^{\frac{1}{n}} = \sqrt[n]{a}$ • $a^{\frac{m}{n}} = (\sqrt[n]{a})^m$ or $\sqrt[n]{a^m}$ • $a^{-\frac{m}{n}} = \frac{1}{a^{\frac{m}{n}}}$	• $121^{\frac{1}{2}} = \sqrt{121} = 11$ • $64^{\frac{1}{3}} = \sqrt[3]{64} = 4$ • $27^{\frac{5}{3}} = \left(\sqrt[3]{27}\right)^5 = 3^5 = 3 \cdot 3 \cdot 3 \cdot 3 \cdot 3 = 243$ • $16^{-\frac{3}{4}} = \frac{1}{16^{\frac{3}{4}}} = \frac{1}{\left(\sqrt[4]{16}\right)^3} = \frac{1}{2^3} = \frac{1}{8}$ • $\left(\sqrt[3]{7xy}\right)^4 = (7xy)^{\frac{4}{3}}$
Properties of integer exponents are true for rational exponents. An expression with rational exponents is simplified when no parentheses appear, no powers are raised to powers, each base occurs once, and no negative or zero exponents appear.	Simplify: $\left(8x^{\frac{1}{3}}y^{-\frac{1}{2}}\right)^{\frac{1}{3}}$. $= 8^{\frac{1}{3}}\left(x^{\frac{1}{3}}\right)^{\frac{1}{3}}\left(y^{-\frac{1}{2}}\right)^{\frac{1}{3}}$ $= 2x^{\frac{1}{9}}y^{-\frac{1}{6}} = \frac{2x^{\frac{1}{9}}}{y^{\frac{1}{6}}}$
Some radical expressions can be simplifed using rational exponents. Rewrite the expression using rational exponents, simplify, and rewrite in radical notation if rational exponents still appear.	• $\sqrt[9]{x^3} = x^{\frac{3}{9}} = x^{\frac{1}{3}} = \sqrt[3]{x}$ • $\sqrt[5]{x^2} \cdot \sqrt[4]{x} = x^{\frac{2}{5}} \cdot x^{\frac{1}{4}} = x^{\frac{2}{5}+\frac{1}{4}}$ $= x^{\frac{8}{20}+\frac{5}{20}} = x^{\frac{13}{20}} = \sqrt[20]{x^{13}}$

Section 10.3 Multiplying and Simplifying Radical Expressions

Definitions and Concepts	Examples
The product rule for radicals can be used to multiply radicals with the same indices: $\sqrt[n]{a} \cdot \sqrt[n]{b} = \sqrt[n]{ab}$.	$\sqrt[3]{7x} \cdot \sqrt[3]{10y^2} = \sqrt[3]{7x \cdot 10y^2} = \sqrt[3]{70xy^2}$
The product rule for radicals can be used to simplify radicals: $\sqrt[n]{ab} = \sqrt[n]{a} \cdot \sqrt[n]{b}$. A radical expression with index n is simplified when its radicand has no factors that are perfect nth powers. To simplify, write the radicand as the product of two factors, one of which is the greatest perfect nth power. Then use the product rule to take the nth root of each factor. If all variables in a radicand are positive, then $\sqrt[n]{a^n} = a$. Some radicals can be simplified after the multiplication is performed.	• Simplify: $\sqrt[3]{54x^7y^{11}}$. $= \sqrt[3]{27 \cdot 2 \cdot x^6 \cdot x \cdot y^9 \cdot y^2}$ $= \sqrt[3]{(27x^6y^9)(2xy^2)}$ $= \sqrt[3]{27x^6y^9} \cdot \sqrt[3]{2xy^2} = 3x^2y^3\sqrt[3]{2xy^2}$ • Assuming positive variables, multiply and simplify: $\sqrt[4]{4x^2y} \cdot \sqrt[4]{4xy^3}$. $= \sqrt[4]{4x^2y \cdot 4xy^3} = \sqrt[4]{16x^3y^4}$ $= \sqrt[4]{16y^4} \cdot \sqrt[4]{x^3} = 2y\sqrt[4]{x^3}$

Definitions and Concepts	Examples

Section 10.4 Adding, Subtracting, and Dividing Radical Expressions

Like radicals have the same indices and radicands. Like radicals can be added or subtracted using the distributive property. In some cases, radicals can be combined once they have been simplified.

$$4\sqrt{18} - 6\sqrt{50}$$
$$= 4\sqrt{9\cdot 2} - 6\sqrt{25\cdot 2} = 4\cdot 3\sqrt{2} - 6\cdot 5\sqrt{2}$$
$$= 12\sqrt{2} - 30\sqrt{2} = -18\sqrt{2}$$

The quotient rule for radicals can be used to simplify radicals:

$$\sqrt[n]{\frac{a}{b}} = \frac{\sqrt[n]{a}}{\sqrt[n]{b}}.$$

$$\sqrt[3]{-\frac{8}{x^{12}}} = \frac{\sqrt[3]{-8}}{\sqrt[3]{x^{12}}} = -\frac{2}{x^4}$$

$\sqrt[3]{x^{12}} = (x^{12})^{\frac{1}{3}} = x^4$

The quotient rule for radicals can be used to divide radicals with the same indices:

$$\frac{\sqrt[n]{a}}{\sqrt[n]{b}} = \sqrt[n]{\frac{a}{b}}.$$

Some radicals can be simplified after the division is performed.

Assuming a positive variable, divide and simplify:

$$\frac{\sqrt[4]{64x^5}}{\sqrt[4]{2x^{-2}}} = \sqrt[4]{32x^{5-(-2)}} = \sqrt[4]{32x^7}$$
$$= \sqrt[4]{16\cdot 2\cdot x^4\cdot x^3} = \sqrt[4]{16x^4}\cdot\sqrt[4]{2x^3}$$
$$= 2x\sqrt[4]{2x^3}.$$

Section 10.5 Multiplying with More Than One Term and Rationalizing Denominators

Radical expressions with more than one term are multiplied in much the same way that polynomials with more than one term are multiplied.

- $\sqrt{5}\left(2\sqrt{6} - \sqrt{3}\right) = 2\sqrt{30} - \sqrt{15}$
- $\left(4\sqrt{3} - 2\sqrt{2}\right)\left(\sqrt{3} + \sqrt{2}\right)$

F O I L

$$= 4\sqrt{3}\cdot\sqrt{3} + 4\sqrt{3}\cdot\sqrt{2} - 2\sqrt{2}\cdot\sqrt{3} - 2\sqrt{2}\cdot\sqrt{2}$$
$$= 4\cdot 3 + 4\sqrt{6} - 2\sqrt{6} - 2\cdot 2$$
$$= 12 + 4\sqrt{6} - 2\sqrt{6} - 4 = 8 + 2\sqrt{6}$$

Radical expressions that involve the sum and difference of the same two terms are called conjugates. Use

$$(A + B)(A - B) = A^2 - B^2$$

to multiply conjugates.

$$\left(8 + 2\sqrt{5}\right)\left(8 - 2\sqrt{5}\right)$$
$$= 8^2 - \left(2\sqrt{5}\right)^2 = 64 - 4\cdot 5$$
$$= 64 - 20 = 44$$

The process of rewriting a radical expression as an equivalent expression without any radicals in the denominator is called rationalizing the denominator. When the denominator contains a single radical with an nth root, multiply the numerator and the denominator by a radical of index n that produces a perfect nth power in the denominator's radicand.

Rationalize the denominator: $\dfrac{7}{\sqrt[3]{2x}}$.

$$= \frac{7}{\sqrt[3]{2x}}\cdot\frac{\sqrt[3]{4x^2}}{\sqrt[3]{4x^2}} = \frac{7\sqrt[3]{4x^2}}{\sqrt[3]{8x^3}} = \frac{7\sqrt[3]{4x^2}}{2x}$$

Section 10.5 Multiplying with More Than One Term and Rationalizing Denominators (continued)

Definitions and Concepts	Examples
If the denominator contains two terms, rationalize the denominator by multiplying the numerator and the denominator by the conjugate of the denominator.	$\frac{13}{5-\sqrt{3}} = \frac{13}{5-\sqrt{3}} \cdot \frac{5+\sqrt{3}}{5+\sqrt{3}}$ $= \frac{13(5+\sqrt{3})}{5^2-(\sqrt{3})^2}$ $= \frac{13(5+\sqrt{3})}{25-3} = \frac{13(5+\sqrt{3})}{22}$

Section 10.6 Radical Equations

Definitions and Concepts	Examples
A radical equation is an equation in which the variable occurs in a radicand. **Solving Radical Equations Containing *n*th Roots** 1. Isolate one radical on one side of the equation. 2. Raise both sides to the *n*th power. 3. Solve the resulting equation. 4. Check proposed solutions in the original equation. Solutions of an equation to an even power that is radical-free, but not the original equation, are called extraneous solutions.	Solve: $\sqrt{6x+13} - 2x = 1$. $\sqrt{6x+13} = 2x+1$ Isolate the radical. $(\sqrt{6x+13})^2 = (2x+1)^2$ Square both sides. $6x+13 = 4x^2+4x+1$ $0 = 4x^2-2x-12$ Subtract $6x+13$ from both sides. $0 = 2(2x^2-x-6)$ Factor out the GCF. $0 = 2(2x+3)(x-2)$ Factor completely. $2x+3=0$ or $x-2=0$ Set variable factors equal to zero. $2x=-3$ $\quad x=2$ Solve for x. $x=-\frac{3}{2}$ Check both proposed solutions. 2 checks, but $-\frac{3}{2}$ is extraneous. The solution is 2 and the solution set is $\{2\}$.

Section 10.7 Complex Numbers

Definitions and Concepts	Examples
The imaginary unit *i* is defined as $i = \sqrt{-1}$, where $i^2 = -1$. The set of numbers in the form $a + bi$ is called the set of complex numbers; a is the real part and b is the imaginary part. If $b = 0$, the complex number is a real number. If $b \neq 0$, the complex number is an imaginary number.	• $\sqrt{-81} = \sqrt{81(-1)} = \sqrt{81}\sqrt{-1} = 9i$ • $\sqrt{-75} = \sqrt{75(-1)} = \sqrt{25\cdot 3}\sqrt{-1} = 5\sqrt{3}i$
To add or subtract complex numbers, add or subtract their real parts and add or subtract their imaginary parts.	$(2-4i)-(7-10i)$ $= 2-4i-7+10i$ $= (2-7)+(-4+10)i = -5+6i$
To multiply complex numbers, multiply as if they are polynomials. After completing the multiplication, replace i^2 with -1. When performing operations with square roots of negative numbers, begin by expressing all square roots in terms of *i*. Then multiply.	F O I L • $(2-3i)(4+5i) = 8+10i-12i-15i^2$ $= 8+10i-12i-15(-1)$ $= 23-2i$ • $\sqrt{-36}\cdot\sqrt{-100} = \sqrt{36(-1)}\cdot\sqrt{100(-1)}$ $= 6i\cdot 10i = 60i^2 = 60(-1) = -60$

Definitions and Concepts	Examples
Section 10.7 Complex Numbers (continued)	
The complex numbers $a + bi$ and $a - bi$ are conjugates. Conjugates can be multiplied using the formula $$(A + B)(A - B) = A^2 - B^2.$$ The multiplication of conjugates results in a real number.	$(3 + 5i)(3 - 5i)$ $= 3^2 - (5i)^2$ $= 9 - 25i^2$ $= 9 - 25(-1) = 34$
To divide complex numbers, multiply the numerator and the denominator by the conjugate of the denominator.	$\frac{5 + 2i}{4 - i} = \frac{5 + 2i}{4 - i} \cdot \frac{4 + i}{4 + i} = \frac{20 + 5i + 8i + 2i^2}{16 - i^2}$ $= \frac{20 + 13i + 2(-1)}{16 - (-1)}$ $= \frac{20 + 13i - 2}{16 + 1}$ $= \frac{18 + 13i}{17} = \frac{18}{17} + \frac{13}{17}i$
To simplify powers of i, rewrite the expression in terms of i^2. Then replace i^2 with -1 and simplify.	Simplify: i^{27}. $i^{27} = i^{26} \cdot i = (i^2)^{13}i$ $= (-1)^{13}i = (-1)i = -i$

CHAPTER 10 REVIEW EXERCISES

10.1 *In Exercises 1–5, find the indicated root, or state that the expression is not a real number.*

1. $\sqrt{81}$

2. $-\sqrt{\frac{1}{100}}$

3. $\sqrt[3]{-27}$

4. $\sqrt[4]{-16}$

5. $\sqrt[5]{-32}$

In Exercises 6–7, find the indicated function values for each function. If necessary, round to two decimal places. If the function value is not a real number and does not exist, so state.

6. $f(x) = \sqrt{2x - 5};\quad f(15), f(4), f\left(\frac{5}{2}\right), f(1)$

7. $g(x) = \sqrt[3]{4x - 8};\quad g(4), g(0), g(-14)$

In Exercises 8–9, find the domain of each square root function.

8. $f(x) = \sqrt{x - 2}$

9. $g(x) = \sqrt{100 - 4x}$

In Exercises 10–15, simplify each expression. Assume that each variable can represent any real number, so include absolute value bars where necessary.

10. $\sqrt{25x^2}$

11. $\sqrt{(x + 14)^2}$

12. $\sqrt{x^2 - 8x + 16}$

13. $\sqrt[3]{64x^3}$

14. $\sqrt[4]{16x^4}$

15. $\sqrt[5]{-32(x + 7)^5}$

10.2 *In Exercises 16–18, use radical notation to rewrite each expression. Simplify, if possible.*

16. $(5xy)^{\frac{1}{3}}$

17. $16^{\frac{3}{2}}$

18. $32^{\frac{4}{5}}$

In Exercises 19–20, rewrite each expression with rational exponents.

19. $\sqrt{7x}$

20. $\left(\sqrt[3]{19xy}\right)^5$

In Exercises 21–22, rewrite each expression with a positive rational exponent. Simplify, if possible.

21. $8^{-\frac{2}{3}}$

22. $3x(ab)^{-\frac{4}{5}}$

In Exercises 23–26, use properties of rational exponents to simplify each expression.

23. $x^{\frac{1}{3}} \cdot x^{\frac{1}{4}}$

24. $\frac{5^{\frac{1}{2}}}{5^{\frac{1}{3}}}$

25. $(8x^6y^3)^{\frac{1}{3}}$

26. $\left(x^{-\frac{2}{3}}y^{\frac{1}{4}}\right)^{\frac{1}{2}}$

In Exercises 27–31, use rational exponents to simplify each expression. If rational exponents appear after simplifying, write the answer in radical notation.

27. $\sqrt[3]{x^9y^{12}}$

28. $\sqrt[9]{x^3y^9}$

29. $\sqrt{x}\cdot\sqrt[3]{x}$

30. $\dfrac{\sqrt[3]{x^2}}{\sqrt[4]{x^2}}$

31. $\sqrt[5]{\sqrt[3]{x}}$

32. The function $f(x) = 350x^{\frac{2}{3}}$ models the expenditures, $f(x)$, in millions of dollars, for the U.S. National Park Service x years after 1985. According to this model, what will expenditures be in 2012?

10.3 *In Exercises 33–35, use the product rule to multiply.*

33. $\sqrt{3x}\cdot\sqrt{7y}$

34. $\sqrt[5]{7x^2}\cdot\sqrt[5]{11x}$

35. $\sqrt[6]{x-5}\cdot\sqrt[6]{(x-5)^4}$

36. If $f(x) = \sqrt{7x^2 - 14x + 7}$, express the function, f, in simplified form. Assume that x can be any real number.

In Exercises 37–39, simplify by factoring. Assume that all variables in a radicand represent positive real numbers.

37. $\sqrt{20x^3}$

38. $\sqrt[3]{54x^8y^6}$

39. $\sqrt[4]{32x^3y^{11}z^5}$

In Exercises 40–43, multiply and simplify, if possible. Assume that all variables in a radicand represent positive real numbers.

40. $\sqrt{6x^3}\cdot\sqrt{4x^2}$

41. $\sqrt[3]{4x^2y}\cdot\sqrt[3]{4xy^4}$

42. $\sqrt[5]{2x^4y^3z^4}\cdot\sqrt[5]{8xy^6z^7}$

43. $\sqrt{x+1}\cdot\sqrt{x-1}$

10.4 *Assume that all variables represent positive real numbers.*

In Exercises 44–47, add or subtract as indicated.

44. $6\sqrt[3]{3} + 2\sqrt[3]{3}$

45. $5\sqrt{18} - 3\sqrt{8}$

46. $\sqrt[3]{27x^4} + \sqrt[3]{xy^6}$

47. $2\sqrt[3]{6} - 5\sqrt[3]{48}$

In Exercises 48–50, simplify using the quotient rule.

48. $\sqrt[3]{\dfrac{16}{125}}$

49. $\sqrt{\dfrac{x^3}{100y^4}}$

50. $\sqrt[4]{\dfrac{3y^5}{16x^{20}}}$

In Exercises 51–54, divide and, if possible, simplify.

51. $\dfrac{\sqrt{48}}{\sqrt{2}}$

52. $\dfrac{\sqrt[3]{32}}{\sqrt[3]{2}}$

53. $\dfrac{\sqrt[4]{64x^7}}{\sqrt[4]{2x^2}}$

54. $\dfrac{\sqrt{200x^3y^2}}{\sqrt{2x^{-2}y}}$

10.5 *Assume that all variables represent positive real numbers.*

In Exercises 55–62, multiply as indicated. If possible, simplify any radical expressions that appear in the product.

55. $\sqrt{3}(2\sqrt{6} + 4\sqrt{15})$

56. $\sqrt[3]{5}(\sqrt[3]{50} - \sqrt[3]{2})$

57. $(\sqrt{7} - 3\sqrt{5})(\sqrt{7} + 6\sqrt{5})$

58. $(\sqrt{x} - \sqrt{11})(\sqrt{y} - \sqrt{11})$

59. $(\sqrt{5} + \sqrt{8})^2$

60. $(2\sqrt{3} - \sqrt{10})^2$

61. $(\sqrt{7} + \sqrt{13})(\sqrt{7} - \sqrt{13})$

62. $(7 - 3\sqrt{5})(7 + 3\sqrt{5})$

In Exercises 63–75, rationalize each denominator. Simplify, if possible.

63. $\dfrac{4}{\sqrt{6}}$

64. $\sqrt{\dfrac{2}{7}}$

65. $\dfrac{12}{\sqrt[3]{9}}$

66. $\sqrt{\dfrac{2x}{5y}}$

67. $\dfrac{14}{\sqrt[3]{2x^2}}$

68. $\sqrt[4]{\dfrac{7}{3x}}$

69. $\dfrac{5}{\sqrt[5]{32x^4y}}$

70. $\dfrac{6}{\sqrt{3} - 1}$

71. $\dfrac{\sqrt{7}}{\sqrt{5} + \sqrt{3}}$

72. $\dfrac{10}{2\sqrt{5} - 3\sqrt{2}}$

73. $\dfrac{\sqrt{x} + 5}{\sqrt{x} - 3}$

74. $\dfrac{\sqrt{7} + \sqrt{3}}{\sqrt{7} - \sqrt{3}}$

75. $\dfrac{2\sqrt{3} + \sqrt{6}}{2\sqrt{6} + \sqrt{3}}$

In Exercises 76–79, rationalize each numerator. Simplify, if possible.

76. $\sqrt{\dfrac{2}{7}}$

77. $\dfrac{\sqrt[3]{3x}}{\sqrt[3]{y}}$

78. $\dfrac{\sqrt{7}}{\sqrt{5} + \sqrt{3}}$

79. $\dfrac{\sqrt{7} + \sqrt{3}}{\sqrt{7} - \sqrt{3}}$

10.6 *In Exercises 80–84, solve each radical equation.*

80. $\sqrt{2x+4}=6$

81. $\sqrt{x-5}+9=4$

82. $\sqrt{2x-3}+x=3$

83. $\sqrt{x-4}+\sqrt{x+1}=5$

84. $(x^2+6x)^{\frac{1}{3}}+2=0$

85. The time, $f(x)$, in seconds, for a free-falling object to fall x feet is modeled by the function

$$f(x)=\sqrt{\frac{x}{16}}.$$

If a worker accidently drops a hammer from a building and it hits the ground after 4 seconds, from what height was the hammer dropped?

86. Out of a group of 50,000 births, the number of people, $f(x)$, surviving to age x is modeled by the function

$$f(x)=5000\sqrt{100-x}.$$

To what age will 20,000 people in the group survive?

10.7 *In Exercises 87–89, express each number in terms of i and simplify, if possible.*

87. $\sqrt{-81}$ **88.** $\sqrt{-63}$

89. $-\sqrt{-8}$

In Exercises 90–99, perform the indicated operation. Write the result in the form $a + bi$.

90. $(7+12i)+(5-10i)$

91. $(8-3i)-(17-7i)$

92. $4i(3i-2)$

93. $(7-5i)(2+3i)$

94. $(3-4i)^2$

95. $(7+8i)(7-8i)$

96. $\sqrt{-8}\cdot\sqrt{-3}$

97. $\frac{6}{5+i}$

98. $\frac{3+4i}{4-2i}$ **99.** $\frac{5+i}{3i}$

In Exercises 100–101, simplify each expression.

100. i^{16} **101.** i^{23}

CHAPTER 10 TEST

Remember to use your Chapter Test Prep Video CD to see the worked-out solutions to the test questions you want to review.

1. Let $f(x)=\sqrt{8-2x}$.

a. Find $f(-14)$.

b. Find the domain of f.

2. Evaluate: $27^{-\frac{4}{3}}$.

3. Simplify: $(25x^{-\frac{1}{2}}y^{\frac{1}{4}})^{\frac{1}{2}}$.

In Exercises 4–5, use rational exponents to simplify each expression. If rational exponents appear after simplifying, write the answer in radical notation.

4. $\sqrt[8]{x^4}$

5. $\sqrt[4]{x}\cdot\sqrt[5]{x}$

In Exercises 6–9, simplify each expression. Assume that each variable can represent any real number.

6. $\sqrt{75x^2}$

7. $\sqrt{x^2-10x+25}$

8. $\sqrt[3]{16x^4y^8}$

9. $\sqrt[5]{-\frac{32}{x^{10}}}$

In Exercises 10–17, perform the indicated operation and, if possible, simplify. Assume that all variables represent positive real numbers.

10. $\sqrt[3]{5x^2}\cdot\sqrt[3]{10y}$

11. $\sqrt[4]{8x^3y}\cdot\sqrt[4]{4xy^2}$

12. $3\sqrt{18}-4\sqrt{32}$

13. $\sqrt[3]{8x^4}+\sqrt[3]{xy^6}$

14. $\frac{\sqrt[3]{16x^8}}{\sqrt[3]{2x^4}}$

15. $\sqrt{3}(4\sqrt{6}-\sqrt{5})$

16. $(5\sqrt{6}-2\sqrt{2})(\sqrt{6}+\sqrt{2})$

17. $(7-\sqrt{3})^2$

In Exercises 18–20, rationalize each denominator. Simplify, if possible. Assume all variables represent positive real numbers.

18. $\sqrt{\frac{5}{x}}$

19. $\frac{5}{\sqrt[3]{5x^2}}$

20. $\frac{\sqrt{2}-\sqrt{3}}{\sqrt{2}+\sqrt{3}}$

In Exercises 21–23, solve each radical equation.

21. $3 + \sqrt{2x - 3} = x$

22. $\sqrt{x + 9} - \sqrt{x - 7} = 2$

23. $(11x + 6)^{\frac{1}{3}} + 3 = 0$

24. The function

$$f(x) = 2.9\sqrt{x} + 20.1$$

models the average height, $f(x)$, in inches, of boys who are x months of age, $0 \le x \le 60$. Find the age at which the average height of boys is 40.4 inches.

25. Express in terms of i and simplify: $\sqrt{-75}$.

In Exercises 26–29, perform the indicated operation. Write the result in the form $a + bi$.

26. $(5 - 3i) - (6 - 9i)$

27. $(3 - 4i)(2 + 5i)$

28. $\sqrt{-9} \cdot \sqrt{-4}$

29. $\dfrac{3 + i}{1 - 2i}$

30. Simplify: i^{35}.

CUMULATIVE REVIEW EXERCISES (CHAPTERS 1–10)

In Exercises 1–5, solve each equation, inequality, or system.

1.
$$\begin{aligned} 2x - y + z &= -5 \\ x - 2y - 3z &= 6 \\ x + y - 2z &= 1 \end{aligned}$$

2. $3x^2 - 11x = 4$

3. $2(x + 4) < 5x + 3(x + 2)$

4. $\dfrac{1}{x + 2} + \dfrac{15}{x^2 - 4} = \dfrac{5}{x - 2}$

5. $\sqrt{x + 2} - \sqrt{x + 1} = 1$

6. Graph the solution set of the system:

$$\begin{aligned} x + 2y &< 2 \\ 2y - x &> 4. \end{aligned}$$

In Exercises 7–15, perform the indicated operations.

7. $\dfrac{8x^2}{3x^2 - 12} \div \dfrac{40}{x - 2}$

8. $\dfrac{x + \dfrac{1}{y}}{y + \dfrac{1}{x}}$

9. $(2x - 3)(4x^2 - 5x - 2)$

10. $\dfrac{7x}{x^2 - 2x - 15} - \dfrac{2}{x - 5}$

11. $7(8 - 10)^3 - 7 + 3 \div (-3)$

12. $\sqrt{80x} - 5\sqrt{20x} + 2\sqrt{45x}$

13. $\dfrac{\sqrt{3} - 2}{2\sqrt{3} + 5}$

14. $(2x^3 - 3x^2 + 3x - 4) \div (x - 2)$

15. $(2\sqrt{3} + 5\sqrt{2})(\sqrt{3} - 4\sqrt{2})$

In Exercises 16–17, factor completely.

16. $24x^2 + 10x - 4$

17. $16x^4 - 1$

18. The amount of light provided by a light bulb varies inversely as the square of the distance from the bulb. The illumination provided is 120 lumens at a distance of 10 feet. How many lumens are provided at a distance of 15 feet?

19. You invested \$6000 in two accounts paying 7% and 9% annual interest, respectively. At the end of the year, the total interest from these investments was \$510. How much was invested at each rate?

20. Although there are 2332 students enrolled in the college, this is 12% fewer students than there were enrolled last year. How many students were enrolled last year?

Thanks to the efficiency of U.S. farms, most Americans enjoy an abundance of inexpensive foods. Nearly one in five U.S. kids ages 9 through 13 say they've already been on a weight-loss diet. By contrast, worldwide, every three seconds a child dies of hunger. If all the world's undernourished people were gathered into a new country, it would be the third most populous nation, just behind China and India.

Source: The State of Food Insecurity in the World, 2000; 2000 BBC News

The growth and decline in the number of U.S. farms is developed algebraically in the technology box at the conclusion of Section 11.3 and in Exercise 58 of Exercise Set 11.3.

CHAPTER 11

Quadratic Equations and Functions

Because we tend to take our food abundance for granted, it might surprise you to know that the number of U.S. farms has declined since the 1920s, as individually owned family farms have been swallowed up by huge agribusinesses owned by corporations. Currently, there are about 1.8 acres of cropland to grow food for each American. If the current trends in population growth and loss of farmland continue, there will be only 0.6 acre per American in the year 2050. This is the rate that currently exists worldwide.

Like the number of farms and the amount of cropland in the United States, many phenomena follow trends that involve growth and decline, or vice versa. In this chapter, you will learn to solve equations and graph functions that provide new ways of looking at and understanding these phenomena.

SECTION 11.1

THE SQUARE ROOT PROPERTY AND COMPLETING THE SQUARE; DISTANCE AND MIDPOINT FORMULAS

Objectives

1. Solve quadratic equations using the square root property.
2. Complete the square of a binomial.
3. Solve quadratic equations by completing the square.
4. Solve problems using the square root property.
5. Find the distance between two points.
6. Find the midpoint of a line segment.

"In the future there will be two kinds of corporations; those that go global, and those that go bankrupt."

C. Michael Armstrong, CEO, AT&T

For better or worse, ours is the era of the multinational corporation. New technology that the multinational corporations control is expanding their power. And their numbers are growing fast. There were approximately 300 multinationals in 1900. By 1970, there were close to 7000, and by 1990 the number had swelled to 30,000. In 2001, more than 65,000 multinational corporations enveloped the world.

In this section, you will learn two new methods for solving quadratic equations. These methods are called the *square root method* and *completing the square*. Using these techniques, you will explore the growth of multinational corporations and make predictions about the number of global corporations in the future.

STUDY TIP

Here is a summary of what we already know about quadratic equations and quadratic functions.

1. A **quadratic equation** in x can be written in the standard form
$$ax^2 + bx + c = 0, \quad a \neq 0.$$
2. Some quadratic equations can be solved by factoring.

Solve: $2x^2 + 7x - 4 = 0.$

$$(2x - 1)(x + 4) = 0$$
$$2x - 1 = 0 \quad \text{or} \quad x + 4 = 0$$
$$2x = 1 \qquad\qquad x = -4$$
$$x = \tfrac{1}{2}$$

The solutions are -4 and $\frac{1}{2}$, and the solution set is $\left\{-4, \frac{1}{2}\right\}$.

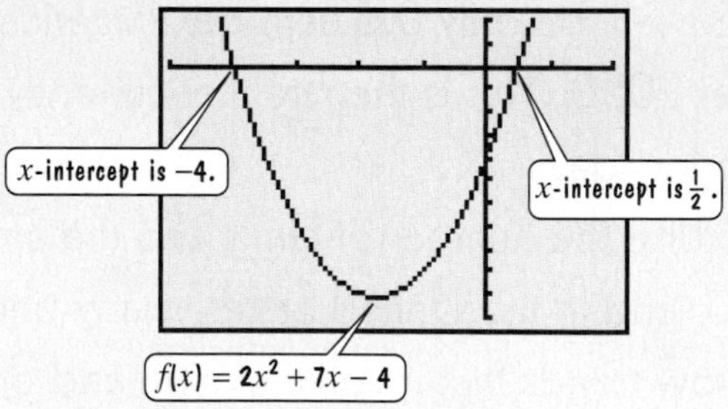

FIGURE 11.1

3. A polynomial function of the form
$$f(x) = ax^2 + bx + c, \quad a \neq 0$$
is a **quadratic function**. Graphs of quadratic functions have cuplike shapes with the same behavior at each end.
4. The real solutions of $ax^2 + bx + c = 0$ correspond to the x-intercepts for the graph of the quadratic function $f(x) = ax^2 + bx + c$. For example, the solutions of the equation $2x^2 + 7x - 4 = 0$ are -4 and $\frac{1}{2}$. Figure 11.1 shows that the solutions appear as x-intercepts on the graph of the quadratic function $f(x) = 2x^2 + 7x - 4$.

Now that we've summarized what we know, let's look at where we go. How do we solve a quadratic equation, $ax^2 + bx + c = 0$, if the trinomial $ax^2 + bx + c$ cannot be factored? Methods other than factoring are needed. In this section, we look at other ways of solving quadratic equations.

1 Solve quadratic equations using the square root property.

The Square Root Property Let's begin with a relatively simple quadratic equation:

$$x^2 = 9.$$

The value of x must be a number whose square is 9. There are two numbers whose square is 9:

$$x = \sqrt{9} = 3 \quad \text{or} \quad x = -\sqrt{9} = -3.$$

Thus, the solutions of $x^2 = 9$ are 3 and -3. This is an example of the **square root property**.

DISCOVER FOR YOURSELF

Solve $x^2 = 9$, or

$$x^2 - 9 = 0,$$

by factoring. What is the advantage of using the square root property?

THE SQUARE ROOT PROPERTY If u is an algebraic expression and d is a nonzero real number, then $u^2 = d$ has exactly two solutions:

$$\text{If } u^2 = d, \quad \text{then } u = \sqrt{d} \text{ or } u = -\sqrt{d}.$$

Equivalently,

$$\text{If } u^2 = d, \quad \text{then } u = \pm\sqrt{d}.$$

Notice that $u = \pm\sqrt{d}$ is a shorthand notation to indicate that $u = \sqrt{d}$ or $u = -\sqrt{d}$. Although we usually read $u = \pm\sqrt{d}$ as "u equals plus or minus the square root of d," we actually mean that u is the positive square root of d or the negative square root of d.

EXAMPLE 1 Solving a Quadratic Equation by the Square Root Property

Solve: $3x^2 = 18$.

SOLUTION To apply the square root property, we need a squared expression by itself on one side of the equation.

$$3x^2 = 18$$

We want x^2 by itself.

We can get x^2 by itself if we divide both sides by 3.

$$3x^2 = 18 \qquad \text{This is the original equation.}$$

$$\frac{3x^2}{3} = \frac{18}{3} \qquad \text{Divide both sides by 3.}$$

$$x^2 = 6 \qquad \text{Simplify.}$$

$$x = \sqrt{6} \quad \text{or} \quad x = -\sqrt{6} \qquad \text{Apply the square root property.}$$

Now let's check these proposed solutions in the original equation.

Check $\sqrt{6}$:

$$3x^2 = 18$$
$$3(\sqrt{6})^2 \stackrel{?}{=} 18$$
$$3 \cdot 6 \stackrel{?}{=} 18$$
$$18 = 18, \quad \text{true}$$

Check $-\sqrt{6}$:

$$3x^2 = 18$$
$$3(-\sqrt{6})^2 \stackrel{?}{=} 18$$
$$3 \cdot 6 \stackrel{?}{=} 18$$
$$18 = 18, \quad \text{true}$$

The solutions are $-\sqrt{6}$ and $\sqrt{6}$. The solution set is $\{-\sqrt{6}, \sqrt{6}\}$ or $\{\pm\sqrt{6}\}$. ■

✔ CHECK POINT 1 Solve: $4x^2 = 28$.

In this section, we will express irrational solutions in simplified radical form, rationalizing denominators when possible.

EXAMPLE 2 Solving a Quadratic Equation by the Square Root Property

Solve: $2x^2 - 7 = 0$.

SOLUTION To solve by the square root property, we isolate the squared expression on one side of the equation.

$$2x^2 - 7 = 0$$

We want x^2 by itself.

$2x^2 - 7 = 0$	This is the original equation.
$2x^2 = 7$	Add 7 to both sides.
$x^2 = \frac{7}{2}$	Divide both sides by 2.
$x = \sqrt{\frac{7}{2}}$ or $x = -\sqrt{\frac{7}{2}}$	Apply the square root property.

Because the proposed solutions are opposites, we can rationalize both denominators at once:

$$\pm\sqrt{\frac{7}{2}} = \pm\frac{\sqrt{7}}{\sqrt{2}}\cdot\frac{\sqrt{2}}{\sqrt{2}} = \pm\frac{\sqrt{14}}{2}.$$

Substitute these values into the original equation and verify that the solutions are $-\frac{\sqrt{14}}{2}$ and $\frac{\sqrt{14}}{2}$. The solution set is $\left\{-\frac{\sqrt{14}}{2}, \frac{\sqrt{14}}{2}\right\}$ or $\left\{\pm\frac{\sqrt{14}}{2}\right\}$. ■

✔ CHECK POINT 2 Solve: $3x^2 - 11 = 0$.

Some quadratic equations have solutions that are imaginary numbers.

EXAMPLE 3 Solving a Quadratic Equation by the Square Root Property

Solve: $9x^2 + 25 = 0$.

SOLUTION

$9x^2 + 25 = 0$ — This is the original equation.

We need to isolate x^2.

$9x^2 = -25$ — Subtract 25 from both sides.

$x^2 = -\frac{25}{9}$ — Divide both sides by 9.

$x = \sqrt{-\frac{25}{9}}$ or $x = -\sqrt{-\frac{25}{9}}$ — Apply the square root property.

$x = \sqrt{\frac{25}{9}}\sqrt{-1}$ $\quad x = -\sqrt{\frac{25}{9}}\sqrt{-1}$

$x = \frac{5}{3}i$ $\quad x = -\frac{5}{3}i$ — $\sqrt{-1} = i$

Because the equation has an x^2-term and no x-term, we can check both proposed solutions, $\pm\frac{5}{3}i$, at once.

Check $\frac{5}{3}i$ and $-\frac{5}{3}i$:

$$9x^2 + 25 = 0$$
$$9\left(\pm\frac{5}{3}i\right)^2 + 25 \stackrel{?}{=} 0$$
$$9\left(\frac{25}{9}i^2\right) + 25 \stackrel{?}{=} 0$$
$$25i^2 + 25 \stackrel{?}{=} 0$$

$i^2 = -1$

$$25(-1) + 25 \stackrel{?}{=} 0$$
$$0 = 0, \quad \text{true}$$

The solutions are $-\frac{5}{3}i$ and $\frac{5}{3}i$. The solution set is $\left\{-\frac{5}{3}i, \frac{5}{3}i\right\}$ or $\left\{\pm\frac{5}{3}i\right\}$. ■

USING TECHNOLOGY

The graph of

$$f(x) = 9x^2 + 25$$

has no x-intercepts. This shows that

$$9x^2 + 25 = 0$$

has no real solutions. Example 3 algebraically establishes that the solutions are imaginary numbers.

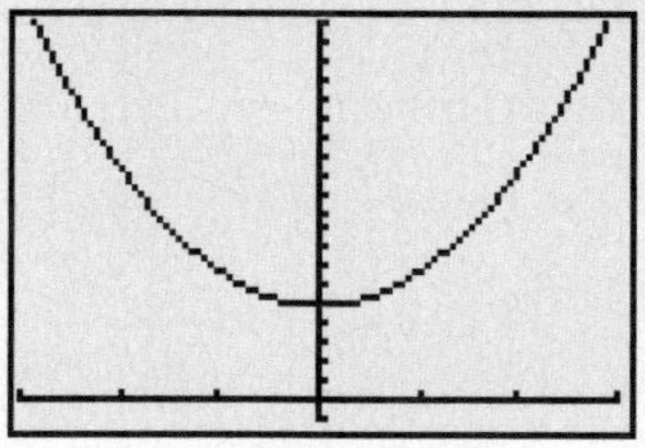

[−3, 3, 1] by [−5, 100, 5]

CHECK POINT 3 Solve: $4x^2 + 9 = 0$.

Can we solve an equation such as $(x - 1)^2 = 5$ using the square root property? Yes. The equation is in the form $u^2 = d$, where u^2, the squared expression, is by itself on the left side.

$$(x - 1)^2 = 5$$

This is u^2 in $u^2 = d$ with $u = x - 1$.

This is d in $u^2 = d$ with $d = 5$.

DISCOVER FOR YOURSELF

Try solving

$$(x - 1)^2 = 5$$

by writing the equation in standard form and factoring. What problem do you encounter?

EXAMPLE 4 Solving a Quadratic Equation by the Square Root Property

Solve by the square root property: $(x - 1)^2 = 5$.

SOLUTION

$$(x - 1)^2 = 5 \quad \text{This is the original equation.}$$

$$x - 1 = \sqrt{5} \quad \text{or} \quad x - 1 = -\sqrt{5} \quad \text{Apply the square root property.}$$

$$x = 1 + \sqrt{5} \qquad x = 1 - \sqrt{5} \quad \text{Add 1 to both sides in each equation.}$$

Check $1 + \sqrt{5}$:

$$(x - 1)^2 = 5$$
$$(1 + \sqrt{5} - 1)^2 \stackrel{?}{=} 5$$
$$(\sqrt{5})^2 \stackrel{?}{=} 5$$
$$5 = 5, \text{ true}$$

Check $1 - \sqrt{5}$:

$$(x - 1)^2 = 5$$
$$(1 - \sqrt{5} - 1)^2 \stackrel{?}{=} 5$$
$$(-\sqrt{5})^2 \stackrel{?}{=} 5$$
$$5 = 5, \text{ true}$$

The solutions are $1 \pm \sqrt{5}$, and the solution set is $\{1 + \sqrt{5}, 1 - \sqrt{5}\}$ or $\{1 \pm \sqrt{5}\}$. ■

✔ **CHECK POINT 4** Solve: $(x - 3)^2 = 10$.

2 Complete the square of a binomial.

Completing the Square We return to the question that opened this section: How do we solve a quadratic equation, $ax^2 + bx + c = 0$, if the trinomial $ax^2 + bx + c$ cannot be factored? We can convert the equation into an equivalent equation that can be solved using the square root property. This is accomplished by **completing the square**.

COMPLETING THE SQUARE If $x^2 + bx$ is a binomial, then by adding $\left(\frac{b}{2}\right)^2$, which is the square of half the coefficient of x, a perfect square trinomial will result. That is,

$$x^2 + bx + \left(\frac{b}{2}\right)^2 = \left(x + \frac{b}{2}\right)^2.$$

EXAMPLE 5 Completing the Square

What term should be added to each binomial so that it becomes a perfect square trinomial? Write and factor the trinomial.

a. $x^2 + 8x$ **b.** $x^2 - 7x$ **c.** $x^2 + \frac{3}{5}x$

SOLUTION To complete the square, we must add a term to each binomial. The term that should be added is the square of half the coefficient of x.

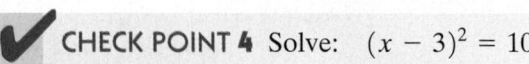

$x^2 + 8x$: Add $\left(\frac{8}{2}\right)^2 = 4^2$. Add 16 to complete the square.

$x^2 - 7x$: Add $\left(\frac{-7}{2}\right)^2$, or $\frac{49}{4}$, to complete the square.

$x^2 + \frac{3}{5}x$: Add $\left(\frac{1}{2} \cdot \frac{3}{5}\right)^2 = \left(\frac{3}{10}\right)^2$. Add $\frac{9}{100}$ to complete the square.

ENRICHMENT ESSAY

Visualizing Completing the Square

This figure, with area $x^2 + 8x$, is not a complete square. The bottom-right corner is missing.

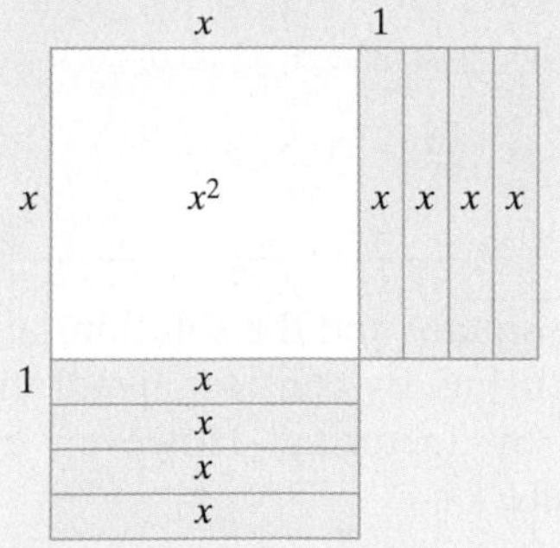

Area: $x^2 + 8x$

Add 16 square units to the missing portion and you, literally, complete the square.

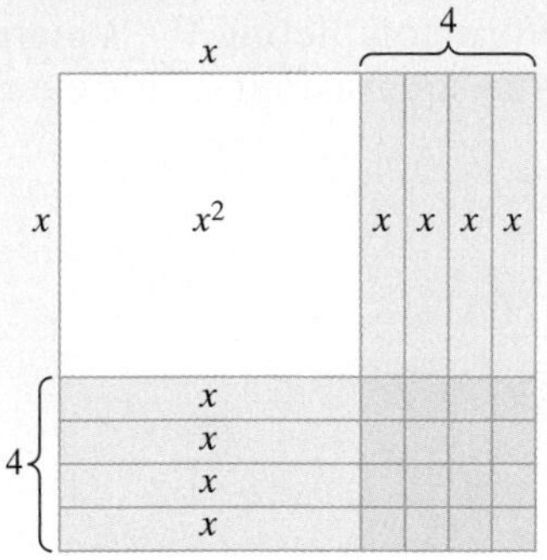

Area: $x^2 + 8x + 16 = (x + 4)^2$

a. The coefficient of the x-term in $x^2 + 8x$ is 8. Half of 8 is 4, and $4^2 = 16$. Add 16. The result is a perfect square trinomial.

$$x^2 + 8x + 16 = (x + 4)^2$$

b. The coefficient of the x-term in $x^2 - 7x$ is -7. Half of -7 is $-\frac{7}{2}$, and $\left(-\frac{7}{2}\right)^2 = \frac{49}{4}$. Add $\frac{49}{4}$. The result is a perfect square trinomial.

$$x^2 - 7x + \frac{49}{4} = \left(x - \frac{7}{2}\right)^2$$

c. The coefficient of the x-term in $x^2 + \frac{3}{5}x$ is $\frac{3}{5}$. Half of $\frac{3}{5}$ is $\frac{1}{2} \cdot \frac{3}{5}$, or $\frac{3}{10}$, and $\left(\frac{3}{10}\right)^2 = \frac{9}{100}$. Add $\frac{9}{100}$. The result is a perfect square trinomial.

$$x^2 + \frac{3}{5}x + \frac{9}{100} = \left(x + \frac{3}{10}\right)^2$$

■

STUDY TIP

You may not be accustomed to factoring perfect square trinomials in which fractions are involved. The constant in the factorization is always half the coefficient of x.

$$x^2 - 7x + \frac{49}{4} = \left(x - \frac{7}{2}\right)^2$$

Half the coefficient of x, -7, is $-\frac{7}{2}$.

$$x^2 + \frac{3}{5}x + \frac{9}{100} = \left(x + \frac{3}{10}\right)^2$$

Half the coefficient of x, $\frac{3}{5}$, is $\frac{3}{10}$.

✔ **CHECK POINT 5** What term should be added to each binomial so that it becomes a perfect square trinomial? Write and factor the trinomial.

a. $x^2 + 10x$

b. $x^2 - 3x$

c. $x^2 + \frac{3}{4}x$.

3 Solve quadratic equations by completing the square.

Solving Quadratic Equations by Completing the Square We can solve any quadratic equation by completing the square. If the coefficient of the x^2-term is one, we add the square of half the coefficient of x to both sides of the equation. **When you add a constant term to one side of the equation to complete the square, be certain to add the same constant to the other side of the equation.** These ideas are illustrated in Example 6.

EXAMPLE 6 Solving a Quadratic Equation by Completing the Square

Solve by completing the square: $x^2 - 6x + 4 = 0$.

SOLUTION We begin by subtracting 4 from both sides. This is done to isolate the binomial $x^2 - 6x$ so that we can complete the square.

$$x^2 - 6x + 4 = 0 \quad \text{This is the original equation.}$$

$$x^2 - 6x = -4 \quad \text{Subtract 4 from both sides.}$$

Next, we work with $x^2 - 6x = -4$ and complete the square. Find half the coefficient of the x-term and square it. The coefficient of the x-term is -6. Half of -6 is -3 and $(-3)^2 = 9$. Thus, we add 9 to both sides of the equation.

$$x^2 - 6x + 9 = -4 + 9 \quad \text{Add 9 to both sides to complete the square.}$$

$$(x - 3)^2 = 5 \quad \text{Factor and simplify.}$$

$$x - 3 = \sqrt{5} \quad \text{or} \quad x - 3 = -\sqrt{5} \quad \text{Apply the square root property.}$$

$$x = 3 + \sqrt{5} \qquad x = 3 - \sqrt{5} \quad \text{Add 3 to both sides in each equation.}$$

The solutions are $3 \pm \sqrt{5}$, and the solution set is $\{3 + \sqrt{5}, 3 - \sqrt{5}\}$ or $\{3 \pm \sqrt{5}\}$.

If you solve a quadratic equation by completing the square and the solutions are rational numbers, the equation can also be solved by factoring. By contrast, quadratic equations with irrational solutions cannot be solved by factoring. However, all quadratic equations can be solved by completing the square.

CHECK POINT 6 Solve by completing the square: $x^2 + 4x - 1 = 0$.

If the coefficient of the x^2-term in a quadratic equation is not 1, you must divide each side of the equation by this coefficient before completing the square. For example, to solve $9x^2 - 6x - 4 = 0$ by completing the square, first divide every term by 9:

$$\frac{9x^2}{9} - \frac{6x}{9} - \frac{4}{9} = \frac{0}{9}$$

$$x^2 - \frac{6}{9}x - \frac{4}{9} = 0$$

$$x^2 - \frac{2}{3}x - \frac{4}{9} = 0.$$

Now that the coefficient of the x^2-term is 1, we can solve by completing the square.

EXAMPLE 7 Solving a Quadratic Equation by Completing the Square

Solve by completing the square: $9x^2 - 6x - 4 = 0$.

SOLUTION

$$9x^2 - 6x - 4 = 0 \quad \text{This is the original equation.}$$

$$x^2 - \frac{2}{3}x - \frac{4}{9} = 0 \quad \text{Divide both sides by 9.}$$

$$x^2 - \frac{2}{3}x = \frac{4}{9} \quad \text{Add } \tfrac{4}{9} \text{ to both sides to isolate the binomial.}$$

$$x^2 - \frac{2}{3}x + \frac{1}{9} = \frac{4}{9} + \frac{1}{9} \quad \text{Complete the square: Half of } -\tfrac{2}{3} \text{ is } -\tfrac{2}{6}, \text{ or } -\tfrac{1}{3}, \text{ and } \left(-\tfrac{1}{3}\right)^2 = \tfrac{1}{9}.$$

$$\left(x - \frac{1}{3}\right)^2 = \frac{5}{9} \quad \text{Factor and simplify.}$$

USING TECHNOLOGY

Obtain a decimal approximation for each solution of $9x^2 - 6x - 4 = 0$, the equation in Example 7.

$$\frac{1+\sqrt{5}}{3} \approx 1.1$$

$$\frac{1-\sqrt{5}}{3} \approx -0.4$$

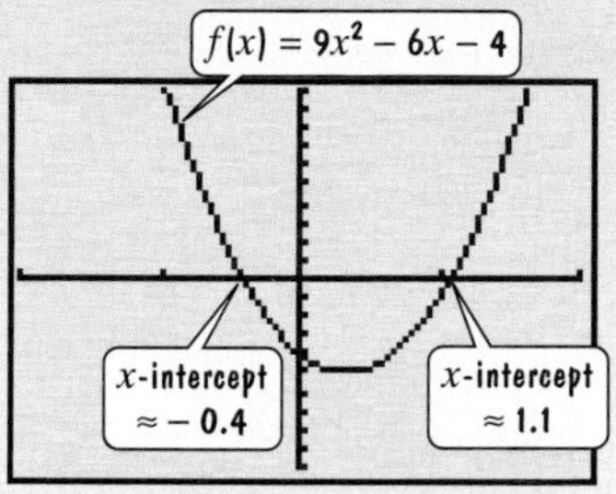

[−2, 2, 1] by [−10, 10, 1]

The x-intercepts of $f(x) = 9x^2 - 6x - 4$ verify the solutions.

$$x - \frac{1}{3} = \sqrt{\frac{5}{9}} \quad \text{or} \quad x - \frac{1}{3} = -\sqrt{\frac{5}{9}}$$ Apply the square root property.

$$x - \frac{1}{3} = \frac{\sqrt{5}}{3} \qquad x - \frac{1}{3} = -\frac{\sqrt{5}}{3}$$ $\sqrt{\frac{5}{9}} = \frac{\sqrt{5}}{\sqrt{9}} = \frac{\sqrt{5}}{3}$

$$x = \frac{1}{3} + \frac{\sqrt{5}}{3} \qquad x = \frac{1}{3} - \frac{\sqrt{5}}{3}$$ Add $\frac{1}{3}$ to both sides and solve for x.

$$x = \frac{1+\sqrt{5}}{3} \qquad x = \frac{1-\sqrt{5}}{3}$$ Express solutions with a common denominator.

The solutions are $\frac{1 \pm \sqrt{5}}{3}$, and the solution set is $\left\{\frac{1 \pm \sqrt{5}}{3}\right\}$. ■

CHECK POINT 7 Solve by completing the square: $2x^2 + 3x - 4 = 0$.

4 Solve problems using the square root property.

Applications We all want a wonderful life with fulfilling work, good health, and loving relationships. And let's be honest: Financial security wouldn't hurt! Achieving this goal depends on understanding how money in a savings account grows in remarkable ways as a result of *compound interest*. **Compound interest** is interest computed on your original investment as well as on any accumulated interest. For example, suppose you deposit \$1000, the principal, in a savings account at a rate of 5%. Table 11.1 shows how the investment grows if the interest earned is automatically added on to the principal.

Table 11.1 Compound Interest on \$1000

Year	Starting Balance	Interest Earned: $I = Pr$	New Balance
1	\$1000	$\$1000 \times 0.05 = \$50$	\$1050
2	\$1050	$\$1050 \times 0.05 = \$52.50$	\$1102.50
3	\$1102.50	$\$1102.50 \times 0.05 \approx \$55.13$	\$1157.63

A faster way to determine the amount, A, in an account subject to compound interest is to use the following formula:

A FORMULA FOR COMPOUND INTEREST Suppose that an amount of money, P, is invested at interest rate r, compounded annually. In t years, the amount, A, or balance, in the account is given by the formula

$$A = P(1 + r)^t.$$

Some compound interest problems can be solved using quadratic equations.

EXAMPLE 8 Solving a Compound Interest Problem

You invested \$1000 in an account whose interest is compounded annually. After 2 years, the amount, or balance, in the account is \$1210. Find the annual interest rate.

SOLUTION We are given that

P (the amount invested) $= \$1000$

t (the time of the investment) $= 2$ years

A (the amount, or balance, in the account) $= \$1210.$

We are asked to find the annual interest rate, r. We substitute the three given values into the compound interest formula and solve for r.

$$A = P(1 + r)^t \quad \text{Use the compound interest formula.}$$

$$1210 = 1000(1 + r)^2 \quad \text{Substitute the given values.}$$

$$\frac{1210}{1000} = (1 + r)^2 \quad \text{Divide both sides by 1000.}$$

$$\frac{121}{100} = (1 + r)^2 \quad \text{Simplify the fraction.}$$

$$1 + r = \sqrt{\frac{121}{100}} \quad \text{or} \quad 1 + r = -\sqrt{\frac{121}{100}} \quad \text{Apply the square root property.}$$

$$1 + r = \frac{11}{10} \qquad 1 + r = -\frac{11}{10} \quad \sqrt{\frac{121}{100}} = \frac{\sqrt{121}}{\sqrt{100}} = \frac{11}{10}$$

$$r = \frac{11}{10} - 1 \qquad r = -\frac{11}{10} - 1 \quad \text{Subtract 1 from both sides and solve for } r.$$

$$r = \frac{1}{10} \qquad r = -\frac{21}{10} \quad \frac{11}{10} - 1 = \frac{11}{10} - \frac{10}{10} = \frac{1}{10} \text{ and } -\frac{11}{10} - 1 = -\frac{11}{10} - \frac{10}{10} = -\frac{21}{10}.$$

Because the interest rate cannot be negative, we reject $-\frac{21}{10}$. Thus, the annual interest rate is $\frac{1}{10} = 0.10 = 10\%$.

We can check this answer using the formula $A = P(1 + r)^t$. If \$1000 is invested for 2 years at 10% interest, compounded annually, the balance in the account is

$$A = \$1000(1 + 0.10)^2 = \$1000(1.10)^2 = \$1210.$$

Because this is precisely the amount given by the problem's conditions, the annual interest rate is, indeed, 10% compounded annually. ■

✔ **CHECK POINT 8** You invested \$3000 in an account whose interest is compounded annually. After 2 years, the amount, or balance, in the account is \$4320. Find the annual interest rate.

The Pythagorean Theorem and the Square Root Property The ancient Greek philosopher and mathematician Pythagoras (approximately 582–500 B.C.) founded a school whose motto was "All is number." Pythagoras is best remembered for his work with the **right triangle**, a triangle with one angle measuring 90°. The side opposite the 90° angle is called the **hypotenuse**. The other sides are called **legs**. Pythagoras found that if he constructed squares on each of the legs, as well as a larger square on the hypotenuse, the sum of the areas of the smaller squares is equal to the area of the larger square. This is illustrated in Figure 11.2.

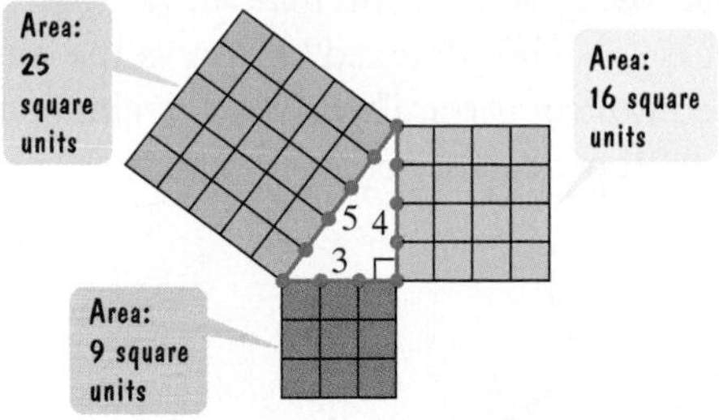

FIGURE 11.2 The area of the large square equals the sum of the areas of the smaller squares.

This relationship is usually stated in terms of the lengths of the three sides of a right triangle and is called the **Pythagorean Theorem**.

THE PYTHAGOREAN THEOREM The sum of the squares of the lengths of the legs of a right triangle equals the square of the length of the hypotenuse.

If the legs have lengths a and b, and the hypotenuse has length c, then

$$a^2 + b^2 = c^2.$$

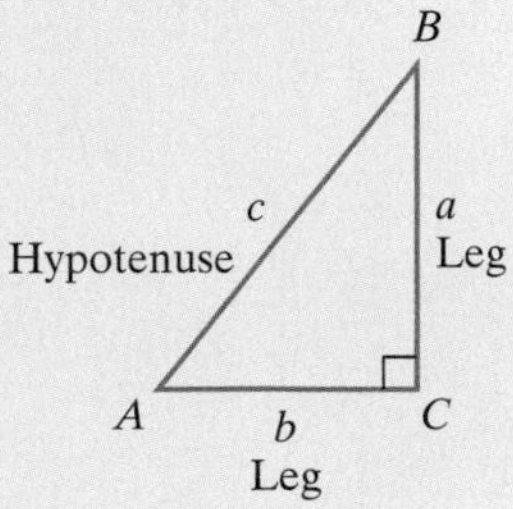

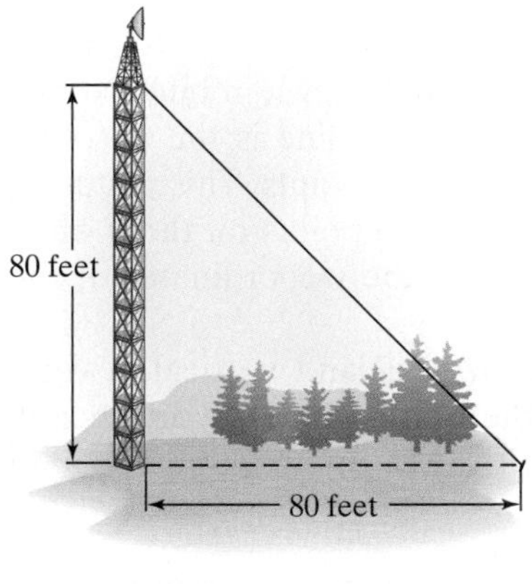

FIGURE 11.3

EXAMPLE 9 Using the Pythagorean Theorem and the Square Root Property

A supporting wire is to be attached to the top of an 80-foot antenna, as shown in Figure 11.3. Because of surrounding trees, the wire must be anchored 80 feet from the base of the antenna. What length of wire is required?

SOLUTION Let w = the wire's length. We can find w in the right triangle using the Pythagorean Theorem.

(Leg)²	plus	(Leg)²	equals	(Hypotenuse)².
80^2	$+$	80^2	$=$	w^2.

We solve this equation using the square root property.

$$6400 + 6400 = w^2 \quad \text{Square 80.}$$

$$12{,}800 = w^2 \quad \text{Add.}$$

$$w = \sqrt{12{,}800} \quad \text{or} \quad w = -\sqrt{12{,}800} \quad \text{Apply the square root property.}$$

$$w = \sqrt{6400}\sqrt{2} \quad \text{or} \quad w = -\sqrt{6400}\sqrt{2} \quad \text{6400 is the greatest perfect square factor of 12,800.}$$

$$w = 80\sqrt{2} \quad \text{or} \quad w = -80\sqrt{2}$$

Because w represents the wire's length, we reject the negative value. Thus, a wire of length $80\sqrt{2}$ feet is needed to support the antenna. Using a calculator, this is approximately 113.1 feet of wire. ■

CHECK POINT 9 A 50-foot supporting wire is to be attached to an antenna. The wire is anchored 20 feet from the base of the antenna. How high up the antenna is the wire attached? Express the answer in simplified radical form. Then find a decimal approximation to the nearest tenth of a foot.

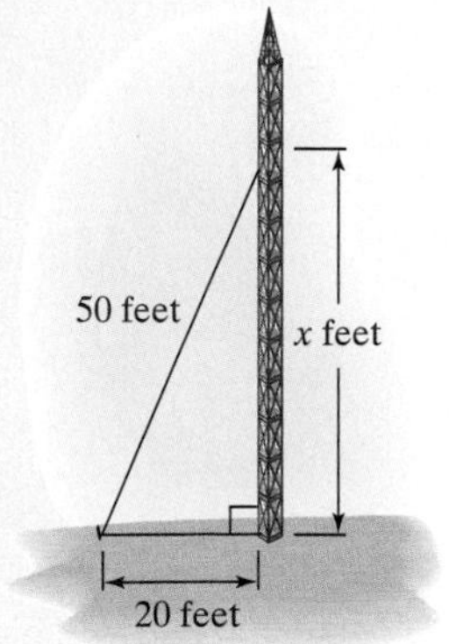

5 Find the distance between two points.

The Distance Formula Using the Pythagorean theorem, we can find the distance between the two points $P_1(x_1, y_1)$ and $P_2(x_2, y_2)$ in the rectangular coordinate system. The two points are illustrated in Figure 11.4.

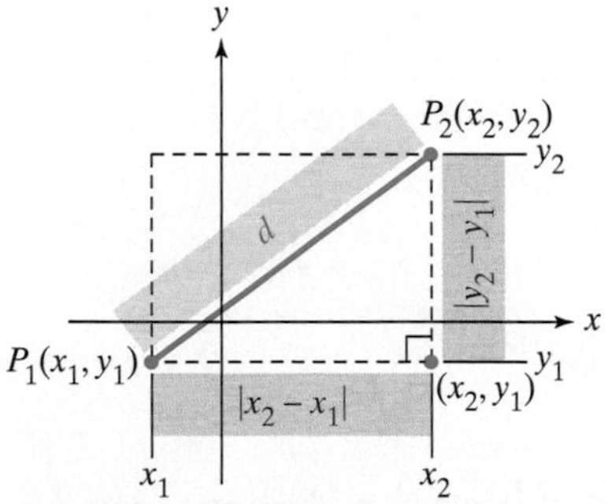

FIGURE 11.4

The distance that we want to find is represented by d and shown in blue. Notice that the distance between two points on the dashed horizontal line is the absolute value of the difference between the x-coordinates of the two points. This distance, $|x_2 - x_1|$, is shown in pink. Similarly, the distance between two points on the dashed vertical line is the absolute value of the difference between the y-coordinates of the two points. This distance, $|y_2 - y_1|$, is also shown in pink.

Because the dashed lines in Figure 11.4 are horizontal and vertical, a right triangle is formed. Thus, we can use the Pythagorean theorem to find the distance d. Squaring the lengths of the triangle's sides results in positive numbers, so absolute value notation is not necessary.

$$d^2 = (x_2 - x_1)^2 + (y_2 - y_1)^2$$ Apply the Pythagorean theorem to the right triangle in Figure 11.4.

$$d = \pm\sqrt{(x_2 - x_1)^2 + (y_2 - y_1)^2}$$ Apply the square root property.

$$d = \sqrt{(x_2 - x_1)^2 + (y_2 - y_1)^2}$$ Because distance is nonnegative, write only the principal square root.

This result is called the **distance formula**.

THE DISTANCE FORMULA The distance, d, between the points (x_1, y_1) and (x_2, y_2) in the rectangular coordinate system is

$$d = \sqrt{(x_2 - x_1)^2 + (y_2 - y_1)^2}.$$

When using the distance formula, it does not matter which point you call (x_1, y_1) and which you call (x_2, y_2).

EXAMPLE 10 Using the Distance Formula

Find the distance between $(-1, -3)$ and $(2, 3)$.

SOLUTION Letting $(x_1, y_1) = (-1, -3)$ and $(x_2, y_2) = (2, 3)$, we obtain

$$d = \sqrt{(x_2 - x_1)^2 + (y_2 - y_1)^2}$$ Use the distance formula.

$$= \sqrt{[2 - (-1)]^2 + [3 - (-3)]^2}$$ Substitute the given values.

$$= \sqrt{(2 + 1)^2 + (3 + 3)^2}$$ Perform subtractions within the grouping symbols.

$$= \sqrt{3^2 + 6^2}$$ Perform the resulting additions.

$$= \sqrt{9 + 36}$$ Square 3 and 6.

$$= \sqrt{45}$$ Add.

$$= 3\sqrt{5} \approx 6.71.$$ $\sqrt{45} = \sqrt{9 \cdot 5} = \sqrt{9}\sqrt{5} = 3\sqrt{5}$

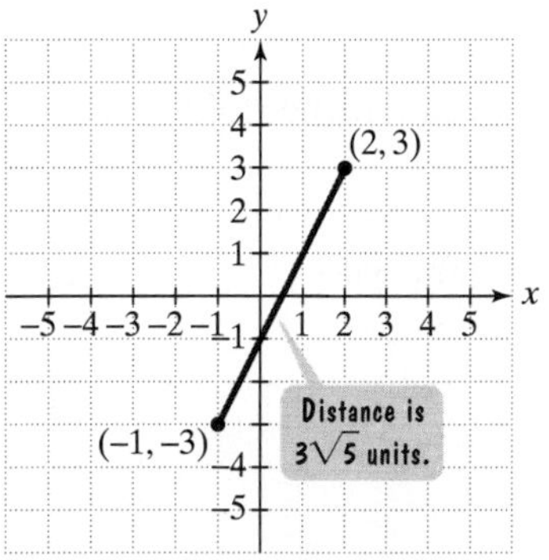

FIGURE 11.5 Finding the distance between two points

The distance between the given points is $3\sqrt{5}$ units, or approximately 6.71 units. The situation is illustrated in Figure 11.5. ■

CHECK POINT 10 Find the distance between $(2, -2)$ and $(5, 2)$.

6 Find the midpoint of a line segment.

The Midpoint Formula The distance formula can be used to derive a formula for finding the midpoint of a line segment between two given points. The formula is given as follows:

THE MIDPOINT FORMULA Consider a line segment whose endpoints are (x_1, y_1) and (x_2, y_2). The coordinates of the segment's midpoint are

$$\left(\frac{x_1 + x_2}{2}, \frac{y_1 + y_2}{2}\right).$$

To find the midpoint, take the average of the two x-coordinates and the average of the two y-coordinates.

STUDY TIP

The midpoint formula requires finding the *sum* of coordinates. By contrast, the distance formula requires finding the *difference* of coordinates:

Midpoint: Sum of coordinates

$$\left(\frac{x_1 + x_2}{2}, \frac{y_1 + y_2}{2}\right)$$

Distance: Difference of coordinates

$$\sqrt{(x_2 - x_1)^2 + (y_2 - y_1)^2}$$

It's easy to confuse the two formulas. Be sure to use addition, not subtraction, when applying the midpoint formula.

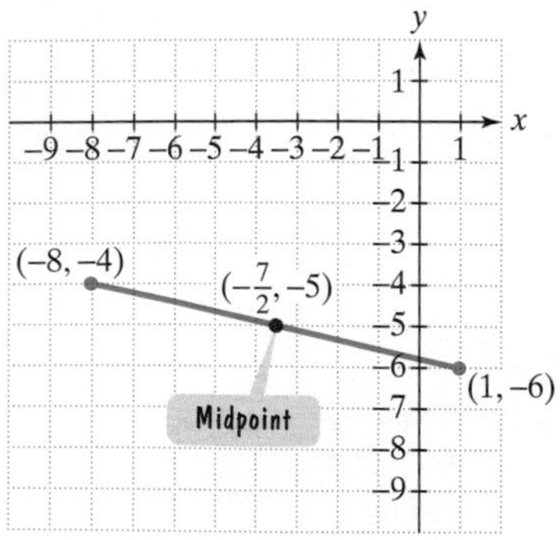

FIGURE 11.6 Finding a line segment's midpoint

EXAMPLE 11 Using the Midpoint Formula

Find the midpoint of the line segment with endpoints $(1, -6)$ and $(-8, -4)$.

SOLUTION To find the coordinates of the midpoint, we average the coordinates of the endpoints.

$$\text{Midpoint} = \left(\frac{1 + (-8)}{2}, \frac{-6 + (-4)}{2}\right) = \left(\frac{-7}{2}, \frac{-10}{2}\right) = \left(-\frac{7}{2}, -5\right)$$

Average the x-coordinates.

Average the y-coordinates.

Figure 11.6 illustrates that the point $\left(-\frac{7}{2}, -5\right)$ is midway between the points $(1, -6)$ and $(-8, -4)$. ■

CHECK POINT 11 Find the midpoint of the line segment with endpoints $(1, 2)$ and $(7, -3)$.

11.1 EXERCISE SET

Student Solutions Manual | CD/Video | PH Math/Tutor Center | MathXL Tutorials on CD | MathXL® | MyMathLab | Interactmath.com

Practice Exercises

In Exercises 1–22, solve each equation by the square root property. If possible, simplify radicals or rationalize denominators. Express imaginary solutions in the form $a + bi$.

1. $3x^2 = 75$

2. $5x^2 = 20$

3. $7x^2 = 42$

4. $8x^2 = 40$

5. $16x^2 = 25$

6. $4x^2 = 49$

7. $3x^2 - 2 = 0$

8. $3x^2 - 5 = 0$

9. $25x^2 + 16 = 0$

10. $4x^2 + 49 = 0$

11. $(x + 7)^2 = 9$

12. $(x + 3)^2 = 64$

13. $(x - 3)^2 = 5$

14. $(x - 4)^2 = 3$

15. $2(x + 2)^2 = 16$

16. $3(x + 2)^2 = 36$

17. $(x - 5)^2 = -9$

18. $(x - 5)^2 = -4$

19. $\left(x + \frac{3}{4}\right)^2 = \frac{11}{16}$

20. $\left(x + \frac{2}{5}\right)^2 = \frac{7}{25}$

21. $x^2 - 6x + 9 = 36$

22. $x^2 - 6x + 9 = 49$

In Exercises 23–34, determine the constant that should be added to the binomial so that it becomes a perfect square trinomial. Then write and factor the trinomial.

23. $x^2 + 2x$

24. $x^2 + 4x$

25. $x^2 - 14x$

26. $x^2 - 10x$

27. $x^2 + 7x$

28. $x^2 + 9x$

29. $x^2 - \frac{1}{2}x$

30. $x^2 - \frac{1}{3}x$

31. $x^2 + \frac{4}{3}x$

32. $x^2 + \frac{4}{5}x$

33. $x^2 - \frac{9}{4}x$

34. $x^2 - \frac{9}{5}x$

In Exercises 35–56, solve each quadratic equation by completing the square.

35. $x^2 + 4x = 32$

36. $x^2 + 6x = 7$

37. $x^2 + 6x = -2$

38. $x^2 + 2x = 5$

39. $x^2 - 8x + 1 = 0$

40. $x^2 + 8x - 5 = 0$

41. $x^2 + 2x + 2 = 0$

42. $x^2 - 4x + 8 = 0$

43. $x^2 + 3x - 1 = 0$

44. $x^2 - 3x - 5 = 0$

45. $x^2 + \frac{4}{7}x + \frac{3}{49} = 0$

46. $x^2 + \frac{6}{5}x + \frac{8}{25} = 0$

47. $x^2 + x - 1 = 0$

48. $x^2 + 3x - 1 = 0$

49. $2x^2 + 3x - 5 = 0$

50. $2x^2 + 5x - 3 = 0$

51. $3x^2 + 6x + 1 = 0$

52. $3x^2 - 6x + 2 = 0$

53. $3x^2 - 8x + 1 = 0$

54. $2x^2 + 3x - 4 = 0$

55. $8x^2 - 4x + 1 = 0$

56. $9x^2 - 6x + 5 = 0$

57. If $f(x) = (x - 1)^2$, find all values of x for which $f(x) = 36$.

58. If $f(x) = (x + 2)^2$, find all values of x for which $f(x) = 25$.

59. If $g(x) = \left(x - \frac{2}{5}\right)^2$, find all values of x for which $g(x) = \frac{9}{25}$.

60. If $g(x) = \left(x + \frac{1}{3}\right)^2$, find all values of x for which $g(x) = \frac{4}{9}$.

61. If $h(x) = 5(x + 2)^2$, find all values of x for which $h(x) = -125$.

62. If $h(x) = 3(x - 4)^2$, find all values of x for which $h(x) = -12$.

In Exercises 63–80, find the distance between each pair of points. If necessary, round answers to two decimal places.

63. $(2, 3)$ and $(14, 8)$

64. $(5, 1)$ and $(8, 5)$

65. $(4, 1)$ and $(6, 3)$

66. $(2, 3)$ and $(3, 5)$

67. $(0, 0)$ and $(-3, 4)$

68. $(0, 0)$ and $(3, -4)$

69. $(-2, -6)$ and $(3, -4)$

70. $(-4, -1)$ and $(2, -3)$

71. $(0, -3)$ and $(4, 1)$

72. $(0, -2)$ and $(4, 3)$

73. $(3.5, 8.2)$ and $(-0.5, 6.2)$

74. $(2.6, 1.3)$ and $(1.6, -5.7)$

75. $(0, -\sqrt{3})$ and $(\sqrt{5}, 0)$

76. $(0, -\sqrt{2})$ and $(\sqrt{7}, 0)$

77. $(3\sqrt{3}, \sqrt{5})$ and $(-\sqrt{3}, 4\sqrt{5})$

78. $(2\sqrt{3}, \sqrt{6})$ and $(-\sqrt{3}, 5\sqrt{6})$

79. $\left(\frac{7}{3}, \frac{1}{5}\right)$ and $\left(\frac{1}{3}, \frac{6}{5}\right)$

80. $\left(-\frac{1}{4}, -\frac{1}{7}\right)$ and $\left(\frac{3}{4}, \frac{6}{7}\right)$

In Exercises 81–92, find the midpoint of the line segment with the given endpoints.

81. $(6, 8)$ and $(2, 4)$

82. $(10, 4)$ and $(2, 6)$

83. $(-2, -8)$ and $(-6, -2)$

84. $(-4, -7)$ and $(-1, -3)$

85. $(-3, -4)$ and $(6, -8)$

86. $(-2, -1)$ and $(-8, 6)$

87. $\left(-\frac{7}{2}, \frac{3}{2}\right)$ and $\left(-\frac{5}{2}, -\frac{11}{2}\right)$

88. $\left(-\frac{2}{5}, \frac{7}{15}\right)$ and $\left(-\frac{2}{5}, -\frac{4}{15}\right)$

89. $(8, 3\sqrt{5})$ and $(-6, 7\sqrt{5})$

90. $(7\sqrt{3}, -6)$ and $(3\sqrt{3}, -2)$

91. $(\sqrt{18}, -4)$ and $(\sqrt{2}, 4)$

92. $(\sqrt{50}, -6)$ and $(\sqrt{2}, 6)$

Practice Plus

93. Three times the square of the difference between a number and 2 is -12. Find the number(s).

94. Three times the square of the difference between a number and 9 is -27. Find the number(s).

In Exercises 95–98, solve the formula for the specified variable. Because each variable is nonnegative, list only the principal square root. If possible, simplify radicals or rationalize denominators.

95. $h = \frac{v^2}{2g}$ for v

96. $s = \frac{kwd^2}{l}$ for d

97. $A = P(1 + r)^2$ for r

98. $C = \frac{kP_1P_2}{d^2}$ for d

In Exercises 99–102, solve each quadratic equation by completing the square.

99. $\frac{x^2}{3} + \frac{x}{9} - \frac{1}{6} = 0$

100. $\frac{x^2}{2} - \frac{x}{6} - \frac{3}{4} = 0$

101. $x^2 - bx = 2b^2$

102. $x^2 - bx = 6b^2$

Application Exercises

In Exercises 103–106, use the compound interest formula

$$A = P(1 + r)^t$$

to find the annual interest rate, r.

103. In 2 years, an investment of \$2000 grows to \$2880.

104. In 2 years, an investment of \$2000 grows to \$2420.

105. In 2 years, an investment of \$1280 grows to \$1445.

106. In 2 years, an investment of \$80,000 grows to \$101,250.

Of the one hundred largest economies in the world, 53 are multinational corporations. In 1970, there were approximately 7000 multinational corporations. By 2001, more than 65,000 corporations enveloped the world. The graph shows this rapid growth from 1970 through 2001, including the starting dates of some notable corporations.

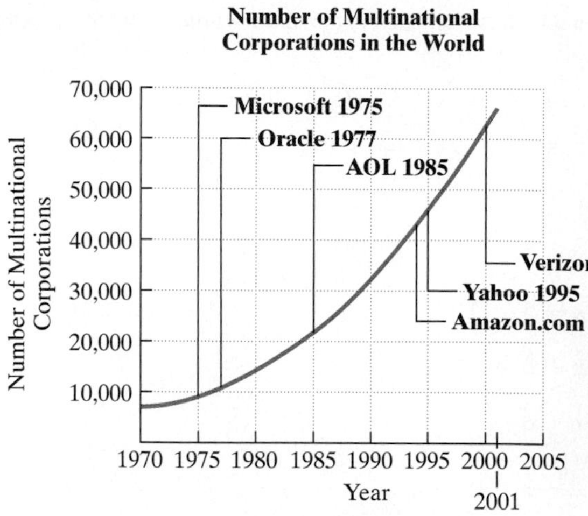

Source: Medard Gabel, *Global Inc.*, The New Press, 2003

The data shown can be modeled by the function

$$f(x) = 62.2x^2 + 7000,$$

where $f(x)$ represents the number of multinational corporations in the world x years after 1970. Use this function and the square root property to solve Exercises 107–108.

107. In which year will there be 92,000 multinational corporations?

108. In which year were there 46,000 multinational corporations? How well does the function model the actual number of corporations for that year shown in the graph?

The function $s(t) = 16t^2$ models the distance, $s(t)$, in feet, that an object falls in t seconds. Use this function and the square root property to solve Exercises 109–110. Express answers in simplified radical form. Then use your calculator and find a decimal approximation to the nearest tenth of a second.

109. A sky diver jumps from an airplane and falls for 4800 feet before opening a parachute. For how many seconds was the diver in a free fall?

110. A sky diver jumps from an airplane and falls for 3200 feet before opening a parachute. For how many seconds was the diver in a free fall?

Use the Pythagorean theorem and the square root property to solve Exercises 111–116. Express answers in simplified radical form. Then find a decimal approximation to the nearest tenth.

111. A rectangular park is 6 miles long and 3 miles wide. How long is a pedestrian route that runs diagonally across the park?

112. A rectangular park is 4 miles long and 2 miles wide. How long is a pedestrian route that runs diagonally across the park?

113. The base of a 30-foot ladder is 10 feet from the building. If the ladder reaches the flat roof, how tall is the building?

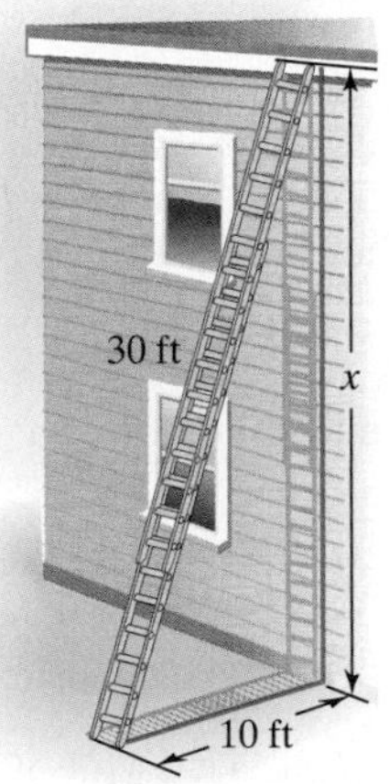

114. The doorway into a room is 4 feet wide and 8 feet high. What is the length of the longest rectangular panel that can be taken through this doorway diagonally?

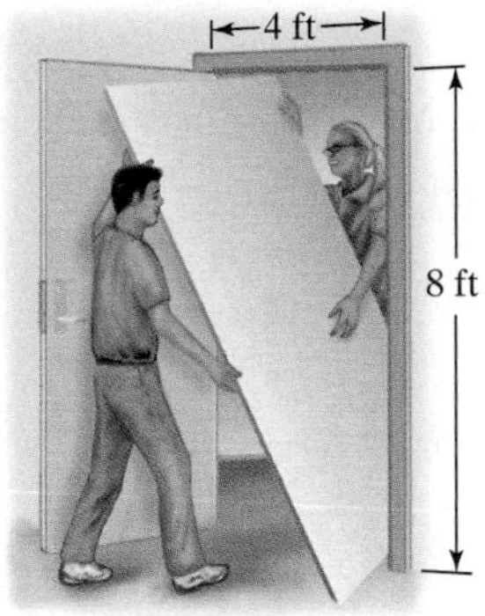

115. A supporting wire is to be attached to the top of a 50-foot antenna. If the wire must be anchored 50 feet from the base of the antenna, what length of wire is required?

116. A supporting wire is to be attached to the top of a 70-foot antenna. If the wire must be anchored 70 feet from the base of the antenna, what length of wire is required?

117. A square flower bed is to be enlarged by adding 2 meters on each side. If the larger square has an area of 196 square meters, what is the length of a side of the original square?

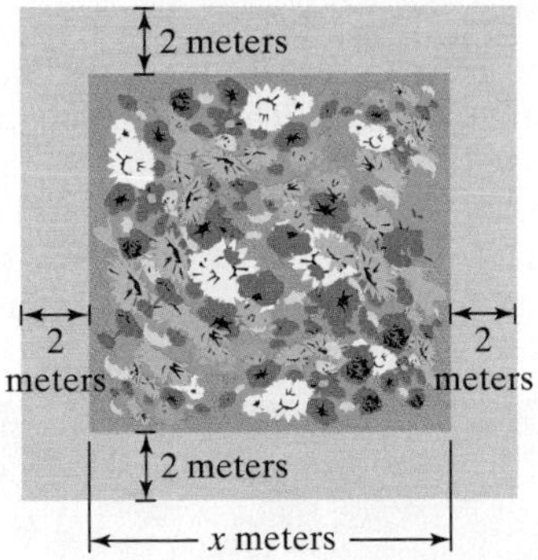

118. A square flower bed is to be enlarged by adding 4 feet on each side. If the larger square has an area of 225 square feet, what is the length of a side of the original square?

119. A rectangular coordinate system with coordinates in miles is placed on the map in the figure shown. Bangkok has coordinates $(-115, 170)$ and Phnom Penh has coordinates $(65, 70)$. How long will it take a plane averaging 400 miles per hour to fly directly from one city to the other? Round to the nearest tenth of an hour. Approximately how many minutes is the flight?

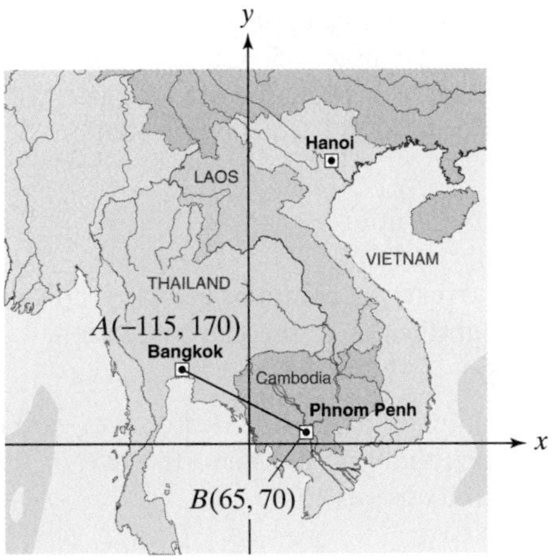

Writing in Mathematics

120. What is the square root property?

121. Explain how to solve $(x - 1)^2 = 16$ using the square root property.

122. Explain how to complete the square for a binomial. Use $x^2 + 6x$ to illustrate your explanation.

123. Explain how to solve $x^2 + 6x + 8 = 0$ by completing the square.

124. What is compound interest?

125. State the Pythagorean theorem.

126. Describe how to find the distance between two points in the rectangular coordinate system.

127. Describe how to find the midpoint of a line segment if its endpoints are known.

128. A basketball player's hang time is the time spent in the air when shooting a basket. The function $f(x) = 4x^2$ models the player's vertical leap, $f(x)$, in feet, in terms of the hang time, x, measured in seconds. Use this function to write a problem that can be solved by the square root property. Then solve the problem. Be realistic: It is unlikely that hang time can exceed 1 second.

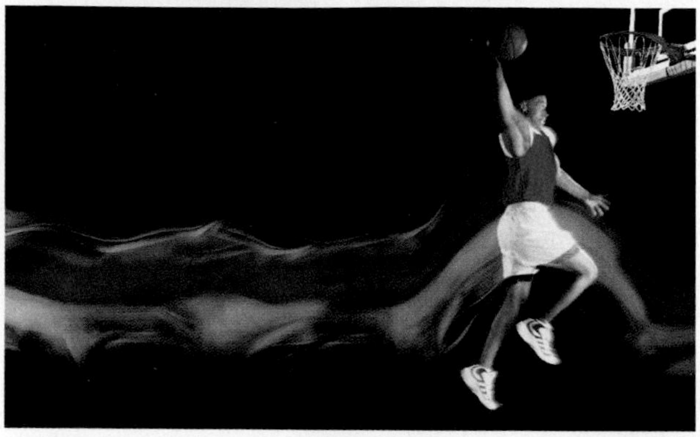

Technology Exercises

129. Use a graphing utility to solve $4 - (x + 1)^2 = 0$. Graph $y = 4 - (x + 1)^2$ in a $[-5, 5, 1]$ by $[-5, 5, 1]$ viewing rectangle. The equation's solutions are the graph's x-intercepts. Check by substitution in the given equation.

130. Use a graphing utility to solve $(x - 1)^2 - 9 = 0$. Graph $y = (x - 1)^2 - 9$ in a $[-5, 5, 1]$ by $[-9, 3, 1]$ viewing rectangle. The equation's solutions are the graph's x-intercepts. Check by substitution in the given equation.

131. Use a graphing utility and x-intercepts to verify any of the real solutions that you obtained for five of the quadratic equations in Exercises 35–56.

Critical Thinking Exercises

132. Which one of the following is true?

a. The equation $(x - 5)^2 = 12$ is equivalent to $x - 5 = 2\sqrt{3}$.

b. In completing the square for $2x^2 - 6x = 5$, we should add 9 to both sides.

c. Although not every quadratic equation can be solved by completing the square, they can all be solved by factoring.

d. The graph of $y = (x - 2)^2 + 3$ cannot have x-intercepts.

133. Solve for y: $\frac{x^2}{a^2} + \frac{y^2}{b^2} = 1$.

134. Solve by completing the square:

$$x^2 + x + c = 0.$$

135. Solve by completing the square:

$$x^2 + bx + c = 0.$$

136. Solve: $x^4 - 8x^2 + 15 = 0$.

137. Show that the points $A(1, 1 + d)$, $B(3, 3 + d)$, and $C(6, 6 + d)$ are collinear (lie along a straight line) by showing that the distance from A to B plus the distance from B to C equals the distance from A to C.

138. Prove the midpoint formula by using the following procedure.

a. Show that the distance between (x_1, y_1) and $\left(\frac{x_1 + x_2}{2}, \frac{y_1 + y_2}{2}\right)$ is equal to the distance between (x_2, y_2) and $\left(\frac{x_1 + x_2}{2}, \frac{y_1 + y_2}{2}\right)$.

b. Use the procedure from Exercise 137 and the distances from part (a) to show that the points (x_1, y_1), $\left(\frac{x_1 + x_2}{2}, \frac{y_1 + y_2}{2}\right)$, and (x_2, y_2) are collinear.

139. Find all points with y-coordinate 2 so that the distance between $(x, 2)$ and $(2, -1)$ is 5.

Review Exercises

140. Simplify: $4x - 2 - 3[4 - 2(3 - x)]$. (Section 1.8, Example 11)

141. Factor: $1 - 8x^3$. (Section 6.4, Example 8)

142. Divide: $(x^4 - 5x^3 + 2x^2 - 6) \div (x - 3)$. (Section 5.6, Example 2 or Example 5)

SECTION 11.2

Objectives

1. Solve quadratic equations using the quadratic formula.
2. Use the discriminant to determine number and type of solutions.
3. Determine the most efficient method to use when solving a quadratic equation.
4. Write quadratic equations from solutions.
5. Use the quadratic formula to solve problems.

THE QUADRATIC FORMULA

America is a nation of immigrants. Since 1820, more than 40 million people have immigrated to the United States from all over the world. They chose to come for various reasons, such as to live in freedom, to practice religion freely, to escape poverty or oppression, and to make better lives for themselves and their children. As a result, a substantial percentage of the U.S. population is foreign-born. The graph in Figure 11.7 shows the percentage of foreign-born Americans from 1930 through 2002.

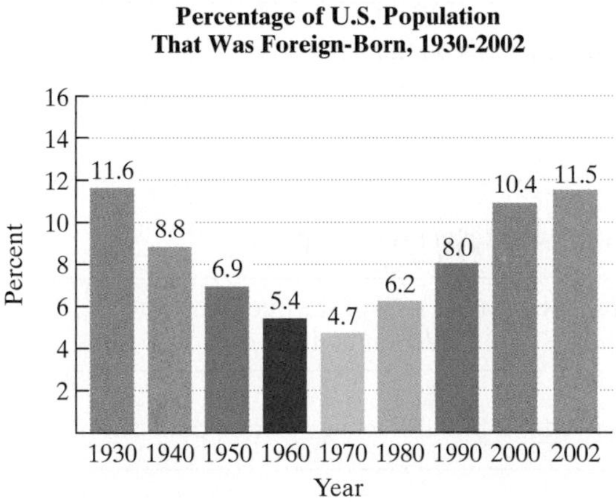

FIGURE 11.7

Source: U.S. Census Bureau

The data shown in Figure 11.7 can be modeled by the quadratic function

$$f(x) = 0.0049x^2 - 0.361x + 11.79,$$

where $f(x)$ is the percentage of the U.S. population that was foreign-born x years after 1930. If the trend shown by the data continues, in which year will 15% of the U.S. population be foreign-born? To answer the question, it is necessary to substitute 15 for $f(x)$ in the given function and solve for x, the number of years after 1930:

$$15 = 0.0049x^2 - 0.361x + 11.79.$$

In this section, we will derive a formula that will enable you to solve quadratic equations more quickly than the method of completing the square. Using this formula, we will work with the function that models the percentage of foreign-born Americans and determine the year in which 15% of our population will be foreign-born.

1 Solve quadratic equations using the quadratic formula.

Solving Quadratic Equations Using the Quadratic Formula We can use the method of completing the square to derive a formula that can be used to solve all quadratic equations. The derivation given here also shows a particular quadratic equation, $3x^2 - 2x - 4 = 0$, to specifically illustrate each of the steps.

Deriving the Quadratic Formula

Standard Form of a Quadratic Equation	Comment	A Specific Example
$ax^2 + bx + c = 0, a > 0$	This is the given equation.	$3x^2 - 2x - 4 = 0$
$x^2 + \frac{b}{a}x + \frac{c}{a} = 0$	Divide both sides by the coefficient of x^2.	$x^2 - \frac{2}{3}x - \frac{4}{3} = 0$
$x^2 + \frac{b}{a}x = -\frac{c}{a}$	Isolate the binomial by adding $-\frac{c}{a}$ on both sides.	$x^2 - \frac{2}{3}x = \frac{4}{3}$
$x^2 + \frac{b}{a}x + \left(\frac{b}{2a}\right)^2 = -\frac{c}{a} + \left(\frac{b}{2a}\right)^2$ (half)^2 $x^2 + \frac{b}{a}x + \frac{b^2}{4a^2} = -\frac{c}{a} + \frac{b^2}{4a^2}$	Complete the square. Add the square of half the coefficient of x to both sides.	$x^2 - \frac{2}{3}x + \left(-\frac{1}{3}\right)^2 = \frac{4}{3} + \left(-\frac{1}{3}\right)^2$ (half)^2 $x^2 - \frac{2}{3}x + \frac{1}{9} = \frac{4}{3} + \frac{1}{9}$
$\left(x + \frac{b}{2a}\right)^2 = -\frac{c}{a}\cdot\frac{4a}{4a} + \frac{b^2}{4a^2}$	Factor on the left side and obtain a common denominator on the right side.	$\left(x - \frac{1}{3}\right)^2 = \frac{4}{3}\cdot\frac{3}{3} + \frac{1}{9}$
$\left(x + \frac{b}{2a}\right)^2 = \frac{-4ac + b^2}{4a^2}$ $\left(x + \frac{b}{2a}\right)^2 = \frac{b^2 - 4ac}{4a^2}$	Add fractions on the right side.	$\left(x - \frac{1}{3}\right)^2 = \frac{12 + 1}{9}$ $\left(x - \frac{1}{3}\right)^2 = \frac{13}{9}$
$x + \frac{b}{2a} = \pm\sqrt{\frac{b^2 - 4ac}{4a^2}}$	Apply the square root property.	$x - \frac{1}{3} = \pm\sqrt{\frac{13}{9}}$
$x + \frac{b}{2a} = \pm\frac{\sqrt{b^2 - 4ac}}{2a}$	Take the square root of the quotient, simplifying the denominator.	$x - \frac{1}{3} = \pm\frac{\sqrt{13}}{3}$
$x = \frac{-b}{2a} \pm \frac{\sqrt{b^2 - 4ac}}{2a}$	Solve for x by subtracting $\frac{b}{2a}$ from both sides.	$x = \frac{1}{3} \pm \frac{\sqrt{13}}{3}$
$x = \frac{-b \pm \sqrt{b^2 - 4ac}}{2a}$	Combine fractions on the right side.	$x = \frac{1 \pm \sqrt{13}}{3}$

The formula shown at the bottom of the left column is called the *quadratic formula*. A similar proof shows that the same formula can be used to solve quadratic equations if a, the coefficient of the x^2-term, is negative.

THE QUADRATIC FORMULA The solutions of a quadratic equation in standard form $ax^2 + bx + c = 0$, with $a \neq 0$, are given by the **quadratic formula**

$$x = \frac{-b \pm \sqrt{b^2 - 4ac}}{2a}.$$

x equals negative b plus or minus the square root of $b^2 - 4ac$, all divided by $2a$.

To use the quadratic formula, write the quadratic equation in standard form if necessary. Then determine the numerical values for a (the coefficient of the x^2-term), b (the coefficient of the x-term), and c (the constant term). Substitute the values of a, b, and c into the quadratic formula and evaluate the expression. The $\pm$ sign indicates that there are two solutions of the equation.

EXAMPLE 1 Solving a Quadratic Equation Using the Quadratic Formula

Solve using the quadratic formula: $8x^2 + 2x - 1 = 0$.

SOLUTION The given equation is in standard form. Begin by identifying the values for a, b, and c.

$$8x^2 + 2x - 1 = 0$$

$a = 8$ $b = 2$ $c = -1$

Substituting these values into the quadratic formula and simplifying gives the equation's solutions.

$$x = \frac{-b \pm \sqrt{b^2 - 4ac}}{2a}$$ Use the quadratic formula.

$$x = \frac{-2 \pm \sqrt{2^2 - 4(8)(-1)}}{2(8)}$$ Substitute the values for a, b, and c: $a = 8$, $b = 2$, and $c = -1$.

$$= \frac{-2 \pm \sqrt{4 - (-32)}}{16}$$ $2^2 - 4(8)(-1) = 4 - (-32)$

$$= \frac{-2 \pm \sqrt{36}}{16}$$ $4 - (-32) = 4 + 32 = 36$

$$= \frac{-2 \pm 6}{16}$$ $\sqrt{36} = 6$

Now we will evaluate this expression in two different ways to obtain the two solutions. On the left, we will *add* 6 to -2. On the right, we will *subtract* 6 from -2.

$$x = \frac{-2 + 6}{16} \quad \text{or} \quad x = \frac{-2 - 6}{16}$$

$$= \frac{4}{16} = \frac{1}{4} \qquad\qquad = \frac{-8}{16} = -\frac{1}{2}$$

The solutions are $-\frac{1}{2}$ and $\frac{1}{4}$, and the solution set is $\left\{-\frac{1}{2}, \frac{1}{4}\right\}$. ■

USING TECHNOLOGY

The graph of the quadratic function

$$y = 8x^2 + 2x - 1$$

has x-intercepts at $-\frac{1}{2}$ and $\frac{1}{4}$. This verifies that $\left\{-\frac{1}{2}, \frac{1}{4}\right\}$ is the solution set of the quadratic equation

$$8x^2 + 2x - 1 = 0.$$

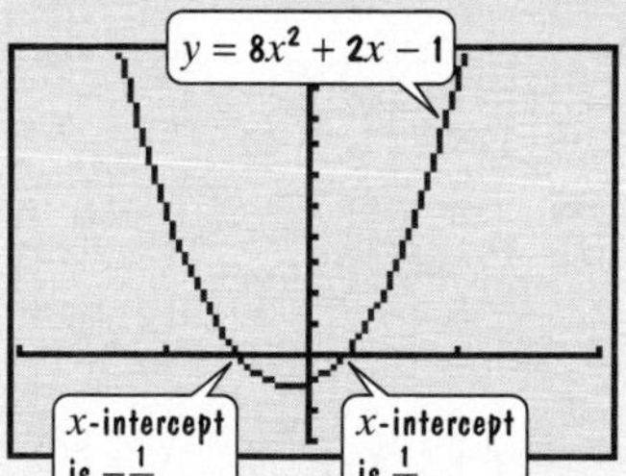

$[-2, 2, 1]$ by $[-3, 10, 1]$

In Example 1, the solutions of $8x^2 + 2x - 1 = 0$ are rational numbers. This means that the equation can also be solved by factoring. The reason that the solutions are rational numbers is that $b^2 - 4ac$, the radicand in the quadratic formula, is 36, which is a perfect square. If a, b, and c are rational numbers, all quadratic equations for which $b^2 - 4ac$ is a perfect square have rational solutions.

✔ **CHECK POINT 1** Solve using the quadratic formula: $2x^2 + 9x - 5 = 0$.

EXAMPLE 2 Solving a Quadratic Equation Using the Quadratic Formula

Solve using the quadratic formula:

$$2x^2 = 4x + 1.$$

SOLUTION The quadratic equation must be in standard form to identify the values for a, b, and c. To move all terms to one side and obtain zero on the right, we subtract $4x + 1$ from both sides. Then we can identify the values for a, b, and c.

$$2x^2 = 4x + 1$$ This is the given equation.

$$2x^2 - 4x - 1 = 0$$ Subtract $4x + 1$ from both sides.

$a = 2$ $b = -4$ $c = -1$

Substituting these values into the quadratic formula and simplifying gives the equation's solutions.

$$x = \frac{-b \pm \sqrt{b^2 - 4ac}}{2a}$$ Use the quadratic formula.

$$x = \frac{-(-4) \pm \sqrt{(-4)^2 - 4(2)(-1)}}{2(2)}$$ Substitute the values for a, b, and c: $a = 2$, $b = -4$, and $c = -1$.

$$= \frac{4 \pm \sqrt{16 - (-8)}}{4}$$ $(-4)^2 - 4(2)(-1) = 16 - (-8)$

$$= \frac{4 \pm \sqrt{24}}{4}$$ $16 - (-8) = 16 + 8 = 24$

$$= \frac{4 \pm 2\sqrt{6}}{4}$$ $\sqrt{24} = \sqrt{4 \cdot 6} = \sqrt{4} \cdot \sqrt{6} = 2\sqrt{6}$

$$= \frac{2(2 \pm \sqrt{6})}{4}$$ Factor out 2 from the numerator.

$$= \frac{2 \pm \sqrt{6}}{2}$$ Divide the numerator and denominator by 2.

The solutions are $\frac{2 \pm \sqrt{6}}{2}$, and the solution set is $\left\{\frac{2 + \sqrt{6}}{2}, \frac{2 - \sqrt{6}}{2}\right\}$ or $\left\{\frac{2 \pm \sqrt{6}}{2}\right\}$. ■

USING TECHNOLOGY

You can use a graphing utility to verify that the solutions of $2x^2 - 4x - 1 = 0$ are $\frac{2 \pm \sqrt{6}}{2}$. Begin by entering $y_1 = 2x^2 - 4x - 1$ in the [Y=] screen. Then evaluate this function at each of the proposed solutions.

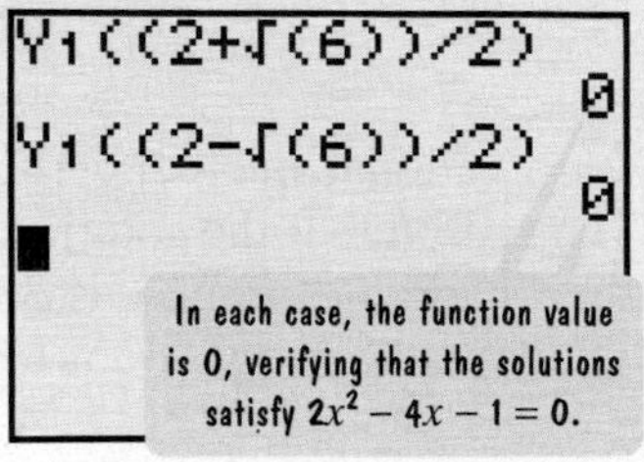
Y1((2+√(6))/2)
0
Y1((2−√(6))/2)
0
In each case, the function value is 0, verifying that the solutions satisfy 2x² − 4x − 1 = 0.

In Example 2, the solutions of $2x^2 = 4x + 1$ are irrational numbers. This means that the equation cannot be solved by factoring. The reason that the solutions are irrational numbers is that $b^2 - 4ac$, the radicand in the quadratic formula, is 24, which is not a perfect square. Notice, too, that the solutions, $\frac{2+\sqrt{6}}{2}$ and $\frac{2-\sqrt{6}}{2}$, are conjugates.

STUDY TIP

Many students use the quadratic formula correctly until the last step, where they make an error in simplifying the solutions. Be sure to factor the numerator before dividing the numerator and the denominator by the greatest common factor.

$$\frac{4 \pm 2\sqrt{6}}{4} = \frac{2\left(2 \pm \sqrt{6}\right)}{4} = \frac{\overset{1}{\cancel{2}}\left(2 \pm \sqrt{6}\right)}{\underset{2}{\cancel{4}}} = \frac{2 \pm \sqrt{6}}{2}$$

You cannot divide just one term in the numerator and the denominator by their greatest common factor.

Incorrect!

$$\frac{\overset{1}{\cancel{4}} \pm 2\sqrt{6}}{\underset{1}{\cancel{4}}} = 1 \pm 2\sqrt{6} \qquad \frac{4 \pm \overset{1}{\cancel{2}}\sqrt{6}}{\underset{2}{\cancel{4}}} = \frac{4 \pm \sqrt{6}}{2}$$

Can all irrational solutions of quadratic equations be simplified? No. The following solutions cannot be simplified:

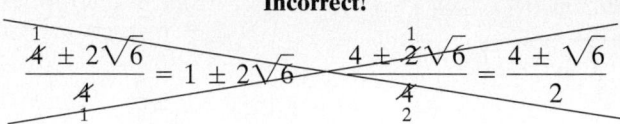

$$\frac{5 \pm 2\sqrt{7}}{2} \quad \text{and} \quad \frac{-4 \pm 3\sqrt{7}}{2}.$$

Other than 1, terms in each numerator have no common factor.

✔ CHECK POINT 2 Solve using the quadratic formula: $2x^2 = 6x - 1$.

EXAMPLE 3 Solving a Quadratic Equation Using the Quadratic Formula

Solve using the quadratic formula:

$$3x^2 + 2 = -4x.$$

SOLUTION Begin by writing the quadratic equation in standard form.

$$3x^2 + 2 = -4x \qquad \text{This is the given equation.}$$

$$3x^2 + 4x + 2 = 0 \qquad \text{Add } 4x \text{ to both sides.}$$

$a = 3$ $b = 4$ $c = 2$

Substituting these values into the quadratic formula and simplifying gives the equation's solutions.

$$x = \frac{-b \pm \sqrt{b^2 - 4ac}}{2a}$$ Use the quadratic formula.

$$x = \frac{-4 \pm \sqrt{4^2 - 4 \cdot 3 \cdot 2}}{2 \cdot 3}$$ Substitute the values for a, b, and c: $a = 3$, $b = 4$, and $c = 2$.

$$= \frac{-4 \pm \sqrt{16 - 24}}{6}$$ Multiply under the radical.

$$= \frac{-4 \pm \sqrt{-8}}{6}$$ Subtract under the radical.

$$= \frac{-4 \pm 2\sqrt{2}i}{6}$$ $\sqrt{-8} = \sqrt{8(-1)} = \sqrt{8}\sqrt{-1} = \sqrt{8}i = \sqrt{4 \cdot 2}i = 2\sqrt{2}i$

$$= \frac{2(-2 \pm \sqrt{2}i)}{6}$$ Factor out 2 from the numerator.

$$= \frac{-2 \pm \sqrt{2}i}{3}$$ Divide the numerator and denominator by 2.

$$= -\frac{2}{3} \pm \frac{\sqrt{2}}{3}i$$ Write in the form $a + bi$.

The solutions are $-\frac{2}{3} \pm \frac{\sqrt{2}}{3}i$, and the solution set is $\left\{-\frac{2}{3} + \frac{\sqrt{2}}{3}i, -\frac{2}{3} - \frac{\sqrt{2}}{3}i\right\}$ or $\left\{-\frac{2}{3} \pm \frac{\sqrt{2}}{3}i\right\}$. ■

USING TECHNOLOGY

The graph of the quadratic function

$$y = 3x^2 + 4x + 2$$

has no x-intercepts. This verifies that the equation in Example 3

$$3x^2 + 2 = -4x, \quad \text{or}$$
$$3x^2 + 4x + 2 = 0$$

has imaginary solutions.

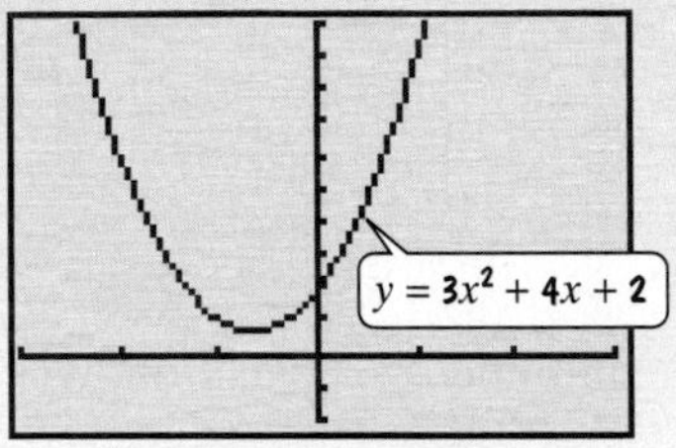

[−3, 3, 1] by [−2, 10, 1]

In Example 3, the solutions of $3x^2 + 2 = -4x$ are imaginary numbers. This means that the equation cannot be solved using factoring. The reason that the solutions are imaginary numbers is that $b^2 - 4ac$, the radicand in the quadratic formula, is -8, which is negative. Notice, too, that the solutions are complex conjugates.

CHECK POINT 3 Solve using the quadratic formula: $3x^2 + 5 = -6x$.

Some rational equations can be solved using the quadratic formula. For example, consider the equation

$$3 + \frac{4}{x} = -\frac{2}{x^2}.$$

The denominators are x and x^2. The least common denominator is x^2. We clear fractions by multiplying both sides of the equation by x^2. Notice that x cannot equal zero.

$$x^2\left(3 + \frac{4}{x}\right) = x^2\left(-\frac{2}{x^2}\right), \quad x \neq 0$$

$$3x^2 + \frac{4}{x} \cdot x^2 = x^2\left(-\frac{2}{x^2}\right)$$ Use the distributive property.

$$3x^2 + 4x = -2$$ Simplify.

By adding 2 to both sides of $3x^2 + 4x = -2$, we obtain the standard form of the quadratic equation:

$$3x^2 + 4x + 2 = 0.$$

This is the equation that we solved in Example 3. The two imaginary solutions are not part of the restriction that $x \neq 0$.

2 Use the discriminant to determine the number and type of solutions.

The Discriminant The quantity $b^2 - 4ac$, which appears under the radical sign in the quadratic formula, is called the **discriminant**. The discriminant of the quadratic equation $ax^2 + bx + c = 0$ determines the number and type of solutions. In Example 1, the discriminant was 36, a positive number that is a perfect square. The equation had two solutions that were rational numbers. In Example 2, the discriminant was 24, a positive number that is not a perfect square. The equation had two solutions that were irrational numbers. Finally, in Example 3, the discriminant was -8, a negative number. The equation had solutions that were imaginary numbers involving i. In this case, the graph of the corresponding quadratic function had no x-intercepts.

These observations are generalized in Table 11.2.

STUDY TIP

Checking irrational and imaginary solutions can be time-consuming. The solutions given by the quadratic formula are always correct, unless you have made a careless error. Checking for computational errors or errors in simplification is sufficient.

Table 11.2 The Discriminant and the Kinds of Solutions to $ax^2 + bx + c = 0$

Discriminant $b^2 - 4ac$	Kinds of Solutions to $ax^2 + bx + c = 0$	Graph of $y = ax^2 + bx + c$
$b^2 - 4ac > 0$	**Two unequal real solutions;** If a, b, and c are rational numbers and the discriminant is a perfect square, the solutions are rational. If the discriminant is not a perfect square, the solutions are irrational conjugates.	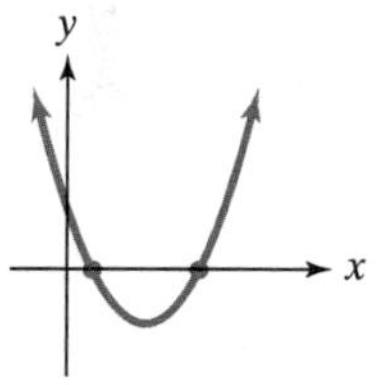 Two x-intercepts
$b^2 - 4ac = 0$	**One real solution (a repeated solution);** If a, b, and c are rational numbers, the repeated solution is also a rational number.	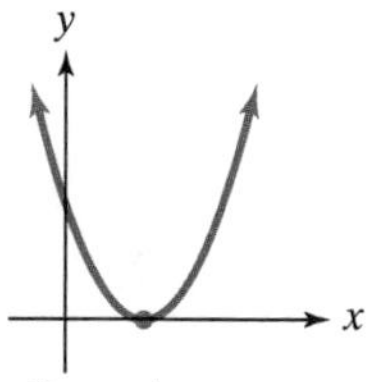 One x-intercept
$b^2 - 4ac < 0$	**No real solution; two imaginary solutions;** The solutions are complex conjugates.	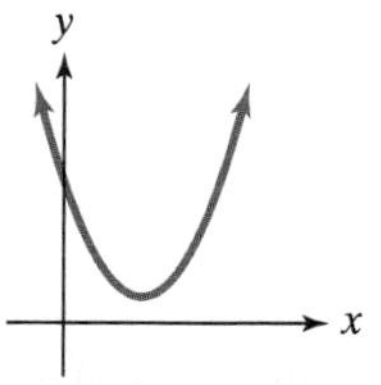 No x-intercepts

EXAMPLE 4 Using the Discriminant

For each equation, compute the discriminant. Then determine the number and type of solutions:

a. $3x^2 + 4x - 5 = 0$ **b.** $9x^2 - 6x + 1 = 0$ **c.** $3x^2 - 8x + 7 = 0$.

SOLUTION Begin by identifying the values for a, b, and c in each equation. Then compute $b^2 - 4ac$, the discriminant.

a. $3x^2 + 4x - 5 = 0$

$a = 3$ $b = 4$ $c = -5$

Substitute and compute the discriminant:

$$b^2 - 4ac = 4^2 - 4 \cdot 3(-5) = 16 - (-60) = 16 + 60 = 76.$$

The discriminant, 76, is a positive number that is not a perfect square. Thus, there are two irrational solutions. (These solutions are conjugates of each other.)

b. $9x^2 - 6x + 1 = 0$

$a = 9$ $b = -6$ $c = 1$

Substitute and compute the discriminant:

$$b^2 - 4ac = (-6)^2 - 4 \cdot 9 \cdot 1 = 36 - 36 = 0.$$

The discriminant, 0, shows that there is only one real solution. This real solution is a rational number.

c. $3x^2 - 8x + 7 = 0$

$a = 3$ $b = -8$ $c = 7$

$$b^2 - 4ac = (-8)^2 - 4 \cdot 3 \cdot 7 = 64 - 84 = -20$$

The negative discriminant, −20, shows that there are two imaginary solutions. (These solutions are complex conjugates of each other.) ■

✔ **CHECK POINT 4** For each equation, compute the discriminant. Then determine the number and type of solutions:

a. $x^2 + 6x + 9 = 0$

b. $2x^2 - 7x - 4 = 0$

c. $3x^2 - 2x + 4 = 0.$

3 Determine the most efficient method to use when solving a quadratic equation.

Determining Which Method to Use All quadratic equations can be solved by the quadratic formula. However, if an equation is in the form $u^2 = d$, such as $x^2 = 5$ or $(2x + 3)^2 = 8$, it is faster to use the square root property, taking the square root of both sides. If the equation is not in the form $u^2 = d$, write the quadratic equation in standard form $(ax^2 + bx + c = 0)$. Try to solve the equation by factoring. If $ax^2 + bx + c$ cannot be factored, then solve the quadratic equation by the quadratic formula.

Because we used the method of completing the square to derive the quadratic formula, we no longer need it for solving quadratic equations. However, we will use completing the square in Chapter 13 to help graph certain kinds of equations.

Table 11.3 summarizes our observations about which technique to use when solving a quadratic equation.

Table 11.3 Determining the Most Efficient Technique to Use when Solving a Quadratic Equation

Description and Form of the Quadratic Equation	Most Efficient Solution Method	Example
$ax^2 + bx + c = 0$ and $ax^2 + bx + c$ can be factored easily.	Factor and use the zero-product principle.	$3x^2 + 5x - 2 = 0$ $(3x - 1)(x + 2) = 0$ $3x - 1 = 0$ or $x + 2 = 0$ $x = \frac{1}{3}$ $\quad x = -2$
$ax^2 + c = 0$ The quadratic equation has no x-term. ($b = 0$)	Solve for x^2 and apply the square root property.	$4x^2 - 7 = 0$ $4x^2 = 7$ $x^2 = \frac{7}{4}$ $x = \pm\frac{\sqrt{7}}{2}$
$u^2 = d$; u is a first-degree polynomial.	Use the square root property.	$(x + 4)^2 = 5$ $x + 4 = \pm\sqrt{5}$ $x = -4 \pm \sqrt{5}$
$ax^2 + bx + c = 0$ and $ax^2 + bx + c$ cannot be factored or the factoring is too difficult.	Use the quadratic formula: $x = \frac{-b \pm \sqrt{b^2 - 4ac}}{2a}$.	$x^2 - 2x - 6 = 0$ $a = 1$ $\quad b = -2$ $\quad c = -6$ $x = \frac{-(-2) \pm \sqrt{(-2)^2 - 4(1)(-6)}}{2(1)}$ $= \frac{2 \pm \sqrt{4 - 4(1)(-6)}}{2(1)}$ $= \frac{2 \pm \sqrt{28}}{2} = \frac{2 \pm \sqrt{4}\sqrt{7}}{2}$ $= \frac{2 \pm 2\sqrt{7}}{2} = \frac{2(1 \pm \sqrt{7})}{2}$ $= 1 \pm \sqrt{7}$

4 Write quadratic equations from solutions.

Writing Quadratic Equations from Solutions Using the zero-product principle, the equation $(x - 3)(x + 5) = 0$ has two solutions, 3 and -5. By applying the zero-product principle in reverse, we can find a quadratic equation that has two given numbers as its solutions.

THE ZERO-PRODUCT PRINCIPLE IN REVERSE If $A = 0$ or $B = 0$, then $AB = 0$.

Jasper Johns, "Zero". © Jasper Johns/Licensed by VAGA, New York, NY.

The special properties of zero make it possible to write a quadratic equation from its solutions.

EXAMPLE 5 Writing Equations from Solutions

Write a quadratic equation with the given solution set:

a. $\left\{-\frac{5}{3}, \frac{1}{2}\right\}$ **b.** $\{-5i, 5i\}$.

SOLUTION

a. Because the solution set is $\left\{-\frac{5}{3}, \frac{1}{2}\right\}$, then

$$x = -\frac{5}{3} \quad \text{or} \quad x = \frac{1}{2}.$$

$$x + \frac{5}{3} = 0 \quad \text{or} \quad x - \frac{1}{2} = 0$$ Obtain zero on one side of each equation.

$$3x + 5 = 0 \quad \text{or} \quad 2x - 1 = 0$$ Clear fractions, multiplying by 3 and 2, respectively.

$$(3x + 5)(2x - 1) = 0$$ Use the zero-product principle in reverse: If A = 0 or B = 0, then AB = 0.

$$6x^2 - 3x + 10x - 5 = 0$$ Use the FOIL method to multiply.

$$6x^2 + 7x - 5 = 0.$$ Combine like terms.

Thus, one equation is $6x^2 + 7x - 5 = 0$. Many other quadratic equations have $\left\{-\frac{5}{3}, \frac{1}{2}\right\}$ for their solution sets. These equations can be obtained by multiplying both sides of $6x^2 + 7x - 5 = 0$ by any nonzero real number.

b. Because the solution set is $\{-5i, 5i\}$, then

$$x = -5i \quad \text{or} \quad x = 5i$$

$$x + 5i = 0 \quad \text{or} \quad x - 5i = 0$$ Obtain zero on one side of each equation.

$$(x + 5i)(x - 5i) = 0$$ Use the zero-product principle in reverse: If A = 0 or B = 0, then AB = 0.

$$x^2 - (5i)^2 = 0$$ Multiply conjugates using $(A + B)(A - B) = A^2 - B^2$.

$$x^2 - 25i^2 = 0$$ $(5i)^2 = 5^2i^2 = 25i^2$

$$x^2 - 25(-1) = 0$$ $i^2 = -1$

$$x^2 + 25 = 0.$$ This is the required equation. ■

CHECK POINT 5 Write a quadratic equation with the given solution set:

a. $\left\{-\frac{3}{5}, \frac{1}{4}\right\}$ **b.** $\{-7i, 7i\}$.

5 Use the quadratic formula to solve problems.

Applications Quadratic equations can be solved to answer questions about variables contained in quadratic functions.

EXAMPLE 6 Making Predictions about the U.S. Foreign-Born Population

The quadratic function

$$f(x) = 0.0049x^2 - 0.361x + 11.79$$

models the percentage of the U.S. population, $f(x)$, that was foreign-born x years after 1930. According to this model, in which year will 15% of the U.S. population be foreign-born?

SOLUTION Because we are interested in when the foreign-born percentage will reach 15%, we substitute 15 for $f(x)$ in the given function. Then we solve for x, the number of years after 1930.

$$f(x) = 0.0049x^2 - 0.361x + 11.79$$ This is the given function.

$$15 = 0.0049x^2 - 0.361x + 11.79$$ Substitute 15 for $f(x)$.

$$0 = 0.0049x^2 - 0.361x - 3.21$$ Subtract 15 from both sides and write the quadratic equation in standard form.

$a = 0.0049$ $b = -0.361$ $c = -3.21$

Because the trinomial on the right side of the equation is not easily factored, if it can be factored at all, we solve using the quadratic formula.

$$x = \frac{-b \pm \sqrt{b^2 - 4ac}}{2a}$$ Use the quadratic formula.

$$= \frac{-(-0.361) \pm \sqrt{(-0.361)^2 - 4(0.0049)(-3.21)}}{2(0.0049)}$$ Substitute the values for a, b, and c: $a = 0.0049$, $b = -0.361$, $c = -3.21$.

$$= \frac{0.361 \pm \sqrt{0.193237}}{0.0098}$$ Perform the indicated computations. Use a calculator to simplify the radicand.

$$x = \frac{0.361 + \sqrt{0.193237}}{0.0098} \quad \text{or} \quad x = \frac{0.361 - \sqrt{0.193237}}{0.0098}$$

$$x \approx 82 \qquad\qquad x \approx -8$$

Reject this solution. The model applies to years **after**, not **before**, 1930.

According to the model, approximately 82 years after 1930, in 2012, 15% of the U.S. population will be foreign-born. ■

USING TECHNOLOGY

A graphing utility's TABLE feature can be used to verify the solution to Example 6.

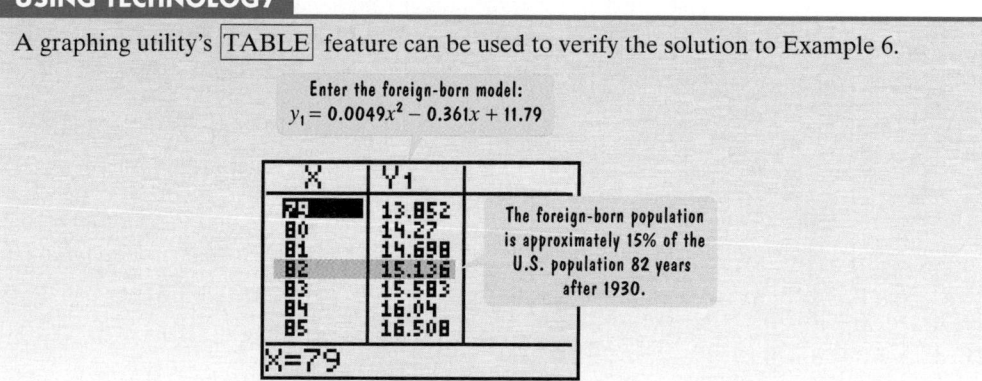

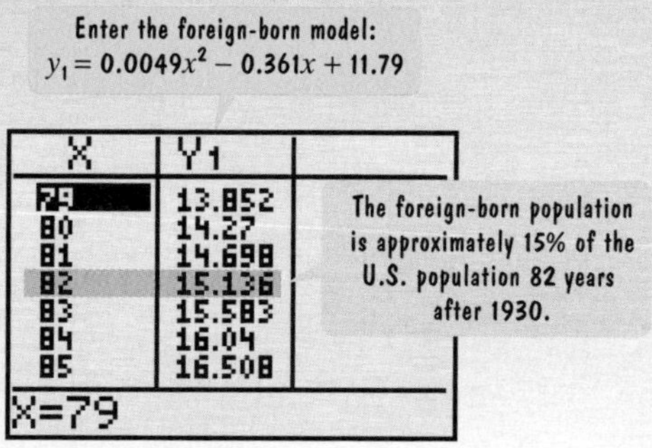

CHECK POINT 6 According to the model in Example 6, in which year(s) was 5.4% of the U.S. population foreign-born? How well does the function describe the situation for that (these) year(s) shown in Figure 11.7 on page 735?

11.2 EXERCISE SET

Student Solutions Manual | CD/Video | PH Math/Tutor Center | MathXL Tutorials on CD | MathXL® | MyMathLab | Interactmath.com

Practice Exercises

In Exercises 1–18, solve each equation using the quadratic formula. Simplify solutions, if possible.

1. $x^2 + 8x + 12 = 0$

2. $x^2 + 8x + 15 = 0$

3. $2x^2 - 7x = -5$

4. $5x^2 + 8x = -3$

5. $x^2 + 3x - 20 = 0$

6. $x^2 + 5x - 10 = 0$

7. $3x^2 - 7x = 3$

8. $4x^2 + 3x = 2$

9. $6x^2 = 2x + 1$

10. $2x^2 = -4x + 5$

11. $4x^2 - 3x = -6$

12. $9x^2 + x = -2$

13. $x^2 - 4x + 8 = 0$

14. $x^2 + 6x + 13 = 0$

15. $3x^2 = 8x - 7$

16. $3x^2 = 4x - 6$

17. $2x(x - 2) = x + 12$

18. $2x(x + 4) = 3x - 3$

In Exercises 19–30, compute the discriminant. Then determine the number and type of solutions for the given equation.

19. $x^2 + 8x + 3 = 0$

20. $x^2 + 7x + 4 = 0$

21. $x^2 + 6x + 8 = 0$

22. $x^2 + 2x - 3 = 0$

23. $2x^2 + x + 3 = 0$

24. $2x^2 - 4x + 3 = 0$

25. $2x^2 + 6x = 0$

26. $3x^2 - 5x = 0$

27. $5x^2 + 3 = 0$

28. $5x^2 + 4 = 0$

29. $9x^2 = 12x - 4$

30. $4x^2 = 20x - 25$

In Exercises 31–46, solve each equation by the method of your choice. Simplify solutions, if possible.

31. $3x^2 - 4x = 4$

32. $2x^2 - x = 1$

33. $x^2 - 2x = 1$

34. $2x^2 + 3x = 1$

35. $(2x - 5)(x + 1) = 2$

36. $(2x + 3)(x + 4) = 1$

37. $(3x - 4)^2 = 16$

38. $(2x + 7)^2 = 25$

39. $\dfrac{x^2}{2} + 2x + \dfrac{2}{3} = 0$

40. $\dfrac{x^2}{3} - x - \dfrac{1}{6} = 0$

41. $(3x - 2)^2 = 10$

42. $(4x - 1)^2 = 15$

43. $\dfrac{1}{x} + \dfrac{1}{x + 2} = \dfrac{1}{3}$

44. $\dfrac{1}{x} + \dfrac{1}{x + 3} = \dfrac{1}{4}$

45. $(2x - 6)(x + 2) = 5(x - 1) - 12$

46. $7x(x - 2) = 3 - 2(x + 4)$

In Exercises 47–60, write a quadratic equation in standard form with the given solution set.

47. $\{-3, 5\}$

48. $\{-2, 6\}$

49. $\left\{-\dfrac{2}{3}, \dfrac{1}{4}\right\}$

50. $\left\{-\dfrac{5}{6}, \dfrac{1}{3}\right\}$

51. $\{-6i, 6i\}$

52. $\{-8i, 8i\}$

53. $\{-\sqrt{2}, \sqrt{2}\}$

54. $\{-\sqrt{3}, \sqrt{3}\}$

55. $\{-2\sqrt{5}, 2\sqrt{5}\}$

56. $\{-3\sqrt{5}, 3\sqrt{5}\}$

57. $\{1 + i, 1 - i\}$

58. $\{2 + i, 2 - i\}$

59. $\{1 + \sqrt{2}, 1 - \sqrt{2}\}$

60. $\{1 + \sqrt{3}, 1 - \sqrt{3}\}$

Practice Plus

Exercises 61–64 describe quadratic equations. Match each description with the graph of the corresponding qudratic function. Each graph is shown in a $[-10, 10, 1]$ *by* $[-10, 10, 1]$ *viewing rectangle.*

61. A quadratic equation whose solution set contains imaginary numbers

62. A quadratic equation whose discriminant is 0

63. A quadratic equation whose solution set is $\{3 \pm \sqrt{2}\}$

64. A quadratic equation whose solution set contains integers

a.

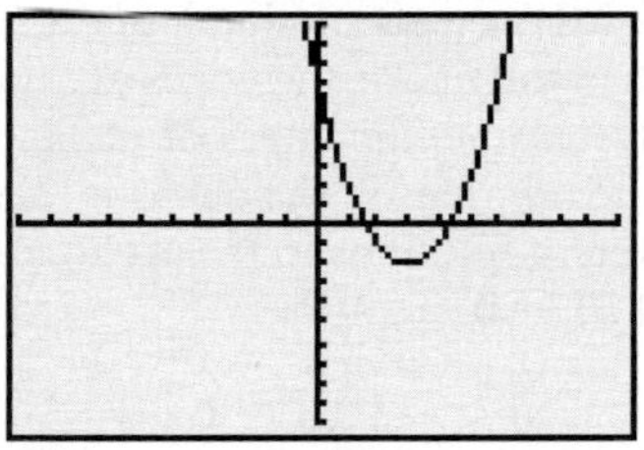

b.

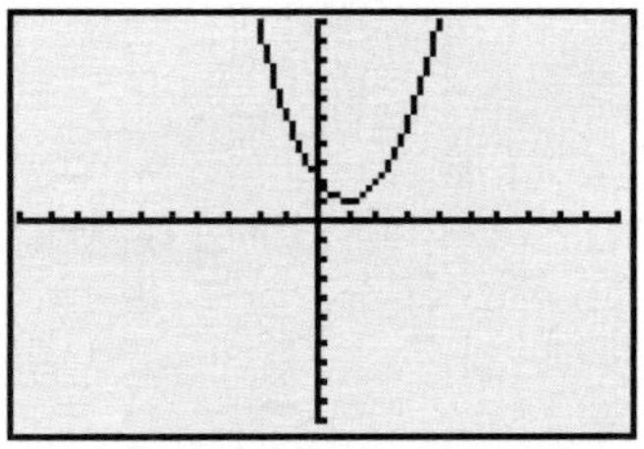

c.

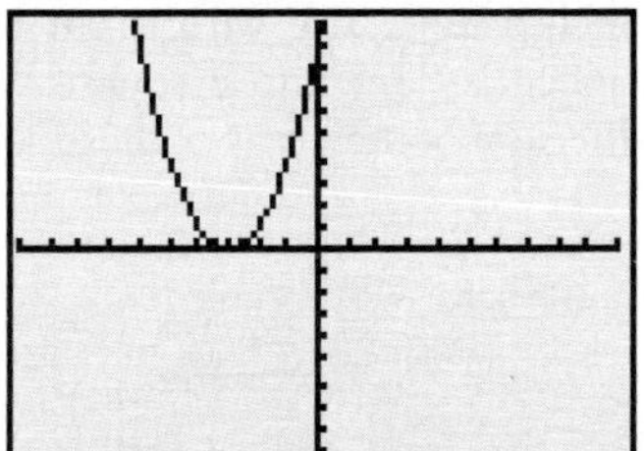

d.

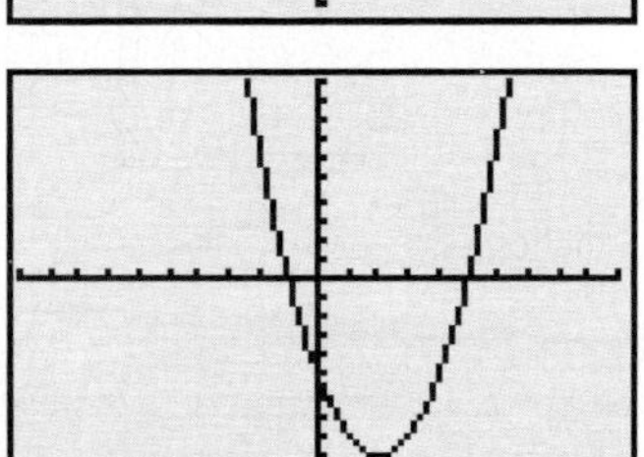

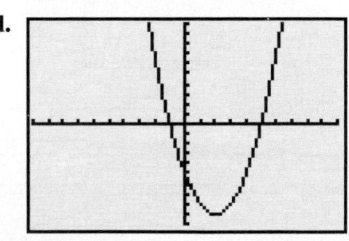

65. When the sum of 6 and twice a positive number is subtracted from the square of the number, 0 results. Find the number.

66. When the sum of 1 and twice a negative number is subtracted from twice the square of the number, 0 results. Find the number.

In Exercises 67–72, solve each equation by the method of your choice.

67. $\dfrac{1}{x^2 - 3x + 2} = \dfrac{1}{x + 2} + \dfrac{5}{x^2 - 4}$

68. $\dfrac{x - 1}{x - 2} + \dfrac{x}{x - 3} = \dfrac{1}{x^2 - 5x + 6}$

69. $\sqrt{2}x^2 + 3x - 2\sqrt{2} = 0$

70. $\sqrt{3}x^2 + 6x + 7\sqrt{3} = 0$

71. $|x^2 + 2x| = 3$

72. $|x^2 + 3x| = 2$

Application Exercises

A driver's age has something to do with his or her chance of getting into a fatal car crash. The bar graph shows the number of fatal vehicle crashes per 100 million miles driven for drivers of various age groups. For example, 25-year-old drivers are involved in 4.1 fatal crashes per 100 million miles driven. Thus, when a group of 25-year-old Americans have driven a total of 100 million miles, approximately 4 have been in accidents in which someone died.

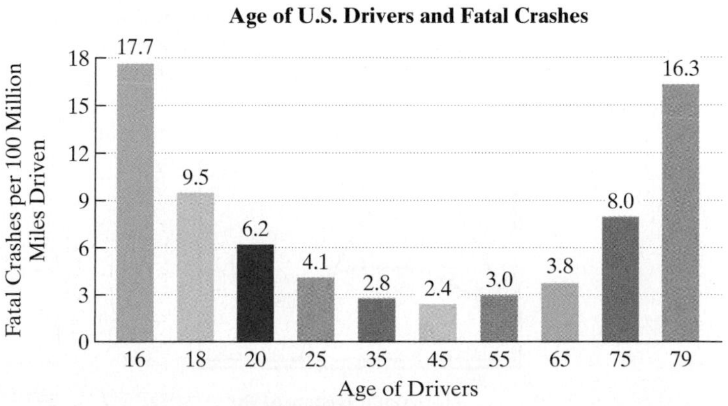

Source: Insurance Institute for Highway Safety

The number of fatal vehicle crashes per 100 million miles, $f(x)$, for drivers of age x can be modeled by the quadratic function

$$f(x) = 0.013x^2 - 1.19x + 28.24.$$

Use the function to solve Exercises 73–74.

73. What age groups are expected to be involved in 3 fatal crashes per 100 million miles driven? How well does the function model the trend in the actual data shown in the bar graph on the previous page?

74. What age groups are expected to be involved in 10 fatal crashes per 100 million miles driven? How well does the function model the trend in the actual data shown in the bar graph on the previous page?

Throwing events in track and field include the shot put, the discus throw, the hammer throw, and the javelin throw. The distance that an athlete can achieve depends on the initial velocity of the object thrown and the angle above the horizontal at which the object leaves the hand.

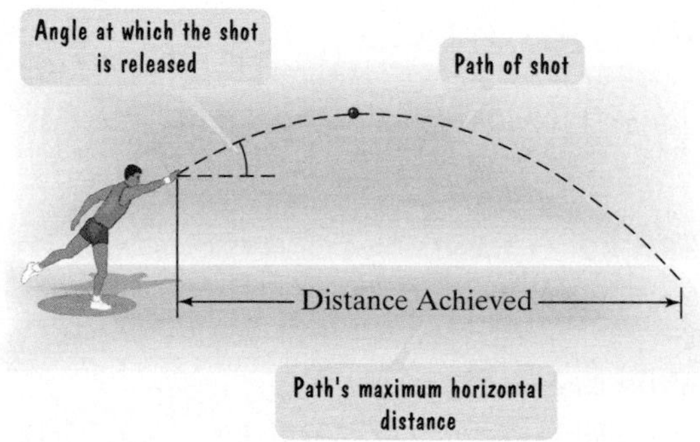

In Exercises 75–76, an athlete whose event is the shot put releases the shot with the same initial velocity, but at different angles.

75. When the shot is released at an angle of 35°, its path can be modeled by the function

$$f(x) = -0.01x^2 + 0.7x + 6.1$$

in which x is the shot's horizontal distance, in feet, and $f(x)$ is its height, in feet. This function is shown by one of the graphs, (a) or (b), in the figure. Use the function to determine the shot's maximum distance. Use a calculator and round to the nearest tenth of a foot. Which graph, (a) or (b), shows the shot's path?

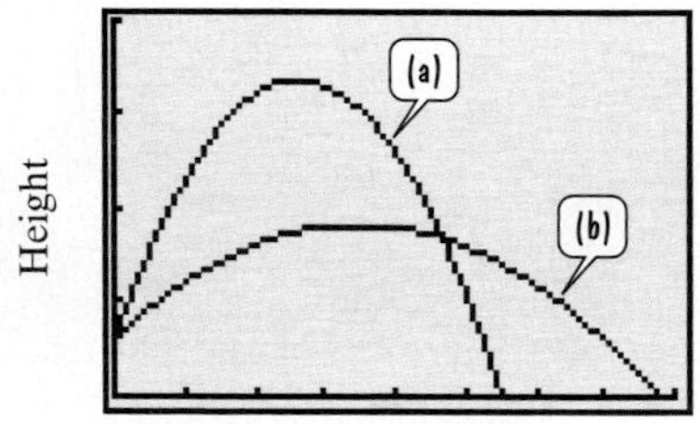

Horizontal Distance
[0, 80, 10] by [0, 40, 10]

76. When the shot is released at an angle of 65°, its path can be modeled by the function

$$f(x) = -0.04x^2 + 2.1x + 6.1$$

in which x is the shot's horizontal distance, in feet, and $f(x)$ is its height, in feet. This function is shown by one of the graphs, (a) or (b), in the figure at the bottom of the previous column. Use the function to determine the shot's maximum distance. Use a calculator and round to the nearest tenth of a foot. Which graph, (a) or (b), shows the shot's path?

77. The length of a rectangle is 4 meters longer than the width. If the area is 8 square meters, find the rectangle's dimensions. Round to the nearest tenth of a meter.

78. The length of a rectangle exceeds twice its width by 3 inches. If the area is 10 square inches, find the rectangle's dimensions. Round to the nearest tenth of an inch.

79. One leg of a right triangle exceeds the shorter leg by 1 inch, and the hypotenuse exceeds the longer leg by 7 inches. Find the lengths of the legs. Round to the nearest tenth of a inch.

80. The hypotenuse of a right triangle is 6 feet long. One leg is 2 feet shorter than the other. Find the lengths of the legs. Round to the nearest tenth of a foot.

81. A rain gutter is made from sheets of aluminum that are 20 inches wide. As shown in the figure, the edges are turned up to form right angles. Determine the depth of the gutter that will allow a cross-sectional area of 13 square inches. Show that there are two different solutions to the problem. Round to the nearest tenth of an inch.

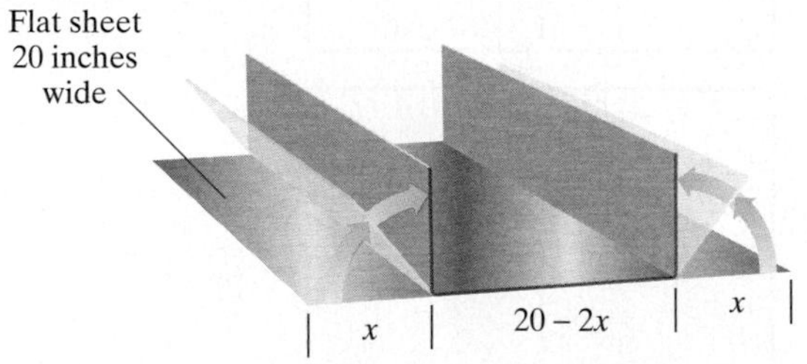

82. A piece of wire is 8 inches long. The wire is cut into two pieces and then each piece is bent into a square. Find the length of each piece if the sum of the areas of these squares is to be 2 square inches.

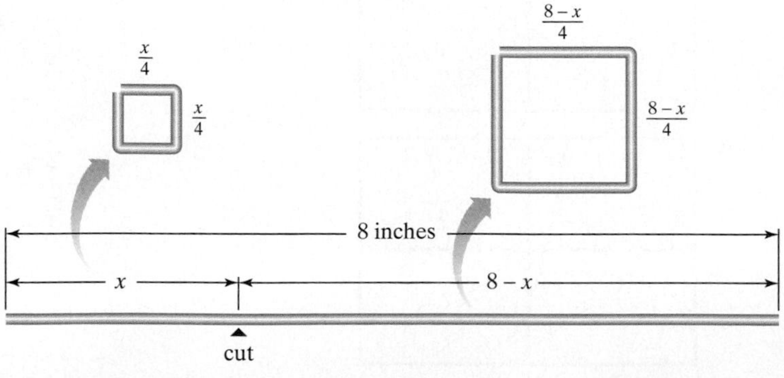

83. Working together, two people can mow a large lawn in 4 hours. One person can do the job alone 1 hour faster than the other person. How long does it take each person working alone to mow the lawn? Round to the nearest tenth of an hour.

84. A pool has an inlet pipe to fill it and an outlet pipe to empty it. It takes 2 hours longer to empty the pool than it does to fill it. The inlet pipe is turned on to fill the pool, but the outlet pipe is accidentally open. Despite this, the pool fills in 8 hours. How long does it take the outlet pipe to empty the pool? Round to the nearest tenth of an hour.

Writing in Mathematics

85. What is the quadratic formula and why is it useful?

86. Without going into specific details for every step, describe how the quadratic formula is derived.

87. Explain how to solve $x^2 + 6x + 8 = 0$ using the quadratic formula.

88. If a quadratic equation has imaginary solutions, how is this shown on the graph of the corresponding quadratic function?

89. What is the discriminant and what information does it provide about a quadratic equation?

90. If you are given a quadratic equation, how do you determine which method to use to solve it?

91. Explain how to write a quadratic equation from its solution set. Give an example with your explanation.

92. Describe the trend shown by the data for the percentage of the U.S. population that was foreign-born from 1930 through 2002. (See Figure 11.7 on page 735.) Do you believe that this trend is likely to continue or might something occur that would make it impossible to extend the model in Example 6 on page 744 into the future? Explain your answer.

Technology Exercises

93. Use a graphing utility to graph the quadratic function related to any five of the quadratic equations in Exercises 19–30. How does each graph illustrate what you determined algebraically using the discriminant?

94. Reread Exercise 81. The cross-sectional area of the gutter is given by the quadratic function

$$f(x) = x(20 - 2x).$$

Graph the function in a $[0, 10, 1]$ by $[0, 60, 5]$ viewing rectangle. Then TRACE along the curve or use the maximum function feature to determine the depth of the gutter that will maximize its cross-sectional area and allow the greatest amount of water to flow. What is the maximum area? Does the situation described in Exercise 81 take full advantage of the sheets of aluminum?

Critical Thinking Exercises

95. Which one of the following is true?

a. The quadratic formula is developed by applying factoring and the zero-product principle to the quadratic equation $ax^2 + bx + c = 0$.

b. In using the quadratic formula to solve the quadratic equation $5x^2 = 2x - 7$, we have $a = 5, b = 2$, and $c = -7$.

c. The quadratic formula cannot be used to solve the equation $x^2 - 9 = 0$.

d. Any quadratic equation that can be solved by completing the square can be solved by the quadratic formula.

96. Solve for t: $s = -16t^2 + v_0t$.

97. A rectangular swimming pool is 12 meters long and 8 meters wide. A tile border of uniform width is to be built around the pool using 120 square meters of tile. The tile is from a discontinued stock (so no additional materials are available) and all 120 square meters are to be used. How wide should the border be? Round to the nearest tenth of a meter. If zoning laws require at least a 2-meter-wide border around the pool, can this be done with the available tile?

98. The area of the shaded region outside the rectangle and inside the triangle is 10 square yards. Find the triangle's height, represented by $2x$. Round to the nearest tenth of a yard.

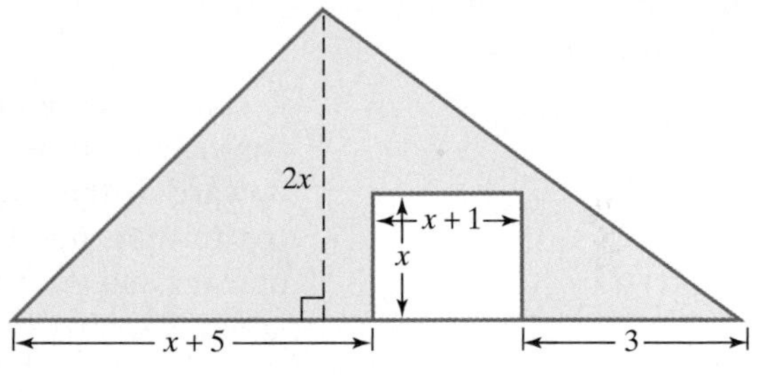

Review Exercises

99. Solve: $|5x + 2| = |4 - 3x|$. (Section 9.3, Example 2)

100. Solve: $\sqrt{2x - 5} - \sqrt{x - 3} = 1$. (Section 10.6, Example 4)

101. Rationalize the denominator: $\dfrac{5}{\sqrt{3} + x}$.

(Section 10.5, Example 5)

SECTION 11.3

Objectives

1. Recognize characteristics of parabolas.
2. Graph parabolas in the form $f(x) = a(x - h)^2 + k$.
3. Graph parabolas in the form $f(x) = ax^2 + bx + c$.
4. Solve problems involving minimizing or maximizing quadratic functions.

QUADRATIC FUNCTIONS AND THEIR GRAPHS

We have a long history of throwing things. Before 400 B.C., the Greeks competed in games that included discus throwing. In the seventeenth century, English soldiers organized cannonball-throwing competitions. In 1827, a Yale University student, disappointed over failing an exam, took out his frustrations at the passing of a collection plate in chapel. Seizing the monetary tray, he flung it in the direction of a large open space on campus. Yale students see this act of frustration as the origin of the Frisbee.

In this section, we study quadratic functions and their graphs. By graphing functions that model the paths of the things we throw, you will be able to determine both the maximum height and the distance of these objects.

1 Recognize characteristics of parabolas.

Graphs of Quadratic Functions The graph of the quadratic function

$$f(x) = ax^2 + bx + c, \quad a \neq 0,$$

is called a **parabola**. Parabolas are shaped like cups, as shown in Figure 11.8. If the coefficient of x^2 (the value of a in $ax^2 + bx + c$) is positive, the parabola opens upward. If the coefficient of x^2 is negative, the parabola opens downward. The **vertex** (or turning point) of the parabola is the lowest point on the graph when it opens upward and the highest point on the graph when it opens downward.

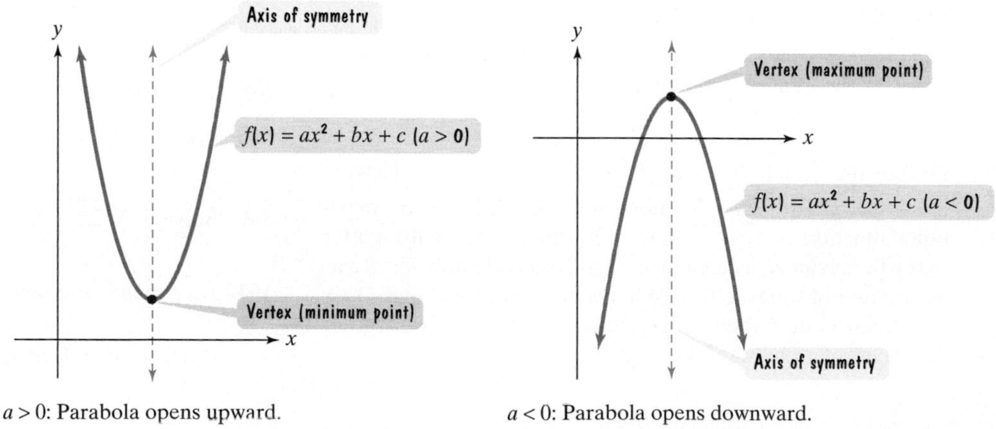

FIGURE 11.8 Characteristics of graphs of quadratic functions

Look at the unusual image of the word "mirror" shown below. The artist, Scott Kim, has created the image so that the two halves of the whole are mirror images of each other. A parabola shares this kind of symmetry, in which a line through the vertex divides the figure in half. Parabolas are symmetric with respect to this line, called the **axis of symmetry**. If a parabola is folded along its axis of symmetry, the two halves match exactly.

2 Graph parabolas in the form $f(x) = a(x - h)^2 + k$.

Graphing Quadratic Functions in the Form $f(x) = a(x - h)^2 + k$ One way to obtain the graph of a quadratic function is to use point plotting. Let's begin by graphing the functions $f(x) = x^2$, $g(x) = 2x^2$, and $h(x) = \frac{1}{2}x^2$ in the same rectangular coordinate system. Select integers for x, starting with -3 and ending with 3. A partial table of coordinates for each function is shown below. The three parabolas are shown in Figure 11.9.

x	$f(x) = x^2$	(x, y) or $(x, f(x))$
-3	$f(-3) = (-3)^2 = 9$	$(-3, 9)$
-2	$f(-2) = (-2)^2 = 4$	$(-2, 4)$
-1	$f(-1) = (-1)^2 = 1$	$(-1, 1)$
0	$f(0) = 0^2 = 0$	$(0, 0)$
1	$f(1) = 1^2 = 1$	$(1, 1)$
2	$f(2) = 2^2 = 4$	$(2, 4)$
3	$f(3) = 3^2 = 9$	$(3, 9)$

x	$g(x) = 2x^2$	(x, y) or $(x, g(x))$
-3	$g(-3) = 2(-3)^2 = 18$	$(-3, 18)$
-2	$g(-2) = 2(-2)^2 = 8$	$(-2, 8)$
-1	$g(-1) = 2(-1)^2 = 2$	$(-1, 2)$
0	$g(0) = 2 \cdot 0^2 = 0$	$(0, 0)$
1	$g(1) = 2 \cdot 1^2 = 2$	$(1, 2)$
2	$g(2) = 2 \cdot 2^2 = 8$	$(2, 8)$
3	$g(3) = 2 \cdot 3^2 = 18$	$(3, 18)$

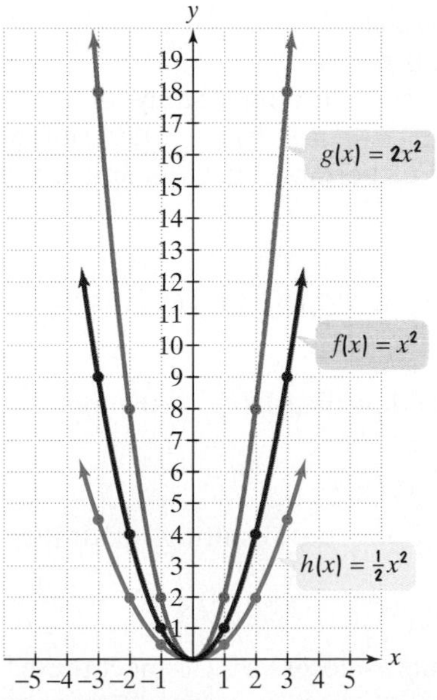

FIGURE 11.9

x	$h(x) = \frac{1}{2}x^2$	(x, y) or $(x, h(x))$
-3	$h(-3) = \frac{1}{2}(-3)^2 = \frac{9}{2}$	$\left(-3, \frac{9}{2}\right)$
-2	$h(-2) = \frac{1}{2}(-2)^2 = 2$	$(-2, 2)$
-1	$h(-1) = \frac{1}{2}(-1)^2 = \frac{1}{2}$	$\left(-1, \frac{1}{2}\right)$
0	$h(0) = \frac{1}{2} \cdot 0^2 = 0$	$(0, 0)$
1	$h(1) = \frac{1}{2} \cdot 1^2 = \frac{1}{2}$	$\left(1, \frac{1}{2}\right)$
2	$h(2) = \frac{1}{2} \cdot 2^2 = 2$	$(2, 2)$
3	$h(3) = \frac{1}{2} \cdot 3^2 = \frac{9}{2}$	$\left(3, \frac{9}{2}\right)$

Can you see that the graphs of f, g, and h all have the same vertex, $(0, 0)$? They also have the same axis of symmetry, the y-axis, or $x = 0$. This is true for all graphs of the form $f(x) = ax^2$. However, the blue graph of $g(x) = 2x^2$ is a narrower parabola than the red graph of $f(x) = x^2$. By contrast, the green graph of $h(x) = \frac{1}{2}x^2$ is a flatter parabola than the red graph of $f(x) = x^2$.

Is there a more efficient method than point plotting to obtain the graph of a quadratic function? The answer is yes. The method is based on comparing graphs of the form $g(x) = a(x - h)^2 + k$ to those of the form $f(x) = ax^2$.

In Figure 11.10(a), the graph of $f(x) = ax^2$ for $a > 0$ is shown in black. The parabola's vertex is $(0, 0)$ and it opens upward. In Figure 11.10(b), the graph of $f(x) = ax^2$ for $a < 0$ is shown in black. The parabola's vertex is $(0, 0)$ and it opens downward.

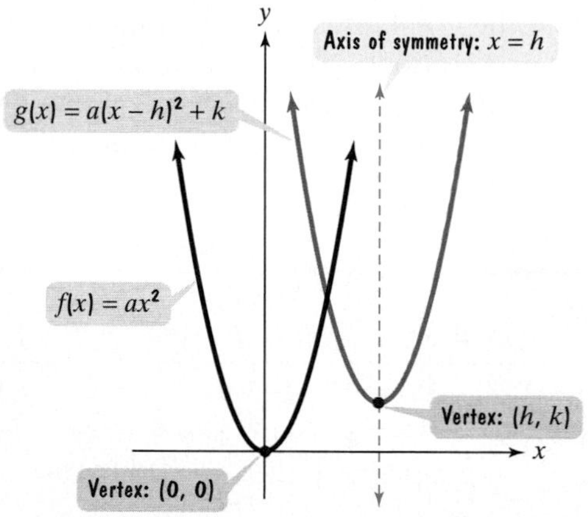

(a) $a > 0$: Parabola opens upward.

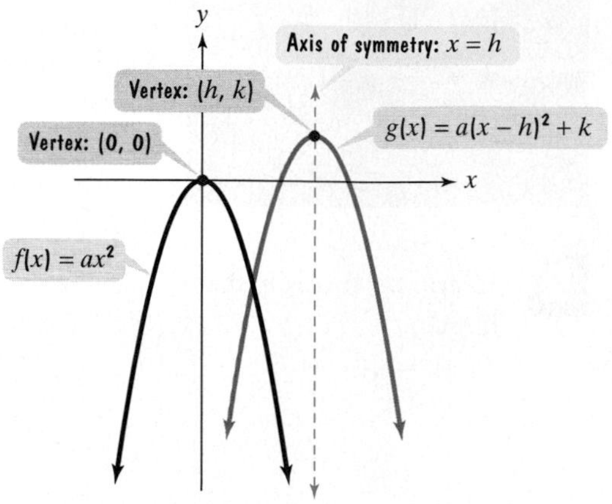

(b) $a < 0$: Parabola opens downward.

FIGURE 11.10 Moving, or shifting, the graph of $f(x) = ax^2$

Figure 11.10 also shows the graph of $g(x) = a(x - h)^2 + k$ in blue. Compare these graphs to those of $f(x) = ax^2$. Observe that h causes a horizontal move, or shift, and k causes a vertical shift of the graph of $f(x) = ax^2$. Consequently, the vertex $(0, 0)$ on the graph of $f(x) = ax^2$ moves to the point (h, k) on the graph of $g(x) = a(x - h)^2 + k$. The axis of symmetry is the vertical line whose equation is $x = h$.

The form of the expression for g is convenient because it immediately identifies the vertex of the parabola as (h, k). The sign of a in $g(x) = a(x - h)^2 + k$ determines whether the parabola opens upward or downward. Furthermore, if $|a|$ is small, the parabola opens more flatly than if $|a|$ is large.

GRAPHING QUADRATIC FUNCTIONS WITH EQUATIONS IN THE FORM $f(x) = a(x - h)^2 + k$ To graph $f(x) = a(x - h)^2 + k$,

1. Determine whether the parabola opens upward or downward. If $a > 0$, it opens upward. If $a < 0$, it opens downward.
2. Determine the vertex of the parabola. The vertex is (h, k).
3. Find any x-intercepts by replacing $f(x)$ with 0. Solve the resulting quadratic equation for x.
4. Find the y-intercept by replacing x with 0.
5. Plot the intercepts and vertex and additional points as necessary. Connect these points with a smooth curve that is shaped like a cup.

EXAMPLE 1 Graphing a Quadratic Function in the Form $f(x) = a(x - h)^2 + k$

Graph the quadratic function $f(x) = -2(x - 3)^2 + 8$.

SOLUTION We can graph this function by following the steps in the preceding box. We begin by identifying values for a, h, and k.

$$f(x) = a(x - h)^2 + k$$

$a = -2$ $h = 3$ $k = 8$

$$f(x) = -2(x - 3)^2 + 8$$

Step 1. **Determine how the parabola opens.** Note that a, the coefficient of x^2, is -2. Thus, $a < 0$; this negative value tells us that the parabola opens downward.

Step 2. **Find the vertex.** The vertex of the parabola is at (h, k). Because $h = 3$ and $k = 8$, the parabola has its vertex at $(3, 8)$.

Step 3. **Find the x-intercepts.** Replace $f(x)$ with 0 in $f(x) = -2(x - 3)^2 + 8$.

$0 = -2(x - 3)^2 + 8$ — Find x-intercepts, setting f(x) equal to 0.

$2(x - 3)^2 = 8$ — Solve for x. Add $2(x - 3)^2$ to both sides of the equation.

$(x - 3)^2 = 4$ — Divide both sides by 2.

$x - 3 = \sqrt{4}$ or $x - 3 = -\sqrt{4}$ — Apply the square root property.

$x - 3 = 2$ $\quad$ $x - 3 = -2$ — $\sqrt{4} = 2$

$x = 5$ $\quad$ $x = 1$ — Add 3 to both sides in each equation.

The x-intercepts are 5 and 1. The parabola passes through $(5, 0)$ and $(1, 0)$.

Step 4. **Find the y-intercept.** Replace x with 0 in $f(x) = -2(x - 3)^2 + 8$.

$$f(0) = -2(0 - 3)^2 + 8 = -2(-3)^2 + 8 = -2(9) + 8 = -10$$

The y-intercept is -10. The parabola passes through $(0, -10)$.

Step 5. **Graph the parabola.** With a vertex at $(3, 8)$, x-intercepts at 5 and 1, and a y-intercept at -10, the graph of f is shown in Figure 11.11. The axis of symmetry is the vertical line whose equation is $x = 3$. ■

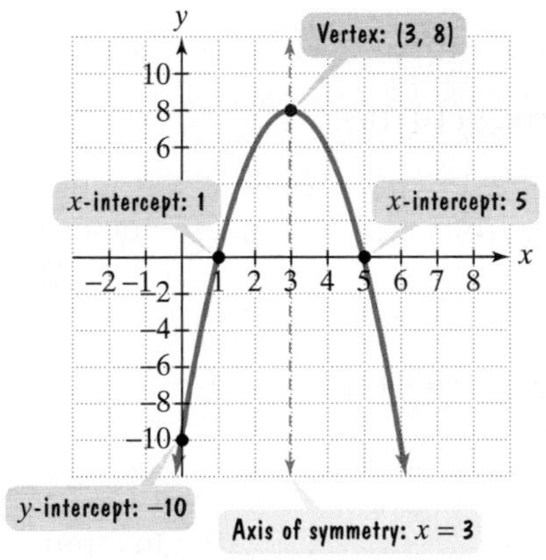

FIGURE 11.11 The graph of $f(x) = -2(x - 3)^2 + 8$

CHECK POINT 1 Graph the quadratic function $f(x) = -(x - 1)^2 + 4$.

EXAMPLE 2 Graphing a Quadratic Function in the Form $f(x) = a(x - h)^2 + k$

Graph the quadratic function $f(x) = (x + 3)^2 + 1$.

SOLUTION We begin by finding values for a, h, and k.

$f(x) = a(x - h)^2 + k$ — Form of quadratic function

$f(x) = (x + 3)^2 + 1$ — Given function

or $f(x) = 1(x - (-3))^2 + 1$

$a = 1$ $h = -3$ $k = 1$

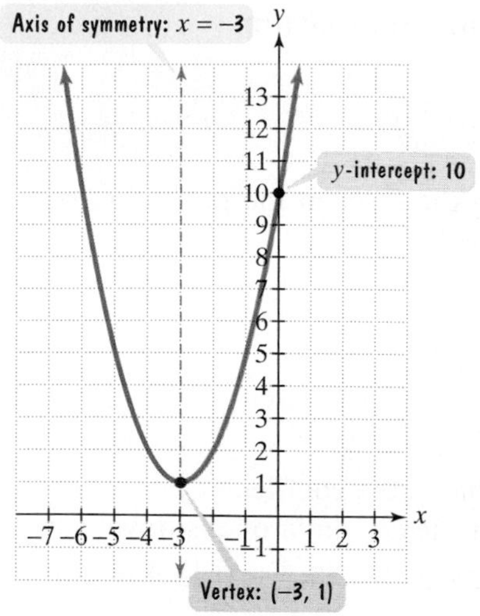

FIGURE 11.12 The graph of $f(x) = (x + 3)^2 + 1$

Step 1. Determine how the parabola opens. Note that a, the coefficient of x^2, is 1. Thus, $a > 0$; this positive value tells us that the parabola opens upward.

Step 2. Find the vertex. The vertex of the parabola is at $(h, k.)$ Because $h = -3$ and $k = 1$, the parabola has its vertex at $(-3, 1)$.

Step 3. Find the x-intercepts. Replace $f(x)$ with 0 in $f(x) = (x + 3)^2 + 1$. Because the vertex is at $(-3, 1)$, which lies above the x-axis, and the parabola opens upward, it appears that this parabola has no x-intercepts. We can verify this observation algebraically.

$$0 = (x + 3)^2 + 1$$ Find possible x-intercepts, setting f(x) equal to 0.

$$-1 = (x + 3)^2$$ Solve for x. Subtract 1 from both sides.

$$x + 3 = \sqrt{-1} \quad \text{or} \quad x + 3 = -\sqrt{-1}$$ Apply the square root property.

$$x + 3 = i \qquad x + 3 = -i$$ $\sqrt{-1} = i$

$$x = -3 + i \qquad x = -3 - i$$ The solutions are $-3 \pm i$.

Because this equation has no real solutions, the parabola has no x-intercepts.

Step 4. Find the y-intercept. Replace x with 0 in $f(x) = (x + 3)^2 + 1$.

$$f(0) = (0 + 3)^2 + 1 = 3^2 + 1 = 9 + 1 = 10$$

The y-intercept is 10. The parabola passes through (0, 10).

Step 5. Graph the parabola. With a vertex at $(-3, 1)$, no x-intercepts, and a y-intercept at 10, the graph of f is shown in Figure 11.12. The axis of symmetry is the vertical line whose equation is $x = -3$. ■

CHECK POINT 2 Graph the quadratic function $f(x) = (x - 2)^2 + 1$.

3 Graph parabolas in the form $f(x) = ax^2 + bx + c$.

Graphing Quadratic Functions in the Form $f(x) = ax^2 + bx + c$ Quadratic functions are frequently expressed in the form $f(x) = ax^2 + bx + c$. How can we identify the vertex of a parabola whose equation is in this form? Completing the square provides the answer to this question.

$$f(x) = ax^2 + bx + c$$

$$= a\left(x^2 + \frac{b}{a}x\right) + c$$ Factor out a from $ax^2 + bx$.

$$= a\left(x^2 + \frac{b}{a}x + \frac{b^2}{4a^2}\right) + c - a\left(\frac{b^2}{4a^2}\right)$$

Complete the square by adding the square of half the coefficient of x.

By completing the square, we added $a \cdot \frac{b^2}{4a^2}$. To avoid changing the function's equation, we must subtract this term.

$$= a\left(x + \frac{b}{2a}\right)^2 + c - \frac{b^2}{4a}$$ Write the trinomial as the square of a binomial and simplify the constant term.

Now let's compare the form of the equation on the bottom of the previous page with a quadratic function in the form $f(x) = a(x - h)^2 + k$.

The form we know how to graph $\quad f(x) = a(x - h)^2 + k$

$h = -\dfrac{b}{2a} \qquad k = c - \dfrac{b^2}{4a}$

Equation under discussion $\quad f(x) = a\left(x - \left(-\dfrac{b}{2a}\right)\right)^2 + c - \dfrac{b^2}{4a}$

The important part of this observation is that h, the x-coordinate of the vertex, is $-\dfrac{b}{2a}$. The y-coordinate can be found by evaluating the function at $-\dfrac{b}{2a}$.

THE VERTEX OF A PARABOLA WHOSE EQUATION IS $f(x) = ax^2 + bx + c$

Consider the parabola defined by the quadratic function $f(x) = ax^2 + bx + c$. The parabola's vertex is $\left(-\dfrac{b}{2a}, f\left(-\dfrac{b}{2a}\right)\right)$.

We can apply our five-step procedure and graph parabolas in the form $f(x) = ax^2 + bx + c$. The only step that is different is how we determine the vertex.

EXAMPLE 3 Graphing a Quadratic Function in the Form $f(x) = ax^2 + bx + c$

Graph the quadratic function $f(x) = -x^2 - 2x + 1$. Use the graph to identify the function's domain and its range.

SOLUTION

Step 1. **Determine how the parabola opens.** Note that a, the coefficient of x^2, is -1. Thus, $a < 0$; this negative value tells us that the parabola opens downward.

Step 2. **Find the vertex.** We know that the x-coordinate of the vertex is $x = -\dfrac{b}{2a}$.

We identify a, b, and c in $f(x) = ax^2 + bx + c$.

$$f(x) = -x^2 - 2x + 1$$

$a = -1 \qquad b = -2 \qquad c = 1$

Substitute the values of a and b into the equation for the x-coordinate:

$$x = -\frac{b}{2a} = -\frac{-2}{2(-1)} = -\left(\frac{-2}{-2}\right) = -1.$$

The x-coordinate of the vertex is -1. We substitute -1 for x in the equation of the function to find the y-coordinate:

$$f(-1) = -(-1)^2 - 2(-1) + 1 = -1 + 2 + 1 = 2.$$

The vertex is $(-1, 2)$.

Step 3. **Find the x-intercepts.** Replace $f(x)$ with 0 in $f(x) = -x^2 - 2x + 1$. We obtain $0 = -x^2 - 2x + 1$. This equation cannot be solved by factoring. We will use the quadratic formula to solve it.

$$-x^2 - 2x + 1 = 0$$

$a = -1$ $b = -2$ $c = 1$

$$x = \frac{-b + \sqrt{b^2 - 4ac}}{2a} = \frac{-(-2) \pm \sqrt{(-2)^2 - 4(-1)(1)}}{2(-1)} = \frac{2 \pm \sqrt{4 - (-4)}}{-2}$$

To locate the x-intercepts, we need decimal approximations. Thus, there is no need to simplify the radical form of the solutions.

$$x = \frac{2 + \sqrt{8}}{-2} \approx -2.4 \quad \text{or} \quad x = \frac{2 - \sqrt{8}}{-2} \approx 0.4$$

The x-intercepts are approximately -2.4 and 0.4. The parabola passes through $(-2.4, 0)$ and $(0.4, 0)$.

Step 4. **Find the y-intercept.** Replace x with 0 in $f(x) = -x^2 - 2x + 1$.

$$f(0) = -0^2 - 2 \cdot 0 + 1 = 1$$

The y-intercept is 1. The parabola passes through $(0, 1)$.

Step 5. **Graph the parabola.** With a vertex at $(-1, 2)$, x-intercepts at -2.4 and 0.4, and a y-intercept at 1, the graph of f is shown in Figure 11.13(a). The axis of symmetry is the vertical line whose equation is $x = -1$.

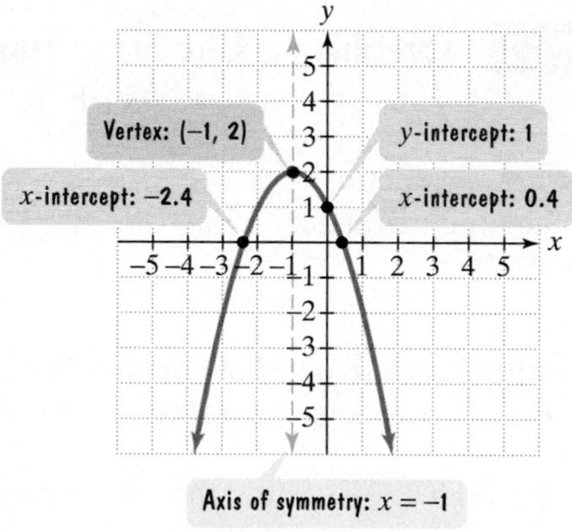

FIGURE 11.13(a) The graph of $f(x) = -x^2 - 2x + 1$

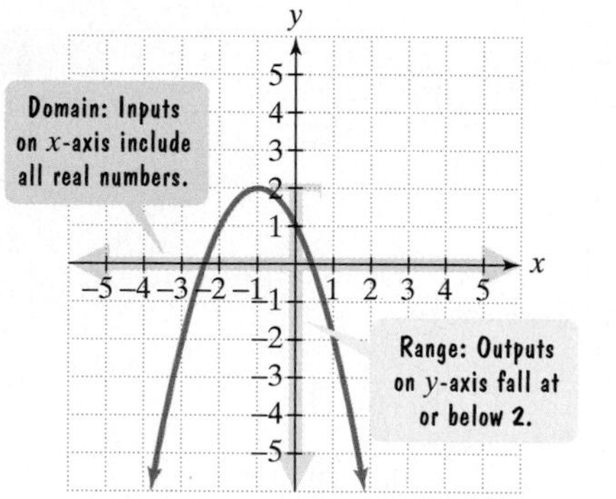

FIGURE 11.13(b) Determining the domain and range of $f(x) = -x^2 - 2x + 1$

Now we are ready to determine the domain and range of $f(x) = -x^2 - 2x + 1$. We can use the parabola, shown again in Figure 11.13(b), to do so. To find the domain, look for all the inputs on the x-axis that correspond to points on the graph. As the graph widens and continues to fall at both ends, can you see that these inputs include all real numbers?

Domain of f is $\{x | x \text{ is a real number}\}$ or $(-\infty, \infty)$.

To find the range, look for all the outputs on the y-axis that correspond to points on the graph. Figure 11.13(b) shows that the parabola's vertex, $(-1, 2)$, is the highest point on the graph. Because the y-coordinate of the vertex is 2, outputs on the y-axis fall at or below 2.

Range of f is $\{y | y \leq 2\}$ or $(-\infty, 2]$. ■

STUDY TIP

The domain of any quadratic function includes all real numbers. If the vertex is the graph's highest point, the range includes all real numbers at or below the y-coordinate of the vertex. If the vertex is the graph's lowest point, the range includes all real numbers at or above the y-coordinate of the vertex.

CHECK POINT 3 Graph the quadratic function $f(x) = -x^2 + 4x + 1$. Use the graph to identify the function's domain and its range.

4 Solve problems involving minimizing or maximizing quadratic functions.

Applications of Quadratic Functions When do the things we throw reach their maximum height and what is that height? How many hours of sleep per night minimize the risk of death? How many units of a product should a business manufacture to maximize its profits? The answers to these questions involve finding the maximum or minimum value of quadratic functions.

Consider the quadratic function $f(x) = ax^2 + bx + c$. If $a > 0$, the parabola opens upward and the vertex is its lowest point. If $a < 0$, the parabola opens downward and the vertex is its highest point. The x-coordinate of the vertex is $-\frac{b}{2a}$. Thus, we can find the minimum or maximum value of f by evaluating the quadratic function at $x = -\frac{b}{2a}$.

MINIMUM AND MAXIMUM: QUADRATIC FUNCTIONS Consider $f(x) = ax^2 + bx + c$.

1. If $a > 0$, then f has a minimum that occurs at $x = -\frac{b}{2a}$. This minimum value is $f\left(-\frac{b}{2a}\right)$.

2. If $a < 0$, then f has a maximum that occurs at $x = -\frac{b}{2a}$. This maximum value is $f\left(-\frac{b}{2a}\right)$.

EXAMPLE 4 Parabolic Paths of a Shot Put

An athlete whose event is the shot put releases the shot with the same initial velocity, but at different angles. Figure 11.14 shows the parabolic paths for shots released at angles of 35° and 65°.

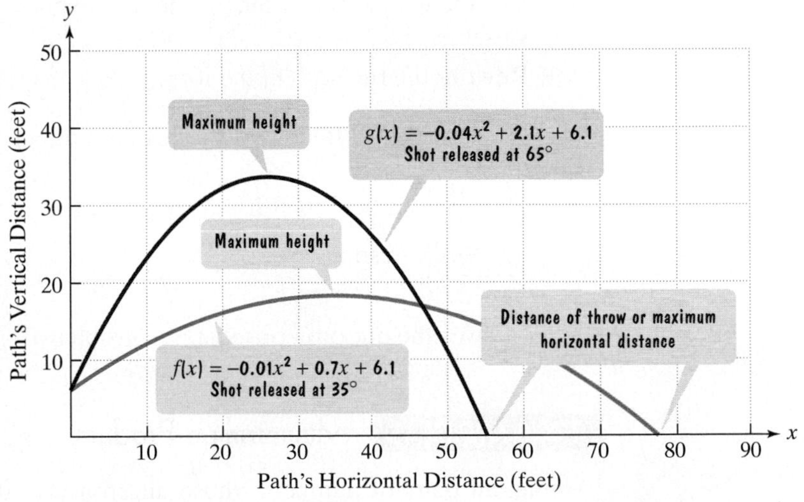

FIGURE 11.14 Two paths of a shot put

When the shot is released at an angle of 35°, its path can be modeled by the function

$$f(x) = -0.01x^2 + 0.7x + 6.1$$

in which x is the shot's horizontal distance, in feet, and $f(x)$ is its height, in feet. What is the maximum height of this shot's path?

SOLUTION The quadratic function is in the form $f(x) = ax^2 + bx + c$, with $a = -0.01$ and $b = 0.7$. Because $a < 0$, the function has a maximum that occurs at $x = -\frac{b}{2a}$.

$$x = -\frac{b}{2a} = -\frac{0.7}{2(-0.01)} = -(-35) = 35$$

This means that the shot's maximum height occurs when its horizontal distance is 35 feet. Can you see how this is shown by the blue graph of f in Figure 11.14 on the previous page? The maximum height of this path is

$$f(35) = -0.01(35)^2 + 0.7(35) + 6.1 = 18.35$$

or 18.35 feet. ■

CHECK POINT 4 Use function g, whose equation and graph are shown in Figure 11.14, to find the maximum height, to the nearest tenth of a foot, when the shot is released at an angle of 65°.

ENRICHMENT ESSAY

Throwing to Achieve Maximum Distance

Although Example 4 focuses on maximum height of a propelled object, the goal of the shot put throw is to achieve the greatest possible distance. If an object is propelled from ground level, an angle of 45° produces the maximum distance for a given initial velocity. An angle slightly less than 45° is better when an object is thrown from above the ground. For the shot in Example 4, released from 6.1 feet above the ground, an angle of approximately 43° produces the maximum distance.

Quadratic functions can also be formed from verbal conditions. Once we have obtained a quadratic function, we can then use the x-coordinate of the vertex to determine its maximum or minimum value. Here is a step-by-step strategy for solving these kinds of problems:

STRATEGY FOR SOLVING PROBLEMS INVOLVING MAXIMIZING OR MINIMIZING QUADRATIC FUNCTIONS

1. Read the problem carefully and decide which quantity is to be maximized or minimized.
2. Use the conditions of the problem to express the quantity as a function in one variable.
3. Rewrite the function in the form $f(x) = ax^2 + bx + c$.
4. Calculate $-\frac{b}{2a}$. If $a > 0$, f has a minimum at $x = -\frac{b}{2a}$. This minimum value is $f\left(-\frac{b}{2a}\right)$. If $a < 0$, f has a maximum at $x = -\frac{b}{2a}$. This maximum value is $f\left(-\frac{b}{2a}\right)$.
5. Answer the question posed in the problem.

EXAMPLE 5 Minimizing a Product

Among all pairs of numbers whose difference is 10, find a pair whose product is as small as possible. What is the minimum product?

SOLUTION

Step 1. **Decide what must be maximized or minimized.** We must minimize the product of two numbers. Calling the numbers x and y, and calling the product P, we must minimize

$$P = xy.$$

Step 2. **Express this quantity as a function in one variable.** In the formula $P = xy$, P is expressed in terms of two variables, x and y. However, because the difference of the numbers is 10, we can write

$$x - y = 10.$$

We can solve this equation for y in terms of x (or vice versa), substitute the result into $P = xy$, and obtain P as a function of one variable.

$$-y = -x + 10$$ Subtract x from both sides of $x - y = 10$.

$$y = x - 10$$ Multiply both sides of the equation by -1 and solve for y

Now we substitute $x - 10$ for y in $P = xy$.

$$P = xy = x(x - 10).$$

Because P is now a function of x, we can write

$$P(x) = x(x - 10).$$

Step 3. **Write the function in the form $f(x) = ax^2 + bx + c$.** We apply the distributive property to obtain

$$P(x) = x(x - 10) = x^2 - 10x$$

$a = 1$ $b = -10$

Step 4. **Calculate $-\frac{b}{2a}$. If $a > 0$, the function has a minimum at this value.** The voice balloons show that $a = 1$ and $b = -10$.

$$x = -\frac{b}{2a} = -\frac{-10}{2(1)} = -(-5) = 5$$

This means that the product, P, of two numbers whose difference is 10 is a minimum when one of the numbers, x, is 5.

Step 5. **Answer the question posed by the problem.** The problem asks for the two numbers and the minimum product. We found that one of the numbers, x, is 5. Now we must find the second number, y.

$$y = x - 10 = 5 - 10 = -5.$$

The number pair whose difference is 10 and whose product is as small as possible is 5, -5. The minimum product is $5(-5)$, or -25. ■

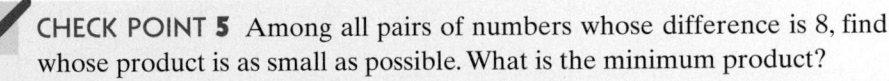

✔ CHECK POINT **5** Among all pairs of numbers whose difference is 8, find a pair whose product is as small as possible. What is the minimum product?

USING TECHNOLOGY

The TABLE feature of a graphing utility can be used to verify our work in Example 5.

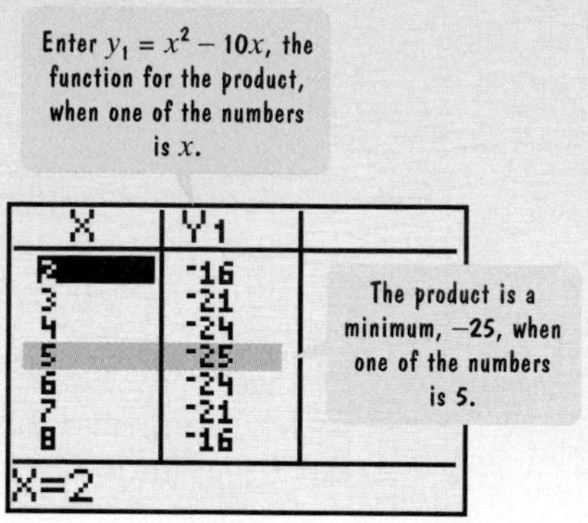
Enter $y_1 = x^2 - 10x$, the function for the product, when one of the numbers is x.
X | Y1
2 | -16
3 | -21
4 | -24
5 | -25
6 | -24
7 | -21
8 | -16
X=2
The product is a minimum, −25, when one of the numbers is 5.

EXAMPLE 6 Maximizing Area

You have 100 yards of fencing to enclose a rectangular region. Find the dimensions of the rectangle that maximize the enclosed area. What is the maximum area?

SOLUTION

Step 1. **Decide what must be maximized or minimized.** We must maximize area. What we do not know are the rectangle's dimensions, x and y.

Step 2. **Express this quantity as a function in one variable.** Because we must maximize area, we have $A = xy$. We need to transform this into a function in which A is represented by one variable. Because you have 100 yards of fencing, the perimeter of the rectangle is 100 yards. This means that

$$2x + 2y = 100.$$

We can solve this equation for y in terms of x, substitute the result into $A = xy$, and obtain A as a function in one variable. We begin by solving for y.

$$2y = 100 - 2x \quad \text{Subtract } 2x \text{ from both sides.}$$

$$y = \frac{100 - 2x}{2} \quad \text{Divide both sides by 2.}$$

$$y = 50 - x \quad \text{Divide each term in the numerator by 2.}$$

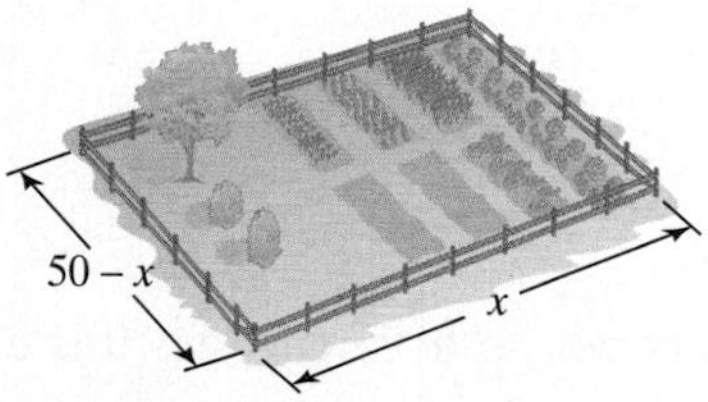
50 − x
x

FIGURE 11.15 What value of x will maximize the rectangle's area?

Now we substitute $50 - x$ for y in $A = xy$.

$$A = xy = x(50 - x)$$

The rectangle and its dimensions are illustrated in Figure 11.15. Because A is now a function of x, we can write

$$A(x) = x(50 - x).$$

This function models the area, $A(x)$, of any rectangle whose perimeter is 100 yards in terms of one of its dimensions, x.

Step 3. **Write the function in the form $f(x) = ax^2 + bx + c$.** We apply the distributive property to obtain

$$A(x) = x(50 - x) = 50x - x^2 = -x^2 + 50x.$$

$a = -1$ $\quad$ $b = 50$

USING TECHNOLOGY

The graph of the area function

$$A(x) = x(50 - x)$$

was obtained with a graphing utility using a $[0, 50, 2]$ by $[0, 700, 25]$ viewing rectangle. The maximum function feature verifies that a maximum area of 625 square yards occurs when one of the dimensions is 25 yards.

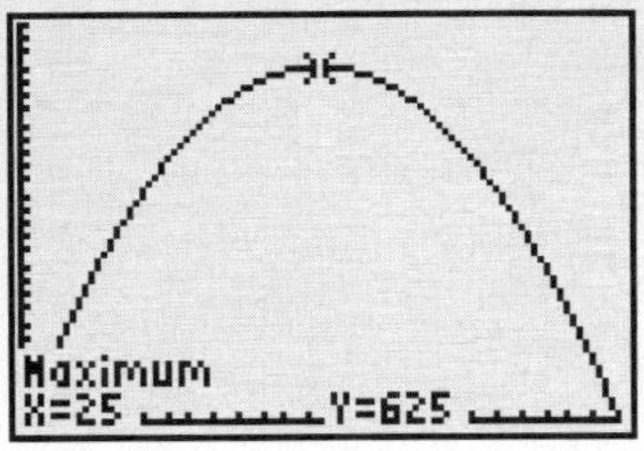

Step 4. **Calculate $-\frac{b}{2a}$. If $a < 0$, the function has a maximum at this value.** The voice balloons on the previous page show that $a = -1$ and $b = 50$.

$$x = -\frac{b}{2a} = -\frac{50}{2(-1)} = 25$$

This means that the area, $A(x)$, of a rectangle with perimeter 100 yards is a maximum when one of the rectangle's dimensions, x, is 25 yards.

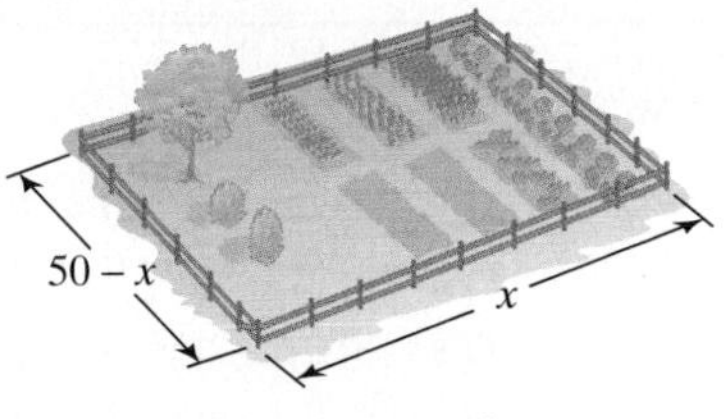

FIGURE 11.15 (repeated)

Step 5. **Answer the question posed by the problem.** We found that $x = 25$. Figure 11.15 shows that the rectangle's other dimension is $50 - x = 50 - 25 = 25$. The dimensions of the rectangle that maximize the enclosed area are 25 yards by 25 yards. The rectangle that gives the maximum area is actually a square with an area of 25 yards · 25 yards, or 625 square yards. ■

CHECK POINT 6 You have 120 feet of fencing to enclose a rectangular region. Find the dimensions of the rectangle that maximize the enclosed area. What is the maximum area?

USING TECHNOLOGY

We've come a long way from the small nation of "embattled farmers" who launched the American Revolution. In the early days of our Republic, 95% of the population was involved in farming. The graph in Figure 11.16 shows the number of farms in the United States from 1850 through 2010 (projected). Because the graph is shaped like a cup, with an increasing number of farms from 1850 to 1910 and a decreasing number of farms from 1910 to 2010, a quadratic function is an appropriate model for the data. You can use the statistical menu of a graphing utility to enter the data in Figure 11.16. We entered the data using (number of decades after 1850, millions of U.S. farms). The data are shown to the right of Figure 11.16.

Number of U. S. Farms, 1850–2010

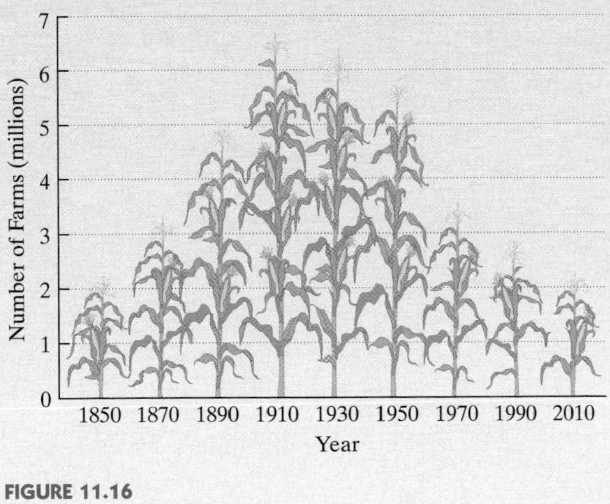

FIGURE 11.16

Source: U. S. Bureau of the Census

Data:

(0, 2.3), (2, 3.3), (4, 5.1),
(6, 6.7), (8, 6.4), (10, 5.8),
(12, 3.6), (14, 2.9), (16, 2.3)

Upon entering the QUADratic REGression program, we obtain the results shown in the screen. Thus, the quadratic function of best fit is approximately

$$f(x) = -0.064x^2 + 0.99x + 2.2,$$

where x represents the number of decades after 1850 and $f(x)$ represents the number of U.S. farms, in millions.

11.3 EXERCISE SET

Student Solutions Manual | CD/Video | PH Math/Tutor Center | MathXL Tutorials on CD | MathXL® | MyMathLab | Interactmath.com

Practice Exercises

In Exercises 1–4, the graph of a quadratic function is given. Write the function's equation, selecting from the following options:

$$f(x) = (x + 1)^2 - 1,\ g(x) = (x + 1)^2 + 1,$$
$$h(x) = (x - 1)^2 + 1,\ j(x) = (x - 1)^2 - 1.$$

1.

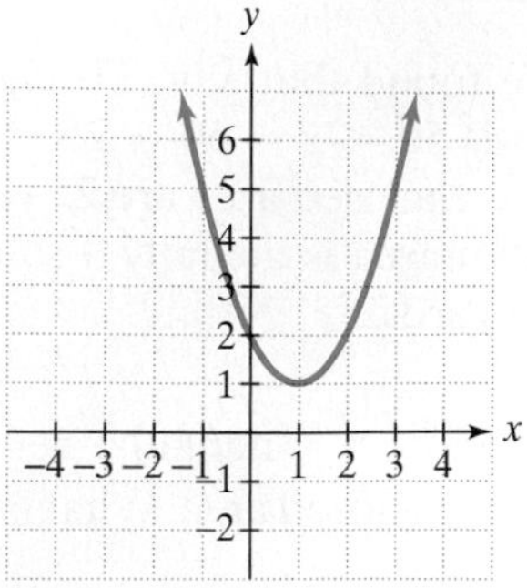

2.

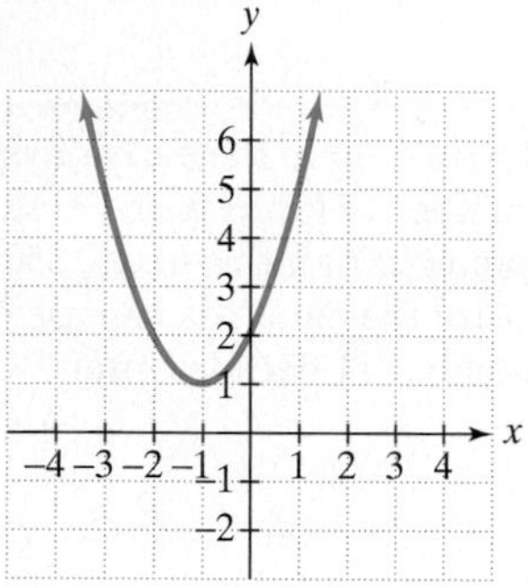

3.

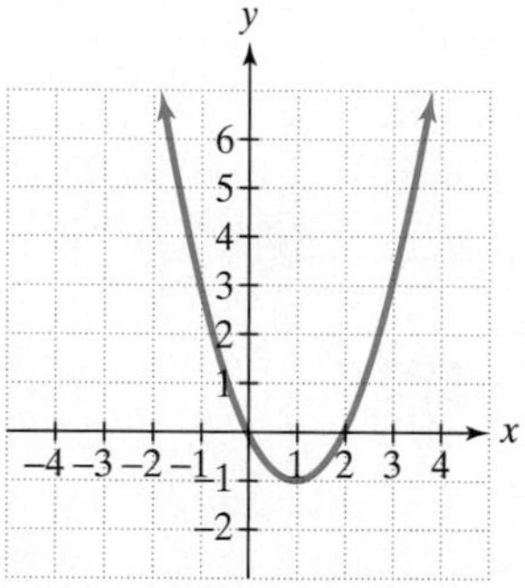

4.

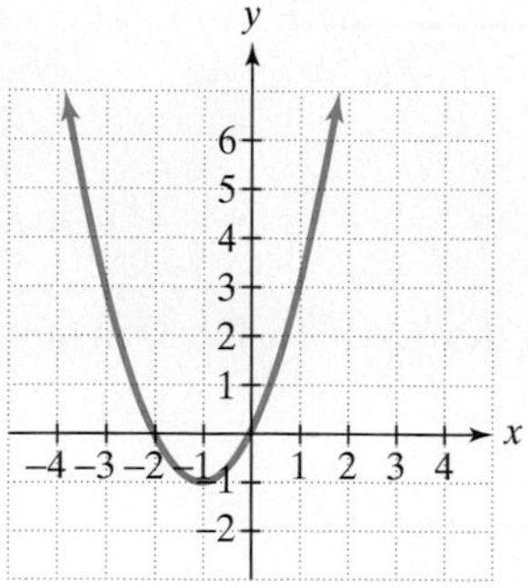

In Exercises 5–8, the graph of a quadratic function is given. Write the function's equation, selecting from the following options:

$$f(x) = x^2 + 2x + 1,\ g(x) = x^2 - 2x + 1,$$
$$h(x) = x^2 - 1,\ j(x) = -x^2 - 1.$$

5.

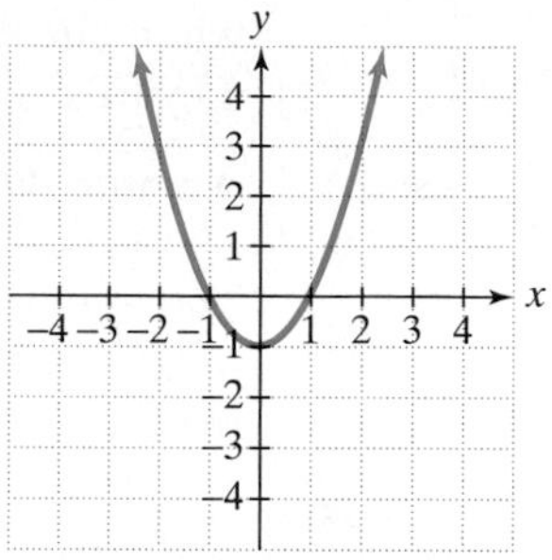

6.

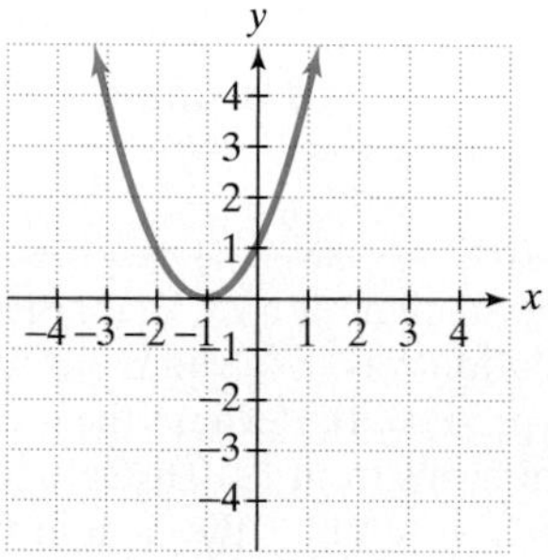

7.

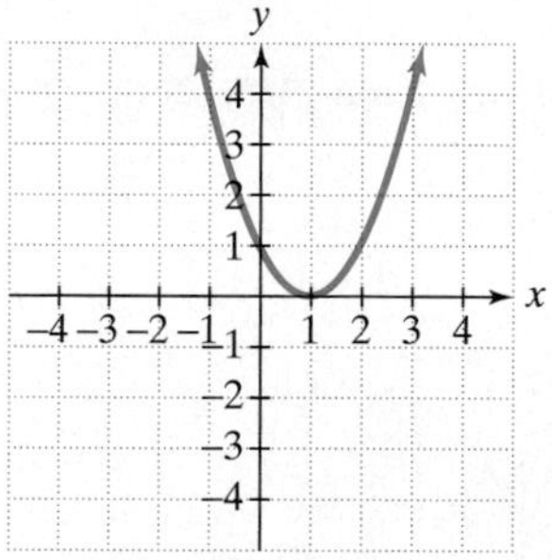

8.

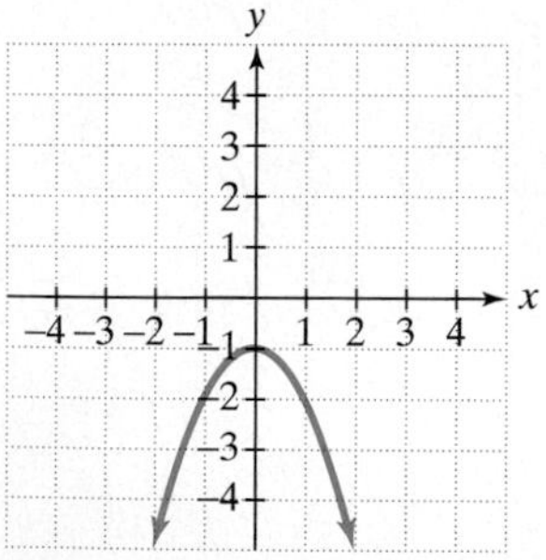

In Exercises 9–16, find the coordinates of the vertex for the parabola defined by the given quadratic function.

9. $f(x) = 2(x - 3)^2 + 1$

10. $f(x) = -3(x - 2)^2 + 12$

11. $f(x) = -2(x + 1)^2 + 5$

12. $f(x) = -2(x + 4)^2 - 8$

13. $f(x) = 2x^2 - 8x + 3$

14. $f(x) = 3x^2 - 12x + 1$

15. $f(x) = -x^2 - 2x + 8$

16. $f(x) = -2x^2 + 8x - 1$

In Exercises 17–34, use the vertex and intercepts to sketch the graph of each quadratic function. Give the equation for the parabola's axis of symmetry. Use the parabola to identify the function's range.

17. $f(x) = (x - 4)^2 - 1$

18. $f(x) = (x - 1)^2 - 2$

19. $f(x) = (x - 1)^2 + 2$

20. $f(x) = (x - 3)^2 + 2$

21. $y - 1 = (x - 3)^2$

22. $y - 3 = (x - 1)^2$

23. $f(x) = 2(x + 2)^2 - 1$

24. $f(x) = \frac{5}{4} - \left(x - \frac{1}{2}\right)^2$

25. $f(x) = 4 - (x - 1)^2$

26. $f(x) = 1 - (x - 3)^2$

27. $f(x) = x^2 - 2x - 3$

28. $f(x) = x^2 - 2x - 15$

29. $f(x) = x^2 + 3x - 10$

30. $f(x) = 2x^2 - 7x - 4$

31. $f(x) = 2x - x^2 + 3$

32. $f(x) = 5 - 4x - x^2$

33. $f(x) = 2x - x^2 - 2$

34. $f(x) = 6 - 4x + x^2$

In Exercises 35–40, determine, without graphing, whether the given quadratic function has a minimum value or a maximum value. Then find the coordinates of the minimum or the maximum point.

35. $f(x) = 3x^2 - 12x - 1$

36. $f(x) = 2x^2 - 8x - 3$

37. $f(x) = -4x^2 + 8x - 3$

38. $f(x) = -2x^2 - 12x + 3$

39. $f(x) = 5x^2 - 5x$

40. $f(x) = 6x^2 - 6x$

Practice Plus

In Exercises 41–44, give the domain and the range of each quadratic function whose graph is described.

41. The vertex is $(-1, -2)$ and the parabola opens up.

42. The vertex is $(-3, -4)$ and the parabola opens down.

43. The maximum point on the parabola is $(10, 3)$.

44. The minimum point on the parabola is $(5, 12)$.

In Exercises 45–48, write an equation of the parabola that has the same shape as the graph of $f(x) = 2x^2$, but with the given point as the vertex.

45. $(5, 3)$

46. $(7, 4)$

47. $(-10, -5)$

48. $(-8, -6)$

In Exercises 49–52, write an equation of the parabola that has the same shape as the graph of $f(x) = 3x^2$ or $g(x) = -3x^2$, but with the given point as a maximum or minimum point.

49. maximum: $(-2, 4)$

50. maximum: $(5, -7)$

51. minimum: $(11, 0)$

52. minimum: $(9, 0)$

Application Exercises

In the United States, HCV, or hepatitis C virus, is four times as widespread as HIV. Few of the nation's three to four million carriers have any idea they are infected. The graph shows the projected mortality, in thousands, from the virus. Use the information in the graph to solve Exercises 53–54.

Projected Hepatitis C Deaths in the U. S.

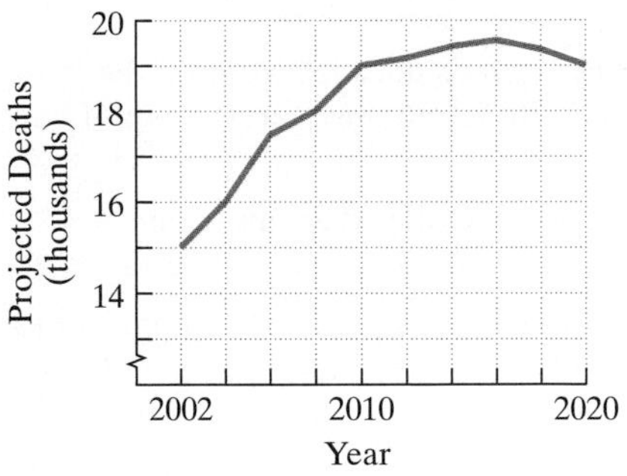

Source: American Journal of Public Health

53. Suppose that a quadratic function is used to model the data shown with ordered pairs representing (number of years after 2002, thousands of hepatitis C deaths). Determine, without obtaining an actual quadratic function that models the data, the approximate coordinates of the vertex for the function's graph. Describe what this means in practical terms. Use the word "maximum" in your description.

54. Why is a quadratic function an appropriate model for the data shown in the graph on the previous page?

55. A person standing close to the edge on the top of a 160-foot building throws a baseball vertically upward. The quadratic function

$$s(t) = -16t^2 + 64t + 160$$

models the ball's height above the ground, $s(t)$, in feet, t seconds after it was thrown.

a. After how many seconds does the ball reach its maximum height? What is the maximum height?

b. How many seconds does it take until the ball finally hits the ground? Round to the nearest tenth of a second.

c. Find $s(0)$ and describe what this means.

d. Use your results from parts (a) through (c) to graph the quadratic function. Begin the graph with $t = 0$ and end with the value of t for which the ball hits the ground.

56. A person standing close to the edge on the top of a 200-foot building throws a baseball vertically upward. The quadratic function

$$s(t) = -16t^2 + 64t + 200$$

models the ball's height above the ground, $s(t)$, in feet, t seconds after it was thrown.

a. After how many seconds does the ball reach its maximum height? What is the maximum height?

b. How many seconds does it take until the ball finally hits the ground? Round to the nearest tenth of a second.

c. Find $s(0)$ and describe what this means.

d. Use your results from parts (a) through (c) to graph the quadratic function. Begin the graph with $t = 0$ and end with the value of t for which the ball hits the ground.

57. The function

$$f(x) = 104.5x^2 - 1501.5x + 6016$$

models the death rate per year per 100,000 males, $f(x)$, for U.S. men who average x hours of sleep each night. How many hours of sleep, to the nearest tenth of an hour, corresponds to the minimum death rate? What is this minimum death rate, to the nearest whole number?

58. The function

$$f(x) = -0.064x^2 + 0.99x + 2.2$$

models the number of U.S. farms, $f(x)$, in millions, x decades after 1850. How many decades after 1850, to the nearest decade, did the number of U.S. farms reach a maximum? How many farms, to the nearest tenth of a million, were there in that year? How well does this model the actual data shown in Figure 11.16 on page 761?

59. Among all pairs of numbers whose sum is 16, find a pair whose product is as large as possible. What is the maximum product?

60. Among all pairs of numbers whose sum is 20, find a pair whose product is as large as possible. What is the maximum product?

61. Among all pairs of numbers whose difference is 16, find a pair whose product is as small as possible. What is the minimum product?

62. Among all pairs of numbers whose difference is 24, find a pair whose product is as small as possible. What is the minimum product?

63. You have 600 feet of fencing to enclose a rectangular plot that borders on a river. If you do not fence the side along the river, find the length and width of the plot that will maximize the area. What is the largest area that can be enclosed?

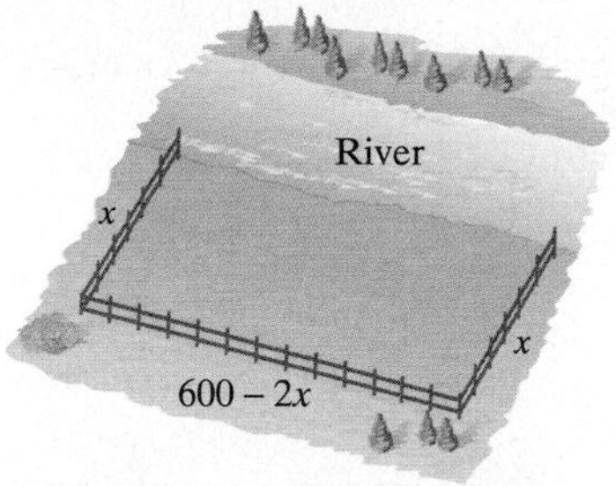

64. You have 200 feet of fencing to enclose a rectangular plot that borders on a river. If you do not fence the side along the river, find the length and width of the plot that will maximize the area. What is the largest area that can be enclosed?

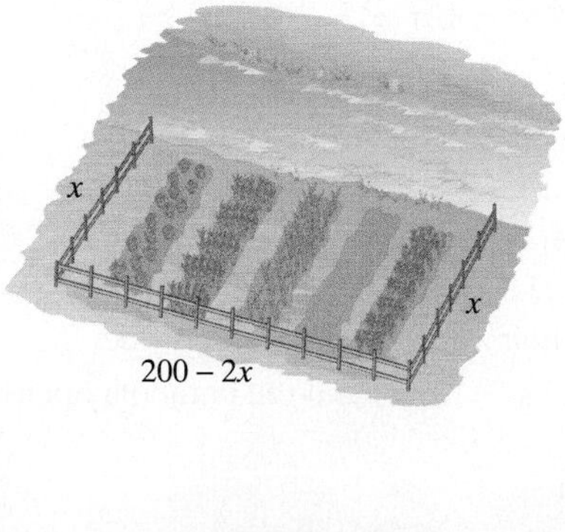

65. You have 50 yards of fencing to enclose a rectangular region. Find the dimensions of the rectangle that maximize the enclosed area. What is the maximum area?

66. You have 80 yards of fencing to enclose a rectangular region. Find the dimensions of the rectangle that maximize the enclosed area. What is the maximum area?

67. A rain gutter is made from sheets of aluminum that are 20 inches wide by turning up the edges to form right angles. Determine the depth of the gutter that will maximize its cross-sectional area and allow the greatest amount of water to flow. What is the maximum cross-sectional area?

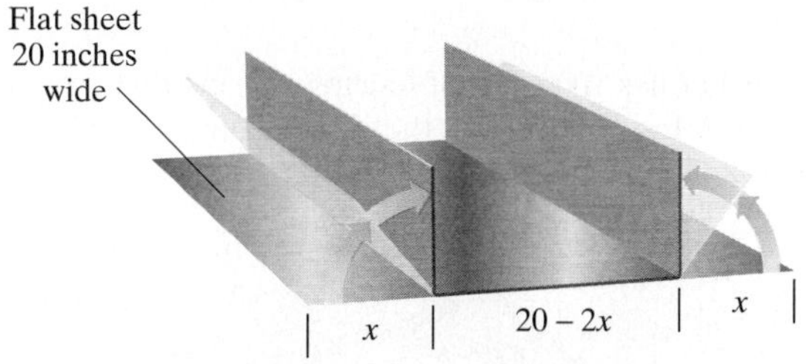

68. A rain gutter is made from sheets of aluminum that are 12 inches wide by turning up the edges to form right angles. Determine the depth of the gutter that will maximize its cross-sectional area and allow the greatest amount of water to flow. What is the maximum cross-sectional area?

In Chapter 9, we saw that the profit, P(x), generated after producing and selling x units of a product is given by the function

$$P(x) = R(x) - C(x),$$

where R and C are the revenue and cost functions, respectively. Use these functions to solve Exercises 69–70.

69. Hunky Beef, a local sandwich store, has a fixed weekly cost of \$525.00, and variable costs for making a roast beef sandwich are \$0.55.

a. Let x represent the number of roast beef sandwiches made and sold each week. Write the weekly cost function, C, for Hunky Beef.

b. The function $R(x) = -0.001x^2 + 3x$ describes the money that Hunky Beef takes in each week from the sale of x roast beef sandwiches. Use this revenue function and the cost function from part (a) to write the store's weekly profit function, P.

c. Use the store's profit function to determine the number of roast beef sandwiches it should make and sell each week to maximize profit. What is the maximum weekly profit?

70. Virtual Fido is a company that makes electronic virtual pets. The fixed weekly cost is \$3000, and variable costs for each pet are \$20.

a. Let x represent the number of virtual pets made and sold each week. Write the weekly cost function, C, for Virtual Fido.

b. The function $R(x) = -x^2 + 1000x$ describes the money that Virtual Fido takes in each week from the sale of x virtual pets. Use this revenue function and the cost function from part (a) to write the weekly profit function, P.

c. Use the profit function to determine the number of virtual pets that should be made and sold each week to maximize profit. What is the maximum weekly profit?

Writing in Mathematics

71. What is a parabola? Describe its shape.

72. Explain how to decide whether a parabola opens upward or downward.

73. Describe how to find a parabola's vertex if its equation is in the form $f(x) = a(x - h)^2 + k$. Give an example.

74. Describe how to find a parabola's vertex if its equation is in the form $f(x) = ax^2 + bx + c$. Use $f(x) = x^2 - 6x + 8$ as an example.

75. A parabola that opens upward has its vertex at $(1, 2)$. Describe as much as you can about the parabola based on this information. Include in your discussion the number of x-intercepts (if any) for the parabola.

Technology Exercises

76. Use a graphing utility to verify any five of your hand-drawn graphs in Exercises 17–34.

77. a. Use a graphing utility to graph $y = 2x^2 - 82x + 720$ in a standard viewing rectangle. What do you observe?

b. Find the coordinates of the vertex for the given quadratic function.

c. The answer to part (b) is $(20.5, -120.5)$. Because the leading coefficient, 2, of the given function is positive, the vertex is a minimum point on the graph. Use this fact to help find a viewing rectangle that will give a relatively complete picture of the parabola. With an axis of symmetry at $x = 20.5$, the setting for x should extend past this, so try Xmin $= 0$ and Xmax $= 30$. The setting for y should include (and probably go below) the y-coordinate of the graph's minimum point, so try Ymin $= -130$. Experiment with Ymax until your utility shows the parabola's major features.

d. In general, explain how knowing the coordinates of a parabola's vertex can help determine a reasonable viewing rectangle on a graphing utility for obtaining a complete picture of the parabola.

In Exercises 78–81, find the vertex for each parabola. Then determine a reasonable viewing rectangle on your graphing utility and use it to graph the parabola.

78. $y = -0.25x^2 + 40x$

79. $y = -4x^2 + 20x + 160$

80. $y = 5x^2 + 40x + 600$

81. $y = 0.01x^2 + 0.6x + 100$

82. The following data show fuel efficiency, in miles per gallon, for all U.S. automobiles in the indicated year.

x (Years after 1940)	y (Average Number of Miles per Gallon for U.S. Automobiles)
1940: 0	14.8
1950: 10	13.9
1960: 20	13.4
1970: 30	13.5
1980: 40	15.9
1990: 50	20.2
2000: 60	22.0

Source: U.S. Department of Transportation

a. Use a graphing utility to draw a scatter plot of the data. Explain why a quadratic function is appropriate for modeling these data.

b. Use the quadratic regression feature to find the quadratic function that best fits the data.

c. Use the equation in part (b) to determine the worst year for automobile fuel efficiency. What was the average number of miles per gallon for that year?

d. Use a graphing utility to draw a scatter plot of the data and graph the quadratic function of best fit on the scatter plot.

Critical Thinking Exercises

83. Which one of the following is true?

a. No quadratic functions have a range of $(-\infty, \infty)$.

b. The vertex of the parabola described by $f(x) = 2(x - 5)^2 - 1$ is at $(5, 1)$.

c. The graph of $f(x) = -2(x + 4)^2 - 8$ has one y-intercept and two x-intercepts.

d. The maximum value of y for the quadratic function $f(x) = -x^2 + x + 1$ is 1.

In Exercises 84–85, find the axis of symmetry for each parabola whose equation is given. Use the axis of symmetry to find a second point on the parabola whose y-coordinate is the same as the given point.

84. $f(x) = 3(x + 2)^2 - 5;\ (-1, -2)$

85. $f(x) = (x - 3)^2 + 2;\ (6, 11)$

In Exercises 86–87, write the equation of each parabola in $f(x) = a(x - h)^2 + k$ form.

86. Vertex: $(-3, -4)$; The graph passes through the point $(1, 4)$.

87. Vertex: $(-3, -1)$; The graph passes through the point $(-2, -3)$.

88. A rancher has 1000 feet of fencing to construct six corrals, as shown in the figure. Find the dimensions that maximize the enclosed area. What is the maximum area?

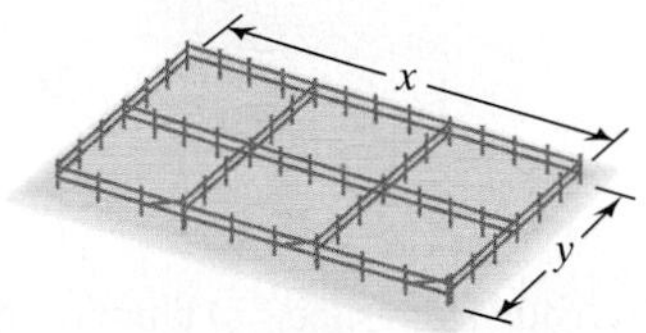

89. The annual yield per lemon tree is fairly constant at 320 pounds when the number of trees per acre is 50 or fewer. For each additional tree over 50, the annual yield per tree for all trees on the acre decreases by 4 pounds due to overcrowding. Find the number of trees that should be planted on an acre to produce the maximum yield. How many pounds is the maximum yield?

Review Exercises

90. Solve: $\frac{2}{x + 5} + \frac{1}{x - 5} = \frac{16}{x^2 - 25}$. (Section 7.6, Example 5)

91. Simplify: $\dfrac{1 + \frac{2}{x}}{1 - \frac{4}{x^2}}$. (Section 7.5, Example 2 or Example 5)

92. Solve the system:

$$\begin{aligned} 2x + 3y &= 6 \\ x - 4y &= 14. \end{aligned}$$

(Section 4.3, Example 2)

MID-CHAPTER CHECK POINT

CHAPTER 11

What You Know: We saw that not all quadratic equations can be solved by factoring. We learned three new methods for solving these equations: the square root property, completing the square, and the quadratic formula. We saw that the discriminant of $ax^2 + bx + c$, namely $b^2 - 4ac$, determines the number and type of the equation's solutions. We graphed quadratic functions using vertices, intercepts, and additional points, as necessary. We learned that the vertex of $f(x) = a(x - h)^2 + k$ is (h, k) and the vertex of $f(x) = ax^2 + bx + c$ is $\left(-\frac{b}{2a}, f\left(-\frac{b}{2a}\right)\right)$. We used the vertex to solve problems that involved minimizing or maximizing quadratic functions.

In Exercises 1–13, solve each equation by the method of your choice. Simplify solutions, if possible.

1. $(3x - 5)^2 = 36$

2. $5x^2 - 2x = 7$

3. $3x^2 - 6x - 2 = 0$

4. $x^2 + 6x = -2$

5. $5x^2 + 1 = 37$

6. $x^2 - 5x + 8 = 0$

7. $2x^2 + 26 = 0$

8. $(2x + 3)(x + 2) = 10$

9. $(x + 3)^2 = 24$

10. $\frac{1}{x^2} - \frac{4}{x} + 1 = 0$

11. $x(2x - 3) = -4$

12. $\frac{x^2}{3} + \frac{x}{2} = \frac{2}{3}$

13. $\frac{2x}{x^2 + 6x + 8} = \frac{x}{x + 4} - \frac{2}{x + 2}$

14. Solve by completing the square: $x^2 + 10x - 3 = 0$.

In Exercises 15–16, find the length (in simplified radical form and rounded to two decimal places) and the midpoint of the line segment with the given endpoints.

15. $(2, -2)$ and $(-2, 2)$

16. $(-5, 8)$ and $(-10, 14)$

In Exercises 17–20, graph the given quadratic function. Give each function's domain and range.

17. $f(x) = (x - 3)^2 - 4$

18. $g(x) = 5 - (x + 2)^2$

19. $h(x) = -x^2 - 4x + 5$

20. $f(x) = 3x^2 - 6x + 1$

In Exercises 21–22, without solving the equation, determine the number and type of solutions.

21. $2x^2 + 5x + 4 = 0$

22. $10x(x + 4) = 15x - 15$

In Exercises 23–24, write a quadratic equation in standard form with the given solution set.

23. $\left\{-\frac{1}{2}, \frac{3}{4}\right\}$

24. $\{-2\sqrt{3}, 2\sqrt{3}\}$

25. A company manufactures and sells bath cabinets. The function

$$P(x) = -x^2 + 150x - 4425$$

models the company's daily profit, $P(x)$, when x cabinets are manufactured and sold per day. How many cabinets should be manufactured and sold per day to maximize the company's profit? What is the maximum daily profit?

26. Among all pairs of numbers whose sum is -18, find a pair whose product is as large as possible. What is the maximum product?

27. The base of a triangle measures 40 inches minus twice the measure of its height. For what measure of the height does the triangle have a maximum area? What is the maximum area?

SECTION 11.4

Objective

1 Solve equations that are quadratic in form.

EQUATIONS QUADRATIC IN FORM

"My husband asked me if we have any cheese puffs. Like he can't go and lift the couch cushion up himself."—Roseanne

Photo by Michael Lutzky. © 2000, *The Washington Post*. Repinted with permission.

How important is it for you to have a clean house? The percentage of people who find this to be quite important varies by age. In the exercise set, you will work with a function that models this phenomenon. Your work will be based on equations that are not quadratic, but that can be written as quadratic equations using an appropriate substitution. Here are some examples:

Given Equation	Substitution	New Equation
$x^4 - 10x^2 + 9 = 0$ or $(x^2)^2 - 10x^2 + 9 = 0$	$t = x^2$	$t^2 - 10t + 9 = 0$
$5x^{\frac{2}{3}} + 11x^{\frac{1}{3}} + 2 = 0$ or $5\left(x^{\frac{1}{3}}\right)^2 + 11x^{\frac{1}{3}} + 2 = 0$	$t = x^{\frac{1}{3}}$	$5t^2 + 11t + 2 = 0$

An equation that is **quadratic in form** is one that can be expressed as a quadratic equation using an appropriate substitution. Both of the preceding given equations are quadratic in form.

1 Solve equations that are quadratic in form.

Equations that are quadratic in form contain an expression to a power, the same expression to that power squared, and a constant term. By letting t equal the expression to the power, a quadratic equation in t will result. Now it's easy. Solve this quadratic equation for t. Finally, use your substitution to find the values for the variable in the given equation. Example 1 shows how this is done.

EXAMPLE 1 Solving an Equation Quadratic in Form

Solve: $x^4 - 10x^2 + 9 = 0$.

SOLUTION Notice that the equation contains an expression to a power, x^2, the same expression to that power squared, x^4 or $(x^2)^2$, and a constant term, 9. We let t equal the expression to the power. Thus,

$$\text{let } t = x^2.$$

USING TECHNOLOGY

The graph of

$$y = x^4 - 10x^2 + 9$$

has x-intercepts at $-3, -1, 1,$ and 3. This verifies that the solution set of

$$x^4 - 10x^2 + 9 = 0$$

is $\{-3, -1, 1, 3\}$.

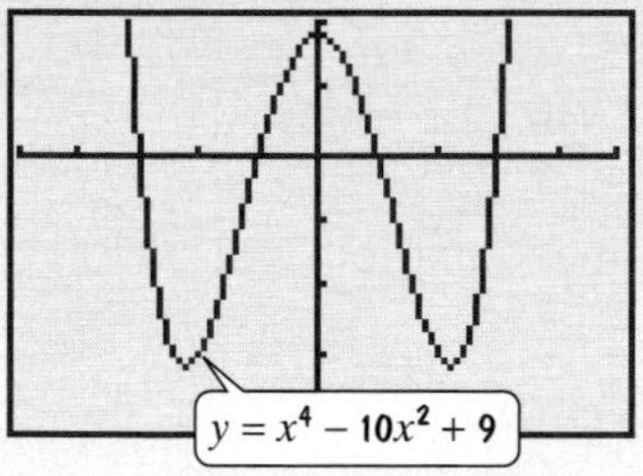

$[-5, 5, 1]$ by $[-20, 10, 5]$

Now we write the given equation as a quadratic equation in t and solve for t.

$x^4 - 10x^2 + 9 = 0$	This is the given equation.
$(x^2)^2 - 10x^2 + 9 = 0$	The given equation contains x^2 and x^2 squared.
$t^2 - 10t + 9 = 0$	Let $t = x^2$. Replace x^2 with t.
$(t - 9)(t - 1) = 0$	Factor.
$t - 9 = 0$ or $t - 1 = 0$	Set each factor equal to zero.
$t = 9$ $\quad$ $t = 1$	Solve for t.

We're not done! Why not? We were asked to solve for x and we have values for t. We use the original substitution, $t = x^2$, to solve for x. Replace t with x^2 in each equation shown.

$x^2 = 9$ or $x^2 = 1$	
$x = \pm\sqrt{9}$ $\quad$ $x = \pm\sqrt{1}$	Apply the square root property.
$x = \pm 3$ $\quad$ $x = \pm 1$	

Substitute these values into the given equation and verify that the solutions are $-3, 3, -1$, and 1. The solution set is $\{-3, 3, -1, 1\}$. You may prefer to express the solution set in numerical order as $\{-3, -1, 1, 3\}$. ■

CHECK POINT 1 Solve: $x^4 - 17x^2 + 16 = 0$.

If checking proposed solutions is not overly cumbersome, you should do so either algebraically or with a graphing utility. The Using Technology box shows a check of the four solutions in Example 1. Are there situations when solving equations quadratic in form where a check is essential? Yes. **If at any point in the solution process both sides of an equation are raised to an even power, a check is required**. Extraneous solutions that are not solutions of the given equation may have been introduced.

EXAMPLE 2 Solving an Equation Quadratic in Form

Solve: $2x - \sqrt{x} - 10 = 0$.

SOLUTION To identify exponents on the terms, let's rewrite $\sqrt{x}$ as $x^{\frac{1}{2}}$. The equation can be expressed as

$$2x^1 - x^{\frac{1}{2}} - 10 = 0.$$

Notice that the equation contains an expression to a power, $x^{\frac{1}{2}}$, the same expression to that power squared, x^1 or $\left(x^{\frac{1}{2}}\right)^2$, and a constant term, -10. We let t equal the expression to the power. Thus,

$$\text{let } t = x^{\frac{1}{2}}.$$

Now we write the given equation as a quadratic equation in t and solve for t.

$2x^1 - x^{\frac{1}{2}} - 10 = 0$	This is the given equation in exponential form.
$2\left(x^{\frac{1}{2}}\right)^2 - x^{\frac{1}{2}} - 10 = 0$	The equation contains $x^{\frac{1}{2}}$ and $x^{\frac{1}{2}}$ squared.
$2t^2 - t - 10 = 0$	Let $t = x^{\frac{1}{2}}$. Replace $x^{\frac{1}{2}}$ with t.
$(2t - 5)(t + 2) = 0$	Factor.
$2t - 5 = 0$ or $t + 2 = 0$	Set each factor equal to zero.
$t = \frac{5}{2}$ $\quad$ $t = -2$	Solve for t.

Use the original substitution, $t = x^{\frac{1}{2}}$, to solve for x. Replace t with $x^{\frac{1}{2}}$ in each of the preceding equations, $t = \frac{5}{2}$ and $t = -2$.

$$x^{\frac{1}{2}} = \frac{5}{2} \quad \text{or} \quad x^{\frac{1}{2}} = -2$$ Replace t with $x^{\frac{1}{2}}$.

$$\left(x^{\frac{1}{2}}\right)^2 = \left(\frac{5}{2}\right)^2 \qquad \left(x^{\frac{1}{2}}\right)^2 = (-2)^2$$ Solve for x by squaring both sides of each equation.

Both sides are raised to even powers. We must check.

$$x = \frac{25}{4} \qquad x = 4$$ Square $\frac{5}{2}$ and -2.

It is essential to check both proposed solutions in the original equation.

Check $\frac{25}{4}$:	**Check 4:**
$2x - \sqrt{x} - 10 = 0$	$2x - \sqrt{x} - 10 = 0$
$2 \cdot \frac{25}{4} - \sqrt{\frac{25}{4}} - 10 \stackrel{?}{=} 0$	$2 \cdot 4 - \sqrt{4} - 10 \stackrel{?}{=} 0$
$\frac{25}{2} - \frac{5}{2} - 10 \stackrel{?}{=} 0$	$8 - 2 - 10 \stackrel{?}{=} 0$
$\frac{20}{2} - 10 \stackrel{?}{=} 0$	$6 - 10 \stackrel{?}{=} 0$
$0 = 0$, true	$-4 = 0$, false

The check indicates that 4 is not a solution. It is an extraneous solution brought about by squaring each side of the equation. The only solution is $\frac{25}{4}$, and the solution set is $\left\{\frac{25}{4}\right\}$. ■

CHECK POINT 2 Solve: $x - 2\sqrt{x} - 8 = 0$.

The equations in Examples 1 and 2 can be solved by methods other than using substitutions.

$$x^4 - 10x^2 + 9 = 0 \qquad 2x - \sqrt{x} - 10 = 0$$

This equation can be solved directly by factoring: $(x^2 - 9)(x^2 - 1) = 0$.

This equation can be solved by isolating the radical term: $2x - 10 = \sqrt{x}$. Then square both sides.

In the examples that follow, solving the equations by methods other than first introducing a substitution becomes increasingly difficult.

EXAMPLE 3 Solving an Equation Quadratic in Form

Solve: $(x^2 - 5)^2 + 3(x^2 - 5) - 10 = 0$.

SOLUTION This equation contains $x^2 - 5$ and $x^2 - 5$ squared. We

$$\text{let} \quad t = x^2 - 5.$$

$(x^2 - 5)^2 + 3(x^2 - 5) - 10 = 0$	This is the given equation.
$t^2 + 3t - 10 = 0$	Let $t = x^2 - 5$.
$(t + 5)(t - 2) = 0$	Factor.
$t + 5 = 0$ or $t - 2 = 0$	Set each factor equal to zero.
$t = -5$ $\qquad$ $t = 2$	Solve for t.

Use the original substitution, $t = x^2 - 5$, to solve for x. Replace t with $x^2 - 5$ in each of the preceding equations.

$x^2 - 5 = -5$ or $x^2 - 5 = 2$	Replace t with $x^2 - 5$.
$x^2 = 0$ $\qquad$ $x^2 = 7$	Solve for x by isolating x^2.
$x = 0$ $\qquad$ $x = \pm\sqrt{7}$	Apply the square root property.

Although we did not raise both sides of an equation to an even power, checking the three proposed solutions in the original equation is a good idea. Do this now and verify that the solutions are $-\sqrt{7}$, 0, and $\sqrt{7}$, and the solution set is $\{-\sqrt{7}, 0, \sqrt{7}\}$. ■

✔ **CHECK POINT 3** Solve: $(x^2 - 4)^2 + (x^2 - 4) - 6 = 0$.

EXAMPLE 4 Solving an Equation Quadratic in Form

Solve: $10x^{-2} + 7x^{-1} + 1 = 0$.

SOLUTION Notice that the equation contains an expression to a power, x^{-1}, the same expression to that power squared, x^{-2} or $(x^{-1})^2$, and a constant term, 1. We let t equal the expression to the power. Thus,

$$\text{let } t = x^{-1}.$$

Now we write the given equation as a quadratic equation in t and solve for t.

$10x^{-2} + 7x^{-1} + 1 = 0$	This is the given equation.
$10(x^{-1})^2 + 7x^{-1} + 1 = 0$	The equation contains x^{-1} and x^{-1} squared.
$10t^2 + 7t + 1 = 0$	Let $t = x^{-1}$.
$(5t + 1)(2t + 1) = 0$	Factor.
$5t + 1 = 0$ or $2t + 1 = 0$	Set each factor equal to zero.
$5t = -1$ $\qquad$ $2t = -1$	Solve each equation for t.
$t = -\frac{1}{5}$ $\qquad$ $t = -\frac{1}{2}$	

Use the original substitution, $t = x^{-1}$, to solve for x. Replace t with x^{-1} in each of the preceding equations, $t = -\frac{1}{5}$ and $t = -\frac{1}{2}$.

$$x^{-1} = -\frac{1}{5} \quad \text{or} \quad x^{-1} = -\frac{1}{2}$$ Replace t with x^{-1}.

$$(x^{-1})^{-1} = \left(-\frac{1}{5}\right)^{-1} \quad (x^{-1})^{-1} = \left(-\frac{1}{2}\right)^{-1}$$ Solve for x by raising both sides of each equation to the -1 power.

$$x = -5 \qquad x = -2$$

$\left(-\frac{1}{5}\right)^{-1} = \frac{1}{-\frac{1}{5}} = -5$ $\qquad$ $\left(-\frac{1}{2}\right)^{-1} = \frac{1}{-\frac{1}{2}} = -2$

We did not raise both sides of an equation to an even power. A check will show that both -5 and -2 are solutions of the original equation. The solution set is $\{-5, -2\}$. ■

CHECK POINT 4 Solve: $2x^{-2} + x^{-1} - 1 = 0$.

EXAMPLE 5 Solving an Equation Quadratic in Form

Solve: $5x^{\frac{2}{3}} + 11x^{\frac{1}{3}} + 2 = 0$.

SOLUTION Notice that the equation contains an expression to a power, $x^{\frac{1}{3}}$, the same expression to that power squared, $x^{\frac{2}{3}}$ or $\left(x^{\frac{1}{3}}\right)^2$, and a constant term, 2. We let t equal the expression to the power. Thus,

$$\text{let } t = x^{\frac{1}{3}}.$$

Now we write the given equation as a quadratic equation in t and solve for t.

$$5x^{\frac{2}{3}} + 11x^{\frac{1}{3}} + 2 = 0$$ This is the given equation.

$$5\left(x^{\frac{1}{3}}\right)^2 + 11\left(x^{\frac{1}{3}}\right) + 2 = 0$$ The given equation contains $x^{\frac{1}{3}}$ and $x^{\frac{1}{3}}$ squared.

$$5t^2 + 11t + 2 = 0$$ Let $t = x^{\frac{1}{3}}$.

$$(5t + 1)(t + 2) = 0$$ Factor.

$$5t + 1 = 0 \quad \text{or} \quad t + 2 = 0$$ Set each factor equal to 0.

$$t = -\tfrac{1}{5} \qquad t = -2$$ Solve for t.

Use the original substitution, $t = x^{\frac{1}{3}}$, to solve for x. Replace t with $x^{\frac{1}{3}}$ in each of the preceding equations.

$$x^{\frac{1}{3}} = -\frac{1}{5} \quad \text{or} \quad x^{\frac{1}{3}} = -2$$ Replace t with $x^{\frac{1}{3}}$.

$$\left(x^{\frac{1}{3}}\right)^3 = \left(-\frac{1}{5}\right)^3 \qquad \left(x^{\frac{1}{3}}\right)^3 = (-2)^3$$ Solve for x by cubing both sides of each equation.

$$x = -\frac{1}{125} \qquad x = -8$$

We did not raise both sides of an equation to an even power. A check will show that both -8 and $-\frac{1}{125}$ are solutions of the original equation. The solution set is $\left\{-8, -\frac{1}{125}\right\}$. ■

CHECK POINT 5 Solve: $3x^{\frac{2}{3}} - 11x^{\frac{1}{3}} - 4 = 0$.

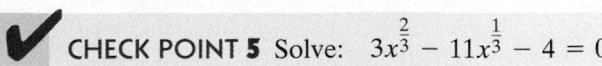

11.4 EXERCISE SET

Student Solutions Manual | CD/Video | PH Math/Tutor Center | MathXL Tutorials on CD | MathXL® | MyMathLab | Interactmath.com

Practice Exercises

In Exercises 1–32, solve each equation by making an appropriate substitution. If at any point in the solution process both sides of an equation are raised to an even power, a check is required.

1. $x^4 - 5x^2 + 4 = 0$

2. $x^4 - 13x^2 + 36 = 0$

3. $x^4 - 11x^2 + 18 = 0$

4. $x^4 - 9x^2 + 20 = 0$

5. $x^4 + 2x^2 = 8$

6. $x^4 + 4x^2 = 5$

7. $x + \sqrt{x} - 2 = 0$

8. $x + \sqrt{x} - 6 = 0$

9. $x - 4x^{\frac{1}{2}} - 21 = 0$

10. $x - 6x^{\frac{1}{2}} + 8 = 0$

11. $x - 13\sqrt{x} + 40 = 0$

12. $2x - 7\sqrt{x} - 30 = 0$

13. $(x - 5)^2 - 4(x - 5) - 21 = 0$

14. $(x + 3)^2 + 7(x + 3) - 18 = 0$

15. $(x^2 - 1)^2 - (x^2 - 1) = 2$

16. $(x^2 - 2)^2 - (x^2 - 2) = 6$

17. $(x^2 + 3x)^2 - 8(x^2 + 3x) - 20 = 0$

18. $(x^2 - 2x)^2 - 11(x^2 - 2x) + 24 = 0$

19. $x^{-2} - x^{-1} - 20 = 0$

20. $x^{-2} - x^{-1} - 6 = 0$

21. $2x^{-2} - 7x^{-1} + 3 = 0$

22. $20x^{-2} + 9x^{-1} + 1 = 0$

23. $x^{-2} - 4x^{-1} = 3$

24. $x^{-2} - 6x^{-1} = -4$

25. $x^{\frac{2}{3}} - x^{\frac{1}{3}} - 6 = 0$

26. $x^{\frac{2}{3}} + 2x^{\frac{1}{3}} - 3 = 0$

27. $x^{\frac{2}{5}} + x^{\frac{1}{5}} - 6 = 0$

28. $x^{\frac{2}{5}} + x^{\frac{1}{5}} - 2 = 0$

29. $2x^{\frac{1}{2}} - x^{\frac{1}{4}} = 1$

30. $2x^{\frac{1}{2}} - 5x^{\frac{1}{4}} = 3$

31. $\left(x - \frac{8}{x}\right)^2 + 5\left(x - \frac{8}{x}\right) - 14 = 0$

32. $\left(x - \frac{10}{x}\right)^2 + 6\left(x - \frac{10}{x}\right) - 27 = 0$

In Exercises 33–38, find the x-intercepts of the given function, f. Then use the x-intercepts to match each function with its graph. [The graphs are labeled (a) through (f).]

33. $f(x) = x^4 - 5x^2 + 4$

34. $f(x) = x^4 - 13x^2 + 36$

35. $f(x) = x^{\frac{1}{3}} + 2x^{\frac{1}{6}} - 3$

36. $f(x) = x^{-2} - x^{-1} - 6$

37. $f(x) = (x + 2)^2 - 9(x + 2) + 20$

38. $f(x) = 2(x + 2)^2 + 5(x + 2) - 3$

a.

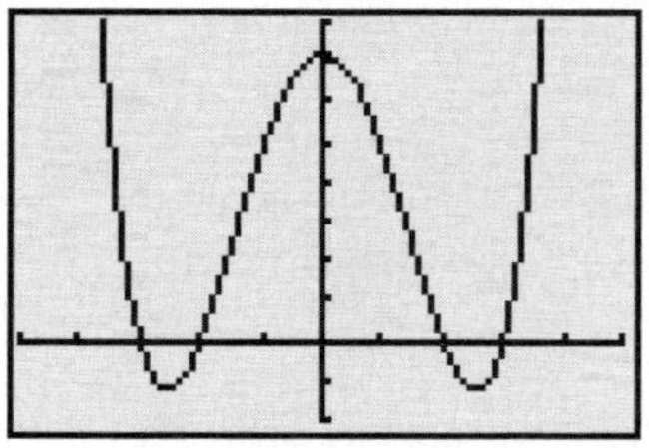

$[-5, 5, 1]$ by $[-10, 40, 5]$

b.

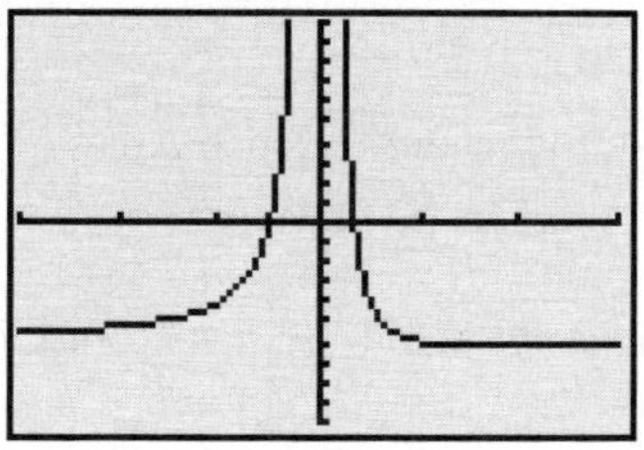

$[-3, 3, 1]$ by $[-10, 10, 1]$

c.

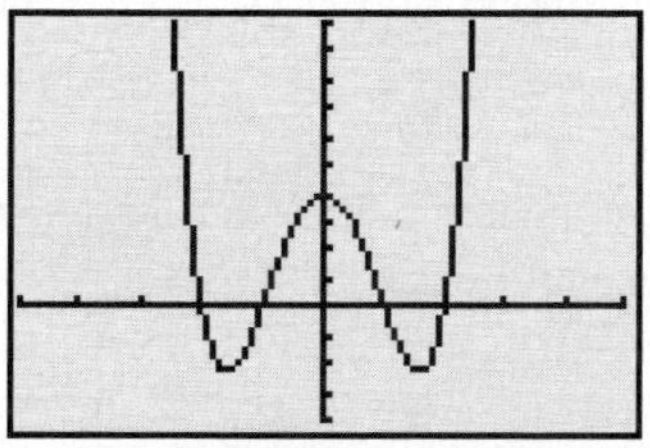

$[-5, 5, 1]$ by $[-4, 10, 1]$

d.

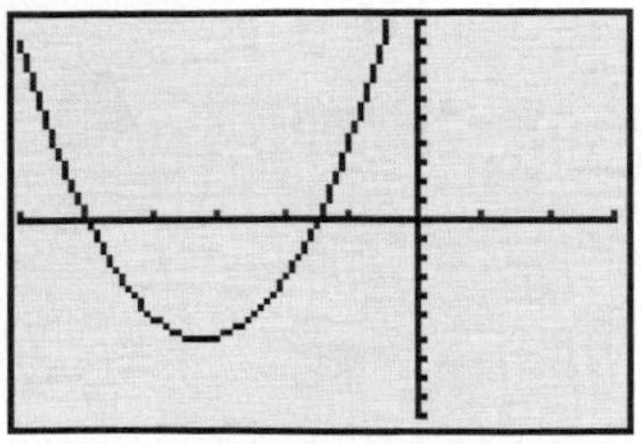

$[-6, 3, 1]$ by $[-10, 10, 1]$

e.

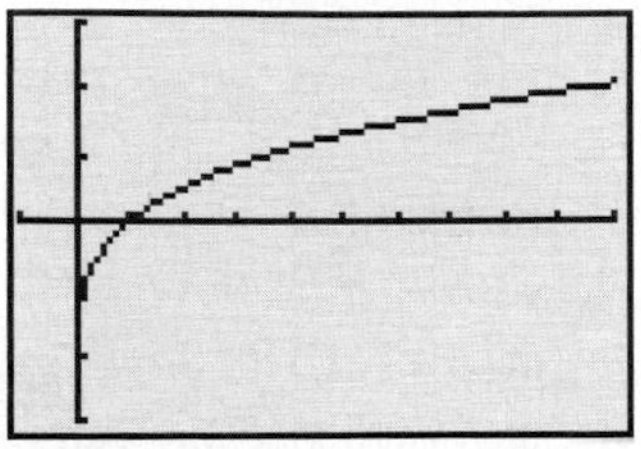

$[-1, 10, 1]$ by $[-3, 3, 1]$

f.

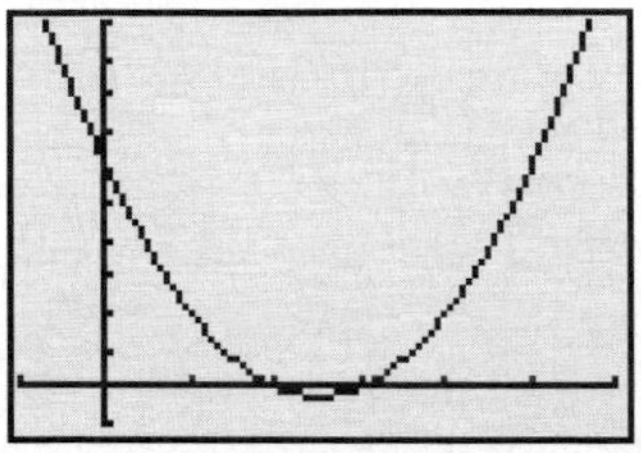

$[-1, 6, 1]$ by $[-1, 10, 1]$

Practice Plus

39. Let $f(x) = (x^2 + 3x - 2)^2 - 10(x^2 + 3x - 2)$. Find all x such that $f(x) = -16$.

40. Let $f(x) = (x^2 + 2x - 2)^2 - 7(x^2 + 2x - 2)$. Find all x such that $f(x) = -6$.

41. Let $f(x) = 3\left(\frac{1}{x} + 1\right)^2 + 5\left(\frac{1}{x} + 1\right)$. Find all x such that $f(x) = 2$.

42. Let $f(x) = 2x^{\frac{2}{3}} + 3x^{\frac{1}{3}}$. Find all x such that $f(x) = 2$.

43. Let $f(x) = \frac{x}{x - 4}$ and $g(x) = 13\sqrt{\frac{x}{x - 4}} - 36$. Find all x such that $f(x) = g(x)$.

44. Let $f(x) = \frac{x}{x - 2} + 10$ and $g(x) = -11\sqrt{\frac{x}{x - 2}}$. Find all x such that $f(x) = g(x)$.

45. Let $f(x) = 3(x - 4)^{-2}$ and $g(x) = 16(x - 4)^{-1}$. Find all x such that $f(x)$ exceeds $g(x)$ by 12.

46. Let $f(x) = 6\left(\frac{2x}{x - 3}\right)^2$ and $g(x) = 5\left(\frac{2x}{x - 3}\right)$. Find all x such that $f(x)$ exceeds $g(x)$ by 6.

Application Exercises

How important is it for you to have a clean house? The bar graph indicates that the percentage of people who find this to be quite important varies by age. The percentage, $P(x)$, who find having a clean house very important can be modeled by the function

$$P(x) = 0.04(x + 40)^2 - 3(x + 40) + 104,$$

where x is the number of years a person's age is above or below 40. Thus, x is positive for people over 40 and negative for people under 40. Use the function to solve Exercises 47–48.

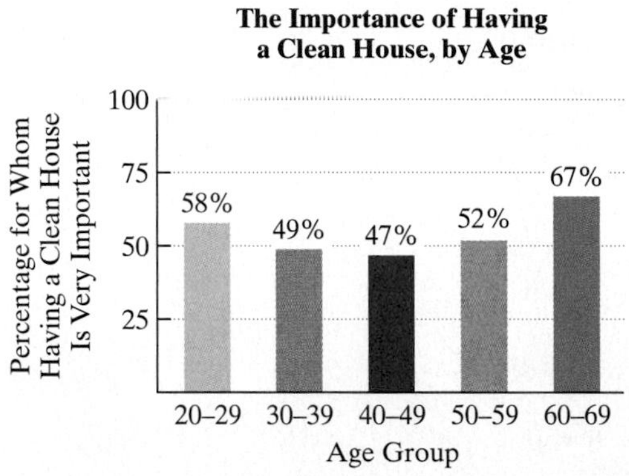

Source: Soap and Detergent Association

47. According to the model, at which ages do 60% of us feel that having a clean house is very important? Substitute 60 for $P(x)$ and solve the quadratic-in-form equation. How well does the function model the data shown in the bar graph?

48. According to the model, at which ages do 50% of us feel that having a clean house is very important? Substitute 50 for $P(x)$ and solve the quadratic-in-form equation. How well does the function model the data shown in the bar graph?

Writing in Mathematics

49. Explain how to recognize an equation that is quadratic in form. Provide two original examples with your explanation.

50. Describe two methods for solving this equation:

$$x - 5\sqrt{x} + 4 = 0.$$

51. In the twenty-first century, collecting and presenting data are big business. Because statisticians both record and influence our behavior, you should always ask yourself if the person or group presenting the data has any special case to make for or against the displayed information. Using this criterion, describe your impressions of the data in the bar graph for Exercises 47–48.

Technology Exercises

52. Use a graphing utility to verify the solutions of any five equations in Exercises 1-32 that you solved algebraically. The real solutions should appear as x-intercepts on the graph of the function related to the given equation.

Use a graphing utility to solve the equations in Exercises 53–60. Check by direct substitution.

53. $x^6 - 7x^3 - 8 = 0$

54. $3(x-2)^{-2} - 4(x-2)^{-1} + 1 = 0$

55. $x^4 - 10x^2 + 9 = 0$

56. $2x + 6\sqrt{x} = 8$

57. $2(x+1)^2 = 5(x+1) + 3$

58. $(x^2 - 3x)^2 + 2(x^2 - 3x) - 24 = 0$

59. $x^{\frac{1}{2}} + 4x^{\frac{1}{4}} = 5$

60. $x^{\frac{2}{3}} - 3x^{\frac{1}{3}} + 2 = 0$

Critical Thinking Exercises

61. Which one of the following is true?

a. If an equation is quadratic in form, there is only one method that can be used to obtain its solution.

b. An equation that is quadratic in form must have a variable factor in one term that is the square of the variable factor in another term.

c. Because x^6 is the square of x^3, the equation $x^6 - 5x^3 + 6x = 0$ is quadratic in form.

d. To solve $x - 9\sqrt{x} + 14 = 0$, we let $\sqrt{t} = x$.

In Exercises 62–64, use a substitution to solve each equation.

62. $x^4 - 5x^2 - 2 = 0$

63. $5x^6 + x^3 = 18$

64. $\sqrt{\frac{x+4}{x-1}} + \sqrt{\frac{x-1}{x+4}} = \frac{5}{2}\left(\text{Let } t = \sqrt{\frac{x+4}{x-1}}.\right)$

Review Exercises

65. Simplify:

$$\frac{2x^2}{10x^3 - 2x^2}.$$

(Section 7.1, Example 3)

66. Divide: $\frac{2+i}{1-i}$. (Section 10.7, Example 5)

67. If $f(x) = \sqrt{x+1}$, find $f(3) - f(24)$. (Section 8.1, Example 3)

SECTION 11.5

POLYNOMIAL AND RATIONAL INEQUALITIES

Objectives

1. Solve polynomial inequalities.
2. Solve rational inequalities.
3. Solve problems modeled by polynomial or rational inequalities.

People are going to live longer in the twenty-first century. This will put added pressure on the Social Security and Medicare systems. The bar graph in Figure 11.17 shows the cost of Medicare, in billions of dollars, through 2005.

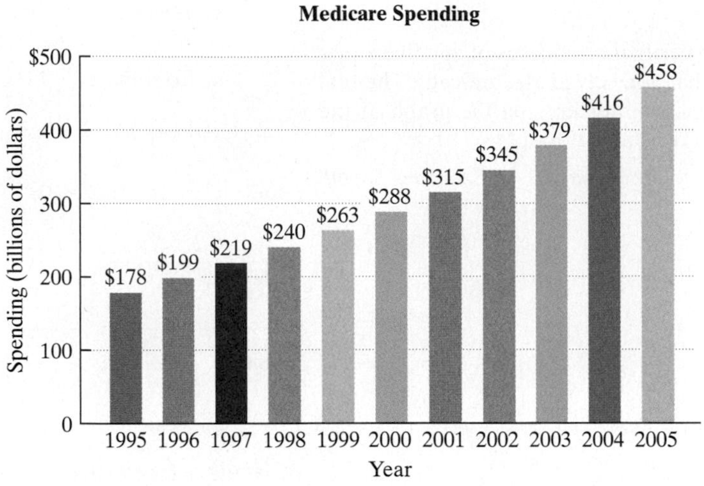

FIGURE 11.17

Source: Congressional Budget Office

Medicare spending, $f(x)$, in billions of dollars, x years after 1995 can be modeled by the quadratic function

$$f(x) = 1.2x^2 + 15.2x + 181.4.$$

To determine in which years Medicare spending will exceed \$500 billion, we must solve the inequality

$$1.2x^2 + 15.2x + 181.4 > 500.$$

Medicare spending exceeds \$500 billion.

We begin by subtracting 500 from both sides. This will give us zero on the right:

$$1.2x^2 + 15.2x + 181.4 - 500 > 500 - 500$$
$$1.2x^2 + 15.2x - 318.6 > 0.$$

The form of this inequality is $ax^2 + bx + c > 0$. Such an inequality is called a *polynomial inequality*.

DEFINITION OF A POLYNOMIAL INEQUALITY A polynomial inequality is any inequality that can be put in one of the forms

$$f(x) < 0, \quad f(x) > 0, \quad f(x) \le 0, \quad \text{or} \quad f(x) \ge 0,$$

where f is a polynomial function.

In this section, we establish the basic techniques for solving polynomial inequalities. We will use these techniques to solve inequalities involving rational functions.

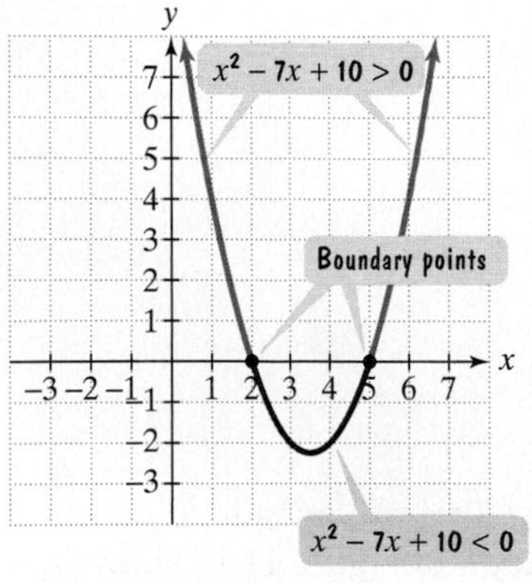

FIGURE 11.18

Solving Polynomial Inequalities Graphs can help us visualize the solutions of polynomial inequalities. For example, the graph of $f(x) = x^2 - 7x + 10$ is shown in Figure 11.18. The x-intercepts, 2 and 5, are **boundary points** between where the graph lies above the x-axis, shown in blue, and where the graph lies below the x-axis, shown in red.

Locating the x-intercepts of a polynomial function, f, is an important step in finding the solution set for polynomial inequalities in the form $f(x) < 0$ or $f(x) > 0$.

We use the x-intercepts of f as boundary points that divide the real number line into intervals. On each interval, the graph of f is either above the x-axis $[f(x) > 0]$ or below the x-axis $[f(x) < 0]$. For this reason, x-intercepts play a fundamental role in solving polynomial inequalities. The x-intercepts are found by solving the equation $f(x) = 0$.

1 Solve polynomial inequalities.

PROCEDURE FOR SOLVING POLYNOMIAL INEQUALITIES

1. Express the inequality in the form

$$f(x) < 0 \quad \text{or} \quad f(x) > 0,$$

where f is a polynomial function

2. Solve the equation $f(x) = 0$. The real solutions are the **boundary points**.

3. Locate these boundary points on a number line, thereby dividing the number line into intervals.

4. Choose one representative number within each interval and evaluate f at that number.

  a. If the value of f is positive, then $f(x) > 0$ for all numbers, x, in the interval.

  b. If the value of f is negative, then $f(x) < 0$ for all numbers, x, in the interval.

5. Write the solution set, selecting the interval(s) that satisfy the given inequality.

This procedure is valid if $<$ is replaced by $\le$ or $>$ is replaced by $\ge$. However, if the inequality involves $\le$ or $\ge$, include the boundary points [the solutions of $f(x) = 0$] in the solution set.

EXAMPLE 1 Solving a Polynomial Inequality

Solve and graph the solution set on a real number line: $2x^2 + x > 15$.

SOLUTION

Step 1. **Express the inequality in the form $f(x) < 0$ or $f(x) > 0$.** We begin by rewriting the inequality so that 0 is on the right side.

$$2x^2 + x > 15 \qquad \text{This is the given inequality.}$$

$$2x^2 + x - 15 > 15 - 15 \qquad \text{Subtract 15 from both sides.}$$

$$2x^2 + x - 15 > 0 \qquad \text{Simplify.}$$

This inequality is equivalent to the one we wish to solve. It is in the form $f(x) > 0$, where $f(x) = 2x^2 + x - 15$.

Step 2. **Solve the equation $f(x) = 0$.** We find the x-intercepts of $f(x) = 2x^2 + x - 15$ by solving the equation $2x^2 + x - 15 = 0$.

$$2x^2 + x - 15 = 0 \qquad \text{This polynomial equation is a quadratic equation.}$$

$$(2x - 5)(x + 3) = 0 \qquad \text{Factor.}$$

$$2x - 5 = 0 \quad \text{or} \quad x + 3 = 0 \qquad \text{Set each factor equal to 0.}$$

$$x = \tfrac{5}{2} \qquad\qquad x = -3 \qquad \text{Solve for x.}$$

The x-intercepts of f are -3 and $\frac{5}{2}$. We will use these x-intercepts as boundary points on a number line.

Step 3. **Locate the boundary points on a number line and separate the line into intervals.** The number line with the boundary points is shown as follows:

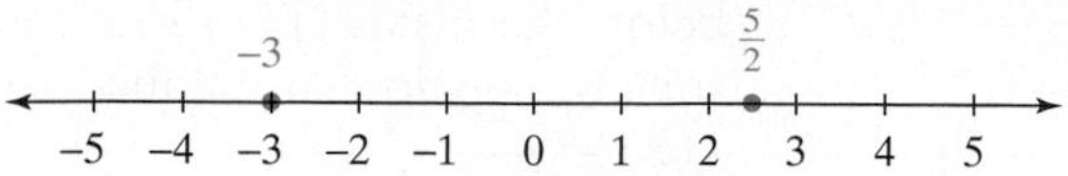

The boundary points divide the number line into three intervals:

$$(-\infty, -3) \quad \left(-3, \tfrac{5}{2}\right) \quad \left(\tfrac{5}{2}, \infty\right).$$

Step 4. **Choose one representative number within each interval and evaluate f at that number.**

Interval	Representative Number	Substitute into $f(x) = 2x^2 + x - 15$	Conclusion
$(-\infty, -3)$	-4	$f(-4) = 2(-4)^2 + (-4) - 15$ $= 13$, positive	$f(x) > 0$ for all x in $(-\infty, -3)$.
$\left(-3, \frac{5}{2}\right)$	0	$f(0) = 2 \cdot 0^2 + 0 - 15$ $= -15$, negative	$f(x) < 0$ for all x in $\left(-3, \frac{5}{2}\right)$.
$\left(\frac{5}{2}, \infty\right)$	3	$f(3) = 2 \cdot 3^2 + 3 - 15$ $= 6$, positive	$f(x) > 0$ for all x in $\left(\frac{5}{2}, \infty\right)$.

Step 5. **Write the solution set, selecting the interval(s) that satisfy the given inequality.** Based on our work in step 4, we see that $f(x) > 0$ for all x in $(-\infty, -3)$ or $\left(\frac{5}{2}, \infty\right)$. Thus, the solution set of the given inequality, $2x^2 + x > 15$, or, equivalently, $2x^2 + x - 15 > 0$, is

$$(-\infty, -3) \cup \left(\tfrac{5}{2}, \infty\right) \text{ or } \left\{x \mid x < -3 \text{ or } x > \tfrac{5}{2}\right\}.$$

The graph of the solution set on a number line is shown as follows:

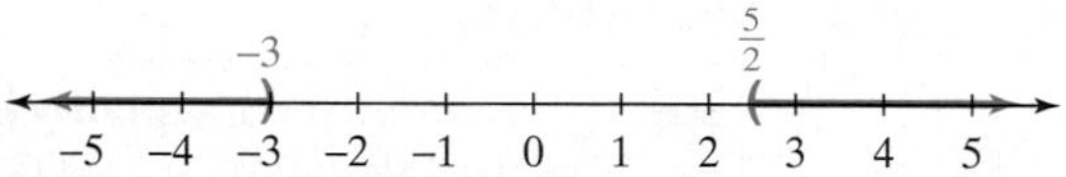

■

USING TECHNOLOGY

The solution set for

$$2x^2 + x > 15$$

or, equivalently,

$$2x^2 + x - 15 > 0$$

can be verified with a graphing utility. The graph of $f(x) = 2x^2 + x - 15$ was obtained using a $[-10, 10, 1]$ by $[-16, 6, 1]$ viewing rectangle. The graph lies above the x-axis, representing $>$, for all x in $(-\infty, -3)$ or $\left(\frac{5}{2}, \infty\right)$.

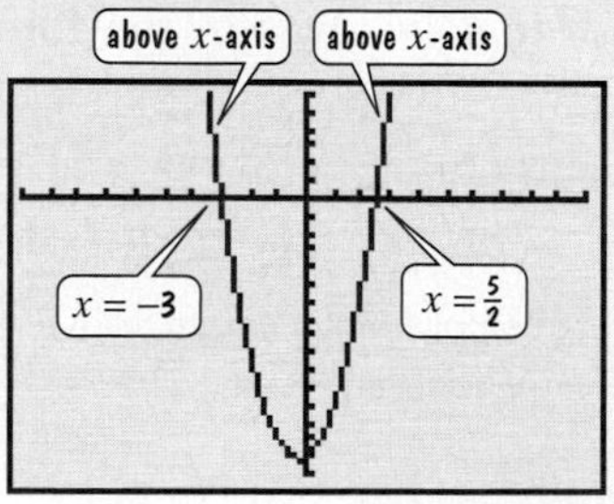

✔ **CHECK POINT 1** Solve and graph the solution set: $x^2 - x > 20$.

EXAMPLE 2 Solving a Polynomial Inequality

Solve and graph the solution set on a real number line: $x^3 + x^2 \leq 4x + 4$.

SOLUTION

Step 1. **Express the inequality in the form $f(x) \leq 0$ or $f(x) \geq 0$.** We begin by rewriting the inequality so that 0 is on the right side.

$x^3 + x^2 \leq 4x + 4$	This is the given inequality.
$x^3 + x^2 - 4x - 4 \leq 4x + 4 - 4x - 4$	Subtract $4x + 4$ from both sides.
$x^3 + x^2 - 4x - 4 \leq 0$	Simplify.

This inequality is equivalent to the one we wish to solve. It is in the form $f(x) \leq 0$, where $f(x) = x^3 + x^2 - 4x - 4$.

USING TECHNOLOGY

The solution set for

$$x^3 + x^2 \le 4x + 4$$

or, equivalently,

$$x^3 + x^2 - 4x - 4 \le 0$$

can be verified with a graphing utility. The graph of $f(x) = x^3 + x^2 - 4x - 4$ lies on or below the x-axis, representing $\le$, for all x in $(-\infty, -2]$ or $[-1, 2]$.

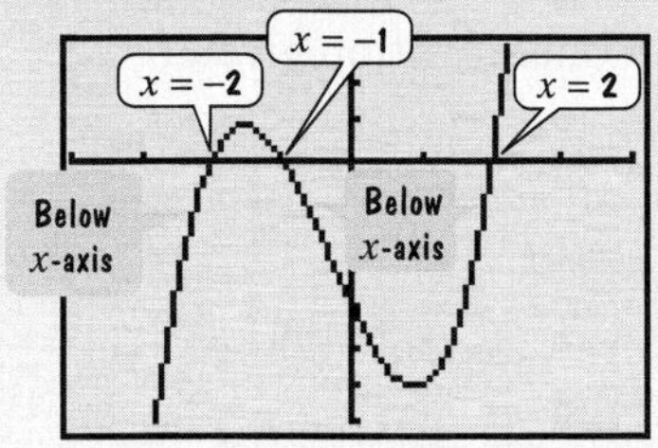

[−4, 4, 1] by [−7, 3, 1]

Step 2. **Solve the equation $f(x) = 0$.** We find the x-intercepts of $f(x) = x^3 + x^2 - 4x - 4$ by solving the equation $x^3 + x^2 - 4x - 4 = 0$.

$$x^3 + x^2 - 4x - 4 = 0$$ This polynomial equation is of degree 3.

$$x^2(x + 1) - 4(x + 1) = 0$$ Factor x^2 from the first two terms and -4 from the last two terms.

$$(x + 1)(x^2 - 4) = 0$$ A common factor of $x + 1$ is factored from the expression.

$$x + 1 = 0 \quad \text{or} \quad x^2 - 4 = 0$$ Set each factor equal to 0.

$$x = -1 \qquad x^2 = 4$$ Solve for x.

$$x = \pm 2$$ Use the square root property.

The x-intercepts of f are -2, -1, and 2. We will use these x-intercepts as boundary points on a number line.

Step 3. **Locate the boundary points on a number line and separate the line into intervals.** The number line with the boundary points is shown as follows:

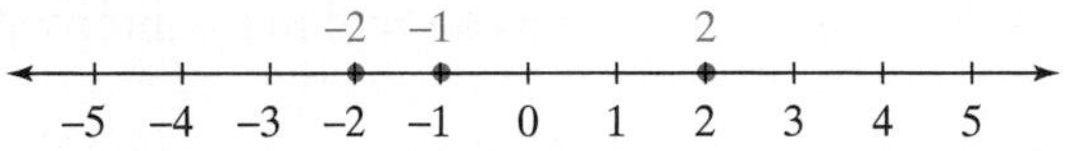

The boundary points divide the number line into four intervals:

$$(-\infty, -2) \quad (-2, -1) \quad (-1, 2) \quad (2, \infty).$$

Step 4. **Choose one representative number within each interval and evaluate f at that number.**

Interval	Representative Number	Substitute into $f(x) = x^3 + x^2 - 4x - 4$	Conclusion
$(-\infty, -2)$	-3	$f(-3) = (-3)^3 + (-3)^2 - 4(-3) - 4$ $= -10$, negative	$f(x) < 0$ for all x in $(-\infty, -2)$.
$(-2, -1)$	-1.5	$f(-1.5) = (-1.5)^3 + (-1.5)^2 - 4(-1.5) - 4$ $= 0.875$, positive	$f(x) > 0$ for all x in $(-2, -1)$.
$(-1, 2)$	0	$f(0) = 0^3 + 0^2 - 4 \cdot 0 - 4$ $= -4$, negative	$f(x) < 0$ for all x in $(-1, 2)$.
$(2, \infty)$	3	$f(3) = 3^3 + 3^2 - 4 \cdot 3 - 4$ $= 20$, positive	$f(x) > 0$ for all x in $(2, \infty)$.

Step 5. **Write the solution set, selecting the interval(s) that satisfy the given inequality.** Based on our work in step 4, we see that $f(x) < 0$ for all x in $(-\infty, -2)$ or $(-1, 2)$. However, because the inequality involves $\le$ (less than or *equal to*), we must also include the solutions of $x^3 + x^2 - 4x - 4 = 0$, namely -2, -1, and 2, in the solution set. Thus, the solution set of the given inequality, $x^3 + x^2 \le 4x + 4$, or, equivalently, $x^3 + x^2 - 4x - 4 \le 0$, is

$$(-\infty, -2] \cup [-1, 2]$$

$$\text{or} \quad \{x \mid x \le -2 \text{ or } -1 \le x \le 2\}.$$

The graph of the solution set on a number line is shown as follows:

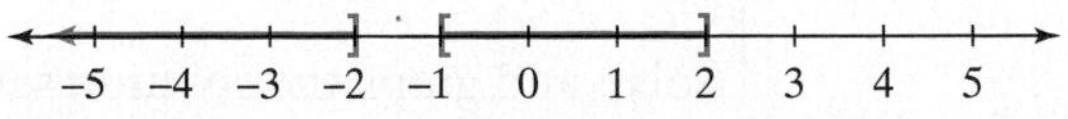

CHECK POINT 2 Solve and graph the solution set on a real number line: $x^3 + 3x^2 \leq x + 3$.

2 Solve rational inequalities.

Solving Rational Inequalities A **rational inequality** is any inequality that can be put in one of the forms

$$f(x) < 0, \quad f(x) > 0, \quad f(x) \leq 0, \quad \text{or} \quad f(x) \geq 0,$$

where f is a rational function. An example of a rational inequality is

$$\frac{3x + 3}{2x + 4} > 0.$$

This inequality is in the form $f(x) > 0$, where f is the rational function given by

$$f(x) = \frac{3x + 3}{2x + 4}.$$

The graph of f is shown in Figure 11.19.

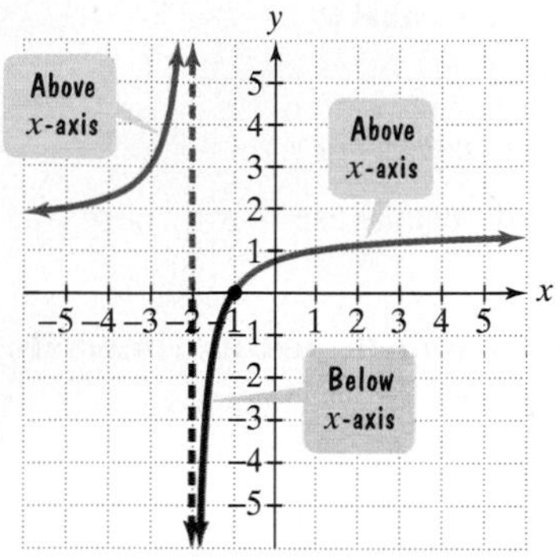

FIGURE 11.19 The graph of $f(x) = \frac{3x + 3}{2x + 4}$

We can find the x-intercept of f by setting the numerator equal to 0:

$$3x + 3 = 0$$
$$3x = -3$$
$$x = -1.$$

f has an x-intercept at −1 and passes through (−1, 0).

We can determine where f is undefined by setting the denominator equal to 0:

$$2x + 4 = 0$$
$$2x = -4$$
$$x = -2.$$

f is undefined at −2. Figure 11.19 shows that the function's vertical asymptote is x = −2.

By setting the numerator and the denominator of f equal to 0, we obtained the solutions -2 and -1. These numbers separate the x-axis into three intervals: $(-\infty, -2)$, $(-2, -1)$, and $(-1, \infty)$. On each interval, the graph of f is either above the x-axis $[f(x) > 0]$ or below the x-axis $[f(x) < 0]$.

Examine the graph in Figure 11.19 carefully. Can you see that it is above the x-axis for all x in $(-\infty, -2)$ or $(-1, \infty)$, shown in blue? Thus, the solution set of $\frac{3x + 3}{2x + 4} > 0$ is $(-\infty, -2) \cup (-1, \infty)$. By contrast, the graph of f lies below the x-axis for all x in $(-2, -1)$, shown in red. Thus, the solution set of $\frac{3x + 3}{2x + 4} < 0$ is $(-2, -1)$.

The first step in solving a rational inequality is to bring all terms to one side, obtaining zero on the other side. Then express the rational function on the nonzero side as a single quotient. The second step is to set the numerator and the denominator of f equal to zero. The solutions of these equations serve as boundary points that separate the real number line into intervals. At this point, the procedure is the same as the one we used for solving polynomial inequalities.

EXAMPLE 3 Solving a Rational Inequality

Solve and graph the solution set: $\frac{x + 1}{x + 3} \geq 2.$

STUDY TIP

Do not begin solving

$$\frac{x+1}{x+3} \geq 2$$

by multiplying both sides by $x + 3$. We do not know if $x + 3$ is positive or negative. Thus, we do not know whether or not to reverse the sense of the inequality.

STUDY TIP

Never include the value that causes a rational function's denominator to equal zero in the solution set of a rational inequality. Division by zero is undefined.

SOLUTION

Step 1. **Express the inequality so that one side is zero and the other side is a single quotient.** We subtract 2 from both sides to obtain zero on the right.

$$\frac{x+1}{x+3} \geq 2 \quad \text{This is the given inequality.}$$

$$\frac{x+1}{x+3} - 2 \geq 0 \quad \text{Subtract 2 from both sides, obtaining 0 on the right.}$$

$$\frac{x+1}{x+3} - \frac{2(x+3)}{x+3} \geq 0 \quad \text{The least common denominator is } x+3. \text{ Express 2 in terms of this denominator.}$$

$$\frac{x+1-2(x+3)}{x+3} \geq 0 \quad \text{Subtract rational expressions.}$$

$$\frac{x+1-2x-6}{x+3} \geq 0 \quad \text{Apply the distributive property.}$$

$$\frac{-x-5}{x+3} \geq 0 \quad \text{Simplify.}$$

This inequality is equivalent to the one we wish to solve. It is in the form $f(x) \geq 0$, where $f(x) = \dfrac{-x-5}{x+3}$.

Step 2. **Set the numerator and the denominator of f equal to zero.** The real solutions are the boundary points.

$$-x - 5 = 0 \qquad x + 3 = 0 \quad \text{Set the numerator and denominator equal to 0. These are the values that make the previous quotient zero or undefined.}$$

$$x = -5 \qquad x = -3 \quad \text{Solve for } x.$$

We will use these solutions as boundary points on a number line.

Step 3. **Locate the boundary points on a number line and separate the line into intervals.** The number line with the boundary points is shown as follows:

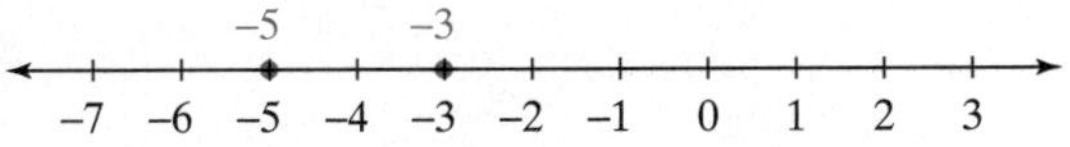

The boundary points divide the number line into three intervals:

$$(-\infty, -5) \quad (-5, -3) \quad (-3, \infty).$$

Step 4. **Choose one representative number within each interval and evaluate f at that number.**

Interval	Representative Number	Substitute into $f(x) = \dfrac{-x-5}{x+3}$	Conclusion
$(-\infty, -5)$	-6	$f(-6) = \dfrac{-(-6)-5}{-6+3} = -\frac{1}{3}$, negative	$f(x) < 0$ for all x in $(-\infty, -5)$.
$(-5, -3)$	-4	$f(-4) = \dfrac{-(-4)-5}{-4+3} = 1$, positive	$f(x) > 0$ for all x in $(-5, -3)$.
$(-3, \infty)$	0	$f(0) = \dfrac{-0-5}{0+3} = -\frac{5}{3}$, negative	$f(x) < 0$ for all x in $(-3, \infty)$.

Step 5. **Write the solution set, selecting the interval(s) that satisfy the given inequality.** Based on our work in step 4, we see that $f(x) > 0$ for all x in $(-5, -3)$. However, because the inequality involves $\geq$ (greater than or *equal to*), we must also include the solution of $f(x) = 0$, namely the value that we obtained when we set the numerator of f equal to zero. Thus, we must include -5 in the solution set. The solution set of the given inequality is

$$[-5, -3) \text{ or } \{x \mid -5 \leq x < -3\}.$$

The graph of the solution set on a number line is shown as follows:

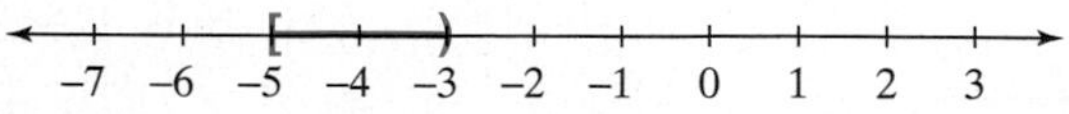

USING TECHNOLOGY

The solution set for

$$\frac{x+1}{x+3} \geq 2$$

or, equivalently,

$$\frac{-x-5}{x+3} \geq 0$$

can be verified with a graphing utility. The graph of $f(x) = \frac{-x-5}{x+3}$ lies on or above the x-axis, representing $\geq$, for all x in $[-5, -3)$.

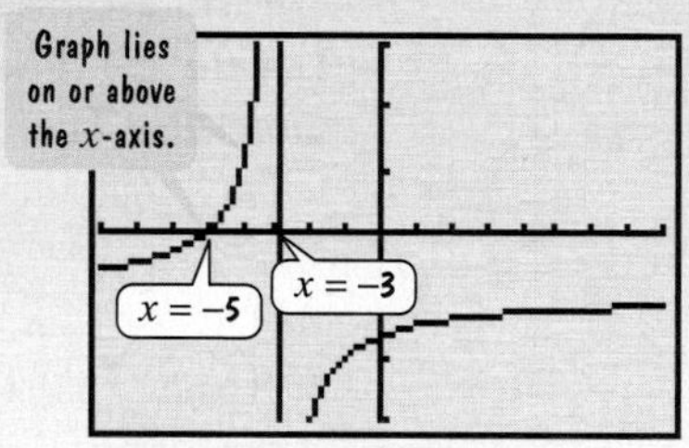

[–8, 8, 1] by [–3, 3, 1]

DISCOVER FOR YOURSELF

Because $(x + 3)^2$ is positive, it is possible so solve

$$\frac{x+1}{x+3} \geq 2$$

by first multiplying both sides by $(x + 3)^2$ (where $x \neq -3$). This will not reverse the sense of the inequality and will clear the fraction. Try using this solution method and compare it to the one shown on pages 780–782.

CHECK POINT 3 Solve and graph the solution set: $\frac{2x}{x+1} \geq 1$.

3 Solve problems modeled by polynomial or rational inequalities.

Applications Polynomial inequalities can be solved to answer questions about variables contained in polynomial functions.

EXAMPLE 4 Modeling the Position of a Free-Falling Object

A ball is thrown vertically upward from the top of the Leaning Tower of Pisa (190 feet high) with an initial velocity of 96 feet per second (Figure 11.20). The function

$$s(t) = -16t^2 + 96t + 190$$

models the ball's height above the ground, $s(t)$, in feet, t seconds after it was thrown. During which time period will the ball's height exceed that of the tower?

FIGURE 11.20 Throwing a ball from the top of the Leaning Tower of Pisa

SOLUTION Using the problem's question and the given model for the ball's height, $s(t) = -16t^2 + 96t + 190$, we obtain a polynomial inequality.

$$-16t^2 + 96t + 190 > 190$$

When will the ball's height | exceed that | of the tower?

$-16t^2 + 96t + 190 > 190$	This is the inequality that models the problem's question. We must find t.
$-16t^2 + 96t > 0$	Subtract 190 from both sides. This inequality is in the form $f(t) > 0$, where $f(t) = -16t^2 + 96t$.
$-16t^2 + 96t = 0$	Solve the equation $f(t) = 0$.
$-16t(t - 6) = 0$	Factor.
$-16t = 0$ or $t - 6 = 0$	Set each factor equal to 0.
$t = 0$ $t = 6$	Solve for t. The boundary points are 0 and 6.

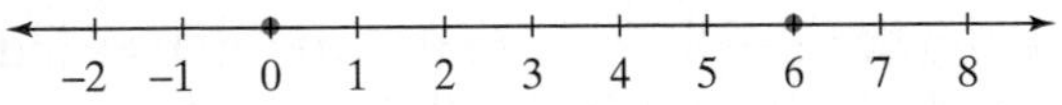

Locate these values on a number line, with $t \geq 0$.

The intervals are $(-\infty, 0)$, $(0, 6)$ and $(6, \infty)$. For our purposes, the mathematical model is useful only from $t = 0$ until the ball hits the ground. (By setting $-16t^2 + 96t + 190$ equal to zero, we find $t \approx 7.57$; the ball hits the ground after approximately 7.57 seconds.) Thus, we use $(0, 6)$ and $(6, 7.57)$ for our intervals.

Interval	Representative Number	Substitute into $f(t) = -16t^2 + 96t$	Conclusion
$(0, 6)$	1	$f(1) = -16 \cdot 1^2 + 96 \cdot 1$ $= 80$, positive	$f(t) > 0$ for all t in $(0, 6)$.
$(6, 7.57)$	7	$f(7) = -16 \cdot 7^2 + 96 \cdot 7$ $= -112$, negative	$f(t) < 0$ for all t in $(6, 7.57)$.

We see that $f(t) > 0$ for all t in $(0, 6)$. This means that the ball's height exceeds that of the tower between 0 and 6 seconds. ■

USING TECHNOLOGY

The graphs of

$$y_1 = -16x^2 + 96x + 190$$

and

$$y_2 = 190$$

are shown in a

$$[0, 8, 1] \text{ by } [0, 360, 36]$$

seconds in motion — height, in feet

viewing rectangle. The graphs show that the ball's height exceeds that of the tower between 0 and 6 seconds.

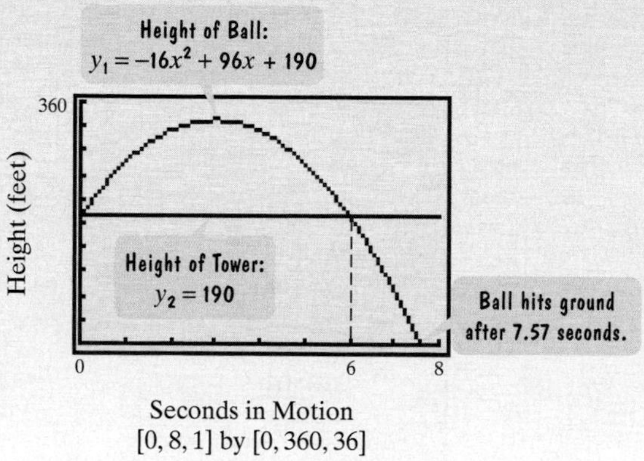

CHECK POINT 4 An object is propelled straight up from ground level with an initial velocity of 80 feet per second. Its height at time t is modeled by

$$s(t) = -16t^2 + 80t,$$

where the height, $s(t)$, is measured in feet and the time, t, is measured in seconds. In which time interval will the object be more than 64 feet above the ground?

11.5 EXERCISE SET

Student Solutions Manual | CD/Video | PH Math/Tutor Center | MathXL Tutorials on CD | MathXL® | MyMathLab | Interactmath.com

Practice Exercises

Solve each polynomial inequality in Exercises 1–36, and graph the solution set on a real number line.

1. $(x - 4)(x + 2) > 0$
2. $(x + 3)(x - 5) > 0$
3. $(x - 7)(x + 3) \leq 0$
4. $(x + 1)(x - 7) \leq 0$
5. $x^2 - 5x + 4 > 0$
6. $x^2 - 4x + 3 < 0$
7. $x^2 + 5x + 4 > 0$
8. $x^2 + x - 6 > 0$
9. $x^2 - 6x + 8 \leq 0$

10. $x^2 - 2x - 3 \geq 0$

11. $3x^2 + 10x - 8 \leq 0$

12. $9x^2 + 3x - 2 \geq 0$

13. $2x^2 + x < 15$

14. $6x^2 + x > 1$

15. $4x^2 + 7x < -3$

16. $3x^2 + 16x < -5$

17. $x^2 - 4x \geq 0$

18. $x^2 + 2x < 0$

19. $2x^2 + 3x > 0$

20. $3x^2 - 5x \leq 0$

21. $-x^2 + x \geq 0$

22. $-x^2 + 2x \geq 0$

23. $x^2 \leq 4x - 2$

24. $x^2 \leq 2x + 2$

25. $x^2 - 6x + 9 < 0$

26. $4x^2 - 4x + 1 \geq 0$

27. $(x - 1)(x - 2)(x - 3) \geq 0$

28. $(x + 1)(x + 2)(x + 3) \geq 0$

29. $x^3 + 2x^2 - x - 2 \geq 0$

30. $x^3 + 2x^2 - 4x - 8 \geq 0$

31. $x^3 - 3x^2 - 9x + 27 < 0$

32. $x^3 + 7x^2 - x - 7 < 0$

33. $x^3 + x^2 + 4x + 4 > 0$

34. $x^3 - x^2 + 9x - 9 > 0$

35. $x^3 \geq 9x^2$

36. $x^3 \leq 4x^2$

Solve each rational inequality in Exercises 37–52 and graph the solution set on a real number line.

37. $\frac{x - 4}{x + 3} > 0$

38. $\frac{x + 5}{x - 2} > 0$

39. $\frac{x + 3}{x + 4} < 0$

40. $\frac{x + 5}{x + 2} < 0$

41. $\frac{-x + 2}{x - 4} \geq 0$

42. $\frac{-x - 3}{x + 2} \leq 0$

43. $\frac{4 - 2x}{3x + 4} \leq 0$

44. $\frac{3x + 5}{6 - 2x} \geq 0$

45. $\frac{x}{x - 3} > 0$

46. $\frac{x + 4}{x} > 0$

47. $\frac{x + 1}{x + 3} < 2$

48. $\frac{x}{x - 1} > 2$

49. $\frac{x + 4}{2x - 1} \leq 3$

50. $\frac{1}{x - 3} < 1$

51. $\frac{x - 2}{x + 2} \leq 2$

52. $\frac{x}{x + 2} \geq 2$

In Exercises 53–56, use the given functions to find all values of x that satisfy the required inequality.

53. $f(x) = 2x^2, g(x) = 5x - 2; f(x) \geq g(x)$

54. $f(x) = 4x^2, g(x) = 9x - 2; f(x) < g(x)$

55. $f(x) = \frac{2x}{x + 1}, g(x) = 1; f(x) < g(x)$

56. $f(x) = \frac{x}{2x - 1}, g(x) = 1; f(x) \geq g(x)$

Practice Plus

Solve each inequality in Exercises 57–62 and graph the solution set on a real number line.

57. $|x^2 + 2x - 36| > 12$

58. $|x^2 + 6x + 1| > 8$

59. $\frac{3}{x+3} > \frac{3}{x-2}$

60. $\frac{1}{x+1} > \frac{2}{x-1}$

61. $\frac{x^2 - x - 2}{x^2 - 4x + 3} > 0$

62. $\frac{x^2 - 3x + 2}{x^2 - 2x - 3} > 0$

In Exercises 63–64, use the graph of the polynomial function to solve each inequality.

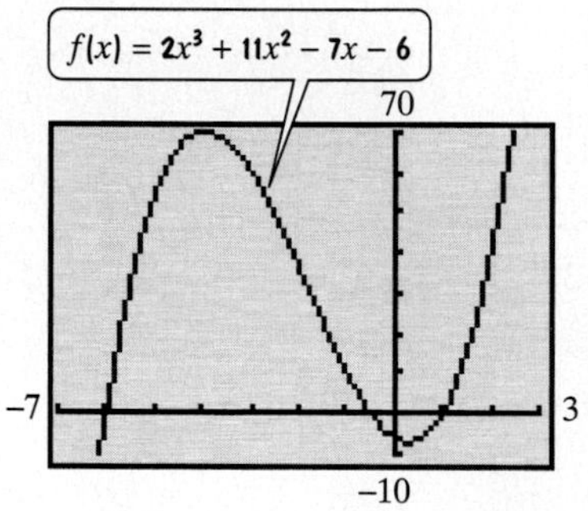

63. $2x^3 + 11x^2 \geq 7x + 6$

64. $2x^3 + 11x^2 < 7x + 6$

In Exercises 65–66, use the graph of the rational function to solve each inequality.

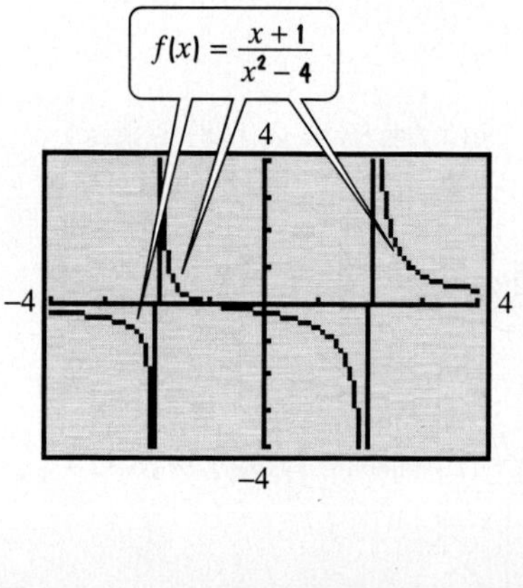

65. $\frac{1}{4(x+2)} \leq -\frac{3}{4(x-2)}$

66. $\frac{1}{4(x+2)} > -\frac{3}{4(x-2)}$

Application Exercises

67. You throw a ball straight up from a rooftop 160 feet high with an initial speed of 48 feet per second. The function

$$s(t) = -16t^2 + 48t + 160$$

models the ball's height above the ground, $s(t)$, in feet, t seconds after it was thrown. During which time period will the ball's height exceed that of the rooftop?

68. Divers in Acapulco, Mexico, dive headfirst from the top of a cliff 87 feet above the Pacific Ocean. The function

$$s(t) = -16t^2 + 8t + 87$$

models a diver's height above the ocean, $s(t)$, in feet, t seconds after leaping. During which time period will the diver's height exceed that of the cliff?

The bar graph in Figure 11.17 on page 776 shows the cost of Medicare, in billions of dollars, through 2005. Using the regression feature of a graphing utility, these data can be modeled by

a linear function, $f(x) = 27x + 163$;
a quadratic function, $g(x) = 1.2x^2 + 15.2x + 181.4$.

In each function, x represents the number of years after 1995. Use these functions to solve Exercises 69–72.

69. The graph indicates that Medicare spending reached \$379 billion in 2003. Find the amount predicted by each of the functions, f and g, for that year. How well do the functions model the value in the graph?

70. The graph indicates that Medicare spending reached \$458 billion in 2005. Find the amount predicted by each of the functions, f and g, for that year. How well do the functions model the value in the graph? Which function serves as a better model for that year?

71. For which years does the quadratic model indicate that Medicare spending will exceed \$536.6 billion?

72. For which years does the quadratic model indicate that Medicare spending will exceed \$629.4 billion?

A company manufactures wheelchairs. The average cost function, $\overline{C}$, of producing x wheelchairs per month is given by

$$\overline{C}(x) = \frac{500{,}000 + 400x}{x}.$$

The graph of the rational function is shown. Use the function to solve Exercises 73–74.

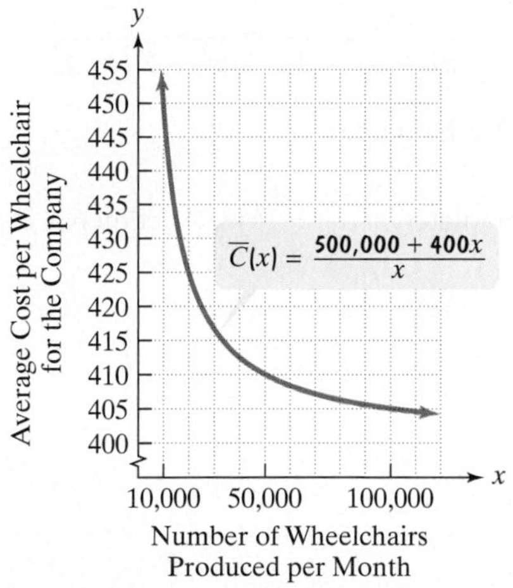

73. Describe the company's production level so that the average cost of producing each wheelchair does not exceed \$425. Use a rational inequality to solve the problem. Then explain how your solution is shown on the graph.

74. Describe the company's production level so that the average cost of producing each wheelchair does not exceed \$410. Use a rational inequality to solve the problem. Then explain how your solution is shown on the graph.

75. The perimeter of a rectangle is 50 feet. Describe the possible length of a side if the area of the rectangle is not to exceed 114 square feet.

76. The perimeter of a rectangle is 180 feet. Describe the possible lengths of a side if the area of the rectangle is not to exceed 800 square feet.

Writing in Mathematics

77. What is a polynomial inequality?

78. What is a rational inequality?

79. Describe similarities and differences between the solutions of

$$(x - 2)(x + 5) \geq 0 \quad \text{and} \quad \frac{x - 2}{x + 5} \geq 0.$$

Technology Exercises

Solve each inequality in Exercises 80–85 using a graphing utility.

80. $x^2 + 3x - 10 > 0$

81. $2x^2 + 5x - 3 \leq 0$

82. $\frac{x - 4}{x - 1} \leq 0$

83. $\frac{x + 2}{x - 3} \leq 2$

84. $\frac{1}{x + 1} \leq \frac{2}{x + 4}$

85. $x^3 + 2x^2 - 5x - 6 > 0$

Critical Thinking Exercises

86. Which one of the following is true?

a. The solution set of $x^2 > 25$ is $(5, \infty)$.

b. The inequality $\frac{x - 2}{x + 3} < 2$ can be solved by multiplying both sides by $x + 3$, resulting in the equivalent inequality $x - 2 < 2(x + 3)$.

c. $(x + 3)(x - 1) \geq 0$ and $\frac{x + 3}{x - 1} \geq 0$ have the same solution set.

d. None of the above statements is true.

87. Write a quadratic inequality whose solution set is $[-3, 5]$.

88. Write a rational inequality whose solution set is $(-\infty, -4) \cup [3, \infty)$.

In Exercises 89–92, use inspection to describe each inequality's solution set. Do not solve any of the inequalities.

89. $(x - 2)^2 > 0$

90. $(x - 2)^2 \leq 0$

91. $(x - 2)^2 < -1$

92. $\frac{1}{(x - 2)^2} > 0$

93. The graphing calculator screen shows the graph of $y = 4x^2 - 8x + 7$.

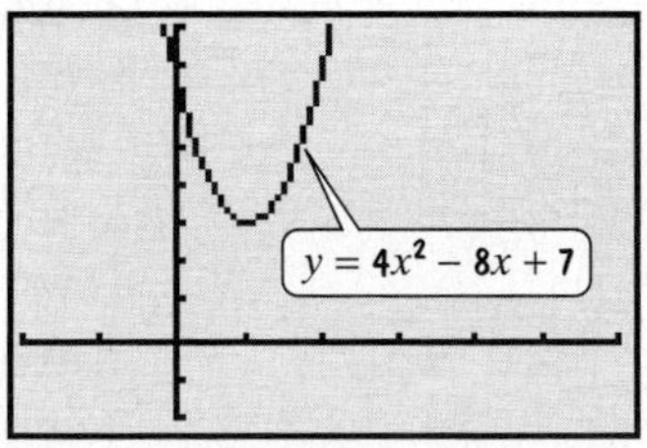

$[-2, 6, 1]$ by $[-2, 8, 1]$

a. Use the graph to describe the solution set for $4x^2 - 8x + 7 > 0$.

b. Use the graph to describe the solution set for $4x^2 - 8x + 7 < 0$.

c. Use an algebraic approach to verify each of your descriptions in parts (a) and (b).

94. The graphing calculator screen shows the graph of $y = \sqrt{27 - 3x^2}$. Write and solve a quadratic inequality that explains why the graph only appears for $-3 \le x \le 3$.

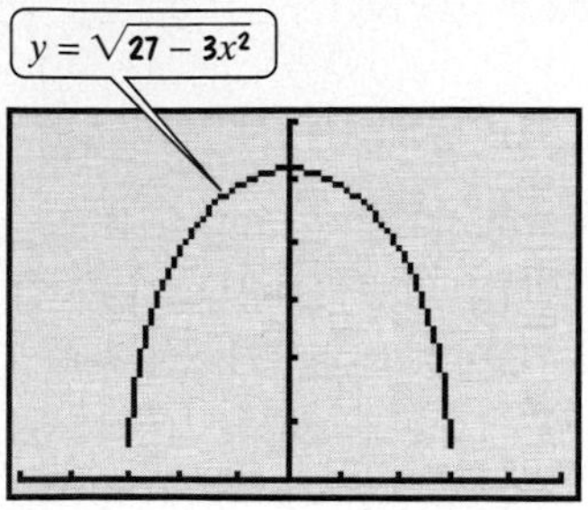

[−5, 5, 1] by [0, 6, 1]

Review Exercises

95. Solve: $\left|\frac{x-5}{3}\right| < 8$. (Section 9.3, Example 3)

96. Divide:

$$\frac{2x+6}{x^2+8x+16} \div \frac{x^2-9}{x^2+3x-4}.$$

(Section 7.2, Example 6)

97. Factor completely: $x^4 - 16y^4$. (Section 6.5, Example 7)

GROUP PROJECT

CHAPTER 11

Throughout the chapter, we have considered functions that model the position of free-falling objects. Any object that is falling, or vertically projected into the air, has its height above the ground, $s(t)$, in feet, after t seconds in motion, modeled by the quadratic function

$$s(t) = -16t^2 + v_o t + s_o,$$

where v_o is the original velocity (initial velocity) of the object, in feet per second, and s_o is the original height (initial height) of the object, in feet, above the ground. In this exercise, group members will be working with this position function. The exercise is appropriate for groups of three to five people.

a. Drop a ball from a height of 3 feet, 6 feet, and 12 feet. Record the number of seconds it takes for the ball to hit the ground.

b. For each of the three initial positions, use the position function to determine the time required for the ball to hit the ground.

c. What factors might result in differences between the times that you recorded and the times indicated by the function?

d. What appears to be happening to the time required for a free-falling object to hit the ground as its initial height is doubled? Verify this observation algebraically and with a graphing utility.

e. Repeat part (a) using a sheet of paper rather than a ball. What differences do you observe? What factor seems to be ignored in the position function?

f. What is meant by the acceleration of gravity and how does this number appear in the position function for a free-falling object?

CHAPTER 11 SUMMARY

Definitions and Concepts	Examples
Section 11.1 The Square Root Property and Completing the Square; Distance and Midpoint Formulas	
The Square Root Property If u is an algebraic expression and d is a real number, then If $u^2 = d$, then $u = \sqrt{d}$ or $u = -\sqrt{d}$. Equivalently, If $u^2 = d$, then $u = \pm\sqrt{d}$.	Solve: $(x-6)^2 = 50$. $x - 6 = \pm\sqrt{50}$ $x - 6 = \pm\sqrt{25 \cdot 2}$ $x - 6 = \pm 5\sqrt{2}$ $x = 6 \pm 5\sqrt{2}$ The solutions are $6 \pm 5\sqrt{2}$ and the solution set is $\{6 \pm 5\sqrt{2}\}$.
Completing the Square If $x^2 + bx$ is a binomial, then by adding $\left(\frac{b}{2}\right)^2$, the square of half the coefficient of x, a perfect square trinomial will result. That is, $x^2 + bx + \left(\frac{b}{2}\right)^2 = \left(x + \frac{b}{2}\right)^2$.	Complete the square: $x^2 + \frac{2}{7}x$. Half of $\frac{2}{7}$ is $\frac{1}{2} \cdot \frac{2}{7} = \frac{1}{7}$ and $\left(\frac{1}{7}\right)^2 = \frac{1}{49}$. $x^2 + \frac{2}{7}x + \frac{1}{49} = \left(x + \frac{1}{7}\right)^2$
Solving Quadratic Equations by Completing the Square 1. If the coefficient of x^2 is not 1, divide both sides by this coefficient. 2. Isolate variable terms on one side. 3. Complete the square by adding the square of half the coefficient of x to both sides. 4. Factor the perfect square trinomial. 5. Solve by applying the square root property.	Solve by completing the square. $2x^2 + 16x - 6 = 0$. $\frac{2x^2}{2} + \frac{16x}{2} - \frac{6}{2} = \frac{0}{2}$ Divide by 2. $x^2 + 8x - 3 = 0$ Simplify. $x^2 + 8x = 3$ Add 3. The coefficient of x is 8. Half of 8 is 4 and $4^2 = 16$. Add 16 to both sides. $x^2 + 8x + 16 = 3 + 16$ $(x+4)^2 = 19$ $x + 4 = \pm\sqrt{19}$ $x = -4 \pm \sqrt{19}$
The Distance Formula The distance, d, between the points (x_1, y_1) and (x_2, y_2) is given by $d = \sqrt{(x_2 - x_1)^2 + (y_2 - y_1)^2}$.	Find the distance between $(-3, -5)$ and $(6, -2)$. $d = \sqrt{[6-(-3)]^2 + [-2-(-5)]^2}$ $= \sqrt{9^2 + 3^2} = \sqrt{81 + 9} = \sqrt{90} = 3\sqrt{10} \approx 9.49$
The Midpoint Formula The midpoint of the line segment whose endpoints are (x_1, y_1) and (x_2, y_2) is the point with coordinates $\left(\frac{x_1 + x_2}{2}, \frac{y_1 + y_2}{2}\right)$.	Find the midpoint of the line segment whose endpoints are $(-3, 6)$ and $(4, 1)$. midpoint $= \left(\frac{-3+4}{2}, \frac{6+1}{2}\right) = \left(\frac{1}{2}, \frac{7}{2}\right)$

Definitions and Concepts	Examples

Section 11.2 The Quadratic Formula

The solutions of a quadratic equation in standard form

$$ax^2 + bx + c = 0, \quad a \neq 0,$$

are given by the quadratic formula

$$x = \frac{-b \pm \sqrt{b^2 - 4ac}}{2a}.$$

Solve using the quadratic formula:

$$2x^2 = 6x - 3.$$

First write the equation in standard form by subtracting $6x$ and adding 3 on both sides.

$$2x^2 - 6x + 3 = 0$$

$a = 2$ $b = -6$ $c = 3$

$$x = \frac{-(-6) \pm \sqrt{(-6)^2 - 4 \cdot 2 \cdot 3}}{2 \cdot 2} = \frac{6 \pm \sqrt{36 - 24}}{4}$$

$$= \frac{6 \pm \sqrt{12}}{4} = \frac{6 \pm \sqrt{4 \cdot 3}}{4} = \frac{6 \pm 2\sqrt{3}}{4}$$

$$= \frac{\cancel{2}(3 \pm \sqrt{3})}{\cancel{2} \cdot 2} = \frac{3 \pm \sqrt{3}}{2}$$

The Discriminant

The discriminant, $b^2 - 4ac$, of the quadratic equation $ax^2 + bx + c = 0$ determines the number and type of solutions.

Discriminant	Solutions
Positive perfect square, with a, b, and c rational numbers	2 rational solutions
Positive and not a perfect square	2 irrational solutions
Zero, with a, b, and c rational numbers	1 rational solution
Negative	2 imaginary solutions

- $2x^2 - 7x - 4 = 0$

$a = 2$ $b = -7$ $c = -4$

$$b^2 - 4ac = (-7)^2 - 4(2)(-4)$$
$$= 49 - (-32) = 49 + 32 = 81$$

Positive perfect square

The equation has 2 rational solutions.

Writing Quadratic Equations from Solutions

The zero-product principle in reverse makes it possible to write a quadratic equation from solutions:

$$\text{If } A = 0 \quad \text{or} \quad B = 0, \quad \text{then} \quad AB = 0.$$

Write a quadratic equation with the solution set $\{-2\sqrt{3}, 2\sqrt{3}\}$.

$$x = -2\sqrt{3} \quad \text{or} \quad x = 2\sqrt{3}$$

$$x + 2\sqrt{3} = 0 \quad \text{or} \quad x - 2\sqrt{3} = 0$$

$$(x + 2\sqrt{3})(x - 2\sqrt{3}) = 0$$

$$x^2 - (2\sqrt{3})^2 = 0$$

$$x^2 - 12 = 0$$

Definitions and Concepts	Examples

Section 11.3 Quadratic Functions and Their Graphs

The graph of the quadratic function

$$f(x) = a(x - h)^2 + k, \quad a \neq 0,$$

is called a parabola. The vertex, or turning point, is (h, k). The graph opens upward if a is positive and downward if a is negative. The axis of symmetry is a vertical line passing through the vertex. The graph can be obtained using the vertex, x-intercepts, if any, (set $f(x)$ equal to zero), and the y-intercept (set $x = 0$).

Graph: $f(x) = -(x + 3)^2 + 1$.

$$f(x) = -1(x - (-3))^2 + 1$$

$a = -1$ $h = -3$ $k = 1$

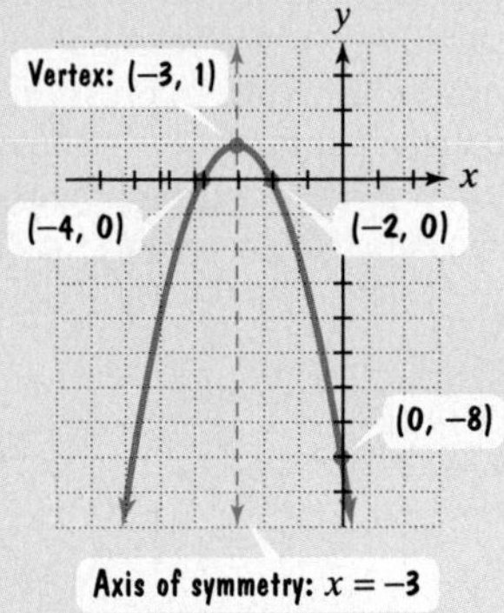

- Vertex: $(h, k) = (-3, 1)$.
- Opens downward because $a < 0$
- x-intercepts: Set $f(x) = 0$.

$$0 = -(x + 3)^2 + 1$$
$$(x + 3)^2 = 1$$
$$x + 3 = \pm\sqrt{1}$$
$$x + 3 = 1 \quad \text{or} \quad x + 3 = -1$$
$$x = -2 \qquad\qquad x = -4$$

- y-intercept: Set $x = 0$.

$$f(0) = -(0 + 3)^2 + 1 = -9 + 1 = -8$$

A parabola whose equation is in the form

$$f(x) = ax^2 + bx + c, \quad a \neq 0,$$

has its vertex at

$$\left(-\frac{b}{2a}, f\left(-\frac{b}{2a}\right)\right).$$

The parabola is graphed as described in the left column on the previous page. The only difference is how we determine the vertex. If $a > 0$, then f has a minimum that occurs at $x = -\frac{b}{2a}$. This minimum value is $f\left(-\frac{b}{2a}\right)$. If $a < 0$, then f has a maximum that occurs at $x = -\frac{b}{2a}$. This maximum value is $f\left(-\frac{b}{2a}\right)$.

Graph: $f(x) = x^2 - 6x + 5$.

$a = 1$ $b = -6$ $c = 5$

- Vertex: $x = -\frac{b}{2a} = -\frac{-6}{2 \cdot 1} = 3$

$f(3) = 3^2 - 6 \cdot 3 + 5 = -4$

Vertex is at $(3, -4)$.

- Opens upward because $a > 0$.
- x-intercepts: Set $f(x) = 0$.

$$x^2 - 6x + 5 = 0$$
$$(x - 1)(x - 5) = 0$$
$$x = 1 \quad \text{or} \quad x = 5$$

- y-intercept: Set $x = 0$.

$$f(0) = 0^2 - 6 \cdot 0 + 5 = 5$$

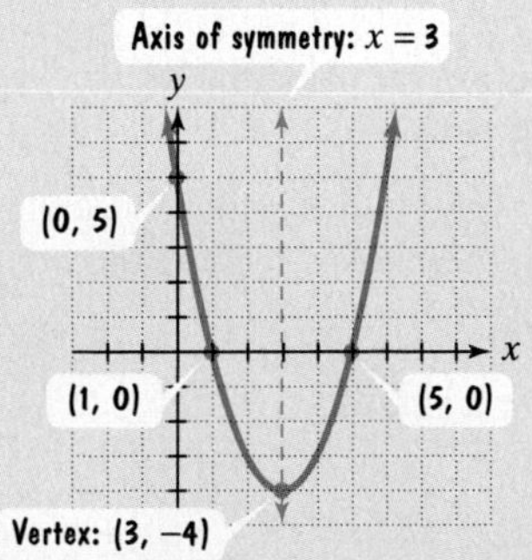

Definitions and Concepts	Examples

Section 11.4 Equations Quadratic in Form

Definitions and Concepts

An equation that is quadratic in form is one that can be expressed as a quadratic equation using an appropriate substitution. These equations contain an expression to a power, the same expression to that power squared, and a constant term: Let $t =$ the expression to the power. If at any point in the solution process both sides of an equation are raised to an even power, a check is required.

Examples

Solve: $x^{\frac{2}{3}} - 3x^{\frac{1}{3}} + 2 = 0.$

$$\left(x^{\frac{1}{3}}\right)^2 - 3x^{\frac{1}{3}} + 2 = 0$$

Let $t = x^{\frac{1}{3}}$.

$$t^2 - 3t + 2 = 0$$

$$(t - 1)(t - 2) = 0$$

$$t - 1 = 0 \quad \text{or} \quad t - 2 = 0$$

$$t = 1 \qquad\qquad t = 2$$

$$x^{\frac{1}{3}} = 1 \qquad\qquad x^{\frac{1}{3}} = 2$$

$$\left(x^{\frac{1}{3}}\right)^3 = 1^3 \qquad\qquad \left(x^{\frac{1}{3}}\right)^3 = 2^3$$

$$x = 1 \qquad\qquad x = 8$$

The solutions are 1 and 8, and the solution set is $\{1, 8\}$.

Section 11.5 Polynomial and Rational Inequalities

Definitions and Concepts

Solving Polynomial Inequalities

1. Express the inequality in the form
$$f(x) < 0 \quad \text{or} \quad f(x) > 0,$$
where f is a polynomial function.
2. Solve the equation $f(x) = 0$. The real solutions are the boundary points.
3. Locate these boundary points on a number line, thereby dividing the number line into intervals.
4. Choose one representative number within each interval and evaluate f at that number.
   a. If the value of f is positive, then $f(x) > 0$ for all x in the interval.
   b. If the value of f is negative, then $f(x) < 0$ for all x in the interval.
5. Write the solution set, selecting the interval(s) that satisfy the given inequality.

This procedure is valid if $<$ is replaced by $\leq$ and $>$ is replaced by $\geq$. In these cases, include the boundary points in the solution set.

Examples

Solve: $2x^2 + x - 6 > 0$.
The form of the inequality is $f(x) > 0$ with $f(x) = 2x^2 + x - 6$. Solve $f(x) = 0$.

$$2x^2 + x - 6 = 0$$

$$(2x - 3)(x + 2) = 0$$

$$2x - 3 = 0 \quad \text{or} \quad x + 2 = 0$$

$$x = \frac{3}{2} \qquad\qquad x = -2$$

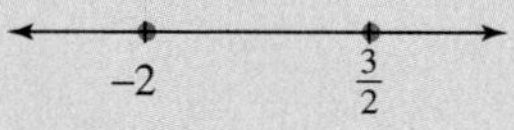

$$f(-3) = 2(-3)^2 + (-3) - 6 = 9, \text{ positive}$$

$$f(x) > 0 \text{ for all } x \text{ in } (-\infty, -2).$$

$$f(0) = 2 \cdot 0^2 + 0 - 6 = -6, \text{ negative}$$

$$f(x) < 0 \text{ for all } x \text{ in } \left(-2, \frac{3}{2}\right).$$

$$f(2) = 2 \cdot 2^2 + 2 - 6 = 4, \text{ positive}$$

$$f(x) > 0 \text{ for all } x \text{ in } \left(\frac{3}{2}, \infty\right).$$

The solution set is $\left\{x \mid x < -2 \text{ or } x > \frac{3}{2}\right\}$ or $(-\infty, -2) \cup \left(\frac{3}{2}, \infty\right)$.

Definitions and Concepts | Examples

Section 11.5 Polynomial and Rational Inequalities (continued)

Solving Rational Inequalities

1. Express the inequality in the form
$$f(x) < 0 \quad \text{or} \quad f(x) > 0,$$
where f is a rational function.
2. Set the numerator and the denominator of f equal to zero. The real zeros are the boundary points.
3. Locate these boundary points on a number line, thereby dividing the number line into intervals.
4. Choose one representative number within each interval and evaluate f at that number.
   - **a.** If the value of f is positive, then $f(x) > 0$ for all x in the interval.
   - **b.** If the value of f is negative, then $f(x) < 0$ for all x in the interval.
5. Write the solution set, selecting the interval(s) that satisfy the given inequality.

This procedure is valid if $<$ is replaced by $\leq$ and $>$ is replaced by $\geq$. In these cases, include any values that make the numerator of f zero. Always exclude any values that make the denominator zero.

Solve: $\dfrac{x}{x+4} \geq 2.$

$$\frac{x}{x+4} - \frac{2(x+4)}{x+4} \geq 0$$
$$\frac{-x-8}{x+4} \geq 0$$

The form of the inequality is $f(x) \geq 0$ with $f(x) = \dfrac{-x-8}{x+4}$.

Set the numerator and the denominator equal to zero.

$$-x - 8 = 0 \qquad x + 4 = 0$$
$$-8 = x \qquad x = -4$$

$$f(-9) = \frac{-(-9)-8}{-9+4} = \frac{1}{-5}, \text{ negative}$$

$f(x) < 0$ for all x in $(-\infty, -8)$.

$$f(-7) = \frac{-(-7)-8}{-7+4} = \frac{-1}{-3} = \frac{1}{3}, \text{ positive}$$

$f(x) > 0$ for all x in $(-8, -4)$

$$f(-3) = \frac{-(-3)-8}{-3+4} = \frac{-5}{1} = -5, \text{ negative}$$

$f(x) < 0$ for all x in $(-4, \infty)$.

Because of $\geq$, include -8, the value that makes the numerator zero, in the solution set.
The solution set is $\{x \mid -8 \leq x < -4\}$ or $[-8, -4)$.

CHAPTER 11 REVIEW EXERCISES

11.1 *In Exercises 1–5, solve each equation by the square root property. If possible, simplify radicals or rationalize denominators. Express imaginary solutions in the form $a + bi$.*

1. $2x^2 - 3 = 125$
2. $3x^2 - 150 = 0$
3. $3x^2 - 2 = 0$
4. $(x - 4)^2 = 18$
5. $(x + 7)^2 = -36$

In Exercises 6–7, determine the constant that should be added to the binomial so that it becomes a perfect square trinomial. Then write and factor the trinomial.

6. $x^2 + 20x$
7. $x^2 - 3x$

In Exercises 8–10, solve each quadratic equation by completing the square.

8. $x^2 - 12x + 27 = 0$
9. $x^2 - 7x - 1 = 0$
10. $2x^2 + 3x - 4 = 0$
11. In 2 years, an investment of \$2500 grows to \$2916. Use the compound interest formula
$$A = P(1 + r)^t$$
to find the annual interest rate, r.
12. The function $W(t) = 3t^2$ models the weight of a human fetus, $W(t)$, in grams, after t weeks, where $0 \leq t \leq 39$. After how many weeks does the fetus weigh 588 grams?

13. A building casts a shadow that is double the length of the building's height. If the distance from the end of the shadow to the top of the building is 300 meters, how high is the building? Express the answer in simplified radical form. Then find a decimal approximation to the nearest tenth of a meter.

In Exercises 14–15, find the distance between each pair of points. If necessary, round answers to two decimal places.

14. $(-2, -3)$ and $(3, 9)$

15. $(-4, 3)$ and $(-2, 5)$

In Exercises 16–17, find the midpoint of the line segment with the given endpoints.

16. $(2, 6)$ and $(-12, 4)$

17. $(4, -6)$ and $(-15, 2)$

11.2 *In Exercises 18–20, solve each equation using the quadratic formula. Simplify solutions, if possible.*

18. $x^2 = 2x + 4$

19. $x^2 - 2x + 19 = 0$

20. $2x^2 = 3 - 4x$

In Exercises 21–23, without solving the given quadratic equation, determine the number and type of solutions.

21. $x^2 - 4x + 13 = 0$

22. $9x^2 = 2 - 3x$

23. $2x^2 + 4x = 3$

In Exercises 24–30, solve each equation by the method of your choice. Simplify solutions, if possible.

24. $3x^2 - 10x - 8 = 0$

25. $(2x - 3)(x + 2) = x^2 - 2x + 4$

26. $5x^2 - x - 1 = 0$

27. $x^2 - 16 = 0$

28. $(x - 3)^2 - 8 = 0$

29. $3x^2 - x + 2 = 0$

30. $\dfrac{5}{x + 1} + \dfrac{x - 1}{4} = 2$

In Exercises 31–33, write a quadratic equation in standard form with the given solution set.

31. $\left\{-\dfrac{1}{3}, \dfrac{3}{5}\right\}$

32. $\{-9i, 9i\}$

33. $\{-4\sqrt{3}, 4\sqrt{3}\}$

34. The bar graph shows the number of convictions of police officers throughout the United States for seven years. The number of police officers convicted of felonies, $f(x)$, x years after 1990 can be modeled by the quadratic function

$$f(x) = 23x^2 - 259x + 816.$$

Convictions of Police Officers

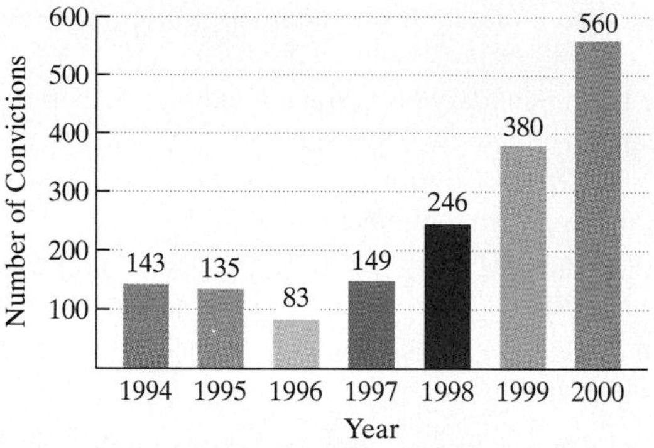

Source: F.B.I.

In which year after 1990 were 1020 police officers convicted of felonies?

35. A baseball is hit by a batter. The function

$$s(t) = -16t^2 + 140t + 3$$

models the ball's height above the ground, $s(t)$, in feet, t seconds after it is hit. How long will it take for the ball to strike the ground? Round to the nearest tenth of a second.

11.3 *In Exercises 36–39, use the vertex and intercepts to sketch the graph of each quadratic function. Give the equation for the parabola's axis of symmetry.*

36. $f(x) = -(x + 1)^2 + 4$

37. $f(x) = (x + 4)^2 - 2$

38. $f(x) = -x^2 + 2x + 3$

39. $f(x) = 2x^2 - 4x - 6$

40. The function

$$f(x) = -0.02x^2 + x + 1$$

models the yearly growth of a young redwood tree, $f(x)$, in inches, with x inches of rainfall per year. How many inches of rainfall per year result in maximum tree growth? What is the maximum yearly growth?

41. A model rocket is launched upward from a platform 40 feet above the ground. The quadratic function

$$s(t) = -16t^2 + 400t + 40$$

models the rocket's height above the ground, $s(t)$, in feet, t seconds after it was launched. After how many seconds does the rocket reach its maximum height? What is the maximum height?

42. Suppose that a quadratic function is used to model the data shown in the graph using

(number of years after 1960, divorce rate per 1000 population).

U. S. Divorce Rate

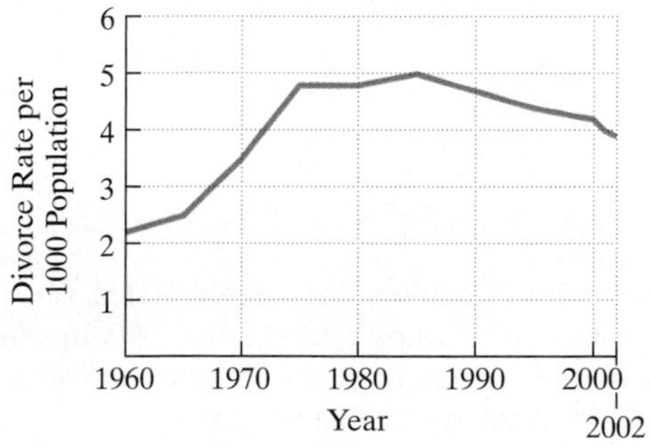

Source: National Center for Health Statistics

Determine, without obtaining an actual quadratic function that models the data, the approximate coordinates of the vertex for the function's graph. Describe what this means in practical terms.

43. A field bordering a straight stream is to be enclosed. The side bordering the stream is not to be fenced. If 1000 yards of fencing material is to be used, what are the dimensions of the largest rectangular field that can be fenced? What is the maximum area?

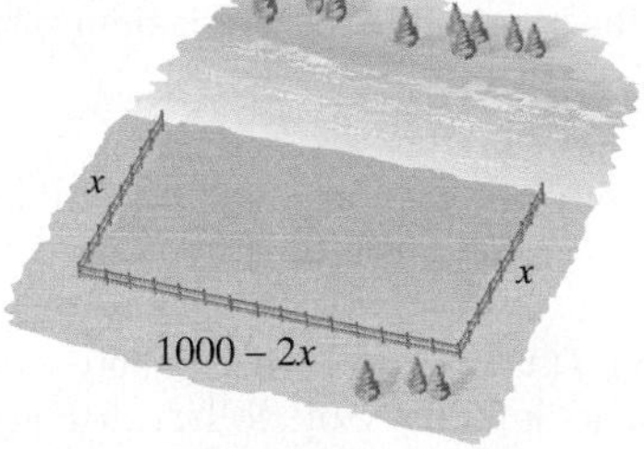

44. Among all pairs of numbers whose difference is 14, find a pair whose product is as small as possible. What is the minimum product?

11.4 *In Exercises 45–50, solve each equation by making an appropriate substitution. When necessary, check proposed solutions.*

45. $x^4 - 6x^2 + 8 = 0$

46. $x + 7\sqrt{x} - 8 = 0$

47. $(x^2 + 2x)^2 - 14(x^2 + 2x) = 15$

48. $x^{-2} + x^{-1} - 56 = 0$

49. $x^{\frac{2}{3}} - x^{\frac{1}{3}} - 12 = 0$

50. $x^{\frac{1}{2}} + 3x^{\frac{1}{4}} - 10 = 0$

11.5 *In Exercises 51–55, solve each inequality and graph the solution set on a real number line.*

51. $2x^2 + 5x - 3 < 0$

52. $2x^2 + 9x + 4 \geq 0$

53. $x^3 + 2x^2 > 3x$

54. $\frac{x - 6}{x + 2} > 0$

55. $\frac{x + 3}{x - 4} \leq 5$

56. A model rocket is launched from ground level. The function

$$s(t) = -16t^2 + 48t$$

models the rocket's height above the ground, $s(t)$, in feet, t seconds after it was launched. During which time period will the rocket's height exceed 32 feet?

57. The function

$$H(x) = \frac{15}{8}x^2 - 30x + 200$$

models heart rate, $H(x)$, in beats per minute, x minutes after a strenuous workout.

a. What is the heart rate immediately following the workout?

b. According to the model, during which intervals of time after the strenuous workout does the heart rate exceed 110 beats per minute? For which of these intervals has model breakdown occurred? Which interval provides a more realistic answer? How did you determine this?

CHAPTER 11 TEST

Remember to use your Chapter Test Prep Video CD to see the worked-out solutions to the test questions you want to review.

Express solutions to all equations in simplified form. Rationalize denominators, if possible.

In Exercises 1–2, solve each equation by the square root property.

1. $2x^2 - 5 = 0$

2. $(x - 3)^2 = 20$

In Exercises 3–4, determine the constant that should be added to the binomial so that it becomes a perfect square trinomial. Then write and factor the trinomial.

3. $x^2 - 16x$

4. $x^2 + \frac{2}{5}x$

5. Solve by completing the square: $x^2 - 6x + 7 = 0$.

6. Use the measurements determined by the surveyor to find the width of the pond. Express the answer in simplified radical form.

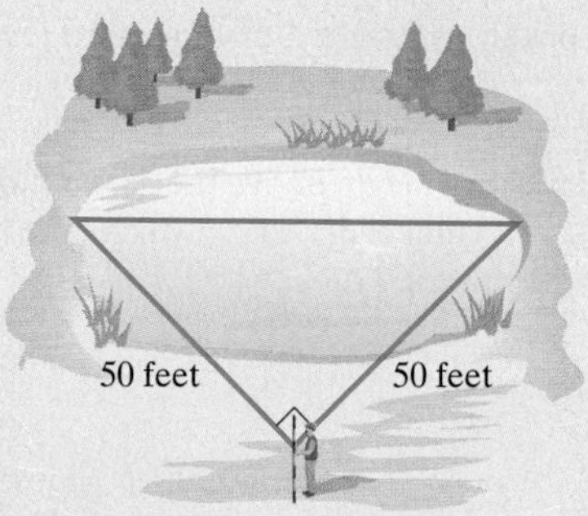

7. Find the distance between $(-1, 5)$ and $(2, -3)$. If necessary, round the answer to two decimal places.

8. Find the midpoint of the line segment whose endpoints are $(-5, -2)$ and $(12, -6)$.

In Exercises 9–10, without solving the given quadratic equation, determine the number and type of solutions.

9. $3x^2 + 4x - 2 = 0$

10. $x^2 = 4x - 8$

In Exercises 11–14, solve each equation by the method of your choice.

11. $2x^2 + 9x = 5$

12. $x^2 + 8x + 5 = 0$

13. $(x + 2)^2 + 25 = 0$

14. $2x^2 - 6x + 5 = 0$

In Exercises 15–16, write a quadratic equation in standard form with the given solution set.

15. $\{-3, 7\}$

16. $\{-10i, 10i\}$

17. The function

$$f(x) = -0.5x^2 + 4x + 19$$

models the number of people in the United States, $f(x)$, in millions, receiving food stamps x years after 1990. In which year(s) were 20 million people receiving food stamps? Use a calculator and round to the nearest year(s).

In Exercises 18–19, use the vertex and intercepts to sketch the graph of each quadratic function. Give the equation for the parabola's axis of symmetry.

18. $f(x) = (x + 1)^2 + 4$

19. $f(x) = x^2 - 2x - 3$

A baseball player hits a pop fly into the air. The function

$$s(t) = -16t^2 + 64t + 5$$

models the ball's height above the ground, $s(t)$, in feet, t seconds after it is hit. Use the function to solve Exercises 20–21.

20. When does the baseball reach its maximum height? What is that height?

21. After how many seconds does the baseball hit the ground? Round to the nearest tenth of a second.

22. The function $f(x) = -x^2 + 46x - 360$ models the daily profit, $f(x)$, in hundreds of dollars, for a company that manufactures x computers daily. How many computers should be manufactured each day to maximize profit? What is the maximum daily profit?

In Exercises 23–25, solve each equation by making an appropriate substitution. When necessary, check proposed solutions.

23. $(2x - 5)^2 + 4(2x - 5) + 3 = 0$

24. $x^4 - 13x^2 + 36 = 0$

25. $x^{\frac{2}{3}} - 9x^{\frac{1}{3}} + 8 = 0$

In Exercises 26–27, solve each inequality and graph the solution set on a real number line.

26. $x^2 - x - 12 < 0$

27. $\dfrac{2x + 1}{x - 3} \le 3$

CUMULATIVE REVIEW EXERCISES (CHAPTERS 1–11)

In Exercises 1–7, solve each equation, inequality, or system.

1. $8 - (4x - 5) = x - 7$

2. $\begin{aligned} 5x + 4y &= 22 \\ 3x - 8y &= -18 \end{aligned}$

3. $\begin{aligned} -3x + 2y + 4z &= 6 \\ 7x - y + 3z &= 23 \\ 2x + 3y + z &= 7 \end{aligned}$

4. $|x - 1| > 3$

5. $\sqrt{x + 4} - \sqrt{x - 4} = 2$

6. $x - 4 \ge 0$ and $-3x \le -6$

7. $2x^2 = 3x - 2$

In Exercises 8–12, graph each function, equation, or inequality in a rectangular coordinate system.

8. $3x = 15 + 5y$

9. $2x - 3y > 6$

10. $f(x) = -\frac{1}{2}x + 1$

11. $f(x) = x^2 + 6x + 8$

12. $f(x) = (x - 3)^2 - 4$

13. If $f(x) = \sqrt[3]{x + 1}$, find $f^{-1}(x)$.

14. Solve for c: $A = \dfrac{cd}{c + d}$.

15. Write the equation of the linear function whose graph contains the point $(-2, 4)$ and is perpendicular to the line whose equation is $2x + y = 10$.

In Exercises 16–20, perform the indicated operations and simplify, if possible.

16. $\dfrac{-5x^3y^7}{15x^4y^{-2}}$

17. $(4x^2 - 5y)^2$

18. $(5x^3 - 24x^2 + 9) \div (5x + 1)$

19. $\dfrac{\sqrt[3]{32xy^{10}}}{\sqrt[3]{2xy^2}}$

20. $\dfrac{x + 2}{x^2 - 6x + 8} + \dfrac{3x - 8}{x^2 - 5x + 6}$

In Exercises 21–22, factor completely.

21. $x^4 - 4x^3 + 8x - 32$

22. $2x^2 + 12xy + 18y^2$

23. The length of a rectangular carpet is 4 feet greater than twice its width. If the area is 48 square feet, find the carpet's length and width.

24. Working alone, you can mow the lawn in 2 hours and your sister can do it in 3 hours. How long will it take you to do the job if you work together?

25. Your motorboat can travel 15 miles per hour in still water. Traveling with the river's current, the boat can cover 20 miles in the same time it takes to go 10 miles against the current. Find the rate of the current.

CHAPTER 12

You've recently taken up weightlifting, recording the maximum number of pounds you can lift at the end of each week. At first your weight limit increases rapidly, but now you notice that this growth is beginning to level off. You wonder about a function that would serve as a mathematical model to predict the number of pounds you can lift as you continue the sport.

This problem appears as Exercise 36 in Exercise Set 12.5 and as the group project on page 862.

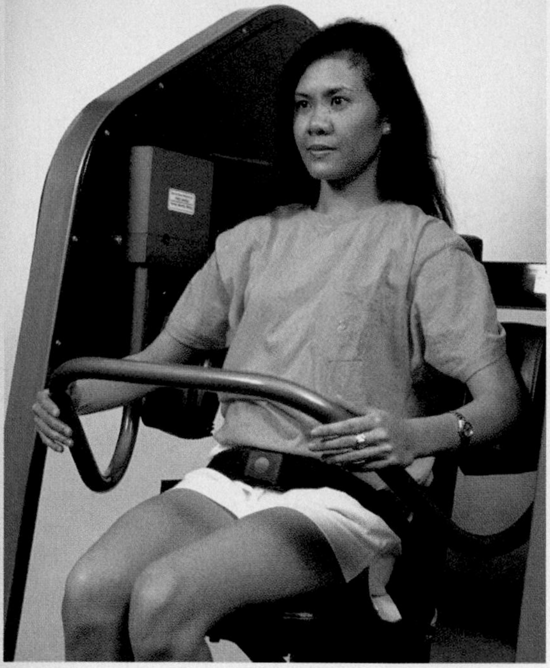

Exponential and Logarithmic Functions

What went wrong on the space shuttle Challenger? Will population growth lead to a future without comfort or individual choice? Can I put aside a small amount of money and have millions for early retirement? Why did I feel I was walking too slowly on my visit to New York City? Why are people in California at far greater risk from drunk drivers than from earthquakes? What is the difference between earthquakes measuring 6 and 7 on the Richter scale? And what can I hope to accomplish in weightlifting?

The functions that you will be learning about in this chapter will provide you with the mathematics for answering these questions. You will see how these remarkable functions enable us to predict the future and rediscover the past.

SECTION 12.1

Objectives

1. Evaluate exponential functions.
2. Graph exponential functions.
3. Evaluate functions with base e.
4. Use compound interest formulas.

EXPONENTIAL FUNCTIONS

The space shuttle *Challenger* exploded approximately 73 seconds into flight on January 28, 1986. The tragedy involved damage to O-rings, which were used to seal the connections between different sections of the shuttle engines. The number of O-rings damaged increases dramatically as temperature falls.

The function

$$f(x) = 13.49(0.967)^x - 1$$

models the number of O-rings expected to fail when the temperature is x°F. Can you see how this function is different from polynomial functions? The variable x is in the exponent. Functions whose equations contain a variable in the exponent are called **exponential functions**. Many real-life situations, including population growth, growth of epidemics, radioactive decay, and other changes that involve rapid increase or decrease, can be described using exponential functions.

> DEFINITION OF THE EXPONENTIAL FUNCTION The **exponential function f with base b** is defined by
>
> $$f(x) = b^x \quad \text{or} \quad y = b^x,$$
>
> where b is a positive constant other than 1 ($b > 0$ and $b \neq 1$) and x is any real number.

Here are some examples of exponential functions:

$f(x) = 2^x$ (Base is 2.) $\qquad g(x) = 10^x$ (Base is 10.) $\qquad h(x) = 3^{x+1}$ (Base is 3.) $\qquad j(x) = \left(\frac{1}{2}\right)^{x-1}$. (Base is $\frac{1}{2}$.)

Each of these functions has a constant base and a variable exponent. By contrast, the following functions are not exponential functions:

$F(x) = x^2$ (Variable is the base and not the exponent.) $\qquad G(x) = 1^x$ (The base of an exponential function must be a positive constant other than 1.) $\qquad H(x) = x^x$. (Variable is both the base and the exponent.)

Why is $G(x) = 1^x$ not classified as an exponential function? The number 1 raised to any power is 1. Thus, the function G can be written as $G(x) = 1$, which is a constant function.

1 Evaluate exponential functions.

You will need a calculator to evaluate exponential expressions. Most scientific calculators have a $\boxed{y^x}$ key. Graphing calculators have a $\boxed{\wedge}$ key. To evaluate expressions of the form b^x, enter the base b, press $\boxed{y^x}$ or $\boxed{\wedge}$, enter the exponent x, and finally press $\boxed{=}$ or $\boxed{\text{ENTER}}$.

EXAMPLE 1 Evaluating an Exponential Function

The exponential function $f(x) = 13.49(0.967)^x - 1$ describes the number of O-rings expected to fail, $f(x)$, when the temperature is x°F. On the morning the *Challenger* was launched, the temperature was 31°F, colder than any previous experience. Find the number of O-rings expected to fail at this temperature.

SOLUTION Because the temperature was 31°F, substitute 31 for x and evaluate the function at 31.

$$f(x) = 13.49(0.967)^x - 1 \quad \text{This is the given function.}$$

$$f(31) = 13.49(0.967)^{31} - 1 \quad \text{Substitute 31 for x.}$$

Use a scientific or graphing calculator to evaluate $f(31)$. Press the following keys on your calculator to do this:

Scientific calculator: $13.49 \boxed{\times} .967 \boxed{y^x} 31 \boxed{-} 1 \boxed{=}$

Graphing calculator: $13.49 \boxed{\times} .967 \boxed{\wedge} 31 \boxed{-} 1 \boxed{\text{ENTER}}$.

The display should be approximately 3.7668627.

$$f(31) = 13.49(0.967)^{31} - 1 \approx 3.8 \approx 4$$

Thus, four O-rings are expected to fail at a temperature of 31°F. ■

✔ **CHECK POINT 1** Use the function in Example 1 to find the number of O-rings expected to fail at a temperature of 60°F. Round to the nearest whole number.

2 Graph exponential functions.

Graphing Exponential Functions We are familiar with expressions involving b^x where x is a rational number. For example,

$$b^{1.7} = b^{\frac{17}{10}} = \sqrt[10]{b^{17}} \quad \text{and} \quad b^{1.73} = b^{\frac{173}{100}} = \sqrt[100]{b^{173}}.$$

However, note that the definition of $f(x) = b^x$ includes all real numbers for the domain x. You may wonder what b^x means when x is an irrational number, such as

$b^{\sqrt{3}}$ or b^{π}. Using closer and closer approximations for $\sqrt{3}$ ($\sqrt{3} \approx 1.73205$), we can think of $b^{\sqrt{3}}$ as the value that has the successively closer approximations

$$b^{1.7}, b^{1.73}, b^{1.732}, b^{1.73205}, \ldots.$$

In this way, we can graph the exponential function with no holes, or points of discontinuity, at the irrational domain values.

EXAMPLE 2 Graphing an Exponential Function

Graph: $f(x) = 2^x$.

SOLUTION We begin by setting up a table of coordinates.

x	$f(x) = 2^x$
-3	$f(-3) = 2^{-3} = \frac{1}{8}$
-2	$f(-2) = 2^{-2} = \frac{1}{4}$
-1	$f(-1) = 2^{-1} = \frac{1}{2}$
0	$f(0) = 2^0 = 1$
1	$f(1) = 2^1 = 2$
2	$f(2) = 2^2 = 4$
3	$f(3) = 2^3 = 8$

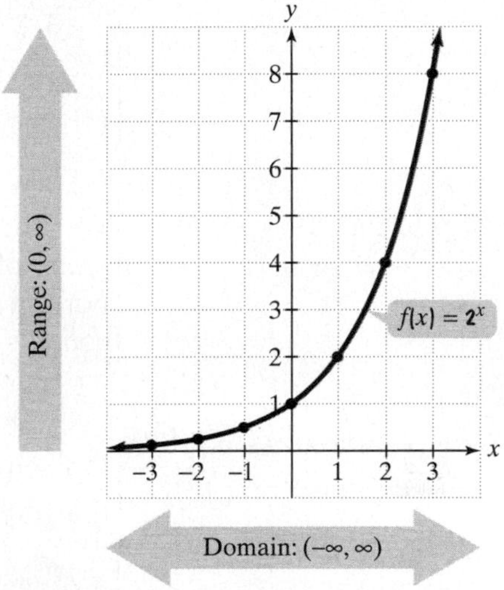

FIGURE 12.1 The graph of $f(x) = 2^x$

We plot these points, connecting them with a continuous curve. Figure 12.1 shows the graph of $f(x) = 2^x$. Observe that the graph approaches, but never touches, the negative portion of the x-axis. Thus, the x-axis is a horizontal asymptote. The range is the set of all positive real numbers. Although we used integers for x in our table of coordinates, you can use a calculator to find additional points. For example, $f(0.3) = 2^{0.3} \approx 1.231$, $f(0.95) = 2^{0.95} \approx 1.932$. The points $(0.3, 1.231)$ and $(0.95, 1.932)$ approximately fit the graph. ■

CHECK POINT 2 Graph: $f(x) = 3^x$.

EXAMPLE 3 Graphing an Exponential Function

Graph: $f(x) = \left(\frac{1}{2}\right)^x$.

SOLUTION We begin by setting up a table of coordinates. We compute the function values by noting that

$$f(x) = \left(\frac{1}{2}\right)^x = (2^{-1})^x = 2^{-x}.$$

x	$f(x) = \left(\frac{1}{2}\right)^x$ or 2^{-x}
-3	$f(-3) = 2^{-(-3)} = 2^3 = 8$
-2	$f(-2) = 2^{-(-2)} = 2^2 = 4$
-1	$f(-1) = 2^{-(-1)} = 2^1 = 2$
0	$f(0) = 2^{-0} = 1$
1	$f(1) = 2^{-1} = \frac{1}{2^1} = \frac{1}{2}$
2	$f(2) = 2^{-2} = \frac{1}{2^2} = \frac{1}{4}$
3	$f(3) = 2^{-3} = \frac{1}{2^3} = \frac{1}{8}$

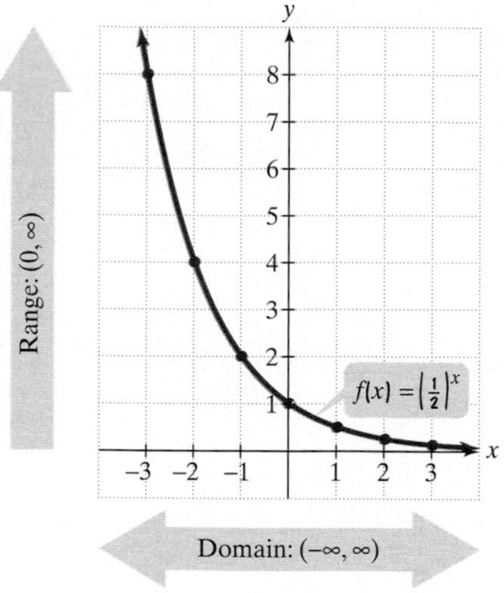

FIGURE 12.2 The graph of $f(x) = \left(\frac{1}{2}\right)^x$

We plot these points, connecting them with a continuous curve. Figure 12.2 shows the graph of $f(x) = \left(\frac{1}{2}\right)^x$. This time the graph approaches, but never touches, the *positive* portion of the x-axis. Once again, the x-axis is a horizontal asymptote. The range consists of the set of all positive real numbers.

Do you notice a relationship between the graphs of $f(x) = 2^x$ and $f(x) = \left(\frac{1}{2}\right)^x$ in Figures 12.1 and 12.2? The graph of $f(x) = \left(\frac{1}{2}\right)^x$ is a mirror image, or reflection, of the graph of $f(x) = 2^x$ about the y-axis.

CHECK POINT 3 Graph: $f(x) = \left(\frac{1}{3}\right)^x$. Note that $f(x) = \left(\frac{1}{3}\right)^x = (3^{-1})^x = 3^{-x}$.

Four exponential functions have been graphed in Figure 12.3. Compare the black and green graphs, where $b > 1$, to those in blue and red, where $b < 1$. When $b > 1$, the value of y increases as the value of x increases. When $b < 1$, the value of y decreases as the value of x increases. Notice that all four graphs pass through $(0, 1)$.

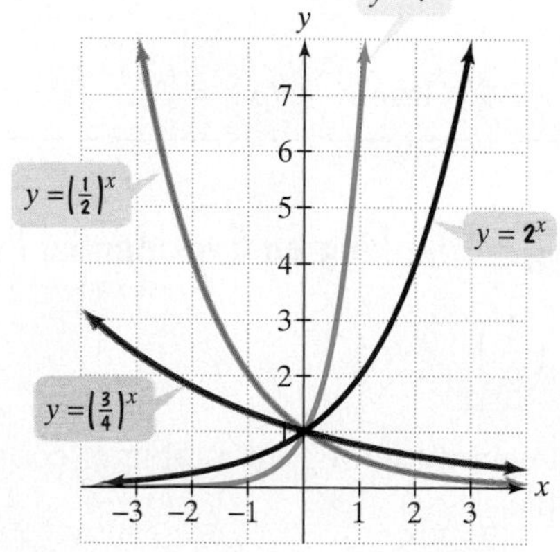

FIGURE 12.3 Graphs of four exponential functions

These graphs illustrate the following general characteristics of exponential functions:

CHARACTERISTICS OF EXPONENTIAL FUNCTIONS OF THE FORM $f(x) = b^x$

1. The domain of $f(x) = b^x$ consists of all real numbers. The range of $f(x) = b^x$ consists of all positive real numbers.
2. The graphs of all exponential functions of the form $f(x) = b^x$ pass through the point $(0, 1)$ because $f(0) = b^0 = 1$ $(b \neq 0)$. The y-intercept is 1.
3. If $b > 1$, $f(x) = b^x$ has a graph that goes up to the right and is an increasing function. The greater the value of b, the steeper the increase.
4. If $0 < b < 1$, $f(x) = b^x$ has a graph that goes down to the right and is a decreasing function. The smaller the value of b, the steeper the decrease.
5. The graph of $f(x) = b^x$ approaches, but does not touch, the x-axis. The x-axis, or $y = 0$, is a horizontal asymptote.

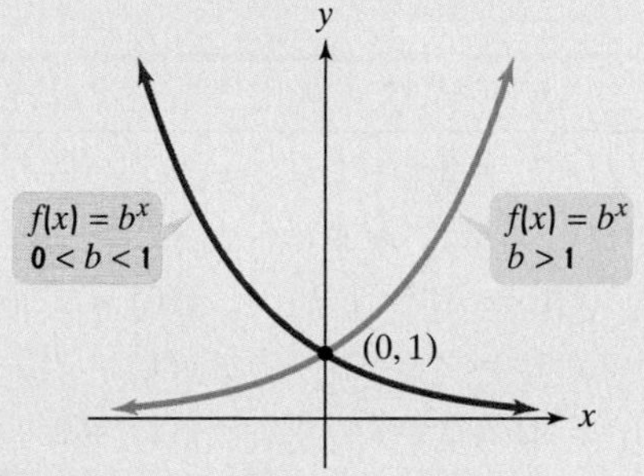

EXAMPLE 4 Graphing Exponential Functions

Graph $f(x) = 3^x$ and $g(x) = 3^{x+1}$ in the same rectangular coordinate system. How is the graph of g related to the graph of f?

SOLUTION We begin by setting up a table showing some of the coordinates for f and g, selecting integers from -2 to 2 for x. Notice that $x + 1$ is the exponent for $g(x) = 3^{x+1}$.

x	$f(x) = 3^x$	$g(x) = 3^{x+1}$
-2	$f(-2) = 3^{-2} = \frac{1}{9}$	$g(-2) = 3^{-2+1} = 3^{-1} = \frac{1}{3}$
-1	$f(-1) = 3^{-1} = \frac{1}{3}$	$g(-1) = 3^{-1+1} = 3^0 = 1$
0	$f(0) = 3^0 = 1$	$g(0) = 3^{0+1} = 3^1 = 3$
1	$f(1) = 3^1 = 3$	$g(1) = 3^{1+1} = 3^2 = 9$
2	$f(2) = 3^2 = 9$	$g(2) = 3^{2+1} = 3^3 = 27$

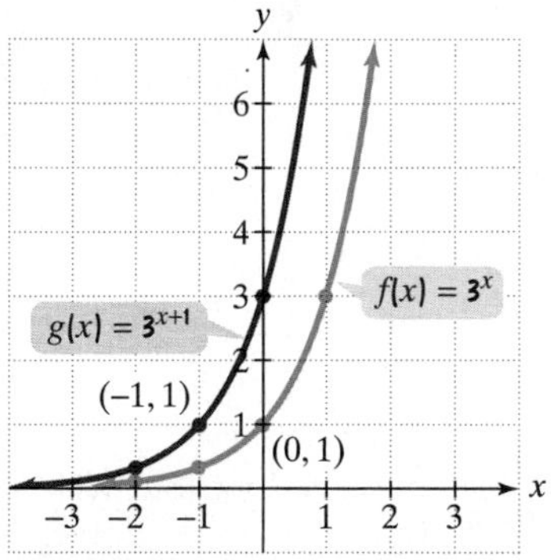

FIGURE 12.4

We plot the points for each function and connect them with a smooth curve. Because of the scale on the y-axis, the points on each function corresponding to $x = 2$ are not shown. Figure 12.4 shows the graphs of $f(x) = 3^x$ and $g(x) = 3^{x+1}$. The graph of g is the graph of f shifted one unit to the left. ■

✔ **CHECK POINT 4** Graph $f(x) = 3^x$ and $g(x) = 3^{x-1}$ in the same rectangular coordinate system. Select integers from -2 to 2 for x. How is the graph of g related to the graph of f?

EXAMPLE 5 Graphing Exponential Functions

Graph $f(x) = 2^x$ and $g(x) = 2^x - 3$ in the same rectangular coordinate system. How is the graph of g related to the graph of f?

SOLUTION We begin by setting up a table showing some of the coordinates for f and g, selecting integers from -2 to 2 for x.

x	$f(x) = 2^x$	$g(x) = 2^x - 3$
-2	$f(-2) = 2^{-2} = \frac{1}{4}$	$g(-2) = 2^{-2} - 3 = \frac{1}{4} - 3 = -2\frac{3}{4}$
-1	$f(-1) = 2^{-1} = \frac{1}{2}$	$g(-1) = 2^{-1} - 3 = \frac{1}{2} - 3 = -2\frac{1}{2}$
0	$f(0) = 2^0 = 1$	$g(0) = 2^0 - 3 = 1 - 3 = -2$
1	$f(1) = 2^1 = 2$	$g(1) = 2^1 - 3 = 2 - 3 = -1$
2	$f(2) = 2^2 = 4$	$g(2) = 2^2 - 3 = 4 - 3 = 1$

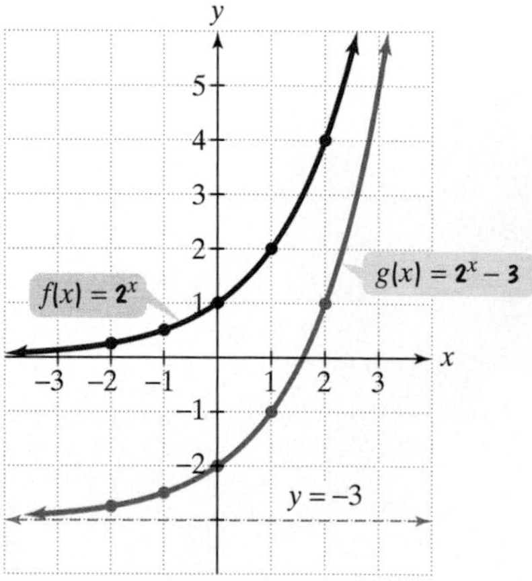

FIGURE 12.5

We plot the points for each function and connect them with a smooth curve. Figure 12.5 shows the graphs of $f(x) = 2^x$ and $g(x) = 2^x - 3$. The graph of g is the graph of f shifted down three units. As a result, $y = -3$ is the horizontal asymptote for g. ■

✔ **CHECK POINT 5** Graph $f(x) = 2^x$ and $g(x) = 2^x + 3$ in the same rectangular coordinate system. Select integers from -2 to 2 for x. How is the graph of g related to the graph of f?

3 Evaluate functions with base e.

The Natural Base e An irrational number, symbolized by the letter e, appears as the base in many applied exponential functions. This irrational number is approximately equal to 2.72. More accurately,

$$e \approx 2.71828\ldots.$$

The number e is called the **natural base**. The function $f(x) = e^x$ is called the **natural exponential function**.

Use a scientific or graphing calculator with an $\boxed{e^x}$ key to evaluate e to various powers. For example, to find e^2, press the following keys on most calculators:

Scientific calculator: 2 $\boxed{e^x}$

Graphing calculator: $\boxed{e^x}$ 2 $\boxed{\text{ENTER}}$.

The display should be approximately 7.389.

$$e^2 \approx 7.389$$

The number e lies between 2 and 3. Because $2^2 = 4$ and $3^2 = 9$, it makes sense that e^2, approximately 7.389, lies between 4 and 9.

Because $2 < e < 3$, the graph of $y = e^x$ is between the graphs of $y = 2^x$ and $y = 3^x$, shown in Figure 12.6.

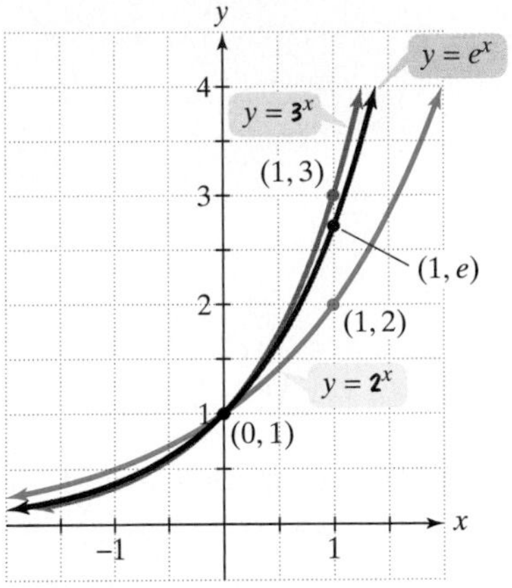

FIGURE 12.6 Graphs of three exponential functions

EXAMPLE 6 World Population

In a report entitled *Resources and Man*, the U.S. National Academy of Sciences concluded that a world population of 10 billion "is close to (if not above) the maximum that an intensely managed world might hope to support with some degree of comfort and individual choice." At the time the report was issued in 1969, world population was approximately 3.6 billion, with a growth rate of 2% per year. The function

$$f(x) = 3.6e^{0.02x}$$

describes world population, $f(x)$, in billions, x years after 1969. Use the function to find world population in the year 2020. Is there cause for alarm?

SOLUTION Because 2020 is 51 years after 1969, we substitute 51 for x in $f(x) = 3.6e^{0.02x}$:

$$f(51) = 3.6e^{0.02(51)}.$$

Perform this computation on your calculator.

Scientific calculator: 3.6 $\boxed{\times}$ $\boxed{(}$ .02 $\boxed{\times}$ 51 $\boxed{)}$ $\boxed{e^x}$ $\boxed{=}$

Graphing calculator: 3.6 $\boxed{\times}$ $\boxed{e^x}$ $\boxed{(}$ .02 $\boxed{\times}$ 51 $\boxed{)}$ $\boxed{\text{ENTER}}$

The display should be approximately 9.9835012. Thus,

$$f(51) = 3.6e^{0.02(51)} \approx 9.98.$$

This indicates that world population in the year 2020 will be approximately 9.98 billion. Because this number is quite close to 10 billion, the given function suggests that there may be cause for alarm. ■

World Population (billions)

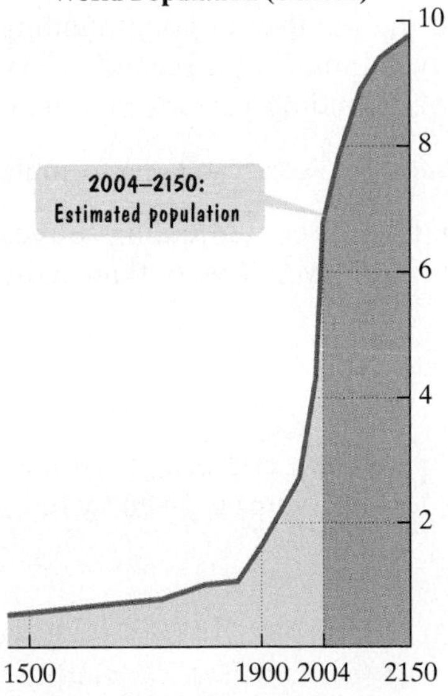

Source: U.N. Population Division

World population in 2002 was approximately 6.3 billion, but the growth rate was no longer 2%. It had slowed down to 1.23%. Using this current growth rate, exponential functions now predict a world population of 7.9 billion in the year 2020. Experts think the population may stabilize at 10 billion after 2200 if the growth rate continues to decline.

CHECK POINT 6 The function $f(x) = 6.3e^{0.0123x}$ describes world population, $f(x)$, in billions, x years after 2002 subject to a growth rate of 1.23% annually. Use the function to predict world population in 2050.

4 Use compound interest formulas.

Compound Interest In Chapter 11, we saw that the amount of money, A, that a principal, P, will be worth after t years at interest rate r, compounded annually, is given by the formula

$$A = P(1 + r)^t.$$

Most savings institutions have plans in which interest is paid more than once a year. If compound interest is paid twice a year, the compounding period is six months. We say that the interest is **compounded semiannually**. When compound interest is paid four times a year, the compounding period is three months and the interest is said to be **compounded quarterly**. Some plans allow for monthly compounding or daily compounding.

In general, when compound interest is paid n times a year, we say that there are $\boldsymbol{n}$ **compounding periods per year**. The formula $A = P(1 + r)^t$ can be adjusted to take into account the number of compounding periods in a year. If there are n compounding periods per year, the formula becomes

$$A = P\left(1 + \frac{r}{n}\right)^{nt}.$$

Some banks use **continuous compounding**, where the number of compounding periods increases infinitely (compounding interest every trillionth of a second, every quadrillionth of a second, etc.). As n, the number of compounding periods in a year, increases without bound, the expression $\left(1 + \frac{1}{n}\right)^n$ approaches e. As a result, the formula for continuous compounding is $A = Pe^{rt}$. Although continuous compounding sounds terrific, it yields only a fraction of a percent more interest over a year than daily compounding.

FORMULAS FOR COMPOUND INTEREST After t years, the balance, A, in an account with principal P and annual interest rate r (in decimal form) is given by the following formulas:

1. For n compoundings per year: $A = P\left(1 + \frac{r}{n}\right)^{nt}$

2. For continuous compounding: $A = Pe^{rt}$.

ENRICHMENT ESSAY

Exploring the Behavior of $\left(1 + \frac{1}{n}\right)^n$ as n Increases without Bound

A table or graphs can be used to show that as n, the number of compounding periods in a year, increases without bound, the expression $\left(1 + \frac{1}{n}\right)^n$ approaches e.

Numerical Approach

n	$\left(1 + \frac{1}{n}\right)^n$
1	2
2	2.25
5	2.48832
10	2.59374246
100	2.704813829
1000	2.716923932
10,000	2.718145927
100,000	2.718268237
1,000,000	2.718280469
1,000,000,000	2.718281827

As n takes on increasingly large values, the expression $\left(1 + \frac{1}{n}\right)^n$ approaches e.

Graphical Approach

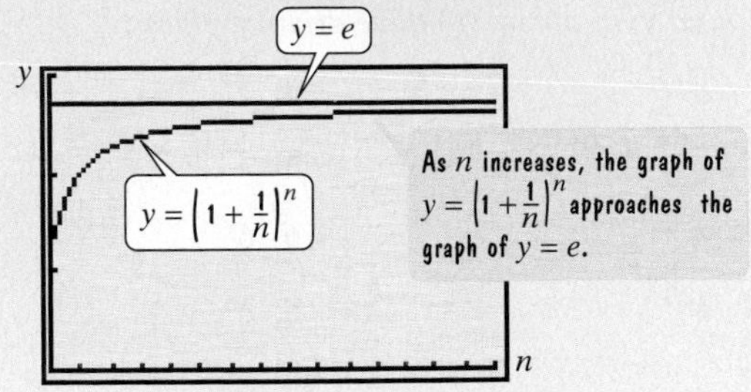

[0, 15, 1] by [0, 3, 1]

EXAMPLE 7 Choosing between Investments

You plan to invest \$8000 for 6 years and you have a choice between two accounts. The first pays 7% per year, compounded monthly. The second pays 6.85% per year, compounded continuously. Which is the better investment?

SOLUTION The better investment is the one with the greater balance in the account after 6 years. Let's begin with the account with monthly compounding. We use the compound interest model with $P = 8000$, $r = 7\% = 0.07$, $n = 12$ (monthly compounding means 12 compoundings per year), and $t = 6$.

$$A = P\left(1 + \frac{r}{n}\right)^{nt} = 8000\left(1 + \frac{0.07}{12}\right)^{12 \cdot 6} \approx 12{,}160.84$$

The balance in this account after 6 years is \$12,160.84.

For the second investment option, we use the model for continuous compounding with $P = 8000$, $r = 6.85\% = 0.0685$, and $t = 6$.

$$A = Pe^{rt} = 8000e^{0.0685(6)} \approx 12{,}066.60$$

The balance in this account after 6 years is \$12,066.60, slightly less than the previous amount. Thus, the better investment is the 7% monthly compounding option. ■

CHECK POINT 7 A sum of \$10,000 is invested at an annual rate of 8%. Find the balance in the account after 5 years subject to **a.** quarterly compounding and **b.** continuous compounding.

12.1 EXERCISE SET

Student Solutions Manual | CD/Video | PH Math/Tutor Center | MathXL Tutorials on CD | MathXL® | MyMathLab | Interactmath.com

Practice Exercises

In Exercises 1–10, approximate each number using a calculator. Round your answer to three decimal places.

1. $2^{3.4}$ **2.** $3^{2.4}$

3. $3^{\sqrt{5}}$ **4.** $5^{\sqrt{3}}$

5. $4^{-1.5}$ **6.** $6^{-1.2}$

7. $e^{2.3}$ **8.** $e^{3.4}$

9. $e^{-0.95}$ **10.** $e^{-0.75}$

In Exercises 11–16, set up a table of coordinates for each function. Select integers from −2 to 2 for x. Then use the table of coordinates to match the function with its graph. [The graphs are labeled (a) through (f).]

11. $f(x) = 3^x$ **12.** $f(x) = 3^{x-1}$

13. $f(x) = 3^x - 1$ **14.** $f(x) = -3^x$

15. $f(x) = 3^{-x}$ **16.** $f(x) = -3^{-x}$

(a)

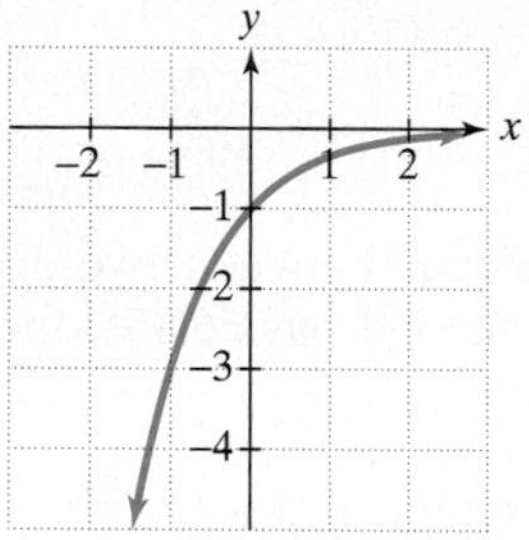

(b)

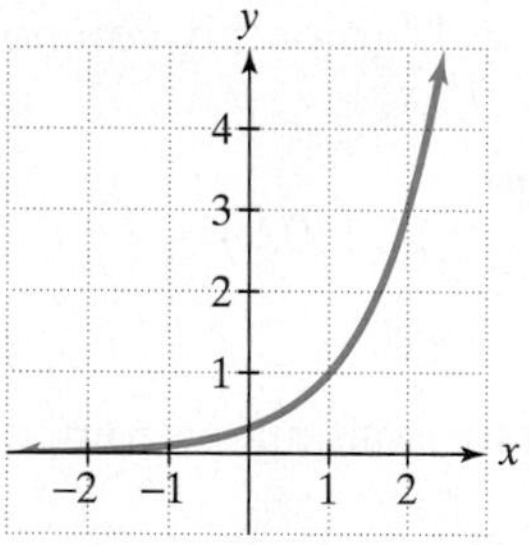

(c)

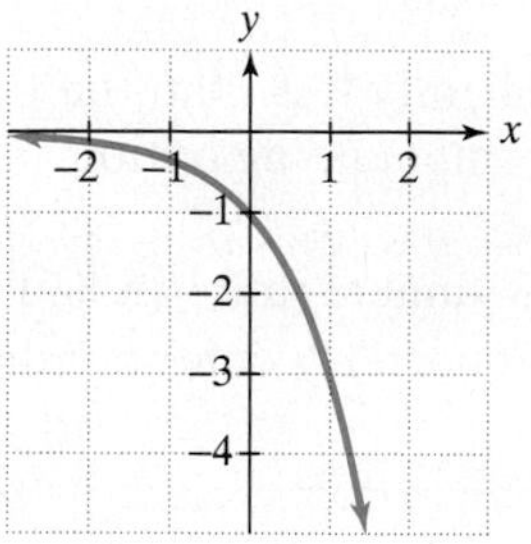

(d)

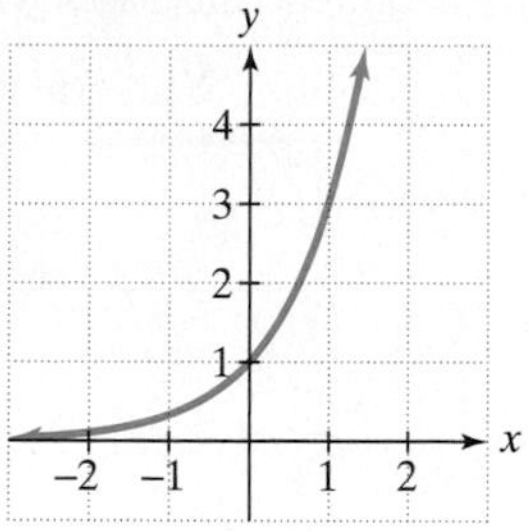

(e)

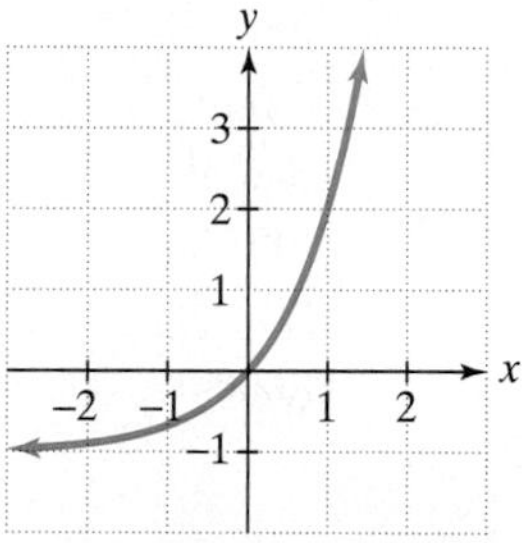

(f)

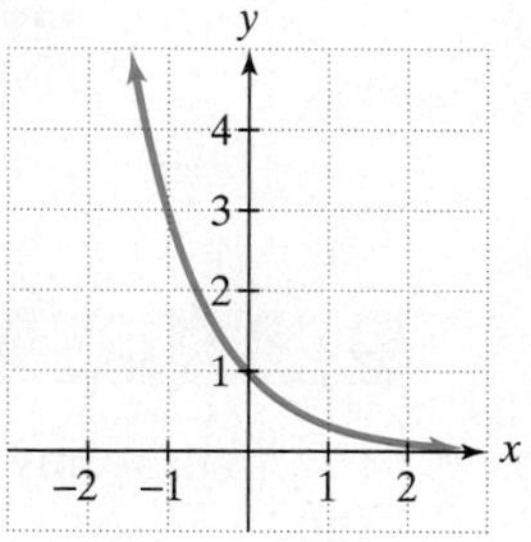

In Exercises 17–24, graph each function by making a table of coordinates. If applicable, use a graphing utility to confirm your hand-drawn graph.

17. $f(x) = 4^x$ **18.** $f(x) = 5^x$

19. $g(x) = \left(\frac{3}{2}\right)^x$ **20.** $g(x) = \left(\frac{4}{3}\right)^x$

21. $h(x) = \left(\frac{1}{2}\right)^x$ **22.** $h(x) = \left(\frac{1}{3}\right)^x$

23. $f(x) = (0.6)^x$ **24.** $f(x) = (0.8)^x$

In Exercises 25–38, graph functions f and g in the same rectangular coordinate system. Select integers from −2 to 2 for x. Then describe how the graph of g is related to the graph of f. If applicable, use a graphing utility to confirm your hand-drawn graphs.

25. $f(x) = 2^x$ and $g(x) = 2^{x+1}$

26. $f(x) = 2^x$ and $g(x) = 2^{x+2}$

27. $f(x) = 2^x$ and $g(x) = 2^{x-2}$

28. $f(x) = 2^x$ and $g(x) = 2^{x-1}$

29. $f(x) = 2^x$ and $g(x) = 2^x + 1$

30. $f(x) = 2^x$ and $g(x) = 2^x + 2$

31. $f(x) = 2^x$ and $g(x) = 2^x - 2$

32. $f(x) = 2^x$ and $g(x) = 2^x - 1$

33. $f(x) = 3^x$ and $g(x) = -3^x$

34. $f(x) = 3^x$ and $g(x) = 3^{-x}$

35. $f(x) = 2^x$ and $g(x) = 2^{x+1} - 1$

36. $f(x) = 2^x$ and $g(x) = 2^{x+1} - 2$

37. $f(x) = 3^x$ and $g(x) = \frac{1}{3} \cdot 3^x$

38. $f(x) = 3^x$ and $g(x) = 3 \cdot 3^x$

Use the compound interest formulas, $A = P\left(1 + \frac{r}{n}\right)^{nt}$ *and* $A = Pe^{rt}$, *to solve Exercises 39–42. Round answers to the nearest cent.*

39. Find the accumulated value of an investment of \$10,000 for 5 years at an interest rate of 5.5% if the money is **a.** compounded semiannually; **b.** compounded monthly; **c.** compounded continuously.

40. Find the accumulated value of an investment of \$5000 for 10 years at an interest rate of 6.5% if the money is **a.** compounded semiannually; **b.** compounded monthly; **c.** compounded continuously.

41. Suppose that you have \$12,000 to invest. Which investment yields the greater return over 3 years: 7% compounded monthly or 6.85% compounded continuously?

42. Suppose that you have \$6000 to invest. Which investment yields the greater return over 4 years: 8.25% compounded quarterly or 8.3% compounded semiannually?

Practice Plus

In Exercises 43–48, use each exponential function's graph to determine the function's domain and range.

43.

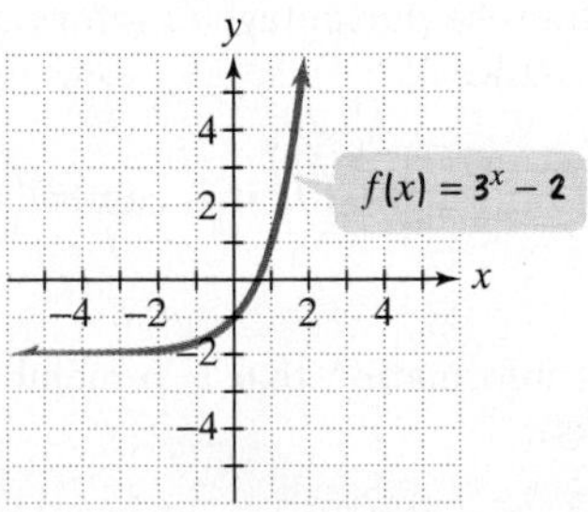

44.

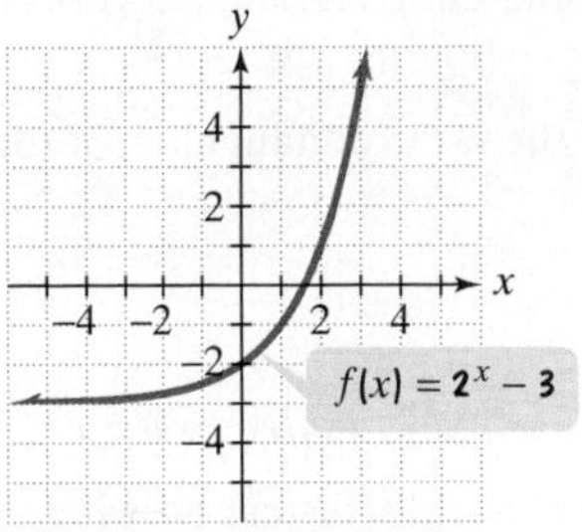

45.

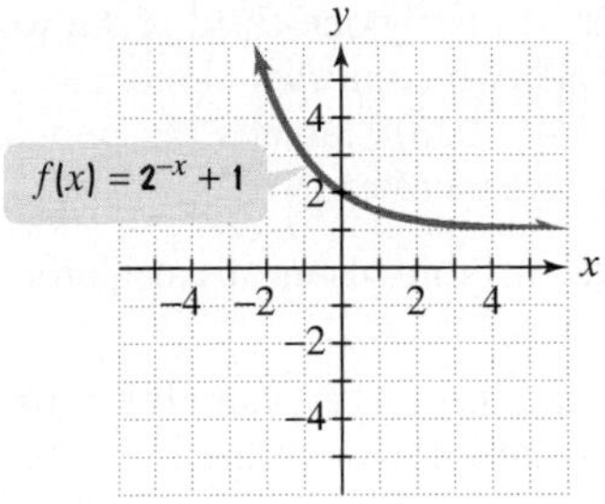

46.

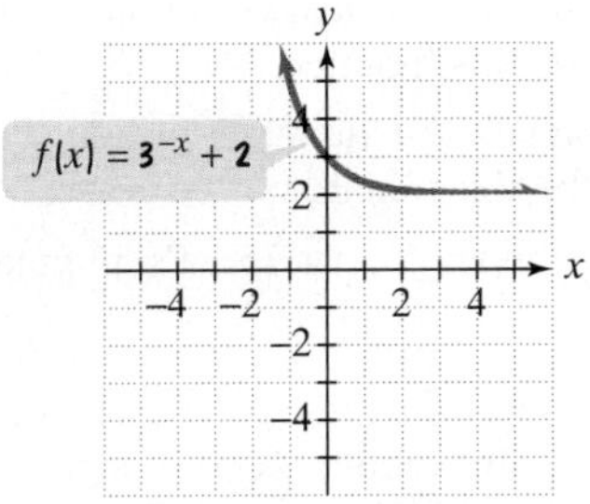

47.

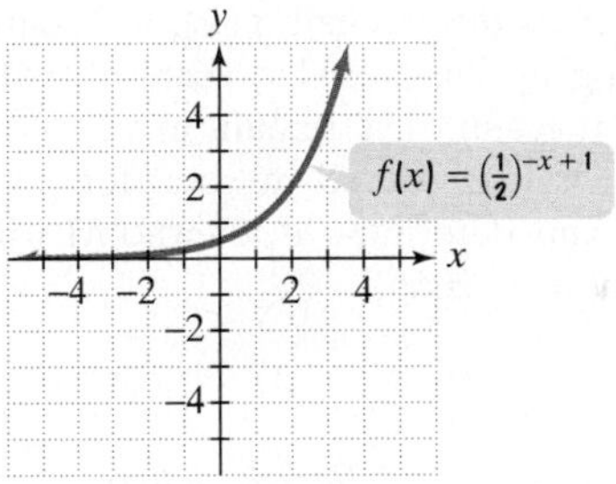

48.

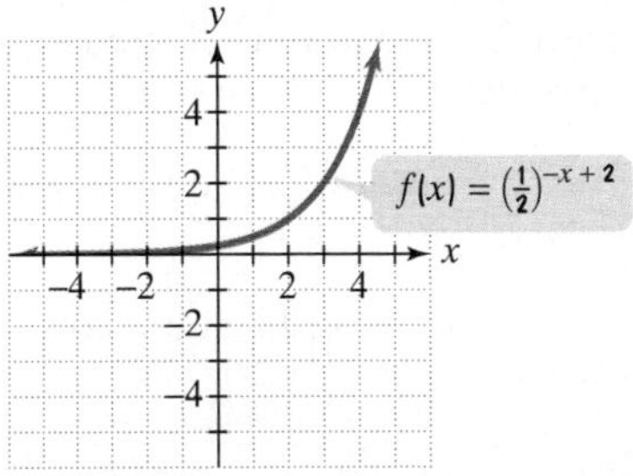

In Exercises 49–50, graph f and g in the same rectangular coordinate system. Then find the point of intersection of the two graphs.

49. $f(x) = 2^x$, $g(x) = 2^{-x}$

50. $f(x) = 2^{x+1}$, $g(x) = 2^{-x+1}$

51. Graph $y = 2^x$ and $x = 2^y$ in the same rectangular coordinate system.

52. Graph $y = 3^x$ and $x = 3^y$ in the same rectangular coordinate system.

Application Exercises

Use a calculator with a $\boxed{y^x}$ *key or a* $\boxed{\wedge}$ *key to solve Exercises 53–56.*

53. India is currently one of the world's fastest-growing countries. By 2040, the population of India will be larger than the population of China; by 2050, nearly one-third of the world's population will live in these two countries alone. The exponential function $f(x) = 574(1.026)^x$ models the population of India, $f(x)$, in millions, x years after 1974.

a. Substitute 0 for x and, without using a calculator, find India's population in 1974.

b. Substitute 27 for x and use your calculator to find India's population, to the nearest million, in the year 2001 as modeled by this function.

c. Find India's population, to the nearest million, in the year 2028 as predicted by this function.

d. Find India's population, to the nearest million, in the year 2055 as predicted by this function.

e. What appears to be happening to India's population every 27 years?

54. The 1986 explosion at the Chernobyl nuclear power plant in the former Soviet Union sent about 1000 kilograms of radioactive cesium-137 into the atmosphere. The function $f(x) = 1000(0.5)^{\frac{x}{30}}$ describes the amount, $f(x)$, in kilograms, of cesium-137 remaining in Chernobyl x years after 1986. If even 100 kilograms of cesium-137 remain in Chernobyl's atmosphere, the area is considered unsafe for human habitation. Find $f(80)$ and determine if Chernobyl will be safe for human habitation by 2066.

The formula $S = C(1 + r)^t$ models inflation, where C = the value today, r = the annual inflation rate, and S = the inflated value t years from now. Use this formula to solve Exercises 55–56. Round answers to the nearest dollar.

55. If the inflation rate is 6%, how much will a house now worth \$65,000 be worth in 10 years?

56. If the inflation rate is 3%, how much will a house now worth \$110,000 be worth in 5 years?

Use a calculator with an $\boxed{e^x}$ *key to solve Exercises 57–63.*

The graph in the next column shows the number of U.S. cellular telephone subscribers from 1995 through 2002. The data can be modeled by the exponential function

$$f(x) = 35.86e^{0.207x},$$

in which $f(x)$ is the number of U.S. cellular telephone subscribers, in millions, x years after 1995. Use this function to solve Exercises 57–58.

Millions of U.S. Cellular Telephone Subscribers

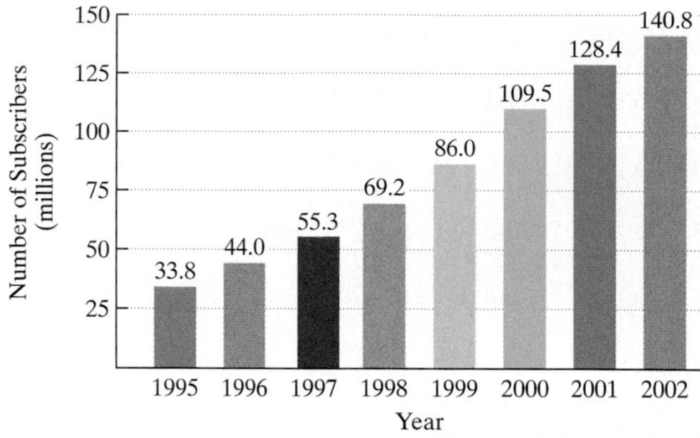

Source: CTIA

57. According to the model, how many cellular telephone subscribers were there in 2001? Round to the nearest tenth of a million. How well does the model describe the actual data shown by the bar graph for this year?

58. According to the model, how many U.S. cellular telephone subscribers will there be in 2007? Round to the nearest tenth of a million. Does this seem realistic or has model breakdown occurred? Explain your answer.

59. In college, we study large volumes of information—information that, unfortunately, we do not often retain for very long. The function

$$f(x) = 80e^{-0.5x} + 20$$

describes the percentage of information, $f(x)$, that a particular person remembers x weeks after learning the information.

a. Substitute 0 for x and, without using a calculator, find the percentage of information remembered at the moment it is first learned.

b. Substitute 1 for x and find the percentage of information that is remembered after 1 week.

c. Find the percentage of information that is remembered after 4 weeks.

d. Find the percentage of information that is remembered after one year (52 weeks).

60. In 1626, Peter Minuit persuaded the Wappinger Indians to sell him Manhattan Island for $24. If the Native Americans had put the $24 into a bank account paying 5% interest, how much would the investment have been worth in the year 2005 if interest were compounded

a. monthly?

b. continuously?

The function

$$f(x) = \frac{90}{1 + 270e^{-0.122x}}$$

models the percentage, $f(x)$, of people x years old with some coronary heart disease. Use this function to solve Exercises 61–62. Round answers to the nearest tenth of a percent.

61. Evaluate $f(30)$ and describe what this means in practical terms.

62. Evaluate $f(70)$ and describe what this means in practical terms.

63. The function

$$N(t) = \frac{30{,}000}{1 + 20e^{-1.5t}}$$

describes the number of people, $N(t)$, who become ill with influenza t weeks after its initial outbreak in a town with 30,000 inhabitants. The horizontal asymptote in the graph indicates that there is a limit to the epidemic's growth.

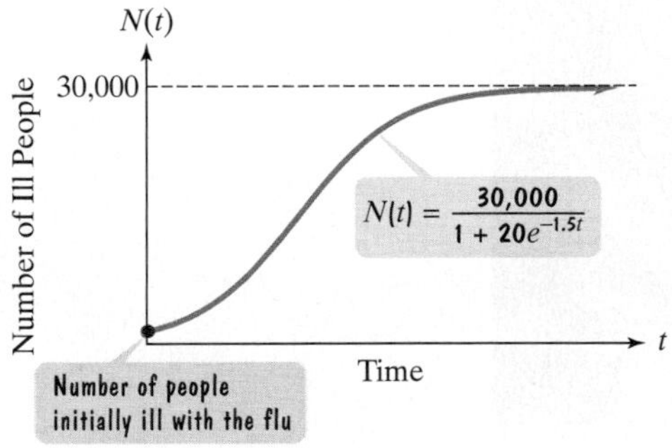

a. How many people became ill with the flu when the epidemic began? (When the epidemic began, $t = 0$.)

b. How many people were ill by the end of the third week?

c. Why can't the spread of an epidemic simply grow indefinitely? What does the horizontal asymptote shown in the graph indicate about the limiting size of the population that becomes ill?

Writing in Mathematics

64. What is an exponential function?

65. What is the natural exponential function?

66. Use a calculator to evaluate $\left(1 + \frac{1}{x}\right)^x$ for $x = 10, 100, 1000,$ 10,000, 100,000, and 1,000,000. Describe what happens to the expression as x increases.

67. Write an example similar to Example 7 on page 807 in which continuous compounding at a slightly lower yearly interest rate is a better investment than compounding n times per year.

68. Describe how you could use the graph of $f(x) = 2^x$ to obtain a decimal approximation for $\sqrt{2}$.

69. In 2002, world population was 6.3 billion, with an annual growth rate of 1.23%. Discuss two factors that would cause this growth rate to slow down over the next ten years.

Technology Exercises

70. Graph $y = 13.49(0.967)^x - 1$, the function for the number of O-rings expected to fail at x°F, in a $[0, 90, 10]$ by $[0, 20, 5]$ viewing rectangle. If NASA engineers had used this function and its graph, is it likely they would have allowed the *Challenger* to be launched when the temperature was 31°F? Explain.

71. You have $10,000 to invest. One bank pays 5% interest compounded quarterly and the other pays 4.5% interest compounded monthly.

a. Use the formula for compound interest to write a function for the balance in each account at any time t, in years.

b. Use a graphing utility to graph both functions in an appropriate viewing rectangle. Based on the graphs, which bank offers the better return on your money?

72. a. Graph $y = e^x$ and $y = 1 + x + \frac{x^2}{2}$ in the same viewing rectangle.

b. Graph $y = e^x$ and $y = 1 + x + \frac{x^2}{2} + \frac{x^3}{6}$ in the same viewing rectangle.

c. Graph $y = e^x$ and $y = 1 + x + \frac{x^2}{2} + \frac{x^3}{6} + \frac{x^4}{24}$ in the same viewing rectangle.

d. Describe what you observe in parts (a)–(c). Try generalizing this observation.

Critical Thinking Exercises

73. Which one of the following is true?

a. As the number of compounding periods increases on a fixed investment, the amount of money in the account over a fixed interval of time will increase without bound.

b. The functions $f(x) = 3^{-x}$ and $g(x) = -3^x$ have the same graph.

c. If $f(x) = 2^x$, then $f(a + b) = f(a) + f(b)$.

d. The functions $f(x) = \left(\frac{1}{3}\right)^x$ and $g(x) = 3^{-x}$ have the same graph.

74. The graphs labeled (a)–(d) in the figure represent $y = 3^x$, $y = 5^x$, $y = \left(\frac{1}{3}\right)^x$, and $y = \left(\frac{1}{5}\right)^x$, but not necessarily in that order. Which is which? Describe the process that enables you to make this decision.

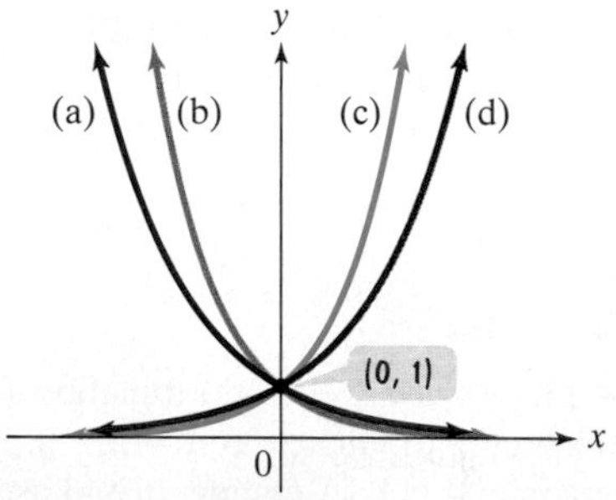

75. The hyperbolic cosine and hyperbolic sine functions are defined by

$$\cosh x = \frac{e^x + e^{-x}}{2} \quad \text{and} \quad \sinh x = \frac{e^x - e^{-x}}{2}.$$

Prove that $(\cosh x)^2 - (\sinh x)^2 = 1$.

Review Exercises

76. Solve for b: $D = \dfrac{ab}{a + b}$.
(Section 7.6, Example 7)

77. Subtract: $\dfrac{2x + 3}{x^2 - 7x + 12} - \dfrac{2}{x - 3}$.
(Section 7.4, Example 7)

78. Solve: $x(x - 3) = 10$.
(Section 11.2, Example 1)

SECTION 12.2

Objectives

1 Change from logarithmic to exponential form.

2 Change from exponential to logarithmic form.

3 Evaluate logarithms.

4 Use basic logarithmic properties.

5 Graph logarithmic functions.

6 Find the domain of a logarithmic function.

7 Use common logarithms.

8 Use natural logarithms.

LOGARITHMIC FUNCTIONS

The earthquake that ripped through northern California on October 17, 1989, measured 7.1 on the Richter scale, killed more than 60 people, and injured more than 2400. Shown here is San Francisco's Marina district, where shock waves tossed houses off their foundations and into the street.

A higher measure on the Richter scale is more devastating than it seems because for each increase in one unit on the scale, there is a tenfold increase in the intensity of an earthquake. In this section, our focus is on the inverse of the exponential function, called the logarithmic function. The logarithmic function will help you to understand diverse phenomena, including earthquake intensity, human memory, and the pace of life in large cities.

STUDY TIP

The inverse of $y = b^x$ is $x = b^y$. Logarithms give us a way to express the inverse function $x = b^y$ for y in terms of x. In case you need to review inverse functions, they are discussed in Section 8.3 on pages 560–566. Here's a summary of what you should already know about functions and their inverses.

1. Only one-to-one functions have inverses that are functions. A function, f, has an inverse function, f^{-1}, if there is no horizontal line that intersects the graph of f at more than one point.
2. If a function is one-to-one, its inverse function can be found by interchanging x and y in the function's equation and solving for y.
3. If $f(a) = b$, then $f^{-1}(b) = a$. The domain of f is the range of f^{-1}. The range of f is the domain of f^{-1}.
4. $f(f^{-1}(x)) = x$ and $f^{-1}(f(x)) = x$.
5. The graph of f^{-1} is the reflection of the graph of f about the line $y = x$.

The Definition of Logarithmic Functions No horizontal line can be drawn that intersects the graph of an exponential function at more than one point. This means that the exponential function is one-to-one and has an inverse. The inverse function of the exponential function with base b is called the *logarithmic function with base b*.

DEFINITION OF THE LOGARITHMIC FUNCTION For $x > 0$ and $b > 0, b \neq 1$,

$$y = \log_b x \text{ is equivalent to } b^y = x.$$

The function $f(x) = \log_b x$ is the **logarithmic function with base b.**

The equations

$$y = \log_b x \quad \text{and} \quad b^y = x$$

are different ways of expressing the same thing. The first equation is in **logarithmic form** and the second equivalent equation is in **exponential form**.

Notice that a **logarithm**, y, **is an exponent**. You should learn the location of the base and exponent in each form.

LOCATION OF BASE AND EXPONENT IN EXPONENTIAL AND LOGARITHMIC FORMS

Logarithmic Form: $y = \log_b x$ (Exponent: y; Base: b)   Exponential Form: $b^y = x$ (Exponent: y; Base: b)

1 Change from logarithmic to exponential form.

EXAMPLE 1 Changing from Logarithmic to Exponential Form

Write each equation in its equivalent exponential form:

a. $2 = \log_5 x$ **b.** $3 = \log_b 64$ **c.** $\log_3 7 = y$.

SOLUTION We use the fact that $y = \log_b x$ means $b^y = x$.

a. $2 = \log_5 x$ means $5^2 = x$.   **b.** $3 = \log_b 64$ means $b^3 = 64$.

Logarithms are exponents.   Logarithms are exponents.

c. $\log_3 7 = y$ or $y = \log_3 7$ means $3^y = 7$. ■

CHECK POINT 1 Write each equation in its equivalent exponential form:

a. $3 = \log_7 x$ **b.** $2 = \log_b 25$ **c.** $\log_4 26 = y$.

2 Change from exponential to logarithmic form.

EXAMPLE 2 Changing from Exponential to Logarithmic Form

Write each equation in its equivalent logarithmic form:

a. $12^2 = x$ **b.** $b^3 = 8$ **c.** $e^y = 9$.

SOLUTION We use the fact that $b^y = x$ means $y = \log_b x$.

a. $12^2 = x$ means $2 = \log_{12} x$. (Logarithms are exponents.)

b. $b^3 = 8$ means $3 = \log_b 8$. (Logarithms are exponents.)

c. $e^y = 9$ means $y = \log_e 9$. ■

CHECK POINT 2 Write each equation in its equivalent logarithmic form:

a. $2^5 = x$ **b.** $b^3 = 27$ **c.** $e^y = 33$.

3 Evaluate logarithms.

Remembering that logarithms are exponents makes it possible to evaluate some logarithms by inspection. The logarithm of x with base b, $\log_b x$, is the exponent to which b must be raised to get x. For example, suppose we want to evaluate $\log_2 32$. We ask, 2 to what power gives 32? Because $2^5 = 32$, we have $\log_2 32 = 5$.

EXAMPLE 3 Evaluating Logarithms

Evaluate:

a. $\log_2 16$ **b.** $\log_3 9$ **c.** $\log_{25} 5$.

SOLUTION

Logarithmic Expression	Question Needed for Evaluation	Logarithmic Expression Evaluated
a. $\log_2 16$	2 to what power gives 16?	$\log_2 16 = 4$ because $2^4 = 16$.
b. $\log_3 9$	3 to what power gives 9?	$\log_3 9 = 2$ because $3^2 = 9$.
c. $\log_{25} 5$	25 to what power gives 5?	$\log_{25} 5 = \frac{1}{2}$ because $25^{\frac{1}{2}} = \sqrt{25} = 5$.

■

CHECK POINT 3 Evaluate:

a. $\log_{10} 100$ **b.** $\log_3 3$ **c.** $\log_{36} 6$.

4 Use basic logarithmic properties.

Basic Logarithmic Properties Because logarithms are exponents, they have properties that can be verified using the properties of exponents.

BASIC LOGARITHMIC PROPERTIES INVOLVING ONE

1. $\log_b b = 1$ because 1 is the exponent to which b must be raised to obtain b. $(b^1 = b)$
2. $\log_b 1 = 0$ because 0 is the exponent to which b must be raised to obtain 1. $(b^0 = 1)$

EXAMPLE 4 Using Properties of Logarithms

Evaluate:

a. $\log_7 7$ **b.** $\log_5 1$.

SOLUTION

a. Because $\log_b b = 1$, we conclude $\log_7 7 = 1$.

b. Because $\log_b 1 = 0$, we conclude $\log_5 1 = 0$.

CHECK POINT 4 Evaluate:

a. $\log_9 9$ **b.** $\log_8 1$.

The inverse of an exponential function is the logarithmic function with the same base. Thus, if $f(x) = b^x$, then $f^{-1}(x) = \log_b x$. In Section 8.3, we saw how inverse functions "undo" one another. In particular,

$$f(f^{-1}(x)) = x \quad \text{and} \quad f^{-1}(f(x)) = x.$$

Applying these relationships to exponential and logarithmic functions, we obtain the following **inverse properties of logarithms**:

INVERSE PROPERTIES OF LOGARITHMS For $b > 0$ and $b \neq 1$,

$$\log_b b^x = x$$ The logarithm with base b of b raised to a power equals that power.

$$b^{\log_b x} = x$$ b raised to the logarithm with base b of a number equals that number.

STUDY TIP

The voice balloons should help you see the "undoing" that takes place between the exponential and logarithmic functions in the inverse properties.

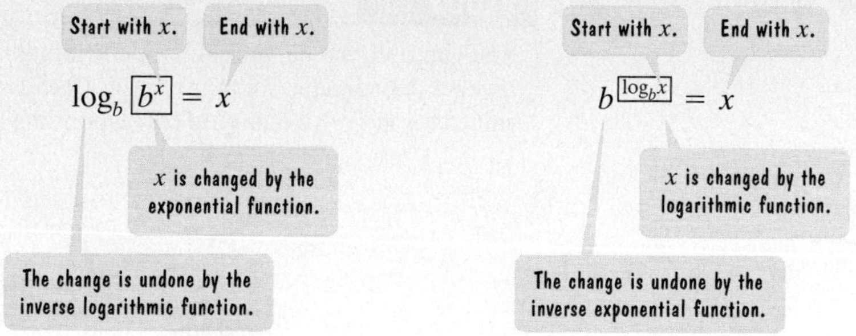
Start with x.
End with x.
log_b b^x = x
x is changed by the exponential function.
The change is undone by the inverse logarithmic function.
Start with x.
End with x.
b^(log_b x) = x
x is changed by the logarithmic function.
The change is undone by the inverse exponential function.

EXAMPLE 5 Using Inverse Properties of Logarithms

Evaluate:

a. $\log_4 4^5$ **b.** $6^{\log_6 9}$.

SOLUTION

a. Because $\log_b b^x = x$, we conclude $\log_4 4^5 = 5$.

b. Because $b^{\log_b x} = x$, we conclude $6^{\log_6 9} = 9$. ■

CHECK POINT 5 Evaluate:

a. $\log_7 7^8$ **b.** $3^{\log_3 17}$.

5 Graph logarithmic functions.

Graphs of Logarithmic Functions How do we graph logarithmic functions? We use the fact that a logarithmic function is the inverse of an exponential function. This means that the logarithmic function reverses the coordinates of the exponential function. It also means that the graph of the logarithmic function is a reflection of the graph of the exponential function about the line $y = x$.

EXAMPLE 6 Graphs of Exponential and Logarithmic Functions

Graph $f(x) = 2^x$ and $g(x) = \log_2 x$ in the same rectangular coordinate system.

SOLUTION We first set up a table of coordinates for $f(x) = 2^x$. Reversing these coordinates gives the coordinates for the inverse function $g(x) = \log_2 x$.

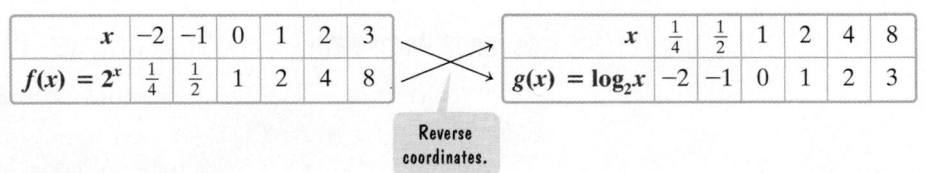

x	-2	-1	0	1	2	3
$f(x) = 2^x$	$\frac{1}{4}$	$\frac{1}{2}$	1	2	4	8

Reverse coordinates.

x	$\frac{1}{4}$	$\frac{1}{2}$	1	2	4	8
$g(x) = \log_2 x$	-2	-1	0	1	2	3

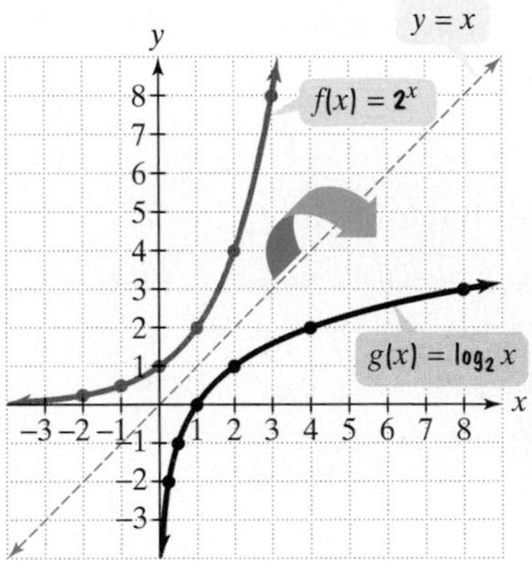

FIGURE 12.7 The graphs of $f(x) = 2^x$ and its inverse function

We now plot the ordered pairs from each table, connecting them with smooth curves. Figure 12.7 shows the graphs of $f(x) = 2^x$ and its inverse function $g(x) = \log_2 x$. The graph of the inverse can also be drawn by reflecting the graph of $f(x) = 2^x$ about the line $y = x$. ■

STUDY TIP

You can obtain a partial table of coordinates for $g(x) = \log_2 x$ without having to obtain and reverse coordinates for $f(x) = 2^x$. Because $g(x) = \log_2 x$ means $2^{g(x)} = x$, we begin with values for $g(x)$ and compute corresponding values for x:

Use $x = 2^{g(x)}$ to compute x. For example, if $g(x) = -2$, $x = 2^{-2} = \frac{1}{2^2} = \frac{1}{4}$.

Start with values for $g(x)$.

x	$\frac{1}{4}$	$\frac{1}{2}$	1	2	4	8
$g(x) = \log_2 x$	-2	-1	0	1	2	3

CHECK POINT 6 Graph $f(x) = 3^x$ and $g(x) = \log_3 x$ in the same rectangular coordinate system.

Figure 12.8 illustrates the relationship between the graph of an exponential function, shown in blue and its inverse, a logarithmic function, shown in red, for bases greater than 1 and for bases between 0 and 1.

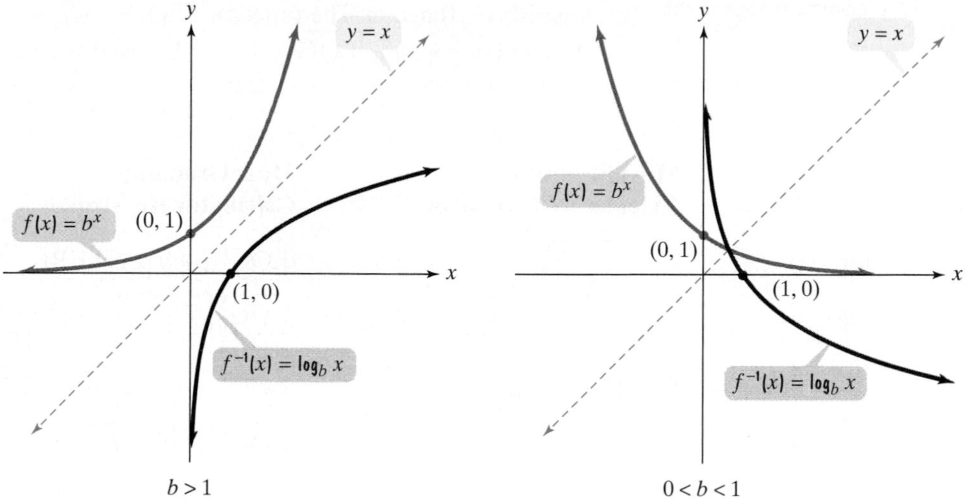

FIGURE 12.8 Graphs of exponential and logarithmic functions

CHARACTERISTICS OF THE GRAPHS OF LOGARITHMIC FUNCTIONS OF THE FORM $f(x) = \log_b x$

- The x-intercept is 1. There is no y-intercept.
- The y-axis is a vertical asymptote.
- If $b > 1$, the function is increasing. If $0 < b < 1$, the function is decreasing.
- The graph is smooth and continuous. It has no sharp corners or gaps.

DISCOVER FOR YOURSELF

Verify each of the four characteristics in the box for the red graphs in Figure 12.8.

6 Find the domain of a logarithmic function.

The Domain of a Logarithmic Function In Section 12.1, we learned that the domain of an exponential function of the form $f(x) = b^x$ includes all real numbers and its range is the set of positive real numbers. Because the logarithmic function reverses the domain and the range of the exponential function, the **domain of a logarithmic function of the form $f(x) = \log_b x$ is the set of all positive real numbers.** Thus, $\log_2 8$ is defined because the value of x in the logarithmic expression, 8, is greater than zero and therefore is included in the domain of the logarithmic function $f(x) = \log_2 x$. However, $\log_2 0$ and $\log_2(-8)$ are not defined because 0 and -8 are not positive real numbers and therefore are excluded from the domain of the logarithmic function $f(x) = \log_2 x$. In general, **the domain of $f(x) = \log_b [g(x)]$ consists of all x for which $g(x) > 0$.**

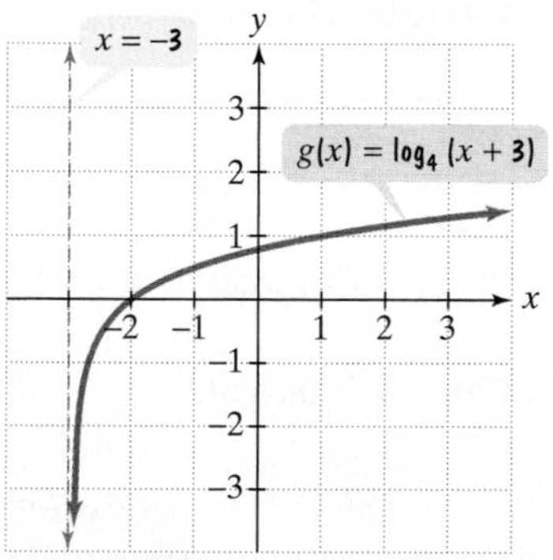

FIGURE 12.9 The domain of $g(x) = \log_4(x + 3)$ is $(-3, \infty)$.

EXAMPLE 7 Finding the Domain of a Logarithmic Function

Find the domain of $g(x) = \log_4(x + 3)$.

SOLUTION The domain of g consists of all x for which $x + 3 > 0$. Solving this inequality for x, we obtain $x > -3$. Thus, the domain of g is $\{x|x > -3\}$ or $(-3, \infty)$. This is illustrated in Figure 12.9. The vertical asymptote is $x = -3$ and all points on the graph of g have x-coordinates that are greater than -3. ■

CHECK POINT 7 Find the domain of $h(x) = \log_4(x - 5)$.

7 Use common logarithms.

Common Logarithms The logarithmic function with base 10 is called the **common logarithmic function**. The function $f(x) = \log_{10} x$ is usually expressed as $f(x) = \log x$. A calculator with a [LOG] key can be used to evaluate common logarithms. Here are some examples:

Logarithm	Most Scientific Calculator Keystrokes	Most Graphing Calculator Keystrokes	Display (or Approximate Display)
$\log 1000$	1000 [LOG]	[LOG] 1000 [ENTER]	3
$\log \frac{5}{2}$	[(] 5 [÷] 2 [)] [LOG]	[LOG] [(] 5 [÷] 2 [)] [ENTER]	0.39794
$\frac{\log 5}{\log 2}$	5 [LOG] [÷] 2 [LOG] [=]	[LOG] 5 [÷] [LOG] 2 [ENTER]	2.32193
$\log(-3)$	3 [+/−] [LOG]	[LOG] [(−)] 3 [ENTER]	[ERROR]

Some graphing calculators display an open parenthesis when the [LOG] key is pressed. In this case, remember to close the set of parentheses after entering the function's domain value: [LOG] 5 [)] [÷] [LOG] 2 [)] [ENTER].

The error message given by many calculators for $\log(-3)$ is a reminder that the domain of the common logarithmic function, $f(x) = \log x$, is the set of positive real numbers.

Many real-life phenomena start with rapid growth and then the growth begins to level off. This type of behavior can be modeled by logarithmic functions.

EXAMPLE 8 Modeling Height of Children

The percentage of adult height attained by a boy who is x years old can be modeled by

$$f(x) = 29 + 48.8 \log(x + 1),$$

where x represents the boy's age and $f(x)$ represents the percentage of his adult height. Approximately what percentage of his adult height has a boy attained at age eight?

SOLUTION We substitute the boy's age, 8, for x and evaluate the function at 8.

$$f(x) = 29 + 48.8 \log(x + 1) \quad \text{This is the given function.}$$

$$f(8) = 29 + 48.8 \log(8 + 1) \quad \text{Substitute 8 for x.}$$

$$= 29 + 48.8 \log 9 \quad \text{Graphing calculator keystrokes:}$$

$$\approx 76 \quad 29 \; [+] \; 48.8 \; [\text{LOG}] \; 9 \; [\text{ENTER}]$$

Thus, an 8-year-old boy has attained approximately 76% of his adult height. ■

CHECK POINT 8 Use the function in Example 8 to answer this question: Approximately what percentage of his adult height has a boy attained at age ten?

The basic properties of logarithms that were listed earlier in this section can be applied to common logarithms.

PROPERTIES OF COMMON LOGARITHMS

General Properties		Common Logarithms
1. $\log_b 1 = 0$		**1.** $\log 1 = 0$
2. $\log_b b = 1$		**2.** $\log 10 = 1$
3. $\log_b b^x = x$	Inverse properties	**3.** $\log 10^x = x$
4. $b^{\log_b x} = x$		**4.** $10^{\log x} = x$

The property $\log 10^x = x$ can be used to evaluate common logarithms involving powers of 10. For example,

$$\log 100 = \log 10^2 = 2, \quad \log 1000 = \log 10^3 = 3, \quad \text{and} \quad \log 10^{7.1} = 7.1.$$

EXAMPLE 9 Earthquake Intensity

The magnitude, R, on the Richter scale of an earthquake of intensity I is given by

$$R = \log\frac{I}{I_0},$$

where I_0 is the intensity of a barely felt zero-level earthquake. The earthquake that destroyed San Francisco in 1906 was $10^{8.3}$ times as intense as a zero-level earthquake. What was its magnitude on the Richter scale?

SOLUTION Because the earthquake was $10^{8.3}$ times as intense as a zero-level earthquake, the intensity, I, is $10^{8.3}I_0$.

$R = \log\frac{I}{I_0}$ This is the formula for magnitude on the Richter scale.

$R = \log\frac{10^{8.3}I_0}{I_0}$ Substitute $10^{8.3}I_0$ for I.

$= \log 10^{8.3}$ Simplify.

$= 8.3$ Use the property $\log 10^x = x$.

San Francisco's 1906 earthquake registered 8.3 on the Richter scale. ■

CHECK POINT 9 Use the formula in Example 9 to solve this problem. If an earthquake is 10,000 times as intense as a zero-level quake ($I = 10{,}000I_0$), what is its magnitude on the Richter scale?

8 Use natural logarithms.

Natural Logarithms The logarithmic function with base e is called the **natural logarithmic function**. The function $f(x) = \log_e x$ is usually expressed as $f(x) = \ln x$, read "el en of x." A calculator with an LN key can be used to evaluate natural logarithms.

Like the domain of all logarithmic functions, the domain of the natural logarithmic function $f(x) = \ln x$ is the set of all positive real numbers. Thus, the domain of $f(x) = \ln[g(x)]$ consists of all x for which $g(x) > 0$.

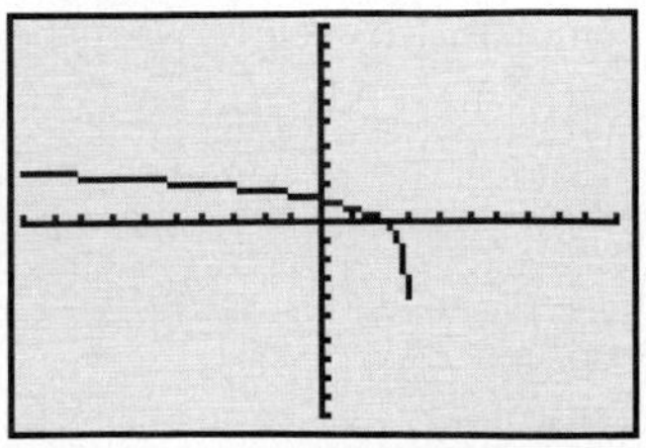

[−10, 10, 1] by [−10, 10, 1]

FIGURE 12.10 The domain of $f(x) = \ln(3 - x)$ is $(-\infty, 3)$.

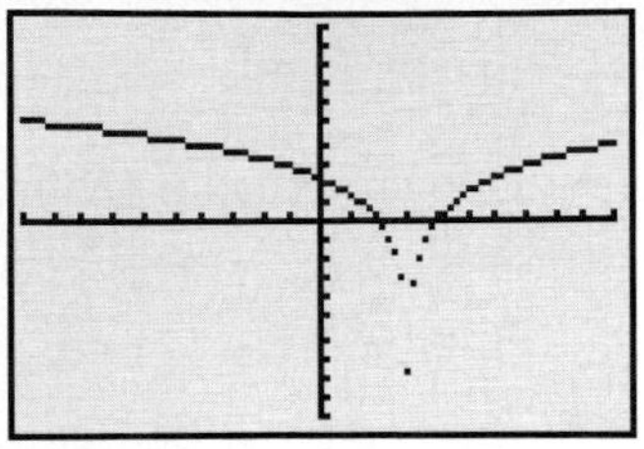

[−10, 10, 1] by [−10, 10, 1]

FIGURE 12.11 3 is excluded from the domain of $g(x) = \ln(x - 3)^2$.

EXAMPLE 10 Finding Domains of Natural Logarithmic Functions

Find the domain of each function:

a. $f(x) = \ln(3 - x)$ **b.** $g(x) = \ln(x - 3)^2$.

SOLUTION

a. The domain of f consists of all x for which $3 - x > 0$. Solving this inequality for x, we obtain $x < 3$. Thus, the domain of f is $\{x|x < 3\}$ or $(-\infty, 3)$. This is verified by the graph in Figure 12.10.

b. The domain of g consists of all x for which $(x - 3)^2 > 0$. It follows that the domain of g is all real numbers except 3. Thus, the domain of g is $\{x|x \neq 3\}$ or $(-\infty, 3) \cup (3, \infty)$. This is shown by the graph in Figure 12.11. To make it more obvious that 3 is excluded from the domain, we used a DOT format. ■

CHECK POINT 10 Find the domain of each function:

a. $f(x) = \ln(4 - x)$

b. $g(x) = \ln x^2$.

The basic properties of logarithms that were listed earlier in this section can be applied to natural logarithms.

PROPERTIES OF NATURAL LOGARITHMS

General Properties		Natural Logarithms
1. $\log_b 1 = 0$		**1.** $\ln 1 = 0$
2. $\log_b b = 1$		**2.** $\ln e = 1$
3. $\log_b b^x = x$	Inverse properties	**3.** $\ln e^x = x$
4. $b^{\log_b x} = x$		**4.** $e^{\ln x} = x$

The property $\ln e^x = x$ can be used to evaluate natural logarithms involving powers of e. For example,

$$\ln e^2 = 2, \quad \ln e^3 = 3, \quad \ln e^{7.1} = 7.1, \quad \text{and} \quad \ln\frac{1}{e} = \ln e^{-1} = -1.$$

EXAMPLE 11 Using Inverse Properties

Use inverse properties to simplify:

a. $\ln e^{7x}$ **b.** $e^{\ln 4x^2}$.

SOLUTION

a. Because $\ln e^x = x$, we conclude that $\ln e^{7x} = 7x$.

b. Because $e^{\ln x} = x$, we conclude that $e^{\ln 4x^2} = 4x^2$. ■

CHECK POINT 11 Use inverse properties to simplify:

a. $\ln e^{25x}$ **b.** $e^{\ln\sqrt{x}}$.

EXAMPLE 12 Modeling the Number of Inmates in U.S. Prisons

The United States has more people in prison, as well as more people in prison per capita, than any other Western industrialized nation. The bar graph in Figure 12.12 shows the number of inmates in U.S. state and federal prisons in seven selected years from 1995 through 2002. The function

$$f(x) = 152.4 \ln x + 1107.7$$

models the number of inmates, $f(x)$, in thousands, x years after 1994. Use the function to find the number of inmates in 2001. How well does this model the actual number shown in Figure 12.12?

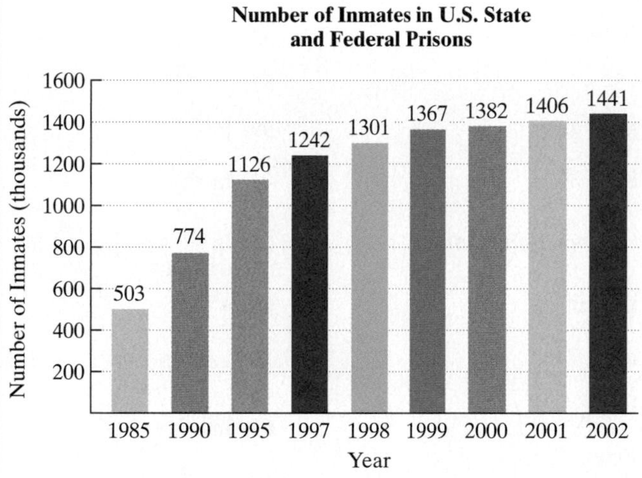

FIGURE 12.12

Source: U.S. Justice Department

SOLUTION We are interested in the U.S. inmate population for 2001. Because 2001 is 7 years after 1994, we substitute 7 for x and evaluate the function at 7.

$$f(x) = 152.4 \ln x + 1107.7 \quad \text{This is the given function.}$$

$$f(7) = 152.4 \ln 7 + 1107.7 \quad \text{Substitute 7 for x.}$$

$$\approx 1404$$

Graphing calculator keystrokes: 152.4 [ln] 7 [+] 1107.7 [ENTER]
On some calculators, a parenthesis is needed after 7.

According to the function, there were approximately 1404 thousand, or 1,404,000, inmates in 2001. Because the number shown in Figure 12.12 is 1406 thousand, the function models the actual data extremely well. ■

CHECK POINT 12 Use the function in Example 12 to predict the number of inmates in 2006.

12.2 EXERCISE SET

Student Solutions Manual | CD/Video | PH Math/Tutor Center | MathXL Tutorials on CD | MathXL® | MyMathLab | Interactmath.com

Practice Exercises

In Exercises 1–8, write each equation in its equivalent exponential form.

1. $4 = \log_2 16$ **2.** $6 = \log_2 64$

3. $2 = \log_3 x$ **4.** $2 = \log_9 x$

5. $5 = \log_b 32$ **6.** $3 = \log_b 27$

7. $\log_6 216 = y$ **8.** $\log_5 125 = y$

In Exercises 9–20, write each equation in its equivalent logarithmic form.

9. $2^3 = 8$ **10.** $5^4 = 625$

11. $2^{-4} = \frac{1}{16}$ **12.** $5^{-3} = \frac{1}{125}$

13. $\sqrt[3]{8} = 2$ **14.** $\sqrt[3]{64} = 4$

15. $13^2 = x$ **16.** $15^2 = x$

17. $b^3 = 1000$ **18.** $b^3 = 343$

19. $7^y = 200$ **20.** $8^y = 300$

In Exercises 21–42, evaluate each expression without using a calculator.

21. $\log_4 16$ **22.** $\log_7 49$

23. $\log_2 64$ **24.** $\log_3 27$

25. $\log_5 \frac{1}{5}$ **26.** $\log_6 \frac{1}{6}$

27. $\log_2 \frac{1}{8}$ **28.** $\log_3 \frac{1}{9}$

29. $\log_7 \sqrt{7}$ **30.** $\log_6 \sqrt{6}$

31. $\log_2 \frac{1}{\sqrt{2}}$ **32.** $\log_3 \frac{1}{\sqrt{3}}$

33. $\log_{64} 8$ **34.** $\log_{81} 9$

35. $\log_5 5$ **36.** $\log_{11} 11$

37. $\log_4 1$ **38.** $\log_6 1$

39. $\log_5 5^7$ **40.** $\log_4 4^6$

41. $8^{\log_8 19}$ **42.** $7^{\log_7 23}$

43. Graph $f(x) = 4^x$ and $g(x) = \log_4 x$ in the same rectangular coordinate system.

44. Graph $f(x) = 5^x$ and $g(x) = \log_5 x$ in the same rectangular coordinate system.

45. Graph $f(x) = \left(\frac{1}{2}\right)^x$ and $g(x) = \log_{\frac{1}{2}} x$ in the same rectangular coordinate system.

46. Graph $f(x) = \left(\frac{1}{4}\right)^x$ and $g(x) = \log_{\frac{1}{4}} x$ in the same rectangular coordinate system.

In Exercises 47–52, find the domain of each logarithmic function.

47. $f(x) = \log_5(x + 4)$

48. $f(x) = \log_5(x + 6)$

49. $f(x) = \log(2 - x)$

50. $f(x) = \log(7 - x)$

51. $f(x) = \ln(x - 2)^2$

52. $f(x) = \ln(x - 7)^2$

In Exercises 53–66, evaluate each expression without using a calculator.

53. $\log 100$ **54.** $\log 1000$

55. $\log 10^7$ **56.** $\log 10^8$

57. $10^{\log 33}$ **58.** $10^{\log 53}$

59. $\ln 1$ **60.** $\ln e$

61. $\ln e^6$ **62.** $\ln e^7$

63. $\ln \frac{1}{e^6}$ **64.** $\ln \frac{1}{e^7}$

65. $e^{\ln 125}$ **66.** $e^{\ln 300}$

In Exercises 67–72, use inverse properties of logarithms to simplify each expression.

67. $\ln e^{9x}$ **68.** $\ln e^{13x}$

69. $e^{\ln 5x^2}$ **70.** $e^{\ln 7x^2}$

71. $10^{\log \sqrt{x}}$ **72.** $10^{\log \sqrt[3]{x}}$

Practice Plus

In Exercises 73–76, write each equation in its equivalent exponential form. Then solve for x.

73. $\log_3(x - 1) = 2$

74. $\log_5(x + 4) = 2$

75. $\log_4 x = -3$

76. $\log_{64} x = \frac{2}{3}$

In Exercises 77–80, evaluate each expression without using a calculator.

77. $\log_3(\log_7 7)$ **78.** $\log_5(\log_2 32)$

79. $\log_2(\log_3 81)$ **80.** $\log(\ln e)$

In Exercises 81–86, match each function with its graph. The graphs are labeled (a) through (f), and each graph is displayed in a $[-5, 5, 1]$ *by* $[-5, 5, 1]$ *viewing rectangle.*

81. $f(x) = \ln(x + 2)$

82. $f(x) = \ln(x - 2)$

83. $f(x) = \ln x + 2$

84. $f(x) = \ln x - 2$

85. $f(x) = \ln(1 - x)$

86. $f(x) = \ln(2 - x)$

(a)

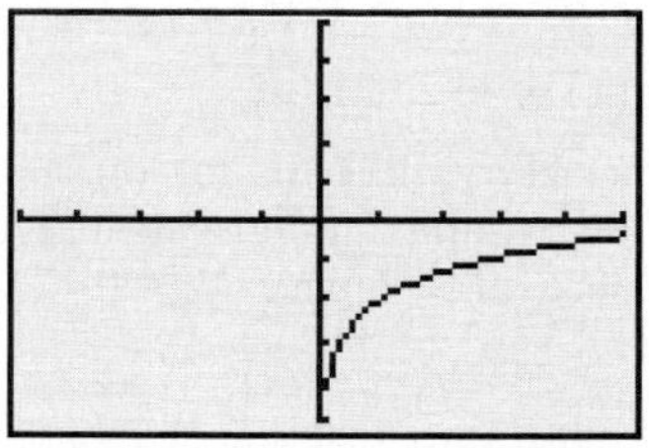

(b)

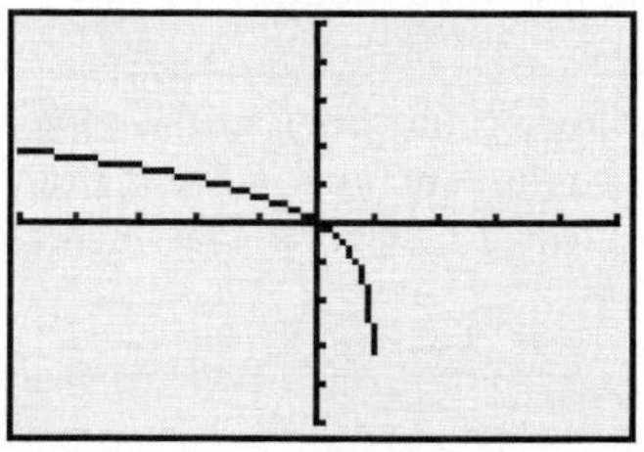

(c)

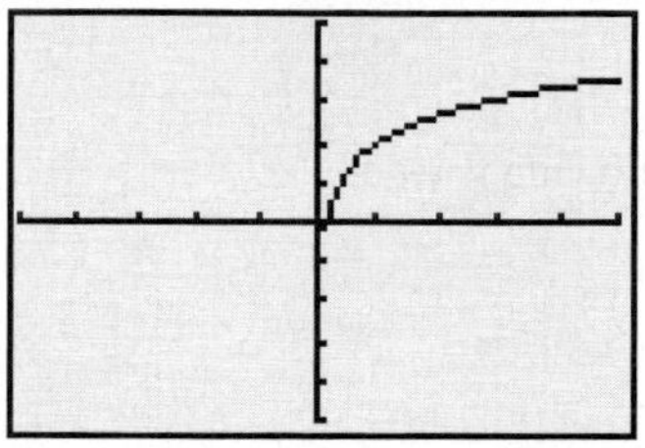

(d)

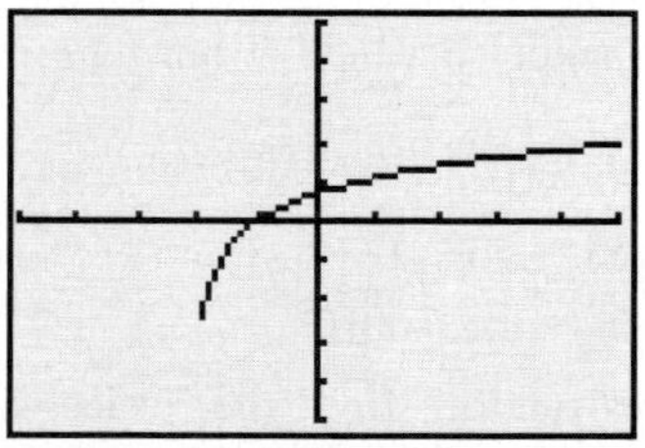

(e)

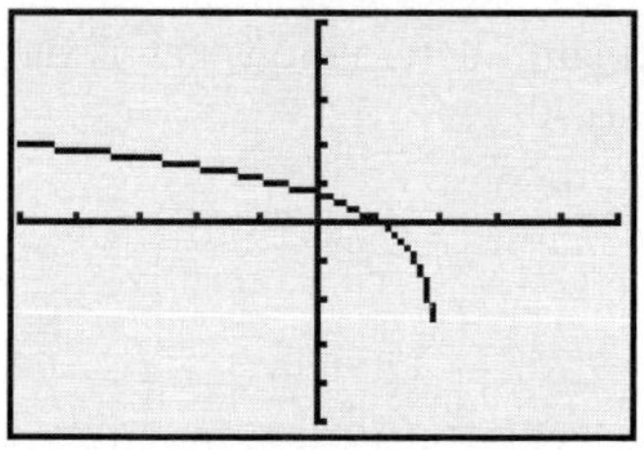

(f)

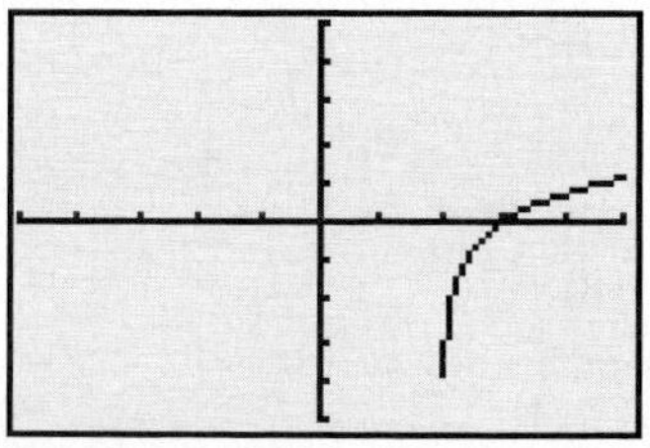

Application Exercises

The percentage of adult height attained by a girl who is x years old can be modeled by

$$f(x) = 62 + 35\log(x - 4),$$

where x represents the girl's age (from 5 to 15) and f(x) represents the percentage of her adult height. Use the formula to solve Exercises 87–88. Round answer to the nearest tenth of a percent.

87. Approximately what percentage of her adult height has a girl attained at age 13?

88. Approximately what percentage of her adult height has a girl attained at age ten?

The bar graph shows the percentage of U.S. companies that performed drug tests on employees or job applicants in five selected years from 1998 through 2003. The function

$$f(x) = -4.9\ln x + 73.8$$

models the percentage of such companies x years after 1997. Use this function to solve Exercises 89–90. Round answers to the nearest percent.

Percentage of U.S. Companies Performing Drug Tests

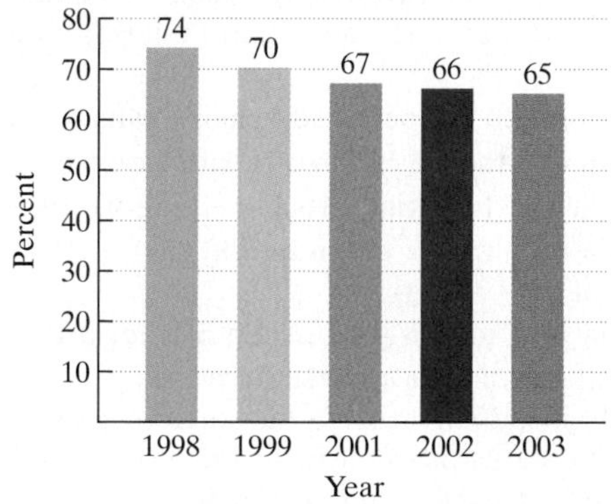

Source: American Management Association

89. Use the function to find the percentage of U.S. companies that performed drug tests in 2003. How well does this model the actual number shown for that year?

90. Use the function to predict the percentage of U.S. companies that will be performing drug tests in 2008.

The loudness level of a sound, D, in decibels, is given by the formula

$$D = 10\log(10^{12}I),$$

where I is the intensity of the sound, in watts per meter^2*. Decibel levels range from 0, a barely audible sound, to 160, a sound resulting in a ruptured eardrum. Use the formula to solve Exercises 91–92.*

91. The sound of a blue whale can be heard 500 miles away, reaching an intensity of 6.3×10^6 watts per meter^2. Determine the decibel level of this sound. At close range, can the sound of a blue whale rupture the human eardrum?

92. What is the decibel level of a normal conversation, 3.2×10^{-6} watt per meter^2?

93. Students in a psychology class took a final examination. As part of an experiment to see how much of the course content they remembered over time, they took equivalent forms of the exam in monthly intervals thereafter. The average score for the group, $f(t)$, after t months was modeled by the function

$$f(t) = 88 - 15 \ln(t + 1), \qquad 0 \le t \le 12.$$

a. What was the average score on the original exam?

b. What was the average score, to the nearest tenth, after 2 months? 4 months? 6 months? 8 months? 10 months? one year?

c. Sketch the graph of f (either by hand or with a graphing utility). Describe what the graph indicates in terms of the material retained by the students.

Writing in Mathematics

94. Describe the relationship between an equation in logarithmic form and an equivalent equation in exponential form.

95. What question can be asked to help evaluate $\log_3 81$?

96. Explain why the logarithm of 1 with base b is 0.

97. Describe the following property using words: $\log_b b^x = x$.

98. Explain how to use the graph of $f(x) = 2^x$ to obtain the graph of $g(x) = \log_2 x$.

99. Explain how to find the domain of a logarithmic function.

100. Logarithmic models are well suited to phenomena in which growth is initially rapid, but then begins to level off. Describe something that is changing over time that can be modeled using a logarithmic function.

101. Suppose that a girl is 4 feet 6 inches at age 10. Explain how to use the function in Exercises 87–88 to determine how tall she can expect to be as an adult.

Technology Exercises

In Exercises 102–105, graph f and g in the same viewing rectangle. Then describe the relationship of the graph of g to the graph of f.

102. $f(x) = \ln x, g(x) = \ln(x + 3)$

103. $f(x) = \ln x, g(x) = \ln x + 3$

104. $f(x) = \log x, g(x) = -\log x$

105. $f(x) = \log x, g(x) = \log(x - 2) + 1$

106. Students in a mathematics class took a final examination. They took equivalent forms of the exam in monthly intervals thereafter. The average score, $f(t)$, for the group after t months is modeled by the human memory function $f(t) = 75 - 10 \log(t + 1)$, where $0 \le t \le 12$. Use a graphing utility to graph the function. Then determine how many months will elapse before the average score falls below 65.

107. In parts (*a*) – (*c*), graph f and g in the same viewing rectangle.

a. $f(x) = \ln(3x), g(x) = \ln 3 + \ln x$

b. $f(x) = \log(5x^2), g(x) = \log 5 + \log x^2$

c. $f(x) = \ln(2x^3), g(x) = \ln 2 + \ln x^3$

d. Describe what you observe in parts (a)–(c). Generalize this observation by writing an equivalent expression for $\log_b(MN)$, where $M > 0$ and $N > 0$.

e. Complete this statement: The logarithm of a product is equal to ______________.

108. Graph each of the following functions in the same viewing rectangle and then place the functions in order from the one that increases most slowly to the one that increases most rapidly.

$$y = x, y = \sqrt{x}, y = e^x, y = \ln x, y = x^x, y = x^2$$

Critical Thinking Exercises

109. Which one of the following is true?

a. $\dfrac{\log_2 8}{\log_2 4} = \dfrac{8}{4}$

b. $\log(-100) = -2$.

c. The domain of $f(x) = \log_2 x$ is $(-\infty, \infty)$.

d. $\log_b x$ is the exponent to which b must be raised to obtain x.

110. Without using a calculator, find the exact value of

$$\frac{\log_3 81 - \log_\pi 1}{\log_{2\sqrt{2}} 8 - \log 0.001}.$$

111. Without using a calculator, find the exact value of $\log_4[\log_3(\log_2 8)]$.

112. Without using a calculator, determine which is the greater number: $\log_4 60$ or $\log_3 40$.

Review Exercises

113. Solve the system:

$$2x = 11 - 5y$$
$$3x - 2y = -12.$$

(Section 4.3, Example 4)

114. Factor completely:

$$6x^2 - 8xy + 2y^2.$$

(Section 6.5, Example 8)

115. Solve: $x + 3 \le -4$ or $2 - 7x \le 16$.
(Section 9.2, Example 6)

SECTION 12.3

Objectives

1. Use the product rule.
2. Use the quotient rule.
3. Use the power rule.
4. Expand logarithmic expressions.
5. Condense logarithmic expressions.
6. Use the change-of-base property.

PROPERTIES OF LOGARITHMS

We all learn new things in different ways. In this section, we consider important properties of logarithms. What would be the most effective way for you to learn these properties? Would it be helpful to use your graphing utility and discover one of these properties for yourself? To do so, work Exercise 107 in Exercise Set 12.2 before continuing. Would the properties become more meaningful if you could see exactly where they come from? If so, you will find details of the proofs of many of these properties in the appendix. The remainder of our work in this chapter will be based on the properties of logarithms that you learn in this section.

1 Use the product rule.

The Product Rule Properties of exponents correspond to properties of logarithms. For example, when we multiply exponential expressions with the same base, we add exponents:

$$b^m \cdot b^n = b^{m+n}.$$

This property of exponents, coupled with an awareness that a logarithm is an exponent, suggests the following property, called the **product rule**:

DISCOVER FOR YOURSELF

We know that $\log 100{,}000 = 5$. Show that you get the same result by writing 100,000 as $1000 \cdot 100$ and then using the product rule. Then verify the product rule by using other numbers whose logarithms are easy to find.

THE PRODUCT RULE Let b, M, and N be positive real numbers with $b \neq 1$.

$$\log_b(MN) = \log_b M + \log_b N$$

The logarithm of a product is the sum of the logarithms.

When we use the product rule to write a single logarithm as the sum of two logarithms, we say that we are **expanding a logarithmic expression**. For example, we can use the product rule to expand $\ln(7x)$:

$$\ln(7x) = \ln 7 + \ln x.$$

The logarithm of a product | is | the sum of the logarithms.

EXAMPLE 1 Using the Product Rule

Use the product rule to expand each logarithmic expression:

a. $\log_4(7 \cdot 5)$ **b.** $\log(10x)$.

SOLUTION

a. $\log_4(7 \cdot 5) = \log_4 7 + \log_4 5$ The logarithm of a product is the sum of the logarithms.

b. $\log(10x) = \log 10 + \log x$ The logarithm of a product is the sum of the logarithms. These are common logarithms with base 10 understood.

$= 1 + \log x$ Because $\log_b b = 1$, then $\log 10 = 1$. ■

CHECK POINT 1 Use the product rule to expand each logarithmic expression:

a. $\log_6(7 \cdot 11)$ **b.** $\log(100x)$.

2 Use the quotient rule.

The Quotient Rule When we divide exponential expressions with the same base, we subtract exponents:

$$\frac{b^m}{b^n} = b^{m-n}.$$

This property suggests the following property of logarithms, called the **quotient rule**:

DISCOVER FOR YOURSELF

We know that $\log_2 16 = 4$. Show that you get the same result by writing 16 as $\frac{32}{2}$ and then using the quotient rule. Then verify the quotient rule using other numbers whose logarithms are easy to find.

THE QUOTIENT RULE Let b, M, and N be positive real numbers with $b \neq 1$.

$$\log_b\left(\frac{M}{N}\right) = \log_b M - \log_b N$$

The logarithm of a quotient is the difference of the logarithms.

When we use the quotient rule to write a single logarithm as the difference of two logarithms, we say that we are **expanding a logarithmic expression**. For example, we can use the quotient rule to expand $\log\frac{x}{2}$:

$$\log\left(\frac{x}{2}\right) = \log x - \log 2.$$

The logarithm of a quotient | is | the difference of the logarithms.

EXAMPLE 2 Using the Quotient Rule

Use the quotient rule to expand each logarithmic expression:

a. $\log_7\left(\frac{19}{x}\right)$ **b.** $\ln\left(\frac{e^3}{7}\right)$.

SOLUTION

a. $\log_7\left(\frac{19}{x}\right) = \log_7 19 - \log_7 x$ — The logarithm of a quotient is the difference of the logarithms.

b. $\ln\left(\frac{e^3}{7}\right) = \ln e^3 - \ln 7$ — The logarithm of a quotient is the difference of the logarithms. These are natural logarithms with base e understood.

$= 3 - \ln 7$ — Because $\ln e^x = x$, then $\ln e^3 = 3$.

CHECK POINT 2 Use the quotient rule to expand each logarithmic expression:

a. $\log_8\left(\frac{23}{x}\right)$ **b.** $\ln\left(\frac{e^5}{11}\right)$.

3 Use the power rule.

The Power Rule When an exponential expression is raised to a power, we multiply exponents:

$$(b^m)^n = b^{mn}.$$

This property suggests the following property of logarithms, called the **power rule**:

THE POWER RULE Let b and M be positive real numbers with $b \neq 1$, and let p be any real number.

$$\log_b M^p = p \log_b M$$

The logarithm of a number with an exponent is the product of the exponent and the logarithm of that number.

When we use the power rule to "pull the exponent to the front," we say that we are **expanding a logarithmic expression**. For example, we can use the power rule to expand $\ln x^2$:

$$\ln x^2 = 2 \ln x.$$

The logarithm of a number with an exponent — is — the product of the exponent and the logarithm of that number.

Figure 12.13 shows the graphs of $y = \ln x^2$ and $y = 2 \ln x$. Are $\ln x^2$ and $2 \ln x$ the same? The graphs illustrate that $y = \ln x^2$ and $y = 2 \ln x$ have different domains. The graphs are only the same if $x > 0$. Thus, we should write

$$\ln x^2 = 2 \ln x \quad \text{for} \quad x > 0.$$

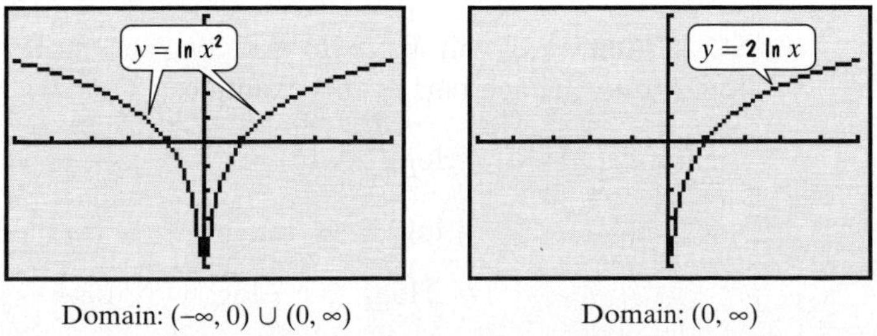

FIGURE 12.13 $\ln x^2$ and $2 \ln x$ have different domains.

When expanding a logarithmic expression, you might want to determine whether the rewriting has changed the domain of the expression. For the rest of this section, assume that all variables and variable expressions represent positive numbers.

4 Expand logarithmic expressions.

EXAMPLE 3 Using the Power Rule

Use the power rule to expand each logarithmic expression:

a. $\log_5 7^4$ **b.** $\ln\sqrt{x}$.

SOLUTION

a. $\log_5 7^4 = 4\log_5 7$ The logarithm of a number with an exponent is the exponent times the logarithm of that number.

b. $\ln\sqrt{x} = \ln x^{\frac{1}{2}}$ Rewrite the radical using a rational exponent.

$= \frac{1}{2}\ln x$ Use the power rule to bring the exponent to the front. ■

CHECK POINT 3 Use the power rule to expand each logarithmic expression:

a. $\log_6 8^9$ **b.** $\ln\sqrt[3]{x}$.

STUDY TIP

The graphs show

$$y_1 = \ln(x+3)$$

and

$$y_2 = \ln x + \ln 3.$$

The graphs are not the same. Thus, we see that

$$\ln(x+3) \neq \ln x + \ln 3.$$

In general,

$$\log_b(M+N) \neq \log_b M + \log_b N.$$

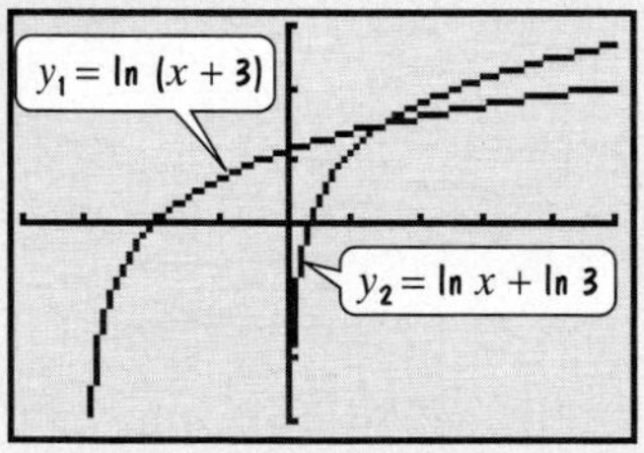

[−4, 5, 1] by [−3, 3, 1]

Try to avoid the following errors:

Incorrect!

$$\log_b(M+N) = \log_b M + \log_b N$$
$$\log_b(M-N) = \log_b M - \log_b N$$
$$\log_b(M \cdot N) = \log_b M \cdot \log_b N$$
$$\log_b\left(\frac{M}{N}\right) = \frac{\log_b M}{\log_b N}$$
$$\frac{\log_b M}{\log_b N} = \log_b M - \log_b N$$

Expanding Logarithmic Expressions It is sometimes necessary to use more than one property of logarithms when you expand a logarithmic expression. Properties for expanding logarithmic expressions are as follows:

PROPERTIES FOR EXPANDING LOGARITHMIC EXPRESSIONS

For $M > 0$ and $N > 0$:

1. $\log_b(MN) = \log_b M + \log_b N$ Product rule
2. $\log_b\left(\frac{M}{N}\right) = \log_b M - \log_b N$ Quotient rule
3. $\log_b M^p = p\log_b M$ Power rule

EXAMPLE 4 Expanding Logarithmic Expressions

Use logarithmic properties to expand each expression as much as possible:

a. $\log_b(x^2\sqrt{y})$ **b.** $\log_6\left(\frac{\sqrt[3]{x}}{36y^4}\right)$.

SOLUTION We will have to use two or more of the properties for expanding logarithms in each part of this example.

a. $\log_b(x^2\sqrt{y}) = \log_b\left(x^2 y^{\frac{1}{2}}\right)$ Use exponential notation.

$= \log_b x^2 + \log_b y^{\frac{1}{2}}$ Use the product rule.

$= 2\log_b x + \frac{1}{2}\log_b y$ Use the power rule.

b. $\log_6\left(\frac{\sqrt[3]{x}}{36y^4}\right) = \log_6\left(\frac{x^{\frac{1}{3}}}{36y^4}\right)$ — Use exponential notation.

$= \log_6 x^{\frac{1}{3}} - \log_6(36y^4)$ — Use the quotient rule.

$= \log_6 x^{\frac{1}{3}} - (\log_6 36 + \log_6 y^4)$ — Use the product rule on $\log_6(36y^4)$.

$= \frac{1}{3}\log_6 x - (\log_6 36 + 4\log_6 y)$ — Use the power rule.

$= \frac{1}{3}\log_6 x - \log_6 36 - 4\log_6 y$ — Apply the distributive property.

$= \frac{1}{3}\log_6 x - 2 - 4\log_6 y$ — $\log_6 36 = 2$ because 2 is the power to which we must raise 6 to get 36. ($6^2 = 36$) ■

CHECK POINT 4 Use logarithmic properties to expand each expression as much as possible:

a. $\log_b(x^4\sqrt[3]{y})$ **b.** $\log_5\left(\frac{\sqrt{x}}{25y^3}\right)$.

5 Condense logarithmic expressions.

STUDY TIP

These properties are the same as those in the box on page 828. The only difference is that we've reversed the sides in each property from the previous box.

Condensing Logarithmic Expressions To **condense a logarithmic expression**, we write the sum or difference of two or more logarithmic expressions as a single logarithmic expression. We use the properties of logarithms to do so:

PROPERTIES FOR CONDENSING LOGARITHMIC EXPRESSIONS

For $M > 0$ and $N > 0$:

1. $\log_b M + \log_b N = \log_b(MN)$ — Product rule
2. $\log_b M - \log_b N = \log_b\left(\frac{M}{N}\right)$ — Quotient rule
3. $p\log_b M = \log_b M^p$ — Power rule

EXAMPLE 5 Condensing Logarithmic Expressions

Write as a single logarithm:

a. $\log_4 2 + \log_4 32$ **b.** $\log(4x - 3) - \log x$.

SOLUTION

a. $\log_4 2 + \log_4 32 = \log_4(2 \cdot 32)$ — Use the product rule.

$= \log_4 64$ — We now have a single logarithm. However, we can simplify.

$= 3$ — $\log_4 64 = 3$ because $4^3 = 64$.

b. $\log(4x - 3) - \log x = \log\left(\frac{4x - 3}{x}\right)$ — Use the quotient rule. ■

CHECK POINT 5 Write as a single logarithm:

a. $\log 25 + \log 4$ **b.** $\log(7x + 6) - \log x$.

Coefficients of logarithms must be 1 before you can condense them using the product and quotient rules. For example, to condense

$$2 \ln x + \ln(x + 1),$$

the coefficient of the first term must be 1. We use the power rule to rewrite the coefficient as an exponent:

1. Use the power rule to make the number in front an exponent.

$$2\ln x + \ln(x + 1) = \ln x^2 + \ln(x + 1) = \ln[x^2(x + 1)].$$

2. Use the product rule. The sum of logarithms with coefficients 1 is the logarithm of the product.

EXAMPLE 6 Condensing Logarithmic Expressions

Write as a single logarithm:

a. $\frac{1}{2}\log x + 4\log(x - 1)$ **b.** $3\ln(x + 7) - \ln x$.

c. $4\log_b x - 2\log_b 6 + \frac{1}{2}\log_b y$.

SOLUTION

a. $\frac{1}{2}\log x + 4\log(x - 1)$

$= \log x^{\frac{1}{2}} + \log(x - 1)^4$ Use the power rule so that all coefficients are 1.

$= \log\left[x^{\frac{1}{2}}(x - 1)^4\right]$ Use the product rule.

$= \log[\sqrt{x}(x - 1)^4]$

b. $3\ln(x + 7) - \ln x$

$= \ln(x + 7)^3 - \ln x$ Use the power rule so that all coefficients are 1.

$= \ln\left[\frac{(x + 7)^3}{x}\right]$ Use the quotient rule.

c. $4\log_b x - 2\log_b 6 + \frac{1}{2}\log_b y$

$= \log_b x^4 - \log_b 6^2 + \log_b y^{\frac{1}{2}}$ Use the power rule so that all coefficients are 1.

$= (\log_b x^4 - \log_b 36) + \log_b y^{\frac{1}{2}}$ This optional step emphasizes the order of operations.

$= \log_b\left(\frac{x^4}{36}\right) + \log_b y^{\frac{1}{2}}$ Use the quotient rule.

$= \log_b\left(\frac{x^4}{36} \cdot y^{\frac{1}{2}}\right)$ Use the product rule.

$= \log_b\left(\frac{x^4\sqrt{y}}{36}\right)$

CHECK POINT 6 Write as a single logarithm:

a. $2 \ln x + \frac{1}{3}\ln(x + 5)$

b. $2 \log(x - 3) - \log x$

c. $\frac{1}{4}\log_b x - 2 \log_b 5 + 10 \log_b y$.

6 Use the change-of-base property.

The Change-of-Base Property We have seen that calculators give the values of both common logarithms (base 10) and natural logarithms (base e). To find a logarithm with any other base, we can use the following change-of-base property:

THE CHANGE-OF-BASE PROPERTY For any logarithmic bases a and b, and any positive number M,

$$\log_b M = \frac{\log_a M}{\log_a b}.$$

The logarithm of M with base b is equal to the logarithm of M with any new base divided by the logarithm of b with that new base.

In the change-of-base property, base b is the base of the original logarithm. Base a is a new base that we introduce. Thus, the change-of-base property allows us to change from base b to *any* new base a, as long as the newly introduced base is a positive number not equal to 1.

The change-of-base property is used to write a logarithm in terms of quantities that can be evaluated with a calculator. Because calculators contain keys for common (base 10) and natural (base e) logarithms, we will frequently introduce base 10 or base e.

Change-of-Base Property	Introducing Common Logarithms	Introducing Natural Logarithms
$\log_b M = \frac{\log_a M}{\log_a b}$	$\log_b M = \frac{\log_{10} M}{\log_{10} b}$	$\log_b M = \frac{\log_e M}{\log_e b}$
a is the new introduced base.	10 is the new introduced base.	e is the new introduced base.

Using the notations for common logarithms and natural logarithms, we have the following results:

THE CHANGE-OF-BASE PROPERTY: INTRODUCING COMMON AND NATURAL LOGARITHMS

Introducing Common Logarithms

$$\log_b M = \frac{\log M}{\log b}$$

Introducing Natural Logarithms

$$\log_b M = \frac{\ln M}{\ln b}$$

EXAMPLE 7 Changing Base to Common Logarithms

Use common logarithms to evaluate $\log_5 140$.

SOLUTION Because $\log_b M = \dfrac{\log M}{\log b}$,

$$\log_5 140 = \frac{\log 140}{\log 5} \approx 3.07.$$

Use a calculator: 140 [LOG] [÷] 5 [LOG] [=] or [LOG] 140 [÷] [LOG] 5 [ENTER]. On some calculators, parentheses are needed after 140 and 5.

This means that $\log_5 140 \approx 3.07$. ■

DISCOVER FOR YOURSELF

Find a reasonable estimate of $\log_5 140$ to the nearest whole number: 5 to what power is 140? Compare your estimate to the value obtained in Example 7.

CHECK POINT 7 Use common logarithms to evaluate $\log_7 2506$.

EXAMPLE 8 Changing Base to Natural Logarithms

Use natural logarithms to evaluate $\log_5 140$.

SOLUTION Because $\log_b M = \dfrac{\ln M}{\ln b}$,

$$\log_5 140 = \frac{\ln 140}{\ln 5} \approx 3.07.$$

Use a calculator: 140 [LN] [÷] 5 [LN] [=] or [LN] 140 [÷] [LN] 5 [ENTER]. On some calculators, parentheses are needed after 140 and 5.

We have again shown that $\log_5 140 \approx 3.07$. ■

CHECK POINT 8 Use natural logarithms to evaluate $\log_7 2506$.

We can use the change-of-base property to graph logarithmic functions with bases other than 10 or e on a graphing utility. For example, Figure 12.14 shows the graphs of

$$y = \log_2 x \quad \text{and} \quad y = \log_{20} x$$

in a [0, 10, 1] by [−3, 3, 1] viewing rectangle. Because $\log_2 x = \dfrac{\ln x}{\ln 2}$ and $\log_{20} x = \dfrac{\ln x}{\ln 20}$, the functions are entered as

$$y_1 = \text{[LN]}x \;\text{[÷]}\;\text{[LN]}2$$
$$\text{and } y_2 = \text{[LN]}x \;\text{[÷]}\;\text{[LN]}20.$$

On some calculators, parentheses are needed after x, 2, and 20.

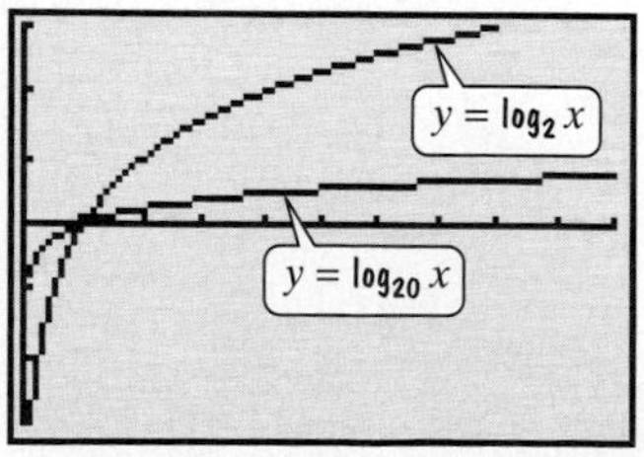

FIGURE 12.14 Using the change-of-base property to graph logarithmic functions

12.3 EXERCISE SET

Student Solutions Manual | CD/Video | PH Math/Tutor Center | MathXL Tutorials on CD | MathXL® | MyMathLab | Interactmath.com

Practice Exercises

In all exercises, assume that all variables and variable expressions represent positive numbers.

In Exercises 1–36, use properties of logarithms to expand each logarithmic expression as much as possible. Where possible, evaluate logarithmic expressions without using a calculator.

1. $\log_5(7 \cdot 3)$

2. $\log_8(13 \cdot 7)$

3. $\log_7(7x)$

4. $\log_9(9x)$

5. $\log(1000x)$

6. $\log(10{,}000x)$

7. $\log_7\left(\frac{7}{x}\right)$

8. $\log_9\left(\frac{9}{x}\right)$

9. $\log\left(\frac{x}{100}\right)$

10. $\log\left(\frac{x}{1000}\right)$

11. $\log_4\left(\frac{64}{y}\right)$

12. $\log_5\left(\frac{125}{y}\right)$

13. $\ln\left(\frac{e^2}{5}\right)$

14. $\ln\left(\frac{e^4}{8}\right)$

15. $\log_b x^3$

16. $\log_b x^7$

17. $\log N^{-6}$

18. $\log M^{-8}$

19. $\ln\sqrt[5]{x}$

20. $\ln\sqrt[7]{x}$

21. $\log_b(x^2y)$

22. $\log_b(xy^3)$

23. $\log_4\left(\frac{\sqrt{x}}{64}\right)$

24. $\log_5\left(\frac{\sqrt{x}}{25}\right)$

25. $\log_6\left(\frac{36}{\sqrt{x+1}}\right)$

26. $\log_8\left(\frac{64}{\sqrt{x+1}}\right)$

27. $\log_b\left(\frac{x^2y}{z^2}\right)$

28. $\log_b\left(\frac{x^3y}{z^2}\right)$

29. $\log\sqrt{100x}$

30. $\ln\sqrt{ex}$

31. $\log\sqrt[3]{\frac{x}{y}}$

32. $\log\sqrt[5]{\frac{x}{y}}$

33. $\log_b\left(\frac{\sqrt{x}y^3}{z^3}\right)$

34. $\log_b\left(\frac{\sqrt[3]{x}y^4}{z^5}\right)$

35. $\log_5\sqrt[3]{\frac{x^2y}{25}}$

36. $\log_2\sqrt[5]{\frac{xy^4}{16}}$

In Exercises 37–60, use properties of logarithms to condense each logarithmic expression. Write the expression as a single logarithm whose coefficient is 1. Where possible, evaluate logarithmic expressions.

37. $\log 5 + \log 2$

38. $\log 250 + \log 4$

39. $\ln x + \ln 7$

40. $\ln x + \ln 3$

41. $\log_2 96 - \log_2 3$

42. $\log_3 405 - \log_3 5$

43. $\log(2x + 5) - \log x$

44. $\log(3x + 7) - \log x$

45. $\log x + 3\log y$

46. $\log x + 7\log y$

47. $\frac{1}{2}\ln x + \ln y$

48. $\frac{1}{3}\ln x + \ln y$

49. $2\log_b x + 3\log_b y$

50. $5\log_b x + 6\log_b y$

51. $5\ln x - 2\ln y$

52. $7\ln x - 3\ln y$

53. $3\ln x - \frac{1}{3}\ln y$

54. $2\ln x - \frac{1}{2}\ln y$

55. $4\ln(x + 6) - 3\ln x$

56. $8\ln(x + 9) - 4\ln x$

57. $3\ln x + 5\ln y - 6\ln z$

58. $4\ln x + 7\ln y - 3\ln z$

59. $\frac{1}{2}(\log_5 x + \log_5 y) - 2\log_5(x + 1)$

60. $\frac{1}{3}(\log_4 x - \log_4 y) + 2\log_4(x + 1)$

In Exercises 61–68, use common logarithms or natural logarithms and a calculator to evaluate to four decimal places.

61. $\log_5 13$

62. $\log_6 17$

63. $\log_{14} 87.5$

64. $\log_{16} 57.2$

65. $\log_{0.1} 17$

66. $\log_{0.3} 19$

67. $\log_\pi 63$

68. $\log_\pi 400$

Practice Plus

In Exercises 69–74, let $\log_b 2 = A$ and $\log_b 3 = C$. Write each expression in terms of A and C.

69. $\log_b \frac{3}{2}$

70. $\log_b 6$

71. $\log_b 8$

72. $\log_b 81$

73. $\log_b \sqrt{\frac{2}{27}}$

74. $\log_b \sqrt{\frac{3}{16}}$

In Exercises 75–78, use properties of logarithms to solve each equation for x.

75. $\log_3 x = \log_3 5 + \log_3 7$

76. $\log_7 x = \log_7 3 + \log_7 5$

77. $\log_5 x = 4\log_5 2 - \log_5 8$

78. $\log_5 x = 5\log_5 2 - \log_5 4$

In Exercises 79–80, use the given functions to find $(f - g)(x)$. Express $f - g$ as a single logarithm whose coefficient is 1. Simplify, if possible.

79. $f(x) = \log x + \log 7 + \log(x^2 - 1)$, $g(x) = \log(x + 1)$

80. $f(x) = \log x + \log 15 + \log(x^2 - 4)$, $g(x) = \log(x + 2)$

Application Exercises

81. The loudness level of a sound can be expressed by comparing the sound's intensity to the intensity of a sound barely audible to the human ear. The formula

$$D = 10(\log I - \log I_0)$$

describes the loudness level of a sound, D, in decibels, where I is the intensity of the sound, in watts per meter², and I_0 is the intensity of a sound barely audible to the human ear.

a. Express the formula so that the expression in parentheses is written as a single logarithm.

b. Use the form of the formula from part (a) to answer this question. If a sound has an intensity 100 times the intensity of a softer sound, how much larger on the decibel scale is the loudness level of the more intense sound?

82. The formula

$$t = \frac{1}{c}[\ln A - \ln(A - N)]$$

describes the time, t, in weeks, that it takes to achieve mastery of a portion of a task, where A is the maximum learning possible, N is the portion of the learning that is to be achieved, and c is a constant used to measure an individual's learning style.

a. Express the formula so that the expression in brackets is written as a single logarithm.

b. The formula is also used to determine how long it will take chimpanzees and apes to master a task. For example, a typical chimpanzee learning sign language can master a maximum of 65 signs. Use the form of the formula from part (a) to answer this question. How many weeks will it take a chimpanzee to master 30 signs if c for that chimp is 0.03?

Writing in Mathematics

83. Describe the product rule for logarithms and give an example.

84. Describe the quotient rule for logarithms and give an example.

85. Describe the power rule for logarithms and give an example.

86. Without showing the details, explain how to condense $\ln x - 2\ln(x + 1)$.

87. Describe the change-of-base property and give an example.

88. Explain how to use your calculator to find $\log_{14} 283$.

89. You overhear a student talking about a property of logarithms in which division becomes subtraction. Explain what the student means by this.

90. Find ln 2 using a calculator. Then calculate each of the following: $1 - \frac{1}{2}$; $1 - \frac{1}{2} + \frac{1}{3}$; $1 - \frac{1}{2} + \frac{1}{3} - \frac{1}{4}$; $1 - \frac{1}{2} + \frac{1}{3} - \frac{1}{4} + \frac{1}{5}$; .... Describe what you observe.

Technology Exercises

91. a. Use a graphing utility (and the change-of-base property) to graph $y = \log_3 x$.

b. Graph $y = 2 + \log_3 x$, $y = \log_3(x + 2)$, and $y = -\log_3 x$ in the same viewing rectangle as $y = \log_3 x$. Then describe the change or changes that need to be made to the graph of $y = \log_3 x$ to obtain each of these three graphs.

92. Graph $y = \log x$, $y = \log(10x)$, and $y = \log(0.1x)$ in the same viewing rectangle. Describe the relationship among the three graphs. What logarithmic property accounts for this relationship?

93. Use a graphing utility and the change-of-base property to graph $y = \log_3 x$, $y = \log_{25} x$, and $y = \log_{100} x$ in the same viewing rectangle.

a. Which graph is on the top in the interval $(0, 1)$? Which is on the bottom?

b. Which graph is on the top in the interval $(1, \infty)$? Which is on the bottom?

c. Generalize by writing a statement about which graph is on top, which is on the bottom, and in which intervals, using $y = \log_b x$ where $b > 1$.

Disprove each statement in Exercises 94–98 by

a. *letting y equal a positive constant of your choice, and*

b. *using a graphing utility to graph the function on each side of the equal sign. The two functions should have different graphs, showing that the equation is not true in general.*

94. $\log(x + y) = \log x + \log y$

95. $\log\frac{x}{y} = \frac{\log x}{\log y}$

96. $\ln(x - y) = \ln x - \ln y$

97. $\ln(xy) = (\ln x)(\ln y)$

98. $\frac{\ln x}{\ln y} = \ln x - \ln y$

Critical Thinking Exercises

99. Which one of the following is true?

a. $\frac{\log_7 49}{\log_7 7} = \log_7 49 - \log_7 7$

b. $\log_b(x^3 + y^3) = 3\log_b x + 3\log_b y$

c. $\log_b(xy)^5 = (\log_b x + \log_b y)^5$

d. $\ln\sqrt{2} = \frac{\ln 2}{2}$

100. Use the change-of-base property to prove that

$$\log e = \frac{1}{\ln 10}.$$

101. If $\log 3 = A$ and $\log 7 = B$, find $\log_7 9$ in terms of A and B.

102. Write as a single term that does not contain a logarithm:

$$e^{\ln 8x^5 - \ln 2x^2}.$$

Review Exercises

103. Graph: $5x - 2y > 10$. (Section 9.4, Example 1)

104. Solve: $x - 2(3x - 2) > 2x - 3$. (Section 9.1, Example 3)

105. Divide and simplify: $\frac{\sqrt[3]{40x^2y^6}}{\sqrt[3]{5xy}}$.

(Section 10.4, Example 5)

MID-CHAPTER CHECK POINT

CHAPTER 12

What You Know: We evaluated and graphed exponential functions $[f(x) = b^x,\ b > 0$ and $b \neq 1]$, including the natural exponential function $[f(x) = e^x,\ e \approx 2.718]$. A function has an inverse that is a function if there is no horizontal line that intersects the function's graph more than once. The exponential function passes this horizontal line test and we called the inverse of the exponential function with base b the logarithmic function with base b. We learned that $y = \log_b x$ is equivalent to $b^y = x$. We evaluated and graphed logarithmic functions, including the common logarithmic function $[f(x) = \log_{10} x$ or $f(x) = \log x]$ and the natural logarithmic function $[f(x) = \log_e x$ or $f(x) = \ln x]$. Finally, we used properties of logarithms to expand and condense logarithmic expressions.

In Exercises 1–4, graph the given function. Give each function's domain and range.

1. $f(x) = 2^x - 3$

2. $f(x) = \left(\frac{1}{3}\right)^x$

3. $f(x) = \log_2 x$

4. $f(x) = \log_2 x + 1$

In Exercises 5–8, find the domain of each function.

5. $f(x) = \log_3(x + 6)$

6. $g(x) = \log_3 x + 6$

7. $h(x) = \log_3(x + 6)^2$

8. $f(x) = 3^{x+6}$

In Exercises 9–19, evaluate each expression without using a calculator. If evaluation is not possible, state the reason.

9. $\log_2 8 + \log_5 25$

10. $\log_3 \frac{1}{9}$

11. $\log_{100} 10$

12. $\log \sqrt[3]{10}$

13. $\log_2(\log_3 81)$

14. $\log_3\left(\log_2 \frac{1}{8}\right)$

15. $6^{\log_6 5}$

16. $\ln e^{\sqrt{7}}$

17. $10^{\log 13}$

18. $\log_{100} 0.1$

19. $\log_\pi \pi^{\sqrt{\pi}}$

In Exercises 20–21, expand and evaluate numerical terms.

20. $\log\left(\frac{\sqrt{xy}}{1000}\right)$

21. $\ln(e^{19}x^{20})$

In Exercises 22–24, write each expression as a single logarithm.

22. $8 \log_7 x - \frac{1}{3}\log_7 y$

23. $7 \log_5 x + 2 \log_5 x$

24. $\frac{1}{2}\ln x - 3 \ln y - \ln(z - 2)$

25. Use the formulas

$$A = P\left(1 + \frac{r}{n}\right)^{nt} \quad \text{and} \quad A = Pe^{rt}$$

to solve this exercise. You plan to invest \$8000 for 3 years at an annual rate of 8%. How much more is the return if the interest is compounded continuously than monthly? Round to the nearest dollar.

SECTION 12.4

Objectives

1. Solve exponential equations.
2. Solve logarithmic equations.
3. Solve applied problems involving exponential and logarithmic equations.

EXPONENTIAL AND LOGARITHMIC EQUATIONS

Is an early retirement awaiting you?

You inherited \$30,000. You'd like to put aside \$25,000 and eventually have over half a million dollars for early retirement. Is this possible? In this section, you will see how techniques for solving equations with variable exponents provide an answer to this question.

1 Solve exponential equations.

Exponential Equations An **exponential equation** is an equation containing a variable in an exponent. Examples of exponential equations include

$$2^{3x-8} = 16, \quad 4^x = 15, \quad \text{and} \quad 40e^{0.6x} = 240.$$

Some exponential equations can be solved by expressing each side of the equation as a power of the same base. All exponential functions are one-to-one—that is, no two different ordered pairs have the same second component. Thus, if b is a positive number other than 1 and $b^M = b^N$, then $M = N$.

SOLVING EXPONENTIAL EQUATIONS BY EXPRESSING EACH SIDE AS A POWER OF THE SAME BASE

If $b^M = b^N$, then $M = N$.

Express each side as a power of the same base. | Set the exponents equal to each other.

EXAMPLE 1 Solving Exponential Equations

Solve: **a.** $2^{3x-8} = 16$ **b.** $16^x = 64$.

SOLUTION In each equation, express both sides as a power of the same base. Then set the exponents equal to each other.

a. Because 16 is 2^4, we express each side of $2^{3x-8} = 16$ in terms of base 2.

$2^{3x-8} = 16$ This is the given equation.

$2^{3x-8} = 2^4$ Write each side as a power of the same base.

$3x - 8 = 4$ If $b^M = b^N$, $b > 0$ and $b \neq 1$, then $M = N$.

$3x = 12$ Add 8 to both sides.

$x = 4$ Divide both sides by 3.

Check 4:

$$2^{3x-8} = 16$$
$$2^{3\cdot4-8} \stackrel{?}{=} 16$$
$$2^4 \stackrel{?}{=} 16$$
$$16 = 16, \text{ true}$$

The solution is 4 and the solution set is $\{4\}$.

b. Because $16 = 4^2$ and $64 = 4^3$, we express each side of $16^x = 64$ in terms of base 4.

$16^x = 64$ This is the given equation.

$(4^2)^x = 4^3$ Write each side as a power of the same base.

$4^{2x} = 4^3$ When an exponential expression is raised to a power, multiply exponents.

$2x = 3$ If two powers of the same base are equal, then the exponents are equal.

$x = \frac{3}{2}$ Divide both sides by 2.

USING TECHNOLOGY

The graphs of

$$y_1 = 2^{3x-8}$$
$$\text{and} \quad y_2 = 16$$

have an intersection point whose x-coordinate is 4. This verifies that $\{4\}$ is the solution set of $2^{3x-8} = 16$.

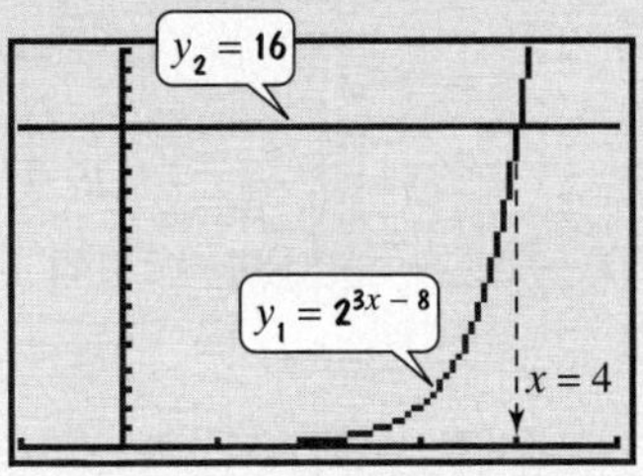

$[-1, 5, 1]$ by $[0, 20, 1]$

DISCOVER FOR YOURSELF

The equation $16^x = 64$ can also be solved by writing each side in terms of base 2. Do this. Which solution method do you prefer?

Check $\frac{3}{2}$:

$$16^x = 64$$
$$16^{\frac{3}{2}} \stackrel{?}{=} 64$$
$$(\sqrt{16})^3 \stackrel{?}{=} 64 \qquad b^{\frac{m}{n}} = (\sqrt[n]{b})^m$$
$$4^3 \stackrel{?}{=} 64$$
$$64 = 64, \quad \text{true}$$

The solution is $\frac{3}{2}$ and the solution set is $\{\frac{3}{2}\}$.

CHECK POINT 1 Solve:

a. $5^{3x-6} = 125$ **b.** $4^x = 32$.

Most exponential equations cannot be rewritten so that each side has the same base. Logarithms are extremely useful in solving these equations. The solution begins with isolating the exponential expression and taking the natural logarithm on both sides. Why can we do this? All logarithmic functions are one-to-one—that is, no two different ordered pairs have the same second component. Thus, if M and N are positive real numbers and $M = N$, then $\log_b M = \log_b N$.

USING NATURAL LOGARITHMS TO SOLVE EXPONENTIAL EQUATIONS

1. Isolate the exponential expression.
2. Take the natural logarithm on both sides of the equation.
3. Simplify using one of the following properties:

$$\ln b^x = x \ln b \quad \text{or} \quad \ln e^x = x.$$

4. Solve for the variable.

EXAMPLE 2 Solving an Exponential Equation

Solve: $4^x = 15$.

SOLUTION Because the exponential expression, 4^x, is already isolated on the left, we begin by taking the natural logarithm on both sides of the equation.

$4^x = 15$ This is the given equation.

$\ln 4^x = \ln 15$ Take the natural logarithm on both sides.

$x \ln 4 = \ln 15$ Use the power rule and bring the variable exponent to the front: $\ln b^x = x \ln b$.

$x = \frac{\ln 15}{\ln 4}$ Solve for x by dividing both sides by ln 4.

We now have an exact value for x. We use the exact value for x in the equation's solution set. Thus, the equation's solution is $\frac{\ln 15}{\ln 4}$ and the solution set is $\left\{\frac{\ln 15}{\ln 4}\right\}$. We can obtain a decimal approximation by using a calculator: $x \approx 1.95$. Because $4^2 = 16$, it seems reasonable that the solution to $4^x = 15$ is approximately 1.95.

DISCOVER FOR YOURSELF

The base that is used when taking the logarithm on both sides of an equation can be any base at all. Solve $4^x = 15$ by taking the common logarithm on both sides. Solve again, this time taking the logarithm with base 4 on both sides. Use the change-of-base property to show that the solutions are the same as the one obtained in Example 2.

CHECK POINT 2 Solve: $5^x = 134$. Find the solution set and then use a calculator to obtain a decimal approximation to two decimal places for the solution.

EXAMPLE 3 Solving an Exponential Equation

Solve: $40e^{0.6x} = 240$.

SOLUTION We begin by dividing both sides by 40 to isolate the exponential expression, $e^{0.6x}$. Then we take the natural logarithm on both sides of the equation.

$$40e^{0.6x} = 240 \quad \text{This is the given equation.}$$

$$e^{0.6x} = 6 \quad \text{Isolate the exponential factor by dividing both sides by 40.}$$

$$\ln e^{0.6x} = \ln 6 \quad \text{Take the natural logarithm on both sides.}$$

$$0.6x = \ln 6 \quad \text{Use the inverse property } \ln e^x = x \text{ on the left.}$$

$$x = \frac{\ln 6}{0.6} \approx 2.99 \quad \text{Divide both sides by 0.6.}$$

Thus, the solution of the equation is $\frac{\ln 6}{0.6} \approx 2.99$. Try checking this approximate solution in the original equation, verifying that $\left\{\frac{\ln 6}{0.6} \approx 2.99\right\}$ is the solution set. ■

CHECK POINT 3 Solve: $7e^{2x} = 63$. Find the solution set and then use a calculator to obtain a decimal approximation to two decimal places for the solution.

2 Solve logarithmic equations.

Logarithmic Equations A **logarithmic equation** is an equation containing a variable in a logarithmic expression. Examples of logarithmic equations include

$$\log_4(x + 3) = 2 \quad \text{and} \quad \ln(2x) = 3.$$

If a logarithmic equation is in the form $\log_b x = c$, we can solve the equation by rewriting it in its equivalent exponential form $b^c = x$. Example 4 illustrates how this is done.

EXAMPLE 4 Solving a Logarithmic Equation

Solve: $\log_4(x + 3) = 2$.

SOLUTION We first rewrite the equation as an equivalent equation in exponential form using the fact that $\log_b x = c$ means $b^c = x$.

$$\log_4(x + 3) = 2 \quad \text{means} \quad 4^2 = x + 3.$$

Logarithms are exponents.

USING TECHNOLOGY

The graphs of

$y_1 = \log_4(x + 3)$ and $y_2 = 2$

have an intersection point whose x-coordinate is 13. This verifies that $\{13\}$ is the solution set for $\log_4(x + 3) = 2$.

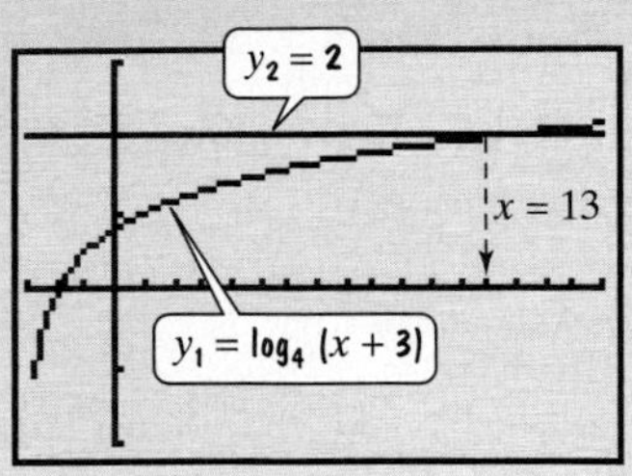

$[-3, 17, 1]$ by $[-2, 3, 1]$

Note:
Because

$$\log_b x = \frac{\ln x}{\ln b}$$

(change-of-base property),

we entered y_1 using

$$y_1 = \log_4(x + 3) = \frac{\ln(x + 3)}{\ln 4}.$$

Now we solve the equivalent equation for x.

$4^2 = x + 3$ — This is the equation equivalent to $\log_4(x + 3) = 2$.

$16 = x + 3$ — Square 4.

$13 = x$ — Subtract 3 from both sides.

Check 13:

$\log_4(x + 3) = 2$ — This is the given logarithmic equation.

$\log_4(13 + 3) \stackrel{?}{=} 2$ — Substitute 13 for x.

$\log_4 16 \stackrel{?}{=} 2$

$2 = 2$, true — $\log_4 16 = 2$ because $4^2 = 16$.

This true statement indicates that the solution is 13 and the solution set is $\{13\}$. ■

✔ **CHECK POINT 4** Solve: $\log_2(x - 4) = 3$.

Logarithmic expressions are defined only for logarithms of positive real numbers. **Always check proposed solutions of a logarithmic equation in the original equation. Exclude from the solution set any proposed solution that produces the logarithm of a negative number or the logarithm of 0.**

To rewrite the logarithmic equation $\log_b x = c$ in the equivalent exponential form $b^c = x$, we need a single logarithm whose coefficient is one. It is sometimes necessary to use properties of logarithms to condense logarithms into a single logarithm. In the next example, we use the product rule for logarithms to obtain a single logarithmic expression on the left side.

EXAMPLE 5 Using the Product Rule to Solve a Logarithmic Equation

Solve: $\log_2 x + \log_2(x - 7) = 3$.

SOLUTION

$\log_2 x + \log_2(x - 7) = 3$ — This is the given equation.

$\log_2[x(x - 7)] = 3$ — Use the product rule to obtain a single logarithm: $\log_b M + \log_b N = \log_b(MN)$.

$2^3 = x(x - 7)$ — $\log_b x = c$ means $b^c = x$.

$8 = x^2 - 7x$ — Evaluate 2^3 on the left and apply the distributive property on the right.

$0 = x^2 - 7x - 8$ — Set the equation equal to 0.

$0 = (x - 8)(x + 1)$ — Factor.

$x - 8 = 0$ or $x + 1 = 0$ — Set each factor equal to 0.

$x = 8$ $\quad$ $x = -1$ — Solve for x.

Check 8:

$\log_2 x + \log_2(x - 7) = 3$

$\log_2 8 + \log_2(8 - 7) \stackrel{?}{=} 3$

$\log_2 8 + \log_2 1 \stackrel{?}{=} 3$

$3 + 0 \stackrel{?}{=} 3$

$3 = 3$, true

Check −1:

$\log_2 x + \log_2(x - 7) = 3$

$\log_2 (-1) + \log_2(-1 - 7) \stackrel{?}{=} 3$

The number −1 does not check. Negative numbers do not have logarithms.

The solution is 8 and the solution set is $\{8\}$. ■

✔ **CHECK POINT 5** Solve: $\log x + \log(x - 3) = 1$.

EXAMPLE 6 Using the Quotient Rule to Solve a Logarithmic Equation

Solve: $\log_4(x+3) - \log_4 x = 2$.

SOLUTION

$$\log_4(x+3) - \log_4 x = 2$$ This is the given equation.

$$\log_4 \frac{x+3}{x} = 2$$ Use the quotient rule to obtain a single logarithm: $\log_b M - \log_b N = \log_b \frac{M}{N}$.

$$4^2 = \frac{x+3}{x}$$ $\log_b x = c$ means $b^c = x$.

$$16 = \frac{x+3}{x}$$ Square 4: $4^2 = 16$.

$$16x = x + 3$$ Multiply both sides by x.

$$15x = 3$$ Subtract x from both sides.

$$x = \frac{3}{15} = \frac{1}{5}$$ Divide both sides by 15.

Check by substituting $\frac{1}{5}$ into the original equation. In this equation, both x and $x + 3$ must be positive. The proposed solution, $\frac{1}{5}$, satisfies these conditions and the solution set is $\{\frac{1}{5}\}$. ■

CHECK POINT 6 Solve: $\log_5(x+1) - \log_5 x = 2$

STUDY TIP

You can also solve $\ln x = 5$, meaning $\log_e x = 5$, by rewriting the equation in exponential form:

$\log_e x = 5$ means

A logarithm is an exponent.

$e^5 = x$.

Equations involving natural logarithms can be solved using the inverse property $e^{\ln x} = x$. For example, to solve

$$\ln x = 5$$

we write both sides of the equation as exponents on base e:

$$e^{\ln x} = e^5.$$

This is called **exponentiating both sides** of the equation. Using the inverse property $e^{\ln x} = x$, we simplify the left side of the equation and obtain the solution:

$$x = e^5.$$

EXAMPLE 7 Solving an Equation with a Natural Logarithm

Solve: $3\ln(2x) = 12$.

SOLUTION

$$3\ln(2x) = 12$$ This is the given equation.

$$\ln(2x) = 4$$ Divide both sides by 3.

$$e^{\ln(2x)} = e^4$$ Exponentiate both sides.

$$2x = e^4$$ Use the inverse property to simplify the left side: $e^{\ln x} = x$.

$$x = \frac{e^4}{2} \approx 27.30$$ Divide both sides by 2.

Check $\frac{e^4}{2}$:

$$3\ln(2x) = 12 \quad \text{This is the given logarithmic equation.}$$

$$3\ln\left[2\left(\frac{e^4}{2}\right)\right] \stackrel{?}{=} 12 \quad \text{Substitute } \frac{e^4}{2} \text{ for } x.$$

$$3\ln e^4 \stackrel{?}{=} 12 \quad \text{Simplify: } \frac{\cancel{2}}{1}\cdot\frac{e^4}{\cancel{2}} = e^4.$$

$$3\cdot 4 \stackrel{?}{=} 12 \quad \text{Because } \ln e^x = x, \text{ we conclude } \ln e^4 = 4.$$

$$12 = 12, \quad \text{true}$$

This true statement indicates that the solution is $\frac{e^4}{2}$ and the solution set is $\left\{\frac{e^4}{2}\right\}$. ■

CHECK POINT 7 Solve: $4\ln 3x = 8$.

3 Solve applied problems involving exponential and logarithmic equations.

Applications Our first applied example provides a mathematical perspective on the old slogan "Alcohol and driving don't mix." In California, where 38% of fatal traffic crashes involve drinking drivers, it is illegal to drive with a blood alcohol concentration of 0.08 or higher. At these levels, drivers may be arrested and charged with driving under the influence.

EXAMPLE 8 Alcohol and Risk of a Car Accident

Medical research indicates that the risk of having a car accident increases exponentially as the concentration of alcohol in the blood increases. The risk is modeled by

$$R = 6e^{12.77x},$$

where x is the blood alcohol concentration and R, given as a percent, is the risk of having a car accident. What blood alcohol concentration corresponds to a 17% risk of a car accident?

SOLUTION For a risk of 17%, we let $R = 17$ in the equation and solve for x, the blood alcohol concentration.

$$R = 6e^{12.77x} \quad \text{This is the given equation.}$$

$$6e^{12.77x} = 17 \quad \text{Substitute 17 for } R \text{ and (optional) reverse the two sides of the equation.}$$

$$e^{12.77x} = \frac{17}{6} \quad \text{Isolate the exponential factor by dividing both sides by 6.}$$

$$\ln e^{12.77x} = \ln\left(\frac{17}{6}\right) \quad \text{Take the natural logarithm on both sides.}$$

$$12.77x = \ln\left(\frac{17}{6}\right) \quad \text{Use the inverse property } \ln e^x = x \text{ on the left side.}$$

$$x = \frac{\ln\left(\frac{17}{6}\right)}{12.77} \approx 0.08 \quad \text{Divide both sides by 12.77.}$$

For a blood alcohol concentration of 0.08, the risk of a car accident is 17%. In many states, it is illegal to drive with this blood alcohol concentration. ■

ENRICHMENT ESSAY

Visualizing the Relationship between Blood Alcohol Concentration and the Risk of a Car Accident

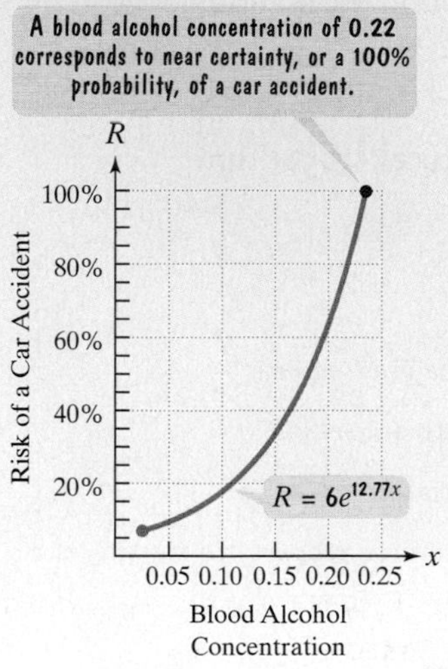

CHECK POINT 8 Use the formula in Example 8 to solve this problem. What blood alcohol concentration corresponds to a 7% risk of a car accident? (In many states, drivers under the age of 21 can lose their licenses for driving at this level.)

Suppose that you inherit \$30,000. Is it possible to invest \$25,000 and have over half a million dollars for early retirement? Our next example illustrates the power of compound interest.

EXAMPLE 9 Revisiting the Formula for Compound Interest

The formula

$$A = P\left(1 + \frac{r}{n}\right)^{nt}$$

describes the accumulated value, A, of a sum of money, P, the principal, after t years at annual percentage rate r (in decimal form) compounded n times a year. How long will it take \$25,000 to grow to \$500,000 at 9% annual interest compounded monthly?

SOLUTION

$$A = P\left(1 + \frac{r}{n}\right)^{nt}$$ This is the given formula.

$$500{,}000 = 25{,}000\left(1 + \frac{0.09}{12}\right)^{12t}$$ A(the desired accumulated value) = 500,000, P(the principal) = 25,000, r(the interest rate) = 9% = 0.09, and n = 12 (monthly compounding).

Our goal is to solve the equation for t. Let's reverse the two sides of the equation and then simplify within parentheses.

$$25{,}000\left(1 + \frac{0.09}{12}\right)^{12t} = 500{,}000$$ Reverse the two sides of the previous equation.

$$25{,}000(1 + 0.0075)^{12t} = 500{,}000$$ Divide within parentheses: $\frac{0.09}{12} = 0.0075$.

$$25{,}000(1.0075)^{12t} = 500{,}000$$ Add within parentheses.

$$(1.0075)^{12t} = 20$$ Divide both sides by 25,000.

$$\ln(1.0075)^{12t} = \ln 20$$ Take the natural logarithm on both sides.

$$12t\ln(1.0075) = \ln 20$$ Use the power rule to bring the exponent to the front: $\ln b^x = x \ln b$.

$$t = \frac{\ln 20}{12 \ln 1.0075}$$ Solve for t, dividing both sides by 12 ln 1.0075.

$$\approx 33.4$$ Use a calculator.

After approximately 33.4 years, the \$25,000 will grow to an accumulated value of \$500,000. If you set aside the money at age 20, you can begin enjoying a life of leisure at about age 53. ■

CHECK POINT 9 How long, to the nearest tenth of a year, will it take \$1000 to grow to \$3600 at 8% annual interest compounded quarterly?

ENRICHMENT ESSAY

Playing Doubles: Interest Rates and Doubling Time

One way to calculate what your savings will be worth at some point in the future is to consider doubling time. The following table shows how long it takes for your money to double at different annual interest rates subject to continuous compounding.

Annual Interest Rate	Years to Double
5%	13.9 years
7%	9.9 years
9%	7.7 years
11%	6.3 years

Of course, the first problem is collecting some money to invest. The second problem is finding a reasonably safe investment with a return of 9% or more.

EXAMPLE 10 The Growth in the Number of U.S. Internet Users

The bar graph in Figure 12.15 shows the number, in millions, of Internet users in the United States from 2000 through 2003. The function

$$f(x) = 34.1 \ln x + 117.7$$

models the number of U.S. Internet users, $f(x)$, in millions, x years after 1999. By which year will there be 185 million Internet users in the United States?

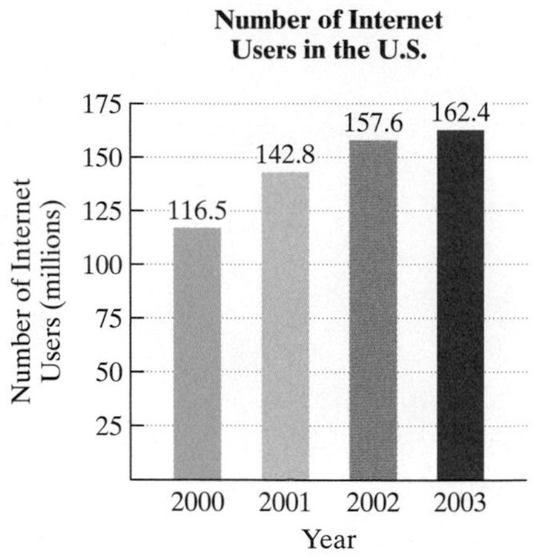

FIGURE 12.15
Source: Jupiter Media

SOLUTION We substitute 185 for $f(x)$ and solve for x, the number of years after 1999.

$f(x) = 34.1 \ln x + 117.7$ This is the given function.

$185 = 34.1 \ln x + 117.7$ Substitute 185 for $f(x)$.

Our goal is to isolate $\ln x$ in the equation $185 = 34.1 \ln x + 117.7$. We can then find x by exponentiating both sides of the equation, using the inverse property $e^{\ln x} = x$.

$34.1 \ln x + 117.7 = 185$ Reverse the two sides of the equation.

$34.1 \ln x = 67.3$ Subtract 117.7 from both sides.

$\ln x = \dfrac{67.3}{34.1}$ Divide both sides by 34.1.

$e^{\ln x} = e^{67.3/34.1}$ Exponentiate both sides.

$x = e^{67.3/34.1}$ $e^{\ln x} = x$

$x \approx 7$ Use a calculator.

Approximately 7 years after 1999, in the year 2006, there will be 185 million Internet users in the United States. ■

CHECK POINT 10 Use the function in Example 10 to find in which year there will be 200 million Internet users in the United States.

12.4 EXERCISE SET

Student Solutions Manual | CD/Video | PH Math/Tutor Center | MathXL Tutorials on CD | MathXL® | MyMathLab | Interactmath.com

Practice Exercises

Solve each exponential equation in Exercises 1–18 by expressing each side as a power of the same base and then equating exponents.

1. $2^x = 64$

2. $3^x = 81$

3. $5^x = 125$

4. $5^x = 625$

5. $2^{2x-1} = 32$

6. $3^{2x+1} = 27$

7. $4^{2x-1} = 64$

8. $5^{3x-1} = 125$

9. $32^x = 8$

10. $4^x = 32$

11. $9^x = 27$

12. $125^x = 625$

13. $3^{1-x} = \frac{1}{27}$

14. $5^{2-x} = \frac{1}{125}$

15. $6^{\frac{x-3}{4}} = \sqrt{6}$

16. $7^{\frac{x-2}{6}} = \sqrt{7}$

17. $4^x = \dfrac{1}{\sqrt{2}}$

18. $9^x = \dfrac{1}{\sqrt[3]{3}}$

Solve each exponential equation in Exercises 19–40 by taking the natural logarithm on both sides. Express the solution set in terms of natural logarithms. Then use a calculator to obtain a decimal approximation, correct to two decimal places, for the solution.

19. $10^x = 3.91$

20. $10^x = 8.07$

21. $e^x = 5.7$

22. $e^x = 0.83$

23. $5^x = 17$

24. $19^x = 143$

25. $5e^x = 25$

26. $9e^x = 99$

27. $3e^{5x} = 1977$

28. $4e^{7x} = 10{,}273$

29. $e^{0.7x} = 13$

30. $e^{0.08x} = 4$

31. $1250e^{0.055x} = 3750$

32. $1250e^{0.065x} = 6250$

33. $30 - (1.4)^x = 0$

34. $135 - (4.7)^x = 0$

35. $e^{1-5x} = 793$

36. $e^{1-8x} = 7957$

37. $7^{x+2} = 410$

38. $5^{x-3} = 137$

39. $2^{x+1} = 5^x$

40. $4^{x+1} = 9^x$

In Exercises 41–62, solve each logarithmic equation. Be sure to reject any value of x that produces the logarithm of a negative number or the logarithm of zero in the original equation.

41. $\log_3 x = 4$

42. $\log_5 x = 3$

43. $\log_2 x = -4$

44. $\log_2 x = -5$

45. $\log_9 x = \frac{1}{2}$

46. $\log_{25} x = \frac{1}{2}$

47. $\log x = 2$

48. $\log x = 3$

49. $\log_4(x + 5) = 3$

50. $\log_5(x - 7) = 2$

51. $\log_3(x - 4) = -3$

52. $\log_7(x + 2) = -2$

53. $\log_4(3x + 2) = 3$

54. $\log_2(4x + 1) = 5$

55. $\log_5 x + \log_5(4x - 1) = 1$

56. $\log_6(x + 5) + \log_6 x = 2$

57. $\log_3(x - 5) + \log_3(x + 3) = 2$

58. $\log_2(x - 1) + \log_2(x + 1) = 3$

59. $\log_2(x + 2) - \log_2(x - 5) = 3$

60. $\log_4(x + 2) - \log_4(x - 1) = 1$

61. $\log(3x - 5) - \log 5x = 2$

62. $\log(2x - 1) - \log x = 2$

Exercises 63–74 involve equations with natural logarithms. Solve each equation by isolating the natural logarithm and exponentiating both sides. Express the answer in terms of e. Then use a calculator to obtain a decimal approximation, correct to two decimal places, for the solution.

63. $\ln x = 2$

64. $\ln x = 3$

65. $\ln x = -3$

66. $\ln x = -4$

67. $5\ln(2x) = 20$

68. $6\ln(2x) = 30$

69. $6 + 2\ln x = 5$

70. $7 + 3\ln x = 6$

71. $\ln\sqrt{x + 3} = 1$

72. $\ln\sqrt{x + 4} = 1$

73. $\ln(x + 1) - \ln x = 1$

74. $\ln(x + 2) - \ln x = 2$

Practice Plus

In Exercises 75–84, solve each equation.

75. $5^{2x} \cdot 5^{4x} = 125$

76. $3^{x+2} \cdot 3^x = 81$

77. $2\log_3(x + 4) = \log_3 9 + 2$

78. $3\log_2(x - 1) = 5 - \log_2 4$

79. $3^{x^2} = 45$

80. $5^{x^2} = 50$

81. $\log_2(x - 6) + \log_2(x - 4) - \log_2 x = 2$

82. $\log_2(x - 3) + \log_2 x - \log_2(x + 2) = 2$

83. $5^{x^2-12} = 25^{2x}$

84. $3^{x^2-12} = 9^{2x}$

Application Exercises

Use the formula $R = 6e^{12.77x}$, where x is the blood alcohol concentration and R, given as a percent, is the risk of having a car accident, to solve Exercises 85–86.

85. What blood alcohol concentration corresponds to certainty, or a 100% risk, of a car accident?

86. What blood alcohol concentration corresponds to a 50% risk of a car accident?

87. The formula $A = 18.9e^{0.0055t}$ models the population of New York State, A, in millions, t years after 2000.
   a. What was the population of New York in 2000?
   b. When will the population of New York reach 19.54 million?

88. The formula $A = 15.9e^{0.0235t}$ models the population of Florida, A, in millions, t years after 2000.
   a. What was the population of Florida in 2000?
   b. When will the population of Florida reach 19.2 million?

In Exercises 89–92, complete the table for a savings account subject to n compoundings yearly $\left[A = P\left(1 + \frac{r}{n}\right)^{nt}\right]$. *Round answers to one decimal place.*

	Amount Invested	Number of Compounding Periods	Annual Interest Rate	Accumulated Amount	Time t in Years
89.	\$12,500	4	5.75%	\$20,000	
90.	\$7250	12	6.5%	\$15,000	
91.	\$1000	360		\$1400	2
92.	\$5000	360		\$9000	4

In Exercises 93–96, complete the table for a savings account subject to continuous compounding ($A = Pe^{rt}$). Round answers to one decimal place.

	Amount Invested	Annual Interest Rate	Accumulated Amount	Time t in Years
93.	\$8000	8%	Double the amount invested	
94.	\$8000		\$12,000	2
95.	\$2350		Triple the amount invested	7
96.	\$17,425	4.25%	\$25,000	

97. The function $f(x) = 2246 - 501.4 \ln x$ models the average price of a new computer, $f(x)$, in dollars, x years after 1995. When was the average price of a new computer \$1270? Round to the nearest year.

98. The function $f(x) = 68.41 + 1.75 \ln x$ models the life expectancy, $f(x)$, in years, for African-American females born x years after 1969. In which birth year is life expectancy 73.7 years? Round to the nearest year.

The function $P(x) = 95 - 30 \log_2 x$ models the percentage, $P(x)$, of students who could recall the important features of a classroom lecture as a function of time, where x represents the number of days that have elapsed since the lecture was given. The figure shows the graph of the function. Use this information to solve Exercises 99–100. Round answers to one decimal place.

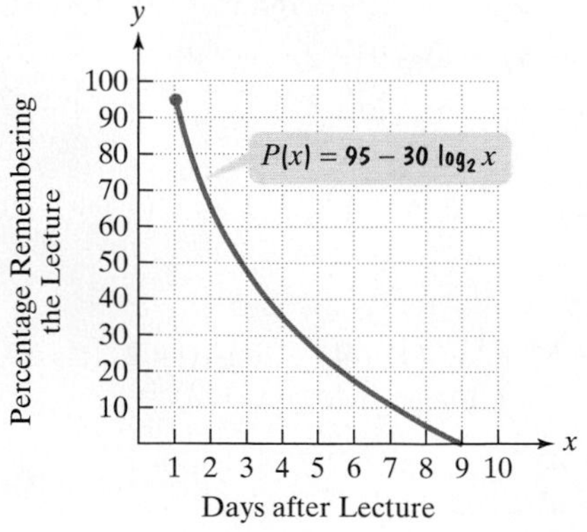

99. After how many days do only half the students recall the important features of the classroom lecture? (Let $P(x) = 50$ and solve for x.) Locate the point on the graph that conveys this information.

100. After how many days have all students forgotten the important features of the classroom lecture? (Let $P(x) = 0$ and solve for x.) Locate the point on the graph that conveys this information.

The pH of a solution ranges from 0 to 14. An acid solution has a pH less than 7. Pure water is neutral and has a pH of 7. Normal, unpolluted rain has a pH of about 5.6. The pH of a solution is given by

$$\text{pH} = -\log x,$$

where x represents the concentration of the hydrogen ions in the solution, in moles per liter. Use the formula to solve Exercises 101–102.

101. An environmental concern involves the destructive effects of acid rain. The most acidic rainfall ever had a pH of 2.4. What was the hydrogen ion concentration? Express the answer as a power of 10 and then round to the nearest thousandth.

102. The figure that follows shows very acidic rain in the northeast United States. What is the hydrogen ion concentration of rainfall with a pH of 4.2? Express the answer as a power of 10 and then round to the nearest hundred-thousandth.

Acid Rain Over Canada and the United States

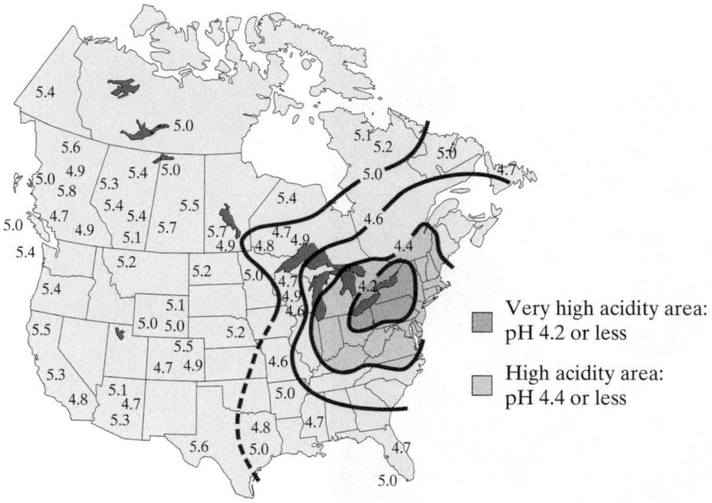

Source: National Atmospheric Program

Writing in Mathematics

103. What is an exponential equation?

104. Explain how to solve an exponential equation when both sides can be written as a power of the same base.

105. Explain how to solve an exponential equation when both sides cannot be written as a power of the same base. Use $3^x = 140$ in your explanation.

106. What is a logarithmic equation?

107. Explain how to solve a logarithmic equation. Use $\log_3(x - 1) = 4$ in your explanation.

108. In many states, a 17% risk of a car accident with a blood alcohol concentration of 0.08 is the lowest level for charging a motorist with driving under the influence. Do you agree with the 17% risk as a cutoff percentage, or do you feel that the percentage should be lower or higher? Explain your answer. What blood alcohol concentration corresponds to what you believe is an appropriate percentage?

Technology Exercises

In Exercises 109–116, use your graphing utility to graph each side of the equation in the same viewing rectangle. Then use the x-coordinate of the intersection point to find the equation's solution set. Verify this value by direct substitution into the equation.

109. $2^{x+1} = 8$

110. $3^{x+1} = 9$

111. $\log_3(4x - 7) = 2$

112. $\log_3(3x - 2) = 2$

113. $\log(x + 3) + \log x = 1$

114. $\log(x - 15) + \log x = 2$

115. $3^x = 2x + 3$

116. $5^x = 3x + 4$

Hurricanes are one of nature's most destructive forces. These low-pressure areas often have diameters of over 500 miles. The function $f(x) = 0.48\ln(x + 1) + 27$ models the barometric air pressure, $f(x)$, in inches of mercury, at a distance of x miles from the eye of a hurricane. Use this function to solve Exercises 117–118.

117. Graph the function in a $[0, 500, 50]$ by $[27, 30, 1]$ viewing rectangle. What does the shape of the graph indicate about barometric air pressure as the distance from the eye increases?

118. Use an equation to answer this question: How far from the eye of a hurricane is the barometric air pressure 29 inches of mercury? Use the TRACE and ZOOM features or the intersect command of your graphing utility to verify your answer.

119. The function $P(t) = 145e^{-0.092t}$ models a runner's pulse, $P(t)$, in beats per minute, t minutes after a race, where $0 \le t \le 15$. Graph the function using a graphing utility. TRACE along the graph and determine after how many minutes the runner's pulse will be 70 beats per minute. Round to the nearest tenth of a minute. Verify your observation algebraically.

120. The function $W(t) = 2600(1 - 0.51e^{-0.075t})^3$ models the weight, $W(t)$, in kilograms, of a female African elephant at age t years. (1 kilogram $\approx$ 2.2 pounds) Use a graphing utility to graph the function. Then TRACE along the curve to estimate the age of an adult female elephant weighing 1800 kilograms.

Critical Thinking Exercises

121. Which one of the following is true?

a. If $\log(x + 3) = 2$, then $e^2 = x + 3$.

b. If $\log(7x + 3) - \log(2x + 5) = 4$, then in exponential form $10^4 = (7x + 3) - (2x + 5)$.

c. If $x = \frac{1}{k} \ln y$, then $y = e^{kx}$.

d. Examples of exponential equations include $10^x = 5.71$, $e^x = 0.72$, and $x^{10} = 5.71$.

122. If \$4000 is deposited into an account paying 3% interest compounded annually and at the same time \$2000 is deposited into an account paying 5% interest compounded annually, after how long will the two accounts have the same balance? Round to the nearest year.

Solve each equation in Exercises 123–125. Check each proposed solution by direct substitution or with a graphing utility.

123. $(\ln x)^2 = \ln x^2$

124. $(\log x)(2 \log x + 1) = 6$

125. $\ln(\ln x) = 0$

Review Exercises

126. Solve: $\sqrt{x+4} - \sqrt{x-1} = 1$
(Section 10.6, Example 4)

127. Solve: $\frac{3}{x+1} - \frac{5}{x} = \frac{19}{x^2 + x}$.
(Section 7.6, Example 4)

128. Simplify: $(-2x^3y^{-2})^{-4}$.
(Section 5.7, Example 6)

SECTION 12.5

EXPONENTIAL GROWTH AND DECAY; MODELING DATA

Objectives

1. Model exponential growth and decay.
2. Model data with exponential and logarithmic functions.
3. Express an exponential model in base e.

The most casual cruise on the Internet shows how people disagree when it comes to making predictions about the effects of the world's growing population. Some argue that there is a recent slowdown in the growth rate, economies remain robust, and famines in North Korea and Ethiopia are aberrations rather than signs of the future. Others say that the 6.3 billion people on Earth is twice as many as can be supported in middle-class comfort, and the world is running out of arable land and fresh water. Debates about entities that are growing exponentially can be approached mathematically: We can create functions that model data and use these functions to make predictions. In this section, we will show you how this is done.

1 Model exponential growth and decay.

Exponential Growth and Decay One of algebra's many applications is to predict the behavior of variables. This can be done with *exponential growth* and *decay models*. With exponential growth or decay, quantities grow or decay at a rate directly proportional to their size. Populations that are growing exponentially grow extremely rapidly as they get larger because there are more adults to have offspring. For example, the **growth rate** for world population is 1.3%, or 0.013. This means that each year world population is 1.3% more than what it was in the previous year.

In 2001, world population was 6.2 billion. Thus, we compute the world population in 2002 as follows:

$$6.2 \text{ billion} + 1.3\% \text{ of } 6.2 \text{ billion} = 6.2 + (0.013)(6.2) = 6.2806.$$

This computation indicates that 6.2806 billion people populated the world in 2002. The 0.0806 billion represents an increase of 80.6 million people from 2001 to 2002, the equivalent of the population of Germany. Using 1.3% as the annual growth rate, world population for 2003 is found in a similar manner:

$$6.2806 + 1.3\% \text{ of } 6.2806 = 6.2806 + (0.013)(6.2806) \approx 6.3622.$$

This computation indicates that approximately 6.3622 billion people populated the world in 2003.

The explosive growth of world population may remind you of the growth of money in an account subject to compound interest. Just as the growth rate for world population is multiplied by the population plus any increase in the population, a compound interest rate is multiplied by your original investment plus any accumulated interest. The balance in an account subject to continuous compounding and world population are special cases of *exponential growth models*.

STUDY TIP

You have seen the formula for exponential growth before, but with different letters. It is the formula for compound interest with continous compounding.

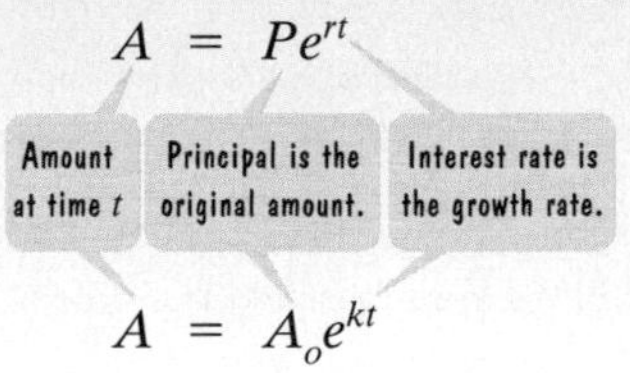

EXPONENTIAL GROWTH AND DECAY MODELS The mathematical model for **exponential growth** or **decay** is given by

$$f(t) = A_0e^{kt} \quad \text{or} \quad A = A_0e^{kt}.$$

- **If $k > 0$, the function models the amount, or size, of a *growing* entity.**
  A_0 is the original amount, or size, of the growing entity at time $t = 0$, A is the amount at time t, and k is a constant representing the growth rate.
- **If $k < 0$, the function models the amount, or size, of a *decaying* entity.**
  A_0 is the original amount, or size, of the decaying entity at time $t = 0$, A is the amount at time t, and k is a constant representing the decay rate.

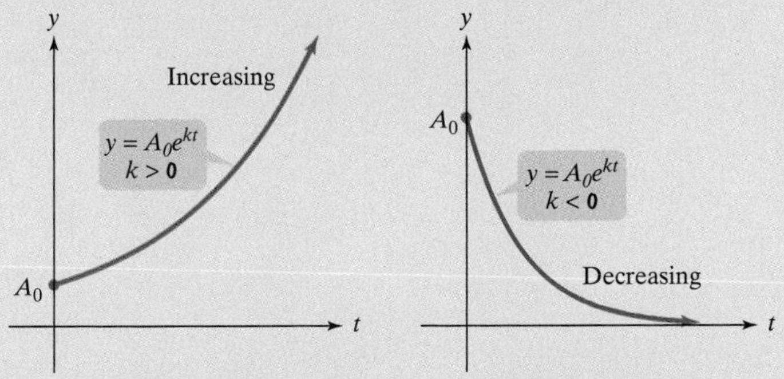

(a) Exponential growth (b) Exponential decay

Sometimes we need to use given data to determine k, the rate of growth or decay. After we compute the value of k, we can use the formula $A = A_0e^{kt}$ to make predictions. This idea is illustrated in our first two examples.

EXAMPLE 1 Modeling the Growth of the U.S. Population

The graph in Figure 12.16 shows the U.S. population, in millions, for five selected years from 1970 through 2003. In 1970, the U.S. population was 203.3 million. By 2003, it had grown to 294 million.

a. Find the exponential growth function that models the data for 1970 through 2003.

b. By which year will the U.S. population reach 315 million?

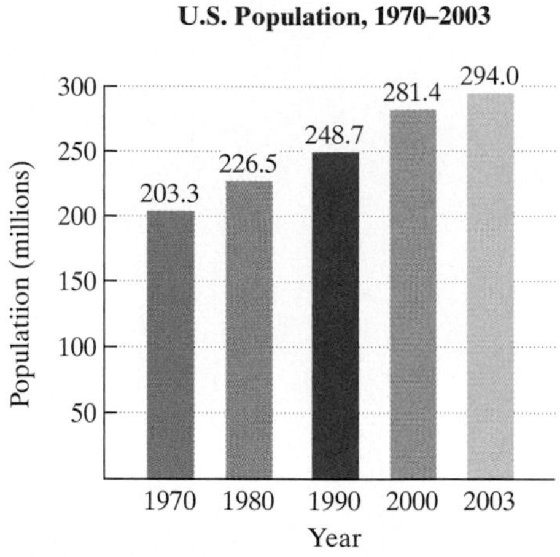

FIGURE 12.16

Source: U.S. Census Bureau

SOLUTION

a. We use the exponential growth model

$$A = A_0e^{kt}$$

in which t is the number of years after 1970. This means that 1970 corresponds to $t = 0$. At that time the U.S. population was 203.3 million, so we substitute 203.3 for A_0 in the growth model:

$$A = 203.3e^{kt}.$$

We are given that 294 million is the population in 2003. Because 2003 is 33 years after 1970, when $t = 33$ the value of A is 294. Substituting these numbers into the growth model will enable us to find k, the growth rate. We know that $k > 0$ because the problem involves growth.

$$A = 203.3e^{kt}$$ Use the growth model with $A_0 = 203.3$.

$$294 = 203.3e^{k \cdot 33}$$ When $t = 33$, $A = 294$. Substitute these numbers into the model.

$$e^{33k} = \frac{294}{203.3}$$ Isolate the exponential factor by dividing both sides by 203.3. We also reversed the sides.

$$\ln e^{33k} = \ln\left(\frac{294}{203.3}\right)$$ Take the natural logarithm on both sides.

$$33k = \ln\left(\frac{294}{203.3}\right)$$ Simplify the left side using $\ln e^x = x$.

$$k = \frac{\ln\left(\frac{294}{203.3}\right)}{33} \approx 0.011$$ Divide both sides by 33 and solve for k. Then use a calculator.

The value of k, approximately 0.011, indicates a growth rate of about 1.1%. We substitute 0.011 for k in the growth model, $A = 203.3e^{kt}$, to obtain the exponential growth function for the U.S. population. It is

$$A = 203.3e^{0.011t},$$

where t is measured in years after 1970.

b. To find the year in which the U.S. population will reach 315 million, substitute 315 for A in the model from part (a) and solve for t.

$$A = 203.3e^{0.011t}$$ This is the model from part (a).

$$315 = 203.3e^{0.011t}$$ Substitute 315 for A.

$$e^{0.11t} = \frac{315}{203.3}$$ Divide both sides by 203.3. We also reversed the sides.

$$\ln e^{0.011t} = \ln\left(\frac{315}{203.3}\right)$$ Take the natural logarithm on both sides.

$$0.011t = \ln\left(\frac{315}{203.3}\right)$$ Simplify on the left using $\ln e^x = x$.

$$t = \frac{\ln\left(\frac{315}{203.3}\right)}{0.011} \approx 40$$ Divide both sides by 33 and solve for t. Then use a calculator.

Because t represents the number of years after 1970, the model indicates that the U.S. population will reach 315 million by 1970 + 40, or in the year 2010. ■

CHECK POINT 1 In 1990, the population of Africa was 643 million and by 2000 it had grown to 813 million.

a. Use the exponential growth model $A = A_0e^{kt}$, in which t is the number of years after 1990, to find the exponential growth function that models the data.

b. By which year will Africa's population reach 2000 million, or two billion?

ENRICHMENT ESSAY

Creating an Inaccurate Picture by Leaving Something Out

On Monday, October 19, 1987, the Dow Jones Industrial Average plunged 508 points, losing 22.6% of its value. The graph shown on the left, which appeared in a major newspaper following "Black Monday" (as it was instantly dubbed), creates the impression that the Dow average had been "bullish" from 1972 through 1987, increasing throughout this period. The graph creates this inaccurate picture by leaving something out. The graph on the right illustrates that the stock market rose and fell sharply over these years. The impressively smooth curve on the left was obtained by plotting only three of the data points. By ignoring most of the data, increases and decreases are not accounted for and the actual behavior of the market over the 15 years leading to "Black Monday" is inaccurately conveyed.

The Dow Jones Industrial Average: 1972–1987

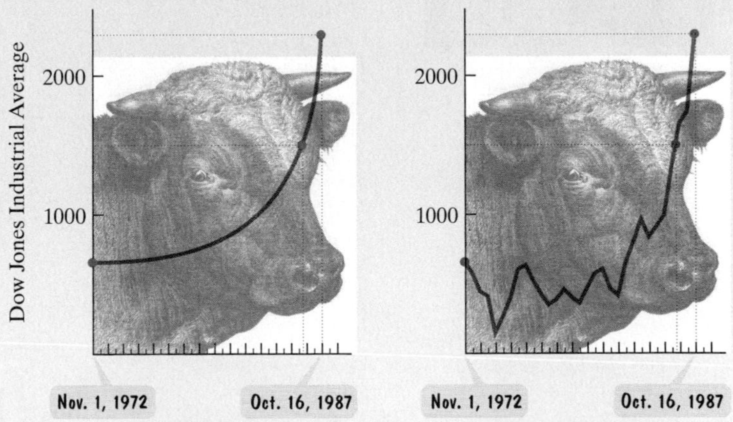

Growth Using Only Three Data Points

Growth and Decline Using All Data

Source: A. K. Dewdney, *200% of Nothing*

In Example 1, we used only two data values, the population for 1970 and the population for 2003, to develop a model for U.S. population growth from 1970 through 2003. By not using data for any other years, have we created a model that inaccurately describes both the existing data and future population projections given by the U.S. Census Bureau? Something else to think about: Is an exponential model the best choice for describing U.S. population growth, or might a linear model provide a better description? We return to these issues in Exercises 39–43 in the exercise set.

ENRICHMENT ESSAY

Carbon Dating and Artistic Development

The artistic community was electrified by the discovery in 1995 of spectacular cave paintings in a limestone cavern in France. Carbon dating of the charcoal from the site showed that the images, created by artists of remarkable talent, were 30,000 years old, making them the oldest cave paintings ever found. The artists seemed to have used the cavern's natural contours to heighten a sense of perspective. The quality of the painting suggests that the art of early humans did not mature steadily from primitive to sophisticated in any simple linear fashion.

Our next example involves exponential decay and its use in determining the age of fossils and artifacts. The method is based on considering the percentage of carbon-14 remaining in the fossil or artifact. Carbon-14 decays exponentially with a *half-life* of approximately 5715 years. The **half-life** of a substance is the time required for half of a given sample to disintegrate. Thus, after 5715 years a given amount of carbon-14 will have decayed to half the original amount. Carbon dating is useful for artifacts or fossils up to 80,000 years old. Older objects do not have enough carbon-14 left to date age accurately.

EXAMPLE 2 Carbon-14 Dating: The Dead Sea Scrolls

a. Use the fact that after 5715 years a given amount of carbon-14 will have decayed to half the original amount to find the exponential decay model for carbon-14.

b. In 1947, earthenware jars containing what are known as the Dead Sea Scrolls were found by an Arab Bedouin herdsman. Analysis indicated that the scroll wrappings contained 76% of their original carbon-14. Estimate the age of the Dead Sea Scrolls.

SOLUTION

a. We begin with the exponential decay model $A = A_0e^{kt}$. We know that $k < 0$ because the problem involves the decay of carbon-14. After 5715 years ($t = 5715$), the amount of carbon-14 present, A, is half the original amount, A_0. Thus, we can substitute $\frac{A_0}{2}$ for A in the exponential decay model. This will enable us to find k, the decay rate.

$$A = A_0e^{kt}$$ Begin with the exponential decay model.

$$\frac{A_0}{2} = A_0e^{k5715}$$ After 5715 years ($t = 5715$), $A = \frac{A_0}{2}$ (because the amount present, A, is half the original amount, A_0).

$$\left(\frac{1}{2}\right) = e^{5715k}$$ Divide both sides of the equation by A_0.

$$\ln\left(\frac{1}{2}\right) = \ln e^{5715k}$$ Take the natural logarithm on both sides.

$$\ln\left(\frac{1}{2}\right) = 5715k$$ Simplify the right side using $\ln e^x = x$.

$$k = \frac{\ln\left(\frac{1}{2}\right)}{5715} \approx -0.000121$$ Divide both sides by 5715 and solve for k.

Substituting for k in the decay model, $A = A_0e^{kt}$, the model for carbon-14 is

$$A = A_0e^{-0.000121t}.$$

b. In 1947, the Dead Sea Scrolls contained 76% of their original carbon-14. To find their age in 1947, substitute $0.76A_0$ for A in the model from part (a) and solve for t.

$$A = A_0e^{-0.000121t}$$ This is the decay model for carbon-14.

$$0.76A_0 = A_0e^{-0.000121t}$$ A, the amount present, is 76% of the original amount, so $A = 0.76A_0$.

$$0.76 = e^{-0.000121t}$$ Divide both sides of the equation by A_0.

$$\ln 0.76 = \ln e^{-0.000121t}$$ Take the natural logarithm on both sides.

$$\ln 0.76 = -0.000121t$$ Simplify the right side using $\ln e^x = x$.

$$t = \frac{\ln 0.76}{-0.000121} \approx 2268$$ Divide both sides by -0.000121 and solve for t.

The Dead Sea Scrolls are approximately 2268 years old plus the number of years between 1947 and the current year. ■

CHECK POINT 2 Strontium-90 is a waste product from nuclear reactors. As a consequence of fallout from atmospheric nuclear tests, we all have a measurable amount of strontium-90 in our bones.

a. Use the fact that after 28 years a given amount of strontium-90 will have decayed to half the original amount to find the exponential decay model for strontium-90.

b. Suppose that a nuclear accident occurs and releases 60 grams of strontium-90 into the atmosphere. How long will it take for strontium-90 to decay to a level of 10 grams?

2 Model data with exponential and logarithmic functions.

The Art of Modeling Throughout this chapter, we have been working with models that were given. However, we can create functions that model data by observing patterns in scatter plots. Figure 12.17 shows scatter plots for data that are exponential or logarithmic.

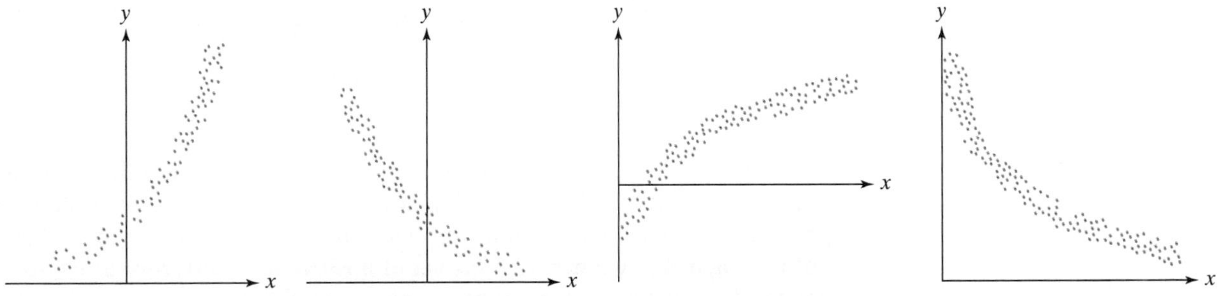

$y = ab^x, a > 0, b > 1$ Exponential | $y = ab^x, 0 < b < 1, a > 0$ Exponential | $y = a + b \ln x, a > 0, b > 0$ Logarithmic | $y = a + b \ln x, a > 0, b < 0$ Logarithmic

FIGURE 12.17 Scatter plots for exponential or logarithmic models

EXAMPLE 3 Choosing a Model for Data

Figure 12.18(a) shows the percentage of U.S. households with televisions that subscribe to cable television. The data are displayed for five selected years from 1980 through 2002. A scatter plot is shown in Figure 12.18(b). What function would be a good choice for modeling the data?

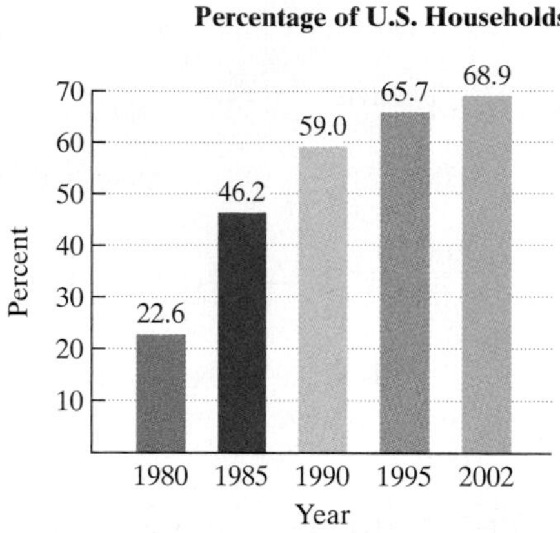

FIGURE 12.18 (a)
Source: Nielsen Media Research

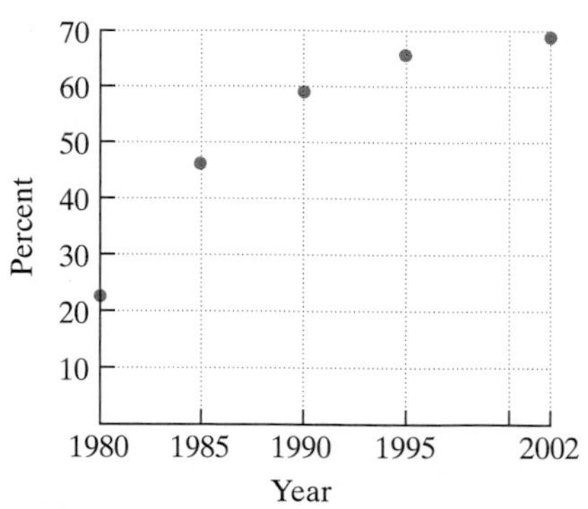

FIGURE 12.18 (b)

SOLUTION Because the data in the scatter plot increase rapidly at first and then begin to level off a bit, the shape suggests that a logarithmic function is a good choice for modeling the data. ■

CHECK POINT 3 Table 12.1 shows the populations of various cities, in thousands, and the average walking speed, in feet per second, of a person living in the city. Create a scatter plot for the data. Based on the scatter plot, what function would be a good choice for modeling the data?

Table 12.1 Population and Walking Speed

Population (thousands)	Walking Speed (feet per second)
5.5	0.6
14	1.0
71	1.6
138	1.9
342	2.2

Source: Mark and Helen Bornstein, "The Pace of Life"

How can we obtain a logarithmic function that models the data for the percentage of U.S. households with cable television shown in Figure 12.18(a)? A graphing utility can be used to obtain a logarithmic model of the form $y = a + b \ln x$. **Because the domain of the logarithmic function is the set of positive numbers, zero must not be a value for x.** What does this mean for our cable television data that begin in the year 1980? We must start values of x after 0. Thus, we'll assign x to represent the number of years after 1979. This gives us the data shown in Table 12.2 at the top of the next page. Using the Logarithmic REGression option, we obtain the equation in Figure 12.19.

Table 12.2

x, Number of Years after 1979	y, Percentage of U.S. Households with Cable TV
1 (1980)	22.6
6 (1985)	46.2
11 (1990)	59.0
16 (1995)	65.7
23 (2002)	68.9

FIGURE 12.19 A logarithmic model for the data in Table 12.2

From Figure 12.19, we see that the logarithmic model of the data, with numbers rounded to three decimal places, is

$$y = 21.779 + 15.202 \ln x.$$

The number r that appears in Figure 12.19 is called the **correlation coefficient** and is a measure of how well the model fits the data. The value of r is such that $-1 \le r \le 1$. A positive r means that as the x-values increase, so do the y-values. A negative r means that as the x-values increase, the y-values decrease. **The closer that r is to -1 or 1, the better the model fits the data.** Because r is approximately 0.996, the model fits the data very well.

EXAMPLE 4 Choosing a Model for Data

Figure 12.20(a) shows world population, in billions, for seven selected years from 1950 through 2003. A scatter plot is shown in Figure 12.20(b). Suggest two functions that would be good choices for modeling the data.

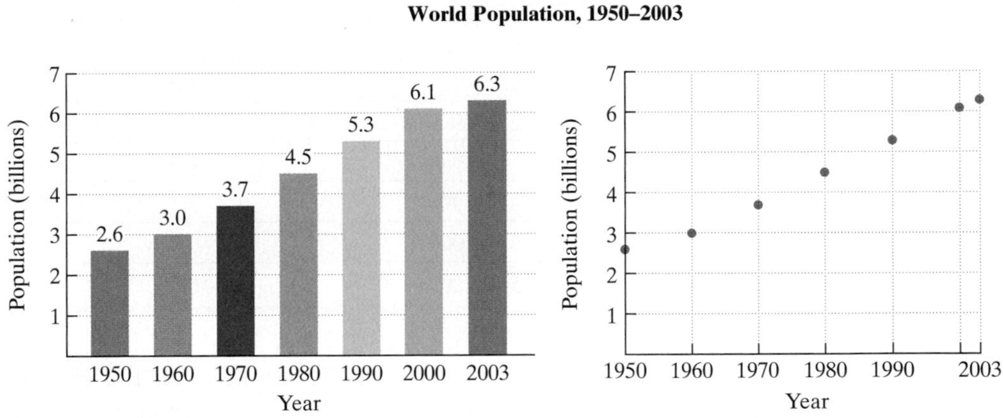

FIGURE 12.20 (a)

Source: U.S. Census Bureau, International Database

FIGURE 12.20 (b)

SOLUTION Because the data in the scatter plot appear to increase more and more rapidly, the shape suggests that an exponential model might be a good choice. Furthermore, we can probably draw a line that passes through or near the seven points. Thus, a linear function would also be a good choice for modeling the data. ■

CHECK POINT 4 Table 12.3 shows the number of international basketball players in the National Basketball Association for six selected years from 1980 through 2003. Create a scatter plot for the data. Based on the scatter plot, what function would be a good choice for modeling the data?

Table 12.3 Growing Diversity of the National Basketball Association

Year (Season's End)	Number of International Players
1980	3
1985	19
1990	25
1995	30
2000	50
2003	65

Source: National Basketball Association

If we choose to model world population shown in Figure 12.20(a) on the previous page with an exponential function, a graphing utility's Exponential REGression option can be used to obtain the function's equation. With this feature, a graphing utility fits the data to an exponential model of the form $y = ab^x$.

Although the domain of the exponential function $y = ab^x$ is the set of all real numbers, some graphing utilities only accept positive values for x. What does this mean for our data for world population that starts in the year 1950? We will start values of x after 0. Thus, we'll assign x to represent the number of years after 1949. This gives us the data shown in Table 12.4. Using the Exponential REGression option, we obtain the equation in Figure 12.21.

Table 12.4

x, Numbers of Years after 1949	y, World Population (billions)
1 (1950)	2.6
11 (1960)	3.0
21 (1970)	3.7
31 (1980)	4.5
41 (1990)	5.3
51 (2000)	6.1
54 (2003)	6.3

FIGURE 12.21 An exponential model for the data in Table 12.4

STUDY TIP

Once you have obtained one or more models for data, you can use a graphing utility's TABLE feature to numerically see how well each model describes the data. Enter the models as y_1, y_2, and so on. Create a table, scroll through the table, and compare the table values given by the models to the actual data.

From Figure 12.21, we see that the exponential model of the data for world population x years after 1949, with numbers rounded to three decimal places, is

$$y = 2.557(1.017)^x.$$

The correlation coefficient, r, is close to 1, indicating that the model fits the data very well.

When using a graphing utility to model data, begin with a scatter plot, drawn either by hand or with the graphing utility, to obtain a general picture for the shape of the data. It might be difficult to determine which model best fits the data—linear, logarithmic, exponential, quadratic, or something else. If necessary, use your graphing utility to fit several models to the data. The best model is the one that yields the value r, the correlation coefficient, closest to 1 or -1. Finding a proper fit for data can be almost as much art as it is mathematics. In this era of technology, the process of creating models that best fit data is one that involves more decision making than computation.

3 Express an exponential model in base e.

Expressing $y = ab^x$ in Base e Graphing utilities display exponential models in the form $y = ab^x$. However, our discussion of exponential growth involved base e. Because of the inverse property $b = e^{\ln b}$, we can rewrite any model in the form $y = ab^x$ in terms of base e.

EXPRESSING AN EXPONENTIAL MODEL IN BASE e

$$y = ab^x \quad \text{is equivalent to} \quad y = ae^{(\ln b)\cdot x}$$

EXAMPLE 5 Rewriting the Model for World Population in Base e

We have seen that the function

$$y = 2.557(1.017)^x$$

models world population, y, in billions, x years after 1949. Rewrite the model in terms of base e.

SOLUTION We use the two equivalent equations shown in the voice balloons to rewrite the model in terms of base e.

$y = ab^x$ | $y = ae^{(\ln b)\cdot x}$

$$y = 2.557(1.017)^x \quad \text{is equivalent to} \quad y = 2.557e^{(\ln 1.017)\cdot x}.$$

Using $\ln 1.017 \approx 0.017$, the exponential growth model for world population, y, in billions, x years after 1949 is

$$y = 2.557e^{0.017x}.$$

In Example 5, we can replace y with A and x with t so that the model has the same letters as those in the exponential growth model $A = A_0e^{kt}$.

$A = A_o\ e^{kt}$ — This is the exponential growth model.

$A = 2.557e^{0.017t}$ — This is the model for world population.

The value of k, 0.017, indicates a growth rate of 1.7%. Although this is an excellent model for the data, we must be careful about making projections about world population using this growth function. Why? World population growth rate is now 1.3%, not 1.7%, so our model will overestimate future populations.

CHECK POINT 5 Rewrite $y = 4(7.8)^x$ in terms of base e. Express the answer in terms of a natural logarithm and then round to three decimal places.

12.5 EXERCISE SET

Student Solutions Manual | CD/Video | PH Math/Tutor Center | MathXL Tutorials on CD | MathXL® | MyMathLab | Interactmath.com

Practice Exercises and Application Exercises

The exponential models describe the population of the indicated country, A, in millions, t years after 2003. Use these models to solve Exercises 1–6.

India $A = 1049.7e^{0.015t}$

Iraq $A = 24.7e^{0.028t}$

Japan $A = 127.2e^{0.001t}$

Russia $A = 144.5e^{-0.004t}$

1. What was the population of Japan in 2003?

2. What was the population of Iraq in 2003?

3. Which country has the greatest growth rate? By what percentage is the population of that country increasing each year?

4. Which country has a decreasing population? By what percentage is the population of that country decreasing each year?

5. When will India's population be 1238 million?

6. When will India's population be 1416 million?

About the size of New Jersey, Israel has seen its population soar to more than 6 million since it was established. With the help of U.S. aid, the country now has a diversified economy rivaling those of other developed Western nations. By contrast, the Palestinians, living under Israeli occupation and a corrupt regime, endure bleak conditions. The graphs show that by 2050, Palestinians in the West Bank, Gaza Strip, and East Jerusalem will outnumber Israelis. Exercises 7–8 involve the projected growth of these two populations.

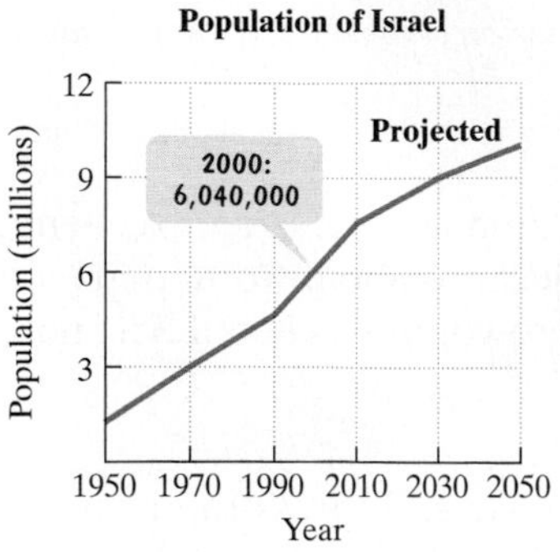

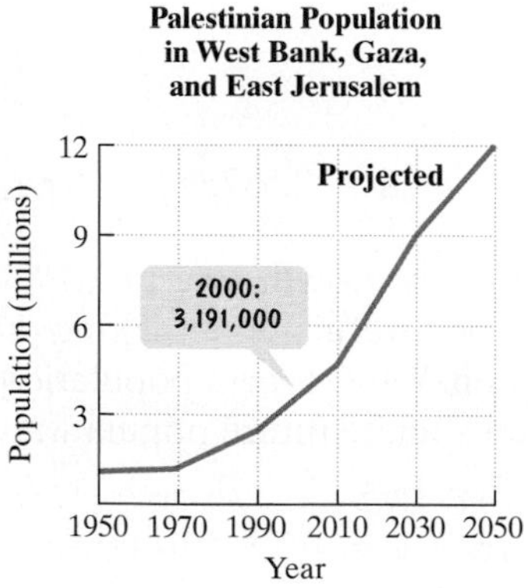

Source: Newsweek

7. a. In 2000, the population of Israel was approximately 6.04 million and by 2050 it is projected to grow to 10 million. Use the exponential growth model $A = A_0e^{kt}$, in which t is the number of years after 2000, to find an exponential growth function that models the data.

b. In which year will Israel's population be 9 million?

8. a. In 2000, the population of the Palestinians in the West Bank, Gaza Strip, and East Jerusalem was approximately 3.2 million and by 2050 it is projected to grow to 12 million. Use the exponential growth model $A = A_0e^{kt}$, in which t is the number of years after 2000, to find the exponential growth function that models the data.

b. In which year will the Palestinian population be 9 million?

An artifact originally had 16 grams of carbon-14 present. The decay model $A = 16e^{-0.000121t}$ describes the amount of carbon-14 present after t years. Use this model to solve Exercises 9–10.

9. How many grams of carbon-14 will be present in 5715 years?

10. How many grams of carbon-14 will be present in 11,430 years?

11. The half-life of the radioactive element krypton-91 is 10 seconds. If 16 grams of krypton-91 are initially present, how many grams are present after 10 seconds? 20 seconds? 30 seconds? 40 seconds? 50 seconds?

12. The half-life of the radioactive element plutonium-239 is 25,000 years. If 16 grams of plutonium-239 are initially present, how many grams are present after 25,000 years? 50,000 years? 75,000 years? 100,000 years? 125,000 years?

Use the exponential decay model for carbon-14, $A = A_0e^{-0.000121t}$, to solve Exercises 13–14.

13. Prehistoric cave paintings were discovered in a cave in France. The paint contained 15% of the original carbon-14. Estimate the age of the paintings.

14. Skeletons were found at a construction site in San Francisco in 1989. The skeletons contained 88% of the expected amount of carbon-14 found in a living person. In 1989, how old were the skeletons?

15. The August 1978 issue of *National Geographic* described the 1964 find of bones of a newly discovered dinosaur weighing 170 pounds, measuring 9 feet, with a 6-inch claw on one toe of each hind foot. The age of the dinosaur was estimated using potassium-40 dating of rocks surrounding the bones.

a. Potassium-40 decays exponentially with a half-life of approximately 1.31 billion years. Use the fact that after 1.31 billion years a given amount of potassium-40 will have decayed to half the original amount to show that the decay model for potassium-40 is given by $A = A_0e^{-0.52912t}$, where t is in billions of years.

b. Analysis of the rocks surrounding the dinosaur bones indicated that 94.5% of the original amount of potassium-40 was still present. Let $A = 0.945A_0$ in the model in part (a) and estimate the age of the bones of the dinosaur.

16. A bird species in danger of extinction has a population that is decreasing exponentially ($A = A_0e^{kt}$). Five years ago the population was at 1400 and today only 1000 of the birds are alive. Once the population drops below 100, the situation will be irreversible. When will this happen?

17. Use the exponential growth model, $A = A_0e^{kt}$, to show that the time it takes a population to double (to grow from A_0 to $2A_0$) is given by $t = \frac{\ln 2}{k}$.

18. Use the exponential growth model, $A = A_0e^{kt}$, to show that the time it takes a population to triple (to grow from A_0 to $3A_0$) is given by $t = \frac{\ln 3}{k}$.

Use the formula $t = \frac{\ln 2}{k}$ that gives the time for a population with a growth rate k to double to solve Exercises 19–20. Express each answer to the nearest whole year.

19. The growth model $A = 4e^{0.007t}$ describes New Zealand's population, A, in millions, t years after 2003.

a. What is New Zealand's growth rate?

b. How long will it take New Zealand to double its population?

20. The growth model $A = 104.9e^{0.017t}$ describes Mexico's population, A, in millions, t years after 2003.

a. What is Mexico's growth rate?

b. How long will it take Mexico to double its population?

Exercises 21–26 present data in the form of tables. For each data set shown by the table,

a. *Create a scatter plot for the data.*

b. *Use the scatter plot to determine whether an exponential function, a logarithmic function, or a linear function is the best choice for modeling the data. (If applicable, in Exercise 45, you will use your graphing utility to obtain these functions.)*

21. **Percent of Miscarriages, by Age**

Woman's Age	Percent of Miscarriages
22	9%
27	10%
32	13%
37	20%
42	38%
47	52%

Source: Time

22. **Number of Countries Connected to the Internet**

Year	Number of Countries Connected to the Internet
1985	11
1991	91
1994	146
1997	195
2002	220

Source: Medard Gabel, *Global Inc.*, 2003

23. **Percentage of U.S. Population Living in Urban Communities of 2500 or More People**

Year	Percent in Urban Communities
1950	56%
1960	69%
1970	73%
1980	75%
2001	76%

Source: U.S. Census Bureau

24. Cumulative Number of Deaths from AIDS in the United States

Year	Cumulative Number of Deaths
1990	122,463
1993	247,151
1995	349,222
1997	409,142
2000	458,551
2002	483,920

Source: Centers for Disease Control

25. Percentage of U.S. Homes with Cellular Telephones

Year	Percent
1999	44%
2000	51%
2001	59%
2002	66%

Source: Telecommunications Association

26. Percentage of U.S. Students Dropping out of High School

Year	Percent
1998	11.8%
1999	11.2%
2000	10.9%
2001	10.7%

Source: U.S. Department of Education

In Exercises 27–30, rewrite the equation in terms of base e. Express the answer in terms of a natural logarithm and then round to three decimal places.

27. $y = 100(4.6)^x$

28. $y = 1000(7.3)^x$

29. $y = 2.5(0.7)^x$

30. $y = 4.5(0.6)^x$

Writing in Mathematics

31. Nigeria has a growth rate of 0.025 or 2.5%. Describe what this means.

32. How can you tell if an exponential model describes exponential growth or exponential decay?

33. Suppose that a population that is growing exponentially increases from 800,000 people in 1997 to 1,000,000 people in 2000. Without showing the details, describe how to obtain the exponential growth function that models the data.

34. What is the half-life of a substance?

35. Describe the shape of a scatter plot that suggests modeling the data with an exponential function.

36. You take up weightlifting and record the maximum number of pounds you can lift at the end of each week. You start off with rapid growth in terms of the weight you can lift from week to week, but then the growth begins to level off. Describe how to obtain a function that models the number of pounds you can lift at the end of each week. How can you use this function to predict what might happen if you continue the sport?

37. Would you prefer that your salary be modeled exponentially or logarithmically? Explain your answer.

38. One problem with all exponential growth models is that nothing can grow exponentially forever. Describe factors that might limit the size of a population.

Technology Exercises

In Example 1 on page 850, we used two data points and an exponential function to model the population of the United States from 1970 through 2003. The data are shown again in the table. Use all five data points to solve Exercises 39–43.

x, Number of Years after 1969	y, U.S. Population (millions)
1 (1970)	203.3
11 (1980)	226.5
21 (1990)	248.7
31 (2000)	281.4
34 (2003)	294.0

39. a. Use your graphing utility's Exponential REGression option to obtain a model of the form $y = ab^x$ that fits the data. How well does the correlation coefficient, r, indicate that the model fits the data?

b. Rewrite the model in terms of base e. By what percentage is the population of the United States increasing each year?

40. Use your graphing utility's Logarithmic REGression option to obtain a model of the form $y = a + b \ln x$ that fits the data. How well does the correlation coefficient, r, indicate that the model fits the data?

41. Use your graphing utility's Linear REGression option to obtain a model of the form $y = ax + b$ that fits the data. How well does the correlation coefficient, r, indicate that the model fits the data?

42. Use your graphing utility's Power REGression option to obtain a model of the form $y = ax^b$ that fits the data. How well does the correlation coefficient, r, indicate that the model fits the data?

43. Use the value of r in Exercises 39–42 to select the two models of best fit. Use each of these models to predict by which year the U.S. population will reach 315 million. How do these answers compare to the year we found in Example 1, namely 2010? If you obtained different years, how do you account for this difference?

44. The figure shows the number of people in the United States age 65 and over, with projected figures for the year 2010 and beyond.

U.S. Population Age 65 and Over

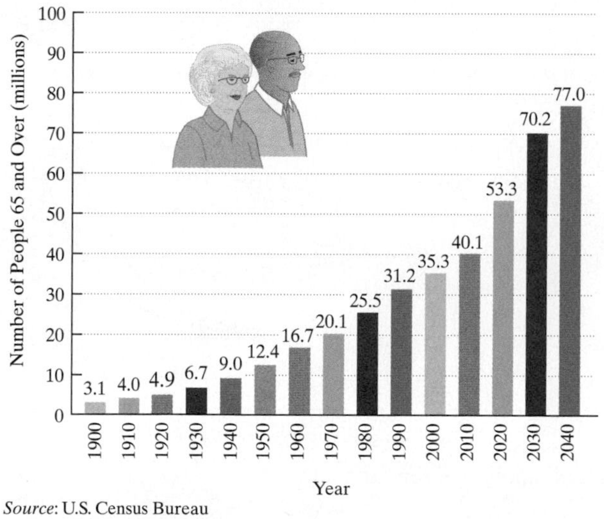

Source: U.S. Census Bureau

a. Let x represent the number of years after 1899 and let y represent the U.S. population age 65 and over, in millions. Use your graphing utility to find the model that best fits the data in the bar graph.

b. Rewrite the model in terms of base e. By what percentage is the 65 and over population increasing each year?

45. In Exercises 21–26, you determined the best choice for the kind of function that modeled the data in the table. For each of these exercises that you worked, use a graphing utility to find the actual function that best fits the data. Then use the model to make a reasonable prediction for a value that exceeds those shown in the table's first column.

Critical Thinking Exercises

46. The exponential growth models describe the population of the indicated country, A, in millions, t years after 2003.

Canada $A = 32.2e^{0.003t}$

Uganda $A = 25.6e^{0.03t}$

According to these models, which one of the following is true?

a. In 2003, Uganda's population was ten times that of Canada's.

b. In 2003, Canada's population exceeded Uganda's by 660,000.

c. In 2012, Uganda's population will exceed Canada's.

d. None of these statements is true.

47. Over a period of time, a hot object cools to the temperature of the surrounding air. This is described mathematically by Newton's Law of Cooling:

$$T = C + (T_0 - C)e^{-kt},$$

where t is the time it takes for an object to cool from temperature T_0 to temperature T, C is the surrounding air temperature, and k is a positive constant that is associated with the cooling object. A cake removed from the oven has a temperature of 210°F and is left to cool in a room that has a temperature of 70°F. After 30 minutes, the temperature of the cake is 140°F. What is the temperature of the cake after 40 minutes?

Review Exercises

48. Divide:

$$\frac{x^2 - 9}{2x^2 + 7x + 3} \div \frac{x^2 - 3x}{2x^2 + 11x + 5}.$$

(Section 7.2, Example 6)

49. Solve: $x^{\frac{2}{3}} + 2x^{\frac{1}{3}} - 3 = 0.$
(Section 11.4, Example 5)

50. Simplify: $6\sqrt{2} - 2\sqrt{50} + 3\sqrt{98}.$
(Section 10.4, Example 2)

GROUP PROJECT

CHAPTER 12

This activity is intended for three or four people who would like to take up weightlifting. Each person in the group should record the maximum number of pounds that he or she can lift at the end of each week for the first 10 consecutive weeks. Use the Logarithmic REGression option of a graphing utility to obtain a model showing the amount of weight that group members can lift from week 1 through week 10. Graph each of the models in the same viewing rectangle to observe similarities and differences among weight–growth patterns of each member. Use the functions to predict the amount of weight that group members will be able to lift in the future. If the group continues to work out together, check the accuracy of these predictions.

CHAPTER 12 SUMMARY

Definitions and Concepts	Examples

Section 12.1 Exponential Functions

The exponential function with base b is defined by $f(x) = b^x$, where $b > 0$ and $b \neq 1$. The graph contains the point $(0, 1)$. When $b > 1$, the graph rises from left to right. When $0 < b < 1$, the graph falls from left to right. The x-axis is a horizontal asymptote. The domain is $(-\infty, \infty)$; the range is $(0, \infty)$. The natural exponential function is $f(x) = e^x$, where $e \approx 2.71828$.

Graph $f(x) = 2^x$ and $g(x) = 2^{x-1}$.

x	$f(x) = 2^x$	$g(x) = 2^{x-1}$
-2	$2^{-2} = \frac{1}{4}$	$2^{-3} = \frac{1}{8}$
-1	$2^{-1} = \frac{1}{2}$	$2^{-2} = \frac{1}{4}$
0	$2^0 = 1$	$2^{-1} = \frac{1}{2}$
1	$2^1 = 2$	$2^0 = 1$
2	$2^2 = 4$	$2^1 = 2$

The graph of g is the graph of f shifted one unit to the right.

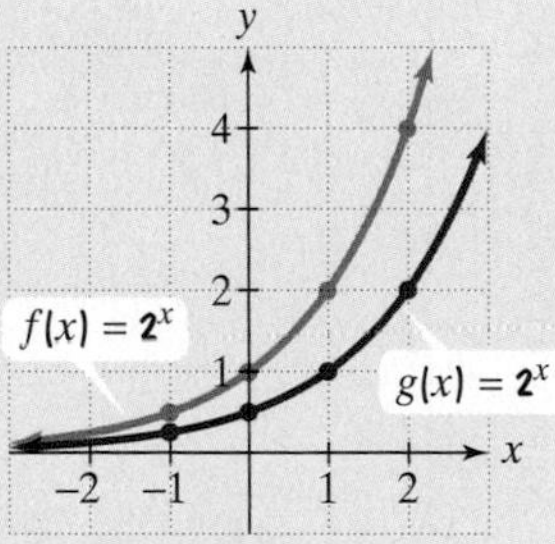

Formulas for Compound Interest
After t years, the balance, A, in an account with principal P and annual interest rate r is given by the following formulas:

1. For n compoundings per year: $A = P\left(1 + \frac{r}{n}\right)^{nt}$
2. For continuous compounding: $A = Pe^{rt}$

Select the better investment for \$4000 over 6 years:

- 6% compounded semiannually

$$A = P\left(1 + \frac{r}{n}\right)^{nt}$$
$$= 4000\left(1 + \frac{0.06}{2}\right)^{2 \cdot 6} \approx \$5703$$

- 5.9% compounded continuously

$$A = Pe^{rt} = 4000e^{0.059(6)} \approx \$5699$$

The first investment is better.

Definitions and Concepts	Examples

Section 12.2 Logarithmic Functions

Definition of the logarithmic function: For $x > 0$ and $b > 0, b \neq 1$, $y = \log_b x$ is equivalent to $b^y = x$. The function $f(x) = \log_b x$ is the logarithmic function with base b. This function is the inverse function of the exponential function with base b.

- Write $\log_2 32 = 5$ in exponential form.

$$2^5 = 32 \qquad y = \log_b x \text{ means } b^y = x.$$

- Write $\sqrt{49} = 7$, or $49^{\frac{1}{2}} = 7$, in logarithmic form.

$$\frac{1}{2} = \log_{49} 7 \qquad b^y = x \text{ means } y = \log_b x.$$

The graph of $f(x) = \log_b x$ can be obtained from $f(x) = b^x$ by reversing coordinates. The graph of $f(x) = \log_b x$ contains the point $(1, 0)$. If $b > 1$, the graph rises from left to right. If $0 < b < 1$, the graph falls from left to right. The y-axis is a vertical asymptote. The domain is $(0, \infty)$; the range is $(-\infty, \infty)$. $f(x) = \log x$ means $f(x) = \log_{10} x$ and is the common logarithmic function. $f(x) = \ln x$ means $f(x) = \log_e x$ and is the natural logarithmic function. The domain of $f(x) = \log_b[g(x)]$ consists of all x for which $g(x) > 0$.

- Graph $f(x) = \log_3 x$.
- Find the domain: $f(x) = \log_6(4 - x)$.

$$4 - x > 0$$
$$4 > x \quad (\text{or } x < 4)$$

The domain is $\{x \mid x < 4\}$ or $(-\infty, 4)$.

Basic Logarithmic Properties

Base b ($b > 0, b \neq 1$)	**Base 10 (Common Logarithms)**	**Base e (Natural Logarithms)**
$\log_b 1 = 0$	$\log 1 = 0$	$\ln 1 = 0$
$\log_b b = 1$	$\log 10 = 1$	$\ln e = 1$
$\log_b b^x = x$	$\log 10^x = x$	$\ln e^x = x$
$b^{\log_b x} = x$	$10^{\log x} = x$	$e^{\ln x} = x$

- $\log_8 1 = 0$ because $\log_b 1 = 0$.
- $\log_4 4 = 1$ because $\log_b b = 1$.
- $\ln e^{8x} = 8x$ because $\ln e^x = x$.
- $e^{\ln \sqrt[3]{x}} = \sqrt[3]{x}$ because $e^{\ln x} = x$.
- $\log_t t^{25} = 25$ because $\log_b b^x = x$.

Definitions and Concepts	Examples

Section 12.3 Properties of Logarithms

Properties of Logarithms For $M > 0$ and $N > 0$:

1. *The Product Rule*: $\log_b(MN) = \log_b M + \log_b N$
2. *The Quotient Rule*: $\log_b\left(\frac{M}{N}\right) = \log_b M - \log_b N$
3. *The Power Rule*: $\log_b M^p = p \log_b M$
4. *The Change-of Base Property*:

The General Property	Introducing Common Logarithms	Introducing Natural Logarithms
$\log_b M = \dfrac{\log_a M}{\log_a b}$	$\log_b M = \dfrac{\log M}{\log b}$	$\log_b M = \dfrac{\ln M}{\ln b}$

- Expand: $\log_3(81x^7)$.

$$= \log_3 81 + \log_3 x^7$$
$$= 4 + 7\log_3 x$$

- Write as a single logarithm: $7\ln x - 4\ln y$.

$$= \ln x^7 - \ln y^4 = \ln\left(\frac{x^7}{y^4}\right)$$

- Evaluate: $\log_6 92$.

$$\log_6 92 = \frac{\ln 92}{\ln 6} \approx 2.5237$$

Section 12.4 Exponential and Logarithmic Equations

An exponential equation is an equation containing a variable in an exponent. Some exponential equations can be solved by expressing both sides as a power of the same base. Then set the exponents equal to each other:

$$\text{If } b^M = b^N, \text{ then } M = N.$$

Solve: $4^{2x-1} = 64$.

$$4^{2x-1} = 4^3$$
$$2x - 1 = 3$$
$$2x = 4$$
$$x = 2$$

The solution is 2 and the solution set is $\{2\}$.

If both sides of an exponential equation cannot be expressed as a power of the same base, isolate the exponential expression and take the natural logarithm on both sides. Simplify using

$$\ln b^x = x\ln b \quad \text{or} \quad \ln e^x = x.$$

Solve: $7^x = 103$.

$$\ln 7^x = \ln 103$$
$$x\ln 7 = \ln 103$$
$$x = \frac{\ln 103}{\ln 7}$$

The solution is $\frac{\ln 103}{\ln 7}$ and the solution set is $\left\{\frac{\ln 103}{\ln 7}\right\}$.

A logarithmic equation is an equation containing a variable in a logarithmic expression. Logarithmic equations in the form $\log_b x = c$ can be solved by rewriting as $b^c = x$. When checking logarithmic equations, reject proposed solutions that produce the logarithm of a negative number or the logarithm of zero in the original equation.

Solve: $\log_2(3x - 1) = 5$.

$$2^5 = 3x - 1$$
$$32 = 3x - 1$$
$$33 = 3x$$
$$11 = x$$

The solution is 11 and the solution set is $\{11\}$.

Equations involving natural logarithms are solved by isolating the natural logarithm with coefficient 1 on one side and exponentiating both sides. Simplify using $e^{\ln x} = x$.

Solve: $3\ln 2x = 15$.

$$\ln 2x = 5$$
$$e^{\ln 2x} = e^5$$
$$2x = e^5$$
$$x = \frac{e^5}{2}$$

The solution is $\frac{e^5}{2}$ and the solution set is $\left\{\frac{e^5}{2}\right\}$.

Definitions and Concepts	Examples

Section 12.5 Exponential Growth and Decay; Modeling Data

Definitions and Concepts

Exponential growth and decay models are given by $A = A_0e^{kt}$ in which t represents time, A_0 is the amount present at $t = 0$, and A is the amount present at time t. If $k > 0$, the model describes growth and k is the growth rate. If $k < 0$, the model describes decay and k is the decay rate. Scatter plots for exponential and logarithmic models are shown in Figure 12.17 on page 853. When using a graphing utility to model data, the closer that the correlation coefficient r is to -1 or 1, the better the model fits the data.

Examples

The 1970 population of the Tokyo, Japan, urban area was 16.5 million: in 2000, it was 26.4 million. Write the exponential growth function that describes the population, in millions, t years after 1970. Begin with $A = A_0e^{kt}$.

$$A = 16.5e^{kt}$$ In 1970 ($t = 0$), population was 16.5 million.

$$26.4 = 16.5e^{k \cdot 30}$$ When $t = 30$ (in 2000), $A = 26.4$.

$$e^{30k} = \frac{26.4}{16.5}$$ Isolate the exponential factor.

$$\ln e^{30k} = \ln\left(\frac{26.4}{16.5}\right)$$ Take the natural logarithm on both sides.

$$30k = \ln\left(\frac{26.4}{16.5}\right) \quad \text{and} \quad k = \frac{\ln\left(\frac{26.4}{16.5}\right)}{30} \approx 0.016$$

The growth function is $A = 16.5e^{0.016t}$.

Growth rate is 0.016 or 1.6%.

Expressing an Exponential Model in Base e

$y = ab^x$ is equivalent to $y = ae^{(\ln b)x}$.

Rewrite in terms of base e: $y = 24(7.2)^x$.

$$y = 24e^{(\ln 7.2)x} \approx 24e^{1.974x}$$

CHAPTER 12 REVIEW EXERCISES

12.1 *In Exercises 1–4, set up a table of coordinates for each function. Select integers from -2 to 2 for x. Then use the table of coordinates to match the function with its graph. [The graphs are labeled (a) through (d).]*

1. $f(x) = 4^x$

2. $f(x) = 4^{-x}$

3. $f(x) = -4^{-x}$

4. $f(x) = -4^{-x} + 3$

a.

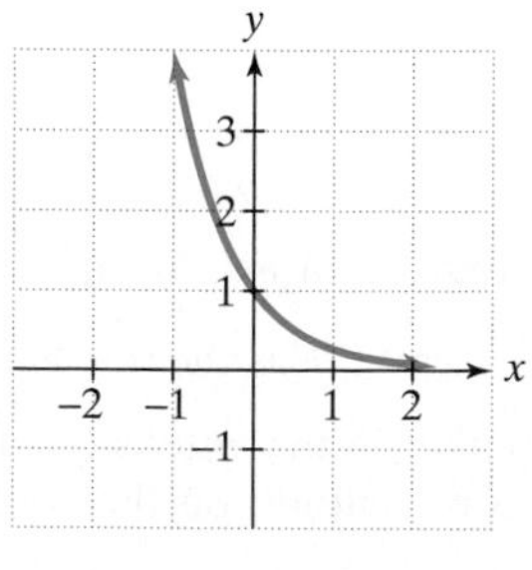

b.

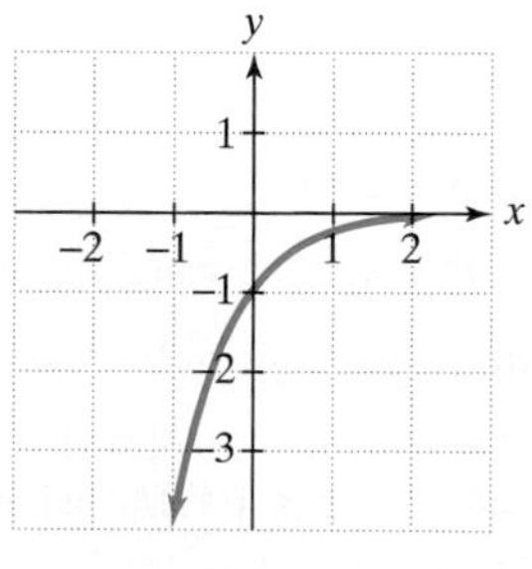

c.

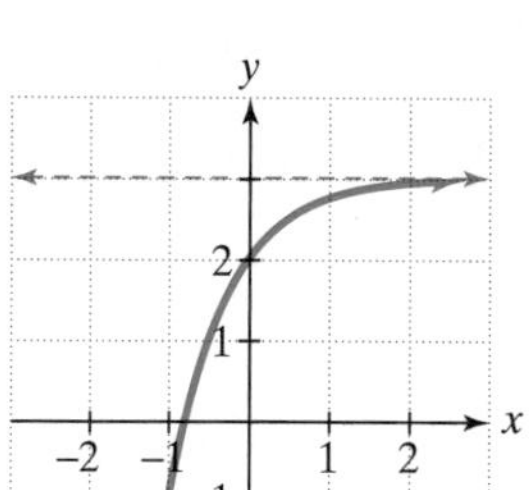

d.

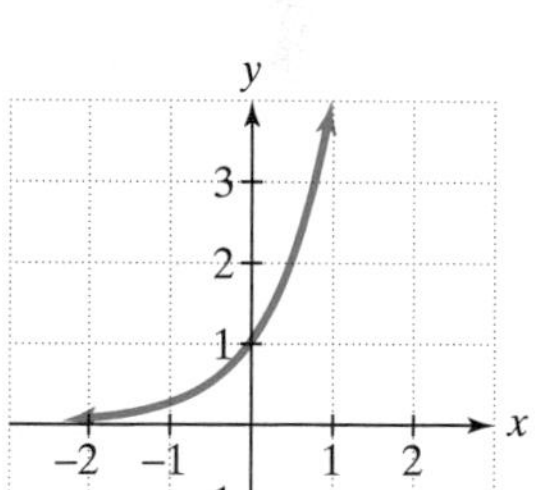

In Exercises 5–8, graph functions f and g in the same rectangular coordinate system. Select integers from -2 to 2 for x. Then describe how the graph of g is related to the graph of f. If applicable, use a graphing utility to confirm your hand-drawn graphs.

5. $f(x) = 2^x$ and $g(x) = 2^{x-1}$

6. $f(x) = 2^x$ and $g(x) = \left(\frac{1}{2}\right)^x$

7. $f(x) = 3^x$ and $g(x) = 3^x - 1$

8. $f(x) = 3^x$ and $g(x) = -3^x$

Use the compound interest formulas

$$A = P\left(1 + \frac{r}{n}\right)^{nt} \quad \text{and} \quad A = Pe^{rt}$$

to solve Exercises 9–10.

9. Suppose that you have \$5000 to invest. Which investment yields the greater return over 5 years: 5.5% compounded semiannually or 5.25% compounded monthly?

10. Suppose that you have \$14,000 to invest. Which investment yields the greater return over 10 years: 7% compounded monthly or 6.85% compounded continuously?

11. A cup of coffee is taken out of a microwave oven and placed in a room. The temperature, T, in degrees Fahrenheit, of the coffee after t minutes is modeled by the function $T = 70 + 130e^{-0.04855t}$. The graph of the function is shown in the figure.

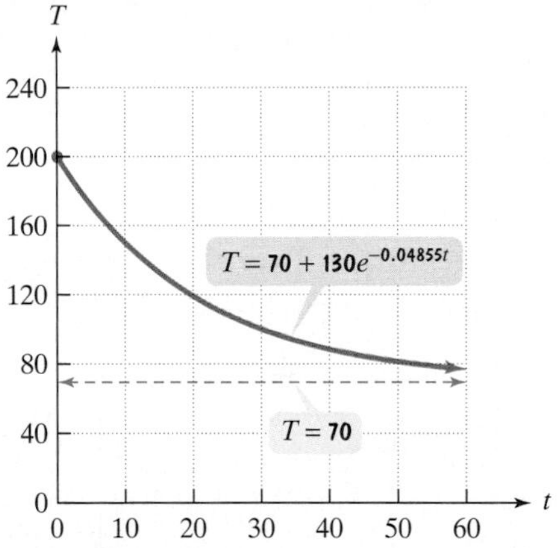

Use the graph to answer each of the following questions.

a. What was the temperature of the coffee when it was first taken out of the microwave?

b. What is a reasonable estimate of the temperature of the coffee after 20 minutes? Use your calculator to verify this estimate.

c. What is the limit of the temperature to which the coffee will cool? What does this tell you about the temperature of the room?

12.2 *In Exercises 12–14, write each equation in its equivalent exponential form.*

12. $\frac{1}{2} = \log_{49} 7$

13. $3 = \log_4 x$

14. $\log_3 81 = y$

In Exercises 15–17, write each equation in its equivalent logarithmic form.

15. $6^3 = 216$

16. $b^4 = 625$

17. $13^y = 874$

In Exercises 18–28, evaluate each expression without using a calculator. If evaluation is not possible, state the reason.

18. $\log_4 64$

19. $\log_5 \frac{1}{25}$

20. $\log_3(-9)$

21. $\log_{16} 4$

22. $\log_{17} 17$

23. $\log_3 3^8$

24. $\ln e^5$

25. $\log_3 \frac{1}{\sqrt{3}}$

26. $\ln \frac{1}{e^2}$

27. $\log \frac{1}{1000}$

28. $\log_3(\log_8 8)$

29. Graph $f(x) = 2^x$ and $g(x) = \log_2 x$ in the same rectangular coordinate system. Use the graphs to determine each function's domain and range.

30. Graph $f(x) = \left(\frac{1}{3}\right)^x$ and $g(x) = \log_{\frac{1}{3}} x$ in the same rectangular coordinate system. Use the graphs to determine each function's domain and range.

In Exercises 31–33, find the domain of each logarithmic function.

31. $f(x) = \log_8(x + 5)$

32. $f(x) = \log(3 - x)$

33. $f(x) = \ln(x - 1)^2$

In Exercises 34–36, use inverse properties of logarithms to simplify each expression.

34. $\ln e^{6x}$

35. $e^{\ln \sqrt{x}}$

36. $10^{\log 4x^2}$

37. On the Richter scale, the magnitude, R, of an earthquake of intensity I is given by $R = \log \frac{I}{I_0}$, where I_0 is the intensity of a barely felt zero-level earthquake. If the intensity of an earthquake is $1000I_0$, what is its magnitude on the Richter scale?

38. Students in a psychology class took a final examination. As part of an experiment to see how much of the course content they remembered over time, they took equivalent forms of the exam in monthly intervals thereafter. The average score, $f(t)$, for the group after t months is modeled by the function $f(t) = 76 - 18\log(t + 1)$, where $0 \le t \le 12$.

a. What was the average score when the exam was first given?

b. What was the average score, to the nearest tenth, after 2 months? 4 months? 6 months? 8 months? one year?

c. Use the results from parts (a) and (b) to graph f. Describe what the shape of the graph indicates in terms of the material retained by the students.

39. The formula

$$t = \frac{1}{c}\ln\left(\frac{A}{A - N}\right)$$

describes the time, t, in weeks, that it takes to achieve mastery of a portion of a task. In the formula, A represents maximum learning possible, N is the portion of the learning that is to be achieved, and c is a constant used to measure an individual's learning style. A 50-year-old man decides to start running as a way to maintain good health. He feels that the maximum rate he could ever hope to achieve is 12 miles per hour. How many weeks will it take before the man can run 5 miles per hour if $c = 0.06$ for this person?

12.3 *In Exercises 40–43, use properties of logarithms to expand each logarithmic expression as much as possible. Where possible, evaluate logarithmic expressions without using a calculator. Assume that all variables represent positive numbers.*

40. $\log_6(36x^3)$

41. $\log_4\left(\frac{\sqrt{x}}{64}\right)$

42. $\log_2\left(\frac{xy^2}{64}\right)$

43. $\ln\sqrt[3]{\frac{x}{e}}$

In Exercises 44–47, use properties of logarithms to condense each logarithmic expression. Write the expression as a single logarithm whose coefficient is 1.

44. $\log_b 7 + \log_b 3$

45. $\log 3 - 3\log x$

46. $3\ln x + 4\ln y$

47. $\frac{1}{2}\ln x - \ln y$

In Exercises 48–49, use common logarithms or natural logarithms and a calculator to evaluate to four decimal places.

48. $\log_6 72{,}348$

49. $\log_4 0.863$

12.4 *Solve each exponential equation in Exercises 50–52 by expressing each side as a power of the same base and then equating exponents.*

50. $2^{4x-2} = 64$

51. $125^x = 25$

52. $9^x = \frac{1}{27}$

Solve each exponential equation in Exercises 53–55 by taking the natural logarithm on both sides. Express the solution in terms of natural logarithms. Then use a calculator to obtain a decimal approximation, correct to two decimal places, for the solution.

53. $8^x = 12{,}143$

54. $9e^{5x} = 1269$

55. $30e^{0.045x} = 90$

In Exercises 56–62, solve each logarithmic equation.

56. $\log_5 x = -3$

57. $\log x = 2$

58. $\log_4(3x - 5) = 3$

59. $\log_2(x + 3) + \log_2(x - 3) = 4$

60. $\log_3(x - 1) - \log_3(x + 2) = 2$

61. $\ln x = -1$

62. $3 + 4\ln 2x = 15$

63. The function $P(x) = 14.7e^{-0.21x}$ models the average atmospheric pressure, $P(x)$, in pounds per square inch, at an altitude of x miles above sea level. The atmospheric pressure at the peak of Mt. Everest, the world's highest mountain, is 4.6 pounds per square inch. How many miles above sea level, to the nearest tenth of a mile, is the peak of Mt. Everest?

64. The amount of carbon dioxide in the atmosphere, measured in parts per million, has been increasing as a result of the burning of oil and coal. The buildup of gases and particles traps heat and raises the planet's temperature, a phenomenon called the greenhouse effect. Carbon dioxide accounts for about half of the warming. The function $f(t) = 364(1.005)^t$ projects carbon dioxide concentration, $f(t)$, in parts per million, t years after 2000. Using the projections given by the function, when will the carbon dioxide concentration be double the preindustrial level of 280 parts per million?

65. The function $W(x) = 0.37 \ln x + 0.05$ models the average walking speed, $W(x)$, in feet per second, of residents in a city whose population is x thousand. Visitors to New York City frequently feel they are moving too slowly to keep pace with New Yorkers' average walking speed of 3.38 feet per second. What is the population of New York City? Round to the nearest thousand.

66. Use the compound interest formula

$$A = P\left(1 + \frac{r}{n}\right)^{nt}$$

to solve this problem. How long, to the nearest tenth of a year, will it take \$12,500 to grow to \$20,000 at 6.5% annual interest compounded quarterly?

Use the compound interest formula

$$A = Pe^{rt}$$

to solve Exercises 67–68.

67. How long, to the nearest tenth of a year, will it take \$50,000 to triple in value at 7.5% annual interest compounded continuously?

68. What interest rate is required for an investment subject to continuous compounding to triple in 5 years?

12.5

69. According to the U.S. Census Bureau, in 1990 there were 22.4 million residents of Hispanic origin living in the United States. By 2000, the number had increased to 35.3 million. The exponential growth function $A = 22.4e^{kt}$ describes the U.S. Hispanic population, A, in millions, t years after 1990.

a. Find k, correct to three decimal places.

b. Use the resulting model to project the Hispanic resident population in 2010.

c. In which year will the Hispanic resident population reach 60 million?

70. Use the exponential decay model for carbon-14, $A = A_0 e^{-0.000121t}$, to solve this exercise. Prehistoric paintings were discovered in the Lascaux cave in France. The paint contained 15% of the original carbon-14. Estimate the age of the paintings at the time of the discovery.

Exercises 71–72 present data in the form of tables. For each data set shown by the table,

a. Create a scatter plot for the data.

b. Use the scatter plot to determine whether an exponential function or a logarithmic function is the best choice for modeling the data.

71. Percentage of the U.S. Population, Ages 25 or More, Completing At Least Four Years of High School

Year	Percent
1970	55.2%
1980	68.6%
1991	78.4%
2002	84.1%

Source: U.S. Census Bureau

72. Projection of U.S. Jobs Moving Overseas

Year	Number of Jobs Moving Overseas (millions)
2003	0.3
2008	1
2010	1.5
2012	2.5
2015	3.3

Source: Forrester Research, Inc.

In Exercises 73 74, rewrite the equation in terms of base e. Express the answer in terms of a natural logarithm and then round to three decimal places.

73. $y = 73(2.6)^x$

74. $y = 6.5(0.43)^x$

75. The figure shows world population projections through the year 2150. The data are from the United Nations Family Planning Program and are based on optimistic or pessimistic expectations for successful control of human population growth. Suppose that you are interested in modeling these data using exponential, logarithmic, linear, and quadratic functions. Which function would you use to model each of the projections? Explain your choices. For the choice corresponding to a quadratic model, would your formula involve one with a positive or negative leading coefficient? Explain.

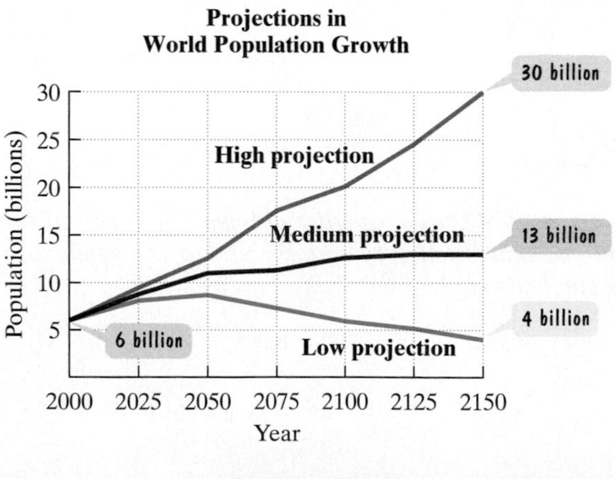

CHAPTER 12 TEST

Remember to use your Chapter Test Prep Video CD to see the worked-out solutions to the test questions you want to review.

1. Graph $f(x) = 2^x$ and $g(x) = 2^{x+1}$ in the same rectangular coordinate system.

2. Use $A = P\left(1 + \frac{r}{n}\right)^{nt}$ and $A = Pe^{rt}$ to solve this problem.

Suppose you have \$3000 to invest. Which investment yields the greater return over 10 years: 6.5% compounded semiannually or 6% compounded continuously? How much more (to the nearest dollar) is yielded by the better investment?

3. Write in exponential form: $\log_5 125 = 3$.

4. Write in logarithmic form: $\sqrt{36} = 6$.

5. Graph $f(x) = 3^x$ and $g(x) = \log_3 x$ in the same rectangular coordinate system. Use the graphs to determine each function's domain and range.

In Exercises 6–8, simplify each expression.

6. $\ln e^{5x}$ **7.** $\log_b b$ **8.** $\log_6 1$

9. Find the domain: $f(x) = \log_5(x - 7)$.

10. On the decibel scale, the loudness of a sound, in decibels, is given by $D = 10 \log \frac{I}{I_0}$, where I is the intensity of the sound, in watts per meter^2, and I_0 is the intensity of a sound barely audible to the human ear. If the intensity of a sound is $10^{12}I_0$, what is its loudness in decibels? (Such a sound is potentially damaging to the ear.)

In Exercises 11–12, use properties of logarithms to expand each logarithmic expression as much as possible. Where possible, evaluate logarithmic expressions without using a calculator.

11. $\log_4(64x^5)$

12. $\log_3\left\{\frac{\sqrt[3]{x}}{81}\right\}$

In Exercises 13–14, write each expression as a single logarithm.

13. $6 \log x + 2 \log y$

14. $\ln 7 - 3 \ln x$

15. Use a calculator to evaluate $\log_{15} 71$ to four decimal places.

In Exercises 16–22, solve each equation.

16. $3^{x-2} = 81$

17. $5^x = 1.4$

18. $400e^{0.005x} = 1600$

19. $\log_{25} x = \frac{1}{2}$

20. $\log_6(4x - 1) = 3$

21. $\log x + \log(x + 15) = 2$

22. $2 \ln 3x = 8$

23. The function

$$A = 82.3e^{-0.002t}$$

models the population of Germany, A, in millions, t years after 2003.

a. What was the population of Germany in 2003?

b. Is the population of Germany increasing or decreasing? Explain.

c. In which year will the population of Germany be 81.5 million?

Use the formulas

$$A = P\left(1 + \frac{r}{n}\right)^{nt} \quad \text{and} \quad A = Pe^{rt}$$

to solve Exercises 24–25.

24. How long, to the nearest tenth of a year, will it take \$4000 to grow to \$8000 at 5% annual interest compounded quarterly?

25. What interest rate is required for an investment subject to continuous compounding to double in 10 years?

26. The 1990 population of Europe was 509 million; in 2000, it was 729 million. Write the exponential growth function that describes the population of Europe, in millions, t years after 1990.

27. Use the exponential decay model for carbon-14, $A = A_0e^{-0.000121t}$, to solve this exercise. Bones of a prehistoric man were discovered and contained 5% of the original amount of carbon-14. How long ago did the man die?

In Exercises 28–31, determine whether the values in each table belong to an exponential function, a logarithmic function, a linear function, or a quadratic function.

28.

x	y
0	3
1	1
2	−1
3	−3
4	−5

29.

x	y
$\frac{1}{3}$	−1
1	0
3	1
9	2
27	3

30.

x	y
0	1
1	5
2	25
3	125
4	625

31.

x	y
0	12
1	3
2	0
3	3
4	12

32. Rewrite $y = 96(0.38)^x$ in terms of base e. Express the answer in terms of a natural logarithm and then round to three decimal places.

CUMULATIVE REVIEW EXERCISES (CHAPTERS 1–12)

In Exercises 1–15, solve each equation, inequality, or system.

1. $9(x - 1) = 1 + 3(x - 2)$

2. $3x + 4y = -7$
$x - 2y = -9$

3. $x - y + 3z = -9$
$2x + 3y - z = 16$
$5x + 2y - z = 15$

4. $7x + 18 \le 9x - 2$

5. $4x - 3 < 13$ and $-3x - 4 \ge 8$

6. $2x + 4 > 8$ or $x - 7 \ge 3$

7. $|2x - 1| < 5$

8. $\left|\frac{2}{3}x - 4\right| = 2$

9. $\frac{4}{x - 3} - \frac{6}{x + 3} = \frac{24}{x^2 - 9}$

10. $\sqrt{x + 4} - \sqrt{x - 3} = 1$

11. $2x^2 = 5 - 4x$

12. $x^{\frac{2}{3}} - 5x^{\frac{1}{3}} + 6 = 0$

13. $2x^2 + x - 6 \le 0$

14. $\log_8 x + \log_8(x + 2) = 1$

15. $5^{2x+3} = 125$

In Exercises 16–20, graph each function, equation, or inequality in a rectangular coordinate system.

16. $x - 3y = 6$

17. $f(x) = \frac{1}{2}x - 1$

18. $3x - 2y > -6$

19. $f(x) = -2(x - 3)^2 + 2$

20. $y = \log_2 x$

In Exercises 21–31, perform the indicated operations, and simplify, if possible.

21. $4[2x - 6(x - y)]$

22. $(-5x^3y^2)(4x^4y^{-6})$

23. $(8x^2 - 9xy - 11y^2) - (7x^2 - 4xy + 5y^2)$

24. $(3x - 1)(2x + 5)$

25. $(3x^2 - 4y)^2$

26. $\frac{3x}{x + 5} - \frac{2}{x^2 + 7x + 10}$

27. $\dfrac{1 - \frac{9}{x^2}}{1 + \frac{3}{x}}$

28. $\dfrac{x^2 - 6x + 8}{3x + 9} \div \dfrac{x^2 - 4}{x + 3}$

29. $\sqrt{5xy} \cdot \sqrt{10x^2y}$

30. $4\sqrt{72} - 3\sqrt{50}$

31. $(5 + 3i)(7 - 3i)$

In Exercises 32–34, factor completely.

32. $81x^4 - 1$

33. $24x^3 - 22x^2 + 4x$

34. $x^3 + 27y^3$

In Exercises 35–38, let $f(x) = x^2 + 3x - 15$ *and* $g(x) = x - 2$. *Find each indicated expression.*

35. $(f - g)(x)$ and $(f - g)(5)$

36. $\left(\dfrac{f}{g}\right)(x)$ and the domain of $\dfrac{f}{g}$

37. $f(g(x))$

38. $g(f(x))$

39. If $f(x) = 7x - 3$, find $f^{-1}(x)$.

40. Divide using synthetic division:

$$(3x^3 - x^2 + 4x + 8) \div (x + 2).$$

41. Solve for R: $I = \dfrac{R}{R + r}$.

42. Write the slope-intercept form of the equation of the line through $(-2, 5)$ and parallel to the line whose equation is $3x + y = 9$.

43. Write as a single logarithm whose coefficient is 1:

$$2 \ln x - \frac{1}{2} \ln y.$$

In Exercises 44–45, find the domain of each function.

44. $f(x) = \dfrac{x - 2}{x^2 - 3x + 2}$

45. $f(x) = \ln(2x - 8)$

46. The price of a computer is reduced by 30% to \$434. What was the original price?

47. The area of a rectangle is 52 square yards. The length of the rectangle is 1 yard longer than 3 times its width. Find the rectangle's dimensions.

48. You invested \$4000 in two stocks paying 12% and 14% annual interest, respectively. At the end of the year, the total interest from these investments was \$508. How much was invested at each rate?

49. Use the formula for continuous compounding, $A = Pe^{rt}$, to solve this problem. What interest rate is required for an investment of \$6000 subject to continuous compounding to grow to \$18,000 in 10 years?

50. The current, I, in amperes, flowing in an electrical circuit varies inversely as the resistance, R, in ohms, in the circuit. When the resistance of an electric percolator is 22 ohms, it draws 5 amperes of current. How much current is needed when the resistance is 10 ohms?

CHAPTER 13

One minute you're in class, enjoying the lecture. Then a sharp pain radiates down your side. The next minute you're being diagnosed with, of all things, a kidney stone. It took your cousin six weeks to recover from kidney stone surgery, but your doctor assures you there is nothing to worry about. A new procedure, based on a curve that looks like the cross section of a football, will dissolve the stone painlessly and let you return to class in a day or two. How can this be?

This problem appears in Section 13.2 in the discussion on applications of ellipses.

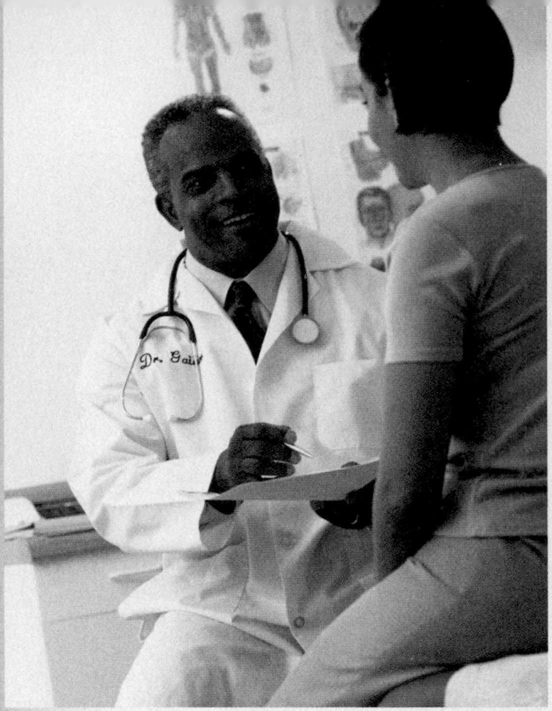

Conic Sections and Systems of Nonlinear Equations

From ripples in water to the path on which humanity journeys through space, certain curves occur naturally throughout the universe. Over 2000 years ago, the ancient Greeks studied these curves, called *conic sections*, without regard to their immediate usefulness simply because studying them elicited ideas that were exciting, challenging, and interesting. The ancient Greeks could not have imagined the applications of these curves in the twenty-first century. Overwhelmed by the choices on satellite television? Blame it on a conic section! In this chapter, we use the rectangular coordinate system to study the conic sections and the mathematics behind their surprising applications.

SECTION 13.1

Objectives

1. Write the standard form of a circle's equation.
2. Give the center and radius of a circle whose equation is in standard form.
3. Convert the general form of a circle's equation to standard form.

THE CIRCLE

It's a good idea to know your way around a circle. Clocks, angles, maps, and compasses are based on circles. Circles occur everywhere in nature: in ripples on water, patterns on a butterfly's wings, and cross sections of trees. Some consider the circle to be the most pleasing of all shapes.

The rectangular coordinate system gives us a unique way of knowing a circle. It enables us to translate a circle's geometric definition into an algebraic equation. We begin with this geometric definition.

DEFINITION OF A CIRCLE A **circle** is the set of all points in a plane that are equidistant from a fixed point, called the **center**. The fixed distance from the circle's center to any point on the circle is called the **radius**.

Figure 13.1 is our starting point for obtaining a circle's equation. We've placed the circle into a rectangular coordinate system. The circle's center is (h, k) and its radius is r. We let (x, y) represent the coordinates of any point on the circle.

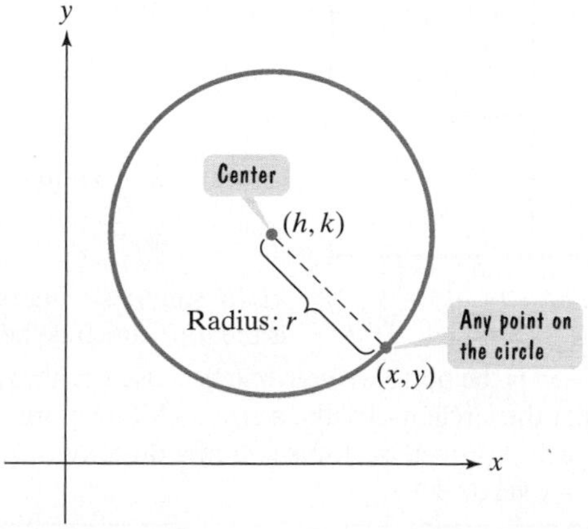

FIGURE 13.1 A circle centered at (h, k) with radius r

What does the geometric definition of a circle tell us about the point (x, y) in Figure 13.1? The point is on the circle if and only if its distance from the center is r. We can use the distance formula to express this idea algebraically:

$$\underbrace{\sqrt{(x-h)^2 + (y-k)^2}}_{\text{The distance between } (x, y) \text{ and } (h, k)} \underbrace{=}_{\text{is always}} \underbrace{r.}_{r.}$$

Squaring both sides of this equation yields the *standard form of the equation of a circle.*

1 Write the standard form of a circle's equation.

THE STANDARD FORM OF THE EQUATION OF A CIRCLE The **standard form of the equation of a circle** with center (h, k) and radius r is

$$(x-h)^2 + (y-k)^2 = r^2.$$

EXAMPLE 1 Finding the Standard Form of a Circle's Equation

Write the standard form of the equation of the circle with center $(0, 0)$ and radius 2. Graph the circle.

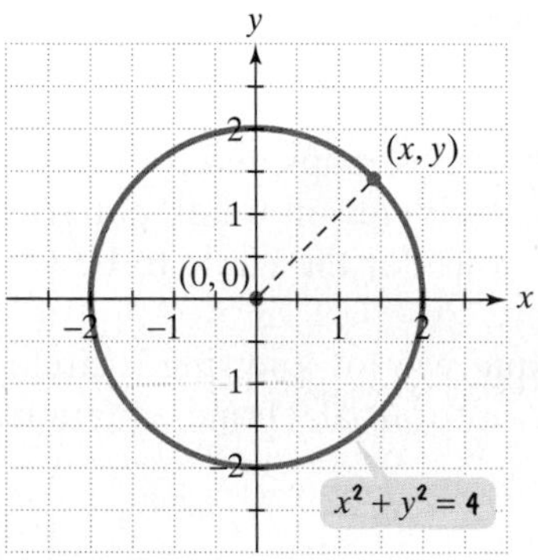

FIGURE 13.2 The graph of $x^2 + y^2 = 4$

SOLUTION The center is $(0, 0)$. Because the center is represented as (h, k) in the standard form of the equation, $h = 0$ and $k = 0$. The radius is 2, so we will let $r = 2$ in the equation.

$(x-h)^2 + (y-k)^2 = r^2$ This is the standard form of a circle's equation.

$(x-0)^2 + (y-0)^2 = 2^2$ Substitute 0 for h, 0 for k, and 2 for r.

$x^2 + y^2 = 4$ Simplify.

The standard form of the equation of the circle is $x^2 + y^2 = 4$. Figure 13.2 shows the graph. ■

CHECK POINT 1 Write the standard form of the equation of the circle with center $(0, 0)$ and radius 4.

USING TECHNOLOGY

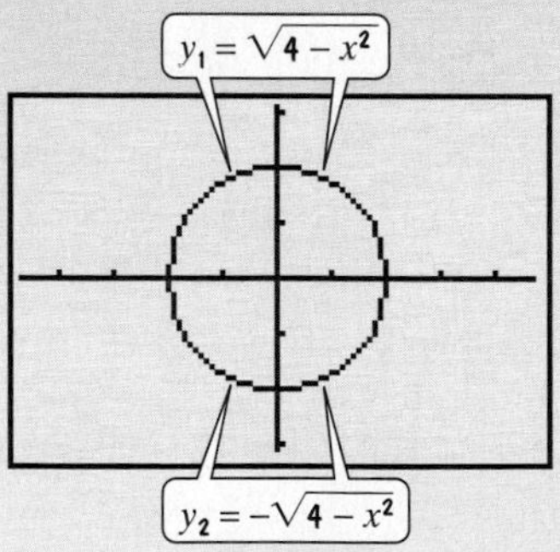

To graph a circle with a graphing utility, first solve the equation for y.

$$x^2 + y^2 = 4$$
$$y^2 = 4 - x^2$$
$$y = \pm\sqrt{4 - x^2}$$

Graph the two equations

$$y_1 = \sqrt{4 - x^2} \quad \text{and} \quad y_2 = -\sqrt{4 - x^2}$$

in the same viewing rectangle. The graph of $y_1 = \sqrt{4 - x^2}$ is the top semicircle because y is always positive. The graph of $y_2 = -\sqrt{4 - x^2}$ is the bottom semicircle because y is always negative. Use a ZOOM SQUARE setting so that the circle looks like a circle. (Many graphing utilities have problems connecting the two semicircles because the segments directly to the left and to the right of the center become nearly vertical.)

Example 1 and Check Point 1 involved circles centered at the origin. The standard form of the equation of all such circles is $x^2 + y^2 = r^2$, where r is the circle's radius. Now, let's consider a circle whose center is not at the origin.

EXAMPLE 2 Finding the Standard Form of a Circle's Equation

Write the standard form of the equation of the circle with center $(-2, 3)$ and radius 4.

SOLUTION The center is $(-2, 3)$. Because the center is represented as (h, k) in the standard form of the equation, $h = -2$ and $k = 3$. The radius is 4, so we will let $r = 4$ in the equation.

$$(x - h)^2 + (y - k)^2 = r^2$$ This is the standard form of a circle's equation.

$$[x - (-2)]^2 + (y - 3)^2 = 4^2$$ Substitute −2 for h, 3 for k, and 4 for r.

$$(x + 2)^2 + (y - 3)^2 = 16$$ Simplify.

The standard form of the equation of the circle is $(x + 2)^2 + (y - 3)^2 = 16$. ■

✔ **CHECK POINT 2** Write the standard form of the equation of the circle with center $(5, -6)$ and radius 10.

2 Give the center and radius of a circle whose equation is in standard form.

EXAMPLE 3 Using the Standard Form of a Circle's Equation to Graph the Circle

Find the center and radius of the circle whose equation is

$$(x - 2)^2 + (y + 4)^2 = 9$$

and graph the equation.

SOLUTION To graph the circle, we have to know its center, (h, k), and its radius, r. We can find the values for h, k, and r by comparing the given equation to the standard form of the equation of a circle, $(x - h)^2 + (y - k)^2 = r^2$.

$$(x - 2)^2 + (y + 4)^2 = 9$$

$$(x - 2)^2 + (y - (-4))^2 = 3^2$$

This is $(x - h)^2$, with $h = 2$. This is $(y - k)^2$, with $k = -4$. This is r^2, with $r = 3$.

We see that $h = 2$, $k = -4$, and $r = 3$. Thus, the circle has center $(h, k) = (2, -4)$ and a radius of 3 units. To graph this circle, first plot the center $(2, -4)$. Because the radius is 3, you can locate at least four points on the circle by going out three units to the right, to the left, up, and down from the center.

The points three units to the right and to the left of $(2, -4)$ are $(5, -4)$ and $(-1, -4)$, respectively. The points three units up and down from $(2, -4)$ are $(2, -1)$ and $(2, -7)$, respectively.

Using these points, we obtain the graph in Figure 13.3. ■

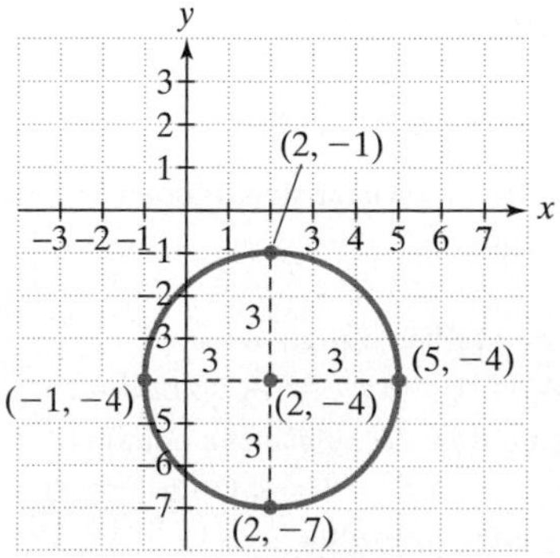

FIGURE 13.3 The graph of $(x - 2)^2 + (y + 4)^2 = 9$

CHECK POINT 3 Find the center and radius of the circle whose equation is

$$(x + 3)^2 + (y - 1)^2 = 4$$

and graph the equation.

If we square $x - 2$ and $y + 4$ in the standard form of the equation of Example 3, we obtain another form for the circle's equation.

$(x - 2)^2 + (y + 4)^2 = 9$	This is the standard form of the equation from Example 3.
$x^2 - 4x + 4 + y^2 + 8y + 16 = 9$	Square $x - 2$ and $y + 4$.
$x^2 + y^2 - 4x + 8y + 20 = 9$	Combine numerical terms and rearrange terms.
$x^2 + y^2 - 4x + 8y + 11 = 0$	Subtract 9 from both sides.

This result suggests that an equation in the form $x^2 + y^2 + Dx + Ey + F = 0$ can represent a circle. This is called the *general form of the equation of a circle*.

THE GENERAL FORM OF THE EQUATION OF A CIRCLE The **general form of the equation of a circle** is

$$x^2 + y^2 + Dx + Ey + F = 0.$$

3 Convert the general form of a circle's equation to standard form.

We can convert the general form of the equation of a circle to the standard form $(x - h)^2 + (y - k)^2 = r^2$. We do so by completing the square on x and y. Let's see how this is done.

EXAMPLE 4 Converting the General Form of a Circle's Equation to Standard Form and Graphing the Circle

Write in standard form and graph: $x^2 + y^2 + 4x - 6y - 23 = 0$.

SOLUTION Because we plan to complete the square on both x and y, let's rearrange the terms so that x-terms are arranged in descending order, y-terms are arranged in descending order, and the constant term appears on the right.

$x^2 + y^2 + 4x - 6y - 23 = 0$	This is the given equation.
$(x^2 + 4x \quad\;) + (y^2 - 6y \quad\;) = 23$	Rewrite in anticipation of completing the square.
$(x^2 + 4x + 4) + (y^2 - 6y + 9) = 23 + 4 + 9$	Complete the square on x: $\frac{1}{2} \cdot 4 = 2$ and $2^2 = 4$, so add 4 to both sides. Complete the square on y: $\frac{1}{2}(-6) = -3$ and $(-3)^2 = 9$, so add 9 to both sides.
$(x + 2)^2 + (y - 3)^2 = 36$	Factor on the left and add on the right.

Remember that numbers added on the left side must also be added on the right side.

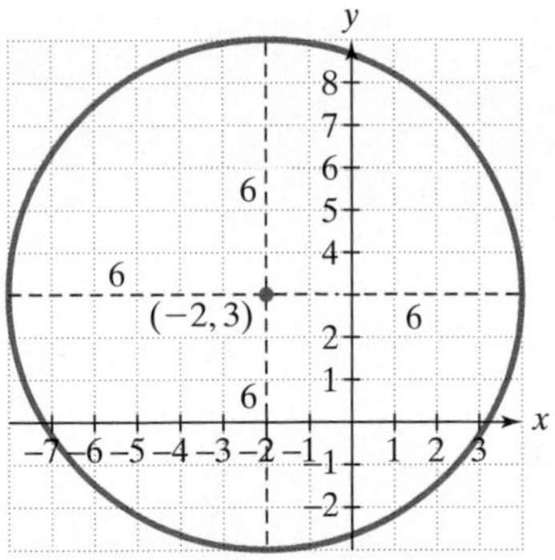

FIGURE 13.4 The graph of $(x + 2)^2 + (y - 3)^2 = 36$

This last equation is in standard form. We can identify the circle's center and radius by comparing this equation to the standard form of the equation of a circle, $(x - h)^2 + (y - k)^2 = r^2$.

$$(x + 2)^2 + (y - 3)^2 = 36$$

$$(x - (-2))^2 + (y - 3)^2 = 6^2$$

This is $(x - h)^2$, with $h = -2$. | This is $(y - k)^2$, with $k = 3$. | This is r^2, with $r = 6$.

We use the center, $(h, k) = (-2, 3)$, and the radius, $r = 6$, to graph the circle. The graph is shown in Figure 13.4. ■

USING TECHNOLOGY

To graph $x^2 + y^2 + 4x - 6y - 23 = 0$, rewrite the equation as a quadratic equation in y.

$$y^2 - 6y + (x^2 + 4x - 23) = 0$$

Now solve for y using the quadratic formula, with $a = 1, b = -6$, and $c = x^2 + 4x - 23$.

$$y = \frac{-b \pm \sqrt{b^2 - 4ac}}{2a} = \frac{-(-6) \pm \sqrt{(-6)^2 - 4 \cdot 1(x^2 + 4x - 23)}}{2 \cdot 1} = \frac{6 \pm \sqrt{36 - 4(x^2 + 4x - 23)}}{2}$$

Because we will enter these equations, there is no need to simplify further. Enter

$$y_1 = \frac{6 + \sqrt{36 - 4(x^2 + 4x - 23)}}{2}$$

and

$$y_2 = \frac{6 - \sqrt{36 - 4(x^2 + 4x - 23)}}{2}.$$

Use a ZOOM SQUARE setting. The graph is shown on the right.

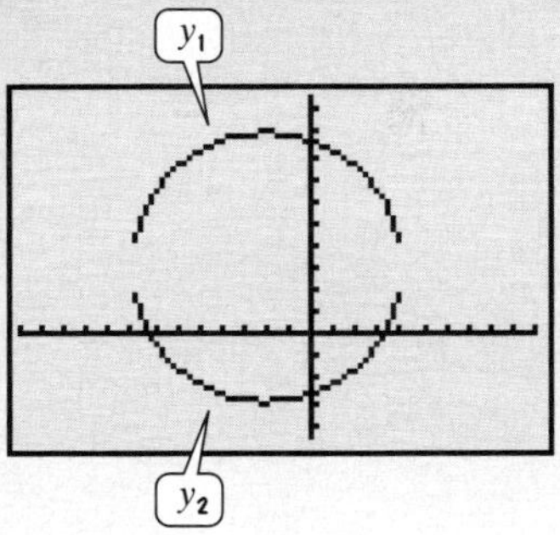

CHECK POINT 4 Write in standard form and graph:

$$x^2 + y^2 + 4x - 4y - 1 = 0.$$

13.1 EXERCISE SET

Student Solutions Manual | CD/Video | PH Math/Tutor Center | MathXL Tutorials on CD | MathXL® | MyMathLab | Interactmath.com

Practice Exercises

In Exercises 1–10, write the standard form of the equation of the circle with the given center and radius.

1. Center $(0, 0)$, $r = 7$

2. Center $(0, 0)$, $r = 8$

3. Center $(3, 2)$, $r = 5$

4. Center $(2, -1)$, $r = 4$

5. Center $(-1, 4)$, $r = 2$

6. Center $(-3, 5)$, $r = 3$

7. Center $(-3, -1)$, $r = \sqrt{3}$

8. Center $(-5, -3)$, $r = \sqrt{5}$

9. Center $(-4, 0)$, $r = 10$

10. Center $(-2, 0)$, $r = 6$

In Exercises 11–18, give the center and radius of the circle described by the equation and graph each equation.

11. $x^2 + y^2 = 16$

12. $x^2 + y^2 = 49$

13. $(x - 3)^2 + (y - 1)^2 = 36$

14. $(x - 2)^2 + (y - 3)^2 = 16$

15. $(x + 3)^2 + (y - 2)^2 = 4$

16. $(x + 1)^2 + (y - 4)^2 = 25$

17. $(x + 2)^2 + (y + 2)^2 = 4$

18. $(x + 4)^2 + (y + 5)^2 = 36$

In Exercises 19–26, complete the square and write the equation in standard form. Then give the center and radius of each circle and graph the equation.

19. $x^2 + y^2 + 6x + 2y + 6 = 0$

20. $x^2 + y^2 + 8x + 4y + 16 = 0$

21. $x^2 + y^2 - 10x - 6y - 30 = 0$

22. $x^2 + y^2 - 4x - 12y - 9 = 0$

23. $x^2 + y^2 + 8x - 2y - 8 = 0$

24. $x^2 + y^2 + 12x - 6y - 4 = 0$

25. $x^2 - 2x + y^2 - 15 = 0$

26. $x^2 + y^2 - 6y - 7 = 0$

Practice Plus

In Exercises 27–30, find the solution set for each system by graphing both of the system's equations in the same rectangular coordinate system and finding points of intersection. Check all solutions in both equations.

27. $x^2 + y^2 = 16$

$x - y = 4$

28. $x^2 + y^2 = 9$

$x - y = 3$

29. $(x - 2)^2 + (y + 3)^2 = 4$

$y = x - 3$

30. $(x - 3)^2 + (y + 1)^2 = 9$

$y = x - 1$

In Exercises 31–34, write the standard form of the equation of the circle with the given graph.

31.

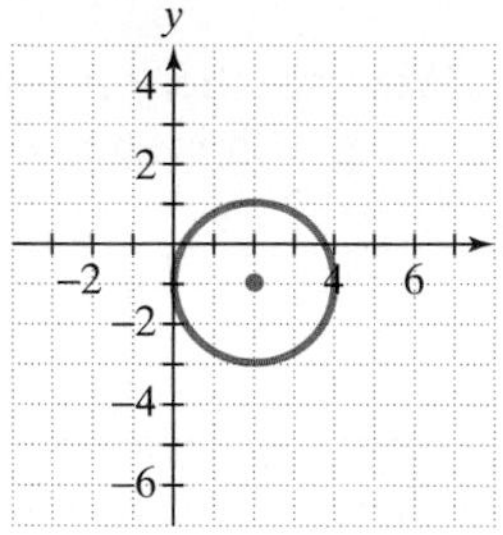

32.

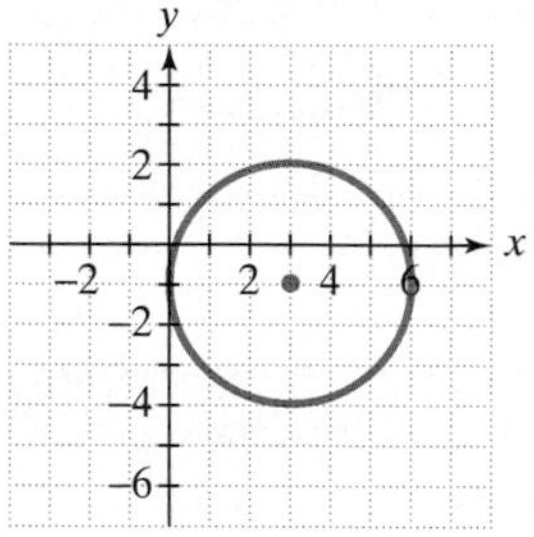

33.

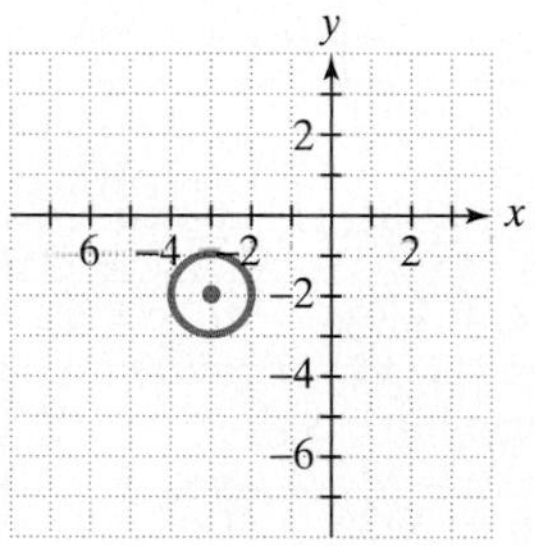

34.

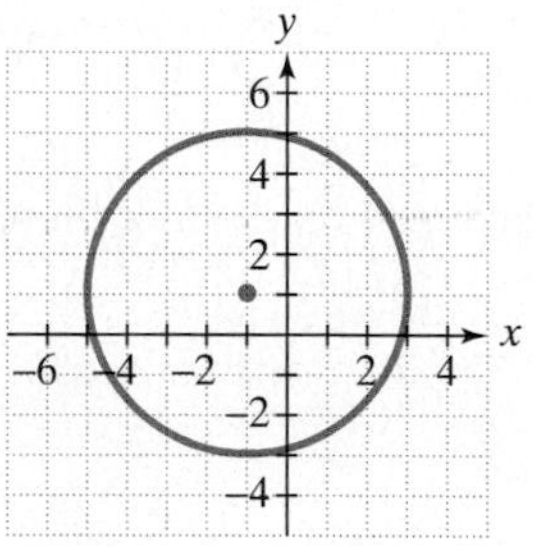

In Exercises 35–36, a line segment through the center of each circle intersects the circle at the points shown.

a. *Find the coordinates of the circle's center.*

b. *Find the radius of the circle.*

c. *Use your answers from parts (a) and (b) to write the standard form of the circle's equation.*

35.

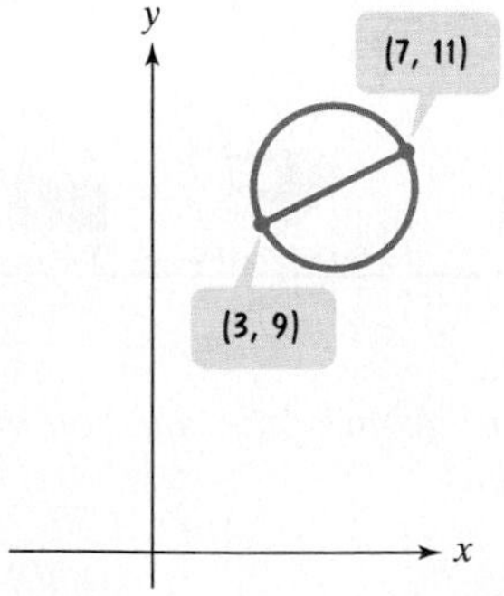

36.

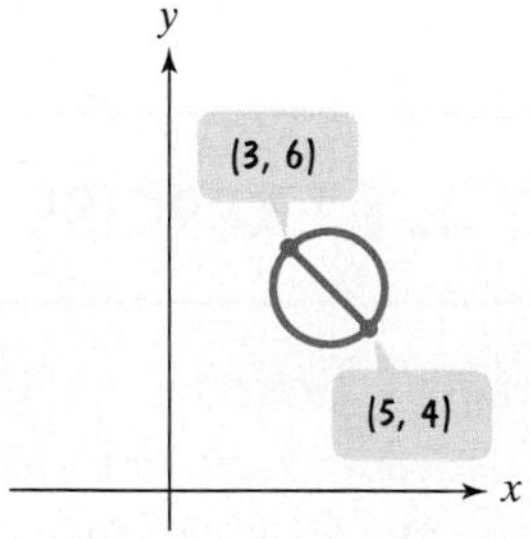

Application Exercises

37. A rectangular coordinate system with coordinates in miles is placed with the origin at the center of Los Angeles. The figure indicates that the University of Southern California is located 2.4 miles west and 2.7 miles south of central Los Angeles. A seismograph on the campus shows that a small earthquake occurred. The quake's epicenter is estimated to be approximately 30 miles from the university. Write the standard form of the equation for the set of points that could be the epicenter of the quake.

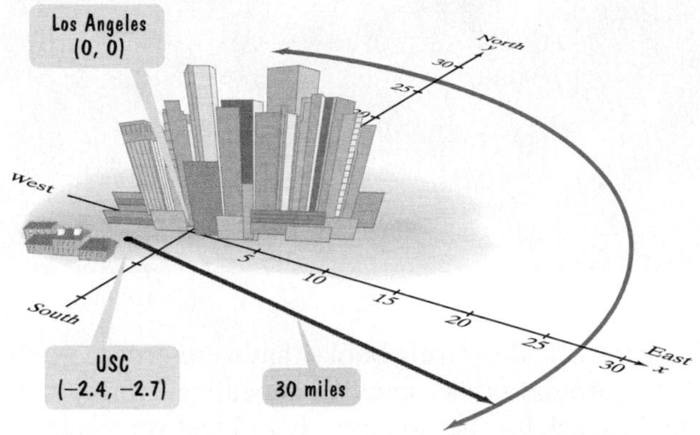

38. The ferris wheel in the figure has a radius of 68 feet. The clearance between the wheel and the ground is 14 feet. The rectangular coordinate system shown has its origin on the ground directly below the center of the wheel. Use the coordinate system to write the equation of the circular wheel.

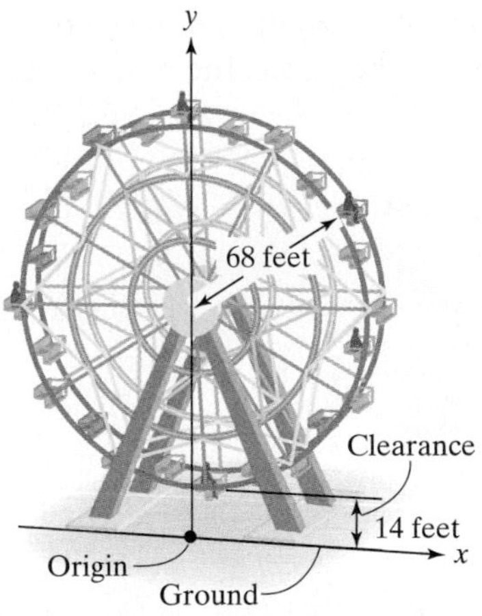

Writing in Mathematics

39. What is a circle? Without using variables, describe how the definition of a circle can be used to obtain a form of its equation.

40. Give an example of a circle's equation in standard form. Describe how to find the center and radius for this circle.

41. How is the standard form of a circle's equation obtained from its general form?

42. Does $(x - 3)^2 + (y - 5)^2 = 0$ represent the equation of a circle? If not, describe the graph of this equation.

43. Does $(x - 3)^2 + (y - 5)^2 = -25$ represent the equation of a circle? What sort of set is the graph of this equation?

Technology Exercises

In Exercises 44–46, use a graphing utility to graph each circle whose equation is given.

44. $x^2 + y^2 = 25$

45. $(y + 1)^2 = 36 - (x - 3)^2$

46. $x^2 + 10x + y^2 - 4y - 20 = 0$

Critical Thinking Exercises

47. Which one of the following is true?

a. The equation of the circle whose center is at the origin with radius 16 is $x^2 + y^2 = 16$.

b. The graph of $(x - 3)^2 + (y + 5)^2 = 36$ is a circle with radius 6 centered at $(-3, 5)$.

c. The graph of $(x - 4) + (y + 6) = 25$ is a circle with radius 5 centered at $(4, -6)$.

d. None of the above is true.

48. Find the area of the doughnut-shaped region bounded by the graphs of $(x - 2)^2 + (y + 3)^2 = 25$ and $(x - 2)^2 + (y + 3)^2 = 36$.

49. A **tangent line** to a circle is a line that intersects the circle at exactly one point. The tangent line is perpendicular to the radius of the circle at this point of contact. Write the point-slope equation of a line tangent to the circle whose equation is $x^2 + y^2 = 25$ at the point $(3, -4)$.

Review Exercises

50. If $f(x) = x^2 - 2$ and $g(x) = 3x + 4$, find $f(g(x))$ and $g(f(x))$. (Section 8.3, Example 1)

51. Solve: $2x = \sqrt{7x - 3} + 3$.
(Section 10.6, Example 3)

52. Solve: $|2x - 5| < 10$.
(Section 9.3, Example 3)

SECTION 13.2

Objectives

1 Graph ellipses centered at the origin.

2 Graph ellipses not centered at the origin.

3 Solve applied problems involving ellipses.

THE ELLIPSE

You took on a summer job driving a truck, delivering books that were ordered online. You're an avid reader, so just being around books sounded appealing. However, now you're feeling a bit shaky driving the truck for the first time. It's 10 feet wide and 9 feet high; compared to your compact car, it feels like you're behind the wheel of a tank. Up ahead you see a sign at the semielliptical entrance to a tunnel: Caution! Tunnel is 10 Feet High at Center Peak. Then you see another sign: Caution! Tunnel is 40 Feet Wide. Will your truck clear the opening of the tunnel's archway?

Mathematics is present in the movements of planets, bridge and tunnel construction, navigational systems used to keep track of a ship's location, manufacture of lenses for telescopes, and even in a procedure for disintegrating kidney stones. The mathematics behind these applications involves conic sections. **Conic sections** are curves that result from the intersection of a right circular cone and a plane. Figure 13.5 illustrates the four conic sections: the circle, the ellipse, the parabola, and the hyperbola.

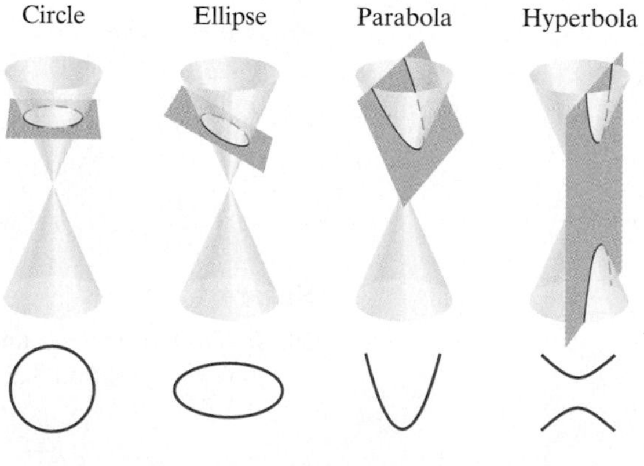

FIGURE 13.5 Obtaining the conic sections by intersecting a plane and a cone

In this section, we study the symmetric oval-shaped curve known as the ellipse. We will use a geometric definition for an ellipse to derive its equation. With this equation, we will determine if your delivery truck will clear the tunnel's entrance.

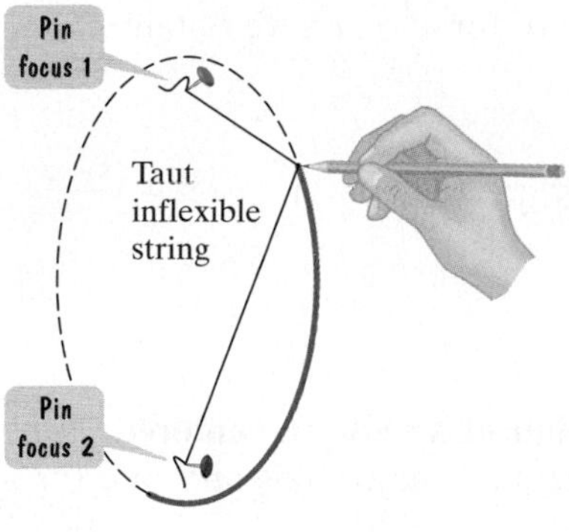

FIGURE 13.6 Drawing an ellipse

Definition of an Ellipse Figure 13.6 illustrates how to draw an ellipse. Place pins at two fixed points, each of which is called a focus (plural: foci). If the ends of a fixed length of string are fastened to the pins and we draw the string taut with a pencil, the path traced by the pencil will be an ellipse. Notice that the sum of the distances of the pencil point from the foci remains constant because the length of the string is fixed. This procedure for drawing an ellipse illustrates its geometric definition.

> DEFINITION OF AN ELLIPSE An **ellipse** is the set of all points, P, in a plane the sum of whose distances from two fixed points, F_1 and F_2, is constant. (See Figure 13.7.) These two fixed points are called the **foci** (plural of **focus**). The midpoint of the segment connecting the foci is the **center** of the ellipse.

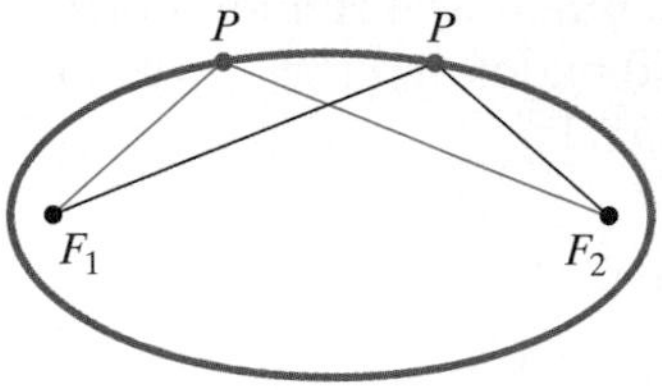

FIGURE 13.7

Figure 13.8 illustrates that an ellipse can be elongated in any direction. We will limit our discussion to ellipses that are elongated horizontally or vertically. The line through the foci intersects the ellipse at two points, called the **vertices** (singular: **vertex**). The line segment that joins the vertices is the **major axis**. Notice that the midpoint of the major axis is the center of the ellipse. The line segment whose endpoints are on the ellipse and that is perpendicular to the major axis at the center is called the **minor axis** of the ellipse.

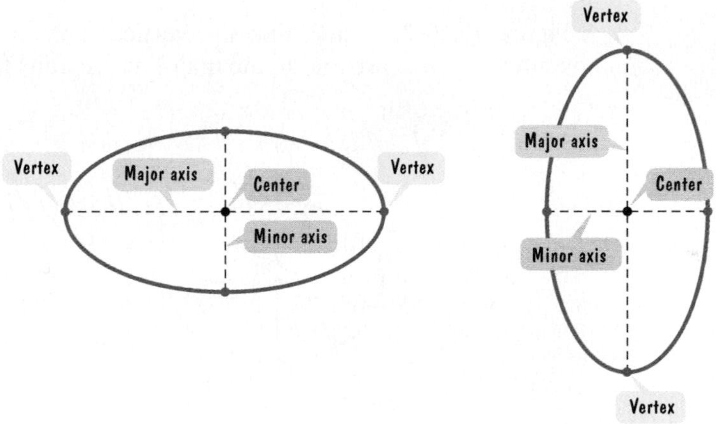

FIGURE 13.8 Horizontal and vertical elongations of an ellipse

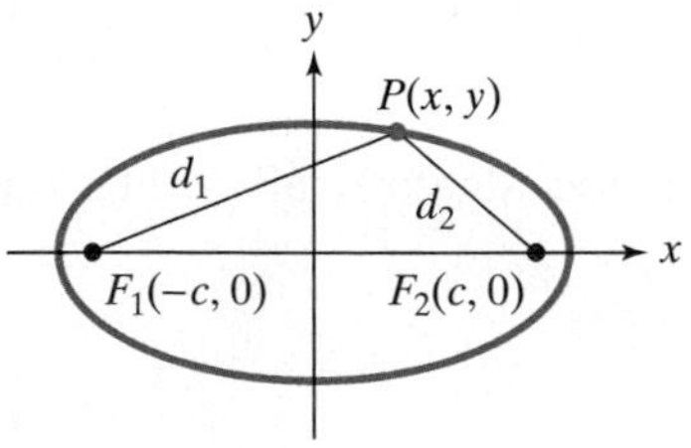

FIGURE 13.9

Standard Form of the Equation of an Ellipse The rectangular coordinate system gives us a unique way of describing an ellipse. It enables us to translate an ellipse's geometric definition into an algebraic equation.

We start with Figure 13.9 to obtain an ellipse's equation. We've placed an ellipse that is elongated horizontally into a rectangular coordinate system. The foci are on the x-axis at $(-c, 0)$ and $(c, 0)$, as in Figure 13.9. In this way, the center of the ellipse is at the origin. We let (x, y) represent the coordinates of any point, P, on the ellipse.

What does the definition of an ellipse tell us about the point (x, y) in Figure 13.9? For any point (x, y) on the ellipse, the sum of the distances to the two foci, $d_1 + d_2$, must be constant. As we shall see, it is convenient to denote this constant by $2a$. Thus, the point (x, y) is on the ellipse if and only if

$$d_1 + d_2 = 2a.$$

$$\sqrt{(x + c)^2 + y^2} + \sqrt{(x - c)^2 + y^2} = 2a \qquad \text{Use the distance formula.}$$

After eliminating radicals and simplifying, we obtain

$$(a^2 - c^2)x^2 + a^2y^2 = a^2(a^2 - c^2).$$

STUDY TIP

The algebraic details behind eliminating the radicals and obtaining the equation shown can be found in Appendix C. There you will find a step-by-step derivation of the ellipse's equation.

To simplify this equation, let $b^2 = a^2 - c^2$. Substituting b^2 for $a^2 - c^2$, we obtain

$$b^2x^2 + a^2y^2 = a^2b^2$$

$$\frac{b^2x^2}{a^2b^2} + \frac{a^2y^2}{a^2b^2} = \frac{a^2b^2}{a^2b^2} \quad \text{Divide both sides by } a^2b^2.$$

$$\frac{x^2}{a^2} + \frac{y^2}{b^2} = 1 \quad \text{Simplify.}$$

This last equation is the **standard form of the equation of an ellipse centered at the origin**. There are two such equations, one for a horizontal major axis and one for a vertical major axis.

STANDARD FORMS OF THE EQUATIONS OF AN ELLIPSE The **standard form of the equation of an ellipse** with center at the origin, and major and minor axes of lengths $2a$ and $2b$ (where a and b are positive, and $a^2 > b^2$) is

$$\frac{x^2}{a^2} + \frac{y^2}{b^2} = 1 \quad \text{or} \quad \frac{x^2}{b^2} + \frac{y^2}{a^2} = 1.$$

Figure 13.10 illustrates that the vertices are on the major axis, a units from the center. The foci are are on the major axis, c units from the center.

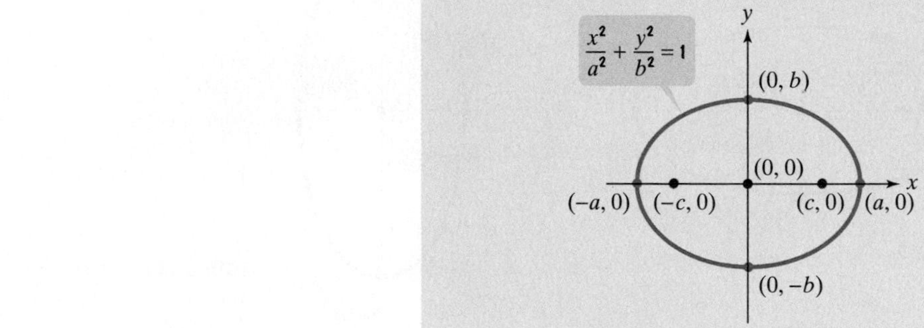

FIGURE 13.10 **(a)** Major axis is horizontal with length $2a$.

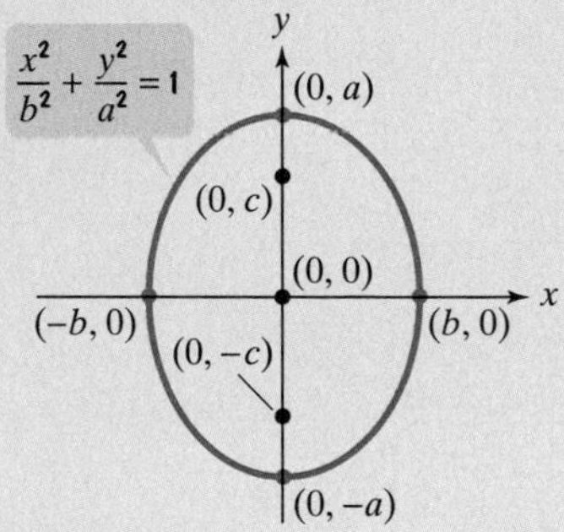

FIGURE 13.10 **(b)** Major axis is vertical with length $2a$.

The intercepts shown in Figure 13.10 can be obtained algebraically. Let's do this for

$$\frac{x^2}{a^2} + \frac{y^2}{b^2} = 1.$$

***x*-intercepts: Set $y = 0$.**	***y*-intercepts: Set $x = 0$.**
$\frac{x^2}{a^2} = 1$	$\frac{y^2}{b^2} = 1$
$x^2 = a^2$	$y^2 = b^2$
$x = \pm a$	$y = \pm b$
x-intercepts are $-a$ and a. The graph passes through $(-a, 0)$ and $(a, 0)$, which are the vertices.	y-intercepts are $-b$ and b. The graph passes through $(0, -b)$ and $(0, b)$.

1 Graph ellipses centered at the origin.

USING TECHNOLOGY

We graph $\frac{x^2}{9} + \frac{y^2}{4} = 1$ with a graphing utility by solving for y.

$$\frac{y^2}{4} = 1 - \frac{x^2}{9}$$

$$y^2 = 4\left(1 - \frac{x^2}{9}\right)$$

$$y = \pm 2\sqrt{1 - \frac{x^2}{9}}$$

Notice that the square root property requires us to define two functions. Enter

$y_1 = 2$ [√] (1 [−] x [∧] 2 [÷] 9)

and

$y_2 = -y_1$.

To see the true shape of the ellipse, use the [ZOOM SQUARE] feature so that one unit on the y-axis is the same length as one unit on the x-axis.

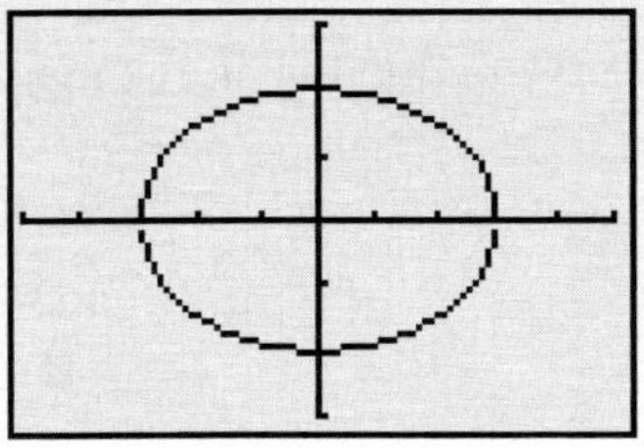

[−5, 5, 1] by [−3, 3, 1]

Using the Standard Form of the Equation of an Ellipse We can use the standard form of an ellipse's equation to graph the ellipse.

EXAMPLE 1 Graphing an Ellipse Centered at the Origin

Graph the ellipse: $\frac{x^2}{9} + \frac{y^2}{4} = 1$.

SOLUTION The given equation is the standard form of an ellipse's equation with $a^2 = 9$ and $b^2 = 4$.

$$\frac{x^2}{9} + \frac{y^2}{4} = 1$$

$a^2 = 9$. This is the larger of the two denominators.

$b^2 = 4$. This is the smaller of the two denominators.

Because the denominator of the x^2-term is greater than the denominator of the y^2-term, the major axis is horizontal. Based on the standard form of the equation, we know that the vertices are $(-a, 0)$ and $(a, 0)$. Because $a^2 = 9$, $a = 3$. Thus, the vertices are $(-3, 0)$ and $(3, 0)$, shown in Figure 13.11.

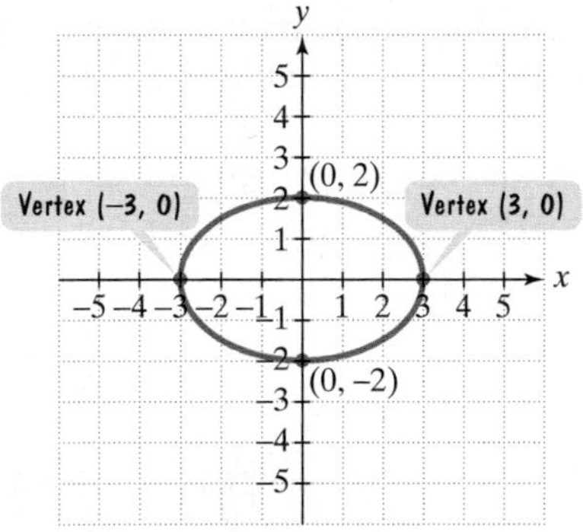

FIGURE 13.11 The graph of $\frac{x^2}{9} + \frac{y^2}{4} = 1$

Now let us find the endpoints of the vertical minor axis. According to the standard form of the equation, these endpoints are $(0, -b)$ and $(0, b)$. Because $b^2 = 4$, $b = 2$. Thus, the endpoints of the minor axis are $(0, -2)$ and $(0, 2)$. They are shown in Figure 13.11.

You can sketch the ellipse in Figure 13.11 by locating the endpoints on the major and minor axes.

$$\frac{x^2}{3^2} + \frac{y^2}{2^2} = 1$$

Endpoints of the major axis are 3 units to the right and left of the center.

Endpoints of the minor axis are 2 units up and down from the center.

■

✔ **CHECK POINT 1** Graph the ellipse: $\frac{x^2}{36} + \frac{y^2}{9} = 1$.

EXAMPLE 2 Graphing an Ellipse Centered at the Origin

Graph the ellipse: $25x^2 + 16y^2 = 400$.

SOLUTION We begin by expressing the equation in standard form. Because we want 1 on the right side, we divide both sides by 400.

$$\frac{25x^2}{400} + \frac{16y^2}{400} = \frac{400}{400}$$

$$\frac{x^2}{16} + \frac{y^2}{25} = 1$$

$b^2 = 16$. This is the smaller of the two denominators.

$a^2 = 25$. This is the larger of the two denominators.

The equation is the standard form of an ellipse's equation with $a^2 = 25$ and $b^2 = 16$. Because the denominator of the y^2-term is greater than the denominator of the x^2-term, the major axis is vertical. Based on the standard form of the equation, we know that the vertices are $(0, -a)$ and $(0, a)$. Because $a^2 = 25$, $a = 5$. Thus, the vertices are $(0, -5)$ and $(0, 5)$, shown in Figure 13.12.

Now let us find the endpoints of the horizontal minor axis. According to the standard form of the equation, these endpoints are $(-b, 0)$ and $(b, 0)$. Because $b^2 = 16$, $b = 4$. Thus, the endpoints of the minor axis are $(-4, 0)$ and $(4, 0)$. They are shown in Figure 13.12.

You can sketch the ellipse in Figure 13.12 by locating the endpoints on the major and minor axes.

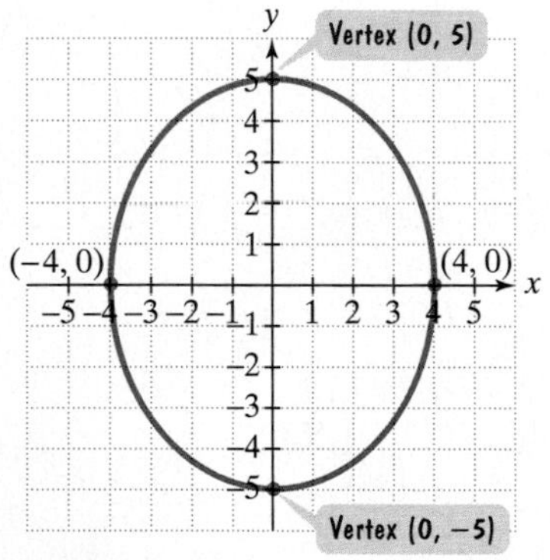

FIGURE 13.12 The graph of $\frac{x^2}{16} + \frac{y^2}{25} = 1$

$$\frac{x^2}{4^2} + \frac{y^2}{5^2} = 1$$

Endpoints of the minor axis are 4 units to the right and left of the center.

Endpoints of the major axis are 5 units up and down from the center.

■

CHECK POINT 2 Graph the ellipse: $16x^2 + 9y^2 = 144$.

2 Graph ellipses not centered at the origin.

Ellipses Centered at (h, k) Horizontal and vertical translations can be used to graph ellipses that are not centered at the origin. Figure 13.13 illustrates that the graphs of

$$\frac{(x-h)^2}{a^2} + \frac{(y-k)^2}{b^2} = 1 \quad \text{and} \quad \frac{x^2}{a^2} + \frac{y^2}{b^2} = 1$$

have the same size and shape. However, the graph of the first equation is centered at (h, k) rather than at the origin.

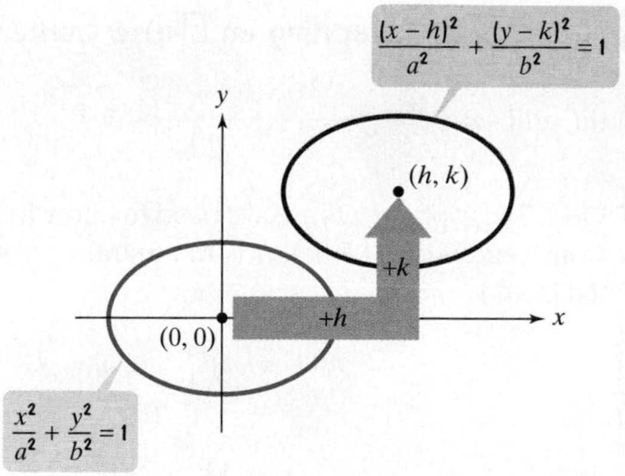

FIGURE 13.13 Translating an ellipse's graph

Table 13.1 gives the standard forms of equations of ellipses centered at (h, k) and shows their graphs.

Table 13.1 Standard Forms of Equations of Ellipses Centered at (h, k)

Equation	Center	Major Axis	Vertices	Graph
$\frac{(x-h)^2}{a^2} + \frac{(y-k)^2}{b^2} = 1$ Endpoints of major axis are a units right and a units left of center. $a^2 > b^2$	(h, k)	Parallel to the x-axis, horizontal	$(h - a, k)$ $(h + a, k)$	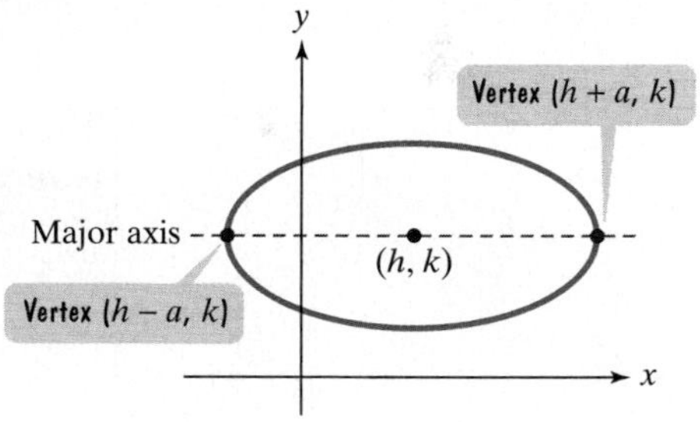
$\frac{(x-h)^2}{b^2} + \frac{(y-k)^2}{a^2} = 1$ $a^2 > b^2$ Endpoints of the major axis are a units above and a units below the center.	(h, k)	Parallel to the y-axis, vertical	$(h, k - a)$ $(h, k + a)$	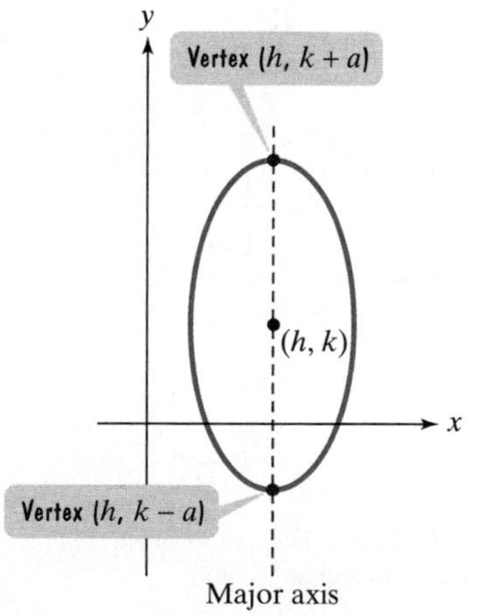

EXAMPLE 3 Graphing an Ellipse Centered at (h, k)

Graph the ellipse: $\frac{(x-1)^2}{4} + \frac{(y+2)^2}{9} = 1.$

SOLUTION To graph the ellipse, we need to know its center, (h, k). In the standard forms of equations centered at (h, k), h is the number subtracted from x and k is the number subtracted from y.

This is $(x-h)^2$ with $h = 1$.

This is $(y-k)^2$ with $k = -2$.

$$\frac{(x-1)^2}{4} + \frac{(y-(-2))^2}{9} = 1$$

We see that $h = 1$ and $k = -2$. Thus, the center of the ellipse, (h, k), is $(1, -2)$. We can graph the ellipse by locating endpoints on the major and minor axes. To do this, we must identify a^2 and b^2.

$$\frac{(x-1)^2}{4} + \frac{(y+2)^2}{9} = 1$$

$b^2 = 4$. This is the smaller of the two denominators.

$a^2 = 9$. This is the larger of the two denominators.

The larger number is under the expression involving y. This means that the major axis is vertical and parallel to the y-axis.

We can sketch the ellipse by locating endpoints on the major and minor axes.

$$\frac{(x-1)^2}{2^2} + \frac{(y+2)^2}{3^2} = 1$$

Endpoints of the minor axis are 2 units to the right and left of the center.

Endpoints of the major axis (the vertices) are 3 units up and down from the center.

We categorize the observations in the voice balloons as follows:

3 units above and below center

For a Vertical Major Axis with Center $(1, -2)$	
Vertices	**Endpoints of Minor Axis**
$(1, -2 + 3) = (1, 1)$	$(1 + 2, -2) = (3, -2)$
$(1, -2 - 3) = (1, -5)$	$(1 - 2, -2) = (-1, -2)$

2 units right and left of center

Using the center and these four points, we can sketch the ellipse shown in Figure 13.14. ■

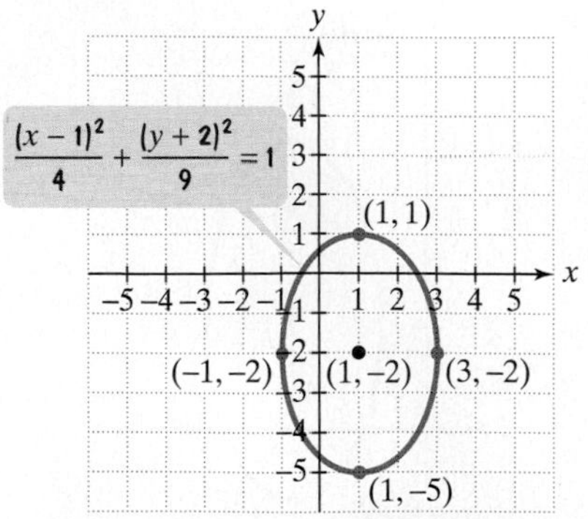

FIGURE 13.14 The graph of an ellipse centered at $(1, -2)$

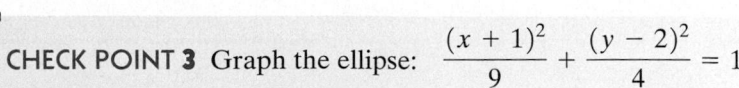

CHECK POINT 3 Graph the ellipse: $\frac{(x+1)^2}{9} + \frac{(y-2)^2}{4} = 1.$

3 Solve applied problems involving ellipses.

Applications Ellipses have many applications. German scientist Johannes Kepler (1571–1630) showed that the planets in our solar system move in elliptical orbits, with the sun at a focus. Earth satellites also travel in elliptical orbits, with Earth at a focus.

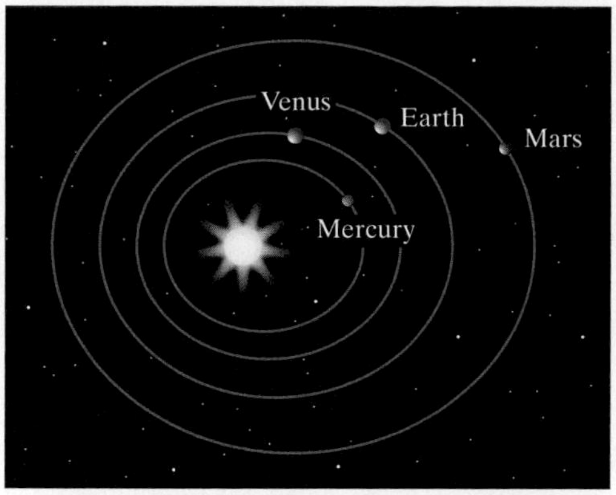

Planets move in elliptical orbits.

One intriguing aspect of the ellipse is that a ray of light or a sound wave emanating from one focus will be reflected from the ellipse to exactly the other focus. A whispering gallery is an elliptical room with an elliptical, dome-shaped ceiling. People standing at the foci can whisper and hear each other quite clearly, while persons in other locations in the room cannot hear them. Statuary Hall in the U.S. Capitol Building is elliptical. President John Quincy Adams, while a member of the House of Representatives, was aware of this acoustical phenomenon. He situated his desk at a focal point of the elliptical ceiling, easily eavesdropping on the private conversations of other House members located near the other focus.

The elliptical reflection principle is used in a procedure for disintegrating kidney stones. The patient is placed within a device that is elliptical in shape. The patient is placed so the kidney is centered at one focus, while ultrasound waves from the other focus hit the walls and are reflected to the kidney stone. The convergence of the ultrasound waves at the kidney stone causes vibrations that shatter it into fragments. The small pieces can then be passed painlessly through the patient's system. The patient recovers in days, as opposed to up to six weeks if surgery is used instead.

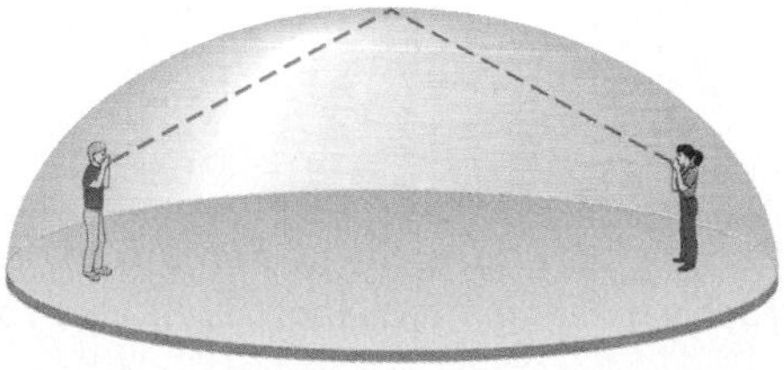

Whispering in an elliptical dome

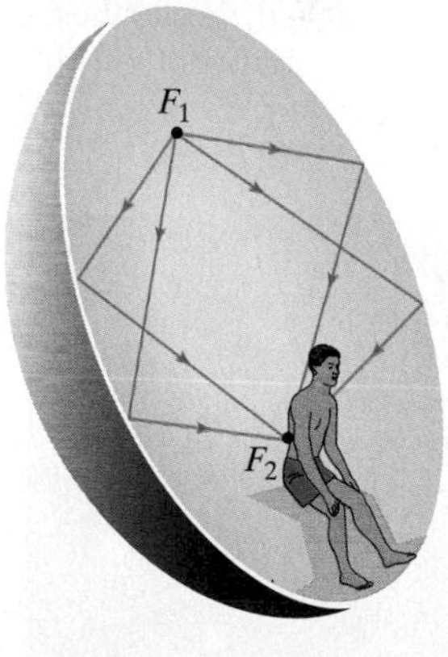

Disintegrating kidney stones

Ellipses are often used for supporting arches of bridges and in tunnel construction. This application forms the basis of our next example.

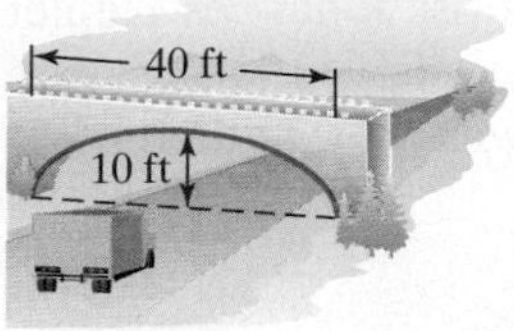

FIGURE 13.15 A semielliptical archway

ENRICHMENT ESSAY

Halley's Comet

Halley's Comet has an elliptical orbit with the sun at one focus. The comet returns every 76.3 years. The first recorded sighting was in 239 B.C. It was last seen in 1986. At that time, spacecraft went close to the comet, measuring its nucleus to be 7 miles long and 4 miles wide. By 2024, Halley's Comet will have reached the farthest point in its elliptical orbit before returning to be next visible from Earth in 2062.

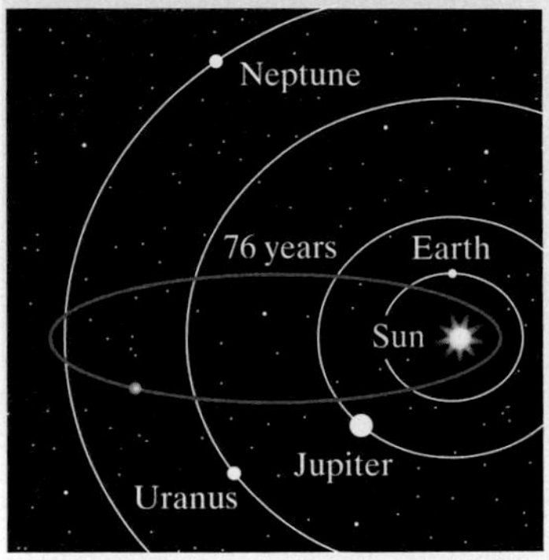

The elliptical orbit of Halley's Comet

EXAMPLE 4 An Application Involving an Ellipse

A semielliptical archway over a one-way road has a height of 10 feet and a width of 40 feet (see Figure 13.15). Your truck has a width of 10 feet and a height of 9 feet. Will your truck clear the opening of the archway?

SOLUTION Because your truck's width is 10 feet, to determine the clearance, we must find the height of the archway 5 feet from the center. If that height is 9 feet or less, the truck will not clear the opening.

In Figure 13.16, we've constructed a coordinate system with the x-axis on the ground and the origin at the center of the archway. Also shown is the truck, whose height is 9 feet.

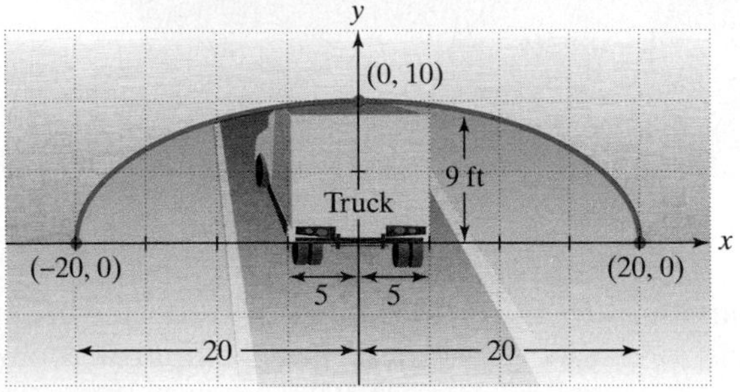

FIGURE 13.16

Using the equation $\frac{x^2}{a^2} + \frac{y^2}{b^2} = 1$, we can express the equation of the blue archway in Figure 13.16 as $\frac{x^2}{20^2} + \frac{y^2}{10^2} = 1$, or $\frac{x^2}{400} + \frac{y^2}{100} = 1$.

As shown in Figure 13.16, the edge of the 10-foot-wide truck corresponds to $x = 5$. We find the height of the archway 5 feet from the center by substituting 5 for x and solving for y.

$$\frac{5^2}{400} + \frac{y^2}{100} = 1 \quad \text{Substitute 5 for } x \text{ in } \frac{x^2}{400} + \frac{y^2}{100} = 1.$$

$$\frac{25}{400} + \frac{y^2}{100} = 1 \quad \text{Square 5.}$$

$$400\left(\frac{25}{400} + \frac{y^2}{100}\right) = 400(1) \quad \text{Clear fractions by multiplying both sides by 400.}$$

$$25 + 4y^2 = 400 \quad \text{Use the distributive property and simplify.}$$

$$4y^2 = 375 \quad \text{Subtract 25 from both sides.}$$

$$y^2 = \frac{375}{4} \quad \text{Divide both sides by 4.}$$

$$y = \sqrt{\frac{375}{4}} \quad \text{Take only the positive square root. The archway is above the } x\text{-axis, so } y \text{ is nonnegative.}$$

$$\approx 9.68 \quad \text{Use a calculator.}$$

Thus, the height of the archway 5 feet from the center is approximately 9.68 feet. Because your truck's height is 9 feet, there is enough room for the truck to clear the archway. ■

CHECK POINT 4 Will a truck that is 12 feet wide and has a height of 9 feet clear the opening of the archway described in Example 4?

13.2 EXERCISE SET

Student Solutions Manual | CD/Video | PH Math/Tutor Center | MathXL Tutorials on CD | MathXL® | MyMathLab | Interactmath.com

Practice Exercises

In Exercises 1–16, graph each ellipse.

1. $\frac{x^2}{16} + \frac{y^2}{4} = 1$ **2.** $\frac{x^2}{25} + \frac{y^2}{16} = 1$

3. $\frac{x^2}{9} + \frac{y^2}{36} = 1$ **4.** $\frac{x^2}{16} + \frac{y^2}{49} = 1$

5. $\frac{x^2}{25} + \frac{y^2}{64} = 1$ **6.** $\frac{x^2}{49} + \frac{y^2}{36} = 1$

7. $\frac{x^2}{49} + \frac{y^2}{81} = 1$ **8.** $\frac{x^2}{64} + \frac{y^2}{100} = 1$

9. $25x^2 + 4y^2 = 100$ **10.** $9x^2 + 4y^2 = 36$

11. $4x^2 + 16y^2 = 64$ **12.** $16x^2 + 9y^2 = 144$

13. $25x^2 + 9y^2 = 225$ **14.** $4x^2 + 25y^2 = 100$

15. $x^2 + 2y^2 = 8$ **16.** $12x^2 + 4y^2 = 36$

In Exercises 17–20, find the standard form of the equation of each ellipse.

17.

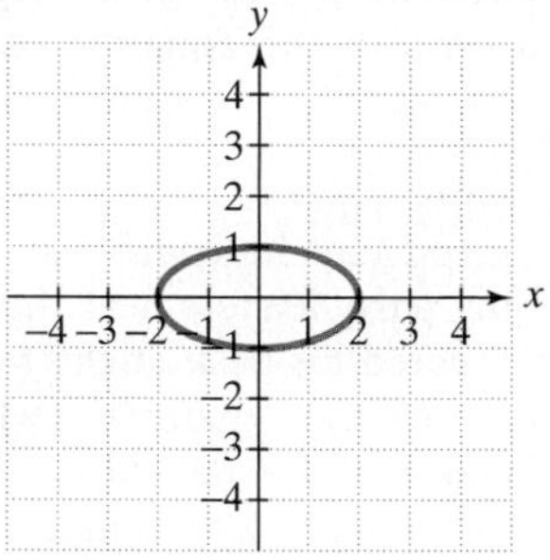

18.

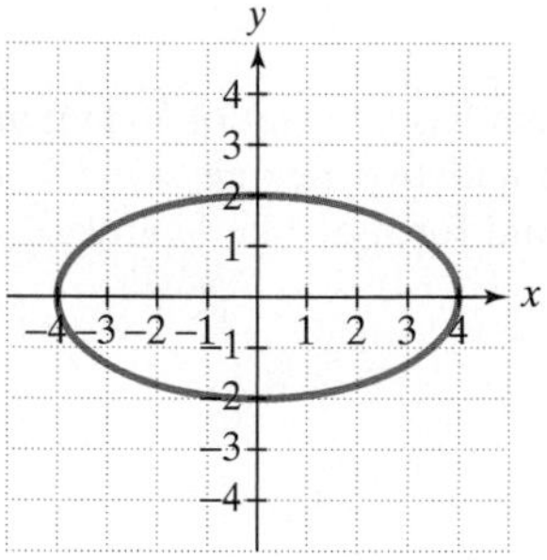

19.

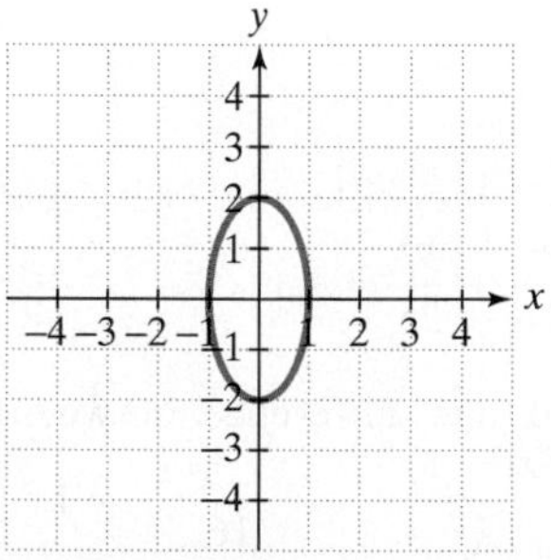

20.

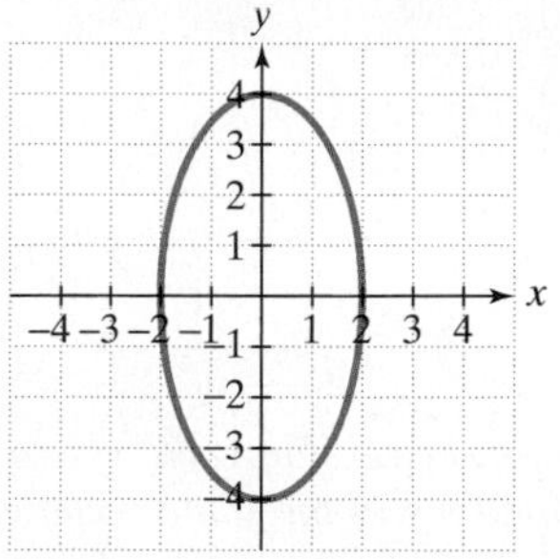

In Exercises 21–32, graph each ellipse.

21. $\frac{(x-2)^2}{9} + \frac{(y-1)^2}{4} = 1$

22. $\frac{(x-1)^2}{16} + \frac{(y+2)^2}{9} = 1$

23. $(x+3)^2 + 4(y-2)^2 = 16$

24. $(x-3)^2 + 9(y+2)^2 = 36$

25. $\frac{(x-4)^2}{9} + \frac{(y+2)^2}{25} = 1$

26. $\frac{(x-3)^2}{9} + \frac{(y+1)^2}{16} = 1$

27. $\frac{x^2}{25} + \frac{(y-2)^2}{36} = 1$

28. $\frac{(x-4)^2}{4} + \frac{y^2}{25} = 1$

29. $\frac{(x+3)^2}{9} + (y-2)^2 = 1$

30. $\frac{(x+2)^2}{16} + (y-3)^2 = 1$

31. $9(x-1)^2 + 4(y+3)^2 = 36$

32. $36(x+4)^2 + (y+3)^2 = 36$

Practice Plus

In Exercises 33–34, find the standard form of the equation of each ellipse.

33.

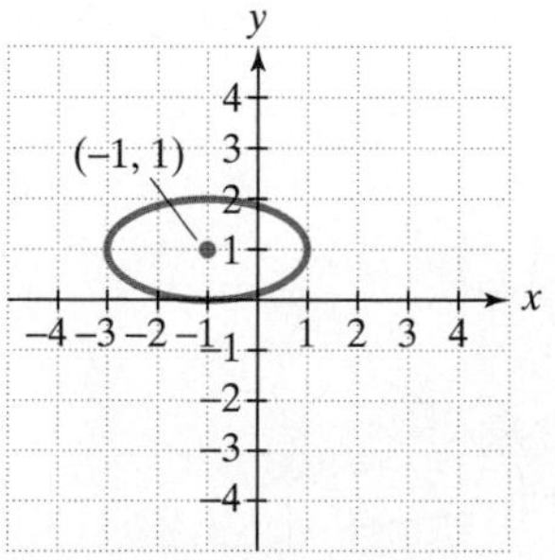

34.

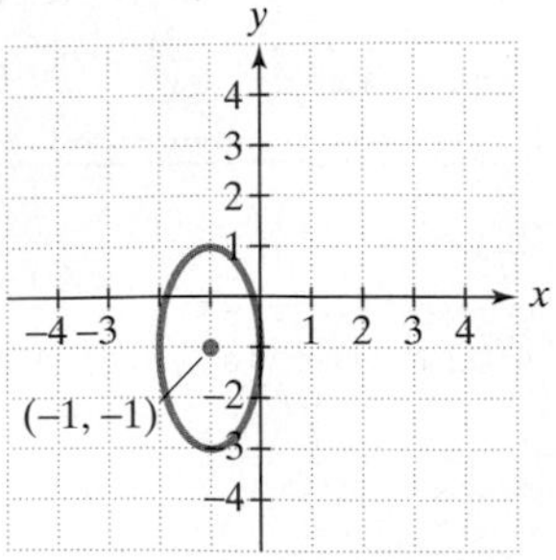

In Exercises 35–40, find the solution set for each system by graphing both of the system's equations in the same rectangular coordinate system and finding points of intersection. Check all solutions in both equations.

35. $x^2 + y^2 = 1$
$x^2 + 9y^2 = 9$

36. $x^2 + y^2 = 25$
$25x^2 + y^2 = 25$

37. $\frac{x^2}{25} + \frac{y^2}{9} = 1$
$y = 3$

38. $\frac{x^2}{4} + \frac{y^2}{36} = 1$
$x = -2$

39. $4x^2 + y^2 = 4$
$2x - y = 2$

40. $4x^2 + y^2 = 4$
$x + y = 3$

In Exercises 41–42, graph each semiellipse.

41. $y = -\sqrt{16 - 4x^2}$

42. $y = -\sqrt{4 - 4x^2}$

Application Exercises

43. Will a truck that is 8 feet wide carrying a load that reaches 7 feet above the ground clear the semielliptical arch on the one-way road that passes under the bridge shown in the figure?

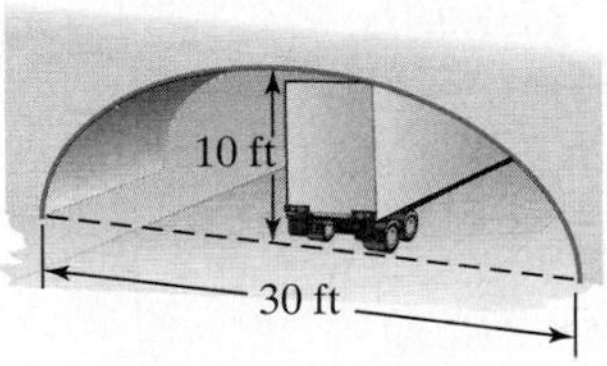

44. A semielliptic archway has a height of 20 feet and a width of 50 feet, as shown in the figure. Can a truck 14 feet high and 10 feet wide drive under the archway without going into the other lane?

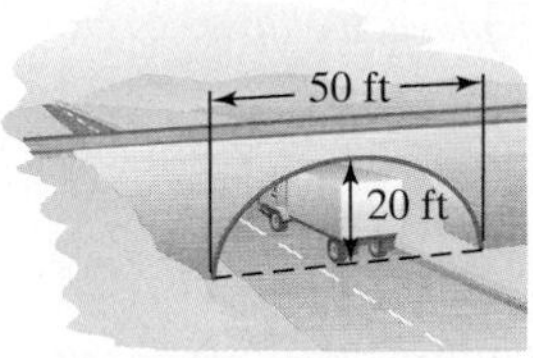

45. The elliptical ceiling in Statuary Hall in the U.S. Capitol Building is 96 feet long and 23 feet tall.

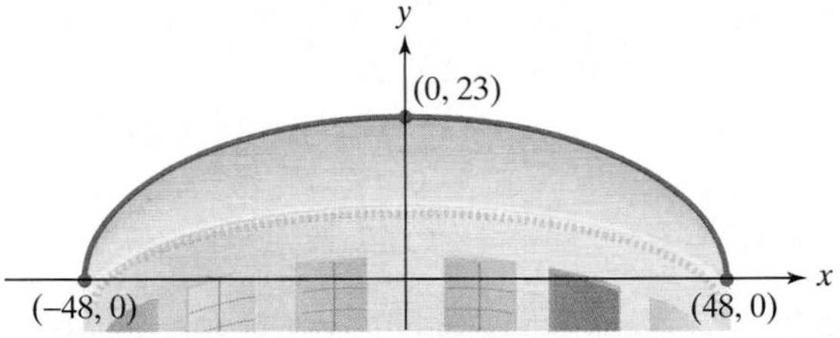

a. Using the rectangular coordinate system in the figure shown, write the standard form of the equation of the elliptical ceiling.

b. John Quincy Adams discovered that he could overhear the conversations of opposing party leaders near the left side of the chamber if he situated his desk at the focus, $(c, 0)$, at the right side of the chamber, where $c^2 = a^2 - b^2$. How far from the center of the ellipse along the major axis did Adams situate his desk? (Round to the nearest foot.)

46. If an elliptical whispering room has a height of 30 feet and a width of 100 feet, where should two people stand if they would like to whisper back and forth and be heard?

Writing in Mathematics

47. What is an ellipse?

48. Describe how to graph $\frac{x^2}{25} + \frac{y^2}{16} = 1$.

49. Describe one similarity and one difference between the graphs of $\frac{x^2}{25} + \frac{y^2}{16} = 1$ and $\frac{x^2}{16} + \frac{y^2}{25} = 1$.

50. Describe one similarity and one difference between the graphs of $\frac{x^2}{25} + \frac{y^2}{16} = 1$ and $\frac{(x-1)^2}{25} + \frac{(y-1)^2}{16} = 1$.

51. An elliptipool is an elliptical pool table with only one pocket. A pool shark places a ball on the table, hits it in what appears to be a random direction, and yet it bounces off the edge, falling directly into the pocket. Explain why this happens.

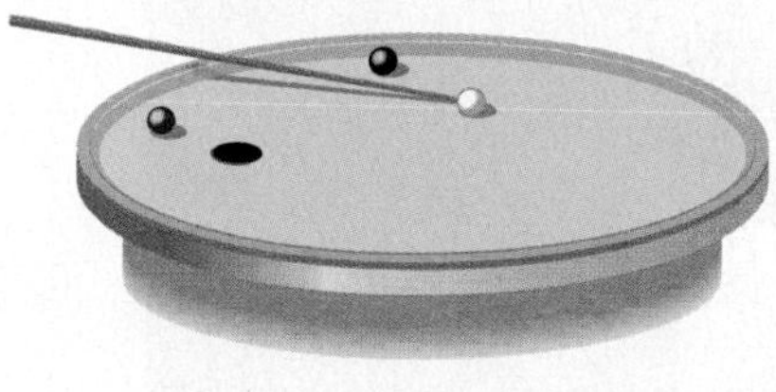

Technology Exercises

52. Use a graphing utility to graph any five of the ellipses that you graphed by hand in Exercises 1–16.

53. Use a graphing utility to graph any three of the ellipses that you graphed by hand in Exercises 21–32. First solve the given equation for y by using the square root property. Enter each of the two resulting equations to produce each half of the ellipse.

Critical Thinking Exercises

54. Which one of the following is true?

a. The graphs of $x^2 + y^2 = 16$ and $\frac{x^2}{4} + \frac{y^2}{9} = 1$ do not intersect.

b. Some ellipses have equations that define y as a function of x.

c. The graph of $\frac{x^2}{9} + \frac{y^2}{4} = 0$ is an ellipse.

d. None of the above is true.

55. Find the standard form of the equation of an ellipse with vertices at $(0, -6)$ and $(0, 6)$, passing through $(2, -4)$.

In Exercises 56–57, convert each equation to standard form by completing the square on x and y. Then graph the ellipse.

56. $9x^2 + 25y^2 - 36x + 50y - 164 = 0$

57. $4x^2 + 9y^2 - 32x + 36y + 64 = 0$

58. An Earth satellite has an elliptical orbit described by

$$\frac{x^2}{(5000)^2} + \frac{y^2}{(4750)^2} = 1.$$

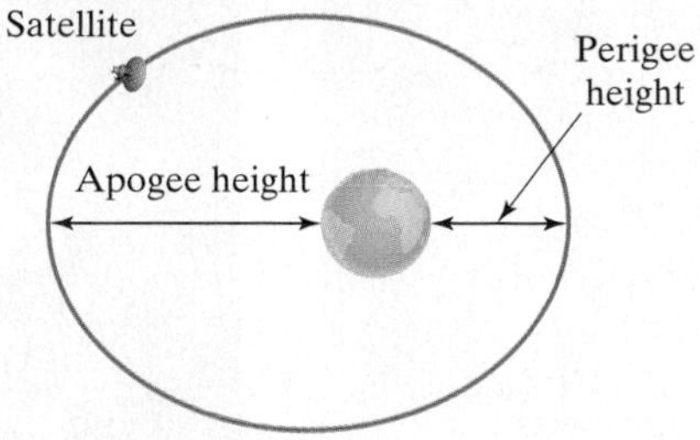

(All units are in miles.) The coordinates of the center of Earth are $(16, 0)$.

a. The perigee of the satellite's orbit is the point that is nearest Earth's center. If the radius of Earth is approximately 4000 miles, find the distance of the perigee above Earth's surface.

b. The apogee of the satellite's orbit is the point that is the greatest distance from Earth's center. Find the distance of the apogee above Earth's surface.

59. The equation of the red ellipse in the figure shown is

$$\frac{x^2}{25} + \frac{y^2}{9} = 1.$$

Write the equation for each circle shown in the figure.

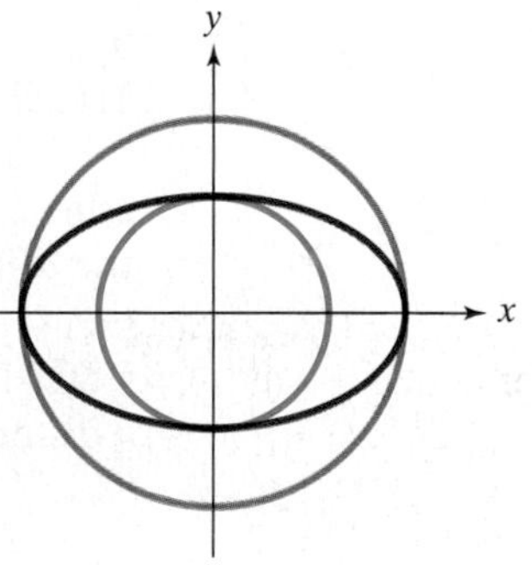

Review Exercises

60. Factor completely: $x^3 + 2x^2 - 4x - 8$. (Section 6.5, Example 4)

61. Simplify: $\sqrt[3]{40x^4y^7}$. (Section 10.3, Example 5)

62. Solve: $\frac{2}{x+2} + \frac{4}{x-2} = \frac{x-1}{x^2-4}$. (Section 7.6, Example 5)

SECTION 13.3

Objectives

1. Locate a hyperbola's vertices.
2. Graph hyperbolas centered at the origin.

THE HYPERBOLA

St. Mary's Cathedral

Conic sections are often used to create unusual architectural designs. The top of St. Mary's Cathedral in San Francisco is a 2135-cubic-foot dome with walls rising 200 feet above the floor and supported by four massive concrete pylons that extend 94 feet into the ground. Cross sections of the roof are parabolas and hyperbolas. In this section, we study the curve with two parts known as the hyperbola.

FIGURE 13.17 Casting hyperbolic shadows

Definition of a Hyperbola Figure 13.17 shows a cylindrical lampshade casting two shadows on a wall. These shadows indicate the distinguishing feature of hyperbolas: Their graphs contain two disjoint parts, called **branches**. Although each branch might look like a parabola, its shape is actually quite different.

The definition of a hyperbola is similar to the definition of an ellipse. For an ellipse, the *sum* of the distances from the foci is a constant. By contrast, for a hyperbola, the *difference* of the distances from the foci is a constant.

DEFINITION OF A HYPERBOLA A **hyperbola** is the set of points in a plane the difference of whose distances from two fixed points, called foci, is constant.

Figure 13.18 illustrates the two branches of a hyperbola. The line through the foci intersects the hyperbola at two points, called the **vertices**. The line segment that joins the vertices is the **transverse axis**. The midpoint of the transverse axis is the **center** of the hyperbola. Notice that the center lies midway between the vertices, as well as midway between the foci.

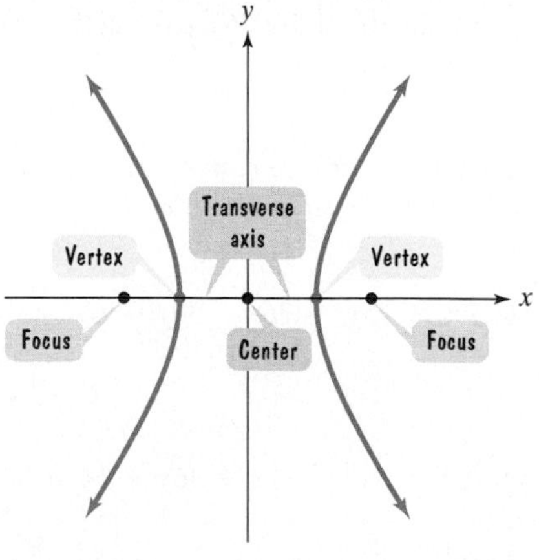

FIGURE 13.18 The two branches of a hyperbola

Standard Form of the Equation of a Hyperbola The rectangular coordinate system enables us to translate a hyperbola's geometric definition into an algebraic equation.

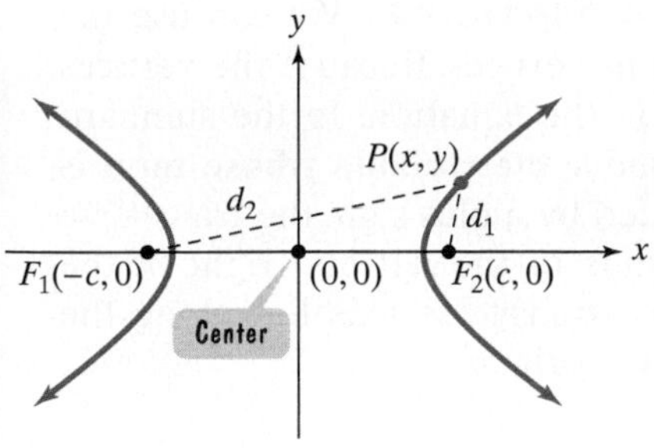

FIGURE 13.19

Figure 13.19 is our starting point for obtaining an equation. We place the foci, F_1 and F_2, on the x-axis at the points $(-c, 0)$ and $(c, 0)$. Note that the center of this hyperbola is at the origin. We let (x, y) represent the coordinates of any point, P, on the hyperbola.

What does the definition of a hyperbola tell us about the point (x, y) in Figure 13.19? For any point (x, y) on the hyperbola, the absolute value of the difference of the distances from the two foci, $|d_2 - d_1|$, must be constant. We denote this constant by $2a$, just as we did for the ellipse. Thus, the point (x, y) is on the hyperbola if and only if

$$|d_2 - d_1| = 2a$$

$$\left|\sqrt{(x + c)^2 + y^2} - \sqrt{(x - c)^2 + y^2}\right| = 2a \qquad \text{Use the distance formula.}$$

After eliminating radicals and simplifying, we obtain

$$(c^2 - a^2)x^2 - a^2y^2 = a^2(c^2 - a^2).$$

For convenience, let $b^2 = c^2 - a^2$. Substituting b^2 for $c^2 - a^2$ in the preceding equation, we obtain

$$b^2x^2 - a^2y^2 = a^2b^2.$$

$$\frac{b^2x^2}{a^2b^2} - \frac{a^2y^2}{a^2b^2} = \frac{a^2b^2}{a^2b^2} \qquad \text{Divide both sides by } a^2b^2.$$

$$\frac{x^2}{a^2} - \frac{y^2}{b^2} = 1. \qquad \text{Simplify.}$$

This last equation is called the **standard form of the equation of a hyperbola centered at the origin**. There are two such equations. The first is for a hyperbola in which the transverse axis lies on the x-axis. The second is for a hyperbola in which the transverse axis lies on the y-axis.

STUDY TIP

When the x^2-term is preceded by a plus sign, the transverse axis is horizontal. When the y^2-term is preceded by a plus sign, the transverse axis is vertical.

STANDARD FORMS OF THE EQUATIONS OF A HYPERBOLA The **standard form of the equation of a hyperbola** with center at the origin is

$$\frac{x^2}{a^2} - \frac{y^2}{b^2} = 1 \quad \text{or} \quad \frac{y^2}{a^2} - \frac{x^2}{b^2} = 1.$$

Figure 13.20 illustrates that for the equation on the left, the transverse axis lies on the x-axis. For the equation on the right, the transverse axis lies on the y-axis. The vertices are a units from the center and the foci are c units from the center.

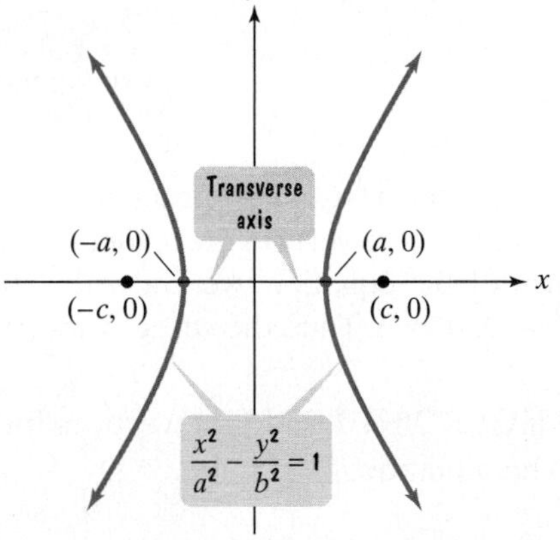

FIGURE 13.20 **(a)** Transverse axis lies on the x-axis.

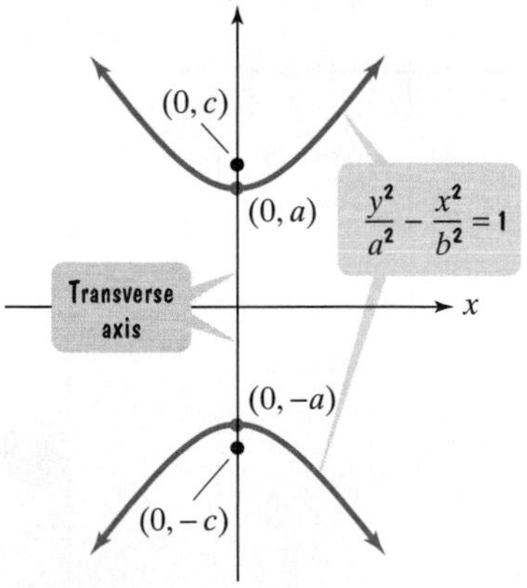

FIGURE 13.20 **(b)** Transverse axis lies on the y-axis.

1 Locate a hyperbola's vertices.

Using the Standard Form of the Equation of a Hyperbola We can use the standard form of the equation of a hyperbola to find its vertices. Because the vertices are a units from the center, begin by identifying a^2 in the equation. In the standard form of a hyperbola's equation, **a^2 is the number under the variable whose term is preceded by a plus sign $(+)$**. If the x^2-term is preceded by a plus sign, the transverse axis lies along the x-axis. Thus, the vertices are a units to the left and right of the origin. If the y^2-term is preceded by a plus sign, the transverse axis lies along the y-axis. Thus, the vertices are a units above and below the origin.

EXAMPLE 1 Finding Vertices from a Hyperbola's Equation

Find the vertices for each of the following hyperbolas with the given equation:

a. $\dfrac{x^2}{16} - \dfrac{y^2}{9} = 1$ **b.** $\dfrac{y^2}{9} - \dfrac{x^2}{16} = 1.$

SOLUTION Both equations are in standard form. We begin by identifying a^2 and b^2 in each equation.

a. The first equation is in the form $\dfrac{x^2}{a^2} - \dfrac{y^2}{b^2} = 1.$

$$\frac{x^2}{16} - \frac{y^2}{9} = 1$$

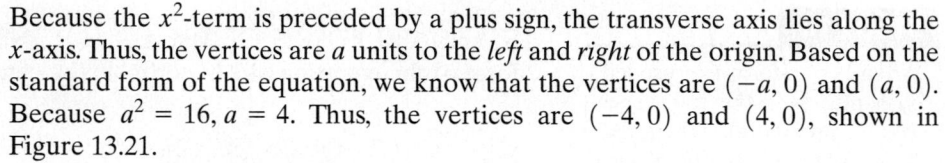

$a^2 = 16$. This is the denominator of the term preceded by a plus sign.

$b^2 = 9$. This is the denominator of the term preceded by a minus sign.

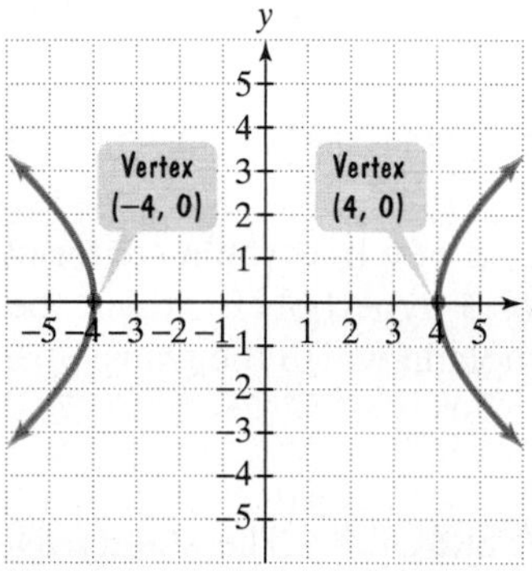

FIGURE 13.21 The graph of $\dfrac{x^2}{16} - \dfrac{y^2}{9} = 1$

Because the x^2-term is preceded by a plus sign, the transverse axis lies along the x-axis. Thus, the vertices are a units to the *left* and *right* of the origin. Based on the standard form of the equation, we know that the vertices are $(-a, 0)$ and $(a, 0)$. Because $a^2 = 16$, $a = 4$. Thus, the vertices are $(-4, 0)$ and $(4, 0)$, shown in Figure 13.21.

b. The second given equation is in the form $\dfrac{y^2}{a^2} - \dfrac{x^2}{b^2} = 1.$

$$\frac{y^2}{9} - \frac{x^2}{16} = 1$$

$a^2 = 9$. This is the denominator of the term preceded by a plus sign.

$b^2 = 16$. This is the denominator of the term preceded by a minus sign.

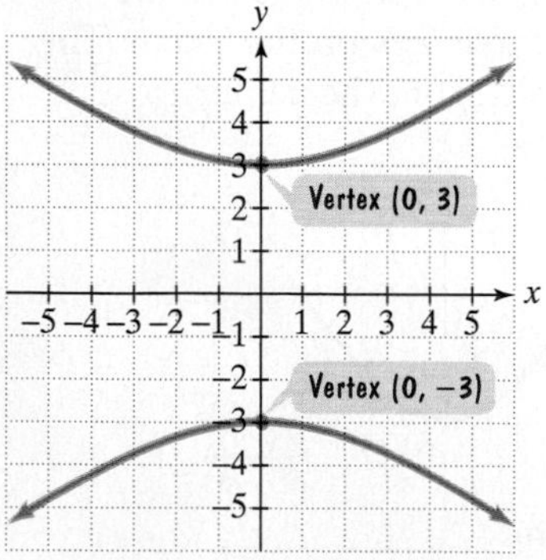

FIGURE 13.22 The graph of $\dfrac{y^2}{9} - \dfrac{x^2}{16} = 1$

Because the y^2-term is preceded by a plus sign, the transverse axis lies along the y-axis. Thus, the vertices are a units *above* and *below* the origin. Based on the standard form of the equation, we know that the vertices are $(0, -a)$ and $(0, a)$. Because $a^2 = 9$, $a = 3$. Thus, the vertices are $(0, -3)$ and $(0, 3)$, shown in Figure 13.22. ■

CHECK POINT 1 Find the vertices for each of the following hyperbolas with the given equation:

a. $\dfrac{x^2}{25} - \dfrac{y^2}{16} = 1$ **b.** $\dfrac{y^2}{25} - \dfrac{x^2}{16} = 1.$

The Asymptotes of a Hyperbola As x and y get larger, the two branches of the graph of a hyperbola approach a pair of intersecting lines, called **asymptotes**. The asymptotes pass through the center of the hyperbola and are helpful in graphing hyperbolas.

Figure 13.23 shows the asymptotes for the graphs of hyperbolas centered at the origin. The asymptotes pass through the corners of a rectangle. Note that the dimensions of this rectangle are $2a$ by $2b$.

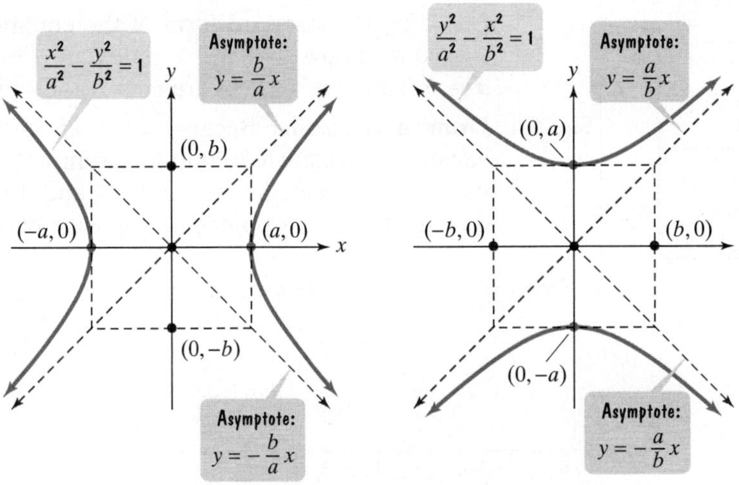

FIGURE 13.23 Asymptotes of a hyperbola

Why are $y = \pm\frac{b}{a}x$ the asymptotes for a hyperbola whose transverse axis is horizontal? The proof can be found in the appendix.

2 Graph hyperbolas centered at the origin.

Graphing Hyperbolas Centered at the Origin Hyperbolas are graphed using vertices and asymptotes.

GRAPHING HYPERBOLAS

1. Locate the vertices.
2. Use dashed lines to draw the rectangle centered at the origin with sides parallel to the axes, crossing one axis at $\pm a$ and the other at $\pm b$.
3. Use dashed lines to draw the diagonals of this rectangle and extend them to obtain the asymptotes.
4. Draw the two branches of the hyperbola by starting at each vertex and approaching the asymptotes.

EXAMPLE 2 Graphing a Hyperbola

Graph the hyperbola: $\frac{x^2}{25} - \frac{y^2}{16} = 1.$

USING TECHNOLOGY

Graph $\frac{x^2}{25} - \frac{y^2}{16} = 1$ by solving for y:

$$y_1 = \frac{\sqrt{16x^2 - 400}}{5}$$

$$y_2 = -\frac{\sqrt{16x^2 - 400}}{5} = -y_1.$$

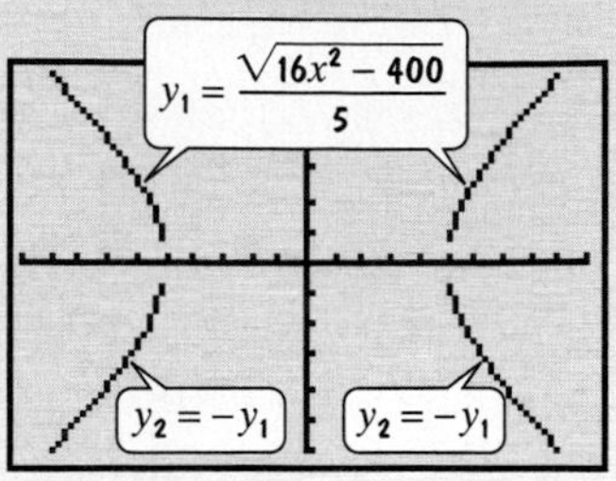

[−10, 10, 1] by [−6, 6, 1]

SOLUTION

Step 1. **Locate the vertices.** The given equation is in the form $\frac{x^2}{a^2} - \frac{y^2}{b^2} = 1$, with $a^2 = 25$ and $b^2 = 16$.

$$\frac{x^2}{25} - \frac{y^2}{16} = 1$$

$a^2 = 25$ $b^2 = 16$

Based on the standard form of the equation with the transverse axis on the x-axis, we know that the vertices are $(-a, 0)$ and $(a, 0)$. Because $a^2 = 25$, $a = 5$. Thus, the vertices are $(-5, 0)$ and $(5, 0)$, shown in Figure 13.24.

Step 2. **Draw a rectangle.** Because $a^2 = 25$ and $b^2 = 16$, $a = 5$ and $b = 4$. We construct a rectangle to find the asymptotes, using -5 and 5 on the x-axis (the vertices are located here) and -4 and 4 on the y-axis. The rectangle passes through these four points, shown using dashed lines in Figure 13.24.

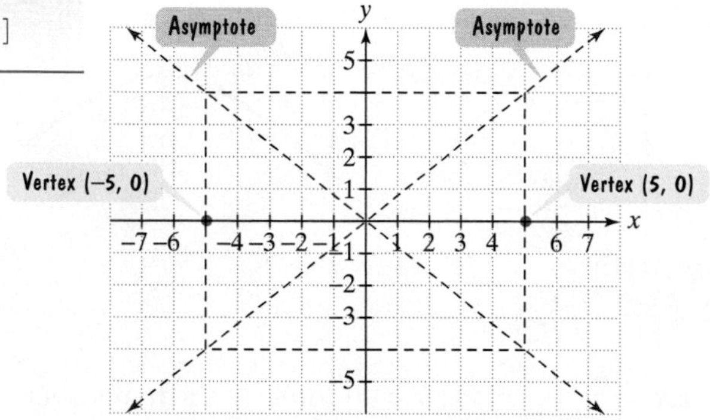

FIGURE 13.24 Preparing to graph $\frac{x^2}{25} - \frac{y^2}{16} = 1$

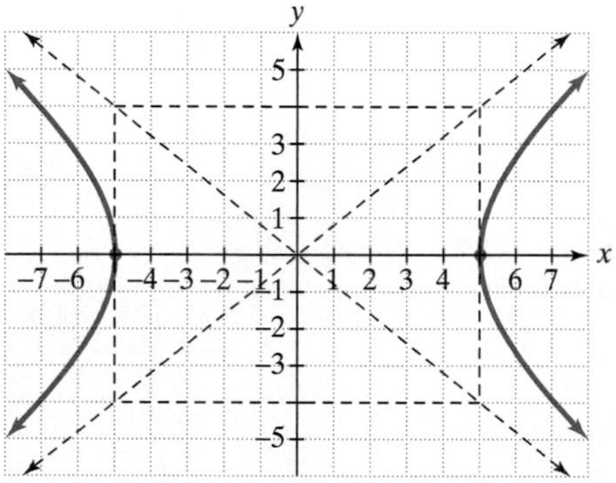

FIGURE 13.25 The graph of $\frac{x^2}{25} - \frac{y^2}{16} = 1$

Step 3. **Draw extended diagonals for the rectangle to obtain the asymptotes.** We draw dashed lines through the opposite corners of the rectangle, shown in Figure 13.24, to obtain the graph of the asymptotes.

Step 4. **Draw the two branches of the hyperbola by starting at each vertex and approaching the asymptotes.** The hyperbola is shown in Figure 13.25. ■

✔ **CHECK POINT 2** Graph the hyperbola: $\frac{x^2}{36} - \frac{y^2}{9} = 1.$

EXAMPLE 3 Graphing a Hyperbola

Graph the hyperbola: $9y^2 - 4x^2 = 36$.

SOLUTION We begin by writing the equation in standard form. The right side should be 1, so we divide both sides by 36.

$$\frac{9y^2}{36} - \frac{4x^2}{36} = \frac{36}{36}$$

$$\frac{y^2}{4} - \frac{x^2}{9} = 1 \quad \text{Simplify. The right side is now 1.}$$

Now we are ready to use our four-step procedure for graphing hyperbolas.

Step 1. **Locate the vertices.** The equation that we obtained, $\frac{y^2}{4} - \frac{x^2}{9} = 1$, is in the form $\frac{y^2}{a^2} - \frac{x^2}{b^2} = 1$, with $a^2 = 4$ and $b^2 = 9$.

$$\frac{y^2}{4} - \frac{x^2}{9} = 1$$

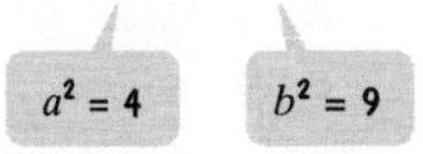

Based on the standard form of the equation with the transverse axis on the y-axis, we know that the vertices are $(0, -a)$ and $(0, a)$. Because $a^2 = 4$, $a = 2$. Thus, the vertices are $(0, -2)$ and $(0, 2)$, shown in Figure 13.26.

Step 2. **Draw a rectangle.** Because $a^2 = 4$ and $b^2 = 9$, $a = 2$ and $b = 3$. We construct a rectangle to find the asymptotes, using -2 and 2 on the y-axis (the vertices are located here) and -3 and 3 on the x-axis. The rectangle passes through these four points, shown using dashed lines in Figure 13.26.

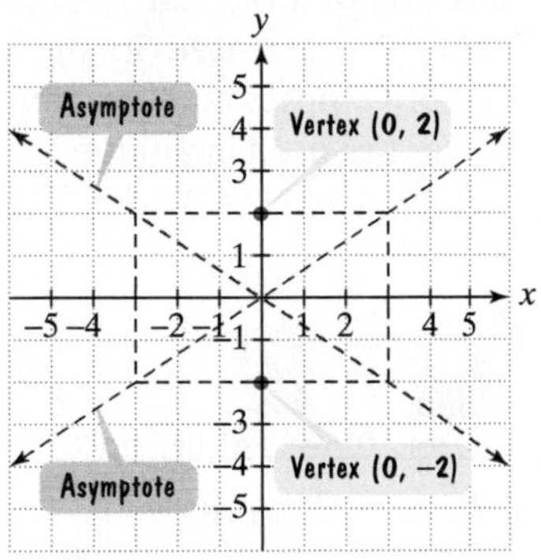

FIGURE 13.26 Preparing to graph $\frac{y^2}{4} - \frac{x^2}{9} = 1$

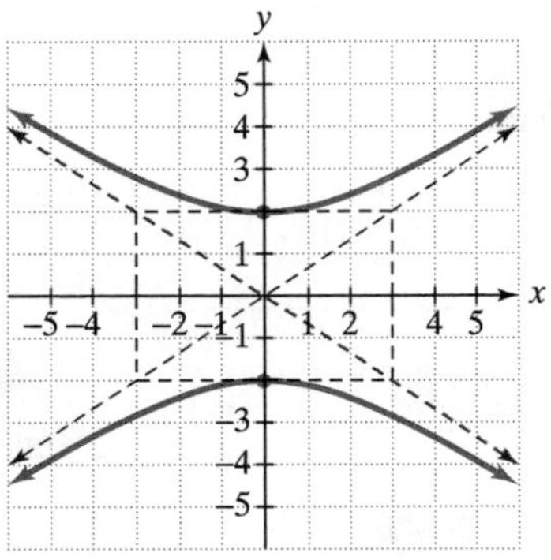

FIGURE 13.27 The graph of $\frac{y^2}{4} - \frac{x^2}{9} = 1$

Step 3. **Draw extended diagonals of the rectangle to obtain the asymptotes.** We draw dashed lines through the opposite corners of the rectangle, shown in Figure 13.26, to obtain the graph of the asymptotes.

Step 4. **Draw the two branches of the hyperbola by starting at each vertex and approaching the asymptotes.** The hyperbola is shown in Figure 13.27. ■

CHECK POINT 3 Graph the hyperbola: $y^2 - 4x^2 = 4$.

Applications Hyperbolas have many applications. When a jet flies at a speed greater than the speed of sound, the shock wave that is created is heard as a sonic boom. The wave has the shape of a cone. The shape formed as the cone hits the ground is one branch of a hyperbola.

Halley's Comet, a permanent part of our solar system, travels around the sun in an elliptical orbit. Other comets pass through the solar system only once, following a hyperbolic path with the sun as a focus.

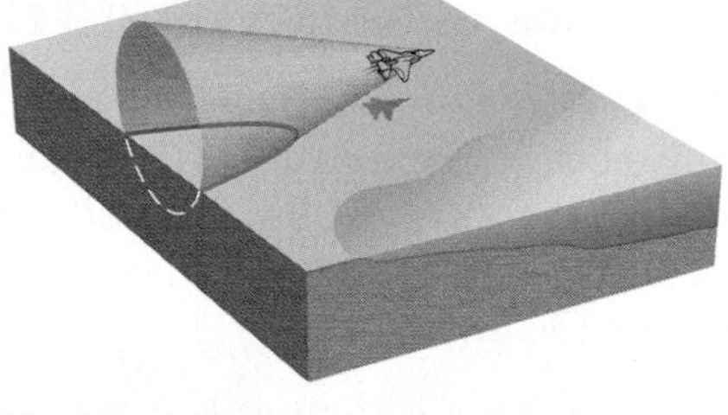

The hyperbolic shape of a sonic boom

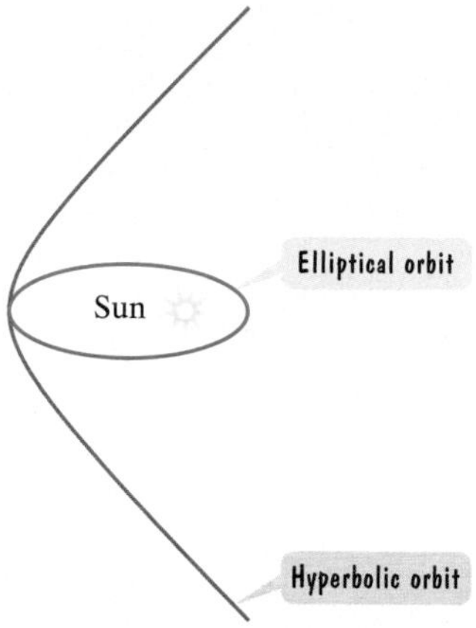

Orbits of comets

Hyperbolas are of practical importance in fields ranging from architecture to navigation. Cooling towers used in the design for nuclear power plants have cross sections that are both ellipses and hyperbolas. Three-dimensional solids whose cross sections are hyperbolas are used in unique architectural creations, including the TWA building at Kennedy Airport in New York City and the St. Louis Science Center Planetarium.

13.3 EXERCISE SET

Student Solutions Manual | CD/Video | PH Math/Tutor Center | MathXL Tutorials on CD | MathXL® | MyMathLab | Interactmath.com

Practice Exercises

In Exercises 1–4, find the vertices of the hyperbola with the given equation. Then match each equation to one of the graphs that are shown and labeled (a)–(d).

1. $\frac{x^2}{4} - \frac{y^2}{1} = 1$

2. $\frac{x^2}{1} - \frac{y^2}{4} = 1$

3. $\frac{y^2}{4} - \frac{x^2}{1} = 1$

4. $\frac{y^2}{1} - \frac{x^2}{4} = 1$

(a)

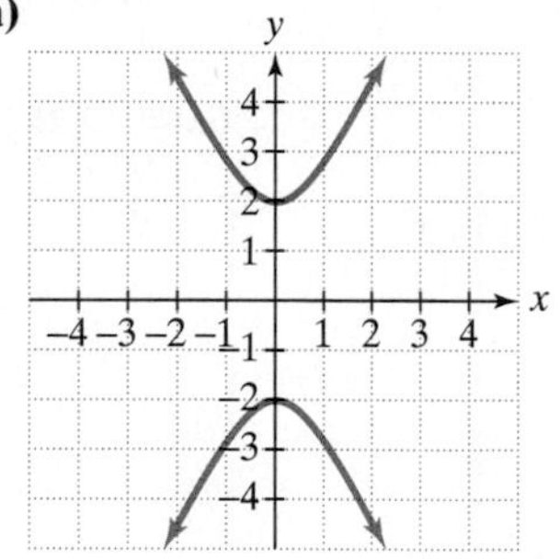

(b)

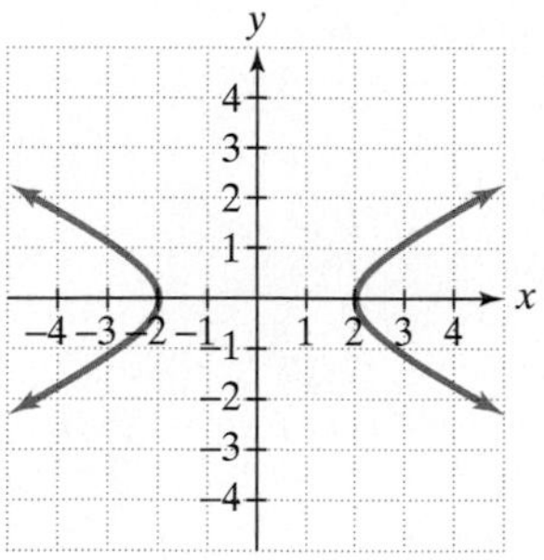

(c)

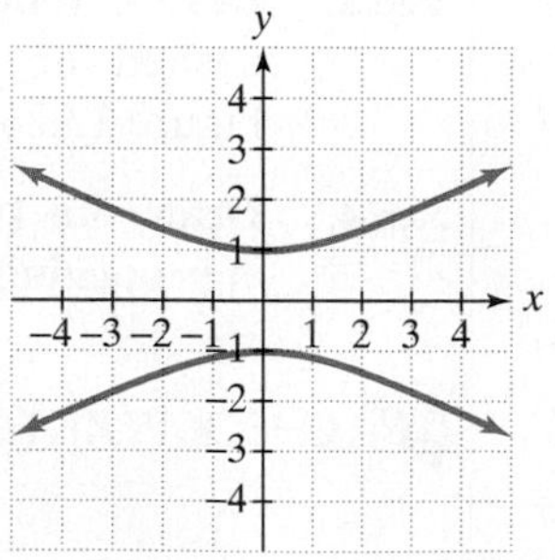

(d)

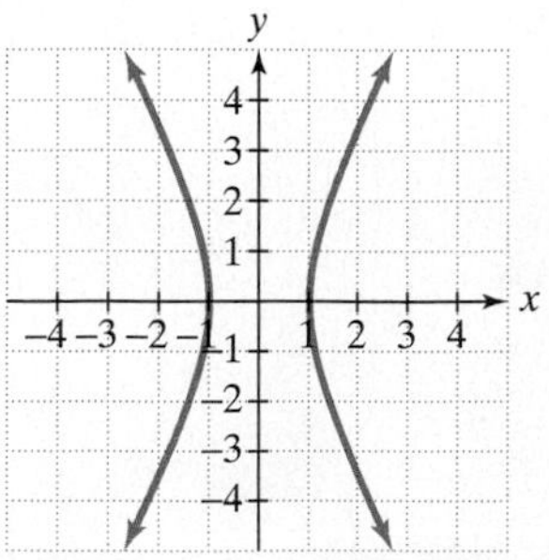

In Exercises 5–18, use vertices and asymptotes to graph each hyperbola.

5. $\frac{x^2}{9} - \frac{y^2}{25} = 1$

6. $\frac{x^2}{16} - \frac{y^2}{25} = 1$

7. $\frac{x^2}{100} - \frac{y^2}{64} = 1$

8. $\frac{x^2}{144} - \frac{y^2}{81} = 1$

9. $\frac{y^2}{16} - \frac{x^2}{36} = 1$

10. $\frac{y^2}{25} - \frac{x^2}{64} = 1$

11. $\frac{y^2}{36} - \frac{x^2}{25} = 1$

12. $\frac{y^2}{100} - \frac{x^2}{49} = 1$

13. $9x^2 - 4y^2 = 36$

14. $4x^2 - 25y^2 = 100$

15. $9y^2 - 25x^2 = 225$

16. $16y^2 - 9x^2 = 144$

17. $4x^2 = 4 + y^2$

18. $25y^2 = 225 + 9x^2$

In Exercises 19–22, find the standard form of the equation of each hyperbola.

19.

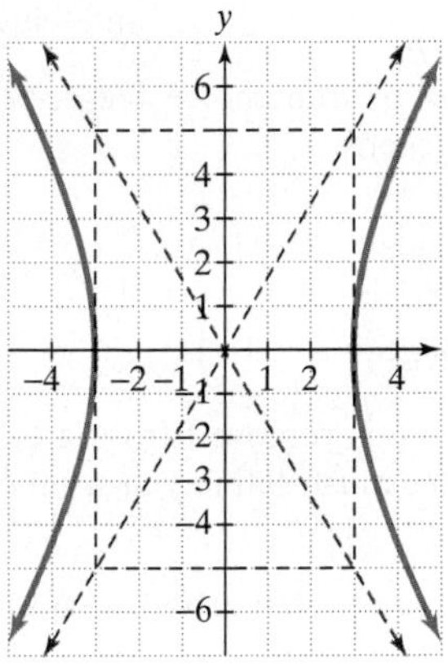

20.

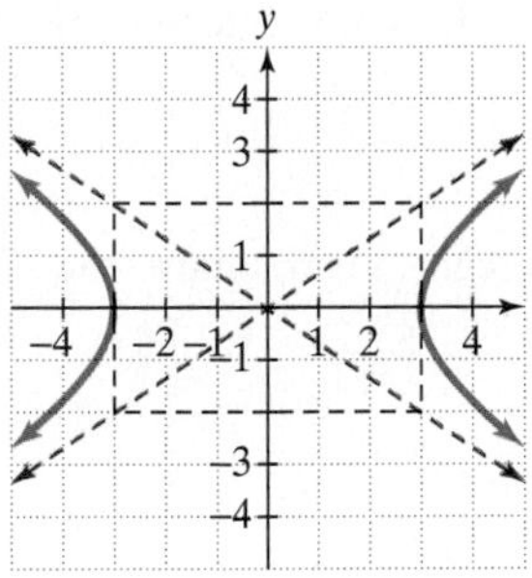

21.

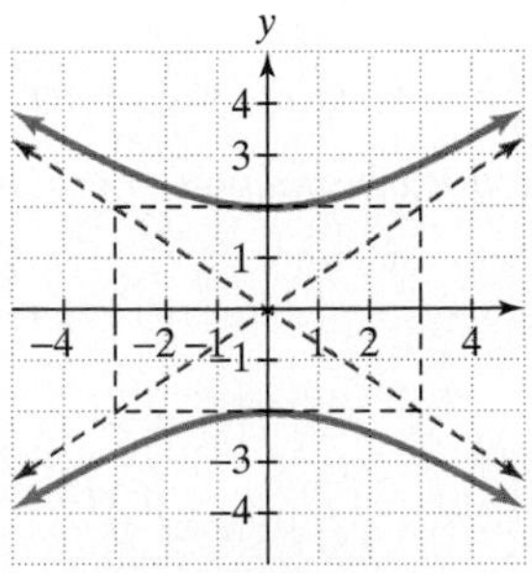

22.

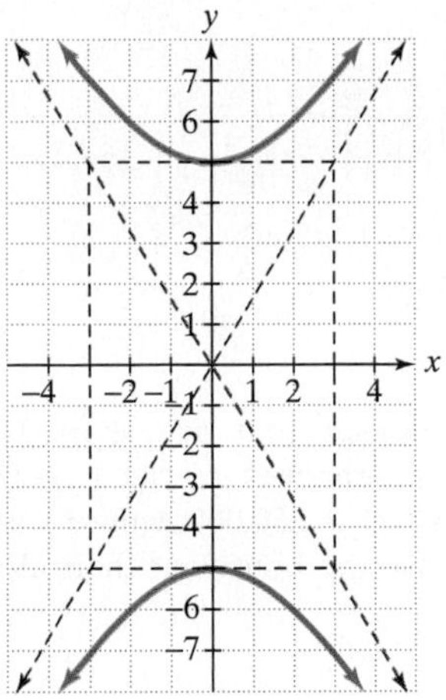

Practice Plus

In Exercises 23–28, graph each relation. Use the relation's graph to determine its domain and range.

23. $\frac{x^2}{9} - \frac{y^2}{16} = 1$

24. $\frac{x^2}{25} - \frac{y^2}{4} = 1$

25. $\frac{x^2}{9} + \frac{y^2}{16} = 1$

26. $\frac{x^2}{25} + \frac{y^2}{4} = 1$

27. $\frac{y^2}{16} - \frac{x^2}{9} = 1$

28. $\frac{y^2}{4} - \frac{x^2}{25} = 1$

In Exercises 29–32, find the solution set for each system by graphing both of the system's equations in the same rectangular coordinate system and finding points of intersection. Check all solutions in both equations.

29. $x^2 - y^2 = 4$

$x^2 + y^2 = 4$

30. $x^2 - y^2 = 9$

$x^2 + y^2 = 9$

31. $9x^2 + y^2 = 9$
$y^2 - 9x^2 = 9$

32. $4x^2 + y^2 = 4$
$y^2 - 4x^2 = 4$

Application Exercises

33. An architect designs two houses that are shaped and positioned like a part of the branches of the hyperbola whose equation is $625y^2 - 400x^2 = 250{,}000$, where x and y are in yards. How far apart are the houses at their closest point?

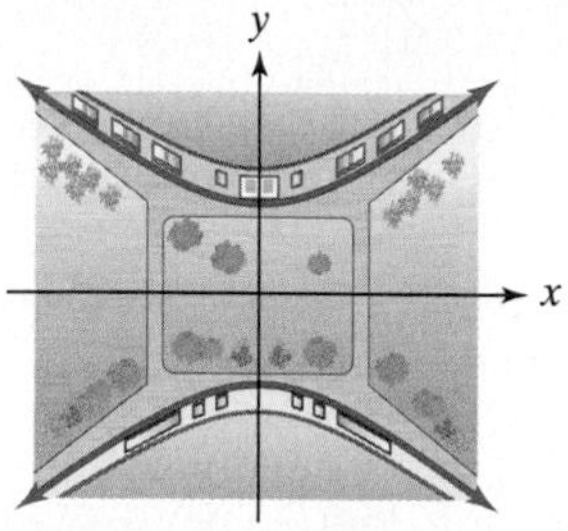

34. Scattering experiments, in which moving particles are deflected by various forces, led to the concept of the nucleus of an atom. In 1911, the physicist Ernest Rutherford (1871–1937) discovered that when alpha particles are directed toward the nuclei of gold atoms, they are eventually deflected along hyperbolic paths, illustrated in the figure. If a particle gets as close as 3 units to the nucleus along a hyperbolic path with an asymptote given by $y = \frac{1}{2}x$, what is the equation of its path?

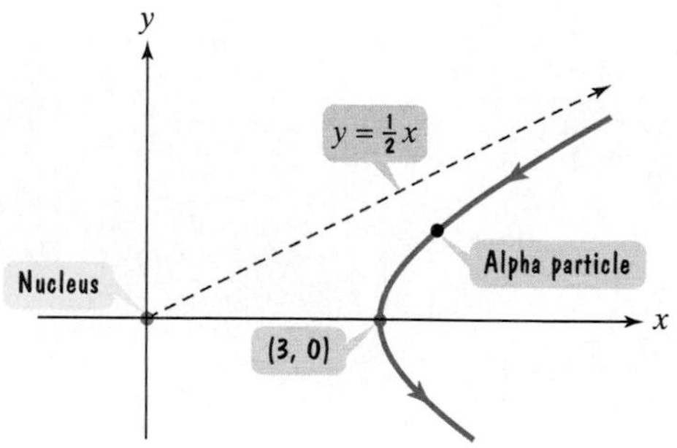

Writing in Mathematics

35. What is a hyperbola?

36. Describe how to graph $\frac{x^2}{9} - \frac{y^2}{1} = 1$.

37. Describe one similarity and one difference between the graphs of $\frac{x^2}{9} - \frac{y^2}{1} = 1$ and $\frac{y^2}{9} - \frac{x^2}{1} = 1$.

38. How can you distinguish an ellipse from a hyperbola by looking at their equations?

39. In 1992, a NASA team began a project called Spaceguard Survey, calling for an international watch for comets that might collide with Earth. Why is it more difficult to detect a possible "doomsday comet" with a hyperbolic orbit than one with an elliptical orbit?

Technology Exercises

40. Use a graphing utility to graph any five of the hyperbolas that you graphed by hand in Exercises 5–18.

41. Use a graphing utility to graph $\frac{x^2}{4} - \frac{y^2}{9} = 0$. Is the graph a hyperbola? In general, what is the graph of $\frac{x^2}{a^2} - \frac{y^2}{b^2} = 0$?

42. Graph $\frac{x^2}{a^2} - \frac{y^2}{b^2} = 1$ and $\frac{x^2}{a^2} - \frac{y^2}{b^2} = -1$ in the same viewing rectangle for values of a^2 and b^2 of your choice. Describe the relationship between the two graphs.

Critical Thinking Exercises

43. Which one of the following is true?

a. If one branch of a hyperbola is removed from a graph, then the branch that remains must define y as a function of x.

b. All points on the asymptotes of a hyperbola also satisfy the hyperbola's equation.

c. The graph of $\frac{x^2}{9} - \frac{y^2}{4} = 1$ does not intersect the line $y = -\frac{2}{3}x$.

d. Two different hyperbolas can never share the same asymptotes.

The graph of

$$\frac{(x-h)^2}{a^2} - \frac{(y-k)^2}{b^2} = 1$$

is the same as the graph of $\frac{x^2}{a^2} - \frac{y^2}{b^2} = 1$, *except the center is at* (h, k) *rather than at the origin. Use this information to graph the hyperbolas in Exercises 44–47.*

44. $\frac{(x-2)^2}{16} - \frac{(y-3)^2}{9} = 1$

45. $\frac{(x+2)^2}{9} - \frac{(y-1)^2}{25} = 1$

46. $(x - 3)^2 - 4(y + 3)^2 = 4$

47. $x^2 - y^2 - 2x - 4y - 4 = 0$

In Exercises 48–49, find the standard form of the equation of the hyperbola satisfying the given conditions.

48. vertices: $(6, 0)$, $(-6, 0)$; asymptotes: $y = 4x$, $y = -4x$

49. vertices: $(0, 7)$, $(0, -7)$; asymptotes: $y = 5x$, $y = -5x$

Review Exercises

50. Use intercepts and the vertex to graph the quadratic function: $y = -x^2 - 4x + 5$. (Section 11.3, Example 3)

51. Solve: $3x^2 - 11x - 4 \geq 0$. (Section 11.5, Example 2)

52. Solve: $\log_4(3x + 1) = 3$. (Section 12.4, Example 4)

MID-CHAPTER CHECK POINT

CHAPTER 13

What You Know: We learned to graph circles with center (h, k) and radius r $[(x-h)^2 + (y-k)^2 = r^2]$. We graphed ellipses centered at (h, k) $\left[\frac{(x-h)^2}{a^2} + \frac{(y-k)^2}{b^2} = 1 \text{ or } \frac{(x-h)^2}{b^2} + \frac{(y-k)^2}{a^2} = 1, a^2 > b^2\right]$ and we saw that the larger denominator (a^2) determines whether the major axis is horizontal or vertical. We used vertices and asymptotes to graph hyperbolas centered at the origin $\left[\frac{x^2}{a^2} - \frac{y^2}{b^2} = 1 \text{ with vertices } (-a, 0) \text{ and } (a, 0) \text{ or } \frac{y^2}{a^2} - \frac{x^2}{b^2} = 1 \text{ with vertices } (0, -a) \text{ and } (0, a)\right]$.

In Exercises 1–4, graph each circle.

1. $x^2 + y^2 = 9$

2. $(x - 3)^2 + (y + 2)^2 = 25$

3. $x^2 + (y - 1)^2 = 4$

4. $x^2 + y^2 - 4x - 2y - 4 = 0$

In Exercises 5–8, graph each ellipse.

5. $\frac{x^2}{25} + \frac{y^2}{4} = 1$

6. $9x^2 + 4y^2 = 36$

7. $\frac{(x-2)^2}{16} + \frac{(y+1)^2}{25} = 1$

8. $\frac{(x+2)^2}{25} + \frac{(y-1)^2}{16} = 1$

In Exercises 9–12, graph each hyperbola.

9. $\frac{x^2}{9} - y^2 = 1$

10. $\frac{y^2}{9} - x^2 = 1$

11. $y^2 - 4x^2 = 16$

12. $4x^2 - 49y^2 = 196$

In Exercises 13–18, graph each equation.

13. $x^2 + y^2 = 4$

14. $x + y = 4$

15. $x^2 - y^2 = 4$

16. $x^2 + 4y^2 = 4$

17. $(x+1)^2 + (y-1)^2 = 4$

18. $x^2 + 4(y - 1)^2 = 4$

SECTION 13.4

Objectives

1. Graph horizontal parabolas.
2. Identify conic sections by their equations.

THE PARABOLA; IDENTIFYING CONIC SECTIONS

At first glance, this image looks like columns of smoke rising from a fire into a starry sky. Those are, indeed, stars in the background, but you are not looking at ordinary smoke columns. These stand almost 6 trillion miles high and are 7000 light-years from Earth—more than 400 million times as far away as the sun.

This NASA photograph is one of a series of stunning images captured from the ends of the universe by the Hubble Space Telescope. The image shows infant star systems the size of our solar system emerging from the gas and dust that shrouded their creation. Using a parabolic mirror 94.5 inches in diameter, the Hubble provided answers to many of the profound mysteries of the cosmos: How big and how old is the universe? How did the galaxies come to exist? Do other Earth-like planets orbit other sun-like stars? In this section, we study parabolas and their applications, including parabolic shapes that gather distant rays of light and focus them into spectacular images.

Definition of a Parabola In Chapter 11, we studied parabolas, viewing them as graphs of quadratic functions in the form

$$y = a(x - h)^2 + k \quad \text{or} \quad y = ax^2 + bx + c.$$

STUDY TIP

Here is a summary of what you should already know about graphing parabolas.

Graphing $y = a(x - h)^2 + k$ and $y = ax^2 + bx + c$

1. If $a > 0$, the graph opens upward. If $a < 0$, the graph opens downward.
2. The vertex of $y = a(x - h)^2 + k$ is (h, k).
3. The x-coordinate of the vertex of $y = ax^2 + bx + c$ is $x = -\frac{b}{2a}$.

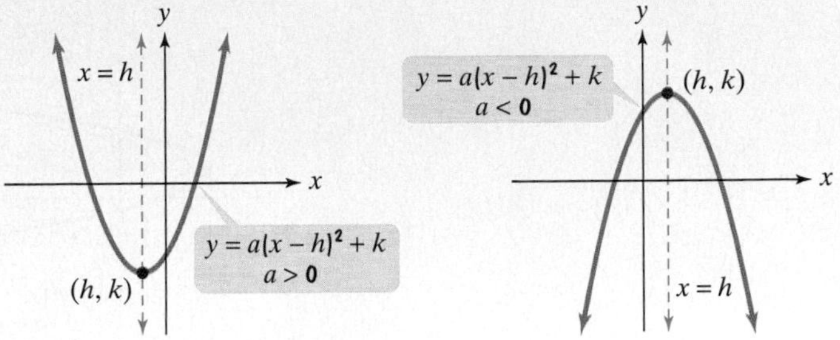

Parabolas can be given a geometric definition that enables us to include graphs that open to the left or to the right. The definitions of ellipses and hyperbolas involved two fixed points, the foci. By contrast, the definition of a parabola is based on one point and a line.

DEFINITION OF A PARABOLA A **parabola** is the set of all points in a plane that are equidistant from a fixed line (the **directrix**) and a fixed point (the **focus**) that is not on the line (see Figure 13.28).

In Figure 13.28, find the line passing through the focus and perpendicular to the directrix. This is the **axis of symmetry** of the parabola. The point of intersection of the parabola with its axis of symmetry is the parabola's **vertex**. Notice that the vertex is midway between the focus and the directrix.

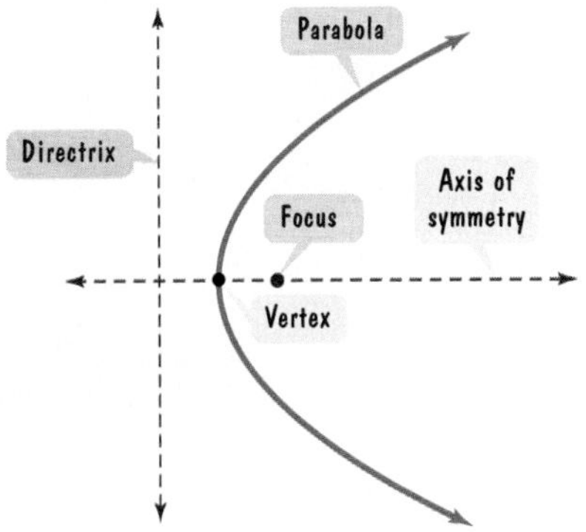

FIGURE 13.28

Parabolas can open to the left, right, upward, or downward. Figure 13.29 shows a basic "family" of four parabolas and their equations. Notice that the red and blue parabolas that open to the left or right are not functions of x because they fail the vertical line test—that is, it is possible to draw vertical lines that intersect these graphs at more than one point.

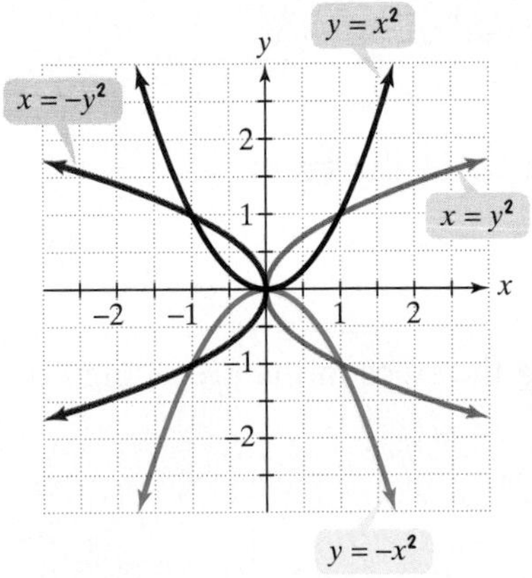

FIGURE 13.29

The equation $x = y^2$ interchanges the variables in the equation $y = x^2$. By interchanging x and y in the two forms of a parabola's equation,

$$y = a(x - h)^2 + k \quad \text{and} \quad y = ax^2 + bx + c,$$

we can obtain the forms of the equations of parabolas that open to the right or to the left.

PARABOLAS OPENING TO THE LEFT OR TO THE RIGHT The graphs of

$$x = a(y - k)^2 + h \quad \text{and} \quad x = ay^2 + by + c$$

are parabolas opening to the left or to the right.

1. If $a > 0$, the graph opens to the right. If $a < 0$, the graph opens to the left.
2. The vertex of $x = a(y - k)^2 + h$ is (h, k).
3. The y-coordinate of the vertex of $x = ay^2 + by + c$ is $y = -\dfrac{b}{2a}$.

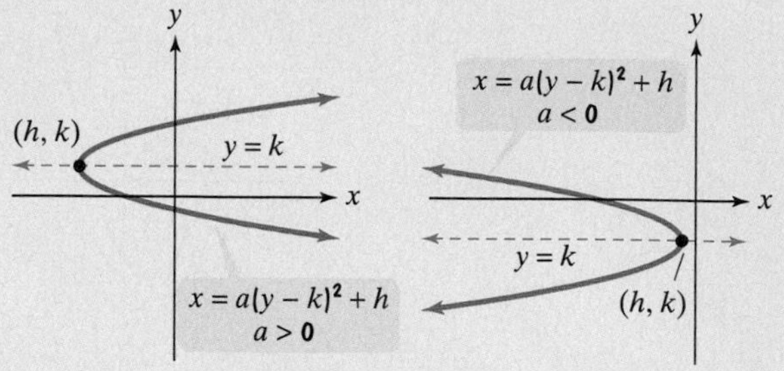

1 Graph horizontal parabolas.

Graphing Parabolas Opening to the Left or the Right Here is a procedure for graphing horizontal parabolas that are not functions. Notice how this procedure is similar to the one that we used in Chapter 11 for graphing vertical parabolas.

GRAPHING HORIZONTAL PARABOLAS To graph $x = a(y - k)^2 + h$ or $x = ay^2 + by + c$,

1. Determine whether the parabola opens to the left or to the right. If $a > 0$, it opens to the right. If $a < 0$, it opens to the left.
2. Determine the vertex of the parabola. The vertex of $x = a(y - k)^2 + h$ is (h, k). The y-coordinate of the vertex of $x = ay^2 + by + c$ is $y = -\dfrac{b}{2a}$. Substitute this value of y into the equation to find the x-coordinate.
3. Find the x-intercept by replacing y with 0.
4. Find any y-intercepts by replacing x with 0. Solve the resulting quadratic equation for y.
5. Plot the intercepts and the vertex. Connect these points with a smooth curve. If additional points are needed to obtain a more accurate graph, select values for y on each side of the axis of symmetry and compute values for x.

EXAMPLE 1 Graphing a Horizontal Parabola in the Form $x = a(y - k)^2 + h$

Graph: $x = -3(y - 1)^2 + 2$.

SOLUTION We can graph this equation by following the steps in the preceding box. We begin by identifying values for a, k, and h.

$$x = a(y - k)^2 + h$$

$a = -3 \quad k = 1 \quad h = 2$

$$x = -3(y - 1)^2 + 2$$

$x = a(y - k)^2 + h$

$a = -3$ | $k = 1$ | $h = 2$

$x = -3(y - 1)^2 + 2$

Identifying a, k, and h (repeated)

Step 1. Determine how the parabola opens. Note that a, the coefficient of y^2, is -3. Thus, $a < 0$; this negative value tells us that the parabola opens to the left.

Step 2. Find the vertex. The vertex of the parabola is at (h, k). Because $k = 1$ and $h = 2$, the parabola has its vertex at $(2, 1)$.

Step 3. Find the x-intercept. Replace y with 0 in $x = -3(y - 1)^2 + 2$.

$$x = -3(0 - 1)^2 + 2 = -3(-1)^2 + 2 = -3(1) + 2 = -1$$

The x-intercept is -1. The parabola passes through $(-1, 0)$.

Step 4. Find the y-intercepts. Replace x with 0 in the given equation.

$$x = -3(y - 1)^2 + 2 \qquad \text{This is the given equation.}$$

$$0 = -3(y - 1)^2 + 2 \qquad \text{Replace } x \text{ with } 0.$$

$$3(y - 1)^2 = 2 \qquad \text{Solve for } y. \text{ Add } 3(y - 1)^2 \text{ to both sides of the equation.}$$

$$(y - 1)^2 = \frac{2}{3} \qquad \text{Divide both sides by 3.}$$

$$y - 1 = \sqrt{\frac{2}{3}} \quad \text{or} \quad y - 1 = -\sqrt{\frac{2}{3}} \qquad \text{Apply the square root property.}$$

$$y = 1 + \sqrt{\frac{2}{3}} \qquad y = 1 - \sqrt{\frac{2}{3}} \qquad \text{Add 1 to both sides in each equation.}$$

$$y \approx 1.8 \qquad y \approx 0.2 \qquad \text{Use a calculator.}$$

The y-intercepts are approximately 1.8 and 0.2. The parabola passes through approximately $(0, 1.8)$ and $(0, 0.2)$.

Step 5. Graph the parabola. With a vertex at $(2, 1)$, an x-intercept at -1, and y-intercepts at approximately 1.8 and 0.2, the graph of the parabola is shown in Figure 13.30. The axis of symmetry is the horizontal line whose equation is $y = 1$.

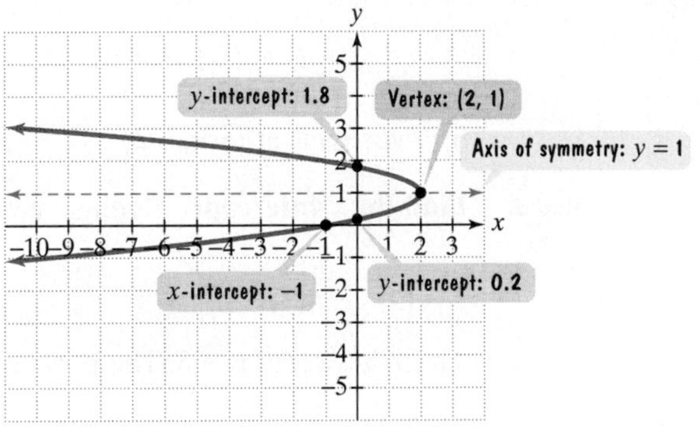

FIGURE 13.30 The graph of $x = -3(y - 1)^2 + 2$

We can possibly improve our graph in Figure 13.30 by finding additional points on the parabola. Choose values of y on each side of the axis of symmetry, $y = 1$. We use $y = 2$ and $y = -1$. Then we compute values for x. The values in the table of coordinates show that $(-1, 2)$ and $(-10, -1)$ are points on the parabola. Locate each point in Figure 13.30.

x	y
-1	2
-10	-1

Choose values for y.

Compute values for x using $x = -3(y - 1)^2 + 2$:

$x = -3(2 - 1)^2 + 2 = -3 \cdot 1^2 + 2 = -1$

$x = -3(-1 - 1)^2 + 2 = -3(-2)^2 + 2 = -10.$

CHECK POINT 1 Graph: $x = -(y-2)^2 + 1$.

EXAMPLE 2 Graphing a Horizontal Parabola in the Form $x = ay^2 + by + c$

Graph: $x = y^2 + 4y - 5$.

SOLUTION

Step 1. **Determine how the parabola opens.** Note that a, the coefficient of y^2, is 1. Thus $a > 0$; this positive value tells us that the parabola opens to the right.

Step 2. **Find the vertex.** We know that the y-coordinate of the vertex is $y = -\frac{b}{2a}$. We identify a, b, and c in $x = ay^2 + by + c$.

$$x = y^2 + 4y - 5$$

a = 1 b = 4 c = −5

Substitute the values of a and b into the equation for the y-coordinate:

$$y = -\frac{b}{2a} = -\frac{4}{2 \cdot 1} = -2.$$

The y-coordinate of the vertex is -2. We substitute -2 for y into the parabola's equation, $x = y^2 + 4y - 5$, to find the x-coordinate:

$$x = (-2)^2 + 4(-2) - 5 = 4 - 8 - 5 = -9.$$

The vertex is at $(-9, -2)$.

Step 3. **Find the x-intercept.** Replace y with 0 in $x = y^2 + 4y - 5$.

$$x = 0^2 + 4 \cdot 0 - 5 = -5$$

The x-intercept is -5. The parabola passes through $(-5, 0)$.

Step 4. **Find the y-intercepts.** Replace x with 0 in the given equation.

$x = y^2 + 4y - 5$	This is the given equation.
$0 = y^2 + 4y - 5$	Replace x with 0.
$0 = (y-1)(y+5)$	Use factoring to solve the quadratic equation.
$y - 1 = 0$ or $y + 5 = 0$	Set each factor equal to 0.
$y = 1$ $\quad$ $y = -5$	Solve.

The y-intercepts are 1 and -5. The parabola passes through $(0, 1)$ and $(0, -5)$.

USING TECHNOLOGY

To graph $x = y^2 + 4y - 5$ using a graphing utility, rewrite the equation as a quadratic equation in y.

$$y^2 + 4y + (-x - 5) = 0$$

a = 1 b = 4 c = −x − 5

Use the quadratic formula to solve for y and enter the resulting equations.

$$y_1 = \frac{-4 + \sqrt{16 - 4(-x-5)}}{2}$$

$$y_2 = \frac{-4 - \sqrt{16 - 4(-x-5)}}{2}$$

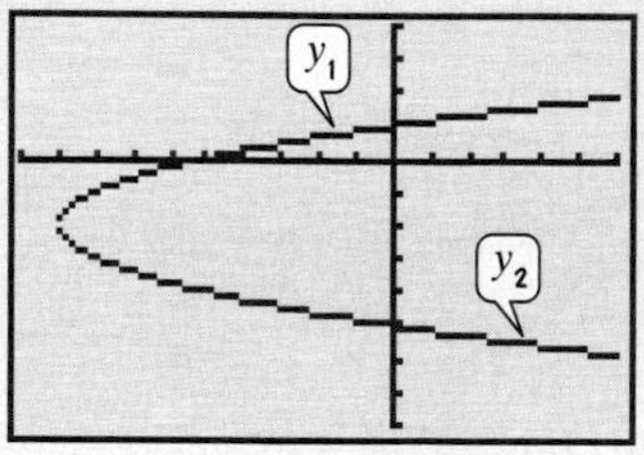

$[-10, 6, 1]$ by $[-8, 4, 1]$

Step 5. **Graph the parabola.** With a vertex at $(-9, -2)$, an x-intercept at -5, and y-intercepts at 1 and -5, the graph of the parabola is shown in Figure 13.31. The axis of symmetry is the horizontal line whose equation is $y = -2$. ■

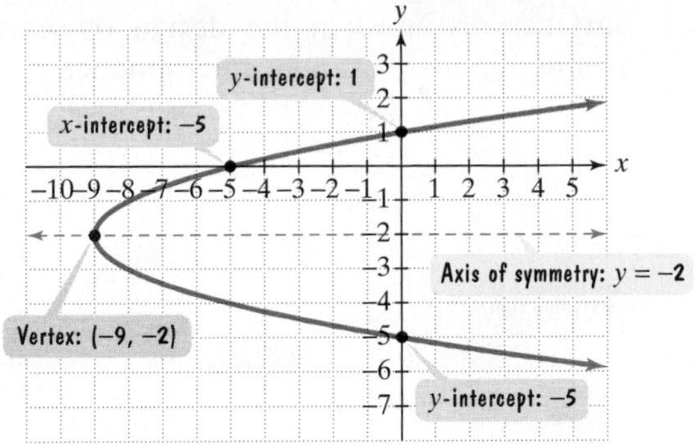

FIGURE 13.31 The graph of $x = y^2 + 4y - 5$

CHECK POINT 2 Graph: $x = y^2 + 8y + 7$.

Applications Parabolas have many applications. Cables hung between structures to form suspension bridges form parabolas. Arches constructed of steel and concrete, whose main purpose is strength, are usually parabolic in shape.

Suspension bridge

Arch bridge

We have seen that comets in our solar system travel in orbits that are ellipses and hyperbolas. Some comets also follow parabolic paths. Only comets with elliptical orbits, such as Halley's Comet, return to our part of the galaxy.

You throw a ball directly upward. As illustrated in Figure 13.32, the height of such a projectile as a function of time is parabolic.

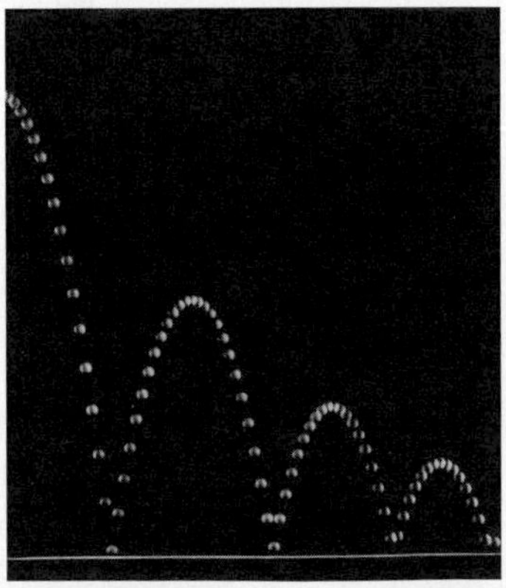

FIGURE 13.32 Multiflash photo showing the parabolic path of a ball thrown into the air
© Berenice Abbott/Commerce Graphics Ltd., Inc.

If a parabola is rotated about its axis of symmetry, a parabolic surface is formed. Figure 13.33(a) shows how a parabolic surface can be used to reflect light. Light originates at the focus. Note how the light is reflected by the parabolic surface, so that the outgoing light is parallel to the axis of symmetry. The reflective properties of parabolic surfaces are used in the design of searchlights [see Figure 13.33(b)], automobile headlights, and parabolic microphones.

ENRICHMENT ESSAY

The Hubble Space Telescope

The Hubble Space Telescope

For decades, astronomers hoped to create an observatory above the atmosphere that would provide an unobscured view of the universe. This vision was realized with the 1990 launching of the Hubble Space Telescope. The telescope initially had blurred vision due to problems with its parabolic mirror. The mirror had been ground two millionths of a meter smaller than design specifications. In 1993, astronauts from the Space Shuttle *Endeavor* equipped the telescope with optics to correct the blurred vision. "A small change for a mirror, a giant leap for astronomy," Christopher J. Burrows of the Space Telescope Science Institute said when clear images from the ends of the universe were presented to the public after the repair mission.

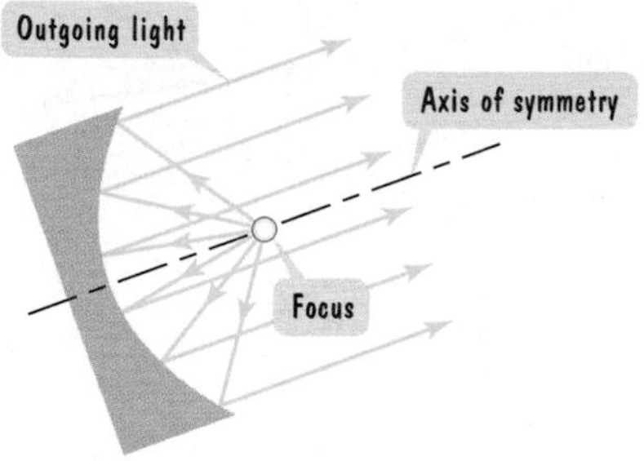

FIGURE 13.33 **(a)** Parabolic surface reflecting light

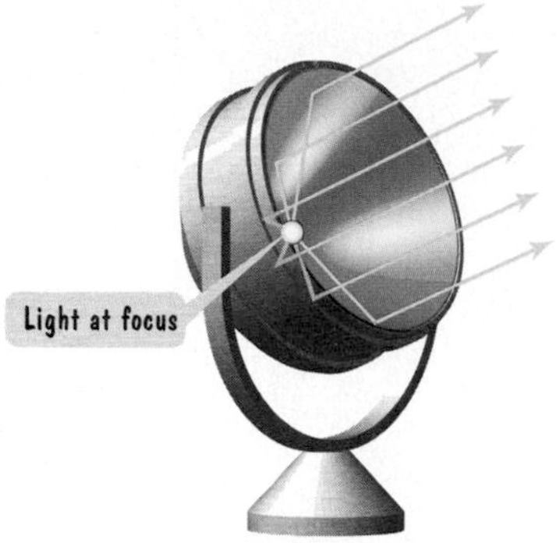

FIGURE 13.33 **(b)** Light from the focus is reflected parallel to the axis of symmetry.

Figure 13.34(a) shows how a parabolic surface can be used to reflect *incoming* light. Note that light rays strike the surface and are reflected *to the focus*. This principle is used in the design of reflecting telescopes, radar, and television satellite dishes. Reflecting telescopes magnify the light from distant stars by reflecting the light from these bodies to the focus of a parabolic mirror [see Figure 13.34(b)].

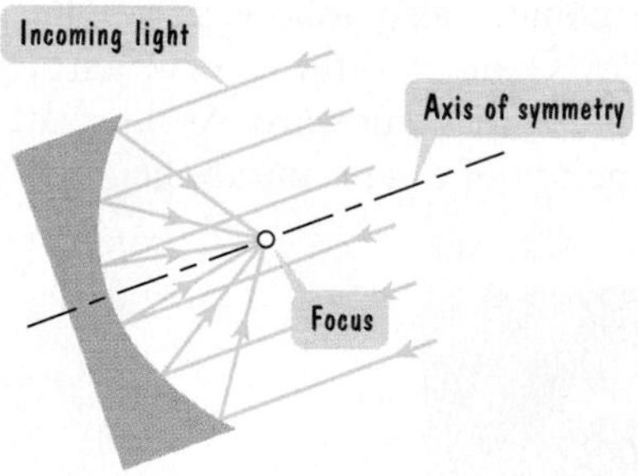

FIGURE 13.34 **(a)** Parabolic surface reflecting incoming light

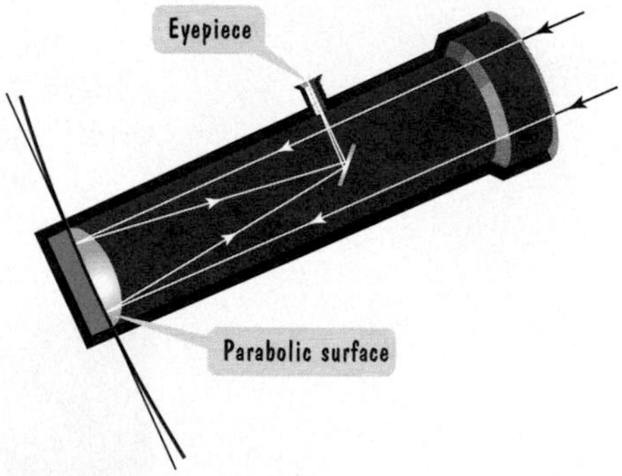

FIGURE 13.34 **(b)** Incoming light rays are reflected to the focus.

2 Identify conic sections by their equations.

Identifying Conic Sections by Their Equations What do the equations of the conic sections in this chapter have in common? They contain x^2-terms, y^2-terms, or both. Furthermore, they do not contain terms with exponents greater than 2. On the next page is a table that shows how to identify conic sections by their equations.

Recognizing Conic Sections from Equations

Conic Section	How to Identify the Equation	Example
Circle	When x^2- and y^2-terms are on the same side, they have the same coefficient.	$x^2 + y^2 = 16$
Ellipse	When x^2- and y^2-terms are on the same side, they have different coefficients of the same sign.	$4x^2 + 16y^2 = 64$ or (dividing by 64) $\frac{x^2}{16} + \frac{y^2}{4} = 1$
Hyperbola	When x^2- and y^2-terms are on the same side, they have coefficients with opposite signs.	$9y^2 - 4x^2 = 36$ or (dividing by 36) $\frac{y^2}{4} - \frac{x^2}{9} = 1$
Parabola	Only one of the variables is squared.	$x = y^2 + 4y - 5$

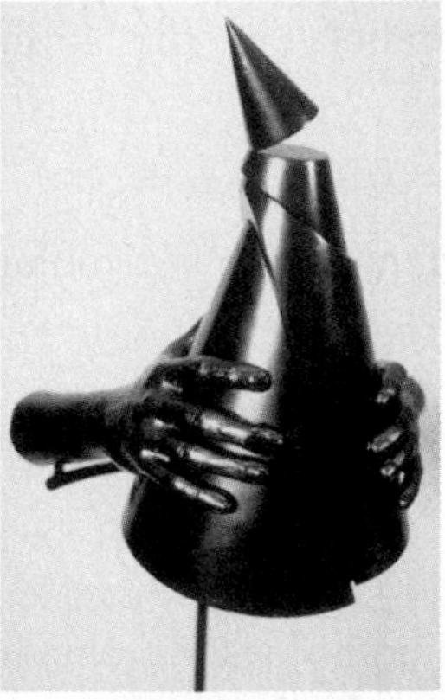

Richard E. Prince "The Cone of Apollonius" (detail), fiberglass, steel, paint, graphite, 51×18×14 in. Collection: Vancouver Art Gallery, Vancouver, Canada. Photo courtesy of Equinox Gallery, Vancouver, Canada.

EXAMPLE 3 Recognizing Equations of Conic Sections

Indicate whether the graph of each equation is a circle, an ellipse, a hyperbola, or a parabola:

a. $4y^2 = 16 - 4x^2$ **b.** $x^2 = y^2 + 9$

c. $x + 7 - 6y = y^2$ **d.** $x^2 = 16 - 16y^2$.

SOLUTION (Throughout the solution, in addition to identifying each equation's graph, we'll also discuss the graph's important features.) If both variables are squared, the graph of the equation is not a parabola. In this case, we collect the x^2- and y^2-terms on the same side of the equation.

a. $4y^2 = 16 - 4x^2$

Both variables, x and y, are squared.

The graph is a circle, an ellipse, or a hyperbola. To see which one it is, we collect the x^2- and y^2-terms on the same side. Add $4x^2$ to both sides. We obtain

$$4x^2 + 4y^2 = 16.$$

Because the coefficients of x^2 and y^2 are the same, namely 4, the equation's graph is a circle. This becomes more obvious if we divide both sides by 4.

$$\frac{4x^2}{4} + \frac{4y^2}{4} = \frac{16}{4}$$ Divide both sides by 4.

$$x^2 + y^2 = 4$$ Simplify.

$$(x - 0)^2 + (y - 0)^2 = 2^2$$ Write in the form $(x - h)^2 + (y - k)^2 = r^2$ with center (h, k) and radius r.

The graph is a circle with center at the origin and radius 2.

b. $x^2 = y^2 + 9$

Both variables, x and y, are squared.

The graph cannot be a parabola. To see if it is a circle, an ellipse, or a hyperbola, we collect the x^2- and y^2-terms on the same side. Subtract y^2 from both sides. We obtain

$$x^2 - y^2 = 9.$$

Because the x^2- and y^2-terms have coefficients with opposite signs, the equation's graph is a hyperbola. This becomes more obvious if we divide both sides by 9 to obtain 1 on the right.

$$\frac{x^2}{9} - \frac{y^2}{9} = 1$$

$a^2 = 9$ $\quad b^2 = 9$

The hyperbola's vertices are $(a, 0)$ and $(-a, 0)$, namely $(3, 0)$. and $(-3, 0)$.

c. $x + 7 - 6y = y^2$

Only one variable, y, is squared.

Because only one variable is squared, the graph of the equation is a parabola. We can express the equation of the horizontal parabola in the form $x = ay^2 + by + c$ by isolating x on the left. We obtain

$$x = y^2 + 6y - 7. \qquad \text{Add } 6y - 7 \text{ to both sides.}$$

Because the coefficient of the y^2-term, 1, is positive, the graph of the horizontal parabola opens to the right.

d. $x^2 = 16 - 16y^2$

Both variables, x and y, are squared.

The graph cannot be a parabola. To see if it is a circle, an ellipse, or a hyperbola, we collect the x^2- and y^2-terms on the same side. Add $16y^2$ to both sides. We obtain

$$x^2 + 16y^2 = 16.$$

Because the x^2- and y^2-terms have different coefficients of the same sign, namely 1 and 16, the equation's graph is an ellipse. This becomes more obvious if we divide both sides by 16 to obtain 1 on the right.

$$\frac{x^2}{16} + \frac{y^2}{1} = 1$$

$a^2 = 16$ $\quad b^2 = 1$

An equation in the form $\frac{x^2}{a^2} + \frac{y^2}{b^2} = 1$ is an ellipse.

The vertices are $(4, 0)$ and $(-4, 0)$. ■

CHECK POINT 3 Identify whether the graph of each equation is a circle, an ellipse, a hyperbola, or a parabola:

a. $x^2 = 4y^2 + 16$

b. $x^2 = 16 - 4y^2$

c. $4x^2 = 16 - 4y^2$

d. $x = -4y^2 + 16y$.

13.4 EXERCISE SET

Student Solutions Manual | CD/Video | PH Math/Tutor Center | MathXL Tutorials on CD | MathXL® | MyMathLab | Interactmath.com

Practice Exercises

In Exercises 1–6, the equation of a horizontal parabola is given. For each equation: Determine how the parabola opens. Find the parabola's vertex. Use your results to identify the equation's graph. [The graphs are labeled (a) through (f).]

1. $x = (y - 2)^2 - 1$

2. $x = (y + 2)^2 - 1$

3. $x = (y + 2)^2 + 1$

4. $x = (y - 2)^2 + 1$

5. $x = -(y - 2)^2 + 1$

6. $x = -(y + 2)^2 + 1$

a.

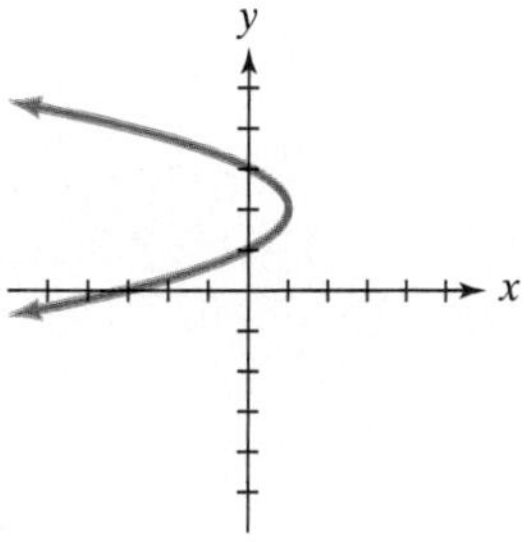

b.

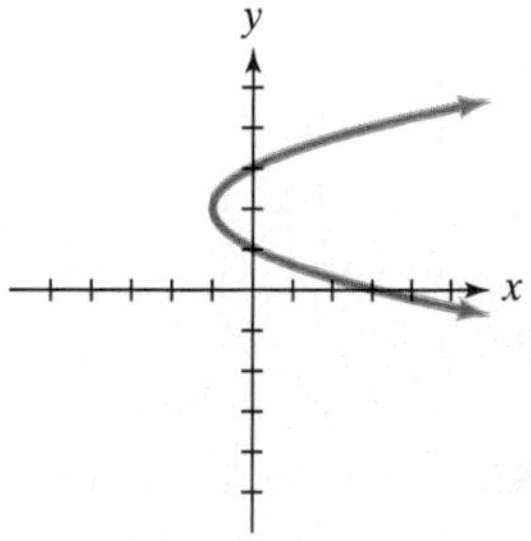

c.

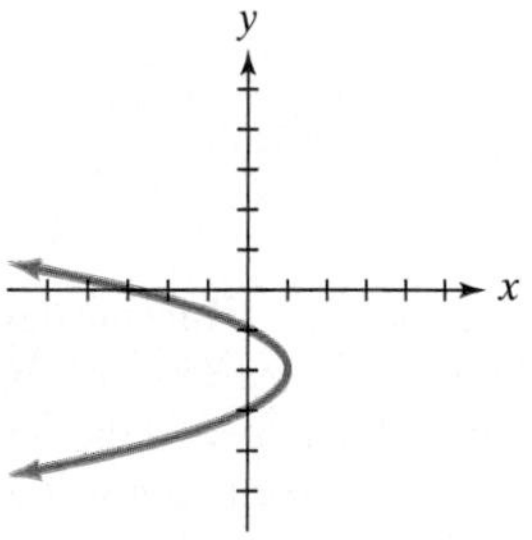

d.

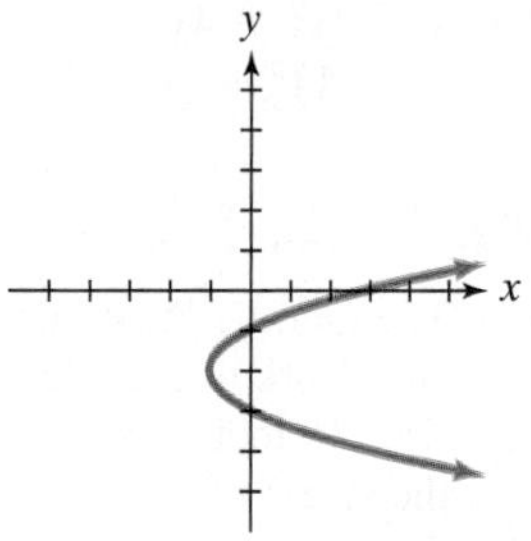

e.

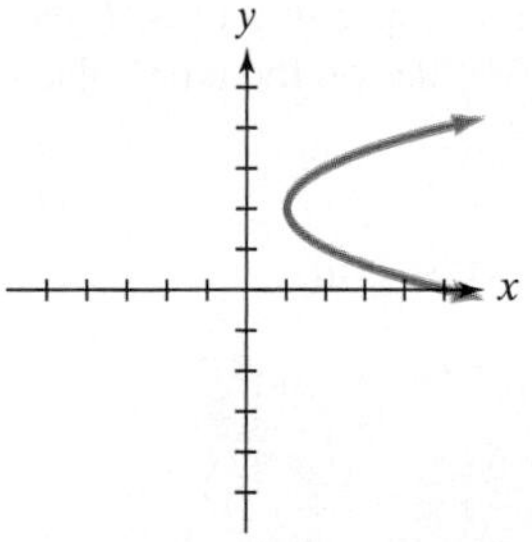

f.

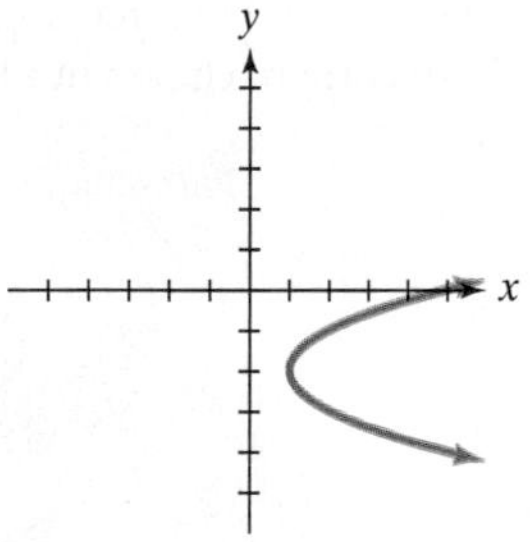

In Exercises 7–18, find the coordinates of the vertex for the horizontal parabola defined by the given equation.

7. $x = 2y^2$

8. $x = 4y^2$

9. $x = (y - 2)^2 + 3$

10. $x = (y - 3)^2 + 4$

11. $x = -4(y + 2)^2 - 1$

12. $x = -2(y + 5)^2 - 1$

13. $x = 2(y - 6)^2$

14. $x = 3(y - 7)^2$

15. $x = y^2 - 6y + 6$

16. $x = y^2 + 6y + 8$

17. $x = 3y^2 + 6y + 7$

18. $x = -2y^2 + 4y + 6$

In Exercises 19–42, use the vertex and intercepts to sketch the graph of each equation. Give the equation for the parabola's axis of symmetry. If needed, find additional points on the parabola by choosing values of y on each side of the axis of symmetry.

19. $x = (y - 2)^2 - 4$

20. $x = (y - 3)^2 - 4$

21. $x = (y - 3)^2 - 5$

22. $x = (y + 2)^2 - 3$

23. $x = -(y - 5)^2 + 4$

24. $x = -(y - 3)^2 + 4$

25. $x = (y - 4)^2 + 1$

26. $x = (y - 2)^2 + 3$

27. $x = -3(y - 5)^2 + 3$

28. $x = -2(y + 6)^2 + 2$

29. $x = -2(y + 3)^2 - 1$

30. $x = -3(y + 1)^2 - 2$

31. $x = \frac{1}{2}(y + 2)^2 + 1$

32. $x = \frac{1}{2}(y + 1)^2 + 2$

33. $x = y^2 + 2y - 3$

34. $x = y^2 - 6y + 8$

35. $x = -y^2 - 4y + 5$

36. $x = -y^2 - 6y + 7$

37. $x = y^2 + 6y$

38. $x = y^2 + 4y$

39. $x = -2y^2 - 4y$

40. $x = -3y^2 - 6y$

41. $x = -2y^2 - 4y + 1$

42. $x = -2y^2 + 4y - 3$

In Exercises 43–54, the equation of a parabola is given. Determine:

a. *if the parabola is horizontal or vertical.*

b. *the way the parabola opens.*

c. *the vertex.*

43. $x = 2(y - 1)^2 + 2$

44. $x = 2(y - 3)^2 + 1$

45. $y = 2(x - 1)^2 + 2$

46. $y = 2(x - 3)^2 + 1$

47. $y = -(x + 3)^2 + 4$

48. $y = -(x + 1)^2 + 4$

49. $x = -(y + 3)^2 + 4$

50. $x = -(y + 1)^2 + 4$

51. $y = x^2 - 4x - 1$

52. $y = x^2 + 6x + 10$

53. $x = -y^2 + 4y + 1$

54. $x = -y^2 - 6y - 10$

In Exercises 55–64, indicate whether the graph of each equation is a circle, an ellipse, a hyperbola, or a parabola.

55. $x - 7 - 8y = y^2$

56. $x - 3 - 4y = 6y^2$

57. $4x^2 = 36 - y^2$

58. $4x^2 = 36 + y^2$

59. $x^2 = 36 + 4y^2$

60. $x^2 = 36 - 4y^2$

61. $3x^2 = 12 - 3y^2$

62. $3x^2 = 27 - 3y^2$

63. $3x^2 = 12 + 3y^2$

64. $3x^2 = 27 + 3y^2$

In Exercises 65–74, indicate whether the graph of each equation is a circle, an ellipse, a hyperbola, or a parabola. Then graph the conic section.

65. $x^2 - 4y^2 = 16$

66. $7x^2 - 7y^2 = 28$

67. $4x^2 + 4y^2 = 16$

68. $7x^2 + 7y^2 = 28$

69. $x^2 + 4y^2 = 16$

70. $4x^2 + y^2 = 16$

71. $x = (y - 1)^2 - 4$

72. $x = (y - 4)^2 - 1$

73. $(x - 2)^2 + (y + 1)^2 = 16$

74. $(x - 1)^2 + (y + 2)^2 = 16$

Practice Plus

In Exercises 75–80, use the vertex and the direction in which the parabola opens to determine the relation's domain and range. Is the relation a function?

75. $x = y^2 + 6y + 5$

76. $x = y^2 - 2y - 5$

77. $y = -x^2 + 4x - 3$

78. $y = -x^2 - 4x + 4$

79. $x = -4(y - 1)^2 + 3$

80. $x = -3(y - 1)^2 - 2$

In Exercises 81–86, find the solution set for each system by graphing both of the system's equations in the same rectangular coordinate system and finding points of intersection. Check all solutions in both equations.

81. $x = (y - 2)^2 - 4$
$y = -\frac{1}{2}x$

82. $x = (y - 3)^2 + 2$
$x + y = 5$

83. $x = y^2 - 3$
$x = y^2 - 3y$

84. $x = y^2 - 5$
$x^2 + y^2 = 25$

85. $x = (y + 2)^2 - 1$
$(x - 2)^2 + (y + 2)^2 = 1$

86. $x = 2y^2 + 4y + 5$
$(x + 1)^2 + (y - 2)^2 = 1$

Application Exercises

87. The George Washington Bridge spans the Hudson River from New York to New Jersey. Its two towers are 3500 feet apart and rise 316 feet above the road. The cable between the towers has the shape of a parabola, and the cable just touches the sides of the road midway between the towers. The parabola is positioned in a rectangular coordinate system with its vertex at the origin. The point (1750, 316) lies on the parabola, as shown.

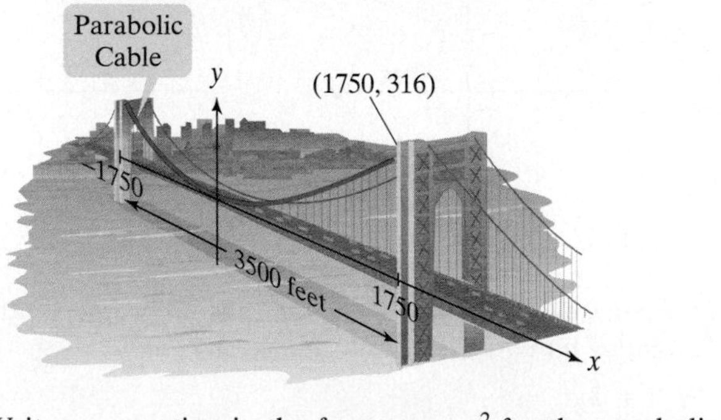

a. Write an equation in the form $y = ax^2$ for the parabolic cable. Do this by substituting 1750 for x and 316 for y and determining the value of a.

b. Use the equation in part (a) to find the height of the cable 1000 feet from a tower. Round to the nearest foot.

88. The towers of the Golden Gate Bridge connecting San Francisco to Marin County are 1280 meters apart and rise 140 meters above the road. The cable between the towers has the shape of a parabola, and the cable just touches the sides of the road midway between the towers. The parabola is positioned in a rectangular coordinate system with its vertex at the origin. The point (640, 140) lies on the parabola, as shown.

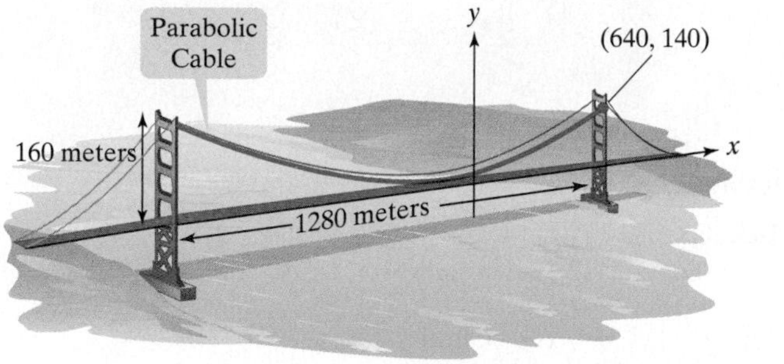

a. Write an equation in the form $y = ax^2$ for the parabolic cable. Do this by substituting 640 for x and 140 for y and determining the value of a.

b. Use the equation in part (a) to find the height of the cable 200 meters from a tower. Round to the nearest meter.

89. A satellite dish is in the shape of a parabolic surface. Signals coming from a satellite strike the surface of the dish and are reflected to the focus, where the receiver is located. The satellite dish shown has a diameter of 12 feet and a depth of 2 feet. The parabola is positioned in a rectangular coordinate system with its vertex at the origin.

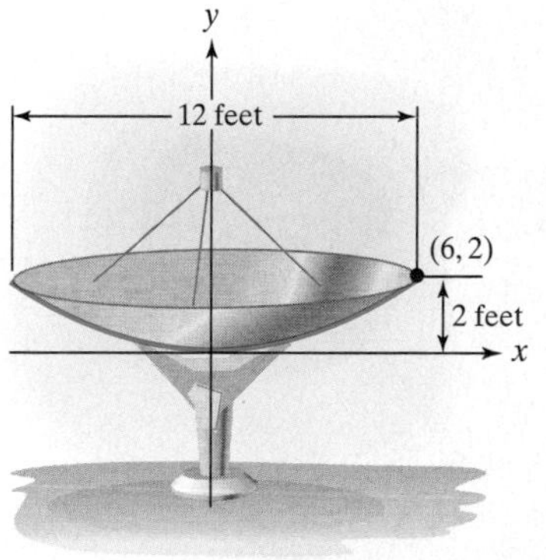

a. Write an equation in the form $y = ax^2$ for the parabola used to shape the dish.

b. The receiver should be placed at the focus $(0, p)$. The value of p is given by the equation $a = \frac{1}{4p}$. How far from the base of the dish should the receiver be placed?

90. An engineer is designing a flashlight using a parabolic reflecting mirror and a light source, as shown. The casting has a diameter of 4 inches and a depth of 2 inches. The parabola is positioned in a rectangular coordinate system with its vertex at the origin.

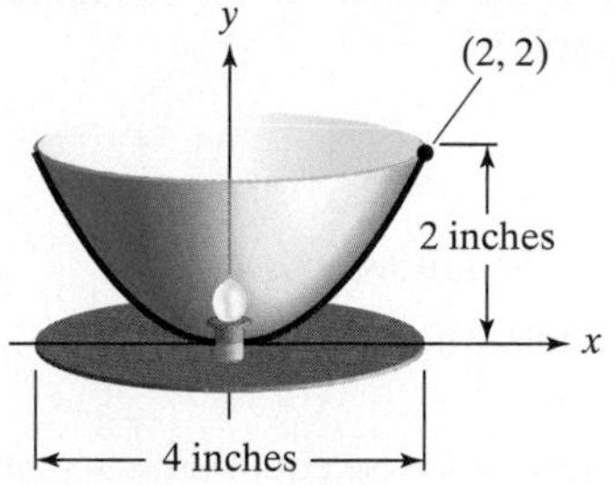

a. Write an equation in the form $y = ax^2$ for the parabola used to shape the mirror.

b. The light source should be placed at the focus $(0, p)$. The value of p is given by the equation $a = \frac{1}{4p}$. Where should the light source be placed relative to the mirror's vertex?

Writing in Mathematics

91. What is a parabola?

92. If you are given an equation of a parabola, explain how to determine if the parabola opens to the right, to the left, upward, or downward.

93. Explain how to use $x = 2(y + 3)^2 - 5$ to find the parabola's vertex.

94. Explain how to use $x = y^2 + 8y + 9$ to find the parabola's vertex.

95. Describe one similarity and one difference between the graphs of $x = 4y^2$ and $x = 4(y - 1)^2 + 2$.

96. How can you distinguish parabolas from other conic sections by looking at their equations?

97. How can you distinguish ellipses from hyperbolas by looking at their equations?

98. How can you distinguish ellipses from circles by looking at their equations?

Technology Exercises

Use a graphing utility to graph the parabolas in Exercises 99–100. Write the given equation as a quadratic equation in y and use the quadratic formula to solve for y. Enter each of the equations to produce the complete graph.

99. $y^2 + 2y - 6x + 13 = 0$

100. $y^2 + 10y - x + 25 = 0$

101. Use a graphing utility to graph any three of the parabolas that you graphed by hand in Exercises 19–42. First solve the given equation for y, possibly using the square root method. Enter each of the two resulting equations to produce the complete graph.

Critical Thinking Exercises

102. Which one of the following is true?

a. The parabola whose equation is $x = 2y - y^2 + 5$ opens to the right.

b. If the parabola whose equation is $x = ay^2 + by + c$ has its vertex at $(3, 2)$ and $a > 0$, then it has no y-intercepts.

c. Some parabolas that open to the right have equations that define y as a function of x.

d. The graph of $x = a(y - k) + h$ is a parabola with vertex at (h, k).

103. Look at the satellite dish shown in Exercise 89. Why must the receiver for a shallow dish be farther from the base of the dish than for a deeper dish of the same diameter?

104. The parabolic arch shown in the figure is 50 feet above the water at the center and 200 feet wide at the base. Will a boat that is 30 feet tall clear the arch 30 feet from the center?

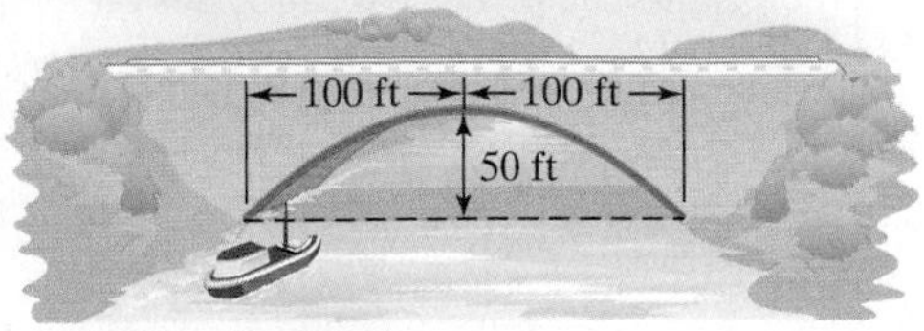

Review Exercises

105. Graph: $f(x) = 2^{1-x}$. (Section 12.1, Example 4)

106. If $f(x) = \frac{1}{3}x - 5$, find $f^{-1}(x)$. (Section 8.3, Example 4)

107. Solve the system:

$$4x - 3y = 12$$
$$3x - 4y = 2.$$

(Section 4.3, Example 3)

SECTION 13.5

Objectives

1. Recognize systems of nonlinear equations in two variables.
2. Solve systems of nonlinear equations by substitution.
3. Solve systems of nonlinear equations by addition.
4. Solve problems using systems of nonlinear equations.

SYSTEMS OF NONLINEAR EQUATIONS IN TWO VARIABLES

Scientists debate the probability that a "doomsday rock" will collide with Earth. It has been estimated that an asteroid, a tiny planet that revolves around the sun, crashes into Earth about once every 250,000 years, and that such a collision would have disastrous results. In 1908, a small fragment struck Siberia, leveling thousands of acres of trees. One theory about the extinction of dinosaurs 65 million years ago involves Earth's collision with a large asteroid and the resulting drastic changes in Earth's climate.

Understanding the path of Earth and the path of a comet is essential to detecting threatening space debris. Orbits about the sun are not described by linear equations in the form $Ax + By = C$. The ability to solve systems that do not contain linear equations provides NASA scientists watching for troublesome asteroids with a way to locate possible collision points with Earth's orbit.

1 Recognize systems of nonlinear equations in two variables.

Systems of Nonlinear Equations and their Solutions A **system of** two **nonlinear equations** in two variables, also called a **nonlinear system**, contains at least one equation that cannot be expressed in the form $Ax + By = C$. Here are two examples:

$$x^2 = 2y + 10$$
$$3x - y = 9$$

Not in the form $Ax + By = C$. The term x^2 is not linear.

$$y = x^2 + 3$$
$$x^2 + y^2 = 9.$$

Neither equation is in the form $Ax + By = C$. The terms x^2 and y^2 are not linear.

A **solution** of a nonlinear system in two variables is an ordered pair of real numbers that satisfies all equations in the system. The **solution set** of the system is the set of all such ordered pairs. As with linear systems in two variables, the solution of a nonlinear system (if there is one) corresponds to the intersection point(s) of the graphs of the equations in the system. Unlike linear systems, the graphs can be circles, ellipses, hyperbolas, parabolas, or anything other than two lines. We will solve nonlinear systems using the substitution method and the addition method.

2 Solve systems of nonlinear equations by substitution.

Eliminating a Variable Using the Substitution Method The substitution method involves converting a nonlinear system into one equation in one variable by an appropriate substitution. The steps in the solution process are exactly the same as those used to solve a linear system by substitution. However, when you obtain an equation in one variable, this equation may not be linear. In our first example, this equation is quadratic.

EXAMPLE 1 Solving a Nonlinear System by the Substitution Method

Solve by the substitution method:

$$x^2 = 2y + 10 \quad \text{(The graph is a parabola.)}$$
$$3x - y = 9. \quad \text{(The graph is a line.)}$$

SOLUTION

Step 1. **Solve one of the equations for one variable in terms of the other.** We begin by isolating one of the variables raised to the first power in either of the equations. By solving for y in the second equation, which has a coefficient of -1, we can avoid fractions.

$3x - y = 9$	This is the second equation in the given system.
$3x = y + 9$	Add y to both sides.
$3x - 9 = y$	Subtract 9 from both sides.

Step 2. **Substitute the expression from step 1 into the other equation.** We substitute $3x - 9$ for y in the first equation.

$$y = \boxed{3x - 9} \qquad x^2 = 2\boxed{y} + 10$$

This gives us an equation in one variable, namely

$$x^2 = 2(3x - 9) + 10.$$

The variable y has been eliminated.

Step 3. Solve the resulting equation containing one variable.

$x^2 = 2(3x - 9) + 10$	This is the equation containing one variable.
$x^2 = 6x - 18 + 10$	Use the distributive property.
$x^2 = 6x - 8$	Combine numerical terms on the right.
$x^2 - 6x + 8 = 0$	Move all terms to one side and set the quadratic equation equal to 0.
$(x - 4)(x - 2) = 0$	Factor.
$x - 4 = 0$ or $x - 2 = 0$	Set each factor equal to 0.
$x = 4$ $\quad$ $x = 2$	Solve for x.

Step 4. Back-substitute the obtained values into the equation from step 1. Now that we have the x-coordinates of the solutions, we back-substitute 4 for x and 2 for x into the equation $y = 3x - 9$.

If x is 4, $\quad y = 3(4) - 9 = 3,\quad$ so $(4, 3)$ is a solution.
If x is 2, $\quad y = 3(2) - 9 = -3,\quad$ so $(2, -3)$ is a solution.

Step 5. Check the proposed solutions in both of the system's given equations. We begin by checking $(4, 3)$. Replace x with 4 and y with 3.

$x^2 = 2y + 10$	$3x - y = 9$	These are the given equations.
$4^2 \stackrel{?}{=} 2(3) + 10$	$3(4) - 3 \stackrel{?}{=} 9$	Let $x = 4$ and $y = 3$.
$16 \stackrel{?}{=} 6 + 10$	$12 - 3 \stackrel{?}{=} 9$	Simplify.
$16 = 16$, true	$9 = 9$, true	True statements result.

The ordered pair $(4, 3)$ satisfies both equations. Thus, $(4, 3)$ is a solution of the system.

Now let's check $(2, -3)$. Replace x with 2 and y with -3 in both given equations.

$x^2 = 2y + 10$	$3x - y = 9$	These are the given equations.
$2^2 \stackrel{?}{=} 2(-3) + 10$	$3(2) - (-3) \stackrel{?}{=} 9$	Let $x = 2$ and $y = -3$.
$4 \stackrel{?}{=} -6 + 10$	$6 + 3 \stackrel{?}{=} 9$	Simplify.
$4 = 4$, true	$9 = 9$, true	True statements result.

The ordered pair $(2, -3)$ also satisfies both equations and is a solution of the system. The solutions are $(4, 3)$ and $(2, -3)$, and the solution set is $\{(4, 3), (2, -3)\}$.

Figure 13.35 shows the graphs of the equations in the system and the solutions as intersection points. ■

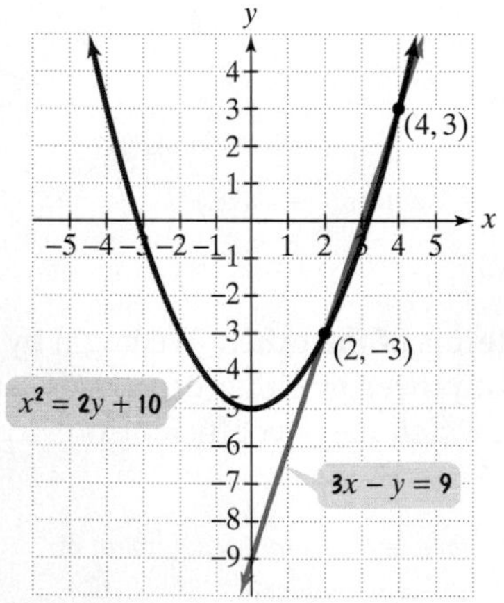

FIGURE 13.35 Points of intersection illustrate the nonlinear system's solutions.

CHECK POINT 1 Solve by the substitution method:

$$x^2 = y - 1$$
$$4x - y = -1.$$

EXAMPLE 2 Solving a Nonlinear System by the Substitution Method

Solve by the substitution method:

$x - y = 3$	(The graph is a line.)
$(x-2)^2 + (y+3)^2 = 4.$	(The graph is a circle.)

SOLUTION Graphically, we are finding the intersection of a line and a circle with center $(2, -3)$ and radius 2.

Step 1. **Solve one of the equations for one variable in terms of the other.** We will solve for x in the linear equation — that is, the first equation. (We could also solve for y.)

$x - y = 3$	This is the first equation in the given system.
$x = y + 3$	Add y to both sides.

Step 2. **Substitute the expression from step 1 into the other equation.** We substitute $y + 3$ for x in the second equation.

$$x = \boxed{y+3} \qquad (\boxed{x} - 2)^2 + (y+3)^2 = 4$$

This gives an equation in one variable, namely

$$(y + 3 - 2)^2 + (y+3)^2 = 4.$$

The variable x has been eliminated.

Step 3. **Solve the resulting equation containing one variable.**

$(y+3-2)^2 + (y+3)^2 = 4$	This is the equation containing one variable.
$(y+1)^2 + (y+3)^2 = 4$	Combine numerical terms in the first parentheses.
$y^2 + 2y + 1 + y^2 + 6y + 9 = 4$	Use the formula $(A+B)^2 = A^2 + 2AB + B^2$ to square $y + 1$ and $y + 3$.
$2y^2 + 8y + 10 = 4$	Combine like terms on the left.
$2y^2 + 8y + 6 = 0$	Subtract 4 from both sides and set the quadratic equation equal to 0.
$2(y^2 + 4y + 3) = 0$	Factor out 2.
$2(y+3)(y+1) = 0$	Factor completely.
$y + 3 = 0$ or $y + 1 = 0$	Set each variable factor equal to 0.
$y = -3 \qquad y = -1$	Solve for y.

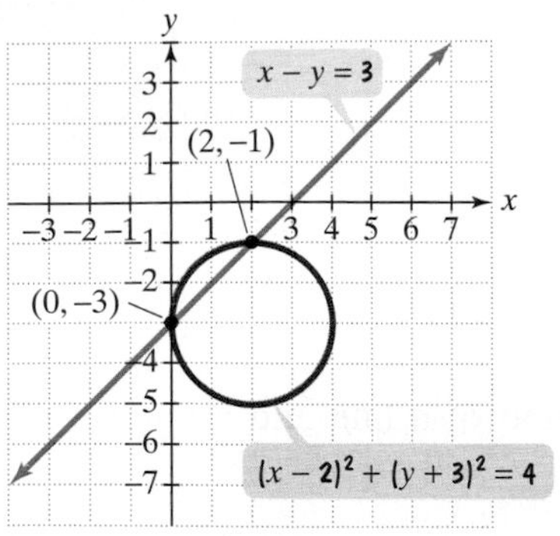

FIGURE 13.36 Points of intersection illustrate the nonlinear system's solutions.

Step 4. **Back-substitute the obtained values into the equation from step 1.** Now that we have the y-coordinates of the solutions, we back-substitute -3 for y and -1 for y in the equation $x = y + 3$.

If $y = -3$: $x = -3 + 3 = 0$, so $(0, -3)$ is a solution.

If $y = -1$: $x = -1 + 3 = 2$, so $(2, -1)$ is a solution.

Step 5. **Check the proposed solution in both of the system's given equations.** Take a moment to show that each ordered pair satisfies both equations. The solutions are $(0, -3)$ and $(2, -1)$, and the solution set of the given system is $\{(0, -3), (2, -1)\}$.

Figure 13.36 shows the graphs of the equations in the system and the solutions as intersection points. ■

CHECK POINT 2 Solve by the substitution method:

$$x + 2y = 0$$
$$(x - 1)^2 + (y - 1)^2 = 5.$$

3 Solve systems of nonlinear equations by addition.

Eliminating a Variable Using the Addition Method In solving linear systems with two variables, we learned that the addition method works well when each equation is in the form $Ax + By = C$. For nonlinear systems, the addition method can be used when each equation is in the form $Ax^2 + By^2 = C$. If necessary, we will multiply either equation or both equations by appropriate numbers so that the coefficients of x^2 or y^2 will have a sum of 0. We then add equations. The sum will be an equation in one variable.

EXAMPLE 3 Solving a Nonlinear System by the Addition Method

Solve the system:

$$4x^2 + y^2 = 13 \quad \text{Equation 1 (The graph is an ellipse.)}$$
$$x^2 + y^2 = 10. \quad \text{Equation 2 (The graph is a circle.)}$$

SOLUTION We can use the same steps that we did when we solved linear systems by the addition method.

Step 1. **Write both equations in the form $Ax^2 + By^2 = C$.** Both equations are already in this form, so we can skip this step.

Step 2. **If necessary, multiply either equation or both equations by appropriate numbers so that the sum of the x^2-coefficients or the sum of the y^2-coefficients is 0.** We can eliminate y^2 by multiplying Equation 2 by -1.

$$4x^2 + y^2 = 13 \xrightarrow{\text{No change}} 4x^2 + y^2 = 13$$
$$x^2 + y^2 = 10 \xrightarrow{\text{Multiply by } -1.} -x^2 - y^2 = -10$$

Steps 3 and 4. Add equations and solve for the remaining variable.

$$\begin{aligned} 4x^2 + y^2 &= 13 \\ \underline{-x^2 - y^2} &= \underline{-10} \\ 3x^2 &= 3 \quad \text{Add equations.} \\ x^2 &= 1 \quad \text{Divide both sides by 3.} \\ x &= \pm 1 \quad \text{Use the square root property: If } x^2 = c\text{, then } x = \pm\sqrt{c}. \end{aligned}$$

Step 5. **Back-substitute and find the values for the other variable.** We must back-substitute each value of x into either one of the original equations. Let's use $x^2 + y^2 = 10$, Equation 2. If $x = 1$,

$$\begin{aligned} 1^2 + y^2 &= 10 \quad \text{Replace } x \text{ with 1 in Equation 2.} \\ y^2 &= 9 \quad \text{Subtract 1 from both sides.} \\ y &= \pm 3. \quad \text{Apply the square root property.} \end{aligned}$$

$(1, 3)$ and $(1, -3)$ are solutions. If $x = -1$,

$$\begin{aligned} (-1)^2 + y^2 &= 10 \quad \text{Replace } x \text{ with } -1 \text{ in Equation 2.} \\ y^2 &= 9 \quad \text{The steps are the same as before.} \\ y &= \pm 3. \end{aligned}$$

$(-1, 3)$ and $(-1, -3)$ are solutions.

Step 6. **Check.** Take a moment to show that each of the four ordered pairs satisfies the given equations, $4x^2 + y^2 = 13$ and $x^2 + y^2 = 10$. The solutions are $(1, 3)$, $(1, -3)$, $(-1, 3)$, and $(-1, -3)$, and the solution set of the given system is $\{(1, 3), (1, -3), (-1, 3), (-1, -3)\}$.

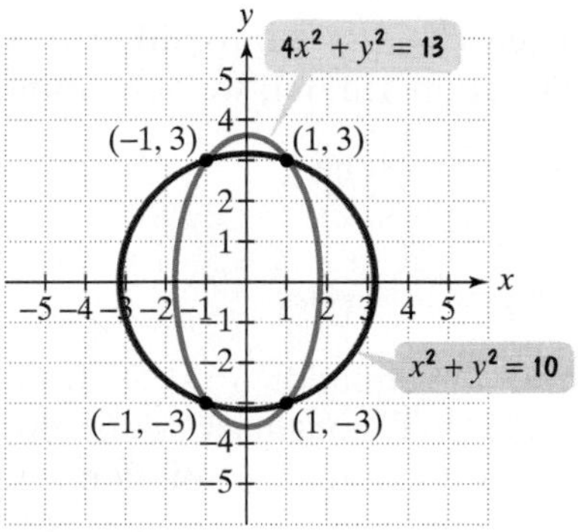

FIGURE 13.37 A system with four solutions

Figure 13.37 shows the graphs of the equations in the system, an ellipse and a circle, and the solutions as intersection points. ■

STUDY TIP

When solving nonlinear systems, extra solutions may be introduced that do not satisfy both equations in the system. Therefore, you should get into the habit of checking all proposed pairs in each of the system's two equations.

✔ **CHECK POINT 3** Solve the system:

$$\begin{aligned} 3x^2 + 2y^2 &= 35 \\ 4x^2 + 3y^2 &= 48. \end{aligned}$$

In solving nonlinear systems, we will include only ordered pairs with real numbers in the solution set. We have seen that each of these ordered pairs corresponds to a point of intersection of the system's graphs.

EXAMPLE 4 Solving a Nonlinear System by the Addition Method

Solve the system:

$$\begin{aligned} y &= x^2 + 3 \quad \text{Equation 1 (The graph is a parabola.)} \\ x^2 + y^2 &= 9. \quad \text{Equation 2 (The graph is a circle.)} \end{aligned}$$

SOLUTION We could use substitution because Equation 1, $y = x^2 + 3$, has y expressed in terms of x, but this would result in a fourth-degree equation. However, we can rewrite Equation 1 by subtracting x^2 from both sides and adding the equations to eliminate the x^2-terms.

Notice how like terms are arranged in columns.

$$\begin{array}{rcrcl} -x^2 & + & y & & = 3 \\ x^2 & & & + y^2 & = 9 \\ \hline & & y & + y^2 & = 12 \end{array}$$

Subtract x^2 from both sides of Equation 1.
This is Equation 2.
Add the equations.

We now solve this quadratic equation.

$$y + y^2 = 12 \quad \text{This is the equation containing one variable.}$$

$$y^2 + y - 12 = 0 \quad \text{Subtract 12 from both sides and write the quadratic equation in standard form.}$$

$$(y + 4)(y - 3) = 0 \quad \text{Factor.}$$

$$y + 4 = 0 \quad \text{or} \quad y - 3 = 0 \quad \text{Set each factor equal to 0.}$$

$$y = -4 \qquad\qquad y = 3 \quad \text{Solve for } y.$$

To complete the solution, we must back-substitute each value of y into either one of the original equations. We will use $y = x^2 + 3$, Equation 1. First, we substitute -4 for y.

$$-4 = x^2 + 3$$

$$-7 = x^2 \quad \text{Subtract 3 from both sides.}$$

Because the square of a real number cannot be negative, the equation $x^2 = -7$ does not have real-number solutions. We will not include the imaginary solutions, $x = \pm\sqrt{-7}$, or $\sqrt{7}i$ and $-\sqrt{7}i$, in the ordered pairs that make up the solution set. Thus, we move on to our other value for y, 3, and substitute this value into Equation 1.

$$y = x^2 + 3 \quad \text{This is Equation 1.}$$

$$3 = x^2 + 3 \quad \text{Back-substitute 3 for } y.$$

$$0 = x^2 \quad \text{Subtract 3 from both sides.}$$

$$0 = x \quad \text{Solve for } x.$$

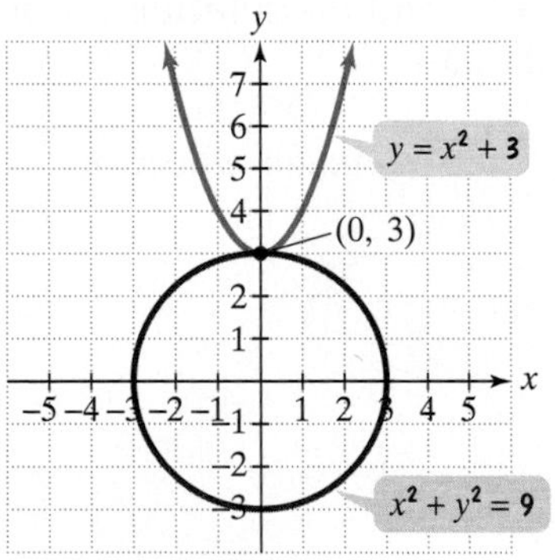

FIGURE 13.38 A system with one real solution.

We showed that if $y = 3$, then $x = 0$. Thus, $(0, 3)$ is the solution with a real ordered pair. Take a moment to show that $(0, 3)$ satisfies the given equations, $y = x^2 + 3$ and $x^2 + y^2 = 9$. The solution set of the system is $\{(0, 3)\}$. Figure 13.38 shows the system's graphs and the solution as an intersection point. ■

CHECK POINT 4 Solve the system:

$$y = x^2 + 5$$

$$x^2 + y^2 = 25.$$

4 Solve problems using systems of nonlinear equations.

Applications Many geometric problems can be modeled and solved by the use of systems of nonlinear equations. We will use our step-by-step strategy for solving problems using mathematical models that are created from verbal models.

EXAMPLE 5 An Application of a Nonlinear System

You have 36 yards of fencing to build the enclosure in Figure 13.39. Some of this fencing is to be used to build an internal divider. If you'd like to enclose 54 square yards, what are the dimensions of the enclosure?

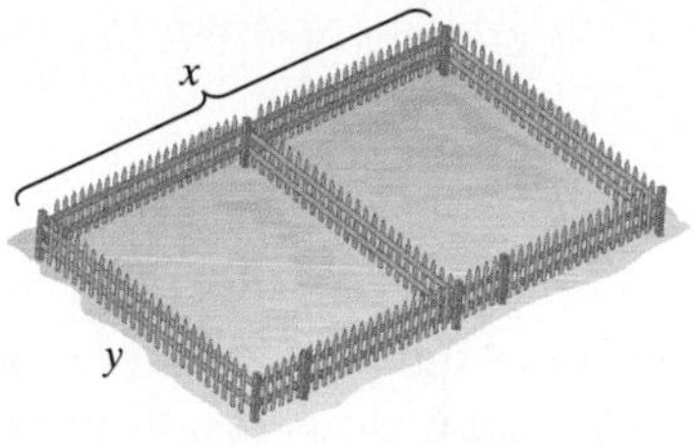

FIGURE 13.39 Building an enclosure

SOLUTION

Step 1. **Use variables to represent unknown quantities.** Let x = the enclosure's length and y = the enclosure's width. These variables are shown in Figure 13.39.

Step 2. **Write a system of equations describing the problem's conditions.** The first condition is that you have 36 yards of fencing.

Fencing along both lengths	plus	fencing along both widths	plus	fencing for the internal divider	equals	36 yards.
$2x$	$+$	$2y$	$+$	y	$=$	36

Adding like terms, we can express the equation that models the verbal conditions for the fencing as $2x + 3y = 36$.

The second condition is that you'd like to enclose 54 square yards. The rectangle's area, the product of its length and its width, must be 54 square yards.

Length	times	width	is	54 square yards.
x	$\cdot$	y	$=$	54

Step 3. **Solve the system and answer the problem's question.** We must solve the system

$$2x + 3y = 36 \quad \text{Equation 1}$$
$$xy = 54. \quad \text{Equation 2}$$

We will use substitution. Because Equation 1 has no coefficients of 1 or -1, we will work with Equation 2 and solve for y. Dividing both sides of $xy = 54$ by x, we obtain

$$y = \frac{54}{x}.$$

Now we substitute $\frac{54}{x}$ for y in Equation 1 and solve for x.

$$2x + 3y = 36 \quad \text{This is Equation 1.}$$
$$2x + 3 \cdot \frac{54}{x} = 36 \quad \text{Substitute } \frac{54}{x} \text{ for } y.$$
$$2x + \frac{162}{x} = 36 \quad \text{Multiply.}$$
$$x\left(2x + \frac{162}{x}\right) = 36 \cdot x \quad \text{Clear fractions by multiplying both sides by } x.$$
$$2x^2 + 162 = 36x \quad \text{Use the distributive property on the left side.}$$
$$2x^2 - 36x + 162 = 0 \quad \text{Subtract } 36x \text{ from both sides and write the quadratic equation in standard form.}$$
$$2(x^2 - 18x + 81) = 0 \quad \text{Factor out 2.}$$
$$2(x - 9)^2 = 0 \quad \text{Factor completely using } A^2 - 2AB + B^2 = (A - B)^2.$$
$$x - 9 = 0 \quad \text{Set the repeated factor equal to zero.}$$
$$x = 9 \quad \text{Solve for } x.$$

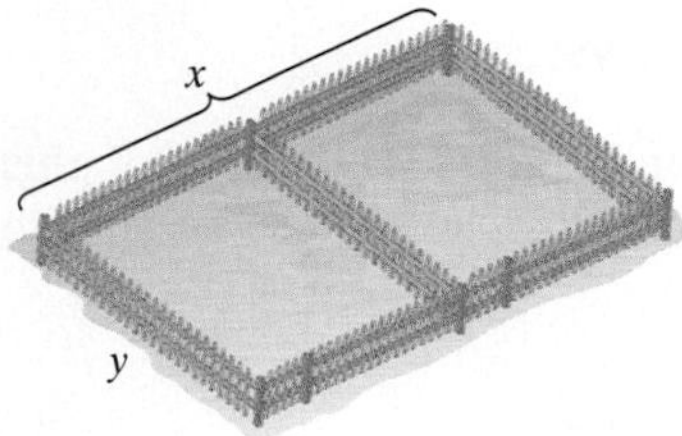

FIGURE 13.39 (repeated)

We back-substitute this value of x into $y = \frac{54}{x}$.

$$\text{If } x = 9, \quad y = \frac{54}{9} = 6.$$

This means that the dimensions of the enclosure in Figure 13.39 are 9 yards by 6 yards.

Step 4. **Check the proposed solution in the original wording of the problem.** Take a moment to check that a length of 9 yards and a width of 6 yards results in 36 yards of fencing and an area of 54 square yards. ■

CHECK POINT 5 Find the length and width of a rectangle whose perimeter is 20 feet and whose area is 21 square feet.

13.5 EXERCISE SET

Student Solutions Manual | CD/Video | PH Math/Tutor Center | MathXL Tutorials on CD | MathXL® | MyMathLab | Interactmath.com

Practice Exercises

In Exercises 1–18, solve each system by the substitution method.

1. $x + y = 2$
$y = x^2 - 4$

2. $x - y = -1$
$y = x^2 + 1$

3. $x + y = 2$
$y = x^2 - 4x + 4$

4. $2x + y = -5$
$y = x^2 + 6x + 7$

5. $y = x^2 - 4x - 10$
$y = -x^2 - 2x + 14$

6. $y = x^2 + 4x + 5$
$y = x^2 + 2x - 1$

7. $x^2 + y^2 = 25$
$x - y = 1$

8. $x^2 + y^2 = 5$
$3x - y = 5$

9. $xy = 6$
$2x - y = 1$

10. $xy = -12$
$x - 2y + 14 = 0$

11. $y^2 = x^2 - 9$
$2y = x - 3$

12. $x^2 + y = 4$
$2x + y = 1$

13. $xy = 3$
$x^2 + y^2 = 10$

14. $xy = 4$
$x^2 + y^2 = 8$

15. $x + y = 1$
$x^2 + xy - y^2 = -5$

16. $x + y = -3$
$x^2 + 2y^2 = 12y + 18$

17. $x + y = 1$
$(x - 1)^2 + (y + 2)^2 = 10$

18. $2x + y = 4$
$(x + 1)^2 + (y - 2)^2 = 4$

In Exercises 19–28, solve each system by the addition method.

19. $x^2 + y^2 = 13$
$x^2 - y^2 = 5$

20. $4x^2 - y^2 = 4$
$4x^2 + y^2 = 4$

21. $x^2 - 4y^2 = -7$
$3x^2 + y^2 = 31$

22. $3x^2 - 2y^2 = -5$
$2x^2 - y^2 = -2$

23. $3x^2 + 4y^2 - 16 = 0$
$2x^2 - 3y^2 - 5 = 0$

24. $16x^2 - 4y^2 - 72 = 0$
$x^2 - y^2 - 3 = 0$

25. $x^2 + y^2 = 25$
$(x - 8)^2 + y^2 = 41$

26. $x^2 + y^2 = 4$
$x^2 + (y - 3)^2 = 9$

27. $y^2 - x = 4$
$x^2 + y^2 = 4$

28. $x^2 - 2y = 8$
$x^2 + y^2 = 16$

In Exercises 29–42, solve each system by the method of your choice.

29. $3x^2 + 4y^2 = 16$
$2x^2 - 3y^2 = 5$

30. $x + y^2 = 4$
$x^2 + y^2 = 16$

31. $2x^2 + y^2 = 18$
$xy = 4$

32. $x^2 + 4y^2 = 20$
$xy = 4$

33. $x^2 + 4y^2 = 20$
$x + 2y = 6$

34. $3x^2 - 2y^2 = 1$
$4x - y = 3$

35. $x^3 + y = 0$
$x^2 - y = 0$

36. $x^3 + y = 0$
$2x^2 - y = 0$

37. $x^2 + (y - 2)^2 = 4$
$x^2 - 2y = 0$

38. $x^2 - y^2 - 4x + 6y - 4 = 0$
$x^2 + y^2 - 4x - 6y + 12 = 0$

39. $y = (x + 3)^2$
$x + 2y = -2$

40. $(x - 1)^2 + (y + 1)^2 = 5$
$2x - y = 3$

41. $x^2 + y^2 + 3y = 22$
$2x + y = -1$

42. $x - 3y = -5$
$x^2 + y^2 - 25 = 0$

In Exercises 43–46, let x represent one number and let y represent the other number. Use the given conditions to write a system of nonlinear equations. Solve the system and find the numbers.

43. The sum of two numbers is 10 and their product is 24. Find the numbers.

44. The sum of two numbers is 20 and their product is 96. Find the numbers.

45. The difference between the squares of two numbers is 3. Twice the square of the first number increased by the square of the second number is 9. Find the numbers.

46. The difference between the squares of two numbers is 5. Twice the square of the second number subtracted from three times the square of the first number is 19. Find the numbers.

Practice Plus

In Exercises 47–52, solve each system by the method of your choice.

47. $2x^2 + xy = 6$
$x^2 + 2xy = 0$

48. $4x^2 + xy = 30$
$x^2 + 3xy = -9$

49. $-4x + y = 12$
$y = x^3 + 3x^2$

50. $-9x + y = 45$
$y = x^3 + 5x^2$

51. $\frac{3}{x^2} + \frac{1}{y^2} = 7$
$\frac{5}{x^2} - \frac{2}{y^2} = -3$

52. $\frac{2}{x^2} + \frac{1}{y^2} = 11$
$\frac{4}{x^2} - \frac{2}{y^2} = -14$

In Exercises 53–54, make a rough sketch in a rectangular coordinate system of the graphs representing the equations in each system.

53. The system, whose graphs are a circle and an ellipse, has two solutions. Both solutions can be represented as points on the x-axis.

54. The system, whose graphs are a line with positive slope and a parabola that is not a function, has two solutions.

Application Exercises

55. A planet follows an elliptical path described by $16x^2 + 4y^2 = 64$. A comet follows the parabolic path $y = x^2 - 4$. Where might the comet intersect the orbiting planet?

56. A system for tracking ships indicates that a ship lies on a hyperbolic path described by $2y^2 - x^2 = 1$. The process is repeated and the ship is found to lie on a hyperbolic path described by $2x^2 - y^2 = 1$. If it is known that the ship is located in the first quadrant of the coordinate system, determine its exact location.

57. Find the length and width of a rectangle whose perimeter is 36 feet and whose area is 77 square feet.

58. Find the length and width of a rectangle whose perimeter is 40 feet and whose area is 96 square feet.

Use the formula for the area of a rectangle and the Pythagorean theorem to solve Exercises 59–60.

59. A small television has a picture with a diagonal measure of 10 inches and a viewing area of 48 square inches. Find the length and width of the screen.

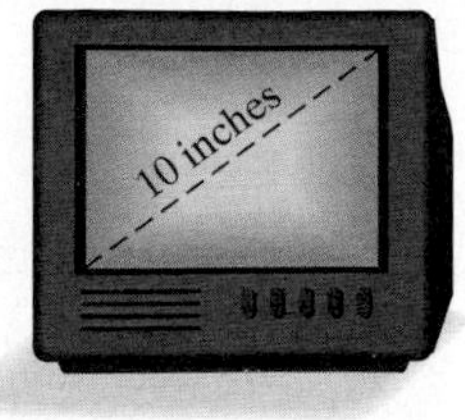

60. The area of a rug is 108 square feet, and the length of its diagonal is 15 feet. Find the length and width of the rug.

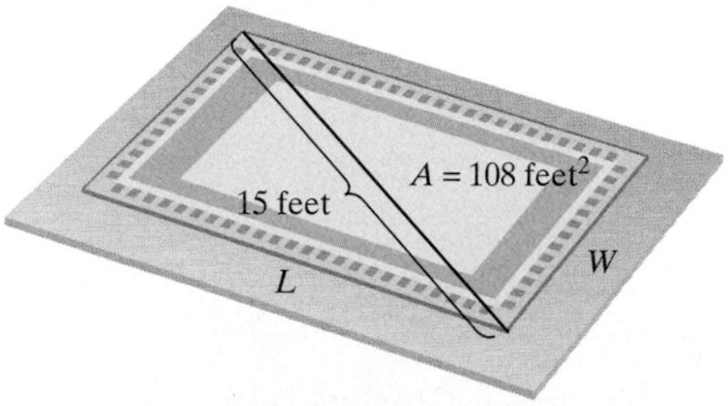

61. The figure shows a square floor plan with a smaller square area that will accommodate a combination fountain and pool. The floor with the fountain-pool area removed has an area of 21 square meters and a perimeter of 24 meters. Find the dimensions of the floor and the dimensions of the square that will accommodate the fountain and pool.

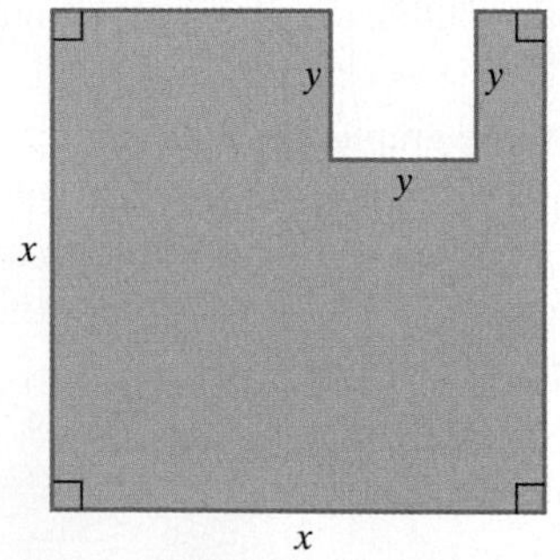

62. The area of the rectangular piece of cardboard shown on the left is 216 square inches. The cardboard is used to make an open box by cutting a 2-inch square from each corner and turning up the sides. If the box is to have a volume of 224 cubic inches, find the length and width of the cardboard that must be used.

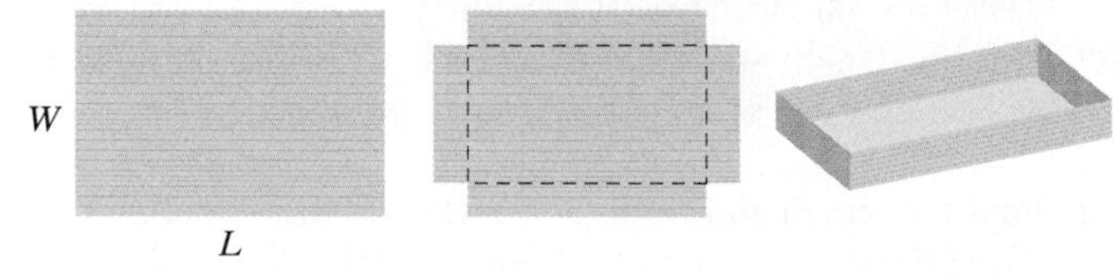

Writing in Mathematics

63. What is a system of nonlinear equations? Provide an example with your description.

64. Explain how to solve a nonlinear system using the substitution method. Use $x^2 + y^2 = 9$ and $2x - y = 3$ to illustrate your explanation.

65. Explain how to solve a nonlinear system using the addition method. Use $x^2 - y^2 = 5$ and $3x^2 - 2y^2 = 19$ to illustrate your explanation.

66. The daily demand and supply models for a carrot cake supplied by a bakery to a convenience store are given by the demand model $N = 40 - 3p$ and the supply model $N = \frac{p^2}{10}$, in which p is the price of the cake and N is the number of cakes sold or supplied each day to the convenience store. Explain how to determine the price at which supply and demand are equal. Then describe how to find how many carrot cakes can be supplied and sold each day at this price.

Technology Exercises

67. Verify your solutions to any five exercises from Exercises 1–42 by using a graphing utility to graph the two equations in the system in the same viewing rectangle. Then use the trace or intersection feature to verify the solutions.

68. Write a system of equations, one equation whose graph is a line and the other whose graph is a parabola, that has no ordered pairs that are real numbers in its solution set. Graph the equations using a graphing utility and verify that you are correct.

Critical Thinking Exercises

69. Which one of the following is true?

a. A system of two equations in two variables whose graphs are a circle and a line can have four real ordered-pair solutions.

b. A system of two equations in two variables whose graphs are a parabola and a circle can have four real ordered-pair solutions.

c. A system of two equations in two variables whose graphs are two circles must have at least two real ordered-pair solutions.

d. A system of two equations in two variables whose graphs are a parabola and a circle cannot have only one real ordered-pair solution.

70. Find a and b in this figure.

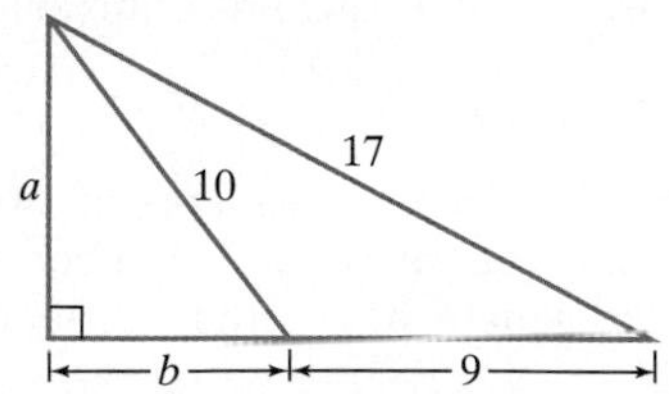

Solve the systems in Exercises 71–72.

71. $\log_y x = 3$
$\log_y(4x) = 5$

72. $\log x^2 = y + 3$
$\log x = y - 1$

Review Exercises

73. Graph: $3x - 2y \le 6$.
(Section 9.4, Example 1)

74. Find the slope of the line passing through $(-2, -3)$ and $(1, 5)$.
(Section 3.3, Example 1)

75. Multiply: $(3x - 2)(2x^2 - 4x + 3)$.
(Section 5.2, Example 7)

GROUP PROJECT

CHAPTER 13

Modeling Planetary Motion

Polish astronomer Nicolaus Copernicus (1473–1543) was correct in stating that planets in our solar system revolve around the sun and not Earth. However, he incorrectly believed that celestial orbits move in perfect circles, calling his system "the ballet of the planets."

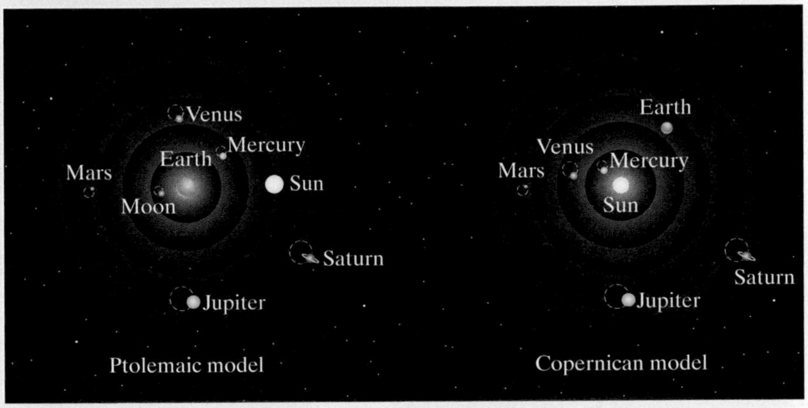

German scientist and mathematician Johannes Kepler (1571–1630) discovered that planets move in elliptical orbits with the sun at one focus. In this exercise, group members will write equations for two of these orbits and use a graphing utility to see what they look like. Use the following information:

Earth's orbit: Length of major axis: 186 million miles
Length of minor axis: 185.8 million miles

Mars's orbit: Length of major axis: 283.5 million miles
Length of minor axis: 278.5 million miles.

a. Group members should write equations in the form $\frac{x^2}{a^2} + \frac{y^2}{b^2} = 1$ for the elliptical orbits of Earth and Mars.

b. Use a graphing utility to graph the two ellipses in the same $[-300, 300, 50]$ by $[-200, 200, 50]$ viewing rectangle. Based on these graphs, explain why early astronomers incorrectly used the Copernican model to describe planetary motion.

c. Describing planetary orbits, Kepler wrote, "The heavenly motions are nothing but a continuous song for several voices, to be perceived by the intellect, not by the ear." Group members should discuss what Kepler meant by this statement.

CHAPTER 13 SUMMARY

Definitions and Concepts	Examples

Section 13.1 The Circle

A circle is the set of all points in a plane that are equidistant from a fixed point, the center. The distance from the center to any point on the circle is the radius.

Standard Form of the Equation of a Circle
The graph of $(x - h)^2 + (y - k)^2 = r^2$ is a circle with center (h, k) and radius r.

General Form of the Equation of a Circle

$$x^2 + y^2 + Dx + Ey + F = 0$$

Convert from the general form to the standard form by completing the square on x and y.

Find the center and radius and graph:

$$(x - 3)^2 + (y + 4)^2 = 16.$$

$$(x - 3)^2 + (y - (-4))^2 = 4^2$$

$h = 3$ $k = -4$ $r = 4$

The center, (h, k), is $(3, -4)$ and the radius, r, is 4.

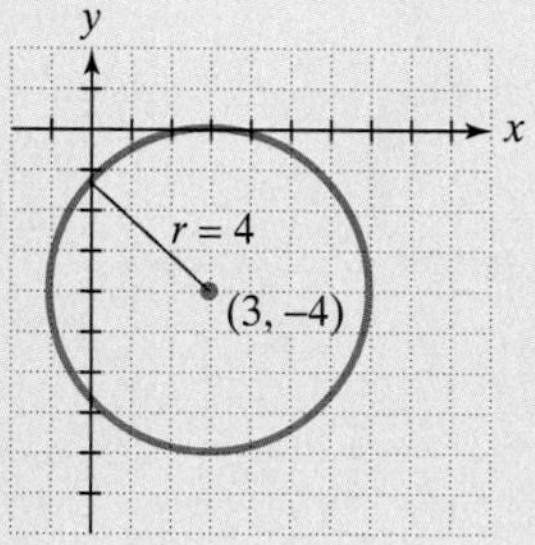

Section 13.2 The Ellipse

An ellipse is the set of all points in a plane the sum of whose distances from two fixed points, the foci, is constant. The midpoint of the segment connecting the foci is the center of the ellipse.

Standard Forms of the Equations of an Ellipse Centered at the Origin $(a^2 > b^2)$

Horizontal with vertices $(a, 0)$ and $(-a, 0)$	Vertical with vertices $(0, a)$ and $(0, -a)$
$\frac{x^2}{a^2} + \frac{y^2}{b^2} = 1$	$\frac{x^2}{b^2} + \frac{y^2}{a^2} = 1$
Endpoints of major axis a units left and right of center; minor axis b units up and down from center.	Endpoints of major axis a units up and down from center; minor axis b units left and right of center.

The equations $\frac{(x-h)^2}{a^2} + \frac{(y-k)^2}{b^2} = 1$ and $\frac{(x-h)^2}{b^2} + \frac{(y-k)^2}{a^2} = 1$ $(a^2 > b^2)$ represent ellipses centered at (h, k).

Graph: $\frac{(x + 2)^2}{9} + \frac{(y + 4)^2}{25} = 1.$

$$\frac{(x - (-2))^2}{9} + \frac{(y - (-4))^2}{25} = 1$$

$b^2 = 9$ $a^2 = 25$

Center, (h, k), is $(-2, -4)$. With $a^2 = 25$, vertices are 5 units above and below the center. With $b^2 = 9$, endpoints of minor axis are 3 units to the left and right of the center.

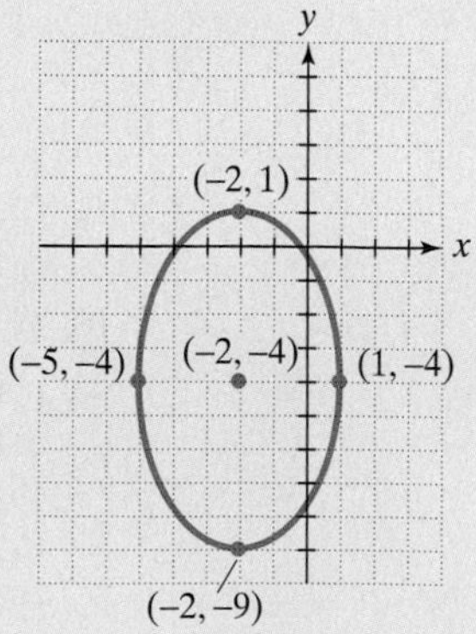

Definitions and Concepts	Examples

Section 13.3 The Hyperbola

A hyperbola is the set of all points in a plane the difference of whose distances from two fixed points, the foci, is constant.

Standard Forms of the Equations of a Hyperbola Centered at the Origin

$$\frac{x^2}{a^2} - \frac{y^2}{b^2} = 1 \qquad \frac{y^2}{a^2} - \frac{x^2}{b^2} = 1$$

Vertices are $(a, 0)$ and $(-a, 0)$. Vertices are $(0, a)$ and $(0, -a)$.

As x and y get larger, the two branches of a hyperbola approach a pair of intersecting lines, called asymptotes. Draw the rectangle centered at the origin with sides parallel to the axes, crossing one axis at $\pm a$ and the other at $\pm b$. Draw the diagonals of this rectangle and extend them to obtain the asymptotes. Draw the two branches of the hyperbola by starting at each vertex and approaching the asymptotes.

Graph: $4x^2 - 9y^2 = 36$.

$$\frac{4x^2}{36} - \frac{9y^2}{36} = \frac{36}{36}$$

$$\frac{x^2}{9} - \frac{y^2}{4} = 1$$

$a^2 = 9$ $\quad$ $b^2 = 4$

Vertices are $(3, 0)$ and $(-3, 0)$. Draw a rectangle using -3 and 3 on the x-axis and -2 and 2 on the y-axis. Its extended diagonals are the asymptotes.

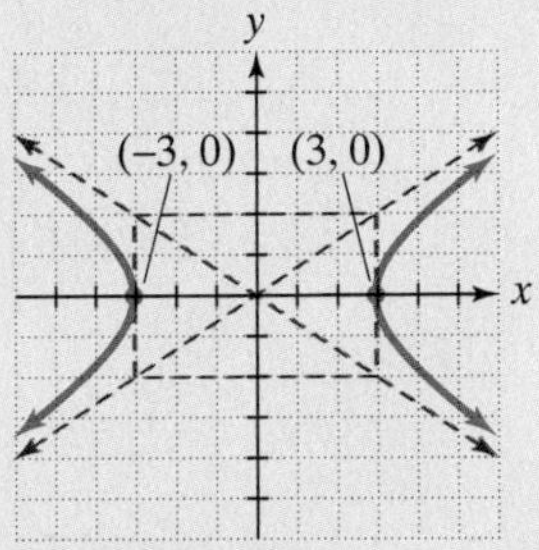

Section 13.4 The Parabola; Identifying Conic Sections

A parabola is the set of all points in a plane that are equidistant from a fixed line, the directrix, and a fixed point, the focus, that is not on the line. The line passing through the focus and perpendicular to the directrix is the axis of symmetry. The point of intersection of the parabola with its axis of symmetry is the vertex.

Equations of Horizontal Parabolas

$x = a(y - k)^2 + h$
Vertex is (h, k).

$x = ay^2 + by + c$
y-coordinate of vertex is $y = -\frac{b}{2a}$.

If $a > 0$, the parabola opens to the right. If $a < 0$, the parabola opens to the left.

Equations of Vertical Parabolas

$y = a(x - h)^2 + k$
Vertex is (h, k).

$y = ax^2 + bx + c$
x-coordinate of vertex is $x = -\frac{b}{2a}$.

If $a > 0$, the parabola opens upward. If $a < 0$, the parabola opens downward.

Find the vertex and graph:

$$x = -(y + 3)^2 + 4.$$

$$x = -1(y - (-3))^2 + 4$$

$a = -1$ $\quad$ $k = -3$ $\quad$ $h = 4$

Parabola opens to the left. Vertex, (h, k), is $(4, -3)$.
x-intercept: Let $y = 0$.

$$x = -(0 + 3)^2 + 4 = -9 + 4 = -5$$

y-intercepts: Let $x = 0$.

$$0 = -(y + 3)^2 + 4$$
$$(y + 3)^2 = 4$$
$$y + 3 = \pm\sqrt{4}$$
$$y = -1 \quad \text{or} \quad y = -5$$

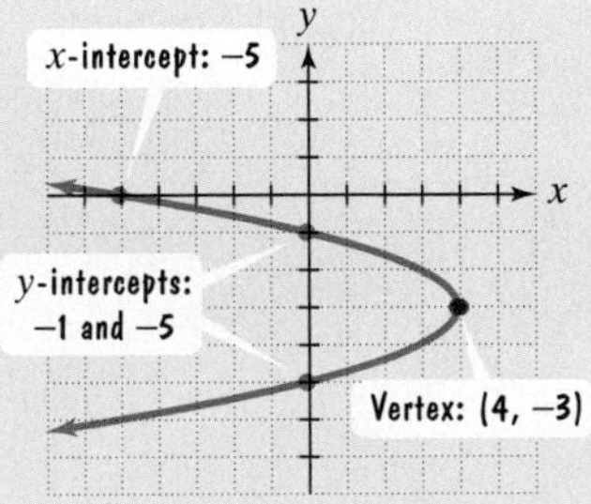

Definitions and Concepts	Examples

Section 13.4 The Parabola; Identifying Conic Sections (continued)

Recognizing Conic Sections from Equations
Conic sections result from the intersection of a cone and a plane. Their equations contain x^2-terms, y^2-terms, or both. For parabolas, only one variable is squared. For circles, ellipses, and hyperbolas, both variables are squared. Collect x^2- and y^2-terms on the same side of the equation to identify the graph:

- Circle: x^2 and y^2 have the same coefficient of the same sign.
- Ellipse: x^2 and y^2 have different coefficients of the same sign.
- Hyperbola: x^2 and y^2 have coefficients with opposite signs.

Examples:

- $9x^2 + 4y^2 = 36$

  Different coefficients of the same sign; ellipse

- $9y^2 = 25x^2 + 225$

  $9y^2 - 25x^2 = 225$

  Coefficients with opposite signs; hyperbola

- $x = -y^2 + 4y$

  Only one variable, y, is squared; parabola

- $\dfrac{x^2}{9} + \dfrac{y^2}{9} = 1$

  $x^2 + y^2 = 9$

  Same coefficient; circle

Section 13.5 Systems of Nonlinear Equations in Two Variables

A system of two nonlinear equations in two variables, x and y, contains at least one equation that cannot be expressed in the form $Ax + By = C$. Systems can be solved by the substitution method or the addition method. Each solution corresponds to a point of intersection of the system's graphs.

Solve:

$$x^2 + y^2 = 25$$

$$x - 3y = -5.$$

Using substitution: $x = 3y - 5$

$$(3y - 5)^2 + y^2 = 25$$

$$9y^2 - 30y + 25 + y^2 = 25$$

$$10y^2 - 30y = 0$$

$$10y(y - 3) = 0$$

$$y = 0 \quad \text{or} \quad y = 3$$

If $y = 0$: $x = 3y - 5 = 3 \cdot 0 - 5 = -5$.

If $y = 3$: $x = 3y - 5 = 3 \cdot 3 - 5 = 4$.

The solutions are $(-5, 0)$ and $(4, 3)$, and the solution set is $\{(-5, 0), (4, 3)\}$.

CHAPTER 13 REVIEW EXERCISES

13.1

In Exercises 1–2, write the standard form of the equation of the circle with the given center and radius.

1. Center $(0, 0)$, $r = 3$

2. Center $(-2, 4)$, $r = 6$

In Exercises 3–6, give the center and radius of each circle and graph its equation.

3. $x^2 + y^2 = 1$

4. $(x + 2)^2 + (y - 3)^2 = 9$

5. $x^2 + y^2 - 4x + 2y - 4 = 0$

6. $x^2 + y^2 - 4y = 0$

13.2 *In Exercises 7–12, graph each ellipse.*

7. $\dfrac{x^2}{36} + \dfrac{y^2}{25} = 1$

8. $\dfrac{x^2}{25} + \dfrac{y^2}{16} = 1$

9. $4x^2 + y^2 = 16$

10. $4x^2 + 9y^2 = 36$

11. $\dfrac{(x-1)^2}{16} + \dfrac{(y+2)^2}{9} = 1$

12. $\dfrac{(x+1)^2}{9} + \dfrac{(y-2)^2}{16} = 1$

13. A semielliptic archway has a height of 15 feet at the center and a width of 50 feet, as shown in the figure. The 50-foot width consists of a two-lane road. Can a truck that is 12 feet high and 14 feet wide drive under the archway without going into the other lane?

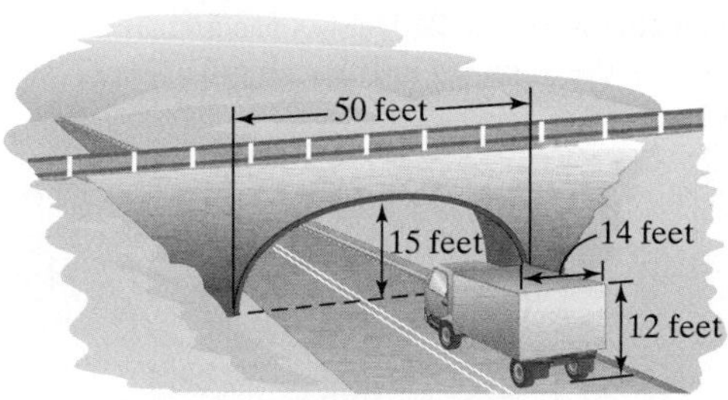

13.3 *In Exercises 14–17, use vertices and asymptotes to graph each hyperbola.*

14. $\dfrac{x^2}{16} - y^2 = 1$

15. $\dfrac{y^2}{16} - x^2 = 1$

16. $9x^2 - 16y^2 = 144$

17. $4y^2 - x^2 = 16$

13.4 *In Exercises 18–21, use the vertex and intercepts to sketch the graph of each equation. Give the equation for the horizontal parabola's axis of symmetry. If needed, find additional points on the parabola by choosing values of y on each side of the axis of symmetry.*

18. $x = (y - 3)^2 - 4$

19. $x = -2(y + 3)^2 + 2$

20. $x = y^2 - 8y + 12$

21. $x = -y^2 - 4y + 6$

In Exercises 22–28, indicate whether the graph of each equation is a circle, an ellipse, a hyperbola, or a parabola.

22. $x + 8y = y^2 + 10$

23. $16x^2 = 32 - y^2$

24. $x^2 = 25 + 25y^2$

25. $x^2 = 4 - y^2$

26. $36y^2 = 576 + 16x^2$

27. $\dfrac{(x+3)^2}{9} + \dfrac{(y-4)^2}{25} = 1$

28. $y = x^2 + 6x + 9$

In Exercises 29–37, indicate whether the graph of each equation is a circle, an ellipse, a hyperbola, or a parabola. Then graph the conic section.

29. $5x^2 + 5y^2 = 180$

30. $4x^2 + 9y^2 = 36$

31. $4x^2 - 9y^2 = 36$

32. $\frac{x^2}{25} + \frac{y^2}{1} = 1$

33. $x + 3 = -y^2 + 2y$

34. $y - 3 = x^2 - 2x$

35. $\frac{(x+2)^2}{16} + \frac{(y-5)^2}{4} = 1$

36. $(x - 3)^2 + (y + 2)^2 = 4$

37. $x^2 + y^2 + 6x - 2y + 6 = 0$

38. An engineer is designing headlight units for automobiles. The unit has a parabolic surface with a diameter of 12 inches and a depth of 3 inches. The situation is illustrated in the figure, where a coordinate system has been superimposed.

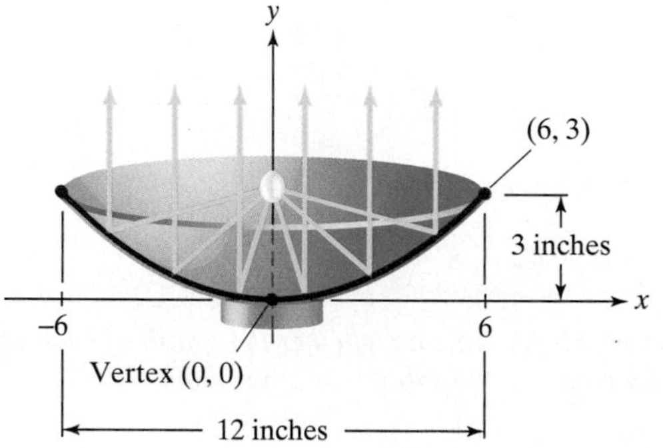

a. Use the point $(6, 3)$ to write an equation in the form $y = ax^2$ for the parabola used to design the headlight.

b. The light source should be placed at the focus $(0, p)$. The value of p is given by the equation $a = \frac{1}{4p}$. Where should the light source be placed? Describe this placement relative to the vertex.

13.5 *In Exercises 39–49, solve each system by the method of your choice.*

39. $5y = x^2 - 1$
$x - y = 1$

40. $y = x^2 + 2x + 1$
$x + y = 1$

41. $x^2 + y^2 = 2$
$x + y = 0$

42. $2x^2 + y^2 = 24$
$x^2 + y^2 = 15$

43. $xy - 4 = 0$
$y - x = 0$

44. $y^2 = 4x$
$x - 2y + 3 = 0$

45. $x^2 + y^2 = 10$
$y = x + 2$

46. $xy = 1$
$y = 2x + 1$

47. $x + y + 1 = 0$
$x^2 + y^2 + 6y - x = -5$

48. $x^2 + y^2 = 13$
$x^2 - y = 7$

49. $2x^2 + 3y^2 = 21$
$3x^2 - 4y^2 = 23$

50. The perimeter of a rectangle is 26 meters and its area is 40 square meters. Find its dimensions.

51. Find the coordinates of all points (x, y) that lie on the line whose equation is $2x + y = 8$, so that the area of the rectangle shown in the figure is 6 square units.

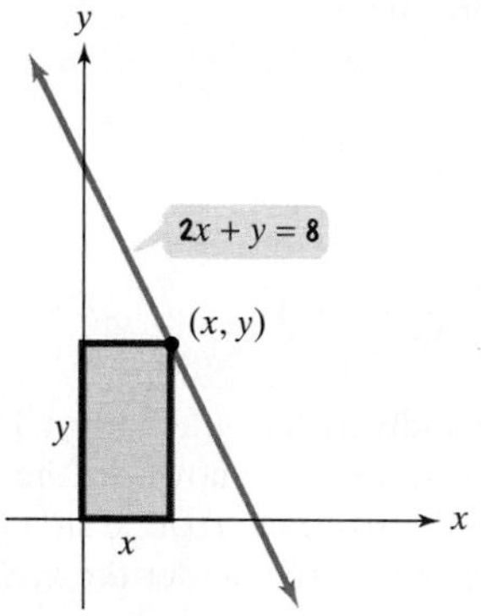

52. Two adjoining square fields with an area of 2900 square feet are to be enclosed with 240 feet of fencing. The situation is represented in the figure. Find the length of each side where a variable appears.

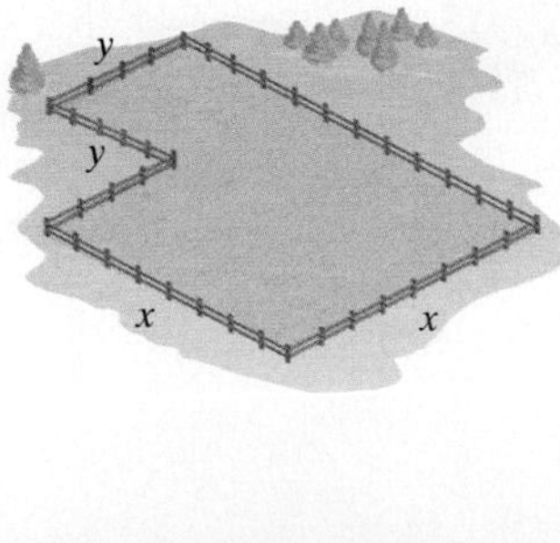

CHAPTER 13 TEST

Remember to use your Chapter Test Prep Video CD to see the worked-out solutions to the test questions you want to review.

1. Write the standard form of the equation of the circle with center $(3, -2)$ and radius 5.

In Exercises 2–3, give the center and radius of each circle.

2. $(x - 5)^2 + (y + 3)^2 = 49$

3. $x^2 + y^2 + 4x - 6y - 3 = 0$

In Exercises 4–5, give the coordinates of the vertex for each parabola.

4. $x = -2(y + 3)^2 + 7$

5. $x = y^2 + 10y + 23$

In Exercises 6–14, indicate whether the graph of each equation is a circle, an ellipse, a hyperbola, or a parabola. Then graph the conic section.

6. $\frac{x^2}{4} - \frac{y^2}{9} = 1$

7. $4x^2 + 9y^2 = 36$

8. $x = (y + 1)^2 - 4$

9. $16x^2 + y^2 = 16$

10. $25y^2 = 9x^2 + 225$

11. $x = -y^2 + 6y$

12. $\frac{(x - 2)^2}{16} + \frac{(y + 3)^2}{9} = 1$

13. $(x + 1)^2 + (y + 2)^2 = 9$

14. $\frac{x^2}{4} + \frac{y^2}{4} = 1$

In Exercises 15–16, solve each system.

15. $x^2 + y^2 = 25$
$x + y = 1$

16. $2x^2 - 5y^2 = -2$
$3x^2 + 2y^2 = 35$

17. The rectangular plot of land shown in the figure is to be fenced along three sides using 39 feet of fencing. No fencing is to be placed along the river's edge. The area of the plot is 180 square feet. What are its dimensions?

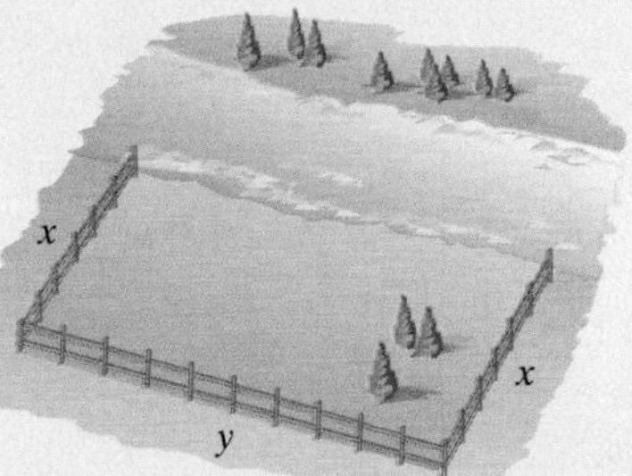

18. A rectangle has a diagonal of 5 feet and a perimeter of 14 feet. Find the rectangle's dimensions.

CUMULATIVE REVIEW EXERCISES (CHAPTERS 1–13)

In Exercises 1–7, solve each equation, inequality, or system.

1. $3x + 7 > 4$ or $6 - x < 1$

2. $x(2x - 7) = 4$

3. $\frac{5}{x - 3} = 1 + \frac{30}{x^2 - 9}$

4. $3x^2 + 8x + 5 < 0$

5. $3^{2x-1} = 81$

6. $30e^{0.7x} = 240$

7. $3x^2 + 4y^2 = 39$
$5x^2 - 2y^2 = -13$

In Exercises 8–11, graph each function, equation, or inequality in a rectangular coordinate system.

8. $f(x) = -\frac{2}{3}x + 4$

9. $3x - y > 6$

10. $x^2 + y^2 + 4x - 6y + 9 = 0$

11. $9x^2 - 4y^2 = 36$

In Exercises 12–15, perform the indicated operations, and simplify, if possible.

12. $-2(3^2 - 12)^3 - 45 \div 9 - 3$

13. $(3x^3 - 19x^2 + 17x + 4) \div (3x - 4)$

14. $\sqrt[3]{4x^2y^5} \cdot \sqrt[3]{4xy^2}$

15. $(2 + 3i)(4 - i)$

In Exercises 16–17, factor completely.

16. $12x^3 - 36x^2 + 27x$

17. $x^3 - 2x^2 - 9x + 18$

18. Find the domain: $f(x) = \sqrt{6 - 3x}$.

19. Rationalize the denominator: $\dfrac{1 - \sqrt{x}}{1 + \sqrt{x}}$.

20. Write as a single logarithm: $\dfrac{1}{3}\ln x + 7\ln y$.

21. Divide using synthetic division:

$$(3x^3 - 5x^2 + 2x - 1) \div (x - 2).$$

22. Write a quadratic equation whose solution set is $\{-2\sqrt{3}, 2\sqrt{3}\}$.

23. Two cars leave from the same place at the same time, traveling in opposite directions. The rate of the faster car exceeds that of the slower car by 10 miles per hour. After 2 hours, the cars are 180 miles apart. Find the rate of each car.

24. Rent-a-Truck charges a daily rental rate of \$39 plus \$0.16 per mile. A competing agency, Ace Truck Rentals, charges \$25 a day plus \$0.24 per mile for the same truck. How many miles must be driven in a day to make the daily cost of both agencies the same? What will be the cost?

25. Three apples and two bananas provide 354 calories, and two apples and three bananas provide 381 calories. Find the number of calories in one apple and one banana.

Something incredible has happened. Your college roommate, a gifted athlete, has been given a six-year contract with a professional baseball team. He will be playing against the likes of Barry Bonds and Sammy Sosa. Management offers him three options. One is a beginning salary of $1,700,000 with annual increases of $70,000 per year starting in the second year. A second option is $1,700,000 the first year with an annual increase of 2% per year beginning in the second year. The third offer involves less money the first year—$1,500,000—but there is an annual increase of 9% yearly after that. Which option offers the most money over the six-year contract?

This problem appears as Exercise 75 in Exercise Set 14.3 and as the group project on page 966.

CHAPTER 14

Sequences, Series, and Probability

We often save for the future by investing small amounts at periodic intervals. To understand how our savings accumulate, we need to understand properties of lists of numbers that are related to each other by a rule. Such lists are called *sequences*. Learning about properties of sequences will show you how to make your financial goals a reality. Your knowledge of sequences will enable you to inform your college roommate of the best of the three appealing offers.

SECTION 14.1

Objectives

1. Find particular terms of a sequence from the general term.
2. Use factorial notation.
3. Use summation notation.

SEQUENCES AND SUMMATION NOTATION

Sequences Many creations in nature involve intricate mathematical designs, including a variety of spirals. For example, the arrangement of the individual florets in the head of a sunflower forms spirals. In some species, there are 21 spirals in the clockwise direction and 34 in the counterclockwise direction. The precise numbers depend on the species of sunflower: 21 and 34, or 34 and 55, or 55 and 89, or even 89 and 144.

This observation becomes more interesting when we consider a sequence of numbers investigated by Leonardo of Pisa, also known as Fibonacci, an Italian mathematician of the thirteenth century. The **Fibonacci sequence** of numbers is an infinite sequence that begins as follows:

$$1, 1, 2, 3, 5, 8, 13, 21, 34, 55, 89, 144, 233, \ldots.$$

The first two terms are 1. Every term thereafter is the sum of the two preceding terms. For example, the third term, 2, is the sum of the first and second terms: $1 + 1 = 2$. The fourth term, 3, is the sum of the second and third terms: $1 + 2 = 3$, and so on. Did you know that the number of spirals in a daisy or a sunflower, 21 and 34, are two Fibonacci numbers? The number of spirals in a pinecone, 8 and 13, and a pineapple, 8 and 13, are also Fibonacci numbers.

We can think of the Fibonacci sequence as a function. The terms of the sequence

$$1, 1, 2, 3, 5, 8, 13, 21, 34, 55, 89, 144, 233, \ldots$$

are the range values for a function whose domain is the set of positive integers.

Domain:	1,	2,	3,	4,	5,	6,	7,	...
	↓	↓	↓	↓	↓	↓	↓	
Range:	1,	1,	2,	3,	5,	8,	13,	...

Thus, $f(1) = 1, f(2) = 1, f(3) = 2, f(4) = 3, f(5) = 5, f(6) = 8, f(7) = 13$, and so on.

The letter a with a subscript is used to represent function values of a sequence, rather than the usual function notation. The subscripts make up the domain of the sequence and they identify the location of a term. Thus, a_1 represents the first term of the sequence, a_2 represents the second term, a_3 the third term, and so on. This notation is shown for the first six terms of the Fibonacci sequence:

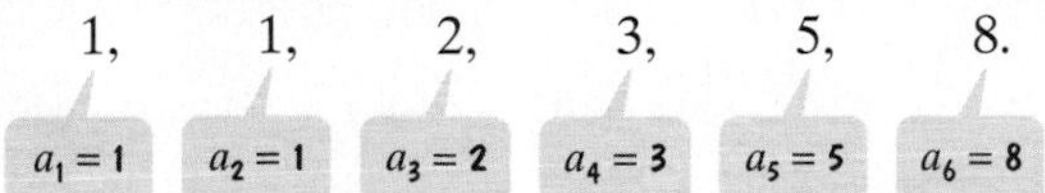

ENRICHMENT ESSAY

Fibonacci Numbers on the Piano Keyboard

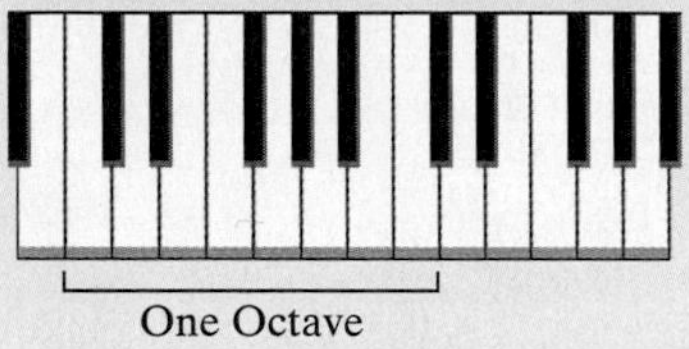

Numbers in the Fibonacci sequence can be found in an octave on the piano keyboard. The octave contains 2 black keys in one cluster and 3 black keys in another cluster, for a total of 5 black keys. It also has 8 white keys, for a total of 13 keys. The numbers 2, 3, 5, 8, and 13 are the third through seventh terms of the Fibonacci sequence.

The notation a_n represents the nth term, or **general term**, of a sequence. The entire sequence is represented by $\{a_n\}$.

DEFINITION OF A SEQUENCE An **infinite sequence** $\{a_n\}$ is a function whose domain is the set of positive integers. The function values, or **terms**, of the sequence are represented by

$$a_1, a_2, a_3, a_4, \ldots, a_n, \ldots.$$

Sequences whose domains consist only of the first n positive integers are called **finite sequences**.

1 Find particular terms of a sequence from the general term.

EXAMPLE 1 Writing Terms of a Sequence from the General Term

Write the first four terms of the sequence whose nth term, or general term, is given:

a. $a_n = 3n + 4$ **b.** $a_n = \frac{(-1)^n}{3^n - 1}$.

SOLUTION

a. We need to find the first four terms of the sequence whose general term is $a_n = 3n + 4$. To do so, we replace n in the formula with 1, 2, 3, and 4.

a_1, 1st term: $3 \cdot 1 + 4 = 3 + 4 = 7$ a_2, 2nd term: $3 \cdot 2 + 4 = 6 + 4 = 10$

a_3, 3rd term: $3 \cdot 3 + 4 = 9 + 4 = 13$ a_4, 4th term: $3 \cdot 4 + 4 = 12 + 4 = 16$

The first four terms are 7, 10, 13, and 16. The sequence defined by $a_n = 3n + 4$ can be written as

$$7, 10, 13, 16, \ldots, 3n + 4, \ldots.$$

b. We need to find the first four terms of the sequence whose general term is $a_n = \frac{(-1)^n}{3^n - 1}$. To do so, we replace each occurrence of n in the formula with 1, 2, 3, and 4.

STUDY TIP

The factor $(-1)^n$ in the general term of a sequence causes the signs of the terms to alternate between positive and negative, depending on whether n is even or odd.

a_1, 1st term: $\frac{(-1)^1}{3^1 - 1} = \frac{-1}{3 - 1} = -\frac{1}{2}$ a_2, 2nd term: $\frac{(-1)^2}{3^2 - 1} = \frac{1}{9 - 1} = \frac{1}{8}$

a_3, 3rd term: $\frac{(-1)^3}{3^3 - 1} = \frac{-1}{27 - 1} = -\frac{1}{26}$ a_4, 4th term: $\frac{(-1)^4}{3^4 - 1} = \frac{1}{81 - 1} = \frac{1}{80}$

The first four terms are $-\frac{1}{2}, \frac{1}{8}, -\frac{1}{26}$, and $\frac{1}{80}$. The sequence defined by $\frac{(-1)^n}{3^n - 1}$ can be written as

$$-\frac{1}{2}, \frac{1}{8}, -\frac{1}{26}, \frac{1}{80}, \ldots, \frac{(-1)^n}{3^n - 1}, \ldots.$$

■

CHECK POINT 1 Write the first four terms of the sequence whose nth term, or general term, is given:

a. $a_n = 2n + 5$ **b.** $a_n = \frac{(-1)^n}{2^n + 1}$.

USING TECHNOLOGY

Graphing utilities can write the terms of a sequence and graph them. For example, to find the first six terms of $\{a_n\} = \left\{\frac{1}{n}\right\}$, enter

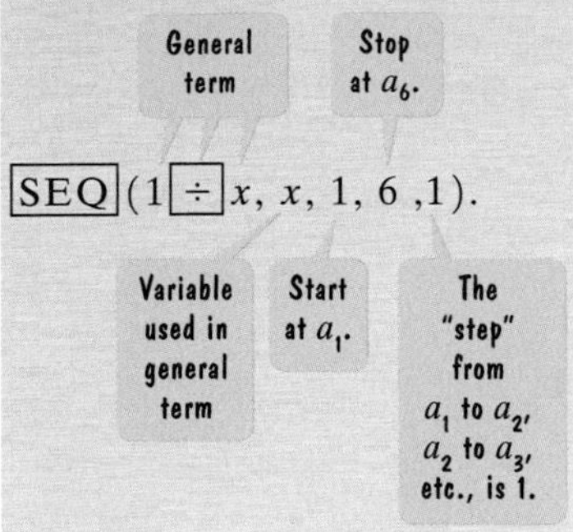

The first few terms of the sequence are shown in the viewing rectangle. By pressing the right arrow key to scroll right, you can see the remaining terms.

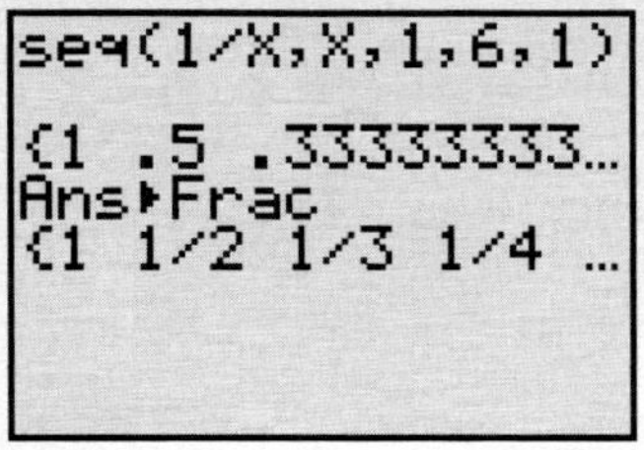

Although sequences are usually named with the letter a, any lowercase letter can be used. For example, the first four terms of the sequence $\{b_n\} = \left\{\left(\frac{1}{2}\right)^n\right\}$ are $b_1 = \frac{1}{2}, b_2 = \frac{1}{4}, b_3 = \frac{1}{8}$, and $b_4 = \frac{1}{16}$.

Because a sequence is a function whose domain is the set of positive integers, the **graph of a sequence** is a set of discrete points. For example, consider the sequence whose general term is $a_n = \frac{1}{n}$. How does the graph of this sequence differ from the graph of the rational function $f(x) = \frac{1}{x}$? The graph of $f(x) = \frac{1}{x}$ is shown in Figure 14.1(a) for positive values of x. To obtain the graph of the sequence $\{a_n\} = \left\{\frac{1}{n}\right\}$, remove all the points from the graph of f except those whose x-coordinates are positive integers. Thus, we remove all points except $(1, 1), \left(2, \frac{1}{2}\right), \left(3, \frac{1}{3}\right), \left(4, \frac{1}{4}\right)$, and so on. The remaining points are the graph of the sequence $\{a_n\} = \left\{\frac{1}{n}\right\}$, shown in Figure 14.1(b). Notice that the horizontal axis is labeled n and the vertical axis is labeled a_n.

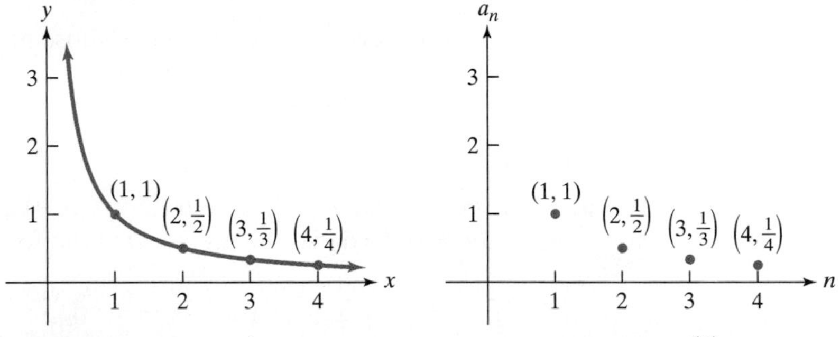

(a) The graph of $f(x) = \frac{1}{x}, x > 0$ (b) The graph of $\{a_n\} = \left\{\frac{1}{n}\right\}$

FIGURE 14.1 Comparing a continuous graph to the graph of a sequence

Factorial Notation Products of consecutive positive integers occur quite often in sequences. These products can be expressed in a special notation, called **factorial notation**.

2 Use factorial notation.

> FACTORIAL NOTATION If n is a positive integer, the notation $n!$ (read "n factorial") is the product of all positive integers from n down through 1.
>
> $$n! = n(n-1)(n-2)\cdots(3)(2)(1)$$
>
> 0! (zero factorial), by definition, is 1.
>
> $$0! = 1$$

The values of $n!$ for the first six positive integers are

$$1! = 1$$
$$2! = 2 \cdot 1 = 2$$
$$3! = 3 \cdot 2 \cdot 1 = 6$$
$$4! = 4 \cdot 3 \cdot 2 \cdot 1 = 24$$
$$5! = 5 \cdot 4 \cdot 3 \cdot 2 \cdot 1 = 120$$
$$6! = 6 \cdot 5 \cdot 4 \cdot 3 \cdot 2 \cdot 1 = 720.$$

Factorials affect only the number or variable that they follow unless grouping symbols appear. For example,

$$2 \cdot 3! = 2(3 \cdot 2 \cdot 1) = 2 \cdot 6 = 12$$

USING TECHNOLOGY

Most calculators have factorial keys. To find 5 factorial, most calculators use one of the following:

Many Scientific Calculators:

5 [x!]

Many Graphing Calculators:

5 [!] [ENTER]

Because $n!$ becomes quite large as n increases, your calculator will display these larger values in scientific notation.

whereas

$$(2 \cdot 3)! = 6! = 6 \cdot 5 \cdot 4 \cdot 3 \cdot 2 \cdot 1 = 720.$$

In this sense, factorials are similar to exponents.

EXAMPLE 2 Finding Terms of a Sequence Involving Factorials

Write the first four terms of the sequence whose nth term is

$$a_n = \frac{2^n}{(n-1)!}.$$

SOLUTION We need to find the first four terms of the sequence. To do so, we replace each n in the formula with 1, 2, 3, and 4.

a_1, 1st term: $$\frac{2^1}{(1-1)!} = \frac{2}{0!} = \frac{2}{1} = 2$$

a_2, 2nd term: $$\frac{2^2}{(2-1)!} = \frac{4}{1!} = \frac{4}{1} = 4$$

a_3, 3rd term: $$\frac{2^3}{(3-1)!} = \frac{8}{2!} = \frac{8}{2 \cdot 1} = 4$$

a_4, 4th term: $$\frac{2^4}{(4-1)!} = \frac{16}{3!} = \frac{16}{3 \cdot 2 \cdot 1} = \frac{16}{6} = \frac{8}{3}$$

The first four terms are 2, 4, 4, and $\frac{8}{3}$. ■

CHECK POINT 2 Write the first four terms of the sequence whose nth term is

$$a_n = \frac{20}{(n+1)!}.$$

3 Use summation notation.

Summation Notation It is sometimes useful to find the sum of the first n terms of a sequence. For example, consider the cost of raising a child born in the United States in 2002 to a middle-income ($39,700–$66,900 per year) family, shown in Table 14.1.

Table 14.1 The Cost of Raising a Child Born in the U.S. in 2002 to a Middle-Income Family

Year	2002	2003	2004	2005	2006	2007	2008	2009	2010
Average Cost	$9230	$9530	$9830	$10,420	$10,750	$11,100	$11,440	$11,810	$12,180
	Child is under 1.	Child is 1.	Child is 2.	Child is 3.	Child is 4.	Child is 5.	Child is 6.	Child is 7.	Child is 8.

Year	2011	2012	2013	2014	2015	2016	2017	2018	2019
Average Cost	$12,440	$12,840	$13,250	$14,750	$15,230	$15,710	$16,520	$17,050	$17,600
	Child is 9.	Child is 10.	Child is 11.	Child is 12.	Child is 13.	Child is 14.	Child is 15.	Child is 16.	Child is 17.

Source: U.S. Department of Agriculture

We can let a_n represent the cost of raising a child in year n, where $n = 1$ corresponds to 2002, $n = 2$ to 2003, $n = 3$ to 2004, and so on. The terms of the finite sequence in Table 14.1 are given as follows:

9230,	9530,	9830,	10,420,	10,750,	11,100,	11,440,	11,810,	12,180,
a_1	a_2	a_3	a_4	a_5	a_6	a_7	a_8	a_9
12,440,	12,840,	13,250,	14,750,	15,230,	15,710,	16,520,	17,050,	17,600.
a_{10}	a_{11}	a_{12}	a_{13}	a_{14}	a_{15}	a_{16}	a_{17}	a_{18}

Why might we want to add the terms of this sequence? We do this to find the total cost of raising a child born in 2002 from birth through age 17. Thus,

$$a_1 + a_2 + a_3 + a_4 + a_5 + a_6 + a_7 + a_8 + a_9 + a_{10} + a_{11} + a_{12} + a_{13} + a_{14} + a_{15} + a_{16} + a_{17} + a_{18}$$
$$= 9230 + 9530 + 9830 + 10{,}420 + 10{,}750 + 11{,}100 + 11{,}440 + 11{,}810 + 12{,}180$$
$$+ 12{,}440 + 12{,}840 + 13{,}250 + 14{,}750 + 15{,}230 + 15{,}710 + 16{,}520 + 17{,}050 + 17{,}600$$
$$= 231{,}680.$$

We see that the total cost of raising a child born in 2002 from birth through age 17 is \$231,680.

There is a compact notation for expressing the sum of the first n terms of a sequence. For example, rather than write

$$a_1 + a_2 + a_3 + a_4 + a_5 + a_6 + a_7 + a_8 + a_9 + a_{10} + a_{11} + a_{12} + a_{13} + a_{14} + a_{15} + a_{16} + a_{17} + a_{18},$$

we can use *summation notation* to express the sum as

$$\sum_{i=1}^{18} a_i.$$

We read this expression as "the sum as i goes from 1 to 18 of a_i." The letter i is called the *index of summation* and is not related to the use of i to represent $\sqrt{-1}$.

You can think of the symbol Σ (the uppercase Greek letter sigma) as an instruction to add up the terms of a sequence.

SUMMATION NOTATION The sum of the first n terms of a sequence is represented by the **summation notation**

$$\sum_{i=1}^{n} a_i = a_1 + a_2 + a_3 + a_4 + \cdots + a_n,$$

where i is the **index of summation,** 1 is the **lower limit of summation,** and n is the **upper limit of summation.**

Any letter can be used for the index of summation. The letters i, j, and k are used commonly. Furthermore, the lower limit of summation can be an integer other than 1.

When we write out a sum that is given in summation notation, we are **expanding the summation**. Example 3 shows how to do this.

EXAMPLE 3 Using Summation Notation

Expand and evaluate the sum:

a. $\sum_{i=1}^{6} (i^2 + 1)$ **b.** $\sum_{k=4}^{7} [(-2)^k - 5]$ **c.** $\sum_{i=1}^{5} 3.$

USING TECHNOLOGY

Graphing utilities can calculate the sum of a sequence. For example, to find the sum of the sequence in Example 3(a), enter

SUM SEQ $(x^2 + 1, x, 1, 6, 1)$.

Then press ENTER; 97 should be displayed. Use this capability to verify Example 3b.

SOLUTION

a. We must replace i in the expression $i^2 + 1$ with all consecutive integers from 1 to 6, inclusive. Then we add.

$$\begin{aligned}\sum_{i=1}^{6}(i^2+1) &= (1^2+1)+(2^2+1)+(3^2+1)+(4^2+1)\\ &\quad +(5^2+1)+(6^2+1)\\ &= 2+5+10+17+26+37\\ &= 97\end{aligned}$$

b. This time the index of summation is k. First we evaluate $(-2)^k - 5$ for all consecutive integers from 4 through 7, inclusive. Then we add.

$$\begin{aligned}\sum_{k=4}^{7}[(-2)^k-5] &= [(-2)^4-5]+[(-2)^5-5]\\ &\quad +[(-2)^6-5]+[(-2)^7-5]\\ &= (16-5)+(-32-5)+(64-5)+(-128-5)\\ &= 11+(-37)+59+(-133)\\ &= -100\end{aligned}$$

c. To find $\sum_{i=1}^{5} 3$, we observe that every term of the sum is 3. The notation $i = 1$ through 5 indicates that we must add the first five terms of a sequence in which every term is 3.

$$\sum_{i=1}^{5} 3 = 3+3+3+3+3 = 15$$

■

CHECK POINT 3 Expand and evaluate the sum:

a. $\sum_{i=1}^{6} 2i^2$

b. $\sum_{k=3}^{5} (2^k - 3)$

c. $\sum_{i=1}^{5} 4$.

For a given sum, we can vary the upper and lower limits of summation, as well as the letter used for the index of summation. By doing so, we can produce different-looking summation notations for the same sum. For example, the sum of the squares of the first four positive integers, $1^2 + 2^2 + 3^2 + 4^2$, can be expressed in a number of equivalent ways:

$$\sum_{i=1}^{4} i^2 = 1^2+2^2+3^2+4^2 = 30$$

$$\begin{aligned}\sum_{i=0}^{3}(i+1)^2 &= (0+1)^2+(1+1)^2+(2+1)^2+(3+1)^2\\ &= 1^2+2^2+3^2+4^2 = 30\end{aligned}$$

$$\begin{aligned}\sum_{k=2}^{5}(k-1)^2 &= (2-1)^2+(3-1)^2+(4-1)^2+(5-1)^2\\ &= 1^2+2^2+3^2+4^2 = 30.\end{aligned}$$

EXAMPLE 4 Writing Sums in Summation Notation

Express each sum using summation notation:

a. $1^3 + 2^3 + 3^3 + \cdots + 7^3$ **b.** $1 + \frac{1}{3} + \frac{1}{9} + \frac{1}{27} + \cdots + \frac{1}{3^{n-1}}.$

SOLUTION In each case, we will use 1 as the lower limit of summation and i for the index of summation.

a. The sum $1^3 + 2^3 + 3^3 + \cdots + 7^3$ has seven terms, each of the form i^3, starting at $i = 1$ and ending at $i = 7$. Thus,

$$1^3 + 2^3 + 3^3 + \cdots + 7^3 = \sum_{i=1}^{7} i^3.$$

b. The sum

$$1 + \frac{1}{3} + \frac{1}{9} + \frac{1}{27} + \cdots + \frac{1}{3^{n-1}}$$

has n terms, each of the form $\frac{1}{3^{i-1}}$, starting at $i = 1$ and ending at $i = n$. Thus,

$$1 + \frac{1}{3} + \frac{1}{9} + \frac{1}{27} + \cdots + \frac{1}{3^{n-1}} = \sum_{i=1}^{n} \frac{1}{3^{i-1}}.$$ ■

CHECK POINT 4 Express each sum using summation notation:

a. $1^2 + 2^2 + 3^2 + \cdots + 9^2$

b. $1 + \frac{1}{2} + \frac{1}{4} + \frac{1}{8} + \cdots + \frac{1}{2^{n-1}}.$

14.1 EXERCISE SET

Student Solutions Manual | CD/Video | PH Math/Tutor Center | MathXL Tutorials on CD | MathXL® | MyMathLab | Interactmath.com

Practice Exercises

In Exercises 1–16, write the first four terms of each sequence whose general term is given.

1. $a_n = 3n + 2$
2. $a_n = 4n - 1$
3. $a_n = 3^n$
4. $a_n = \left(\frac{1}{3}\right)^n$
5. $a_n = (-3)^n$
6. $a_n = \left(-\frac{1}{3}\right)^n$
7. $a_n = (-1)^n(n + 3)$
8. $a_n = (-1)^{n+1}(n + 4)$
9. $a_n = \frac{2n}{n + 4}$
10. $a_n = \frac{3n}{n + 5}$
11. $a_n = \frac{(-1)^{n+1}}{2^n - 1}$
12. $a_n = \frac{(-1)^{n+1}}{2^n + 1}$
13. $a_n = \frac{n^2}{n!}$
14. $a_n = \frac{(n + 1)!}{n^2}$

15. $a_n = 2(n+1)!$

16. $a_n = -2(n-1)!$

In Exercises 17–30, find each indicated sum.

17. $\sum_{i=1}^{6} 5i$

18. $\sum_{i=1}^{6} 7i$

19. $\sum_{i=1}^{4} 2i^2$

20. $\sum_{i=1}^{5} i^3$

21. $\sum_{k=1}^{5} k(k+4)$

22. $\sum_{k=1}^{4} (k-3)(k+2)$

23. $\sum_{i=1}^{4} \left(-\frac{1}{2}\right)^i$

24. $\sum_{i=2}^{4} \left(-\frac{1}{3}\right)^i$

25. $\sum_{i=5}^{9} 11$

26. $\sum_{i=3}^{7} 12$

27. $\sum_{i=0}^{4} \frac{(-1)^i}{i!}$

28. $\sum_{i=0}^{4} \frac{(-1)^{i+1}}{(i+1)!}$

29. $\sum_{i=1}^{5} \frac{i!}{(i-1)!}$

30. $\sum_{i=1}^{5} \frac{(i+2)!}{i!}$

In Exercises 31–42, express each sum using summation notation. Use 1 as the lower limit of summation and i for the index of summation.

31. $1^2 + 2^2 + 3^2 + \cdots + 15^2$

32. $1^4 + 2^4 + 3^4 + \cdots + 12^4$

33. $2 + 2^2 + 2^3 + \cdots + 2^{11}$

34. $5 + 5^2 + 5^3 + \cdots + 5^{12}$

35. $1 + 2 + 3 + \cdots + 30$

36. $1 + 2 + 3 + \cdots + 40$

37. $\frac{1}{2} + \frac{2}{3} + \frac{3}{4} + \cdots + \frac{14}{14+1}$

38. $\frac{1}{3} + \frac{2}{4} + \frac{3}{5} + \cdots + \frac{16}{16+2}$

39. $4 + \frac{4^2}{2} + \frac{4^3}{3} + \cdots + \frac{4^n}{n}$

40. $\frac{1}{9} + \frac{2}{9^2} + \frac{3}{9^3} + \cdots + \frac{n}{9^n}$

41. $1 + 3 + 5 + \cdots + (2n-1)$

42. $a + ar + ar^2 + \cdots + ar^{n-1}$

In Exercises 43–48, express each sum using summation notation. Use a lower limit of summation of your choice and k for the index of summation.

43. $5 + 7 + 9 + 11 + \cdots + 31$

44. $6 + 8 + 10 + 12 + \cdots + 32$

45. $a + ar + ar^2 + \cdots + ar^{12}$

46. $a + ar + ar^2 + \cdots + ar^{14}$

47. $a + (a+d) + (a+2d) + \cdots + (a+nd)$

48. $(a+d) + (a+d^2) + \cdots + (a+d^n)$

Practice Plus

In Exercises 49–56, use the graphs of $\{a_n\}$ and $\{b_n\}$ to find each indicated sum.

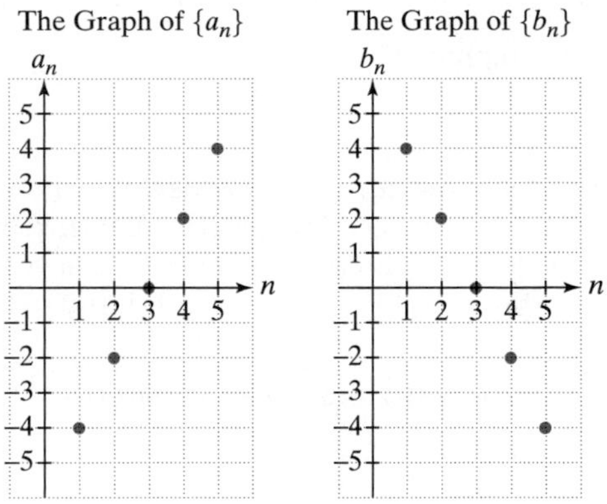

49. $\sum_{i=1}^{5} (a_i^2 + 1)$

50. $\sum_{i=1}^{5} (b_i^2 - 1)$

51. $\sum_{i=1}^{5} (2a_i + b_i)$

52. $\sum_{i=1}^{5} (a_i + 3b_i)$

53. $\sum_{i=4}^{5} \left(\frac{a_i}{b_i}\right)^2$

54. $\sum_{i=4}^{5} \left(\frac{a_i}{b_i}\right)^3$

55. $\sum_{i=1}^{5} a_i^2 + \sum_{i=1}^{5} b_i^2$

56. $\sum_{i=1}^{5} a_i^2 - \sum_{i=3}^{5} b_i^2$

Application Exercises

57. The bar graph shows the number of people in the United States who lived below the poverty level from 1995 through 2001. Let a_n represent the number of people, in millions, living below the poverty level in year n, where $n = 1$ corresponds to 1995, $n = 2$ to 1996, and so on.

Number of People in the U.S. Living below the Poverty Level

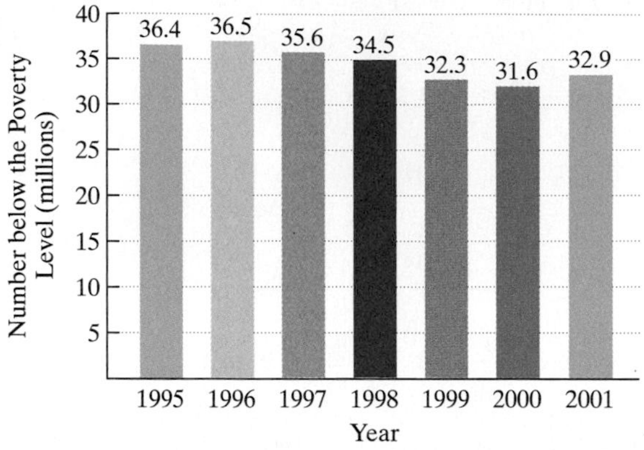

Source: Bureau of the Census

a. Find $\sum_{i=1}^{7} a_i$.

b. Find $\frac{\sum_{i=1}^{7} a_i}{7}$. Round to one decimal place. What does this number represent?

58. The bar graph shows the number of deaths by firearms in the United States from 1994 through 2000. Let a_n represent the number of deaths by firearms, in thousands, in year n, where $n = 1$ corresponds to 1994, $n = 2$ to 1995, and so on.

Number of Deaths by Firearms in the U.S.

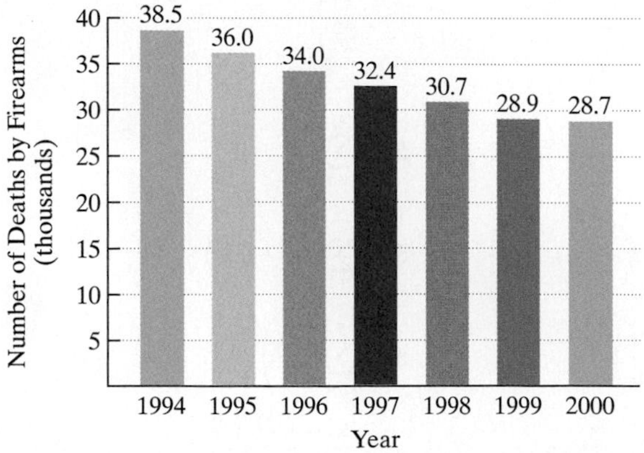

Source: Centers for Disease Control and Prevention

a. Find $\sum_{i=1}^{7} a_i$. What does this number represent?

b. Find $\frac{\sum_{i=1}^{7} a_i}{7}$. Round to one decimal place. What does this number represent?

The graph shows the millions of welfare recipients in the United States who received cash assistance from 1993 through 2000. In Exercises 59–60, consider a sequence whose general term, a_n, represents the millions of Americans receiving cash assistance n years after 1992.

Welfare Recipients in the U.S.

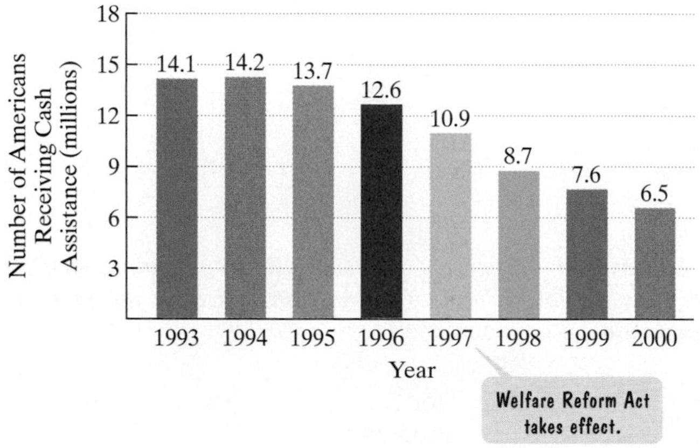

Source: Thomas R. Dye, *Politics in America*, Prentice Hall

59. a. Use the numbers given in the graph to find and interpret $\frac{1}{8}\sum_{i=1}^{8} a_i$.

b. The finite sequence whose general term is $a_n = -1.23n + 16.55$, where $n = 1, 2, 3, \ldots, 8$, models the millions of Americans receiving cash assistance, a_n, n years after 1992. Use the model to find $\frac{1}{8}\sum_{i=1}^{8} a_i$. Does this seem reasonable in terms of the actual sum in part (a), or has model breakdown occurred?

60. a. Use the numbers given in the graph to find and interpret $\frac{1}{8}\sum_{i=1}^{8} a_i$.

b. The finite sequence whose general term is $a_n = -0.11n^2 - 0.22n + 14.88$, where $n = 1, 2, 3, \ldots, 8$, models the millions of Americans receiving cash assistance, a_n, n years after 1992. Use the model to find $\frac{1}{8}\sum_{i=1}^{8} a_i$. Does this seem reasonable in terms of the actual sum in part (a), or has model breakdown occurred?

61. A deposit of \$6000 is made in an account that earns 6% interest compounded quarterly. The balance in the account after n quarters is given by the sequence

$$a_n = 6000\left(1 + \frac{0.06}{4}\right)^n, \quad n = 1, 2, 3, \ldots.$$

Find the balance in the account after five years. Round to the nearest cent.

62. A deposit of \$10,000 is made in an account that earns 8% interest compounded quarterly. The balance in the account after n quarters is given by the sequence

$$a_n = 10{,}000\left(1 + \frac{0.08}{4}\right)^n, \quad n = 1, 2, 3, \ldots.$$

Find the balance in the account after six years. Round to the nearest cent.

Writing in Mathematics

63. What is a sequence? Give an example with your description.

64. Explain how to write terms of a sequence if the formula for the general term is given.

65. What does the graph of a sequence look like? How is it obtained?

66. Explain how to find $n!$ if n is a positive integer.

67. What is the meaning of the symbol Σ? Give an example with your description.

68. You buy a new car for \$24,000. At the end of n years, the value of your car is given by the sequence

$$a_n = 24{,}000\left(\frac{3}{4}\right)^n, \quad n = 1, 2, 3, \ldots.$$

Find a_5 and write a sentence explaining what this value represents. Describe the nth term of the sequence in terms of the value of your car at the end of each year.

69. It is estimated that 4 to 6 million people in the United States have overwhelming physical, psychological, and social problems that make it impossible for them to work. (*Source*: Thomas R. Dye, *Politics in America*, Prentice Hall) Describe what this means in terms of projecting the model in Exercise 59(b) into the first decade of the new millennium. In writing your answer, use the model and be as specific as possible.

Technology Exercises

70. Use the SEQ (sequence) capability of a graphing utility to verify the terms of the sequences you obtained for any five sequences from Exercises 1–16.

71. Use the SUM SEQ (sum of the sequence) capability of a graphing utility to verify any five of the sums you obtained in Exercises 17–30.

72. As n increases, the terms of the sequence

$$a_n = \left(1 + \frac{1}{n}\right)^n$$

get closer and closer to the number e (where $e \approx 2.7183$). Use a calculator to find $a_{10}, a_{100}, a_{1000}, a_{10{,}000}$, and $a_{100{,}000}$, comparing these terms to the decimal approximation for e.

Many graphing utilities have a sequence-graphing mode that plots the terms of a sequence as points on a rectangular coordinate system. Consult your manual; if your graphing utility has this capability, use it to graph each of the sequences in Exercises 73–76. What appears to be happening to the terms of each sequence as n gets larger?

73. $a_n = \dfrac{n}{n+1}$; n: [0, 10, 1] by a_n: [0, 1, 0.1]

74. $a_n = \dfrac{100}{n}$; n: [0, 1000, 100] by a_n: [0, 1, 0.1]

75. $a_n = \dfrac{2n^2 + 5n - 7}{n^3}$; n: [0, 10, 1] by a_n: [0, 2, 0.2]

76. $a_n = \dfrac{3n^4 + n - 1}{5n^4 + 2n^2 + 1}$; n: [0, 10, 1] by a_n: [0, 1, 0.1]

Critical Thinking Exercises

77. Which one of the following is true?

a. $\sum_{i=1}^{2} (-1)^i 2^i = 0$

b. $\sum_{i=1}^{2} a_i b_i = \sum_{i=1}^{2} a_i \sum_{i=1}^{2} b_i$

c. $\sum_{i=1}^{4} 3i + \sum_{i=1}^{4} 4i = \sum_{i=1}^{4} 7i$

d. $\sum_{i=0}^{6} (-1)^i (i+1)^2 = \sum_{j=1}^{7} (-1)^j j^2$

In Exercises 78–85, find a general term, a_n, for each sequence. More than one answer may be possible.

78. $1, \frac{1}{2}, \frac{1}{3}, \frac{1}{4}, \ldots$

79. $1, 4, 9, 16, \ldots$

80. $-1, 1, -1, 1, \ldots$

81. $1 \cdot 3, 2 \cdot 4, 3 \cdot 5, 4 \cdot 6, \ldots$

82. $\frac{3}{2}, \frac{4}{3}, \frac{5}{4}, \frac{6}{5}, \ldots$

83. $5, 7, 9, 11, \ldots$

84. $\frac{4}{1}, \frac{9}{2}, \frac{16}{3}, \frac{25}{4}, \ldots$

85. $4, -8, 16, -32, \ldots$

86. Evaluate without using a calculator: $\frac{600!}{599!}$.

In Exercises 87–88, rewrite each expression as a polynomial in standard form.

87. $\frac{(n+4)!}{(n+2)!}$

88. $\frac{n!}{(n-3)!}$

In Exercises 89–90, expand and write the answer as a single logarithm with a coefficient of 1.

89. $\sum_{i=1}^{4} \log(2i)$

90. $\sum_{i=2}^{4} 2i \log x$

91. If $a_1 = 7$ and $a_n = a_{n-1} + 5$ for $n \geq 2$, write the first four terms of the sequence.

Review Exercises

92. Simplify: $\sqrt[3]{40x^4y^7}$. (Section 10.3, Example 5)

93. Factor: $27x^3 - 8$. (Section 6.4, Example 8)

94. Solve: $\frac{6}{x} + \frac{6}{x+2} = \frac{5}{2}$.

(Section 7.6, Example 3)

SECTION 14.2

Objectives

1. Find the common difference for an arithmetic sequence.
2. Write terms of an arithmetic sequence.
3. Use the formula for the general term of an arithmetic sequence.
4. Use the formula for the sum of the first n terms of an arithmetic sequence.

ARITHMETIC SEQUENCES

Your grandmother and her financial counselor are looking at options in case an adult residential facility is needed in the future. The good news is that your grandmother's total assets are $400,000. The bad news is that yearly adult residential community costs average $58,730, increasing by $1800 each year. In this section, we will see how sequences can be used to describe your grandmother's situation and help her to identify realistic options.

Arithmetic Sequences A mathematical model for the average annual salaries of major league baseball players generates the following data:

Year	1996	1997	1998	1999	2000	2001	2002
Salary	1,076,865	1,304,152	1,531,439	1,758,726	1,986,013	2,213,300	2,440,587

From 1996 to 1997, salaries increased by \$1,304,152 − \$1,076,865 = \$227,287. From 1997 to 1998, salaries increased by \$1,531,439 − \$1,304,152 = \$227,287. If we make these computations for each year, we find that the yearly salary increase is \$227,287. The sequence of annual salaries shows that each term after the first, 1,076,865, differs from the preceding term by a constant amount, namely 227,287. The sequence of annual salaries

$$1{,}076{,}865,\ 1{,}304{,}152,\ 1{,}531{,}439,\ 1{,}758{,}726,\ 1{,}986{,}013, \ldots$$

is an example of an *arithmetic sequence.*

DEFINITION OF AN ARITHMETIC SEQUENCE An **arithmetic sequence** is a sequence in which each term after the first differs from the preceding term by a constant amount. The difference between consecutive terms is called the **common difference** of the sequence.

1 Find the common difference for an arithmetic sequence.

The common difference, d, is found by subtracting any term from the term that directly follows it. In the following examples, the common difference is found by subtracting the first term from the second term, $a_2 - a_1$.

Arithmetic Sequence	**Common Difference**
1,076,865, 1,304,152, 1,531,439, 1,758,726, ...	$d = 1{,}304{,}152 - 1{,}076{,}865 = 227{,}287$
$-5, -2, 1, 4, 7, \ldots$	$d = -2 - (-5) = -2 + 5 = 3$
$8, 3, -2, -7, -12, \ldots$	$d = 3 - 8 = -5$

Figure 14.2 shows the graphs of the last two arithmetic sequences in our list. The common difference for the increasing sequence in Figure 14.2(a) is 3. The common difference for the decreasing sequence in Figure 14.2(b) is −5.

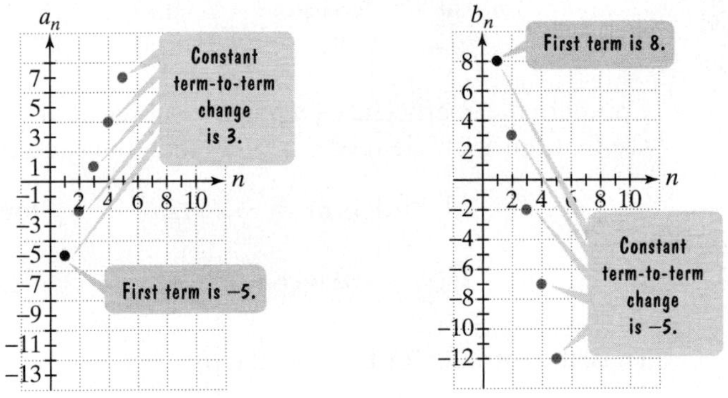

FIGURE 14.2 (a) The graph of $\{a_n\} = -5, -2, 1, 4, 7, \ldots$ (b) The graph of $\{b_n\} = 8, 3, -2, -7, -12, \ldots$

The graph of each arithmetic sequence in Figure 14.2 forms a set of discrete points lying on a straight line. This illustrates that **an arithmetic sequence is a linear function whose domain is the set of positive integers.**

If the first term of an arithmetic sequence is a_1, each term after the first is obtained by adding d, the common difference, to the previous term.

2 Write terms of an arithmetic sequence.

EXAMPLE 1 Writing the Terms of an Arithmetic Sequence Using the First Term and the Common Difference

Write the first six terms of the arithmetic sequence with first term 6 and common difference −2.

SOLUTION To find the second term, we add −2 to the first term, 6, giving 4. For the next term, we add −2 to 4, and so on.

$$\begin{aligned} a_1 \text{ (first term)} &= 6 \\ a_2 \text{ (second term)} &= 6 + (-2) = 4 \\ a_3 \text{ (third term)} &= 4 + (-2) = 2 \\ a_4 \text{ (fourth term)} &= 2 + (-2) = 0 \\ a_5 \text{ (fifth term)} &= 0 + (-2) = -2 \\ a_6 \text{ (sixth term)} &= -2 + (-2) = -4 \end{aligned}$$

The first six terms are

$$6, 4, 2, 0, -2, \text{ and } -4.$$

CHECK POINT 1 Write the first six terms of the arithmetic sequence with first term 100 and common difference −30.

3 Use the formula for the general term of an arithmetic sequence.

The General Term of an Arithmetic Sequence Consider an arithmetic sequence whose first term is a_1 and whose common difference is d. We are looking for a formula for the general term, a_n. Let's begin by writing the first six terms. The first term is a_1. The second term is $a_1 + d$. The third term is $a_1 + d + d$, or $a_1 + 2d$. Thus, we start with a_1 and add d to each successive term. The first six terms are

$$a_1, \quad a_1 + d, \quad a_1 + 2d, \quad a_1 + 3d, \quad a_1 + 4d, \quad a_1 + 5d.$$

a_1, first term — a_2, second term — a_3, third term — a_4, fourth term — a_5, fifth term — a_6, sixth term

Compare the coefficient of d and the subscript of a denoting the term number. Can you see that the coefficient of d is 1 less than the subscript of a denoting the term number?

$$a_3\text{: third term} = a_1 + 2d \qquad a_4\text{: fourth term} = a_1 + 3d$$

2 is one less than 3. — 3 is one less than 4.

Thus, the formula for the nth term is

$$a_n\text{: } n\text{th term} = a_1 + (n - 1)d.$$

$n - 1$ is one less than n.

GENERAL TERM OF AN ARITHMETIC SEQUENCE The nth term (the general term) of an arithmetic sequence with first term a_1 and common difference d is

$$a_n = a_1 + (n - 1)d.$$

EXAMPLE 2 Using the Formula for the General Term of an Arithmetic Sequence

Find the eighth term of the arithmetic sequence whose first term is 4 and whose common difference is −7.

SOLUTION To find the eighth term, a_8, we replace n in the formula with 8, a_1 with 4, and d with −7.

$$a_n = a_1 + (n - 1)d$$
$$a_8 = 4 + (8 - 1)(-7) = 4 + 7(-7) = 4 + (-49) = -45$$

The eighth term is −45. We can check this result by writing the first eight terms of the sequence:

$$4, -3, -10, -17, -24, -31, -38, -45.$$

■

CHECK POINT 2 Find the ninth term of the arithmetic sequence whose first term is 6 and whose common difference is −5.

EXAMPLE 3 Using an Arithmetic Sequence to Model Teachers' Earnings

According to the National Education Association, teachers in the United States earned an average of $30,532 in 1990. This amount has increased by approximately $1472 per year.

a. Write a formula for the nth term of the arithmetic sequence that describes teachers' average earnings n years after 1989.

b. How much will U.S. teachers earn, on average, by the year 2010?

SOLUTION

a. We can express teachers' earnings by the following arithmetic sequence:

30,532,	32,004,	33,476,	30,948, . . .
a_1: earnings in 1990, 1 year after 1989	a_2: earnings in 1991, 2 years after 1989	a_3: earnings in 1992, 3 years after 1989	a_4: earnings in 1993, 4 years after 1989

In this sequence, a_1, the first term, represents the amount teachers earned in 1990. Each subsequent year this amount increases by $1472, so $d = 1472$. We use the formula for the general term of an arithmetic sequence to write the nth term of the sequence that describes teachers' earnings n years after 1989.

$a_n = a_1 + (n - 1)d$	This is the formula for the general term of an arithmetic sequence.
$a_n = 30{,}532 + (n - 1)1472$	$a_1 = 30{,}532$ and $d = 1472$.
$a_n = 30{,}532 + 1472n - 1472$	Distribute 1472 to each term in parentheses.
$a_n = 1472n + 29{,}060$	Simplify.

Thus, teachers' earnings n years after 1989 can be described by $a_n = 1472n + 29{,}060$.

b. Now we need to find teachers' earnings in 2010. The year 2010 is 21 years after 1989: That is, $2010 - 1989 = 21$. Thus, $n = 21$. We substitute 21 for n in $a_n = 1472n + 29{,}060$.

$$a_{21} = 1472 \cdot 21 + 29{,}060 = 59{,}972$$

The 21st term of the sequence is 59,972. Therefore, U.S. teachers are predicted to earn an average of $59,972 by the year 2010. ■

CHECK POINT 3 According to the U.S. Census Bureau, new one-family houses sold for an average of \$159,000 in 1995. This average sales price has increased by approximately \$9700 per year.

a. Write a formula for the nth term of the arithmetic sequence that describes the average cost of new one-family houses n years after 1994.

b. How much will new one-family houses cost, on average, by the year 2010?

4 Use the formula for the sum of the first n terms of an arithmetic sequence.

The Sum of the First n Terms of an Arithmetic Sequence The sum of the first n terms of an arithmetic sequence, denoted by S_n, and called the ***n*th partial sum**, can be found without having to add up all the terms. Let

$$S_n = a_1 + a_2 + a_3 + \cdots + a_n$$

be the sum of the first n terms of an arithmetic sequence. Because d is the common difference between terms, S_n can be written forward and backward as follows.

Forward: Start with the first term, a_1. Keep adding d.

Backward: Start with the last term, a_n. Keep subtracting d.

$$\begin{aligned} S_n &= a_1 + (a_1 + d) + (a_1 + 2d) + \cdots + a_n \\ S_n &= a_n + (a_n - d) + (a_n - 2d) + \cdots + a_1 \\ \hline 2S_n &= (a_1 + a_n) + (a_1 + a_n) + (a_1 + a_n) + \cdots + (a_1 + a_n) \end{aligned}$$

Add the two equations.

Because there are n sums of $(a_1 + a_n)$ on the right side, we can express this side as $n(a_1 + a_n)$. Thus, the last equation can be written as follows:

$$2S_n = n(a_1 + a_n).$$

$$S_n = \frac{n}{2}(a_1 + a_n)$$

Solve for S_n, dividing both sides by 2.

We have proved the following result:

THE SUM OF THE FIRST n TERMS OF AN ARITHMETIC SEQUENCE The sum, S_n, of the first n terms of an arithmetic sequence is given by

$$S_n = \frac{n}{2}(a_1 + a_n),$$

in which a_1 is the first term and a_n is the nth term.

To find the sum of the terms of an arithmetic sequence using $S_n = \frac{n}{2}(a_1 + a_n)$, we need to know the first term, a_1, the last term, a_n, and the number of terms, n. The following examples illustrate how to use this formula.

EXAMPLE 4 Finding the Sum of n Terms of an Arithmetic Sequence

Find the sum of the first 100 terms of the arithmetic sequence: 1, 3, 5, 7, ...

SOLUTION By finding the sum of the first 100 terms of 1, 3, 5, 7, . . . , we are finding the sum of the first 100 odd numbers. To find the sum of the first 100 terms, S_{100}, we replace n in the formula with 100.

$$S_n = \frac{n}{2}(a_1 + a_n)$$

$$S_{100} = \frac{100}{2}(a_1 + a_{100})$$

The first term, a_1, is 1. We must find a_{100}, the 100th term.

We use the formula for the general term of a sequence to find a_{100}. The common difference, d, of 1, 3, 5, 7, . . . , is 2.

$$a_n = a_1 + (n - 1)d$$ This is the formula for the nth term of an arithmetic sequence. Use it to find the 100th term.

$$a_{100} = 1 + (100 - 1) \cdot 2$$ Substitute 100 for n, 2 for d, and 1 (the first term) for a_1.

$$= 1 + 99 \cdot 2$$

$$= 1 + 198 = 199$$

Now we are ready to find the sum of the 100 terms 1, 3, 5, 7, . . . , 199.

$$S_n = \frac{n}{2}(a_1 + a_n)$$ Use the formula for the sum of the first n terms of an arithmetic sequence. Let $n = 100$, $a_1 = 1$, and $a_{100} = 199$.

$$S_{100} = \frac{100}{2}(1 + 199) = 50(200) = 10{,}000$$

The sum of the first 100 odd numbers is 10,000. Equivalently, the 100th partial sum of the sequence 1, 3, 5, 7, . . . is 10,000. ■

CHECK POINT 4 Find the sum of the first 15 terms of the arithmetic sequence: 3, 6, 9, 12, ...

USING TECHNOLOGY

To find

$$\sum_{i=1}^{25} (5i - 9)$$

on a graphing utility, enter:

SUM SEQ $(5x - 9, x, 1, 25, 1)$.

Then press ENTER.

```
sum(seq(5X-9,X,1
,25,1))
            1400
```

EXAMPLE 5 Using S_n to Evaluate a Summation

Find the following sum: $\sum_{i=1}^{25} (5i - 9)$.

SOLUTION

$$\sum_{i=1}^{25} (5i - 9) = (5 \cdot 1 - 9) + (5 \cdot 2 - 9) + (5 \cdot 3 - 9) + \cdots + (5 \cdot 25 - 9)$$

$$= -4 + 1 + 6 + \cdots + 116$$

By evaluating the first three terms and the last term, we see that $a_1 = -4$; d, the common difference, is $1 - (-4)$ or 5; and a_{25}, the last term, is 116.

$$S_n = \frac{n}{2}(a_1 + a_n)$$ Use the formula for the sum of the first n terms of an arithmetic sequence. Let $n = 25$, $a_1 = -4$, and $a_{25} = 116$.

$$S_{25} = \frac{25}{2}(-4 + 116) = \frac{25}{2}(112) = 1400$$

Thus,

$$\sum_{i=1}^{25} (5i - 9) = 1400.$$

■

CHECK POINT 5 Find the following sum: $\sum_{i=1}^{30} (6i - 11)$.

EXAMPLE 6 Modeling Total Nursing Home Costs over a Six-Year Period

Your grandmother has assets of \$400,000. One option that she is considering involves an adult residential community for a six-year period beginning in 2006. The model

$$a_n = 1800n + 58{,}730$$

describes yearly adult residential community costs n years after 2005. Does your grandmother have enough to pay for the facility?

SOLUTION We must find the sum of an arithmetic sequence. The first term of the sequence corresponds to the facility's costs in the year 2006. The last term corresponds to costs in the year 2011. Because the model describes costs n years after 2005, $n = 1$ describes the year 2006 and $n = 6$ describes the year 2011.

$$a_n = 1800n + 58{,}730$$ This is the given formula for the general term of the sequence.

$$a_1 = 1800 \cdot 1 + 58{,}730 = 60{,}530$$ Find a_1 by replacing n with 1.

$$a_6 = 1800 \cdot 6 + 58{,}730 = 69{,}530$$ Find a_6 by replacing n with 6.

The first year the facility will cost \$60,530. By year six, the facility will cost \$69,530. Now we must find the sum of the costs for all six years. We focus on the sum of the first six terms of the arithmetic sequence

$$\underbrace{60{,}530}_{a_1},\ \underbrace{62{,}330}_{a_2},\ \ldots,\ \underbrace{69{,}530}_{a_6}.$$

We find this sum using the formula for the sum of the first n terms of an arithmetic sequence. We are adding 6 terms: $n = 6$. The first term is 60,530: $a_1 = 60{,}530$. The last term—that is, the sixth term—is 69,530: $a_6 = 69{,}530$.

$$S_n = \frac{n}{2}(a_1 + a_n)$$

$$S_6 = \frac{6}{2}(60{,}530 + 69{,}530) = 3(130{,}060) = 390{,}180$$

Total adult residential community costs for your grandmother are predicted to be \$390,180. Because your grandmother's assets are \$400,000, she has enough to pay for the facility for the six-year period. ■

CHECK POINT 6 In Example 6, how much would it cost for the adult residential community for a ten-year period beginning in 2006?

14.2 EXERCISE SET

Student Solutions Manual CD/Video PH Math/Tutor Center MathXL Tutorials on CD MathXL® MyMathLab Interactmath.com

Practice Exercises

In Exercises 1–6, find the common difference for each arithmetic sequence.

1. 2, 6, 10, 14, . . .
2. 3, 8, 13, 18, . . .
3. −7, −2, 3, 8, . . .
4. −10, −4, 2, 8, . . .
5. 714, 711, 708, 705, . . .
6. 611, 606, 601, 596, . . .

In Exercises 7–16, write the first six terms of each arithmetic sequence with the given first term, a_1, and common difference, d.

7. $a_1 = 200, d = 20$
8. $a_1 = 300, d = 50$
9. $a_1 = -7, d = 4$
10. $a_1 = -8, d = 5$
11. $a_1 = 300, d = -90$
12. $a_1 = 200, d = -60$
13. $a_1 = \frac{5}{2}, d = -\frac{1}{2}$
14. $a_1 = \frac{3}{4}, d = -\frac{1}{4}$
15. $a_1 = -0.4, d = -1.6$
16. $a_1 = -0.3, d = -1.7$

In Exercises 17–24, use the formula for the general term (the nth term) of an arithmetic sequence to find the indicated term of each sequence with the given first term, a_1, and common difference, d.

17. Find a_6 when $a_1 = 13, d = 4$.
18. Find a_{16} when $a_1 = 9, d = 2$.
19. Find a_{50} when $a_1 = 7, d = 5$.
20. Find a_{60} when $a_1 = 8, d = 6$.
21. Find a_{200} when $a_1 = -40, d = 5$.
22. Find a_{150} when $a_1 = -60, d = 5$.
23. Find a_{60} when $a_1 = 35, d = -3$.
24. Find a_{70} when $a_1 = -32, d = 4$.

In Exercises 25–34, write a formula for the general term (the nth term) of each arithmetic sequence. Then use the formula for a_n to find a_{20}, the 20th term of the sequence.

25. 1, 5, 9, 13, . . .
26. 2, 7, 12, 17, . . .
27. 7, 3, −1, −5, . . .
28. 6, 1, −4, −9, . . .
29. −20, −24, −28, −32, . . .
30. −70, −75, −80, −85, . . .
31. $a_1 = -\frac{1}{3}, d = \frac{1}{3}$
32. $a_1 = -\frac{1}{4}, d = \frac{1}{4}$
33. $a_1 = 4, d = -0.3$
34. $a_1 = 5, d = -0.2$
35. Find the sum of the first 20 terms of the arithmetic sequence: 4, 10, 16, 22, ...
36. Find the sum of the first 25 terms of the arithmetic sequence: 7, 19, 31, 43, ...
37. Find the sum of the first 50 terms of the arithmetic sequence: −10, −6, −2, 2, ...
38. Find the sum of the first 50 terms of the arithmetic sequence: −15, −9, −3, 3, ...
39. Find $1 + 2 + 3 + 4 + \cdots + 100$, the sum of the first 100 natural numbers.
40. Find $2 + 4 + 6 + 8 + \cdots + 200$, the sum of the first 100 positive even integers.
41. Find the sum of the first 60 positive even integers.
42. Find the sum of the first 80 positive even integers.
43. Find the sum of the even integers between 21 and 45.
44. Find the sum of the odd integers between 30 and 54.

For Exercises 45–50, write out the first three terms and the last term. Then use the formula for the sum of the first n terms of an arithmetic sequence to find the indicated sum.

45. $\sum_{i=1}^{17} (5i + 3)$
46. $\sum_{i=1}^{20} (6i - 4)$
47. $\sum_{i=1}^{30} (-3i + 5)$
48. $\sum_{i=1}^{40} (-2i + 6)$
49. $\sum_{i=1}^{100} 4i$
50. $\sum_{i=1}^{50} (-4i)$

Practice Plus

Use the graphs of the arithmetic sequences $\{a_n\}$ and $\{b_n\}$ to solve Exercises 51–58.

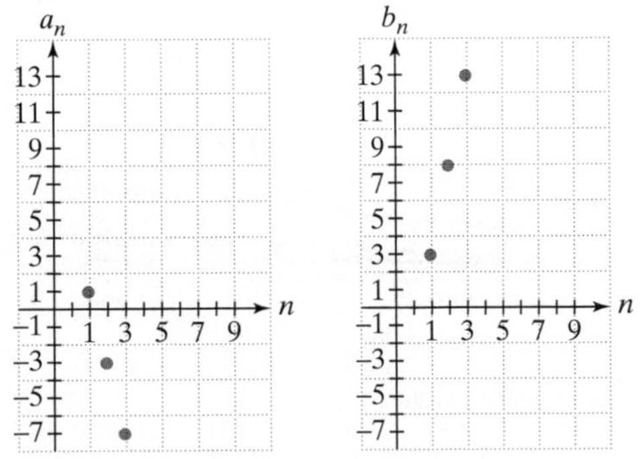

In Exercises 51–58, be sure to refer to the graphs on the previous page.

51. Find $a_{14} + b_{12}$.

52. Find $a_{16} + b_{18}$.

53. If $\{a_n\}$ is a finite sequence whose last term is -83, how many terms does $\{a_n\}$ contain?

54. If $\{b_n\}$ is a finite sequence whose last term is 93, how many terms does $\{b_n\}$ contain?

55. Find the difference between the sum of the first 14 terms of $\{b_n\}$ and the sum of the first 14 terms of $\{a_n\}$.

56. Find the difference between the sum of the first 15 terms of $\{b_n\}$ and the sum of the first 15 terms of $\{a_n\}$.

57. Write a linear function $f(x) = mx + b$, whose domain is the set of positive integers, that represents $\{a_n\}$.

58. Write a linear function $g(x) = mx + b$, whose domain is the set of positive integers, that represents $\{b_n\}$.

Use a system of two equations in two variables, a_1 and d, to solve Exercises 59–60.

59. Write a formula for the general term (the nth term) of the arithmetic sequence whose second term, a_2, is 4 and whose sixth term, a_6, is 16.

60. Write a formula for the general term (the nth term) of the arithmetic sequence whose third term, a_3, is 7 and whose eighth term, a_8, is 17.

Application Exercises

The graph shows pounds of various food groups consumed per year by the average American. Exercises 61–64 involve developing arithmetic sequences that model the data. In Exercises 63–64, models will vary.

Per Capita Consumption of Various Food Groups

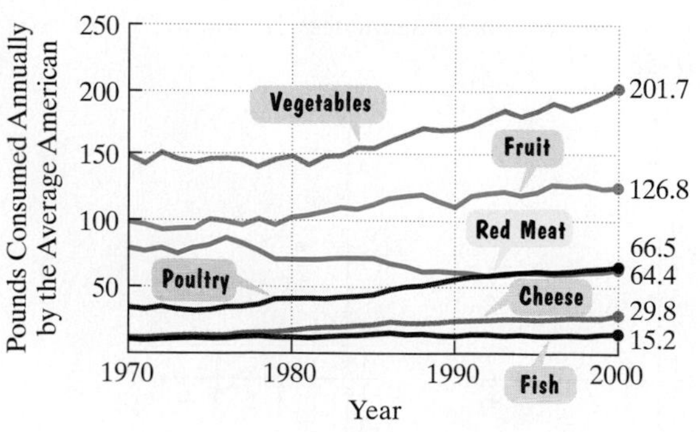

Source: U.S. Department of Agriculture

61. The graph shows that the average American consumed 150 pounds of vegetables in 1970. On average, this amount has increased by approximately 1.7 pounds per person per year.

a. Write a formula for the nth term of the arithmetic sequence that describes the number of pounds of vegetables consumed annually by the average American n years after 1969.

b. How many pounds of vegetables will be consumed by the average American in 2009?

62. The graph shows that the average American consumed 100 pounds of fruit in 1970. On average, this amount has increased by approximately 0.9 pound per person per year.

a. Write a formula for the nth term of the arithmetic sequence that describes the number of pounds of fruit consumed annually by the average American n years after 1969.

b. How many pounds of fruit will be consumed by the average American in 2009?

63. a. Use the data shown to write a formula for the nth term of the arithmetic sequence that describes the number of pounds of cheese consumed annually by the average American n years after 1969.

b. How many pounds of cheese will be consumed by the average American in 2006?

64. Use the data shown for fish, poultry, or red meats, and repeat both parts of Exercise 63.

65. Company A pays \$24,000 yearly with raises of \$1600 per year. Company B pays \$28,000 yearly with raises of \$1000 per year. Which company will pay more in year 10? How much more?

66. Company A pays \$23,000 yearly with raises of \$1200 per year. Company B pays \$26,000 yearly with raises of \$800 per year. Which company will pay more in year 10? How much more?

The bar graph shows the average cost of tuition, fees, and room and board at public and private colleges in the United States for four academic years. Use this information to solve Exercises 67–68.

The Cost of Higher Education

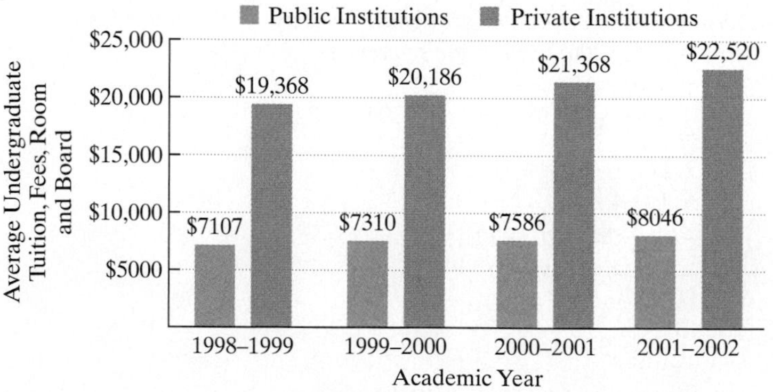

Source: U.S. Department of Education

67. **a.** Use the numbers shown in the bar graph on the previous page to find the total cost of higher education at a public college for a four-year period, beginning with the 1998–1999 academic year and ending with the 2001–2002 academic year.

b. The model

$$a_n = 309n + 6739$$

describes the cost of higher education at a public college in academic year n, where $n = 1$ corresponds to 1998–1999, $n = 2$ to 1999–2000, and so on. Use this model and the formula for S_n to find the total cost of a higher education at a public college for a four-year period, beginning with the 1998–1999 academic year and ending with the 2001–2002 academic year. How well does the model describe the actual sum that you obtained in part (a)?

68. **a.** Use the numbers shown in the bar graph on the previous page to find the total cost of higher education at a private college for a four-year period, beginning with the 1998–1999 academic year and ending with the 2001–2002 academic year.

b. The model

$$a_n = 1064n + 18{,}201$$

describes the cost of higher education at a private college in academic year n, where $n = 1$ corresponds to 1998–1999, $n = 2$ to 1999–2000, and so on. Use this model and the formula for S_n to find the total cost of a higher education at a private college for a four-year period, beginning with the 1998–1999 academic year and ending with the 2001–2002 academic year. How well does the model describe the actual sum that you obtained in part (a)?

69. Use one of the models in Exercises 67–68 and the formula for S_n to find the total cost of your undergraduate education. How well does the model describe your anticipated costs?

70. A company offers a starting yearly salary of \$33,000 with raises of \$2500 per year. Find the total salary over a ten-year period.

71. You are considering two job offers. Company A will start you at \$19,000 a year and guarantee a raise of \$2600 per year. Company B will start you at a higher salary, \$27,000 a year, but will only guarantee a raise of \$1200 per year. Find the total salary that each company will pay over a ten-year period. Which company pays the greater total amount?

72. A theater has 30 seats in the first row, 32 seats in the second row, increasing by 2 seats each row for a total of 26 rows. How many seats are there in the theater?

73. A section in a stadium has 20 seats in the first row, 23 seats in the second row, increasing by 3 seats each row for a total of 38 rows. How many seats are in this section of the stadium?

Writing in Mathematics

74. What is an arithmetic sequence? Give an example with your explanation.

75. What is the common difference in an arithmetic sequence?

76. Explain how to find the general term of an arithmetic sequence.

77. Explain how to find the sum of the first n terms of an arithmetic sequence without having to add up all the terms.

Technology Exercises

78. Use the [SEQ] (sequence) capability of a graphing utility and the formula you obtained for a_n to verify the value you found for a_{20} in any five exercises from Exercises 25–34.

79. Use the capability of a graphing utility to calculate the sum of a sequence to verify any five of your answers to Exercises 45–50.

Critical Thinking Exercises

80. Give examples of two different arithmetic sequences whose fourth term, a_4, is 10.

81. In the sequence 21,700, 23,172, 24,644, 26,116, ..., which term is 314,628?

82. A *degree-day* is a unit used to measure the fuel requirements of buildings. By definition, each degree that the average daily temperature is below 65°F is 1 degree-day. For example, a temperature of 42°F constitutes 23 degree-days. If the average temperature on January 1 was 42°F and fell 2°F for each subsequent day up to and including January 10, how many degree-days are included from January 1 to January 10?

83. Show that the sum of the first n positive odd integers,

$$1 + 3 + 5 + \cdots + (2n - 1),$$

is n^2.

Review Exercises

84. Solve: $\log(x^2 - 25) - \log(x + 5) = 3$.

(Section 12.4, Example 6)

85. Solve: $x^2 + 3x \le 10$.

(Section 11.5, Example 1)

86. Solve for P: $A = \dfrac{Pt}{P + t}$.

(Section 7.6, Example 7)

SECTION 14.3

Objectives

1 Find the common ratio of a geometric sequence.

2 Write terms of a geometric sequence.

3 Use the formula for the general term of a geometric sequence.

4 Use the formula for the sum of the first n terms of a geometric sequence.

5 Find the value of an annuity.

6 Use the formula for the sum of an infinite geometric series.

GEOMETRIC SEQUENCES AND SERIES

Here we are at the closing moments of a job interview. You're shaking hands with the manager. You managed to answer all the tough questions without losing your poise, and now you've been offered a job. As a matter of fact, your qualifications are so terrific that you've been offered two jobs—one just the day before, with a rival company in the same field! One company offers \$30,000 the first year, with increases of 6% per year for four years after that. The other offers \$32,000 the first year, with annual increases of 3% per year after that. Over a five-year period, which is the better offer?

If salary raises amount to a certain percent each year, the yearly salaries over time form a geometric sequence. In this section, we investigate geometric sequences and their properties. After studying the section, you will be in a position to decide which job offer to accept: You will know which company will pay you more over five years.

Geometric Sequences Figure 14.3 shows a sequence in which the number of squares is increasing. From left to right, the number of squares is 1, 5, 25, 125, and 625. In this sequence, each term after the first, 1, is obtained by multiplying the preceding term by a constant amount, namely 5. This sequence of increasing number of squares is an example of a *geometric sequence*.

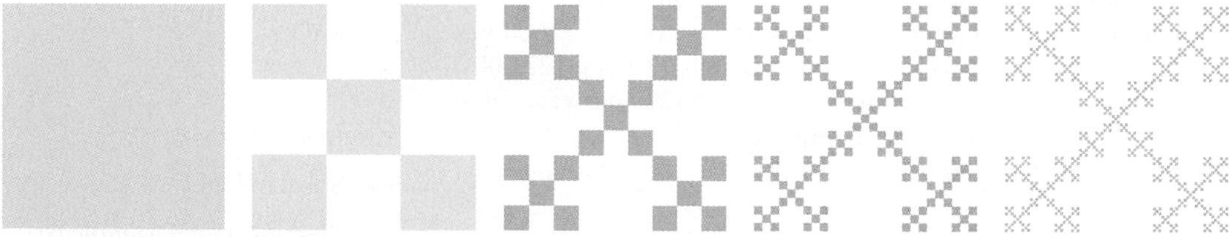

FIGURE 14.3 A geometric sequence of squares

DEFINITION OF A GEOMETRIC SEQUENCE A **geometric sequence** is a sequence in which each term after the first is obtained by multiplying the preceding term by a fixed nonzero constant. The amount by which we multiply each time is called the **common ratio** of the sequence.

1 Find the common ratio of a geometric sequence.

The common ratio, r, is found by dividing any term after the first term by the term that directly precedes it. In the following examples, the common ratio is found by dividing the second term by the first term, $\frac{a_2}{a_1}$.

Geometric sequence	Common ratio
$1, 5, 25, 125, 625, \ldots$	$r = \frac{5}{1} = 5$
$4, 8, 16, 32, 64, \ldots$	$r = \frac{8}{4} = 2$
$6, -12, 24, -48, 96, \ldots$	$r = \frac{-12}{6} = -2$
$9, -3, 1, -\frac{1}{3}, \frac{1}{9}, \ldots$	$r = \frac{-3}{9} = -\frac{1}{3}$

STUDY TIP

When the common ratio of a geometric sequence is negative, the signs of the terms alternate.

Figure 14.4 shows a partial graph of the first geometric sequence in our list. The graph forms a set of discrete points lying on the exponential function $f(x) = 5^{x-1}$. This illustrates that **a geometric sequence with a positive common ratio other than 1 is an exponential function whose domain is the set of positive integers.**

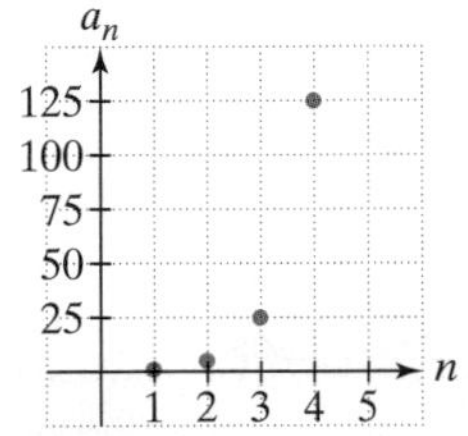

FIGURE 14.4 The graph of $\{a_n\} = 1, 5, 25, 125, \ldots$

2 Write terms of a geometric sequence.

How do we write out the terms of a geometric sequence when the first term and the common ratio are known? We multiply the first term by the common ratio to get the second term, multiply the second term by the common ratio to get the third term, and so on.

EXAMPLE 1 Writing the Terms of a Geometric Sequence

Write the first six terms of the geometric sequence with first term 6 and common ratio $\frac{1}{3}$.

SOLUTION The first term is 6. The second term is $6 \cdot \frac{1}{3}$, or 2. The third term is $2 \cdot \frac{1}{3}$, or $\frac{2}{3}$. The fourth term is $\frac{2}{3} \cdot \frac{1}{3}$, or $\frac{2}{9}$, and so on. The first six terms are

$$6, 2, \frac{2}{3}, \frac{2}{9}, \frac{2}{27}, \text{ and } \frac{2}{81}.$$

CHECK POINT 1 Write the first six terms of the geometric sequence with first term 12 and common ratio $\frac{1}{2}$.

3 Use the formula for the general term of a geometric sequence.

The General Term of a Geometric Sequence Consider a geometric sequence whose first term is a_1 and whose common ratio is r. We are looking for a formula for the general term, a_n. Let's begin by writing the first six terms. The first term is a_1. The second term is $a_1 r$. The third term is $a_1 r \cdot r$, or $a_1 r^2$. The fourth term is $a_1 r^2 \cdot r$, or $a_1 r^3$, and so on. Starting with a_1 and multiplying each successive term by r, the first six terms are

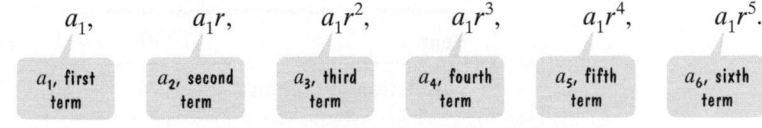

$$a_1, \quad a_1 r, \quad a_1 r^2, \quad a_1 r^3, \quad a_1 r^4, \quad a_1 r^5.$$

a_1	$a_1 r$	$a_1 r^2$	$a_1 r^3$	$a_1 r^4$	$a_1 r^5$
a_1, first term	a_2, second term	a_3, third term	a_4, fourth term	a_5, fifth term	a_6, sixth term

Compare the exponent on r and the subscript of a denoting the term number. Can you see that the exponent on r is 1 less than the subscript of a denoting the term number?

$$a_3\text{: third term} = a_1r^2 \qquad a_4\text{: third term} = a_1r^3$$

2 is one less than 3.   3 is one less than 4.

Thus, the formula for the nth term is

$$a_n = a_1r^{n-1}.$$

$n-1$ is one less than n.

GENERAL TERM OF A GEOMETRIC SEQUENCE The nth term (the general term) of a geometric sequence with first term a_1 and common ratio r is

$$a_n = a_1r^{n-1}.$$

STUDY TIP

Be careful with the order of operations when evaluating

$$a_1r^{n-1}.$$

First find r^{n-1}. Then multiply the result by a_1.

EXAMPLE 2 Using the Formula for the General Term of a Geometric Sequence

Find the eighth term of the geometric sequence whose first term is -4 and whose common ratio is -2.

SOLUTION To find the eighth term, a_8, we replace n in the formula with 8, a_1 with -4, and r with -2.

$$a_n = a_1r^{n-1}$$
$$a_8 = -4(-2)^{8-1} = -4(-2)^7 = -4(-128) = 512$$

The eighth term is 512. We can check this result by writing the first eight terms of the sequence:

$$-4, 8, -16, 32, -64, 128, -256, 512.$$ ■

CHECK POINT 2 Find the seventh term of the geometric sequence whose first term is 5 and whose common ratio is -3.

In Chapter 12, we studied exponential functions of the form $f(x) = b^x$ and the explosive exponential growth of world population. In our next example, we consider Florida's geometric population growth. Because a geometric sequence is an exponential function whose domain is the set of positive integers, geometric and exponential growth mean the same thing.

EXAMPLE 3 Geometric Population Growth

The population of Florida from 1990 through 1997 is shown in the following table:

Year	1990	1991	1992	1993	1994	1995	1996	1997
Population in millions	12.94	13.20	13.46	13.73	14.00	14.28	14.57	14.86

ENRICHMENT ESSAY

Geometric Population Growth

Economist Thomas Malthus (1766–1834) predicted that population growth would increase as a geometric sequence and food production would increase as an arithmetic sequence. He concluded that eventually population would exceed food production. If two sequences, one geometric and one arithmetic, are increasing, the geometric sequence will eventually overtake the arithmetic sequence, regardless of any head start that the arithmetic sequence might initially have.

a. Show that the population is increasing geometrically.

b. Write the general term for the geometric sequence describing population growth for Florida n years after 1989.

c. Estimate Florida's population, in millions, for the year 2000.

SOLUTION

a. First, we divide the population for each year by the population in the preceding year.

$$\frac{13.20}{12.94} \approx 1.02, \quad \frac{13.46}{13.20} \approx 1.02, \quad \frac{13.73}{13.46} \approx 1.02$$

Continuing in this manner, we will keep getting approximately 1.02. This means that the population is increasing geometrically with $r \approx 1.02$. In this situation, the common ratio is the growth rate, indicating that the population of Florida in any year shown in the table is approximately 1.02 times the population the year before.

b. The sequence of Florida's population growth is

$$12.94, 13.20, 13.46, 13.73, 14.00, 14.28, 14.57, 14.86, \ldots.$$

Because the population is increasing geometrically, we can find the general term of this sequence using

$$a_n = a_1 r^{n-1}.$$

In this sequence, $a_1 = 12.94$ and [from part (a)] $r \approx 1.02$. We substitute these values into the formula for the general term. This gives the general term for the geometric sequence describing Florida's population n years after 1989.

$$a_n = 12.94(1.02)^{n-1}$$

c. We can use the formula for the general term, a_n, in part (b) to estimate Florida's population for the year 2000. The year 2000 is 11 years after 1989—that is, $2000 - 1989 = 11$. Thus, $n = 11$. We substitute 11 for n in $a_n = 12.94(1.02)^{n-1}$.

$$a_{11} = 12.94(1.02)^{11-1} = 12.94(1.02)^{10} \approx 15.77$$

The formula indicates that Florida had a population of approximately 15.77 million in the year 2000. According to the U.S. Census Bureau, Florida's population in 2000 was 15.98 million. Our geometric sequence models the actual population fairly well. ■

CHECK POINT 3 Write the general term for the geometric sequence

$$3, 6, 12, 24, 48, \ldots.$$

Then use the formula for the general term to find the eighth term.

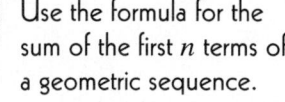

4 Use the formula for the sum of the first n terms of a geometric sequence.

The Sum of the First n Terms of a Geometric Sequence The sum of the first n terms of a geometric sequence, denoted by S_n, and called the ***n*th partial sum**, can be found without having to add up all the terms. Recall that the first n terms of a geometric sequence are

$$a_1, a_1r, a_1r^2, \ldots, a_1r^{n-2}, a_1r^{n-1}.$$

We proceed as follows:

$$S_n = a_1 + a_1r + a_1r^2 + \cdots + a_1r^{n-2} + a_1r^{n-1}$$ S_n is the sum of the first n terms of the sequence.

$$rS_n = a_1r + a_1r^2 + a_1r^3 + \cdots + a_1r^{n-1} + a_1r^n$$ Multiply both sides of the equation by r.

$$S_n - rS_n = a_1 - a_1r^n$$ Subtract the second equation from the first equation.

$$S_n(1 - r) = a_1(1 - r^n)$$ Factor out S_n on the left and a_1 on the right.

$$S_n = \frac{a_1(1 - r^n)}{1 - r}.$$ Solve for S_n by dividing both sides by $1 - r$ (assuming that $r \neq 1$).

We have proved the following result:

THE SUM OF THE FIRST n TERMS OF A GEOMETRIC SEQUENCE The sum, S_n, of the first n terms of a geometric sequence is given by

$$S_n = \frac{a_1(1 - r^n)}{1 - r}$$

in which a_1 is the first term and r is the common ratio ($r \neq 1$).

STUDY TIP

If the common ratio is 1, the geometric sequence is

$$a_1, a_1, a_1, a_1, \ldots.$$

The sum of the first n terms of this sequence is na_1:

$$S_n = \underbrace{a_1 + a_1 + a_1 + \cdots + a_1}_{\text{There are } n \text{ terms.}}$$
$$= na_1.$$

To find the sum of the terms of a geometric sequence, we need to know the first term, a_1, the common ratio, r, and the number of terms, n. The following examples illustrate how to use this formula.

EXAMPLE 4 Finding the Sum of the First n Terms of a Geometric Sequence

Find the sum of the first 18 terms of the geometric sequence: $2, -8, 32, -128, \ldots$.

SOLUTION To find the sum of the first 18 terms, S_{18}, we replace n in the formula with 18.

$$S_n = \frac{a_1(1 - r^n)}{1 - r}$$

$$S_{18} = \frac{a_1(1 - r^{18})}{1 - r}$$

The first term, a_1, is 2. We must find r, the common ratio.

We can find the common ratio by dividing the second term of $2, -8, 32, -128, \ldots$ by the first term.

$$r = \frac{a_2}{a_1} = \frac{-8}{2} = -4$$

Now we are ready to find the sum of the first 18 terms of 2, −8, 32, −128, ...

$$S_n = \frac{a_1(1 - r^n)}{1 - r}$$ Use the formula for the sum of the first n terms of a geometric sequence.

$$S_{18} = \frac{2[1 - (-4)^{18}]}{1 - (-4)}$$ a_1 (the first term) = 2, $r = -4$, and $n = 18$ because we want the sum of the first 18 terms.

$$= -27{,}487{,}790{,}694$$ Use a calculator.

The sum of the first 18 terms is −27,487,790,694. Equivalently, this number is the 18th partial sum of the sequence 2, −8, 32, −128, ... ■

CHECK POINT 4 Find the sum of the first nine terms of the geometric sequence: 2, −6, 18, −54, ...

EXAMPLE 5 Using S_n to Evaluate a Summation

Find the following sum: $\sum_{i=1}^{10} 6 \cdot 2^i$.

SOLUTION Let's write out a few terms in the sum.

$$\sum_{i=1}^{10} 6 \cdot 2^i = 6 \cdot 2 + 6 \cdot 2^2 + 6 \cdot 2^3 + \cdots + 6 \cdot 2^{10}$$

Do you see that each term after the first is obtained by multiplying the preceding term by 2? To find the sum of the 10 terms ($n = 10$), we need to know the first term, a_1, and the common ratio, r. The first term is $6 \cdot 2$ or 12: $a_1 = 12$. The common ratio is 2.

$$S_n = \frac{a_1(1 - r^n)}{1 - r}$$ Use the formula for the sum of the first n terms of a geometric sequence.

$$S_{10} = \frac{12(1 - 2^{10})}{1 - 2}$$ a_1 (the first term) = 12, $r = 2$, and $n = 10$ because we are adding ten terms.

$$= 12{,}276$$ Use a calculator.

Thus,

$$\sum_{i=1}^{10} 6 \cdot 2^i = 12{,}276.$$ ■

USING TECHNOLOGY

To find

$$\sum_{i=1}^{10} 6 \cdot 2^i$$

on a graphing utility, enter

[SUM] [SEQ] $(6 \times 2^x, x, 1, 10, 1)$.

Then press [ENTER].

```
sum(seq(6*2^X,X,
1,10,1))
           12276
```

CHECK POINT 5 Find the following sum: $\sum_{i=1}^{8} 2 \cdot 3^i$.

Some of the exercises in the previous exercise set involved situations in which salaries increased by a fixed amount each year. A more realistic situation is one in which salary raises increase by a certain percent each year. Example 6 shows how such a situation can be described using a geometric sequence.

EXAMPLE 6 Computing a Lifetime Salary

A union contract specifies that each worker will receive a 5% pay increase each year for the next 30 years. One worker is paid \$20,000 the first year. What is this person's total lifetime salary over a 30-year period?

SOLUTION The salary for the first year is \$20,000. With a 5% raise, the second-year salary is computed as follows:

$$\text{Salary for year 2} = 20{,}000 + 20{,}000(0.05) = 20{,}000(1 + 0.05) = 20{,}000(1.05).$$

Each year, the salary is 1.05 times what it was in the previous year. Thus, the salary for year 3 is 1.05 times 20,000(1.05), or $20{,}000(1.05)^2$. The salaries for the first five years are given in the table.

Yearly Salaries

Year 1	Year 2	Year 3	Year 4	Year 5	...
20,000	20,000(1.05)	$20{,}000(1.05)^2$	$20{,}000(1.05)^3$	$20{,}000(1.05)^4$	...

The numbers in the bottom row form a geometric sequence with $a_1 = 20{,}000$ and $r = 1.05$. To find the total salary over 30 years, we use the formula for the sum of the first n terms of a geometric sequence, with $n = 30$.

$$S_n = \frac{a_1(1 - r^n)}{1 - r}$$

$$S_{30} = \frac{20{,}000[1 - (1.05)^{30}]}{1 - 1.05}$$

Total salary over 30 years

$$= \frac{20{,}000[1 - (1.05)^{30}]}{-0.05}$$

$$\approx 1{,}328{,}777 \qquad \text{Use a calculator.}$$

The total salary over the 30-year period is approximately \$1,328,777. ■

CHECK POINT 6 A job pays a salary of \$30,000 the first year. During the next 29 years, the salary increases by 6% each year. What is the total lifetime salary over the 30-year period?

5 Find the value of an annuity.

Annuities The compound interest formula

$$A = P(1 + r)^t$$

gives the future value, A, after t years, when a fixed amount of money, P, the principal, is deposited in an account that pays an annual interest rate r (in decimal form) compounded once a year. However, money is often invested in small amounts at periodic intervals. For example, to save for retirement, you might decide to place \$1000 into an Individual Retirement Account (IRA) at the end of each year until you retire. An **annuity** is a sequence of equal payments made at equal time periods. An IRA is an example of an annuity.

Suppose P dollars is deposited into an account at the end of each year. The account pays an annual interest rate, r, compounded annually. At the end of the first year, the account contains P dollars. At the end of the second year, P dollars is deposited again. At the time of this deposit, the first deposit has received interest earned during the second year. The **value of the annuity** is the sum of all deposits made plus all interest paid. Thus, the value of the annuity after two years is

$$P + P(1 + r).$$

Deposit of P dollars at end of second year

First-year deposit of P dollars with interest earned for a year

The value of the annuity after three years is

$$P \quad + \quad P(1 + r) \quad + \quad P(1 + r)^2.$$

Deposit of P dollars at end of third year

Second-year deposit of P dollars with interest earned for a year

First-year deposit of P dollars with interest earned over two years

The value of the annuity after t years is

$$P + P(1 + r) + P(1 + r)^2 + P(1 + r)^3 + \cdots + P(1 + r)^{t-1}.$$

Deposit of P dollars at end of year t

First-year deposit of P dollars with interest earned over $t - 1$ years

This is the sum of the terms of a geometric sequence with first term P and common ratio $1 + r$. We use the formula

$$S_n = \frac{a_1(1 - r^n)}{1 - r}$$

to find the sum of the terms:

$$S_t = \frac{P[1 - (1 + r)^t]}{1 - (1 + r)} = \frac{P[1 - (1 + r)^t]}{-r} = P\frac{(1 + r)^t - 1}{r}.$$

This formula gives the value of an annuity after t years if interest is compounded once a year. We can adjust the formula to find the value of an annuity if equal payments are made at the end of each of n yearly compounding periods.

VALUE OF AN ANNUITY: INTEREST COMPOUNDED n TIMES PER YEAR If P is the deposit made at the end of each compounding period for an annuity at r percent annual interest compounded n times per year, the value, A, of the annuity after t years is

$$A = P\frac{\left(1 + \frac{r}{n}\right)^{nt} - 1}{\frac{r}{n}}.$$

EXAMPLE 7 Determining the Value of an Annuity

To save for retirement, you decide to deposit \$1000 into an IRA at the end of each year for the next 30 years. If the interest rate is 10% per year compounded annually, find the value of the IRA after 30 years.

SOLUTION The annuity involves 30 year-end deposits of $P = \$1000$. The interest rate is 10%: $r = 0.10$. Because the deposits are made once a year and the interest is compounded once a year, $n = 1$. The number of years is 30: $t = 30$. We replace the variables in the formula for the value of an annuity with these numbers.

$$A = P\frac{\left(1 + \frac{r}{n}\right)^{nt} - 1}{\frac{r}{n}}$$

$$A = 1000\frac{\left(1 + \frac{0.10}{1}\right)^{1\cdot 30} - 1}{\frac{0.10}{1}} \approx 164{,}494$$

The value of the IRA at the end of 30 years is approximately \$164,494. ■

CHECK POINT 7 If \$3000 is deposited into an IRA at the end of each year for 40 years and the interest rate is 10% per year compounded annually, find the value of the IRA after 40 years.

6 Use the formula for the sum of an infinite geometric series.

Geometric Series An infinite sum of the form

$$a_1 + a_1r + a_1r^2 + a_1r^3 + \cdots + a_1r^{n-1} + \cdots$$

with first term a_1 and common ratio r is called an **infinite geometric series**. How can we determine which infinite geometric series have sums and which do not? We look at what happens to r^n as n gets larger in the formula for the sum of the first n terms of this series, namely

$$S_n = \frac{a_1(1 - r^n)}{1 - r}.$$

If r is any number between -1 and 1, that is, $-1 < r < 1$, the term r^n approaches 0 as n gets larger. For example, consider what happens to r^n for $r = \frac{1}{2}$:

$$\left(\frac{1}{2}\right)^1 = \frac{1}{2} \quad \left(\frac{1}{2}\right)^2 = \frac{1}{4} \quad \left(\frac{1}{2}\right)^3 = \frac{1}{8} \quad \left(\frac{1}{2}\right)^4 = \frac{1}{16} \quad \left(\frac{1}{2}\right)^5 = \frac{1}{32} \quad \left(\frac{1}{2}\right)^6 = \frac{1}{64}.$$

These numbers are approaching 0 as n gets larger.

Take another look at the formula for the sum of the first n terms of a geometric sequence.

$$S_n = \frac{a_1(1 - r^n)}{1 - r}$$

If $-1 < r < 1$, r^n approaches 0 as n gets larger.

Let us replace r^n with 0 in the formula for S_n. This change gives us a formula for the sum of an infinite geometric series with a common ratio between -1 and 1.

THE SUM OF AN INFINITE GEOMETRIC SERIES If $-1 < r < 1$ (equivalently, $|r| < 1$), then the sum of the infinite geometric series

$$a_1 + a_1 r + a_1 r^2 + a_1 r^3 + \cdots$$

in which a_1 is the first term and r is the common ratio is given by

$$S = \frac{a_1}{1 - r}.$$

If $|r| \geq 1$, the infinite series does not have a sum.

To use the formula for the sum of an infinite geometric series, we need to know the first term and the common ratio. For example, consider

$$\frac{1}{2} + \frac{1}{4} + \frac{1}{8} + \frac{1}{16} + \frac{1}{32} + \cdots.$$

First term, a_1, is $\frac{1}{2}$.

Common ratio, r, is $\frac{a_2}{a_1}$.
$r = \frac{1}{4} \div \frac{1}{2} = \frac{1}{4} \cdot 2 = \frac{1}{2}$

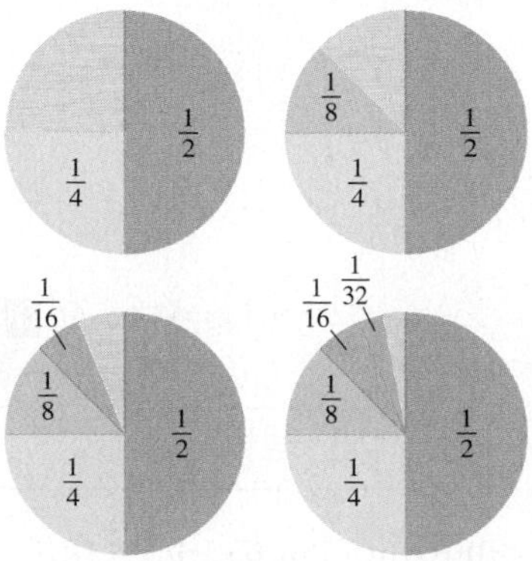

FIGURE 14.5 The sum $\frac{1}{2} + \frac{1}{4} + \frac{1}{8} + \frac{1}{16} + \frac{1}{32} + \cdots$ is approaching 1.

With $r = \frac{1}{2}$, the condition that $|r| < 1$ is met, so the infinite geometric series has a sum given by $S = \frac{a_1}{1 - r}$. The sum of the series is found as follows:

$$\frac{1}{2} + \frac{1}{4} + \frac{1}{8} + \frac{1}{16} + \frac{1}{32} + \cdots = \frac{a_1}{1 - r} = \frac{\frac{1}{2}}{1 - \frac{1}{2}} = \frac{\frac{1}{2}}{\frac{1}{2}} = 1.$$

Thus, the sum of the infinite geometric series is 1. Notice how this is illustrated in Figure 14.5. As more terms are included, the sum is approaching the area of one complete circle.

EXAMPLE 8 Finding the Sum of an Infinite Geometric Series

Find the sum of the infinite geometric series: $\frac{3}{8} - \frac{3}{16} + \frac{3}{32} - \frac{3}{64} + \cdots$.

SOLUTION Before finding the sum, we must find the common ratio.

$$r = \frac{a_2}{a_1} = \frac{-\frac{3}{16}}{\frac{3}{8}} = -\frac{3}{16} \cdot \frac{8}{3} = -\frac{1}{2}$$

Because $r = -\frac{1}{2}$, the condition that $|r| < 1$ is met. Thus, the infinite geometric series has a sum.

$$S = \frac{a_1}{1 - r}$$ This is the formula for the sum of an infinite geometric series. Let $a_1 = \frac{3}{8}$ and $r = -\frac{1}{2}$.

$$= \frac{\frac{3}{8}}{1 - \left(-\frac{1}{2}\right)} = \frac{\frac{3}{8}}{\frac{3}{2}} = \frac{3}{8} \cdot \frac{2}{3} = \frac{1}{4}$$

Thus, the sum of $\frac{3}{8} - \frac{3}{16} + \frac{3}{32} - \frac{3}{64} + \cdots$ is $\frac{1}{4}$. Put in an informal way, as we continue to add more and more terms, the sum is approximately $\frac{1}{4}$. ■

CHECK POINT 8 Find the sum of the infinite geometric series: $3 + 2 + \frac{4}{3} + \frac{8}{9} + \cdots$.

We can use the formula for the sum of an infinite geometric series to express a repeating decimal as a fraction in lowest terms.

EXAMPLE 9 Writing a Repeating Decimal as a Fraction

Express $0.\overline{7}$ as a fraction in lowest terms.

SOLUTION

$$0.\overline{7} = 0.7777\ldots = \frac{7}{10} + \frac{7}{100} + \frac{7}{1000} + \frac{7}{10{,}000} + \cdots$$

Observe that $0.\overline{7}$ is an infinite geometric series with first term $\frac{7}{10}$ and common ratio $\frac{1}{10}$. Because $r = \frac{1}{10}$, the condition that $|r| < 1$ is met. Thus, we can use our formula to find the sum. Therefore,

$$0.\overline{7} = \frac{a_1}{1-r} = \frac{\frac{7}{10}}{1-\frac{1}{10}} = \frac{\frac{7}{10}}{\frac{9}{10}} = \frac{7}{10} \cdot \frac{10}{9} = \frac{7}{9}.$$

An equivalent fraction for $0.\overline{7}$ is $\frac{7}{9}$. ■

CHECK POINT 9 Express $0.\overline{9}$ as a fraction in lowest terms.

Infinite geometric series have many applications, as illustrated in Example 10.

EXAMPLE 10 Tax Rebates and the Multiplier Effect

A tax rebate that returns a certain amount of money to taxpayers can have a total effect on the economy that is many times this amount. In economics, this phenomenon is called the **multiplier effect**. Suppose, for example, that the government reduces taxes so that each consumer has \$2000 more income. The government assumes that each person will spend 70% of this (=\$1400). The individuals and businesses receiving this \$1400 in turn spend 70% of it (=\$980), creating extra income for other people to spend, and so on. Determine the total amount spent on consumer goods from the initial \$2000 tax rebate.

SOLUTION The total amount spent is given by the infinite geometric series

$$1400 + 980 + 686 + \cdots.$$

70% of 1400 | 70% of 980

The first term is 1400: $a_1 = 1400$. The common ratio is 70%, or 0.7: $r = 0.7$. Because $r = 0.7$, the condition that $|r| < 1$ is met. Thus, we can use our formula to find the sum. Therefore,

$$1400 + 980 + 686 + \cdots = \frac{a_1}{1-r} = \frac{1400}{1-0.7} \approx 4667.$$

This means that the total amount spent on consumer goods from the initial \$2000 rebate is approximately \$4667. ■

CHECK POINT 10 Rework Example 10 and determine the total amount spent on consumer goods with a \$1000 tax rebate and 80% spending down the line.

14.3 EXERCISE SET

Student Solutions Manual | CD/Video | PH Math/Tutor Center | MathXL Tutorials on CD | MathXL® | MyMathLab | Interactmath.com

Practice Exercises

In Exercises 1–8, find the common ratio for each geometric sequence.

1. 5, 15, 45, 135, ...
2. 5, 10, 20, 40, ...
3. −15, 30, −60, 120, ...
4. −2, 6, −18, 54, ...
5. $3, \frac{9}{2}, \frac{27}{4}, \frac{81}{8}, \ldots$
6. $4, \frac{8}{3}, \frac{16}{9}, \frac{32}{27}, \ldots$
7. 4, −0.4, 0.04, −0.004, ...
8. 7, −0.7, 0.07, −0.007, ...

In Exercises 9–16, write the first five terms of each geometric sequence with the given first term, a_1, and common ratio, r.

9. $a_1 = 2, r = 3$
10. $a_1 = 2, r = 4$
11. $a_1 = 20, r = \frac{1}{2}$
12. $a_1 = 24, r = \frac{1}{3}$
13. $a_1 = -4, r = -10$
14. $a_1 = -3, r = -10$
15. $a_1 = -\frac{1}{4}, r = -2$
16. $a_1 = -\frac{1}{16}, r = -4$

In Exercises 17–24, use the formula for the general term (the nth term) of a geometric sequence to find the indicated term of each sequence with the given first term, a_1, and common ratio, r.

17. Find a_8 when $a_1 = 6, r = 2$.
18. Find a_8 when $a_1 = 5, r = 3$.
19. Find a_{12} when $a_1 = 5, r = -2$.
20. Find a_{12} when $a_1 = 4, r = -2$.
21. Find a_6 when $a_1 = 6400, r = -\frac{1}{2}$.
22. Find a_6 when $a_1 = 8000, r = -\frac{1}{2}$.
23. Find a_8 when $a_1 = 1{,}000{,}000, r = 0.1$.
24. Find a_8 when $a_1 = 40{,}000, r = 0.1$.

In Exercises 25–32, write a formula for the general term (the nth term) of each geometric sequence. Then use the formula for a_n to find a_7, the seventh term of the sequence.

25. 3, 12, 48, 192, ...
26. 3, 15, 75, 375, ...
27. $18, 6, 2, \frac{2}{3}, \ldots$
28. $12, 6, 3, \frac{3}{2}, \ldots$
29. 1.5, −3, 6, −12, ...
30. $5, -1, \frac{1}{5}, -\frac{1}{25}, \ldots$
31. 0.0004, −0.004, 0.04, −0.4. ...
32. 0.0007, −0.007, 0.07, −0.7, ...

Use the formula for the sum of the first n terms of a geometric sequence to solve Exercises 33–38.

33. Find the sum of the first 12 terms of the geometric sequence: 2, 6, 18, 54 ....
34. Find the sum of the first 12 terms of the geometric sequence: 3, 6, 12, 24, ...
35. Find the sum of the first 11 terms of the geometric sequence: 3, −6, 12, −24, ...
36. Find the sum of the first 11 terms of the geometric sequence: 4, −12, 36, −108, ...
37. Find the sum of the first 14 terms of the geometric sequence: $-\frac{3}{2}, 3, -6, 12, \ldots$

38. Find the sum of the first 14 terms of the geometric sequence: $-\frac{1}{24}, \frac{1}{12}, -\frac{1}{6}, \frac{1}{3}, \ldots$.

In Exercises 39–44, find the indicated sum. Use the formula for the sum of the first n terms of a geometric sequence.

39. $\sum_{i=1}^{8} 3^i$

40. $\sum_{i=1}^{6} 4^i$

41. $\sum_{i=1}^{10} 5 \cdot 2^i$

42. $\sum_{i=1}^{7} 4(-3)^i$

43. $\sum_{i=1}^{6} \left(\frac{1}{2}\right)^{i+1}$

44. $\sum_{i=1}^{6} \left(\frac{1}{3}\right)^{i+1}$

In Exercises 45–52, find the sum of each infinite geometric series.

45. $1 + \frac{1}{3} + \frac{1}{9} + \frac{1}{27} + \cdots$

46. $1 + \frac{1}{4} + \frac{1}{16} + \frac{1}{64} + \cdots$

47. $3 + \frac{3}{4} + \frac{3}{4^2} + \frac{3}{4^3} + \cdots$

48. $5 + \frac{5}{6} + \frac{5}{6^2} + \frac{5}{6^3} + \cdots$

49. $1 - \frac{1}{2} + \frac{1}{4} - \frac{1}{8} + \cdots$

50. $3 - 1 + \frac{1}{3} - \frac{1}{9} + \cdots$

51. $\sum_{i=1}^{\infty} 26(-0.3)^{i-1}$

52. $\sum_{i=1}^{\infty} 51(-0.7)^{i-1}$

In Exercises 53–58, express each repeating decimal as a fraction in lowest terms.

53. $0.\overline{5} = \frac{5}{10} + \frac{5}{100} + \frac{5}{1000} + \frac{5}{10{,}000} + \cdots$

54. $0.\overline{1} = \frac{1}{10} + \frac{1}{100} + \frac{1}{1000} + \frac{1}{10{,}000} + \cdots$

55. $0.\overline{47} = \frac{47}{100} + \frac{47}{10{,}000} + \frac{47}{1{,}000{,}000} + \cdots$

56. $0.\overline{83} = \frac{83}{100} + \frac{83}{10{,}000} + \frac{83}{1{,}000{,}000} + \cdots$

57. $0.\overline{257}$

58. $0.\overline{529}$

In Exercises 59–64, the general term of a sequence is given. Determine whether the sequence is arithmetic, geometric, or neither. If the sequence is arithmetic, find the common difference; if it is geometric, find the common ratio.

59. $a_n = n + 5$

60. $a_n = n - 3$

61. $a_n = 2^n$

62. $a_n = \left(\frac{1}{2}\right)^n$

63. $a_n = n^2 + 5$

64. $a_n = n^2 - 3$

Practice Plus

In Exercises 65–70, let

$$\{a_n\} = -5, 10, -20, 40, \ldots,$$
$$\{b_n\} = 10, -5, -20, -35, \ldots,$$

and

$$\{c_n\} = -2, 1, -\tfrac{1}{2}, \tfrac{1}{4}, \ldots.$$

65. Find $a_{10} + b_{10}$.

66. Find $a_{11} + b_{11}$.

67. Find the difference between the sum of the first 10 terms of $\{a_n\}$ and the sum of the first 10 terms of $\{b_n\}$.

68. Find the difference between the sum of the first 11 terms of $\{a_n\}$ and the sum of the first 11 terms of $\{b_n\}$.

69. Find the product of the sum of the first 6 terms of $\{a_n\}$ and the sum of the infinite series containing all the terms of $\{c_n\}$.

70. Find the product of the sum of the first 9 terms of $\{a_n\}$ and the sum of the infinite series containing all the terms of $\{c_n\}$.

In Exercises 71–72, find a_2 and a_3 for each geometric sequence.

71. $8, a_2, a_3, 27$

72. $2, a_2, a_3, -54$

Application Exercises

Use the formula for the general term (the nth term) of a geometric sequence to solve Exercises 73–76.

In Exercises 73–74, suppose you save \$1 the first day of a month, \$2 the second day, \$4 the third day, and so on. That is, each day you save twice as much as you did the day before.

73. What will you put aside for savings on the fifteenth day of the month?

74. What will you put aside for savings on the thirtieth day of the month?

75. A professional baseball player signs a contract with a beginning salary of \$3,000,000 for the first year and an annual increase of 4% per year beginning in the second year. That is, beginning in year 2, the athlete's salary will be 1.04 times what it was in the previous year. What is the athlete's salary for year 7 of the contract? Round to the nearest dollar.

76. You are offered a job that pays \$30,000 for the first year with an annual increase of 5% per year beginning in the second year. That is, beginning in year 2, your salary will be 1.05 times what it was in the previous year. What can you expect to earn in your sixth year on the job? Round to the nearest dollar.

77. The population of California from 1990 through 1997 is shown in the following table.

Year	1990	1991	1992	1993
Population in millions	29.76	30.15	30.54	30.94
Year	1994	1995	1996	1997
Population in millions	31.34	31.75	32.16	32.58

a. Divide the population for each year by the population in the preceding year. Round to three decimal places and show that the population of California is increasing geometrically.

b. Write the general term of the geometric sequence describing population growth for California n years after 1989.

c. Use your model from part (b) to estimate California's population, in millions, for the year 2000. According to the U.S. Census Bureau, California's population in 2000 was 33.87 million. How well does your geometric sequence model the actual population?

78. The population of Texas from 1990 through 1997 is shown in the following table.

Year	1990	1991	1992	1993
Population in millions	16.99	17.35	17.71	18.08
Year	1994	1995	1996	1997
Population in millions	18.46	18.85	19.25	19.65

a. Divide the population for each year by the population in the preceding year. Round to three decimal places and show that the population of Texas is increasing geometrically.

b. Write the general term of the geometric sequence describing population growth for Texas n years after 1989.

c. Use your model from part (b) to estimate Texas's population, in millions, for the year 2000. According to the U.S. Census Bureau, Texas's population in 2000 was 20.85 million. How well does your geometric sequence model the actual population?

Use the formula for the sum of the first n terms of a geometric sequence to solve Exercises 79–84.

In Exercises 79–80, you save \$1 the first day of a month, \$2 the second day, \$4 the third day, continuing to double your savings each day.

79. What will your total savings be for the first 15 days?

80. What will your total savings be for the first 30 days?

81. A job pays a salary of \$24,000 the first year. During the next 19 years, the salary increases by 5% each year. What is the total lifetime salary over the 20-year period? Round to the nearest dollar.

82. You are investigating two employment opportunities. Company A offers \$30,000 the first year. During the next four years, the salary is guaranteed to increase by 6% per year. Company B offers \$32,000 the first year, with guaranteed annual increases of 3% per year after that. Which company offers the better total salary for a five-year contract? By how much? Round to the nearest dollar.

83. A pendulum swings through an arc of 20 inches. On each successive swing, the length of the arc is 90% of the previous length.

$$20,\quad 0.9(20),\quad 0.9^2(20),\quad 0.9^3(20),\ldots$$

1st swing | 2nd swing | 3rd swing | 4th swing

After 10 swings, what is the total length of the distance the pendulum has swung? Round to the nearest hundredth of an inch.

84. A pendulum swings through an arc of 16 inches. On each successive swing, the length of the arc is 96% of the previous length.

$$16,\quad 0.96(16),\quad (0.96)^2(16),\quad (0.96)^3(16),\ldots$$

1st swing | 2nd swing | 3rd swing | 4th swing

After 10 swings, what is the total length of the distance the pendulum has swung? Round to the nearest hundredth of an inch.

Use the formula for the value of an annuity to solve Exercises 85–88. Round answers to the nearest dollar.

85. To save for retirement, you decide to deposit \$2500 into an IRA at the end of each year for the next 40 years. If the interest rate is 9% per year compounded annually, find the value of the IRA after 40 years.

86. You decide to deposit \$100 at the end of each month into an account paying 8% interest compounded monthly to save for your child's education. How much will you save over 16 years?

87. You contribute \$600 at the end of each quarter to a Tax Sheltered Annuity (TSA) paying 8% annual interest compounded quarterly. Find the value of the TSA after 18 years.

88. To save for a new home, you invest \$500 per month in a mutual fund with an annual rate of return of 10% compounded monthly. How much will you have saved after four years?

Use the formula for the sum of an infinite geometric series to solve Exercises 89–91.

89. A new factory in a small town has an annual payroll of \$6 million. It is expected that 60% of this money will be spent in the town by factory personnel. The people in the town who receive this money are expected to spend 60% of what they receive in the town, and so on. What is the total of all this spending, called the *total economic impact* of the factory, on the town each year?

90. How much additional spending will be generated by a \$10 billion tax rebate if 60% of all income is spent?

91. If the shading process shown in the figure is continued indefinitely, what fractional part of the largest square will eventually be shaded?

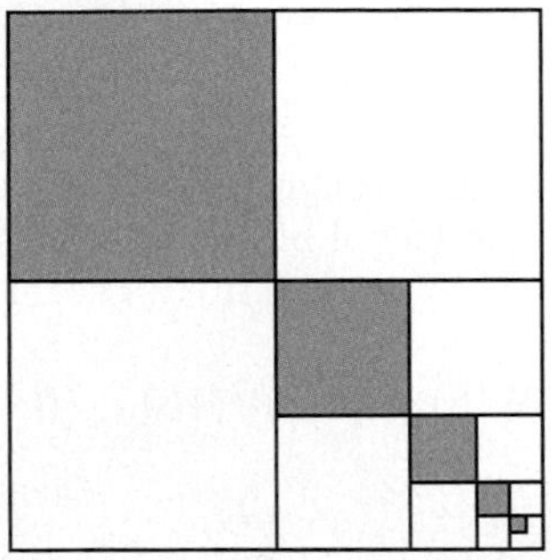

Writing in Mathematics

92. What is a geometric sequence? Give an example with your explanation.

93. What is the common ratio in a geometric sequence?

94. Explain how to find the general term of a geometric sequence.

95. Explain how to find the sum of the first n terms of a geometric sequence without having to add up all the terms.

96. What is an annuity?

97. What is the difference between a geometric sequence and an infinite geometric series?

98. How do you determine if an infinite geometric series has a sum? Explain how to find the sum of such an infinite geometric series.

99. Would you rather have \$10,000,000 and a brand new BMW or 1¢ today, 2¢ tomorrow, 4¢ on day 3, 8¢ on day 4, 16¢ on day 5, and so on, for 30 days? Explain.

100. For the first 30 days of a flu outbreak, the number of students on your campus who become ill is increasing. Which is worse: the number of students with the flu is increasing arithmetically or is increasing geometrically? Explain your answer.

Technology Exercises

101. Use the SEQ (sequence) capability of a graphing utility and the formula you obtained for a_n to verify the value you found for a_7 in any three exercises from Exercises 25–32.

102. Use the capability of a graphing utility to calculate the sum of a sequence to verify any three of your answers to Exercises 39–44.

In Exercises 103–104, use a graphing utility to graph the function. Determine the horizontal asymptote for the graph of f and discuss its relationship to the sum of the given series.

	Function	Series
103.	$f(x) = \dfrac{2\left[1 - \left(\frac{1}{3}\right)^x\right]}{1 - \frac{1}{3}}$	$2 + 2\left(\frac{1}{3}\right) + 2\left(\frac{1}{3}\right)^2 + 2\left(\frac{1}{3}\right)^3 + \cdots$
104.	$f(x) = \dfrac{4[1 - (0.6)^x]}{1 - 0.6}$	$4 + 4(0.6) + 4(0.6)^2 + 4(0.6)^3 + \cdots$

Critical Thinking Exercises

105. Which one of the following is true?

a. The sequence $2, 6, 24, 120, \ldots$ is an example of a geometric sequence.

b. The sum of the geometric series $\frac{1}{2} + \frac{1}{4} + \frac{1}{8} + \cdots + \frac{1}{512}$ can only be estimated without knowing precisely what terms occur between $\frac{1}{8}$ and $\frac{1}{512}$.

c. $10 - 5 + \frac{5}{2} - \frac{5}{4} + \cdots = \dfrac{10}{1 - \frac{1}{2}}$

d. If the nth term of a geometric sequence is $a_n = 3(0.5)^{n-1}$, the common ratio is $\frac{1}{2}$.

106. In a pest-eradication program, sterilized male flies are released into the general population each day. Ninety percent of those flies will survive a given day. How many flies should be released each day if the long-range goal of the program is to keep 20,000 sterilized flies in the population?

107. You are now 25 years old and would like to retire at age 55 with a retirement fund of \$1,000,000. How much should you deposit at the end of each month for the next 30 years in an IRA paying 10% annual interest compounded monthly to achieve your goal? Round to the nearest dollar.

Review Exercises

108. Simplify: $\sqrt{28} - 3\sqrt{7} + \sqrt{63}$.
(Section 10.4, Example 2)

109. Solve: $2x^2 = 4 - x$.
(Section 11.2, Example 2)

110. Rationalize the denominator: $\dfrac{6}{\sqrt{3} - \sqrt{5}}$.
(Section 10.5, Example 5)

MID-CHAPTER CHECK POINT

CHAPTER 14

What You Know: We learned that a sequence is a function whose domain is the set of positive integers. In an arithmetic sequence, each term after the first differs from the preceding term by a constant, the common difference, d. In a geometric sequence, each term after the first is obtained by multiplying the preceding term by a nonzero constant, the common ratio, r. We found the general term of arithmetic sequences $[a_n = a_1 + (n-1)d]$ and geometric sequences $[a_n = a_1 r^{n-1}]$ and used these formulas to find particular terms. We determined the sum of the first n terms of arithmetic sequences $\left[S_n = \frac{n}{2}(a_1 + a_n)\right]$ and geometric sequences $\left[S_n = \frac{a_1(1-r^n)}{1-r}\right]$. Finally, we determined the sum of an infinite geometric series, $a_1 + a_1 r + a_1 r^2 + a_1 r^3 + \cdots$, if $-1 < r < 1 \left(S = \frac{a_1}{1-r}\right)$.

In Exercises 1–3, write the first five terms of each sequence. Assume that d represents the common difference of an arithmetic sequence and r represents the common ratio of a geometric sequence.

1. $a_n = (-1)^{n+1}\dfrac{n}{(n-1)!}$

2. $a_1 = 5, d = -3$

3. $a_1 = 5, r = -3$

In Exercises 4–6, write a formula for the general term (the nth term) of each sequence. Then use the formula to find the indicated term.

4. $2, 6, 10, 14, \ldots; a_{20}$

5. $3, 6, 12, 24, \ldots; a_{10}$

6. $\frac{3}{2}, 1, \frac{1}{2}, 0, \ldots; a_{30}$

7. Find the sum of the first ten terms of the sequence:

$$5, 10, 20, 40, \ldots.$$

8. Find the sum of the first 50 terms of the sequence:

$$-2, 0, 2, 4, \ldots.$$

9. Find the sum of the first ten terms of the sequence:

$$-20, 40, -80, 160, \ldots.$$

10. Find the sum of the first 100 terms of the sequence:

$$4, -2, -8, -14, \ldots.$$

In Exercises 11–14, find each indicated sum.

11. $\sum_{i=1}^{4} (i+4)(i-1)$

12. $\sum_{i=1}^{50} (3i-2)$

13. $\sum_{i=1}^{6} \left(\frac{3}{2}\right)^i$

14. $\sum_{i=1}^{\infty} \left(-\frac{2}{5}\right)^{i-1}$

15. Express $0.\overline{45}$ as a fraction in lowest terms.

16. Express the sum using summation notation. Use i for the index of summation.

$$\frac{1}{3} + \frac{2}{4} + \frac{3}{5} + \cdots + \frac{18}{20}$$

17. A skydiver falls 16 feet during the first second of a dive, 48 feet during the second second, 80 feet during the third second, 112 feet during the fourth second, and so on. Find the distance that the skydiver falls during the 15th second and the total distance the skydiver falls in 15 seconds.

18. If the average value of a house increases 10% per year, how much will a house costing \$120,000 be worth in 10 years? Round to the nearest dollar.

SECTION 14.4

Objectives

1. Recognize patterns in binomial expansions.
2. Evaluate a binomial coefficient.
3. Expand a binomial raised to a power.
4. Find a particular term in a binomial expansion.

THE BINOMIAL THEOREM

Galaxies are groupings of billions of stars bound together by gravity. Some galaxies, such as the Centaurus galaxy shown here, are elliptical in shape.

Is mathematics discovered or invented? For example, planets revolve in elliptical orbits. Does that mean that the ellipse is out there, waiting for the mind to discover it? Or do people create the definition of an ellipse just as they compose a song? And is it possible for the same mathematics to be discovered/invented by independent researchers separated by time, place, and culture? This is precisely what occurred when mathematicians attempted to find efficient methods for raising binomials to higher and higher powers, such as

$$(x + 2)^3, (x + 2)^4, (x + 2)^5, (x + 2)^6,$$

and so on. In this section, we study higher powers of binomials and a method first discovered/invented by great minds in Eastern and in Western cultures working independently.

1 Recognize patterns in binomial expansions.

Patterns in Binomial Expansions When we write out the *binomial expression* $(a + b)^n$, where n is a positive integer, several patterns begin to appear.

$$\begin{aligned}
(a + b)^1 &= a + b \\
(a + b)^2 &= a^2 + 2ab + b^2 \\
(a + b)^3 &= a^3 + 3a^2b + 3ab^2 + b^3 \\
(a + b)^4 &= a^4 + 4a^3b + 6a^2b^2 + 4ab^3 + b^4 \\
(a + b)^5 &= a^5 + 5a^4b + 10a^3b^2 + 10a^2b^3 + 5ab^4 + b^5
\end{aligned}$$

DISCOVER FOR YOURSELF

Each expanded form of the binomial expression is a polynomial. Study the five polynomials above and answer the following questions:

1. For each polynomial, describe the pattern for the exponents on a. What is the largest exponent on a? What happens to the exponent on a from term to term?
2. Describe the pattern for the exponents on b. What is the exponent on b in the first term? What is the exponent on b in the second term? What happens to the exponent on b from term to term?
3. Find the sum of the exponents on the variables in each term for the polynomials in the five rows. Describe the pattern.
4. How many terms are there in the polynomials on the right in relation to the power of the binomial?

How many of the following patterns were you able to discover?

Expansions of Binomials

$(a + b)^1 = a + b$

$(a + b)^2 = a^2 + 2ab + b^3$

$(a + b)^3 = a^3 + 3a^2b + 3ab^2 + b^3$

$(a + b)^4 = a^4 + 4a^3b + 6a^2b^2 + 4ab^3 + b^4$

$(a + b)^5 = a^5 + 5a^4b + 10a^3b^2 + 10a^2b^3 + 5ab^4 + b^5$

1. The first term in the expansion of $(a + b)^n$ is a^n. The exponents on a decrease by 1 in each successive term.

2. The exponents on b in the expansion of $(a + b)^n$ increase by 1 in each successive term. In the first term, the exponent on b is 0. (Because $b^0 = 1$, b is not shown in the first term.) The last term is b^n.

3. The sum of the exponents on the variables in any term in the expansion of $(a + b)^n$ is equal to n.

4. The number of terms in the polynomial expansion is one greater than the power of the binomial, n. There are $n + 1$ terms in the expanded form of $(a + b)^n$.

Using these observations, the variable parts of the expansion of $(a + b)^6$ are

$$a^6, \quad a^5b, \quad a^4b^2, \quad a^3b^3, \quad a^2b^4, \quad ab^5, \quad b^6.$$

The first term is a^6, with the exponents on a decreasing by 1 in each successive term. The exponents on b increase from 0 to 6, with the last term being b^6. The sum of the exponents in each term is equal to 6.

We can generalize from these observations to obtain the variable parts of the expansion of $(a + b)^n$. They are

$$a^n, \quad a^{n-1}b, \quad a^{n-2}b^2, \quad a^{n-3}b^3, \ldots, \quad ab^{n-1}, \quad b^n.$$

Exponents on a are decreasing by 1. Exponents on b are increasing by 1.

Sum of exponents: $n - 1 + 1 = n$

Sum of exponents: $n - 3 + 3 = n$

Sum of exponents: $1 + n - 1 = n$

Let's now establish a pattern for the coefficients of the terms in the binomial expansion. Notice that each row in the figure shown below begins and ends with 1. Any other number in the row can be obtained by adding the two numbers immediately above it.

Coefficients for $(a + b)^1$:	1 1
Coefficients for $(a + b)^2$:	1 2 1
Coefficients for $(a + b)^3$:	1 3 3 1
Coefficients for $(a + b)^4$:	1 4 6 4 1
Coefficients for $(a + b)^5$:	1 5 10 10 5 1

This triangular array of coefficients is called **Pascal's triangle**. If we continue with the sixth row, the first and last numbers are 1. Each of the other numbers is obtained by finding the sum of the two closest numbers above it in the fifth row.

1 1

1 2 1

1 3 3 1

1 4 6 4 1

1 5 10 10 5 1

1 + 5   5 + 10   10 + 10   10 + 5   5 + 1

1 6 15 20 15 6 1

STUDY TIP

We have not shown the number in the top row of Pascal's triangle on the right. The top row is *row zero* because it corresponds to $(a + b)^0 = 1$. With row zero, the triangle appears as

1

1 1

1 2 1

1 3 3 1

1 4 6 4 1

etc.

We can use the numbers in the sixth row and the variable parts we found above to write the expansion for $(a + b)^6$. It is

$$(a + b)^6 = a^6 + 6a^5b + 15a^4b^2 + 20a^3b^3 + 15a^2b^4 + 6ab^5 + b^6.$$

2 Evaluate a binomial coefficient.

Binomial Coefficients Pascal's triangle becomes cumbersome when a binomial is raised to a relatively large power. Therefore, the coefficients in a binomial expansion are instead given in terms of factorials. The coefficients are written in a special notation, which we define next.

DEFINITION OF A BINOMIAL COEFFICIENT $\binom{n}{r}$ For nonnegative integers n and r, with $n \geq r$, the expression $\binom{n}{r}$ (read "n above r") is called a **binomial coefficient** and is defined by

$$\binom{n}{r} = \frac{n!}{r!(n-r)!}.$$

The symbol ${}_nC_r$ is often used in place of $\binom{n}{r}$ to denote binomial coefficients.

Can you see that the definition of a binomial coefficient involves a fraction with factorials in the numerator and the denominator? When evaluating such an expression, try to reduce the fraction before performing the multiplications. For example, consider $\frac{26!}{21!}$. Rather than writing out 26! as the product of all integers from 26 down to 1, we can express 26! as

$$26! = 26 \cdot 25 \cdot 24 \cdot 23 \cdot 22 \cdot 21!.$$

In this way, we can divide both the numerator and the denominator by the common factor, 21!.

$$\frac{26!}{21!} = \frac{26 \cdot 25 \cdot 24 \cdot 23 \cdot 22 \cdot \cancel{21!}}{\cancel{21!}} = 26 \cdot 25 \cdot 24 \cdot 23 \cdot 22 = 7{,}893{,}600$$

EXAMPLE 1 Evaluating Binomial Coefficients

Evaluate:

a. $\binom{6}{2}$ **b.** $\binom{3}{0}$ **c.** $\binom{9}{3}$ **d.** $\binom{4}{4}$.

SOLUTION In each case, we apply the definition of the binomial coefficient.

a. $\binom{6}{2} = \frac{6!}{2!(6-2)!} = \frac{6!}{2!4!} = \frac{6 \cdot 5 \cdot \cancel{4!}}{2 \cdot 1 \cdot \cancel{4!}} = 15$

b. $\binom{3}{0} = \frac{3!}{0!(3-0)!} = \frac{\cancel{3!}}{0!\cancel{3!}} = \frac{1}{1} = 1$

Remember that 0! = 1.

c. $\binom{9}{3} = \frac{9!}{3!(9-3)!} = \frac{9!}{3!6!} = \frac{9 \cdot 8 \cdot 7 \cdot \cancel{6!}}{3 \cdot 2 \cdot 1 \cdot \cancel{6!}} = 84$

d. $\binom{4}{4} = \frac{4!}{4!(4-4)!} = \frac{\cancel{4!}}{\cancel{4!}0!} = \frac{1}{1} = 1$

USING TECHNOLOGY

Graphing utilities can compute binomial coefficients. For example, to find $\binom{6}{2}$, many utilities require the sequence

6 [nCr] 2 [ENTER].

The graphing utility will display 15. Consult your manual and verify the other evaluations in Example 1.

CHECK POINT 1 Evaluate:

a. $\binom{6}{3}$ **b.** $\binom{6}{0}$ **c.** $\binom{8}{2}$ **d.** $\binom{3}{3}$.

3 Expand a binomial raised to a power.

The Binomial Theorem If we use binomial coefficients and the pattern for the variable part of each term, a formula called the **Binomial Theorem** can be used to expand any positive integral power of a binomial.

A FORMULA FOR EXPANDING BINOMIALS: THE BINOMIAL THEOREM

For any positive integer n,

$$(a+b)^n = \binom{n}{0}a^n + \binom{n}{1}a^{n-1}b + \binom{n}{2}a^{n-2}b^2 + \binom{n}{3}a^{n-3}b^3 + \cdots + \binom{n}{n}b^n$$

$$= \sum_{r=0}^{n}\binom{n}{r}a^{n-r}b^r.$$

EXAMPLE 2 Using the Binomial Theorem

Expand: $(x+2)^4$.

SOLUTION We use the Binomial Theorem

$$(a+b)^n = \binom{n}{0}a^n + \binom{n}{1}a^{n-1}b + \binom{n}{2}a^{n-2}b^2 + \binom{n}{3}a^{n-3}b^3 + \cdots + \binom{n}{n}b^n$$

to expand $(x+2)^4$. In $(x+2)^4$, $a = x$, $b = 2$, and $n = 4$. In the expansion, powers of x are in descending order, starting with x^4. Powers of 2 are in ascending order, starting with 2^0. (Because $2^0 = 1$, a 2 is not shown in the first term.) The sum of the exponents on x and 2 in each term is equal to 4, the exponent in the expression $(x+2)^4$.

$$(x+2)^4 = \binom{4}{0}x^4 + \binom{4}{1}x^3 \cdot 2 + \binom{4}{2}x^2 \cdot 2^2 + \binom{4}{3}x \cdot 2^3 + \binom{4}{4}2^4$$

These binomial coefficients are evaluated using $\binom{n}{r} = \frac{n!}{r!(n-r)!}$.

$$= \frac{4!}{0!4!}x^4 + \frac{4!}{1!3!}x^3 \cdot 2 + \frac{4!}{2!2!}x^2 \cdot 4 + \frac{4!}{3!1!}x \cdot 8 + \frac{4!}{4!0!} \cdot 16$$

$\frac{4!}{2!2!} = \frac{4 \cdot 3 \cdot 2!}{2! \cdot 2 \cdot 1} = \frac{12}{2} = 6$

Take a few minutes to verify the other factorial evaluations.

$$= 1 \cdot x^4 + 4x^3 \cdot 2 + 6x^2 \cdot 4 + 4x \cdot 8 + 1 \cdot 16$$

$$= x^4 + 8x^3 + 24x^2 + 32x + 16$$

■

CHECK POINT 2 Expand: $(x+1)^4$.

ENRICHMENT ESSAY

The Universality of Mathematics

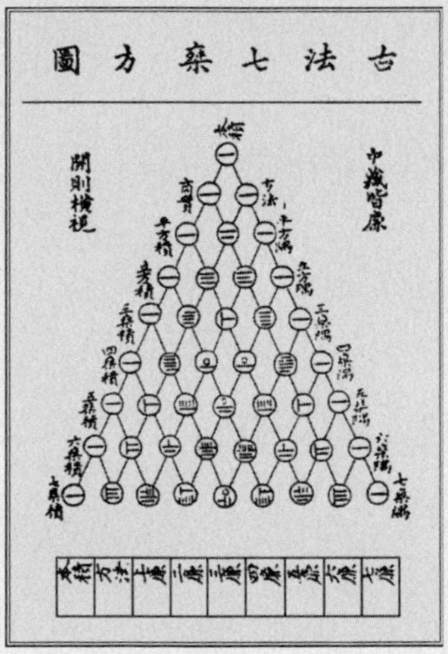

"Pascal's" triangle, credited to French mathematician Blaise Pascal (1623–1662), appeared in a Chinese document written in 1303. The Binomial Theorem was known in Eastern cultures prior to its discovery in Europe. The same mathematics is often discovered/invented by independent researchers separated by time, place, and culture.

EXAMPLE 3 Using the Binomial Theorem

Expand: $(2x - y)^5$.

SOLUTION Because the Binomial Theorem involves the addition of two terms raised to a power, we rewrite $(2x - y)^5$ as $[2x + (-y)]^5$. We use the Binomial Theorem

$$(a + b)^n = \binom{n}{0}a^n + \binom{n}{1}a^{n-1}b + \binom{n}{2}a^{n-2}b^2 + \binom{n}{3}a^{n-3}b^3 + \cdots + \binom{n}{n}b^n$$

to expand $[2x + (-y)]^5$. In $[2x + (-y)]^5$, $a = 2x$, $b = -y$, and $n = 5$. In the expansion, powers of $2x$ are in descending order, starting with $(2x)^5$. Powers of $-y$ are in ascending order, starting with $(-y)^0$. [Because $(-y)^0 = 1$, a $-y$ is not shown in the first term.] The sum of the exponents on $2x$ and $-y$ in each term is equal to 5, the exponent in the expression $(2x - y)^5$.

$$(2x - y)^5 = [2x + (-y)]^5$$

$$= \binom{5}{0}(2x)^5 + \binom{5}{1}(2x)^4(-y) + \binom{5}{2}(2x)^3(-y)^2 + \binom{5}{3}(2x)^2(-y)^3 + \binom{5}{4}(2x)(-y)^4 + \binom{5}{5}(-y)^5$$

Evaluate binomial coefficients using $\binom{n}{r} = \frac{n!}{r!(n-r)!}$.

$$= \frac{5!}{0!5!}(2x)^5 + \frac{5!}{1!4!}(2x)^4(-y) + \frac{5!}{2!3!}(2x)^3(-y)^2 + \frac{5!}{3!2!}(2x)^2(-y)^3 + \frac{5!}{4!1!}(2x)(-y)^4 + \frac{5!}{5!0!}(-y)^5$$

$\frac{5!}{2!3!} = \frac{5 \cdot 4 \cdot 3!}{2 \cdot 1 \cdot 3!} = 10$
Take a few minutes to verify the other factorial evaluations.

$$= 1(2x)^5 + 5(2x)^4(-y) + 10(2x)^3(-y)^2 + 10(2x)^2(-y)^3 + 5(2x)(-y)^4 + 1(-y)^5$$

Raise both factors in these parentheses to the indicated powers.

$$= 1(32x^5) + 5(16x^4)(-y) + 10(8x^3)(-y)^2 + 10(4x^2)(-y)^3 + 5(2x)(-y)^4 + 1(-y)^5$$

Now raise $-y$ to the indicated powers.

$$= 1(32x^5) + 5(16x^4)(-y) + 10(8x^3)y^2 + 10(4x^2)(-y^3) + 5(2x)y^4 + 1(-y^5)$$

Multiplying factors in each of the six terms gives us the desired expansion:

$$(2x - y)^5 = 32x^5 - 80x^4y + 80x^3y^2 - 40x^2y^3 + 10xy^4 - y^5.$$ ■

CHECK POINT 3 Expand: $(x - 2y)^5$.

4 Find a particular term in a binomial expansion.

Finding a Particular Term in a Binomial Expansion By observing the terms in the formula for expanding binomials, we can find a formula for finding a particular term without writing the entire expansion.

1st term: $\binom{n}{0}a^nb^0$ 2nd term: $\binom{n}{1}a^{n-1}b^1$ 3rd term: $\binom{n}{2}a^{n-2}b^2$

The exponent on b is 1 less than the term number.

Based on the observation in the bottom voice balloon, the $(r + 1)$st term of the expansion of $(a + b)^n$ is the term that contains b^r.

FINDING A PARTICULAR TERM IN A BINOMIAL EXPANSION The $(r + 1)$st term of the expansion of $(a + b)^n$ is

$$\binom{n}{r} a^{n-r} b^r.$$

EXAMPLE 4 Finding a Single Term of a Binomial Expansion

Find the fourth term in the expansion of $(3x + 2y)^7$.

SOLUTION The fourth term in the expansion of $(3x + 2y)^7$ contains $(2y)^3$. To find the fourth term, first note that $4 = 3 + 1$. Equivalently, the fourth term of $(3x + 2y)^7$ is the $(3 + 1)$st term. Thus, $r = 3$, $a = 3x$, $b = 2y$, and $n = 7$. The fourth term is

$$\binom{7}{3}(3x)^{7-3}(2y)^3 = \binom{7}{3}(3x)^4(2y)^3 = \frac{7!}{3!(7-3)!}(3x)^4(2y)^3.$$

Use the formula for the $(r + 1)$st term of $(a + b)^n$: $\binom{n}{r}a^{n-r}b^r$.

We use $\binom{n}{r} = \frac{n!}{r!(n-r)!}$ to evaluate $\binom{7}{3}$.

Now we need to evaluate the factorial expression and raise $3x$ and $2y$ to the indicated powers. We obtain

$$\frac{7!}{3!4!}(81x^4)(8y^3) = \frac{7 \cdot 6 \cdot 5 \cdot \cancel{4!}}{3 \cdot 2 \cdot 1 \cdot \cancel{4!}}(81x^4)(8y^3) = 35(81x^4)(8y^3) = 22{,}680x^4y^3.$$

The fourth term of $(3x + 2y)^7$ is $22{,}680x^4y^3$. ■

CHECK POINT 4 Find the fifth term in the expansion of $(2x + y)^9$.

14.4 EXERCISE SET

Student Solutions Manual | CD/Video | PH Math/Tutor Center | MathXL Tutorials on CD | MathXL® | MyMathLab | Interactmath.com

Practice Exercises

In Exercises 1–8, evaluate the given binomial coefficient.

1. $\binom{8}{3}$ **2.** $\binom{7}{2}$ **3.** $\binom{12}{1}$

4. $\binom{11}{1}$ **5.** $\binom{6}{6}$ **6.** $\binom{15}{2}$

7. $\binom{100}{2}$ **8.** $\binom{100}{98}$

In Exercises 9–30, use the Binomial Theorem to expand each binomial and express the result in simplified form.

9. $(x + 2)^3$

10. $(x + 4)^3$

11. $(3x + y)^3$

12. $(x + 3y)^3$

13. $(5x - 1)^3$

14. $(4x - 1)^3$

15. $(2x + 1)^4$

16. $(3x + 1)^4$

17. $(x^2 + 2y)^4$

18. $(x^2 + y)^4$

19. $(y - 3)^4$

20. $(y - 4)^4$

21. $(2x^3 - 1)^4$

22. $(2x^5 - 1)^4$

23. $(c + 2)^5$

24. $(c + 3)^5$

25. $(x - 1)^5$

26. $(x - 2)^5$

27. $(3x - y)^5$

28. $(x - 3y)^5$

29. $(2a + b)^6$

30. $(a + 2b)^6$

In Exercises 31–38, write the first three terms in each binomial expansion, expressing the result in simplified form.

31. $(x + 2)^8$

32. $(x + 3)^8$

33. $(x - 2y)^{10}$

34. $(x - 2y)^9$

35. $(x^2 + 1)^{16}$

36. $(x^2 + 1)^{17}$

37. $(y^3 - 1)^{20}$

38. $(y^3 - 1)^{21}$

In Exercises 39–46, find the term indicated in each expansion.

39. $(2x + y)^6$; third term

40. $(x + 2y)^6$; third term

41. $(x - 1)^9$; fifth term

42. $(x - 1)^{10}$; fifth term

43. $(x^2 + y^3)^8$; sixth term

44. $(x^3 + y^2)^8$; sixth term

45. $\left(x - \frac{1}{2}\right)^9$; fourth term

46. $\left(x + \frac{1}{2}\right)^8$; fourth term

Practice Plus

In Exercises 47–50, use the Binomial Theorem to expand each expression and write the result in simplified form.

47. $(x^3 + x^{-2})^4$

48. $(x^2 + x^{-3})^4$

49. $\left(x^{\frac{1}{3}} - x^{-\frac{1}{3}}\right)^3$

50. $\left(x^{\frac{2}{3}} - \frac{1}{\sqrt[3]{x}}\right)^3$

Exercises 51–52 involve expressions containing i, where $i = \sqrt{-1}$. Expand each expression and use powers of i to simplify the result.

51. $(-1 + \sqrt{3}i)^3$

52. $(-1 - \sqrt{3}i)^3$

In Exercises 53–54, find $\dfrac{f(x + h) - f(x)}{h}$ *and simplify.*

53. $f(x) = x^4 + 7$

54. $f(x) = x^5 + 8$

55. Find the middle term in the expansion of $\left(\dfrac{3}{x} + \dfrac{x}{3}\right)^{10}$.

56. Find the middle term in the expansion of $\left(\dfrac{1}{x} - x^2\right)^{12}$.

Application Exercises

Bariatrics is the field of medicine that deals with the overweight. Bariatric surgery closes off a large part of the stomach. As a result, patients eat less and have a diminished appetite. Celebrities like pop singer Carnie Wilson and the Today *show's weatherman Al Roker have become no-longer-larger-than-life walking billboards for the operation. The figure shows the number of bariatric surgeries from 1992 through 2002.*

The function

Bariatric Surgery in the United States

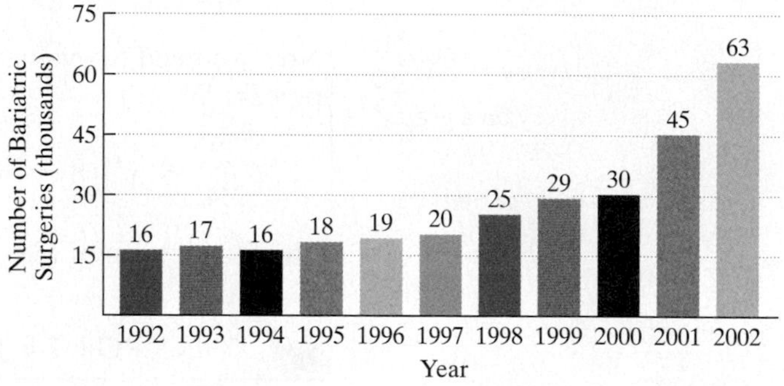

Source: American Society for Bariatric Surgery

$$f(x) = 0.12x^3 - x^2 + 3x + 15$$

models the number of bariatric surgeries, f(x), in thousands, x years after 1992. Use this function to solve Exercises 57–58.

57. **a.** How can we adjust the function so that $x = 0$ corresponds to 1995 rather than 1992? We shift the graph of f 3 units to the left. We obtain $g(x) = f(x + 3)$. Use the Binomial Theorem to express g in descending powers of x.

b. Find $f(5)$ and $g(2)$. How well do these function values model the number shown in the bar graph?

58. **a.** How can we adjust the function so that $x = 0$ corresponds to 1997 rather than 1992? We shift the graph of f 5 units to the left. We obtain $g(x) = f(x + 5)$. Use the Binomial Theorem to express g in descending powers of x.

b. Find $f(7)$ and $g(2)$. How well do these function values model the number shown in the bar graph?

Writing in Mathematics

59. Describe the pattern on the exponents on a in the expansion of $(a + b)^n$.

60. Describe the pattern on the exponents on b in the expansion of $(a + b)^n$.

61. What is true about the sum of the exponents on a and b in any term in the expansion of $(a + b)^n$?

62. How do you determine how many terms there are in a binomial expansion?

63. What is Pascal's triangle? How do you find the numbers in any row of the triangle?

64. Explain how to evaluate $\binom{n}{r}$. Provide an example with your explanation.

65. Explain how to use the Binomial Theorem to expand a binomial. Provide an example with your explanation.

66. Explain how to find a particular term in a binomial expansion without having to write out the entire expansion.

67. Are there situations in which it is easier to use Pascal's triangle than binomial coefficients? Describe these situations.

Technology Exercises

68. Use the [nCr] key on a graphing utility to verify your answers in Exercises 1–8.

In Exercises 69–70, graph each of the functions in the same viewing rectangle. Describe how the graphs illustrate the Binomial Theorem.

69. $f_1(x) = (x + 2)^3$

$f_2(x) = x^3$

$f_3(x) = x^3 + 6x^2$

$f_4(x) = x^3 + 6x^2 + 12x$

$f_5(x) = x^3 + 6x^2 + 12x + 8$

Use a $[-10, 10, 1]$ by $[-30, 30, 10]$ viewing rectangle.

70. $f_1(x) = (x + 1)^4$

$f_2(x) = x^4$

$f_3(x) = x^4 + 4x^3$

$f_4(x) = x^4 + 4x^3 + 6x^2$

$f_5(x) = x^4 + 4x^3 + 6x^2 + 4x$

$f_6(x) = x^4 + 4x^3 + 6x^2 + 4x + 1$

Use a $[-5, 5, 1]$ by $[-30, 30, 10]$ viewing rectangle.

In Exercises 71–73, use the Binomial Theorem to find a polynomial expansion for each function. Then use a graphing utility and an approach similar to the one in Exercises 69 and 70 to verify the expansion.

71. $f_1(x) = (x - 1)^3$

72. $f_1(x) = (x - 2)^4$

73. $f_1(x) = (x + 2)^6$

74. Graphing utilities capable of symbolic manipulation, such as the TI-92, will expand binomials. On the TI-92, to expand $(3a - 5b)^{12}$, input the following:

[EXPAND] ((3a [−] 5b) [∧] 12) [ENTER].

Use a graphing utility with this capability to verify any five of the expansions you performed by hand in Exercises 9–30.

Critical Thinking Exercises

75. Which one of the following is true?

a. The binomial expansion for $(a + b)^n$ contains n terms.

b. The Binomial Theorem can be written in condensed form as $(a + b)^n = \sum_{r=0}^{n} \binom{n}{r} a^{n-r}b^r$.

c. The sum of the binomial coefficients in $(a + b)^n$ cannot be 2^n.

d. There are no values of a and b such that $(a + b)^4 = a^4 + b^4$.

76. Use the Binomial Theorem to expand and then simplify the result: $(x^2 + x + 1)^3$. [*Hint*: Write $x^2 + x + 1$ as $x^2 + (x + 1)$].

77. Find the term in the expansion of $(x^2 + y^2)^5$ containing x^4 as a factor.

Review Exercises

78. If $f(x) = x^2 + 2x + 3$, find $f(a + 1)$. (Section 8.1, Example 3)

79. If $f(x) = x^2 + 5x$ and $g(x) = 2x - 3$, find $f(g(x))$ and $g(f(x))$. (Section 8.3, Example 1)

80. Subtract: $\dfrac{x}{x + 3} - \dfrac{x + 1}{2x^2 - 2x - 24}$.

(Section 7.4, Example 7)

SECTION 14.5

COUNTING PRINCIPLES, PERMUTATIONS, AND COMBINATIONS

Objectives

1. Use the Fundamental Counting Principle.
2. Use the permutations formula.
3. Distinguish between permutation problems and combination problems.
4. Use the combinations formula.

Have you ever imagined what your life would be like if you won the lottery? What changes would you make? Before you fantasize about becoming a person of leisure with a staff of obedient elves, think about this: The probability of winning top prize in the lottery is about the same as the probability of being struck by lightning. There are millions of possible number combinations in lottery games and only one way of winning the grand prize. Determining the probability of winning involves calculating the chance of getting the winning combination from all possible outcomes. In this section, we begin preparing for the surprising world of probability by looking at methods for counting possible outcomes.

1 Use the Fundamental Counting Principle.

The Fundamental Counting Principle It's early morning, you're groggy, and you have to select something to wear for your 8 A.M. class. (What *were* you thinking of when you signed up for a class at that hour?!) Fortunately, your "lecture wardrobe" is rather limited—just two pairs of jeans to choose from (one blue, one black), three T-shirts to choose from (one beige, one yellow, and one blue), and two pairs of sneakers to select from (one black pair, one red pair). Your possible outfits are shown in Figure 14.6.

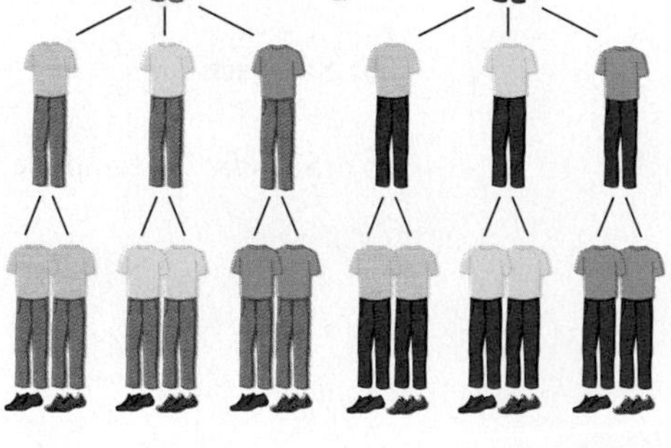

FIGURE 14.6 Selecting a wardrobe

The number of possible ways of playing the first four moves on each side in a game of chess is 318,979,564,000.

The **tree diagram**, so named because of its branches, shows that you can form 12 outfits from your two pairs of jeans, three T-shirts, and two pairs of sneakers. Notice that the number of outfits can be obtained by multiplying the number of choices for jeans, 2, the number of choices for the T-shirts, 3, and the number of choices for the sneakers, 2:

$$2 \cdot 3 \cdot 2 = 12.$$

We can generalize this idea to any two or more groups of items—not just jeans, T-shirts, and sneakers—with the **Fundamental Counting Principle:**

THE FUNDAMENTAL COUNTING PRINCIPLE The number of ways in which a series of successive things can occur is found by multiplying the number of ways in which each thing can occur.

For example, if you own 30 pairs of jeans, 20 T-shirts, and 12 pairs of sneakers, you have

$$30 \cdot 20 \cdot 12 = 7200$$

choices for your wardrobe!

EXAMPLE 1 Options in Planning a Course Schedule

Next semester you are planning to take three courses—math, English, and humanities. Based on time blocks and highly recommended professors, there are 8 sections of math, 5 of English, and 4 of humanities that you find suitable. Assuming no scheduling conflicts, how many different three-course schedules are possible?

SOLUTION This situation involves making choices with three groups of items.

Math	English	Humanities
8 choices	5 choices	4 choices

We use the Fundamental Counting Principle to find the number of three-course schedules. Multiply the number of choices for each of the three groups:

$$8 \cdot 5 \cdot 4 = 160.$$

Thus, there are 160 different three-course schedules. ■

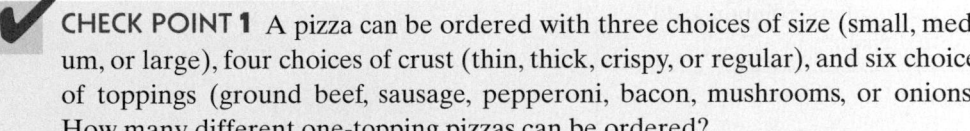

CHECK POINT 1 A pizza can be ordered with three choices of size (small, medium, or large), four choices of crust (thin, thick, crispy, or regular), and six choices of toppings (ground beef, sausage, pepperoni, bacon, mushrooms, or onions). How many different one-topping pizzas can be ordered?

EXAMPLE 2 A Multiple-Choice Test

You are taking a multiple-choice test that has ten questions. Each of the questions has four answer choices, with one correct answer per question. If you select one of these four choices for each question and leave nothing blank, in how many ways can you answer the questions?

SOLUTION This situation involves making choices with ten questions.

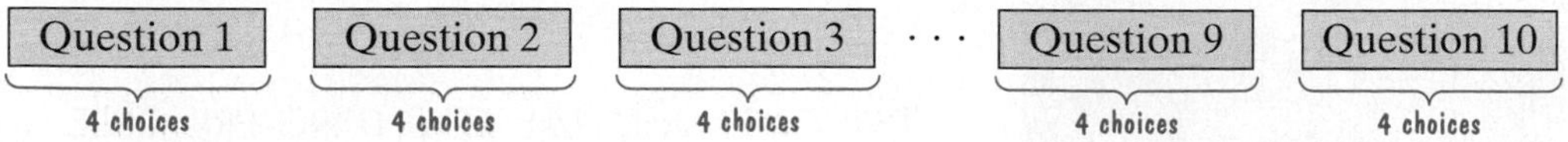

We use the Fundamental Counting Principle to determine the number of ways that you can answer the questions on the test. Multiply the number of choices, 4, for each of the ten questions.

$$4 \cdot 4 \cdot 4 \cdot 4 \cdot 4 \cdot 4 \cdot 4 \cdot 4 \cdot 4 \cdot 4 = 4^{10} = 1{,}048{,}576$$

Thus, you can answer the questions in 1,048,576 different ways. ■

ENRICHMENT ESSAY

Running Out of Telephone Numbers

By the year 2020, portable telephones used for business and pleasure will all be videophones. At that time, the U.S. population is expected to be 323 million. Faxes, beepers, cell phones, computer phone lines, and business lines may result in certain areas running out of phone numbers. Solution: Add more digits!

Are you surprised that there are over one million ways of answering a ten-question multiple-choice test? Of course, there is only one way to answer the test and receive a perfect score. The probability of guessing your way into a perfect score involves calculating the chance of getting a perfect score, just one way, from all 1,048,576 possible outcomes. In short, prepare for the test and do not rely on guessing!

✔ **CHECK POINT 2** You are taking a multiple-choice test that has six questions. Each of the questions has three answer choices, with one correct answer per question. If you select one of these three choices for each question and leave nothing blank, in how many ways can you answer the questions?

EXAMPLE 3 Telephone Numbers in the United States

Telephone numbers in the United States begin with three-digit area codes followed by seven-digit local telephone numbers. Area codes and local telephone numbers cannot begin with 0 or 1. How many different telephone numbers are possible?

SOLUTION This situation involves making choices with ten groups of items.

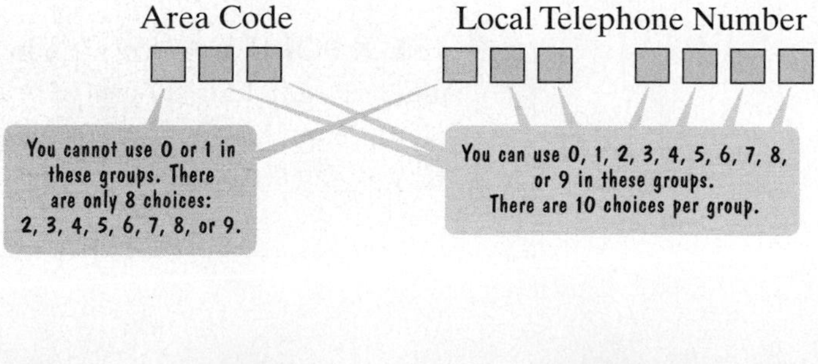

Here are the numbers of choices for each of the ten groups of items:

Area Code	**Local Telephone Number**	
8 10 10	8 10 10	10 10 10 10 .

We use the Fundamental Counting Principle to determine the number of different telephone numbers that are possible. The total number of telephone numbers possible is

$$8 \cdot 10 \cdot 10 \cdot 8 \cdot 10 \cdot 10 \cdot 10 \cdot 10 \cdot 10 \cdot 10 = 6{,}400{,}000{,}000.$$

There are six billion four hundred million different telephone numbers that are possible. ■

CHECK POINT 3 License plates in a particular state display two letters followed by three numbers, such as AT-887 or BB-013. How many different license plates can be manufactured?

2 Use the permutations formula.

Permutations You are the coach of a little league baseball team. There are 13 players on the team (and lots of parents hovering in the background, dreaming of stardom for their little "Barry Bonds"). You need to choose a batting order having 9 players. The order makes a difference, because, for instance, if bases are loaded and "Little Barry" is fourth or fifth at bat, his possible home run will drive in three additional runs. How many batting orders can you form?

You can choose any of 13 players for the first person at bat. Then you will have 12 players from which to choose the second batter, then 11 from which to choose the third batter, and so on. The situation can be shown as follows:

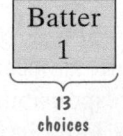

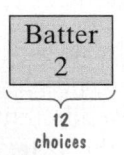

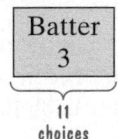

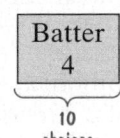

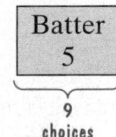

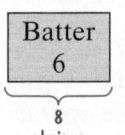

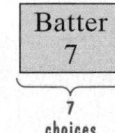

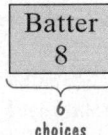

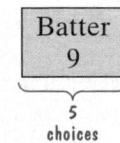

Batter 1	Batter 2	Batter 3	Batter 4	Batter 5	Batter 6	Batter 7	Batter 8	Batter 9
13 choices	12 choices	11 choices	10 choices	9 choices	8 choices	7 choices	6 choices	5 choices

We use the Fundamental Counting Principle to find the number of batting orders. The total number of batting orders is

$$13 \cdot 12 \cdot 11 \cdot 10 \cdot 9 \cdot 8 \cdot 7 \cdot 6 \cdot 5 = 259{,}459{,}200.$$

Nearly 260 million batting orders are possible for your 13-player little league team. Each batting order is called a *permutation* of 13 players taken 9 at a time. The number of permutations of 13 players taken 9 at a time is 259,459,200.

A **permutation** is an ordered arrangement of items that occurs when

- No item is used more than once. (Each of the 9 players in the batting order bats exactly once.)
- The order of arrangement makes a difference.

We can obtain a formula for finding the number of permutations of 13 players taken 9 at a time by rewriting our computation:

$$13 \cdot 12 \cdot 11 \cdot 10 \cdot 9 \cdot 8 \cdot 7 \cdot 6 \cdot 5$$

$$= \frac{13 \cdot 12 \cdot 11 \cdot 10 \cdot 9 \cdot 8 \cdot 7 \cdot 6 \cdot 5 \cdot \boxed{4 \cdot 3 \cdot 2 \cdot 1}}{\boxed{4 \cdot 3 \cdot 2 \cdot 1}} = \frac{13!}{4!} = \frac{13!}{(13-9)!}.$$

Thus, the number of permutations of 13 things taken 9 at a time is $\frac{13!}{(13-9)!}$. The special notation ${}_{13}P_9$ is used to replace the phrase "the number of permutations of 13 things taken 9 at a time." Using this new notation, we can write

$$ {}_{13}P_9 = \frac{13!}{(13-9)!}. $$

The numerator of this expression is the number of items, 13 team members, expressed as a factorial: 13! The denominator is also a factorial. It is the factorial of the difference between the number of items, 13, and the number of items in each permutation, 9 batters: $(13 - 9)!$.

The notation ${}_nP_r$ means the **number of permutations of n things taken r at a time.** We can generalize from the situation in which 9 batters were taken from 13 players. By generalizing, we obtain the following formula for the number of permutations if r items are taken from n items.

STUDY TIP

Because all permutation problems are also Fundamental Counting problems, they can be solved using the formula for ${}_nP_r$ or using the Fundamental Counting Principle.

PERMUTATIONS OF n THINGS TAKEN r AT A TIME The number of possible permutations if r items are taken from n items is

$$ {}_nP_r = \frac{n!}{(n-r)!}. $$

EXAMPLE 4 Using the Formula for Permutations

You and 19 of your friends have decided to form an Internet marketing consulting firm. The group needs to choose three officers—a CEO, an operating manager, and a treasurer. In how many ways can those offices be filled?

SOLUTION Your group is choosing $r = 3$ officers from a group of $n = 20$ people (you and 19 friends). The order in which the officers are chosen matters because the CEO, the operating manager, and the treasurer each have different responsibilities. Thus, we are looking for the number of permutations of 20 things taken 3 at a time. We use the formula

$$ {}_nP_r = \frac{n!}{(n-r)!} $$

with $n = 20$ and $r = 3$.

$$ {}_{20}P_3 = \frac{20!}{(20-3)!} = \frac{20!}{17!} = \frac{20 \cdot 19 \cdot 18 \cdot 17!}{17!} = \frac{20 \cdot 19 \cdot 18 \cdot \cancel{17!}}{\cancel{17!}} = 20 \cdot 19 \cdot 18 = 6840 $$

Thus, there are 6840 different ways of filling the three offices. ■

USING TECHNOLOGY

Graphing utilities have a menu item for calculating permutations, usually labeled $\boxed{{}_nP_r}$. For example, to find ${}_{20}P_3$, the keystrokes are

20 $\boxed{{}_nP_r}$ 3 $\boxed{\text{ENTER}}$.

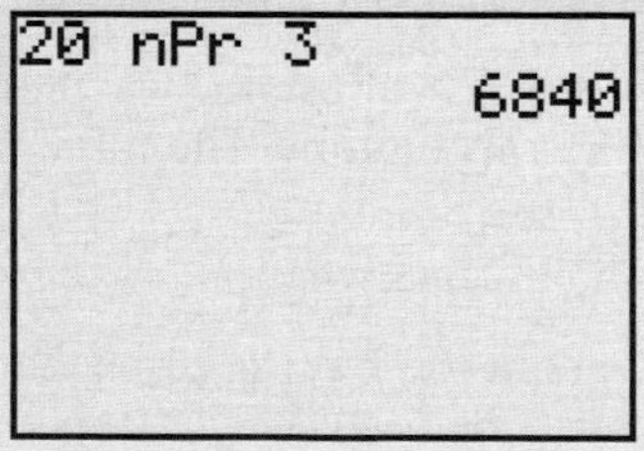

If you are using a scientific calculator, check your manual for the location of the menu item for calculating permutations and the required keystrokes.

CHECK POINT 4 A corporation has seven members on its board of directors. In how many different ways can it elect a president, vice-president, secretary, and treasurer?

EXAMPLE 5 Using the Formula for Permutations

You need to arrange seven of your favorite books along a small shelf. How many different ways can you arrange the books, assuming that the order of the books makes a difference to you?

ENRICHMENT ESSAY

How to Pass the Time for $2\frac{1}{2}$ Million Years

If you were to arrange 15 different books on a shelf and it took you one minute for each permutation, the entire task would take 2,487,965 years.

Source: Isaac Asimov's *Book of Facts*.

SOLUTION Because you are using all seven of your books in every possible arrangement, you are arranging $r = 7$ books from a group of $n = 7$ books. Thus, we are looking for the number of permutations of 7 things taken 7 at a time. We use the formula

$$_nP_r = \frac{n!}{(n-r)!}$$

with $n = 7$ and $r = 7$.

$$_7P_7 = \frac{7!}{(7-7)!} = \frac{7!}{0!} = \frac{7!}{1} = 5040$$

Thus, you can arrange the books in 5040 ways. There are 5040 different possible permutations. ■

CHECK POINT 5 In how many ways can 6 books be lined up along a shelf?

3 Distinguish between permutation problems and combination problems.

Combinations As the twentieth century drew to a close, *Time* magazine presented a series of special issues on the most influential people of the century. In their issue on heroes and icons (June 14, 1999), they discussed a number of people whose careers became more profitable after their tragic deaths, including Marilyn Monroe, James Dean, Jim Morrison, Kurt Cobain, and Selena.

Imagine that you ask your friends the following question: "Of these five people, which three would you select to be included in a documentary featuring the best of their work?" You are not asking your friends to rank their three favorite artists in any kind of order—they should merely select the three to be included in the documentary.

One friend answers, "Jim Morrison, Kurt Cobain, and Selena." Another responds, "Selena, Kurt Cobain, and Jim Morrison." These two people have the same artists in their group of selections, even if they are named in a different order. We are interested *in which artists are named, not the order in which they are named*, for the documentary. Because the items are taken without regard to order, this is not a permutation problem. No ranking of any sort is involved.

Marilyn Monroe, actress (1927–1962)

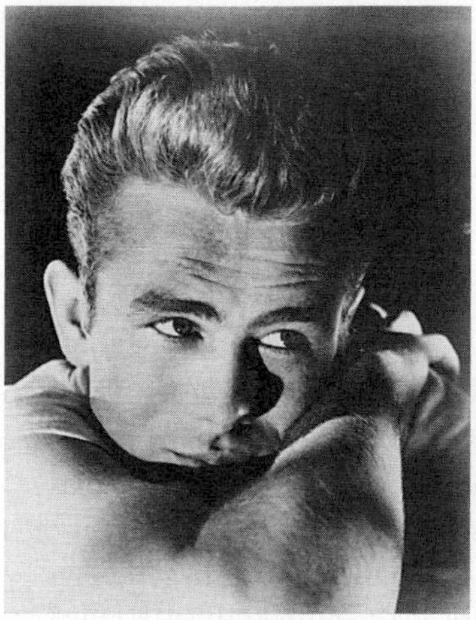

James Dean, actor (1931–1955)

Jim Morrison, musician and lead singer of the Doors (1943–1971)

Kurt Cobain, musician and front man for Nirvana (1967–1994)

Selena, musician of Tejano music (1971–1995)

Later on, you ask your roommate which three artists she would select for the documentary. She names Marilyn Monroe, James Dean, and Selena. Her selection is different from those of your two other friends because different entertainers are cited.

Mathematicians describe the group of artists given by your roommate as a *combination*. A **combination** of items occurs when

- The items are selected from the same group (the five stars who died young and tragically).
- No item is used more than once. (You may adore Selena, but your three selections cannot be Selena, Selena, and Selena.)
- The order of items makes no difference. (Morrison, Cobain, Selena is the same group in the documentary as Selena, Cobain, Morrison.)

Do you see the difference between a permutation and a combination? A permutation is an ordered arrangement of a given group of items. A combination is a group of items taken without regard to their order. **Permutation** problems involve situations in which **order matters**. **Combination** problems involve situations in which the **order** of items **makes no difference**.

EXAMPLE 6 Distinguishing between Permutations and Combinations

For each of the following problems, determine whether the problem is one involving permutations or combinations. (It is not necessary to solve the problem.)

a. Six students are running for student government president, vice-president, and treasurer. The student with the greatest number of votes becomes the president, the second highest vote-getter becomes vice-president, and the student who gets the third largest number of votes will be treasurer. How many different outcomes are possible for these three positions?

b. Six people are on the board of supervisors for your neighborhood park. A three-person committee is needed to study the possibility of expanding the park. How many different committees could be formed from the six people?

c. Baskin-Robbins offers 31 different flavors of ice cream. One of their items is a bowl consisting of three scoops of ice cream, each a different flavor. How many such bowls are possible?

SOLUTION

a. Students are choosing three student government officers from six candidates. The order in which the officers are chosen makes a difference because each of the offices (president, vice-president, treasurer) is different. Order matters. This is a problem involving permutations.

b. A three-person committee is to be formed from the six-person board of supervisors. The order in which the three people are selected does not matter because they are not filling different roles on the committee. Because order makes no difference, this is a problem involving combinations.

c. A three-scoop bowl of three different flavors is to be formed from Baskin-Robbin's 31 flavors. The order in which the three scoops of ice cream are put into the bowl is irrelevant. A bowl with chocolate, vanilla, and strawberry is exactly the same as a bowl with vanilla, strawberry, and chocolate. Different orderings do not change things, and so this is a problem involving combinations. ■

✔ CHECK POINT 6 For each of the following problems, explain if the problem is one involving permutations or combinations. (It is not necessary to solve the problem.)

a. How many ways can you select 6 free videos from a list of 200 videos?

b. In a race in which there are 50 runners and no ties, in how many ways can the first three finishers come in?

4 Use the combinations formula.

A Formula for Combinations We have seen that the notation $_nP_r$ means the number of permutations of n things taken r at a time. Similarly, the notation $_nC_r$ **means the number of combinations of n things taken r at a time.**

We can develop a formula for $_nC_r$ by comparing permutations and combinations. Consider the letters A, B, C, and D. The number of permutations of these four letters taken three at a time is

$$_4P_3 = \frac{4!}{(4-3)!} = \frac{4!}{1!} = \frac{4 \cdot 3 \cdot 2 \cdot 1}{1} = 24.$$

Here are the 24 permutations:

ABC,	ABD,	ACD,	BCD,
ACB,	ADB,	ADC,	BDC,
BAC,	BAD,	CAD,	CBD,
BCA,	BDA,	CDA,	CDB,
CAB,	DAB,	DAC,	DBC,
CBA,	DBA,	DCA,	DCB.
This column contains only one combination, ABC.	This column contains only one combination, ABD.	This column contains only one combination, ACD.	This column contains only one combination, BCD.

Because the order of items makes no difference in determining combinations, each column of six permutations represents one combination. There are a total of four combinations:

ABC, ABD, ACD, BCD.

Thus, $_4C_3 = 4$: The number of combinations of 4 things taken 3 at a time is 4. With 24 permutations and only four combinations, there are 6, or 3!, times as many permutations as there are combinations.

In general, there are $r!$ times as many permutations of n things taken r at a time as there are combinations of n things taken r at a time. Thus, we find the number of combinations of n things taken r at a time by dividing the number of permutations of n things taken r at a time by $r!$.

$$_nC_r = \frac{_nP_r}{r!} = \frac{\frac{n!}{(n-r)!}}{r!} = \frac{n!}{(n-r)!r!}$$

STUDY TIP

The number of combinations if r items are taken from n items cannot be found using the Fundamental Counting Principle and requires the use of the formula shown on the right.

COMBINATIONS OF n THINGS TAKEN r AT A TIME The number of possible combinations if r items are taken from n items is

$$_nC_r = \frac{n!}{(n-r)!r!}.$$

Notice that the formula for ${}_nC_r$ is the same as the formula for the binomial coefficient $\binom{n}{r}$.

USING TECHNOLOGY

Graphing utilities have a menu item for calculating combinations, usually labeled ${}_nC_r$. For example, to find ${}_8C_3$, the keystrokes on most graphing utilities are

8 [${}_nC_r$] 3 [ENTER].

If you are using a scientific calculator, check your manual to see whether there is a menu item for calculating combinations.

If you use your calculator's factorial key to find $\frac{8!}{5!3!}$, be sure to enclose the factorials in the denominator with parentheses

8 [!] [÷] [(] 5 [!] [×] 3 [!] [)]

pressing [=] or [ENTER] to obtain the answer.

EXAMPLE 7 Using the Formula for Combinations

A three-person committee is needed to study ways of improving public transportation. How many committees could be formed from the eight people on the board of supervisors?

SOLUTION The order in which the three people are selected does not matter. This is a problem of selecting $r = 3$ people from a group of $n = 8$ people. We are looking for the number of combinations of eight things taken three at a time. We use the formula

$$ {}_nC_r = \frac{n!}{(n-r)!r!} $$

with $n = 8$ and $r = 3$.

$$ {}_8C_3 = \frac{8!}{(8-3)!3!} = \frac{8!}{5!3!} = \frac{8 \cdot 7 \cdot 6 \cdot 5!}{5! \cdot 3 \cdot 2 \cdot 1} = \frac{8 \cdot 7 \cdot 6 \cdot \cancel{5!}}{\cancel{5!} \cdot 3 \cdot 2 \cdot 1} = 56 $$

Thus, 56 committees of three people each can be formed from the eight people on the board of supervisors. ■

✔ **CHECK POINT 7** From a group of 10 physicians, in how many ways can four people be selected to attend a conference on acupuncture?

EXAMPLE 8 Using the Formula for Combinations

In poker, a person is dealt 5 cards from a standard 52-card deck. The order in which you are dealt the 5 cards does not matter. How many different 5-card poker hands are possible?

SOLUTION Because the order in which the 5 cards are dealt does not matter, this is a problem involving combinations. We are looking for the number of combinations of $n = 52$ cards drawn $r = 5$ at a time. We use the formula

$$ {}_nC_r = \frac{n!}{(n-r)!r!} $$

with $n = 52$ and $r = 5$.

$$ {}_{52}C_5 = \frac{52!}{(52-5)!5!} = \frac{52!}{47!5!} = \frac{52 \cdot 51 \cdot 50 \cdot 49 \cdot 48 \cdot \cancel{47!}}{\cancel{47!} \cdot 5 \cdot 4 \cdot 3 \cdot 2 \cdot 1} = 2{,}598{,}960 $$

Thus, there are 2,598,960 different 5-card poker hands possible. It surprises many people that more than 2.5 million 5-card hands can be dealt from a mere 52 cards. ■

FIGURE 14.7 A royal flush

If you are a card player, it does not get any better than to be dealt the 5-card poker hand shown in Figure 14.7. This hand is called a *royal flush*. It consists of an ace, king, queen, jack, and 10, all of the same suit: all hearts, all diamonds, all clubs, or all spades. The probability of being dealt a royal flush involves calculating the number of ways of being dealt such a hand: just 4 of all 2,598,960 possible hands. In the next section, we move from counting possibilities to computing probabilities.

CHECK POINT 8 How many different 4-card hands can be dealt from a deck that has 16 different cards?

14.5 EXERCISE SET

Student Solutions Manual | CD/Video | PH Math/Tutor Center | MathXL Tutorials on CD | MathXL® | MyMathLab | Interactmath.com

Practice Exercises

In Exercises 1–8, use the formula for ${}_nP_r$ to evaluate each expression.

1. ${}_9P_4$ **2.** ${}_7P_3$

3. ${}_8P_5$ **4.** ${}_{10}P_4$

5. ${}_6P_6$ **6.** ${}_9P_9$

7. ${}_8P_0$ **8.** ${}_6P_0$

In Exercises 9–16, use the formula for ${}_nC_r$ to evaluate each expression.

9. ${}_9C_5$ **10.** ${}_{10}C_6$

11. ${}_{11}C_4$ **12.** ${}_{12}C_5$

13. ${}_7C_7$ **14.** ${}_4C_4$

15. ${}_5C_0$ **16.** ${}_6C_0$

In Exercises 17–20, does the problem involve permutations or combinations? Explain your answer. (It is not necessary to solve the problem.)

17. A medical researcher needs 6 people to test the effectiveness of an experimental drug. If 13 people have volunteered for the test, in how many ways can 6 people be selected?

18. Fifty people purchase raffle tickets. Three winning tickets are selected at random. If first prize is \$1000, second prize is \$500, and third prize is \$100, in how many different ways can the prizes be awarded?

19. How many different four-letter passwords can be formed from the letters A, B, C, D, E, F, and G if no repetition of letters is allowed?

20. Fifty people purchase raffle tickets. Three winning tickets are selected at random. If each prize is \$500, in how many different ways can the prizes be awarded?

Practice Plus

In Exercises 21–28, evaluate each expression.

21. $\dfrac{{}_7P_3}{3!} - {}_7C_3$ **22.** $\dfrac{{}_{20}P_2}{2!} - {}_{20}C_2$

23. $1 - \dfrac{{}_3P_2}{{}_4P_3}$ **24.** $1 - \dfrac{{}_5P_3}{{}_{10}P_4}$

25. $\dfrac{{}_7C_3}{{}_5C_4} - \dfrac{98!}{96!}$ **26.** $\dfrac{{}_{10}C_3}{{}_6C_4} - \dfrac{46!}{44!}$

27. $\dfrac{{}_4C_2 \cdot {}_6C_1}{{}_{18}C_3}$ **28.** $\dfrac{{}_5C_1 \cdot {}_7C_2}{{}_{12}C_3}$

Application Exercises

Use the Fundamental Counting Principle to solve Exercises 29–40.

29. The model of the car you are thinking of buying is available in nine different colors and three different styles (hatchback, sedan, or station wagon). In how many ways can you order the car?

30. A popular brand of pen is available in three colors (red, green, or blue) and four writing tips (bold, medium, fine, or micro). How many different choices of pens do you have with this brand?

31. An ice cream store sells two drinks (sodas or milk shakes), in four sizes (small, medium, large, or jumbo), and five flavors (vanilla, strawberry, chocolate, coffee, or pistachio). In how many ways can a customer order a drink?

32. A restaurant offers the following lunch menu.

Main Course	Vegetables	Beverages	Desserts
Ham	Potatoes	Coffee	Cake
Chicken	Peas	Tea	Pie
Fish	Green beans	Milk	Ice cream
Beef		Soda	

If one item is selected from each of the four groups, in how many ways can a meal be ordered? Describe two such orders.

33. You are taking a multiple-choice test that has five questions. Each of the questions has three answer choices, with one correct answer per question. If you select one of these three choices for each question and leave nothing blank, in how many ways can you answer the questions?

34. You are taking a multiple-choice test that has eight questions. Each of the questions has three answer choices, with one correct answer per question. If you select one of these three choices for each question and leave nothing blank, in how many ways can you answer the questions?

35. In the original plan for area codes in 1945, the first digit could be any number from 2 through 9, the second digit was either 0 or 1, and the third digit could be any number except 0. With this plan, how many different area codes were possible?

36. How many different four-letter radio station call letters can be formed if the first letter must be W or K?

37. Six performers are to present their comedy acts on a weekend evening at a comedy club. One of the performers insists on being the last stand-up comic of the evening. If this performer's request is granted, how many different ways are there to schedule the appearances?

38. Five singers are to perform at a night club. One of the singers insists on being the last performer of the evening. If this singer's request is granted, how many different ways are there to schedule the appearances?

39. In the *Cambridge Encyclopedia of Language* (Cambridge University Press, 1987), author David Crystal presents five sentences that make a reasonable paragraph regardless of their order. The sentences are as follows:

- Mark had told him about the foxes.
- John looked out the window.
- Could it be a fox?
- However, nobody had seen one for months.
- He thought he saw a shape in the bushes.

How many different five-sentence paragraphs can be formed if the paragraph begins with "He thought he saw a shape in the bushes" and ends with "John looked out of the window"?

40. A television programmer is arranging the order that five movies will be seen between the hours of 6 P.M. and 4 A.M. Two of the movies have a G rating and they are to be shown in the first two time blocks. One of the movies is rated NC-17 and it is to be shown in the last of the time blocks, from 2 A.M. until 4 A.M. Given these restrictions, in how many ways can the five movies be arranged during the indicated time blocks?

Use the formula for $_nP_r$ to solve Exercises 41–48.

41. A club with ten members is to choose three officers—president, vice-president, and secretary-treasurer. If each office is to be held by one person and no person can hold more than one office, in how many ways can those offices be filled?

42. A corporation has ten members on its board of directors. In how many different ways can it elect a president, vice-president, secretary, and treasurer?

43. For a segment of a radio show, a disc jockey can play 7 records. If there are 13 records to select from, in how many ways can the program for this segment be arranged?

44. Suppose you are asked to list, in order of preference, the three best movies you have seen this year. If you saw 20 movies during the year, in how many ways can the three best be chosen and ranked?

45. In a race in which six automobiles are entered and there are no ties, in how many ways can the first three finishers come in?

46. In a production of *West Side Story*, eight actors are considered for the male roles of Tony, Riff, and Bernardo. In how many ways can the director cast the male roles?

47. Nine bands have volunteered to perform at a benefit concert, but there is only enough time for five of the bands to play. How many lineups are possible?

48. How many arrangements can be made using four of the letters of the word COMBINE if no letter is to be used more than once?

Use the formula for $_nC_r$ to solve Exercises 49–56.

49. An election ballot asks voters to select three city commissioners from a group of six candidates. In how many ways can this be done?

50. A four-person committee is to be elected from an organization's membership of 11 people. How many different committees are possible?

51. Of 12 possible books, you plan to take 4 with you on vacation. How many different collections of 4 books can you take?

52. There are 14 standbys who hope to get seats on a flight, but only 6 seats are available on the plane. How many different ways can the 6 people be selected?

53. You volunteer to help drive children at a charity event to the zoo, but you can fit only 8 of the 17 children present in your van. How many different groups of 8 children can you drive?

54. Of the 100 people in the U.S. Senate, 18 serve on the Foreign Relations Committee. How many ways are there to select Senate members for this committee (assuming party affiliation is not a factor in selection)?

55. To win at LOTTO in the state of Florida, one must correctly select 6 numbers from a collection of 53 numbers (1 through 53). The order in which the selection is made does not matter. How many different selections are possible?

56. To win in the New York State lottery, one must correctly select 6 numbers from 59 numbers. The order in which the selection is made does not matter. How many different selections are possible?

In Exercises 57–66, solve by the method of your choice.

57. In a race in which six automobiles are entered and there are no ties, in how many ways can the first four finishers come in?

58. A book club offers a choice of 8 books from a list of 40. In how many ways can a member make a selection?

59. A medical researcher needs 6 people to test the effectiveness of an experimental drug. If 13 people have volunteered for the test, in how many ways can 6 people be selected?

60. Fifty people purchase raffle tickets. Three winning tickets are selected at random. If first prize is \$1000, second prize is \$500, and third prize is \$100, in how many different ways can the prizes be awarded?

61. From a club of 20 people, in how many ways can a group of three members be selected to attend a conference?

62. Fifty people purchase raffle tickets. Three winning tickets are selected at random. If each prize is \$500, in how many different ways can the prizes be awarded?

63. How many different four-letter passwords can be formed from the letters A, B, C, D, E, F, and G if no repetition of letters is allowed?

64. Nine comedy acts will perform over two evenings. Five of the acts will perform on the first evening and the order in which the acts perform is important. How many ways can the schedule for the first evening be made?

65. Using 15 flavors of ice cream, how many cones with three different flavors can you create if it is important to you which flavor goes on the top, middle, and bottom?

66. Baskin-Robbins offers 31 different flavors of ice cream. One of their items is a bowl consisting of three scoops of ice cream, each a different flavor. How many such bowls are possible?

Writing in Mathematics

67. Explain the Fundamental Counting Principle.

68. Write an original problem that can be solved using the Fundamental Counting Principle. Then solve the problem.

69. What is a permutation?

70. Describe what $_nP_r$ represents.

71. Write a word problem that can be solved by evaluating $_7P_3$.

72. What is a combination?

73. Explain how to distinguish between permutation and combination problems.

74. Write a word problem that can be solved by evaluating $_7C_3$.

Technology Exercises

75. Use a graphing utility with an $\boxed{_nP_r}$ menu item to verify your answers in Exercises 1–8.

76. Use a graphing utility with an $\boxed{_nC_r}$ menu item to verify your answers in Exercises 9–16.

Critical Thinking Exercises

77. Which one of the following is true?

a. The number of ways to choose four questions out of ten questions on an essay test is $_{10}P_4$.

b. If $r > 1$, $_nP_r$ is less than $_nC_r$.

c. $_7P_3 = 3!_7C_3$

d. The number of ways to pick a winner and first runner-up in a piano recital with 20 contestants is $_{20}C_2$.

78. Five men and five women line up at a checkout counter in a store. In how many ways can they line up if the first person in line is a woman and the people in line alternate woman, man, woman, man, and so on?

79. How many four-digit odd numbers less than 6000 can be formed using the digits 2, 4, 6, 7, 8, and 9?

80. A mathematics exam consists of 10 multiple-choice questions and 5 open-ended problems in which all work must be shown. If an examinee must answer 8 of the multiple-choice questions and 3 of the open-ended problems, in how many ways can the questions and problems be chosen?

Review Exercises

81. If $f(x) = x^2 + 2x - 5$ and $g(x) = 4x - 1$, find $(f \circ g)(x)$. (Section 8.3. Example 1)

82. Solve: $|2x - 5| > 3$. (Section 9.3, Example 4)

83. Give the center and radius. Then graph the equation:

$$x^2 + y^2 - 2x + 4y - 4 = 0.$$

(Section 13.1, Example 4)

SECTION 14.6

Objectives

1. Compute empirical probability.
2. Compute theoretical probability.
3. Find the probability that an event will not occur.
4. Find the probability of one event or a second event occurring.
5. Find the probability of one event and a second event occurring.

PROBABILITY

Table 14.2 The Hours of Sleep Americans Get on a Typical Night

Hours of Sleep	Number of Americans, in millions
4 or less	11.36
5	25.56
6	71.00
7	85.20
8	76.68
9	8.52
10 or more	5.68

Total: 284.00

Source: Discovery Health Media

How many hours of sleep do you typically get each night? Table 14.2 indicates that 71 million out of 284 million Americans are getting six hours of sleep on a typical night. The *probability* of an American getting six hours of sleep on a typical night is $\frac{71}{284}$. This fraction can be reduced to $\frac{1}{4}$, or expressed as 0.25, or 25%. Thus, 25% of Americans get six hours of sleep each night.

We find a probability by dividing one number by another. Probabilities are assigned to an *event*, such as getting six hours of sleep on a typical night. Events that are certain to occur are assigned probabilities of 1, or 100%. For example, the probability that a given individual will eventually die is 1. Regrettably, taxes and death are always certain! By contrast, if an event cannot occur, its probability is 0. For example, the probability that Elvis will return from the dead and serenade us with one final reprise of "Heartbreak Hotel" is 0.

Probabilities of events are expressed as numbers ranging from 0 to 1, or 0% to 100%. The closer the probability of a given event is to 1, the more likely it is that the event will occur. The closer the probability of a given event is to 0, the less likely it is that the event will occur.

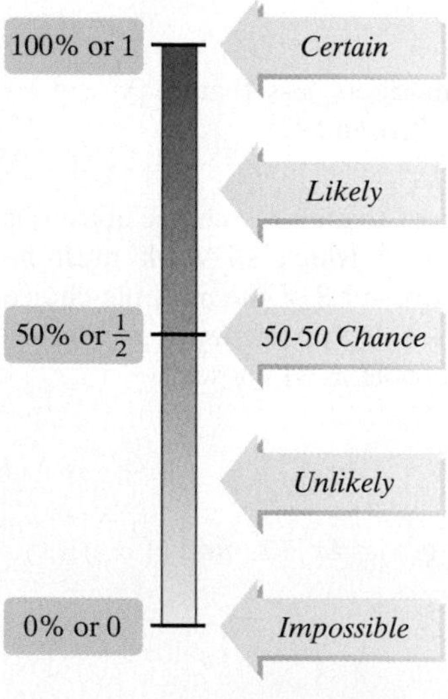

Possible Values for Probabilities

Empirical Probability Empirical probability applies to situations in which we observe how frequently an event occurs. We use the following formula to compute the empirical probability of an event:

1 Compute empirical probability.

COMPUTING EMPIRICAL PROBABILITY The **empirical probability** of event E, denoted by $P(E)$, is

$$P(E) = \frac{\text{observed number of times } E \text{ occurs}}{\text{total number of observed occurrences}}.$$

EXAMPLE 1 Empirical Probabilities with Real-World Data

When women turn 40, their gynecologists typically remind them that it is time to undergo mammography screening for breast cancer. The data in Table 14.3 are based on 100,000 U.S. women, ages 40 to 50, who participated in mammography screening.

Table 14.3 Mammography Screening on 100,000 U.S. Women, Ages 40 to 50

	Breast Cancer	No Breast Cancer
Positive Mammogram	720	6944
Negative Mammogram	80	92,256

$720 + 6944 = 7664$ women have positive mammograms.

$80 + 92{,}256 = 92{,}336$ women have negative mammograms.

$720 + 80 = 800$ women have breast cancer.

$6944 + 92{,}256 = 99{,}200$ women do not have breast cancer.

Source: Gerd Gigerenzer, *Calculated Risks*, Simon and Schuster, 2002

a. Use Table 14.3 to find the probability that a woman aged 40 to 50 has breast cancer.

b. Among women without breast cancer, find the probability of a positive mammogram.

c. Among women with positive mammograms, find the probability of not having breast cancer.

SOLUTION

a. We begin with the probability that a woman aged 40 to 50 has breast cancer. The probability of having breast cancer is the number of women with breast cancer divided by the total number of women.

$$P(\text{breast cancer}) = \frac{\text{number of women with breast cancer}}{\text{total number of women}}$$

$$= \frac{800}{100{,}000} = \frac{1}{125} = 0.008$$

The empirical probability that a woman aged 40 to 50 has breast cancer is $\frac{1}{125}$, or 0.008.

b. Now, we find the probability of a positive mammogram among women without breast cancer. Thus, we restrict the data to women without breast cancer:

	No Breast Cancer
Positive Mammogram	6944
Negative Mammogram	92,256

Within the restricted data, the probability of a positive mammogram is the number of women with positive mammograms divided by the total number of women.

	No Breast Cancer
Positive Mammogram	6944
Negative Mammogram	92,256

Part (b) table (repeated)

$$P(\text{positive mammogram}) = \frac{\text{number of women with positive mammograms}}{\text{total number of women in the restricted data}}$$

$$= \frac{6944}{6944 + 92{,}256} = \frac{6944}{99{,}200} = 0.07$$

This is the total number of women without breast cancer.

Among women without breast cancer, the empirical probability of a positive mammogram is $\frac{6944}{99{,}200}$, or 0.07.

c. Now, we find the probability of not having breast cancer among women with positive mammograms. Thus, we restrict the data to women with positive mammograms:

	Breast Cancer	No Breast Cancer
Positive Mammogram	720	6944

Within the restricted data, the probability of not having breast cancer is the number of women with no breast cancer divided by the total number of women.

$$P(\text{no breast cancer}) = \frac{\text{number of women with no breast cancer}}{\text{total number of women in the restricted data}}$$

$$= \frac{6944}{720 + 6944} = \frac{6944}{7664} \approx 0.906$$

This is the total number of women with positive mammograms.

Among women with positive mammograms, the probability of not having breast cancer is $\frac{6944}{7664}$, or approximately 0.906. ■

CHECK POINT 1 Use the data in Table 14.3 to answer this exercise. Express probabilities as fractions and as decimals to three decimal places.

a. Find the probability that a woman aged 40 to 50 has a positive mammogram.

b. Among women with breast cancer, find the probability of a positive mammogram.

c. Among women with positive mammograms, find the probability of having breast cancer.

2 Compute theoretical probability.

Theoretical Probability You toss a coin. Although it is equally likely to land either heads up, denoted by H, or tails up, denoted by T, the actual outcome is uncertain. Any occurrence for which the outcome is uncertain is called an **experiment**. Thus, tossing a coin is an example of an experiment. The set of all possible outcomes of an experiment is the **sample space** of the experiment, denoted by S. The sample space for the coin-tossing experiment is

$$S = \{H, T\}.$$

Lands heads up — Lands tails up

We can define an event more formally using these concepts. An **event**, denoted by E, is any subcollection, or subset, of a sample space. For example, the subset $E = \{T\}$ is the event of landing tails up when a coin is tossed.

Theoretical probability applies to situations like this, in which the sample space only contains equally-likely outcomes, all of which are known. To calculate the theoretical probability of an event, we divide the number of outcomes resulting in the event by the number of outcomes in the sample space.

COMPUTING THEORETICAL PROBABILITY If an event E has $n(E)$ equally-likely outcomes and its sample space S has $n(S)$ equally-likely outcomes, the **theoretical probability** of event E, denoted by $P(E)$, is

$$P(E) = \frac{\text{number of outcomes in event } E}{\text{number of outcomes in sample space } S} = \frac{n(E)}{n(S)}.$$

The sum of the theoretical probabilities of all possible outcomes in the sample space is 1.

How can we use this formula to compute the probability of a coin landing tails up? We use the following sets:

$$E = \{T\}. \qquad S = \{H, T\}.$$

This is the event of landing tails up.

This is the sample space with all equally-likely outcomes.

The probability of a coin landing tails up is

$$P(E) = \frac{n(E)}{n(S)} = \frac{1}{2}.$$

FIGURE 14.8 Outcomes when a die is rolled

Theoretical probability applies to many games of chance, including dice rolling, lotteries, card games, and roulette. The next example deals with the experiment of rolling a die. Figure 14.8 illustrates that when a die is rolled, there are six equally-likely outcomes. The sample space can be shown as

$$S = \{1, 2, 3, 4, 5, 6\}.$$

EXAMPLE 2 Computing Theoretical Probability

A die is rolled. Find the probability of getting a number less than 5.

SOLUTION The sample space of equally-likely outcomes is $S = \{1, 2, 3, 4, 5, 6\}$. There are six outcomes in the sample space, so $n(S) = 6$.

We are interested in the probability of getting a number less than 5. The event of getting a number less than 5 can be represented by

$$E = \{1, 2, 3, 4\}.$$

There are four outcomes in this event, so $n(E) = 4$.

The probability of rolling a number less than 5 is

$$P(E) = \frac{n(E)}{n(S)} = \frac{4}{6} = \frac{2}{3}.$$

■

CHECK POINT 2 A die is rolled. Find the probability of getting a number greater than 4.

EXAMPLE 3 Computing Theoretical Probability

Two ordinary six-sided dice are rolled. What is the probability of getting a sum of 8?

SOLUTION Each die has six equally-likely outcomes. By the Fundamental Counting Principle, there are $6 \cdot 6$, or 36, equally-likely outcomes in the sample space. That is, $n(S) = 36$. The 36 outcomes are shown below as ordered pairs. The five ways of rolling a sum of 8 appear in the green highlighted diagonal.

First Die \ Second Die	⚀	⚁	⚂	⚃	⚄	⚅
⚀	(1, 1)	(1, 2)	(1, 3)	(1, 4)	(1, 5)	(1, 6)
⚁	(2, 1)	(2, 2)	(2, 3)	(2, 4)	(2, 5)	(2, 6)
⚂	(3, 1)	(3, 2)	(3, 3)	(3, 4)	(3, 5)	(3, 6)
⚃	(4, 1)	(4, 2)	(4, 3)	(4, 4)	(4, 5)	(4, 6)
⚄	(5, 1)	(5, 2)	(5, 3)	(5, 4)	(5, 5)	(5, 6)
⚅	(6, 1)	(6, 2)	(6, 3)	(6, 4)	(6, 5)	(6, 6)

$$\begin{aligned} S = \{&(1,1), (1,2), (1,3), (1,4),\\ &(1,5), (1,6), (2,1), (2,2),\\ &(2,3), (2,4), (2,5), (2,6),\\ &(3,1), (3,2), (3,3), (3,4),\\ &(3,5), (3,6), (4,1), (4,2),\\ &(4,3), (4,4), (4,5), (4,6),\\ &(5,1), (5,2), (5,3), (5,4),\\ &(5,5), (5,6), (6,1), (6,2),\\ &(6,3), (6,4), (6,5), (6,6)\}\end{aligned}$$

The phrase "getting a sum of 8" describes the event

$$E = \{(6,2), (5,3), (4,4), (3,5), (2,6)\}.$$

This event has 5 outcomes, so $n(E) = 5$. Thus, the probability of getting a sum of 8 is

$$P(E) = \frac{n(E)}{n(S)} = \frac{5}{36}.$$

■

CHECK POINT 3 What is the probability of getting a sum of 5 when two six-sided dice are rolled?

Computing Theoretical Probability without Listing an Event and the Sample Space In some situations, we can compute theoretical probability without having to write out each event and each sample space. For example, suppose you are dealt one card from a standard 52-card deck, illustrated in Figure 14.9. The deck has four suits: Hearts and diamonds are red, and clubs and spades are black. Each suit has 13 different face values—A(ace), 2, 3, 4, 5, 6, 7, 8, 9, 10, J(jack), Q(queen), and K(king). Jacks, queens, and kings are called **picture cards** or **face cards**.

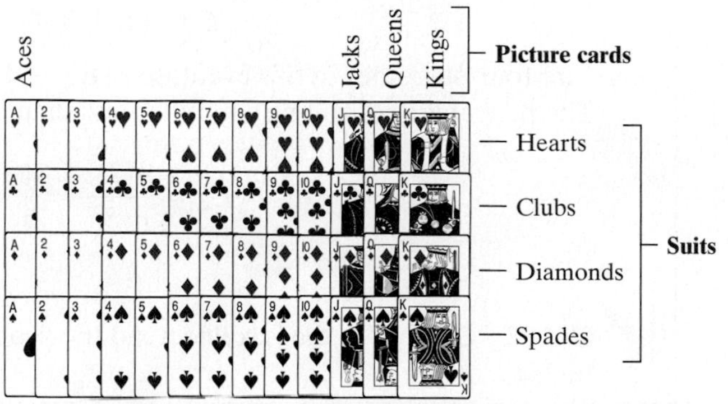

FIGURE 14.9 A standard 52-card bridge deck

EXAMPLE 4 Probability and a Deck of 52 Cards

You are dealt one card from a standard 52-card deck. Find the probability of being dealt a heart.

SOLUTION Let E be the event of being dealt a heart. Because there are 13 hearts in the deck, the event of being dealt a heart can occur in 13 ways. The number of outcomes in event E is 13: $n(E) = 13$. With 52 cards in the deck, the total number of possible ways of being dealt a single card is 52. The number of outcomes in the sample space is 52: $n(S) = 52$. The probability of being dealt a heart is

$$P(E) = \frac{n(E)}{n(S)} = \frac{13}{52} = \frac{1}{4}.$$

CHECK POINT 4 If you are dealt one card from a standard 52-card deck, find the probability of being dealt a king.

State lotteries keep 50 cents on the dollar, resulting in $10 billion a year for public funding.

If your state has a lottery drawing each week, the probability that someone will win the top prize is relatively high. If there is no winner this week, it is virtually certain that eventually someone will be graced with millions of dollars. So, why are you so unlucky compared to this undisclosed someone? In Example 5, we provide an answer to this question, using the counting principles discussed in Section 14.5.

EXAMPLE 5 Probability and Combinations: Winning the Lottery

Florida's lottery game, LOTTO, is set up so that each player chooses six different numbers from 1 to 53. If the six numbers chosen match the six numbers drawn randomly, the player wins (or shares) the top cash prize. (As of this writing, the top cash prize has ranged from $7 million to $106.5 million.) With one LOTTO ticket, what is the probability of winning this prize?

SOLUTION Because the order of the six numbers does not matter, this is a situation involving combinations. Let E be the event of winning the lottery with one ticket. With one LOTTO ticket, there is only one way of winning. Thus, $n(E) = 1$. The sample space is the set of all possible six-number combinations. We can use the combinations formula

$${}_nC_r = \frac{n!}{(n-r)!r!}$$

to find the total number of possible combinations. We are selecting $r = 6$ numbers from a collection of $n = 53$ numbers.

$${}_{53}C_6 = \frac{53!}{(53-6)!6!} = \frac{53!}{47!6!} = \frac{53 \cdot 52 \cdot 51 \cdot 50 \cdot 49 \cdot 48 \cdot \cancel{47!}}{\cancel{47!} \cdot 6 \cdot 5 \cdot 4 \cdot 3 \cdot 2 \cdot 1} = 22{,}957{,}480$$

There are nearly 23 million number combinations possible in LOTTO. If a person buys one LOTTO ticket, the probability of winning is

$$P(E) = \frac{n(E)}{n(S)} = \frac{1}{22{,}957{,}480} \approx 0.0000000436.$$

The probability of winning the top prize with one LOTTO ticket is $\frac{1}{22{,}957{,}480}$, or about 1 in 23 million.

ENRICHMENT ESSAY

Comparing the Probability of Dying to the Probability of Winning Florida's LOTTO

As a healthy nonsmoking 30-year-old, your probability of dying this year is approximately 0.001. Divide this probability by the probability of winning LOTTO with one ticket:

$$\frac{0.001}{0.0000000436} \approx 22{,}936.$$

A healthy 30-year-old is nearly 23,000 times more likely to die this year than to win Florida's lottery.

In 2003, Americans spent nearly 19 billion dollars on lotteries set up by revenue-hungry states. If a person buys, say 5000 different tickets in Florida's LOTTO, that person has selected 5000 different combinations of the six numbers. The probability of winning is

$$\frac{5000}{22,957,480} \approx 0.000218.$$

The chances of winning top prize are about 218 in a million. At $1 per LOTTO ticket, it is highly probable that our LOTTO player will be $5000 poorer.

CHECK POINT 5 People lose interest when they do not win at games of chance, including Florida's LOTTO. With drawings twice weekly instead of once, the game described in Example 5 was brought in to bring back lost players and increase ticket sales. The original LOTTO was set up so that each player chose six different numbers from 1 to 49, rather than from 1 to 53, with a lottery drawing only once a week. With one LOTTO ticket, what was the probability of winning the top cash prize in Florida's original LOTTO? Express the answer as a fraction and as a decimal correct to ten places.

3 Find the probability that an event will not occur.

Probability of an Event Not Occurring If we know $P(E)$, the probability of an event E, we can determine the probability that the event will not occur, denoted by $P(\text{not } E)$. Because the sum of the probabilities of all possible outcomes in any situation is 1,

$$P(E) + P(\text{not } E) = 1.$$

We now solve this equation for $P(\text{not } E)$, the probability that event E will not occur, by subtracting $P(E)$ from both sides. The resulting formula is given in the following box:

THE PROBABILITY OF AN EVENT NOT OCCURRING The probability that an event E will not occur is equal to 1 minus the probability that it will occur.

$$P(\text{not } E) = 1 - P(E)$$

EXAMPLE 6 The Probability of an Event Not Occurring

The graph in Figure 14.10 shows the distribution, by branch and gender, of the 1.37 million, or 1370 thousand, active-duty personnel in the U.S. military in 2001. Numbers are given in thousands and rounded to the nearest ten thousand. If one person is randomly selected from the U.S. military and the distribution is the same as it was in 2001, find the probability that this person is not in the Army.

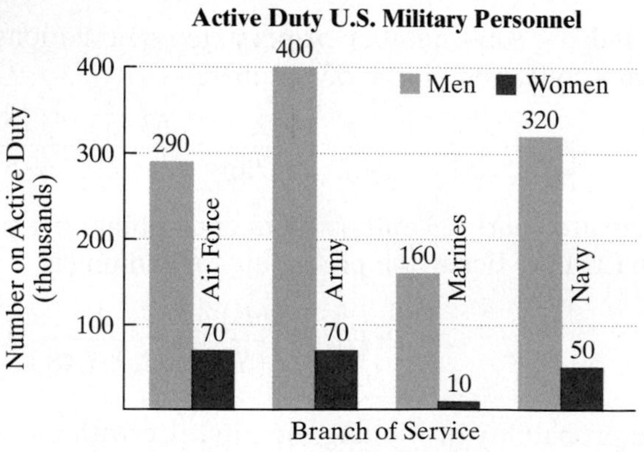

Source: U.S. Defense Department

FIGURE 14.10

SOLUTION We begin by finding the probability that the selected person *is* in the Army.

$$P(\text{Army}) = \frac{\text{number of people in the Army}}{\text{total number of people in the U.S. military}}$$

$$= \frac{400 + 70}{1370}$$

The graph shows 400 thousand men and 70 thousand women in the Army.

This number was given, but can be obtained by adding the eight numbers above the bars.

$$= \frac{470}{1370} = \frac{47}{137}$$

Thus,

$$P(\text{not in Army}) = 1 - P(\text{Army}) = 1 - \frac{47}{137} = \frac{137}{137} - \frac{47}{137} = \frac{90}{137}.$$

The probability that a person selected from the U.S. military is not in the Army is $\frac{90}{137}$. ■

CHECK POINT 6 Use the graph in Figure 14.10. If one person is randomly selected from the U.S. military, find the probability that this person is not in the Marines.

4 Find the probability of one event or a second event occurring.

Or **Probabilities with Mutually Exclusive Events** Suppose that you randomly select one card from a deck of 52 cards. Let A be the event of selecting a king and let B be the event of selecting a queen. Only one card is selected, so it is impossible to get both a king and a queen. The outcomes of selecting a king and a queen cannot occur simultaneously. They are called *mutually exclusive events*. If it is impossible for any two events, A and B, to occur simultaneously, they are said to be **mutually exclusive**. If A and B are mutually exclusive events, the probability that either A or B will occur is determined by adding their individual probabilities.

OR **PROBABILITIES WITH MUTUALLY EXCLUSIVE EVENTS** If A and B are mutually exclusive events, then

$$P(A \text{ or } B) = P(A) + P(B).$$

Using set notation, $P(A \cup B) = P(A) + P(B)$.

EXAMPLE 7 The Probability of Either of Two Mutually Exclusive Events Occurring

If one card is randomly selected from a deck of cards, what is the probability of selecting a king or a queen?

SOLUTION We find the probability that either of these mutually exclusive events will occur by adding their individual probabilities.

$$P(\text{king or queen}) = P(\text{king}) + P(\text{queen}) = \frac{4}{52} + \frac{4}{52} = \frac{8}{52} = \frac{2}{13}$$

The probability of selecting a king or a queen is $\frac{2}{13}$. ■

CHECK POINT 7 If you roll a single, six-sided die, what is the probability of getting either a 4 or a 5?

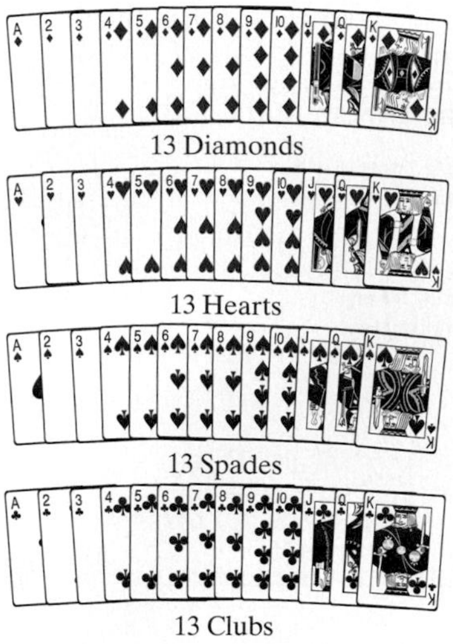

FIGURE 14.11 A deck of 52 cards

***Or* Probabilities with Events That Are Not Mutually Exclusive** Consider the deck of 52 cards shown in Figure 14.11. Suppose that these cards are shuffled and you randomly select one card from the deck. What is the probability of selecting a diamond or a picture card (jack, queen, king)? Begin by adding their individual probabilities.

$$P(\text{diamond}) + P(\text{picture card}) = \frac{13}{52} + \frac{12}{52}$$

There are 13 diamonds in the deck of 52 cards.

There are 12 picture cards in the deck of 52 cards.

However, this sum is not the probability of selecting a diamond or a picture card. The problem is that there are three cards that are *simultaneously* diamonds and picture cards, shown in Figure 14.12. The events of selecting a diamond and selecting a picture card are not mutually exclusive. It is possible to select a card that is both a diamond and a picture card.

FIGURE 14.12 Three diamonds are picture cards.

The situation is illustrated in the diagram in Figure 14.13. Why can't we find the probability of selecting a diamond or a picture card by adding their individual probabilities? The diagram shows that three of the cards, the three diamonds that are picture cards, get counted twice when we add the individual probabilities. First the three cards get counted as diamonds and then they get counted as picture cards. In order to avoid the error of counting the three cards twice, we need to subtract the probability of getting a diamond and a picture card, $\frac{3}{52}$, as follows:

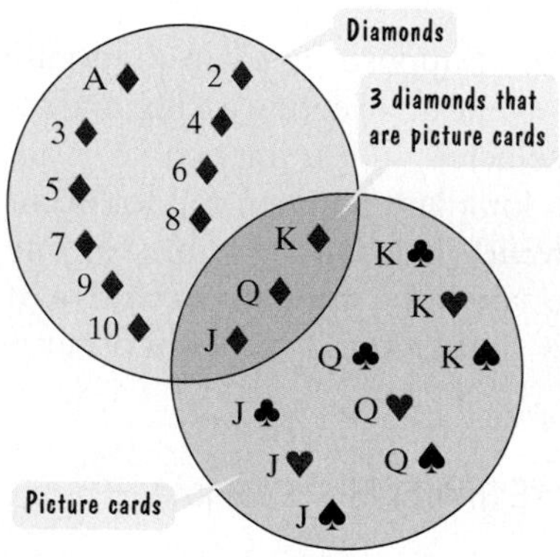

FIGURE 14.13

$P(\text{diamond or picture card})$

$$= P(\text{diamond}) + P(\text{picture card}) - P(\text{diamond and picture card})$$

$$= \frac{13}{52} + \frac{12}{52} - \frac{3}{52} = \frac{13 + 12 - 3}{52} = \frac{22}{52} = \frac{11}{26}.$$

Thus, the probability of selecting a diamond or a picture card is $\frac{11}{26}$.

In general, if A and B are events that are not mutually exclusive, the probability that A or B will occur is determined by adding their individual probabilities and then subtracting the probability that A and B occur simultaneously.

OR PROBABILITIES WITH EVENTS THAT ARE NOT MUTUALLY EXCLUSIVE

If A and B are not mutually exclusive events, then

$$P(A \text{ or } B) = P(A) + P(B) - P(A \text{ and } B).$$

Using set notation,

$$P(A \cup B) = P(A) + P(B) - P(A \cap B).$$

EXAMPLE 8 An *Or* Probability with Events That Are Not Mutually Exclusive

FIGURE 14.14 It is equally probable that the pointer will land on any one of the eight regions.

Figure 14.14 illustrates a spinner. It is equally probable that the pointer will land on any one of the eight regions, numbered 1 through 8. If the pointer lands on a borderline, spin again. Find the probability that the pointer will stop on an even number or a number greater than 5.

SOLUTION It is possible for the pointer to land on a number that is both even and greater than 5. Two of the numbers, 6 and 8, are even and greater than 5. These events are not mutually exclusive. The probability of landing on a number that is even or greater than 5 is calculated as follows:

$$P\left(\begin{matrix}\text{even or}\\ \text{greater than 5}\end{matrix}\right) = P(\text{even}) + P(\text{greater than 5}) - P\left(\begin{matrix}\text{even and}\\ \text{greater than 5}\end{matrix}\right)$$

$$= \frac{4}{8} + \frac{3}{8} - \frac{2}{8}$$

Four of the eight numbers, 2, 4, 6, and 8, are even.

Three of the eight numbers, 6, 7, and 8, are greater than 5.

Two of the eight numbers, 6 and 8, are even and greater than 5.

$$= \frac{4+3-2}{8} = \frac{5}{8}.$$

The probability that the pointer will stop on an even number or a number greater than 5 is $\frac{5}{8}$. ■

✔ **CHECK POINT 8** Use Figure 14.14 to find the probability that the pointer will stop on an odd number or a number less than 5.

EXAMPLE 9 An *Or* Probability with Real-World Data

Earlier in this section, we saw a graph showing the distribution, by branch and gender, of active-duty personnel in the U.S. military. The data are shown again in Table 14.4. If one person is randomly selected from the U.S. military, find the probability that this person is in the Army or is a woman.

Table 14.4 Active-Duty U.S. Military Personnel, in Thousands

	Air Force	Army	Marine Corps	Navy	Total
Male	290	400	160	320	1170
Female	70	70	10	50	200
Total	360	470	170	370	1370

Total male: 290 + 400 + 160 + 320 = 1170

Total female: 70 + 70 + 10 + 50 = 200

Total Air Force · Total Army · Total Marines · Total Navy · Total on active duty

Source: U.S. Defense Department

SOLUTION It is possible to select a person who is both in the Army and is a woman. Thus, these events are not mutually exclusive.

Table 14.4 (repeated)

	Air Force	Army	Marine Corps	Navy	Total
Male	290	400	160	320	1170
Female	70	70	10	50	200
Total	360	470	170	370	1370

Source: U.S. Defense Department

$$P(\text{Army or woman}) = P(\text{Army}) + P(\text{woman}) - P(\text{Army and woman})$$

$$= \frac{470}{1370} + \frac{200}{1370} - \frac{70}{1370}$$

Of the 1370 (thousand) personnel, 470 are in the Army – 400 men and 70 women.

Of the 1370 personnel, 200 are women – 70 Air Force + 70 Army + 10 Marines + 50 Navy.

Of the 1370 personnel, 70 are Army women.

$$= \frac{470 + 200 - 70}{1370} = \frac{600}{1370} = \frac{60}{137}$$

The probability that a person selected from the U.S. military is in the Army or is a woman is $\frac{60}{137}$. ■

CHECK POINT 9 Use Table 14.4. If one person is randomly selected from the U.S. military, find the probability that this person is in the Navy or is a man.

5 Find the probability of one event and a second event occurring.

***And* Probabilities with Independent Events** Suppose that you toss a fair coin two times in succession. The outcome of the first toss, heads or tails, does not affect what happens when you toss the coin a second time. For example, the occurrence of tails on the first toss does not make tails more likely or less likely to occur on the second toss. The repeated toss of a coin produces *independent events* because the outcome of one toss does not influence the outcome of others. Two events are **independent events** if the occurrence of either of them has no effect on the probability of the other.

If two events are independent, we can calculate the probability of the first occurring and the second occurring by multiplying their probabilities.

***AND* PROBABILITIES WITH INDEPENDENT EVENTS** If A and B are independent events, then

$$P(A \text{ and } B) = P(A) \cdot P(B).$$

EXAMPLE 10 Independent Events on a Roulette Wheel

Figure 14.15 shows a U.S. roulette wheel that has 38 numbered slots (1 through 36, 0, and 00). Of the 38 compartments, 18 are black, 18 are red, and two are green. A play has the dealer spin the wheel and a small ball in opposite directions. As the ball slows to a stop, it can land with equal probability on any one of the 38 numbered slots. Find the probability of red occurring on two consecutive plays.

FIGURE 14.15 A U.S. roulette wheel

SOLUTION The wheel has 38 equally-likely outcomes and 18 are red. Thus, the probability of red occurring on a play is $\frac{18}{38}$, or $\frac{9}{19}$. The result that occurs on each play is independent of all previous results. Thus,

$$P(\text{red and red}) = P(\text{red}) \cdot P(\text{red}) = \frac{9}{19} \cdot \frac{9}{19} = \frac{81}{361} \approx 0.224.$$

The probability of red occurring on two consecutive plays is $\frac{81}{361}$. ■

Some roulette players incorrectly believe that if red occurs on two consecutive plays, then another color is "due." Because the events are independent, the outcomes of previous spins have no effect on any other spins.

CHECK POINT 10 Find the probability of green occurring on two consecutive plays on a roulette wheel.

The *and* rule for independent events can be extended to cover three or more events. Thus, if A, B, and C are independent events, then

$$P(A \text{ and } B \text{ and } C) = P(A) \cdot P(B) \cdot P(C).$$

EXAMPLE 11 Independent Events in a Family

The picture in the margin shows a family that has had nine girls in a row. Find the probability of this occurrence.

SOLUTION If two or more events are independent, we can find the probability of them all occurring by multiplying their probabilities. The probability of a baby girl is $\frac{1}{2}$, so the probability of nine girls in a row is $\frac{1}{2}$ used as a factor nine times.

$$P(\text{nine girls in a row}) = \frac{1}{2}\cdot\frac{1}{2}\cdot\frac{1}{2}\cdot\frac{1}{2}\cdot\frac{1}{2}\cdot\frac{1}{2}\cdot\frac{1}{2}\cdot\frac{1}{2}\cdot\frac{1}{2}$$

$$= \left(\frac{1}{2}\right)^9 = \frac{1}{512}$$

The probability of a run of nine girls in a row is $\frac{1}{512}$. (If another child is born into the family, this event is independent of the other nine, and the probability of a girl is still $\frac{1}{2}$.) ■

CHECK POINT 11 Find the probability of a family having four boys in a row.

14.6 EXERCISE SET

Student Solutions Manual | CD/Video | PH Math/Tutor Center | MathXL Tutorials on CD | MathXL® | MyMathLab | Interactmath.com

Practice and Application Exercises

The table shows the breakdown of the 89 thousand single parents on active duty in the U.S. military in 2002. All numbers are in thousands and rounded to the nearest thousand. Use the data in the table to solve Exercises 1–10.

Single Parents on Active Duty in the U.S. Military, in Thousands

	Army	Navy	Marine Corps	Air Force	Total
Male	26	23	5	12	66
Female	10	6	1	6	23
Total	36	29	6	18	89

Total male: $26 + 23 + 5 + 12 = 66$

Total female: $10 + 6 + 1 + 6 = 23$

Total Army; Total Navy; Total Marines; Total Air Force; Total on active duty

Source: U.S. Defense Department

(In Exercises 1-10, be sure to refer to the table on the bottom of page 1001.) Find the probability that a randomly selected single parent in the U.S. military is

1. female.
2. male.
3. in the Army.
4. in the Navy.
5. a woman in the Air Force.
6. a man in the Marine Corps.
7. Among single parents in the Air Force, find the probability of selecting a woman.
8. Among single parents in the Marine Corps, find the probability of selecting a man.
9. Among the female single parents in the military, find the probability of selecting a woman in the Air Force.
10. Among the male single parents in the military, find the probability of selecting a man in the Marine Corps.

In Exercises 11–16, a die is rolled. Find the probability of getting

11. a 4.
12. a 5.
13. an odd number.
14. a number greater than 3.
15. a number greater than 4.
16. a number greater than 7.

In Exercises 17–20, you are dealt one card from a standard 52 card deck. Find the probability of being dealt

17. a queen.
18. a diamond.
19. a picture card.
20. a card greater than 3 and less than 7.

In Exercises 21–22, a fair coin is tossed two times in succession. The sample space of equally-likely outcomes is {HH, HT, TH, TT}. Find the probability of getting

21. two heads.
22. the same outcome on each toss.

In Exercises 23–24, you select a family with three children. If M represents a male child and F a female child, the sample space of equally-likely outcomes is {MMM, MMF, MFM, MFF, FMM, FMF, FFM, FFF}. Find the probability of selecting a family with

23. at least one male child.
24. at least two female children.

In Exercises 25–26, a single die is rolled twice. The 36 equally-likely outcomes are shown as follows:

	Second Roll ⚀	⚁	⚂	⚃	⚄	⚅
First Roll ⚀	(1, 1)	(1, 2)	(1, 3)	(1, 4)	(1, 5)	(1, 6)
⚁	(2, 1)	(2, 2)	(2, 3)	(2, 4)	(2, 5)	(2, 6)
⚂	(3, 1)	(3, 2)	(3, 3)	(3, 4)	(3, 5)	(3, 6)
⚃	(4, 1)	(4, 2)	(4, 3)	(4, 4)	(4, 5)	(4, 6)
⚄	(5, 1)	(5, 2)	(5, 3)	(5, 4)	(5, 5)	(5, 6)
⚅	(6, 1)	(6, 2)	(6, 3)	(6, 4)	(6, 5)	(6, 6)

Find the probability of getting

25. two numbers whose sum is 4.
26. two numbers whose sum is 6.
27. To play the California lottery, a person has to correctly select 6 out of 51 numbers, paying \$1 for each six-number selection. If you pick six numbers that are the same as the ones drawn by the lottery, you win mountains of money. What is the probability that a person with one combination of six numbers will win? What is the probability of winning if 100 different lottery tickets are purchased?
28. A state lottery is designed so that a player chooses six numbers from 1 to 30 on one lottery ticket. What is the probability that a player with one lottery ticket will win? What is the probability of winning if 100 different lottery tickets are purchased?

Exercises 29–30 involve a deck of 52 cards. If necessary, refer to the picture of a deck of cards, Figure 14.11 on page 998.

29. A poker hand consists of five cards.
  a. Find the total number of possible five-card poker hands.
  b. A diamond flush is a five-card hand consisting of all diamonds. Find the number of possible diamond flushes.
  c. Find the probability of being dealt a diamond flush.
30. If you are dealt 3 cards from a shuffled deck of 52 cards, find the probability that all 3 cards are picture cards.

The table shows the educational attainment of the U.S. population, ages 25 and over. Use the data in the table, expressed in millions, to solve Exercises 31–36.

Educational Attainment, in Millions, of the U.S. Population, Ages 25 and Over

	Less Than 4 Years High School	4 Years High School Only	Some College (Less than 4 years)	4 Years College (or More)	Total
Male	14	25	20	23	82
Female	15	31	24	22	92
Total	29	56	44	45	174

Source: U.S. Census Bureau

Find the probability that a randomly selected American, aged 25 or over

31. has not completed four years (or more) of college.

32. has not completed four years of high school.

33. has completed four years of high school only or less than four years of college.

34. has completed less than four years of high school or four years of high school only.

35. has completed four years of high school only or is a man.

36. has completed four years of high school only or is a woman.

In Exercises 37–42, you are dealt one card from a 52-card deck. Find the probability that

37. you are not dealt a king.

38. you are not dealt a picture card.

39. you are dealt a 2 or a 3.

40. you are dealt a red 7 or a black 8.

41. you are dealt a 7 or a red card.

42. you are dealt a 5 or a black card.

In Exercises 43–44, it is equally probable that the pointer on the spinner shown will land on any one of the eight regions, numbered 1 through 8. If the pointer lands on a borderline, spin again.

Find the probability that the pointer will stop on

43. an odd number or a number less than 6.

44. an odd number or a number greater than 3.

Use this information to solve Exercises 45–46. The mathematics department of a college has 8 male professors, 11 female professors, 14 male teaching assistants, and 7 female teaching assistants. If a person is selected at random from the group, find the probability that the selected person is

45. a professor or a male.

46. a professor or a female.

In Exercises 47–50, a single die is rolled twice. Find the probability of rolling

47. a 2 the first time and a 3 the second time.

48. a 5 the first time and a 1 the second time.

49. an even number the first time and a number greater than 2 the second time.

50. an odd number the first time and a number less than 3 the second time.

51. If you toss a fair coin six times, what is the probability of getting all heads?

52. If you toss a fair coin seven times, what is the probability of getting all tails?

53. The probability that South Florida will be hit by a major hurricane (category 4 or 5) in any single year is $\frac{1}{16}$. (*Source*: National Hurricane Center)

a. What is the probability that South Florida will be hit by a major hurricane two years in a row?

b. What is the probability that South Florida will be hit by a major hurricane in three consecutive years?

c. What is the probability that South Florida will not be hit by a major hurricane in the next ten years?

d. What is the probability that South Florida will be hit by a major hurricane at least once in the next ten years?

Writing in Mathematics

54. Describe the difference between theoretical probability and empirical probability.

55. Give an example of an event whose probability must be determined empirically rather than theoretically.

56. Write a probability word problem whose answer is one of the following fractions: $\frac{1}{6}$ or $\frac{1}{4}$ or $\frac{1}{3}$.

57. Explain how to find the probability of an event not occurring. Give an example.

58. What are mutually exclusive events? Give an example of two events that are mutually exclusive.

59. Explain how to find *or* probabilities with mutually exclusive events. Give an example.

60. Give an example of two events that are not mutually exclusive.

61. Explain how to find *or* probabilities with events that are not mutually exclusive. Give an example.

62. Explain how to find *and* probabilities with independent events. Give an example.

63. The president of a large company with 10,000 employees is considering mandatory cocaine testing for every employee. The test that would be used is 90% accurate, meaning that it will detect 90% of the cocaine users who are tested, and that 90% of the nonusers will test negative. This also means that the test gives 10% false positive. Suppose that 1% of the employees actually use cocaine. Find the probability that someone who tests positive for cocaine use is, indeed, a user.

Hint: Find the following probability fraction:

$$\frac{\text{the number of employees who test positive and are cocaine users}}{\text{the number of employees who test positive}}.$$

This fraction is given by

$$\frac{90\% \text{ of } 1\% \text{ of } 10{,}000}{\text{the number who test positive who actually use cocaine plus the number who test positive who do not use cocaine}}.$$

What does this probability indicate in terms of the percentage of employees who test positive who are not actually users? Discuss these numbers in terms of the issue of mandatory drug testing. Write a paper either in favor of or against mandatory drug testing, incorporating the actual percentage accuracy for such tests.

Critical Thinking Exercises

64. The target in the figure shown contains four squares. If a dart thrown at random hits the target, find the probability that it will land in a yellow region.

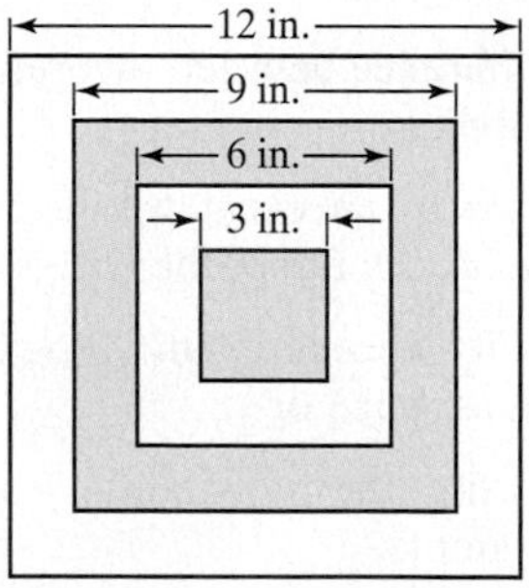

65. Suppose that it is a week in which the cash prize in Florida's LOTTO is promised to exceed \$50 million. If a person purchases 22,957,480 tickets in LOTTO at \$1 per ticket (all possible combinations), isn't this a guarantee of winning the lottery? Because the probability in this situation is 1, what's wrong with doing this?

66. Some three-digit numbers, such as 101 and 313, read the same forward and backward. If you select a number from all three-digit numbers, find the probability that it will read the same forward and backward.

67. In a class of 50 students, 29 are Democrats, 11 are business majors, and 5 of the business majors are Democrats. If one student is randomly selected from the class, find the probability of choosing

a. a Democrat who is not a business major.

b. a student who is neither a Democrat nor a business major.

68. One New Year's Eve, the probability of a person driving while intoxicated or having a driving accident is 0.35. If the probability of driving while intoxicated is 0.32 and the probability of having a driving accident is 0.09, find the probability of a person having a driving accident while intoxicated.

Review Exercises

69. Graph: $4x^2 + 25y^2 = 100$. (Section 13.2, Example 2)

70. Solve: $\log_2(x + 5) + \log_2(x - 1) = 4$. (Section 12.4, Example 5)

71. Divide $x^3 + 5x^2 + 3x - 10$ by $x + 2$.

(Section 5.6, Example 4)

GROUP PROJECT

CHAPTER 14

Group members serve as a financial team analyzing the three options given to the professional baseball player described in the chapter opener on page 933. As a group, determine which option provides the most amount of money over the six-year contract and which provides the least. Describe one advantage and one disadvantage to each option.

CHAPTER 14 SUMMARY

Definitions and Concepts	Examples

Section 14.1 Sequences and Summation Notation

An infinite sequence $\{a_n\}$ is a function whose domain is the set of positive integers. The function values, or terms, are represented by

$$a_1, a_2, a_3, a_4, \ldots, a_n, \ldots.$$

General term: $a_n = \frac{(-1)^n}{n^3}$.

$$a_1 = \frac{(-1)^1}{1^3} = -1, \quad a_2 = \frac{(-1)^2}{2^3} = \frac{1}{8},$$
$$a_3 = \frac{(-1)^3}{3^3} = -\frac{1}{27}, \quad a_4 = \frac{(-1)^4}{4^3} = \frac{1}{64}$$

First four terms are $-1, \frac{1}{8}, -\frac{1}{27}$, and $\frac{1}{64}$.

Factorial Notation

$$n! = n(n-1)(n-2)\cdots(3)(2)(1) \quad \text{and} \quad 0! = 1$$

$$6! = 6\cdot5\cdot4\cdot3\cdot2\cdot1 = 720$$
$$3! = 3\cdot2\cdot1 = 6$$

Summation Notation

$$\sum_{i=1}^{n} a_i = a_1 + a_2 + a_3 + a_4 + \cdots + a_n$$

In the summation shown here, i is the index of summation, n is the upper limit of summation, and 1 is the lower limit of summation.

$$\sum_{i=3}^{7} (i^2 - 4)$$
$$= (3^2 - 4) + (4^2 - 4) + (5^2 - 4) + (6^2 - 4) + (7^2 - 4)$$
$$= (9 - 4) + (16 - 4) + (25 - 4) + (36 - 4) + (49 - 4)$$
$$= 5 + 12 + 21 + 32 + 45$$
$$= 115$$

Section 14.2 Arithmetic Sequences

In an arithmetic sequence, each term after the first differs from the preceding term by a constant, the common difference. Subtract any term from the term that directly follows to find the common difference.

General Term of an Arithmetic Sequence
The nth term (the general term) of an arithmetic sequence with first term a_1 and common difference d is

$$a_n = a_1 + (n-1)d.$$

Find the general term and the tenth term:

$$3, 7, 11, 15, \ldots.$$

$a_1 = 3$ $\quad$ $d = 7 - 3 = 4$

Using $a_n = a_1 + (n-1)d$,

$$a_n = 3 + (n-1)4 = 3 + 4n - 4 = 4n - 1.$$

The general term is $a_n = 4n - 1$.
The tenth term is $a_{10} = 4\cdot10 - 1 = 39$.

The Sum of the First n Terms of an Arithmetic Sequence
The sum, S_n, of the first n terms of an arithmetic sequence is given by

$$S_n = \frac{n}{2}(a_1 + a_n)$$

in which a_1 is the first term and a_n is the nth term.

Find the sum of the first ten terms:

$$2, 5, 8, 11, \ldots.$$

$a_1 = 2$ $\quad$ $d = 5 - 2 = 3$

First find a_{10}, the 10th term. Using $a_n = a_1 + (n-1)d$,

$$a_{10} = 2 + (10-1)\cdot3 = 2 + 9\cdot3 = 29.$$

Find the sum of the first ten terms using

$$S_n = \frac{n}{2}(a_1 + a_n).$$

$$S_{10} = \frac{10}{2}(a_1 + a_{10}) = 5(2 + 29) = 5(31) = 155$$

Definitions and Concepts / Examples

Section 14.3 Geometric Sequences and Series

In a geometric sequence, each term after the first is obtained by multiplying the preceding term by a nonzero constant, the common ratio. Divide any term after the first by the term that directly precedes it to find the common ratio.

General Term of a Geometric Sequence
The nth term (the general term) of a geometric sequence with first term a_1 and common ratio r is

$$a_n = a_1 r^{n-1}.$$

Find the general term and the ninth term:

$$12, -6, 3, -\frac{3}{2}, \ldots.$$

$a_1 = 12$ $\quad$ $r = \frac{-6}{12} = -\frac{1}{2}$

Using $a_n = a_1 r^{n-1}$,

$$a_n = 12\left(-\frac{1}{2}\right)^{n-1} \text{ is the general term.}$$

The ninth term is

$$a_9 = 12\left(-\frac{1}{2}\right)^{9-1} = 12\left(-\frac{1}{2}\right)^{8} = \frac{12}{256} = \frac{3}{64}.$$

The Sum of the First n Terms of a Geometric Sequence
The sum, S_n, of the first n terms of a geometric sequence is given by

$$S_n = \frac{a_1(1 - r^n)}{1 - r}$$

in which a_1 is the first term and r is the common ratio ($r \neq 1$).

Find $\sum_{i=1}^{8} 4 \cdot 3^i$

$$= 4 \cdot 3 + 4 \cdot 3^2 + 4 \cdot 3^3 + \cdots + 4 \cdot 3^8.$$

$a_1 = 12$ $\quad$ $r = \frac{4 \cdot 3^2}{4 \cdot 3} = 3$

Using $S_n = \dfrac{a_1(1 - r^n)}{1 - r}$,

$$S_8 = \frac{12(1 - 3^8)}{1 - 3} = 39{,}360.$$

The Sum of an Infinite Geometric Series
If $-1 < r < 1$ (equivalently, $|r| < 1$), then the sum of the infinite geometric series

$$a_1 + a_1 r + a_1 r^2 + a_1 r^3 + \cdots$$

in which a_1 is the first term and r is the common ratio is given by

$$S = \frac{a_1}{1 - r}.$$

If $|r| \geq 1$, the infinite series does not have a sum.

Find the sum:

$$6 + \frac{6}{3} + \frac{6}{3^2} + \frac{6}{3^3} + \cdots.$$

$a_1 = 6$ $\quad$ $r = \frac{1}{3}$

Using $S = \dfrac{a_1}{1 - r}$, the sum is

$$S = \frac{6}{1 - \frac{1}{3}} = \frac{6}{\frac{2}{3}} = 6 \cdot \frac{3}{2} = 9.$$

Section 14.4 The Binomial Theorem

Definition of a Binomial Coefficient

$$\binom{n}{r} = \frac{n!}{r!(n - r)!}$$

$$\binom{8}{3} = \frac{8!}{3!(8 - 3)!} = \frac{8!}{3!5!}$$

$$= \frac{8 \cdot 7 \cdot 6 \cdot \cancel{5!}}{3 \cdot 2 \cdot 1 \cdot \cancel{5!}} = 56$$

Definitions and Concepts	Examples

Section 14.4 The Binomial Theorem (continued)

A Formula for Expanding Binomials: The Binomial Theorem
For any positive integer n,

$$(a+b)^n = \binom{n}{0}a^n + \binom{n}{1}a^{n-1}b + \binom{n}{2}a^{n-2}b^2 + \binom{n}{3}a^{n-3}b^3 + \cdots + \binom{n}{n}b^n.$$

Expand: $(3x - y)^4 = [3x + (-y)]^4$.

$$= \binom{4}{0}(3x)^4 + \binom{4}{1}(3x)^3(-y) + \binom{4}{2}(3x)^2(-y)^2 + \binom{4}{3}(3x)^1(-y)^3 + \binom{4}{4}(-y)^4$$

$$= 1\cdot 81x^4 + 4\cdot 27x^3(-y) + 6\cdot 9x^2y^2 + 4\cdot 3x(-y^3) + 1\cdot y^4$$

$$= 81x^4 - 108x^3y + 54x^2y^2 - 12xy^3 + y^4$$

Finding a Particular Term in a Binomial Expansion
The $(r+1)$st term in the expansion of $(a+b)^n$ is

$$\binom{n}{r}a^{n-r}b^r.$$

The 8th term, or $(7+1)$st term $(r=7)$, of $(x+2y)^{10}$ is

$$\binom{10}{7}x^{10-7}(2y)^7$$

$$= \binom{10}{7}x^3(2y)^7 = 120x^3\cdot 128y^7$$

$$= 15{,}360x^3y^7.$$

Section 14.5 Counting Principles, Permutations, and Combinations

The Fundamental Counting Principle
The number of ways in which a series of successive things can occur is found by multiplying the number of ways in which each thing can occur.

How many ways can 6 applicants fill three different positions?

$$6\cdot 5\cdot 4 = 120 \text{ ways}$$

Permutations
A permutation from a group of items occurs when no item is used more than once and the order of arrangement makes a difference.
Permutations Formula: The number of possible permutations if r items are taken from n items is ${}_nP_r = \dfrac{n!}{(n-r)!}$.

How many ways can 6 applicants fill three different positions?

$${}_6P_3 = \frac{6!}{(6-3)!} = \frac{6!}{3!} = \frac{6\cdot 5\cdot 4\cdot 3!}{3!}$$

$$= 6\cdot 5\cdot 4 = 120 \text{ ways}$$

Combinations
A combination from a group of items occurs when no item is used more than once and the order of items makes no difference.
Combinations Formula: The number of possible combinations if r items are taken from n items is ${}_nC_r = \dfrac{n!}{(n-r)!r!}$.

How many different sets of 3 books can be selected from 6 books?

$${}_6C_3 = \frac{6!}{(6-3)!3!} = \frac{6!}{3!3!}$$

$$= \frac{6\cdot 5\cdot 4\cdot 3!}{3\cdot 2\cdot 1\cdot 3!} = \frac{6\cdot 5\cdot 4}{3\cdot 2\cdot 1}$$

$$= 20 \text{ ways}$$

Definitions and Concepts	Examples

Section 14.6 Probability

Empirical Probability
Empirical probability applies to situations in which we observe the frequency of occurrence of an event. The empirical probability of event E is

$$P(E) = \frac{\text{observed number of times } E \text{ occurs}}{\text{total number of observed occurrences}}.$$

Teachers in U.S. Catholic High Schools

Total	Religious	Lay
47,730	4149	43,581

Source: National Catholic Education Association

The probability that a randomly selected U.S. Catholic high school teacher belongs to a religious order is

$$\frac{4149}{47{,}730} \approx 0.087.$$

Theoretical Probability
Theoretical probability applies to situations in which the sample space of all equally-likely outcomes is known. The theoretical probability of event E is

$$P(E) = \frac{\text{number of outcomes in event } E}{\text{number of outcomes in sample space } S} = \frac{n(E)}{n(S)}.$$

Probability of an event not occurring: $P(\text{not } E) = 1 - P(E)$.

A die is rolled.

$$S = \{1, 2, 3, 4, 5, 6\}$$

Probability of getting a number greater than 4 $(E = \{5, 6\})$ is $\frac{2}{6} = \frac{1}{3}$.

Probability of not getting a number greater than 4 is $1 - \frac{1}{3} = \frac{2}{3}$.

***Or* Probabilities**
If it is impossible for events A and B to occur simultaneously, the events are mutually exclusive.
If A and B are mutually exclusive events, then
$P(A \text{ or } B) = P(A) + P(B)$.
If A and B are not mutually exclusive events, then
$P(A \text{ or } B) = P(A) + P(B) - P(A \text{ and } B)$.

A die is rolled: $S = \{1, 2, 3, 4, 5, 6\}$.
Probability (2 or 5)

$$= P(2) + P(5) = \frac{1}{6} + \frac{1}{6} = \frac{2}{6} = \frac{1}{3}.$$

Probability (even or greater than 3)

$$= P(\text{even}) + P(>3) - P(\text{even and} > 3)$$

$$= \frac{3}{6} + \frac{3}{6} - \frac{2}{6} = \frac{4}{6} = \frac{2}{3}.$$

This event is {4, 6}.

***And* Probabilities**
Two events are independent if the occurrence of either of them has no effect on the probability of the other.
If A and B are independent events, then

$$P(A \text{ and } B) = P(A) \cdot P(B).$$

The probability of a succession of independent events is the product of each of their probabilities.

A quiz contains six multiple-choice questions. Each question has four answer choices, with one correct answer per question. If you guess at every question, the probability of answering all correctly is

$$\frac{1}{4} \cdot \frac{1}{4} \cdot \frac{1}{4} \cdot \frac{1}{4} = \frac{1}{256}.$$

CHAPTER 14 REVIEW EXERCISES

14.1 *In Exercises 1–4, write the first four terms of each sequence whose general term is given.*

1. $a_n = 7n - 4$

2. $a_n = (-1)^n \dfrac{n+2}{n+1}$

3. $a_n = \dfrac{1}{(n-1)!}$

4. $a_n = \dfrac{(-1)^{n+1}}{2^n}$

In Exercises 5–6, find each indicated sum.

5. $\sum_{i=1}^{5}(2i^2 - 3)$ **6.** $\sum_{i=0}^{4}(-1)^{i+1}i!$

In Exercises 7–8, express each sum using summation notation. Use i for the index of summation.

7. $\frac{1}{3} + \frac{2}{4} + \frac{3}{5} + \cdots + \frac{15}{17}$

8. $4^3 + 5^3 + 6^3 + \cdots + 13^3$

14.2 *In Exercises 9–11, write the first six terms of each arithmetic sequence.*

9. $a_1 = 7, d = 4$

10. $a_1 = -4, d = -5$

11. $a_1 = \frac{3}{2}, d = -\frac{1}{2}$

In Exercises 12–14, use the formula for the general term (the nth term) of an arithmetic sequence to find the indicated term of each sequence.

12. Find a_6 when $a_1 = 5, d = 3$.

13. Find a_{12} when $a_1 = -8, d = -2$.

14. Find a_{14} when $a_1 = 14, d = -4$.

In Exercises 15–18, write a formula for the general term (the nth term) of each arithmetic sequence. Then use the formula for a_n to find a_{20}, the 20th term of the sequence.

15. $-7, -3, 1, 5, \ldots$

16. $a_1 = 200, d = -20$

17. $a_1 = -12, d = -\frac{1}{2}$

18. $15, 8, 1, -6, \ldots$

19. Find the sum of the first 22 terms of the arithmetic sequence: $5, 12, 19, 26, \ldots$.

20. Find the sum of the first 15 terms of the arithmetic sequence: $-6, -3, 0, 3, \ldots$.

21. Find $3 + 6 + 9 + \cdots + 300$, the sum of the first 100 positive multiples of 3.

In Exercises 22–24, use the formula for the sum of the first n terms of an arithmetic sequence to find the indicated sum.

22. $\sum_{i=1}^{16}(3i + 2)$ **23.** $\sum_{i=1}^{25}(-2i + 6)$

24. $\sum_{i=1}^{30}(-5i)$

25. The graph shows the changing pattern of work in the United States from 1900 through 2000.

The Changing Pattern of Work in the United States

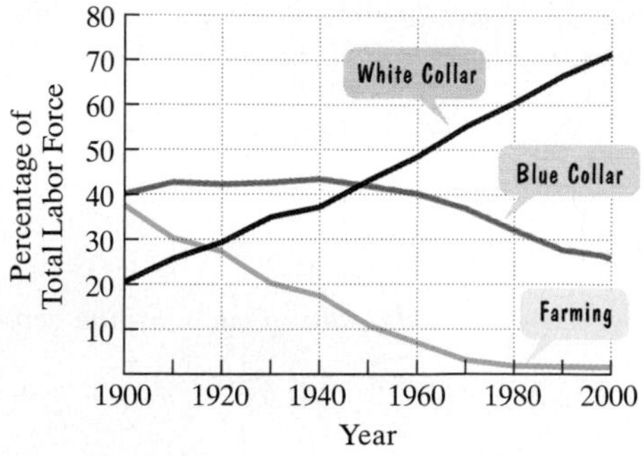

Source: U.S. Department of Labor

In 1900, 20% of the total labor force was comprised of white-collar workers. On average, this increased by approximately 0.52% per year since then.

a. Write a formula for the *n*th term of the arithmetic sequence that describes the percentage of white-collar workers in the labor force *n* years after 1899.

b. Use the model to predict the percentage of white-collar workers in the labor force by the year 2010.

26. A company offers a starting salary of \$31,500 with raises of \$2300 per year. Find the total salary over a ten-year period.

27. A theater has 25 seats in the first row and 35 rows in all. Each successive row contains one additional seat. How many seats are in the theater?

14.3 *In Exercises 28–31, write the first five terms of each geometric sequence.*

28. $a_1 = 3, r = 2$

29. $a_1 = \frac{1}{2}, r = \frac{1}{2}$

30. $a_1 = 16, r = -\frac{1}{4}$

31. $a_1 = -5, r = -1$

In Exercises 32–34, use the formula for the general term (the nth term) of a geometric sequence to find the indicated term of each sequence.

32. Find a_7 when $a_1 = 2, r = 3$.

33. Find a_6 when $a_1 = 16, r = \frac{1}{2}$.

34. Find a_5 when $a_1 = -3, r = 2$.

In Exercises 35–37, write a formula for the general term (the nth term) of each geometric sequence. Then use the formula for a_n to find a_8, the eighth term of the sequence.

35. $1, 2, 4, 8, \ldots$

36. $100, 10, 1, \frac{1}{10}, \ldots$

37. $12, -4, \frac{4}{3}, -\frac{4}{9}, \ldots$

38. Find the sum of the first 15 terms of the geometric sequence: $5, -15, 45, -135, \ldots$.

39. Find the sum of the first 7 terms of the geometric sequence: $8, 4, 2, 1, \ldots$.

In Exercises 40–42, use the formula for the sum of the first n terms of a geometric sequence to find the indicated sum.

40. $\sum_{i=1}^{6} 5^i$

41. $\sum_{i=1}^{7} 3(-2)^i$

42. $\sum_{i=1}^{5} 2\left(\frac{1}{4}\right)^{i-1}$

In Exercises 43–46, find the sum of each infinite geometric series.

43. $9 + 3 + 1 + \frac{1}{3} + \cdots$

44. $2 - 1 + \frac{1}{2} - \frac{1}{4} + \cdots$

45. $-6 + 4 - \frac{8}{3} + \frac{16}{9} - \cdots$

46. $\sum_{i=1}^{\infty} 5(0.8)^i$

In Exercises 47–48, express each repeating decimal as a fraction in lowest terms.

47. $0.\overline{6}$

48. $0.\overline{47}$

49. The population of Iraq from 1998 through 2001 is shown in the following table.

Year	1998	1999	2000	2001
Population in Millions	19.96	20.72	21.51	22.33

Source: U.N. Population Division

a. Show that Iraq's population is increasing geometrically.

b. Write the general term of the geometric sequence describing population growth for Iraq n years after 1997.

c. Estimate Iraq's population, in millions, for the year 2008.

50. A job pays \$32,000 for the first year with an annual increase of 6% per year beginning in the second year. What is the salary in the sixth year? What is the total salary paid over this six-year period? Round answers to the nearest dollar.

51. You decide to deposit \$200 at the end of each month into an account paying 10% interest compounded monthly to save for your child's education. How much will you save over 18 years? Round to the nearest dollar.

52. A factory in an isolated town has an annual payroll of \$4 million. It is estimated that 70% of this money is spent within the town, that people in the town receiving this money will again spend 70% of what they receive in the town, and so on. What is the total of all this spending in the town each year?

14.4 *In Exercises 53–54, evaluate the given binomial coefficient.*

53. $\binom{11}{8}$

54. $\binom{90}{2}$

In Exercises 55–58, use the Binomial Theorem to expand each binomial and express the result in simplified form.

55. $(2x + 1)^3$

56. $(x^2 - 1)^4$

57. $(x + 2y)^5$

58. $(x - 2)^6$

In Exercises 59–60, write the first three terms in each binomial expansion, expressing the result in simplified form.

59. $(x^2 + 3)^8$

60. $(x - 3)^9$

In Exercises 61–62, find the term indicated in each expansion.

61. $(x + 2)^5$; fourth term

62. $(2x - 3)^6$; fifth term

14.5 *In Exercises 63–66, evaluate each expression.*

63. ${}_8P_3$

64. ${}_9P_5$

65. ${}_8C_3$

66. ${}_{13}C_{11}$

In Exercises 67–73, solve by the method of your choice.

67. A popular brand of pen comes in red, green, blue, or black ink. The writing tip can be chosen from extra bold, bold, regular, fine, or micro. How many different choices of pens do you have with this brand?

68. A stock can go up, go down, or stay unchanged. How many possibilities are there if you own five stocks?

69. A club with 15 members is to choose four officers —president, vice-president, secretary, and treasurer. In how many ways can these offices be filled?

70. How many different ways can a director select 4 actors from a group of 20 actors to attend a workshop on performing in rock musicals?

71. From the 20 CDs that you've bought during the past year, you plan to take 3 with you on vacation. How many different sets of three CDs can you take?

72. How many different ways can a director select from 20 male actors and cast the roles of Mark, Roger, Angel, and Collins in the musical *Rent*?

73. In how many ways can five airplanes line up for departure on a runway?

14.6 *The table shows differences in political ideology, by education, for a random sample of U.S. voters. (The ratios for each group's ideologies are sourced from voting patterns in the 2000 U.S. election. The frequencies shown are hypothetical.) Use the data to solve Exercises 74–79. Express probabilities as simplified fractions.*

	Liberal	Moderate	Conservative
High School only	7	35	13
College	10	15	20

Find the probability that a randomly selected person from this group

74. is liberal.

75. is not conservative.

76. is moderate or conservative.

77. is conservative or attended college.

78. Among people with a conservative ideology, find the probability of selecting a person who attended high school only.

79. Among people who attended college, find the probability of selecting a person with a liberal ideology.

In Exercises 80–81, a die is rolled. Find the probability of

80. getting a number less than 5.

81. getting a number less than 3 or greater than 4.

In Exercises 82–83, you are dealt one card from a 52-card deck. Find the probability of

82. getting an ace or a king.

83. getting a queen or a red card.

In Exercises 84–86, it is equally probable that the pointer on the spinner shown will land on any one of the six regions, numbered 1 through 6, and colored as shown. If the pointer lands on a borderline, spin again. Find the probability of

84. not stopping on yellow.

85. stopping on red or a number greater than 3.

86. stopping on green on the first spin and stopping on a number less than 4 on the second spin.

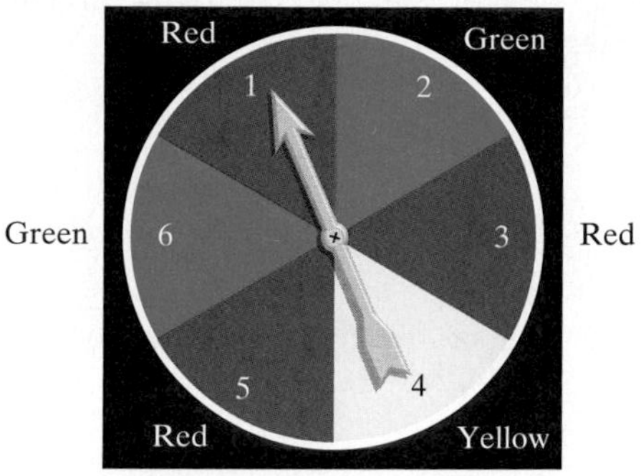

87. A lottery game is set up so that each player chooses five different numbers from 1 to 20. If the five numbers match the five numbers drawn in the lottery, the player wins (or shares) the top cash prize. What is the probability of winning the prize

a. with one lottery ticket?

b. with 100 different lottery tickets?

88. What is the probability of a family having five boys born in a row?

89. The probability of a flood in any given year in a region prone to floods is 0.2.

a. What is the probability of a flood two years in a row?

b. What is the probability of a flood for three consecutive years?

c. What is the probability of no flooding for four consecutive years?

CHAPTER 14 TEST

Remember to use your Chapter Test Prep Video CD to see the worked-out solutions to the test questions you want to review.

1. Write the first five terms of the sequence whose general term is $a_n = \dfrac{(-1)^{n+1}}{n^2}$.

2. Find the indicated sum: $\sum_{i=1}^{5} (i^2 + 10)$.

3. Express the sum using summation notation. Use i for the index of summation.

$$\frac{2}{3} + \frac{3}{4} + \frac{4}{5} + \cdots + \frac{21}{22}$$

In Exercises 4–5, write a formula for the general term (the nth term) of each sequence. Then use the formula to find the twelfth term of the sequence.

4. 4, 9, 14, 19, . . .

5. 16, 4, 1, $\frac{1}{4}$, . . .

In Exercises 6–7, use the formula for the sum of the first n terms of an arithmetic sequence.

6. Find the sum of the first ten terms of the arithmetic sequence: $-7, -14, -21, -28, \ldots$.

7. Find $\sum_{i=1}^{20}(3i - 4)$.

In Exercises 8–9, use the formula for the sum of the first n terms of a geometric sequence.

8. Find the sum of the first ten terms of the geometric sequence: $7, -14, 28, -56, \ldots$.

9. Find $\sum_{i=1}^{15}(-2)^i$.

10. Find the sum of the infinite geometric series:

$$4 + \frac{4}{2} + \frac{4}{2^2} + \frac{4}{2^3} + \cdots.$$

11. Express $0.\overline{73}$ in fractional notation.

12. A job pays $30,000 for the first year with an annual increase of 4% per year beginning in the second year. What is the total salary paid over an eight-year period? Round to the nearest dollar.

13. Evaluate: $\binom{9}{2}$.

14. Use the Binomial Theorem to expand and simplify: $(x^2 - 1)^5$.

15. Use the Binomial Theorem to write the first three terms in the expansion and simplify: $(x + y^2)^8$.

16. A human resource manager has 11 applicants to fill three different positions. Assuming that all applicants are equally qualified for any of the three positions, in how many ways can this be done?

17. From the ten books that you've recently bought but not read, you plan to take four with you on vacation. How many different sets of four books can you take?

18. How many seven-digit local telephone numbers can be formed if the first three digits are 279?

A class is collecting data on eye color and gender. They organize the data they collected into the table shown. Numbers in the table represent the number of students from the class that belong to each of the categories. Use the data to solve Exercises 19–22. Express probabilities as simplified fractions.

	Brown	Blue	Green
Male	22	18	10
Female	18	20	12

Find the probability that a randomly selected student from this class

19. does not have brown eyes.

20. has brown eyes or blue eyes.

21. is female or has green eyes.

22. Among the students with blue eyes, find the probability of selecting a male.

23 A lottery game is set up so that each player chooses six different numbers from 1 to 15. If the six numbers match the six numbers drawn in the lottery, the player wins (or shares) the top cash prize. What is the probability of winning the prize with 50 different lottery tickets?

24. One card is randomly selected from a deck of 52 cards. Find the probability of selecting a black card or a picture card.

25. A group of students consists of 10 male freshmen, 15 female freshmen, 20 male sophomores, and 5 female sophomores. If one person is randomly selected from the group, find the probability of selecting a freshman or a female.

26 A quiz consisting of four multiple-choice questions has four answer choices for each question, with one correct answer per question. If a person guesses at every question, what is the probability of answering *all* questions correctly?

27. If the spinner shown is spun twice, find the probability that the pointer lands on red on the first spin and blue on the second spin.

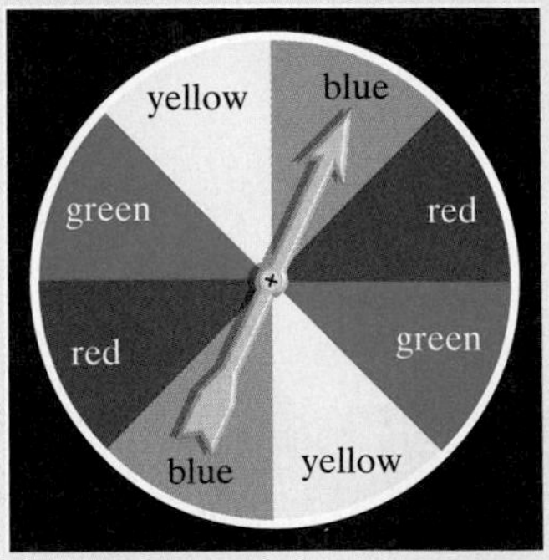

CUMULATIVE REVIEW EXERCISES (CHAPTERS 1–14)

In Exercises 1–10, solve each equation, inequality, or system.

1. $\sqrt{2x+5} - \sqrt{x+3} = 2$

2. $(x-5)^2 = -49$

3. $x^2 + x > 6$

4. $6x - 3(5x+2) = 4(1-x)$

5. $\dfrac{2}{x-3} - \dfrac{3}{x+3} = \dfrac{12}{x^2-9}$

6. $3x + 2 < 4$ and $4 - x > 1$

7. $\begin{aligned} 3x - 2y + z &= 7 \\ 2x + 3y - z &= 13 \\ x - y + 2z &= -6 \end{aligned}$

8. $\log_9 x + \log_9(x-8) = 1$

9. $\begin{aligned} 2x^2 - 3y^2 &= 5 \\ 3x^2 + 4y^2 &= 16 \end{aligned}$

10. $\begin{aligned} 2x^2 - y^2 &= -8 \\ x - y &= 6 \end{aligned}$

In Exercises 11–15, graph each function, equation or inequality in a rectangular coordinate system.

11. $f(x) = (x+2)^2 - 4$

12. $y < -3x + 5$

13. $f(x) = 3^{x-2}$

14. $\dfrac{x^2}{16} + \dfrac{y^2}{4} = 1$

15. $x^2 - y^2 = 9$

In Exercises 16–19, perform the indicated operations and simplify, if possible.

16. $\dfrac{2x+1}{x-5} - \dfrac{4}{x^2-3x-10}$

17. $\dfrac{\dfrac{1}{x-1}+1}{\dfrac{1}{x+1}-1}$

18. $\dfrac{6}{\sqrt{5}-\sqrt{2}}$

19. $8\sqrt{45} + 2\sqrt{5} - 7\sqrt{20}$

20. Rationalize the denominator: $\dfrac{5}{\sqrt[3]{2x^2y}}$.

21. Factor completely: $5ax + 5ay - 4bx - 4by$.

22. Write as a single logarithm: $5 \log x - \dfrac{1}{2}\log y$.

23. Solve for p: $\dfrac{1}{p} + \dfrac{1}{q} = \dfrac{1}{f}$.

24. Find the distance between $(6, -1)$ and $(-3, -4)$. Round to two decimal places.

25. Find the indicated sum: $\displaystyle\sum_{i=2}^{5} (i^3 - 4)$.

26. Find the sum of the first 30 terms of the arithmetic sequence: 2, 6, 10, 14, ...

27. Express $0.\overline{3}$ as a fraction in lowest terms.

28. Use the Binomial Theorem to expand and simplify: $(2x - y^3)^4$.

In Exercises 29–31, find the domain of each function.

29. $f(x) = \dfrac{2}{x^2+2x-15}$

30. $f(x) = \sqrt{2x-6}$

31. $f(x) = \ln(1-x)$

32. The length of a rectangular garden is 2 feet more than twice its width. If 22 feet of fencing is needed to enclose the garden, what are its dimensions?

33. With a 6% raise, you will earn \$19,610 annually. What is your salary before this raise?

34. The function $F(t) = 1 - k\ln(t+1)$ models the fraction of people, $F(t)$, who remember all the words in a list of nonsense words t hours after memorizing them. After 3 hours, only half the people could remember all the words. Determine the value of k and then predict the fraction of people in the group who will remember all the words after 6 hours. Round to three decimal places and then express the fraction with a denominator of 1000.

APPENDIX A

Objectives

1 Write the augmented matrix for a linear system.

2 Perform matrix row operations.

3 Use matrices to solve systems.

4 Use matrices to identify inconsistent and dependent systems.

MATRIX SOLUTIONS TO LINEAR SYSTEMS

Yes, we overindulged, but it was delicious. Anyway, a few hours of moderate activity and we'll just burn off those extra calories. The chart below should help. It shows that the number of calories burned per hour depends on our weight. Four hours of tennis and we'll be as good as new!

How Fast You Burn Off Calories

	Weight (pounds)					
	110	**132**	**154**	**176**	**187**	**209**
Activity	**Calories Burned per Hour**					
Housework	175	210	245	285	300	320
Cycling	190	215	245	270	280	295
Tennis	335	380	425	470	495	520
Watching TV	60	70	80	85	90	95

The 24 numbers inside the red brackets are arranged in four rows and six columns. This rectangular array of 24 numbers, arranged in rows and columns and placed in brackets, is an example of a **matrix** (plural: **matrices**). The numbers inside the brackets are called **elements** of the matrix. Matrices are used to display information and to solve systems of linear equations.

1 Write the augmented matrix for a linear system.

Solving Linear Systems by Using Matrices A matrix gives us a shortened way of writing a system of equations. The first step in solving a system of linear equations using matrices is to write the *augmented matrix*. An **augmented matrix** has a vertical bar separating the columns of the matrix into two groups. The coefficients of each variable are placed to the left of the vertical line and the constants are placed to the right. If any variable is missing, its coefficient is 0. Here are two examples:

System of Linear Equations	**Augmented Matrix**
$\begin{aligned} x + 3y &= 5 \\ 2x - y &= -4 \end{aligned}$	$\left[\begin{array}{cc\|c} 1 & 3 & 5 \\ 2 & -1 & -4 \end{array}\right]$
$\begin{aligned} x + 2y - 5z &= -19 \\ y + 3z &= 9 \\ z &= 4 \end{aligned}$	$\left[\begin{array}{ccc\|c} 1 & 2 & -5 & -19 \\ 0 & 1 & 3 & 9 \\ 0 & 0 & 1 & 4 \end{array}\right]$

Notice how the second augmented matrix contains 1s down the diagonal from upper left to lower right and 0s below the 1s. This arrangement makes it easy to find the solution of the system of equations, as Example 1 shows.

EXAMPLE 1 Solving a System Using a Matrix

Write the solution for a system of equations represented by the matrix

$$\left[\begin{array}{ccc|c} 1 & 2 & -5 & -19 \\ 0 & 1 & 3 & 9 \\ 0 & 0 & 1 & 4 \end{array}\right].$$

SOLUTION The system represented by the given matrix is

$$\left[\begin{array}{ccc|c} 1 & 2 & -5 & -19 \\ 0 & 1 & 3 & 9 \\ 0 & 0 & 1 & 4 \end{array}\right] \rightarrow \begin{aligned} 1x + 2y - 5z &= -19 \\ 0x + 1y + 3z &= 9 \\ 0x + 0y + 1z &= 4 \end{aligned}.$$

This system can be simplified as follows.

$$\begin{aligned} x + 2y - 5z &= -19 && \text{Equation 1} \\ y + 3z &= 9 && \text{Equation 2} \\ z &= 4 && \text{Equation 3} \end{aligned}$$

The value of z is known. We can find y by back-substitution.

$$\begin{aligned} y + 3z &= 9 && \text{Equation 2} \\ y + 3(4) &= 9 && \text{Substitute 4 for } z. \\ y + 12 &= 9 && \text{Multiply.} \\ y &= -3 && \text{Subtract 12 from both sides.} \end{aligned}$$

With values for y and z, we can now use back-substitution to find x.

$$\begin{aligned} x + 2y - 5z &= -19 && \text{Equation 1} \\ x + 2(-3) - 5(4) &= -19 && \text{Substitute } -3 \text{ for } y \text{ and 4 for } z. \\ x - 6 - 20 &= -19 && \text{Multiply.} \\ x - 26 &= -19 && \text{Add.} \\ x &= 7 && \text{Add 26 to both sides.} \end{aligned}$$

We see that $x = 7$, $y = -3$, and $z = 4$. The solution is $(7, -3, 4)$ and the solution set for the system is $\{(7, -3, 4)\}$. ■

CHECK POINT 1 Write the solution for a system of equations represented by the matrix

$$\left[\begin{array}{ccc|c} 1 & -1 & 1 & 8 \\ 0 & 1 & -12 & -15 \\ 0 & 0 & 1 & 1 \end{array}\right].$$

Our goal in solving a linear system using matrices is to produce a matrix with 1s down the diagonal from upper left to lower right, called the **main diagonal**, and 0s below the 1s. In general, the matrix will be one of the following forms.

This is the desired form for systems with two equations. $\left[\begin{array}{cc|c} 1 & a & b \\ 0 & 1 & c \end{array}\right]$

$\left[\begin{array}{ccc|c} 1 & a & b & c \\ 0 & 1 & d & e \\ 0 & 0 & 1 & f \end{array}\right]$ This is the desired form for systems with three equations.

The last row of these matrices gives us the value of one variable. The other variables can then be found by back-substitution.

2 Perform matrix row operations.

A matrix with 1s down the main diagonal and 0s below the 1s is said to be in **row-echelon form**. How do we produce a matrix in this form? We use **row operations** on the augmented matrix. These row operations are just like what you did when solving a linear system by the addition method. The difference is that we no longer write the variables, usually represented by x, y, and z.

MATRIX ROW OPERATIONS The following row operations produce matrices that represent systems with the same solution:

1. Two rows of a matrix may be interchanged. This is the same as interchanging two equations in the linear system.
2. The elements in any row may be multiplied by a nonzero number. This is the same as multiplying both sides of an equation by a nonzero number.
3. The elements in any row may be multiplied by a nonzero number, and these products may be added to the corresponding elements in any other row. This is the same as multiplying an equation by a nonzero number and then adding equations to eliminate a variable.

Two matrices are **row equivalent** if one can be obtained from the other by a sequence of row operations.

STUDY TIP

When performing the row operation

$$kR_i + R_j$$

you use row i to find the products. However, **elements in row i do not change. It is the elements in row j that change:** Add k times the elements in row i to the corresponding elements in row j. Replace elements in row j by these sums.

Each matrix row operation in the preceding box can be expressed symbolically as follows:

1. Interchange the elements in the ith and jth rows: $R_i \leftrightarrow R_j$.
2. Multiply each element in the ith row by k: kR_i.
3. Add k times the elements in row i to the corresponding elements in row j: $kR_i + R_j$.

EXAMPLE 2 Performing Matrix Row Operations

Use the matrix

$$\left[\begin{array}{rrr|r} 3 & 18 & -12 & 21 \\ 1 & 2 & -3 & 5 \\ -2 & -3 & 4 & -6 \end{array}\right]$$

and perform each indicated row operation:

a. $R_1 \leftrightarrow R_2$ **b.** $\frac{1}{3}R_1$ **c.** $2R_2 + R_3$.

STUDY TIP

As you read the solution, keep looking back at the given matrix

$$\left[\begin{array}{rrr|r} 3 & 18 & -12 & 21 \\ 1 & 2 & -3 & 5 \\ -2 & -3 & 4 & -6 \end{array}\right].$$

SOLUTION

a. The notation $R_1 \leftrightarrow R_2$ means to interchange the elements in row 1 and row 2. This results in the row-equivalent matrix

$$\left[\begin{array}{rrr|r} 1 & 2 & -3 & 5 \\ 3 & 18 & -12 & 21 \\ -2 & -3 & 4 & -6 \end{array}\right].$$

This was row 2; now it's row 1.

This was row 1; now it's row 2.

b. The notation $\frac{1}{3}R_1$ means to multiply each element in row 1 by $\frac{1}{3}$. This results in the row-equivalent matrix

$$\left[\begin{array}{ccc|c} \frac{1}{3}(3) & \frac{1}{3}(18) & \frac{1}{3}(-12) & \frac{1}{3}(21) \\ 1 & 2 & -3 & 5 \\ -2 & -3 & 4 & -6 \end{array}\right] = \left[\begin{array}{rrr|r} 1 & 6 & -4 & 7 \\ 1 & 2 & -3 & 5 \\ -2 & -3 & 4 & -6 \end{array}\right].$$

c. The notation $2R_2 + R_3$ means to add 2 times the elements in row 2 to the corresponding elements in row 3. Replace the elements in row 3 by these sums. First, we find 2 times the elements in row 2, namely, 1, 2, −3 and 5:

$$2(1) \text{ or } 2, \qquad 2(2) \text{ or } 4, \qquad 2(-3) \text{ or } -6, \qquad 2(5) \text{ or } 10.$$

Now we add these products to the corresponding elements in row 3. Although we use row 2 to find the products, row 2 does not change. It is the elements in row 3 that change, resulting in the row-equivalent matrix

$$\left[\begin{array}{ccc|c} 3 & 18 & -12 & 21 \\ 1 & 2 & -3 & 5 \\ -2+2=0 & -3+4=1 & 4+(-6)=-2 & -6+10=4 \end{array}\right] = \left[\begin{array}{ccc|c} 3 & 18 & -12 & 21 \\ 1 & 2 & -3 & 5 \\ 0 & 1 & -2 & 4 \end{array}\right].$$

■

CHECK POINT 2 Use the matrix

$$\left[\begin{array}{ccc|c} 4 & 12 & -20 & 8 \\ 1 & 6 & -3 & 7 \\ -3 & -2 & 1 & -9 \end{array}\right]$$

and perform each indicated row operation:

a. $R_1 \leftrightarrow R_2$

b. $\frac{1}{4}R_1$

c. $3R_2 + R_3$.

3 Use matrices to solve systems.

The process that we use to solve linear systems using matrix row operations is often called **Gaussian elimination**, after the German mathematician Carl Friedrich Gauss (1777–1855). Here are the steps used in solving systems with matrices:

SOLVING LINEAR SYSTEMS USING MATRICES

1. Write the augmented matrix for the system.
2. Use matrix row operations to simplify the matrix to one with 1s down the main diagonal from upper left to lower right, and 0s below the 1s.
3. Write the system of linear equations corresponding to the matrix in step 2 and use back-substitution to find the system's solution.

EXAMPLE 3 Using Matrices to Solve a Linear System

Use matrices to solve the system:

$$\begin{aligned} 4x - 3y &= -15 \\ x + 2y &= -1. \end{aligned}$$

SOLUTION

Step 1. Write the augmented matrix for the system.

Linear System	**Augmented Matrix**
$\begin{aligned} 4x - 3y &= -15 \\ x + 2y &= -1 \end{aligned}$	$\left[\begin{array}{cc\|c} 4 & -3 & -15 \\ 1 & 2 & -1 \end{array}\right]$

Step 2. **Use matrix row operations to simplify the matrix to one with 1s down the diagonal from upper left to lower right, and 0s below the 1s.** Our goal is to obtain a matrix of the form

$$\left[\begin{array}{cc|c} 1 & a & b \\ 0 & 1 & c \end{array}\right].$$

Our first step in achieving this goal is to get 1 in the top position of the first column.

We want 1 in this position.
$$\left[\begin{array}{cc|c} 4 & -3 & -15 \\ 1 & 2 & -1 \end{array}\right]$$

To get 1 in this position, we interchange rows 1 and 2: $R_1 \leftrightarrow R_2$.

$$\left[\begin{array}{cc|c} 1 & 2 & -1 \\ 4 & -3 & -15 \end{array}\right]$$
This was row 2; now it's row 1.
This was row 1; now it's row 2.

Now we want 0 below the 1 in the first column.

We want 0 in this position.
$$\left[\begin{array}{cc|c} 1 & 2 & -1 \\ 4 & -3 & -15 \end{array}\right]$$

Let's get a 0 where there is now a 4. If we multiply the top row of numbers by -4 and add these products to the second row of numbers, we will get 0 in this position: $-4R_1 + R_2$. *We change only row 2.*

$-4R_1 + R_2$
$$\left[\begin{array}{cc|c} 1 & 2 & -1 \\ -4(1)+4 & -4(2)+(-3) & -4(-1)+(-15) \end{array}\right] = \left[\begin{array}{cc|c} 1 & 2 & -1 \\ 0 & -11 & -11 \end{array}\right]$$
row 1 element, row 2 element, row 1 element, row 2 element, row 1 element, row 2 element

We move on to the second column. We want 1 in the second row, second column.

We want 1 in this position.
$$\left[\begin{array}{cc|c} 1 & 2 & -1 \\ 0 & -11 & -11 \end{array}\right]$$

To get 1 in the desired position, we multiply -11 by its multiplicative inverse, or reciprocal, $-\frac{1}{11}$. Therefore, we multiply all the numbers in the second row by $-\frac{1}{11}$: $-\frac{1}{11}R_2$.

$-\frac{1}{11}R_2$
$$\left[\begin{array}{cc|c} 1 & 2 & -1 \\ -\frac{1}{11}(0) & -\frac{1}{11}(-11) & -\frac{1}{11}(-11) \end{array}\right] = \left[\begin{array}{cc|c} 1 & 2 & -1 \\ 0 & 1 & 1 \end{array}\right]$$

We now have the desired matrix with 1s down the main diagonal and 0 below the 1.

Step 3. **Write the system of linear equations corresponding to the matrix in step 2 and use back-substitution to find the system's solution.** The system represented by the matrix in step 2 is

$$\left[\begin{array}{cc|c} 1 & 2 & -1 \\ 0 & 1 & 1 \end{array}\right] \rightarrow \begin{aligned} 1x + 2y &= -1 \\ 0x + 1y &= 1 \end{aligned} \quad \text{or} \quad \begin{aligned} x + 2y &= -1 \\ y &= 1. \end{aligned}$$

We immediately see that the value for y is 1. To find x, we back-substitute 1 for y in the first equation.

$$x + 2y = -1 \quad \text{Equation 1}$$
$$x + 2 \cdot 1 = -1 \quad \text{Substitute 1 for } y.$$
$$x = -3 \quad \text{Solve for } x.$$

With $x = -3$ and $y = 1$, the proposed solution is $(-3, 1)$. Take a moment to show that $(-3, 1)$ satisfies both equations. The solution is $(-3, 1)$ and the solution set is $\{(-3, 1)\}$. ■

CHECK POINT 3 Use matrices to solve the system:

$$\begin{aligned} 2x - y &= -4 \\ x + 3y &= 5. \end{aligned}$$

When solving a system of linear equations in three variables, use matrix row operations in the following order to obtain a matrix with 1s down the main diagonal and 0s below the 1s:

$$\left[\begin{array}{ccc|c} 1 & * & * & * \\ * & * & * & * \\ * & * & * & * \end{array}\right] \rightarrow \left[\begin{array}{ccc|c} 1 & * & * & * \\ 0 & * & * & * \\ 0 & * & * & * \end{array}\right] \rightarrow \left[\begin{array}{ccc|c} 1 & * & * & * \\ 0 & 1 & * & * \\ 0 & * & * & * \end{array}\right] \rightarrow \left[\begin{array}{ccc|c} 1 & * & * & * \\ 0 & 1 & * & * \\ 0 & 0 & * & * \end{array}\right] \rightarrow \left[\begin{array}{ccc|c} 1 & * & * & * \\ 0 & 1 & * & * \\ 0 & 0 & 1 & * \end{array}\right]$$

Get 1 in the upper left-hand corner.

Use the 1 in the first column to get 0s below it.

Get 1 in the second row, second column position.

Use the 1 in the second column to get 0 below it.

Get 1 in the third row, third column position.

EXAMPLE 4 Using Matrices to Solve a Linear System

Use matrices to solve the system:

$$\begin{aligned} 3x + y + 2z &= 31 \\ x + y + 2z &= 19 \\ x + 3y + 2z &= 25. \end{aligned}$$

SOLUTION

Step 1. **Write the augmented matrix for the system.**

Linear System

$$\begin{aligned} 3x + y + 2z &= 31 \\ x + y + 2z &= 19 \\ x + 3y + 2z &= 25 \end{aligned}$$

Augmented Matrix

$$\left[\begin{array}{ccc|c} 3 & 1 & 2 & 31 \\ 1 & 1 & 2 & 19 \\ 1 & 3 & 2 & 25 \end{array}\right]$$

Step 2. **Use matrix row operations to simplify the matrix to one with 1s down the diagonal from upper left to lower right, and 0s below the 1s.** Our first step in achieving this goal is to get 1 in the top position of the first column.

We want 1 in this position.

$$\left[\begin{array}{ccc|c} 3 & 1 & 2 & 31 \\ 1 & 1 & 2 & 19 \\ 1 & 3 & 2 & 25 \end{array}\right]$$

To get 1 in this position, we interchange rows 1 and 2: $R_1 \leftrightarrow R_2$. (We could also interchange rows 1 and 3 to attain our goal.)

$$\left[\begin{array}{ccc|c} 1 & 1 & 2 & 19 \\ 3 & 1 & 2 & 31 \\ 1 & 3 & 2 & 25 \end{array}\right]$$

This was row 2; now it's row 1.

This was row 1; now it's row 2.

Now we want to get 0s below the 1 in the first column.

We want 0 in these positions.

$$\left[\begin{array}{ccc|c} 1 & 1 & 2 & 19 \\ 3 & 1 & 2 & 31 \\ 1 & 3 & 2 & 25 \end{array}\right]$$

To get a 0 where there is now a 3, multiply the top row of numbers by -3 and add these products to the second row of numbers: $-3R_1 + R_2$. To get a 0 where there is now a 1, multiply the top row of numbers by -1 and add these products to the third row of numbers: $-1R_1 + R_3$. Although we are using row 1 to find the products, the numbers in row 1 do not change.

$-3R_1 + R_2$, $-1R_1 + R_3$

$$\left[\begin{array}{ccc|c} 1 & 1 & 2 & 19 \\ -3(1)+3 & -3(1)+1 & -3(2)+2 & -3(19)+31 \\ -1(1)+1 & -1(1)+3 & -1(2)+2 & -1(19)+25 \end{array}\right] = \left[\begin{array}{ccc|c} 1 & 1 & 2 & 19 \\ 0 & -2 & -4 & -26 \\ 0 & 2 & 0 & 6 \end{array}\right]$$

We move on to the second column. We want 1 in the second row, second column.

We want 1 in this position.

$$\left[\begin{array}{ccc|c} 1 & 1 & 2 & 19 \\ 0 & -2 & -4 & -26 \\ 0 & 2 & 0 & 6 \end{array}\right]$$

To get 1 in the desired position, we multiply -2 by its reciprocal, $-\frac{1}{2}$. Therefore, we multiply all the numbers in the second row by $-\frac{1}{2}$: $-\frac{1}{2}R_2$.

$-\frac{1}{2}R_2$

$$\left[\begin{array}{ccc|c} 1 & 1 & 2 & 19 \\ -\frac{1}{2}(0) & -\frac{1}{2}(-2) & -\frac{1}{2}(-4) & -\frac{1}{2}(-26) \\ 0 & 2 & 0 & 6 \end{array}\right] = \left[\begin{array}{ccc|c} 1 & 1 & 2 & 19 \\ 0 & 1 & 2 & 13 \\ 0 & 2 & 0 & 6 \end{array}\right].$$

We want 0 in this position.

We are not yet done with the second column. The voice balloon shows that we want to get a 0 where there is now a 2. If we multiply the second row of numbers by -2 and add these products to the third row of numbers, we will get 0 in this position: $-2R_2 + R_3$. Although we are using the numbers in row 2 to find the products, the numbers in row 2 do not change.

$-2R_2 + R_3$

$$\left[\begin{array}{ccc|c} 1 & 1 & 2 & 19 \\ 0 & 1 & 2 & 13 \\ -2(0)+0 & -2(1)+2 & -2(2)+0 & -2(13)+6 \end{array}\right] = \left[\begin{array}{ccc|c} 1 & 1 & 2 & 19 \\ 0 & 1 & 2 & 13 \\ 0 & 0 & -4 & -20 \end{array}\right]$$

We move on to the third column. We want 1 in the third row, third column.

We want 1 in this position.

$$\left[\begin{array}{ccc|c} 1 & 1 & 2 & 19 \\ 0 & 1 & 2 & 13 \\ 0 & 0 & -4 & -20 \end{array}\right]$$

To get 1 in the desired position, we multiply -4 by its reciprocal, $-\frac{1}{4}$. Therefore, we multiply all the numbers in the third row by $-\frac{1}{4}$: $-\frac{1}{4}R_3$.

$-\frac{1}{4}R_3$

$$\left[\begin{array}{ccc|c} 1 & 1 & 2 & 19 \\ 0 & 1 & 2 & 13 \\ -\frac{1}{4}(0) & -\frac{1}{4}(0) & -\frac{1}{4}(-4) & -\frac{1}{4}(-20) \end{array}\right] = \left[\begin{array}{ccc|c} 1 & 1 & 2 & 19 \\ 0 & 1 & 2 & 13 \\ 0 & 0 & 1 & 5 \end{array}\right].$$

We now have the desired matrix with 1s down the main diagonal and 0s below the 1s.

USING TECHNOLOGY

Most graphing utilities can convert an augmented matrix to one with 1s down the diagonal from upper left to lower right, and 0s below the 1s. Recall that this is called row-echelon form. However, row-echelon form is not unique. Your graphing utility might give a row-echelon form different from the one you obtained by hand. However, all row-echelon forms for a given system's augmented matrix produce the same solution to the system. Enter the augmented matrix and name it A. Then use the REF (row-echelon form) command on matrix A.

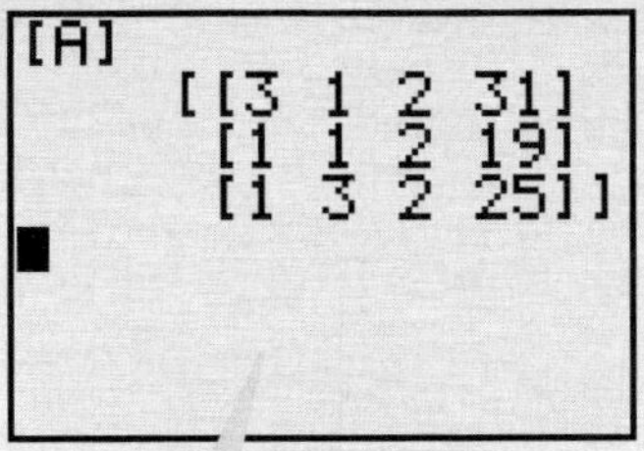

This is the augmented matrix in Example 4

This matrix is equivalent to the row-echelon form we obtained by hand. It produces the same solution: $z = 5$, $y = 3$, and $x = 6$.

Step 3. Write the system of linear equations corresponding to the matrix in step 2 and use back-substitution to find the system's solution. The system represented by the matrix in step 2 is

$$\left[\begin{array}{ccc|c} 1 & 1 & 2 & 19 \\ 0 & 1 & 2 & 13 \\ 0 & 0 & 1 & 5 \end{array}\right] \rightarrow \begin{aligned} 1x + 1y + 2z &= 19 \\ 0x + 1y + 2z &= 13 \\ 0x + 0y + 1z &= 5 \end{aligned} \quad \text{or} \quad \begin{aligned} x + y + 2z &= 19 \quad (1) \\ y + 2z &= 13. \quad (2) \\ z &= 5 \quad (3) \end{aligned}$$

We immediately see from equation (3) that the value for z is 5. To find y, we back-substitute 5 for z in the second equation.

$$\begin{aligned} y + 2z &= 13 && \text{Equation (2)} \\ y + 2(5) &= 13 && \text{Substitute 5 for } z. \\ y &= 3 && \text{Solve for } y. \end{aligned}$$

Finally, back-substitute 3 for y and 5 for z in the first equation.

$$\begin{aligned} x + y + 2z &= 19 && \text{Equation (1)} \\ x + 3 + 2(5) &= 19 && \text{Substitute 3 for } y \text{ and 5 for } z. \\ x + 13 &= 19 && \text{Multiply and add.} \\ x &= 6 && \text{Subtract 13 both sides.} \end{aligned}$$

The solution of the original system is $(6, 3, 5)$ and the solution set is $\{(6, 3, 5)\}$. Check to see that the solution satisfies all three equations in the given system. ■

CHECK POINT 4 Use matrices to solve the system:

$$\begin{aligned} 2x + y + 2z &= 18 \\ x - y + 2z &= 9 \\ x + 2y - z &= 6. \end{aligned}$$

Modern supercomputers are capable of solving systems with more than 600,000 variables. The augmented matrices for such systems are huge, but the solution using matrices is exactly like what we did in Example 4. Work with the augmented matrix, one column at a time. First, get 1 in the desired position. Then get 0s below the 1.

4 Use matrices to identify inconsistent and dependent systems.

Inconsistent Systems and Systems with Dependent Equations When solving a system using matrices, you might obtain a matrix with a row in which the numbers to the left of the vertical bar are all zeros, but a nonzero number number appears on the right. In such a case, the system is inconsistent and has no solution. For example, a system of equations that yields the following matrix is an inconsistent system:

$$\left[\begin{array}{cc|c} 1 & -2 & 3 \\ 0 & 0 & -4 \end{array}\right].$$

The second row of the matrix represents the equation $0x + 0y = -4$, which is false for all values of x and y.

If you obtain a matrix in which a 0 appears across an entire row, the system contains dependent equations and has infinitely many solutions. This row of zeros represents $0x + 0y = 0$ or $0x + 0y + 0z = 0$. These equations are satisfied by infinitely many ordered pairs or triples.

A EXERCISE SET

Student Solutions Manual | CD/Video | PH Math/Tutor Center | MathXL Tutorials on CD | MathXL® | MyMathLab | Interactmath.com

Practice Exercises

In Exercises 1–8, write the system of linear equations represented by the augmented matrix. Use x, y, and, if necessary, z, for the variables. Once the system is written, use back-substitution to find its solution.

1. $\left[\begin{array}{cc|c} 1 & -3 & 11 \\ 0 & 1 & -3 \end{array}\right]$

2. $\left[\begin{array}{cc|c} 1 & 3 & 5 \\ 0 & 1 & 2 \end{array}\right]$

3. $\left[\begin{array}{cc|c} 1 & -3 & 1 \\ 0 & 1 & -1 \end{array}\right]$

4. $\left[\begin{array}{cc|c} 1 & 2 & 13 \\ 0 & 1 & 4 \end{array}\right]$

5. $\left[\begin{array}{ccc|c} 1 & 0 & -4 & 5 \\ 0 & 1 & -12 & 13 \\ 0 & 0 & 1 & -\frac{1}{2} \end{array}\right]$

6. $\left[\begin{array}{ccc|c} 1 & 2 & 1 & 0 \\ 0 & 1 & 0 & -2 \\ 0 & 0 & 1 & 3 \end{array}\right]$

7. $\left[\begin{array}{ccc|c} 1 & \frac{1}{2} & 1 & \frac{11}{2} \\ 0 & 1 & \frac{3}{2} & 7 \\ 0 & 0 & 1 & 4 \end{array}\right]$

8. $\left[\begin{array}{ccc|c} 1 & 1 & 0 & 3 \\ 0 & 1 & \frac{3}{2} & -2 \\ 0 & 0 & 1 & 0 \end{array}\right]$

In Exercises 9–22, perform each matrix row operation and write the new matrix.

9. $\left[\begin{array}{cc|c} 2 & 2 & 5 \\ 1 & -\frac{3}{2} & 5 \end{array}\right] R_1 \leftrightarrow R_2$

10. $\left[\begin{array}{cc|c} -6 & 9 & 4 \\ 1 & -\frac{3}{2} & 4 \end{array}\right] R_1 \leftrightarrow R_2$

11. $\left[\begin{array}{cc|c} -6 & 8 & -12 \\ 3 & 5 & -2 \end{array}\right] -\frac{1}{6}R_1$

12. $\left[\begin{array}{cc|c} -2 & 3 & -10 \\ 4 & 2 & 5 \end{array}\right] -\frac{1}{2}R_1$

13. $\left[\begin{array}{cc|c} 1 & -3 & 5 \\ 2 & 6 & 4 \end{array}\right] -2R_1 + R_2$

14. $\left[\begin{array}{cc|c} 1 & -3 & 1 \\ 2 & 1 & -5 \end{array}\right] -2R_1 + R_2$

15. $\left[\begin{array}{cc|c} 1 & -\frac{3}{2} & \frac{7}{2} \\ 3 & 4 & 2 \end{array}\right] -3R_1 + R_2$

16. $\left[\begin{array}{cc|c} 1 & -\frac{2}{5} & \frac{3}{4} \\ 4 & 2 & -1 \end{array}\right] -4R_1 + R_2$

17. $\left[\begin{array}{ccc|c} 2 & -6 & 4 & 10 \\ 1 & 5 & -5 & 0 \\ 3 & 0 & 4 & 7 \end{array}\right] \frac{1}{2}R_1$

18. $\left[\begin{array}{ccc|c} 3 & -12 & 6 & 9 \\ 1 & -4 & 4 & 0 \\ 2 & 0 & 7 & 4 \end{array}\right] \frac{1}{3}R_1$

19. $\left[\begin{array}{ccc|c} 1 & -3 & 2 & 0 \\ 3 & 1 & -1 & 7 \\ 2 & -2 & 1 & 3 \end{array}\right] -3R_1 + R_2$

20. $\left[\begin{array}{ccc|c} 1 & -1 & 5 & -6 \\ 3 & 3 & -1 & 10 \\ 1 & 3 & 2 & 5 \end{array}\right] -3R_1 + R_2$

21. $\left[\begin{array}{ccc|c} 1 & 1 & -1 & 6 \\ 2 & -1 & 1 & -3 \\ 3 & -1 & -1 & 4 \end{array}\right] \begin{array}{c} -2R_1 + R_2 \\ \text{and} \\ -3R_1 + R_3 \end{array}$

22. $\left[\begin{array}{ccc|c} 1 & 2 & 1 & 2 \\ -2 & -1 & 2 & 5 \\ 1 & 3 & -2 & -8 \end{array}\right] \begin{array}{c} 2R_1 + R_2 \\ \text{and} \\ -1R_1 + R_3 \end{array}$

In Exercises 23–46, solve each system using matrices. If there is no solution or if there are infinitely many solutions and a system's equations are dependent, so state.

23. $x + y = 6$
$x - y = 2$

24. $x + 2y = 11$
$x - y = -1$

25. $2x + y = 3$
$x - 3y = 12$

26. $3x - 5y = 7$
$x - y = 1$

27. $5x + 7y = -25$
$11x + 6y = -8$

28. $3x - 5y = 22$
$4x - 2y = 20$

29. $4x - 2y = 5$
$-2x + y = 6$

30. $-3x + 4y = 12$
$6x - 8y = 16$

31. $\begin{aligned} x - 2y &= 1 \\ -2x + 4y &= -2 \end{aligned}$

32. $\begin{aligned} 3x - 6y &= 1 \\ 2x - 4y &= \frac{2}{3} \end{aligned}$

33. $\begin{aligned} x + y - z &= -2 \\ 2x - y + z &= 5 \\ -x + 2y + 2z &= 1 \end{aligned}$

34. $\begin{aligned} x - 2y - z &= 2 \\ 2x - y + z &= 4 \\ -x + y - 2z &= -4 \end{aligned}$

35. $\begin{aligned} x + 3y \quad &= 0 \\ x + y + z &= 1 \\ 3x - y - z &= 11 \end{aligned}$

36. $\begin{aligned} 3y - z &= -1 \\ x + 5y - z &= -4 \\ -3x + 6y + 2z &= 11 \end{aligned}$

37. $\begin{aligned} 2x + 2y + 7z &= -1 \\ 2x + y + 2z &= 2 \\ 4x + 6y + z &= 15 \end{aligned}$

38. $\begin{aligned} 3x + 2y + 3z &= 3 \\ 4x - 5y + 7z &= 1 \\ 2x + 3y - 2z &= 6 \end{aligned}$

39. $\begin{aligned} x + y + z &= 6 \\ x \quad - z &= -2 \\ y + 3z &= 11 \end{aligned}$

40. $\begin{aligned} x + y + z &= 3 \\ -y + 2z &= 1 \\ -x \quad + z &= 0 \end{aligned}$

41. $\begin{aligned} x - y + 3z &= 4 \\ 2x - 2y + 6z &= 7 \\ 3x - y + 5z &= 14 \end{aligned}$

42. $\begin{aligned} 3x - y + 2z &= 4 \\ -6x + 2y - 4z &= 1 \\ 5x - 3y + 8z &= 0 \end{aligned}$

43. $\begin{aligned} x - 2y + z &= 4 \\ 5x - 10y + 5z &= 20 \\ -2x + 4y - 2z &= -8 \end{aligned}$

44. $\begin{aligned} x - 3y + z &= 2 \\ 4x - 12y + 4z &= 8 \\ -2x + 6y - 2z &= -4 \end{aligned}$

45. $\begin{aligned} x + y &= 1 \\ y + 2z &= -2 \\ 2x - z &= 0 \end{aligned}$

46. $\begin{aligned} x + 3y &= 3 \\ y + 2z &= -8 \\ x - z &= 7 \end{aligned}$

Practice Plus

In Exercises 47–48, write the system of linear equations represented by the augmented matrix. Use w, x, y, and z, for the variables. Once the system is written, use back-substitution to find its solution, (w, x, y, z).

47. $\left[\begin{array}{cccc|c} 1 & -1 & 1 & 1 & 3 \\ 0 & 1 & -2 & -1 & 0 \\ 0 & 0 & 1 & 6 & 17 \\ 0 & 0 & 0 & 1 & 3 \end{array}\right]$

48. $\left[\begin{array}{cccc|c} 1 & 2 & -1 & 0 & 2 \\ 0 & 1 & 1 & -2 & -3 \\ 0 & 0 & 1 & -1 & -2 \\ 0 & 0 & 0 & 1 & 3 \end{array}\right]$

In Exercises 49–50, perform each matrix row operation and write the new matrix.

49. $\left[\begin{array}{cccc|c} 1 & -1 & 1 & 1 & 3 \\ 0 & 1 & -2 & -1 & 0 \\ 2 & 0 & 3 & 4 & 11 \\ 5 & 1 & 2 & 4 & 6 \end{array}\right] \begin{array}{l} \\ \\ -2R_1 + R_3 \\ -5R_1 + R_4 \end{array}$

50. $\left[\begin{array}{cccc|c} 1 & -5 & 2 & -2 & 4 \\ 0 & 1 & -3 & -1 & 0 \\ 3 & 0 & 2 & -1 & 6 \\ -4 & 1 & 4 & 2 & -3 \end{array}\right] \begin{array}{l} \\ \\ -3R_1 + R_3 \\ 4R_1 + R_4 \end{array}$

In Exercises 51–52, solve each system using matrices. You will need to use matrix row operations to obtain matrices like those in Exercises 47 and 48, with 1s down the main diagonal and 0s below the 1s. Express the solution as (w, x, y, z).

51. $\begin{aligned} w + x + y + z &= 4 \\ 2w + x - 2y - z &= 0 \\ w - 2x - y - 2z &= -2 \\ 3w + 2x + y + 3z &= 4 \end{aligned}$

52. $\begin{aligned} w + x + y + z &= 5 \\ w + 2x - y - 2z &= -1 \\ w - 3x - 3y - z &= -1 \\ 2w - x + 2y - z &= -2 \end{aligned}$

APPENDIX B

Objectives

1 Evaluate a second-order determinant.

2 Solve a system of linear equations in two variables using Cramer's rule.

3 Evaluate a third-order determinant.

4 Solve a system of linear equations in three variables using Cramer's rule.

5 Use determinants to identify inconsistent and dependent systems.

DETERMINANTS AND CRAMER'S RULE

A portion of Charles Babbage's unrealized difference engine

As cyberspace absorbs more and more of our work, play, shopping, and socializing, where will it all end? Which activities will still be offline in 2025?

Our technologically transformed lives can be traced back to the English inventor Charles Babbage (1791–1871). Babbage knew of a method for solving linear systems called *Cramer's rule*, in honor of the Swiss geometer Gabriel Cramer (1704–1752). Cramer's rule was simple, but involved numerous multiplications for large systems. Babbage designed a machine, called the "difference engine," that consisted of toothed wheels on shafts for performing these multiplications. Despite the fact that only one-seventh of the functions ever worked, Babbage's invention demonstrated how complex calculations could be handled mechanically. In 1944, scientists at IBM used the lessons of the difference engine to create the world's first computer.

Those who invented computers hoped to relegate the drudgery of repeated computation to a machine. In this section, we look at a method for solving linear systems that played a critical role in this process. The method uses arrays of numbers, called *determinants*. As with matrix methods, solutions are obtained by writing down the coefficients and constants of a linear system and performing operations with them.

1 Evaluate a second-order determinant.

The Determinant of a 2 × 2 Matrix A matrix of **order $m \times n$** has m rows and n columns. If $m = n$, a matrix has the same number of rows as columns and is called a **square matrix**. Associated with every square matrix is a real number, called its **determinant**. The determinant for a 2×2 square matrix is defined as follows:

STUDY TIP

To evaluate a second-order determinant, find the difference of the product of the two diagonals.

$$\begin{vmatrix} a_1 & b_1 \\ a_2 & b_2 \end{vmatrix} = a_1b_2 - a_2b_1$$

DEFINITION OF THE DETERMINANT OF A 2 × 2 MATRIX The determinant of the matrix $\begin{bmatrix} a_1 & b_1 \\ a_2 & b_2 \end{bmatrix}$ is denoted by $\begin{vmatrix} a_1 & b_1 \\ a_2 & b_2 \end{vmatrix}$ and is defined by

$$\begin{vmatrix} a_1 & b_1 \\ a_2 & b_2 \end{vmatrix} = a_1b_2 - a_2b_1.$$

We also say that the **value** of the **second-order determinant** $\begin{vmatrix} a_1 & b_1 \\ a_2 & b_2 \end{vmatrix}$ is $a_1b_2 - a_2b_1$.

Example 1 illustrates that the determinant of a matrix may be positive or negative. The determinant can also have 0 as its value.

EXAMPLE 1 Evaluating the Determinant of a 2 × 2 Matrix

Evaluate the determinant of each of the following matrices:

a. $\begin{bmatrix} 5 & 6 \\ 7 & 3 \end{bmatrix}$ **b.** $\begin{bmatrix} 2 & 4 \\ -3 & -5 \end{bmatrix}$.

DISCOVER FOR YOURSELF

Write and then evaluate three determinants, one whose value is positive, one whose value is negative, and one whose value is 0.

SOLUTION We multiply and subtract as indicated.

a. $\begin{vmatrix} 5 & 6 \\ 7 & 3 \end{vmatrix} = 5 \cdot 3 - 7 \cdot 6 = 15 - 42 = -27$ The value of the second-order determinant is −27.

b. $\begin{vmatrix} 2 & 4 \\ -3 & -5 \end{vmatrix} = 2(-5) - (-3)(4) = -10 + 12 = 2$ The value of the second-order determinant is 2. ■

CHECK POINT 1 Evaluate the determinant of each of the following matrices:

a. $\begin{bmatrix} 10 & 9 \\ 6 & 5 \end{bmatrix}$ **b.** $\begin{bmatrix} 4 & 3 \\ -5 & -8 \end{bmatrix}$.

2 Solve a system of linear equations in two variables using Cramer's rule.

Solving Systems of Linear Equations in Two Variables Using Determinants

Determinants can be used to solve a linear system in two variables. In general, such a system appears as

$$a_1x + b_1y = c_1$$
$$a_2x + b_2y = c_2.$$

Let's first solve this system for x using the addition method. We can solve for x by eliminating y from the equations. Multiply the first equation by b_2 and the second equation by $-b_1$. Then add the two equations:

$$\begin{aligned} a_1x + b_1y = c_1 &\xrightarrow{\text{Multiply by } b_2.} & a_1b_2x + b_1b_2y &= c_1b_2 \\ a_2x + b_2y = c_2 &\xrightarrow{\text{Multiply by } -b_1.} & -a_2b_1x - b_1b_2y &= -c_2b_1 \\ & \text{Add:} & (a_1b_2 - a_2b_1)x &= c_1b_2 - c_2b_1 \\ & & x &= \frac{c_1b_2 - c_2b_1}{a_1b_2 - a_2b_1} \end{aligned}$$

Because

$$\begin{vmatrix} c_1 & b_1 \\ c_2 & b_2 \end{vmatrix} = c_1b_2 - c_2b_1 \quad \text{and} \quad \begin{vmatrix} a_1 & b_1 \\ a_2 & b_2 \end{vmatrix} = a_1b_2 - a_2b_1$$

we can express our answer for x as the quotient of two determinants:

$$x = \frac{\begin{vmatrix} c_1 & b_1 \\ c_2 & b_2 \end{vmatrix}}{\begin{vmatrix} a_1 & b_1 \\ a_2 & b_2 \end{vmatrix}}.$$

Similarly, we could use the addition method to solve our system for y, again expressing y as the quotient of two determinants. This method of using determinants to solve the linear system, called **Cramer's rule**, is summarized in the box.

SOLVING A LINEAR SYSTEM IN TWO VARIABLES USING DETERMINANTS

Cramer's Rule

If

$$a_1x + b_1y = c_1$$
$$a_2x + b_2y = c_2$$

then

$$x = \frac{\begin{vmatrix} c_1 & b_1 \\ c_2 & b_2 \end{vmatrix}}{\begin{vmatrix} a_1 & b_1 \\ a_2 & b_2 \end{vmatrix}} \quad \text{and} \quad y = \frac{\begin{vmatrix} a_1 & c_1 \\ a_2 & c_2 \end{vmatrix}}{\begin{vmatrix} a_1 & b_1 \\ a_2 & b_2 \end{vmatrix}},$$

where

$$\begin{vmatrix} a_1 & b_1 \\ a_2 & b_2 \end{vmatrix} \neq 0.$$

Here are some helpful tips when solving

$$a_1x + b_1y = c_1$$
$$a_2x + b_2y = c_2$$

using determinants:

1. Three different determinants are used to find x and y. The determinants in the denominators for x and y are identical. The determinants in the numerators for x and y differ. In abbreviated notation, we write

$$x = \frac{D_x}{D} \quad \text{and} \quad y = \frac{D_y}{D}, \text{ where } D \neq 0.$$

2. The elements of D, the determinant in the denominator, are the coefficients of the variables in the system.

$$D = \begin{vmatrix} a_1 & b_1 \\ a_2 & b_2 \end{vmatrix}$$

3. D_x, the determinant in the numerator of x, is obtained by replacing the x-coefficients in D, a_1 and a_2, with the constants on the right sides of the equations, c_1 and c_2.

$$D = \begin{vmatrix} a_1 & b_1 \\ a_2 & b_2 \end{vmatrix} \quad \text{and} \quad D_x = \begin{vmatrix} c_1 & b_1 \\ c_2 & b_2 \end{vmatrix}$$

Replace the column with a_1 and a_2 with the constants c_1 and c_2 to get D_x.

4. D_y, the determinant in the numerator for y, is obtained by replacing the y-coefficients in D, b_1 and b_2, with the constants on the right sides of the equations, c_1 and c_2.

$$D = \begin{vmatrix} a_1 & b_1 \\ a_2 & b_2 \end{vmatrix} \quad \text{and} \quad D_y = \begin{vmatrix} a_1 & c_1 \\ a_2 & c_2 \end{vmatrix}$$

Replace the column with b_1 and b_2 with the constants c_1 and c_2 to get D_y.

EXAMPLE 2 Using Cramer's Rule to Solve a Linear System

Use Cramer's rule to solve the system:

$$5x - 4y = 2$$
$$6x - 5y = 1.$$

SOLUTION Because

$$x = \frac{D_x}{D} \quad \text{and} \quad y = \frac{D_y}{D},$$

we will set up and evaluate the three determinants D, D_x, and D_y.

1. D, the determinant in both denominators, consists of the x- and y-coefficients.

$$D = \begin{vmatrix} 5 & -4 \\ 6 & -5 \end{vmatrix} = (5)(-5) - (6)(-4) = -25 + 24 = -1$$

Because this determinant is not zero, we continue to use Cramer's rule to solve the system.

2. D_x, the determinant in the numerator for x, is obtained by replacing the x-coefficients in D, 5 and 6, by the constants on the right sides of the equations, 2 and 1.

$$D_x = \begin{vmatrix} 2 & -4 \\ 1 & -5 \end{vmatrix} = (2)(-5) - (1)(-4) = -10 + 4 = -6$$

3. D_y, the determinant in the numerator for y, is obtained by replacing the y-coefficients in D, -4 and -5, by the constants on the right sides of the equations, 2 and 1.

$$D_y = \begin{vmatrix} 5 & 2 \\ 6 & 1 \end{vmatrix} = (5)(1) - (6)(2) = 5 - 12 = -7$$

4. Thus,

$$x = \frac{D_x}{D} = \frac{-6}{-1} = 6 \quad \text{and} \quad y = \frac{D_y}{D} = \frac{-7}{-1} = 7.$$

As always, the ordered pair $(6, 7)$ should be checked by substituting these values into the original equations. The solution is $(6, 7)$ and the solution set is $\{(6, 7)\}$. ■

✔ **CHECK POINT 2** Use Cramer's rule to solve the system:

$$5x + 4y = 12$$
$$3x - 6y = 24.$$

3 Evaluate a third-order determinant.

The Determinant of a 3 × 3 Matrix The determinant for a 3 × 3 matrix is defined in terms of second-order determinants:

DEFINITION OF THE DETERMINANT OF A 3 × 3 MATRIX A third-order determinant is defined by

Subtract. Add.

$$\begin{vmatrix} a_1 & b_1 & c_1 \\ a_2 & b_2 & c_2 \\ a_3 & b_3 & c_3 \end{vmatrix} = a_1\begin{vmatrix} b_2 & c_2 \\ b_3 & c_3 \end{vmatrix} - a_2\begin{vmatrix} b_1 & c_1 \\ b_3 & c_3 \end{vmatrix} + a_3\begin{vmatrix} b_1 & c_1 \\ b_2 & c_2 \end{vmatrix}$$

Each a on the right comes from the first column.

Here are some tips that should be helpful when evaluating the determinant of a 3 × 3 matrix:

EVALUATING THE DETERMINANT OF A 3 × 3 MATRIX

1. Each of the three terms in the definition contains two factors—a numerical factor and a second-order determinant.
2. The numerical factor in each term is an element from the first column of the third-order determinant.
3. The minus sign precedes the second term.
4. The second-order determinant that appears in each term is obtained by crossing out the row and the column containing the numerical factor.

$$a_1\begin{vmatrix} b_2 & c_2 \\ b_3 & c_3 \end{vmatrix} - a_2\begin{vmatrix} b_1 & c_1 \\ b_3 & c_3 \end{vmatrix} + a_3\begin{vmatrix} b_1 & c_1 \\ b_2 & c_2 \end{vmatrix}$$

$$\downarrow \qquad\qquad \downarrow \qquad\qquad \downarrow$$

$$\begin{vmatrix} a_1 & b_1 & c_1 \\ a_2 & b_2 & c_2 \\ a_3 & b_3 & c_3 \end{vmatrix} \quad \begin{vmatrix} a_1 & b_1 & c_1 \\ a_2 & b_2 & c_2 \\ a_3 & b_3 & c_3 \end{vmatrix} \quad \begin{vmatrix} a_1 & b_1 & c_1 \\ a_2 & b_2 & c_2 \\ a_3 & b_3 & c_3 \end{vmatrix}$$

The **minor** of an element is the determinant that remains after deleting the row and column of that element. For this reason, we call this method **expansion by minors**.

EXAMPLE 3 Evaluating the Determinant of a 3 × 3 Matrix

Evaluate the determinant of the following matrix:

$$\begin{bmatrix} 4 & 1 & 0 \\ -9 & 3 & 4 \\ -3 & 8 & 1 \end{bmatrix}.$$

SOLUTION We know that each of the three terms in the determinant contains a numerical factor and a second-order determinant. The numerical factors are from the first column of the given matrix. They are shown in red in the following matrix:

$$\begin{bmatrix} 4 & 1 & 0 \\ -9 & 3 & 4 \\ -3 & 8 & 1 \end{bmatrix}.$$

We find the minor for each numerical factor by deleting the row and column of that element:

$$\begin{bmatrix} 4 & 1 & 0 \\ -9 & 3 & 4 \\ -3 & 8 & 1 \end{bmatrix} \quad \begin{bmatrix} 4 & 1 & 0 \\ -9 & 3 & 4 \\ -3 & 8 & 1 \end{bmatrix} \quad \begin{bmatrix} 4 & 1 & 0 \\ -9 & 3 & 4 \\ -3 & 8 & 1 \end{bmatrix}$$

The minor for 4 is $\begin{vmatrix} 3 & 4 \\ 8 & 1 \end{vmatrix}$.

The minor for −9 is $\begin{vmatrix} 1 & 0 \\ 8 & 1 \end{vmatrix}$.

The minor for −3 is $\begin{vmatrix} 1 & 0 \\ 3 & 4 \end{vmatrix}$.

Now we have three numerical factors, 4, −9, and −3, and three second-order determinants. We multiply each numerical factor by its second-order determinant to find the three terms of the third-order determinant:

$$4\begin{vmatrix} 3 & 4 \\ 8 & 1 \end{vmatrix}, \quad -9\begin{vmatrix} 1 & 0 \\ 8 & 1 \end{vmatrix}, \quad -3\begin{vmatrix} 1 & 0 \\ 3 & 4 \end{vmatrix}.$$

Based on the preceding definition, we subtract the second term from the first term and add the third term:

Don't forget to supply the minus sign.

$$\begin{vmatrix} 4 & 1 & 0 \\ -9 & 3 & 4 \\ -3 & 8 & 1 \end{vmatrix} = 4\begin{vmatrix} 3 & 4 \\ 8 & 1 \end{vmatrix} - (-9)\begin{vmatrix} 1 & 0 \\ 8 & 1 \end{vmatrix} - 3\begin{vmatrix} 1 & 0 \\ 3 & 4 \end{vmatrix}$$

$= 4(3\cdot1 - 8\cdot4) + 9(1\cdot1 - 8\cdot0) - 3(1\cdot4 - 3\cdot0)$ — Begin evaluating the three second-order determinants.

$= 4(3 - 32) + 9(1 - 0) - 3(4 - 0)$ — Multiply within parentheses.

$= 4(-29) + 9(1) - 3(4)$ — Subtract within parentheses.

$= -116 + 9 - 12$ — Multiply.

$= -119$ — Add and subtract as indicated. ■

USING TECHNOLOGY

A graphing utility can be used to evaluate the determinant of a matrix. Enter the matrix and call it A. Then use the determinant command. The screen below verifies our result in Example 3.

```
[A]
        [[4  1 0]
         [-9 3 4]
         [-3 8 1]]
det([A])
            -119
```

CHECK POINT 3 Evaluate the determinant of the following matrix:

$$\begin{bmatrix} 2 & 1 & 7 \\ -5 & 6 & 0 \\ -4 & 3 & 1 \end{bmatrix}.$$

4 Solve a system of linear equations in three variables using Cramer's rule.

Solving Systems of Linear Equations in Three Variables Using Determinants

Cramer's rule can be applied to solving systems of linear equations in three variables. The determinants in the numerator and denominator of all variables are third-order determinants.

SOLVING THREE EQUATIONS IN THREE VARIABLES USING DETERMINANTS

Cramer's Rule

If

$$\begin{aligned} a_1x + b_1y + c_1z &= d_1 \\ a_2x + b_2y + c_2z &= d_2 \\ a_3x + b_3y + c_3z &= d_3 \end{aligned}$$

then

$$x = \frac{D_x}{D}, \quad y = \frac{D_y}{D}, \quad \text{and} \quad z = \frac{D_z}{D}.$$

These four third-order determinants are given by

$$D = \begin{vmatrix} a_1 & b_1 & c_1 \\ a_2 & b_2 & c_2 \\ a_3 & b_3 & c_3 \end{vmatrix}$$

These are the coefficients of the variables x, y, and z, $D \neq 0$.

$$D_x = \begin{vmatrix} d_1 & b_1 & c_1 \\ d_2 & b_2 & c_2 \\ d_3 & b_3 & c_3 \end{vmatrix}$$

Replace x-coefficients in D with the constants on the right of the three equations.

$$D_y = \begin{vmatrix} a_1 & d_1 & c_1 \\ a_2 & d_2 & c_2 \\ a_3 & d_3 & c_3 \end{vmatrix}$$

Replace y-coefficients in D with the constants on the right of the three equations.

$$D_z = \begin{vmatrix} a_1 & b_1 & d_1 \\ a_2 & b_2 & d_2 \\ a_3 & b_3 & d_3 \end{vmatrix}.$$

Replace z-coefficients in D with the constants on the right of the three equations.

EXAMPLE 4 Using Cramer's Rule to Solve a Linear System in Three Variables

Use Cramer's rule to solve:

$$\begin{aligned} x + 2y - z &= -4 \\ x + 4y - 2z &= -6 \\ 2x + 3y + z &= 3. \end{aligned}$$

SOLUTION Because

$$x = \frac{D_x}{D}, \quad y = \frac{D_y}{D}, \quad \text{and} \quad z = \frac{D_z}{D},$$

we need to set up and evaluate four determinants.

Step 1. Set up the determinants.

1. D, the determinant in all three denominators, consists of the x-, y-, and z-coefficients.

$$D = \begin{vmatrix} 1 & 2 & -1 \\ 1 & 4 & -2 \\ 2 & 3 & 1 \end{vmatrix}$$

$$\begin{aligned} x + 2y - z &= -4 \\ x + 4y - 2z &= -6 \\ 2x + 3y + z &= 3 \end{aligned}$$

The system we are solving (repeated)

2. D_x, the determinant in the numerator for x, is obtained by replacing the x-coefficients in D, 1, 1, and 2, with the constants on the right sides of the equations, −4, −6, and 3.

$$D_x = \begin{vmatrix} -4 & 2 & -1 \\ -6 & 4 & -2 \\ 3 & 3 & 1 \end{vmatrix}$$

3. D_y, the determinant in the numerator for y, is obtained by replacing the y-coefficients in D, 2, 4, and 3, with the constants on the right sides of the equations, −4, −6, and 3.

$$D_y = \begin{vmatrix} 1 & -4 & -1 \\ 1 & -6 & -2 \\ 2 & 3 & 1 \end{vmatrix}$$

4. D_z, the determinant in the numerator for z, is obtained by replacing the z-coefficients in D, −1, −2, and 1, with the constants on the right sides of the equations, −4, −6, and 3.

$$D_z = \begin{vmatrix} 1 & 2 & -4 \\ 1 & 4 & -6 \\ 2 & 3 & 3 \end{vmatrix}$$

Step 2. Evaluate the four determinants.

$$D = \begin{vmatrix} 1 & 2 & -1 \\ 1 & 4 & -2 \\ 2 & 3 & 1 \end{vmatrix} = 1\begin{vmatrix} 4 & -2 \\ 3 & 1 \end{vmatrix} - 1\begin{vmatrix} 2 & -1 \\ 3 & 1 \end{vmatrix} + 2\begin{vmatrix} 2 & -1 \\ 4 & -2 \end{vmatrix}$$

$$= 1(4 + 6) - 1(2 + 3) + 2(-4 + 4)$$

$$= 1(10) - 1(5) + 2(0) = 5$$

Using the same technique to evaluate each determinant, we obtain

$$D_x = -10, \quad D_y = 5, \quad \text{and} \quad D_z = 20.$$

Step 3. Substitute these four values and solve the system.

$$x = \frac{D_x}{D} = \frac{-10}{5} = -2$$

$$y = \frac{D_y}{D} = \frac{5}{5} = 1$$

$$z = \frac{D_z}{D} = \frac{20}{5} = 4$$

The ordered triple $(-2, 1, 4)$ can be checked by substitution into the original three equations. The solution is $(-2, 1, 4)$ and the solution set is $\{(-2, 1, 4)\}$. ■

STUDY TIP

To find D_x, D_y, and D_z, you'll need to apply the evaluation process for a 3 × 3 determinant three times. The values of D_x, D_y, and D_z cannot be obtained from the numbers that occur in the computation of D.

CHECK POINT 4 Use Cramer's rule to solve the system:

$$\begin{aligned} 3x - 2y + z &= 16 \\ 2x + 3y - z &= -9 \\ x + 4y + 3z &= 2. \end{aligned}$$

5 Use determinants to identify inconsistent and dependent systems.

Cramer's Rule with Inconsistent and Dependent Systems If D, the determinant in the denominator, is 0, the variables described by the quotient of determinants are not real numbers. However, when $D = 0$, this indicates that the system is inconsistent or contains dependent equations. This gives rise to the following two situations:

DETERMINANTS: INCONSISTENT AND DEPENDENT SYSTEMS

1. If $D = 0$ and at least one of the determinants in the numerator is not 0, then the system is inconsistent. The solution set is $\varnothing$.
2. If $D = 0$ and all the determinants in the numerators are 0, then the equations in the system are dependent. The system has infinitely many solutions.

Although we have focused on applying determinants to solve linear systems, they have other applications, some of which we consider in the exercise set that follows.

B EXERCISE SET

Student Solutions Manual | CD/Video | PH Math/Tutor Center | MathXL Tutorials on CD | MathXL® | MyMathLab | Interactmath.com

Practice Exercises

Evaluate each determinant in Exercises 1–10.

1. $\begin{vmatrix} 5 & 7 \\ 2 & 3 \end{vmatrix}$

2. $\begin{vmatrix} 4 & 8 \\ 5 & 6 \end{vmatrix}$

3. $\begin{vmatrix} -4 & 1 \\ 5 & 6 \end{vmatrix}$

4. $\begin{vmatrix} 7 & 9 \\ -2 & -5 \end{vmatrix}$

5. $\begin{vmatrix} -7 & 14 \\ 2 & -4 \end{vmatrix}$

6. $\begin{vmatrix} 1 & -3 \\ -8 & 2 \end{vmatrix}$

7. $\begin{vmatrix} -5 & -1 \\ -2 & -7 \end{vmatrix}$

8. $\begin{vmatrix} \frac{1}{5} & \frac{1}{6} \\ -6 & 5 \end{vmatrix}$

9. $\begin{vmatrix} \frac{1}{2} & \frac{1}{2} \\ \frac{1}{8} & -\frac{3}{4} \end{vmatrix}$

10. $\begin{vmatrix} \frac{2}{3} & \frac{1}{3} \\ -\frac{1}{2} & \frac{3}{4} \end{vmatrix}$

For Exercises 11–26, use Cramer's rule to solve each system or to determine that the system is inconsistent or contains dependent equations.

11. $x + y = 7$
$x - y = 3$

12. $2x + y = 3$
$x - y = 3$

13. $12x + 3y = 15$
$2x - 3y = 13$

14. $x - 2y = 5$
$5x - y = -2$

15. $4x - 5y = 17$
$2x + 3y = 3$

16. $3x + 2y = 2$
$2x + 2y = 3$

17. $x - 3y = 4$
$3x - 4y = 12$

18. $2x - 9y = 5$
$3x - 3y = 11$

19. $3x - 4y = 4$
$2x + 2y = 12$

20. $3x = 7y + 1$
$2x = 3y - 1$

21. $2x = 3y + 2$
$5x = 51 - 4y$

22. $y = -4x + 2$
$2x = 3y + 8$

23. $3x = 2 - 3y$
$2y = 3 - 2x$

24. $x + 2y - 3 = 0$
$12 = 8y + 4x$

25. $4y = 16 - 3x$
$6x = 32 - 8y$

26. $2x = 7 + 3y$
$4x - 6y = 3$

Evaluate each determinant in Exercises 27–32.

27. $\begin{vmatrix} 3 & 0 & 0 \\ 2 & 1 & -5 \\ 2 & 5 & -1 \end{vmatrix}$

28. $\begin{vmatrix} 4 & 0 & 0 \\ 3 & -1 & 4 \\ 2 & -3 & 5 \end{vmatrix}$

29. $\begin{vmatrix} 3 & 1 & 0 \\ -3 & 4 & 0 \\ -1 & 3 & -5 \end{vmatrix}$

30. $\begin{vmatrix} 2 & -4 & 2 \\ -1 & 0 & 5 \\ 3 & 0 & 4 \end{vmatrix}$

31. $\begin{vmatrix} 1 & 1 & 1 \\ 2 & 2 & 2 \\ -3 & 4 & -5 \end{vmatrix}$

32. $\begin{vmatrix} 1 & 2 & 3 \\ 2 & 2 & -3 \\ 3 & 2 & 1 \end{vmatrix}$

In Exercises 33–40, use Cramer's rule to solve each system.

33. $\begin{aligned} x + y + z &= 0 \\ 2x - y + z &= -1 \\ -x + 3y - z &= -8 \end{aligned}$

34. $\begin{aligned} x - y + 2z &= 3 \\ 2x + 3y + z &= 9 \\ -x - y + 3z &= 11 \end{aligned}$

35. $\begin{aligned} 4x - 5y - 6z &= -1 \\ x - 2y - 5z &= -12 \\ 2x - y &= 7 \end{aligned}$

36. $\begin{aligned} x - 3y + z &= -2 \\ x + 2y &= 8 \\ 2x - y &= 1 \end{aligned}$

37. $\begin{aligned} x + y + z &= 4 \\ x - 2y + z &= 7 \\ x + 3y + 2z &= 4 \end{aligned}$

38. $\begin{aligned} 2x + 2y + 3z &= 10 \\ 4x - y + z &= -5 \\ 5x - 2y + 6z &= 1 \end{aligned}$

39. $\begin{aligned} x + 2z &= 4 \\ 2y - z &= 5 \\ 2x + 3y &= 13 \end{aligned}$

40. $\begin{aligned} 3x + 2z &= 4 \\ 5x - y &= -4 \\ 4y + 3z &= 22 \end{aligned}$

Practice Plus

In Exercises 41–42, evaluate each determinant.

41. $\begin{vmatrix} \begin{vmatrix} 3 & 1 \\ -2 & 3 \end{vmatrix} & \begin{vmatrix} 7 & 0 \\ 1 & 5 \end{vmatrix} \\ \begin{vmatrix} 3 & 0 \\ 0 & 7 \end{vmatrix} & \begin{vmatrix} 9 & -6 \\ 3 & 5 \end{vmatrix} \end{vmatrix}$

42. $\begin{vmatrix} \begin{vmatrix} 5 & 0 \\ 4 & -3 \end{vmatrix} & \begin{vmatrix} -1 & 0 \\ 0 & -1 \end{vmatrix} \\ \begin{vmatrix} 7 & -5 \\ 4 & 6 \end{vmatrix} & \begin{vmatrix} 4 & 1 \\ -3 & 5 \end{vmatrix} \end{vmatrix}$

In Exercises 43–44, write the system of linear equations for which Cramer's rule yields the given determinants.

43. $D = \begin{vmatrix} 2 & -4 \\ 3 & 5 \end{vmatrix}, \quad D_x = \begin{vmatrix} 8 & -4 \\ -10 & 5 \end{vmatrix}$

44. $D = \begin{vmatrix} 2 & -3 \\ 5 & 6 \end{vmatrix}, \quad D_x = \begin{vmatrix} 8 & -3 \\ 11 & 6 \end{vmatrix}$

In Exercises 45–48, solve each equation for x.

45. $\begin{vmatrix} -2 & x \\ 4 & 6 \end{vmatrix} = 32$

46. $\begin{vmatrix} x + 3 & -6 \\ x - 2 & -4 \end{vmatrix} = 28$

47. $\begin{vmatrix} 1 & x & -2 \\ 3 & 1 & 1 \\ 0 & -2 & 2 \end{vmatrix} = -8$

48. $\begin{vmatrix} 2 & x & 1 \\ -3 & 1 & 0 \\ 2 & 1 & 4 \end{vmatrix} = 39$

Application Exercises

Determinants are used to find the area of a triangle whose vertices are given by three points in a rectangular coordinate system. The area of a triangle with vertices (x_1, y_1), (x_2, y_2), and (x_3, y_3) is

$$\text{Area} = \pm\frac{1}{2}\begin{vmatrix} x_1 & y_1 & 1 \\ x_2 & y_2 & 1 \\ x_3 & y_3 & 1 \end{vmatrix}.$$

where the $\pm$ symbol indicates that the appropriate sign should be chosen to yield a positive area. Use this information to work Exercises 49–50.

49. Use determinants to find the area of the triangle whose vertices are $(3, -5)$, $(2, 6)$, and $(-3, 5)$.

50. Use determinants to find the area of the triangle whose vertices are $(1, 1)$, $(-2, -3)$, and $(11, -3)$.

Determinants are used to show that three points lie on the same line (are collinear). If

$$\begin{vmatrix} x_1 & y_1 & 1 \\ x_2 & y_2 & 1 \\ x_3 & y_3 & 1 \end{vmatrix} = 0,$$

then the points (x_1, y_1), (x_2, y_2), and (x_3, y_3) are collinear. If the determinant does not equal 0, then the points are not collinear. Use this information to work Exercises 51–52.

51. Are the points $(3, -1)$, $(0, -3)$, and $(12, 5)$ collinear?

52. Are the points $(-4, -6)$, $(1, 0)$, and $(11, 12)$ collinear?

Determinants are used to write an equation of a line passing through two points. An equation of the line passing through the distinct points (x_1, y_1) and (x_2, y_2) is given by

$$\begin{vmatrix} x & y & 1 \\ x_1 & y_1 & 1 \\ x_2 & y_2 & 1 \end{vmatrix} = 0.$$

Use this information to work Exercises 53–54.

53. Use the determinant to write an equation for the line passing through $(3, -5)$ and $(-2, 6)$. Then expand the determinant, expressing the line's equation in slope-intercept form.

54. Use the determinant to write an equation for the line passing through $(-1, 3)$ and $(2, 4)$. Then expand the determinant, expressing the line's equation in slope-intercept form.

APPENDIX C

WHERE DID THAT COME FROME? SELECTED PROOFS

Section 12.3 Properties of Logarithms

The Product Rule

Let b, M, and N be positive real numbers with $b \neq 1$.

$$\log_b(MN) = \log_b M + \log_b N$$

Proof. We begin by letting $\log_b M = R$ and $\log_b N = S$.
Now we write each logarithm in exponential form.

$$\log_b M = R \quad \text{means} \quad b^R = M.$$
$$\log_b N = S \quad \text{means} \quad b^S = N.$$

By substituting and using a property of exponents, we see that

$$MN = b^R b^S = b^{R+S}$$

Now we change $MN = b^{R+S}$ to logarithmic form.

$$MN = b^{R+S} \quad \text{means} \quad \log_b(MN) = R + S.$$

Finally, substituting $\log_b M$ for R and $\log_b N$ for S gives us

$$\log_b(MN) = \log_b M + \log_b N,$$

the property that we wanted to prove.
The quotient and power rules for logarithms are proved using similar procedures.

The Change-of-Base Property

For any logarithmic bases a and b, and any positive number M,

$$\log_b M = \frac{\log_a M}{\log_a b}.$$

Proof. To prove the change-of-base property, we let x equal the logarithm on the left side:

$$\log_b M = x.$$

Now we rewrite this logarithm in exponential form.

$$\log_b M = x \quad \text{means} \quad b^x = M.$$

Because b^x and M are equal, the logarithms with base a for each of these expressions must be equal. This means that

$$\log_a b^x = \log_a M$$

$$x \log_a b = \log_a M \qquad \text{Apply the power rule for logarithms on the left side.}$$

$$x = \frac{\log_a M}{\log_a b} \qquad \text{Solve for } x \text{ by dividing both sides by } \log_a b.$$

In our first step, we let x equal $\log_b M$. Replacing x on the left side by $\log_b M$ gives us

$$\log_b M = \frac{\log_a M}{\log_a b},$$

which is the change-of-base property.

Section 13.2 The Ellipse

The Standard Form of the Equation of an Ellipse with a Horizontal Major Axis Centered at the Origin

Proof. Refer to Figure A.1.

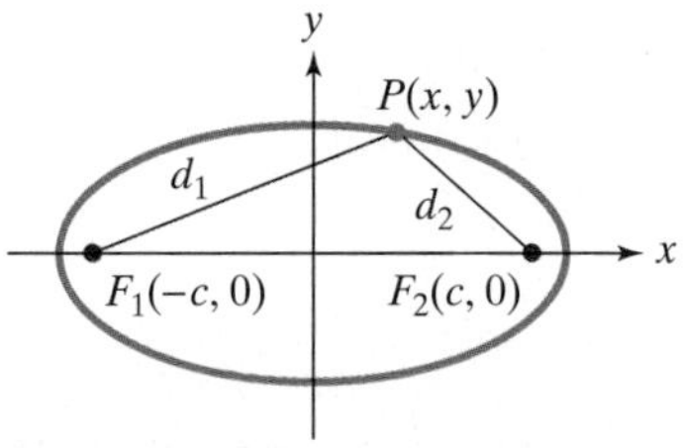

FIGURE A.1

$$d_1 + d_2 = 2a$$
This sum of the distances from P to the foci equals a constant, $2a$.

$$\sqrt{(x+c)^2 + y^2} + \sqrt{(x-c)^2 + y^2} = 2a$$
Use the distance formula.

$$\sqrt{(x+c)^2 + y^2} = 2a - \sqrt{(x-c)^2 + y^2}$$
Isolate a radical.

$$(x+c)^2 + y^2 = 4a^2 - 4a\sqrt{(x-c)^2 + y^2} + (x-c)^2 + y^2$$
Square both sides.

$$x^2 + 2cx + c^2 + y^2 = 4a^2 - 4a\sqrt{(x-c)^2 + y^2} + x^2 - 2cx + c^2 + y^2$$
Square $x + c$ and $x - c$.

$$4cx - 4a^2 = -4a\sqrt{(x-c)^2 + y^2}$$
Simplify and isolate the radical.

$$cx - a^2 = -a\sqrt{(x-c)^2 + y^2}$$
Divide both sides by 4.

$$(cx - a^2)^2 = a^2[(x-c)^2 + y^2]$$
Square both sides.

$$c^2x^2 - 2a^2cx + a^4 = a^2(x^2 - 2cx + c^2 + y^2)$$
Square $cx - a^2$ and $x - c$.

$$c^2x^2 - 2a^2cx + a^4 = a^2x^2 - 2a^2cx + a^2c^2 + a^2y^2$$
Use the distributive property.

$$c^2x^2 + a^2 = a^2x^2 + a^2c^2 + a^2y^2$$
Add $2a^2cx$ to both sides.

$$c^2x^2 - a^2x^2 - a^2y^2 = a^2c^2 - a^4$$
Rearrange the terms.

$$(c^2 - a^2)x^2 - a^2y^2 = a^2(c^2 - a^2)$$
Factor out x^2 and a^2, respectively.

$$(a^2 - c^2)x^2 + a^2y^2 = a^2(a^2 - c^2)$$
Multiply both sides by -1.

Refer to the discussion on page 882 and let $b^2 = a^2 - c^2$ in the preceding equation.

$$b^2x^2 + a^2y^2 = a^2b^2$$

$$\frac{x^2}{a^2} + \frac{y^2}{b^2} = 1$$
Divide both sides by a^2b^2.

Section 13.3 The Hyperbola

The Asymptotes of a Hyperbola Centered at the Origin

The hyperbola

$$\frac{x^2}{a^2} - \frac{y^2}{b^2} = 1$$

with a horizontal transverse axis has the two asymptotes

$$y = \frac{b}{a}x \quad \text{and} \quad y = -\frac{b}{a}x.$$

Proof. Begin by solving the hyperbola's equation for y.

$$\frac{x^2}{a^2} - \frac{y^2}{b^2} = 1$$ This is the standard form of the equation of a hyperbola.

$$\frac{y^2}{b^2} = \frac{x^2}{a^2} - 1$$ We isolate the term involving y^2 to solve for y.

$$y^2 = \frac{b^2x^2}{a^2} - b^2$$ Multiply both sides by b^2.

$$y^2 = \frac{b^2x^2}{a^2}\left(1 - \frac{a^2}{x^2}\right)$$ Factor out $\frac{b^2x^2}{a^2}$ on the right. Verify that this result is correct by multiplying using the distributive property and obtaining the previous step.

$$y = \pm\sqrt{\frac{b^2x^2}{a^2}\left(1 - \frac{a^2}{x^2}\right)}$$ Solve for y using the square root property: If $u^2 = d$, then $u = \pm\sqrt{d}$.

$$y = \pm\frac{b}{a}x\sqrt{1 - \frac{a^2}{x^2}}$$ Simplify.

As x increases or decreases without bound, the value of $\frac{a^2}{x^2}$ approaches 0. Consequently, the value of y can be approximated by

$$y = \pm\frac{b}{a}x.$$

This means that the lines whose equations are $y = \frac{b}{a}x$ and $y = -\frac{b}{a}x$ are asymptotes for the graph of the hyperbola.

Answers to Selected Exercises

CHAPTER 1

Section 1.1

Check Point Exercises

1. $\frac{21}{8}$ **2.** $1\frac{2}{3}$ **3.** $2 \cdot 2 \cdot 3 \cdot 3$ **4. a.** $\frac{2}{3}$ **b.** $\frac{7}{4}$ **c.** $\frac{13}{15}$ **d.** $\frac{1}{5}$ **5. a.** $\frac{8}{33}$ **b.** $\frac{18}{5}$ or $3\frac{3}{5}$ **c.** $\frac{2}{7}$ **d.** $\frac{51}{10}$ or $5\frac{1}{10}$ **6. a.** $\frac{10}{3}$ or $3\frac{1}{3}$ **b.** $\frac{2}{9}$ **c.** $\frac{3}{2}$ or $1\frac{1}{2}$ **7. a.** $\frac{5}{11}$ **b.** $\frac{2}{3}$ **c.** $\frac{9}{4}$ or $2\frac{1}{4}$ **8.** $\frac{14}{21}$ **9. a.** $\frac{11}{10}$ or $1\frac{1}{10}$ **b.** $\frac{7}{12}$ **c.** $\frac{5}{4}$ or $1\frac{1}{4}$ **10.** $\frac{53}{60}$ **11.** $\frac{29}{96}$ **12.** $\frac{1}{4}$

Exercise Set 1.1

1. $\frac{19}{8}$ **3.** $\frac{38}{5}$ **5.** $\frac{135}{16}$ **7.** $4\frac{3}{5}$ **9.** $8\frac{4}{9}$ **11.** $35\frac{11}{20}$ **13.** $2 \cdot 11$ **15.** $2 \cdot 2 \cdot 5$ **17.** prime **19.** $2 \cdot 2 \cdot 3 \cdot 3$ **21.** $2 \cdot 2 \cdot 5 \cdot 7$ **23.** prime **25.** $3 \cdot 3 \cdot 3 \cdot 3$ **27.** $2 \cdot 2 \cdot 2 \cdot 2 \cdot 3 \cdot 5$ **29.** $\frac{5}{8}$ **31.** $\frac{5}{6}$ **33.** $\frac{7}{10}$ **35.** $\frac{2}{5}$ **37.** $\frac{22}{25}$ **39.** $\frac{60}{43}$ **41.** $\frac{2}{15}$ **43.** $\frac{21}{88}$ **45.** $\frac{36}{7}$ **47.** $\frac{1}{12}$ **49.** $\frac{15}{14}$ **51.** 6 **53.** $\frac{15}{16}$ **55.** $\frac{9}{5}$ **57.** $\frac{5}{9}$ **59.** 3 **61.** $\frac{7}{10}$ **63.** $\frac{1}{2}$ **65.** 6 **67.** $\frac{6}{11}$ **69.** $\frac{2}{3}$ **71.** $\frac{5}{4}$ **73.** $\frac{1}{6}$ **75.** 2 **77.** $\frac{7}{10}$ **79.** $\frac{9}{10}$ **81.** $\frac{19}{24}$ **83.** $\frac{7}{18}$ **85.** $\frac{7}{12}$ **87.** $\frac{41}{80}$ **89.** $1\frac{5}{12}$ or $\frac{17}{12}$ **91.** $\frac{3a}{20}$ **93.** $\frac{20}{x}$ **95.** $\frac{4}{15}$ **97. a.** $\frac{11}{20}$ **b.** 240 **99.** $\frac{2}{5}$ **101.** $1\frac{1}{5}$ cups **103.** $\frac{23}{20}$ mi; $\frac{7}{20}$ mi farther **105.** 38 mi **117.** d

Section 1.2

Check Point Exercises

1. a. -500 **b.** -282 **2.**

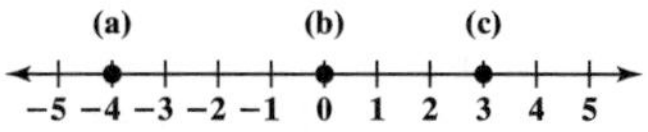

3.

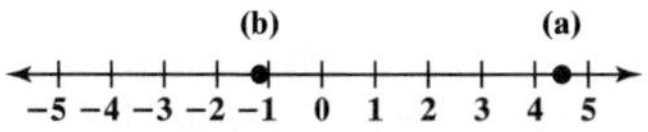

4. a. 0.375 **b.** $0.\overline{45}$ **5. a.** $\sqrt{9}$ **b.** $0, \sqrt{9}$ **c.** $-9, 0, \sqrt{9}$ **d.** $-9, -1.3, 0, 0.\overline{3}, \sqrt{9}$ **e.** $\frac{\pi}{2}, \sqrt{10}$ **f.** $-9, -1.3, 0, 0.\overline{3}, \frac{\pi}{2}, \sqrt{9}, \sqrt{10}$ **6. a.** $>$ **b.** $<$ **c.** $<$ **d.** $<$ **7. a.** true **b.** true **c.** false **8. a.** 4 **b.** 6 **c.** $\sqrt{2}$

Exercise Set 1.2

1. -20 **3.** 8 **5.** -3000 **7.** -4 billion **9–19.**

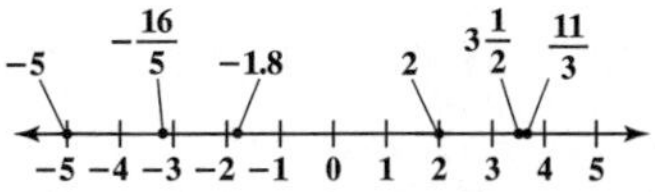

21. 0.75 **23.** 0.35 **25.** 0.875 **27.** $0.\overline{81}$ **29.** -0.5 **31.** $-0.8\overline{3}$ **33. a.** $\sqrt{100}$ **b.** $0, \sqrt{100}$ **c.** $-9, 0, \sqrt{100}$ **d.** $-9, -\frac{4}{5}, 0, 0.25, 9.2, \sqrt{100}$ **e.** $\sqrt{3}$ **f.** $-9, -\frac{4}{5}, 0, 0.25, \sqrt{3}, 9.2, \sqrt{100}$ **35. a.** $\sqrt{64}$ **b.** $0, \sqrt{64}$ **c.** $-11, 0, \sqrt{64}$ **d.** $-11, -\frac{5}{6}, 0, 0.75, \sqrt{64}$ **e.** $\sqrt{5}, \pi$ **f.** $-11, -\frac{5}{6}, 0, 0.75, \sqrt{5}, \pi, \sqrt{64}$ **37.** 0 **39.** Answers will vary; $\frac{1}{2}$ is an example. **41.** Answers will vary; 6 is an example. **43.** Answers will vary; π is an example. **45.** $<$ **47.** $>$ **49.** $>$ **51.** $<$ **53.** $>$ **55.** $<$ **57.** $<$ **59.** $>$ **61.** $>$ **63.** true **65.** true **67.** true **69.** false **71.** 6 **73.** 7 **75.** $\frac{5}{6}$ **77.** $\sqrt{11}$ **79.** $>$ **81.** $=$ **83.** $<$ **85.** $=$ **87.** -2 **89.** 1997, 2002; budget deficit **103.** c **105.** -7 and -6 **107.** 1.732; 1 and 2 **109.** -0.414; -1 and 0

Section 1.3

Check Point Exercises

1.

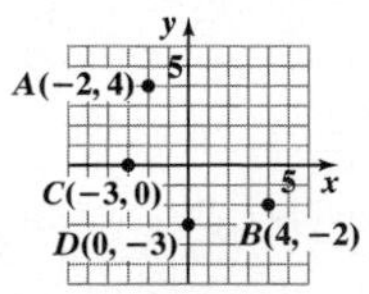

2. $E(-4, -2), F(-2, 0), G(6, 0)$ **3.** $B(8, 200)$; After 8 sec, the watermelon is 200 ft above ground. **4.** $D(8.8, 0)$; After approximately 8.8 sec, the watermelon is 0 ft above ground. Equivalently, the watermelon splatters on the ground after approximately 8.8 sec. **5.** 30 thousand **6. a.** 22 **b.** United States, France, Canada, Sweden

Exercise Set 1.3

1.
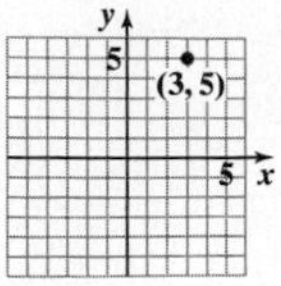

3.
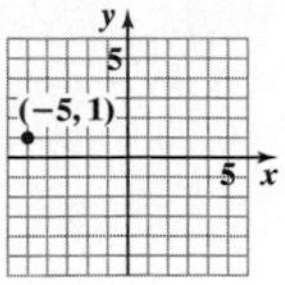

5.
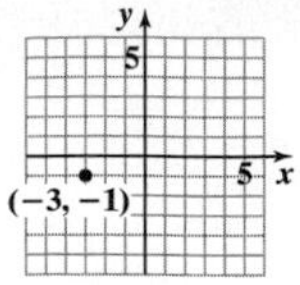

7.
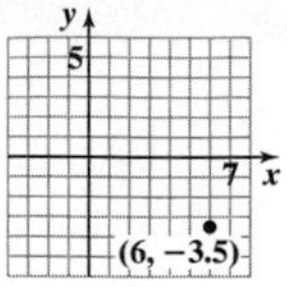

9–23.
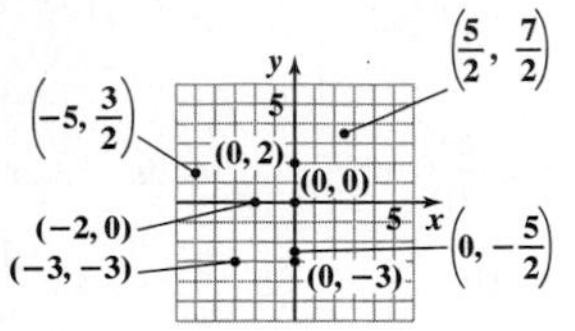

25. $(5, 2)$ **27.** $(-6, 5)$ **29.** $(-2, -3)$ **31.** $(5, -3)$ **33.** I and II **35.** I and III

37. Answers will vary; examples are $(-2, 2)$, $(-1, 1)$, $(1, 1)$, $(2, 2)$.

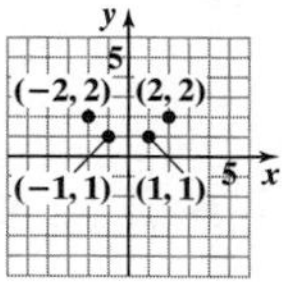

39. $4\frac{1}{2}$ or $\frac{9}{2}$ **41.** $(2, 7)$; The football is 7 ft above ground when it is 2 yd from the quarterback. **43.** $(6, 9.25)$ **45.** 12 ft; 15 yd

47. $(91, 125)$; In 1991, 125 thousand acres were used for cultivation.

49. 2001; 25 thousand acres **51.** 1991 and 1992 **53.** 7.1 million barrels per day **55.** 1990; 7.2 million barrels per day

57. 1970; 9.8 million barrels per day **59.** 1.7 million barrels per day **61.** \$120 million **63.** *No Way Out*, *The Postman*, *Dragonfly*

65. 48 yr old **67.** approx 17 yr **69.** \$22 thousand **71.** \$7 thousand **73.** 270; 14 **75.** 2002 **77.** 1972, 1982, 1992 **79.** $\frac{1}{20}$

89. a.
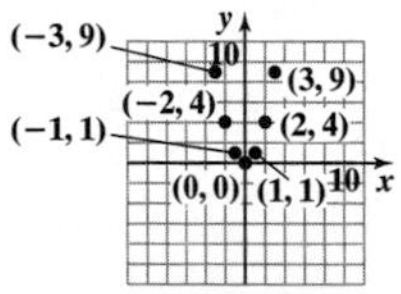

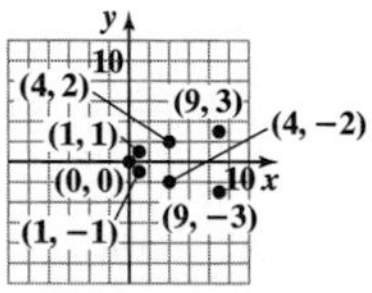

b. A; B; Points line up vertically when x-coordinates are associated with more than one y-coordinate. **91.** $\frac{23}{20}$

92. $<$ **93.** 5.83

Section 1.4

Check Point Exercises

1. 226.5; In 1980, the population of the United States was 226.5 million. **2. a.** 3 terms **b.** 6 **c.** 11 **d.** $6x$ and $2x$
3. a. $14 + x$ **b.** $y7$ **4. a.** $17 + 5x$ **b.** $x5 + 17$ **5. a.** $20 + x$ or $x + 20$ **b.** $30x$ **6.** $12 + x$ or $x + 12$ **7.** $5x + 15$
8. $24y + 42$ **9. a.** $10x$ **b.** $5a$ **10. a.** $18x + 10$ **b.** $14x + 8y$ **11.** $25x + 21$ **12.** $38x + 23y$ **13.** 108 beats/min

Exercise Set 1.4

1. 15 **3.** 80 **5.** 46 **7.** 35 **9.** 25 **11. a.** 2 **b.** 3 **c.** 5 **d.** no **13. a.** 3 **b.** 1 **c.** 2 **d.** yes; x and $5x$
15. a. 3 **b.** 4 **c.** 1 **d.** no **17.** $4 + y$ **19.** $3x + 5$ **21.** $5y + 4x$ **23.** $5(3 + x)$ **25.** $x9$ **27.** $x + 6y$ **29.** $x7 + 23$
31. $(x + 3)5$ **33.** $(7 + 5) + x = 12 + x$ **35.** $(7 \cdot 4)x = 28x$ **37.** $3x + 15$ **39.** $16x + 24$ **41.** $4 + 2r$ **43.** $5x + 5y$
45. $3x - 6$ **47.** $8x - 10$ **49.** $\frac{5}{2}x - 6$ **51.** $8x + 28$ **53.** $6x + 18 + 12y$ **55.** $15x - 10 + 20y$ **57.** $17x$ **59.** $8a$ **61.** $14 + x$
63. $11y + 3$ **65.** $9x + 1$ **67.** $14a + 14$ **69.** $15x + 6$ **71.** $15x + 2$ **73.** $41a + 24b$ **75.** Commutative property of addition
77. Associative property of addition **79.** false **81.** true **83.** 300; You can stay in the sun for 300 min without burning with a number 15 lotion.
85. 8805; In 2002, credit card debt per U.S. household was \$8805; fairly well. **87. a.** $0.38x + 0.02$ **b.** 1.92 million; very well
101. c **103. a.** \$50.50; \$5.50; \$1.00 **b.** No; When producing 2000 clocks, the average cost is \$3. This is more than the \$1.50 maximum price at which the manufacturer can sell the clocks.

104. $0.\overline{4}$ **105.**
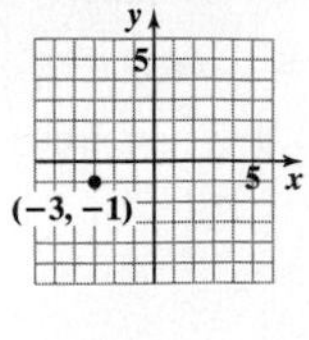

106. $\frac{1}{5}$

Mid-Chapter 1 Check Point

1. $24a + 1$ **2.** $\frac{1}{6}$ **3.** $\frac{1}{2}$ **4.** $65x + 16$ **5.** $\frac{25}{66}$ **6.** $\frac{2}{3}$ **7.** $\frac{1}{5}$ **8.** $\frac{25}{27}$ **9.** $\frac{37}{18}$ or $2\frac{1}{18}$ **10.** IV

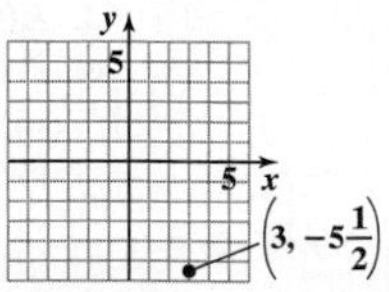

11. $(x + 3)5$ **12.** $5(3 + x)$ **13.** $5x + 15$ **14.** 7% **15.** 2002, 2003, 2004 **16.** $<$

17. $-11, -\frac{3}{7}, 0, 0.45, \sqrt{25}$ **18.** 13 **19.** 4

20. (67; 20); In 1967, 20% of grades of undergraduate college students were A's. **21.** 1970; 23% **22.** 40% **23.** 19.3 **24.** $0.\overline{09}$ **25.** $56x - 80 + 24y$

Section 1.5

Check Point Exercises

1. $4 + (-7) = -3$

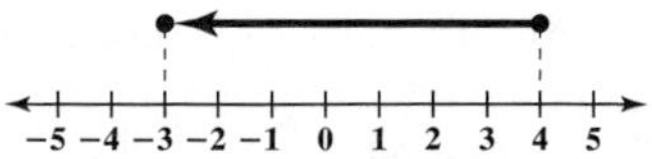

2. a. $-1 + (-3) = -4$

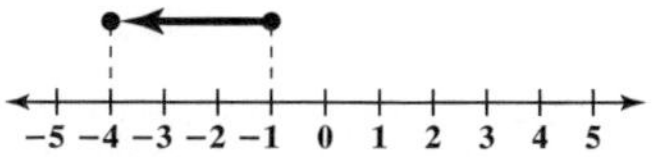

b. $-5 + 3 = -2$

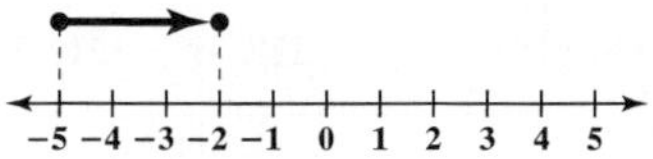

3. a. -35 **b.** -1.5 **c.** $-\frac{5}{6}$ **4. a.** -13 **b.** 1.2 **c.** $-\frac{1}{2}$ **5. a.** $-17x$ **b.** $-7y + 6z$ **c.** $20 - 20x$ **6.** The water level is down 3 ft at the end of 5 months.

Exercise Set 1.5

1. 4

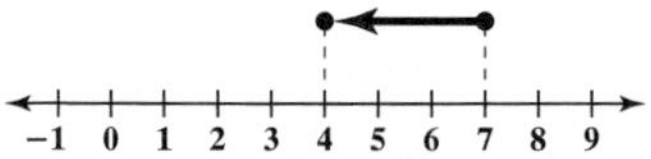

3. -7

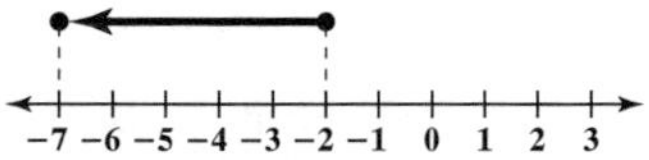

5. -4

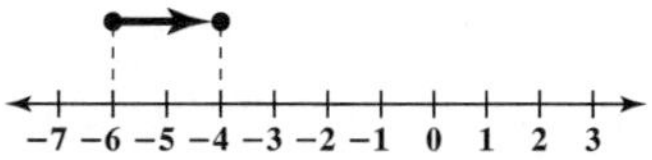

7. 0

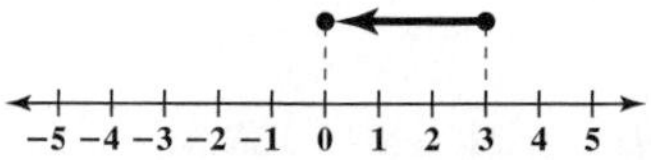

9. -7 **11.** 0 **13.** -60 **15.** -18 **17.** -1.3 **19.** -1 **21.** -5 **23.** 4 **25.** -3 **27.** -1.5 **29.** -5.7 **31.** $\frac{3}{10}$ **33.** $\frac{1}{8}$ **35.** $-\frac{43}{35}$ **37.** -8 **39.** 62 **41.** 8 **43.** -21 **45.** 22.1 **47.** $-8x$ **49.** $13y$ **51.** $-23a$ **53.** $1 - 6x$ **55.** $-4 + 6b$ **57.** $-2x - 3y$ **59.** $20x - 6$ **61.** $24 + 3y$ **63.** $47 - 33a$ **65.** 12 **67.** -30 **69.** $>$ **71.** The high temperature was 44°. **73.** The elevation is 600 ft below sea level. **75.** The temperature at 4:00 P.M. was 3°F. **77.** The ball was at the 25-yard line at the end of the fourth play. **79.** \$455 billion deficit **89.** d **91.** $-18y$ **94.** true **95. a.** $\sqrt{4}$ **b.** $0, \sqrt{4}$ **c.** $-6, 0, \sqrt{4}$ **d.** $-6, 0, 0.\overline{7}, \sqrt{4}$ **e.** $-\pi, \sqrt{3}$ **f.** $-6, -\pi, 0, 0.\overline{7}, \sqrt{3}, \sqrt{4}$ **96.** IV

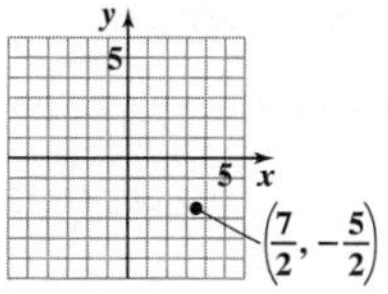

Section 1.6

Check Point Exercises

1. a. -8 **b.** 9 **c.** -5 **2. a.** 9.2 **b.** $-\frac{14}{15}$ **c.** 7π **3.** 15 **4.** $-6, 4a, -7ab$ **5. a.** $4 - 7x$ **b.** $-9x + 4y$ **6.** 19,763 m

Exercise Set 1.6

1. a. -12 **b.** $5 + (-12)$ **3. a.** 7 **b.** $5 + 7$ **5.** 6 **7.** -6 **9.** 23 **11.** 11 **13.** -11 **15.** -38 **17.** 0 **19.** 0 **21.** 26 **23.** -13 **25.** 13 **27.** $-\frac{2}{7}$ **29.** $\frac{4}{5}$ **31.** -1 **33.** $-\frac{3}{5}$ **35.** $\frac{3}{4}$ **37.** $\frac{1}{4}$ **39.** 7.6 **41.** -2 **43.** 2.6 **45.** 0 **47.** 3π **49.** 13π **51.** 19 **53.** -3 **55.** -15 **57.** 0 **59.** -52 **61.** -187 **63.** $\frac{7}{6}$ **65.** -4.49 **67.** $-\frac{3}{8}$ **69.** $-3x, -8y$ **71.** $12x, -5xy, -4$ **73.** $-6x$ **75.** $4 - 10y$ **77.** $5 - 7a$ **79.** $-4 - 9b$ **81.** $24 + 11x$ **83.** $3y - 8x$ **85.** 9 **87.** $\frac{7}{8}$ **89.** -2 **91.** 19,757 ft **93.** 1039 thousand jobs **95.** 268 thousand jobs **97.** 21°F **99.** 3°F **101.** The maximum concentration is 0.05 and it occurs during the 3rd hour. **103.** 0.015 **105.** from 0 to 3 hr **113.** 711 yr

117.

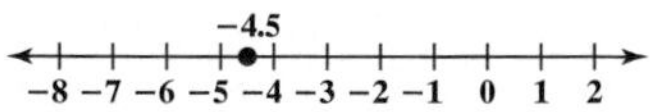

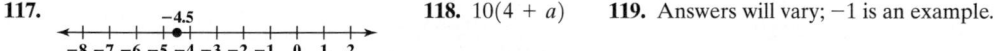

118. $10(4 + a)$ **119.** Answers will vary; -1 is an example.

Section 1.7

Check Point Exercises

1. a. -40 **b.** $-\frac{4}{21}$ **c.** 36 **d.** 5.5 **e.** 0 **2. a.** 24 **b.** -30 **3. a.** $\frac{1}{7}$ **b.** 8 **c.** $-\frac{1}{6}$ **d.** $-\frac{13}{7}$ **4. a.** -4 **b.** 8 **5. a.** 8 **b.** $-\frac{8}{15}$ **c.** -7.3 **d.** 0 **6. a.** $-20x$ **b.** $10x$ **c.** $-b$ **d.** $-21x + 28$ **e.** $-7y + 6$ **7.** $-y - 26$ **8. a.** \$330 **b.** \$60 **c.** \$33

Exercise Set 1.7

1. −45 **3.** 24 **5.** −21 **7.** 19 **9.** 0 **11.** −12 **13.** 9 **15.** $\frac{12}{35}$ **17.** $-\frac{14}{27}$ **19.** −3.6 **21.** 0.12 **23.** 30 **25.** −72 **27.** 24 **29.** −27 **31.** 90 **33.** 0 **35.** $\frac{1}{4}$ **37.** 5 **39.** $-\frac{1}{10}$ **41.** $-\frac{5}{2}$ **43. a.** $-32 \cdot \left(\frac{1}{4}\right)$ **b.** −8 **45. a.** $-60 \cdot \left(-\frac{1}{5}\right)$ **b.** 12 **47.** −3 **49.** −7 **51.** 30 **53.** 0 **55.** undefined **57.** −5 **59.** −12 **61.** 6 **63.** 0 **65.** undefined **67.** −4.3 **69.** $\frac{5}{6}$ **71.** $-\frac{16}{9}$ **73.** −1 **75.** −15 **77.** $-10x$ **79.** $3y$ **81.** $9x$ **83.** $-4x$ **85.** $-b$ **87.** $3y$ **89.** $-8x + 12$ **91.** $6x - 12$ **93.** $-2y + 5$ **95.** $y - 14$ **97.** $4(-10) + 8 = -32$ **99.** $(-9)(-3) - (-2) = 29$ **101.** $\frac{-18}{-15 + 12} = 6$ **103.** $-6 - \left(\frac{12}{-4}\right) = -3$ **105.** 6.3 million; fairly well, but slight overestimation **107. a.** 11 Latin words **b.** 11 Latin words; it is the same. **109. a.** \$2,000,000 (or \$200 tens of thousands) **b.** \$8,000,000 (or \$800 tens of thousands) **c.** Cost increases. **119.** b **121.** $5x$ **123.** $\frac{x}{12}$ **127.** $1.144x + 2.5$ **129.** −9 **130.** −3 **131.** 2

Section 1.8

Check Point Exercises

1. a. 36 **b.** −64 **c.** 1 **d.** −1 **2. a.** $21x^2$ **b.** $8x^3$ **c.** cannot simplify **3.** 15 **4.** 32 **5. a.** 36 **b.** 12 **6.** $\frac{3}{4}$ **7.** −40 **8.** −31 **9.** $\frac{5}{7}$ **10.** −5 **11.** $7x^2 + 15$ **12.** For 40-year-old drivers, there are 200 accidents per 50 million miles driven. **13.** 30°C

Exercise Set 1.8

1. 81 **3.** 64 **5.** 16 **7.** −64 **9.** 625 **11.** −625 **13.** −100 **15.** $19x^2$ **17.** $15x^3$ **19.** $9x^4$ **21.** $-x^2$ **23.** x^3 **25.** cannot be simplified **27.** 0 **29.** 25 **31.** 27 **33.** 12 **35.** 5 **37.** 45 **39.** −24 **41.** 300 **43.** 0 **45.** −32 **47.** 64 **49.** 30 **51.** $\frac{4}{3}$ **53.** 2 **55.** 2 **57.** 3 **59.** 88 **61.** −60 **63.** −36 **65.** 14 **67.** $-\frac{3}{4}$ **69.** $-\frac{9}{40}$ **71.** $-\frac{37}{36}$ or $-1\frac{1}{36}$ **73.** 24 **75.** 28 **77.** 9 **79.** −7 **81.** $15x - 27$ **83.** $15 - 3y$ **85.** $16y - 25$ **87.** $-2x^2 - 9$ **89.** $-10 - (-2)^3 = -2$ **91.** $[2(7 - 10)]^2 = 36$ **93.** $x - (5x + 8) = -4x - 8$ **95.** $5(x^3 - 4) = 5x^3 - 20$ **97.** 135 beats/min; (40, 135) on blue graph **99.** \$287.4 billion; fairly well **101.** 65 thousand; fairly well, but slight overestimation **103.** 20°C **105.** −30°C **113.** $-\frac{79}{4}$ **115.** $\left(2 \cdot 5 - \frac{1}{2} \cdot 10\right) \cdot 9 = 45$ **117.** Answers will vary. **119.** 6 **120.** −24 **121.** Answers will vary; −3 is an example.

Review Exercises

1. $\frac{23}{7}$ **2.** $\frac{64}{11}$ **3.** $1\frac{8}{9}$ **4.** $5\frac{2}{5}$ **5.** $2 \cdot 2 \cdot 3 \cdot 5$ **6.** $3 \cdot 3 \cdot 7$ **7.** prime **8.** $\frac{5}{11}$ **9.** $\frac{8}{15}$ **10.** $\frac{21}{50}$ **11.** $\frac{8}{3}$ **12.** $\frac{1}{4}$ **13.** $\frac{2}{3}$ **14.** $\frac{29}{18}$ **15.** $\frac{37}{60}$ **16.** $\frac{5}{12}$

17.

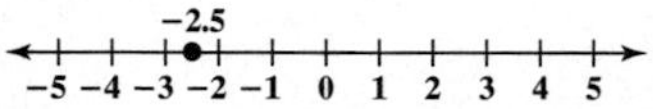

18.

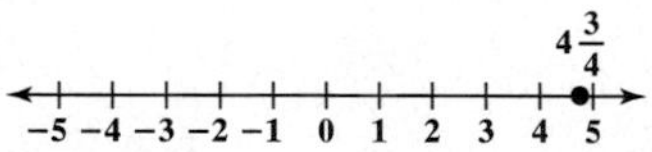

19. 0.625 **20.** $0.\overline{27}$

21. a. $\sqrt{81}$ **b.** 0, $\sqrt{81}$ **c.** −17, 0, $\sqrt{81}$ **d.** −17, $-\frac{9}{13}$, 0, 0.75, $\sqrt{81}$ **e.** $\sqrt{2}$, π **f.** −17, $-\frac{9}{13}$, 0, 0.75, $\sqrt{2}$, π, $\sqrt{81}$ **22.** Answers will vary; −2 is an example. **23.** Answers will vary; $\frac{1}{2}$ is an example. **24.** Answers will vary; π is an example. **25.** < **26.** > **27.** > **28.** < **29.** false **30.** true **31.** 58 **32.** 2.75

33. IV

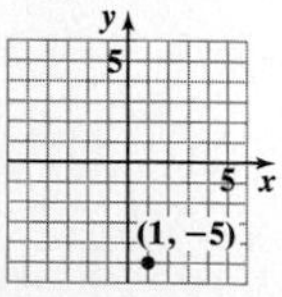

34. IV

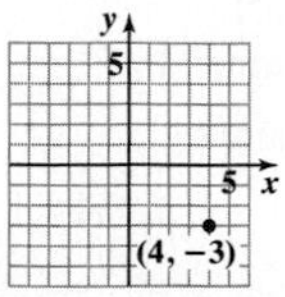

35. I

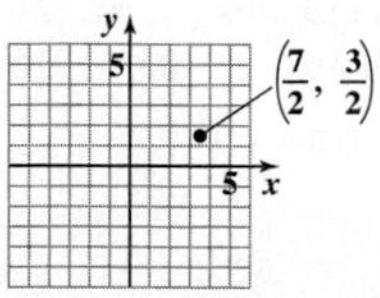

36. II

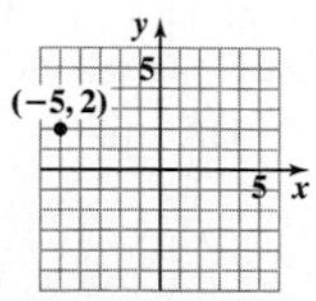

37. $A(5, 6)$; $B(-3, 0)$; $C(-5, 2)$; $D(-4, -2)$; $E(0, -5)$; $F(3, -1)$ **38.** 65 **39.** 80 **40.** 1989–1993 **41.** 1981–1985 **42.** 1977 **43.** 85 **44.** Finland, Germany, Austria **45.** 73 **46.** 40 **47.** $13y + 7$ **48.** $(x + 7)9$ **49.** $(6 + 4) + y = 10 + y$ **50.** $(7 \cdot 10)x = 70x$ **51.** $24x - 12 + 30y$ **52.** $7a + 2$ **53.** $28x + 19$ **54.** 1800; The sale price at 25% off is \$1800 for a \$2400 computer.

55. 2

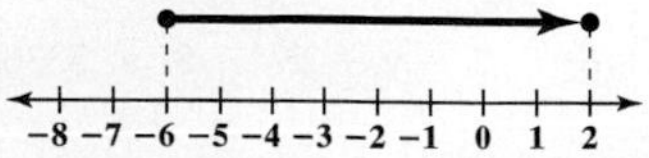

56. −3 **57.** $-\frac{11}{20}$ **58.** −7 **59.** $5y - 4x$ **60.** $40 - 2y$ **61.** 800 ft below sea level

62. The reservoir's level at the end of 5 months is 23 ft. **63.** $9 + (-13)$ **64.** 4 **65.** $-\frac{6}{5}$ **66.** -1.5 **67.** -7 **68.** -3 **69.** $-5 - 8a$ **70.** 27,150 ft **71.** 84 **72.** $-\frac{3}{11}$ **73.** -120 **74.** -9 **75.** undefined **76.** 2 **77.** $3x$ **78.** $-x - 1$ **79.** 36 **80.** -36 **81.** -32 **82.** $6x^3$ **83.** cannot be simplified **84.** -16 **85.** -16 **86.** 10 **87.** -2 **88.** 17 **89.** -88 **90.** 14 **91.** $-\frac{20}{3}$ **92.** $-\frac{2}{5}$ **93.** 6 **94.** 10 **95.** $28a - 20$ **96.** $6y - 12$ **97.** 13 lb; (4, 13) **98.** 16 lb; (6, 16) **99.** 805 million; fairly well, but slight overestimation **100.** Sales increase from 1998 to 2000 and decrease from 2000 to 2002.

Chapter 1 Test

1. 4 **2.** -11 **3.** -51 **4.** $\frac{1}{5}$ **5.** $-\frac{35}{6}$ or $-5\frac{5}{6}$ **6.** -5 **7.** 1 **8.** -4 **9.** -32 **10.** 1 **11.** $4x + 4$ **12.** $13x - 19y$ **13.** $10 - 6x$ **14.** $-7, -\frac{4}{5}, 0, 0.25, \sqrt{4}, \frac{22}{7}$ **15.** $>$ **16.** 12.8 **17.** II;

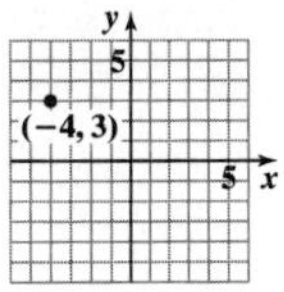

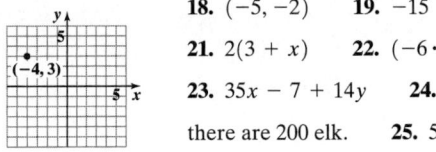

18. $(-5, -2)$ **19.** -15 **20.** 150 **21.** $2(3 + x)$ **22.** $(-6 \cdot 4)x = -24x$ **23.** $35x - 7 + 14y$ **24.** (30, 200); After 30 yr, there are 200 elk. **25.** 50 **26.** 725 million **27.** 16 sec **28.** 17,030 ft **29.** \$135 thousand **30.** \$134.7 thousand

CHAPTER 2

Section 2.1

Check Point Exercises

1. a. not a solution **b.** solution **2.** 17 or $\{17\}$ **3.** 2.29 or $\{2.29\}$ **4.** $\frac{1}{4}$ or $\left\{\frac{1}{4}\right\}$ **5.** 13 or $\{13\}$ **6.** 12 or $\{12\}$ **7.** 11 or $\{11\}$ **8.** 2100 words

Exercise Set 2.1

1. $\{23\}$ **3.** $\{-20\}$ **5.** $\{-16\}$ **7.** $\{-12\}$ **9.** $\{4\}$ **11.** $\{-11\}$ **13.** $\{2\}$ **15.** $\left\{-\frac{17}{12}\right\}$ **17.** $\left\{\frac{21}{4}\right\}$ **19.** $\left\{-\frac{11}{20}\right\}$ **21.** $\{4.3\}$ **23.** $\left\{-\frac{21}{4}\right\}$ **25.** $\{18\}$ **27.** $\left\{\frac{9}{10}\right\}$ **29.** $\{-310\}$ **31.** $\{4.3\}$ **33.** $\{0\}$ **35.** $\{11\}$ **37.** $\{5\}$ **39.** $\{-13\}$ **41.** $\{6\}$ **43.** $\{-12\}$ **45.** $x = \triangle + \square$ **47.** $\triangle - \square = x$ **49.** $x - 12 = -2$; 10 **51.** $\frac{2}{5}x - 8 = \frac{7}{5}x$; -8 **53.** \$1700 **55.** 525,000 deaths **57.** \$17 billion **65.** Answers will vary; example: $x - 100 = -101$. **67.** $\{2.7529\}$ **68.** II

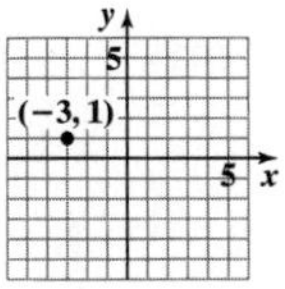

69. -12 **70.** $6 - 9x$

Section 2.2

Check Point Exercises

1. 36 or $\{36\}$ **2. a.** 21 or $\{21\}$ **b.** -4 or $\{-4\}$ **c.** -3.1 or $\{-3.1\}$ **3. a.** 24 or $\{24\}$ **b.** -16 or $\{-16\}$ **4. a.** -5 or $\{-5\}$ **b.** 3 or $\{3\}$ **5.** 6 or $\{6\}$ **6.** -10 or $\{-10\}$ **7.** 6 or $\{6\}$ **8.** 2020

Exercise Set 2.2

1. $\{30\}$ **3.** $\{-33\}$ **5.** $\{7\}$ **7.** $\{-9\}$ **9.** $\left\{-\frac{7}{2}\right\}$ **11.** $\{6\}$ **13.** $\left\{-\frac{3}{4}\right\}$ **15.** $\{0\}$ **17.** $\{18\}$ **19.** $\{-8\}$ **21.** $\{-17\}$ **23.** $\{47\}$ **25.** $\{45\}$ **27.** $\{-5\}$ **29.** $\{5\}$ **31.** $\{6\}$ **33.** $\{-1\}$ **35.** $\{-2\}$ **37.** $\left\{\frac{9}{4}\right\}$ **39.** $\{-6\}$ **41.** $\{-3\}$ **43.** $\{-3\}$ **45.** $\{4\}$ **47.** $\left\{-\frac{3}{2}\right\}$ **49.** $\{2\}$ **51.** $\{-4\}$ **53.** $\{-6\}$ **55.** $x = \square \cdot \triangle$ **57.** $-\triangle = x$ **59.** $6x = 10$; $\frac{5}{3}$ **61.** $\frac{x}{-9} = 5$; -45 **63.** 10 sec **65.** 1502.2 mph **67.** 60 yd **73.** d **75.** Answers will vary; example: $\frac{5}{4}x = -20$. **77.** $\{6.5\}$ **78.** 100 **79.** -100 **80.** 3

Section 2.3

Check Point Exercises

1. 6 or $\{6\}$ **2.** 2 or $\{2\}$ **3.** 5 or $\{5\}$ **4.** -2 or $\{-2\}$ **5.** inconsistent, no solution; $\{\ \}$ or $\varnothing$ **6.** all real numbers or $\mathbb{R}$ **7.** 124 lb

Exercise Set 2.3

1. $\{3\}$ **3.** $\{-1\}$ **5.** $\{4\}$ **7.** $\{4\}$ **9.** $\left\{\frac{7}{2}\right\}$ **11.** $\{-3\}$ **13.** $\{6\}$ **15.** $\{8\}$ **17.** $\{4\}$ **19.** $\{1\}$ **21.** $\{-4\}$ **23.** $\{5\}$ **25.** $\{6\}$ **27.** $\{1\}$ **29.** $\{-57\}$ **31.** $\{-10\}$ **33.** $\{18\}$ **35.** $\left\{\frac{7}{4}\right\}$ **37.** $\{1\}$ **39.** $\{24\}$ **41.** $\{-6\}$ **43.** $\{20\}$ **45.** $\{-7\}$ **47.** no solution; { } or ∅ **49.** all real numbers; ℝ **51.** $\left\{\frac{2}{3}\right\}$ **53.** all real numbers; ℝ **55.** no solution; { } or ∅ **57.** $\{0\}$ **59.** no solution; { } or ∅ **61.** $\{0\}$ **63.** $\left\{\frac{4}{3}\right\}$ **65.** $x = □\$ - □\triangle$ **67.** 240 **69.** $\frac{x}{5} + \frac{x}{3} = 16$; 30 **71.** $\frac{3x}{4} - 3 = \frac{x}{2}$; 12 **73.** 85 mph **75.** 3.7; point (3.7, 10) on low-humor graph **77.** 409.2 ft **83.** c **85.** $\{3\}$ **87.** $\{-4.2\}$ **89.** < **90.** < **91.** -10

Section 2.4

Check Point Exercises

1. $l = \frac{A}{w}$ **2.** $l = \frac{P - 2w}{2}$ **3.** $m = \frac{T - D}{p}$ **4.** $x = 15 + 12y$ **5.** 2.3% **6. a.** 0.67 **b.** 2.5 **7.** 4.5 **8.** 15 **9.** 36% **10.** 35% **11. a.** \$1152 **b.** 4% decrease

Exercise Set 2.4

1. $r = \frac{d}{t}$ **3.** $P = \frac{I}{rt}$ **5.** $r = \frac{C}{2\pi}$ **7.** $m = \frac{E}{c^2}$ **9.** $m = \frac{y - b}{x}$ **11.** $p = \frac{T - D}{m}$ **13.** $b = \frac{2A}{h}$ **15.** $n = 5M$ **17.** $c = 4F - 160$ **19.** $a = 2A - b$ **21.** $r = \frac{S - P}{Pt}$ **23.** $b = \frac{2A}{h} - a$ **25.** $x = \frac{C - By}{A}$ **27.** 89% **29.** 0.2% **31.** 478% **33.** 10,000% **35.** 0.27 **37.** 0.634 **39.** 1.7 **41.** 0.03 **43.** 0.005 **45.** 6 **47.** 7.2 **49.** 5 **51.** 170 **53.** 20% **55.** 12% **57.** 60% **59.** 75% **61.** $x = \frac{y}{a + b}$ **63.** $x = \frac{y - 5}{a - b}$ **65.** $x = \frac{y}{c + d}$ **67.** $x = \frac{y + C}{A + B}$ **69. a.** $z = 3A - x - y$ **b.** 96% **71. a.** $t = \frac{d}{r}$ **b.** 2.5 hr **73.** 408 **75.** 2,369,200 **77.** 59% **79.** 12.5% **81.** \$9 **83. a.** \$1008 **b.** \$17,808 **85. a.** \$103.20 **b.** \$756.80 **87.** 15% **89.** no; 2% loss **97.** d **99.** 1.5 sec and 60 ft **100.** $\{12\}$ **101.** $\{20\}$ **102.** $0.7x$

Mid-Chapter 2 Check Point

1. $\{16\}$ **2.** $\{-3\}$ **3.** $C = \frac{825H}{E}$ **4.** 8.4 **5.** $\{30\}$ **6.** $\left\{-\frac{8}{9}\right\}$ **7.** $r = \frac{S}{2\pi h}$ **8.** 40 **9.** $\{20\}$ **10.** $\{-1\}$ **11.** $x = \frac{By + C}{A}$ **12.** no solution; { } or ∅ **13.** 2008 **14.** $\{40\}$ **15.** 12.5% **16.** $\left\{-\frac{6}{5}\right\}$ **17.** 25% **18.** all real numbers; ℝ

Section 2.5

Check Point Exercises

1. a. $4x + 6$ **b.** $\frac{x - 4}{9}$ **2.** 12 **3.** MRI scan: \$575; Acupuncture: \$112 **4.** pages 96 and 97 **5.** 32; 4 mi **6.** 40 ft wide and 120 ft long **7.** \$940

Exercise Set 2.5

1. $x + 7$ **3.** $25 - x$ **5.** $9 - 4x$ **7.** $\frac{83}{x}$ **9.** $2x + 40$ **11.** $9x - 93$ **13.** $8(x + 14)$ **15.** $x + 60 = 410$; 350 **17.** $x - 23 = 214$; 237 **19.** $7x = 126$; 18 **21.** $\frac{x}{19} = 5$; 95 **23.** $4 + 2x = 56$; 26 **25.** $5x - 7 = 178$; 37 **27.** $x + 5 = 2x$; 5 **29.** $2(x + 4) = 36$; 14 **31.** $9x = 3x + 30$; 5 **33.** $\frac{3x}{5} + 4 = 34$; 50 **35.** *Titanic*: \$200 million; *Waterworld*: \$175 million **37.** Liberals: 39.6%; Conservatives: 17.6% **39.** pages 314 and 315 **41.** Rolling Stones: \$121 million; Springsteen: \$120 million **43.** 32 and 34 **45.** 800 mi **47.** 6 months **49.** 50 yd wide and 200 yd long **51.** 160 ft wide and 360 ft long **53.** 12 ft long and 4 ft high **55.** \$400 **57.** \$67,000 **59.** \$14,500 **61.** 11 hr **67.** d **69.** 5 ft 7 in. **71.** The uncle is 60 years old and the woman is 20 years old. **73.** -20 **74.** 0 **75.** $w = \frac{3V}{lh}$

Section 2.6

Check Point Exercises

1. 12 ft **2.** 400π ft$^2 \approx 1256$ ft^2; 40π ft ≈ 126 ft **3.** large pizza **4.** 2 times **5.** No; About 32 more cubic inches are needed. **6.** 120°, 40°, 20° **7.** 60°

Exercise Set 2.6

1. 18 m; 18 m^2 **3.** 56 in.^{2} **5.** 91 m^2 **7.** 50 ft **9.** 8 ft **11.** 50 cm **13.** 16π cm$^2 \approx 50$ cm^2; 8π cm ≈ 25 cm **15.** 36π yd$^2 \approx 113$ yd^2; 12π yd ≈ 38 yd **17.** 7 in.; 14 in. **19.** 36 in.^{3} **21.** 150π cm$^3 \approx 471$ cm^3 **23.** 972π cm$^3 \approx 3054$ cm^3 **25.** 48π m$^3 \approx 151$ m^3 **27.** $h = \frac{V}{\pi r^2}$ **29.** 9 times **31.** 50°, 50°, 80° **33.** $4x = 76$; $3x + 4 = 61$; $2x + 5 = 43$; 76°, 61°, 43°

35. 40°, 80°, 60° **37.** 32° **39.** 2° **41.** 48° **43.** 90° **45.** 75° **47.** 135° **49.** 50° **51.** 72 m^2 **53.** 70.5 cm^2 **55.** 168 cm^3 **57.** \$698.18 **59.** large pizza **61.** \$2262 **63.** approx 19.7 ft **65.** 21,000 yd^3 **67.** the can with diameter of 6 in. and height of 5 in. **69.** Yes, the water tank is a little over one cubic foot too small. **81.** 2.25 times **83.** Volume increases 8 times. **85.** 35° **86.** $s = \frac{P - b}{2}$ **87.** $\{8\}$ **88.** 0

Section 2.7

Check Point Exercises

1. a. **b.** **c.** 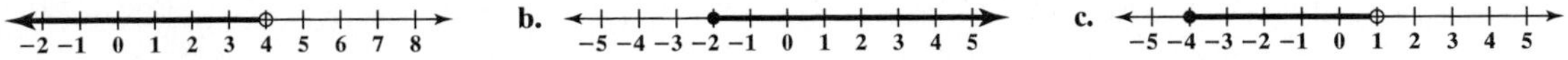

2. $\{x|x < 3\}$ 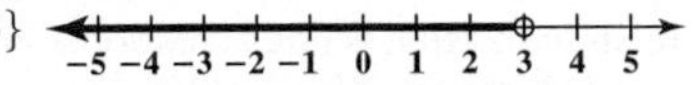 **3.** $\{x|x \geq -2\}$ 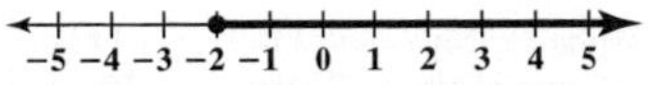

4. a. $\{x|x < 8\}$ 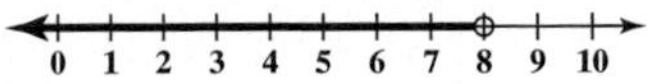 **b.** $\{x|x > -3\}$ 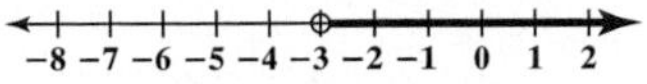

5. $\{y|y \geq 4\}$ 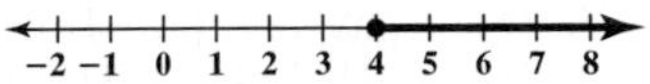 **6.** $\{x|x \geq 1\}$ 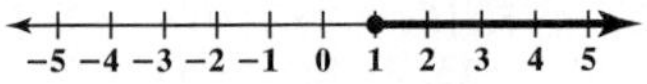

7. $\{x|x \geq 1\}$ 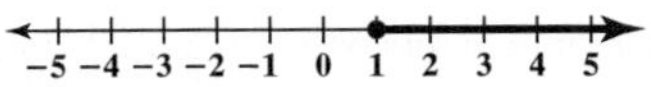 **8.** no solution; ∅ **9.** $\{x|x \text{ is a real number}\}$ or $\mathbb{R}$ **10.** at least 83%

Exercise Set 2.7

1. **3.** **5.** 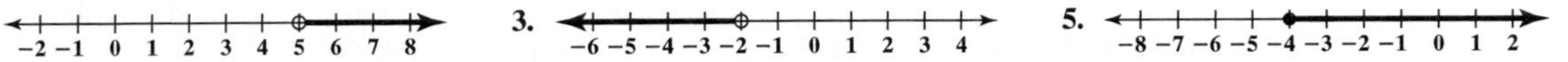

7. **9.** **11.** 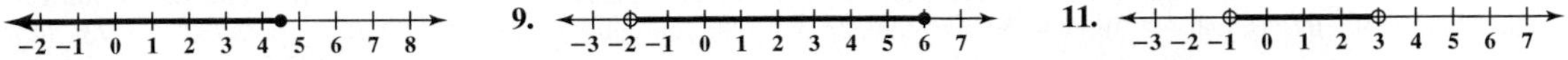

13. $\{x|x > -2\}$ **15.** $\{x|x \geq 4\}$ **17.** $\{x|x \geq 3\}$

19. $\{x|x > 7\}$ 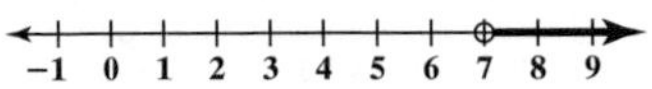 **21.** $\{x|x \leq 6\}$ 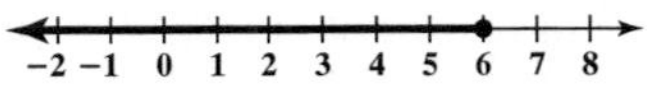

23. $\{y|y < 2\}$ 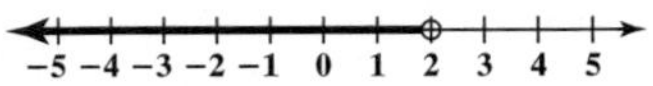 **25.** $\{x|x \leq 3\}$ 

27. $\{x|x < 16\}$ 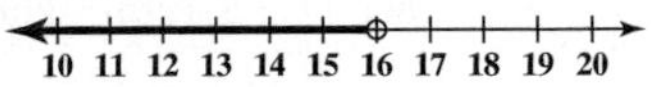 **29.** $\{x|x > 4\}$ 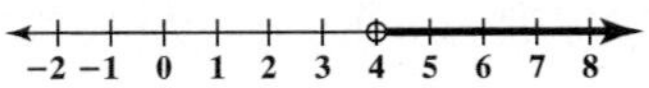

31. $\left\{x \middle| x > \frac{7}{6}\right\}$ 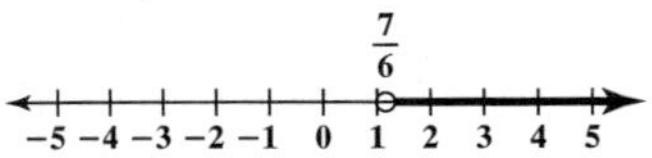 **33.** $\left\{y \middle| y \leq -\frac{3}{8}\right\}$ 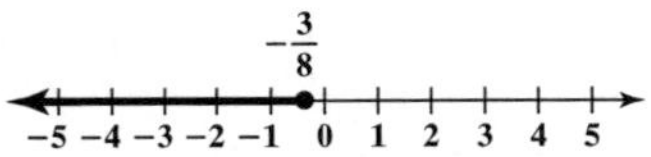

35. $\{y|y > 0\}$  **37.** $\{x|x < 8\}$ 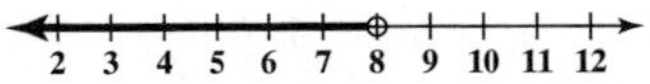

39. $\{x|x > -6\}$ 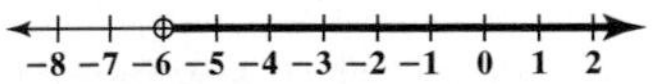 **41.** $\{x|x < 5\}$ 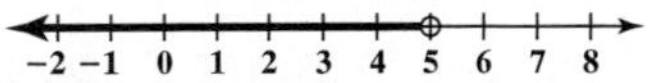

43. $\{x|x \geq -7\}$ 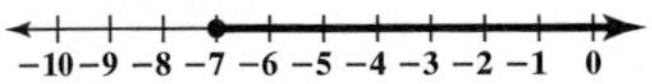 **45.** $\{x|x > -5\}$ 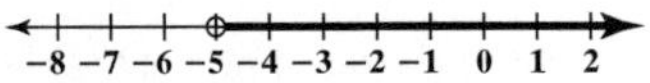

47. $\{x|x \leq -5\}$ 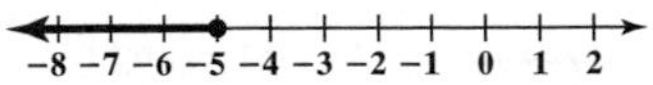 **49.** $\{x|x < 3\}$ 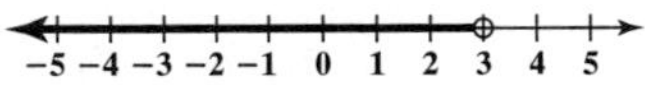

51. $\left\{y \middle| y \geq -\frac{1}{8}\right\}$ 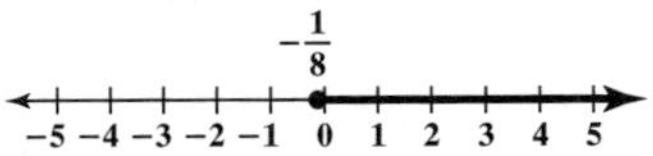 **53.** $\{x|x > -4\}$ 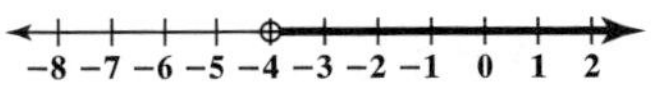

55. $\{x|x > 5\}$ 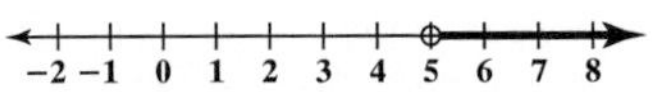 **57.** $\{x|x < 5\}$ 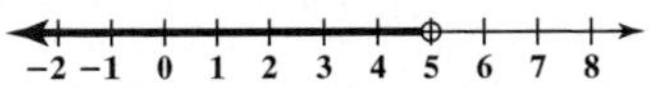

59. $\{x|x \geq -2\}$ 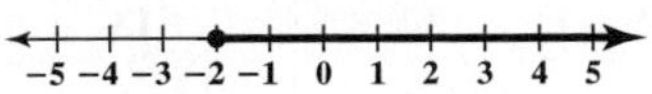 **61.** $\{x|x > -3\}$ 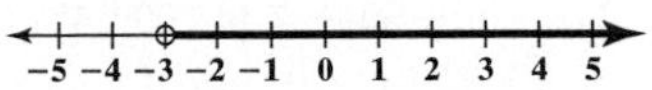

63. $\{x|x \geq 4\}$ 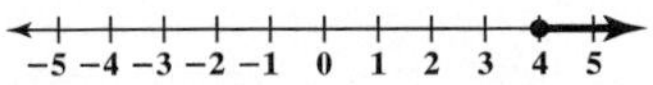 **65.** $\left\{x \middle| x > \frac{11}{3}\right\}$ 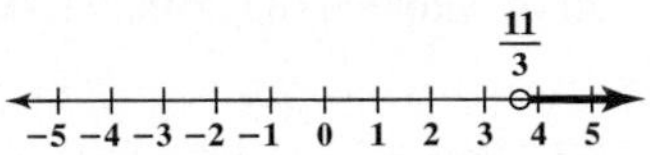

67. $\{y|y > 2\}$ 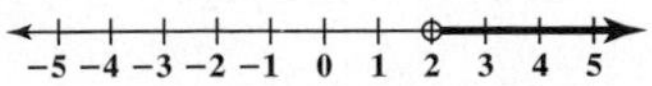 **69.** $\{y|y < 2\}$ 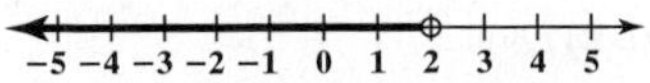

71. $\{x|x < 3\}$ 

73. $\left\{x \middle| x > \frac{5}{3}\right\}$ 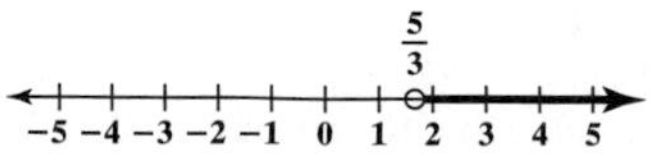

75. $\{x|x \geq 9\}$ 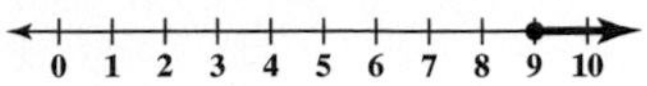

77. $\{x|x < -6\}$ 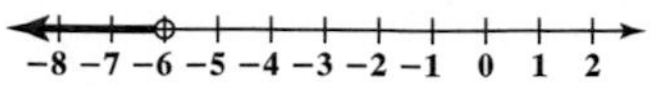

79. no solution; ∅ **81.** $\{x|x$ is a real number$\}$ or ℝ **83.** no solution; ∅ **85.** $\{x|x$ is a real number$\}$ or ℝ **87.** $\{x|x \leq 0\}$ **89.** $x > \frac{b-a}{3}$ **91.** $\frac{y-b}{m} \geq x$ **93.** x is between −2 and 2; $|x| < 2$ **95.** x is greater than 2 or less than −2; $|x| > 2$ **97.** Denmark, Netherlands, Norway **99.** Japan, Mexico **101.** Netherlands, Norway, Canada, U.S. **103.** 20 yr; from 2009 onward **105. a.** at least 96 **b.** if you get less than 66 on the final **107.** up to 1280 mi **109.** up to 29 bags of cement **111.** An open dot indicates an endpoint that is not a solution and a closed dot indicates an endpoint that is a solution. **117.** more than 720 mi **119.** $\{x|x < 0.4\}$ **121.** 20 **122.** length: 11 in.; width: 6 in. **123.** $\{4\}$

Review Exercises

1. $\{32\}$ **2.** $\{-22\}$ **3.** $\{12\}$ **4.** $\{-22\}$ **5.** $\{5\}$ **6.** $\{80\}$ **7.** $\{-56\}$ **8.** $\{11\}$ **9.** $\{4\}$ **10.** $\{-15\}$ **11.** $\{-12\}$ **12.** $\{-25\}$ **13.** $\{10\}$ **14.** $\{6\}$ **15.** $\{-5\}$ **16.** $\{-10\}$ **17.** $\{2\}$ **18.** $\{1\}$ **19.** 16 yr; 2006 **20.** $\{-1\}$ **21.** $\{12\}$ **22.** $\{-13\}$ **23.** $\{-3\}$ **24.** $\{-10\}$ **25.** $\{2\}$ **26.** $\{2\}$ **27.** no solution or ∅ **28.** all real numbers or ℝ **29.** 20 yr old **30.** $r = \frac{I}{P}$ **31.** $h = \frac{3V}{B}$ **32.** $w = \frac{P-2l}{2}$ **33.** $B = 2A - C$ **34.** $m = \frac{T-D}{P}$ **35.** 72% **36.** 0.35% **37.** 0.65 **38.** 1.5 **39.** 0.03 **40.** 9.6 **41.** 200 **42.** 48% **43.** 100% **44.** 40% **45.** 12.5% **46.** no; 1% **47. a.** $h = 7r$ **b.** 5 ft 3 in. **48.** 350 gallons **49.** 10 **50.** New York: 55 days; Los Angeles: 213 days **51.** pages 46 and 47 **52.** Streisand: 49 albums; Madonna: 47 albums **53.** 9 yr; 2012 **54.** 18 checks **55.** length: 150 yd; width: 50 yd **56.** \$240 **57.** 32.5 ft^2 **58.** 50 cm^2 **59.** 135 yd^2 **60.** 20π m ≈ 63 m; 100π m^2 ≈ 314 m^2 **61.** 6 ft **62.** 156 ft^2 **63.** \$1890 **64.** medium pizza **65.** 60 cm^3 **66.** 128π yd^3 ≈ 402 yd^3 **67.** 288π m^3 ≈ 905 m^3 **68.** 4800 m^3 **69.** 16 fish **70.** $x = 30, 3x = 90, 2x = 60$; 30°, 60°, 90° **71.** 85°, 35°, 60° **72.** 33° **73.** 105° **74.** 57.5° **75.** 45° and 135° **76.** 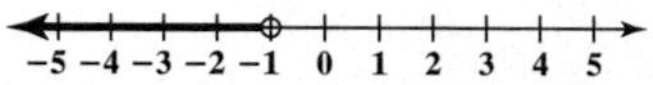

77. 

78. $\{x|x > 4\}$
79. $\{x|x \leq -3\}$ **80.** $\{x|x < 4\}$ 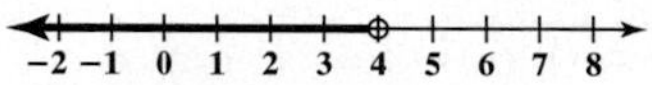

81. $\{x|x > -8\}$ 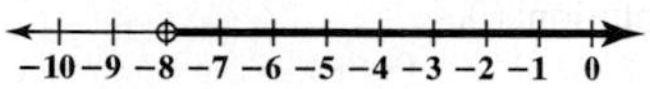

82. $\{x|x \geq -3\}$ 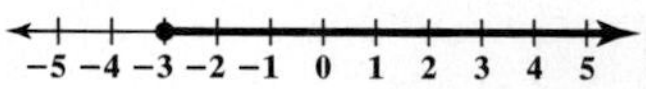

83. $\{x|x > 6\}$ 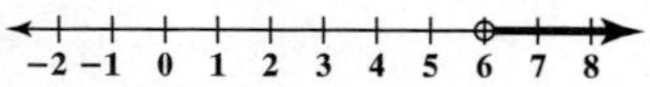

84. $\{x|x \geq 4\}$ 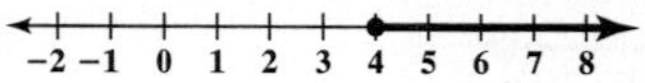

85. $\{x|x \leq 2\}$ 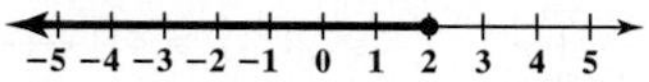

86. $\{x|x$ is a real number$\}$ or ℝ
87. no solution; ∅ **88.** at least 64 **89.** no more than 99 min

Chapter 2 Test

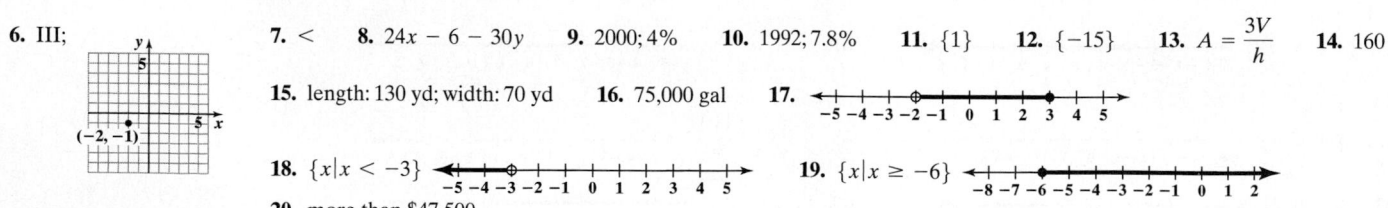

1. $\left\{\frac{9}{2}\right\}$ **2.** $\{-5\}$ **3.** $\left\{-\frac{4}{3}\right\}$ **4.** $\{2\}$ **5.** $\{-20\}$ **6.** $\left\{-\frac{5}{3}\right\}$ **7.** 60 yr; 2020 **8.** $h = \frac{V}{\pi r^2}$ **9.** $w = \frac{P-2l}{2}$ **10.** 8.4 **11.** 150 **12.** 5% **13.** 63 **14.** fitness trainer: \$50,950; teacher: \$28,080 **15.** 600 min **16.** length: 150 yd; width: 75 yd **17.** \$35 **18.** 517 m^2 **19.** 525 $in.^2$ **20.** 18 $in.^3$ **21.** 175π cm^3 ≈ 550 cm^3 **22.** \$650 **23.** 14 ft **24.** 126°, 42°, 12° **25.** 53° **26.** 

27. 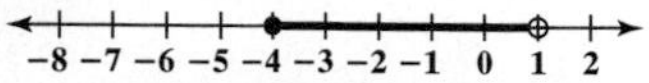

28. $\{x|x \leq -1\}$ **29.** $\{x|x < -6\}$ 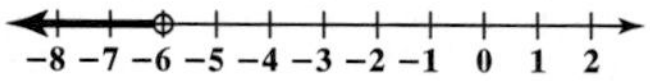

30. $\{x|x \leq -3\}$ 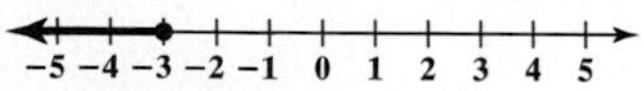

31. $\{x|x > 7\}$ 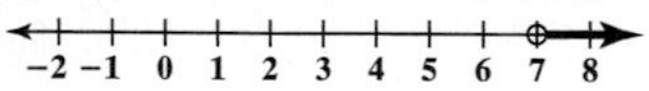

32. at least 92 **33.** widths greater than 8 inches

Cumulative Review Exercises (Chapters 1–2)

1. −4 **2.** −2 **3.** −128 **4.** $-103 - 20x$ **5.** $-4, -\frac{1}{3}, 0, \sqrt{4}, 1063$

6. III; 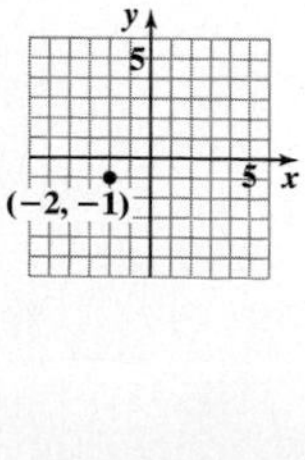

7. < **8.** $24x - 6 - 30y$ **9.** 2000; 4% **10.** 1992; 7.8% **11.** $\{1\}$ **12.** $\{-15\}$ **13.** $A = \frac{3V}{h}$ **14.** 160 **15.** length: 130 yd; width: 70 yd **16.** 75,000 gal **17.** 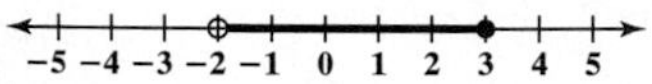

18. $\{x|x < -3\}$ 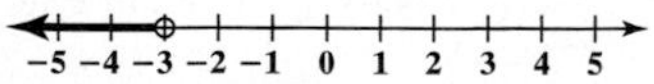

19. $\{x|x \geq -6\}$ 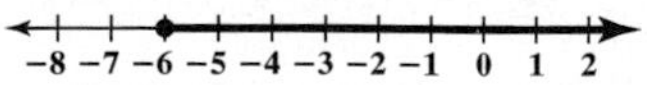

20. more than \$47,500

CHAPTER 3

Section 3.1

Check Point Exercises

1. a. solution **b.** not a solution **2.** $(-2, -4), (-1, -1), (0, 2), (1, 5)$, and $(2, 8)$

3.

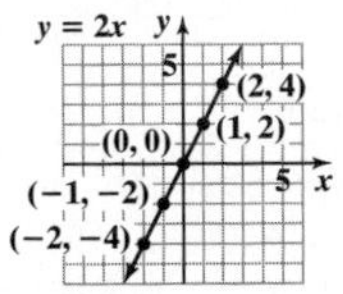

4.

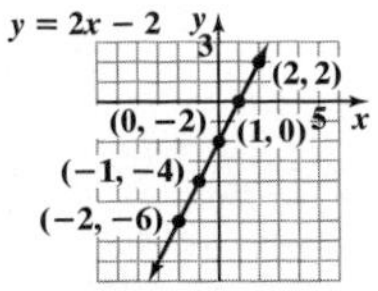

5.

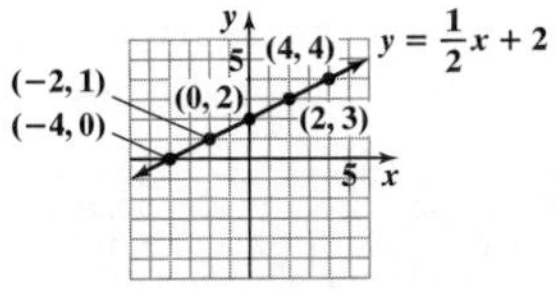

6.

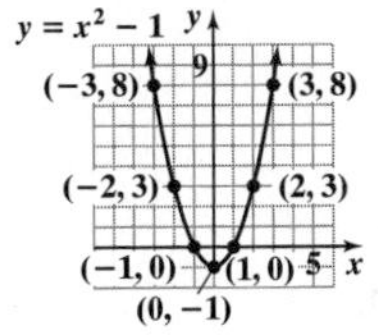

7. a.

x	$y = 2x$: (x, y)
0	(0, 0)
2	(2, 4)
4	(4, 8)
6	(6, 12)
8	(8, 16)
10	(10, 20)
12	(12, 24)

x	$y = 10 + x$: (x, y)
0	(0, 10)
2	(2, 12)
4	(4, 14)
6	(6, 16)
8	(8, 18)
10	(10, 20)
12	(12, 22)

b.

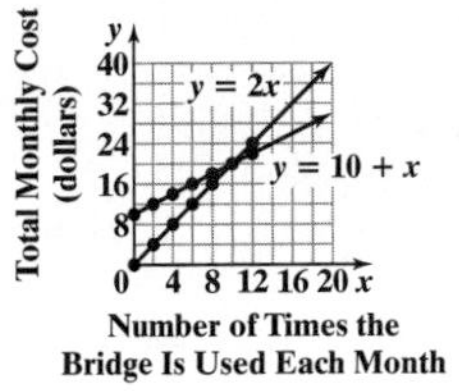

c. (10, 20); If the bridge is used 10 times in a month, the total monthly cost without the coupon book is the same as the monthly cost with the coupon book, namely \$20.

Exercise Set 3.1

1. (2, 3) and (3, 2) are not solutions; $(-4, -12)$ is a solution. **3.** $(-5, -20)$ is not a solution; (0, 0) and $(9, -36)$ are solutions. **5.** $(2, -2)$ is not a solution; (0, 6) and $(-3, 0)$ are solutions. **7.** (0, 5) is not a solution; $(-5, 6)$ and $(10, -3)$ are solutions. **9.** $\left(1, \frac{1}{3}\right)$ is not a solution; (0, 0) and $\left(2, -\frac{2}{3}\right)$ are solutions. **11.** (3, 4) and $(0, -4)$ are not solutions; (4, 7) is a solution.

13.

x	(x, y)
−2	(−2, −24)
−1	(−1, −12)
0	(0, 0)
1	(1, 12)
2	(2, 24)

15.

x	(x, y)
−2	(−2, 20)
−1	(−1, 10)
0	(0, 0)
1	(1, −10)
2	(2, −20)

17.

x	(x, y)
−2	(−2, −21)
−1	(−1, −13)
0	(0, −5)
1	(1, 3)
2	(2, 11)

19.

x	(x, y)
−2	(−2, 13)
−1	(−1, 10)
0	(0, 7)
1	(1, 4)
2	(2, 1)

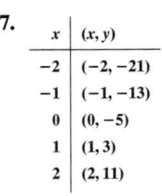

21.

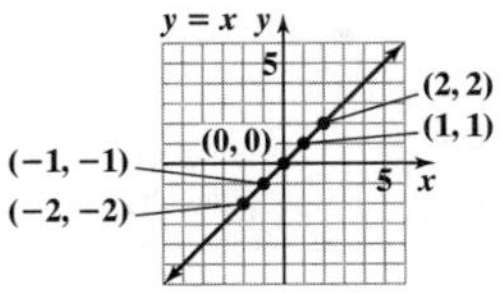

23.

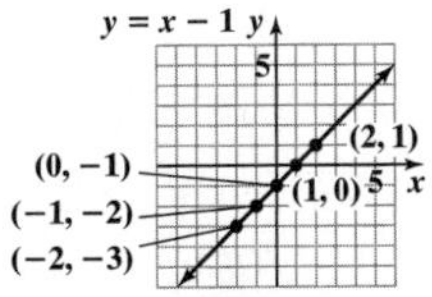

25.

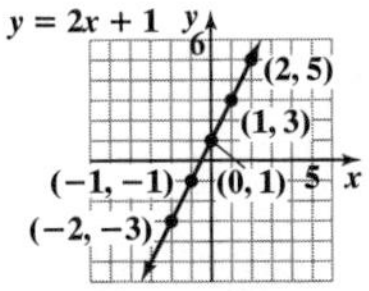

27.

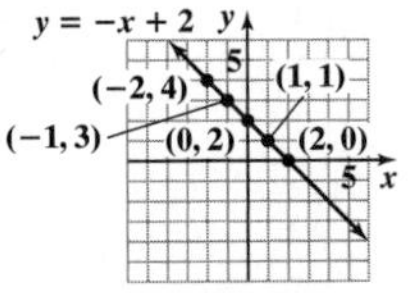

29.

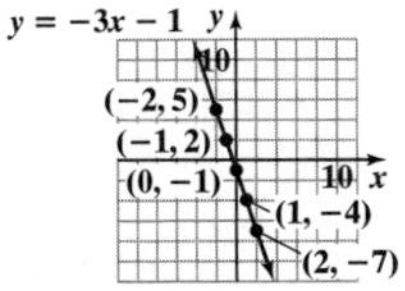

31.

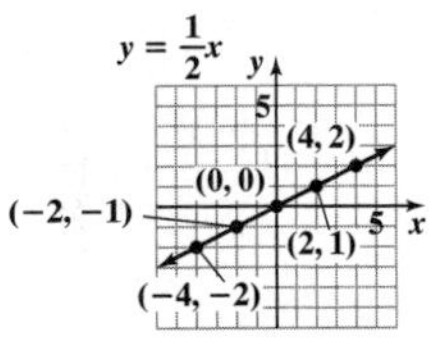

33.

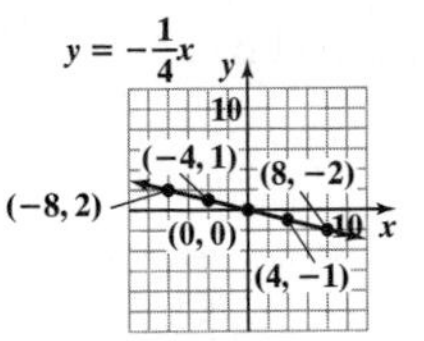

35.

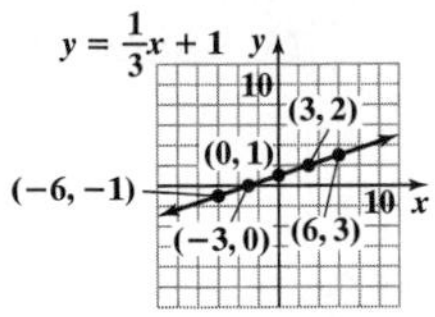

37.

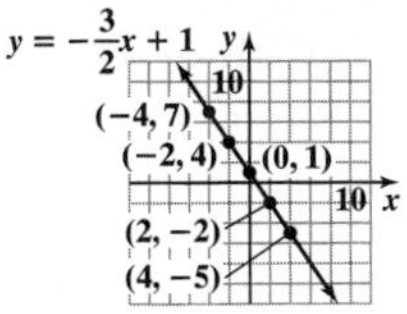

39.

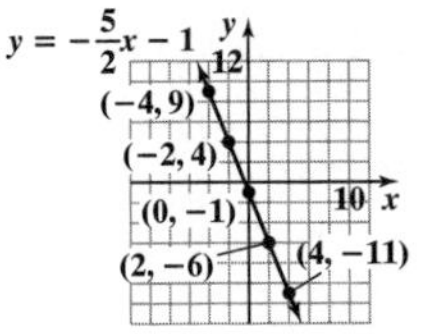

41.

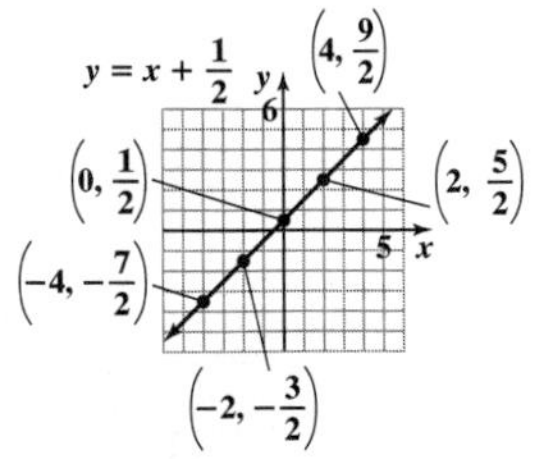

43.

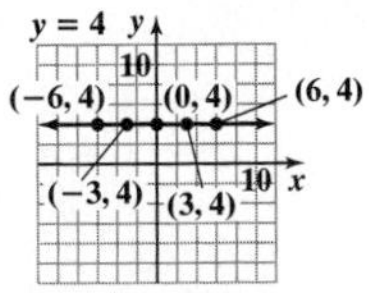

45.

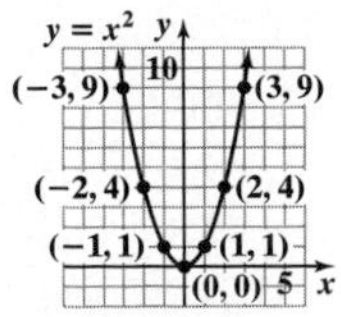

47.

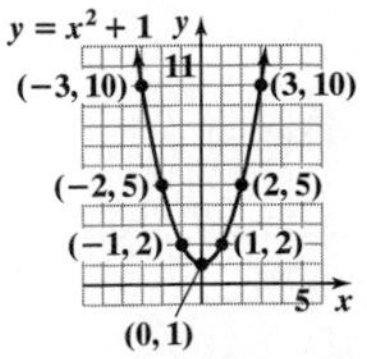

49.

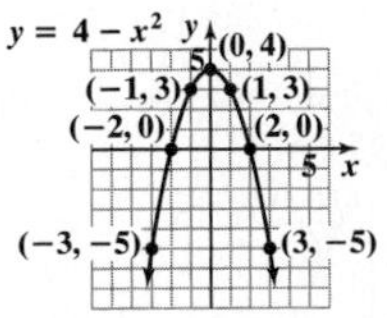

51.

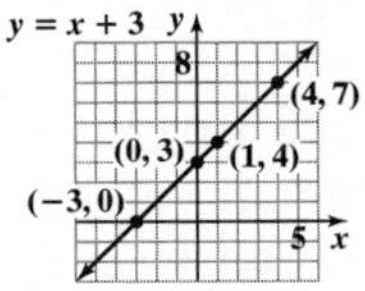

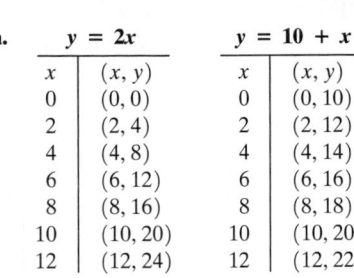

53.
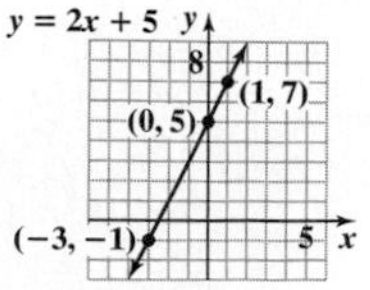

55. a. $8x + 6y = 14.50$ **b.** \$1.25

57. a. African Americans

x	(x, y)
0	(0, 5.2)
10	(10, 7.6)
20	(20, 10)
30	(30, 12.4)

b. Whites

x	(x, y)
0	(0, 3.4)
10	(10, 5.6)
20	(20, 7.8)
30	(30, 10)

c. Hispanics

x	(x, y)
0	(0, 4.2)
10	(10, 5.5)
20	(20, 6.8)
30	(30, 8.1)

59. a. 40; Both rental companies have the same cost when the truck is used for 40 miles. **b.** 55
c. 54; Both rental companies have the same cost of \$54 when the truck is used for 40 miles.

61. a.

x	(x, y)
0	(0, 30,000)
10	(10, 30,500)
20	(20, 31,000)
30	(30, 31,500)
40	(40, 32,000)

b.
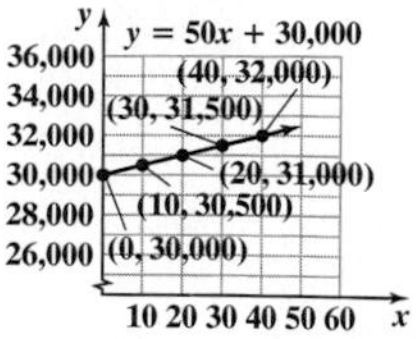

67.
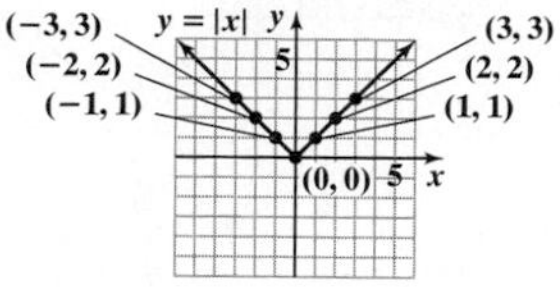

69. a. (0, 0.6), (1, 0.3), (2, 0.2), (3, 0.3), (4, 0.6), (5, 1.1) **b.** between 10:00 A.M. and noon

71.
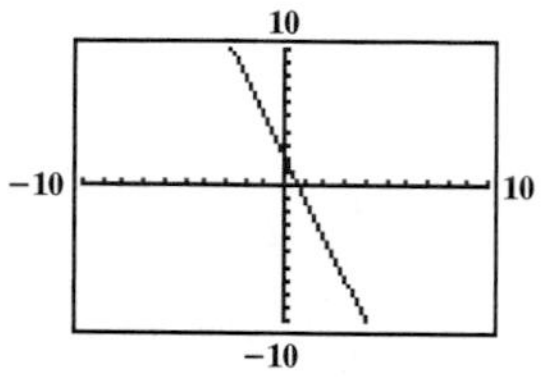

Answers will vary.

73.
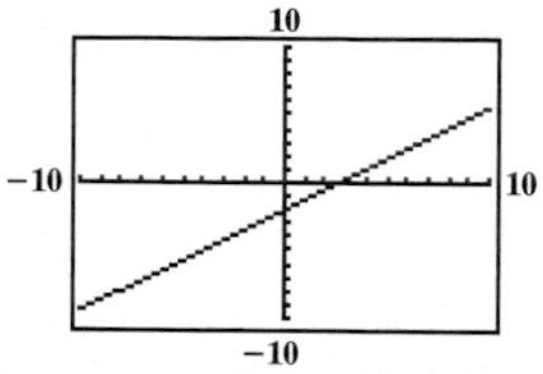

Answers will vary.

75.
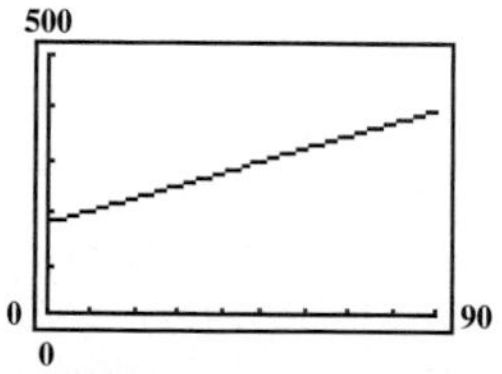

U.S. population increases.

76. 2 **77.** 1 **78.** $h = \frac{3V}{A}$

Section 3.2

Check Point Exercises

1. a. x-intercept is -3; y-intercept is 5. **b.** y-intercept is 4; no x-intercept. **c.** x-intercept is 0; y-intercept is 0. **2.** 3 **3.** -4

4.
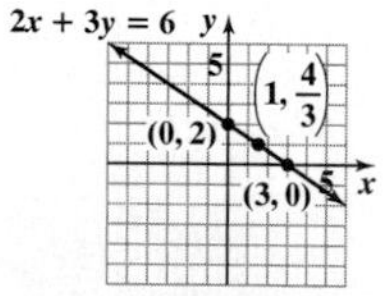

5.
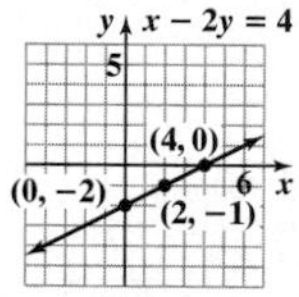

6.
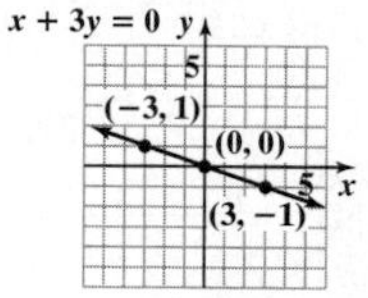

7.
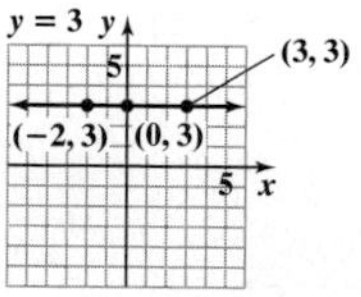

8.
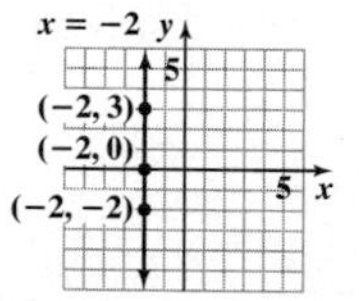

Exercise Set 3.2

1. a. 3 **b.** 4 **3. a.** -4 **b.** -2 **5. a.** 0 **b.** 0 **7. a.** no x-intercept **b.** -2 **9.** x-intercept is 10; y-intercept is 4.
11. x-intercept is $\frac{15}{2}$, or $7\frac{1}{2}$; y-intercept is -5. **13.** x-intercept is 8; y-intercept is $-\frac{8}{3}$, or $-2\frac{2}{3}$. **15.** x-intercept is 0; y-intercept is 0.
17. x-intercept is $-\frac{11}{2}$, or $-5\frac{1}{2}$; y-intercept is $\frac{11}{3}$, or $3\frac{2}{3}$.

19.
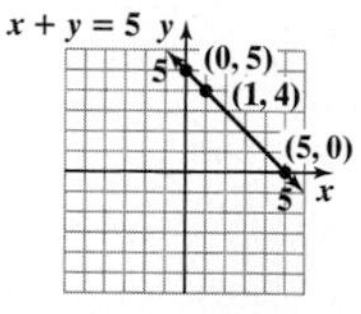

21.
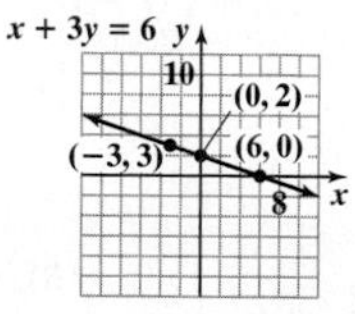

23.
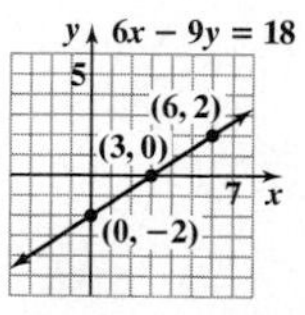

25.
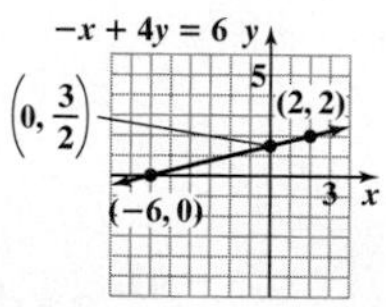

27.
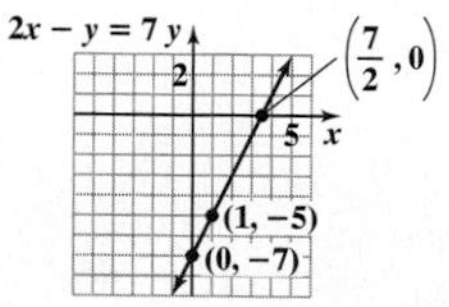

29.
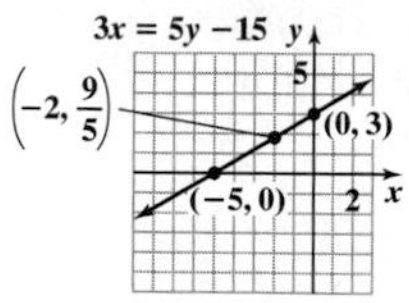

31.
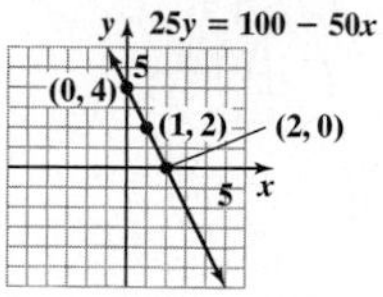

33.
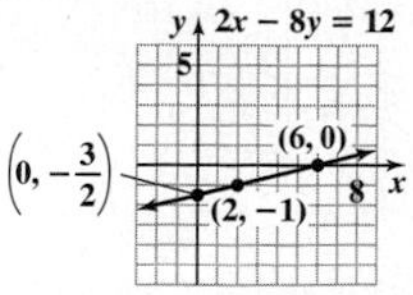

35.
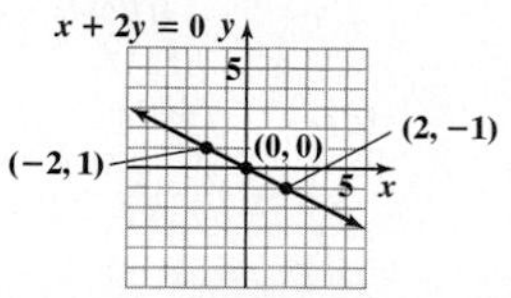

37.
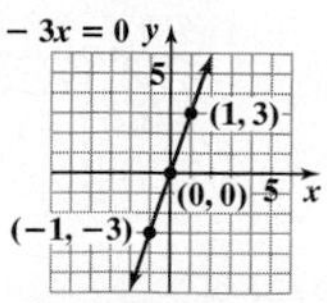

39. 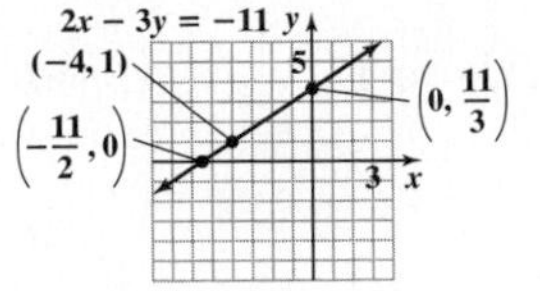

41. $y = 3$ **43.** $x = -3$ **45.** $y = 0$ **47.** 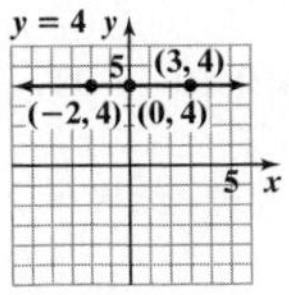

49. 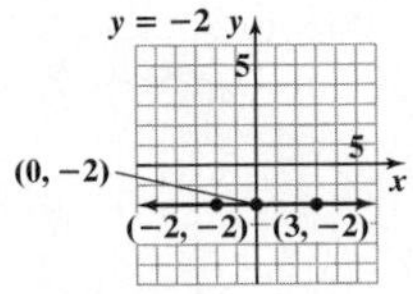

51. 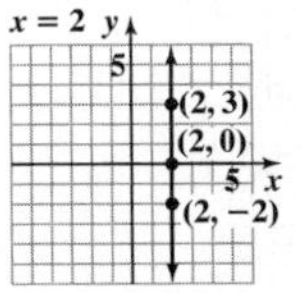

53. 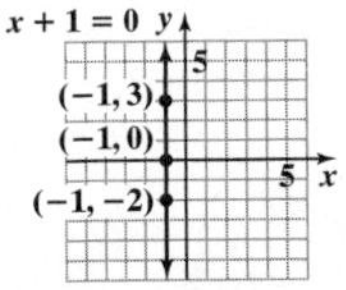

55. 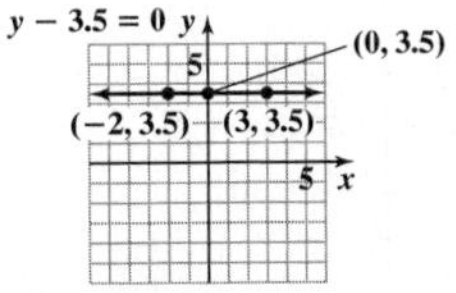

57. 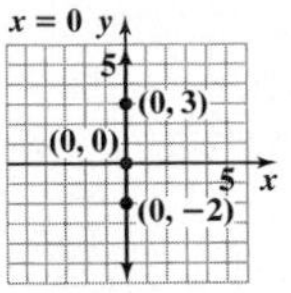

59. 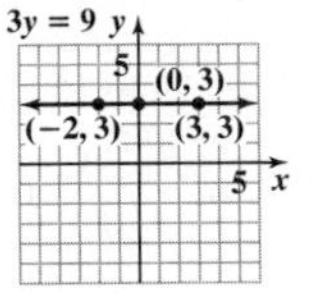

61. 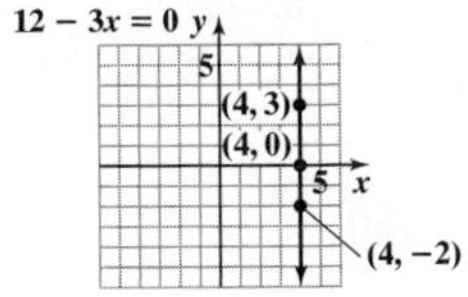

63. Exercise 4 **65.** Exercise 7 **67.** Exercise 1 **69.** 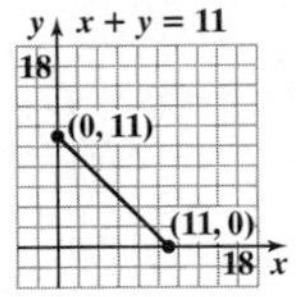

71. from 3 to 12 sec **73.** 45; The vulture was 45 m above the ground when the observation started.
75. 12, 13, 14, 15, 16; The vulture is on the ground at this time.

77. 8:00 A.M. to 11:00 A.M. **79.** 11:00 A.M. to 1:00 P.M. **81. a.** 80; In 1994, carbonated beverages had 80% of the market share. **b.** very well **c.** 140; In 2134, there will be no consumption of carbonated beverages. Model breakdown: model extended too far into the future. **93.** $y = 6$
95. $-6; 3$ **97.** 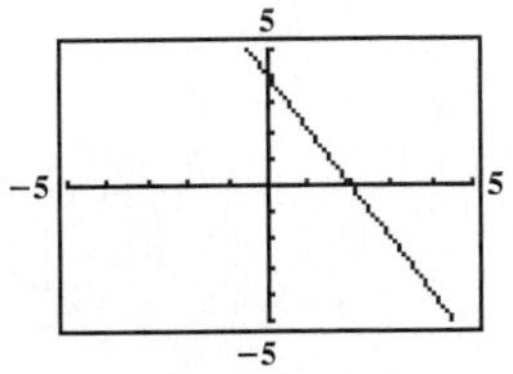
x-intercept: 2; y-intercept: 4
99. 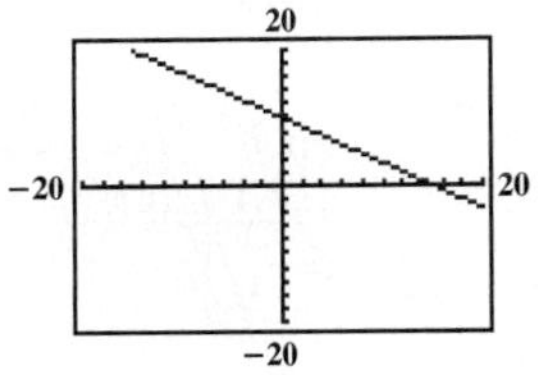
x-intercept: 15; y-intercept: 10
101. 13.4 **102.** $4x + 5$
103. 

Section 3.3

Check Point Exercises

1. a. 6 **b.** $-\frac{7}{5}$ **2. a.** 0 **b.** undefined **3.** Both slopes equal 2, so the lines are parallel. **4.** $\frac{2}{15} \approx 0.13$; The slope indicates that the number of U.S. men living alone is projected to increase by 0.13 million each year. The rate of change is 0.13 million men per year.

Exercise Set 3.3

1. $\frac{3}{4}$; rises **3.** $\frac{1}{4}$; rises **5.** 0; horizontal **7.** -5; falls **9.** undefined; vertical **11.** $\frac{1}{2}$ **13.** $-\frac{1}{3}$

15. $-\frac{1}{2}$ **17.** $-\frac{2}{3}$ **19.** 0 **21.** undefined **23.** parallel **25.** not parallel

27. 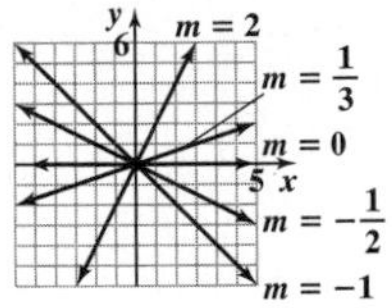

29. Slopes of corresponding opposite sides are equal: $-\frac{2}{5}$ and $\frac{4}{3}$. **31.** 2 **33.** collinear
35. 250; The amount spent online per U.S. online household increased by $250 each year from 1999 to 2001.
37. -0.4; The percentage of men is decreasing by 0.4% per year. **39.** 2900; Mean income is increasing $2900 per year.
41. 0.40; Cost increases by $0.40 per mile driven. **43.** pitch $= \frac{1}{3}$ **45.** 8.3% **53.** b

55. b_2, b_1, b_4, b_3 **57.** -3 **59.** $\frac{3}{4}$ **61.** 12 in. and 24 in. **62.** -42 **63.** $\{x | x \le 4\}$; 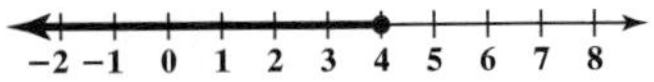

Section 3.4

Check Point Exercises

1. a. 5; -3 **b.** $\frac{2}{3}$; 4 **c.** -7; 6 **2.** 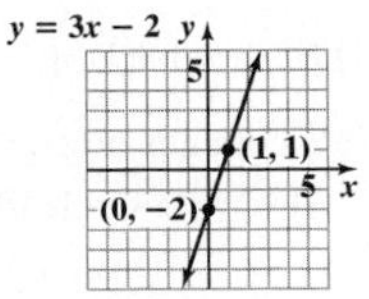

3. 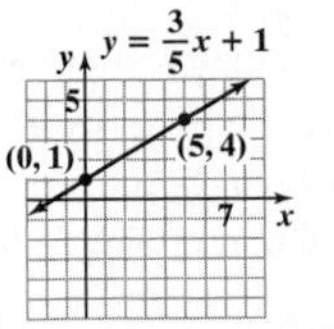

4. 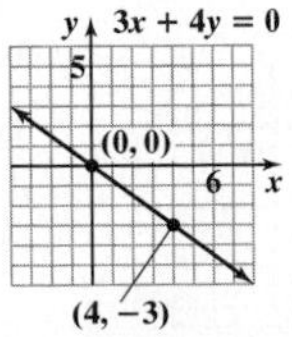

Exercise Set 3.4

1. 3; 2 **3.** 3; −5 **5.** $-\frac{1}{2}$; 5 **7.** 7; 0 **9.** 0; 10 **11.** −1; 4 **13.** $y = 5x + 7$; 5; 7 **15.** $y = -x + 6$; −1; 6 **17.** $y = -6x$; −6; 0 **19.** $y = 2x$; 2; 0 **21.** $y = -\frac{2}{7}x$; $-\frac{2}{7}$; 0 **23.** $y = -\frac{3}{2}x + \frac{3}{2}$; $-\frac{3}{2}$; $\frac{3}{2}$ **25.** $y = \frac{3}{4}x - 3$; $\frac{3}{4}$; −3

27.

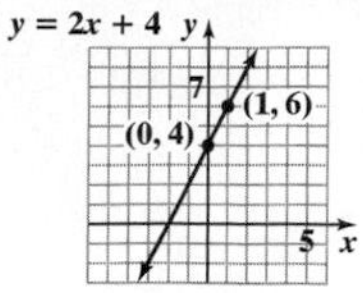

29.

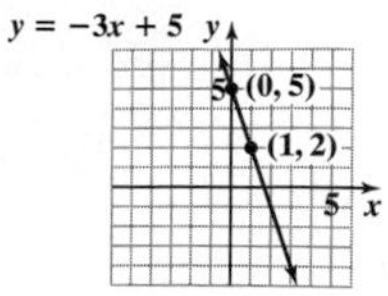

31.

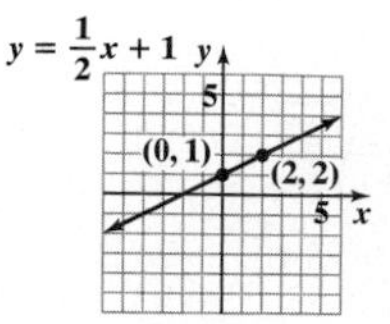

33.

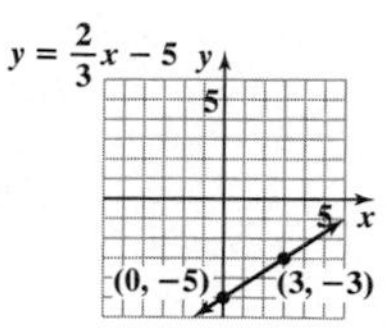

35.

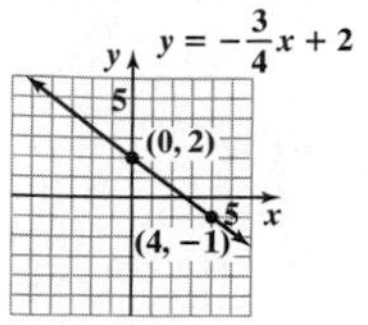

37.

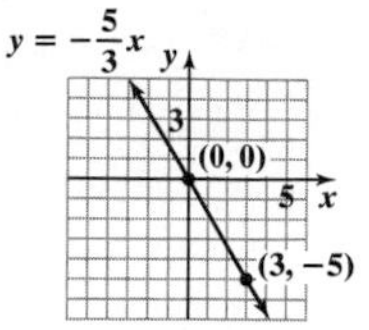

39. a. $y = -3x$ **b.** −3; 0 **c.**

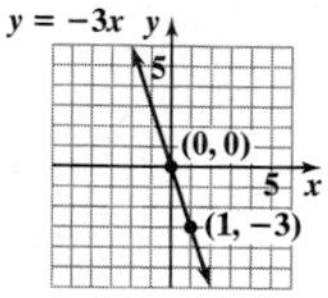

41. a. $y = \frac{4}{3}x$ **b.** $\frac{4}{3}$; 0 **c.**

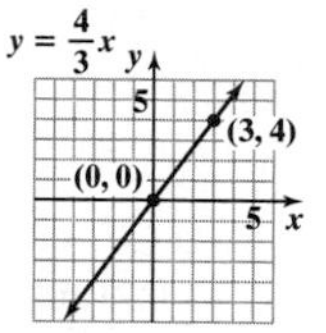

43. a. $y = -2x + 3$ **b.** −2; 3 **c.**

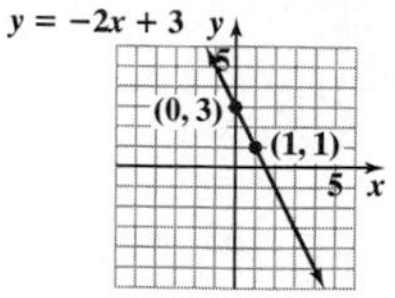

45. a. $y = -\frac{7}{2}x + 7$ **b.** $-\frac{7}{2}$; 7 **c.**

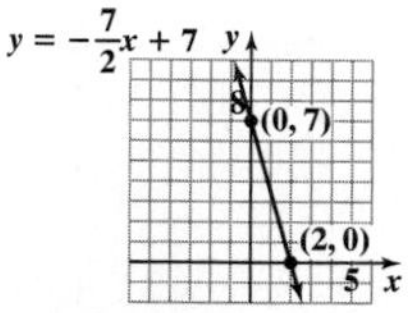

47.

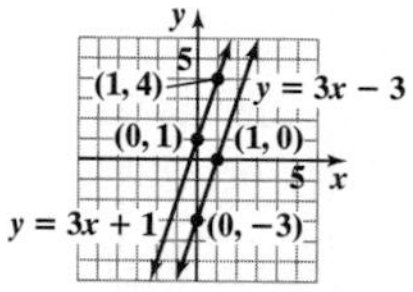

parallel; The slopes are equal.

49.

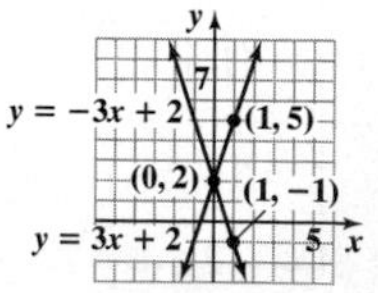

not parallel; The slopes are not equal.

51.

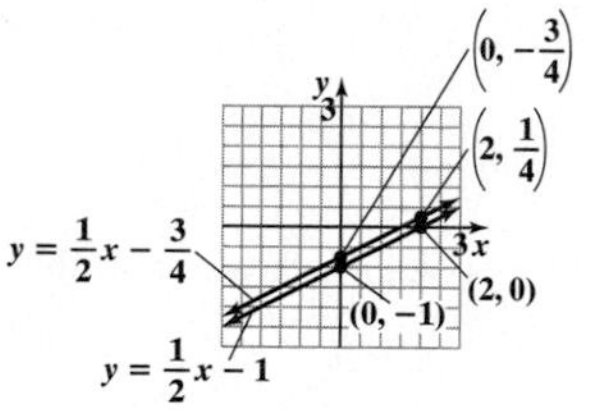

parallel; The slopes are equal.

53. $y = -3x + 5$ **55.** $y = -x + 2$ **57.** $y = x$ **59. a.** 10; The number of endangered species in the United States has been increasing by 10 species per year. **b.** 166; In 1980, there were 166 endangered animal species in the United States. **c.** 166; 216; 266; 316; 366; 396; Answers will vary. **61. a.** 21; In 1997, 21 million people in sub-Saharan Africa were living with AIDS. **b.** 1.6; Number of people living with AIDS in sub-Saharan Africa is increasing by 1.6 million per year. **c.** $y = 1.6x + 21$ **d.** 35.4 million **67.** $F = \frac{9}{5}C + 32$ **69.** 8 **70.** 0 **71.** 56

Mid-Chapter 3 Check Point

1. a. 4 **b.** 2 **c.** $-\frac{1}{2}$ **2. a.** −5 **b.** no y-intercept **c.** undefined slope **3. a.** 0 **b.** 0 **c.** $\frac{3}{5}$

4.

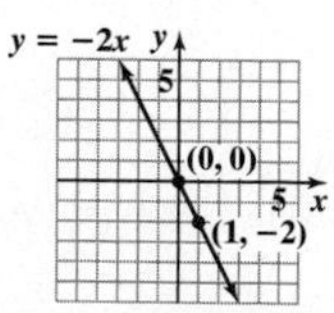

5.

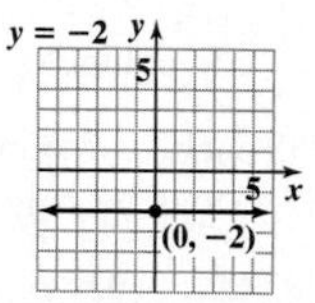

6.

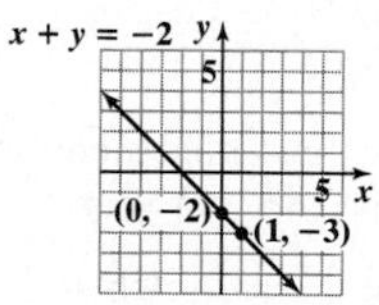

7.

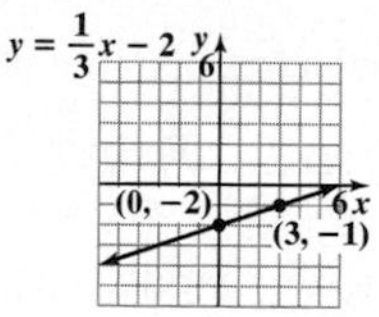

8.

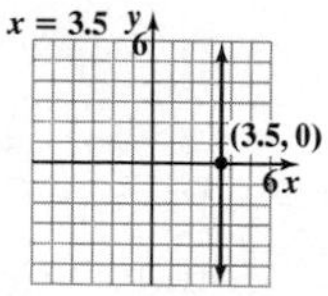

9.

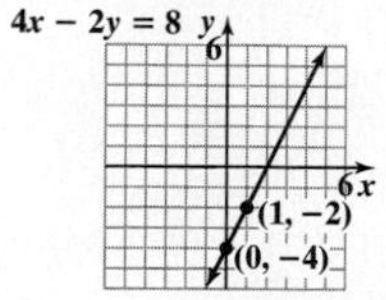

10.

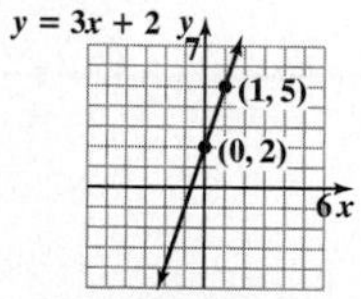

11.

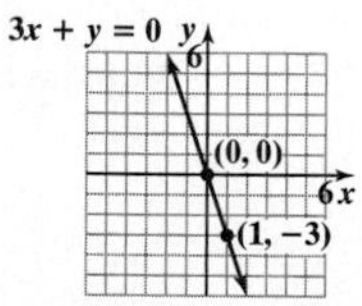

12.

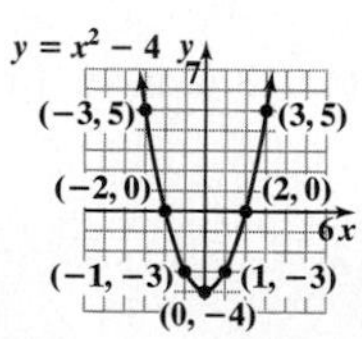

13.

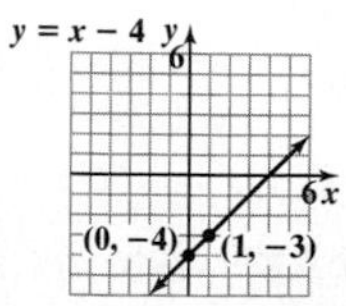

14.

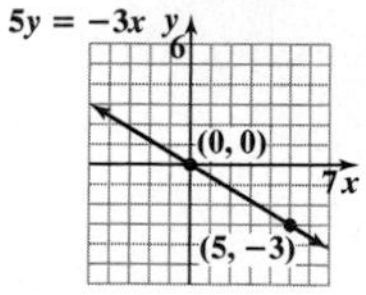

15.

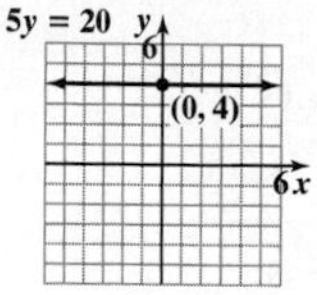

16. $m = \frac{5}{2}$; y-intercept is −5. **17.** parallel; Both have slopes of $\frac{4}{5}$. **18. a.** 33 **b.** In 1995, 33% of U.S. colleges offered distance learning. **c.** 7.8 **d.** For the years 1995 through 2002, the percentage of colleges that offered distance learning increased at a rate of 7.8% per year.

Section 3.5

Check Point Exercises

1. $y + 5 = 6(x - 2)$; $y = 6x - 17$ **2. a.** $y + 1 = -5(x + 2)$, or $y + 6 = -5(x + 1)$ **b.** $y = -5x - 11$
3. $y - 5 = 3(x + 2)$; $y = 3x + 11$ or $f(x) = 3x + 11$ **4.** 3 **5.** $y = 0.28x + 27.2$; 41.2

Exercise Set 3.5

1. $y - 5 = 3(x - 2)$; $y = 3x - 1$ **3.** $y - 6 = 5(x + 2)$; $y = 5x + 16$ **5.** $y + 2 = -8(x + 3)$; $y = -8x - 26$
7. $y - 0 = -12(x + 8)$; $y = -12x - 96$ **9.** $y + 2 = -1\left(x + \frac{1}{2}\right)$; $y = -x - \frac{5}{2}$ **11.** $y - 0 = \frac{1}{2}(x - 0)$; $y = \frac{1}{2}x$
13. $y + 2 = -\frac{2}{3}(x - 6)$; $y = -\frac{2}{3}x + 2$ **15.** $y - 2 = 2(x - 1)$, or $y - 10 = 2(x - 5)$; $y = 2x$
17. $y - 0 = 1(x + 3)$, or $y - 3 = 1(x - 0)$; $y = x + 3$ **19.** $y + 1 = 1(x + 3)$, or $y - 4 = 1(x - 2)$; $y = x + 2$
21. $y + 1 = \frac{5}{7}(x + 4)$, or $y - 4 = \frac{5}{7}(x - 3)$; $y = \frac{5}{7}x + \frac{13}{7}$ **23.** $y + 1 = 0(x + 3)$, or $y + 1 = 0(x - 4)$; $y = -1$
25. $y - 4 = 1(x - 2)$, or $y - 0 = 1(x + 2)$; $y = x + 2$ **27.** $y - 0 = 8\left(x + \frac{1}{2}\right)$, or $y - 4 = 8(x - 0)$; $y = 8x + 4$ **29. a.** 5 **b.** $-\frac{1}{5}$
31. a. -7 **b.** $\frac{1}{7}$ **33. a.** $\frac{1}{2}$ **b.** -2 **35. a.** $-\frac{2}{5}$ **b.** $\frac{5}{2}$ **37. a.** -4 **b.** $\frac{1}{4}$ **39. a.** $-\frac{1}{2}$ **b.** 2 **41. a.** $\frac{2}{3}$ **b.** $-\frac{3}{2}$
43. a. undefined **b.** 0 **45.** $y - 2 = 2(x - 4)$; $y = 2x - 6$ **47.** $y - 4 = -\frac{1}{2}(x - 2)$; $y = -\frac{1}{2}x + 5$
49. $y + 10 = -4(x + 8)$; $y = -4x - 42$ **51.** $y + 3 = -5(x - 2)$; $y = -5x + 7$ **53.** $y - 2 = \frac{2}{3}(x + 2)$; $y = \frac{2}{3}x + \frac{10}{3}$
55. $y + 7 = -2(x - 4)$; $y = -2x + 1$ **57.** $y = 3x - 2$ **59.** $y = -5x - 20$ **61.** $y = 5$ **63.** $y = -\frac{1}{2}x + 1$ **65.** $y = -\frac{2}{3}x - 2$ **67.** $-\frac{A}{B}$
69. a. $y - 162 = 1(x - 2)$, or $y - 168 = 1(x - 8)$ **b.** $y = x + 160$ **c.** 180 lb
71. Answers will vary. Example: $y - 3 = -\frac{2}{3}(x - 12)$; $y = -\frac{2}{3}x + 11$; 6.3.

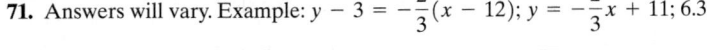

73. a–d. Answers will vary.

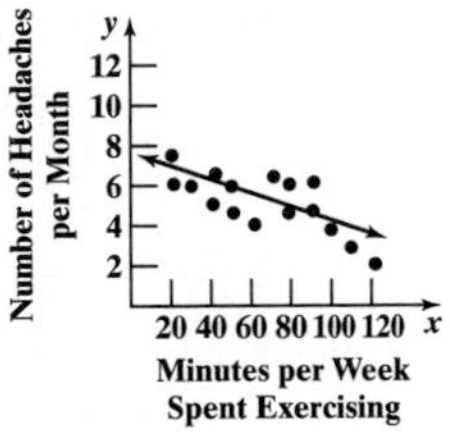

77. c

81. a.

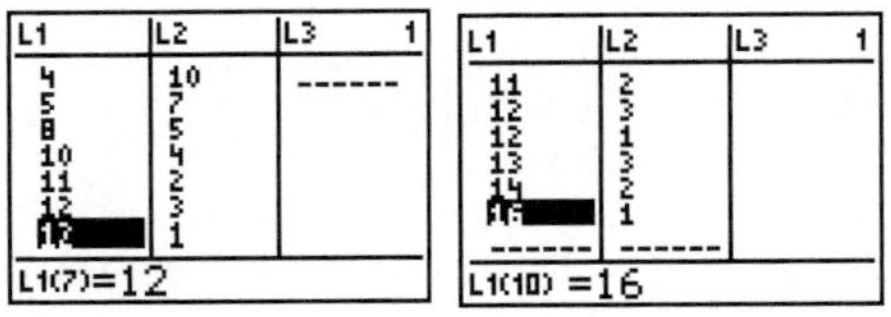

b.

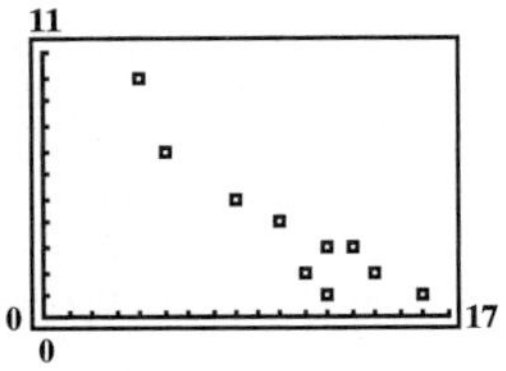

c. $a = -0.6867924528$; $b = 11.01132075$; $r = -0.9214983162$;

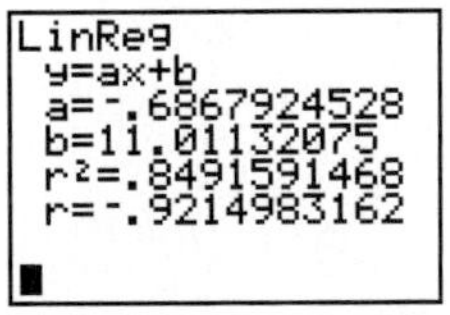

d.

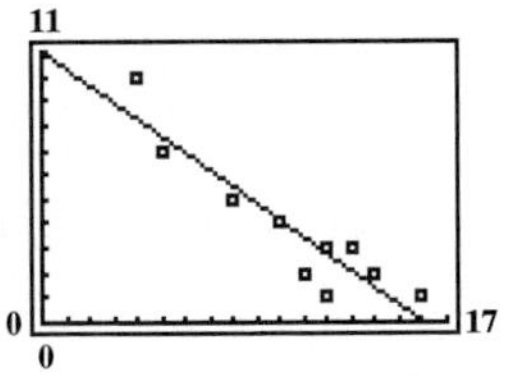

82. at most 12 sheets of paper **83.** 1, $\sqrt{4}$ **84.**

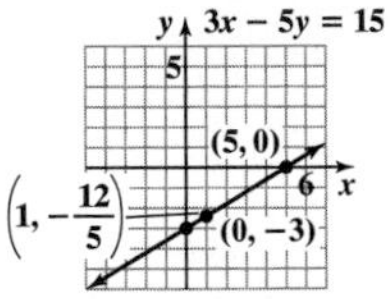

Review Exercises

1. $(-3, 3)$ is not a solution; $(0, 6)$ and $(1, 9)$ are solutions. **2.** $(0, 4)$ and $(-1, 15)$ are not solutions; $(4, 0)$ is a solution.

3. a.

x	(x, y)
-2	$(-2, -7)$
-1	$(-1, -5)$
0	$(0, -3)$
1	$(1, -1)$
2	$(2, 1)$

b.

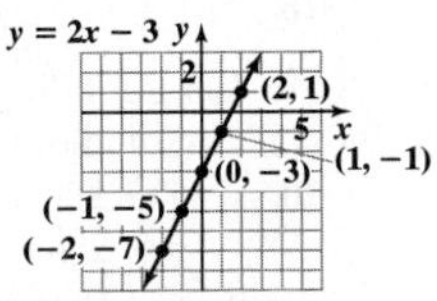

4. a.

x	(x, y)
-2	$(-2, 0)$
-1	$\left(-1, \frac{1}{2}\right)$
0	$(0, 1)$
1	$\left(1, \frac{3}{2}\right)$
2	$(2, 2)$

b.

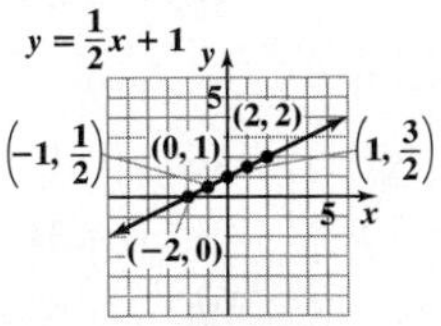

5.

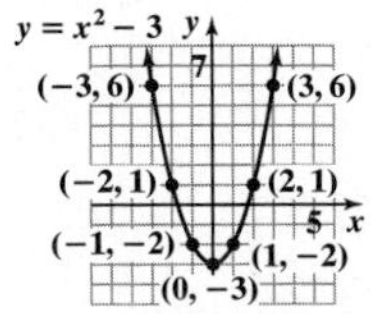

6. a.

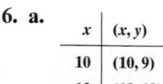

x	(x, y)
10	(10, 9)
12	(12, 19)
14	(14, 29)
16	(16, 39)

b. Answers will vary.

7. a. -2 **b.** -4

8. a. no x-intercept **b.** 2

9. a. 0 **b.** 0

10.

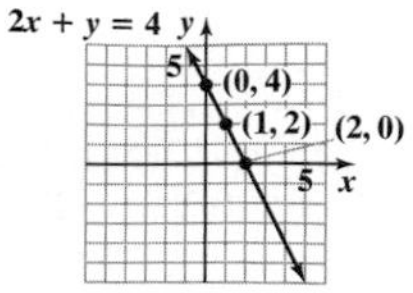

11.

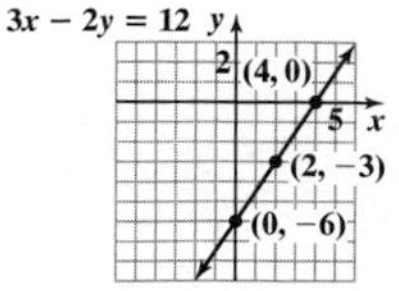

12.

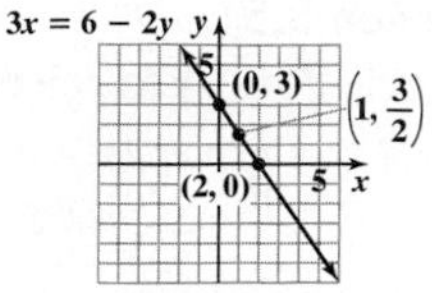

13.

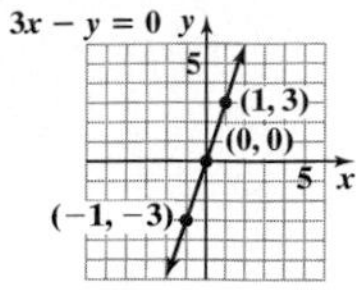

14.

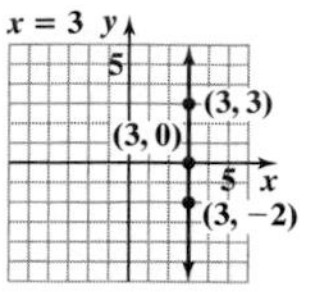

15.

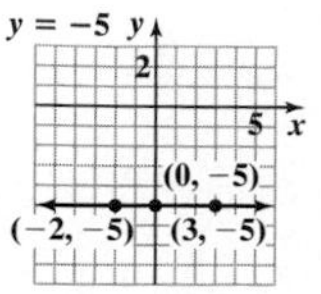

16.

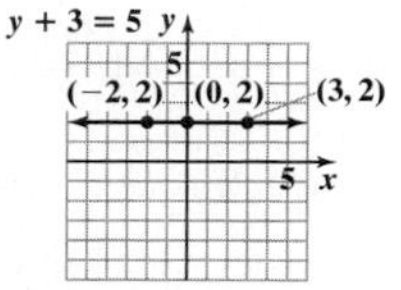

17.

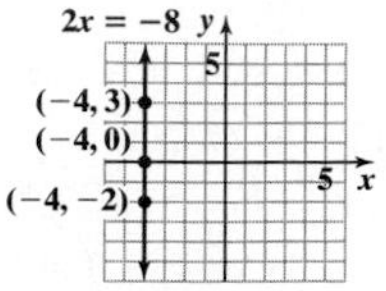

18. a. 5:00 P.M.; -4°F **b.** 8:00 P.M.; 16°F **c.** 4 and 6; At 4:00 P.M. and 6:00 P.M, the temperature was 0°F. **d.** 12; At noon, the temperature was 12°F. **e.** The temperature stayed the same, 12°F. **19.** $-\frac{1}{2}$; falls **20.** 3; rises **21.** 0; horizontal **22.** undefined; vertical **23.** $\frac{3}{5}$ **24.** undefined **25.** $-\frac{1}{3}$ **26.** 0 **27.** not parallel **28.** parallel **29. a.** $m = -44$; The number of new AIDS diagnoses decreased at a rate of 44 each year from 1999 to 2001. **b.** $m = 909$; The number of new AIDS diagnoses increased at a rate of 909 each year from 2001 to 2003. **c.** $m = 432.5$; yes; Answers will vary. **30.** 5; -7 **31.** -4; 6 **32.** 0; 3 **33.** $-\frac{2}{3}$; 2

34.

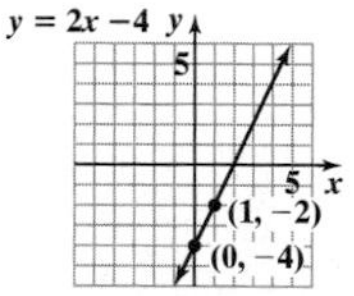

35.

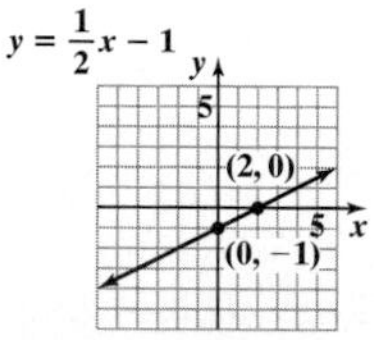

36.

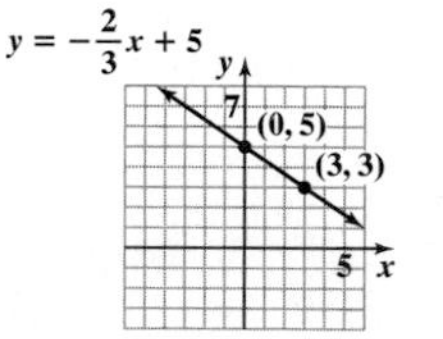

37.

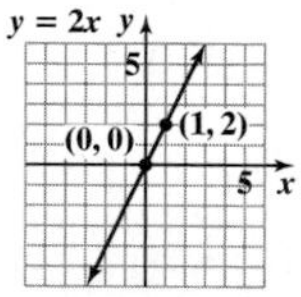

38.

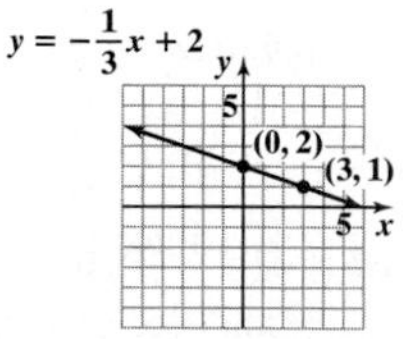

39.

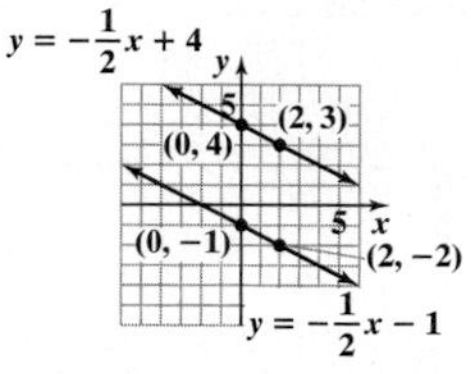

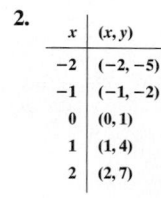

Yes, they are parallel since both have slopes of $-\frac{1}{2}$ and different y-intercepts.

40. a. 25; In 1990, the average age of U.S. Hispanics was 25.
b. 0.3; The average age for U.S. whites increased at a rate of about 0.3 each year from 1990 to 2000.
c. $y = 0.3x + 35$ **d.** 41 years old

41. $y - 7 = 6(x + 4)$; $y = 6x + 31$

42. $y - 4 = 3(x - 3)$, or $y - 1 = 3(x - 2)$; $y = 3x - 5$

43. $y + 7 = -3(x - 4)$; $y = -3x + 5$

44. $y - 6 = -3(x + 2)$; $y = -3x$

45. a. $y - 1.5 = 0.95(x - 1)$, or $y - 3.4 = 0.95(x - 3)$ **b.** $y = 0.95x + 0.55$ **c.** $10.05 billion

Chapter 3 Test

1. $(-2, 1)$ is not a solution; $(0, -5)$ and $(4, 3)$ are solutions.

2.

x	(x, y)
−2	(−2, −5)
−1	(−1, −2)
0	(0, 1)
1	(1, 4)
2	(2, 7)

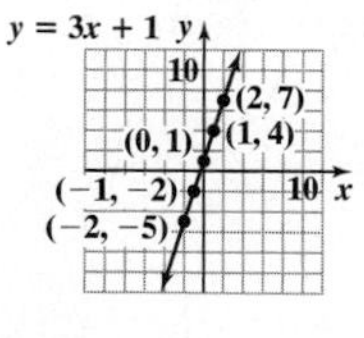

3.

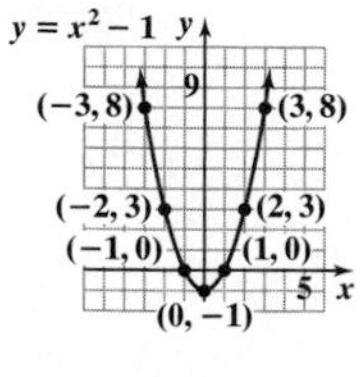

4. a. 2 **b.** -3

5.

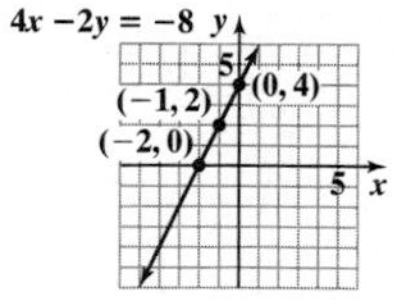

6.

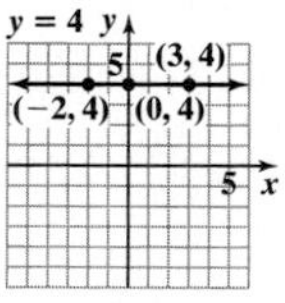

7. 3; rises **8.** undefined; vertical **9.** $\frac{3}{2}$ **10.** parallel **11.** -1; 10 **12.** -2; 6

13.

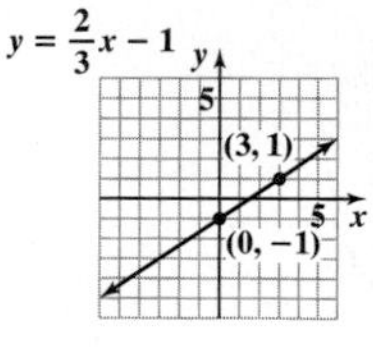

14.

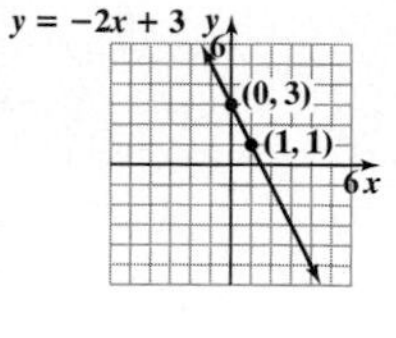

15. $y - 4 = -2(x + 1)$; $y = -2x + 2$

16. $y - 1 = 3(x - 2)$, or $y + 8 = 3(x + 1)$; $y = 3x - 5$

17. $y - 3 = 2(x + 2)$; $y = 2x + 7$

18. $y + 4 = -\frac{1}{2}(x - 6)$; $y = -\frac{1}{2}x - 1$

19. 106; Spending per pupil increased at a rate of about $106 each year.

20. a. $y - 2.3 = 0.3(x - 1)$ or $y - 3.2 = 0.3(x - 4)$ **b.** $y = 0.3x + 2$ **c.** 4.4 million

Cumulative Review Exercises (Chapters 1–3)

1. 2 **2.** $4 - 6x$ **3.** $\sqrt{5}$ **4.** $\{19\}$ **5.** $\left\{\frac{5}{4}\right\}$ **6.** $x = \frac{y - b}{m}$ **7.** 800 **8.** 40 mph **9.** $\{x|x \le -2\}$ 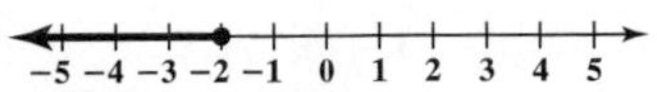

10. $\{x|x < 0\}$ 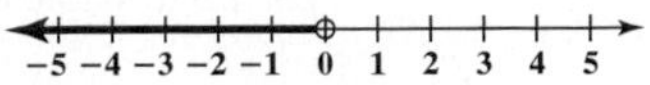

11. 39 **12.** $<$ **13.** 6° F **14.** 2015 **15.** width: 53 m; lenght: 120 m

16. 200 lb **17.** 6 hr **18.** 40°, 60°, 80° **19.** 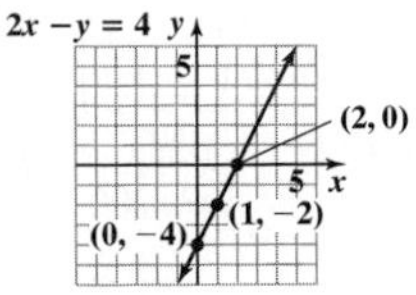

20. 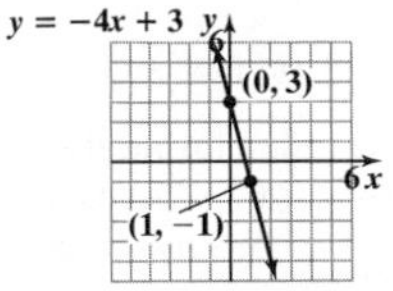

CHAPTER 4

Section 4.1

Check Point Exercises

1. a. solution **b.** not a solution **2.** (1, 4) or $\{(1, 4)\}$ **3.** (3, 3) or $\{(3, 3)\}$ **4.** no solution or ∅
5. infinitely many solutions; $\{(x, y)|x + y = 3\}$ or $\{(x, y)|2x + 2y = 6\}$

Exercise Set 4.1

1. solution **3.** solution **5.** not a solution **7.** solution **9.** not a solution **11.** $\{(4, 2)\}$ **13.** $\{(-1, 2)\}$ **15.** $\{(3, 0)\}$
17. $\{(1, 0)\}$ **19.** $\{(-1, 4)\}$ **21.** $\{(2, 4)\}$ **23.** $\{(2, -1)\}$ **25.** no solution; ∅ **27.** $\{(-2, 6)\}$
29. infinitely many solutions; $\{(x, y)|x - 2y = 4\}$ or $\{(x, y)|2x - 4y = 8\}$ **31.** $\{(2, 3)\}$ **33.** no solution; ∅
35. infinitely many solutions; $\{(x, y)|x - y = 0\}$ or $\{(x, y)|y = x\}$ **37.** $\{(2, 4)\}$ **39.** no solution; ∅ **41.** no solution; ∅
43. $m = \frac{1}{2}$; $b = -3$ and -5; no solution **45.** $m = -\frac{1}{2}$ and 3; $b = 4$; one solution **47.** $m = 3$; $b = -6$; infinite number of solutions
49. $m = -3$; $b = 0$ and 1; no solution **51. a.** Answers will vary. Example: (1996, 41); Mothers 30 years and older had 41 thousand births in 1996.
b. There are more births to mothers 30 years and older. **61.** c **67.** $\{(6, -1)\}$ **69.** $\{(3, 0)\}$ **71.** $\{(2, -1)\}$ **73.** $\{(-4, 4)\}$
74. −12 **75.** 6 **76.** 27

Section 4.2

Check Point Exercises

1. (3, 2) or $\{(3, 2)\}$ **2.** (1, −2) or $\{(1, -2)\}$ **3.** no solution or ∅
4. infinitely many solutions; $\{(x, y)|y = 3x - 4\}$ or $\{(x, y)|9x - 3y = 12\}$ **5.** \$30; 400 units

Exercise Set 4.2

1. $\{(1, 3)\}$ **3.** $\{(5, 1)\}$ **5.** $\{(2, 1)\}$ **7.** $\{(-1, 3)\}$ **9.** $\{(4, 5)\}$ **11.** $\left\{\left(-\frac{2}{5}, -\frac{11}{5}\right)\right\}$ **13.** no solution or ∅
15. infinitely many solutions; $\{(x, y)|y = 3x - 5\}$ or $\{(x, y)|21x - 35 = 7y\}$ **17.** $\{(0, 0)\}$ **19.** $\left\{\left(\frac{17}{7}, -\frac{8}{7}\right)\right\}$ **21.** no solution; ∅
23. $\left\{\left(\frac{43}{5}, -\frac{1}{5}\right)\right\}$ **25.** $\{(200, 700)\}$ **27.** $\{(7, 3)\}$ **29.** $\{(-1, -1)\}$ **31.** $\{(5, 4)\}$ **33.** $x + y = 81$, $y = x + 41$; 20 and 61
35. $x - y = 5$, $4x = 6y$; 10 and 15 **37.** $x - y = 1$, $x + 2y = 7$; 2 and 3 **39.** (2, 8) **41. a.** 6500 sold; 6200 supplied **b.** \$50; 6250 tickets
43. 2700 gal **45.** 2032; about 0.6 deaths per 1000 live births or less than one death per 1000 live births **53.** $x = 1, y = -3, z = 5$

55. 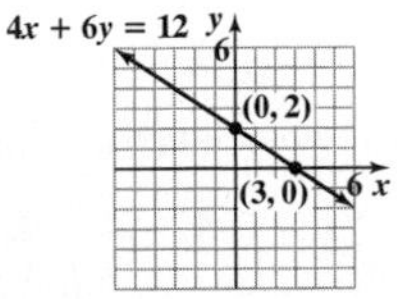

56. 12 **57.** $-73, 0, \frac{3}{1}$

Section 4.3

Check Point Exercises

1. (7, −2) or $\{(7, -2)\}$ **2.** (6, 2) or $\{(6, 2)\}$ **3.** (2, −1) or $\{(2, -1)\}$ **4.** $\left(\frac{60}{17}, -\frac{11}{17}\right)$ or $\left\{\left(\frac{60}{17}, -\frac{11}{17}\right)\right\}$ **5.** no solution or ∅
6. infinitely many solutions; $\{(x, y)|x - 5y = 7\}$ or $\{(x, y)|3x - 15y = 21\}$

Exercise Set 4.3

1. $\{(4, -7)\}$ **3.** $\{(3, 0)\}$ **5.** $\{(-3, 5)\}$ **7.** $\{(3, 1)\}$ **9.** $\{(2, 1)\}$ **11.** $\{(-2, 2)\}$ **13.** $\{(-7, -1)\}$ **15.** $\{(2, 0)\}$
17. $\{(1, -2)\}$ **19.** $\{(-1, 1)\}$ **21.** $\{(3, 1)\}$ **23.** $\{(-5, -2)\}$ **25.** $\left\{\left(\frac{11}{12}, -\frac{7}{6}\right)\right\}$ **27.** $\left\{\left(\frac{23}{16}, \frac{3}{8}\right)\right\}$ **29.** no solution; $\varnothing$
31. infinitely many solutions; $\{(x, y) | x + 3y = 2\}$ or $\{(x, y) | 3x + 9y = 6\}$ **33.** no solution; $\varnothing$ **35.** $\left\{\left(\frac{1}{2}, -\frac{1}{2}\right)\right\}$
37. infinitely many solutions; $\{(x, y) | x = 5 - 3y\}$ or $\{(x, y) | 2x + 6y = 10\}$ **39.** $\left\{\left(\frac{1}{3}, 1\right)\right\}$ **41.** $\{(-10, 21)\}$ **43.** $\{(0, 1)\}$ **45.** $\{(2, -1)\}$
47. $\{(1, -3)\}$ **49.** $\{(4, 3)\}$ **51.** no solution; $\varnothing$ **53.** infinitely many solutions; $\{(x, y) | 2(x + 2y) = 6\}$ or $\{(x, y) | 3(x + 2y - 3) = 0\}$
55. $\{(3, 2)\}$ **57.** $\{(-1, 2)\}$ **59.** $\{(-1, 0)\}$ **61.** $\{(3, 1)\}$ **63.** dependent **65.** 2002; 48% for and 48% against
73. $x = -a - b$; $y = -2a - b$ **77.** $\{(5, 1)\}$ **79.** 10 **80.** II **81.** 26

Mid-Chapter 4 Check Point

1. $\{(2, 0)\}$ **2.** $\{(1, 1)\}$ **3.** no solution or $\varnothing$ **4.** $\{(5, 8)\}$ **5.** $\{(2, -1)\}$ **6.** $\{(2, 9)\}$ **7.** $\{(-2, -3)\}$
8. $\{(10, -1)\}$ **9.** no solution or $\varnothing$ **10.** $\{(-12, -1)\}$ **11.** $\{(-7, -23)\}$ **12.** $\left\{\left(\frac{16}{17}, -\frac{12}{17}\right)\right\}$
13. infinitely many solutions; $\{(x, y) | y - 2x = 7\}$ or $\{(x, y) | 4x = 2y - 14\}$ **14.** $\{(7, 11)\}$ **15.** $\{(6, 10)\}$

Section 4.4

Check Point Exercises

1. A bustard weighs 46 lb; a condor weighs 27 lb. **2.** A Quarter Pounder has 420 cal; a Whopper with cheese has 589 cal. **3.** 17.5 yr; $24,250
4. $3150 at 9%; $1850 at 11% **5.** 12% solution: 100 oz; 20% solution: 60 oz **6.** boat: 35 mph; current: 7 mph

Exercise Set 4.4

1. $x + y = 17, x - y = -3$; 7 and 10 **3.** $3x - y = -1, x + 2y = 23$; 3 and 10 **5.** 5.8 million lb of potato chips; 4.6 million lb of tortilla chips
7. pan pizza: 1120 calories; beef burrito: 430 calories **9.** scrambled eggs: 366 mg; Double Beef Whopper: 175 mg **11.** sweater: $12; shirt: $10
13. a. 300 min; $35 **b.** plan B; Answers will vary. **15.** $600 of merchandise; $580 **17.** 26 yr; 2011; college grads: $1158; high school grads: $579
19. 2 servings of macaroni and 4 servings of broccoli **21.** $2000 at 6% and $5000 at 8% **23.** first fund: $8000; second fund: $6000
25. $17,000 at 12%; $3000 at a 5% loss **27.** California: 100 gal; French: 100 gal **29.** 18-karat gold: 96 grams; 12-karat gold: 204 grams
31. cheaper candy: 30 lb; more expensive candy: 45 lb **33.** 8 nickels and 7 dimes **35.** plane: 130 mph; wind: 30 mph
37. crew: 6 km/hr; current: 2 km/hr **39.** in still water: 4.5 mph; current: 1.5 mph **45.** 10 birds and 20 lions
47. There are 5 people downstairs and 7 people upstairs. **50.** $-\frac{1}{4}$ **51.** $-\frac{11}{20}$ **52.**

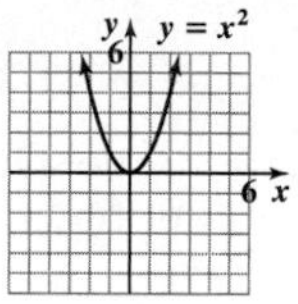

Section 4.5

Check Point Exercises

1. $(-1) - 2(-4) + 3(5) = 22; 2(-1) - 3(-4) - 5 = 5; 3(-1) + (-4) - 5(5) = -32$ **2.** $(1, 4, -3)$ or $\{(1, 4, -3)\}$
3. $(4, 5, 3)$ or $\{(4, 5, 3)\}$ **4.** $y = 3x^2 - 12x + 13$

Exercise Set 4.5

1. not a solution **3.** solution **5.** $\{(2, 3, 3)\}$ **7.** $\{(2, -1, 1)\}$ **9.** $\{(1, 2, 3)\}$ **11.** $\{(3, 1, 5)\}$ **13.** $\{(1, 0, -3)\}$ **15.** $\{(1, -5, -6)\}$
17. no solution or $\varnothing$ **19.** infinitely many solutions; dependent equations **21.** $\left\{\left(\frac{1}{2}, \frac{1}{3}, -1\right)\right\}$ **23.** $y = 2x^2 - x + 3$ **25.** $y = 2x^2 + x - 5$
27. 7, 4, and 5 **29.** $\{(4, 8, 6)\}$ **31.** $y = -\frac{3}{4}x^2 + 6x - 11$ **33.** $\left(\frac{8}{a}, -\frac{3}{b}, -\frac{5}{c}\right)$ **35. a.** (0, 13.6), (70, 4.7), (102, 11.5)
b. $0a + 0b + c = 13.6$; $4900a + 70b + c = 4.7$; $10{,}404a + 102b + c = 11.5$ **37. a.** $y = -16x^2 + 40x + 200$ **b.** 0; After 5 seconds, the ball hits the ground.
39. Carnegie: $100 billion; Vanderbilt: $96 billion; Gates: $60 billion **41.** $1200 at 8%; $2000 at 10%; $3500 at 12%
43. 200 $8 tickets; 150 $10 tickets; 50 $12 tickets **45.** 4 oz of food A; 0.5 oz of food B; 1 oz of food C
55. c **57.** 60°, 55°, 65° **59.**

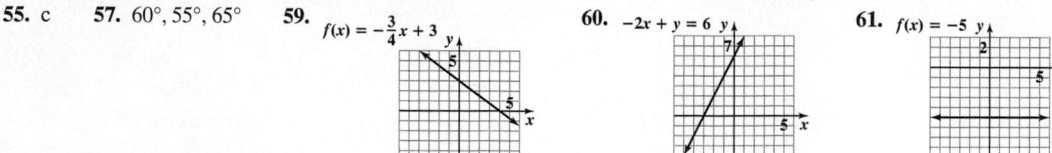

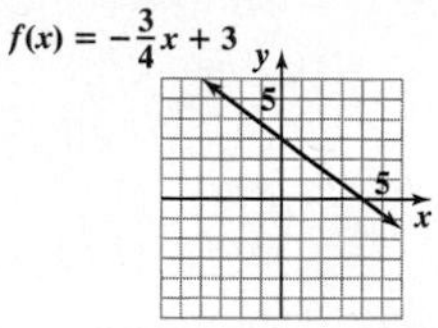

60.

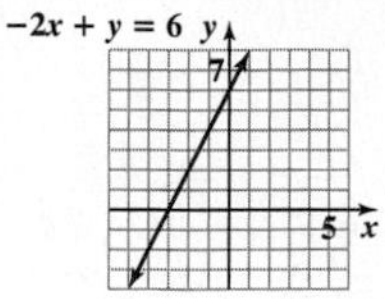

61.

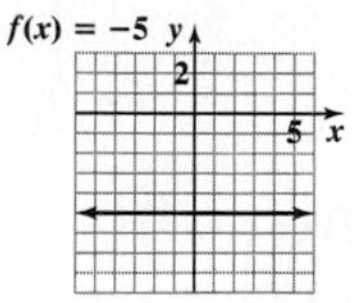

Review Exercises

1. solution **2.** not a solution **3.** no; $(-1, 3)$ does not satisfy $2x + y = -5$. **4.** $\{(4, -2)\}$ **5.** $\{(3, -2)\}$ **6.** $\{(2, 0)\}$ **7.** $\{(2, 1)\}$ **8.** $\{(4, -1)\}$ **9.** no solution; $\varnothing$ **10.** infinitely many solutions; $\{(x, y)|2x - 4y = 8\}$ or $\{(x, y)|x - 2y = 4\}$ **11.** no solution; $\varnothing$ **12.** $\{(-2, -6)\}$ **13.** $\{(2, 5)\}$ **14.** no solution; $\varnothing$ **15.** $\{(2, -1)\}$ **16.** $\{(5, 4)\}$ **17.** $\{(-2, -1)\}$ **18.** $\{(1, -4)\}$ **19.** $\{(20, -21)\}$ **20.** infinitely many solutions; $\{(x, y)|4x + y = 5\}$ or $\{(x, y)|12x + 3y = 15\}$ **21.** no solution; $\varnothing$ **22.** $\{(4, 18)\}$ **23.** $\left\{\left(-1, -\frac{1}{2}\right)\right\}$ **24.** \$12.50; 250 copies **25.** $\{(2, 4)\}$ **26.** $\{(-1, -1)\}$ **27.** $\{(2, -1)\}$ **28.** $\{(3, 2)\}$ **29.** $\{(2, 1)\}$ **30.** $\{(0, 0)\}$ **31.** $\left\{\left(\frac{17}{7}, -\frac{15}{7}\right)\right\}$ **32.** no solution; $\varnothing$ **33.** infinitely many solutions; $\{(x, y)|3x - 4y = -1\}$ or $\{(x, y)|-6x + 8y = 2\}$ **34.** $\{(4, -2)\}$ **35.** $\{(-8, -6)\}$ **36.** $\{(-4, 1)\}$ **37.** $\left\{\left(\frac{5}{2}, 3\right)\right\}$ **38.** $\{(3, 2)\}$ **39.** $\left\{\left(\frac{1}{2}, -2\right)\right\}$ **40.** no solution; $\varnothing$ **41.** $\{(3, 7)\}$ **42.** Japan: 73.6 yr; Switzerland: 72.8 yr **43.** gorilla: 485 lb; orangutan: 165 lb **44.** shrimp: 42 mg; scallops: 15 mg **45.** 9 ft by 5 ft **46.** room: \$80; car: \$60 **47.** 200 min; \$25 **48.** \$2500 at 4%; \$6500 at 7% **49.** 10 ml of 34%; 90 ml of 4% **50.** plane: 630 mph; wind: 90 mph **51.** no **52.** $\{(0, 1, 2)\}$ **53.** $\{(2, 1, -1)\}$ **54.** infinitely many solutions; dependent equations **55.** $y = 3x^2 - 4x + 5$ **56.** per dog: \$129; per horse: \$97; per cat: \$81

Chapter 4 Test

1. solution **2.** not a solution **3.** $\{(2, 4)\}$ **4.** $\{(2, 4)\}$ **5.** $\{(1, -3)\}$ **6.** $\{(2, -3)\}$ **7.** no solution; $\varnothing$ **8.** $\{(-1, 4)\}$ **9.** $\{(-4, 3)\}$ **10.** infinitely many solutions; $\{(x, y)|3x - 2y = 2\}$ or $\{(x, y)|-9x + 6y = -6\}$ **11.** Mary: 2.6%; Patricia: 1.1% **12.** 500 min; \$40 **13.** \$2000 at 6% and \$7000 at 7% **14.** 6% solution: 12 oz; 9% solution: 24 oz **15.** boat: 14 mph; current: 2 mph **16.** $(1, 3, 2)$ or $\{(1, 3, 2)\}$

Cumulative Review Exercises (Chapters 1–4)

1. -36 **2.** $17x - 11$ **3.** $\{-6\}$ **4.** $\{20\}$ **5.** $t = \frac{A - P}{Pr}$ **6.** $\{x|x > 2\}$;

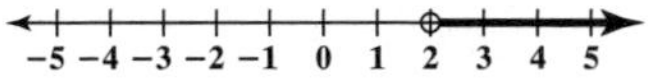

7.

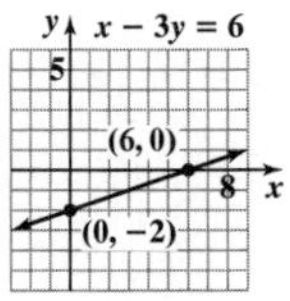

8.

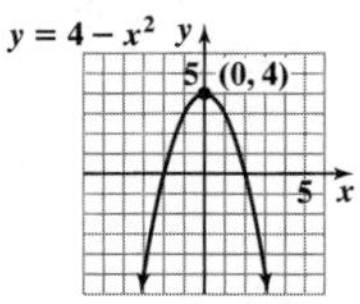

9.

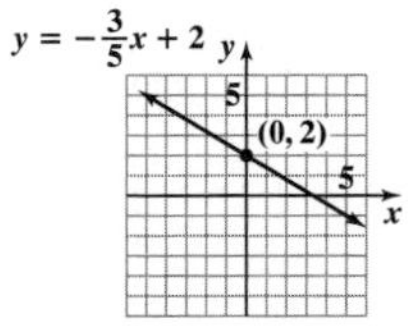

10. $\{(0, -2)\}$ **11.** $\left\{\left(\frac{3}{2}, -2\right)\right\}$ **12.** 1 **13.** $y - 6 = -4(x + 1)$; $y = -4x + 2$ **14.** 10 ft **15.** pen: \$0.80; pad: \$1.20 **16.** $-93, 0, \frac{7}{1}, \sqrt{100}$ **17.** 20% **18.** one computer; The percentage of households with one computer is increasing faster than the percentage for multiple computers. **19.** 8 yr, 2005 **20.** 2012

CHAPTER 5

Section 5.1

Check Point Exercises

1. $5x^3 + 4x^2 - 8x - 20$ **2.** $5x^3 + 4x^2 - 8x - 20$ **3.** $7x^2 + 11x + 4$ **4.** $7x^3 + 3x^2 + 12x - 8$ **5.** $3y^3 - 10y^2 - 11y - 8$ **6.** approximately 4 per thousand; approximately $(40, 4)$

Exercise Set 5.1

1. binomial, 1 **3.** binomial, 3 **5.** monomial, 2 **7.** monomial, 0 **9.** trinomial, 2 **11.** trinomial, 4 **13.** binomial, 3 **15.** monomial, 23 **17.** $-8x + 13$ **19.** $12x^2 + 15x - 9$ **21.** $10x^2 - 12x$ **23.** $5x^2 - 3x + 13$ **25.** $4y^3 + 10y^2 + y - 2$ **27.** $3x^3 + 2x^2 - 9x + 7$ **29.** $-2y^3 + 4y^2 + 13y + 13$ **31.** $-3y^6 + 8y^4 + y^2$ **33.** $10x^3 + 1$ **35.** $-\frac{2}{5}x^4 + x^3 - \frac{1}{8}x^2$ **37.** $0.01x^5 + x^4 - 0.1x^3 + 0.3x + 0.33$ **39.** $11y^3 - 3y^2$ **41.** $-2x^2 - x + 1$ **43.** $-\frac{1}{4}x^4 - \frac{7}{15}x^3 - 0.3$ **45.** $-y^3 + 8y^2 - 3y - 14$ **47.** $-5x^3 - 6x^2 + x - 4$ **49.** $7x^4 - 2x^3 + 4x - 2$ **51.** $8x^2 + 7x - 5$ **53.** $9x^3 - 4.9x^2 + 11.1$ **55.** $-2x - 10$ **57.** $-5x^2 - 9x - 12$ **59.** $-5x^2 - x$ **61.** $-4x^2 - 4x - 6$ **63.** $-2y - 6$ **65.** $6y^3 + y^2 + 7y - 20$ **67.** $n^3 + 2$ **69.** $y^6 - y^3 - y^2 + y$ **71.** $26x^4 + 9x^2 + 6x$ **73.** $\frac{5}{7}x^3 - \frac{9}{20}x$ **75.** $4x + 6$ **77.** $10x^2 - 7$ **79.** $-4y^2 - 7y + 5$ **81.** $9x^3 + 11x^2 - 8$ **83.** $-y^3 + 8y^2 + y + 14$ **85.** $7x^4 - 2x^3 + 3x^2 - x + 2$ **87.** $0.05x^3 + 0.02x^2 + 1.02x$ **89.** $x^2 + 12x$ **91.** $y^2 - 19y + 16$ **93.** $2x^3 + 3x^2 + 7x - 5$ **95.** $-10y^3 + 2y^2 + y + 3$ **97.** 5 billion, 7.25 billion, 17 billion, 25.25 billion, and 5 billion; 3 days; 4 days **99.** 2029 cigarettes; Answers will vary. **101. a.** 73% **b.** 72.8% **c.** 73.3%; Answers will vary. **103. a.** 42 human years **b.** 41.8 human years **105.** 3 dog years **117.** $-3x^2 - x - 2$ **120.** -10 **121.** 5.6 **122.** $\{-4\}$

Section 5.2

Check Point Exercises

1. a. 2^6 or 64 **b.** x^{10} **c.** y^8 **d.** y^9 **2. a.** 3^{20} **b.** x^{90} **c.** $(-5)^{21}$ **3. a.** $16x^4$ **b.** $-64y^6$ **4. a.** $70x^3$ **b.** $-20x^9$ **5. a.** $3x^2 + 15x$ **b.** $30x^5 - 12x^3 + 18x^2$ **6. a.** $x^2 + 9x + 20$ **b.** $10x^2 - 29x - 21$ **7.** $5x^3 - 18x^2 + 7x + 6$ **8.** $6x^5 - 19x^4 + 22x^3 - 8x^2$

Exercise Set 5.2

1. x^{18} **3.** y^{12} **5.** x^{11} **7.** 7^{19} **9.** 6^{90} **11.** x^{45} **13.** $(-20)^9$ **15.** $8x^3$ **17.** $25x^2$ **19.** $16x^6$ **21.** $16y^{24}$ **23.** $-32x^{35}$ **25.** $14x^2$ **27.** $24x^3$ **29.** $-15y^7$ **31.** $\frac{1}{8}a^5$ **33.** $-48x^7$ **35.** $4x^2 + 12x$ **37.** $x^2 - 3x$ **39.** $2x^2 - 12x$ **41.** $-12y^2 - 20y$ **43.** $4x^3 + 8x^2$ **45.** $2y^4 + 6y^3$ **47.** $6y^4 - 8y^3 + 14y^2$ **49.** $6x^4 + 8x^3$ **51.** $-2x^3 - 10x^2 + 6x$ **53.** $12x^4 - 3x^3 + 15x^2$ **55.** $x^2 + 8x + 15$ **57.** $2x^2 + 9x + 4$ **59.** $x^2 - 2x - 15$ **61.** $x^2 - 2x - 99$ **63.** $2x^2 + 3x - 20$ **65.** $\frac{3}{16}x^2 + \frac{11}{4}x - 4$ **67.** $x^3 + 3x^2 + 5x + 3$ **69.** $y^3 - 6y^2 + 13y - 12$ **71.** $2a^3 - 9a^2 + 19a - 15$ **73.** $x^4 + 3x^3 + 5x^2 + 7x + 4$ **75.** $4x^4 - 4x^3 + 6x^2 - \frac{17}{2}x + 3$ **77.** $x^4 + x^3 + x^2 + 3x + 2$ **79.** $x^3 + 3x^2 - 37x + 24$ **81.** $2x^3 - 9x^2 + 27x - 27$ **83.** $2x^4 + 9x^3 + 6x^2 + 11x + 12$ **85.** $12z^4 - 14z^3 + 19z^2 - 22z + 8$ **87.** $21x^5 - 43x^4 + 38x^3 - 24x^2$ **89.** $4y^6 - 2y^5 - 6y^4 + 5y^3 - 5y^2 + 8y - 3$ **91.** $x^4 + 6x^3 - 11x^2 - 4x + 3$ **93.** $2x - 2$ **95.** $15x^5 + 42x^3 - 8x^2$ **97.** $2y^3$ **99.** $16y + 32$ **101.** $2x^2 + 7x - 15$ ft² **103. a.** $(2x + 1)(x + 2)$ **b.** $2x^2 + 5x + 2$ **c.** $(2x + 1)(x + 2) = 2x^2 + 5x + 2$ **115.** $8x + 16$ **117.** $-8x^4$ **118.** $\{x \mid x < -1\}$ **119.**

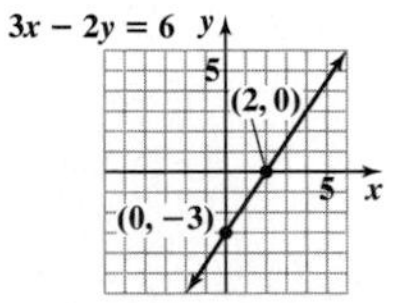

120. $-\frac{2}{3}$

Section 5.3

Check Point Exercises

1. $x^2 + 11x + 30$ **2.** $28x^2 - x - 15$ **3.** $6x^2 - 22x + 20$ **4. a.** $49y^2 - 64$ **b.** $16x^2 - 25$ **c.** $4a^6 - 9$ **5. a.** $x^2 + 20x + 100$ **b.** $25x^2 + 40x + 16$ **6. a.** $x^2 - 18x + 81$ **b.** $49x^2 - 42x + 9$

Exercise Set 5.3

1. $x^2 + 10x + 24$ **3.** $y^2 - 4y - 21$ **5.** $2x^2 + 7x - 15$ **7.** $4y^2 - y - 3$ **9.** $10x^2 - 9x - 9$ **11.** $12y^2 - 43y + 35$ **13.** $-15x^2 - 32x + 7$ **15.** $6y^2 - 28y + 30$ **17.** $15x^4 - 47x^2 + 28$ **19.** $-6x^2 + 17x - 10$ **21.** $x^3 + 5x^2 + 3x + 15$ **23.** $8x^5 + 40x^3 + 3x^2 + 15$ **25.** $x^2 - 9$ **27.** $9x^2 - 4$ **29.** $9r^2 - 16$ **31.** $9 - r^2$ **33.** $25 - 49x^2$ **35.** $4x^2 - \frac{1}{4}$ **37.** $y^4 - 1$ **39.** $r^6 - 4$ **41.** $1 - y^8$ **43.** $x^{20} - 25$ **45.** $x^2 + 4x + 4$ **47.** $4x^2 + 20x + 25$ **49.** $x^2 - 6x + 9$ **51.** $9y^2 - 24y + 16$ **53.** $16x^4 - 8x^2 + 1$ **55.** $49 - 28x + 4x^2$ **57.** $4x^2 + 2x + \frac{1}{4}$ **59.** $16y^2 - 2y + \frac{1}{16}$ **61.** $x^{16} + 6x^8 + 9$ **63.** $x^3 - 1$ **65.** $x^2 - 2x + 1$ **67.** $9y^2 - 49$ **69.** $12x^4 + 3x^3 + 27x^2$ **71.** $70y^2 + 2y - 12$ **73.** $x^4 + 2x^2 + 1$ **75.** $x^4 + 3x^2 + 2$ **77.** $x^4 - 16$ **79.** $4 - 12x^5 + 9x^{10}$ **81.** $\frac{3}{16}x^4 + 7x^2 - 96$ **83.** $x^2 + 2x + 1$ **85.** $4x^2 - 9$ **87.** $6x + 22$ **89.** $16x^4 - 72x^2 + 81$ **91.** $16x^4 - 1$ **93.** $x^3 + 6x^2 + 12x + 8$ **95.** $x^2 + 6x + 9 - y^2$ **97.** $(x + 1)(x + 2)$ yd² **99.** 56 yd²; (6, 56) **101.** $(x^2 + 4x + 4)$ in² **109.** $(x - 10)$ and $(x + 2)$ **111.** $x^2 + 2x$ **113.** Change $x^2 + 2x + 4$ to $x^2 + 4x + 4$. **115.** Graphs coincide. **116.** $(2, -1)$ **117.** $(1, 1)$ **118.** $\left\{\left(\frac{3}{2}, -1, \frac{1}{2}\right)\right\}$

Section 5.4

Check Point Exercises

1. -9 **2.** polynomial degree: 9;

Term	Coefficient	Degree
$8x^4y^5$	8	9
$-7x^3y^2$	-7	5
$-x^2y$	-1	3
$-5x$	-5	1
11	11	0

3. $2x^2y + 2xy - 4$ **4.** $3x^3 + 2x^2y + 5xy^2 - 10$ **5.** $60x^5y^5$ **6.** $60x^5y^7 - 12x^3y^3 + 18xy^2$ **7. a.** $21x^2 - 25xy + 6y^2$ **b.** $4x^2 + 16xy + 16y^2$ **8. a.** $36x^2y^4 - 25x^2$ **b.** $x^3 - y^3$

Exercise Set 5.4

1. 1 **3.** −47 **5.** −6 **7.** polynomial degree: 9;

Term	Coefficient	Degree
x^3y^2	1	5
$-5x^2y^7$	−5	9
$6y^2$	6	2
−3	−3	0

9. $7x^2y - 4xy$ **11.** $2x^2y + 13xy + 13$ **13.** $-11x^4y^2 - 11x^2y^2 + 2xy$ **15.** $-5x^3 + 8xy - 9y^2$ **17.** $x^4y^2 + 8x^3y + y - 6x$ **19.** $5x^3 + x^2y - xy^2 - 4y^3$ **21.** $-3x^2y^2 + xy^2 + 5y^2$ **23.** $8a^2b^4 + 3ab^2 + 8ab$ **25.** $-30x + 37y$ **27.** $40x^3y^2$ **29.** $-24x^5y^9$ **31.** $45x^2y + 18xy^2$ **33.** $50x^3y^2 - 15xy^3$ **35.** $28a^3b^5 + 8a^2b^3$ **37.** $-a^2b + ab^2 - b^3$ **39.** $7x^2 + 38xy + 15y^2$ **41.** $2x^2 + xy - 21y^2$ **43.** $15x^2y^2 + xy - 2$ **45.** $4x^2 + 12xy + 9y^2$ **47.** $x^2y^2 - 6xy + 9$ **49.** $x^4 + 2x^2y^2 + y^4$ **51.** $x^4 - 4x^2y^2 + 4y^4$ **53.** $9x^2 - y^2$ **55.** $a^2b^2 - 1$ **57.** $x^2 - y^4$ **59.** $9a^4b^2 - a^2$ **61.** $9x^2y^4 - 16y^2$ **63.** $a^3 - ab^2 + a^2b - b^3$ **65.** $x^3 + 4x^2y + 4xy^2 + y^3$ **67.** $x^3 - 4x^2y + 4xy^2 - y^3$ **69.** $x^2y^2 - a^2b^2$ **71.** $x^6y + x^4y + x^4 + 2x^2 + 1$ **73.** $x^4y^4 - 6x^2y^2 + 9$ **75.** $x^2 + 2xy + y^2 - 1$ **77.** $3x^2 + 8xy + 5y^2$ **79.** $2xy + y^2$ **81.** $x^{12}y^{12} - 2x^6y^6 + 1$ **83.** $x^4y^4 - 18x^2y^2 + 81$ **85.** $x^2 - y^2 - 2yz - z^2$ **87.** no; need 120 more board feet **89.** 192 ft **91.** 0 ft; The ball hits the ground. **93.** 2.5 to 6 sec **95.** (2, 192) **97.** 2.5 sec; 196 ft **101.** c **103.** $5x^2 - 16y^2$ **105.** $W = \frac{2R - L}{3}$ **106.** 3.8 **107.** $y - 5 = 3(x + 2)$; $y = 3x + 11$

Mid-Chapter 5 Check Point

1. $-55x^4y^6$ **2.** $6x^2y^3$ **3.** $12x^2 - x - 35$ **4.** $-x + 12$ **5.** $2x^3 - 11x^2 + 17x - 5$ **6.** $x^2 - x - 4$ **7.** $64x^2 - 48x + 9$ **8.** $70x^9$ **9.** $x^4 - 4$ **10.** $x^4 + 4x^2 + 4$ **11.** $18a^2 - 11ab - 10b^2$ **12.** $70x^5 - 14x^3 + 21x^2$ **13.** $5a^2b^3 + 2ab - b^2$ **14.** $18y^2 - 50$ **15.** $2x^3 - x^2 + x$ **16.** $10x^2 - 5xy - 3y^2$ **17.** $-4x^5 + 7x^4 - 10x + 23$ **18.** $x^3 + 27y^3$ **19.** $10x^7 - 5x^4 + 8x^3 - 4$ **20.** $y^2 - 12yz + 36z^2$ **21.** $-21x^2 + 7$

Section 5.5

Check Point Exercises

1. a. 5^8 **b.** x^7 **c.** y^{19} **2. a.** 1 **b.** 1 **c.** −1 **d.** 20 **e.** 1 **3. a.** $\frac{x^2}{25}$ **b.** $\frac{x^{12}}{8}$ **c.** $\frac{16a^{40}}{b^{12}}$ **4. a.** $-2x^8$ **b.** $\frac{1}{5}$ **c.** $3x^5y^3$ **5.** $-5x^7 + 2x^3 - 3x$ **6.** $5x^6 - \frac{7}{5}x + 2$ **7.** $3x^6y^4 - xy + 10$

Exercise Set 5.5

1. 3^{15} **3.** x^4 **5.** y^8 **7.** $5^3 \cdot 2^4$ **9.** $x^{75}y^{40}$ **11.** 1 **13.** 1 **15.** −1 **17.** 100 **19.** 1 **21.** 0 **23.** −2 **25.** $\frac{x^2}{9}$ **27.** $\frac{x^6}{64}$ **29.** $\frac{4x^6}{25}$ **31.** $-\frac{64}{27a^9}$ **33.** $-\frac{32a^{35}}{b^{20}}$ **35.** $\frac{x^8y^{12}}{16z^4}$ **37.** $3x^5$ **39.** $-2x^{20}$ **41.** $-\frac{1}{2}y^3$ **43.** $\frac{7}{5}y^{12}$ **45.** $6x^5y^4$ **47.** $-\frac{1}{2}x^{12}$ **49.** $\frac{9}{7}$ **51.** $-\frac{1}{10}x^8y^9z^4$ **53.** $5x^4 + x^3$ **55.** $2x^3 - x^2$ **57.** $y^6 - 9y + 1$ **59.** $-8x^2 + 5x$ **61.** $6x^3 + 2x^2 + 3x$ **63.** $3x^3 - 2x^2 + 10x$ **65.** $4x - 6$ **67.** $-6z^2 - 2z$ **69.** $4x^2 + 3x - 1$ **71.** $5x^4 - 3x^2 - x$ **73.** $-9x^3 + \frac{9}{2}x^2 - 10x + 5$ **75.** $4xy + 2x - 5y$ **77.** $-4x^5y^3 + 3xy + 2$ **79.** $4x^2 - x + 6$ **81.** $-xy^2$ **83.** $y + 5$ **85.** $3x^{12n} - 6x^{9n} + 2$ **87. a.** $\frac{6t^4 - 207t^3 + 2128t^2 - 6622t + 15{,}220}{28t^4 - 711t^3 + 5963t^2 - 1695t + 27{,}424}$ **b.** No; the divisor is not a monomial. **97.** $18x^8 - 27x^6 + 36x^4$ **99.** $\frac{3x^{14} - 6x^{12} - 9x^7}{-3x^7}$ **100.** 20.3 **101.** 0.875 **102.**

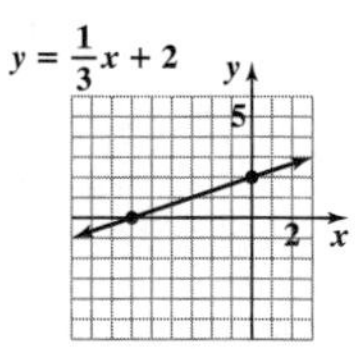

Section 5.6

Check Point Exercises

1. $x + 5$ **2.** $4x - 3 - \frac{3}{2x + 3}$ **3.** $x^2 + x + 1$ **4.** $2x^2 + 7x + 14 + \frac{21x - 10}{x^2 - 2x}$ **5.** $x^2 - 2x - 3$

Exercise Set 5.6

1. $x + 4$ **3.** $2x + 5$ **5.** $x - 2$ **7.** $2y + 1$ **9.** $x - 5 + \frac{14}{x + 2}$ **11.** $y + 3 + \frac{4}{y + 2}$ **13.** $x^2 - 5x + 2$ **15.** $6y - 1$ **17.** $2a + 3$ **19.** $y^2 - y + 2$ **21.** $3x + 5 - \frac{5}{2x - 5}$ **23.** $x^2 + 2x + 8 + \frac{13}{x - 2}$ **25.** $2y^2 + y + 1 + \frac{6}{2y + 3}$ **27.** $2y^2 - 3y + 2 + \frac{1}{3y + 2}$ **29.** $9x^2 + 3x + 1$ **31.** $y^3 - 9y^2 + 27y - 27$ **33.** $2y + 4 + \frac{4}{2y - 1}$ **35.** $y^3 + y^2 - y - 1 + \frac{4}{y - 1}$

37. $4x^2 + 3x - 8 + \frac{18}{x^2 + 3}$ **39.** $5x^2 + x + 3 + \frac{x + 7}{3x^2 - 1}$ **41.** $2x + 5$ **43.** $3x - 8 + \frac{20}{x + 5}$ **45.** $4x^2 + x + 4 + \frac{3}{x - 1}$
47. $6x^4 + 12x^3 + 22x^2 + 48x + 93 + \frac{187}{x - 2}$ **49.** $x^3 - 10x^2 + 51x - 260 + \frac{1300}{5 + x}$ **51.** $3x^2 + 3x - 3$ **53.** $x^4 + x^3 + 2x^2 + 2x + 2$
55. $x^3 + 4x^2 + 16x + 64$ **57.** $2x^4 - 7x^3 + 15x^2 - 31x + 64 - \frac{129}{x + 2}$ **59.** $x^3 - x^2y + xy^2 - y^3 + \frac{2y^4}{x + y}$ **61.** $3x^2 + 2x - 1$
63. $4x - 7 + \frac{4x + 8}{x^2 + x + 1}$ **65.** $x^3 + x^2 - x - 3 + \frac{-x + 5}{x^2 - x + 2}$ **67.** $4x^2 + 5xy - y^2$ **69.** $x^2 + 2x + 3$ **71.** $x^2 + 2x + 3$ units
73. a. $\frac{30{,}000x^3 - 30{,}000}{x - 1}$ **b.** $30{,}000x^2 + 30{,}000x + 30{,}000$ **c.** \$94,575 **79.** b **81.** -3 **83.** Graphs coincide.
85. $2x + 3$ should be $2x + 23 + \frac{130}{x - 5}$. **87.** $x^2 - 2x + 3$ should be $x^2 + 2x + 3$. **88.** $\{(5, 3)\}$ **89.** 1.2 **90.** -6

Section 5.7

Check Point Exercises

1. a. $\frac{1}{6^2} = \frac{1}{36}$ **b.** $\frac{1}{5^3} = \frac{1}{125}$ **c.** $\frac{1}{(-3)^4} = \frac{1}{81}$ **d.** $-\frac{1}{3^4} = -\frac{1}{81}$ **e.** $\frac{1}{8^1} = \frac{1}{8}$ **2. a.** $\frac{7^2}{2^3} = \frac{49}{8}$ **b.** $\frac{5^2}{4^2} = \frac{25}{16}$ **c.** $\frac{y^2}{7}$ **d.** $\frac{y^8}{x^1} = \frac{y^8}{x}$
3. $\frac{1}{x^{10}}$ **4. a.** $\frac{1}{x^8}$ **b.** $\frac{15}{x^6}$ **c.** $-\frac{2}{y^6}$ **5.** $\frac{36}{x^3}$ **6.** $\frac{1}{x^{20}}$ **7. a.** 7,400,000,000 **b.** 0.000003017 **8. a.** 7.41×10^9 **b.** 9.2×10^{-8}
9. a. 6×10^{10} **b.** 2.1×10^{11} **c.** 6.4×10^{-5} **10.** 5.2×10^5 mi

Exercise Set 5.7

1. $\frac{1}{8^2} = \frac{1}{64}$ **3.** $\frac{1}{5^3} = \frac{1}{125}$ **5.** $\frac{1}{(-6)^2} = \frac{1}{36}$ **7.** $-\frac{1}{6^2} = -\frac{1}{36}$ **9.** $\frac{1}{4^1} = \frac{1}{4}$ **11.** $\frac{1}{2^1} + \frac{1}{3^1} = \frac{1}{2} + \frac{1}{3} = \frac{5}{6}$ **13.** $3^2 = 9$ **15.** $(-3)^2 = 9$
17. $\frac{8^2}{2^3} = 8$ **19.** $\frac{4^2}{1^2} = 16$ **21.** $\frac{5^3}{3^3} = \frac{125}{27}$ **23.** $\frac{x^5}{6}$ **25.** $\frac{y^1}{x^8} = \frac{y}{x^8}$ **27.** $3 \cdot (-5)^3 = -375$ **29.** $\frac{1}{x^5}$ **31.** $\frac{8}{x^3}$ **33.** $\frac{1}{x^6}$ **35.** $\frac{1}{y^{99}}$
37. $\frac{3}{z^5}$ **39.** $-\frac{4}{x^4}$ **41.** $-\frac{1}{3a^3}$ **43.** $\frac{7}{5w^8}$ **45.** $\frac{1}{x^5}$ **47.** $\frac{1}{y^{11}}$ **49.** $\frac{16}{x^2}$ **51.** $216y^{17}$ **53.** $\frac{1}{x^6}$ **55.** $\frac{1}{16x^{12}}$ **57.** $\frac{x^2}{9}$ **59.** $-\frac{y^3}{8}$
61. $\frac{2x^6}{5}$ **63.** x^8 **65.** $16y^6$ **67.** $\frac{1}{y^2}$ **69.** $\frac{1}{y^{50}}$ **71.** $\frac{1}{a^{12}b^{15}}$ **73.** $\frac{a^8}{b^{24}}$ **75.** $\frac{4}{x^4}$ **77.** $\frac{y^9}{x^6}$ **79.** 870 **81.** 923,000 **83.** 3.4
85. 0.79 **87.** 0.0215 **89.** 0.000786 **91.** 3.24×10^4 **93.** 2.2×10^8 **95.** 7.13×10^2 **97.** 6.751×10^3 **99.** 2.7×10^{-3}
101. 2.02×10^{-5} **103.** 5×10^{-3} **105.** 3.14159×10^0 **107.** 6×10^5 **109.** 1.6×10^9 **111.** 3×10^4 **113.** 3×10^6 **115.** 3×10^{-6}
117. 9×10^4 **119.** 2.5×10^6 **121.** 1.25×10^8 **123.** 8.1×10^{-7} **125.** 2.5×10^{-7} **127.** 1 **129.** $\frac{y}{16x^8z^6}$ **131.** $\frac{1}{x^{12}y^{16}z^{20}}$
133. $\frac{x^{18}y^6}{4}$ **135.** 2.5×10^{-3} **137.** 8×10^{-5} **139.** 9.2×10^3 **141.** 2.5×10^{-16} **143.** 6×10^8 **145.** 3.99×10^{11}
147. \$$4 \times 10^{11}$ **149.** \$$3.48 \times 10^{10}$ **151.** 1.25 sec **159.** b **167.** $\{x \mid x < 2\}$ **168.** 5 **169.** $0, \sqrt{16}$

Review Exercises

1. binomial, 4 **2.** trinomial, 2 **3.** monomial, 1 **4.** $8x^3 + 10x^2 - 20x - 4$ **5.** $13y^3 - 8y^2 + 7y - 5$ **6.** $11y^2 - 4y - 4$
7. $8x^4 - 5x^3 + 6$ **8.** $-14x^4 - 13x^2 + 16x$ **9.** $7y^4 - 5y^3 + 3y^2 - y - 4$ **10.** $3x^2 - 7x + 9$ **11.** $10x^3 - 9x^2 + 2x + 11$
12. 1451; 1451 per 100,000 is the death rate for men averaging 10 hr of sleep each night. **13.** x^{23} **14.** y^{14} **15.** x^{100}
16. $100y^2$ **17.** $-64x^{30}$ **18.** $50x^4$ **19.** $-36y^{11}$ **20.** $30x^{12}$ **21.** $21x^3 + 63x$ **22.** $20x^5 - 55x^4$ **23.** $-21y^4 + 9y^3 - 18y^2$
24. $16y^8 - 20y^7 + 2y^5$ **25.** $x^3 - 2x^2 - 13x + 6$ **26.** $12y^3 + y^2 - 21y + 10$ **27.** $3y^3 - 17y^2 + 41y - 35$
28. $8x^4 + 8x^3 - 18x^2 - 20x - 3$ **29.** $x^2 + 8x + 12$ **30.** $6y^2 - 7y - 5$ **31.** $4x^4 - 14x^2 + 6$ **32.** $25x^2 - 16$
33. $49 - 4y^2$ **34.** $y^4 - 1$ **35.** $x^2 + 6x + 9$ **36.** $9y^2 + 24y + 16$ **37.** $y^2 - 2y + 1$ **38.** $25y^2 - 20y + 4$
39. $x^4 + 8x^2 + 16$ **40.** $x^4 - 16$ **41.** $x^4 - x^2 - 20$ **42.** $x^2 + 7x + 12$ **43.** $x^2 + 50x + 600$ yd² **44.** 28

45. polynomial degree: 5;

Term	Coefficient	Degree
$4x^2y$	4	3
$9x^3y^2$	9	5
$-17x^4$	-17	4
-12	-12	0

46. $-x^2 - 17xy + 5y^2$ **47.** $2x^3y^2 + x^2y - 6x^2 - 4$ **48.** $-35x^6y^9$ **49.** $15a^3b^5 - 20a^2b^3$ **50.** $3x^2 + 16xy - 35y^2$
51. $36x^2y^2 - 31xy + 3$ **52.** $9x^2 + 30xy + 25y^2$ **53.** $x^2y^2 - 14xy + 49$ **54.** $49x^2 - 16y^2$ **55.** $a^3 - b^3$ **56.** 6^{30} **57.** x^{15}
58. 1 **59.** -1 **60.** 400 **61.** $\frac{x^{12}}{8}$ **62.** $\frac{81}{16y^{24}}$ **63.** $-5y^6$ **64.** $8x^7y^3$ **65.** $3x^3 - 2x + 6$ **66.** $-6x^5 + 5x^4 + 8x^2$
67. $9x^2y - 3x - 6y$ **68.** $2x + 7$ **69.** $x^2 - 3x + 5$ **70.** $x^2 + 5x + 2 + \frac{7}{x - 7}$ **71.** $y^2 + 3y + 9$ **72.** $2x^2 + 3x - 1$

73. $4x^2 - 7x + 5 - \frac{4}{x + 1}$ **74.** $3x^3 + 6x^2 + 10x + 10$ **75.** $x^3 - 4x^2 + 16x - 64 + \frac{272}{x + 4}$ **76.** $\frac{1}{7^2} = \frac{1}{49}$ **77.** $\frac{1}{(-4)^3} = -\frac{1}{64}$

78. $\frac{1}{2^1} + \frac{1}{4^1} = \frac{1}{2} + \frac{1}{4} = \frac{3}{4}$ **79.** $5^2 = 25$ **80.** $\frac{5^3}{2^3} = \frac{125}{8}$ **81.** $\frac{1}{x^6}$ **82.** $\frac{6}{y^2}$ **83.** $\frac{30}{x^5}$ **84.** x^8 **85.** $81y^2$ **86.** $\frac{1}{y^{19}}$

87. $\frac{x^3}{8}$ **88.** $\frac{1}{x^6}$ **89.** y^{20} **90.** 23,000 **91.** 0.00176 **92.** 0.9 **93.** 7.39×10^7 **94.** 6.2×10^{-4} **95.** 3.8×10^{-1} **96.** 3.8×10^0
97. 9×10^3 **98.** 5×10^4 **99.** 1.6×10^{-3} **100.** 1000 nanosec **101.** 1.26×10^{10} people

Chapter 5 Test

1. trinomial, 2 **2.** $13x^3 + x^2 - x - 24$ **3.** $5x^3 + 2x^2 + 2x - 9$ **4.** $-35x^{11}$ **5.** $48x^5 - 30x^3 - 12x^2$ **6.** $3x^3 - 10x^2 - 17x - 6$
7. $6y^2 - 13y - 63$ **8.** $49x^2 - 25$ **9.** $x^4 + 6x^2 + 9$ **10.** $25x^2 - 30x + 9$ **11.** 30 **12.** $2x^2y^3 + 3xy + 12y^2$
13. $12a^2 - 13ab - 35b^2$ **14.** $4x^2 + 12xy + 9y^2$ **15.** $-5x^{12}$ **16.** $3x^3 - 2x^2 + 5x$ **17.** $x^2 - 2x + 3 + \frac{1}{2x + 1}$ **18.** $3x^3 - 4x^2 + 7$
19. $\frac{1}{10^2} = \frac{1}{100}$ **20.** $4^3 = 64$ **21.** $-27x^6$ **22.** $\frac{4}{x^5}$ **23.** $-\frac{21}{x^6}$ **24.** $16y^4$ **25.** $\frac{x^8}{25}$ **26.** $\frac{1}{x^{15}}$ **27.** 0.00037
28. 7.6×10^6 **29.** 1.23×10^{-2} **30.** 2.1×10^8 **31.** $x^2 + 10x + 16$

Cumulative Review Exercises (Chapters 1–5)

1. $\left\{\frac{35}{9}\right\}$ **2.** $\{-128\}$ **3.** 15,050 ft **4.** $-\frac{2}{3}$ **5.** $-\frac{5}{3}$ **6.** 4 ft by 10 ft **7.** $\{x | x \geq 6\}$;

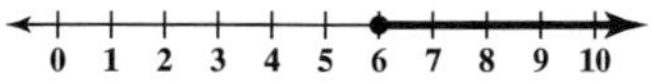

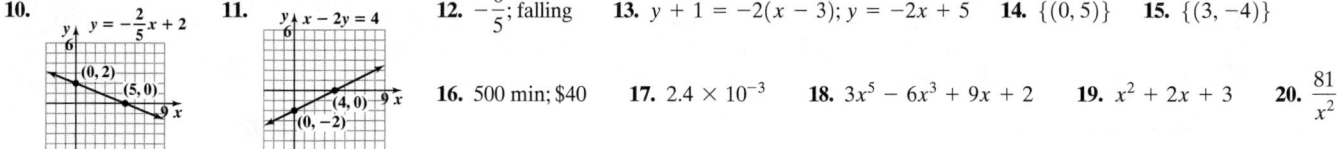

8. \$3400 at 12%; \$2600 at 14% **9.** 15 l of 70%; 5 l of 30%;

10.

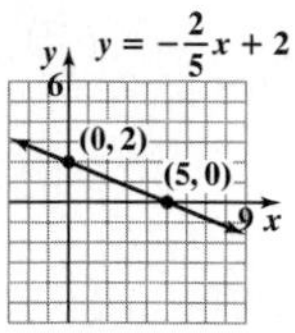

11.

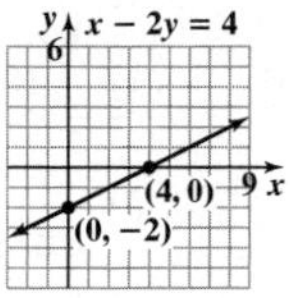

12. $-\frac{6}{5}$; falling **13.** $y + 1 = -2(x - 3)$; $y = -2x + 5$ **14.** $\{(0, 5)\}$ **15.** $\{(3, -4)\}$
16. 500 min; \$40 **17.** 2.4×10^{-3} **18.** $3x^5 - 6x^3 + 9x + 2$ **19.** $x^2 + 2x + 3$ **20.** $\frac{81}{x^2}$

CHAPTER 6

Section 6.1

Check Point Exercises

1. a. $3x^2$ **b.** $4x^2$ **c.** x^2y **2.** $6(x^2 + 3)$ **3.** $5x^2(5 + 7x)$ **4.** $3x^3(5x^2 + 4x - 9)$ **5.** $2xy(4x^2y - 7x + 1)$
6. a. $(x + 1)(x^2 + 7)$ **b.** $(y + 4)(x - 7)$ **7.** $(x + 5)(x^2 + 2)$ **8.** $(y + 3)(x - 5)$

Exercise Set 6.1

1. Answers will vary; 3 examples are: $(2x)(4x^2)$, $(4x)(2x^2)$, and $(8x)(x^2)$. **3.** Answers will vary; 3 examples are: $(-4x^3)(3x^2)$, $(2x^2)(-6x^3)$, and $(-3)(4x^5)$. **5.** Answers will vary; 3 examples are: $(6x^2)(6x^2)$, $(-2x)(-18x^3)$, and $(4x^3)(9x)$. **7.** 4 **9.** $4x$ **11.** $2x^3$ **13.** $3y$
15. xy **17.** $4x^4y^3$ **19.** $8(x + 1)$ **21.** $4(y - 1)$ **23.** $5(x + 6)$ **25.** $6(5x - 2)$ **27.** $x(x + 5)$ **29.** $6(3y^2 + 2)$
31. $7x^2(2x + 3)$ **33.** $y(13y - 25)$ **35.** $9y^4(1 + 3y^2)$ **37.** $4x^2(2 - x^2)$ **39.** $4(3y^2 + 4y - 2)$ **41.** $3x^2(3x^2 + 6x + 2)$
43. $50y^2(2y^3 - y + 2)$ **45.** $5x(2 - 4x + x^2)$ **47.** cannot be factored **49.** $3xy(2x^2y + 3)$ **51.** $10xy(3xy^2 - y + 2)$
53. $8x^2y(4xy - 3x - 2)$ **55.** $(x + 5)(x + 3)$ **57.** $(x + 2)(x - 4)$ **59.** $(y + 6)(x - 7)$ **61.** $(x + y)(3x - 1)$
63. $(3x + 1)(4x + 1)$ **65.** $(5x + 4)(7x^2 + 1)$ **67.** $(x + 2)(x + 4)$ **69.** $(x - 5)(x + 3)$ **71.** $(x^2 + 5)(x - 2)$
73. $(x^2 + 2)(x - 1)$ **75.** $(y + 5)(x + 9)$ **77.** $(y - 1)(x + 5)$ **79.** $(x - 2y)(3x + 5y)$ **81.** $(3x - 2)(x^2 - 2)$ **83.** $(x - a)(x - b)$
85. $6x^2yz(4xy^2z^2 + 5y + 3z)$ **87.** $(x^3 - 4)(1 + 3y)$ **89.** $2x^2(x + 1)(2x^3 - 3x - 4)$ **91.** $(x - 1)(3x^4 + x^2 + 5)$
93. $36x^2 - 4\pi x^2$; $4x^2(9 - \pi)$ **95. a.** 48 ft **b.** $16x(4 - x)$ **c.** 48 ft; yes; no; Answers will vary. **97.** $x^3 - 2$ **105.** d
109. $-3(x - 2)$ should be $-3(x + 2)$. **111.** $x(x + 2) + 1$ should be $(x + 2)(x + 1)$. **112.** $x^2 + 17x + 70$ **113.** $\{(-3, -2)\}$
114. $y - 2 = 1(x + 7)$, or $y - 5 = 1(x + 4)$; $y = x + 9$

Section 6.2

Check Point Exercises

1. $(x + 2)(x + 3)$ **2.** $(x - 2)(x - 4)$ **3.** $(x + 5)(x - 2)$ **4.** $(y - 9)(y + 3)$ **5.** cannot factor over the integers; prime
6. $(x - 3y)(x - y)$ **7.** $2x(x - 4)(x + 7)$

Exercise Set 6.2

1. $(x + 6)(x + 1)$ **3.** $(x + 2)(x + 5)$ **5.** $(x + 1)(x + 10)$ **7.** $(x - 4)(x - 3)$ **9.** $(x - 6)(x - 6)$ **11.** $(y - 3)(y - 5)$ **13.** $(x + 5)(x - 2)$ **15.** $(y + 13)(y - 3)$ **17.** $(x - 5)(x + 3)$ **19.** $(x - 4)(x + 2)$ **21.** prime **23.** $(y - 4)(y - 12)$ **25.** prime **27.** $(w - 32)(w + 2)$ **29.** $(y - 5)(y - 13)$ **31.** $(r + 3)(r + 9)$ **33.** prime **35.** $(x + 6y)(x + y)$ **37.** $(x - 3y)(x - 5y)$ **39.** $(x - 6y)(x + 3y)$ **41.** $(a - 15b)(a - 3b)$ **43.** $3(x + 2)(x + 3)$ **45.** $4(y - 2)(y + 1)$ **47.** $10(x - 10)(x + 6)$ **49.** $3(x - 2)(x - 9)$ **51.** $2r(r + 2)(r + 1)$ **53.** $4x(x + 6)(x - 3)$ **55.** $2r(r + 8)(r - 4)$ **57.** $y^2(y + 10)(y - 8)$ **59.** $x^2(x - 5)(x + 2)$ **61.** $2w^2(w - 16)(w + 3)$ **63.** $15x(y - 1)(y + 4)$ **65.** $x^3(x - y)(x + 4y)$ **67.** $2x^2(y - 15z)(y - z)$ **69.** $(a + b)(x + 5)(x - 4)$ **71.** $(x + 0.3)(x + 0.2)$ **73.** $\left(x - \frac{1}{5}\right)\left(x - \frac{1}{5}\right)$ **75.** $-(x + 8)(x - 5)$ **77. a.** $-16(t - 2)(t + 1)$ **b.** 0; yes; After 2 seconds, you hit the water. **83.** c **85.** 3, 4 **87.** $x(x + 1)(x + 2)$; the product of three consecutive integers **89.** correctly factored **91.** $(x + 1)(x - 1)$ should be $(x - 1)(x - 1)$. **93.** $2x^2 - x - 6$ **94.** $9x^2 + 15x + 4$ **95.** $\{13\}$

Section 6.3

Check Point Exercises

1. $(5x - 4)(x - 2)$ **2.** $(3x - 1)(2x + 7)$ **3.** $(3x - y)(x - 4y)$ **4.** $(3x + 5)(x - 2)$ **5.** $(2x - 1)(4x - 3)$ **6.** $y^2(5y + 3)(y + 2)$

Exercise Set 6.3

1. $(2x + 3)(x + 1)$ **3.** $(3x + 1)(x + 4)$ **5.** $(2x + 3)(x + 4)$ **7.** $(5y - 1)(y - 3)$ **9.** $(3y + 4)(y - 1)$ **11.** $(3x - 2)(x + 5)$ **13.** $(3x - 1)(x - 7)$ **15.** $(5y - 1)(y - 3)$ **17.** $(3x - 2)(x - 5)$ **19.** $(3w - 4)(2w - 1)$ **21.** $(8x + 1)(x + 4)$ **23.** $(5x - 2)(x + 7)$ **25.** $(7y - 3)(2y + 3)$ **27.** prime **29.** $(5z - 3)(5z - 3)$ **31.** $(3y + 1)(5y - 2)$ **33.** prime **35.** $(5y - 1)(2y + 9)$ **37.** $(4x + 1)(2x - 1)$ **39.** $(3y - 1)(3y - 2)$ **41.** $(5x + 8)(4x - 1)$ **43.** $(2x + y)(x + y)$ **45.** $(3x + 2y)(x + y)$ **47.** $(2x - 3y)(x - 3y)$ **49.** $(2x - 3y)(3x + 2y)$ **51.** $(3x - 2y)(5x + 7y)$ **53.** $(2a + 5b)(a + b)$ **55.** $(3a - 2b)(5a + 3b)$ **57.** $(3x - 4y)(4x - 3y)$ **59.** $2(2x + 3)(x + 5)$ **61.** $3(3x + 4)(x - 2)$ **63.** $2(2y - 5)(y + 3)$ **65.** $3(3y - 4)(y + 5)$ **67.** $x(3x + 1)(x + 1)$ **69.** $x(2x - 5)(x + 1)$ **71.** $3y(3y - 1)(y - 4)$ **73.** $5z(6z + 1)(2z + 1)$ **75.** $3x^2(5x - 3)(x - 2)$ **77.** $x^3(2x - 3)(5x - 1)$ **79.** $3(2x + 3y)(x - 2y)$ **81.** $2(2x - y)(3x + 4y)$ **83.** $2y(4x - 7)(x + 6)$ **85.** $2b(2a - 7b)(3a - b)$ **87.** $10(y + 1)(x + 1)(3x - 2)$ **89.** $-4y^4(8x + 3)(x - 1)$ **91. a.** $(2x + 1)(x - 3)$ **b.** $(2y + 3)(y - 2)$ **93.** $(x - 2)(3x - 2)(x - 1)$ **95. a.** $x^2 + 3x + 2$ **b.** $(x + 2)(x + 1)$ **c.** $x^2 + 3x + 2 = (x + 2)(x + 1)$ **101.** a **103.** 5, 7, −5, −7 **105.** $(2x^n + 1)(x^n - 4)$ **106.** $81x^2 - 100$ **107.** $16x^2 + 40xy + 25y^2$ **108.** $x^3 + 8$

Mid-Chapter 6 Check Point

1. $x^4(x + 1)$ **2.** $(x + 9)(x - 2)$ **3.** $x^2y(y^2 - y + 1)$ **4.** prime **5.** $(7x - 1)(x - 3)$ **6.** $(x^2 + 3)(x + 5)$ **7.** $x(2x - 1)(x - 5)$ **8.** $(x - 4)(y - 7)$ **9.** $(x - 15y)(x - 2y)$ **10.** $(5x + 2)(5x - 7)$ **11.** $2(8x - 3)(x - 4)$ **12.** $(3x + 7y)(x + y)$

Section 6.4

Check Point Exercises

1. a. $(x + 9)(x - 9)$ **b.** $(6x + 5)(6x - 5)$ **2. a.** $(5 + 2x^5)(5 - 2x^5)$ **b.** $(10x + 3y)(10x - 3y)$ **3. a.** $2x(3x + 1)(3x - 1)$ **b.** $18(2 + x)(2 - x)$ **4.** $(9x^2 + 4)(3x + 2)(3x - 2)$ **5. a.** $(x + 7)^2$ **b.** $(x - 3)^2$ **c.** $(4x - 7)^2$ **6.** $(2x + 3y)^2$ **7.** $(x + 3)(x^2 - 3x + 9)$ **8.** $(1 - y)(1 + y + y^2)$ **9.** $(5x + 2)(25x^2 - 10x + 4)$

Exercise Set 6.4

1. $(x + 5)(x - 5)$ **3.** $(y + 1)(y - 1)$ **5.** $(2x + 3)(2x - 3)$ **7.** $(5 + x)(5 - x)$ **9.** $(1 + 7x)(1 - 7x)$ **11.** $(3 + 5y)(3 - 5y)$ **13.** $(x^2 + 3)(x^2 - 3)$ **15.** $(7y^2 + 4)(7y^2 - 4)$ **17.** $(x^5 + 3)(x^5 - 3)$ **19.** $(5x + 4y)(5x - 4y)$ **21.** $(x^2 + y^5)(x^2 - y^5)$ **23.** $(x^2 + 4)(x + 2)(x - 2)$ **25.** $(4x^2 + 9)(2x + 3)(2x - 3)$ **27.** $2(x + 3)(x - 3)$ **29.** $2x(x + 6)(x - 6)$ **31.** prime **33.** $3x(x^2 + 9)$ **35.** $2(3 + y)(3 - y)$ **37.** $3y(y + 4)(y - 4)$ **39.** $2x(3x + 1)(3x - 1)$ **41.** $(x + 1)^2$ **43.** $(x - 7)^2$ **45.** $(x - 1)^2$ **47.** $(x + 11)^2$ **49.** $(2x + 1)^2$ **51.** $(5y - 1)^2$ **53.** prime **55.** $(x + 7y)^2$ **57.** $(x - 6y)^2$ **59.** prime **61.** $(4x - 5y)^2$ **63.** $3(2x - 1)^2$ **65.** $x(3x + 1)^2$ **67.** $2(y - 1)^2$ **69.** $2y(y + 7)^2$ **71.** $(x + 1)(x^2 - x + 1)$ **73.** $(x - 3)(x^2 + 3x + 9)$ **75.** $(2y - 1)(4y^2 + 2y + 1)$ **77.** $(3x + 2)(9x^2 - 6x + 4)$ **79.** $(xy - 4)(x^2y^2 + 4xy + 16)$ **81.** $y(3y + 2)(9y^2 - 6y + 4)$ **83.** $2(3 - 2y)(9 + 6y + 4y^2)$ **85.** $(4x + 3y)(16x^2 - 12xy + 9y^2)$ **87.** $(5x - 4y)(25x^2 + 20xy + 16y^2)$ **89.** $\left(5x + \frac{2}{7}\right)\left(5x - \frac{2}{7}\right)$ **91.** $y\left(y - \frac{1}{10}\right)\left(y^2 + \frac{y}{10} + \frac{1}{100}\right)$ **93.** $x(0.5 + x)(0.5 - x)$ **95.** $(x + 6)(x - 4)$ **97.** $(x - 3)(x + 1)^2$ **99.** $x^2 - 25 = (x + 5)(x - 5)$ **101.** $x^2 - 16 = (x + 4)(x - 4)$ **107.** b **109.** $(x + y)(x - y + 3)$ **111.** $(2x^n + 3)^2$ **113.** 6, −6 **115.** $(4x + 3)(4x - 3)$ should be $(2x + 3)(2x - 3)$. **117.** $(4x - 1)^2$ should be $(2x - 1)^2$. **119.** $80x^9y^{14}$ **120.** $-4x^2 + 3$ **121.** $2x + 5$

Section 6.5

Check Point Exercises

1. $5x^2(x + 3)(x - 3)$ **2.** $4(x - 6)(x + 2)$ **3.** $4x(x^2 + 4)(x + 2)(x - 2)$ **4.** $(x - 4)(x + 3)(x - 3)$ **5.** $3x(x - 5)^2$ **6.** $2x^7(x + 3)(x^2 - 3x + 9)$ **7.** $3y(x^2 + 4y^2)(x + 2y)(x - 2y)$ **8.** $3x(2x + 3y)^2$

Exercise Set 6.5

1. $5x(x + 2)(x - 2)$ **3.** $7x(x^2 + 1)$ **5.** $5(x - 3)(x + 2)$ **7.** $2(x^2 + 9)(x + 3)(x - 3)$ **9.** $(x + 2)(x + 3)(x - 3)$ **11.** $3x(x - 4)^2$ **13.** $2x^2(x + 1)(x^2 - x + 1)$ **15.** $2x(3x + 4)$ **17.** $2(y - 8)(y + 7)$ **19.** $7y^2(y + 1)^2$ **21.** prime **23.** $2(4y + 1)(2y - 1)$ **25.** $r(r - 25)$ **27.** $(2w + 5)(2w - 1)$ **29.** $x(x + 2)(x - 2)$ **31.** prime **33.** $(9y + 4)(y + 1)$ **35.** $(y + 2)(y + 2)(y - 2)$ **37.** $(4y + 3)^2$ **39.** $4y(y - 5)(y - 2)$ **41.** $y(y^2 + 9)(y + 3)(y - 3)$ **43.** $5a^2(2a + 3)(2a - 3)$ **45.** $(4y - 1)(3y - 2)$ **47.** $(3y + 8)(3y - 8)$ **49.** prime **51.** $(2y + 3)(y + 5)(y - 5)$ **53.** $2r(r + 17)(r - 2)$ **55.** $2x^3(2x + 1)(2x - 1)$ **57.** $3(x^2 + 81)$ **59.** $x(x + 2)(x^2 - 2x + 4)$ **61.** $2y^2(y - 1)(y^2 + y + 1)$ **63.** $2x(3x + 4y)$ **65.** $(y - 7)(x + 3)$ **67.** $(x - 4y)(x + y)$ **69.** $12a^2(6ab^2 + 1 - 2a^2b^2)$ **71.** $3(a + 6b)(a + 3b)$ **73.** $3x^2y(4x + 1)(4x - 1)$ **75.** $b(3a + 2)(2a - 1)$ **77.** $7xy(x^2 + y^2)(x + y)(x - y)$ **79.** $2xy(5x - 2y)(x - y)$ **81.** $2b(x + 11)^2$ **83.** $(5a + 7b)(3a - 2b)$ **85.** $2xy(9x - 2y)(2x - 3y)$ **87.** $(y - x)(a + b)(a - b)$ **89.** $ax(3x + 7)(3x - 2)$ **91.** $y(9x^2 + y^2)(3x + y)(3x - y)$ **93.** $(x + 1)(5x - 6)(2x + 1)$ **95.** $(x^2 + 6)(6x^2 - 1)$ **97.** $(x - 7 + 2a)(x - 7 - 2a)$ **99.** $(x + 4 + 5a)(x + 4 - 5a)$ **101.** $y(y^2 + 1)(y^4 - y^2 + 1)$ **103.** $16(4 + t)(4 - t)$ **105.** $\pi b^2 - \pi a^2$; $\pi(b + a)(b - a)$ **109.** d **111.** $5y^2(y - 1)(y + 2)(y - 2)$ **113.** $(x - 5)^2$ **115.** $(4x - 3)^2$ should be $(2x - 3)^2$. **117.** correctly factored

119. $2(x + 5)(x^2 + 1)$ should be $2(x + 5)(x + 1)(x - 1)$. **120.** $(3x + 4)(3x - 4)$ **121.**

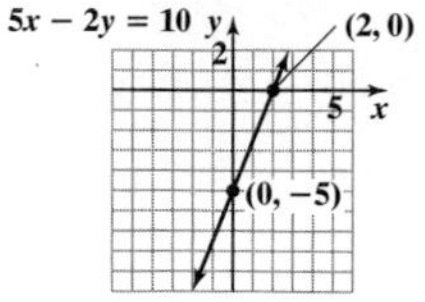

122. 20°, 60°, 100°

Section 6.6

Check Point Exercises

1. $-\frac{1}{2}$ and 4, or $\left\{-\frac{1}{2}, 4\right\}$ **2.** 5 and 1, or $\{1, 5\}$ **3.** 0 and $\frac{1}{2}$, or $\left\{0, \frac{1}{2}\right\}$ **4.** 5 or $\{5\}$ **5.** $-\frac{5}{4}$ and $\frac{5}{4}$, or $\left\{-\frac{5}{4}, \frac{5}{4}\right\}$ **6.** -2 and 9, or $\{-2, 9\}$ **7.** 1 sec and 2 sec; (1, 192) and (2, 192) **8.** length: 9 ft; width: 6 ft

Exercise Set 7.6

1. $\{-7, 0\}$ **3.** $\{-4, 6\}$ **5.** $\left\{-\frac{4}{5}, 9\right\}$ **7.** $\left\{-\frac{9}{2}, 4\right\}$ **9.** $\{-5, -3\}$ **11.** $\{-3, 5\}$ **13.** $\{-3, 7\}$ **15.** $\{-8, -1\}$ **17.** $\{-4, 0\}$ **19.** $\{0, 5\}$ **21.** $\{0, 4\}$ **23.** $\left\{0, \frac{5}{2}\right\}$ **25.** $\left\{-\frac{5}{3}, 0\right\}$ **27.** $\{-2\}$ **29.** $\{6\}$ **31.** $\left\{\frac{3}{2}\right\}$ **33.** $\left\{-\frac{1}{2}, 4\right\}$ **35.** $\left\{-2, \frac{9}{5}\right\}$ **37.** $\{-7, 7\}$ **39.** $\left\{-\frac{5}{2}, \frac{5}{2}\right\}$ **41.** $\left\{-\frac{5}{9}, \frac{5}{9}\right\}$ **43.** $\{-3, 7\}$ **45.** $\left\{-\frac{5}{2}, \frac{3}{2}\right\}$ **47.** $\{-6, 3\}$ **49.** $\left\{-2, -\frac{3}{2}\right\}$ **51.** $\{4\}$ **53.** $\left\{-\frac{5}{2}\right\}$ **55.** $\left\{\frac{3}{8}\right\}$ **57.** $\{-3, -2, 4\}$ **59.** $\{-6, 0, 6\}$ **61.** $\{-2, -1, 0\}$ **63.** $\{2, 8\}$ **65.** $\{4, 5\}$ **67.** 5 sec; Each tick represents one second. **69.** 2 sec; (2, 276) **71.** $\frac{1}{2}$ sec and 4 sec **73.** 1995; (15, 1100) **75.** 10 yr **77.** (10, 7250) **79.** 10 teams **81.** length: 15 yd; width: 12 yd **83.** base: 5 cm; height: 6 cm **85. a.** $4x^2 + 44x$ **b.** 3 ft **89.** d **91.** -4, 1, and 4 **93.** 4 and 1 **95.** a **97.** b **99.** -3 and 2 **101.** 1 **104.** $\{(2, 4, 4)\}$ **105.** $\frac{4}{x^6}$ **106.** -2

Review Exercises

1. $15(2x - 3)$ **2.** $4x(3x^2 + 4x - 100)$ **3.** $5x^2y(6x^2 + 3x + 1)$ **4.** $5(x + 3)$ **5.** $(7x^2 - 1)(x + y)$ **6.** $(x^2 + 2)(x + 3)$ **7.** $(x + 1)(y + 4)$ **8.** $(x^2 + 5)(x + 1)$ **9.** $(x - 2)(y + 4)$ **10.** $(x - 2)(x - 1)$ **11.** $(x - 5)(x + 4)$ **12.** $(x + 3)(x + 16)$ **13.** $(x - 4y)(x - 2y)$ **14.** prime **15.** $(x + 17y)(x - y)$ **16.** $3(x + 4)(x - 2)$ **17.** $3x(x - 11)(x - 1)$ **18.** $(x + 5)(3x + 2)$ **19.** $(y - 3)(5y - 2)$ **20.** $(2x + 5)(2x - 3)$ **21.** prime **22.** $2(2x + 3)(2x - 1)$ **23.** $x(2x - 9)(x + 8)$ **24.** $4y(3y + 1)(y + 2)$ **25.** $(2x - y)(x - 3y)$ **26.** $(5x + 4y)(x - 2y)$ **27.** $(2x + 1)(2x - 1)$ **28.** $(9 + 10y)(9 - 10y)$ **29.** $(5a + 7b)(5a - 7b)$ **30.** $(z^2 + 4)(z + 2)(z - 2)$ **31.** $2(x + 3)(x - 3)$ **32.** prime **33.** $x(3x + 1)(3x - 1)$ **34.** $2x(3y + 2)(3y - 2)$ **35.** $(x + 11)^2$ **36.** $(x - 8)^2$ **37.** $(3y + 8)^2$ **38.** $(4x - 5)^2$ **39.** prime **40.** $(6x + 5y)^2$ **41.** $(5x - 4y)^2$ **42.** $(x - 3)(x^2 + 3x + 9)$ **43.** $(4x + 1)(16x^2 - 4x + 1)$ **44.** $2(3x - 2y)(9x^2 + 6xy + 4y^2)$ **45.** $y(3x + 2)(9x^2 - 6x + 4)$ **46.** $(a + 3)(a - 3)$ **47.** $(a + 2b)(a - 2b)$ **48.** $A^2 + 2A + 1 = (A + 1)^2$ **49.** $x(x - 7)(x - 1)$ **50.** $(5y + 2)(2y + 1)$ **51.** $2(8 + y)(8 - y)$ **52.** $(3x + 1)^2$ **53.** $4x^3(5x^4 - 9)$ **54.** $(x - 3)^2(x + 3)$ **55.** prime **56.** $x(2x + 5)(x + 7)$ **57.** $3x(x - 5)^2$ **58.** $3x^2(x - 2)(x^2 + 2x + 4)$ **59.** $4y^2(y + 3)(y - 3)$ **60.** $5(x + 7)(x - 3)$ **61.** prime **62.** $2x^3(5x - 2)(x - 4)$ **63.** $(10y + 7)(10y - 7)$ **64.** $9x^4(x - 2)$ **65.** $(x^2 + 1)(x + 1)(x - 1)$ **66.** $2(y - 2)(y^2 + 2y + 4)$ **67.** $(x + 4)(x^2 - 4x + 16)$ **68.** $(3x - 2)(2x + 5)$ **69.** $3x^2(x + 2)(x - 2)$ **70.** $(x - 10)(x + 9)$ **71.** $(5x + 2y)(5x + 3y)$ **72.** $x(x + 5)(x^2 - 5x + 25)$ **73.** $2y(4y + 3)(4y + 1)$ **74.** $2(y - 4)^2$ **75.** $(x + 5y)(x - 7y)$ **76.** $(x + y)(x + 7)$ **77.** $(3x + 4y)^2$ **78.** $2x^2y(x + 1)(x - 1)$ **79.** $(10y + 7z)(10y - 7z)$ **80.** prime **81.** $3x^2y^2(x + 2y)(x - 2y)$ **82.** $\{0, 12\}$ **83.** $\left\{-\frac{9}{4}, 7\right\}$ **84.** $\{-7, 2\}$ **85.** $\{-4, 0\}$ **86.** $\left\{-8, \frac{1}{2}\right\}$ **87.** $\{-4, 8\}$ **88.** $\{-8, 7\}$ **89.** $\{7\}$ **90.** $\left\{-\frac{10}{3}, \frac{10}{3}\right\}$ **91.** $\{-5, -2\}$ **92.** $\left\{\frac{1}{3}, 7\right\}$ **93.** 2 sec **94.** width: 5 ft; length: 8 ft **95.** 11 m by 11 m

Chapter 6 Test

1. $(x-3)(x-6)$ **2.** $(x-7)^2$ **3.** $5y^2(3y-1)(y-2)$ **4.** $(x^2+3)(x+2)$ **5.** $x(x-9)$ **6.** $x(x+7)(x-1)$ **7.** $2(7x-3)(x+5)$ **8.** $(5x+3)(5x-3)$ **9.** $(x+2)(x^2-2x+4)$ **10.** $(x+3)(x-7)$ **11.** prime **12.** $3y(2y+1)(y+1)$ **13.** $4(y+3)(y-3)$ **14.** $4(2x+3)^2$ **15.** $2(x^2+4)(x+2)(x-2)$ **16.** $(6x-7)^2$ **17.** $(7x-1)(x-7)$ **18.** $(x^2-5)(x+2)$ **19.** $3y(2y+3)(2y-5)$ **20.** $(y-5)(y^2+5y+25)$ **21.** $5(x-3y)(x+2y)$ **22.** $\{-6, 4\}$ **23.** $\left\{-\frac{1}{3}, 2\right\}$ **24.** $\{-2, 8\}$ **25.** $\left\{0, \frac{7}{2}\right\}$ **26.** $\left\{-\frac{9}{4}, \frac{9}{4}\right\}$ **27.** $\left\{-1, \frac{6}{5}\right\}$ **28.** $x^2-4=(x+2)(x-2)$ **29.** 6 sec **30.** width: 5 ft; length: 11 ft

Cumulative Review Exercises (Chapters 1–6)

1. -48 **2.** $\{0\}$ **3.** $\{12\}$ **4.** $\{x \mid x < -2\}$;

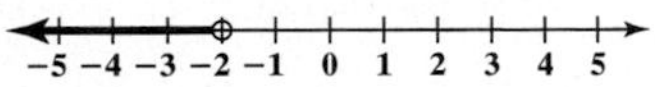

5. 38°, 38°, 104° **6.** \$150

7.

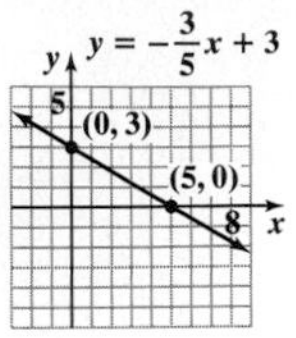

8. $y+4=5(x-2)$, or $y-1=5(x-3)$; $y=5x-14$ **9.** $\{(2,-2,-1)\}$ **10.** $\{(3,-1)\}$ **11.** $\left\{\left(-\frac{2}{13}, \frac{23}{13}\right)\right\}$ **12.** $-\frac{13}{40}$ **13.** $2x^4-x^3+3x+9$ **14.** $6x^2+17xy-45y^2$ **15.** $2x^2+5x-3-\frac{2}{3x-5}$ **16.** 7.1×10^{-3} **17.** $(3x+2)(x+3)$ **18.** $y(y^2+4)(y+2)(y-2)$ **19.** $(2x+3)^2$ **20.** width: 4 ft; length: 6 ft

CHAPTER 7

Section 7.1

Check Point Exercises

1. a. $x=5$ **b.** $x=-7$ and $x=4$ **2.** $\frac{x+4}{3x}$ **3.** $\frac{x^2}{7}$ **4.** $\frac{x-1}{x+1}$ **5.** $-\frac{3x+7}{4}$

Exercise Set 7.1

1. $x=0$ **3.** $x=8$ **5.** $x=4$ **7.** $x=-9$ and $x=2$ **9.** $x=\frac{17}{3}$ and $x=-3$ **11.** $x=-4$ and $x=3$ **13.** defined for all real numbers **15.** $y=-1$ and $y=\frac{3}{4}$ **17.** $y=-5$ and $y=5$ **19.** defined for all real numbers **21.** $2x$ **23.** $\frac{x-3}{5}$ **25.** $\frac{x-4}{2x}$ **27.** $\frac{1}{x-3}$ **29.** $-\frac{5}{x-3}$ **31.** 3 **33.** $\frac{1}{x-5}$ **35.** $\frac{2}{3}$ **37.** $\frac{1}{x-3}$ **39.** $\frac{4}{x-2}$ **41.** $\frac{y-1}{y+9}$ **43.** $\frac{y-3}{y-2}$ **45.** cannot be simplified **47.** $\frac{x+6}{x-6}$ **49.** x^2+1 **51.** x^2+2x+4 **53.** $\frac{x-4}{x+4}$ **55.** cannot be simplified **57.** cannot be simplified **59.** -1 **61.** -1 **63.** cannot be simplified **65.** -2 **67.** $-\frac{2}{x}$ **69.** $-x-1$ **71.** $-y-3$ **73.** $-\frac{1}{x}$ **75.** $\frac{x-y}{2x-y}$ **77.** $\frac{x-6}{x^2+3x+9}$ **79.** $\frac{3+y}{3-y}$ **81.** $\frac{y+3}{x+3}$ **83.** $\frac{2}{1-2x}$ **85. a.** It costs \$86.67 million to inoculate 40% of the population. It costs \$520 million to inoculate 80% of the population. It costs \$1170 million to inoculate 90% of the population. **b.** $x=100$ **c.** The cost keeps rising; No amount of money will be enough to inoculate 100% of the population. **87.** 400 mg **89. a.** \$300 **b.** \$125 **c.** decrease **91.** 1.5 mg per liter; (3, 1.5) **93.** 20.8; no **95. a.** $\frac{-0.4t+14.2}{3.7t+257.4}$ **b.** 0.04; 4000 **c.** fairly well; Answers will vary. **105.** $\frac{x^2-x-6}{x+2}$ **107.** $2x^2-1$ should be $2x+1$. **110.** $\frac{3}{10}$ **111.** $\frac{1}{6}$ **112.** $\{(4, 2)\}$

Section 7.2

Check Point Exercises

1. $\frac{9x-45}{2x+8}$ **2.** $\frac{3}{8}$ **3.** $\frac{x+2}{9}$ **4.** $-\frac{5(x+1)}{7x(2x-3)}$ **5.** $\frac{(x+3)(x+7)}{x-4}$ **6.** $\frac{x+3}{x-5}$ **7.** $\frac{y+1}{5y(y^2+1)}$

Exercise Set 7.2

1. $\frac{4x-20}{9x+27}$ **3.** $\frac{4x}{x+5}$ **5.** $\frac{4}{5}$ **7.** $\frac{4}{9}$ **9.** 1 **11.** $\frac{x+5}{x}$ **13.** $\frac{2}{y}$ **15.** $\frac{y+2}{y+4}$ **17.** $4(y+3)$ **19.** $\frac{x-1}{x+2}$ **21.** $\frac{x^2+2x+4}{3x}$ **23.** $\frac{x-2}{x-1}$ **25.** $-\frac{2}{x(x+1)}$ **27.** $-\frac{y-10}{y-7}$ **29.** $(x-y)(x+y)$ **31.** $\frac{4(x+y)}{3(x-y)}$ **33.** $\frac{3x}{35}$ **35.** $\frac{1}{4}$ **37.** 10 **39.** $\frac{7}{9}$ **41.** $\frac{3}{4}$ **43.** $\frac{(x-2)^2}{x}$ **45.** $\frac{(y-4)(y^2+4)}{y-1}$ **47.** $\frac{y}{3}$ **49.** $\frac{2(x+3)}{3}$ **51.** $\frac{x-5}{2}$ **53.** $\frac{y^2+1}{y^2}$ **55.** $\frac{y+1}{y-7}$ **57.** 6 **59.** $\frac{2}{3}$ **61.** $\frac{(x+y)^2}{32(x-y)^2}$

63. $\frac{y(x+y)}{(x+1)^2}$ **65.** $\frac{y+3}{y-3}$ **67.** $\frac{5(x-5)}{4}$ **69.** $\frac{(x+y)^2}{x-y}$ **71.** $\frac{2}{y^2-by+b^2}$ **73.** $\frac{125x}{100-x}$ **79.** numerator: x; denominator: $x-4$

81. $(3x-y)(3x+y)$ **83.** correct answer **85.** $2x-1$ should be $2x+1$. **86.** $\{x|x>18\}$ **87.** $3(x-7)(x+2)$ **88.** $\left\{-5, \frac{1}{2}\right\}$

Section 7.3

Check Point Exercises

1. $x+2$ **2.** $\frac{x-5}{x+5}$ **3. a.** $\frac{3x+5}{x+7}$ **b.** $3x-4$ **4.** $\frac{3y+2}{y-4}$ **5.** $x+3$ **6.** $\frac{-4x^2+12x}{x^2-2x-9}$

Exercise Set 7.3

1. $\frac{9x}{13}$ **3.** $\frac{3x}{5}$ **5.** $\frac{x+3}{2}$ **7.** $\frac{6}{x}$ **9.** $\frac{7}{3x}$ **11.** $\frac{9}{x+3}$ **13.** $\frac{5x+5}{x-3}$ **15.** 2 **17.** $\frac{2y+3}{y-5}$ **19.** $\frac{7}{5y}$ **21.** $\frac{2}{x+2}$ **23.** $\frac{x-2}{x-3}$ **25.** $\frac{3x-4}{5x-4}$ **27.** 1 **29.** 1 **31.** $\frac{x}{2x-1}$ **33.** $-\frac{1}{y}$ **35.** $\frac{y+3}{3y+8}$ **37.** $\frac{3y+2}{y-3}$ **39.** $\frac{2}{x-3}$ **41.** $\frac{3x+7}{x-6}$ **43.** $\frac{3x+1}{3x-4}$ **45.** $x+2$ **47.** 0 **49.** $\frac{11}{x-1}$ **51.** $\frac{12}{x+3}$ **53.** $\frac{y+1}{y-1}$ **55.** $\frac{x-2}{x-7}$ **57.** $\frac{2x-4}{x^2-25}$ **59.** 1 **61.** $\frac{2}{x+y}$ **63.** $\frac{x-3}{x-1}$ **65.** $\frac{4b}{4b-3}$ **67.** $\frac{y}{y-5}$

69. $-\frac{1}{c+d}$ **71.** $\frac{3y^2+8y-5}{(y+1)(y-4)}$ **73. a.** $\frac{100W}{L}$ **b.** round **75.** 10 m **81.** d **83.** $x-2$ **85.** $20x-6$ **87.** $a+12$

89. correct answer **91.** $x+4$ should be $x-4$. **92.** $\frac{31}{45}$ **93.** $(9x^2+1)(3x+1)(3x-1)$ **94.** $3x^2-7x-5$

Section 7.4

Check Point Exercises

1. $30x^2$ **2.** $(x+3)(x-3)$ **3.** $7x(x+4)(x+4)$ or $7x(x+4)^2$ **4.** $\frac{9+14x}{30x^2}$ **5.** $\frac{6x+6}{(x+3)(x-3)}$ **6.** $-\frac{5}{x+5}$ **7.** $-\frac{5+y}{5y}$ **8.** $\frac{x-15}{(x+5)(x-5)}$

Exercise Set 7.4

1. $120x^2$ **3.** $30x^5$ **5.** $(x-3)(x+1)$ **7.** $7y(y+2)$ **9.** $(x+4)(x-4)$ **11.** $y(y+3)(y-3)$ **13.** $(y+1)(y-1)(y-1)$ **15.** $(x-5)(x+4)(2x-1)$ **17.** $\frac{3x+5}{x^2}$ **19.** $\frac{37}{18x}$ **21.** $\frac{8x+7}{2x^2}$ **23.** $\frac{6x+1}{x}$ **25.** $\frac{2+9x}{x}$ **27.** $\frac{x+1}{2}$ **29.** $\frac{7x-20}{x(x-5)}$ **31.** $\frac{5x+1}{(x-1)(x+2)}$ **33.** $\frac{11y+15}{4y(y+5)}$ **35.** $-\frac{7}{x+7}$ **37.** $\frac{3x-55}{(x+5)(x-5)}$ **39.** $\frac{x^2+6x}{(x-4)(x+4)}$ **41.** $\frac{y+12}{(y+3)(y-3)}$ **43.** $\frac{7x-10}{(x-1)(x-1)}$ **45.** $\frac{9y}{4(y-5)}$ **47.** $\frac{8y+16}{y(y+4)}$ **49.** $\frac{17}{(x-3)(x-4)}$ **51.** $\frac{7x-1}{(x+1)(x+1)(x-1)}$ **53.** $\frac{x^2-x}{(x+3)(x-2)(x+5)}$ **55.** $\frac{y^2+8y+4}{(y+4)(y+1)(y+1)}$ **57.** $\frac{2x^2-4x+34}{(x+3)(x-5)}$ **59.** $\frac{5-3y}{2y(y-1)}$ **61.** $-\frac{x^2}{(x+3)(x-3)}$ **63.** $\frac{2}{y+5}$ **65.** $\frac{4x-11}{x-3}$ **67.** $\frac{3}{y+1}$ **69.** $\frac{-x^2+7x+3}{(x-3)(x+2)}$ **71.** $\frac{x^2+2x-1}{(x+1)(x-1)(x+2)}$ **73.** $\frac{-y^2+8y+9}{15y^2}$ **75.** $\frac{x^2-2x-6}{3(x+2)(x-2)}$ **77.** $\frac{-y-2}{(y+1)(y-1)}$ **79.** $\frac{x+2xy-y}{xy}$ **81.** $\frac{5x+2y}{(x+y)(x-y)}$ **83.** $\frac{-x+6}{(x+2)(x-2)}$ **85.** $\frac{6x+5}{(x-5)(x+5)(x-6)}$ **87.** $\frac{9}{(x-3)(x^2+3x+9)}$ **89.** $\frac{3}{y-3}$ **91.** $-\frac{22y}{(x+3y)(x+y)(x-3y)}$ **93.** $\frac{D}{5}$ **95.** $\frac{D}{24}$ **97.** no **99.** 5 yr **101.** $\frac{4x^2+14x}{(x+3)(x+4)}$ **105.** $\frac{3}{x+5}$ should be $\frac{5+2x}{5x}$. **109.** $\frac{y^3+y^2-10y-2}{(y-1)(y-1)(y+3)}$ **111.** $\frac{3x-1}{x^2(x-1)}$ **113.** $6x^2-11x-35$

114. 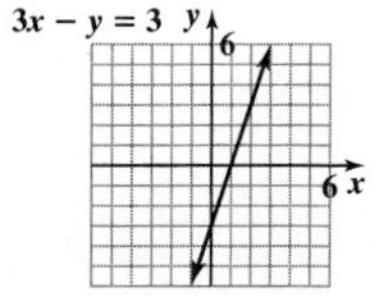

115. $y = x - 1$

Mid-Chapter 7 Check Point

1. $x=-2$ and $x=4$ **2.** $\frac{x-2}{2x+1}$ **3.** $\frac{-3}{y-2}$ or $\frac{3}{2-y}$ **4.** $\frac{2}{w}$ **5.** $\frac{4}{x+4}$ **6.** $\frac{4}{(x-2)^2}$ **7.** $\frac{x+5}{x-2}$ **8.** $\frac{x+2}{x-4}$ **9.** $\frac{2x+1}{(x+2)(x+3)(x-1)}$ **10.** $\frac{9-x}{x-5}$ **11.** $-\frac{2y-1}{3y}$ or $\frac{1-2y}{3y}$ **12.** $\frac{-y^2}{(y+1)(y+2)}$ **13.** $\frac{w+1}{(w-3)(w+5)}$ **14.** $\frac{2z^2-3z+15}{(z+3)(z-3)(z+1)}$ **15.** $\frac{3z^2+5z+3}{(3z-1)^2}$ **16.** $\frac{3x-1}{(x+7)(x-3)}$ **17.** x **18.** $\frac{2x+1}{(x+1)(x-2)(x+3)}$ **19.** $\frac{5x-5y}{x}$ **20.** $\frac{x+3}{x+5}$

Section 7.5

Check Point Exercises

1. $\frac{11}{5}$ **2.** $\frac{2x-1}{2x+1}$ **3.** $y-x$ **4.** $\frac{11}{5}$ **5.** $\frac{2x-1}{2x+1}$ **6.** $y-x$

Exercise Set 7.5

1. $\frac{9}{10}$ **3.** $\frac{18}{23}$ **5.** $-\frac{4}{5}$ **7.** $\frac{3-4x}{3+4x}$ **9.** $\frac{7x-2}{5x+1}$ **11.** $\frac{2y+3}{y-7}$ **13.** $\frac{4-6y}{4+3y}$ **15.** $x-5$ **17.** $\frac{x}{x-1}$ **19.** $-\frac{1}{y}$ **21.** $\frac{xy+2}{x}$ **23.** $\frac{y+x}{x^2y^2}$ **25.** $\frac{x^2+y}{y(y+1)}$ **27.** $\frac{1}{y}$ **29.** $\frac{4-x}{5x-3}$ **31.** $\frac{2y}{y-3}$ **33.** $\frac{1}{x+3}$ **35.** $\frac{x^2-5x+3}{x^2-7x+2}$ **37.** $\frac{2y^2+3xy}{y^2+2x}$ **39.** $-\frac{6}{5}$ **41.** $\frac{1-x}{(x-3)(x+6)}$ **43.** $\frac{1}{y(y+5)}$ **45.** $\frac{1}{x-1}$ **47.** $\frac{x+1}{2x+1}$ **49.** $\frac{2r_1r_2}{r_1+r_2}$; $34\frac{2}{7}$ mph **55.** It cubes x. **57.** $\frac{5y+3}{(y-3)(y+1)}$ **59.** 2 should be $1+x$. **61.** $2x(x-5)^2$ **62.** $\{-2\}$ **63.** x^3+y^3

Section 7.6

Check Point Exercises

1. 4 or $\{4\}$ **2.** 3 or $\{3\}$ **3.** -3 and -2, or $\{-3, -2\}$ **4.** 24 or $\{24\}$ **5.** no solution or $\emptyset$ **6.** 75% **7.** $x = \frac{yz}{y-z}$

Exercise Set 7.6

1. $\{12\}$ **3.** $\{0\}$ **5.** $\{-2\}$ **7.** $\{6\}$ **9.** $\{-2\}$ **11.** $\{2\}$ **13.** $\{15\}$ **15.** $\{4\}$ **17.** $\{-6, 1\}$ **19.** $\{-5, 5\}$ **21.** $\{-3, 3\}$ **23.** $\{-8, 1\}$ **25.** $\{-1, 16\}$ **27.** $\{3\}$ **29.** no solution or $\emptyset$ **31.** $\left\{\frac{2}{3}, 4\right\}$ **33.** no solution or $\emptyset$ **35.** $\{1\}$ **37.** $\left\{\frac{1}{5}\right\}$ **39.** $\{-3\}$ **41.** no solution or $\emptyset$ **43.** $\left\{-6, \frac{1}{2}\right\}$ **45.** $P_1 = \frac{P_2V_2}{V_1}$ **47.** $f = \frac{pq}{q+p}$ **49.** $r = \frac{A-P}{P}$ **51.** $m_1 = \frac{Fd^2}{Gm_2}$ **53.** $x = \overline{x} + zs$ **55.** $R = \frac{E-Ir}{I}$ **57.** $f_1 = \frac{ff_2}{f_2-f}$ **59.** $\{-3\}$ **61.** $\frac{-6x-18}{(x-5)(x+4)}$ **63.** $\{1, 5\}$ **65.** $\{3\}$ **67.** 10,000 wheelchairs **69.** 50% **71.** 5 yr old **73.** either 50 or approx 67 cases; (50, 350) or $\left(66\frac{2}{3}, 350\right)$ **75.** 10 more hits **83.** b **85.** no solution **87.** $\{8\}$ **89.** $\{-3, -2\}$ **90.** $(x^3-3)(x+2)$ **91.** $-\frac{12}{x^8}$ **92.** $-20x+55$

Section 7.7

Check Point Exercises

1. 2 mph **2.** $2\frac{2}{3}$ hr or 2 hr 40 min **3.** \$1500 **4.** 720 deer **5.** 32 in. **6.** 32 yd

Exercise Set 7.7

1. Walking rate is 6 mph; car rate is 9 mph. **3.** Downhill rate is 10 mph; uphill rate is 6 mph. **5.** Water's current is 5 mph. **7.** Walking rate is 3 mph; jogging rate is 6 mph. **9.** Still water rate is 10 mph. **11.** It will take about 8.6 min, which is enough time. **13.** It will take about 171.4 hr, which is not enough time. **15.** It will take 2.4 hr, or 2 hr 24 min. **17.** \$1800 **19.** 20,489 fur seal pups **21.** \$950 **23.** 154.1 in. **25.** 5 in. **27.** 6 m **29.** 16 in. **31.** 16 ft **43.** 30 mph **45.** 7 hr **47.** 2.5 hr **49.** $(5x+9)(5x-9)$ **50.** $\{6\}$ **51.**

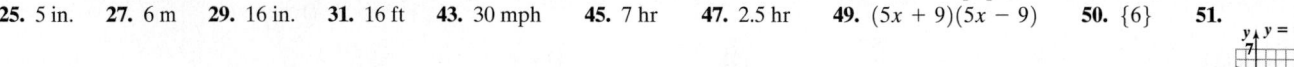

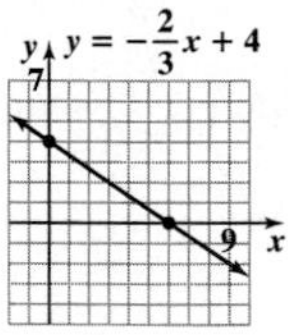

Section 7.8

Check Point Exercises

1. 66 gal **2.** 512 cycles per second **3.** 24 min **4.** 96π cubic feet

Exercise Set 7.8

1. 156 **3.** 30 **5.** $\frac{5}{6}$ **7.** 240 **9.** 50 **11.** $x = kyz$; $y = \frac{x}{kz}$ **13.** $x = \frac{kz^3}{y}$; $y = \frac{kz^3}{x}$ **15.** $x = \frac{kyz}{\sqrt{w}}$; $y = \frac{x\sqrt{w}}{kz}$ **17.** $x = kz(y+w)$; $y = \frac{x-kzw}{kz}$ **19.** $x = \frac{kz}{y-w}$; $y = \frac{xw+kz}{x}$ **21.** 5.4 ft **23.** 80 in. **25.** about 607 lb **27.** 32° **29.** 90 milliroentgens per hour **31.** This person has a BMI of 24.4 and is not overweight. **33.** 1800 Btu **35.** $\frac{1}{4}$ of what it was originally **37. a.** $C = \frac{kP_1P_2}{d^2}$ **b.** $k \approx 0.02$; $C = \frac{0.02\,P_1P_2}{d^2}$ **c.** approximately 39,813 daily phone calls

39. a.

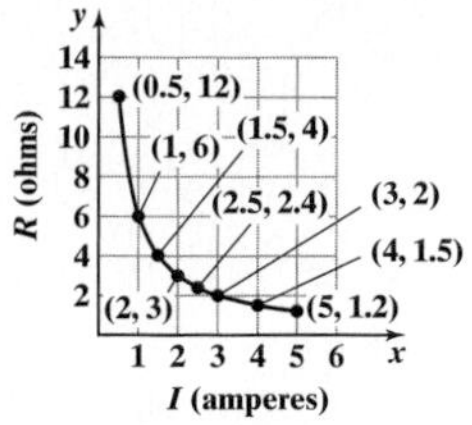

b. Current varies inversely as resistance. **c.** $R = \frac{6}{I}$

45. z varies directly as the square root of x and inversely as the square of y.

49. The wind pressure is 4 times more destructive.

51. Distance is increased by $\sqrt{50}$, or about 7.07, for the space telescope.

52. $\left\{\frac{16}{3}\right\}$ **53.** $9x^2 - 6x + 4 - \frac{16}{3x + 2}$ **54.** $6x(x - 5)(x + 4)$

Review Exercises

1. $x = 4$ **2.** $x = 2$ and $x = -5$ **3.** $x = 1$ and $x = 2$ **4.** defined for all real numbers **5.** $\frac{4x}{3}$ **6.** $x + 2$ **7.** x^2 **8.** $\frac{x - 3}{x - 6}$ **9.** $\frac{x - 5}{x + 7}$ **10.** $\frac{y}{y + 2}$ **11.** cannot be simplified **12.** $-2(x + 3y)$ **13.** $\frac{x - 2}{4}$ **14.** $\frac{5}{2}$ **15.** $\frac{x + 3}{x + 2}$ **16.** $\frac{2y - 1}{5}$ **17.** $\frac{-3(y + 1)}{5y(2y - 3)}$ **18.** $\frac{x - 1}{4}$ **19.** $\frac{2}{x(x + 1)}$ **20.** $\frac{1}{7(y + 3)}$ **21.** $(y - 3)(y - 6)$ **22.** $\frac{8}{x - y}$ **23.** 4 **24.** 4 **25.** $3x - 5$ **26.** $3y + 4$ **27.** $\frac{4}{x - 2}$ **28.** $\frac{2x + 5}{x - 3}$ **29.** $36x^3$ **30.** $x^2(x - 1)^2$ **31.** $(x + 3)(x + 1)(x + 7)$ **32.** $\frac{14x + 15}{6x^2}$ **33.** $\frac{7x + 2}{x(x + 1)}$ **34.** $\frac{7x + 25}{(x + 3)^2}$ **35.** $\frac{3}{y - 2}$ **36.** $-\frac{y}{y - 1}$ **37.** $\frac{x^2 + y^2}{xy}$ **38.** $\frac{3x^2 - x}{(x + 1)^2(x - 1)}$ **39.** $\frac{5x^2 - 3x}{(x + 1)(x - 1)}$ **40.** $\frac{4}{(x + 2)(x - 2)(x - 3)}$ **41.** $\frac{2x + 13}{x + 3}$ **42.** $\frac{y - 4}{3y}$ **43.** $\frac{7}{2}$ **44.** $\frac{1}{x - 1}$ **45.** $x + y$ **46.** $\frac{3}{x}$ **47.** $\frac{3x}{x - 4}$ **48.** $\{12\}$ **49.** $\{-1\}$ **50.** $\{-6, 1\}$ **51.** no solution or $\varnothing$ **52.** $\{5\}$ **53.** $\{-6, 2\}$ **54.** $\{5\}$ **55.** $\{-5\}$ **56.** 3 yr **57.** 30% **58.** $C = R - nP$ **59.** $T_1 = \frac{P_1V_1T_2}{P_2V_2}$ **60.** $P = \frac{A}{rT + 1}$ **61.** $R = \frac{R_1R_2}{R_2 + R_1}$ **62.** $n = \frac{IR}{E - Ir}$ **63.** 4 mph **64.** Slower car's rate is 20 mph; faster car's rate is 30 mph. **65.** 4 hr **66.** 324 teachers **67.** 287 trout **68.** 5 ft **69.** $12\frac{1}{2}$ ft **70.** \$4935 **71.** 1600 ft **72.** 440 vibrations per second **73.** 112 decibels **74.** 16 hr **75.** 800 cubic feet

Chapter 7 Test

1. $x = -9$ and $x = 4$ **2.** $\frac{x + 3}{x - 2}$ **3.** $\frac{4x}{x + 1}$ **4.** $\frac{x - 4}{2}$ **5.** $\frac{x}{x + 3}$ **6.** $\frac{2x + 6}{x + 1}$ **7.** $\frac{5}{(y - 1)(y - 3)}$ **8.** $2y$ **9.** $\frac{2}{y + 3}$ **10.** $\frac{x^2 + 2x + 15}{(x + 3)(x - 3)}$ **11.** $\frac{8x - 14}{(x + 2)(x - 1)(x - 3)}$ **12.** $\frac{-x - 1}{x - 3}$ **13.** $\frac{x + 2}{x - 1}$ **14.** $\frac{11}{(x - 3)(x - 4)}$ **15.** $\frac{4}{y + 4}$ **16.** $\frac{3x - 3y}{x}$ **17.** $\frac{5x + 5}{2x + 1}$ **18.** $\frac{y - x}{y}$ **19.** 6 or $\{6\}$ **20.** -8 or $\{-8\}$ **21.** 0 and 2, or $\{0, 2\}$ **22.** $a = \frac{Rs}{s - R}$ or $a = -\frac{RS}{R - s}$ **23.** Water's current is 2 mph. **24.** 12 min **25.** 6000 tule elk **26.** 3.2 in. **27.** 52.5 amp

Cumulative Review Exercises (Chapters 1–7)

1. $\{2\}$ **2.** $\{x|x < 2\}$ **3.** $\{-6, 3\}$ **4.** $\{-6\}$ **5.** $\{(3, 3)\}$ **6.** $\{(-2, 2)\}$

7.

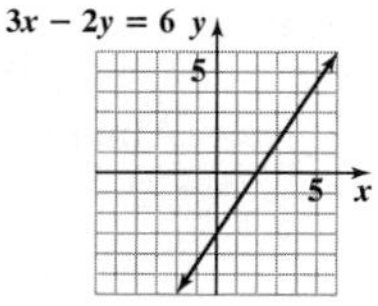

8.

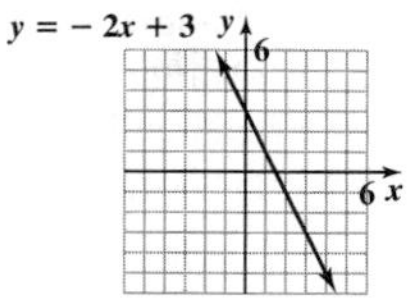

9.

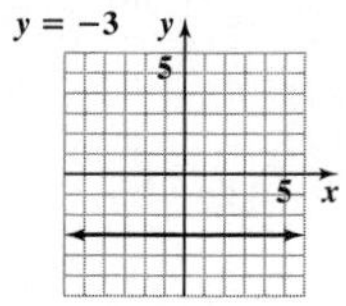

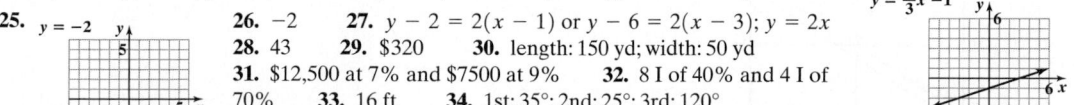

10. -19 **11.** $8x^9$ **12.** $\frac{1 - 2x}{4x - 1}$ **13.** $(4x - 1)(x - 3)$ **14.** $(2x - 5)^2$ **15.** $3(x + 5)(x - 5)$ **16.** $-x^2 + 4x + 8$ **17.** $-2x^4 + 3x^2 - 1$ **18.** $\frac{3x^2 + 6x + 16}{(x - 2)(x + 3)}$ **19.** \$1225 at 5% and \$2775 at 9% **20.** 17 in. and 51 in.

Mid-Textbook Check Point

1. 4 or $\{4\}$ **2.** 10 or $\{10\}$ **3.** $\{x|x \leq 7\}$ **4.** $(-3, 4)$ or $\{(-3, 4)\}$ **5.** $(3, 4)$ or $\{(3, 4)\}$ **6.** -8 or $\{-8\}$ **7.** -3 and -2, or $\{-3, -2\}$ **8.** $\frac{4}{x^9}$ **9.** 31 **10.** $10x^2 - 9x + 4$ **11.** $21x^2 - 23x - 20$ **12.** $25x^2 - 20x + 4$ **13.** $x^3 + y^3$ **14.** $\frac{x + 4}{3x^3}$ **15.** $\frac{-7x}{(x + 3)(x - 1)(x - 4)}$ **16.** $\frac{x}{5}$ **17.** $(2x + 7)(2x - 7)$ **18.** $(x + 3)(x + 1)(x - 1)$ **19.** $2(x - 3)(x + 7)$ **20.** $x(x^2 + 4)(x + 2)(x - 2)$ **21.** $x(x - 5)^2$ **22.** $(x - 2)(x^2 + 2x + 4)$

23.

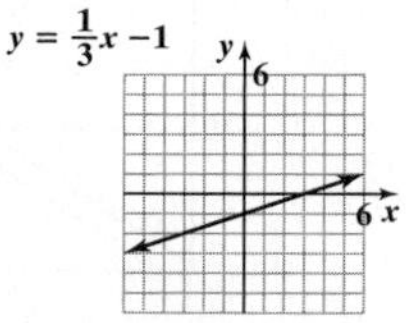

24.

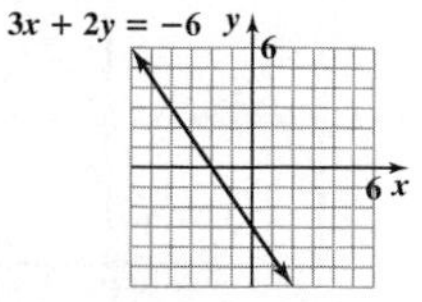

25.

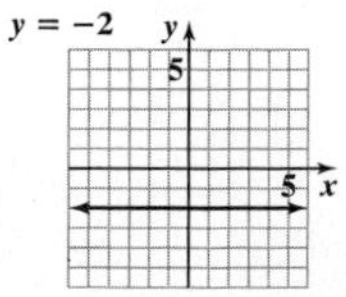

26. -2 **27.** $y - 2 = 2(x - 1)$ or $y - 6 = 2(x - 3)$; $y = 2x$ **28.** 43 **29.** \$320 **30.** length: 150 yd; width: 50 yd **31.** \$12,500 at 7% and \$7500 at 9% **32.** 8 l of 40% and 4 l of 70% **33.** 16 ft **34.** 1st: 35°; 2nd: 25°; 3rd: 120° **35.** TV: \$350; stereo: \$370 **36.** length: 11m; width: 5 m

CHAPTER 8

Section 8.1

Check Point Exercises

1. domain: $\{5, 10, 15, 20, 25\}$; range: $\{12.8, 16.2, 18.9, 20.7, 21.8\}$ **2. a.** not a function **b.** function **3. a.** 29 **b.** 65 **c.** 46 **d.** $6a + 6h + 9$ **4. a.** function **b.** function **c.** not a function **5. a.** 16 **b.** 8 **6. a.** Domain $= \{x|-2 \le x \le 1\}$; Range $= \{y|0 \le y \le 3\}$ **b.** Domain $= \{x|-2 < x \le 1\}$; Range $= \{y|-1 \le y < 2\}$ **c.** Domain $= \{x|-3 \le x < 0\}$; Range $= \{y|y = -3, -2, -1\}$

Exercise Set 8.1

1. function; domain: $\{1, 3, 5\}$; range: $\{2, 4, 5\}$ **3.** not a function; domain: $\{3, 4\}$; range: $\{4, 5\}$ **5.** function; domain: $\{-3, -2, -1, 0\}$; range: $\{-3, -2, -1, 0\}$ **7.** not a function; domain: $\{1\}$; range: $\{4, 5, 6\}$ **9. a.** 1 **b.** 6 **c.** -7 **d.** $2a + 1$ **e.** $a + 3$ **11. a.** -2 **b.** -17 **c.** 0 **d.** $12b - 2$ **e.** $3b + 10$ **13. a.** 5 **b.** 8 **c.** 53 **d.** 32 **e.** $48b^2 + 5$ **15. a.** -1 **b.** 26 **c.** 19 **d.** $2b^2 + 3b - 1$ **e.** $50a^2 + 15a - 1$ **17. a.** $\frac{3}{4}$ **b.** -3 **c.** $\frac{11}{8}$ **d.** $\frac{13}{9}$ **e.** $\frac{2a + 2h - 3}{a + h - 4}$ **f.** Denominator would be zero. **19.** function **21.** not a function **23.** function **25.** not a function **27.** -4 **29.** 4 **31.** 0 **33.** 2 **35.** 2 **37.** -2 **39.** Domain $= \{x|0 \le x < 5\}$; Range $= \{y|-1 \le y < 5\}$ **41.** Domain $= \{x|x \ge 0\}$; Range $= \{y|y \ge 1\}$ **43.** Domain $= \{x|-2 \le x \le 6\}$; Range $= \{y|-2 \le y \le 6\}$ **45.** Domain $= \{x|x$ is a real number$\}$; Range $= \{y|y \le -2\}$ **47.** Domain $= \{x|x = -5, -2, 0, 1, 3\}$; Range $= \{y|y = 2\}$ **49.** -2; 10 **51.** -38 **53.** $-2x^3 - 2x$ **55. a.** -1 **b.** 7 **c.** 19 **d.** 112 **57.** Domain $= \{x|x \ne 1\}$; Range $= \{y|y \ne 0\}$ **59. a.** {(EL, 1%), (L, 7%), (SL, 11%), (M, 52%), (SC, 13%), (C, 13%), (EC, 3%)} **b.** Yes; Each ideology corresponds to exactly one percentage. **c.** {(1%, EL), (7%, L), (11%, SL), (52%, M), (13%, SC), (13%, C), (3%, EC)} **d.** No; 13% in the domain corresponds to two ideologies in the range, SC and C **61.** 5.22; In 2000, there were 5.22 million women enrolled in U.S. colleges; (2000, 5.22). **63.** 1.4; In 2004, there were 1.4 million more women than men enrolled in U.S. colleges. **65.** 440; For 20-year-old drivers, there are 440 accidents per 50 million miles driven; (20, 440). **67.** $x = 45$; $y = 190$; The minimum number of accidents is 190 per 50 million miles driven and is attributed to 45-year-old drivers. **69.** 3.1; In 1960, Jewish Americans made up about 3.1% of the U.S. population. **71.** 19 and 64; In 1919 and in 1964, Jewish Americans made up about 3% of the U.S. population. **73.** 1940; 3.7% **75.** Each year corresponds to only one percentage. **77.** 0.83; It costs $0.83 to mail a 3-ounce first-class letter. **79.** $0.60 **91.** 3 **93.** $f(2) = 6$; $f(3) = 9$; $f(4) = 12$; no **94.** 0 **95.** $\frac{y^{10}}{9x^4}$ **96.** -15

Section 8.2

Check Point Exercises

1. a. $\{x|x$ is a real number$\}$ **b.** $\{x|x$ is a real number and $x \ne -5\}$ **2. a.** $3x^2 + 6x + 6$ **b.** 78 **3. a.** $\frac{5}{x} + \frac{7}{x - 8}$ **b.** $\{x|x$ is a real number and $x \ne 0$ and $x \ne 8\}$ **4. a.** 23 **b.** $x^2 - 3x - 3$; 1 **c.** $x^3 + x^2 - 6x$; -24 **d.** $\frac{x^2 - 2x}{x + 3}$; $\frac{7}{2}$

Exercise Set 8.2

1. $\{x|x$ is a real number$\}$ **3.** $\{x|x$ is a real number and $x \ne -4\}$ **5.** $\{x|x$ is a real number and $x \ne 3\}$ **7.** $\{x|x$ is a real number and $x \ne 5\}$ **9.** $\{x|x$ is a real number and $x \ne -7$ and $x \ne 9\}$ **11. a.** $5x - 5$ **b.** 20 **13. a.** $3x^2 + x - 5$ **b.** 75 **15. a.** $2x^2 - 2$ **b.** 48 **17.** $\{x|x$ is a real number$\}$ **19.** $\{x|x$ is a real number and $x \ne 5\}$ **21.** $\{x|x$ is a real number and $x \ne 0$ and $x \ne 5\}$ **23.** $\{x|x$ is a real number and $x \ne 2$ and $x \ne -3\}$ **25.** $\{x|x$ is a real number and $x \ne 2\}$ **27.** $\{x|x$ is a real number$\}$ **29.** $x^2 + 3x + 2$; 20 **31.** 0 **33.** $x^2 + 5x - 2$; 48 **35.** -8 **37.** $-x^3 - 2x^2 + 8x$; 0 **39.** -135 **41.** $\frac{x^2 + 4x}{2 - x}$; 5 **43.** -1 **45.** $\{x|x$ is a real number$\}$ **47.** $\{x|x$ is a real number and $x \ne 2\}$ **49.** 5 **51.** -1 **53.** $\{x|-4 \le x \le 3\}$

55.

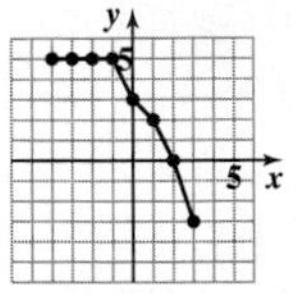

57. -4 **59.** -4 **61.** $\{x|x = 0, 1, 2, \ldots, 7\}$ **63. a.** $(B - D)(x) = 4565x + 1{,}578{,}296$; change in U.S. population **b.** $(B - D)(6) = 1{,}605{,}686$; In 2001, U.S. population increased by 1,605,686. **c.** 1,609,508; fairly well **65.** $h(x)$ or total world population in year x **67.** 5.9 billion **69.** $-200{,}000$; 0; 200,000; The company has a loss of $200,000, breaks even, and a profit of $200,000 when 20,000, 30,000, and 40,000 radios are produced, respectively. **75.** Per capita costs of Medicare are increasing more and more rapidly over time.

77.

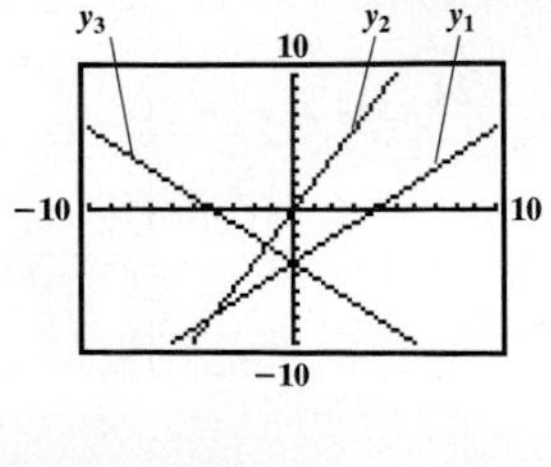

79.

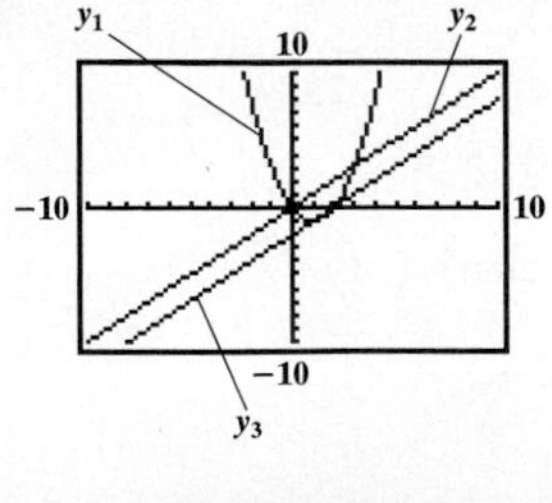

81. d **83.** $\{(-1, 2)\}$ **84.** 7 **85.** $6b + 8$

Mid-Chapter 8 Check Point

1. not a function; Domain: $\{1, 2\}$; Range: $\{-6, 4, 6\}$ **2.** function; Domain: $\{0, 2, 3\}$; Range: $\{1, 4\}$ **3.** function; Domain: $\{x|-2 \le x < 2\}$; Range: $\{y|0 \le y \le 3\}$ **4.** not a function; Domain: $\{x|-3 < x \le 4\}$; Range: $\{y|-1 \le y \le 2\}$ **5.** not a function; Domain: $\{-2, -1, 0, 1, 2\}$; Range: $\{-2, -1, 1, 3\}$ **6.** function; Domain: $\{x|x \le 1\}$; Range: $\{y|y \ge -1\}$ **7.** No vertical line intersects the graph of f more than once. **8.** 3 **9.** -2 **10.** -6 and 2 **11.** $\{x|x \text{ is a real number}\}$ **12.** $\{y|y \le 4\}$ **13.** $\{x|x \text{ is a real number}\}$ **14.** $\{x|x \text{ is a real number and } x \ne -2 \text{ and } x \ne 2\}$ **15.** 23 **16.** 23 **17.** $a^2 - 5a - 3$ **18.** $x^2 - 5x + 3$; 17 **19.** $x^2 - x + 13$; 33 **20.** $-2x^3 + x^2 - x - 40$; -36 **21.** $\dfrac{x^2 - 3x + 8}{-2x - 5}$; 12 **22.** $\left\{x \middle| x \text{ is a real number and } x \ne -\dfrac{5}{2}\right\}$

Section 8.3

Check Point Exercises

1. a. $5x^2 + 1$ **b.** $25x^2 + 60x + 35$ **2.** $f(g(x)) = 7\left(\dfrac{x}{7}\right) = x$; $g(f(x)) = \dfrac{7x}{7} = x$

3. $f(g(x)) = 4\left(\dfrac{x + 7}{4}\right) - 7 = x + 7 - 7 = x$; $g(f(x)) = \dfrac{(4x - 7) + 7}{4} = \dfrac{4x}{4} = x$ **4.** $f^{-1}(x) = \dfrac{x - 7}{2}$

5. (b) and (c) **6.**

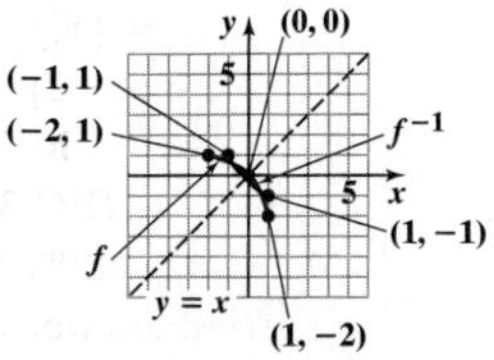

Exercise Set 8.3

1. a. $2x + 14$ **b.** $2x + 7$ **c.** 18 **3. a.** $2x + 5$ **b.** $2x + 9$ **c.** 9 **5. a.** $20x^2 - 11$ **b.** $80x^2 - 120x + 43$ **c.** 69 **7. a.** $x^4 - 4x^2 + 6$ **b.** $x^4 + 4x^2 + 2$ **c.** 6 **9. a.** $\sqrt{x - 1}$ **b.** $\sqrt{x} - 1$ **c.** 1 **11. a.** x **b.** x **c.** 2 **13. a.** x **b.** x **c.** 2 **15.** $f(g(x)) = x$; $g(f(x)) = x$; inverses **17.** $f(g(x)) = x$; $g(f(x)) = x$; inverses **19.** $f(g(x)) = \dfrac{5x - 56}{9}$; $g(f(x)) = \dfrac{5x - 4}{9}$; not inverses **21.** $f(g(x)) = x$; $g(f(x)) = x$; inverses **23.** $f(g(x)) = x$; $g(f(x)) = x$; inverses **25. a.** $f^{-1}(x) = x - 3$ **b.** $f(f^{-1}(x)) = (x - 3) + 3 = x$ and $f^{-1}(f(x)) = (x + 3) - 3 = x$ **27. a.** $f^{-1}(x) = \dfrac{x}{2}$ **b.** $f(f^{-1}(x)) = 2\left(\dfrac{x}{2}\right) = x$ and $f^{-1}(f(x)) = \dfrac{2x}{2} = x$ **29. a.** $f^{-1}(x) = \dfrac{x - 3}{2}$ **b.** $f(f^{-1}(x)) = 2\left(\dfrac{x - 3}{2}\right) + 3 = x$ and $f^{-1}(f(x)) = \dfrac{(2x + 3) - 3}{2} = x$ **31. a.** $f^{-1}(x) = \dfrac{1}{x}$ **b.** $f(f^{-1}(x)) = \dfrac{1}{\frac{1}{x}} = x$ and $f^{-1}(f(x)) = \dfrac{1}{\frac{1}{x}} = x$

33. a. $f^{-1}(x) = \dfrac{3x + 1}{x - 2}$ **b.** $f(f^{-1}(x)) = \dfrac{2\left(\dfrac{3x + 1}{x - 2}\right) + 1}{\left(\dfrac{3x + 1}{x - 2}\right) - 3} = x$ and $f^{-1}(f(x)) = \dfrac{3\left(\dfrac{2x + 1}{x - 3}\right) + 1}{\left(\dfrac{2x + 1}{x - 3}\right) - 2} = x$ **35.** no inverse **37.** no inverse

39. inverse function **41.**

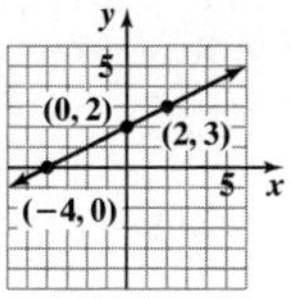

43.

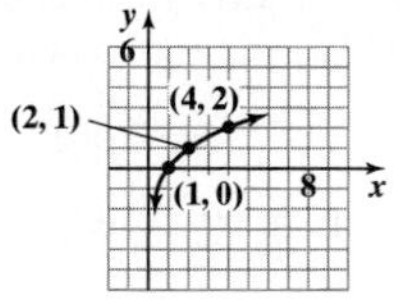

45. 5 **47.** 1 **49.** 2 **51.** 1 **53.** -6 **55.** -7 **57.** 3 **59.** 11

61. a. f represents the price after a \$400 discount and g represents the price after a 25% discount (75% of the regular price).
b. $0.75x - 400$; $f \circ g$ represents an additional \$400 discount on a price that has already been reduced by 25%.
c. $0.75(x - 400) = 0.75x - 300$; $g \circ f$ represents an additional 25% discount on a price that has already been reduced \$400.
d. $f \circ g$; $0.75x - 400 < 0.75x - 300$, so $f \circ g$ represents the lower price after the two discounts.
e. $f^{-1}(x) = x + 400$; f^{-1} represents the regular price, since the value of x here is the price after a \$400 discount.

63. a. f: {(Zambia, 4.2), (Colombia, 4.5), (Poland, 3.3), (Italy, 3.3), (U.S., 2.5)}
b. {(4.2, Zambia), (4.5, Colombia), (3.3, Poland), (3.3, Italy), (2.5, U.S.)}; no; The input 3.3 is associated with two outputs, Poland and Italy.

65. a. No horizontal line intersects the graph of f in more than one point.
b. $f^{-1}(0.25)$, or approximately 15, represents the number of people who would have to be in the room so that the probability of two sharing a birthday would be 0.25; $f^{-1}(0.5)$, or approximately 23, represents the number of people so that the probability would be 0.5; $f^{-1}(0.7)$, or approximately 30, represents the number of people so that the probability would be 0.7.

73.

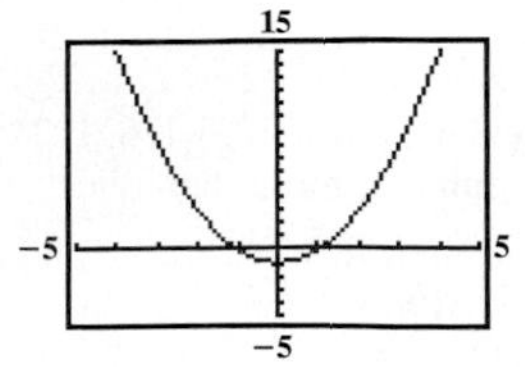

no inverse function

75.

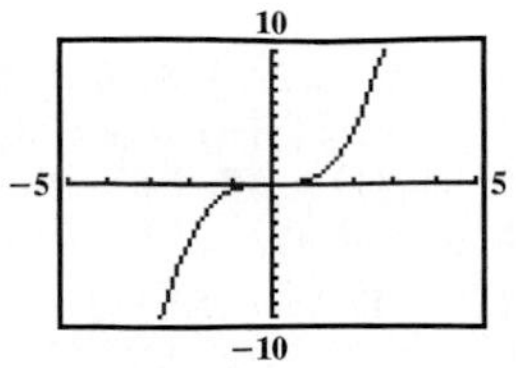

inverse function

77.

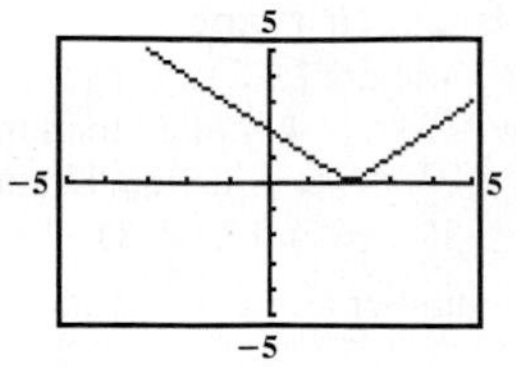

no inverse function

79.

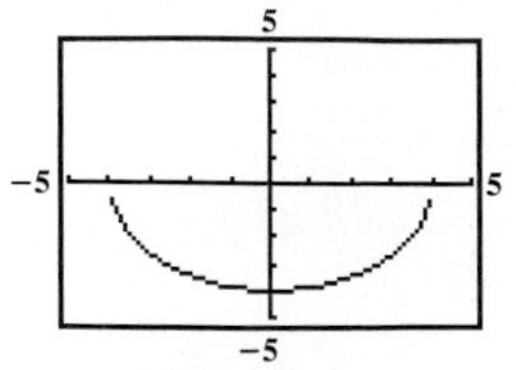

no inverse function

81.

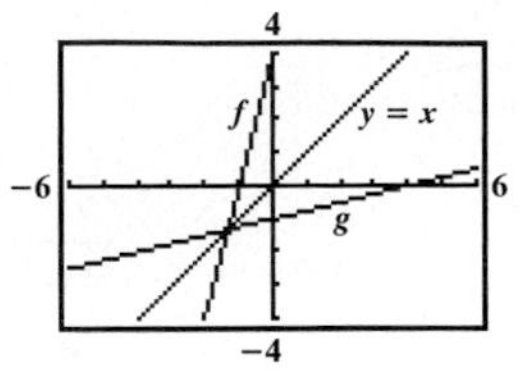

inverses

83.

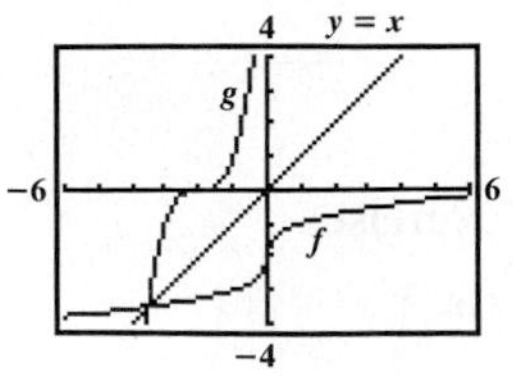

inverses

85. Answers will vary; Examples are $f(x) = \sqrt{x+5}$ and $g(x) = 3x^2$.

87. $f(f(x)) = \dfrac{3\left(\frac{3x-2}{5x-3}\right) - 2}{5\left(\frac{3x-2}{5x-3}\right) - 3} = \dfrac{3(3x-2) - 2(5x-3)}{5(3x-2) - 3(5x-3)} = \dfrac{9x - 6 - 10x + 6}{15x - 10 - 15x + 9} = \dfrac{-x}{-1} = x$ **89.** 5×10^8 **90.** $x^2 + 9x + 16 + \dfrac{35}{x-2}$ **91.** $\left\{\left(\frac{4}{5}, \frac{9}{5}\right)\right\}$

Review Exercises

1. function: domain: $\{3, 4, 5\}$; range: $\{10\}$ **2.** function: domain: $\{1, 2, 3, 4\}$; range: $\{12, 100, \pi, -6\}$ **3.** not a function; domain: $\{13, 15\}$; range: $\{14, 16, 17\}$ **4. a.** -5 **b.** 16 **c.** -75 **d.** $14a - 5$ **e.** $7a + 9$ **5. a.** 2 **b.** 52 **c.** 70 **d.** $3b^2 - 5b + 2$ **e.** $48a^2 - 20a + 2$ **6.** not a function **7.** function **8.** function **9.** not a function **10.** not a function **11.** function **12.** -3 **13.** -2 **14.** 3 **15.** $\{x \mid -3 \le x < 5\}$ **16.** $\{y \mid -5 \le y \le 0\}$ **17. a.** For each time, there is only one height. **b.** 0; The vulture was on the ground after 15 seconds. **c.** 45 m **d.** 7 and 22; After 7 and 22 seconds, the vulture's height is 20 meters. **e.** Answers will vary. **18.** $\{x \mid x$ is a real number$\}$ **19.** $\{x \mid x$ is a real number and $x \ne -8\}$ **20.** $\{x \mid x$ is a real number and $x \ne 5\}$ **21. a.** $6x - 4$ **b.** 14 **22. a.** $5x^2 + 1$ **b.** 46 **23.** $\{x \mid x$ is a real number and $x \ne 4\}$ **24.** $\{x \mid x$ is a real number and $x \ne -6$ and $x \ne -1\}$ **25.** $x^2 - x - 5; 1$ **26.** 1 **27.** $x^2 - 3x + 5; 3$ **28.** 9 **29.** $x^3 - 7x^2 + 10x; -120$ **30.** $\dfrac{x^2 - 2x}{x - 5}; -8$ **31.** $\{x \mid x$ is a real number$\}$ **32.** $\{x \mid x$ is a real number and $x \ne 5\}$ **33. a.** $16x^2 - 8x + 4$ **b.** $4x^2 + 11$ **c.** 124 **34. a.** $\sqrt{x+1}$ **b.** $\sqrt{x} + 1$ **c.** 2 **35.** $f(g(x)) = x - \dfrac{7}{10}$; $g(f(x)) = x - \dfrac{7}{6}$; not inverses **36.** $f(g(x)) = x$; $g(f(x)) = x$; inverses **37. a.** $f^{-1}(x) = \dfrac{x+3}{4}$ **b.** $f(f^{-1}(x)) = 4\left(\dfrac{x+3}{4}\right) - 3 = x$ and $f^{-1}(f(x)) = \dfrac{(4x-3)+3}{4} = x$ **38. a.** $f^{-1}(x) = -\dfrac{1}{x}$ **b.** $f(f^{-1}(x)) = -\dfrac{1}{\left(-\frac{1}{x}\right)} = x$ and $f^{-1}(f(x)) = -\dfrac{1}{\left(-\frac{1}{x}\right)} = x$ **39.** inverse function **40.** no inverse function **41.** inverse function **42.** no inverse function **43.**

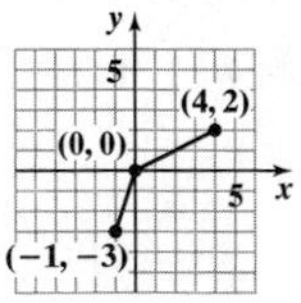

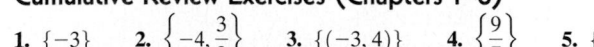

Chapter 8 Test

1. function; domain: $\{1, 3, 5, 6\}$; range: $\{2, 4, 6\}$ **2.** not a function; domain: $\{2, 4, 6\}$; range: $\{1, 3, 5, 6\}$ **3.** $3a + 10$ **4.** 285. function **6.** not a function **7.** -3 **8.** -2 and 3 **9.** $\{x \mid x$ is a real number$\}$ **10.** $\{y \mid y \le 3\}$ **11.** $\{x \mid x$ is a real number and $x \ne 10\}$ **12.** $x^2 + 5x + 2; 26$ **13.** $x^2 + 3x - 2; -4$ **14.** $x^3 + 6x^2 + 8x; -15$ **15.** $\dfrac{x^2 + 4x}{x + 2}; 3$ **16.** $\{x \mid x$ is a real number and $x \ne -2\}$ **17.** $(f \circ g)(x) = 9x^2 - 3x$; $(g \circ f)(x) = 3x^2 + 3x - 1$ **18.** $f^{-1}(x) = \dfrac{x+7}{5}$ **19. a.** No horizontal line intersects the graph of f in more than one point. **b.** 2000 **c.** $f^{-1}(2000)$ represents the income, \$80 thousand, of a family that gives \$2000 to charity.

Cumulative Review Exercises (Chapters 1–8)

1. $\{-3\}$ **2.** $\left\{-4, \frac{3}{2}\right\}$ **3.** $\{(-3, 4)\}$ **4.** $\left\{\frac{9}{7}\right\}$ **5.** $\{-5\}$ **6.** $x^2 - 3x - 1; -1$ **7.** $-\dfrac{2}{x^4}$ **8.** 4 **9.** $\dfrac{3}{x}$ **10.** $-\dfrac{2+x}{3x+1}$ **11.** 5 **12.** $(x-7)(x-11)$ **13.** $x(x+5)(x-5)$ **14.** $6x^2 - 7x + 2$ **15.** $8x^3 - 27$ **16.** $\dfrac{1}{x-1}$ **17.** $\dfrac{5x-1}{4x-3}$ **18.** $\{(3, 2, 4)\}$

19.

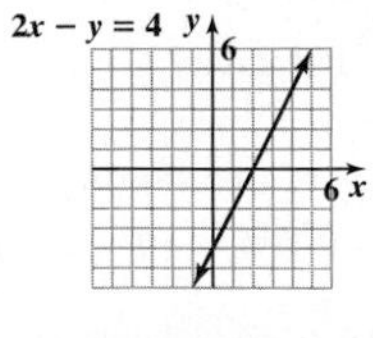

20.

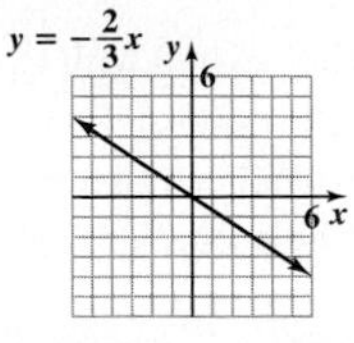

21. function; domain: $\{1, 2, 3, 4, 6\}$; range: $\{5\}$ **22.** $\left[\frac{8}{3}\right.$ **23.** $y + 3 = 5(x + 2)$; $y = 5x + 7$ **24.** 2.1×10^{-5} **25.** $\{x \mid x$ is a real number and $x \ne 15\}$

CHAPTER 9

Section 9.1

Check Point Exercises

1. a. 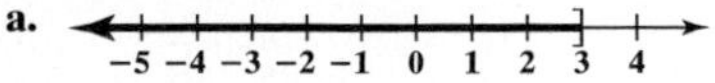

b. 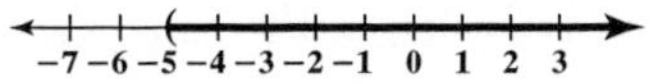

c. 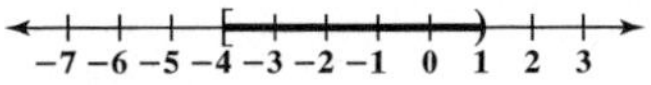

2. a. $\{x|-2 \le x < 5\}$ 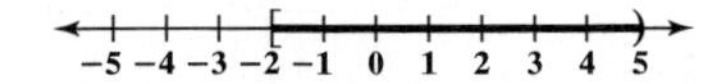

b. $\{x|1 \le x \le 3.5\}$ 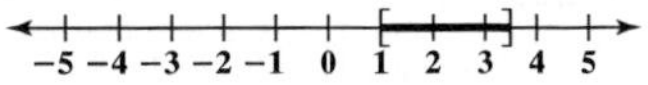

c. $\{x|x < -1\}$ 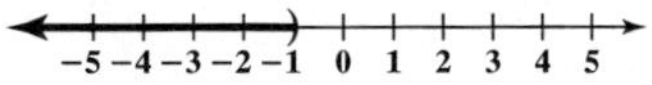

3. $\{x|x \ge -9\}$ or $[-9, \infty)$ 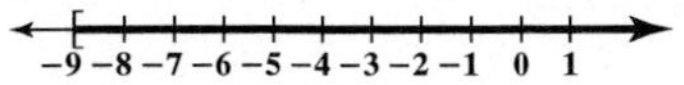

4. $\{x|x > 13\}$ or $(13, \infty)$ 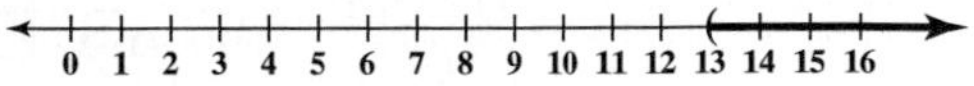

5. a. $P(x) = 125x - 160{,}000$ **b.** more than 1280 units

6. a. $C(x) = 300{,}000 + 30x$ **b.** $R(x) = 80x$ **c.** $P(x) = 50x - 300{,}000$ **d.** more than 6000 pairs

Exercise Set 9.1

1. $\{x|1 < x \le 6\}$ 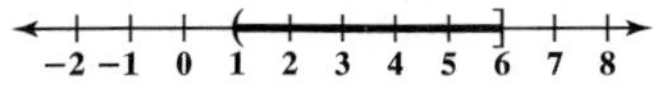

3. $\{x|-5 \le x < 2\}$ 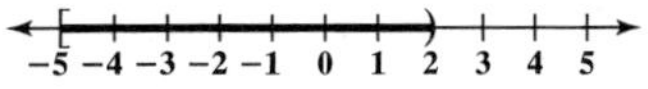

5. $\{x|-3 \le x \le 1\}$ 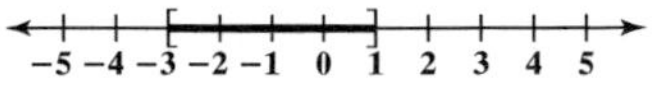

7. $\{x|x > 2\}$ 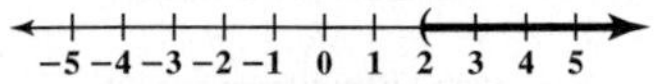

9. $\{x|x \ge -3\}$ 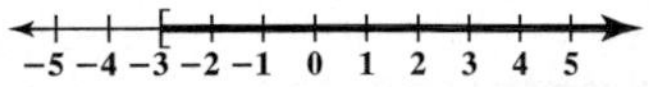

11. $\{x|x < 3\}$ 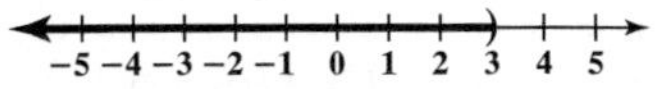

13. $\{x|x < 5.5\}$ 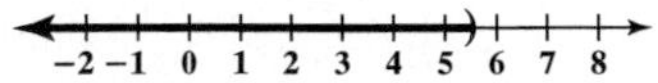

15. $\{x|x < 3\}$ or $(-\infty, 3)$ 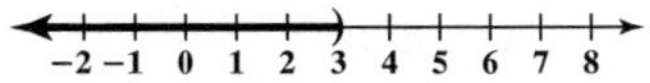

17. $\{x|x \ge 7\}$ or $[7, \infty)$ 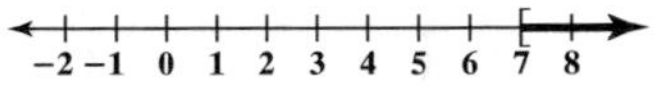

19. $\{x|x \le -4\}$ or $(-\infty, -4]$ 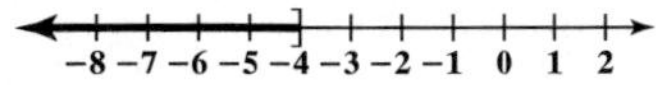

21. $\left\{x \middle| x \le -\frac{2}{5}\right\}$ or $\left(-\infty, -\frac{2}{5}\right]$ 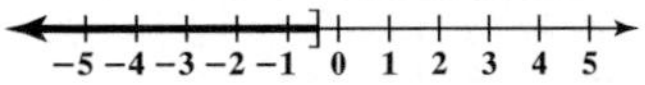

23. $\{x|x \ge 0\}$ or $[0, \infty)$ 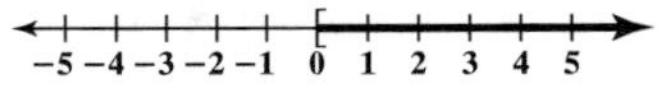

25. $\{x|x < 1\}$ or $(-\infty, 1)$ 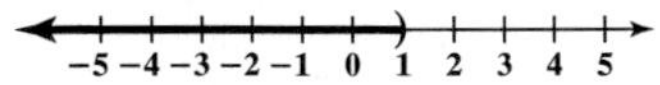

27. $\{x|x \ge 6\}$ or $[6, \infty)$ 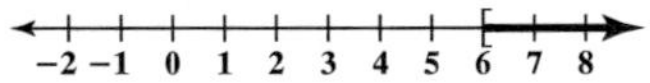

29. $\{x|x \ge -6\}$ or $[-6, \infty)$ 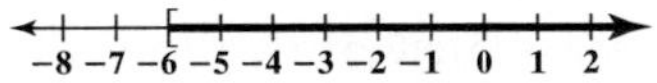

31. $\{x|x < -6\}$ or $(-\infty, -6)$ 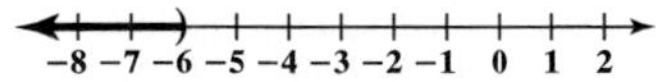

33. $\{x|x \ge 13\}$ or $[13, \infty)$ 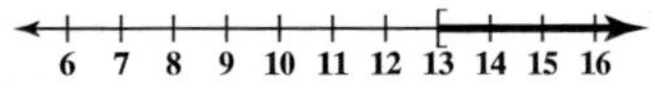

35. $\{y|y > -6\}$ or $(-6, \infty)$ 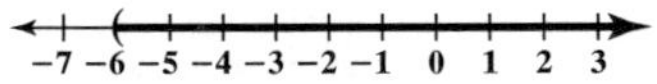

37. $\{x|x \ge -1\}$ or $[-1, \infty)$ 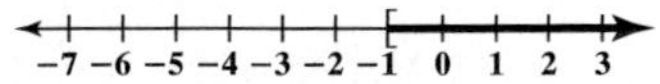

39. $\{y|y < -2\}$ or $(-\infty, -2)$ 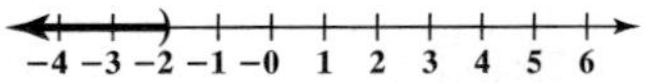

41. $\{x|x < 5\}$ or $(-\infty, 5)$
43. $\left\{x \middle| x \ge -\frac{4}{7}\right\}$ or $\left[-\frac{4}{7}, \infty\right)$

45. a. $P(x) = 17x - 25{,}500$ **b.** more than 1500 units **47. a.** $P(x) = 140x - 70{,}000$ **b.** more than 500 units

49. $\{x|x < 2\}$ or $(-\infty, 2)$ 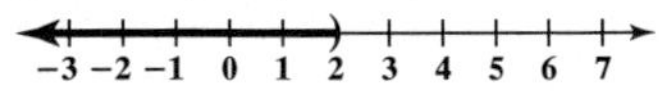

51. $x < \frac{c - b}{a}$ **53.** $\{x|x \le -3\}$ or $(-\infty, -3]$ **55.** $\{x|x > -1.4\}$ or $(-1.4, \infty)$ **57.** $(0, 4)$
59. intimacy ≥ passion or passion ≤ intimacy **61.** commitment > passion or passion < commitment
63. 9; after 3 years **65.** voting years after 2006

67. $\{t|t > 175\}$ or $(175, \infty)$; The women's speed skating times will be less than the men's after the year 2075.
69. a. $C(x) = 18{,}000 + 20x$ **b.** $R(x) = 80x$ **c.** $P(x) = 60x - 18{,}000$ **d.** more than 300 canoes
71. a. $C(x) = 30{,}000 + 2500x$ **b.** $R(x) = 3125x$ **c.** $P(x) = 625x - 30{,}000$ **d.** more than 48 sold-out performances
73. more than 6250 tapes **75.** more than 300 minutes **83.** 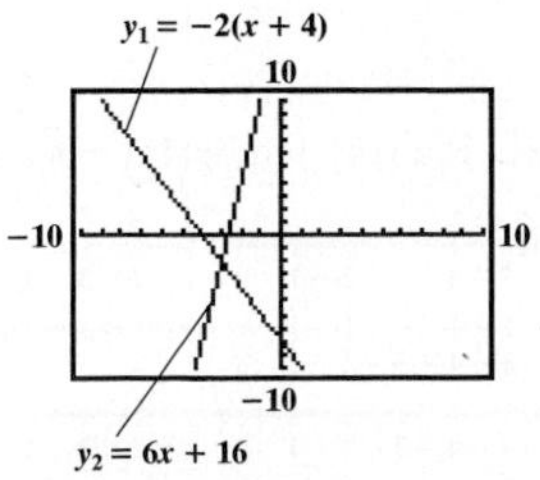

$\{x|x < -3\}$ or $(-\infty, -3)$

85. ∅ **87. a.** plan A: $4 + 0.10x$; plan B: $2 + 0.15x$ **b.**

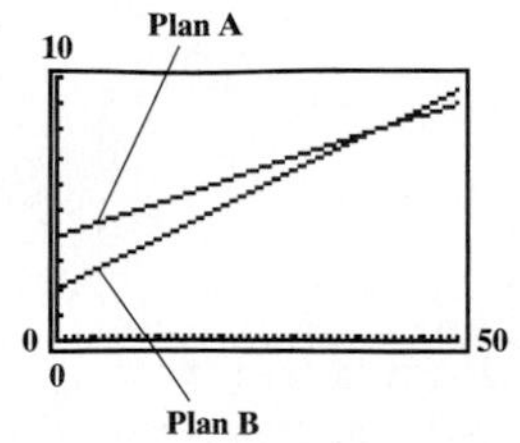

c–d. more than 40 checks per month
89. Since $x > y$, $y - x < 0$. Thus, when both sides were multiplied by $y - x$, the sense of the inequality should have been changed.
91. 29 **92.** $(-1, -1, 2)$ **93.** $(5x + 9)(5x - 9)$

Section 9.2

Check Point Exercises

1. $\{3, 7\}$ **2.** $\{x | x < 1\}$ or $(-\infty, 1)$ **3.** ∅ **4.** $\{x | -1 \le x < 4\}$ or $[-1, 4)$;

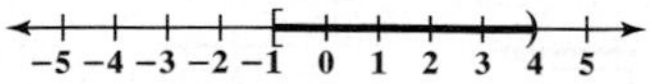

5. $\{3, 4, 5, 6, 7, 8, 9\}$ **6.** $\{x | x \le 1 \text{ or } x > 3\}$ or $(-\infty, 1] \cup (3, \infty)$ **7.** $\{x | x \text{ is a real number}\}$ or $\mathbb{R}$ or $(-\infty, \infty)$

Exercise Set 9.2

1. $\{2, 4\}$ **3.** ∅ **5.** ∅

7. $\{x | x > 6\}$; $(6, \infty)$

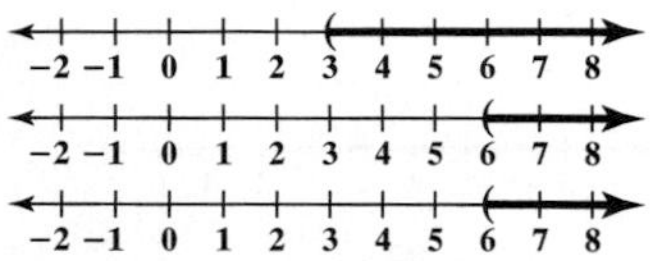

9. $\{x | x \le 1\}$; $(-\infty, 1]$

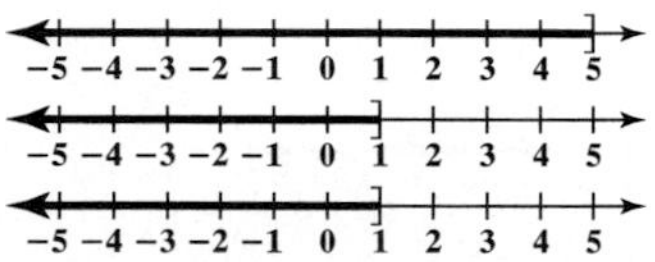

11. $\{x | -1 \le x < 2\}$; $[-1, 2)$

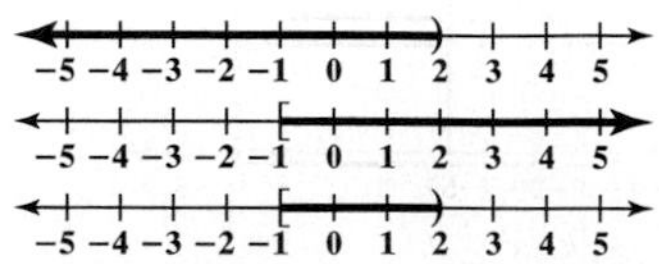

13. ∅

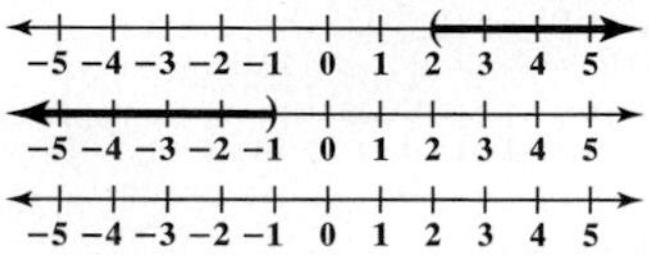

15. $\{x | -6 < x < -4\}$; $(-6, -4)$

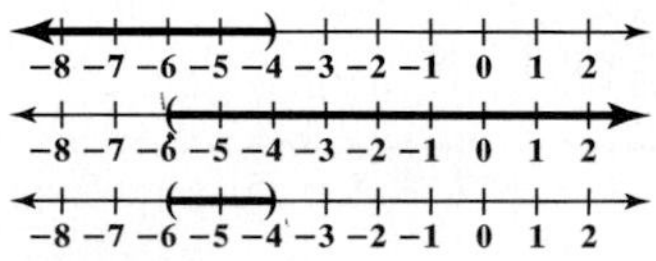

17. $\{x | -3 < x \le 6\}$; $(-3, 6]$

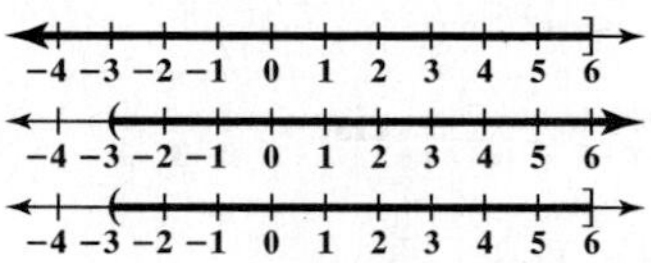

19. $\{x | 2 < x < 5\}$; $(2, 5)$

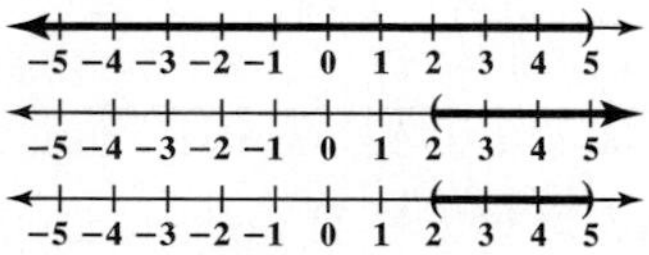

21. ∅

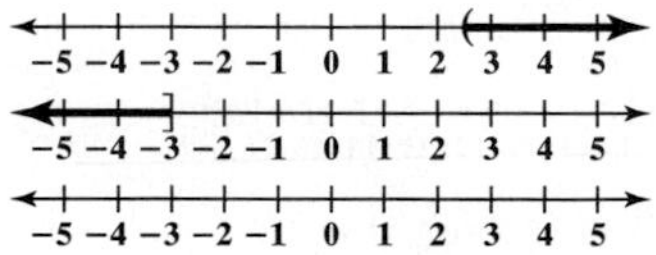

23. $\{x | 0 \le x < 2\}$; $[0, 2)$

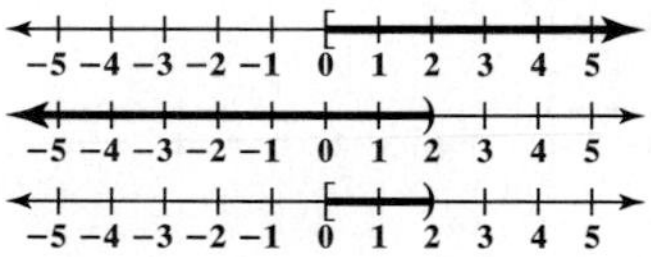

25. $\{x | 3 < x < 5\}$; $(3, 5)$

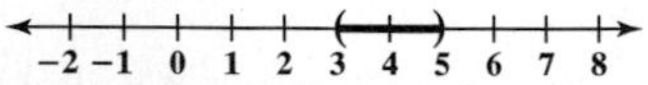

27. $\{x | -1 \le x < 3\}$; $[-1, 3)$

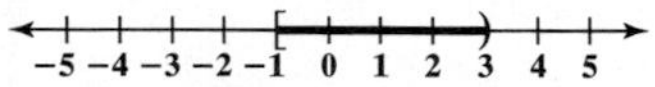

29. $\{x | -5 < x \le -2\}$; $(-5, -2]$

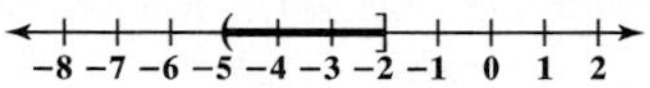

31. $\{x | 3 \le x < 6\}$; $[3, 6)$

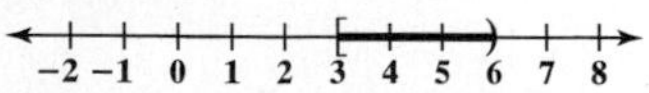

33. $\{1, 2, 3, 4, 5\}$

35. $\{1, 2, 3, 4, 5, 6, 7, 8, 10\}$

37. $\{a, e, i, o, u\}$

39. $\{x | x > 3\}$; $(3, \infty)$

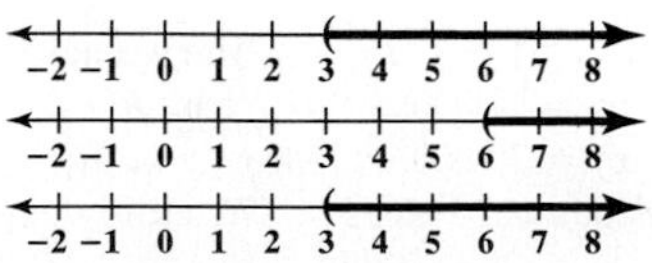

41. $\{x | x \le 5\}$; $(-\infty, 5]$

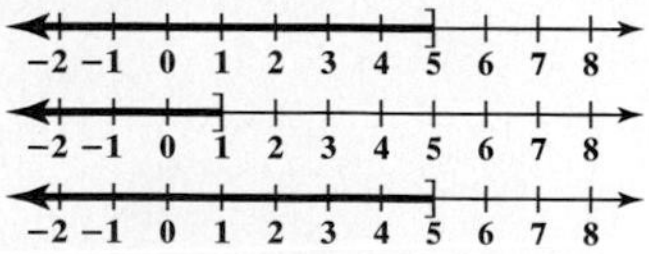

43. $\{x | x \text{ is a real number}\}$; $(-\infty, \infty)$

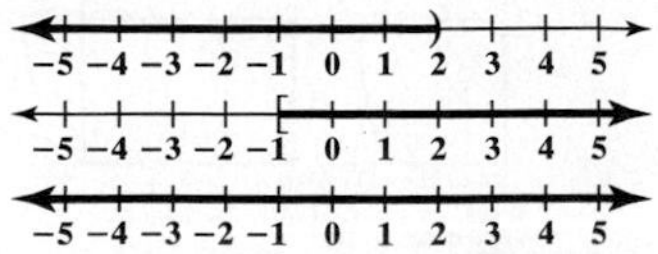

45. $\{x | x < -1 \text{ or } x \ge 2\}$; $(-\infty, -1) \cup [2, \infty)$

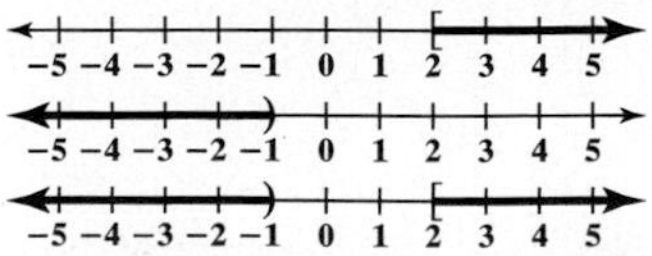

47. $\{x|x < -3 \text{ or } x > 4\}$; $(-\infty, -3) \cup (4, \infty)$

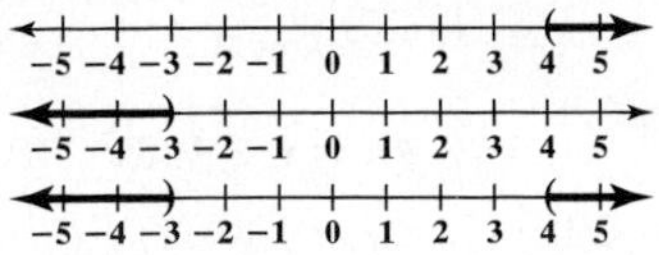

49. $\{x|x \le 1 \text{ or } x \ge 3\}$; $(-\infty, 1] \cup [3, \infty)$

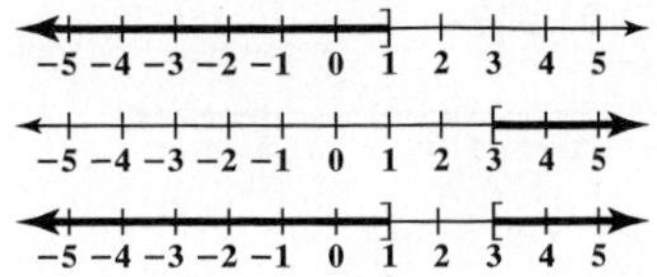

51. $\{x|x \text{ is a real number}\}$; $(-\infty, \infty)$

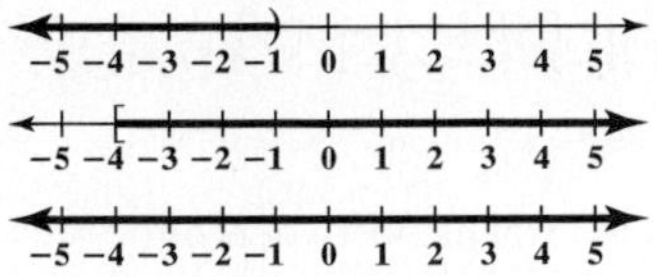

53. $\{x|x < 2\}$; $(-\infty, 2)$

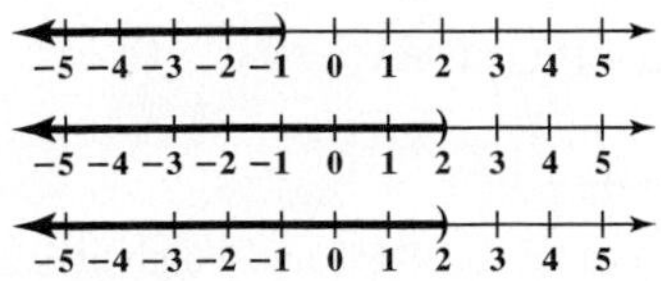

55. $\{x|x > 4\}$ or $(4, \infty)$ **57.** $\{x|x < 0 \text{ or } x > 6\}$ or $(-\infty, 0) \cup (6, \infty)$ **59.** $\frac{b-c}{a} < x < \frac{b+c}{a}$ **61.** $\{x|-1 \le x \le 3\}$ or $[-1, 3]$ **63.** $\{x|-1 < x < 3\}$ or $(-1, 3)$ **65.** $\{x|-1 \le x < 2\}$ or $[-1, 2)$ **67.** $\{-3, -2, -1\}$ **69.** spatial-temporal toys, sports equipment, and toy cars and trucks **71.** dollhouses, spatial-temporal toys, sports equipment, and toy cars and trucks **73.** none of the toys **75.** between 80 and 110 minutes, inclusive

77. $[76, 126)$; If the highest grade is 100, then $[76, 100]$. **79.** more than 3 and less than 15 crossings per 3-month period

89. $\{x|-6 < x < 2\}$ or $(-6, 2)$

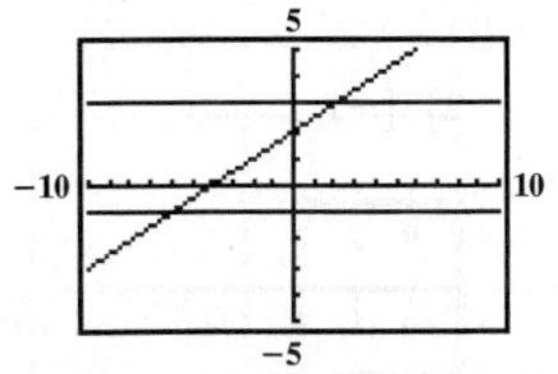

91. $\{x|-3 \le x \le 2\}$ or $[-3, 2]$

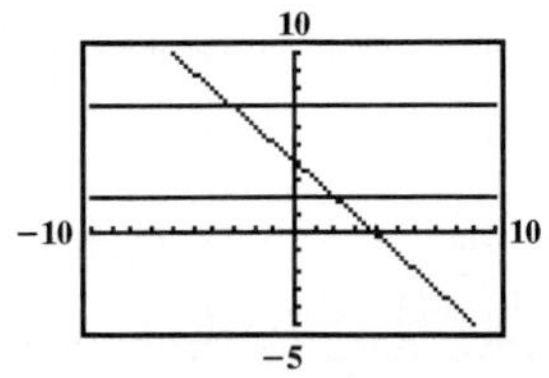

93. $(-1, 5]$ **95.** $[-1, \infty)$ **97.** $(-1, 4]$ **99.** $-x^2 + 5x - 9$; -15 **100.** $f(x) = -\frac{1}{2}x + 4$ **101.** $17 - 2x$

Section 9.3

Check Point Exercises

1. -2 and 3, or $\{-2, 3\}$ **2.** $\frac{4}{3}$ and 10, or $\left\{\frac{4}{3}, 10\right\}$

3. $\{x|-3 < x < 7\}$ or $(-3, 7)$

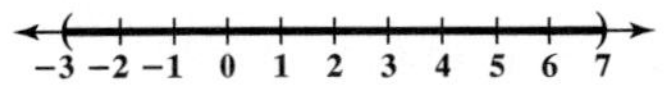

4. $\{x|x \le 1 \text{ or } x \ge 4\}$ or $(-\infty, 1] \cup [4, \infty)$

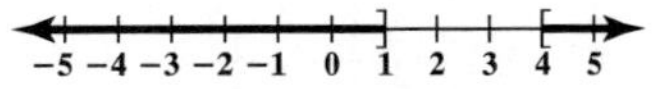

5. $\{x|8.1 \le x \le 13.9\}$ or $[8.1, 13.9]$; The percentage of children in the population who think that not being able to do everything they want is a bad thing is between a low of 8.1% and a high of 13.9%.

Exercise Set 9.3

1. $\{-8, 8\}$ **3.** $\{-5, 9\}$ **5.** $\{-3, 4\}$ **7.** $\{-1, 2\}$ **9.** $\varnothing$ **11.** $\{-3\}$ **13.** $\{-11, -1\}$ **15.** $\{-3, 4\}$ **17.** $\left\{-\frac{13}{3}, 5\right\}$ **19.** $\left\{-\frac{2}{5}, \frac{2}{5}\right\}$ **21.** $\varnothing$ **23.** $\varnothing$ **25.** $\left\{\frac{1}{2}\right\}$ **27.** $\left\{\frac{3}{4}, 5\right\}$ **29.** $\left\{\frac{5}{3}, 3\right\}$ **31.** $\{0\}$ **33.** $\{4\}$ **35.** $\{4\}$ **37.** $\{-1, 15\}$

39. $\{x|-3 < x < 3\}$ or $(-3, 3)$

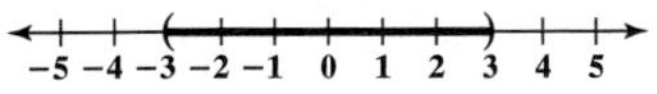

41. $\{x|1 < x < 3\}$ or $(1, 3)$

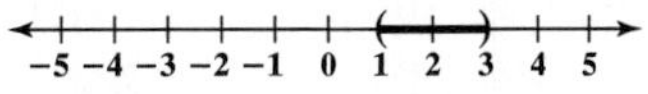

43. $\{x|-3 \le x \le -1\}$ or $[-3, -1]$

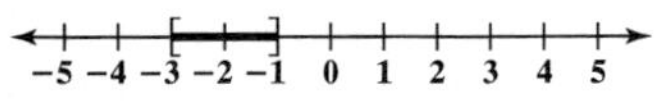

45. $\{x|-1 < x < 7\}$ or $(-1, 7)$

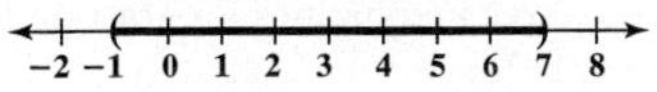

47. $\{x|x < -3 \text{ or } x > 3\}$ or $(-\infty, -3) \cup (3, \infty)$

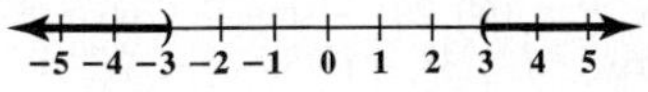

49. $\{x|x < -4 \text{ or } x > -2\}$ or $(-\infty, -4) \cup (-2, \infty)$

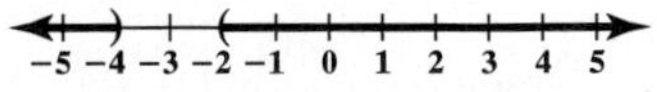

51. $\{x|x \le 2 \text{ or } x \ge 6\}$ or $(-\infty, 2] \cup [6, \infty)$

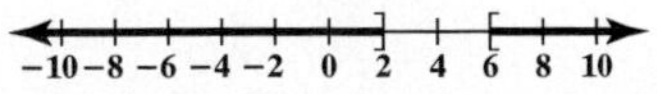

53. $\left\{x \middle| x < \frac{1}{3} \text{ or } x > 5\right\}$ or $\left(-\infty, \frac{1}{3}\right) \cup (5, \infty)$

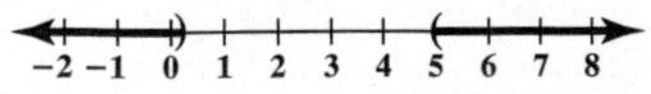

55. $\{x|-5 \le x \le 3\}$ or $[-5, 3]$

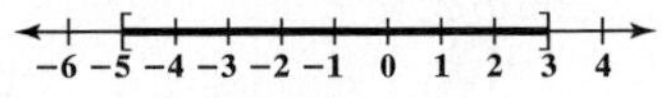

57. $\{x|-6 < x < 0\}$ or $(-6, 0)$

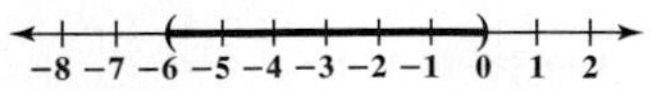

59. $\{x|x \le -5 \text{ or } x \ge 3\}$ or $(-\infty, -5] \cup [3, \infty)$

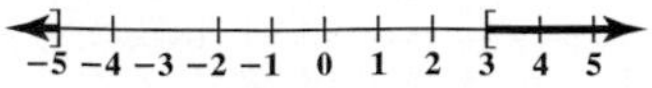

61. $\{x|x < -3 \text{ or } x > 12\}$ or $(-\infty, -3) \cup (12, \infty)$

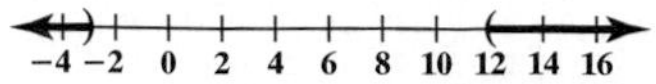

63. $\varnothing$

65. $\{x|x \text{ is a real number}\}$ or $\mathbb{R}$ or $(-\infty, \infty)$

67. $\{x|-9 \le x \le 5\}$ or $[-9, 5]$

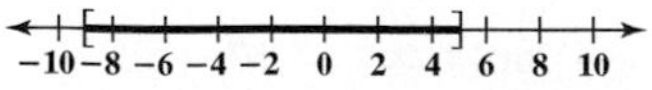

69. $\{x|x < 1 \text{ or } x > 2\}$ or $(-\infty, 1) \cup (2, \infty)$

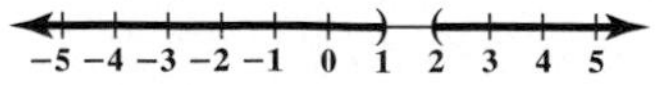

71. $-\frac{3}{2}$ and 4 **73.** −7 and −2 **75.** $\left\{x \middle| -\frac{7}{3} \le x \le 1\right\}$ or $\left[-\frac{7}{3}, 1\right]$ **77.** $\{x|x < -1 \text{ or } x > 4\}$ or $(-\infty, -1) \cup (4, \infty)$

79. $\left(-\infty, -\frac{1}{3}\right] \cup [3, \infty)$ **81.** $\left\{x \middle| \frac{-c-b}{a} < x < \frac{c-b}{a}\right\}$ **83.** 3 and 5 **85.** $\{x|-2 \le x \le 1\}$ or $[-2, 1]$

87. $\{x|58.6 \le x \le 61.8\}$ or $[58.6, 61.8]$; The percentage of the U.S. population that watched M*A*S*H is between a low of 58.6% and a high of 61.8%.; 1.6% **89.** $\{T|50 \le T \le 64\}$ or $[50, 64]$; The monthly average temperature for San Francisco, CA is between a low of 50°F and a high of 64°F. **91.** $\{x|8.59 \le x \le 8.61\}$ or $[8.59, 8.61]$; A machine part that is supposed to be 8.6 centimeters is acceptable between a low of 8.59 and a high of 8.61 centimeters. **93.** If the number of outcomes that result in heads is 41 or less or 59 or more, then the coin is unfair.

103. −8 and 0

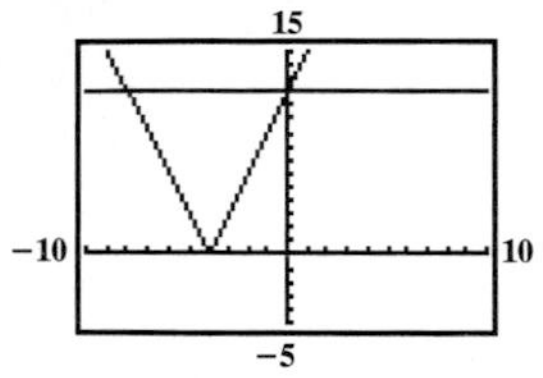

105. $\{x|-4 < x < 1\}$ or $(-4, 1)$

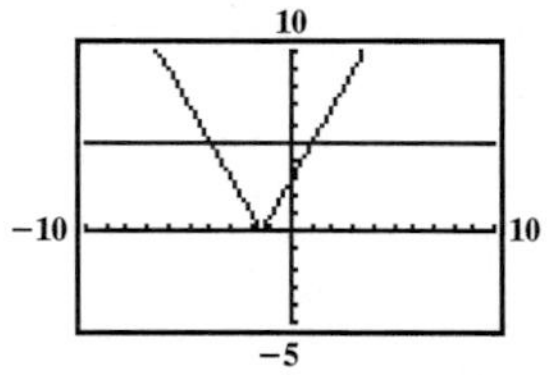

107. $\varnothing$

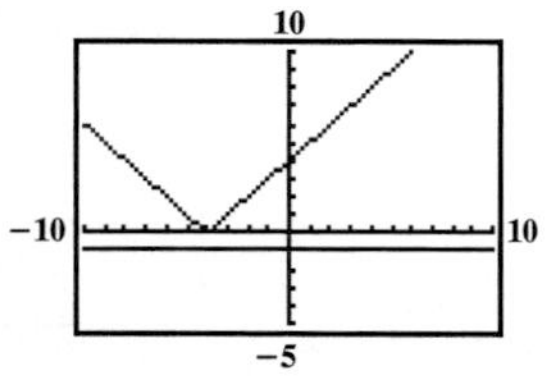

109. $\{x|x < 2 \text{ or } x > 6\}$ or $(-\infty, 2) \cup (6, \infty)$

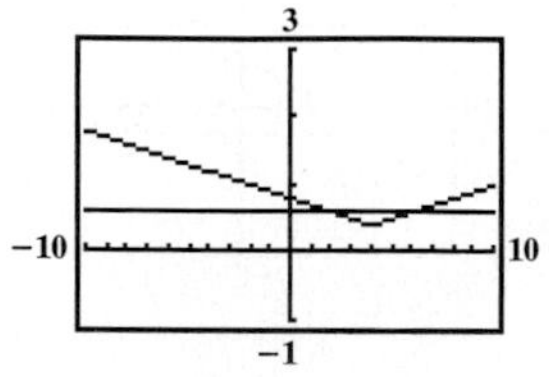

113. a. $|x - 4| < 3$ **b.** $|x - 4| \ge 3$ **115.** 1

116.

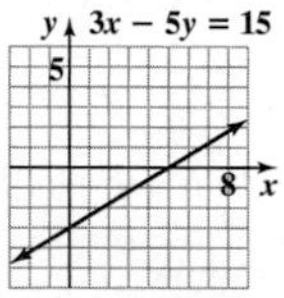

117.

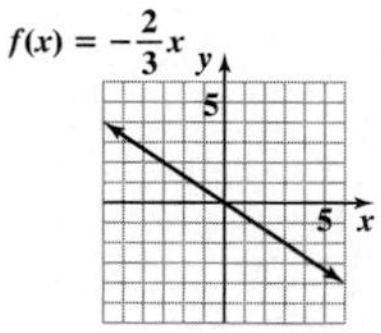

118.

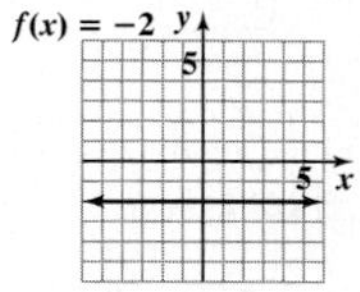

Mid-Chapter 9 Check Point

1. $\{x|x \le -4\}; (-\infty, -4]$ **2.** $\{x|3 \le x < 5\}; [3, 5)$ **3.** $\left\{\frac{1}{2}, 3\right\}$ **4.** $\{x|x < -1\}; (-\infty, -1)$

5. $\{x|x < -9 \text{ or } x > -5\}; (-\infty, -9) \cup (-5, \infty)$ **6.** $\left\{x \middle| -\frac{2}{3} \le x \le 2\right\}; \left[-\frac{2}{3}, 2\right]$ **7.** $\left\{\frac{1}{2}, \frac{13}{4}\right\}$ **8.** $\{x|-4 < x \le -2\}; (-4, -2]$

9. $\{x|x \text{ is real number}\}; (-\infty, \infty)$ **10.** $\{x|x \le -3\}; (-\infty, -3]$ **11.** $\{x|x < -3\}; (-\infty, -3)$

12. $\left\{x \middle| x < -1 \text{ or } x > -\frac{1}{5}\right\}; (-\infty, -1) \cup \left(-\frac{1}{5}, \infty\right)$ **13.** $\{x|x \le -10 \text{ or } x \ge 2\}; (-\infty, -10] \cup [2, \infty)$ **14.** $\left\{x \middle| x < -\frac{5}{3}\right\}; \left(-\infty, -\frac{5}{3}\right)$

15. $\{x|x > 4\}; (4, \infty)$ **16.** $\{-2, 7\}$ **17.** $\varnothing$ **18. a.** $C(x) = 60{,}000 + 0.18x$ **b.** $R(x) = 0.30x$ **c.** $P(x) = 0.12x - 60{,}000$ **d.** at least 750,000 discs each month **19.** no more than 80 miles per day **20.** [49%, 99%) **21.** at least $120,000

Section 9.4

Check Point Exercises

1.

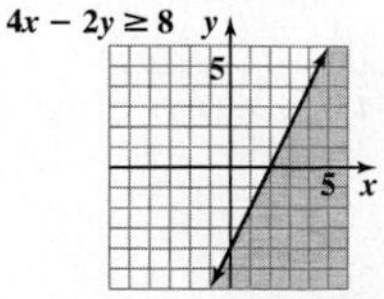

2.

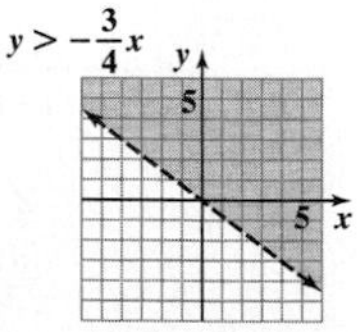

3. a.

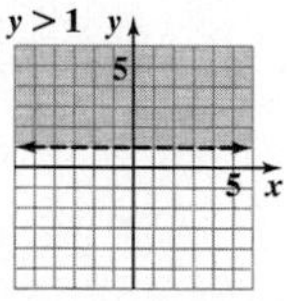

b.

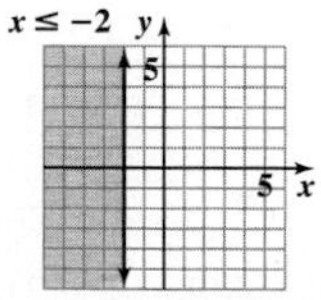

4.

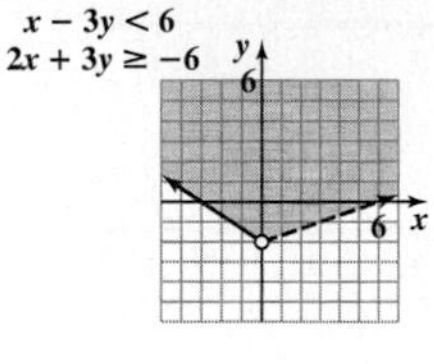

5.

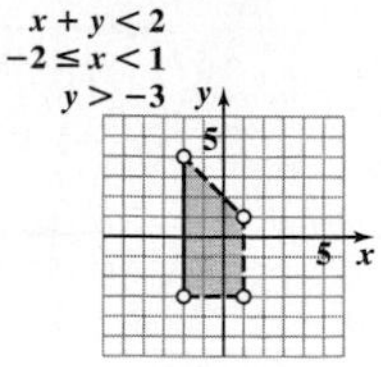

6. $B = (60, 20)$; Using $T = 60$ and $P = 20$, each of the three inequalities for grasslands is true: $60 \ge 35$, true; $5(60) - 7(20) \ge 70$, true; $3(60) - 35(20) \le -140$, true.

Exercise Set 9.4

1. $x + y \geq 3$
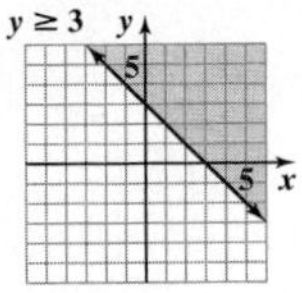

3. $x - y < 5$
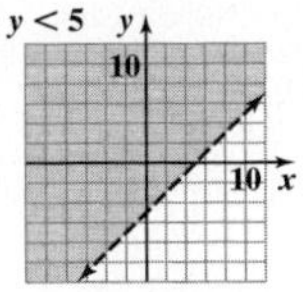

5. $x + 2y > 4$
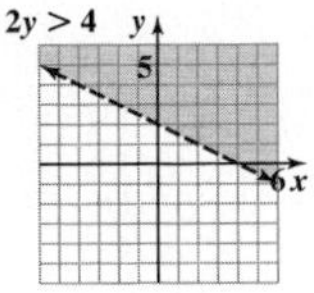

7. $3x - y \leq 6$
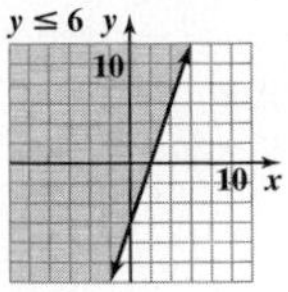

9. $\frac{x}{2} + \frac{y}{3} < 1$
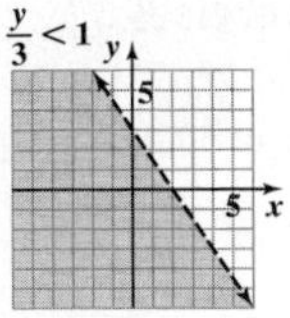

11. $y > \frac{1}{3}x$
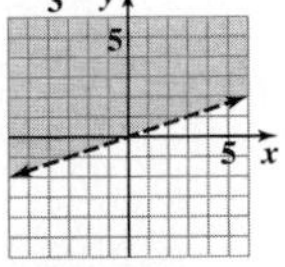

13. $y \leq 3x + 2$
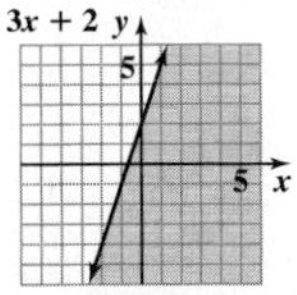

15. $y < -\frac{1}{4}x$
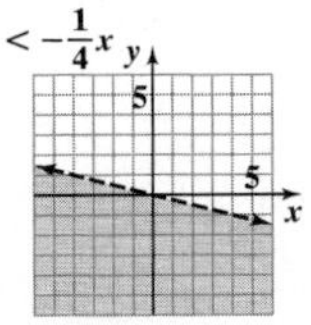

17. $x \leq 2$
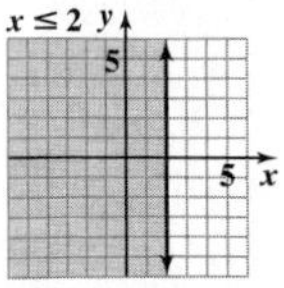

19. $y > -4$
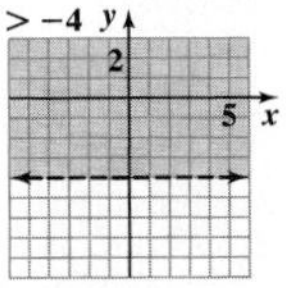

21. $y \geq 0$
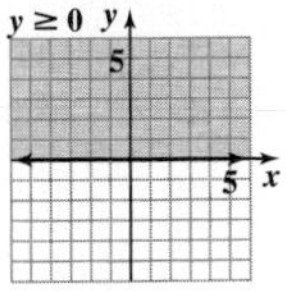

23. $3x + 6y \leq 6$
$2x + y \leq 8$
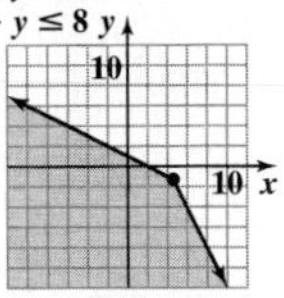

25. $2x - 5y \leq 10$
$3x - 2y > 6$
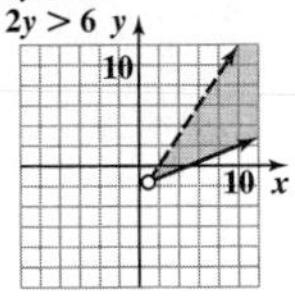

27. $y > 2x - 3$
$y < -x + 6$
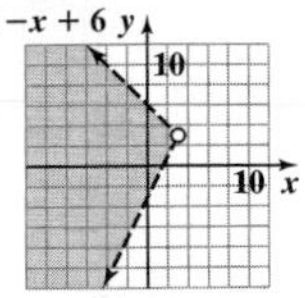

29. $x + 2y \leq 4$
$y \geq x - 3$
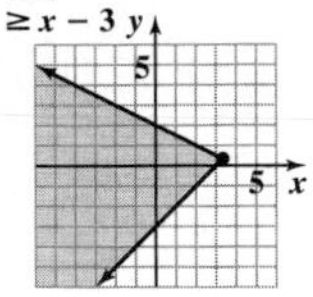

31. $x \leq 2$
$y \geq -1$
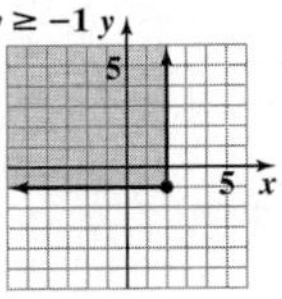

33. $-2 \leq x < 5$
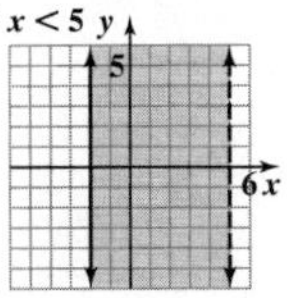

35. $x - y \leq 1$
$x \geq 2$
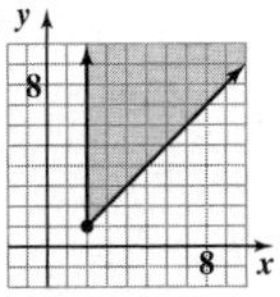

37. $\varnothing$

39. $x + y > 4$
$x + y > -1$
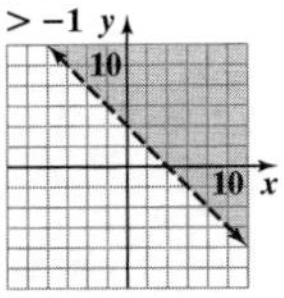

41. $x - y \leq 2$
$x \geq -2$
$y \leq 3$
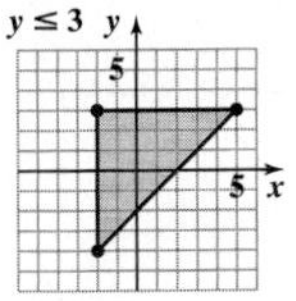

43. $x \geq 0$
$y \geq 0$
$2x + 5y \leq 10$
$3x + 4y \leq 12$
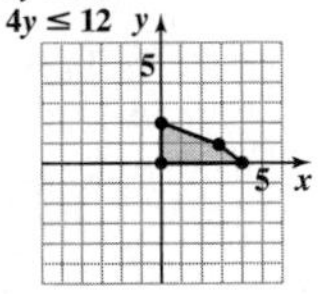

45. $3x + y \leq 6$
$2x - y \leq -1$
$x \geq -2$
$y \leq 4$
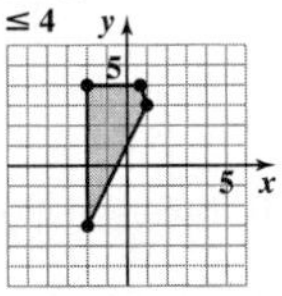

47. $y \geq -2x + 4$

$y \geq -2x + 4$
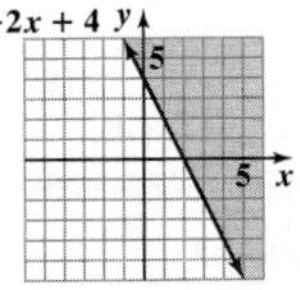

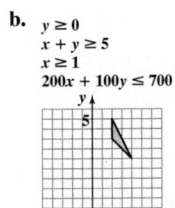

49. $x + y \leq 4$ and $3x + y \leq 6$

$x + y \leq 4$
$3x + y \leq 6$
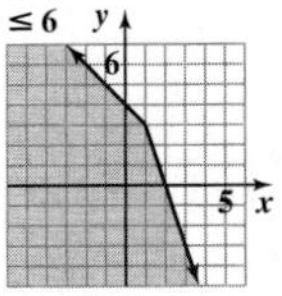

51. $-2 \leq x \leq 2; -3 \leq y \leq 3$

$-2 \leq x \leq 2$
$-3 \leq y \leq 3$
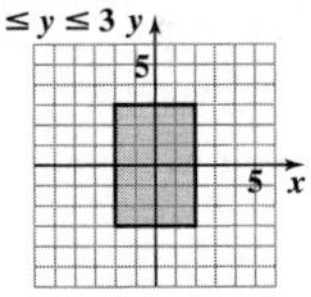

53. $y > \frac{3}{2}x - 2$ or $y < 4$
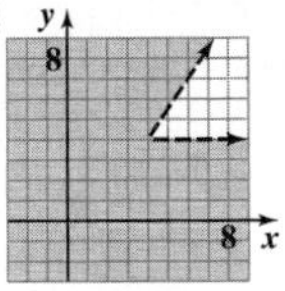

55. no solution **57.** infinitely many solutions

59. a. $A = (20, 150)$; A 20-year-old with a heart rate of 150 beats per minute is within the target range.
b. $10 \leq 20 \leq 70$, true; $150 \geq 0.7(220 - 20)$, true; $150 \leq 0.8(220 - 20)$, true

61. $10 \leq a \leq 70$; $H \geq 0.6(220 - a)$; $H \leq 0.7(220 - a)$

63. a. $y \geq 0$; $x + y \geq 5$; $x \geq 1$; $200x + 100y \leq 700$

b. $y \geq 0$
$x + y \geq 5$
$x \geq 1$
$200x + 100y \leq 700$
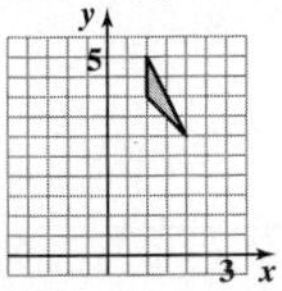

c. 2 nights

75.
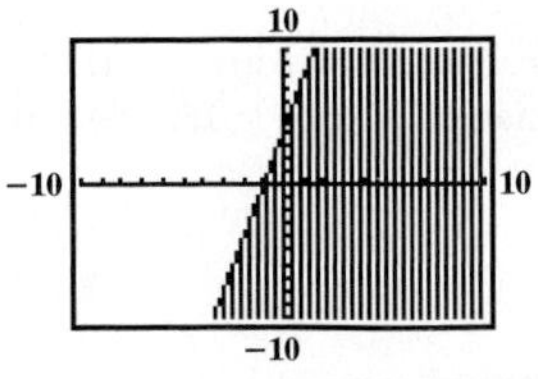

77.
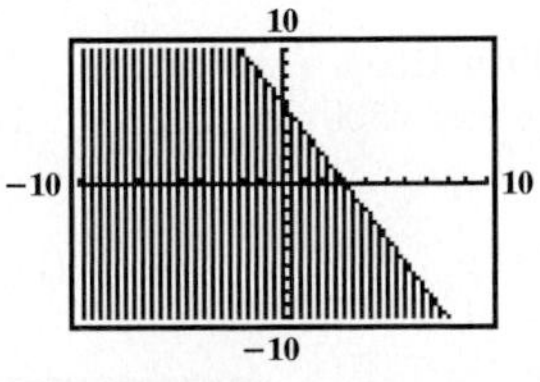

83. $x \geq -2$; $y > -1$ **85.** Answers will vary. **87.** $\{(3, 1)\}$ **88.** $\{(2, 4)\}$ **89.** $2x^4(x + 5y)^2$

Section 9.5

Check Point Exercises

1. $z = 25x + 55y$ **2.** $x + y \le 80$ **3.** $30 \le x \le 80, 10 \le y \le 30; z = 25x + 55y, x + y \le 80, 30 \le x \le 80, 10 \le y \le 30$
4. 50 bookshelves and 30 desks; \$2900 **5.** 30

Exercise Set 9.5

1. (1, 2): 17; (2, 10): 70; (7, 5): 65; (8, 3): 58; maximum: 70; minimum: 17 **3.** (0, 0): 0; (0, 8): 400; (4, 9): 610; (8, 0): 320; maximum: 610; minimum: 0

5. a.

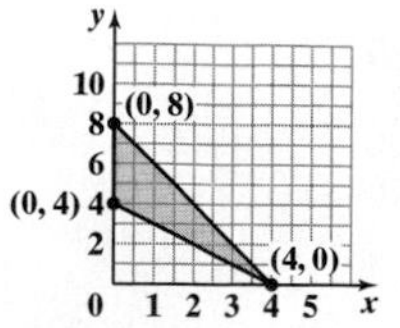

b. (0, 4): 8; (0, 8): 16; (4, 0): 12

c. maximum: 16; at (0, 8)

7. a.

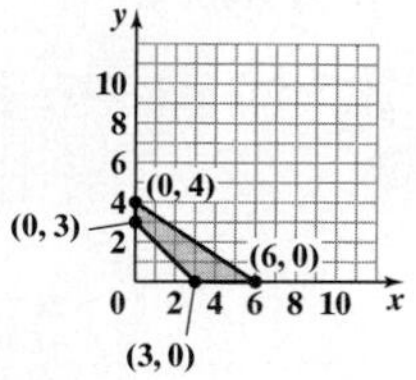

b. (0, 3): 3; (0, 4): 4; (6, 0): 24; (3, 0): 12

c. maximum: 24; at (6, 0)

9. a.

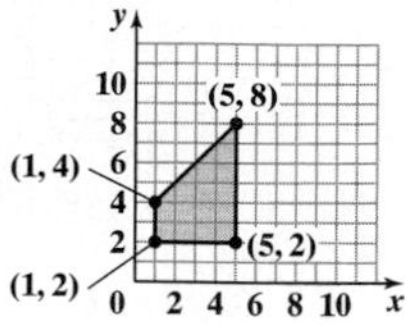

b. (1, 2): −1; (1, 4): −5; (5, 8): −1; (5, 2): 11

c. maximum: 11; at (5, 2)

11. a.

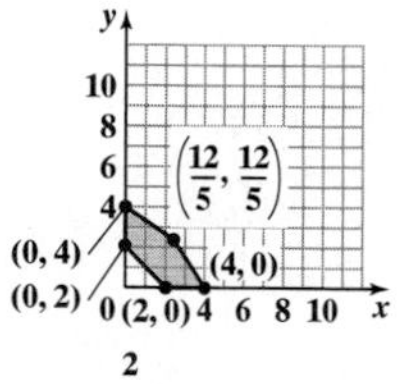

b. (0, 2): 4; (0, 4): 8; $\left(\frac{12}{5}, \frac{12}{5}\right): \frac{72}{5} = 14.4$; (4, 0): 16; (2, 0): 8

c. maximum: 16; at (4, 0)

13. a.

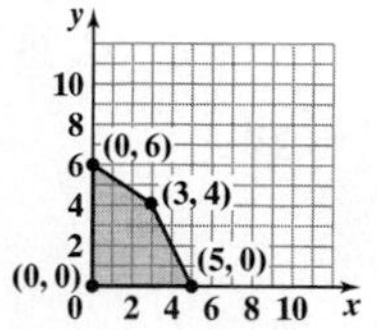

b. (0, 0): 0; (0, 6): 72; (3, 4): 78; (5, 0): 50

c. maximum: 78; at (3, 4)

15. a. $z = 125x + 200y$

b. $x \le 450$; $y \le 200$; $600x + 900y \le 360{,}000$

c.

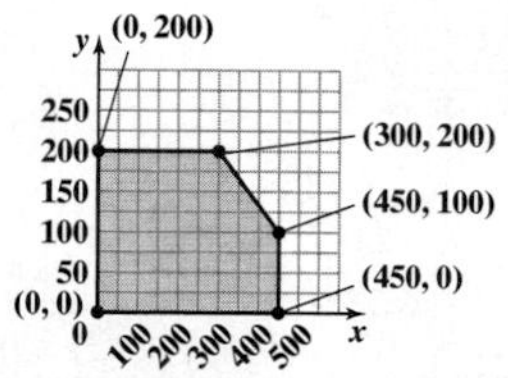

d. 0; 40,000; 77,500; 76,250; 56,250
e. 300; 200; 77,500

17. 40 of model A and 0 of model B **19.** 300 boxes of food and 200 boxes of clothing **21.** 100 parents and 50 students
23. 10 Boeing 727s and 42 Falcon 20s **29.** \$5000 in stocks and \$5000 in bonds **31.** $\{6\}$ **32.** $\frac{x + 9}{x - 15}$ **33.** 10

Review Exercises

1. $\{x|-2 < x \le 3\}$

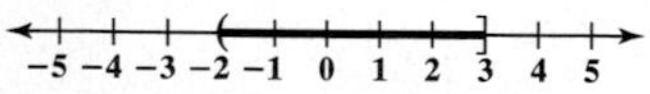

2. $\{x|-1.5 \le x \le 2\}$

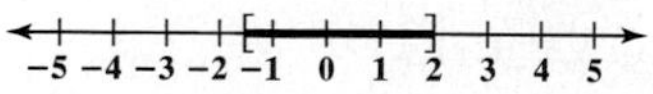

3. $\{x|x > -1\}$

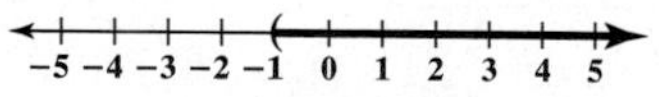

4. $\{x|x \ge -2\}; [-2, \infty)$

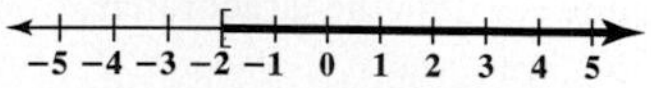

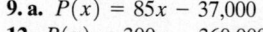

5. $\left\{x \middle| x \ge \frac{3}{5}\right\}; \left[\frac{3}{5}, \infty\right)$

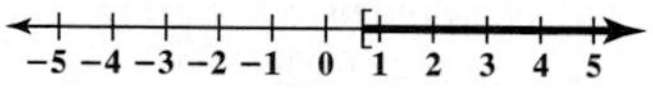

6. $\left\{x \middle| x < -\frac{21}{2}\right\}; \left(-\infty, -\frac{21}{2}\right)$

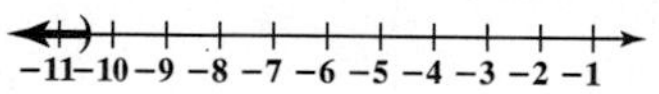

7. $\{x|x > -3\}; (-3, \infty)$

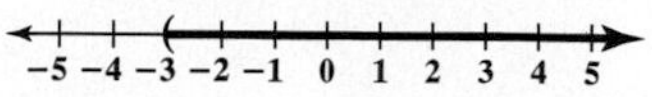

8. $\{x|x \le -2\}; (-\infty, -2]$

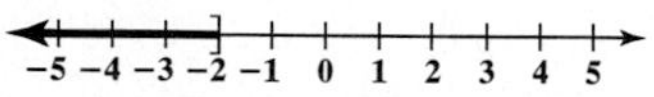

9. a. $P(x) = 85x - 37{,}000$ **b.** more than 4200 toaster ovens **10.** $C(x) = 360{,}000 + 850x$ **11.** $R(x) = 1150x$
12. $P(x) = 300x - 360{,}000$ **13.** more than 4200 computers **14.** more than 50 checks **15.** more than \$13,500 in sales **16.** $\{a, c\}$
17. $\{a\}$ **18.** $\{a, b, c, d, e\}$ **19.** $\{a, b, c, d, f, g\}$

20. $\{x|x \le 3\}; (-\infty, 3]$

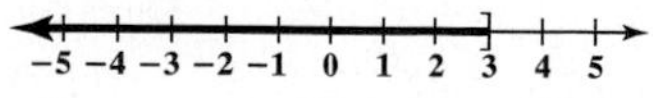

21. $\{x|x < 6\}; (-\infty, 6)$

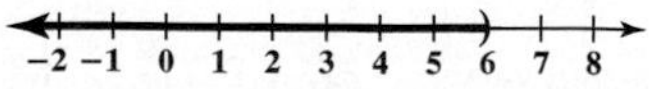

22. $\{x|6 < x < 8\}; (6, 8)$

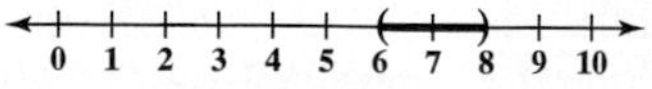

23. $\{x \mid x \leq 1\}$; $(-\infty, 1]$

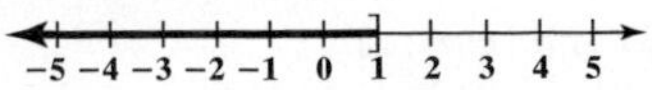

24. $\varnothing$ **25.** $\{x \mid x < 1 \text{ or } x > 2\}$; $(-\infty, 1) \cup (2, \infty)$

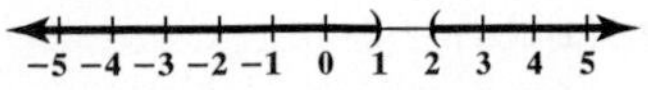

26. $\{x \mid x \leq -4 \text{ or } x > 2\}$; $(-\infty, -4] \cup (2, \infty)$

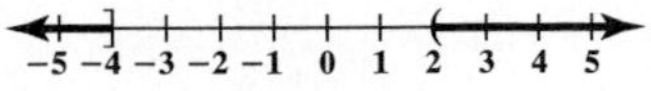

27. $\{x \mid x < -2\}$; $(-\infty, -2)$

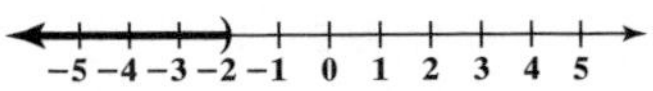

28. $\{x \mid x \text{ is a real number}\}$; $(-\infty, \infty)$

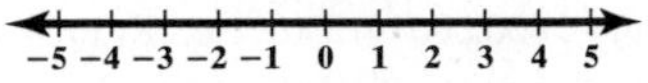

29. $\{x \mid -5 < x \leq 2\}$; $(-5, 2]$

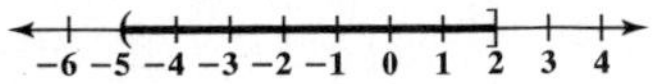

30. $\left\{x \,\middle|\, -\frac{3}{4} \leq x \leq 1\right\}$; $\left[-\frac{3}{4}, 1\right]$

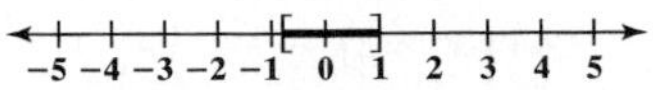

31. $[81, 131)$; If the highest grade is 100, then $[81, 100]$.

32. $\{-4, 3\}$ **33.** $\varnothing$ **34.** $\left\{-\frac{11}{2}, \frac{23}{2}\right\}$ **35.** $\left\{-4, -\frac{6}{11}\right\}$

36. $\{x \mid -9 \leq x \leq 6\}$; $[-9, 6]$

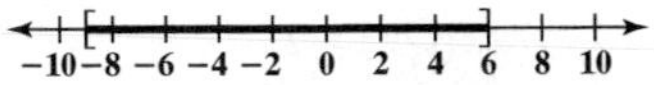

37. $\{x \mid x < -6 \text{ or } x > 0\}$; $(-\infty, -6) \cup (0, \infty)$

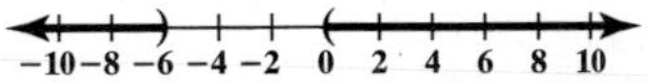

38. $\{x \mid -3 < x < -2\}$; $(-3, -2)$

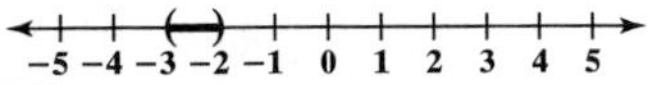

39. $\varnothing$ **40.** Approximately 90% of the population sleeps between 5.5 hours and 7.5 hours daily, inclusive.

41. $3x - 4y > 12$

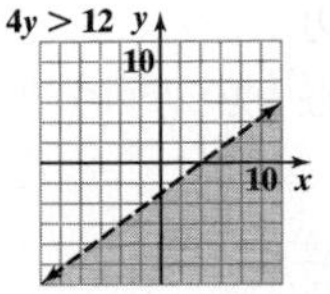

42. $x - 3y \leq 6$

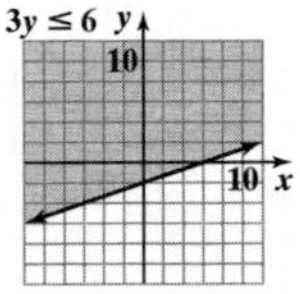

43. $y \leq -\frac{1}{2}x + 2$

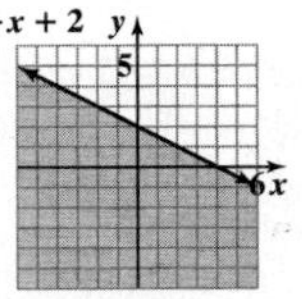

44. $y > \frac{3}{5}x$

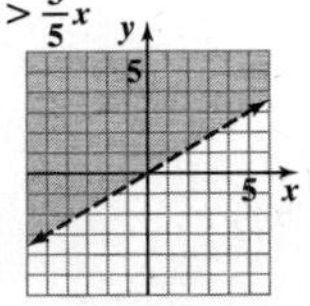

45. $x \leq 2$

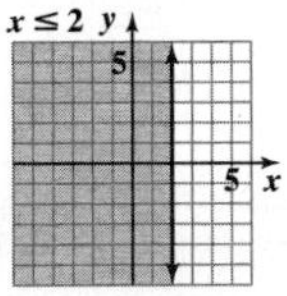

46. $y > -3$

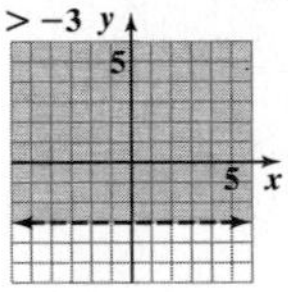

47. $2x - y \leq 4$, $x + y \geq 5$

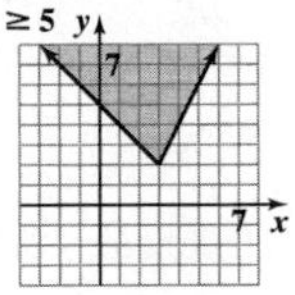

48. $y < -x + 4$, $y > x - 4$

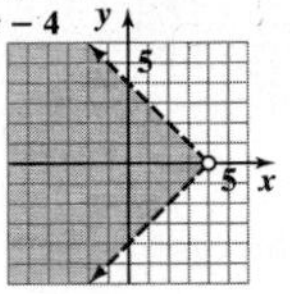

49. $-3 \leq x < 5$

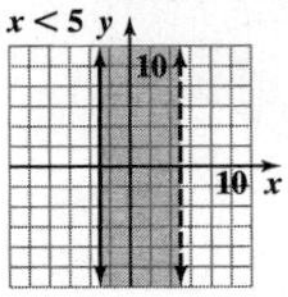

50. $-2 < y \leq 6$

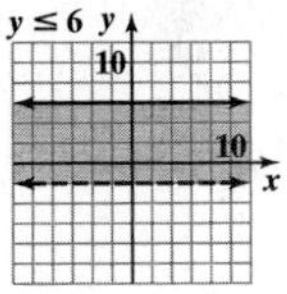

51. $x \geq 3$, $y \leq 0$

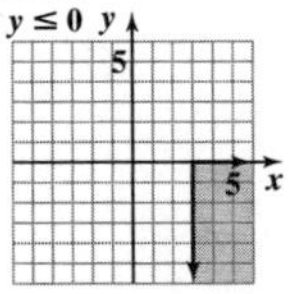

52. $2x - y > -4$, $x \geq 0$

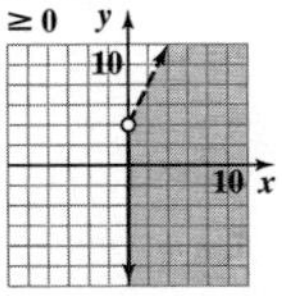

53. $x + y \leq 6$, $y \geq 2x - 3$

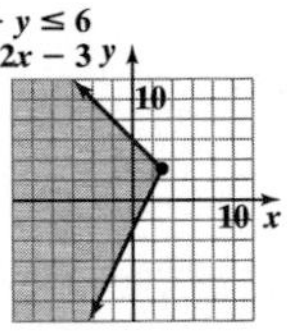

54. $3x + 2y \geq 4$, $x - y \leq 3$, $x \geq 0, y \geq 0$

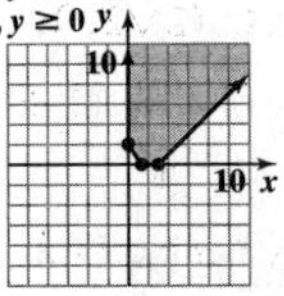

55. $\varnothing$

56. $\left(\frac{1}{2}, \frac{1}{2}\right)$: $\frac{5}{2}$; $(2, 2)$: 10; $(4, 0)$: 8; $(1, 0)$: 2; maximum: 10; minimum: 2

57.

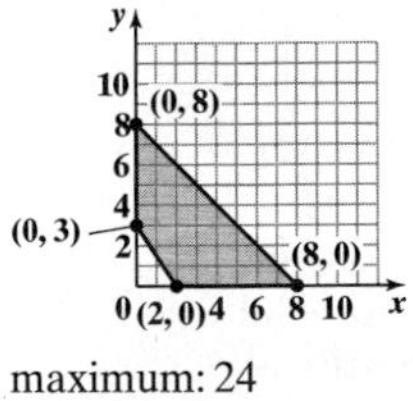

maximum: 24

58.

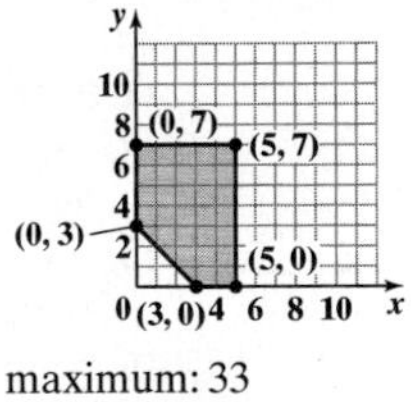

maximum: 33

59.

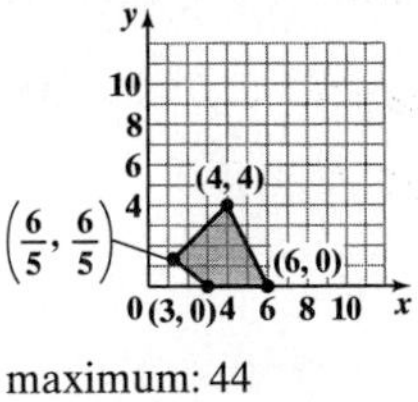

maximum: 44

60. a. $z = 500x + 350y$ **b.** $x + y \leq 200$; $x \geq 10$; $y \geq 80$

c.

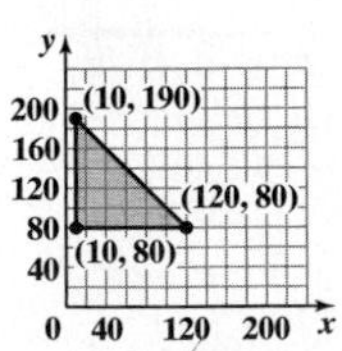

d. $(10, 80)$: 33,000; $(10, 190)$: 71,500; $(120, 80)$: 88,000

e. 120; 80; 88,000 **61.** 480 of model A and 240 of model B

Chapter 9 Test

1. $\{x \mid -3 \le x < 2\}$

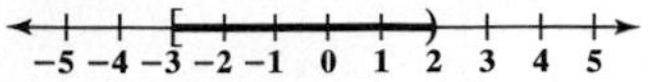

2. $\{x \mid x \le -1\}$

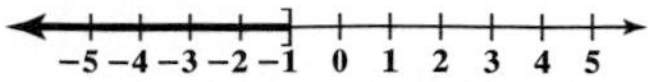

3. $\{x \mid x \le 12\}$; $(-\infty, 12]$

4. $\left\{x \middle| x \ge \frac{21}{8}\right\}$; $\left[\frac{21}{8}, \infty\right)$

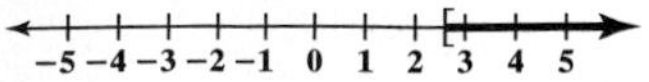

5. a. $C(x) = 600{,}000 + 200x$ **b.** $R(x) = 450x$
c. $P(x) = 250x - 60{,}000$
d. more than 240 desks

6. $\{4, 6\}$ **7.** $\{2, 4, 6, 8, 10, 12, 14\}$

8. $\{x \mid -2 < x < -1\}$; $(-2, -1)$

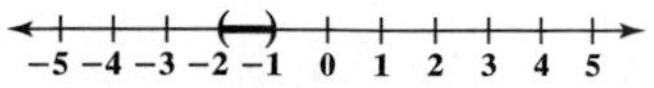

9. $\{x \mid x \ge -2\}$; $[-2, \infty)$

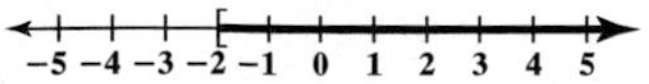

10. $\{x \mid x < 4\}$; $(-\infty, 4)$

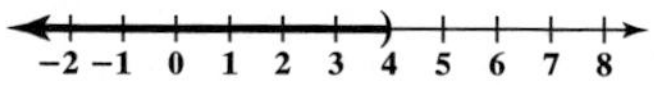

11. $\{x \mid x \le -4 \text{ or } x > 2\}$; $(-\infty, -4] \cup (2, \infty)$

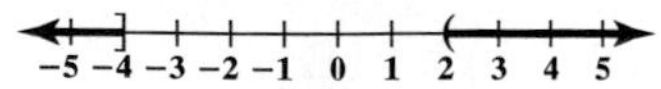

12. $\left\{x \middle| -7 \le x < \frac{13}{2}\right\}$; $\left[-7, \frac{13}{2}\right)$

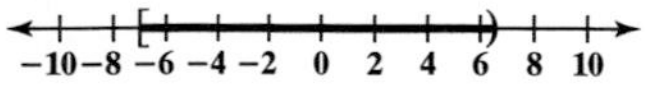

13. $\left\{-2, \frac{4}{5}\right\}$ **14.** $\left\{-\frac{8}{5}, 7\right\}$

15. $\{x \mid -3 < x < 4\}$; $(-3, 4)$

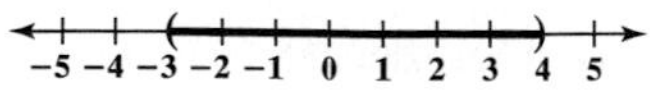

16. $\{x \mid x \le -1 \text{ or } x \ge 4\}$; $(-\infty, -1] \cup [4, \infty)$

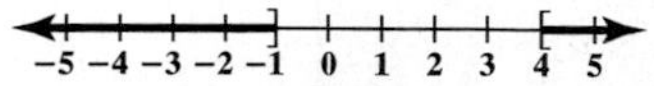

17. $\{b \mid b < 90.6 \text{ or } b > 106.6\}$ or $(-\infty, 90.6) \cup (106.6, \infty)$; Hypothermia: Body temperature below 90.6°F; Hyperthermia: Body temperature above 106.6°F

18.

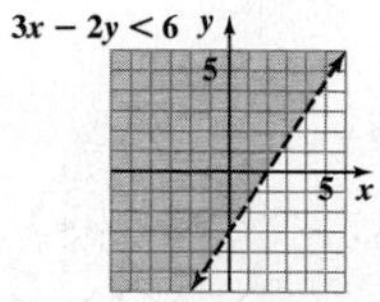

19.

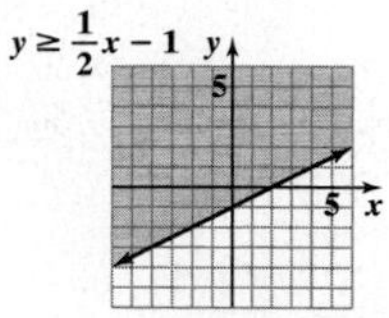

20.

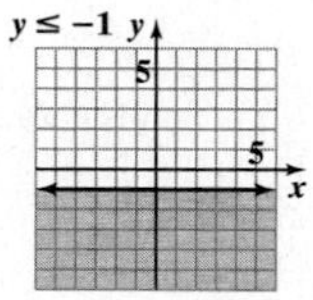

21.

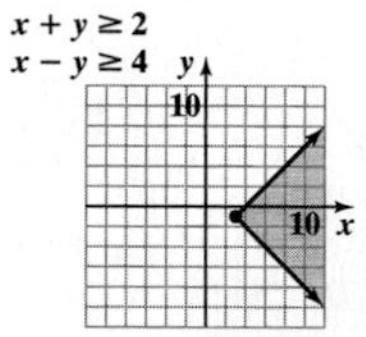

22.

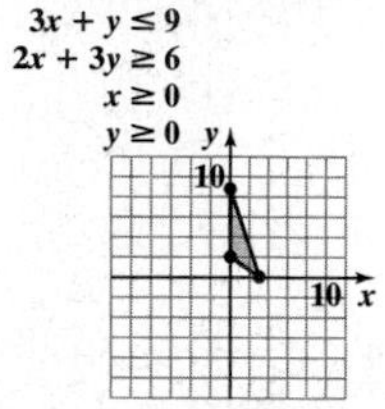

23.

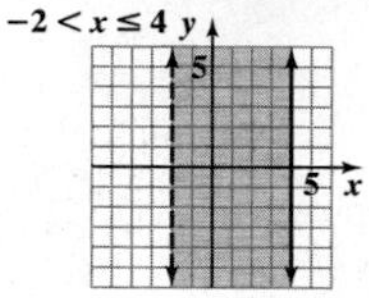

24. maximum: 26 **25.** 50 regular and 100 deluxe; \$35,000

Cumulative Review Exercises (Chapters 1–9)

1. $\{-1\}$ **2.** $\{8\}$ **3.** $-\frac{2y^7}{3x^5}$ **4.** $22; 4a^2 - 6a + 4$ **5.** $2x^2 + x + 2; 12$ **6.** $f(x) = -\frac{1}{2}x + 4$

7.

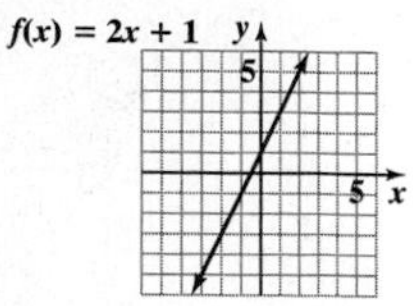

8.

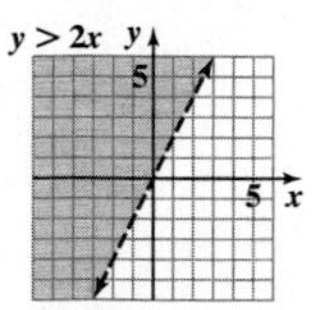

9.

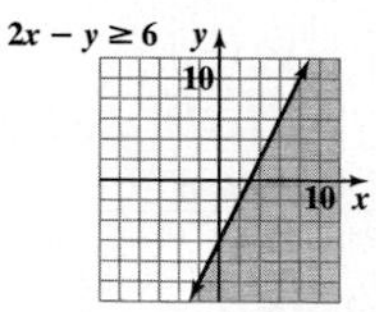

10.

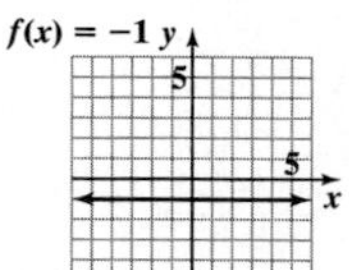

11. $(-4, 2, -1)$ **12.** $f^{-1}(x) = 3x + 12$
13. $f(g(x)) = 3x^2 + 12x + 11$; $g(f(x)) = 3x^2 + 1$
14. 46 rooms with kitchen facilities and 14 without kitchen facilities
15. a. and b. are functions.

16. $\{x \mid x \ge -7\}$; $[-7, \infty)$

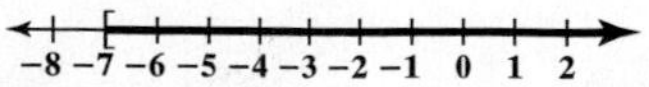

17. $\{x \mid x < -6\}$; $(-\infty, -6)$

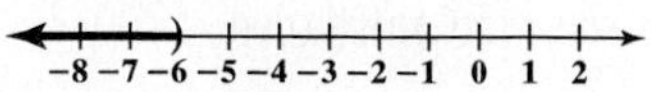

18. $\{x \mid x \le 3 \text{ or } x \ge 5\}$; $(-\infty, 3] \cup [5, \infty)$

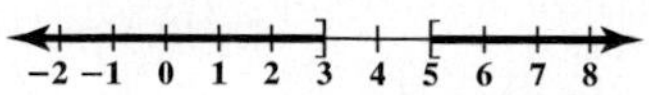

19. $\{x \mid -10 \le x \le 7\}$; $[-10, 7]$

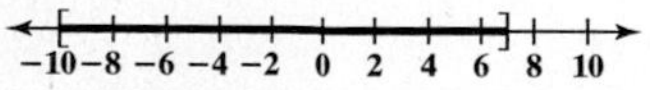

20. $\left\{x \middle| x < \frac{1}{3} \text{ or } x > 5\right\}$; $\left(-\infty, \frac{1}{3}\right) \cup (5, \infty)$

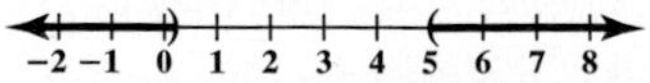

CHAPTER 10

Section 10.1

Check Point Exercises

1. a. 8 **b.** -7 **c.** $\frac{4}{5}$ **d.** 0.09 **e.** 5 **f.** 7 **2. a.** 4 **b.** $-\sqrt{24} \approx -4.90$ **3.** $\{x|x \geq 3\}$ or $[3, \infty)$
4. 53.9%; This is the percent given for 2001, not 2002, so the model does not provide a totally accurate description of the data.
5. a. 7 **b.** $|x+8|$ **c.** $|7x^5|$ or $7|x^5|$ **d.** $|x-3|$ **6. a.** 3 **b.** -2 **7.** $-3x$ **8. a.** 2 **b.** -2 **c.** not a real number **d.** -1
9. a. $|x+6|$ **b.** $3x-2$ **c.** 8

Exercise Set 10.1

1. 6 **3.** -6 **5.** not a real number **7.** $\frac{1}{5}$ **9.** $-\frac{3}{4}$ **11.** 0.9 **13.** -0.2 **15.** 3 **17.** 1 **19.** not a real number
21. 4; 1; 0; not a real number **23.** -5; $-\sqrt{5} \approx -2.24$; -1; not a real number **25.** 4; 2; 1; 6 **27.** $\{x|x \geq 3\}$ or $[3, \infty)$; c
29. $\{x|x \geq -5\}$ or $[-5, \infty)$; d **31.** $\{x|x \leq 3\}$ or $(-\infty, 3]$; e **33.** 5 **35.** 4 **37.** $|x-1|$ **39.** $|6x^2|$ or $6x^2$
41. $-|10x^3|$ or $-10|x^3|$ **43.** $|x+6|$ **45.** $-|x-4|$ **47.** 3 **49.** -3 **51.** $\frac{1}{5}$ **53.** 3; 2; -1; -4 **55.** -2; 0; 2
57. 1 **59.** 2 **61.** -2 **63.** not a real number **65.** -1 **67.** not a real number **69.** -4 **71.** 2 **73.** -2
75. x **77.** $|y|$ **79.** $-2x$ **81.** -5 **83.** 5 **85.** $|x+3|$ **87.** $-2(x-1)$

89.

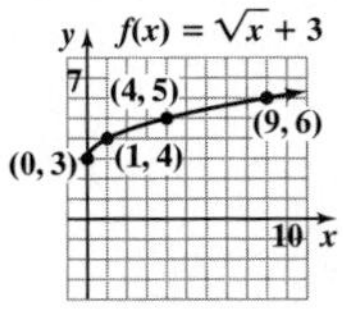

Domain: $\{x|x \geq 0\} = [0, \infty)$;
Range: $\{y|y \geq 3\} = [3, \infty)$

91.

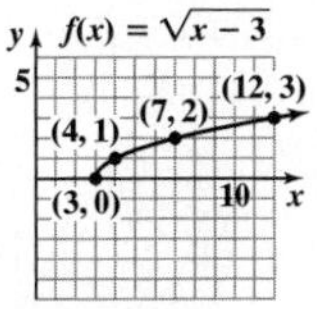

Domain: $\{x|x \geq 3\} = [3, \infty)$;
Range: $\{y|y \geq 0\} = [0, \infty)$

93. $\{x|x < 15\} = (-\infty, 15)$
95. $\{x|1 \leq x < 3\} = [1, 3)$
97. 3
99. 40.2 in.; The model describes actual data well.
101. 70 mph; The officer should not believe the motorist.; Answers will vary.

115.

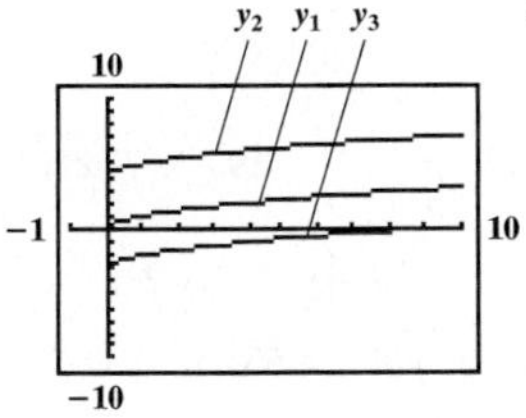

; Answers will

117.

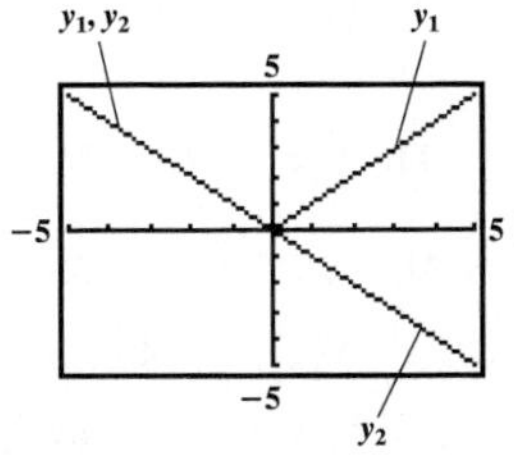

a. $\{x|x \leq 0\}$ **b.** $\{x|x > 0\}$
119. Answers will vary; an example is $f(x) = \sqrt{15 - 3x}$.
121. $|(2x+3)^5|$
123. The graph of h is the graph of f shifted left by 3 units.

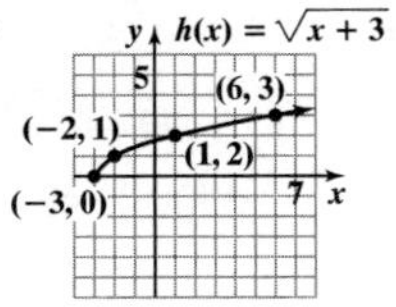

124. $7x + 30$ **125.** $\frac{x^8}{9y^6}$ **126.** $\left\{x \middle| x < -\frac{7}{3} \text{ or } x > 5\right\}$ or $\left(-\infty, -\frac{7}{3}\right) \cup (5, \infty)$

Section 10.2

Check Point Exercises

1. a. 5 **b.** -2 **c.** $\sqrt[4]{5xy^2}$ **2. a.** $(5xy)^{1/4}$ **b.** $\left(\frac{a^3b}{2}\right)^{1/5}$ **3. a.** $(\sqrt[3]{8})^4 = 16$ **b.** $(\sqrt{25})^3 = 125$ **c.** $-(\sqrt[4]{81})^3 = -27$ **4. a.** $6^{4/3}$ **b.** $(2xy)^{7/5}$
5. a. $\frac{1}{100^{1/2}} = \frac{1}{10}$ **b.** $\frac{1}{8^{1/3}} = \frac{1}{2}$ **c.** $\frac{1}{32^{3/5}} = \frac{1}{8}$ **d.** $\frac{1}{(3xy)^{5/9}}$ **6.** 25° **7. a.** $7^{5/6}$ **b.** $\frac{5}{x}$ **c.** $9.1^{3/10}$ **d.** $\frac{y^{1/12}}{x^{1/5}}$
8. a. $\sqrt{x}$ **b.** $2a^4$ **c.** $\sqrt[4]{x^2y}$ **d.** $\sqrt[6]{x}$ **e.** $\sqrt[6]{x}$

Exercise Set 10.2

1. $\sqrt{49} = 7$ **3.** $\sqrt[3]{-27} = -3$ **5.** $-\sqrt[4]{16} = -2$ **7.** $\sqrt[3]{xy}$ **9.** $\sqrt[5]{2xy^3}$ **11.** $(\sqrt{81})^3 = 729$ **13.** $(\sqrt[3]{125})^2 = 25$ **15.** $(\sqrt[5]{-32})^3 = -8$ **17.** $(\sqrt[3]{27})^2 + (\sqrt[4]{16})^3 = 17$ **19.** $\sqrt[7]{(xy)^4}$ **21.** $7^{1/2}$ **23.** $5^{1/3}$ **25.** $(11x)^{1/5}$ **27.** $x^{3/2}$ **29.** $x^{3/5}$ **31.** $(x^2y)^{1/5}$ **33.** $(19xy)^{3/2}$ **35.** $(7xy^2)^{5/6}$ **37.** $2xy^{2/3}$ **39.** $\frac{1}{49^{1/2}} = \frac{1}{7}$ **41.** $\frac{1}{27^{1/3}} = \frac{1}{3}$ **43.** $\frac{1}{16^{3/4}} = \frac{1}{8}$ **45.** $\frac{1}{8^{2/3}} = \frac{1}{4}$ **47.** $\left(\frac{27}{8}\right)^{1/3} = \frac{3}{2}$ **49.** $\frac{1}{(-64)^{2/3}} = \frac{1}{16}$ **51.** $\frac{1}{(2xy)^{7/10}}$ **53.** $\frac{5x}{z^{1/3}}$ **55.** 3 **57.** 4 **59.** $x^{5/6}$ **61.** $x^{3/5}$ **63.** $\frac{1}{x^{5/12}}$ **65.** 25 **67.** $\frac{1}{y^{1/6}}$ **69.** $32x$ **71.** $5x^2y^3$ **73.** $\frac{x^{1/4}}{y^{3/10}}$ **75.** 3 **77.** $27y^{2/3}$ **79.** $\sqrt[4]{x}$ **81.** $2a^2$ **83.** x^2y^3 **85.** x^6y^6 **87.** $\sqrt[5]{3y}$ **89.** $\sqrt[3]{4a^2}$ **91.** $\sqrt[3]{x^2y}$ **93.** $\sqrt[6]{2^5}$ or $\sqrt[6]{32}$ **95.** $\sqrt[10]{x^9}$ **97.** $\sqrt[12]{a^{10}b^7}$ **99.** $\sqrt[20]{x}$ **101.** $\sqrt{y}$ **103.** $\sqrt[8]{x}$ **105.** $\sqrt[4]{x^2y}$ **107.** $\sqrt[12]{2x}$ **109.** x^9y^{15} **111.** $\sqrt[4]{a^3b^3}$ **113.** $x^{2/3} - x$ **115.** $x + 2x^{1/2} - 15$ **117.** $2x^{1/2}(3 + x)$ **119.** $15x^{1/3}(1 - 4x^{2/3})$ **121.** $\frac{x^2}{7y^{3/2}}$ **123.** $\frac{x^3}{y^2}$ **125.** 58 species of plants **127.** about 1872 calories per day **129. a.** $C(v) = 35.74 - 35.74v^{4/25}$ **b.** $C(25) \approx -24$; When the air temperature is 0°F and the wind speed is 25 miles per hour, the windchill temperature is −24°. **c.** the point (25, −24) **131. a.** $L + 1.25S^{1/2} - 9.8D^{1/3} \le 16.296$ **b.** eligible

143. simplified correctly

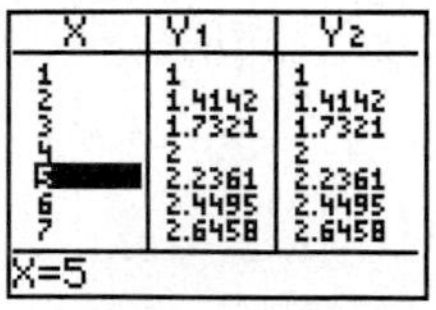

145. Right side should be $x^{1/2}$.

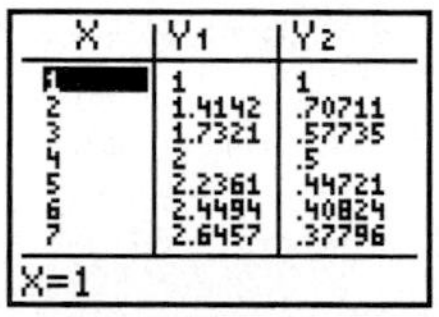

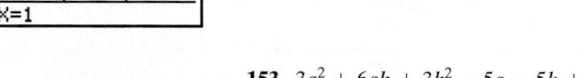

147. 8 years old **149.** 61 **151.** $y = -2x + 11$ or $f(x) = -2x + 11$ **152.**

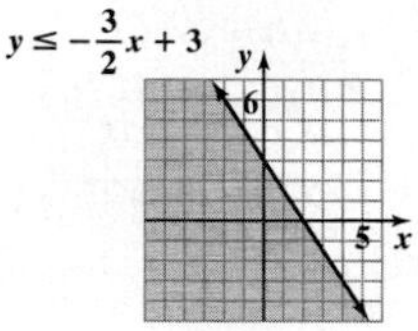

153. $3a^2 + 6ah + 3h^2 - 5a - 5h + 4$

Section 10.3

Check Point Exercises

1. a. $\sqrt{55}$ **b.** $\sqrt{x^2 - 16}$ **c.** $\sqrt[3]{60}$ **d.** $\sqrt[7]{12x^4}$ **2. a.** $4\sqrt{5}$ **b.** $2\sqrt[3]{5}$ **c.** $2\sqrt[4]{2}$ **d.** $10|x|\sqrt{2y}$
3. $f(x) = \sqrt{3}|x - 2|$ **4.** $x^4y^5z\sqrt{xyz}$ **5.** $2x^3y^4\sqrt[3]{5xy^2}$ **6.** $2x^2z\sqrt[5]{x^2y^2z^3}$ **7. a.** $2\sqrt{3}$ **b.** $100\sqrt[3]{4}$ **c.** $2x^2y\sqrt[4]{2}$

Exercise Set 10.3

1. $\sqrt{15}$ **3.** $\sqrt[3]{18}$ **5.** $\sqrt[4]{33}$ **7.** $\sqrt{33xy}$ **9.** $\sqrt[5]{24x^4}$ **11.** $\sqrt{x^2 - 9}$ **13.** $\sqrt[6]{(x - 4)^5}$ **15.** $\sqrt{x}$ **17.** $\sqrt[4]{\frac{3x}{7y}}$ **19.** $\sqrt[7]{77x^5y^3}$ **21.** $5\sqrt{2}$ **23.** $3\sqrt{5}$ **25.** $5\sqrt{3x}$ **27.** $2\sqrt[3]{2}$ **29.** $3x$ **31.** $-2y\sqrt[3]{2x^2}$ **33.** $6|x + 2|$ **35.** $2(x + 2)\sqrt[3]{4}$ **37.** $|x - 1|\sqrt{3}$ **39.** $x^3\sqrt{x}$ **41.** $x^4y^4\sqrt{y}$ **43.** $4x\sqrt{3x}$ **45.** $y^2\sqrt[3]{y^2}$ **47.** $x^4y\sqrt[3]{x^2z}$ **49.** $3x^2y^2\sqrt[3]{3x^2}$ **51.** $(x + y)\sqrt[3]{(x + y)^2}$ **53.** $y^3\sqrt[5]{y^2}$ **55.** $2xy^3\sqrt[5]{2xy^2}$ **57.** $2x^2\sqrt[4]{5x^2}$ **59.** $(x - 3)^2\sqrt[4]{(x - 3)^2}$ or $(x - 3)^2\sqrt{x - 3}$ **61.** $2\sqrt{6}$ **63.** $5\sqrt{2xy}$ **65.** $6x$ **67.** $10xy\sqrt{2y}$ **69.** $60\sqrt{2}$ **71.** $2\sqrt[3]{6}$ **73.** $2x^2\sqrt{10x}$ **75.** $5xy^4\sqrt[3]{x^2y^2}$ **77.** $2xyz\sqrt[4]{x^2z^3}$ **79.** $2xy^2z\sqrt[5]{2y^3z}$ **81.** $(x - y)^2\sqrt[3]{(x - y)^2}$ **83.** $-6x^3y^3\sqrt[3]{2yz^2}$ **85.** $-6y^2\sqrt[5]{2x^3y}$ **87.** $-6x^3y^3\sqrt{2}$ **89.** $-12x^3y^4\sqrt[4]{x^3}$ **91.** $xy^2\sqrt[5]{16y^3} + 2x^2y^3$ **93.** $6\sqrt{3}$ miles; 10.4 miles **95.** $8\sqrt{3}$ ft per sec; 14 ft per sec **97. a.** $\frac{7.644}{2\sqrt[4]{2}} = \frac{3.822}{\sqrt[4]{2}}$ **b.** 3.21 liters of blood per minute per square meter; (32, 3.21)

105. Graphs are the same; simplification is correct.

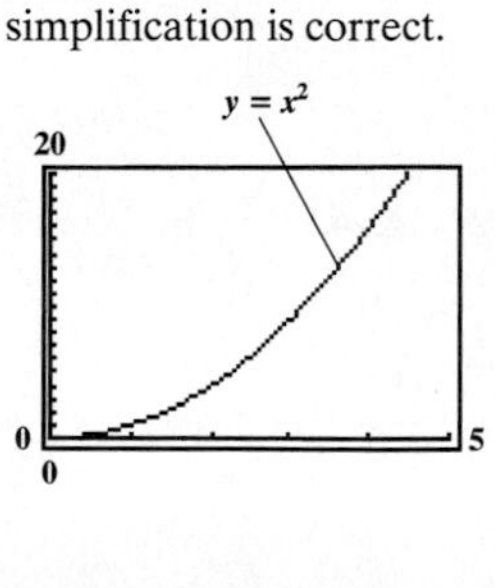

107. Graphs are not the same; $\sqrt{3x^2 - 6x + 3} = |x - 1|\sqrt{3}$.

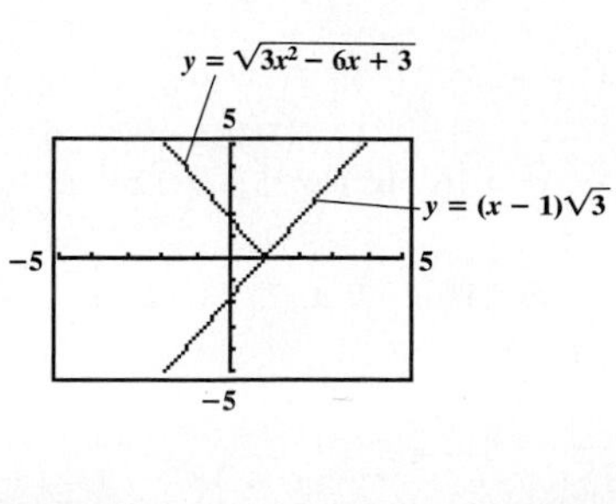

109. d

111. The number must be multiplied by 27.

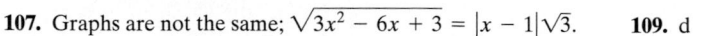

113.

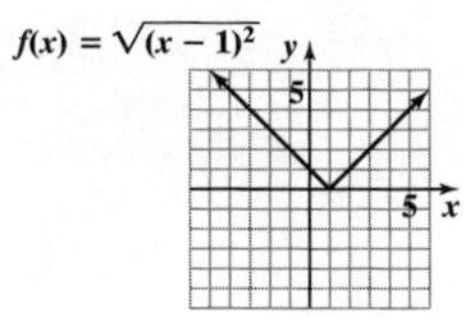

114. $\{x|5 \le x \le 11\}$ or $[5, 11]$ **115.** $\{(2, -4)\}$ **116.** $(4x - 3)(16x^2 + 12x + 9)$

Section 10.4

Check Point Exercises

1. a. $10\sqrt{13}$ **b.** $(21 - 6x)\sqrt[3]{7}$ **c.** $5\sqrt[4]{3x} + 2\sqrt[3]{3x}$ **2. a.** $21\sqrt{5}$ **b.** $-12\sqrt{3x}$ **c.** cannot be simplified

3. a. $-9\sqrt[3]{3}$ **b.** $(5 + 3xy)\sqrt[3]{x^2y}$ **4. a.** $\frac{2\sqrt{3}}{5}$ **b.** $\frac{3x\sqrt{x}}{y^5}$ **c.** $\frac{2y^2\sqrt[3]{y}}{x^4}$ **5. a.** $2x^2\sqrt{5}$ **b.** $\frac{5\sqrt{xy}}{2}$ **c.** $2x^2y$

Exercise Set 10.4

1. $11\sqrt{5}$ **3.** $7\sqrt[3]{6}$ **5.** $2\sqrt[5]{2}$ **7.** $\sqrt{13} + 2\sqrt{5}$ **9.** $7\sqrt{5} + 2\sqrt[3]{x}$ **11.** $4\sqrt{3}$ **13.** $19\sqrt{3}$ **15.** $6\sqrt{2x}$ **17.** $13\sqrt[3]{2}$ **19.** $(9x + 1)\sqrt{5x}$

21. $7y\sqrt[3]{2x}$ **23.** $(3x - 2)\sqrt[3]{2x}$ **25.** $4\sqrt{x - 2}$ **27.** $5x\sqrt[3]{xy^2}$ **29.** $\frac{\sqrt{11}}{2}$ **31.** $\frac{\sqrt[3]{19}}{3}$ **33.** $\frac{x}{6y^4}$ **35.** $\frac{2x\sqrt{2x}}{5y^3}$ **37.** $\frac{x\sqrt[3]{x}}{2y}$ **39.** $\frac{x^2\sqrt[3]{50x^2}}{3y^4}$

41. $\frac{y\sqrt[4]{9y^2}}{x^2}$ **43.** $\frac{2x^2\sqrt[5]{2x^3}}{y^4}$ **45.** $2\sqrt{2}$ **47.** 2 **49.** $3x$ **51.** x^2y **53.** $2x^2\sqrt{5}$ **55.** $4a^5b^5$ **57.** $3\sqrt{xy}$ **59.** $2xy$ **61.** $2x^2y^2\sqrt[4]{y^2}$ or $2x^2y^2\sqrt{y}$

63. $\sqrt[3]{x + 3}$ **65.** $\sqrt[3]{a^2 - ab + b^2}$ **67.** $\frac{43\sqrt{2}}{35}$ **69.** $-11xy\sqrt{2x}$ **71.** $25x\sqrt{2x}$ **73.** $7x\sqrt{3xy}$ **75.** $3x\sqrt[3]{5xy}$ **77.** $\left(\frac{f}{g}\right)(x) = 4x\sqrt{x}$; Domain of $\frac{f}{g} = \{x|x > 0\}$ or $(0, \infty)$ **79.** $\left(\frac{f}{g}\right)(x) = 2x\sqrt[3]{2x}$; Domain of $\frac{f}{g} = \{x|x$ is a real number and $x \neq 0\}$ or $(-\infty, 0) \cup (0, \infty)$

81. a. $R_f\frac{\sqrt{c^2 - v^2}}{\sqrt{c^2}} = R_f\sqrt{\frac{c^2 - v^2}{c^2}} = R_f\sqrt{\frac{c^2}{c^2} - \frac{v^2}{c^2}} = R_f\sqrt{1 - \frac{v^2}{c^2}} = R_f\sqrt{1 - \left(\frac{v}{c}\right)^2}$ **b.** $R_f\sqrt{1 - \left(\frac{c}{c}\right)^2} = 0$; Aging rate is 0 relative to a friend on Earth; A person moving at the speed of light does not age relative to a friend on Earth. **83.** $P = 18\sqrt{5}$ ft ; $A = 100$ sq ft

93. Graphs are not the same.; $\sqrt{16x} - \sqrt{9x} = \sqrt{x}$.

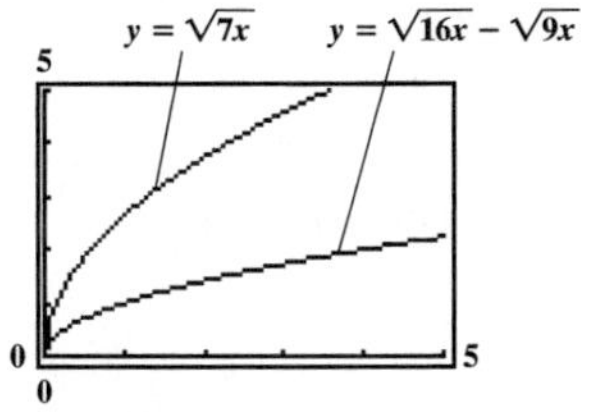

95. d **97.** $-\frac{31\sqrt{5}}{12}$ **99.** $\{0\}$ **100.** $(x - 2y)(x - 6y)$ **101.** $\frac{3x^2 + 8x + 6}{(x + 3)^2(x + 2)}$

Mid-Chapter 10 Check Point

1. 13 **2.** $2x^2y^3\sqrt{2xy}$ **3.** $5\sqrt[3]{4x^2}$ **4.** $12x\sqrt[3]{2x}$ **5.** 1 **6.** $4xy^{1/12}$ **7.** $-\sqrt{3}$ **8.** $\frac{5x\sqrt{5x}}{y^2}$ **9.** $\sqrt[4]{x^3}$ **10.** $3x\sqrt[3]{2x^2}$ **11.** $2\sqrt[3]{10}$

12. $\frac{x^2}{y^4}$ **13.** $x^{7/12}$ **14.** $x\sqrt[3]{y^2}$ **15.** $(x - 2)\sqrt[7]{(x - 2)^2}$ **16.** $2x^2y^4\sqrt[4]{2x^3y}$ **17.** $14\sqrt[3]{2}$ **18.** $x\sqrt[7]{x^2y^2}$ **19.** $\frac{1}{25}$ **20.** $\sqrt[6]{32}$

21. $\frac{4x}{y^2}$ **22.** $8x^2y^3\sqrt{y}$ **23.** $-10x^3y\sqrt{6y}$ **24.** $\{x|x \le 6\} = (-\infty, 6]$ **25.** $\{x|x$ is a real number$\} = (-\infty, \infty)$

Section 10.5

Check Point Exercises

1. a. $x\sqrt{6} + 2\sqrt{15}$ **b.** $y - \sqrt[3]{7y}$ **c.** $36 - 18\sqrt{10}$ **2. a.** $11 + 2\sqrt{30}$ **b.** 1 **c.** $a - 7$ **3. a.** $\frac{\sqrt{21}}{7}$ **b.** $\frac{\sqrt[3]{6}}{3}$

4. a. $\frac{\sqrt{14xy}}{7y}$ **b.** $\frac{\sqrt[3]{3xy^2}}{3y}$ **c.** $\frac{3\sqrt[5]{4x^3y}}{y}$ **5.** $12\sqrt{3} - 18$ **6.** $\frac{3\sqrt{5} + 3\sqrt{2} + \sqrt{35} + \sqrt{14}}{3}$ **7.** $\frac{1}{\sqrt{x + 3} + \sqrt{x}}$

Exercise Set 10.5

1. $x\sqrt{2} + \sqrt{14}$ **3.** $7\sqrt{6} - 6$ **5.** $12\sqrt{2} - 6$ **7.** $\sqrt[3]{12} + 4\sqrt[3]{10}$ **9.** $2x\sqrt[3]{2} - \sqrt[3]{x^2}$ **11.** $32 + 11\sqrt{2}$ **13.** $34 - 15\sqrt{5}$

15. $117 - 36\sqrt{7}$ **17.** $\sqrt{6} + \sqrt{10} + \sqrt{21} + \sqrt{35}$ **19.** $\sqrt{6} - \sqrt{10} - \sqrt{21} + \sqrt{35}$ **21.** $-48 + 7\sqrt{6}$ **23.** $8 + 2\sqrt{15}$

25. $3x - 2\sqrt{3xy} + y$ **27.** -44 **29.** -71 **31.** 6 **33.** $6 - 5\sqrt{x} + x$ **35.** $\sqrt[3]{x^2} + \sqrt[3]{x} - 20$ **37.** $2x^2 + x\sqrt[3]{y^2} - y\sqrt[3]{y}$ **39.** $\frac{\sqrt{10}}{5}$

41. $\frac{\sqrt{11x}}{x}$ **43.** $\frac{3\sqrt{3y}}{y}$ **45.** $\frac{\sqrt[3]{4}}{2}$ **47.** $3\sqrt[3]{2}$ **49.** $\frac{\sqrt[3]{18}}{3}$ **51.** $\frac{4\sqrt[3]{x^2}}{x}$ **53.** $\frac{\sqrt[3]{2y}}{y}$ **55.** $\frac{7\sqrt[3]{4x}}{2x}$ **57.** $\frac{\sqrt[3]{2x^2y}}{xy}$ **59.** $\frac{3\sqrt[4]{x^3}}{x}$

61. $\frac{3\sqrt[5]{4x^2}}{x}$ **63.** $x\sqrt[5]{8x^3y}$ **65.** $\frac{3\sqrt{3y}}{xy}$ **67.** $-\frac{5a^2\sqrt{3ab}}{b^2}$ **69.** $\frac{\sqrt{2mn}}{2m}$ **71.** $\frac{3\sqrt[4]{x^3y}}{x^2y}$ **73.** $-\frac{6\sqrt[3]{xy}}{x^2y^3}$ **75.** $8\sqrt{5} - 16$ **77.** $\frac{13\sqrt{11} + 39}{2}$

79. $3\sqrt{5} - 3\sqrt{3}$ **81.** $\frac{a + \sqrt{ab}}{a - b}$ **83.** $25\sqrt{2} + 15\sqrt{5}$ **85.** $4 + \sqrt{15}$ **87.** $\frac{x - 2\sqrt{x} - 3}{x - 9}$ **89.** $\frac{3\sqrt{6} + 4}{2}$ **91.** $\frac{4\sqrt{xy} + 4x + y}{y - 4x}$ **93.** $\frac{3}{\sqrt{6}}$

95. $\frac{2x}{\sqrt[3]{2x^2y}}$ **97.** $\frac{x - 9}{x - 3\sqrt{x}}$ **99.** $\frac{a - b}{a - 2\sqrt{ab} + b}$ **101.** $\frac{1}{\sqrt{x + 5} + \sqrt{x}}$ **103.** $\frac{1}{(x + y)(\sqrt{x} - \sqrt{y})}$ **105.** $\frac{3\sqrt{2}}{2}$ **107.** $-2\sqrt[3]{25}$ **109.** $\frac{7\sqrt{6}}{6}$

111. $7\sqrt{3} - 7\sqrt{2}$ **113.** 0 **115.** 6 **117.** 51; In 2001, approximately 51% of U.S. households were online.; Answer models data well. **119.** 6.5%

121. approximately 6.5%; This answer models the actual yearly percentage increase well. **123. a.** $\frac{15.92}{\sqrt{t + h} + \sqrt{t}}$ **b.** $\frac{15.92}{2\sqrt{t}}$ or $\frac{7.96}{\sqrt{t}}$ **c.** 3.2%

125. $P = 8\sqrt{2}$ in.; $A = 7$ sq in. **127.** $\frac{7\sqrt{2 \cdot 2 \cdot 3}}{6} = \frac{7 \cdot 2\sqrt{3}}{6} = \frac{14\sqrt{3}}{6} = \frac{7\sqrt{3}}{3} = \frac{7}{3}\sqrt{3}$

139.

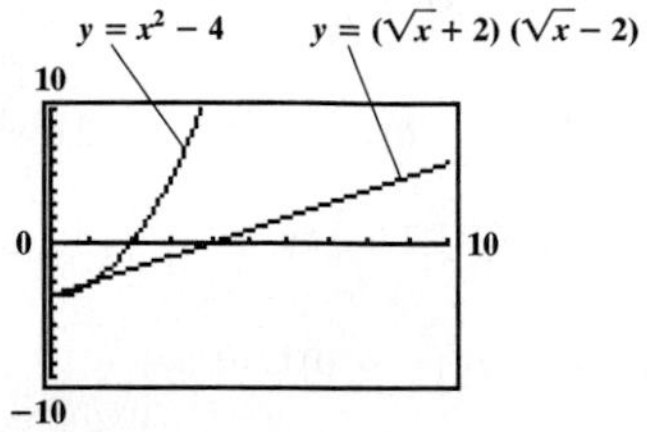

Graphs are not the same; $(\sqrt{x} + 2)(\sqrt{x} - 2) = x - 4$ for $x \geq 0$.

141.

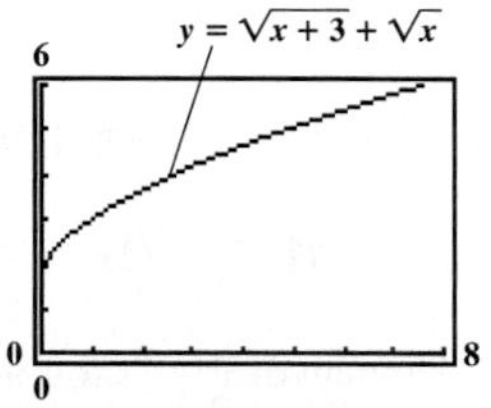

Graphs are the same; Simplification is correct.

143. $\frac{45}{7}$ or $\left\{\frac{45}{7}\right\}$

145. $\frac{5\sqrt{2} + 3\sqrt{3} - 4\sqrt{6} + 2}{23}$

146. $\frac{2x + 7}{x^2 - 4}$

147. $\{x|0 \leq x \leq 2\}$ or $[0, 2]$

148. yes; no

Section 10.6

Check Point Exercises

1. 20 or $\{20\}$ **2.** no solution or $\varnothing$ **3.** -1 and 3, or $\{-1, 3\}$ **4.** 4 or $\{4\}$ **5.** -12 or $\{-12\}$ **6.** 2009; faster

Exercise Set 10.6

1. $\{6\}$ **3.** $\{17\}$ **5.** $\varnothing$ **7.** $\{8\}$ **9.** $\{0, 3\}$ **11.** $\{1, 3\}$ **13.** $\{3, 7\}$ **15.** $\{9\}$ **17.** $\{8\}$ **19.** $\{35\}$ **21.** $\{16\}$ **23.** $\{2\}$ **25.** $\{2, 6\}$

27. $\left\{\frac{5}{2}\right\}$ **29.** $\{5\}$ **31.** $\varnothing$ **33.** $\{2, 6\}$ **35.** $\{0, 10\}$ **37.** $\{8\}$ **39.** 4 **41.** -7 **43.** $V = \frac{\pi r^2 h}{3}$ or $\frac{1}{3}\pi r^2 h$ **45.** $l = \frac{8t^2}{\pi^2}$ **47.** 8

49. 9 **51.** High: 66.0°; Low: 61.5° **53.** Using f: 2014; Using g: 2010 **55.** 36 years old; (36, 40,000) **57.** 27 sq mi **59.** 149 million km

69.

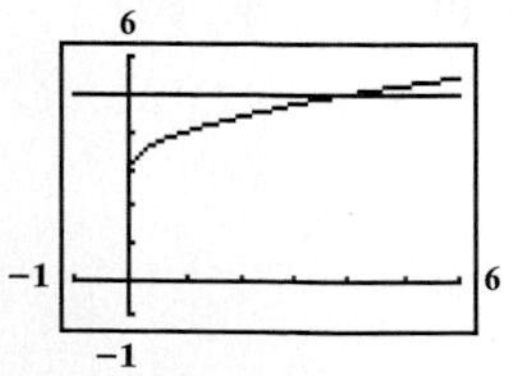

4 or $\{4\}$

71.

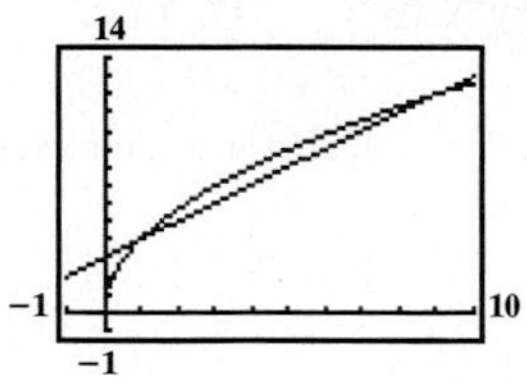

1 and 9, or $\{1, 9\}$

73. c **75.** 81 or $\{81\}$ **77.** 129 or $\{129\}$

78. $4x^3 - 15x^2 + 47x - 142 + \frac{425}{x + 3}$

79. $\frac{x + 2}{2(x + 3)}$

80. $(y - 3 + 5x)(y - 3 - 5x)$

Section 10.7

Check Point Exercises

1. a. $8i$ **b.** $\sqrt{11}i$ **c.** $4\sqrt{3}i$ **2. a.** $8 + i$ **b.** $-10 + 10i$ **3. a.** $63 + 14i$ **b.** $58 - 11i$

4. $-\sqrt{35}$ **5.** $\frac{18}{25} + \frac{26}{25}i$ **6.** $-\frac{1}{2} - \frac{3}{4}i$ **7. a.** 1 **b.** i **c.** $-i$

Exercise Set 10.7

1. $10i$ **3.** $\sqrt{23}i$ **5.** $3\sqrt{2}i$ **7.** $3\sqrt{7}i$ **9.** $-6\sqrt{3}i$ **11.** $5 + 6i$ **13.** $15 + \sqrt{3}i$ **15.** $-2 - 3\sqrt{2}i$ **17.** $8 + 3i$ **19.** $8 - 2i$ **21.** $5 + 3i$ **23.** $-1 - 7i$ **25.** $-2 + 9i$ **27.** $8 + 15i$ **29.** $-14 + 17i$ **31.** $9 + 5\sqrt{3}i$ **33.** $-6 + 10i$ **35.** $-21 - 15i$ **37.** $-35 - 14i$ **39.** $7 + 19i$ **41.** $-1 - 31i$ **43.** $3 + 36i$ **45.** 34 **47.** 34 **49.** 11 **51.** $-5 + 12i$ **53.** $21 - 20i$ **55.** $-\sqrt{14}$ **57.** -6 **59.** $-5\sqrt{7}$ **61.** $-2\sqrt{6}$ **63.** $\frac{3}{5} - \frac{1}{5}i$ **65.** $1 + i$ **67.** $\frac{28}{25} + \frac{21}{25}i$ **69.** $-\frac{12}{13} + \frac{18}{13}i$ **71.** $0 + i$ or i **73.** $\frac{3}{10} - \frac{11}{10}i$ **75.** $\frac{11}{13} - \frac{16}{13}i$ **77.** $-\frac{23}{58} + \frac{43}{58}i$ **79.** $0 - \frac{7}{3}i$ or $-\frac{7}{3}i$ **81.** $-\frac{5}{2} - 4i$ **83.** $-\frac{7}{3} + \frac{4}{3}i$ **85.** -1 **87.** $-i$ **89.** -1 **91.** 1 **93.** i **95.** 1 **97.** $-i$ **99.** 0 **101.** $-11 - 5i$ **103.** $-5 + 10i$ **105.** $0 + 47i$ or $47i$ **107.** $1 - i$ **109.** 0 **111.** $10 + 10i$ **113.** $\frac{20}{13} + \frac{30}{13}i$ **115.** $(47 + 13i)$ volts **117.** $(5 + \sqrt{15}i) + (5 - \sqrt{15}i) = 10$; $(5 + \sqrt{15}i)(5 - \sqrt{15}i) = 25 - 15i^2 = 25 + 15 = 40$ **131.** $\sqrt{-9} + \sqrt{-16} = 3i + 4i = 7i$ **133.** d **135.** $\frac{6}{5} + 0i$ or $\frac{6}{5}$ **137.** $\frac{x^2}{y^2}$ **138.** $x = \frac{yz}{y - z}$ **139.** -18

Review Exercises

1. 9 **2.** $-\frac{1}{10}$ **3.** -3 **4.** not a real number **5.** -2 **6.** 5; 1.73; 0; not a real number **7.** 2; −2; −4 **8.** $\{x \mid x \geq 2\}$ or $[2, \infty)$ **9.** $\{x \mid x \leq 25\}$ or $(-\infty, 25]$ **10.** $5|x|$ **11.** $|x + 14|$ **12.** $|x - 4|$ **13.** $4x$ **14.** $2|x|$ **15.** $-2(x + 7)$ **16.** $\sqrt[3]{5xy}$ **17.** $(\sqrt{16})^3 = 64$ **18.** $(\sqrt[5]{32})^4 = 16$ **19.** $(7x)^{1/2}$ **20.** $(19xy)^{5/3}$ **21.** $\frac{1}{8^{2/3}} = \frac{1}{4}$ **22.** $\frac{3x}{(ab)^{4/5}} = \frac{3x}{\sqrt[5]{(ab)^4}}$ **23.** $x^{7/12}$ **24.** $5^{1/6}$ **25.** $2x^2y$ **26.** $\frac{y^{1/8}}{x^{1/3}}$ **27.** x^3y^4 **28.** $y\sqrt[3]{x}$ **29.** $\sqrt[6]{x^5}$ **30.** $\sqrt[6]{x}$ **31.** $\sqrt[15]{x}$ **32.** \$3150 million **33.** $\sqrt{21xy}$ **34.** $\sqrt[5]{77x^3}$ **35.** $\sqrt[6]{(x - 5)^5}$ **36.** $f(x) = \sqrt{7}|x - 1|$ **37.** $2x\sqrt{5x}$ **38.** $3x^2y^2\sqrt[3]{2x^2}$ **39.** $2y^2z\sqrt[4]{2x^3y^3z}$ **40.** $2x^2\sqrt{6x}$ **41.** $2xy\sqrt[3]{2y^2}$ **42.** $xyz^2\sqrt[5]{16y^4z}$ **43.** $\sqrt{x^2 - 1}$ **44.** $8\sqrt[3]{3}$ **45.** $9\sqrt{2}$ **46.** $(3x + y^2)\sqrt[3]{x}$ **47.** $-8\sqrt[3]{6}$ **48.** $\frac{2}{5}\sqrt[3]{2}$ **49.** $\frac{x\sqrt{x}}{10y^2}$ **50.** $\frac{y\sqrt[4]{3y}}{2x^5}$ **51.** $2\sqrt{6}$ **52.** $2\sqrt[3]{2}$ **53.** $2x\sqrt[4]{2x}$ **54.** $10x^2\sqrt{xy}$ **55.** $6\sqrt{2} + 12\sqrt{5}$ **56.** $5\sqrt[3]{2} - \sqrt[3]{10}$ **57.** $-83 + 3\sqrt{35}$ **58.** $\sqrt{xy} - \sqrt{11x} - \sqrt{11y} + 11$ **59.** $13 + 4\sqrt{10}$ **60.** $22 - 4\sqrt{30}$ **61.** -6 **62.** 4 **63.** $\frac{2\sqrt{6}}{3}$ **64.** $\frac{\sqrt{14}}{7}$ **65.** $4\sqrt[3]{3}$ **66.** $\frac{\sqrt{10xy}}{5y}$ **67.** $\frac{7\sqrt[3]{4x}}{x}$ **68.** $\frac{\sqrt[4]{189x^3}}{3x}$ **69.** $\frac{5\sqrt[5]{xy^4}}{2xy}$ **70.** $3\sqrt{3} + 3$ **71.** $\frac{\sqrt{35} - \sqrt{21}}{2}$ **72.** $10\sqrt{5} + 15\sqrt{2}$ **73.** $\frac{x + 8\sqrt{x} + 15}{x - 9}$ **74.** $\frac{5 + \sqrt{21}}{2}$ **75.** $\frac{3\sqrt{2} + 2}{7}$ **76.** $\frac{2}{\sqrt{14}}$ **77.** $\frac{3x}{\sqrt[3]{9x^2y}}$ **78.** $\frac{7}{\sqrt{35} + \sqrt{21}}$ **79.** $\frac{2}{5 - \sqrt{21}}$ **80.** $\{16\}$ **81.** $\varnothing$ **82.** $\{2\}$ **83.** $\{8\}$ **84.** $\{-4, -2\}$ **85.** 256 feet **86.** 84 years old **87.** $9i$ **88.** $3\sqrt{7}i$ **89.** $-2\sqrt{2}i$ **90.** $12 + 2i$ **91.** $-9 + 4i$ **92.** $-12 - 8i$ **93.** $29 + 11i$ **94.** $-7 - 24i$ **95.** $113 + 0i$ or 113 **96.** $-2\sqrt{6} + 0i$ or $-2\sqrt{6}$ **97.** $\frac{15}{13} - \frac{3}{13}i$ **98.** $\frac{1}{5} + \frac{11}{10}i$ **99.** $\frac{1}{3} - \frac{5}{3}i$ **100.** 1 **101.** $-i$

Chapter 10 Test

1. a. 6 **b.** $\{x \mid x \leq 4\}$ or $(-\infty, 4]$ **2.** $\frac{1}{81}$ **3.** $\frac{5\sqrt[8]{y}}{\sqrt[4]{x}}$ **4.** $\sqrt{x}$ **5.** $\sqrt[20]{x^9}$ **6.** $5|x|\sqrt{3}$ **7.** $|x - 5|$ **8.** $2xy^2\sqrt[3]{xy^2}$ **9.** $-\frac{2}{x^2}$ **10.** $\sqrt[3]{50x^2y}$ **11.** $2x\sqrt[4]{2y^3}$ **12.** $-7\sqrt{2}$ **13.** $(2x + y^2)\sqrt[3]{x}$ **14.** $2x\sqrt[3]{x}$ **15.** $12\sqrt{2} - \sqrt{15}$ **16.** $26 + 6\sqrt{3}$ **17.** $52 - 14\sqrt{3}$ **18.** $\frac{\sqrt{5x}}{x}$ **19.** $\frac{\sqrt[3]{25x}}{x}$ **20.** $-5 + 2\sqrt{6}$ **21.** 6 or $\{6\}$ **22.** 16 or $\{16\}$ **23.** -3 or $\{-3\}$ **24.** 49 months **25.** $5\sqrt{3}i$ **26.** $-1 + 6i$ **27.** $26 + 7i$ **28.** $-6 + 0i$ or -6 **29.** $\frac{1}{5} + \frac{7}{5}i$ **30.** $-i$

Cumulative Review Exercises (Chapters 1–10)

1. $\{(-2, -1, -2)\}$ **2.** $\left\{-\frac{1}{3}, 4\right\}$ **3.** $\left\{x \,\middle|\, x > \frac{1}{3}\right\}$ or $\left(\frac{1}{3}, \infty\right)$ **4.** $\left\{\frac{3}{4}\right\}$ **5.** $\{-1\}$

6. $x + 2y < 2$
$2y - x > 4$

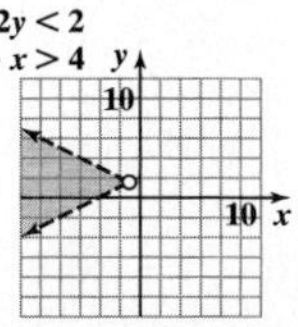

7. $\frac{x^2}{15(x + 2)}$ **8.** $\frac{x}{y}$ **9.** $8x^3 - 22x^2 + 11x + 6$ **10.** $\frac{5x - 6}{(x - 5)(x + 3)}$ **11.** -64 **12.** 0 **13.** $-\frac{16 - 9\sqrt{3}}{13}$ **14.** $2x^2 + x + 5 + \frac{6}{x - 2}$ **15.** $-34 - 3\sqrt{6}$ **16.** $2(3x + 2)(4x - 1)$ **17.** $(4x^2 + 1)(2x + 1)(2x - 1)$ **18.** about 53 lumens **19.** \$1500 at 7% and \$4500 at 9% **20.** 2650 students

CHAPTER 11

Section 11.1

Check Point Exercises

1. $\pm\sqrt{7}$ or $\{\pm\sqrt{7}\}$ **2.** $\pm\frac{\sqrt{33}}{3}$ or $\left\{\pm\frac{\sqrt{33}}{3}\right\}$ **3.** $\pm\frac{3}{2}i$ or $\left\{\pm\frac{3}{2}i\right\}$ **4.** $3 \pm \sqrt{10}$ or $\{3 \pm \sqrt{10}\}$ **5. a.** $25; x^2 + 10x + 25 = (x + 5)^2$ **b.** $\frac{9}{4}; x^2 - 3x + \frac{9}{4} = \left(x - \frac{3}{2}\right)^2$ **c.** $\frac{9}{64}; x^2 + \frac{3}{4}x + \frac{9}{64} = \left(x + \frac{3}{8}\right)^2$ **6.** $-2 \pm \sqrt{5}$ or $\{-2 \pm \sqrt{5}\}$ **7.** $\frac{-3 \pm \sqrt{41}}{4}$ or $\left\{\frac{-3 \pm \sqrt{41}}{4}\right\}$ **8.** 20% **9.** $10\sqrt{21}$ ft; 45.8 ft **10.** 5 units **11.** $\left(4, -\frac{1}{2}\right)$

Exercise Set 11.1

1. $\{\pm 5\}$ **3.** $\{\pm\sqrt{6}\}$ **5.** $\left\{\pm\frac{5}{4}\right\}$ **7.** $\left\{\pm\frac{\sqrt{6}}{3}\right\}$ **9.** $\left\{\pm\frac{4}{5}i\right\}$ **11.** $\{-10, -4\}$ **13.** $\{3 \pm \sqrt{5}\}$ **15.** $\{-2 \pm 2\sqrt{2}\}$ **17.** $\{5 \pm 3i\}$ **19.** $\left\{\frac{-3 \pm \sqrt{11}}{4}\right\}$ **21.** $\{-3, 9\}$ **23.** $1; x^2 + 2x + 1 = (x + 1)^2$ **25.** $49; x^2 - 14x + 49 = (x - 7)^2$ **27.** $\frac{49}{4}; x^2 + 7x + \frac{49}{4} = \left(x + \frac{7}{2}\right)^2$ **29.** $\frac{1}{16}; x^2 - \frac{1}{2}x + \frac{1}{16} = \left(x - \frac{1}{4}\right)^2$ **31.** $\frac{4}{9}; x^2 + \frac{4}{3}x + \frac{4}{9} = \left(x + \frac{2}{3}\right)^2$ **33.** $\frac{81}{64}; x^2 - \frac{9}{4}x + \frac{81}{64} = \left(x - \frac{9}{8}\right)^2$ **35.** $\{-8, 4\}$ **37.** $\{-3 \pm \sqrt{7}\}$ **39.** $\{4 \pm \sqrt{15}\}$ **41.** $\{-1 \pm i\}$ **43.** $\left\{\frac{-3 \pm \sqrt{13}}{2}\right\}$ **45.** $\left\{-\frac{3}{7}, -\frac{1}{7}\right\}$ **47.** $\left\{\frac{-1 \pm \sqrt{5}}{2}\right\}$ **49.** $\left\{-\frac{5}{2}, 1\right\}$ **51.** $\left\{\frac{-3 \pm \sqrt{6}}{3}\right\}$ **53.** $\left\{\frac{4 \pm \sqrt{13}}{3}\right\}$ **55.** $\left\{\frac{1}{4} \pm \frac{1}{4}i\right\}$ **57.** $-5, 7$ **59.** $-\frac{1}{5}, 1$ **61.** $-2 \pm 5i$ **63.** 13 units **65.** $2\sqrt{2}$ or 2.83 units **67.** 5 units **69.** $\sqrt{29}$ or 5.39 units **71.** $4\sqrt{2}$ or 5.66 units **73.** $2\sqrt{5}$ or 4.47 units **75.** $2\sqrt{2}$ or 2.83 units **77.** $\sqrt{93}$ or 9.64 units **79.** $\sqrt{5}$ or 2.24 units **81.** $(4, 6)$ **83.** $(-4, -5)$ **85.** $\left(\frac{3}{2}, -6\right)$ **87.** $(-3, -2)$ **89.** $(1, 5\sqrt{5})$ **91.** $(2\sqrt{2}, 0)$ **93.** $2 \pm 2i$ **95.** $v = \sqrt{2gh}$ **97.** $r = \frac{\sqrt{AP}}{P} - 1$ **99.** $\frac{-1 \pm \sqrt{19}}{6}$ or $\left\{\frac{-1 \pm \sqrt{19}}{6}\right\}$ **101.** $-b$ and $2b$, or $\{-b, 2b\}$ **103.** 20% **105.** 6.25% **107.** approximately 2007 **109.** $10\sqrt{3}$ sec; 17.3 sec **111.** $3\sqrt{5}$ mi; 6.7 mi **113.** $20\sqrt{2}$ ft, or 28.3 ft **115.** $50\sqrt{2}$ ft; 70.7 ft **117.** 10 m **119.** 0.5 hour or 30 minutes

129.

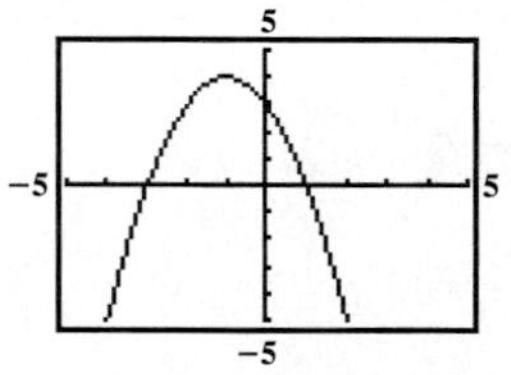

-3 and 1, or $\{-3, 1\}$

133. $y = \pm\frac{b\sqrt{a^2 - x^2}}{a}$ **135.** $\frac{-b \pm \sqrt{b^2 - 4c}}{2}$ or $\left\{\frac{-b \pm \sqrt{b^2 - 4c}}{2}\right\}$ **137.** The distance from A to B is $2\sqrt{2}$ and the distance from B to C is $3\sqrt{2}$. The distance from A to C is $5\sqrt{2}$, which is equal to $2\sqrt{2} + 3\sqrt{2}$. **139.** $(-2, 2)$ and $(6, 2)$ **140.** $4 - 2x$ **141.** $(1 - 2x)(1 + 2x + 4x^2)$ **142.** $x^3 - 2x^2 - 4x - 12 - \frac{42}{x - 3}$

Section 11.2

Check Point Exercises

1. -5 and $\frac{1}{2}$, or $\left\{-5, \frac{1}{2}\right\}$ **2.** $\frac{3 \pm \sqrt{7}}{2}$ or $\left\{\frac{3 \pm \sqrt{7}}{2}\right\}$ **3.** $-1 \pm \frac{\sqrt{6}}{3}i$ or $\left\{-1 \pm \frac{\sqrt{6}}{3}i\right\}$ **4. a.** 0; one real solution **b.** 81; two rational solutions **c.** -44; two imaginary solutions that are complex conjugates **5. a.** $20x^2 + 7x - 3 = 0$ **b.** $x^2 + 49 = 0$ **6.** 1960 and 1974; very well

Exercise Set 11.2

1. $\{-6, -2\}$ **3.** $\left\{1, \frac{5}{2}\right\}$ **5.** $\left\{\frac{-3 \pm \sqrt{89}}{2}\right\}$ **7.** $\left\{\frac{7 \pm \sqrt{85}}{6}\right\}$ **9.** $\left\{\frac{1 \pm \sqrt{7}}{6}\right\}$ **11.** $\left\{\frac{3}{8} \pm \frac{\sqrt{87}}{8}i\right\}$ **13.** $\{2 \pm 2i\}$ **15.** $\left\{\frac{4}{3} \pm \frac{\sqrt{5}}{3}i\right\}$ **17.** $\left\{-\frac{3}{2}, 4\right\}$ **19.** 52; two irrational solutions **21.** 4; two rational solutions **23.** -23; two imaginary solutions

25. 36; two rational solutions **27.** −60; two imaginary solutions **29.** 0; one (repeated) rational solution **31.** $\left\{-\frac{2}{3}, 2\right\}$

33. $\{1 \pm \sqrt{2}\}$ **35.** $\left\{\frac{3 \pm \sqrt{65}}{4}\right\}$ **37.** $\left\{0, \frac{8}{3}\right\}$ **39.** $\left\{\frac{-6 \pm 2\sqrt{6}}{3}\right\}$

41. $\left\{\frac{2 \pm \sqrt{10}}{3}\right\}$ **43.** $\{2 \pm \sqrt{10}\}$ **45.** $\left\{1, \frac{5}{2}\right\}$ **47.** $x^2 - 2x - 15 = 0$ **49.** $12x^2 + 5x - 2 = 0$

51. $x^2 + 36 = 0$ **53.** $x^2 - 2 = 0$ **55.** $x^2 - 20 = 0$ **57.** $x^2 - 2x + 2 = 0$ **59.** $x^2 - 2x - 1 = 0$ **61.** b **63.** a **65.** $1 + \sqrt{7}$

67. $\left\{\frac{-1 \pm \sqrt{21}}{2}\right\}$ **69.** $\left\{-2\sqrt{2}, \frac{\sqrt{2}}{2}\right\}$ **71.** $\{-3, 1, -1 \pm \sqrt{2}i\}$

73. 33-year-olds and 58-year-olds; The function models the actual data well. **75.** 77.8 ft; (b) **77.** 5.5 m by 1.5 m **79.** 17.6 in. and 18.6 in.

81. 9.3 in. and 0.7 in. **83.** 7.5 hr and 8.5 hr **95.** d **97.** 2.4 m; yes **99.** $\left\{-3, \frac{1}{4}\right\}$

100. $\{3, 7\}$ **101.** $\frac{5\sqrt{3} - 5x}{3 - x^2}$

Section 11.3

Check Point Exercises

1.

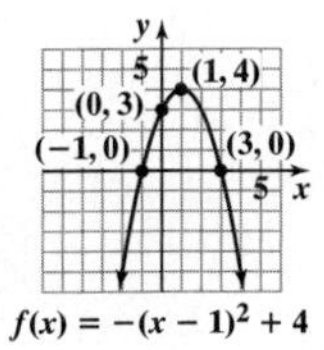

2.

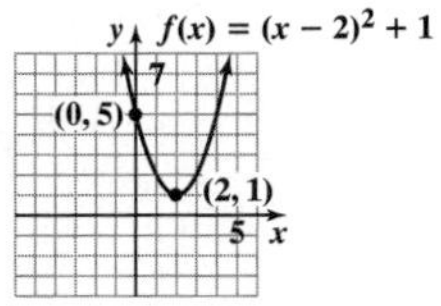

3.

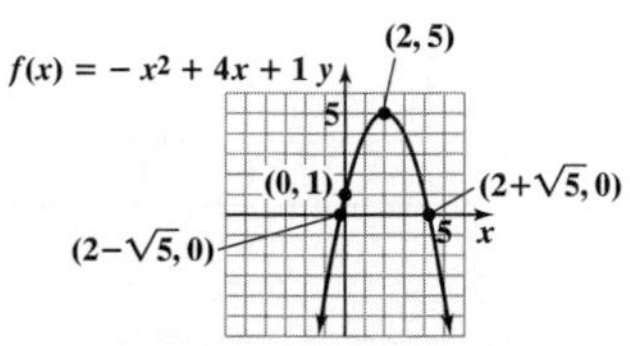

Domain = $\{x|x$ is a real number$\}$ or $(-\infty, \infty)$; Range = $\{y|y \le 5\}$ or $(-\infty, 5]$

4. 33.7 ft **5.** 4, −4; −16
6. 30 ft by 30 ft; 900 sq ft

Exercise Set 11.3

1. $h(x) = (x - 1)^2 + 1$ **3.** $j(x) = (x - 1)^2 - 1$ **5.** $h(x) = x^2 - 1$ **7.** $g(x) = x^2 - 2x + 1$

9. $(3, 1)$ **11.** $(-1, 5)$ **13.** $(2, -5)$ **15.** $(-1, 9)$

17.

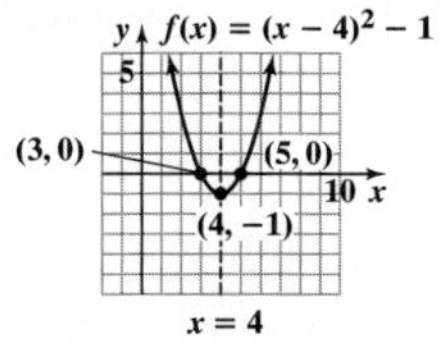

$x = 4$; $\{y|y \ge -1\}$ or $[-1, \infty)$

19.

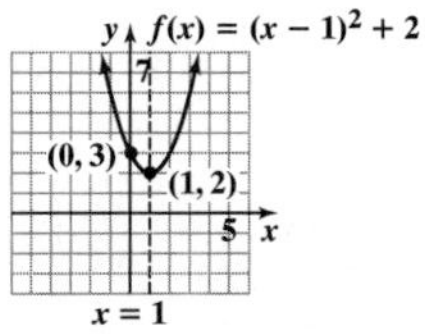

$x = 1$; $\{y|y \ge 2\}$ or $[2, \infty)$

21.

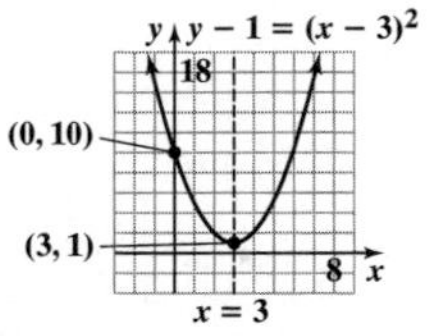

$x = 3$; $\{y|y \ge 1\}$ or $[1, \infty)$

23.

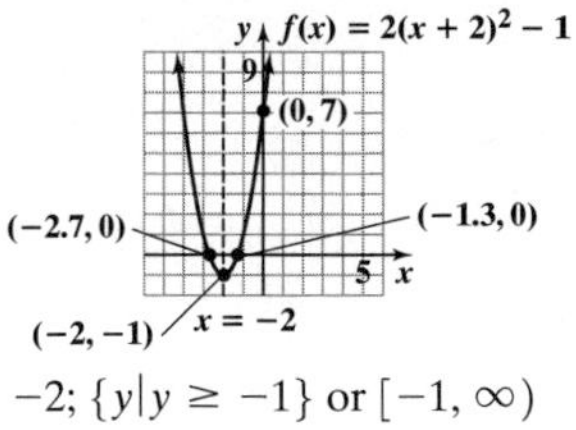

$x = -2$; $\{y|y \ge -1\}$ or $[-1, \infty)$

25.

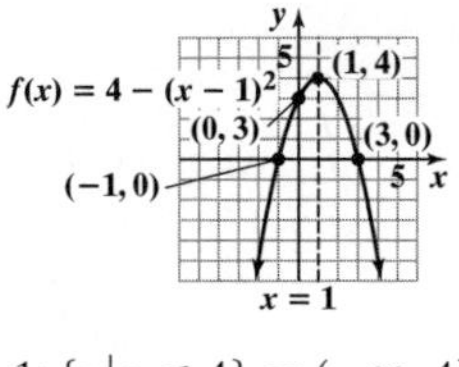

$x = 1$; $\{y|y \le 4\}$ or $(-\infty, 4]$

27.

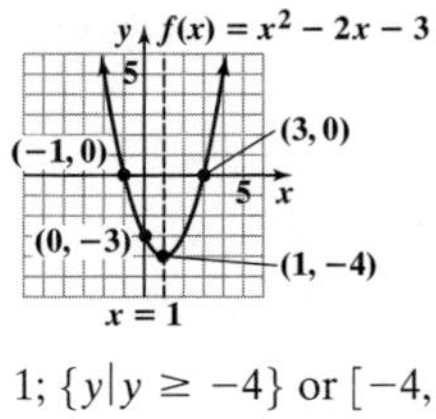

$x = 1$; $\{y|y \ge -4\}$ or $[-4, \infty)$

29.

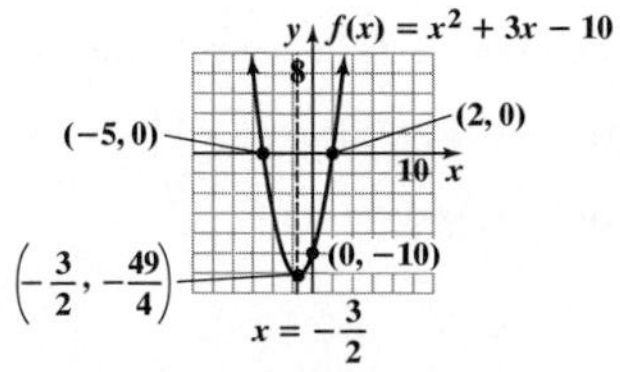

$x = -\frac{3}{2}$; $\left\{y|y \ge -\frac{49}{4}\right\}$ or $\left[-\frac{49}{4}, \infty\right)$

31.

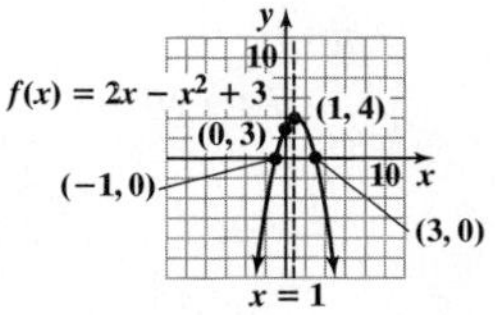

$x = 1$; $\{y|y \le 4\}$ or $(-\infty, 4]$

33.

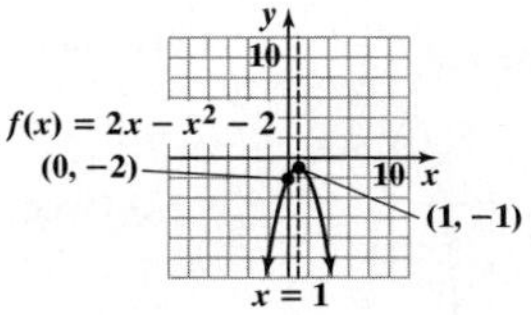

$x = 1$; $\{y|y \le -1\}$ or $(-\infty, -1]$

35. minimum; $(2, -13)$ **37.** maximum; $(1, 1)$ **39.** minimum; $\left(\frac{1}{2}, -\frac{5}{4}\right)$

41. Domain $= \{x|x$ is a real number$\}$ or $(-\infty, \infty)$; Range $= \{y|y \geq -2\}$ or $[-2, \infty)$

43. Domain $= \{x|x$ is a real number$\}$ or $(-\infty, \infty)$; Range $= \{y|y \leq 3\}$ or $(-\infty, 3]$ **45.** $f(x) = 2(x - 5)^2 + 3$ **47.** $f(x) = 2(x + 10)^2 - 5$

49. $f(x) = -3(x + 2)^2 + 4$ **51.** $f(x) = 3(x - 11)^2$ **53.** (14, 19.5); In 2016, the projected mortality will reach a maximum of 19,500.

55. a. 2 sec; 224 ft

b. 5.7 sec

c. 160; 160 feet is the height of the building.

d.

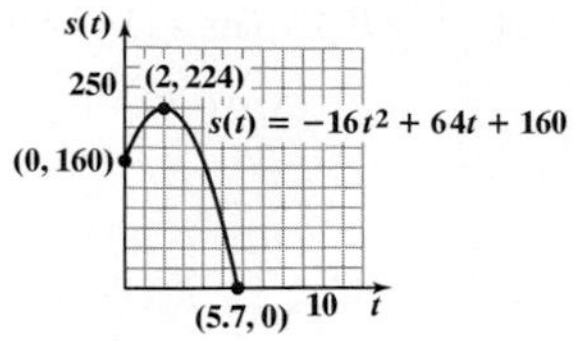

57. 7.2 hr; 622 per 100,000 males

59. 8 and 8; 64

61. 8, −8; −64

63. length: 300 ft; width: 150 ft; maximum area: 45,000 sq ft

65. 12.5 yd by 12.5 yd; 156.25 sq yd

67. 5 in.; 50 sq in.

69. a. $C(x) = 525 + 0.55x$

b. $P(x) = -0.001x^2 + 2.45x - 525$

c. 1225 sandwiches; $975.63

77. a.

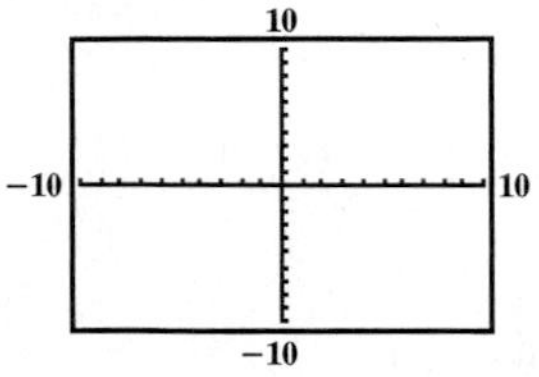

79. (2.5, 185)

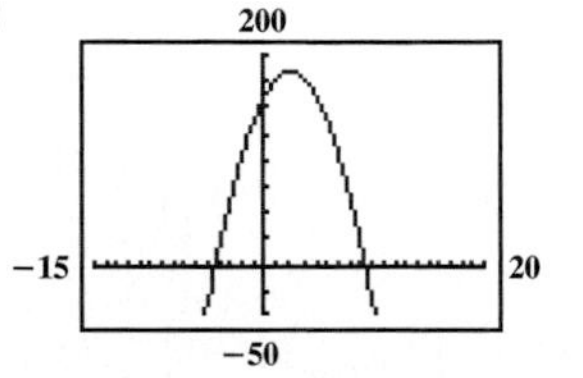

b. $(20.5, -120.5)$

c.

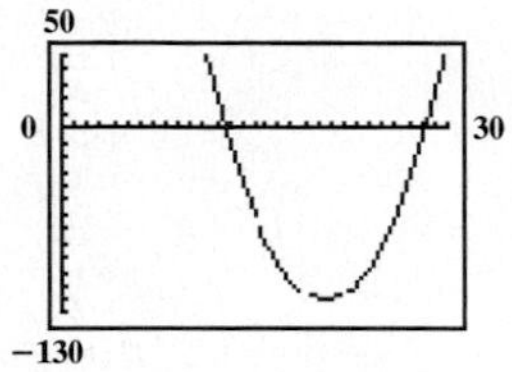

d. Answers will vary.

81. $(-30, 91)$

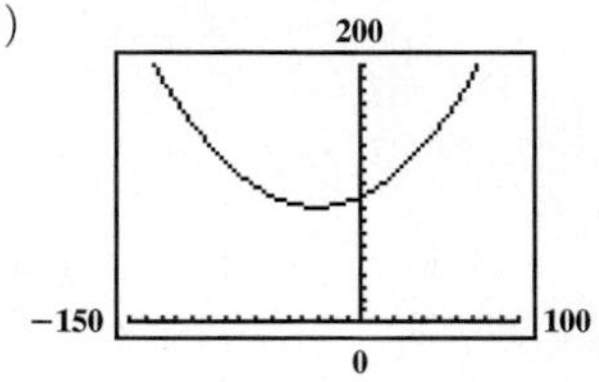

83. a **85.** $x = 3$; $(0, 11)$

87. $f(x) = -2(x + 3)^2 - 1$

89. 65 trees; 16,900 lb

90. $\{7\}$

91. $\dfrac{x}{x - 2}$

92. $(6, -2)$ or $\{(6, -2)\}$

Mid-Chapter 11 Check Point

1. $\left\{-\frac{1}{3}, \frac{11}{3}\right\}$ **2.** $\left\{-1, \frac{7}{5}\right\}$ **3.** $\left\{\frac{3 \pm \sqrt{15}}{3}\right\}$ **4.** $\{-3 \pm \sqrt{7}\}$ **5.** $\left\{\pm\frac{6\sqrt{5}}{5}\right\}$ **6.** $\left\{\frac{5}{2} \pm \frac{\sqrt{7}}{2}i\right\}$ **7.** $\{\pm\sqrt{13}i\}$ **8.** $\left\{-4, \frac{1}{2}\right\}$

9. $\{-3 \pm 2\sqrt{6}\}$ **10.** $\{2 \pm \sqrt{3}\}$ **11.** $\left\{\frac{3}{4} \pm \frac{\sqrt{23}}{4}i\right\}$ **12.** $\left\{\frac{-3 \pm \sqrt{41}}{4}\right\}$ **13.** $\{-2, 4\}$ **14.** $\{-5 \pm 2\sqrt{7}\}$

15. $4\sqrt{2} \approx 5.66$ units; $(0, 0)$ **16.** $\sqrt{61} \approx 7.81$ units; $\left(-\frac{15}{2}, 11\right)$

17.

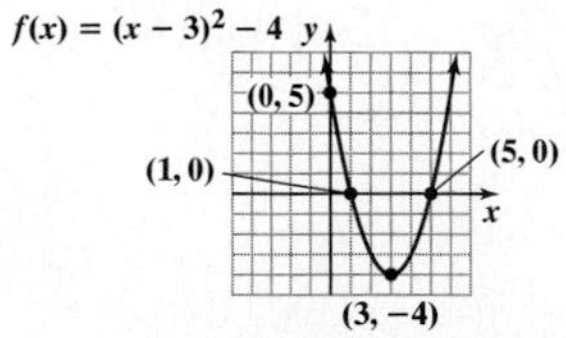

Domain $= \{x|x$ is a real number$\}$ or $(-\infty, \infty)$; Range $= \{y|y \geq -4\}$ or $[-4, \infty)$

18.

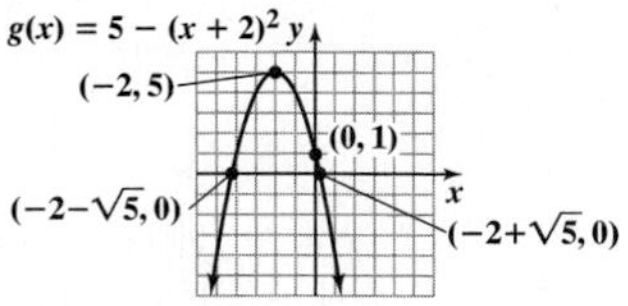

Domain $= \{x|x$ is a real number$\}$ or $(-\infty, \infty)$; Range $= \{y|y \leq 5\}$ or $(-\infty, 5]$

19.

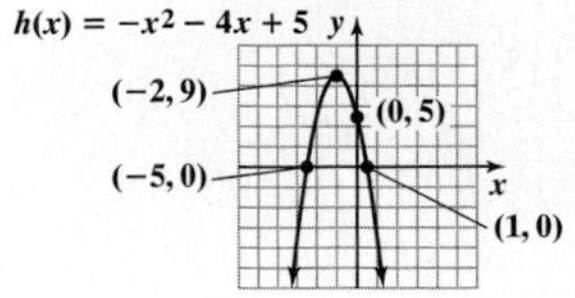

Domain $= \{x|x$ is a real number$\}$ or $(-\infty, \infty)$; Range $= \{y|y \leq 9\}$ or $(-\infty, 9]$

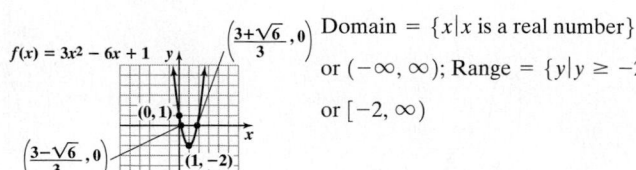

20.

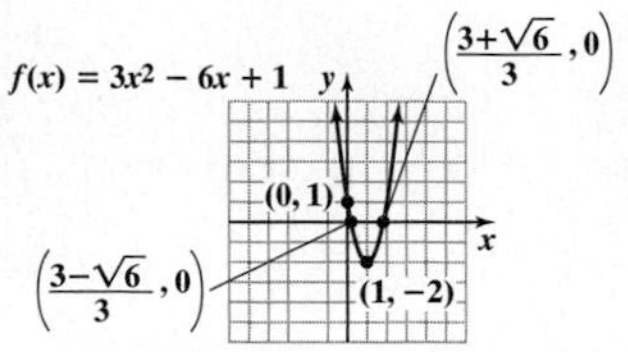

Domain $= \{x|x$ is a real number$\}$ or $(-\infty, \infty)$; Range $= \{y|y \geq -2\}$ or $[-2, \infty)$

21. two imaginary solutions **22.** two rational solutions **23.** $8x^2 - 2x - 3 = 0$ **24.** $x^2 - 12 = 0$ **25.** 75 cabinets per day; $1200

26. −9, −9; 81 **27.** 10 in.; 100 sq in.

Section 11.4

Check Point Exercises

1. $-4, 4, -1$, and 1, or $\{-4, 4, -1, 1\}$ **2.** 16 or $\{16\}$ **3.** $-\sqrt{6}, -1, 1$, and $\sqrt{6}$, or $\{-\sqrt{6}, -1, 1, \sqrt{6}\}$ **4.** -1 and 2, or $\{-1, 2\}$ **5.** $-\frac{1}{27}$ and 64, or $\left\{-\frac{1}{27}, 64\right\}$

Exercise Set 11.4

1. $\{-2, 2, -1, 1\}$ **3.** $\{-3, 3, -\sqrt{2}, \sqrt{2}\}$ **5.** $\{-2i, 2i, -\sqrt{2}, \sqrt{2}\}$ **7.** $\{1\}$ **9.** $\{49\}$ **11.** $\{25, 64\}$ **13.** $\{2, 12\}$ **15.** $\{-\sqrt{3}, 0, \sqrt{3}\}$ **17.** $\{-5, -2, -1, 2\}$ **19.** $\left\{-\frac{1}{4}, \frac{1}{5}\right\}$ **21.** $\left\{\frac{1}{3}, 2\right\}$ **23.** $\left\{\frac{-2 \pm \sqrt{7}}{3}\right\}$ **25.** $\{-8, 27\}$ **27.** $\{-243, 32\}$ **29.** $\{1\}$ **31.** $\{-8, -2, 1, 4\}$ **33.** $-2, 2, -1$, and 1; c **35.** 1; e **37.** 2 and 3; f **39.** $-5, -4, 1$, and 2 **41.** $-\frac{3}{2}$ and $-\frac{1}{3}$ **43.** $\frac{64}{15}$ and $\frac{81}{20}$ **45.** $\frac{5}{2}$ and $\frac{25}{6}$ **47.** ages 20 and 55; The function models the data well. **53.** $\{-1, 2\}$ **55.** $\{-1, 1, -3, 3\}$ **57.** $\left\{-\frac{3}{2}, 2\right\}$ **59.** $\{1\}$ **61.** b **63.** $\left\{\sqrt[3]{-2}, \frac{\sqrt[3]{225}}{5}\right\}$ **65.** $\frac{1}{5x - 1}$ **66.** $\frac{1}{2} + \frac{3}{2}i$ **67.** -5

Section 11.5

Check Point Exercises

1. $\{x|x < -4 \text{ or } x > 5\}$ or $(-\infty, -4) \cup (5, \infty)$

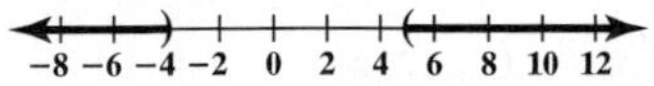

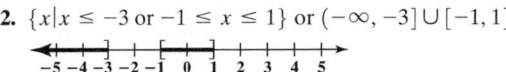

2. $\{x|x \le -3 \text{ or } -1 \le x \le 1\}$ or $(-\infty, -3] \cup [-1, 1]$

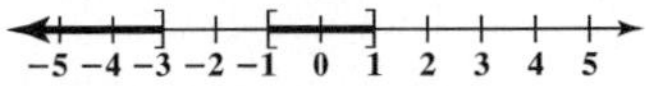

3. $\{x|x < -1 \text{ or } x \ge 1\}$ or $(-\infty, -1) \cup [1, \infty)$

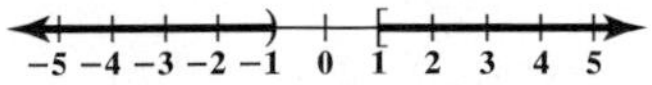

4. between 1 and 4 seconds, excluding $t = 1$ and $t = 4$

Exercise Set 11.5

1. $\{x|x < -2 \text{ or } x > 4\}$ or $(-\infty, -2) \cup (4, \infty)$

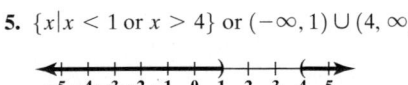

3. $\{x|-3 \le x \le 7\}$ or $[-3, 7]$

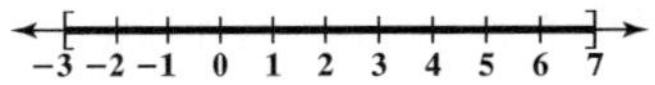

5. $\{x|x < 1 \text{ or } x > 4\}$ or $(-\infty, 1) \cup (4, \infty)$

7. $\{x|x < -4 \text{ or } x > -1\}$ or $(-\infty, -4) \cup (-1, \infty)$

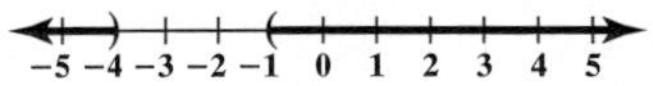

9. $\{x|2 \le x \le 4\}$ or $[2, 4]$

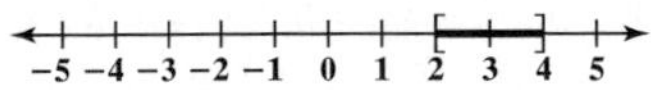

11. $\left\{x\middle|-4 \le x \le \frac{2}{3}\right\}$ or $\left[-4, \frac{2}{3}\right]$

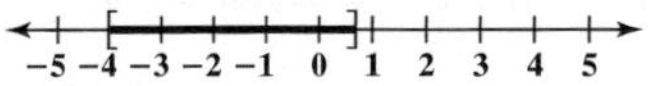

13. $\left\{x\middle|-3 < x < \frac{5}{2}\right\}$ or $\left(-3, \frac{5}{2}\right)$

15. $\left\{x\middle|-1 < x < -\frac{3}{4}\right\}$ or $\left(-1, -\frac{3}{4}\right)$

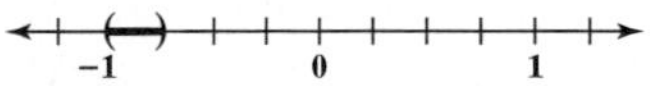

17. $\{x|x \le 0 \text{ or } x \ge 4\}$ or $(-\infty, 0] \cup [4, \infty)$

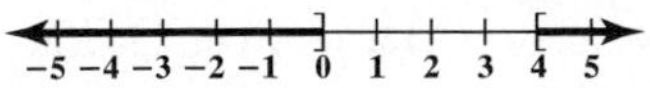

19. $\left\{x\middle|x < -\frac{3}{2} \text{ or } x > 0\right\}$ or $\left(-\infty, -\frac{3}{2}\right) \cup (0, \infty)$

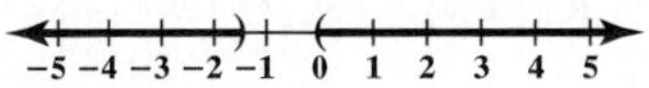

21. $\{x|0 \le x \le 1\}$ or $[0, 1]$

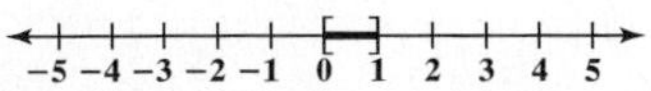

23. $\{x|2 - \sqrt{2} \le x \le 2 + \sqrt{2}\}$ or $[2 - \sqrt{2}, 2 + \sqrt{2}]$

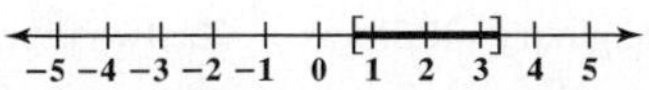

25. no solution or $\varnothing$

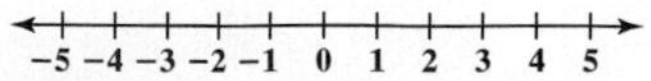

27. $\{x|1 \le x \le 2 \text{ or } x \ge 3\}$ or $[1, 2] \cup [3, \infty)$

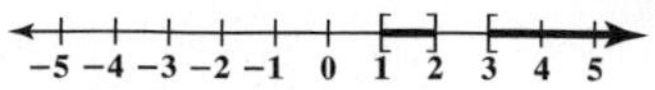

29. $\{x|-2 \le x \le -1 \text{ or } x \ge 1\}$ or $[-2, -1] \cup [1, \infty)$

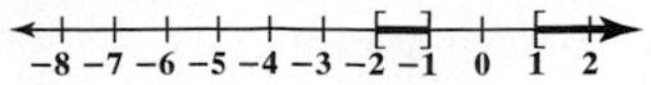

31. $\{x|x < -3\}$ or $(-\infty, -3)$

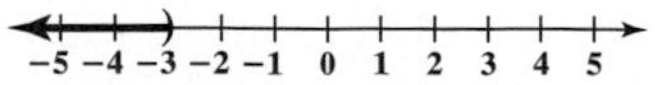

33. $\{x|x > -1\}$ or $(-1, \infty)$

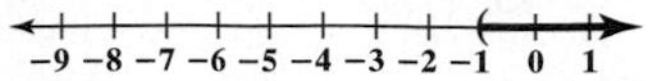

35. $\{x|x = 0 \text{ or } x \ge 9\}$ or $\{0\} \cup [9, \infty)$

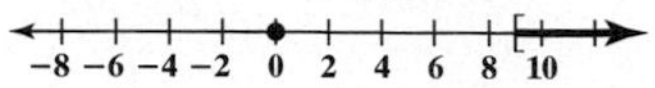

37. $\{x|x < -3 \text{ or } x > 4\}$ or $(-\infty, -3) \cup (4, \infty)$

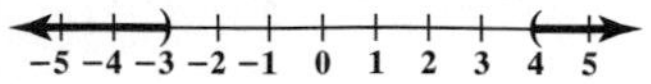

39. $\{x|-4 < x < -3\}$ or $(-4, -3)$

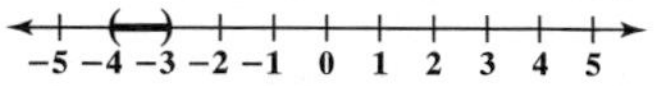

41. $\{x|2 \le x < 4\}$ or $[2, 4)$

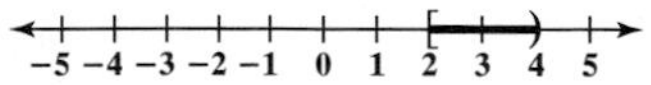

43. $\left\{x \middle| x < -\frac{4}{3} \text{ or } x \ge 2\right\}$ or $\left(-\infty, -\frac{4}{3}\right) \cup [2, \infty)$

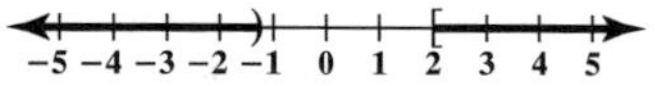

45. $\{x|x < 0 \text{ or } x > 3\}$ or $(-\infty, 0) \cup (3, \infty)$

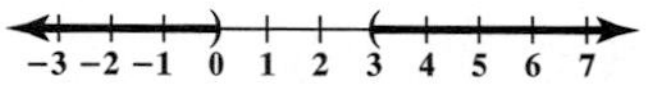

47. $\{x|x < -5 \text{ or } x > -3\}$ or $(-\infty, -5) \cup (-3, \infty)$

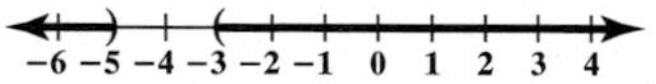

49. $\left\{x \middle| x < \frac{1}{2} \text{ or } x \ge \frac{7}{5}\right\}$ or $\left(-\infty, \frac{1}{2}\right) \cup \left[\frac{7}{5}, \infty\right)$

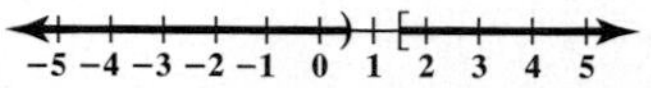

51. $\{x|x \le -6 \text{ or } x > -2\}$ or $(-\infty, -6] \cup (-2, \infty)$

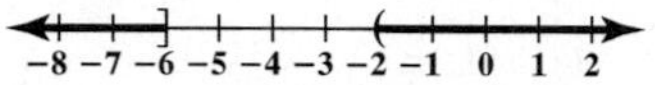

53. $\left\{x \middle| x \le \frac{1}{2} \text{ or } x \ge 2\right\}$ or $\left(-\infty, \frac{1}{2}\right] \cup [2, \infty)$

55. $\{x|-1 < x < 1\}$ or $(-1, 1)$

57. $\{x|x < -8 \text{ or } -6 < x < 4 \text{ or } x > 6\}$ or $(-\infty, -8) \cup (-6, 4) \cup (6, \infty)$

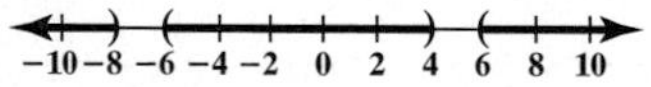

59. $\{x|-3 < x < 2\}$ or $(-3, 2)$

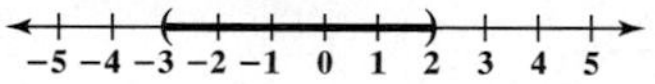

61. $\{x|x < -1 \text{ or } 1 < x < 2 \text{ or } x > 3\}$ or $(-\infty, -1) \cup (1, 2) \cup (3, \infty)$

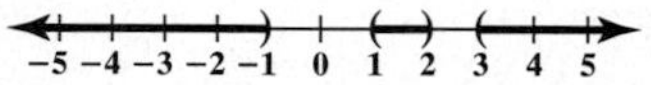

63. $\left\{x \middle| -6 \le x \le -\frac{1}{2} \text{ or } x \ge 1\right\}$ or $\left[-6, -\frac{1}{2}\right] \cup [1, \infty)$

65. $\{x|x < 2 \text{ or } -1 \le x < 2\}$ or $(-\infty, 2) \cup [-1, 2)$ **67.** between 0 and 3 seconds, excluding $t = 0$ and $t = 3$ **69.** f: \$379 billion; g: \$379.8 billion; The functions model the value in the graph well. **71.** after 2007 **73.** The company's production level must be at least 20,000 wheelchairs per month. For values of x greater than or equal to 20,000, the graph lies on or below the line $y = 425$. **75.** The length of the shorter side cannot exceed 6 ft. **81.** $\left\{x \middle| -3 \le x \le \frac{1}{2}\right\}$ or $\left[-3, \frac{1}{2}\right]$ **83.** $\{x|x < 3 \text{ or } x \ge 8\}$ or $(-\infty, 3) \cup [8, \infty)$ **85.** $\{x|-3 < x < -1 \text{ or } x > 2\}$ or $(-3, -1) \cup (2, \infty)$ **87.** Answers will vary; example: $x^2 - 2x - 15 \le 0$. **89.** $\{x|x < 2 \text{ or } x > 2\}$ or $(-\infty, 2) \cup (2, \infty)$ **91.** $\varnothing$ **93. a.** all real numbers **b.** empty set **95.** $\{x|-19 < x < 29\}$ or $(-19, 29)$ **96.** $\frac{2(x-1)}{(x+4)(x-3)}$ **97.** $(x^2 + 4y^2)(x + 2y)(x - 2y)$

Review Exercises

1. $\{\pm 8\}$ **2.** $\{\pm 5\sqrt{2}\}$ **3.** $\left\{\pm \frac{\sqrt{6}}{3}\right\}$ **4.** $\{4 \pm 3\sqrt{2}\}$ **5.** $\{-7 \pm 6i\}$ **6.** 100; $x^2 + 20x + 100 = (x + 10)^2$ **7.** $\frac{9}{4}$; $x^2 - 3x + \frac{9}{4} = \left(x - \frac{3}{2}\right)^2$ **8.** $\{3, 9\}$ **9.** $\left\{\frac{7 \pm \sqrt{53}}{2}\right\}$ **10.** $\left\{\frac{-3 \pm \sqrt{41}}{4}\right\}$ **11.** 8% **12.** 14 weeks **13.** $60\sqrt{5}$ m; 134.2 m **14.** 13 units **15.** $2\sqrt{2} \approx 2.83$ units **16.** $(-5, 5)$ **17.** $\left(-\frac{11}{2}, -2\right)$ **18.** $\{1 \pm \sqrt{5}\}$ **19.** $\{1 \pm 3\sqrt{2}i\}$ **20.** $\left\{\frac{-2 \pm \sqrt{10}}{2}\right\}$ **21.** two imaginary solutions **22.** two rational solutions

23. two irrational solutions **24.** $\left\{-\frac{2}{3}, 4\right\}$ **25.** $\{-5, 2\}$ **26.** $\left\{\frac{1 \pm \sqrt{21}}{10}\right\}$ **27.** $\{-4, 4\}$ **28.** $\{3 \pm 2\sqrt{2}\}$ **29.** $\left\{\frac{1}{6} \pm \frac{\sqrt{23}}{6}i\right\}$

30. $\{4 \pm \sqrt{5}\}$ **31.** $15x^2 - 4x - 3 = 0$ **32.** $x^2 + 81 = 0$ **33.** $x^2 - 48 = 0$ **34.** 2002 **35.** 8.8 sec

36.

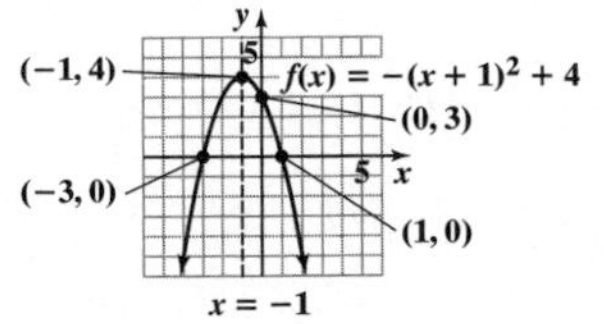

37.

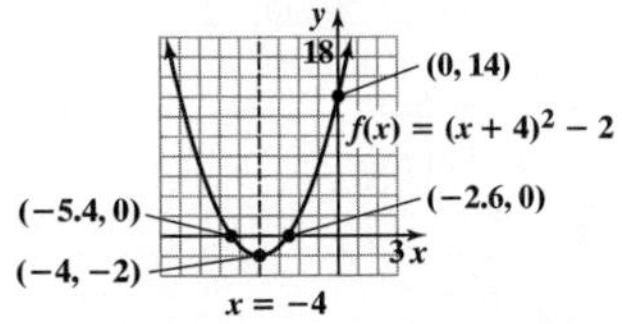

38.

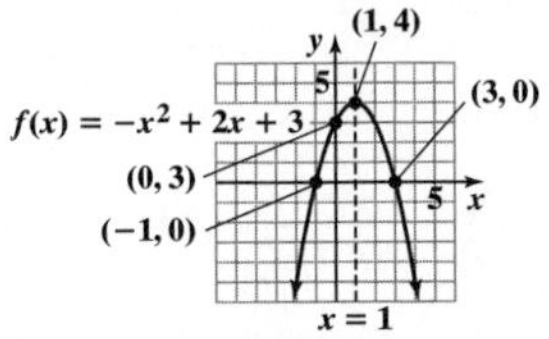

39.

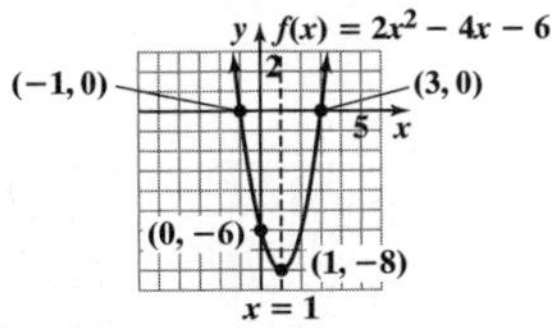

40. 25 in. of rainfall per year; 13.5 in. of growth **41.** 12.5 sec; 2540 feet **42.** (25, 5); The maximum divorce rate of about 5 per 1000 population was in 1985. **43.** 250 yd by 500 yd; 125,000 sq yard **44.** −7 and 7; −49 **45.** $\{-\sqrt{2}, \sqrt{2}, -2, 2\}$

46. $\{1\}$ **47.** $\{-5, -1, 3\}$ **48.** $\left\{-\frac{1}{8}, \frac{1}{7}\right\}$ **49.** $\{-27, 64\}$ **50.** $\{16\}$

51. $\left\{x \middle| -3 < x < \frac{1}{2}\right\}$ or $\left(-3, \frac{1}{2}\right)$;

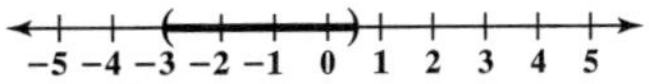

52. $\left\{x \middle| x \le -4 \text{ or } x \ge -\frac{1}{2}\right\}$ or $(-\infty, -4] \cup \left[-\frac{1}{2}, \infty\right)$;

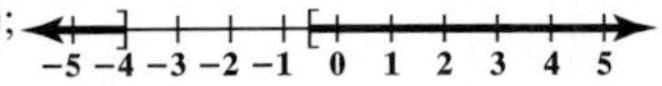

53. $\{x|-3 < x < 0 \text{ or } x > 1\}$ or $(-3, 0) \cup (1, \infty)$;

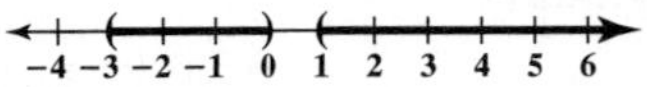

54. $\{x|x < -2 \text{ or } x > 6\}$ or $(-\infty, -2) \cup (6, \infty)$;

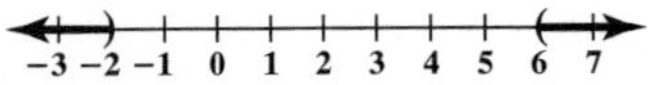

55. $\left\{x \middle| x < 4 \text{ or } x \ge \frac{23}{4}\right\}$ or $(-\infty, 4) \cup \left[\frac{23}{4}, \infty\right)$;

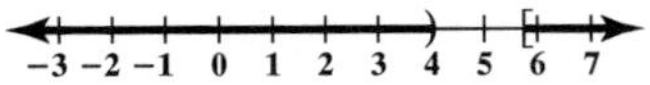

56. between 1 and 2 seconds, excluding $t = 1$ and $t = 2$
57. a. 200 beats per minute **b.** between 0 and 4 minutes and more than 12 minutes after the workout; between 0 and 4 minutes; Answers will vary.

Chapter 11 Test

1. $\pm\frac{\sqrt{10}}{2}$ or $\left\{\pm\frac{\sqrt{10}}{2}\right\}$ **2.** $3 \pm 2\sqrt{5}$ or $\{3 \pm 2\sqrt{5}\}$ **3.** 64; $x^2 - 16x + 64 = (x - 8)^2$ **4.** $\frac{1}{25}$; $x^2 + \frac{2}{5}x + \frac{1}{25} = \left(x + \frac{1}{5}\right)^2$ **5.** $3 \pm \sqrt{2}$ or $\{3 \pm \sqrt{2}\}$

6. $50\sqrt{2}$ ft **7.** $\sqrt{73}$ or 8.54 units **8.** $\left(\frac{7}{2}, -4\right)$ **9.** two irrational solutions **10.** two imaginary solutions **11.** −5 and $\frac{1}{2}$, or $\left\{-5, \frac{1}{2}\right\}$

12. $-4 \pm \sqrt{11}$ or $\{-4 \pm \sqrt{11}\}$ **13.** $-2 \pm 5i$ or $\{-2 \pm 5i\}$ **14.** $\frac{3}{2} \pm \frac{1}{2}i$ or $\left\{\frac{3}{2} \pm \frac{1}{2}i\right\}$ **15.** $x^2 - 4x - 21 = 0$ **16.** $x^2 + 100 = 0$

17. 1990 and 1998

18.

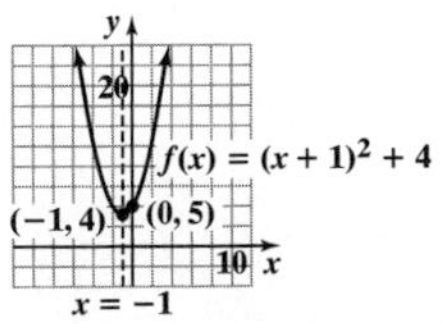

19.

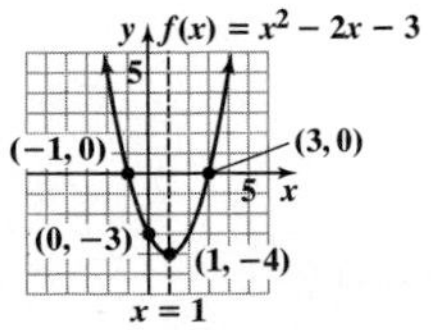

20. after 2 sec; 69 ft **21.** 4.1 sec **22.** 23 computers; \$169 hundreds or \$16,900
23. 1 and 2, or $\{1, 2\}$ **24.** −3, 3, −2, and 2, or $\{-3, 3, -2, 2\}$
25. 1 and 512, or $\{1, 512\}$
26. $\{x|-3 < x < 4\}$ or $(-3, 4)$

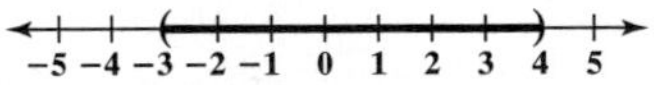

27. $\{x|x < 3 \text{ or } x \ge 10\}$ or $(-\infty, 3) \cup [10, \infty)$

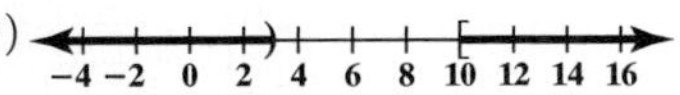

Cumulative Review Exercises (Chapters 1–11)

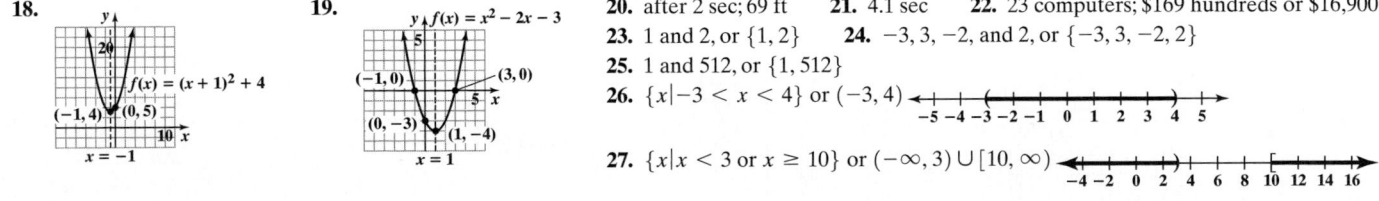

1. $\{4\}$ **2.** $\{(2, 3)\}$ **3.** $\{(2, 0, 3)\}$ **4.** $\{x|x < -2 \text{ or } x > 4\}$ or $(-\infty, -2) \cup (4, \infty)$ **5.** $\{5\}$ **6.** $\{x|x \ge 4\}$ or $[4, \infty)$ **7.** $\left\{\frac{3}{4} \pm \frac{\sqrt{7}}{4}i\right\}$

8. 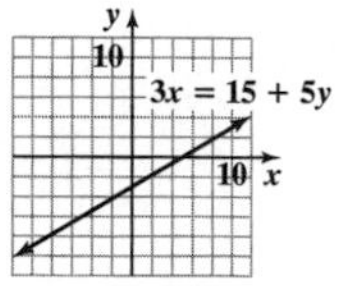

9. 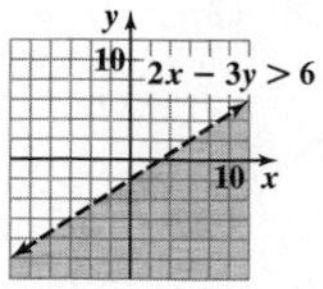

10. 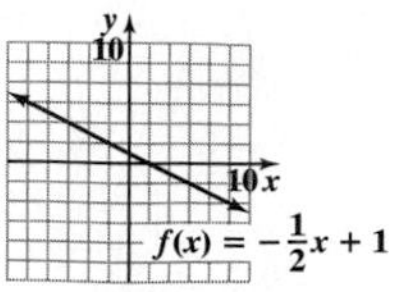

11. 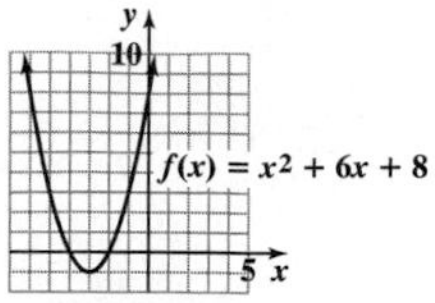

12. 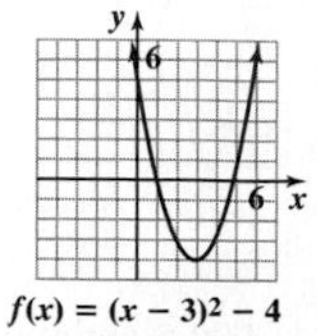

13. $f^{-1}(x) = x^3 - 1$ **14.** $c = \dfrac{Ad}{d - A}$ **15.** $y = \dfrac{1}{2}x + 5$ or $f(x) = \dfrac{1}{2}x + 5$ **16.** $-\dfrac{y^9}{3x}$ **17.** $16x^4 - 40x^2y + 25y^2$ **18.** $x^2 - 5x + 1 + \dfrac{8}{5x + 1}$ **19.** $2y^2\sqrt[3]{2y^2}$ **20.** $\dfrac{4x - 13}{(x - 3)(x - 4)}$ **21.** $(x - 4)(x + 2)(x^2 - 2x + 4)$ **22.** $2(x + 3y)^2$ **23.** length: 12 ft; width: 4 ft **24.** $\dfrac{6}{5}$ hr or 1 hr and 12 min **25.** 5 mph

CHAPTER 12

Section 12.1

Check Point Exercises

1. one **2.** 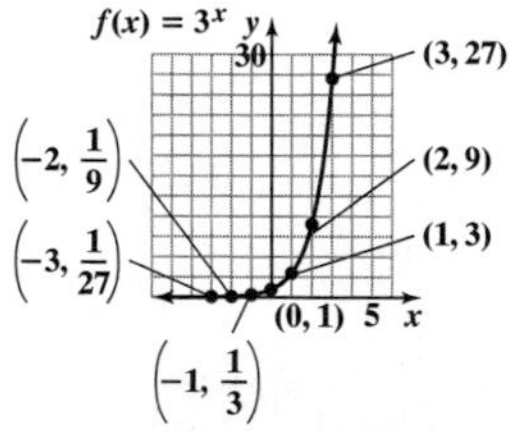

3. 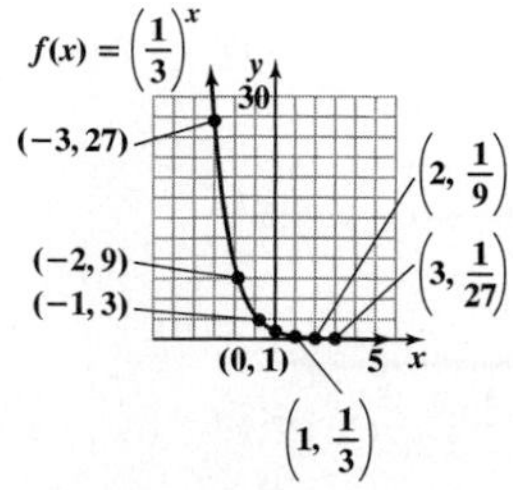

4. 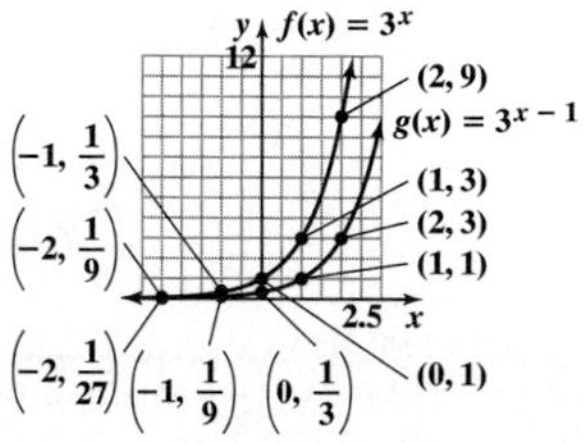

The graph of g is the graph of f shifted 1 unit to the right.

5. 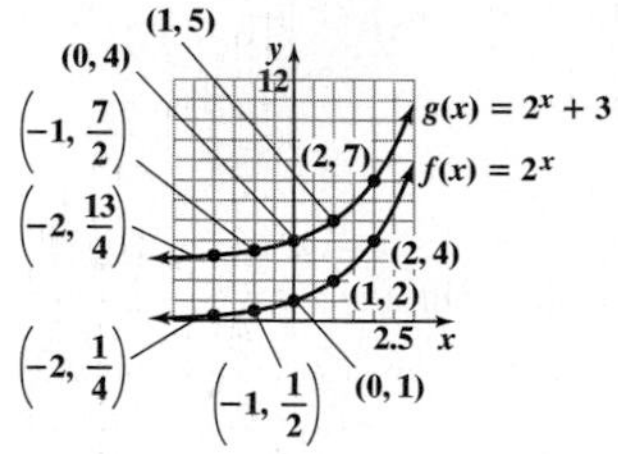

The graph of g is the graph of f shifted up 3 units.

6. approximately 11.37 billion
7. a. $14,859.47
b. $14,918.25

Exercise Set 12.1

1. 10.556 **3.** 11.665 **5.** 0.125 **7.** 9.974 **9.** 0.387

11. ; d

x	$f(x)$
−2	$\frac{1}{9}$
−1	$\frac{1}{3}$
0	1
1	3
2	9

13. ; e

x	$f(x)$
−2	$-\frac{8}{9}$
−1	$-\frac{2}{3}$
0	0
1	2
2	8

15. ; f

x	$f(x)$
−2	9
−1	3
0	1
1	$\frac{1}{3}$
2	$\frac{1}{9}$

17.

x	$f(x)$
−2	$\frac{1}{16}$
−1	$\frac{1}{4}$
0	1
1	4
2	16

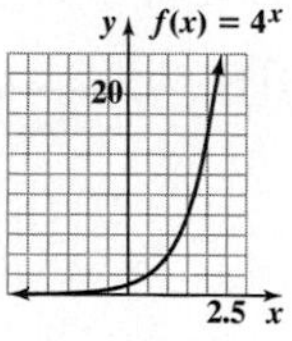

19.

x	$g(x)$
−2	$\frac{4}{9}$
−1	$\frac{2}{3}$
0	1
1	$\frac{3}{2}$
2	$\frac{9}{4}$

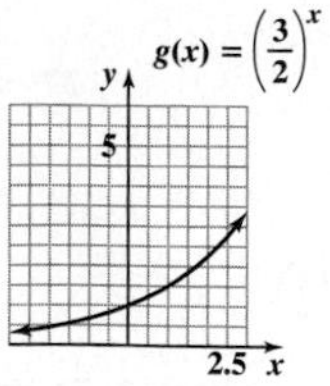

21.

x	$h(x)$
-2	4
-1	2
0	1
1	$\frac{1}{2}$
2	$\frac{1}{4}$

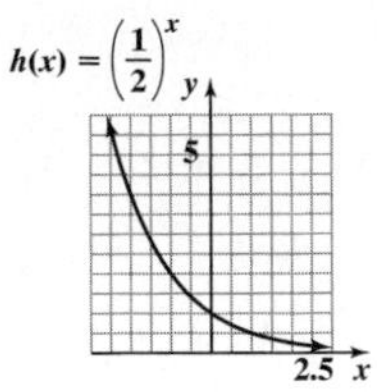

23.

x	$f(x)$
-2	2.78
-1	1.67
0	1
1	0.6
2	0.36

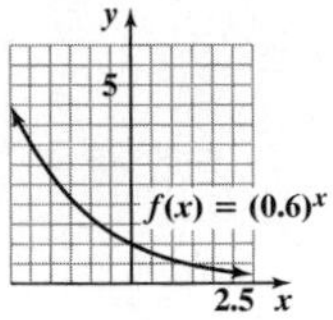

25.

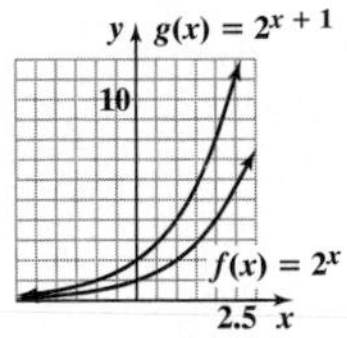

The graph of g is the graph of f shifted 1 unit to the left.

27.

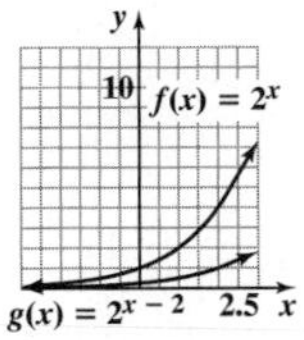

The graph of g is the graph of f shifted 2 units to the right.

29.

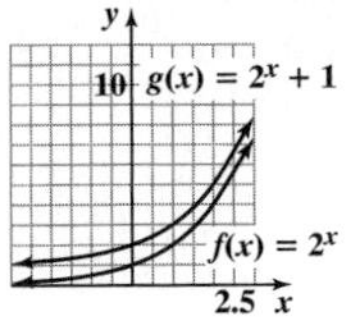

The graph of g is the graph of f shifted up 1 unit.

31.

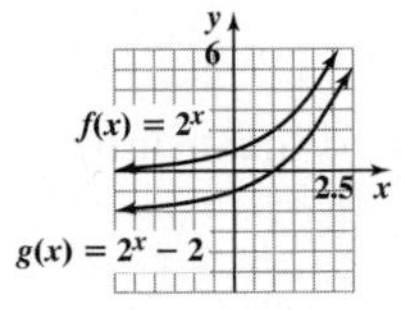

The graph of g is the graph of f shifted down 2 units.

33.

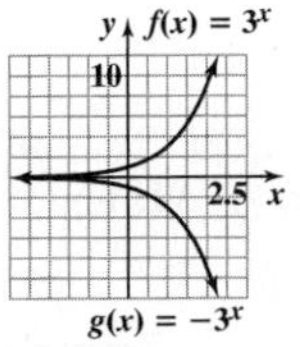

The graph of g is a reflection of the graph of f across the x-axis.

35.

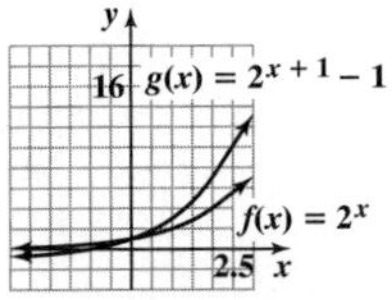

The graph of g is the graph of f shifted 1 unit to the left and 1 unit down.

37.

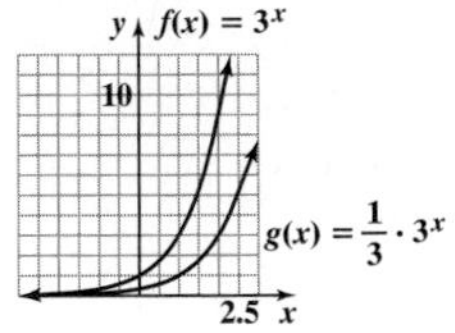

The graph of g is the graph of f stretched vertically by a factor of $\frac{1}{3}$.

39. **a.** \$13,116.51 **b.** \$13,157.04 **c.** \$13,165.31
41. 7% compounded monthly
43. Domain $= \{x|x$ is a real number$\}$ or $(-\infty, \infty)$; Range $= \{y|y > -2\}$ or $(-2, \infty)$
45. Domain $= \{x|x$ is a real number$\}$ or $(-\infty, \infty)$; Range $= \{y|y > 1\}$ or $(1, \infty)$
47. Domain $= \{x|x$ is a real number$\}$ or $(-\infty, \infty)$; Range $= \{y|y > 0\}$ or $(0, \infty)$

49. $(0, 1)$

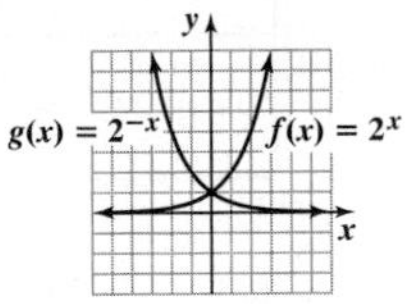

51.

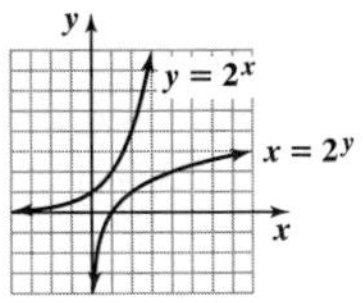

53. **a.** 574 million **b.** 1148 million **c.** 2295 million **d.** 4590 million **e.** It appears to double. **55.** \$116,405 **57.** 124.2 million; fairly well **59.** **a.** 100% **b.** about 68.5% **c.** about 30.8% **d.** about 20% **61.** 11.3; About 11.3% of 30-year-olds have some coronary heart disease. **63.** **a.** about 1429 people **b.** about 24,546 people **c.** The number of ill people cannot exceed the population.; The asymptote indicates that the number of ill people will not exceed 30,000, the population of the town.

71. **a.** $f(t) = 10{,}000\left(1 + \frac{0.05}{4}\right)^{4t}$; $f(t) = 10{,}000\left(1 + \frac{0.045}{12}\right)^{12t}$

b.

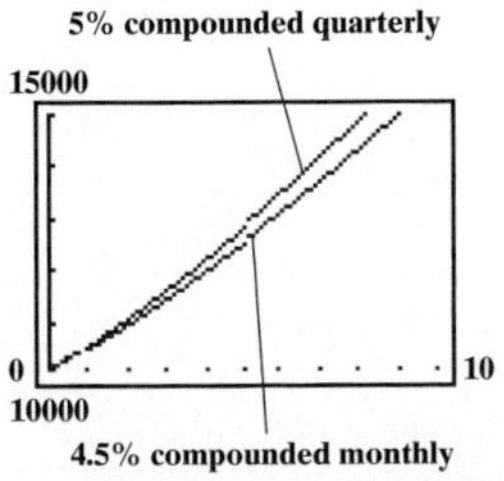

The bank that pays 5% interest compounded quarterly

73. d

75. $(\cosh x)^2 - (\sinh x)^2 = \left(\frac{e^x + e^{-x}}{2}\right)^2 - \left(\frac{e^x - e^{-x}}{2}\right)^2$

$$= \frac{e^{2x} + 2e^xe^{-x} + e^{-2x}}{4} - \left(\frac{e^{2x} - 2e^xe^{-x} + e^{-2x}}{4}\right) = \frac{4e^xe^{-x}}{4} = \frac{4}{4} = 1$$

76. $b = \dfrac{Da}{a - D}$

77. $\dfrac{11}{(x-3)(x-4)}$

78. $\{-2, 5\}$

Section 12.2

Check Point Exercises

1. a. $7^3 = x$ **b.** $b^2 = 25$ **c.** $4^y = 26$ **2. a.** $5 = \log_2 x$ **b.** $3 = \log_b 27$ **c.** $y = \log_e 33$ **3. a.** 2 **b.** 1 **c.** $\frac{1}{2}$ **4. a.** 1 **b.** 0

5. a. 8 **b.** 17 **6.**

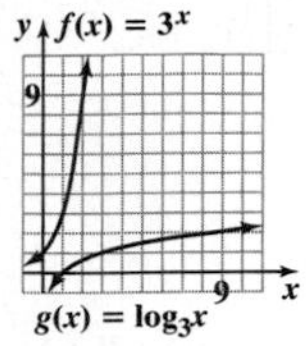

7. $\{x|x > 5\}$ or $(5, \infty)$ **8.** approximately 80% **9.** 4

10. a. $\{x|x < 4\}$ or $(-\infty, 4)$ **b.** $\{x|x \neq 0\}$ or $(-\infty, 0) \cup (0, \infty)$

11. a. $25x$ **b.** $\sqrt{x}$

12. 1486 thousand or 1,486,000

Exercise Set 12.2

1. $2^4 = 16$ **3.** $3^2 = x$ **5.** $b^5 = 32$ **7.** $6^y = 216$ **9.** $\log_2 8 = 3$ **11.** $\log_2 \frac{1}{16} = -4$ **13.** $\log_8 2 = \frac{1}{3}$ **15.** $\log_{13} x = 2$

17. $\log_b 1000 = 3$ **19.** $\log_7 200 = y$ **21.** 2 **23.** 6 **25.** -1 **27.** -3 **29.** $\frac{1}{2}$ **31.** $-\frac{1}{2}$ **33.** $\frac{1}{2}$ **35.** 1 **37.** 0 **39.** 7 **41.** 19

43.

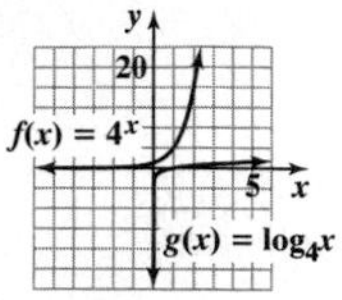

45.

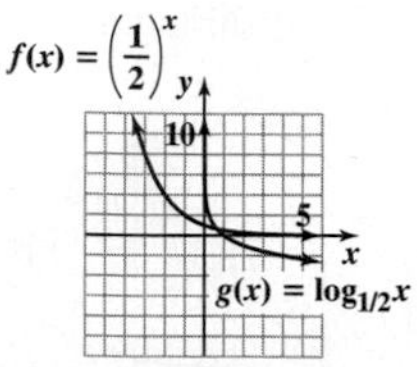

47. $\{x|x > -4\}$ or $(-4, \infty)$ **49.** $\{x|x < 2\}$ or $(-\infty, 2)$ **51.** $\{x|x \neq 2\}$ or $(-\infty, 2) \cup (2, \infty)$ **53.** 2 **55.** 7 **57.** 33 **59.** 0 **61.** 6 **63.** -6 **65.** 125 **67.** $9x$ **69.** $5x^2$ **71.** $\sqrt{x}$ **73.** $3^2 = x - 1$; 10 or $\{10\}$ **75.** $4^{-3} = x$; $\frac{1}{64}$ or $\left\{\frac{1}{64}\right\}$ **77.** 0 **79.** 2 **81.** d **83.** c **85.** b **87.** approximately 95.4% **89.** 65%; extremely well (gives the actual percent) **91.** approximately 188 decibels; yes

93. a. 88
b. 71.5; 63.9; 58.8; 55.0; 52.0; 49.5
c.

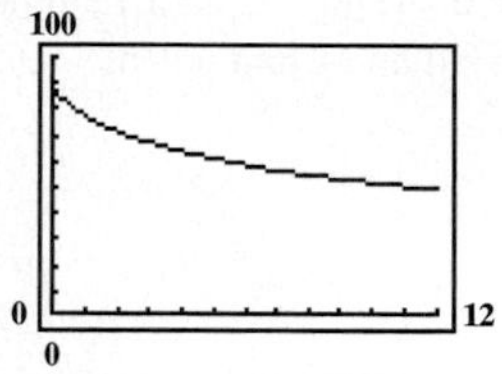

The students remembered less of the material over time.

103.

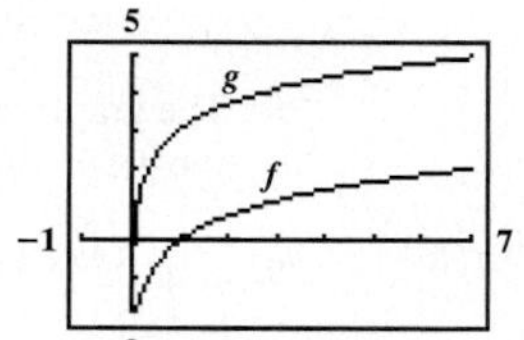

The graph of g is the graph of f shifted up 3 units.

105.

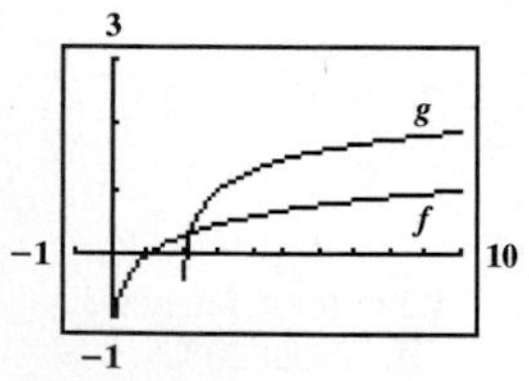

The graph of g is the graph of f shifted 2 units to the right and 1 unit up.

107. a.

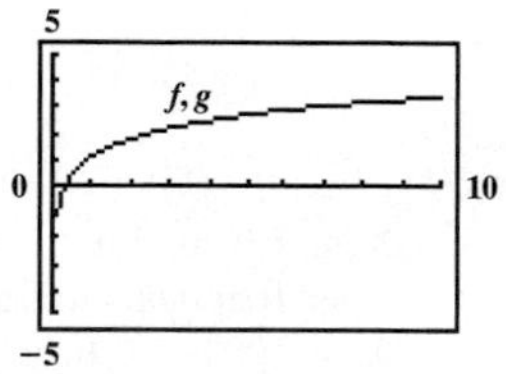

b.

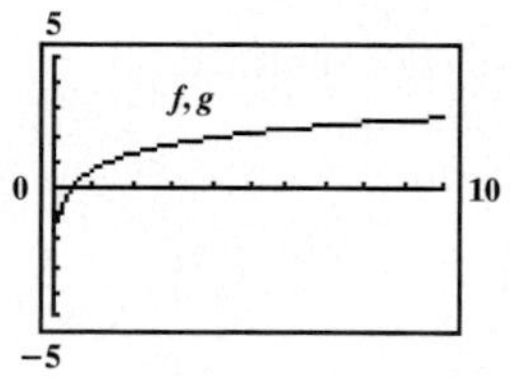

c.

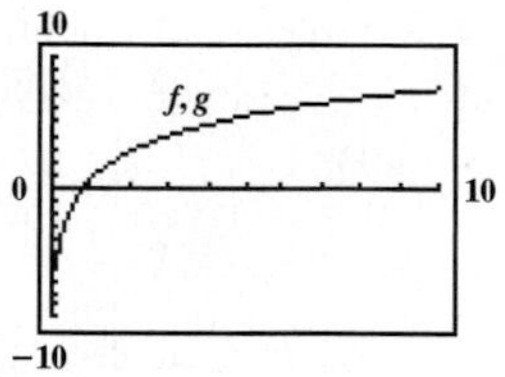

d. In each case, the graphs of f and g are the same.; $\log_b(MN) = \log_b M + \log_b N$
e. the sum of the logarithms of the factors

109. d **111.** 0 **113.** $\{(-2, 3)\}$ **114.** $2(3x - y)(x - y)$ **115.** $\{x|x \leq -7 \text{ or } x \geq -2\}$ or $(-\infty, -7] \cup [-2, \infty)$

Section 12.3

Check Point Exercises

1. a. $\log_6 7 + \log_6 11$ **b.** $2 + \log x$ **2. a.** $\log_8 23 - \log_8 x$ **b.** $5 - \ln 11$ **3. a.** $9 \log_6 8$ **b.** $\frac{1}{3} \ln x$

4. a. $4 \log_b x + \frac{1}{3} \log_b y$ **b.** $\frac{1}{2} \log_5 x - 2 - 3 \log_5 y$ **5. a.** $\log 100 = 2$ **b.** $\log\left(\frac{7x + 6}{x}\right)$

6. a. $\ln (x^2 \sqrt[3]{x + 5})$ **b.** $\log\left[\frac{(x - 3)^2}{x}\right]$ **c.** $\log_b\left(\frac{\sqrt[4]{x}\, y^{10}}{25}\right)$ **7.** $\frac{\log 2506}{\log 7} \approx 4.02$ **8.** $\frac{\ln 2506}{\ln 7} \approx 4.02$

Exercise Set 12.3

1. $\log_5 7 + \log_5 3$ **3.** $1 + \log_7 x$ **5.** $3 + \log x$ **7.** $1 - \log_7 x$ **9.** $\log x - 2$ **11.** $3 - \log_4 y$ **13.** $2 - \ln 5$ **15.** $3 \log_b x$ **17.** $-6 \log N$ **19.** $\frac{1}{5} \ln x$ **21.** $2 \log_b x + \log_b y$ **23.** $\frac{1}{2} \log_4 x - 3$ **25.** $2 - \frac{1}{2} \log_6(x + 1)$ **27.** $2 \log_b x - \log_b y - 2 \log_b z$ **29.** $1 + \frac{1}{2} \log x$ **31.** $\frac{1}{3} \log x - \frac{1}{3} \log y$ **33.** $\frac{1}{2} \log_b x + 3 \log_b y - 3 \log_b z$ **35.** $\frac{2}{3} \log_5 x + \frac{1}{3} \log_5 y - \frac{2}{3}$ **37.** $\log 10 = 1$ **39.** $\ln (7x)$ **41.** $\log_2 32 = 5$ **43.** $\log\left(\frac{2x + 5}{x}\right)$ **45.** $\log (xy^3)$ **47.** $\ln (y\sqrt{x})$ **49.** $\log_b (x^2y^3)$ **51.** $\ln\left(\frac{x^5}{y^2}\right)$ **53.** $\ln\left(\frac{x^3}{\sqrt[3]{y}}\right)$ **55.** $\ln\left[\frac{(x + 6)^4}{x^3}\right]$ **57.** $\ln\left(\frac{x^3y^5}{z^6}\right)$ **59.** $\log_5\left[\frac{\sqrt{xy}}{(x + 1)^2}\right]$ **61.** 1.5937 **63.** 1.6944 **65.** −1.2304 **67.** 3.6193 **69.** $C - A$ **71.** $3A$ **73.** $\frac{1}{2}A - \frac{3}{2}C$ **75.** 35 or {35} **77.** 2 or {2} **79.** $(f - g)(x) = \log [7x(x - 1)]$ **81. a.** $D = 10 \log\left(\frac{I}{I_0}\right)$ **b.** 20 decibels

91. a. & b.

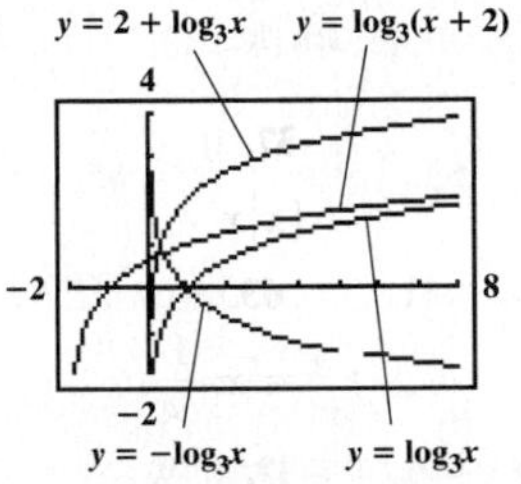

$y = \log_3 x$ is shifted up 2 units to obtain $y = 2 + \log_3 x$, $y = \log_3 x$ is shifted to the left 2 units to obtain $y = \log_3 (x + 2)$, and $y = \log_3 x$ is reflected across the x-axis to obtain $y = -\log_3 x$.

93.

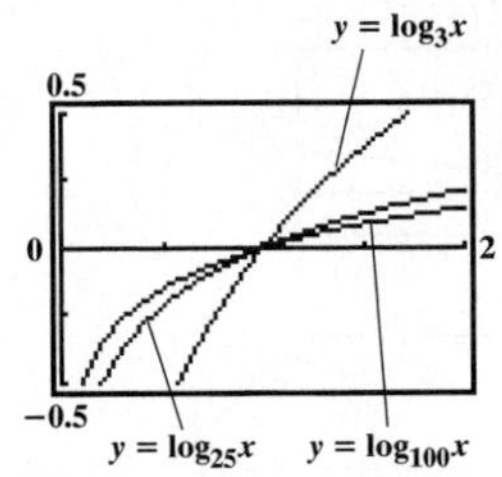

a. $y = \log_{100} x$ is on the top and $y = \log_3 x$ is on the bottom.
b. $y = \log_3 x$ is on the top and $y = \log_{100} x$ is on the bottom.
c. If $y = \log_b x$ is graphed for two different values of b, the graph of the one with the larger base will be on top in the interval $(0, 1)$ and the one with the smaller base will be on top in the interval $(1, \infty)$.

99. d **101.** $\frac{2A}{B}$ **103.**

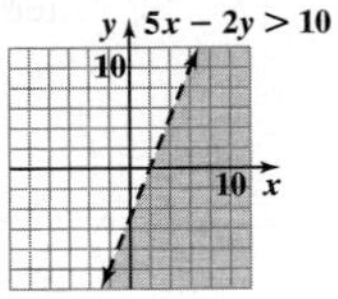

104. $\{x|x < 1\}$ or $(-\infty, 1)$ **105.** $2y\sqrt[3]{xy^2}$

Mid-Chapter 12 Check Point

1.

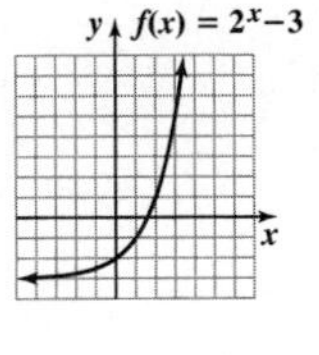

Domain: $\{x|x$ is a real number$\}$ or $(-\infty, \infty)$; Range: $\{y|y > -3\}$ or $(-3, \infty)$

2.

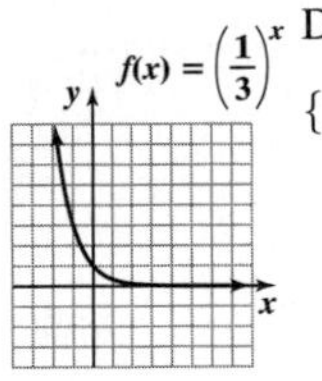

Domain: $\{x|x$ is a real number$\}$ or $(-\infty, \infty)$; Range: $\{y|y > 0\}$ or $(0, \infty)$

3.

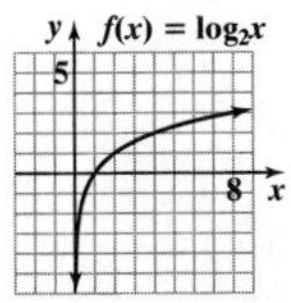

Domain: $\{x|x > 0\}$ or $(0, \infty)$; Range: $\{y|y$ is a real number$\}$ or $(-\infty, \infty)$

4.

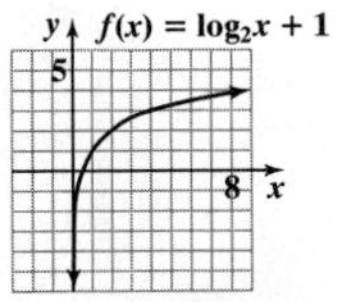

Domain: $\{x|x > 0\}$ or $(0, \infty)$; Range: $\{y|y$ is a real number$\}$ or $(-\infty, \infty)$

5. $\{x|x > -6\}$ or $(-6, \infty)$ **6.** $\{x|x > 0\}$ or $(0, \infty)$ **7.** $\{x|x \neq -6\}$ or $(-\infty, -6) \cup (-6, \infty)$ **8.** $\{x|x$ is a real number$\}$ or $(-\infty, \infty)$ **9.** 5 **10.** −2 **11.** $\frac{1}{2}$ **12.** $\frac{1}{3}$ **13.** 2 **14.** Evaluation not possible; $\log_2 \frac{1}{8} = -3$ and $\log_3(-3)$ is undefined. **15.** 5 **16.** $\sqrt{7}$ **17.** 13 **18.** $-\frac{1}{2}$ **19.** $\sqrt{\pi}$ **20.** $\frac{1}{2} \log x + \frac{1}{2} \log y - 3$ **21.** $19 + 20 \ln x$ **22.** $\log_7\left(\frac{x^8}{\sqrt[3]{y}}\right)$ **23.** $\log_5 x^9$ **24.** $\ln\left[\frac{\sqrt{x}}{y^3(z - 2)}\right]$ **25.** $8

Section 12.4

Check Point Exercises

1. a. 3 or $\{3\}$ **b.** $\frac{5}{2}$ or $\left\{\frac{5}{2}\right\}$ **2.** $\frac{\ln 134}{\ln 5} \approx 3.04$ or $\left\{\frac{\ln 134}{\ln 5} \approx 3.04\right\}$ **3.** $\frac{\ln 9}{2} = \ln 3 \approx 1.10$ or $\{\ln 3 \approx 1.10\}$ **4.** 12 or $\{12\}$

5. 5 or $\{5\}$ **6.** $\frac{1}{24}$ or $\left\{\frac{1}{24}\right\}$ **7.** $\frac{e^2}{3}$ or $\left\{\frac{e^2}{3}\right\}$ **8.** blood alcohol concentration of 0.01 **9.** 16.2 years **10.** 2010

Exercise Set 12.4

1. $\{6\}$ **3.** $\{3\}$ **5.** $\{3\}$ **7.** $\{2\}$ **9.** $\left\{\frac{3}{5}\right\}$ **11.** $\left\{\frac{3}{2}\right\}$ **13.** $\{4\}$ **15.** $\{5\}$ **17.** $\left\{-\frac{1}{4}\right\}$ **19.** $\left\{\frac{\ln 3.91}{\ln 10} \approx 0.59\right\}$

21. $\{\ln 5.7 \approx 1.74\}$ **23.** $\left\{\frac{\ln 17}{\ln 5} \approx 1.76\right\}$ **25.** $\{\ln 5 \approx 1.61\}$ **27.** $\left\{\frac{\ln 659}{5} \approx 1.30\right\}$ **29.** $\left\{\frac{\ln 13}{0.7} \approx 3.66\right\}$ **31.** $\left\{\frac{\ln 3}{0.055} \approx 19.97\right\}$

33. $\left\{\frac{\ln 30}{\ln 1.4} \approx 10.11\right\}$ **35.** $\left\{\frac{1 - \ln 793}{5} \approx -1.14\right\}$ **37.** $\left\{\frac{\ln 410}{\ln 7} - 2 \approx 1.09\right\}$ **39.** $\left\{\frac{\ln 2}{\ln 5 - \ln 2} \approx 0.76\right\}$ **41.** $\{81\}$ **43.** $\left\{\frac{1}{16}\right\}$

45. $\{3\}$ **47.** $\{100\}$ **49.** $\{59\}$ **51.** $\left\{\frac{109}{27}\right\}$ **53.** $\left\{\frac{62}{3}\right\}$ **55.** $\left\{\frac{5}{4}\right\}$ **57.** $\{6\}$ **59.** $\{6\}$ **61.** $\varnothing$ **63.** $\{e^2 \approx 7.39\}$ **65.** $\{e^{-3} \approx 0.05\}$

67. $\left\{\frac{e^4}{2} \approx 27.30\right\}$ **69.** $\{e^{-1/2} \approx 0.61\}$ **71.** $\{e^2 - 3 \approx 4.39\}$ **73.** $\left\{\frac{1}{e - 1} \approx 0.58\right\}$ **75.** $\left\{\frac{1}{2}\right\}$ **77.** $\{5\}$ **79.** $\left\{\pm\sqrt{\frac{\ln 45}{\ln 3}} \approx \pm 1.86\right\}$

81. $\{12\}$ **83.** $\{-2, 6\}$ **85.** a blood alcohol concentration of approximately 0.22 **87. a.** 18.9 million **b.** 2006
89. 8.2 **91.** 16.8% **93.** 8.7 **95.** 15.7% **97.** 2002 **99.** about 2.8 days; (2.8, 50) **101.** $10^{-2.4}$ or 0.004 mole per liter

109. 2 or $\{2\}$

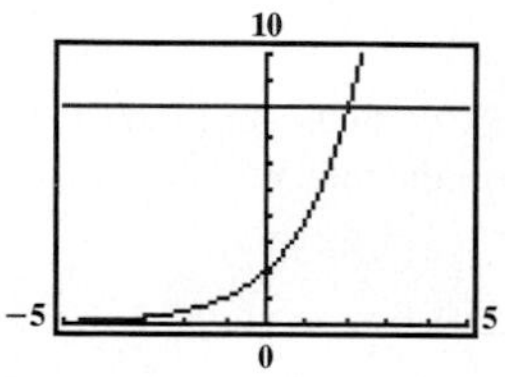

111. 4 or $\{4\}$

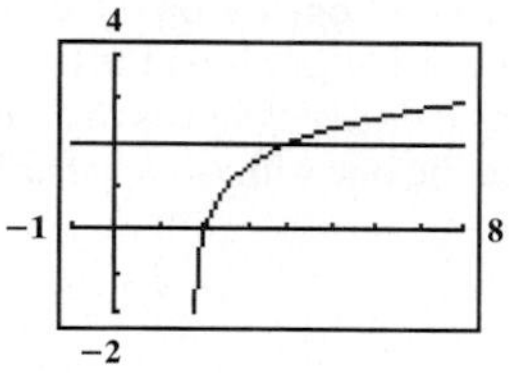

113. 2 or $\{2\}$

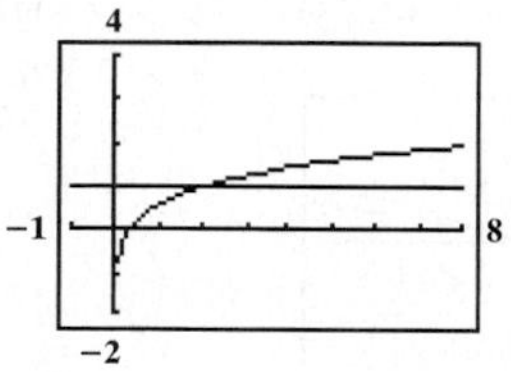

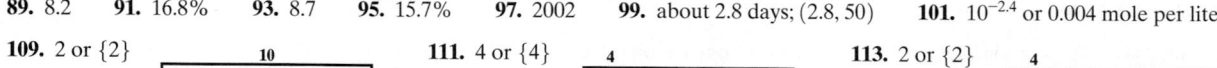

115. -1.39 and 1.69, or $\{-1.39, 1.69\}$

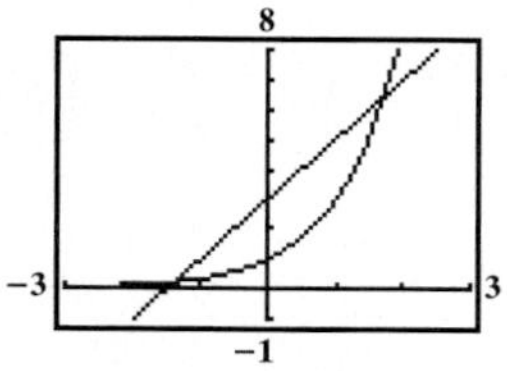

117.

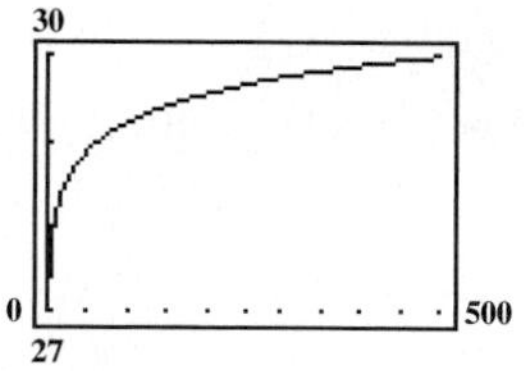

The barometric air pressure increases as the distance from the eye increases.

119.

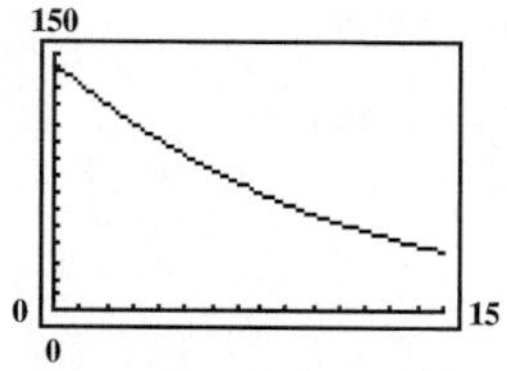

The runner's pulse will be 70 beats per minute after about 7.9 minutes.

121. c **123.** 1 and e^2, or $\{1, e^2\}$ **125.** e or $\{e\}$ **126.** $\{5\}$ **127.** $\{-12\}$ **128.** $\frac{y^8}{16x^{12}}$

Section 12.5

Check Point Exercises

1. a. $A = 643e^{0.023t}$ **b.** 2039 **2. a.** $A = A_0e^{-0.0248t}$ **b.** about 72.2 years

3.

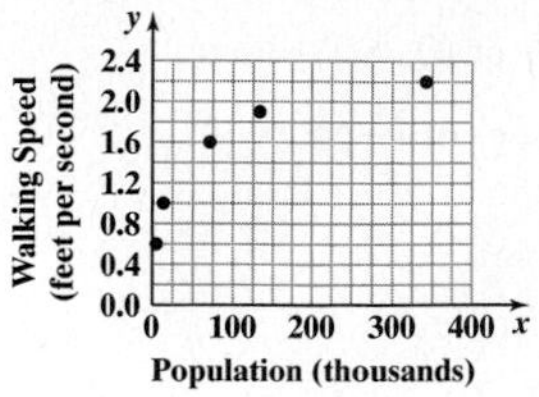

logarithmic function

4.

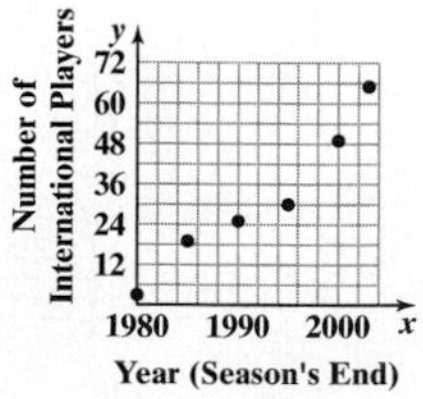

exponential function **5.** $y = 4e^{(\ln 7.8)x}$; $y = 4e^{2.054x}$

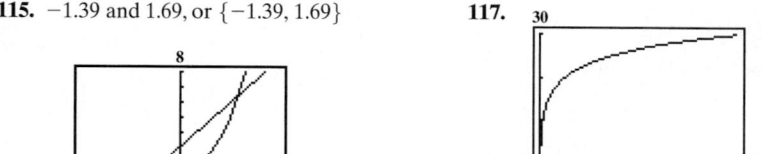

Exercise Set 12.5

1. 127.2 million **3.** Iraq; 2.8% **5.** 2014 **7. a.** $A = 6.04e^{0.01t}$ **b.** 2040 **9.** approximately 8 grams
11. 8 grams after 10 seconds; 4 grams after 20 seconds; 2 grams after 30 seconds; 1 gram after 40 seconds; 0.5 gram after 50 seconds

13. approximately 15,679 years old **15. a.** $\frac{1}{2} = e^{1.31k}$ yields $k = \dfrac{\ln\left(\frac{1}{2}\right)}{1.31} \approx -0.52912.$ **b.** about 0.1069 billion or 106,900,000 years old

17. $2A_0 = A_0e^{kt}$; $2 = e^{kt}$; $\ln 2 = \ln e^{kt}$; $\ln 2 = kt$; $\dfrac{\ln 2}{k} = t$ **19. a.** 0.7% **b.** about 99 years

21.

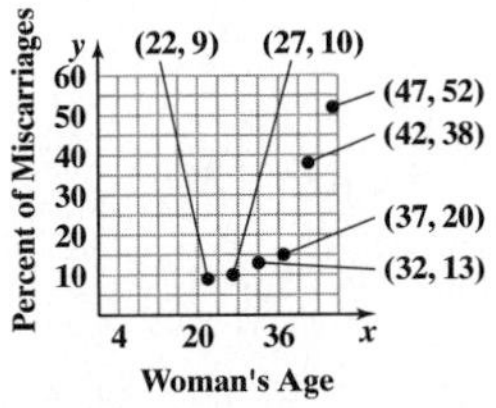

exponential function

23.

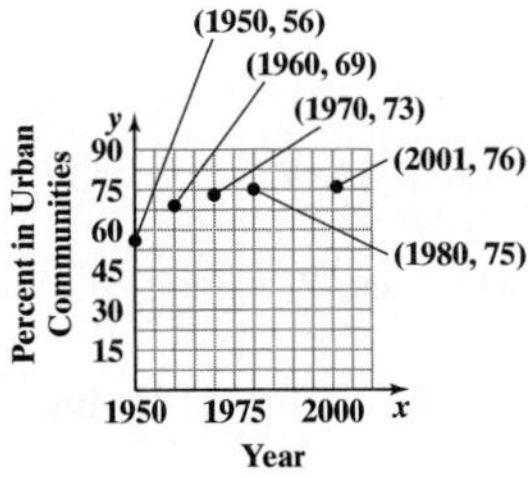

logarithmic function

25.

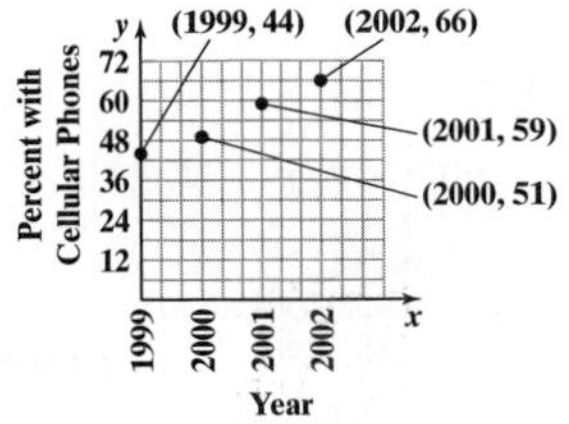

linear function

27. $y = 100e^{(\ln 4.6)x}$; $y = 100e^{1.526x}$ **29.** $y = 2.5e^{(\ln 0.7)x}$; $y = 2.5e^{-0.357x}$ **39. a.** $y = 200.2(1.011)^x$; $r \approx 0.998$; Since r is close to 1, the model fits the data well. **b.** $y = 200.2e^{(\ln 1.011)x}$; $y = 200.2e^{0.0109x}$; by approximately 1% **41.** $y = 2.714x + 197.586$; $r \approx 0.995$; Since r is close to 1, the model fits the data well. **43.** $y = 200.2(1.011)^x$; $y = 2.714x + 197.586$; using exponential, by 2010; using linear, by 2012; Answers will vary.
45. Models will vary. Examples are given. Predictions will vary. For Exercise 21: $y = 1.402(1.078)^x$; For Exercise 23: $y = 56.269 + 5.274 \ln x$; For Exercise 25: $y = 7.4x + 43.9$ **47.** about 126°F **48.** $\dfrac{x+5}{x}$ **49.** $\{-27, 1\}$ **50.** $17\sqrt{2}$

Review Exercises

1.

x	$f(x)$; d
−2	$\frac{1}{16}$
−1	$\frac{1}{4}$
0	1
1	4
2	16

2.

x	$f(x)$; a
−2	16
−1	4
0	1
1	$\frac{1}{4}$
2	$\frac{1}{16}$

3.

x	$f(x)$; b
−2	−16
−1	−4
0	−1
1	$-\frac{1}{4}$
2	$-\frac{1}{16}$

4.

x	$f(x)$; c
−2	−13
−1	−1
0	2
1	$\frac{11}{4}$
2	$\frac{47}{16}$

5.

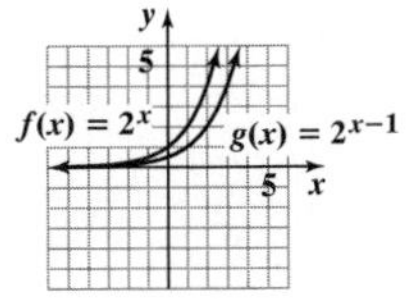

The graph of g is the graph of f shifted to the right 1 unit.

6.

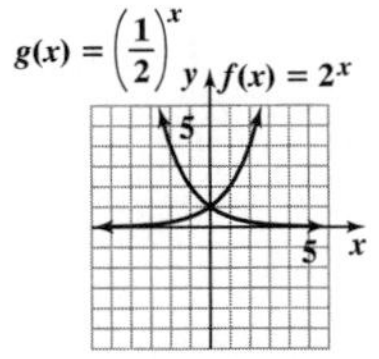

The graph of g is the reflection of f across the y-axis.

7.

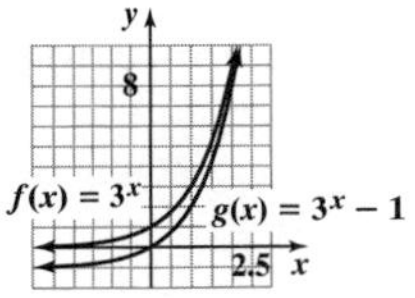

The graph of g is the graph of f shifted down 1 unit.

8.

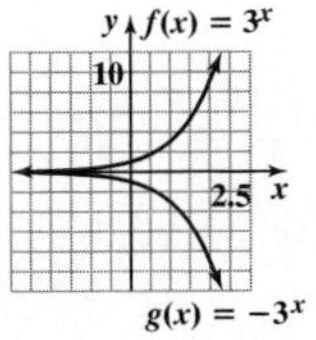

The graph of g is the reflection of f across the x-axis.

9. 5.5% compounded semiannually
10. 7% compounded monthly
11. a. 200°F **b.** about 119°F **c.** 70°F; The temperature of the room is 70°F.
12. $49^{1/2} = 7$ **13.** $4^3 = x$ **14.** $3^y = 81$ **15.** $\log_6 216 = 3$ **16.** $\log_b 625 = 4$
17. $\log_{13} 874 = y$ **18.** 3 **19.** −2 **20.** −9 is not in the domain of $y = \log_3 x$.
21. $\frac{1}{2}$ **22.** 1 **23.** 8 **24.** 5 **25.** $-\frac{1}{2}$ **26.** −2 **27.** −3 **28.** 0

29. Domain of $f = \{x|x \text{ is a real number}\}$ or $(-\infty, \infty)$;
Range of $f = \{y|y > 0\}$ or $(0, \infty)$;
Domain of $g = \{x|x > 0\}$ or $(0, \infty)$;
Range of $g = \{y|y \text{ is a real number}\}$ or $(-\infty, \infty)$

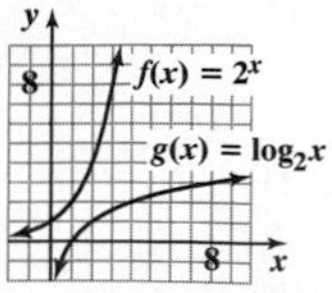

30. Domain of $f = \{x|x \text{ is a real number}\}$ or $(-\infty, \infty)$;
Range of $f = \{y|y > 0\}$ or $(0, \infty)$;
Domain of $g = \{x|x > 0\}$ or $(0, \infty)$;
Range of $g = \{y|y \text{ is a real number}\}$ or $(-\infty, \infty)$

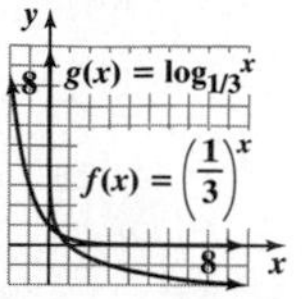

31. $\{x|x > -5\}$ or $(-5, \infty)$ **32.** $\{x|x < 3\}$ or $(-\infty, 3)$ **33.** $\{x|x \neq 1\}$ or $(-\infty, 1) \cup (1, \infty)$ **34.** $6x$ **35.** $\sqrt{x}$ **36.** $4x^2$ **37.** 3

38. a. 76
b. 67.4 after 2 months; 63.4 after 4 months; 60.8 after 6 months; 58.8 after 8 months; 55.9 after one year
c.

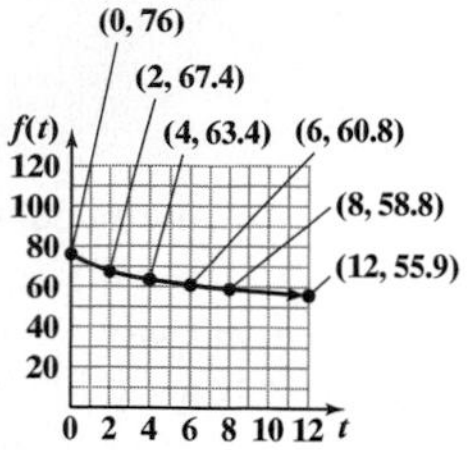

As time increases, the amount of material retained by the students decreases.

39. about 9 weeks **40.** $2 + 3\log_6 x$ **41.** $\frac{1}{2}\log_4 x - 3$ **42.** $\log_2 x + 2\log_2 y - 6$ **43.** $\frac{1}{3}\ln x - \frac{1}{3}$ **44.** $\log_b 21$ **45.** $\log\left(\frac{3}{x^3}\right)$ **46.** $\ln(x^3y^4)$ **47.** $\ln\left(\frac{\sqrt{x}}{y}\right)$ **48.** 6.2448 **49.** -0.1063 **50.** $\{2\}$ **51.** $\left\{\frac{2}{3}\right\}$ **52.** $\left\{-\frac{3}{2}\right\}$ **53.** $\left\{\frac{\ln 12{,}143}{\ln 8} \approx 4.52\right\}$ **54.** $\left\{\frac{\ln 141}{5} \approx 0.99\right\}$ **55.** $\left\{\frac{\ln 3}{0.045} \approx 24.41\right\}$ **56.** $\left\{\frac{1}{125}\right\}$ **57.** $\{100\}$ **58.** $\{23\}$ **59.** $\{5\}$ **60.** $\varnothing$ **61.** $\left\{\frac{1}{e}\right\}$ **62.** $\left\{\frac{e^3}{2}\right\}$ **63.** 5.5 mi **64.** approximately 2086 **65.** approximately 8103 thousand or 8,103,000 **66.** 7.3 years

67. 14.6 years **68.** about 22% **69. a.** $k = 0.045$ **b.** about 55.1 million **c.** 2012 **70.** 15,679 years old

71.

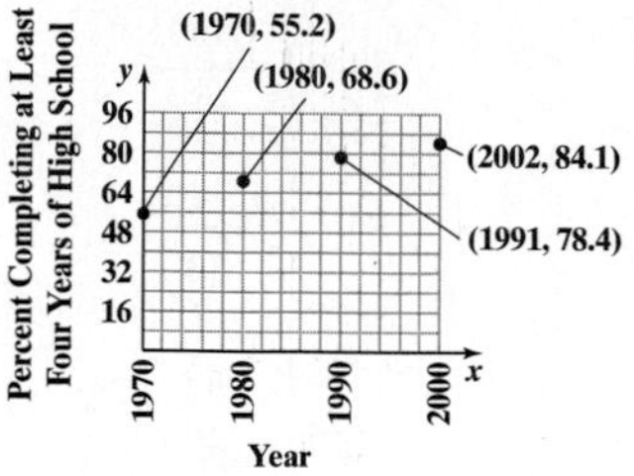

logarithmic function

72.

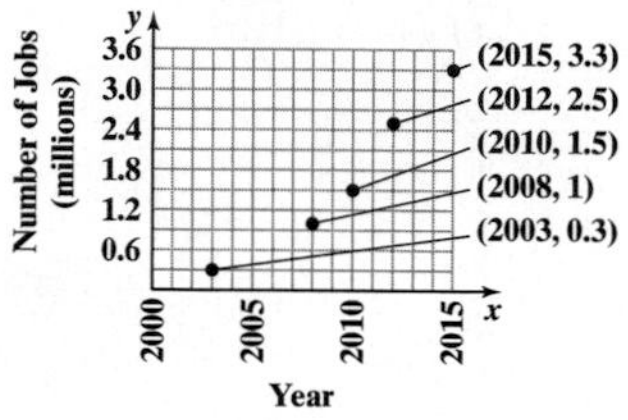

exponential function

73. $y = 73e^{(\ln 2.6)x}$; $y = 73e^{0.956x}$ **74.** $y = 6.5e^{(\ln 0.43)x}$; $y = 6.5e^{-0.844x}$ **75.** Answers will vary.

Chapter 12 Test

1.

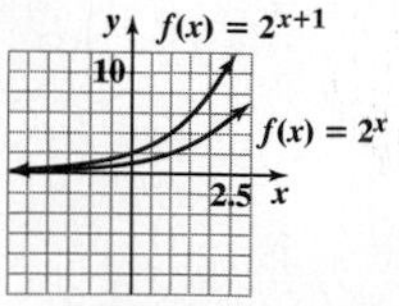

2. 6.5% compounded semiannually **3.** $5^3 = 125$ **4.** $\log_{36} 6 = \frac{1}{2}$

5.

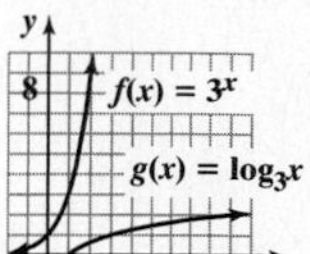

Domain of $f = \{x|x \text{ is a real number}\}$ or $(-\infty, \infty)$;
Range of $f = \{y|y > 0\}$ or $(0, \infty)$; Domain of $g = \{x|x > 0\}$ or $(0, \infty)$;
Range of $g = \{y|y \text{ is a real number}\}$ or $(-\infty, \infty)$

6. $5x$ **7.** 1 **8.** 0 **9.** $\{x|x > 7\}$ or $(7, \infty)$ **10.** 120 decibels **11.** $3 + 5\log_4 x$ **12.** $\frac{1}{3}\log_3 x - 4$ **13.** $\log(x^6y^2)$ **14.** $\ln\left(\frac{7}{x^3}\right)$ **15.** 1.5741 **16.** 6 or $\{6\}$ **17.** $\frac{\ln 1.4}{\ln 5}$ or $\left\{\frac{\ln 1.4}{\ln 5}\right\}$ **18.** $\frac{\ln 4}{0.005}$ or $\left\{\frac{\ln 4}{0.005}\right\}$ **19.** 5 or $\{5\}$

20. $\frac{217}{4}$ or $\left\{\frac{217}{4}\right\}$ **21.** 5 or $\{5\}$ **22.** $\frac{e^4}{3}$ or $\left\{\frac{e^4}{3}\right\}$ **23. a.** 82.3 million **b.** decreasing; The growth rate, -0.002, is negative. **c.** 2008

24. 13.9 years **25.** about 6.9% **26.** $A = 509e^{0.036t}$ **27.** about 24,758 years ago **28.** linear **29.** logarithmic **30.** exponential

31. quadratic **32.** $y = 96e^{(\ln 0.38)x}$; $y = 96e^{-0.968x}$

Cumulative Review Exercises (Chapters 1–12)

1. $\left\{\frac{2}{3}\right\}$ **2.** $\{(-5, 2)\}$ **3.** $\{(1, 4, -2)\}$ **4.** $\{x|x \geq 10\}$ or $[10, \infty)$ **5.** $\{x|x \leq -4\}$ or $(-\infty, -4]$ **6.** $\{x|x > 2\}$ or $(2, \infty)$ **7.** $\{x|-2 < x < 3\}$ or $(-2, 3)$ **8.** $\{3, 9\}$ **9.** $\varnothing$ **10.** $\{12\}$ **11.** $\left\{\frac{-2 \pm \sqrt{14}}{2}\right\}$ **12.** $\{8, 27\}$ **13.** $\left\{x \middle| -2 \leq x \leq \frac{3}{2}\right\}$ or $\left[-2, \frac{3}{2}\right]$ **14.** $\{2\}$ **15.** $\{0\}$

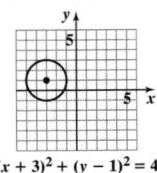

16.

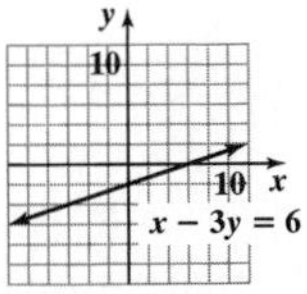

17.

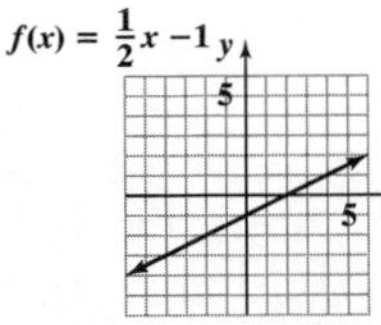

18.

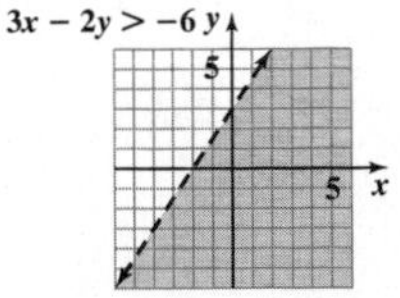

19.

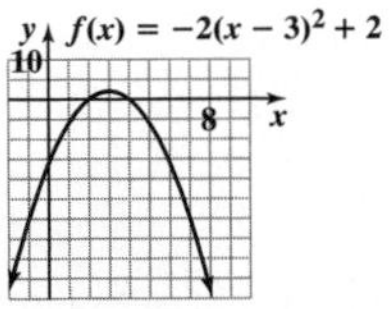

20.

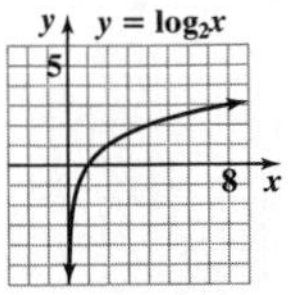

21. $-16x + 24y$ **22.** $-\frac{20x^7}{y^4}$ **23.** $x^2 - 5xy - 16y^2$ **24.** $6x^2 + 13x - 5$ **25.** $9x^4 - 24x^2y + 16y^2$ **26.** $\frac{3x^2 + 6x - 2}{(x + 5)(x + 2)}$ **27.** $\frac{x - 3}{x}$ **28.** $\frac{x - 4}{3x + 6}$ **29.** $5xy\sqrt{2x}$ **30.** $9\sqrt{2}$ **31.** $44 + 6i$ **32.** $(9x^2 + 1)(3x + 1)(3x - 1)$ **33.** $2x(4x - 1)(3x - 2)$ **34.** $(x + 3y)(x^2 - 3xy + 9y^2)$ **35.** $x^2 + 2x - 13; 22$ **36.** $x + 5$; $\{x|x$ is a real number and $x \neq 2\}$ or $(-\infty, 2) \cup (2, \infty)$ **37.** $x^2 - x - 17$ **38.** $x^2 + 3x - 17$ **39.** $f^{-1}(x) = \frac{x + 3}{7}$ **40.** $3x^2 - 7x + 18 - \frac{28}{x + 2}$ **41.** $R = -\frac{Ir}{I - 1}$ or $R = \frac{Ir}{1 - I}$ **42.** $y = -3x - 1$ or $f(x) = -3x - 1$ **43.** $\ln\left(\frac{x^2}{\sqrt{y}}\right)$ **44.** $\{x|x$ is a real number and $x \neq 1$ and $x \neq 2\}$ or $(-\infty, 1) \cup (1, 2) \cup (2, \infty)$ **45.** $\{x|x > 4\}$ or $(4, \infty)$ **46.** \$620 **47.** 13 yd by 4 yd **48.** \$2600 at 12% and \$1400 at 14% **49.** approximately 11% **50.** 11 amps

CHAPTER 13

Section 13.1

Check Point Exercises

1. $x^2 + y^2 = 16$ **2.** $(x - 5)^2 + (y + 6)^2 = 100$ **3.** center: $(-3, 1)$; radius: 2 units

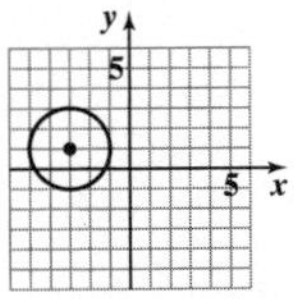

$(x + 3)^2 + (y - 1)^2 = 4$

4. $(x + 2)^2 + (y - 2)^2 = 9$

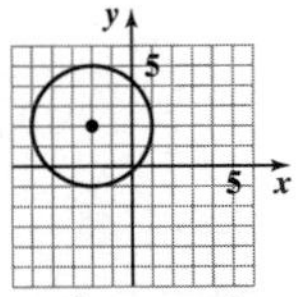

$(x + 2)^2 + (y - 2)^2 = 9$

Exercise Set 13.1

1. $x^2 + y^2 = 49$ **3.** $(x - 3)^2 + (y - 2)^2 = 25$ **5.** $(x + 1)^2 + (y - 4)^2 = 4$ **7.** $(x + 3)^2 + (y + 1)^2 = 3$ **9.** $(x + 4)^2 + y^2 = 100$

11. center: $(0, 0)$; $r = 4$

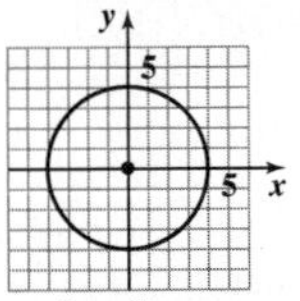

$x^2 + y^2 = 16$

13. center: $(3, 1)$; $r = 6$

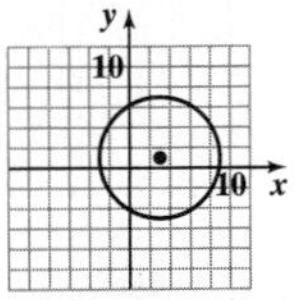

$(x - 3)^2 + (y - 1)^2 = 36$

15. center: $(-3, 2)$; $r = 2$

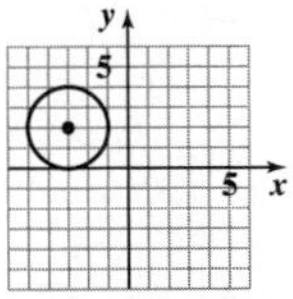

$(x + 3)^2 + (y - 2)^2 = 4$

17. center: $(-2, -2)$; $r = 2$

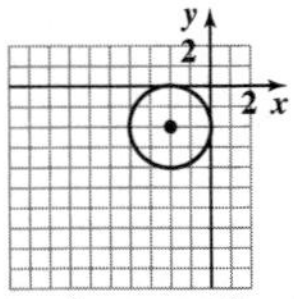

$(x + 2)^2 + (y + 2)^2 = 4$

19. $(x + 3)^2 + (y + 1)^2 = 4$; center: $(-3, -1)$; $r = 2$

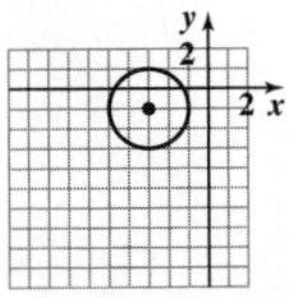

$x^2 + y^2 + 6x + 2y + 6 = 0$

21. $(x - 5)^2 + (y - 3)^2 = 64$; center: $(5, 3)$; $r = 8$

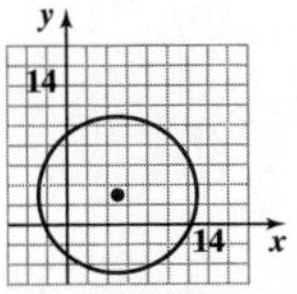

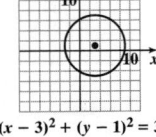

$x^2 + y^2 - 10x - 6y - 30 = 0$

23. $(x+4)^2+(y-1)^2=25$; center: $(-4,1)$; $r=5$

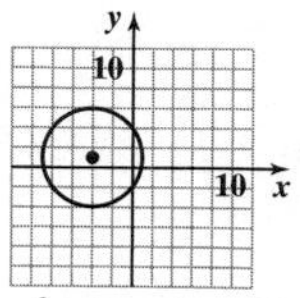

$x^2+y^2+8x-2y-8=0$

25. $(x-1)^2+y^2=16$; center: $(1,0)$; $r=4$

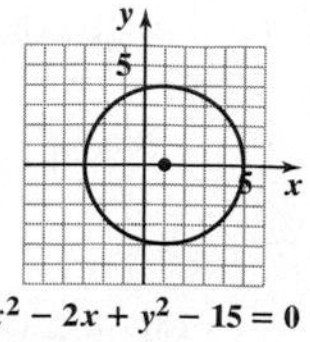

$x^2-2x+y^2-15=0$

27. $x^2+y^2=16$
$x-y=4$

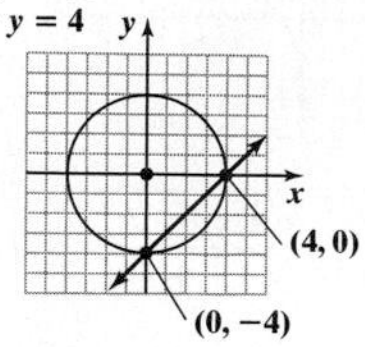

$\{(0,-4),(4,0)\}$

29. $(x-2)^2+(y+3)^2=4$
$y=x-3$

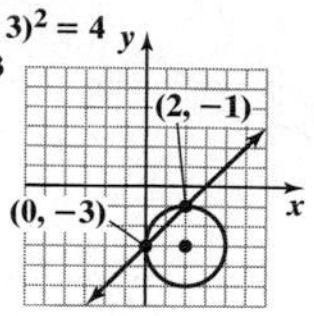

$\{(0,-3),(2,-1)\}$

31. $(x-2)^2+(y+1)^2=4$ **33.** $(x+3)^2+(y+2)^2=1$ **35. a.** $(5,10)$ **b.** $\sqrt{5}$ **c.** $(x-5)^2+(y-10)^2=5$

37. $(x+2.4)^2+(y+2.7)^2=900$ **45.**

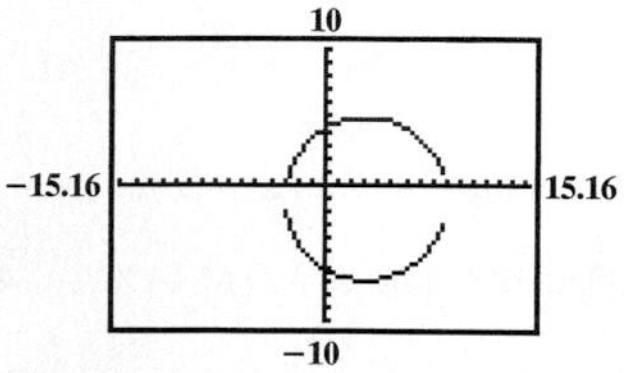

47. d

49. $y+4=\dfrac{3}{4}(x-3)$

50. $f(g(x))=9x^2+24x+14$; $g(f(x))=3x^2-2$ **51.** $\{4\}$ **52.** $\left\{x\,\middle|\,-\dfrac{5}{2}<x<\dfrac{15}{2}\right\}$ or $\left(-\dfrac{5}{2},\dfrac{15}{2}\right)$

Section 13.2

Check Point Exercises

1.

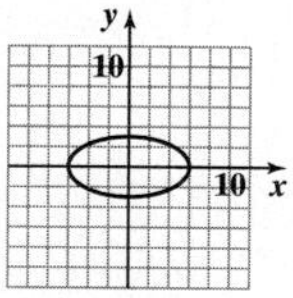

$\dfrac{x^2}{36}+\dfrac{y^2}{9}=1$

2.

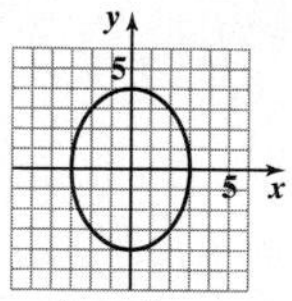

$16x^2+9y^2=144$

3.

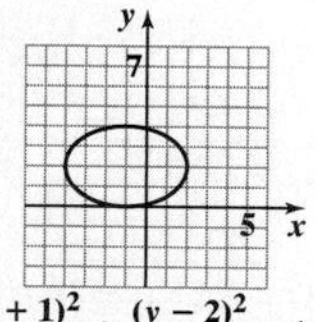

$\dfrac{(x+1)^2}{9}+\dfrac{(y-2)^2}{4}=1$

4. Yes, the height of the archway 6 feet from the center is approximately 9.54 feet.

Exercise Set 13.2

1.

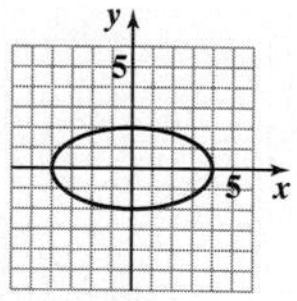

$\dfrac{x^2}{16}+\dfrac{y^2}{4}=1$

3.

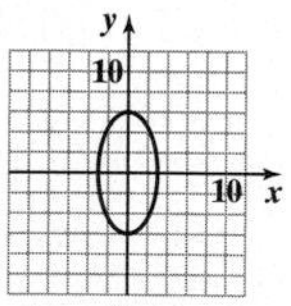

$\dfrac{x^2}{9}+\dfrac{y^2}{36}=1$

5.

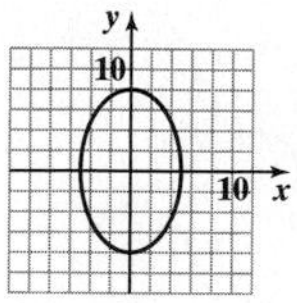

$\dfrac{x^2}{25}+\dfrac{y^2}{64}=1$

7.

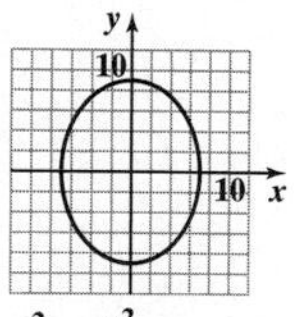

$\dfrac{x^2}{49}+\dfrac{y^2}{81}=1$

9.

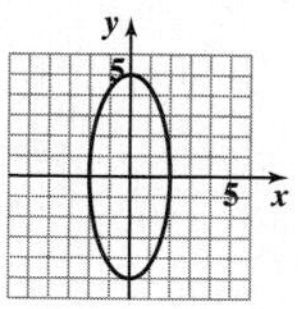

$25x^2+4y^2=100$

11.

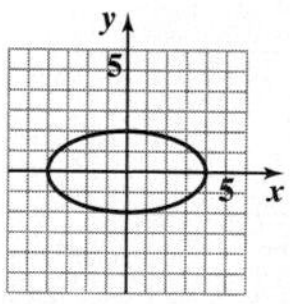

$4x^2+16y^2=64$

13.

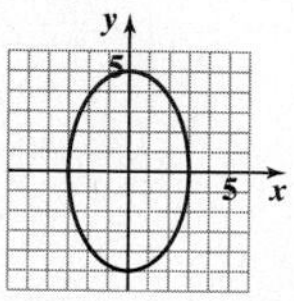

$25x^2+9y^2=225$

15.

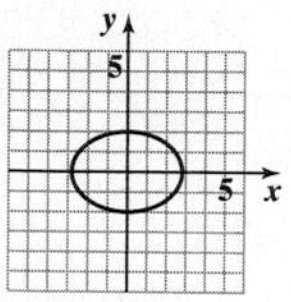

$x^2+2y^2=8$

17. $\dfrac{x^2}{4}+\dfrac{y^2}{1}=1$ or $\dfrac{x^2}{4}+y^2=1$ **19.** $\dfrac{x^2}{1}+\dfrac{y^2}{4}=1$ or $x^2+\dfrac{y^2}{4}=1$

21. 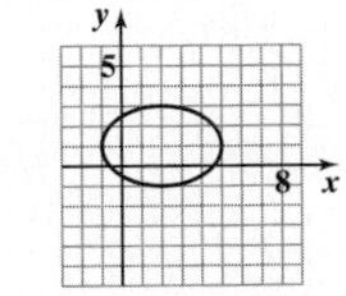
$\frac{(x-2)^2}{9} + \frac{(y-1)^2}{4} = 1$

23. 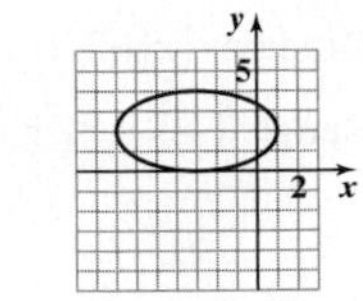
$(x+3)^2 + 4(y-2)^2 = 16$

25. 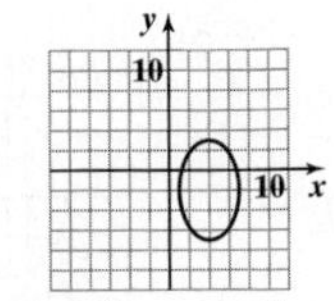
$\frac{(x-4)^2}{9} + \frac{(y+2)^2}{25} = 1$

27. 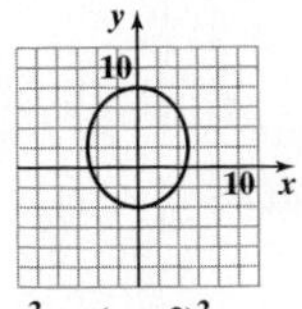
$\frac{x^2}{25} + \frac{(y-2)^2}{36} = 1$

29. 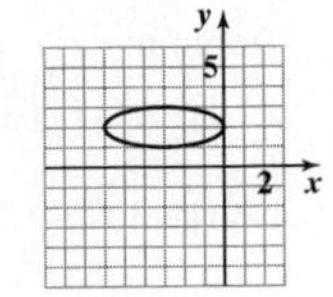
$\frac{(x+3)^2}{9} + (y-2)^2 = 1$

31. 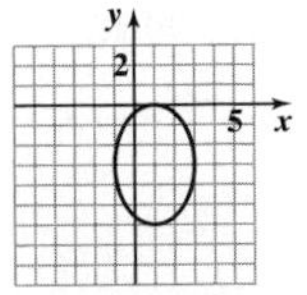
$9(x-1)^2 + 4(y+3)^2 = 36$

33. $\frac{(x+1)^2}{4} + \frac{(y-1)^2}{1} = 1$

35. $\{(0,-1), (0,1)\}$

37. $\{(0,3)\}$

39. $\{(0,-2), (1,0)\}$

41. 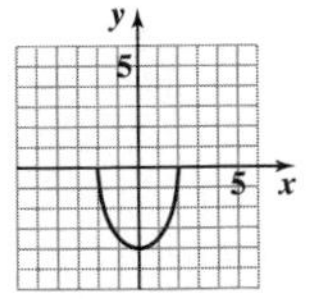
$y = -\sqrt{16-4x^2}$

43. Yes, the height of the archway 4 feet from the center is approximately 9.64 feet.

45. **a.** $\frac{x^2}{48^2} + \frac{y^2}{23^2} = 1$ or $\frac{x^2}{2304} + \frac{y^2}{529} = 1$ **b.** approximately 42 feet

55. $\frac{x^2}{\frac{36}{5}} + \frac{y^2}{36} = 1$

57. 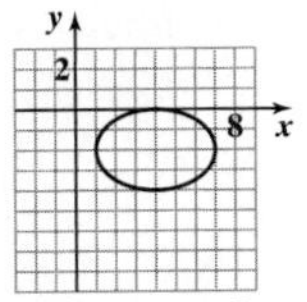
$\frac{(x-4)^2}{9} + \frac{(y+2)^2}{4} = 1$

59. small circle: $x^2 + y^2 = 9$; large circle: $x^2 + y^2 = 25$

60. $(x+2)(x+2)(x-2)$ or $(x+2)^2(x-2)$

61. $2xy^2\sqrt[3]{5xy}$

62. $\{-1\}$

Section 13.3

Check Point Exercises

1. **a.** $(-5, 0)$ and $(5, 0)$ **b.** $(0, -5)$ and $(0, 5)$

2. 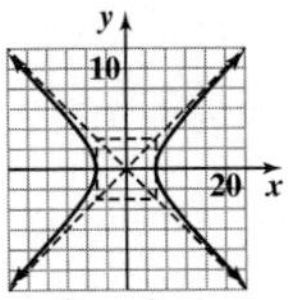
$\frac{x^2}{36} - \frac{y^2}{9} = 1$

3. 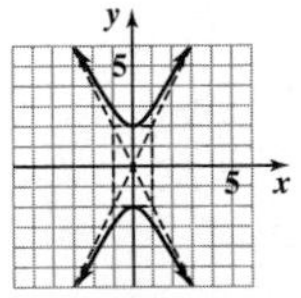
$y^2 - 4x^2 = 4$

Exercise Set 13.3

1. $(-2, 0)$ and $(2, 0)$; b **3.** $(0, -2)$ and $(0, 2)$; a

5. 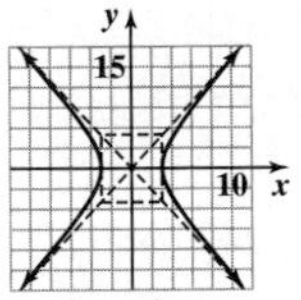
$\frac{x^2}{9} - \frac{y^2}{25} = 1$

7. 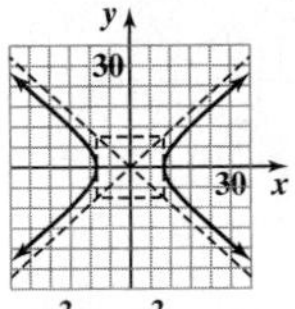
$\frac{x^2}{100} - \frac{y^2}{64} = 1$

9. 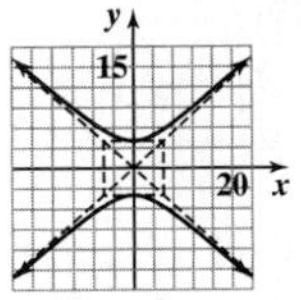
$\frac{y^2}{16} - \frac{x^2}{36} = 1$

11. 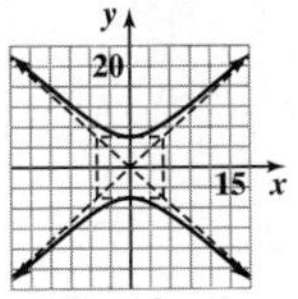
$\frac{y^2}{36} - \frac{x^2}{25} = 1$

13. $9x^2 - 4y^2 = 36$ 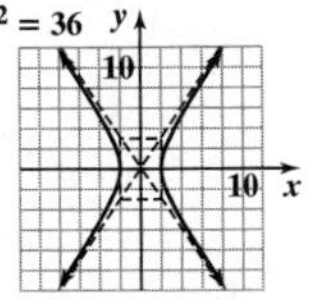

15. 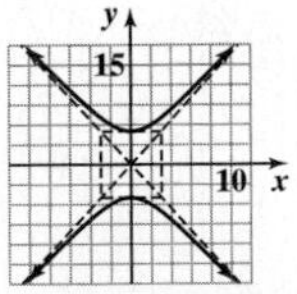
$9y^2 - 25x^2 = 225$

17. 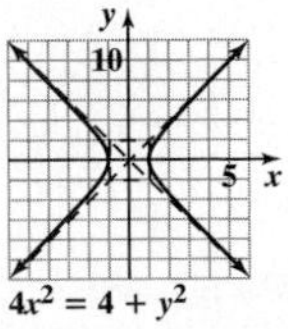
$4x^2 = 4 + y^2$

19. $\frac{x^2}{9} - \frac{y^2}{25} = 1$

21. $\frac{y^2}{4} - \frac{x^2}{9} = 1$

23.
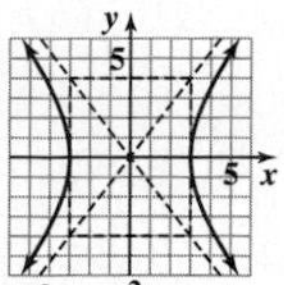

$\frac{x^2}{9} - \frac{y^2}{16} = 1$

Domain: $\{x|x \le -3 \text{ or } x \ge 3\}$ or $(-\infty, -3] \cup [3, \infty)$; Range: $\{y|y \text{ is a real number}\}$ or $(-\infty, \infty)$

25.
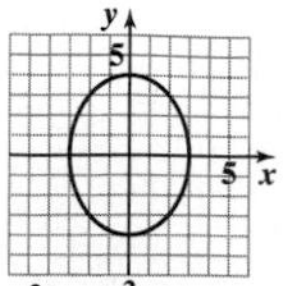

$\frac{x^2}{9} + \frac{y^2}{16} = 1$

Domain: $\{x|-3 \le x \le 3\}$ or $[-3, 3]$; Range: $\{y|-4 \le y \le 4\}$ or $[-4, 4]$

27.
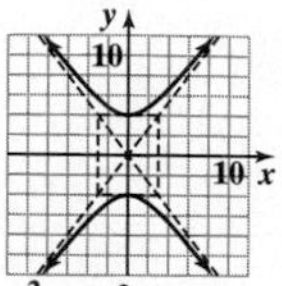

$\frac{y^2}{16} - \frac{x^2}{9} = 1$

Domain: $\{x|x \text{ is a real number}\}$ or $(-\infty, \infty)$; Range: $\{y|y \le -4 \text{ or } y \ge 4\}$ or $(-\infty, -4] \cup [4, \infty)$

29. $\{(-2, 0), (2, 0)\}$ **31.** $\{(0, -3), (0, 3)\}$ **33.** 40 yd

41.
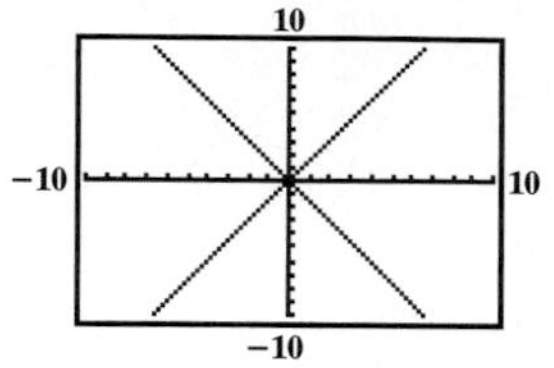

no; two lines; $y = \frac{b}{a}x$ and $y = -\frac{b}{a}x$

43. c **45.**
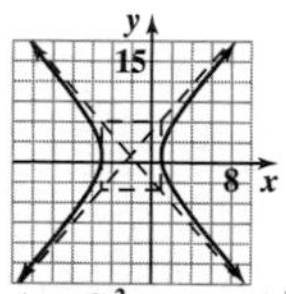

$\frac{(x + 2)^2}{9} - \frac{(y - 1)^2}{25} = 1$

47.
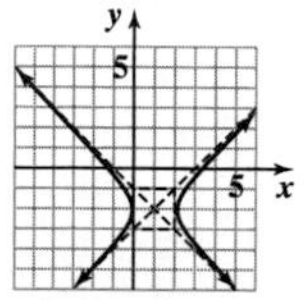

$x^2 - y^2 - 2x - 4y - 4 = 0$

49. $\dfrac{y^2}{49} - \dfrac{x^2}{\frac{49}{25}} = 1$

50.
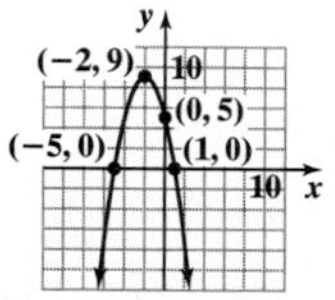

$y = -x^2 - 4x + 5$

51. $\left\{x \middle| x \le -\frac{1}{3} \text{ or } x \ge 4\right\}$ or $\left(-\infty, -\frac{1}{3}\right] \cup [4, \infty)$ **52.** $\{21\}$

Mid-Chapter 13 Check Point

1.
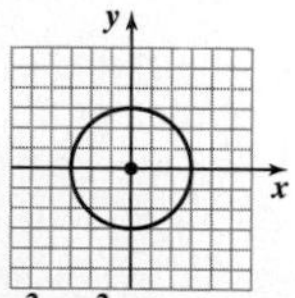

$x^2 + y^2 = 9$

2.
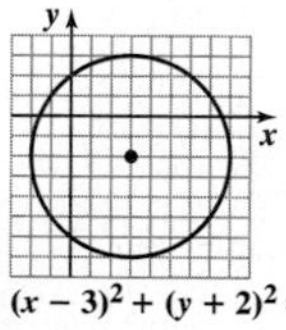

$(x - 3)^2 + (y + 2)^2 = 25$

3.
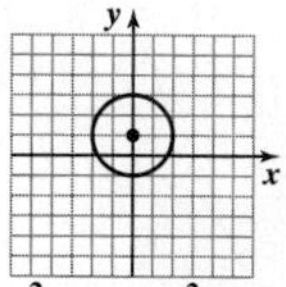

$x^2 + (y - 1)^2 = 4$

4.
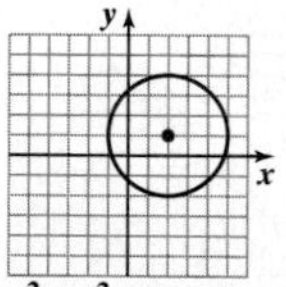

$x^2 + y^2 - 4x - 2y - 4 = 0$

5.
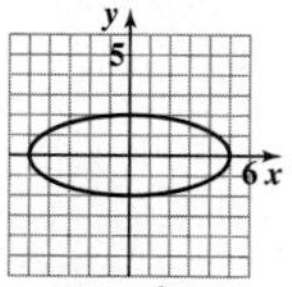

$\frac{x^2}{25} + \frac{y^2}{4} = 1$

6.
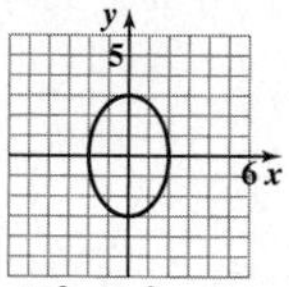

$9x^2 + 4y^2 = 36$

7.
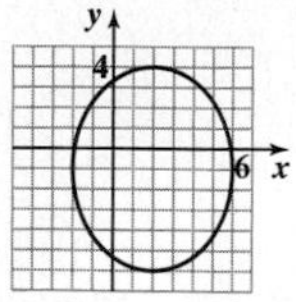

$\frac{(x - 2)^2}{16} + \frac{(y + 1)^2}{25} = 1$

8.
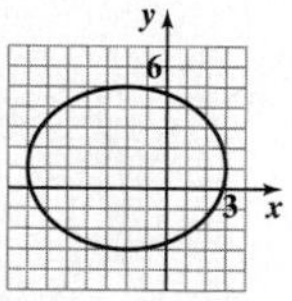

$\frac{(x + 2)^2}{25} + \frac{(y - 1)^2}{16} = 1$

9.
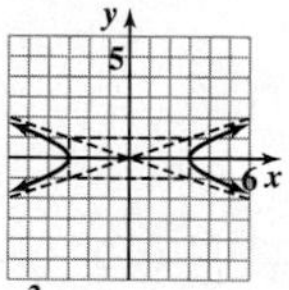

$\frac{x^2}{9} - y^2 = 1$

10.
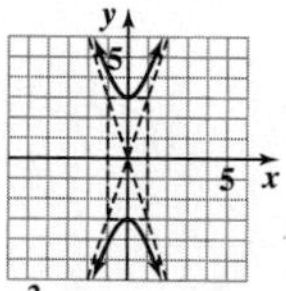

$\frac{y^2}{9} - x^2 = 1$

11.
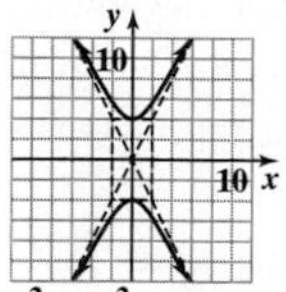

$y^2 - 4x^2 = 16$

12.
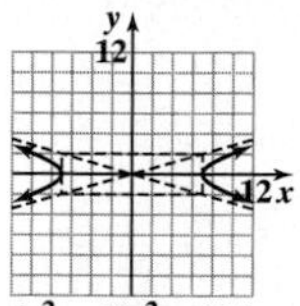

$4x^2 - 49y^2 = 196$

13.
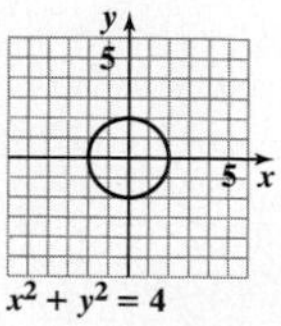

$x^2 + y^2 = 4$

14.
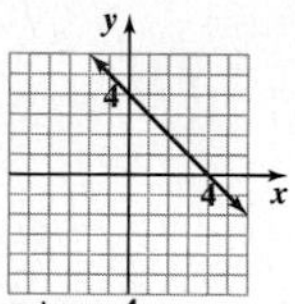

$x + y = 4$

15.
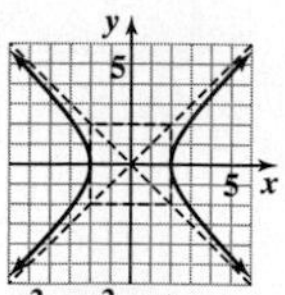

$x^2 - y^2 = 4$

16.
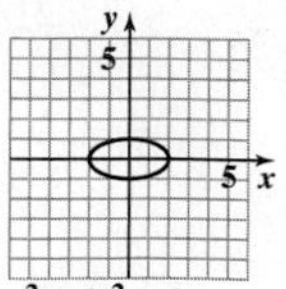

$x^2 + 4y^2 = 4$

17.
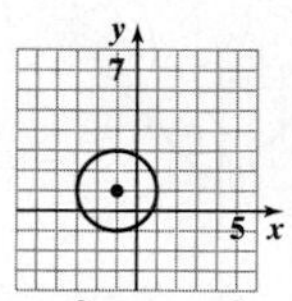

$(x + 1)^2 + (y - 1)^2 = 4$

18.
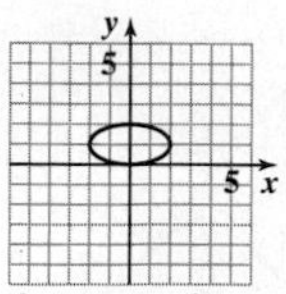

$x^2 + 4(y - 1)^2 = 4$

Section 13.4

Check Point Exercises

1.

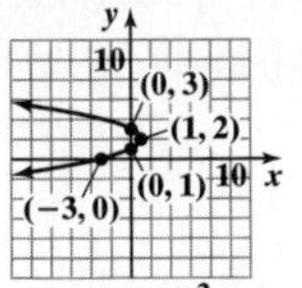

$x = -(y - 2)^2 + 1$

2.

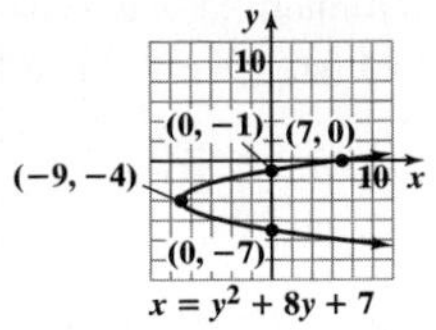

$x = y^2 + 8y + 7$

3. a. hyperbola **b.** ellipse **c.** circle **d.** parabola

Exercise Set 13.4

1. opens to right; $(-1, 2)$; b **3.** opens to right; $(1, -2)$; f **5.** opens to left; $(1, 2)$; a

7. $(0, 0)$ **9.** $(3, 2)$ **11.** $(-1, -2)$ **13.** $(0, 6)$ **15.** $(-3, 3)$ **17.** $(4, -1)$

19.

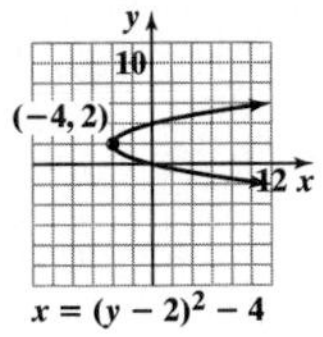

$x = (y - 2)^2 - 4$

axis of symmetry: $y = 2$

21.

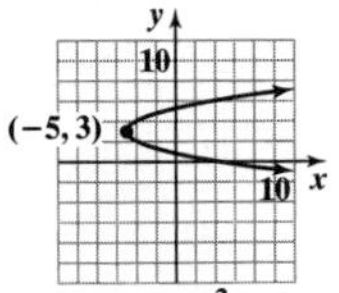

$x = (y - 3)^2 - 5$

axis of symmetry: $y = 3$

23.

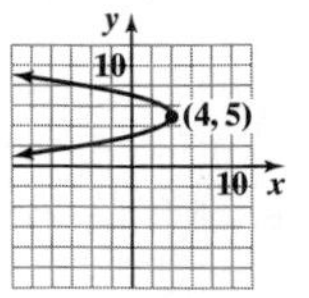

$x = -(y - 5)^2 + 4$

axis of symmetry: $y = 5$

25.

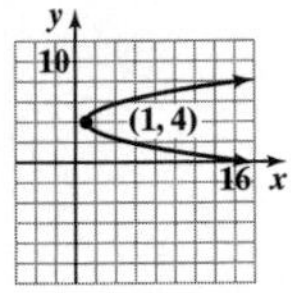

$x = (y - 4)^2 + 1$

axis of symmetry: $y = 4$

27.

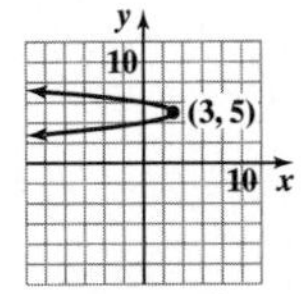

$x = -3(y - 5)^2 + 3$

axis of symmetry: $y = 5$

29.

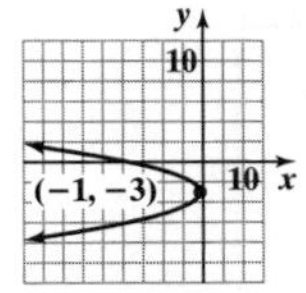

$x = -2(y + 3)^2 - 1$

axis of symmetry: $y = -3$

31.

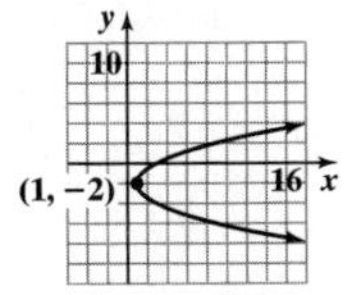

$x = \frac{1}{2}(y + 2)^2 + 1$

axis of symmetry: $y = -2$

33.

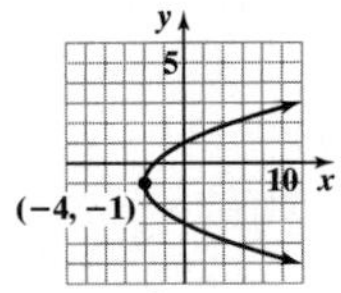

$x = y^2 + 2y - 3$

axis of symmetry: $y = -1$

35.

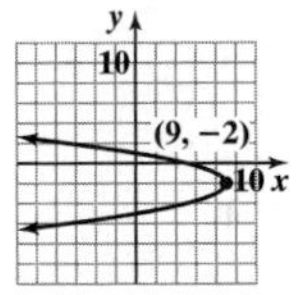

$x = -y^2 - 4y + 5$

axis of symmetry: $y = -2$

37.

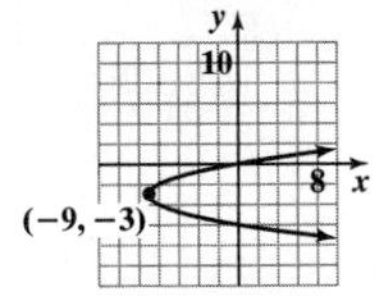

$x = y^2 + 6y$

axis of symmetry: $y = -3$

39.

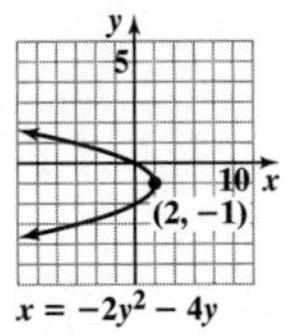

$x = -2y^2 - 4y$

axis of symmetry: $y = -1$

41.

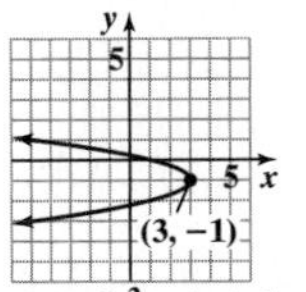

$x = -2y^2 - 4y + 1$

axis of symmetry: $y = -1$

43. a. horizontal **b.** to the right **c.** $(2, 1)$ **45. a.** vertical **b.** upward **c.** $(1, 2)$ **47. a.** vertical **b.** downward **c.** $(-3, 4)$
49. a. horizontal **b.** to the left **c.** $(4, -3)$ **51. a.** vertical **b.** upward **c.** $(2, -5)$ **53. a.** horizontal **b.** to the left **c.** $(5, 2)$
55. parabola **57.** ellipse **59.** hyperbola **61.** circle **63.** hyperbola **65.** hyperbola

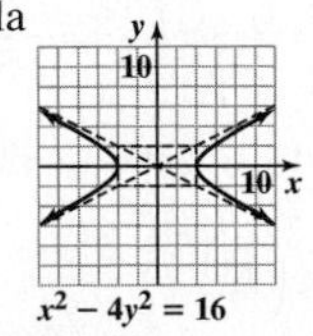

$x^2 - 4y^2 = 16$

67. circle

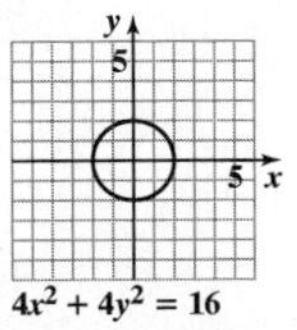

$4x^2 + 4y^2 = 16$

69. ellipse

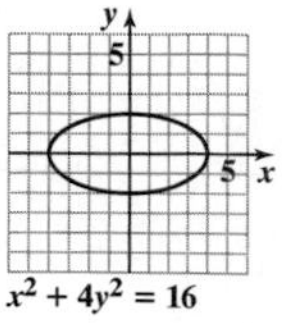

71. parabola

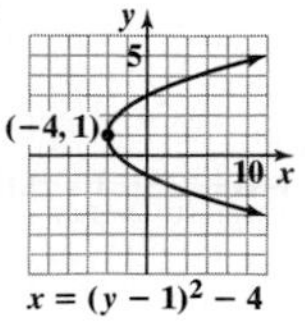

73. circle

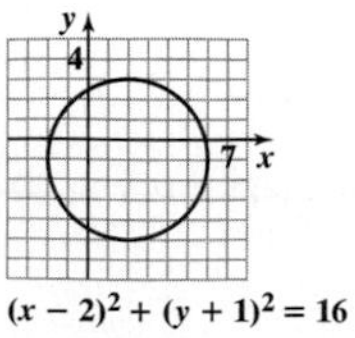

75. Domain: $\{x|x \geq -4\}$ or $[-4, \infty)$; Range: $\{y|y$ is a real number$\}$ or $(-\infty, \infty)$; not a function
77. Domain: $\{x|x$ is a real number$\}$ or $(-\infty, \infty)$; Range: $\{y|y \leq 1\}$ or $(-\infty, 1]$; function

79. Domain: $\{x|x \leq 3\}$ or $(-\infty, 3]$; Range: $\{y|y$ is a real number$\}$ or $(-\infty, \infty)$; not a function

81. $\{(-4, 2), (0, 0)\}$

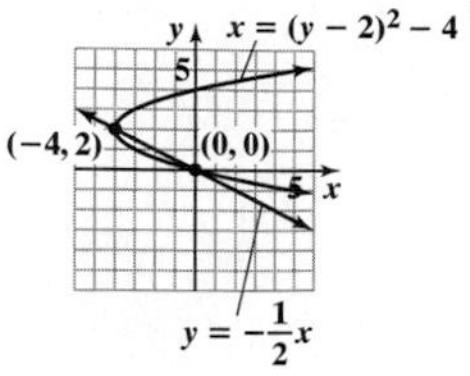

83. $\{(-2, 1)\}$

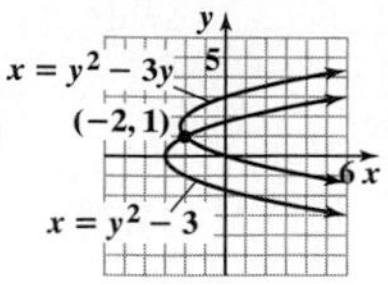

85. ∅

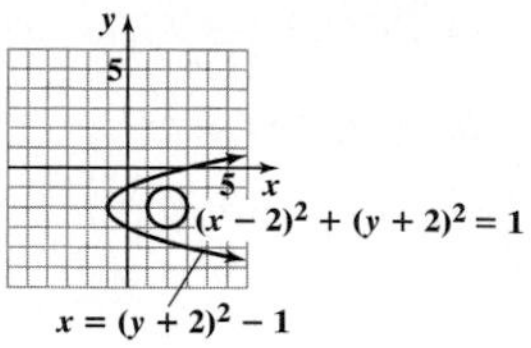

87. a. $y = 0.0001032x^2$ **b.** 58 ft **89. a.** $y = \frac{1}{18}x^2$ **b.** 4.5 ft

99. $y^2 + 2y + (-6x + 13) = 0$; $y = -1 \pm \sqrt{6x - 12}$

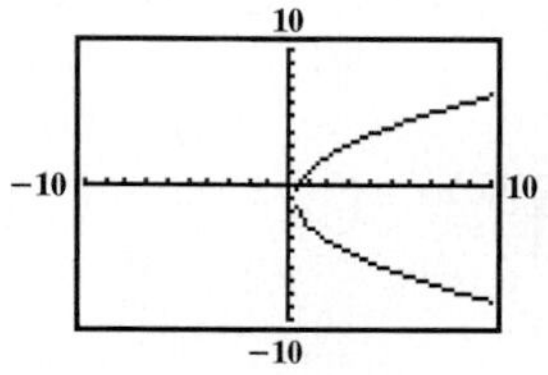

105.

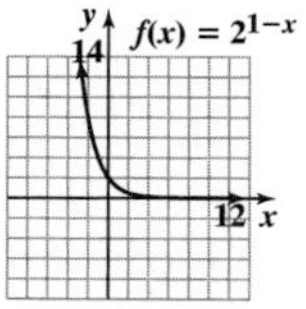

106. $f^{-1}(x) = 3x + 15$
107. $\{(6, 4)\}$

Section 13.5

Check Point Exercises

1. (0, 1) and (4, 17), or $\{(0, 1), (4, 17)\}$ **2.** $(2, -1)$ and $\left(-\frac{6}{5}, \frac{3}{5}\right)$, or $\left\{(2, -1), \left(-\frac{6}{5}, \frac{3}{5}\right)\right\}$
3. $(3, 2), (-3, 2), (3, -2)$, and $(-3, -2)$, or $\{(3, 2), (-3, 2), (3, -2), (-3, -2)\}$
4. (0, 5) or $\{(0, 5)\}$ **5.** length: 7 feet; width: 3 feet

Exercise Set 13.5

1. $\{(2, 0), (-3, 5)\}$ **3.** $\{(2, 0), (1, 1)\}$ **5.** $\{(-3, 11), (4, -10)\}$ **7.** $\{(-3, -4), (4, 3)\}$ **9.** $\left\{(2, 3), \left(-\frac{3}{2}, -4\right)\right\}$
11. $\{(3, 0), (-5, -4)\}$ **13.** $\{(3, 1), (-1, -3), (1, 3), (-3, -1)\}$ **15.** $\{(4, -3), (-1, 2)\}$ **17.** $\{(4, -3), (0, 1)\}$
19. $\{(3, 2), (-3, 2), (3, -2), (-3, -2)\}$ **21.** $\{(3, 2), (-3, 2), (3, -2), (-3, -2)\}$ **23.** $\{(2, 1), (-2, 1), (2, -1), (-2, -1)\}$
25. $\{(3, 4), (3, -4)\}$ **27.** $\{(0, 2), (0, -2), (-1, -\sqrt{3}), (-1, -\sqrt{3})\}$ **29.** $\{(2, 1), (-2, 1), (2, -1), (-2, -1)\}$
31. $\{(-1, -4), (1, 4), (2\sqrt{2}, \sqrt{2}), (-2\sqrt{2}, -\sqrt{2})\}$ **33.** $\{(4, 1), (2, 2)\}$ **35.** $\{(0, 0), (-1, 1)\}$ **37.** $\{(0, 0), (2, 2), (-2, 2)\}$
39. $\left\{(-4, 1), \left(-\frac{5}{2}, \frac{1}{4}\right)\right\}$ **41.** $\left\{(-2, 3), \left(\frac{12}{5}, -\frac{29}{5}\right)\right\}$ **43.** $x + y = 10$; $xy = 24$; 6 and 4
45. $x^2 - y^2 = 3$; $2x^2 + y^2 = 9$; 2 and 1, −2 and 1, −2 and −1, or 2 and −1 **47.** $\{(2, -1), (-2, 1)\}$ **49.** $\{(2, 20), (-2, 4), (-3, 0)\}$
51. $\left\{\left(-1, -\frac{1}{2}\right), \left(-1, \frac{1}{2}\right), \left(1, -\frac{1}{2}\right), \left(1, \frac{1}{2}\right)\right\}$

53. Answers will vary. example:

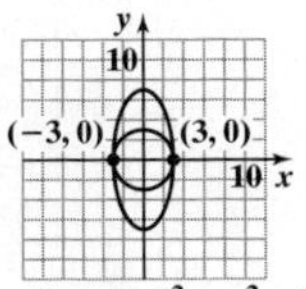

Circle : $x^2 + y^2 = 9$

Ellipse : $\frac{x^2}{9} + \frac{y^2}{49} = 1$

55. $(0, -4), (2, 0), (-2, 0)$
57. length: 11 feet; width: 7 feet
59. length: 8 inches; width: 6 inches
61. large square: 5 meters by 5 meters; small square: 2 meters by 2 meters
69. b
71. $(8, 2)$ or $\{(8, 2)\}$

73. $3x - 2y \le 6$

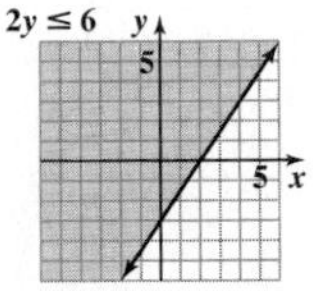

74. $m = \frac{8}{3}$

75. $6x^3 - 16x^2 + 17x - 6$

Review Exercises

1. $x^2 + y^2 = 9$ **2.** $(x + 2)^2 + (y - 4)^2 = 36$

3. center: $(0, 0)$; $r = 1$

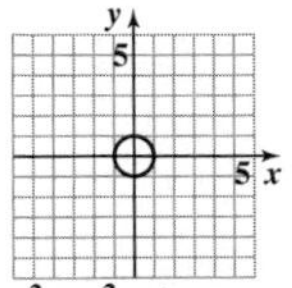
$x^2 + y^2 = 1$

4. center: $(-2, 3)$; $r = 3$

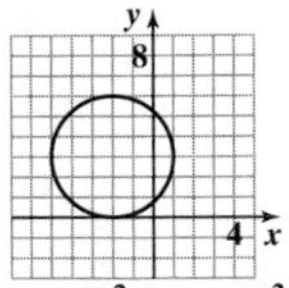
$(x + 2)^2 + (y - 3)^2 = 9$

5. center: $(2, -1)$; $r = 3$

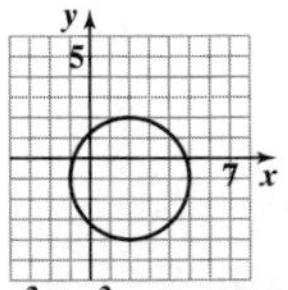
$x^2 + y^2 - 4x + 2y - 4 = 0$

6. center: $(0, 2)$; $r = 2$

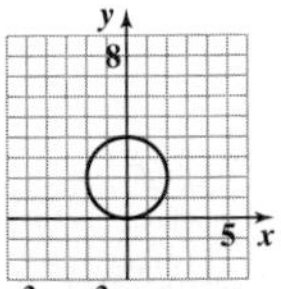
$x^2 + y^2 - 4y = 0$

7.

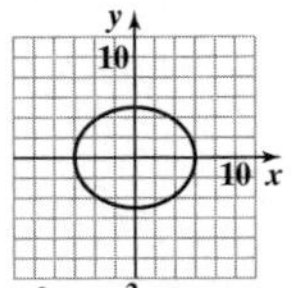
$\frac{x^2}{36} + \frac{y^2}{25} = 1$

8.

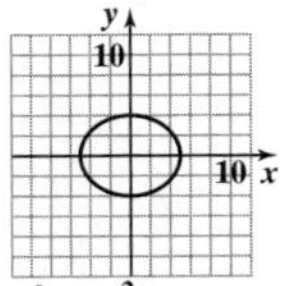
$\frac{x^2}{25} + \frac{y^2}{16} = 1$

9.

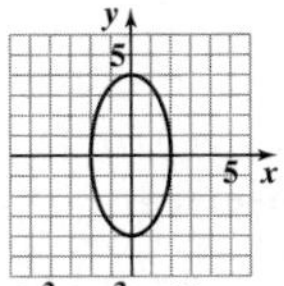
$4x^2 + y^2 = 16$

10.

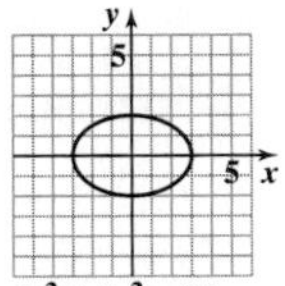
$4x^2 + 9y^2 = 36$

11.

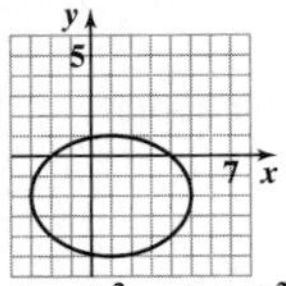
$\frac{(x - 1)^2}{16} + \frac{(y + 2)^2}{9} = 1$

12.

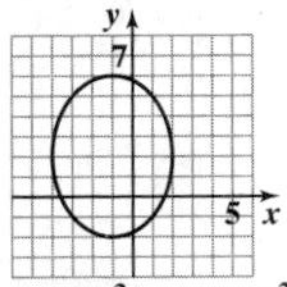
$\frac{(x + 1)^2}{9} + \frac{(y - 2)^2}{16} = 1$

13. Yes, the height of the archway 14 feet from the center is approximately 12.43 feet.

14.

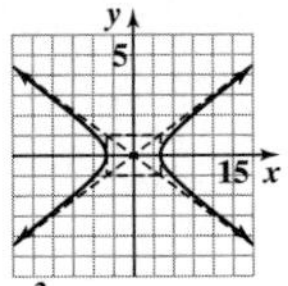
$\frac{x^2}{16} - y^2 = 1$

15.

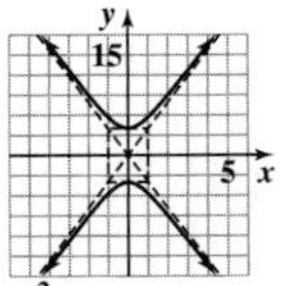
$\frac{y^2}{16} - x^2 = 1$

16.

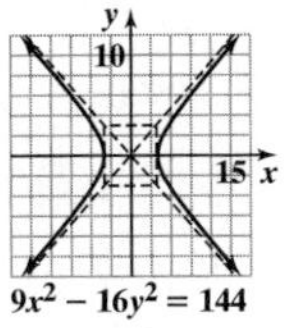
$9x^2 - 16y^2 = 144$

17.

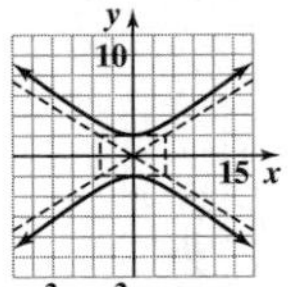
$4y^2 - x^2 = 16$

18.

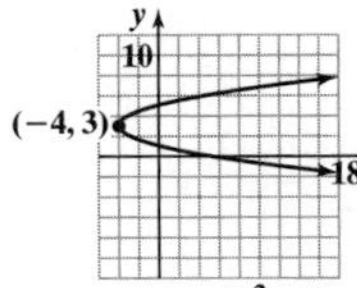

$x = (y - 3)^2 - 4$

axis of symmetry: $y = 3$

19.

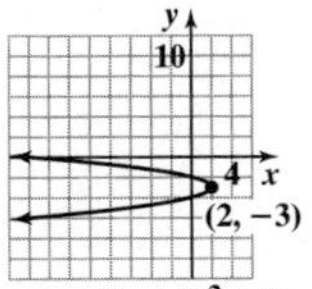

$x = -2(y + 3)^2 + 2$

axis of symmetry: $y = -3$

20.

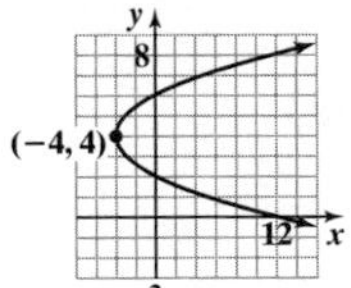

$x = y^2 - 8y + 12$

axis of symmetry: $y = 4$

21.

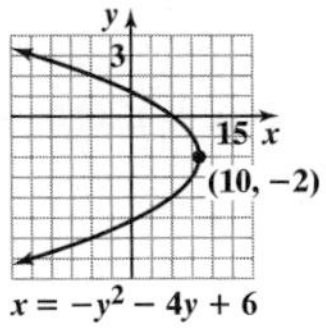

$x = -y^2 - 4y + 6$

axis of symmetry: $y = -2$

22. parabola **23.** ellipse **24.** hyperbola **25.** circle **26.** hyperbola **27.** ellipse **28.** parabola

29. circle

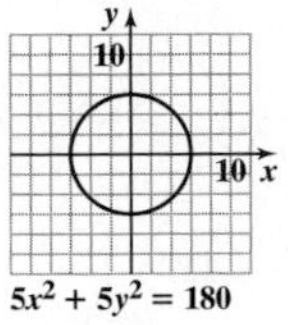
$5x^2 + 5y^2 = 180$

30. ellipse

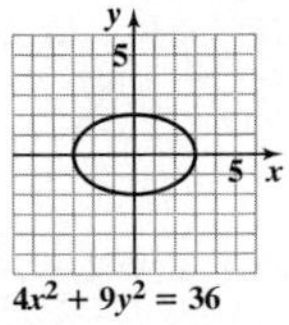
$4x^2 + 9y^2 = 36$

31. hyperbola

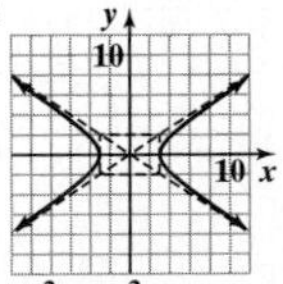
$4x^2 - 9y^2 = 36$

32. ellipse

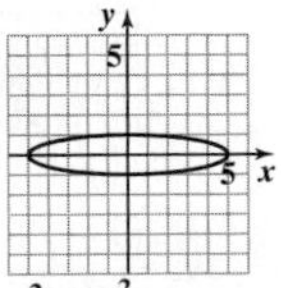
$\frac{x^2}{25} + \frac{y^2}{1} = 1$

33. parabola

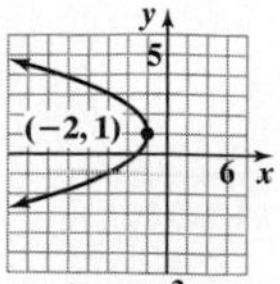

$x + 3 = -y^2 + 2y$

34. parabola

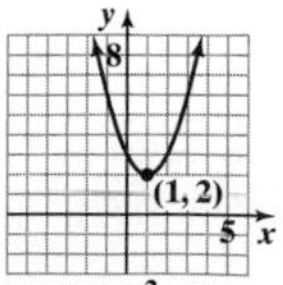

$y - 3 = x^2 - 2x$

35. ellipse

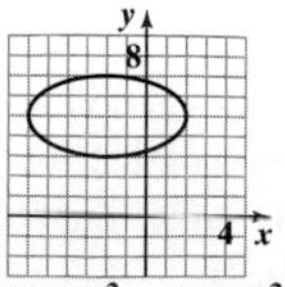

$\frac{(x+2)^2}{16} + \frac{(y-5)^2}{4} = 1$

36. circle

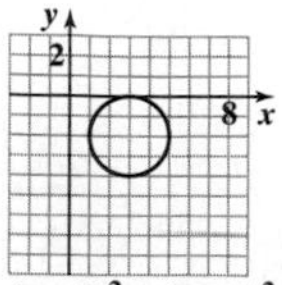

$(x-3)^2 + (y+2)^2 = 4$

37. circle

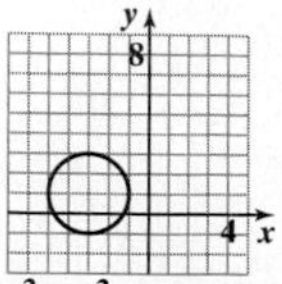

$x^2 + y^2 + 6x - 2y + 6 = 0$

38. a. $y = \frac{1}{12}x^2$ **b.** $(0, 3)$; 3 inches above the vertex **39.** $\{(1, 0), (4, 3)\}$ **40.** $\{(0, 1), (-3, 4)\}$

41. $\{(-1, 1), (1, -1)\}$ **42.** $\{(3, \sqrt{6}), (-3, \sqrt{6}), (3, -\sqrt{6}), (-3, -\sqrt{6})\}$ **43.** $\{(2, 2), (-2, -2)\}$

44. $\{(1, 2), (9, 6)\}$ **45.** $\{(-3, -1), (1, 3)\}$ **46.** $\left\{(-1, -1), \left(\frac{1}{2}, 2\right)\right\}$ **47.** $\left\{(0, -1), \left(\frac{5}{2}, -\frac{7}{2}\right)\right\}$

48. $\{(3, 2), (-3, 2), (2, -3), (-2, -3)\}$ **49.** $\{(3, 1), (-3, 1), (3, -1), \text{and } (-3, -1)\}$ **50.** 8 meters by 5 meters

51. $(1, 6)$ and $(3, 2)$ **52.** x: 46 ft; y: 28 ft or x: 50 ft; y: 20 ft

Chapter 13 Test

1. $(x - 3)^2 + (y + 2)^2 = 25$ **2.** center: $(5, -3)$; $r = 7$ **3.** center: $(-2, 3)$; $r = 4$ **4.** $(7, -3)$ **5.** $(-2, -5)$

6. hyperbola

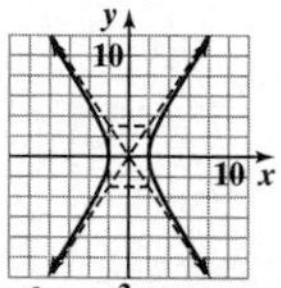

$\frac{x^2}{4} - \frac{y^2}{9} = 1$

7. ellipse

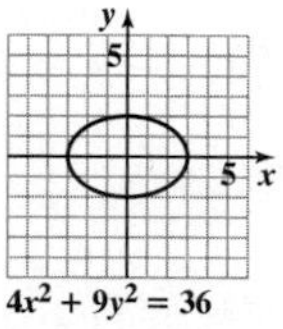

$4x^2 + 9y^2 = 36$

8. parabola

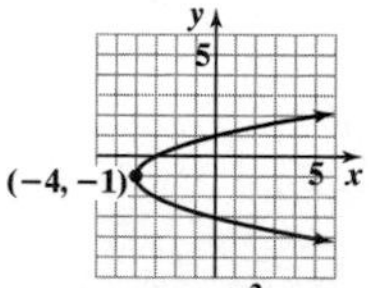

$x = (y+1)^2 - 4$

9. ellipse

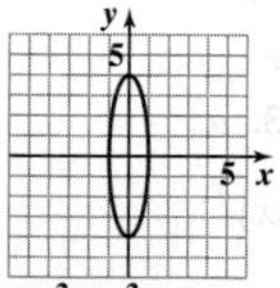

$16x^2 + y^2 = 16$

10. hyperbola

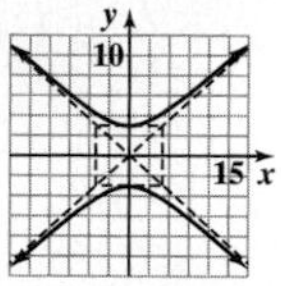

$25y^2 = 9x^2 + 225$

11. parabola

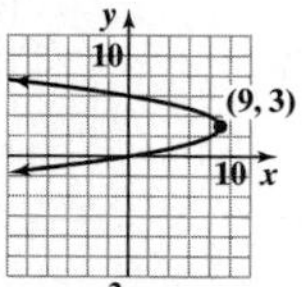

$x = -y^2 + 6y$

12. ellipse

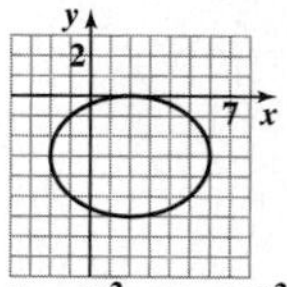

$\frac{(x-2)^2}{16} + \frac{(y+3)^2}{9} = 1$

13. circle

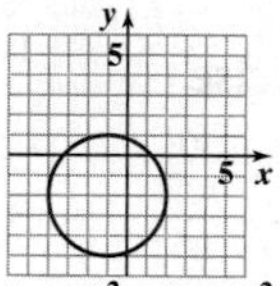

$(x+1)^2 + (y+2)^2 = 9$

14. circle

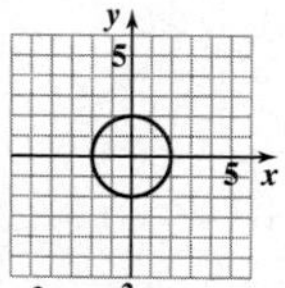

$\frac{x^2}{4} + \frac{y^2}{4} = 1$

15. $(4, -3)$ and $(-3, 4)$, or $\{(4, -3), (-3, 4)\}$

16. $(3, 2), (-3, 2), (3, -2)$, and $(-3, -2)$, or $\{(3, 2), (-3, 2), (3, -2), (-3, -2)\}$

17. 15 feet by 12 feet or 24 feet by 7.5 feet

18. 4 feet by 3 feet

Cumulative Review Exercises (Chapters 1–13)

1. $\{x | x > -1\}$ or $(-1, \infty)$ **2.** $\left\{-\frac{1}{2}, 4\right\}$ **3.** $\{2\}$ **4.** $\left\{x \middle| -\frac{5}{3} < x < -1\right\}$ or $\left(-\frac{5}{3}, -1\right)$ **5.** $\left\{\frac{5}{2}\right\}$ **6.** $\left\{\frac{\ln 8}{0.7} \approx 2.97\right\}$

7. $\{(1, 3), (-1, 3), (1, -3), (-1, -3)\}$

8.

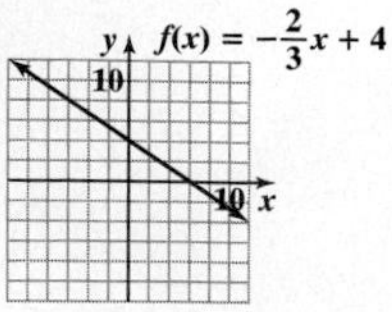

9.

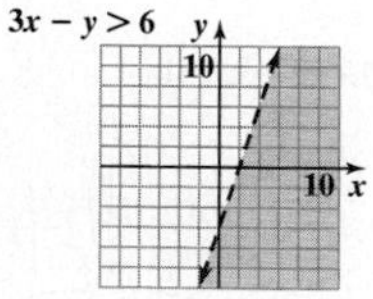

10.

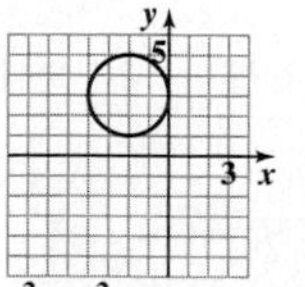

$x^2 + y^2 + 4x - 6y + 9 = 0$

11.

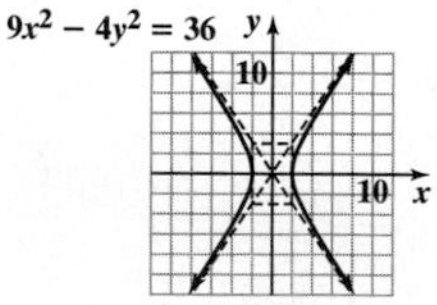

12. 46 **13.** $x^2 - 5x - 1$

14. $2xy^2\sqrt[3]{2y}$ **15.** $11 + 10i$

16. $3x(2x - 3)^2$

17. $(x - 2)(x + 3)(x - 3)$

18. $\{x | x \le 2\}$ or $(-\infty, 2]$ **19.** $\frac{(1 - \sqrt{x})^2}{1 - x}$ or $\frac{1 - 2\sqrt{x} + x}{1 - x}$ **20.** $\ln(x^{1/3}y^7)$ **21.** $3x^2 + x + 4 + \frac{7}{x - 2}$ **22.** $x^2 - 12 = 0$

23. faster car: 50 mph; slower car: 40 mph **24.** 175 miles; \$67 **25.** apple: 60 calories; banana: 87 calories

CHAPTER 14

Section 14.1

Check Point Exercises

1. a. 7, 9, 11, 13 **b.** $-\frac{1}{3}, \frac{1}{5}, -\frac{1}{9}, \frac{1}{17}$ **2.** $10, \frac{10}{3}, \frac{5}{6}, \frac{1}{6}$ **3. a.** $2(1)^2 + 2(2)^2 + 2(3)^2 + 2(4)^2 + 2(5)^2 + 2(6)^2 = 182$
b. $(2^3 - 3) + (2^4 - 3) + (2^5 - 3) = 47$ **c.** $4 + 4 + 4 + 4 + 4 = 20$ **4. a.** $\sum_{i=1}^{9} i^2$ **b.** $\sum_{i=1}^{n} \frac{1}{2^{i-1}}$

Exercise Set 14.1

1. 5, 8, 11, 14 **3.** 3, 9, 27, 81 **5.** −3, 9, −27, 81 **7.** −4, 5, −6, 7 **9.** $\frac{2}{5}, \frac{2}{3}, \frac{6}{7}, 1$ **11.** $1, -\frac{1}{3}, \frac{1}{7}, -\frac{1}{15}$ **13.** $1, 2, \frac{3}{2}, \frac{2}{3}$
15. 4, 12, 48, 240 **17.** 105 **19.** 60 **21.** 115 **23.** $-\frac{5}{16}$ **25.** 55 **27.** $\frac{3}{8}$ **29.** 15 **31.** $\sum_{i=1}^{15} i^2$ **33.** $\sum_{i=1}^{11} 2^i$
35. $\sum_{i=1}^{30} i$ **37.** $\sum_{i=1}^{14} \frac{i}{i+1}$ **39.** $\sum_{i=1}^{n} \frac{4^i}{i}$ **41.** $\sum_{i=1}^{n} (2i - 1)$ **43.** Answers will vary; examples are: $\sum_{k=1}^{14} (2k + 3)$ or $\sum_{k=2}^{15} (2k + 1)$.
45. Answers will vary; an example is: $\sum_{k=0}^{12} ar^k$. **47.** Answers will vary; an example is: $\sum_{k=0}^{n} (a + kd)$. **49.** 45 **51.** 0 **53.** 2 **55.** 80
57. a. 239.8 **b.** 34.3; From 1995 through 2001, the average number of people living below the poverty level each year was approximately 34.3 million.
59. a. 11.0375; From 1993 through 2000, the average number of welfare recipients each year was 11.0375 million. **b.** 11.015; This is a reasonable model.
61. \$8081.13 **73.** As n gets larger, the terms get closer to 1. **75.** As n gets larger, the terms get closer to 0. **77.** c **79.** $a_n = n^2$
81. $a_n = n(n + 2)$ **83.** $a_n = 2n + 3$ **85.** $a_n = (-2)^{n+1}$ **87.** $n^2 + 7n + 12$ **89.** $\log 2 + \log 4 + \log 6 + \log 8 = \log 384$
91. 7, 12, 17, 22 **92.** $2xy^2\sqrt[3]{5xy}$ **93.** $(3x - 2)(9x^2 + 6x + 4)$ **94.** $\left\{-\frac{6}{5}, 4\right\}$

Section 14.2

Check Point Exercises

1. 100, 70, 40, 10, −20, −50 **2.** −34 **3. a.** $a_n = 9700n + 149{,}300$ **b.** \$304,500 **4.** 360 **5.** 2460 **6.** \$686,300

Exercise Set 14.2

1. 4 **3.** 5 **5.** −3 **7.** 200, 220, 240, 260, 280, 300 **9.** −7, −3, 1, 5, 9, 13 **11.** 300, 210, 120, 30, −60, −150 **13.** $\frac{5}{2}, 2, \frac{3}{2}, 1, \frac{1}{2}, 0$
15. −0.4, −2, −3.6, −5.2, −6.8, −8.4 **17.** 33 **19.** 252 **21.** 955 **23.** −142 **25.** $a_n = 4n - 3$; $a_{20} = 77$
27. $a_n = 11 - 4n$; $a_{20} = -69$ **29.** $a_n = -4n - 16$; $a_{20} = -96$ **31.** $a_n = \frac{1}{3}n - \frac{2}{3}$; $a_{20} = 6$ **33.** $a_n = 4.3 - 0.3n$; $a_{20} = -1.7$ **35.** 1220
37. 4400 **39.** 5050 **41.** 3660 **43.** 396 **45.** $8 + 13 + 18 + \cdots + 88 = 816$ **47.** $2 + (-1) + (-4) + \cdots + (-85) = -1245$
49. $4 + 8 + 12 + \cdots + 400 = 20{,}200$ **51.** 7 **53.** 22 **55.** 847 **57.** $f(x) = -4x + 5$ **59.** $a_n = 3n - 2$ **61. a.** $a_n = 1.7n + 148.3$
b. 216.3 lb **63.** Answers will vary. **65.** company A; \$1400 **67. a.** \$30,049 **b.** \$30,046; very well
71. Company A: \$307,000; Company B: \$324,000; Company B pays the greater total amount. **73.** 2869 seats **81.** 200th term
83. $S_n = \frac{n}{2}[1 + (2n - 1)] = \frac{n}{2}(2n) = n^2$ **84.** {1005} **85.** $\{x \mid -5 \le x \le 2\}$ or $[-5, 2]$ **86.** $P = \frac{At}{t - A}$

Section 14.3

Check Point Exercises

1. $12, 6, 3, \frac{3}{2}, \frac{3}{4}, \frac{3}{8}$ **2.** 3645 **3.** $a_n = 3(2)^{n-1}$; $a_8 = 384$ **4.** 9842 **5.** 19,680 **6.** approximately \$2,371,746
7. approximately \$1,327,778 **8.** 9 **9.** $\frac{1}{1}$ or 1 **10.** \$4000

Exercise Set 14.3

1. $r = 3$ **3.** $r = -2$ **5.** $r = \frac{3}{2}$ **7.** $r = -0.1$ **9.** 2, 6, 18, 54, 162 **11.** $20, 10, 5, \frac{5}{2}, \frac{5}{4}$ **13.** −4, 40, −400, 4000, −40,000
15. $-\frac{1}{4}, \frac{1}{2}, -1, 2, -4$ **17.** $a_8 = 768$ **19.** $a_{12} = -10{,}240$ **21.** $a_6 = -200$ **23.** $a_8 = 0.1$ **25.** $a_n = 3(4)^{n-1}$; $a_7 = 12{,}288$

27. $a_n = 18\left(\frac{1}{3}\right)^{n-1}; a_7 = \frac{2}{81}$ **29.** $a_n = 1.5(-2)^{n-1}; a_7 = 96$ **31.** $a_n = 0.0004(-10)^{n-1}; a_7 = 400$ **33.** 531,440 **35.** 2049
37. $\frac{16{,}383}{2}$ or 8191.5 **39.** 9840 **41.** 10,230 **43.** $\frac{63}{128}$ **45.** $\frac{3}{2}$ **47.** 4 **49.** $\frac{2}{3}$ **51.** 20 **53.** $\frac{5}{9}$ **55.** $\frac{47}{99}$ **57.** $\frac{257}{999}$
59. arithmetic; $d = 1$ **61.** geometric; $r = 2$ **63.** neither **65.** 2435 **67.** 2280 **69.** -140 **71.** $a_2 = 12, a_3 = 18$ **73.** \$16,384
75. approximately \$3,795,957 **77. a.** 1.013 for each division **b.** $a_n = 29.76(1.013)^{n-1}$ **c.** approximately 33.86 million; very well **79.** \$32,767
81. approximately \$793,583 **83.** approximately 130.26 in. **85.** approximately \$844,706 **87.** approximately \$94,834 **89.** \$9 million **91.** $\frac{1}{3}$

103.

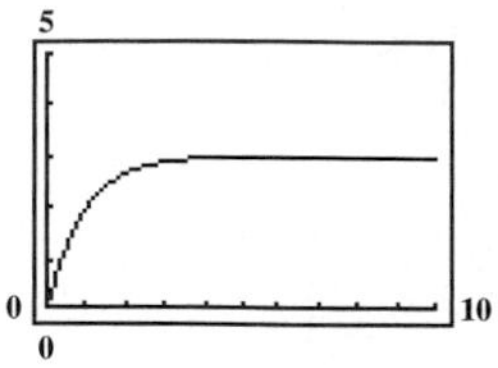

horizontal asymptote: $y = 3$; sum of series: 3

105. d
107. approximately \$442
108. $2\sqrt{7}$
109. $\left\{\frac{-1 \pm \sqrt{33}}{4}\right\}$
110. $-3(\sqrt{3} + \sqrt{5})$

Mid-Chapter 14 Check Point

1. $1, -2, \frac{3}{2}, -\frac{2}{3}, \frac{5}{24}$ **2.** 5, 2, −1, −4, −7 **3.** 5, −15, 45, −135, 405 **4.** $a_n = 4n - 2; a_{20} = 78$ **5.** $a_n = 3(2)^{n-1}; a_{10} = 1536$
6. $a_n = -\frac{1}{2}n + 2; a_{30} = -13$ **7.** 5115 **8.** 2350 **9.** 6820 **10.** −29,300 **11.** 44 **12.** 3725 **13.** $\frac{1995}{64}$
14. $\frac{5}{7}$ **15.** $\frac{5}{11}$ **16.** Answers will vary. An example is $\sum_{i=1}^{18} \frac{i}{i+2}$. **17.** 464 ft; 3600 ft **18.** \$311,249

Section 14.4

Check Point Exercises

1. a. 20 **b.** 1 **c.** 28 **d.** 1 **2.** $x^4 + 4x^3 + 6x^2 + 4x + 1$ **3.** $x^5 - 10x^4y + 40x^3y^2 - 80x^2y^3 + 80xy^4 - 32y^5$ **4.** $4032x^5y^4$

Exercise Set 14.4

1. 56 **3.** 12 **5.** 1 **7.** 4950 **9.** $x^3 + 6x^2 + 12x + 8$ **11.** $27x^3 + 27x^2y + 9xy^2 + y^3$ **13.** $125x^3 - 75x^2 + 15x - 1$
15. $16x^4 + 32x^3 + 24x^2 + 8x + 1$ **17.** $x^8 + 8x^6y + 24x^4y^2 + 32x^2y^3 + 16y^4$ **19.** $y^4 - 12y^3 + 54y^2 - 108y + 81$
21. $16x^{12} - 32x^9 + 24x^6 - 8x^3 + 1$ **23.** $c^5 + 10c^4 + 40c^3 + 80c^2 + 80c + 32$ **25.** $x^5 - 5x^4 + 10x^3 - 10x^2 + 5x - 1$
27. $243x^5 - 405x^4y + 270x^3y^2 - 90x^2y^3 + 15xy^4 - y^5$ **29.** $64a^6 + 192a^5b + 240a^4b^2 + 160a^3b^3 + 60a^2b^4 + 12ab^5 + b^6$ **31.** $x^8 + 16x^7 + 112x^6$
33. $x^{10} - 20x^9y + 180x^8y^2$ **35.** $x^{32} + 16x^{30} + 120x^{28}$ **37.** $y^{60} - 20y^{57} + 190y^{54}$ **39.** $240x^4y^2$ **41.** $126x^5$ **43.** $56x^6y^{15}$
45. $-\frac{21}{2}x^6$ **47.** $x^{12} + 4x^7 + 6x^2 + \frac{4}{x^3} + \frac{1}{x^8}$ **49.** $x - 3x^{1/3} + \frac{3}{x^{1/3}} - \frac{1}{x}$ **51.** 8 **53.** $4x^3 + 6x^2h + 4xh^2 + h^3$ **55.** 252
57. a. $g(x) = 0.12x^3 + 0.08x^2 + 0.24x + 18.24$ **b.** $f(5) = 20; g(2) = 20$; They give the exact value of the number shown.

69.

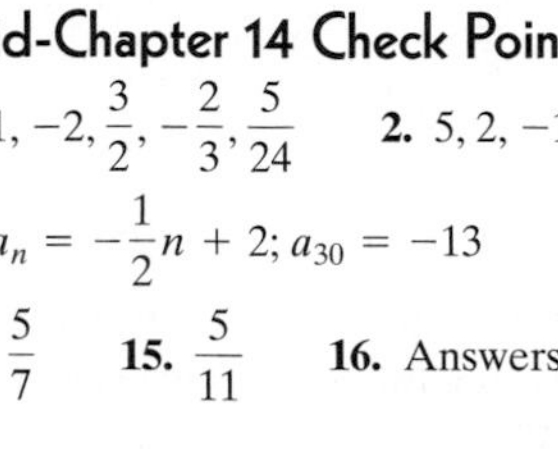

f_2, f_3, and f_4 are approaching $f_1 = f_5$.

71. $x^3 - 3x^2 + 3x - 1$
73. $x^6 + 12x^5 + 60x^4 + 160x^3 + 240x^2 + 192x + 64$
75. b **77.** $10x^4y^6$
78. $a^2 + 4a + 6$
79. $f(g(x)) = 4x^2 - 2x - 6; g(f(x)) = 2x^2 + 10x - 3$
80. $\frac{2x^2 - 9x - 1}{2(x-4)(x+3)}$

Section 14.5

Check Point Exercises

1. 72 **2.** 729 **3.** 676,000 **4.** 840 **5.** 720 **6. a.** combinations **b.** permutations **7.** 210 **8.** 1820

Exercise Set 14.5

1. 3024 **3.** 6720 **5.** 720 **7.** 1 **9.** 126 **11.** 330 **13.** 1 **15.** 1 **17.** combinations **19.** permutations **21.** 0 **23.** $\frac{3}{4}$
25. −9499 **27.** $\frac{3}{68}$ **29.** 27 **31.** 40 **33.** 243 **35.** 144 **37.** 120 **39.** 6 **41.** 720 **43.** 8,648,640 **45.** 120 **47.** 15,120 **49.** 20
51. 495 **53.** 24,310 **55.** 22,957,480 **57.** 360 **59.** 1716 **61.** 1140 **63.** 840 **65.** 2730
77. c **79.** 144 **81.** $(f \circ g)(x) = 16x^2 - 6$ **82.** $\{x \mid x < 1 \text{ or } x > 4\}$ or $(-\infty, 1) \cup (4, \infty)$ **83.**

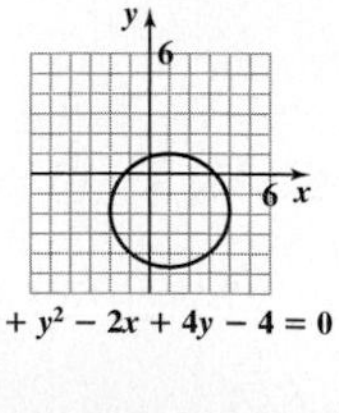

$x^2 + y^2 - 2x + 4y - 4 = 0$

Section 14.6

Check Point Exercises

1. a. $\frac{7664}{100{,}000} \approx 0.077$ **b.** $\frac{720}{800} = 0.9$ **c.** $\frac{720}{7664} \approx 0.094$ **2.** $\frac{1}{3}$ **3.** $\frac{1}{9}$ **4.** $\frac{1}{13}$ **5.** $\frac{1}{13{,}983{,}816} \approx 0.0000000715$ **6.** $\frac{120}{137}$ **7.** $\frac{1}{3}$ **8.** $\frac{3}{4}$ **9.** $\frac{122}{137}$ **10.** $\frac{1}{361} \approx 0.00277$ **11.** $\frac{1}{16}$

Exercise Set 14.6

1. $\frac{23}{89}$ **3.** $\frac{36}{89}$ **5.** $\frac{6}{89}$ **7.** $\frac{1}{3}$ **9.** $\frac{6}{23}$ **11.** $\frac{1}{6}$ **13.** $\frac{1}{2}$ **15.** $\frac{1}{3}$ **17.** $\frac{1}{13}$ **19.** $\frac{3}{13}$ **21.** $\frac{1}{4}$ **23.** $\frac{7}{8}$ **25.** $\frac{1}{12}$ **27.** $\frac{1}{18{,}009{,}460} \approx 0.0000000555$; $\frac{5}{900{,}473} \approx 0.00000555$ **29. a.** 2,598,960 **b.** 1287 **c.** $\frac{1287}{2{,}598{,}960} \approx 0.000495$ **31.** $\frac{43}{58}$ **33.** $\frac{50}{87}$ **35.** $\frac{113}{174}$ **37.** $\frac{12}{13}$ **39.** $\frac{2}{13}$ **41.** $\frac{7}{13}$ **43.** $\frac{3}{4}$ **45.** $\frac{33}{40}$ **47.** $\frac{1}{36}$ **49.** $\frac{1}{3}$ **51.** $\frac{1}{64}$ **53. a.** $\frac{1}{256}$ **b.** $\frac{1}{4096}$ **c.** $\left(\frac{15}{16}\right)^{10} \approx 0.524$ **d.** $1 - \left(\frac{15}{16}\right)^{10} \approx 0.476$ **63.** $\frac{1}{12}$; Answers will vary. **67. a.** $\frac{12}{25}$ **b.** $\frac{3}{10}$

69.

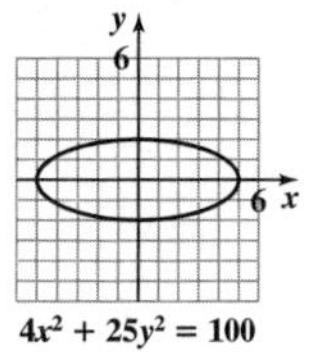

70. 3 or {3} **71.** $x^2 + 3x - 3 - \frac{4}{x+2}$

Chapter 14 Review Exercises

1. 3, 10, 17, 24 **2.** $-\frac{3}{2}, \frac{4}{3}, -\frac{5}{4}, \frac{6}{5}$ **3.** $1, 1, \frac{1}{2}, \frac{1}{6}$ **4.** $\frac{1}{2}, -\frac{1}{4}, \frac{1}{8}, -\frac{1}{16}$ **5.** 95 **6.** −20 **7.** Answers will vary; An example is $\sum_{i=1}^{15} \frac{i}{i+2}$. **8.** Answers will vary; Examples are $\sum_{i=4}^{13} i^3$ or $\sum_{i=1}^{10} (i+3)^3$. **9.** 7, 11, 15, 19, 23, 27 **10.** −4, −9, −14, −19, −24, −29 **11.** $\frac{3}{2}, 1, \frac{1}{2}, 0, -\frac{1}{2}, -1$ **12.** 20 **13.** −30 **14.** −38 **15.** $a_n = 4n - 11; a_{20} = 69$ **16.** $a_n = 220 - 20n; a_{20} = -180$ **17.** $a_n = -\frac{23}{2} - \frac{1}{2}n; a_{20} = -\frac{43}{2}$ **18.** $a_n = 22 - 7n; a_{20} = -118$ **19.** 1727 **20.** 225 **21.** 15,150 **22.** 440 **23.** −500 **24.** −2325 **25. a.** $a_n = 0.52n + 19.48$ **b.** 77.2% **26.** $418,500 **27.** 1470 seats **28.** 3, 6, 12, 24, 48 **29.** $\frac{1}{2}, \frac{1}{4}, \frac{1}{8}, \frac{1}{16}, \frac{1}{32}$ **30.** $16, -4, 1, -\frac{1}{4}, \frac{1}{16}$ **31.** −5, 5, −5, 5, −5 **32.** $a_7 = 1458$ **33.** $a_6 = \frac{1}{2}$ **34.** $a_5 = -48$ **35.** $a_n = 1(2)^{n-1}$ or $a_n = 2^{n-1}$; $a_8 = 128$ **36.** $a_n = 100\left(\frac{1}{10}\right)^{n-1}$; $a_8 = \frac{1}{100{,}000} = 0.00001$ **37.** $a_n = 12\left(-\frac{1}{3}\right)^{n-1}$; $a_8 = -\frac{4}{729}$ **38.** 17,936,135 **39.** $\frac{127}{8}$ or 15.875 **40.** 19,530 **41.** −258 **42.** $\frac{341}{128}$ **43.** $\frac{27}{2}$ **44.** $\frac{4}{3}$ **45.** $-\frac{18}{15}$ **46.** 20 **47.** $\frac{2}{3}$ **48.** $\frac{47}{99}$ **49. a.** Dividing the population for each year by the population in the preceding year gives approximately 1.038 for each division. **b.** $a_n = 19.96(1.038)^{n-1}$ **c.** approximately 27.92 million **50.** $42,823; $223,210 **51.** $120,113 **52.** $\$9\frac{1}{3}$ million **53.** 165 **54.** 4005 **55.** $8x^3 + 12x^2 + 6x + 1$ **56.** $x^8 - 4x^6 + 6x^4 - 4x^2 + 1$ **57.** $x^5 + 10x^4y + 40x^3y^2 + 80x^2y^3 + 80xy^4 + 32y^5$ **58.** $x^6 - 12x^5 + 60x^4 - 160x^3 + 240x^2 - 192x + 64$ **59.** $x^{16} + 24x^{14} + 252x^{12}$ **60.** $x^9 - 27x^8 + 324x^7$ **61.** $80x^2$ **62.** $4860x^2$ **63.** 336 **64.** 15,120 **65.** 56 **66.** 78 **67.** 20 **68.** 243 **69.** 32,760 **70.** 4845 **71.** 1140 **72.** 116,280 **73.** 120 **74.** $\frac{17}{100}$ **75.** $\frac{67}{100}$ **76.** $\frac{83}{100}$ **77.** $\frac{29}{50}$ **78.** $\frac{13}{33}$ **79.** $\frac{2}{9}$ **80.** $\frac{2}{3}$ **81.** $\frac{2}{3}$ **82.** $\frac{2}{13}$ **83.** $\frac{7}{13}$ **84.** $\frac{5}{6}$ **85.** $\frac{5}{6}$ **86.** $\frac{1}{6}$ **87. a.** $\frac{1}{15{,}504} \approx 0.0000645$ **b.** $\frac{25}{3876} \approx 0.00645$ **88.** $\frac{1}{32}$ **89. a.** 0.04 **b.** 0.008 **c.** 0.4096

Chapter 14 Test

1. $1, -\frac{1}{4}, \frac{1}{9}, -\frac{1}{16}, \frac{1}{25}$ **2.** 105 **3.** Answers will vary; Examples are $\sum_{i=2}^{21} \frac{i}{i+1}$ or $\sum_{i=1}^{20} \frac{i+1}{i+2}$. **4.** $a_n = 5n - 1; a_{12} = 59$ **5.** $a_n = 16\left(\frac{1}{4}\right)^{n-1}$; $a_{12} = \frac{1}{262{,}144}$ **6.** −385 **7.** 550 **8.** −2387 **9.** −21,846 **10.** 8 **11.** $\frac{73}{99}$ **12.** $276,427 **13.** 36 **14.** $x^{10} - 5x^8 + 10x^6 - 10x^4 + 5x^2 - 1$ **15.** $x^8 + 8x^7y^2 + 28x^6y^4$ **16.** 990 **17.** 210 **18.** 10,000 **19.** $\frac{3}{5}$ **20.** $\frac{39}{50}$ **21.** $\frac{3}{5}$ **22.** $\frac{9}{19}$ **23.** $\frac{10}{1001}$ **24.** $\frac{8}{13}$ **25.** $\frac{3}{5}$ **26.** $\frac{1}{256}$ **27.** $\frac{1}{16}$

Cumulative Review Exercises (Chapters 1–14)

1. $\{22\}$ **2.** $\{5 \pm 7i\}$ **3.** $\{x|x < -3 \text{ or } x > 2\}$ or $(-\infty, -3) \cup (2, \infty)$ **4.** $\{-2\}$ **5.** $\varnothing$

6. $\left\{x \middle| x < \frac{2}{3}\right\}$ or $\left(-\infty, \frac{2}{3}\right)$ **7.** $\{(4, 0, -5)\}$ **8.** $\{9\}$ **9.** $\{(2, 1), (-2, 1), (2, -1), (-2, -1)\}$ **10.** $\{(-14, -20), (2, -4)\}$

11.
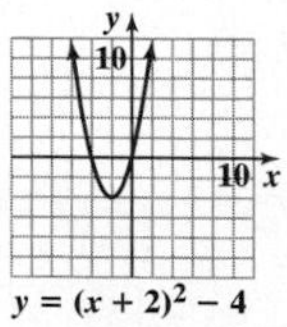

12.
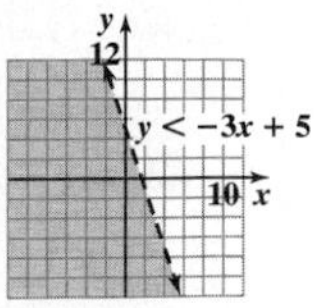

13.
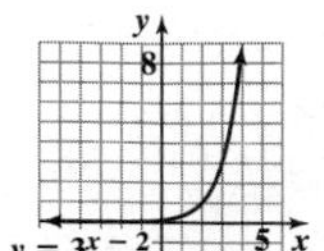

14.
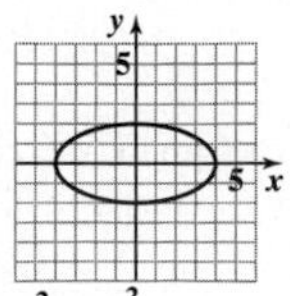

15.
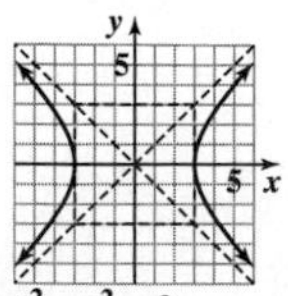

16. $\frac{2x^2 + 5x - 2}{(x - 5)(x + 2)}$ **17.** $-\frac{x + 1}{x - 1}$ **18.** $2\sqrt{5} + 2\sqrt{2}$

19. $12\sqrt{5}$ **20.** $\frac{5\sqrt[3]{4xy^2}}{2xy}$ **21.** $(x + y)(5a - 4b)$ **22.** $\log\left(\frac{x^5}{\sqrt{y}}\right)$ **23.** $p = \frac{qf}{q - f}$ **24.** $3\sqrt{10} \approx 9.49$ units **25.** 208 **26.** 1800

27. $\frac{1}{3}$ **28.** $16x^4 - 32x^3y^3 + 24x^2y^6 - 8xy^9 + y^{12}$ **29.** $\{x|x \neq -5 \text{ and } x \neq 3\}$ or $(-\infty, -5) \cup (-5, 3) \cup (3, \infty)$ **30.** $\{x|x \geq 3\}$ or $[3, \infty)$

31. $\{x|x < 1\}$ or $(-\infty, 1)$ **32.** 8 ft by 3 ft **33.** \$18,500 **34.** $k = \frac{1}{2 \ln 4} \approx 0.3607$; about 0.298 or $\frac{298}{1000}$

Appendix A

Check Point Exercises

1. $(4, -3, 1)$ or $\{(4, -3, 1)\}$ **2. a.** $\left[\begin{array}{rrr|r} 1 & 6 & -3 & 7 \\ 4 & 12 & -20 & 8 \\ -3 & -2 & 1 & -9 \end{array}\right]$ **b.** $\left[\begin{array}{rrr|r} 1 & 3 & -5 & 2 \\ 1 & 6 & -3 & 7 \\ -3 & -2 & 1 & -9 \end{array}\right]$ **c.** $\left[\begin{array}{rrr|r} 4 & 12 & -20 & 8 \\ 1 & 6 & -3 & 7 \\ 0 & 16 & -8 & 12 \end{array}\right]$

3. $(-1, 2)$ or $\{(-1, 2)\}$ **4.** $(5, 2, 3)$ or $\{(5, 2, 3)\}$

Appendix A Exercise Set

1. $x - 3y = 11$; $y = -3$; $(2, -3)$ **3.** $x - 3y = 1$; $y = -1$; $(-2, -1)$ **5.** $x - 4z = 5$; $y - 12z = 13$; $z = -\frac{1}{2}$; $\left(3, 7, -\frac{1}{2}\right)$

7. $x + \frac{1}{2}y + z = \frac{11}{2}$; $y + \frac{3}{2}z = 7$; $z = 4$; $(1, 1, 4)$ **9.** $\left[\begin{array}{rr|r} 1 & -\frac{3}{2} & 5 \\ 2 & 2 & 5 \end{array}\right]$ **11.** $\left[\begin{array}{rr|r} 1 & -\frac{4}{3} & 2 \\ 3 & 5 & -2 \end{array}\right]$ **13.** $\left[\begin{array}{rr|r} 1 & -3 & 5 \\ 0 & 12 & -6 \end{array}\right]$ **15.** $\left[\begin{array}{rr|r} 1 & -\frac{3}{2} & \frac{7}{2} \\ 0 & \frac{17}{2} & -\frac{17}{2} \end{array}\right]$

17. $\left[\begin{array}{rrr|r} 1 & -3 & 2 & 5 \\ 1 & 5 & -5 & 0 \\ 3 & 0 & 4 & 7 \end{array}\right]$ **19.** $\left[\begin{array}{rrr|r} 1 & -3 & 2 & 0 \\ 0 & 10 & -7 & 7 \\ 2 & -2 & 1 & 3 \end{array}\right]$ **21.** $\left[\begin{array}{rrr|r} 1 & 1 & -1 & 6 \\ 0 & -3 & 3 & -15 \\ 0 & -4 & 2 & -14 \end{array}\right]$ **23.** $\{(4, 2)\}$ **25.** $\{(3, -3)\}$ **27.** $\{(2, -5)\}$ **29.** no solution or $\varnothing$ **31.** infinitely many solutions; dependent equations

33. $\{(1, -1, 2)\}$ **35.** $\{(3, -1, -1)\}$ **37.** $\{(1, 2, -1)\}$ **39.** $\{(1, 2, 3)\}$ **41.** no solution or $\varnothing$ **43.** infinitely many solutions; dependent equations
45. $\{(-1, 2, -2)\}$ **47.** $w - x + y + z = 3$; $x - 2y - z = 0$; $y + 6z = 17$; $z = 3$; $(2, 1, -1, 3)$

49. $\left[\begin{array}{rrrr|r} 1 & -1 & 1 & 1 & 3 \\ 0 & 1 & -2 & -1 & 0 \\ 0 & 2 & 1 & 2 & 5 \\ 0 & 6 & -3 & -1 & -9 \end{array}\right]$ **51.** $\{(1, 2, 3, -2)\}$

Appendix B

Check Point Exercises

1. a. -4 **b.** -17 **2.** $(4, -2)$ or $\{(4, -2)\}$ **3.** 80 **4.** $(2, -3, 4)$ or $\{(2, -3, 4)\}$

Appendix B Exercise Set

1. 1 **3.** -29 **5.** 0 **7.** 33 **9.** $-\frac{7}{16}$ **11.** $\{(5, 2)\}$ **13.** $\{(2, -3)\}$ **15.** $\{(3, -1)\}$ **17.** $\{(4, 0)\}$ **19.** $\{(4, 2)\}$ **21.** $\{(7, 4)\}$
23. inconsistent; no solution or $\varnothing$ **25.** dependent equations; infinitely many solutions **27.** 72 **29.** -75 **31.** 0 **33.** $\{(-5, -2, 7)\}$
35. $\{(2, -3, 4)\}$ **37.** $\{(3, -1, 2)\}$ **39.** $\{(2, 3, 1)\}$ **41.** -42 **43.** $2x - 4y = 8$; $3x + 5y = 10$ **45.** -11 **47.** 4 **49.** 28 sq units **51.** yes

53. $\begin{vmatrix} x & y & 1 \\ 3 & -5 & 1 \\ -2 & 6 & 1 \end{vmatrix} = 0$; $y = -\frac{11}{5}x + \frac{8}{5}$

APPLICATIONS INDEX

A

B

C

D

E

F

G

H

I

J

K

L

R

U

V

W

Y

SUBJECT INDEX

D

E

F

G

H

I

J

K

L

Q

R

T

PHOTO CREDITS

CHAPTER 1 **CO** Nita Winter Photography **p. 2** Photograph by Richard Strauss, Smithsonian Institution, Collection of the Supreme Court of the United States **p. 17** SuperStock, Inc. **p. 26** Allan Tannenbaum/The Image Works **p. 28** S.S. Archives/Shooting Star International Photo Agency **p. 30** Globe Photos, Inc. **p. 37** Jake Rajs/Getty Images, Inc.-Stone/Allstock **p. 46** Brian S. Wilcox/ImageState/International Stock Photography Ltd. **p. 51** SuperStock, Inc. **p. 59** Getty Images, Inc.-Stone/Allstock **p. 66** Bob Daemmrich/Stock Boston **p. 78** (left) Steve Smith/Getty Images, Inc.-Taxi; (right) Peter Cade/Getty Images, Inc.-Stone/Allstock

CHAPTER 2 **CO** Warner Bros. TV/Bright/Kauffman/Crane Pro/Picture Desk, Inc./Kobal Collection **p. 103** A. Ramu/Stock Boston **p. 111** (left) Robert Daly/Getty Images, Inc.-Stone/Allstock; (right) Dan Bosler/Getty Images, Inc.-Stone/Allstock **p. 117** Faidley/Agliolo/International Stock/Photo Network **p. 118** John Dominis/Index Stock Imagery, Inc. **p. 120** Squeak Carnwath "Equations" 1981, oil on cotton canvas 96 in. h × 72 in. w. **p. 125** Kurt Amsler/Vandystadt/Getty Images, Inc.-Allsport Photography **p. 126** Erlendur Berg/SuperStock, Inc. **p. 138** Chuck Savage/Corbis/Stock Market **p. 143** Picture Desk, Inc./Kobal Collection **p. 146** 20th Century Fox/Shooting Star International Photo Agency/Paramount Pictures Corporation, Inc. **p. 148** United Media/United Feature Syndicate, Inc. PEANUTS reprinted by permission of United Feature Syndicate, Inc. **p. 149** © Manfred Kage/Peter Arnold, Inc. **p. 150** Benjamin Shearn/Getty Images, Inc.-Taxi **p. 151** J. Silver/SuperStock, Inc. **p. 161** Pascal Quittemelle/Stock Boston

CHAPTER 3 **CO** Daniel J. Cox/Getty Images, Inc.-Stone/Allstock **p. 185** Hans Neleman/Neleman Inc. **p. 191** Simon Bruty/Getty Images, Inc.-Allsport Photography **p. 194** Melvin L. Prueitt **p. 197** Todd A. Gipstein/Corbis/Bettmann **p. 203** Bob Winsett/ Corbis/Bettman **p. 208** (left) Paul Avis/Getty Images, Inc.-Taxi (right) Bob Schatz/Getty Images, Inc.-Stone/Allstock **p. 209** Carol Simowitz/San Francisco Convention and Visitors Bureau **p. 212** Porterfield/Chickering/Photo Researchers, Inc. **p. 216** Reuters/Landov LLC **p. 226** Culver Pictures, Inc.

CHAPTER 4 **CO** (left) Brown Brothers; (right) AP Wide World Photos **p. 246** Dave Martin/AP Wide World Photos **p. 252** AP Wide World Photos **p. 255** Paul Sakuma/AP Wide World Photos **p. 263** Karen Ducey/AP Wide World Photos **p. 271** Pierre Tremblay/Masterfile Corporation **p. 273** Chris Everard/Getty Images, Inc.-Stone/Allstock **p. 285** United Media/United Feature Syndicate, Inc. **p. 286** (top) Courtesy of the Lilly Library, Indiana University, Bloomington, Indiana **p. 286** (bottom) David W. Hamilton/Getty Images, Inc.-Image Bank

CHAPTER 5 **CO** Tim Davis/Getty Images, Inc.-Stone/Allstock **p. 306** Purdue News Service/AP Wide World Photos **p. 315** Mehau Kulyk/Science Photo Library/Photo Researchers, Inc. **p. 322** Barbara Penoyar/Getty Images, Inc.-Photodisc **p. 325** Photofest **p. 334** Dale Wilson/Masterfile Corporation **p. 342** Reed Saxon/AP Wide World Photos **p. 350** (left) Townsend P. Dickinson/The Image Works; (right)Townsend P. Dickinson/The Image Works **p. 362** Robert Brenner/PhotoEdit **p. 371** David Weintraub/Photo Researchers, Inc.

CHAPTER 6 **CO** F. Stuart Westmorland/Photo Researchers, Inc. **p. 383** Gabe Palmer/Mug Shots/Corbis/Stock Market **p. 391** Tom Sanders/ImageState/International Stock Photography Ltd. **p. 399** Hood Museum of Art **p. 407** Jay Brousseau/Getty Images, Inc.-Image Bank **p. 414** Orion Pictures Corp./Photofest **p. 416** Robert F. Blitzer **p. 423** Stephen G. Maka/Index Stock Imagery, Inc.

CHAPTER 7 **CO** Dale Wilson/Masterfile Corporation **p. 442** Steve Vidler/SuperStock, Inc. **p. 452** Bebeto Matthews/AP Wide World Photos **p. 466** LWA-Dann Tardif/Corbis/Stock Market **p. 478** (left) Robert F. Blitzer. Art by Henry L. Small; (right) Robert F. Blitzer. Art by Henry L. Small **p. 485** Edouard Berne/Getty Images, Inc.-Stone/Allstock **p. 495** Doc White/TCL/Getty Images, Inc.-Taxi **p. 498** Bob Daemmrich/The Image Works **p. 501** Frans Lanting/Minden Pictures **p. 505** The Granger Collection **p. 507** Stephen Simpson/Getty Images, Inc.-Taxi **p. 509** Gerald French/Corbis/Bettmann **p. 513** David Madison/Duomo Photography Incorporated **p. 514** Duomo Photography Incorporated **p. 515** UPI/Corbis/Bettmann

CHAPTER 8 **CO** Chris Salvo/Getty Images, Inc.-Taxi **p. 531** Photofest **p. 546** General Motors Corp. Used with permission, GM Media Archives. **p. 557** Chuck Savage/Corbis/Stock Market

CHAPTER 9 **CO** Peter Griffith **p. 579** (left) Esbin-Anderson/ Photo Network; (right) Richard Falco/Black Star **p. 587** Laura Rauch/AP/Wide World Photos **p. 592** (left) Sue Ann Miller/Getty Images, Inc.-Stone/Allstock; (right) Gary Walts/The Image Works **p. 601** CBS-TV/Picture Desk, Inc./Kobal Collection **p. 610** Frank Clarkson/Getty Images, Inc.-Liaison **p. 621** AP/Wide World Photos **p. 622** Pascal Parrot/Corbis/Sygma

CHAPTER 10 **CO** Rick Rusing/eStock Photography LLC **p. 638** Reuters/Getty Images, Inc.-Hulton Archive Photos **p. 650** Peter Arnold, Inc. © Fred Bavendam/Peter Arnold, Inc. **p. 655** David Pollack/Corbis/Bettmann **p. 662** Greg Heins/Addison Gallery of American Art. **p. 670** Kauko Helavuo/Getty Images, Inc.-Image Bank **p. 678** United Media/United Feature Syndicate, Inc. PEANUTS reprinted by permission of United Feature Syndicate, Inc. **p. 683** John G. Ross/Photo Researchers, Inc. **p. 685** Alan Thornton/Getty Images, Inc.-Stone/Allstock **p. 690** (left) Victor Ramos/ImageState/International Stock Photography Ltd.; (right) Shoot-ze/Masterfile Corporation **p. 700** The Cartoon Bank © 2003 Roz Chast from Cartoonbank. com. All rights reserved. **p. 704** Stamp from the private collection of Professor C. M. Lang, photography by Gary J. Shulfer, University of Wisconsin, Stevens Point. "Germany: #5"; Scott Standard Postage Stamp Catalogue, Scott Pub. Co., Sidney, Ohio. **p. 706** R.F. Voss "29-Fold M-set Seahorse" Computer-generated image. © 1990 Richard F. Voss/IBM Research.

CHAPTER 11 **CO** John & Lisa Merrill/Getty Images, Inc.-Stone/Allstock **p. 718** Les Stone/The Image Works **p. 734** © Steve Chenn/CORBIS **p. 735** © Bettmann/CORBIS **p. 744** Christie's Images Inc. **p. 750** Joe McBride/Getty Images, Inc.-Stone/Allstock **p. 751** (left) Simon Bruty/Getty Images, Inc.-Allsport Photography **p. 751** (right) © 1981Scott Kim, scottkim.com. All rights reserved. **p. 758** Stephanie Sinclair/Corbis/Bettmann **p. 768** Michael Lutzky © 2000, The Washington Post Writers Group. Reprinted with permission. **p. 775** © Jay Syverson/CORBIS

CHAPTER 12 **CO** Anthony Neste **p. 799** Bruce Weaver/AP/Wide World Photos **p. 812** David Weintraub/Photo Researchers, Inc. **p. 825** Ron Chapple/ Getty Images, Inc.-Taxi **p. 836** (left) Gary Kufner/Sharpshooters/ Corbis; (right) Gary Kufner/Sharpshooters/ Corbis **p. 848** Bullit Marquez/AP/Wide World Photos **p. 852** Jean-Marie Chauvet/Corbis/Sygma

CHAPTER 13 **CO** SuperStock, Inc. **p. 873** Skip Moody/Dembinsky Photo Associates **p. 880** © Kevin Fleming/CORBIS **p. 888** David Austen **p. 892** Andrea Pistolesi/Getty Images, Inc.-Image Bank **p. 902** Jeff Hester and Paul Scowen (Arizona State University), and NASA **p. 907** © Berenice Abbott/Commerce Graphics Ltd., Inc. **p. 908** Space Telescope Science Institute **p. 909** Equinox Gallery **p. 914** Index Stock Imagery, Inc.

CHAPTER 14 **CO** Reuters/Colin Braley/Corbis/Bettmann **p. 934** Dick Morton **p. 944** David Young-Wolff/PhotoEdit **p. 954** SuperStock, Inc. **p. 957** Richard Lord/The Image Works **p. 960** (bills) U.S. Bureau of Engraving and Printing **p. 970** Dr. Rudolph Schild/Science Photo Library/Photo Researchers, Inc. **p. 973** From Science and Civilization in China, Vol. 3, pg. 135, Fig. "Pascal's triangle as depicted by Chu Shih Chieh," by Joseph Needham, 1959. Reprinted with the permission of Cambridge University Press. **p. 978** L. Schwartzward/Corbis/Sygna. **p. 979** SuperStock, Inc. **p. 981** Getty Images, Inc.-PhotoDisc **p. 983** (Marilyn Monroe) Picture Desk, Inc./Kobal Collection (James Dean) Imapress/Globe Photos, Inc. (Jim Morrison) Michael Ochs Archives.com (Kurt Cobain) S.I.N./Corbis/Bettmann (Selena) AP/Wide World Photos **p. 990** Kwame Zikomo/SuperStock, Inc. **p. 995** © Damon Higgins/*The Palm Beach Post* **p. 1001** UPI/Corbis/Bettmann

APPENDIX A **p. 1014** Steven Needham/Envision Stock Photography, Inc.

APPENDIX B **p. 1024** David Parker/Science Museum/Science Photo Library/Photo Researchers, Inc.

INTRODUCTORY & INTERMEDIATE ALGEBRA FOR COLLEGE STUDENTS 2E, CHAPTER TEST PREP VIDEO CDs

Robert Blitzer
0-13-149259-4
© 2006 Pearson Education, Inc.
Pearson Prentice Hall
Pearson Education, Inc.
Upper Saddle River, NJ 07458
Pearson Prentice Hall™ is a trademark of Pearson Education, Inc.

YOU SHOULD CAREFULLY READ THE TERMS AND CONDITIONS BEFORE USING THE CD-ROM PACKAGE. USING THIS CD-ROM PACKAGE INDICATES YOUR ACCEPTANCE OF THESE TERMS AND CONDITIONS.

Pearson Education, Inc. provides this program and licenses its use. You assume responsibility for the selection of the program to achieve your intended results, and for the installation, use, and results obtained from the program. This license extends only to use of the program in the United States or countries in which the program is marketed by authorized distributors.

LICENSE GRANT

You hereby accept a nonexclusive, nontransferable, permanent license to install and use the program ON A SINGLE COMPUTER at any given time. You may copy the program solely for backup or archival purposes in support of your use of the program on the single computer. You may not modify, translate, disassemble, decompile, or reverse engineer the program, in whole or in part.

TERM

The License is effective until terminated. Pearson Education, Inc. reserves the right to terminate this License automatically if any provision of the License is violated. You may terminate the License at any time. To terminate this License, you must return the program, including documentation, along with a written warranty stating that all copies in your possession have been returned or destroyed.

LIMITED WARRANTY

THE PROGRAM IS PROVIDED "AS IS" WITHOUT WARRANTY OF ANY KIND, EITHER EXPRESSED OR IMPLIED, INCLUDING, BUT NOT LIMITED TO, THE IMPLIED WARRANTIES OR MERCHANTABILITY AND FITNESS FOR A PARTICULAR PURPOSE. THE ENTIRE RISK AS TO THE QUALITY AND PERFORMANCE OF THE PROGRAM IS WITH YOU. SHOULD THE PROGRAM PROVE DEFECTIVE, YOU (AND NOT PEARSON EDUCATION, INC. OR ANY AUTHORIZED DEALER) ASSUME THE ENTIRE COST OF ALL NECESSARY SERVICING, REPAIR, OR CORRECTION.

NO ORAL OR WRITTEN INFORMATION OR ADVICE GIVEN BY PEARSON EDUCATION, INC., ITS DEALERS, DISTRIBUTORS, OR AGENTS SHALL CREATE A WARRANTY OR INCREASE THE SCOPE OF THIS WARRANTY. SOME STATES DO NOT ALLOW THE EXCLUSION OF IMPLIED WARRANTIES, SO THE ABOVE EXCLUSION MAY NOT APPLY TO YOU. THIS WARRANTY GIVES YOU SPECIFIC LEGAL RIGHTS AND YOU MAY ALSO HAVE OTHER LEGAL RIGHTS THAT VARY FROM STATE TO STATE.

Pearson Education, Inc. does not warrant that the functions contained in the program will meet your requirements or that the operation of the program will be uninterrupted or error-free. However, Pearson Education, Inc. warrants the CD-ROM(s) on which the program is furnished to be free from defects in material and workmanship under normal use for a period of ninety (90) days from the date of delivery to you as evidenced by a copy of your receipt. The program should not be relied on as the sole basis to solve a problem whose incorrect solution could result in injury to person or property. If the program is employed in such a manner, it is at the user's own risk and Pearson Education, Inc. explicitly disclaims all liability for such misuse.

LIMITATION OF REMEDIES

Pearson Education, Inc.'s entire liability and your exclusive remedy shall be:

1. the replacement of any CD-ROM not meeting Pearson Education, Inc.'s "LIMITED WARRANTY" and that is returned to Pearson Education, or
2. if Pearson Education is unable to deliver a replacement CD-ROM that is free of defects in materials or workmanship, you may terminate this agreement by returning the program.

IN NO EVENT WILL PEARSON EDUCATION, INC. BE LIABLE TO YOU FOR ANY DAMAGES, INCLUDING ANY LOST PROFITS, LOST SAVINGS, OR OTHER INCIDENTAL OR CONSEQUENTIAL DAMAGES ARISING OUT OF THE USE OR INABILITY TO USE SUCH PROGRAM EVEN IF PEARSON EDUCATION, INC. OR AN AUTHORIZED DISTRIBUTOR HAS BEEN ADVISED OF THE POSSIBILITY OF SUCH DAMAGES, OR FOR ANY CLAIM BY ANY OTHER PARTY.

SOME STATES DO NOT ALLOW FOR THE LIMITATION OR EXCLUSION OF LIABILITY FOR INCIDENTAL OR CONSEQUENTIAL DAMAGES, SO THE ABOVE LIMITATION OR EXCLUSION MAY NOT APPLY TO YOU.

GENERAL

You may not sublicense, assign, or transfer the license of the program. Any attempt to sublicense, assign or transfer any of the rights, duties, or obligations hereunder is void.

This Agreement will be governed by the laws of the State of New York.

Should you have any questions concerning this Agreement, you may contact Pearson Education, Inc. by writing to:
ESM Media Development
Higher Education Division
Pearson Education, Inc.
1 Lake Street
Upper Saddle River, NJ 07458

Should you have any questions concerning technical support, you may write to:
New Media Production
Higher Education Division
Pearson Education, Inc.
1 Lake Street
Upper Saddle River, NJ 07458

YOU ACKNOWLEDGE THAT YOU HAVE READ THIS AGREEMENT, UNDERSTAND IT, AND AGREE TO BE BOUND BY ITS TERMS AND CONDITIONS. YOU FURTHER AGREE THAT IT IS THE COMPLETE AND EXCLUSIVE STATEMENT OF THE AGREEMENT BETWEEN US THAT SUPERSEDES ANY PROPOSAL OR PRIOR AGREEMENT, ORAL OR WRITTEN, AND ANY OTHER COMMUNICATIONS BETWEEN US RELATING TO THE SUBJECT MATTER OF THIS AGREEMENT.

System Requirements

- Windows:
Pentium II 300 MHz processor
Windows 98, NT, 2000, ME, or XP
64 MB RAM (128 MB RAM required for Windows XP)
4.3 available hard drive space (optional—for minimum QuickTime installation)
800 × 600 resolution
8x or faster CD-ROM drive
QuickTime 6.x
Sound Card

- Macintosh:
Power PC G3 233 MHz or better
Mac OS 9.x or 10.x
64 MB RAM
10 MB available hard drive space for Mac OS 9, 19 MB on OS X (optional—if QuickTime installation is needed)
800 × 600 resolution
8x or faster CD-ROM drive
QuickTime 6.x

Support Information

If you are having problems with this software, call (800) 677-6337 between 8:00 a.m. and 8:00 p.m. EST, Monday through Friday, and 5:00 p.m. through Midnight EST on Sundays. You can also get support by filling out the web form located at : http://247.prenhall.com/mediaform

Our technical staff will need to know certain things about your system in order to help us solve your problems more quickly and efficiently. If possible, please be at your computer when you call for support. You should have the following information ready:

- Textbook ISBN
- CD-Rom/Diskette ISBN
- corresponding product and title
- computer make and model
- Operating System (Windows or Macintosh) and Version
- RAM available
- hard disk space available
- Sound card? Yes or No
- printer make and model
- network connection
- detailed description of the problem, including the exact wording of any error messages.

NOTE: Pearson does not support and/or assist with the following:

- third-party software (i.e. Microsoft including Microsoft Office Suite, Apple, Borland, etc.)
- homework assistance
- Textbooks and CD-Rom's purchased used are not supported and are non-replaceable. To purchase a new CD-Rom contact Pearson Individual Order Copies at 1-800-282-0693

Definitions, Rules, and Formulas

The Real Numbers

Natural Numbers: $\{1, 2, 3, \ldots\}$

Whole Numbers: $\{0, 1, 2, 3, \ldots\}$

Integers: $\{\ldots, -3, -2, -1, 0, 1, 2, 3, \ldots\}$

Rational Numbers: $\{\frac{a}{b} \mid a$ and b are integers, $b \neq 0\}$

Irrational Numbers: $\{x \mid x$ is real and not rational$\}$

Basic Rules of Algebra

Commutative: $a + b = b + a$; $ab = ba$

Associative: $(a + b) + c = a + (b + c)$;
$(ab)c = a(bc)$

Distributive: $a(b + c) = ab + ac$; $a(b - c) = ab - ac$

Identity: $a + 0 = a$; $a \cdot 1 = a$

Inverse: $a + (-a) = 0$; $a \cdot \frac{1}{a} = 1 (a \neq 0)$

Multiplication Properties: $(-1)a = -a$;
$(-1)(-a) = a$; $a \cdot 0 = 0$; $(-a)(b) = (a)(-b) = -ab$;
$(-a)(-b) = ab$

Set-Builder Notation, Interval Notation, and Graphs

$(a, b) = \{x \mid a < x < b\}$ 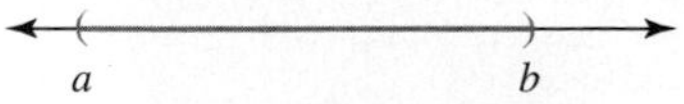

$[a, b) = \{x \mid a \le x < b\}$ 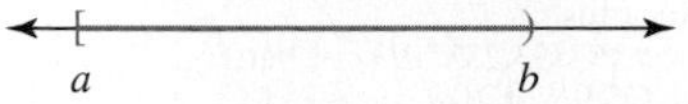

$(a, b] = \{x \mid a < x \le b\}$ 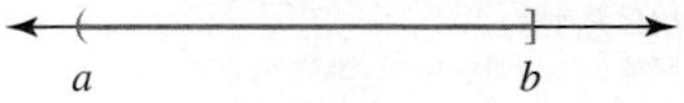

$[a, b] = \{x \mid a \le x \le b\}$ 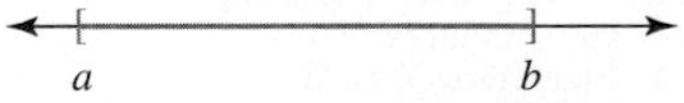

$(-\infty, b) = \{x \mid x < b\}$ 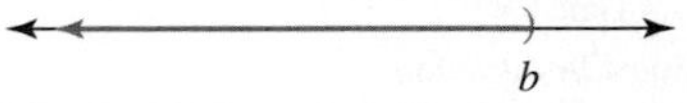

$(-\infty, b] = \{x \mid x \le b\}$ 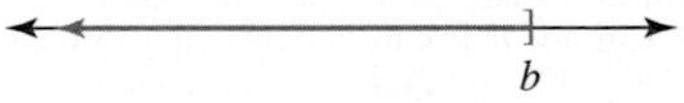

$(a, \infty) = \{x \mid x > a\}$ 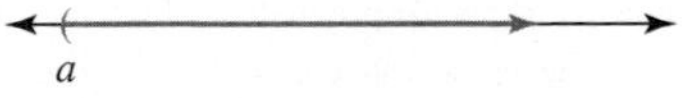

$[a, \infty) = \{x \mid x \ge a\}$ 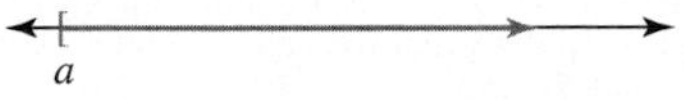

$(-\infty, \infty) = \{x \mid x$ is a real number$\} = \{x \mid x \in R\}$

Slope Formula

$$\text{slope } (m) = \frac{\text{change in } y}{\text{change in } x} = \frac{y_2 - y_1}{x_2 - x_1} \quad (x_1 \neq x_2)$$

Equations of Lines

1. *Slope-intercept form:* $y = mx + b$
   m is the line's slope and b is its y-intercept.
2. *Standard form:* $Ax + By = C$
3. *Point-slope form:* $y - y_1 = m(x - x_1)$
   m is the line's slope and (x_1, y_1) is a fixed point on the line.
4. *Horizontal line parallel to the x-axis:* $y = b$
5. *Vertical line parallel to the y-axis:* $x = a$

Systems of Equations

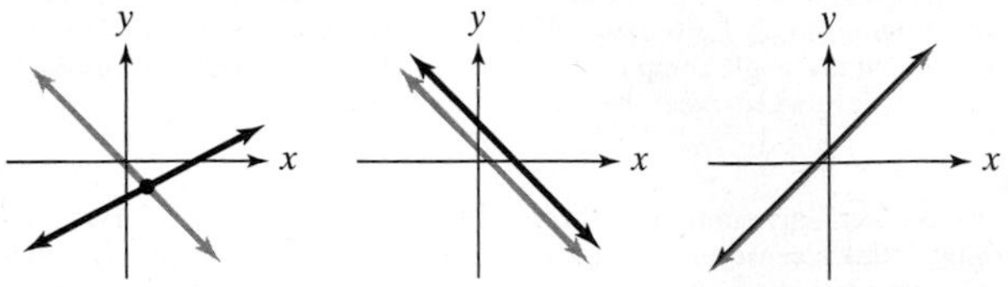

One solution: consistent | No solution: inconsistent | Infinitely many solutions: dependent and consistent

A system of linear equations may be solved: (a)graphically, (b) by the substitution method, (c) by the addition or elimination method, (d) by matrices, or (e) by determinants.

$$\begin{vmatrix} a_1 b_1 \\ a_2 b_2 \end{vmatrix} = a_1 b_2 - a_2 b_1$$

Cramer's Rule:

Given a system of a equations of the form

$$\begin{aligned} a_1 x + b_1 y &= c_1 \\ a_2 x + b_2 y &= c_2 \end{aligned}, \quad \text{then } x = \frac{\begin{vmatrix} c_1 b_1 \\ c_2 b_2 \end{vmatrix}}{\begin{vmatrix} a_1 b_1 \\ a_2 b_2 \end{vmatrix}} \text{ and } y = \frac{\begin{vmatrix} a_1 c_1 \\ a_2 c_2 \end{vmatrix}}{\begin{vmatrix} a_1 b_1 \\ a_2 b_2 \end{vmatrix}}.$$

Absolute Value

1. $|x| = \begin{cases} x & \text{if } x \ge 0 \\ -x & \text{if } x < 0 \end{cases}$
2. If $|x| = c$, then $x = c$ or $x = -c$. $(c > 0)$
3. If $|x| < c$, then $-c < x < c$. $(c > 0)$
4. If $|x| > c$, then $x < -c$ or $x > c$. $(c > 0)$

Special Factorizations

1. *Difference of two squares:*
   $$A^2 - B^2 = (A + B)(A - B)$$
2. *Perfect square trinomials:*
   $$A^2 + 2AB + B^2 = (A + B)^2$$
   $$A^2 - 2AB + B^2 = (A - B)^2$$

3. *Sum of two cubes:*

$$A^3 + B^3 = (A + B)(A^2 - AB + B^2)$$

4. *Difference of two cubes:*

$$A^3 - B^3 = (A - B)(A^2 + AB + B^2)$$

Variation

English Statement	*Equation*
y varies directly as x.	$y = kx$
y varies directly as x^n.	$y = kx^n$
y varies inversely as x.	$y = \frac{k}{x}$
y varies inversely as x^n.	$y = \frac{k}{x^n}$
y varies jointly as x and z.	$y = kxz$

Exponents

Definitions of Rational Exponents

1. $a^{\frac{1}{n}} = \sqrt[n]{a}$
2. $a^{\frac{m}{n}} = (\sqrt[n]{a})^m$ or $\sqrt[n]{a^m}$
3. $a^{-\frac{m}{n}} = \frac{1}{a^{\frac{m}{n}}}$

Properties of Rational Exponents

If m and n are rational exponents, and a and b are real numbers for which the following expressions are defined, then

1. $b^m \cdot b^n = b^{m+n}$
2. $\frac{b^m}{b^n} = b^{m-n}$
3. $(b^m)^n = b^{mn}$
4. $(ab)^n = a^n b^n$
5. $\left(\frac{a}{b}\right)^n = \frac{a^n}{b^n}$

Radicals

1. If n is even, then $\sqrt[n]{a^n} = |a|$.
2. If n is odd, then $\sqrt[n]{a^n} = a$.
3. The product rule: $\sqrt[n]{a} \cdot \sqrt[n]{b} = \sqrt[n]{ab}$
4. The quotient rule: $\frac{\sqrt[n]{a}}{\sqrt[n]{b}} = \sqrt[n]{\frac{a}{b}}$

Complex Numbers

1. The imaginary unit i is defined as

$$i = \sqrt{-1}, \quad \text{where} \quad i^2 = -1.$$

The set of numbers in the form $a + bi$ is called the set of complex numbers. If $b = 0$, the complex number is a real number. If $b \neq 0$, the complex number is an imaginary number.

2. The complex numbers $a + bi$ and $a - bi$ are conjugates. Conjugates can be multiplied using the formula

$$(A + B)(A - B) = A^2 - B^2.$$

The multiplication of conjugates results in a real number.

3. To simplify powers of i, rewrite the expression in terms of i^2. Then replace i^2 with -1 and simplify.

Quadratic Equations and Functions

1. The solutions of a quadratic equation in standard form

$$ax^2 + bx + c = 0, \quad a \neq 0,$$

are given by the quadratic formula

$$x = \frac{-b \pm \sqrt{b^2 - 4ac}}{2a}.$$

2. The discriminant, $b^2 - 4ac$, of the quadratic equation $ax^2 + bx + c = 0$ determines the number and type of solutions.

Discriminant	**Solutions**
Positive perfect square with a, b, and c rational numbers	2 rational solutions
Positive and not a perfect square	2 irrational solutions
Zero, with a, b, and c rational numbers	1 rational solution
Negative	2 imaginary solutions

3. The graph of the quadratic function

$$f(x) = a(x - h)^2 + k, \quad a \neq 0,$$

is called a parabola. The vertex, or turning point, is (h, k). The graph opens upward if a is positive and downward if a negative. The axis of symmetry is a vertical line passing through the vertex. The graph can be obtained using the vertex, x-intercepts, if any, [set $f(x)$ equal to zero], and the y-intercept (set $x = 0$).

4. A parabola whose equation is in the form

$$f(x) = ax^2 + bx + c, \quad a \neq 0,$$

has its vertex at

$$\left(-\frac{b}{2a}, f\left(-\frac{b}{2a}\right)\right).$$

If $a > 0$, then f has a minimum that occurs at $x = -\frac{b}{2a}$. If $a < 0$, then f has a maximum that occurs at $x = -\frac{b}{2a}$.

Definitions, Rules, and Formulas (continued)

Exponential and Logarithmic Functions

1. Exponential Function: $f(x) = b^x, b > 0, b \neq 1$
Graphs:

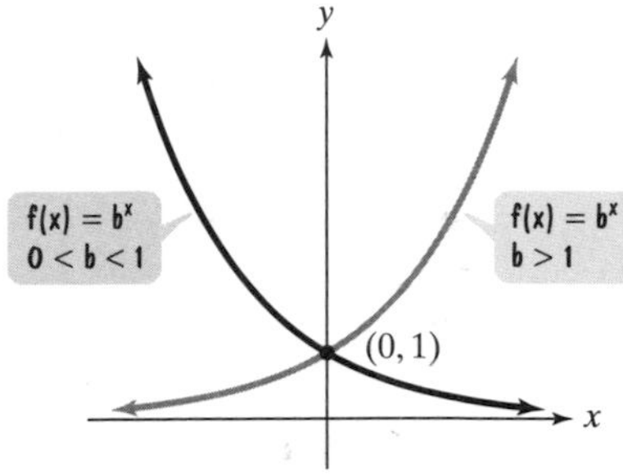

2. Logarithmic Function: $f(x) = \log_b x, b > 0, b \neq 1$
$y = \log_b x$ is equivalent to $x = b^y$.

Graphs:

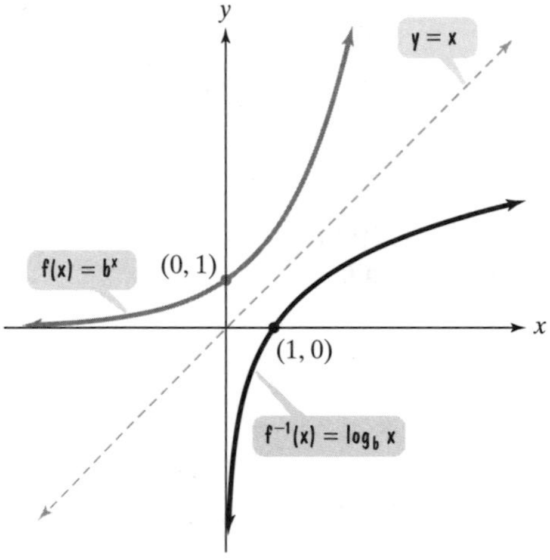

$b > 1$

3. Properties of Logarithms

a. $\log_b(MN) = \log_b M + \log_b N$

b. $\log_b\left(\frac{M}{N}\right) = \log_b M - \log_b N$

c. $\log_b M^p = p\log_b M$

d. $\log_b M = \frac{\log_a M}{\log_a b} = \frac{\ln M}{\ln b} = \frac{\log M}{\log b}$

e. $\log_b b^x = x; \log 10^x = x; \ln e^x = x$

f. $b^{\log_b x} = x; 10^{\log x} = x; e^{\ln x} = x$

Distance and Midpoint Formulas

1. The distance from (x_1, y_1) to (x_2, y_2) is

$$\sqrt{(x_2 - x_1)^2 + (y_2 - y_1)^2}.$$

2. The midpoint of the line segment with endpoints (x_1, y_1) and (x_2, y_2) is

$$\left(\frac{x_1 + x_2}{2}, \frac{y_1 + y_2}{2}\right).$$

Conic Sections

Circle

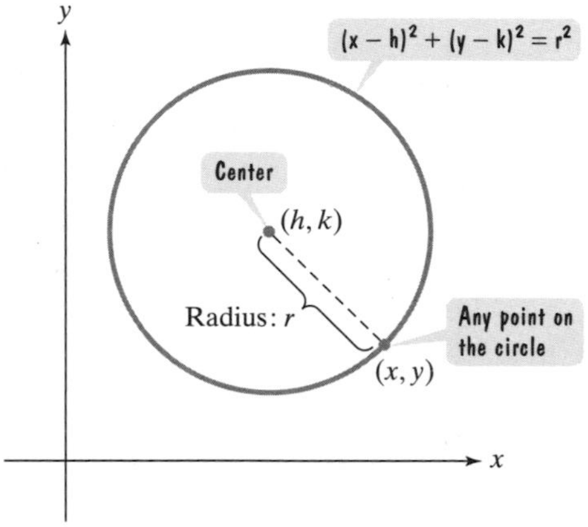

Ellipse

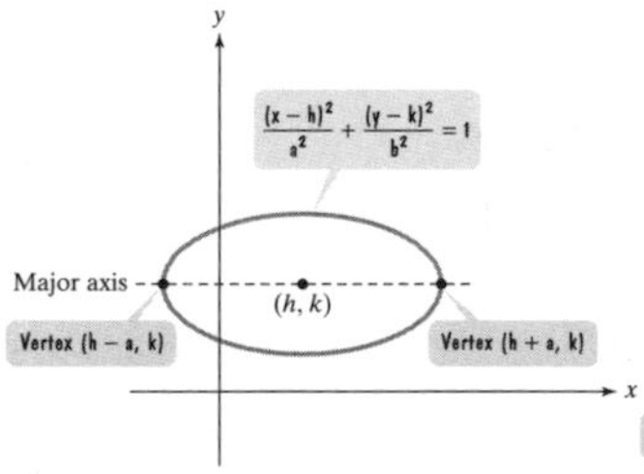

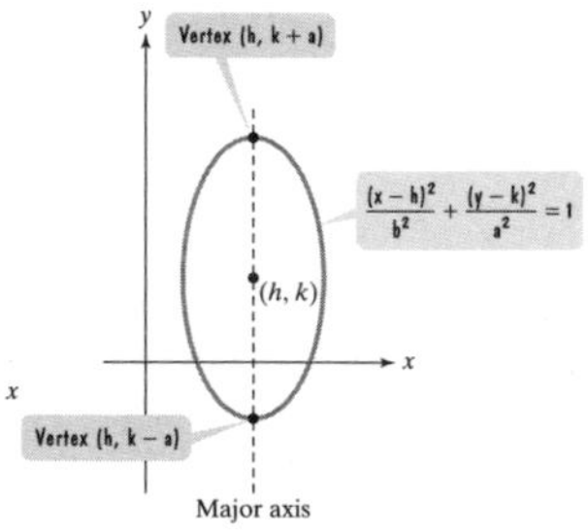

Hyperbola

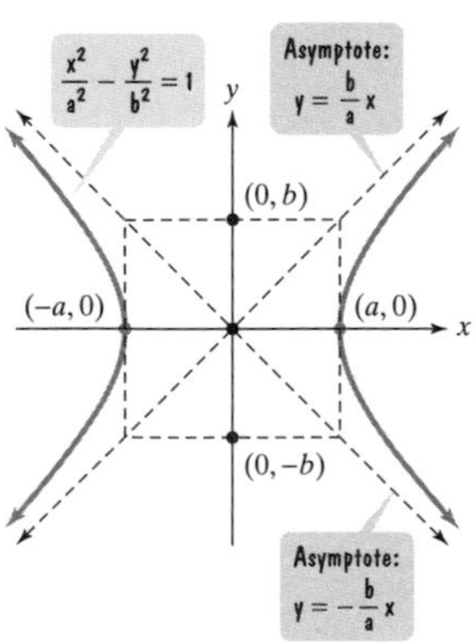

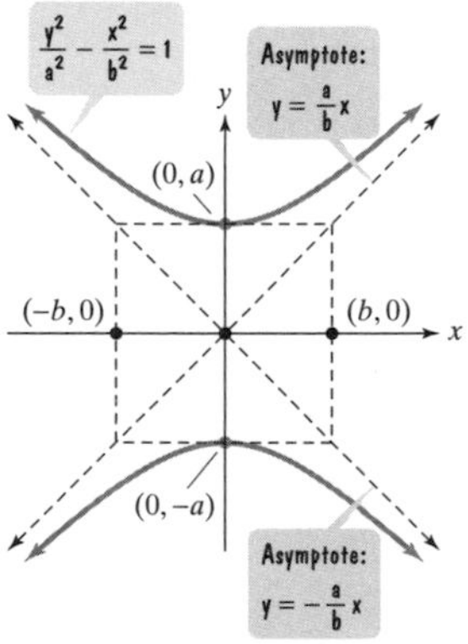